Recommended Dietary Allowances (RDA) and Adequate Intakes (AI) for Vitamins

Age (yr)	Thiamin RDA (mg/day)	Riboflavin RDA (mg/day)	Niacin RDA (mg/day)[a]	Biotin AI (µg/day)	Pantothenic acid AI (mg/day)	Vitamin B_6 RDA (mg/day)	Folate RDA (µg/day)[b]	Vitamin B_{12} RDA (µg/day)	Choline AI (mg/day)	Vitamin C RDA (mg/day)	Vitamin A RDA (µg/day)[c]	Vitamin D RDA (IU/day)[d]	Vitamin E RDA (mg/day)[e]	Vitamin K AI (µg/day)
Infants														
0–0.5	0.2	0.3	2	5	1.7	0.1	65	0.4	125	40	400	400 (10 µg)	4	2.0
0.5–1	0.3	0.4	4	6	1.8	0.3	80	0.5	150	50	500	400 (10 µg)	5	2.5
Children														
1–3	0.5	0.5	6	8	2	0.5	150	0.9	200	15	300	600 (15 µg)	6	30
4–8	0.6	0.6	8	12	3	0.6	200	1.2	250	25	400	600 (15 µg)	7	55
Males														
9–13	0.9	0.9	12	20	4	1.0	300	1.8	375	45	600	600 (15 µg)	11	60
14–18	1.2	1.3	16	25	5	1.3	400	2.4	550	75	900	600 (15 µg)	15	75
19–30	1.2	1.3	16	30	5	1.3	400	2.4	550	90	900	600 (15 µg)	15	120
31–50	1.2	1.3	16	30	5	1.3	400	2.4	550	90	900	600 (15 µg)	15	120
51–70	1.2	1.3	16	30	5	1.7	400	2.4	550	90	900	600 (15 µg)	15	120
>70	1.2	1.3	16	30	5	1.7	400	2.4	550	90	900	800 (20 µg)	15	120
Females														
9–13	0.9	0.9	12	20	4	1.0	300	1.8	375	45	600	600 (15 µg)	11	60
14–18	1.0	1.0	14	25	5	1.2	400	2.4	400	65	700	600 (15 µg)	15	75
19–30	1.1	1.1	14	30	5	1.3	400	2.4	425	75	700	600 (15 µg)	15	90
31–50	1.1	1.1	14	30	5	1.3	400	2.4	425	75	700	600 (15 µg)	15	90
51–70	1.1	1.1	14	30	5	1.5	400	2.4	425	75	700	600 (15 µg)	15	90
>70	1.1	1.1	14	30	5	1.5	400	2.4	425	75	700	800 (20 µg)	15	90
Pregnancy														
≤18	1.4	1.4	18	30	6	1.9	600	2.6	450	80	750	600 (15 µg)	15	75
19–30	1.4	1.4	18	30	6	1.9	600	2.6	450	85	770	600 (15 µg)	15	90
31–50	1.4	1.4	18	30	6	1.9	600	2.6	450	85	770	600 (15 µg)	15	90
Lactation														
≤18	1.4	1.6	17	35	7	2.0	500	2.8	550	115	1200	600 (15 µg)	19	75
19–30	1.4	1.6	17	35	7	2.0	500	2.8	550	120	1300	600 (15 µg)	19	90
31–50	1.4	1.6	17	35	7	2.0	500	2.8	550	120	1300	600 (15 µg)	19	90

NOTE: For all nutrients, values for infants are AI. The glossary on the inside back cover defines units of nutrient measure.

[a]Niacin recommendations are expressed as niacin equivalents (NE), except for recommendations for infants younger than 6 months, which are expressed as preformed niacin.

[b]Folate recommendations are expressed as dietary folate equivalents (DFE).

[c]Vitamin A recommendations are expressed as retinol activity equivalents (RAE).

[d]Vitamin D recommendations are expressed as cholecalciferol and assume an absence of adequate exposure to sunlight.

[e]Vitamin E recommendations are expressed as α-tocopherol.

Recommended Dietary Allowances (RDA) and Adequate Intakes (AI) for Minerals

Age (yr)	Sodium AI (mg/day)	Chloride AI (mg/day)	Potassium AI (mg/day)	Calcium RDA (mg/day)	Phosphorus RDA (mg/day)	Magnesium RDA (mg/day)	Iron RDA (mg/day)	Zinc RDA (mg/day)	Iodine RDA (µg/day)	Selenium RDA (µg/day)	Copper RDA (µg/day)	Manganese AI (mg/day)	Fluoride AI (mg/day)	Chromium AI (µg/day)	Molybdenum RDA (µg/day)
Infants															
0–0.5	120	180	400	200	100	30	0.27	2	110	15	200	0.003	0.01	0.2	2
0.5–1	370	570	700	260	275	75	11	3	130	20	220	0.6	0.5	5.5	3
Children															
1–3	1000	1500	3000	700	460	80	7	3	90	20	340	1.2	0.7	11	17
4–8	1200	1900	3800	1000	500	130	10	5	90	30	440	1.5	1.0	15	22
Males															
9–13	1500	2300	4500	1300	1250	240	8	8	120	40	700	1.9	2	25	34
14–18	1500	2300	4700	1300	1250	410	11	11	150	55	890	2.2	3	35	43
19–30	1500	2300	4700	1000	700	400	8	11	150	55	900	2.3	4	35	45
31–50	1500	2300	4700	1000	700	420	8	11	150	55	900	2.3	4	35	45
51–70	1300	2000	4700	1000	700	420	8	11	150	55	900	2.3	4	30	45
>70	1200	1800	4700	1200	700	420	8	11	150	55	900	2.3	4	30	45
Females															
9–13	1500	2300	4500	1300	1250	240	8	8	120	40	700	1.6	2	21	34
14–18	1500	2300	4700	1300	1250	360	15	9	150	55	890	1.6	3	24	43
19–30	1500	2300	4700	1000	700	310	18	8	150	55	900	1.8	3	25	45
31–50	1500	2300	4700	1000	700	320	18	8	150	55	900	1.8	3	25	45
51–70	1300	2000	4700	1200	700	320	8	8	150	55	900	1.8	3	20	45
>70	1200	1800	4700	1200	700	320	8	8	150	55	900	1.8	3	20	45
Pregnancy															
≤18	1500	2300	4700	1300	1250	400	27	12	220	[illegible]	[illegible]	[illegible]	[illegible]	[illegible]	[illegible]
19–30	1500	2300	4700	1000	700	350	27	11	220	[illegible]	[illegible]	[illegible]	[illegible]	[illegible]	[illegible]
31–50	1500	2300	4700	1000	700	360	27	11	220	[illegible]	[illegible]	[illegible]	[illegible]	[illegible]	[illegible]
Lactation															
≤18	1500	2300	5100	1300	1250	360	10	13	290	[illegible]	[illegible]	[illegible]	[illegible]	[illegible]	[illegible]
19–30	1500	2300	5100	1000	700	310	9	12	290	[illegible]	[illegible]	[illegible]	[illegible]	[illegible]	[illegible]
31–50	1500	2300	5100	1000	700	320	9	12	290	[illegible]	[illegible]	[illegible]	[illegible]	[illegible]	[illegible]

NOTE: For all nutrients, values for infants are AI. The glossary on the inside back cover defines units of nutrient measure.

Tolerable Upper Intake Levels (UL) for Vitamins

Age (yr)	Niacin (mg/day)[a]	Vitamin B_6 (mg/day)	Folate (µg/day)[a]	Choline (mg/day)	Vitamin C (mg/day)	Vitamin A (IU/day)[b]	Vitamin D (IU/day)	Vitamin E (mg/day)[c]
Infants								
0–0.5	—	—	—	—	—	600	1000 (25 µg)	—
0.5–1	—	—	—	—	—	600	1500 (38 µg)	—
Children								
1–3	10	30	300	1000	400	600	2500 (63 µg)	200
4–8	15	40	400	1000	650	900	3000 (75 µg)	300
9–13	20	60	600	2000	1200	1700	4000 (100 µg)	600
Adolescents								
14–18	30	80	800	3000	1800	2800	4000 (100 µg)	800
Adults								
19–70	35	100	1000	3500	2000	3000	4000 (100 µg)	1000
>70	35	100	1000	3500	2000	3000	4000 (100 µg)	1000
Pregnancy								
≤18	30	80	800	3000	1800	2800	4000 (100 µg)	800
19–50	35	100	1000	3500	2000	3000	4000 (100 µg)	1000
Lactation								
≤18	30	80	800	3000	1800	2800	4000 (100 µg)	800
19–50	35	100	1000	3500	2000	3000	4000 (100 µg)	1000

[a]The UL for niacin and folate apply to synthetic forms obtained from supplements, fortified foods, or a combination of the two.
[b]The UL for vitamin A applies to the preformed vitamin only.
[c]The UL for vitamin E applies to any form of supplemental α-tocopherol, fortified foods, or a combination of the two.

Tolerable Upper Intake Levels (UL) for Minerals

Age (yr)	Sodium (mg/day)	Chloride (mg/day)	Calcium (mg/day)	Phosphorus (mg/day)	Magnesium (mg/day)[d]	Iron (mg/day)	Zinc (mg/day)	Iodine (µg/day)	Selenium (µg/day)	Copper (µg/day)	Manganese (mg/day)	Fluoride (mg/day)	Molybdenum (µg/day)	Boron (mg/day)	Nickel (mg/day)	Vanadium (mg/day)
Infants																
0–0.5	—	—	1000	—	—	40	4	—	45	—	—	0.7	—	—	—	—
0.5–1	—	—	1500	—	—	40	5	—	60	—	—	0.9	—	—	—	—
Children																
1–3	1500	2300	2500	3000	65	40	7	200	90	1000	2	1.3	300	3	0.2	—
4–8	1900	2900	2500	3000	110	40	12	300	150	3000	3	2.2	600	6	0.3	—
9–13	2200	3400	3000	4000	350	40	23	600	280	5000	6	10	1100	11	0.6	—
Adolescents																
14–18	2300	3600	3000	4000	350	45	34	900	400	8000	9	10	1700	17	1.0	—
Adults																
19–50	2300	3600	2500	4000	350	45	40	1100	400	10,000	11	10	2000	20	1.0	1.8
51–70	2300	3600	2000	4000	350	45	40	1100	400	10,000	11	10	2000	20	1.0	1.8
>70	2300	3600	2000	3000	350	45	40	1100	400	10,000	11	10	2000	20	1.0	1.8
Pregnancy																
≤18	2300	3600	3000	3500	350	45	34	900	400	8000	9	10	1700	17	1.0	—
19–50	2300	3600	2500	3500	350	45	40	1100	400	10,000	11	10	2000	20	1.0	—
Lactation																
≤18	2300	3600	3000	4000	350	45	34	900	400	8000	9	10	1700	17	1.0	—
19–50	2300	3600	2500	4000	350	45	40	1100	400	10,000	11	10	2000	20	1.0	—

[d]The UL for magnesium applies to synthetic forms obtained from supplements or drugs only.

NOTE: An upper Limit was not established for vitamins and minerals not listed and for those age groups listed with a dash (—) because of a lack of data, not because these nutrients are safe to consume at any level of intake. All nutrients can have adverse effects when intakes are excessive.

SOURCE: Adapted with permission from the *Dietary Reference Intakes for Calcium and Vitamin D*, © 2011 by the National Academies of Sciences, Courtesy of the National Academies Press, Washington, D.C.

NUTRITION FOR HEALTH AND HEALTH CARE

fifth edition

Linda Kelly DeBruyne
Kathryn Pinna

Australia • Brazil • Japan • Korea • Mexico • Singapore • Spain • United Kingdom • United States

Nutrition for Health and Health Care, **Fifth Edition**
Linda Kelly DeBruyne, Kathryn Pinna

Publisher: Yolanda Cossio
Acquisitions Editor: Peggy Williams
Developmental Editor: Elesha Feldman
Editorial Assistant: Kellie Petruzzelli
Media Editor: Miriam Myers
Brand Manager: Jennifer N. Levanduski
Market Development Manager: Tom Ziolkowski
Content Project Manager: Carol Samet
Art Director: John Walker
Manufacturing Planner: Karen Hunt
Rights Acquisitions Specialist: Roberta Broyer
Production Service: Lachina Publishing Services
Photo Researcher: Q2A/Bill Smith
Text Researcher: Melissa Tomaselli
Copy Editor: Lachina Publishing Services
Text Designer: Diane Beasley
Cover Designer: John Walker
Cover Image: Gettyimages/Anthony Harvie, Gettyimages/Peter Dazeley
Compositor: Lachina Publishing Services

Library of Congress Control Number: 2013932311

ISBN-13: 978-1-133-59911-1

ISBN-10: 1-133-59911-7

Wadsworth
20 Davis Drive
Belmont, CA 94002-3098
USA

Cengage Learning is a leading provider of customized learning solutions with office locations around the globe, including Singapore, the United Kingdom, Australia, Mexico, Brazil, and Japan. Locate your local office at **www.cengage.com/global**.

Cengage Learning products are represented in Canada by Nelson Education, Ltd.

To learn more about Wadsworth, visit **www.cengage.com/Wadsworth**

Purchase any of our products at your local college store or at our preferred online store **www.CengageBrain.com**.

Printed in the United States of America
2 3 4 5 6 7 17 16 15 14

To my grandson, Ryder Koa DeBruyne. You are my blue sky and my sunny day. Aloha little man.

LINDA KELLY DEBRUYNE

To my mother, Tina C. Pinna, for her unwavering love and support, and to David L. Stone, for keeping the music in my life.

KATHRYN PINNA

About the Authors

LINDA KELLY DEBRUYNE, M.S., R.D., received her B.S. in 1980 and her M.S. in 1982 in nutrition and food science at Florida State University. She is a founding member of Nutrition and Health Associates, an information resource center in Tallahassee, Florida, where her specialty areas are life cycle nutrition and fitness. Her other publications include the textbooks *Nutrition and Diet Therapy* and *Health: Making Life Choices* and a multi-media CD-ROM called *Nutrition Interactive*. As a consultant for a group of Tallahassee pediatricians, she teaches infant nutrition classes to parents. She is a registered dietitian and maintains a professional membership in the Academy of Nutrition and Dietetics.

KATHRYN PINNA, Ph.D., R.D., received her M.S. and Ph.D. in nutrition from the University of California at Berkeley. She has taught nutrition, food science, and human biology courses in the San Francisco Bay area for over 25 years and currently teaches nutrition classes at City College of San Francisco. She has also worked as an outpatient dietitian, Internet consultant, and freelance writer. Her other publications include the textbooks *Understanding Normal and Clinical Nutrition* and *Nutrition and Diet Therapy*. She is a registered dietitian and a member of the American Society for Nutrition and the Academy of Nutrition and Dietetics.

Brief Contents

Brief Contents

Contents

Case Studies

How to Features

Preface

Whether for the prevention of disease or appropriate treatment of illness, health professionals and patients rank nutrition among their most serious concerns. Moreover, medical personnel are often called upon to answer questions about foods and diets or provide nutrition care. This fifth edition of *Nutrition for Health and Health Care* provides a solid foundation in nutrition science and the role of nutrition in clinical care. Although much of the material has been written for nursing students and is relevant to nursing care, this textbook can be useful for students of other health-related professions, including nursing assistants, physician assistants, dietitians, dietary technicians, and health educators.

Each chapter of this textbook includes essential nutrition concepts along with practical information for addressing nutrition concerns and solving nutrition problems. The introductory chapters (Chapters 1 and 2) provide an overview of the nutrients and nutrition recommendations and describe the process of digestion and absorption. Chapters 3 through 5 introduce the attributes and functions of carbohydrates, lipids, and protein and explain how appropriate intakes of these nutrients support health. Chapters 6 and 7 introduce the concepts of energy balance and weight management and describe the health effects of overweight, underweight, and eating disorders. Chapters 8 and 9 introduce the vitamins and minerals, describing their roles in the body, appropriate intakes, and food sources. Chapters 10 through 12 explain how nutrient needs change throughout the life cycle. Chapters 13 and 14 explore how health professionals can use information from nutrition assessments to identify and address a patient's dietary needs. The remaining chapters (Chapters 15–23) examine nutrition therapy and its role in the prevention and treatment of common medical conditions.

SPECIAL FEATURES

Students of nutrition often begin a nutrition course with some practical knowledge of nutrition; after all, they may purchase food, read food labels, and be familiar with common nutrition problems such as obesity or lactose intolerance. After just a few weeks of class, however, the nutrition student realizes that nutrition is a biological and chemical science with a fair amount of new terminology and new concepts to learn. This book contains abundant pedagogy to help students master the subject matter. Within each chapter, definitions and notes in the margins clarify nutrition information, remind readers of previously defined terms, and provide cross-references. How To skill boxes help readers work through calculations or give practical suggestions for applying nutrition advice. The Nursing Diagnosis feature enables nursing students to correlate nutrition care with nursing care. Review Notes summarize the information following each major heading; these summaries can be used to preview or review key chapter concepts. The Self Check at the end of each chapter provides questions to help review chapter information. Each chapter concludes with a Nutrition on the Net feature, which lists websites relevant to the topics covered in the chapter.

In the life cycle and clinical chapters, Case Studies guide readers in applying nutrition therapy to patient care. Diet-Drug Interaction boxes in the clinical chapters identify important nutrient-drug and food-drug interactions. Clinical Applications throughout the text encourage readers to practice mathematical calculations, synthesize information from previous chapters, or understand how dietary adjustments affect patients. Nutrition Assessment Checklists remind readers of assessment parameters relevant to specific stages of the life cycle or medical problems.

The Nutrition in Practice sections that follow the chapters explore issues of current interest, advanced topics, or specialty areas such as dental health or dialysis. Examples of topics covered include foodborne illness, the glycemic index, vegetarian diets, alcohol in health and disease, nutritional genomics, the metabolic syndrome, and childhood obesity and chronic disease.

APPENDIXES

The appendixes support the book with a wealth of information on the nutrient contents of thousands of foods, Canadian nutrient recommendations and food choices, U.S. nutrient intake recommendations, the exchange system, physical activity and energy requirements, nutrition assessments, enteral formulas, aids to calculations, and answers to Self Check questions.

NEW TO THIS EDITION

Due to the rapid pace of nutrition research, staying current is a primary concern. Each chapter of this book has been substantially updated to reflect advances in research and clinical practice since the fourth edition. In addition, we have made the following changes:

Chapter 1

- Included *Healthy People 2020* nutrition and weight status objectives
- Introduced Dietary Guidelines for Americans, 2010
- Introduced a new MyPlate figure and discussion
- Added definitions of solid fats, added sugars, and nutrient profiling
- Added a new figure on label health claims

Chapter 2

- Enhanced and clarified the GI tract figure, adding labels to describe the function of each part of the system and moving descriptions of carbohydrate, fat, and protein digestion and absorption to each respective chapter
- Mentioned the Food Safety Modernization Act in the Nutrition in Practice
- Reorganized and shortened the foodborne illness table to include the pathogens that cause most foodborne illnesses

Chapter 3

- Added a new figure showing glycogen and starch molecules and their branching
- Reorganized and enhanced the discussion of blood glucose regulation
- Added a brief discussion of and a table on carbohydrate digestion
- Added a discussion of sugar and heart disease
- Emphasized the term *nonnutritive sweeteners* rather than *artificial sweeteners* as per Academy of Nutrition and Dietetics position paper
- Added a new figure of sources of added sugars from the 2010 *Dietary Guidelines*
- Added a new "How to" box on reducing intake of added sugars

Chapter 4

- Added a brief discussion and definition of conjugated linoleic acid
- Added a brief discussion of and table on lipid digestion and absorption
- Greater emphasis on and explanation of solid fats
- Added a new figure of sources of solid fats in U.S. diets
- Added a table of solid fats on food labels
- Added a new figure about replacing saturated fat with unsaturated fat in the Nutrition in Practice

Chapter 5

- Added a brief section and table on protein digestion and absorption
- Enhanced the discussion of proteins as antibodies
- Reorganized the section on protein deficiency
- Introduced and defined the new WHO term *severe acute malnutrition* and defined *chronic malnutrition*
- Added a table comparing severe acute malnutrition and chronic malnutrition
- Enhanced the glossary of terms describing vegetarian diets
- Added the USDA Food Patterns' recommended weekly amounts of protein foods for vegetarians and vegans

Chapter 6

- Updated the discussion of energy balance to explain the shortcomings of the traditional 3500 kcal per pound rule
- Added risks of underweight to coincide with risks of overweight
- Enhanced and updated the table comparing popular diets
- Added a new table called "Tips for Identifying Fad Diets and Weight-Loss Scams"
- Added a brief discussion of protein and kcalorie restriction

Chapter 7

- Introduced and defined screen time, food deserts, satiation, and satiety
- Added a new section on neighborhood obstacles to physical activity and healthy food choices
- Streamlined obesity drug discussion and simplified and updated the list of FDA-approved obesity drugs
- Separated and enhanced tables of weight-loss and weight-gain strategies
- Addressed new thinking and research about the degree of energy restriction required to achieve weight loss over time

Chapter 8

- Revised and updated information on vitamin D as per the 2011 DRI
- Added a new table comparing fat-soluble and water-soluble vitamins
- Figures of good sources of vitamins (A, folate, and C) updated for current USDA nutrient data
- Added a new table in the Nutrition in Practice called "Tips for Consuming Phytochemicals"

Chapter 9

- Simplified the table of major and trace minerals
- Added a new margin list of top 10 sodium food sources
- Added a section on potassium and hypertension and noted this mineral's status as a *Dietary Guidelines* "nutrient of concern"
- Addressed the U.S. Preventive Task Force findings on use of calcium supplements for fracture protection
- Added information on magnesium and disease
- Added information on iron excess and oxidative stress
- Added information on hepcidin, the iron-regulating hormone

Chapter 10

- Enhanced the discussion of essential fatty acids in breast milk
- Added information on breastfeeding and reduced risk of SIDS
- Added a new table of the benefits of breastfeeding
- Personalized the table of strategies for successful breastfeeding

Chapter 11

- Added a new table providing tips for picky eaters
- Enhanced the table of food skills for preschoolers, adding developmental milestones
- Added a new figure to highlight MyPlate resources for children
- Described the new National School Lunch/Breakfast requirements
- Added a new table of physical activities for children and adolescents
- Updated and simplified the "How to" on protecting against lead toxicity

Chapter 12

- Added a new table comparing signs of Alzheimer's and typical age-related changes
- Added the definition of health care communities from the Academy of Nutrition and Dietetics
- Added a brief discussion of the obesity/food insecurity paradox
- Enhanced the "How to" on meal planning to stretch food dollars and reduce waste

Chapter 13

- Reorganized the table on the criteria for identifying malnutrition risk
- Included a discussion of the multiple-pass method for conducting a 24-hour dietary recall interview

Chapter 14

- Introduced the use of indirect calorimetry for determining RMR
- Updated names for diets in the "Modified Diets" section to more closely match current dietetics terminology
- Moved the discussion about improving food intake from the "Modified Diets" section to the "Foodservice" section
- In the "Foodservice" section, introduced the room service model adopted at many hospitals and shortened the discussion about marking selective menus
- In the "Diet-Drug Interactions" section, modified some paragraphs and added some common names for drugs in addition to their generic names; moved discussions on isoniazid and corticosteroids to sections that better reflect their effects on nutrition status
- In the Nutrition in Practice, modified the table listing examples of herb-drug interactions

Chapter 15

- Introduced a new section about oral supplements at the beginning of the chapter
- Reorganized the enteral nutrition section to address tube feeding candidates and tube feeding routes before describing enteral formulas
- Simplified the presentation of tube feeding initiation and advancement
- In the Nutrition in Practice on inborn errors, revised the introductory paragraph about PKU, removed the figure related to PKU, and added a table listing examples of inborn errors that are related to defects in nutrient metabolism

Chapter 16

- Added a photo showing pressure sores
- Simplified the discussion about the inflammatory process
- Updated the discussion about the clinical effects of altering omega-6 and omega-3 fatty acid intakes based on recent analyses
- Added the concept of hypocaloric feedings for obese critical care patients to the section on energy needs in acute illness
- Updated the Diet-Drug Interactions box with current drug treatments for COPD

Chapter 17

- Modified the table of suggestions for managing dry mouth
- Modified the discussion of the causes and signs of dysphagia, expanded the discussion about food thickeners, and updated the "How to" about improving the acceptance of mechanically altered foods
- Rewrote the paragraph about causes of vomiting
- Added a table listing suggestions for preventing oral diseases to the Nutrition in Practice about oral health

Chapter 18

- Revised or reorganized the tables describing laxatives and bulk-forming agents and foods that cause intestinal gas
- Revised the sections on acute and chronic pancreatitis, cystic fibrosis, inflammatory bowel diseases, short bowel syndrome, and diverticular disease of the colon

Chapter 19

- Modified the introduction of fatty liver
- Modified the descriptions of the different types of hepatitis viruses
- Modified some sections on cirrhosis complications and the table on the stages of hepatic encephalopathy
- Revised the sections on the medical treatment and nutrition therapy for cirrhosis

- In the Nutrition in Practice about alcohol, expanded the discussion about the benefits of alcohol for chronic illness risk reduction

Chapter 20

- Revised the discussion about complications of diabetes and included a new figure that outlines the acute effects of insulin insufficiency
- Added a definition for continuous glucose monitoring to the section on evaluating diabetes treatment
- Reorganized the discussion about body weight concerns: the topic of weight gain in type 1 diabetes was moved to the insulin therapy section, and weight loss for type 2 diabetes was moved to the nutrition therapy section
- In the nutrition therapy section, added a paragraph about macronutrient distribution, shortened the discussion about fat intake, and modified the paragraph related to micronutrient supplementation
- Revised the section about physical activity and diabetes management, the discussion about pregnancy in type 1 or type 2 diabetes, and several sections in the Nutrition in Practice on metabolic syndrome

Chapter 21

- Reorganized and revised several sections on the causes of atherosclerosis; modified some sections on the risk factors and treatment of hypertension
- Expanded and modified the discussion of CHD risk assessment
- Modified several sections related to therapeutic lifestyle changes, adding a new discussion about cholesterol and egg intakes

Chapter 22

- Modified the sections related to the nephrotic syndrome, uremia, and the uremic syndrome
- Clarified/updated some sections related to the nutrition therapy for chronic kidney disease to reflect the current recommendations
- Added a new table listing foods high in phosphorus and a table describing an appropriate menu for a person with chronic kidney disease; eliminated the table on the dietary guidelines following a kidney transplant and a table listing foods high in purines
- Added a new section about the medical treatment of kidney stones

Chapter 23

- Replaced the paragraph about high-fat diets and cancer risk with a discussion about alcohol consumption and cancer risk
- Modified the section about hematopoietic stem cell transplantation, added a section about biological therapies for cancer, and modified the section about medications used for treating anorexia and wasting
- Expanded the "How to" about increasing energy and protein in meals
- Replaced the section about nutrition therapy for bone marrow transplant patients with a section about the low-microbial diet for individuals with suppressed immunity
- Modified the introductory paragraphs in the HIV/AIDS section
- Expanded the table listing antiretroviral drugs to include recently approved medications for HIV infection
- Reorganized and revised the section on nutrition therapy for HIV infection, and included a discussion about weight management and overweight/obesity

INSTRUCTOR TOOLS

A number of helpful ancillary materials are available for instructors, including a Power Lecture DVD-ROM that includes figures and photos from the text as well as PowerPoint lecture presentations, a test bank in Word and ExamView formats, and an instructor's manual. A printed version of the instructor's manual and test bank is also available.

STUDENT ANCILLARIES

A printed study guide for students provides numerous review exercises including multiple-choice, true/false, fill-in, matching, and discussion questions; math exercises; and case study problems to help students master chapter concepts. The book's CourseMate website, accessible via CengageBrain.com, provides additional study tools. The new MindTap Reader, a fully interactive online e-reader, seamlessly integrates content, rich media assets, and robust note-taking.

We hope that as you discover the many fascinating aspects of nutrition science and nutrition therapy, you will enthusiastically apply the concepts in both your personal and professional life.

Linda Kelly DeBruyne
Kathryn Pinna

Acknowledgments

Among the most difficult words to write are those that express the depth of our gratitude to the many dedicated people whose efforts have made this book possible. We extend special thanks to Fran Webb and Sharon Rolfes for their expertise and valuable contributions throughout this text. Thanks also to Sylvia Smith and Chris Black for their careful attention to the numerous details involved in production; John Walker and Diane Beasley for designing these pages; Roberta Broyer, Melissa Tomaselli, and Chris Arena for their help in securing permissions and selecting photographs; and the folks at Axxya for compiling the food composition appendix. We are indebted to our editorial team, Peggy Williams and Elesha Feldman, and to our production manager, Carol Samet, for seeing this project through from start to finish. We also thank those who helped to prepare the many supplements available with the text: Lynn Earle of Old Dominion University, Diane Kraft of Alvernia University, Phyllis Magaletto of Cochran School of Nursing, Jillann Neely of Onondaga Community College, Tania Rivera of Florida International University, Karen Schmitz of Madonna University, Lynn Thomas of the University of South Carolina School of Medicine, and Elesha Feldman and Miriam Myers, who managed the preparation of the print and media supplements. We also acknowledge Tom Ziolkowski for his marketing efforts. To the many others involved in designing, indexing, typesetting, dummying, and marketing, we offer our thanks. We are especially grateful to our associates, family, and friends for their continued encouragement and support and to the following reviewers who offered excellent suggestions for improving the text:

John W. Carbone, *Eastern Michigan University*
Barbara A. Caton, *Missouri State University—West Plains*
Lori P. Enriquez, *Drexel University*
Diane Woznicki Kraft, *Alvernia University*

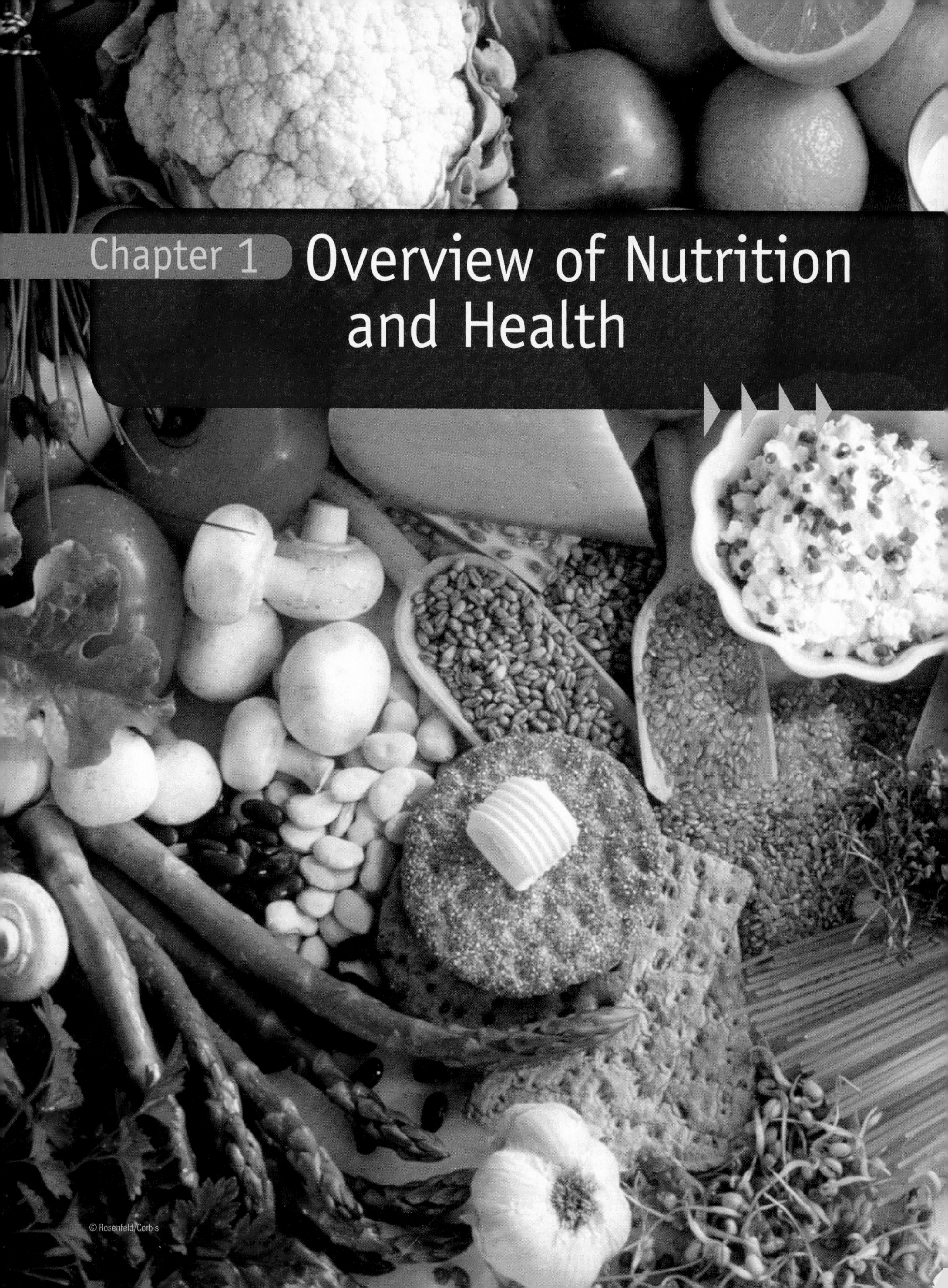

Chapter 1 Overview of Nutrition and Health

© Rosenfeld/Corbis

EVERY DAY, SEVERAL TIMES A DAY, YOU MAKE CHOICES THAT WILL either improve your **health** or harm it. Each choice may influence your health only a little, but when these choices are repeated over years and decades, their effects become significant.

The choices people make each day affect not only their physical health but also their **wellness**—all the characteristics that make a person strong, confident, and able to function well with family, friends, and others. People who consistently make poor lifestyle choices, on a daily basis, increase their risks of developing diseases. Figure 1-1 shows how a person's health can fall anywhere along a continuum, from maximum wellness on the one end to total failure to function (death) on the other.

As nurses or other health care professionals, when you take responsibility for your own health by making daily choices and practicing behaviors that enhance your well-being, you prepare yourself physically, mentally, and emotionally to meet the demands of your profession. As health care professionals, however, you have a responsibility to your clients (■) as well as to yourselves. You have unique opportunities to make your clients aware of the benefits of positive health choices and behaviors, to show them how to change their behaviors and make daily choices to enhance their own health, and to serve as role models for those behaviors.

This text focuses on how nutrition choices affect health and disease. The early chapters introduce the basics of nutrition to support good health. The later chapters emphasize medical nutrition therapy and its role in supporting health and treating diseases and symptoms.

■ Nurses generally use either *client* or *patient* when referring to an individual under their care. The first 12 chapters of this text emphasize the nutrition concerns of people in good health; therefore, the term *client* is used in these chapters.

health: a range of states with physical, mental, emotional, spiritual, and social components. At a minimum, health means freedom from physical disease, mental disturbances, emotional distress, spiritual discontent, social maladjustment, and other negative states. At a maximum, health means *wellness*.

wellness: maximum well-being; the top range of health states; the goal of the person who strives toward realizing his or her full potential physically, mentally, emotionally, spiritually, and socially.

FIGURE 1-1 The Health Line

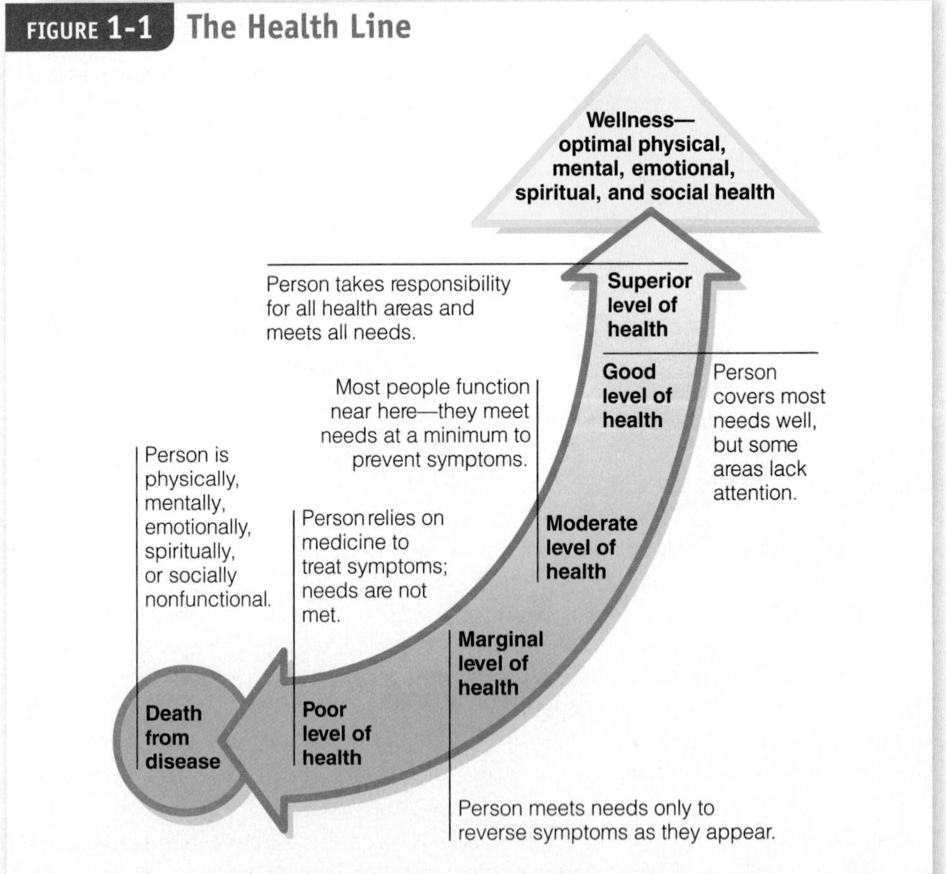
Wellness—optimal physical, mental, emotional, spiritual, and social health
Superior level of health
Person takes responsibility for all health areas and meets all needs.
Good level of health
Person covers most needs well, but some areas lack attention.
Most people function near here—they meet needs at a minimum to prevent symptoms.
Moderate level of health
Marginal level of health
Person meets needs only to reverse symptoms as they appear.
Poor level of health
Person relies on medicine to treat symptoms; needs are not met.
Death from disease
Person is physically, mentally, emotionally, spiritually, or socially nonfunctional.

No matter how well you maintain your health today, you may still be able to improve tomorrow. Likewise, a person who is well today can slip by failing to maintain health-promoting habits.

Food Choices

Sound **nutrition** throughout life does not ensure good health and long life, but it can certainly help to tip the balance in their favor. Nevertheless, most people choose foods for reasons other than their nourishing value. Even people who claim to choose foods primarily for the sake of health or nutrition will admit that other factors also influence their food choices. Because food choices become an integral part of people's lifestyles, they sometimes find it difficult to change their eating habits. Health care professionals who help clients make diet changes must understand the dynamics of food choices because people will alter their eating habits only if their preferences are honored. Developing **cultural competence** is an important aspect of honoring individual preferences, especially for health care professionals who help clients to achieve a nutritious diet.[1]

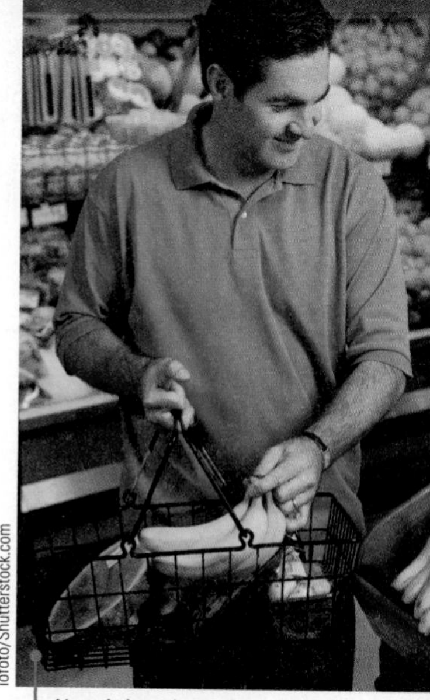
iofoto/Shutterstock.com

Nutrition is only one of the many factors that influence people's food choices.

Preference Why do people like certain foods? One reason, of course, is their preference for certain tastes. Some tastes are widely liked, such as the sweetness of sugar and the zest of salt. Research suggests that genetics influence people's taste preferences, a finding that may eventually have implications for clinical nutrition.[2] For example, sensitivity to bitter taste is an inheritable trait. People born with great sensitivity to bitter tastes tend to avoid foods with bitter flavors such as broccoli, cabbage, brussels sprouts, spinach, and grapefruit juice. These foods, as well as many other fruits and vegetables, contain **bioactive food components—phytochemicals** and nutrients—that may reduce the risk of cancer. (■) Thus, the role that genetics may play in food selection is gaining importance in cancer research.

■ Nutrition in Practice 8 addresses phytochemicals and their role in disease prevention.

Habit Sometimes habit dictates people's food choices. People eat a sandwich for lunch or drink orange juice at breakfast simply because they have always done so.

Associations People also like foods with happy associations—foods eaten in the midst of warm family gatherings on traditional holidays or given to them as children by someone who loved them. By the same token, people can attach intense and unalterable dislikes to foods that they ate when they were sick or that were forced on them when they weren't hungry.

Ethnic Heritage and Regional Cuisines Every country, and every region of a country, has its own typical foods and ways of combining them into meals. The **foodways** of North America reflect the many different cultural and ethnic backgrounds of its inhabitants. Many foods with foreign origins are familiar items on North American menus: tacos, egg rolls, lasagna, sushi, and gyros, to name a few. Still others, such as spaghetti and croissants, are almost staples in the "American diet." North American regional cuisines like Cajun and TexMex blend the traditions of several cultures. Table 1-1 (p. 4) presents selected **ethnic diets** and food choices.

Values Food choices may reflect people's environmental ethics, religious beliefs, and political views. By choosing to eat some foods or to avoid others, people make statements that reflect their values. For example, people may select only foods that come in containers that can be reused or recycled. A concerned consumer may boycott fruit or vegetables picked by migrant workers who have been exploited. People may buy vegetables from local farmers to save the fuel and environmental costs of foods shipped from far away. Labels on some foods carry statements or symbols—known as ecolabels—that imply that the foods have been produced in ways that are considered environmentally favorable.

Religion also influences many people's food choices. Jewish law sets forth an extensive set of dietary rules. Many Christians forgo meat on Fridays during Lent, the period prior to Easter. In Islamic dietary laws, permitted or lawful foods are called *halal*. Other faiths prohibit some dietary practices and promote others. Diet planners can foster sound nutrition practices only if they respect and honor each person's values.

nutrition: the science of foods and the nutrients and other substances they contain, and of their ingestion, digestion, absorption, transport, metabolism, interaction, storage, and excretion. A broader definition includes the study of the environment and of human behavior as it relates to these processes.

cultural competence: an awareness and acceptance of one's own and others' cultures combined with the skills needed to interact effectively with people of diverse cultures.

bioactive food components: compounds in foods (either nutrients or phytochemicals) that alter physiological processes in the body.

phytochemicals (FIGH-toe-CHEM-ih-cals): compounds in plants that confer color, taste, and other characteristics. Some phytochemicals are bioactive food components in functional foods. Nutrition in Practice 8 provides details.

foodways: the eating habits and culinary practices of a people, region, or historical period.

ethnic diets: foodways and cuisines typical of national origins, races, cultural heritages, or geographic locations.

TABLE 1-1 Selected Ethnic Cuisines and Food Choices

	Grains	Vegetables	Fruits	Protein Foods	Milk
Asian © Becky Luigart-Stayner/Corbis	Millet, rice, rice or wheat noodles	Amaranth, baby corn, bamboo shoots, bok choy, cabbages, mung bean sprouts, scallions, seaweed, snow peas, straw mushrooms, water chestnuts, wild yam	Kumquats, loquats, lychee, mandarin oranges, melons, pears, persimmon, plums	Pork, poultry, fish and other seafood, squid, soybeans, tofu, duck eggs, cashews, peanuts	Soy milk
Mediterranean Photodisc, Inc./Getty Images	Bulgur, couscous, focaccia, Italian bread, pastas, pita pocket bread, polenta, rice	Cucumbers, eggplant, grape leaves, onions, peppers, tomatoes	Dates, figs, grapes, lemons, melons, olives, raisins	Beef, gyros, lamb, pork, sausage, chicken, fish and other seafood, fava beans, lentils, almonds, walnuts	Feta, goat, mozzarella, parmesan, provolone, and ricotta cheeses; yogurt
Mexican Mitch Hrdlicka/Photodisc/Getty Images	Taco shells, tortillas (corn or flour), rice	Cactus, cassava, chayote, chilies, corn, jicama, onions, tomatoes, tomato salsa, yams	Avocado, bananas, guava, lemons, limes, mango, oranges, papaya, plantain	Beef, chorizo, chicken, fish, refried beans, eggs	Cheese, flan (caramel custard)

© Cengage Learning

Monkey Business Images/Shutterstock.com

Ethnic meals and family gatherings nourish the spirit as well as the body.

Social Interaction Social interaction is another powerful influence on people's food choices. Meals are social events, and the sharing of food is part of hospitality. It is often considered rude to refuse food or drink being shared by a group or offered by a host. Food brings people together for many different reasons: to celebrate a holiday or special event, to renew an old friendship, to make new friends, to conduct business, and many more. Sometimes food is used to influence or impress someone. For example, a business executive invites a prospective new client out to dinner in hopes of edging out the competition. In each case, for whatever the purpose, food plays an integral part of the social interaction.

Emotional State Some people cannot eat when they are emotionally upset. Others may eat in response to a variety of emotional stimuli—for example, to relieve boredom or depression or to calm anxiety. A depressed person may choose to eat rather than to call a friend. A person who has returned home from an exciting evening out may unwind with a late-night snack. Eating in response to emotions can easily lead to overeating and obesity but may be appropriate at times. For example, sharing food at times of bereavement serves both the giver's need to provide comfort and the receiver's need to be cared for and to interact with others as well as to take nourishment.

Availability, Convenience, and Economy The influence of these factors on people's food selections is clear. You cannot eat foods if they are not available, if you cannot get to the grocery store, if you do not have the time or skill to prepare them, or if you cannot afford them. Consumers who value convenience frequently eat out, bring home ready-to-eat meals, cook meals ahead at storefront meal preparation sites, or have food delivered.[3] Whether decisions based on convenience meet a person's nutrition needs depends on the choices made. Eating a banana or a candy bar may be equally convenient, but the fruit provides more vitamins and minerals and less sugar and fat.

Rising food costs have shifted some consumers' priorities and changed their shopping habits. They are less likely to buy higher priced convenience foods and more likely to prepare home-cooked meals.[4] Those who frequently prepare their own meals eat fast food less often and are more likely to meet dietary guidelines for fat, calcium, fruits, vegetables, and whole grains. Not surprisingly, when eating out, consumers choose low-cost fast-food outlets over more expensive fine-dining restaurants.[5]

Some people have jobs that keep them away from home for days at a time, require them to conduct business in restaurants or at conventions, or involve hectic schedules that allow little or no time for meals at home. For these people, the kinds of restaurants available to them and the cost of eating out so often may limit food choices.

Age Age influences people's food choices. Infants, for example, depend on others to choose foods for them. Older children also rely on others but become more active in selecting foods that taste sweet and are familiar to them and rejecting those whose taste or texture they dislike. In contrast, the links between taste preferences and food choices in adults are less direct than in children. Adults often choose foods based on health concerns such as body weight. Indeed, adults may avoid sweet or familiar foods because of such concerns.

Body Weight and Image Sometimes people select certain foods and supplements that they believe will improve their physical appearance and avoid those they believe might be detrimental. Such decisions can be beneficial when based on sound nutrition and fitness knowledge but may undermine good health when based on fads or carried to extremes. (■)

Medical Conditions Sometimes medical conditions and their treatments (including medications) limit the foods a person can select. For example, a person with heart disease might need to adopt a diet low in certain types of fats. The chemotherapy needed to treat cancer can interfere with a person's appetite or limit food choices by causing vomiting. Allergy to certain foods can also limit choices. The second half of this text discusses how diet can be modified to accommodate different medical conditions.

Health and Nutrition Finally, of course, many consumers make food choices they believe will improve their health.[6] Food manufacturers and restaurant chefs have responded to scientific findings linking health with nutrition by offering an abundant selection of health-promoting foods and beverages. Foods that provide health benefits beyond their nutrient contributions are called **functional foods**.[7] Whole foods—as natural and familiar as oatmeal or tomatoes—are the simplest functional foods. In other cases, foods have been modified through fortifications, enrichment, or enhancement. Examples of these functional foods include orange juice fortified with calcium to build strong bones, bread enriched with folate to promote normal fetal development, and margarine enhanced with a plant sterol to lower blood cholesterol. (■)

Consumers typically welcome new foods into their diets, provided that these foods are reasonably priced, clearly labeled, easy to find in the grocery store, and convenient to prepare. These foods must also taste good—as good as the traditional choices. Of course, a person need not eat any "special" foods to enjoy a healthy diet; many "regular" foods provide numerous health benefits as well. In fact, foods such as whole grains; vegetables and legumes; fruits; meats, seafood, poultry, eggs, nuts, and seeds; and milk products are among the healthiest choices a person can make.

■ Eating disorders are the topic of Nutrition in Practice 7.

■ Nutrition in Practice 8 offers more discussion of functional foods.

functional foods: whole or modified foods that contain bioactive food components believed to provide health benefits, such as reduced disease risks, beyond the benefits that their nutrients contribute. All whole foods are functional in some ways because they provide at least some needed substances, but certain foods stand out as rich sources of bioactive food components.

Review Notes

- A person selects foods for many different reasons.
- Food choices influence health—both positively and negatively. Individual food selections neither make nor break a diet's healthfulness, but the balance of foods selected over time can make an important difference to health.
- In the interest of health, people are wise to think "nutrition" when making their food choices.

The Nutrients

You are a collection of molecules that move. All these moving parts are arranged in patterns of extraordinary complexity and order—cells, tissues, and organs. Although the arrangement remains constant, the parts are continually changing, using **nutrients** and energy derived from nutrients.

Almost any food you eat is composed of dozens or even hundreds of different kinds of materials. Spinach, for example, is composed mostly of water (95 percent), and most of its solid materials are the compounds carbohydrates, fats (properly called lipids), and proteins. If you could remove these materials, you would find a tiny quantity of minerals, vitamins, and other compounds.

■ *Metabolism* is the set of processes by which nutrients are rearranged into body structures or broken down to yield energy.

SIX CLASSES OF NUTRIENTS

Water, carbohydrates, fats, proteins, vitamins, and minerals are the six classes of nutrients commonly found in spinach and other foods. Some of the other materials in foods, such as the pigments and other phytochemicals, are not nutrients but may still be important to health. The body can make some nutrients for itself, at least in limited quantities, but it cannot make them all, and it makes some in insufficient quantities to meet its needs. Therefore, the body must obtain many nutrients from foods. The nutrients that foods must supply are called **essential nutrients**.

■ Food energy can also be measured in kilojoules (kJ). The kilojoule is the international unit of energy. One kcalorie equals 4.2 kJ.

nutrients: substances obtained from food and used in the body to provide energy and structural materials and to serve as regulating agents to promote growth, maintenance, and repair. Nutrients may also reduce the risks of some diseases.

essential nutrients: nutrients a person must obtain from food because the body cannot make them for itself in sufficient quantities to meet physiological needs.

organic: carbon containing. The four organic nutrients are carbohydrate, fat, protein, and vitamins.

energy-yielding nutrients: the nutrients that break down to yield energy the body can use. The three energy-yielding nutrients are carbohydrate, protein, and fat.

inorganic: not containing carbon or pertaining to living things.

calories: units in which energy is measured. Food energy is measured in **kilocalories** (1000 calories equal 1 kilocalorie), abbreviated **kcalories** or kcal. One kcalorie is the amount of heat necessary to raise the temperature of 1 kilogram (kg) of water 1°C. The scientific use of the term *kcalorie* is the same as the popular use of the term *calorie*.

Carbohydrates, Fats, and Proteins

Four of the six classes of nutrients (carbohydrates, fats, proteins, and vitamins) contain carbon, which is found in all living things. They are therefore **organic** (meaning, literally, "alive"). During metabolism, three of these four (carbohydrates, fats, and proteins) provide energy the body can use. (■) These **energy-yielding nutrients** continually replenish the energy you spend daily. Carbohydrates and fats meet most of the body's energy needs; proteins make a significant contribution only when other fuels are unavailable.

Vitamins, Minerals, and Water

Vitamins are organic but do not provide energy to the body. They facilitate the release of energy from the three energy-yielding nutrients. In contrast, minerals and water are **inorganic** nutrients. Minerals yield no energy in the human body, but, like vitamins, they help to regulate the release of energy, among their many other roles. As for water, it is the medium in which all of the body's processes take place.

KCALORIES: A MEASURE OF ENERGY

The amount of energy that carbohydrates, fats, and proteins release can be measured in **calories**—tiny units of energy so small that a single apple provides tens of thousands of them. To ease calculations, energy is expressed in 1000-calorie metric units known as **kilocalories** (shortened to **kcalories**, but commonly called "calories"). When you read in popular books or magazines that an apple provides "100 calories," understand that it means 100 kcalories. This book uses the term *kcalorie* and its abbreviation *kcal* throughout, as do other scientific books and journals. (■) kCalories are not constituents

HOW TO Calculate the Energy a Food Provides

To calculate the energy available from a food, multiply the number of grams of carbohydrate, protein, and fat by 4, 4, and 9, respectively. Then add the results together. For example, one slice of bread with 1 tablespoon of peanut butter on it contains 16 grams of carbohydrate, 7 grams of protein, and 9 grams of fat:

$$16 \text{ g carbohydrate} \times 4 \text{ kcal/g} = 64 \text{ kcal}$$

$$7 \text{ g protein} \times 4 \text{ kcal/g} = 28 \text{ kcal}$$

$$9 \text{ g fat} \times 9 \text{ kcal/g} = 81 \text{ kcal}$$

$$\text{Total} = 173 \text{ kcal}$$

From this information, you can calculate the percentage of kcalories each of the energy nutrients contributes to the total.

To determine the percentage of kcalories from fat, for example, divide the 81 fat kcalories by the total 173 kcalories:

$$81 \text{ fat kcal} \div 173 \text{ total kcal} = 0.468 \text{ (rounded to 0.47)}$$

Then multiply by 100 to get the percentage:

$$0.47 \times 100 = 47\%$$

Dietary recommendations that urge people to limit fat intake to 20 to 35 percent of kcalories refer to the day's total energy intake, not to individual foods. Still, if the proportion of fat in each food choice throughout a day exceeds 35 percent of kcalories, then the day's total surely will, too. Knowing that this snack provides 47 percent of its kcalories from fat alerts a person to the need to make lower-fat selections at other times that day.

of foods; they are a measure of the energy foods provide. The energy a food provides depends on how much carbohydrate, fat, and protein the food contains.

Carbohydrate yields 4 kcalories of energy from each gram, and so does protein. Fat yields 9 kcalories per gram. Thus, fat has a greater **energy density** than either carbohydrate or protein. (■) If you know how many grams of each nutrient a food contains, you can derive the number of kcalories potentially available from the food. Simply multiply the carbohydrate grams times 4, the protein grams times 4, and the fat grams times 9, and add the results together (the accompanying "How to" describes how to calculate the energy a food provides).

Energy Nutrients in Foods

Most foods contain mixtures of all three energy-yielding nutrients, although foods are sometimes classified by their predominant nutrient. To speak of meat as "a protein" or of bread as "a carbohydrate," however, is inaccurate. Each is rich in a particular nutrient, but a protein-rich food such as beef contains a lot of fat along with the protein, and a carbohydrate-rich food such as cornbread also contains fat (corn oil) and protein. Only a few foods are exceptions to this rule, the common ones being sugar (which is pure carbohydrate) and oil (which is pure fat).

Energy Storage in the Body

The body first uses the energy-yielding nutrients to build new compounds and fuel metabolic and physical activities. Excesses are then rearranged into storage compounds, primarily body fat, and put away for later use. Thus, if you take in more energy than you expend, whether from carbohydrate, fat, or protein, the result is an increase in energy stores and weight gain. Similarly, if you take in less energy than you expend, the result is a decrease in energy stores and weight loss.

Alcohol, Not a Nutrient

One other substance contributes energy: alcohol. The body derives energy from alcohol at the rate of 7 kcalories per gram. Alcohol is not a nutrient, however, because it cannot support the body's growth, maintenance, or repair. Nutrition in Practice 19 discusses alcohol's effects on nutrition.

■ Chapter 7 comes back to energy density with regard to weight management.

energy density: a measure of the energy a food provides relative to the amount of food (kcalories per gram).

Review Notes

- Foods provide nutrients—substances that support the growth, maintenance, and repair of the body's tissues.
- The six classes of nutrients are water, carbohydrates, fats, proteins, vitamins, and minerals.
- Vitamins, minerals, and water do not yield energy; instead they facilitate a variety of activities in the body.
- Foods rich in the energy-yielding nutrients (carbohydrates, fats, and proteins) provide the major materials for building the body's tissues and yield energy the body can use or store.
- Energy is measured in kcalories.

■ Appendix B presents nutrient recommendations developed by the World Health Organization.

Dietary Reference Intakes (DRI): a set of values for the dietary nutrient intakes of healthy people in the United States and Canada. These values are used for planning and assessing diets.

Recommended Dietary Allowances (RDA): a set of values reflecting the average daily amounts of nutrients considered adequate to meet the known nutrient needs of practically all healthy people in a particular life stage and gender group; a goal for dietary intake by individuals.

Adequate Intakes (AI): a set of values that are used as guides for nutrient intakes when scientific evidence is insufficient to determine an RDA.

requirement: the lowest continuing intake of a nutrient that will maintain a specified criterion of adequacy.

deficient: in regard to nutrient intake, describes the amount below which almost all healthy people can be expected, over time, to experience deficiency symptoms.

Estimated Average Requirements (EAR): the average daily nutrient intake levels estimated to meet the requirements of half of the healthy individuals in a given age and gender group; used in nutrition research and policymaking and as the basis on which RDA values are set.

Nutrient Recommendations

Nutrient recommendations are used as standards to evaluate healthy people's energy and nutrient intakes. (■) Nutrition experts use the recommendations to assess nutrient intakes and to guide people on amounts to consume. Individuals can use them to decide how much of a nutrient they need to consume.

DIETARY REFERENCE INTAKES

Defining the amounts of energy, nutrients, and other dietary components that best support health is a huge task. Nutrition experts have produced a set of standards that define the amounts of energy, nutrients, other dietary components, and physical activity that best support health. These recommendations are called **Dietary Reference Intakes (DRI)** and reflect the collaborative efforts of scientists in both the United States and Canada.*[8] The inside front covers of this book present the DRI values.

Setting Nutrient Recommendations: RDA and AI

One advantage of the DRI is that they apply to the diets of individuals. The DRI committee offers two sets of values to be used as nutrient intake goals by individuals: a set called the **Recommended Dietary Allowances (RDA)** and a set called **Adequate Intakes (AI)**.

Based on solid experimental evidence and other reliable observations, the RDA are the foundation of the DRI. The AI values are based on less extensive scientific findings and rely more heavily on scientific judgment. The committee establishes an AI value whenever scientific evidence is insufficient to generate an RDA.[9] To see which nutrients have an AI and which have an RDA, turn to the inside front cover.

In the last several decades, abundant new research has linked nutrients in the diet with the promotion of health and the prevention of chronic diseases. An advantage of the DRI is that, where appropriate, they take into account disease prevention as well as an adequate nutrient intake. For example, the RDA for calcium is based on intakes thought to reduce the likelihood of osteoporosis-related fractures later in life.

To ensure that the vitamin and mineral recommendations meet the needs of as many people as possible, the recommendations are set near the top end of the range of the population's estimated average requirements (see Figure 1-2). Small amounts above the daily **requirement** do no harm, whereas amounts below the requirement may lead to health problems. When people's intakes are consistently **deficient**, their nutrient stores decline, and over time this decline leads to deficiency symptoms and poor health.

FIGURE 1-2 Nutrient Intake Recommendations

Nutrients
EAR[a]
RDA
Number of people
20 30 40 50 60 70
Daily requirements (units/day)

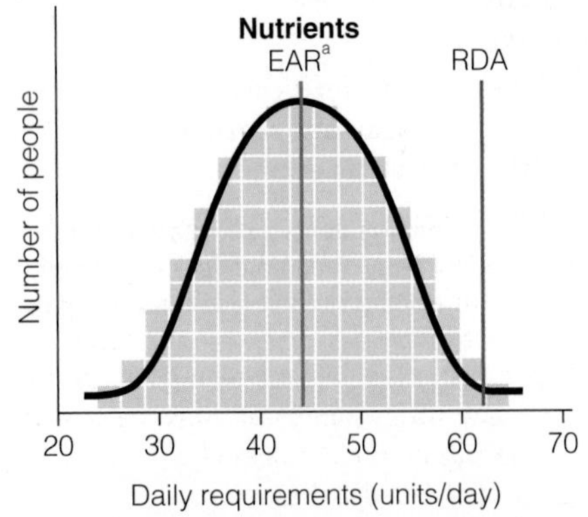

The nutrient intake recommendations are set high enough to cover nearly everyone's requirements (the boxes represent people).

[a]Estimated Average Requirement

Facilitating Nutrition Research and Policy: EAR

In addition to the RDA and AI, the DRI committee has established another set of values: **Estimated Average Requirements (EAR)**. These values establish average requirements for given life stage and gender groups that researchers and nutrition policymakers use in their work. Nutrition scientists may use the EAR as standards in research. Public health officials may use them to assess nutrient intakes of populations and make recommendations. The EAR values form the scientific basis on which the RDA are set.

Establishing Safety Guidelines: UL

The DRI committee also establishes upper limits of intake for nutrients posing a hazard when consumed in excess. These values, the **Tolerable Upper Intake Levels (UL)**, are indispensable to consumers who take supplements. Consumers need

*The DRI reports are produced by the Food and Nutrition Board, Institute of Medicine of the National Academies, with active involvement of scientists from Canada.

to know how much of a nutrient is too much. The UL are also of value to public health officials who set allowances for nutrients that are added to foods and water. The UL values are listed on the inside front cover.

Using Nutrient Recommendations Each of the four DRI categories serves a unique purpose. For example, the EAR are most appropriately used to develop and evaluate nutrition programs for *groups* such as schoolchildren or military personnel. The RDA (or AI, if an RDA is not available) can be used to set goals for *individuals*. The UL help to keep nutrient intakes below the amounts that increase the risk of toxicity. With these understandings, professionals can use the DRI for a variety of purposes.

In addition to understanding the unique purposes of the DRI, it is important to keep their uses in perspective. Consider the following:

- The values are recommendations for safe intakes, not minimum requirements; except for energy, they include a generous margin of safety. Figure 1-3 presents an accurate view of how a person's nutrient needs fall within a range, with marginal and danger zones both below and above the range.
- The values reflect daily intakes to be achieved on average, over time. They assume that intakes will vary from day to day, and they are set high enough to ensure that body nutrient stores will meet nutrient needs during periods of inadequate intakes lasting a day or two for some nutrients and up to a month or two for others.
- The values are chosen in reference to specific indicators of nutrient adequacy, such as blood nutrient concentrations, normal growth, and reduction of certain chronic diseases or other disorders when appropriate, rather than prevention of deficiency symptoms alone.
- The recommendations are designed to meet the needs of most healthy people. Medical problems alter nutrient needs, as later chapters describe.
- The recommendations are specific for people of both genders as well as various ages and stages of life: infants, children, adolescents, men, women, pregnant women, and lactating women.

Tolerable Upper Intake Levels (UL): a set of values reflecting the highest average daily nutrient intake levels that are likely to pose no risk of toxicity to almost all healthy individuals in a particular life stage and gender group. As intake increases above the UL, the potential risk of adverse health effects increases.

Estimated Energy Requirement (EER): the dietary energy intake level that is predicted to maintain energy balance in a healthy adult of a defined age, gender, weight, and physical activity level consistent with good health.

FIGURE 1-3 Naive versus Accurate View of Nutrient Intakes

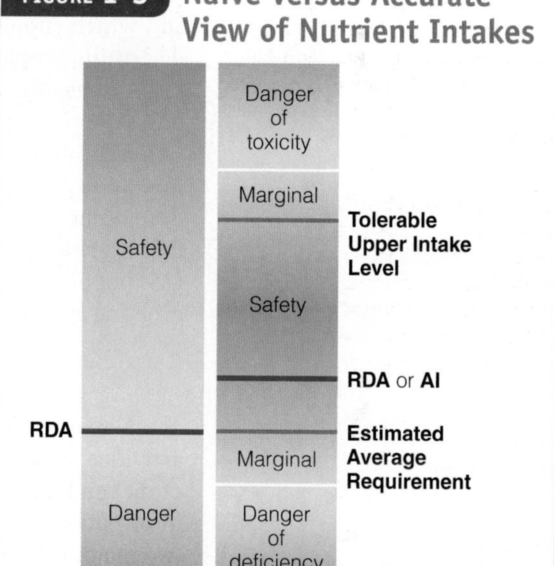

The RDA or AI for a given nutrient represents a point that lies within a range of appropriate and reasonable intakes between toxicity and deficiency. Both of these recommendations are high enough to provide reserves in times of short-term dietary inadequacies but not so high as to approach toxicity. Nutrient intakes above or below this range may be equally harmful.

Setting Energy Recommendations In contrast to the vitamin and mineral recommendations, the recommendation for energy, called the **Estimated Energy Requirement (EER)**, is not generous because excess energy cannot be excreted and is eventually stored as body fat. Rather, the key to the energy recommendation is balance. For a person who has a body weight, body composition, and physical activity level consistent with good health, energy intake from food should match energy expenditure, so the person achieves energy balance. Enough energy is needed to sustain a healthy, active life, but too much energy leads to obesity. The EER is therefore set at a level of energy intake predicted to maintain energy balance in a healthy adult of a defined age, gender, weight, height, and physical activity level.* Another difference between the requirements for other nutrients and those for energy is that each person has an obvious indicator of whether energy intake is inadequate, adequate, or excessive: body weight. Because *any* amount of energy in excess of need leads to weight gain, the DRI committee did not set a Tolerable Upper Intake Level.

*The EER for children, pregnant women, and lactating women includes energy needs associated with the deposition of tissue or the secretion of milk at rates consistent with good health.

ACCEPTABLE MACRONUTRIENT DISTRIBUTION RANGES (AMDR)

As noted earlier, the DRI committee considers prevention of chronic disease as well as nutrient adequacy when establishing recommendations. To that end, the committee established healthy ranges of intakes for the energy-yielding nutrients—carbohydrate, fat, and protein—known as **Acceptable Macronutrient Distribution Ranges (AMDR)**. Each of these three energy-yielding nutrients contributes to a person's total energy (kcalorie) intake, and those contributions vary in relation to each other. The DRI committee has determined that a diet that provides the energy-yielding nutrients in the following proportions provides adequate energy and nutrients and reduces the risk of chronic disease:

- 45 to 65 percent of kcalories from carbohydrate
- 20 to 35 percent of kcalories from fat
- 10 to 35 percent of kcalories from protein

National Nutrition Surveys

How do nutrition experts know whether people are meeting nutrient recommendations? The dietary reference intakes and other major reports that examine the relationships between diet and health depend on information collected from nutrition surveys. Researchers use nutrition surveys to learn which foods people are eating and which supplements they are taking, to assess people's nutritional health, and to determine people's knowledge, attitudes, and behaviors about nutrition and how these relate to health. The resulting wealth of information can be used for a variety of purposes. For example, Congress uses this information to establish public policy on nutrition education, assess food assistance programs, and regulate the food supply. The food industry uses the information to guide decisions in public relations and product development. Scientists use the information to establish research priorities. One of the first nutrition surveys, taken before World War II, suggested that up to a third of the U.S. population might be eating poorly. Programs to correct **malnutrition** have been evolving ever since.

COORDINATING NUTRITION SURVEY DATA

The National Nutrition Monitoring program coordinates the many nutrition-related activities of various federal agencies. All major reports that examine the contribution of diet and nutrition status to the health of the people of the United States depend on information collected and coordinated by this national program. The integration of two major national surveys provides comprehensive data efficiently. One portion of the survey collects data on the kinds and amounts of foods people eat.* Researchers then calculate the energy and nutrients in the foods and compare the amounts consumed with standards such as the DRI. The other portion of the survey examines the people themselves, using nutrition assessment methods.† The data provide valuable information on several nutrition-related conditions such as growth retardation, heart disease, and nutrient deficiencies. These data also provide the basis for developing and monitoring national health goals.

Acceptable Macronutrient Distribution Ranges (AMDR): ranges of intakes for the energy-yielding nutrients that provide adequate energy and nutrients and reduce the risk of chronic disease.

malnutrition: any condition caused by deficient or excess energy or nutrient intake or by an imbalance of nutrients.

Healthy People: a national public health initiative under the jurisdiction of the U.S. Department of Health and Human Services (DHHS) that identifies the most significant preventable threats to health and focuses efforts toward eliminating them.

*This survey was formerly called the Continuing Survey of Food Intakes by Individuals (CSFII), conducted by the U.S. Department of Agriculture (USDA).

†This survey is known as the National Health and Nutrition Examination Survey (NHANES).

NATIONAL HEALTH GOALS

Healthy People is a program that identifies the nation's health priorities and guides policies that promote health and prevent disease. At the start of each decade, the program sets goals for improving the nation's health during the following 10 years. Nutrition is one of 38 topic areas of Healthy People 2020, each with numerous objectives. Table 1-2 lists the nutrition and weight status objectives for 2020.

TABLE 1-2 Healthy People 2020 Nutrition and Weight Status Objectives

- Increase the proportion of adults who are at a healthy weight
- Reduce the proportion of adults who are obese
- Reduce iron deficiency among young children and females of childbearing age
- Reduce iron deficiency among pregnant females
- Reduce the proportion of children and adolescents who are overweight or obese
- Increase the contribution of fruits to the diets of the population aged 2 years and older
- Increase the variety and contribution of vegetables to the diets of the population aged 2 years and older
- Increase the contribution of whole grains to the diets of the population aged 2 years and older
- Reduce consumption of saturated fat in the population aged 2 years and older
- Reduce consumption of sodium in the population aged 2 years and older
- Increase consumption of calcium in the population aged 2 years and older
- Increase the proportion of worksites that offer nutrition or weight management classes or counseling
- Increase the proportion of physician office visits that include counseling or education related to nutrition or weight
- Eliminate very low food security among children in U.S. households
- Prevent inappropriate weight gain in youth and adults
- Increase the proportion of primary care physicians who regularly measure the body mass index of their patients
- Reduce consumption of kcalories from solid fats and added sugars in the population aged 2 years and older
- Increase the number of states that have state-level policies that incentivize food retail outlets to provide foods that are encouraged by the *Dietary Guidelines*
- Increase the number of states with nutrition standards for foods and beverages provided to preschool-aged children in childcare
- Increase the percentage of schools that offer nutritious foods and beverages outside of school meals

Source: www.healthypeople.gov.

Review Notes

- The Dietary Reference Intakes (DRI) are a set of nutrient intake values that can be used to plan and evaluate dietary intakes for healthy people.
- The Estimated Energy Requirement (EER) defines the energy intake level needed to maintain energy balance in a healthy adult of a defined age, gender, weight, height, and physical activity level.
- The Acceptable Macronutrient Distribution Ranges (AMDR) define the proportions contributed by carbohydrate, fat, and protein to a healthy diet.
- Nutrition surveys measure people's food consumption and evaluate the nutrition status of populations.
- Information gathered from nutrition surveys serves as the basis for many major diet and nutrition reports, including Healthy People.

■ A nutritious diet has the following characteristics:

- Adequacy
- Balance
- kCalorie (energy) control
- Nutrient density
- Moderation
- Variety

■ Balance in the diet helps to ensure adequacy.

overnutrition: overconsumption of food energy or nutrients sufficient to cause disease or increased susceptibility to disease; a form of malnutrition.

undernutrition: underconsumption of food energy or nutrients severe enough to cause disease or increased susceptibility to disease; a form of malnutrition.

chronic diseases: diseases characterized by slow progression, long duration, and degeneration of body organs due in part to such personal lifestyle elements as poor food choices, smoking, alcohol use, and lack of physical activity.

eating pattern: customary intake of foods and beverages over time.

adequacy: the characteristic of a diet that provides all the essential nutrients, fiber, and energy necessary to maintain health and body weight.

balance: the dietary characteristic of providing foods in proportion to one another and in proportion to the body's needs.

Dietary Guidelines and Food Guides

Today, government authorities are as much concerned about **overnutrition** as they once were about **undernutrition**. Research confirms that dietary excesses, especially of energy, sodium, certain fats, and alcohol, contribute to many **chronic diseases**, including heart disease, cancer, stroke, diabetes, and liver disease.[10] Only two common lifestyle habits have more influence on health than a person's choice of diet: smoking and other tobacco use, and excessive drinking of alcohol. Table 1-3 lists the leading causes of death in the United States; notice that three of the top four are nutrition related (and related to tobacco use). Note, however, that although diet is a powerful influence on these diseases, they cannot be prevented by a healthy diet alone; genetics, physical activity, age, gender, and other factors also play a role. Within the range set by genetic inheritance, however, disease development is strongly influenced by the foods a person chooses to eat.

Sound nutrition does not depend on the selection of any one food. Instead, it depends on the overall **eating pattern**—the combination of many different foods and beverages at numerous meals over days, months, and years.[11] So how can health care professionals help people select foods to create an eating pattern that supplies all the needed nutrients in amounts consistent with good health? The principle is simple enough: encourage clients to eat a variety of foods that supply all the nutrients the body needs. In practice, how do people do this? It helps to keep in mind that a nutritious diet achieves six basic ideals.

TABLE 1-3 Leading Causes of Death in the United States

The diseases in bold italics are nutrition related.

1. ***Heart disease***
2. ***Cancers***
3. Chronic lung diseases
4. ***Strokes***
5. Accidents
6. Alzheimer's disease
7. ***Diabetes mellitus***
8. Kidney disease
9. Pneumonia and influenza
10. Suicide

Source: S.L. Murphy and coauthors, Deaths: Preliminary data for 2010, *National Vital Statistics Reports* 60 (2012):1–68.

DIETARY IDEALS

A nutritious diet has six characteristics. (■) The first, **adequacy**, was already addressed in the earlier discussion on the DRI. An adequate diet has enough energy and enough of every nutrient (as well as fiber) to meet the needs of healthy people. Second is **balance**: the food choices do not overemphasize one nutrient or food type at the expense of another. (■)

The essential minerals calcium and iron illustrate the importance of dietary balance. Meat is rich in iron but poor in calcium. Conversely, milk is rich in calcium but poor in iron. Use some meat for iron; use some milk for calcium; and save some space for other foods, too, because a diet consisting of milk and meat alone would not be adequate. For other nutrients, people need to consume other protein foods, whole grains, vegetables, and fruit.

The third characteristic is **kcalorie (energy) control**: the foods provide the amount of energy needed to maintain a healthy body weight—not more, not less. The key to kcalorie control is to select foods that deliver the most nutrients for the least food energy. This fourth characteristic is known as **nutrient density**. Consider foods containing calcium, for example. You can get about 300 milligrams of calcium from either 1½ ounces of cheddar cheese or 1 cup of fat-free milk, but the cheese delivers about twice as much food energy (kcalories) as the milk. The fat-free milk, then, is twice as calcium dense as the cheddar cheese; it offers the same amount of calcium for half the kcalories. Both foods are excellent choices for adequacy's sake alone, but to achieve adequacy while controlling kcalories, (■) the fat-free milk is the better choice. (Alternatively, a person could select a low-fat cheddar cheese providing kcalories comparable to fat-free milk.)

Just as a financially responsible person pays for rent, food, clothes, and tuition on a limited budget, healthy people obtain iron, calcium, and all the other essential nutrients on a limited energy (kcalorie) allowance. Success depends on getting many nutrients for each kcalorie "dollar." For example, a can of cola and handful of grapes may both provide about the same number of kcalories, but grapes deliver many more nutrients. A person who makes nutrient-dense choices, such as fruit instead of cola, can meet daily nutrient needs on a lower energy budget. Such choices support good health.

Foods that are notably low in nutrient density—such as potato chips, candy, and colas—are sometimes called **empty-kcalorie foods**. The kcalories these foods provide are called "empty" because they deliver energy (from added sugars, solid fats, or both) with little or no protein, vitamins, or minerals.

The concept of nutrient density is relatively simple when examining the contributions of one nutrient to a food or diet. With respect to calcium, milk ranks high and meats rank low. With respect to iron, meats rank high and milk ranks low. But which food is more nutritious? Answering that question is a more complex task because we need to consider several nutrients—those that may harm health and those that may be beneficial.[12] Ranking foods based on their overall nutrient composition is known as **nutrient profiling**. Researchers have yet to agree on an ideal way to rate foods based on the nutrient profile, but when they do, nutrient profiling will be quite useful in helping consumers identify nutritious foods and plan healthy diets.[13]

The fifth characteristic of a nutritious diet is **moderation**. (■) Foods rich in fat and sugar often provide enjoyment and energy but relatively few nutrients. In addition, they promote weight gain when eaten in excess. A person who practices moderation eats such foods only on occasion and regularly selects foods low in **solid fats** and **added sugars**, a practice that automatically improves nutrient density. Returning to the example of cheddar cheese and fat-free milk, the milk not only offers more calcium for less energy, but it contains far less fat than the cheese.

Finally, the sixth characteristic of a nutritious diet is **variety**: the foods chosen differ from one day to the next. A diet may have all the virtues just described and still lack variety if a person eats the same foods day after day. People should select foods from each of the food groups daily and vary their choices within each food group from day to day for a couple of reasons. First, different foods within the same group contain different arrays of nutrients. Among the fruits, for example, strawberries are especially rich in vitamin C while apricots are rich in vitamin A. Variety improves nutrient adequacy. Second, no food is guaranteed entirely free of substances that, in excess, could be harmful. The strawberries might contain trace amounts of one contaminant, the apricots another. By alternating fruit choices, a person will ingest very little of either contaminant.

DIETARY GUIDELINES FOR AMERICANS

What should a person eat to stay healthy? The answers can be found in the *Dietary Guidelines for Americans*. These guidelines translate the *nutrient* recommendations of the DRI into *food* recommendations.[14] *The Dietary Guidelines for Americans 2010* provide evidence-based advice to help people attain and maintain a healthy weight,

■ Nutrient density promotes adequacy and kcalorie control.

■ Moderation contributes to adequacy, balance, and kcalorie control.

kcalorie (energy) control: management of food energy intake.

nutrient density: a measure of the nutrients a food provides relative to the energy it provides. The more nutrients and the fewer kcalories, the higher the nutrient density.

empty-kcalorie foods: a popular term used to denote foods that contribute energy but lack protein, vitamins, and minerals.

nutrient profiling: ranking foods based on their nutrient composition.

moderation: the provision of enough, but not too much, of a substance.

solid fats: fats that are not usually liquid at room temperature; commonly found in most foods derived from animals and vegetable oils that have been hydrogenated. Solid fats typically contain more saturated and *trans* fats than most oils (Chapter 4 provides more details).

added sugars: sugars, syrups, and other kcaloric sweeteners that are added to foods during processing or preparation or at the table. Added sugars do not include the naturally occurring sugars found in fruits and milk products.

variety (dietary): consumption of a wide selection of foods within and among the major food groups (the opposite of monotony).

reduce the risk of chronic diseases, and promote overall health through diet and physical activity.[15] Table 1-4 presents the key recommendations of the *Dietary Guidelines for Americans 2010*, clustered into four major topic areas. The first area focuses on balancing kcalories to manage a healthy body weight by improving eating habits and engaging in regular physical activity. The second area advises people to reduce their intakes of such foods and food components as sodium, solid fats (and the saturated fats, *trans* fats, and cholesterol they contain), added sugars, refined grain products, and

TABLE 1-4 Key Recommendations of the *Dietary Guidelines for Americans 2010*

Balancing kCalories to Manage Weight

- Prevent and/or reduce overweight and obesity through improved eating and physical activity behaviors.
- Control total kcalorie intake to manage body weight. For people who are overweight or obese, this will mean consuming fewer kcalories from foods and beverages.
- Increase physical activity and reduce time spent in sedentary behaviors.
- Maintain appropriate kcalorie balance during each stage of life—childhood, adolescence, adulthood, pregnancy and breastfeeding, and older age.

Foods and Food Components to Reduce

- Reduce daily sodium intake to less than 2300 milligrams and further reduce intake to 1500 milligrams among persons who are 51 and older and those of any age who are African American or have hypertension, diabetes, or chronic kidney disease.
- Consume less than 10 percent of kcalories from saturated fatty acids by replacing them with monounsaturated and polyunsaturated fatty acids.
- Consume less than 300 milligrams per day of dietary cholesterol.
- Keep *trans* fatty acid consumption as low as possible by limiting foods that contain synthetic sources of *trans* fats, such as partially hydrogenated oils, and by limiting other solid fats.
- Reduce the intake of kcalories from solid fats and added sugars.
- Limit the consumption of foods that contain refined grains, especially refined grain foods that contain solid fats, added sugars, and sodium.
- If alcohol is consumed, it should be consumed in moderation—up to one drink per day for women and two drinks per day for men—and only by adults of legal drinking age.

Foods and Nutrients to Increase

- Increase vegetable and fruit intake.
- Eat a variety of vegetables, especially dark green and red and orange vegetables and beans and peas.
- Consume at least half of all grains as whole grains. Increase whole-grain intake by replacing refined grains with whole grains.
- Increase intake of fat-free or low-fat milk and milk products, such as milk, yogurt, cheese, or fortified soy beverages.
- Choose a variety of protein foods, which include seafood, lean meat and poultry, eggs, beans and peas, soy products, and unsalted nuts and seeds.
- Increase the amount and variety of seafood consumed by choosing seafood in place of some meat and poultry.
- Replace protein foods that are higher in solid fats with choices that are lower in solid fats and kcalories and/or are sources of oils.
- Use oils to replace solid fats where possible.
- Choose foods that provide more potassium, dietary fiber, calcium, and vitamin D, which are nutrients of concern in American diets. These foods include vegetables, fruits, whole grains, and milk and milk products.

Building Healthy Eating Patterns

- Select an eating pattern that meets nutrient needs over time at an appropriate kcalorie level.
- Account for all foods and beverages consumed and assess how they fit within a total healthy eating pattern.
- Follow food safety recommendations when preparing and eating foods to reduce the risk of foodborne illnesses.

Note: These guidelines are intended for adults and healthy children ages 2 and older.

Source: Dietary Guidelines for Americans 2010, available at www.dietaryguidelines.gov.

alcoholic beverages (for those who partake). The third area encourages consumers to select a variety of fruits and vegetables, whole grains, low-fat milk products, and protein foods (including seafood). The fourth area helps consumers build healthy eating patterns that meet energy and nutrient needs while reducing the risk of foodborne illnesses. Together, the *Dietary Guidelines for Americans 2010* point the way toward longer, healthier, and more active lives.

■ Chapter 6 offers a discussion of body mass index (BMI).

Review Notes

- A well-planned diet delivers adequate nutrients, a balanced array of nutrients, and an appropriate amount of energy.
- A well-planned diet is based on nutrient-dense foods, moderate in substances that can be detrimental to health, and varied in its selections.
- The *Dietary Guidelines* apply these principles, offering practical advice on how to eat for good health.

aerobic physical activity: activity in which the body's large muscles move in a rhythmic manner for a sustained period of time. Aerobic activity, also called *endurance activity*, improves cardiorespiratory fitness. Brisk walking, running, swimming, and bicycling are examples.

moderate-intensity physical activity: physical activity that requires some increase in breathing and/or heart rate and expends 3.5 to 7 kcalories per minute. Walking at a speed of 3 to 4.5 miles per hour (about 15 to 20 minutes to walk one mile) is an example.

vigorous-intensity physical activity: physical activity that requires a large increase in breathing and/or heart rate and expends more than 7 kcalories per minute. Walking at a very brisk pace (>4.5 miles per hour) or running at a pace of at least 5 miles per hour are examples.

fitness: the characteristics that enable the body to perform physical activity; more broadly, the ability to meet routine physical demands with enough reserve energy to rise to a physical challenge; or the body's ability to withstand stress of all kinds.

FITNESS GUIDELINES

The *Dietary Guidelines for Americans* (see Table 1-4) also emphasize the benefits of increasing physical activity and reducing sedentary activities to achieve or maintain a healthy body weight and reduce the risk of chronic disease. What does a person have to do to reap the health rewards of physical activity? The *Physical Activity Guidelines for Americans, 2008* (see Table 1-5) specify the minimum amount of **aerobic physical activity** people need to gain substantial *health* benefits.[16] The guidelines recommend longer or shorter times for activity depending on whether the activity is **moderate-intensity physical activity** or **vigorous-intensity physical activity**. For clarity and effectiveness, a minimum length of 10 minutes for short bouts of aerobic physical activity is recommended.[17] Of course, more time and greater intensity bring even greater health benefits—such as maintaining a healthy body weight (BMI of 18.5 to 24.9) and further reducing the risk of chronic diseases. (■)

In addition to providing health benefits, physical activity helps to develop and maintain **fitness**. Table 1-6 (p.16) presents the American College of Sports Medicine (ACSM) guidelines for physical activity.[18] The kinds and amounts of physical activity

TABLE 1-5 *Physical Activity Guidelines for Americans, 2008*

- For substantial health benefits, adults should do at least 150 minutes (2 hours and 30 minutes) a week of moderate-intensity aerobic physical activity, or 75 minutes (1 hour and 15 minutes) a week of vigorous-intensity aerobic activity, or an equivalent combination of moderate- and vigorous-intensity aerobic activity. Aerobic activity should be performed in episodes of at least 10 minutes, and, preferably, it should be spread throughout the week.
- For additional and more extensive health benefits, adults should increase their aerobic physical activity to 300 minutes (5 hours) a week of moderate-intensity, or 150 minutes a week of vigorous-intensity aerobic physical activity, or an equivalent combination of moderate- and vigorous-intensity activity. Additional health benefits are gained by engaging in physical activity beyond this amount.
- Adults should also do muscle-strengthening activities that are moderate or high intensity and involve all major muscle groups on 2 or more days a week, as these activities provide additional health benefits.

Source: "Key Guidelines for Adults" from http://www.health.gov/paguidelines/guidelines/summary.aspx.

TABLE 1-6 American College of Sports Medicine's Guidelines for Physical Activity

	Cardiorespiratory	Strength	Flexibility
	Steve Cole/Photodisc/ Getty Images	© David Hanover Photography	© David Hanover Photography
Type of activity	Aerobic activity that uses large-muscle groups and can be maintained continuously	Resistance activity that is performed at a controlled speed and through a full range of motion	Stretching activity that uses the major muscle groups
Frequency	5 to 7 days per week	2 or more nonconsecutive days per week	2 to 7 days per week
Intensity	Moderate (equivalent to walking at a pace of 3 to 4 miles per hour)[a]	Enough to enhance muscle strength and improve body composition	Enough to feel tightness or slight discomfort
Duration	At least 30 minutes per day	8 to 12 repetitions of 8 to 10 different exercises (minimum)	2 to 4 repetitions of 15 to 30 seconds per muscle group
Examples	Running, cycling, swimming, inline skating, rowing, power walking, cross-country skiing, kickboxing, jumping rope; sports activities such as basketball, soccer, racquetball, tennis, volleyball	Pull-ups, push-ups, weight lifting, pilates	Yoga

[a]For those who prefer vigorous-intensity aerobic activity such as walking at a very brisk pace (>4.5 mph) or running (5 mph), a minimum of 20 minutes per day, 3 days per week is recommended.

Source: American College of Sports Medicine position stand, Quantity and quality of exercise for developing and maintaining cardiorespiratory, musculoskeletal, and neuromuscular fitness in apparently healthy adults: Guidance for prescribing exercise, *Medicine & Science in Sports & Exercise* 43 (2011): 1334–1359; W.L. Haskell and coauthors, Physical activity and public health: Updated recommendation for adults from the American College of Sports Medicine and the American Heart Association, *Medicine & Science in Sports & Exercise* 39 (2007): 1423–1434.

that improve physical fitness also provide still greater health benefits (substantially lower risk of premature death compared with those who are inactive and improved body composition, for example).[19]

Extensive evidence confirms that regular physical activity promotes health and reduces the risk of developing a number of diseases.[20] Despite an increasing awareness of the health benefits that physical activity confers, however, less than half of adults in the United States meet physical activity recommendations.[21] A sedentary lifestyle is linked to the major degenerative diseases—heart disease, cancer, stroke, diabetes, and hypertension—the primary killers of adults in developed countries.[22] Therefore, one of the most important challenges health care professionals face is to motivate more people to become physically active. To motivate others, health care professionals must first become more physically active themselves, thereby enhancing their own health. Second, they can include regular physical activity as a component of therapy for their clients. As a person becomes physically fit, the health of the entire body improves. In general, physically fit people enjoy:

- *Restful sleep.* Rest and sleep occur naturally after periods of physical activity. During rest, the body repairs injuries, disposes of wastes generated during activity, and builds new physical structures.
- *Nutritional health.* Physical activity expends energy and thus allows people to eat more food. If they choose wisely, active people will consume more nutrients and be less likely to develop nutrient deficiencies.
- *Optimal body composition.* A balanced program of physical activity limits body fat and increases or maintains lean tissue. Thus, physically active people have relatively less body fat than sedentary people at the same body weight.[23]

- *Optimal bone density.* Weight-bearing physical activity builds bone strength and protects against osteoporosis.[24]
- *Resistance to colds and other infectious diseases.* Fitness enhances immunity.[*25]
- *Low risks of some types of cancers.* Lifelong physical activity may help to protect against colon cancer, breast cancer, and some other cancers.[26]
- *Strong circulation and lung function.* Physical activity that challenges the heart and lungs strengthens the circulatory system.
- *Low risk of cardiovascular disease.* Physical activity lowers blood pressure, slows resting pulse rate, and lowers blood cholesterol, thus reducing the risks of heart attacks and strokes.[27] Some research suggests that physical activity may reduce the risk of cardiovascular disease in another way as well—by reducing intra-abdominal fat stores.[28]
- *Low risk of type 2 diabetes.* Physical activity normalizes glucose tolerance.[29] Regular physical activity reduces the risk of developing type 2 diabetes and benefits those who already have the condition.
- *Reduced risk of gallbladder disease.* Regular physical activity reduces the risk of gallbladder disease—perhaps by facilitating weight control and lowering blood lipid levels.[30]
- *Low incidence and severity of anxiety and depression.* Physical activity may improve one's mood and enhance the quality of life by reducing depression and anxiety.[31]
- *Strong self-image.* The sense of achievement that comes from meeting physical challenges promotes self-confidence.
- *Long life and high quality of life in the later years.* Active people live longer, healthier lives than sedentary people do.[32] Even as little as 15 minutes a day of moderate-intensity activity can add years to a person's life.[33] In addition to extending longevity, physical activity supports independence and mobility in later life by reducing the risk of falls and minimizing the risk of injury should a fall occur.[34]

VILevi/Shutterstock.com

Physical activity helps you look good, feel good, and have fun, and it brings many long-term health benefits as well.

Review Notes

- Regular physical activity promotes health and reduces risk of chronic disease.
- The *Physical Activity Guidelines, 2008* recommend at least 30 minutes of physical activity each day for health benefits and 60 minutes or more for maintaining body weight and still greater health benefits.
- The ACSM has issued recommendations for physical activity to develop and maintain physical fitness.

THE USDA FOOD PATTERNS

To help people achieve the goals set forth by the *Dietary Guidelines for Americans* (see Table 1-4), the USDA provides a **food group plan**—the **USDA Food Patterns**—that builds a diet from categories of foods that are similar in vitamin and mineral content. (■) Thus, each group provides a set of nutrients that differs somewhat from the nutrients supplied by the other groups. Selecting foods from each of the groups eases the task of creating an adequate and balanced diet.

Figure 1-4 (on pp. 18–19) presents the USDA Food Patterns. The USDA Food Patterns assign foods to five major food groups and recommend daily amounts of foods from each group to meet nutrient needs. (■) The judicious use of oils is also described in the USDA Food Patterns. In addition to presenting the food groups, Figure 1-4 indicates the most notable nutrients of each group, the serving equivalents, and the foods within each group. (■)

■ The DASH Eating Plan, presented in Chapter 21, is another dietary pattern that meets the goals of the *Dietary Guidelines for Americans*.

■ Five food groups:
- Fruits
- Vegetables
- Grains
- Protein foods
- Milk and milk products

■ Chapter 11 provides a food guide for children, and Appendix B presents Canada's food group plan, *Eating Well with Canada's Food Guide*.

food group plan: a diet-planning tool that sorts foods into groups based on nutrient content and then specifies that people should eat certain amounts of food from each group.

USDA Food Patterns: the USDA's food group plan for ensuring dietary adequacy that assigns foods to five major food groups.

*Moderate physical activity can stimulate immune function. Intense, vigorous, prolonged activity such as marathon running, however, may compromise immune function.

FIGURE 1-4 USDA Food Patterns: Food Groups and Subgroups

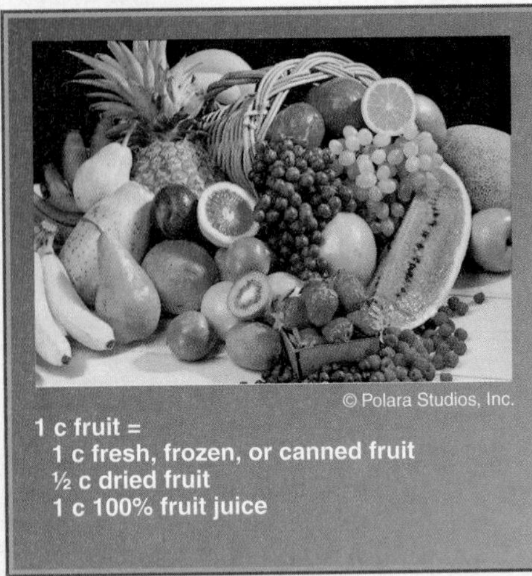

© Polara Studios, Inc.

1 c fruit =
1 c fresh, frozen, or canned fruit
½ c dried fruit
1 c 100% fruit juice

Fruits contribute folate, vitamin A, vitamin C, potassium, and fiber.

Consume a variety of fruits, and choose whole or cut-up fruits more often than fruit juice.

Apples, apricots, avocados, bananas, blueberries, cantaloupe, cherries, grapefruit, grapes, guava, honeydew, kiwi, mango, nectarines, oranges, papaya, peaches, pears, pineapples, plums, raspberries, strawberries, tangerines, watermelon; dried fruit (dates, figs, prunes, raisins); 100% fruit juices

Limit these fruits that contain solid fats and/or added sugars:
Canned or frozen fruit in syrup; juices, punches, ades, and fruit drinks with added sugars; fried plantains

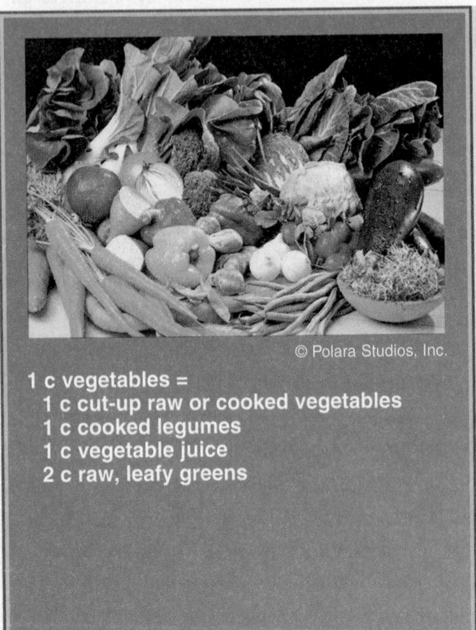

© Polara Studios, Inc.

1 c vegetables =
1 c cut-up raw or cooked vegetables
1 c cooked legumes
1 c vegetable juice
2 c raw, leafy greens

Vegetables contribute folate, vitamin A, vitamin C, vitamin K, vitamin E, magnesium, potassium, and fiber.

Consume a variety of vegetables each day, and choose from all five subgroups several times a week.

Dark-green vegetables: Broccoli and leafy greens such as arugula, beet greens, bok choy, collard greens, kale, mustard greens, romaine lettuce, spinach, turnip greens, watercress

Red and orange vegetables: Carrots, carrot juice, pumpkin, red bell peppers, sweet potatoes, tomatoes, tomato juice, vegetable juice, winter squash (acorn, butternut)

Legumes: Black beans, black-eyed peas, garbanzo beans (chickpeas), kidney beans, lentils, navy beans, pinto beans, soybeans and soy products such as tofu, split peas, white beans

Starchy vegetables: Cassava, corn, green peas, hominy, lima beans, potatoes

Other vegetables: Artichokes, asparagus, bamboo shoots, bean sprouts, beets, brussels sprouts, cabbages, cactus, cauliflower, celery, cucumbers, eggplant, green beans, green bell peppers, iceberg lettuce, mushrooms, okra, onions, seaweed, snow peas, zucchini

Limit these vegetables that contain solid fats and/or added sugars:
Baked beans, candied sweet potatoes, coleslaw, french fries, potato salad, refried beans, scalloped potatoes, tempura vegetables

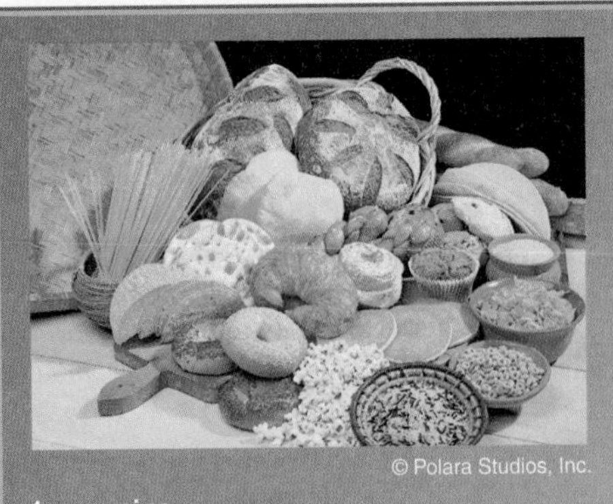

© Polara Studios, Inc.

1 oz grains =
1 slice bread
½ c cooked rice, pasta, or cereal
1 oz dry pasta or rice
1 c ready-to-eat cereal
3 c popped popcorn

Grains contribute folate, niacin, riboflavin, thiamin, iron, magnesium, selenium, and fiber.

Make most (at least half) of the grain selections whole grains.

Whole grains: amaranth, barley, brown rice, buckwheat, bulgur, cornmeal, millet, oats, quinoa, rye, wheat, wild rice and whole-grain products such as breads, cereals, crackers, and pastas; popcorn

Enriched refined products: bagels, breads, cereals, pastas (couscous, macaroni, spaghetti), pretzels, white rice, rolls, tortillas

Limit these grains that contain solid fats and/or added sugars:
Biscuits, cakes, cookies, cornbread, crackers, croissants, doughnuts, fried rice, granola, muffins, pastries, pies, presweetened cereals, taco shells

continued

continued

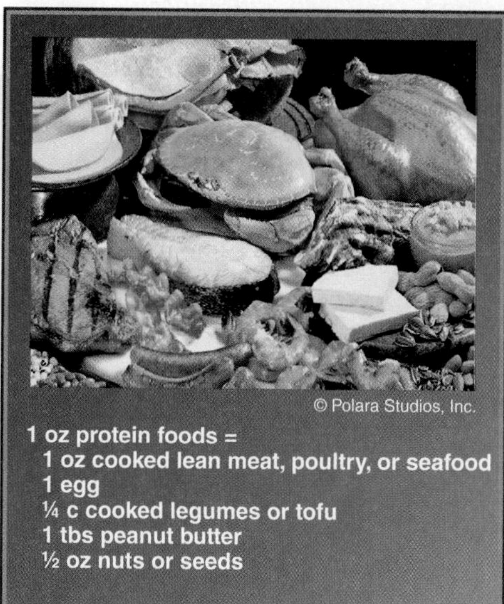

© Polara Studios, Inc.

1 oz protein foods =
1 oz cooked lean meat, poultry, or seafood
1 egg
¼ c cooked legumes or tofu
1 tbs peanut butter
½ oz nuts or seeds

Protein foods contribute protein, essential fatty acids, niacin, thiamin, vitamin B_6, vitamin B_{12}, iron, magnesium, potassium, and zinc.

Choose a variety of protein foods from the three subgroups, including seafood in place of meat or poultry twice a week.

Seafood: Fish (catfish, cod, flounder, haddock, halibut, herring, mackerel, pollock, salmon, sardines, sea bass, snapper, trout, tuna), shellfish (clams, crab, lobster, mussels, oysters, scallops, shrimp)

Meats, poultry, eggs: Lean or low-fat meats (fat-trimmed beef, game, ham, lamb, pork, veal), poultry (no skin), eggs

Nuts, seeds, soy products: Unsalted nuts (almonds, cashews, filberts, pecans, pistachios, walnuts), seeds (flaxseeds, pumpkin seeds, sesame seeds, sunflower seeds), legumes, soy products (textured vegetable protein, tofu, tempeh), peanut butter, peanuts

Limit these protein foods that contain solid fats and/or added sugars:
Bacon; baked beans; fried meat, seafood, poultry, eggs, or tofu; refried beans; ground beef; hot dogs; luncheon meats; marbled steaks; poultry with skin; sausages; spare ribs

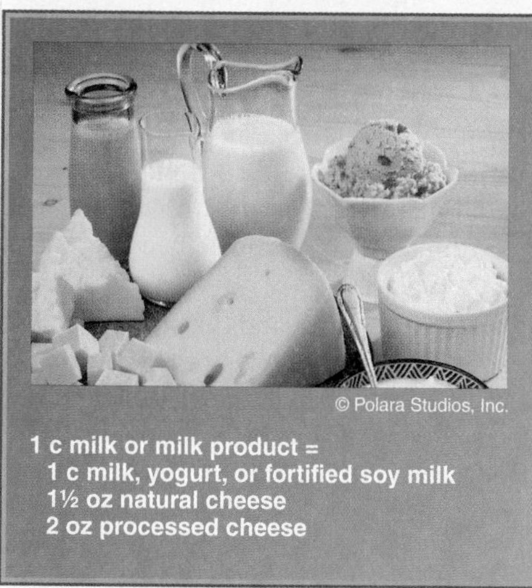

© Polara Studios, Inc.

1 c milk or milk product =
1 c milk, yogurt, or fortified soy milk
1½ oz natural cheese
2 oz processed cheese

Milk and milk products contribute protein, riboflavin, vitamin B_{12}, calcium, potassium, and, when fortified, vitamin A and vitamin D.

Make fat-free or low-fat choices. Choose other calcium-rich foods if you don't consume milk.

Fat-free or 1% low-fat milk and fat-free or 1% low-fat milk products such as buttermilk, cheeses, cottage cheese, yogurt; fat-free fortified soy milk

Limit these milk products that contain solid fats and/or added sugars:
2% reduced-fat milk and whole milk; 2% reduced-fat and whole-milk products such as cheeses, cottage cheese, and yogurt; flavored milk with added sugars such as chocolate milk, custard, frozen yogurt, ice cream, milk shakes, pudding, sherbet; fortified soy milk

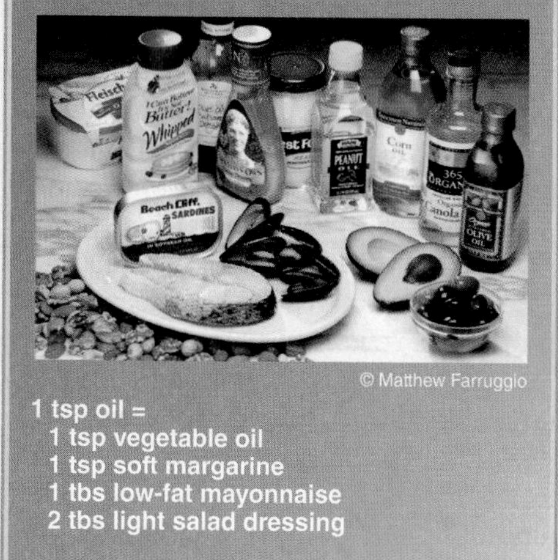

© Matthew Farruggio

1 tsp oil =
1 tsp vegetable oil
1 tsp soft margarine
1 tbs low-fat mayonnaise
2 tbs light salad dressing

Oils are not a food group, but are featured here because they contribute vitamin E and essential fatty acids.

Use oils instead of solid fats, when possible.

Liquid vegetable oils such as canola, corn, flaxseed, nut, olive, peanut, safflower, sesame, soybean, sunflower oils; mayonnaise, oil-based salad dressing, soft *trans*-free margarine; unsaturated oils that occur naturally in foods such as avocados, fatty fish, nuts, olives, seeds (flaxseeds, sesame seeds), shellfish

Limit these solid fats:
Butter, animal fats, stick margarine, shortening

TABLE 1-7 USDA Food Patterns: Recommended Daily Amounts from Each Food Group

Food Group	1600 kcal	1800 kcal	2000 kcal	2200 kcal	2400 kcal	2600 kcal	2800 kcal	3000 kcal
Fruits	1½ c	1½ c	2 c	2 c	2 c	2 c	2½ c	2½ c
Vegetables	2 c	2½ c	2½ c	3 c	3 c	3½ c	3½ c	4 c
Grains	5 oz	6 oz	6 oz	7 oz	8 oz	9 oz	10 oz	10 oz
Protein foods	5 oz	5 oz	5½ oz	6 oz	6½ oz	6½ oz	7 oz	7 oz
Milk	3 c	3 c	3 c	3 c	3 c	3 c	3 c	3 c
Oils	5 tsp	5 tsp	6 tsp	6 tsp	7 tsp	8 tsp	8 tsp	10 tsp
Discretionary kcalories	121 kcal	161 kcal	258 kcal	266 kcal	330 kcal	362 kcal	395 kcal	459 kcal

Recommended Daily Food Amounts All food groups offer valuable nutrients, and people should make selections from each group daily. Table 1-7 specifies the amounts of food needed from each group daily to create a healthful diet for several energy (kcalorie) levels. (■) A person needing 2000 kcalories a day, for example, would select 2 cups of fruit; 2½ cups of vegetables; 6 ounces of grain foods; 5½ ounces of protein foods; and 3 cups of milk or milk products.* Additionally, a small amount of unsaturated oil, such as vegetable oil or the oils of nuts, olives, or fatty fish, is required to supply needed nutrients. Estimated daily kcalorie needs for sedentary and active men and women are shown in Table 1-8.

■ Chapter 6 explains how to determine energy needs.

A portion of grains is 1 ounce, yet most bagels today weigh 4 ounces or more—meaning that a single bagel can easily supply four or more portions of grains, not one, as many people assume.

All vegetables provide an array of vitamins, fiber, and the mineral potassium, but some vegetables are especially good sources of certain nutrients and beneficial phytochemicals. For this reason, the vegetable group is sorted into five subgroups. The dark green vegetables deliver the B vitamin folate; the red and orange vegetables provide vitamin A; legumes supply iron and protein; the starchy vegetables contribute carbohydrate energy; and the other vegetables fill in the gaps and add more of these same nutrients.

In a 2000-kcalorie diet, then, the recommended 2½ cups of daily vegetables should be varied among the subgroups over a week's time. In other words, eating 2½ cups of potatoes or even nutrient-rich spinach every day for seven days does *not* meet the recommended vegetable intakes. Potatoes and spinach make excellent choices when consumed in balance with vegetables from the other subgroups. One way to help ensure selections for all of the subgroups is to eat vegetables of various colors—for example, green broccoli, orange sweet potatoes, black beans, yellow corn, and white cauliflower. Intakes of vegetables are appropriately averaged over a week's time—it isn't necessary to include every subgroup every day.

For similar reasons, the protein foods group is sorted into three subgroups. Perhaps most notably, each of these subgroups contributes a different assortment of fats. Table 1-9 presents the recommended *weekly* amounts for each of the subgroups for vegetables and protein foods.

Notable Nutrients As Figure 1-4 notes, each food group contributes key nutrients. This feature provides flexibility in diet planning because a person can select any

* Milk and milk products also can be referred to as dairy products.

TABLE 1-8 Estimated Daily kCalorie Needs for Adults

	Sedentary[a]	Active[b]
Women		
19–30 yr	2000	2400
31–50 yr	1800	2200
51+ yr	1600	2100
Men		
19–30 yr	2500	3000
31–50 yr	2300	2900
51+ yr	2100	2600

[a]*Sedentary* describes a lifestyle that includes only the activities typical of day-to-day life.

[b]*Active* describes a lifestyle that includes physical activity equivalent to walking more than 3 miles per day at a rate of 3 to 4 miles per hour, in addition to the activities typical of day-to-day life. kCalorie values for active people reflect the midpoint of the range appropriate for age and gender, but within each group older adults may need fewer kcalories and younger adults may need more. In addition to gender, age, and activity level, energy needs vary with height and weight (see Chapter 6).

food from a food group (or its subgroup) and receive similar nutrients. For example, a person can choose milk, cheese, or yogurt and receive the same key nutrients. Importantly, foods provide not only these key nutrients, but small amounts of other nutrients and phytochemicals as well.

Legumes contribute the same key nutrients—notably protein, iron, and zinc—as meats, poultry, and seafood. They are also excellent sources of fiber, folate, and potassium, which are commonly found in vegetables. To encourage frequent consumption

legumes (lay-GYOOMS, LEG-yooms): plants of the bean and pea family with seeds that are rich in protein compared with other plant-derived foods.

TABLE 1-9 USDA Food Patterns: Recommended Weekly Amounts from the Vegetable and Protein Foods Subgroups

Vegetable Subgroups	1600 kcal	1800 kcal	2000 kcal	2200 kcal	2400 kcal	2600 kcal	2800 kcal	3000 kcal
Dark green	1½ c	1½ c	1½ c	2 c	2 c	2½ c	2½ c	2½ c
Red and orange	4 c	5½ c	5½ c	6 c	6 c	7 c	7 c	7½ c
Legumes	1 c	1½ c	1½ c	2 c	2 c	2½ c	2½ c	3 c
Starchy	4 c	5 c	5 c	6 c	6 c	7 c	7 c	8 c
Other	3½ c	4 c	4 c	5 c	5 c	5½ c	5½ c	7 c
Protein Foods Subgroups								
Seafood	8 oz	8 oz	8 oz	9 oz	10 oz	10 oz	11 oz	11 oz
Meats, poultry, eggs	24 oz	24 oz	26 oz	29 oz	31 oz	31 oz	34 oz	34 oz
Nuts, seeds, soy products	4 oz	4 oz	4 oz	4 oz	5 oz	5 oz	5 oz	5 oz

Note: Table 1-7 specifies the recommended amounts of total vegetables and protein foods per *day*. This table shows those amounts dispersed among five vegetable and three protein foods subgroups per *week*.

of these nutrient-rich foods, legumes are included as a subgroup of both the vegetable group and the protein foods group. Thus, legumes can be counted in either the vegetable group or the protein foods group.[35] In general, people who regularly eat meat, poultry, and seafood count legumes as a vegetable, and vegetarians and others who seldom eat meat, poultry, or seafood count legumes in the protein foods group.

The USDA Food Patterns encourage greater consumption from certain food groups to provide the nutrients most often lacking in the diets of Americans. (■) In general, most people need to eat:

- More vegetables, fruits, whole grains, seafood, and fat-free or low-fat milk and milk products.
- Less sodium, saturated fat, *trans* fat, and cholesterol, and *fewer* refined grains and foods and beverages with solid fats and added sugars.

Nutrient-Dense Choices A healthy eating pattern emphasizes nutrient-dense options within each food group. By consistently selecting nutrient-dense foods, a person can obtain all the nutrients needed and still keep kcalories under control. In contrast, eating foods that are low in nutrient density makes it difficult to get enough nutrients without exceeding energy needs and gaining weight. For this reason, consumers should select low-fat foods from each group and foods without added fats or sugars—for example, fat-free milk instead of whole milk, baked chicken without the skin instead of hot dogs, green beans instead of french fries, orange juice instead of fruit punch, and whole-wheat bread instead of biscuits. Notice that Figure 1-4 indicates which foods *within each group* contain solid fats and/or added sugars and therefore should be limited. Oil is a notable exception: even though oil is pure fat and therefore rich in kcalories, a small amount of oil from sources such as nuts, fish, or vegetable oils is necessary every day to provide nutrients lacking from other foods. Consequently, these high-fat foods are listed among the nutrient-dense foods (see Nutrition in Practice 4 to learn why).

■ Nutrients of concern:
- Dietary fiber
- Vitamin D
- Calcium
- Potassium

discretionary kcalories: the kcalories remaining in a person's energy allowance after consuming enough nutrient-dense foods to meet all nutrient needs for a day.

Discretionary kCalories At each kcalorie level, people who consistently choose nutrient-dense foods may be able to meet their nutrient needs without consuming their full allowance of kcalories. The difference between the kcalories needed to supply nutrients and those needed to maintain weight might be considered **discretionary kcalories** (see Figure 1-5).

Discretionary kcalories allow a person to choose whether to:

- Eat additional nutrient-dense foods, such as an extra serving of skinless chicken or a second ear of corn.
- Select a few foods with fats or added sugars, such as reduced-fat milk or sweetened cereal.
- Add a little fat or sugar to foods, such as butter or jelly on toast.
- Consume some alcohol. (Nutrition in Practice 19 explains why this may not be a good choice for some individuals.)

FIGURE 1-5 Discretionary kCalories in a 2000-kCalorie Diet

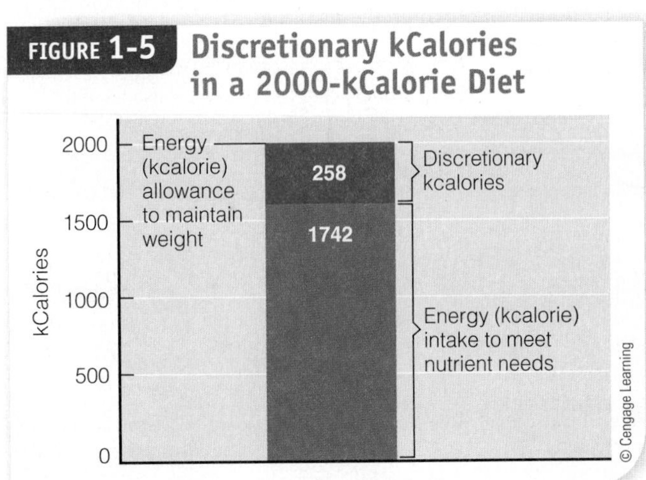

Alternatively, a person wanting to lose weight might choose to:

- *Not* use discretionary kcalories.

Serving Equivalents Recommended daily amounts for fruits, vegetables, and milk are measured in cups and those for grains and protein foods, in ounces. Figure 1-4 provides the equivalent measures for foods that are not readily measured in cups and ounces. For example, 1 ounce of grains is considered equivalent to 1 slice of bread or ½ cup of cooked rice.

Consumers using the USDA Food Patterns can learn how standard serving sizes compare with their personal

portion sizes by determining the answers to questions such as these: What portion of a cup is a small handful of raisins? Is a "helping" of mashed potatoes more or less than a half cup? How many ounces of cereal do you typically pour into the bowl? How many ounces is the steak at your favorite restaurant? How many cups of milk does your glass hold? The margin offers some tips for estimating portion sizes. (■)

Mixtures of Foods Some foods—such as casseroles, soups, and sandwiches—fall into two or more food groups. With a little practice, users can learn to divide these foods into food groups. From the USDA Food Patterns point of view, a taco represents four different food groups: the taco shell from the grains group; the onions, lettuce, and tomatoes from the vegetable group; the ground beef from the protein foods group; and the cheese from the milk group.

Vegetarian Food Guide Vegetarian diets are plant-based eating patterns that rely mainly on grains, vegetables, legumes, fruits, seeds, and nuts. Some vegetarian diets include eggs, milk products, or both. People who do not eat meats or milk products can still use the USDA Food Patterns to create an adequate diet.[36] The subgroups for protein foods have been reorganized to eliminate meats, poultry, and seafood (see Table NP5-1 on p. 138). The other food groups and the recommended daily amounts for each food group remain the same. Nutrition in Practice 5 defines vegetarian terms and provides details on planning healthy vegetarian diets.

Ethnic Food Choices People can use the USDA Food Patterns and still enjoy a diverse array of culinary styles by sorting ethnic foods into their appropriate food groups. For example, a person eating Mexican foods would find tortillas in the grains group, jicama in the vegetable group, and guava in the fruit group. Table 1-1 (p. 4) features ethnic food choices.

■ For quick and easy estimates, visualize each portion as being about the size of a common object:

- ¼ c dried fruit or nuts = a golf ball
- 1 c fruit or vegetables = a baseball
- 3 oz meat = a deck of cards
- 1 oz cheese = 4 stacked dice
- ½ c ice cream = a racquetball
- 2 tbs peanut butter = a ping pong ball

portion size: the quantity of food served or eaten at one meal or snack; *not* a standard amount.

MYPLATE

The USDA created an educational tool called MyPlate to illustrate the five food groups and remind consumers to make healthy food choices. The MyPlate icon, shown in Figure 1-6, divides a plate into four sections, each representing a food group—fruits, vegetables, grains, and protein foods. The sections vary in size, indicating the relative proportion each food group contributes to a healthy diet. A circle next to the plate represents the milk group (dairy).

The MyPlate icon does not stand alone as an educational tool. A wealth of information can be found at the website (chooseMyPlate.gov). The USDA's MyPlate online suite of information makes applying the USDA Food Patterns easier. Consumers can create a personal profile to estimate kcalorie needs and determine the kinds and amounts of foods they need to eat each day based on their height, weight, age, gender, and activity level. Information is also available for children, pregnant and lactating women, and vegetarians. In addition to creating a personal plan, consumers can find daily tips to help them improve their diet and increase physical activity. A key message of the website is to enjoy food, but eat less by avoiding oversized portions.

FIGURE 1-6 MyPlate

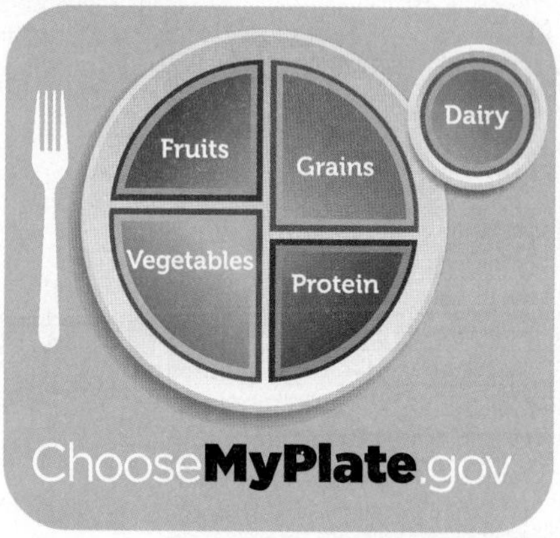

Note that vegetables and fruits occupy half the plate and that the grains portion is slightly larger than the portion of protein foods.

Source: USDA, www.choosemyplate.gov.

Review Notes

- Food group plans such as the USDA Food Patterns help consumers select the types and amounts of foods to provide adequacy, balance, and variety in the diet.
- Each food group contributes key nutrients, a feature that provides flexibility in diet planning.
- MyPlate is an educational tool used to illustrate the five food groups.

Food Labels

Today, consumers know more about the links between diet and disease than they did in the past, and they are demanding still more information on disease prevention. Many people rely on food labels to help them select foods with less saturated fat, *trans* fat, cholesterol, and sodium and more vitamins, minerals, and dietary fiber.[37] Food labels appear on virtually all packaged foods, and posters or brochures provide similar nutrition information for fresh fruits, vegetables, and seafoods. Under a new ruling, packages of meat cuts, ground meats, and poultry also must display nutrition information.[38] Figure 1-7 illustrates the requirements for label information. A few foods need not carry nutrition labels: those contributing few nutrients, such as plain coffee, tea, and spices; foods produced by small businesses; and foods prepared and sold in the same establishment.

Household and metric measures:
- 1 teaspoon (tsp) = 5 milliliters (mL)
- 1 tablespoon (tbs) = 15 mL
- 1 cup (c) = 240 mL
- 1 fluid ounce (fl oz) = 30 mL
- 1 ounce (oz) = 28 grams (g)

THE INGREDIENT LIST

All packaged foods must list *all* ingredients on the label in descending order of predominance by weight. Knowing that the first ingredient predominates by weight, consumers can glean much information. Compare these products, for example:

- A beverage powder that contains "sugar, citric acid, natural flavors . . ." versus a juice that contains "water, tomato concentrate, concentrated juices of carrots, celery. . . ."
- A cereal that contains "puffed milled corn, sugar, corn syrup, molasses, salt . . ." versus one that contains "100 percent rolled oats. . . ."

In each comparison, consumers can tell that the second product is the more nutrient dense.

FIGURE 1-7 Example of a Food Label

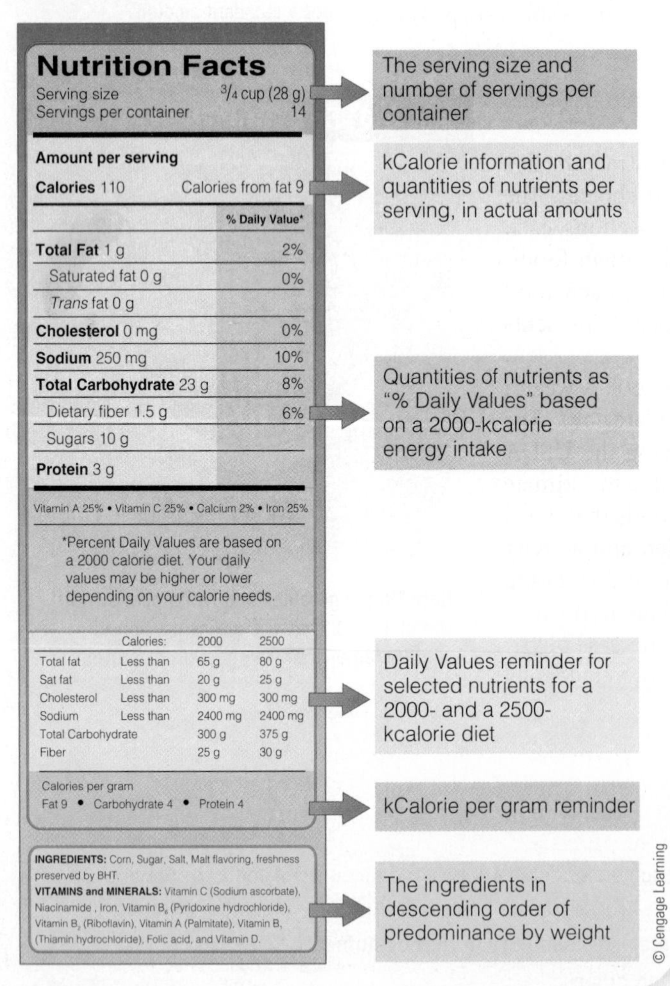

NUTRITION FACTS PANEL

The Food and Drug Administration (FDA) requires food labels to include key nutrition facts. The "Nutrition Facts" panel provides such information as serving sizes, Daily Values, and nutrient quantities.

Serving Sizes Because labels present nutrient information per serving, they must identify the size of a serving. The FDA has established specific serving sizes for various foods and requires that all labels for a given product use the same serving size. For example, the serving size for all ice creams is ½ cup and for all beverages, 8 fluid ounces. This facilitates comparison shopping. Consumers can see at a glance which brand has more or fewer kcalories or grams of fat, for example. Standard serving sizes are expressed in both common household measures, such as cups, and metric measures, such as milliliters, to accommodate users of both types of measures (see the margin).

When examining the nutrition information on a food label, consumers need to consider how the serving size compares with the actual quantity eaten. If it is not the same, they will need to adjust the quantities accordingly. For example, if the serving size is four cookies and you only eat two, then you need to cut the nutrient and kcalorie values in half; similarly, if you eat eight cookies, then you need to double the values. Notice, too, that small bags or individually wrapped items, such as chips or candy bars, may contain more than a single serving. The number of servings per container is listed just below the serving size.

The Daily Values To help consumers evaluate the information found on labels, the FDA created a set of nutrient standards called the **Daily Values** specifically for use on food labels. The Daily Values do two things: they set adequacy standards for nutrients that are desirable in the diet such as protein, vitamins, minerals, and fiber, and they also set moderation standards for other nutrients that must be limited, such as fat, saturated fat, cholesterol, and sodium.

The "% Daily Value" column on a label provides a ballpark estimate of how individual foods contribute to the total diet. It compares key nutrients in a serving of food with the daily goals of a person consuming 2000 kcalories. Most labels list, at the bottom, Daily Values for both a 2000-kcalorie and a 2500-kcalorie diet, but the "% Daily Value" column on all labels applies only to a 2000-kcalorie diet. Although the Daily Values are based on a 2000-kcalorie diet, people's actual energy intakes vary widely; some people need fewer kcalories, and some people need many more. This makes the Daily Values most useful for comparing one food with another and less useful as nutrient intake targets for individuals. By examining a food's general nutrient profile, however, a person can determine whether the food contributes "a little" or "a lot" of a nutrient, whether it contributes "more" or "less" than another food, and how well it fits into the consumer's overall diet.

Nutrient Quantities In addition to the serving size and the servings per container, the FDA requires that the "Nutrition Facts" panel on a label present nutrient information in two ways—in quantities (such as grams) and as percentages of the Daily Values. The Nutrition Facts panel must provide the nutrient amount, percent Daily Value, or both for the following:

Daily Values: reference values developed by the FDA specifically for use on food labels.

- Total food energy (kcalories)
- Food energy from fat (kcalories)
- Total fat (grams and percent Daily Value)
- Saturated fat (grams and percent Daily Value)
- *Trans* fat (grams)
- Cholesterol (milligrams and percent Daily Value)
- Sodium (milligrams and percent Daily Value)
- Total carbohydrate, including starch, sugar, and fiber (grams and percent Daily Value)
- Dietary fiber (grams and percent Daily Value)
- Sugars (grams), including both those naturally present in and those added to the food
- Protein (grams)

The labels must also present nutrient content information as a percentage of the Daily Values for the following vitamins and minerals:

- Vitamin A
- Vitamin C
- Iron
- Calcium

The FDA developed the Daily Values for use on food labels because comparing nutrient amounts against a standard helps make them meaningful to consumers. A person might wonder, for example, whether 1 milligram of iron or calcium is a little or a lot. As Table 1-10 shows, the Daily Value for iron is 18 milligrams, so 1 milligram of iron is enough to take notice of: it is more than 5 percent. But the Daily Value for calcium on food labels is 1000 milligrams, so 1 milligram of calcium is a negligible amount.

TABLE 1-10 Daily Values for Food Labels

Food labels must present the "% Daily Value" for these nutrients.

Food Component	Daily Value	Calculation Factors
Fat	65 g	30% of kcal
Saturated fat	20 g	10% of kcal
Cholesterol	300 mg	—
Carbohydrate (total)	300 g	60% of kcal
Fiber	25 g	11.5 g per 1000 kcal
Protein	50 g	10% of kcal
Sodium	2400 mg	—
Potassium	3500 mg	—
Vitamin C	60 mg	—
Vitamin A	1500 µg	—
Calcium	1000 mg	—
Iron	18 mg	—

Note: Daily Values were established for adults and children 4 years old and older. The values for energy-yielding nutrients are based on 2000 kcalories a day. For fiber, the Daily Value was rounded up from 23.

CLAIMS ON LABELS

In addition to the Nutrition Facts panel, consumers may find various claims on labels. These claims include nutrient claims, health claims, and structure-function claims.

Nutrient Claims

Have you noticed phrases such as "good source of fiber" on a box of cereal or "rich in calcium" on a package of cheese? These and other **nutrient claims** may be used on labels only if the claims meet FDA definitions, which include the conditions under which each term can be used. For example, in addition to having less than 2 milligrams of cholesterol, a "cholesterol-free" product may not contain more than 2 grams of saturated fat and *trans* fat combined per serving. Table 1-11 defines nutrient terms on food labels, including criteria for foods described as "low," "reduced," and "free."

nutrient claims: statements that characterize the quantity of a nutrient in a food.

TABLE 1-11 Terms Used on Food Labels

General Terms
free: "nutritionally trivial" and unlikely to have a physiological consequence; synonyms include *without, no,* and *zero*. A food that does not contain a nutrient naturally may make such a claim but only as it applies to all similar foods (for example, "applesauce, a fat-free food").
good source of: the product provides between 10 and 19 percent of the Daily Value for a given nutrient per serving.
healthy: a food that is low in fat, saturated fat, cholesterol, and sodium and that contains at least 10 percent of the Daily Values for vitamin A, vitamin C, iron, calcium, protein, or fiber.
high: 20 percent or more of the Daily Value for a given nutrient per serving; synonyms include *rich in* or *excellent source.*
less: at least 25 percent less of a given nutrient or kcalories than the comparison food (see individual nutrients); synonyms include *fewer* and *reduced.*
light or lite: one-third fewer kcalories than the comparison food; 50 percent or less of the fat or sodium than the comparison food; any use of the term other than as defined must specify what it is referring to (for example, "light in color" or "light in texture").
low: an amount that would allow frequent consumption of a food without exceeding the Daily Value for the nutrient. A food that is naturally low in a nutrient may make such a claim but only as it applies to all similar foods (for example, "fresh cauliflower, a low-sodium food"); synonyms include *little, few,* and *low source of.*
more: at least 10 percent more of the Daily Value for a given nutrient than the comparison food; synonyms include *added* and *extra.*
organic (on food labels): at least 95 percent of the product's ingredients have been grown and processed according to USDA regulations defining the use of fertilizers, herbicides, insecticides, fungicides, preservatives, and other chemical ingredients.
Energy
kcalorie-free: fewer than 5 kcalories per serving.
low kcalorie: 40 kcalories or less per serving.
reduced kcalorie: at least 25 percent fewer kcalories per serving than the comparison food.
Fat and Cholesterol[a]
percent fat free: may be used only if the product meets the definition of *low fat* or *fat free* and must reflect the amount of fat in 100 grams (for example, a food that contains 2.5 grams of fat per 50 grams can claim to be "95 percent fat free").
fat free: less than 0.5 gram of fat per serving (and no added fat or oil); synonyms include *zero-fat, no-fat,* and *nonfat.*

continued

continued

Fat and Cholesterol[a]
low fat: 3 grams or less fat per serving.
less fat: at least 25 percent less fat than the comparison food.
saturated fat free: less than 0.5 gram of saturated fat and 0.5 gram of *trans* fat per serving.
low saturated fat: 1 gram or less saturated fat and less than 0.5 gram of *trans* fat per serving.
less saturated fat: at least 25 percent less saturated fat and *trans* fat combined than the comparison food.
***trans* fat free:** less than 0.5 gram of *trans* fat and less than 0.5 gram of saturated fat per serving.
cholesterol-free: less than 2 milligrams cholesterol per serving and 2 grams or less saturated fat and *trans* fat combined per serving.
low cholesterol: 20 milligrams or less cholesterol per serving and 2 grams or less saturated fat and *trans* fat combined per serving.
less cholesterol: at least 25 percent less cholesterol than the comparison food (reflecting a reduction of at least 20 milligrams per serving), and 2 grams or less saturated fat and *trans* fat combined per serving.
extra lean: less than 5 grams of fat, 2 grams of saturated fat and *trans* fat combined, and 95 milligrams of cholesterol per serving and per 100 grams of meat, poultry, and seafood.
lean: less than 10 grams of fat, 4.5 grams of saturated fat and *trans* fat combined, and 95 milligrams of cholesterol per serving and per 100 grams of meat, poultry, and seafood. For mixed dishes such as burritos and sandwiches, less than 8 grams of fat, 3.5 grams of saturated fat, and 80 milligrams of cholesterol per reference amount customarily consumed.
Carbohydrates: Fiber and Sugar
high fiber: 5 grams or more fiber per serving. A high-fiber claim made on a food that contains more than 3 grams fat per serving and per 100 grams of food must also declare total fat.
sugar-free: less than 0.5 gram of sugar per serving.
Sodium
sodium-free and **salt-free:** less than 5 milligrams of sodium per serving.
low sodium: 140 milligrams or less per serving.
very low sodium: 35 milligrams or less per serving.

[a]Foods containing more than 13 grams total fat per serving or per 50 grams of food must indicate those contents immediately after a cholesterol claim. As you can see, all cholesterol claims are prohibited when the food contains more than 2 grams saturated fat and *trans* fat combined per serving.

Some descriptions *imply* that a food contains, or does not contain, a nutrient. Implied claims are prohibited unless they meet specified criteria. For example, a claim that a product "contains no oil" implies that the food contains no fat. If the product is truly fat free, then it may make the no-oil claim, but if it contains another source of fat, such as butter, it may not.

Health Claims

Until recently, the FDA held manufacturers to the highest standards of scientific evidence before allowing them to place **health claims** on food labels.[39] When a label stated, "diets low in sodium may reduce the risk of high blood pressure," for example, consumers could be sure that the FDA had examined much scientific

health claims: statements that characterize the relationship between a nutrient or other substance in food and a disease or health-related condition.

TABLE 1-12 Reliable Health Claims on Food Labels

• Calcium and reduced risk of osteoporosis
• Sodium and reduced risk of hypertension
• Dietary saturated fat and cholesterol and reduced risk of coronary heart disease
• Dietary fat and reduced risk of cancer
• Fiber-containing grain products, fruits, and vegetables and reduced risk of cancer
• Fruits, vegetables, and grain products that contain fiber, particularly soluble fiber, and reduced risk of coronary heart disease
• Fruits and vegetables and reduced risk of cancer
• Folate and reduced risk of neural tube defects
• Sugar alcohols and reduced risk of tooth decay
• Soluble fiber from whole oats and from psyllium seed husk and reduced risk of heart disease
• Soy protein and reduced risk of heart disease
• Whole grains and reduced risk of heart disease and certain cancers
• Plant sterol and plant stanol esters and reduced risk of heart disease
• Potassium and reduced risk of hypertension and stroke

evidence and found substantial support for the claim. Such reliable health claims still appear on food labels, and they have a high degree of scientific validity (see Table 1-12).

Today, however, the FDA also allows other claims backed by weaker evidence to appear on labels. These are *qualified* claims, in the sense that labels bearing them must also state the strength of the scientific evidence backing them up. Unfortunately, most people are not knowledgeable enough to distinguish between scientifically reliable claims those that are best ignored.[40]

Structure-Function Claims

Unlike health claims, which require food manufacturers to collect scientific evidence and petition the FDA, **structure-function claims** can be made without any FDA approval. Product labels can claim to "slow aging," "improve memory," and "build strong bones" without any proof. The only criterion for a structure-function claim is that it must not mention a disease or symptom. Unfortunately, structure-function claims can be deceptively similar to health claims. Consider these statements:

- "May reduce the risk of heart disease."
- "Promotes a healthy heart."

Although most consumers do not distinguish between these two types of claims, the first is a health claim that requires FDA approval, whereas the second is an unproven, but legal, structure-function claim. Figure 1-8 compares the three types of label claims.

structure-function claims: statements that describe how a product may affect a structure or function of the body; for example, "calcium builds strong bones." Structure-function claims do not require FDA authorization.

FIGURE 1-8 Label Claims

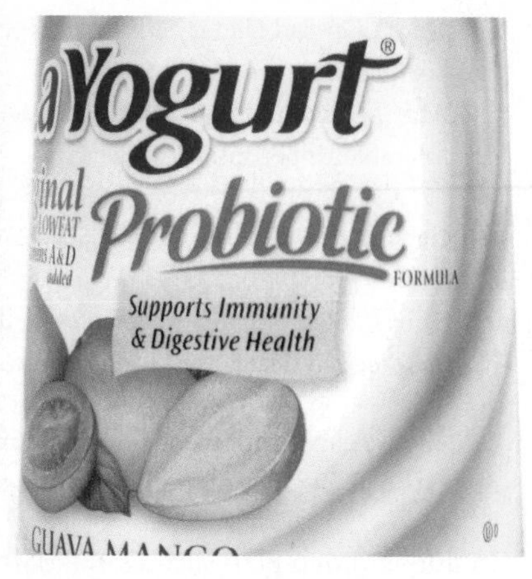

© Same Kolich/Bill Smith Group/Cengage Learning

Nutrient claims characterize the level of a nutrient in the food—for example, "fat free" or "less sodium."

Health claims characterize the relationship of a food or food component to a disease or health-related condition—for example, "soluble fiber from oatmeal daily in a diet low in saturated fat and cholesterol may reduce the risk of heart disease" or "a diet low in total fat may reduce the risk of some cancers."

Structure/function claims describe the effect that a substance has on the structure or function of the body and do not make reference to a disease—for example, "supports immunity and digestive health" or "calcium builds strong bones."

Review Notes

- Food labels list the ingredients, the serving size, the number of kcalories provided, and the key nutrient quantities in a food—information consumers need to select foods that will help them meet their nutrition and health goals.
- Daily Values are a set of nutrient standards created by the FDA for use on food labels.
- Reliable health claims are backed by the highest standards of scientific evidence.

Self Check

1. When people eat the foods typical of their families or geographic area, their choices are influenced by:
   a. occupation.
   b. nutrition.
   c. emotional state.
   d. ethnic heritage or regional cuisine.

2. The energy-yielding nutrients are:
   a. fats, minerals, and water.
   b. minerals, proteins, and vitamins.
   c. carbohydrates, fats, and vitamins.
   d. carbohydrates, fats, and proteins.

3. The inorganic nutrients are:
   a. proteins and fats.
   b. vitamins and minerals.
   c. minerals and water.
   d. vitamins and proteins.

4. Alcohol is not a nutrient because:
   a. the body derives no energy from it.
   b. it is organic.
   c. it is converted to body fat.
   d. it does not contribute to the body's growth or repair.

5. The nutrient standards in use today include all of the following except:
   a. Recommended Dietary Allowances (RDA).
   b. Adequate Intakes (AI).
   c. Daily Minimum Requirements (DMR).
   d. Tolerable Upper Intake Levels (UL).

6. Which of the following is consistent with the *Dietary Guidelines for Americans*?
   a. Limit intakes of fruits, vegetables, and whole grains.
   b. Increase physical activity and reduce time spent in sedentary activities.
   c. Choose a diet with plenty of whole-milk products.
   d. Eat an abundance of foods to ensure nutrient adequacy.

7. In a food group plan such as the USDA Food Patterns, foods within a given food group are similar in their contents of:
   a. energy.
   b. proteins and fibers.
   c. vitamins and minerals.
   d. carbohydrates and fats.

8. A slice of apple pie supplies 350 kcalories with 3 grams of fiber; an apple provides 80 kcalories and the same 3 grams of fiber. This is an example of:
   a. kcalorie control.
   b. nutrient density.
   c. variety.
   d. essential nutrients.

9. According to the USDA Food Patterns, which of the following vegetables should be limited?
   a. carrots
   b. avocados
   c. baked beans
   d. potatoes

10. Food labels list ingredients in:
    a. alphabetical order.
    b. ascending order of predominance by weight.
    c. descending order of predominance by weight.
    d. manufacturer's order of preference.

Answers to these questions appear in Appendix H. For more chapter review: Access an interactive eBook, chapter-specific interactive learning tools, including flashcards, quizzes, videos, and more in your Nutrition CourseMate, accessed through CengageBrain.com.

Clinical Applications

1. Make a list of the foods and beverages you've consumed in the past two days. Look at each item on your list and consider why you chose the particular food or beverage you did. Did you eat cereal for breakfast because that's what you always eat (habit), or because it was the easiest, quickest food to prepare (convenience)? Did you put fat-free milk on the cereal because you want to control your energy intake (nutrition)? In going down your list, you may be surprised to discover exactly why you chose certain foods.

2. As a nurse, you can uncover clues about a client's food choices by paying close attention. You may be surprised to discover why a client chooses certain foods, but you can then use this knowledge to serve the best interests of the client. For example, an elderly, undernourished widower may eat the same sandwich for lunch every day. In talking with the client, you discover that this is what he and his wife fixed together each day. Consider ways you might be able to help the client learn to eat other foods and vary his choices.

3. Using the list of foods and beverages from exercise #1, compare your day's intake with the USDA Food Patterns. Did you vary your choices within each food group? Did your intake match the daily recommended amounts from each group? If not, list some changes you could have made to meet the recommendations.

Nutrition on the Net

For further study of the topics in this chapter, access these websites.

- Search for "nutrition" at the U.S. government health information site:
  www.healthfinder.gov
- Review the Dietary Reference Intakes:
  www.nap.edu
- Review nutrient recommendations from the Food and Agriculture Organization and the World Health Organization:
  www.fao.org and www.who.int
- Learn more about the *Dietary Guidelines for Americans*:
  www.dietaryguidelines.gov
- View Canadian information on nutrition guidelines and food labels:
  www.hc-sc.gc.ca
- Visit MyPlate:
  www.choosemyplate.gov
- See food pyramids for various ethnic groups at Oldways Preservation and Exchange Trust:
  www.oldwayspt.org
- View the Healthy People objectives for the nation:
  www.healthypeople.gov
- Learn more about food labeling from the Food and Drug Administration:
  www.fda.gov/food

Notes

1. K. Stein, Navigating cultural competency: In preparation for an expected standard in 2010, *Journal of the American Dietetic Association* 110 (2010): S13–S20.
2. E. R. Grimm and N. I. Steinle, Genetics of eating behavior: Established and emerging concepts, *Nutrition Reviews* 69 (2011): 52–60; J. E. Hayes and R. S. Keast, Two decades of supertasting: Where do we stand? *Physiology and Behavior* 104 (2011): 1072–1074; J. R. Krebs, The gourmet ape: Evolution and human food preferences, *American Journal of Clinical Nutrition* 90 (2009): 707S–711S; Q. Y. Chen and coauthors, Perceptual variation in umami taste and polymorphisms in *TAS1R* taste receptor genes, *American Journal of Clinical Nutrition* 90 (2009): 770S–779S; B. J. Tepper, Nutritional implications of genetic taste variation: The role of PROP sensitivity and other taste phenotypes, *Annual Review of Nutrition* 28 (2008): 367–388.
3. J. E. Tillotson, Fast food—through the ages: Part 1, *Nutrition Today* 43 (2008): 70–74.
4. J. Tillotson, American's food shopping in today's lousy economy (part 3), *Nutrition Today* 44 (2009): 265–268.
5. J. Tillotson, Americans' food shopping in today's lousy economy (part 2), *Nutrition Today* 44 (2009): 218–221.
6. International Food Information Council Foundation, *2011 Food & Health Survey*, www.foodinsight.org, accessed October 16, 2012.
7. Position of the American Dietetic Association: Functional foods, *Journal of the American Dietetic Association* 109 (2009): 735–746.
8. Committee on Dietary Reference Intakes, *Dietary Reference Intakes for Calcium and Vitamin D* (Washington, DC: National Academies Press, 2011); Committee on the Scientific Evaluation of Dietary Reference Intakes, *Food and Nutrition Board, Institute of Medicine, Dietary Reference Intakes for Water, Potassium, Sodium, Chloride, and Sulfate* (Washington, DC: National Academies Press, 2004); Committee on the Scientific Evaluation of Dietary Reference Intakes, *Food and Nutrition Board, Institute of Medicine, Dietary Reference Intakes for Energy, Carbohydrate, Fiber, Fat, Fatty Acids, Cholesterol, Protein, and Amino Acids* (Washington, DC: National Academies Press, 2005); Committee on the Scientific Evaluation of Dietary Reference Intakes, Food and Nutrition Board, Institute of Medicine, *Dietary Reference Intakes for Vitamin A, Vitamin K, Arsenic, Boron, Chromium, Copper, Iodine, Iron, Manganese, Molybdenum, Nickel, Silicon, Vanadium, and Zinc* (Washington, DC: National Academy Press, 2001); Committee on the Scientific Evaluation of Dietary Reference Intakes, Food and Nutrition Board, Institute of Medicine, *Dietary Reference Intakes for Vitamin C, Vitamin E, Selenium, and Carotenoids* (Washington, DC: National Academy Press, 2000); Committee on the Scientific Evaluation of Dietary Reference Intakes, Food and Nutrition Board, Institute of Medicine, *Dietary Reference Intakes for Thiamin, Riboflavin, Niacin, Vitamin B6, Folate, Vitamin B12, Pantothenic Acid, Biotin, and Choline* (Washington, DC: National Academy Press, 1998); Committee on the Scientific Evaluation of Dietary Reference Intakes, Food and Nutrition Board, Institute of Medicine, *Dietary Reference Intakes for Calcium, Phosphorus, Magnesium, Vitamin D, and Fluoride* (Washington, DC: National Academy Press, 1997).
9. Institute of Medicine, *Dietary Reference Intakes: The Essential Guide to Nutrient Requirements*, eds., J. J. Otten, J. P. Hellwig, and L. D. Meyers (Washington, DC: National Academies Press, 2006), pp. 5–18.
10. V. L. Roger and coauthors, Heart disease and stroke statistics—2012 update: A report from the American Heart Association, *Circulation* 125 (2012): e12–e230; F. Magkos and coauthors, Management of the metabolic syndrome and type 2 diabetes through lifestyle modification, *Annual Review of Nutrition* 29 (2009): 223–256; S. S. Gidding and coauthors, Implementing American Heart Association pediatric and adult nutrition guidelines: A scientific statement from the American Heart Association Nutrition Committee of the Council on Nutrition, Physical Activity and Metabolism, Council on Cardiovascular Disease in the Young, Council on Arteriosclerosis, Thrombosis and Vascular Biology, Council on Cardiovascular Nursing, Council on Epidemiology and Prevention, and Council for High Blood Pressure Research, *Circulation* 119 (2009): 1161–1175; A. Galimanis and coauthors, Lifestyle and stroke risk: A review, *Current Opinion in Neurology* 22 (2009): 60–68; World Cancer Research Fund/American Institute for Cancer Research, *Food, Nutrition, Physical Activity, and the Prevention of Cancer: A Global Perspective* (Washington, DC: World Cancer Research Fund/American Institute for Cancer Research, 2007), pp. 66–196.
11. Position of the American Dietetic Association, Total diet approach to communicating food and nutrition information, *Journal of the American Dietetic Association* 107 (2007): 1224–1232.
12. A. Drewnowski and coauthors, Achieve better health with nutrient-rich foods, *Nutrition Today* 47 (2012): 23–29; G. D. Miller and coauthors, It is time for a positive approach to dietary guidance using nutrient density as a basic principle, *Journal of Nutrition* 139 (2009): 1198–1202.
13. Drewnowski and coauthors, 2012; C. Kapica, The science behind current nutrition profiling systems to promote consumer intake of nutrient-dense foods, *American Journal of Clinical Nutrition* 91 (2010): entire supplement; N. Darmon and coauthors, Nutrient profiles discriminate between foods according to their contribution to nutritionally adequate diets: A validation study using linear programming and the SAIN, LIM system, *American Journal of Clinical Nutrition* 89 (2009): 1227–1236.
14. D. Mozaffarian and D. S. Ludwig, Dietary guidelines in the 21st century: A time for food, *Journal of the American Medical Association* 304 (2010): 681–682.
15. U.S. Department of Agriculture and U.S. Department of Health and Human Services, *Dietary Guidelines for Americans, 2010,* available at www.dietaryguidelines.gov.

16. Centers for Disease Control and Prevention, www.cdc.gov/physical activity/everyone, site updated March 30, 2011.
17. American College of Sports Medicine position stand, Quantity and quality of exercise for developing and maintaining cardiorespiratory, musculoskeletal, and neuromotor fitness in apparently healthy adults: Guidance for prescribing exercise, *Medicine and Science in Sports and Exercise* 43 (2011): 1334–1359; W. L. Haskell and coauthors, Physical activity and public health: Updated recommendation for adults from the American College of Sports Medicine and the American Heart Association, *Medicine and Science in Sports and Exercise* 39 (2007): 1423–1434.
18. American College of Sports Medicine position stand, 2011; Haskell and coauthors, 2007.
19. Centers for Disease Control and Prevention, www.cdc.gov/physicalactivity/everyone, 2011; Haskell and coauthors, 2007.
20. M. Hamer and coauthors, Physical activity and cardiovascular mortality risk: Possible protective mechanisms, *Medicine and Science in Sports and Exercise* 44 (2012): 84–88; Centers for Disease Control and Prevention, www.cdc.gov/physicalactivity/everyone, 2011; J. Sattelmair and coauthors, Dose response between physical activity and risk of coronary heart disease: A meta-analysis, *Circulation* 124 (2011): 789–795; A. K. Chomistek and coauthors, Vigorous physical activity, mediating biomarkers, and risk of myocardial infarction, *Medicine and Science in Sports and Exercise* 43 (2011): 1884–1890; H.-K. Na and S. Oliynyk, Effects of physical activity on cancer prevention, *Annals of the New York Academy of Sciences* 1229 (2011): 176–183.
21. J. S. Schiller and coauthors, Summary health statistics for U.S. adults: National Health Interview Survey, 2010, *Vital and Health Statistics* 10 (252), 2012, 1–207.
22. C. E. Matthews and coauthors, Amount of time spent in sedentary behaviors and cause-specific mortality in US adults, *American Journal of Clinical Nutrition* 95 (2012): 437–445; A. Grøntved and F. B. Hu, Television viewing and risk of type 2 diabetes, cardiovascular disease, and all-cause mortality, *Journal of the American Medical Association* 305 (2011): 2448–2455; T. Y. Warren and coauthors, Sedentary behaviors increase risk of cardiovascular disease mortality in men, *Medicine and Science in Sports and Exercise* 42 (2010): 879–885; A. V. Patel and coauthors, Leisure time spent sitting in relation to total mortality in prospective cohort of adults, *American Journal of Epidemiology* 172 (2010): 419–429; P. T. Katzmarzyk and coauthors, Sitting time and mortality from all causes, cardiovascular disease, and cancer, *Medicine and Science in Sports and Exercise* 41 (2009): 998–1005.
23. American College of Sports Medicine, Position paper; Appropriate physical activity intervention strategies for weight loss and prevention of weight regain for adults, *Medicine and Science in Sports and Exercise* 41 (2009): 459–471; K. S. Vimaleswaran and coauthors, Physical activity attenuates the body mass index-increasing influence of genetic variation in the *FTO* gene, *American Journal of Clinical Nutrition* 90 (2009): 425-428.
24. R. S. Rector and coauthors, Lean body mass and weight-bearing activity in the prediction of bone mineral density in physically active men, *Journal of Strength and Conditioning* 23 (2009): 427–435; A. Guadalupe-Grau and coauthors, Exercise and bone mass in adults, *Sports Medicine* 39 (2009): 439–468.
25. J. Romeo and coauthors, Physical activity, immunity and infection, *Proceedings of the Nutrition Society* 69 (2010): 390–399.
26. World Cancer Research Fund/American Institute for Cancer Research, *Continuous Update Project Interim Report Summary: Food, Nutrition, Physical Activity, and the Prevention of Colorectal Cancer* (Washington, DC: AICR, 2011); X. Sui and coauthors, Influence of cardiorespiratory fitness on lung cancer mortality, *Medicine and Science in Sports and Exercise* 42 (2010): 872–878; J. B. Peel and coauthors, A prospective study of cardiorespiratory fitness and breast cancer mortality, *Medicine and Science in Sports and Exercise* 41 (2009): 742–748; S. Y. Pan and M. DesMeules, Energy intake, physical activity energy balance, and cancer: Epidemiologic evidence, *Methods in Molecular Biology* 472 (2009): 191–215; World Cancer Research Fund/American Institute for Cancer Research, *Food, Nutrition, Physical Activity, and the Prevention of Cancer*, 2007, pp. 244–321.
27. Hamer and coauthors, Physical activity and cardiovascular mortality risk, 2012; N. T. Attinian and coauthors, Interventions to promote physical activity and dietary lifestyle changes for cardiovascular risk factor reduction in adults: A scientific statement from the American Heart Association, *Circulation* 122 (2010): 406–441; N. L. Chase and coauthors, The association of cardiorespiratory fitness and physical activity with incidence of hypertension in men, *American Journal of Hypertension* 22 (2009): 417–424; P. T. Williams, Reduced diabetic, hypertensive, and cholesterol medication use with walking, *Medicine and Science in Sports and Exercise* 40 (2008): 433–443.
28. T. S. Church and coauthors, Changes in weight, waist circumference and compensatory responses with different doses of exercise among sedentary, overweight postmenopausal women, *PLoS ONE* 4 (2009): e4515; B. A. Irving and coauthors, Effect of exercise training intensity on abdominal visceral fat and body composition, *Medicine and Science in Sports and Exercise* 40 (2008): 1863–1872; M. Fogelhom, How physical activity can work? *International Journal of Pediatric Obesity* 3 (2008): 10–14.
29. American College of Sports Medicine, American Diabetes Association: Joint position statement, Exercise and type 2 diabetes, *Medicine and Science in Sports and Exercise* 42 (2010): 2282–2301; J. Ralph and coauthors, Low-intensity exercise reduces the prevalence of hyperglycemia in type 2 diabetes, *Medicine and Science in Sports and Exercise* 42 (2010): 219–225.
30. P. J. Banim and coauthors, Physical activity reduces the risk of symptomatic gallstones: A prospective cohort study, *European Journal of Gastroenterology and Hepatology* 22 (2010): 983–988; P. T. Williams, Independent effects of cardiorespiratory fitness, vigorous physical activity, and body mass index on clinical gallbladder disease risk, *American Journal of Gastroenterology* 103 (2008): 2239–2247.
31. J. C. Sieverdes and coauthors, Association between leisure-time physical activity and depressive symptoms in men, *Medicine and Science in Sports and Exercise* 44 (2012): 260–265; D.B. Nelson and coauthors, Effect of physical activity on menopausal symptoms among urban women, *Medicine and Science in Sports and Exercise* 40 (2008): 50–58.
32. L. B. Yates and coauthors, Exceptional longevity in men, *Archives of Internal Medicine* 168 (2008): 284–290; P. Kokkinos and coauthors, Exercise capacity and mortality in black and white men, *Circulation* 117 (2008): 614–622.
33. C. P. Wen and coauthors, Minimum amount of physical activity for reduced mortality and extended life expectancy: A prospective cohort study, *Lancet* 378 (2011): 1244–1253.
34. M. E. Nelson and coauthors, Physical activity and public health in older adults: Recommendation from the American College of Sports Medicine and the American Heart Association, *Medicine and Science in Sports and Exercise* 39 (2007): 1435–1445; N. Takeshima and coauthors, Functional fitness gain varies in older adults depending on exercise mode, *Medicine and Science in Sports and Exercise* 39 (2007): 2036–2043.
35. U.S. Department of Agriculture, Vegetables: Peas and beans are unique foods, www.choosemyplate.gov/food-groups/vegetables-beans-peas.html, accessed May 8, 2012.
36. Position of the American Dietetic Association: Vegetarian diets, *Journal of the American Dietetic Association* 109 (2009): 1266–1282.
37. N. J. Ollberding, R. L. Wolf, and I. Contento, Food label use and its relation dietary intake among US adults, *Journal of the American Dietetic Association* 110 (2010): 1233–1237; D. Schorr and coauthors, Nutrition facts you can't miss: The evolution of front-of-pack labeling, *Nutrition Today* 45 (2010): 22–32; J. L. Lewis and coauthors, Food label use and awareness of nutritional information and recommendations among persons with chronic disease, *American Journal of Clinical Nutrition* 90 (2009): 1351–1357.
38. R. M. Bliss, Nutrient data in time for the new year, *Agricultural Research* January 2012, pp. 20–21.
39. C. L. Taylor and V. L. Wilkening, How the nutrition food label was developed, part 2: The purpose and promise of nutrition claims, *Journal of the American Dietetic Association* 108 (2008): 618–623.
40. Position of the American Dietetic Association, Functional foods, *Journal of the American Dietetic Association* 109 (2009); 735–746.

Nutrition in Practice

Finding the Truth about Nutrition

Nutrition and health receive so much attention on television, on the radio, in the popular press, and on the Internet that it is easy to be overwhelmed with inconsistent, unclear information.[41] More than two-thirds of U.S. consumers are interested in the relationship between nutrition and health, but the majority of them agree that such information is often confusing and conflicting.[42] Determining whether nutrition information is accurate can be a challenging task. It is also an important task because nutrition affects a person both professionally and personally.

A person watches a nutrition report on television and then reads a conflicting report in the newspaper. Why do nutrition news reports and claims for nutrition products seem to contradict each other so often?

The problem of conflicting messages arises for several reasons:

- Popular media, often faced with tight deadlines and limited time or space to report new information, rush to present the latest "breakthrough" in a headline or a 60-second spot. They can hardly help omitting important facts about the study or studies that the "breakthrough" is based on.
- Despite tremendous advances in the past few decades, scientists still have much to learn about the human body and nutrition. Scientists themselves often disagree on their first tentative interpretations of new research findings, yet these are the very findings that the public hears most about.
- The popular media often broadcast preliminary findings in hopes of grabbing attention and boosting readership or television ratings.
- Commercial promoters turn preliminary findings into advertisements for products or supplements long before the findings have been validated—or disproved. The scientific process requires many experiments or trials to confirm a new finding. Seldom do promoters wait as long as they should to make their claims.
- Promoters are aware that consumers like to try new products or treatments even though they probably will not withstand the tests of time and scientific scrutiny.

So how can a person tell what claims to believe?

Valid nutrition information derives from scientific research, which has the following characteristics:

- Scientists test their ideas by conducting properly designed scientific experiments. They report their methods and procedures in detail so that other scientists can verify the findings through replication.
- Scientists recognize the inadequacy of personal testimonials.

Glossary of Nutrition Terms Associated with Nutrition Experts

Academy of Nutrition and Dietetics: the professional organization of dietitians in the United States; formerly the American Dietetic Association. The Canadian equivalent is Dietitians of Canada, which operates similarly.

dietetic technicians: persons who have completed a minimum of an associate's degree from an accredited college or university and an approved dietetic technician program. A **dietetic technician, registered (DTR)** has also passed a national examination and maintains registration through continuing professional education.

dietetics: the application of nutrition principles to achieve and maintain optimal human health.

nutritionists: all registered dietitians are nutritionists, but not all nutritionists are registered dietitians. Some state licensing boards set specific qualifications for holding the title. For states that regulate this title, the definition varies from state to state. To obtain some "nutritionist" credentials requires little more than a payment.

registered dietitians (RDs): food and nutrition experts who have earned a minimum of a bachelor's degree from an accredited university or college after completing a program of coursework approved by the Academy of Nutrition and Dietetics (or Dietitians of Canada). The dietitians must serve in an approved, supervised, internship or coordinated program to practice the necessary skills, pass the registration examination, and maintain competency through continuing education. Many states require licensing for practicing dietitians. Licensed dietitians (LDs) have met all *state* requirements to offer nutrition advice.

- Scientists who use animals in their research do not apply their findings directly to human beings.
- Scientists may use specific segments of the population in their research. When they do, they are careful not to generalize the findings to all people.
- Scientists report their findings in respected scientific journals. Their work must survive a screening review by their peers before it is accepted for publication.

With each report from scientists, the field of nutrition changes a little—each finding contributes another piece to the whole body of knowledge. Table NP1-1 features warning signs of nutrition quackery to help consumers distinguish valid from misleading nutrition information.

Because nutrition misinformation harms the health and economic status of consumers, the **Academy of Nutrition and Dietetics** (formerly the American Dietetic Association) works with health care professionals and educators to present sound nutrition information to the public and to actively confront nutrition misinformation.[43] Table NP1-2 offers a list of credible sources of nutrition information.

TABLE NP1-1 Warning Signs of Nutrition Quackery

1. Quick and easy fixes. Even proven treatments take time to be effective.
2. Personal testimonials. Hearsay is the weakest form of scientific validity.
3. One product does it all. No one product can treat every disease and condition.
4. Natural. Natural is not necessarily safer or better. Any product strong enough to be effective is strong enough to cause side effects.
5. Time-tested or latest innovation. Such findings would be widely publicized and accepted by health professionals.
6. Satisfaction guaranteed. Marketers of fraudulent products may make generous promises, but consumers won't be able to collect on them.
7. Paranoid accusations. These claims suggest that health professionals and legitimate drug manufacturers are conspiring with each other to promote drug companies' products for financial gain.
8. Meaningless medical jargon. Phony terms hide the lack of scientific proof.
9. Too good to be true. If it sounds too good to be true, it probably isn't true.

What about nutrition and health information found on the Internet? How does a person know whether the websites are reliable?

With hundreds of millions of websites on the Internet, searching for nutrition and health information can be daunting. The Internet offers no guarantee of the accuracy of the information found there, and much of it is pure fiction. Websites must be evaluated for their accuracy, just like every other source. Table NP1-3 provides clues to identifying reliable nutrition information sites and lists some credible sites.

One of the most trustworthy sites used by scientists and others is the National Library of Medicine's PubMed (www.ncbi.nlm.nih.gov/pubmed), which provides free access to more than 10 million abstracts (short descriptions) of research papers published in scientific journals around the world. Many abstracts provide links to websites where full articles are available.

Promoters of fraudulent "health" products use the Internet as a primary means to sell their wares. Agencies such as the Food and Drug Administration (FDA) take action against fraudulent marketing of supplements and health products on the Internet.[44] The latest actions target unscrupulous companies that use the Internet to promote products to the most vulnerable consumers—those with diseases such as cancer or AIDS. Of greatest concern are those products that not only make false promises but also are potentially dangerous. For example, herbal products touted as safe treatments for serious illnesses such as cancer may interact with and impair the effectiveness of medications. The FDA advises consumers to be suspicious of:

- Claims that a product is "natural" or "nontoxic." "Natural" or "nontoxic" does not always mean safe.
- Claims that a product is a "scientific breakthrough," "miraculous cure," "secret ingredient," or "ancient remedy."
- Claims that a product cures a wide range of illnesses.
- Claims that use impressive-sounding medical terms.
- Claims of a "money-back" guarantee.

Consumers with questions or suspicions about fraud can contact the FDA on the Internet at www.FDA.gov or by telephone at (888) INFO-FDA.

Everyone seems to be giving advice on nutrition. How can a person tell whom to listen to?

Registered dietitians (RDs) and nutrition professionals with advanced degrees (M.S., Ph.D.) are experts (see the glossary on p. 33). These professionals are probably in the best position to answer a person's nutrition questions. On the other hand, a "**nutritionist**" may be an expert or a quack, depending on which state the person practices in.

TABLE NP1-2 Credible Sources of Nutrition Information

Government agencies, volunteer associations, consumer groups, and professional organizations provide consumers with reliable health and nutrition information. Credible sources of nutrition information include:

• Nutrition and food science departments at a university or community college	
• Local agencies such as the health department or County Cooperative Extension Service	
• Government resources such as:	
• Centers for Disease Control and Prevention (CDC)	• www.cdc.gov
• Department of Agriculture (USDA)	• www.usda.gov
• Department of Health and Human Services (DHHS)	• www.hhs.gov
• *Dietary Guidelines for Americans*	• www.dietaryguidelines.gov
• Food and Drug Administration (FDA)	• www.fda.gov
• Health Canada	• http://www.hc-sc.gc.ca/fn-an/index-eng.php
• Healthy People	• www.healthypeople.gov
• Let's Move!	• www.letsmove.gov
• MyPlate	• www.choosemyplate.gov
• National Institutes of Health	• www.nih.gov
• *Physical Activity Guidelines for Americans*	• www.health.gov/paguidelines
• Volunteer health agencies such as:	
• American Cancer Society	• www.cancer.org
• American Diabetes Association	• www.diabetes.org
• American Heart Association	• www.americanheart.org
• Reputable consumer groups such as:	
• American Council on Science and Health	• www.acsh.org
• Federal Citizen Information Center	• www.usa.gov
• International Food Information Council	• www.foodinsight.org
• Professional health organizations such as:	
• Academy of Nutrition and Dietetics	• www.eatright.org
• American Medical Association	• www.ama-assn.org
• Dietitians of Canada	• www.dietitians.ca
• Journals such as:	
• *American Journal of Clinical Nutrition*	• www.ajcn.org
• *Journal of the Academy of Nutrition and Dietetics*	• www.adajournal.org
• *New England Journal of Medicine*	• www.nejm.org
• *Nutrition Reviews*	• www.ilsi.org

Some states require people who use this title to meet strict standards. In other states, a "nutritionist" may be any individual who claims a career connection with the nutrition field. There is no accepted national definition for the term *nutritionist*.

Other purveyors of nutrition information may also lack credentials. A health food store owner may be in the nutrition business simply because it is a lucrative market. The owner may have a background in business or sales and no education in nutrition at all. Such a person is not qualified to provide nutrition information to customers. For accurate nutrition information, seek out a trained professional with a college education in nutrition—an expert in the field of **dietetics**.

What about nurses and other health care professionals?

All members of the health care team share responsibility for helping each client to achieve optimal health, but the registered dietitian is usually the primary nutrition expert. Each of the other team members has a related specialty. Some physicians are specialists in clinical nutrition and are also experts in the field. Other physicians, nurses, and **dietetic technicians** often assist dietitians in providing nutrition information and may help to administer direct nutrition care. Nurses play central roles in client care management and client relationships. Visiting nurses and home health care nurses may become intimately

TABLE NP1-3 Evaluating the Reliability of Websites

To determine whether an Internet site offers reliable nutrition information, answer the following questions.
• **Who is responsible for the site?** Clues can be found in the three-letter "tag" that follows the dot in the site's name. For example, "gov" and "edu" indicate government and university sites, respectively, which are usually reliable sources of information.
• **Do the names and credentials of information providers appear? Is an editorial board identified?** Many legitimate sources provide e-mail addresses or other ways to obtain more information about the site and the information providers behind it.
• **Are links with other reliable information sites provided?** Reputable organizations almost always provide links with other similar sites because they want you to know of other experts in their area of knowledge. Caution is needed when you evaluate a site by its links, however. Anyone, even a quack, can link a web page to a reputable site without the organization's permission. Doing so may give the quack's site the appearance of legitimacy, just the effect for which the quack is hoping.
• **Is the site updated regularly?** Nutrition information changes rapidly, and sites should be updated often.
• **Is the site selling a product or service?** Commercial sites may provide accurate information, but they also may not. Their profit motive increases the risk of bias.
• **Does the site charge a fee to gain access to it?** Many academic and government sites offer the best information, usually for free. Some legitimate sites do charge fees, but before paying up, check the free sites. Chances are good you will find what you are looking for without paying.
• **Some credible websites include** • National Council Against Health Fraud www.ncahf.org • Stephen Barrett's Quackwatch www.quackwatch.com

involved in clients' nutrition care at home, teaching them both theory and cooking techniques. Physical therapists can provide individualized exercise programs related to nutrition—for example, to help control obesity. Social workers may provide practical and emotional support.

What roles might these other health care professionals play in nutrition care?

Some of the responsibilities of the health care professional might be:

- Helping people understand why nutrition is important to them.
- Answering questions about food and diet.
- Explaining to clients how modified diets work.
- Collecting information about clients that may influence their nutritional health.
- Identifying clients at risk for poor nutrition status (see Chapter 13) and recommending or taking appropriate action.
- Recognizing when clients need extra help with nutrition problems (in such cases, the problems should be referred to a dietitian or physician).

Health care professionals may routinely perform these nutrition-related tasks:

- Obtaining diet histories.
- Taking weight and height measurements.
- Feeding clients who cannot feed themselves.
- Recording what clients eat or drink.
- Observing clients' responses and reactions to foods.
- Helping clients mark menus.
- Monitoring weight changes.
- Monitoring food and drug interactions.
- Encouraging clients to eat.
- Assisting clients at home in planning their diets and managing their kitchen chores.
- Alerting the physician or dietitian when nutrition problems are identified.
- Charting actions taken and communicating on these matters with other professionals as needed.

Thus, although the dietitian assumes the primary role as the nutrition expert on a health care team, other health care professionals play important roles in administering nutrition care.

Notes

1. D. Quagliani and M. Hermann, Practice paper of the Academy of Nutrition and Dietetics: Communicating accurate food and nutrition information, *Journal of the Academy of Nutrition and Dietetics* 112 (2012): 759.
2. D. Schor and coauthors, Nutrition facts you can't miss: The evolution of front-of-pack labeling, *Nutrition Today* 45 (2010): 22–32.
3. Quagliani and Hermann, 2012.
4. Food and Drug Administration, *FDA 101: Health Fraud Awareness*, FDA Consumer Health Information, May 2009, available at www.fda.gov.

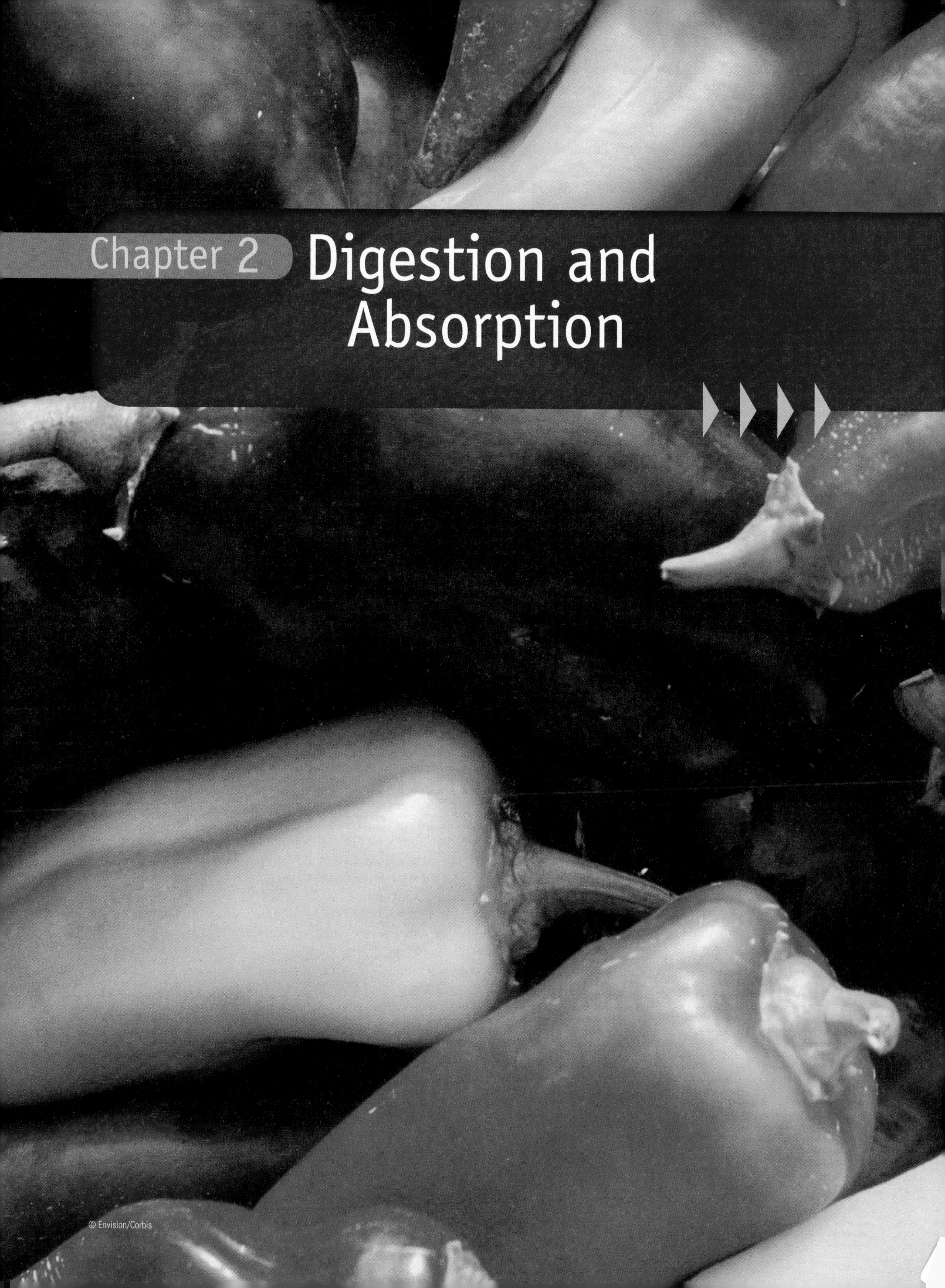

Chapter 2 Digestion and Absorption

THE BODY'S ABILITY TO TRANSFORM THE FOODS A PERSON EATS INTO the nutrients that fuel the body's work is quite remarkable. Yet most people probably give little, if any, thought to all the body does with food once it is eaten. This chapter offers the reader the opportunity to learn how the body digests, absorbs, and transports the nutrients and how it excretes the unwanted substances in foods.

One of the beauties of the digestive tract is that it is selective. Materials that are nutritive for the body are broken down into particles that can be absorbed into the bloodstream. Most of the nonnutritive materials are left undigested and pass out the other end of the digestive tract.

Anatomy of the Digestive Tract

The **gastrointestinal (GI) tract** is a flexible muscular tube extending from the mouth to the anus. Figure 2-1 on p. 39 shows the gastrointestinal tract and its associated organs. The accompanying glossary defines GI anatomical terms. In a sense, the human body surrounds the GI tract. Only when a nutrient or other substance passes through the cells of the digestive tract wall does it actually enter the body.

gastrointestinal (GI) tract: the digestive tract. The principal organs are the stomach and intestines.

gastro = stomach

digestion: the process by which complex food particles are broken down to smaller absorbable particles.

THE DIGESTIVE ORGANS

The process of **digestion** begins in the **mouth**. As you chew, your teeth crush and soften the food, while saliva mixes with the food mass and moistens it for comfortable swallowing. Saliva also helps dissolve the food so that you can taste it; only particles in solution can react with taste buds.

Glossary of GI Terms

These terms are listed in order from start to end of the digestive system.

mouth: the oral cavity containing the tongue and teeth.

pharynx (FAIR-inks): the passageway leading from the nose and mouth to the larynx and esophagus, respectively.

epiglottis (epp-ih-GLOTT-iss): cartilage in the throat that guards the entrance to the trachea and prevents fluid or food from entering it when a person swallows.

- *epi* = upon (over)
- *glottis* = back of tongue

esophagus (ee-SOFF-ah-gus): the food pipe; the conduit from the mouth to the stomach.

sphincter (SFINK-ter): a circular muscle surrounding, and able to close, a body opening. Sphincters are found at specific points along the GI tract and regulate the flow of food particles.

- *sphincter* = band (binder)

esophageal (ee-SOF-a-GEE-al) sphincter: a sphincter muscle at the upper or lower end of the esophagus. The *lower esophageal sphincter* is also called the *cardiac sphincter*.

stomach: a muscular, elastic, saclike portion of the digestive tract that grinds and churns swallowed food, mixing it with acid and enzymes to form chyme.

pyloric (pie-LORE-ic) sphincter: the circular muscle that separates the stomach from the small intestine and regulates the flow of partially digested food into the small intestine; also called *pylorus* or *pyloric valve*.

- *pylorus* = gatekeeper

small intestine: a 10-foot length of small-diameter intestine that is the major site of digestion of food and absorption of nutrients. Its segments are the duodenum, jejunum, and ileum.

duodenum (doo-oh-DEEN-um, doo-ODD-num): the top portion of the small intestine (about "12 fingers' breadth long" in ancient terminology).

- *duodecim* = twelve

jejunum (je-JOON-um): the first two-fifths of the small intestine beyond the duodenum.

ileum (ILL-ee-um): the last segment of the small intestine.

gallbladder: the organ that stores and concentrates bile. When it receives the signal that fat is present in the duodenum, the gallbladder contracts and squirts bile through the bile duct into the duodenum.

pancreas: a gland that secretes digestive enzymes and juices into the duodenum. (The pancreas also secretes hormones that help to maintain glucose homeostasis into the blood.)

ileocecal (ill-ee-oh-SEEK-ul) valve: the sphincter separating the small and large intestines.

large intestine or colon (COAL-un): the lower portion of intestine that completes the digestive process. Its segments are the ascending colon, the transverse colon, the descending colon, and the sigmoid colon.

- *sigmoid* = shaped like the letter S (sigma in Greek)

appendix: a narrow blind sac extending from the beginning of the colon that stores lymph cells.

rectum: the muscular terminal part of the intestine, extending from the sigmoid colon to the anus.

anus (AY-nus): the terminal outlet of the GI tract.

The tongue allows you not only to taste food but also to move food around the mouth, facilitating chewing and swallowing. When you swallow a mouthful of food, it passes through the **pharynx**, a short tube that is shared by both the **digestive system** and the respiratory system.

digestive system: all the organs and glands associated with the ingestion and digestion of food.

bolus (BOH-lus): the portion of food swallowed at one time.

Mouth to the Esophagus Once a mouthful of food has been chewed and swallowed, it is called a **bolus**. Each bolus first slides across your **epiglottis**, bypassing the entrance to your lungs. During each swallow, the epiglottis closes off your trachea, the air passageway to the lungs, so that you do not choke.

Esophagus to the Stomach The **esophagus** has a **sphincter** muscle at each end. During a swallow, the upper **esophageal sphincter** opens. The bolus then slides down the esophagus, which conducts it through the diaphragm to the **stomach**. The lower

FIGURE 2-1 The Digestive System

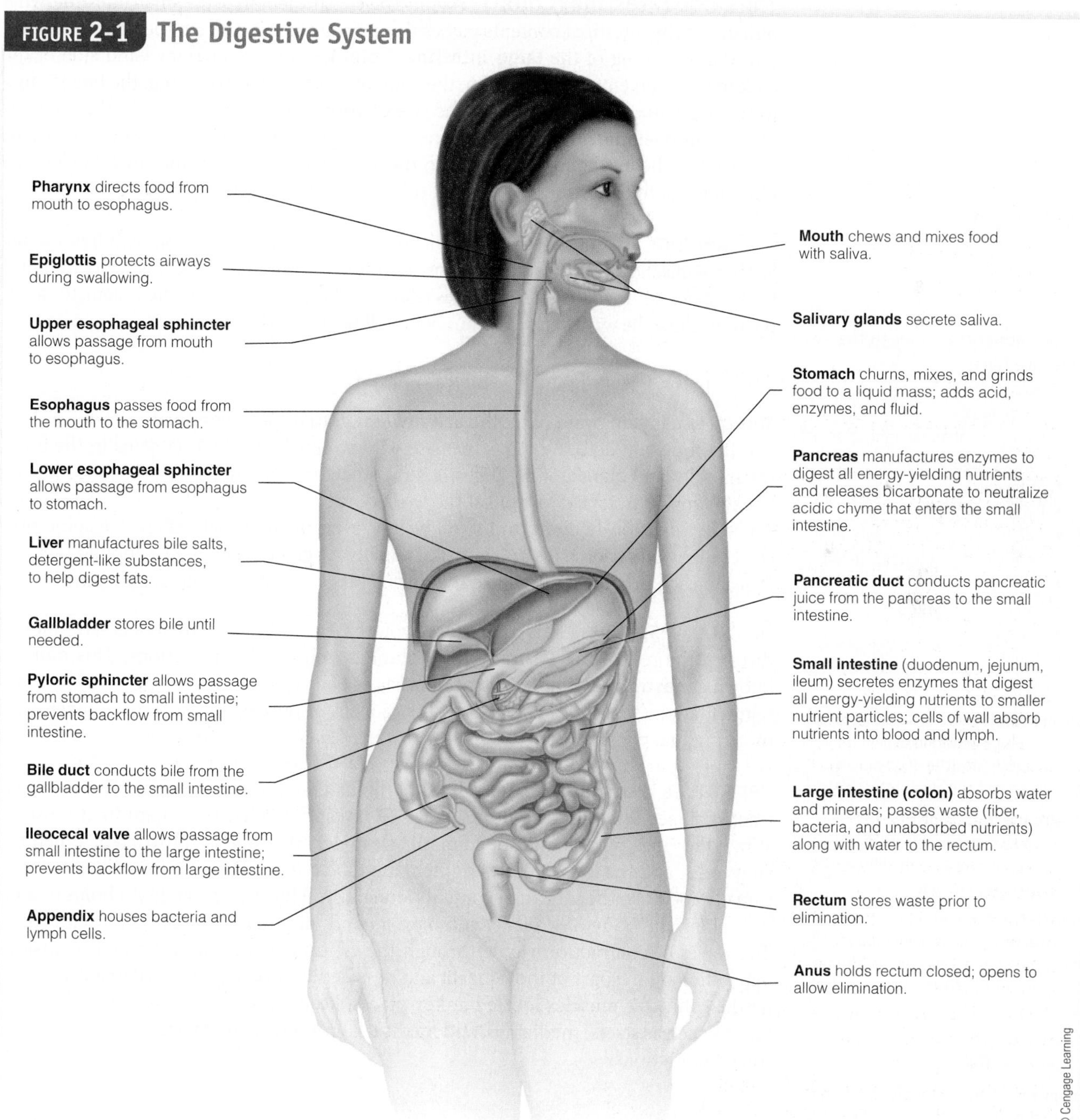
Pharynx directs food from mouth to esophagus.
Epiglottis protects airways during swallowing.
Upper esophageal sphincter allows passage from mouth to esophagus.
Esophagus passes food from the mouth to the stomach.
Lower esophageal sphincter allows passage from esophagus to stomach.
Liver manufactures bile salts, detergent-like substances, to help digest fats.
Gallbladder stores bile until needed.
Pyloric sphincter allows passage from stomach to small intestine; prevents backflow from small intestine.
Bile duct conducts bile from the gallbladder to the small intestine.
Ileocecal valve allows passage from small intestine to the large intestine; prevents backflow from large intestine.
Appendix houses bacteria and lymph cells.
Mouth chews and mixes food with saliva.
Salivary glands secrete saliva.
Stomach churns, mixes, and grinds food to a liquid mass; adds acid, enzymes, and fluid.
Pancreas manufactures enzymes to digest all energy-yielding nutrients and releases bicarbonate to neutralize acidic chyme that enters the small intestine.
Pancreatic duct conducts pancreatic juice from the pancreas to the small intestine.
Small intestine (duodenum, jejunum, ileum) secretes enzymes that digest all energy-yielding nutrients to smaller nutrient particles; cells of wall absorb nutrients into blood and lymph.
Large intestine (colon) absorbs water and minerals; passes waste (fiber, bacteria, and unabsorbed nutrients) along with water to the rectum.
Rectum stores waste prior to elimination.
Anus holds rectum closed; opens to allow elimination.
© Cengage Learning

esophageal sphincter closes behind the bolus so that it cannot slip back. The stomach retains the bolus for a while, adds juices to it (gastric juices are discussed on p. 43), and transforms it into a semiliquid mass called **chyme**. Then, bit by bit, the stomach releases the chyme through another sphincter, the **pyloric sphincter**, which opens into the **small intestine** and then closes after the chyme passes through.

The Small Intestine At the beginning of the small intestine, the chyme passes by an opening from the common bile duct, which secretes digestive fluids into the small intestine from two organs outside the GI tract—the **gallbladder** and the **pancreas**. The chyme travels on down the small intestine through its three segments—the **duodenum**, the **jejunum**, and the **ileum**. Together, the segments amount to a total of about 10 feet of tubing coiled within the abdomen.* Digestion is completed within the small intestine.

The Large Intestine (Colon) Having traveled the length of the small intestine, what remains of the intestinal contents passes through another sphincter, the **ileocecal valve**, into the beginning of the **large intestine (colon)** in the lower right-hand side of the abdomen. Upon entering the colon, the contents pass another opening: the one leading to the **appendix**, a blind sac about the size of your little finger. Normally, the contents bypass this opening, however, and travel up the right-hand side of the abdomen, across the front to the left-hand side, down to the lower left-hand side, and finally below the other folds of the intestines to the back side of the body above the **rectum**.

The Rectum As the intestinal contents pass to the rectum, the colon withdraws water, leaving semisolid waste. The strong muscles of the rectum hold back this waste until it is time to defecate. Then the rectal muscles relax, and the last sphincter in the system, the **anus**, opens to allow the wastes to pass. Thus, food follows the path shown in the margin. (■)

■ The path of food through the digestive tract:

- Mouth
- Esophagus
- Lower esophageal sphincter (or cardiac sphincter)
- Stomach
- Pyloric sphincter
- Duodenum (common bile duct enters here), jejunum, ileum
- Ileocecal valve
- Large intestine (colon)
- Rectum
- Anus

THE INVOLUNTARY MUSCLES AND THE GLANDS

You are usually unaware of all the activity that goes on between the time you swallow and the time you defecate. As is the case with so much else that happens in the body, the muscles and **glands** of the digestive tract meet internal needs without your having to exert any conscious effort to get the work done.

People consciously chew and swallow, but even in the mouth there are some processes over which you have no control. The salivary glands secrete just enough saliva to moisten each mouthful of food so that it can pass easily down your esophagus.

Gastrointestinal Motility Once you have swallowed, materials are moved through the rest of the GI tract by involuntary muscular contractions. This motion, known as **gastrointestinal motility**, consists of two types of movement, peristalsis and segmentation (see Figure 2-2). Peristalsis propels, or pushes; segmentation mixes, with more gradual pushing.

Peristalsis **Peristalsis** begins when the bolus enters the esophagus. The entire GI tract is ringed with circular muscles, which are surrounded by longitudinal muscles. When the rings tighten and the long muscles relax, the tube is constricted. When the rings relax and the long muscles tighten, the tube bulges. These actions alternate continually and push the intestinal contents along. If you have ever watched a bolus of food pass along the body of a snake, you have a good picture of how these muscles work. The waves of contraction ripple through the GI tract at varying rates and intensities depending on the part of the GI tract and on whether food is present. Peristalsis, aided by the sphincter muscles located at key places, keeps things moving along. However, factors such as stress, medicines, and medical conditions may interfere with normal GI tract contractions.[1]

chyme (KIME): the semiliquid mass of partly digested food expelled by the stomach into the duodenum (the top portion of the small intestine).

gastrointestinal motility: spontaneous motion in the digestive tract accomplished by involuntary muscular contractions.

peristalsis (peri-STALL-sis): successive waves of involuntary muscular contractions passing along the walls of the GI tract that push the contents along.

peri = around
stellein = wrap

segmentation: a periodic squeezing or partitioning of the intestine by its circular muscles that both mixes and slowly pushes the contents along.

*The small intestine is almost two and a half times shorter in living adults than it is at death, when muscles are relaxed and elongated.

FIGURE 2-2 Peristalsis and Segmentation

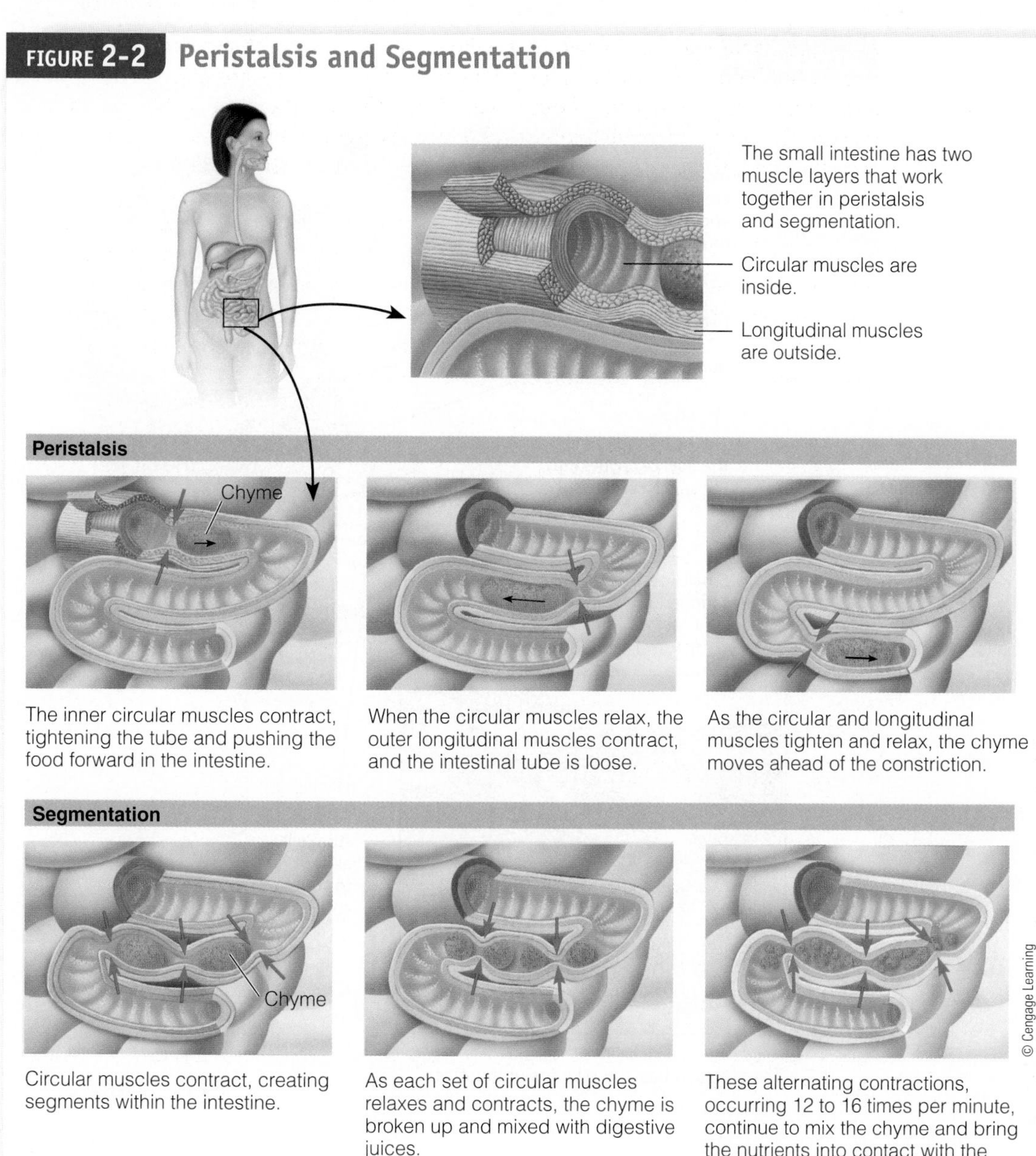

The small intestine has two muscle layers that work together in peristalsis and segmentation.
Circular muscles are inside.
Longitudinal muscles are outside.
Peristalsis
Chyme
The inner circular muscles contract, tightening the tube and pushing the food forward in the intestine.
When the circular muscles relax, the outer longitudinal muscles contract, and the intestinal tube is loose.
As the circular and longitudinal muscles tighten and relax, the chyme moves ahead of the constriction.
Segmentation
Chyme
Circular muscles contract, creating segments within the intestine.
As each set of circular muscles relaxes and contracts, the chyme is broken up and mixed with digestive juices.
These alternating contractions, occurring 12 to 16 times per minute, continue to mix the chyme and bring the nutrients into contact with the intestinal lining for absorption.

Segmentation The intestines not only push but also periodically squeeze their contents as if a string tied around the intestines were being pulled tight. This motion, called **segmentation**, forces the contents back a few inches, mixing them and promoting close contact with the digestive juices and the absorbing cells of the intestinal walls before letting the contents slowly move along again.

Liquefying Process Besides forcing the intestinal contents along, the muscles of the GI tract help to liquefy them to chyme so that the digestive juices will have access to all their nutrients. The mouth initiates this liquefying process by chewing, adding saliva, and stirring with the tongue to reduce the food to a coarse mash suitable for swallowing. The stomach then further mixes and kneads the food.

Stomach Action The stomach has the thickest walls and strongest muscles of all the GI tract organs. In addition to circular and longitudinal muscles, the stomach has a third layer of diagonal muscles that also alternately contract and relax (see Figure 2-3 on p. 42). These three sets of muscles work to force the chyme downward, but the pyloric sphincter usually remains tightly closed so that the stomach's contents are

FIGURE 2-3 Stomach Muscles

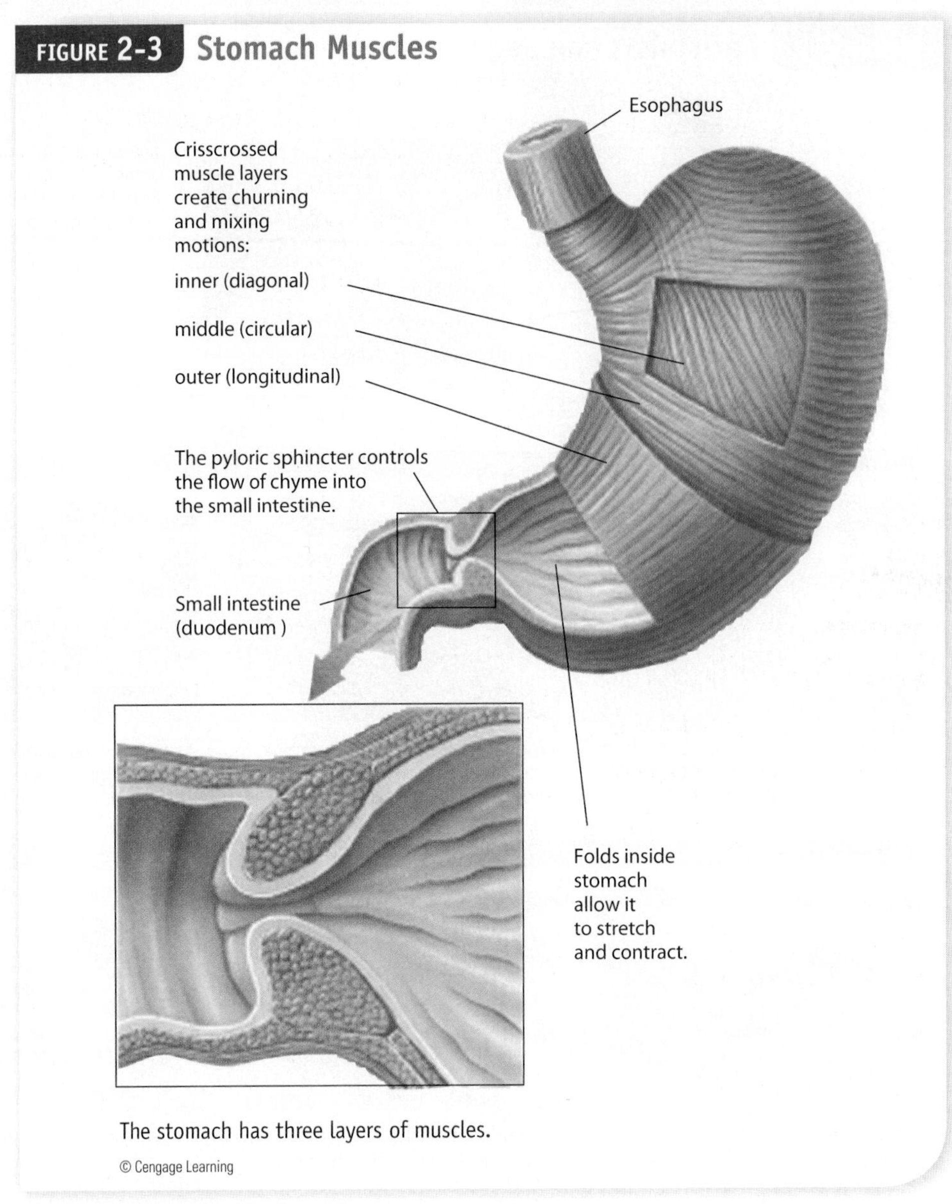

Esophagus
Crisscrossed muscle layers create churning and mixing motions:
inner (diagonal)
middle (circular)
outer (longitudinal)
The pyloric sphincter controls the flow of chyme into the small intestine.
Small intestine (duodenum)
Folds inside stomach allow it to stretch and contract.

The stomach has three layers of muscles.

thoroughly mixed and squeezed before being released. Meanwhile, the gastric glands are adding juices. When the chyme is thoroughly liquefied, the pyloric sphincter opens briefly, about three times a minute, to allow small portions through. At this point, the intestinal contents no longer resemble food in the least.

Review Notes

- As Figure 2-1 shows, food enters the mouth and travels down the esophagus and through the lower esophageal sphincter to the stomach, then through the pyloric sphincter to the small intestine, on through the ileocecal valve to the large intestine, and past the appendix to the rectum, exiting through the anus.
- The wavelike contractions of peristalsis and the periodic squeezing of segmentation keep things moving at a reasonable pace.
- The mouth begins the process of liquefying food by chewing and adding saliva to reduce food to a coarse mash for swallowing. The stomach then further mixes and kneads the food.

The Process of Digestion

One person eats nothing but vegetables, fruits, and nuts; another, nothing but meat, milk, and potatoes. How is it that both people wind up with essentially the same body composition? It all comes down to the body rendering food—whatever it is to start with—into the basic units that make up carbohydrate, fat, and protein. The body absorbs these units and builds its tissues from them.

To digest food, five different body organs secrete digestive juices: the salivary glands, the stomach, the small intestine, the liver (via the gallbladder), and the pancreas. These secretions enter the GI tract at various points along the way, bringing an abundance of water and a variety of enzymes. (■) Each of the juices has a turn to mix with the food and promote its breakdown to small units that can be absorbed into the body. The accompanying glossary defines some of the digestive glands and their juices.

DIGESTION IN THE MOUTH

Digestion of carbohydrate begins in the mouth, where the **salivary glands** secrete **saliva**, which contains water, salts, and enzymes (including salivary **amylase**) that break the bonds in the chains of starch. Saliva also protects the tooth surfaces and linings of the mouth, esophagus, and stomach from attack by molecules that might harm them. The enzymes in the mouth do not, for the most part, affect the fats, proteins, vitamins, minerals, and fiber that are present in the foods people eat.

■ Enzymes are formally introduced in Chapter 5, but for now a simple definition will suffice. An *enzyme* is a protein that facilitates a chemical reaction—making a compound, breaking down a compound, changing the arrangement of a compound, or exchanging parts. Enzymes themselves are not changed by the reactions they facilitate.

DIGESTION IN THE STOMACH

Gastric juice, secreted by the **gastric glands**, is composed of water, enzymes, and **hydrochloric acid**. The acid is so strong that it burns the throat if it happens to reflux into the upper esophagus and mouth. The strong acidity of the stomach prevents bacterial

Glossary of Digestive Glands and Their Secretions

These terms are listed in order from the beginning of the digestive tract to the end.

glands: cells or groups of cells that secrete materials for special uses in the body. Glands may be *exocrine* (EKS-oh-crin) *glands*, secreting their materials "out" (into the digestive tract or onto the surface of the skin), or *endocrine* (EN-doe-crin) *glands*, secreting their materials "in" (into the blood).

- *exo* = outside
- *endo* = inside
- *krine* = to separate

salivary glands: exocrine glands that secrete saliva into the mouth.

saliva: the secretion of the salivary glands. The principal enzyme is salivary amylase.

amylase (AM-uh-lace): an enzyme that splits amylose (a form of starch). Amylase is a carbohydrase. The ending *-ase* indicates an enzyme; the root tells what it digests. Other examples: *protease, lipase.*

gastric glands: exocrine glands in the stomach wall that secrete gastric juice into the stomach.

- *gastro* = stomach

gastric juice: the digestive secretion of the gastric glands containing a mixture of water, hydrochloric acid, and enzymes. The principal enzymes are pepsin (acts on proteins) and lipase (acts on emulsified fats).

hydrochloric acid (HCl): an acid composed of hydrogen and chloride atoms; normally produced by the gastric glands.

mucus (MYOO-cuss): a mucopolysaccharide (a relative of carbohydrate) secreted by cells of the stomach wall that protects the cells from exposure to digestive juices (and other destructive agents). The cellular lining of the stomach wall with its coat of mucus is known as the *mucous membrane.* (The noun is *mucus*; the adjective is *mucous.*)

pepsin: a protein-digesting enzyme (gastric protease) in the stomach. It circulates as a precursor, pepsinogen, and is converted to pepsin by the action of stomach acid.

intestinal juice: the secretion of the intestinal glands; contains enzymes for the digestion of carbohydrate and protein and a minor enzyme for fat digestion.

liver: the organ that manufactures bile. (The liver's other functions are described in Chapter 19.)

bile: an emulsifier that prepares fats and oils for digestion; made by the liver, stored in the gallbladder, and released into the small intestine when needed.

pancreatic (pank-ree-AT-ic) juice: the exocrine secretion of the pancreas, containing enzymes for the digestion of carbohydrate, fat, and protein. Juice flows from the pancreas into the small intestine through the pancreatic duct. The pancreas also has an endocrine function, the secretion of insulin and other hormones.

bicarbonate: an alkaline secretion of the pancreas; part of the pancreatic juice. (Bicarbonate also occurs widely in all cell fluids.)

growth and kills most bacteria that enter the body with food. You might expect that the stomach's acid would attack the stomach itself, but the cells of the stomach wall secrete **mucus**, a thick, slimy, white polysaccharide that coats and protects the stomach's lining.

The major digestive event in the stomach is the initial breakdown of proteins. Other than being crushed and mixed with saliva in the mouth, nothing happens to protein until it comes in contact with the gastric juices in the stomach. There, the acid helps to uncoil (denature) the protein's tangled strands so that the stomach enzymes can attack the bonds. Both the enzyme **pepsin** and the stomach acid itself act as catalysts in the process. Minor events are the digestion of some fat by a gastric lipase, the digestion of sucrose (to a very small extent) by the stomach acid, and the attachment of a protein carrier to vitamin B_{12}.

The stomach enzymes work most efficiently in the stomach's strong acid, but salivary amylase, which is swallowed with food, does not work in acid this strong. Consequently, the digestion of starch gradually ceases as the acid penetrates the bolus. In fact, salivary amylase becomes just another protein to be digested.

DIGESTION IN THE SMALL AND LARGE INTESTINES

By the time food leaves the stomach, digestion of all three energy-yielding nutrients has begun, but the process gains momentum in the small intestine. There, the pancreas and the liver contribute additional digestive juices through the duct leading into the duodenum, and the small intestine adds **intestinal juice**. These juices contain digestive enzymes, bicarbonate, and bile.

■ Mayonnaise, made from vinegar and oil, would separate as other vinegar-and-oil salad dressings do if food chemists did not blend in a third ingredient—an emulsifier. The emulsifier mixes well with the fatty oil and the watery vinegar. In the case of mayonnaise, the emulsifier is lecithin from egg yolks.

emulsifier: a substance that mixes with both fat and water and that disperses the fat in the water, forming an emulsion.

Digestive Enzymes **Pancreatic juice** contributes enzymes that digest fats, proteins, and carbohydrates. Glands in the intestinal wall also secrete digestive enzymes. (Review the glossary of digestive glands and their secretions on p. 43 for details.)

Bicarbonate The pancreatic juice also contains sodium **bicarbonate**, which neutralizes the acidic chyme as it enters the small intestine. From this point on, the contents of the digestive tract are neutral or slightly alkaline. The enzymes from both the intestine and the pancreas work best in this environment.

Bile **Bile** is secreted continuously by the **liver** and is concentrated and stored in the gallbladder. The gallbladder squirts bile into the duodenum whenever fat arrives there. Bile is not an enzyme but an **emulsifier** that brings fats into suspension in water (see Figure 2-4). (■) After the fats are emulsified, enzymes can work on them, and they

FIGURE 2-4 Emulsification of Fat by Bile

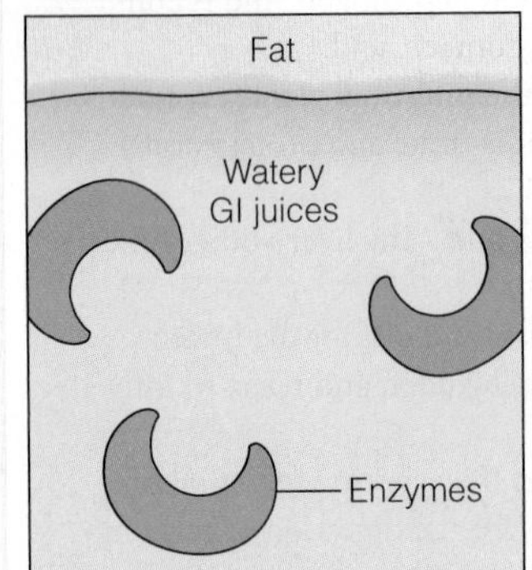

In the stomach, the fat and watery GI juices tend to separate. The enzymes are in the water and can't get at the fat.

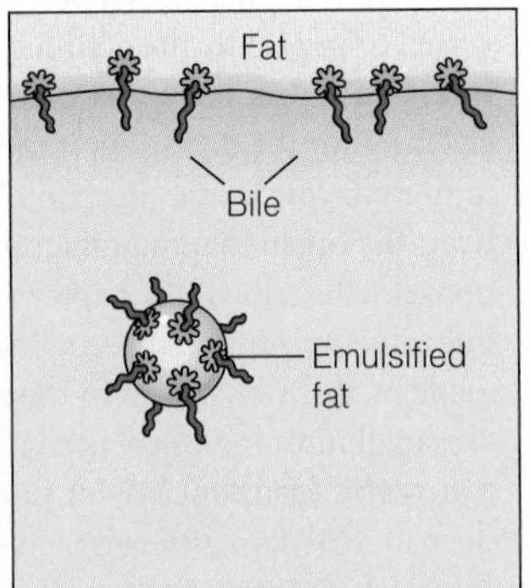

When fat enters the small intestine, the gallbladder secretes bile. Bile has an affinity for both fat and water, so it can bring the fat into the water.

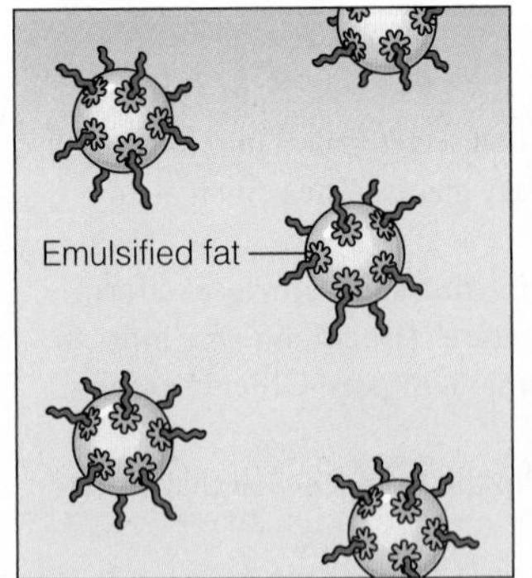

Bile's emulsifying action converts large fat globules into small droplets that repel each other.

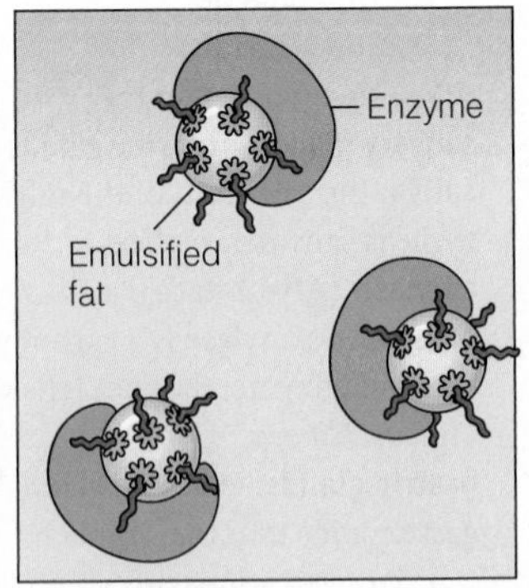

After emulsification, the enzymes have easy access to the fat droplets.

Like bile, detergents are emulsifiers and work the same way, which is why they are effective at removing grease spots from clothes. Molecule by molecule, the grease is dissolved out of the spot and suspended in water, where it can be rinsed away.

can be absorbed. Thanks to all these secretions, all three energy-yielding nutrients are digested in the small intestine.

The Rate of Digestion The rate of digestion of the energy nutrients depends on the contents of the meal. If the meal is high in simple sugars, digestion proceeds fairly rapidly. On the other hand, if the meal is rich in fat, digestion is slower.

Protective Factors The intestines contain bacteria that produce a variety of vitamins, including biotin and vitamin K (although bacteria alone cannot meet the need for these vitamins). The GI bacteria also protect people from infections. Provided that the normal **intestinal flora** are thriving, infectious bacteria have a hard time getting established and launching an attack on the system. (■) In addition, the small intestine and the entire GI tract manufacture and maintain a strong arsenal of defenses against foreign invaders. Several different types of defending cells are present there and confer specific immunity against intestinal diseases.

The Final Stage The story of how food is broken down into nutrients that can be absorbed is now nearly complete. The three energy-yielding nutrients—carbohydrate, fat, and protein—are disassembled to basic building blocks before they are absorbed. Most of the other nutrients—vitamins, minerals, and water—are absorbed as they are. Undigested residues, such as some fibers, are not absorbed but continue through the digestive tract as a semisolid mass that stimulates the tract's muscles, helping them remain strong and able to perform peristalsis efficiently. Fiber also retains water, keeping the stools soft, and carries some bile acids, sterols, and fat out of the body. Drinking plenty of water in conjunction with eating foods high in fiber supplies fluid for the fiber to take up. This is the basis for the recommendation to drink water and eat fiber-rich foods to relieve constipation.

The process of absorbing the nutrients into the body is discussed in the next section. For the moment, let us assume that the digested nutrients simply disappear from the GI tract as they are ready. Virtually all nutrients are gone by the time the contents of the GI tract reach the end of the small intestine. Little remains but water, a few salts and body secretions, and undigested materials such as fiber. These enter the large intestine (colon).

In the colon, intestinal bacteria degrade some of the fiber to simpler compounds. The colon itself retrieves from its contents the materials that the body is designed to recycle—water and dissolved salts. The waste that is finally excreted has little or nothing of value left in it. The body has extracted all that it can use from the food.

Review Notes

- Digestive enzymes secreted by the salivary glands, stomach, pancreas, and small intestine break down macronutrients into absorbable components.
- Bile produced by the liver and delivered by the gallbladder emulsifies fats to prepare them for digestion.

The Absorptive System

Within three or four hours after you have eaten a meal, your body must find a way to absorb millions of molecules one by one. The absorptive system is ingeniously designed to accomplish this task.

THE SMALL INTESTINE

Most absorption takes place in the small intestine. The small intestine is a tube about 10 feet long and about an inch across, yet it provides a surface comparable in area to a tennis court. When nutrient molecules make contact with this surface, they are absorbed and carried off to the liver and other parts of the body.

Villi and Microvilli How does the intestine manage to provide such a large absorptive surface area? Its inner surface looks smooth, but viewed through a microscope, it turns out to be wrinkled into hundreds of folds. Each fold is covered with thousands of fingerlike projections called **villi**. The villi are as numerous as the hairs on velvet fabric. A single villus, magnified still more, turns out to be composed of several hundred cells, each covered with microscopic hairs called **microvilli** (see Figure 2-5 on p. 46).

■ The influence of probiotics on intestinal health is the topic of Nutrition in Practice 18.

intestinal flora: the bacterial inhabitants of the GI tract.

flora = plant growth

villi (VILL-ee or VILL-eye): fingerlike projections from the folds of the small intestine. The singular form is **villus**.

villus = shaggy hair

microvilli (MY-cro-VILL-ee or MY-cro-VILL-eye): tiny, hairlike projections on each cell of every villus that can trap nutrient particles and transport them into the cells. The singular form is **microvillus**.

FIGURE 2-5 The Small Intestinal Villi

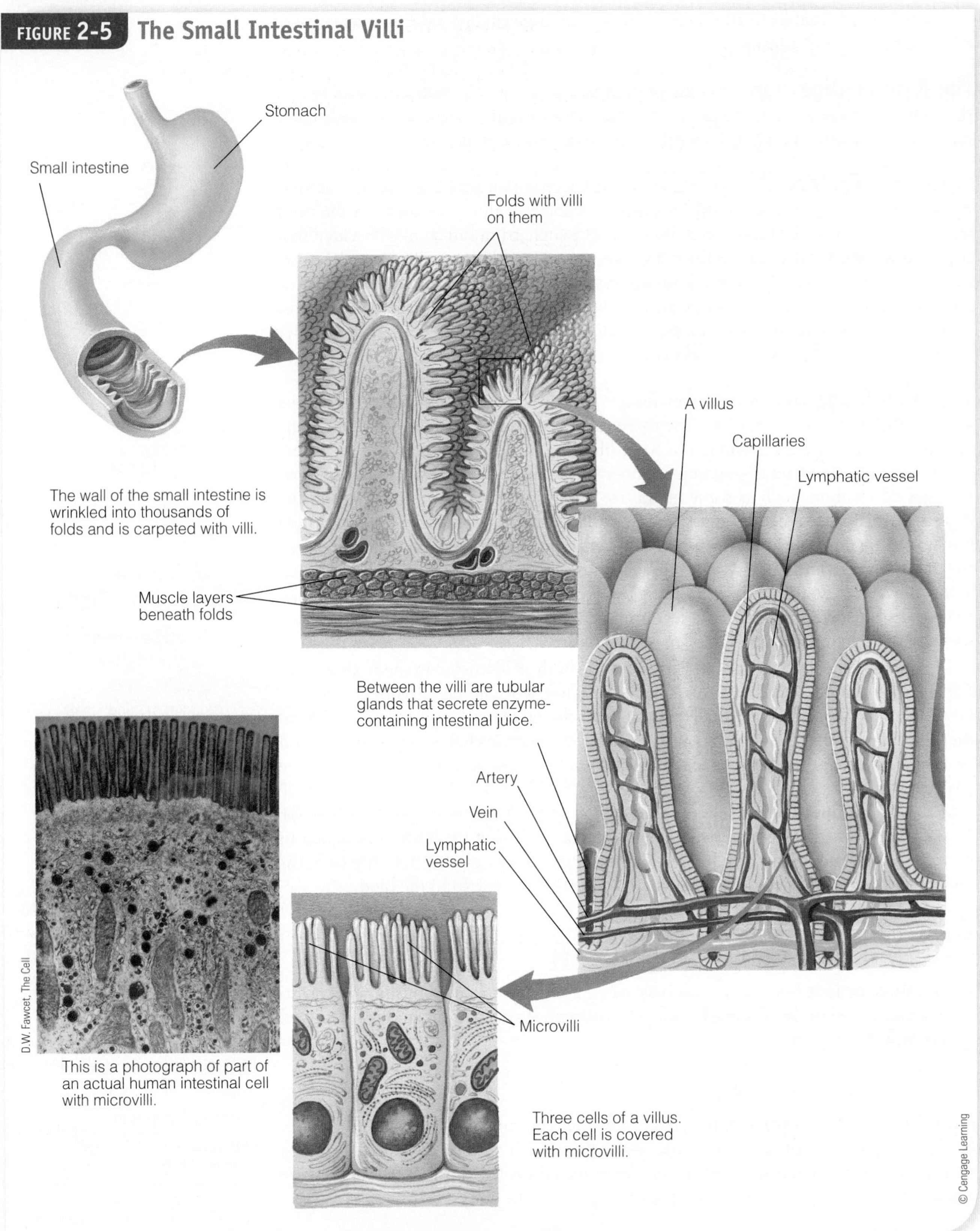

Stomach
Small intestine
Folds with villi on them
The wall of the small intestine is wrinkled into thousands of folds and is carpeted with villi.
Muscle layers beneath folds
A villus
Capillaries
Lymphatic vessel
Between the villi are tubular glands that secrete enzyme-containing intestinal juice.
Artery
Vein
Lymphatic vessel
Microvilli
This is a photograph of part of an actual human intestinal cell with microvilli.
D.W. Fawcet, The Cell
Three cells of a villus. Each cell is covered with microvilli.
© Cengage Learning

The villi are in constant motion. A thin sheet of muscle lines each villus so that it can wave, squirm, and wiggle like the tentacles of a sea anemone. Any nutrient molecule small enough to be absorbed is trapped among the microvilli and drawn into a cell beneath them. Some partially digested nutrients are caught in the microvilli, digested further by enzymes there, and then absorbed into the cells.

Specialization in the Intestinal Tract As you can see, the intestinal tract is beautifully designed to perform its functions. A further refinement of the system is that the cells of successive portions of the tract are specialized to absorb different nutrients. The nutrients that are ready for absorption early are absorbed near the top of the tract; those that take longer to be digested are absorbed farther down. The rate at which the nutrients travel through the GI tract is finely adjusted to maximize their availability to the appropriate absorptive segment of the tract when they are ready. The lowly "gut" turns out to be one of the most elegantly designed organ systems in the body.

The Myth of "Food Combining" Some popular fad diets advocate the idea that people should not eat certain food combinations (for example, fruit and meat) at the same meal because the digestive system cannot handle more than one task at a time. This is a myth. The art of "food combining" (which actually emphasizes "food separating") is based on this idea, and it represents faulty logic and a gross underestimation of the body's capabilities. In fact, the opposite is often true: foods eaten together can enhance each other's use by the body. For example, vitamin C in a pineapple or citrus fruit can enhance the absorption of iron from a meal of chicken and rice or other iron-containing foods. Many other instances of mutually beneficial interactions are presented in later chapters.

ABSORPTION OF NUTRIENTS

Once a molecule has entered a cell in a villus, the next step is to transmit it to a destination elsewhere in the body by way of the body's two transport systems—the bloodstream and the **lymphatic system**. As Figure 2-5 shows, both systems supply vessels to each villus. Through these vessels, the nutrients leave the cell and enter either the **lymph** or the blood. In either case, the nutrients end up in the blood, at least for a while. The water-soluble nutrients (and the smaller products of fat digestion) are released directly into the bloodstream by way of the capillaries, but the larger fats and the fat-soluble vitamins find direct access into the capillaries impossible because these nutrients are insoluble in water (and blood is mostly water). They require some packaging before they are released.

The intestinal cells assemble the products of fat digestion into larger molecules called **triglycerides**. These triglycerides, fat-soluble vitamins (when present), and other large lipids (cholesterol and the phospholipids) are then packaged for transport. They cluster together with special proteins to form **chylomicrons**, one kind of **lipoproteins** (lipoproteins are described beginning on p. 49). Finally, the cells release the chylomicrons into the lymphatic system. They can then glide through the lymph spaces until they arrive at a point of entry into the bloodstream near the heart. Thus, some materials from the GI tract initially enter the lymphatic system but soon reach the bloodstream.

Review Notes

- The many folds and villi of the small intestine dramatically increase its surface area, facilitating nutrient absorption.
- Nutrients pass through the cells of the villi and enter either the blood (if they are water soluble or small fat fragments) or the lymph (if they are fat soluble).

Transport of Nutrients

Once a nutrient has entered the bloodstream or the lymphatic system, it may be transported to any part of the body, from the tips of the toes to the roots of the hair, where it becomes available to any of the cells. The circulatory systems are arranged to deliver nutrients wherever they are needed.

lymphatic system: a loosely organized system of vessels and ducts that conveys the products of digestion toward the heart.

lymph (LIMF): the body fluid found in lymphatic vessels. Lymph consists of all the constituents of blood except red blood cells.

triglycerides (try-GLISS-er-rides): one of the main classes of lipids: the chief form of fat in foods and the major storage form of fat in the body; composed of glycerol with three fatty acids attached.

tri = three
glyceride = a compound of glycerol

chylomicrons (kye-lo-MY-crons): the lipoproteins that transport lipids from the intestinal cells into the body. The cells of the body remove the lipids they need from the chylomicrons, leaving chylomicron remnants to be picked up by the liver cells.

lipoproteins: clusters of lipids associated with proteins that serve as transport vehicles for lipids in the lymph and blood.

THE VASCULAR SYSTEM

The vascular or blood circulatory system is a closed system of vessels through which blood flows continuously in a figure eight, with the heart serving as a pump at the crossover point. On each loop of the figure eight, blood travels a simple route: heart to arteries to capillaries to veins to heart.

The routing of the blood through the digestive system is different, however. The blood is carried to the digestive system (as it is to all organs) by way of an **artery**, which (as in all organs) branches into **capillaries** to reach every cell. Blood leaving the digestive system, however, goes by way of a **vein**. The **hepatic portal vein** directs blood not back to the heart but to another organ—the liver. This vein *again* branches into a network of small blood vessels (*sinusoids*) so that every cell of the liver has access to the newly absorbed nutrients that the blood is carrying. Blood leaving the liver then *again* collects into a vein, called the **hepatic vein**, which returns the blood to the heart. The route is thus heart to arteries to capillaries (in intestines) to hepatic portal vein to sinusoids (in liver) to hepatic vein to heart. (■)

■ The artery that delivers oxygen-rich blood from the heart and lungs to the liver is the *hepatic artery*.

An anatomist studying this system knows there must be a reason for this special arrangement. The liver is located in the circulation system at the point where it will have the first chance at most of the materials absorbed from the GI tract. In fact, the liver is the body's major metabolic organ (see Figure 2-6) and must prepare the absorbed nutrients for use by the rest of the body. Furthermore, the liver stands as a gatekeeper to waylay intruders that might otherwise harm the heart or brain. Chapter 19 offers more information about this noble organ.

THE LYMPHATIC SYSTEM

The lymphatic system is a one-way route for fluids to travel from tissue spaces into the blood. The lymphatic system has no pump; instead, lymph is squeezed from one portion of the body to another like water in a sponge, as muscles contract and create pressure here and there. Ultimately, the lymph collects in a large duct behind the heart. This duct terminates in a vein that conducts the lymph into the heart. (■)

■ The duct that conveys lymph toward the heart is the *thoracic* (thor-ASS-ic) *duct*. The *subclavian vein* connects this duct with the right upper chamber of the heart, providing a passageway by which lymph can be returned to the vascular system.

artery: a vessel that carries blood away from the heart.

capillaries: small vessels that branch from an artery. Capillaries connect arteries to veins. Oxygen, nutrients, and waste materials are exchanged across capillary walls.

vein: a vessel that carries blood back to the heart.

hepatic portal vein: the vein that collects blood from the GI tract and conducts it to capillaries in the liver.

portal = gateway

hepatic vein: the vein that collects blood from the liver capillaries and returns it to the heart.

hepatic = liver

very-low-density lipoproteins (VLDL): the type of lipoproteins made primarily by liver cells to transport lipids to various tissues in the body; composed primarily of triglycerides.

low-density lipoproteins (LDL): the type of lipoproteins derived from VLDL as cells remove triglycerides from them. LDL carry cholesterol and triglycerides from the liver to the cells of the body and are composed primarily of cholesterol.

high-density lipoproteins (HDL): the type of lipoproteins that transport cholesterol back to the liver from peripheral cells; composed primarily of protein.

TRANSPORT OF LIPIDS: LIPOPROTEINS

Within the circulatory system, lipids always travel from place to place bundled with protein, that is, as lipoproteins. When physicians measure a person's blood lipid profile, they are interested in both the types of fat present (such as triglycerides and cholesterol) and the types of lipoproteins that carry them.

VLDL, LDL, and HDL As mentioned earlier, chylomicrons transport newly absorbed (*diet-derived*) lipids from the intestinal cells to the rest of the body. As chylomicrons circulate through the body, cells remove their lipid contents, so the chylomicrons get smaller and smaller. The liver picks up these chylomicron remnants. When necessary, the liver can assemble different lipoproteins, which are known as **very-low-density lipoproteins (VLDL)**. As the body's cells remove triglycerides from the VLDL, the proportions of their lipid and protein contents shift. As this occurs, VLDL become cholesterol-rich **low-density lipoproteins (LDL)**. Cholesterol returning to the liver for metabolism or excretion from other parts of the body is packaged in lipoproteins known as **high-density lipoproteins (HDL)**. HDL are synthesized primarily in the liver.

The density of lipoproteins varies according to the proportion of lipids and protein they contain. The more lipids in the lipoprotein molecule, the lower the density; the more protein, the higher the density. Both LDL and HDL carry lipids around in the blood, but LDL are larger, lighter, and filled with more lipid; HDL are smaller, denser, and packaged with more protein. LDL deliver cholesterol and triglycerides from the liver to the tissues; HDL scavenge excess cholesterol from the tissues and return it to the liver for metabolism or disposal. Figure 2-7 (p. 50) shows the relative sizes and compositions of the lipoproteins.

Health Implications of LDL and HDL The distinction between LDL and HDL has implications for the health of the heart and blood vessels. Elevated LDL concentrations in the blood are associated with a high risk of heart disease, and elevated HDL concentrations are associated with a low risk.[2] These associations explain why some people refer to LDL as "bad" cholesterol and HDL as "good" cholesterol. Keep in mind, though, that there is only *one* kind of cholesterol molecule; the differences between LDL and HDL reflect *proportions* of lipids and proteins within them—not the type of cholesterol. Factors that improve the LDL-to-HDL ratio include:

- Weight management (see Chapter 7).
- Polyunsaturated or monounsaturated, instead of saturated, fatty acids in the diet (see Chapter 4).
- Soluble fibers (see Chapter 3).
- Physical activity.

Lipoproteins and heart disease are discussed in Chapter 21.

FIGURE 2-6 The Liver and Its Circulatory System

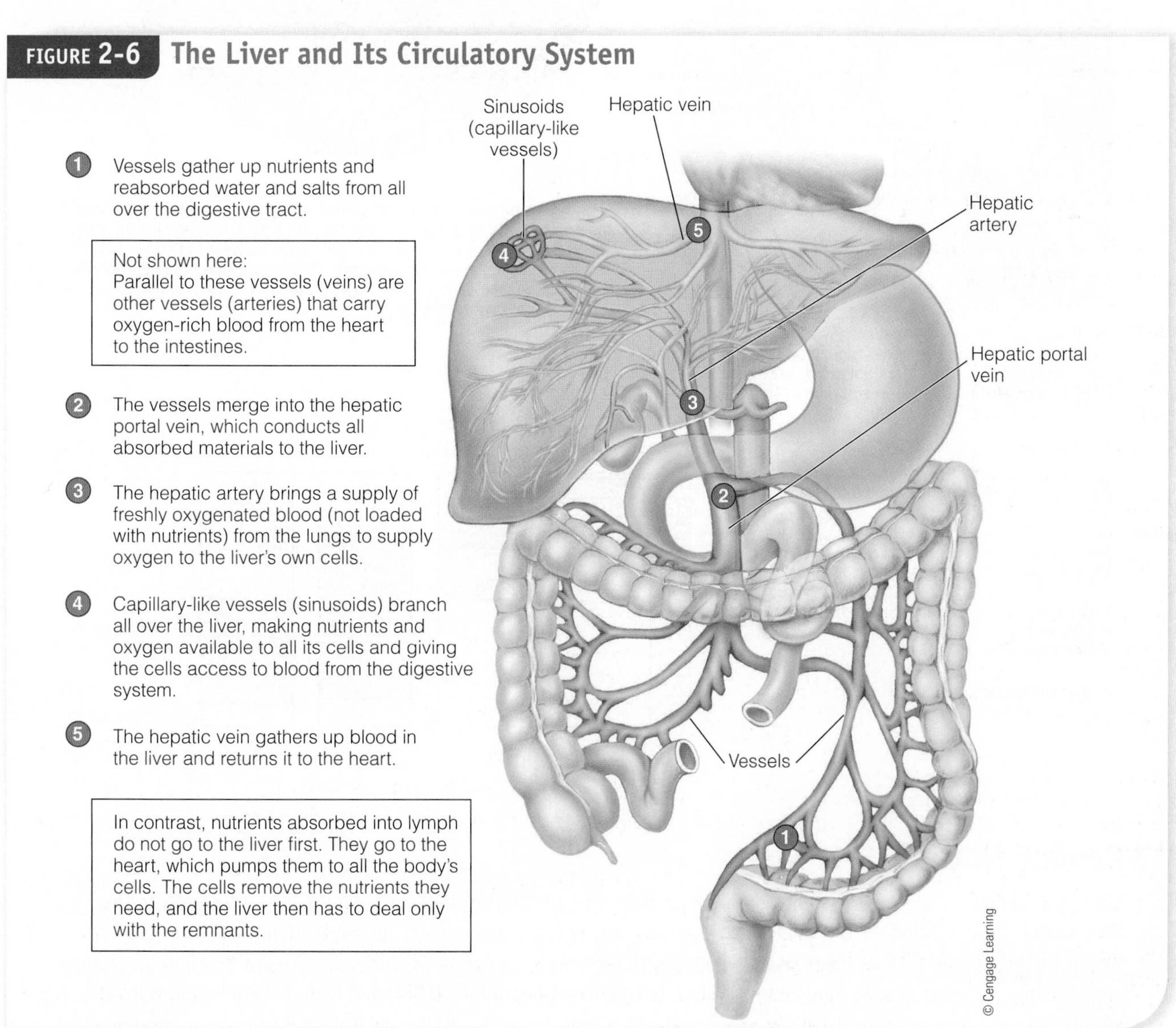

1 Vessels gather up nutrients and reabsorbed water and salts from all over the digestive tract.
Not shown here: Parallel to these vessels (veins) are other vessels (arteries) that carry oxygen-rich blood from the heart to the intestines.
2 The vessels merge into the hepatic portal vein, which conducts all absorbed materials to the liver.
3 The hepatic artery brings a supply of freshly oxygenated blood (not loaded with nutrients) from the lungs to supply oxygen to the liver's own cells.
4 Capillary-like vessels (sinusoids) branch all over the liver, making nutrients and oxygen available to all its cells and giving the cells access to blood from the digestive system.
5 The hepatic vein gathers up blood in the liver and returns it to the heart.
In contrast, nutrients absorbed into lymph do not go to the liver first. They go to the heart, which pumps them to all the body's cells. The cells remove the nutrients they need, and the liver then has to deal only with the remnants.
Sinusoids (capillary-like vessels)
Hepatic vein
Hepatic artery
Hepatic portal vein
Vessels

FIGURE 2-7 The Lipoproteins

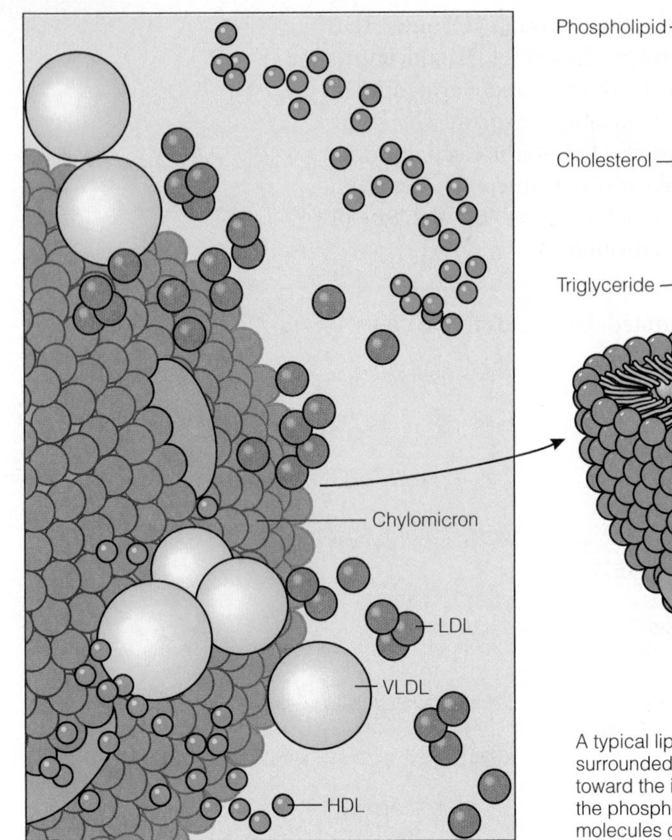

This solar system of lipoproteins shows their relative sizes. Notice how large the fat-filled chylomicron is compared with the others and how the others get progressively smaller as their proportion of fat declines and protein increases.

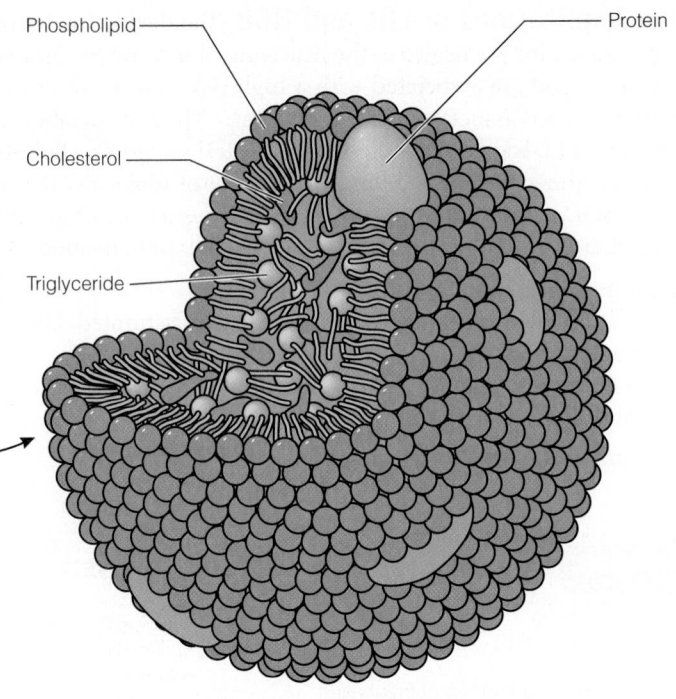

A typical lipoprotein contains an interior of triglycerides and cholesterol surrounded by phospholipids. The phospholipids' fatty acid "tails" point toward the interior, where the lipids are. Proteins near the outer ends of the phospholipids cover the structure. This arrangement of hydrophobic molecules on the inside and hydrophilic molecules on the outside allows lipids to travel through the watery fluids of the blood.

Chylomicrons contain so little protein and so much triglyceride that they are the lowest in density.

Very-low-density lipoproteins (VLDL) are half triglycerides, accounting for their low density.

Low-density lipoproteins (LDL) are half cholesterol, accounting for their implication in heart disease.

High-density lipoproteins (HDL) are half protein, accounting for their high density.

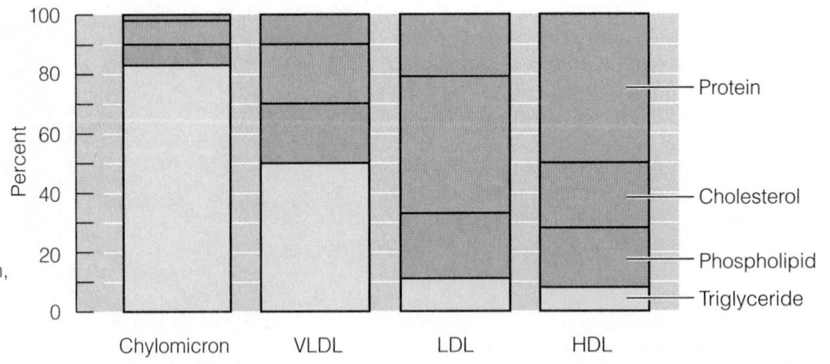

Review Notes

- Nutrients leaving the digestive system via the blood are routed directly to the liver before being transported to the body's cells. Those leaving via the lymphatic system eventually enter the vascular system but bypass the liver at first.
- Within the circulatory system, lipids travel bundled with proteins as lipoproteins. Different types of lipoproteins include chylomicrons, very-low-density lipoproteins (VLDL), low-density lipoproteins (LDL), and high-density lipoproteins (HDL).
- Elevated blood concentrations of LDL are associated with a high risk of heart disease. Elevated HDL are associated with a low risk of heart disease.

The System at Its Best

The GI tract is the first organ in the body to deal with the nutrients that will ultimately maintain the health and nutrition status of the whole body. The intricate architecture of the GI tract makes it sensitive and responsive to conditions in its environment. One condition indispensable to its performance is its own good health. Such lifestyle factors as sleep, physical activity, state of mind, and nutrition affect GI tract health. Adequate sleep allows for repair and maintenance of tissue. Physical activity promotes healthy muscle tone and may protect against cancer of the colon.[3] Mental state profoundly affects digestion and absorption through the activity of nerves and hormones that help regulate these processes. A relaxed, peaceful attitude during a meal enhances digestion and absorption.

Self Check

1. Once food is swallowed, it travels through the digestive tract in this order:
   a. esophagus, stomach, large intestine, liver.
   b. esophagus, stomach, small intestine, large intestine.
   c. small intestine, stomach, esophagus, large intestine.
   d. small intestine, large intestine, stomach, esophagus.

2. Once chyme travels the length of the small intestine, it passes through the ileocecal valve at the beginning of the:
   a. large intestine.
   b. stomach.
   c. esophagus.
   d. jejunum.

3. The periodic squeezing or partitioning of the intestine by its circular muscles that both mixes and slowly pushes the contents along is known as:
   a. secretion.
   b. absorption.
   c. peristalsis.
   d. segmentation.

4. An enzyme in saliva begins the digestion of:
   a. starch.
   b. vitamins.
   c. protein.
   d. minerals.

5. Bile is:
   a. an enzyme that splits starch.
   b. an alkaline secretion of the pancreas.
   c. an emulsifier made by the liver that prepares fats and oils for digestion.
   d. a stomach secretion containing water, hydrochloric acid, and the enzymes pepsin and lipase.

6. Which nutrient passes through the large intestine mostly unabsorbed?
   a. Fiber
   b. Vitamins
   c. Minerals
   d. Starch

7. The two major nutrient transport systems in the body are:
   a. LDL and HDL.
   b. digestion and absorption.
   c. lipoproteins and chylomicrons.
   d. the vascular and lymphatic systems.

8. Within the circulatory system, lipids always travel from place to place bundled with proteins as:
   a. microvilli.
   b. chylomicrons.
   c. lipoproteins.
   d. phospholipids.

9. Elevated LDL concentrations in the blood are associated with:
   a. a high-protein diet.
   b. a low risk of diabetes.
   c. too much physical activity.
   d. a high risk of heart disease.

10. Three factors that improve the LDL-to-HDL ratio include:
    a. polyunsaturated fat, rest, and dietary HDL.
    b. antioxidants, insoluble fibers, and dietary HDL.
    c. saturated fat, antioxidants, and insoluble fibers.
    d. weight control, soluble fibers, and physical activity.

Answers to these questions can be found in Appendix H. For more chapter review: Access an interactive eBook, chapter-specific interactive learning tools, including flashcards, quizzes, videos, and more in your Nutrition CourseMate, accessed through CengageBrain.com.

Clinical Applications

1. People who experience malabsorption frequently have the most difficulty digesting fat. Considering the differences in fat, carbohydrate, and protein digestion and absorption, can you offer an explanation?

2. How might you explain the importance of dietary fiber to a client who frequently experiences constipation?

Nutrition on the Net

For further study of the topics in this chapter, access these websites.

- Visit the Center for Digestive Health and Nutrition: www.gihealth.com
- Visit the Patients section of the American College of Gastroenterology: http://gi.org//

Notes

1. A. Stengel and Y. Taché, Neuroendocrine control of the gut during stress: Corticotropin-releasing factor signaling pathways in the spotlight, *Annual Review of Physiology* 71 (2009): 219–239; P. Holzer, Opioid receptors in the gastrointestinal tract, *Regulatory Peptides* 155 (2009): 11–17; J. López-Herce, Gastrointestinal complications in critically ill patients: What differs between adults and children? *Current Opinion in Clinical Nutrition and Metabolic Care* 12 (2009): 180–185.
2. A. K. Chhatriwalla and coauthors, Low levels of low-density lipoprotein cholesterol and blood pressure and progression of coronary atherosclerosis, *Journal of the American College of Cardiology* 53 (2009): 1110–1115; A. H. Lichtenstein and coauthors, Diet and lifestyle recommendations revision 2006: A scientific statement from the American Heart Association Nutrition Committee, *Circulation* 114 (2006): 82–96; Expert Panel on Detection, Evaluation, and Treatment of High Blood Cholesterol in Adults (Adult Treatment Panel III), *Third Report of the National Cholesterol Education Program (NCEP)*, NIH publication no. 02-5215 (Bethesda, MD: National Heart, Lung, and Blood Institute, 2002) pp. II-1–II-61.
3. L. H. Kushi and coauthors, American Cancer Society guidelines on nutrition and physical activity for cancer prevention, *CA Cancer Journal for Clinicians* 62 (2012): 30-67; World Cancer Research Fund/American Institute for Cancer Research, *Food, Nutrition, Physical Activity, and the Prevention of Cancer: A Global Perspective* (Washington, DC: American Institute for Cancer Research, 2007), pp. 244–321.

Nutrition in Practice

Food Safety

The Food and Drug Administration (FDA) lists **foodborne illness** as the leading food safety concern in the United States because **outbreaks** of food poisoning far outnumber episodes of any other kind of food contamination. The **CDC (Centers for Disease Control and Prevention)** estimates that 48 million cases of foodborne illnesses occur each year in the United States.[1] More than 100,000 people become so sick as to need hospitalization. For some 3000 people each year, the symptoms (Table NP2-1) are so severe as to cause death. Most vulnerable are pregnant women; very young, very old, sick, or malnourished people; and those with a weakened immune system (as in AIDS).[2] By taking the proper precautions, people can minimize their chances of contracting foodborne illnesses. The accompanying glossary defines related terms.

What is foodborne illness?

Foodborne illness can be caused by either an infection or an intoxication. Table NP2-2 (p. 54) summarizes the **pathogens** responsible for 90 percent of foodborne illnesses, related hospitalizations, and deaths, along with food sources, general symptoms, and prevention methods.

TABLE NP2-1 Symptoms of Foodborne Illness

Get medical help when these symptoms occur:
• Bloody diarrhea
• Diarrhea lasting more than three days
• Prolonged vomiting that prevents keeping liquids down and can lead to dehydration
• Difficulty breathing
• Difficulty swallowing
• Double vision
• Fever lasting more than 24 hours
• Headache accompanied by muscle stiffness and fever
• Numbness, muscle weakness, and tingling sensations in the skin
• Rapid heart rate, fainting, and dizziness

What is the difference between foodborne infections and food intoxications?

Foodborne infections are caused by eating foods contaminated by infectious microbes. Among foodborne

Glossary

CDC (Centers for Disease Control and Prevention): a branch of the Department of Health and Human Services that is responsible for, among other things, monitoring foodborne diseases (www.cdc.gov).

cross-contamination: the contamination of food by bacteria that occurs when the food comes into contact with surfaces previously touched by raw meat, poultry, or seafood.

foodborne illness: illness transmitted to human beings through food and water, caused by either an infectious agent (foodborne infection) or a poisonous substance (foodborne intoxication); commonly known as **food poisoning**.

Hazard Analysis Critical Control Points (HACCP): a systematic plan to identify and correct potential microbial hazards in the manufacturing, distribution, and commercial use of food products; commonly referred to as "HASS-ip."

outbreaks: two or more cases of a similar illness resulting from the ingestion of a common food.

pasteurization: heat processing of food that inactivates some, but not all, microorganisms in the food; not a sterilization process. Bacteria that cause spoilage are still present.

pathogens (PATH-oh-jens): microorganisms capable of producing disease.

sushi: vinegar-flavored rice and seafood, typically wrapped in seaweed and stuffed with colorful vegetables. Some sushi is stuffed with raw fish; other varieties contain cooked seafood.

traveler's diarrhea: nausea, vomiting, and diarrhea caused by consuming food or water contaminated by any of several organisms, most commonly, *E. coli*, *Shigella*, *Campylobacter jejuni*, and *Salmonella*.

TABLE NP2-2 The Major Microbes of Foodborne Illnesses

Organism Name	Most Frequent Food Sources	Onset and General Symptoms	Prevention Methods[a]
Foodborne Infections			
Campylobacter (KAM-pee-loh-BAK-ter) bacterium	Raw and undercooked poultry, unpasteurized milk, contaminated water	Onset: 2 to 5 days. Diarrhea, vomiting, abdominal cramps, fever; sometimes bloody stools; lasts 2 to 10 days.	Cook foods thoroughly; use pasteurized milk; use sanitary food-handling methods.
Escherichia (esh-uh-REEK-ee-uh) ***coli*** (KOH-lye) bacterium (including Shiga toxin-producing strains)[b]	Undercooked ground beef, unpasteurized milk and juices, raw fruits and vegetables, contaminated water, and person-to-person contact	Onset: 1 to 8 days. Severe bloody diarrhea, abdominal cramps, vomiting; lasts 5 to 10 days.	Cook ground beef thoroughly; use pasteurized milk; use sanitary food-handling methods; use treated, boiled, or bottled water.
Norovirus	Person-to-person contact; raw foods, salads, sandwiches	Onset: 1 to 2 days. Vomiting; lasts 1 to 2 days.	Use sanitary food-handling methods.
Listeria (lis-TER-ee-AH) bacterium	Unpasteurized milk; fresh soft cheeses; luncheon meats, hot dogs	Onset: 1 to 21 days. Fever, muscle aches; nausea, vomiting, blood poisoning, complications in pregnancy, and meningitis (stiff neck, severe headache, and fever).	Use sanitary food-handling methods; cook foods thoroughly; use pasteurized milk.
Clostridium (claw-STRID-ee-um) ***perfringens*** (per-FRINGE-enz) bacterium	Meats and meat products stored at between 120°F and 130°F	Onset: 8 to 16 hours. Abdominal pain, diarrhea, nausea; lasts 1 to 2 days.	Use sanitary food-handling methods; use pasteurized milk; cook foods thoroughly; refrigerate foods promptly and properly.
Salmonella (sal-moh-NEL-ah) bacteria (>2300 types)	Raw or undercooked eggs, meats, poultry, raw milk and other dairy products, shrimp, frog legs, yeast, coconut, pasta, and chocolate	Onset: 1 to 3 days. Fever, vomiting, abdominal cramps, diarrhea; lasts 4 to 7 days; can be fatal.	Use sanitary food-handling methods; use pasteurized milk; cook foods thoroughly; refrigerate foods promptly and properly.
Toxoplasma (TOK-so-PLAZ-ma) ***gondii*** parasite	Raw or undercooked meat; contaminated water; unpasteurized goat's milk; contact with infected cat feces	Onset: 7 to 21 days. Swollen glands, fever, headache, muscle pain, stiff neck.	Use sanitary food-handling methods; cook foods thoroughly.

continued

infections, norovirus and *Salmonella* are the leading causes of hospitalizations and deaths.[3] Pathogens commonly enter the GI tract in contaminated foods such as undercooked poultry and unpasteurized milk. Symptoms generally include abdominal cramps, fever, vomiting, and diarrhea.

Norovirus illustrates the importance of personal hygiene.[4] Norovirus is passed in the stool and vomit of infected people. Thus, infected people who do not wash their hands adequately can pass the virus directly to other people, or they can pass it indirectly by way of contaminated food or water. Outbreaks in the United States are often linked to food touched by infected food handlers or to person-to-person contact in day care centers, in nursing homes, and on cruise ships.[5] People can also be infected with norovirus by eating raw shellfish such as oysters and clams that are grown in sewage-contaminated waters.

continued

Organism Name	Most Frequent Food Sources	Onset and General Symptoms	Prevention Methods[a]
Foodborne Intoxications			
Clostridium (claw-STRID-ee-um) ***botulinum*** (bot-chew-LINE-um) bacterium produces botulin toxin, responsible for causing botulism	Anaerobic environment of low acidity (canned corn, peppers, green beans, soups, beets, asparagus, mushrooms, ripe olives, spinach, tuna, chicken, chicken liver, liver pâté, luncheon meats, ham, sausage, stuffed eggplant, lobster, and smoked and salted fish)	Onset: 4 to 36 hours. Nervous system symptoms, including double vision, inability to swallow, speech difficulty, and progressive paralysis of the respiratory system; often fatal; leaves prolonged symptoms in survivors.	Use proper canning methods for low-acid foods; refrigerate homemade garlic and herb oils; avoid commercially prepared foods with leaky seals or with bent, bulging, or broken cans. Do not give infants honey because it may contain spores of *Clostridium botulinum*, which is a common source of infection for infants.
Staphylococcus (STAF-il-oh-KOK-us) ***aureus*** bacterium produces staphylococcal toxin	Toxin produced in improperly refrigerated meats; egg, tuna, potato, and macaroni salads; cream-filled pastries	Onset: 1 to 6 hours. Diarrhea, nausea, vomiting, abdominal cramps, fever; lasts 1 to 2 days.	Use sanitary food-handling methods; cook food thoroughly; refrigerate foods promptly and properly; use proper home-canning methods.

Note: Travelers' diarrhea is most commonly caused by *E. coli, Campylobacter jejuni, Shigella,* and *Salmonella.*
[a]Table NP2-3 on pp. 57–59 provides more details on the proper handling, cooking, and refrigeration of foods.
[b]0157, 0145, and other Shiga toxin-producing strains.

Food intoxications are caused by eating foods containing natural toxins or, more likely, microbes that produce toxins. The most common food toxin is produced by *Staphylococcus aureus*; it affects more than 1 million people each year. Less common, but more infamous, is *Clostridium botulinum*, an organism that produces a deadly toxin in anaerobic conditions such as improperly canned (especially home-canned) foods and homemade garlic or herb-flavored oils stored at room temperature. The botulism toxin paralyzes muscles, making it difficult to see, speak, swallow, and breathe. Because death can occur within 24 hours of onset, botulism demands immediate medical attention. Even then, survivors may suffer the effects for months or years.

How do people get foodborne illness?

Transmission of foodborne illness has changed as our food supply and lifestyles have changed.[6] In the past, foodborne illness was caused by one person's error in a small setting, such as improperly refrigerated egg salad at a family picnic, and affected only a few victims. Today, we are eating more foods prepared and packaged by others. Consequently, when a food manufacturer or restaurant chef makes an error, foodborne illness can quickly affect many people. An estimated 80 percent of reported foodborne illnesses are caused by errors in a commercial setting, such as the improper **pasteurization** of milk at a large dairy.

In 2009, *Salmonella* was found in peanut butter that had been used in more than 2100 products made by more than 200 companies; another *Salmonella* outbreak in 2010 led to the recall of 500 million eggs from two farms. In 2011, a cantaloupe farm had to recall more than 300,000 cases of fruit when *Listeria* poisoning killed 29 people and made 139 others sick. These incidents and others focus the national spotlight on two important safety issues: disease-causing organisms are commonly found in foods, and safe food-handling practices can minimize harm from most of these foodborne pathogens.

What kinds of programs are in place to help keep foods safe?

To improve the safety of the U.S. food supply, the Food Safety Modernization Act (FSMA) was signed into law in 2011. It has been called "historic" because it shifts the focus of FDA activities from reacting *after* people become ill to preventing foodborne illness in the first place.[7] The new law stresses prevention at food-processing facilities; provides the FDA with greater enforcement, inspection, and recall authorities; and affords the FDA greater oversight of imported foods.

In addition, the U.S. Department of Agriculture (USDA), the FDA, and the food-processing industries have developed and implemented programs to control foodborne illness.* For example, USDA inspectors examine meat-processing plants every day to ensure that these facilities meet government standards. Seafood, egg, produce, and processed food facilities are inspected less often, but all food producers must use a **Hazard Analysis Critical Control Points (HACCP)** plan to help prevent foodborne illnesses at their source. Each slaughterhouse, packer, distributor, and transporter of susceptible foods must identify "critical control points" that pose a risk of contamination and implement verifiable procedures to eliminate or minimize the risk. The HACCP system has proved a remarkable success for domestic products, but such programs do not apply to imported foods.

An estimated $2 trillion worth of products are imported into the United States from more than 150 countries each year. Many countries cooperate with the FDA and have adopted many of the safe food-handling practices used in the United States, but some imported foods come from countries with little or no regulatory oversight. To help consumers distinguish between imported and domestic foods, certain foods—including fish, shellfish, meats, fruits, vegetables, and some nuts—must display a Country of Origin Label specifying where they were produced.[8]

Importantly, the implementation of the FSMA strengthens the FDA's ability to safeguard imported foods.[9] Under the FSMA, the FDA is permitted to inspect foreign facilities. If a food producer in another country does not allow the FDA to inspect its facility, the FDA can refuse to allow food from that facility into the United States. The FSMA also requires importers to verify the food safety practices of their suppliers and adds new checks on imported foods.

Are foods bought in grocery stores and foods eaten in restaurants safe?

Canned and packaged foods sold in grocery stores are easily controlled, but rare accidents do happen. Batch numbering makes it possible to recall contaminated foods through public announcements via the Internet, newspapers, television, and radio. In the grocery store, consumers can buy items before the "sell by" date and inspect the safety seals and wrappers of packages. A broken seal, bulging can lid, or mangled package fails to protect the consumer against microbes, insects, spoilage, or even vandalism.

State and local health regulations provide guidelines on the cleanliness of facilities and the safe preparation of foods for restaurants, cafeterias, and fast-food establishments. Even so, consumers should take these actions to help prevent foodborne illnesses when dining out:

- Wash hands with hot, soapy water before meals.
- Expect clean tabletops, dinnerware, utensils, and food preparation areas.
- Expect cooked foods to be served piping hot and salads to be fresh and cold.
- Refrigerate take-home items within two hours and use leftovers within three to four days.

Improper handling of foods can occur anywhere along the line from commercial manufacturers to large supermarkets to small restaurants to private homes. Maintaining a safe food supply requires everyone's efforts.

What can people do to protect themselves from foodborne illness?

Whether microbes multiply and cause illness depends, in part, on a few key food-handling behaviors in the kitchen—whether the kitchen is in your home, a school cafeteria, a gourmet restaurant, or a canned goods manufacturer. Figure NP2-1 summarizes the four simple things that can help most to prevent foodborne illness:

- *Clean.* Keep a clean, safe kitchen by washing hands and surfaces often. Wash countertops, cutting boards, sponges, and utensils in hot, soapy water before and after each step of food preparation. To reduce bacterial contamination on hands, wash hands with soap and warm water; if soap and water are not available, use an alcohol-based sanitizing gel.[10]
- *Separate.* Avoid cross-contamination by keeping raw eggs, meat, poultry, and seafood separate from other foods. Wash all utensils and surfaces (such as cutting boards or platters) that have been in contact with these foods with hot, soapy water before using them again. Bacteria inevitably left on the surfaces from the raw meat can recontaminate the cooked meat or other

FIGURE NP2-1 Fight Bac!

Four ways to keep food safe. The Fight Bac! website is at www.fightbac.org.

*In addition to HACCP, other programs initiated under the Food Safety Initiative include FoodNet, PulseNet, the Environmental Health specialists Network (EHS-Net), and Fight BAC!

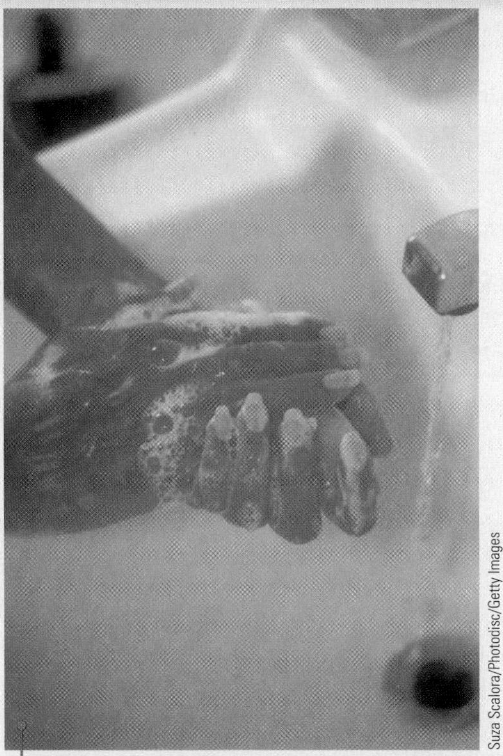

Suza Scalora/Photodisc/Getty Images

Wash your hands with warm water and soap for at least 20 seconds before preparing or eating food to reduce the chance of microbial contamination.

foods—a problem known as **cross-contamination**. Washing raw eggs, meat, and poultry is not recommended as the extra handling increases the risk of cross-contamination.

- *Cook.* Keep hot foods hot by cooking to proper temperatures. Foods need to cook long enough to reach internal temperatures that will kill microbes, and maintain adequate temperatures to prevent bacterial growth until the foods are served.
- *Chill.* Keep cold foods cold by refrigerating promptly. Go directly home upon leaving the grocery store and immediately unpack foods into the refrigerator or freezer upon arrival. After a meal, refrigerate any leftovers immediately.

Unfortunately, consumers commonly fail to follow these simple food-handling recommendations. See Table NP2-3 for additional food safety tips.

What precautions need to be taken when preparing meat and poultry?

Figure NP2-2 presents label instructions for the safe handling of meat and poultry and two types of USDA seals. Meats and poultry contain bacteria and provide a moist, nutrient-rich environment that favors microbial growth. Ground meat is especially susceptible because it receives more handling than other kinds of meat and

FIGURE NP2-2 Meat and Poultry Safety, Grading, and Inspection Seals

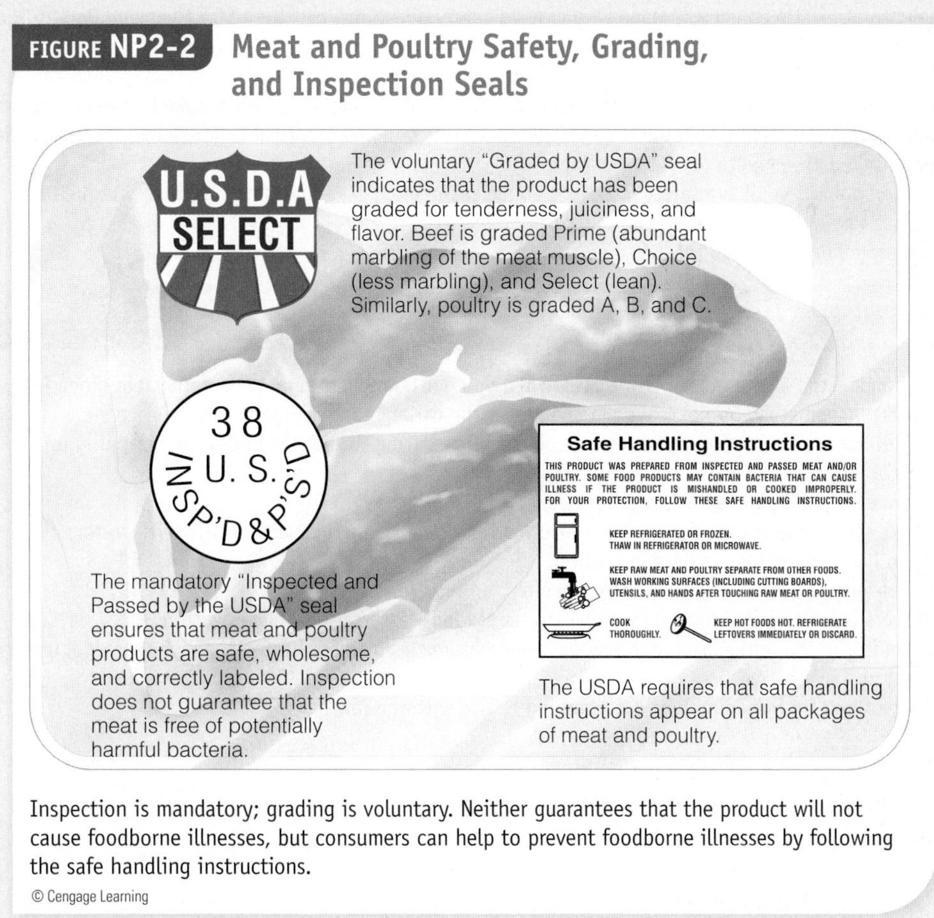

U.S.D.A
SELECT
The voluntary "Graded by USDA" seal indicates that the product has been graded for tenderness, juiciness, and flavor. Beef is graded Prime (abundant marbling of the meat muscle), Choice (less marbling), and Select (lean). Similarly, poultry is graded A, B, and C.
38
U. S.
INSP'D & P'S'D
The mandatory "Inspected and Passed by the USDA" seal ensures that meat and poultry products are safe, wholesome, and correctly labeled. Inspection does not guarantee that the meat is free of potentially harmful bacteria.
Safe Handling Instructions
THIS PRODUCT WAS PREPARED FROM INSPECTED AND PASSED MEAT AND/OR POULTRY. SOME FOOD PRODUCTS MAY CONTAIN BACTERIA THAT CAN CAUSE ILLNESS IF THE PRODUCT IS MISHANDLED OR COOKED IMPROPERLY. FOR YOUR PROTECTION, FOLLOW THESE SAFE HANDLING INSTRUCTIONS.
KEEP REFRIGERATED OR FROZEN. THAW IN REFRIGERATOR OR MICROWAVE.
KEEP RAW MEAT AND POULTRY SEPARATE FROM OTHER FOODS. WASH WORKING SURFACES (INCLUDING CUTTING BOARDS), UTENSILS, AND HANDS AFTER TOUCHING RAW MEAT OR POULTRY.
COOK THOROUGHLY.
KEEP HOT FOODS HOT. REFRIGERATE LEFTOVERS IMMEDIATELY OR DISCARD.
The USDA requires that safe handling instructions appear on all packages of meat and poultry.

Inspection is mandatory; grading is voluntary. Neither guarantees that the product will not cause foodborne illnesses, but consumers can help to prevent foodborne illnesses by following the safe handling instructions.

TABLE NP2-3 Strategies to Prevent Foodborne Illnesses

Most foodborne illnesses can be prevented by following four simple rules: clean, separate, cook, and chill.

Clean

- Wash fruits and vegetables in a clean sink with a scrub brush and warm water; store washed and unwashed produce separately.
- Use hot, soapy water to wash hands, utensils, dishes, nonporous cutting boards, and countertops before handling food and between tasks when working with different foods. Use a bleach solution on cutting boards (one capful per gallon of water).
- Cover cuts with clean bandages before food preparation; dirty bandages carry harmful microorganisms.
- Mix foods with utensils, not hands; keep hands and utensils away from mouth, nose, and hair.
- Anyone may be a carrier of bacteria and should avoid coughing or sneezing over food. A person with a skin infection or infectious disease should not prepare food.
- Wash or replace sponges and towels regularly.
- Clean up food spills and crumb-filled crevices.

Separate

- Wash all surfaces that have been in contact with raw meats, poultry, eggs, fish, and shellfish before reusing.
- Serve cooked foods on a clean plate with a clean utensil. Separate raw foods from those that have been cooked.
- Don't use marinade that was in contact with raw meat for basting or sauces.

Cook

- When cooking meats or poultry, use a thermometer to test the internal temperature. Insert the thermometer between the thigh and the body of a turkey or into the thickest part of other meats, making sure the tip of the thermometer is not in contact with bone or the pan. Cook to the temperature indicated for that particular meat (see Figure NP2-3 on p. 60); cook hamburgers to at least medium well done. If you have safety questions, call the USDA Meat and Poultry Hotline: (800) 535-4555.
- Cook stuffing separately, or stuff poultry just prior to cooking.
- Do not cook large cuts of meat or turkey in a microwave oven; it leaves some parts undercooked while overcooking others.
- Cook eggs before eating them (soft-boiled for at least 3½ minutes; scrambled until set, not runny; fried for at least 3 minutes on one side and 1 minute on the other).
- Cook seafood thoroughly. If you have safety questions about seafood, call the FDA hotline: (800) FDA-4010.
- When serving foods, maintain temperatures at 140°F or higher.
- Heat leftovers thoroughly to at least 165°F.

Chill

- When running errands, stop at the grocery store last. When you get home, refrigerate the perishable groceries (such as meats and dairy products) immediately. Do not leave perishables in the car any longer than it takes for ice cream to melt.
- Put packages of raw meat, fish, or poultry on a plate before refrigerating to prevent juices from dripping on food stored below.
- Buy only foods that are solidly frozen in store freezers.
- Keep cold foods at 40° F or less; keep frozen foods at 0°F or less (keep a thermometer in the refrigerator).
- Marinate meats in the refrigerator, not on the counter.
- Look for "Keep Refrigerated" or "Refrigerate After Opening" on food labels.
- Refrigerate leftovers promptly; use shallow containers to cool foods faster; use leftovers within 3 to 4 days.
- Thaw meats or poultry in the refrigerator, not at room temperature. If you must hasten thawing, use cool water (changed every 30 minutes) or a microwave oven.
- Freeze meat, fish, or poultry immediately if not planning to use within a few days.

continued

continued

In General

- Do not reuse disposable containers; use nondisposable containers or recycle instead.
- Do not taste food that is suspect. "If in doubt, throw it out."
- Throw out foods with danger-signaling odors. Be aware, though, that most food-poisoning bacteria are odorless, colorless, and tasteless.
- Do not buy or use items that have broken seals or mangled packaging; such containers cannot protect against microbes, insects, spoilage, or even vandalism. Check safety seals, buttons, and expiration dates.
- Follow label instructions for storing and preparing packaged and frozen foods; throw out foods that have been thawed or refrozen.
- Discard foods that are discolored, moldy, or decayed or that have been contaminated by insects or rodents.

For Specific Food Items

- *Canned goods.* Carefully discard food from cans that leak or bulge so that other people and animals will not accidentally ingest it; before canning, seek professional advice from the USDA National Institute of Food and Agriculture (check your phone book under U.S. government listings, or go online: www.csrees.usda.gov).
- *Milk and cheeses.* Use only pasteurized milk and milk products. Aged cheeses, such as cheddar and Swiss, do well for an hour or two without refrigeration, but they should be refrigerated or stored in an ice chest for longer periods.
- *Eggs.* Use clean eggs with intact shells. Do not eat eggs, even pasteurized eggs, raw; raw eggs are commonly found in Caesar salad dressing, eggnog, cookie dough, hollandaise sauce, and key lime pie. Cook eggs until whites are firmly set and yolks begin to thicken.
- *Honey.* Honey may contain dormant bacterial spores, which can awaken in the human body to produce botulism. In adults, this poses little hazard, but infants younger than 1 year of age should never be fed honey. Honey can accumulate enough toxin to kill an infant; it has been implicated in several cases of sudden infant death. (Honey can also be contaminated with environmental pollutants picked up by the bees.)
- *Mayonnaise.* Commercial mayonnaise may actually help a food to resist spoilage because of the acid content. Still, keep it refrigerated after opening.
- *Mixed salads.* Mixed salads of chopped ingredients spoil easily because they have extensive surface area for bacteria to invade, and they have been in contact with cutting boards, hands, and kitchen utensils that easily transmit bacteria to food (regardless of their mayonnaise content). Chill them well before, during, and after serving.
- *Picnic foods.* Choose foods that last without refrigeration, such as fresh fruits and vegetables, breads and crackers, and canned spreads and cheeses that can be opened and used immediately. Pack foods cold, layer ice between foods, and keep foods out of water.
- *Seafood.* Buy only fresh seafood that has been properly refrigerated or iced. Cooked seafood should be stored separately from raw seafood to avoid cross-contamination.

has more surface exposed to bacterial contamination. Consumers cannot detect the harmful bacteria in or on meat. For safety's sake, cook meat thoroughly, using a thermometer to test the internal temperature (see Figure NP2-3 on p. 60).

How can a person enjoy seafood safely?

Most seafood available in the United States and Canada is safe, but eating it undercooked or raw can cause severe illnesses—hepatitis, worms, parasites, viral intestinal disorders, and other diseases.* Rumor has it that freezing fish will make it safe to eat raw, but this is only partly true. Commercial freezing will kill mature parasitic worms, but only cooking can kill all worm eggs and other microorganisms that can cause illness. For safety's sake, all seafood should be cooked until it is opaque. Even **sushi** can be safe to eat when chefs combine cooked seafood and other ingredients into delicacies.

Eating raw oysters can be dangerous for anyone, but people with liver disease and weakened immune systems are most vulnerable. At least 10 species of bacteria found in raw oysters can cause serious illness and even death. Raw oysters may also carry the hepatitis A virus, which can cause liver disease. Some hot sauces can kill many of these bacteria but not the virus; alcohol may also protect

*Diseases caused by toxins from the sea include ciguatera poisoning, scombroid poisoning, and paralytic and neurotoxic shellfish poisoning.

some people against some oyster-borne illnesses but not enough to guarantee protection (or to recommend drinking alcohol). Pasteurization of raw oysters—holding them at a specified temperature for a specified time—holds promise for killing bacteria without cooking the oyster or altering its texture or flavor.

As population density increases along the shores of seafood-harvesting waters, pollution inevitably invades the sea life there. Preventing seafood-borne illness is in large part a task of controlling water pollution. To help ensure a safe seafood market, the FDA requires processors to adopt food safety practices based on the HACCP system mentioned earlier.

Chemical pollution and microbial contamination lurk not only in the water but also in the boats and warehouses where seafood is cleaned, prepared, and refrigerated. Seafood is one of the most perishable foods: time and temperature are critical to its freshness and flavor. To keep seafood as fresh as possible, people in the industry "keep it cold, keep it clean, and keep it moving." Wise consumers eat it cooked.

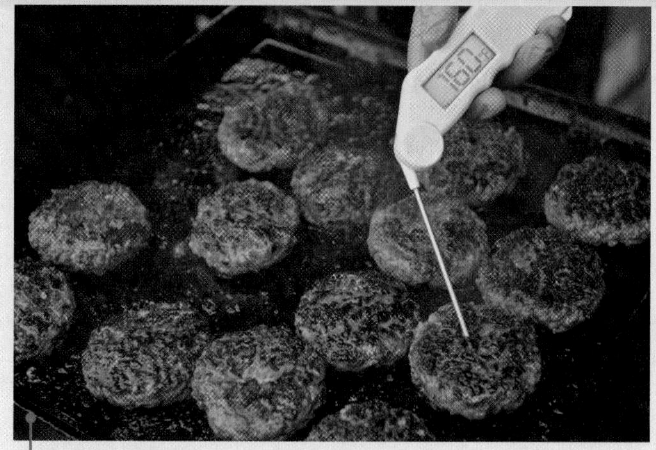

Cook hamburgers to 160°F; color alone cannot determine doneness. Some burgers will turn brown before reaching 160°F, whereas others may retain some pink color, even when cooked to 175°F.

© Charles Stirling/Alamy

FIGURE NP2-3 Recommended Safe Temperatures (Fahrenheit)

- Well-done meats
- 170° — Stuffing, all poultry, including ground chicken and turkey; reheat leftovers
- 165° — Medium-done meats, raw eggs, egg dishes, ground meats (beef, veal, lamb, pork)
- 160° — Beef, pork, lamb, and veal (steaks, roasts, and chops): Allow to rest at least 3 minutes. During the 3 minutes after meat is removed from the heat source, its temperature remains constant or continues to rise, which destroys pathogens.
- 145°
- 140° — Hold hot foods
- DANGER ZONE: Do not keep foods between 40°F and 140°F for more than 2 hours or for more than 1 hour when the air temperature is greater than 90°F.
- 40° — Refrigerator temperatures
- 0° — Freezer temperatures

Bacteria multiply rapidly at temperatures between 40°F and 140°F. Cook foods to the temperatures shown on this thermometer and hold them at 140°F or higher.

Eating raw seafood is a risky proposition.

David Chasey/Photodisc/Getty Images

Do foods that are unsafe to eat smell bad?

Fresh food generally smells fresh. Not all types of food poisoning are detectable by odor, but some bacterial wastes produce "off" odors. Food with an abnormal odor is spoiled. Throw it out or, if it was recently purchased, return it to the grocery store. Do not taste it. Table NP2-4 lists safe refrigerator storage times for selected foods.

Local health departments and the USDA Extension Service can provide additional information about food safety. Should precautions fail and mild foodborne illness develop, drink clear liquids to replace fluids lost through vomiting and diarrhea. If serious foodborne illness is suspected, first call a physician. Then wrap the remainder of the suspected food and label the container so that the food cannot be mistakenly eaten, place it in the refrigerator, and hold it for possible inspection by health authorities.

How can a person defend against foodborne illness when traveling to foreign countries?

People who travel to other countries have a 50-50 chance of contracting a foodborne illness, commonly described as **traveler's diarrhea**. Like many other foodborne illnesses, travelers' diarrhea is a sometimes serious, always annoying bacterial infection of the digestive tract. The risk is high because some countries' cleanliness standards for food and water may be lower than those in the United States and Canada. Also, every region's microbes are different, and while people are immune to those in their own neighborhoods, they have had no chance to develop immunity to the pathogens in places they are visiting for the first time. In addition to the food safety tips outlined on pp. 58–59, precautions while traveling include:

- Wash hands frequently with soap and hot water, especially before handling food or eating. Use sanitizing gel or hand wipes regularly.
- Eat only well-cooked and hot or canned foods. Eat raw fruits or vegetables only if washed in purified water and peeled with clean hands.
- Use purified, bottled water for drinking, making ice cubes, and brushing teeth. Alternatively, use disinfecting tablets or boil water.
- Refuse dairy products that have not been pasteurized and refrigerated properly.
- Travel with antidiarrheal medication in case efforts to avoid illness fail.

To sum up these recommendations, "Boil it, cook it, peel it, or forget it."

TABLE NP2-4 Safe Refrigerator Storage Times (≤40°F)

Storage Time	Foods
1 to 2 Days	Raw ground meats, breakfast or other raw sausages, raw fish or poultry; gravies
3 to 5 Days	Raw steaks, roasts, or chops; cooked meats, poultry, vegetables, and mixed dishes; lunchmeats (packages opened); mayonnaise salads (chicken, egg, pasta, tuna); fresh vegetables (spinach, green beans, tomatoes)
1 Week	Hard-cooked eggs, bacon or hot dogs (opened packages); smoked sausages or seafood; milk, cottage cheese
1 to 2 Weeks	Yogurt; carrots, celery, lettuce
2 to 4 Weeks	Fresh eggs (in shells); lunchmeats, bacon, or hot dogs (packages unopened); dry sausages (pepperoni, hard salami); most aged and processed cheeses (Swiss, brick)
2 Months	Mayonnaise (opened jar); most dry cheeses (Parmesan, Romano)

Notes

1. M. T. Osterholm, Foodborne disease in 2011—The rest of the story, *New England Journal of Medicine* 364 (2011): 889–891; Centers for Disease Control and Prevention, Press release: New estimates more precise, December 15, 2010, available at www.cdc.gov/media/pressrel/2010/r101215.html.
2. B. M. Lund and S. J. O'Brien, The occurrence and prevention of foodborne disease in vulnerable people, *Foodborne Pathogens and Disease* 8 (2011): 961–973.
3. L. H. Gould and coauthors, Surveillance for foodborne disease outbreaks—United States, 2008, *Morbidity and Mortality Weekly Report* 60 (2011): 1197–1202.
4. Centers for Disease Control and Prevention, Surveillance for foodborne disease outbreaks—United States, 2006, *Morbidity and Mortality Weekly Report* 58 (2009): 609–614.
5. L. Verhoef and coauthors, Emergence of new norovirus variants on spring cruise ships and prediction of winter epidemics, *Emerging Infectious Diseases* 14 (2008): 238–243.
6. Position of the American Dietetic Association: Food and water safety, *Journal of the American Dietetic Association* 109 (2009): 1449–1460.
7. K. Stewart and L. O. Gostin, Food and Drug Administration regulation of food safety, *Journal of the American Medical Association* 306 (2011): 88–89.
8. Department of Agriculture, Mandatory Country of Origin Labeling of beef, pork, lamb, chicken, goat meat, wild and farm-raised fish and shellfish, perishable agricultural commodities, peanuts, pecans, ginseng, and macadamia nuts, *Federal Register* 74 (2009): 2658–2707.
9. Stewart and Gostin, Food and Drug Administration regulation of food safety, 2011.
10. Centers for Disease Control and Prevention, Handwashing: Clean hands save lives, www.cdc.gov/handwashing, accessed October 18, 2012.

Chapter 3 Carbohydrates

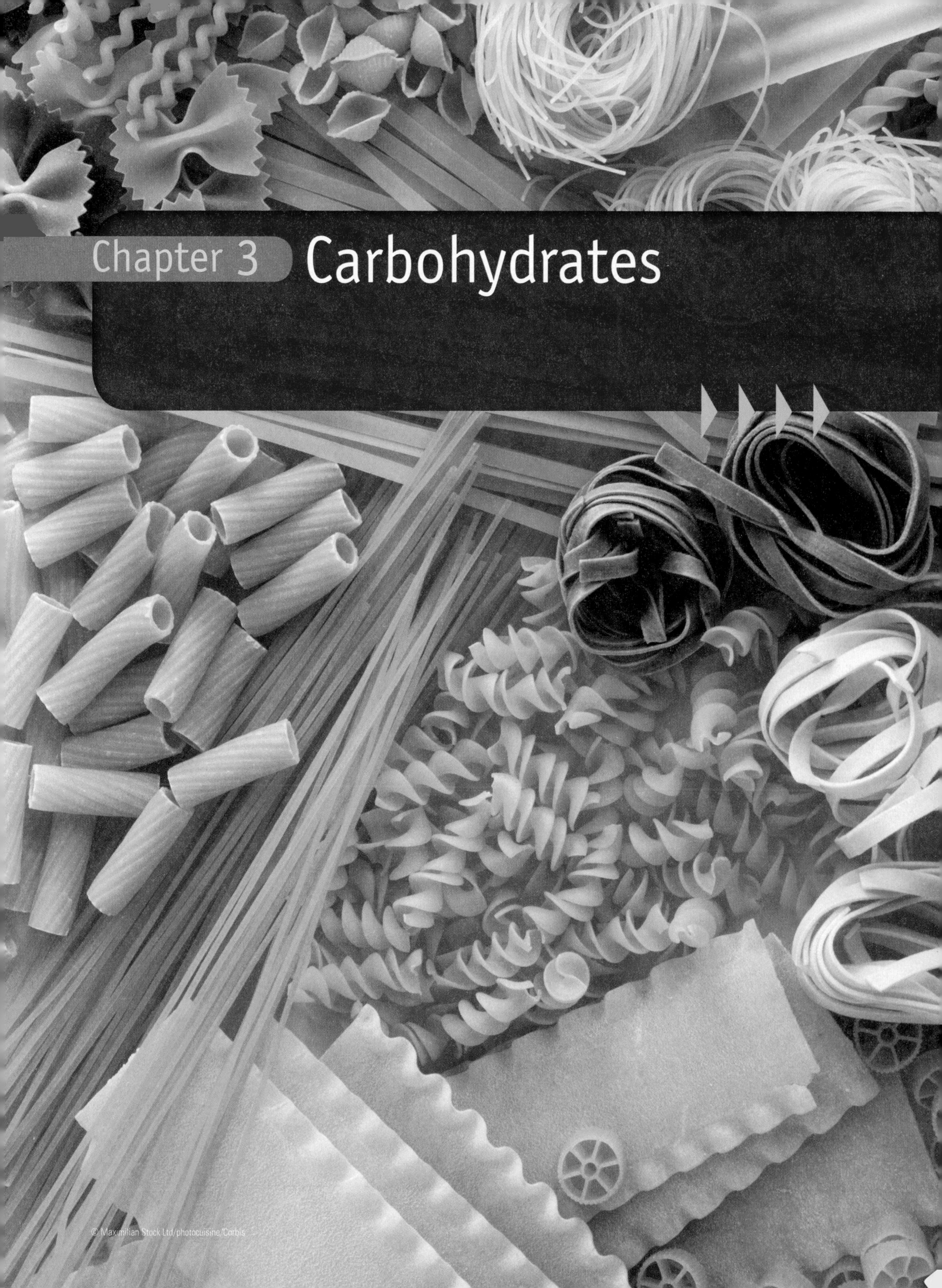

MOST PEOPLE WOULD LIKE TO FEEL GOOD ALL THE TIME. PART OF THE secret of feeling good is replenishing the body's energy supply with food. Carbohydrate is the preferred energy source for many of the body's functions. As long as carbohydrate is available, the human brain depends exclusively on it as an energy source. Athletes eat a "high-carb" diet to store as much muscle fuel as possible, and dietary recommendations urge people to eat carbohydrate-rich foods for better health. Many people, however, mistakenly think of carbohydrate-rich foods as "fattening" and avoid them. In truth, people who wish to lose fat, maintain lean tissue, and stay healthy can best do so by being physically active, paying close attention to portion sizes, and designing an eating pattern based on foods that supply carbohydrate in balance with other energy nutrients.[1] Most unrefined plant foods—whole grains, vegetables, legumes, and fruits—provide ample carbohydrate and fiber with little or no fat. Milk is the only animal-derived food that contains significant amounts of carbohydrate.

This chapter on carbohydrates is the first of three on the energy-yielding nutrients. Fats are the topic of Chapter 4 and protein is featured in Chapter 6. Nutrition in Practice 19 addresses one other contributor of energy to the human diet, alcohol. Alcohol, of course, has well-known undesirable side effects when used in excess.

carbohydrates: energy nutrients composed of monosaccharides.

carbo = carbon
hydrate = water

monosaccharides (mon-oh-SACK-uh-rides): single sugar units.

mono = one
saccharide = sugar

disaccharides (dye-SACK-uh-rides): pairs of sugar units bonded together.

di = two

polysaccharides: long chains of monosaccharide units arranged as starch, glycogen, or fiber.

poly = many

glucose: a monosaccharide; the sugar common to all disaccharides and polysaccharides; also called *blood sugar* or *dextrose*.

fructose: a monosaccharide; sometimes known as *fruit sugar*.

fruct = fruit

galactose: a monosaccharide; part of the disaccharide lactose.

The Chemist's View of Carbohydrates

The dietary **carbohydrates** include the sugars, starch, and fiber. Chemists describe the sugars as:

- **Monosaccharides** (single sugars).
- **Disaccharides** (double sugars).

Starch and fiber are:*

- **Polysaccharides**—compounds composed of chains of monosaccharide units.

All of these carbohydrates are composed of the single sugar **glucose** and other compounds that are much like glucose in composition and structure. Figure 3-1 shows the chemical structure of glucose.

MONOSACCHARIDES

Three monosaccharides are important in nutrition: glucose, **fructose**, and **galactose**. All three monosaccharides have the same number and kinds of atoms but in different arrangements.

Glucose

Most cells depend on glucose for their fuel to some extent, and the cells of the brain and the rest of the nervous system depend almost exclusively on glucose for their energy. The body can obtain this glucose from carbohydrates. To function optimally, the body must maintain blood glucose within limits that allow the cells to nourish themselves. A later section describes blood glucose regulation.

Fructose

Fructose is the sweetest of the sugars. Fructose occurs naturally in fruits, in honey, and as part of table sugar. However, most fructose is consumed in sweet beverages such as soft drinks, in ready-to-eat cereals, and in other products sweetened with **high-fructose corn syrup** or other **added sugars**. Glucose and fructose are the most common monosaccharides in nature.

Galactose

The third single sugar, galactose, occurs mostly as part of lactose, a disaccharide also known as milk sugar. During digestion, galactose is freed as a single sugar.

Polara Studios, Inc.

Grains, vegetables, legumes, fruits, and milk offer ample carbohydrate.

*Monosaccharides and disaccharides (sugars) are sometimes called *simple carbohydrates*, and the polysaccharides (starch and fiber) are sometimes called *complex carbohydrates*.

DISACCHARIDES

In disaccharides, pairs of single sugars are linked together. Three disaccharides are important in nutrition: maltose, sucrose, and lactose. All three contain glucose as one of their single sugars. As Table 3-1 shows, the other monosaccharide is either another glucose (in maltose), fructose (in sucrose), or galactose (in lactose). The shapes of the sugars in Table 3-1 reflect their chemical structures as drawn on paper.

FIGURE 3-1 Chemical Structure of Glucose

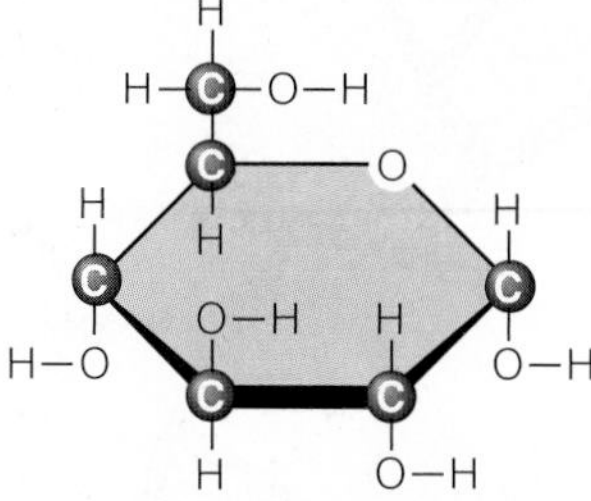

On paper, the structure of glucose has to be drawn flat, but in nature the five carbons and oxygen are roughly in a plane, with the H, OH, and CH_2OH extending out above and below it.

Sucrose

Sucrose (table, or white, sugar) is the most familiar of the three disaccharides and is what people mean when they speak of "sugar." This sugar is usually obtained by refining the juice from sugar beets or sugarcane to provide the brown, white, and powdered sugars available in the supermarket, but it occurs naturally in many fruits and vegetables. When a person eats a food containing sucrose, enzymes in the digestive tract split the sucrose into its glucose and fructose components.

Lactose

Lactose is the principal carbohydrate of milk. Most human infants are born with the digestive enzymes necessary to split lactose into its two monosaccharide parts, glucose and galactose, so as to absorb it. Breast milk thus provides a simple, easily digested carbohydrate that meets an infant's energy needs; many formulas do, too, because they are made from milk. (■)

■ Many people lose the ability to digest lactose after infancy. This condition, known as lactose intolerance, is discussed in Chapter 18.

Maltose

The third disaccharide, **maltose**, is a plant sugar that consists of two glucose units. Maltose is produced whenever starch breaks down—as happens in plants when they break down their stored starch for energy and start to sprout and in human beings during carbohydrate digestion.

POLYSACCHARIDES

Unlike the sugars, which contain the three monosaccharides—glucose, fructose, and galactose—in different combinations, the polysaccharides are composed almost entirely of glucose (and, in some cases, other monosaccharides). Three types of polysaccharides are important in nutrition: glycogen, starch, and fibers.

Glycogen is a storage form of energy for human beings and animals; starch plays that role in plants; and fibers provide structure in stems, trunks, roots, leaves, and skins of plants. Both glycogen and starch are built entirely of glucose units; fibers are composed of a variety of monosaccharides and other carbohydrate derivatives.

Glycogen

Glycogen molecules, which are also made of chains of glucose, are more highly branched than starch molecules (see the left side of Figure 3-2 on p. 66). Glycogen is found in meats only to a limited extent and not at all in plants.* For this reason, glycogen is not a significant food source of carbohydrate, but it does play an important role in the body. The human body stores much of its glucose as glycogen in the liver and muscles.

TABLE 3-1 The Major Sugars

Monosaccharides	Disaccharides
Glucose	Sucrose (glucose + fructose)
Fructose	Lactose (glucose + galactose)
Galactose (found mostly as part of lactose)	Maltose (glucose + glucose)

*Glycogen in animal muscles rapidly breaks down after slaughter.

high-fructose corn syrup: a widely used commercial kcaloric sweetener made by adding enzymes to cornstarch to convert a portion of its glucose molecules into sweet-tasting fructose.

added sugars: sugars, syrups, and other kcaloric sweeteners that are added to foods during processing, preparation, or at the table. Added sugars do not include the naturally occurring sugars found in fruits and milk products. Also called *carbohydrate sweeteners,* they include glucose, fructose, high-fructose corn syrup, concentrated fruit juice, and other sweet carbohydrates. Also defined in Chapter 1.

sucrose: a disaccharide composed of glucose and fructose; commonly known as *table sugar, beet sugar,* or *cane sugar.*

sucro = sugar

lactose: a disaccharide composed of glucose and galactose; commonly known as *milk sugar.*

lact = milk

maltose: a disaccharide composed of two glucose units; sometimes known as *malt sugar.*

glycogen (GLY-co-gen): a polysaccharide composed of glucose, made and stored by liver and muscle tissues of human beings and animals as a storage form of glucose. Glycogen is not a significant food source of carbohydrate and is not counted as one of the polysaccharides in foods.

FIGURE 3-2 Glycogen and Starch Compared

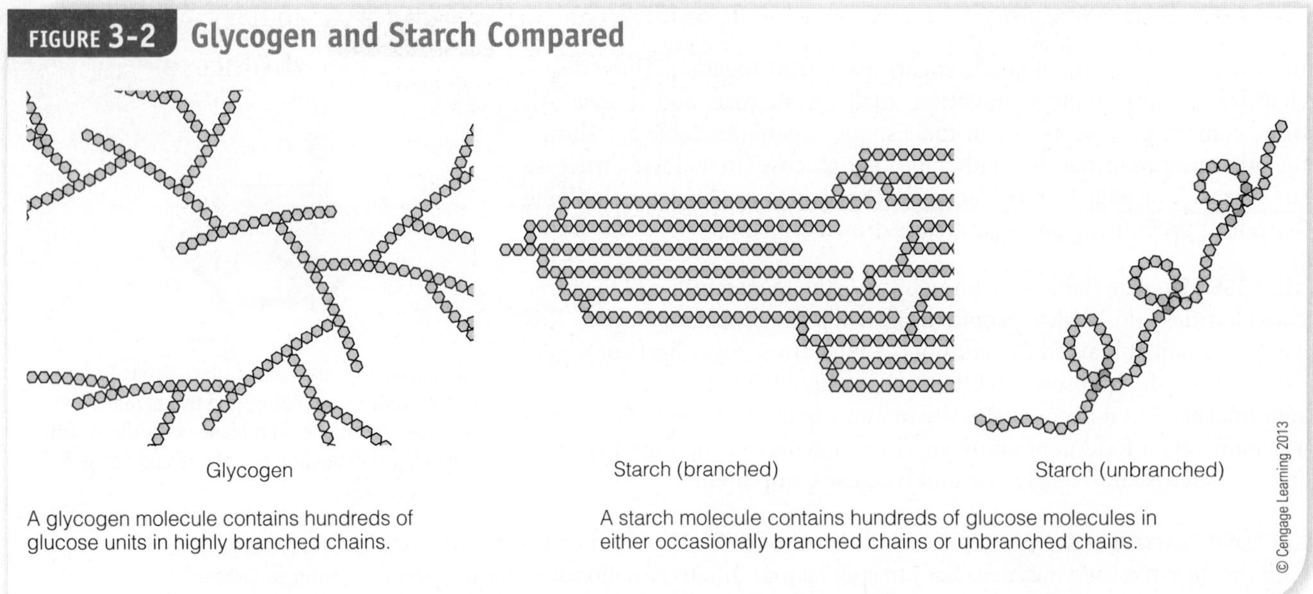

A glycogen molecule contains hundreds of glucose units in highly branched chains.

A starch molecule contains hundreds of glucose molecules in either occasionally branched chains or unbranched chains.

Starch **Starch** is a long, straight or branched chain of hundreds or thousands of glucose units linked together (see the middle and right side of Figure 3-2). These giant molecules are packed side by side in grains such as rice or wheat, in root crops and tubers such as yams and potatoes, and in legumes such as peas and beans. When a person eats the plant, the body splits the starch into glucose units (■) and uses the glucose for energy.

All starchy foods come from plants. Grains are the richest food source of starch. In most human societies, people depend on a staple grain for much of their food energy: rice in Asia; wheat in Canada, the United States, and Europe; corn in much of Central and South America; and millet, rye, barley, and oats elsewhere. A second important source of starch is the legume (bean and pea) family. Legumes include peanuts and "dry" beans such as butter beans, kidney beans, "baked" beans, black-eyed peas (cowpeas), chickpeas (garbanzo beans), and soybeans. Root vegetables (tubers) such as potatoes and yams are a third major source of starch, and in many non-Western societies, they are the primary starch sources. Grains, legumes, and tubers not only are rich in starch but may also contain abundant dietary fiber, protein, and other nutrients.

■ The short chains of glucose units that result from the breakdown of starch are known as *dextrins*. The word sometimes appears on food labels because dextrins can be used as thickening agents in foods.

Fibers **Dietary fibers** are the structural parts of plants and thus are found in all plant-derived foods—vegetables, fruits, **whole grains**, and legumes. Most dietary fibers are polysaccharides—chains of sugars—just as starch is, but in fibers the sugar units are held together by bonds that human digestive enzymes cannot break. Consequently, most dietary fibers pass through the body, providing little or no energy for its use. Figure 3-3 illustrates the difference in the bonds that link glucose molecules together in starch and those found in the plant fiber cellulose. In addition to cellulose, fibers include the polysaccharides hemicellulose, pectins, gums, and mucilages, as well as the nonpolysaccharide lignins.

starch: a plant polysaccharide composed of glucose and digestible by human beings.

dietary fibers: a general term denoting in plant foods the polysaccharides cellulose, hemicellulose, pectins, gums and mucilages, as well as the nonpolysaccharide lignins, which are not digested by human digestive enzymes, although some are digested by GI tract bacteria.

whole grains: grains or foods made from them that contain all the essential parts and naturally occurring nutrients of the entire grain seed (except the inedible husk).

FIGURE 3-3 The Bonds of Starch and Cellulose Molecules Compared (Small Segments)

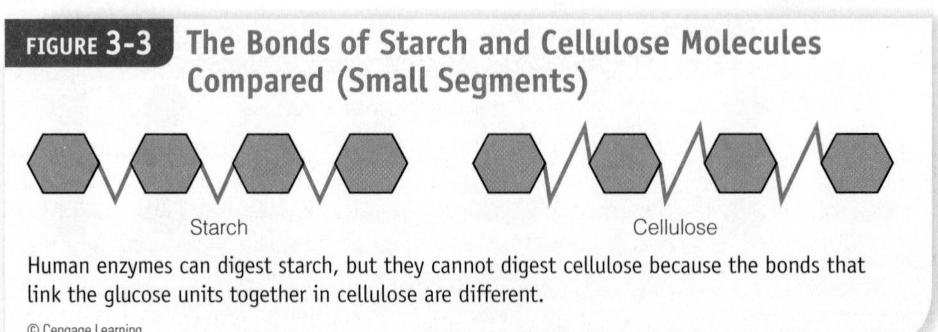

Human enzymes can digest starch, but they cannot digest cellulose because the bonds that link the glucose units together in cellulose are different.

Cellulose is the main constituent of plant cell walls, so it is found in all vegetables, fruits, and legumes. Hemicellulose is the main constituent of cereal fibers. Pectins are abundant in vegetables and fruits, especially citrus fruits and apples. The food industry uses pectins to thicken jelly and keep salad dressing from separating. Gums and mucilages have similar structures and are used as additives or stabilizers by the food industry. Lignins are the tough, woody parts of plants; few foods people eat contain much lignin.

A few starches are classified as fibers. Known as **resistant starches**, these starches escape digestion and absorption in the small intestine. Starch may resist digestion for several reasons, including the individual's efficiency in digesting starches and the food's physical properties. Resistant starch is common in whole or partially milled grains, legumes, raw potatoes, and unripe bananas. Cooked potatoes, pasta, and rice that have been chilled also contain resistant starch. Similar to some fibers, resistant starch may support a healthy colon.[2]

Although cellulose and other dietary fibers are not broken down by human enzymes, some fibers can be digested by bacteria in the human digestive tract. Bacterial **fermentation** of fibers can generate some absorbable products that can yield energy when metabolized. Food fibers, therefore, can contribute some energy (1.5 to 2.5 kcalories per gram), depending on the extent to which they break down in the body.

Fibers are divided into two general groups by their chemical and physical properties.* In the first group are fibers that dissolve in water (**soluble fibers**). These form gels (are **viscous**) and are more readily digested by bacteria in the human large intestine (are easily fermented). Commonly found in barley, legumes, fruits, oats, and vegetables, these fibers are often associated with lower risks of chronic diseases (as discussed in a later section). In foods, soluble fibers add a pleasing consistency; for example, pectin puts the gel in jelly, and gums are added to salad dressings and other foods to thicken them.

Other fibers do not dissolve in water (**insoluble fibers**), do not form gels (are not viscous), and are less readily fermented. Insoluble fibers, such as cellulose and many hemicelluloses, are found in the outer layers of whole grains (bran), the strings of celery, the hulls of seeds, and the skins of corn kernels. These fibers retain their structure and rough texture even after hours of cooking. In the body, they aid the digestive system by easing elimination.[3]

Starch- and fiber-rich foods are the foods to emphasize.

Review Notes

- Carbohydrate is the body's preferred energy source. Six sugars are important in nutrition: the three monosaccharides (glucose, fructose, and galactose) and the three disaccharides (sucrose, lactose, and maltose).
- The three disaccharides are pairs of monosaccharides; each contains glucose paired with one of the three monosaccharides. The polysaccharides (chains of monosaccharides) are glycogen, starches, and fibers.
- Both glycogen and starch are storage forms of glucose—glycogen in the body and starch in plants—and both yield energy for human use.
- The dietary fibers also contain glucose (and other monosaccharides), but their bonds cannot be broken by human digestive enzymes, so they yield little, if any, energy.

Digestion and Absorption of Carbohydrates

The ultimate goal of digestion and absorption of sugars and starches is to break them into small molecules—chiefly glucose—that the body can absorb and use. The large starch molecules require extensive breakdown; the disaccharides need only be broken once and the monosaccharides not at all. Most fiber passes intact through the small

*The DRI committee has proposed these fiber definitions: the term *dietary fibers* refers to naturally occurring fibers in intact foods, and *functional fibers* refers to added fibers that have health benefits; *total fiber* refers to the sum of fibers from both sources.

resistant starches: starches that escape digestion and absorption in the small intestine of healthy people.

fermentation: the anaerobic (without oxygen) breakdown of carbohydrates by microorganisms that releases small organic compounds along with carbon dioxide and energy.

soluble fibers: indigestible food components that readily dissolve in water and often impart gummy or gel-like characteristics to foods. An example is pectin from fruit, which is used to thicken jellies.

viscous: having a gel-like consistency.

insoluble fibers: the tough, fibrous structures of fruits, vegetables, and grains; indigestible food components that do not dissolve in water.

■ Diabetes, a disorder characterized by elevated blood glucose, is the topic of Chapter 20.

homeostasis (HOME-ee-oh-STAY-sis): the maintenance of constant internal conditions (such as chemistry, temperature, and blood pressure) by the body's control system.

homeo = the same
stasis = staying

insulin: a hormone secreted by the pancreas in response to (among other things) high blood glucose. It promotes cellular glucose uptake for use or storage.

glucagon (GLOO-ka-gon): a hormone that is secreted by special cells in the pancreas in response to low blood glucose concentration and that elicits release of glucose from storage.

intestine to the large intestine. There, bacteria digest many soluble fibers to produce short-chain fatty acids, which are rapidly absorbed by the large intestine. Table 3-2 provides the details.

Regulation of Blood Glucose

If blood glucose falls below normal, a person may become dizzy and weak; if it rises substantially above normal, the person may become fatigued. (■) Left untreated, fluctuations to the extremes—either high or low—can be fatal. Blood glucose **homeostasis** is regulated primarily by two hormones: **insulin**, which moves glucose from the blood into the cells, and **glucagon**, which brings glucose out of storage when blood glucose falls (as occurs between meals).

Insulin After a meal, as blood glucose rises, the pancreas is the first organ to respond. It releases the hormone insulin, which signals body tissues to take up surplus glucose. Muscle tissue responds to insulin by taking up excess blood glucose and using it to make glycogen. The liver takes up excess blood glucose, too, but it needs no help from insulin to do so. Instead, the liver cells respond to insulin by speeding up their glycogen production. Adipose (fat) tissue also responds to insulin by both taking up

TABLE 3-2 Carbohydrate Digestion and Absorption

Sugar and Starch	Fiber
Mouth and salivary glands The salivary glands secrete saliva into the mouth to moisten the food. The salivary enzyme amylase begins digestion: Starch —(Amylase)→ Small polysaccharides, maltose	**Mouth** The mechanical action of the mouth crushes and tears fiber in food and mixes it with saliva to moisten it for swallowing.
Stomach Stomach acid inactivates salivary enzymes, halting starch digestion.	**Stomach** Fiber is not digested, and it delays gastric emptying.
Small intestine and pancreas The pancreas produces an amylase that is released through the pancreatic duct into the small intestine: Starch —(Pancreatic amylase)→ Small polysaccharides, maltose Then disaccharidase enzymes on the surface of the small intestinal cells hydrolyze the disaccharides into monosaccharides: Maltose —(Maltase)→ Glucose + Glucose Sucrose —(Sucrase)→ Fructose + Glucose Lactose —(Lactase)→ Galactose + Glucose Intestinal cells absorb these monosaccharides.	**Small intestine** Fiber is not digested but passes intact through to the large intestine. It delays absorption of some nutrients. **Large intestine** Bacterial enzymes digest many soluble fibers, producing short-chain fatty acids that are preferentially and immediately absorbed by the large intestine. Soluble fibers —(Bacterial enzymes)→ Short-chain fatty acids, gas Most insoluble fibers remain intact, retain some water, and bind substances such as bile, cholesterol, and some minerals, carrying them out of the body.

blood glucose and slowing its release of the fat stored within its cells.[4] Simply put, insulin regulates blood glucose by:

- Facilitating blood glucose uptake by the muscles and adipose tissue.
- Stimulating glycogen synthesis in the liver.

The muscles hoard two-thirds of the body's total glycogen to ensure that glucose, a critical fuel for physical activity, is available for muscular work. The brain stores a tiny fraction of the total, thought to provide an emergency glucose reserve sufficient to fuel the brain for an hour or two in severe glucose deprivation. The liver stores the remainder and is generous with its glycogen, making it available as blood glucose for the brain or other tissues when the supply runs low. Without carbohydrate from food to replenish it, the liver glycogen stores can be depleted in less than a day. Balanced meals and snacks, eaten on a regular schedule, help the body to maintain its blood glucose. Meals with starch and soluble fiber combined with some protein and a little fat slow digestion so that glucose enters the blood gradually in an ongoing, steady rate.

The Release of Glucose from Glycogen The glycogen molecule is highly branched with hundreds of ends bristling from each molecule's surface (review this structure in Figure 3-2 on p. 66). When blood glucose starts to fall too low, the hormone glucagon is released into the bloodstream and triggers the breakdown of liver glycogen to single glucose molecules. Enzymes in liver cells respond to glucagon by attacking a multitude of glycogen ends simultaneously to release a surge of glucose into the blood for use by all the body's cells. Thus, the highly branched structure of glycogen uniquely suits the purpose of releasing glucose on demand.

Health Effects of Sugars and Alternative Sweeteners

Fiber-rich carbohydrate foods such as vegetables, whole grains, legumes, and fruits should predominate in people's diets; the **naturally occurring sugars** in these foods and in milk are acceptable because they are accompanied by many nutrients. In contrast, concentrated sweets such as candy, cola beverages and other soft drinks, cookies, pies, cakes, and other foods with added sugars add kcalories, but few, if any, other nutrients or fiber. The *Dietary Guidelines for Americans* offer clear advice on added sugars: reduce intake.[5] People who want to limit their use of sugar may choose from two sets of alternative sweeteners: sugar alcohols and nonnutritive sweeteners.

naturally occurring sugars: sugars that are not added to a food but are present as its original constituents, such as the sugars of fruit or milk.

SUGARS

Recent decades have seen a dramatic upward trend in consumption of added sugars. All kinds of sugary foods and beverages taste delicious, cost little money, and are constantly available, making overconsumption extremely likely. Though people are adding less sugar in the kitchen, food manufacturers are adding plenty to foods during processing. Soft drinks and other sugar-sweetened beverages are the main source of added sugars in the diets of U.S. consumers (see Figure 3-4 on p. 70).[6] Most people can afford only a little added sugar in their diets if they are to meet nutrient needs within kcalorie limits.

© Polara Studios, Inc.

Sugary soft drinks are the leading source of added sugars in the United States; cakes, cookies, pies, and other baked goods come next; and sweetened fruit drinks and punches follow closely behind.

The U.S. population is not alone in increasing sugar consumption. The trend is a worldwide phenomenon. In response, the World Health Organization has also taken a stand on sugar intake: consume no more than 10 percent of total kcalories from added sugars.*

The increase in sugar consumption has raised many questions about sugar's effects on health. In moderate amounts, sugars add pleasure to meals without harming health.[7] In excess, however, sugars can be detrimental, and the average American diet

*The World Health Organization uses the term *free sugars* to mean all monosaccharides and disaccharides added to foods by the manufacturer, cook, or consumer plus the naturally occurring sugars in honey, syrups, and fruit juices.

FIGURE 3-4 Sources of Added Sugars in the U.S. Diet[a]

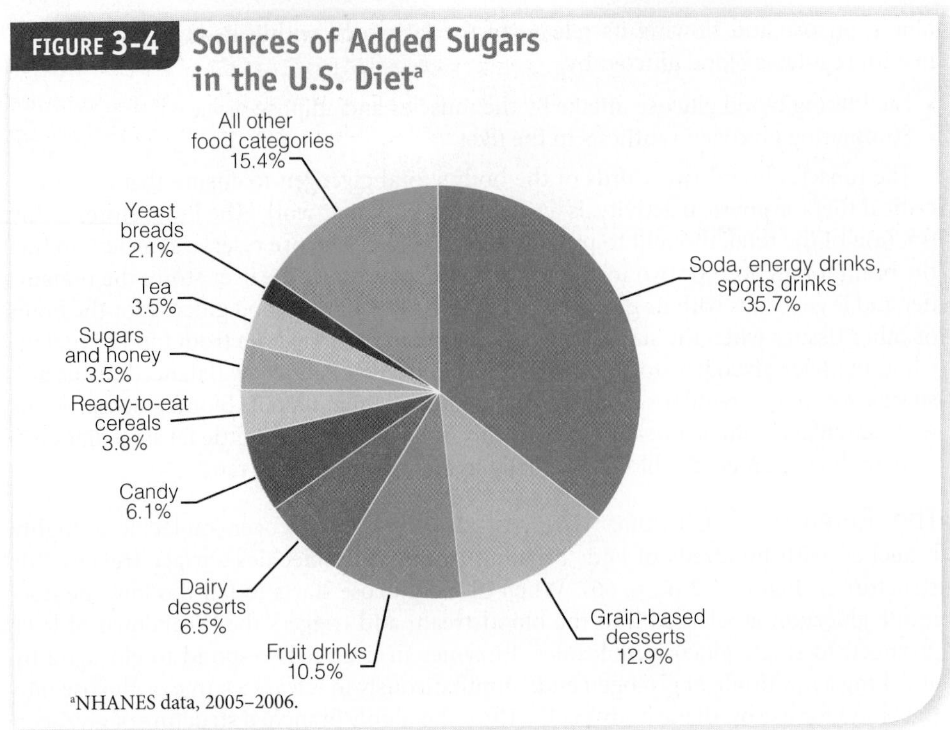

[a]NHANES data, 2005–2006.

Source: U.S. Department of Agriculture and U.S. Department of Health and Human Services, *Dietary Guidelines for Americans 2010*, available at www.dietaryguidelines.gov. Figure 3-6, p. 29.

currently delivers excessive amounts.[8] Mounting evidence links high intakes of added sugars with obesity and other chronic diseases.[9] Sugars can also contribute to nutrient deficiencies by supplying energy (kcalories) without providing nutrients and to tooth decay, or **dental caries**.

Sugar and Obesity Over the past several decades, as obesity rates increased sharply, consumption of added sugars reached an all-time high—largely because high-fructose corn syrup use, especially in beverages, surged.[10] High-fructose corn syrup is composed of fructose and glucose in a ratio of about 50:50. Compared with sucrose, high-fructose corn syrup is less expensive, easier to use, and more stable. In addition to being used in beverages, high-fructose corn syrup sweetens candies, baked goods, and hundreds of other foods. The use of high-fructose corn syrup sweetener parallels unprecedented increases in the incidence of obesity, but does this mean that the increasing sugar intakes are responsible for the increase in body fat and its associated health problems?[11] Excess sugar in the diet may be associated with more fat on the body.[12] When they are eaten in excess of need, energy from added sugars contributes to body fat stores, raising the risk of weight gain.[13] When total energy intake is controlled, however, *moderate* amounts of sugar do not *cause* obesity. Thus, to the extent that sugar contributes to an excessive energy intake, it can play a role in the development of obesity.

The liquid form of sugar in soft drinks makes it especially easy to overconsume kcalories. Swallowing liquid kcalories requires little effort. The sugar kcalories of sweet beverages also cost less than many other energy sources, and they are widely available. Also, beverages are energy-dense, providing more than 150 kcalories per 12-ounce can, and many people drink several cans a day. The convenience, economy, availability, and flavors of sugary foods and beverages encourage overconsumption.

Limiting selections of foods and beverages high in added sugars can be an effective weight-loss strategy, especially for people whose excess kcalories come primarily from added sugars. Replacing a can of cola with a glass of water every day, for example, can help a person lose a pound (or at least not gain a pound) in one month.[14] That may not sound like much, but it adds up to more than 10 pounds a year, for very little effort.

dental caries: the gradual decay and disintegration of a tooth.

Sugar and Heart Disease Added sugars also influence the balance between the body's fat-making and fat-clearing mechanisms, a balance that plays critical roles in the development of heart disease.[15] Fructose stimulates the body's fat-making pathways and impairs its fat-clearing pathways in ways that could lead to an unhealthy buildup of blood lipids (triglycerides, discussed in Chapter 5).[16]

To determine whether added sugars raise blood lipids, researchers studied more than 6000 healthy adults. They observed that people with higher intakes of added sugars had blood lipid values indicating an increased risk of heart disease.[17] In another study, young adults who consumed more sugar-sweetened beverages had greater abdominal fatness, more harmful blood lipids, and higher blood pressure than those who drank fewer.[18] It may not take an unrealistic amount of added sugars to cause this effect. As little as the equivalent of one or two fructose- or sucrose-sweetened soft drinks a day consumed for only a few weeks significantly changed blood lipids in ways that may pose risks to the heart and arteries.[19]

Sugar and Nutrient Deficiencies Empty-kcalorie foods that contain lots of added sugars such as cakes, candies, and sodas provide the body with glucose and energy, but few, if any, other nutrients. By comparison, foods such as whole grains, vegetables, legumes, and fruits that contain some natural sugars and lots of starches and fibers also provide protein, vitamins, and minerals.

A person spending 200 kcalories of a day's energy allowance on a 16-ounce soda gets little of value for those kcalories. In contrast, a person using 200 kcalories on three slices of whole-wheat bread gets 9 grams of protein, 6 grams of fiber, plus several of the B vitamins with those kcalories. For the person who wants something sweet, a reasonable compromise might be two slices of bread with a teaspoon of jam on each. The amount of sugar a person can afford to eat depends on how many discretionary kcalories are available beyond those needed to deliver indispensable vitamins and minerals.

By following the USDA Food Patterns and making careful food selections, a typical adult can obtain all the needed nutrients within an allowance of about 1500 kcalories. Some people have more generous energy allowances. For example, an active teenage boy may need as many as 3000 kcalories a day. If he eats mostly nutritious foods, then he may have discretionary kcalories available for cola beverages and other "extras." In contrast, an inactive older woman who is limited to fewer than 1500 kcalories a day can afford to eat only the most nutrient-dense foods—with few, or no, discretionary kcalories available.

Added sugars contribute to nutrient deficiencies by displacing nutrients. For nutrition's sake, the appropriate attitude to take is not that sugar is "bad" and must be avoided, but that nutritious foods must come first. If nutritious foods crowd sugar out of the diet, that is fine—but not the other way around. As always, balance, variety, and moderation guide healthy food choices.

Sugar and Dental Caries Does sugar contribute to dental caries? The evidence says yes. Any carbohydrate-containing food, including bread, bananas, or milk, as well as sugar, can support bacterial growth in the mouth. These bacteria produce the acid that eats away tooth enamel. Of major importance is the length of time the food stays in the mouth. This, in turn, depends on the composition of the food, how sticky the food is, how often a person eats the food, and especially whether the teeth are brushed afterward. Total sugar intake still plays a major role in caries incidence; populations whose diets provide no more than 10 percent of kcalories from sugar have a low prevalence of dental caries. The *Dietary Guidelines for Americans* recommend a combined approach to prevent dental caries—practicing good oral hygiene, drinking fluoridated water, and reducing the amount of time sugars and starches are in the mouth. Nutrition in Practice 17 discusses nutrition and oral health.

Recommended Sugar Intakes Moderate sugar intakes are not harmful and make eating more enjoyable, but the average U.S. intake is well above moderate. Thus, the *Dietary Guidelines for Americans* urge people to "reduce the intake of kcalories from added sugars." These added sugar kcalories (and those from solid fats and alcohol) are considered discretionary kcalories—and most people need to limit their intake. The USDA Food Patterns suggest about 8 teaspoons of sugar, about the amount in one 12-ounce soft drink, in a nutrient-dense 2200-kcalorie eating pattern (the margin lists other amounts). (■) The USDA Food Patterns recommendations represent about 5 to 10 percent of the day's total energy intake. As noted earlier, the World Health Organization agrees that people should restrict their consumption of added sugars to 10 percent or less of total energy. The accompanying "How To" provides strategies for reducing the intake of added sugars.

The DRI committee did not set a Tolerable Upper Intake Level for added sugars, but, as mentioned, excessive intakes can interfere with sound nutrition and good health. Few people can eat lots of sugary treats and still meet all of their nutrient needs without exceeding their kcalorie allowance. Instead, the DRI committee suggests that added sugars should account for no more than 25 percent of the day's total energy intake.[20] One out of eight in the U.S. population exceeds this rather high maximum intake.[21] For a person consuming 2000 kcalories a day, 25 percent represents 500 kcalories from added sugars—quite a lot of sugar. (■) Perhaps an athlete in training whose energy needs are high can afford the added sugars from sports drinks without compromising nutrient intake, but most people would do better following recommendations to limit added sugar consumption to less than 10 percent of the day's total energy intake. Added sugars contribute an average of about 16 percent of the total energy in the typical American diet.[22]

■ The USDA Food Patterns suggest:*
- 4 tsp for 1600 kcal
- 5 tsp for 1800 kcal
- 8 tsp for 2000 kcal
- 8 tsp for 2200 kcal
- 10 tsp for 2400 kcal

■ For perspective, each of these sources of concentrated sugars provides about 500 kcalories:
- 40 oz cola
- 80 oz sports drink
- ½ cup honey
- 125 jelly beans
- 23 marshmallows
- 30 tsp sugar

Recognizing Sugars People often fail to recognize sugar in all its forms and so do not realize how much they consume. To help your clients estimate their sugar intakes, tell them to treat all of the following concentrated sweets as equivalent to 1 teaspoon of white sugar (4 grams of carbohydrate):

- 1 teaspoon brown sugar, candy, jam, jelly, any corn sweetener, syrup, honey, molasses, or maple sugar or syrup
- 1 tablespoon ketchup
- 1½ ounces carbonated soft drink

These portions of sugar all provide about the same number of kcalories. Some are closer to 10 kcalories (for example, 14 kcalories for molasses), whereas some are more than 20 kcalories (22 kcalories for honey), so an average figure of 16 kcalories is an acceptable approximation. The accompanying glossary presents the multitude of names that denote sugar on food labels.

HOW TO Reduce Intakes of Added Sugars

- Use less table sugar when preparing meals and at the table.
- Use your sugar kcalories to sweeten nutrient-dense foods (such as oatmeal) instead of consuming empty-kcalorie foods and beverages (such as candy and soda).
- Replace empty-kcalorie-rich regular sodas, sports drinks, energy drinks, and fruit drinks with water, fat-free milk, 100% fruit juice, or unsweetened tea or coffee.
- Use sweet spices such as cinnamon, nutmeg, allspice, or clove.
- Select fruit for dessert. Eat less cake, cookies, ice cream, other desserts, and candy. If you do eat these foods, have a small portion.
- Warm up sweet foods before serving (heat enhances sweet tastes).
- Read the Nutrition Facts on labels to choose foods with less sugar. Compare the unsweetened version of a food (such as corn flakes) with the sweetened version (such as frosted corn flakes) to estimate the quantity of added sugars.
- Read the ingredients list to identify foods with little or no added sugars. A food is likely to be high in added sugars if its ingredient list starts with any of the sugars named in the glossary on p. 73, or if it includes several of them.

*The amounts of added sugars suggested here are *not* specific recommendations for amounts of added sugars to consume but rather represent the amounts that can be included in the diet at each kcalorie level.

Glossary of Added Sugars

brown sugar: refined white sugar with molasses added; 95 percent pure sucrose.

concentrated fruit juice sweetener: a concentrated sugar syrup made from dehydrated, deflavored fruit juice, commonly grape juice; used to sweeten products that can then claim to be "all fruit."

confectioner's sugar: finely powdered sucrose; 99.9 percent pure.

corn sweeteners: corn syrup and sugar solutions derived from corn.

corn syrup: a syrup, mostly glucose, partly maltose, produced by the action of enzymes on cornstarch. It may be dried and used as *corn syrup solids.*

dextrose, anhydrous dextrose: forms of glucose.

evaporated cane juice: raw sugar from which impurities have been removed.

granulated sugar: white sugar.

high-fructose corn syrup (HFCS): a widely used commercial kcaloric sweetener made by adding enzymes to cornstarch to convert a portion of its glucose molecules into sweet-tasting fructose.

honey: a concentrated solution primarily composed of glucose and fructose; produced by enzymatic digestion of the sucrose in nectar by bees.

invert sugar: a mixture of glucose and fructose formed by the splitting of sucrose in an industrial process. Sold only in liquid form and sweeter than sucrose, invert sugar forms during certain cooking procedures and works to prevent crystallization of sucrose in soft candies and sweets.

malt syrup: a sweetener made from sprouted barley.

maple syrup: a concentrated solution of sucrose derived from the sap of the sugar maple tree, mostly sucrose. This sugar was once common but is now usually replaced by sucrose and artificial maple flavoring.

molasses: a thick, brown syrup left over from the refining of sucrose from sugarcane. The major nutrient in molasses is iron, a contaminant from the machinery used in processing it.

raw sugar: the first crop of crystals harvested during sugar processing. Raw sugar cannot be sold in the United States because it contains too much filth (dirt, insect fragments, and the like). Sugar sold as "raw sugar" is actually evaporated cane juice.

turbinado (ter-bih-NOD-oh) sugar: raw sugar from which the filth has been washed; legal to sell in the United States.

white sugar: granulated sucrose or "table sugar," produced by dissolving, concentrating, and recrystallizing raw sugar; 99.9 percent pure.

People often ask: What is the difference between honey and white sugar? Is honey, by virtue of being natural, more nutritious? Honey, like white sugar, contains glucose and fructose. The difference is that, in white sugar, the glucose and fructose are bonded together in pairs, whereas in honey some of them are paired and some are free single sugars. When you eat either white sugar or honey, though, your body breaks all of the sugars apart into single sugars. It ultimately makes no difference, then, whether you eat single sugars linked together, as in white sugar, or the same sugars unlinked, as in honey; they will end up as single sugars in your body.

Honey does contain a few vitamins and minerals but not many. Honey is denser than crystalline sugar, too, so it provides more energy per spoonful. Table 3-3 (p. 74) shows that honey and white sugar are similar nutritionally—and both fall short of milk, legumes, fruits, grains, and vegetables. Honey may offer some health benefits, however: It seems to relieve nighttime coughing in children and reduce the severity of mouth ulcers in cancer patients undergoing chemotherapy or radiation.

Some sugar sources are more nutritious than others, though. Consider a fruit such as an orange. The orange provides the same sugars and about the same energy as a tablespoon of sugar or honey, but the packaging makes a big difference in nutrient density. The sugars of the orange are diluted in a large volume of fluid that contains valuable vitamins and minerals, and the flesh and skin of the orange are supported by fibers that also offer health benefits. A tablespoon of honey offers no such bonuses. Of course, a cola beverage, containing many teaspoons of sugar, offers no advantages either.

You receive about the same amount and kinds of sugars from an orange as from a tablespoon of honey, but the packaging makes a big nutrition difference.

TABLE 3-3 Sample Nutrients in Sugars and Other Foods

The indicated portion of any of these foods provides approximately 100 kcalories. Notice that—for a similar number of kcalories and grams of carbohydrate—milk, legumes, fruits, grains, and vegetables offer more of the other nutrients than do the sugars.

	Size of 100 kCal Portion	Carbohydrate (g)	Protein (g)	Calcium (mg)	Iron (mg)	Vitamin A (µg)	Vitamin C (mg)
Foods							
Milk, 1% low-fat	1 c	12	8	300	0.1	144	2
Kidney beans	½ c	20	7	30	1.6	0	2
Apricots	6	24	2	30	1.1	554	22
Bread, whole wheat	1½ slices	20	4	30	1.9	0	0
Broccoli, cooked	2 c	20	12	188	2.2	696	148
Sugars							
Sugar, white	2 tbs	24	0	trace	trace	0	0
Molasses, blackstrap	2½ tbs	28	0	343	12.6	0	0.1
Cola beverage	1 c	26	0	6	trace	0	0
Honey	1½ tbs	26	trace	2	0.2	0	trace

Review Notes

- In moderation, sugars pose no major health threat except for an increased risk of dental caries.
- Excessive sugar intakes may displace needed nutrients and fiber and may contribute to obesity.
- A person deciding to limit daily sugar intake should recognize that it is the added sugars in concentrated sweets, which are high in kcalories and relatively lacking in other nutrients, that should be restricted. Sugars that occur naturally in fruits, vegetables, and milk are acceptable.

sugar alcohols: sugarlike compounds in the chemical family *alcohol* derived from fruits or manufactured from carbohydrates; sugar alcohols are absorbed more slowly than other sugars, are metabolized differently, and do not elevate the risk of dental caries. Examples are maltitol, mannitol, sorbitol, isomalt, lactitol, and xylitol.

nutritive sweeteners: sweeteners that yield energy, including both the sugars and the sugar alcohols.

ALTERNATIVE SWEETENERS: SUGAR ALCOHOLS

The **sugar alcohols** are carbohydrates, but they trigger a lower glycemic response and yield slightly less energy (2 to 3 kcalories per gram) than sucrose (4 kcalories per gram) because they are not absorbed completely.[23] The sugar alcohols are sometimes called **nutritive sweeteners** because they do yield some energy. One exception, erythritol, cannot be metabolized by human enzymes and so is kcalorie free.

The sugar alcohols occur naturally in fruits and vegetables; they are also used by manufacturers to provide sweetness and bulk to cookies, sugarless gum, hard candies, and jams and jellies. Unlike sucrose, sugar alcohols are fermented in the large intestine by intestinal bacteria. Consequently, side effects such as gas, abdominal discomfort, and diarrhea make the sugar alcohols less attractive than the nonnutritive sweeteners.

The advantage of using sugar alcohols is that they do not contribute to dental caries. Bacteria in the mouth metabolize sugar alcohols much more slowly than sucrose, thereby inhibiting the production of acids that promote caries formation. They are therefore valuable in chewing gums, breath mints, and other products that people keep in their mouths awhile. The Food and Drug Administration (FDA) allows food labels to carry a health claim (see p. 28 in Chapter 1) about the relationship between sugar alcohols and the nonpromotion of dental caries as long as certain FDA criteria, including those for sugar-free status, are met. Figure 3-5 presents labeling information for products using sugar alternatives.

ALTERNATIVE SWEETENERS: NONNUTRITIVE SWEETENERS

The **nonnutritive sweeteners** sweeten with minimal or no carbohydrate or energy. The human taste buds perceive many of them as extremely sweet, so just tiny amounts are added to foods to achieve the desired sweet taste.[24] The FDA endorses nonnutritive sweeteners as safe for use over a lifetime within **Acceptable Daily Intake (ADI)** levels. Like the sugar alcohols, nonnutritive sweeteners make foods taste sweet without promoting tooth decay. Table 3-4 (p. 76) offers details about nonnutritive sweeteners, including ADI levels.

Nonnutritive Sweeteners and Weight Management

Whether the use of nonnutritive sweeteners promotes weight loss or improves health by reducing total kcalorie intakes is not known with certainty.[25] Some research even suggests that their use may *promote* weight gain through unknown mechanisms; these are topics of current research.

When people reduce their energy intakes by replacing sugar in their diets with nonnutritive sweeteners and then compensate for the reduced energy at later meals, energy intake may stay the same or increase. Using nonnutritive sweeteners will not automatically lower energy intake; to successfully control energy intake, a person needs to make informed diet and activity decisions throughout the day.

nonnutritive sweeteners: synthetic or natural food additives that offer sweet flavor but with negligible or no calories per serving; also called *artificial sweeteners, intense sweeteners, noncaloric sweeteners,* and *very-low-calorie sweeteners.*

Acceptable Daily Intake (ADI): the amount of a nonnutritive sweetener that individuals can safely consume each day over the course of a lifetime without adverse effect. It includes a 100-fold safety factor.

FIGURE 3-5 Sugar Alternatives on Food Labels

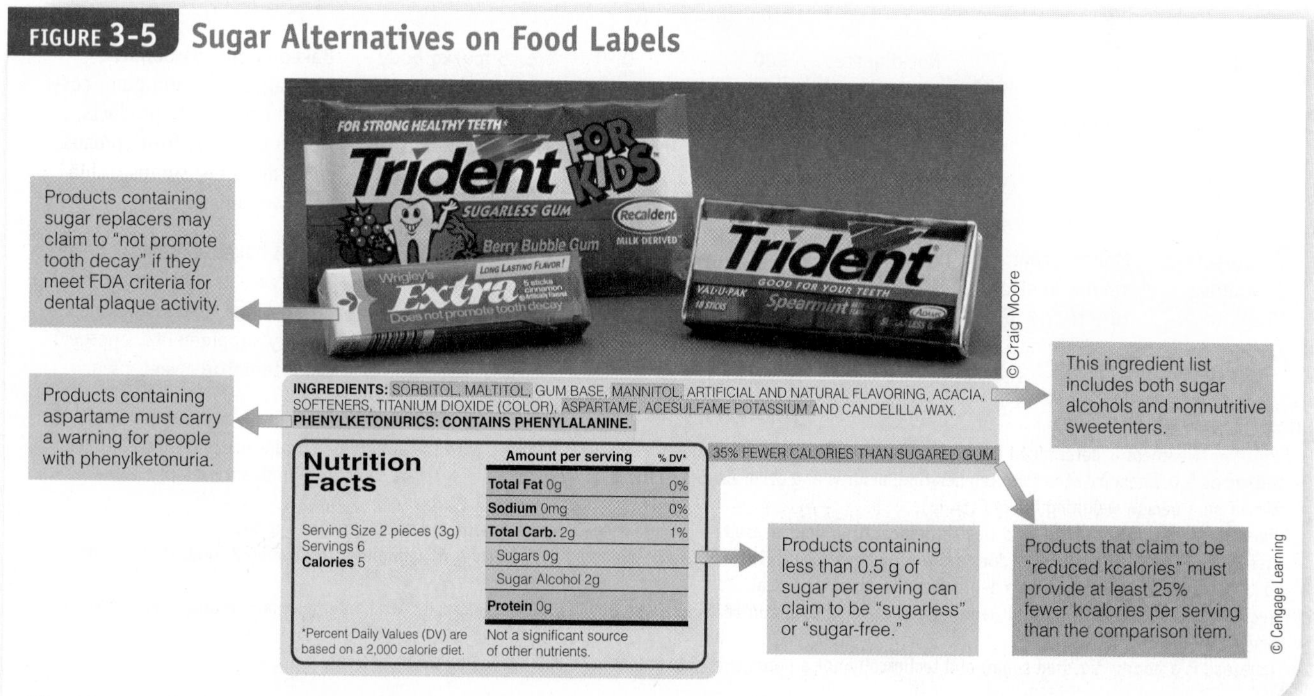

FOR STRONG HEALTHY TEETH
Trident FOR KIDS
SUGARLESS GUM
Berry Bubble Gum
Recaldent
MILK DERIVED
Wrigley's
Extra
Long Lasting Flavor!
Does not promote tooth decay
Trident
GOOD FOR YOUR TEETH
VAL-U-PAK
Spearmint
© Craig Moore
Products containing sugar replacers may claim to "not promote tooth decay" if they meet FDA criteria for dental plaque activity.
Products containing aspartame must carry a warning for people with phenylketonuria.
INGREDIENTS: SORBITOL, MALTITOL, GUM BASE, MANNITOL, ARTIFICIAL AND NATURAL FLAVORING, ACACIA, SOFTENERS, TITANIUM DIOXIDE (COLOR), ASPARTAME, ACESULFAME POTASSIUM AND CANDELILLA WAX.
PHENYLKETONURICS: CONTAINS PHENYLALANINE.
This ingredient list includes both sugar alcohols and nonnutritive sweetenters.
Nutrition Facts
Serving Size 2 pieces (3g)
Servings 6
Calories 5
Amount per serving
% DV*
Total Fat 0g 0%
Sodium 0mg 0%
Total Carb. 2g 1%
Sugars 0g
Sugar Alcohol 2g
Protein 0g
*Percent Daily Values (DV) are based on a 2,000 calorie diet.
Not a significant source of other nutrients.
35% FEWER CALORIES THAN SUGARED GUM.
Products containing less than 0.5 g of sugar per serving can claim to be "sugarless" or "sugar-free."
Products that claim to be "reduced kcalories" must provide at least 25% fewer kcalories per serving than the comparison item.
© Cengage Learning

TABLE 3-4 U.S.-Approved Nonnutritive Sweeteners

Sweetener	Chemical Composition	Digestion/ Absorption	Sweetness Relative to Sucrose[a]	Energy (kcal/g)	Acceptable Daily Intake (ADI) and (Estimated Equivalent[b])	Approved Uses
Acesulfame potassium or acesulfame-K (Sunette, Sweet One)	Potassium salt	Not digested or absorbed	200	0	15 mg/kg body weight[c] (30 cans diet soda)	General use, except in meat and poultry. Tabletop sweeteners. Heat stable.
Aspartame (NutraSweet, Equal, others)	Amino acids (phenylalanine and aspartic acid) and a methyl group	Digested and absorbed	180	4[d]	50 mg/kg body weight[e] (18 cans diet soda)	General use in all foods and beverages; warning to population with PKU. Degrades when heated.
Luo han guo	Curcurbine, glycosides from monk fruit extract	Digested and absorbed	150-300	1	No ADI determined	GRAS[f]; general use as a food ingredient and table top sweetener.
Neotame	Aspartame with an additional side group attached	Not digested or absorbed	7000	0	18 mg/day	General use, except in meat and poultry.
Saccharin (SugarTwin, Sweet'N Low, others)	Benzoic sulfimide	Rapidly absorbed and excreted	300	0	5 mg/kg body weight (10 packets of sweetener)	Tabletop sweeteners, wide range of foods, beverages, cosmetics, and pharmaceutical products.
Stevia (Sweetleaf, Truvia, PurVia)	Glyosides found in the leaves of the *Steviarebaudiana* herb	Digested and absorbed	200-300	0	4 mg/kg body weight	GRAS[f]; tabletop sweeteners, a variety of foods and beverages.
Sucralose (Splenda)	Sucrose with Cl atoms instead of OH groups	Not digested or absorbed	600	0	5 mg/kg body weight (6 cans diet soda)	Baked goods, carbonated beverages, chewing gum, coffee and tea, dairy products, frozen desserts, fruit spreads, salad dressing, syrups, tabletop sweeteners.
Tagatose[g] (Nutralose, Nutrilatose, Tagatesse)	Monosaccharide similar in structure to fructose; naturally occurring or derived from lactose	Not well absorbed	0.9	1.5	7.5 g/day	GRAS[f]; bakery products, beverages, cereals, chewing gum, confections, dairy products, dietary supplements, energy bars, tabletop sweeteners.

[a]Relative sweetness is determined by comparing the approximate sweetness of a sugar substitute with the sweetness of pure sucrose, which has been defined as 1.0. Chemical structure, temperature, acidity, and other flavors of the foods in which the substance occurs all influence relative sweetness.
[b]Based on a person weighing 70 kg (154 lb).
[c]Recommendations from the World Health Organization limit acesulfame-K intake to 9 mg per kilogram of body weight per day.
[d]Aspartame provides 4 kcal/g, as does protein, but because so little is used, its energy contribution is negligible. In powdered form, it is sometimes mixed with lactose, however, so a 1-g packet may provide 4 kcal.
[e]Recommendations from the World Health Organization and in Europe and Canada limit aspartame intake to 40 mg per kilogram of body weight per day.
[f]Generally recognized as safe.
[g]Tagatose is a poorly digested sugar, and technically not a nonnutritive sweetener.

Safety of Nonnutritive Sweeteners Through the years, questions have emerged about the safety of nonnutritive sweeteners, but these issues have since been resolved. For example, early research indicating that large quantities of saccharin caused bladder tumors in laboratory animals was later shown to be inapplicable to humans. Common sense dictates that consuming large amounts of saccharin is probably not safe, but consuming moderate amounts poses no known hazard.

Aspartame, a sweetener made from two amino acids (phenylalanine and aspartic acid) is one of the most thoroughly studied food additives ever approved, and no scientific evidence supports the Internet stories that accuse it of causing disease.[26] However, aspartame's phenylalanine base poses a threat to those with the inherited disease phenylketonuria (PKU). People with PKU cannot dispose of phenylalanine efficiently (see Nutrition in Practice 15). Food labels warn people with PKU of the presence of phenylalanine in aspartame-sweetened foods (see Figure 3-5). In addition, foods and drinks containing nonnutritive sweeteners have no place in the diets of even healthy infants or toddlers.

Review Notes

- Two types of alternative sweeteners are sugar alcohols and nonnutritive sweeteners.
- Sugar alcohols are carbohydrates, but they yield slightly less energy than sucrose.
- Sugar alcohols do not contribute to dental caries.
- The FDA endorses the use of nonnutritive sweeteners within ADI levels as safe over a lifetime.
- The nonnutritive sweeteners sweeten with minimal or no carbohydrate and energy.
- Like the sugar alcohols, nonnutritive sweeteners do not promote tooth decay.

Health Effects of Starch and Dietary Fibers

Despite dietary recommendations that people should eat generous servings of starch- and fiber-rich carbohydrate foods for their health, many people still believe that carbohydrate is the "fattening" component of foods. Gram for gram, carbohydrates contribute fewer kcalories to the body than do dietary fats, so a moderate diet based on starch- and fiber-rich carbohydrate foods is likely to be lower in kcalories than a diet based on high-fat foods.

For health's sake, most people should increase their intakes of carbohydrate-rich foods such as whole grains, vegetables, legumes, and fruits—foods noted for their starch, fiber, and naturally occurring sugars.[27] In addition, most people should also limit their intakes of foods high in added sugars and the types of fats associated with heart disease (see Chapter 4). A diet that emphasizes whole grains, vegetables, legumes, and fruits is almost invariably moderate in food energy, low in fats that can harm health, and high in dietary fiber, vitamins, and minerals. All these factors working together can help reduce the risks of obesity, cancer, cardiovascular disease, diabetes, dental caries, gastrointestinal disorders, and malnutrition.

CARBOHYDRATES: DISEASE PREVENTION AND RECOMMENDATIONS

Fiber-rich carbohydrate foods benefit health in many ways. Foods such as whole grains, legumes, vegetables, and fruits supply valuable vitamins, minerals, and phytochemicals, along with abundant dietary fiber and little or no fat. The following paragraphs describe some of the health benefits of diets that emphasize a variety of these foods each day.

Heart Disease Diets rich in whole grains, legumes, and vegetables, especially those rich in whole grains, may protect against heart disease and stroke by lowering blood pressure, improving blood lipids, and reducing inflammation.[28] Such diets are generally low in saturated fat, *trans* fat, and cholesterol and high in dietary fibers, vegetable proteins, and phytochemicals—all factors associated with a lower risk of heart disease. (■) Foods rich in soluble fibers (such as oat bran, barley, and legumes) lower blood cholesterol by binding cholesterol compounds and carrying them out of the body with the feces.[29] High-fiber foods may also lower blood cholesterol indirectly by displacing fatty, cholesterol-raising foods from the diet. Even when dietary fat intake is low, research shows that high intakes of soluble fiber exert separate and significant blood cholesterol–lowering effects.

Diabetes High-fiber foods—and especially whole grains—play a key role in reducing the risk of **type 2 diabetes** (see Chapter 20).[30] The soluble fibers of foods such as oats and legumes can help regulate the blood glucose following a carbohydrate-rich meal. Soluble fibers trap nutrients and delay their transit through the digestive tract, slowing glucose absorption and preventing the glucose surge and rebound often associated with diabetes onset.

The term **glycemic response** refers to how quickly glucose is absorbed after a person eats, how high blood glucose rises, and how quickly it returns to normal. Slow absorption, a modest rise in blood glucose, and a smooth return to normal are desirable (a low glycemic response). Fast absorption, a surge in blood glucose, and an overreaction that plunges glucose below normal are less desirable (a high glycemic response). Different foods have different effects on blood glucose. The **glycemic index**, a method of classifying foods according to their potential to raise blood glucose, is the topic of Nutrition in Practice 3.

GI Health Soluble and insoluble fibers, along with ample fluid intake, may enhance the health of the large intestine. The healthier the intestinal walls, the better they can block absorption of unwanted constituents. Soluble fibers help to maintain normal colonic bacteria necessary for intestinal health.[31] Insoluble fibers that both enlarge and soften stools such as cellulose (in cereal brans, fruits, and vegetables) ease elimination for the rectal muscles and thereby alleviate or prevent constipation and hemorrhoids.

Some fibers (again, such as cereal bran) help keep the contents of the intestinal tract moving easily. This action helps prevent compaction of the intestinal contents, which could obstruct the appendix and permit bacteria to invade and infect it. In addition, fibers stimulate the muscles of the gastrointestinal (GI) tract so that they retain their strength and resist bulging out in places, as occurs in diverticulosis.[32] Insoluble fiber seems to be most beneficial in lowering the risk of diverticulosis, which is described in Chapter 18.

Cancer Many studies show that, as people increase their dietary fiber intakes, their risk for colon cancer declines.[33] A recent meta-analysis using data from several studies exposed a strong, linear inverse association between dietary fiber and colon cancer.[34] People who ate the most fiber (24 grams per day) reduced their risk of colon and rectal cancer by almost 30 percent compared with those who ate the least (10 grams per day). Mid-range intakes (18 grams per day) reduced the risk by 20 percent. Importantly, fiber from food but not from supplements demonstrates this association, possibly because fiber supplements lack the nutrients and phytochemicals of whole foods that may also help to protect against cancers.

All plant foods—vegetables, fruits, and whole-grain products—have attributes that may reduce the risks of colon and rectal cancers. Their fiber dilutes, binds, and rapidly removes potential cancer-causing agents from the colon. In addition, the colon's bacteria ferment soluble fibers, forming small fatlike molecules that lower the pH. (■) These small fatlike molecules activate cancer-killing enzymes and inhibit inflammation in the colon.[35]

■ The role of saturated fat, *trans* fat, and cholesterol in heart disease is discussed in Chapter 4. The role of vegetable proteins in heart disease is presented in Chapter 5. The benefits of phytochemicals in disease prevention are presented in Nutrition in Practice 8.

■ pH is the unit of measure expressing a substance's acidity or alkalinity. (Chapter 5 provides a more detailed definition.)

type 2 diabetes: the type of diabetes that accounts for 90 to 95 percent of diabetes cases and usually results from insulin resistance coupled with insufficient insulin secretion.

glycemic response: the extent to which a food raises the blood glucose concentration and elicits an insulin response.

glycemic index: a method of classifying foods according to their potential for raising blood glucose.

Other processes may also be at work. As research progresses, cancer experts recommend that fiber in the diet come from 5 to 9 half-cup servings of vegetables and fruit daily, along with generous portions of whole grains and legumes.

Weight Management Fiber-rich foods tend to be low in fat and added sugars and therefore prevent weight gain and promote weight loss by delivering less energy per bite.[36] In addition, fibers absorb water from the digestive juices; as they swell, they create feelings of fullness, delay hunger, and reduce food intake.[37] Soluble fibers may be especially useful for appetite control. In a recent study, soluble fiber from barley shifted the body's mix of appetite-regulating hormones toward reducing food intake.[38] By whatever mechanism, as populations eat more refined low-fiber foods and concentrated sweets, body fat stores creep up.[39] In contrast, people who eat three or more whole grain servings each day tend to have lower body and abdominal fatness over time.[40]

Commercial weight-loss products often contain bulk-inducing fibers such as methylcellulose, but pure fiber compounds are not advised. High-fiber foods not only add bulk to the diet but are economical, are nutritious, and supply health-promoting phytochemicals—benefits that no purified fiber preparation can match. Figure 3-6 summarizes fibers and their health benefits.

Harmful Effects of Excessive Fiber Intake Despite fiber's benefits to health, when too much fiber is consumed, some minerals may bind to it and be excreted with it without becoming available for the body to use. When mineral intake is adequate, however, a reasonable intake of high-fiber foods does not seem to compromise mineral balance.

People with marginal intakes who eat mostly high-fiber foods may not be able to take in enough food to meet energy or nutrient needs. The malnourished, the elderly, and young children adhering to all-plant (vegan) diets are especially vulnerable to this problem. Fibers also carry water out of the body and can cause dehydration. Advise

FIGURE 3-6 Characteristics, Sources, and Health Effects of Fibers

People who eat these foods . . .		obtain these types of fibers . . .	with these actions in the body . . .	and these probable health benefits . . .
		Viscous, soluble, more fermentable		
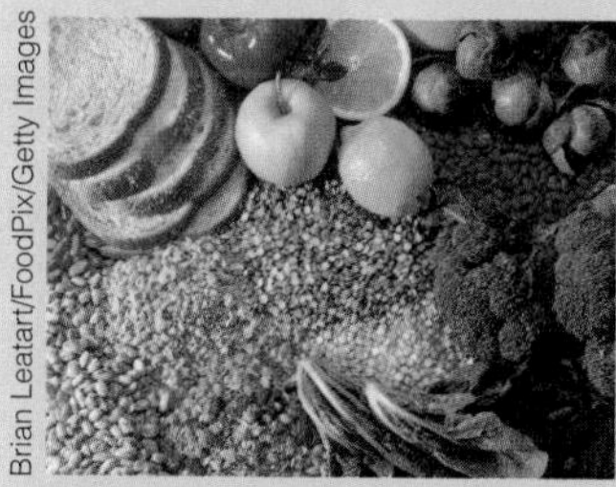 Brian Leatart/FoodPix/Getty Images	• Barley, oats, oat bran, rye, fruits (apples, citrus), legumes (especially young green peas and black-eyed peas), seaweeds, seeds and husks, many vegetables, fibers used as food additives	• Gums • Pectins • Psyllium[a] • Some hemicellulose	• Lower blood cholesterol by binding bile • Slow glucose absorption • Slow transit of food through upper GI tract • Hold moisture in stools, softening them • Yield small fat molecules after fermentation that the colon can use for energy • Increase satiety	• Lower risk of heart disease • Lower risk of diabetes • Lower risk of colon and rectal cancer • Increased satiety, and may help with weight management
		Nonviscous, insoluble, less fermentable		
 Polara Studios, Inc.	• Brown rice, fruits, legumes, seeds, vegetables (cabbage, carrots, Brussels sprouts), wheat bran, whole grains, extracted fibers used as food additives	• Cellulose • Lignins • Resistant starch • Hemicellulose	• Increase fecal weight and speed fecal passage through colon • Provide bulk and feelings of fullness	• Alleviate constipation • Lower risk of diverticulosis, hemorrhoids, and appendicitis • Lower risk of colon and rectal cancer

[a]Psyllium, a soluble fiber derived from seeds, is used as a laxative and food additive.

clients to add an extra glass or two of water to go along with the fiber added to their diets. Athletes may want to avoid bulky, fiber-rich foods just prior to competition.

Recommended Intakes of Starches and Fibers The DRI committee advises that carbohydrates should contribute about half (45 to 65 percent) of the energy requirement. A person consuming 2000 kcalories a day should therefore obtain 900 to 1300 kcalories' worth of carbohydrate, or between 225 and 325 grams. (■) This amount is more than adequate to meet the RDA for carbohydrate, which is set at 130 grams per day based on the average minimum amount of glucose used by the brain.[41] (■)

When it established the Daily Values that appear on food labels, the FDA used a guideline of 60 percent of kcalories in setting the Daily Value for carbohydrate at 300 grams per day. (■) For most people, this means increasing total carbohydrate intake. To this end, the *Dietary Guidelines for Americans* encourage people to choose fiber-rich whole grains, vegetables, fruits, and legumes daily.

Recommendations for fiber encourage the same foods just mentioned: whole grains, vegetables, fruits, and legumes, which also provide vitamins, minerals, and phytochemicals. The FDA set the Daily Value for fiber at 25 grams, rounding up from the recommended 11.5 grams per 1000 kcalories for a 2000-kcalorie intake. (■) The DRI recommendation is slightly higher, at 14 grams per 1000-kcalorie intake—roughly 25 to 35 grams of dietary fiber daily. (■) These recommendations are about two times higher than the usual intake in the United States.[42]

As health care professionals, you can advise your clients that an effective way to add dietary fiber while lowering fat is to substitute plant sources of proteins (legumes) for some of the animal sources of protein (meats and cheeses) in the diet. Another way to add fiber is to encourage clients to consume the recommended amounts of fruits and vegetables each day. People choosing high-fiber foods are wise to seek out a variety of fiber sources and to drink extra fluids to help the fiber do its job. Many foods provide fiber in varying amounts, as Figure 3-7 shows.

As mentioned earlier, too much fiber is no better than too little. The World Health Organization recommends an upper limit of 40 grams of dietary fiber a day.

■ 45% of 2000 kcal:
2000 x 0.45 = 900 kcal
900 kcal ÷ 4 kcal/g = 225 g

65% of 2000 kcal:
2000 x 0.65 = 1300 kcal
1300 kcal ÷ 4 kcal/g = 325 g

■ RDA for carbohydrate:
130 g/day
45 to 65% of energy intake

■ Daily Value:
300 g carbohydrate (based on 60% of 2000-kcal diet)

■ Daily Values:
25 g fiber (based on 11.5 g/1000 kcal)

■ Fiber AI:
14 g/1000 kcal/day

Men:
19–50 yr: 38 g/day
51+ yr: 30 g/day

Women:
19–50 yr: 25 g/day
51+ yr: 21 g/day

Reminder: An *Adequate Intake (AI)* is used as a guide for nutrient intake when an RDA cannot be established (see Chapter 1).

Review Notes

- A diet rich in starches and dietary fibers helps prevent heart disease, diabetes, GI disorders, and possibly some types of cancer. It also supports efforts to manage body weight.
- For these reasons, recommendations urge people to eat plenty of whole grains, vegetables, legumes, and fruits—enough to provide 45 to 65 percent of the daily energy from carbohydrate and 14 grams of fiber per 1000 kcalories.

CARBOHYDRATES: FOOD SOURCES

A day's meals based on the USDA Food Patterns not only meet carbohydrate recommendations but provide abundant fiber, too. Grains, vegetables, fruits, and legumes deliver dietary fiber and are noted for their valuable energy-yielding starches and dilute sugars. Each class of foods makes its own typical carbohydrate contribution. The USDA Food Patterns in Chapter 1 can help you and your clients choose carbohydrate-rich foods.

Grains Most foods in this group—a slice of whole-wheat bread, half an English muffin or bagel, a 6-inch tortilla, or ½ cup of rice, pasta, or cooked cereal—provide about 15 grams of carbohydrate, mostly as starch.* Be aware that some foods in this group, especially snack crackers and baked goods such as biscuits, croissants, and muffins, contain added sugars, solid fats, and sodium. When selecting from the grain group,

*Gram values in this section are adapted from *Choose Your Foods: Exchange Lists for Diabetes.*

FIGURE 3-7 Fiber in Selected Foods

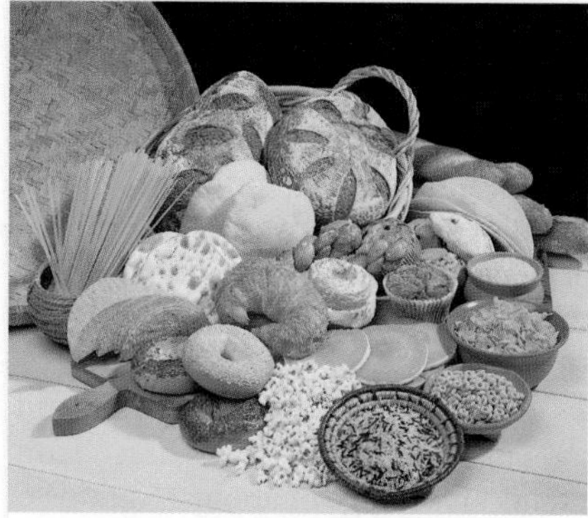

Grains

Whole-grain products provide 1 to 2 g of fiber or more per serving:

- 1 slice whole-wheat or rye bread (1 g).
- 1 slice pumpernickel bread (2 g).
- 1 oz ready-to-eat cereal (100% bran cereals contain 10 g or more).
- 1/2 c cooked barley, bulgur, grits, oatmeal (2 to 3 g).

Vegetables

Most vegetables contain 2 to 3 g of fiber per serving:

- 1 c raw bean sprouts.
- 1/2 c cooked broccoli, brussels sprouts, cabbage, carrots, cauliflower, collards, corn, eggplant, green beans, green peas, kale, mushrooms, okra, parsnips, potatoes, pumpkin, spinach, sweet potatoes, swiss chard, winter squash.
- 1/2 c chopped raw carrots, peppers.

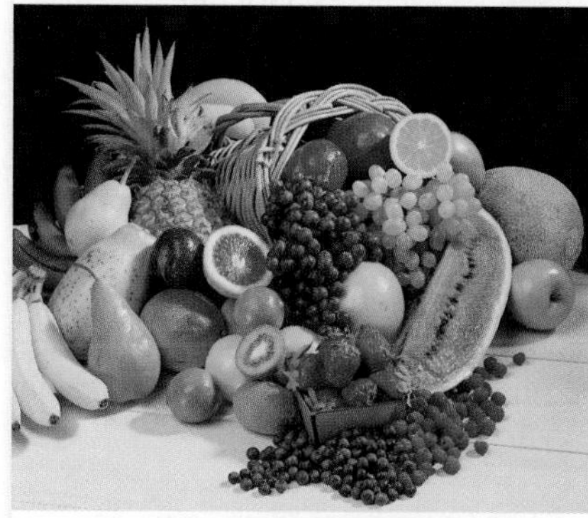

Fruits

Fresh, frozen, and dried fruits have about 2 g of fiber per serving:

- 1 medium apple, banana, kiwi, nectarine, orange, pear.
- 1/2 c applesauce, blackberries, blueberries, raspberries, strawberries.
- Fruit juices contain very little fiber.

Legumes and Nuts

Many legumes provide about 8 g of fiber per serving:

- 1/2 c cooked baked beans, black beans, black-eyed peas, kidney beans, navy beans, pinto beans.

Some legumes provide about 5 g of fiber per serving:

- 1/2 c cooked garbanzo beans, great northern beans, lentils, lima beans, split peas.

Most nuts and seeds provide 1 to 3 g of fiber per serving:

- 1 oz almonds, cashews, hazelnuts, peanuts, pecans, pumpkin seeds, sunflower seeds.

limit refined grains and be sure at least half of the foods chosen are whole-grain products. People who eat more whole grains tend to have healthier diets.[43]

Vegetables Some vegetables are major contributors of starch in the diet. Just a small white or sweet potato or ½ cup of cooked dry beans, corn, peas, plantain, or winter squash provides 15 grams of carbohydrate, as much as in a slice of bread, though as a mixture of sugars and starch. One-half cup of carrots, okra, onions, tomatoes, cooked greens, or most other nonstarchy vegetables or a cup of salad greens provides about 5 grams as a mixture of starch and sugars. Each of these foods also contributes a little protein, some fiber, and no fat.

Fruits The size of a typical serving of fruit varies depending on the form of the fruit: ½ cup of juice; a small banana, apple, or orange; ½ cup of most canned or fresh fruit; or ¼ cup of dried fruit. A typical fruit serving contains an average of about 15 grams of carbohydrate, mostly as sugars, including the fruit sugar fructose. Fruits vary greatly in their water and fiber contents; therefore, their sugar concentrations vary also. No more than one-half of the day's fruit should come from juice. With the exception of avocado, which is high in fat, the fruits contain insignificant amounts of fat and protein.

Milk and Milk Products One cup of milk or yogurt or the equivalent (1 cup of buttermilk, ⅓ cup of dry milk powder, or ½ cup of evaporated milk) provides a generous 12 grams of carbohydrate. Among cheeses, cottage cheese provides about 6 grams of carbohydrate per cup, whereas most other types contain little, if any, carbohydrate. These foods also contribute high-quality protein as well as several important vitamins and minerals. Calcium-fortified soy beverages are options for providing calcium and about the same amount of carbohydrate as milk. All milk products vary in fat content, an important consideration in choosing among them; Chapter 4 provides the details.

Cream and butter, although dairy products, are not equivalent to milk because they contain little or no carbohydrate and insignificant amounts of the other nutrients important in milk. They are appropriately placed with the solid fats and added sugars.

Protein Foods With two exceptions, foods of this group provide almost no carbohydrate to the diet. The exceptions are nuts, which provide a little starch and fiber along with their abundant fat, and dry beans, which are excellent sources of both starch and fiber. Just ½ cup of beans provides 15 grams of carbohydrate, an amount equal to the richest carbohydrate sources. Among sources of fiber, beans and other legumes are outstanding, providing as much as 8 grams in ½ cup. The carbohydrate content of a diet can be determined by using a nutrient composition table such as that found in Appendix A, the exchange list system described in Chapter 20, or a computer diet analysis program.

Review Notes

- Grains, vegetables, fruits, and legumes contribute dietary fiber and energy-yielding starches and dilute sugars to people's diets. One cup of milk or yogurt or the equivalent contributes a generous 12 grams of carbohydrate as well as protein and other important nutrients.
- Food labels list grams of total carbohydrate and also provide separate listings of grams of fiber and sugar.

CARBOHYDRATES: FOOD LABELS AND HEALTH CLAIMS

Food labels list the amount, in grams, of total carbohydrate—including starch, fibers, and sugars—per serving. Fiber and sugar grams are also listed separately. (With this information, consumers can calculate starch grams by subtracting the grams of fibers and sugars from the total carbohydrate.) Sugars on the Nutrition Facts panel of a food label reflect both added sugars and those that occur naturally in foods. Total carbohydrate and dietary fiber are also expressed as "% Daily Values" for a person consuming 2000 kcalories; there is no Daily Value for sugars.

The FDA authorizes four health claims on food labels concerning fiber-rich carbohydrate foods. One is for "fiber-containing grain products, fruits, and vegetables and reduced risk of cancer." Another is for "fruits, vegetables, and grain products that contain fiber, and reduced risk of coronary heart disease." A third is for "soluble fiber from whole oats and from psyllium seed husk and reduced risk of coronary heart disease," and a fourth is for "whole grains and reduced risk of heart disease and certain cancers." Chapter 1 describes the criteria foods must meet to bear these health claims.

Self Check

1. Which of the following foods is *not* a good source of carbohydrates?
   a. Plain yogurt
   b. Steak
   c. Brown rice
   d. Green peas

2. Polysaccharides include:
   a. galactose, starch, and glycogen.
   b. starch, glycogen, and fiber.
   c. lactose, maltose, and glycogen.
   d. sucrose, fructose, and glucose.

3. The chief energy source of the body is:
   a. sucrose.
   b. starch.
   c. glucose.
   d. fructose.

4. The primary form of stored glucose in animals is:
   a. glycogen.
   b. cellulose.
   c. starch.
   d. lactose.

5. The polysaccharide that helps form the cell walls of plants is:
   a. cellulose.
   b. starch.
   c. glycogen.
   d. lactose.

6. Which of the following terms on a food label may denote sugar?
   a. Corn syrup
   b. Aspartame
   c. Xylitol
   d. Cellulose

7. The two types of alternative sweeteners are:
   a. saccharin and cyclamate.
   b. sugar alcohols and nonnutritive sweeteners.
   c. sorbitol and xylitol.
   d. sucrose and fructose.

8. A diet high in carbohydrate-rich foods such as whole grains, vegetables, fruits, and legumes is:
   a. most likely low in fat.
   b. most likely low in fiber.
   c. most likely poor in vitamins and minerals.
   d. most likely disease promoting.

9. A fiber-rich diet may help to prevent or control:
   a. diabetes.
   b. heart disease.
   c. constipation.
   d. all of the above.

10. The DRI fiber recommendation is:
    a. 10 grams per 1000 kcalories.
    b. 15 to 25 grams per day.
    c. 14 grams per 1000 kcalories.
    d. 40 to 55 grams per day.

Answers to these questions can be found in Appendix H. For more chapter review: Access an interactive eBook, chapter-specific interactive learning tools, including flashcards, quizzes, videos, and more in your Nutrition CourseMate, accessed through CengageBrain.com.

Clinical Applications

1. Considering the health benefits of carbohydrate-rich foods, especially those that provide starch and fiber, what suggestions would you offer to a client who reports the following:
   - Eats only three servings of refined, sugary breads or cereals each day.
   - Eats one serving of vegetables (usually french fries) each day.
   - Drinks fruit juice once a day but never eats fruit.
   - Eats cheese at least twice a day but does not drink milk.
   - Eats large servings of meat at least twice a day.
   - Eats hard candy two or three times a day.

Nutrition on the Net

For further study of the topics in this chapter, access these websites.

- Search for "artificial sweeteners" at the U.S. government health information site:
  www.healthfinder.gov
- Search for "sugars" and "fiber" at the International Food Information Council site:
  www.foodinsight.org
- Learn more about dental caries from the American Dental Association and the National Institute of Dental and Craniofacial Research:
  www.ada.org and www.nidcr.nih.gov
- Learn more about diabetes from the American Diabetes Association, the Canadian Diabetes Association, and the National Diabetes Information Clearinghouse:
  www.diabetes.org, www.diabetes.ca, and http://diabetes.niddk.nih.gov

Notes

1. U.S. Department of Agriculture and U.S. Department of Health and Human Services, *Dietary Guidelines for Americans* 2010, www.dietaryguidelines.gov; A. T. Merchant and coauthors, Carbohydrate intake and overweight and obesity among healthy adults, *Journal of the American Dietetic Association* 109 (2009): 1165–1172; Standing Committee on the Scientific Evaluation of Dietary Reference Intakes, Food and Nutrition Board, Institute of Medicine, *Dietary Reference Intakes for Energy, Carbohydrate, Fiber, Fat, Fatty Acids, Cholesterol, Protein, and Amino Acids* (Washington, DC: National Academies Press, 2005), pp. 265–338.
2. S. S. Dronamraju and coauthors, Cell kinetics and gene expression changes in colorectal cancer patients give resistant starch: A randomised controlled trial, *Gut* 58 (2009): 413–420; K. C. Maki and coauthors, Beneficial effects of resistant starch on laxation in healthy adults, *International Journal of Food Sciences and Nutrition* 139 (2009): 296–305; S. J. D. O'Keefe and coauthors, Products of the colonic microbiota mediate the effects of diet on colon cancer risk, *Journal of Nutrition* 139 (2009): 2044–2048.
3. K. Raninen and coauthors, Dietary fiber type reflects physiological functionality: Comparison of grain fiber, insulin, and polydextrose, *Nutrition Reviews* 69 (2011): 9–21; Position of the American Dietetic Association: Health implications of dietary fiber, *Journal of the American Dietetic Association* 108 (2008): 1716–1731.
4. F. Magkos, S. Wang, and B. Mittendorfer, Metabolic actions of insulin in men and women, *Nutrition* 28 (2010): 686–691.
5. U.S. Department of Agriculture and U. S. Department of Health and Human Services, *Dietary Guidelines for Americans 2010*, www.dietaryguidelines.gov.
6. R. K. Johnson and coauthors, Dietary sugars intake and cardiovascular health: A scientific statement from the American Heart Association, *Circulation* 120 (2009): 1011–1020.
7. S. W. Rizkalla, health implications of fructose consumption: A review of recent data, *Nutrition and Metabolism* 7 (2010): 82-98.
8. Johnson and coauthors, 2009.
9. K. L. Stanhope, Role of fructose-containing sugars in the epidemics of obesity and metabolic syndrome, *Annual Review of Medicine* 63 (2012): 19.1–19.15; J. L. Sievenpiper and coauthors, Effect of fructose on body weight in controlled feeding trials: A systematic review and meta-analysis, *Annals of Internal Medicine* 156 (2012): 291–304; L. de Koning and coauthors, sugar-sweetened and artificially sweetened beverage consumption and risk of type 2 diabetes in men, *American Journal of Clinical Nutrition* 93 (2011): 1321–1327; R. D. Mattes and coauthors, Nutritively sweetened beverage consumption and body weight: A systematic review and meta-analysis of randomized experiments, *Obesity Reviews* 12 (2011): 346–365; K. J. Duffey and coauthors, Drinking caloric beverages increases the risk of adverse cardiometabolic outcomes in the Coronary Artery Risk Development in Young Adults (CARDIA) Study, *American Journal of Clinical Nutrition* 92 (2010): 954-959; V. S. Malik and coauthors, Sugar-sweetened beverages and risk of metabolic syndrome and type 2 diabetes; A meta-analysis, *Diabetes Care* 33 (2010): 2477–2483; F. B. Hu and V. S. Malik, Sugar-sweetened beverages and risk of obesity and type 2 diabetes: Epidemiologic evidence, *Physiology and Behavior* 100 (2010): 47–54; Johnson and coauthors, 2009.
10. O. I. Bermudez and X. Gao, Greater consumption of sweetened beverages and added sugars is associated with obesity among US young adults, *Annals of Nutrition and Metabolism* 57 (2010): 211–218; S. N. Bleich and coauthors, Increasing consumption of sugar-sweetened beverages among US adults: 1988–1994 to 1999–2004, *American Journal of Clinical Nutrition* 89 (2009): 372–381.
11. K. L. Stanhope, 2012; Sievenpiper and coauthors, 2012; Mattes and coauthors, 2011; Johnson and coauthors, 2009.
12. Bermudez and Gao, 2010; Hu and Malik, 2010; V. S. Malik and coauthors, Sugar-sweetened beverages, obesity, type 2 diabetes mellitus, and cardiovascular disease risk, *Circulation* 121 (2010): 1356-1364.
13. L. Trappy and coauthors, Fructose and metabolic diseases: New findings, new questions, *Nutrition* 26 (2010): 1044–1049.
14. D. F. Tate and coauthors, Replacing caloric beverages with water or diet beverages for weight loss in adults: Main results of the Choose Healthy Options Consciously Everyday (CHOICE) randomized clinical trial, *American Journal of Clinical Nutrition* 95 (2012); 555–563; L. Chen and coauthors, Reduction in consumption of sugar-sweetened beverages is associated with weight loss: The PREMIER trial, *American Journal of Clinical Nutrition* 89 (2009): 1299–1306.
15. L. de Koning and coauthors, Sweetened beverages consumption, incident coronary heart disease and biomarkers of risk in men, *Circulation* 125 (2012): 1735–1741; T. J. Angelopoulos and coauthors, The effect of high-fructose corn syrup consumption on triglycerides and uric acid, *Journal of Nutrition* 139 (2009): 1242S–1245S; K. L. Teff and coauthors, Endocrine and metabolic effects of consuming fructose- and glucose-sweetened beverages with meals in obese men and women: Influence of insulin resistance on plasma triglyceride responses, *Journal of Clinical Endocrinology and Metabolism* 94 (2009): 1652–1659.
16. Stanhope, 2012; M. J. Dekker and coauthors, Fructose: A highly lipogenic nutrient implicated in insulin resistance, hepatic steatosis, and the metabolic syndrome, *American Journal of Physiology, Endocrinology and Metabolism* 299 (2010): E685–E694.
17. J. A. Welsh and coauthors, Caloric sweetener consumption and dyslipidemia among US adults, *Journal of the American Medical Association* 303 (2010): 1490–1497.
18. K. J. Duffy and coauthors, Drinking caloric beverages increases the risk of adverse cardiometabolic outcomes in the Coronary Artery Risk Development in Young Adults (CARDIA) Study, *American Journal of Clinical Nutrition* 92 (2010): 954–959.
19. I. Aeberli and coauthors, Low to moderate sugar-sweetened beverage consumption impairs glucose and lipid metabolism and promotes inflammation in healthy young men: A randomized controlled trial, *American Journal of Clinical Nutrition* 94 (2011): 479–485; K. L. Stanhope, Consumption of fructose and high fructose corn syrup increase postprandial triglycerides, LDL-cholesterol, and apolipoprotein-B in young men and women, *Journal of Clinical Endocrinology and Metabolism* 96 (2011): E1596–E1605; J. P. Bantle, Dietary fructose and metabolic syndrome and diabetes, *Journal of Nutrition* 139 (2009): 1263S–1268S.
20. Standing Committee on the Scientific Evaluation of Dietary Reference Intakes, 2005, p. 770.

21. B. P. Marriott and coauthors, Intake of added sugars and selected nutrients in the United States, National Health and Nutrition Examination Survey (NHANES) 2003–2006, *Critical Reviews in Food Science and Nutrition* 50 (2010): 228–258.
22. *Dietary Guidelines for Americans* 2010.
23. Position of the Academy of Nutrition and Dietetics: Use of nutritive and nonnutritive sweeteners, *Journal of the Academy of Nutrition and Dietetics* 112 (2012): 739-758.
24. Position of the Academy of Nutrition and Dietetics: Use of nutritive and nonnutritive sweeteners, 2012.
25. C. Gardner and coauthors, Nonnutritive sweeteners: Current use and health perspectives, *Diabetes Care* 35 (2012): 1798–1808.
26. Position of the Academy of Nutrition and Dietetics: Use of nutritive and nonnutritive sweeteners, 2012.
27. S. Chuang and coauthors, Fiber intake and total and cause-specific mortality in the European Prospective Investigation into Cancer and Nutrition cohort, *American Journal of Clinical Nutrition* 96 (2012): 164–174.
28. R. J. J. van de Laar and coauthors, Lower lifetime dietary fiber intake is associated with carotid artery stiffness: The Amsterdam Growth and Health Longitudinal Study, *American Journal of Clinical Nutrition* 96 (2012): 14–23; K. Raninen and coauthors, Dietary fiber type reflects physiological functionality: Comparison of grain fiber, inulin, and polydextrose, *Nutrition Reviews* 69 (2011): 9-21; A. Fardet, New hypotheses for the health-protective mechanisms of whole-grain cereals: What is beyond fibre? *Nutrition Research Reviews* 23 (2010): 65-134; T. M. S. Wolever and coauthors, Physicochemical properties of oat β glucan influence its ability to reduce serum LDL cholesterol in humans: A randomized clinical trial, *American Journal of Clinical Nutrition* 92 (2010): 723–732; J. W. Anderson and coauthors, Health benefits of dietary fiber, *Nutrition Reviews* 67 (2009): 188–205; Position of the American Dietetic Association, Health implications of dietary fiber, 2008.
29. Raninen and coauthors, 2011; J. A. Nettleton and coauthors, Dietary patterns and incident cardiovascular disease in the MultiEthnic Study of Atherosclerosis, *American Journal of Clinical Nutrition* 90 (2009): 647–654; A. Mente and coauthors, A systematic review of the evidence supporting a causal link between dietary factors and coronary heart disease, *Archives of Internal Medicine* 169 (2009): 659–669; Position of the American Dietetic Association, Health implications of dietary fiber, 2008.
30. Raninen and coauthors, 2011; J. A. Nettleton and coauthors, Interactions of dietary whole-grain intake with fasting glucose- and insulin-related genetic loci in individuals of European descent: A meta-analysis of 14 cohort studies, *Diabetes Care* 33 (2010): 2684-2691; V. Vuksan and coauthors, Fiber facts: Benefits and recommendations for individuals with type 2 diabetes, *Current Diabetes Reports* 9 (2009): 405–411; H. Kim and coauthors, Glucose and insulin responses to whole grain breakfasts varying in soluble fiber, beta-glucan: A dose response study in obese women with increased risk for insulin resistance, *European Journal of Nutrition* 48 (2009): 170–175.
31. A. M. Brownawell and coauthors, Prebiotics and the health benefits of fiber: Current regulatory status, future research, and goals, *Journal of Nutrition* 142 (2012): 962–974; J. E. Ravikoff and J. R. Korzenik, The role of fiber indiverticular disease, *Journal of Clinical Gastroenterology* 45 (2011): S7–S11.
32. S. Tarleton and J. K. Dibaise, Low-residue diet in diverticular disease: Putting an end to a myth, *Nutrition in Clinical Practice* 26 (2011): 137–142; A. Rocco and coauthors, Treatment options for uncomplicated diverticular disease of the colon, *Journal of Clinical Gastroenterology* 43 (2009): 803–808.
33. D. Aune and coauthors, Dietary fibre, whole grains, and risk of colorectal cancer: Systematic review and dose-response meta-analysis of prospective studies, *British Medical Journal* 343 (2011), doi: 10.1136/bmj.d6617; C. C. Dahm and coauthors, Dietary fiber and colorectal cancer risk: A nested case-control study using food diaries, *Journal of the National Cancer Institute* 102 (2010): 614–626; L. B. Sansbury and coauthors, The effect of strict adherence to a high-fiber, high-fruit and -vegetable, and low-fat eating pattern on adenoma recurrence, *American Journal of Epidemiology* 170 (2009): 576–584.
34. Dahm and coauthors, 2010.
35. M. H. Pan and coauthors, Molecular mechanisms for chemoprevention of colorectal cancer by natural dietary compounds, *Molecular Nutrition and Food Research* 55 (2011): 32–45.
36. L. A. Tucker and K. S. Thomas, Increasing total fiber intake reduces of weight and fat gains in women, *Journal of Nutrition* 139 (2009): 567–581.
37. N. Schroeder and coauthors, Influence of whole grain barley, whole grain wheat, and refined rice-based foods on short-term satiety and energy intake, *Appetite* 53 (2009): 363–369; M. Lyly and coauthors, Fiber in beverages can enhance perceived satiety, *European Journal of Nutrition* 48 (2009): 251–258; K. R. Juvonen and coauthors, Viscosity of oat bran-enriched beverages influences gastrointestinal hormonal responses in healthy humans, *Journal of Nutrition* 139 (2009): 461–466; R. A. Samra and G. H. Anderson, Insoluble cereal fiber reduces appetite and short-term food intake and glycemic response to food consumed 75 min later by healthy men, *American Journal of Clinical Nutrition* 86 (2007): 972–979.
38. P. Vitaglione and coauthors, β-Glucan-enriched bread reduces energy intake and modifies plasma ghrelin and peptide YY concentrations in the short term, *Appetite* 53 (2009): 338–344.
39. H. Du and coauthors, Dietary fiber and subsequent changes in body weight and waist circumference in European men and women, *American Journal of Clinical Nutrition* 91 (2010): 329–336.
40. M. Kristensen and coauthors, Whole grain compared with refined wheat decreases the percentage of body fat following a 12-week, energy-restricted dietary intervention in postmenopausal women, *Journal of Nutrition* 142 (2012): 710–716; S. S. Jonnalagadda and coauthors, Putting the whole grain puzzle together: Health benefits associated with whole grains—Summary of American Society for Nutrition 2010 Satellite symposium, *Journal of Nutrition* 141 (2011): 1011S–1022S; K. A. Harris and P. M. Kris-Etherton, Effects of whole grains on coronary heart disease risk, *Current Atherosclerosis Reports* 12 (2010): 368–376.
41. Standing Committee on the Scientific Evolution of Dietary Reference Intakes, 2005, p. 265.
42. Position of the American Dietetic Association, 2008.
43. C. E. O'Neil and coauthors, Consumption of whole grains is associated with improved diet quality and nutrient intake in children and adolescents: The National Health and Nutrition Examination Survey 1999–2004, *Public Health Nutrition* 14 (2011): 347–355; C. E. O'Neil and coauthors, Whole-grain consumption is associated with diet quality and nutrient intake in adults: The National Health and Nutrition Examination Survey, 1999–2004, *Journal of the American Dietetic Association* 110 (2010): 1461–1468.

Nutrition in Practice

The Glycemic Index in Nutrition Practice

Carbohydrate-rich foods vary in the degree to which they elevate both blood glucose and insulin concentrations. Chapter 3 introduced the *glycemic index (GI),* a ranking of carbohydrate foods based on their glycemic effect after ingestion. The glycemic index may be of interest to people with diabetes who must regulate their blood glucose to protect their health. In diabetes treatment, however, the total amount of carbohydrate is more important than the type of carbohydrate consumed.[1] Thus, dietetics experts have debated the usefulness of the glycemic index for diabetes treatment. Despite some controversy, however, the American Diabetes Association encourages low-glycemic foods that are rich in fiber and other nutrients.[2] Furthermore, because some recent research shows that a low-glycemic index diet can improve blood glucose control in type 2 diabetes, the use of low-glycemic diets for this purpose may be gaining credibility.[3]

Researchers are also trying to determine whether low-GI diets may be helpful for improving risk factors for a number of other chronic diseases.[4] This Nutrition in Practice will describe the factors that contribute to a food's glycemic effect and the results of research studies that have examined the potential benefits of selecting mainly low-GI foods.

How is the glycemic index measured?

The glycemic index is essentially a measure of how quickly the carbohydrate in a food is digested and absorbed. Although testing methods vary to some degree, the most common protocol is to feed the test food—which contains a measured quantity of digestible carbohydrate—to research subjects and then measure blood glucose levels for two or three hours after the feeding. The increase in blood glucose over the two- or three-hour period is then compared to the blood glucose rise after an identical amount of digestible carbohydrate is ingested from a reference food such as pure glucose or white bread. Figure NP3-1 illustrates the difference in the blood glucose response to a low-GI food and a high-GI food. The blood glucose curve displays the surge in blood glucose above normal fasting levels after the food is consumed and the subsequent fall over several hours. Table NP3-1 lists the GIs of various carbohydrate-containing foods, arranged from highest to lowest within each food group listed.

FIGURE NP3-1 Blood Glucose Response to High-GI and Low-GI Foods

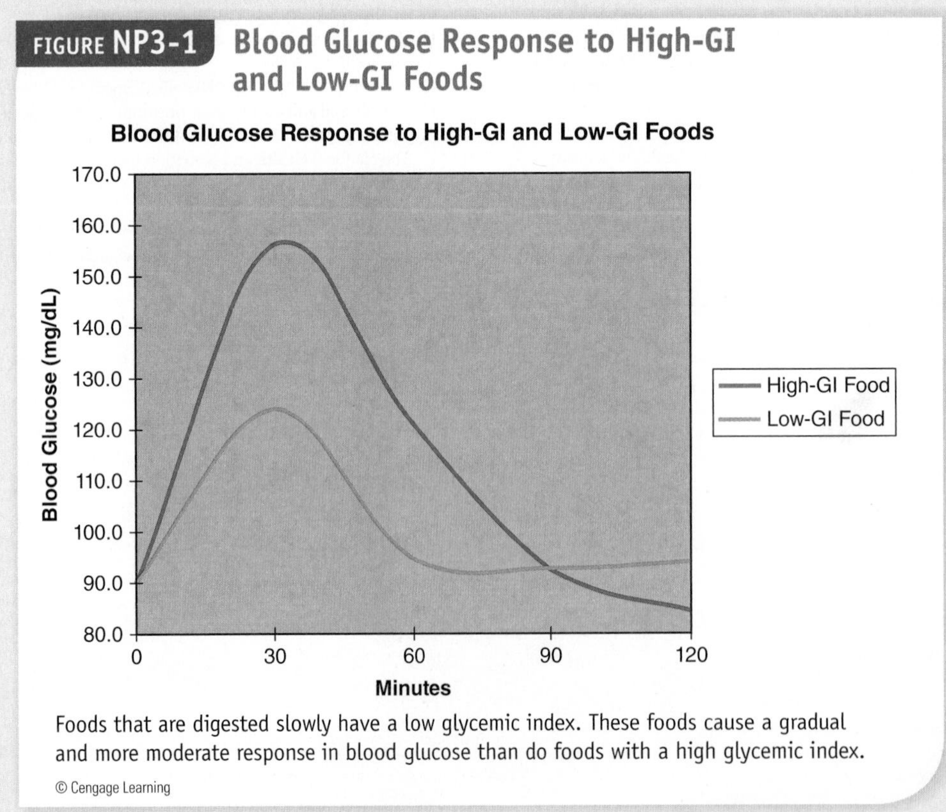

Foods that are digested slowly have a low glycemic index. These foods cause a gradual and more moderate response in blood glucose than do foods with a high glycemic index.

TABLE NP3-1 Glycemic Index (GI) of Selected Foods[a]

Value	GI
High	≥70
Medium	56–69
Low	≤55

Food Item	Glycemic Index
Grains	
Cornflakes	81
Instant oatmeal, cooked	79
White bread, enriched	75
Whole wheat bread	74
White rice	73
Bagel, white	69
Brown rice	68
Couscous	65
Popcorn	65
Bran flakes	63
Oatmeal, cooked	55
Spaghetti, white, boiled	49
Spaghetti, whole meal, boiled	48
Corn tortilla	46
Oat bran bread; 50% oat bran	44
Barley	28
Milk products	
Ice cream	51
Yogurt, fruit	41
Whole milk	39
Nonfat milk	37
Soy milk	34
Legumes	
Lentils	32
Garbanzo beans	28
Kidney beans	24
Soy beans	16

Food Item	Glycemic Index
Vegetables	
Russet potato, baked	111
Potato, instant mash	87
Potato, boiled	78
Pumpkin, boiled	64
Potato, french fries	63
Sweet potato, boiled	63
Taro	53
Sweet corn	52
Green peas, boiled	51
Carrots, boiled	39
Fruits	
Watermelon	76
Pineapple	59
Bananas	51
Mangoes	51
Orange juice	50
Oranges	43
Apple juice	41
Apples	36
Snack Foods/Beverages	
Fruit punch	67
Soft drink/soda	59
Chocolate	56
Potato chips	40
Sugars	
Sucrose	65
Honey	61
Fructose	15

[a]Reference food: glucose = 100

Source: F. S. Atkinson, K. Foster-Powell, and J. C. Brand-Miller, International tables of glycemic index and glycemic load values: 2008, *Diabetes Care* (2008): 2281–2283.

The *amount* of carbohydrate consumed also influences the glycemic response. A food's total glycemic effect—expressed as the *glycemic load (GL)*—is the product of its GI and the amount of available carbohydrate from the portion consumed, divided by 100. For example, if the GI of sweet potato is 60, a 100-gram serving containing about 20 grams of carbohydrates would have a GL of 12. The GL is used to standardize GI values to the carbohydrate content and portion size of a food or a meal.[5]

What factors influence a food's glycemic effect?

Table NP3-1 shows that starchy foods such as bread and potatoes tend to have high GI values, whereas many fruits and legumes have low GI values. The main factors that influence the GI value of a food include the following:

- *Starch structure.* Starch is present in foods as either a straight chain or branched chain of glucose molecules. Whereas digestion of the branched form tends to release glucose quickly, the straight chain is resistant to digestion. Thus, foods that contain mainly the branched form of starch tend to raise blood glucose levels more quickly and have a high GI value. Due to the subtle differences in starch among foods, different species of the same foods can have substantially different GI values; for example, current GI values for rice range from low (38 for parboiled white rice) to high (85 for Japanese sushi-style white rice).[6]
- *Fiber content.* Certain types of dietary fibers (primarily soluble fibers) increase the viscosity of chyme, slowing the passage of food in the stomach and upper intestine and making it more difficult for enzymes to digest the food. Therefore, foods such as beans, fruits, and vegetables, which contain soluble fibers, tend to have lower GI values.
- *Presence of fat and protein.* The fat in foods tends to slow stomach emptying, thus reducing the rate of digestion and absorption; hence, the presence of fat usually reduces a food's GI value. The protein in foods can also influence the GI because protein promotes insulin secretion, increasing the rate at which glucose is taken up from the blood.
- *Food processing.* The manner in which a food is processed and cooked influences the interactions among starch, protein, and fiber and thus affects the final GI value. For example, both pasta and bread are prepared from wheat flour, but pasta (cooked *al dente*) has a lower GI because the starch granules in pasta are surrounded by a sturdy protein barrier that hampers starch digestion. Cooking the pasta for longer periods can break down its structure and raise the GI value. As another example, the GI values for oatmeal vary according to the size and thickness of the oats used to prepare it: oatmeal prepared from steel-cut oats has a lower GI value than oatmeal prepared from quick oats. This is because the steel-cut oats are solid particles of grain, whereas "quick oats" are small, thin flakes.
- *Mixture of foods in a meal.* Because foods are rarely consumed in isolation, the GI value of an individual food may be less important than the combination of foods consumed at a meal. For example, in a cheese sandwich, the high GI of the bread is lowered by the addition of fat and protein in the cheese.
- *Individual glucose tolerance.* Cellular responses to insulin vary; thus, individual variability affects the glycemic response to foods. Persons with diabetes or prediabetes exhibit higher blood glucose levels after ingesting carbohydrate foods than do healthy individuals.[7]

What evidence suggests that a low-GI diet may influence chronic disease risk?

Some research shows that low-GI diets may reduce the risks of developing diabetes, heart disease, and obesity and help individuals lose weight.[8] Other studies, however, do not support such findings.[9] Studies are often difficult to interpret because low-GI foods often provide abundant soluble fiber, and soluble fiber slows glucose absorption, sustains feelings of fullness, and improves blood lipids. Therefore, it could be that soluble fiber, and not the low-GI diet, is responsible for any reported effects.[10] Because of mixed findings so far, health practitioners do not routinely recommend that patients consume low-GI diets to prevent or treat disease. An abundance of ongoing research

Noah Armonn/Shutterstock.com

Pasta cooked *al dente* has a lower glycemic index than many other starchy foods.

to reveal specific relationships between low-GI diets and chronic disease risk, however, may change such thinking. Examples of research include the following:

- *Diabetes prevention.* Some researchers have proposed that a high glycemic load can increase the body's demand for insulin and eventually reduce pancreatic function, resulting in inadequate insulin secretion. Indeed, results of several studies suggest that low-GI diets might prevent or delay the onset of type 2 diabetes in those at risk.[11]
- *Heart disease risk.* Although some research suggests that low-GI diets may improve blood lipid levels, other research shows no consistent effects of low-GI diets on heart disease risk.[12]
- *Appetite and weight loss.* Research shows that low-GI foods can slow the after-meal rise in blood glucose, supporting the hypothesis that such foods may promote satiety and suppress hunger.[13] Research results are inconsistent, however, on whether low-GI diets assist in weight loss.[14]

Given the mixed results of research studies on chronic disease prevention, are there any benefits associated with consuming low-GI foods?

Yes, if the low-GI foods are nutrient-dense, high-fiber foods. Not all low-GI foods meet these criteria: note that cakes, cookies, and candy bars may have a low GI due to their high fat content. Thus, a food's GI should be considered along with other nutrient criteria when assessing the health benefits.

The GI can be a helpful tool for choosing the most healthful food from a food group. For example, low-GI breakfast cereals tend to be high in fiber and low in added sugars, whereas high-GI cereals tend to be those that contain refined flours and significant amounts of added sugars. In other words, low-GI foods are often wholesome foods that have been minimally processed.

In general, should people avoid consuming high-GI foods?

Some people assume that starchy foods such as breads and potatoes should be avoided due to their high GI values. As mentioned earlier, these foods are rarely consumed in isolation, and their GI values are reduced in a mixed meal. For example, breads often have a GI greater than 70, but adding cheese or peanut butter reduces the GI to 55 or 59, respectively. Also worth considering is that GI values often vary considerably. For example, published values for white potatoes range from 24 to 101, and many samples have values in the mid-50s.[15] For these reasons and others, more studies are needed to confirm whether the GI is practical or beneficial for healthy people.

Given the complexity of the GI, what are the current recommendations?

The potential benefits associated with consuming low-GI diets are still under investigation. As discussed earlier, people with type 2 diabetes may benefit from limiting high-GI foods—those that produce too great a rise, or too sudden a fall, in blood glucose. Additional research is needed to justify the use of diets based on the GI for preventing or treating diseases such as heart disease, obesity, or other medical problems.

At present, many nutrition scientists advocate consuming a plant-based diet that contains minimally processed grains, legumes, vegetables, and fruits. Such a diet would include abundant fiber and limited amounts of solid fats and added sugars. Undoubtedly, meals consisting of these foods would tend to have low or medium GI values.

Notes

1. American Diabetes Association, Position statement: Nutrition recommendations and interventions for diabetes, *Diabetes Care* 31 (2008): S61–S78.
2. American Diabetes Association, Standards of medical care in diabetes—2011, *Diabetes Care* 34 (2011): S11–S61; American Diabetes Association, 2008.
3. A. Pande, G. Krishnamoorthy, and N. D. Moulick, Hypoglycaemic and hypolipidaemic effects of low GI and medium GL Indian diets in type 2 diabetics for a period of 4 weeks: A prospective study, *International Journal of Food Sciences and Nutrition* 63 (2012): 649–658; J. Brand-Miller and A. E. Buyken, The glycemic index issue, *Current Opinion in Lipidology* 23 (2012): 62–67; T. P. J. Solomon and coauthors, A low-glycemic index diet combined with exercise reduces insulin resistance, postprandial hyperinsulinemia, and glucose-dependent insulinotropic polypeptide responses in obese, prediabetic humans, *American Journal of Clinical Nutrition* 92 (2010): 1359–1368; D. Thomas and E. J. Elliott, Low glycaemic index, or low glycaemic load, diets for diabetes mellitus, *Cochrane Database of Systematic Reviews*, January 21, 2009, CD006296; Low glycaemic index diet and disposition index in type 2 diabetes (the Canadian trial of carbohydrates in diabetes): A randomized controlled trial, *Diabetologia* 51 (2009): 1607–1615.
4. Brand-Miller and Buyken, The glycemic index issue, 2012; I. Krog-Mikkelsen and coauthors, A low glycemic index diet does not affect postprandial energy metabolism but decreases postprandial insulinemia and increases fullness ratings in healthy women, *Journal of Nutrition* 141 (2011): 1679–1684; G. Radulian and coauthors, Metabolic effects of low glycaemic index diets, *Nutrition Journal* 8 (2009): 5.
5. S. Vega-López and S. N. Mayol-Kreiser, Use of the glycemic index for weight loss and glycemic control: A review of recent evidence, *Current Diabetes Reports* 9 (2009): 379–388.
6. F. S. Atkinson, K. Foster-Powell, and J. C. Brand-Miller, International tables of glycemic index and glycemic load values: 2008, *Diabetes Care* 31 (2008): 2281–2283.
7. American Diabetes Association, Diagnosis and classification of diabetes mellitus, *Diabetes Care* 34 (2011): S62–S69.
8. Brand-Miller and Buyken, The glycemic index issue, 2012; M. L. Neuhouser and coauthors, A low-glycemic load diet reduces serum C-reactive protein and modestly increases adiponectin in overweight and obese adults, *Journal of Nutrition* 142 (2012): 369–374; O. Gögebakan and coauthors, Effects of weight loss and long-term weight maintenance with diets varying in protein

and glycemic index on cardiovascular risk factors: The diet, obesity, and genes (DiOGenes) study: A randomized, controlled trial, *Circulation* 124 (2011): 2829–2838; G. M. Turner-McGrievy and coauthors, Decreases in dietary glycemic index are related to weight loss among individuals following therapeutic diets for type 2 diabetes, *Journal of Nutrition* 141 (2011): 1469–1474; I. Sluijs and coauthors, Carbohydrate quantity and quality and risk of type 2 diabetes in the Europena Prospective Investigation into Cancer and Nutrition—Netherlands (EPIC-NL) study, *American Journal of Clinical Nutrition* 92 (2010): 905–911; Solomon and coauthors, 2010; Radulian and coauthors, 2009.

9. Vega-López and Mayol-Kreiser, 2009; J. M. Shikany and coauthors, Effects of low-and high-glycemic index/glycemic load diets on coronary heart disease risk factors in overweight/obese men, *Metabolism* 58 (2009): 1793–1801; N. R. Sahyoun and coauthors, Dietary glycemic index and glycemic load and the risk of type 2 diabetes in older adults, *American Journal of Clinical Nutrition* 87 (2008): 126–131.
10. G. Livesey and H. Tagami, Interventions to lower the glycemic response to carbohydrate foods with a low-viscosity fiber (resistant maltodextrin): Meta-analysis of randomized controlled trials, *American Journal of Clinical Nutrition* 89 (2009): 114–125.
11. Brand-Miller and Buyken, The glycemic index issue, 2012; Radulian and coauthors, 2009.
12. L. M. Goff and coauthors, Low glycaemic index diets and blood lipids: A systematic review and meta-analysis of randomized controlled trials, *Nutrition, Metabolism, and Cardiovascular Diseases,* 2012 [Epub ahead of print], available at www.ncbi.nlm.nih.gov/pubmed/22841185; M. L. Wheeler and coauthors, Macronutrients, food groups, and eating patterns in the management of diabetes: A systematic review of the literature, 2010, *Diabetes Care* 35 (2012): 434-445; E. Denova-Gutierrez and coauthors, Dietary glycemic index, dietary glycemic load, blood lipids, and coronary heart disease, *Journal of Nutrition and Metabolism*, 2010, doi:10.1155/2010/170680; C. E. Finley and coauthors, Glycemic index, glycemic load, and prevalence of the metabolic syndrome in the Cooper Center Longitudinal Study, *Journal of the American Dietetic Association* 110 (2010): 1820–1829; Shikany and coauthors, Effects of low- and high-glycemic index/glycemic load diets on coronary heart disease risk factors in overweight/obese men, 2009.
13. Krog-Mikkelsen and coauthors, 2011; R. C. Reynolds and coauthors, Effect of the glycemic index of carbohydrates on day-long (10 h) profiles of plasma glucose, insulin, cholecystokinin and ghrelin, *European Journal of Clinical Nutrition* 63 (2009): 872–878.
14. Turner-McGrievy and coauthors, 2011; Vega-Lopez and Mayol-Kreiser, 2009; M. A. Mendez and coauthors, Glycemic load, glycemic index, and body mass index in Spanish adults, *American Journal of Clinical Nutrition* 89 (2009): 316–322.
15. Atkinson, Foster-Powell, and Brand-Miller, 2008.

Chapter 4 Lipids

MOST PEOPLE KNOW THAT TOO MUCH FAT, ESPECIALLY CERTAIN KINDS of fat, in the diet imposes health risks, but they may be surprised to learn that too little does, too. People in the United States, however, are more likely to eat too much fat than too little.

Fat is a member of the class of compounds called **lipids**. The lipids in foods and in the human body include triglycerides (**fats** and **oils**), phospholipids, and sterols.

Roles of Body Fat

Lipids perform many tasks in the body, but, most importantly, they provide energy. A constant flow of energy is so vital to life that, in a pinch, any other function is sacrificed to maintain it. Chapter 3 described one safeguard against such an emergency—the stores of glycogen in the liver that provide glucose to the blood whenever the supply runs short. The body's stores of glycogen are limited, however. In contrast, the body's capacity to store fat for energy is virtually unlimited due to the fat-storing cells of the **adipose tissue**. The fat cells of the adipose tissue readily take up and store fat, growing in size as they do so. Fat cells are more than just storage depots, however; fat cells secrete hormones that help to regulate the appetite and influence other body functions.[1] Figure 4-1 shows a fat cell.

The fat stored in fat cells supplies 60 percent of the body's ongoing energy needs during rest. The fat embedded in muscle tissue shares with muscle glycogen the task of providing energy when the muscles are active. During some types of physical activity or prolonged periods of food deprivation, (■) fat stores may make an even greater energy contribution. The brain and nerves, however, need their energy as glucose, and,

FIGURE 4-1 A Fat Cell

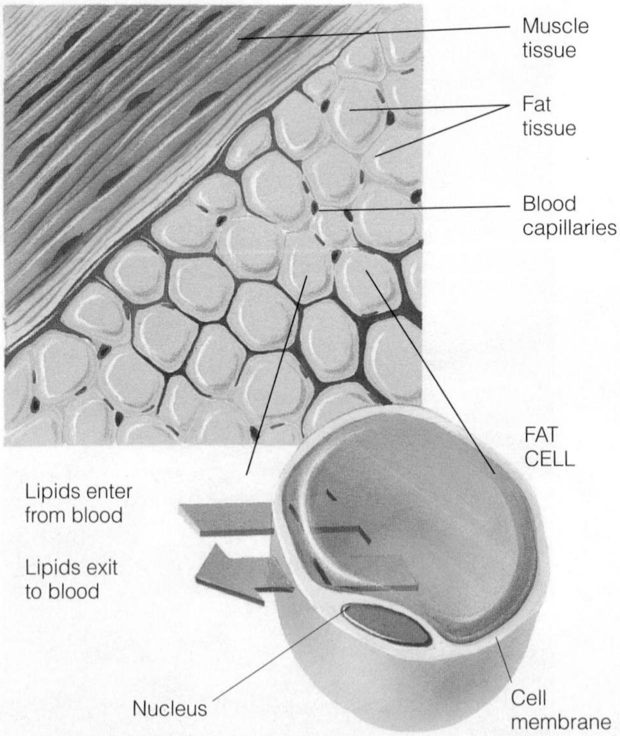

Within the fat, or adipose, cell, lipid is stored in a droplet. This droplet can greatly enlarge, and the fat cell membrane will expand to accommodate its swollen contents.

■ Chapter 6 discusses fat use during fasting.

lipids: a family of compounds that includes triglycerides (fats and oils), phospholipids, and sterols. Lipids are characterized by their insolubility in water.

fats: lipids that are solid at room temperature (70°F or 21°C).

oils: lipids that are liquid at room temperature (70°F or 21°C).

adipose tissue: the body's fat, which consists of masses of fat-storing cells called adipose cells.

TABLE 4-1 The Functions of Fats in the Body

- *Energy stores.* Fats are the body's chief form of stored energy.
- *Muscle fuel.* Fats provide much of the energy to fuel muscular work.
- *Padding.* Fat pads inside the body cavity protect the internal organs from shock.
- *Insulation.* Fats insulate against temperature extremes by forming a fat layer under the skin.
- *Cell membranes.* Fats form the major material of cell membranes.
- *Raw materials.* Fats are converted to other compounds, such as hormones, bile, and vitamin D, as needed.

as explained in Chapter 6, fat is an inefficient source of glucose. After a long period of glucose deprivation (during fasting or starvation), brain and nerve cells develop the ability to derive about half of their energy from a special form of fat known as **ketones**, but they still require glucose as well. This means that people wanting to lose weight need to eat a certain minimum amount of carbohydrate to meet their energy needs, even when they are limiting their food intakes.

In addition to supplying energy, fat serves other roles in the body. Natural oils in the skin provide a radiant complexion; in the scalp, they help nourish the hair and make it glossy. The layer of fat beneath the skin insulates the body from extremes of temperature. A pad of hard fat beneath each kidney protects it from being jarred and damaged, even during a motorcycle ride on a bumpy road. The soft fat in a woman's breasts protects her mammary glands from heat and cold and cushions them against shock. The phospholipids and the sterol cholesterol are cell membrane constituents that help maintain the structure and health of all cells. Table 4-1 summarizes the major functions of fats in the body.

ketones (KEY-tones): acidic, water-soluble compounds produced by the liver during the breakdown of fat when carbohydrate is not available; technically known as *ketone bodies*.

triglycerides (try-GLISS-er-rides): one of the main classes of lipids; the chief form of fat in foods and the major storage form of fat in the body; composed of glycerol with three fatty acids attached.

tri = three
glyceride = a compound of glycerol

fatty acids: organic compounds composed of a chain of carbon atoms with hydrogen atoms attached and an acid group at one end.

glycerol (GLISS-er-ol): an organic compound, three carbons long, that can form the backbone of triglycerides and phospholipids.

Review Notes

- Lipids in the body not only serve as energy reserves but also protect the body from temperature extremes, cushion the vital organs, and provide the major material of cell membranes.

The Chemist's View of Lipids

The diverse and vital functions that lipids perform in the body reveal why eating too little fat can be harmful. As mentioned earlier, though, too much fat in the diet seems to be the greater problem for most people. To understand both the beneficial and harmful effects that fats exert on the body, a closer look at the structure and function of members of the lipid family is in order.

TRIGLYCERIDES

When people talk about fat—for example, "I'm too fat" or "That meat is fatty"—they are usually referring to triglycerides. Among lipids, **triglycerides** predominate—both in the diet and in the body. The name *triglyceride* almost explains itself: three (*tri*) **fatty acids** attached to a **glycerol** "backbone." Figure 4-2 (p. 94) shows how three fatty acids combine with glycerol to make a triglyceride.

Body fat supplies much of the fuel that muscles need to do their work.

FIGURE 4-2 Triglyceride Formation

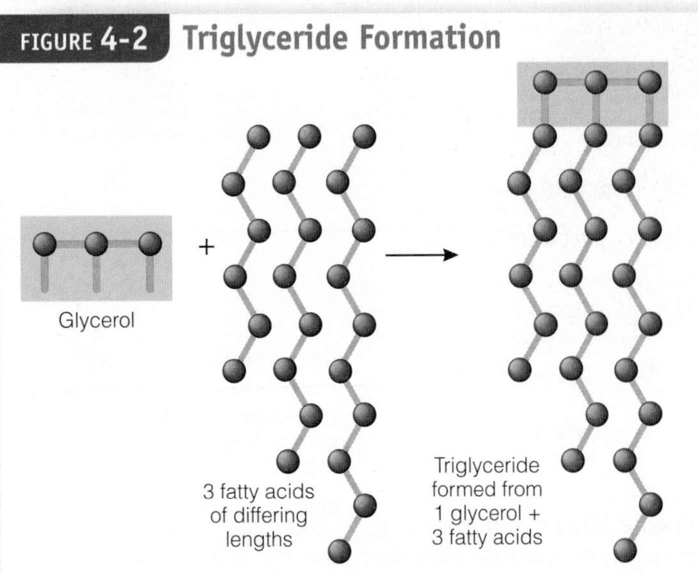

Glycerol, a small, water-soluble compound, plus three fatty acids, equals a triglyceride.

FATTY ACIDS

When energy from any energy-yielding nutrient is to be stored as fat, the nutrient is first broken into small fragments. Then the fragments are linked together into chains known as fatty acids. The fatty acids are then packaged, three at a time, with glycerol to make triglycerides.

Chain Length and Saturation Fatty acids may differ from one another in two ways—in chain length and in degree of saturation. The chain length refers to the number of carbons in a fatty acid. Saturation also refers to its chemical structure—specifically, to the number of hydrogen atoms the carbons in the fatty acid are holding. If every available carbon is filled to capacity with hydrogen atoms, the chain is called a **saturated fatty acid**. A saturated fatty acid is fully loaded with hydrogen atoms and has only single bonds between the carbons. The first zigzag structure in Figure 4-3 represents a saturated fatty acid.

Unsaturated Fatty Acids In some fatty acids, including most of those in plants and fish, hydrogen atoms are missing from the fatty acid chains. The places where the hydrogen atoms are missing are called points of unsaturation, and a chain containing such points is called an **unsaturated fatty acid**. An unsaturated fatty acid has at least one double bond between its carbons. If there is one point of unsaturation, the chain is a **monounsaturated fatty acid**. The second structure in Figure 4-3

FIGURE 4-3 Three Types of Fatty Acids

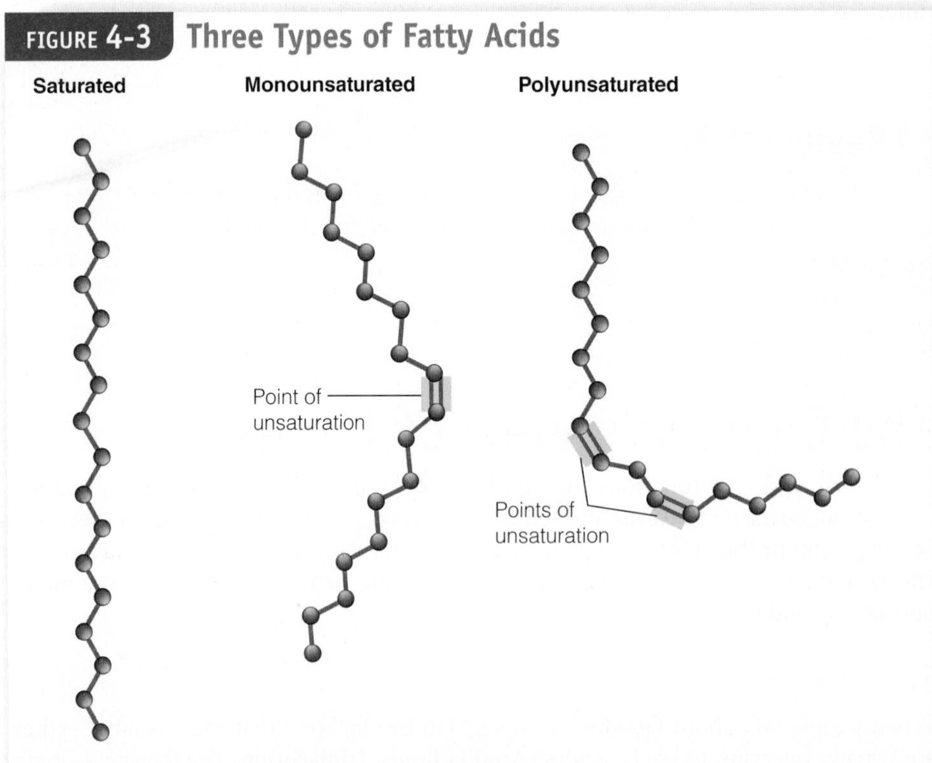

The more carbon atoms in a fatty acid, the longer it is. The more hydrogen atoms attached to those carbons, the more saturated the fatty acid is.

saturated fatty acid: a fatty acid carrying the maximum possible number of hydrogen atoms (having no points of unsaturation).

unsaturated fatty acid: a fatty acid with one or more points of unsaturation where hydrogen atoms are missing (includes monounsaturated and polyunsaturated fatty acids).

monounsaturated fatty acid (MUFA): a fatty acid that has one point of unsaturation; for example, the oleic acid found in olive oil.

is an example. If there are two or more points of unsaturation, then the fatty acid is a **polyunsaturated fatty acid** (see the third structure in Figure 4-3).

Hard and Soft Fat A triglyceride can contain any combination of fatty acids—long chain or short chain and saturated, monounsaturated, or polyunsaturated. The degree of saturation of the fatty acids in a fat influences the health of the body (discussed in a later section) and the characteristics of foods. Fats that contain the shorter chain or the more unsaturated fatty acids are softer at room temperature and melt more readily. A comparison of three fats—lard (which comes from pork), chicken fat, and safflower oil—illustrates these differences: lard is the most saturated and the hardest; chicken fat is less saturated and somewhat soft; and safflower oil, which is the most unsaturated, is a liquid at room temperature.

Stability Saturation also influences stability. Fats can become **rancid** when exposed to oxygen. Polyunsaturated fatty acids spoil most readily because their double bonds are unstable. The **oxidation** of unsaturated fats produces a variety of compounds that smell and taste rancid; saturated fats are more resistant to oxidation and thus less likely to become rancid. Other types of spoilage can occur due to microbial growth, however.

Manufacturers can protect fat-containing products against rancidity in three ways—none of them perfect. First, products may be sealed airtight and refrigerated—an expensive and inconvenient storage system. Second, manufacturers may add **antioxidants** to compete for the oxygen and thus protect the oil (examples are the additives **BHA** and **BHT** and vitamins C and E).* Third, manufacturers may saturate some or all of the points of unsaturation by adding hydrogen atoms—a process known as hydrogenation.

Hydrogenation **Hydrogenation** offers two advantages: it protects against oxidation (thereby prolonging shelf life) and also alters the texture of foods by increasing the solidity of fats. When partially hydrogenated, vegetable oils become spreadable margarine. Hydrogenated fats make pie crusts flaky and puddings creamy. A disadvantage is that hydrogenation makes polyunsaturated fats more saturated. Consequently, any health advantages of using polyunsaturated fats instead of saturated fats are lost with hydrogenation.

***Trans*-Fatty Acids** Another disadvantage of hydrogenation is that some of the molecules that remain unsaturated after processing change shape from *cis* to *trans*. In nature, most unsaturated fatty acids are *cis*-fatty acids—meaning that the hydrogen atoms next to the double bonds are on the same side of the carbon chain. Only a few fatty acids in nature (notably a small percentage of those found in milk and meat products) are ***trans*-fatty acids**—meaning that the hydrogen atoms next to the double bonds are on opposite sides of the carbon chain (see Figure 4-4). These arrangements result in different configurations for the fatty acids, and this difference affects function: in the body, *trans*-fatty acids behave more like saturated fats, increasing blood cholesterol and the risk of heart disease (as a later section describes).[2]

FIGURE 4-4 ***Cis*- and *Trans*-Fatty Acids Compared**

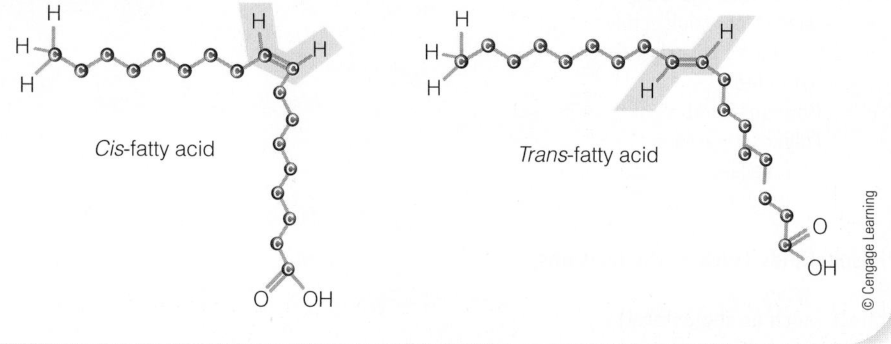

Cis-fatty acid

Trans-fatty acid

© Cengage Learning

*BHA is butylated hydroxyanisole; BHT is butylated hydroxytolvene.

polyunsaturated fatty acids (PUFA): fatty acids with two or more points of unsaturation. For example, linoleic acid has two such points, and linolenic acid has three. Thus, polyunsaturated *fat* is composed of triglycerides containing a high percentage of PUFA.

rancid: the term used to describe fats when they have deteriorated, usually by oxidation. Rancid fats often have an "off" odor.

oxidation (OKS-ee-day-shun): the process of a substance combining with oxygen.

antioxidants: as a food additive, preservatives that delay or prevent rancidity of foods and other damage to food caused by oxygen.

BHA, BHT: preservatives commonly used to slow the development of "off" flavors, odors, and color changes caused by oxidation.

hydrogenation (high-dro-gen-AY-shun): a chemical process by which hydrogen atoms are added to monounsaturated or polyunsaturated fats to reduce the number of double bonds, making the fats more saturated (solid) and more resistant to oxidation (protecting against rancidity). Hydrogenation produces *trans*-fatty acids.

***trans*-fatty acids:** fatty acids in which the hydrogen atoms next to the double bond are on opposite sides of the carbon chain.

Researchers are trying to determine whether the health effects of naturally occurring *trans* fats differ from those of commercially created *trans* fats.[3] In any case, the important distinction is that intake of naturally occurring *trans* fatty acids is typically low. At current levels of consumption, naturally occurring *trans* fats are unlikely to have adverse effects on blood lipids. The naturally occurring *trans* fatty acid **conjugated linoleic acid** may even have health benefits.[4]

Essential Fatty Acids Using carbohydrate, fat, or protein, the human body can synthesize all the fatty acids it needs except for two—**linoleic acid** and **linolenic acid**. Both linoleic acid and linolenic acid are polyunsaturated fatty acids. Because they cannot be made from other substances in the body, they must be obtained from food and are therefore called **essential fatty acids**. Linoleic acid and linolenic acid are found in small amounts in plant oils, and the body readily stores them, making deficiencies unlikely. From both of these essential fatty acids, the body makes important substances that help regulate a wide range of body functions: blood pressure, clot formation, blood lipid concentration, the immune response, the inflammatory response to injury, and many others.[5] These two essential nutrients also serve as structural components of cell membranes.

Linoleic Acid: An Omega-6 Fatty Acid Linoleic acid is an **omega-6 fatty acid**, (■) found in the seeds of plants and in the oils produced from the seeds. Any diet that contains vegetable oils, seeds, nuts, and whole-grain foods provides enough linoleic acid to meet the body's needs. Researchers have long known and appreciated the importance of the omega-6 fatty acid family.

Linolenic Acid and Other Omega-3 Fatty Acids Linolenic acid belongs to a family of polyunsaturated fatty acids known as **omega-3 fatty acids**, a family that also includes **EPA** and **DHA**. EPA and DHA are found primarily in fish oils. As mentioned, the human body cannot make linolenic acid, but given dietary linolenic acid, it can make EPA and DHA, although the process is slow.

The importance of omega-3 fatty acids has been recognized since the 1980s, and research continues to unveil impressive roles for EPA and DHA in metabolism and disease prevention. The brain has a high content of DHA, and both EPA and DHA are needed for normal brain development.[6] DHA is also especially active in the rods and cones of the retina of the eye.[7] Today, researchers know that these omega-3 fatty acids are essential for normal growth and development and that they may play an important role in the prevention and treatment of heart disease.[8]

PHOSPHOLIPIDS

Up to now, this discussion has focused on one class of lipids, the triglycerides (fats and oils), and their component parts, the fatty acids (see Table 4-2). Two

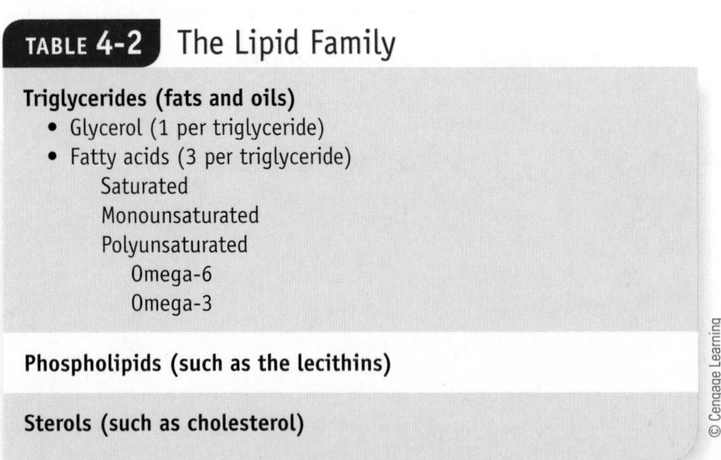

TABLE 4-2 The Lipid Family

Triglycerides (fats and oils) • Glycerol (1 per triglyceride) • Fatty acids (3 per triglyceride) Saturated Monounsaturated Polyunsaturated Omega-6 Omega-3
Phospholipids (such as the lecithins)
Sterols (such as cholesterol)

■ Chemists use the term *omega*, the last letter of the Greek alphabet, to refer to the position of the last double bond in a fatty acid.

conjugated linoleic acid: a collective term for several fatty acids that have the same chemical formulas as linoleic acid but with different configurations.

linoleic acid, linolenic acid: polyunsaturated fatty acids that are essential for human beings.

essential fatty acids: fatty acids that the body requires but cannot make and so must be obtained through the diet.

omega-6 fatty acid: a polyunsaturated fatty acid with its endmost double bond six carbons back from the end of its carbon chain; long recognized as important in nutrition. Linoleic acid is an example.

omega-3 fatty acids: polyunsaturated fatty acids in which the endmost double bond is three carbons back from the end of the carbon chain; relatively newly recognized as important in nutrition. Linolenic acid is an example.

EPA, DHA: omega-3 fatty acids made from linolenic acid. The full name for EPA is *eicosapentaenoic* (EYE-cosa-PENTA-ee-NO-ick) *acid*. The full name for DHA is *docosahexaenoic* (DOE-cosa-HEXA-ee-NO-ick) *acid*.

other classes of lipids, the **phospholipids** and sterols, make up only 5 percent of the lipids in the diet, but they are nevertheless worthy of attention. Among the phospholipids, the lecithins are of particular interest.

Structure of Phospholipids Like the triglycerides, the **lecithins** and other phospholipids have a backbone of glycerol; they differ from the triglycerides in having only two fatty acids attached to the glycerol. In place of the third fatty acid, they have a phosphate group (a phosphorus-containing acid) and a molecule of **choline** or a similar compound. The fatty acids make the phospholipids soluble in fat; the phosphate group enables them to dissolve in water. Such versatility benefits the food industry, which uses phospholipids as emulsifiers to mix fats with water in such products as mayonnaise and candy bars. (■)

Phospholipids in Foods In addition to the phospholipids used by the food industry as emulsifiers, phospholipids are also found naturally in foods. The richest food sources of lecithin are eggs, liver, soybeans, wheat germ, and peanuts.

Roles of Phospholipids Lecithins and other phospholipids are important constituents of cell membranes. They also act as emulsifiers in the body, helping to keep other fats in solution in the watery blood and body fluids. In addition, some phospholipids generate signals inside the cells in response to hormones, such as insulin, to help alter body conditions.

STEROLS

Sterols are large, complex molecules consisting of interconnected rings of carbon. Cholesterol is the most familiar sterol, but others, such as vitamin D and the sex hormones (for example, testosterone), are important, too.

Sterols in Foods Foods derived from both plants and animals contain sterols, but only those from animals—meats, eggs, fish, poultry, and dairy products—contain significant amounts of cholesterol. Organ meats, such as liver and kidneys, and eggs are richest in cholesterol; cheeses and meats have less. Shellfish contain many sterols but much less cholesterol than was previously thought.

Sterols other than cholesterol are naturally found in plants. Being structurally similar to cholesterol, plant sterols interfere with cholesterol absorption. Food manufacturers have fortified foods such as margarine with plant sterols, creating a functional food that helps to reduce blood cholesterol.

Cholesterol Synthesis Like the lecithins, cholesterol can be made by the body, so it is not an essential nutrient. Right now, as you read, your liver is manufacturing cholesterol from fragments of carbohydrate, protein, and fat. Most of the body's cholesterol ends up in the membranes of cells, where it performs vital structural and metabolic functions.

Cholesterol's Two Routes in the Body After it is made, cholesterol leaves the liver by two routes:

1. It may be incorporated into bile, stored in the gallbladder, and delivered to the intestine. (■)
2. It may travel, via the bloodstream, to all the body's cells.

The bile that is made from cholesterol in the liver is released into the intestine to aid in the digestion and absorption of fat. After bile does its job, most of it is absorbed and reused by the body; the rest is excreted in the feces.

Cholesterol Excreted While bile is in the intestine, some of it may be trapped by soluble fibers or by some medications, which carry it out of the body in feces. The excretion of bile reduces the total amount of cholesterol remaining in the body.

■ Reminder: Emulsifiers are substances that mix with both fat and water and are able to disperse the fat in the water, forming an emulsion.

■ Reminder: Bile is a compound made by the liver from cholesterol and stored in the gallbladder. Bile prepares fat for digestion.

phospholipids: one of the three main classes of lipids; compounds that are similar to triglycerides but have *choline* (or another compound) and a phosphorus-containing acid in place of one of the fatty acids.

lecithins: one type of phospholipid.

choline: a nutrient that can be made in the body from an amino acid.

sterols: one of the main classes of lipids; includes cholesterol, vitamin D, and the sex hormones (such as testosterone).

Cholesterol Transport Recall from Chapter 2 that some cholesterol, packaged with other lipids and protein, leaves the liver via the arteries and is transported to the body tissues by the blood. (■) These packages of lipids and proteins are called lipoproteins. As the lipoproteins travel through the body, tissues can extract lipids from them. Cholesterol can be harmful to the body when it forms deposits in the artery walls. These deposits contribute to **atherosclerosis**, a disease that can cause heart attacks and strokes.

Review Notes

- Table 4-2 summarizes the members of the lipid family.
- The predominant lipids both in foods and in the body are triglycerides, which have glycerol backbones with three fatty acids attached.
- Fatty acids vary in the length of their carbon chains and their degree of saturation. Those that are fully loaded with hydrogen atoms are saturated; those that are missing hydrogen atoms and therefore have double bonds are unsaturated (monounsaturated or polyunsaturated).
- Most triglycerides contain more than one type of fatty acid.
- Fatty acid saturation affects the physical characteristics and storage properties of fats.
- Hydrogenation, which makes polyunsaturated fats more saturated, gives rise to *trans*-fatty acids, altered fatty acids that may have health effects similar to those of saturated fatty acids.
- Linoleic acid and linolenic acid are essential nutrients. In addition to serving as structural parts of cell membranes, they make powerful substances that help regulate blood pressure, blood clot formation, and the immune response.
- Phospholipids, including the lecithins, have a unique chemical structure that allows them to be soluble in both water and fat.
- In the body, phospholipids are major constituents of cell membranes; the food industry uses phospholipids as emulsifiers.
- Sterols include cholesterol, bile, vitamin D, and the sex hormones.
- Only animal-derived foods contain significant amounts of cholesterol.

Digestion and Absorption of Lipids

The goal of fat digestion is to dismantle triglycerides into small molecules that the body can absorb and use—namely **monoglycerides**, fatty acids, and glycerol. Table 4-3 provides the details.

■ Both the intestine and the liver make lipoproteins. Chapter 2 tells the story of lipid transport.

Health Effects and Recommended Intakes of Fats

Of all the dietary factors related to chronic diseases prevalent in developed countries, high intakes of certain fats are by far the most significant. The person who chooses a diet too high in saturated fats or *trans* fats invites the risk of **cardiovascular disease (CVD)**, and heart disease is the number one killer of adults in the United States and Canada. As for cancer, evidence is less compelling than for heart disease, but it does suggest that a diet high in certain kinds of fat is associated with a greater-than-average risk of developing some types of cancer.[9] Conversely, some research suggests that omega-3 fatty acids from fish may protect against some cancers.[10] Nutrition and cancer is a topic of Chapter 23.

Obesity carries serious risks to health. A diet high in energy-rich fatty foods makes it easy for people to exceed their energy needs and encourages unneeded weight gain.[11]

atherosclerosis (ath-er-oh-scler-OH-sis): a type of artery disease characterized by accumulations of lipid-containing material on the inner walls of the arteries (see Chapter 21).

monoglycerides: molecules of glycerol with one fatty acid attached. A molecule of glycerol with two fatty acids attached is a *diglyceride*.

cardiovascular disease (CVD): a general term for all diseases of the heart and blood vessels (see Chapter 21).

TABLE 4-3 Fat Digestion and Absorption

Mouth and salivary glands

Some hard fats begin to melt as they reach body temperature. The sublingual salivary gland in the base of the tongue secretes lingual lipase. The degree of hydrolysis by lingual lipase is slight for most fats but may be appreciable for milk fats.

Stomach

The stomach's churning action mixes fat with water and acid. A gastric lipase accesses and hydrolyzes (only a very small amount of) fat.

Small intestine and pancreas

Cholecystokinin (CCK) signals the gallbladder to release bile (via the common bile duct):

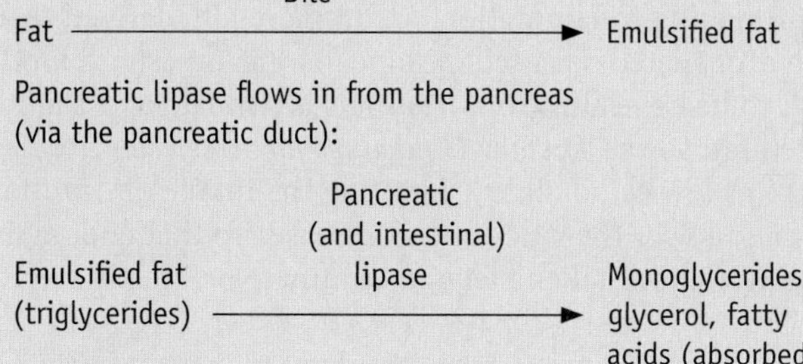

Large intestine

Some fat and cholesterol, trapped in fiber, exit in feces.

An increasing waistline, in turn, often increases blood triglycerides, which can indicate an increased risk of heart disease and other chronic diseases.[12] The health risks of obesity are described in Chapter 6, and Chapter 7 focuses on weight management. The links between diet and disease are the focus of much research. Some points about fats and heart health are presented here because they underlie dietary recommendations concerning fats. Nutrition and heart disease is the topic of Chapter 21.

FATS AND HEART HEALTH

As noted earlier and described in Chapter 2, cholesterol travels in the blood within lipoproteins. Two of the lipoproteins, LDL and HDL, play major roles with regard to heart health and are the focus of most recommendations made for reducing the risk of heart disease. A high blood LDL cholesterol concentration is a predictor of the likelihood of suffering a fatal heart attack or stroke, and the higher the LDL, the earlier the episode is expected to occur. Conversely, high HDL cholesterol signifies a *lower* disease risk.[13]

Most people realize that elevated blood cholesterol is an important risk factor for heart disease. Most people may not realize, though, that cholesterol in *food* is not the main influential factor in raising *blood* cholesterol.

Saturated Fats and Blood Cholesterol

The main dietary factors associated with elevated blood LDL cholesterol are high saturated fat and high *trans* fat intakes.*[14] High LDL cholesterol levels increase the risk of heart disease because high LDL concentrations promote the uptake of cholesterol in the blood vessel walls. (■)

Solid fats, introduced in Chapter 1, are foods or ingredients in foods (such as shortening in cakes or pies) that provide abundant saturated fat, *trans* fat, and/or cholesterol and many kcalories. The current American diet delivers excessive amounts of solid

■ Nutrition in Practice 4 examines various types of fats and their roles in supporting or harming heart health.

solid fats: fats that are not usually liquid at room temperature; commonly found in most foods derived from animals and vegetable oils that have been hydrogenated. Solid fats typically contain more saturated and *trans* fats than most oils. Also defined in Chapter 1.

*It should be noted that not all saturated fatty acids have the same cholesterol-raising effect. Stearic acid, an 18-carbon fatty acid, does not seem to raise blood cholesterol.

fats—representing an average of almost one-fifth of the day's total kcalories.[15] The easiest way to lower saturated fat, then, is to limit solid fats in the diet. Figure 4-5 shows that grain-based desserts, pizza, cheese, and processed and fatty meats are major providers of solid fats. Solid fats from animal sources contribute a great deal of the saturated fat in most people's diets. Some vegetable fats (coconut oil, palm kernel oil, and palm oil) and hydrogenated fats such as shortening or stick margarine provide smaller amounts of saturated fats.

Importantly, replacing dietary saturated fats with added sugars and refined starches is often counterproductive.[16] The best diet for health not only replaces saturated fats with polyunsaturated and monounsaturated oils (as discussed in a later section), but also is adequate, balanced, kcalorie-controlled, and based on mostly nutrient-dense whole foods.

Trans-Fatty Acids and Blood Cholesterol Consuming commercially derived *trans* fat poses a risk to the health of the heart and arteries by raising LDL and lowering HDL cholesterol, and by producing inflammation.[17] Commercially derived *trans* fats are found in the partially hydrogenated oils used in some margarines, snack foods, and prepared desserts. The risk to heart health from *trans* fats is similar to or slightly greater than that from saturated fat, so the *Dietary Guidelines for Americans* suggest that people keep *trans* fat intake as low as possible.[18] Limiting the intake of *trans* fats can improve blood cholesterol and lower the risk of heart disease. To that end, many restaurants and food manufacturers have taken steps to eliminate or greatly reduce *trans* fats in foods.[19]

For example, margarine makers have reformulated their products to contain much less *trans* fat. (■) Soft or liquid varieties are made from unhydrogenated oils, which are mostly

■ The words *hydrogenated vegetable oil* or *shortening* in an ingredients list indicate *trans*-fatty acids in the product.

FIGURE 4-5 Sources of Solid Fats in the U.S. Diet

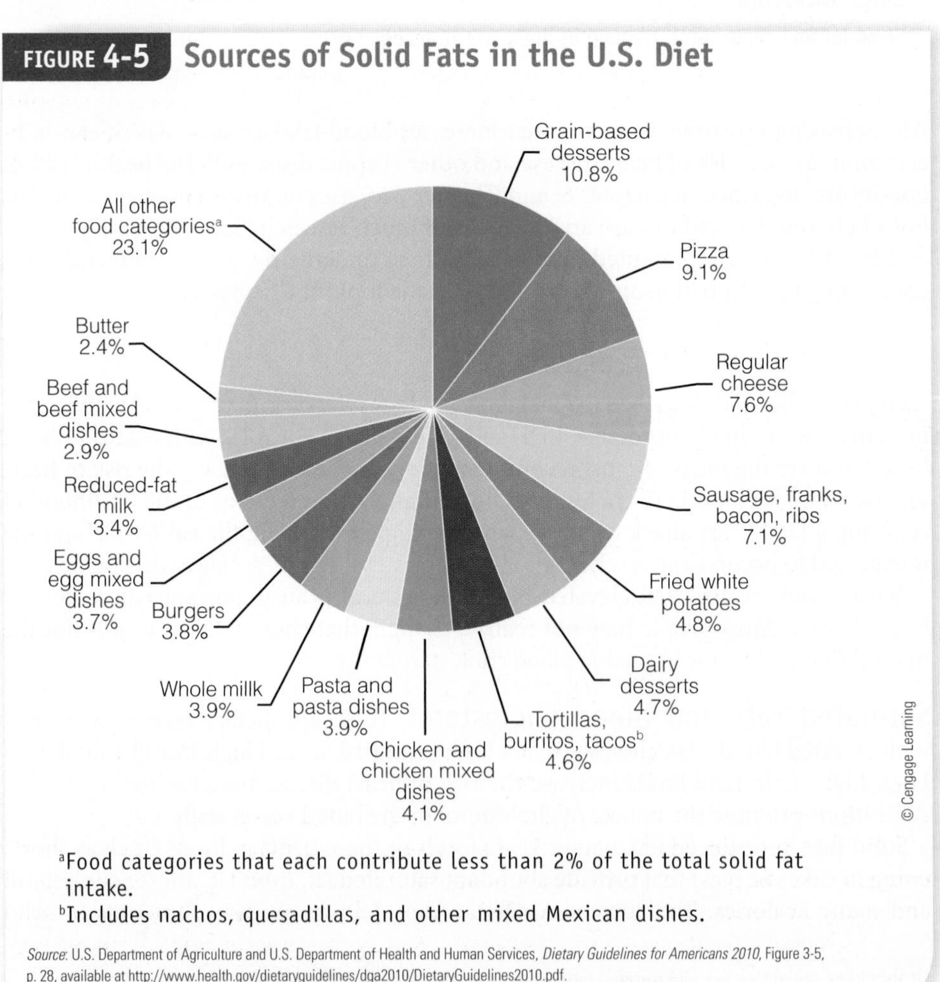

[a]Food categories that each contribute less than 2% of the total solid fat intake.
[b]Includes nachos, quesadillas, and other mixed Mexican dishes.

Source: U.S. Department of Agriculture and U.S. Department of Health and Human Services, *Dietary Guidelines for Americans 2010*, Figure 3-5, p. 28, available at http://www.health.gov/dietaryguidelines/dga2010/DietaryGuidelines2010.pdf.

unsaturated and so are less likely to elevate blood cholesterol than the saturated fats of butter. Some margarines contain olive oil, omega-3 fatty acids, or plant sterols (mentioned earlier), making these products preferable to butter and other margarines for the heart.*

In the past, most commercially fried foods, from doughnuts to chicken, delivered a sizeable amount of *trans* fats to consumers. (■) Today, newly formulated commercial oils and fats perform the same tasks as the previously used hydrogenated fats but with fewer *trans*-fatty acids. Some new fats, however, merely substitute saturated fats—which pose well-established risks to heart health—for *trans* fats. When reformulating their products, food companies must consider not only the fat composition, but also the taste, texture, cost, and availability of materials. No health benefits can be expected when saturated fats replace *trans* fats in the diet.

■ Major food sources of *trans* fats:
- Cakes, cookies, pies, doughnuts, and crackers
- Hard margarine
- Deep-fried foods (vegetable shortening)
- Snack chips

Dietary Cholesterol and Blood Cholesterol Although its effect is not as strong as that of saturated fat or *trans* fat, dietary cholesterol may contribute to elevated blood cholesterol in some people. Less clear is its role heart disease.[20] The *Dietary Guidelines* recommend limiting dietary cholesterol to less than 300 milligrams per day for healthy people (less than 200 milligrams for some people with or at high risk of heart disease). On average, women take in about 240 milligrams a day and men take in about 350 milligrams.[21] Foods providing the greatest share of cholesterol to the U.S. diet are eggs and egg dishes, chicken and chicken dishes, beef and beef dishes, and all types of beef burgers. (■)

■ The cholesterol content of one egg is about 210 milligrams.

In healthy people, evidence suggests no association between consuming one egg per day and increased risk of heart disease.[22] For individuals with or at high risk of heart disease, however, consuming one egg per day may worsen or accelerate the progression of heart disease.

■ Major sources of monounsaturated fats:
- Olive oil, canola oil, peanut oil
- Avocados

■ Major sources of polyunsaturated fats:
- Vegetable oils (sunflower, sesame, soy, corn)
- Nuts and seeds

Monounsaturated Fatty Acids and Blood Cholesterol Replacing saturated and *trans* fats with monounsaturated fat such as olive oil may be an effective dietary strategy to prevent heart disease. (■) The lower rates of heart disease among people in the Mediterranean region of the world are often attributed to their liberal use of olive oil, a rich source of monounsaturated fatty acids.[23] Olive oil also delivers valuable phytochemicals that help to protect against heart disease.[24] Nutrition in Practice 4 examines the role of olive oil and other fats in supporting or harming heart health.

Polyunsaturated Fatty Acids, Blood Cholesterol, and Heart Disease Risk Polyunsaturated fatty acids (PUFA) of the omega-6 and omega-3 families are potent protectors against heart disease. The primary omega-6 fatty acid, linoleic acid, which is found in vegetable oils such as corn and sunflower oil, exerts most of its beneficial effect by lowering both total blood cholesterol and LDL cholesterol.[25] (■)

The omega-3 fatty acids EPA and DHA, which are found mainly in fatty fish, exert their beneficial effects by influencing the function of both the heart and blood vessels. Specifically, EPA and DHA protect heart health by:[26]

- Lowering blood triglycerides.
- Preventing blood clots.
- Protecting against irregular heartbeats.
- Lowering blood pressure.
- Defending against inflammation.

The primary member of the omega-3 family, linolenic acid, may benefit heart health as well, but evidence for this effect is much less certain than for EPA and DHA. Table 4-4 (p. 102) names sources of omega-6 and omega-3 fatty acids.

Grilling or broiling fish, instead of frying them, preserves their beneficial omega-3 fatty acids while adding little or no saturated fat.

The *Dietary Guidelines* recommend choosing 8 to 12 ounces of a variety of seafood each week, or about one of every five protein servings, to provide an average

*Two brand names of margarines with plant sterols currently on the market are *Benecol* and *Take Control*.

TABLE 4-4 Food Sources of Omega-3 and Omega-6 Fatty Acids

Omega-3	
Linolenic acid	Oils (canola, flaxseed, soybean, walnut, wheat germ, liquid or soft margarine made from canola or soybean oil) Nuts and seeds (flaxseeds, walnuts, soybeans) Vegetables (soybeans)
EPA and DHA	Human milk Fish and seafood: *>500 mg per 3.5-oz serving:* European seabass (bronzini), herring (Atlantic and Pacific), mackerel, oyster (Pacific wild), salmon (wild and farmed), sardines, toothfish (includes Chilian seabass), trout (wild and farmed) *150–500 mg per 3.5-oz serving:* black bass, catfish (wild and farmed), clam, cod (Atlantic), crab (Alaskan king), croakers, flounder, haddock, hake, halibut, oyster (eastern and farmed), perch, scallop, shrimp (mixed varieties), sole, swordfish, tilapia (farmed) *<150 mg per 3.5-oz serving:* cod (pacific), grouper, lobster, mahi-mahi, monkfish, red snapper, skate, triggerfish, tuna, wahoo
Omega-6	
Linoleic acid	Seeds, nuts, vegetable oils (corn, cottonseed, safflower, sesame, soybean, sunflower), poultry fat

Source for fish data: K. L. Weaver and coauthors, The content of favorable and unfavorable polyunsaturated fatty acids found in commonly eaten fish, *Journal of the American Dietetic Association* 108 (2008): 1178–1185; P. M. Kris-Etherton, W. S. Harris, and L. J. Appel, Fish consumption, fish oil, omega-3 fatty acids, and cardiovascular disease, *Circulation* 106 (2002): 2747–2757.

of 250 milligrams of EPA and DHA per day along with the beneficial array of other nutrients that seafood provides.[27] For example, fish, low in saturated fat and high in protein, contributes not only EPA and DHA but also the mineral selenium, a nutrient of concern for heart health (see Chapter 9).[28]

Greater heart health benefits can be expected when fish is grilled, baked, or broiled, partly because the varieties prepared this way often contain more EPA and DHA than species used for fried fish in fast-food restaurants and frozen products.[29] Additionally, benefits are attained by avoiding commercial frying fats, which may be laden with *trans* fat and saturated fat. Further benefits arise when fish replaces high-fat meats or other foods rich in saturated fats in several meals each week.

Some species of fish and shellfish, however, may contain significant levels of mercury or other environmental contaminants. Most healthy people can safely consume most species of ocean fish several times a week, but for some, the risks are greater. Women who may become pregnant, pregnant and lactating women, and children are more sensitive to contaminants than others, but even they can benefit from consuming safer fish varieties within recommended limits (see Chapter 10 for details).

For everyone, consuming a variety of different types of fish to minimize exposure to any single toxin that may accumulate in a favored species is a good idea. The margin lists the species most heavily contaminated with mercury and those that are low in mercury. (■) Consumers should check local advisories to determine the safety of freshwater fish caught by family and friends.

■ • Fish most heavily contaminated with mercury: king mackerel, shark, swordfish, and tilefish (also called golden bass or golden snapper).
• Fish or shellfish lower in mercury: catfish, pollock, salmon, sardines, and canned light tuna. Canned albacore ("white") tuna generally contains more mercury than light tuna.

Omega-3 Supplements Fish, not fish oil supplements, is the preferred source of omega-3 fatty acids. High intakes of omega-3 polyunsaturated fatty acids may increase bleeding time, interfere with wound healing, raise LDL cholesterol, and suppress immune function. Evidence is mixed for people with heart disease—several studies show hopeful results, while others reveal no benefits from supplements.[30] Because supplements pose risks, such as excessive bleeding, those taking daily fish oil supplements need medical supervision.[31] The benefits and risks from EPA and DHA illustrate an important concept in nutrition: too much of a nutrient is often as harmful as too little.

RECOMMENDATIONS

Some fat in the diet is essential for good health. The *Dietary Guidelines for Americans* recommend that a portion of each day's total fat come from raw oils. (■) Oils are naturally present in foods such as nuts, avocados, and seafood. In addition, many commonly used oils such as olive, peanut, safflower, soybean, and sunflower oils are extracted from plants. When choosing oils, alternate among the various types to obtain the benefits different oils offer. Peanut and safflower oils are especially rich in vitamin E. Olive oil contributes naturally occurring antioxidant phytochemicals with potential heart benefits (see Nutrition in Practice 4), and canola oil is rich in monounsaturated and essential fatty acids.

Defining the exact amount of fat, saturated fat, or cholesterol that benefits health or begins to harm health, however, is not possible. For this reason, no RDA or UL has been set.[32] Instead, the DRI and *Dietary Guidelines* suggest a diet that is low in saturated fat, *trans* fat, and cholesterol and provides 20 to 35 percent of the daily energy intake from fat. (■) These recommendations recognize that diets with up to 35 percent of kcalories from fat can be compatible with good health if energy intake is reasonable and saturated and *trans* fat intakes are low. When total fat intake exceeds 35 percent of kcalories, saturated fat intakes increase to unhealthy levels.[33] Fat and oil intakes below 20 percent of kcalories increase the risk of inadequate essential fatty acid intakes. The FDA established Daily Values for food labels using 30 percent of energy intake as the guideline for fat. (■)

Part of the allowance for total fat provides for the essential fatty acids—linoleic acid and linolenic acid. Recommendations suggest that linoleic acid provide 5 to 10 percent of the daily energy intake and linolenic acid, 0.6 to 1.2 percent. (■)

Recommendations urge people to eat diets that are low in saturated fat, *trans* fat, and cholesterol. (■) Specifically, consume less than 10 percent of kcalories from saturated fat, keep *trans* fat intakes as low as possible, and consume less than 300 milligrams of cholesterol each day.[34] To help consumers meet these goals, the FDA established Daily Values for food labels using 10 percent of energy intake for saturated fat; the Daily Value for cholesterol is 300 milligrams, regardless of energy intake. (■) There is no Daily Value for *trans* fat.

■ An adequate intake of the needed fat-soluble nutrients can be ensured by a small daily intake of oil:
- 1600-kcalorie diet = 22 g (5 tsp)
- 2000-kcalorie diet = 27 g (6 tsp)
- 2400-kcalorie diet = 31 g (7 tsp)
- 2800-kcalorie diet = 36 g (8 tsp)

■ DRI for fat:
- 20 to 35% of energy intake

■ Daily Value:
- 65 g fat (based on 30% of 2000 kcal)

■ Linoleic acid (omega-6) AI:
- 5 to 10% of energy intake

Men:
- 19–50 yr: 17 g/day
- 51+ yr: 14 g/day

Women:
- 19–50 yr: 12 g/day
- 51+ yr: 11 g/day

Linolenic acid (omega-3) AI:
- 0.6 to 1.2% of energy intake
- Men: 1.6 g/day
- Women: 1.1 g/day

■ *Dietary Guidelines for Americans*: Consume less than 10% of kcalories from saturated fatty acids and less than 300 mg/day of cholesterol, and keep *trans*-fatty acid consumption as low as possible.

■ Daily Values:
- 20 g saturated fat (based on 10% of 2000 kcal)
- 300 mg cholesterol

Review Notes

- High intakes of saturated or *trans* fats contribute to heart disease, obesity, and other health problems.
- High blood cholesterol, specifically, poses a risk of heart disease, and high intakes of saturated fat contribute most to high blood cholesterol. High intakes of *trans*-fatty acids also appear to raise blood cholesterol. Cholesterol in foods presents less of a risk.
- Polyunsaturated fatty acids of the omega-6 and omega-3 families protect against heart disease.
- When monounsaturated fat such as olive oil replaces saturated and *trans* fats in the diet, the risk of heart disease may be lessened.
- Though some fat in the diet is necessary, health authorities recommend a diet moderate in total fat and low in saturated fat, *trans* fat, and cholesterol.

Fats in Foods

Fats are important in foods as well as in the body. Many of the compounds that give foods their flavors and aromas are found in fats and oils. The delicious aromas associated with bacon, ham, and other meats, as well as with onions being sautéed, come from fats. Fats also influence the texture of many foods, enhancing smoothness, creaminess, moistness, or crispness. In addition, four vitamins—A, D, E, and K—are soluble in fat. When the fat is removed from a food, many fat-soluble compounds, including these vitamins, are also removed. Table 4-5 summarizes the roles of fats in foods.

Fats are also an important part of most people's ethnic or national cuisines. Each culture has its own favorite food sources of fats and oils. In Canada, canola oil (also known as rapeseed oil) is widely used. In the Mediterranean area, Greeks, Italians, and Spaniards rely heavily on olive oil. Both canola oil and olive oil are rich sources of monounsaturated fatty acids. Asians use the polyunsaturated oil of soybeans. Jewish people traditionally employ chicken fat. Everywhere in North America, butter and margarine are widely used.

FINDING THE FATS IN FOODS

The remainder of this chapter and the Nutrition in Practice show you how to choose fats wisely with the goals of providing optimal health and pleasure in eating. To achieve such goals, you need to know which foods offer unsaturated oils that provide the essential fatty acids and which foods are loaded with solid fats—the saturated and *trans* fats. Also important for many people is learning to control portion sizes, particularly portions of fatty foods that can pack hundreds of kcalories into just a few bites.

Keep in mind that, whether solid or liquid, essential or nonessential, all fats bring the same abundant kcalories to the diet and excesses contribute to body fat stores. According to the *Dietary Guidelines for Americans,* no benefits can be expected when oil is added to an already fat-rich eating pattern. The following amounts of these fats contain about 5 grams of pure fat, providing 45 kcalories and negligible protein and carbohydrate:

- 1 teaspoon of oil or shortening
- 1½ teaspoons of mayonnaise, butter, or margarine
- 1 tablespoon of regular salad dressing, cream cheese, or heavy cream
- 1½ tablespoons of sour cream

The solid fat of some foods, such as the rim of fat on a steak, is visible (and therefore identifiable and removable). Other solid fats, such as those in candy, cheeses, coconut, hamburger, homogenized milk, and lunch meats, are invisible (and therefore easily missed or ignored). Equally hidden are the solid fats blended into biscuits, cakes, cookies, chip dips, ice cream, mixed dishes, pastries, sauces, and creamy soups and in fried foods. Invisible fats supply the majority of solid fats in the U.S. diet.[35]

Milk and Milk Products

Milk products go by different names that reflect their varying fat contents. (■) A cup of homogenized whole milk contains the protein and carbohydrate of fat-free milk, but in addition it contains about 80 extra kcalories from butterfat, a solid fat. A cup of reduced-fat (2 percent fat) milk falls between whole milk and fat-free, with 45 kcalories from fat.

Note that cream and butter do not appear in the milk and milk products group. Milk and yogurt are rich in calcium and protein,

■ 1 c whole milk:
- 8 g fat
- 5 g saturated fat
- 24 mg cholesterol

1 c reduced-fat milk:
- 5 g fat
- 2 g saturated fat
- 20 mg cholesterol

1 c low-fat milk:
- 2 g fat
- 1.5 g saturated fat
- 10 mg cholesterol

1 c fat-free milk:
- 0 g fat
- 0 g saturated fat
- 5 mg cholesterol

TABLE 4-5 The Functions of Fats in Foods

• *Nutrient.* Food fats provide essential fatty acids and other raw materials.
• *Energy.* Food fats provide a concentrated energy source.
• *Transport.* Fats carry fat-soluble vitamins A, D, E, and K along with some phytochemicals and assist in their absorption.
• *Sensory appeal.* Fats contribute to the taste and smell of foods.
• *Appetite.* Fats stimulate the appetite.
• *Satiety.* Fats contribute to feelings of fullness.
• *Texture.* Fats make fried foods crisp and other foods tender.

but cream and butter are not. Cream and butter are solid fats, as are whipped cream, sour cream, and cream cheese. Other cheeses, grouped with milk products, vary in their fat contents and are major contributors of saturated fat in people's diets.

Protein Foods Meats conceal a good deal of the fat—and much of the solid fat—that people consume. To help "see" the fat in meats, it is useful to think of them in four categories according to their fat contents: very lean, lean, medium-fat, and high-fat meats (as the Exchange Lists in Appendix C do). Meats in all four categories contain about equal amounts of protein, but their fat, saturated fat, and kcalorie amounts vary significantly. Table 1-11 on pp. 26 and 27 in Chapter 1 provided definitions for common terms used to describe the fat contents of meats.

The USDA Food Patterns suggest that most adults limit their intake of protein foods to about 5 to 7 ounces per day. For comparison, the smallest fast-food hamburger weighs about 3 ounces. A steak served in a restaurant often runs 8, 12, or 16 ounces, more than a whole day's meat allowance. You may have to weigh a serving or two of meat to see how much you are eating.

People think of meat as protein food, but calculation of its nutrient content reveals a surprising fact. A big (4-ounce) fast-food hamburger sandwich contains 24 grams of protein and 18 grams of fat, 7 of them saturated fat. Because protein offers 4 kcalories per gram and fat offers 9, the sandwich provides 96 kcalories from protein but 162 kcalories from fat. Hot dogs, fried chicken sandwiches, and fried fish sandwiches also provide hundreds of fat kcalories, mostly from invisible solid fat. Because so much meat fat is hidden from view, meat eaters can easily and unknowingly consume a great many grams of solid fat from this source.

When choosing beef or pork, look for lean cuts named *loin* or *round* from which the fat can be trimmed, and eat small portions. Chicken and turkey flesh are naturally lean, but commercial processing and frying add solid fats, especially in "patties," "nuggets," "fingers," and "wings." Chicken wings are mostly skin, and a chicken stores most of its fat just under its skin. The tastiest wing snacks have also been fried in cooking fat (often a hydrogenated, saturated type with *trans*-fatty acids); smothered with a buttery, spicy sauce; and then dipped in blue cheese dressing, making wings an extraordinarily high-fat snack. People who snack on wings may want to plan on eating low-fat foods at several other meals to balance them out.

Vegetables, Fruits, and Grains Choosing vegetables, fruits, whole grains, and legumes also helps lower the saturated fat, cholesterol, and total fat content of the diet. Most vegetables and fruits naturally contain little or no fat; avocados and olives are exceptions, but most of their fat is unsaturated, which is not harmful to heart health. Most grains contain only small amounts of fat. Some refined grain *products* such as fried taco shells, croissants, and biscuits are high in saturated fat, so consumers need to read food labels. Similarly, many people add butter, margarine, or cheese sauce to grains and vegetables, which raises their saturated and *trans* fat contents. Because fruits are often eaten without added fat, a diet that includes several servings of fruit daily can help a person meet the dietary recommendations for fat.

A diet rich in vegetables, fruits, whole grains, and legumes offers abundant vitamin C, folate, vitamin A, vitamin E, and dietary fiber—all important in supporting health. Consequently, such a diet protects against disease both by reducing saturated fat, cholesterol, and total fat and by increasing nutrients. It also provides valuable phytochemicals that help defend against heart disease.

CUTTING SOLID FATS AND CHOOSING UNSATURATED FATS

Meeting today's lipid guidelines can be challenging. Reducing intakes of saturated and *trans* fatty acids, for example, involves identifying food sources of these fatty acids—that is, foods that contain solid fats. Then, replacing them appropriately involves

TABLE 4-6 Solid Fat Ingredients Listed on Food Labels

Solid Fat Ingredients
• Beef fat (tallow)
• Butter
• Chicken fat
• Coconut oil
• Cream
• Hydrogenated oil
• Milk fat
• Palm kernel oil; palm oil
• Partially hydrogenated oil
• Pork fat (lard)
• Shortening
• Stick margarine

Source: U.S. Department of Agriculture and U. S. Department of Health and Human Services, *Dietary Guidelines for Americans 2010*, available at www.dietaryguidelines.gov.

identifying unsaturated oils. To help simplify these tasks, the *Dietary Guidelines* suggest:

1. Selecting the most nutrient-dense foods from all food groups. Solid fats and high-kcalorie choices can be found in every food group.
2. Consuming fewer and smaller portions of foods and beverages that contain solid fats.
3. Replacing solid fats with liquid oils whenever possible.
4. Checking Nutrition Facts labels and selecting foods with little saturated fat and no *trans* fat.

Such advice is easily dispensed but not easily followed. The first step in doing so is learning which foods contain heavy doses of solid fats. Table 4-6 lists some terms that indicate solid fats in a food label ingredients list.

The accompanying "How to" offers strategies for making heart-healthy choices, food group by food group. The best diet for heart health is also rich in fruits, vegetables, nuts, and whole grains that offer many health advantages by supplying abundant nutrients, fiber, and phytochemicals.

Fats and kCalories Removing fat from food also removes energy, as Figure 4-6 shows. A pork chop with the fat trimmed to within a half-inch of the lean provides 290 kcalories; with the fat trimmed off completely, it supplies 174 kcalories. A baked potato with butter and sour cream (1 tablespoon each) has 315 kcalories; a plain baked potato

FIGURE 4-6 Cutting Fat Cuts kCalories—and Saturated Fat

Pork chop with fat (290 kcal, 24 g fat, 9 g saturated fat)

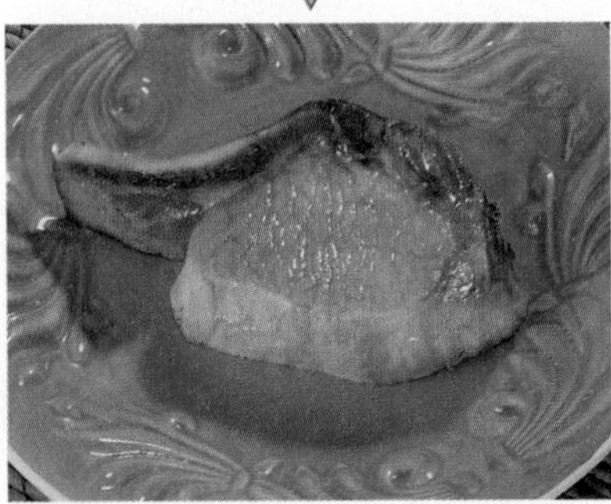

Pork chop with fat trimmed off (174 kcal, 8 g fat, 3 g saturated fat)

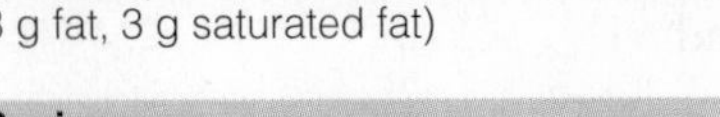

Savings: 116 kcal, 16 g fat, 6 g saturated fat

Potato with 1 tbs butter and 1 tbs sour cream (315 kcal, 14 g fat, 9 g saturated fat)

Plain potato (188 kcal, <1 g fat, 0 g saturated fat)

Savings: 127 kcal, 13 g fat, 9 g saturated fat

Whole milk, 1 c (150 kcal, 8 g fat, 5 g saturated fat)

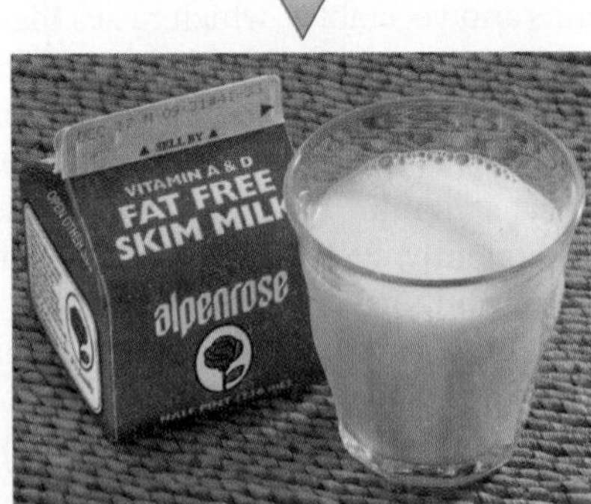

Fat-free milk, 1 c (90 kcal, <1 g fat, <1 g saturated fat)

Savings: 60 kcal, 7 g fat, 4 g saturated fat

has 188 kcalories. The single most effective step you can take to reduce the energy value of a food is to eat it with less fat.

Choosing Unsaturated Fats When a person does eat fats, those to choose are the unsaturated ones. Remember, the softer a fat is, the more unsaturated it is. Generally speaking, vegetable and fish oils are rich in polyunsaturates, olive oil and canola oil

HOW TO Make Heart-Healthy Choices—by Food Group

Breads and Cereals

- Select whole-grain breads, cereals, and crackers that are low in saturated and *trans* fat (for example, bagels instead of croissants).
- Prepare pasta with a tomato sauce instead of a cheese or cream sauce.

Vegetables and Fruits

- Enjoy the natural flavor of steamed vegetables (without butter) for dinner and fruits for dessert.
- Eat at least two vegetables (in addition to a salad) with dinner.
- Snack on raw vegetables or fruits instead of high-fat items like potato chips.
- Buy frozen vegetables without sauce.

Milk and Milk Products

- Switch from whole milk to reduced-fat, from reduced-fat to low-fat, and from low-fat to fat-free (nonfat).
- Use fat-free and low-fat cheeses (such as part-skim ricotta and low-fat mozzarella) instead of regular cheeses.
- Use fat-free or low-fat yogurt or sour cream instead of regular sour cream.
- Use evaporated fat-free milk instead of cream.
- Enjoy fat-free frozen yogurt, sherbet, or ice milk instead of ice cream.

Meat and Legumes

- Fat adds up quickly, even with lean meat; limit intake to about 6 ounces (cooked weight) daily.
- Eat at least two servings of fish per week (particularly fish such as mackerel, lake trout, herring, sardines, and salmon).
- Choose fish, poultry, or lean cuts of pork or beef; look for unmarbled cuts named *round* or *loin* (eye of round, top round, bottom round, round tip, tenderloin, sirloin, center loin, and top loin).
- Trim the fat from pork and beef; remove the skin from poultry.
- Grill, roast, broil, bake, stir-fry, stew, or braise meats; don't fry. When possible, place food on a rack so that fat can drain.
- Use lean ground turkey or lean ground beef in recipes; brown ground meats without added fat, then drain off fat.
- Select tuna, sardines, and other canned meats packed in water; rinse oil-packed items with hot water to remove much of the fat.
- Fill kabob skewers with lots of vegetables and slivers of meat; create main dishes and casseroles by combining a little meat, fish, or poultry with whole-grain pasta or brown rice and generous amounts of vegetables.
- Use legumes often.
- Eat a meatless meal or two daily.
- Use egg substitutes in recipes instead of whole eggs, or use two egg whites in place of each whole egg.

Fats and Oils

- Use butter or stick margarine sparingly; select soft margarines instead of hard margarines.
- Use fruit butters, reduced-kcalorie margarines, or butter replacers instead of butter.
- Use low-fat or fat-free mayonnaise and salad dressing instead of regular.
- Limit use of lard and meat fat.
- Limit use of products made with coconut oil, palm kernel oil, and palm oil (read labels on bakery goods, processed foods, microwave popcorn, and nondairy creamers).
- Reduce use of hydrogenated shortenings and stick margarines and products that contain them (read labels on crackers, cookies, and other commercially prepared baked goods); use vegetable oils instead.

Miscellaneous

- Use a nonstick pan, or coat the pan lightly with vegetable oil.
- Refrigerate soups and stews; when the fat solidifies, remove it before reheating.
- Use wine; lemon, orange, or tomato juice; herbs; spices; fruits; or broth instead of butter or margarine when cooking.
- Stir-fry in a small amount of oil; add moisture and flavor with broth, tomato juice, or wine.
- Use variety to enhance enjoyment of the meal: vary colors, textures, and temperatures—hot cooked versus cool raw foods—and use garnishes to complement food.
- Omit high-fat meat gravies and cheese sauces.
- Order pizzas with lots of vegetables, a little lean meat, and half the cheese.

Source: Adapted from *Expert Panel on Detection, Evaluation, and Treatment of High Blood Cholesterol in Adults (Adult Treatment Panel III), Third Report of the National Cholesterol Education Program (NCEP)*, NIH publication no. 02-5215 (Bethesda, MD: National Heart, Lung, and Blood Institute, 2002), pp. V-25–V-27.

At room temperature, unsaturated fats (such as those found in oil), are usually liquid, whereas saturated fats (such as those found in butter) are solid.

are rich in monounsaturates, and the harder fats—animal fats—are more saturated (see Figure 4-7). Remember, however, that vegetable fat or vegetable oil doesn't always mean unsaturated fat. Both coconut oil and palm oil, for example, which are often used in nondairy creamers, are saturated fats, and both raise blood cholesterol.

Don't Overdo Fat Restriction Some people actually manage to eat too *little* fat—to their detriment. Among them are young women and men with eating disorders, described in Nutrition in Practice 7. As a practical guideline, it is wise to include the equivalent of at least a teaspoon of fat in every meal.

Fat Replacers Today, consumers can choose from thousands of fat-reduced products. Many bakery goods, lunch meats, cheeses, spreads, frozen desserts, and other products made with **fat replacers** offer less than half a gram of fat, saturated fat, and *trans* fat in a serving. Some of these products contain **artificial fats**, and others use conventional ingredients in unconventional ways to reduce fats and kcalories. Among the latter, manufacturers can:

- Add water or whip air into foods.
- Add fat-free milk to creamy foods.
- Use lean meats and soy protein to replace high-fat meats.
- Bake foods instead of frying them.

Common food ingredients such as fibers, sugars, or proteins can also take the place of fats in some foods. Products made from sugars or proteins still provide kcalories but far

FIGURE 4-7 Comparison of Dietary Fats

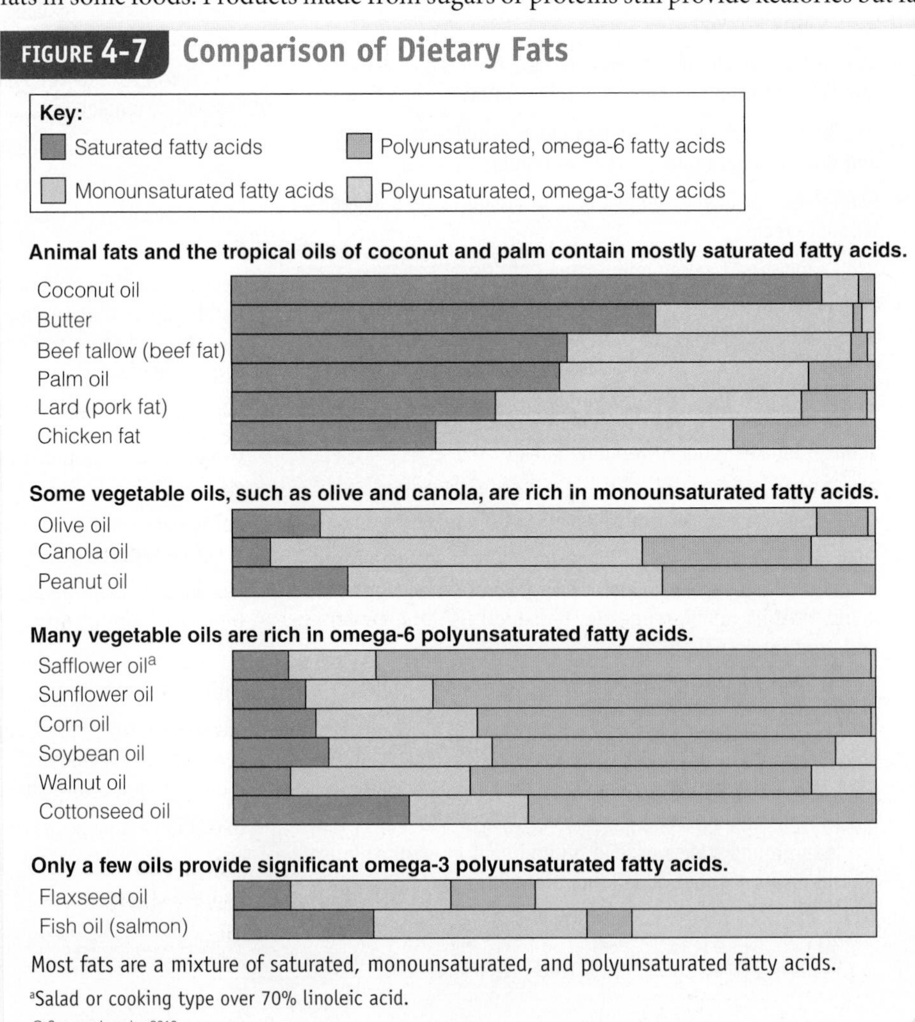

Most fats are a mixture of saturated, monounsaturated, and polyunsaturated fatty acids.

[a]Salad or cooking type over 70% linoleic acid.

fat replacers: ingredients that replace some or all of the functions of fat in foods and may or may not provide energy.

artificial fats: zero-energy fat replacers that are chemically synthesized to mimic the sensory and cooking qualities of naturally occurring fats but are totally or partially resistant to digestion.

fewer kcalories from fats. Manufactured fat replacers consist of chemical derivatives of carbohydrate, protein, or fat, or modified versions of foods rich in those constituents.

A familiar example of an artificial fat that has been approved for use in snack foods such as potato chips, crackers, and tortilla chips is **olestra**. Olestra's chemical structure is similar to that of a regular fat (a triglyceride) but with important differences. A triglyceride is composed of a glycerol molecule with three fatty acids attached, whereas olestra is made of a sucrose molecule with six to eight fatty acids attached. Enzymes in the digestive tract cannot break the bonds of olestra, so unlike sucrose or fatty acids, olestra passes through the system unabsorbed.

The FDA's evaluation of olestra's safety addressed two questions. First, is olestra toxic? Research on both animals and humans supports the safety of olestra as a partial replacement for dietary fats and oils, with no reports of cancer or birth defects. Second, does olestra affect either nutrient absorption or the health of the digestive tract? When olestra passes through the digestive tract unabsorbed, it binds with the fat-soluble vitamins A, D, E, and K and carries them out of the body, robbing the person of these valuable nutrients. To compensate for these losses, the FDA requires the manufacturer to fortify olestra with vitamins A, D, E, and K. Saturating olestra with these vitamins does not make the product a good source of vitamins, but it does block olestra's ability to bind with the vitamins from other foods. An asterisk in the ingredients list informs consumers that these added vitamins are "dietarily insignificant."

Consumers need to keep in mind that low-fat and fat-free foods still deliver kcalories. Decades ago, consumers hailed the arrival of artificial sweeteners as a weight-loss wonder, but in reality, kcalories saved by using artificial sweeteners were readily replaced by kcalories from other foods. Alternatives to fat can help to lower energy intake and support weight loss only when they actually *replace* fat and energy in the diet.

Read Food Labels Labels list total fat, saturated fat, *trans* fat, and cholesterol contents of foods in addition to fat kcalories per serving. Because each package provides information for a single serving and serving sizes are standardized, consumers can easily compare similar products.

Review Notes

- Fats in foods contribute to sensory appeal—enhancing the flavor, aroma, and texture of foods.
- Fats in foods deliver fat-soluble vitamins, energy, and essential fatty acids.
- While some fat in the diet is necessary, the *Dietary Guidelines for Americans* recommend limiting intakes of saturated fat and *trans* fat.
- Fats added to foods during preparation or at the table are a major source of fat in the diet.
- The choice between whole and fat-free milk products can make a big difference to the fat, saturated fat, and cholesterol content of a diet.
- Meats account for a large proportion of hidden solid fats in many people's diets.
- Most people consume more meat than is recommended.
- Most vegetables and fruits naturally contain little or no fat.
- Grain products such as croissants and biscuits can be high in saturated fat, so consumers need to read food labels to learn which foods in this group contain fats.
- Consumers today can choose from an array of fat-reduced products, and many bakery goods and other foods made with fat replacers offer less than a half a gram of fat, saturated fat, and *trans* fat in a serving.
- Some products use artificial fats such as olestra, while others use conventional ingredients such as water or fat-free milk to reduce fat and kcalories.
- Food labels list total fat, saturated fat, cholesterol, and *trans* fat, as well as fat kcalories per serving.

olestra: a synthetic fat made from sucrose and fatty acids that provides zero kcalories per gram; also known as *sucrose polyester*.

Chapters 3 and 4 have looked briefly at the two major energy fuels in the body—carbohydrate and fat. When used for energy, each has desirable characteristics. The glucose derived from carbohydrate is needed by the brain and nerve tissues and is easily used for energy in other cells. Fat is a particularly useful fuel because the body stores it efficiently and in generous amounts. Chapter 5 looks at protein, a nutrient that can be used as fuel, but whose primary role is to provide machinery for getting things done.

Self Check

1. Three classes of lipids in the body are:
a. triglycerides, fatty acids, and cholesterol.
b. triglycerides, phospholipids, and sterols.
c. fatty acids, phospholipids, and cholesterol.
d. glycerol, fatty acids, and triglycerides.

2. A triglyceride consists of:
a. three glycerols attached to a lipid.
b. three fatty acids attached to a glucose.
c. three fatty acids attached to a glycerol.
d. three phospholipids attached to a cholesterol.

3. A fatty acid that has the maximum possible number of hydrogen atoms is known as a(n):
a. saturated fatty acid.
b. monounsaturated fatty acid.
c. PUFA.
d. essential fatty acid.

4. The difference between *cis*- and *trans*-fatty acids is:
a. the number of double bonds.
b. the length of their carbon chains.
c. the location of the first double bond.
d. the configuration around the double bond.

5. Essential fatty acids:
a. are used to make substances that regulate blood pressure, among other functions.
b. can be made from carbohydrates.
c. include lecithin and cholesterol.
d. cannot be found in commonly eaten foods.

6. Lecithins and other phospholipids in the body function as:
a. emulsifiers.
b. enzymes.
c. temperature regulators.
d. shock absorbers.

7. To minimize saturated fat intake and lower the risk of heart disease, most people need to:
a. eat less meat.
b. select fat-free milk.
c. use nonhydrogenated margarines and cooking oils such as olive oil or canola oil.
d. All of the above

8. To include omega-3 fatty acids in the diet, the *Dietary Guidelines for Americans* recommends eating:
a. cholesterol-free margarine.
b. fish oil supplements.
c. hydrogenated margarine.
d. 8 to 12 ounces of seafood per week.

9. Some examples of foods with hidden solid fats are:
a. cheese, lettuce, and fruit juices.
b. fried foods, sauces, dips, and lunch meats.
c. fish, rice, and potatoes.
d. baked potatoes, vegetables, and fruits.

10. Generally speaking, vegetable and fish oils are rich in:
a. polyunsaturated fat.
b. saturated fat.
c. cholesterol.
d. *trans*-fatty acids.

Answers to these questions are found in Appendix H. For more chapter review: Access an interactive eBook, chapter-specific interactive learning tools, including flashcards, quizzes, videos, and more in your Nutrition CourseMate, accessed through CengageBrain.com.

Clinical Applications

1. The connection between the overconsumption of saturated and *trans* fats and chronic diseases (obesity, diabetes, cancer, and cardiovascular disease) underscores the importance of being alert to a client's fat intake. What advice would you offer a client who reports the following?
   - Eats two or more 6-ounce servings of meat each day.
   - Drinks whole milk and eats regular cheddar cheese each day.
   - Eats four to five servings of breads and cereals each day, including a bagel with cream cheese for breakfast, a bologna sandwich on white bread for lunch, and biscuits or cornbread with butter to accompany dinner.
   - Eats one serving of fruit each day and eats vegetables only on occasion.
2. Make a list of foods, beverages, and seasonings that your client can substitute for foods high in saturated fat.

Nutrition on the Net

For further study of the topics in this chapter, access these websites.

- Search for cholesterol and dietary fat at the U.S. government health information site: www.healthfinder.gov
- Search for "fat" at the International Food Information Council site: www.foodinsight.org

Notes

1. C. Stryjecki and D. M. Mutch, Fatty acid-gene interactions, adipokines, and obesity, *European Journal of Clinical Nutrition* 65 (2011): 285–297; N. Ouchi and coauthors, Adipokines in inflammation and metabolic disease, *Nature Reviews. Immunology* 11 (2011): 85–97; Y. Deng and P. E. Scherer, Adipokines as novel biomarkers and regulators of the metabolic syndrome, *Annals of the New York Academy of Sciences* 1212 (2010): E1–E19.G. Govindarajan, M. A. Alpert, and L. Tejwani, Endocrine and metabolic effects of fat: Cardiovascular implications, *American Journal of Medicine* 121 (2008): 366–370; T. Yamauchi and T. Kadowaki, Physiological and pathophysiological roles of adiponectin and adiponectin receptors in the integrated regulation of metabolic and cardiovascular diseases, *International Journal of Obesity* 32 (2008): S13–S18; T. Yamada and H. Katagiri, Avenues of communication between the brain and tissues/organs involved in energy homeostasis, *Endocrine Journal* 54 (2007): 497–505.
2. V. Remig and coauthors, *Trans* fats in America: A review of their use, consumption, health implications, and regulation, *Journal of the American Dietetic Association* 110 (2010): 585–592.
3. E. Lacroix and coauthors, Randomized controlled study of the effect of a butter naturally enriched in *trans* fatty acids on blood lipids in healthy women, *American Journal of Clinical Nutrition* 95 (2012): 318–325; S. Gebauer and coauthors, Effects of ruminant trans fatty acids on cardiovascular disease and cancer: A comprehensive review of epidemiological, clinical, and mechanistic studies, *Advances in Nutrition* 2 (2011): 332–354; I. A. Brouwer, A. J. Wanders, and M. B. Katan, Effect of animal and industrial trans fatty acids on HDL and LDL cholesterol levels in humans—A quantitative review, *PLoS ONE* 5 (2010): e9434.
4. S. W. Ing and M. A. Belury, Impact of conjugated linoleic acid on bone physiology: Proposed mechanism involving inhibition of adipogenesis, *Nutrition Reviews* 69 (2011): 123–131; L. A. Smit, A. Baylin, and H. Campos, Conjugated linoleic acid in adipose tissue and risk of myocardial infarction, *American Journal of Clinical Nutrition* 92 (2010): 34–40.
5. G. H. Johnson and K. Fritsche, Effect of dietary linoleic acid on markers of inflammation in healthy persons: A systematic review of randomized controlled trials, *Journal of the Academy of Nutrition and Dietetics* 112 (2012): 1029–1041; R. J. Deckelbaum and C. Torrejon, The omega-3 fatty acid nutritional landscape: Health benefits and sources, *Journal of Nutrition* 142 (2012): 587S–591S; W. S. Harris and coauthors, Omega-6 fatty acids and risk for cardiovascular disease, A science advisory from the American Heart Association Nutrition Subcommittee of the Council on Nutrition, Physical Activity, and Metabolism; Council on Cardiovascular Nursing; and Council on Epidemiology and Prevention, *Circulation* 119 (2009): 902–907.
6. N. G. Bazan, M. F. Molina, and W. C. Gordon, Docosahexaenoic acid signalolipidomics in nutrition: Significance in aging, neuroinflammation, macular degeneration, Alzheimer's, and other neurodegenerative diseases, *Annual Review of Nutrition* 31 (2011): 321–351; R. K. McNamara and coauthors, Docosahexaenoic acid supplementation increases prefrontal cortex activation during sustained attention in healthy boys: A placebo-controlled, dose-ranging, functional magnetic resonance imaging study, *American Journal of Clinical Nutrition* 91 (2010): 1060–1067.
7. Bazan, Molina, and Gordon, 2011; E. E. Birch and coauthors, The DIAMOND (DHA Intake and Measurement of Neural Development) Study: A double-masked, randomized controlled clinical trial of the maturation of infant visual acuity as a function of the dietary level of docosahexaenoic acid, *American Journal of Clinical Nutrition* 91 (2010): 848–859.
8. D. Mozaffarian and J. H. Y. Wu, Omega-3 fatty acids and cardiovascular disease: Effects on risk factors, molecular pathways, and clinical events, *Journal of the American College of Cardiology* 58 (2011): 2047–2067; M. N. Di Minno and coauthors, Exploring newer cardioprotective strategies: ω-3 fatty acids in perspective, *Thrombosis and Haemostasis* 104 (2010): 664–680; P. J. Smith and coauthors, Association between n-3 fatty acid consumption and ventricular ectopy after myocardial infarction, *American Journal of Clinical Nutrition* 89 (2009): 1315–1320.
9. T. Psaltopoulou and coauthors, Olive oil intake is inversely related to cancer prevalence: A systematic review and a meta-analysis of 13800 patients and 23340 controls in 19 observational studies, *Lipids in Health and Disease* 10 (2011): 127–143; M. Solanas and coauthors, Dietary olive oil and corn oil differentially affect experimental breast cancer through distinct modulation of the p21 Ras signaling and the proliferation-apoptosis balance, *Carcinogenesis* 31 (2010): 871–879; J. Y. Lee, L. Zhao, and D. H. Hwang, Modulation of pattern recognition receptor-mediated inflammation and risk of chronic diseases by dietary fatty acids, *Nutrition Reviews* 68 (2009): 38–61.

10. T. M. Brasky and coauthors, Specialty supplements and breast cancer risk in the VITamins And Lifestyle (VITAL) Cohort, *Cancer Epidemiology, Biomarkers, and Prevention* 19 (2010); 1696–1708; M. Gerber, Background review paper on total fat, fatty acid intake and cancers, *Annals of Nutrition and Metabolism* 55 (2009): 140–161.
11. G. A. Bray, Is dietary fat important? *American Journal of Clinical Nutrition* 93 (2011): 481–482.
12. M. Miller and coauthors, Triglycerides and cardiovascular disease: A scientific statement from the American Heart Association, *Circulation* 123 (2011): 2292–2333.
13. J. Heineckem, HDL and cardiovascular-disease risk: Time for a new approach? *New England Journal of Medicine* 364 (2011): 170–171.
14. S. J. Baum and coauthors, Fatty acids in cardiovascular health and disease: A comprehensive update, *Journal of Clinical Lipidology* 6 (2012): 216–234; A. Astrup, the role of reducing intakes of saturated fats in the prevention of cardiovascular disease: Where does the evidence stand in 2010? *American Journal of Clinical Nutrition* 93 (2011): 684-688.
15. U.S. Department of Agriculture, U.S. Department of Health and Human Services, *Dietary Guidelines for Americans 2010*, available at www.dietaryguidelines.gov.
16. Baum and coauthors, 2012; K. Zelman, The great fat debate: A closer look at the controversy—Questioning the validity of age-old dietary guidance, *Journal of the American Dietetic Association* 111 (2011): 655–658.
17. I. A. Brouwer, A. J. Wanders, and M. B. Katan, Effect of animal and industrial *trans* fatty acids on HDL and LDL cholesterol levels in humans—A quantitative review, *PLoS One* 5 (2010): e9434; S. K. Wallace and D. Mozaffarian, Trans-fatty acids and nonlipid risk factors, *Current Atherosclerosis Reports* 11 (2009): 423–433; D. Mozaffarian, A. Aro, and W. C. Willett, Health effects of trans-fatty acids: Experimental and observational evidence, *European Journal of Clinical Nutrition* 63 (2009): S5–S21.
18. U.S. Department of Agriculture, U.S. Department of Health and Human Services, *Dietary Guidelines for Americans 2010*, available at www.dietaryguidelines.gov.
19. V. Bemig and coauthors, *Trans* fats in America; A review of their use, consumption, health implications, and regulation, *Journal of the American Dietetic Association* 110 (2010): 585–592.
20. A. M. Brownawell and M. C. Falk, Where science and public health policy intersect, *Nutrition Reviews* 68 (2010): 355–364; M. L. Fernandez and M. Calle, Revisiting dietary cholesterol recommendations: Does the evidence support a limit of 300 mg/d?, *Current Atherosclerosis Reports* 12 (2010): 377–383.
21. U.S. Department of Agriculture, U.S. Department of Health and Human Services, *Dietary Guidelines for Americans 2010*, available at www.dietaryguidelines.gov.
22. M. R. Flock and P. M. Kris-Etherton, Dietary Guidelines for Americans 2010: Implications for cardiovascular disease, *Current Atherosclerosis Reports* 13 (2011): 499–507.
23. O. Castañer and coauthors, Protection of LDL from oxidation by olive oil polyphenols is associated with a downregulation of CD40-ligand expression and its downstream products in vivo in humans, *American Journal of Clinical Nutrition* 95 (2012): 1238–1244; S. Cicerale, L. J. Lucas, and R. S. Keast, Antimicrobial, antioxidant and anti-inflammatory phenolic activities in extra virgin olive oil, *Current Opinion in Biotechnology* 23 (2012): 129–135; C. Samieri and coauthors, Olive oil consumption, plasma oleic acid, and stroke incidence: The Three-City Study, *Neurology* 77 (2011): 418–425; S. Cicerale and coauthors, Chemistry and health of olive oil phenolics, *Critical Reviews in Food Science and Nutrition* 49 (2009): 218–236.
24. Castañer and coauthors, 2012; Cicerale, Lucas, and Keast, 2012; Cicerale and coauthors, 2009.
25. Harris and coauthors, 2009.
26. Baum and coauthors, 2012; Flock and Kris-Etherton, 2011; Mozaffarian and Wu, 2011; N. D. Riediger and coauthors, A systemic review of the roles of n-3 fatty acids in health and disease, *Journal of the American Dietetic Association* 109 (2009): 668–679; P. P. Dimitrow and M. Jawien, Pleiotropic, cardioprotective effects of omega-3 polyunsaturated fatty acids, *Mini Reviews in Medicinal Chemistry* 9 (2009): 1030–1039.
27. U.S. Department of Agriculture, U.S. Department of Health and Human Services, *Dietary Guidelines for Americans 2010*, available at www.dietaryguidelines.gov.
28. C. Berr and coauthors, Increased selenium intake in elderly high-fish consumers may account for health benefits previously ascribed to omega-3 fatty acids, *Journal of Nutrition, Health, and Aging* 13 (2009): 14–18.
29. R. J. Belin and coauthors, Fish intake and the risk of incident heart failure: The Women's Health Initiative, *Circulation Heart Failure* 4 (2011): 404–413.
30. P. C. Calder, Mechanisms of action of (n-3) fatty acids, *Journal of Nutrition* 142 (2012): 592S–599S; S. M. Kwak and coauthors, Efficacy of omega-3 fatty acid supplements (eicosapentaenoic acid and docosahexaenoic acid) in the secondary prevention of cardiovascular disease: a meta-analysis of randomized, double-blind, placebo-controlled trials, *Archives of Internal Medicine,* 172 (2012): 686–694; V. A. Andreeva and coauthors, B vitamin and/or ω-3 fatty acid supplementation and cancer, *Archives of Internal Medicine* 172 (2012): 540–547; Flock and Kris-Etherton, 2011.
31. American Heart Association, Fish and omega-3 fatty acids, available at www.heart.org, updated September 2010.
32. Standing Committee on the Scientific Evaluation of Dietary Reference Intakes, 2005, pp. 769–770.
33. Standing Committee on the Scientific Evaluation of Dietary Reference Intakes, 2005, pp. 797–802.
34. U.S. Department of Agriculture, U.S. Department of Health and Human Services, *Dietary Guidelines for Americans 2010*, available at www.dietaryguidelines.gov.
35. U.S. Department of Agriculture, U.S. Department of Health and Human Services, *Dietary Guidelines for Americans 2010*, available at www.dietaryguidelines.gov.

Nutrition in Practice

Figuring Out Fats

To consumers, advice about dietary fat appears to change almost daily. "Eat less fat." "Eat more fatty fish." "Give up butter—use margarine instead." "Give up margarine—replace it with olive oil." "Steer clear of saturated fats." "Seek out omega-3." "Stay away from *trans* fats." "Stick with mono- and polyunsaturated fats." No wonder some people feel confused about dietary fat. This Nutrition in Practice begins with a look at the dietary guidelines for fat intake. It continues by identifying which foods provide which fats. It closes with strategies to help consumers choose the right amounts of the right kinds of fats for a healthy diet.

Why do today's fat messages seem to change constantly and become more confusing?

The confusion stems in part from the complexities of fat and in part from the nature of recommendations. As Chapter 4 explained, "dietary fat" refers to several kinds of fats; some fats support health, whereas others damage it, and foods typically provide a mixture of fats in varying proportions. It has taken researchers decades to sort through the relationships among the various kinds of fat and their roles in supporting or harming health. Translating these research findings into dietary recommendations is a challenging process. Too little information can mislead consumers, but too much detail can overwhelm them. As scientific understanding has grown, recommendations have evolved to become less general and more specific. Recommendations may seem to "change constantly and become more confusing," but in fact they are becoming more meaningful.

How exactly have dietary recommendations for fat changed to become more meaningful for consumers?

Dietary recommendations for fat have shifted away from reducing total fat, in general, to limiting saturated and *trans* fats, specifically, along with a greater emphasis on kcalorie control to maintain a healthy body weight.[1] For decades, health experts urged consumers to limit total fat intake to 30 percent or less of energy intake. This advice was straightforward—cut the fat and improve your health. Health experts recognized that saturated fats and *trans* fats were the ones that raise blood cholesterol, but they reasoned that when total fat was limited, saturated and *trans* fat intake would decline as well. People were simply advised to cut back on all fat so that they would cut back on saturated and *trans* fat. Such advice may have oversimplified the message and unnecessarily restricted total fat.

But low-fat diets have been recommended for years to help people manage weight and reduce the risk of heart disease. Are you saying that low-fat diets are no longer recommended?

Low-fat diets remain a key recommendation in treatment plans for people with elevated blood lipids or heart disease and therefore are important in nutrition.[2] As for healthy people, evidence from around the world has led researchers to change population-wide recommendations from a "low-fat" to a "wise-fat" approach. Several problems accompany low-fat diets. For one, many people find low-fat diets difficult to maintain over time. For another, low-fat diets are not necessarily low-kcalorie diets; if energy intake exceeds energy needs, weight gain follows, and obesity brings a host of health problems, including heart disease. For another, diets high in refined carbohydrates, even if low in fat, can cause blood triglycerides to rise and HDL to fall, a deleterious combination for heart health.[3] Finally, taken to the extreme, a low-fat diet may exclude fatty fish, nuts, seeds, and vegetable oils—all valuable sources of many essential fatty acids, phytochemicals, vitamins, and minerals. Importantly, the fats from these sources protect against heart disease, as later sections explain.

How have today's recommendations for fat been revised?

Today, health experts have revised dietary recommendations to acknowledge that not all fats have damaging health consequences.[4] In fact, higher intakes of some kinds of fats (for example, the omega-3 fatty acids) support good health. Instead of urging people to cut back on all fats, current recommendations suggest carefully replacing the "bad" saturated and *trans* fats with the "good" unsaturated fats and enjoying these fats within kcalorie limits.[5] The goal is to create an eating pattern moderate in kcalories that provides enough of the fats that support good health, but not too much of those that harm health. (Turn to pp. 99–102 for a review of the health consequences of each type of fat.)

With these findings and goals in mind, the committee writing the *Dietary Guidelines for Americans 2010* concluded that a diet containing up to 35 percent of total kcalories from fat, but reduced in saturated fat and *trans* fat and moderate in energy, is compatible with low rates of heart disease, diabetes, obesity, and cancer.[6]

How can people distinguish between the fats in foods that support health and those that might harm it?

Asking consumers to limit their total fat intake was less than perfect advice, but it was straightforward—find the fat and cut back. Asking consumers to keep their intakes of saturated fats, *trans* fats, and cholesterol low and to use monounsaturated and polyunsaturated fats instead may be more on target with heart health, but it also makes diet planning more complicated. To make appropriate selections, consumers must first learn which foods contain which fats. For example, avocados, bacon, walnuts, potato chips, and mackerel are all high-fat foods, yet some of these foods have detrimental effects on heart health when consumed in excess, and others seem neutral or even beneficial.

Is there evidence to clarify why some high-fat foods are compatible with a heart-healthy diet and others are not?

Yes. The traditional diets of Greece and other countries in the Mediterranean region are exemplary in their use of "good" fats, especially olives and olive oil. A classic study of the world's people, the Seven Countries Study, found that death rates from heart disease were strongly associated with diets high in saturated fats, but only weakly linked with total fat.[7] In fact, the two countries with the highest fat intakes, Finland and the Greek island of Crete, had the highest (Finland) and lowest (Crete) rates of heart disease deaths. In both countries, the people consumed 40 percent or more of their kcalories from fat. Clearly, a high-fat diet was not the primary problem, so researchers refocused their attention on the type of fat. They found that the Cretes ate diets high in olive oil but low in saturated fat (less than 10 percent of kcalories), a pattern they linked with relatively low disease risks. Many studies that followed yielded similar results—people who follow "Mediterranean-type" eating patterns have low rates of heart disease, some cancers, and other chronic diseases, and their life expectancy is high.[8] Unfortunately, many busy Mediterranean people today, especially the young, are trading labor-intensive traditional diets for convenient and fast Western-style foods. At the same time, their health advantages are rapidly disappearing.[9]

When olive oil replaces saturated fats, such as those of butter, coconut oil or palm oil, hydrogenated stick margarine, lard, or shortening, it may offer numerous health benefits.[10] Olive oil helps to protect against heart disease by:

- Lowering total and LDL cholesterol and not lowering HDL cholesterol or raising triglycerides.[11]
- Reducing blood-clotting factors.[12]
- Providing antioxidant phytochemicals that may reduce LDL cholesterol's vulnerability to oxidation. (see Nutrition in Practice 8).[13]
- Lowering blood pressure.[14]
- Interfering with the inflammatory process related to disease progression.[15]

The phytochemicals of olives captured in extra virgin olive oil, and not its monounsaturated fatty acids, seem responsible for these potential effects.[16] When processors lighten olive oils to make them more appealing to consumers, they strip away the intensely flavored phytochemicals of the olives, thus diminishing not only the bitter flavor of the oils, but also their potential for protecting the health of the heart.

Importantly, olive oil is not a magic potion; drizzling it on foods does not make them healthier. Like other fats, olive oil delivers 9 kcalories per gram, which can contribute to weight gain in people who fail to balance their energy intake with their energy output. Its role in a healthy diet is to *replace* the saturated fats.

Other vegetable oils, such as canola oil or safflower oil, in their liquid unhydrogenated states are also generally low in saturated fats and high in unsaturated fats. Such oils, when they replace solid, saturated fats in the diet, may help to preserve heart health.

Good, olive oil may help protect against heart disease; are there other food fats that may also be protective?

Possibly so. People who eat an ounce of nuts on two or more days a week appear to have lower risks of sudden death from heart events than those consuming no nuts.[17] Even in women with diabetes, whose risks are high, nuts

Matthew Farruggio

Olives and their oil may benefit heart health.

(5 ounces per week) or peanut butter (5 tbs per week) were associated with reduced heart disease risk.[18] While as little as 2 ounces of nuts per week were linked with a detectable benefit, higher intakes were associated with greater benefits, with about an 8 percent reduction in cardiovascular risk for each additional weekly serving.

The nuts under study are those commonly eaten in the United States: almonds, Brazil nuts, cashews, hazelnuts, macadamia nuts, pecans, pistachios, walnuts, even peanuts, and, as mentioned, peanut butter.[19] On average, these nuts contain mostly monounsaturated fat (59 percent), some polyunsaturated fat (27 percent), and little saturated fat (14 percent).

Research has shown a benefit from walnuts and almonds in particular. In study after study, walnuts, when substituted for other fats in the diet, produce favorable effects on blood lipids—even in people with elevated total and LDL cholesterol.[20] Results are similar for almonds.[21]

Studies on peanuts, macadamia nuts, pecans, and pistachios follow suit, indicating that including nuts may be a wise strategy against heart disease. Nuts may protect against heart disease because they provide:

- Monounsaturated and polyunsaturated fats in abundance but few saturated fats.
- Fiber, vegetable protein, essential fatty acids, and other valuable nutrients, including the antioxidant vitamin E.
- Phytochemicals that act as antioxidants (see Nutrition in Practice 8).
- Plant sterols.

In addition to their heart benefits, nuts may also benefit other body organs—people who frequently consume nuts and other healthy fats suffer fewer gallbladder problems.[22]

Tree nuts and peanuts once had no place in a low-fat or low-kcalorie diet—and for good reason. Nuts provide up to 80 percent of their kcalories from fat, and a quarter cup (about an ounce) of mixed nuts provides more than 200 kcalories. Despite this, people who regularly eat nuts tend to be leaner, not fatter, and generally have smaller waistlines than others.[23] No one can yet say, however, whether nut eaters owe their leaner physiques to an overall health-conscious lifestyle or whether nuts might be extra satiating and thus reduce food and kcalorie intake at other meals.[24] In any case, replacing potato chips or chocolate candy snacks with nuts certainly improves nutrition by reducing intakes of saturated fats and increasing intakes of vitamin E and other nutrients.[25]

When designing a test diet, researchers must carefully use nuts *instead of*, not in addition to, other fat sources (such as meat, potato chips, or oil) to keep kcalories constant. If you decide to snack on nuts, you should probably do the same thing and use them to replace other fats in your diet. Remember that nuts, although not associated with obesity or weight gain in research, provide substantially more kcalories per bite than crunchy raw vegetables, for example.

Matthew Farruggio

Stay mindful of kcalories when snacking on nuts.

What about fish? I have a friend whose doctor told her that eating fish is good for the heart. Is this true?

Yes. The preceding chapter made clear that fish oils hold the potential to improve health, and particularly the health of the heart. Research studies have provided strong evidence that increasing omega-3 fatty acids in the diet supports heart health and lowers the risk of death from heart disease.[26] For this reason, the American Heart Association and other authorities recommend including two fatty fish servings (see Table 4-4 on p. 102) a week in a heart-healthy diet. People who eat some fish each week can lower their risks of heart attack and stroke.

Fish is the best source of EPA and DHA in the diet, but it is also a major source of mercury and other environmental contaminants. Most fish contain at least trace amounts of mercury, but tilefish, swordfish, king mackerel, and shark have especially high levels. Freshwater fish may contain PCBs and other pollutants, so local advisories warn sport fishers of species that can pose problems. The chapter listed safer species of fish. To minimize risks while obtaining fish benefits, vary your choices among fatty fish species often.

If olive oil, nuts, and fatty fish are protective against heart disease, which fats are harmful?

The number one dietary determinant of LDL cholesterol is saturated fat. Each 1 percent increase in energy from saturated fatty acids in the diet may produce a 2 percent jump in heart disease risk by elevating blood LDL cholesterol. Conversely, reducing saturated fat intake by 1 percent can be expected to produce a 2 percent drop in heart disease risk by the same mechanism. Even a 2 percent drop in LDL represents a significant improvement for the health of the heart.[27] Like saturated fats, *trans* fats also raise heart disease risk by elevating LDL cholesterol. A heart-healthy eating pattern limits foods rich in these two types of fat.

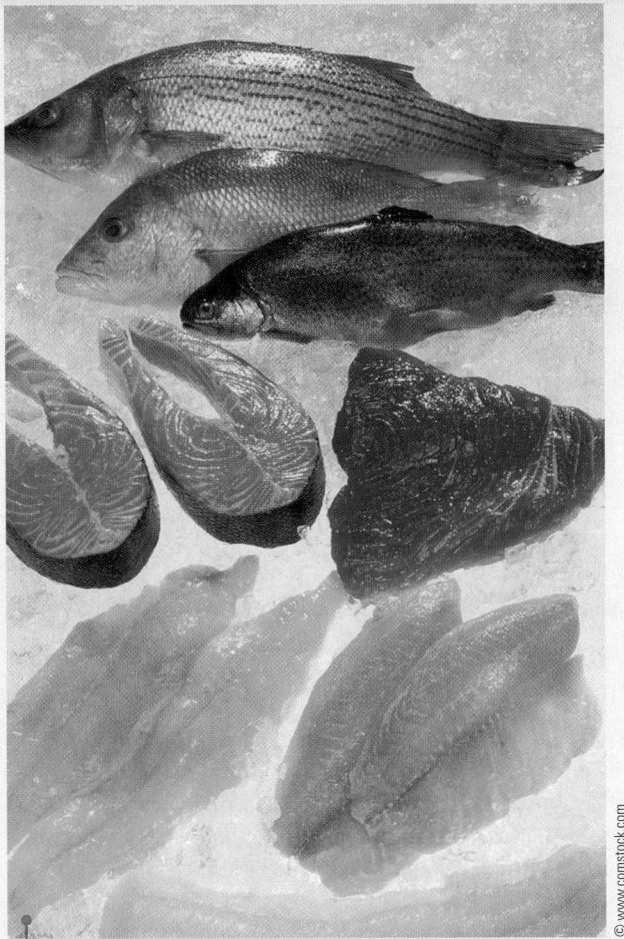

Fish is a good source of the omega-3 fatty acids.

Which foods are highest in saturated and *trans* fats?

The major sources of saturated fats in the U.S. diet are fatty meats, whole-milk products, tropical oils, and products made from any of these foods. To limit saturated fat intake, consumers must choose carefully among these high-fat foods. More than a third of the fat in most meats is saturated. Similarly, more than half of the fat is saturated in whole milk and other high-fat dairy products, such as cheese, butter, cream, half-and-half, cream cheese, sour cream, and ice cream. Consumers rarely use the tropical oils of palm, palm kernel, and coconut in the kitchen, but these oils are used heavily by food manufacturers and so are commonly found in many commercially prepared foods.

When choosing meats, milk products, and commercially prepared foods, look for those lowest in saturated fat. Labels provide a useful guide for comparing products in this regard, and Appendix A lists the saturated fat in several thousand foods.

Even with careful selections, a nutritionally adequate diet will provide some saturated fat. Zero saturated fat is not possible even when experts design menus with the mission to minimize saturated fat.[28] Eating patterns based on fruits, vegetables, legumes, nuts, soy products, and whole grains can, and often do, deliver less saturated fat than diets that depend heavily on animal-derived foods, however.

As for *trans* fats, this chapter explained that solid shortening and margarine are made from vegetable oil that has been hardened through hydrogenation. This process both saturates some of the unsaturated fatty acids and introduces *trans*-fatty acids. Many convenience foods contain *trans* fats. Table NP4-1 summarizes which foods provide which fats. Substituting unsaturated fats for saturated fats at each meal and snack can help protect against heart disease. Figure NP4-1 compares two meals and shows how such substitutions can lower saturated fat and raise unsaturated fat—even when total fat and kcalories remain the same.

So it seems that some fats are "good" and others are "bad" from the body's point of view. Is that right?

The saturated and *trans* fats do indeed seem mostly bad for the health of the heart. Aside from providing energy, which unsaturated fats can do equally well, saturated and *trans* fats bring no indispensable benefits to the body. Furthermore, no harm can come from consuming diets low in them.

In contrast, the unsaturated fats are mostly good for the health of the heart when consumed in moderation. To date, their one proven fault seems to be that they, like all fats, provide abundant energy to the body and so may promote obesity if they drive kcalorie intakes higher than energy needs.[29] Obesity, in turn, often begets many body ills, as Chapter 6 makes clear.

When judging foods by their fatty acids, keep in mind that the fat in foods is a mixture of "good" and "bad," providing both saturated and unsaturated fatty acids. Even predominantly monounsaturated olive oil delivers some saturated fat. Consequently, even when a person chooses foods with mostly unsaturated fats, saturated fat can still add up if total fat is high. For this reason, fat must be kept below 35 percent of total kcalories if the diet is to be moderate in saturated fat.

Additionally, food manufacturers may come to the assistance of consumers wishing to avoid the health threats from saturated and *trans* fats. Some companies now make margarine without *trans* fats, and many snack manufacturers have reduced the saturated and *trans* fats in some products and offer snack foods in 100-kcalorie packages. Other companies are likely to follow if consumers respond favorably.

Eating a balanced diet based on vegetables, fruits, and legumes is a good idea, as is *replacing* saturated fats such as butter, shortening, and meat fat with unsaturated fats like olive oil and the oils from nuts and fish. These foods provide vitamins, minerals, and phytochemicals—all valuable in protecting the body's health. To further protect health, you may want to reduce fats from convenience foods and fast foods; choose small portions of meats, fish, and poultry; and include fresh foods from all the groups each day. Take care to select portion sizes that will best meet your energy needs. Also, be physically active each day.

TABLE NP4-1 Food Sources of Various Fatty Acids

Healthful Fatty Acids

Monounsaturated	Omega-6 Polyunsaturated	Omega-3 Polyunsaturated
• Avocado • Oils (canola, olive, peanut, sesame) • Nuts (almonds, cashews, filberts, hazelnuts, macadamia nuts, peanuts, pecans, pistachios) • Olives • Peanut butter • Seeds (sesame)	• Margarine (nonhydrogenated) • Oils (corn, cottonseed, safflower, soybean) • Nuts (pine nuts, walnuts) • Mayonnaise • Salad dressing • Seeds (pumpkin, sunflower)	• Fatty fish (herring, mackerel, salmon, tuna) • Flaxseed • Nuts (walnuts)

Harmful Fatty Acids

Saturated	*Trans*
• Bacon, butter, lard • Cheese, whole milk products • Chocolate, coconut • Cream, half-and-half, cream cheese, sour cream • Meats • Oil (coconut, palm, palm kernel) • Shortening	• Fried foods (hydrogenated shortening) • Margarine (hydrogenated or partially hydrogenated) • Nondairy creamers • Many fast foods • Shortening • Commercial baked goods (including doughnuts, cakes, cookies) • Many snack foods (including microwave popcorn, chips, crackers)

FIGURE NP4-1 Two Meals Compared: Replacing Saturated Fat with Unsaturated Fat

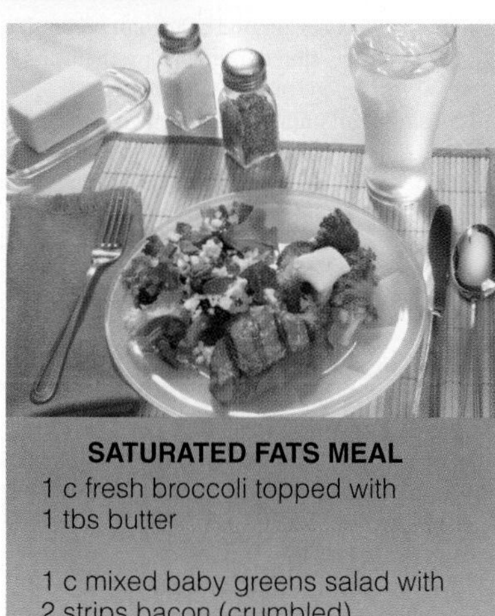

SATURATED FATS MEAL
1 c fresh broccoli topped with
1 tbs butter
1 c mixed baby greens salad with
2 strips bacon (crumbled)
1 oz blue cheese crumble
1 tbs light Italian dressing
4 oz grilled steak
Energy = 600 kcal

To lower saturated fat and raise monounsaturated and polyunsaturated fats . . .

UNSATURATED FATS MEAL
1 c fresh broccoli sautéed in
1 tbs olive oil
1 c mixed baby greens salad with
½ avocado
2 tbs sunflower seeds
1 tbs light Italian dressing
4 oz grilled salmon
Energy = 600 kcal

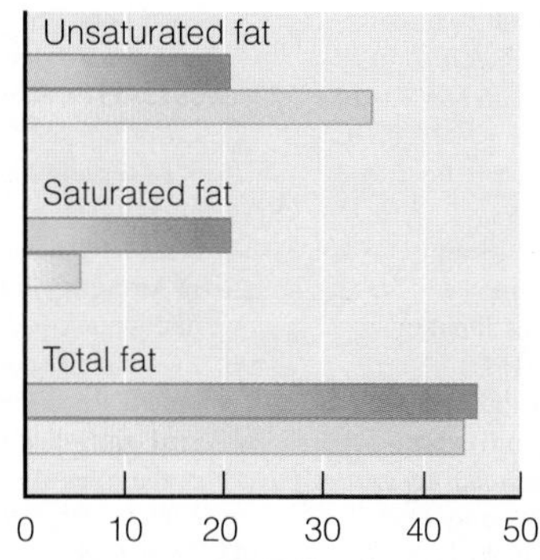

Examples of ways to replace saturated fats with unsaturated fats include sautéing vegetables in olive oil instead of butter, garnishing salads with avocado and sunflower seeds instead of bacon and blue cheese, and eating salmon instead of steak. Each of these meals provides roughly the same number of kcalories and grams of fat, but the one on the left has almost four times as much saturated fat and only half as many omega-3 fatty acids.

Both photos: © Matthew Farrugio

Notes

1. U.S. Department of Agriculture and U.S. Department of Health and Human Services, *Dietary Guidelines for Americans 2010,* available at www.dietaryguidelines.gov; A. Astrup and coauthors, The role of reducing intakes of saturated fat in the prevention of cardiovascular disease: Where does the evidence stand in 2010? *American Journal of Clinical Nutrition* 93 (2011): 684–688.
2. M.R. Flock and P. M. Kris-Etherton, Dietary Guidelines for Americans 2010: Implications for cardiovascular disease, *Current Atherosclerosis Reports* 13 (2011): 499–507; S. S. Gidding and coauthors, Implementing American Heart Association Pediatric and Adult Nutrition Guidelines: A scientific statement from the American Heart Association Nutrition Committee of the Council on Nutrition, Physical Activity and Metabolism, Council on Cardiovascular Disease in the Young, Council on arteriosclerosis, Thrombosis and Vascular Biology, council on Cardiovascular Nursing, Council on Epidemiology and Prevention, and Council for High Blood Pressure Research, *Circulation* 119 (2009): 1161–1175.
3. T. J. Angelopoulos and coauthors, The effect of high-fructose corn syrup consumption on triglycerides and uric acid, *Journal of Nutrition* 139 (2009): 1242S–1245S; K. L. Teff and coauthors, Endocrine and metabolic effects of consuming fructose- and glucose-sweetened beverages with meals in obese men and women: Influcence of insulin resistance on plasma triglyceride responses, *Journal of Clinical Endocrinology and Metabolism* 94 (2009): 1562–1569.
4. Flock and Kris-Etherton, 2011; S. S. Gidding and coauthors, Implementing American Heart Association Pediatric and Adult Nutrition Guidelines: A scientific statement from the American Heart Association Nutrition Committee of the Council on Nutrition, Physical Activity and Metabolism, Council on Cardiovascular Disease in the Young, Council on Arteriosclerosis, Thrombosis and Vascular biology, Council on Cardiovascular Nursing, Council on Epidemiology and Prevention, and Council for High Blood Pressure Research, *Circulation* 119 (2009): 1161–1175; M. U. Jakobsen and coauthors, Major types of dietary fat and risk of coronary heart disease: A pooled analysis of 11 cohort studies, *American Journal of Clinical Nutrition* 89 (2009): 1425–1432.
5. U.S. Department of Agriculture and U.S. Department of Health and Human Services, *Dietary Guidelines for Americans 2010,* available at www.dietaryguidelines.gov; Standing Committee on the Scientific Evaluation of Dietary Reference Intakes, Food and Nutrition Board, Institute of Medicine, *Dietary Reference Intakes for Energy, Carbohydrate, Fiber, Fat, Fatty Acids, Cholesterol, Protein, and Amino Acids* (Washington, DC: National Academies Press, 2005).
6. U.S. Department of Agriculture and U.S. Department of Health and Human Services, *Dietary Guidelines for Americans 2010,* available at www.dietaryguidelines.gov.
7. A. Keys, *Seven Countries: A Multivariate Analysis of Death and Coronary Heart Disease* (Cambridge, MA: Harvard University Press, 1980).
8. H. Gardener and coauthors, Mediterranean-style diet and risk of ischemic stroke, myocardial infarction, and vascular death: the Northern Manhattan Study, *American Journal of Clinical Nutrition* 94 (2011): 1458–1464; A. Trichopoulou, C. Bamia, and D. Trichopoulos, Anatomy of health effects of Mediterranean diet: Greek EPIC prospective cohort study, *British Medical Journal* 338 (2009): b2337; F. Sofi, The Mediterranean diet revisited: Evidence of its effectiveness grows, *Current Opinion in Cardiology* 24 (2009): 442–446; M. C. Sparling and J. J. B. Anderson, The Mediterranean diet and cardiovascular diseases, *Nutrition Today* 44 (2009): 124–133.
9. L. M. León-Muñoz and coauthors, Adherence to the Mediterranean Diet Pattern has declined in Spanish adults, *Journal of Nutrition* 142(2012): 1843–1850; F. Bellisle, Infrequently asked questions about the Mediterranean diet, *Public Health Nutrition* 12 (2009): 1644–1647.
10. O. Castañer and coauthors, Protection of LDL from oxidation by olive oil polyphenols is associated with a downregulation of CD40-ligand expression and its downstream products *in vivo* in humans, *American Journal of Clinical Nutrition* 95 (2012): 1238–1244; S. Cicerale, L. J. Lucas, and R. S. Keast, Antimicrobial, antioxidant and anti-inflammatory phenolic activities in extra virgin olive oil, *Current Opinion in Biotechnology* 23 (2012): 129–135; C. Samieri and coauthors, Olive oil consumption, plasma oleic acid, and stroke incidence: The Three-City Study, *Neurology* 77 (2011): 418–425; S. Cicerale and coauthors, Chemistry and health of olive oil phenolics, *Critical Reviews in Food Science and Nutrition* 49 (2009): 218–236.
11. S. Cicerale, L. Lucas, and R. Keast, Biological activities of phenolic compounds present in virgin olive oil, *International Journal of Molecular Sciences* 11 (2010): 458–479.
12. S. Granados-Principal and coauthors, Hydroxytyrosol: From laboratory investigations to future clinical trials, *Nutrition Reviews* 68 (2010): 191–206.
13. Cicerale, Lucas, and Keast, 2012.
14. R. Estruch, Anti-inflammatory effects of the Mediterranean diet: The experience of the PREDIMED study, *Proceedings of the Nutrition Society* 69 (2010): 333–340; J. S. Perona and coauthors, Evaluation of the effect of dietary virgin olive oil on blood pressure and lipid composition of serum and low-density lipoprotein in elderly type 2 diabetic subjects, *Journal of Agricultural and Food Chemistry* 57 (2009): 11427–11433.
15. Castañer and coauthors, 2012; L. Lucas, A. Russell, and R. Keast, Molecular mechanisms of inflammation. Anti-inflammatory benefits of virgin olive oil and the phenolic compound oleocanthal, *Current Pharmacological Design* 17 (2011): 754–768; Cicerale, Lucas, and Keast, 2010.
16. C. Degirolamo and L. L. Rudel, Dietary monounsaturated fatty acids appear not to provide cardioprotection, *Current Atherosclerosis Reports* 12 (2010): 391–396; Cicerale, Lucas, and Keast, 2010.
17. E. Ross, Health benefits of nut consumption, *Nutrients* 2 (2010): 652–682; J. Sabaté and Y. Ang, Nuts and health outcomes: New epidemiologic evidence, *American Journal of Clinical Nutrition* 89 (2009): 1643S–1488S.
18. T. Y. Li and coauthors, Regular consumption of nuts is associated with a lower risk of cardiovascular disease in women with type 2 diabetes, *Journal of Nutrition* 139 (2009): 1333–1338.
19. M. L. Dreher, Pistachio nuts: Composition and potential health benefits, *Nutrition Reviews* 70 (2012): 234–240; Sabaté and Ang, 2009; S. Rajaram and coauthors, Walnuts and fatty fish influence different serum lipid fractions in normal to mildly hyperlipidemic individuals: A randomized controlled study, *American Journal of Clinical Nutrition* 89 (2009): 1657S–1663S.
20. N. R. Damasceno and coauthors, Crossover study of diets enriched with virgin olive oil, walnuts, or almonds. Effects on lipids and other cardiovascular risk markers, *Nutrition, Metabolism and Cardiovascular Disease* 21 (2011): S14–S20; S. Rajaram and coauthors, Walnuts and fatty fish influence different serum lipid fractions in normal to mildly hyperlipidemic individuals: A randomized controlled study, *American Journal of Clinical Nutrition* 89 (2009): 1657S–1663S; D. K. Banel and F. B. Hu, Effects of walnut consumption on blood lipids and other cardiovascular risk factors: A meta-analysis and systematic review, *American Journal of Clinical Nutrition* 90 (2009): 56–63.
21. C. E. Berryman and coauthors, Effects of almond consumption on the reduction of LDL-cholesterol: A discussion of potential mechanisms and future research directions, *Nutrition Reviews* 69 (2011): 171–185; S. C. Li and coauthors, Almond consumption improved glycemic control and lipid profiles in patients with type 2 diabetes mellitus, *Metabolism* 60 (2011): 474–479; P. M. Kris-Etherton, W. Karmally, and R. Ramakrishnan, Almonds lower LDL cholesterol, *Journal of the American Dietetic Association* 109 (2009): 1521–1522; O. J. Phung and coauthors, Almonds have a neutral effect on serum lipid profile: A meta-analysis of randomized trials, *Journal of the American Dietetic Association* 109 (2009): 865–873.
22. Sabaté and Ang, 2009.
23. P. Casa-Agustench and coauthors, Cross-sectional association of nut intake with adiposity in a Mediterranean population, *Nutrition, Metabolism, and Cardiovascular Diseases* 21 (2011): 518–525; M. Bes-Rastrollo and coauthors, Prospective study of nut consumption, long-term weight change, and obesity risk in women, *American Journal of Clinical Nutrition* 89 (2009): 1913–1919; Sabaté and Ang, 2009.
24. Casa-Agustench and coauthors, 2011.
25. S. L. Tey and coauthors, Nuts improve diet quality compared to other energy-dense snacks while maintaining body weight, *Journal of Nutrition and Metabolism* 2011, doi: 10.1155/2011/357350.
26. D. Mozaffarian and J. H. Y. Wu, Omega-3 fatty acids and cardiovascular disease, *Journal of the American College of Cardiology* 58 (2011): 2047–2067; Rajaram and coauthors, Walnuts and fatty fish influence different serum lipid fractions in normal to mildly hyperlipidemic individuals: A randomized controlled study, 2009; N. D. Riediger and coauthors, A systemic review of the roles of n-3 fatty acids in health and disease, *Journal of the American Dietetic Association* 109 (2009): 668–679.
27. *Third Report of the National Cholesterol Education Program (NCEP) Expert Panel on Detection, Evaluation, and Treatment of High Blood Cholesterol in Adults (Adult Treatment Panel III)*, NIH publication no. 025215 (Bethesda, MD: National Heart, Lung, and Blood Institute, 2002), p. V-8.
28. Standing Committee on the Scientific Evaluation of Dietary Reference Intakes, 2005, p. 835.
29. Standing Committee on the Scientific Evaluation of Dietary Reference Intakes, 2005, pp. 796–797.

Chapter 5 Protein

THE VERSATILITY OF PROTEINS IN THE BODY IS REMARKABLE. PROTEINS help your muscles to contract, your blood to clot, and your eyes to see. They keep you alive and well by facilitating chemical reactions and defending against infections. Without them, your bones, skin, and hair would have no structure. No new living tissue can be built without them. No wonder they were named *proteins,* meaning "of prime importance."

The Chemist's View of Proteins

Proteins are chemical compounds that contain the same atoms as carbohydrates and lipids—carbon (C), hydrogen (H), and oxygen (O)—but proteins are different in that they also contain nitrogen (N) atoms. These nitrogen atoms give the name *amino* (nitrogen containing) to the amino acids that form the links in the chains we call proteins.

THE STRUCTURE OF PROTEINS

About 20 different **amino acids** may appear in proteins.* All amino acids share a common chemical "backbone," and it is these backbones that are linked together to form proteins. Each amino acid also carries a side group, which varies from one amino acid to another (see Figure 5-1). The side group makes the amino acids differ in size, shape, and electrical charge. The side groups on amino acids are what make proteins so varied in comparison with either carbohydrates or lipids.

Protein Chains

The 20 amino acids can be linked end to end in a virtually infinite variety of sequences to form proteins. When two amino acids bond together, the resulting structure is known as a **dipeptide**. Three amino acids bonded together form a **tripeptide**. As additional amino acids join the chain, the structure becomes a **polypeptide**. Most proteins are a few dozen to several hundred amino acids long.

Protein Shapes

Polypeptide chains twist into complex shapes. Each amino acid has special characteristics that attract it to, or repel it from, the surrounding fluids and other amino acids. Because of these interactions, polypeptide chains fold and intertwine into intricate coils (see Figure 5-2) and other shapes. The amino acid sequence of a protein determines the specific way the chain will fold.

Protein Functions

The dramatically different shapes of proteins enable them to perform different tasks in the body. Some, such as hemoglobin in the blood (see Figure 5-3), are globular in shape; some are hollow balls that can carry and store materials within them; and some, such as those that form tendons, are more than 10 times as long as they are wide, forming stiff, sturdy, rodlike structures.

FIGURE 5-1 Amino Acid Structure and Examples of Amino Acids

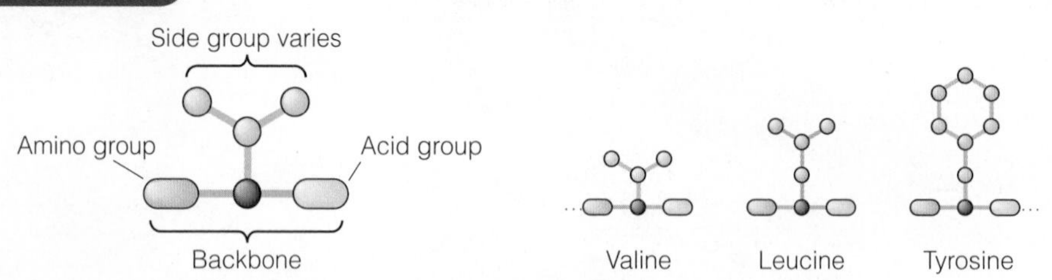

All amino acids have a "backbone" made of an amino acid group (which contains nitrogen) and an acid group. The side group varies from one amino acid to the next. Note that the side group is a unique structure that differentiates one amino acid from another.

*Besides the 20 common amino acids, which can all be components of proteins, others occur individually (for example, ornithine).

proteins: compounds made from strands of amino acids composed of carbon, hydrogen, oxygen, and nitrogen atoms. Some amino acids also contain sulfur atoms.

amino (a-MEEN-oh) acids: building blocks of protein. Each contains an amino group, an acid group, a hydrogen atom, and a distinctive side group, all attached to a central carbon atom.

amino = containing nitrogen

dipeptide: two amino acids bonded together.

di = two
peptide = amino acid

tripeptide: three amino acids bonded together.

tri = three

polypeptide: 10 or more amino acids bonded together. An intermediate strand of between 4 and 10 amino acids is an *oligopeptide*.

poly = many
oligo = few

FIGURE 5-2 The Coiling and Folding of a Protein Molecule

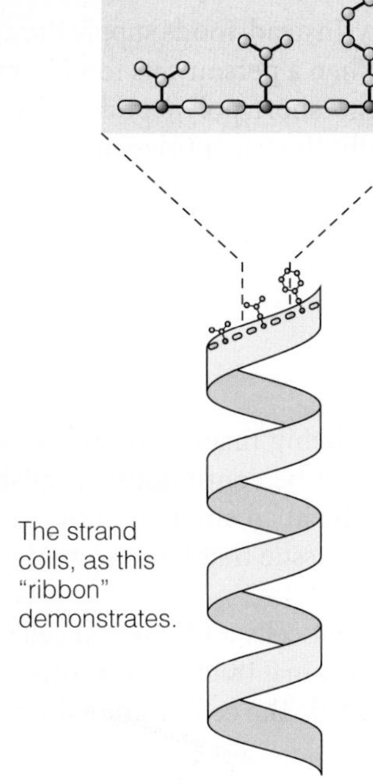
The strand
coils, as this
"ribbon"
demonstrates.

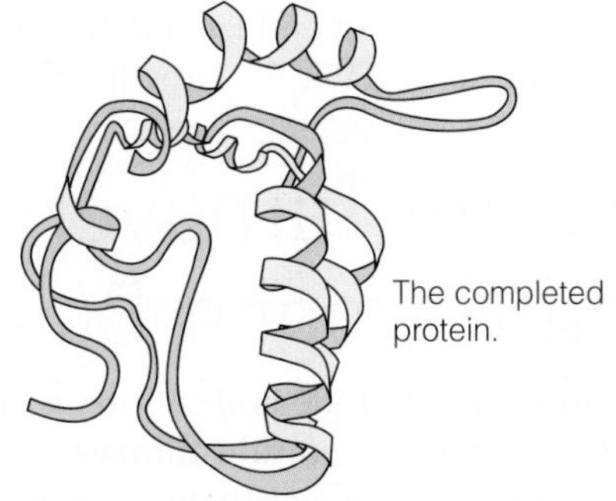
A portion of a strand of amino acids.
The completed protein.

Once coiled and folded, the protein may be functional as is, or it may need to join with other proteins or add a vitamin or mineral to become active.

The strand of amino acids takes on a spring-like shape as their side groups variously attract and repel each other.

© Cengage Learning

■ Some researchers refer to essential amino acids as *indispensable* and to nonessential amino acids as *dispensable*.

nonessential amino acids: amino acids that the body can synthesize.

essential amino acids: amino acids that the body cannot synthesize in amounts sufficient to meet physiological need. Nine amino acids are known to be essential for human adults:

histidine (HISS-tuh-deen)
isoleucine (eye-so-LOO-seen)
leucine (LOO-seen)
lysine (LYE-seen)
methionine (meh-THIGH-oh-neen)
phenylalanine (fen-il-AL-uh-neen)
threonine (THREE-oh-neen)
tryptophan (TRIP-toe-fane, TRIP-toe-fan)
valine (VAY-leen)

conditionally essential amino acid: an amino acid that is normally nonessential but must be supplied by the diet in special circumstances when the need for it becomes greater than the body's ability to produce it.

NONESSENTIAL AND ESSENTIAL AMINO ACIDS

More than half of the amino acids are **nonessential amino acids**, meaning that the body can make them for itself. Proteins in foods usually deliver these amino acids, but it is not essential that they do so. There are other amino acids that the body cannot make at all, however, and some that it cannot make fast enough to meet its needs. The proteins in foods must supply these nine amino acids to the body; they are therefore called **essential amino acids**. (■)

Under special circumstances, a nonessential amino acid can become essential. For example, the body normally makes tyrosine (a nonessential amino acid) from the essential amino acid phenylalanine. If the diet fails to supply enough phenylalanine or if the body cannot make the conversion for some reason (as happens in the inherited disease phenylketonuria), then tyrosine becomes a **conditionally essential amino acid**.

Review Notes

- Chemically speaking, proteins are more complex than carbohydrates or lipids; proteins are made of some 20 different nitrogen-containing amino acids, 9 of which the body cannot make (they are essential).
- Each amino acid contains a central carbon atom with an amino group, an acid group, a hydrogen atom, and a unique side group attached to it.
- The distinctive sequence of amino acids in each protein determines its shape and function.

FIGURE 5-3 The Structure of Hemoglobin

Four highly folded polypeptide chains form the globular hemoglobin protein.

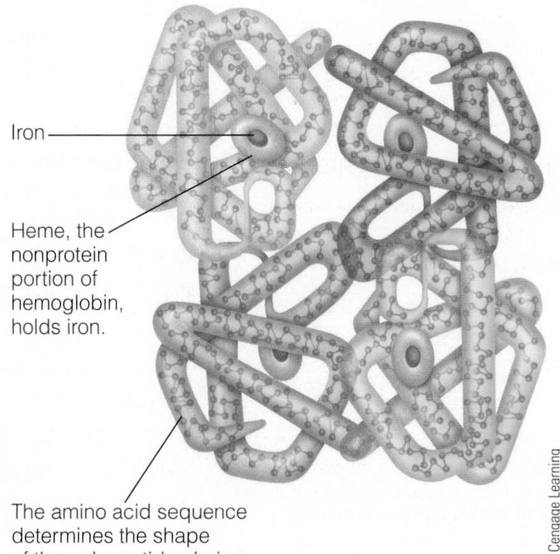
Iron
Heme, the nonprotein portion of hemoglobin, holds iron.
The amino acid sequence determines the shape of the polypeptide chain.

© Cengage Learning

Protein Digestion and Absorption

Proteins in foods do not become body proteins directly. Instead, foods supply the amino acids from which the body makes its own proteins. When a person eats foods containing protein, enzymes break the long polypeptides into short polypeptides, the short polypeptides into tripeptides and dipeptides, and finally, the tripeptides and dipeptides into amino acids. Table 5-1 provides the details.

Protein Turnover and Nitrogen Balance

Within each cell of the body, proteins are continually being made and broken down, a process known as **protein turnover**. Amino acids must be continuously available to build the proteins of new tissues. The new tissues may be in an embryo, in the muscles of an athlete in training, in a growing child, in the scar tissue that heals wounds, or in new hair and nails.

Less obvious is the protein that helps replace worn-out cells and internal cell structures. For example, the millions of cells that line the intestinal tract live for three to five days; they are constantly being shed and must be replaced. The cells of the skin die and rub off, and new ones grow from underneath.

PROTEIN TURNOVER

When proteins break down, their component amino acids are liberated within the cells or released into the bloodstream. Some of these amino acids are promptly recycled into other proteins. By reusing amino acids to build proteins, the body conserves and

TABLE 5-1 Protein Digestion and Absorption

Mouth and Salivary Glands

Chewing and crushing moisten protein-rich foods and mix them with saliva to be swallowed.

Stomach

Hydrochloric acid (HCl) uncoils protein strands and activates stomach enzymes:

Protein —(Pepsin, HCl)→ Smaller polypeptides

Small Intestine and Pancreas

Pancreatic and small intestinal enzymes split polypeptides further:

Polypeptides —(Pancreatic and intestinal proteases)→ Tripeptides, dipeptides, amino acids

Then enzymes on the surface of the small intestinal cells hydrolyze these peptides and the cells absorb them:

Peptides —(Intestinal tripeptidases and dipeptidases)→ Amino acids (absorbed)

protein turnover: the continuous breakdown and synthesis of body proteins involving the recycling of amino acids.

recycles a valuable commodity. Other amino acids are stripped of their nitrogen and used for energy. Each day, about a quarter of the body's available amino acids are irretrievably broken down and used for energy. For this reason, amino acids from food are needed each day to support the new growth and maintenance of cells.

NITROGEN BALANCE

Researchers use **nitrogen balance** studies to estimate protein requirements. (■) In healthy adults, protein synthesis balances with protein degradation, and nitrogen intake from protein in food balances with nitrogen excretion in the urine, feces, and sweat. When nitrogen intake equals nitrogen output, a person is in nitrogen equilibrium, or zero nitrogen balance.

■ Nitrogen balance:
- Nitrogen equilibrium (zero nitrogen balance): N in = N out
- Positive nitrogen balance: N in > N out
- Negative nitrogen balance: N in < N out

If the body synthesizes more than it degrades and adds protein, nitrogen status becomes positive. Nitrogen status is positive in growing infants, children, and adolescents; pregnant women; and people recovering from protein deficiency or illness; their nitrogen intake exceeds their nitrogen output. They are retaining protein in new tissues as they add blood, bone, skin, and muscle to their bodies.

If the body degrades more than it synthesizes and loses protein, nitrogen status becomes negative. Nitrogen status is negative in people who are starving or suffering other severe stresses such as burns, injuries, infections, and fever; their nitrogen output exceeds their nitrogen intake. During these times, the body loses nitrogen as it breaks down muscle and other body proteins for energy.

Growing children end each day with more bone, blood, muscle, and skin cells than they had at the beginning of the day.

Review Notes

- The process by which proteins are continually being made and broken down is known as protein turnover.
- The body needs dietary amino acids to grow new cells and to replace worn-out ones.
- When nitrogen intake equals nitrogen output, a person is in nitrogen equilibrium, or zero nitrogen balance.

Roles of Body Proteins

What distinguishes you chemically from any other human being are minute differences in your particular body proteins (enzymes, antibodies, and others). These differences are determined by your proteins' amino acid sequences, which are written into the genes you inherited from your parents and ancestors. The genes direct the making of all the body's proteins.

The human body has more than 20,000 genes that code for hundreds of thousands of proteins. Relatively few proteins have been studied in detail, although this number is growing rapidly with the surge in knowledge gained from sequencing the human genome. (■) Only a few of the many roles proteins play are described here, but these should serve to illustrate proteins' versatility, uniqueness, and importance.

■ The *human genome* is the full set of chromosomes, including all of the genes and associated DNA. Nutritional genomics is the topic of Nutrition in Practice 13.

nitrogen balance: the amount of nitrogen consumed (N in) as compared with the amount of nitrogen excreted (N out) in a given period of time. The laboratory scientist can estimate the protein in a sample of food, body tissue, or excreta by measuring the nitrogen in it.

As Structural Components A great deal of the body's protein is found in muscle tissue, which allows the body to move. The amino acids of muscle protein can also be released when the need is dire, as in starvation. These amino acids are integral parts of the muscle structure, and their loss exacts a cost in functional protein, as a later section makes clear.

Other structural proteins confer shape and strength on bones, teeth, tendons, cartilage, blood vessels, and other tissues. These proteins exist in a stable form and are more resistant to breakdown than are the proteins of muscles.

As Enzymes **Enzymes** are catalysts that are essential to all life processes. Enzymes in the cells of plants or animals put together the pairs of sugars that make disaccharides and the long strands of sugars that make starch, cellulose, and glycogen. Enzymes also dismantle these compounds to free their constituent parts and release energy. Enzymes also assemble and disassemble lipids, assemble all other compounds that the body makes, and disassemble all compounds that the body can use for building tissue and other metabolic work. It is enzymes that put amino acids together to make needed proteins, too. In other words, these proteins can even make other proteins. As Figure 5-4 shows, enzymes themselves are not altered by the reactions they facilitate.

The protein story moves in a circle. To follow the circle in nutrition, start with a person eating food proteins. The food proteins are broken down by digestive enzymes, proteins themselves, into amino acids. The amino acids enter the cells of the body, where other proteins (enzymes) put the amino acids together in long chains whose sequences are specified by the genes. The chains fold and twist back on themselves to form proteins, and some of these proteins become enzymes themselves. Some of these enzymes break apart compounds; others put compounds together. Day by day, in billions of reactions, these processes repeat themselves, and life goes on. Only living systems can achieve such self-renewal. A toaster cannot produce another toaster; a car cannot fix a broken-down car. Only living creatures and the parts they are composed of—the cells—can duplicate and repair themselves.

As Transporters A large group of proteins specialize in transporting other substances, such as lipids, vitamins, and minerals, around the body. To do their jobs, those substances must move from place to place within the blood, into and out of cells, or around the cellular interiors. Two familiar examples: the protein hemoglobin carries oxygen from the lungs to the cells, and the lipoproteins transport lipids in the watery blood.

As Regulators of Fluid and Electrolyte Balance Proteins help maintain the body's **fluid and electrolyte balance**. As Figure 5-5 shows, the body's fluids are contained in three major body compartments: (1) the spaces inside the blood vessels, (2) the spaces within the cells, and (3) the spaces between the cells (the interstitial spaces outside the blood vessels). Fluids flow back and forth between these compartments, and proteins in the fluids, together with minerals, (■) help to maintain the needed distribution of these fluids.

■ Minerals are helper nutrients. The attraction of protein and mineral particles to water is due to osmotic pressure (see Chapter 9).

enzymes: protein catalysts. A catalyst is a compound that facilitates chemical reactions without itself being changed in the process.

fluid and electrolyte balance: maintenance of the necessary amounts and types of fluid and minerals in each compartment of the body fluids.

FIGURE 5-4 Enzyme Action

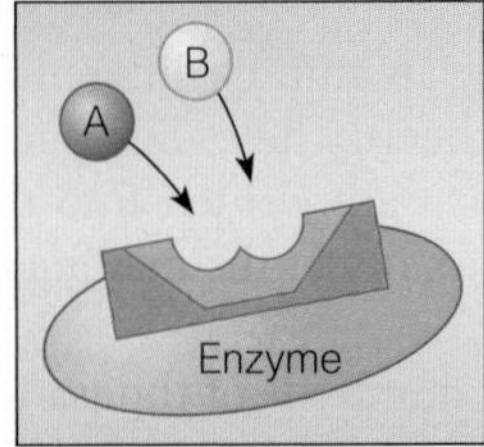

The separate compounds, A and B, are attracted to the enzyme's active site, making a reaction likely.

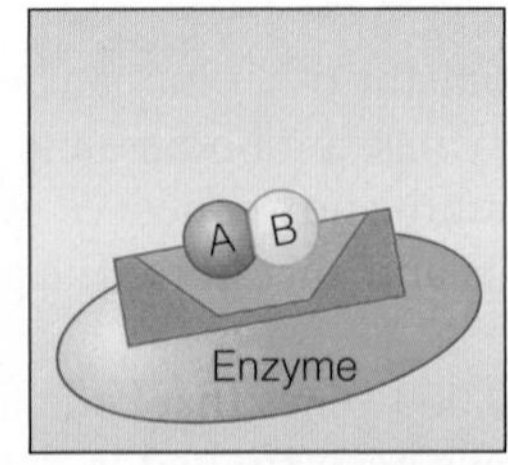

The enzyme forms a complex with A and B.

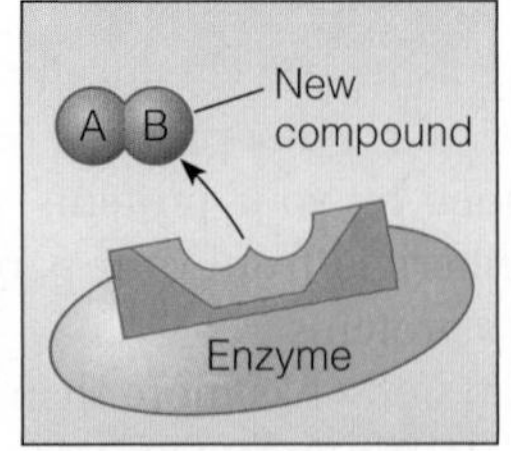

The enzyme is unchanged, but A and B have formed a new compound, AB.

Each enzyme facilitates a specific chemical reaction. In this diagram, an enzyme enables two compounds to make a more complex structure, but the enzyme itself remains unchanged.

Proteins are able to help determine the distribution of fluids in living systems for two reasons: first, proteins cannot pass freely across the membranes that separate the body compartments, and second, they are attracted to water. A cell that "wants" a certain amount of water in its interior space cannot move the water around directly, but it can manufacture proteins, and these proteins will hold water. Thus, the cell can use proteins to help regulate the distribution of water indirectly. Similarly, the body makes proteins for the blood and the interstitial (intercellular) spaces. These proteins help maintain the fluid volume in those spaces. Excess fluid accumulation in the interstitial spaces is called **edema**.

Not only is the quantity of the body fluids vital to life, but so is their composition. Special transport proteins in the membranes of cells continuously transfer substances into and out of cells to maintain balance. For example, sodium is concentrated outside the cells, and potassium is concentrated inside. The balance of these two minerals is critical to nerve transmission and muscle contraction. Any disturbance in this balance triggers a major medical emergency. Such imbalances can cause irregular heartbeats, kidney failure, muscular weakness, and even death.

FIGURE 5-5 One Cell and Its Associated Fluids

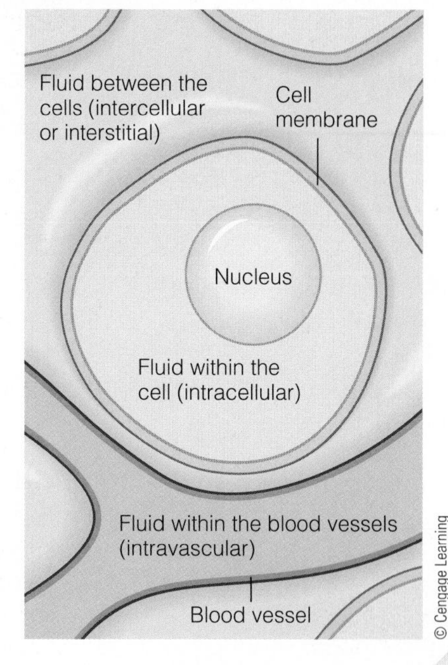

As Regulators of Acid–Base Balance

Proteins also help maintain the balance between **acids** and **bases** within the body's fluids. Normal body processes continually produce acids and bases, which must be carried by the blood to the kidneys and lungs for excretion. The blood must do this without upsetting its own **acid–base balance**. Blood **pH** is one of the most tightly controlled conditions in the body. If the blood becomes too acidic, vital proteins may undergo **denaturation**, losing their shape and ability to function. A similar situation arises when there is an excess of base. These imbalances are known as **acidosis** and **alkalosis**, respectively, and both can be fatal. Figure 5-6 (p. 126) shows the normal and abnormal pH ranges of body fluids, as well as the pHs of some common substances.

Proteins such as albumin in blood help to prevent acid–base imbalances. In a sense, the proteins protect one another by gathering up extra acid (hydrogen) ions when there are too many in the surrounding medium and by releasing them when there are too few. By accepting and releasing hydrogen ions, proteins act as **buffers**, maintaining the acid–base balance of the blood and body fluids.

As Antibodies

Proteins also defend the body against disease. A virus—whether it is one that causes flu, smallpox, measles, or the common cold—enters the cells and multiplies there. One virus may produce 100 replicas of itself within an hour or so. Each replica can then burst out and invade 100 different cells, soon yielding 10,000 viruses, which invade 10,000 cells. Left free to do their worst, they will soon overwhelm the body with disease.

Fortunately, when the body detects these invading **antigens**, it manufactures **antibodies**, giant protein molecules designed specifically to combat them. The antibodies work so swiftly and efficiently that in a healthy individual, most diseases never get started. Without sufficient protein, though, the body cannot maintain its army of antibodies to resist infectious diseases.

Each antibody is designed to destroy a specific antigen. Once the body has manufactured antibodies against a particular antigen (such as the measles virus), it "remembers" how to make them. Consequently, the next time the body encounters that same antigen, it produces antibodies even more quickly. In other words, the body develops a molecular memory, known as **immunity**. (Chapter 11 describes food allergies—the immune system's response to food antigens.)

As Hormones

The blood also carries messenger molecules known as **hormones**, and *some* hormones are proteins. (Recall that some hormones are sterols, members of the lipid family.) Among the proteins that act as hormones are glucagon and insulin. Hormones have many profound effects, which will become evident in subsequent chapters.

edema (eh-DEEM-uh): the swelling of body tissue caused by leakage of fluid from the blood vessels and accumulation of the fluid in the interstitial spaces.

acids: compounds that release hydrogen ions in a solution.

bases: compounds that accept hydrogen ions in a solution.

acid–base balance: the balance maintained between acid and base concentrations in the blood and body fluids.

pH: the concentration of hydrogen ions. The lower the pH, the stronger the acid. Thus, pH 2 is a strong acid; pH 6 is a weak acid; pH 7 is neutral; and a pH above 7 is alkaline.

denaturation (dee-nay-cher-AY-shun): the change in a protein's shape brought about by heat, acid, or other agents. Past a certain point, denaturation is irreversible.

acidosis: acid accumulation in the blood and body fluids; depresses the central nervous system and can lead to disorientation and, eventually, coma.

alkalosis: excessive base in the blood and body fluids.

buffers: compounds that can reversibly combine with hydrogen ions to help keep a solution's acidity or alkalinity constant.

FIGURE 5-6 The pH Scale

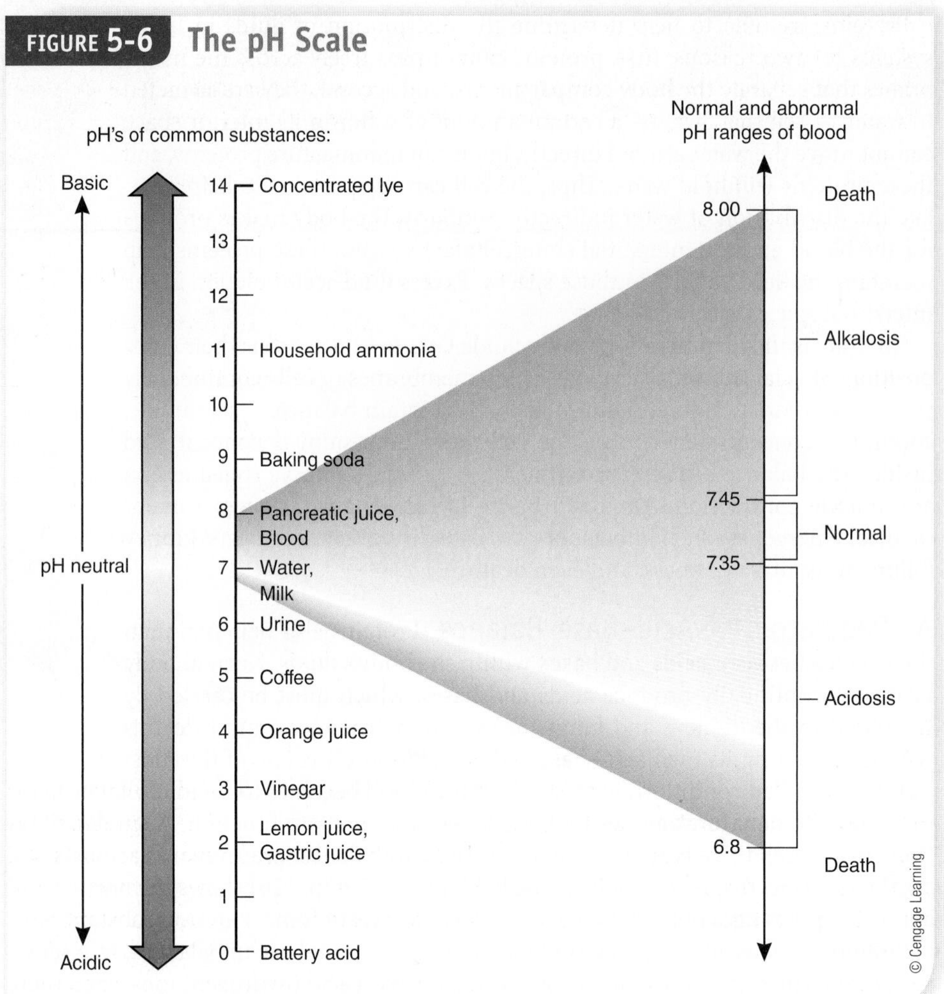

antigens: substances that elicit the formation of antibodies or an inflammation reaction from the immune system. A bacterium, a virus, a toxin, and a protein in food that causes allergy are all examples of antigens.

antibodies: large proteins of the blood and body fluids, produced in response to invasion of the body by unfamiliar molecules (mostly proteins) called *antigens*. Antibodies inactivate the invaders and so protect the body.

anti = against

immunity: the body's ability to defend itself against diseases.

hormones: chemical messengers. Hormones are secreted by a variety of glands in the body in response to altered conditions. Each travels to one or more target tissues or organs and elicits specific responses to restore normal conditions.

urea (yoo-REE-uh): the principal nitrogen-excretion product of protein metabolism; generated mostly by removal of amine groups from unneeded amino acids or from the sacrifice of amino acids to meet a need for energy.

As a Source of Energy and Glucose

Without energy, cells die; without glucose, the brain and nervous system falter. Even though amino acids are needed to do the work that only they can perform—build vital proteins—they will be sacrificed to provide energy and glucose during times of starvation or insufficient carbohydrate intake.[1] When glucose or fatty acids are limited, cells are forced to use amino acids for energy and glucose. When amino acids are degraded for energy or converted into glucose, their nitrogen-containing amine groups are stripped off and used elsewhere or are incorporated by the liver into **urea** and sent to the kidneys for excretion in the urine. The fragments that remain are composed of carbon, hydrogen, and oxygen, as are carbohydrate and fat, and can be used to build glucose or fatty acids or can be metabolized like them.

The body does not make a specialized storage form of protein as it does for carbohydrate and fat. Glucose is stored as glycogen in the liver and muscles, fat as triglycerides in the adipose tissue, but body protein is available only as the working and structural components of the tissues. When the need arises, the body dismantles its tissue proteins and uses them for energy. Thus, over time, energy deprivation (starvation) always incurs wasting of lean body tissue as well as fat loss.

Review Notes

- The list of protein functions discussed here and summarized in Table 5-2 is by no means exhaustive. Nevertheless, it does give some sense of the immense variety of proteins and their importance in the body.

TABLE 5-2 Summary of Functions of Proteins

Structural components. Proteins form integral parts of most body tissues and confer shape and strength on bones, skin, tendons, and other tissues. Structural proteins of muscles allow movement.
Enzymes. Proteins facilitate chemical reactions.
Transporters. Proteins transport substances such as lipids, vitamins, minerals, and oxygen around the body.
Fluid and electrolyte balance. Proteins help to maintain the distribution and composition of various body fluids.
Acid-base balance. Proteins help maintain the acid-base balance of body fluids by acting as buffers.
Antibodies. Proteins inactivate disease-causing agents, thus protecting the body.
Hormones. Proteins regulate body processes. Some, but not all, hormones are proteins.
Energy and glucose. Proteins provide some fuel, and glucose if needed, for the body's energy needs.
Other. The protein fibrin creates blood clots; the protein collagen forms scars; the protein opsin participates in vision.

Protein and Health

During the time that scientists have been studying nutrition, no nutrient has been more intensely scrutinized than protein. As you know by now, it is indispensable to life. And it should come as no surprise that protein deficiency can have devastating effects on people's health. But, as with the other nutrients, overconsumption of protein can be harmful, too, so this section also discusses the consequences of protein excess.

PROTEIN DEFICIENCY

In protein deficiency, when the diet supplies too little protein or lacks a specific essential amino acid relative to the others (a limiting amino acid), the body slows its synthesis of proteins while increasing its breakdown of body tissue protein to liberate the amino acids it needs to build other proteins of critical importance. When these proteins are not available to perform their roles, many of the body's life-sustaining activities come to a halt. The most recognizable consequences of protein deficiency include slow growth in children, impaired brain and kidney functions, weakened immune defenses, and impaired nutrient absorption from the digestive tract.

SEVERE ACUTE MALNUTRITION

The term *protein-energy malnutrition (PEM)* has traditionally been used to describe the condition that develops when the diet delivers too little protein, too little energy, or both. (■) The causes and consequences are complex, but clearly, such malnutrition reflects insufficient food intake. Importantly, not only are protein and energy inadequate, but so are many, if not all, of the vitamins and minerals. For this reason, the term **severe acute malnutrition (SAM)** is now used to describe severely malnourished infants and children. The next sections describe the symptoms of **kwashiorkor** and **marasmus**, two clinical expressions of SAM that can appear individually or together in a starving child.[2]

The form of malnutrition manifested in a child's condition depends partly on the nature of the food shortage. About 10 percent of the world's children suffer from SAM, which most often occurs when food suddenly becomes unavailable, such as in drought or war. These children are often underweight for their height (described as *wasting*).

■ PEM can be a consequence of many different conditions. PEM has been recognized in people with many chronic diseases such as cancer and AIDS and in those experiencing severe stresses such as burns or extensive infections (see Chapter 16). The problems associated with PEM and illness are described throughout later chapters.

severe acute malnutrition (SAM): malnutrition caused by recent severe food restriction; characterized in children by underweight for height (wasting).

kwashiorkor (kwash-ee-OR-core or kwash-ee-or-CORE): severe malnutrition characterized by failure to grow and develop, edema, changes in the pigmentation of the hair and skin, fatty liver, anemia, and apathy.

marasmus (ma-RAZZ-mus): severe malnutrition characterized by poor growth, dramatic weight loss, loss of body fat and muscle, and apathy.

The starving child faces this threat to life by engaging in as little activity as possible—not even crying for food. The body gathers all its forces to meet the crisis, so it cuts down on any expenditure of energy not needed for the heart, lungs, and brain to function. Digestive enzymes are in short supply, the digestive tract lining deteriorates, and absorption fails. The child cannot assimilate what little food is eaten.

Each year, 7.6 million children younger than age 5, as many as 5 children *every minute,* die as a result of SAM.[3] Most of them do not starve to death—they die from the diarrhea and dehydration that accompany infections.

Kwashiorkor Kwashiorkor is a Ghanaian word meaning a "sickness that infects the first child when the second child is born." When a mother who has been nursing her first child bears a second child, she weans the first child and puts the second one on the breast. The first child, suddenly switched from nutrient-dense, protein-rich breast milk to a starchy, protein-poor gruel, soon begins to sicken and die.

With too little nutritious food, concentrations of the blood protein albumin fall, causing fluids to shift out of the blood and into the tissues—edema. Some muscle wasting may occur, but it may not be apparent because the child's face, limbs, and abdomen become swollen with edema—a distinguishing feature of kwashiorkor. Infections with parasites commonly further expand the enlarged abdomen. Other effects of too little dietary protein include loss of hair color because melanin, a dark pigment, is made from the amino acid tyrosine; in addition, patchy and scaly skin develops, often with sores that fail to heal.

Marasmus Marasmus, appropriately named from the Greek word meaning "dying away," reflects a prolonged, unrelenting deprivation of food observed in children living in impoverished nations. Children living in poverty simply do not have enough food. They subsist on diluted cereal drinks that supply scant energy and protein of low quality; such food can barely sustain life, much less support growth. Consequently, marasmic children look like little old people—just skin and bones.

Without adequate nutrition, muscles, including the heart muscle, waste and weaken. Because the brain normally grows to almost its full adult size within the first two years of life, marasmus impairs brain development and learning ability. Reduced synthesis of key hormones leads to a metabolism so slow that body temperature drops below

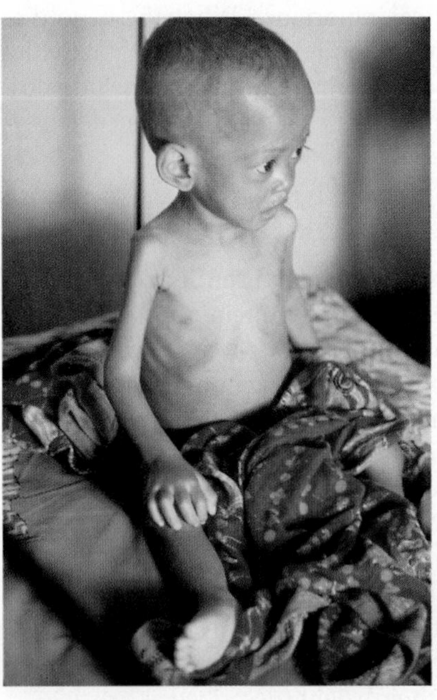
© FAO/7752/F. Botts

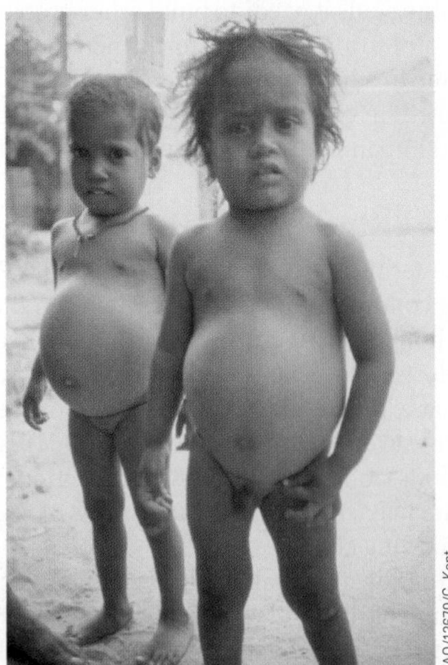
FAO/13670/G. Kent

In the photo on the left, the extreme loss of muscle and fat characteristic of marasmus is apparent in the child's matchstick arms and legs. In contrast, the edema and fatty liver characteristic of kwashiorkor are apparent in the swollen bellies of the children in the photo on the right.

TABLE 5-3 Identifying Characteristics of Severe Acute Malnutrition and Chronic Malnutrition

	Severe Acute Malnutrition	Chronic Malnutrition
Food deprivation	Current or recent	Long-term
Physical features	Rapid weight loss Wasting (underweight for height; small upper arm circumference) Edema	Minimal height gains Stunting (short for age)
World prevalence of children younger than age 5	5 to 15%	20 to 50%

Note: Vitamin and mineral deficiencies are common in both of these types of malnutrition.

normal. There is little or no fat under the skin to insulate against cold. Some hospital workers find that the primary need of marasmic children is to be clothed, covered, and kept warm.

CHRONIC MALNUTRITION

A much greater number, 25 percent of children worldwide, live with **chronic malnutrition**.[4] These children have enough food to survive but not to thrive. They are short for their age—they stop growing because they chronically lack the nutrients required for normal growth (described as stunting). These stunted children may be no larger at age 4 than at age 2. Table 5-3 compares key features of SAM with those of chronic malnutrition.

Rehabilitation Ideally, optimal breastfeeding and improved complementary feedings would prevent malnutrition and save the lives of children. Should mild to moderate malnutrition occur, it can be quickly remedied with supplemental foods in the community. SAM, on the other hand, requires hospitalization, which demands intensive nursing care, diet, and medication.[5]

Experts assure us that we possess the knowledge, technology, and resources to end the hunger that leads to SAM. Programs that have involved the local people in the process of identifying the problem and devising its solution have met with some success. Until those who have the food, technology, and resources make fighting hunger a priority, however, the war on hunger will not be won.

PROTEIN EXCESS

While many of the world's people struggle to obtain enough food and enough protein to survive, in the developed nations protein is so abundant that problems of protein excess are observed. Overconsumption of protein offers no benefits and may pose health risks for the heart and weakened kidneys.[6]

Heart Disease Protein itself is not known to contribute to heart disease and mortality but some of its food sources may do so.[7] Selecting too many animal-derived protein foods, such as fatty red meats, processed meats, and fat-containing milk products, adds a burden of fat kcalories and saturated fat to the diet and crowds out fruits, vegetables, legumes, and whole grains. Consequently, it is not surprising that people who eat substantial amounts of high-fat meats—particularly processed meats such as lunch meats and hot dogs—have higher body weights and a greater risk of obesity, heart disease, and diabetes than those who eat less.[8] As the Nutrition in Practice points out, people who substitute vegetable protein for animal protein lower their risk of dying from heart disease.[9]

chronic malnutrition: malnutrition caused by long-term food deprivation; characterized in children by short height for age (stunting).

Kidney Disease Excretion of the end products of protein metabolism depends, in part, on an adequate fluid intake and healthy kidneys. A high protein intake increases the work of the kidneys but does not appear to damage healthy kidneys or cause kidney disease.[10] In people with chronic kidney disease, however, a high-protein diet may accelerate the kidneys' decline. One of the most effective ways to slow the progression of kidney disease is to restrict dietary protein (see Chapter 22).

Review Notes

- Children suffering from severe acute malnutrition (SAM) may be underweight for their height (wasting), while those experiencing chronic malnutrition are short for their age (stunting).
- Kwashiorkor and marasmus are clinical expressions of malnutrition.
- Excesses of protein offer no advantage; in fact, overconsumption of certain protein-rich foods may contribute to health problems.

PROTEIN AND AMINO ACID SUPPLEMENTS

Why do people take protein or amino acid supplements? Athletes often take them when trying to build muscle. Dieters may take them to spare their bodies' protein while losing weight. Some women take them to strengthen their fingernails. People take individual amino acid supplements, too—to cure herpes, to improve sleep, to lose weight, and to relieve pain and depression. Do protein and amino acid supplements really do these things? Probably not. Are they safe? Not always.

Protein Supplements Though protein supplements are popular with athletes, well-fed athletes do not need them. Dietary protein is necessary for building muscle tissue, and consuming protein in conjunction with exercise helps muscles build new proteins.[11] Protein supplements, however, do not improve athletic performance beyond the gains from well-timed meals of ordinary foods.

Weight-loss dieters may benefit from consistently consuming protein-rich foods because protein often satisfies the appetite. Research is ongoing to determine whether sufficient protein content of a meal may help to prolong feelings of fullness or delay the urge to eat.[12] However, extra protein from powders, pills, or beverages is unlikely to dampen the appetite further, although it contributes unneeded kcalories—the wrong effect for weight loss. Evidence does not support taking protein supplements for weight loss, and common sense opposes it.

Amino Acid Supplements Enthusiastic popular reports have led to widespread use of individual amino acids. One such amino acid is lysine, promoted to prevent or relieve the infections that cause herpes sores on the mouth or genital organs. Lysine does not cure herpes infections. Whether it reduces outbreaks or even whether it is safe is unknown because scientific studies are lacking.

Tryptophan supplements are advertised to relieve pain, depression, and insomnia. Tryptophan plays a role as a precursor for the brain neurotransmitter serotonin, an important regulator of sleep, appetite, mood, and sensory perception. The DRI committee concludes that high doses of tryptophan may induce sleepiness, but they may also cause side effects, such as nausea and skin disorders.

The body is designed to handle whole proteins best. It breaks them into manageable pieces (dipeptides and tripeptides) and then splits these, a few at a time, simultaneously releasing them into the blood. This slow bit-by-bit assimilation is ideal because groups of chemically similar amino acids compete for the carriers that absorb them into the blood. An excess of one amino acid can produce such a demand for a carrier that it limits the absorption of another amino acid, creating a temporary imbalance.[13]

The DRI committee reviewed the available research on amino acids, but with next to no safety research in existence, the committee was unable to set Tolerable Upper Intake Levels for supplemental doses. Until research becomes available, no level of amino acid supplementation can be assumed to be safe for all people. Growth or altered metabolism makes the following groups of people especially likely to suffer harm from amino acid supplements:

- All women of childbearing age
- Pregnant or lactating women
- Infants, children, and adolescents
- Elderly people
- People with inborn errors of metabolism that affect their bodies' handling of amino acids
- Smokers
- People on low-protein diets
- People with chronic or acute mental or physical illnesses who take amino acids without medical supervision

A known side effect of these products is digestive disturbances: amino acids in concentrated supplements cause excess water to flow into the digestive tract, causing diarrhea. Anyone considering taking amino acid supplements should be cautious not to exceed levels normally found in foods.[14]

PROTEIN RECOMMENDATIONS AND INTAKES

The committee that established the RDA states that a generous daily protein allowance for a healthy adult is 0.8 gram per kilogram (2.2 pounds) of healthy body weight. (■) The RDA covers the needs for replacing worn-out tissue, so it increases for larger people; it also covers the needs for building new tissue during growth, so is slightly higher for infants, children, and pregnant and lactating women.

In setting the RDA, the committee assumes that the protein eaten will be of good quality, that it will be consumed together with adequate energy from carbohydrate and fat, and that other nutrients in the diet will be adequate. The committee also assumes that the RDA will be applied only to healthy individuals with no unusual alteration of protein metabolism.

Most people assume that Americans eat too much protein. Research demonstrates that a median protein intake for U.S. adult males is about 16 percent of total kcalories, an amount that falls directly within the DRI suggested range of between 10 and 35 percent of kcalories.[15] Women, children, and some elderly people may typically take in less protein—13 to 15 percent. A small percentage of adolescent girls and elderly women consume insufficient protein or barely enough to meet their needs.[16]

■ RDA for protein (adult) = 0.8 g/kg
- 10 to 35 percent of energy intake

To calculate your protein need:
1. Find your body weight in pounds (or kilograms).
2. Convert pounds to kilograms, if necessary (pounds divided by 2.2).
3. Multiply kilograms by 0.8 to find total grams of protein recommended.

For example:
1. Weight = 150 lb
2. 150 lb ÷ 2.2 lb/kg = 68 kg
3. 68 kg ÷ 0.8 g/kg = 54 g

Review Notes

- Normal, healthy people do not need amino acid or protein supplements.
- Optimally, an adult's diet will be adequate in energy from carbohydrate and fat and will deliver at least 0.8 grams of protein per kilogram of healthy body weight each day.

Protein in Foods

In the United States and Canada, where nutritious foods are abundant, most people easily obtain enough protein to receive all the amino acids that they need. In countries where food is scarce and the people eat only marginal amounts of protein-rich foods, however, the *quality* of the protein becomes crucial.

PROTEIN QUALITY

The protein quality of the diet determines, in large part, how well children grow and how well adults maintain their health. Put simply, **high-quality proteins** provide enough of all the essential amino acids needed to support the body's work, and low-quality proteins don't. Two factors influence protein quality—the protein's digestibility and its amino acid composition.

Digestibility As explained earlier, proteins must be digested before they can provide amino acids. **Protein digestibility** depends on such factors as the protein's source and the other foods eaten with it. The digestibility of most animal proteins is high (90 to 99 percent); plant proteins are less digestible (70 to 90 percent for most, but more than 90 percent for soy).

Amino Acid Composition To make proteins, cells must have all the needed amino acids available simultaneously. The liver can produce any nonessential amino acid that may be in short supply so that the cells can continue linking amino acids into protein strands. If an essential amino acid is missing, however, a cell must dismantle its own proteins to obtain it. Therefore, to prevent protein breakdown, dietary protein must supply at least the nine essential amino acids plus enough nitrogen-containing amino groups and energy for the synthesis of the others. If the diet supplies too little of any essential amino acid, protein synthesis will be limited. The body makes whole proteins only; if one amino acid is missing, the others cannot form a "partial" protein. An essential amino acid that is available in the shortest supply relative to the amount needed to support protein synthesis is called a **limiting amino acid**.

high-quality proteins: dietary proteins containing all the essential amino acids in relatively the same amounts that human beings require. They may also contain nonessential amino acids.

protein digestibility: a measure of the amount of amino acids absorbed from a given protein intake.

limiting amino acid: an essential amino acid that is present in dietary protein in the shortest supply relative to the amount needed for protein synthesis in the body.

complementary proteins: two or more proteins whose amino acid assortments complement each other in such a way that the essential amino acids missing from one are supplied by the other.

High-Quality Proteins A high-quality protein contains all the essential amino acids in amounts adequate for human use; it may or may not contain all the others. Generally, proteins derived from animal foods (meat, seafood, poultry, cheese, eggs, and milk and milk products) are high quality, although gelatin is an exception. Proteins derived from plant foods (legumes, grains, nuts, seeds, and vegetables) tend to be limiting in one or more essential amino acids. Some plant proteins are notoriously low quality—for example, corn protein. Others are high quality—for example, soy protein. As discussed in Nutrition in Practice 5, the educated vegetarian can design a diet that is adequate in protein by choosing a variety of legumes, whole grains, nuts, and vegetables. Table 5-4 lists the protein contents of foods based on the food groups of the USDA Food Patterns in Chapter 1. Fruits are not included in Table 5-4 because they contribute only small amounts of protein.

© Polara Studios, Inc.

Vegetarians obtain their protein from whole grains, legumes, nuts, vegetables, and, in some cases, eggs and milk products.

Complementary Proteins If the body does not receive all the essential amino acids it needs, the supply of essential amino acids will dwindle until body organs are compromised. Obtaining enough essential amino acids presents no problem to people who regularly eat high-quality proteins, such as those of meat, seafood, poultry, cheese, eggs, milk, and many soybean products. The proteins of these foods contain ample amounts of all the essential amino acids. An equally sound choice is to eat two different protein foods from plants so that each supplies the amino acids missing in the other. In this strategy, the two protein-rich foods are combined to yield **complementary proteins** (see Figure 5-7)—proteins containing all the essential amino acids in amounts sufficient to support health. The two proteins need not even be eaten together, as long as the day's meals supply them both, and the diet provides enough energy and total protein from a variety of sources.

TABLE 5-4 Protein-Containing Foods

Milk and Milk Products

Each of the following provides about 8 g of protein:
- 1 c milk, buttermilk, or yogurt (choose low-fat or fat-free)
- 1 oz regular cheese (for example, cheddar or Swiss; choose low-fat)
- ¼ c cottage cheese (choose low-fat or fat-free)

Protein Foods

Each of the following provides about 7 g of protein:
- 1 oz meat, poultry, or fish (choose lean meats to limit saturated fat intake)
- ½ c legumes (navy beans, pinto beans, black beans, lentils, soybeans, and other dried beans and peas)
- 1 egg
- ½ c tofu (soybean curd)
- 2 tbs peanut butter
- 1 to 2 oz nuts or seeds

Grains

Each of the following provides about 3 g of protein:
- 1 slice of bread
- ½ c cooked rice, pasta, cereals, or other grain foods

Vegetables

Each of the following provides about 2 g of protein:
- ½ c cooked vegetables
- 1 c raw vegetables

PROTEIN SPARING

Dietary protein—no matter how high the quality—will not be used efficiently and will not support growth when energy from carbohydrate and fat is lacking. The body assigns top priority to meeting its energy need and, if necessary, will break down protein to meet this need. After stripping off and excreting the nitrogen from the amino acids, the body will use the remaining carbon skeletons in much the same way it uses those from glucose or fat. A major reason why people must have ample carbohydrate and fat in the diet is to prevent this wasting of protein. (■)

FIGURE 5-7 Complementary Proteins

	Ile	Lys	Met	Trp
Legumes	✓	✓		
Grains			✓	✓
Together	✓	✓	✓	✓

In general, legumes provide plenty of isoleucine (Ile) and lysine (Lys) but fall short in methionine (Met) and tryptophan (Trp). Grains have the opposite strengths and weaknesses, making them a perfect match for legumes.

PROTEIN ON FOOD LABELS

All food labels must state the *quantity* of protein in grams. The "percent Daily Value" for protein is not mandatory on all labels, but it is required whenever a food makes a protein claim or is intended for consumption by children younger than age four.* (■) Whenever the Daily Value percentage is declared, researchers must determine the *quality* of the protein. Thus, when a % Daily Value is stated for protein, it reflects both quantity and quality.

■ Reminder: Carbohydrate and fat allow amino acids to be used to build body proteins. This is known as the *protein-sparing effect* of carbohydrate and fat.

■ Daily Value:
- 50 g protein (based on 10 percent of 2000 kcal)

*For labeling purposes, the Daily Values for protein are as follows: for infants, 14 grams; for children younger than age four, 16 grams; for older children and adults, 50 grams; for pregnant women, 60 grams; and for lactating women, 65 grams.

Review Notes

- A diet inadequate in any of the essential amino acids limits protein synthesis.
- The best guarantee of amino acid adequacy is to eat foods containing high-quality proteins or combinations of foods containing complementary proteins so that each can supply the amino acids missing in the other.
- Vegetarians who consume no foods of animal origin can meet their protein needs by eating a variety of whole grains, legumes, seeds, nuts, and vegetables.
- Ample carbohydrate and fat in the diet allow amino acids to be used to build body proteins.
- All food labels must state the quantity of protein in grams.

Self Check

1. Proteins are chemically different from carbohydrates and fats because they also contain:
a. iron.
b. sodium.
c. nitrogen.
d. phosphorus.

2. The basic building blocks for protein are:
a. side groups.
b. amino acids.
c. glucose units.
d. saturated bonds.

3. Enzymes are proteins that, among other things:
a. defend the body against disease.
b. regulate fluid and electrolyte balance.
c. facilitate chemical reactions by changing themselves.
d. help assemble disaccharides into starch, cellulose, or glycogen.

4. Functions of proteins in the body include:
a. supplying omega-3 fatty acids for growth, lowering serum cholesterol, and helping with weight control.
b. supplying fiber to aid digestion, digesting cellulose, and providing the main fuel source for muscles.
c. protecting organs against shock, helping the body use carbohydrate efficiently, and providing triglycerides.
d. serving as structural components, supplying hormones to regulate body processes, and maintaining fluid and electrolyte balance.

5. The swelling of body tissue caused by the leakage of fluid from the blood vessels into the interstitial spaces is called:
a. edema.
b. anemia.
c. acidosis.
d. sickle-cell anemia.

6. Major proteins in the blood that protect against bacteria and other disease agents are called:
a. acids.
b. buffers.
c. antigens.
d. antibodies.

7. Marasmus can be distinguished from kwashiorkor because in marasmus:
a. only adults are victims.
b. the cause is usually an infection.
c. severe wasting of body fat and muscle are evident.
d. the limbs, face, and belly swell with edema.

8. The RDA for protein for a healthy adult is ___ gram(s) per kilogram of appropriate body weight for height.
a. 0.5
b. 0.8
c. 1.1
d. 1.4

9. Generally speaking, from which of the following foods are complete proteins derived?
a. Milk, gelatin, and soy
b. Rice, potatoes, and eggs
c. Meats, fish, and poultry
d. Vegetables, grains, and fruits

10. An incomplete protein lacks one or more:
a. hydrogen bonds.
b. essential fatty acids.
c. saturated fatty acids.
d. essential amino acids.

Answers to these questions can be found in Appendix H. For more chapter review: Access an interactive eBook, chapter-specific interactive learning tools, including flashcards, quizzes, videos, and more in your Nutrition CourseMate, accessed through CengageBrain.com.

Clinical Applications

1. Considering the health effects of too little dietary protein, what suggestions would you have for a teenage girl who reports the following information about her food intake:
   - She never eats any meat or other animal-derived foods because she is a vegan. On a typical day, she consumes toast and juice for breakfast; chips, a soft drink, and a piece of fruit for lunch; and a small serving of plain pasta with tomato sauce or steamed vegetables for dinner, along with a glass of water or tea.
   - She takes amino acid supplements because a friend told her that the only way to get amino acids if she doesn't eat meat is to take them as supplements.
2. Considering the health effects of excess dietary protein, what advice would you have for a college athlete who tells you he wants to bulk up his muscles and reports the following information about his food intake:
   - He eats large servings of meat (usually red meat) at least twice a day. He drinks whole milk two or three times a day and eats eggs and bacon for breakfast almost every day.
   - He avoids breads, cereals, and pasta in order to save room for protein-rich foods such as meat, milk, and eggs.
   - He eats a piece of fruit once in a while but seldom eats vegetables because they are too time-consuming to prepare.

Nutrition on the Net

For further study of the topics of this chapter, access these websites.

- To learn more about protein in foods, visit the Academy of Nutrition and Dietetics (AND) site: www.eatright.org
- For more on vegetarian diets, explore the resources indexed by Medline Plus: www.nlm.nih.gov/medlineplus/vegetariandiet.html
- Search among thousands of current scientific and medical abstracts for any topic related to protein at: www.ncbi.nlm.nih.gov/PubMed/
- Learn more about protein-energy malnutrition and world hunger from the World Health Organization's Nutrition Programme: www.who.int

Notes

1. Standing Committee on the Scientific Evaluation of Dietary Reference Intakes, Food and Nutrition Board, Institute of Medicine, *Dietary Reference Intakes for Energy, Carbohydrate, Fiber, Fat, Fatty Acids, Cholesterol, Protein, and Amino Acids* (Washington, DC: National Academies Press, 2005), p. 605.
2. T. E. Forrester and coauthors, Prenatal factors contribute to the emergence of kwashiorkor or marasmus in severe undernutrition: Evidence for the predictive adaptation model, *PLoS ONE* 7 (2012): e35907; World Health Organization/UNICEF, *WHO Child Growth Standards and the Identification of Severe Acute Malnutrition in Infants and Children* (Geneva: WHO Press, 2009), available at www.who.int/nutrition/publications/severemalnutrition/9789241598163_eng.pdf.
3. World Health Organization, 10 Facts on Child Health, October 2011, www.who.int/features/factfiles/child_health2/en/index.html.
4. P. Svedberg, How many people are malnourished, *Annual Review of Nutrition* 31 (2011): 263–283.
5. D. R. Brewster, Inpatient management of severe malnutrition: Time for a change in protocol and practice, *Annals of Tropical Paediatrics* 31 (2011): 97–107.
6. H. Frank and coauthors, Effect of short term high-protein compared with normal-protein diets on renal hemodynamics and associated variables in healthy young men, *American Journal of Clinical Nutrition* 90 (2009): 1509–1516; Standing Committee on the Scientific Evaluation of Dietary Reference Intakes, 2005, p. 694.
7. A. Pan and coauthors, Red meat consumption and mortality: Results from 2 prospective cohort studies, *Archives of Internal Medicine* 172 (2012): 555–563; P. M. Clifton, Protein and coronary heart disease: The role of different protein sources, *Current Atherosclerosis Reports* 13 (2011): 493–498; A. M. Bernstein and coauthors, Major dietary protein sources and risk of coronary heart disease in women, *Circulation* 122 (2010): 876–883.
8. Pan and coauthors, 2012; A. M. Fretts and coauthors, Associations of processed meat and unprocessed red meat intake with incident diabetes: The Strong Heart Family Study, *American Journal of Clinical Nutrition* 95 (2012): 752–758; Y. Wang and M. A. Beydoun, Meat consumption is associated with obesity and central obesity among US adults, *International Journal of Obesity* 33 (2009): 621–628.
9. Clifton, 2011.
10. H. Frank and coauthors, Effect of short-term high-protein compared with normal protein diets on renal hemodynamics and associated variables in healthy young men, *American Journal of Clinical Nutrition* 90 (2009): 1509–1516.
11. A. R. Josse and coauthors, Body composition and strength changes in women with milk and resistance exercise, *Medicine and Science in Sports and Exercise* 42 (2010): 1122–1130.
12. A. D. Blatt and coauthors, Increasing the protein content of meals and its effect on daily energy intake, *Journal of the American Dietetic Association* 111 (2011): 290–294; S. Pombo-Rodrigues and coauthors, The effects of consuming eggs for lunch on satiety and subsequent food intake, *International Journal of Food Sciences and Nutrition* 62 (2011): 593–599.
13. G. Wu, Amino acids: Metabolism, functions, and nutrition, *Amino Acids*, 37 (2009): 1–17.
14. *Dietary Reference Intakes—The Essential Guide to Nutrient Requirements* (Washington, DC: National Academies Press, 2006), p. 152.
15. V. L. Fulgoni III, Current protein intake in America: Analysis of the National Health and Nutrition Examination survey, 2003–2004, *American Journal of Clinical Nutrition* 87 (2008): 1554S–1557S.
16. Fulgoni III, 2008.

Nutrition in Practice

Vegetarian Diets

Eating patterns all along the continuum of dietary choices—from one end, where people eat no foods of animal origin, to the other end, where they eat generous quantities of meat every day—can support or compromise nutritional health. The quality of the diet depends not on whether it consists of all plant foods or centers on meat but on whether the eater's food choices are based on sound nutrition principles: adequacy of nutrient intakes, balance and variety of foods chosen, appropriate energy intake, and moderation in intakes of substances such as saturated fat, *trans* fat, added sugars, sodium, and alcohol that are harmful when consumed in excess. As mentioned in Chapter 5, however, because vegetarian diets exclude at least some animal-derived foods, they are usually lower in saturated fat and cholesterol than many meat-based diets.

The health benefits of a primarily vegetarian diet seem to have encouraged many people to eat more plant-based meals. U.S. sales of vegetarian foods have increased during the last decade, as has the number of people who are reducing their consumption of meat. The popular press sometimes refers to individuals who eat small amounts of meat, seafood, or poultry from time to time as "flexitarians."

People who choose to exclude meat and other animal-derived foods from their diets today do so for many of the same reasons the Greek philosopher Pythagoras cited in the sixth century B.C.: physical health, ecological responsibility, and philosophical concerns. They might also cite world hunger issues, concern for animal welfare, or religious beliefs as motivating factors. Whatever their reasons, vegetarians and health professionals who work with them should be aware of the nutrition and health implications of vegetarian diets.

Vegetarians generally are categorized not by their motivations but by the foods they choose to exclude (see the accompanying glossary). Because **vegetarian** diets vary in both the types and the amounts of animal-derived foods they include, these differences must be considered when evaluating the health status of vegetarians.

Are vegetarian diets nutritionally sound?

The Academy of Nutrition and Dietetics takes the position that well-planned vegetarian diets offer nutrition and health benefits to adults in general.[1] Research suggests that meat-eating adults who switch to vegetarian diets reduce their risks of heart disease, hypertension, diabetes, some types of cancer, and obesity.[2]

What should be my main concerns when planning a nutritionally sound vegetarian diet?

A vegetarian diet planner faces the same task as other diet planners—obtaining a variety of foods that provide all the needed nutrients within an energy allowance that maintains a healthy body weight. The challenge is to do so using at least one less food group. Because all vegetarians omit meat and some omit other animal-derived foods, protein, the nutrient that meat is famous for, merits some discussion here.

Isn't protein a problem in vegetarian diets?

No, protein is not the problem it was once thought to be in vegetarian diets. People who include animal-derived

Glossary of Terms Used to Describe Vegetarian Diets

fruitarian: includes only raw or dried fruits, seeds, and nuts in the diet.

lacto-ovo vegetarian: includes dairy products, eggs, vegetables, grains, legumes, fruits, and nuts; excludes meat, poultry, and seafood.

lacto-vegetarian: includes dairy products, vegetables, grains, legumes, fruits, and nuts; excludes meat, poultry, seafood, and eggs.

macrobiotic diet: a vegan diet composed mostly of whole grains, beans, and certain vegetables; taken to extremes, macrobiotic diets can compromise nutrient status.

ovo-vegetarian: includes eggs, vegetables, grains, legumes, fruits, and nuts; excludes meat, poultry, seafood, and milk products.

partial vegetarian: a term sometimes used to mean an eating style that includes seafood, poultry, eggs, dairy products, vegetables, grains, legumes, fruits, and nuts; excludes or strictly limits certain meats, such as red meat. Also called *semi-vegetarian.*

vegan: includes only food from plant sources: vegetables, grains, legumes, fruits, seeds, and nuts; also called *strict vegetarian.*

vegetarian: includes plant-based foods and eliminates some or all animal-derived foods.

foods such as milk and eggs in their diets need not worry at all about protein deficiency. Even for those who choose a **vegan** eating pattern—which includes only plant-based foods—protein intakes are usually satisfactory as long as energy intakes are adequate and protein sources are varied.[3] A mixture of proteins from whole grains, legumes, seeds, nuts, and vegetables can provide adequate amounts of high-quality protein. An advantage of many vegetarian sources of protein is that they are generally lower in saturated fat than meat and are often high in fiber and richer in some vitamins and minerals.

Some vegetarians may use meat replacements made of textured vegetable protein (soy protein). These foods are formulated to look and taste like meat, fish, or poultry. Many of these products are fortified to provide the vitamins and minerals found in animal sources of protein. Some products, however, may fall short of providing the nutrients of meat, and they may be high in salt, sugar, or other additives. Labels list all of the ingredients in such foods. A wise vegetarian learns to use a variety of whole, unrefined foods often and commercially prepared foods less frequently. Vegetarians may also use soybeans in other forms, such as plain tofu (soybean curd), edamame (cooked green soybeans), or soy flour, to bolster protein intake without consuming unwanted salt, sugar, or other additives.

What sorts of food energy intakes do vegetarian diets provide?

Researchers find that vegetarians as a group are closer to a healthy body weight than nonvegetarians.[4] Because obesity impairs health in a number of ways, vegetarians therefore have a health advantage. Vegetarian diets tend to be high in starch- and fiber-rich carbohydrates and low in saturated fat, characteristics that are consistent with current dietary recommendations aimed at reducing the incidence of obesity and other chronic diseases in this country.

Not all vegetarians fit the average pattern, though. Obesity can be a concern for vegetarians who include whole milk, eggs, and cheese in their diets. They can easily consume excess saturated fat and food energy and so must be careful to select fat-free and low-fat milk and milk products and to avoid relying too heavily on these foods in general.

In contrast, people who exclude all animal-derived foods (vegans) may have trouble obtaining *enough* food energy. This is especially true for children. Vegan diets can fail to provide food energy sufficient to support the growth of a child within a bulk of food small enough for the child to eat. Frequent meals of fortified breads, cereals, or pastas with legumes, nuts, nut butters, and sources of unsaturated fats can help to meet protein and energy needs in a smaller volume at each sitting.[5] The MyPlate resources, introduced in Chapter 1, include tips for planning vegetarian diets using an adaptation of the USDA Food Patterns.

How can vegetarians and health professionals who plan vegetarian meals best utilize the USDA Food Patterns?

The recommended daily amounts from the food groups are the same for both vegetarians and nonvegetarians (see Table 1-7 on p. 20). The USDA Food Patterns are flexible enough that a variety of people can use them: people who have adopted various vegetarian diets, those who want to make the transition to a vegetarian diet, and those who simply want to include more plant-based meals in their diets. Selections from within the food groups may differ, of course. For example, the milk group features fortified soy milks for those who do not use milk, cheese, or yogurt. When selecting from the vegetable and fruit groups, vegetarians should emphasize particularly good sources of calcium and iron, respectively. Some green leafy vegetables, for example, provide almost five times as much calcium per serving as other vegetables. Similarly, dried fruits deserve special notice in the fruit group because they deliver six times as much iron as other fruits. The protein foods group includes eggs (for those who use them), legumes, soy products, nuts, and seeds. Table NP 5-1 (p. 138) provides recommended *weekly* amounts of protein food subgroups for both vegetarians and vegans.

Most vegetarians easily obtain large quantities of the nutrients that are abundant in plant foods: carbohydrate, fiber, thiamin, folate, vitamin B_6, vitamin C, vitamin A, and vitamin E. Well-planned vegetarian eating patterns help to ensure adequate intakes of the main nutrients vegetarian diets might otherwise lack: protein, iron, zinc, calcium, vitamin B_{12}, vitamin D, and omega-3 fatty acids. Table NP5-2 (p. 139) presents good vegetarian sources of these key nutrients. For example, the use of vegetable oils rich in unsaturated fats provides essential omega-3 fatty acids. To ensure adequate intakes of vitamin B_{12}, vitamin D, and calcium, vegetarians need to select fortified foods or use supplements daily.

Tell me about vitamins and minerals. Does a person eating a vegetarian diet need to take vitamin supplements?

That depends on the kind of vegetarian diet. The diet of **lacto-ovo vegetarians** can be adequate in all vitamins, but for vegans, several vitamins may be a problem. One such vitamin is B_{12}. Because vitamin B_{12} occurs only in animal-derived foods, regular use of vitamin B_{12}–fortified foods, such as fortified soy and rice beverages, some breakfast cereals, and meat replacements, or supplements, is necessary to

TABLE NP5-1 USDA Food Patterns: Recommended Weekly Amounts of Protein Foods for Vegetarians and Vegans

The daily amounts for protein foods are the same for both vegetarians and nonvegetarians, but the subgroups and weekly amounts for protein foods differ. The recommended daily amounts from each of the other food groups—fruits, vegetables, grains, and milk products—are the same (see Table 1-7).

Protein Foods	1600 kcal	1800 kcal	2000 kcal	2200 kcal	2400 kcal	2600 kcal	2800 kcal	3000 kcal
Daily Amounts	5 oz	5 oz	5½ oz	6 oz	6½ oz	6½ oz	7 oz	7 oz
Vegetarian Subgroups								
Eggs	4 oz	4 oz	4 oz	4 oz	5 oz	5 oz	5 oz	6 oz
Legumes	9 oz	9 oz	10 oz	10 oz	11 oz	11 oz	12 oz	12 oz
Soy products	11 oz	11 oz	12 oz	13 oz	14 oz	14 oz	15 oz	15 oz
Nuts and seeds	12 oz	12 oz	13 oz	15 oz	16 oz	16 oz	17 oz	17 oz
Vegan Subgroups								
Legumes	12 oz	12 oz	13 oz	15 oz	16 oz	16 oz	17 oz	17 oz
Soy products	9 oz	9 oz	10 oz	11 oz	11 oz	11 oz	12 oz	12 oz
Nuts and seeds	14 oz	14 oz	15 oz	17 oz	18 oz	18 oz	20 oz	20 oz

Note: Total recommended amounts for legumes include the sum of both the vegetables and protein foods. An ounce-equivalent of legumes in the protein foods group is ¼ cup. For a 2000-kcal vegan diet, that's 2½ cups legumes for protein foods plus 1½ cups legumes for vegetables (see Table 1-9), or about 4 cups legumes weekly.

prevent deficiency.[6] Women who have adhered to all-plant diets for many years are especially likely to have low vitamin B_{12} stores. Pregnant vegan women, whose needs for vitamin B_{12} are especially high, find it virtually impossible to maintain adequate vitamin B_{12} status without taking supplements or including a reliable food source of the nutrient.

People who stop eating animal-derived foods containing vitamin B_{12} may take several years to develop deficiency symptoms because the body recycles much of its vitamin B_{12}, reabsorbing it over and over again. Even when the body fails to absorb vitamin B_{12}, deficiency may take up to three years to develop because the body conserves its supply. For pregnant and lactating women, obtaining vitamin B_{12} is critical to prevent serious deficiency-related disorders in infants who do not receive sufficient vitamin B_{12}.[7] All vegan mothers must be sure to take the appropriate supplements or to use vitamin B_{12}–fortified products.

Do vegans, who do not drink vitamin D-fortified cow's milk, get enough vitamin D?

Overall, vegetarians are similar to nonvegetarians in their vitamin D status; factors such as taking supplements, skin color, and sun exposure have a greater influence on vitamin D than diet.[8] People who do not use vitamin D-fortified foods and do not receive enough exposure to sunlight to synthesize adequate vitamin D may need supplements to fend off bone loss. Of particular concern are infants, children, and older adults in northern climates during winter months.

So, on a vegan diet, vitamin B_{12} and vitamin D can be problems if a person is not careful. What about minerals?

For *all* vegetarians, not just the vegan, two minerals may be of concern—iron and zinc. The iron in plant foods such as legumes, dark green leafy vegetables, iron-fortified cereals, and whole-grain breads and cereals is not as absorbable as that in meat. For this reason, the iron recommendation for adult vegetarian men, premenopausal women, and adolescent girls is almost double the recommendation for meat eaters of the same gender and age.[9]

Fortunately, the body seems to adapt to a vegetarian diet by absorbing iron more efficiently. Furthermore, iron absorption is enhanced by vitamin C, and vegetarians typically eat many vitamin C–rich fruits and vegetables.

TABLE NP5-2 Vegetarian Sources of Key Nutrients

Nutrients	Food Groups					
	Grains	Vegetables	Fruits	Legumes and Other Protein-Rich Foods	Milk or Soy Milk	Oils
Protein	Whole grains[a]			Eggs (for ovo-vegetarians)	Milk, cheese, yogurt (for lacto-vegetarians) Soy milk, soy yogurt, soy cheeses	
Iron	Fortified cereals, enriched and whole grains	Dark green leafy vegetables (spinach, turnip greens)	Dried fruits (apricots, prunes, raisins)	Legumes (black-eyed peas, kidney beans, lentils), soy products		
Zinc	Fortified cereals, whole grains			Legumes (garbanzo beans, kidney beans, navy beans), nuts, seeds (pumpkin seeds)	Milk, cheese, yogurt (for lacto-vegetarians) Soy milk, soy yogurt, soy cheeses	
Calcium	Fortified cereals	Dark green leafy vegetables (bokchoy, broccoli, collard greens, kale, mustard greens, turnip greens, watercress)	Fortified juices, figs	Fortified soy products, nuts (almonds), seeds (sesame seeds)	Milk, cheese, yogurt (for lacto-vegetarians) Fortified soy milk, fortified soy yogurt, fortified soy cheese	
Vitamin B_{12}	Fortified cereals			Eggs (for ovo-vegetarians) Fortified soy products	Milk, cheese, yogurt (for lacto-vegetarians) Fortified soy milk, fortified soy yogurt, fortified soy cheese	
Vitamin D	Fortified cereals				Milk, cheese, yogurt (for lacto-vegetarians) Fortified soy milk, fortified soy yogurt, fortified soy cheese	
Omega-3 Fatty Acids		Marine algae and its oils		Flaxseed, walnuts, soybeans Fortified margarine Fortified eggs (for ovo-vegetarians)		Flaxseed oil, walnut oil, soybean oil

[a]As Chapter 5 explains, many plant proteins do not contain all the essential amino acids in the amounts and proportions needed by human beings. To improve protein quality, vegetarians can eat grains and legumes together, for example, although it is not necessary if protein intake is varied and energy intake is sufficient.

Consequently, vegetarians suffer no more iron deficiency than other people do.

Zinc is similar to iron in that meat is its richest food source, and zinc from plant sources is not as well absorbed. In addition, phytates, fiber, and calcium, which are common in vegetarian diets, interfere with zinc absorption. The zinc needs of vegetarians and the effects of mineral binders are subjects of intensive study at the present time. While research continues, vegetarians are advised to eat varied diets that include legumes such as navy beans and kidney beans, zinc-enriched cereals, and whole-grain breads well leavened with yeast, which improves the availability of their minerals. For those who include seafood in their diets, oysters, crabmeat, and shrimp are rich in zinc.

What about calcium for the vegan?

Good thinking. Yes, calcium is of concern. The milk-drinking vegetarian is protected from deficiency, but the vegan must find other sources of calcium. Some good calcium sources are regular and ample servings of dark green leafy vegetables such as kale and collard; legumes; calcium-fortified foods such as breakfast cereals, soy milk, and orange juice; some nuts such as almonds; and certain seeds such as sesame seeds. The choices should be varied because binders in some of these foods may hinder calcium absorption. The vegan is urged to use calcium-fortified soy milk in ample quantities regularly. This is especially important for children. Infant formula based on soy is fortified with calcium and can easily be used in food preparation, even for adults.

Do vegetarian diets provide adequate amounts of the essential fatty acids?

Vegetarian diets typically provide enough of the essential fatty acids linoleic acid and linolenic acid, but they lack a dietary source of EPA and DHA.[10] Fatty fish and DHA-fortified eggs and other products can provide EPA and DHA, but fortified sources that ultimately derive from fish are unacceptable for vegans. Alternatively, certain marine algae and their oils provide DHA, and vegans can select foods fortified with such oils, which are listed among the ingredients on a food's label. A vegetarian's daily diet should include small amounts of flaxseed, walnuts, and their oils, as well as soybeans and canola oil to provide essential fatty acids.

Are there any other health advantages to the vegetarian diet?

Yes. Vegetarian protein foods are often higher in fiber, richer in certain vitamins and minerals, and lower in fat—especially saturated fat—than meats. Vegetarians can enjoy a nutritious diet low in saturated fat provided that they limit foods such as butter, cream cheese, and sour cream. If vegetarians follow the guidelines presented here and plan carefully, they can support their health as well as, or perhaps better than, nonvegetarians.

Abundant evidence supports the idea that vegetarians may actually be healthier than meat eaters. Informed vegetarians are not only more likely to be at the desired weights for their heights, but they are also more likely to have lower blood cholesterol levels, lower rates of certain kinds of cancer, better digestive function, and more. Even among people who are health conscious, generally vegetarians experience fewer deaths from cardiovascular disease than meat eaters do. Because many vegetarians also abstain from smoking and the consumption of alcohol, dietary practices alone probably do not account for all the aspects of improved health. Clearly, however, they contribute significantly to it.

Notes

1. Position of the American Dietetic Association, Vegetarian diets, *Journal of the American Dietetic Association* 109 (2009): 1266–1282.
2. B. J. Pettersen and coauthors, Vegetarian diets and blood pressure among white subjects: Results from the Adventist Health Study-2 (AHS-2), *Public Health Nutrition* 15 (2012): 1909–1916; N. S. Rizzo and coauthors, Vegetarian dietary patterns are associated with a lower risk of metabolic syndrome, *Diabetes Care* 34 (2011): 1225–1227; S. Tonstad and coauthors, Vegetarian diets and incidence of diabetes in the Adventist Health Study-2, *Nutrition, Metabolism, and Cardiovascular Diseases,* 2011, epub ahead of print; B. Farmer and coauthors, A vegetarian dietary pattern as nutrient-dense approach to weight management: An analysis of the national Health and Nutrition Examination Survey 1999–2004, *Journal of the American Dietetic Association* 111 (2011): 819–827; W. J. Craig, Nutrition concerns and health effects of vegetarian diets, *Nutrition in Clinical Practice* 25 (2010): 613–620; G. E. Fraser, Vegetarian diets: What do we know of their effects on common chronic diseases? *American Journal of Clinical Nutrition* 89 (2009): 1607S–1612S; N. D. Barnard and coauthors, A low-fat vegan diet and a conventional diabetes diet in the treatment of type 2 diabetes: A randomized, controlled, 74-wk clinical trial, *American Journal of Clinical Nutrition* 89 (2009): 1588S–1596S.
3. Position of the American Dietetic Association, 2009.
4. J. Sabate and M. Wien, Vegetarian diets and childhood obesity prevention, *American Journal of Clinical Nutrition* 91 (2010): 1525S–1529S; S. Tonstad and coauthors, Type of vegetarian diet, body weight, and prevalence of type 2 diabetes, *Diabetes Care* 32 (2009): 791–796; Fraser, 2009; R. Robinson-O'Brien and coauthors, Adolescent and young adult vegetarianism: Better dietary intake and weight outcomes but increased risk of disordered eating behaviors, *Journal of the American Dietetic Association* 109 (2009): 648–655.
5. Position of the American Dietetic Association, 2009.
6. Position of the American Dietetic Association, 2009.
7. M. R. Pepper and M. M. Black, B12 in fetal development, *Seminars in Cell and Developmental Biology* 22 (2011): 619–623.
8. J. Chan, K. Jaceldo-Siegl, and Gary E. Fraser, Serum 25-hydroxyvitamin D status of vegetarians, partial vegetarians, and nonvegetarians: The Adventist Health Study, *American Journal of Clinical Nutrition* 89 (2009): 1686S–1692S.
9. Standing Committee on the Scientific Evaluation of Dietary Reference Intakes, Food and Nutrition Board, Institute of Medicine, *Dietary Reference Intakes for Vitamin A, Vitamin K, Arsenic, Boron, Chromium, Copper, Iodine, Iron, Manganese, Molybdenum, Nickel, Silicon, Vanadium, and Zinc* (Washington, DC: National Academy Press, 2001), pp. 9–45.
10. I. Mangat, Do vegetarians have to eat fish for optimal cardiovascular protection? *American Journal of Clinical Nutrition* 89 (2009): 1597–1601.

Chapter 6 Energy Balance and Body Composition

THE BODY MANAGES ITS ENERGY SUPPLY WITH AMAZING PRECISION.

Consider that most people maintain their weight within about a 10- to 20-pound range throughout their lives. How do they do this? How does the body manage excess energy? And how does it manage to do without food for prolonged periods—as when someone is starving or fasting? The answers to these questions lie in our understanding of how the body adapts to energy imbalances, the focus of the first part of this chapter. The chapter then describes energy balance, body composition, and the health risks associated with too much or too little body fat. The next chapter offers strategies toward solving the problems of too much or too little body fat.

Energy Imbalance

The average person takes in and expends close to a million kcalories a year while maintaining a stable weight for years on end. In other words, the body's energy is in balance. Many people, however, eat too much or exercise too little and get fat; others eat too little or exercise too much and get thin. This section examines the two extremes of energy imbalance—feasting and fasting.

FEASTING

When people consume more energy than they expend, much of the excess is stored as body fat. Fat can be made from an excess of any energy-yielding nutrient. In addition, excess energy from alcohol is also stored as fat. Alcohol has also been shown to slow down the body's use of fat for fuel, causing more fat to be stored, much of it as abdominal fat tissue.[1] Alcohol therefore is fattening, both through the kcalories it provides and through its effects on fat metabolism. The fat cells of the adipose tissue enlarge as they fill with fat, as Figure 6-1 shows.

Excess Carbohydrate

Surplus carbohydrate (glucose) is first stored as glycogen in the liver and muscles, but the glycogen-storing cells have a limited capacity. Once glycogen stores are filled, most of the additional carbohydrate is burned for energy, displacing the body's use of fat for energy and allowing body fat to accumulate. Thus, excess carbohydrate can contribute to obesity.

Excess Fat

Surplus dietary fat contributes more directly to the body's fat stores. After a meal, fat is routed to the body's adipose tissue, where it is stored until needed for energy. Thus, excess fat from food easily adds to body fat.

Excess Protein

Surplus protein may also contribute to body fat. If not needed to build body protein (as in response to physical activity) or to meet energy needs, amino

FIGURE 6-1 Fat Cell Enlargement

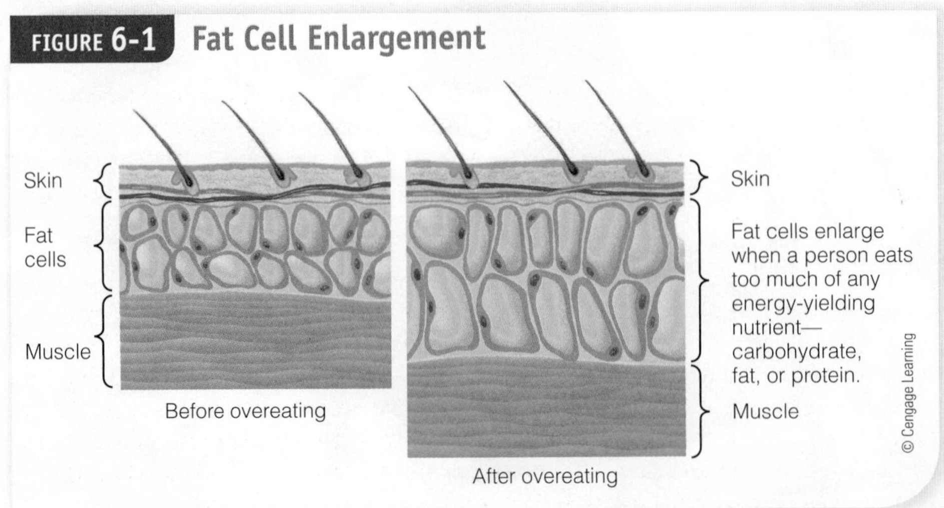

Skin
Fat cells
Muscle
Before overeating
Skin
Fat cells enlarge when a person eats too much of any energy-yielding nutrient—carbohydrate, fat, or protein.
Muscle
After overeating
© Cengage Learning

FIGURE 6-2 Feasting

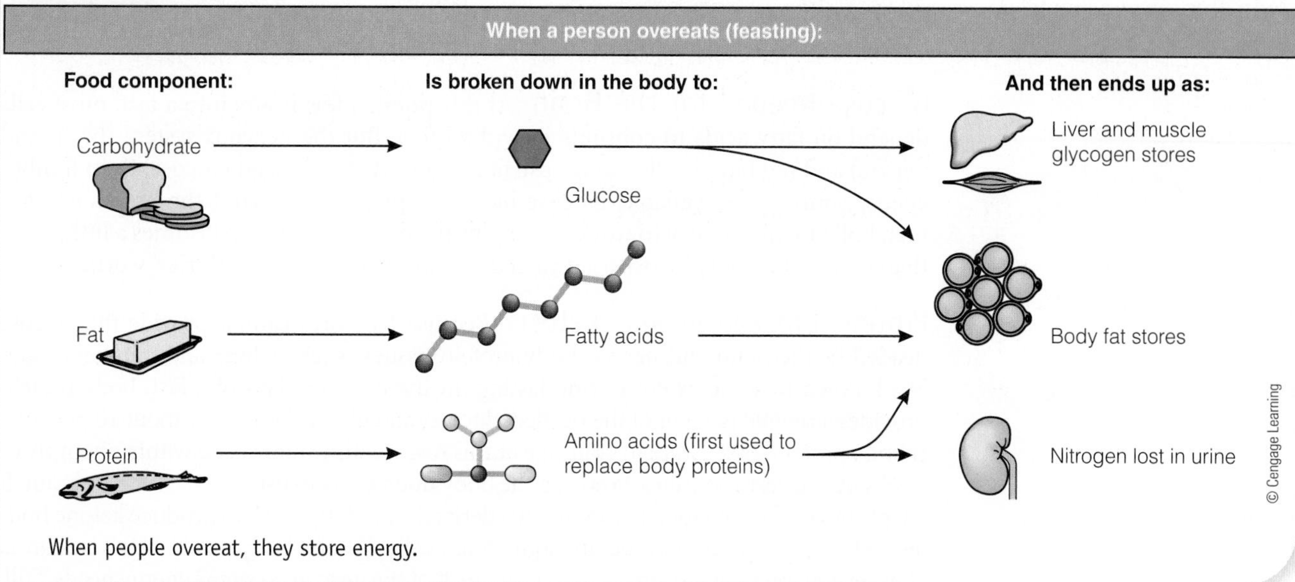

When people overeat, they store energy.

acids will lose their nitrogens and be converted, through intermediates, to triglycerides. These, too, swell the fat cells and add to body weight. Figure 6-2 shows the metabolic events of feasting.

Review Notes

- Too much food, too little physical activity, or both encourage body fat accumulation.
- A net excess of energy is almost all stored in the body as fat in adipose (fat) tissue.
- Alcohol both delivers kcalories and encourages storage of body fat.
- Once glycogen stores are filled, excess carbohydrate is used for energy, displacing the body's use of fat for energy and allowing fat to accumulate.
- Fat from food is particularly easy for the body to store as adipose tissue.
- If not needed to build body protein or to meet energy needs, excess protein can be converted to fat.
- In short, excess energy intake from carbohydrate, fat, protein, and alcohol leads to storage of body fat.

THE ECONOMICS OF FASTING

The body expends energy all the time. Even when a person is asleep and totally relaxed, the cells of many organs are hard at work. In fact, this cellular work, which maintains all life processes, represents about two-thirds of the total energy a sedentary person expends in a day. (The other one-third is the work that a person's muscles do voluntarily during waking hours.)

Energy Deficit The body's top priority is to meet the energy needs for this ongoing cellular activity. Its normal way of doing so is by periodic refueling, that is, by eating several times a day. When food is not available, the body uses fuel reserves from its own tissues. If people voluntarily choose not to eat, we say they are fasting; if they have no choice (as in a famine), we say they are starving. The body, however, makes no distinction between the two—metabolically, fasting and starving are identical. In either case, the body is forced to switch to a wasting metabolism, drawing on its stores of carbohydrate and fat and, within a day or so, on its vital protein tissues as well.

Glycogen Used First As fasting begins, glucose from the liver's glycogen stores and fatty acids from the body's adipose tissue flow into the cells to fuel their work. Within a day, liver glycogen is exhausted, and most of the glucose is used up. Low blood glucose concentrations serve as a signal to promote further fat breakdown.

Glucose Needed for the Brain At this point, a few hours into a fast, most cells depend on fatty acids to continue providing fuel. But the nervous system (brain and nerves) and red blood cells cannot use fatty acids; they still need glucose. Even if other energy sources are available, glucose has to be present to permit the brain's energy-metabolizing machinery to work. Normally, the nervous system consumes a little more than half of the total glucose used each day—about 400 to 600 kcalories' worth.

Protein Breakdown and Ketosis Because fat stores cannot provide the glucose needed by the brain and nerves, body protein tissues (such as liver and muscle) always break down to some extent during fasting. In the first few days of a fast, body protein provides about 90 percent of the needed glucose, and glycerol provides about 10 percent.* If body protein losses were to continue at this rate, death would ensue within about three weeks. As the fast continues, however, the body finds a way to use its fat to fuel the brain. It adapts by condensing together fragments derived from fatty acids to produce ketone bodies, (■) which can serve as fuel for some brain cells. Ketone body production rises until, after several weeks of fasting, it is meeting much of the nervous system's energy needs. Still, many areas of the brain rely exclusively on glucose, and body protein continues to be sacrificed to produce it. (■) Figure 6-3 shows the metabolic events that occur during fasting.

Slowed Metabolism As fasting continues and the nervous system shifts to partial dependence on ketone bodies for energy, the body simultaneously reduces its energy output (metabolic rate) and conserves both fat and lean tissue. Because of the slowed metabolism, energy use falls to a bare minimum.

Hazards of Fasting The body's adaptations to fasting are sufficient to maintain life for a long period. Mental alertness need not be diminished. Even physical energy may remain unimpaired for a surprisingly long time. Still, fasting is not without its hazards. Among the many changes that take place in the body are:

- Wasting of lean tissues.
- Impairment of disease resistance.
- Lowering of body temperature.
- Disturbances of the body's fluid and electrolyte balances.

For the person who wants to lose weight, fasting is not the best way to go. The body's lean tissue continues to be degraded, sometimes amounting to as much as 50 percent of the weight lost over the first week. Over the long term, a diet only moderately restricted in energy promotes primarily *fat* loss and the retention of more lean tissue than a severely restrictive fast.

■ Reminder: *Ketone bodies* are acidic, water-soluble compounds produced by the liver during the breakdown of fat when carbohydrate is not available. Small amounts of ketone bodies are normally produced during energy metabolism, but when their blood concentration rises, they spill into the urine. The combination of a high blood concentration of ketone bodies (*ketonemia*) and ketone bodies in the urine (*ketonuria*) is called *ketosis*.

■ In fasting, muscle and lean tissues give up protein to supply amino acids for conversion to glucose. This glucose, with ketone bodies produced from fat, fuels the brain's activities.

Review Notes

- When fasting, the body makes a number of adaptations: increasing the breakdown of fat to provide energy for most of the cells, using glycerol and amino acids to make glucose for the red blood cells and central nervous system, producing ketones to fuel the brain, and slowing metabolism.
- All of these measures conserve energy and minimize losses.
- Over the long term, a diet moderately restricted in energy promotes primarily fat loss and the retention of lean tissue.

*The small glycerol portion of a triglyceride can yield glucose, but glycerol represents only about 5 percent of the weight of a triglyceride molecule. Thus, fat is an inefficient source of glucose. About 95 percent of a triglyceride (the fatty acids attached to the glycerol) cannot be converted to glucose.

FIGURE 6-3 Fasting

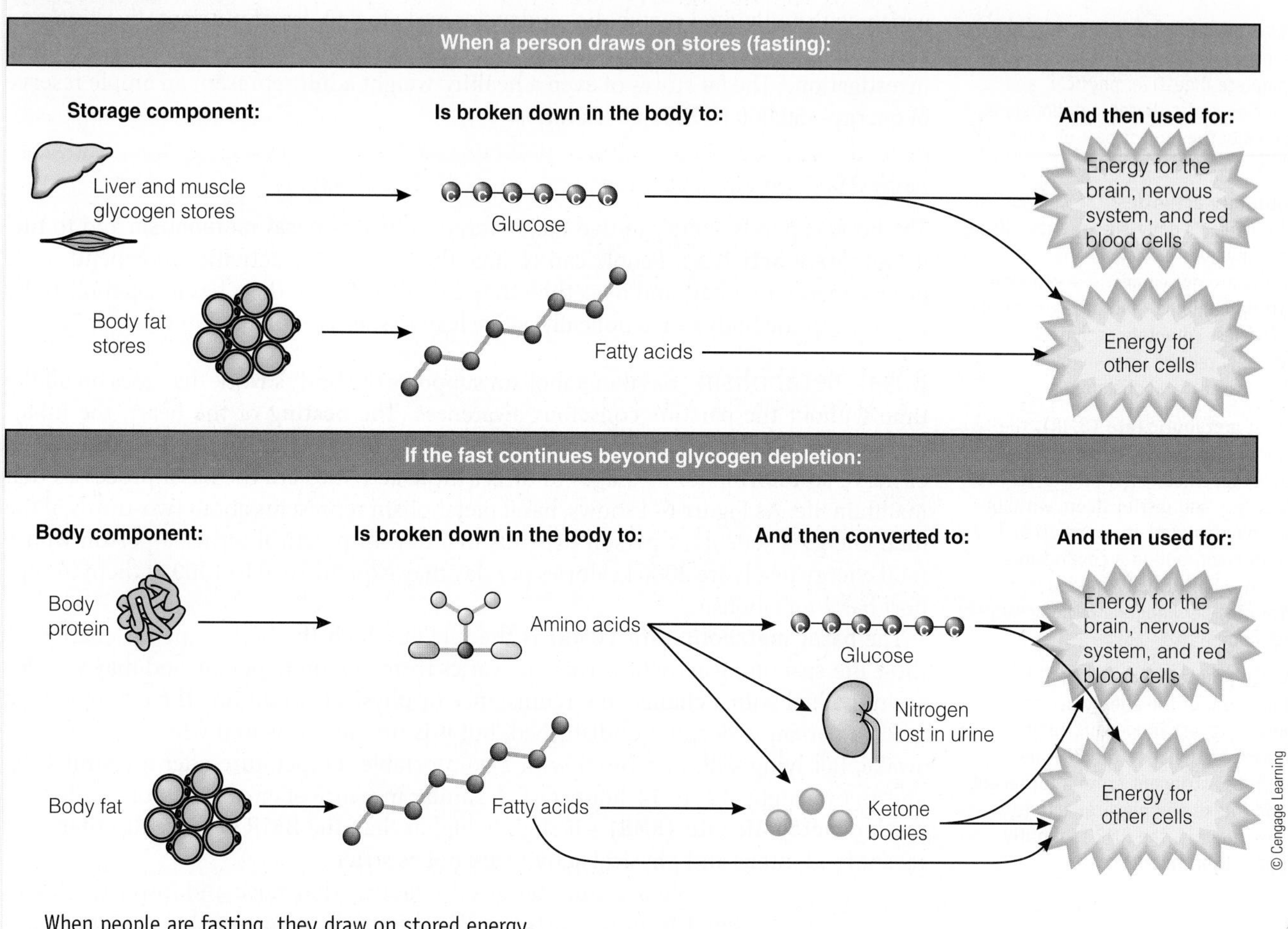

When people are fasting, they draw on stored energy.

Energy Balance

If a person maintains a healthy weight over time, the person is in energy balance. Food energy intake equals energy expenditure: deposits of fat made at one time have been compensated for by withdrawals made at another. In other words, the body uses fat as a savings account for energy. But, unlike money, having more fat is not better; there is an optimum.

A day's energy balance can be stated like this: change in energy stores equals the food energy taken in (kcalories) minus the energy spent on metabolism and physical activities (kcalories). More simply:

Change in energy stores = energy in (kcalories) − energy out (kcalories).

ENERGY IN

The energy in food and beverages is the only contributor to the "energy in" side of the energy balance equation. Before you can decide how much food will supply the energy you or one of your clients needs in a day, you must first become familiar with the amounts of energy in foods and beverages. One way to do so is to look up the kcalories provided by various foods and beverages in the Table of Food Composition (Appendix A). Alternatively, computer programs can readily provide this information.

Food composition data would reveal that an apple provides about 70 kcalories from carbohydrate and a candy bar supplies about 250 kcalories, mostly from fat and carbohydrate. You may have heard that for each 3500 kcalories you eat in excess of

basal metabolism: the energy needed to maintain life when a person is at complete digestive, physical, and emotional rest. Basal metabolism is normally the largest part of a person's daily energy expenditure.

voluntary activities: the component of a person's daily energy expenditure that involves conscious and deliberate muscular work—walking, lifting, climbing, and other physical activities. Voluntary activities normally require less energy in a day than basal metabolism does.

basal metabolic rate (BMR): the rate of energy use for metabolism under specified conditions: after a 12-hour fast and restful sleep, without any physical activity or emotional excitement, and in a comfortable setting. It is usually expressed as kcalories per kilogram of body weight per hour.

resting metabolic rate (RMR): a measure of the energy use of a person at rest in a comfortable setting—similar to the BMR but with less stringent criteria for recent food intake and physical activity. Consequently, the RMR is slightly higher than the BMR.

expenditures, you store 1 pound of body fat—a general rule that has previously been used for mathematical estimations. Keep in mind, however, that this number can vary widely with individual metabolic tendencies and efficiencies of nutrient digestion and absorption. Currently, the dynamics of energy storage is a topic of intense scientific investigation.[2] The fat stores of even a healthy-weight adult represent an ample reserve of energy—50,000 to 200,000 kcalories.

ENERGY OUT

The body expends energy in two major ways: to fuel its **basal metabolism** and to fuel its **voluntary activities**. People can change their voluntary activities to expend more or less energy in a day, and over time they can also change their basal metabolism by building up the body's metabolically active lean tissue, as explained in Chapter 7.

Basal Metabolism

Basal metabolism supports the body's work that goes on all the time without the person's conscious awareness. The beating of the heart, the inhaling and exhaling of air, the maintenance of body temperature, and the transmission of nerve and hormonal messages to direct these activities are the basal processes that maintain life. As Figure 6-4 shows, basal metabolism represents about two-thirds of the total energy a sedentary person expends in a day. In practical terms, a person whose total energy needs are 2000 kcalories per day may expend 1000 to 1300 of them to support basal metabolism.

The **basal metabolic rate (BMR)** is the rate at which the body expends energy for these life-sustaining activities. This rate varies from person to person and may vary for an individual with a change in circumstance or physical condition. The rate is slowest when a person is sleeping undisturbed, but it is usually measured when the person is awake, but lying still, in a room with a comfortable temperature after a restful sleep and an overnight (12- to 14-hour) fast. A similar measure of energy output—called the **resting metabolic rate (RMR)**—is slightly higher than the BMR because its criteria for recent food intake and physical activity are not as strict.

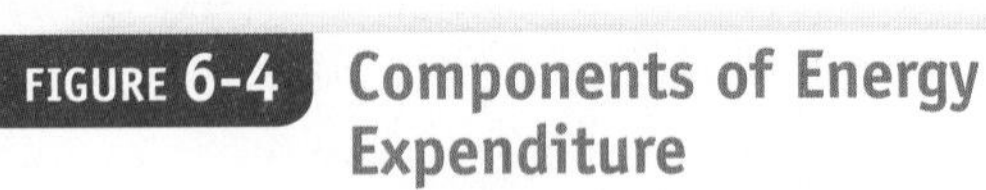

FIGURE 6-4 Components of Energy Expenditure

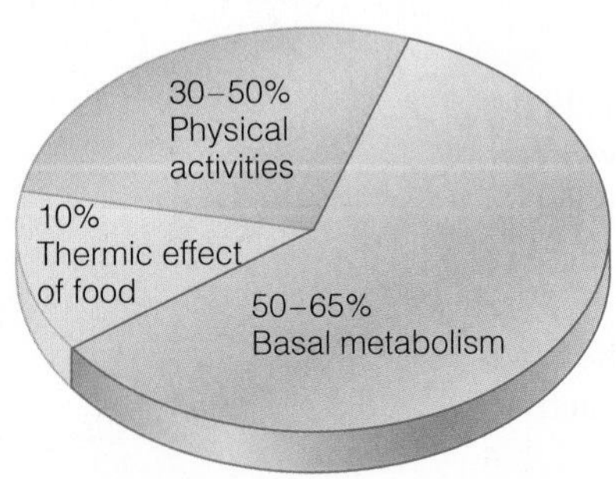

The amount of energy spent in a day differs for each individual, but in general, basal metabolism is the largest component of energy expenditure, and the thermic effect of food is the smallest. The amount spent in voluntary physical activities has the greatest variability, depending on a person's activity patterns. For a sedentary person, physical activities may account for less than half as much energy as basal metabolism, whereas an extremely active person may expend as much on activity as for basal metabolism.

Table 6-1 summarizes the factors that raise and lower the BMR. For the most part, the BMR is highest in people who are growing (children, adolescents, and pregnant women) and in those with considerable lean body mass (physically fit people and males). One way to increase the BMR, then, is to maximize lean body tissue by participating regularly in endurance and strength-building activities. The BMR is also fast in people who are tall and so have a large surface area for their weight, in people with fever or under stress, in people taking certain medications, and in people with highly active thyroid glands. The BMR slows down with a loss of lean body mass and during fasting and malnutrition.

Energy for Physical Activities

The number of kcalories spent on voluntary activities depends on three factors: muscle mass, body weight, and activity. The larger the muscle mass required for the activity and the heavier the weight of the body part being moved, the more kcalories are spent. The activity's duration, frequency, and intensity also influence energy costs: the longer, the more frequent, and the more intense the activity, the more kcalories are expended. Table 6-2 (pp. 148–149) gives average energy expenditures for various activities.

Energy to Manage Food

When food is taken into the body, many cells that have been dormant become active. The muscles that move the food through the intestinal tract speed up their rhythmic contractions, and the cells that manufacture and secrete digestive juices begin their tasks. All these and other cells need extra energy

TABLE 6-1 Factors That Affect the BMR

Factor	Effect on BMR
Age	Lean body mass diminishes with age, slowing the BMR.[a]
Height	In tall, thin people, the BMR is higher.[b]
Growth	In children, adolescents, and pregnant women, the BMR is higher.
Body composition (gender)	The more lean tissue, the higher the BMR (which is why males usually have a higher BMR than females). The more fat tissue, the lower the BMR.
Fever	Fever raises the BMR.[c]
Stresses	Stresses (including many diseases and certain drugs) raise the BMR.
Environmental temperature	Both heat and cold raise the BMR.
Fasting/starvation	Fasting/starvation lowers the BMR.[d]
Malnutrition	Malnutrition lowers the BMR.
Hormones (gender)	The thyroid hormone thyroxin, for example, can speed up or slow down the BMR.[e] Premenstrual hormones slightly raise the BMR.
Smoking	Nicotine increases energy expenditure.
Caffeine	Caffeine increases energy expenditure.
Sleep	BMR is lowest when sleeping.

[a]The BMR begins to decrease in early adulthood (after growth and development cease) at a rate of about 2 percent a decade. A reduction in voluntary activity as well brings the total decline in energy expenditure to 5 percent a decade.
[b]If two people weigh the same, the taller, thinner person will have the faster metabolic rate, reflecting the greater skin surface, through which heat is lost by radiation, in proportion to the body's volume.
[c]Fever raises the BMR by 7 percent for each degree Fahrenheit.
[d]Prolonged starvation reduces the total amount of metabolically active lean tissue in the body, although the decline occurs sooner and to a greater extent than body losses alone can explain. More likely, the neural and hormonal changes that accompany fasting are responsible for changes in the BMR.
[e]The thyroid gland releases hormones that travel to the cells and influence cellular metabolism. Thyroid hormone activity can speed up or slow down the rate of metabolism by as much as 50 percent.

as they participate in the digestion, absorption, and metabolism of food. This cellular activity produces heat and is known as the **thermic effect of food**. The thermic effect of food is generally thought to represent about 10 percent of the total food energy taken in. For purposes of rough estimates, though, the thermic effect of food is not always included.

ESTIMATING ENERGY REQUIREMENTS

In estimating energy requirements, the DRI committee developed equations that consider how the following factors influence energy expenditure: (■)

- *Gender.* Women generally have a lower BMR than men, in large part because men typically have more lean body mass. In addition, menstrual hormones influence the BMR in women, raising it just prior to menstruation. Two sets of energy equations—one for men and one for women—were developed to accommodate the influence of gender on energy expenditure.
- *Growth.* The BMR is high in people who are growing. For this reason, pregnant and lactating women, infants, children, and adolescents have their own sets of energy equations.

■ Note that Table 6-1 above listed these factors among those that influence BMR and, consequently, energy expenditure.

thermic effect of food: an estimation of the energy required to process food (digest, absorb, transport, metabolize, and store ingested nutrients).

TABLE 6-2 Energy Spent on Various Activities

The values listed in this table reflect both the energy spent in physical activity and the amount used for BMR. To calculate kcalories spent per minute of activity for your own body weight, multiply kcal/lb/min by your exact weight and then multiply that number by the number of minutes spent in the activity. For example, if you weigh 142 pounds, and you want to know how many kcalories you spent doing 30 minutes of vigorous aerobic dance: 0.062 × 142 = 8.8 kcalories per minute; 8.8 × 30 (minutes) = 264 total kcalories spent.

Activity	kcal/lb/ min
Aerobic dance (vigorous)	0.062
Basketball (vigorous, full court)	0.097
Bicycling	
13 mph	0.045
15 mph	0.049
17 mph	0.057
19 mph	0.076
21 mph	0.090
23 mph	0.109
25 mph	0.139
Canoeing, flat water, moderate pace	0.045
Cross-country skiing	0.104
8 mph	
Golf (carrying clubs)	0.045
Handball	0.078
Horseback riding (trot)	0.052
Rowing (vigorous)	0.097
Running	
5 mph	0.061
6 mph	0.074
7.5 mph	0.094
9 mph	0.103
10 mph	0.114
11 mph	0.131
Soccer (vigorous)	0.097
Studying	0.011
Swimming	
20 yd/min	0.032
45 yd/min	0.058
50 yd/min	0.070
Table tennis (skilled)	0.045
Tennis (beginner)	0.032
Walking (brisk pace)	
3.5 mph	0.035
4.5 mph	0.048

continued

Activity	kcal/lb/ min
Weight lifting	
light-to-moderate effort	0.024
vigorous effort	0.048
Wheelchair basketball	0.084
Wheeling self in wheelchair	0.030
Wii games	
bowling	0.021
boxing	0.021
tennis	0.022

- *Age.* The BMR declines during adulthood as lean body mass diminishes. Physical activities tend to decline as well, bringing the average reduction in energy expenditure to about 5 percent per decade. The decline in the BMR that occurs when a person becomes less active reflects the loss of lean body mass and may be prevented with ongoing physical activity. Because age influences energy expenditure, it is also factored into the energy equations.
- *Physical activity.* Using individual values for various physical activities (as in Table 6-2) is time consuming and impractical for estimating the energy needs of a population. Instead, various activities are clustered according to the typical intensity of a day's efforts (Appendix D provides details).
- *Body composition and body size.* The BMR is high in people who are tall and so have a large surface area. Similarly, the more a person weighs, the more energy is expended on basal metabolism. For these reasons, the energy equations include a factor for both height and weight.

As just explained, energy needs vary among individuals depending on such factors as gender, growth, age, physical activity, and body composition. Even when two people are similarly matched, however, their energy needs will still differ because of genetic differences. Perhaps one day genetic research will reveal how to estimate requirements for each individual. For now, the "How to" box (p. 150) provides instructions on calculating estimated energy requirements using the DRI equations and physical activity factors. (■)

Physical activity expends energy and benefits health in many ways.

Review Notes

- A person takes in energy from food and, on average, expends most of it on basal metabolic activities, some of it on physical activities, and about 10 percent on the thermic effect of food.
- Because energy requirements vary from person to person, such factors as age, gender, and weight must be considered when calculating energy expended on basal metabolism, and the intensity and duration of the activity must be taken into account when calculating expenditures on physical activities.

■ Appendix D presents tables that provide a shortcut to estimating total energy expenditure.

HOW TO Estimate Energy Requirements

To determine your estimated energy requirements (EER), use the appropriate equation (of the two options below), inserting your age in years, weight (wt) in kilograms, height (ht) in meters, and physical activity (PA) factor from the accompanying table. (To convert pounds to kilograms, divide by 2.2; to convert inches to meters, divide by 39.37.)

- For men 19 years and older:

$$EER = [662 - (9.53 \times age)] + PA \times [(15.91 \times wt) + (539.6 \times ht)]$$

- For women 19 years and older:

$$EER = [354 - (6.91 \times age)] + PA \times [(9.36 \times wt) + (726 \times ht)]$$

For example, consider an active 30-year-old male who is 5 feet 11 inches tall and weighs 178 pounds. First, he converts his weight from pounds to kilograms and his height from inches to meters, if necessary:

$$178 \text{ lb} \div 2.2 = 80.9 \text{ kg}$$

$$71 \text{ in} \div 39.37 = 1.8 \text{ m}$$

Next, he considers his level of daily physical activity and selects the appropriate PA factor from the accompanying table (in this example, 1.25 for an active male). Then, he inserts his age, PA factor, weight, and height into the appropriate equation:

$$EER = [662 - (9.53 \times 30)] + 1.25 \times [(15.91 \times 80.9) + (539.6 \times 1.8)]$$

(A reminder: do calculations within the parentheses first, and multiply before adding or subtracting.) He calculates

$$EER = [662 - (9.53 \times 30)] + 1.25 \times (1287 + 971)$$

$$EER = [662 - (9.53 \times 30)] + (1.25 \times 2258)$$

$$EER = 662 - 286 + 2823$$

$$EER = 3199$$

The estimated energy requirement for an active 30-year-old male who is 5 feet 11 inches tall and weighs 178 pounds is about 3200 kcalories/day. His actual requirement probably falls within a range of 200 kcalories above and below this estimate. (■)

Physical Activity (PA) Factors for EER Equations

	Men	Women	Physical Activity
Sedentary	1.0	1.0	Typical daily living activities
Low active	1.11	1.12	Plus 30 to 60 minutes moderate activity
Active	1.25	1.27	Plus ≥ 60 minutes moderate activity
Very active	1.48	1.45	Plus ≥ 60 minutes moderate activity and 60 minutes vigorous activity or 120 minutes moderate activity

Note: Moderate activity is equivalent to walking at 3½ to 4½ miles per hour.

Body Weight and Body Composition

The body's weight reflects its composition—the proportions of its bone, muscle, fat, fluid, and other tissue. All of these body components can vary in quantity and quality: the bones can be dense or porous, the muscles can be well developed or underdeveloped, fat can be abundant or scarce, and so on. By far the most variable tissue, though, is body fat. For health's sake, weight management efforts should focus on eating and activity habits to improve body composition.

DEFINING HEALTHY BODY WEIGHT

How much should a person weigh? How can a person know if her weight is appropriate for her height and age? How can a person know if his weight is jeopardizing his health? Questions such as these seem so simple, yet the answers can be complex—and different depending on whom you ask.

■ For *most* people, the energy requirement falls within these ranges:
(Men) EER ±200 kcal; (Women) EER ±160 kcal.

For almost all people, the actual energy requirement falls within these larger ranges:
(Men) EER ±400 kcal; (Women) EER ±320 kcal.

The Criterion of Fashion

When asking the question, "What is an ideal body weight?" people often mistakenly turn to fashion for the answer. Without a doubt, our society sets unrealistic ideals for body weight, especially for women.[3] Magazines, movies, and television all convey the message that to be thin is to be beautiful and happy. As a result, the media have a great influence on the weight concerns and dieting patterns of people of all ages but most tragically on young, impressionable children and adolescents.[4]

Importantly, perceived body image has little to do with actual body weight or size. People of all shapes, sizes, and ages—including extremely thin fashion models with anorexia nervosa and fitness instructors with ideal body composition—have learned to be unhappy with their "overweight" bodies. Such dissatisfaction can lead to damaging behaviors, such as starvation diets, diet pill abuse, and failure to seek health care. The first step toward making healthy changes may be self-acceptance. Keep in mind that fashion is fickle; the body shapes that our society values change with time and, furthermore, differ from those valued by other societies. The standards defining "ideal" are subjective and frequently have little in common with health. Table 6-3 offers some tips for adopting health as an ideal.

The Criterion of Health

Even if our society were to accept fat as beautiful, obesity would still be a major risk factor for several life-threatening diseases, as discussed later in the chapter. For this reason, the most important criterion for determining how much a person should weigh and how much body fat a person needs is not appearance but good health and longevity. A range of healthy body weights has been identified using a common measure of weight and height—the body mass index.

Body Mass Index

The **body mass index (BMI)** describes relative weight for height: (■)

$$\text{BMI} = \frac{\text{weight (kg)}}{\text{height (m)}^2} \quad \text{or} \quad \frac{\text{weight (lb)}}{\text{height (in)}^2} \times 703$$

Weight classifications based on BMI are presented in the table on the inside back cover. Notice that a healthy weight falls between a BMI of 18.5 and 24.9, with **underweight** below 18.5, **overweight** above 25, and **obese** above 30. Figure 6-5 (p. 152) presents body shapes associated with various BMI values. Most people with a BMI within the healthy weight range have few of the health risks typically associated with too-low or too-high body weight. Risks increase as BMI falls below 18.5 or rises above 24.9 (see Figure 6-6 on p. 152), reflecting the reality that both underweight and overweight impair health status.

The BMI values are most accurate in assessing degrees of obesity and are less useful for evaluating nonobese people's body fatness. BMI values fail to reveal two valuable

■ • To convert pounds to kilograms, divide by 2.2.
• To convert inches to meters, divide by 39.37.

body mass index (BMI): an index of a person's weight in relation to height; determined by dividing the weight (in kilograms) by the square of the height (in meters).

underweight: body weight lower than the weight range that is considered healthy; BMI below 18.5.

overweight: body weight greater than the weight range that is considered healthy; BMI 25.0 to 29.9.

obese: having too much body fat with adverse health effects; BMI 30 or more.

TABLE 6-3 Tips for Accepting a Healthy Body Weight

- Value yourself and others for human attributes other than body weight. Realize that prejudging people by weight is as harmful as prejudging them by race, religion, or gender.
- Use positive, nonjudgmental descriptions of your body.
- Accept positive comments from others.
- Focus on your whole self including your intelligence, social grace, and professional and scholastic achievements.
- Accept that no magic diet exists.
- Stop dieting to lose weight. Adopt a lifestyle of healthy eating and physical activity permanently.
- Follow the USDA Food Patterns. Never restrict food intake below the minimum levels that meet nutrient needs.
- Become physically active, not because it will help you get thin but because it will make you feel good and enhance your health.
- Seek support from loved ones. Tell them of your plan for a healthy life in the body you have been given.
- Seek professional counseling, *not* from a weight-loss counselor but from someone who can help you make gains in self-esteem without weight as a factor.

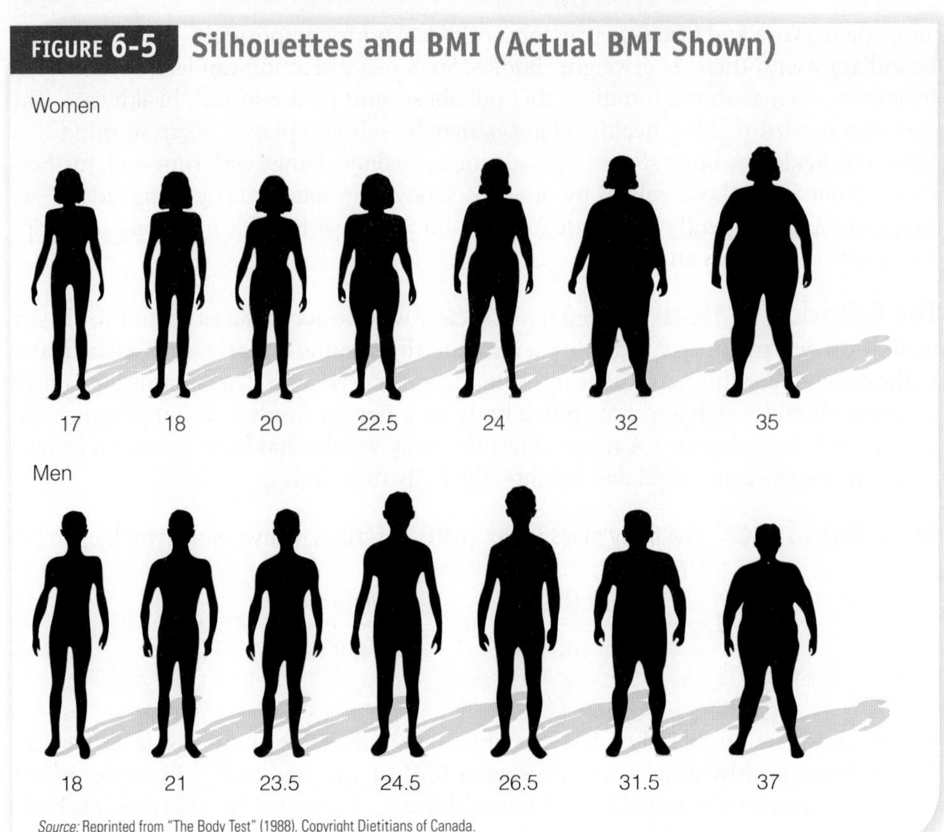

FIGURE 6-5 Silhouettes and BMI (Actual BMI Shown)

Source: Reprinted from "The Body Test" (1988). Copyright Dietitians of Canada.

■ The inflammatory response is described in detail in Chapter 16.

visceral fat: fat stored within the abdominal cavity in association with the internal abdominal organs, as opposed to fat stored directly under the skin (subcutaneous fat); also called *intra-abdominal fat.*

central obesity: excess fat around the trunk of the body; also called *abdominal fat* or *upper-body fat.*

adipokines (AD-ih-poh-kynz): protein hormones made and released by adipose tissue (fat) cells.

inflammation: an immunological response to cellular injury characterized by an increase in white blood cells.

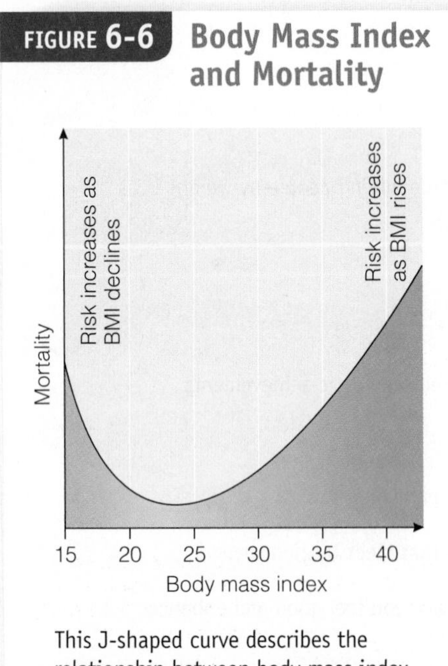

FIGURE 6-6 Body Mass Index and Mortality

This J-shaped curve describes the relationship between body mass index (BMI) and mortality and shows that both underweight and overweight present risks of a premature death.

pieces of information used in assessing disease risk: they don't reveal how much of the weight is fat, and they don't indicate where the fat is located. To obtain these data, measures of body composition are needed.

BODY COMPOSITION

For many people, being overweight compared with the standard means that they are over*fat*. This is not the case, though, for athletes with dense bones and well-developed muscles; they may be over*weight* but carry little body fat. Conversely, inactive people may seem to have acceptable weights but still carry too much body fat. In addition, among some racial and ethnic groups, BMI values may not precisely identify overweight and obesity. African American people of all ages may have more lean tissue per pound of body weight than Asians or Caucasians, for example.[5] Thus, a diagnosis of obesity or overweight requires a BMI value *plus* some measure of body composition and fat distribution.

Central Obesity

The distribution of fat on the body may influence health as much as, or more than, the total fat alone. **Visceral fat** that is stored deep within the central abdominal area of the body is referred to as **central obesity** or upper body fat (see Figure 6-7). Much research supports the widely held belief that central obesity—significantly and independently of BMI—contributes to heart disease and related deaths.[6] Some research, however, casts doubt on this premise and suggests that *all* types of obesity are linked to heart disease and central obesity is no riskier than other shapes.[7]

One possible explanation for why fat in the abdomen may increase the risk of disease involves **adipokines**, hormones released by adipose tissue. Adipokines help to regulate **inflammation** (■) and energy metabolism in the tissues.[8]

FIGURE 6-7 Abdominal Fat

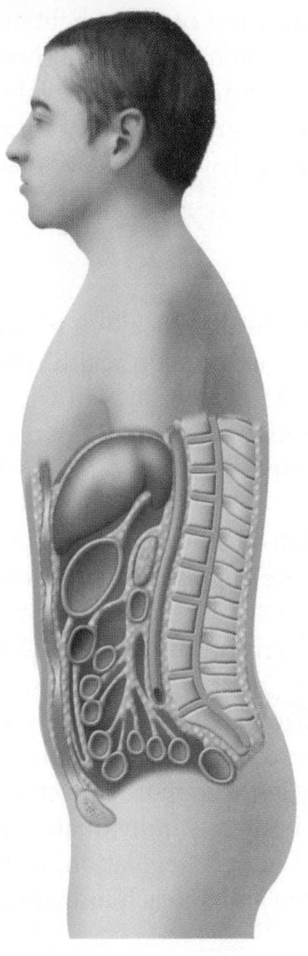

In healthy-weight people, some fat is stored around the organs of the abdomen.

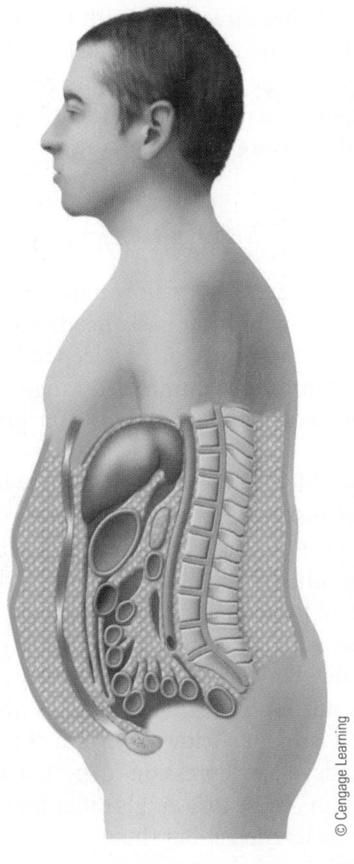

In overweight people, excess abdominal fat increases the risks of diseases.

In central obesity, a shift occurs in the balance of adipokines, favoring those that increase both inflammation and insulin resistance of tissues.[9] The resulting chronic inflammation and insulin resistance contribute to diabetes, atherosclerosis (a cause of heart disease), and other chronic diseases.[10] (■)

Visceral fat creates the "apple" profile of central obesity. **Subcutaneous fat** around the hips and thighs creates more of a "pear" profile (see Figure 6-8). Visceral fat is common in women past menopause and even more common in men. Even when total body fat is similar, men have more visceral fat than either premenopausal or postmenopausal women. For those women with visceral fat, the risks of cardiovascular disease and mortality are increased, just as they are for men. Interestingly, smokers tend to have more visceral fat than nonsmokers even though they typically have a lower BMI.[11] Two other factors that may affect body fat distribution are intakes of alcohol and physical activity. Moderate-to-high alcohol consumption may favor central obesity.[12] In contrast, regular physical activity seems to prevent abdominal fat accumulation.[13]

Waist Circumference

A person's **waist circumference** is a good indicator of fat distribution and central obesity (see Appendix E). In general, women with a waist circumference greater than 35 inches and men with a waist circumference greater than 40 inches have a high risk of central obesity–related health problems.

■ Insulin resistance is a central feature of the metabolic syndrome, which is discussed in Nutrition in Practice 20.

subcutaneous fat: fat stored directly under the skin.

sub = beneath
cutaneous = skin

waist circumference: a measurement used to assess a person's abdominal fat.

FIGURE 6-8 "Apple" and "Pear" Body Shapes Compared

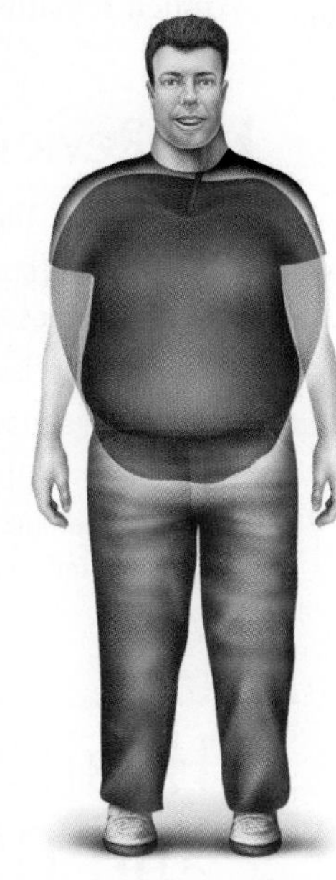

Upper-body fat is more common in men than in women and is closely associated with heart disease, stroke, diabetes, hypertension, and some types of cancer.

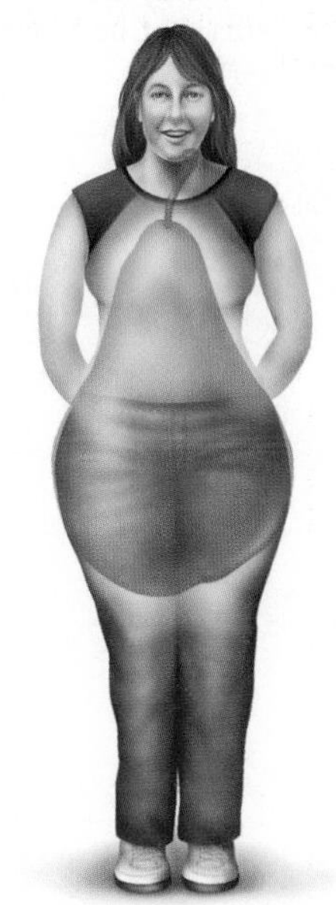

Lower-body fat is more common in women than in men and is not usually associated with chronic diseases.

Popular articles sometimes call bodies with upper-body fat "apples" and those with lower-body fat, "pears."

© Rick Schaff

At 6 feet 4 inches tall and 250 pounds, this runner would be considered overweight by most standards. Yet he is clearly not overfat.

Skinfold Measures Skinfold measurements provide an accurate estimate of total body fat and a fair assessment of the fat's location. About half of the fat in the body lies directly beneath the skin, so the thickness of this subcutaneous fat is assumed to reflect total body fat. Measures taken from central-body sites (around the abdomen) better reflect changes in fatness than those taken from upper sites (arm and back). A skilled assessor can obtain an accurate **skinfold measure** and then compare the measurement with standards (see Appendix E).

HOW MUCH BODY FAT IS TOO MUCH?

People often ask exactly how much fat is too fat for health. Ideally, a person has enough fat to meet basic needs but not so much as to incur health risks. The ideal amount of body fat depends partly on the person. A man with a BMI within the recommended range may have between 18 and 24 percent body fat; a woman, because of her greater quantity of essential fat, 23 to 31 percent.

Many athletes have a lower percentage of body fat—just enough fat to provide fuel, insulate and protect the body, assist in nerve impulse transmissions, and support normal hormone activity, but not so much as to burden the muscles. For athletes, then, body fat might be 5 to 10 percent for men and 15 to 20 percent for women.

For an Alaskan fisherman, a higher-than-average percentage of body fat is probably beneficial because fat helps prevent heat loss in cold weather. A woman starting a pregnancy needs sufficient body fat to support conception and fetal growth. Below a certain threshold for body fat, individuals may become infertile, develop depression, experience abnormal hunger regulation, or be unable to keep warm. These thresholds differ for each function and for each individual; much remains to be learned about them.

Review Notes

- Clearly, the most important criterion of appropriate fatness is health.
- Current standards for body weight are based on the body mass index (BMI), which describes a person's weight in relation to height.
- Health risks increase with a BMI below 18.5 or above 24.9.
- Central obesity, in which excess fat is distributed around the trunk of the body, may present greater health risks than excess fat distributed on the lower body.
- Researchers use a number of techniques to assess body composition, including waist circumference and skinfold measures.

Health Risks of Underweight and Obesity

As mentioned earlier and shown in Figure 6-6 (p. 152), health risks increase as BMI falls below 18.5 or rises above 24.9. People who are extremely underweight or extremely obese carry higher risks of early death than those whose weights fall within the healthy, or even the slightly overweight, range.[14] These mortality risks decline with age. Independently of BMI, factors such as smoking habits raise health risks, and physical fitness lowers them.

Health Risks of Underweight Some underweight people enjoy an active, healthy life, but others are underweight because of malnutrition, smoking habits, substance

skinfold measure: a clinical estimate of total body fatness in which the thickness of a fold of skin on the back of the arm (over the triceps muscle), below the shoulder blade (subscapular), or in other places is measured with a caliper.

abuse, or illnesses. Weight and fat measures alone would not reveal these underlying causes, but a complete assessment that includes a diet and medical history, physical examination, and biochemical analysis would.

Underweight people, especially older adults, may be unable to preserve lean tissue when fighting a wasting disease such as cancer. Overly thin people are also at a disadvantage in the hospital, where nutrient status can easily deteriorate if they have to go without food for days at a time while undergoing tests or surgery.[15] Underweight women often develop menstrual irregularities and become infertile. Underweight and significant weight loss are also associated with osteoporosis and bone fractures. For all these reasons, underweight people may benefit from enough of a weight gain to provide an energy reserve and protective amounts of all the nutrients that can be stored.

An extreme underweight condition known as anorexia nervosa is sometimes seen in young people who exercise unreasonable self-denial in order to control their weight. Anorexia nervosa is a major eating disorder seen in our society today. Eating disorders are the subject of Nutrition in Practice 7.

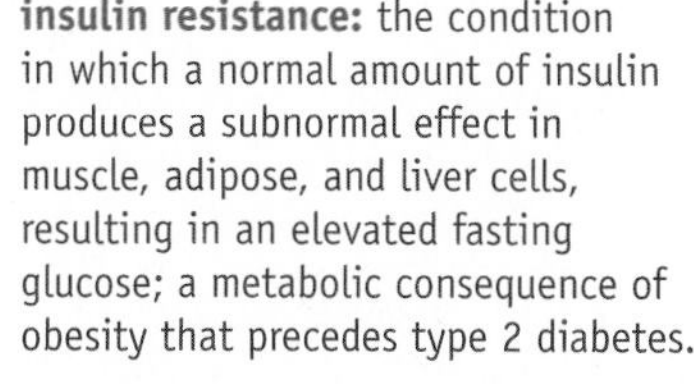
insulin resistance: the condition in which a normal amount of insulin produces a subnormal effect in muscle, adipose, and liver cells, resulting in an elevated fasting glucose; a metabolic consequence of obesity that precedes type 2 diabetes.

Health Risks of Overweight and Obesity

Despite our nation's preoccupation with body image and weight loss, the prevalence of overweight and obesity continues to rise dramatically (see Figure 6-9).[16] During the previous three decades, obesity increased in every state, in both genders, and across all ages, races, and education levels. Today, 68 percent of U.S. adults are overweight (BMI of 25 or greater) or dangerously obese (BMI of 30 or greater).[17] Obesity rates are expected to continue rising into the foreseeable future, although at a somewhat slower rate than in the past.[18] The problem reaches around the globe, in urban and rural areas alike.[19] In short, obesity is a major public health problem that is becoming more prevalent.

The growing prevalence of overweight and obesity is a matter of concern because both conditions present risks to health. Indeed, the health risks of obesity are so numerous that it has been declared a disease. Excess weight contributes to up to half of all cases of hypertension, thereby increasing the risk of heart attack and stroke. Obesity raises blood pressure in part by altering kidney function, increasing blood volume, and promoting blood vessel damage through insulin resistance.[20] Often weight loss alone can normalize the blood pressure of an overweight person.

Excess body weight also increases the risk of type 2 diabetes. Most adults with type 2 diabetes are overweight or obese, and obesity itself can directly cause some degree of **insulin resistance**.[21] Diabetes (type 2) is three times more likely to develop in an obese person than in a nonobese person. Furthermore, the person with type 2 diabetes often has central obesity.[22] Central-body fat cells appear to be larger and more insulin resistant than lower-body fat cells.

In addition to diabetes and hypertension, other risks threaten obese adults. Among them are high blood lipids, cardiovascular disease, sleep apnea (abnormal cessation of breathing during sleep), osteoarthritis, abdominal hernias, some cancers, varicose veins, gout, gallbladder disease, kidney stones, respiratory problems (including Pickwickian syndrome, a breathing blockage linked with sudden death), nonalcoholic fatty liver disease, complications in pregnancy and surgery, flat feet, and even a high accident rate.[23] Each year these obesity-related illnesses cost our nation billions of dollars. The cost in terms of lives is also great. People with lifelong obesity are twice as likely to die prematurely as others. In the United States, obesity is second only to tobacco use as the most significant cause of preventable death.[24]

FIGURE 6-9 The Increasing Prevalence of Obesity among U.S. Adults by State

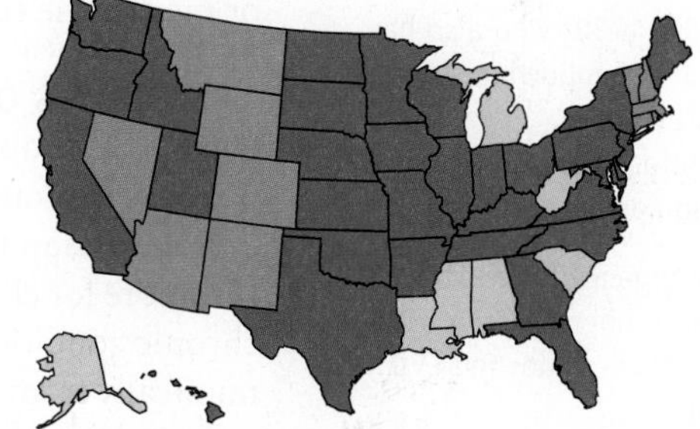
1998: Most states had prevalence rates less than 20 percent.

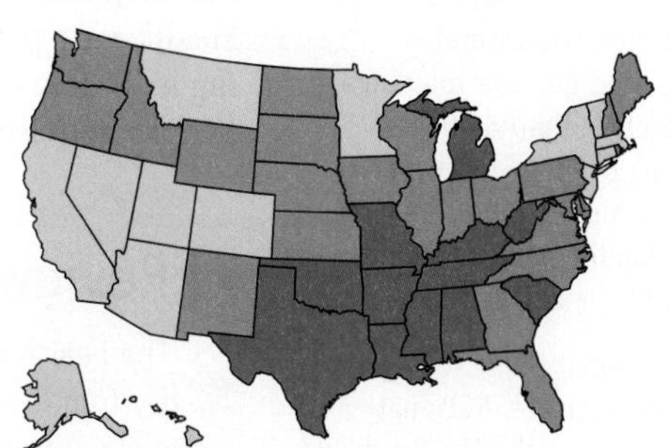
2010: No state had a prevalence rate less than 20 percent, and most states had prevalence rates greater than 25 percent, with twelve states reporting prevalence rates greater than or equal to 30 percent.

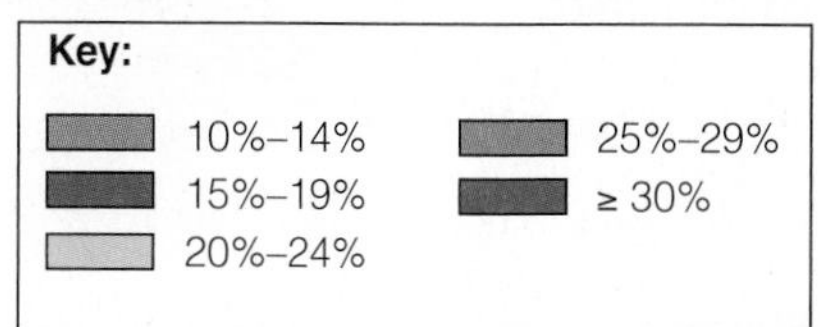

Source: www.cdc.nccdphp/dnpa/obesity/trend/maps/index.htm

Some overweight people, however, remain healthy and live long lives despite their body fatness. Genetic inheritance, smoking habits, and level of physical activity may help to explain why some overweight individuals stay well while others become ill. Overweight but fit people have lower risks than normal-weight, unfit ones, for example.[25] To help in identifying those most at risk, obesity experts have developed guidelines for health practitioners, described next.

National Guidelines for Identifying Those at Risk from Obesity The U.S. guidelines for identifying and evaluating the risks to health from overweight and obesity rely on three indicators. The first indicator is a person's BMI. As a general guideline, overweight for adults is defined as BMI of 25.0 through 29.9, and obesity is defined as BMI equal to or greater than 30.

The second indicator is waist circumference, which, as discussed earlier, reflects the degree of abdominal fatness in proportion to body fatness. Women with a waist circumference greater than 35 inches and men with a waist circumference greater than 40 inches are at greater risk of type 2 diabetes, hypertension, and cardiovascular disease than women or men with waist circumferences equal to, or below, these measures. In other words, waist circumference is an independent predictor of disease risk.

The third indicator is the person's disease risk profile. The disease risk profile takes into account life-threatening diseases, family history, and common risk factors for chronic diseases (such as blood lipid profile). The higher the BMI, the greater the waist circumference, and the more risk factors, the greater the urgency to treat obesity. People who have one or more of the diseases listed in the margin, (■) or three or more of the cardiovascular disease risk factors listed, have a very high risk for disease complications and mortality that requires aggressive treatment to manage the disease or modify the risk factors.

■ The National Heart, Lung, and Blood Institute states that aggressive treatment is urgently needed for a clinically obese person (BMI ≥ 30) who also has any one of the following:

- Heart disease
- Diabetes (type 2)
- Sleep apnea

The same urgency for treatment exists for an obese person with any *three* of the following CVD risk factors:

- Hypertension
- Cigarette smoking
- High LDL cholesterol
- Low HDL cholesterol
- Impaired glucose tolerance
- Age older than 45 years (men) or 55 years (women)
- Heart disease of an immediate family member before age 55 (male) or 65 (female)

Source: National Heart, Lung, and Blood Institute, National Institutes of Health, *The Practical Guide: Identification, Evaluation, and Treatment of Overweight and Obesity in Adults,* NIH publication no. 00-4084 (Washington, DC: Government Printing Office, 2000).

Other Risks of Obesity Although some obese people seem to escape health problems, few in our society can avoid the social and economic handicaps. Our society places enormous value on thinness. Obese people are less sought after for romance, less often hired, and less often admitted to college.[26] They pay higher insurance premiums and more for clothing. This is especially true for women. In contrast, people with other chronic conditions such as asthma, diabetes, and epilepsy do not differ socially or economically from healthy non-overweight people.

Prejudice defines people by their appearance rather than by their abilities and characters. Obese people suffer emotional pain when others treat them with insensitivity, hostility, and contempt, and they may internalize a sense of guilt and self-deprecation. Health-care professionals, even dietitians, can be among the offenders without realizing it. To free our society of its obsession with body fatness and prejudice against obese people, activists are promoting respect for individuals of all body weights.

Review Notes

- The health risks of underweight include infertility (women), bone loss, and inability to preserve lean tissue when fighting a wasting disease such as cancer; risks of obesity include chronic diseases such as hypertension, type 2 diabetes, and heart disease. Both underweight and obesity increase the risk of premature death.
- Guidelines for identifying the health risks of overweight and obesity are based on a person's BMI, waist circumference, and disease risk profile.
- Obesity also incurs social, economic, and psychological risks.

Self Check

1. As carbohydrate and fat stores are depleted during fasting or starvation, the body then uses ___ as its fuel source.
   a. alcohol
   b. protein
   c. glucose
   d. triglycerides

2. When carbohydrate is not available to provide energy for the brain, as in starvation, the body produces ketone bodies from:
   a. glucose.
   b. glycerol.
   c. fatty acid fragments.
   d. amino acids.

3. Three hazards of fasting are:
   a. water weight loss, decrease in mental alertness, and wasting of lean tissue.
   b. water weight gain, impairment of disease resistance, and lowering of body temperature.
   c. water weight gain, decrease in mental alertness, and impairment of disease resistance.
   d. wasting of lean tissue, impairment of disease resistance, and disturbances of the body's salt and water balance.

4. Two activities that contribute to the basal metabolic rate are:
   a. walking and running.
   b. maintenance of heartbeat and running.
   c. maintenance of body temperature and walking.
   d. maintenance of heartbeat and body temperature.

5. Three factors that affect the body's basal metabolic rate are:
   a. height, weight, and energy intake.
   b. age, body composition, and height.
   c. fever, body composition, and altitude.
   d. weight, fever, and environmental temperature.

6. The largest component of energy expenditure is:
   a. basal metabolism.
   b. physical activity.
   c. indirect calorimetry.
   d. thermic effect of food.

7. Which of the following reflects height and weight?
   a. Body mass index
   b. Central obesity
   c. Waist circumference
   d. Body composition

8. The BMI range that correlates with the fewest health risks is:
   a. 16.5 to 20.9.
   b. 18.5 to 24.9.
   c. 25.5 to 30.9.
   d. 30.5 to 34.9.

9. The profile of central obesity is sometimes referred to as a(n):
   a. beer.
   b. pear.
   c. apple.
   d. potato.

10. Which of the following health risks is *not* associated with being overweight?
    a. Hypertension
    b. Heart disease
    c. Type 1 diabetes
    d. Gallbladder disease

Answers to these questions can be found in Appendix H. For more chapter review: Access an interactive eBook, chapter-specific interactive learning tools, including flashcards, quizzes, videos, and more in your Nutrition CourseMate, accessed through CengageBrain.com.

Clinical Applications

1. Compare the energy a person might spend on various physical activities. Refer to Table 6-2 on pp. 148–149, and compute how much energy a person who weighs 142 pounds would spend doing each of the following activities. An example using aerobic dance has been provided for you. You may want to compare various activities based on your own weight.

   30 minutes of vigorous aerobic dance:

   0.062 kcal/lb/min × 142 lb = 8.8 kcal/min.

   8.8 kcal/min × 30 min = 264 kcal.

   a. 2 hours of golf, carrying clubs.
   b. 20 minutes running at 9 mph.
   c. 45 minutes of swimming at 20 yd/min.
   d. 1 hour of walking at 3.5 mph.

2. Using the "How to" on p. 150 as a guide, determine your Estimated Energy Requirement (EER).

Answers

1. (a) 0.045 kcal/lb/min × 142 lb = 6.4 kcal/min, 6.4 kcal/min × 120 min = 768 kcal; (b) 0.103 kcal/lb/min × 142 lb = 14.6 kcal/min, 14.6 kcal/min × 20 min = 292 kcal; (c) 0.032 kcal/lb/min × 142 lb = 4.5 kcal/min, 4.5 kcal/min × 45 = 202.5 kcal; (d) 0.035 kcal/lb/min × 142 lb = 5 kcal/min, 5 kcal/min × 60 min = 300 kcal.

Nutrition on the Net

For further study of the topics in this chapter, access these websites.

- Obtain food composition data from the USDA Nutrient Data Laboratory: **http://www.ars.usda.gov/nutrientdata**
- Search for "obesity" and "weight control" at the U.S. government health information site: **www.healthfinder.gov**
- Review the Clinical Guidelines on the Identification, Evaluation, and Treatment of Overweight and Obesity in Adults: **www.nhlbi.nih.gov/guidelines/obesity/ob_home.htm**
- Learn about the 10,000 Steps Program at Shape Up America: **www.shapeup.org**
- Visit the special web pages and interactive applications for Healthy Weight: **www.nhlbi.nih.gov/health/prof/heart/index.htm#obesity**
- Find helpful information on achieving and maintaining a healthy weight from the Calorie Control Council: **www.caloriecontrol.org**
- Learn how to end size discrimination and improve the quality of life for fat people from the National Association to Advance Fat Acceptance: **www.naafaonline.com/dev2/**
- Consider ways to live a healthy life at any weight: **www.bodypositive.com**

Notes

1. M. M. Bergmann and coauthors, The association of lifetime alcohol use with measures of abdominal and general adiposity in a large-scale European cohort, *European Journal of Clinical Nutrition* 65 (2011): 1079–1087.
2. S. B. Hymsfield and coauthors, Energy content of weight loss: Kinetic features during voluntary caloric restriction, *Metabolism* 61 (2012): 937–943; K. D. Hall and coauthors, Energy balance and its components: Implications for body weight regulation, *American Journal of Clinical Nutrition* 95 (2012): 989–994; K. D. Hall and coauthors, Quantification of the effect energy imbalance on bodyweight, *Lancet* 378 (2011): 826–837.
3. J. B. Martin, The development of ideal body image perceptions in the United States, *Nutrition Today* 45 (2010): 98–110.
4. D. Anschutz and coauthors, Watching your weight? The relations between watching soaps and music television and body dissatisfaction and restrained eating in young girls, *Psychology and Health* 24 (2009): 1035–1050; M. J. Hogan and V. C. Strasburger, Body image, eating disorders, and the media, *Adolescent Medicine: State of the Art Reviews* 19 (2009): 521–546.
5. M. Heo and coauthors, Percentage of body fat cutoffs by sex, age, and race-ethnicity in the US adult population from NHANES 1999–2004, *American Journal of Clinical Nutrition* 95 (2012): 594–602; H. R. Hull and coauthors, Fat-free mass index: Changes and race/ethnic differences in adulthood, *International Journal of Obesity (London)* 35 (2011): 121–127; N. Farah and coauthors, Comparison in maternal body composition between Caucasian Irish and Indian women, *Journal of Obstetrics and Gynecology* 31 (2011): 483–485; A. Liu and coauthors, Ethnic differences in the relationship between body mass index and percentage of body fat among Asian children from different backgrounds, *British Journal of Nutrition* 106 (2011): 1390–1397.

6. B. J. Arsenault and coauthors, Physical inactivity, abdominal obesity and risk of coronary heart disease in apparently healthy men and women, *International Journal of Obesity* 34 (2010): 340–347; E. J. Jacobs and coauthors, Waist circumference and all-cause mortality in a large US cohort, *Archives of Internal Medicine* 170 (2010): 1293–1301; S. S. Dhaliwal and T. A. Welborn, Central obesity and multivariable cardiovascular risk as assessed by the Framingham prediction scores, *American Journal of Cardiology* 103 (2009): 1403–1407; J. P. Reis and coauthors, Overall obesity and abdominal adiposity as predictors of mortality in U.S. white and black adults, *Annals of Epidemiology* 19 (2009): 134–142; E. B. Levitan and coauthors, Adiposity and incidence of heart failure hospitalization and mortality: A population-based prospective study, *Circulation Heart Failure* 2 (2009): 202–208.
7. D. Wormser and coauthors, Separate and combined associations of body-mass index and abdominal adiposity with cardiovascular disease: Collaborative analysis of 58 prospective studies, *The Lancet* 377 (2011): 1085–1095; A. E. Taylor and coauthors, comparison of the associations of body mass index and measures of central adiposity and fat mass with coronary heart disease, diabetes, and all-cause mortality: A study using data from 4 UK cohorts, *American Journal of Clinical Nutrition* 91 (2010): 547–556; K. M. Flegal and B. I. Graubard, Estimates of excess deaths associated with body mass index and other anthropometric variables, *American Journal of Clinical Nutrition* 89 (2009): 1213–1219.
8. Y. Deng and P. E. Scherer, Adipokines as novel biomarkers and regulators of the metabolic syndrome, *Annals of the New York Academy of Sciences* 1212 (2010): E1–E19.
9. W. V. Brown and coauthors, Obesity: Why be concerned? *American Journal of Medicine* 122 (2009): S4–S11; J. Korner, S. C. Woods, and K. A. Woodworth, Regulation of energy homeostasis and health consequences in obesity, *American Journal of Medicine* 122 (2009): S12–S18.
10. Brown and coauthors, 2009; Komer, Woods, and Woodworth, 2009.
11. E. W. Demerath, Causes and consequences of human variation in visceral adiposity, *American Journal of Clinical Nutrition* 91 (2010): 1–2; E. A. Molenaar and coauthors, Association of lifestyle factors with abdominal subcutaneous and visceral adiposity: The Framingham Heart Study, *Diabetes Care* 32 (2009): 505–510.
12. Molenaar and coauthors, 2009.
13. U. Ekelund and coauthors, Physical activity and gain in abdominal adiposity and body weight: Prospective cohort study in 288,498 men and women, *American Journal of Clinical Nutrition* 93 (2011): 826–835; B. J. Arsenault and coauthors, Physical inactivity, abdominal obesity and risk of coronary heart disease in apparently healthy men and women, *International Journal of Obesity* 34 (2010): 340–347; Y. Kim and S. Lee, Physical activity and abdominal obesity in youth, *Applied Physiology, Nutrition and Metabolism* 34 (2009): 571–581.
14. Flegal and Graubard, 2009.
15. O. Bouillanne and coauthors, Fat mass protects hospitalized elderly persons against morbidity and mortality, *American Journal of Clinical Nutrition* 90 (2009): 505–510.
16. C. M. Apovian, The causes, prevalence, and treatment of obesity revisited in 2009: What have we learned so far? *American Journal of Clinical Nutrition* 91 (2010): 277S–279S; K. M. Flegal and coauthors, Prevalence and trends in obesity among US adults, 1999–2008, *Journal of the American Medical Association* 303 (2010): 235–241.
17. Flegal and coauthors, 2010.
18. E. A. Finkelstein and coauthors, Obesity and severe obesity forecasts through 2030, *American Journal of Preventive Medicine* 42 (2012): 563–570; Flegal and coauthors, 2010.
19. B. M. Popkin, L. S. Adair, and S. W. Ng, Global nutrition transition and pandemic of obesity in developing countries, *Nutrition Reviews* 70 (2012): 3–21; B. A. Swinburn and coauthors, The global obesity pandemic: Shaped by global drivers and local environments, *Lancet* 378 (2011): 804–814; B. M. Popkin, Recent dynamics suggest selected countries catching up to U.S. obesity, *American Journal of Clinical Nutrition* 91 (2010): 284S–288S.
20. C. W. Mende, Obesity and hypertension: A common coexistence, *Journal of Clinical Hypertension* 14 (2012): 137–138; J. Redon and coauthors, Mechanisms of hypertension in the cardiometabolic syndrome, *Journal of Hypertension* 27 (2009): 441–451; F. W. Visser and coauthors, Rise in extracellular fluid volume during high sodium depends on BMI in healthy men, *Obesity (Silver Spring)* 17 (2009): 1684–1688.
21. M. L. Biggs and coauthors, Association between adiposity in midlife and older age and risk of diabetes in older adults, *Journal of the American Medical Association* 303 (2010): 2504–2512; American Diabetes Association, Diagnosis and classification of diabetes mellitus, *Diabetes Care* 32 (2009): S62–S67.
22. The InterAct Consortium, Long-term risk of incident type 2 diabetes and measures of overall and regional obesity: The EPIC-InterAct Case-Cohort Study, *PLos Medicine* 9 (2012): e1001230, doi:10.1371/journal.pmed.1001230.
23. C. Eheman and coauthors, Annual Report to the Nation on the status of cancer, 1975–2008, featuring cancers associated with excess weight and lack of sufficient physical activity, *Cancer* 118 (2012): 2338–2366; World Cancer Research Fund/American Institute for Cancer Research, Continuous Update Project, Colorectal Cancer Report 2010 Summary, *Food, Nutrition, Physical Activity, and the Prevention of Colorectal Cancer* (Washington, DC: American Institute of Cancer Research, 2011); L. W. York, S. Puthalapattu, and G. Y. Wu, Nonalcoholic fatty liver disease and low-carbohydrate diets, *Annual Review of Nutrition* 29 (2009): 365–379; G. Whitlock and coauthors, Body-mass index and cause-specific mortality in 900,000 adults: Collaborative analyses of 57 prospective studies, *Lancet* 373 (2009): 1083–1096; C. J. Lavie, R. V. Milani, and H. O. Ventura, Obesity and cardiovascular disease: Risk factor, paradox, and impact of weight loss, *Journal of the American College of Cardiology* 53 (2009): 1925–1932.
24. H. Jia and E. I. Lubetkin, Trends in quality-adjusted life-years lost contributed by smoking and obesity, *American Journal of Preventive Medicine* 38 (2010): 138–144.
25. M. Hamer and E. Stamatakis, Metabolically healthy obesity and risk of all-cause and cardiovascular disease mortality, *Journal of Clinical Endocrinology and Metabolism* 97 (2012): 2482–2488; D. Lee and coauthors, Long-term effects of changes in cardiorespiratory fitness and body mass index on all-cause and cardiovascular disease mortality in men, *Circulation* 124 (2011): 2483–2490; D. E. Larson-Meyer and coauthors, Caloric restriction with or without exercise: The fitness versus fatness debate, *Medicine and Science in Sports and Exercise* 42 (2010): 152–159.
26. M. S. Argugete, J. L. Edman, and A. Yates, Romantic interest in obese college students, *Eating Behaviors* 10 (2009): 143–145.

Nutrition in Practice

Fad Diets

To paraphrase William Shakespeare, "a fad diet by any other name would still be a fad diet." And the names are legion: the Atkins Diet, the Cheater's Diet, the South Beach Diet, the Zone Diet.* Year after year, "new and improved" diets appear on bookstore shelves and circulate among friends. People of all sizes eagerly try the best diet ever on the market, hoping that this one will really work. Sometimes fad diets seem to work for a while, but more often than not, their success is short-lived. Then another fad diet takes the spotlight. Here's how Dr. K. Brownell, an obesity researcher at Yale University, describes this phenomenon: "When I get calls about the latest diet fad, I imagine a trick birthday cake candle that keeps lighting up and we have to keep blowing it out."†

Why don't health professionals speak out against the relentless promotion of fad diets?

Realizing that many fad diets do not offer a safe and effective plan for weight loss, health professionals speak out, but they never get the candle blown out permanently. New fad diets can keep making outrageous claims because no one requires their advocates to prove what they say. Fad diet promoters do not have to conduct credible research on the benefits or dangers of their diets. They can simply make recommendations and then later, if questioned, search for bits and pieces of research that support the conclusions they have already reached. That's backwards. Diet and health recommendations should *follow* years of sound research that has been reviewed by panels of scientists *before* being offered to the public.

AP Photo/Evan Vucci

*The Academy of Nutrition and Dietetics offers evaluations of popular diets for your review. Look for Popular Diet Reviews at their website, www.eatright.org/media.

†Dr. K. Brownell, an obesity researcher at Yale University.

How can the promoters of fad diets get away with exaggerated claims?

Because anyone can publish anything—in books or on the Internet—peddlers of fad diets can make unsubstantiated statements that fall far short of the truth but sound impressive to the uninformed. They often offer distorted bits of legitimate research. They may start with one or more actual facts but then leap from one erroneous conclusion to the next. Anyone who wants to believe these claims has to wonder how the thousands of scientists working on obesity research over the past century could possibly have missed such obvious connections.

Each popular diet features a unique way to make losing weight fast and easy. Which one works best?

Fad diets come in almost as many shapes and sizes as the people who search them out. Some restrict fats or carbohydrates, some limit portion sizes, some focus on food combinations, and some claim that a person's genetic type or blood type determines the foods best suited to manage weight and prevent disease. Despite claims that each new diet is "unique" in its approach to weight loss, most fad diets are designed to ensure a low energy intake. The "magic feature" that best supports weight loss is to limit energy intake to less than energy expenditure. Most of the sample menu plans, especially in the early stages, deliver an average of 1200 kcalories per day. Total kcalories tend to be low simply because food intake is so limited. Table NP6-1 compares some of today's more popular diets.

What is the appeal of fad diets?

Probably the greatest appeal of some fad diets is that they tend to ignore current diet recommendations. Foods such as meats and milk products that need to be selected carefully to limit saturated fat can now be eaten with abandon. Whole grains, legumes, vegetables, and fruits that should be eaten in abundance can now be bypassed. For some people, this is a dream come true: steaks without the potatoes, ribs without the coleslaw, and meatballs without

TABLE NP6-1 Popular Diets Compared

Diet	Major Premise Promoted	Strong Point(s)	Weak Point(s)
The 4-Hour Body	• Less is more, and small, simple changes produce long-lasting effects.	• Quick results.	• Restricts carbohydrates. • All fruit and milk (except cottage cheese) are excluded, and vegetables are limited. • Encourages eating the same small meals repeatedly. • Recommends a weekly binge.
The 17 Day Diet	• Changing the way you eat every few days creates "body confusion," which prevents metabolism from settling into homeostasis.	• You can boost metabolism by "eating clean," which means no sugar, no processed food, and no fried foods. • Quick results. • Prevents boredom by alternating between cycles. • Fairly well-balanced diet that promotes healthy eating.	• No scientific evidence that changing the diet creates "body confusion." • No individualized kcalorie goals. • Promotes its own processed foods.
Atkins Diet	• People are overweight or obese because they have metabolic imbalances caused by eating too many carbohydrates; by restricting carbohydrates, these imbalances can be corrected. • You can lose weight without lowering kcalorie intake.	• Quick, short-term weight loss is achieved.	• Restricts carbohydrates to a level that induces ketosis, which can cause nausea, lightheadedness, and fatigue and can worsen existing medical problems such as kidney disease. • A diet high in fat such as Atkins can increase the risk of heart disease and some cancers.
Cheater's Diet	• Successful weight loss depends on eliminating boredom and allowing indulgences. • Cheating on weekends "stokes your metabolism."	• Meals are proportioned one-half fruit or vegetables, one-fourth lean protein, and one-fourth whole grains. • Encourages as much exercise as possible.	• No scientific data on cheating boosting metabolism or supporting weight loss.
Cinch!	• A nutrient-dense diet composed mainly of plant-based foods will help you lose weight and lower the risk of disease.	• Plant-based, nutrient-dense diet. • Stresses the importance of exercise.	• A little confusing and dense with facts.

continued

continued

Diet	Major Premise Promoted	Strong Point(s)	Weak Point(s)
The Dukan Diet	• A high-protein, low-kcalorie diet promotes rapid weight loss and will keep it off for good.	• Encourages daily exercise, moderate salt intake, and lifelong weight management. • Provides a highly structured plan.	• Restricts carbohydrates to a level that induces ketosis, which can cause nausea, lightheadedness, and fatigue and can worsen medical problems such as kidney disease. • Not suited for vegetarians and others who prefer not to emphasize animal proteins.
Glucose Revolution	• Low-glycemic index foods satisfy hunger, control blood glucose, and promote weight loss.	• Emphasizes fiber-rich vegetables, legumes, fruits, and whole grains. • Minimizes saturated fat intake.	• Difficult to know the glycemic index of some foods.
New Sonoma Diet	• Enjoying portion-controlled Coastal California style foods supports weight loss and promotes good health.	• Emphasizes nutrient-dense foods. • Limits processed foods.	• No individualized kcalorie plans.
Ornish Diet	• By strictly limiting fat (both animal and vegetable), you eat fewer kcalories without eating less food.	• High-fiber, low-fat foods in this plan can lower blood cholesterol and blood pressure.	• So little fat that essential fatty acids may be lacking. • Limits fish, nuts, and olive oil, which may protect against heart disease.
South Beach Diet	• Eating "good carbohydrates" such as vegetables, whole-wheat pastas, and brown rice will maintain satiety and resist cravings for "bad carbohydrates" such as white rice and potatoes.	• Encourages consumption of vegetables, lean meats, and fish, and the use of unsaturated oils when cooking. • Restricts fatty meats and cheeses as well as sweets.	• Starchy carbohydrates and all fruits are completely excluded during the first two weeks.
Ultimate Weight Solution Diet	• Foods that require great effort to prepare and eat are nutrient-dense; eating these kinds of foods (raw vegetables, vegetable soups, whole grains, beans, meats, poultry, and fish) will lead to weight loss. • Foods that take little effort to prepare and eat provide excess kcalories relative to nutrients; eating these kinds of foods (fast foods, puddings, high-kcalorie convenience foods, processed foods) leads to uncontrolled eating and weight gain.	• Encourages consumption of lean meats and fish; whole grains; vegetables; fruit; and low-fat milk, yogurt, and cheese. • Restricts fatty meats and cheeses as well as sweets. • Encourages exercise.	• Confusing as to exactly what to eat or how much.
Zone Diet	• Eating the correct proportions of carbohydrates, fat, and protein leads to hormonal balance, weight loss, disease prevention, and increased vitality.	• Promotes weight loss because it is a low-kcalorie diet.	• The diet is rigid, restrictive, and complicated, making it difficult for most people to follow accurately. • The overblown health claims of the diet's proponents are based on misinterpreted science and remain unsubstantiated.

the pasta. Who can resist the promise of weight loss while eating freely from a list of favorite foods?

Dieters are also lured into fad diets by sophisticated—yet often erroneous—explanations of the metabolic consequences of eating certain foods. Terms such as *eicosanoids* or *adipokines* are scattered about, often intimidating readers into believing that the authors must be right given their brilliance in understanding the body.

With over half of our nation's adults overweight and many more concerned about their weight, weight-loss books and products are a $33 billion-a-year business. Even a plan that offers only minimal weight-loss success easily attracts a following. Table NP6-2 presents some tips for identifying fad diets and weight-loss scams.

Are fad diets adequate?

When food choices are limited, nutrient intakes may be inadequate. To help shore up some of these inadequacies, fad diets often recommend a dietary supplement. Conveniently, many of the companies selling fad diets also sell these supplements, often at an inflated price. As Nutrition in Practice 9 explains, however, foods offer many more health benefits than any supplement can provide.[1] Quite simply, if the diet is inadequate, it needs to be improved, not supplemented.

Are the diets effective?

If fad diets were entirely ineffective, consumers would eventually stop pursuing them. Obviously, this is not the case. Similarly, if the diets were especially effective, then consumers who tried them would lose weight, and the obesity problem would be solved. Clearly, this is not happening either. As mentioned earlier, most fad diets succeed in contriving ways to limit kcalorie intakes, and so produce weight loss (at least in the short term). Studies demonstrate, however, that fad diets are particularly ineffective for weight-loss maintenance—people may drop some weight, but they quickly gain it back.[2] Straightforward kcalorie deficit is the real key to weight loss—not the elimination of protein, carbohydrate, or fat, or the still unidentified metabolic mechanisms proposed by many fad diets.[3]

For example, diet promoters make much of research showing that high-protein, low-carbohydrate diets produce a little more weight loss than balanced diets over the first few months of dieting. However, in the long run, any low-kcalorie diet produces about the same degree of loss.[4] In addition, most people cannot sustain such a diet and they quickly return to their original or an even higher weight.[5]

Protein nutrition during kcalorie restriction deserves attention, however. A meal with too little protein may not

TABLE NP6-2 Tips for Identifying Fad Diets and Weight-Loss Scams

It may be a fad diet or weight-loss scam if it:
• Sounds too good to be true.
• Recommends using a single food consistently as the key to the program's success.
• Promises quick and easy weight loss with no effort. "Lose weight while you sleep!"
• Eliminates an entire food group such as grains or milk and milk products.
• Guarantees an unrealistic outcome in an unreasonable time period. "Lose 10 pounds in 2 days!"
• Bases evidence for its effectiveness on anecdotal stories.
• Requires that you buy special products that are not readily available in the marketplace at affordable prices.
• Specifies a proportion for the energy nutrients that falls outside the recommended ranges—carbohydrate (45 to 65 percent), fat (20 to 35 percent), and protein (10 to 35 percent).
• Claims to alter your genetic code or reset your metabolism.
• Fails to mention potential risks or additional costs.
• Promotes products or procedures that have not been proven safe and effective.
• Neglects plans for weight maintenance following weight loss.

produce enough satiety, the satisfaction of feeling full after a meal, to prevent between-meal hunger. Diets providing more than 35 percent of total kcalories as protein (the upper end of the DRI range), however, are no more effective for producing long-term weight loss than more balanced kcalorie-controlled diets.[6] Eating the normal amount of lean protein-rich foods while reducing carbohydrate- and fat-containing foods automatically reduces kcalories and shifts the balance towards a higher percentage of energy from protein.[7]

Many people seem confused about what kinds of dietary changes they need to make to lose weight. Isn't it helpful to follow a plan?

Yes. Most people need specific instructions and examples to make dietary changes. Popular diets offer dieters a plan. The user doesn't have to decide what foods to eat, how to prepare them, or how much to eat. Unfortunately, these instructions serve short-term weight-loss needs only. Over the long run, people tire of the monotony of diet plans, and most revert to their previous patterns.

By now you can see that the major drawback of most fad weight-loss schemes is that they fail to create lifestyle changes needed to support long-term weight maintenance and improve health. For that, people must do the hard work of learning the facts about nutrition and weight loss, setting their own realistic goals and then devising a lifestyle plan that can achieve them. A balanced, kcalorie-restricted diet with sufficient physical activity to support it may not be the shortcut that most people wish for, but it ensures nutrient adequacy and provides the best chance of long-term success.

Some currently popular diet plans offer a sensible approach to weight loss and healthy eating. The challenge is sorting the fad diets from the healthy options.

Chapter 7 describes reasonable approaches to weight management and concludes that the ideal diet is one you can live with for the rest of your life. Keep that criterion in mind when you evaluate the next "latest and greatest weight-loss diet" that comes along.

Notes

1. D. R. Jacobs, M. D. Gross, and L. C. Tapsell, Food Synergy: An operational concept for understanding nutrition, *American Journal of Clinical Nutrition* 89 (2009): 1543S–1548S.
2. J. M. Nicklas and coauthors, Successful weight loss among obese U.S., adults, *American Journal of Preventive Medicine* 42 (2012): 481–485.
3. W. S. Yancy and coauthors, A randomized trial of a low-carbohydrate diet vs orlistat plus a low-fat diet for weight loss, *Archives of Internal Medicine* 170 (2010): 136-145; F. M. Sacks and coauthors, Comparison of weight-loss diets with different compositions of fat, protein, and carbohydrates, *New England Journal of Medicine* 360 (2009): 859–873.
4. R. J. de Souza and coauthors, Effects of 4 weight-loss diets differing in fat, protein, and carbohydrate on fat mass, lean mass, visceral adipose tissue, and hepatic fat: Results from the POUNDS LOST trial, *American Journal of Clinical Nutrition* 95 (2012): 614–625.
5. de Souza and coauthors, 2012; Sacks and coauthors, 2009.
6. *The Dietary Guidelines for Americans 2010*, www.dietaryguidelines.gov.
7. A. M. Johnstone, Safety and efficacy of high-protein diets for weight loss, *Proceedings of the Nutrition Society* 71 (2012): 339–349.

Chapter 7 Weight Management

ARE YOU PLEASED WITH YOUR BODY WEIGHT? IF YOU ANSWERED YES, you are a rare individual. Nearly all people in our society think they should weigh more or less (mostly less) than they do. Usually, their primary reason is appearance, but they often perceive, correctly, that their weight is also related to physical health. Chapter 6 addressed the health risks of being overweight or underweight.

Overweight and underweight both result from energy imbalance. The simple picture is as follows. Overweight people have consumed more food energy than they have expended and have banked the surplus in their body fat. To reduce body fat, overweight people need to expend more energy than they take in from food. In contrast, underweight people have consumed too little food energy to support their activities and so have depleted their bodies' fat stores and possibly some of their lean tissues as well. To gain weight, they need to take in more food energy than they expend.

This chapter's missions are to present strategies for solving the problems of excessive and deficient body fatness and to point out how appropriate body composition, once achieved, can be maintained. The chapter emphasizes overweight and obesity, partly because they have been more intensively studied and partly because they represent a major health problem in the United States and a growing concern worldwide.

Causes of Obesity

Henceforth, this chapter will use the term *obesity* to refer to excess body fat. Excess body fat accumulates when people take in more food energy than they expend. Why do they do this? Is it genetic? Metabolic? Psychological? Behavioral? All of these? Most likely, obesity has many interrelated causes.

GENETICS AND WEIGHT

A person's genetic makeup influences the body's tendency to consume or store too much energy or to expend too little.[1] Evidence that genes influence eating behavior and body composition comes from family, twin, and adoption studies.[2] Adopted children tend to be more similar in weight to their biological parents than to their adoptive parents.[3] Studies of twins yield similar findings: compared with fraternal twins, identical twins are twice as likely to weigh the same.[4] Genes clearly influence a person's tendency to gain weight or stay lean.[5] The risk of obesity is two to three times higher for a person with a family history of obesity than for a person without such history. Despite such findings, however, only 1 to 5 percent of obesity cases can be explained by a single gene mutation.[6] Furthermore, although genomics researchers have identified hundreds of genes with possible roles in obesity development, they have not yet identified genetic causes of common obesity.[7]

Exceptionally complex relationships exist among the many genes related to energy metabolism and obesity, and they each interact with environmental factors, too. Although an individual's genetic inheritance may make obesity likely, it will not necessarily develop unless given a push by environmental factors that encourage energy consumption and discourage energy expenditure.[8] The following sections describe research involving proteins that might help explain appetite control, energy regulation, and obesity development.

■ Genes instruct cells to make proteins, and each protein performs a unique function.

lipoprotein lipase (LPL): an enzyme mounted on the surface of fat cells (and other cells) that hydrolyzes triglycerides in the blood into fatty acids and glycerol for absorption into the cells. There they are metabolized or reassembled for storage.

leptin: a hormone produced by fat cells under the direction of the (*ob*) gene. It decreases appetite and increases energy expenditure.

leptos = thin

Lipoprotein Lipase Some of the research investigating genetic influence on obesity focuses on the enzyme **lipoprotein lipase (LPL)**, which promotes fat storage in fat cells and muscle cells. People with high LPL activity are especially efficient at storing fat. Obese people generally have much more LPL activity in their fat cells than lean people do (their muscle cell LPL activity is similar, though). This high LPL activity makes fat storage especially efficient. Consequently, even modest excesses in energy intake have a more dramatic impact on obese people than on lean people.

Leptin Researchers have identified a gene in humans called the obesity *(ob)* gene. The obesity gene codes for the protein **leptin**. (■) Leptin is a hormone primarily produced and secreted by the fat cells in proportion to the amount of fat stored. A gain in body fatness stimulates the production of leptin, which, by way of the **hypothalamus**, suppresses the appetite, increases energy expenditure, and produces fat loss.[9] Fat loss produces the opposite effect—suppression of leptin production, increased appetite, and decreased energy expenditure. As the accompanying photo shows, mice with a defective obesity gene do not produce leptin and can weigh up to three times as much as normal mice. When injected with leptin, the mice lose weight. (Because leptin is a protein, it would be destroyed during digestion if given orally; consequently, it must be given by injection.)

Courtesy Amgen, Inc. Photo © John Sholtis/ The Rockefeller University

The mouse on the left is genetically obese—it lacks the gene for producing leptin. The mouse on the right is *also* genetically obese, but because it receives leptin, it eats less, expends more energy, and is less obese than it would be had it not received the leptin.

Although it is extremely rare, researchers have identified a genetic deficiency of leptin in human beings as well.[10] An error in the gene that codes for leptin was discovered in two extremely obese children whose blood levels of leptin were barely detectable. Without leptin, the children had little appetite control; they were constantly hungry and ate considerably more than their siblings or peers. Given daily injections of leptin, these children lost a substantial amount of weight, confirming leptin's role in regulating appetite and body weight.

Most obese people do not have leptin deficiency, however. In fact, most obese people produce plenty of leptin, but they fail to respond to it, a condition called *leptin resistance.*[11] Researchers speculate that blood leptin rises in an effort to suppress appetite and inhibit fat storage when fat cells are ample. Obese people with elevated leptin concentrations may be resistant to its satiating effect. The absence of or resistance to leptin in obesity parallels the scenario of insulin in diabetes: some people have an insulin deficiency (type 1), whereas many others have elevated insulin but are resistant to its glucose-storing effect (type 2).

Ghrelin Another protein known as **ghrelin** works in the opposite direction of leptin. Ghrelin is synthesized and secreted primarily by the stomach cells but works in the hypothalamus to promote a positive energy balance by stimulating appetite and promoting efficient energy storage.[12] The role ghrelin plays in regulating food intake and body weight is the subject of much intense research.[13] Pharmaceutical companies are eager to develop products that mimic ghrelin to treat wasting conditions, as well as products that oppose ghrelin's action to treat obesity.[14]

Ghrelin powerfully triggers the desire to eat. Blood levels of ghrelin typically rise before and fall after a meal in proportion to the kcalories ingested—reflecting the hunger and satiety that precede and follow eating. In general, fasting blood levels correlate inversely with body weight: lean people have high ghrelin levels and obese people have low levels.

Ghrelin fights to maintain a stable body weight. On average, ghrelin levels are high whenever the body is in negative energy balance, as occurs during low-kcalorie diets, for example. This response may help explain why weight loss is so difficult to maintain. Ghrelin levels decline again whenever the body is in positive energy balance, as occurs with weight gains.

Fat Cell Development When "energy in" exceeds "energy out," much of the excess energy is stored in the fat cells of adipose tissue. The amount of fat in adipose tissue reflects both the *number* and the *size* of the fat cells.* The number of fat cells increases most rapidly during the growing years of late childhood and early puberty. After growth ceases, fat cell number may continue to increase whenever energy balance is positive.[15] Obese people have more fat cells than healthy-weight people; their fat cells are also larger.

Fat cells can also expand in size. After they reach their maximum size, more cells can develop to store more fat. Thus, obesity develops when a person's fat cells increase in number, in size, or quite often both. Figure 7-1 (p. 168) illustrates fat cell development.

With fat loss, the size of the fat cells shrinks, but their number cannot. For this reason, people with extra fat cells may tend to regain lost weight rapidly. Prevention of

*Obesity due to an increase in the *number* of fat cells is *hyperplastic obesity.* Obesity due to an increase in the *size* of fat cells is *hypertrophic obesity.*

hypothalamus (high-po-THAL-ah-mus): a brain center that controls activities such as maintenance of water balance, regulation of body temperature, and control of appetite.

ghrelin (GRELL-in): a hormone produced primarily by the stomach cells. It signals the hypothalamus of the brain to stimulate appetite and food intake.

FIGURE 7-1 Fat Cell Development

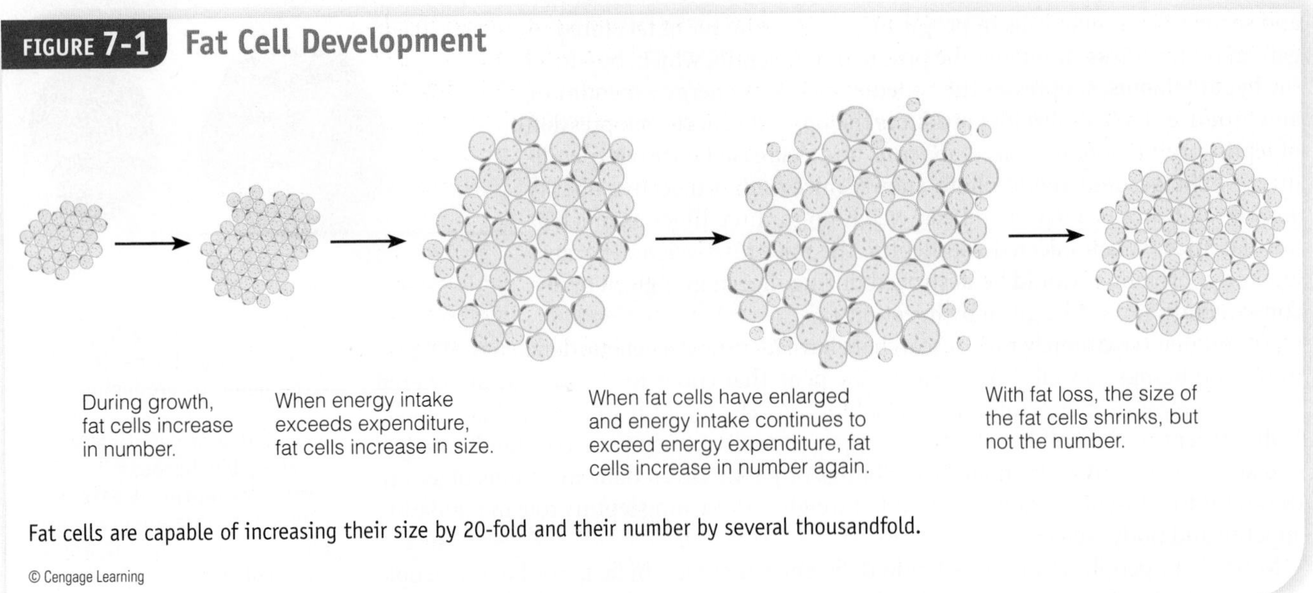

During growth, fat cells increase in number.
When energy intake exceeds expenditure, fat cells increase in size.
When fat cells have enlarged and energy intake continues to exceed energy expenditure, fat cells increase in number again.
With fat loss, the size of the fat cells shrinks, but not the number.

Fat cells are capable of increasing their size by 20-fold and their number by several thousandfold.

obesity is most critical, then, during the growing years of childhood and adolescence when fat cells increase in number most profoundly.

Set-Point Theory One popular theory of why a person may store too much fat is the **set-point theory**. The set-point theory proposes that body weight, like body temperature, is physiologically regulated.[16] Researchers have noted that many people who lose weight quickly regain it all. This suggests that somehow the body chooses a preferred weight and defends that weight by regulating eating behaviors and hormonal actions. After weight gains or losses, the body adjusts its metabolism to restore the original weight.

Energy expenditure increases after weight gain and decreases after weight loss. These changes in energy expenditure are greater than those predicted based on body composition alone, and they help to explain why it can be difficult for an underweight person to maintain weight gains and an overweight person to maintain weight losses. An individual's set point for body weight may be adjustable, shifting over the life span in response to physiological changes and to genetic, dietary, and other factors.

ENVIRONMENTAL STIMULI

As discussed earlier, genetic factors play a partial role in determining a person's susceptibility to obesity, but they do not fully explain obesity. Obesity rates have risen dramatically during recent decades, but the human gene pool has remained unchanged. The environment must therefore play a role as well. Obesity reflects the interaction between genes and the environment.[17] The *environment* includes all of the circumstances that people encounter daily that push them toward fatness or thinness. Over the past four decades, the demand for physical activity has decreased as the abundance of food has increased.[18]

Overeating People may overeat in response to stimuli in their surroundings—primarily, the availability of many delectable foods. Most people in the United States find high-kcalorie foods readily available, relatively inexpensive, heavily advertised, and reasonably tasty. Food is available everywhere, all the time—thanks largely to fast food. Our highways are lined with fast-food restaurants, and convenience stores and service stations offer fast food as well. Fast food is available in malls, airports, and even schools. Fast food is convenient, and it's available morning, noon, and night—and all times in between. With around-the-clock access to rich palatable foods, we eat more and more often than in decades past—and energy intakes have risen accordingly.[19]

set-point theory: the theory that the body tends to maintain a certain weight by means of its own internal controls.

Most alarming are the extraordinarily large serving sizes and ready-to-go meals offered in supersize combinations. Eating large portion sizes multiple times a day accounts for much of the weight increase seen over the decades.[20] People buy the large sizes and combinations, perceiving them to be a good value, but then they eat more than they need. Research shows that people eat more if they're served more.[21] "Portion distortion" is a term used to describe the perception that large portions are the appropriate amounts to eat at a single sitting.[22] And portion sizes of virtually all foods and beverages (and even sizes of plates, glasses, and utensils) have increased markedly in the past several decades, most notably at fast-food restaurants. The increase in portion sizes parallels the growing prevalence of overweight and obesity in the United States, beginning in the 1970s, rising sharply in the 1980s, and continuing today.

Research suggests that fast-food consumption contributes significantly to the development of obesity.[23] Fast food is often energy-dense food, which increases energy intake, BMI, and body fatness. The combination of large portions and energy-dense foods strikes a double blow. Reducing portion sizes is helpful, but the real kcalorie savings come from lowering the energy density. Satisfying portions of foods with low energy density such as fruits and vegetables can help with weight loss.

Consumers' health would benefit from restaurants providing appropriate portion sizes and offering more fruits, vegetables, legumes, and whole grains. In an effort to help consumers make healthier choices, national legislation now requires restaurants with 20 or more locations to provide menu listings of an item's kcalories, grams of saturated fat, and milligrams of sodium.[24] Once such information displays become standard throughout the country, perhaps restaurants will have greater incentives to offer healthier choices.

hunger: the physiological need to eat, experienced as a drive to obtain food; an unpleasant sensation that demands relief.

appetite: the psychological desire to eat; a learned motivation that is experienced as a pleasant sensation that accompanies the sight, smell, or thought of appealing foods.

satiation (say-she-AY-shun): the feeling of satisfaction and fullness that occurs during a meal and halts eating. Satiation determines how much food is consumed during a meal.

satiety: the feeling of fullness and satisfaction that occurs after a meal and inhibits eating until the next meal. Satiety determines how much time passes between meals.

Learned Behavior Psychological stimuli also trigger inappropriate eating behaviors in some people. Appropriate eating behavior is a response to **hunger**. Hunger is a drive programmed into people by their heredity. **Appetite**, in contrast, is learned and can lead people to ignore hunger or to over-respond to it. Hunger is physiological, whereas appetite is psychological, and the two do not always coincide.

During the course of a meal, as food enters the GI tract and hunger diminishes, **satiation** occurs. As receptors in the stomach stretch and hormones become active, the person begins to feel full. The response is satiation, which prompts the person to stop eating.

After a meal, the feeling of **satiety** continues to suppress hunger and allow a person to not eat again for a while. Whereas *satiation* tells us to "stop eating," *satiety* reminds us to "not start eating again." As mentioned above, people can override these signals, especially when presented with favorite foods or stressful situations.

Food behavior is intimately connected to deep emotional needs such as the primitive fear of starvation. Yearnings, cravings, and addictions with profound psychological significance can express themselves in people's eating behavior. An emotionally insecure person might eat rather than call a friend and risk rejection. Another person might eat to relieve boredom or to ward off depression.

Physical Inactivity The possible causes of obesity mentioned so far all relate to the input side of the energy equation. What about output? People may be obese, not because they eat too much, but because they move too little—both in purposeful exercise and in the activities of daily life.[25] Obese people observed closely are often seen to eat less than lean people, but they are sometimes so extraordinarily inactive that they still manage to accumulate an energy surplus. Reducing their

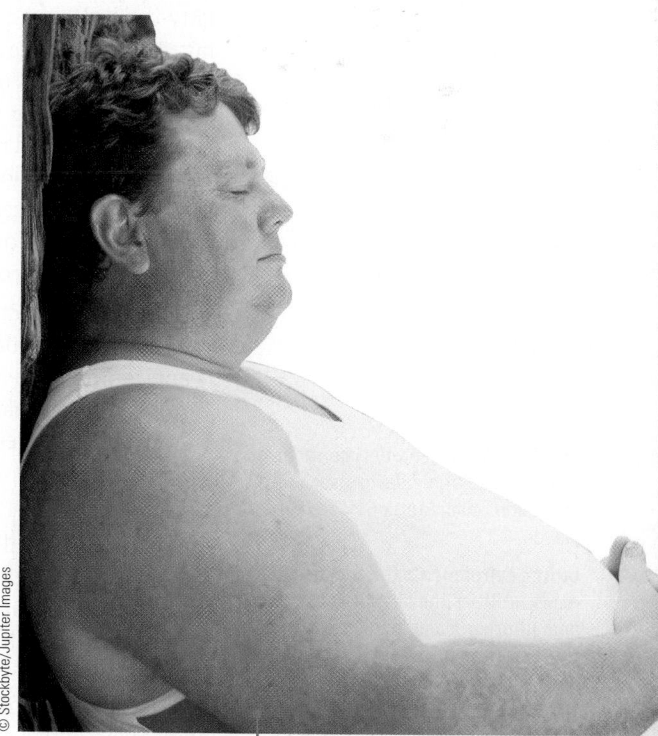

Lack of physical activity fosters obesity.

food intake further would jeopardize health and incur nutrient deficiencies. Physical activity, then, is a necessary component of nutritional health. People must be physically active if they are to eat enough food to deliver all the nutrients needed without unhealthy weight gain.

Our environment, however, fosters inactivity. In turn, inactivity contributes to obesity and poor health.[26] Sedentary **screen time** has all but replaced outdoor activity for many people. In addition, most people work at sedentary jobs. One hundred years ago, 30 percent of the energy used in farm and factory work came from muscle power; today, only 1 percent does. Modern technology has replaced physical activity at home, at work, at school, and in transportation. The more time spent sitting still, the higher the risk of dying from heart disease and other causes.[27]

Health experts urge people to "take the stairs instead of the elevator" or "walk or bike to work." These are excellent suggestions: climbing stairs provides an impromptu workout, and people who walk or ride a bicycle for transportation most often meet their needs for physical activity. Physical activity strategies for weight loss and weight maintenance are offered in a later section of this chapter.

Neighborhood Obstacles to Physical Activity and Healthy Foods

Some aspects of the **built environment**, including buildings, sidewalks, and transportation opportunities, can discourage physical activity. For example, most stairwells of modern buildings are inconvenient, isolated, and unsafe. Roadways often lack sidewalks, crosswalks, or lanes marked for bicycles. The air on roadways can be dangerously high in carbon monoxide gas and other pollutants from gasoline engine emissions. Hot and cold weather also pose hazards for outdoor commuters. In contrast, those with access to health-promoting foods and built environments more easily make healthy choices.[28] Safe, affordable biking and walking areas and public exercise facilities help maintain health and body leanness.

In addition, residents of many low-income urban and rural areas lack access to even a single supermarket.[29] Often overweight and lacking transportation, residents of these so-called **food deserts** have limited access to the affordable, fresh, nutrient-dense foods they need.[30] Instead, they shop at local convenience stores and fast-food places, where they can purchase mostly refined packaged sweets and starches, sugary soft drinks, fatty canned meats, or fast foods, and they often have an eating pattern that predicts nutrient deficiencies and excesses along with high rates of obesity and type 2 diabetes.[31]

In truth, in most neighborhoods across the United States, the most accessible, affordable, and tempting foods and beverages are high-kcalorie, good-tasting, inexpensive fare from fast-food restaurants, and it takes a great deal of attention, planning, and time to "go against the flow" to keep kcalorie intakes reasonable. Today, only about a third of the population succeeds in doing so. The prestigious National Academies' Institute of Medicine has put forth these national goals as most likely to slow or reverse the obesity epidemic and improve the nation's health:

- Make physical activity an integral and routine part of American life.
- Make healthy foods and beverages available everywhere—create food and beverage environments to make healthy food and beverage choices the routine, easy choice.
- Advertise and market what matters for a healthy life.
- Activate employers and health-care professionals.
- Strengthen schools as the heart of health.[32]

Accomplishing any one of these goals on its own might speed up progress in preventing obesity, but if all these goals are realized, their effects will be powerful allies in the nationwide struggle to regain control over weight and health. Such changes require leaders at all levels and citizenry across all sectors of society to work toward one goal: improving the health of the nation.

Until these changes occur, the best way for most people to attain a healthy body composition boils down to overcoming obstacles to making dietary changes and increasing physical activity. Later sections focus on these areas, while the next section delves into the details of obesity treatment.

screen time: sedentary time spent using an electronic device, such as a television, computer, or video game player.

built environment: the buildings, roads, utilities, homes, fixtures, parks, and all other manufactured entities that form the physical characteristics of a community.

food deserts: urban and rural low-income areas with limited access to affordable and nutritious foods.

Review Notes

- Genetics, fat cell development, set point, overeating, and inactivity all offer possible, but still incomplete, explanations of obesity.
- Most likely, obesity has not one cause but different causes and combinations of causes in different people.
- The easy availability of large portions of energy-dense foods is a major environmental factor thought to be contributing to the obesity epidemic.
- Built environments that lack opportunities for safe physical activity and access to fresh, nutrient-dense foods are also linked with obesity.

Obesity Treatment: Who Should Lose?

Millions of U.S. adults are trying to lose weight at any given time. Some of these people do not even need to lose weight. Others may benefit from weight loss but are not successful. Relatively few people succeed in losing weight, and even fewer succeed permanently.

Many people assume that every overweight person can achieve slenderness and should pursue that goal. Consider, however, that most overweight people cannot become slender. People vary in their weight tendencies just as they vary in their potentials for height and degrees of health. The question of whether a person should lose weight depends on many factors: the extent of overweight, age, health, and genetics, to name a few. Weight-loss advice, then, does not apply equally to all overweight people. Some people may risk more in the process of losing weight than in remaining overweight. Others may reap significant health benefits with even modest weight loss.

Review Notes

- Weight-loss advice does not apply equally to all overweight people.
- Some people may risk more through misguided efforts to lose weight than by remaining overweight, whereas others may benefit from even a modest weight loss.

Inappropriate Obesity Treatments

The risks people incur in attempting to lose weight often depend on how they go about it. Weight-loss plans and obesity treatments abound—some are adequate, but many are ineffective and possibly dangerous. Fad diets are the topic of Nutrition in Practice 6. This section addresses other inappropriate obesity interventions. Aggressive approaches to obesity for those obese people who face high risks of medical problems and must lose weight rapidly are discussed in the next section. Reasonable approaches to overweight for those seeking safe, gradual weight losses are saved for the last part of the obesity treatment discussion.

OVER-THE-COUNTER WEIGHT-LOSS PRODUCTS

Millions of people in the United States use over-the-counter (OTC) weight-loss products, believing them to be safe. Most of the people who use such products are women, especially young overweight women, but almost 10 percent are of normal weight. Promoters and marketers of weight-loss products make all kinds of claims for their products with only one intention—profit. Such claims as "eat all you want and lose weight," "take three pills before bedtime and watch the fat disappear," "blocks carbs," "blocks fat," and many more lure people into believing that maybe this time a product will really work.

In an investigation of OTC weight-loss pills, powders, and other "dietary supplements," the FDA found that an alarming number of products illegally contained prescription medications. Strong diuretics, unproven experimental drugs, psychotropic drugs used to treat mental illnesses, and even drugs deemed unsafe and so banned from U.S. markets were among those discovered, and all pose serious health risks.[33]

In their search for weight-loss magic, some consumers turn to "natural" herbal products and dietary supplements, even though few have proved to be effective.[34] People falsely believe that "natural" herbs are not harmful to the body, but many herbs contain toxins. Belladonna and hemlock are infamous examples, but many lesser-known herbs, such as sassafras, contain toxins as well. Furthermore, because herbs are marketed as "dietary supplements," manufacturers need not present scientific evidence of their safety or effectiveness to the FDA before marketing them. Evidence about their safety is gathered only through reports of consumers who sicken or die after using the remedies.

A now familiar example is ephedra (also called ma huang), an herb that showed promise as a weight-loss drug in preliminary studies. Immediately, ephedra-containing products for dieters and athletes flooded the market. Many consumers of these products reported ill effects including cardiac arrest, abnormal heartbeats, hypertension, strokes, and seizures; the supplements have been linked to some deaths as well. For this reason, the FDA has banned the sale of dietary supplements containing ephedra and its active constituent, ephedrine.* (■)

OTC weight-loss pills, powders, herbs, and other "dietary supplements" are not associated with successful weight loss and maintenance.[35] Nutrition in Practice 14 explores the possible benefits and potential dangers of herbal products and other alternative therapies. Anyone using dietary supplements for weight loss should first consult a physician.

OTHER GIMMICKS

Other gimmicks don't help with weight loss either. Hot baths do not speed up metabolism so that pounds can be lost in hours. Steam and sauna baths do not melt the fat off the body, although they may dehydrate people so that they lose water weight. Brushes, sponges, wraps, creams, and massages intended to move, burn, or break up fat do nothing of the kind.

Review Note

- Efforts to lose weight via unwise techniques, such as use of over-the-counter supplements and herbal products, can be physically and psychologically damaging.

Aggressive Treatments of Obesity

■ Ephedrine is an amphetamine-like substance extracted from the Chinese herb ma huang.

For some obese people, the medical problems caused by obesity demand treatment approaches that may, themselves, incur some risks. The health benefits to be gained by weight loss, however, may make these risks worth taking.

OBESITY DRUGS

■ Drugs may be an option for people with all of the following conditions:

- Unable to achieve adequate weight loss with diet and exercise
- BMI ≥ 30, or BMI ≥ 27 with weight-related health problems
- No medical contraindications

Several prescription medications for weight loss have been tried over the years. When used as part of a long-term, comprehensive weight-loss program, medications can help with modest weight loss.[36] (■) Because weight regain commonly occurs with the discontinuation of drug therapy, treatment is long term—and the long-term use of

*Ma huang (ephedrine) is illegal in Canada.

TABLE 7-1 FDA-Approved Drugs for Weight Loss

Product	Action	Side Effects
Orlistat (OR-leh-stat), trade names: Alli, Xenical	Inhibits pancreatic lipase activity in the GI tract, thus blocking digestion and absorption of dietary fat and limiting energy intake	Cramping, diarrhea, gas, frequent bowel movements, reduced absorption of fat-soluble vitamins; rare cases of liver injury
Phentermine (FEN-ter-mean), diethylpropion (DYE-eth-ill-PRO-pee-on), phendimetrazine (FEN-dye-MEH-tra-zeen)	Enhances the release of the neurotransmitter norepinephrine, which suppresses appetite	Increased blood pressure and heart rate, insomnia, nervousness, dizziness, headache
Lorcaserin hydrochloride, trade name: Belviq (BELL-veek)	Interacts with brain serotonin receptors to increase satiety	Headache, dizziness, fatigue, nausea, dry mouth, and constipation; low blood glucose in people with diabetes; serotonin syndrome, including agitation, confusion, fever, loss of coordination, rapid or irregular heart rate, shivering, seizures, and unconsciousness; cannot be safely used by pregnant or lactating women or people with heart valve problems; high doses cause hallucinations
Qsymia (kyoo-sim-EE-uh)	Combines phentermine (an appetite suppressant) and topiramate (a seizure/migraine medication) that makes food seem less appealing and increases feelings of fullness	Increased heart rate; can cause birth defects if taken in the first weeks or months of pregnancy; may worsen glaucoma or hyperthyroidism; may interact with other medications

Note: Weight-loss drugs are most effective when taken as directed and used in combination with a reduced-kcalorie diet and increased physical activity.

medications poses risks. Medical experts do not yet know whether a person would benefit more from maintaining a 50-pound excess or from taking a medication for a decade to keep the 50 pounds off.

The challenge, then, is to develop an effective drug—or more likely, a combination of drugs—that can be used over time without adverse side effects or the potential for abuse. Weight-loss drugs should be prescribed in tandem with a healthy diet and physical activity program. Table 7-1 presents FDA-approved drugs used to treat obesity.

SURGERY

The prevalence of **clinically severe obesity** is increasing at an incredibly rapid rate.[37] At this level of obesity, lifestyle changes and modest weight losses can improve disease risks a little, but the most effective treatment is surgery.[38] Two procedures, **gastric bypass** and **gastric banding**, have gained wide acceptance. Both procedures limit food intake by effectively reducing the capacity of the stomach. In addition, gastric bypass suppresses hunger by reducing production of gastrointestinal hormones. The results can be dramatic: greater than 90 percent of surgical patients achieve a weight loss of more than 50 percent of their excess body weight, and much of the loss may be maintained over time.[39] More long-term studies are needed, but surgery with weight loss often brings immediate and lasting improvements in blood lipids, diabetes, sleep apnea, heart disease, and hypertension.[40]

Laparoscopic weight-loss surgery techniques are now used to perform gastric bypass and other weight-loss surgeries. Laparoscopic weight-loss surgery produces significant, long-term weight loss, shortens recovery, and is less invasive than open surgery.

clinically severe obesity: a BMI of 40 or greater, or a BMI of 35 or greater with one or more serious conditions such as hypertension. Another term used to describe the same condition is *morbid obesity*.

gastric bypass: surgery that restricts stomach size and reroutes food from the stomach to the lower part of the small intestine; creates a chronic, lifelong state of malabsorption by preventing normal digestion and absorption of nutrients.

gastric banding: a surgical means of producing weight loss by restricting stomach size with a constricting band.

laparoscopic weight-loss surgery: a weight-loss surgery procedure in which surgeons gain access to the abdomen via several small incisions. A tiny video camera is inserted through one of the incisions and surgical instruments through the others. The surgeons watch their work on a large-screen monitor.

The long-term safety and effectiveness of gastric surgery depend, in large part, on compliance with dietary instructions. Common immediate postsurgical complications include infections, nausea, vomiting, and dehydration; in the long term, vitamin and mineral deficiencies and psychological problems are common. Lifelong medical supervision is necessary for those who choose the surgical route, but in suitable candidates, the health benefits of weight loss may prove worth the risks.[41]

Review Note

- Obese people with high risks of medical problems may need aggressive treatment, including drugs or surgery.

Reasonable Strategies for Weight Loss

Efforts to combat obesity must integrate healthy eating patterns, physical activities, supportive environments, and psychosocial support.[42] Successful weight-loss strategies embrace small changes, moderate losses, and reasonable goals.[43] People who lose 10 to 20 pounds in a year by consistently choosing nutrient-dense foods and engaging in regular physical activity are much more likely to maintain the loss and reap health benefits than if they were to lose more weight in less time by adopting the latest fad diet. In keeping with this philosophy, the *Dietary Guidelines for Americans* advise those who need to lose weight to "consume fewer kcalories from foods and beverages, increase physical activity, and reduce time spent in sedentary behaviors." Modest weight loss, even when a person is still overweight, can improve control of diabetes and reduce the risks of heart disease by lowering blood pressure and blood cholesterol, especially for those with central obesity.[44]

Of course, the same eating and activity habits that improve health often lead to a healthier body weight and composition as well. Successful weight loss, then, is defined not by pounds lost but by health gained. People less concerned with disease risks may prefer to set goals for personal fitness, such as being able to play with children or climb stairs without becoming short of breath.

Whether the goal is health or fitness, weight-loss expectations need to be reasonable. Unreachable targets ensure frustration and failure. Setting reasonable goals helps to achieve the desired result in managing weight. For example, obese people who must reduce their weight to lower their disease risks might set three broad goals:

1. Reduce body weight by about 10 percent over half a year's time.
2. Maintain a lower body weight over the long term.
3. At a minimum, prevent further weight gain.

Such goals may be achieved or even exceeded, providing a sense of accomplishment instead of disappointment.

After all, excess weight takes years to accumulate. Losing excess body fat also takes time, along with patience and perseverance. The person must adopt healthy eating patterns, engage in physical activity, create a supportive environment, and seek out behavioral and social support; continue these behaviors for at least six months for initial weight loss; and then continue all of it for a lifetime to maintain the losses.[45] Setbacks are a given, and according to recent evidence, the size of the kcalorie deficit required to lose a pound of weight initially may be smaller than that required later on. In other words, weight loss is hard at first, and then it may get harder.[46]

Courtesy of the author

A healthy body contains enough lean tissue to support health and the right amount of fat to meet body needs.

A HEALTHFUL EATING PLAN

Contrary to the claims of many fad diets, no particular eating plan is magical, and no specific food must be either included or avoided for weight management. In designing an eating pattern, people need only consider foods that they like or can learn to like, that are available, and that are within their means.

A Realistic Energy Intake The main characteristic of a weight-loss diet is that it provides less energy than the person needs to maintain present body weight. If food energy is restricted too severely, dieters may not receive sufficient nutrients and may lose lean tissue. Rapid weight loss usually means excessive loss of lean tissue, a lower BMR, and a rapid weight gain to follow. Restrictive eating may also set in motion the unhealthy behaviors of eating disorders (described in Nutrition in Practice 7).

More than a decade ago, obesity experts from the National Institutes of Health recommended that reductions in energy intake should be based on a person's BMI.[47] Dieters with a BMI of 35 or greater are encouraged to reduce their usual daily kcalorie intakes by about 500 to 1,000 kcalories. People with a BMI between 27 and 35 should reduce energy intake by 300 to 500 kcalories a day. Based on the assumption that a 3,500-kcalorie deficit will consistently produce a pound of weight loss, these recommendations are predicted to produce a weight loss of about 1 to 2 pounds per week while retaining lean tissue.[48] However, this assumption has been called into question.[49]

Recent research suggests that the kcalorie deficit required to lose a pound of weight may increase as weight loss progresses.[50] Weight loss may proceed rapidly for some weeks or months but will eventually slow down. The following factors contribute to a decline in the rate of loss:

- Metabolism may slow in response to a lower kcalorie intake and loss of metabolically active lean tissue.
- Less energy may be expended in physical activity as body weight diminishes.

In addition, the composition of weight loss itself may affect the rate of loss. Body weight lost early in dieting may be composed of a greater percentage of water and lean tissue, which contains fewer kcalories per pound as compared with later losses that appear to be composed mostly of fat.[51] This may mean that dieters should expect a slowdown in weight loss as they progress past the initial phase.

Newer schemes for determining the kcalorie deficit required for weight loss reflect changes in energy use by the body over time. These require extensive calculations for accuracy and are most efficiently applied with the power of computer programs.* A panel of experts recommended this shortcut, however: "every permanent 10-kcalorie change in energy intake per day will lead to an eventual weight change of 1 pound when the body weight reaches a new steady state. It will take nearly 1 year to achieve 50 percent and 3 years to achieve 95 percent of this weight loss."[52†]

Most people can lose weight safely on an eating pattern providing approximately 1200 kcalories per day for women and 1600 kcalories per day for men.[53] Table 7-2 suggests daily food amounts from which to build balanced 1200- to 1600-kcalorie diets.

Some people skip meals, typically breakfast, in an effort to reduce energy intake, but research

TABLE 7-2 Daily Amounts from Each Food Group for 1200- to 1600-kCalorie Diets

Food Group	1200 kCalories	1400 kCalories	1600 kCalories
Fruit	1 c	1½ c	1½ c
Vegetables	1½ c	1½ c	2 c
Grains	4 oz	5 oz	5 oz
Protein foods	3 oz	4 oz	5 oz
Milk	2½ c	2½ c	3 c
Oils	4 tsp	4 tsp	5 tsp

*Two such programs are available at http://bwsimulator.niddk.nih.gov and at http://www.pbrc.edu/the-research/tools/weight-loss-predictor.

†Source: K.D. Hall and coauthors, Energy balance and its components: Implications for body weight regulation, *American Journal of Clinical Nutrition* 95 (2012): 989–994.

suggests such a strategy may be counterproductive. Breakfast frequency is inversely associated with obesity—that is, people who frequently eat breakfast have lower BMI values than those who tend to skip breakfast.[54] Furthermore, when people eat breakfast, overall diet quality is better and daily energy density is lower—two factors that support healthy body weight.

Nutritional Adequacy Nutritional adequacy is difficult to achieve on fewer than 1200 kcalories a day, and most healthy adults need never consume any less than that. A plan that provides an adequate intake supports a healthier and more successful weight loss than a restrictive plan that creates feelings of starvation and deprivation, which can lead to an irresistible urge to binge.

Take a look at the 1200-kcalorie diet in Table 7-2. Such an intake would allow most people to lose weight and still meet their nutrient needs with careful, nutrient-dense food selections. Healthy eating patterns for weight loss should provide all of the needed nutrients in the form of fresh fruits and vegetables; low-fat milk products or substitutes; legumes; small amounts of lean meats, seafood, poultry, or meat alternatives; nuts; and whole grains.[55] These foods are necessary for adequate protein, carbohydrate, fiber, vitamins, and minerals and are generally associated with leanness. They are also best for managing weight.

Wholesome, high-fiber, unprocessed or lightly processed foods offer bulk and satiety for fewer kcalories than smooth, quickly consumed refined foods. Thus, choosing whole grains and fiber-rich vegetables in place of most refined grains and added sugars benefits both weight and nutrition.

Choose fats sensibly by avoiding most solid fats and including enough unsaturated oils (details in Nutrition in Practice 4) to support health but not so much as to oversupply kcalories. Nuts provide unsaturated fat and protein, and people who regularly eat nuts often maintain a healthy body weight.[56] Lean meats or other low-fat protein sources also play important roles in weight loss and provide satiety. Sufficient protein foods may also help to preserve lean tissue, including muscle tissue, during weight loss.[57]

A dietary supplement providing vitamins and minerals—especially iron and calcium for women—at or below 100 percent of the Daily Values can help people following low-kcalorie eating patterns to achieve nutrient adequacy. A person who plans resolutely to include all of the foods from each group needed each day will be satisfied, be well-nourished, and have little appetite left for high-kcalorie treats.

Small Portions As mentioned earlier, large portion sizes increase energy intakes, and the huge helpings served by restaurants and sold in packages are the enemies of people striving to manage their weight. Similarly, big bowls, plates, utensils, and drinking glasses encourage people to take and consume larger portions. Small plates and bowls, tall and thin drinking glasses, and luncheon-sized plates have the opposite effect.

For health's sake, overweight people may need to learn to eat less food at each meal—one piece of chicken for dinner instead of two, a teaspoon of butter on the vegetables instead of a tablespoon, and one cookie for dessert instead of six. Chew foods slowly and thoroughly. The goal is to eat enough food for energy, nutrients, and pleasure, but not more. This amount should leave a person feeling satisfied—not necessarily full. Keep in mind that even fat-free and low-fat foods can deliver a lot of kcalories when a person eats large quantities.

People who have difficulty making low-kcalorie selections or controlling portion sizes may find it easier to use prepared meal plans. Prepared meals that provide low-kcalorie, nutritious meals or snacks can support weight loss while easing the task of diet planning.[58] Ideally, those using a prepared meal plan will also receive counsel from a registered dietitian to learn how to select appropriately from conventional food choices as well.

Lower Energy Density As discussed earlier, to lower energy intake, people can choose smaller portion sizes, or they can reduce the energy density of the foods they eat. Research shows that eating satisfying portions of foods that are low in energy density (such as fruits, vegetables, and broth-based soups) maintains satiety while reducing energy intake.[59] Foods containing substantial water or fiber and those low in fat help to

FIGURE 7-2 Energy Density

Selecting grapes with their high water content instead of raisins increases the volume and cuts the energy intake in half.

Even at the same weight and similar serving sizes, the fiber-rich broccoli delivers twice the fiber of the potatoes for about one-fourth the energy.

By selecting the water-packed tuna (on the right) instead of the oil-packed tuna, a person can enjoy the same amount for fewer kcalories.

Decreasing the energy density (kcal/g) of foods allows a person to eat satisfying portions while still reducing energy intake. To lower energy density, select foods high in water or fiber and low in fat.

Matthew Farruggio (all)

lower a meal's energy density (see Figure 7-2). The Clinical Applications feature at the end of this chapter describes how to calculate the energy density of foods.

Sugar and Alcohol A person trying to achieve or maintain a healthy weight needs to limit intakes of added sugars and alcohol, as well as fat. Including them for pleasure on occasion is compatible with health as long as most daily choices are of nutrient-dense foods.

Meal Spacing Three meals a day is standard in our society, but no law says you can't have four or five—just be sure they are smaller, of course. People who eat small, frequent meals can be as successful at weight loss and maintenance as those who eat three.[60] Make sure that mild hunger, not appetite, is prompting you to eat. Eat regularly, and eat before you become extremely hungry.

Adequate Water Learn to satisfy thirst with water. Water helps with weight management in several ways. For one, foods with high water content (such as broth-based soups) increase fullness, reduce hunger, and consequently reduce energy intake. For another, drinking a large glass of water before a meal may ease hunger, fill the stomach, and reduce energy intake.[61] Importantly, water adds no kcalories. The average U.S. diet delivers an estimated 75 to 150 kcalories a day from sweetened beverages. Simply replacing nutrient-poor, energy-dense beverages with water could save a person up to 15 pounds a year. Water also helps the GI tract adapt to a high-fiber diet.

PHYSICAL ACTIVITY

The best approach to weight management includes physical activity.[62] Either energy restriction or physical activity alone can produce some weight loss. Clearly, however, the combination is most effective.[63] People who combine diet and physical activity are more likely to lose more fat, retain more muscle, and regain less weight than those who only diet.[64] Table 1-5 in Chapter 1 presented the *Physical Activity Guidelines for Americans, 2008,* which specify the minimum amount of physical activity people need to gain general *health* benefits. Table 7-3 (p. 178) presents physical activity strategies from the American College of Sports Medicine developed specifically to prevent weight gain and promote at least modest weight loss.[65]

Even without weight loss, physical activity may also help counteract some of the negative effects of excess body weight on health.[66] For example, physical activity reduces abdominal obesity, and this change improves blood pressure, insulin resistance, and fitness of the heart and lungs, even without weight loss.[67]

TABLE 7-3 Physical Activity Strategies for Weight Management
A negative energy balance incurred through physical activity will result in weight loss, and the larger the negative energy balance, the greater the weight loss.
• Weight gain prevention and augmented weight loss occur with at least 2 hours and 30 minutes (150 minutes) per week of at least moderate-intensity physical activity.
• Greater weight loss and improved weight maintenance after weight loss occur with more than 4 hours and 10 minutes (>250 minutes) per week of at least moderate-intensity physical activity.
• Both aerobic (endurance) and muscle-strengthening (resistance) physical activities are beneficial, but kcalorie restriction must accompany resistance training to achieve weight loss.

Source: Adapted from American College of Sports Medicine, Position stand: Appropriate physical activity intervention strategies for weight loss and prevention of weight regain for adults, *Medicine and Science in Sports and Exercise* 41 (2009): 459–471.

Energy Expenditure Physical activity directly increases energy output by the muscles and cardiovascular system. Table 6-2 in Chapter 6 (pp. 148–149) shows how much energy each of several activities uses. The number of kcalories spent in an activity depends on body weight, intensity, and duration. For example, a 150-pound person who walks 3.5 miles in 60 minutes expends about 315 kcalories. That same person running 3 miles in 30 minutes uses a similar amount. By comparison, a 200-pound person running 3 miles in 30 minutes expends an additional 125 kcalories or so (about 444 kcalories total). The goal is to expend as much energy as time allows. The greater the energy deficit created by physical activity, the greater the fat loss. A word of caution, however: people who reward themselves with high-kcalorie foods for "good behavior" can easily negate any kcalorie deficits incurred with physical activity.

BMR Although the BMR is elevated in the hours after physical activity, this effect requires a sustained high-intensity workout beyond the level achievable by most weight-loss seekers.[68] Over the long term, however, a person who engages in daily vigorous activity gradually develops more lean tissue, which is more active metabolically than fat tissue. Metabolic rate rises accordingly, and this makes a contribution toward continued weight loss or maintenance.

Appetite Control Physical activity also helps to control appetite. People think that exercising will make them want to eat, but this is not entirely true. Active people do have healthy appetites, but appetite is suppressed immediately following a workout and satiation during meals is heightened.[69] The reasons are unclear, but physical activity may help to control appetite by altering the levels of appetite-regulating hormones.[70]

■ Benefits of physical activity in a weight-management program:
- Improved body composition
- Favorable effects on disease risks
- Short-term increase in energy expenditure (from exercise and from a slight rise in BMR)
- Long-term increase (slight) in BMR
- Appetite control
- Stress reduction and control of stress eating
- Physical, and therefore psychological, well-being
- Improved self-esteem

Psychological Benefits Physical activity helps especially to curb the inappropriate appetite that prompts a person to eat when bored, anxious, or depressed. Weight-management programs encourage people to go out and be active when they're tempted to eat but are not really hungry.

Physical activity also helps to reduce stress. Because stress itself is a cue to inappropriate eating behavior for many people, activity can help here, too.

Activity offers still more psychological advantages. The fit person looks and feels healthy, and high self-esteem accompanies these benefits. High self-esteem tends to support a person's resolve to persist in a weight-control effort, rounding out a beneficial cycle. (■)

resistance training: the use of free weights or weight machines to provide resistance for developing muscle strength, power, and endurance; also called *weight training*. A person's own body weight may also be used to provide resistance as when a person does push-ups, pull-ups, or abdominal crunches.

Choosing Activities What kind of physical activity is best? For health, a combination of moderate to vigorous aerobic physical activity along with **resistance training** at a safe level provides benefits. However, any physical activity is better than being sedentary.[71]

People seeking to lose weight should choose activities that they enjoy and are willing to do regularly. Health care professionals frequently advise people who want to manage their body weight and lose fat to engage in activities of low-to-moderate intensity for a long duration, such as an hour-long fast-paced walk. The reasoning behind such advice is that people exercising at low to moderate intensity are likely to stick with their activity for longer times and are less likely to injure themselves. People who regularly engage in more *vigorous* physical activities (fast bicycling or endurance running, for example), however, have less body fat than those who engage in moderately intense activities. The conditioned body that is adapted to strenuous and prolonged aerobic activity uses more fat all day long, not just during activity. The bottom line on physical activity and weight and/or fat loss seems to be that total energy expenditure is the main factor, regardless of how a person does it.

In addition to activities such as walking or cycling, there are hundreds of ways to incorporate energy-expending activities into daily routines: take the stairs instead of the elevator, walk to the neighbor's apartment instead of making a phone call, and rake the leaves instead of using a blower. These activities burn only a few kcalories each, but over a year's time they become significant.

Spot Reducing

People sometimes ask about "spot reducing." Unfortunately, no one part of the body gives up fat in preference to another. Fat cells all over the body release fat in response to demand, and the fat is then used by whatever muscles are active. No exercise can remove the fat from any one particular area—and, incidentally, neither can a massage machine that claims to break up fat on trouble spots.

Physical activity can help with trouble spots in another way, though. Strengthening muscles in a trouble area can help to improve their tone; stretching to gain flexibility can help with posture problems. Thus, cardiorespiratory endurance, strength, and flexibility workouts all have a place in fitness programs.

behavior modification: the changing of behavior by the manipulation of *antecedents* (cues or environmental factors that trigger behavior), the behavior itself, and *consequences* (the penalties or rewards attached to behavior).

BEHAVIOR AND ATTITUDE

Behavior-modification therapy provides ways to overcome barriers to making dietary changes and increasing physical activity. Behavior-modification therapy does more than help people decide which behaviors to change: it also teaches them how to change. Behavior and attitude are important supporting factors in achieving and maintaining appropriate body weight and composition. Changing the behaviors of overeating and underexercising that lead to, and perpetuate, obesity requires time and effort. A person must commit to take action.

Becoming Aware of Behaviors

A person who is aware of all the behaviors that create a problem has a head start on developing a solution. First, the person needs to establish a baseline (a record of present eating and physical activity behaviors) against which to measure future progress.[72] It is best to keep a diary (see Figure 7-3 on p. 180) that includes the time and place of meals and snacks, the type and amount of foods eaten, the persons present when food is eaten, and a description of the individual's feelings when eating. The diary should also record physical activities: the kind, the intensity level, the duration, and the person's feelings about them. These entries will help the individual identify possible behaviors to change.

In this era of technology, many companies have developed weight-loss applications for smartphones and other mobile devices to help users manage their daily food and physical activity behaviors. Applications include diet analysis tools that can track eating habits, scanning devices that can quickly enter food data, customized activity and meal plans that can be sent to users, and support programs that deliver encouraging messages and helpful tips. Social media sites allow users to upload progress reports and receive texts. Using these applications can help a person become more aware of behaviors that lead to weight gains and losses.

Being active—even if overweight—is healthier than being sedentary. Weight loss itself, however, best reduces disease risks.

Making Small Changes

Behavior modification strategies focus on learning desired eating and exercise behaviors and eliminating unwanted behaviors. With so many possible behavior changes, a person can feel overwhelmed. Start with small time-specific goals for each behavior—for example, "I'm going to take a 30-minute walk after dinner every evening" instead of "I'm going to run

FIGURE 7-3 Food and Activity Diary

Time	Place	Activity or food eaten	People present	Mood
10:30–10:40	School vending machine	6 peanut butter crackers and 12 oz. cola	by myself	Starved
12:15–12:30	Restaurant	Sub sandwich and 12 oz. cola	friends	relaxed & friendly
3:00–3:45	Gym	Weight training	work out partner	tired
4:00–4:10	Snack bar	Small frozen yogurt	by myself	OK

The entries in a food and activity diary should include the times and places of meals and snacks, the types and amounts of foods eaten, and a description of the individual's feelings when eating. The diary should also record physical activities: the kind, the intensity level, the duration, and the person's feelings about them.

a marathon someday." Practice desired behaviors until they become routine. Using a reward system seems to effectively support weight-loss efforts. The "How to" describes behavioral strategies to support weight management. A particularly attractive feature of these strategies is that they do not involve blaming oneself or putting oneself down—an important element in fostering self-esteem.

Cognitive Skills Behavior therapists often teach **cognitive skills**, or new ways of thinking, to help overweight people solve problems and correct false thinking that can undermine healthy eating behaviors. Thinking habits are as important for achieving a healthy body weight as eating and activity habits, and they can be changed.[73] A paradox of making a change is that it takes belief in oneself and honoring of oneself to lay the foundation for changing that self. That is, self-acceptance predicts success, while self-loathing predicts failure. "Positive self-talk" is a concept worth cultivating—many people succeed because their mental dialogue supports, rather than degrades, their efforts. Negative thoughts ("I'm not getting thin anyway, so what is the use of continuing?") should be viewed in light of empirical evidence ("my starting weight: 174 pounds; today's weight: 163 pounds").

Take credit for new behaviors; be aware of any physical improvements, too, such as lower blood pressure, or less painful knees, even without a noticeable change in pant size. Finally, remember to enjoy your emerging fit and healthy self.

Personal Attitude For many people, overeating and being overweight may have become an integral part of their identity. Changing diet and activity behaviors without attention to a person's self-concept invites failure.

Many people overeat to cope with the stresses of life. To break out of that pattern, they must first identify the particular stressors that trigger their urges to overeat. Then, when faced with these situations, they must learn to practice problem-solving skills. When the problems that trigger the urge to overeat are dealt with in alternative ways, people may find that they eat less. The message is that sound emotional health supports the ability to take care of health in all ways—including nutrition, weight management, and fitness.

cognitive skills: as taught in behavior therapy, changes to conscious thoughts with the goal of improving adherence to lifestyle modifications; examples are problem-solving skills or the correction of false negative thoughts, termed *cognitive restructuring*.

HOW TO Apply Behavior Modification to Manage Body Fatness

1. Eliminate inappropriate cues:
   - Do not buy problem foods.
   - Eat only in one room at the designated time.
   - Shop when not hungry.
   - Replace large plates, cups, and utensils with smaller ones.
   - Avoid vending machines, fast-food restaurants, and convenience stores.
   - Turn off television, video games, and computers or measure out appropriate food portions to eat during entertainment.
2. Suppress the cues you cannot eliminate:
   - Serve individual plates; do not serve "family style."
   - Remove food from the table after eating a meal to enjoy company and ambience without excess food to trigger overeating.
   - Create obstacles to consuming problem foods—wrap them and freeze them, making them less quickly accessible.
   - Control deprivation; plan and eat regular meals.
   - Plan the time spent in sedentary activities, such as watching television or using a computer—do not use these activities just to fill time.
3. Strengthen cues to appropriate behaviors:
   - Choose to dine with companions who make appropriate food choices.
   - Store appropriate foods in convenient spots in the refrigerator.
   - Learn appropriate portion sizes.
   - Plan appropriate snacks.
   - Keep sports and play equipment by the door.
4. Repeat desired behaviors:
   - Slow down eating—put down utensils between bites.
   - Always use utensils.
   - Leave some food on your plate.
   - Move more—shake a leg, pace, stretch often.
   - Join groups of active people and participate.
5. Arrange negative consequences for negative behavior:
   - Ask that others respond neutrally to your deviations (make no comments—even negative attention is a reward).
   - If you slip, do not punish yourself.
6. Reward yourself personally and immediately for positive behaviors:
   - Buy tickets to sports events, movies, concerts, or other nonfood amusement.
   - Indulge in a new small purchase.
   - Get a massage; buy some flowers.
   - Take a hot bath; read a good book.
   - Treat yourself to a lesson in a new active pursuit such as horseback riding, handball, or tennis.
   - Praise yourself; visit friends.
   - Nap; relax.

WEIGHT MAINTENANCE

Finally, be aware that it can be hard to maintain weight loss. Millions of people have experienced the frustration of achieving a desired change in weight only to see their hard work visibly slip away in a seemingly never-ending cycle of weight loss and weight regain. Disappointment, frustration, and self-condemnation are common in people who have slipped back to their original weight or even higher.

A key to weight maintenance is accepting it as a lifelong endeavor and not a goal to be achieved and then forgotten. People who maintain their weight loss continue to employ the behaviors that reduced their weight in the first place. They cultivate habits of people who maintain a healthy weight, such as eating low-kcalorie meals (averaging 1800 kcalories per day) and being physically active. They must also control susceptibility to overeating regardless of the initial weight-loss method.[74] Those who maintain weight losses over time also generally:

- Believe they have the ability to control their weight, an attribute known as **self-efficacy**.
- Eat breakfast every day.
- Average about one hour of physical activity per day.
- Monitor body weight about once a week.
- Maintain consistent lower-kcalorie eating patterns.
- Quickly address small **lapses** to prevent small gains from turning into major ones.
- Watch less than 10 hours of television per week.
- Eat high-fiber food, particularly whole grains, vegetables, and fruit, and consume sufficient water each day.
- Cultivate and honor realistic expectations regarding body size and shape.

self-efficacy: a person's belief in his or her ability to succeed in an undertaking.

lapses: periods of returning to old habits.

Review Notes

- A person who adopts a lifelong "eating plan for good health" rather than a "diet for weight loss" will be more likely to keep the lost weight off. Table 7-4 offers strategies for successful weight management.
- Reducing daily energy intake, choosing foods of low energy density, and eating smaller portions are important dietary strategies for weight management.
- Physical activity should be an integral part of a weight-management program.
- Physical activity can increase energy expenditure, improve body composition, help control appetite, reduce stress and stress eating, and enhance physical and psychological well-being.
- Behavior modification provides ways to overcome barriers to successful weight management.

TABLE 7-4 Weight-Loss Strategies

Food	Activities
• To maintain weight, consume foods and drinks to meet, not exceed, kcalorie needs. To lose weight, energy out should exceed energy in by about 500 kcalories/day. • Emphasize foods with a low energy density and a high nutrient density; make legumes, whole grains, vegetables, and fruits central to your eating pattern. • Eat slowly. • Drink water before you eat and while you eat; drink plenty of water throughout the day. • Track food and kcalorie intake. • Plan ahead to make better food choices. • Limit kcalorie intake from solid fats and added sugars. • Reduce portions, especially of high-kcalorie foods. • Cook and eat more meals at home, instead of eating out. When eating out, think about choosing healthy options.	• Limit screen time. • Increase physical activity. • Choose moderate- or vigorous-intensity physical activities. • Avoid inactivity. Some physical activity is better than none. • Slowly build up the amount of physical activity you choose.

Strategies for Weight Gain

Underweight is far less prevalent than overweight, affecting about 2 percent of U.S. adults.[75] Whether the underweight person needs to gain weight is a question of health and, like weight loss, a highly individual matter. People who are healthy at their present weight may stay there; there are no compelling reasons to try to gain weight. Those who are thin because of malnourishment or illness, however, might benefit from a diet that supports weight gain. (■) Medical advice can help make the distinction.

■ Eating disorders are the subject of the Nutrition in Practice that follows this chapter.

Some people are unalterably thin by reasons of genetics or early physical influences. Those who wish to gain weight for appearance's sake or to improve athletic performance should be aware that a healthful weight can be achieved only through physical activity, particularly strength training, combined with a high energy intake. Eating many high-kcalorie foods can bring about weight gain, but it will be mostly fat, and this can be as detrimental to health as being slightly underweight. In an athlete, such a weight gain can impair performance. Therefore, in weight gain, as in weight loss, physical activity and energy intake are essential components of a sound plan.

Physical Activity to Build Muscles The person who wants to gain weight should use resistance training primarily. As activity is increased, energy intake must be increased to support that activity. Eating extra food will then support a gain of both muscle and fat.

Energy-Dense Foods Energy-dense foods (the very ones eliminated from a successful weight-loss diet) hold the key to weight gain. Pick the highest-kcalorie items from each food group—that is, milk shakes instead of fat-free milk, peanut butter instead of lean meat, avocados instead of cucumbers, and whole-wheat muffins instead of whole-wheat bread. Because fat contains more than twice as many kcalories per teaspoon as sugar does, fat adds kcalories without adding much bulk.

Be aware that health experts recommend a moderate-fat diet for the general U.S. population because the general population is overweight and at risk for heart disease. Consumption of excessive fat is not healthy for most people, of course, but may be essential for an underweight individual who needs to gain weight. An underweight person who is physically active and eating a nutritionally adequate diet can afford a few extra kcalories from fat. For health's sake, it is wise to select foods with monounsaturated and polyunsaturated fats instead of those with saturated or *trans* fats: for example, sautéing vegetables in olive oil instead of butter or hydrogenated margarine.

Three Meals Daily People wanting to gain weight should eat at least three healthy meals a day. Many people who are underweight have simply been too busy (sometimes for months) to eat enough to gain or maintain weight. Therefore, they need to make meals a priority and plan them in advance. Taking time to prepare and eat each meal can help, as can learning to eat more food within the first 20 minutes of a meal before you begin to feel full. Another suggestion is to eat meaty appetizers or the main course first and leave the soup or salad until later.

Large Portions It is also important for the underweight person to learn to eat more food at each meal: have two sandwiches for lunch instead of one, drink milk from a larger glass, and eat cereal from a larger bowl. Expect to feel full. Most underweight individuals are accustomed to small quantities of food. When they begin eating significantly more, they feel uncomfortable. This is normal and passes over time.

Extra Snacks Because a substantially higher energy intake is needed each day, in addition to eating more food at each meal, it is necessary to eat more frequently. Between-meal snacking offers a solution. For example, a student might make three sandwiches in the morning and eat them between classes in addition to the day's three regular meals.

Juice and Milk Beverages provide an easy way to increase energy intake. Consider that 6 cups of cranberry juice add almost 1000 kcalories to the day's intake. kCalories can be added to milk by mixing in powdered milk or packets of instant breakfast.

For people who are underweight due to illness, concentrated liquid formulas are often recommended because a weak person can swallow them easily. A registered dietitian can recommend high-protein, high-kcalorie formulas to help the underweight person maintain or gain weight. Used in addition to regular meals, these formulas can help considerably.

Review Notes

- Both the incidence of underweight and the health problems associated with it are less prevalent than overweight and its associated problems.
- To gain weight, a person must train physically and increase energy intake by selecting energy-dense foods, eating regular meals, taking larger portions, and consuming extra snacks and beverages. Table 7-5 offers a summary of weight-gain strategies.

TABLE 7-5 Weight-Gain Strategies

In General:

- Eat enough to store more energy than you expend—at least 500 extra kcalories a day.
- Exercise to build muscle.
- Be patient. Weight gain takes time (1 pound per month would be reasonable).
- Choose energy-dense foods most often.
- Eat at least three meals a day, and add snacks between meals.
- Choose large portions and expect to feel full.
- Drink kcaloric fluids—juice, chocolate milk, sweet coffee drinks, sweet iced tea.

In Addition:

- Cook and bake often—delicious cooking aromas whet the appetite.
- Invite others to the table—companionship often boosts eating.
- Make meals interesting—try new vegetables and fruit, add crunchy nuts or creamy avocado, and explore the flavors of herbs and spices.
- Keep a supply of favorite snacks, such as trail mix or granola bars, handy for grabbing.
- Control stress and relax. Enjoy your food.

Self Check

1. Two causes of obesity in humans are:
   a. set-point theory and BMI.
   b. genetics and physical inactivity.
   c. genetics and low-carbohydrate diets.
   d. mineral imbalances and fat cell imbalance.

2. The protein produced by the fat cells under the direction of the *ob* gene is called:
   a. leptin.
   b. orlistat.
   c. sibutramine.
   d. lipoprotein lipase.

3. All of the following describe the behavior of fat cells *except:*
   a. the number decreases when fat is lost from the body.
   b. the storage capacity for fat depends on both cell number and cell size.
   c. the size is larger in obese people than in normal-weight people.
   d. the number increases most rapidly during the growth years and tapers off when adult status is reached.

4. The obesity theory that suggests the body chooses to be at a specific weight is the:
   a. fat cell theory.
   b. enzyme theory.
   c. set-point theory.
   d. external cue theory.

5. The biggest problem associated with the use of prescription drugs in the treatment of obesity is:
   a. cost.
   b. the necessity for long-term use.
   c. ineffectiveness.
   d. adverse side effects.

6. A nutritionally sound weight-loss diet might restrict daily energy intake to create a:
   a. 1000-kcalorie-per-month deficit.
   b. 500-kcalorie-per-month deficit.
   c. 500-kcalorie-per-day deficit.
   d. 3500-kcalorie-per-day deficit.

7. What is the best approach to weight loss?
   a. Avoid foods containing carbohydrates.
   b. Eliminate all fats from the diet, and decrease water intake.
   c. Greatly increase protein intake to prevent body protein loss.
   d. Reduce daily energy intake, and increase energy expenditure.

8. Physical activity does *not* help a person to:
   a. lose weight.
   b. lose fat in trouble spots.
   c. retain muscle.
   d. maintain weight loss.

9. Suggestions to change behaviors for successful weight control include:
   a. shop only when hungry.
   b. eat in front of the television for distraction.
   c. learn appropriate portion sizes.
   d. eat quickly.

10. Which strategy would *not* help an underweight person to gain weight?
   a. Exercise.
   b. Drink plenty of water.
   c. Eat snacks between meals.
   d. Eat large portions of foods.

Answers to these questions can be found in Appendix H. For more chapter review: Access an interactive eBook, chapter-specific interactive learning tools, including flashcards, quizzes, videos, and more in your Nutrition CourseMate, accessed through CengageBrain.com.

Clinical Applications

1. Chapter 1 discussed the nutrient density of foods—their nutrient contribution per kcalorie. Another way to evaluate foods is to consider their energy density—their energy contribution per gram:
   - A carrot weighing 72 grams delivers 31 kcalories.
   - To calculate the energy density, divide kcalories by grams: 31 kcalories divided by 72 grams = 0.43 kcalories per gram.

2. Do the same for french fries weighing 50 grams and contributing 167 kcalories: 167 kcalories divided by 50 grams = _____ kcalories per gram.
   - The more kcalories per gram, the greater the energy density.
   - Which food is more energy dense? The conclusion is no surprise, but understanding the mathematics may offer valuable insight into the concept of energy density.
   - French fries are more energy dense, providing 3.34 kcalories per gram. They provide more energy per gram—and per bite.

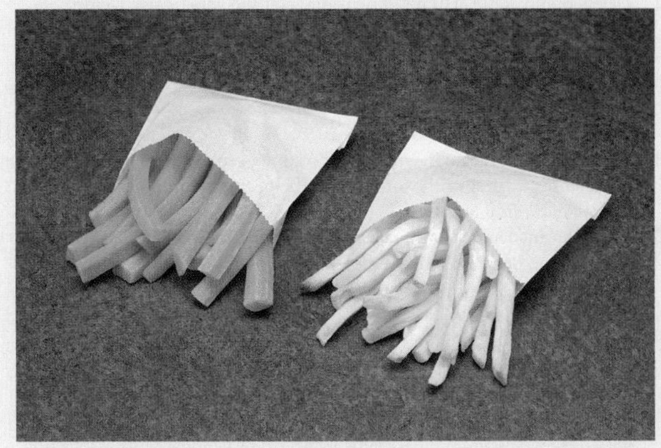
Matthew Farruggio

3. Considering a food's energy density is especially useful in planning diets for weight management. Foods with a high energy density can help with weight gain, whereas those with a low energy density can help with weight loss. Give some examples of foods that you might suggest for a client who wants to gain weight and some that might be appropriate for a client who is trying to lose weight.

Nutrition on the Net

For further study of the topics in this chapter, access these websites.

- Information on a variety of obesity topics is available at the Obesity Society website:
  www.obesity.org/
- Search for obesity at:
  www.eatright.org/
- Explore the Weight Management section of the U.S. government site:
  www.nutrition.gov
- Learn about weight control and the WIN program from the Weight-Control Information Network:
  www.win.niddk.nih.gov
- Learn about the drugs used for weight loss from the Center for Drug Evaluation and Research:
  www.fda.gov/Drugs/default.htm
- Visit weight-loss support groups, such as Take Off Pounds Sensibly (TOPS), Overeaters Anonymous (OA), and Weight Watchers:
  www.tops.org, www.oa.org, and www.weightwatchers.com
- Read the latest materials on obesity diagnosis and treatment here:
  www.nhlbi.nih.gov/guidelines/obesity/ob_home.htm
- Find helpful information on achieving and maintaining a healthy weight from the Calorie Control Council:
  www.caloriecontrol.org

Notes

1. J. Cecil and coauthors, Obesity and eating behavior in children and adolescents: Contribution of common gene polymorphisms, *International Review of Psychiatry* 24 (2012): 200–210; C. E. Elks and coauthors, Adult obesity susceptibility variants are associated with greater childhood weight gain and a faster tempo of growth: The 1946 British Birth Cohort Study, *American Journal of Clinical Nutrition* 95 (2012): 1150–1156; K. E. North and coauthors, Genetic epidemiology of BMI and body mass change from adolescence to young adulthood, *Obesity (Silver Spring)* 18 (2010): 1474–1476; M. de Krom and coauthors, Genetic variation and effects on human eating behavior, *Annual Review of Nutrition* 29 (2009): 283–304.
2. J. Hebebrand and A. Hinney, Environmental and genetic risk factors in obesity, *Child and Adolescent Psychiatric Clinics of North America* 18 (2009): 83–94.
3. K. Silventhoinen and coauthors, The genetic and environment influences on childhood obesity: A systematic review of twin and adoption studies, *International Journal of Obesity* 34 (2010): 29–40.
4. J. Wardle and coauthors, Evidence for a strong genetic influence on childhood adiposity despite the force of the obesogenic environment, *American Journal of Clinical Nutrition* 87 (2008): 398–404.
5. J. Naukkarinen and coauthors, Causes and consequences of obesity: The contribution of recent twin studies, *International Journal of Obesity* 36 (2012): 1017–1024; M. M. Hetherington and J. E. Cecil, Gene-environment interactions in obesity, *Forum of Nutrition* 63 (2010): 195–203.
6. C. Bouchard, Defining the genetic architecture of the predisposition to obesity: A challenging but not insurmountable task, *American Journal of Clinical Nutrition* 91 (2010): 5–6.
7. Li Shengxu and coauthors, Cumulative effects and predictive value of common obesity-susceptibility variants identified by genome-wide association studies, *American Journal of Clinical Nutrition* 91 (2010): 184–190.
8. Hetherington and Cecil, 2010.
9. J. M. Friedman, Leptin at 14 y of age: An ongoing story, *American Journal of Clinical Nutrition* 89 (2009): 973S–979S.
10. I. S. Farooqi and S. O'Rahilly, Leptin: A pivotal regulator of human homeostasis, *American Journal of Clinical Nutrition* 89 (2009): 980S–984S.
11. Friedman, 2009.
12. C. Delporte, Recent advances in potential clinical application of ghrelin in obesity, *Journal of Obesity* 2012, doi: 10.1155/2012/535624; T. R. Castañeda and coauthors, Ghrelin in the regulation of body weight and metabolism, *Frontiers in Neuroendocrinology* 31 (2010): 44–60.
13. Delporte, 2012; Castañeda and coauthors, 2010.
14. E. Egecioglu and coauthors, Hedonic and incentive signals for body weight control, *Reviews in Endocrine and Metabolic Disorders,* 12 (2011): 141–151; Castañeda and coauthors, 2010.
15. Y. D. Tchoukalova and coauthors, Regional differences in cellular mechanisms of adipose tissue gain with overfeeding, *Proceedings of the National Academy of Sciences of the United States of America,* 107 (2010): 18226–18231.
16. J. R. Speakman and coauthors, Set points, settling points and some alternative models: Theoretical options to understand how genes and environment combine to regulate body adiposity, *Disease Models and Mechanisms* 4 (2011): 733–745; M. M. Farias, A. M. Cuevas, and F. Rodriguez, Set-point theory and obesity, *Metabolic Syndrome and Related Disorders* 9 (2011): 85–89.
17. Hetherington and Cecil, 2010; D. Heber, An integrative view of obesity, *American Journal of Clinical* 91 (2010): 280S–283S; L. Qi and Y. A. Cho, Gene-environment interaction and obesity, *Nutrition Reviews* 66 (2008): 684–694.
18. B. A. Swinburn, G. Sacks, and E. Ravussin, Increased food energy supply is more than sufficient to explain the US epidemic of obesity, *American Journal of Clinical Nutrition* 90 (2009): 1453–1456; B. A. Swinburn and coauthors, Estimating the changes in energy flux that characterizes the rise in obesity prevalence, *American Journal of Clinical Nutrition* 89 (2009): 1723–1728.
19. B. M. Popkin and K. J. Duffey, Does hunger and satiety drive eating anymore? Increasing eating occasions and decreasing time between eating occasions in the United States, *American Journal of Clinical Nutrition* 91 (2010): 1342–1347; K. J. Duffey and coauthors, Regular consumption from fast food establishments relative to other restaurants is differentially associated with metabolic outcomes in young adults, *Journal of Nutrition* 139 (2009): 2113–2118.
20. K. J. Duffey and B. M. Popkin, Energy density, portion size, and eating occasions: Contributions to increased energy intake in the United States, 1977–2006, *PLoS Medicine* 8 (2011): e1001050.
21. B. J. Rolls, Plenary Lecture 1: Dietary strategies for the prevention and treatment of obesity, *Proceedings of the Nutrition Society* 69 (2010): 70–79; I. H. Steenhuis and W. M. Vermeer, Portion size: Review and framework for interventions, *International Journal of Behavioral Nutrition and Physical Activity* 6 (2009): 58.
22. Position of the American Dietetic Association, Weight management, *Journal of the American Dietetic Association* 109 (2009): 330–346.
23. Duffey and coauthors, 2009.
24. L. Matt, National restaurant menu labeling, legislation: Public nutrition education and professional opportunities, *Journal of the American Dietetic Association* 111 (2011): S7; K. Stein, A national approach to restaurant menu labeling: The Patient Protection and Affordable Health Care Act, section 4205, *Journal of the American Dietetic Association* 110 (2010): 1280–1286.

25. J. S. Schiller and coauthors, Summary health statistics for U.S. adults: National Health Interview Survey, *Vital and Health Statistics* 10 (2012): 1–217.
26. P. T. Katzmarzyk and coauthors, Sitting time and mortality from all causes, cardiovascular disease, and cancer, *Medicine and Science in Sports and Exercise* 41 (2009): 998–1005; S. Mandic and coauthors, Characterizing differences in mortality at the low end of the fitness spectrum, *Medicine and Science in Sports and Exercise* 41 (2009): 1573–1579; American College of Sports Medicine, Position stand: Appropriate physical activity intervention strategies for weight loss and prevention of weight regain for adults, *Medicine and Science in Sports and Exercise* 41 (2009): 459–471; P. Gordon-Larsen and coauthors, Fifteen-year longitudinal trends in walking patterns and their impact on weight change, *American Journal of Clinical Nutrition* 89 (2009): 19–26.
27. H. P. van der Ploeg and coauthors, Sitting time and all-cause mortality risk in 222,497 Australian adults, *Archives of Internal Medicine* 172 (2012): 494–500; Katzmarzyk and coauthors, 2009.
28. J. F. Salis, Role of built environments in physical activity, obesity, and cardiovascular disease, *Circulation* 125 (2012): 729–737; D. Ding and coauthors, Neighborhood environment and physical activity among youth: A review, *American Journal of Preventive Medicine* 41 (2011): 442–455.
29. J. Beaulac, E. Kristjansson, and S. Cummins, A systematic review of food deserts, 1966–2007, *Preventing Chronic Disease: Public Health Research, Practice, and Policy*, 6 (2009), epub available at www.cdc.gov.
30. J. L. Blitstein, J. Snider, and W. D. Evans, Perceptions of the food shopping environment are associated with greater consumption of fruits and vegetables, *Public Health Nutrition* 21 (2012): 1–6; M. I. Larson, M. T. Story, and M. C. Nelson, Neighborhood environments: Disparities in access to healthy foods in the U.S., *American Journal of Preventive Medicine* 36 (2009): 74–81.
31. I. N. Bezerra, C. Curioni, and R. Sichieri, Association between eating out of home and body weight, *Nutrition Reviews* 70 (2012): 65–79; A. M. Fretts and coauthors, Associations of processed meat and unprocessed red meat intake with incident diabetes: The Strong Heart Family Study, *American Journal of Clinical Nutrition* 95 (2012): 752–758; H. J. Song and coauthors, Understanding a key feature of urban food stores to develop nutrition intervention, *Journal of Hunger & Environmental Nutrition* 7 (2012): 77–90.
32. Institute of Medicine (U.S.) Committee on Accelerating Progress in Obesity Prevention, *Accelerating Progress in Obesity Prevention: Solving the Weight of the Nation* (Washington, DC: National Academies Press, 2012), available at www.nap.edu.
33. U.S. Food and Drug Administration, Questions and answers about FDA's initiative against contaminated weight loss products, April 30, 2009, available at http://www.fda.gov/Drugs/ResourcesForYou/Consumers/QuestionsAnswers/ucm136187.htm.
34. S. Hasani-Ranjbar and coauthors, A systematic review of the efficacy and safety of herbal medicines used in the treatment of obesity, *World Journal of Gastroenterology* 15 (2009): 3073–3085.
35. J. M. Nicklas and coauthors, Successful weight loss among obese U.S. adults, *American Journal of Preventive Medicine* 42 (2012): 481–485; D. Laddu and coauthors, A review of evidence-based strategies to treat obesity in adults, *Nutrition in Clinical Practice* 26 (2011): 512–525.
36. Laddu and coauthors, 2011.
37. E. A. Finkelstein and coauthors, Obesity and severe obesity forecasts through 2030, *American Journal of Preventive Medicine* 42 (2012): 563–570.
38. A. Nagle, Bariatric surgery—A surgeon's perspective, *Journal of the American Dietetic Association* 110 (2010): 520–523; G. L. Blackburn, S. Wollner, and S. B. Heymsfield, Lifestyle interventions for the treatment of class III obesity: A primary target for nutrition medicine in the obesity epidemic, *American Journal of Clinical Nutrition* 91 (2010): 289S–292S.
39. R. Padwal and coauthors, Bariatric surgery: A systematic review of the clinical and economic evidence, *Journal of General Internal Medicine* 26 (2011): 1183–1194.
40. M. Attiah and coauthors, Durability of Roux-en-Y gastric bypass surgery: A meta-regression study, *Annals of Surgery* 256 (2012): 251–254; L. Sjöström and coauthors, Bariatric surgery and long-term cardiovascular events, *Journal of the American Medical Association* 307 (2012): 56–65; P. R. Schauer and coauthors, Bariatric surgery versus intensive medical therapy in obese patients with diabetes, *New England Journal of Medicine* 366 (2012): 1567–1576; The Endocrine Society, Evaluating the benefits of treating type 2 diabetes with bariatric surgery, An Endocrine Society statement to providers on study findings related to medical versus surgical treatment of obese patients with type 2 diabetes, March 2012, available at www.endo-society.org/advocacy/index.cfm.
41. M. K. Robinson, Surgical Treatment of obesity—Weighing the facts, *New England Journal of Medicine* 361 (2009): 520–521.
42. Laddu and coauthors, 2011; D. Herber, An integrative view of obesity, *American Journal of Clinical Nutrition* 91 (2010): 280S–283S.
43. Position of the American Dietetic Association, Weight management, 2009; J. O. Hill, Can a small-changes approach help address the obesity epidemic? A report of the Joint Task Force of the American Society for Nutrition, Institute of Food Technologists, and International Food Information Council, *American Journal of Clinical Nutrition* 89 (2009): 477–484.
44. D. R. Jacobs and coauthors, Association of 1-y change in diet pattern with cardiovascular disease risk factors and adipokines: Results from the 1-y randomized Oslo Diet and Exercise Study, *American Journal of Clinical Nutrition* 89 (2009): 509–517.
45. S.F. Kirk and coauthors, Effective weight management practice; A review of the lifestyle intervention evidence, *International Journal of Obesity* 36 (2012): 178–185.
46. S. B. Heymsfield and coauthors, Energy content of weight loss: Kinetic features during voluntary caloric restriction, *Metabolism* 61 (2012): 937–943.
47. National Heart, Lung and Blood Institute, National Institutes of Health, *The Practical Guide: Identification, Evaluation, and Treatment of Overweight and Obesity in Adults,* NIH publication no. 00-4084 (Washington, DC: Government Printing Office, 2000).
48. Position of the American Dietetic Association, 2009.
49. K. D. Hall and coauthors, Energy balance and its components: Implications for body weight regulation, *American Journal of Clinical Nutrition* 95 (2012): 989–995.
50. Hymsfield and coauthors, 2012.
51. Hymsfield and coauthors, 2012.
52. Hall and coauthors, 2012.
53. National Heart, Lung, and Blood Institute, National Institutes of Health, 2000, pp. 26–27.
54. P. Deshmukh-Taskar and coauthors, The relationship of breakfast skipping and type of breakfast consumed with overweight/obesity, abdominal obesity, other cardiometabolic risk factors and the metabolic syndrome in young adults. The National Health and Nutrition Examination Survey (NHANES): 1999–2006, *Public Health Nutrition* October 3, 2012, e-pub ahead of print; L. Dubois and coauthors, Breakfast skipping is associated with differences in meal patterns, macronutrient intakes and overweight among pre-school children, *Public Health Nutrition* 12 (2009): 19–28.
55. C. M. Shay and coauthors, Food and nutrient intakes and their associations with lower BMI in middle-aged US adults: The International Study of Macro-/Micronutrients and Blood Pressure, *American Journal of Clinical Nutrition* 96 (2012): 483–491; M. Kristensen and coauthors, Whole grain compared with refined wheat decreases the percentage of body fat following a 12-week, energy restricted dietary intervention in postmenopausal women, *Journal of Nutrition* 142 (2012): 710–716.
56. M. Bes-Rastrollo and coauthors, Prospective study of nut consumption, long-term weight change, and obesity risk in women, *American Journal of Clinical Nutrition* 89 (2009): 1913–1919.
57. J. W. Carbone, J. P. McClung, and S. M. Pasiakos, Skeletal muscle responses to negative energy balance: Effects of dietary protein, *Advances in Nutrition* 3 (2012): 119–126.
58. C. L. Rock and coauthors, Effect of a free prepared meal and incentivized weight loss program on weight loss and weight loss maintenance in obese and overweight women: A randomized controlled trial, *Journal of the American Medical Association* 304 (2010): 1803–1810; Position of the American Dietetic Association, 2009.
59. R. Perez-Escamilla and coauthors, Dietary energy density and body weight in adults and children: A systematic review, *Journal of the Academy of Nutrition and Dietetics* 112 (2012): 671–684; B. J. Rolls, Plenary lecture 1: Dietary strategies for the prevention and treatment of obesity, *Proceeding of the Nutrition Society* 3 (2009): 1–10.
60. M. A. Palmer, S. Capra, and S. K. Baines, Association between eating frequency, weight, and health, *Nutrition Reviews* 67 (2009): 379–390.
61. M. C. Daniels and B. M. Popkin, Impact of water intake on energy intake and weight status: A systematic review, *Nutrition Reviews* 68 (2010): 505–521.
62. Nicklas and coauthors, 2012; B. H. Goodpaster and coauthors, Effects of diet and physical activity interventions on weight loss and cardiometabolic risk factors in severely obese adults: A randomized study, *Journal of the American Medical Association* 304 (2010): 1795–1802; A. L. Hankinson and coauthors, Maintaining a high physical activity level over 20 years and weight gain, *Journal of the American Medical Association* 304 (2010):

2603–2610; D. E. Larson-Meyer and coauthors, Caloric restriction with or without exercise: The fitness versus fatness debate, *Medicine and Science in Sports and Exercise* 42 (2010): 152–159; American College of Sports Medicine, 2009 American Dietetic Association, 2009.
63. A. E. Field and coauthors, Weight control behaviors and subsequent weight change among adolescents and young adult females, *American Journal of Clinical Nutrition* 91 (2010): 147–153; American College of Sports Medicine, 2009.
64. B. J. Nicklas and coauthors, Effect of exercise intensity on abdominal fat loss during calorie restriction in overweight and obese postmenopausal women: A randomized, controlled trial, *American Journal of Clinical Nutrition* 89 (2009): 1043–1052; American College of Sports Medicine, 2009.
65. American College of Sports Medicine, 2009.
66. Larson-Meyer and coauthors, 2010; F. Magkos and coauthors, Management of the Metabolic Syndrome and type 2 diabetes through lifestyle modification, *Annual Review of Nutrition* 29 (2009): 223–256.
67. M. Hamer and G. O'Donovan, Cardiorespiratory fitness and metabolic risk factors in obesity, *Current Opinion in Lipidology* 21 (2010): 1–7.
68. A. M. Knab and coauthors, A 45-minute vigorous exercise bout increases metabolic rate for 14 hours, *Medicine and Science in Sports and Exercise* 43 (2011): 1642–1648.
69. K. Deighton, J. C. Zahra, and D. J. Stensel, Appetite, energy intake and resting metabolic responses to 60 min treadmill running performed in a fasted versus a postprandial state, *Appetite* 58 (2012): 946–954; N. A. King and coauthors, Dual process action of exercise on appetite control: Increase in orexigenic drive but improvement in meal-induced satiety, *American Journal of Clinical Nutrition* 90 (2009): 921–927.
70. D. Stensel, Exercise, appetite and appetite-regulating hormones: Implications for food intake and weight control, *Annals of Nutrition and Metabolism* 57 (2010): 36–42.
71. U.S. Department of Agriculture and U.S. Department of Health and Human Services, *2008 Physical Activity Guidelines for Americans*, available at www.health.gov/paguidelines/default.aspx.
72. L. E. Burke, J. Wang, and M. A. Sevick, Self-monitoring in weight loss: A systematic review of the literature, *Journal of the American Dietetic Association* 111 (2011): 92–102.
73. N. R. Reyes and coauthors, similarities and differences between weight loss maintainers and regainers: A qualitative analysis, *Journal of the Academy of Nutrition and Dietetics* 112 (2012): 499–505.
74. D. S. Bond, Weight-loss maintenance in successful weight losers: Surgical vs non-surgical methods, *International Journal of Obesity (London)* 33 (2009): 173–180.
75. C. D. Fryar and C. L. Ogden, prevalence of underweight among adults aged 20 and over; United States, 1960–1962 through 2007–2010, 2012, www.cdc.gov/nchs/data/hestat/underweight_adult_07_08/underweight_adult_07_08.htm.

Nutrition in Practice

Eating Disorders

The exact number of people in the United States afflicted with some form of **eating disorder** (see the accompanying glossary for the relevant terms) is unknown because many cases are never reported.[1] An estimated 6 percent of females and 3 percent of males have **anorexia nervosa**, **bulimia nervosa**, or **binge eating disorder**. Many more suffer from other related conditions that do not meet the strict criteria for anorexia nervosa, bulimia nervosa, or binge eating disorder but still imperil a person's well-being. Characteristics of disordered eating, such as restrained eating, binge eating, purging, fear of fatness, and distortion of body image, are common, especially among young middle-class girls. The incidence and prevalence of eating disorders in young people has increased steadily since the 1950s.[2] Most alarming is the rising prevalence of eating disorders at progressively younger ages. In most other societies, these behaviors and attitudes are much less prevalent.

Why do so many young people in our society suffer from eating disorders?

Most experts agree that the causes are multifactorial: sociocultural, psychological, genetic, and probably also neurochemical.[3] However, excessive pressure to be thin in our society is at least partly to blame. When low body weight becomes an important goal, people begin to view normal, healthy body weight as too fat. Healthy people then take unhealthy actions to lose weight. Severe restriction of food intake can create intense stress and extreme hunger that leads to binges.[4] Research confirms this theory, showing that unhealthy or dangerous diets often precede binge eating in adolescent girls.[5] Energy restriction followed by bingeing can set in motion a pattern of **weight cycling**, which may make weight loss and maintenance more difficult over time. Importantly, healthful dieting and physical activity in overweight adolescents do not appear to trigger eating disorders.

People who attempt extreme weight loss are dissatisfied with their bodies to begin with; they may also be depressed or suffer social anxiety.[6] As weight loss becomes increasingly difficult, psychological problems worsen, and the likelihood of developing full-blown eating disorders increases.

People with anorexia nervosa suffer from an extreme preoccupation with weight loss that seriously endangers their health and even their lives. People with bulimia engage in episodes of binge eating alternating with periods of severe dieting or self-starvation. Some bulimics also follow binge eating with self-induced vomiting, laxative abuse, or diuretic abuse in an attempt to undo the perceived damage caused by the binge.

Glossary

amenorrhea (ay-MEN-oh-REE-ah): the absence of or cessation of menstruation. Primary amenorrhea is menarche delayed beyond 16 years of age. Secondary amenorrhea is the absence of three to six consecutive menstrual cycles.

anorexia nervosa: an eating disorder characterized by a refusal to maintain a minimally normal body weight, self-starvation to the extreme, and a disturbed perception of body weight and shape; seen (usually) in adolescent girls and young women.

anorexia = without appetite

nervosa = of nervous origin

binge eating disorder: an eating disorder whose criteria are similar to those of bulimia nervosa, excluding purging or other compensatory behaviors.

bulimia (byoo-LEEM-ee-uh) nervosa: recurring episodes of binge eating combined with a morbid fear of becoming fat, usually followed by self-induced vomiting or purging.

cathartic: a strong laxative.

cognitive therapy: psychological therapy aimed at changing undesirable behaviors by changing underlying thought processes contributing to these behaviors. In anorexia nervosa, a goal is to replace false beliefs about body weight, eating, and self-worth with health-promoting beliefs.

disordered eating: eating behaviors that are neither normal nor healthy, including restrained eating, fasting, binge eating, and purging.

eating disorder: a disturbance in eating behavior that jeopardizes a person's physical and psychological health.

emetic (em-ETT-ic): an agent that causes vomiting.

female athlete triad: a potentially fatal triad of medical problems: disordered eating, amenorrhea, and osteoporosis.

stress fractures: bone damage or breaks caused by stress on bone surfaces during exercise.

weight cycling: repeated rounds of weight loss and subsequent regain, with reduced ability to lose weight with each attempt; also called *yo-yo dieting*.

Are there other groups, besides girls and young women, who are vulnerable to anorexia nervosa and bulimia nervosa?

Yes. Athletes and dancers are at special risk for eating disorders.[7] They may severely restrict energy intakes in an attempt to enhance performance or appearance or to meet weight guidelines of a sport. In reality, severe energy restriction causes a loss of lean tissue that impairs physical performance and imposes a risk of eating disorders. Risk factors for eating disorders among athletes include:

- Young age (adolescence)
- Pressure to excel in a sport
- Focus on achieving or maintaining an "ideal" body weight or body fat percentage
- Participation in sports or competitions that judge performance on aesthetic appeal such as gymnastics, figure skating, or dance
- Unhealthy, unsupervised weight-loss dieting at an early age

Female athletes are most vulnerable, but males—especially dancers, wrestlers, skaters, jockeys, and gymnasts—suffer from eating disorders, too, and their numbers may be increasing.[8]

Male athletes and dancers with eating disorders often deny having them because they mistakenly believe that such disorders afflict only women. Under the same pressures as female athletes, males may also develop eating disorders. They skip meals, restrict fluids, practice in plastic suits, or train in heated rooms to lose a quick 4 to 7 pounds.[9] Wrestlers, especially, must "make weight" to compete in the lowest possible weight class to face smaller opponents. Conversely, male athletes may suffer weight-*gain* problems. When young men with low self-esteem internalize unrealistically bulky male body images, they can become dissatisfied with their own healthy bodies and practice unhealthy behaviors, even steroid drug abuse.

Even among athletes, however, women are most susceptible to developing eating disorders. Many female athletes appear healthy but in fact may develop the three interrelated components of the **female athlete triad**: **disordered eating**, **amenorrhea** (the absence of three or more consecutive menstrual cycles), and osteoporosis (see Figure NP7-1).

How does the female athlete triad develop?

Many athletic women engage in self-destructive eating behaviors (disordered eating) because they and their coaches have adopted unsuitable weight standards. An athlete's body must be heavier for height than a nonathlete's body because the athlete's bones and muscles are denser. Weight standards that may be appropriate for others are inappropriate for athletes. Techniques such as skinfold measures yield more useful information about body composition.

As for amenorrhea, its prevalence among premenopausal women in the United States is about 2 to 5 percent overall, but among female athletes it may be as high as 66 percent. Amenorrhea is *not* a normal adaptation to strenuous physical training: it is a symptom of something going wrong.[10] Amenorrhea is characterized by low blood estrogen, infertility, and often bone mineral losses.

In general, weight-bearing physical activity, dietary calcium, and the hormone estrogen protect against the bone loss of osteoporosis, but in women with disordered eating and amenorrhea, strenuous activity may impair bone health.[11] Vigorous training combined with low food energy intakes disrupts metabolic and hormonal balances. These disturbances compromise bone health, greatly increasing the risks of **stress fractures**. Stress fractures, a serious form of bone injury, commonly occur among dancers and other competitive athletes with amenorrhea, low calcium intakes, and disordered eating. Many underweight young athletes have bones like those of postmenopausal women, and they may never recover their lost bone even after diagnosis and treatment—which makes prevention critical. Young athletes should be encouraged to consume at least 1300 milligrams of calcium each day, to eat nutrient-dense foods, and to obtain enough food energy to support weight gain and the energy expended in physical activity. Nutrition is critical to bone recovery.

What can be done to prevent eating disorders in athletes and dancers?

To prevent eating disorders in athletes and dancers, both the performers and their coaches must be educated about links between inappropriate body weight ideals, improper weight-loss techniques, eating disorder development, adequate nutrition, and safe weight-control methods. Coaches and dance instructors should never encourage unhealthy weight loss to qualify for competition or to conform with distorted artistic ideals. Frequent weighings can

FIGURE NP7-1 The Female Athlete Triad

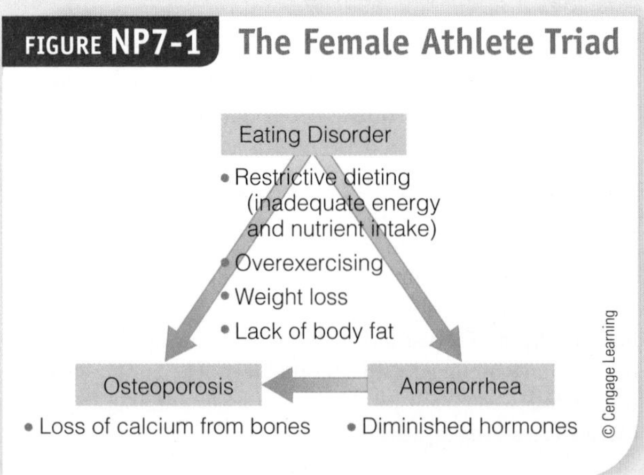

push young people who are striving to lose weight into a cycle of starving to confront the scale and then bingeing uncontrollably afterward. The erosion of self-esteem that accompanies these events can interfere with the normal psychological development of the teen years and set the stage for serious problems later on.

Table NP7-1 provides some suggestions to help athletes and dancers protect themselves against developing eating disorders. The next sections describe eating disorders that anyone, athlete or nonathlete, may experience.

Who is at risk for anorexia nervosa?

Most anorexia nervosa victims are females who come from middle- or upper-class families. The person with anorexia nervosa is often a perfectionist who works hard to please her parents. She may identify so strongly with her parents' ideals and goals for her that she sometimes feels she has no identity of her own. She is respectful of authority but sometimes feels like a robot, and she may act that way, too: polite but controlled, rigid, and unspontaneous. She earnestly desires to control her own destiny, but she feels controlled by others. When she does not eat, she gains control.

How does a person know when dieting is going too far?

When a person loses weight to well below the average for her height, becoming too slim, and still doesn't stop, she has gone too far. Regardless of how thin she is, she looks in the mirror and sees herself as fat. Central to the diagnosis of anorexia nervosa is a distorted body image that overestimates body fatness. Anorexia nervosa resembles an addiction. The characteristic behavior is obsessive and compulsive. Before drawing conclusions about someone who is extremely thin or who eats very little, remember that diagnosis of anorexia nervosa requires professional assessment.

TABLE NP7-1 Tips for Combating Eating Disorders

The following guidelines may be useful in combating eating disorders:

- Never restrict food intakes to below the amounts suggested for adequacy by the USDA Food Patterns.
- Eat frequently. People often do not eat frequent meals because of time constraints, but eating can be incorporated into other activities, such as snacking while studying or commuting. The person who eats frequently never gets so hungry as to allow hunger to dictate food choices.
- If not at a healthy weight, establish a reasonable weight goal based on a healthy body composition.
- Allow a reasonable time to achieve the goal. A reasonable loss of excess fat can be achieved at the rate of about 1 percent of body weight per week.
- Learn to recognize media image biases, and reject ultra-thin standards for beauty. Shift focus to health, competencies, and human interactions; bring behaviors in line with those beliefs.

Specific guidelines for athletes and dancers:

- Replace weight-based goals with performance-based goals.
- Remember that eating disorders impair physical performance. Seek confidential help in obtaining treatment if needed.
- Restrict weight-loss activities to the off-season.
- Focus on proper nutrition as an important facet of your training—as important as proper technique.

People with anorexia nervosa see themselves as fat, even when they are dangerously underweight.

What is the harm in being very thin?

Anorexia nervosa damages the body much as starvation does. In young people, growth ceases, and normal development falters. They lose so much lean tissue that basal metabolic rate slows. Bones weaken, too—osteoporosis develops in about a third of those with anorexia nervosa.[12] Additionally, the heart pumps inefficiently and irregularly, the heart muscle becomes weak and thin, the heart chambers diminish in size, and the blood pressure falls. Minerals that help to regulate the heartbeat become unbalanced. Many deaths in people with anorexia nervosa are due to heart failure. Kidneys often fail as well.[13]

Starvation brings other physical consequences: loss of brain tissue, impaired immune response, anemia, and a loss of digestive function that worsens malnutrition. Digestive functioning becomes sluggish, the stomach empties slowly, and the lining of the intestinal tract shrinks. The ailing digestive tract fails to sufficiently digest any food the victim may eat. The pancreas slows its production of digestive enzymes. The person may suffer from diarrhea, further worsening malnutrition.

What kind of treatment helps people with anorexia nervosa?

Treatment of anorexia nervosa requires a multidisciplinary approach that addresses two sets of issues and behaviors: those relating to food and weight, and those involving relationships with oneself and others. Teams of physicians, nurses, psychiatrists, family therapists, and dietitians work together to treat people with anorexia nervosa. The expertise of a registered dietitian is essential because an appropriate, individually crafted diet is crucial for normalizing body weight, and nutrition counseling is indispensable.[14] Seldom are clients willing to eat for themselves, but if they are, chances are they can recover without other interventions.

Professionals classify clients based on the risks posed by the degree of malnutrition present.* Clients with low risks may benefit from family counseling, **cognitive therapy**, behavior modification, and nutrition guidance; those with greater risks may also need other forms of psychotherapy and supplemental formulas to provide extra nutrients and energy.

Sometimes, intensive behavior management treatment in a live-in facility can help to normalize food intake and exercise.[15] When starvation leads to severe underweight (less than 75 percent of ideal body weight), high medical risks ensue, and patients require hospitalization. They must be stabilized and carefully fed to forestall death. However, involuntary feeding through a tube can cause psychological trauma and may not be necessary in all cases. Antidepressants and other drugs are commonly prescribed but are often ineffective in anorexia nervosa.

*Indicators of malnutrition include a low percentage of body fat, low blood proteins, and impaired immune response.

Stopping weight loss is a first goal; establishing regular eating patterns is next. Because body weight is low and fear of weight gain is high, initial food intake may be small—1200 kcalories per day can be an achievement. A variety of high-energy foods and beverages can help deliver needed kcalories and nutrients.[16] Even after recovery, however, energy intakes and eating behaviors may not fully return to normal.

Almost half of women who are treated can successfully maintain their body weight at just below a healthy weight; at that weight, many of them begin menstruating again. The other half have poor or fair treatment outcomes, relapse into abnormal eating behaviors, or die. Anorexia nervosa has one of the highest mortality rates among psychiatric disorders—most commonly from cardiac complications or by suicide.[17]

How does bulimia nervosa differ from anorexia nervosa?

Bulimia nervosa is distinct from anorexia nervosa and is more prevalent. More men suffer from bulimia nervosa than from anorexia, but bulimia is still more common in women. The secretive nature of bulimic behaviors makes recognition of the problem difficult, but once it is recognized, diagnosis is based on such criteria as frequency of binge eating episodes and inappropriate compensatory behaviors (at least once a week for three months) and distorted body image.[18]

The typical person with bulimia is well educated, in her early twenties, and close to ideal body weight. She is a high achiever but emotionally insecure. She experiences considerable social anxiety and has difficulty establishing personal relationships. She is sometimes depressed and often exhibits impulsive behavior.

Like the person with anorexia nervosa, the person with bulimia spends much time thinking about her body weight and food. Her preoccupation with food manifests itself in secretive binge eating episodes followed by self-induced vomiting, fasting, or the use of laxatives or diuretics. Such behaviors typically begin in late adolescence after a long series of various unsuccessful weight-reduction diets. People with bulimia commonly follow a pattern of restrictive dieting interspersed with bulimic behaviors and experience weight fluctuations of more than 10 pounds over short periods of time.

Unlike the person with anorexia nervosa, the person with bulimia is aware of the consequences of her behavior, feels that it is abnormal, and is deeply ashamed of it. She feels inadequate and unable to control her eating, so she tends to be passive and to look to others for confirmation of her sense of self-worth. When she is rejected, either in reality or in her imagination, her bulimia becomes worse. If her depression deepens, she may seek solace in drug or alcohol abuse or other addictive behaviors. Clinical depression is common in people with bulimia nervosa, and the rates of substance abuse are high.

What exactly is binge eating?

Binge eating is unlike normal eating, and the food is not consumed for its nutritional value. The binge eater has a compulsion to eat. A typical binge occurs periodically, is done in secret, usually at night, and lasts an hour or more. A binge frequently follows a period of rigid dieting, so the binge eating is accelerated by hunger. During a binge, the person with bulimia may consume a thousand kcalories or more. The food typically contains little fiber or water, has a smooth texture, and is high in sugar and fat, so it is easy to consume vast amounts rapidly with little chewing.

What are the consequences of binge eating and purging?

After a binge, the person may use a **cathartic**—a strong laxative that can injure the lower intestinal tract. Or the person may induce vomiting, using an **emetic**—a drug intended as first aid for poisoning.

On first glance, purging seems to offer a quick and easy solution to the problems of unwanted kcalories and body weight. Many people perceive such behavior as neutral or even positive, when, in fact, bingeing and purging have serious physical consequences. Fluid and electrolyte imbalances caused by vomiting or diarrhea can lead to abnormal heart rhythms and injury to the kidneys. Vomiting causes irritation and infection of the pharynx, esophagus, and salivary glands; erosion of the teeth; and dental caries. The esophagus may rupture or tear, as may the stomach. Overuse of emetics depletes potassium concentrations and can lead to death by heart failure.

What is the treatment for bulimia nervosa?

As for people with anorexia nervosa, a team approach provides the most effective treatment for people with bulimia nervosa. Bulimia nervosa is easier to treat than anorexia nervosa in many respects because it seems to be more of a chosen behavior. People with bulimia know that their behavior is abnormal, and many are willing to try to cooperate.

The goal of the dietary plan to treat bulimia is to help the client gain control, establish regular eating patterns, and restore nutritional health. Energy intake should not be severely restricted—hunger can be a trigger for a binge. The person needs to learn to eat a quantity of nutritious food sufficient to nourish her body and to satisfy hunger (at least 1600 kcalories per day). Table NP7-2 offers some ways to begin correcting bulimia nervosa. Most people diagnosed with bulimia nervosa recover within 5 to 10 years, with or without treatment, but treatment probably speeds the recovery process.

Anorexia nervosa and bulimia nervosa are distinct eating disorders, yet they sometimes overlap. Anorexia victims may purge, and victims of both conditions share an overconcern with body weight and the tendency to drastically undereat. The two disorders can also appear in the same person, or one can lead to the other. Other people have eating disorders that fall short of anorexia nervosa or bulimia nervosa but share some of their features, such as fear of body fatness. One such condition is binge eating disorder.

How does binge eating disorder differ from bulimia nervosa?

Up to half of all people who restrict eating to lose weight periodically binge without purging, including about one-third of obese people. Obesity itself, however, does not constitute an eating disorder.

TABLE NP7-2 Diet Strategies for Combating Bulimia Nervosa

Planning Principles

- Plan meals and snacks; record plans in a food diary prior to eating.
- Plan meals and snacks that require eating at the table and using utensils.
- Refrain from finger foods.
- Refrain from "dieting" or skipping meals.

Nutrition Principles

- Eat a well-balanced diet and regularly timed meals consisting of a variety of foods.
- Include raw vegetables, salad, or raw fruit at meals to prolong eating times.
- Choose whole-grain, high-fiber breads, pasta, rice, and cereals to increase bulk.
- Consume adequate fluid, particularly water.

Other Tips

- Choose foods that provide protein and fat for satiety and bulky, fiber-rich carbohydrates for immediate feelings of fullness.
- Try including soups and other water-rich foods for satiety.
- Consume the amounts of food specified in the USDA Food Patterns (p. 20).
- For convenience (and to reduce temptation), select foods that naturally divide into portions. Select one potato, rather than rice or pasta that can be overloaded onto the plate; purchase yogurt and cottage cheese in individual containers; look for small packages of precut steak or chicken; choose frozen dinners with measured portions.
- Include 30 minutes of physical activity every day—exercise may be an important tool in defeating bulimia.

Clinicians note differences between people with bulimia nervosa and those with binge eating disorder. People with binge eating disorder consume less during a binge, rarely purge, and exert less restraint during times of dieting. Similarities also exist, including feeling out of control, disgusted, depressed, embarrassed, or guilty because of their self-perceived gluttony. Binge eating behavior responds more readily to treatment than other eating disorders, and resolving such behaviors can be a first step to authentic weight control. Successful treatment also improves physical health, mental health, and the chances of breaking the cycle of rapid weight losses and gains.

Chapter 6 describes how our society sets unrealistic ideals for body weight, especially in women, and devalues those who do not conform to them. Anorexia nervosa and bulimia nervosa are not a form of rebellion against these unreasonable expectations but rather the exaggerated acceptance of them. Body dissatisfaction is a primary factor in the development of eating disorders.[19] Perhaps a person's best defense against these disorders is to learn to appreciate his or her own uniqueness.

Notes

1. Position of the American Dietetic Association, Nutrition intervention in the treatment of eating disorders, *Journal of the American Dietetic Association* 111 (2011): 1236–1241.
2. D. S. Rosen and the Committee on Adolescence, Clinical report—Identification and management of eating disorders in children and adolescents, *Pediatrics* 126 (2010): 1240–1253.
3. T. K. Clarke, A. R. Weiss, and W. H. Berrettini, The genetics of anorexia nervosa, *Clinical Pharmacology and Therapeutics* 91 (2012): 181–188; R. Calati and coauthors, The 5-HTTLPR polymorphism and eating disorders: A meta-analysis, *International Journal of Eating Disorders* 44 (2011): 191–199; S. E. Mazzeo and C. M. Bulik, Environmental and genetic risk factors for eating disorders: What the clinician needs to know, *Child and Adolescent Psychiatric Clinics of North America* 18 (2009): 67–82; S. S. O'Sullivan, A. H. Evans, and A. J. Lees, Dopamine dysregulation syndrome: An overview of its epidemiology, mechanisms, and management, *CNS Drugs* 23 (2009): 157–170; T. D. Müller and coauthors, Leptin-mediated neuroendocrine alterations in anorexia nervosa: Somatic and behavioral implications, *Child and Adolescent Psychiatric Clinics of North America* 18 (2009): 117–129; V. Costarelli, M. Demerzi, and D. Stamou, Disordered eating attitudes in relation body image and emotional intelligence in young women, *Journal of Human Nutrition and Dietetics* 22 (2009): 239–245.
4. K. D. Carr, Food scarcity, neuroadaptations, and the pathogenic potential of dieting in an unnatural ecology: Binge eating and drug abuse, *Psychology and Behavior* 104 (2011): 515–524.
5. D. Neumark-Sztainer and coauthors, Dieting and disordered eating behaviors from adolescence to young adulthood: Findings from a 10-year longitudinal study, *Journal of the American Dietetic Association* 111 (2011): 1004–1011.
6. M. L. Norris and coauthors, An examination of medical and psychological morbidity in adolescent males with eating disorders, *Eating Disorders* 20 (2012): 405–415; U. Pauli-Pott and coauthors, Links between psychopathological symptoms and disordered eating behaviors in overweight/obese youths, *International Journal of Eating Disorders* September 18, 2012, doi: 10.1002/eat.22055; R. D. Grave, Eating disorders: Progress and challenges, *European Journal of Internal Medicine* 22 (2011): 153–160.
7. C. A. Hincapie and J. D. Cassidy, Disordered eating, menstrual disturbances, and low bone mineral density in dancers: A systematic review, *Archives of Physical Medicine and Rehabilitation* 91 (2010): 1777–1789; American College of Sports Medicine, Position stand: The female athlete triad, *Medicine and Science in Sports and Exercise* 39 (2007): 1867–1882.
8. J. B. Martin, The development of ideal body image perceptions in the United States, *Nutrition Today* 45 (2010): 98–110.
9. G. G. Artioli and coauthors, Prevalence, magnitude, and methods of rapid weight loss among judo competitors, *Medicine and Science in Sports and Exercise* 42 (2010): 436–442.
10. C. J. Rosen and A. Klibanski, Bone, fat, and body composition: Evolving concepts in the pathogenesis of osteoporosis, *American Journal of Medicine* 122 (2009): 409–414.
11. M. T. Barrack and coauthors, Physiologic and behavioral indicators of energy deficiency in female adolescent runners with elevated bone turnover, *American Journal of Clinical Nutrition* 92 (2010): 652–659; Position of the American Dietetic Association, Dietitians of Canada, and the American College of Sports Medicine, Nutrition and Athletic Performance, *Journal of the American Dietetic Association* 109 (2009): 509–527.
12. A. P. Winston, The clinical biochemistry of anorexia nervosa, *Annals of Clinical Biochemistry* 49 (2012): 132–143; P. S. Mehler, B. S. Cleary, and J. L. Gaudiani, Osteoporosis in anorexia nervosa, *Eating Disorders* 19 (2011): 194–202.
13. Winston, 2012.
14. Position of the American Dietetic Association, 2011.
15. E. Attia and B. T. Walsh, Behavioral management for anorexia nervosa, *New England Journal of Medicine* 360 (2009): 500–506.
16. J. E. Schebendach and coauthors, Dietary energy density and diet variety as predictors of outcome in anorexia nervosa, *American Journal of Clinical Nutrition* 87 (2008): 810–816.
17. Grave, 2011.
18. Position of the American Dietetic Association, 2011.
19. Martin, 2010.

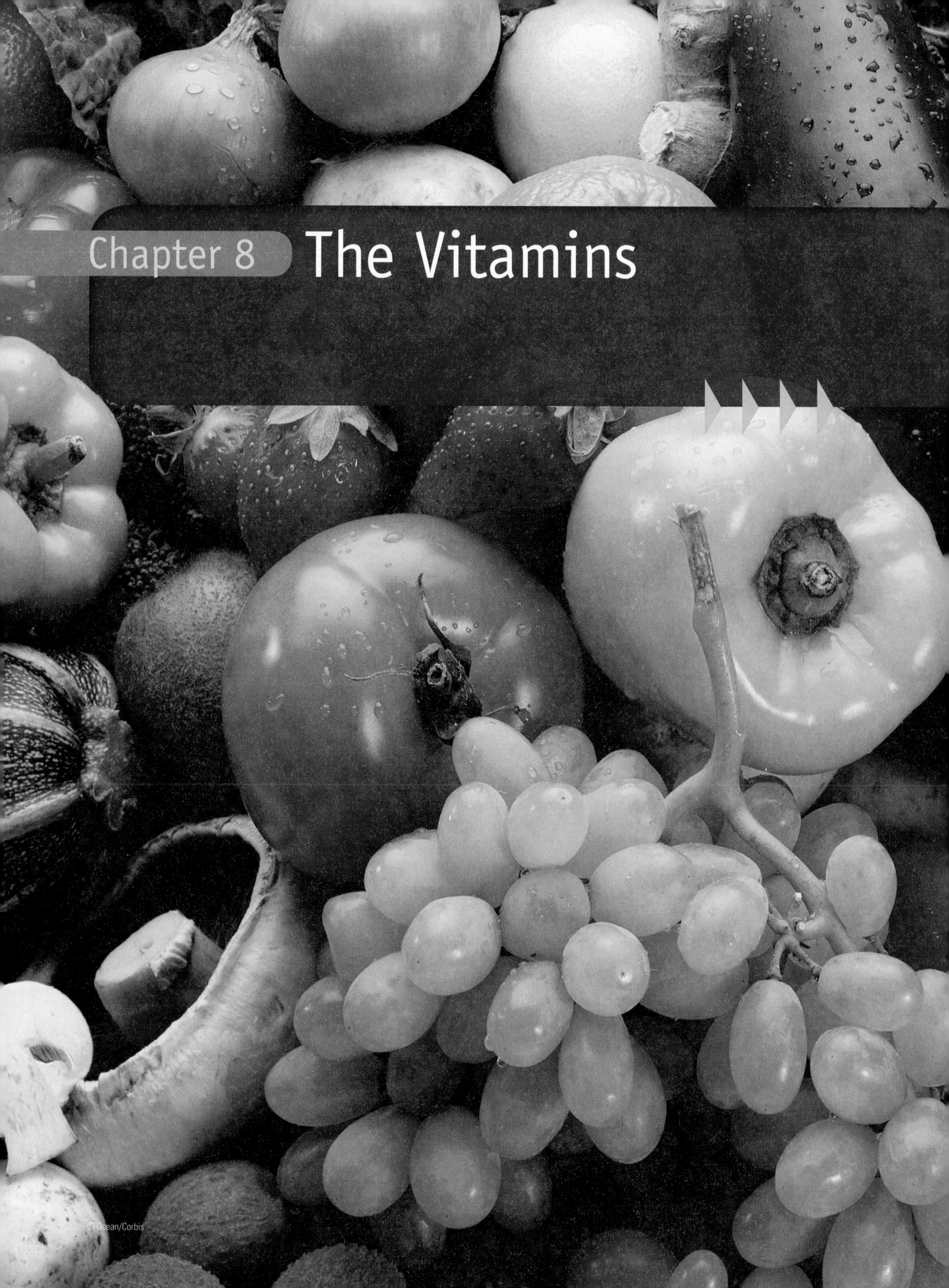

Chapter 8 The Vitamins

EARLIER CHAPTERS FOCUSED PRIMARILY ON THE ENERGY-YIELDING nutrients—carbohydrate, fat, and protein. This chapter and the next one discuss the nutrients everyone thinks of when nutrition is mentioned—the vitamins and minerals.

The Vitamins—An Overview

The **vitamins** occur in foods in much smaller quantities than do the energy-yielding nutrients, and they themselves contribute no energy to the body. Instead, they serve mostly as facilitators of body processes. They are a powerful group of substances, as their absence attests: Vitamin A deficiency can cause blindness, a lack of niacin can cause dementia, and a lack of vitamin D can retard bone growth. The consequences of deficiencies are so dire and the effects of restoring the needed nutrients so dramatic that people spend billions of dollars each year on **dietary supplements** to cure many different ailments. Vitamins certainly contribute to sound nutritional health, but supplements do not cure all ills. Furthermore, vitamin supplements do not offer the many benefits that come from vitamin-rich foods. The only disease a vitamin will *cure* is the one caused by a deficiency of that vitamin. The vitamins' roles in supporting optimal health extend far beyond preventing deficiency diseases, however. Emerging evidence supports the role of vitamin-rich *foods*, but not vitamin supplements, as protective against cancer and heart disease.[1]

A child once defined vitamins as "what, if you don't eat, you get sick." The description is both insightful and accurate. A more prosaic definition is that vitamins are potent, essential, nonkcaloric, organic nutrients needed from foods in trace amounts to perform specific functions that promote growth, reproduction, and the maintenance of health and life. The vitamins differ from carbohydrates, fats, and proteins in the following ways:

- *Structure.* Vitamins are individual units; they are not linked together (as are molecules of glucose or amino acids).
- *Function.* Vitamins do not yield energy when metabolized; many of them do, however, assist the enzymes that participate in the release of energy from carbohydrates, fats, and proteins.
- *Dietary intakes.* The amounts of vitamins people ingest daily from foods and the amounts they require are measured in *micrograms* (μg) or *milligrams* (mg), rather than grams (g).

The vitamins are similar to the energy-yielding nutrients, though, in that they are vital to life, organic, and available from foods.

As the individual vitamins were discovered, they were named or given letters, numbers, or both. This led to the confusion that still exists today. This chapter uses the names shown in Table 8-1; alternative names are given in Tables 8-4 and 8-5, which appear later in the chapter.

Bioavailability

The amount of vitamins available from foods depends on two factors: the quantity provided by a food and the amount absorbed and used by the body (the vitamin's **bioavailability**). Researchers analyze foods to determine their vitamin contents and publish the results in tables of food composition such as Appendix A. Determining the bioavailability of a vitamin is more difficult because it depends on many factors, including:

- Efficiency of digestion and time of transit through the GI tract.
- Previous nutrient intake and nutrition status.
- Other foods consumed at the same time.
- The method of food preparation (raw or cooked, for example).
- Source of the nutrient (naturally occurring, synthetic, or fortified).

This chapter and the next describe factors that inhibit or enhance the absorption of individual vitamins and minerals. Experts consider these factors when estimating recommended intakes.

vitamins: essential, nonkcaloric, organic nutrients needed in tiny amounts in the diet.

dietary supplements: products that are added to the diet and contain any of the following ingredients: a vitamin, a mineral, an herb or other botanical, an amino acid, a metabolite, a constituent, or an extract.

bioavailability: the rate and extent to which a nutrient is absorbed and used.

Precursors Some of the vitamins are available from foods in inactive forms known as **precursors**, or provitamins. Once inside the body, the precursor is converted to the active form of the vitamin. Thus, in measuring a person's vitamin intake, it is important to count both the amount of the actual vitamin and the potential amount available from its precursors. Tables 8-4 and 8-5 later in the chapter specify which vitamins have precursors.

Organic Nature Fresh foods naturally contain vitamins, but because they are organic, (■) vitamins can be readily destroyed during processing. Therefore, processed foods should be used sparingly, and fresh foods must be handled with care during storage and in cooking. Prolonged heating may destroy much of the thiamin in food. Because riboflavin can be destroyed by the ultraviolet rays of the sun or by fluorescent light, foods stored in transparent glass containers are most likely to lose riboflavin. Oxygen destroys vitamin C, so losses occur when foods are cut, processed, and stored. Table 8-2 summarizes ways to minimize nutrient losses in the kitchen.

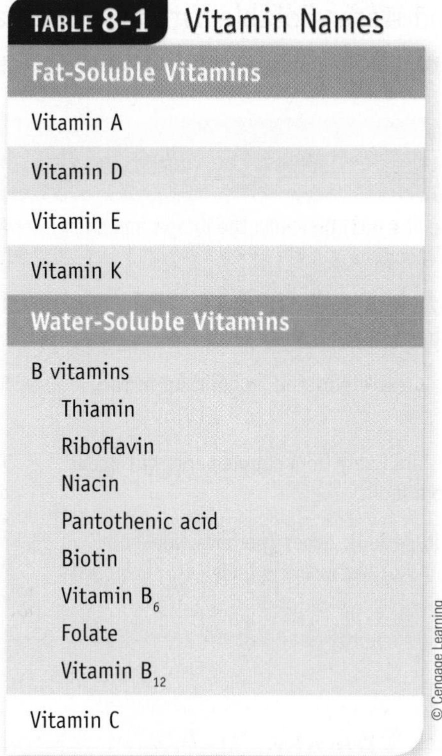

TABLE 8-1 Vitamin Names

Fat-Soluble Vitamins
Vitamin A
Vitamin D
Vitamin E
Vitamin K
Water-Soluble Vitamins
B vitamins
Thiamin
Riboflavin
Niacin
Pantothenic acid
Biotin
Vitamin B_6
Folate
Vitamin B_{12}
Vitamin C

Solubility Vitamins fall naturally into two classes—fat soluble and water soluble. The water-soluble vitamins are the B vitamins and vitamin C; the fat-soluble ones are vitamins A, D, E, and K. The solubility of a vitamin confers on it many characteristics and determines how it is absorbed, transported, stored, and excreted (see Table 8-3 on p. 198). This discussion of vitamins begins with the fat-soluble vitamins.

TABLE 8-2 Minimizing Nutrient Losses

Each of these tactics saves a small percentage of the vitamins in foods, but repeated each day this can add up to significant amounts in a year's time.

Prevent Enzymatic Destruction

- Refrigerate most fruits, vegetables, and juices to slow breakdown of vitamins.

Protect from Light and Air

- Store milk and enriched grain products in opaque containers to protect riboflavin.
- Store cut fruits and vegetables in the refrigerator in airtight wrappers; reseal opened juice containers before refrigerating.

Prevent Heat Destruction or Losses in Water

- Wash intact fruits and vegetables before cutting or peeling to prevent vitamin losses during washing.
- Cook fruits and vegetables in a microwave oven, or quickly stir fry, or steam them over a small amount of water to preserve heat-sensitive vitamins and to prevent vitamin loss in cooking water. Recapture dissolved vitamins by using cooking water for soups, stews, or gravies.
- Avoid high temperatures and long cooking times.

■ Organic nutrients contain carbon.

precursors: compounds that can be converted into other compounds; with regard to vitamins, compounds that can be converted into active vitamins; also known as *provitamins*.

TABLE 8-3 Fat-Soluble and Water-Soluble Vitamins Compared

While each vitamin has unique functions and features, a few generalizations about the fat-soluble and water-soluble vitamins can aid understanding.

	Fat-Soluble Vitamins: Vitamins A, D, E, and K	Water-Soluble Vitamins: B Vitamins and Vitamin C
Absorption	Absorbed like fats, first into the lymph and then into the blood.	Absorbed directly into the blood.
Transport and storage	Must travel with protein carriers in watery body fluids; stored in the liver or fatty tissues.	Travel freely in watery fluids; most are not stored in the body.
Excretion	Not readily excreted; tend to build up in the tissues.	Readily excreted in the urine.
Toxicity	Toxicities are likely from supplements but occur rarely from food.	Toxicities are unlikely but possible with high doses from supplements.
Requirement	Needed in periodic doses (perhaps weekly or even monthly) because the body can draw on its stores.	Needed in frequent doses (perhaps every 1 to 3 days) because the body does not store most of them to any extent.

Review Notes

- Vitamins are essential, nonkcaloric, organic nutrients that are needed in trace amounts in the diet to help facilitate body processes.
- The amount of vitamins available from foods depends on the quantity provided by a food and the amount absorbed and used by the body—the vitamins' bioavailability.
- Vitamin precursors in foods are converted into active vitamins in the body.
- Vitamins can be readily destroyed during processing.
- The water-soluble vitamins are the B vitamins and vitamin C; the fat-soluble vitamins are vitamins A, D, E, and K.

The Fat-Soluble Vitamins

The fat-soluble vitamins—A, D, E, and K—usually occur together in the fats and oils of foods, and the body absorbs them in the same way it absorbs lipids. Therefore, any condition that interferes with fat absorption can precipitate a deficiency of the fat-soluble vitamins. Once absorbed, fat-soluble vitamins are stored in the liver and fatty tissues until the body needs them. They are not readily excreted, and, unlike most of the water-soluble vitamins, they can build up to toxic concentrations.

The capacity to store fat-soluble vitamins affords a person some flexibility in dietary intake. When blood concentrations begin to decline, the body can retrieve the vitamins from storage. Thus, a person need not eat a day's allowance of each fat-soluble vitamin every day but need only make sure that, over time, average daily intakes approximate recommended intakes. In contrast, most water-soluble vitamins must be consumed more regularly because the body does not store them to any great extent.

vitamin A: a fat-soluble vitamin. Its three chemical forms are *retinol* (the alcohol form), *retinal* (the aldehyde form), and *retinoic acid* (the acid form).

beta-carotene: a vitamin A precursor made by plants and stored in human fat tissue; an orange pigment.

VITAMIN A AND BETA-CAROTENE

Vitamin A has the distinction of being the first fat-soluble vitamin to be recognized. More than a century later, vitamin A and its plant-derived precursor, **beta-carotene**, continue to intrigue researchers with their diverse roles and profound effects on health.

Vitamin A is a versatile vitamin, with roles in gene expression, vision, cell differentiation (thereby maintaining the health of body linings and skin), immunity, and reproduction and growth.[2] Three different forms of vitamin A are active in the body: retinol, retinal, and retinoic acid. Each form of vitamin A performs specific tasks. Retinol supports reproduction and is the major transport and storage form of the vitamin; the cells convert retinol to retinal or retinoic acid as needed. Retinal is active in vision, and retinoic acid acts as a hormone, regulating cell differentiation, growth, and embryonic development. A special transport protein, **retinol-binding protein (RBP)**, picks up retinol from the liver, where it is stored, and carries it in the blood. (■)

■ Measurement of the blood concentration of RBP is a sensitive test of vitamin A status.

■ Photosensitive cells of the retina:
- Rods contain the rhodopsin pigment and respond to faint light.
- Cones contain the iodopsin pigment and function in color vision.

retinol-binding protein (RBP): the specific protein responsible for transporting retinol.

cornea (KOR-nee-uh): the hard, transparent membrane covering the outside of the eye.

retina (RET-in-uh): the layer of light-sensitive nerve cells lining the back of the inside of the eye; consists of rods and cones.

pigment: a molecule capable of absorbing certain wavelengths of light so that it reflects only those that we perceive as a certain color.

rhodopsin (ro-DOP-sin): a light-sensitive pigment of the retina; contains the retinal form of vitamin A and the protein opsin.

hod = red (pigment)
opsin = visual protein

opsin (OP-sin): the protein portion of the visual pigment molecule.

night blindness: the slow recovery of vision after exposure to flashes of bright light at night; an early symptom of vitamin A deficiency.

Vitamin A's Role in Gene Expression

Vitamin A exerts considerable influence on an array of body functions through its interaction with genes—hundreds of genes are regulated by the retinoic acid form of the vitamin.[3] Genes direct the synthesis of proteins, including enzymes, and enzymes perform the metabolic work of the tissues (see Chapter 5). Hence, factors that influence gene expression also affect the metabolic activities of the tissues and the health of the body.

Researchers have long known that simply possessing the genetic equipment needed to make a particular protein does not guarantee that the protein will be produced, any more than owning a car guarantees you a trip across town. To make the journey, you must also use the right key to trigger the events that start up its engine or turn it off at the appropriate time. Some dietary components, including the retinoic acid form of vitamin A, are now known to be such keys—they help to activate or deactivate certain genes and thus affect the production of specific proteins.[4]

Vitamin A's Role in Vision

Vitamin A plays two indispensable roles in the eye. It helps maintain a healthy, crystal-clear outer window, the **cornea**; and it participates in light detection at the **retina**. Some of the photosensitive cells (■) of the retina contain **pigment** molecules called **rhodopsin**; each rhodopsin molecule is composed of a protein called **opsin** bonded to a molecule of the retinal form of vitamin A. When light passes through the cornea of the eye and strikes the retina, rhodopsin responds by changing shape and becoming bleached. In turn, this initiates the signal that conveys the sensation of sight to the optic center in the brain. Figure 8-1 shows vitamin A's site of action inside the eye.

When vitamin A is lacking, the eye has difficulty adapting to changing light levels. At night, after the eye has adapted to darkness, a flash of bright light is followed by a brief delay before the eye can see again. This lag in the recovery of night vision is known as **night blindness**. Because night blindness is easy to test, it aids in the diagnosis of vitamin A deficiency. Night blindness is only a symptom, however, and may indicate a condition other than vitamin A deficiency.

FIGURE 8-1 Vitamin A's Role in Vision

As light enters the eye, pigments within the cells of the retina absorb the light.

Light energy

Cornea

Retina cells (rods and cones)

Eye

Nerve impulses to the brain

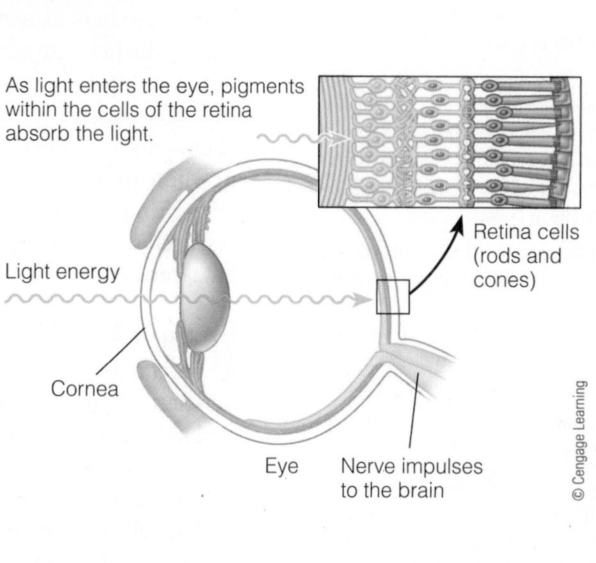

Vitamin A's Role in Protein Synthesis and Cell Differentiation

The role that vitamin A plays in vision is undeniably important, but only one-thousandth of the body's vitamin A is in the retina. Much more is in the skin and the linings of organs, where it works behind the scenes at the genetic level to promote protein synthesis and cell **differentiation**. The process of cell differentiation allows each type of cell to mature so that it is capable of performing a specific function.

All body surfaces, both inside and out, are covered by layers of cells known as **epithelial cells**. The **epithelial tissue** on the outside of the body is, of course, the skin. In the eye, epithelial tissue covers the outermost layer of the cornea, where it blocks the passage of foreign materials such as dust, water, and bacteria. The epithelial tissues inside the

■ Nutrition during pregnancy is discussed in Chapter 10.

■ A placebo is an inactive substance used in research studies.

■ Reminder: *Phytochemicals* are compounds in plants that confer color, taste, and other characteristics. Some phytochemicals are bioactive food components in functional foods. Phytochemicals and functional foods are the topic of Nutrition in Practice 8.

■ The progressive blindness caused by vitamin A deficiency is called *xerophthalmia* (zer-off-THAL-mee-uh).

xero = dry
ophthalm = eye

■ An early sign of xerophthalmia is *xerosis* (drying of the cornea); the last and most severe stage is *keratomalacia* (kerr-uh-to-mal-AY-shuh), or total blindness.

malacia = softening, weakening

differentiation: the development of specific functions different from those of the original.

epithelial (ep-i-THEE-lee-ul) cells: cells on the surface of the skin and mucous membranes.

epithelial tissue: tissue composing the layers of the body that serve as selective barriers between the body's interior and the environment (examples are the cornea, the skin, the respiratory lining, and the lining of the digestive tract).

mucous membrane: membrane composed of mucus-secreting cells that lines the surfaces of body tissues. (Reminder: *Mucus* is the smooth, slippery substance secreted by these cells.)

antioxidant (anti-OX-ih-dant): a compound that protects other compounds from oxygen by itself reacting with oxygen. Oxidation is a potentially damaging effect of normal cell chemistry involving oxygen.

anti = against
oxy = oxygen

free radicals: highly reactive chemical forms that can cause destructive changes in nearby compounds, sometimes setting up a chain reaction.

body include the linings of the mouth, stomach, and intestines; the linings of the lungs and the passages leading to them; the lining of the bladder; the linings of the uterus and vagina; and the linings of the eyelids and sinus passageways. The epithelial tissues on the inside of the body must be kept smooth. To ensure that they are, the epithelial cells on their surfaces secrete a smooth, slippery substance (mucus) that coats the tissues and protects them from invasive microorganisms and other harmful particles. The **mucous membrane** that lines the stomach also shields its cells from digestion by gastric juices. Vitamin A, by way of its role in cell differentiation, helps to maintain the integrity of the epithelial cells.

Vitamin A's Role in Immunity Vitamin A has gained a reputation as an "anti-infective" vitamin because so many of the body's defenses against infection depend on an adequate supply.[5] Much research supports the need for vitamin A in the regulation of the genes involved in immunity. Without sufficient vitamin A, these genetic interactions produce an altered response to infection that weakens the body's defenses.

Vitamin A's Role in Reproduction, Growth, and Development Vitamin A is crucial to normal reproduction and growth. In men, vitamin A participates in sperm development, and, in women, vitamin A promotes normal fetal growth and development.[6] During pregnancy, vitamin A is transferred to the fetus and is essential to the development of the nervous system, lungs, heart, kidneys, skeleton, eyes, and ears. (■)

Beta-Carotene's Role as an Antioxidant For many years scientists believed beta-carotene to be of interest solely as a vitamin A precursor. Eventually, though, researchers began to recognize that beta-carotene is an extremely effective **antioxidant** in the body. Antioxidants are compounds that protect other compounds (such as lipids in cell membranes) from attack by oxygen. Oxygen triggers the formation of compounds known as **free radicals** that can start chain reactions in cell membranes. If left uncontrolled, these chain reactions can damage cell structures and impair cell functions. Oxidative and free-radical damage to cells is suspected of instigating some early stages of cancer and heart disease.[7] Research has identified links between oxidative damage and the development of many other diseases, including age-related blindness, Alzheimer's disease, arthritis, cataracts, diabetes, and kidney disease.[8]

Studies of populations suggest that people whose diets are low in foods rich in beta-carotene have higher incidences of certain types of cancer than those whose diets contain generous amounts of such foods. Based on findings that beta-carotene in foods may protect against cancer, researchers designed a study to determine the effects of beta-carotene *supplements* on the incidence of lung cancer among smokers. The researchers expected to see a beneficial effect, but instead they found that smokers taking the beta-carotene supplements suffered a *greater* incidence of lung cancer than those taking placebos. (■) Beta-carotene is one of many **dietary antioxidants** present in foods—others include vitamin E, vitamin C, the mineral selenium, and many phytochemicals. (■) Dietary antioxidants are just one class of a complex array of constituents in whole foods that seem to benefit health synergistically.[9] Until more is known, eating beta-carotene–rich foods, not supplements, is in the best interests of health. Based on research so far, the DRI committee has not established a recommended intake value for beta-carotene.

Vitamin A Deficiency Up to a year's supply of vitamin A can be stored in the body, 90 percent of it in the liver. If a healthy adult were to stop eating vitamin A–rich foods, deficiency symptoms would not begin to appear until after stores were depleted, which would take one to two years. Then, however, the consequences would be profound and severe. Table 8-4, later in this chapter, lists some of them.

In vitamin A deficiency, cell differentiation and maturation are impaired. The epithelial cells flatten and begin to produce **keratin**—the hard, inflexible protein of hair and nails. In the eye, this process leads to drying and hardening of the cornea, which may progress to permanent blindness. (■)(■) Vitamin A deficiency is the major cause of preventable blindness in children worldwide, causing as many as half a million children to lose their sight every year.[10] As many as 250 million children worldwide endure less severe forms of vitamin A deficiency, making them vulnerable to infectious diseases. Routine vitamin A supplementation and food fortification can be life-saving interventions.[11]

All body surfaces, both inside and out, maintain their integrity with the help of vitamin A. When vitamin A is lacking, cells of the skin harden and flatten, making it dry, rough, scaly, and hard. An accumulation of keratin makes a lump around each hair **follicle** (keratinization). (■)

In the mouth, a vitamin A deficiency results in drying and hardening of the salivary glands, making them susceptible to infection. Secretions of mucus in the stomach and intestines are reduced, hindering normal digestion and absorption of nutrients. Infections of other mucous membranes also become likely.

Vitamin A's role in maintaining the body's defensive barriers may partially explain the relationship between vitamin A deficiency and susceptibility to infection. In developing countries around the world, measles is a devastating infectious disease, killing nearly 380 children each day.[12] Deaths are usually due to related infections such as pneumonia and severe diarrhea. Providing large doses of vitamin A reduces the risk of dying from these infections.

The evidence that vitamin A reduces the severity of measles and measles-related infections and diarrhea has prompted the World Health Organization (WHO) and the United Nations International Children's Emergency Fund (UNICEF) to make control of vitamin A deficiency a major goal in their quest to improve child survival throughout the developing world. They recommend routine vitamin A supplementation for all children with measles in areas where vitamin A deficiency is a problem or where the measles death rate is high. The American Academy of Pediatrics recommends vitamin A supplementation for certain groups of measles-infected infants and children in the United States.

Vitamin A Toxicity

Vitamin A toxicity is a real possibility when people consume concentrated amounts of **preformed vitamin A** in foods derived from animals, fortified foods, or supplements. Plant foods contain the vitamin only as beta-carotene, its inactive precursor form. The precursor does not convert to active vitamin A rapidly enough to cause toxicity.

Overdoses of vitamin A damage the same body systems that exhibit symptoms in vitamin A deficiency (see Table 8-4 later in the chapter). Children are most vulnerable to vitamin A toxicity because they need less vitamin A and are more sensitive to overdoses. The availability of breakfast cereals, instant meals, fortified milk, and chewable candy-like vitamins, each containing 100 percent or more of the recommended daily intake of vitamin A, makes it possible for a well-meaning parent to provide several times the daily allowance of the vitamin to a child within a few hours.

Excessive vitamin A also poses a **teratogenic** risk.[13] Excessive vitamin A during pregnancy can injure the spinal cord and other tissues of the developing fetus, increasing the risk of birth defects.[14] The Tolerable Upper Intake Level of 3000 micrograms for women of childbearing age is based on the teratogenic effect of vitamin A.[15]

Excessive amounts of vitamin A over the years may weaken the bones and contribute to fractures and osteoporosis (see the later discussion of vitamin D for more on osteoporosis).[16] More research is needed to clarify the relationship between vitamin A intake and bone health, but such findings suggest that most people should not take vitamin A supplements. Even multivitamin supplements provide more vitamin A than most people need. (■)

Certain vitamin A relatives are available by prescription as acne treatments. When applied directly to the skin surface, these preparations help relieve the symptoms of acne. Taking massive doses of vitamin A internally will *not* cure acne, however, and may cause the symptoms itemized in Table 8-4, later in the chapter. In most cases, foods are a better choice than supplements for needed nutrients. The best way to ensure a safe vitamin A intake is to eat generous servings of vitamin A–rich foods.

■ The accumulation of the hard material keratin around each hair follicle is *follicular hyperkeratosis*.

■ Multivitamin supplements typically provide:
- 750 µg (2500 IU)
- 1500 µg (5000 IU)

For perspective, the RDA for vitamin A is 700 µg for women and 900 µg for men. The IU (*international unit*) is often used to express the amount of vitamins in foods and on supplement labels.

dietary antioxidants: compounds typically found in plant foods that significantly decrease the adverse effects of oxidation on living tissues. The major antioxidant vitamins are vitamin E, vitamin C, and beta-carotene.

keratin (KERR-uh-tin): a water-insoluble protein; the normal protein of hair and nails. Keratin-producing cells may replace mucus-producing cells in vitamin A deficiency.

follicle (FOLL-i-cul): a group of cells in the skin from which a hair grows.

preformed vitamin A: vitamin A in its active form.

teratogenic (ter-AT-oh-jen-ik): causing abnormal fetal development and birth defects.

terato = monster
genic = to produce

FIGURE 8-2 Symptom of Beta-Carotene Excess—Discoloration of the Skin

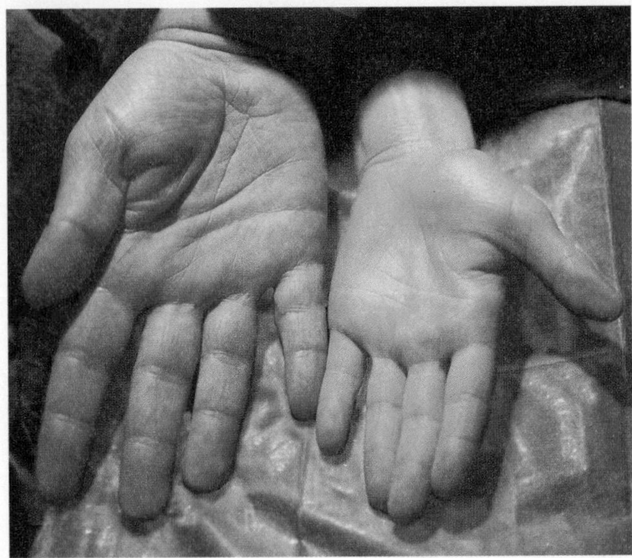

The hand on the right shows the skin discoloration that occurs from excess beta-carotene. (The hand on the left belongs to someone else and is shown for comparison.)

Beta-Carotene Conversion and Toxicity

Nutrition scientists do not use micrograms to specify the quantity of beta-carotene in foods. Instead, they use a value known as **retinol activity equivalents (RAE)**, (■) which express the amount of retinol the body actually derives from a plant food after conversion. The body can make one unit of retinol from about 12 units of beta-carotene.

As mentioned earlier, beta-carotene from plant foods is not converted to the active form of vitamin A rapidly enough to be hazardous. It has, however, been known to turn people bright yellow if they eat too much. (■) Beta-carotene builds up in the fat just beneath the skin and imparts a yellowish cast (see Figure 8-2). Additionally, as discussed earlier, overconsumption of beta-carotene from *supplements* may be harmful, especially to smokers.

Vitamin A in Foods Preformed vitamin A is found only in foods of animal origin. The richest sources of vitamin A are liver and fish oil, but milk, cheese, and fortified cereals are also good sources. Healthy people can eat vitamin A–rich foods—with the possible exception of liver—in large amounts without risking toxicity. Eating liver once every week or so is enough. Butter and eggs also provide some vitamin A to the diet.

Because vitamin A is fat soluble, it is lost when milk is skimmed. To compensate, reduced-fat, low-fat, and fat-free milks are often fortified with vitamin A. Margarine is also usually fortified so as to provide the same amount of vitamin A as butter. Figure 8-3 shows a sampling of the richest food sources of both preformed vitamin A and beta-carotene.

Fast-food meals often lack vitamin A. When fast-food restaurants offer salads with cheese, carrots, and other vitamin A–rich foods, the nutritional quality of their meals greatly improves.

Beta-Carotene in Foods Many foods from plants contain beta-carotene, the orange pigment responsible for the bold colors of many fruits and vegetables. Carrots, sweet potatoes, pumpkins, cantaloupe, and apricots are all rich sources, and their bright orange color enhances the eye appeal of the plate. Another colorful group, *dark* green vegetables, such as spinach, other greens, and broccoli, owe their color to both chlorophyll and beta-carotene. The orange and green pigments together impart a deep, murky green color to the vegetables. Other colorful vegetables, such as iceberg lettuce, beets, and sweet corn, can fool you into thinking they contain beta-carotene, but these foods derive their color from other pigments and are poor sources of beta-carotene. As for "white" plant foods such as rice and potatoes, they have little or none. Recommendations to eat *dark* green or *deep* orange vegetables and fruits at least every other day help people to meet their vitamin A needs.

VITAMIN D

Vitamin D (■) is different from all the other nutrients in that the body can synthesize it in significant quantities with the help of sunlight. Therefore, in a sense, vitamin D is not an essential nutrient. Given enough sun, people need no vitamin D from foods.

Vitamin D's Metabolic Conversions The liver manufactures a vitamin D precursor, (■) which migrates to the skin, where it is converted to a second precursor

■ 1 μg RAE = 1 μg retinol = 12 μg beta-carotene from food

■ Yellowing of the skin caused by excess carotene in the blood is known as *carotenemia* (KAR-oh-teh-NEE-me-ah). Carotenemia can be distinguished from jaundice because the mucous membranes lining the eyelids do not turn yellow as they do in jaundice.

■ Vitamin D comes in many forms, but the two most important in the diet are a plant version called vitamin D_2, or *ergocalciferol* (ER-go-kal-SIF-er-ol), and an animal version called vitamin D_3, or *cholecalciferol* (KO-lee-kal-SIF-er-ol).

■ The precursor of vitamin D made in the liver is 7-dehydrocholesterol, which is made from cholesterol. This is one of the body's many "good" uses for cholesterol.

retinol activity equivalents (RAE): a measure of vitamin A activity; the amount of retinol that the body will derive from a food containing preformed retinol or its precursor beta-carotene.

FIGURE 8-3 Good Sources of Vitamin A and Beta-Carotene[a]

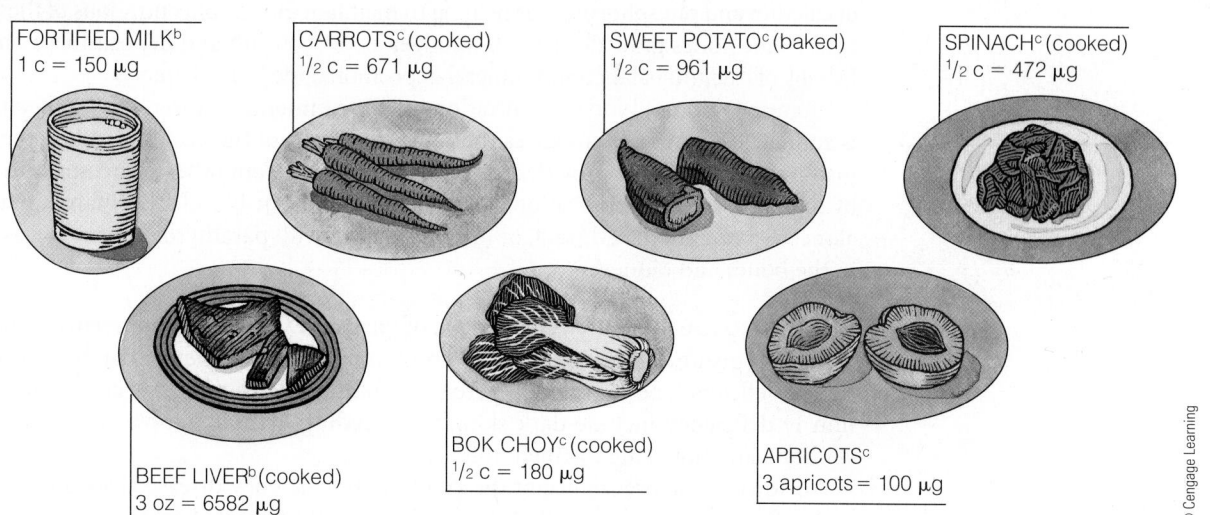

FORTIFIED MILK[b]
1 c = 150 µg

CARROTS[c] (cooked)
1/2 c = 671 µg

SWEET POTATO[c] (baked)
1/2 c = 961 µg

SPINACH[c] (cooked)
1/2 c = 472 µg

BEEF LIVER[b] (cooked)
3 oz = 6582 µg

BOK CHOY[c] (cooked)
1/2 c = 180 µg

APRICOTS[c]
3 apricots = 100 µg

© Cengage Learning

[a]These foods provide 10 percent or more of the vitamin A Daily Value in a serving. For a 2000-kcalorie diet, the DV is 900 µg/day.
[b]This food contains preformed vitamin A.
[c]This food contains beta-carotene.

with the help of the sun's ultraviolet rays. Next, the liver and then the kidneys alter the second precursor to produce the active vitamin. (■) Whether made from sunlight or obtained from food, vitamin D requires the same two conversions by the liver and kidneys to become active. The biological activity of the active vitamin is 500- to 1000-fold greater than that of its precursor. Diseases that affect either the liver or the kidneys may impair the transformations of precursor vitamin D to active vitamin D and therefore produce symptoms of vitamin D deficiency.

■ The final, active vitamin is calcitriol (1-25 dihydroxyvitamin D).

Vitamin D's Actions Although known as a vitamin, vitamin D is actually a hormone—a compound manufactured by one organ of the body that has effects on another. The best-known vitamin D target organs are the small intestine, the kidneys, and the bones, but scientists have discovered more than 30 other vitamin D target tissues, including the brain, the pancreas, the skin, the reproductive organs, and cells of the immune system.[17] In many cases, vitamin D enhances or suppresses the activity of genes that regulate cell growth. As such, it may be valuable in treating a number of diseases.

Research suggests that a deficit of vitamin D is associated with an increased risk of high blood pressure; cardiovascular diseases; some common cancers; infections such as tuberculosis; inflammatory conditions; autoimmune diseases such as type 1 diabetes, rheumatoid arthritis, and multiple sclerosis; and even premature death.[18] Even so, evidence does not support taking vitamin D supplements to prevent diseases (except those caused by deficiency), and experts recommend against it.[19] The well-established vitamin D roles, however, concern calcium balance and the bones during growth and throughout life, and these form the basis of the recommended intakes for vitamin D, discussed in a later section.

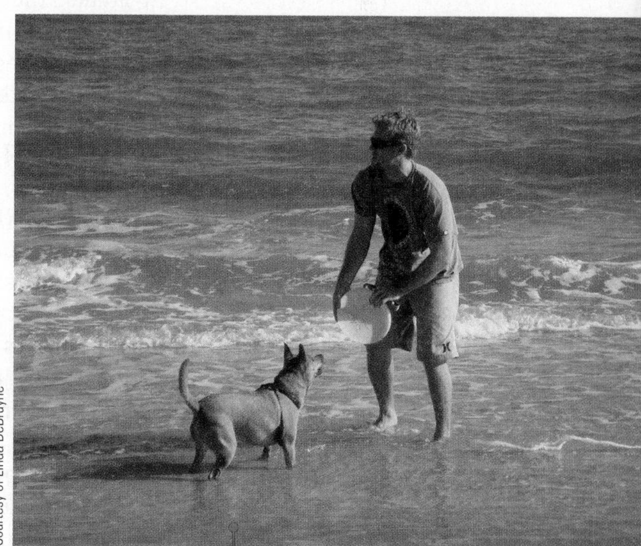
Courtesy of Linda DeBruyne

Sunlight promotes vitamin D synthesis in the skin. Exposure to the sun should be moderate, however; excessive exposure may cause skin cancer.

Vitamin D's Roles in Bone Growth Vitamin D is a member of a large, cooperative bone-making and maintenance team composed of nutrients and other compounds, including vitamins A, C, and K; the hormones parathormone (parathyroid hormone) and calcitonin; the protein collagen; and the minerals calcium, phosphorus, magnesium, and fluoride. (■) Many of their interactions take place at the genetic level in ways that are under investigation. Vitamin D's special role in bone health is to assist in the absorption of calcium and phosphorus, thus helping to maintain blood concentrations of these minerals.[20] The bones grow denser and stronger as they absorb and deposit these minerals. Details of calcium balance and mineral deposition appear in Chapter 9.

Vitamin D raises blood concentrations of bone minerals in three ways. When the diet is sufficient, vitamin D enhances their absorption from the GI tract. When the diet is insufficient, vitamin D provides the needed minerals from other sources: reabsorption by the kidneys and mobilization from the bones into the blood. The vitamin may work alone, as it does in the GI tract, or in combination with parathyroid hormone, as it does in the bones and kidneys.

Vitamin D Deficiency Overt signs of vitamin D deficiency are relatively rare, but vitamin D insufficiency is remarkably common.[21] Almost 10 percent of the U.S. population is deficient, and another 25 percent is marginal.[22] Factors that contribute to vitamin D deficiency include dark skin, breastfeeding without supplementation, lack of sunlight, and not using fortified milk.

Worldwide, the prevalence of the vitamin D–deficiency disease **rickets** is extremely high, affecting more than half of the children in countries such as Mongolia, Tibet, and the Netherlands. In the United States, more than 58 million children are reported to be vitamin D deficient or insufficient, but rickets itself is uncommon.[23] When it occurs, black children and adolescents—especially females and overweight teens—are most likely to be affected.[24] To prevent rickets and support optimal bone health, the DRI committee recommends that all infants, children, and adolescents consume the recommended 15 micrograms (600 IU) of vitamin D each day.[25] (■) In rickets, the bones fail to calcify normally, causing growth retardation and skeletal abnormalities. The bones become so weak that they bend when they have to support the body's weight (see Figure 8-4). A child with rickets who is old enough to walk characteristically develops bowed legs, often the most obvious sign of the disease. Another sign is the beaded ribs that result from the poorly formed attachments of the bones to the cartilage.

Adolescents, who often abandon vitamin D–fortified milk in favor of soft drinks, may also prefer indoor pastimes such as computer games to outdoor activities during daylight hours. Such teens often lack vitamin D and so fail to develop the bone density needed to prevent bone loss in later life.

In adults, the poor mineralization of bone results in the painful bone disease **osteomalacia**. Any failure to synthesize adequate vitamin D or obtain enough from foods sets the stage for a loss of calcium from the bones, which can result in fractures secondary to **osteoporosis** (reduced bone density). The simple act of taking a vitamin D supplement could easily save the life of an elderly person who might otherwise suffer dangerous bone fractures and falls.[26]

Vitamin D Toxicity Vitamin D clearly illustrates how nutrients in optimal amounts support health, but both inadequacies and excesses cause harm. Vitamin D is among the vitamins most likely to have toxic effects when consumed in excess. Excess vitamin D raises the concentration of blood calcium. Excess blood calcium tends to precipitate in the soft tissues and form stones, especially in the kidneys, where calcium is concentrated in an effort to excrete it. Calcification may also harden the blood vessels and is especially dangerous in the major arteries of the brain, heart, and lungs, where it can cause death.

The amounts of vitamin D made by the skin and found in foods are well within safe limits, but supplements containing the vitamin in concentrated form should be kept out of the reach of children and used cautiously, if at all, by adults. The DRI committee has set a Tolerable Upper Intake Level for vitamin D at 50 micrograms per day (2000 IU on supplement labels).

Vitamin D from the Sun Most of the world's population relies on natural exposure to sunlight to maintain adequate vitamin D nutrition. The sun imposes no risk of

■ Key bone nutrients:

- Vitamin D, vitamin K, vitamin A
- Calcium, phosphorus, magnesium, fluoride

■ 1 μg of vitamin D is equivalent to 40 IU of vitamin D.

rickets: the vitamin D–deficiency disease in children.

osteomalacia (os-tee-oh-mal-AY-shuh): a bone disease characterized by softening of the bones. Symptoms include bending of the spine and bowing of the legs. The disease occurs most often in adults with renal failure or malabsorption disorders.

osteo = bone
mal = bad (soft)

osteoporosis (os-tee-oh-pore-OH-sis): literally, porous bones; reduced density of the bones, also known as *adult bone loss*.

FIGURE 8-4 Vitamin D–Deficiency Symptoms—Bowed Legs and Beaded Ribs of Rickets

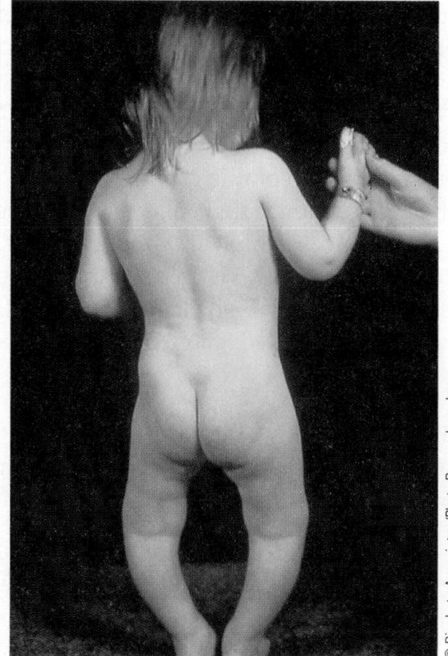

Bowed legs. In rickets, the poorly formed long bones of the legs bend outward as weight-bearing activities such as walking begin.

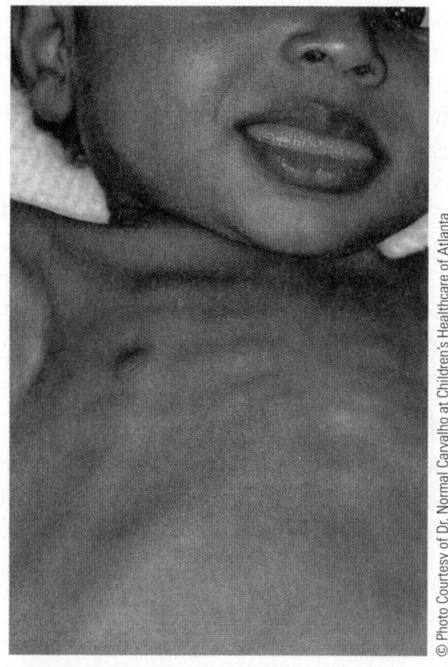

Beaded ribs. In rickets, a series of "beads" develop where the cartilages and bones attach.

vitamin D toxicity. Prolonged exposure to sunlight degrades the vitamin D precursor in the skin, preventing its conversion to the active vitamin. Even lifeguards on southern beaches are safe from vitamin D toxicity from the sun.

Prolonged exposure to sunlight has other undesirable consequences, however, such as premature wrinkling of the skin and skin cancers. Sunscreens help reduce the risks of these outcomes, but sunscreens with sun protection factors (SPF) of 8 and above also retard vitamin D synthesis. Still, even with an SPF 15–30 sunscreen, sufficient vitamin D synthesis can be obtained in 10 to 20 minutes of sun exposure. A strategy to avoid this dilemma is to apply sunscreen after enough time has elapsed to provide sufficient vitamin D. For most people, exposing hands, face, and arms on a clear summer day for 5 to 10 minutes, a few times a week, should be sufficient to maintain vitamin D nutrition. Dark-skinned people require longer exposure than light-skinned people, but by three hours, vitamin D synthesis in heavily pigmented skin arrives at the same plateau as in fair skin after 30 minutes.

Latitude, season, and time of day also have dramatic effects on vitamin D synthesis and status. (■) Heavy clouds, smoke, or smog block the ultraviolet (UV) rays of the sun that promote vitamin D synthesis. Differences in skin pigmentation, latitude, and smog may account for the finding that African American people, especially those in northern, smoggy cities, are most likely to be vitamin D deficient and develop rickets. Vitamin D deficiency is especially prevalent in the winter and in the Arctic and Antarctic regions of the world.[27] To ensure an adequate vitamin D status, supplements may be needed. The body's vitamin D supplies from summer synthesis alone are insufficient to meet winter needs.

The ultraviolet rays from tanning lamps and tanning booths may also stimulate vitamin D synthesis, but the hazards outweigh any possible benefits. The Food and Drug Administration (FDA) warns that, if the lamps are not properly filtered, people using tanning booths risk burns, damage to the eyes and blood vessels, and skin cancer.

■ Factors that may limit sun exposure and, therefore, vitamin D synthesis:

- Geographic location
- Season of the year
- Time of day
- Air pollution
- Clothing
- Tall buildings
- Indoor living
- Sunscreens

Vitamin D in Foods Only a few animal foods, notably, eggs, liver, butter, some fatty fish, and fortified milk, supply significant amounts of vitamin D. For those who use margarine in place of butter, fortified margarine is a significant source. Infant formulas are fortified with vitamin D in amounts adequate for daily intake as long as infants consume at least one liter (1000 milliliters) or one quart (32 ounces) of formula. Breast milk is low in vitamin D, so vitamin D supplements (10 micrograms daily) are recommended for infants who are breastfed exclusively and for those who do not receive at least 1000 milliliters of vitamin D–fortified formula per day.[28] These sources, plus any exposure to the sun, provide infants with more than enough of this vitamin.

The fortification of milk with vitamin D is the best guarantee that children will meet their vitamin D needs and underscores the importance of milk in children's diets. Unlike milk, cheese and yogurt are not fortified with vitamin D. Vegans, and especially their children, may have low vitamin D intakes because few fortified plant sources exist. Exceptions include margarine and some soy milks. In the United States, breakfast cereals may be fortified with vitamin D, as their labels indicate.

Vitamin D Recommendations In setting the recently revised dietary recommendations, the DRI committee assumed that no vitamin D was available from skin synthesis.[29] Advancing age increases the risk of vitamin D deficiency, so intake recommendations increase with age (see the margin). (■)

VITAMIN E

Almost a century ago, researchers discovered a compound in vegetable oils necessary for reproduction in rats. The compound was named **tocopherol**, which means "offspring." Eventually, the compound was named vitamin E. When chemists isolated four tocopherol compounds, they designated them by the first four letters of the Greek alphabet: alpha, beta, gamma, and delta. Of these, alpha-tocopherol is the gold standard for vitamin E activity; recommended intakes are based on it. Table 8-4 later in the chapter summarizes important information about vitamin E.

Vitamin E as an Antioxidant Like beta-carotene, vitamin E is a fat-soluble antioxidant. It protects other substances from oxidation (■) by being oxidized itself. If there is plenty of vitamin E in the membranes of cells exposed to an oxidant, chances are this vitamin will take the brunt of any oxidative attack, protecting the lipids and other vulnerable components of the membranes. Vitamin E is especially effective in preventing the oxidation of the polyunsaturated fatty acids (PUFA), but it protects all other lipids (for example, vitamin A) as well.

Vitamin E exerts an especially important antioxidant effect in the lungs, where the cells are exposed to high concentrations of oxygen. Vitamin E also protects the lungs from air pollutants that are strong oxidants.

Vitamin E may also offer protection against heart disease by protecting low-density lipoproteins (LDL) from oxidation and reducing inflammation.[30] The oxidation of LDL encourages the development of atherosclerosis; thus, in theory, halting LDL oxidation should reduce atherosclerosis. The results of controlled clinical studies in which human beings were given vitamin E supplements, however, have been disappointing.[31] Several leading vitamin E researchers point out that clinical trials of vitamin E supplementation differ in many important ways, such as the selection of subjects, the source of the vitamin, the dose of the vitamin, and the outcomes studied. Such differences may partly explain the inconsistent findings.

Further research that addresses such issues may clarify the relationship between vitamin E and heart disease. In the mean time, the American Heart Association supports the consumption of antioxidant-rich fruits and vegetables, as well as whole grains and nuts, to reduce the risk of heart disease.[32]

Vitamin E Deficiency When blood concentrations of vitamin E fall below a certain critical level, the red blood cells tend to break open and spill their contents, probably because the PUFA in their membranes oxidize. This classic vitamin E–deficiency

■ Vitamin D RDA:
- Adults (19–70 yr): 15 µg/day
- Adults (>70 yr): 20 µg/day

■ Reminder: *Oxidation* is a type of chemical reaction, so named because oxygen is one of the agents that often brings it about.

tocopherol (tuh-KOFF-er-ol): a general term for several chemically related compounds, one of which has vitamin E activity.

symptom, known as **erythrocyte hemolysis**, is seen in premature infants born before the transfer of vitamin E from the mother to the fetus that takes place in the last weeks of pregnancy. Vitamin E treatment corrects **hemolytic anemia**.

The few symptoms of vitamin E deficiency that have been observed in adults include loss of muscle coordination and reflexes with impaired movement, vision, and speech. Vitamin E treatment corrects all of these symptoms.

In adults, vitamin E deficiency is usually associated with diseases, notably those that cause malabsorption of fat. These include diseases of the liver, gallbladder, and pancreas, as well as various hereditary diseases involving digestion and use of nutrients. (■)

On rare occasions, vitamin E deficiencies develop in people without diseases. Most likely, such deficiencies occur after years of eating diets extremely low in fat; using fat substitutes, such as diet margarines and salad dressings, as the only sources of fat; or consuming diets composed of highly processed or "convenience" foods. Extensive heating in the processing of foods destroys vitamin E.

■ Nutrition and upper GI disorders are discussed in Chapter 17.

Nutrition and lower GI disorders are discussed in Chapter 18.

Nutrition and liver disorders are discussed in Chapter 19.

erythrocyte (er-REETH-ro-cite) hemolysis (he-MOLL-uh-sis): rupture of the red blood cells, caused by vitamin E deficiency.

erythro = red
cyte = cell
hemo = blood
lysis = breaking

hemolytic (HE-moh-LIT-ick) anemia: the condition of having too few red blood cells as a result of erythrocyte hemolysis.

Vitamin E Toxicity

Vitamin E in foods is safe to consume. Reports of vitamin E toxicity symptoms are rare across a broad range of intakes. Vitamin E supplement use has increased in recent years, however, as its antioxidant action against disease has been recognized. As a result, signs of toxicity are now known or suspected, although vitamin E toxicity is not nearly as common, and its effects are not as serious, as vitamin A or vitamin D toxicity. Extremely high doses of vitamin E interfere with the blood-clotting action of vitamin K and enhance the action of anticoagulant medications, leading to hemorrhage.

The pooled results from 67 experiments involving almost a quarter-million people suggested that taking vitamin E supplements may slightly increase mortality in both healthy and sick people.[33] Other studies find no effect or a slight decrease in mortality among certain groups.[34] To err on the safe side, people who use vitamin E supplements should probably keep their dosages low, not to exceed the UL of 1000 milligrams of alpha-tocopherol per day. The UL for vitamin E is more than 65 times greater than the recommended intake for adults (15 milligrams).

Vitamin E in Foods

Vitamin E is widespread in foods. Much of the vitamin E in the diet comes from vegetable oils and the products made from them, such as margarine, salad dressings, and shortenings. Wheat germ oil is especially rich in vitamin E. Other sources of the vitamin include whole grains, fruits, vegetables, and nuts. Because vitamin E is readily destroyed by heat processing and oxidation, fresh or lightly processed foods are the best sources of this vitamin.

Prior to 2000, values for the vitamin E in food reflected all of the different tocopherols and were expressed in "milligrams of tocopherol equivalents." These measures overestimated the amount of alpha-tocopherol. To estimate the alpha-tocopherol content of foods with values stated in tocopherol equivalents, multiply by 0.8.

VITAMIN K

Vitamin K has long been known for its role in blood clotting, where its presence can make the difference between life and death. Vitamin K appropriately gets it name from the Danish word *koagulation* (coagulation, or "clotting"). The vitamin also participates in the synthesis of several bone proteins. Without vitamin K, the bones produce an abnormal protein that cannot effectively bind to the minerals that normally form bones.[35] An adequate intake of vitamin K helps to decrease bone turnover and protect against fractures.[36] Vitamin K supplements seem ineffective against bone loss, however, and more research is needed to clarify the links between vitamin K and bone health.[37] Vitamin K is also under investigation for roles in heart disease prevention.[38]

Vegetable oils, some nuts and seeds such as almonds and sunflower seeds, and wheat germ are rich in vitamin E.

Blood Clotting

At least 13 different proteins and the mineral calcium are involved in making blood clots. Vitamin K is essential for the activation of several of these proteins, among them prothrombin, the precursor of the enzyme thrombin

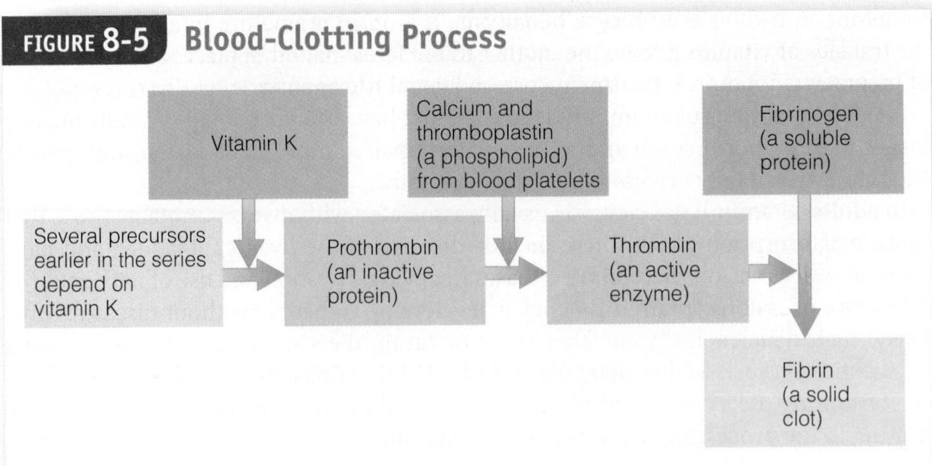
FIGURE 8-5 Blood-Clotting Process

When blood is exposed to air, foreign substances, or secretions from injured tissues, platelets (small, cell-like structures in the blood) release a phospholipid known as thromboplastin. Thromboplastin catalyzes the conversion of the inactive protein prothrombin to the active enzyme thrombin. Thrombin then catalyzes the conversion of the precursor protein fibrinogen to the active protein fibrin that forms the clot.

■ Reminder: The bacterial inhabitants of the digestive tract are known as the *intestinal flora*.

flora = plant inhabitants

hemorrhagic (hem-oh-RAJ-ik) disease: the vitamin K–deficiency disease in which blood fails to clot.

sterile: free of microorganisms such as bacteria.

(see Figure 8-5). When any of the blood-clotting factors is lacking, **hemorrhagic disease** results. If an artery or vein is cut or broken, bleeding goes unchecked. Note, though, that hemorrhaging is not always caused by a vitamin K deficiency.

Intestinal Synthesis Like vitamin D, vitamin K can be obtained from a nonfood source. Bacteria in the intestinal tract (■) synthesize vitamin K that the body can absorb, but people cannot depend on this source alone for their vitamin K.

Vitamin K Deficiency Vitamin K deficiency is rare, but it may occur in two circumstances. First, it may arise in conditions of fat malabsorption. Second, some medications interfere with vitamin K's synthesis and action in the body: antibiotics kill the vitamin K–producing bacteria in the intestine, and anticoagulant medications interfere with vitamin K metabolism and activity. When vitamin K deficiency does occur, it can be fatal.

Vitamin K for Newborns Newborn infants present a unique case of vitamin K nutrition. An infant is born with a **sterile** digestive tract, and some weeks pass before the vitamin K–producing bacteria become fully established in the infant's intestines. At the same time, plasma prothrombin concentrations are low (this helps prevent blood clotting during the stress of birth, which might otherwise be fatal). A single dose of vitamin K, usually in a water-soluble form, is given at birth to prevent hemorrhagic disease in the newborn.[39]

Notable food sources of vitamin K include green vegetables such as collards, spinach, bib lettuce, brussels sprouts, and cabbage and vegetable oils such as soybean oil and canola oil.

Vitamin K Toxicity Vitamin K toxicity is rare, and no adverse effects have been reported with high intakes. Therefore, a Tolerable Upper Intake Level has not been established. High doses of vitamin K can reduce the effectiveness of anticoagulant medications used to prevent blood clotting. People taking these medications should eat vitamin K–rich foods in moderation and keep their intakes consistent from day to day.

Vitamin K in Foods Many foods contain ample amounts of vitamin K, notably, green leafy vegetables, members of the cabbage family, and some vegetable oils. (■) Other vegetables such as iceberg lettuce and green beans provide smaller amounts.

■ Vitamin K AI:
- Men: 120 µg/day
- Women: 90 µg/day

Review Notes

- The fat-soluble vitamins are vitamins A, D, E, and K.
- Vitamin A is essential to gene expression, vision, cell differentiation and integrity of epithelial tissues, immunity, and reproduction and growth.
- Vitamin A deficiency can cause blindness, sickness, and death and is a major problem worldwide.
- Overdoses of vitamin A are possible and dangerous.
- Vitamin D raises calcium and phosphorus levels in the blood. A deficiency can cause rickets in children or osteomalacia in adults.
- Vitamin D is the most toxic of all the vitamins.
- People exposed to the sun make vitamin D in their skin; fortified milk is an important food source.
- Vitamin E acts as an antioxidant in cell membranes and is especially important in the lungs, where cells are exposed to high concentrations of oxygen.
- Vitamin E may protect against heart disease, but the evidence is not conclusive yet.
- Vitamin E deficiency is rare in healthy human beings. The vitamin is widely distributed in plant foods.
- Vitamin K is necessary for blood to clot and for bone health.
- The bacterial inhabitants of the digestive tract produce vitamin K, but people need vitamin K from foods as well.
- Dark green, leafy vegetables are good sources of vitamin K.

Table 8-4 offers a complete summary of the fat-soluble vitamins.

The Water-Soluble Vitamins

The B vitamins and vitamin C are the water-soluble vitamins. These vitamins, found in the watery compartments of foods, are distributed into water-filled compartments of the body. They are easily absorbed into the bloodstream and are just as easily excreted if their blood concentrations rise too high. Thus, the water-soluble vitamins are less likely to reach toxic concentrations in the body than are the fat-soluble vitamins. Foods never deliver excessive amounts of the water-soluble vitamins, but the large doses concentrated in vitamin supplements can reach toxic levels.

THE B VITAMINS

Despite advertisements that claim otherwise, the B vitamins do not give people energy. Carbohydrate, fat, and protein—the *energy-yielding* nutrients—are used for fuel. The B vitamins *help* the body use that fuel but do not serve as fuel themselves.

Coenzymes The eight B vitamins were listed in Table 8-1. Each is part of an enzyme helper known as a **coenzyme**. Some B vitamins have other important functions in the body as well, but the roles these vitamins play as parts of coenzymes are the best understood. A coenzyme is a small molecule that combines with an enzyme to make it active. With the coenzyme in place, a substance is attracted to the enzyme, and the reaction proceeds instantaneously. Figure 8-6 (p. 211) illustrates coenzyme action.

coenzyme (co-EN-zime): a small molecule that works with an enzyme to promote the enzyme's activity. Many coenzymes contain B vitamins as part of their structure.

co = with

TABLE 8-4 The Fat-Soluble Vitamins—A Summary

Vitamin Name	Chief Functions	Deficiency Symptoms	Toxicity Symptoms	Significant Sources
Vitamin A (Retinol, retinal, retinoic acid; main precursor is beta-carotene)	Vision, maintenance of cornea, epithelial cells, mucous membranes, skin; bone and tooth growth; reproduction; regulation of gene expression; immunity	Infectious diseases, night blindness, blindness (xerophthalmia), keratinization	*Chronic:* reduced bone mineral density, liver abnormalities, birth defects *Acute (single large dose or short-term):* blurred vision, nausea, vomiting, vertigo; increase of pressure inside skull; headache; muscle incoordination	*Retinol:* milk and milk products; eggs; liver *Beta-carotene:* spinach and other dark, leafy greens; broccoli; deep orange fruits (apricots, cantaloupe) and vegetables (carrots, winter squashes, sweet potatoes, pumpkin)
Vitamin D (Calciferol, cholecalciferol, dihydroxy vitamin D; precursor is cholesterol)	Mineralization of bones (raises blood calcium and phosphorus by increasing absorption from digestive tract, withdrawing calcium from bones, stimulating retention by kidneys)	Rickets, osteomalacia	Calcium imbalance (calcification of soft tissues and formation of stones)	Synthesized in the body with the help of sunshine; fortified milk, margarine, butter, and cereals; eggs; liver; fatty fish (salmon, sardines)
Vitamin E (Alpha-tocopherol, tocopherol)	Antioxidant (stabilization of cell membranes, regulation of oxidation reactions, protection of polyunsaturated fatty acids [PUFA] and vitamin A)	Erythrocyte hemolysis, nerve damage	Hemorrhagic effects	Polyunsaturated plant oils (margarine, salad dressings, shortenings), green and leafy vegetables, wheat germ, whole-grain products, nuts, seeds
Vitamin K (Phylloquinone, menaquinone, naphthoquinone)	Synthesis of blood-clotting proteins and bone proteins	Hemorrhage	None known	Synthesized in the body by GI bacteria; green, leafy vegetables; cabbage-type vegetables; vegetable oils

Active forms of five of the B vitamins—thiamin, riboflavin, niacin, pantothenic acid, and biotin—participate in the release of energy from carbohydrate, fat, and protein. A coenzyme containing vitamin B_6 assists enzymes that metabolize amino acids. Folate and vitamin B_{12} help cells to multiply. Among these cells are the red blood cells and the cells lining the GI tract—cells that deliver energy to all the others.

The eight B vitamins play many specific roles in helping the enzymes to perform thousands of different molecular conversions in the body. They must be present in every cell continuously for the cells to function as they should. As for vitamin C, its primary role, discussed later, is as an antioxidant.

B Vitamin Deficiencies In academic and clinical discussions of the vitamins, different sets of deficiency symptoms are ascribed to each individual vitamin. Such clear-cut symptoms, however, are found only in laboratory animals that have been fed contrived diets that lack just one nutrient. In reality, a deficiency of any single B vitamin seldom shows up in isolation because people do not eat nutrients one by one; they eat foods containing mixtures of many nutrients. If a major class of foods is missing from the diet, all of the nutrients delivered by those foods will be lacking to various extents.

In only two cases have dietary deficiencies associated with single B vitamins been observed on a large scale in human populations. Diseases have been named for these deficiency states. One of them, **beriberi**, was first observed in Southeast Asia when the custom of polishing rice became widespread. Rice contributed 80 percent of the energy intake of the people in these areas, and rice bran was their principal source of thiamin. When the bran was removed to make the rice whiter, beriberi spread like wildfire.

The niacin-deficiency disease, **pellagra**, became widespread in the southern United States in the early part of the 20th century among people who subsisted on a low-protein diet with corn as a staple grain. This diet was unusual in that it supplied neither enough niacin nor enough tryptophan, its amino acid precursor, to make the niacin intake adequate.

Even in the cases of beriberi and pellagra, the deficiencies were probably not pure. When foods were provided containing the one vitamin known to be needed, other vitamins that may have been in short supply came as part of the package.

Major deficiency diseases such as pellagra and beriberi no longer occur in the United States and Canada, but more subtle deficiencies of nutrients, including the B vitamins, are sometimes observed. When they do occur, it is usually in people whose food choices are poor because of poverty, ignorance, illness, or poor health habits such as alcohol abuse.

FIGURE 8-6 Coenzyme Action

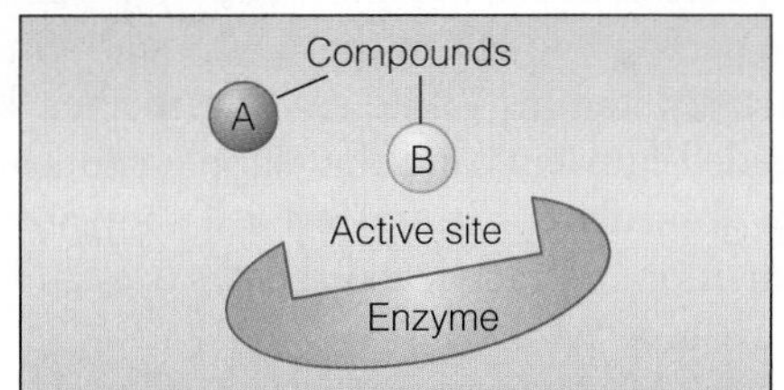

Without the coenzyme, compounds A and B do not respond to the enzyme.

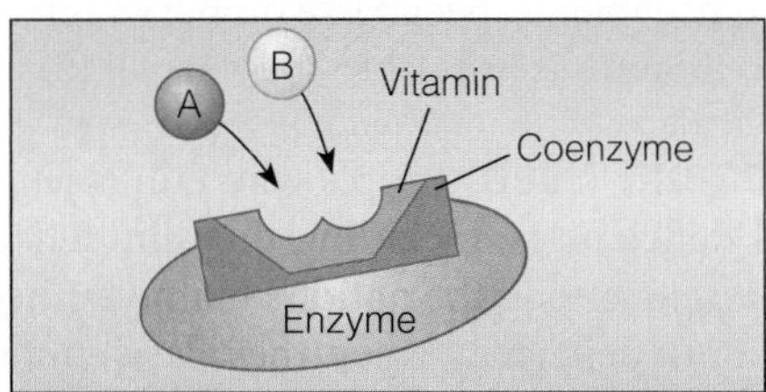

With the coenzyme in place, compounds A and B are attracted to the active site on the enzyme, and they react.

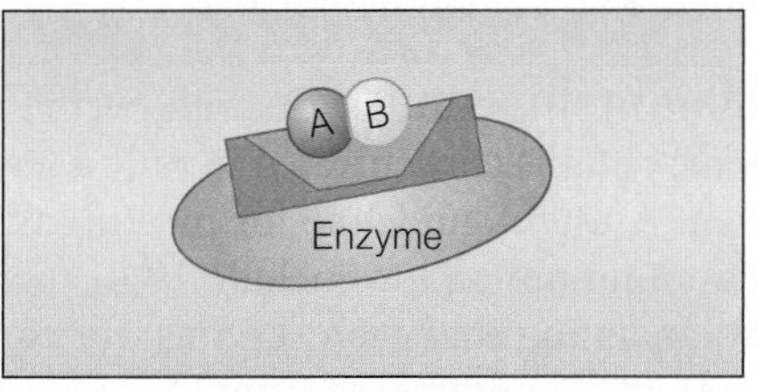

The reaction is completed with the formation of a new product. In this case the product is AB.

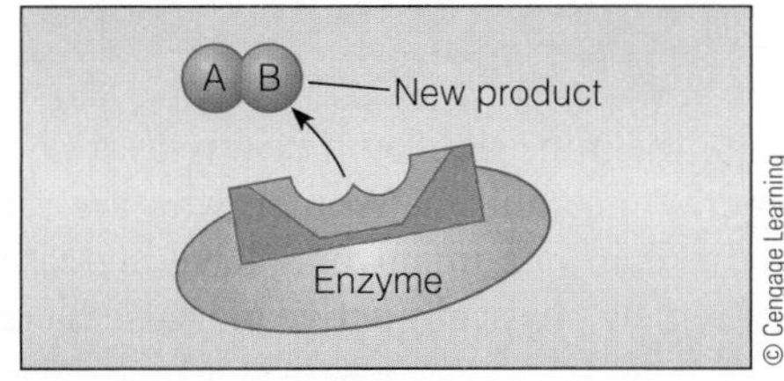

The product AB is released.

Interdependent Systems

Table 8-5, at the end of this chapter, sums up a few of the better-established facts about B vitamin deficiencies. A look at the table will make another generalization possible. Different body systems depend on these vitamins to different extents. Processes in nerves and in their responding tissues, the muscles, depend heavily on glucose metabolism and hence on thiamin, so paralysis sets in when this vitamin is lacking. But thiamin is important in all cells, not just in nerves and muscles. Similarly, because the red blood cells and GI tract cells divide the most rapidly, two of the first symptoms of a deficiency of folate are a type of anemia and GI tract deterioration—but again, all systems depend on folate, not just these. The list of symptoms in Table 8-5, later in the chapter, is far from complete.

B Vitamin Enrichment of Foods

If the staple food of a region is made from **refined grain**, vitamin B deficiencies are especially likely. One way to protect people from deficiencies is to add nutrients to their staple food, a process known as **fortification** or **enrichment**. (■) The enrichment of refined breads and cereals has drastically reduced the incidence of iron and B vitamin deficiencies.

■ Note: The terms *fortified* and *enriched* may be used interchangeably.

beriberi: the thiamin-deficiency disease; characterized by loss of sensation in the hands and feet, muscular weakness, advancing paralysis, and abnormal heart action.

pellagra (pell-AY-gra): the niacin-deficiency disease. Symptoms include the "4 Ds": diarrhea, dermatitis, dementia, and, ultimately, death.

pellis = skin
agra = seizure

refined grain: a product from which the bran, germ, and husk have been removed, leaving only the endosperm.

fortification: the addition to a food of nutrients that were either not originally present or present in insignificant amounts. Fortification can be used to correct or prevent a widespread nutrient deficiency, to balance the total nutrient profile of a food, or to restore nutrients lost in processing.

enrichment: the addition to a food of nutrients to meet a specified standard. In the case of refined bread or cereal, five nutrients have been added: thiamin, riboflavin, niacin, and folate in amounts approximately equivalent to, or higher than, those originally present and iron in amounts to alleviate the prevalence of iron-deficiency anemia.

© Polara Studios, Inc.

Nutritious foods such as pork, legumes, sunflower seeds, and enriched and whole-grain breads are valuable sources of thiamin.

■ Severe thiamin deficiency in alcohol abusers is called *Wernicke-Korsakoff syndrome*. Symptoms include disorientation, loss of short-term memory, jerky eye movements, and staggering gait.

■ Thiamin RDA:
- Men: 1.2 mg/day
- Women: 1.1 mg/day

■ Riboflavin RDA:
- Men: 1.3 mg/day
- Women: 1.1 mg/day

The preceding discussion has shown both the great importance of the B vitamins in promoting normal, healthy functioning of all body systems and the severe consequences of deficiency. Now you may want to know how to be sure you and your clients are getting enough of these vital nutrients. The next sections present information on each B vitamin. While reading further, keep in mind that *foods* can provide all the needed nutrients and that supplements are a poor second choice. Some supplements are absurdly costly, but even if they are inexpensive, most people don't need them. Nutrition in Practice 9 discusses uses and choices of supplements in more detail.

THIAMIN

All cells use thiamin, which plays a critical role in their energy metabolism. Thiamin also occupies a special site on nerve cell membranes. Consequently, as mentioned earlier, thiamin is critical to the normal functioning of the nerves and muscles.

Thiamin Need People who fail to eat enough food to meet energy needs risk nutrient deficiencies, including thiamin deficiency. Inadequate thiamin intakes have been reported among the nation's malnourished and homeless people. Similarly, people risk thiamin deficiency when they derive most of their energy from empty-kcalorie foods and beverages. Alcohol is a good example. (■) It contributes energy but provides few, if any, nutrients and often displaces food. In addition, alcohol impairs thiamin absorption and enhances thiamin excretion in the urine, doubling the risk of deficiency. Many alcoholics are thiamin deficient.

Thiamin in Foods Thiamin occurs in small quantities in virtually all nutritious foods, but it is concentrated in only a few foods, of which pork is the most commonly eaten. A useful guideline for meeting thiamin needs is to keep empty-kcalorie foods to a minimum and to include 10 or more different servings of nutritious foods each day, assuming that each serving will contribute, on the average, about 10 percent of needs. (■) Foods chosen from the bread and cereal group should be either whole grain or enriched. Thiamin is not stored in the body to any great extent, so daily intake is important.

RIBOFLAVIN

Like thiamin, riboflavin serves as a coenzyme in many reactions, most notably in energy metabolism. Women who are carrying more than one fetus or breastfeeding more than one infant may have increased needs for riboflavin. Individuals who are extremely physically active may also have increased riboflavin needs.

Riboflavin Deficiency and Toxicity When thiamin is deficient, riboflavin may be lacking too, but its deficiency symptoms, such as cracks at the corners of the mouth and sore throat, may go undetected because those of thiamin are more severe. A diet that remedies riboflavin deficiency invariably contains some thiamin and so clears up both deficiencies. Excesses of riboflavin appear to cause no harm, and no UL has been established.

Riboflavin in Foods Unlike thiamin, riboflavin is not evenly distributed among the food groups. (■) The major contributors of riboflavin to people's diets are milk and milk products, followed by enriched breads, cereals, and other grain products. Green vegetables (broccoli, turnip greens, asparagus, and spinach) and meats are also contributors. The riboflavin richness of milk and milk products is a good reason to include these foods in every day's meals. No other commonly eaten food can make such a substantial contribution. People who omit milk and milk products from their diets can substitute generous servings of dark green, leafy vegetables. Among the meats, liver and heart are the richest sources, but all lean meats, as well as eggs, offer some riboflavin.

© Polara Studios, Inc.

Milk and milk products supply much (about 50 percent) of the riboflavin in people's diets, but meats, eggs, green vegetables, and enriched and whole-grain breads and cereals are good sources, too.

Effects of Light Riboflavin is light sensitive; the ultraviolet rays of the sun or of fluorescent lamps can destroy it, as can irradiation. For this reason, milk is often sold in cardboard or opaque plastic containers to protect the riboflavin in the milk from light. In contrast, riboflavin is heat stable, so ordinary cooking does not destroy it.

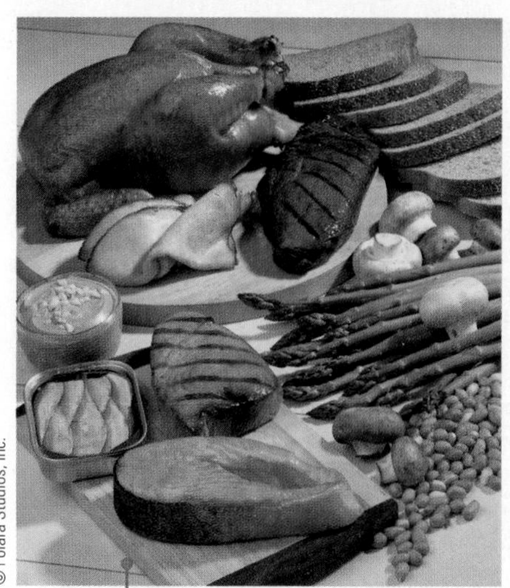

Niacin-rich foods include meat, fish, poultry, and peanut butter, as well as enriched breads and cereals and a few vegetables.

NIACIN

Like thiamin and riboflavin, niacin participates in the energy metabolism of every body cell. Niacin is unique among the B vitamins in that the body can make it from protein. The amino acid tryptophan can be converted to niacin in the body: 60 milligrams of tryptophan yield 1 milligram of niacin. Recommended intakes are therefore stated in **niacin equivalents (NE)**, (■) reflecting the body's ability to convert tryptophan to niacin. (■)

Naturally occurring niacin from foods causes no harm. Certain forms of niacin supplements taken in doses three to four times the dietary recommendation or larger cause "niacin flush," a dilation of the capillaries of the skin with perceptible tingling that, if intense, can be painful.[40] The Tolerable Upper Intake Level (35 milligrams NE) is based on flushing as the critical adverse effect.

Niacin Used as a Medication

Physicians sometimes use diet and large doses of a form of niacin (nicotinic acid) to lower blood cholesterol in the treatment of atherosclerosis. When used this way, niacin leaves the realm of nutrition to become a pharmacological agent, a drug. (■) As with any medication, self-dosing with niacin is ill advised; large doses may injure the liver and produce some symptoms of diabetes.[41]

Niacin in Foods

Meat, poultry, fish, legumes, and enriched and whole grains contribute about half the niacin equivalents most people consume. Among the vegetables, mushrooms, asparagus, and potatoes are the richest niacin sources. Niacin is less vulnerable to losses during food preparation and storage than other water-soluble vitamins. Being fairly heat resistant, niacin can withstand reasonable cooking times, but like other water-soluble vitamins, it will leach into cooking water.

■ A food containing 1 mg of niacin and 60 mg of tryptophan contains the niacin equivalent of 2 mg, or 2 mg NE.

■ Niacin RDA:

- Men: 16 mg NE/day
- Women: 14 mg NE/day

■ When a normal dose of a nutrient clears up a deficiency condition, the effect is a *physiological* one. When a large dose of a nutrient overwhelms a body system and acts like a drug, the effect is a *pharmacological* one.

PANTOTHENIC ACID AND BIOTIN

Two other B vitamins—pantothenic acid and biotin—are also important in energy metabolism. Pantothenic acid was first recognized as a substance that stimulates growth. It is a component of a key enzyme that makes possible the release of energy from the energy nutrients. Pantothenic acid is involved in more than 100 different steps in the synthesis of lipids, neurotransmitters, steroid hormones, and hemoglobin. Biotin plays an important role in metabolism as a coenzyme that carries carbon dioxide. Emerging evidence indicates that biotin participates in other processes such as gene expression and cell signaling and in the structure of DNA-binding proteins in the cell nucleus.[42]

Pantothenic Acid and Biotin in Foods

Both pantothenic acid (■) and biotin (■) are more widespread in foods than the other vitamins discussed so far. There seems to be no danger that people who consume a variety of foods will suffer deficiencies. Claims that pantothenic acid and biotin are needed in pill form to prevent or cure disease conditions are at best unfounded and at worst intentionally misleading.

■ Pantothenic acid AI:

- Adults: 5 mg/day

■ Biotin AI:

- Adults: 30 μg/day

Biotin Deficiency

Biotin deficiencies are rare but have been reported in adults fed artificially by vein without biotin supplementation. Researchers can induce biotin deficiency in animals or human beings by feeding them raw egg whites, which contain a protein that binds biotin and prevents its absorption. (■)

■ The protein *avidin* in egg whites binds biotin.

VITAMIN B_6

A surge of research interest in the past several decades has not only revealed new knowledge about vitamin B_6 but has also raised new questions. Most recently, research interest has centered on a possible role for vitamin B_6 in the treatment of disease.[43]

niacin equivalents (NE): the amount of niacin present in food, including the niacin that can theoretically be made from tryptophan, its precursor, present in the food.

© Polara Studios, Inc.

Most protein-rich foods such as meat, fish, and poultry provide ample vitamin B_6; some vegetables and fruits are good sources, too.

Metabolic Roles of Vitamin B_6 Vitamin B_6 has long been known to play roles in protein and amino acid metabolism. In the cells, vitamin B_6 helps convert one kind of amino acid, which the cells have in abundance, to other nonessential amino acids that the cells lack. It also aids in the conversion of the amino acid tryptophan to niacin and plays important roles in the synthesis of hemoglobin and neurotransmitters, the communication molecules of the brain. Vitamin B_6 also assists in releasing stored glucose from glycogen and thus contributes to the regulation of blood glucose.

Vitamin B_6 Deficiency Vitamin B_6 deficiency is expressed in general symptoms, such as weakness, depression, confusion, and irritability. Other symptoms include a greasy, flaky dermatitis; anemia; and, in advanced cases, convulsions. A shortage of vitamin B_6 may also weaken the immune response. Some evidence links low vitamin B_6 intakes with increased risk of some cancers and cardiovascular disease; more research is needed to clarify these associations.[44]

Vitamin B_6 Toxicity For years it was believed that vitamin B_6, like other water-soluble vitamins, could not reach toxic concentrations in the body. Toxic effects of vitamin B_6 became known when a physician reported them in women who had been taking more than 2 *grams* of vitamin B_6 daily (20 times the current UL of 100 *milligrams*) for two months or more, attempting to cure premenstrual syndrome. The first symptom of toxicity was numb feet; then the women lost sensation in their hands; then they became unable to walk. The women recovered after they discontinued the supplements.

Vitamin B_6 Recommendations Because vitamin B_6 coenzymes play many roles in amino acid metabolism, previous RDA were expressed in terms of protein intakes; the current RDA (■) for vitamin B_6, however, is not. Research does not support claims that large doses of vitamin B_6 enhance muscle strength or physical endurance.

■ Vitamin B_6 RDA:
- Adults (19–50): 1.3 mg/day

Vitamin B_6 in Foods The richest food sources of vitamin B_6 are protein-rich meat, fish, and poultry. Potatoes, a few other vegetables, and some fruits are good sources, too. Foods lose vitamin B_6 when heated.

FOLATE

The B vitamin folate is active in cell division. During periods of rapid growth and cell division, such as pregnancy and adolescence, folate needs increase, and deficiency is especially likely. When a deficiency occurs, the replacement of the rapidly dividing cells of the blood and the GI tract falters. Not surprisingly, then, two of the first symptoms of a folate deficiency are a type of anemia and GI tract deterioration (see Table 8-5 later in the chapter).

Folate, Alcohol, and Drugs Of all the vitamins, folate appears to be the most vulnerable to interactions with alcohol and other drugs. As Nutrition in Practice 19 describes, alcohol-addicted people risk folate deficiency because alcohol impairs folate's absorption and increases its excretion. Furthermore, as people's alcohol intakes rise, their folate intakes decline. Many medications, including aspirin, oral contraceptives, and anticonvulsants, also impair folate status. Smoking exerts a negative effect on folate status as well.

■ The two main types of neural tube defects are *spina bifida* (literally, "split spine") and *anencephaly* ("no brain"). Chapter 10 includes a figure of a neural tube defect and further discussion.

neural tube defects (NTD): malformations of the brain, spinal cord, or both that occur during embryonic development.

Folate and Neural Tube Defects Research studies confirm the importance of folate in preventing **neural tube defects (NTD)**. (■) The brain and spinal cord develop from the neural tube, and defects in its orderly formation during the early weeks of pregnancy may result in various central nervous system disorders and death. Folate supplements taken before conception and continued throughout the first trimester of pregnancy can prevent NTD. For this reason, all women of childbearing age who are capable of becoming pregnant should consume 400 micrograms (0.4 milligrams) of folate daily from supplements, fortified foods, or both, *in addition* to eating folate-rich foods.

Folate status improves more with supplementation or fortification than with a dietary intake that meets recommendations. (■) Neural tube defects arise early in pregnancy before most women realize they are pregnant, and most women eat too few fruits and vegetables to supply even half the folate needed to prevent NTD.[45] For these reasons, the FDA mandated that enriched grain products (flour, cornmeal, pasta, and rice) be fortified with an especially absorbable synthetic form of folate, folic acid. (■)

Fortification has improved folate status in women of childbearing age and lowered the number of neural tube defects that occur each year.[46] Folate fortification also raises safety concerns, however. High doses of folate can complicate the diagnosis of vitamin B_{12} deficiency, as discussed later. Other suspected but unconfirmed potential harms from high blood folic acid levels include suppression of normal immune functioning and increased cancer risks.[47] The DRI committee set a Tolerable Upper Intake Level of 1000 micrograms per day from fortified foods or supplements. Except for individuals who take supplements with more than 400 micrograms of folic acid, few people (less than 3 percent of the U.S. population) exceed the Tolerable Upper Intake Level for folic acid.[48] Thus, researchers conclude that the current level of fortification of the food supply appears to be safe.

■ Folate RDA:
- Adults: 400 µg/day

■ Bread products, flour, corn grits, and pasta must be fortified with 140 µg folate per 100 g of food (about ½ c cooked food or 1 slice of bread).

Folate in Foods As Figure 8-7 shows, the best food sources of folate are liver, legumes, beets, and leafy green vegetables (the vitamin's name suggests the word *foliage*). Among the fruits, oranges, orange juice, and cantaloupe are the best sources. With fortification, grain products are good sources of folate, too. Heat and oxidation during cooking and storage can destroy up to half of the folate in foods.

The difference in absorption between naturally occurring food folate and synthetic folate that enriches foods and is added to supplements necessitated a new unit of measurement for folate: the **dietary folate equivalents**, or **DFE**. The DFE convert all forms of folate into units that are equivalent to the folate in foods. Most food labels and tables of food composition express folate values in micrograms, however, so the accompanying "How to" describes how to estimate dietary folate equivalents.

dietary folate equivalents (DFE): the amount of folate available to the body from naturally occurring sources, fortified foods, and supplements, accounting for differences in bioavailability from each source.

FIGURE 8-7 Good Sources of Folate[a]

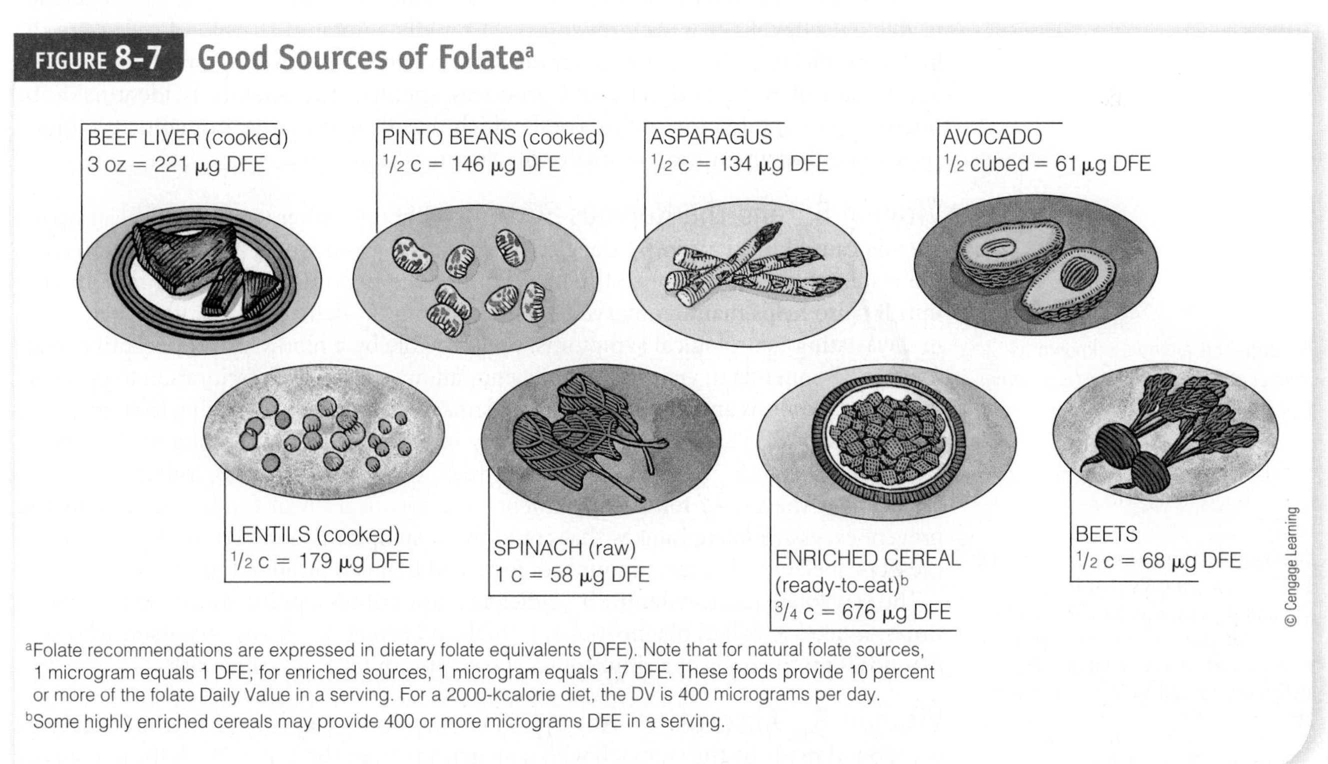

[a]Folate recommendations are expressed in dietary folate equivalents (DFE). Note that for natural folate sources, 1 microgram equals 1 DFE; for enriched sources, 1 microgram equals 1.7 DFE. These foods provide 10 percent or more of the folate Daily Value in a serving. For a 2000-kcalorie diet, the DV is 400 micrograms per day.

[b]Some highly enriched cereals may provide 400 or more micrograms DFE in a serving.

HOW TO Estimate Dietary Folate Equivalents

Folate is expressed in terms of dietary folate equivalents (DFE) because synthetic folate from supplements and fortified foods is absorbed at almost twice (1.7 times) the rate of naturally occurring folate from other foods. Use the following equation to calculate:

$$\text{DFE} = \mu\text{g food folate} + (1.7 \times \mu\text{g synthetic folate})$$

Consider, for example, a pregnant woman who takes a supplement and eats a bowl of fortified corn flakes, two slices of fortified bread, and one cup of fortified pasta. From the supplement and fortified foods, she obtains synthetic folate:

Supplement	100 µg folate
Fortified corn flakes	100 µg folate
Fortified bread	40 µg folate
Fortified pasta	60 µg folate
	300 µg folate

To calculate the DFE, multiply the amount of synthetic folate by 1.7:

$$300\ \mu\text{g} \times 1.7 = 510\ \mu\text{g DFE}$$

Now add the naturally occurring folate from the other foods in her diet—in this example, another 90 micrograms of folate.

$$510\ \mu\text{g DFE} + 90\ \mu\text{g} = 600\ \mu\text{g DFE}$$

Notice that if we had not converted synthetic folate from supplements and fortified foods to DFE, this woman's intake would appear to fall short of the 600 µg recommended for pregnancy (300 µg + 90 µg = 390 µg). But as this example shows, her intake does meet the recommendation. At this time, supplement and fortified food labels list folate in micrograms only, not micrograms DFE, making such calculations necessary.

VITAMIN B_{12}

Vitamin B_{12} and folate share a special relationship: vitamin B_{12} assists folate in cell division. Their roles intertwine, but each performs a specific task that the other cannot accomplish.

Vitamin B_{12}, Folate, and Cell Division Vitamin B_{12} (in coenzyme form) stands by to accept carbon groups from folate as folate removes them from other compounds. The passing of these carbon groups from folate to vitamin B_{12} regenerates the active form of folate so that it can continue its dismantling tasks. In the absence of vitamin B_{12}, folate is trapped in its inactive, metabolically useless form, unable to do its job. When folate is either trapped due to a vitamin B_{12} deficiency or unavailable due to a deficiency of folate itself, cells that are growing most rapidly—notably, the blood cells—are the first to be affected. Thus, a deficiency of either nutrient—vitamin B_{12} or folate—impairs maturation of the blood cells and produces anemia. The anemia is identifiable by microscopic examination of the blood, which reveals many large, immature red blood cells. (■) Either vitamin B_{12} or folate will clear up the anemia.

■ Large-cell anemia is known as *macrocytic* or *megaloblastic anemia*.

macro = large
cyte = cell
mega = large
blast = immature cell

Vitamin B_{12} and the Nervous System Although either vitamin will clear up the anemia caused by vitamin B_{12} deficiency, if folate is given when vitamin B_{12} is needed, the result is disastrous, not to the blood but to the nervous system. The reason: vitamin B_{12} also helps maintain nerve fibers. A vitamin B_{12} deficiency can ultimately result in devastating neurological symptoms, undetectable by a blood test. A deceptive folate "cure" of the anemia in vitamin B_{12} deficiency allows the nerve deterioration to progress, leading to paralysis and permanent nerve damage. Evidence is mounting to suggest that even a marginal vitamin B_{12} deficiency may impair mental functioning in the elderly, worsening dementia.[49] This interaction between folate and vitamin B_{12} raises safety concerns about the use of folate supplements and fortification of foods.[50] In an effort to prevent excessive folate intakes that could mask symptoms of a vitamin B_{12} deficiency, the FDA specifies the exact amounts of folic acid that can be added to enriched foods.

The way folate masks vitamin B_{12} deficiency underlines a point already made several times: it takes a skilled diagnostician to make a correct diagnosis. A person who self-diagnoses on the basis of a single observed symptom takes a serious risk.

intrinsic factor: a substance secreted by the stomach cells that binds with vitamin B_{12} in the small intestine to aid in the absorption of vitamin B_{12}. Anemia that reflects a vitamin B_{12} deficiency caused by lack of intrinsic factor is known as *pernicious anemia*.

intrinsic = on the inside

Vitamin B_{12} Absorption Absorption of vitamin B_{12} requires an **intrinsic factor**, a compound made by the stomach with instructions from the genes. With the help of the stomach's acid to liberate vitamin B_{12} from the food proteins that bind it, intrinsic factor attaches to the vitamin and the complex is absorbed into the bloodstream.

A few people have an inherited defect in the gene for an intrinsic factor, which makes vitamin B_{12} absorption abnormal beginning in mid-adulthood. Many others lose the ability to produce enough stomach acid and intrinsic factor to allow efficient absorption of vitamin B_{12} in later life.* In these cases, vitamin B_{12} must be supplied by injection to bypass the defective absorptive system. The anemia of the vitamin B_{12} deficiency caused by lack of intrinsic factor is known as **pernicious anemia**.

Vitamin B_{12} in Foods A unique characteristic of vitamin B_{12} is that it is found almost exclusively in foods derived from animals. People who eat meat are guaranteed an adequate intake, (■) and lacto-ovo vegetarians (who consume milk, cheese, and eggs) are also protected from deficiency. It is a myth, however, that fermented soy products, such as miso (a soybean paste), or sea algae, such as spirulina, provide vitamin B_{12} in its active form. Extensive research shows that the amounts of vitamin B_{12} listed on the labels of these plant products are inaccurate and misleading because the vitamin B_{12} in these products occurs in an inactive, unavailable form. Vegans must take vitamin B_{12} supplements or find other sources of active vitamin B_{12}. Some loss of vitamin B_{12} occurs when foods are heated in microwave ovens.

Vitamin B_{12} Deficiency in Vegans Vegans are at special risk for undetected vitamin B_{12} deficiency for two reasons: first, they receive none in their diets, and second, they consume large amounts of folate in the vegetables they eat. Because the body can store many times the amount of vitamin B_{12} used each day, a deficiency may take years to develop in a new vegetarian. When a deficiency does develop, though, it may progress to a dangerous extreme because the deficiency of vitamin B_{12} may be masked by the high folate intake.

Worldwide, vitamin B_{12} deficiency among vegetarians is a growing problem.[51] A pregnant or lactating vegetarian woman who eats no foods of animal origin should be aware that her infant can develop a vitamin B_{12} deficiency, even if the mother appears healthy. Breastfed infants born to vegan mothers with low concentrations of vitamin B_{12} in their breast milk can develop severe neurological symptoms such as seizures and cognitive problems.

NON-B VITAMINS

Other compounds are sometimes inappropriately called B vitamins because, like the true B vitamins, they serve as coenzymes in metabolism. Even if they were essential, however, supplements would be unnecessary because these compounds are abundant in foods.

Inositol, Choline, and Carnitine Among the non–B vitamins are a trio of substances known as inositol, choline, and carnitine. Researchers are exploring the possibility that these substances may be essential. Thus far, only choline has been assigned an Adequate Intake value. (■)

Other Non–B Vitamins Other substances have also been mistaken for essential nutrients. They include para-aminobenzoic acid (PABA), bioflavonoids (vitamin P or hesperidin), and ubiquinone. Other names you may hear are "vitamin B_{15}" (a hoax) and "vitamin B_{17}" (laetrile, a fake cancer-curing drug and not a vitamin by any stretch of the imagination). There is, however, one other water-soluble vitamin of great interest and importance—vitamin C.

■ Vitamin B_{12} RDA:
- Adults: 2.4 µg/day

■ Choline AI:
- Men: 550 mg/day
- Women: 425 mg/day

pernicious (per-NISH-us) anemia: a blood disorder that reflects a vitamin B_{12} deficiency caused by lack of intrinsic factor and characterized by large, immature red blood cells and damage to the nervous system (*pernicious* means "highly injurious or destructive").

*The condition is atrophic gastritis (a-TROH-fik gas-TRY-tis), a chronic inflammation of the stomach accompanied by a diminished size and functioning of the stomach's mucous membrane and glands.

VITAMIN C

More than three hundred years ago, any man who joined the crew of a seagoing ship knew he had only half a chance of returning alive—not because he might be slain by pirates or die in a storm, but because he might contract the dread disease **scurvy**. Then, a physician with the British navy found that citrus fruits could cure the disease, and thereafter, all ships were required to carry lime juice for every sailor. (This is why British sailors are still called "limeys" today.) In the 1930s, the antiscurvy factor in citrus fruits was isolated from lemon juice and named **ascorbic acid**. Today, hundreds of millions of vitamin C pills are produced in pharmaceutical laboratories.

Metabolic Roles of Vitamin C Vitamin C's action defies a simple, tidy description. It plays many important roles in the body, and its modes of action differ in different situations.

Vitamin C's Role in Collagen Formation The best-understood action of vitamin C is its role in helping to form **collagen**, the single most important protein of connective tissue. Collagen serves as the matrix on which bone is formed, the material of scars, and an important part of the "glue" that attaches one cell to another. This latter function is especially important in the artery walls, which must expand and contract with each beat of the heart, and in the walls of the capillaries, which are thin and fragile. Vitamin C also plays a role in the production of carnitine, important for transporting fatty acids within cells.

Vitamin C as an Antioxidant Vitamin C is also an important antioxidant.[52] Recall that the antioxidants beta-carotene and vitamin E protect fat-soluble substances from oxidizing agents; vitamin C protects water-soluble substances the same way. By being oxidized itself, vitamin C regenerates already-oxidized substances such as iron and copper to their original, active form. In the intestines, it protects iron from oxidation and so enhances iron absorption. In the cells and body fluids, it helps to protect other molecules, including the fat-soluble compounds vitamin A, vitamin E, and the polyunsaturated fatty acids.

Vitamin C in Amino Acid Metabolism Vitamin C is also involved in the metabolism of several amino acids. Some of these amino acids end up being used to make substances of great importance in body functioning, among them the neurotransmitter norepinephrine and the hormone thyroxine.

Role of Stress During stress, the adrenal glands release large quantities of vitamin C together with the stress hormones epinephrine and norepinephrine. The vitamin's exact role in the stress reaction remains unclear, but physical stresses (described in a later section) raise vitamin C needs.

Vitamin C as a Possible Antihistamine Newspaper headlines touting vitamin C as a cure for colds have appeared frequently over the years. Some research suggests that vitamin C (2 grams per day for two weeks) may reduce the severity and duration of cold and allergy symptoms by reducing blood histamine concentrations. In other words, vitamin C acts as an antihistamine. If further research confirms vitamin C's antihistamine effect, its use may permit people to rely less heavily on antihistamine drugs when suffering from cold and allergy symptoms.

Vitamin C's Role in Cancer Prevention and Treatment The role of vitamin C in the prevention and treatment of cancer is still being studied.[53] Evidence to date indicates that foods containing vitamin C probably protect against cancer of the esophagus.[54] The correlation may reflect not just an association with vitamin C but the broader benefits of a diet rich in fruits and vegetables and low in fat. It does not support the taking of vitamin C supplements to prevent or treat cancer.

scurvy: the vitamin C–deficiency disease.

ascorbic acid: one of the two active forms of vitamin C. Many people refer to vitamin C by this name.

a = without
scorbic = having scurvy

collagen: the characteristic protein of connective tissue.

kolla = glue
gennan = produce

Vitamin C Deficiency When intake of vitamin C is inadequate, the body's vitamin C pool dwindles, and the blood vessels show the first deficiency signs. The gums around the teeth begin to bleed easily, and capillaries under the skin break spontaneously, producing pinpoint hemorrhages. As vitamin C concentrations continue to fall, the symptoms of scurvy appear. Muscles, including the heart muscle, may degenerate. The skin becomes rough, brown, scaly, and dry. Wounds fail to heal because scar tissue will not form without collagen. Bone rebuilding falters; the ends of the long bones become softened, malformed, and painful; and fractures occur. The teeth may become loose in the jawbone and fall out. Anemia and infections are common. Sudden death is likely, perhaps because of massive bleeding into the joints and body cavities.

It takes only 10 or so milligrams of vitamin C a day to prevent scurvy, and not much more than that to cure it. Once diagnosed, scurvy is readily reversible with moderate doses, in the neighborhood of 100 milligrams per day. Such an intake is easily achieved by including vitamin C–rich foods in the diet.

Vitamin C Toxicity The easy availability of vitamin C in pill form and the publication of books recommending vitamin C to prevent everything from the common cold to life-threatening cancer have led thousands of people to take megadoses of vitamin C. (■) Not surprisingly, instances of vitamin C causing harm have surfaced.

Some of the suspected toxic effects of vitamin C megadoses have not been confirmed, but others have been seen often enough to warrant concern. Nausea, abdominal cramps, and diarrhea are often reported. Several instances of interference with medical regimens are known. Large amounts of vitamin C excreted in the urine obscure the results of tests used to detect diabetes. People taking anticoagulants (■) may unwittingly counteract the effect of these medications if they also take massive doses of vitamin C. Vitamin C megadoses can also enhance iron absorption too much, resulting in iron overload (see Chapter 9).

People with sickle-cell anemia may be especially vulnerable to megadoses of vitamin C. Those who have a tendency toward **gout**, as well as those who have a genetic abnormality that alters the way they metabolize vitamin C, are more prone to forming kidney stones if they take megadoses of vitamin C.

Recommended Intakes of Vitamin C The vitamin C RDA is 90 milligrams for men and 75 milligrams for women. These amounts are far higher than the 10 milligrams per day needed to prevent the symptoms of scurvy. In fact, they are close to the amount at which the body's pool of vitamin C is full to overflowing: about 100 milligrams per day.

Special Needs for Vitamin C As is true of all nutrients, unusual circumstances may raise vitamin C needs. Among the stresses known to do so are infections; burns; surgery; extremely high or low temperatures; toxic doses of heavy metals, such as lead, mercury, and cadmium; and the chronic use of certain medications, including aspirin, barbiturates, and oral contraceptives. Smoking, too, has adverse effects on vitamin C status. Cigarette smoke contains oxidants, which deplete this potent antioxidant. Accordingly, the vitamin C recommendation for smokers is set high, at 125 milligrams for men and 110 milligrams for women.

Safe Limits Few instances warrant the taking of more than 100 to 300 milligrams of vitamin C a day. The risks may not be great for adults who dose themselves with 1 to 2 grams a day, but those taking more than 2 grams, and especially those taking more than 3 grams per day, should be aware of the distinct possibility of harm.[55]

Vitamin C in Foods The inclusion of intelligently selected fruits and vegetables in the daily diet guarantees a generous intake of vitamin C. Even those who wish to ingest amounts well above the RDA can easily meet their goals by eating certain foods (see Figure 8-8 on p. 220). Citrus fruits are rightly famous for their vitamin C contents. Certain

■ Doses of 10 to 30 or more times the recommended intake of a nutrient are termed *megadoses*. In the case of vitamin C, current recommendations are 75 mg/day for women and 90 mg/day for men. The Tolerable Upper Intake Level for vitamin C is 2000 mg/day.

■ The anticoagulants with which vitamin C interferes are warfarin and dicumarol.

gout (GOWT): a metabolic disease in which crystals of uric acid precipitate in the joints.

FIGURE 8-8 Good Sources of Vitamin C[a]

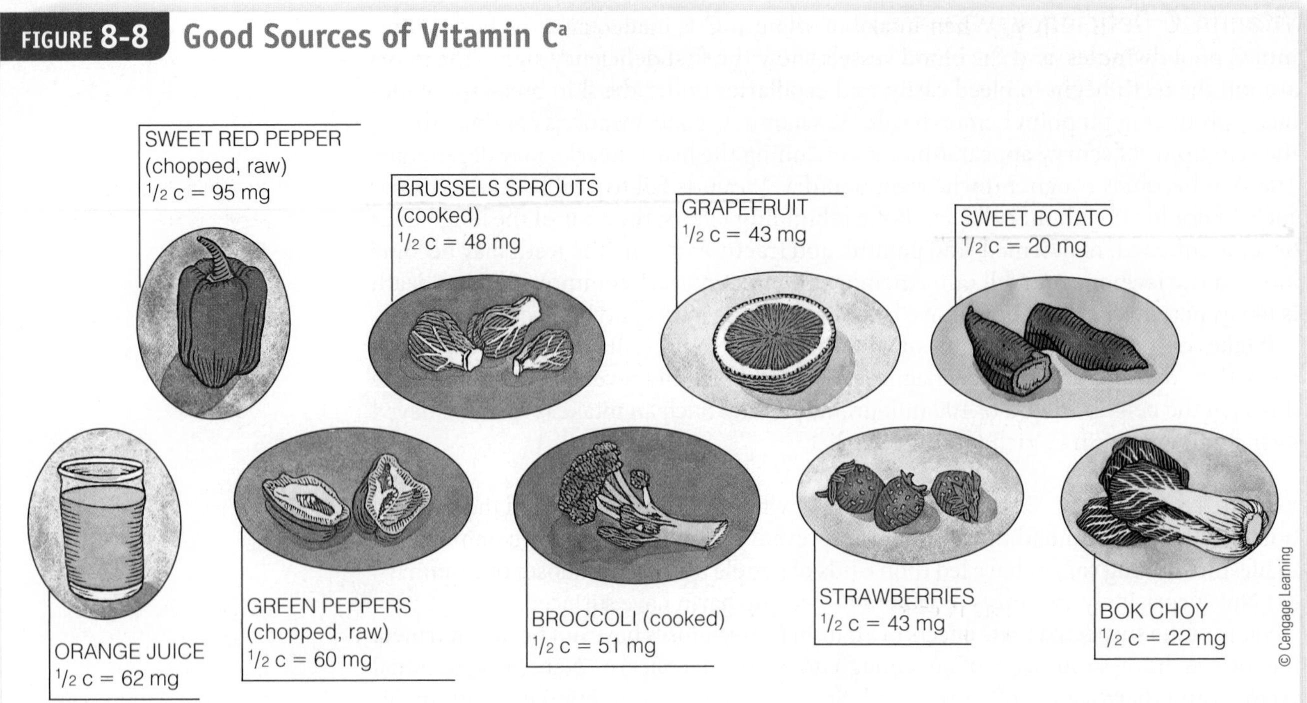

[a]These foods provide 10 percent or more of the vitamin C Daily Value in a serving. For a 2000-kcalorie diet, the DV is 60 mg/day.

other fruits and vegetables are also rich sources: cantaloupe, strawberries, broccoli, and brussels sprouts. No animal foods other than organ meats, such as chicken liver and kidneys, contain vitamin C. The humble potato is an important source of vitamin C in Western countries, where potatoes are eaten so frequently that they make substantial vitamin C contributions overall. They provide about 20 percent of all the vitamin C in the average diet. Vitamin C in foods is easily oxidized, so store cut produce and juices in airtight containers.

Vitamin C and Iron Absorption Eating foods containing vitamin C at the same meal with foods containing iron can double or triple the absorption of iron from those foods. (■) This strategy is highly recommended for women and children, whose energy intakes are not large enough to guarantee that they will get enough iron from the foods they eat.

Review Notes

- The B vitamins and vitamin C are the water-soluble vitamins.
- Each B vitamin is part of an enzyme helper known as a coenzyme.
- As parts of coenzymes, the B vitamins assist in the release of energy from glucose, amino acids, and fats and help in many other body processes.
- Folate and vitamin B12 are important in cell division.
- Vitamin C's primary role is as an antioxidant.
- Historically, famous B vitamin–deficiency diseases are beriberi (thiamin) and pellagra (niacin). The vitamin C–deficiency disease is known as scurvy.

Table 8-5 summarizes functions, deficiency and toxicity symptoms, and food sources of the water-soluble vitamins.

■ Iron is discussed in Chapter 9.

TABLE 8-5 The Water-Soluble Vitamins—A Summary

Vitamin Name	Chief Functions	Deficiency Symptoms	Toxicity Symptoms	Significant Sources
Thiamin (Vitamin B_1)	Part of a coenzyme used in energy metabolism	Beriberi (edema or muscle wasting), anorexia and weight loss, neurological disturbances, muscular weakness, heart enlargement and failure	None reported	Enriched, fortified, or whole-grain products; pork
Riboflavin (Vitamin B_2)	Part of coenzymes used in energy metabolism	Inflammation of the mouth, skin, and eyelids; sensitivity to light; sore throat	None reported	Milk products; enriched, fortified, or whole-grain products; liver
Niacin (Nicotinic acid, nicotinamide, niacinamide, vitamin B_3; precursor is dietary tryptophan, an amino acid)	Part of coenzymes used in energy metabolism	Pellagra (diarrhea, dermatitis, and dementia)	Niacin flush, liver damage, impaired glucose tolerance	Milk, eggs, meat, poultry, fish, whole-grain and enriched breads and cereals, nuts, and all protein-containing foods
Biotin	Part of a coenzyme used in energy metabolism	Skin rash, hair loss, neurological disturbances	None reported	Widespread in foods; GI bacteria synthesis
Pantothenic acid	Part of a coenzyme used in energy metabolism	Digestive and neurological disturbances	None reported	Widespread in foods
Vitamin B_6 (Pyridoxine, pyridoxal, pyridoxamine)	Part of coenzymes used in amino acid and fatty acid metabolism	Scaly dermatitis, depression, confusion, convulsions, anemia	Nerve degeneration, skin lesions	Meats, fish, poultry, potatoes, legumes, non-citrus fruits, fortified cereals, liver, soy products
Folate (Folic acid, folacin, pteroylglutamic acid)	Activates vitamin B_{12}; helps synthesize DNA for new cell growth	Anemia; smooth, red tongue; mental confusion; elevated homocysteine	Masks vitamin B_{12} deficiency	Fortified grains, leafy green vegetables, legumes, seeds, liver
Vitamin B_{12} (Cobalamin)	Activates folate; helps synthesize DNA for new cell growth; protects nerve cells	Anemia; nerve damage and paralysis	None reported	Foods derived from animals (meat, fish, poultry, shellfish, milk, cheese, eggs), fortified cereals
Vitamin C (Ascorbic acid)	Synthesis of collagen, carnitine, hormones, neurotransmitters; antioxidant	Scurvy (bleeding gums, pinpoint hemorrhages, abnormal bone growth, and joint pain)	Diarrhea, GI distress	Citrus fruits, cabbage-type vegetables, dark green vegetables (such as bell peppers and broccoli), cantaloupe, strawberries, lettuce, tomatoes, potatoes, papayas, mangoes

Self Check

1. Which of the following vitamins are fat soluble?
   a. Vitamins B, C, and E
   b. Vitamins B, C, D, and E
   c. Vitamins A, C, E, and K
   d. Vitamins A, D, E, and K

2. Which of the following describes fat-soluble vitamins?
   a. They include thiamin, vitamin A, and vitamin K.
   b. They cannot be stored to any great extent and so must be consumed daily.
   c. Toxic levels can be reached by consuming citrus fruits and vegetables.
   d. They can be stored in the liver and fatty tissues and can reach toxic concentrations.

3. Night blindness and susceptibility to infection are the result of a deficiency of which vitamin?
   a. Niacin
   b. Vitamin C
   c. Vitamin A
   d. Vitamin K

4. Good sources of vitamin D include:
   a. eggs, fortified milk, and sunlight.
   b. citrus fruits, sweet potatoes, and spinach.
   c. leafy, green vegetables, cabbage, and liver.
   d. breast milk, polyunsaturated plant oils, and citrus fruits.

5. Which of the following describes water-soluble vitamins?
   a. They include vitamins D and E.
   b. They are frequently toxic.
   c. They are stored extensively in tissues.
   d. They are easily absorbed and excreted.

6. A coenzyme is:
   a. a fat-soluble vitamin.
   b. an energy-yielding nutrient.
   c. a source of vitamin K.
   d. a molecule that combines with an enzyme to make it active.

7. Good food sources of folate include:
   a. citrus fruits, dairy products, and eggs.
   b. liver, legumes, and leafy, green vegetables.
   c. dark green vegetables, corn, and cabbage.
   d. potatoes, broccoli, and whole-wheat bread.

8. Which vitamin is present only in foods of animal origin?
   a. Riboflavin
   b. Pantothenic acid
   c. Vitamin B_{12}
   d. The inactive form of vitamin A

9. Which of the following nutrients is an antioxidant that protects water-soluble substances from oxidizing agents?
   a. Beta-carotene
   b. Thiamin
   c. Vitamin C
   d. Vitamin D

10. Eating foods containing vitamin C at the same meal can increase the absorption of which mineral?
    a. Iron
    b. Calcium
    c. Magnesium
    d. Folate

Answers to these questions appear in Appendix H. For more chapter review: Access an interactive eBook, chapter-specific interactive learning tools, including flashcards, quizzes, videos, and more in your Nutrition CourseMate, accessed through CengageBrain.com.

Clinical Applications

1. How might a vitamin deficiency weaken a client's resistance to disease?

2. Pull together information from Chapter 1 about the different food groups and the significant sources of vitamins shown in the photos and figures throughout this chapter. Consider which vitamins might be lacking in the diet of a client who reports the following:
   - Dislikes leafy, green vegetables
   - Never uses milk, milk products, or cheese
   - Follows a very low-fat diet
   - Eats a fruit or vegetable once a day

   What additional information would help you pinpoint problems with vitamin intake?

Nutrition on the Net

For further study of the topics in this chapter, access these websites. Be aware that many websites on the Internet are peddling vitamin supplements, not accurate information.

- Search for "vitamins" at the Academy of Nutrition and Dietetics: www.eatright.org
- Visit the World Health Organization to learn about "vitamin deficiencies" around the world: www.who.int
- Search for "vitamins" at the U.S. government health information site: www.healthfinder.gov
- Learn more about neural tube defects from the Spina Bifida Association of America: www.sbaa.org
- Read about Dr. Joseph Goldberger and his groundbreaking discovery linking pellagra to diet by searching for his name at: www.nih.gov or www.pbs.org
- Learn how fruits and vegetables support a healthy diet rich in vitamins from the Fruits & Veggies: More Matters program: www.fruitsandveggiesmorematters.org

Notes

1. S. T. Mayne, L. M. Ferrucci, and B. Cartmel, Lessons learned from randomized clinical trials of micronutrient supplementation for cancer prevention, *Annual Review of Nutrition* 32 (2012): 369–390; L. Franzini and coauthors, Food selection based on high total antioxidant capacity improves endothelial function in a low cardiovascular risk population, *Nutrition, Metabolism, and Cardiovascular Diseases* 22 (2012): 50–57.
2. J. C. Saari, Vitamin A metabolism in rod and cone visual cycles, *Annual Review of Nutrition* 32 (2012): 125–145; M. Clagett-Dame and D. Knutson, Vitamin A in reproduction and development, *Nutrients* 3 (2011): 385–428; J. von Lintig, Colors with functions: Elucidating the biochemical and molecular basis of carotenoid metabolism, *Annual Review of Nutrition* 30 (2010): 35–56; N. Noy, Between death and survival: Retinoic acid in regulation of apoptosis, *Annual Review of Nutrition* 30 (2010): 201–217; S. M. Ahmad and coauthors, Markers of innate immune function are associated with vitamin A stores in men, *Journal of Nutrition* 139 (2009): 377–385; M. A. Marzinke and coauthors, Calmin expression in embryos and the adult brain, and its regulation by all-trans retinoic acid, *Developmental Dynamics* 239 (2009): 610–619.
3. A. S. Kato and R. Fujiki, Transcriptional controls by nuclear fat-soluble vitamin receptors through chromatin reorganization, *Bioscience, Biotechnology, and Biochemistry* 75 (2011): 410–413; A. C. Ross and R. Zolfaghari, Cytochrome P450s in the regulation of cellular retinoic acid metabolism, *Annual Review of Nutrition* 31 (2011): 65–87.
4. Noy, 2010.
5. Y. Yang and coauthors, Effects of vitamin A deficiency on mucosal immunity and response to intestinal infection in rats, *Nutrition* 27 (2011): 227–232; Ahmad and coauthors, 2009; C. Trottier and coauthors, Retinoids inhibit measles virus through a type I IFN-dependent bystander effect, *FASEB Journal* 23 (2009): 3203–3212.
6. Clagett-Dame and Knutson, 2011.
7. L. Franzini and coauthors, Food selection based on high total antioxidant capacity improves endothelial function in a low cardiovascular risk population, *Nutrition, Metabolism, and Cardiovascular Diseases,* 22 (2012): 50–57; M. Goodman and coauthors, Clinical trials of antioxidants as cancer prevention agents: Past, present, and future, *Free Radical Biology and Medicine* 51 (2011): 1068–1084; E. Herrera and coauthors, Aspects of antioxidant foods and supplements in health and disease, *Nutrition Reviews* 67 (2009): S140–S144; L. C. Yong and coauthors, High dietary antioxidant intakes are associated with decreased chromosome translocation frequency in airline pilots, *American Journal of Clinical Nutrition* 90 (2009): 1402–1410.
8. M. Ramamoorthy and coauthors, Sporadic Alzheimer Disease fibroblasts display an oxidative stress phenotype, *Free Radical Biology and Medicine* 53 (2012): 1371–1380; L. Ho and coauthors, Reducing the genetic risk of age-related macular degeneration with dietary antioxidants, zinc, and ω-3 fatty acids, *Archives of Ophthalmology* 129 (2011): 758–766; R. L. Roberts, J. Green, and B. Lewis, Lutein and zeaxanthin in eye and skin health, *Clinical Dermatology* 27 (2009): 195–201; Herrera and coauthors, 2009.
9. D. R. Jacobs, M. D. Gross, and L. C. Tapsell, Food synergy: An operational concept for understanding nutrition, *American Journal of Clinical Nutrition* 89 (2009): 1543S–1548S.
10. World Health Organization, Micronutrient deficiencies: Vitamin A deficiency, available at www.who.int/nutrition/topics/vad/en; Standing Committee on the Scientific Evaluation of Dietary Reference Intakes, *Dietary Reference Intakes for Vitamin A, Vitamin D, Arsenic, Boron, Chromium, Copper, Iodine, Iron, Manganese, Molybdenum, Nickel, Silicon, Vanadium, and Zinc* (Washington, DC: National Academy Press, 2001), pp. 95–97.
11. E. Mayo-Wilson and coauthors, Vitamin A supplements for preventing mortality, illness, and blindness in children aged under 5: Systematic review and meta-analysis, *British Medical Journal* 343 (2011): d5094; S. A. Abrams and DC Hilmers, Postnatal vitamin A supplementation in developing countries: An intervention whose time has come? *Pediatrics* 122 (2008): 180–181.
12. World Health Organization, Measles, April 2012, available at www.who.int/mediacentre/factsheets.
13. Standing Committee on the Scientific Evaluation of Dietary Reference Intakes, 2001, pp. 128–129.
14. M. M. G. Ackermans and coauthors, Vitamin A and clefting: Putative biological mechanisms, *Nutrition Reviews* 69 (2011): 613–624; J. Zhao and coauthors, Retinoic acid downregulates microRNAs to induce abnormal development of spinal cord in spina bifida rat model, *Child's Nervous System* 24 (2008): 485–492.
15. Standing Committee on the Scientific Evaluation of Dietary Reference Intakes, 2001, pp. 126–133.
16. S. L. Morgan, Nutrition and bone: It is more than calcium and vitamin D, *Women's Health* 5 (2009): 727–737.
17. C. C. Sung and coauthors, Role of vitamin D in insulin resistance, *Journal of Biomedicine and Biotechnology* 2012: 634195.
18. G. J. Fung and coauthors, Vitamin D intake is inversely related to risk of developing metabolic syndrome in African American and white men and women over 20 y: The Coronary Artery Risk Development in Young Adults Study, *American Journal of Clinical Nutrition* 96 (2012): 24–29; I. Laaski, Vitamin D and respiratory infections in adults, *Proceedings of the Nutrition Society* 71 (2012): 90–97; Y. Liss and W. H. Frishman, Vitamin D: A cardioprotective agent? *Cardiology in Review* 20 (2012): 38–44; A. Zittermann and coauthors, Vitamin D deficiency and mortality risk in the general population: A meta-analysis of prospective cohort studies, *American Journal of Clinical Nutrition* 95 (2012): 91–100; C. D. Davis and J. A. Milner, Nutrigenomics, vitamin D and cancer prevention, *Journal of Nutrigenetics and Nutrigenomics* 4 (2011): 1–11; R. Jorde and G. Grimnes, Vitamin D and metabolic health with special reference to the effect of vitamin D on serum lipids, *Progress in Lipid Research* 50 (2011): 303–312; M. Hewison, Vitamin D and innate and adaptive immunity, *Vitamins & Hormones* 86 (2011): 23–62; E. M. Mowry, Vitamin D: Evidence for its role as a prognostic factor in multiple sclerosis, *Journal of Neurological Science* 311 (2011): 19–22; M. H. Hopkins and coauthors, Effects of supplemental vitamin D and calcium on biomarkers of inflammation in colorectal adenoma patients: A randomized, controlled clinical trial, *Cancer Prevention Research* 4 (2011): 1645–1654; K. Luong and L. T. Nguyen, Impact of vitamin D in the treatment of tuberculosis, *American Journal of Medical Sciences* 341 (2011): 493–498; A. E. Millen and coauthors, Vitamin D status and early age-related macular degeneration in postmenopausal women, *Archives of Ophthalmology* 129 (2011): 481–489; N. Parekh, Protective role of vitamin D against age-related macular degeneration: A hypothesis, *Topics in Clinical Nutrition* 25 (2010): 290–301; A. G. Pittas and coauthors, Systematic review: Vitamin D and cardiometabolic outcomes, *Annals of Internal Medicine* 152

(2010): 307–314; S. Cheng and coauthors, Adiposity, cardiometabolic risk, and vitamin status: The Framingham heart study, *Diabetes* 59 (2010): 242–248; C. F. Garland and coauthors, Vitamin D for cancer prevention: Global perspective, *Annals of Epidemiology* 19 (2009): 468–483; A. Zitterman, J. Gummert, and J. Börgermann, Vitamin D deficiency and mortality, *Current Opinion in Clinical Nutrition and Metabolic Care* 12 (2009): 634–639.

19. U.S. Preventive Services Task Force, Vitamin D and calcium supplementation to prevent cancer and osteoporotic fractures, Draft recommendations, August 2012, available at www.uspreventiveservicestaskforce.org/uspstf12/vitamind/vitdart.htm; M. Chung and coauthors, Vitamin D with or without calcium supplementation for prevention of cancer and fractures: An updated meta-analysis for the U. S. Preventive Services Task Force, *Annals of Internal Medicine* 155 (2011): 827–838; C. McGreevy and coauthors, New insights about vitamin D and cardiovascular disease, *Annals of Internal Medicine* 155 (2011): 820–826.
20. Committee on Dietary Reference Intakes, *Dietary Reference Intakes for Calcium and Vitamin D* (Washington, DC: National Academies Press, 2011), pp. 75–124.
21. C. J. Rosen, Vitamin D insufficiency, *New England Journal of Medicine* 364 (2011): 248–254.
22. A. C. Looker and coauthors, Vitamin D status: United States, 2001–2006, *NCHS Data Brief* 59 (2011): 1–8.
23. J. Kumar and coauthors, Prevalence and associations of 25-hydroxyvitamin D deficiency in US children: NHANES 2001–2004, *Pediatrics* 124 (2009): e362–e370.
24. S. Saintonge, H. Bang, and L. M. Gerber, Implications of a new definition of vitamin D deficiency in a multiracial US adolescent population: The National Health and Nutrition Examination Survey III, *Pediatrics* 123 (2009): 797–803.
25. Committee on Dietary Reference Intakes, *Dietary Reference Intakes for Calcium and Vitamin D*, 2011, pp. 362–402.
26. U.S. Preventive Services Task Force, Vitamin D and calcium supplementation to prevent cancer and osteoporotic fractures, Draft recommendations, 2012; V. A. Moyer and the U.S. Preventive Services Task Force, Prevention of falls in community-dwelling older adults: U.S. Preventive Services Task Force Recommendation Statement, *Annals of Internal Medicine* 157 (2012): 197–204; H. A. Bischoff-Ferrari and coauthors, A pooled analysis of vitamin D dose requirements for fracture prevention, *New England Journal of Medicine* 367 (2012): 40–49; P. Lips and coauthors, Once-weekly dose of 8400 IU vitamin D3 compared with placebo: Effects on neuromuscular function and tolerability in older adults with vitamin D insufficiency, *American Journal of Clinical Nutrition* 91 (2010): 985–991.
27. S. Sharma and coauthors, Vitamin D deficiency and disease risk among aboriginal Arctic populations, *Nutrition Reviews* 69 (2011): 468–478; L. M. Hall and coauthors, Vitamin D intake needed to maintain target serum 25-hydroxyvitamin D concentrations in participants with low sun exposure and dark skin pigmentation is substantially higher than current recommendations, *Journal of Nutrition* 140 (2010): 542–550; S. M. Smith and coauthors, Vitamin D supplementation during Antarctic winter, *American Journal of Clinical Nutrition* 89 (2009): 1092–1098.
28. C. L. Wagner, F. R. Greer, and the Section on Breastfeeding and Committee on Nutrition, Prevention of rickets and vitamin D deficiency in infants, children, and adolescents, *Pediatrics* 122 (2008): 1142–1152.
29. Committee on Dietary Reference Intakes, 2011, pp. 362–402.
30. M. G. Traber and J. F. Stevens, Vitamins C and E: Beneficial effects from a mechanistic perspective, *Free Radical Biology and Medicine* 51 (2011): 1000–1013.
31. G. Riccioni and coauthors, Carotenoids and vitamins C and E in the prevention of cardiovascular disease, *International Journal for Vitamin and Nutrition Research* 82 (2012): 15–26; B. J. Wilcox, J. D. Curb, and B. L. Rodriguez, Antioxidants in cardiovascular health and disease: Key lessons from epidemiologic studies, *American Journal of Cardiology* 101 (2009): 75D–86D.
32. S. S. Gidding and coauthors, Implementing American Heart Association pediatric and adult nutrition guidelines: A scientific statement from the American Heart Association Nutrition Committee of the Council on Nutrition, Physical Activity and Metabolism, Council on Cardiovascular Disease in the Young, Council on Arteriosclerosis, Thrombosis and Vascular Biology, Council on Cardiovascular Nursing, Council on Epidemiology and Prevention, and Council for High Blood Pressure Research, *Circulation* 119 (2009): 1161–1175.
33. G. Bjelakovic and coauthors, Antioxidant supplements for prevention of mortality in healthy participants and patients with various diseases, *Cochrane Database of Systematic Reviews*, April 16, 2008: CD007176.
34. G. Pocobelli and coauthors, Use of supplements of multivitamins, vitamin C and vitamin E in relation to mortality, *American Journal of Epidemiology* 170 (2009): 472–483.
35. H. Ahmadieh and A. Arabi, Vitamins and bone health: Beyond calcium and vitamin D, *Nutrition Reviews* 69 (2011): 584–598; S. L. Booth, Roles for vitamin K beyond coagulation, *Annual Review of Nutrition* 29 (2009): 89–110.
36. Booth, 2009.
37. C. M. Gundberg, J. B. Lian, and S. L. Booth, Vitamin K-dependent carboxylation of osteocalcin; Friend or foe? *Advances in Nutrition* 3 (2012): 149–157.
38. M. K. Shea and coauthors, Vitamin K supplementation and progression of coronary artery calcium in older men and women, *American Journal of Clinical Nutrition* 89 (2009): 1799–1807.
39. G. Lippi and M. Franchini, Vitamin K in neonates: Facts and myths, *Blood Transfusion* 9 (2011): 4–9.
40. D. MacKay, J. Hathcock, and E. Guarneri, Niacin: Chemical forms, bioavailability, and health effects, *Nutrition Reviews* 70 (2012): 357–366.
41. MacKay, Hathcock, and Guarneri, 2012.
42. C. A. Perry and M. A. Caudill, Biotin: Critical for fetal growth and development yet often overlooked, *Nutrition Today* 47 (2012): 79–85.
43. J. Shen and coauthors, Association of vitamin B-6 status with inflammation, oxidative stress, and chronic inflammatory conditions: The Boston Puerto Rican Health Study, *American Journal of Clinical Nutrition* 91 (2010): 337–342.
44. S. C. Larsson, N. Orsini, and A. Wolk, Vitamin B6 and risk of colorectal cancer: A meta-analysis of prospective studies, *Journal of the American Medical Association* 303 (2010): 1077–1083; J. Shen and coauthors, 2010; J. H. Page and coauthors, Plasma vitamin B(6) and risk of myocardial infarction in women, *Circulation* 120 (2009): 649–655.
45. R. L. Bailey and coauthors, Total folate and folic acid intake from foods and dietary supplements in the United States: 2003–2006, *American Journal of Clinical Nutrition* 91 (2010): 231–237.
46. Centers for Disease Control and Prevention, Folic acid data and statistics, available at www.cdc.gov/ncbddd/folicacid/data.html, updated July 7, 2010.
47. J. B. Mason, Folate consumption and cancer risk: A confirmation and some reassurance, but we're not out of the woods quite yet, *American Journal of Clinical Nutrition* 94 (2011): 965–966; M. Ebbing and coauthors, Cancer incidence and mortality after treatment with folic acid and vitamin B12, *Journal of the American Medical Association* 302 (2009): 2119–2126; U. C. Ericson and coauthors, Increased breast cancer risk at high plasma folate concentrations among women with the MTHFR 677T allele, *American Journal of Clinical Nutrition* 90 (2009): 1380–1389.
48. Bailey and coauthors, 2010; Q. Yang and coauthors, Folic acid source, usual intake, and folate and vitamin B-12 status in US adults: National Health and Nutrition Examination Survey (NHANES) 2003–2006, *American Journal of Clinical Nutrition* 91 (2010): 64–72.
49. J. G. Walker and coauthors, Oral folic acid and vitamin B-12 supplementation to prevent cognitive decline in community-dwelling older adults with depressive symptoms—The Beyond Ageing Project: A randomized controlled trial, *American Journal of Clinical Nutrition* 95 (2012): 194–203; E. Moore and coauthors, Cognitive impairment and vitamin B12: A review, *International Psychogeriatrics* 24 (2012): 541–556; Y. Minn and coauthors, Sequential involvement of the nervous system in subacute combined degeneration, *Yonsei Medical Journal* 53 (2012): 276–278; L. Feng and coauthors, Vitamin B-12, apolipoprotein E genotype, and cognitive performance in community-living older adults: Evidence of a gene-micronutrient interaction, *American Journal of Clinical Nutrition* 89 (2009): 1263–1268.
50. J. W. Miller and coauthors, Metabolic evidence of vitamin B-12 deficiency, including high homocysteine and methylmalonic acid and low holotranscobalamin, is more pronounced in older adults with elevated plasma folate, *American Journal of Clinical Nutrition* 90 (2009): 1586–1592.
51. I. Elmadfa and I. Singer, Vitamin B-12 and homocysteine status among vegetarians, *American Journal of Clinical Nutrition* 89 (2009): 1693S–1698S.
52. M. F. Garcia-Saura and coauthors, Nitroso-redox status and vascular function in marginal and severe ascorbate deficiency, *Antioxidants and Redox Signaling* 17 (2012): 937–950.
53. A. C. Mamede and coauthors, Cytotoxicity of ascorbic acid in a human colorectal adenocarcinoma cell line (WiDr): In vitro and in vivo studies, *Nutrition and Cancer* 64 (2012): 1049–1057; M. G. Traber and J. F. Stevens, Vitamins C and E: Beneficial effects from a mechanistic perspective, *Free Radical Biology and Medicine* 51 (2011): 1000–1013.
54. World Cancer Research Fund and American Institute for Cancer Research, *Food, Nutrition, Physical Activity, and the Prevention of Cancer: A Global Perspective* (Washington, DC: American Institute for Cancer Research, 2007), pp. 253–258.
55. Standing Committee on the Scientific Evaluation of Dietary Reference Intakes, *Dietary Reference Intakes for Vitamin C, Vitamin E, Selenium, and Carotenoids* (Washington, DC: National Academy Press, 2000), p. 155.

Nutrition in Practice

Phytochemicals and Functional Foods

The wisdom of the familiar advice, "Eat your vegetables; they're good for you," stands on firmer scientific ground today than ever before as population studies around the world suggest that diets rich in vegetables and fruits protect against heart disease, cancer, and other chronic diseases.[1] We now know that the "goodness" of vegetables, fruits, and other whole foods such as legumes and grains comes not only from the nutrients they contain but also from the **phytochemicals** that they offer.[2] Phytochemicals often act as **bioactive food components**, food constituents with the ability to alter body processes. (Terms are defined in the accompanying glossary.)

Vegetables, fruits, and other whole foods are the simplest examples of foods now known as **functional foods**. Functional foods provide health benefits beyond basic nutrition by altering one or more physiological processes. Modified foods, such as those that have been fortified, enriched, or enhanced with nutrients, phytochemicals, herbs, or other food components, also are functional foods.[3] Functional foods that fit this description include orange juice fortified with calcium, folate-enriched cereal, beverages with herbal additives, and margarine enhanced with **plant sterols**. This Nutrition in Practice begins with a look at the evidence concerning the effectiveness and safety of a few selected phytochemicals in the simplest of functional foods—vegetables, fruits, and other whole foods. Then, the discussion turns to examine the most controversial of functional foods—novel foods to which

Glossary of Phytochemical and Functional Food Terms

antioxidants (anti-OX-ih-dants): compounds that protect other compounds from damaging reactions involving oxygen by themselves reacting with oxygen (*anti* means "against"; *oxy* means "oxygen"); *oxidation* is a potentially damaging effect of normal cell chemistry involving oxygen.

bioactive food components: compounds in foods, either nutrients or phytochemicals, that alter physiological processes in the body. Also defined in Chapter 1.

carotenoids (kah-ROT-eh-noyds): pigments commonly found in plants and animals, some of which have vitamin A activity. The carotenoid with the greatest vitamin A activity is beta-carotene.

flavonoids (FLAY-von-oyds): a common and widespread group of phytochemicals, with more than 6000 identified members; physiologic effects may include antioxidant, antiviral, anticancer, and other activities. Flavonoids are yellow pigments in foods; *flavus* means "yellow."

flaxseed: small brown seed of the flax plant; used in baking, cereals, and other foods. Valued in nutrition as a source of fiber, lignans, and the omega-3 fatty acid linolenic acid.

functional foods: whole or modified foods that contain bioactive food components believed to provide health benefits, such as reduced disease risks, beyond the benefits that their nutrients contribute. All whole foods are functional in some ways because they provide at least some needed substances, but certain foods stand out as rich sources of bioactive food components. Also defined in Chapter 1.

genistein (GEN-ih-steen): a phytoestrogen found primarily in soybeans that both mimics and blocks the action of estrogen in the body.

lignans: phytochemicals present in flaxseed, but not flaxseed oil, that are converted to phytoestrogens by intestinal bacteria and are under study as possible anticancer agents.

lutein (LOO-teen): a plant pigment of yellow hue; a phytochemical believed to play roles in eye functioning and health.

lycopene (LYE-koh-peen): a pigment responsible for the red color of tomatoes and other red-hued vegetables; a phytochemical that may act as an antioxidant in the body.

organosulfur compounds: a large group of phytochemicals containing the mineral sulfur. Organosulfur phytochemicals are responsible for the pungent flavors and aromas of foods belonging to the onion, leek, chive, shallot, and garlic family and are thought to stimulate cancer defenses in the body.

phytochemicals (FIGH-toe-CHEM-ih-cals): compounds in plants that confer color, taste, and other characteristics. Some phytochemicals are bioactive food components in functional foods. Also defined in Chapter 1.

phytoestrogens (FIGHT-toe-ESS-troh-gens): phytochemicals structurally similar to human estrogen. Phytoestrogens weakly mimic or modulate estrogen in the body.

plant sterols: phytochemicals that resemble cholesterol in structure but that lower blood cholesterol by interfering with cholesterol absorption in the intestine. Plant sterols include sterol esters and stanol esters. Formerly called *phytosterols*.

resveratrol (rez-VER-ah-trol): a phytochemical of grapes under study for potential health benefits.

tofu: white curd made of soybeans, popular in Asian cuisines, and considered to be a functional food.

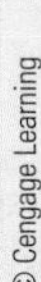

phytochemicals have been added to promote health. How these foods fit into a healthy diet is still unclear.[4]

What are phytochemicals, and what do they do?

Phytochemicals are bioactive compounds found in plants. In foods, phytochemicals impart tastes, aromas, colors, and other characteristics. They give hot peppers their burning sensation, garlic and onions their pungent flavor, chocolate its bitter tang, and tomatoes their dark red color. In the body, phytochemicals can have profound physiological effects—acting as antioxidants, mimicking hormones, stimulating or inhibiting enzymes, interfering with DNA replication, destroying bacteria, and binding physically to cell walls. Any of these actions may suppress the development of diseases, depending in part on how genetic factors interact with the phytochemicals.[5] Notably, cancer and heart disease are linked to processes involving oxygen compounds in the body, and **antioxidants** are thought to oppose these actions.[6] Table NP8-1 introduces the names, possible physiological effects, and food sources of some of the better-known phytochemicals.

TABLE NP8-1 Phytochemicals—Possible Health Effects and Food Sources

Chemical Name	Possible Effects	Food Sources
Alkylresorcinols (phenolic lipids)	May contribute to the protective effect of grains in reducing the risks of diabetes, heart disease, and some cancers	Whole-grain wheat and rye
Allicin (organosulfur compound)	Antimicrobial that may reduce ulcers; may lower blood cholesterol	Chives, garlic, leeks, onions
Capsaicin	Modulates blood clotting, possibly reducing the risk of fatal clots in heart and artery disease	Hot peppers
Carotenoids (include beta-carotene, lycopene, lutein, and hundreds of related compounds)	Act as antioxidants, possibly reducing risks of cancer and other diseases	Deeply pigmented fruits and vegetables (apricots, broccoli, cantaloupe, carrots, pumpkin, spinach, sweet potatoes, tomatoes)
Curcumin	Acts as an antioxidant and anti-inflammatory agent; may reduce blood clot formation; may inhibit enzymes that activate carcinogens	Tumeric, a yellow-colored spice
Flavonoids (include flavones, flavonols, isoflavones, catechins, and others)	Act as antioxidants; scavenge carcinogens; bind to nitrates in the stomach, preventing conversion to nitrosamines; inhibit cell proliferation	Berries, black tea, celery, citrus fruits, green tea, olives, onions, oregano, grapes, purple grape juice, soybeans and soy products, vegetables, whole wheat and other grains, wine
Genistein and daidzein (isoflavones)	Phytoestrogens that inhibit cell replication in GI tract; may reduce or elevate risk of breast, colon, ovarian, prostate, and other estrogen-sensitive cancers; may reduce cancer cell survival; may reduce risk of osteoporosis	Soybeans, soy flour, soy milk, tofu, textured vegetable protein, other legume products
Indoles (organosulfur compounds)	May trigger production of enzymes that block DNA damage from carcinogens; may inhibit estrogen action	Cruciferous vegetables such as broccoli, brussels sprouts, cabbage, cauliflower; horseradish, mustard greens, kale

continued

Chemical Name	Possible Effects	Food Sources
Isothiocyanates (organosulfur compounds that include sulforaphane)	Act as antioxidants; inhibit enzymes that activate carcinogens; activate enzymes that detoxify carcinogens; may reduce risk of breast cancer, prostate cancer	Cruciferous vegetables such as broccoli, brussels sprouts, cabbage, cauliflower; horseradish, mustard greens, kale
Lignans	Phytoestrogens that block estrogen activity in cells, possibly reducing the risk of cancer of the breast, colon, ovaries, and prostate	Flaxseed, whole grains
Monoterpenes (including limonene)	May trigger enzyme production to detoxify carcinogens; inhibit cancer promotion and cell proliferation	Citrus fruit peels and oils
Phenolic acids	May trigger enzyme production to make carcinogens water soluble, facilitating excretion	Coffee beans, fruits (apples, blueberries, cherries, grapes, oranges, pears, prunes), oats, potatoes, soybeans
Phytic acid	Binds to minerals, preventing free-radical formation, possibly reducing cancer risk	Whole grains
Resveratrol	Acts as antioxidant; may inhibit cancer growth; reduces inflammation, LDL oxidation, and blood clot formation	Red wine, peanuts, grapes, raspberries
Saponins (glucosides)	May interfere with DNA replication, preventing cancer cells from multiplying; stimulate immune response	Alfalfa sprouts, other sprouts, green vegetables, potatoes, tomatoes
Tannins	Act as antioxidants; may inhibit carcinogen activation and cancer promotion	Black-eyed peas, grapes, lentils, red and white wine, tea

Why are phytochemicals receiving so much attention these days, and what are some examples of those in the spotlight?

Diets rich in whole grains, legumes, vegetables, and fruits seem to be protective against heart disease and cancer, but identifying *the* specific foods or components of foods that are responsible is difficult. Scientists are conducting extensive research studies to discover phytochemical connections to disease prevention, but, so far, solid evidence is generally lacking. Some of the likeliest candidates include **flavonoids** and **carotenoids** (including **lycopene**).

What are flavonoids, and in which foods are they found?

Flavonoids, a large group of phytochemicals known for their health-promoting qualities, are found in whole grains, soy, vegetables, fruits, herbs, spices, teas, chocolate, nuts, olive oil, and red wine. Flavonoids are powerful antioxidants that may help to protect LDL against oxidation, minimize inflammation, and reduce blood platelet stickiness, thereby slowing the progression of atherosclerosis and making blood clots less likely.[7] Whereas an abundance of flavonoid-containing *foods* in the diet may lower the risks of chronic diseases, no claims can be made for flavonoids themselves as the protective factor, particularly when they are extracted from foods and sold as supplements. In fact, purified flavonoids may even be harmful.[8]

Flavonoids impart a bitter taste to foods, so manufacturers often refine away the natural flavonoids to please consumers, who usually prefer milder flavors. For example, for white grape juice or white wine, manufacturers remove the red, flavonoid-rich grape skins to lighten the flavor and color of the product, while greatly reducing its beneficial flavonoid content. One such flavonoid in purple grape juice and red wine, **resveratrol**, seems to hold promise as a disease fighter, but the amount present in wine or a serving of grape juice may be too small to benefit human health.[9]

What about carotenoids?

In addition to flavonoids, fruits and vegetables are rich in carotenoids—the red and yellow pigments of plants. Some carotenoids, such as beta-carotene, are vitamin A precursors. Some research suggests that a diet rich in carotenoids is associated with a lower risk of hypertension and heart disease.[10] Among the carotenoids that may defend against heart disease as well as stroke is lycopene, although findings are somewhat inconsistent.[11] Lycopene may also protect against certain types of cancer.[12]

What is lycopene, and what foods contain it?

Lycopene is a red pigment with powerful antioxidant activity found in guava, papaya, pink grapefruit, tomatoes (especially cooked tomatoes and tomato products), and watermelon. More than 80 percent of the lycopene consumed in the United States comes from tomato products such as tomato sauce, tomato juice, and catsup. Around the world, people who eat five or more tomato-containing meals per week are less likely to suffer from cancers of the esophagus, prostate, or stomach than those who avoid tomatoes. Lycopene is a leading candidate for this protective effect.

Theoretically, the potent antioxidant capability of lycopene may play a role in its action against cancer, but research suggests that several other mechanisms may underlie lycopene's possible anticancer activity.[13] The FDA concludes that no or very little solid evidence links lycopene or tomato consumption with reduced cancer risks. Tomatoes contain many other phytochemicals and nutrients that may contribute to the beneficial health effects of eating tomatoes and tomato products.

Do foods contain other phytochemicals that may help to protect people from cancer or other diseases?

Foods contain thousands of different phytochemicals, and so far only a few have been researched at all. There are still many questions about the phytochemicals that have been studied and only tentative answers about their roles in human health. For example, compared with people in the West, Asians living in Asia consume far more soybeans and soy products such as **tofu**, and they suffer less frequently from heart disease and certain cancers.* Women in Asia also suffer less from problems arising in menopause, the midlife drop in blood estrogen and cessation of menstruation, such as sensations of heat ("hot flashes") and loss of minerals from the bones. When Asians living in the United States adopt Western diets and habits, however, they experience diseases and symptoms at the same rates as native Westerners.[14]

In research, evidence concerning soy and heart health seems promising.[15] Soy's cholesterol-like plant sterols theoretically could inhibit cholesterol absorption in the intestine, and thus lower blood cholesterol.[16] A meta-analysis revealed a significant blood cholesterol–lowering effect from soy foods, attributable partly to soy's metabolic effects on the body and partly to the replacement of saturated fat–rich meats and dairy foods with soy foods in the diet.[17]

Cancers of the breast, colon, and prostate can be estrogen-sensitive—meaning that they grow when exposed to estrogen.[18] In addition to plant sterols, soy contains **phytoestrogens**, chemical relatives of human estrogen that may mimic or oppose its effects.[19] Girls who eat soy foods during childhood and adolescence may have reduced breast cancer risk as young adults.[20] A study of women in China suggests a somewhat better outcome for breast cancer among soy consumers.[21] Studies that include U.S. women report no effect or mixed results.[22] Clearly, more research is needed before conclusions may be drawn about soy intake and cancer risk.

As for menopause, no consistent findings indicate that soy phytoestrogens can eliminate hot flashes, and in some studies, soy intake accompanied a greater incidence.[23] Some evidence does suggest that soy foods may help to preserve bone density after menopause, but supplements of isolated soy phytoestrogens fail to do so.[24]

Low doses of one soy phytoestrogen, **genistein**, appear to speed up division of breast cancer cells in laboratory cultures and in mice, whereas high doses seem to do the opposite.[25] However, it seems unlikely that moderate intakes of soy foods cause harm.[26] Still under study is whether dietary soy phytoestrogens have similar effects on cancer cells in living people, but it seems unlikely that moderate intakes of soy foods would do so.[27] If they did, soy-eating cultures would have higher, not lower, incidences of these cancers.

The opposing actions of phytoestrogens should raise a red flag against taking supplements, especially by people who have had cancer or have close relatives with cancer. The American Cancer Society recommends that breast cancer survivors and those under treatment for breast cancer should consume only moderate amounts of soy foods as part of a healthy plant-based diet and should not intentionally ingest very high levels of soy products.

Other foods under study for potential health benefits include **flaxseed** and its oil. Flaxseed is found as a whole seed or ground meal, or as flaxseed oil. Flaxseed is of interest for its possible benefits to heart health because it is a good source of soluble fiber, and it is the richest known source of both the omega-3 fatty acid linolenic acid and **lignans**, compounds converted into biologically active phytoestrogens by bacteria that normally reside in the human intestine.[28] Flaxseed oil, though rich in linolenic acid, does not contain fiber or lignans. Large quantities of flaxseed can cause digestive distress, and severe allergic reactions to flaxseed have been reported.

*Among the cancers occurring less often in Asia are breast, colon, and prostate cancers.

What about other phytochemical supplements?

Even when people don't make healthy food choices, taking supplements of purified phytochemicals is not the way to go. Phytochemicals can alter body functions, sometimes powerfully. Researchers are just beginning to understand how a handful of phytochemicals work, and what is current today may change tomorrow. Foods deliver thousands of bioactive food components, all within a food matrix that maximizes their availability and effectiveness.[29] The body is equipped to handle phytochemicals in diluted form, mixed with all of the other constituents of foods, but it does not adapt well to phytochemicals in concentrated form.[30] The best way to reap the benefits of phytochemicals is by eating foods, not taking supplements (see Figure NP8-1).

How do whole foods compare with processed foods that have been enriched with phytochemicals?

Good question. The American food supply is being transformed by a proliferation of functional foods—foods claimed to provide health benefits beyond those of the traditional nutrients. Virtually all whole foods have some special value in supporting health and are therefore functional foods. Cranberries may protect against urinary tract

FIGURE NP8-1 An Array of Phytochemicals in a Variety of Fruits and Vegetables

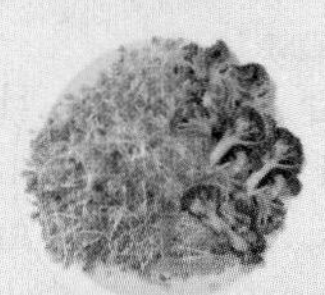

Broccoli and broccoli sprouts contain an abundance of the cancer-fighting phytochemical sulforaphane.

An apple a day—rich in flavonoids—may protect against lung cancer.

The phytoestrogens of soybeans seem to starve cancer cells and inhibit tumor growth; the phytosterols may lower blood cholesterol and protect cardiac arteries.

Garlic, with its abundant organosulfur compounds, may lower blood cholesterol and protect against stomach cancer.

The phytochemical resveratrol found in grapes (and nuts) protects against cancer by inhibiting cell growth and against heart disease by limiting clot formation and inflammation.

The ellagic acid of strawberries may inhibit certain types of cancer.

Tomatoes, with their abundant lycopene, may defend against cancer by protecting DNA from oxidative damage.

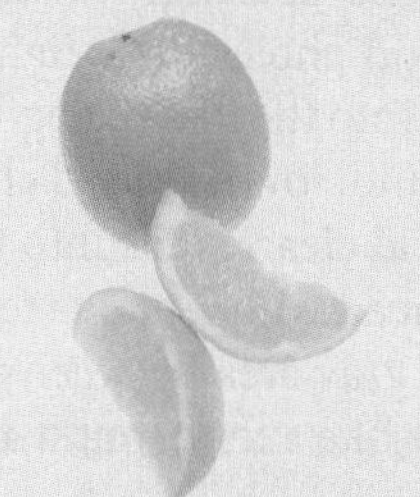

The monoterpenes of citrus fruits (and cherries) may inhibit cancer growth.

The flavonoids in black tea may protect against heart disease, whereas those in green tea may defend against cancer.

The flavonoids in cocoa and chocolate defend against oxidation and reduce the tendency of blood to clot.

Spinach and other colorful vegetables contain the carotenoids lutein and zeaxanthin, which help protect the eyes against macular degeneration.

Flaxseed, the richest source of lignans, may prevent the spread of cancer.

Blueberries, a rich source of flavonoids, improve memory in animals.

infections because cranberries contain a phytochemical that dislodges bacteria from the tract.[31] Cooked tomatoes, as mentioned, provide lycopene, along with **lutein** (an antioxidant associated with healthy eye function), vitamin C (an antioxidant vitamin), and many other healthful attributes. This has not stopped food manufacturers from trying to create functional foods as well. As consumer demand for healthful foods continues to grow, so will the development of functional foods.[32]

What are some examples of manufactured functional foods?

Many processed foods become functional foods when they are fortified with nutrients (calcium-fortified orange juice to support bone health, for example). In other cases, processed foods are enhanced with bioactive food components (margarine blended with a plant sterol to lower cholesterol, for example). The creation of some novel manufactured functional foods raises the question—is it a food or a drug?

Isn't the distinction between a food and a drug pretty clear?

Not too long ago, most of us could agree on what was a food and what was a drug. Today, functional foods blur the distinctions. They have characteristics similar to both foods and drugs but do not fit neatly into either category. Consider the margarine example below.

Eating nonhydrogenated margarine sparingly instead of butter generously may lower blood cholesterol slightly over several months and clearly falls into the food category. Taking a statin drug, on the other hand, lowers blood cholesterol significantly within weeks and clearly falls into the drug category. But margarine enhanced with a plant sterol that lowers blood cholesterol is in a gray area between the two.* The margarine looks and tastes like a food, but it acts like a drug.

What are the health advantages and disadvantages of eating manufactured functional foods?

To achieve a desired health effect, which is the better choice: to eat a food designed to affect some body function or simply to adjust the diet? Does it make more sense to use a margarine enhanced with a plant sterol that lowers blood cholesterol or simply to limit the amount of butter eaten? Is it smarter to eat eggs enriched with omega-3 fatty acids or to restrict egg consumption? Might functional foods offer a sensible solution for improving our nation's health—if done correctly? Perhaps so, but there is a problem with functional foods: the food industry is moving too fast for either scientists or the Food and Drug Administration (FDA) to keep up. Consumers were able to buy soup with St. John's wort that claimed to enhance mood and fruit juice with echinacea that was supposed to fight colds while scientists were still conducting their studies on these ingredients. Research to determine the safety and effectiveness of these substances is still in progress. Until this work is complete, consumers are on their own in finding the answers to the following questions:

- *Does it work?* Research is generally lacking, and findings are often inconclusive.
- *How much does it contain?* Food labels are not required to list the quantities of added phytochemicals. Even if they were, consumers have no standard for comparison and cannot deduce whether the amounts listed are a little or a lot. Most importantly, until research is complete, food manufacturers do not know what amounts (if any) are most effective—or most toxic.
- *Is it safe?* Functional foods can act like drugs. They contain ingredients that can alter body functions and cause allergies, drug interactions, drowsiness, and other side effects. Yet, unlike drug labels, food labels do not provide instructions for the dosage, frequency, or duration of treatment.
- *Has the FDA issued warnings about any of the ingredients?* Check the FDA's website (www.fda.gov) to find out.
- *Is it healthy?* Adding phytochemicals to a food does not magically make it a healthy choice. A candy bar may be fortified with phytochemicals, but it is still made mostly of sugar and fat.

Critics suggest that the designation "functional foods" may be nothing more than a marketing tool. After all, even the most experienced researchers cannot yet identify the perfect combination of nutrients and phytochemicals

Functional foods currently on the market promise to "enhance mood," "promote relaxation and good karma," "increase alertness," and "improve memory," among other claims.

*Margarine products that lower blood cholesterol contain either sterol esters—from vegetable oils, soybeans, and corn—or stanol esters from wood pulp.

to support optimal health. Yet manufacturers are freely experimenting with various concoctions as if they possessed that knowledge. Is it okay for them to sprinkle phytochemicals on fried snack foods and label them "functional," thus implying health benefits?

What is the final word regarding phytochemicals and functional foods?

Nature has elegantly designed foods to provide us with a complex array of dozens of nutrients and thousands of additional compounds that may benefit health—most of which we have yet to identify or understand. Over the years, we have taken those foods and first deconstructed them and then reconstructed them in an effort to "improve" them. With new scientific understandings of how nutrients—and the myriad of other compounds in foods—interact with genes, we may someday be able to design *specific* eating patterns to meet the *exact* health needs of *each* individual.[33] Indeed, our knowledge of the human genome and of human nutrition may well merge to allow specific recommendations for individuals based on their predisposition to diet-related diseases.

If the present trend continues, then someday physicians may be able to prescribe the perfect foods to enhance a person's health, and farmers will be able to grow them. In the mean time, however, it seems clear that a moderate approach to phytochemicals and functional foods is warranted. People who eat the recommended amounts of a variety of fruits and vegetables may cut their risk of many diseases by as much as half. Replacing some meat with soy foods and other legumes may also lower heart disease and cancer risks. Beneficial constituents are widespread among foods. Take a no-nonsense approach where your health is concerned: choose a wide variety of whole grains, legumes, fruits, and vegetables in the context of an adequate, balanced, and varied diet, and receive all of the health benefits that these foods offer. Table NP8-2 offers some tips for consuming the whole foods known to provide phytochemicals.

Nature offers a variety of functional foods that provide us with many health benefits.

TABLE NP8-2 Tips for Consuming Phytochemicals

- *Eat more fruit.* The average U.S. diet provides little more than ½ cup fruit per day. Remember to choose juices and raw, dried, or cooked fruits and vegetables at mealtimes as well as for snacks. Choose dried fruit in place of candy.
- *Increase vegetable portions to meet recommendations.* Choose 1 cup of cut-up raw or cooked vegetables, or 2 cups of raw, leafy greens.
- *Use herbs and spices.* Cookbooks offer ways to include parsley, basil, garlic, hot peppers, oregano, and other beneficial seasonings.
- *Replace some meat with grains, legumes, and vegetables.* Oatmeal, soy meat replacer, or grated carrots mixed with ground meat and seasonings make a luscious, nutritious meat loaf, for example.
- *Add grated vegetables.* Carrots in chili or meatballs and celery, mushrooms, and squash in spaghetti sauce or other sauces add phytochemicals without greatly changing the taste of the food.
- *Try new foods.* Try a new fruit, vegetable, or whole grain each week. Walk through vegetable aisles and visit farmers' markets. Read recipes. Try tofu, fortified soy milk, or soybeans in cooking.

Notes

1. L. M. Oude Griep, and coauthors, Colors of fruit and vegetables and 10-year incidence of stroke, *Stroke* 42 (2011): 3190–3195; F. L. Crowe and coauthors, Fruit and vegetable intake and mortality from ischaemic heart disease: Results from the European Prospective Investigation into Cancer and Nutrition (EPIC)-Heart study, *European Heart Journal* 32 (2011): 1235–1243; T. J. Key, Fruit and vegetables and cancer risk, *British Journal of Cancer* (2010), doi:10.1038/sj.bjc.6606032; S. S. Gidding and coauthors, Implementing American Heart Association pediatric and adult nutrition guidelines: A scientific statement from the American Heart Association Nutrition Committee of the Council on Nutrition, Physical Activity and Metabolism, Council on Cardiovascular Disease in the Young, Council on Arteriosclerosis, Thrombosis, and Vascular Biology, Council on Cardiovascular Nursing,

Council on Epidemiology and Prevention, and Council for High Blood Pressure Research, *Circulation* 119 (2009): 1161–11745; F. Sofi, The Mediterranean diet revisited: Evidence of its effectiveness grows, *Current Opinion in Cardiology* 24 (2009): 442–446; J. A. Nettleton and coauthors, Dietary patterns and incident cardiovascular disease in the Multi-Ethnic Study of Atherosclerosis, *American Journal of Clinical Nutrition* 90 (2009): 647–654; F. J. B. van Duijnhoven and coauthors, Fruit, vegetables, and colorectal cancer risk: The European Prospective Investigation into Cancer and Nutrition, *American Journal of Clinical Nutrition* 89 (2009): 1441–1452.

2. M. M. Murphy and coauthors, Phytonutrient intake by adults in the United States in relation to fruit and vegetable consumption, *Journal of the Academy of Nutrition and Dietetics* 112 (2012): 222–229; N. P. Gullett and coauthors, Cancer prevention with natural compounds, *Seminars in Oncology* 37 (2010): 258–281; J. M. Matés and coauthors, Anticancer antioxidant regulatory functions of phytochemicals, *Current Medicinal Chemistry* 18 (2011): 2315–2338; B. B. Aggarwal, Targeting inflammation-induced obesity and metabolic diseases by curcumin and other nutraceuticals, *Annual Review of Nutrition* 30 (2010): 173–199.
3. Position of the American Dietetic Association, Functional foods, *Journal of the American Dietetic Association* 109 (2009): 735–746.
4. G. Williamson and coauthors, Functional foods for health promotion: State-of-the-science on dietary flavonoids. Extended abstracts from the 12th Annual Conference on Functional Foods for Health Promotion, April, 2009, *Nutrition Reviews* 67 (2009): 736–743; L. R. Ferguson, Nutrigenomics approaches to functional foods, *Journal of the American Dietetic Association* 109 (2009): 452–458.
5. J. W. Lampe, Interindividual differences in response to plant-based diets: Implications for cancer risk, *American Journal of Clinical Nutrition* 89 (2009): 1553S–1557S; S. Jew, S. S. AbulMweis, and P. J. Jones, Evolution of the human diet: Linking our ancestral diet to modern functional foods as a means of chronic disease prevention, *Journal of Medicinal Foods* 12 (2009): 925–934.
6. L. Franzini and coauthors, Food selection based on high total antioxidant capacity improves endothelial function in a low cardiovascular risk population, *Nutrition Metabolism and Cardiovascular Diseases* 22 (2012): 50–57.
7. M. L. McCullough and coauthors, Flavonoid intake and cardiovascular disease mortality in a prospective cohort of US adults, *American Journal of Clinical Nutrition* 95 (2012): 454–464; G. Chiva-Blanch and coauthors, Differential effects of polyphenols and alcohol of red wine on the expression of adhesion molecules and inflammatory cytokines related to atherosclerosis: A randomized clinical trial, *American Journal of Clinical Nutrition* 95 (2012): 326–334; A. Jennings and coauthors, Higher anthocyanin intake is associated with lower arterial stiffness and central blood pressure in women, *American Journal of Clinical Nutrition* 96 (2012): 781–788; A. Basu, M. Rhone, and T. J. Lyons, Berries: Emerging impact on cardiovascular health, *Nutrition Reviews* 68 (2010): 168–177; Williamson and coauthors, 2009; M. Monagas and coauthors, Effect of cocoa powder on the modulation of inflammatory biomarkers in patients at high risk of cardiovascular disease, *American Journal of Clinical Nutrition* 90 (2009): 1144–1150.
8. S. Egert and G. Rimbach, Which sources of flavonoids: Complex diets or dietary supplements? *Advances in Nutrition* 2 (2011): 8–14.
9. J. K. Aluyen and coauthors, Resveratrol: Potential as anticancer agent, *Journal of Dietary Supplements* 9 (2012): 45–56; J. M. Smoliga, J. A. Baur, and H. A. Hausenblas, Resveratrol and health—A comprehensive review of human clinical trials, *Molecular Nutrition and Food Research* 55 (2011): 1129–1141; Y. Shukla and R. Singh, Resveratrol and cellular mechanisms of cancer prevention, *Annals of the New York Academy of Sciences*, 1215 (2011): 1–8; J. M. Wu, T. C. Hsieh, and Z. Wang, Cardioprotection by resveratrol: A review of effects/targets in cultured cells and animal tissues, *American Journal of Cardiovascular Disease* 1 (2011): 38–47; M. D. Knutson and C. Leeuwenburgh, Resveratrol and novel potent activators of SIRT1: Effects on aging and age-related diseases, *Nutrition Reviews* 66 (2008): 591–596.
10. J. Karppi and coauthors, Low serum lycopene and β-carotene increase risk of acute myocardial infarction in men, *European Journal of Public Health*, December 7, 2011, epub ahead of print; C. Li and coauthors, Serum alpha-carotene concentrations and risk of death among US adults, *Archives of Internal Medicine* 171 (2011): 507–515; A. Hozawa and coauthors, Circulating carotenoid concentrations and incident hypertension: The Coronary Artery Risk Development in Young Adults (CARDIA) Study, *Journal of Hypertension* 27 (2009): 237–242; G. Riccioni, Carotenoids and cardiovascular disease, *Current Atherosclerosis Reports* 11 (2009): 434–439.
11. F. Thies and coauthors, Effect of a tomato-rich diet on markers of cardiovascular disease risk in moderately overweight, disease-free, middle-aged adults: A randomized controlled trial, *American Journal of Clinical Nutrition* 95 (2012): 1013–1022; J. Karppi and coauthors, Serum lycopene decreases the risk of stroke in men, *Neurology* 79 (2012): 1540–1547; A. Mordente and coauthors, Lycopene and cardiovascular diseases: An update, *Current Medicinal Chemistry* 18 (2011): 1146–1163.
12. A. J. Teodoro and coauthors, Effect of lycopene on cell viability and cell cycle progression in human cancer cell lines, *Cancer Cell International* 12 (2012): 36; Y. Sharoni and coauthors, The role of lycopene and its derivatives in the regulation of transcription systems: Implications for cancer prevention, *American Journal of Clinical Nutrition* 96 (2012): 1173S–1178S; J. Talvas and coauthors, Differential effects of lycopene consumed in tomato paste and lycopene in the form of a purified extract on target genes of cancer prostatic cells, *American Journal of Clinical Nutrition* 91 (2010): 1716–1724.
13. Teodoro and coauthors, 2012; Sharoni and coauthors, The role of lycopene and its derivatives in the regulation of transcription systems, 2012; Talvas and coauthors, 2010.
14. N. Mehrotra, S. Gaur, and A. Petrova, Health care practices of the foreign born Asian Indians in the United States. A community based survey, *Journal of Community Health* 37 (2012): 328–334.
15. D. J. Jenkins and coauthors, Soy protein reduces serum cholesterol by both intrinsic and food displacement mechanisms, *Journal of Nutrition* 140 (2010): 2302S–2311S.
16. S. B. Racette and coauthors, Dose effects of dietary plant sterols on cholesterol metabolism: A controlled feeding study, *American Journal of Clinical Nutrition* 91 (2010): 32–38.
17. Jenkins and coauthors, 2010.
18. P. L. de Souza and coauthors, Clinical pharmacology of isoflavones and its relevance for potential prevention of prostate cancer, *Nutrition Reviews* 68 (2010): 542–555; X. O. Shu and coauthors, Soy food intake and breast cancer survival, *Journal of the American Medical Association* 302 (2009): 2437–2443.
19. J. H. van Ee, Soy constituents: Modes of action in low-density lipoprotein management, *Nutrition Reviews* 67 (2009): 222–234.
20. S. A. Lee and coauthors, Adolescent and adult soy food intake and breast cancer risk: Results from the Shanghai Women's Health Study, *American Journal of Clinical Nutrition* 89 (2009): 1920–1926; M. Messina and A. H. Wu, Perspectives on the soy–breast cancer relation, *American Journal of Clinical Nutrition* 89 (2009): 1673S–1679S.
21. Shu and coauthors, 2009.
22. S. J. Nechuta and coauthors, Soy food intake after diagnosis of breast cancer and survival: An in-depth analysis of combined evidence from cohort studies of U.S. and Chinese women, *American Journal of Clinical Nutrition* 96 (2012): 123–132; S. A. Khan and coauthors, Soy isoflavones supplementation for breast cancer risk reduction: A randomized phase II trial, *Cancer Prevention Research* 5 (2012): 309–319.
23. S. Levis and coauthors, Soy isoflavones in the prevention of menopausal bone loss and menopausal symptoms, *Archives of Internal Medicine* 171 (2011): 1363–1369; North American Menopause Society, The role of soy isoflavones in menopausal health: Report of the North American Menopause Society/Wulf H. Utian Translational Science Symposium in Chicago, IL (October 2010), *Menopause* 18 (2011): 732–753.
24. D. L. Alekel and coauthors, The Soy Isoflavones for Reducing Bone Loss (SIRBL) Study: A 3-y randomized controlled trial in postmenopausal women, *American Journal of Clinical Nutrition* 91 (2010): 218–230.
25. R. Bosviel and coauthors, Can soy phytoestrogens decrease DNA methylation in BRCA1 and BRCA2 oncosuppressor genes in breast cancer? *Omics* 16 (2012): 235–244; T. T. Rajah and coauthors, Physiological concentrations of genistein and 17β-estradiol inhibit MDA-MB231 breast cancer cell growth by increasing BAX/BC1-2 and reducing pERK 1/2, *Anticancer Research* 32 (2012): 1181–1191.
26. C. K. Taylor and coauthors, The effect of genistein aglycone on cancer and cancer risk: A review of in vitro, preclinical, and clinical studies, *Nutrition Reviews* 67 (2009): 398–414.
27. Taylor and coauthors, 2009.
28. A. Pan and coauthors, Meta-analysis of the effects of flaxseed interventions on blood lipids, *American Journal of Clinical Nutrition* 90 (2009): 288–297.
29. D. R. Jacobs, M. D. Gross, and L. C. Tapsell, Food synergy: An operational concept for understanding nutrition, *American Journal of Clinical Nutrition* 89 (2009): 1543S–1548S.
30. Egert and Rimbach, 2011.
31. C. Wang and coauthors, Cranberry-containing products for prevention of urinary tract infections in susceptible populations, *Archives of Internal Medicine* 172 (2012): 988–996; H. Shmuely and coauthors, Cranberry components for the therapy of infectious disease, *Current Opinion in Biotechnology* 23 (2012): 148–152; D. R. Guay, Cranberry and urinary tract infections, *Drugs* 69 (2009): 775–807; J. Jass and G. Reid, Effect of cranberry drink on bacterial adhesion in vitro and vaginal microbiota in healthy females, *Canadian Journal of Urology* 16 (2009): 4901–4907.
32. Position of the American Dietetic Association, 2009.
33. L. R. Ferguson, Nutrigenomics approaches to functional foods, *Journal of the American Dietetic Association* 109 (2009): 452–458.

Chapter 9 Water and the Minerals

THE BODY'S WATER CANNOT BE CONSIDERED SEPARATELY FROM THE minerals dissolved in it. A person can drink pure water, but in the body, water mingles with minerals to become fluids in which all life processes take place. This chapter begins by discussing the body's fluids and their chief minerals. The focus then shifts to other functions of the minerals.

Water and Body Fluids

water balance: the balance between water intake and water excretion that keeps the body's water content constant.

dehydration: the loss of water from the body that occurs when water output exceeds water input. The symptoms progress rapidly from thirst, to weakness, to exhaustion and delirium, and end in death if not corrected.

water intoxication: the rare condition in which body water contents are too high. The symptoms may include confusion, convulsion, coma, and even death in extreme cases.

Water constitutes about 60 percent of an adult's body weight and a higher percentage of a child's. Every cell in the body is bathed in a fluid of the exact composition that is best for that cell. The body fluids bring to each cell the ingredients it requires and carry away the end products of the life-sustaining reactions that take place within the cell's boundaries. Without water, cells quickly die. The water in the body fluids:

- Carries nutrients and waste products throughout the body.
- Maintains the structure of large molecules such as proteins and glycogen.
- Participates in metabolic reactions.
- Serves as the solvent for minerals, vitamins, amino acids, glucose, and many other small molecules so that they can participate in metabolic activities.
- Maintains blood volume.
- Aids in the regulation of normal body temperature, as the evaporation of sweat from the skin removes excess heat from the body.
- Acts as a lubricant and cushion around joints and inside the eyes, spinal cord, and amniotic sac surrounding a fetus in the womb.

To support these and other vital functions, the body actively regulates its **water balance**.

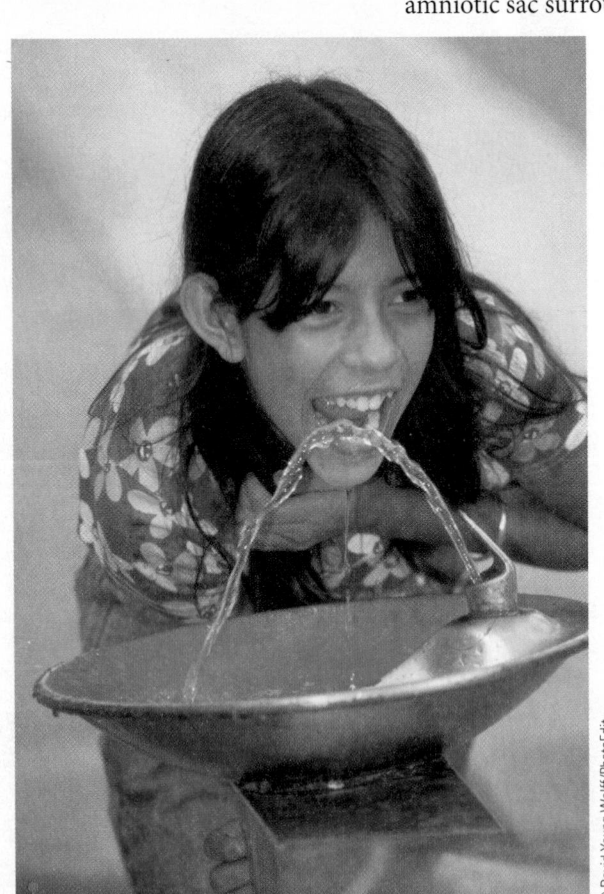

© David Young Wolff/PhotoEdit

Water is the most indispensable nutrient of all.

WATER BALANCE

The cells themselves regulate the composition and amounts of fluids within and surrounding them. The entire system of cells and fluids remains in a delicate but firmly maintained state of dynamic equilibrium. Imbalances such as **dehydration** (see Table 9-1) and **water intoxication** can occur, but the body quickly restores the balance to normal if it can. The body controls both water intake and water excretion to maintain water equilibrium.

Water Intake Regulation

The body can survive for only a few days without water. In healthy people, thirst and satiety govern water intake.[1] Thirst is finely adjusted to ensure a water intake that meets the body's needs. When the blood becomes too concentrated (having lost water but not salt and other dissolved substances), the mouth becomes dry, and the brain center known as the **hypothalamus** initiates drinking behavior.

Thirst lags behind the lack of water. A water deficiency that develops slowly can switch on drinking behavior in time to prevent serious dehydration, but a deficiency that develops quickly may not. Also, thirst itself does not remedy a water deficiency; a person must pay attention to the thirst signal and take the time to get a drink. With aging, thirst sensations may diminish. Dehydration can threaten elderly people who do not develop the habit of drinking water regularly.

TABLE 9-1 Signs of Mild and Severe Dehydration

Mild Dehydration (Loss of <5% Body Weight)	Severe Dehydration (Loss of >5% Body Weight)
• Thirst	• Pale skin
• Sudden weight loss	• Bluish lips and fingertips
• Rough, dry skin	• Confusion; disorientation
• Dry mouth, throat, body linings	• Rapid, shallow breathing
• Rapid pulse	• Weak, rapid, irregular pulse
• Low blood pressure	• Thickening of blood
• Lack of energy; weakness	• Shock; seizures
• Impaired kidney function	• Coma; death
• Reduced quantity of urine; concentrated urine	
• Decreased mental functioning	
• Decreased muscular work and athletic performance	
• Fever or increased internal temperature	
• Fainting	

Source: Based on Standing Committee on the Scientific Evaluation of Dietary Reference Intakes, Food and Nutrition Board, Institute of Medicine, Dietary Reference Intakes: Water, Potassium, Sodium, Chloride, and Sulfate (Washington, DC: National Academies Press, 2005), pp. 90–122.

Water intoxication, on the other hand, is rare but can occur with excessive water consumption and kidney disorders that reduce urine production. The symptoms may include severe headache, confusion, convulsions, and even death in extreme cases. Excessive water ingestion (several gallons) within a few hours dilutes the sodium concentration of the blood and contributes to a dangerous condition known as **hyponatremia**.

Water Excretion Regulation Water excretion is regulated by the brain and the kidneys. The cells of the brain's hypothalamus, which monitor blood salts, stimulate the **pituitary gland** to release **antidiuretic hormone (ADH)** whenever the salts are too concentrated, or the blood volume or blood pressure is too low. ADH stimulates the kidneys to reabsorb water rather than excrete it. Thus, the more water you need, the less you excrete.

If too much water is lost from the body, blood volume and blood pressure fall. Cells in the kidneys respond to the low blood pressure by releasing an enzyme. (■) Through a complex series of events involving the hormone **aldosterone**, this enzyme also causes the kidneys to retain more water. Again, the effect is that, when more water is needed, less is excreted.

Minimum Water Needed These mechanisms can maintain water balance only if a person drinks enough water. The body must excrete a minimum of about 500 milliliters (■) each day as urine—enough to carry away the waste products generated by a day's metabolic activities. Above this amount, excretion adjusts to balance intake, so the more a person drinks, the more dilute the urine becomes. In addition to urine, some water is lost from the lungs as vapor, some is excreted in feces, and some evaporates from the skin. A person's water losses from all of these routes total about 2½ liters (about 2½ quarts) a day on the average. Table 9-2 (p. 236) shows how fluid intake and output naturally balance out.

■ The enzyme renin (REN-in), released by the kidneys in response to low blood pressure, aids the kidneys in retaining water through the renin-angiotensin mechanism.

■ 500 mL = about ½ qt

hypothalamus (high-poh-THAL-ah-mus): a brain center that controls activities such as maintenance of water balance, regulation of body temperature, and control of appetite.

hyponatremia (HIGH-po-na-TREE-me-ah): a decreased concentration of sodium in the blood.

pituitary (pit-TOO-ih-tary) gland: in the brain, the "king gland" that regulates the operation of many other glands.

antidiuretic hormone (ADH): a hormone released by the pituitary gland in response to high salt concentrations in the blood. The kidneys respond by reabsorbing water.

aldosterone (al-DOS-ter-own): a hormone secreted by the adrenal glands that stimulates the reabsorption of sodium by the kidneys; also regulates chloride and potassium concentrations.

TABLE 9-2 Water Balance

Water Sources	Amount (mL)	Water Losses	Amount (mL)
Liquids	550 to 1500	Kidneys (urine)	500 to 1400
Foods	700 to 1000	Skin (sweat)	450 to 900
Metabolic water	200 to 300	Lungs (breath)	350
		GI tract (feces)	150
Total	1450 to 2800	Total	1450 to 2800

Water Recommendations and Sources Water needs vary greatly depending on the foods a person eats, the environmental temperature and humidity, the person's activity level, and other factors. Accordingly, a general water requirement is difficult to establish. In the past, recommendations for adults were expressed in proportion to the amount of energy expended under normal environmental conditions. For the person who expends about 2000 kcalories a day, this works out to 2 to 3 liters, or about 8 to 12 cups. This recommendation is in line with the Adequate Intake (AI) for *total* water (■) set by the DRI committee. Total water includes not only drinking water but also water in other beverages and in foods.[2]

Because a wide range of water intakes will prevent dehydration and its harmful consequences, the AI is based on average intakes. Strenuous physical activity and heat stress can increase water needs considerably, however.[3] In general, you can tell from the color of the urine whether a person needs more water. Pale yellow urine reflects appropriate dilution.

The obvious dietary sources of water are water itself and other beverages, but nearly all foods also contain water. Water constitutes up to 95 percent of the volume of most fruits and vegetables and at least 50 percent of many meats and cheeses (see Table 9-3). The energy nutrients in foods also give up water during metabolism.

Which beverages are best? Any beverage can readily meet the body's fluid needs, but those with few or no kcalories do so without contributing to weight gain. Given that obesity is a major health problem and that beverages contribute more than 20 percent of the total energy intake in the United States, water is the best choice for most people.[4] Other choices include tea, coffee, nonfat and low-fat milk and soy milk, artificially sweetened beverages, fruit and vegetable juices, sports drinks, and, lastly, sweetened nutrient-poor beverages. By far, carbonated soft drinks are chosen most often, but this choice often crowds more nutritious beverages out of the diet, and the regular sugar-sweetened varieties provide many empty kcalories from added sugars.

People often ask whether caffeine-containing beverages such as coffee, tea, or soda can help to meet water needs. People who drink caffeinated beverages lose a little more fluid than when they drink water because caffeine acts as a mild diuretic. The DRI committee considered such findings in making its recommendations for water intake and concluded that "caffeinated beverages contribute to the daily total water intake similar to that contributed by non-caffeinated beverages."[5] In other words, it doesn't seem to matter whether people rely on caffeine-containing beverages or other beverages to meet their fluid needs.

In contrast, alcohol should probably not be used to meet fluid needs. As Nutrition in Practice 19 explains, alcohol acts as a diuretic, and it has many adverse effects on health and nutrition status.

■ AI for *total* water:
- Men: 3.7 L/day
- Women: 2.7 L/day

TABLE 9-3 Percentage of Water in Selected Foods

100%	Water, diet soft drinks, seltzer (unflavored), plain tea
90–99%	Fat-free milk, black coffee, Gatorade, gelatin dessert, strawberries, watermelon, grapefruit, tomato, lettuce, cabbage, celery, spinach, broccoli, cucumber, summer squash, clear broth
80–89%	Fruit juice, sugar-sweetened soft drinks, whole milk, yogurt, low-fat cottage cheese, apples, grapes, oranges, carrots, egg white, cooked oatmeal
70–79%	Shrimp, bananas, instant pudding, corn, potatoes, avocados, ricotta cheese
60–69%	Pasta, macaroni and cheese, cooked rice, legumes, salmon, lean steak, chicken breast, ice cream, low-kcalorie mayonnaise
50–59%	Ground beef, pork chop, pork sausage, hot dogs, feta cheese
40–49%	Pizza, cheeseburger
30–39%	Cheddar cheese, bagels, bread
20–29%	Pepperoni sausage, cake, biscuits
10–19%	Butter, margarine, raisins, regular mayonnaise
1–9%	Crackers, ready-to-eat cereals, pretzels, taco shells, peanut butter, nuts
0%	Oils, white sugar, meat fats, shortening

■ Exceptions: A compound in which the positive ions are hydrogen ions (H^+) is an acid (example: hydrochloric acid, or H^+Cl^-); a compound in which the negative ions are hydroxyl ions (OH^-) is a base (example: potassium hydroxide, or K^+OH^-).

■ The simple statement that water follows salt describes the force that chemists call *osmosis*.

salts: compounds composed of charged particles (ions). An example of a salt is potassium chloride (K^+Cl^-).

electrolyte: a salt that dissolves in water and dissociates into charged particles called ions.

electrolyte solutions: solutions that can conduct electricity.

FLUID AND ELECTROLYTE BALANCE

When mineral **salts** dissolve in water, they separate (dissociate) into charged particles known as ions, which can conduct electricity. For this reason, a salt that dissociates in water is known as an **electrolyte**. (■) The body fluids, which contain water and partly dissociated salts, are **electrolyte solutions**.

The body's electrolytes are vital to the life of the cells and therefore must be closely regulated to help maintain the appropriate distribution of body fluids. The major minerals form salts that dissolve in the body fluids; the cells direct where these salts go; and the movement of the salts determines where the fluids flow because water follows salt. (■) Cells use this force to move fluids back and forth across their membranes. Thanks to the electrolytes, water can be held in compartments where it is needed.

Proteins in the cell membranes move ions into or out of the cells. These protein pumps tend to concentrate sodium and chloride outside cells and potassium and other ions inside. By maintaining specific amounts of sodium outside and potassium inside, cells can regulate the exact amounts of water inside and outside their boundaries.

Healthy kidneys regulate the body's sodium, as well as its water, with remarkable precision. The intestinal tract absorbs sodium readily, and it travels freely in the blood, but the kidneys excrete unneeded amounts. The kidneys actually filter all of the sodium out of the blood; then they return to the bloodstream the exact amount the body needs to retain. Thus, the body's total electrolytes remain constant, while the urinary electrolytes fluctuate according to what is eaten.

Water follows salt. Notice the beads of "sweat," formed on the right-hand slices of eggplant, which were sprinkled with salt. Cellular water moves across each cell's membrane (water-permeable divider) toward the higher concentration of salt (dissolved particles) on the surface.

In some cases, the body's mechanisms for maintaining fluid and electrolyte balances cannot compensate for a sudden loss of large amounts of fluid and electrolytes. Vomiting, diarrhea, heavy sweating, fever, burns, wounds, and the like may incur great fluid and electrolyte losses, precipitating an emergency that demands medical intervention. (■)

ACID–BASE BALANCE

The body uses ions not only to help maintain water balance but also to regulate the acidity (pH) of its fluids. Like proteins, electrolyte mixtures in the body fluids protect the body against changes in acidity by acting as **buffers**—substances that can accommodate excess acids or bases.

The body's buffer systems serve as a first line of defense against changes in the fluids' acid–base balance. The lungs, skin, gastrointestinal (GI) tract, and kidneys provide other defenses. Of these organ systems, the kidneys play the primary role in maintaining acid–base balance. Thus, disorders of the kidneys impair the body's ability to regulate its acid–base balance, as well as its fluid and electrolyte balances.

■ The body's responses to severe stress and trauma are discussed in Chapter 16.

buffers: compounds that can reversibly combine with hydrogen ions to help keep a solution's acidity or alkalinity constant. Also defined in Chapter 5.

major minerals: essential mineral nutrients required in the adult diet in amounts greater than 100 milligrams per day.

trace minerals: essential mineral nutrients required in the adult diet in amounts less than 100 milligrams per day.

Review Notes

- Water makes up about 60 percent of the body's weight.
- Water helps transport nutrients and waste products throughout the body, participates in metabolic reactions, acts as a solvent, assists in maintaining blood volume and body temperature, acts as a lubricant and cushion around joints, and serves as a shock absorber.
- To maintain water balance, intake from liquids, foods, and metabolism must equal losses from the kidneys, skin, lungs, and feces.
- Electrolytes help maintain the appropriate distribution of body fluids and acid–base balance.

The Major Minerals

Table 9-4 lists the major minerals and the nine essential trace minerals. Other trace minerals are recognized as essential nutrients for some animals but have not been proved to be required for human beings. Figure 9-1 shows the amounts of the major minerals found in the body, and for comparison, some of the trace minerals. As you can see, the most prevalent minerals are calcium and phosphorus, the chief minerals of bone. The distinction between the **major minerals** and the **trace minerals** does not mean that one group is more important than the other. A deficiency of the few micrograms of iodine needed daily is just as serious as a deficiency of the several hundred milligrams of calcium. The major minerals are so named because they are present, and needed, in larger amounts in the body than the trace minerals.

Although all the major minerals influence the body's fluid balance, sodium, chloride, and potassium are most noted for that role. For this reason, these three minerals are discussed first. Each major mineral also plays other specific roles in the body. Sodium, potassium, calcium, and magnesium are critical to nerve transmission and muscle contractions. Phosphorus and magnesium are involved in energy metabolism. Calcium, phosphorus, and magnesium contribute to the structure of the bones. Sulfur helps determine the shape of proteins. Table 9-7, shown later in the chapter, provides a summary of information about the major minerals.

TABLE 9-4 The Major and Trace Minerals

Major Minerals	Trace Minerals
Calcium	Chromium
Chloride	Copper
Magnesium	Fluoride
Phosphorus	Iodine
Potassium	Iron
Sodium	Manganese
Sulfur	Molybdenum
	Selenium
	Zinc

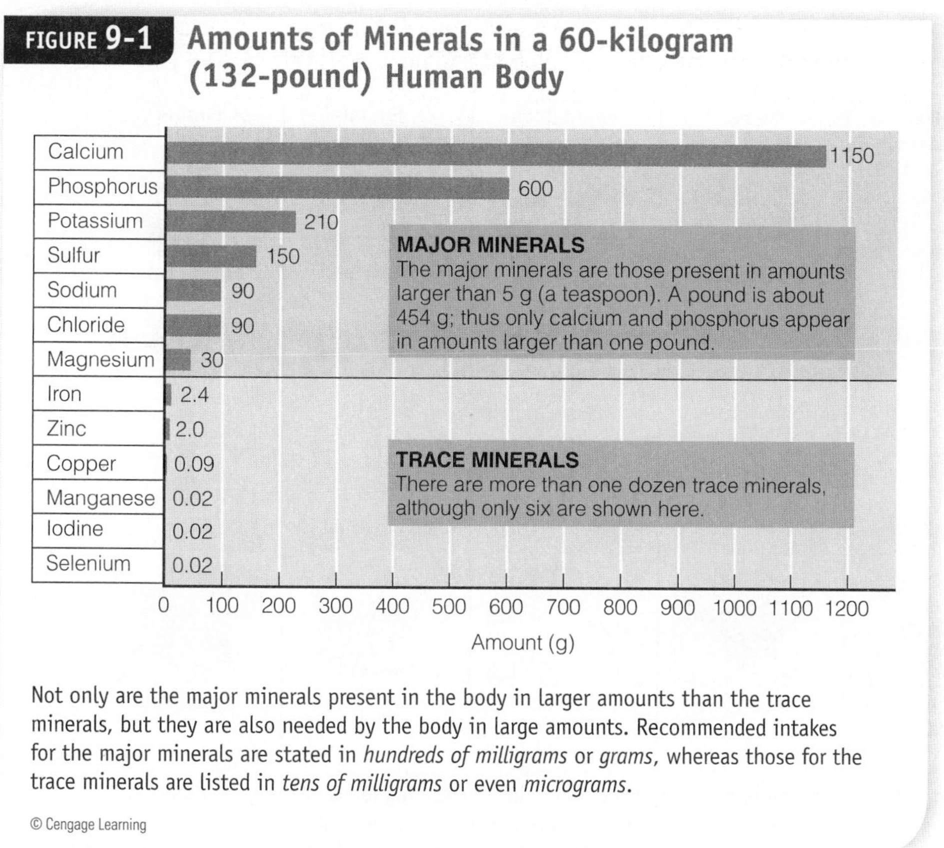

Not only are the major minerals present in the body in larger amounts than the trace minerals, but they are also needed by the body in large amounts. Recommended intakes for the major minerals are stated in *hundreds of milligrams* or *grams*, whereas those for the trace minerals are listed in *tens of milligrams* or even *micrograms*.

SODIUM

Sodium is the principal electrolyte in the **extracellular fluid** (the fluid outside the cells) and the primary regulator of the extracellular fluid volume. When the blood concentration of sodium rises, as when a person eats salted foods, thirst prompts the person to drink water until the appropriate sodium-to-water ratio is restored. Sodium also helps maintain acid–base balance and is essential to muscle contraction and nerve transmission. Too much sodium, however, can contribute to high blood pressure (hypertension).

Sodium Recommendations and Food Sources

Diets rarely lack sodium, and even when intakes are low, the body adapts by reducing sodium losses in urine and sweat, thus making deficiencies unlikely. Sodium recommendations (■) are set low enough to protect against high blood pressure but high enough to allow an adequate intake of other nutrients. Because high sodium intakes correlate with high blood pressure, the Tolerable Upper Intake Level (UL) for adults is set at 2300 milligrams per day, slightly lower than the Daily Value used on food labels (2400 milligrams). The UL corresponds to about 1 teaspoon of salt (sodium chloride). (■) Three groups of people encompassing the majority of U.S. adults—all those age 51 and older, all African Americans, and everyone with hypertension, diabetes, or chronic kidney disease—are urged to limit their sodium intakes to no more than 1500 milligrams of sodium per day for the sake of their blood pressure and heart health.[6]

Today, the average U.S. sodium intake is more than 3400 mg per day, an amount that far exceeds the UL.[7] People who eat mostly processed and fast foods have the highest sodium intakes, whereas those who eat mostly whole, unprocessed foods, such as fresh fruits and vegetables, have the lowest intakes. In fact, about three-fourths of the sodium in people's diets comes from salt added to foods by manufacturers. Figure 9-2 (p. 240) shows that processed foods contain not only more sodium but also less potassium than their less processed counterparts.

■ Sodium AI:
- 1500 mg/day (19–50 yr)
- 1300 mg/day (51–70 yr)
- 1200 mg/day (>70 yr)

■ Salt (sodium chloride) is about 40 percent sodium.
- 1 g salt contributes about 400 mg sodium.
- 6 g salt = 1 tsp.
- 1 tsp salt contributes about 2300 mg sodium.

extracellular fluid: fluid residing outside the cells; includes the fluid between the cells (*interstitial fluid*), plasma, and the water of structures such as the skin and bones. Extracellular fluid accounts for about one-third of the body's water.

FIGURE 9-2 What Processing Does to the Sodium and Potassium Contents of Foods

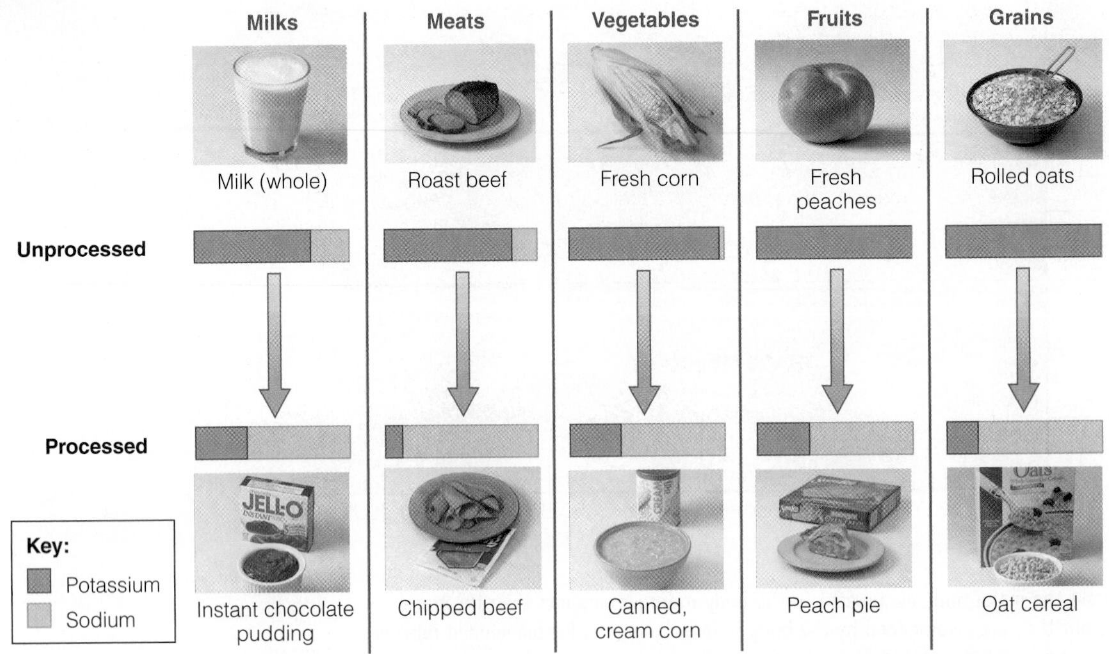

People who eat foods high in salt often happen to be eating fewer potassium-containing foods at the same time. Notice how potassium is lost and sodium is gained as foods become more processed, causing the potassium-to-sodium ratio to fall dramatically. Even when potassium is not lost, the addition of sodium still lowers the potassium-to-sodium ratio.

Many experts today are calling for reductions of sodium in the food supply to give consumers more low-salt options to choose from.[8] Reducing the sodium content in processed foods could prevent an estimated 100,000 deaths and save up to $24 billion in health care costs in the United States annually.[9] The margin lists the 10 common foods that are the top sodium providers.[10](■)

■ Top contributors of sodium in the diet:

- Breads and rolls
- Cold cuts and cured meats
- Pizza
- Fresh and processed poultry
- Soups
- Sandwiches (including cheeseburgers)
- Cheese
- Pasta dishes
- Meat mixtures (including meatloaf)
- Salty snacks (including popcorn, chips, and pretzels)

Sodium and Blood Pressure High intakes of salt among the world's people correlate with high rates of **hypertension**, heart disease, and stroke.[11] Over time, a high-salt diet may damage the linings of blood vessels in ways that make hypertension likely to develop.[12] One-third of U.S. adults have hypertension and the rate among African American adults is one of the highest in the world—42 percent.[13] An additional 30 percent of U.S. adults have **prehypertension.** (■)

■ Chapter 21 presents more details about blood pressure and lists standard ranges for blood pressure readings.

In many people, the relationship between salt intake and blood pressure is direct—as chronic sodium intakes increase, blood pressure rises with them in a stepwise fashion.[14] Once hypertension sets in, the risk of death from stroke and heart disease climbs steeply.

Because reducing salt intake causes no harm and diminishes the risk of hypertension and heart disease, the *Dietary Guidelines for Americans* advise limiting daily *sodium* intakes to less than 2300 milligrams (approximately 1 teaspoon of *salt*). Higher intakes seem to be well tolerated in most healthy people, however. The accompanying "How to" offers strategies for cutting salt (and, therefore, sodium) intake.

A proven eating pattern that can help people to reduce their sodium and increase potassium intakes, and thereby often reduce their blood pressure, is DASH (Dietary Approaches to Stop Hypertension).[15] The DASH approach emphasizes potassium-rich fruits and vegetables and fat-free or low-fat milk products; includes whole grains, nuts, poultry, and fish; and calls for reduced intakes of red and processed meats, sweets, and

hypertension: high blood pressure.

prehypertension: blood pressure values that predict hypertension.

HOW TO Cut Salt Intake

Most people eat more salt (and, therefore, sodium) than they need. Some people can lower their blood pressure by avoiding highly salted foods and removing the salt shaker from the table. Foods eaten without salt may seem less tasty at first, but with repetition, people can learn to enjoy the natural flavors of ... gies to cut salt intake include:

- ... foods.
- ... ed salt.
- ... -free spices such as basil, bay ..., mint, oregano, pepper, rosemary, ... negar; or wine.
- ... table; taste foods before adding ...
- ... for sodium. (See Table 1-11 on ... scribe the sodium contents of ...
- ... ucts when available.

Use these foods sparingly:

- Foods prepared in brine, such as pickles, olives, and sauerkraut.
- Salty or smoked meats, such as bologna, corned or chipped beef, bacon, frankfurters, ham, lunch meats, salt pork, sausage, and smoked tongue.
- Salty or smoked fish, such as anchovies, caviar, salted and dried cod, herring, sardines, and smoked salmon.
- Snack items such as potato chips, pretzels, salted popcorn, salted nuts, and crackers.
- Condiments such as bouillon cubes; seasoned salts; MSG; soy, teriyaki, Worcestershire, and barbeque sauces; prepared horseradish, catsup, and mustard.
- Cheeses, especially processed types.
- Canned and instant soups.

sugar-containing beverages. The DASH diet in combination with a reduced sodium intake is even more effective at lowering blood pressure than either strategy alone. Chapter 21 offers a complete discussion of hypertension and the dietary recommendations for its prevention and treatment.

Many Americans have much to gain in terms of cardiovascular health and nothing to lose from cutting back on salt as part of an overall lifestyle strategy to reduce blood pressure. Physical activity should also be part of that lifestyle because regular moderate exercise reliably lowers blood pressure.

CHLORIDE

The chloride ion is the major negative ion of the extracellular fluids, where it occurs primarily in association with sodium. Like sodium, chloride is critical to maintaining fluid, electrolyte, and acid–base balances in the body. In the stomach, the chloride ion is part of hydrochloric acid, which maintains the strong acidity of the gastric fluids.

Salt is a major food source of chloride, and, as with sodium, processed foods are a major contributor of this mineral to people's diets. Because salt contains a higher proportion of chloride (by weight) than sodium, chloride recommendations (■) are slightly higher than, but still equivalent to, those of sodium. In other words, ¾ teaspoon of salt will deliver some sodium and more chloride, and still meet the AI for both.

POTASSIUM

Outside the body's cells, sodium is the principle positively charged ion. *Inside* the cells, potassium takes the role of the principal positively charged ion. Potassium plays a major role in maintaining fluid and electrolyte balance and cell integrity. During nerve impulse transmission and muscle contraction, potassium and sodium briefly trade places across the cell membrane. The cell then quickly pumps them back into place. Controlling potassium distribution is a high priority for the body because it affects many aspects of homeostasis, including maintaining a steady heartbeat. The sudden deaths that occur with fasting, eating disorders, severe diarrhea, or severe malnutrition in children may be due to heart failure caused by potassium loss.

Potassium Deficiency and Toxicity

Potassium deficiency is characterized by an increase in blood pressure, salt sensitivity, kidney stones, and bone turnover. As deficiency progresses, symptoms include irregular heartbeats, muscle weakness, and

■ Chloride AI:
- 2300 mg/day (19–50 yr)
- 2000 mg/day (51–70 yr)
- 1800 mg/day (>70 yr)

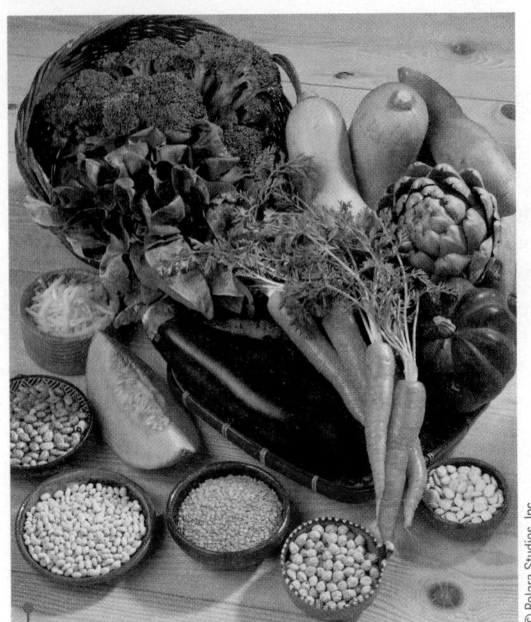
© Polara Studios, Inc.

Fresh fruits and vegetables provide potassium in abundance.

glucose intolerance. Potassium deficiency results more often from excessive losses than from deficient intakes. Deficiency arises in abnormal conditions such as diabetic acidosis, dehydration, or prolonged vomiting or diarrhea; potassium deficiency can also result from the regular use of certain medications, including **diuretics**, **steroids**, and **cathartics**.

Potassium toxicity does not result from overeating foods high in potassium; therefore, a Tolerable Upper Intake Level was not set. Toxicity can result from overconsumption of potassium salts or supplements and from certain diseases or medications. Given more potassium than the body needs, the kidneys accelerate their excretion. If the GI tract is bypassed, however, and potassium is injected directly into a vein, it can stop the heart.

Potassium Recommendations and Food Sources In healthy people, almost any reasonable diet provides enough potassium to prevent the dangerously low blood potassium that indicates a severe deficiency. Potassium is abundant inside all living cells, both plant and animal, and because cells remain intact until foods are processed, the richest sources of potassium are *fresh* foods of all kinds—especially fruits and vegetables. A typical U.S. eating pattern, however, with its low intakes of fruits and vegetables, provides far less potassium than the recommended intake.[16] (■) Although blood potassium may remain normal on such a diet, chronic diseases are more likely to occur.

Potassium and Hypertension Low potassium intakes, especially when combined with high sodium intakes, raise blood pressure and increase the risk of death from heart disease and stroke.[17] In contrast, eating patterns with ample potassium, particularly when low in sodium, appear to both prevent and correct hypertension. This effect earns potassium its status as a *Dietary Guidelines* nutrient of concern.[18] Recall that the DASH eating pattern described earlier is used to lower blood pressure and emphasizes potassium-rich foods such as fruits and vegetables.

CALCIUM

Calcium occupies more space in this chapter than any other major mineral. Other minerals are revisited later in this book, where their key roles in heart disease and kidney disease are discussed. Calcium, though, deserves emphasis here in the normal nutrition part of the book because an adequate intake of calcium early in life helps grow a healthy skeleton and prevent bone disease in later life.

Calcium Roles in the Body Calcium owns the distinction of being the most abundant mineral in the body. Ninety-nine percent of the body's calcium is stored in the bones (and teeth), where it plays two important roles. First, it is an integral part of bone structure. Second, it serves as a calcium bank available to the body fluids should a drop in blood calcium occur.

Calcium in Bone As bones begin to form, calcium salts form crystals on a matrix of the protein collagen. As the crystals become denser, they give strength and rigidity to the maturing bones. As a result, the long leg bones of children can support their weight by the time they have learned to walk. Figure 9-3 shows the lacy network of calcium-containing crystals in the bone.

Many people have the idea that bones are inert, like rocks. Not so. Bones continuously gain and lose minerals in an ongoing process of remodeling. Growing children gain more bone than they lose, and healthy adults maintain a reasonable balance. When withdrawals substantially exceed deposits, however, problems such as osteoporosis develop.

■ Potassium AI:
- Adults: 4700 mg

diuretics (dye-yoo-RET-ics): medications that promote the excretion of water through the kidneys. Not all diuretics increase the urinary loss of potassium. Some, called potassium-sparing diuretics, are less likely to result in a potassium deficiency (see Chapter 21).

steroids (STARE-oids): medications used to reduce tissue inflammation, to suppress the immune response, or to replace certain steroid hormones in people who cannot synthesize them.

cathartics (ca-THART-ics): strong laxatives.

From birth to approximately age 20, the bones are actively growing by modifying their length, width, and shape (see Figure 9-4). This rapid growth phase overlaps with the next period of peak bone mass development, which occurs between the ages of 12 and 30. During this period, skeletal mass increases. Bones grow thicker and denser by remodeling, a maintenance and repair process involving the loss of existing bone and the deposition of new bone. In the final phase, which begins between 30 and 40 years of age and continues throughout the remainder of life, bone loss exceeds new bone formation.

Calcium in Body Fluids The 1 percent of the body's calcium that circulates in the fluids as ionized calcium is vital to life. It plays these major roles:

- Regulates the transport of ions across cell membranes and is particularly important in nerve transmission.
- Helps maintain normal blood pressure.
- Plays an essential role in the clotting of blood.
- Is essential for muscle contraction and therefore for the heartbeat.
- Allows secretion of hormones, digestive enzymes, and neurotransmitters.
- Activates cellular enzymes that regulate many processes.

Because of its importance, blood calcium is tightly controlled.

Calcium Balance Whenever blood calcium rises too high, a system of hormones and vitamin D promotes its deposit into bone. Whenever blood calcium falls too low, the regulatory system (■) acts in three locations to raise it:

1. The small intestine absorbs more calcium.
2. The bones release more calcium into the blood.
3. The kidneys excrete less calcium.

Thus, blood calcium rises to normal.

FIGURE 9-3 Cross Section of Bone

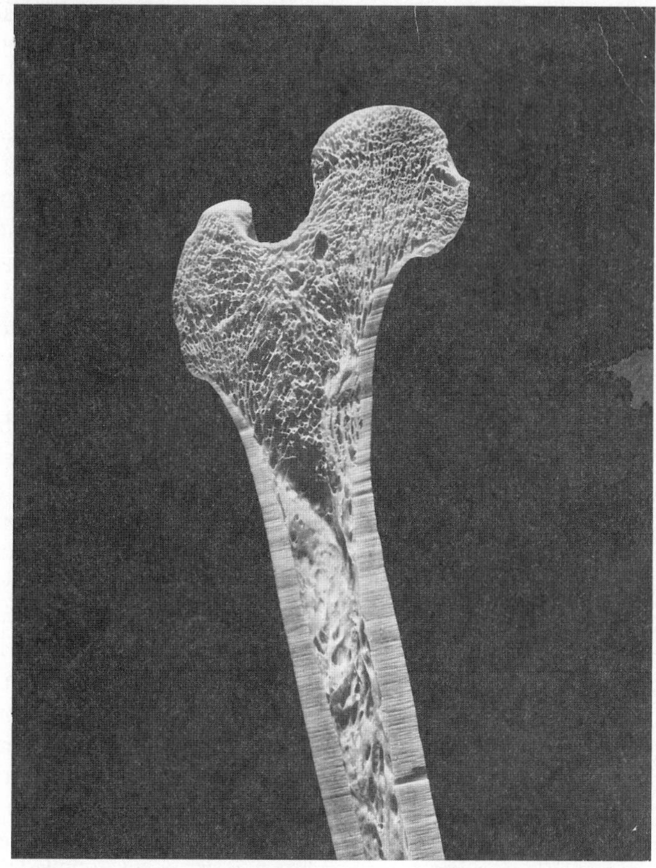

The lacy structural elements are trabeculae (tra-BECK-you-lee), which can be drawn on to replenish blood calcium.

Courtesy of Gjon Mili

■ The regulators are hormones from the thyroid and parathyroid glands, as well as vitamin D. One hormone, parathormone, raises blood calcium. Another hormone, calcitonin, lowers blood calcium by inhibiting release of calcium from bone. The hormonelike vitamin D raises blood calcium by acting at the three sites listed.

FIGURE 9-4 Phases of Bone Development throughout Life

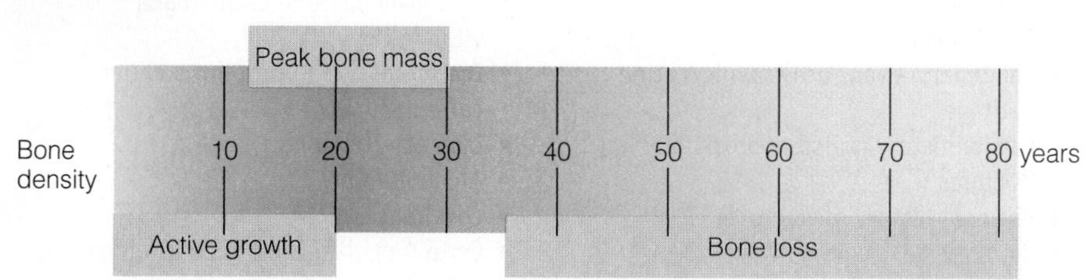

The active growth phase occurs from birth to approximately age 20. The next phase of peak bone mass development occurs between the ages of 12 and 30. The final phase, when bone resorption exceeds formation, begins between the ages of 30 and 40 and continues through the remainder of life.

The calcium stored in bone provides a nearly inexhaustible source of calcium for the blood. Even in a calcium deficiency, blood calcium remains normal. Blood calcium changes only in response to abnormal regulatory control, not to diet.

Although a chronic *dietary* deficiency of calcium or a chronic deficiency due to poor absorption does not change blood calcium, it does deplete the calcium in the bones. Because this is an important concept, we repeat: it is the bones, not the blood, that are robbed by calcium deficiency.

Calcium and Osteoporosis As mentioned earlier, bone mass peaks at the time of skeletal maturity (about age 30), and a high peak bone mass is the best protection against later age-related bone loss and fracture. Adequate calcium nutrition during the growing years is essential to achieving optimal peak bone mass.[19] Following menopause, women may lose up to 20 percent of their bone mass, as may middle-aged and older men. When bone loss has reached such an extreme that bones fracture under even common, everyday stresses, the condition is known as osteoporosis. (■) An estimated 44 million people in the United States, most of them women older than 50, have or are developing osteoporosis.[20] Men, however, are not immune to osteoporosis. Each year, 1.5 million people—30 percent of them men—break a hip, leg, arm, hand, ankle, or other bone as a result of osteoporosis.

Both genetic and environmental factors contribute to osteoporosis; Table 9-5 summarizes these risk factors. Osteoporosis is more prevalent in women than men for several reasons. First, women consume less dietary calcium than men do. Second, at all ages, women's bone mass is lower than men's because women generally have smaller bodies. Finally, women often lose more bone, particularly in the 6 to 8 years following menopause when the hormone estrogen diminishes.[21]

In addition to calcium, many other minerals and vitamins, including phosphorus, magnesium, fluoride, and vitamin D, help to form and stabilize the structure of bones. Any or all of these elements are needed to prevent bone loss. The first, most obvious lines of defense, however, are to maintain a lifelong adequate intake of calcium and to "exercise it into place." Physical activity supports bone growth during adolescence and may protect the bones later on.[22] Weight-bearing physical activity, such as walking, running, dancing, and weight training, prompts the bones to deposit minerals. It has long been known that, when people are confined to bed, both their muscles and their bones lose strength. Muscle strength and bone strength go together: when muscles work, they pull on the bones, and both are stimulated to grow stronger.

TABLE 9-5 Risk Factors for Osteoporosis

Nonmodifiable	Modifiable
• Female gender • Older age • Small frame • Caucasian, Asian, or Hispanic/Latino heritage • Family history of osteoporosis or fractures • Personal history of fractures • Estrogen deficiency in women (amenorrhea or menopause, especially early or surgically induced); testosterone deficiency in men	• Sedentary lifestyle • Diet inadequate in calcium and vitamin D • Diet excessive in protein, sodium, caffeine • Cigarette smoking • Alcohol abuse • Low body weight • Certain medications, such as glucocorticoids and anticonvulsants

■ Reminder: *Osteoporosis* is a condition characterized by reduced density of the bones. The bones become porous and fragile and fracture easily.

Calcium and Disease Prevention Other roles for calcium are emerging as well. Calcium may protect against hypertension.[23] Considering the success of the DASH diet in lowering blood pressure, restricting sodium to treat hypertension may be narrow advice. The DASH diet is not particularly low in sodium, but it is rich in calcium, as well as in potassium and magnesium. As mentioned earlier, the DASH diet, together with a reduced sodium intake, is more effective at lowering blood pressure than either strategy alone. Some research also suggests that calcium may protect against colon and rectal cancers.[24] Calcium from low-fat milk and milk products (but not from supplements) has been linked with healthy body weight in some, but not all, studies.[25]

Calcium Recommendations As mentioned earlier, blood calcium concentration does not reflect calcium status. Calcium recommendations (■) are therefore based on balance studies, which measure daily intake and excretion. An optimal calcium intake reflects the amount needed to retain the most calcium. The more calcium retained, the greater the bone density (within genetic limits) and, potentially, the lower the risk of osteoporosis. Calcium recommendations during adolescence are set high (1300 milligrams) to help ensure that the skeleton will be strong and dense. Between the ages of 19 and 50, recommendations are lowered slightly, and for women over 50 and all adults over age 70, recommendations are raised again to minimize bone loss. Many people in the United States have calcium intakes below current recommendations.[26] Findings about the effectiveness of calcium supplements in reducing fractures in older women, however, are inconclusive or negative.[27] Because adverse effects such as kidney stone formation are possible with high supplemental doses, a UL has been established (see inside front cover).

Calcium in Foods Calcium is found most abundantly in a single food group—milk and milk products. For this reason, dietary recommendations advise daily consumption of low-fat or fat-free milk products. A cup of milk offers about 300 milligrams of calcium, so an adult who drinks 3 cups of milk a day (or eats the equivalent in yogurt) is well on the way to meeting daily calcium needs (see Table 9-6). The other dairy food that contains comparable amounts of calcium is cheese. One slice of cheese (1 ounce) contains about two-thirds as much calcium as a cup of milk. Cottage cheese, however, contains much less. Figure 9-5 (p. 246) shows foods that are rich in calcium, and the "How to" (p. 246) suggests ways of adding calcium to meals.

Some foods offer large amounts of calcium because of fortification. Calcium-fortified juice, high-calcium milk (milk with extra calcium added), and calcium-fortified cereals are examples. Some calcium-rich mineral waters may also be a useful calcium source. The calcium from mineral water may be as absorbable as that from milk but accompanied by zero kcalories.

TABLE 9-6 Suggested Minimum Daily Fluid Milk Intakes

Young children (4–8 years of age)	2½ cups
Older children and adolescents	3 cups
Adults	3 cups
Pregnant or lactating women	3 cups
Women past menopause	3 cups

■ Calcium RDA:
- Adults (19–50 yr): 1000 mg/day
- Men (51–70): 1000 mg/day
- Women (51–70): 1200 mg/day
- Adults (>70): 1200 mg/day

FIGURE 9-5 Good Sources of Calcium[a]

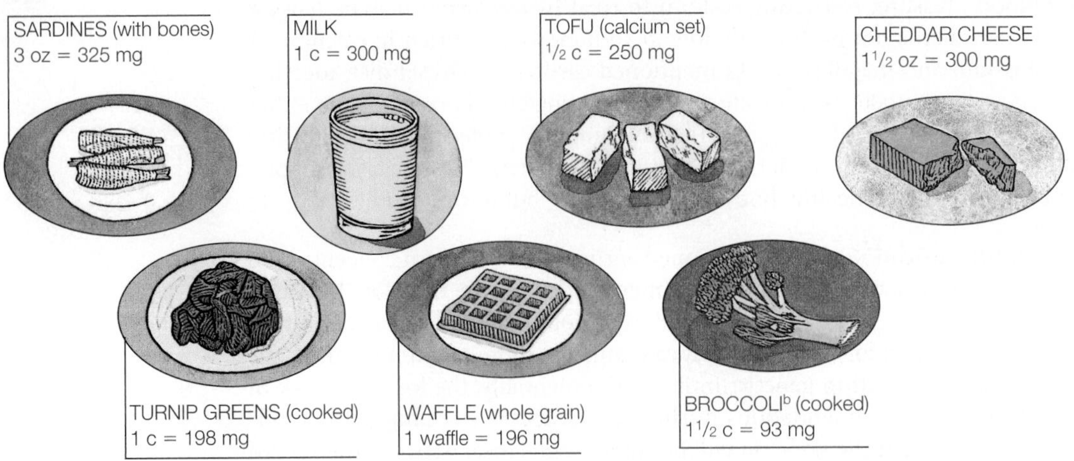

[a]These foods provide 10 percent or more of the calcium Daily Value in a serving. For a 2000-kcalorie diet, the DV is 1000 mg/day.
[b]Broccoli, kale, and some other cooked green leafy vegetables are important sources of bioavailable calcium. Almonds also supply calcium. Other greens, such as spinach and chard, contain calcium in an unabsorbable form. Some calcium-rich mineral waters may also be good sources.

HOW TO Add Calcium to Daily Meals

For those who tolerate milk, many cooks slip extra calcium into meals by sprinkling a tablespoon or two of fat-free dry milk into almost everything. The added kcalorie value is small, and changes to the taste and texture of the dish are practically nil. Yet each 2 tablespoons adds about 100 extra milligrams of calcium and moves people closer to meeting the recommendation to obtain 3 cups of milk each day. Here are some more tips for including calcium-rich foods in your meals.

At Breakfast

- Choose calcium-fortified orange or vegetable juice.
- Serve tea or coffee, hot or iced, with milk.
- Choose cereals, hot or cold, with milk.
- Cook hot cereals with milk instead of water; then mix in 2 tablespoons of fat-free dry milk.
- Make muffins or quick breads with milk and extra fat-free powdered milk.
- Add milk to scrambled eggs.
- Moisten cereals with flavored yogurt.

At Lunch

- Add low-fat cheeses to sandwiches, burgers, or salads.
- Use a variety of green vegetables, such as watercress or kale, in salads and on sandwiches.
- Drink fat-free milk or calcium-fortified soy milk as a beverage or in a smoothie.
- Drink calcium-rich mineral water as a beverage (studies suggest significant calcium absorption).
- Marinate cabbage shreds or broccoli spears in low-fat Italian dressing for an interesting salad that provides calcium.
- Choose coleslaw over potato and macaroni salads.
- Mix the mashed bones of canned salmon into salmon salad or patties.
- Eat sardines with their bones.
- Stuff potatoes with broccoli and low-fat cheese.
- Try pasta such as ravioli stuffed with low-fat ricotta cheese instead of meat.
- Sprinkle Parmesan cheese on pasta salads.

At Supper

- Toss a handful of thinly sliced green vegetables, such as kale or young turnip greens, with hot pasta; the greens wilt pleasingly in the steam of the freshly cooked pasta.
- Serve a green vegetable every night and try new ones—how about kohlrabi? It tastes delicious when cooked like broccoli.
- Learn to stir-fry Chinese cabbage and other Asian foods.
- Try tofu (the calcium-set kind); this versatile food has inspired whole cookbooks devoted to creative uses.
- Add fat-free powdered milk to almost anything—meatloaf, sauces, gravies, soups, stuffings, casseroles, blended beverages, puddings, quick breads, cookies, brownies. Be creative.
- Choose frozen yogurt, ice milk, or custards for dessert.

Among the vegetables, beet greens, bok choy (a Chinese cabbage), broccoli, kale, mustard greens, rutabaga, and turnip greens provide some available calcium. So do collard greens, green cabbage, kohlrabi, parsley, and watercress. Some dark green, leafy vegetables—notably, spinach and Swiss chard—appear to be calcium rich but actually provide very little, if any, calcium to the body. These foods contain **binders** that prevent calcium absorption. The presence of calcium binders does not make spinach an inferior food. Spinach is also rich in iron, beta-carotene, riboflavin, and dozens of other nutrients and phytochemicals.

Aided by vitamin D, the body is able to regulate its absorption of calcium by altering its production of the calcium-binding protein. More of this protein is made if more calcium is needed. Infants and children absorb up to 60 percent of the calcium they ingest, and pregnant women, about 50 percent. Other adults, who are not growing, absorb about 25 to 30 percent.[28]

People may think that taking a calcium supplement is preferable to getting calcium from food, but foods offer important fringe benefits. For example, drinking 3 cups of milk fortified with vitamins A and D will supply substantial amounts of other nutrients. Furthermore, the vitamin D and possibly other nutrients in the milk enhance calcium absorption. Some people absorb calcium better from milk and milk products than from even the most absorbable supplements. The National Institutes of Health concludes that foods are the best sources of calcium and recommends supplements only when intake from food is insufficient.

PHOSPHORUS

Phosphorus is the second most abundant mineral in the body. About 85 percent of it is found combined with calcium in the crystals of the bones and teeth. As part of one of the body's buffer systems (phosphoric acid), phosphorus is also found in all body tissues. Phosphorus is a part of DNA and RNA, the genetic material present in every cell. Thus, phosphorus is necessary for all growth. Phosphorus also plays many key roles in the transfer of energy that occurs during cellular metabolism. Phosphorus-containing lipids (phospholipids) help transport other lipids in the blood. Phospholipids are also principal components of cell membranes.

Animal protein is the best source of phosphorus because the mineral is so abundant in the cells of animals. Milk and cheese are also rich sources. Diets that provide adequate energy and protein also supply adequate phosphorus. (■) Dietary deficiencies are rare. A summary of facts about phosphorus appears in Table 9-7 later in the chapter.

MAGNESIUM

Magnesium barely qualifies as a major mineral. Only about 1 ounce of magnesium is present in the body of a 130-pound person, more than half of it in the bones. Most of the rest is in the muscles, heart, liver, and other soft tissues, with only 1 percent in the body fluids. Bone magnesium seems to be a reservoir to ensure that some will be on hand for vital reactions regardless of recent dietary intake.

Magnesium is critical to the operation of hundreds of enzymes and other cellular functions.[29] It acts in all the cells of the soft tissues, where it forms part of the protein-making machinery and is necessary for the release of energy. Magnesium and calcium work together for proper functioning of the muscles: calcium promotes contraction, and magnesium helps relax the muscles afterward. Magnesium is also critical to normal heart function.[30] Like many other nutrients, magnesium supports the normal functioning of the immune system and inflammatory response.[31]

Magnesium Deficiency

Magnesium deficiency can result from inadequate intake, vomiting, diarrhea, alcohol abuse, or malnutrition; in people who have been fed nutritionally incomplete fluids intravenously for too long after surgery; or in people using diuretics. Magnesium deficiency symptoms include a low blood calcium level, muscle cramps, and seizures. Magnesium deficiency is thought to cause the hallucinations commonly experienced during withdrawal from alcohol intoxication. In addition,

■ Phosphorus RDA:
- Adults: 700 mg/day

binders: chemical compounds in foods that combine with nutrients (especially minerals) to form complexes the body cannot absorb. Examples include *phytates* and *oxalates*.

FIGURE 9-6 Good Sources of Magnesium[a]

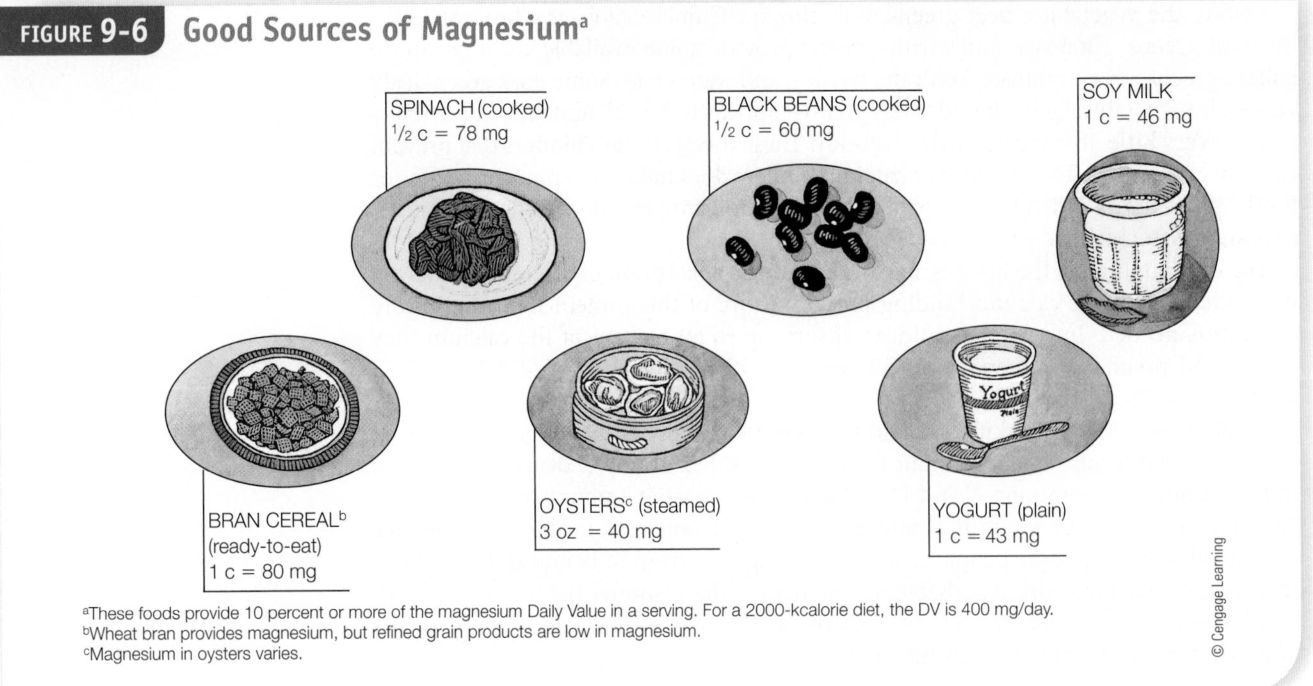

[a]These foods provide 10 percent or more of the magnesium Daily Value in a serving. For a 2000-kcalorie diet, the DV is 400 mg/day.
[b]Wheat bran provides magnesium, but refined grain products are low in magnesium.
[c]Magnesium in oysters varies.

magnesium deficiency may worsen inflammation associated with many chronic diseases and may increase the risk of stroke and sudden death by heart failure, even in otherwise healthy people.[32] Low intakes of magnesium may have adverse effects on bone metabolism and increase the risk for osteoporosis as well.[33]

Magnesium Toxicity Magnesium toxicity is rare, but it can be fatal. Toxicity occurs only with high intakes from nonfood sources such as supplements or magnesium salts. Accidental poisonings may occur in children with access to medicine cabinets and in older adults who abuse magnesium-containing laxatives, antacids, and other medications. The consequences include diarrhea, abdominal cramps, and, in severe cases, acid–base imbalance and potassium depletion.[34]

Magnesium Intakes and Food Sources Although almost half the U.S. population has magnesium intakes below those recommended, deficiency symptoms are rare in healthy people.[35] (■) In some areas of the country, the water naturally contains both calcium and magnesium. Known as "hard" water, this water can contribute significantly to magnesium intakes.

Magnesium-rich food sources include dark green, leafy vegetables; nuts; legumes; whole-grain breads and cereals; seafood; chocolate; and cocoa (see Figure 9-6). Magnesium is easily lost from foods during processing, so unprocessed foods are the best choices.

■ Magnesium RDA:
- Men (19–30 yr): 400 mg/day
  (31 and older): 420 mg/day
- Women (19–30 yr): 310 mg/day
  (31 and older): 320 mg/day

SULFATE

Sulfate is the oxidized form of sulfur as it exists in food and water. The body requires sulfate for the synthesis of many important sulfur-containing compounds. Sulfur-containing amino acids play an important role in helping to shape strands of protein. (■) The particular shape of a protein enables it to do its specific job, such as enzyme work. Skin, hair, and nails contain some of the body's more rigid proteins, which have high sulfur contents.

There is no recommended intake for sulfur, and no deficiencies are known. Only a person who lacks protein to the point of severe deficiency will lack the sulfur-containing amino acids.

■ The sulfur-containing amino acids are methionine and cysteine. Cysteine in one part of a protein chain can bind to cysteine in another part of the chain by way of a sulfur-sulfur bridge, thus helping to stabilize the protein structure.

Review Notes

- All of the major minerals influence the body's fluid balance, but sodium, chloride, and potassium are most noted for this role.
- Excess sodium in the diet contributes to high blood pressure.
- Most of the body's calcium is in the bones, where it provides a rigid structure and a reservoir of calcium for the blood.
- Magnesium is critical to the operation of hundreds of enzymes and other cellular functions.

Table 9-7 (p. 250) offers a summary of the major minerals and their functions.

The Trace Minerals

Figure 9-1, earlier in this chapter, shows how tiny the quantities of trace minerals in the human body are. If you could remove all of them from your body, you would have only a bit of dust, hardly enough to fill a teaspoon. Yet each of the trace minerals performs some vital role for which no substitute will do. A deficiency of any of them can be fatal, and an excess of many can be equally deadly. Table 9-8, at the end of the chapter, provides a summary of the trace minerals.

IRON

Every living cell—both plant and animal—contains iron. Most of the iron in the body is a component of the proteins **hemoglobin** in red blood cells and **myoglobin** in muscle cells. The iron in both hemoglobin and myoglobin helps them carry and hold oxygen and then release it. Hemoglobin in the blood carries oxygen from the lungs to tissues throughout the body. Myoglobin holds oxygen for the muscles to use when they contract. As part of many enzymes, iron is vital to the processes by which cells generate energy. Iron is also needed to make new cells, amino acids, hormones, and neurotransmitters.

The special provisions the body makes for iron's handling show that it is a precious mineral to be tightly hoarded. For example, when a red blood cell dies, the liver saves the iron and returns it to the bone marrow, which uses it to build new red blood cells. The body does lose iron from the digestive tract, in nail and hair trimmings, and in shed skin cells, but only in tiny amounts.[36] Bleeding, however, can cause significant iron loss from the body.

Special measures are needed to maintain an appropriate iron balance in the body. Iron is a powerful oxidant that generates free-radical reactions. Free radicals increase oxidative stress and inflammation associated with diseases such as diabetes, heart disease, and cancer.[37] To ensure that enough iron is available to meet the body's needs and yet guard against excessive and damaging levels, special proteins transport and store the body's iron supply, and its absorption is tightly regulated.[38] Because the body has no active means of excreting excess iron, regulation of absorption plays a critical role in iron homeostasis.[39]

Normally, only about 10 to 15 percent of dietary iron is absorbed, but if the body's supply is diminished or if the need increases for any reason (such as pregnancy), absorption increases. The body makes several provisions for absorbing iron. A special protein in the intestinal cells captures iron and holds it in reserve for release into the body as needed; another protein transfers the iron to a special iron carrier in the blood. The blood protein (**transferrin**) carries the iron to tissues throughout the body. When more iron is needed, more of these special proteins are produced so that more than the usual amount of iron can be absorbed and carried. If there is a surplus of iron, special storage proteins (■) in the liver, bone marrow, and other organs store it.

■ The storage proteins are ferritin (FERR-i-tin) and hemosiderin (heem-oh-SID-er-in).

hemoglobin: the oxygen-carrying protein of the red blood cells.

hemo = blood
globin = globular protein

myoglobin: the oxygen-carrying protein of the muscle cells.

myo = muscle

transferrin (trans-FERR-in): the body's iron-carrying protein.

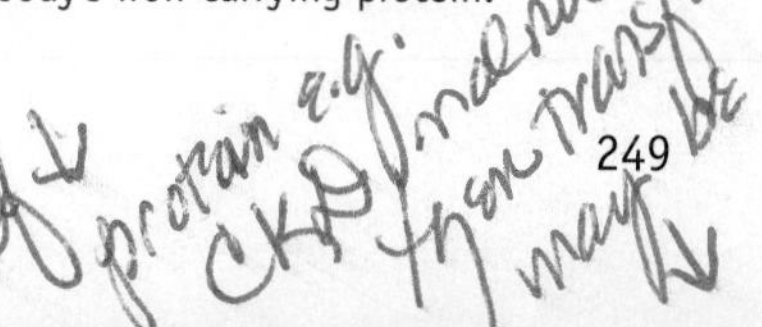

TABLE 9-7 The Major Minerals—A Summary

Mineral Name	Chief Functions in the Body	Deficiency Symptoms	Toxicity Symptoms	Significant Sources
Sodium	With chloride and potassium (electrolytes), maintains cells' normal fluid balance and acid–base balance in the body. Also critical to nerve impulse transmission and muscle contraction.	Muscle cramps, mental apathy, loss of appetite	Hypertension	Salt, soy sauce, processed foods
Chloride	Part of the hydrochloric acid found in the stomach and necessary for proper digestion.	Does not occur under normal circumstances	Normally harmless (the gas chlorine is a poison but evaporates from water); can cause vomiting	Salt, soy sauce; moderate quantities in whole, unprocessed foods; large amounts in processed foods
Potassium	Facilitates reactions, including the making of protein; the maintenance of fluid and electrolyte balance; the support of cell integrity; the transmission of nerve impulses; and the contraction of muscles, including the heart.	Moderate deficiency: elevated blood pressure, increased salt sensitivity, increased risk of kidney stones, increased bone turnover. Severe deficiency: cardiac arrhythmias, muscle weakness, glucose intolerance.	Causes muscular weakness; triggers vomiting; if given into a vein, can stop the heart	All whole foods: meats, milk, fruits, vegetables, grains, legumes
Calcium	The principal mineral of bones and teeth. Also acts in normal muscle contraction and relaxation, nerve functioning, blood clotting, blood pressure, and immune defenses.	Stunted growth in children; adult bone loss (osteoporosis)	Constipation; increased risk of urinary stone formation and kidney dysfunction; interference with absorption of other minerals	Milk and milk products, oysters, small fish (with bones), tofu (bean curd), greens, legumes
Phosphorus	Involved in the mineralization of bones and teeth. Important in cells' genetic material, in cell membranes as phospholipids, in energy transfer, and in buffering systems.	Phosphorus deficiency unknown	Calcification of nonskeletal tissues, particularly the kidneys	Foods from animal sources (meat, fish, poultry, eggs, milk)
Magnesium	Another factor involved in bone mineralization, the building of protein, enzyme action, normal muscular contraction, and transmission of nerve impulses.	Low blood calcium; muscle cramps; confusion; if extreme, seizures, bizarre movements, hallucinations, and difficulty in swallowing; in children, growth failure.	From nonfood sources only; diarrhea, nausea, and abdominal cramps; acid–base imbalance; potassium depletion	Nuts, legumes, whole grains, dark green vegetables, seafoods, chocolate, cocoa
Sulfate	A component of certain amino acids; part of the vitamins biotin and thiamin and the hormone insulin; stabilizes protein shape by forming sulfur–sulfur bridges.	None known; protein deficiency would occur first	Would occur only if sulfur amino acids were eaten in excess; this (in animals) depresses growth	All protein-containing foods (meats, fish, poultry, eggs, milk, legumes, nuts)

At an extraordinary pace researchers are discovering genes that participate in the regulation of the body's iron.[40] Along the way, new information about how the body controls the uptake, storage, and distribution of iron is emerging. The hormone **hepcidin**, produced by the liver, is central to the regulation of iron balance.[41] Hepcidin helps to maintain blood iron within the normal range by limiting absorption from the small intestine and controlling release from the liver, spleen, and bone marrow.[42] Many details are known about this process, but simply described, hepcidin works in an elegant feedback system to control blood iron:

- Elevated levels of iron in the blood (and liver) trigger hepcidin secretion, which reduces iron absorption and inhibits the release of stored iron, thereby reducing the blood iron concentration.
- Low levels of iron in the blood suppress hepcidin secretion, which permits increased iron absorption and mobilizes storage iron, raising the blood iron concentrations.

Thus, the body adjusts to changing iron needs and iron availability in the diet.[43]

Iron Deficiency Worldwide, **iron deficiency** is the most common nutrient deficiency, with **iron-deficiency anemia** affecting more than 1.6 billion people—almost half of them preschool children and pregnant women.[44] In the United States, iron deficiency is less prevalent, but it still affects about 14 percent of toddlers and 10 percent of adolescent girls and women of childbearing age.[45] Iron deficiency is also more prevalent among overweight children and adolescents compared with those who are normal weight.[46] The association between iron deficiency and obesity has yet to be explained, but researchers are currently examining the relationships between the inflammation that develops with excess body fat and reduced iron absorption.[47]

Some stages of life demand more iron but provide less, making deficiency likely.[48] Women are especially prone to iron deficiency during their reproductive years because of blood losses during menstruation. Pregnancy places further iron demands on women: iron is needed to support the added blood volume, the growth of the fetus, and blood loss during childbirth. Infants (six months or older) and young children receive little iron from their high-milk diets, yet they need extra iron to support their rapid growth and brain development. The rapid growth of adolescence, especially for males, and the blood losses of menstruation for females also demand extra iron that a typical teen diet may not provide. To summarize, an adequate iron intake is especially important during these stages of life:

- Women in their reproductive years.
- Pregnant women.
- Infants and toddlers.
- Adolescents.

Causes of Iron Deficiency The cause of iron deficiency is usually inadequate intake resulting from an ignorance of which foods to choose, a sheer lack of food altogether, or a high consumption of iron-poor foods. In the Western world, high sugar and fat intakes are often associated with low iron intakes. Blood loss is the primary nonnutritional cause, especially in poor regions of the world where parasitic infections of the GI tract may lead to blood loss.

Assessment of Iron Deficiency Iron deficiency develops in stages. (■) This section provides a brief overview of how to detect these stages, and Appendix E provides more details. In the first stage of iron deficiency, iron stores diminish, as do levels of ferritin, an iron-storing protein. Measures of serum ferritin (in the blood) reflect iron stores and are most valuable in assessing iron status at this earliest stage.

The second stage of iron deficiency is characterized by a decrease in transport iron: levels of serum iron fall, and levels of the iron-carrying protein transferrin *increase* (an adaptation that enhances iron absorption). Together, these two measures can determine the severity of the deficiency—the more transferrin and the less iron in the blood,

■ Stages of iron deficiency:
- Iron stores diminish.
- Transport iron decreases.
- Hemoglobin production declines.

hepcidin (HEP-sid-in): a hormone secreted by the liver in response to elevated blood iron. Hepcidin reduces iron's absorption from the intestine and its release from storage.

iron deficiency: the condition of having depleted iron stores.

iron-deficiency anemia: a blood iron deficiency characterized by small, pale red blood cells; also called *microcytic hypochromic anemia.*

micro = small
cytic = cells
hypo = too little
chrom = color

the more advanced the deficiency is. Transferrin saturation—the percentage of transferrin that is saturated with iron—decreases as iron stores decline.

The third stage of iron deficiency occurs when the lack of iron limits hemoglobin production. Now the hemoglobin precursor, **erythrocyte protoporphyrin**, begins to accumulate as hemoglobin and **hematocrit** values decline.

Hemoglobin and hematocrit tests are easy, quick, and inexpensive, so they are the tests most commonly used in evaluating iron status. Their usefulness is limited, however, because they are late indicators of iron deficiency. Furthermore, other nutrient deficiencies and medical conditions can influence their values.

Iron Deficiency and Anemia

Iron deficiency and iron-deficiency anemia are not the same: people may be iron deficient without being anemic. The term *iron deficiency* refers to depleted body iron stores without regard to the degree of depletion or to the presence of anemia. The term *iron-deficiency anemia* refers to the severe depletion of iron stores that results in a low hemoglobin concentration. In iron-deficiency anemia, red blood cells are pale and small (see Figure 9-7). They can't carry enough oxygen from the lungs to the tissues. Without adequate iron, energy metabolism in the cells falters. The result is fatigue, weakness, headaches, apathy, pallor, and poor resistance to cold temperatures. Because hemoglobin is the bright red pigment of the blood, the skin of a fair person who is anemic may become noticeably pale. In a dark-skinned person, the tongue and eye lining, normally pink, will be very pale.

The fatigue that accompanies iron-deficiency anemia differs from the tiredness a person experiences from a simple lack of sleep. People with anemia feel fatigue only when they exert themselves. Iron supplementation can, over time, relieve the fatigue and improve the body's response to physical activity.

Less severe iron deficiency produces symptoms, too. Long before the red blood cells are affected and anemia is diagnosed, a developing iron deficiency affects behavior.[49] Even at slightly lowered iron levels, energy metabolism is impaired and neurotransmitter synthesis is altered, reducing physical work capacity and mental productivity.[50] Without the physical energy and mental alertness to work, plan, think, play, sing, or learn, people simply do these things less. They have no obvious deficiency signs; they just appear unmotivated, apathetic, and less physically fit. Children deprived of iron become irritable, restless, and unable to pay attention. (■) These symptoms are among the first to appear when the body's iron begins to fall and among the first to disappear when iron status is restored.

Iron Deficiency and Pica

A curious behavior seen in some iron-deficient (and sometimes zinc-deficient) people, especially in women and children of low-income groups, is **pica**—the craving for and consumption of ice, chalk, starch, and other

■ The effects of iron deficiency on children's behavior are revisited in Chapter 11.

erythrocyte protoporphyrin (PRO-toh-PORE-fe-rin): a precursor to hemoglobin.

hematocrit (hee-MAT-oh-crit): measurement of the volume of the red blood cells packed by centrifuge in a given volume of blood.

pica (PIE-ka): a craving for nonfood substances; also known as *geophagia* (jee-oh-FAY-jee-uh) when referring to clay-eating behavior.

picus = woodpecker or magpie
geo = earth
phagein = to eat

FIGURE 9-7 Normal and Anemic Blood Cells

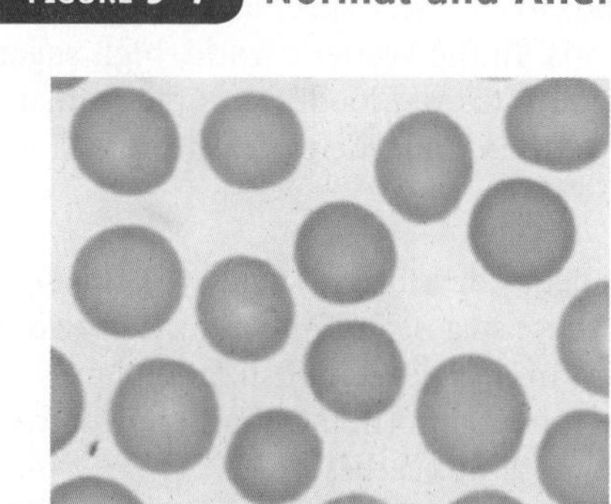

Normal red blood cells. Both size and color are normal.

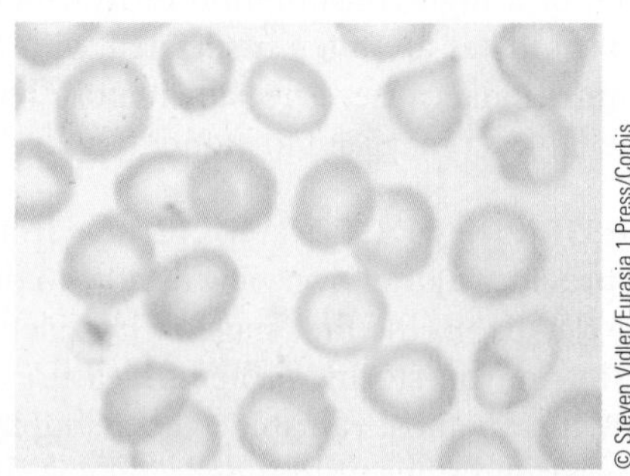

Blood cells in iron-deficiency anemia. These cells are small and pale because they contain less hemoglobin.

nonfood substances. These substances contain no iron and cannot remedy a deficiency; in fact, clay actually inhibits iron absorption, which may explain the iron deficiency that accompanies such behavior. Pica is poorly understood. Its cause is unknown, but researchers hypothesize that it may be motivated by hunger, nutrient deficiencies, or an attempt to protect against toxins or microbes.[51]

■ Binding proteins in the intestinal cells (*mucosal ferritin* and *mucosal transferrin*) capture and hold unneeded iron to be shed with the cells, thereby forming a mucosal block to iron absorption.

Caution on Self-Diagnosis Low hemoglobin may reflect an inadequate iron intake, and, if it does, the physician may prescribe iron supplements. Any nutrient deficiency, disease, or agent that interferes with hemoglobin synthesis, disrupts hemoglobin function, or causes a loss of red blood cells can precipitate anemia, however.

Feeling fatigued, weak, and apathetic is thus a sign that something is wrong, but it does not indicate that a person should take iron supplements; it means that the person should consult a physician. In fact, taking iron supplements may be the worst possible thing a person can do because such supplements can mask a serious medical condition, such as hidden bleeding from cancer or an ulcer. Furthermore, a person can waste precious time by not seeking treatment. Remember, don't self-diagnose.

■ Iron RDA:
- Men (19 and older): 8 mg/day
- Women (19–50 yr): 18 mg/day
- Women (>50 yr): 8 mg/day

Iron Overload Normally, the body protects itself against absorbing too much iron by setting up a block in the intestinal cells. (■) The system can be overwhelmed, however, resulting in **iron overload**. Once considered rare, iron overload has emerged as an important disorder of iron metabolism and regulation.

Iron overload, known as **hemochromatosis**, is usually caused by a genetic failure to prevent unneeded iron in the diet from being absorbed.[52] Other causes of iron overload include repeated blood transfusions, massive doses of supplementary iron, and other rare metabolic disorders.

iron overload: toxicity from excess iron.

hemochromatosis (heem-oh-crome-a-TOH-sis): iron overload characterized by deposits of iron-containing pigment in many tissues, with tissue damage. Hemochromatosis is usually caused by a hereditary defect in iron absorption.

Some of the signs and symptoms of iron overload are similar to those of iron deficiency: apathy, lethargy, and fatigue. Therefore, taking iron supplements before assessing iron status is clearly unwise; hemoglobin tests alone would fail to make the distinction because excess iron accumulates in storage. Iron overload assessment tests measure transferrin saturation and serum ferritin.

Iron overload is characterized by free-radical tissue damage, especially in iron-storing organs such as the liver.[53] Infections are likely because bacteria thrive on iron-rich blood.[54] Symptoms are most severe in alcohol abusers because alcohol damages the intestine, further impairing its defenses against absorbing excess iron. Untreated hemochromatosis aggravates the risk of diabetes, liver cancer, heart disease, and arthritis.

Iron overload is much more common in men than in women and is twice as prevalent among men as iron deficiency. The widespread fortification of foods with iron makes it difficult for people with hemochromatosis to follow an iron-restricted diet.

Iron Poisoning The rapid ingestion of massive amounts of iron can cause sudden death. Iron-containing supplements can easily cause accidental poisonings in young children.[55] As few as five iron tablets have caused death in a child. Keep iron-containing supplements out of children's reach. If you suspect iron poisoning, call the nearest poison center or a physician immediately.

© Tim Hill/Alamy

This chili dinner provides iron and MFP factor from meat, iron from legumes, and vitamin C from tomatoes. The combination of heme iron, nonheme iron, MFP factor, and vitamin C helps to achieve maximum iron absorption.

Iron Recommendations The average eating pattern in the United States provides only about 6 to 7 milligrams of iron for every 1000 kcalories. (■) Men need 8 milligrams of iron each day; most men easily take in more than 2000 kcalories, so a man can meet his iron needs without special effort. The recommendation for women during childbearing years, however, is 18 milligrams. Because women have higher iron needs and typically consume fewer than 2000 kcalories per day, they have trouble achieving appropriate iron intakes. On the average, women receive only 12 to 13 milligrams of iron per day, not enough until after menopause. To meet their iron needs from foods, premenopausal women need to select iron-rich foods at every meal. Vegetarians, because vegetable sources of iron are poorly absorbed, should aim

FIGURE 9-8 Good Sources of Iron[a]

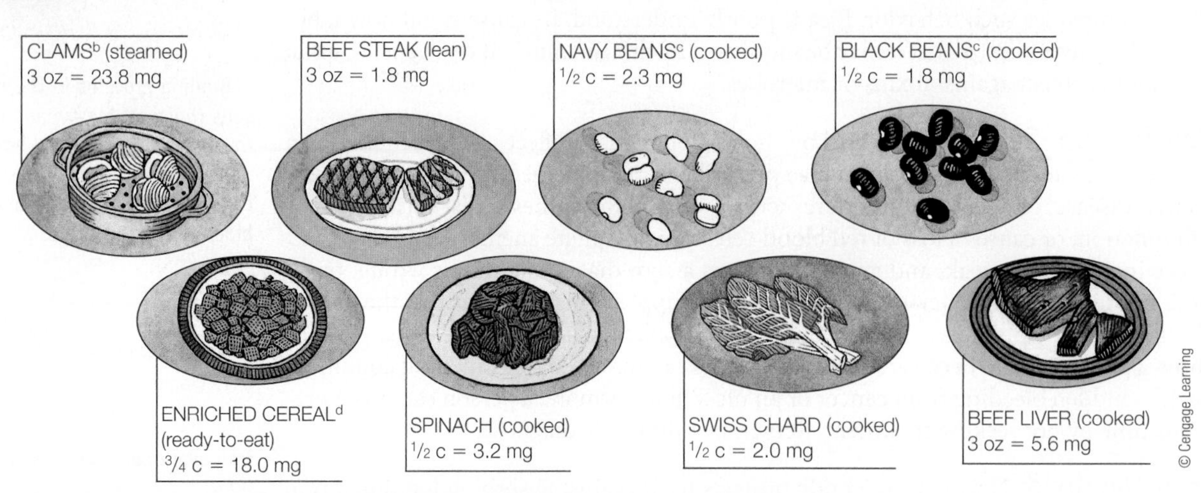

[a]These foods provide 10 percent or more of the iron Daily Value in a serving. For a 2000-kcalorie diet, the DV is 18 mg/day.
Note: Dried figs contain 0.6 mg per 1/4 cup; raisins contain 0.8 mg per 1/4 cup.
[b]Some clams may contain less, but most types are iron-rich foods.
[c]Legumes contain phytates that reduce iron absorption.
[d]Enriched cereals vary widely in iron content.

for 1.8 times the normal requirement. (■) Vegetarian diets are the topic of Nutrition in Practice 5.

Iron in Foods Iron occurs in two forms in foods, one of which is up to 10 times more absorbable than the other. The more absorbable form is heme iron, which is bound into the iron-carrying proteins hemoglobin and myoglobin in meat, poultry, and fish. (■) Heme iron contributes a small portion of the iron consumed by most people, but it is absorbed at a fairly constant rate of about 23 percent. The less absorbable form is nonheme iron, found in meats and also in plant foods. People absorb nonheme iron at a lower rate (2 to 20 percent); its absorption depends on several dietary factors and iron stores. Most of the iron people consume is nonheme iron from vegetables, grains, eggs, meat, fish, and poultry. Figure 9-8 shows the iron amounts found in usual serving sizes of different foods.

Iron absorption from foods can be maximized by two substances that enhance iron absorption: MFP factor and vitamin C. Meat, fish, and poultry contain a factor (MFP factor) other than heme that promotes the absorption of iron. MFP factor even enhances the absorption of nonheme iron from other foods eaten at the same time. Vitamin C eaten in the same meal also doubles or triples nonheme iron absorption. Additionally, cooking with iron skillets can contribute iron to the diet. Some substances impair iron absorption; they include the **tannins** of tea and coffee, the calcium in milk, and the **phytates** that accompany fiber in legumes and whole-grain cereals.[56] The accompanying "How to" offers suggestions for obtaining adequate iron.

■ To calculate the RDA for vegetarians, multiply by 1.8:

- 8 mg/day × 1.8 = 14 mg/day (vegetarian men)
- 18 mg/day × 1.8 = 32 mg/day (vegetarian women, 19–50 yr)

■ About 40 percent of the iron in meat, fish, and poultry is bound into molecules of heme (HEEM), the iron-holding part of the hemoglobin and myoglobin proteins. This heme iron is much more absorbable than nonheme iron.

tannins: compounds in tea (especially black tea) and coffee that bind iron.

phytates: nonnutrient components of grains, legumes, and seeds. Phytates can bind minerals such as iron, zinc, calcium, and magnesium in insoluble complexes in the intestine, and the body excretes them unused.

ZINC

Zinc is a versatile trace mineral necessary for the activation of more than 50 different enzymes.[57] These zinc-requiring enzymes perform tasks in the eyes, liver, kidneys, muscles, skin, bones, and male reproductive organs. Zinc works with the enzymes that make genetic material; manufacture heme; digest food; metabolize carbohydrate, protein, and fat; liberate vitamin A from storage in the liver; and dispose of damaging free radicals.[58] Zinc also interacts with platelets in blood clotting, affects thyroid hormone function, assists in immune function, and affects behavior and learning performance.[59] Zinc is needed to produce the active form of vitamin A in visual pigments and is essential to wound healing, taste perception, the making of sperm, and fetal development. When zinc deficiency occurs, it impairs all these and other functions.

HOW TO Add Iron to Daily Meals

The following set of guidelines can be used for planning an iron-rich diet:

- *Grains.* Use only whole-grain, enriched, and fortified products (iron is one of the enrichment nutrients).
- *Vegetables.* The dark green, leafy vegetables are good sources of vitamin C and iron. Eat vitamin C–rich vegetables often to enhance absorption of the iron from foods eaten with them.
- *Fruits.* Dried fruits, such as raisins, apricots, peaches, and prunes, are high in iron. Eat vitamin C–rich fruits often with iron-containing foods.
- *Milk and milk products.* Do not overdo foods from the milk group; they are poor sources of iron. But do not omit them either because they are rich in calcium. Drink fat-free milk to free kcalories to be invested in iron-rich foods.
- *Protein foods.* Meat, fish, and poultry are excellent iron sources. Include legumes frequently. One cup of peas or beans can supply up to 7 milligrams of iron.

The body's handling of zinc differs from that of iron but with some interesting similarities. For example, like iron, extra zinc that enters the body is held within the intestinal cells, and only the amount needed is released into the bloodstream. As with iron, zinc status influences the percentage of zinc absorbed from the diet; if more is needed, more is absorbed.

Zinc's main transport vehicle in the blood is the protein albumin. This may account for observations that serum zinc concentrations decline in conditions that lower plasma albumin concentrations—for example, pregnancy and malnutrition.

Zinc Deficiency Zinc deficiency in human beings was first reported in the 1960s in studies of growing children and male adolescents in Egypt, Iran, and Turkey. Their diets were typically low in zinc and high in fiber and phytates (which impair zinc absorption). The zinc deficiency was marked by dwarfism, or severe growth retardation, and arrested sexual maturation—symptoms that were responsive to zinc supplementation.

Since that time, zinc deficiency has been recognized elsewhere and is known to affect more than growth. It drastically impairs immune function, causes loss of appetite, and, during pregnancy, may lead to growth and developmental disorders.[60] Zinc deficiency is a substantial contributor to illness throughout the developing world and is responsible for almost half a million deaths each year.[61] A detailed list of symptoms of zinc deficiency is presented later in Table 9-8.

Pronounced zinc deficiency is not widespread in developed countries, but deficiencies do occur in the most vulnerable groups of the U.S. population—pregnant women, young children, the elderly, and the poor. Even mild zinc deficiency can result in metabolic changes such as impaired immune response, abnormal taste, and abnormal dark adaptation (zinc is required to produce the active form of vitamin A, retinal, in visual pigments).

Some people are at greater risk of zinc deficiency than others. Pregnant teenagers need zinc for their own growth as well as for the developing fetus. Vegetarians whose diets emphasize whole grains, legumes, and other plant foods may also be at risk. These foods, though rich in zinc, also contain phytate, a potent inhibitor of zinc absorption.[62] Protein enhances zinc absorption, but most plant-protein foods also contain phytate. The DRI committee suggests that the dietary zinc requirement for vegetarians who exclude all animal-derived foods may be as much as 50 percent greater than the RDA, but so far evidence is insufficient to establish zinc recommendations based on the presence of other food components or nutrients.[63] Vegetarians who include cheese, eggs, or other animal protein in their diet absorb more zinc than those who exclude these foods.

Zinc Toxicity Zinc can be toxic if consumed in large enough quantities. A high zinc intake is known to produce copper-deficiency anemia by inducing the intestinal cells to synthesize large amounts of a protein (metallothionein) that captures copper in a nonabsorbable form. Accidental consumption of high levels of zinc can cause vomiting,

FIGURE 9-9 Good Sources of Zinc[a]

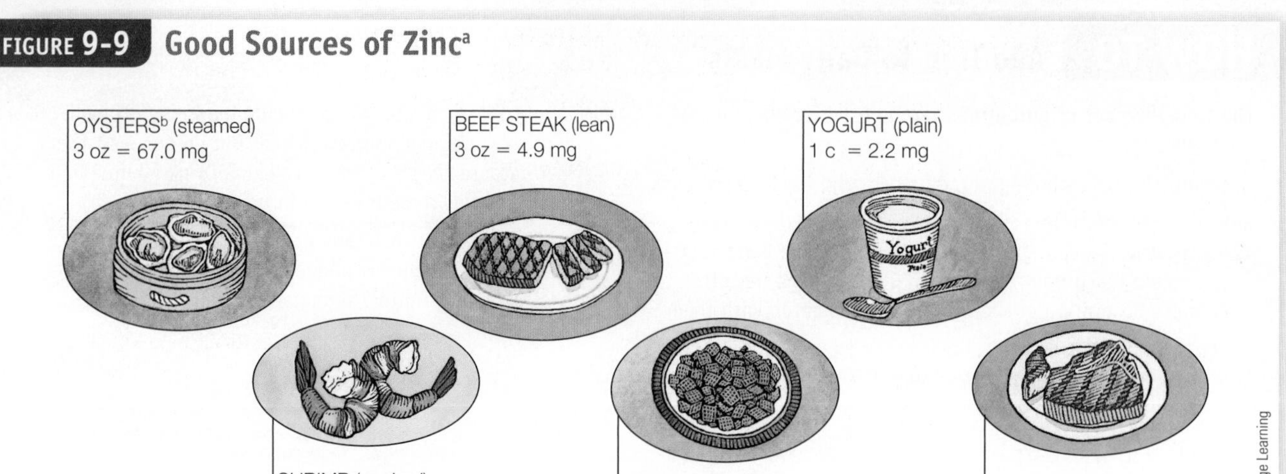

[a]These foods provide 10 percent or more of the zinc Daily Value in a serving. For a 2000-kcalorie diet, the DV is 15 mg/day.
[b]Some oysters contain more or less than this amount, but all types are zinc-rich foods.
[c]Enriched cereals vary widely in zinc content.

diarrhea, headaches, exhaustion, and other symptoms (see Table 9-8, later in the chapter). Large doses can even be fatal. The Tolerable Upper Intake Level for zinc for adults is 40 milligrams per day.

Zinc Recommendations and Food Sources Most people in the United States have zinc intakes that approximate recommendations.[64] (■) Zinc is most abundant in foods high in protein, such as shellfish (especially oysters), meats, poultry, and milk products. In general, two ordinary servings a day of animal protein provide most of the zinc a healthy person needs. Legumes and whole-grain products are good sources of zinc if eaten in large quantities. For infants, breast milk is a good source of zinc, which is more efficiently absorbed from human milk than from cow's milk. Commercial infant formulas are fortified with zinc, of course. Figure 9-9 shows zinc-rich foods.

Zinc supplements are not recommended except for an accurately diagnosed zinc deficiency or when needed for use as a medication to displace other ions in unusual medical circumstances. Normally, it should be possible to obtain enough zinc from the diet. Zinc from cold-relief lozenges, sprays, and gels may sometimes shorten the duration of a cold, but they can upset the stomach and contribute supplemental zinc to the body.[65]

SELENIUM

Selenium is an essential trace mineral that functions as an antioxidant nutrient, working primarily as a part of proteins—most notably, the enzyme glutathione peroxidase.[66] Glutathione peroxidase and vitamin E work in tandem. Glutathione peroxidase prevents free-radical formation, thus blocking the damaging chain reaction before it begins; if free radicals do form, and a chain reaction starts, vitamin E halts it. Selenium-containing enzymes are necessary for the proper functioning of the iodine-containing thyroid hormones that regulate metabolism.[67]

Selenium and Cancer The question of whether selenium protects against the development of certain cancers, particularly prostate cancer, is under intense investigation.[68] Adequate *blood* selenium seems protective against cancers of the prostate, colon, and other sites.[69] Many questions remain unanswered regarding selenium status, intake, and cancer, however. For example, the range of selenium status and the levels of

■ Zinc RDA:
- Men: 11 mg/day
- Women: 8 mg/day

intakes most beneficial to reducing cancer risk have not been established.[70] Given the potential for harm from excess selenium and the lack of conclusive evidence, recommendations to take selenium supplements to prevent cancer would be premature.

Selenium Deficiency Selenium deficiency is associated with a heart disease in children and young women living in regions of China where the soil and foods lack selenium. Although the primary cause of this heart disease is probably a virus, selenium deficiency appears to predispose people to it, and adequate selenium seems to prevent it. The heart disease is named *Keshan disease* for one of the provinces of China where it was studied.

Selenium Toxicity Because high doses of selenium are toxic, a UL has been set (see inside front cover). Selenium toxicity causes vomiting, diarrhea, loss of hair and nails, and lesions of the skin and nervous system.[71]

Selenium Recommendations and Intakes Anyone who eats a normal diet composed mostly of unprocessed foods need not worry about meeting selenium recommendations. (■) Selenium is widely distributed in foods such as meats and shellfish and in vegetables and grains grown on selenium-rich soil. Some regions in the United States and Canada produce crops on selenium-poor soil, but people are protected from deficiency because they eat selenium-rich meat and supermarket foods transported from other regions. Eating as few as two Brazil nuts a day effectively improves selenium status.

■ Selenium RDA:
- Adults: 55 µg/day

■ Iodine RDA: 150 µg/day

goiter (GOY-ter): an enlargement of the thyroid gland due to an iodine deficiency, malfunction of the gland, or overconsumption of a thyroid antagonist. Goiter caused by iodine deficiency is sometimes called *simple goiter*.

goitrogen (GOY-troh-jen): a substance that enlarges the thyroid gland and causes *toxic goiter*. Goitrogens occur naturally in such foods as cabbage, kale, brussels sprouts, cauliflower, broccoli, and kohlrabi.

cretinism (CREE-tin-ism): an iodine-deficiency disease characterized by mental and physical retardation.

IODINE

Traces of the iodine ion (called iodide) are indispensible to life. In the GI tract, iodine from foods becomes iodide. This chapter uses the term *iodine* when referring to the nutrient in foods, and *iodide* when referring to it in the body. Iodide occurs in the body in minuscule amounts, but its principal role in human nutrition is well known, and the amount needed is well established. Iodide is an integral part of the thyroid hormones, which regulate body temperature, metabolic rate, reproduction, growth, the making of blood cells, nerve and muscle function, and more.

Iodine Deficiency When the iodide concentration in the blood is low, the cells of the thyroid gland enlarge in an attempt to trap as many particles of iodide as possible. If the gland enlarges until it is visible, the swelling is called a **goiter**. People with iodine deficiency this severe become sluggish and may gain weight. Goiter afflicts about 200 million people the world over, many of them in South America, Asia, and Africa. In most cases of goiter, the cause is iodine deficiency, but some people have goiter because they overconsume foods of the cabbage family and others that contain an antithyroid substance (**goitrogen**) whose effect is not counteracted by dietary iodine.

Goiter may be the earliest and most obvious sign of iodine deficiency, but the most tragic and prevalent damage occurs in the brain. Iodine deficiency is the most common cause of *preventable* mental retardation and brain damage in the world. Children with even a mild iodine deficiency typically have goiters and perform poorly in school. With sustained treatment, however, mental performance in the classroom as well as thyroid function improves.[72] Programs to provide iodized salt to the world's iodine-deficient areas now prevent much misery and suffering worldwide.[73]

A severe iodine deficiency during pregnancy causes the extreme and irreversible mental and physical retardation known as **cretinism**. A child with cretinism may have an IQ as low as 20 (100 is normal) and a face and body with many abnormalities. Cretinism can be averted if the pregnant woman's deficiency is detected and treated in time.[74]

In iodine deficiency, the thyroid gland enlarges—a condition known as simple goiter.

Iodine Toxicity Excessive intakes of iodine can enlarge the thyroid gland, just as deficiencies can. Average intakes in the United States are slightly above the recommended intake of 150 micrograms (■) but still below the Tolerable Upper Intake Level of 1100 micrograms per day for an adult.[75] Iodine intakes of young women, however, barely meet their need.[76]

Iodine Sources The ocean is the world's major source of iodine. In coastal areas, seafood, water, and even iodine-containing sea mist are important iodine sources.* Further inland, the amount of iodine in the diet is variable and generally reflects the amount present in the soil in which plants are grown or on which animals graze. In the United States and Canada, the use of iodized salt has largely wiped out the iodine deficiency that once was widespread. In the United States, you have to read the label to find out whether salt is iodized; in Canada, all table salt is iodized.

COPPER

The body contains about 100 milligrams of copper. About one-fourth is in the muscles; one-fourth is in the liver, brain, and blood; and the rest is in the bones, kidneys, and other tissues. The primary function of copper in the body is to serve as a constituent of enzymes. The copper-containing enzymes have diverse metabolic roles: they catalyze the formation of hemoglobin, help manufacture the protein collagen, inactivate histamine, degrade serotonin, assist in the healing of wounds, and help maintain the sheaths around nerve fibers. One of copper's most vital roles is to help cells use iron. Like iron, copper is needed in many reactions related to respiration and energy metabolism. One copper-dependent enzyme helps to control damage from free-radical activity in the tissues.†

Copper Deficiency Copper deficiency is rare but not unknown. It has been seen in premature infants and malnourished infants. High intakes of zinc interfere with copper absorption and can lead to deficiency.

Copper Toxicity Some genetic disorders create a copper toxicity. Copper toxicity from foods, however, is unlikely. The Tolerable Upper Intake Level for copper is set at 10,000 micrograms per day.

Copper Recommendations and Food Sources The RDA for copper is 900 micrograms per day, (■) which is slightly below the average intake for adults in the United States.[77] The best food sources of copper are organ meats, legumes, whole grains, seafood, nuts, and seeds.

MANGANESE

The human body contains a tiny 20 milligrams of manganese, mostly in the bones and metabolically active organs such as the liver, kidneys, and pancreas. Manganese is a cofactor for many enzymes, helping to facilitate dozens of different metabolic processes. Deficiencies of manganese have not been noted in people, but toxicity may be severe. Miners who inhale large quantities of manganese dust on the job over prolonged periods show many symptoms of a brain disease, along with abnormalities in appearance and behavior. The Tolerable Upper Intake Level for manganese is 11 milligrams per day.

Manganese requirements are low, (■) and plant foods such as nuts, whole grains, and leafy green vegetables contain significant amounts of this trace mineral. Deficiencies are therefore unlikely.

FLUORIDE

Fluoride's primary role in health is the prevention of dental caries throughout life.[78] Only a trace of fluoride occurs in the human body, but it is important to the mineralization of the bones and teeth. When bones and teeth become mineralized, first a crystal called hydroxyapatite forms from calcium and phosphorus. Then fluoride replaces the hydroxy portion of hydroxyapatite, forming **fluorapatite**. During development, fluorapatite enlarges crystals in bones and teeth, decreasing their susceptibility to demineralization and making the teeth more resistant to decay. Once the teeth have erupted, the topical application of fluoride by way of toothpaste or mouth rinse continues to exert a caries-reducing effect.

■ Copper RDA:
- Adults: 900 µg/day

■ Manganese AI:
- Men: 2.3 mg/day
- Women: 1.8 mg/day

fluorapatite (floor-APP-uh-tite): the stabilized form of bone and tooth crystal, in which fluoride has replaced the hydroxy portion of hydroxyapatite.

*Iodine in sea mist combines with particles in the air or with water. Iodine may then enter the soil or surface water or land on plants when these particles fall to the ground or when it rains.

†The enzyme is superoxide dismutase.

Fluoride Deficiency Where fluoride is lacking in the water supply, the incidence of dental decay is high. Fluoridation of water is thus recommended as an important public health measure. Those fortunate enough to have had sufficient fluoride during the tooth-forming years of infancy and childhood are protected throughout life from dental decay. Dental problems are of great concern because they can lead to a multitude of other health problems affecting the whole body. Based on the accumulated evidence of its beneficial effects, water fluoridation has been endorsed by nearly 100 national and international organizations including the National Institute of Dental Health, the Academy of Nutrition and Dietetics, the American Medical Association, the National Cancer Institute, and the Centers for Disease Control and Prevention (CDC).[79] In fact, the CDC named water fluoridation as one of the 10 most important public health measures of the 20th century. About 70 percent of the U.S. population served by public water systems receives optimal levels of fluoride (about 0.7 milligram per liter). Most bottled waters lack fluoride.

FIGURE 9-10 **Fluoride Toxicity Symptom—The Mottled Teeth of Fluorosis**

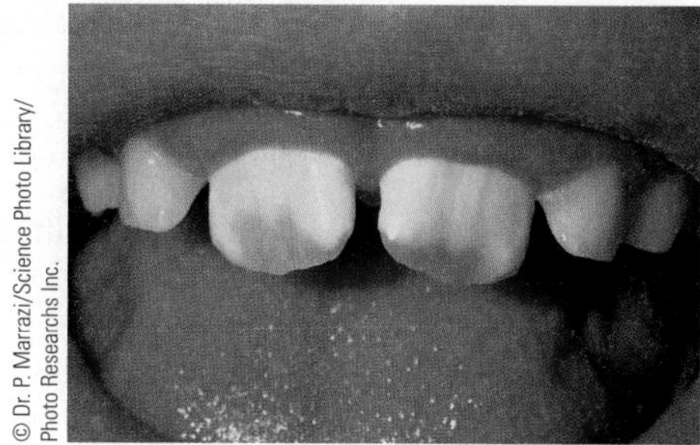

Fluoride Sources All normal diets include some fluoride, but drinking water; processed soft drinks and fruit juice made with fluoridated water; and fluoride toothpastes, gels, and oral rinses are the most common fluoride sources in the United States. (■) Fish and tea may supply substantial amounts as well.

In some areas, the natural fluoride concentration in water is high, and too much fluoride can damage teeth, causing **fluorosis**. For this reason, a Tolerable Upper Intake Level (■) has been established. In mild cases, the teeth develop small white specks; in severe cases, the enamel becomes pitted and permanently stained (see Figure 9-10). Fluorosis occurs only during tooth development and cannot be reversed, making its prevention a high priority. (■)

CHROMIUM

Chromium is an essential mineral that participates in carbohydrate and lipid metabolism. Chromium enhances the activity of the hormone insulin.[80] (■) When chromium is lacking, a diabetes-like condition with elevated blood glucose and impaired glucose tolerance, insulin response, and glucagon response may develop. Some research findings suggest that chromium supplements improve glucose or insulin responses in diabetes, but these relationships are uncertain.[81]

Chromium deficiency is unlikely, given the small amount of chromium required (■) and its presence in a variety of foods. The more refined foods people eat, however, the less chromium they obtain from their diets. Unrefined foods such as liver, brewer's yeast, whole grains, nuts, and cheeses are the best sources.

OTHER TRACE MINERALS

An RDA has been established for one other trace mineral, molybdenum. (■) **Molybdenum** functions as a working part of several metal-containing enzymes, some of which are giant proteins. Deficiencies or toxicities of molybdenum are unknown.

Several other trace minerals are known or suspected to contribute to the health of the body. Nickel is recognized as important for the health of many body tissues. Nickel deficiencies harm the liver and other organs. Silicon is involved in the formation of bones and collagen. Cobalt is found in the large vitamin B_{12} molecule. Boron influences the activity of many enzymes and may play a key role in bone health, brain activities, and immune response. The future may reveal that other trace minerals also play key roles. Even arsenic—famous as the deadly poison in many murder mysteries and known to be a carcinogen—may turn out to be an essential nutrient in tiny quantities.

■ Fluoride AI:
- Men: 4 mg/day
- Women: 3 mg/day

■ Fluoride Tolerable Upper Intake Level:
- Adults: 10 mg/day

■ To prevent fluorosis:
- Monitor the fluoride content of the local water supply.
- Supervise children younger than six when they brush their teeth (to ensure that they don't swallow the toothpaste), and use only a pea-size amount of toothpaste.
- Use fluoride supplements only as prescribed by a physician.

■ Small organic compounds that enhance insulin's activity are called glucose tolerance factors (GTF). Some glucose tolerance factors contain chromium.

■ Chromium AI:
- Men (19–50 yr): 35 µg/day (51 and older): 30 µg/day
- Women (19–50 yr): 25 µg/day (51 and older): 20 µg/day

■ Molybdenum RDA: 45 µg/day

fluorosis (floor-OH-sis): mottling of the tooth enamel from ingestion of too much fluoride during tooth development.

molybdenum (mo-LIB-duh-num): a trace element.

Review Notes

- The body requires trace minerals in tiny amounts, and they function in similar ways—assisting enzymes all over the body.
- Eating a diet that consists of a variety of foods is the best way to ensure an adequate intake of these important nutrients.
- Many dietary factors, including the trace minerals themselves, affect the absorption and availability of these nutrients.

Table 9-8 offers a summary of facts about trace minerals in the body.

TABLE 9-8 The Trace Minerals—A Summary

Mineral Name	Chief Functions in the Body	Deficiency Symptoms	Toxicity Symptoms	Significant Sources
Iron	Part of the protein hemoglobin, which carries oxygen in the blood; part of the protein myoglobin in muscles, which makes oxygen available for muscle contraction; necessary for the utilization of energy	Anemia: weakness, pallor, headaches, reduced work productivity, inability to concentrate, impaired cognitive function (children), lowered cold tolerance	Iron overload: infections, liver injury, possible increased risk of heart attack, acidosis, bloody stools, shock	Red meats, fish, poultry, shellfish, eggs, legumes, dried fruits
Zinc	Part of the hormone insulin and many enzymes; involved in making genetic material and proteins, immune reactions, transport of vitamin A, taste perception, wound healing, the making of sperm, and normal fetal development	Growth retardation, delayed sexual maturation, impaired immune function, loss of taste, poor wound healing, eye and skin lesions	Loss of appetite, impaired immunity, low HDL, nausea, vomiting, diarrhea, headaches, copper and iron deficiencies	Protein-containing foods: meats, fish, shellfish, poultry, grains, vegetables
Selenium	Assists a group of enzymes that break down reactive chemicals that harm cells	Predisposition to heart disease characterized by cardiac tissue becoming fibrous (uncommon)	Nausea, abdominal pain, nail and hair changes, nerve damage	Seafoods, organ meats, other meats, whole grains and vegetables (depending on soil content)
Iodine	A component of two thyroid hormones, which help to regulate growth, development, and metabolic rate	Goiter, cretinism	Depressed thyroid activity; goiterlike thyroid enlargement	Iodized salt; seafood; bread; plants grown in most parts of the country and animals fed those plants
Copper	Necessary for the absorption and use of iron in the formation of hemoglobin; part of several enzymes	Anemia, bone abnormalities	Vomiting, diarrhea, liver damage	Organ meats, seafood, nuts, seeds, whole grains, drinking water
Manganese	Facilitator, with enzymes, of many cell processes; bone formation	Rare	Nervous system disorders	Nuts, whole grains, leafy vegetables, tea

continued

continued

Mineral Name	Chief Functions in the Body	Deficiency Symptoms	Toxicity Symptoms	Significant Sources
Fluoride	An element involved in the formation of bones and teeth; helps to make teeth resistant to decay	Susceptibility to tooth decay	Fluorosis (pitting and discoloration of teeth); skeletal fluorosis (weak, malformed bones)	Drinking water (if fluoride containing or fluoridated), tea, seafood
Chromium	Enhances insulin action and may improve glucose tolerance	Diabetes-like condition marked by an inability to use glucose normally	None reported	Meats, whole grains

Self Check

1. Which of the following body structures helps to regulate thirst?
   a. Brainstem
   b. Cerebellum
   c. Optic nerve
   d. Hypothalamus

2. Which of the following is *not* a function of water in the body?
   a. Lubricant
   b. Source of energy
   c. Maintains protein structure
   d. Participant in chemical reactions

3. Two situations in which a person may experience fluid and electrolyte imbalances are:
   a. vomiting and burns.
   b. diarrhea and cuts.
   c. broken bones and fever.
   d. heavy sweating and excessive carbohydrate intake.

4. Three-fourths of the sodium in people's diets comes from:
   a. fresh meats.
   b. home-cooked foods.
   c. frozen vegetables and meats.
   d. salt added to food by manufacturers.

5. Which mineral is critical to keeping the heartbeat steady and plays a major role in maintaining fluid and electrolyte balance?
   a. Sodium
   b. Calcium
   c. Potassium
   d. Magnesium

6. The two best ways to prevent age-related bone loss and fracture are to:
   a. take calcium supplements and estrogen.
   b. participate in aerobic activity and drink eight glasses of milk daily.
   c. eat a diet low in fat and salt and refrain from smoking.
   d. maintain a lifelong adequate calcium intake and engage in weight-bearing physical activity.

7. Three good food sources of calcium are:
   a. milk, sardines, and broccoli.
   b. spinach, yogurt, and sardines.
   c. cottage cheese, spinach, and tofu.
   d. Swiss chard, mustard greens, and broccoli.

8. Foods high in iron that help prevent or treat anemia include:
   a. green peas and cheese.
   b. dairy foods and fresh fruits.
   c. homemade breads and most fresh vegetables.
   d. meat and dark green, leafy vegetables.

9. Two groups of people who are especially at risk for zinc deficiency are:
   a. Asians and children.
   b. infants and teenagers.
   c. smokers and athletes.
   d. pregnant adolescents and vegetarians.

10. A deficiency of ___ is one of the world's most common preventable causes of mental retardation.
   a. zinc
   b. iodine
   c. selenium
   d. magnesium

Answers to these questions appear in Appendix H. For more chapter review: Access an interactive eBook, chapter-specific interactive learning tools, including flashcards, quizzes, videos, and more in your Nutrition CourseMate, accessed through CengageBrain.com.

Clinical Applications

1. Pull together information from Chapter 1 about the different food groups and the significant sources of minerals shown or discussed in this chapter. Consider which minerals might be lacking (or excessive) in the diet of a client who reports the following:
   - Relies on highly processed foods, snack foods, and fast foods as mainstays of the diet.
   - Never uses milk, milk products, or cheese.
   - Dislikes leafy green vegetables.
   - Never eats meat, fish, poultry, or even meat alternates such as legumes.

   What additional information would help you pinpoint problems with mineral intake?

Nutrition on the Net

For further study of the topics in this chapter, access these websites.

- Find information about mineral supplements: http://ods.od.nih.gov/
- Search for "minerals" at the Academy of Nutrition and Dietetics site: www.eatright.org
- Find tips and recipes for including more milk in the diet: www.milkmustache.com/
- Learn about the benefits of calcium from the National Dairy Council: www.nationaldairycouncil.org
- Search for the individual minerals by name at the U.S. government health information site: www.healthfinder.org
- Learn more about iron overload from the Iron Overload Diseases Association: www.ironoverload.org
- Learn more about iodine deficiency and thyroid disease from the American Thyroid Association: www.thyroid.org

Notes

1. B. M. Popkin, K. E. D'Anci, and I. H. Rosenberg, Water, hydration, and health, *Nutrition Reviews* 68 (2010): 439–458.
2. Standing Committee on the Scientific Evaluation of Dietary Reference Intakes, Food Nutrition Board, Institute of Medicine, *Dietary Reference Intakes for Water, Potassium, Sodium, Chloride, and Sulfate* (Washington, DC: National Academies Press, 2005), p. 73.
3. F. Péronnet, Healthy hydration for physical activity, *Nutrition Today* 45 (2010): S41–S44; K. M. Kolasa, C. J. Lackey, and A. C. Grandjean, Hydration and health promotion, *Nutrition Today* 44 (2009): 190–201.
4. D. F. Tate and coauthors, Replacing caloric beverages with water or diet beverages for weight loss in adults: Main results of the Choose Healthy Options Consciously Everyday (CHOICE) randomized clinical trial, *American Journal of Clinical Nutrition* 95 (2012): 555–563.
5. Standing Committee on the Scientific Evaluation of Dietary Reference Intakes, 2005, pp. 133–134.
6. U.S. Department of Agriculture and U.S. Department of Health and Human Services, *Dietary Guidelines for Americans* 2010, available at www.dietaryguidelines.gov.
7. U.S. Department of Agriculture and U.S. Department of Health and Human Services, *Dietary Guidelines for Americans* 2010, available at www.dietaryguidelines.gov; Centers for Disease Control and Prevention, Application of Lower Sodium Intake Recommendations to Adults—United States, 1999–2006, *Morbidity and Mortality Weekly Report* 58 (2009): 281–283.
8. Centers for Disease Control and Prevention, Where's the sodium? *Vital Signs*, February 2012, available at www.cdc.gov/VitalSigns/Sodium/index.html.
9. K. Bibbins-Domingo and coauthors, Projected effect of dietary salt reductions on future cardiovascular disease, *New England Journal of Medicine* 362 (2010): 590–599; Institute of Medicine (US) Committee on Strategies

to Reduce Sodium Intake, *Strategies to Reduce Sodium Intake in the United States* (Washington, DC: National Academies Press, 2010).

10. Centers for Disease Control and Prevention, Vital signs: Food categories contributing the most to sodium consumption—United States, 2007–2008, *Morbidity and Mortality Weekly Report* 61 (2012): 92–98.
11. H. Gardener and coauthors, Dietary sodium and risk of stroke in the Northern Manhattan Study, *Stroke* 43 (2012): 1200–1205; R. Takachi and coauthors, Consumption of sodium and salted foods in relation to cancer and cardiovascular disease: The Japan Public Health Center-based prospective study, *American Journal of Clinical Nutrition* 91 (2010): 456–464; F. J. He and G. A. MacGregor, A comprehensive review on salt and health and current experience of worldwide salt reduction programmes, *Journal of Human Hypertension* 23 (2009): 363–384.
12. J. P. Forman and coauthors, Association between sodium intake and change in uric acid, urine albumin excretion, and the risk of developing hypertension, *Circulation* 125 (2012): 3108–3116.
13. V. L. Roger and coauthors, Heart disease and stroke statistics—2012 update: A report from the American Heart Association, *Circulation* 125 (2012): e12–e230.
14. Forman and coauthors, 2012; Standing Committee on the Scientific Evaluation of Dietary Reference Intakes, 2005, pp. 269–272.
15. J. A. Blumenthal and coauthors, Effects of the DASH diet alone and in combination with exercise and weight loss on blood pressure and cardiovascular biomarkers in men and women with high blood pressure: The ENCORE study, *Archives of Internal Medicine* 170 (2010): 126–135.
16. Standing Committee on the Scientific Evaluation of Dietary Reference Intakes, 2005, pp. 186–268.
17. H. Gardener and coauthors, Dietary sodium and risk of stroke in the Northern Manhattan Study, *Stroke* 43 (2012): 1200–1205; L. J. Appel and coauthors, The importance of population-wide sodium reduction as a means to prevent cardiovascular disease and stroke: A call to action from the American Heart Association, *Circulation* 123 (2011): 1138–1143; Q. Yang and coauthors, Sodium and potassium intake and mortality among US adults: Prospective data from the third National Health and Nutrition Examination Survey, *Archives of Internal Medicine* 171 (2011): 1183–1191; M. C. Houston, The importance of potassium in managing hypertension, *Current Hypertension Reports* 13 (2011): 309–317; C. J. Rodriguez and coauthors, Association of sodium and potassium intake with left ventricular mass: Coronary artery risk development in young adults, *Hypertension* 58 (2011): 410–416.
18. U.S. Department of Agriculture and U.S. Department of Health and Human Services, *Dietary Guidelines for Americans* 2010, available at www.dietaryguidelines.gov.
19. R. P. Heaney, Diet, osteoporosis, and fracture prevention: The totality of the evidence, in A. Bendich and R. J. Deckelbaum, eds., *Preventive Nutrition: A Comprehensive Guide for Health Professionals,* 4th ed. (New York: Humana Press, 2010), pp. 443–469.
20. National Osteoporosis Foundation, *Osteoporosis: Fast Facts* (2010), available at www.nof.org/osteoporosis/diseasefacts.htm.
21. J. C. Lo, S. A. Burnett-Bowie, and J. S. Finkelstein, Bone and the perimenopause, *Obstetrics & Gynecology Clinics of North America* 38 (2011): 503–517; B. Frenkel and coauthors, Regulation of adult bone turnover by sex steroids, *Journal of Cellular Physiology* 224 (2010): 305–310.
22. K. F. Janz and coauthors, Early physical activity provides sustained bone health benefits later in childhood, *Medicine and Science in Sports and Exercise* 42 (2010): 1072–1078; A. Guadalupe-Grau and coauthors, Exercise and bone mass in adults, *Sports Medicine* 39 (2009): 439–468.
23. J. Kaluza and coauthors, Dietary calcium and magnesium intake and mortality: A prospective study of men, *American Journal of Epidemiology* 171 (2010): 801–807; I. R. Reid and coauthors, Effects of calcium supplementation on lipids, blood pressure, and body composition in healthy older men: A randomized controlled trial, *American Journal of Clinical Nutrition* 91 (2010): 131–139; V. Centeno and coauthors, Molecular mechanisms triggered by low-calcium diets, *Nutrition Research Reviews* 22 (2009): 163–174.
24. M. Huncharek, J. Muscat, and B. Kupelnick, Colorectal cancer risk and dietary intake of calcium, vitamin D, and dairy products: A meta-analysis of 26,335 cases from 60 observational studies, *Nutrition and Cancer* 61 (2009): 47–69; Y. Park and coauthors, Dairy food, calcium, and risk of cancer in the NIH-AARP Diet and Health Study, *Archives of Internal Medicine* 169 (2009): 391–401.
25. J. A. Gilbert and coauthors, Milk supplementation facilitates appetite control in obese women during weight loss: A randomized, single-blind, placebo-controlled trial, *British Journal of Nutrition* 105 (2011): 133–143; D. R. Shahar and coauthors, Dairy calcium intake, serum vitamin D, and successful weight loss, *American Journal of Clinical Nutrition* 92 (2010): 1017–1022; R. P. Heaney and K. Rafferty, Preponderance of the evidence: An example from the issue of calcium intake and body composition, *Nutrition Reviews* 67 (2009): 32–39.
26. A. Moshfegh and coauthors, *What We Eat in America, NHANES 2005–2006: Usual Nutrient Intakes from Food and Water Compared to 1997 Dietary Reference Intakes for Vitamin D, Calcium, Phosphorus, and Magnesium* (Beltsville, Md.: USDA, 2009).
27. M. Chung and coauthors, Vitamin D with or without calcium supplementation for prevention of cancer and fractures: An updated meta-analysis for the U.S. Preventive Services Task Force, *Annals of Internal Medicine* 155 (2011): 827–838; Committee on Dietary Reference Intakes, *Dietary Reference Intakes for Calcium and Vitamin D* (Washington, DC: National Academies Press, 2011), pp. 410–411.
28. Committee on Dietary Reference Intakes, *Dietary Reference Intakes for Calcium and Vitamin D* (Washington, DC: National Academies Press, 2011), pp. 38–40.
29. A. M. Romani, Cellular magnesium homeostasis, *Archives of Biochemistry and Biophysics* 512 (2011): 1–23.
30. M. Shechter, Magnesium and cardiovascular system, *Magnesium Research* 23 (2010): 60–72.
31. A. Rosanoff, C. M. Weaver, and R. K. Rude, Suboptimal magnesium status in the United States: Are the health consequences underestimated? *Nutrition Reviews* 70 (2012): 153–164; J. Sugimoto and coauthors, Magnesium decreases inflammatory cytokine production: A novel innate immunomodulatory mechanism, *Journal of Immunology* 188 (2012): 6338–6346; W. B. Weglicki, Hypomagnesemia and inflammation: Clinical and basic aspects, *Annual Review of Nutrition* 32 (2012): 55–71.
32. S. C. Larsson, N. Orsini, and A. Wolk, Dietary magnesium intake and risk of stroke: A meta-analysis of prospective studies, *American Journal of Clinical Nutrition* 95 (2012): 362–366; S. E. Chiuve and coauthors, Plasma and dietary magnesium and risk of sudden cardiac death in women, *American Journal of Clinical Nutrition* 93 (2011): 253–260; F. H. Nielsen, Magnesium, inflammation, and obesity in chronic disease, *Nutrition Reviews* 68 (2010): 333–340.
33. Rosanoff, Weaver, and Rude, 2012; R. K. Rude, F. R. Singer, and H. E. Gruber, Skeletal and hormonal effects of magnesium deficiency, *Journal of the American College of Nutrition* 28 (2009): 131–141.
34. *Dietary Reference Intakes: The Essential Guide to Nutrient Requirements,* eds., J. J. Otten, J. P. Hellwig, and L. D. Meyers (Washington, DC: National Academies Press, 2006), pp. 341–349.
35. Moshfegh and coauthors, 2009.
36. J. R. Hunt, C. A. Zito, and L. K. Johnson, Body iron excretion by healthy men and women, *American Journal of Clinical Nutrition* 89 (2009): 1792–1798.
37. M. Wessling-Resnick, Iron homeostasis and the inflammatory response, *Annual Review of Nutrition* 30 (2010): 105–122; Q. Liu and coauthors, Role of iron deficiency and overload in the pathogenesis of diabetes and diabetic complications, *Current Medicinal Chemistry* 16 (2009): 113–129.
38. G. J. Anderson and F. Wang, Essential but toxic: Controlling the flux of iron in the body, *Clinical and Experimental Pharmacology and Physiology* 39 (2012): 719–724.
39. M. D. Knutson, Iron-sensing proteins that regulate hepcidin and enteric iron absorption, *Annual Review of Nutrition* 30 (2010): 149–171.
40. Wessling-Resnick, 2010; Knutson, 2010.
41. L. Tussing-Humphreys and coauthors, Rethinking iron regulation and assessment in iron deficiency, anemia of chronic disease, and obesity: Introducing hepcidin, *Journal of the Academy of Nutrition and Dietetics* 112 (2012): 391–400; Knutson, 2010; M. D. Knutson, Into the matrix: Regulation of the iron regulatory hormone hepcidin by matriptase-2, *Nutrition Reviews* 67 (2009): 284–288; M. F. Young and coauthors, Serum hepcidin is significantly associated with iron absorption from food and supplemental sources in healthy young women, *American Journal of Clinical Nutrition* 89 (2009): 533–538.
42. T. Ganz, Hepcidin and iron regulation: 10 years later, *Blood* 117 (2011): 4425–4433; Knutson, 2010.
43. Knutson, 2010.
44. S. R. Lynch, Why nutritional iron deficiency persists as a worldwide problem, *Journal of Nutrition* 141 (2011): 763S–768S; *Worldwide prevalence of anaemia 1993–2005: WHO Global Database on Anaemia*, published 2008, available at www.who.org.
45. M. E. Cogswell and coauthors, Assessment of iron deficiency in US preschool children and nonpregnant females of childbearing age: National Health and Nutrition Examination survey 2003–2006, *American Journal of Clinical Nutrition* 89 (2009): 1334–1342.
46. Tussing-Humphreys and coauthors, 2012; A. C. Cepeda-Lopez and coauthors, Sharply higher rates of iron deficiency in obese Mexican women

and children are predicted by obesity-related inflammation rather than by differences in dietary iron intake, *American Journal of Clinical Nutrition* 93 (2011): 975–983; L. M. Tussing-Humphreys and coauthors, Excess adiposity, inflammations, and iron-deficiency in female adolescents, *Journal of the American Dietetic Association* 109 (2009): 297–302.

47. Tussing-Humphreys and coauthors, 2012; Wessling-Resnick, 2010; J. P. McClung and J. P. Karl, Iron deficiency and obesity: The contribution of inflammation and diminished iron absorption, *Nutrition Reviews* 67 (2009): 100–104; E. M. del Giudice and coauthors, Hepcidin in obese children as a potential mediator of the association between obesity and iron deficiency, *Journal of Endocrinology and Metabolism* 94 (2009): 5102–5107.
48. N. Milman, Anemia: Still a major health problem in many parts of the world, *Annals of Hematology* 90 (2011): 369–377.
49. L. E. Murray-Kolb, Iron status and neuropsychological consequences in women of reproductive age: What do we know and where are we headed? *Journal of Nutrition* 141 (2011): 747S–755S; K. Kordas, Iron, lead, and children's behavior and cognition, *Annual Review of Nutrition* 30 (2010): 123–148.
50. B. Lozoff, Early iron deficiency has brain and behavior effects consistent with dopaminergic dysfunction, *Journal of Nutrition* 141 (2011): 740S–746S.
51. S. L. Young, Pica in pregnancy: New ideas about an old condition, *Annual Review of Nutrition* 30 (2010): 403–422.
52. P. Brissot and coauthors, Molecular diagnosis of genetic iron-overload disorders, *Expert Review of Molecular Diagnostics* 10 (2010): 755–763.
53. G. A. Ramm and R. G. Ruddell, Iron homeostasis, hepatocellular injury, and fibrogenesis in hemochromatosis: The role of inflammation in a noninflammatory liver disease, *Seminars in Liver Disease* 30 (2010): 271–287; S. Lekawanvijit and N. Chattipakorn, Iron overload thalassemic cardiomyopathy: Iron status assessment and mechanisms of mechanical and electrical disturbance due to iron toxicity, *Canadian Journal of Cardiology* 25 (2009): 213–218.
54. Wessling-Resnick, 2010.
55. A. C. Bronstein and coauthors, 2009 Annual Report of the American Association of Poison Control Centers' National Poison Data System (NPDS): 27th Annual Report, *Clinical Toxicology* 28 (2010): 979–1178.
56. Standing Committee on the Scientific Evaluation of Dietary Reference Intakes, Food and Nutrition Board, National Institute of Health, *Dietary Reference Intakes for Vitamin A, Vitamin K, Arsenic, Boron, Chromium, Copper, Iodine, Iron, Manganese, Molybdenum, Nickel, Silicon, Vanadium, and Zinc* (Washington, DC: National Academies Press, 2001), pp. 311–316.
57. J. C. King, Zinc: An essential but elusive nutrient, *American Journal of Clinical Nutrition* 94 (2011): 679S–684S.
58. S. G. Bell and B. L. Vallee, The metallothionein/thionein system: An oxidoreductive metabolic zinc link, *ChemBioChem* 10 (2009): 55–62; Y. Song and coauthors, Zinc deficiency affects DNA damage, oxidative stress, antioxidant defenses, and DNA repair in rats, *Journal of Nutrition* 139 (2009): 1626–1631.
59. H. Haase and L. Rink, Functional significance of zinc-related signaling pathways in immune cells, *Annual Review of Nutrition* 29 (2009): 133–152; G. A. Kandhro and coauthors, Effect of zinc supplementation on the zinc level in serum and urine and their relation to thyroid hormone profile in male and female goitrous patients, *Clinical Nutrition* 28 (2009): 162–168.
60. S. Y. Hess and J. C. King, Effects of maternal zinc supplementation on pregnancy and lactation outcomes, *Food and Nutrition Bulletin* 30 (2009): S60–S78.
61. C. L. Fischer Walker, M. Ezzati, and R. E. Black, Global and regional child mortality and burden of disease attributable to zinc deficiency, *European Journal of Clinical Nutrition* 63 (2009): 591–597.
62. K. M. Hambidge and coauthors, Zinc bioavailability and homeostasis, *American Journal of Clinical Nutrition* 91 (2010): 1478S–1483S.
63. Standing Committee on the Scientific Evaluation of Dietary Reference Intakes, 2001, pp. 479–480.
64. Standing Committee on the Scientific Evaluation of Dietary Reference Intakes, 2001, p. 442.
65. M. Science and coauthors, Zinc for the treatment of the common cold: A systematic review and meta-analysis of randomized controlled trials, *Canadian Medical Association* 184 (2012): E551–E561; FDA, Warnings on three Zicam intranasal zinc products, *For Consumers,* June 2009, available at www.fda.gov/forconsumers/ConsumerUpdates/ucm166931.htm.
66. F. P. Bellinger and coauthors, Regulation and function of selenoproteins in human disease, *Biochemical Journal* 422 (2009): 11–22.
67. D. L. St. Germain, V. A. Galton, and A. Hernandez, Minireview: Defining the roles of the iodothyronine deiodinases: Current concepts and challenges, *Endocrinology* 150 (2009): 1097–1107.
68. R. Hurst and coauthors, Selenium and prostate cancer: Systematic review and meta-analysis, *American Journal of Clinical Nutrition* 96 (2012): 111–122; M. P. Rayman, Selenium and human health, *Lancet* 379 (2012): 1256–1268; C. D. Davis, R. A. Tsuji, and J. A. Milner, Selenoproteins and cancer prevention, *Annual Review of Nutrition* 32 (2012): 73–95; S. M. Lippman and coauthors, Effect of selenium and vitamin E on risk of prostate cancer and other cancers: The Selenium and Vitamin E Cancer Prevention Trial (SELECT), *Journal of the American Medical Association* 301 (2009): 39–51; N. Facompre and K. El-Bayoumy, Potential stages for prostate cancer prevention with selenium: Implications for cancer survivors, *Cancer Research* 69 (2009): 2699–2703.
69. Hurst and coauthors, 2012; Rayman, 2012; Davis, Tsuji, and Milner, 2012.
70. Hurst and coauthors, 2012; N. Facompre and K. El-Bayoumy, Potential stages for prostate cancer prevention with selenium: Implications for cancer survivors, *Cancer Research* 69 (2009): 2699–2703.
71. B. M. Aldosary and coauthors, Case series of selenium toxicity from a nutritional supplement, *Clinical Toxicology* 50 (2012): 57–64.
72. R. C. Gordon and coauthors, Iodine supplementation improves cognition in mildly iodine-deficient children, *American Journal of Clinical Nutrition* 90 (2009): 1264–1271.
73. GAIN-UNICEF Universal Salt Iodization Partnership Program, 2011, available at www.gainhealth.gov/programs/usi.
74. M. B. Zimmermann, Iodine deficiency in pregnancy and the effects of maternal iodine supplementation on the offspring: A review, *American Journal of Clinical Nutrition* 89 (2009): 668S–672S.
75. Standing Committee on the Scientific Evaluation of Dietary Reference Intakes, 2001, pp. 258–290.
76. Centers for Disease Control and Prevention, *Second National Report on Biochemical Indicators of Diet and Nutrition in the U.S. Population, 2012, Executive Summary*, p. 9, available at www.cdc.gov/nutritionreport.
77. Standing Committee on the Scientific Evaluation of Dietary Reference Intakes, 2001, pp. 224–257.
78. Position of the Academy of Nutrition and Dietetics, The impact of fluoride on health, *Journal of the Academy of Nutrition and Dietetics* 112 (2012): 1443–1453.
79. Position of the Academy of Nutrition and Dietetics, 2012.
80. Standing Committee on the Scientific Evaluation of Dietary Reference Intakes, 2001, pp. 197–223.
81. Z. Q. Wang and W. T. Cefalu, Current concepts about chromium supplementation in type 2 diabetes and insulin resistance, *Current Diabetes Reports* 10 (2010): 145–151.

Nutrition in Practice

Vitamin and Mineral Supplements

At least half of U.S. adults collectively spend tens of *billions* of dollars a year on dietary supplements.[1] This trend is accelerating as scientists discover more and more links between nutrition and disease prevention. Most people take a daily multivitamin and mineral pill to make up for dietary shortfalls; others take single nutrient supplements in an attempt to ward off diseases. In many cases, taking supplements is a costly but harmless practice; sometimes, it is both costly and harmful to health.[2] The main message of this Nutrition in Practice is that most healthy people can get the nutrients they need from foods. Supplements cannot substitute for a healthy diet. For some people, however, certain nutrient supplements may be desirable. In some cases, they can correct deficiencies; in others, they can reduce the risks of disease.

Do foods really contain enough vitamins and minerals to supply all that most people need?

Emphatically, yes, for both healthy adults and children who choose a variety of foods. The USDA Food Patterns described in Chapter 1 are the guide to follow to achieve adequate intakes. People who meet their nutrient needs from foods, rather than supplements, have little risk of deficiency or toxicity.

Do some people need supplements?

Yes, some people may suffer marginal nutrient deficiencies due to illness, alcohol or drug addiction, or other conditions that limit food intake.[3] People who may benefit from nutrient supplements in amounts consistent with the RDA include the following:

- People with specific nutrient deficiencies need specific nutrient supplements.
- People whose energy intakes are particularly low (fewer than 1600 kcalories per day) need multivitamin and mineral supplements.
- Vegetarians who eat all-plant diets (vegans) and older adults with atrophic gastritis need vitamin B_{12}.
- People who have lactose intolerance or milk allergies or who otherwise do not consume enough milk products to forestall extensive bone loss need calcium.
- People in certain stages of the life cycle who have increased nutrient requirements need specific nutrient supplements. For example, infants may need vitamin D, iron, and fluoride; women who are capable of becoming pregnant and pregnant women need folate and iron; and the elderly may benefit from some of the vitamins and minerals in a balanced supplement (they may choose poor diets, have trouble chewing, or absorb or metabolize nutrients less efficiently; see Chapter 12).
- People who have inadequate intakes of milk or milk products, limited sun exposure, or heavily pigmented skin may need vitamin D.
- People who have diseases, infections, or injuries or who have undergone surgery that interferes with the intake, absorption, metabolism, or excretion of nutrients may need specific nutrient supplements. (The increased metabolic needs associated with these severe stresses are discussed in Chapter 16.)
- People taking medications that interfere with the body's use of specific nutrients may need specific nutrient supplements.

© Tom Carter/PhotoEdit

Except for people in these circumstances, most adults can normally get all the nutrients they need by eating a varied diet of nutrient-dense foods. Whenever a health care professional finds a person's diet inadequate, the right corrective step is to improve the person's food choices and eating patterns, not to begin supplementation.

Why do so many people take supplements?

People frequently take supplements for mistaken reasons, such as "They give me energy" or "They make me strong." Other invalid reasons why people may take supplements include:

- Their feeling of insecurity about the nutrient content of the food supply.
- Their belief that extra vitamins and minerals will help them cope with stress.
- Their belief that supplements can enhance athletic performance or build lean body tissue without physical work.
- Their desire to prevent, treat, or cure symptoms or diseases ranging from the common cold to cancer.

In study after study, well-nourished people are the ones found to be taking supplements, adding greater amounts of nutrients to already sufficient intakes.[4] People with low nutrient intakes from food generally do not take supplements. In addition, little relationship exists between the nutrients people need and the ones they take in supplements. In fact, an argument against supplements is that they may lull people into a false sense of security. A person might eat irresponsibly, thinking, "My supplement will cover my needs."

Are there other arguments against taking supplements?

Yes, there are several arguments against taking supplements. First, foods rarely cause nutrient imbalances or toxicities, but supplements can. The higher the dose, the greater the risk of harm. People's tolerances for high doses of nutrients vary, just as their risks of deficiencies do. Amounts that some can tolerate may be harmful for others, and no one knows who falls where along the spectrum. The Tolerable Upper Intake Levels of the DRI (see inside front cover) answer the question "How much is too much?" by defining the highest amount that appears safe for *most* healthy people. A few sensitive people may experience toxicities at lower doses, however.

Second, supplement users are more likely to have excessive intakes of certain nutrients—notably iron, zinc, vitamin A, and niacin. The true extent of supplement toxicity in this country is unknown, but many adverse effects are reported each year from vitamins, minerals, essential oils, herbs, and other supplements.[5] Only a few alert health care professionals can recognize toxicity, even when it is acute. When it is chronic, with the effects developing subtly and progressing slowly, it often goes unrecognized. In view of the potential hazards, some authorities believe supplements should bear warning labels, advising consumers that large doses may be toxic.

Toxic overdoses of vitamins and minerals in children are more readily recognized and, unfortunately, fairly common. Fruit-flavored, chewable vitamins shaped like cartoon characters entice young children to eat them like candy in amounts that can cause poisoning. High-potency iron supplements (30 milligrams of iron or more per tablet) are especially toxic and are the leading cause of accidental ingestion fatalities among children.

Third, FDA recently identified more than 140 dietary supplements sold on the U.S. market that were contaminated with pharmaceutical drugs, such as steroid hormones and stimulants. Toxic plant material, toxic heavy metals, bacteria, and other substances have also shown up in a wide variety of dietary supplements. Even some children's chewable vitamins have contained appreciable lead, a destructive heavy metal for young children.[6] Even though hazardous products are quickly removed from the market upon discovery, many others remain on store shelves because current regulations make the supplement market difficult to monitor and control.[7] Plain multivitamin and mineral supplements from reputable sources, without herbs or add-ons, often test free from contamination.

A fourth argument against taking supplements arises when people who are ill come to believe that high doses of vitamins or minerals can be therapeutic. Not only can high doses be toxic, but the person may take them instead of seeking medical help.

A final argument against taking supplements is that the body absorbs nutrients best from foods that dilute and disperse them among other substances to facilitate their absorption and use by the body.[8] Taken in pure, concentrated form, nutrients are likely to interfere with one another's absorption or with the absorption of other nutrients from foods eaten at the same time. Such effects are particularly well known among the minerals. For example, zinc hinders copper and calcium absorption, iron hinders zinc absorption, and calcium hinders magnesium and iron absorption. Among vitamins, vitamin C supplements *enhance* iron absorption, making iron overload likely in susceptible people. These and other interactions present drawbacks to supplement use.

Do antioxidant supplements prevent cancer?

Again, it is better advice to eat a nutrient-dense diet. Some evidence exists linking high intakes of antioxidant-rich fruits and vegetables with good health and disease prevention.[9] More than 200 population studies have examined the effects of fruits and vegetables on cancer risk, and many show that people who eat more of these foods are less likely to develop certain cancers. Findings from other types of studies, such as intervention studies and clinical trials, however, show weaker associations between fruit and vegetable intake and reduced risk of cancer. Some researchers speculate that fruit and vegetable intake may play a smaller role in total cancer protection than previously thought.[10] However, some fruits and vegetables do contain protective factors for specific cancers. For example, research suggests dietary fiber protects against colorectal cancer.[11] When research combines many different fruits and vegetables, specific protective factors may not stand out. Many experts agree that the antioxidant vitamins in these foods are probably important protective factors, but they also note that other constituents of fruits and vegetables (see Table NP8-1 on p. 226–227) certainly have not been ruled out as contributing factors.

The way to apply this information is to eat nutritious foods. Taking antioxidant supplements instead of making needed lifestyle changes may sound appealing, but evidence does not support a role for supplements against chronic diseases.[12] In some cases, supplements may even be harmful.[13]

When a person needs a vitamin-mineral supplement, what kind should be used?

Take your health care professional's advice, if it is offered. If you are selecting a supplement yourself, a single, balanced vitamin-mineral supplement with no added extras such as herbs should suffice. Choose the kind that provides all the nutrients in amounts less than, equal to, or very close to the RDA (remember, you get some nutrients from foods). For those who require a higher dose, such as young women who need supplemental folate in the childbearing years, choose a supplement with just the needed nutrient or in combination with a reasonable dose of others. Avoid any preparations that, in a daily dose, provide more than the recommended intake of vitamin A, vitamin D, or any mineral or more than the Tolerable Upper Intake Level for any nutrient. In addition, avoid the following:

- High doses of iron (more than 10 milligrams per day), except for menstruating women. People who menstruate need more iron, but people who don't, don't.
- "Organic" or "natural" preparations with added substances. They are no better than standard types, but they cost more.
- "High-potency" or "therapeutic dose" supplements. More is not better.
- Substances not needed in human nutrition such as inositol and carnitine. These particular ingredients won't harm you, but they reveal a marketing strategy that makes the whole mix suspect.

As for price, be aware that local or store brands may be just as good as or better than nationally advertised brands. If they are less expensive, it may be because the price does not have to cover the cost of national advertising. Finally, be aware that if you see a USP symbol on the label, it means that the manufacturer has voluntarily paid an independent laboratory to test the product and affirm that it contains the ingredients listed and will dissolve or disintegrate in the digestive tract to make the ingredients available for absorption. The symbol does not imply that the supplement has been tested for safety or effectiveness with regard to health, however.

Can supplement labels help consumers make informed choices?

Yes, to some extent. To enable consumers to make more informed choices about nutrient supplements, the FDA, with the encouragement of the Academy of Nutrition and Dietetics, published labeling regulations for supplements. The Dietary Supplement Health and Education Act subjects supplements to the same general labeling requirements that apply to foods. Specifically:

This symbol means that a supplement contains the nutrients stated and that it will dissolve in the digestive system—the symbol does not guarantee safety or health advantages.

Credit to come

- Nutrition labeling for dietary supplements is required. The nutrition panel on supplements is called "Supplement Facts" (see Figure NP9-1). The Supplement Facts panel lists the quantity and the percentage of the Daily Value for each nutrient in the supplement. Ingredients that have no Daily Value—for example, sugars and gelatin—appear in a list below the Supplement Facts panel.
- Labels may make nutrient claims (such as "high" or "low") according to specific criteria (for example, "an excellent source of vitamin C").
- Labels may make health claims that are supported by significant scientific agreement and are not brand specific (for example, "folate protects against neural tube defects").

FIGURE NP9-1 An Example of a Supplement Label

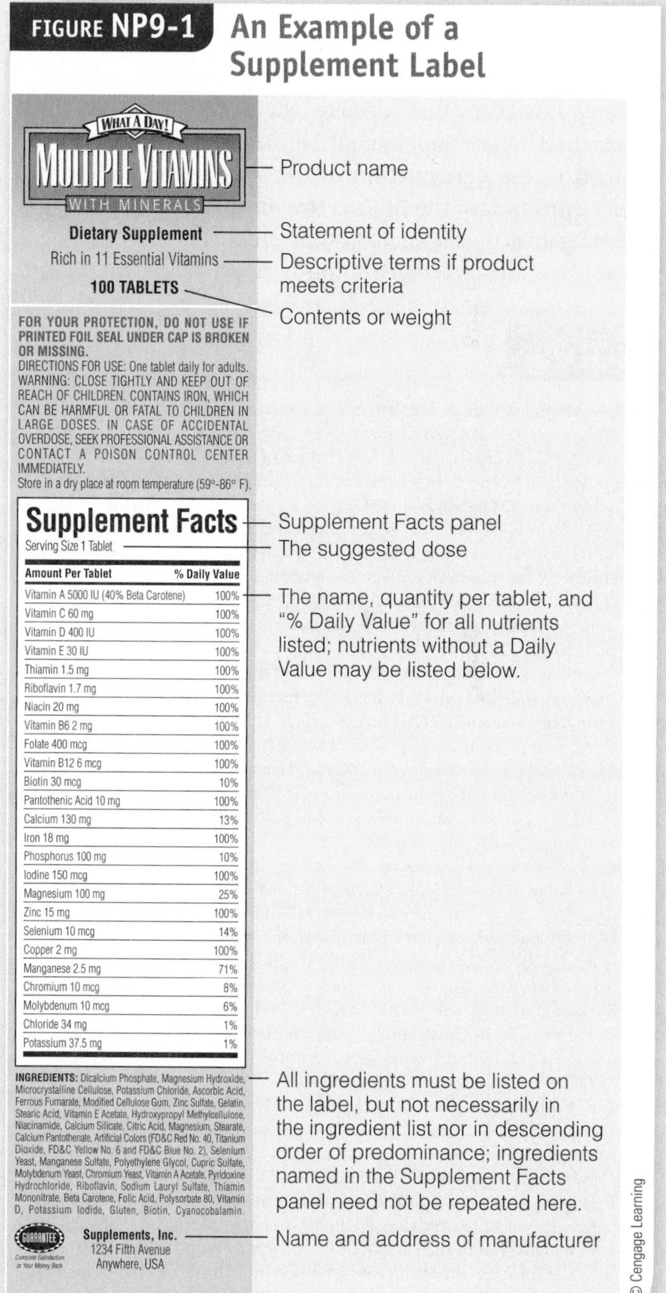

- Labels may claim that the lack of a nutrient can cause a deficiency disease, but if they do, they must also include the prevalence of that deficiency disease in the United States.
- Labels may claim to diagnose, treat, cure, or relieve common complaints such as menstrual cramps or memory loss, but they may *not* make claims about specific diseases (except as noted previously).
- Labels may make structure-function claims (see Chapter 1) about the role a nutrient plays in the body, explain how the nutrient performs its function, and indicate that consuming the nutrient is associated with general well-being. These claims must be accompanied by an FDA disclaimer statement: "This statement has not been evaluated by the Food and Drug Administration. This product is not intended to diagnose, treat, cure, or prevent any disease."

Note, however, that despite these requirements, in effect, the Dietary Supplement Health and Education Act resulted in the deregulation of the supplement industry. Under current law, the FDA is responsible only for taking action against unsafe dietary supplements already on the market. No advance registration or approval by the FDA is needed before a manufacturer can put a supplement on store shelves.[14] To act against unsafe supplements, the FDA must receive manufacturers' reports concerning serious adverse health effects reported to them by consumers. Manufacturers do list contact information on supplement labels for this purpose, but many symptoms of adverse reactions are easily mistaken for something else—stomach flu or headache or fatigue. Consumers can also report adverse reactions from supplements directly to the FDA via its hotline or website, but most people are unaware of these options (see the note on this page).*

People in developed nations are far more likely to suffer from *overnutrition* and poor lifestyle choices than from nutrient deficiencies. People wish that swallowing vitamin pills would boost their health. The truth—that they need to improve their eating and exercise habits—is harder to swallow.

Don't waste time and money trying to single out a few nutrients to take as supplements. Invest energy in eating a wide variety of fruits and vegetables in generous quantities, along with the recommended daily amounts of whole grains, lean meats, and milk products every day, and take supplements only when they are truly needed.

Notes

1. J. Gajcje and coauthors, Dietary supplement use among U.S. adults has increased since NHANES III (1988–1994), *National Center for Health Statistics: Data Brief* 61 (2011): 1–8; Position of the American Dietetic Association: Nutrient Supplementation, *Journal of the American Dietetic Association* 109 (2009): 2073–2085.
2. D. B. McCormick, Vitamin/mineral supplements: Of questionable benefit for the general population, *Nutrition Reviews* 68 (2010): 207–213.
3. Position of the American Dietetic Association, 2009.
4. R. L. Bailey and coauthors, Examination of vitamin intakes among US adults by dietary supplement use, *Journal of the Academy of Nutrition and Dietetics* 112 (2012): 657–663;Y. A. Shakur and coauthors, A comparison of micronutrient inadequacy and risk of high micronutrient intakes among vitamin and mineral supplement users and nonusers in Canada, *Journal of Nutrition* 142 (2012): 534–540; R. L. Bailey and coauthors, Dietary supplement use is associated with higher intakes of minerals from food sources, *American Journal of Clinical Nutrition* 94 (2011): 1376–1381; R. M. Bliss, Monitoring the population's food and supplement intakes, March 1, 2012, available at www.ars.usda.gov/is/pr/2012/120301.htm; McCormick, 2010.
5. A. C. Bronstein and coauthors, 2007 Annual report of the American Association of Poison Control Centers' national poisoning and exposure database, *Clinical Toxicology* 46 (2008): 927–1057.
6. W. R. Mindak and coauthors, Lead in women's and children's vitamins, *Journal of Agricultural and Food Chemistry* 56 (2008): 6892–6896.
7. P. A. Cohen, American roulette—Contaminated dietary supplements, *New England Journal of Medicine* 361 (2009): 1523–1525.
8. D. R. Jacobs, M. D. Gross, and L. C. Tapsell, Food synergy: An operational concept for understanding nutrition, *American Journal of Clinical Nutrition* 89 (2009): 1543S–1548S.
9. L. H. Kushi and coauthors, American Cancer Society guidelines on nutrition and physical activity for cancer prevention: Reducing the risk of cancer with healthy food choices and physical activity, *CA: Cancer Journal for Clinicians* 62 (2012): 30–67; L. Djousse, J. A. Driver, and J. M. Gazianao, Relation between modifiable lifestyle factors and lifetime risk of heart failure, *Journal of the American Medical Association* 302 (2009): 394–400; F. J. B. van Duijnhoven and coauthors, Fruit, vegetables, and colorectal cancer risk: The European Prospective Investigation into Cancer and Nutrition, *American Journal of Clinical Nutrition* 89 (2009): 1441–1452.
10. P. Boffetta and coauthors, Fruit and vegetable intake and overall cancer risk in the European Prospective Investigation into Cancer and Nutrition (EPIC), *Journal of the National Cancer Institute* 102 (2010): 529–537; T. J. Key, Fruit and vegetables and cancer risk, *British Journal of Cancer* 104 (2010): 6–11.
11. D. Aune and coauthors, Dietary fibre, whole grains, and risk of colorectal cancer: Systematic review and dose-response meta-analysis of prospective studies, *British Medical Journal* 343 (2011): d6617; C. C. Dahm and coauthors, Dietary fiber and colorectal cancer risk: A nested case-control study using food diaries, *Journal of the National Cancer Institute* 102 (2010): 614–626.
12. M. E. Martinez and coauthors, Dietary supplements and cancer prevention: Balancing potential benefits against proven harms, *Journal of the National Cancer Institute* 104 (2012): 732–739; M. G. O'Doherty and coauthors, Effect of supplementation with B vitamins and antioxidants on levels of asymmetric dimethylarginine (ADMA) and C-reactive protein (CRP): A double-blind, randomized, factorial design, placebo-controlled trial, *European Journal of Nutrition* 49 (2010): 483–492; G. J. Hankey and VITATOPS Trial Study Group, B vitamins in patients with recent transient ischaemic attack or stroke in the VITAmins TO Prevent Stroke (VITATOPS) trial: A randomized, double-blind, parallel, placebo-controlled trial, *The Lancet Neurology* 9 (2010): 855–865; S. Czernichow and coauthors, Effects of long-term antioxidant supplementation and association of serum antioxidant concentrations with risk of metabolic syndrome in adults, *American Journal of Clinical Nutrition* 90 (2009): 329–335; A. M. Hill, J. A. Fleming, and P. M. Kris-Etherton, The role of diet and nutritional supplements in preventing and treating cardiovascular disease, *Current Opinion in Cardiology* 24 (2009): 433–441; M. L. Neuhouser and coauthors, Multivitamin use and risk of cancer and cardiovascular disease in the Women's Health Initiative Cohorts, *Archives of Internal Medicine* 169 (2009): 294–304; G. Pocobelli and coauthors, Use of supplements of multivitamins, vitamin C, and vitamin E in relation to mortality, *American Journal of Epidemiology* 170 (2009): 472–483.
13. M. P. Rayman, Selenium and human health, *Lancet* 379 (2012): 1256–1268; G. Bjelakovic and C. Gluud, Vitamin and mineral supplement use in relation to all-cause mortality in the Iowa Women's Health Study, *Archives of Internal Medicine* 171 (2011): 1633–1634.
14. U.S. Food and Drug Administration, Center for Food Safety and Applied Nutrition, *Dietary Supplements*, retrieved from www.fda.gov/Food/DietarySupplements/default.htm, site updated October 9, 2012.

*Consumers should report suspected harm from dietary supplements to their health providers or to the FDA's MedWatch program at (800) FDA-1088 or on the Internet at www.fda.gov/medwatch.

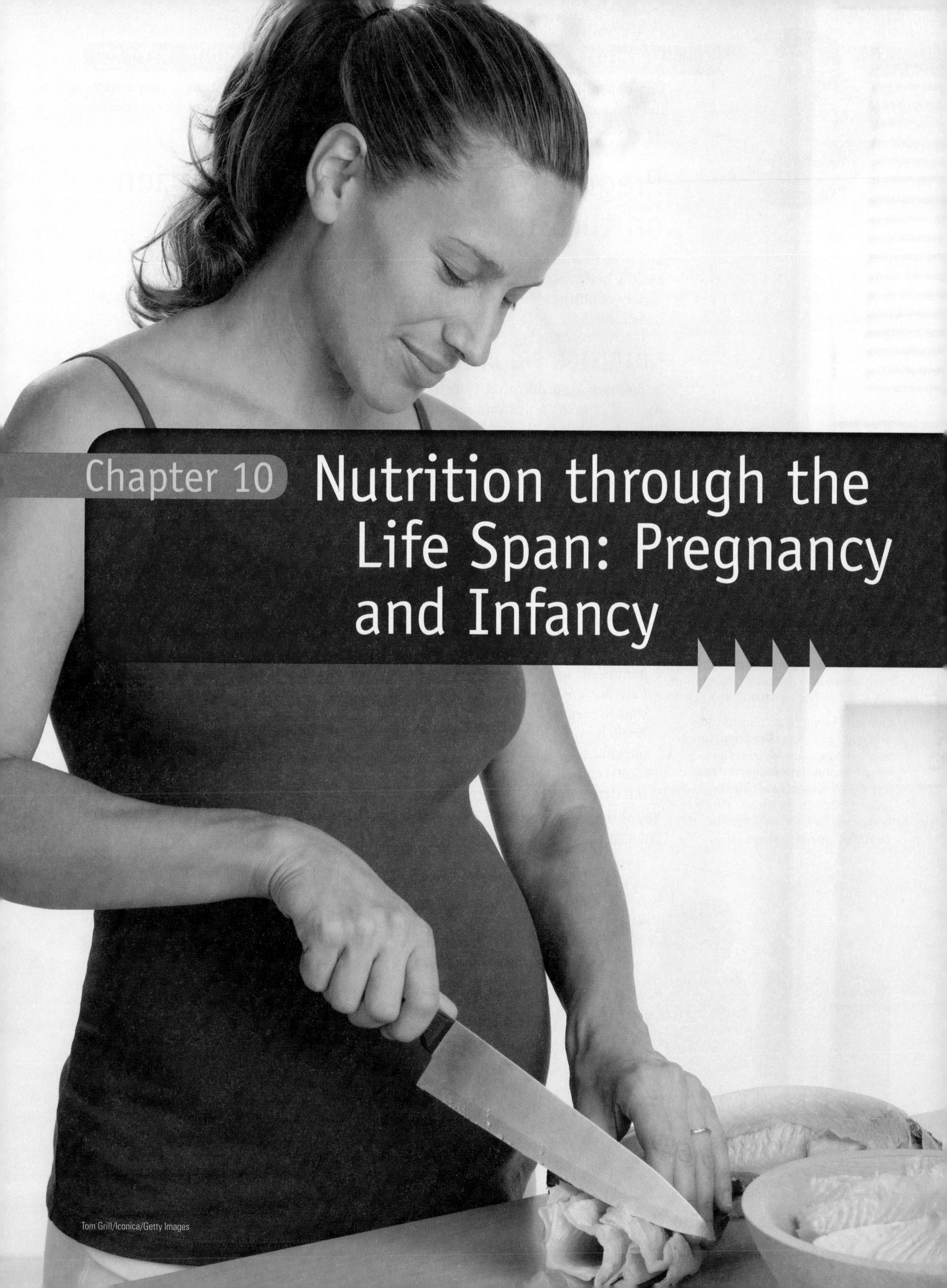

Chapter 10 Nutrition through the Life Span: Pregnancy and Infancy

Tom Grill/Iconica/Getty Images

ALL PEOPLE NEED THE SAME NUTRIENTS, BUT THE AMOUNTS THEY NEED vary depending on their stage of life. This chapter focuses on nutrition in preparation for, and support of, pregnancy, lactation, and infancy. The next two chapters address the needs of children, adolescents, and older adults.

Pregnancy: The Impact of Nutrition on the Future

The woman who enters pregnancy with full nutrient stores, sound eating habits, and a healthy body weight has done much to ensure an optimal pregnancy. Then, if she eats a variety of nutrient-dense foods during pregnancy, her own and her infant's health will benefit further.

NUTRITION PRIOR TO PREGNANCY

A discussion on nutrition prior to pregnancy must, by its nature, focus mainly on women. A man's nutrition may affect his **fertility** and possibly the genetic contributions he makes to his children, but nutrition exerts its primary influence through the woman. Her body provides the environment for the growth and development of a new human being. Full nutrient stores *before* pregnancy are essential both to conception and to healthy infant development during pregnancy. In the early weeks of pregnancy, before many women are even aware that they are pregnant, significant developmental changes occur that depend on a woman's nutrient stores. In preparation for a healthy pregnancy, a woman can establish the following habits:

- *Achieve and maintain a healthy body weight.* Both underweight and overweight women, and their newborns, face increased risks of complications.
- *Choose an adequate and balanced diet.* Malnutrition reduces fertility and impairs the early development of an infant should a woman become pregnant.
- *Be physically active.* A woman who wants to be physically active *when* she is pregnant needs to become physically active *beforehand.*
- *Receive regular medical care.* Regular health care visits can help ensure a healthy start to pregnancy.
- *Avoid harmful influences.* Both maternal and paternal ingestion of, or exposure to, harmful substances (such as cigarettes, alcohol, drugs, or environmental contaminants) can cause miscarriage or abnormalities, alter genes or their expression, and interfere with fertility.[1]

Young adults who nourish and protect their bodies do so not only for their own sakes but also for future generations.

fertility: the capacity of a woman to produce a normal ovum periodically and of a man to produce normal sperm; the ability to reproduce.

low birthweight (LBW): a birthweight less than 5½ lb (2500 g); indicates probable poor health in the newborn and poor nutrition status of the mother during pregnancy. Optimal birthweight for a full-term infant is 6.8 to 7.9 lb (about 3100 to 3600 g).

Low-birthweight infants are of two different types. Some are **premature;** they are born early and are of a weight **appropriate for gestational age (AGA).** Others have suffered growth failure in the uterus; they may or may not be born early, but they are **small for gestational age (SGA).**

© Stockbytes/Jupiter Images

Both parents can prepare in advance for a healthy pregnancy.

PREPREGNANCY WEIGHT

Appropriate weight prior to pregnancy benefits pregnancy outcome. Being either underweight or overweight (see p. 281) presents medical risks during pregnancy and childbirth. Underweight women are therefore advised to gain weight before becoming pregnant and overweight women to lose excess weight.

Underweight Infant birthweight correlates with prepregnancy weight and weight gain during pregnancy and is the most potent single predictor of the infant's future health and survival. An underweight woman has a high risk of having a **low-birthweight** infant, especially if she is unable to gain sufficient weight during pregnancy.[2] Compared with normal-weight infants, low-birthweight infants are more likely to contract diseases and nearly 40 times more likely to die in the first month of life. Impaired growth and development during pregnancy may have long-term health effects as well. Research suggests that, when nutrient supplies fail to meet demands, permanent adaptations take place that may make obesity or chronic diseases such as heart disease and hypertension more likely in later life.[3] Other potential problems of low birthweight may include lower adult IQ and other brain impairments, short stature, and educational disadvantages.[4] Underweight women are therefore advised to gain weight before becoming pregnant and to strive to gain adequately during pregnancy.

Nutritional deficiency, coupled with low birthweight, is the underlying cause of more than half of all the deaths worldwide of children younger than five years of age. In the United States, the infant mortality rate in 2009 was slightly less than 6.5 deaths per 1000 live births.[5] This rate, though higher than that of some other developed countries, has seen a significant steady decline over the past two decades and stands as a tribute to public health efforts aimed at reducing infant deaths (see Figure 10-1).

Not all cases of low birthweight reflect poor nutrition. Heredity, disease conditions, smoking, and drug use (including alcohol) during pregnancy all contribute.[6] Even with optimal nutrition and health during pregnancy, some women give birth to small infants for unknown reasons. But poor nutrition is the major factor in low birthweight—and an avoidable one, as later sections make clear.[7]

FIGURE 10-1 Infant Mortality Decline over Time

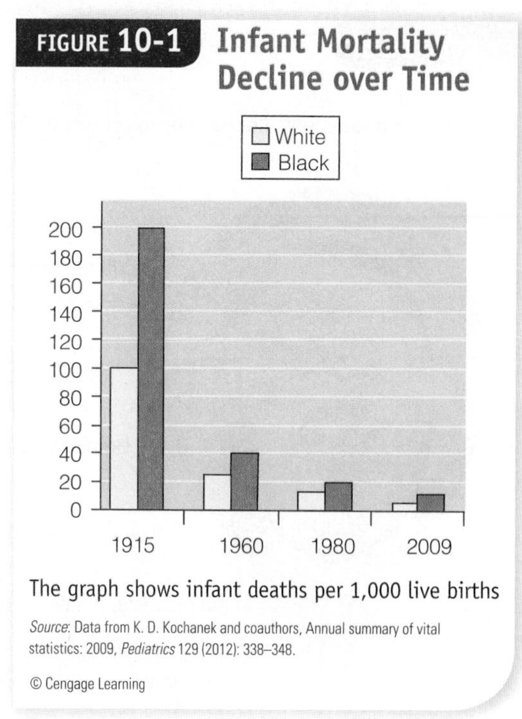

The graph shows infant deaths per 1,000 live births

Source: Data from K. D. Kochanek and coauthors, Annual summary of vital statistics: 2009, *Pediatrics* 129 (2012): 338–348.

Overweight and Obesity

Obese women are also urged to strive for healthy weights before pregnancy. Infants born to obese women are more likely to be large for gestational age, weighing more than 9 pounds.[8] Problems associated with a high birthweight include increases in the likelihood of a difficult labor and delivery, birth trauma, and **cesarean section**.[9] Consequently, these infants have a greater risk of poor health and death than infants of normal weight. Infants of obese mothers may be twice as likely to be born with a neural tube defect, too. Folate's role has been examined, but a more likely explanation seems to be poor glycemic control.[10] Obese women themselves are likely to suffer gestational diabetes, hypertension, and complications during and infections after the birth.[11] (■) In addition, both overweight and obese women have a greater risk of giving birth to infants with heart defects and other abnormalities.[12] An appropriate goal for the obese woman who wishes to become pregnant is to strive to attain a healthy prepregnancy body weight so as to minimize her medical risks and those of her future child.

■ Neural tube defects and gestational diabetes are discussed in later sections.

cesarean (see-ZAIR-ee-un) section: surgical childbirth, in which the infant is taken through an incision in the woman's abdomen.

uterus (YOO-ter-us): the womb, the muscular organ within which the infant develops before birth.

placenta (pla-SEN-tuh): an organ that develops inside the uterus early in pregnancy, in which maternal and fetal blood circulate in close proximity and exchange materials. The fetus receives nutrients and oxygen across the placenta; the mother's blood picks up carbon dioxide and other waste materials to be excreted via her lungs and kidneys.

gestation: the period of about 40 weeks (three trimesters) from conception to birth; the term of a pregnancy.

umbilical (um-BIL-ih-cul) cord: the ropelike structure through which the fetus's veins and arteries reach the placenta; the route of nourishment and oxygen into the fetus and the route of waste disposal from the fetus.

amniotic (am-nee-OTT-ic) sac: the "bag of waters" in the uterus in which the fetus floats.

HEALTHY SUPPORT TISSUES

A major reason that the mother's prepregnancy nutrition is so crucial is that it determines whether her **uterus** will be able to support the growth of a healthy **placenta** during the first month of **gestation**. The placenta is both a supply depot and a waste-removal system for the fetus. If the placenta works perfectly, the fetus wants for nothing; if it doesn't, no alternative source of sustenance is available, and the fetus will fail to thrive. Figure 10-2 (p. 272) shows the placenta, a mass of tissue in which maternal and fetal blood vessels intertwine and exchange materials. The two bloods never mix, but the barrier between them is notably thin. Across this thin barrier, nutrients and oxygen move from the mother's blood into the fetus's blood, and wastes move out of the fetal blood to be excreted by the mother. Thus, by way of the placenta, the mother's digestive tract, respiratory system, and kidneys serve the needs of the fetus as well as her own. The fetus has these organ systems, but they do not yet function. The **umbilical cord** is the pipeline from the placenta to the fetus. The **amniotic sac** surrounds and cradles the fetus, cushioning it with fluids.

The placenta is an active metabolic organ with many responsibilities of its own. It actively gathers up hormones, nutrients, and protein molecules such as antibodies and transfers them into the fetal bloodstream.[13] The placenta also produces a broad range

FIGURE 10-2 The Placenta

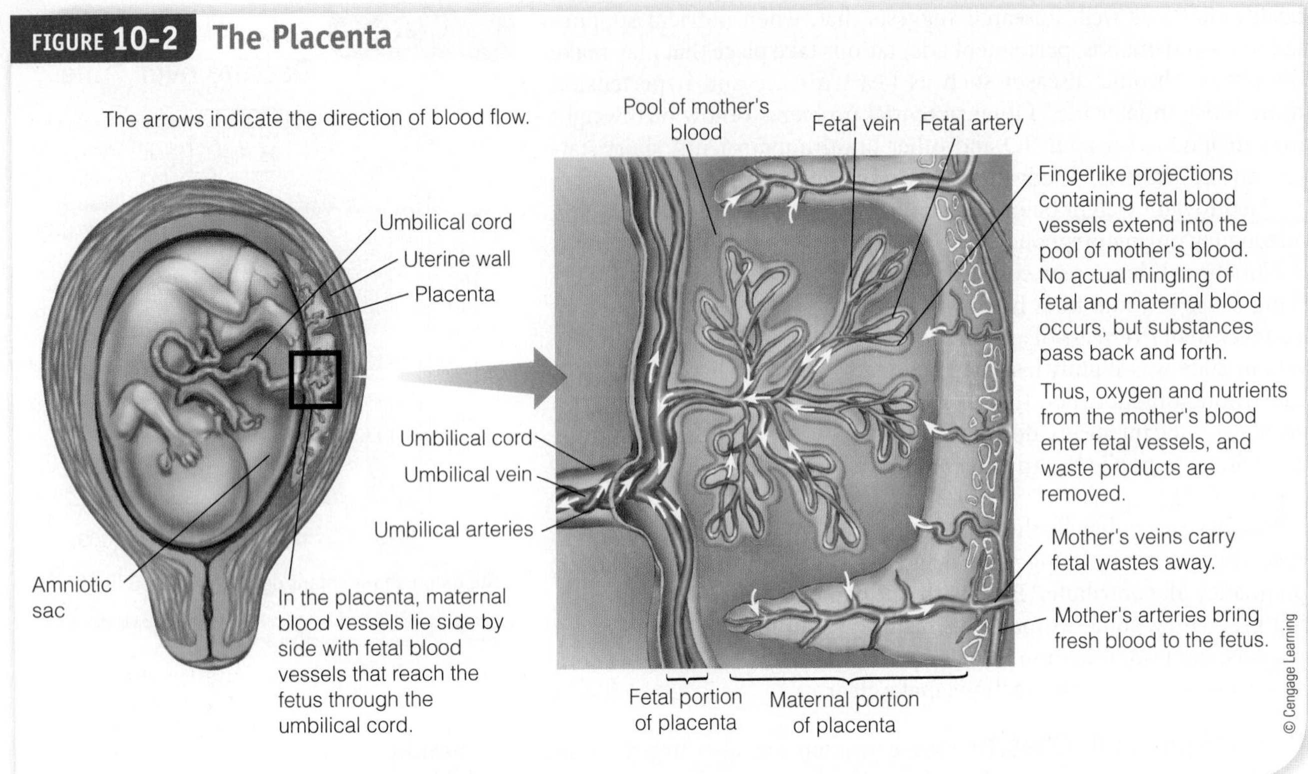

The arrows indicate the direction of blood flow.
Umbilical cord
Uterine wall
Placenta
Amniotic sac
In the placenta, maternal blood vessels lie side by side with fetal blood vessels that reach the fetus through the umbilical cord.
Pool of mother's blood
Fetal vein
Fetal artery
Fingerlike projections containing fetal blood vessels extend into the pool of mother's blood. No actual mingling of fetal and maternal blood occurs, but substances pass back and forth.
Thus, oxygen and nutrients from the mother's blood enter fetal vessels, and waste products are removed.
Mother's veins carry fetal wastes away.
Mother's arteries bring fresh blood to the fetus.
Umbilical cord
Umbilical vein
Umbilical arteries
Fetal portion of placenta
Maternal portion of placenta

of hormones that act in many ways to maintain pregnancy and prepare the mother's breasts for **lactation**. A healthy placenta is essential for the developing fetus to attain its full potential.

lactation: production and secretion of breast milk for the purpose of nourishing an infant.

ovum (OH-vum): the female reproductive cell, capable of developing into a new organism upon fertilization; commonly referred to as an egg.

zygote (ZY-goat): the product of the union of ovum and sperm; a fertilized ovum.

blastocyst (BLASS-toe-sist): the developmental stage of the zygote when it is about five days old and ready for implantation.

implantation: the stage of development in which the blastocyst embeds itself in the wall of the uterus and begins to develop; occurs during the first two weeks after conception.

embryo (EM-bree-oh): the developing infant from two to eight weeks after conception.

fetus (FEET-us): the developing infant from eight weeks after conception until its birth.

Review Notes

- Adequate nutrition before pregnancy establishes physical readiness and nutrient stores to support fetal growth.
- Both underweight and overweight women should strive for appropriate body weights before pregnancy.
- Newborns who weigh less than 5½ pounds face greater health risks than normal-weight infants.
- The healthy development of the placenta depends on adequate nutrition before pregnancy.

THE EVENTS OF PREGNANCY

The newly fertilized **ovum** is called a **zygote**. It begins as a single cell and rapidly divides to become a **blastocyst**. During the first week, the blastocyst floats down into the uterus where it will embed itself in the inner uterine wall—a process known as **implantation**. Minimal growth in size takes place at this time, but it is a crucial period in development. Adverse influences such as smoking, drug abuse, and malnutrition at this time lead to failure to implant or to abnormalities such as neural tube defects that can cause the loss of the developing embryo, possibly before the woman knows she is pregnant.

The Embryo and Fetus

During the next six weeks of development, the **embryo** registers astonishing physical changes (see Figure 10-3). At eight weeks, the **fetus** has a

FIGURE 10-3 Stages of Embryonic and Fetal Development

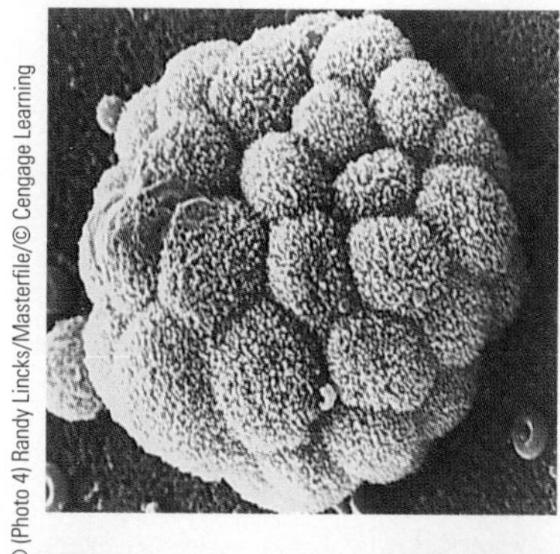

(1) A newly fertilized ovum is called a zygote and is about the size of the period at the end of this sentence. Less than one week after fertilization these cells have rapidly divided multiple times to become a blastocyst ready for implantation.

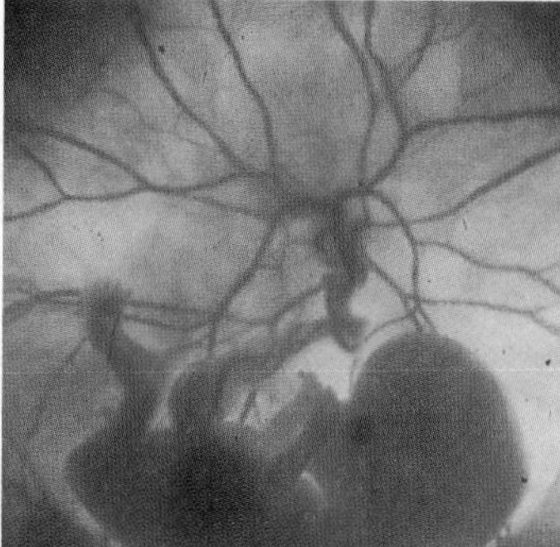

(3) A fetus after 11 weeks of development is just over an inch long. Notice the umbilical cord and blood vessels connecting the fetus with the placenta.

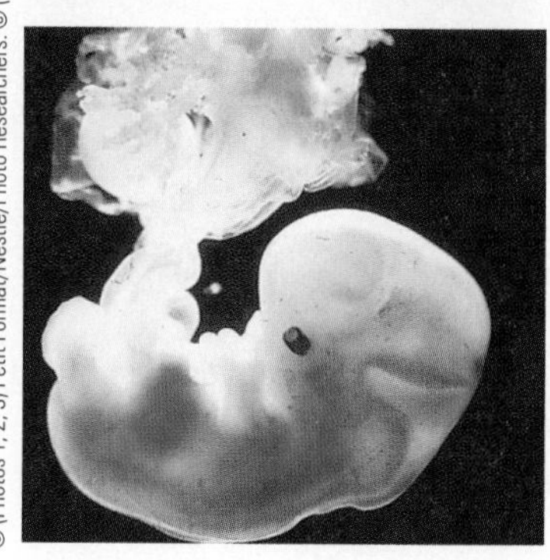

(2) After implantation, the placenta develops and begins to provide nourishment to the developing embryo. An embryo five weeks after fertilization is about 1/2 inch long.

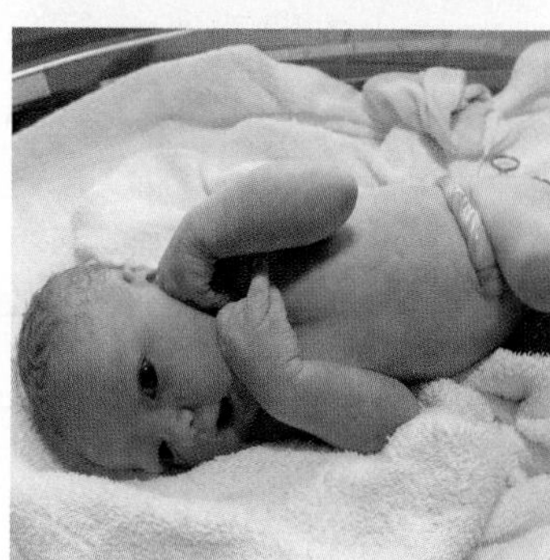

(4) A newborn infant after nine months of development measures close to 20 inches in length. The average birthweight is about 7 1/2 pounds. From eight weeks to term, this infant grew 20 times longer and 50 times heavier.

© (Photos 1, 2, 3) Petit Format/Nestle/Photo Researchers. © (Photo 4) Randy Lincks/Masterfile/© Cengage Learning

complete central nervous system, a beating heart, a fully formed digestive system, well-defined fingers and toes, and the beginnings of facial features.

In the last seven months of pregnancy, the fetal period, the fetus grows 50 times heavier and 20 times longer. Critical periods of cell division and development occur in organ after organ. Most successful pregnancies last 38 to 42 weeks and produce a healthy infant weighing between 6.8 and 7.9 pounds.[14] The 40 or so weeks of pregnancy are divided into thirds, each of which is called a **trimester**.

A Note about Critical Periods Each organ and tissue type grows with its own characteristic pattern and timing. The development of each takes place only at a certain time—the **critical period**. Whatever nutrients and other environmental conditions are necessary during this period must be supplied on time if the organ is to reach its full potential. If the development of an organ is limited during a critical period, recovery is impossible. For example, the fetus's heart and brain are well developed at 14 weeks; the lungs, 10 weeks later. Therefore, early malnutrition impairs the heart and brain; later malnutrition impairs the lungs.

The effects of malnutrition during critical periods of pregnancy are seen in defects of the nervous system of the embryo (explained later), in the child's poor dental health, and in the adolescent's and adult's vulnerability to infections and possibly higher risks of diabetes, hypertension, stroke, or heart disease.[15] The effects of malnutrition during critical periods are irreversible: abundant and nourishing food, consumed after the critical time, cannot remedy harm already done.

Table 10-1 (p. 274) identifies characteristics of a **high-risk pregnancy**. A woman with none of these factors is said to have a **low-risk pregnancy**. The more factors that apply, the higher the risk. All pregnant women, especially those in high-risk categories, need prenatal medical care, including dietary advice.

trimester: a period representing one-third of the term of gestation. A trimester is about 13 to 14 weeks.

critical period: a finite period during development in which certain events occur that will have irreversible effects on later developmental stages; usually a period of rapid cell division.

high-risk pregnancy: a pregnancy characterized by risk factors that make it likely the birth will be surrounded by problems such as premature delivery, difficult birth, retarded growth, birth defects, and early infant death.

low-risk pregnancy: a pregnancy characterized by factors that make it likely the birth will be normal and the infant healthy.

TABLE 10-1 High-Risk Pregnancy Factors

- Prepregnancy BMI either <18.5 or >25
- Insufficient or excessive pregnancy weight gain
- Nutrient deficiencies or toxicities; eating disorders
- Poverty, lack of family support, low level of education, limited food available
- Smoking, alcohol, or other drug use
- Teens, especially 15 years or younger; women 35 years or older
- Many previous pregnancies (three or more to mothers younger than age 20; four or more to mothers age 20 or older)
- Short or long intervals between pregnancies (<18 months or >59 months)
- Previous history of problems
- Twins or triplets
- Low- or high-birthweight infants
- Development of gestational hypertension
- Development of gestational diabetes
- Diabetes; hypertension; heart, respiratory, and kidney disease; certain genetic disorders; special diets and medications

Review Notes

- Placental development, implantation, and early critical periods of embryonic and fetal development depend on maternal nutrition before and during pregnancy.
- The effects of malnutrition during critical periods are irreversible.

NUTRIENT NEEDS DURING PREGNANCY

Nutrient needs during pregnancy increase more for certain nutrients than for others. (■) Figure 10-4 shows the percentage increase in nutrient intakes recommended for pregnant women compared with nonpregnant women. To meet the high nutrient demands of pregnancy, a woman must make careful food choices, but her body will also do its part by maximizing nutrient absorption and minimizing losses.

Energy, Carbohydrate, Protein, and Fat Energy needs vary with the progression of pregnancy. In the first trimester, the pregnant woman needs no additional energy, but as pregnancy progresses, her energy needs rise. She requires an additional 340 kcalories daily during the second trimester and an extra 450 kcalories each day during the third trimester.[16] Well-nourished pregnant women meet these demands for more energy in several ways: some eat more food, some reduce their activity, and some store less of their food energy as fat.[17] A woman can easily meet the need for extra kcalories by selecting more nutrient-dense foods from the five food groups. Table 1-7 (on p. 20) provides suggested eating patterns for several kcalorie levels, and Table 10-2 (p. 276) offers a sample menu for pregnant and lactating women.

If a woman chooses less nutritious options such as sugary soft drinks or fatty snack foods to meet her energy needs, she will undoubtedly come up short on nutrients. The increase in the need for nutrients is even greater than that for energy, so the mother-to-be should choose nutrient-dense foods such as whole-grain breads and cereals, legumes, dark green vegetables, citrus fruits, low-fat milk and milk products, and lean meats, fish, poultry, and eggs.

Ample carbohydrate (ideally, 175 grams or more per day and certainly no less than 135 grams) is necessary to fuel the fetal brain and spare the protein needed for fetal growth. Fiber in carbohydrate-rich foods such as whole grains, vegetables, and fruit can help alleviate the constipation that many pregnant women experience.

The protein RDA for pregnancy is 25 grams per day higher than for nonpregnant women. Pregnant women can easily meet their protein needs by selecting meats,

NURSING DIAGNOSIS

Imbalanced nutrition: less than body requirements applies to clients who lack nutrition knowledge or have too little money to purchase food.

■ Nutrient and energy intake recommendations for pregnant women are listed on the inside front cover.

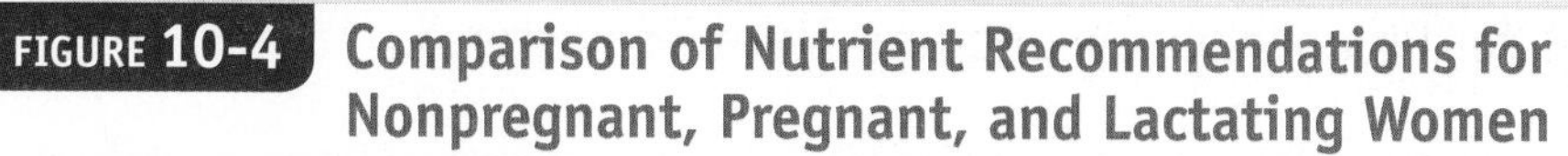

FIGURE 10-4 Comparison of Nutrient Recommendations for Nonpregnant, Pregnant, and Lactating Women

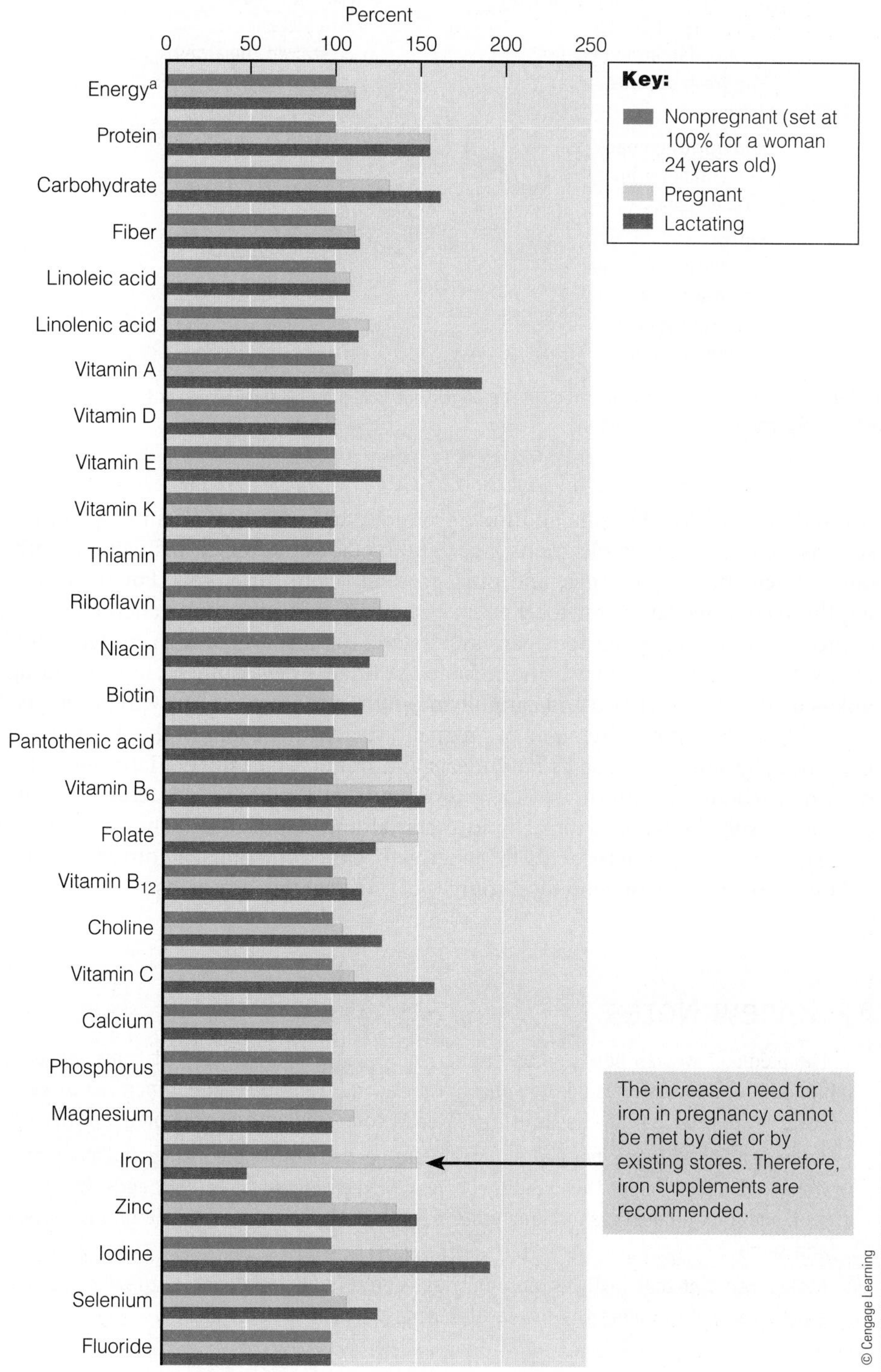

[a]Energy allowance during pregnancy is for second trimester; energy allowance during the third trimester is slightly higher; no additional allowance is provided during the first trimester. Energy allowance during lactation is for the first six months; energy allowance during the second six months is slightly higher.

TABLE 10-2 Daily Food Choices for Pregnancy (Second and Third Trimesters) and Lactation

Food Group	Amounts	Sample Menu	
Fruits	2 c	**Breakfast** 1 whole-wheat English muffin 2 tbs peanut butter 1 c low-fat vanilla yogurt ½ c fresh strawberries 1 c orange juice **Midmorning snack** ½ c cranberry juice 1 oz pretzels **Lunch** Sandwich (tuna salad on whole-wheat bread) ½ carrot (sticks) 1 c low-fat milk	**Dinner** Chicken cacciatore 3 oz chicken ½ c stewed tomatoes 1 c rice ½ c summer squash 1½ c salad (spinach, mushrooms, carrots) 1 tbs salad dressing 1 slice Italian bread 2 tsp soft margarine 1 c low-fat milk
Vegetables	3 c		
Grains	8 oz		
Protein foods	6½ oz		
Milk	3 c		

Note: This sample meal plan provides about 2500 kcalories (55 percent from carbohydrate, 20 percent from protein, and 25 percent from fat) and meets most of the vitamin and mineral needs of pregnant and lactating women.

seafood, poultry, low-fat milk and milk products, and protein-containing plant foods such as legumes, tofu, whole grains, nuts, and seeds. Some vegetarian women limit or omit protein-rich meats, eggs, and milk products from their diets. For them, meeting the recommendation for food energy each day and including generous servings of protein-containing plant foods are imperative. Because use of high-protein supplements during pregnancy may be harmful to the infant's development, it is discouraged unless medically prescribed and carefully monitored to treat fetal growth problems.[18]

The high nutrient requirements of pregnancy leave little room in the diet for excess fat, especially solid fats such as fatty meats and butter. The essential fatty acids, however, are particularly important to the growth and development of the fetus.[19] The brain contains a substantial amount of lipid material and depends heavily on long-chain omega-3 and omega-6 fatty acids for its growth, function, and structure.[20] (See Table 4-4 on p. 102 for a list of good food sources of the essential fatty acids.)

Review Notes

- The pregnant woman needs no additional energy intake during the first trimester, an increase of 340 kcalories per day during the second trimester, and an increase of 450 kcalories per day during the third trimester as compared with her nonpregnant needs.
- An additional 25 grams per day of protein are needed during pregnancy, but pregnant women can easily meet this need by choosing meats, poultry, seafood, eggs, low-fat milk and milk products, and plant-protein foods such as legumes, whole grains, nuts, and seeds.
- A balanced diet that includes more nutrient-dense foods from each of the five food groups can help to meet the increased nutrient demands of pregnancy.

neural tube: the embryonic tissue that later forms the brain and spinal cord.

neural tube defect (NTD): a serious central nervous system birth defect that often results in lifelong disability or death.

anencephaly (AN-en-SEF-a-lee): an uncommon and always fatal type of neural tube defect; characterized by the absence of a brain.

an = not (without)
encephalus = brain

Of Special Interest: Folate and Vitamin B_{12} The vitamins famous for their roles in cell reproduction—folate and vitamin B_{12}—are needed in large amounts during pregnancy. New cells are laid down at a tremendous pace as the fetus grows and develops. At the same time, because the mother's blood volume increases, the number of her red blood cells must rise, requiring more cell division and therefore more vitamins.

To accommodate these needs, the recommendation for folate during pregnancy (■) increases from 400 to 600 micrograms a day.

As described in Chapter 8, folate plays an important role in preventing neural tube defects. To review, the early weeks of pregnancy are a critical period for the formation and closure of the **neural tube** that will later develop to form the brain and spinal cord. By the time a woman suspects she is pregnant, usually around the sixth week of pregnancy, the embryo's neural tube normally has closed. A **neural tube defect (NTD)** (■) occurs when the tube fails to close properly. (■) Each year in the United States, an estimated 3000 pregnancies are affected by an NTD.[21] The two most common types of NTD are anencephaly (no brain) and spina bifida (split spine).

In **anencephaly**, the upper end of the neural tube fails to close. Consequently, the brain is either missing or fails to develop. Pregnancies affected by anencephaly often end in miscarriage; infants born with anencephaly die shortly after birth.

Spina bifida is characterized by incomplete closure of the spinal cord and its bony encasement (see Figure 10-5). The membranes covering the spinal cord and sometimes the cord itself may protrude from the spine as a sac. Spina bifida often produces paralysis in varying degrees, depending on the extent of spinal cord damage. Mild cases may not be noticed. Moderate cases may involve curvature of the spine, muscle weakness, mental handicaps, and other ills, while severe cases can lead to death.

To reduce the risk of neural tube defects, women who are capable of becoming pregnant are advised to obtain 400 micrograms of folic acid daily from supplements, fortified foods, or both, *in addition* to eating folate-rich foods (see Table 10-3 on p. 278). The DRI committee recommends intake of synthetic folate, called folic acid, in supplements and fortified food because it is absorbed better than the folate naturally present in foods. Foods that naturally contain folate are still important, however, because they

spina (SPY-nah) bifida (BIFF-ih-dah): one of the most common types of neural tube defects; characterized by the incomplete closure of the spinal cord and its bony encasement.

spina = spine
bifida = split

■ Folate RDA during pregnancy:
- 600 µg/day

■ Reminder: Neural tube defects (NTD) are malformations of the brain, spinal cord, or both during embryonic development.

■ A pregnancy affected by a neural tube defect can occur in any woman, but these factors make it more likely:[22]
- A previous pregnancy affected by a neural tube defect
- Maternal diabetes
- Maternal use of certain anti-seizure medications
- Mutations in folate-related enzymes
- Maternal obesity

FIGURE 10-5 Spina Bifida—A Neural Tube Defect

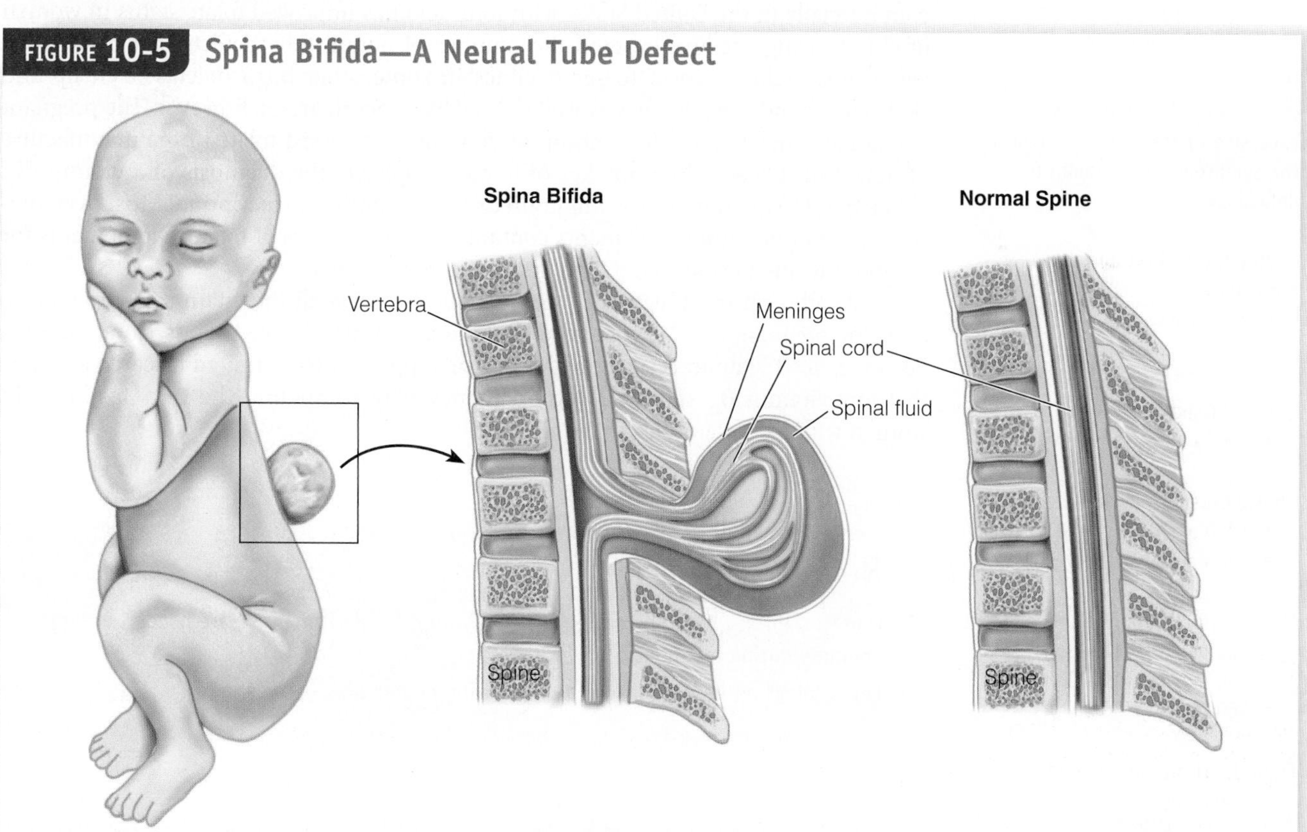

Spina bifida, a common neural tube defect, occurs when the vertebrae of the spine fail to close around the spinal cord, leaving it unprotected. The B vitamin folate—consumed prior to and during pregnancy—helps prevent spina bifida and other neural tube defects.

TABLE 10-3 Rich Folate Sources[a]

Natural Folate Sources	Fortified Folate Sources
Liver (3 oz) 221 µg DFE	Highly enriched ready-to-eat cereals (¾ c) 680 µg DFE
Lentils (½ c) 179 µg DFE	Pasta, cooked (1 c) 154 µg DFE (average value)
Chickpeas or pinto beans (½ c) 145 µg DFE	Rice, cooked (1 c) 153 µg DFE
Asparagus (½ c) 134 µg DFE	Bagel (1 small whole) 156 µg DFE
Spinach (1 c raw) 58 µg DFE	Waffles, frozen (2) 78 µg DFE
Avocado (½ c) 61 µg DFE	Bread, white (1 slice) 48 µg DFE
Orange juice (1 c) 74 µg DFE	
Beets (½ c) 68 µg DFE	

[a]Folate amounts for these and 2000 other foods are listed in the Table of Food Composition in Appendix A.
[b]Folate in cereals varies; read the Nutrition Facts panel of the label.

contribute to folate intakes while providing other needed vitamins, minerals, fiber, and phytochemicals.

The enrichment of grain products (cereal, grits, pasta, rice, bread, and the like) sold commercially in the United States with folic acid has improved folate status in women of childbearing age and lowered the number of neural tube defects that occur each year.[23] Researchers expect to see declines in some other birth defects (cleft lip and cleft palate) and miscarriages as well.[24] A safety concern arises, however. The pregnant woman also needs a greater amount of vitamin B_{12} to assist folate in the manufacture of new cells. Because high intakes of folate complicate the diagnosis of a vitamin B_{12} deficiency, (■) quantities of 1 milligram or more require a prescription. Most over-the-counter multivitamin supplements contain 400 micrograms of folate; supplements for pregnant women usually contain at least 800 micrograms.

People who eat meat, eggs, or dairy products receive all the vitamin B_{12} they need, even for pregnancy. (■) Those who exclude all animal-derived foods from the diet, however, need vitamin B_{12}-fortified foods or supplements.[25] Limited research suggests that low vitamin B_{12} status during pregnancy may compound the risks for NTD in women with low folate status.[26]

■ Chapter 8 describes how excessive folate intakes can mask the symptoms of a vitamin B_{12} deficiency.

■ Vitamin B_{12} RDA during pregnancy:
- 2.6 µg/day

■ Calcium RDA during pregnancy:
- 1300 mg/day (14–18 yr)
- 1000 mg/day (19–50 yr)

Phosphorus RDA during pregnancy:
- 1250 mg/day (14–18 yr)
- 700 mg/day (19–50 yr)

Magnesium RDA during pregnancy:
- 400 mg/day (14–18 yr)
- 350 mg/day (19–30 yr)
- 360 mg/day (31–50 yr)

Fluoride AI during pregnancy:
- 3 mg/day

■ Three cups of milk a day will supply 900 mg of calcium. For other food sources of calcium, see Chapter 9.

Review Notes

- Due to their key roles in cell reproduction, folate and vitamin B_{12} are needed in large amounts during pregnancy.
- Folate plays an important role in preventing neural tube defects.

Vitamin D and Calcium for Bones Vitamin D and the minerals involved in building the skeleton—calcium, phosphorus, magnesium, and fluoride—are in great demand during pregnancy. (■) Insufficient intakes may produce abnormal fetal bone growth and tooth development. (■)

Vitamin D plays a vital role in calcium absorption and utilization. Consequently, severe maternal vitamin D deficiency interferes with normal calcium metabolism, which, in rare cases, may cause rickets in the infant.[27] Regular exposure to sunlight and consumption of vitamin D–fortified milk are usually sufficient to provide the recommended amount of vitamin D during pregnancy (15 μg/day), which is the same as for nonpregnant women.[28] The vitamin D in **prenatal supplements** helps to protect many, but not all, pregnant women from inadequate intakes.[29]

Intestinal absorption of calcium doubles early in pregnancy, when the mother's bones store the mineral. Later, as the fetal bones begin to calcify, a dramatic shift of calcium across the placenta occurs. Whether the calcium added to the mother's bones early in pregnancy is withdrawn to build the fetus's bones later is unclear.[30] In the final weeks of pregnancy, more than 300 milligrams a day are transferred to the fetus.[31] Recommendations to ensure an adequate calcium intake during pregnancy are aimed at conserving the mother's bone mass while supplying fetal needs.

Typically, young women in the United States take in too little calcium. Of particular importance, pregnant women under age 25, whose own bones are still actively depositing minerals, should strive to meet the recommendation for calcium by increasing their intakes of milk, cheese, yogurt, and other calcium-rich foods. The USDA Food Patterns suggest consuming 3 cups per day of fat-free or low-fat milk or the equivalent in milk products. Less preferred, but still acceptable, is a daily supplement of 600 milligrams of calcium. The RDA for calcium intake is the same for nonpregnant and pregnant women in the same age group. Women who exclude milk products need calcium-fortified foods such as soy milk, orange juice, and cereals. Read the labels: contents vary and products fortified with both calcium and vitamin D are recommended.

Review Notes

- Maternal vitamin D deficiency interferes with calcium metabolism in the infant.
- All pregnant women, but especially those who are younger than 25 years of age, need to pay special attention to calcium to ensure adequate intakes.

Iron The body conserves iron especially well during pregnancy: menstruation ceases, and absorption of iron increases up to threefold due to a rise in the blood's iron-absorbing and iron-carrying protein transferrin. Still, iron needs are so high that stores dwindle during pregnancy. (■) To help improve the iron status of women before and during pregnancy, the *Dietary Guidelines for Americans 2010* recommend that all women capable of becoming pregnant do these three things:

- Choose foods that supply heme iron (meat, seafood, and poultry), which is most readily absorbed.
- Choose additional sources, such as iron-rich eggs, vegetables, and legumes.
- Consume foods that enhance iron absorption, such as vitamin C-rich fruits and vegetables.

The developing fetus draws heavily on the mother's iron stores to create stores of its own to last through the first four to six months after birth. The transfer of significant amounts of iron to the fetus is regulated by the placenta, which gives the iron needs of the fetus priority over those of the mother.[32] Even a woman with inadequate iron stores transfers a considerable amount of iron to the fetus. In addition, blood losses are inevitable at birth, especially during a delivery by cesarean section, further draining the mother's iron supply.

Few women enter pregnancy with adequate iron stores, so a daily iron supplement is recommended early in pregnancy, if not before.[33] Women who enter pregnancy with iron-deficiency anemia have a greater-than-normal risk of delivering low-birthweight or preterm infants.[34] When a low hemoglobin or hematocrit level is confirmed by a repeat test, more than the standard iron dose of 30 milligrams may be prescribed. To

■ Iron RDA during pregnancy:
- 27 mg/day

In pregnancy, hemoglobin values of 12 g/dL are not unusual, and 11 g is where the line defining "too low" is often drawn. Appendix E discusses more sensitive measures of iron status.

prenatal supplements: nutrient supplements specifically designed to provide the nutrients needed during pregnancy, particularly folate, iron, and calcium, without excesses or unneeded constituents.

enhance iron absorption, the supplement should be taken between meals and with liquids other than milk, coffee, or tea, which inhibit iron absorption.

Zinc Zinc is required for protein synthesis and cell development during pregnancy. (■) Typical zinc intakes of pregnant women are lower than recommendations but, fortunately, zinc absorption increases when intakes are low. Large doses of iron can interfere with zinc absorption and metabolism, but most prenatal supplements supply the right balance of these minerals for pregnancy. Zinc is abundant in protein-rich foods such as shellfish, meat, and nuts.

■ Zinc RDA during pregnancy:
- 12 mg/day (≤18 yr)
- 11 mg/day (19–50 yr)

■ To provide certain key nutrients for pregnancy, lactation, and growth, WIC offers vouchers for:
- Baby foods
- Eggs, dried and canned beans and peas, tuna fish, peanut butter
- Fruits, vegetables, and their juices
- Iron-fortified cereals
- Milk and cheese
- Soy-based beverages and tofu
- Whole-wheat bread and other whole-grain products
- Iron-fortified formula

Review Notes

- Iron supplements are recommended for pregnant women.
- Large doses of iron can interfere with zinc absorption and metabolism, but most prenatal supplements supply the right balance of these minerals.

Nutrient Supplements A healthy pregnancy and optimal infant development depend on the mother's diet.[35] Pregnant women who make wise food choices can meet most of their nutrient needs, with the possible exception of iron. Even so, physicians often recommend daily multivitamin-mineral supplements for pregnant women. These prenatal supplements typically provide more folate, iron, and calcium than regular supplements (see Figure 10-6). Prenatal supplements are especially beneficial for women who do not eat adequately and for those in high-risk groups: women carrying twins or triplets, cigarette smokers, and alcohol or drug abusers.[36] For these women, prenatal supplements may be of some help in reducing the risks of preterm delivery, low infant birthweights, and birth defects.[37] Be aware, however, that supplements cannot prevent the vast majority of fetal harm from tobacco, alcohol, and drugs, which continues unopposed, as later sections explain.

FIGURE 10-6 Example of a Prenatal Supplement Label

Prenatal Vitamins

Supplement Facts
Serving Size 1 Tablet

Amount Per Tablet	% Daily Value for Pregnant/ Lactating Women
Vitamin A 4000 IU	50%
Vitamin C 100 mg	167%
Vitamin D 400 IU	100%
Vitamin E 11 IU	37%
Thiamin 1.84 mg	108%
Riboflavin 1.7 mg	85%
Niacin 18 mg	90%
Vitamin B6 2.6 mg	104%
Folate 800 mcg	100%
Vitamin B12 4 mcg	50%
Calcium 200 mg	15%
Iron 27 mg	150%
Zinc 25 mg	167%

INGREDIENTS: calcium carbonate, microcrystalline cellulose, dicalcium phosphate, ascorbic acid, ferrous fumarate, zinc oxide, acacia, sucrose ester, niacinamide, modified cellulose gum, di-alpha tocopheryl acetate, hydroxypropyl methylcellulose, hydroxypropyl cellulose, artificial colors (FD&C blue no. 1 lake, FD&C red no. 40 lake, FD&C yellow no. 6 lake, titanium dioxide), polyethylene glycol, starch, pyridoxine hydrochloride, vitamin A acetate, riboflavin, thiamin mononitrate, folic acid, beta carotene, cholecalciferol, maltodextrin, gluten, cyanocobalamin, sodium bisulfite.

Notice that vitamin A is reduced to guard against birth defects, while extra amounts of folate, iron, and other nutrients are provided to meet the specific needs of pregnant women.

Review Notes

- Physicians often recommend daily multivitamin-mineral supplements for pregnant women.
- Women most likely to benefit from prenatal supplements during pregnancy include those who do not eat adequately, those carrying twins or triplets, and those who smoke cigarettes or are alcohol or drug abusers.

FOOD ASSISTANCE PROGRAMS

Women of limited financial means may eat diets too low in calcium, iron, vitamins A and C, and protein. Often, they and their children need help in obtaining food and benefit from nutrition counseling. At the federal level, the **Special Supplemental Nutrition Program for Women, Infants, and Children (WIC)** provides vouchers redeemable for nutritious foods, (■) nutrition education, and referrals to health and social services to low-income pregnant and lactating women and their children.[38] For infants given infant formula, WIC also provides iron-fortified formula. WIC encourages mothers to breastfeed their infants, however, and offers incentives to those who do.

More than 9 million people—most of them infants and young children—receive WIC benefits each month. Participation in the

WIC program benefits both the nutrient status and the growth and development of infants and children. WIC participation during pregnancy can effectively reduce iron deficiency, infant mortality, low birthweight, and maternal and newborn medical costs.

The Supplemental Nutrition Assistance Program (SNAP) provides a debit card that can also help to stretch the low-income pregnant woman's grocery dollars. In addition, many communities and organizations such as the Academy of Nutrition and Dietetics provide educational services and materials, including nutrition, food budgeting, and shopping information.

Review Notes

- Food assistance programs such as WIC can provide nutritious food for pregnant women of limited financial means.
- Participation in WIC during pregnancy can reduce iron deficiency, infant mortality, low birthweight, and maternal and newborn medical costs.

WEIGHT GAIN

Women must gain weight during pregnancy—fetal and maternal well-being depend on it. Ideally, a woman will have begun her pregnancy at a healthy weight, but even more importantly, she will gain within the recommended weight range based on her prepregnancy body mass index (BMI) as shown in Table 10-4. Pregnancy weight gains within the recommended ranges are associated with fewer surgical births, a greater number of healthy birthweights, and other positive outcomes for both mothers and infants, but many women do not gain within these ranges.[39] Among women in the United States, excessive weight gain during pregnancy is more prevalent than inadequate weight gain. To improve pregnancy outcomes, researchers and health care providers are placing greater emphasis on preventing excessive weight gains during pregnancy than in the recent past.[40]

Weight loss during pregnancy is not recommended.[41] Even obese women are advised to gain between 11 and 20 pounds for the best chance of delivering a healthy infant.[42] Ideally, overweight women will achieve a healthy body weight before becoming pregnant, avoid excessive weight gain during pregnancy, and postpone weight loss until after childbirth.[43]

TABLE 10-4 Recommended Weight Gains Based on Prepregnancy Weight

Prepregnancy Weight	Recommended Weight Gain	
	For Single Birth	For Twin Birth
Underweight (BMI <18.5)	28–40 lb (12.5–18.0 kg)	Insufficient data to make recommendation
Healthy weight (BMI 18.5–24.9)	25–35 lb (11.5–16.0 kg)	37–54 lb (17–25 kg)
Overweight (BMI 25.0–29.9)	15–25 lb (7.0–11.5 kg)	31–50 lb (14–23 kg)
Obese (BMI ≥30)	11–20 lb (5–9 kg)	25–42 lb (11–19 kg)

Source: Institute of Medicine, *Weight Gain During Pregnancy: Reexamining the Guidelines.* Reprinted with permission from The National Academies Press, Copyright © 2009, National Academy of Sciences.

Special Supplemental Nutrition Program for Women, Infants, and Children (WIC): a high-quality, cost-effective health care and nutrition services program administered by the U.S. Department of Agriculture for low-income women, infants, and children who are nutritionally at risk. WIC provides supplemental foods, nutrition education, and referrals to health care and other social services.

According to current recommendations, pregnant adolescents who are still growing themselves should strive for gains at the upper end of the target range. Current concerns about obesity raise questions as to whether such recommendations are always appropriate. Compared with older mothers, for example, the risk of lifetime weight retention after pregnancy may be far greater for young adolescents.

The ideal weight gain pattern for a woman who begins pregnancy at a healthy weight is 3½ pounds during the first trimester and 1 pound per week thereafter. (■) If a woman gains more than is recommended early in pregnancy, she should not restrict her energy intake later on in order to lose weight. A sudden, large weight gain is a danger signal, however, because it may indicate the onset of preeclampsia (discussed later in the chapter).

The weight the pregnant woman gains is nearly all lean tissue: placenta, uterus, blood, milk-producing glands, and the fetus itself (see Figure 10-7). The fat she gains is needed later for lactation. Physical activity can help a pregnant woman cope with the extra weight, as a later section explains.

WEIGHT LOSS AFTER PREGNANCY

The pregnant woman loses some weight at delivery. In the following weeks, she loses more as her blood volume returns to normal and she gets rid of accumulated fluids. The typical woman does not, however, return to her prepregnancy weight. In general, the more weight a woman gains beyond the needs of pregnancy, the more she retains—mostly as body fat.[44] Even with an average weight gain during pregnancy, most women tend to retain a few pounds with each pregnancy. When those few pounds become 7 or more and BMI increases by a unit or more, complications such as diabetes and hypertension in future pregnancies as well as chronic diseases later in life can increase—even for women who are not overweight. Women who achieve a healthy weight prior to the first pregnancy and maintain it between pregnancies best avoid the cumulative weight gain that threatens health later on.

FIGURE 10-7 Components of Weight Gain during Pregnancy

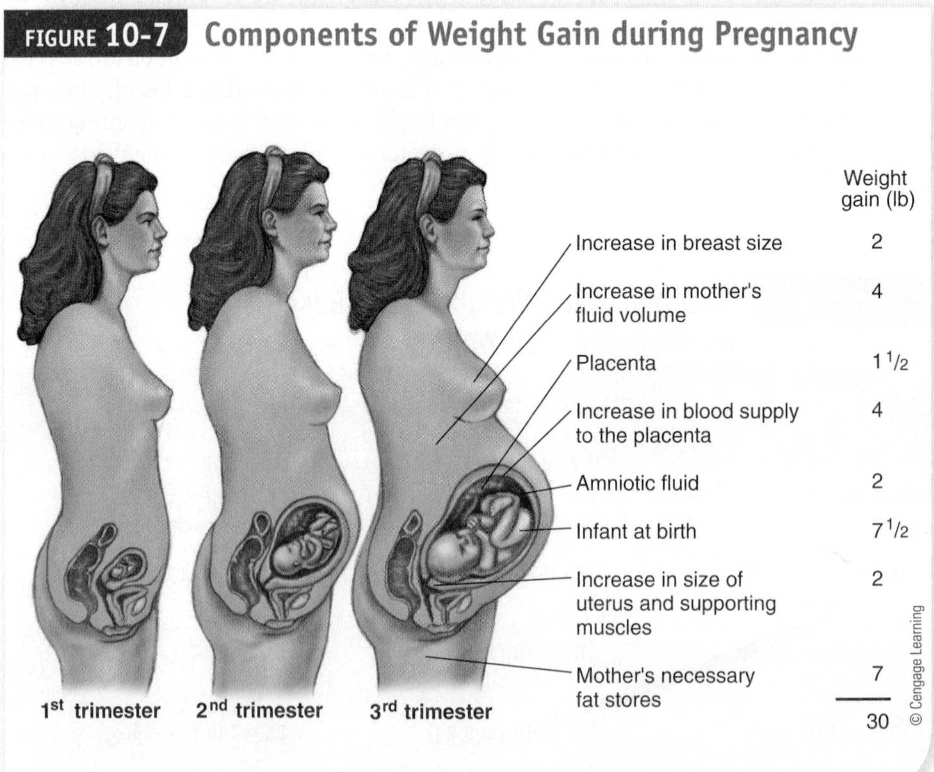

■ A prenatal weight-gain grid (Appendix E) plots the rate of weight gain during pregnancy.

Review Notes

- Appropriate (adequate but not excessive) weight gain is essential for a healthy pregnancy.
- A woman's prepregnancy BMI, her own nutrient needs, and the number of fetuses she is carrying help to determine appropriate weight gain.

PHYSICAL ACTIVITY

An active, physically fit woman experiencing a normal pregnancy can and should continue to exercise throughout pregnancy, adjusting the intensity and duration as the pregnancy progresses. Staying active during the course of a normal, healthy pregnancy improves the fitness of the mother-to-be, facilitates labor, helps to prevent or manage gestational diabetes and gestational hypertension, and reduces psychological stress.[45] Women who remain active report fewer discomforts throughout their pregnancies and retain habits that help in losing excess weight and regaining fitness after the birth.[46]

Pregnant women should choose "low-impact" activities and avoid sports in which they might fall or be hit by other people or objects. Swimming and water aerobics are particularly beneficial because they allow the body to remain cool and move freely with the water's support, thus reducing back pain.

As is true for everyone, the frequency, duration, and intensity of the activity affect the likelihood of the benefits or risks. (■) A few guidelines are offered in Figure 10-8. Several of the guidelines are aimed at preventing excessively high internal body temperatures and dehydration, both of which can harm fetal development. To this end, a pregnant woman should also stay out of saunas, steam rooms, and hot whirlpools. A pregnant woman with a medical condition or pregnancy complication should undergo a thorough evaluation by her health care provider before engaging in physical activity.

■ *Physical Activity Guidelines for Americans*

- Healthy women who are not already highly active or doing vigorous-intensity activity should get at least 150 minutes (2 hours and 30 minutes) of moderate-intensity aerobic activity per week during pregnancy and the postpartum period. Preferably, this activity should be spread throughout the week.
- Pregnant women who habitually engage in vigorous-intensity aerobic activity or are highly active can continue physical activity during pregnancy and the postpartum period, provided that they remain healthy and discuss with their health care provider how and when activity should be adjusted over time.

FIGURE 10-8 Guidelines for Physical Activity during Pregnancy

DO

- Do exercise regularly (most, if not all, days of the week).
- Do warm up with 5 to 10 minutes of light activity.
- Do 30 minutes or more of moderate physical activity.
- Do cool down with 5 to 10 minutes of slow activity and gentle stretching.
- Do drink water before, after, and during exercise.
- Do eat enough to support the additional needs of pregnancy plus exercise.
- Do rest adequately.

© Tracy Frankel / The Image Bank/Getty

Pregnant women can enjoy the benefits of physical activity.

DON'T

- Don't exercise vigorously after long periods of inactivity.
- Don't exercise in hot, humid weather.
- Don't exercise when sick with fever.
- Don't exercise while lying on your back after the first trimester of pregnancy or stand motionless for prolonged periods.
- Don't exercise if you experience any pain or discomfort.
- Don't participate in activities that may harm the abdomen or involve jerky, bouncy movements.
- Don't scuba dive.

Review Notes

- Physically fit women can continue physical activity throughout pregnancy but should choose activities wisely.
- Pregnant women should avoid sports in which they might fall or be hit by other people or objects and take care not to become dehydrated or overheated.

COMMON NUTRITION-RELATED CONCERNS OF PREGNANCY

Food sensitivities, nausea, heartburn, and constipation are common during pregnancy. A few simple strategies can help alleviate maternal discomforts (see Table 10-5).

Food Cravings and Aversions

Some women develop cravings for, or aversions to, certain foods and beverages during pregnancy. Individual **food cravings** during pregnancy do not seem to reflect real physiological needs. In other words, a woman who craves pickles does not necessarily need salt. Similarly, cravings for ice cream are common during pregnancy but do not signify a calcium deficiency. **Food aversions** and cravings that arise during pregnancy are probably due to hormone-induced changes in taste and sensitivities to smells, and they quickly disappear after the birth.

Nonfood Cravings

Some pregnant women develop cravings for and ingest nonfood items such as laundry starch, clay, soil, or ice—a practice known as pica.[47] Pica may be practiced for cultural reasons that reflect a society's folklore; it is especially common among African American women. Pica is often associated with iron deficiency, but whether iron deficiency leads to pica or pica leads to iron deficiency is unclear.[48] Eating clay or soil may interfere with iron absorption and displace iron-rich foods from the diet. Furthermore, if the soil or clay contains environmental contaminants such as lead or parasites, health and nutrition suffer.

TABLE 10-5 Strategies to Alleviate Maternal Discomforts

To alleviate the nausea of pregnancy:
- On waking, arise slowly.
- Eat dry toast or crackers.
- Chew gum or suck hard candies.
- Eat small, frequent meals whenever hunger strikes.
- Avoid foods with offensive odors.
- When nauseated, do not drink citrus juice, water, milk, coffee, or tea.

To prevent or alleviate constipation:
- Eat foods high in fiber.
- Exercise daily.
- Drink at least eight glasses of liquids a day.
- Respond promptly to the urge to defecate.
- Use laxatives only as prescribed by a physician; avoid mineral oil—it carries needed fat-soluble vitamins out of the body.

To prevent or relieve heartburn:
- Relax and eat slowly.
- Chew food thoroughly.
- Eat small, frequent meals.
- Drink liquids between meals.
- Avoid spicy or greasy foods.
- Sit up while eating.
- Wait an hour after eating before lying down.
- Wait two hours after eating before exercising.

food cravings: deep longings for particular foods.

food aversions: strong desires to avoid particular foods.

Morning Sickness The nausea of "morning" (actually, anytime) sickness seems unavoidable and may even be a welcome sign of a healthy pregnancy because it arises from the hormonal changes of early pregnancy. The problem typically peaks at 9 weeks gestation and resolves within a month or two.[49] Many women complain that odors, especially cooking smells, make them sick. Thus, minimizing odors may alleviate morning sickness for some women. Traditional strategies for quelling nausea are listed in Table 10-5, but little evidence exists to support such advice.[50] Some women do best by simply eating what they desire whenever they feel hungry. Morning sickness can be persistent, however. If morning sickness interferes with normal eating for more than a week or two, the woman should seek medical advice to prevent nutrient deficiencies.

> ***NURSING DIAGNOSIS***
>
> *Nausea* applies to clients who are pregnant.

Heartburn Heartburn, a burning sensation in the lower esophagus near the heart, is common during pregnancy and is also benign. As the growing fetus puts increasing pressure on the woman's stomach, acid may back up and create a burning sensation in her throat. Tips to relieve heartburn are also listed in Table 10-5.

Constipation As the hormones of pregnancy alter muscle tone and the thriving infant crowds intestinal organs, an expectant mother may complain of constipation, another harmless but annoying condition. A high-fiber diet, physical activity, and plentiful fluids will help relieve this condition. Also, responding promptly to the urge to defecate can help. Laxatives should be used only as prescribed by a physician.

> ***NURSING DIAGNOSIS***
>
> *Risk for constipation* applies to clients who are pregnant.

Review Notes

- Food cravings typically do not reflect physiological needs.
- Pica is the ingestion of nonfood items such as laundry starch, clay, soil, or ice. In some cases, pica can be harmful to health and nutrition.
- The nausea, heartburn, and constipation that sometimes accompany pregnancy can usually be alleviated with a few simple strategies.

PROBLEMS IN PREGNANCY

Just as adequate nutrition and normal weight gain support the health of the mother and growth of the fetus, maternal diseases can have an adverse effect. If discovered early, many diseases can be controlled—another reason why early prenatal care is recommended. Some nutrition measures can help alleviate the most common problems encountered during pregnancy.

Preexisting Diabetes Pregnancy presents special challenges for the management of diabetes. Insulin-induced hypoglycemia has a more rapid onset during pregnancy and is a danger to the mother, especially in those with type 1 diabetes. Women with type 2 diabetes often start pregnancy with insulin resistance and obesity, making optimal glycemic control difficult. The risks of diabetes during pregnancy depend on how well it is managed before, during, and after. Excellent glycemic control in the first trimester and throughout the pregnancy is associated with the lowest frequency of maternal, fetal, and newborn complications.[51] Without proper management, women face high infertility rates, and those who do conceive may experience episodes of severe hypoglycemia or hyperglycemia, preterm labor, and pregnancy-related hypertension. Infants may be large, suffer physical and mental abnormalities, and experience other complications such as severe hypoglycemia or respiratory distress, both of which can be fatal. Signs of fetal health problems are apparent even when maternal glucose is above normal but still below the diagnosis of diabetes. Ideally, a woman will receive the prenatal care needed to achieve glucose control before conception and continued glucose control throughout pregnancy. For optimal long-term outcomes, continuation of intensified diabetes management after pregnancy is in the best interest of the mother's health.

Gestational Diabetes

Some women are prone to developing a pregnancy-related form of diabetes, **gestational diabetes**. Gestational diabetes usually resolves after the infant is born, but some women go on to develop diabetes (type 2) later in life, especially if they are overweight.[52] Gestational diabetes can lead to fetal or infant sickness or death. When it is identified early and managed properly, however, the most serious risks fall dramatically. More commonly, gestational diabetes leads to surgical birth and high infant birthweight. To ensure that the problems of gestational diabetes are dealt with promptly, physicians screen for the risk factors listed in the margin (■) and test all pregnant women for glucose intolerance.[53] Chapter 20 provides information about medical nutrition therapy for gestational diabetes.

■ Risk factors for gestational diabetes:

- Age 25 or older
- BMI ≥25 or excessive weight gain
- Complications in previous pregnancies, including gestational diabetes or high birthweight
- Prediabetes or symptoms of diabetes
- Family history of diabetes
- Hispanic American, African American, Native American, Asian American, Pacific Islander

Hypertension

Hypertension (blood pressure ≥140/90 millimeters mercury) complicates pregnancy and affects its outcome in different ways, depending on how severe it becomes. Hypertension during pregnancy is classified as **chronic hypertension** or **gestational hypertension**.[54] Chronic hypertension can be a preexisting condition that develops before a woman becomes pregnant. In women whose prepregnancy blood pressure is unknown, diagnosis of chronic hypertension is based on the presence of sustained hypertension before 20 weeks of gestation.[55] In contrast, gestational hypertension develops after the 20th week of gestation. In women with gestational hypertension, blood pressure usually returns to normal during the first few weeks after childbirth.

Both types of hypertension pose risks to the mother and fetus. In addition to the health risks normally imposed by hypertension (heart attack and stroke), high blood pressure increases the risks of growth restriction, preterm birth, and separation of the placenta from the wall of the uterus before the birth.[56] Both chronic hypertension and gestational hypertension also increase the risk of preeclampsia.

■ Warning signs of preeclampsia:

- Hypertension
- Protein in the urine
- Upper abdominal pain
- Severe and constant headaches
- Swelling, especially of the face
- Dizziness
- Blurred vision
- Sudden weight gain (1 lb/day)

Preeclampsia

Preeclampsia is a condition characterized not only by high blood pressure but also by protein in the urine.[57] (■) Preeclampsia usually occurs with first pregnancies and almost always appears after 20 weeks' gestation.[58] Symptoms typically regress within 48 hours of delivery. Because delivery is the only known cure, preeclampsia is a leading cause of indicated preterm delivery and accounts for about 15 percent of infants who are growth restricted.[59]

Preeclampsia affects almost all of the woman's organs—the circulatory system, liver, kidneys, and brain. If it progresses, she may experience seizures; when this occurs, the condition is called **eclampsia**. Maternal mortality during pregnancy is rare in developed countries, but eclampsia is the most common cause. Preeclampsia demands prompt medical attention. Treatment focuses on regulating blood pressure and preventing seizures.

gestational diabetes: glucose intolerance with first onset or first recognition during pregnancy.

chronic hypertension: in pregnant women, hypertension that is present and documented before pregnancy; in women whose prepregnancy blood pressure is unknown, the presence of sustained hypertension before 20 weeks of gestation.

gestational hypertension: high blood pressure that develops in the second half of pregnancy and usually resolves after childbirth.

preeclampsia (PRE-ee-KLAMP-see-ah): a condition characterized by hypertension and protein in the urine.

eclampsia (eh-KLAMP-see-ah): a severe complication during pregnancy in which seizures occur.

Review Notes

- Conditions such as gestational diabetes, hypertension, and preeclampsia can threaten the health and life of both mother and infant.
- Maternal diseases require medical and nutrition treatment.

NURSING DIAGNOSIS

Ineffective health maintenance applies to clients who lack knowledge regarding basic health practices.

PRACTICES TO AVOID

A general guideline for the pregnant woman is to eat a normal, healthy diet and practice moderation. A woman's daily choices during pregnancy take on enormous importance. Forewarned, pregnant women can choose to abstain from or avoid potentially harmful practices.

Cigarette Smoking One practice to be avoided during pregnancy is cigarette smoking. A surgeon general's warning states that parental smoking can kill an otherwise healthy fetus or newborn. Unfortunately, an estimated 10 to 12 percent of pregnant women in the United States smoke, and rates are even higher for unmarried women and those who have not graduated from high school.[60] Constituents of cigarette smoke, such as nicotine, carbon monoxide, arsenic, and cyanide, are toxic to a fetus.[61] Research shows that smoking during pregnancy can cause damage to fetal chromosomes, which could lead to developmental defects or diseases such as cancer. Smoking also restricts the blood supply to the growing fetus and so limits the delivery of oxygen and nutrients and the removal of wastes. It slows fetal growth, can reduce brain size, and may impair the intellectual and behavioral development of the child later in life. Smoking during pregnancy damages fetal blood vessels, an effect that is still apparent at the age of five years.[62]

A mother who smokes is more likely to have a complicated birth and a low-birthweight infant. The more a mother smokes, the smaller her infant will be. Of all preventable causes of low birthweight in the United States, smoking has the greatest impact. Smoking during pregnancy interferes with fetal lung development and increases the risks of respiratory infections and childhood asthma.[63] Sudden infant death syndrome (SIDS), the unexplained death that sometimes occurs in an otherwise healthy infant, has been linked to the mother's cigarette smoking during pregnancy.[64] Even in women who do not smoke, exposure to **environmental tobacco smoke** (**ETS**, or secondhand smoke) during pregnancy increases the risk of low birthweight and the likelihood of SIDS.

In addition to the harms already described, cigarette (and cigar) smoking adversely affects the pregnant woman's nutrition status and thus impairs fetal nutrition and development. Smokers tend to have lower intakes of dietary fiber, vitamin A, beta-carotene, folate, and vitamin C. The margin (■) lists complications of smoking during pregnancy.

Alternatives to smoking—such as using snuff, chewing tobacco, or nicotine-replacement therapy—are not safe during pregnancy.[65] Any woman who uses nicotine in any form and is considering pregnancy or who is already pregnant needs to quit.

Medicinal Drugs and Herbal Supplements Medicinal drugs taken during pregnancy can cause serious birth defects. Pregnant women should not take over-the-counter drugs or any other medications without consulting their physicians, who must weigh the benefits against the risks.

Some pregnant women mistakenly consider herbal supplements to be safe alternatives to medicinal drugs and take them to relieve nausea, promote water loss, alleviate depression, help them sleep, or for other reasons. Some herbal products may be safe, but very few have been tested for safety or effectiveness during pregnancy. Pregnant women should stay away from herbal supplements, teas, or other products unless their safety during pregnancy has been ascertained.[66] Nutrition in Practice 14 offers more information about herbal supplements and other alternative therapies.

Drugs of Abuse Drugs of abuse such as cocaine easily cross the placenta and impair fetal growth and development. Furthermore, such drugs are responsible for preterm births, low-birthweight infants, and sudden infant deaths. If these newborns survive, central nervous system damage is evident: their cries, sleep, and behaviors early in life are abnormal, and their cognitive development later in life is impaired.[67] They may be hypersensitive or underaroused; infants who test positive for drugs suffer the greatest effects of toxicity and withdrawal.[68] Their childhood growth continues, but at a slow rate.

Environmental Contaminants Infants and young children of pregnant women exposed to environmental contaminants such as lead and mercury show signs of impaired mental and psychomotor development. During pregnancy, lead readily moves across the placenta, inflicting severe damage on the developing fetal nervous system. In

■ Complications associated with smoking during pregnancy:

- Fetal growth restriction
- Low birthweight
- Preterm birth
- Complications at birth (prolonged final stage of labor)
- Premature separation of the placenta
- Vaginal bleeding
- Spontaneous abortion
- Fetal death
- Sudden infant death syndrome
- Middle ear diseases
- Cardiac and respiratory diseases

environmental tobacco smoke (ETS): the combination of exhaled smoke (mainstream smoke) and smoke from lighted cigarettes, pipes, or cigars (sidestream smoke) that enters the air and may be inhaled by other people.

addition, infants exposed to even low levels of lead during gestation weigh less at birth and consequently struggle to survive. For these reasons, it is particularly important that pregnant women consume foods and beverages free of contamination. Dietary calcium can help to defend against lead toxicity by reducing its absorption.

Mercury is a contaminant of concern as well. As discussed in Chapter 4, fatty fish are a good source of omega-3 fatty acids, but some fish contain large amounts of the pollutant mercury, which can impair fetal growth and harm the developing brain and nervous system. Because the benefits of moderate seafood consumption seem to outweigh the risks, pregnant (and lactating) women need reliable information on which fish are safe to eat.[69] The *Dietary Guidelines for Americans* advise pregnant and lactating women to do the following:

- Avoid shark, swordfish, king mackerel, and tilefish (also called golden snapper or golden bass).
- Limit average weekly consumption to 12 ounces (cooked or canned) of seafood *or* to 6 ounces (cooked or canned) of white (albacore) tuna.

Supplements of fish oil are not recommended because they may contain concentrated toxins and because their effects on pregnancy remain unknown.

Foodborne Illness The vomiting and diarrhea caused by many foodborne illnesses can leave a pregnant woman exhausted and dangerously dehydrated. Particularly threatening, however, is **listeriosis**, which can cause miscarriage, stillbirth, or severe brain or other infections to fetuses and newborns. Pregnant women are about 20 times more likely than other healthy adults to get listeriosis. A woman with listeriosis may develop symptoms such as fever, vomiting, and diarrhea in about 12 hours after eating a contaminated food, and serious symptoms may develop a week to six weeks later. A blood test can reliably detect listeriosis, and antibiotics given promptly to the pregnant sufferer can often prevent infection of the fetus or newborn. The margin lists tips to prevent listeriosis. (■) Nutrition in Practice 2 includes precautions to minimize the risks of other common foodborne illnesses.

■ To prevent listeriosis:
- Use only pasteurized juices and dairy products; do not eat soft cheeses such as feta, brie, Camembert, Panela, "queso blanco," "queso fresco," and blue-veined cheeses such as Roquefort; do not drink raw (unpasteurized) milk or eat foods that contain it.
- Do not eat hot dogs or luncheon or deli meats unless heated until steaming hot.
- Thoroughly cook meat, poultry, eggs, and seafood.
- Wash all fruits and vegetables.
- Avoid refrigerated (not canned) patés or smoked seafood, or any fish labeled "nova-style," "lox," or "kippered."

Vitamin-Mineral Megadoses Many vitamins and minerals are toxic when taken in excess. Excessive vitamin A is particularly infamous for its role in fetal malformations of the cranial nervous system. Intakes before the seventh week of pregnancy appear to be the most damaging. For this reason, vitamin A supplements are not given during pregnancy, unless there is specific evidence of deficiency, which is rare.

Restrictive Dieting Restrictive dieting, even for short periods, can be hazardous during pregnancy. Low-carbohydrate diets or fasts that cause ketosis deprive the growing fetal brain of needed glucose and may impair cognitive development. Such diets are also likely to be deficient in other nutrients vital to fetal growth. Regardless of prepregnancy weight, pregnant women need an adequate diet to support healthy fetal development.

Sugar Substitutes Artificial sweeteners have been studied extensively and found to be acceptable during pregnancy if used within the FDA's guidelines (see Chapter 3).[70] Still, it would be prudent for pregnant women to use sweeteners in moderation and within an otherwise nutritious and well-balanced diet. Women with phenylketonuria should not use aspartame, as Chapter 3 explains.

listeriosis: a serious foodborne infection that can cause severe brain infection or death in a fetus or newborn; caused by the bacterium *Listeria monocytogenes*, which is found in soil and water.

Caffeine Caffeine crosses the placenta, and the fetus has only a limited ability to metabolize it. Research studies have not proved that caffeine (even in high doses) causes birth defects in human infants (as it does in animals), but limited evidence suggests that heavy use—intake equaling three or more cups of coffee a day—increases the risk of miscarriage and fetal death.[71] Depending on the quantities consumed and the mother's metabolism, caffeine may also interfere with fetal growth.[72] In light of this evidence, the most sensible course is to limit caffeine consumption to the equivalent of one cup of coffee or two 12-ounce cola beverages a day. The amounts of caffeine in foods and beverages are listed in Appendix A.

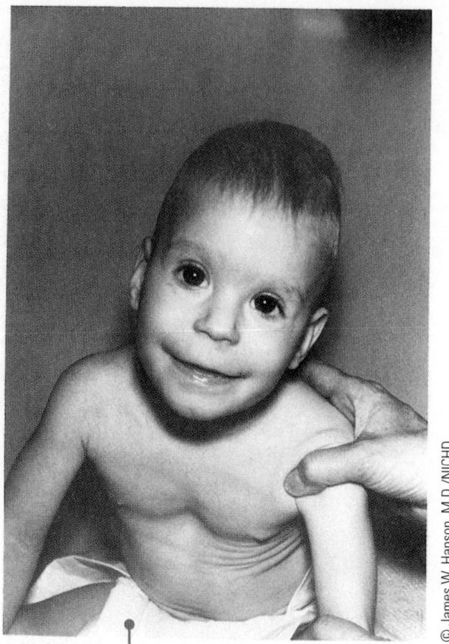

© James W. Hanson, M.D./NICHD

These facial traits are typical of fetal alcohol syndrome, caused by drinking alcohol during pregnancy—low nasal bridge, short eyelid opening, underdeveloped groove in the center of the upper lip, small midface, short nose, and small head circumference.

Alcohol Drinking alcohol during pregnancy threatens the fetus with irreversible brain damage, growth retardation, mental retardation, facial abnormalities, vision abnormalities, and many more health problems—a spectrum of symptoms known as **fetal alcohol spectrum disorders**, or **FASD**. Children at the most severe end of the spectrum (those with all of the symptoms) are defined as having **fetal alcohol syndrome (FAS)**.[73] The fetal brain is extremely vulnerable to a glucose or oxygen deficit, and alcohol causes both by disrupting placental functioning. The lifelong mental retardation and other tragedies of FAS can be prevented by abstaining from drinking alcohol during pregnancy. Once the damage is done, however, the child remains impaired.

The accompanying photo shows the facial abnormalities of FAS, which are easy to depict. A visual picture of the internal harm is impossible, but that damage seals the fate of the child. An estimated 5 to 20 of every 10,000 children are victims of FAS, making it one of the leading known preventable causes of mental retardation in the world.[74]

Despite alcohol's potential for harm, 1 out of 8 pregnant women drinks alcohol sometime during her pregnancy; 1 out of 75 pregnant women reports binge drinking (four or more drinks on one occasion).[75] Almost half of all pregnancies are unintended, and many are conceived during a binge-drinking episode.

For women who know they are pregnant and choose to drink alcohol, the question is how much alcohol is too much. Even one drink a day threatens neurological development and behaviors. Low birthweight is reported among infants born to women who drink 1 ounce (two drinks) of alcohol per day during pregnancy, and FAS is also known to occur with as few as two drinks a day. Birth defects have been reliably observed among the children of women who drink 2 ounces (four drinks) of alcohol daily during pregnancy. The most severe impact is likely to occur in the first two months, possibly before the woman is aware that she is pregnant.

Even when a child does not develop full FAS, prenatal exposure to alcohol can lead to less severe, but nonetheless serious, mental and physical problems. The cluster of mental problems is known as **alcohol-related neurodevelopmental disorder (ARND)**, and the physical malformations are referred to as **alcohol-related birth defects (ARBD)**. Some of these children show no outward sign of impairment, but others are short in stature or display subtle facial abnormalities. Many perform poorly in school and in social interactions and suffer a subtle form of brain damage. Mood disorders and problem behaviors, such as aggression, are common.

For every child diagnosed with full-blown FAS, many more with FASD go undiagnosed until problems develop in the preschool years. Upon reaching adulthood, such children are ill equipped for employment, relationships, and the other facets of life most adults take for granted. Anyone exposed to alcohol before birth may always respond differently to it, and also to certain drugs, than if no exposure had occurred, making addictions likely.

The American Academy of Pediatrics takes the position that women should stop drinking as soon as they *plan* to become pregnant. Researchers have looked for a "safe" alcohol intake limit during pregnancy and have found none. Their conclusion: abstinence from alcohol is the best policy for pregnant women.

For pregnant women who have already drunk alcohol, the advice is "stop now." A woman who has drunk heavily during the first two-thirds of her pregnancy can still prevent some organ damage by stopping heavy drinking during the third trimester.

fetal alcohol spectrum disorders (FASD): a spectrum of physical, behavioral, and cognitive disabilities caused by prenatal alcohol exposure.

fetal alcohol syndrome (FAS): the cluster of symptoms seen in an infant or child whose mother consumed excessive alcohol during her pregnancy. FAS includes, but is not limited to, brain damage, growth retardation, mental retardation, and facial abnormalities.

alcohol-related neurodevelopmental disorder (ARND): a condition caused by prenatal alcohol exposure that is diagnosed when there is a confirmed history of substantial, regular maternal alcohol intake or heavy episodic drinking and behavioral, cognitive, or central nervous system abnormalities known to be associated with alcohol exposure.

alcohol-related birth defects (ARBD): a condition caused by prenatal alcohol exposure that is diagnosed when there is a history of substantial, regular maternal alcohol intake or heavy episodic drinking and birth defects known to be associated with alcohol exposure.

Review Notes

- Smoking during pregnancy delivers toxins to the fetus, damages DNA, restricts fetal growth, and limits delivery of oxygen and nutrients and the removal of wastes.
- Drinking alcohol during pregnancy threatens the fetus with irreversible brain damage, growth retardation, mental retardation, facial abnormalities, vision abnormalities, and many more health problems—a spectrum of symptoms known as fetal alcohol spectrum disorders, or FASD.
- Abstaining from smoking and other drugs, including alcohol, limiting intake of foods known to contain unsafe levels of contaminants such as mercury, taking precautions against foodborne illness, avoiding large doses of nutrients, refraining from restrictive dieting, using artificial sweeteners in moderation, and limiting caffeine use are recommended during pregnancy.

ADOLESCENT PREGNANCY

The number of infants born to teenage mothers has steadily declined during the last 50 years. Despite the long-term decline, however, the U. S. teen birth rate is still one of the highest among industrialized nations.[76] In 2010, more than 367,000 infants were born to teenage mothers.

A pregnant adolescent presents a special case of intense nutrient needs. Young teenage girls have a hard enough time meeting nutrient needs for their own rapid growth and development, let alone those of pregnancy. Many teens enter pregnancy deficient in vitamins B_{12} and D, folate, calcium, and iron, which increases the risk of impaired fetal growth.[77] Pregnant adolescents are less likely to receive early prenatal care and are more likely to smoke during pregnancy—two factors that predict low birthweight and infant death.[78]

NURSING DIAGNOSIS

Imbalanced nutrition: less than body requirements applies to clients with reported food intakes less than the RDA and those who lack nutrition knowledge.

NURSING DIAGNOSIS

Risk for delayed development applies to clients who are young, are pregnant, and lack prenatal care.

The rates of stillbirths, preterm births, and low-birthweight infants are high for teenagers—both for teen moms and for teen dads. Their greatest risk, though, is death of the infant: mothers younger than 16 years of age bear more infants who die within the first year than do women in any other age group. These factors combine to make adolescent pregnancy a major public health problem.

Adequate nutrition is an indispensable component of prenatal care for adolescents and can substantially improve the outlook for both mother and infant. To support the needs of both mother and fetus, a pregnant teenager with a BMI in the normal range is encouraged to gain about 35 pounds to reduce the likelihood of a low-birthweight infant. As mentioned earlier, however, compared with older mothers, the lifetime risk of postpartum weight retention in young adolescents may be far greater.[79] Research shows that women who give birth during adolescence have higher body weights, BMIs, and percent body fat later on than adolescents who do not experience pregnancy.[80] Researchers agree that optimal weight gain recommendations for pregnant adolescents need focused attention. Meanwhile, pregnant and lactating adolescents would do well to follow the eating pattern presented in Table 1-7 (on p. 20), making sure to choose a kcalorie level high enough to support adequate, but not excessive, weight gain.

Review Notes

- Pregnant adolescents have extraordinarily high nutrient needs and an increased likelihood of problem pregnancies.
- Proper nutrition and adequate weight gain are especially important in reducing the risk of poor pregnancy outcome in adolescents.

Breastfeeding

A woman who decides to breastfeed provides her infant with a full array of nutrients and protective factors to support optimal health and development.

The American Academy of Pediatrics (AAP) recommends that infants receive breast milk for at least the first 12 months of life and beyond for as long as mutually desired by mother and child.[81] The Academy of Nutrition and Dietetics (AND) advocates breastfeeding for the nutritional health it confers on the infant as well as for the physiological, social, economic, and other benefits it offers the mother (see Table 10-6).[82] The AAP and the AND recognize **exclusive breastfeeding** for 6 months and breastfeeding with complementary foods for at least 12 months as an optimal feeding pattern for infants.[83] Breast milk's unique nutrient composition and protective factors promote optimal infant health and development. The only acceptable alternative to breast milk is iron-fortified formula. Adequate nutrition of the mother supports successful lactation, and without it, lactation is likely to falter or fail. Health care professionals play an important role in providing encouragement and accurate information on breastfeeding.

NUTRITION DURING LACTATION

By continuing to eat nutrient-dense foods, not restricting weight gain unduly, and enjoying ample food and fluid at frequent intervals throughout lactation, the mother who chooses to breastfeed her infant will be nutritionally prepared to do so. An inadequate diet does not support the stamina, patience, and self-confidence that nursing an infant demands. Figure 10-4 (on p. 275) shows how a lactating woman's nutrient needs differ from those of a nonpregnant woman, and Table 10-2 (on p. 276) presents a sample menu that meets those needs.

TABLE 10-6 Benefits of Breastfeeding

For Infants

- Provides the appropriate composition and balance of nutrients with high bioavailability
- Provides hormones that promote physiological development
- Improves cognitive development
- Protects against a variety of infections and illnesses
- May protect against some chronic diseases—such as diabetes (both types), obesity, atherosclerosis, asthma, and hypertension—later in life
- Protects against food allergies
- Supports healthy weight
- Reduces the risk of SIDS

For Mothers

- Contracts the uterus
- Delays the return of regular ovulation, thus lengthening birth intervals (this is not, however, a dependable method of contraception)
- Conserves iron stores (by prolonging amenorrhea)
- May protect against breast and ovarian cancer and reduce the risk of diabetes (type 2)

Other

- Cost and time savings from not needing medical treatment for childhood illnesses or leaving work to care for sick infants
- Cost and time savings from not needing to purchase and prepare formula (even after adjusting for added foods in the diet of a lactating mother)
- Environmental savings for society from not needing to manufacture, package, and ship formula and dispose of the packaging

exclusive breastfeeding: an infant's consumption of human milk with no supplementation of any type (no water, no juice, no nonhuman milk, and no foods) except for vitamins, minerals, and medications.

Energy A nursing woman produces about 25 ounces of milk a day, with considerable variation from woman to woman and in the same woman from time to time, depending primarily on the infant's demand for milk. Producing this milk costs a woman almost 500 kcalories per day above her regular need during the first six months of lactation. To meet this energy need, the woman is advised to eat foods providing an extra 330 kcalories each day. (■) The other 170 kcalories can be drawn from the fat stores she accumulated during pregnancy. The food energy consumed by the nursing mother should carry with it abundant nutrients. Severe energy restriction hinders milk production and can compromise the mother's health.

Weight Loss After the birth of the infant, many women actively try to lose the extra weight and body fat they accumulated during pregnancy. How much weight a woman retains after pregnancy depends on her gestational weight gain and the duration and intensity of breastfeeding. Many women who follow recommendations for gestational weight gain and breastfeeding can readily return to prepregnancy weight by six months. Neither the quality nor the quantity of breast milk is adversely affected by moderate weight loss, and infants grow normally.

Women often choose to be physically active to lose weight and improve fitness, and this is compatible with breastfeeding and infant growth.[84] A gradual weight loss (1 pound per week) is safe and does not reduce milk output. Too large an energy deficit, however, especially soon after birth, will inhibit lactation.

Vitamins and Minerals A question often raised is whether a mother's milk may lack a nutrient if she fails to get enough in her diet. The answer differs from one nutrient to the next, but, in general, nutritional deprivation of the mother reduces the *quantity,* not the *quality,* of her milk. Women can produce milk with adequate protein, carbohydrate, fat, folate, and most minerals, even when their own supplies are limited. For these nutrients, milk quality is maintained at the expense of maternal stores. This is most evident in the case of calcium: dietary calcium has no effect on the calcium concentration of breast milk, but maternal bones lose some of their density during lactation if calcium intakes are inadequate.[85] Such losses are generally made up quickly when lactation ends, and breastfeeding has no long-term harmful effects on women's bones. The nutrients in breast milk that are most likely to decline in response to prolonged inadequate intakes are the vitamins—especially vitamins B_6, B_{12}, A, and D. Vitamin supplementation of undernourished women appears to help normalize the vitamin concentrations in their milk and may be beneficial.

Water The volume of breast milk produced depends on how much milk the infant demands, not on how much fluid the mother drinks. The nursing mother is nevertheless advised to drink plenty of liquids each day (about 13 cups) to protect herself from dehydration. (■) To help themselves remember to drink enough liquid, many women make a habit of drinking a glass of milk, juice, or water each time the infant nurses as well as at mealtimes.

Particular Foods Foods with strong or spicy flavors (such as onions or garlic) may alter the flavor of breast milk. A sudden change in the taste of the milk may annoy some infants, whereas familiar flavors may enhance enjoyment. Flavors imparted to breast milk by the mother's diet can influence the infant's later food preferences.[86] A mother who is breastfeeding her infant is advised to eat whatever nutritious foods she chooses. Then, if a particular food seems to cause the infant discomfort, she can try eliminating that food from her diet for a few days and see if the problem goes away.

Current evidence does not support a major role for maternal dietary restrictions during lactation to prevent or delay the onset of food allergy in infants.[87] Infants who develop symptoms of food allergy, however, may be more comfortable if the mother's diet excludes the most common offenders—cow's milk, eggs, fish, peanuts, and tree nuts. Generally, infants with a strong family history of food allergies benefit from breastfeeding.[88]

■ Energy requirement during lactation:

- First 6 months: +330 kcal/day
- Second 6 months: +400 kcal/day

■ The DRI recommendation for *total* water intake during lactation is 3.8 L/day. This includes 3.1 L or about 13 c as total beverages, including drinking water.

CONTRAINDICATIONS TO BREASTFEEDING

Some substances impair maternal milk production or enter the breast milk and interfere with infant development. Some medical conditions prohibit breastfeeding.

NURSING DIAGNOSIS

Interrupted breastfeeding applies to clients with contraindications to breastfeeding.

Alcohol Alcohol easily enters breast milk and can adversely affect the production, volume, composition, and ejection of breast milk as well as overwhelm an infant's immature alcohol-degrading system.[89] Alcohol concentration in breast milk peaks within one hour after ingestion of even moderate amounts (equivalent to a can of beer). It may alter the taste of the milk to the disapproval of the nursing infant, who may, in protest, drink less milk than normal.

Tobacco and Caffeine About half the women who quit smoking during pregnancy relapse after delivery.[90] Lactating women who smoke produce less milk, and milk with a lower fat content, than do nonsmokers.[91] Consequently, infants of smokers gain less weight than infants of nonsmokers.

A lactating woman who smokes not only transfers nicotine and other chemicals to her infant via her breast milk but may also expose the infant to sidestream smoke. Infants who are "smoked over" experience a wide array of health problems—poor growth, hearing impairment, vomiting, breathing difficulties, and even unexplained death.[92] Health care professionals should actively discourage smoking by lactating women.

Excessive caffeine can make an infant jittery and wakeful. As during pregnancy, caffeine consumption during lactation should be moderate.

Medications and Illicit Drugs If a nursing mother must take medication that is secreted in breast milk and is known to affect the infant, then breastfeeding must be put off for the duration of treatment. Meanwhile, the flow of milk can be sustained by pumping the breasts and discarding the milk. Many prescription medications do not reach nursing infants in sufficient quantities to affect them adversely and so have no impact on breastfeeding. Other drugs are not at all compatible with breastfeeding, either because they are secreted into the milk and can harm the infant or because they suppress lactation.[93] A nursing mother should consult with the prescribing physician before taking medicines or even herbal supplements—herbs may have unpredictable effects on breastfeeding infants. Breastfeeding is also contraindicated if the mother uses illicit drugs. Breast milk can deliver such high doses of illicit drugs as to cause irritability, tremors, hallucinations, and even death in infants.

Many women wonder about using oral contraceptives during lactation. One type that combines the hormones estrogen and progestin seems to suppress milk output, lower the nitrogen content of the milk, and shorten the duration of breastfeeding. In contrast, progestin-only pills have no effect on breast milk or breastfeeding and are considered appropriate for lactating women.[94]

Maternal Illness If a woman has an ordinary cold, she can go on nursing without worry. If susceptible, the infant will catch it from her anyway, and, thanks to immunological protection, a breastfed baby may be less susceptible than a formula-fed infant would be. A woman who has active, infectious tuberculosis should not breastfeed, but she can resume or begin once she has been treated and it is documented that she is no longer infectious.[95]

The human immunodeficiency virus (HIV), responsible for causing AIDS, can be passed from an infected mother to her infant during pregnancy, at birth, or through breast milk, especially during the early months of breastfeeding. In developed countries such as the United States, where safe alternatives are available, HIV-positive women should not breastfeed their infants.[96]

Throughout the world, breastfeeding prevents millions of infant deaths each year. In developing countries, where the feeding of inappropriate or contaminated formulas causes more than 1 million infant deaths each year, breastfeeding can be critical to infant survival.[97] Thus, the question of whether HIV-infected women in developing countries should breastfeed comes down to a delicate balance between risks and benefits. For HIV-positive women in developing countries, the most appropriate infant-feeding option depends on individual circumstances, including the health status of the mother and the local situation, as well as the health services, counseling, and support available. The World Health Organization (WHO) recommends exclusive breastfeeding for infants of HIV-infected women for the first six months of life unless replacement feeding is acceptable, feasible, affordable, sustainable, and safe for mothers and their infants before that time.[98] Alternatively, HIV-exposed infants may be protected by receiving antiretroviral treatment while being breastfed.

Review Notes

- The lactating woman needs enough energy and nutrients to produce about 25 ounces of milk a day. She also needs extra fluid.
- Alcohol, smoking, caffeine, and drugs may reduce milk production or enter breast milk and impair infant development.
- Some maternal illnesses are incompatible with breastfeeding.

Nutrition of the Infant

Early nutrition affects later development, and early feeding sets the stage for eating habits that will influence nutrition status for a lifetime. Trends change, and experts argue about the fine points, but properly nourishing an infant is relatively simple, overall. The infant initially drinks only breast milk or formula but later begins to eat some foods, as appropriate. Common sense in the selection of infant foods and a nurturing, relaxed environment go far to promote an infant's health and well-being. The remainder of this discussion is devoted to feeding the infant and identifying the nutrients most often deficient in infant diets.

NUTRIENT NEEDS DURING INFANCY

An infant grows faster during the first year than ever again, as Figure 10-9 shows. The growth of infants and children directly reflects their nutritional well-being and is an important parameter in assessing their nutrition status. Health care professionals use growth charts to evaluate the growth and development of children from birth to 20 years of age (see Appendix E).

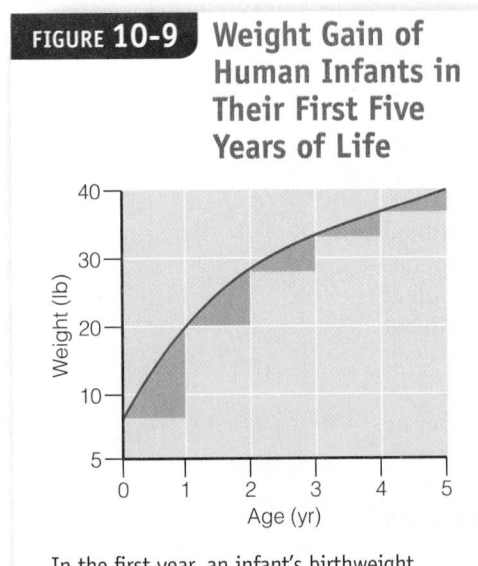

FIGURE 10-9 Weight Gain of Human Infants in Their First Five Years of Life

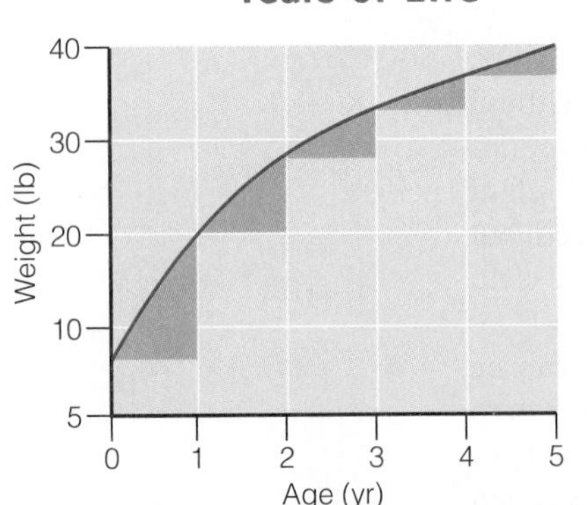

In the first year, an infant's birthweight may triple, but over the following several years, the rate of weight gain gradually diminishes.

Nutrients to Support Growth An infant's birthweight doubles by about five months of age and triples by the age of one year, typically reaching 20 to 25 pounds. (Consider that if an adult, starting at 120 pounds, were to do this, the person's weight would increase to 360 pounds in a single year.) The infant's length changes more slowly than weight, increasing about 10 inches from birth to one year. By the end of the first year, the growth rate slows considerably. An infant typically gains less than 10 pounds during the second year and grows about 5 inches in height. At the age of 2, healthy children have attained approximately half of their adult height.

Not only do infants grow rapidly, but, in proportion to body weight, their basal metabolic rate is remarkably high—about twice that of an adult. The rapid growth and metabolism of the infant demand an ample supply of all the nutrients. Of special importance during infancy are the energy nutrients and the vitamins and minerals critical to the growth process, such as vitamin A, vitamin D, and calcium.

Because they are small, infants need smaller *total* amounts of these nutrients than adults do, but as a percentage of body weight, infants need more than twice as much of most nutrients. Infants require about 100 kcalories per kilogram of body weight per day; most adults require fewer than 40 (see Table 10-7). Figure 10-10 compares a five-month-old infant's needs (per unit of body weight) with those of an adult man. You can see that differences in vitamin D and iodine, for instance, are extraordinary. Around six months of age, energy needs begin to increase less rapidly as the growth rate begins to slow, but some of the energy saved by slower growth is spent on increased activity. When their growth slows, infants spontaneously reduce their energy intakes. Parents should expect their infants to adjust their food intakes downward when appropriate and should not force or coax them to eat more.

TABLE 10-7 Infant and Adult Heart Rate, Respiration Rate, and Energy Needs Compared

	Infants	Adults
Heart rate (beats/minute)	120–140	70–80
Respiration rate (breaths/minute)	20–40	15–20
Energy needs (kcal/body weight)	45/lb (100/kg)	<18/lb (<40/kg)

FIGURE 10-10 Nutrient Recommendations for a Five-Month-Old Infant and an Adult Male Compared on the Basis of Body Weight

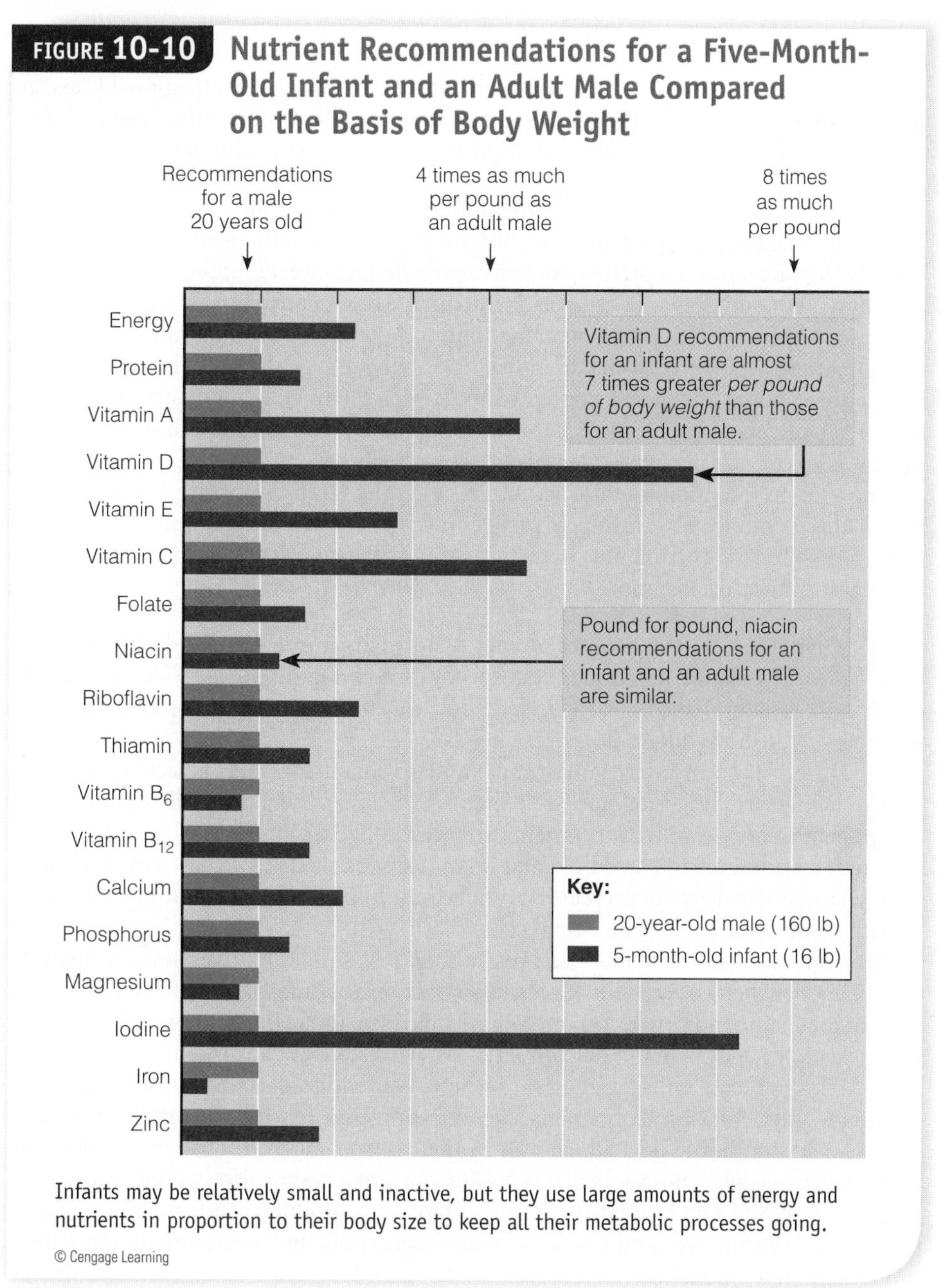

Infants may be relatively small and inactive, but they use large amounts of energy and nutrients in proportion to their body size to keep all their metabolic processes going.

Linda DeBruyne

After six months of age, energy saved by slower growth is spent on increased activity.

Water One of the most important nutrients for infants, as for everyone, is water. The younger a child is, the more of the child's body weight is water. Breast milk or infant formula normally provides enough water to replace fluid losses in a healthy infant. If the environmental temperature is extremely high, however, infants need supplemental water.[99] Because proportionately more of an infant's body water than an adult's is between the cells and in the vascular space, this water is easy to lose. Consequently, conditions that cause rapid fluid loss, such as vomiting or diarrhea, require an electrolyte solution designed for infants.

Review Notes

- An infant's birthweight doubles by about five months of age and triples by one year.
- Infants' rapid growth and development depend on adequate nutrient supplies, including water from breast milk and formula.

BREAST MILK

Breast milk excels as a source of nutrients for the young infant. With the possible exception of vitamin D (discussed later), breast milk provides all the nutrients a healthy infant needs for the first six months of life.[100] It provides many other health benefits as well.

Frequency and Duration of Breastfeeding Breast milk is more easily and completely digested than formula, so breastfed infants usually need to eat more frequently than formula-fed infants do. During the first few weeks, approximately 8 to 12 feedings a day—on demand, as soon as the infant shows early signs of hunger such as increased alertness, activity, or suckling motions—promote optimal milk production and infant growth. Crying is a late indicator of hunger.[101] An infant who nurses every two to three hours and sleeps contentedly between feedings is adequately nourished. As the infant gets older, stomach capacity enlarges and the mother's milk production increases, allowing for longer intervals between feedings.

NURSING DIAGNOSIS

Effective breastfeeding applies to clients with effective mother/infant communication patterns and an infant who is content after feeding.

Even though the infant obtains about half the milk from the breast during the first two or three minutes of suckling, breastfeeding is encouraged for about 10 to 15 minutes on each breast. The infant's suckling, as well as the complete removal of milk from the breast, stimulates lactation.

Energy Nutrients The balance of energy nutrients in breast milk differs dramatically from the balance recommended for adults (see Figure 10-11). Yet, for infants, breast milk is nature's most nearly perfect food, illustrating clearly that people at different stages of life have different nutrient needs.

The carbohydrate in breast milk (and standard infant formula) is lactose. In addition to being easily digested, lactose enhances calcium absorption. The carbohydrate component of breast milk also contains abundant oligosaccharides, which are present only in trace amounts in cow's milk and infant formula made from cow's milk. Human milk oligosaccharides help protect the infant from infection by preventing the binding of pathogens to the infant's intestinal cells.[102]

Human breast milk contains less protein than cow's milk, but this smaller quantity is actually beneficial because it places less stress on the infant's immature kidneys to excrete the major end product of protein metabolism, urea. The protein in breast milk is largely **alpha-lactalbumin**, which is efficiently digested and absorbed.

The lipids in breast milk—and infant formula—provide the main source of energy in the infant's diet. Breast milk contains a generous proportion of the essential fatty acids linoleic acid and linolenic acid, as well as their longer-chain derivatives arachidonic acid and docosahexaenoic acid (DHA). Most formulas today also contain added arachidonic acid and DHA (read the label). Infants can produce some arachidonic acid and DHA from linoleic and linolenic acid, respectively, but some infants may need more than they can make.

alpha-lactalbumin (lackt-AL-byoo-min): the chief protein in human breast milk, as **casein** (CAY-seen) is the chief protein in cow's milk.

FIGURE 10-11 **Percentages of Energy-Yielding Nutrients in Breast Milk and in Recommended Adult Diets**

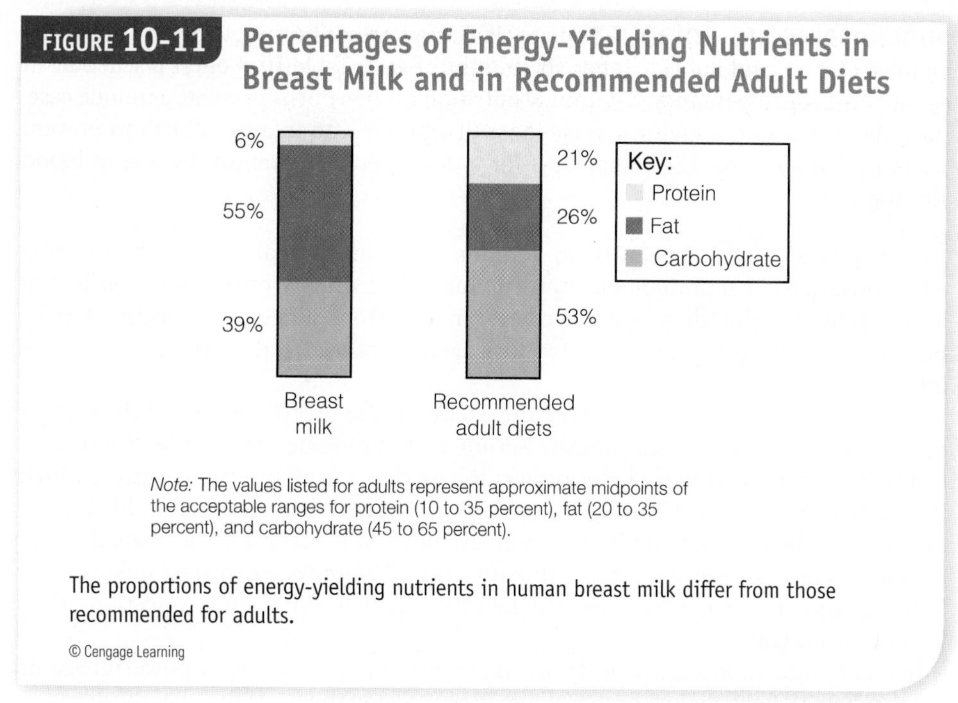
6%
55%
39%
21%
26%
53%
Key:
Protein
Fat
Carbohydrate
Breast milk
Recommended adult diets
Note: The values listed for adults represent approximate midpoints of the acceptable ranges for protein (10 to 35 percent), fat (20 to 35 percent), and carbohydrate (45 to 65 percent).

The proportions of energy-yielding nutrients in human breast milk differ from those recommended for adults.

As Chapter 4 mentioned, DHA is the most abundant fatty acid in the brain and is also present in the retina of the eye. DHA accumulation in the brain is greatest during fetal development and early infancy.[103] Research has focused on the visual and mental development of breastfed infants and infants fed standard formula with and without DHA added.[104] Results of studies for visual acuity development in term infants are mixed. When infants are fed formula with DHA added, only about half of the studies show higher visual acuity compared with controls.[105] Factors such as the amount of DHA provided, the sources of the DHA, and the sensitivity of different measures for visual acuity may have contributed to the mixed outcomes. A smaller number of studies have examined the effects of DHA status during fetal and infant development on cognitive function. Some of the evidence from these studies suggests that DHA supplementation during development can influence certain measures of cognitive function.[106] Studies that follow children beyond infancy, when more detailed tests of cognitive function can be employed, are needed to confirm the importance of DHA for optimal infant and child development.

Vitamins and Minerals With the exception of vitamin D, the vitamin content of the breast milk of a well-nourished mother is ample. Even vitamin C, for which cow's milk is a poor source, is supplied generously. The concentration of vitamin D in breast milk is low, however, and vitamin D deficiency impairs bone mineralization.[107] Vitamin D deficiency is most likely in infants who are not exposed to sunlight daily, have darkly pigmented skin, and receive breast milk without vitamin D supplementation.[108] Reports of infants in the United States developing the vitamin D–deficiency disease rickets and recommendations by the AAP to keep infants under six months of age out of direct sunlight have prompted revisions in vitamin D guidelines. The AAP currently recommends a vitamin D supplement for all infants who are breastfed exclusively and for any infants who do not receive at least 1 liter (1000 milliliters) or 1 quart (32 ounces) of vitamin D–fortified formula daily.[109] Despite such recommendations, most infants in the United States are consuming inadequate amounts of vitamin D.[110]

As for minerals, the calcium content of breast milk is ideal for infant bone growth, and the calcium is well absorbed. Breast milk is also low in sodium. The limited amount of iron in breast milk is highly absorbable, and its zinc, too, is absorbed better than from cow's milk, thanks to the presence of a zinc-binding protein.

Supplements for Infants Pediatricians may prescribe supplements containing vitamin D, iron, and fluoride (after six months of age). Table 10-8 offers a schedule of supplements during infancy. Vitamin K nutrition for newborns presents a unique case: the AAP recommends giving a single dose of vitamin K to infants at birth to prevent uncontrolled bleeding. (See Chapter 8 for a description of vitamin K's role in blood clotting.)

Immunological Protection In addition to its nutritional benefits, breast milk offers unsurpassed immunological protection.[111] Not only is breast milk sterile, but it also actively fights disease and protects infants from illnesses. Protective factors include antiviral agents, anti-inflammatory agents, antibacterial agents, and infection inhibitors.

During the first two or three days after delivery, the breasts produce **colostrum**, a premilk substance containing mostly serum with antibodies and white blood cells. Colostrum (like breast milk) helps protect the newborn from infections against which the mother has developed immunity—precisely those that the infant is most likely to be exposed to. The maternal antibodies in colostrum and breast milk inactivate disease-causing bacteria within the infant's digestive tract before they can start infections.[112] This explains, in part, why breastfed infants have fewer intestinal infections than formula-fed infants.

In addition to antibodies, colostrum and breast milk provide other powerful agents (■) that help to fight against bacterial infection. Among them are the oligosaccharides, described earlier, that prevent pathogens from binding to intestinal cells. Also present are **bifidus factors**, which favor the growth of the "friendly" bacterium *Lactobacillus bifidus* in the infant's digestive tract, so that other, harmful bacteria cannot become established. An iron-binding protein in breast milk, **lactoferrin**, keeps bacteria from getting the iron they need to grow, helps absorb iron into the infant's bloodstream, and kills some bacteria directly. The protein **lactadherin** in breast milk binds to, and inhibits replication of, the virus that causes most infant diarrhea. In addition, a growth factor that is present in breast milk stimulates the development and maintenance of the infant's digestive tract and its protective factors. Several breast milk enzymes such as lipase also help protect the infant against infection.

■ Protective factors in breast milk:
- Antibodies
- Oligosaccharides
- Bifidus factors
- Lactoferrin
- Lactadherin
- Growth factor
- Lipase enzyme

colostrum (co-LAHS-trum): a milk-like secretion from the breasts that is rich in protective factors. Colostrum is present during the first day or so after delivery, before milk appears.

bifidus (BIFF-id-us, by-FEED-us) factors: factors in colostrum and breast milk that favor the growth of the "friendly" bacterium *Lactobacillus* (lack-toe-ba-SILL-us) *bifidus* in the infant's intestinal tract. These bacteria prevent other, less desirable intestinal inhabitants from flourishing.

lactoferrin (lack-toe-FERR-in): protein in breast milk that binds iron and keeps it from supporting the growth of the infant's intestinal bacteria.

lactadherin (lack-tad-HAIR-in): a protein in breast milk that attacks diarrhea-causing viruses.

TABLE 10-8 Supplements for Full-Term Infants

	Vitamin D[a]	Iron[b]	Fluoride[c]
Breastfed infants:			
Birth to six months of age	√		
Six months to one year	√	√	√
Formula-fed infants:			
Birth to six months of age			
Six months to one year	√	√	

[a]Vitamin D supplements are recommended for all infants who are exclusively breastfed and for any infants who do not receive at least 1 liter (1000 milliliters) or 1 quart (32 ounces) of vitamin D–fortified formula per day.
[b]All infants six months of age need additional iron, preferably in the form of iron-fortified infant cereal and/or infant meats. Formula-fed infants need iron-fortified infant formula.
[c]At six months of age, breastfed infants and formula-fed infants who receive ready-to-use formulas (these are prepared with water low in fluoride) or formula mixed with water that contains little or no fluoride (less than 0.3 ppm) need supplements.

Source: Based on Committee on Nutrition, American Academy of Pediatrics, *Pediatric Nutrition Handbook*, 6th ed., ed. R. E. Kleinman (Elk Grove Village, Ill.: American Academy of Pediatrics, 2009).

Breastfeeding also protects against other common illnesses of infancy such as middle ear infection and respiratory illness.[113] Breast milk offers protection against the development of allergies as well. Compared with formula-fed infants, breastfed infants have a lower incidence of allergic reactions such as asthma, wheezing, and skin rash.[114] This protection is especially noticeable among infants with a family history of allergies. Breastfeeding also reduces the risk of sudden infant death syndrome (SIDS).[115] This protective effect is stronger when breastfeeding is exclusive, but any amount of breast milk for any duration is protective against SIDS. In one study, exclusively breastfeeding an infant for the first month reduced the risk of SIDS by half compared to never breastfeeding.[116] Breast milk may also offer protection against the development of cardiovascular disease, but more well-conducted research, using consistent and precise definitions of breastfeeding, is needed to confirm this effect.[117]

Clearly, breast milk is a very special substance. Nutrition in Practice 10 offers suggestions for successful breastfeeding.

Other Potential Benefits Breastfeeding may offer some protection against excessive weight gain later, although findings are inconsistent.[118] For example, some research suggests that the longer the duration of breastfeeding, the lower the risk of overweight in childhood, while other research conflicts with such findings.[119] Researchers note that many other factors—socioeconomic status, other infant and child feeding practices, and especially the mother's weight—strongly predict a child's body weight.[120]

Many studies suggest a beneficial effect of breastfeeding on later intelligence, but when subjected to strict standards of methodology (for example, large sample size and appropriate intelligence testing), the evidence is less convincing.[121] Nevertheless, the possibility that breastfeeding may positively affect later intelligence is intriguing. It may be that some specific component of breast milk, such as DHA, contributes to brain development or that certain factors associated with the feeding process itself promote the intellect.[122] Most likely, a combination of factors is involved. More large, well-controlled studies are needed to confirm the effects, if any, of breastfeeding on later intelligence.

The Case Study below presents a woman who is four months pregnant. Answering the questions offers practice in thinking through some of the issues related to pregnancy and breastfeeding.

INFANT FORMULA

Breastfeeding offers many benefits to both mother and infant, and it should be encouraged whenever possible. The mother who has decided to use formula, however, should be supported in her choice just as the breastfeeding mother should be. She can offer the same closeness, warmth, and stimulation during feedings as the breastfeeding mother can.

Case Study: Woman in Her First Pregnancy

Ellen Cassidy is a 24-year-old woman who is four months pregnant. This is her first pregnancy, and she is eager to learn how to feed herself during pregnancy as well as her infant after birth. She is 5 feet 3 inches tall and currently weighs 150 pounds. Her prepregnancy weight was 148 pounds. Ellen is very concerned about her 2-pound weight gain.

1. Consult the BMI table (inside back cover), and, using the "Healthy Weight" section, find a healthy weight in the middle of the range appropriate for a woman of Ellen's height.
2. Do you think that Ellen's weight at the start of her pregnancy was appropriate for her height? Why or why not? Should Ellen be concerned about her 2-pound weight gain? Why or why not?
3. What advice should you give Ellen about her weight gain during pregnancy? What other dietary advice would you give her?
4. Discuss methods of infant feeding with Ellen and describe some of the advantages breastfeeding would offer her. What advice will you give Ellen if she decides to breastfeed?
5. What are the advantages of formula feeding? What information should Ellen have about formula feeding?

The infant thrives on infant formula offered with affection.

Many mothers choose to breastfeed at first but **wean** their children within the first 1 to 12 months. If infants are younger than a year of age, mothers must wean them onto *infant formula*, (■) not onto plain cow's milk of any kind—whole, reduced-fat, low-fat, or fat-free.

Infant Formula Composition Manufacturers can prepare formulas from cow's milk in such a way that they do not differ significantly from human milk in nutrient content. Figure 10-12 illustrates the energy nutrient balance of breast milk, infant formula, and cow's milk. Notice the higher protein concentration of cow's milk, which stresses an infant's kidneys. The AAP recommends that all formula-fed infants receive iron-fortified infant formulas.[123] Low-iron formulas have no role in infant feeding. Use of iron-fortified formulas has risen in recent decades and is credited with the decline of iron-deficiency anemia in U.S. infants.

Infant Formula Standards National and international standards have been set for the nutrient contents of infant formulas. U.S. standards are based on AAP recommendations, and the FDA mandates quality control procedures to ensure that these standards are met. All standard formulas are therefore nutritionally similar. Small differences in nutrient content are sometimes confusing but usually unimportant.

Special Formulas Standard cow's milk–based formulas are inappropriate for some infants. Special formulas have been designed to meet the dietary needs of infants with specific conditions such as prematurity or inherited diseases. Most infants allergic to milk protein can drink formulas based on soy protein.[124] Soy formulas also use cornstarch and sucrose instead of lactose and so are recommended for infants with lactose intolerance as well. They are also useful as an alternative to milk-based formulas for vegan families. Some infants who are allergic to cow's milk protein may also be allergic to soy protein. For these infants, special formulas based on hydrolyzed protein are available.

Risks of Formula Feeding Infant formulas contain no protective antibodies for infants, but, in general, vaccinations, purified water, and clean environments in developed countries help protect infants from infections. Formulas can be prepared safely by following the rules of proper food handling and by using water that is free of contamination. Of particular concern is lead-contaminated water, a major source of lead poisoning in infants. Because the first water drawn from the tap each day is highest in

■ Formula preparation:

- Liquid concentrate (moderately expensive, relatively easy)—mix with equal part water.
- Powdered formula (least expensive, lightest for travel)—follow label directions.
- Ready-to-feed (easiest, most expensive)—pour directly into clean bottles.

wean: to gradually replace breast milk with infant formula or other foods appropriate to an infant's diet.

FIGURE 10-12 Percentages of Energy-Yielding Nutrients in Breast Milk, Infant Formula, and Cow's Milk

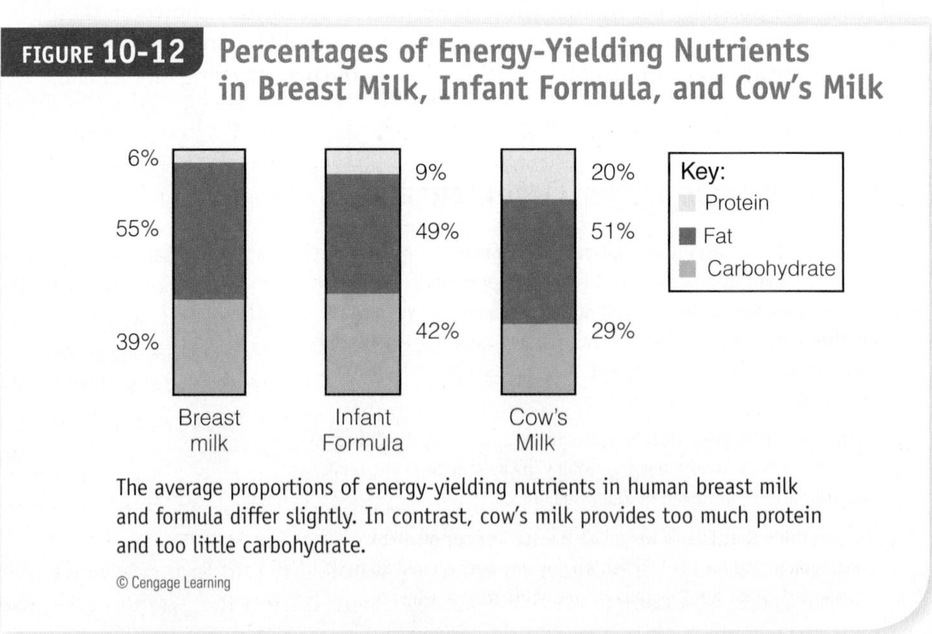

The average proportions of energy-yielding nutrients in human breast milk and formula differ slightly. In contrast, cow's milk provides too much protein and too little carbohydrate.

lead, a person living in a house with old, lead-soldered plumbing should let the water run a few minutes before drinking or using it to prepare formula or food.

In developing countries and in poor areas of the United States, formula may be unavailable, overdiluted in an attempt to save money, or prepared with contaminated water. Overdilution of formula can cause malnutrition and growth failure. Contaminated formula often causes infections leading to diarrhea, dehydration, and malabsorption. Wherever sanitation is poor, breastfeeding should be encouraged over feeding formula. Breast milk is sterile, and its antibodies enhance an infant's resistance to disease.

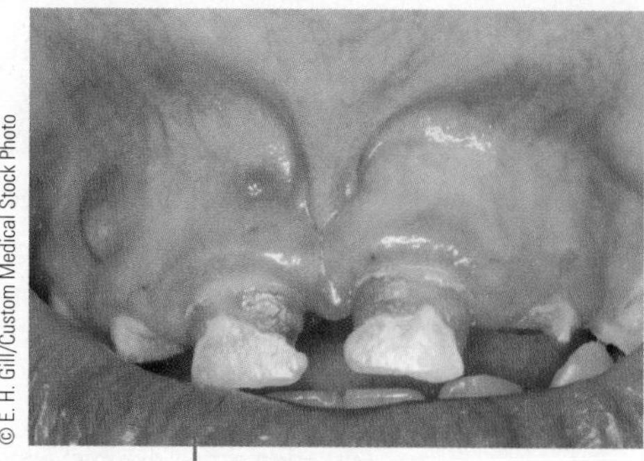

Nursing bottle tooth decay, an extreme example. The teeth have decayed all the way to the gum line.

Nursing Bottle Tooth Decay Dentists advise against putting an infant to bed with a bottle. Salivary flow, which normally cleanses the mouth, diminishes as the infant falls asleep. Prolonged sucking on a bottle of formula, milk, or juice bathes the upper teeth in a carbohydrate-rich fluid that nourishes decay-producing bacteria. (The tongue covers and protects most of the lower teeth, but they, too, may be affected.) The result is extensive and rapid tooth decay. To prevent **nursing bottle tooth decay**, no child should be put to bed with a bottle as a pacifier.

THE TRANSITION TO COW'S MILK

The AAP advises that cow's milk is not appropriate during the first year.[125] In some infants, particularly those younger than six months of age, cow's milk causes intestinal bleeding, which can lead to iron deficiency.[126] Cow's milk is also a poor source of iron. Consequently, it both causes iron loss and fails to replace iron. Furthermore, the bioavailability of iron from infant cereal and other foods is reduced when cow's milk replaces breast milk or iron-fortified formula during the first year. Compared with breast milk or iron-fortified formula, cow's milk is higher in calcium and lower in vitamin C, characteristics that reduce iron absorption. In short, cow's milk is a poor choice during the first year of life; infants need breast milk or iron-fortified infant formula.

Once the infant is obtaining at least two-thirds of total daily food energy from a balanced mixture of cereals, vegetables, fruits, and other foods (after 12 months of age), reduced-fat or low-fat cow's milk (in the context of an overall diet that supplies 30 percent of kcalories from fat) is an acceptable and recommended accompanying beverage.[127] After the age of 2, a transition to fat-free milk can take place, but care should be taken to avoid excessive restriction of dietary fat.

INTRODUCING FIRST FOODS

Changes in the body organs during the first year affect the infant's readiness to accept solid foods. Until the child is several months old, the immature stomach and intestines can digest milk sugar (lactose) but not starch. This is one of the many reasons why breast milk and formula are such good foods for an infant; they provide simple, easily digested carbohydrate that supplies energy for the infant's growth and activity.

When to Introduce Solid Food The AAP supports exclusive breastfeeding for approximately six months but recognizes that infants are often developmentally ready to accept complementary foods between four and six months of age.[128] The main purpose of introducing solid foods is to provide needed nutrients that are no longer supplied adequately by breast milk or formula alone.[129] The foods chosen must be those that the infant is developmentally capable of handling both physically and metabolically. The exact timing depends on the individual infant's needs and developmental readiness (see Table 10-9 on p. 302), which vary from infant to infant because of differences in growth rates, activities, and environmental conditions. In short, the addition of foods to an infant's diet should be governed by three considerations: the infant's nutrient needs, the infant's physical readiness to handle different forms of foods, and the need to detect and control allergic reactions. With respect to nutrient needs, the nutrients needed earliest are iron and then vitamin C.

nursing bottle tooth decay: extensive tooth decay due to prolonged tooth contact with formula, milk, fruit juice, or other carbohydrate-rich liquid offered to an infant in a bottle.

TABLE 10-9 Infant Development and Recommended Foods

Age (mo)	Feeding Skill	Foods Introduced into the Diet
0–4	Turns head toward any object that brushes cheek. Initially swallows using back of tongue; gradually begins to swallow using front of tongue as well. Strong reflex (extrusion) to push food out during first 2 to 3 months.	Feed breast milk or infant formula.
4–6	Extrusion reflex diminishes, and the ability to swallow nonliquid foods develops. Indicates desire for food by opening mouth and leaning forward. Indicates satiety or disinterest by turning away and leaning back. Sits erect with support at 6 months. Begins chewing action. Brings hand to mouth. Grasps objects with palm of hand.	Begin iron-fortified cereal mixed with breast milk, formula, or water. Begin pureed meats, legumes, vegetables, and fruits.
6–8	Able to self-feed finger foods. Develops pincher (finger to thumb) grasp. Begins to drink from cup.	Begin textured vegetables and fruits. Begin unsweetened, diluted fruit juices from cup.
8–10	Begins to hold own bottle. Reaches for and grabs food and spoon. Sits unsupported.	Begin breads and cereals from table. Begin yogurt. Begin pieces of soft, cooked vegetables and fruit from table. Gradually begin finely cut meats, fish, casseroles, cheese, eggs, and mashed legumes.
10–12	Begins to master spoon but still spills some.	Add variety. Gradually increase portion sizes.[a]

Note: Because each stage of development builds on the previous stage, the foods from an earlier stage continue to be included in all later stages.
[a]Portion sizes for infants and young children are smaller than those for an adult. For example, a grain serving might be ½ slice of bread instead of 1 slice, or ¼ cup rice instead of ½ cup.

Source: Adapted in part from Committee on Nutrition, American Academy of Pediatrics, *Pediatric Nutrition Handbook*, 6th ed., ed. R. E. Kleinman (Elk Grove Village, Ill.: American Academy of Pediatrics, 2009), pp. 113–142.

Foods to Provide Iron and Vitamin C Rapid growth demands iron. At about four to six months, the infant begins to need more iron than body stores plus breast milk or iron-fortified formula can provide. In addition to breast milk or iron-fortified formula, infants can receive iron from iron-fortified cereals and, once they readily accept solid foods, from meat or meat alternates such as legumes. Iron-fortified cereals contribute a significant amount of iron to an infant's diet, but the iron's bioavailability is poor.[130] Caregivers can enhance iron absorption from iron-fortified cereals by serving vitamin C–rich foods with meals.

The best sources of vitamin C are fruits and vegetables (see pp. 219–220 in Chapter 8). It has been suggested that infants who are introduced to fruits before vegetables may develop a preference for sweets and find the vegetables less palatable, but there is no evidence to support offering these foods in a particular order.[131] Fruit juice is a source of vitamin C, but excessive juice intake can lead to diarrhea in infants and young children.[132] Furthermore, too much fruit juice contributes excessive kcalories and displaces other nutrient-rich foods. AAP recommendations limit juice consumption for infants and young children (one to six years of age) to between 4 and 6 ounces per day.[133] Fruit juices should be diluted and served in a cup, not a bottle, once the infant is six months of age or older.

© Polara Studios, Inc.

Foods such as iron-fortified cereals and formulas, mashed legumes, and strained meats provide iron.

Physical Readiness for Solid Foods The ability to swallow solid food develops at around four to six months, and food offered by spoon helps to develop swallowing ability. At eight months to a year, an infant can sit up, can handle finger foods, and begins to teethe. At that time, hard crackers and other hard finger foods may be introduced to promote the development of manual dexterity and control of the jaw muscles. These feedings must occur under the watchful eye of an adult because the infant can also choke on such foods.

Some parents want to feed solids at an earlier age, on the theory that "stuffing the baby" at bedtime promotes sleeping through the night. There is no proof for this theory. On average, infants start to sleep through the night at about the same age (three to four months) regardless of when solid foods are introduced.

Allergy-Causing Foods To prevent allergy and to facilitate its prompt identification should it occur, experts recommend introducing single-ingredient foods, one at a time, in small portions, and waiting three to five days before introducing the next new food.[134] For example, rice cereal is usually the first cereal introduced because it is the least allergenic. When it is clear that rice cereal is not causing an allergy, another grain, perhaps barley or oats, is introduced. Wheat cereal is offered last because it is the most common offender. If a cereal causes an allergic reaction such as a skin rash, digestive upset, or respiratory discomfort, it should be discontinued before introducing the next food. Chapter 11 offers more information about food allergies.

Choice of Infant Foods Infant foods should be selected to provide variety, balance, and moderation. Commercial baby foods offer a wide variety of palatable, nutritious foods in a safe and convenient form. Parents or caregivers should not feed directly from the jar—remove the infant's portion and place it in a dish so as not to contaminate the leftovers that will be stored in the jar. Homemade infant foods can be as nutritious as commercially prepared ones, as long as the cook minimizes nutrient losses during preparation. Ingredients for homemade foods should be fresh, whole foods without added salt, sugar, or seasonings. Pureed food can be frozen in ice cube trays, providing convenient-sized blocks of food that can be thawed, warmed, and fed to the infant. To guard against foodborne illnesses, hands and equipment must be kept clean.

Foods to Omit Sweets of any kind (including baby food "desserts") have no place in an infant's diet. The added food energy conveys few, if any, nutrients to support growth and contributes to obesity. Canned vegetables are also inappropriate for infants; they often contain too much sodium. Honey and corn syrup should never be fed to infants because of the risk of botulism. Infants and young children are vulnerable to foodborne illnesses. An infant's caregiver must be on guard against food poisoning and take precautions against it as described in Nutrition in Practice 2.

Infants and even young children cannot safely chew and swallow any of the foods listed in the margin; (■) they can easily choke on these foods, a risk not worth taking.[135] Nonfood items of small size should always be kept out of the infant's reach to prevent choking.

■ To prevent choking, do not give infants or young children the following:

- Gum
- Popcorn
- Large raw apple slices
- Whole grapes
- Whole cherries
- Raw celery
- Raw carrots
- Whole beans
- Hot dog slices
- Sausage sticks or slices
- Hard or gel-type candies
- Marshmallows
- Nuts
- Peanut butter

Keep these nonfood items out of their reach:

- Coins
- Balloons
- Small balls
- Pen tops
- Other items of similar size

Foods at One Year At one year of age, reduced-fat or low-fat cow's milk can become a primary source of most of the nutrients an infant needs; 2 to 3 cups a day meet those needs sufficiently. More milk than this displaces iron-rich foods and can lead to the iron-deficiency anemia known as milk anemia. (■) Other foods—meat and meat alternates, iron-fortified cereals, whole-grain or enriched bread, fruits, and vegetables—should be supplied in variety and in amounts sufficient to round out total energy needs. Ideally, a one-year-old will sit at the table, eat many of the same foods everyone else eats, and drink liquids from a cup—not a bottle. Table 10-10 (p. 304) shows a sample menu that meets a one-year-old's requirements.

■ Reminder: *Milk anemia* develops when an excessive milk intake displaces iron-rich foods from the diet.

Jose Luis Pelaez, Inc./Jupiter Images

Ideally, a one-year-old eats many of the same healthy foods as the rest of the family.

LOOKING AHEAD

Probably the most important single measure to undertake during the first year is to encourage eating habits that will support continued normal weight as the child grows. This means introducing a variety of nutritious foods in an inviting way, not forcing the infant to finish the bottle or the baby food jar, avoiding concentrated sweets and empty-kcalorie foods, and encouraging physical activity. Parents should avoid teaching infants to seek food as a reward, to expect food as comfort for unhappiness, or to associate food deprivation with punishment. Normal dental development is also promoted by supplying nutritious foods, avoiding sweets, and discouraging the association of food with reward or comfort. Oral health is the subject of Nutrition in Practice 17.

MEALTIMES

The nurturing of a young child involves more than nutrition. Those who care for young children are responsible for providing not only nutritious foods, milk, and water but also a safe, loving, secure environment in which the children may grow and develop. The person feeding a one-year-old should be aware that exploring and experimenting are normal and desirable behaviors at this time in a child's life. The child is developing a sense of autonomy that, if allowed to develop, will lay the foundation for later assertiveness in choosing when and how much to eat and when to stop eating. In light of the developmental and nutrient needs of one-year-olds, and in the face of their often contrary and willful behavior, a few feeding guidelines may be helpful:

- *Discourage unacceptable behavior (such as standing at the table or throwing food) by removing the child from the table to wait until later to eat.* Be consistent and firm, not punitive. For example, instead of saying "You make me mad when you don't sit down," say "The fruit salad tastes good—please sit down and eat some with me." The child will soon learn to sit and eat.

TABLE 10-10 Sample Menu for a One-Year-Old

Breakfast	1 scrambled egg 1 slice whole-wheat toast ½ c reduced-fat milk
Morning snack	½ c yogurt ¼ c fruit[a]
Lunch	½ grilled cheese sandwich: 1 slice whole-wheat bread with 1 slice cheese ½ c vegetables[b] (steamed carrots) ¼ c 100% fruit juice
Afternoon snack	½ c fruit[a] ½ c toasted oat cereal
Dinner	1 oz chopped meat or ¼ c well-cooked mashed legumes ½ c rice or pasta ½ c vegetables[b] (chopped broccoli) ½ c reduced-fat milk

Note: This sample menu provides about 1000 kcalories.
[a]Include citrus fruits, melons, and berries.
[b]Include dark green, leafy, and deep yellow vegetables.

- *Let young children explore and enjoy food.* This may mean eating with fingers for a while. Learning to use a spoon will come in time. Children who are allowed to touch, mash, and smell their food while exploring it are more likely to accept it.
- *Don't force food on children.* Rejecting new foods is normal, and acceptance is more likely as children become familiar with new foods through repeated opportunities to taste them. Instead of saying "You cannot go outside to play until you taste your carrots," say "You can try the carrots again another time."
- *Provide nutritious foods, and let children choose which ones, and how much, they will eat.* Gradually, they will acquire a taste for different foods.
- *Limit sweets.* Infants and young children have little room for empty-kcalorie foods in their daily energy allowance. Do not use sweets as a reward for eating meals.
- *Don't turn the dining table into a battleground.* Make mealtimes enjoyable. Teach healthy food choices and eating habits in a pleasant environment. Mealtimes are not the time to fight, argue, or scold.

These recommendations reflect a spirit of tolerance that best serves the emotional and physical interests of the infant. This attitude, carried throughout childhood, helps the child to develop a healthy relationship with food. The Nutrition Assessment Checklist helps to identify nutrition-related factors that may help prevent or correct potential problems in pregnant women and infants. Details of nutrition assessment are presented in Chapter 13.

Review Notes

- The primary food for infants during the first 12 months is either breast milk or iron-fortified formula.
- In addition to nutrients, breast milk also offers immunological protection.
- At about four to six months, infants should gradually begin eating solid foods.
- The addition of foods to an infant's diet should be governed by three considerations: the infant's nutrient needs, the infant's physical readiness to handle different forms of foods, and the need to detect and control allergic reactions.
- By the time infants are one year old, they are drinking from a cup and eating many of the same foods as the rest of the family.

Nutrition Assessment Checklist for Pregnant Women and Infants

MEDICAL HISTORY

Check the medical record for:

- ❍ Alcohol or illicit drug abuse
- ❍ Chronic diseases
- ❍ Gestational diabetes
- ❍ History of previous pregnancies (number, intervals, outcomes, multiple births, and gestational age/birthweights)
- ❍ Hypertension
- ❍ Neural tube defect in an infant born previously
- ❍ Preeclampsia

Note risk factors for complications during pregnancy, including:

- ❍ Cigarette smoking
- ❍ Food faddism
- ❍ Lactose intolerance
- ❍ Low socioeconomic status
- ❍ Significant or prolonged vomiting
- ❍ Very young or old age
- ❍ Weight-loss dieting

Note any complaints of:

- ❍ Constipation
- ❍ Heartburn
- ❍ Morning sickness

MEDICATIONS

For pregnant women who are using drug therapy for medical conditions, note:

- ❍ Potential for contraindication to breastfeeding
- ❍ GI tract side effects that might reduce food intake or change nutrient needs

DIETARY INTAKE

For all pregnant women, especially those considered at risk nutritionally, assess the diet for:

- ❍ Total energy
- ❍ Protein
- ❍ Calcium, phosphorus, magnesium, iron, and zinc
- ❍ Folate and vitamin B_{12}
- ❍ Vitamin D

For infants, note:

- ❍ Method of feeding (breastfeeding, formula, or both)
- ❍ Frequency and duration of breastfeeding
- ❍ Amount of infant formula
- ❍ Practice of putting infant to bed with bottle
- ❍ Solid foods the infant is fed, if any
- ❍ Amount of food the infant is fed

ANTHROPOMETRIC DATA

Measure baseline height and weight:

- ❍ Prepregnancy weight
- ❍ Infant birthweight

Reassess weight at each medical checkup and determine whether gains are appropriate. Note:

- ❍ Weight gain during pregnancy
- ❍ Gestational age
- ❍ Weight, length, and head circumference of infants

LABORATORY TESTS

Monitor the following laboratory tests for pregnant women:

- ❍ Hemoglobin, hematocrit, or other tests of iron status
- ❍ Blood glucose

Monitor the following laboratory tests for infants:

- ❍ Blood glucose of infants born to mothers with gestational diabetes
- ❍ Results of tests for inborn errors

PHYSICAL SIGNS

Blood pressure measurement is a routine measurement in physical exams but is especially important for pregnant women.

Look for physical signs of:

- ❍ Iron deficiency
- ❍ Edema
- ❍ Protein-energy malnutrition
- ❍ Folate deficiency

Self-Check

1. The most important single predictor of an infant's future health and survival is:

a. the infant's birthweight.
b. the infant's iron status at birth.
c. the mother's weight at delivery.
d. the mother's prepregnancy weight.

2. A mother's prepregnancy nutrition is important to a healthy pregnancy because it determines the development of:

a. the largest baby possible.
b. adequate maternal iron stores.
c. an adequate fat supply for the mother.
d. healthy support tissues—the placenta, amniotic sac, umbilical cord, and uterus.

3. A pregnant woman needs an extra 450 calories above the allowance for nonpregnant women during which trimester(s)?

a. First
b. Second
c. Third
d. First, second, and third

4. Two nutrients needed in large amounts during pregnancy for rapid cell proliferation are:

a. vitamin B_{12} and vitamin C.
b. calcium and vitamin B_6.
c. folate and vitamin B_{12}.
d. copper and zinc.

5. For a woman who is at the appropriate weight for height and is carrying a single fetus, the recommended weight gain during pregnancy is:

a. 40 to 60 pounds.
b. 25 to 35 pounds.
c. 10 to 20 pounds.
d. 20 to 40 pounds.

6. Rewards of physical activity during pregnancy may include:

a. weight loss.
b. decreased incidence of pica.
c. relief from morning sickness.
d. reduced stress and easier labor.

7. Breast milk is recommended for the first 12 months of life because it offers complete nutrition and ___ to the infant.
   a. fluoride
   b. fructose
   c. immunological protection
   d. pica

8. An acceptable substitute for breast milk during the first year is:
   a. low-fat cow's milk.
   b. apple juice.
   c. water.
   d. iron-fortified infant formula.

9. The addition of solid foods to an infant's diet should be governed by all of the following considerations *except*:
   a. the infant's nutrient needs.
   b. the need to detect and control allergies.
   c. the infant's physical readiness to handle different forms of foods.
   d. the infant's sleep duration and patterns.

10. During the first year of life, the most important step to undertake to encourage healthy eating is to:
   a. give food as a reward or for comfort.
   b. give sweets as a reward for eating vegetables.
   c. introduce a variety of nutritious foods in an inviting way.
   d. restrict fat to less than 30 percent of energy intake.

Answers to these questions can be found in Appendix H. For more chapter review: Access an interactive eBook, chapter-specific interactive learning tools, including flashcards, quizzes, videos, and more in your Nutrition CourseMate, accessed through CengageBrain.com.

Clinical Applications

1. Consider the different factors in a pregnant woman's history that can affect her nutrition status and the outcome of her pregnancy. Describe what steps you would take to remedy potential problems for the following clients:
   a. A 15-year-old adolescent of low socioeconomic status is in her first trimester of pregnancy. She began the pregnancy at a normal, healthy weight, but her weight gain during pregnancy so far has been less than expected. Her favorite beverages are soft drinks; her favorite foods are French fries and boxed macaroni and cheese.
   b. A lactose-intolerant, 22-year-old pregnant woman has been eating a vegan diet for the past year or so. She began the pregnancy slightly underweight (BMI = 18.4), but her weight gain has been adequate and consistent during the four months of her pregnancy. She complains of feeling tired all the time.

2. What information would you give to a pregnant woman who is considering breastfeeding her infant but isn't quite sure why she should?

3. The parents of a two-month-old infant have been told by the child's grandparents that they should introduce solid foods to help the baby sleep through the night. What advice would you give the parents?

Nutrition on the Net

For further study of the topics in this chapter, access these websites.

- Visit the "Pregnancy week by week" section of the Mayo Clinic site:
  www.mayoclinic.com
- Learn more about having a healthy infant and about birth defects from the March of Dimes:
  www.marchofdimes.com
- Learn more about neural tube defects from the Spina Bifida Association of America:
  www.sbaa.org
- Search for "birth defects," "pregnancy," "adolescent pregnancy," "maternal and infant health," and "breastfeeding" at the U.S. government health information site:
  www.healthfinder.gov
- Search for "pregnancy" at the Academy of Nutrition and Dietetics site:
  www.eatright.org
- Learn more about the WIC program:
  www.fns.usda.gov/wic/
- Visit the American College of Obstetricians and Gynecologists:
  www.acog.org
- Learn more about gestational diabetes from the American Diabetes Association:
  www.diabetes.org
- Learn more about breastfeeding from La Leche League International:
  www.llli.org/
- Read *New Mother's Guide to Breastfeeding* at the American Academy of Pediatrics site:
  www.aap.org

Notes

1. J. C. Sadeu and coauthors, Alcohol, drugs, caffeine, tobacco, and environmental contaminant exposure: Reproductive health consequences and clinical implications, *Critical Reviews in Toxicology* 40 (2010): 633–652.
2. T. A. Simas and coauthors, Prepregnancy weight, gestational weight gain, and risk of growth affected neonates, *Journal of Women's Health* 21 (2012): 410–417.
3. J. L. Tarry-Adkins and S. E. Ozanne, Mechanisms of early life programming: Current knowledge and future directions, *American Journal of Clinical Nutrition* 94 (2011): 1765S–1771S; G. C. Burdge and K. A. Lillycrop, Nutrition, epigenetics, and developmental plasticity: Implications for understanding human disease, *Annual Review of Nutrition* 30 (2010): 315–339; C. Bouchard, Childhood obesity: Are genetic differences involved? *American Journal of Clinical Nutrition* 89 (2009): 1494S–1501S; M. E. Symonds, T. Stephenson, and H. Budge, Early determinants of cardiovascular disease: The role of early diet in later blood pressure control, *American Journal of Clinical Nutrition* 89 (2009): 1518S–1522S.
4. N. Wesglas-Kuperus and coauthors, Intelligence of very preterm or very low birthweight infants in young adulthood, *Archives of Disease in Childhood: Fetal and Neonatal Edition* 94 (2009): F196–200; D. S. Alam, Prevention of low birthweight, *Nestle Nutrition Workshop Series: Pediatric Program* 63 (2009): 209–225.
5. K. D. Kochanek and coauthors, Annual summary of vital statistics: 2009, *Pediatrics* 129 (2012): 338–348.
6. L. M. McCowan and coauthors, Spontaneous preterm birth and small for gestational age infants in women who stop smoking early in pregnancy: Prospective cohort study, *British Medical Journal* 338 (2009): b1081.
7. N. Kozuki, A. C. Lee, and J. Katz, Moderate to severe, but not mild, maternal anemia is associated with increased risk of small-for-gestational-age outcomes, *Journal of Nutrition* 142 (2012): 358–362; Alam, 2009; Position of the American Dietetic Association: Nutrition and lifestyle for a healthy pregnancy outcome, *Journal of the American Dietetic Association* 108 (2008): 553–561.
8. R. Retnakaran and coauthors, Effect of maternal weight, adipokines, glucose intolerance and lipids on infant birth weight among women without gestational diabetes mellitus, *Canadian Medical Association Journal*, 184 (2012): 1353–1360.
9. A. A. Mamun and coauthors, Associations of maternal pre-pregnancy obesity and excess pregnancy weight gains with adverse pregnancy outcomes and length of hospital stay, *BMC Pregnancy and Childbirth* 11 (2011): 62.
10. Position of the American Dietetic Association and American Society for Nutrition, Obesity, reproduction, and pregnancy outcomes, *Journal of the American Dietetic Association* 109 (2009): 918–927.
11. Mamun and coauthors, 2011; Position of the American Dietetic Association and American Society for Nutrition, 2009.
12. J. L. Mills and coauthors, Maternal obesity and congenital heart defects: A population-based study, *American Journal of Clinical Nutrition* 91 (2010): 1543–1549; K. J. Stothard and coauthors, Maternal overweight and obesity and the risk of congenital anomalies: A systematic review and meta-analysis, *Journal of the American Medical Association* 301 (2009): 636–650.
13. M. Desforges and C. P. Sibley, Placental nutrient supply and fetal growth, *The International Journal of Developmental Biology* 54 (2010): 377–390.
14. Position of the American Dietetic Association, 2008.
15. A. F. M. van Abeelen and coauthors, Survival effects of prenatal famine exposure, *American Journal of Clinical Nutrition* 95 (2012): 179–183; Tarry-Adkins and Ozanne, 2011; M. L. de Gusmão Correia and coauthors, Developmental origins of health and disease: Experimental and human evidence of fetal programming for metabolic syndrome, *Journal of Human Hypertension* 26 (2012): 405–419; L. C. Schulz, The Dutch hunger winter and the developmental origins of health and disease, *Proceedings of the National Academy of Sciences* 107 (2010): 16757–16758; S. H. Zeisel, Epigenetic mechanisms for nutrition determinants of later health outcomes, *American Journal of Clinical Nutrition* 89 (2009): 1488S–1493S.
16. Standing Committee on the Scientific Evaluation of Dietary Reference Intakes, Food and Nutrition Board, Institute of Medicine, *Dietary Reference Intakes for Energy, Carbohydrate, Fiber, Fat, Fatty Acids, Cholesterol, Protein, and Amino Acids* (Washington, DC: National Academies Press, 2005), pp. 185–194.
17. M. L. Blumfield and coauthors, Systematic review and meta-analysis of energy and macronutrient intakes during pregnancy in developed countries, *Nutrition Reviews* 70 (2012): 322–336.
18. L. D. Brown and coauthors, Maternal amino acid supplementation for intrauterine growth restriction, *Frontiers in Bioscience* 3 (2011): 428–444.
19. P. Haggarty, Fatty acid supply to the human fetus, *Annual Review of Nutrition* 30 (2010): 237–255.
20. A. S. de Souza, F. S. Fernandes, and M. Das Gracas Tavares do Carmo, Effects of maternal malnutrition and postnatal nutritional rehabilitation on brain fatty acids, learning, and memory, *Nutrition Reviews* 69 (2011): 132–144.
21. Centers for Disease Control and Prevention, Folic acid data and statistics, available at www.cdc.gov/ncbddd/folicacid/data/html.
22. U.S. Preventive Services Task Force, Folic acid for the prevention of neural tube defects: U. S. Preventive Services Task Force recommendation statement, *Annals of Internal Medicine* 150 (2009): 626–631.
23. Centers for Disease Control and Prevention, available at www.cdc.gov/ncbddd/folicacid/data/html.
24. S. H. Blanton and coauthors, Folate pathway and nonsyndromic cleft lip and palate, *Birth Defects Research. Part A, Clinical and Molecular Teratology* 91 (2011): 50–60.
25. I. Elmadfa and I. Singer, Vitamin B-12 and homocysteine status among vegetarians: A global perspective, *American Journal of Clinical Nutrition* 89 (2009): 1693S–1698S.
26. A. M. Molloy and coauthors, Maternal vitamin B12 status and risk of neural tube defects in a population with high neural tube defect prevalence and no folic acid fortification, *Pediatrics* 123 (2009): 917–923.
27. P. M. Brannon and M. F. Picciano, Vitamin D in Pregnancy and lactation in humans, *Annual Review of Nutrition* 31 (2011): 89–115; C. L. Wagner and F. R. Greer, and the Section on Breastfeeding and Committee on Nutrition, Prevention of rickets and vitamin D deficiency in infants, children, and adolescents, *Pediatrics* 122 (2008): 1142–1152.
28. Committee on Dietary Reference Intakes, *Dietary Reference Intakes for Calcium and Vitamin D* (Washington, DC: National Academies Press, 2011).
29. A. Merewood and coauthors, Widespread vitamin D deficiency in urban Massachusetts newborns and their mothers, *Pediatrics* 125 (2010): 640–647.
30. Committee on Dietary Reference Intakes, 2011, pp. 242–250.
31. A. N. Hacker, E. B. Fung, and J. C. King, Role of calcium during pregnancy: Maternal and fetal needs, *Nutrition Reviews* 70 (2012): 397–409.
32. H. J. McArdle and coauthors, Role of the placenta in regulation of fetal iron status, *Nutrition Reviews* 69 (2011): S17–S22.
33. T. O. Scholl, Maternal iron status: Relation to fetal growth, length of gestation, and iron endowment of the neonate, *Nutrition Reviews* 69 (2011): S23–S29.
34. Scholl, 2011.
35. S. H. Zeisel, Is maternal diet supplementation beneficial? Optimal development of infant depends on mother's diet, *American Journal of Clinical Nutrition* 89 (2009): 685S–687S.
36. Position of the American Dietetic Association, 2008.
37. J. M. Catov and coauthors, Periconceptional multivitamin use and risk of preterm or small-for-gestational-age births in the Danish National Birth Cohort, *American Journal of Clinical Nutrition* 94 (2011): 906–912.
38. WIC, The Special Supplemental Nutrition Program for Women, Infants, and Children, available at www.fns.usda.gov/fns.
39. Position of the American Dietetic Association and American Society for Nutrition, 2009; S. Y. Chu and coauthors, Gestational weight gain by body mass index among US women delivering live births, 2004–2005: Fueling future obesity, *American Journal of Obstetrics and Gynecology* 200 (2009): 271, e1–e7.
40. Position of the American Dietetic Association and American Society for Nutrition, 2009.
41. Position of the American Dietetic Association and American Society for Nutrition, 2009.
42. Institute of Medicine, *Weight Gain during Pregnancy: Reexamining the Guidelines* (Washington, DC: The National Academies Press, 2009).
43. J. H. Cohen and H. Kim, Sociodemographic and health characteristics associated with attempting weight loss during pregnancy, *Preventing Chronic Disease* 6 (2009): A07.
44. A. A. Mamun and coauthors, Association of excess weight gain during pregnancy with long-term maternal overweight and obesity: Evidence from 21 y postpartum follow-up, *American Journal* of *Clinical Nutrition* 91 (2010): 1336–1341; Position of the American Dietetic Association and American Society for Nutrition, 2009.

45. K. Melzer and coauthors, Physical activity and pregnancy: Cardiovascular adaptations, recommendations and pregnancy outcomes, *Sports Medicine* 40 (2010): 493–507; Position of the American Dietetic Association, 2008.
46. S. M. Ruchat and coauthors, Nutrition and exercise reduce excessive weight gain in normal-weight pregnant women, *Medicine and Science in Sports and Exercise* 44 (2012): 1419–1426; M. F. Mottola and coauthors, Nutrition and exercise prevent excess weight gain in overweight pregnant women, *Medicine and Science in Sports and Exercise* 42 (2010): 265–272.
47. S. L. Young, Pica in pregnancy: New ideas about an old condition, *Annual Review of Nutrition* 30 (2010): 403–422.
48. Young, 2010.
49. J. R. Niebyl, Nausea and vomiting in pregnancy, *New England Journal of Medicine* 363 (2010): 1544–1550.
50. A. Matthews and coauthors, Interventions for nausea and vomiting in early pregnancy, *Cochrane Database of Systemic Reviews* 8 (2010): CD007575.
51. J. L. Kitzmiller and coauthors, Preconception care for women with diabetes and prevention of major congenital malformations, *Birth Defects Research Part A: Clinical and Molecular Teratology* 88 (2010): 791–803.
52. Position statement, Standards of medical care in diabetes—2011, *Diabetes Care* 34 (2011): S11–S61; Y. Yogev and G. H. Visser, Obesity, gestational diabetes and pregnancy outcome, *Seminars in Fetal and Neonatal Medicine* 14 (2009): 77–84.
53. American Diabetes Association, Diagnosis and classification of diabetes mellitus, *Diabetes Care* 34 (2011): S62–S69.
54. P. E. Marik, Hypertensive disorders of pregnancy, *Postgraduate Medicine* 121 (2009): 69–76.
55. B. M. Sibai, Caring for women with hypertension, *Journal of the American Medical Association* 298 (2007): 1566–1568.
56. E. W. Seely and J. Ecker, Chronic hypertension in pregnancy, *New England Journal of Medicine* 365 (2011): 439–446.
57. H. Xu and coauthors, Role of nutrition in the risk of preeclampsia, *Nutrition Reviews* 67 (2009): 639–657; Position of the American Dietetic Association, 2008.
58. Position of the American Dietetic Association, 2008.
59. Xu and coauthors, 2009.
60. V. T. Tong and coauthors, Trends in smoking before, during, and after pregnancy—Pregnancy Risk Assessment Monitoring System (PRAMS), United States, 31 sites, 2000–2005, *Morbidity and Mortality Weekly Report* 58 (2009): SS-4, pp. 1–31.
61. E. Stephan-Blanchard and coauthors, Perinatal nicotine/smoking exposure and carotid chemoreceptors during development, June 25, 2012, http://dx.doi.org/1016/j.resp.2012.06.023; J. M. Rogers, Tobacco and pregnancy: Overview of exposures and effects, *Birth Defects Research* (Part C) 84 (2008): 1–15.
62. C. C. Geerts and coauthors, Parental smoking and vascular damage in their 5-year-old children, *Pediatrics* 129 (2010): 45–54.
63. H. Burke and coauthors, Prenatal and passive smoke exposure and incidence of asthma and wheeze: Systematic review and meta-analysis, *Pediatrics* 129 (2012): 735–744; A. Bjerg and coauthors, A strong synergism of low birth weight and prenatal smoking on asthma in schoolchildren, *Pediatrics* 127 (2011): e905–e912.
64. F. L. Trachtenberg and coauthors, Risk factor changes for sudden infant death syndrome after initiation of back-to-sleep campaign, *Pediatrics* 129 (2012): 630–638; H. C. Kinney and B. T. Thach, The sudden infant death syndrome, *New England Journal of Medicine* 361 (2009): 795–805.
65. A. Gunnerbeck and coauthors, Relationship of maternal snuff use and cigarette smoking with neonatal apnea, *Pediatrics* 128 (2011): 503–509.
66. Position of the American Dietetic Association, 2008.
67. J. P. Ackerman, T. Riggings, and M. M. Black, A review of the effects of prenatal cocaine exposure among school-aged children, *Pediatrics* 125 (2010): 554–565.
68. M. O'Donnell and coauthors, Increasing prevalence of neonatal withdrawal syndrome: Population study of maternal factors and child protection involvement, *Pediatrics* 123 (2009): e614–e621.
69. A. Bloomingdale and coauthors, A qualitative study of fish consumption during pregnancy, *American Journal of Clinical Nutrition* 92 (2010): 1234–1240.
70. Position of the Academy of Nutrition and Dietetics, Use of nutritive and nonnutritive sweeteners, *Journal of the Academy of Nutrition and Dietetics* 112 (2012): 739–758; Position of the American Dietetic Association, 2008.
71. R. Bakker and coauthors, Maternal caffeine intake, blood pressure, and the risk of hypertensive complications during pregnancy: The Generation R study, *American Journal of Hypertension* 24 (2011): 421–428; E. Maslova and coauthors, Caffeine consumption during pregnancy and risk of preterm birth: A meta-analysis, *American Journal of Clinical Nutrition* 92 (2010): 1120–1132; DC. Greenwood and coauthors, Caffeine intake during pregnancy, late miscarriage and stillbirth, *European Journal of Epidemiology* 25 (2010): 275–280.
72. R. Bakker and coauthors, Maternal caffeine intake from coffee and tea, fetal growth, and the risks of adverse birth outcomes: The Generation R Study, *American Journal of Clinical Nutrition* 91 (2010): 1691–1698.
73. E. P. Riley, M. A. Infante, and K. R. Warren, Fetal alcohol spectrum disorders: An overview, *Neuropsychological Review* 21 (2011): 73–80; K. L. Jones and coauthors, Fetal alcohol spectrum disorders: Extending the range of structural defects, *American Journal of Medical Genetics* 152A (2010): 2731–2735.
74. Centers for Disease Control, *Fetal Alcohol Spectrum* Disorders, www.cdc.gov/ncbddd/fasd/data.html; C. H. Denny and coauthors, Alcohol use among pregnant and nonpregnant women of childbearing age—United States, 1991–2005, *Morbidity and Mortality Weekly Report* 58 (2009): 529–532.
75. C. M. Marchetta and coauthors, Alcohol use and binge drinking among women of childbearing age—United States, 2006–2010, *Morbidity and Mortality Weekly Report* 61 (2012): 534–538; Denny and coauthors, 2009.
76. B. E. Hamilton and S. J. Ventura, Birth rates for U.S. teenagers reach historic lows for all age and ethnic groups, *NCHS Data Brief*, no. 89 (Hyattsville, MD: National Center for Health statistics, 2012).
77. P. N. Baker and coauthors, A prospective study of micronutrient status in adolescent pregnancy, *American Journal of Clinical Nutrition* 89 (2009): 1114–1124; S. Saintonge, H. Bang, and L. M. Gerber, Implications of a new definition of vitamin D deficiency in a multiracial US adolescent population: The National Health and Nutrition Examination Survey III, *Pediatrics* 123 (2009): 797–803.
78. Tong and coauthors, 2009.
79. B. E. Gould Rothberg and coauthors, Gestational weight gain and subsequent postpartum weight loss among young, low-income, ethnic minority women, *American Journal of Obstetrics and Gynecology* 204 (2011): e1–e11; M. M. Thame and coauthors, Weight retention within the puerperium in adolescents: A risk factor for obesity? *Public Health Nutrition* 13 (2010): 283–288.
80. E. P. Gunderson and coauthors, Longitudinal study of growth and adiposity in parous compared with nulligravid adolescents, *Archives of Pediatrics and Adolescent Medicine* 163 (2009): 349–356.
81. American Academy of Pediatrics, Policy statement: Breastfeeding and the use of human milk, *Pediatrics* 129 (2012): e827–e841, available at www.pediatrics.org/content/129/3/e827.full; Breastfeeding, in R. E. Kleinman, ed., Committee on Nutrition, American Academy of Pediatrics, *Pediatric Nutrition Handbook,* 6th ed. (Elk Grove Village, IL: American Academy of Pediatrics, 2009), pp. 29–59.
82. Position of the American Dietetic Association: Promoting and supporting breastfeeding, *Journal of the American Dietetic Association* 109 (2009): 1926–1942.
83. American Academy of Pediatrics, 2012; Position of the American Dietetic Association: Promoting and supporting breastfeeding, 2009.
84. A. J. Daly and coauthors, Maternal exercise and growth in breastfed infants: A meta-analysis of randomized controlled trials, *Pediatrics* 130 (2012): 108–114.
85. Committee on Dietary Reference Intakes, *Dietary Reference Intakes for Calcium and Vitamin D* (Washington, DC: National Academies Press, 2011), pp. 256–257.
86. J. A. Mennella and J. C. Trabulsi, Complementary foods and flavor experiences: Setting the foundation, *Annals of Nutrition and Metabolism* 60 (2012): 40–50; G. Beauchanp and J. A. Mennella, Flavor perception in human infants: Development and functional significance, *Digestion* 83 (2011): 1–6.
87. F. R. Greer, S. H. Sicherer, A. Wesley Burks, and the Committee on Nutrition and Section on Allergy and Immunology, Effects of early nutritional interventions on the development of atopic disease in infants and children: The role of maternal dietary restriction, breastfeeding, timing of introduction of complementary foods, and hydrolyzed formulas, *Pediatrics* 121 (2008): 183–191.
88. Greer, Sicherer, Wesley Burks, and the Committee on Nutrition and Section on Allergy and Immunology, 2008.
89. American Academy of Pediatrics, 2012.
90. Tong and coauthors, 2009.

91. P. Bachour and coauthors, Effects of smoking mother's age, body mass index, and parity number on lipid, protein, and secretory immunoglobulin A concentration of human milk, *Breastfeeding Medicine*, 7 (2012): 179–188.
92. G. Liebrechts-Akkerman and coauthors, Postnatal parental smoking: An important risk factor for SIDS, *European Journal of Pediatrics* 170 (2011): 1281–1291; G Yilmaz and coauthors, Effect of passive smoking on growth and infection rates of breast-fed and non-breast-fed infants, *Pediatrics International* 51 (2009): 352–358.
93. American Academy of Pediatrics, 2012.
94. N. Kapp, K. Curtis, and K. Nanda, Progestogen-only contraceptive use among breastfeeding women: A systematic review, *Contraception* 82 (2010): 17–37.
95. American Academy of Pediatrics, 2012.
96. American Academy of Pediatrics, 2012; P. L. Havens, L. M. Mofenson, and the Committee on Pediatric AIDS, Evaluation and management of the infant exposed to HIV-1 in the United States, *Pediatrics* 123 (2009): 175–187.
97. A. Koyanagi and coauthors, Effect of early exclusive breastfeeding on morbidity among infants born to HIV-negative mothers in Zimbabwe, *American Journal of Clinical Nutrition* 89 (2009): 1375–1382.
98. World Health Organization, *HIV and Infant Feeding*, available at http://www.who.int/child_adolescent_health/topics/prevention_care/child/nutrition/hivif/en/; M. W. Kline, Early exclusive breastfeeding: Still the cornerstone of child survival, *American Journal of Clinical Nutrition* 89 (2009): 1281–1282.
99. Formula feeding of term infants, in R. E. Kleinman, ed., Committee on Nutrition, American Academy of Pediatrics, *Pediatric Nutrition Handbook*, 6th ed. (Elk Grove Village, IL: American Academy of Pediatrics, 2009), pp. 61–78.
100. American Academy of Pediatrics, 2012.
101. Breastfeeding, 2009.
102. S. M. Donovan, Human milk oligosaccharides—The plot thickens, *British Journal of Nutrition* 101 (2009): 1267–1269.
103. Fat and fatty acids, in R. E. Kleinman, ed., *Pediatric Nutrition Handbook*, 6th ed. (Elk Grove Village, IL: American Academy of Pediatrics, 2009), pp. 357–386; S. E. Carlson, Docosahexaenoic acid supplementation in pregnancy and lactation, *American Journal of Clinical Nutrition* 89 (2009): 678S–684S.
104. S. J. Meldrum and coauthors, Achieving definitive results in long-chain polyunsaturated fatty acid supplementation trials of term infants: Factors for consideration, *Nutrition Reviews* 69 (2011): 205–214; L. G. Smithers, R. A. Gibson, and M. Makrides, Maternal supplementation with docosahexaenoic acid during pregnancy does not affect early visual development in the infant; A randomized controlled trial, *American Journal of Clinical Nutrition* 93 (2011): 1293–1299; E. E. Birch and coauthors, The DIAMOND (DHA Intake and Measurement of Neural Development) Study: A double-masked, randomized controlled clinical trial of the maturation of infant visual acuity as a function the dietary level of docosahexaenoic acid, *American Journal of Clinical Nutrition* 91 (2010): 848–859; S. E. Carlson, Early determinants of development: A lipid perspective, *American Journal of Clinical Nutrition* 89 (2009): 1523S–1529S.
105. Carlson, 2009.
106. M. Guxens and coauthors, Breastfeeding, long-chain polyunsaturated fatty acids in colostrums, and infant mental development, *Pediatrics* 128 (2011): e880–e889; Carlson, 2009.
107. S. A. Abrams, What are the risks and benefits to increasing dietary bone minerals and vitamin D intake in infants and small children? *Annual Review of Nutrition* 31 (2011): 285–297.
108. Fat-soluble vitamins, in R. E. Kleinman, ed., Committee on Nutrition, American Academy of Pediatrics, *Pediatric Nutrition Handbook*, 6th ed. (Elk Grove Village, IL: American Academy of Pediatrics, 2009), pp. 461–474.
109. Wagner, and Greer, and the Section on Breastfeeding and Committee on Nutrition, 2008.
110. C. G. Perrine and coauthors, Adherence to vitamin D recommendations among US infants, *Pediatrics* 125 (2010): 627–632.
111. American Academy of Pediatrics, 2012; Breastfeeding, 2009; Position of the American Dietetic Association, Promoting and supporting breastfeeding, 2009.
112. American Academy of Pediatrics, 2012; A. Walker, Breast milk as the gold standard for protective nutrients, *Journal of Pediatrics* 156 (2010): 53–57.
113. American Academy of Pediatrics, 2012; Breastfeeding, 2009; Position of the American Dietetic Association, 2009.
114. American Academy of Pediatrics, 2012.
115. F. R. Hauck, and coauthors, Breastfeeding and reduced risk of sudden infant death syndrome: A meta-analysis, *Pediatrics* 128 (2011): 103–110.
116. M. M. Vennemann and coauthors, Does breastfeeding reduce the risk of sudden infant death syndrome? *Pediatrics* 123 (2009): e406–e410.
117. I. Labayen and coauthors, Association of exclusive breastfeeding duration and fibrinogen levels in childhood and adolescence: The European Youth Heart Study, *Archives of Pediatric and Adolescent Medicine* 166 (2012): 56–61; C. G. Owen, P. H. Whincup, and D. G. Cook, Breast-feeding and cardiovascular risk factors and outcomes in later life: Evidence from epidemiological studies, *Proceedings of the Nutrition Society* 70 (2011): 478–484; M. S. Fewtrell, Breast-feeding and later risk of CVD and obesity: Evidence from randomized trials, *Proceedings of the Nutrition Society* 70 (2011): 472–477; D. A. Leon and G. Ronalds, Breast-feeding influences on later life—Cardiovascular disease, *Advances in Experimental Medicine and Biology* 639 (2009): 153–166.
118. K. Casazza, J. R. Fernandez, and D. B. Allison, Modest protective effects of breast-feeding on obesity, *Nutrition Today* 47 (2012): 33–38; L. Shields and coauthors, Breastfeeding and obesity at 21 years: A cohort study, *Journal of Clinical Nursing* 19 (2010): 1612–1617; L. Schack-Nielsen and coauthors, Late introduction of complementary feeding, rather than duration breastfeeding, may protect against adult overweight, *American Journal of Clinical Nutrition* 91 (2010): 619–627; L. Twells and L. A. Newhook, Can exclusive breastfeeding reduce the likelihood of childhood obesity in some regions of Canada? *Canadian Journal of Public Health* 101 (2010): 36–39; P. Chivers and coauthors, Body mass index, adiposity rebound and early feeding in a longitudinal cohort (Raine Study), *International Journal of Obesity* 34 (2010): 1169–1176; B. Koletzko and coauthors, Can infant feeding choices modulate later obesity risk? *American Journal of Clinical Nutrition* 89 (2009): 1502S–1508S.
119. Chivers and coauthors, 2010; Shields and coauthors, 2010.
120. Casazza, Fernandez, and Allison, 2012; K. L. Whitaker and coauthors, Comparing maternal and paternal intergenerational transmission of obesity risk in a large population-based sample, *American Journal of Clinical Nutrition* 91 (2010): 1560–1567; R. Li, S. B. Fein, and L. M. Grummer-Strawn, Do infants fed from bottles lack self-regulation of milk intake compared with directly breastfed infants? *Pediatrics* 125 (2010): e1386–e1393.
121. W. Jedrychowski and coauthors, Effect of exclusive breastfeeding on the development of children's cognitive function in the Krakow prospective birth cohort study, *European Journal of Pediatrics* 171 (2012): 151–158; M. A. Quigley and coauthors, Breastfeeding is associated with improved child cognitive development: A population-based cohort study, *Journal of Pediatrics* 160 (2012): 25–32; C. McCrory and R. Layte, The effect of breastfeeding on children's educational test scores at nine years of age: Results of an Irish cohort study, *Social Science and Medicine* 72 (2011); 1515–1521.
122. M. Guxens and coauthors, Breastfeeding, long-chain polyunsaturated fatty acids in colostrum, and infant mental development *Pediatrics* 128 (2011): e880–e889.
123. Formula feeding of term infants, 2009.
124. Formula feeding of term infants, 2009.
125. Formula feeding of term infants, 2009.
126. E. E. Ziegler, Consumption of cow's milk as a cause of iron deficiency in infants and toddlers, *Nutrition Reviews* 69 (2011): S37–S42.
127. Expert Panel on Integrated Guidelines for Cardiovascular Health and Risk Reduction in children and Adolescents, Summary report, *Pediatrics* 128 (2011): S213–S256.
128. Complementary feeding, in R. E. Kleinman, ed., Committee on Nutrition, American Academy of Pediatrics, *Pediatric Nutrition Handbook*, 6th ed. (Elk Grove Village, IL: American Academy of Pediatrics, 2009), pp. 113–142.
129. H. Przyrembel, Timing of introduction of complementary food: Short- and long-term health consequences, *Annals of Nutrition and Metabolism* 60 (2012): 8–20.
130. Iron, in R. E. Kleinman, ed., Committee on Nutrition, American Academy of Pediatrics, *Pediatric Nutrition Handbook*, 6th ed. (Elk Grove Village, IL: American Academy of Pediatrics, 2009), pp. 403–422.
131. Complementary feeding, 2009.
132. Feeding the child, in R. E. Kleinman, ed., Committee on Nutrition, American Academy of Pediatrics, *Pediatric Nutrition Handbook*, 6th ed. (Elk Grove Village, IL: American Academy of Pediatrics, 2009), pp. 145–174.
133. Feeding the child, 2009.
134. Complementary feeding, 2009.
135. American Academy of Pediatrics, Policy statement—Prevention of choking among children, *Pediatrics* 125 (2010): 601–607.

Nutrition in Practice

Encouraging Successful Breastfeeding

As discussed in Chapter 10, breastfeeding offers benefits to both mother and infant. The American Academy of Pediatrics (AAP), the Academy of Nutrition and Dietetics, and the Canadian Paediatric Society all advocate breastfeeding as the preferred means of infant feeding.[1] Promotion of breastfeeding is an integral part of the WIC program's nutrition education component. National efforts to promote breastfeeding seem to be working, at least to some extent: the percentage of infants who were ever breastfed rose from 60 percent among those born in 1994 to 77 percent among infants born in 2009.[2] Despite this encouraging trend, by six months, breastfeeding rates fall to 47 percent. The AAP and many other health organizations recommend exclusive breastfeeding for the first six months of life.[3] Exclusive breastfeeding is defined as an infant's consumption of human milk with no supplementation of any kind (no water, no other type of milk, no juice, and no other foods) except for vitamins, minerals, and medications. The AAP recommends that breastfeeding continue for at least a year and thereafter for as long as mutually desired.[4] In the United States, the prevalence of exclusive breastfeeding at six months is low (about 15 percent) and only about one in five infants is still breastfeeding at one year of age.[5] Increasing the rates of breastfeeding initiation and duration is one of the goals of *Healthy People 2020*:

> Increase the proportion of infants who are breastfed ever, at 6 months, and at 1 year. Increase the proportion of infants who are breastfed exclusively through 3 months, and through 6 months.[6]

Despite the trend toward increasing breastfeeding, the percentage of mothers choosing to breastfeed their infants and continuing to do so still falls short of goals.

Why don't more women choose to breastfeed their infants?

Many experts cite two major deterrents: infant formula manufacturers' public advertising and promotion of their products and the medical community's failure to encourage breastfeeding. Infant formula manufacturers spend millions of dollars each year marketing their products, often claiming that formula "is like breast milk." Such advertising efforts seem to be working. According to one survey, one out of four people of various ages, races, and socioeconomic backgrounds agrees with the statement "infant formula is as good as breast milk." Infant formula is an appropriate substitute for breast milk when breastfeeding is specifically contraindicated, but for most infants, the benefits of breast milk outweigh those of formula. Despite medical evidence of the benefits of breast milk, medical practice is not always supportive.[7]

As an example of the medical lack of encouragement, some hospitals routinely separate mother and infant soon after birth. The child's first feeding then comes from the bottle rather than the breast. Furthermore, many hospitals send new mothers home with free samples of infant formula. The World Health Organization opposes this practice because it sends a misleading message that medical authorities favor infant formula over breast milk for infants. Even in hospitals where women are encouraged to breastfeed and are supported in doing so, little, if any, assistance is available after hospital discharge when many breastfeeding women still need assistance. Up to half of mothers who initially breastfeed their infants stop within three months—seemingly due to lack of knowledge and support.

Women who receive early and repeated breastfeeding information and support breastfeed their infants longer than other women do. Information and instruction are especially important during the *prenatal* period when most women decide whether to breastfeed or to feed formula.[8] Nurses and other health care professionals can play a crucial role in encouraging successful breastfeeding by offering women adequate, accurate information about breastfeeding that permits them to make informed choices. Table NP10-1 (p. 312) lists tips for successful breastfeeding.

If breastfeeding is a natural process, what do mothers need to learn?

Although lactation is an automatic physiological process, breastfeeding requires some learning. This learning is most successful in a supportive environment. It begins with preparatory steps taken before the infant is born.

What are these preparatory steps?

Toward the end of pregnancy and throughout lactation, a woman who intends to breastfeed should stop using soap and lotions on her breasts. The natural secretions of the breasts themselves lubricate the nipple area best. A woman who plans to breastfeed should also acquire at least two nursing bras before her infant is born. The bras should

TABLE NP10-1 Tips for Successful Breastfeeding

- Learn about the benefits of breastfeeding.
- Initiate breastfeeding within 1 hour of birth.
- Ask a health care professional to explain how to breastfeed and how to maintain lactation.
- Give newborn infants no food or drink other than breast milk, unless medically indicated.
- Breastfeed on demand.
- Give no artificial nipples or pacifiers to breastfeeding infants.[a]
- Find breastfeeding support groups, books, or websites to help troubleshoot breastfeeding problems.

[a]Compared with nonusers, infants who use pacifiers breastfeed less frequently and stop breastfeeding at a younger age.

provide good support and have drop-flaps so that either breast can be freed for nursing.

How soon after birth should breastfeeding start?

As soon as possible. Immediately after the delivery, for a short period, the infant is intensely alert and intent on suckling. This is the ideal time for the first breastfeeding and facilitates successful lactation.

What does the new mother need to know to continue breastfeeding her infant successfully?

She needs to learn how to relax and position herself so that she and the infant will be comfortable and the infant can breathe freely while nursing. She also needs to understand that infants have a **rooting reflex** that makes them turn toward any touch on the face. (The accompanying glossary defines this and other relevant terms.) Consequently, she should touch the infant's cheek to her nipple so that the infant will turn the right way to nurse. The mother can then place four fingers under the breast and her thumb on top to support the breast and present the nipple to the infant. The mother's fingers and thumb should be behind the areola, the colored ring around the nipple, so as not to interfere with the infant latching onto the breast (see Figure NP10-1). With the breast supported, the mother tickles the infant's lips with the breast until the infant's mouth opens wide. The mother can then gently bring the infant forward onto the breast. The nipple must rest well back on the infant's tongue so that the infant's gums will squeeze on the glands that release the milk and swallowing will be effortless. To break the suction, if necessary, the mother can slip a finger between the infant's mouth and her breast.

NURSING DIAGNOSIS

Ineffective breastfeeding applies to clients whose infants are unable to attach to the maternal breast correctly.

Does it hurt to have the infant sucking so hard on the breast?

Breastfeeding should not be painful if the infant is positioned correctly. The mother has a **letdown reflex** that forces milk to the front of her breast when the infant begins to nurse, virtually propelling the milk into the infant's mouth. Letdown is necessary for the infant to obtain milk easily, and the mother needs to relax for letdown to occur. The mother who assumes a comfortable position in an environment without interruptions will find it easiest to relax.

How long should the infant be allowed to nurse at each feeding?

Even though the infant obtains about half the milk from the breast during the first 2 or 3 minutes of suckling, and 80 to 90 percent of it within 4 minutes, the infant should be encouraged to breastfeed on the first breast for as long as he or she wishes before being offered the second breast. The suckling itself, as well as the complete removal of milk from the breast, stimulates the mammary glands to produce milk for the next nursing session. Successive sessions should start on alternate breasts to ensure that each breast is emptied regularly. This pattern maintains the same supply and demand for each breast and thus prevents either breast from overfilling.

Infants should be fed "on demand" and not be held to a rigid schedule. The breastfed infant may average 8 to 12 feedings per 24-hour period during the first month or so. Once the mother's milk supply is well established and the infant's capacity has increased, the intervals between feedings will become longer.

Glossary of Breastfeeding Terms

engorgement: overfilling of the breasts with milk.
letdown reflex: the reflex that forces milk to the front of the breast when the infant begins to nurse.
mastitis: infection of a breast.
rooting reflex: a reflex that causes an infant to turn toward whichever cheek is touched, in search of a nipple.

FIGURE NP10-1 Infant's Grasp on Mother's Breast

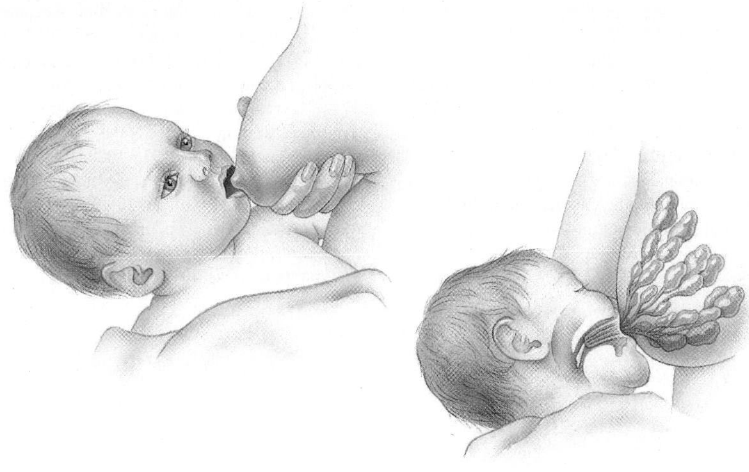

The mother supports the breast with her fingers and thumb behind the areola to present the nipple to the infant. Once the infant latches onto the breast, the infant's lips and gums pump the areola, releasing milk from the mammary glands into the milk ducts that lie beneath the areola.

What if a mother wants to skip one or two feedings daily—for example, because she works outside the home?

The mother can express breast milk into a bottle ahead of time, freeze the breast milk, and, when needed, substitute the expressed breast milk for a nursing session. Breast milk can be kept refrigerated for 48 hours or frozen. Frozen milk can be kept for one month in a freezer attached to a refrigerator or for three to six months in a zero-degree deep freezer.[9]

The mother can hand express her breast milk or use one of several different breast pumps available. The bicycle-horn type of manual breast pump is difficult to keep clean and is not recommended. Cylinder-type manual pumps or electric breast pumps are safer and are also more efficient. Alternatively, a mother can substitute formula for those missed feedings and continue to breastfeed at other times.

What about problems associated with breastfeeding such as sore nipples or infection of the breast?

Most problems associated with breastfeeding can be resolved. Many mothers experience sore nipples during the initial days of breastfeeding. Sore nipples need to be treated kindly, but nursing can continue. Improper feeding position is a frequent cause of sore nipples: the mother should make sure the infant is taking the entire nipple and part of the areola onto the tongue. She should nurse on the less sore breast first to get letdown going while the infant is suckling hardest; then she can switch to the sore breast. Between times, she should expose her nipples to light and air to heal them.

Before lactation is well established, when the schedule changes, or when a feeding is missed, the breasts may become full and hard—an uncomfortable condition known as **engorgement**. The infant cannot grasp an engorged nipple and so cannot provide relief by nursing. A gentle massage or warming the breasts with a cloth soaked in warm water or in a shower helps to initiate letdown and to release some of the accumulated milk; then the mother can pump out some of her milk and allow the infant to nurse.

Infection of the breast, known as **mastitis**, is best managed by continuing to breastfeed. By drawing off the milk, the infant helps to relieve pressure in the infected area. The infant is safe because the infection is between the milk-producing glands, not inside them.

Even if everything is going smoothly, the nursing mother should ideally have enough help and support so that she can rest in bed a few hours each day for the first week or so. Successful breastfeeding requires the support of all those who care. This, plus adequate nutrition, ample fluids, fresh air, and physical activity, will do much to enhance the well-being of mother and infant.

Notes

1. American Academy of Pediatrics, Policy statement: Breastfeeding and the use of human milk, *Pediatrics* 129 (2012): e827–e841, available at www.pediatrics.org/content/129/3/e827.full; Breastfeeding, in R. E. Kleinman, ed., *Pediatric Nutrition Handbook,* 6th ed. (Elk Grove Village, IL: American Academy of Pediatrics, 2009), pp. 29–59; Position of the American Dietetic Association: Promoting and supporting breastfeeding, *Journal of the American Dietetic Association* 109 (2009): 1926–1942.
2. Centers for Disease Control and Prevention, *Breastfeeding report card—United States, 2012,* available at www.cdc.gov/breastfeeding/data/reportcard.htm.
3. American Academy of Pediatrics, 2012; Position of the American Dietetic Association: Promoting and supporting breastfeeding, 2009.
4. American Academy of Pediatrics, 2012.
5. R. Jones and coauthors, Factors associated with exclusive breastfeeding in the United States, *Pediatrics* 128 (2011): 1117–1125.
6. U. S. Department of Health and Human Services, *Healthy People, 2020,* available at www.healthypeople.gov.
7. C. Perrine and coauthors, Vital signs: Hospital practices to support breastfeeding—United States, 2007 and 2009, *Morbidity and Mortality Weekly Report* 60 (2011): 1020–1025.
8. A. Avery and coauthors, Confident commitment is a key factor for sustained breastfeeding, *Birth* 36 (2009): 141–148.
9. Breastfeeding beyond infancy, in J. Y. Meek and W. Yu, eds., *American Academy of Pediatrics, New Mother's Guide to Breastfeeding* (New York: Bantam Books, 2011), pp. 158–183.

Chapter 11 Nutrition through the Life Span: Childhood and Adolescence

Redius Images/Jupiter Images

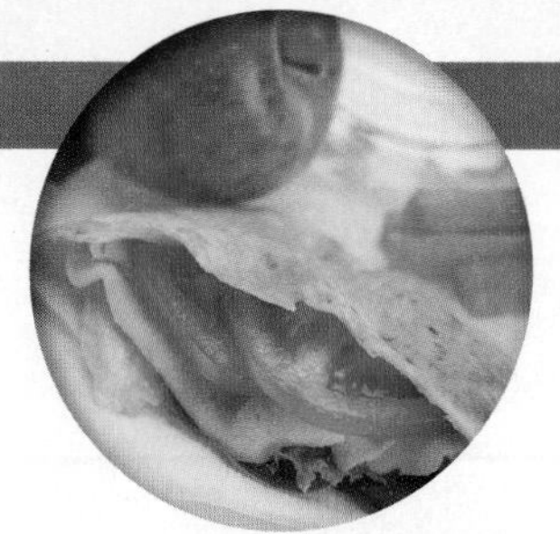

NUTRIENT NEEDS CHANGE THROUGHOUT LIFE, DEPENDING ON RATES OF growth, activity, and many other factors. Nutrient needs also vary from individual to individual, but generalizations are possible and useful. Sound nutrition throughout childhood promotes normal growth and development; facilitates academic and physical performance; and helps prevent obesity, diabetes, heart disease, cancer, and other chronic diseases in adulthood. As children enter the teen years, a foundation built by years of eating nutritious foods best prepares them to meet the upcoming demands of rapid growth.

Nutrition during Childhood

After the age of one, growth rate slows, but the body continues to change dramatically (see Figure 11-1). At one, infants have just learned to stand and toddle; by two, they walk confidently and are learning to run, jump, and climb. Nutrition and physical activity have helped them prepare for these new accomplishments by adding to the mass and density of their bone and muscle tissue. Thereafter, their bones continue to grow longer and their muscles to gain size and strength, though unevenly and more slowly, until adolescence.

ENERGY AND NUTRIENT NEEDS

Children's appetites begin to diminish around the first birthday, consistent with the slowed growth rate. Thereafter, the appetite fluctuates. At times children seem to be insatiable, and at other times they seem to live on air and water. Parents and other caregivers need not worry: given an ample selection of nutritious foods at regular intervals, internal appetite regulation in healthy, normal-weight children guarantees that their food energy intakes will be right for each stage of growth.

Ideally, children accumulate stores of nutrients before adolescence. Then, when they take off on the adolescent growth spurt and their nutrient intakes cannot keep pace with the demands of rapid growth, they can draw on the nutrient stores accumulated earlier. This is especially true of calcium; the denser the bones are in childhood, the better prepared they will be to support teen growth and still withstand the inevitable bone losses of later life.[1] Consequently, the way children eat influences their nutritional health during childhood, during their teen years, and for the rest of their lives.

FIGURE 11-1 Body Shape of a One-Year-Old and a Two-Year-Old Compared

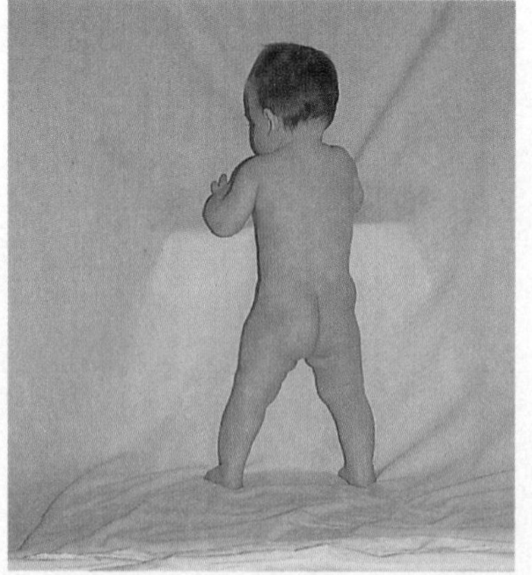

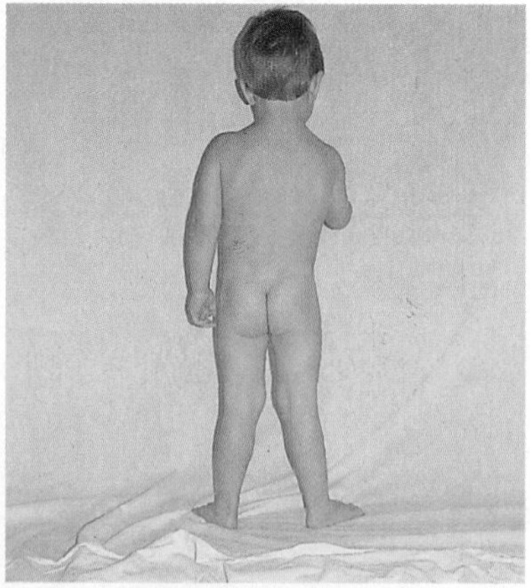

The body shape of a one-year-old (left) changes dramatically by age two (right). The two-year-old has lost much of the baby fat; the muscles (especially in the back, buttocks, and legs) have firmed and strengthened; and the leg bones have lengthened.

Children's Appetites Many people mistakenly believe that they must "make" their children eat the right amounts of food, and children's erratic appetites often reinforce this belief.[2] Although children's food energy intakes vary widely from meal to meal, total daily energy intake remains remarkably constant. If children eat less at one meal, they typically eat more at the next, and vice versa.

Parents do, however, need to help children choose the right foods, and, with overweight children, they may need to help more, as described later. Overweight children may not adjust their energy intakes appropriately and may disregard appetite-regulation signals and eat in response to external cues, such as television commercials.[3]

NURSING DIAGNOSIS

Risk for imbalanced nutrition—more than body requirements applies to clients who eat in response to external cues.

Energy Individual children's energy needs vary widely, depending on their growth and physical activity. A one-year-old child needs approximately 800 kcalories a day; an active six-year-old needs twice as many kcalories a day. By age 10, an active child needs about 2000 kcalories a day. Total energy needs increase gradually with age, but energy needs per kilogram of body weight actually decline. Physically active children of any age need more energy because they expend more, and inactive children can become obese even when they eat less food than the average. Unfortunately, our nation's children are becoming less and less active; child care programs and schools would serve our children well by offering more activities to promote physical fitness.[4]

Some children, notably those adhering to a vegan diet, may have difficulty meeting their energy needs. Grains, vegetables, and fruits provide plenty of fiber, adding bulk, but may provide too few kcalories to support growth. Soy products, other legumes, and nut or seed butters offer more concentrated sources of energy to support optimal growth and development.[5]

NURSING DIAGNOSIS

Imbalanced nutrition—less than body requirements applies to clients who have nutrient intakes insufficient to meet metabolic needs.

Carbohydrate and Fiber Carbohydrate recommendations are based on glucose use by the brain. After one year of age, brain glucose use remains fairly constant and is within the adult range. Carbohydrate recommendations for children after one year are therefore the same as for adults (see inside front cover).[6]

Fiber recommendations (■) derive from adult intakes shown to reduce the risk of coronary heart disease and are based on energy intakes. Consequently, fiber recommendations for younger children with low energy intakes are less than those for older ones with high energy intakes.[7]

Fat and Fatty Acids No RDA for total fat has been established, but the DRI committee recommends a fat intake of 30 to 40 percent of energy for children 1 to 3 years of age and 25 to 35 percent for children 4 to 18 years of age.[8] However, as long as children's energy intakes are adequate, fat intakes below 30 percent of total energy do not impair growth.[9] Children who eat low-fat diets, however, tend to have low intakes of some vitamins and minerals. Recommended intakes of the essential fatty acids are based on average intakes (see inside front cover).

Protein Like energy needs, total protein needs increase slightly with age but actually decline slightly when the child's body weight is considered (see inside front cover). Protein recommendations must address the requirements for maintaining nitrogen balance, the quality of protein consumed, and the added needs of growth.

Vitamins and Minerals The vitamin and mineral needs of children increase with age (see inside front cover). A balanced diet of nutritious foods can meet children's needs for these nutrients, with the notable exception of iron, and possibly vitamin D. Iron-deficiency anemia, a major problem worldwide, is prevalent among U.S. and Canadian children, especially toddlers one to three years of age.[10] During the second year of life, toddlers progress from a diet of iron-rich infant foods such as breast milk, iron-fortified formula, and iron-fortified infant cereal to a diet of adult foods and iron-poor cow's milk. In addition, their appetites often fluctuate—some become finicky about the foods they eat, and others prefer milk and juice to solid foods. These situations can interfere with children eating iron-rich foods at a critical time for brain growth and development.

■ Fiber recommendations for children:

Age (yr)	DRI (g/day)	
1–3	19	
4–8	25	
9–13	Boys	31
	Girls	26
14–18	Boys	38
	Girls	26

TABLE 11-1 Iron-Rich Foods Children Like[a]

Breads, cereals, and grains
- Cream of wheat (¼ c)
- Fortified dry cereals (1 oz)[b]
- Noodles, rice, or barley (½ c)
- Tortillas (1 flour or whole wheat, 2 corn)
- Whole-wheat, enriched, or fortified bread (1 slice)

Vegetables
- Cooked snow peas (½ c)
- Cooked mushrooms (½ c)
- Green peas (½ c)
- Mixed vegetable juice (1 c)

Fruits
- Canned plums (3 plums)
- Cooked dried apricots (¼ c)
- Dried peaches (4 halves)
- Raisins (1 tbs)

Meats and legumes
- Bean dip (¼ c)
- Lean chopped roast beef or cooked ground beef (1 oz)
- Liverwurst on crackers (½ oz)
- Meat casseroles (½ c)
- Mild chili or other bean/meat dishes (¼ c)
- Peanut butter and jelly sandwich (½ sandwich)
- Sloppy joes (½ sandwich)

[a]Each serving provides at least 1 milligram of iron. Vitamin C–rich foods included with these snacks increase iron absorption.
[b]Some fortified breakfast cereals contain more than 10 milligrams of iron per half-cup serving (read the labels).

To prevent iron deficiency, children's foods must deliver 7 to 10 milligrams of iron per day. To achieve this goal, snacks and meals should include iron-rich foods, and milk intake should be reasonable so that it will not displace lean meats, fish, poultry, eggs, legumes, and whole-grain or enriched products. That means 2½ cups of milk per day up to age eight, increasing to 3 cups per day from age nine on. After age one, as long as the overall diet supplies 30 percent of total kcalories from fat, reduced-fat or low-fat milk can be used instead of whole milk.[11] The saved kcalories can be invested in iron-rich foods such as lean meats, fish, poultry, eggs, and legumes. Whole-grain or enriched breads and cereals also contribute iron. Table 11-1 lists iron-rich foods children like.

The DRI committee recently revised their recommendations for vitamin D intakes for healthy Americans.[12] Children typically obtain most of their vitamin D from fortified milk (2.5 micrograms per 1 cup serving) and dry cereals (1 microgram per 1/2 cup serving). Children who do not meet their RDA (15 micrograms) from these sources should receive a vitamin D supplement.[13] Remember that sunlight is also a source of vitamin D, especially in warm climates and warm seasons.

Supplements

With the exception of specific recommendations for fluoride, iron, and vitamin D during infancy and childhood, the AAP and other professional groups agree that well-nourished children do not need vitamin and mineral supplements. Despite this, many children and adolescents take supplements.[14] Ironically, children with poor nutrient intakes typically do not receive supplements, whereas those who do take them usually receive extra nutrients they do not need.[15] Furthermore, researchers are still studying the safety of supplement use by children. The Federal Trade Commission has warned parents about giving children supplements advertised to prevent or cure childhood illnesses such as colds, ear infections, or asthma. Dietary supplements on the market today include many herbal products that have not been tested for safety and effectiveness in children.

Food Patterns for Children

To provide all the needed nutrients, a child's meals and snacks should include a variety of foods from each food group—in amounts suited to the child's appetite and needs. Table 11-2 provides USDA Food Patterns for several kcalorie

TABLE 11-2 USDA Food Patterns: Recommended Daily Amounts for Each Food Group (1000 to 1800 kCalories)

Food Group	1000 kcal	1200 kcal	1400 kcal	1600 kcal	1800 kcal
Fruit	1 c	1 c	1½ c	1½ c	1½ c
Vegetables	1 c	1½ c	1½ c	2 c	2½ c
Grains	3 oz	4 oz	5 oz	5 oz	6 oz
Protein Foods	2 oz	3 oz	4 oz	5 oz	5 oz
Milk	2 c	2½ c	2½ c	3 c	3 c

levels. (■) Estimated daily kcalorie needs for active and sedentary children of various ages are shown in Table 11-3. MyPlate online resources for preschoolers (2 to 5 years) translate the eating patterns into messages that can help parents ensure that the foods they provide meet their child's needs. For children older than 5 (6 to 11 years), the site provides an interactive "Blast Off" nutrition teaching game and other resources for teachers, parents, and children themselves (Figure 11-2). These guidelines and resources also stress the importance of balancing kcalorie intake with kcalorie expenditure through adequate physical activity to promote growth without increasing the chances of developing obesity. Childhood obesity is the topic of a later section in this chapter.

TABLE 11-3 Estimated Daily kCalorie Needs for Children

Children	Sedentary[a]	Active[b]
2 to 3 yr	1000	1400
Females		
4 to 8 yr	1200	1800
9 to 13 yr	1400	2200
Males		
4 to 8 yr	1200	2000
9 to 13 yr	1600	2600

[a]*Sedentary* describes a lifestyle that includes only the activities typical of day-to-day life.
[b]*Active* describes a lifestyle that includes at least 60 minutes per day of moderate physical activity (equivalent to walking more than 3 miles per day at 3 to 4 miles per hour) in addition to the activities of day-to-day life.

Children whose diets follow the patterns presented in Table 11-2 meet their nutrient needs fully, but few children eat according to these recommendations. One analysis of the quality of children's diets found that most children (up to 88 percent) between two and nine years of age have diets that need substantial improvement.[16] A comprehensive survey, called the Feeding Infants and Toddlers Study (FITS), assessed the food and nutrient intakes of more than 3000 infants and toddlers.[17] The survey found that fruit and vegetable intakes of infants and toddlers are limited, and, in fact, about 25 percent of two- and three-year-old children did not eat a single serving of fruits or vegetables in a day.[18] White potatoes were the most commonly consumed vegetable, often in fried form. Of great concern is the finding that more than 80 percent of preschoolers consumed one or more low-nutrient, energy-dense beverages, desserts, and snack foods in a day. Parents and caregivers of infants and toddlers thus need to offer a much greater variety of nutrient-dense vegetables and fruits at meals and snacks to help ensure adequate nutrition. Among other nutrition concerns for U.S. children are inadequate intakes of vitamin E, potassium, and fiber, and excessive intakes of sodium and saturated fat.[19]

Children's Food Choices The childhood years are the parents' best chance to influence their child's food choices.[20] Appropriate eating habits and attitudes toward food, developed in childhood, can help future adults emerge with healthy habits that reduce risks of chronic diseases in later life. The challenge is to deliver nutrients in the form of meals and snacks that are both nutritious and appealing so that children will learn to enjoy a variety of health-promoting foods.

■ For kcalorie levels over 1800, see Table 1-7 on p. 20.

FIGURE 11-2 MyPlate Resources for Children

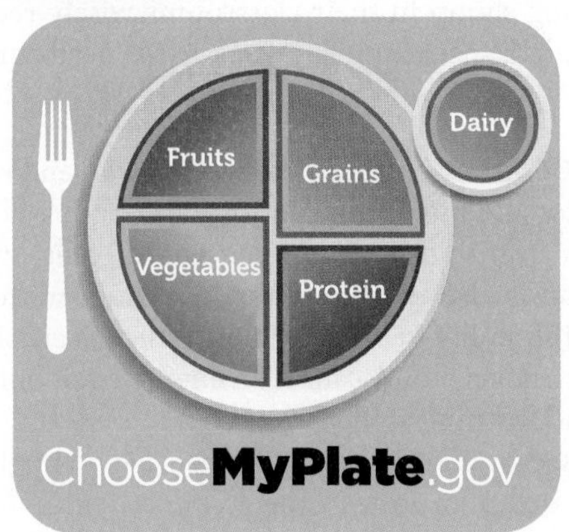

Abundant MyPlate resources for preschool children and older children can be found at http://www.choosemyplate.gov/.

U.S. Department of Agriculture

Candy, cola, and other concentrated sweets must be limited in children's diets. If such foods are permitted in large quantities, the only possible outcomes are nutrient deficiencies, obesity, or both. The *Dietary Guidelines for Americans 2010* recommend limiting intake of added sugars and solid fats to 5 to 15 percent of daily kcalories, but children and adolescents consume approximately 16 percent of total kcaloric intake from added sugars alone.[21]

Children can't be trusted to choose nutritious foods on the basis of taste alone; the preference for sweets is innate, and children naturally gravitate to them. Overweight children, especially, need help in selecting nutrient-dense foods that will meet their nutrient needs within their energy allowances. Underweight children or active, healthy-weight children can enjoy higher-kcalorie foods, but these should still be nutritious. Examples are ice cream and pudding in the milk group and whole-grain or enriched pancakes and crackers in the bread group.

HUNGER AND MALNUTRITION IN CHILDREN

Most children in the United States and Canada have access to regular meals, but hunger and malnutrition do appear in certain circumstances. Children in very low-income families, for example, are more likely to be hungry and malnourished. More than 16 million U.S. children are hungry at least some of the time and are living in poverty.[22] Nutrition in Practice 12 examines the causes and consequences of hunger in the United States.

Hunger and Behavior Both short-term and long-term hunger exert negative effects on behavior and health. Short-term hunger, such as when a child misses a meal, impairs the child's ability to pay attention and to be productive. Hungry children are irritable, apathetic, and uninterested in their environment. Long-term hunger impairs growth and immune defenses. Food assistance programs such as the WIC program (discussed in Chapter 10) and the School Breakfast and National School Lunch Programs (discussed later in this chapter) are designed to protect against hunger and improve the health of children.[23]

Children who eat no breakfast are more likely to be overweight, perform poorly in tasks requiring concentration, have shorter attention spans, and achieve lower test scores than their well-fed peers.[24] A nutritious breakfast is a central feature of a diet that meets the needs of children and supports their healthy growth and development.[25] Children who skip breakfast typically do not make up the deficits at later meals—they simply have lower intakes of energy, vitamins, and minerals than those who eat breakfast. Malnourished children are particularly vulnerable. Common sense dictates that it is unreasonable to expect anyone to learn and perform without fuel. For the child who hasn't had breakfast, the morning's lessons may be lost altogether. Even if a child has eaten breakfast, discomfort from hunger may become distracting by late morning. Teachers aware of the late-morning slump in their classrooms wisely request that midmorning snacks be provided; snacks improve classroom performance all the way to lunchtime.

Healthy, well-nourished children are alert in the classroom and energetic at play.

Iron Deficiency and Behavior Iron deficiency has well-known and widespread effects on children's behavior and intellectual performance.[26] In addition to carrying oxygen in the blood, iron transports oxygen within cells, which use it for energy metabolism. Iron is also used to make neurotransmitters—most notably, those that regulate the ability to pay attention, which is crucial to learning. Consequently, iron deficiency not only causes an energy crisis but also directly impairs attention span and learning ability.

Iron deficiency is often diagnosed by a quick, easy, inexpensive hemoglobin or hematocrit test that detects a deficit of iron in the *blood.* A child's *brain,* however, is sensitive to low iron concentrations long before the blood effects appear. Iron deficiency lowers the "motivation to persist in intellectually challenging tasks" and impairs overall intellectual performance. Anemic children perform poorly on tests and are disruptive in the classroom; iron supplementation improves learning and memory. When combined with other nutrient deficiencies, iron-deficiency anemia has synergistic effects that are especially detrimental to learning. Furthermore, children who had iron-deficiency anemia *as infants* continue to perform poorly as they grow older, even if their iron status improves.[27] The long-term damaging effects on mental development make prevention and treatment of iron deficiency during infancy and early childhood a high priority.

Other Nutrient Deficiencies A child with any of several nutrient deficiencies may be irritable, aggressive, and disagreeable, or sad and withdrawn. Such a child may be labeled "hyperactive," "depressed," or "unlikable," when in fact these traits may be due to simple, even marginal, malnutrition. Though parents and medical practitioners often overlook the possibility that malnutrition may account for these abnormalities, any departure from normal, healthy appearance and behavior is a sign of possible poor nutrition (see Table 11-4). In any such case, inspection of the child's diet by a registered dietitian or other qualified health care professional is in order. Any suspicion of dietary inadequacies, *no matter what other causes may be implicated*, should prompt steps to correct those inadequacies immediately.

TABLE 11-4 Physical Signs of Health and Malnutrition in Children

	Well-Nourished	Malnourished	Possible Nutrient Deficiencies
Hair	Shiny, firm in the scalp	Dull, brittle, dry, loose; falls out	Protein
Eyes	Bright, clear; pink membranes; adjust easily to light	Pale membranes; spots; redness; adjust slowly to darkness	Vitamin A, the B vitamins, zinc, and iron
Teeth and gums	No pain or caries, gums firm, teeth bright	Missing, discolored, decayed teeth; gums bleed easily and are swollen and spongy	Minerals and vitamin C
Face	Clear complexion without dryness or scaliness	Off-color, scaly, flaky, cracked skin	Protein, vitamin A, and iron
Glands	No lumps	Swollen at front of neck, cheeks	Protein and iodine
Tongue	Red, bumpy, rough	Sore, smooth, purplish, swollen	B vitamins
Skin	Smooth, firm, good color	Dry, rough, spotty; "sandpaper" feel or sores; lack of fat under skin	Protein, essential fatty acids, vitamin A, B vitamins, and vitamin C
Nails	Firm, pink	Spoon-shaped, brittle, ridged	Iron
Internal systems	Regular heart rhythm, heart rate, and blood pressure; no impairment of digestive function, reflexes, or mental status	Abnormal heart rate, heart rhythm, or blood pressure; enlarged liver, spleen; abnormal digestion; burning, tingling of hands, feet; loss of balance, coordination; mental confusion, irritability, fatigue	Protein and minerals
Muscles and bones	Muscle tone; posture, long bone development appropriate for age	"Wasted" appearance of muscles; swollen bumps on skull or ends of bones; small bumps on ribs; bowed legs or knock-knees	Protein, minerals, and vitamin D

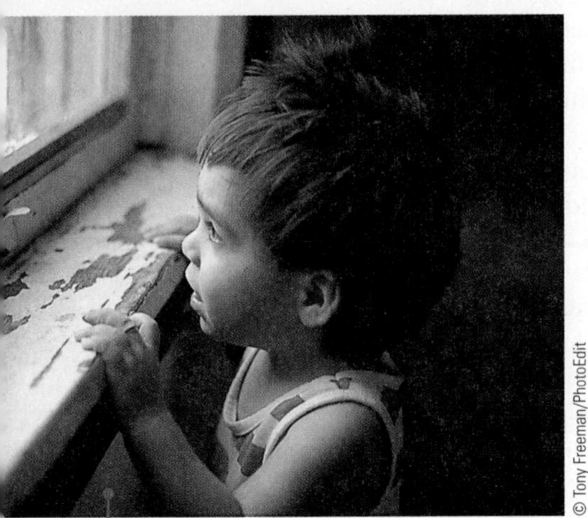
© Tony Freeman/PhotoEdit

Paint is the primary source of lead in children's lives.

LEAD POISONING IN CHILDREN

Children who are malnourished are vulnerable to lead poisoning. They absorb more lead if their stomachs are empty; if they have low intakes of calcium, zinc, vitamin C, or vitamin D; and, of greatest concern because it is so common, if they have an iron deficiency. Iron deficiency weakens the body's defenses against lead absorption, and lead poisoning can cause iron deficiency.[28] Circumstances associated with both iron deficiency and lead poisoning are a low socioeconomic status and a lack of immunizations against infectious diseases. Another common factor is pica—a craving for nonfood items. Many children with lead poisoning eat dirt or chips of old paint, two common sources of lead.

The anemia brought on by lead poisoning may be mistaken for a simple iron deficiency and therefore may be incorrectly treated. Like iron deficiency, mild lead toxicity has nonspecific symptoms, including diarrhea, irritability, and fatigue. Adding iron to the diet does not reverse the symptoms; exposure to lead must stop and treatment for lead poisoning must begin. With further exposure, the symptoms become more pronounced, and children develop learning disabilities and behavior problems. Still more severe lead toxicity can cause irreversible nerve damage, paralysis, mental retardation, and death.

NURSING DIAGNOSIS

Risk for poisoning applies to children who have dangerous products placed within their reach.

Research suggests that childhood lead exposure disrupts normal brain development—a finding that may partially explain the impaired cognitive and behavioral abilities of lead-exposed children. For six years, researchers measured blood lead levels at intervals in young children who lived in lead-contaminated houses.[29] Years later, brain images revealed that the higher the blood lead concentrations during childhood, the smaller the brain size as a young adult. The brains of boys were more affected than the brains of girls.

Approximately half a million children between the ages of one and five in the United States have blood lead levels above 5 micrograms per deciliter, the level at which the Centers for Disease Control and Prevention recommend public health actions be initiated.[30] (■) Lead toxicity in young children results from their own behaviors and activities—putting their hands in their mouths, playing in dirt and dust, and chewing on nonfood items.[31] Unfortunately, the body readily absorbs lead during times of rapid growth and hoards it possessively thereafter. Lead is not easily excreted and accumulates mainly in the bones but also in the brain, teeth, and kidneys. Tragically, a child's neuromuscular system is also maturing during these first few years of life. No wonder children with elevated lead levels experience impairment of balance, motor development, and the relaying of nerve messages to and from the brain. Deficits in intellectual development are only partially reversed when lead levels decline.

NURSING DIAGNOSIS

Delayed growth and development applies to clients who have altered physical growth, decreased response time, and delays in performing skills typical of their age group.

Federal laws mandating reductions in leaded gasolines, lead-based solder, and other products over the past four decades have helped to reduce the amounts of lead in food and in the environment in the United States. As a consequence, the prevalence of lead toxicity in children has declined dramatically for most of the United States. Nevertheless, lead exposure is still a threat in certain communities. The accompanying "How to" presents strategies for defending children against lead toxicity.

FOOD ALLERGY

Food allergy is frequently blamed for physical and behavioral abnormalities in children, but only 4 to 8 percent of children younger than four years of age are diagnosed with true food allergies.[32] Food allergies diminish with age, until in adulthood they affect less than 4 percent of the population. The prevalence of food allergy, especially peanut allergy, is on the rise, however.[33] Reasons for an increase in peanut allergy are not yet clear, but possible contributing factors include genetics, food preparation methods (roasting peanuts at very high temperatures makes them more allergenic), and exposure to medicinal skin creams containing peanut oil.

■ The Centers for Disease Control and Prevention recently reduced the threshold of lead exposure in a child from 10 micrograms per deciliter of blood to 5 micrograms.

food allergy: an adverse reaction to food that involves an immune response; also called *food-hypersensitivity reactions*.

A true food allergy occurs when fractions of a food protein or other large molecule are absorbed into the blood and elicit an immunologic response. (Recall that proteins are normally dismantled in the digestive tract to amino acids that are absorbed without

HOW TO Protect against Lead Toxicity

Strategies to prevent lead poisoning:

- Ask a pediatrician whether your child should be tested for lead poisoning.
- Clean floors, window frames, window sills, and other surfaces regularly. Use a mop or sponge with warm water and a general all-purpose cleaner.
- Prevent children from putting dirty or old painted objects in their mouths, and make sure children wash their hands before eating.
- Feed children balanced, timely meals with ample iron and calcium.
- Wash children's bottle, pacifiers, and toys often.
- Do not use lead-contaminated water to make infant formula.
- Have your water tested for lead. If the cold water hasn't been used for more than a few hours, let it run for 15 to 30 seconds before drinking it.
- Be aware that other countries do not have the same regulations protecting consumers against lead. Children have been poisoned by eating crayons made in China, putting toys from China in their mouths, and drinking fruit juice canned in Mexico.

such a reaction.) The body's immune system reacts to these large food molecules as it does to other antigens—by producing antibodies or other defensive agents. (■)

Detecting Allergy Allergies may have one or two components. They always involve antibodies; they may or may not involve symptoms.* Therefore, allergies cannot be diagnosed from symptoms alone. The National Institute of Allergy and Infectious Diseases (NIAID) has developed clinical guidelines for the diagnosis and management of food allergy.[34] Even symptoms exactly like those of an allergy may not be caused by an allergy. The NIAID recommends that food allergy should be considered when an individual—especially a young child—experiences symptoms such as skin rash, respiratory difficulties, vomiting, diarrhea, or anaphylactic shock (described later) within minutes to hours of eating food.

Diagnosis of food allergy requires medical testing and food challenges. Once a food allergy has been diagnosed, the required treatment is strict elimination of the offending food. Children with allergies, like all children, need all their nutrients, so it is important to include other foods that offer the same nutrients as the omitted foods.[35] Nutritional counseling and growth monitoring are recommended for all children with food allergies.[36]

Immediate and Delayed Reactions Allergic reactions to food may be immediate or delayed. In either case, the antigen interacts immediately with the immune system, but symptoms may appear within minutes or after several (up to 24) hours. Identifying the food that causes an immediate allergic reaction is easy because symptoms correlate closely with the time of eating the food. Identifying the food that caused a delayed reaction is more difficult because the symptoms may not appear until a day after the offending food was eaten; by this time, many other foods will have been eaten, too, complicating the picture.

Anaphylactic Shock The life-threatening food allergy reaction known as **anaphylactic shock** is most often caused by peanuts, tree nuts, milk, eggs, wheat, soybeans, fish, or shellfish. Among these foods, peanuts, milk, and shellfish most often cause problems in children.[37] Children are more likely to outgrow allergies to eggs, milk, wheat, and soy than allergies to peanuts, tree nuts, fish, and shellfish.[38] Peanut allergies cause more life-threatening reactions than do all other food allergies combined. Research is currently under way to help those with peanut allergies tolerate small doses, thus saving lives and minimizing reactions.[39] Families of children with a life-threatening food allergy and school personnel who supervise those children must guard them against any exposure to the allergen. The child must learn to identify which foods pose

■ Reminder: *Antigens* are substances foreign to the body that elicit the formation of antibodies or an inflammation reaction from immune system cells. Food antigens are usually glycoproteins (large proteins with glucose molecules attached). *Antibodies* are large proteins that are produced in response to antigens and then inactivate the antigens.

anaphylactic (AN-ah-feh-LAC-tic) shock: a life-threatening whole-body allergic reaction to an offending substance.

*A person may produce antibodies without having any symptoms (known as *asymptomatic allergy*) or may produce antibodies and have symptoms (known as *symptomatic allergy*).

These eight normally wholesome foods—milk, shellfish, fish, peanuts, tree nuts, eggs, wheat, and soybeans (and soy products)—may cause life-threatening symptoms in people with allergies.

a problem and then to use refusal skills for all foods that may contain the allergen.

Parents of allergic children can pack safe foods for lunches and snacks and ask school officials to strictly enforce a "no swapping" policy in the lunchroom. The child must be able to recognize the symptoms of impending anaphylactic shock, (■) such as a tingling of the tongue, throat, or skin or difficulty breathing. Any person with food allergies severe enough to cause anaphylactic shock should wear a medical alert bracelet or necklace. Finally, the responsible child and the school staff should be prepared to administer injections of **epinephrine**, which prevents anaphylaxis after exposure to the allergen.[40] Many preventable deaths occur each year when people with food allergies accidentally ingest the allergen but have no epinephrine available.

Technology may soon offer new solutions. New drugs are being developed that may interfere with the immune response that causes allergic reactions. Also, through genetic engineering, scientists may one day banish allergens from peanuts, soybeans, and other foods to make them safer.

■ Symptoms of impending anaphylactic shock:

- Tingling sensation in mouth
- Swelling of the tongue and throat
- Irritated, reddened eyes
- Difficulty breathing, asthma
- Hives, swelling, rashes
- Vomiting, abdominal cramps, diarrhea
- Drop in blood pressure
- Loss of consciousness

Food Labeling Food labels must identify any common allergens present in plain language, using the names of the eight most common allergy-causing foods.[41] For example, a food containing "textured vegetable protein" must say "soy" on its label. Similarly, "casein" must be identified as "milk," and so forth. Food producers must also prevent cross-contamination during production and clearly label the foods in which it is likely to occur. For example, equipment used for making peanut butter must be scrupulously clean before being used to pulverize cashew nuts for cashew butter to protect unsuspecting cashew butter consumers from peanut allergens.

Other Adverse Reactions to Foods Not all **adverse reactions** to foods are food allergies, although even physicians may describe them as such. Signs of adverse reactions to foods include stomachaches, headaches, rapid pulse rate, nausea, wheezing, hives, bronchial irritation, coughs, and other such discomforts. Among the causes may be reactions to chemicals in foods, such as the flavor enhancer monosodium glutamate (MSG), the natural laxative in prunes, or the mineral sulfur; digestive diseases, obstructions, or injuries; enzyme deficiencies, such as lactose intolerance; and even psychological aversions. These reactions involve symptoms but no antibody production. Therefore, they are **food intolerances**, not allergies.

Pesticides on produce may also cause adverse reactions. Pesticides that were applied in the fields may linger on the foods. Health risks from pesticide exposure may be low for healthy adults, but children are vulnerable. Therefore, government agencies have set a **tolerance level** for each pesticide by first identifying foods that children commonly eat in large amounts and then considering the effects of pesticide exposure during each developmental stage.

Food Dislikes Parents are advised to watch for signs of food dislikes and take them seriously. Children's food aversions may be the result of nature's efforts to protect them from allergic or other adverse reactions. Test for allergies, and then apply nutrition knowledge conscientiously in deciding how to alter the diet.

HYPERACTIVITY

Hyperactivity affects behavior and learning in about 9 percent of young school-aged children.[42] Left untreated, it can interfere with a child's social development and ability to learn. Treatment focuses on relieving the symptoms and controlling the associated problems. Physicians often manage hyperactivity through behavior modification, special educational techniques, and psychological counseling. In many cases, they prescribe medication.[43]

epinephrine: one of the stress hormones secreted whenever emergency action is needed; prescribed therapeutically to relax the bronchioles during allergy or asthma attacks.

adverse reactions: unusual responses to food (including intolerances and allergies).

food intolerances: adverse reactions to foods or food additives that do not involve the immune system.

tolerance level: the maximum amount of residue permitted in a food when a pesticide is used according to the label directions.

hyperactivity: inattentive and impulsive behavior that is more frequent and severe than is typical of others of a similar age; professionally called **attention deficit/hyperactivity disorder (ADHD)**.

Case Study Boy with Disruptive Behavior

Freddie Willis is a six-year-old boy who seldom sits still, often misbehaves, and is frequently sick. Freddie's eating habits are erratic and poor, as is his appetite. He often misses breakfast because he is too tired to get up in time to eat before school. By midmorning, Freddie is irritable and disruptive in the classroom. At lunchtime, he trades the peanut butter and banana sandwich his mother packed in his lunchbox for a piece of cake. After school he hurries home to watch television while he eats his favorite snack—cola and potato chips. At dinnertime, Freddie picks at his food because he isn't very hungry. Later on, when it's time for bed, Freddie complains that he's hungry. His parents let him stay up to have a bowl of cereal (the kind with marshmallows) before he finally falls asleep.

1. What factors in Freddie's daily routine might be contributing to his restless behavior?
2. Discuss some changes in diet that might improve Freddie's health and disposition.

Research on hyperactivity has focused on several nutritional factors as possible causes or treatments.[44] Parents of hyperactive children often blame sugar as the cause and mistakenly believe that simply eliminating candy and other sweet treats will solve the problem. Despite such speculation, which has been based on personal anecdotes, studies have consistently found no convincing evidence that sugar causes hyperactivity or worsens behavior. Restricting dietary sugar will not cure hyperactivity.

Food additives have also been blamed for hyperactivity and other behavior problems in children, but scientific evidence to substantiate the connection has been elusive.[45] Limited research suggests that food additives such as artificial colors or sodium benzoate preservative (or both) may exacerbate hyperactive symptoms such as inattention and impulsivity. Additional studies are needed to confirm the findings and to determine which additives studied might be responsible for negative behaviors. A Food and Drug Administration (FDA) review determined that evidence linking color additives to hyperactivity is lacking.[46] The FDA did not rule out the possibility that some food additives, including food colorings, may aggravate hyperactivity and other behavioral problems in some susceptible children.

Children can become excitable, rambunctious, and unruly as a result of a desire for attention, lack of sleep, overstimulation, watching too much television or playing too many computer games, too much caffeine from colas or chocolate, or a lack of physical activity. Such behaviors may suggest that more consistent care is needed. It helps to insist on regular hours of sleep, regular mealtimes, and regular outdoor activity. The accompanying Case Study offers an opportunity to think about these issues in relation to a specific child.

Review Notes

- Children's appetites and nutrient needs reflect their stage of growth.
- Long-term hunger and malnutrition impair growth and health.
- Short-term hunger exerts more subtle effects on children's health and behavior—such as poor academic performance.
- Iron deficiency is widespread and has many physical and behavioral consequences.
- Lead toxicity is prevalent among young children and can have irreversible effects on health and behavior.
- True food allergies are somewhat rare in children, and children can outgrow some food allergies.
- Some allergies, however, can cause dangerous, life-threatening reactions in both children and adults.
- "Hyper" behavior is not caused by poor nutrition; misbehavior may reflect inconsistent care.

CHILDHOOD OBESITY

The number of overweight children has increased dramatically over the past three decades. Like their parents, children in the United States are becoming fatter. An estimated 32 percent of U.S. children and adolescents 2 to 19 years of age are overweight, and 17 percent of these children are obese.[47] Based on data from the BMI-for-age growth charts, children and adolescents are categorized as *overweight* above the 85th percentile and as *obese* at the 95th percentile and above.[48] There are exceptions to the use of the 85th and 95th percentile cutoff points. For older adolescents, BMI at the 95th percentile is higher than 30, the adult obesity cutoff point. Therefore, obesity in young people is defined as BMI at the 95th percentile or BMI of 30 or greater, whichever is lower. For children younger than two years of age, BMI values are not available. For this age group, weight-for-height values above the 95th percentile are classified as overweight. Figures E-5A and E-5B in Appendix E present the BMI charts for children and adolescents, indicating cutoff points for obesity and overweight.

The Expert Committee of the American Medical Association recommends a third cutoff point (99th percentile) to define severe obesity in childhood. Unfortunately, severe obesity in children is becoming more prevalent.[49] Many of these children have multiple risk factors for cardiovascular disease and a high risk of severe obesity in adulthood.[50] The special risks and treatment needs of severely obese children need to be recognized.

The problem of obesity in children is especially troubling because overweight children have the potential of becoming obese adults with all the social, economic, and medical ramifications that often accompany obesity. They have additional problems, too, arising from differences in their growth, physical health, and psychological development. In trying to explain the rise in childhood obesity, researchers point to both genetic and environmental factors.

Genetic and Environmental Factors

Parental obesity predicts an early increase in a young child's BMI, and it more than doubles the chances that a young child will become an obese adult.[51] Children with neither parent obese have a less than 10 percent chance of becoming obese in adulthood, whereas overweight teens with at least one obese parent have a greater than 80 percent chance of being obese adults. The chances of an obese child becoming an obese adult grow greater as the child grows older.[52] The link between parental and child obesity reflects both genetic and environmental factors (as described in Chapter 7).

Diet and physical inactivity are the two strongest environmental factors explaining why children are heavier today than they were 40 or so years ago. As the prevalence of childhood obesity throughout the United States has more than doubled for young children and more than tripled for children 6 to 11 years of age and adolescents, the society our children live in has changed considerably. In many families today, both parents work outside the home and work longer hours; more emphasis is placed on convenience foods and foods eaten away from home; meal choices at school are more diverse and often less nutritious; sedentary activities such as watching television and playing computer games occupy much of children's free time; and opportunities for physical activity and outdoor play both during and after school have declined.[53] All of these factors—and many others—influence children's eating and activity patterns.

Children learn food behaviors from their families, and research confirms the significant roles parents play in teaching their children about healthy food choices, providing nutrient-dense foods, and serving as role models.[54] When parents eat fruits and vegetables frequently, their children do, too.[55] The more fruits and vegetables children eat, the more vitamins, minerals, and fiber, and the less saturated fat, in their diets.

In children 2 to 18 years of age, about 40 percent of total energy intake comes from solid fats and added sugars—in other words, empty kcalories.[56] About half of these empty kcalories are contributed by six specific foods: soda, fruit drinks, dairy desserts (ice cream, frozen yogurt, sorbet, sherbet, pudding, and custard), grain desserts (cakes, cookies, pies, cobblers, donuts, and granola bars), pizza, and whole milk. Not surprisingly, when researchers ask, "Do children's food choices today provide more kcalories than those of 40 years ago?" the answer is, "Yes."

As Chapter 3 discussed, as the prevalence of obesity among both children and adults has surged over the past four decades, so has the consumption of added sugars and, especially, high-fructose corn syrup—the easily consumed, energy-dense liquid sugar added to soft drinks.[57] Each 12-ounce can of soft drink provides the equivalent of about 10 teaspoons of sugar and 150 kcalories. More than half of school-age children consume at least one soft drink each day at school; adolescent males consume the most—four or more cans daily. Research shows that consumption of sugar-sweetened beverages such as soft drinks is associated with increased energy intake and body weight.[58]

No doubt, the tremendous increase in soft drink consumption plays a role, but much of the obesity epidemic can be explained by lack of physical activity. Children have become more sedentary, and sedentary children are more often overweight.[59] Television watching (■) may contribute most to physical inactivity. Children 8 to 18 years of age spend an average of 4.5 hours per day watching television.[60] Longer television time is linked to overweight in children.[61] A child who spends more than an hour or two each day in front of a television, computer monitor, or other media can become overweight even while eating fewer kcalories than a more active child.

Children who have television sets in their bedrooms spend more time watching TV, spend less time being physically active, and are more likely to be overweight than children who do not have televisions in their rooms.[62] Children who watch a great deal of television are most likely to be overweight and least likely to eat family meals or fruits and vegetables. They often snack on the nutrient-poor, energy-dense foods that are advertised.[63] The average child sees an estimated 40,000 TV commercials a year—many peddling foods high in sugar, saturated fat, and salt such as sugar-coated breakfast cereals, candy bars, chips, fast foods, and carbonated beverages. More than half of all food advertisements are aimed specifically at children and market their products as fun and exciting. Not surprisingly, the more time children spend watching television, the more they request these advertised foods and beverages—and they get them about half of the time. The most popular foods and beverages are marketed to children and adolescents on the Internet as well, using "advergaming" (advertised product as part of a game), cartoon characters or "spokes-characters," and designated children's areas.[64]

The physically inactive time spent watching television (and playing video and computer games) is second only to time spent sleeping. These activities use no more energy than resting, displace participation in more vigorous activities, and foster snacking on high-fat foods.[65] Compared to sedentary screen-time activities, playing active video games does expend a little more energy, but not enough to count toward the 60 minutes of moderate-to-vigorous physical activity recommended for children.[66] Simply reducing the amount of time spent watching television (and playing video games) can improve a child's BMI. The American Academy of Pediatrics (AAP) now recommends no television viewing before two years of age and thereafter limiting television and video time to two hours per day as a strategy to help prevent childhood obesity.[67]

Growth Overweight children develop a characteristic set of physical traits. They typically begin puberty earlier and so grow taller than their peers at first, but then they stop growing at a shorter height. They develop greater bone and muscle mass in response to the demand of having to carry more weight—both fat and lean weight. Consequently, they appear "stocky" even when they lose their excess fat.

Physical Health Like overweight adults, overweight children display a blood lipid profile indicating that atherosclerosis is beginning to develop—high levels of total cholesterol, triglycerides, and LDL cholesterol. Overweight children also tend to have high blood pressure; in fact, obesity is a leading cause of pediatric hypertension.[68] Their risks for developing type 2 diabetes and respiratory diseases (such as asthma) are also exceptionally high.[69] These relationships between childhood obesity and chronic diseases are discussed fully in Nutrition in Practice 11.

■ TV fosters obesity because it:

- Requires no energy beyond basal metabolism.
- Replaces vigorous activities.
- Encourages snacking.
- Promotes a sedentary lifestyle.

Playing sedentary video games influences children's activity patterns similarly.

Psychological Development In addition to the physical consequences, childhood obesity brings a host of emotional and social problems. Because people frequently judge others on appearance more than on character, overweight children are often victims of prejudice. Many suffer discrimination by adults and rejection by their peers. They may have poor self-images, a sense of failure, and a passive approach to life. Television shows, which are a major influence in children's lives, often portray the fat person as the bumbling misfit. Overweight children may come to accept this negative stereotype in themselves and in others, which can lead to additional emotional and social problems. Researchers investigating children's reactions to various body types find that both normal-weight and underweight children respond unfavorably to overweight bodies.

Prevention and Treatment of Obesity Medical science has worked wonders in preventing or curing many of even the most serious childhood diseases, but obesity remains a challenge. Once excess fat has been stored, it is challenging to lose. In light of all this, parents are encouraged to make major efforts to prevent childhood obesity, starting at birth, or to begin treatment early—before adolescence. The Expert Committee of the American Medical Association recommends specific eating and physical activity behaviors to prevent obesity for all children (see Table 11-5).

Treatment of obesity must consider the many aspects of the problem and possible solutions. The main goal of obesity treatment is to improve long-term physical health through permanent healthy lifestyle habits. The most successful approach integrates diet, physical activity, psychological support, and behavioral changes. As a first step, the Expert Committee recommends that overweight and obese children and their families adopt the same healthy eating and activity behaviors presented in Table 11-5 for obesity prevention. The goal for overweight and obese children is to improve BMI. If

TABLE 11-5 Recommended Eating and Physical Activity Behaviors to Prevent Obesity

The Expert Committee of the American Medical Association recommends the following healthy habits for children 2 to 18 years of age to help prevent childhood obesity:
• Limit consumption of sugar-sweetened beverages, such as soft drinks and fruit-flavored punches.
• Eat the recommended amounts of fruits and vegetables every day (2 to 4.5 cups per day based on age).
• Learn to eat age-appropriate portions of foods.
• Eat foods low in energy density such as those high in fiber and/or water and modest in fat.
• Eat a nutritious breakfast every day.
• Eat a diet rich in calcium.
• Eat a diet balanced in recommended proportions for carbohydrate, fat, and protein.
• Eat a diet high in fiber.
• Eat together as a family as often as possible.
• Limit the frequency of restaurant meals.
• Limit television watching or other screen time to no more than two hours per day, and do not have televisions or computers in sleeping areas.
• Engage in at least 60 minutes of moderate to vigorous physical activity every day.

Source: Based on S. E. Barlow, Expert Committee recommendations regarding the prevention, assessment, and treatment of child and adolescent overweight and obesity: Summary report, *Pediatrics* 120 (2007): S164–S192.

the child's BMI does not improve after several months, the Expert Committee recommends increasing the intensity of the treatment. The level of intensity depends on treatment response, age, degree of obesity, health risks, and the family's readiness to change. Advanced treatment involves close follow-up monitoring by a health care provider and greater support and structure for the child.

Diet The initial goal for overweight children is to reduce the rate of weight gain, that is, to maintain weight as the child grows taller.[70] Continued growth will then accomplish the desired change in BMI. Weight loss is usually not recommended because diet restriction can interfere with growth and development. Intervention for some overweight children with accompanying medical conditions may warrant weight loss, but this treatment requires an individualized approach based on the degree of overweight and severity of the medical conditions. Dietary strategies begin with those listed in Table 11-5 and progress to more structured family meal plans when necessary. For example, the child or the parent may be instructed to keep detailed logs of dietary intake.

Physical Activity The many benefits of physical activity are well known but often are not enough to motivate overweight people, especially children. Yet regular vigorous activity can improve a child's weight, body composition, and physical fitness. The *Physical Activity Guidelines for Americans 2008* make specific recommendations for children that can be used as a guide for how much activity children need. (■) Ideally, parents will limit sedentary activities and encourage at least one hour of daily physical activity to promote strong skeletal, muscular, and cardiovascular development and instill in their children the desire to be physically active throughout life. Opportunities to be physically active can include team, individual, and recreational activities (see Table 11-6 on p. 330). Most importantly, parents need to set a good example. Physical activity is a natural and lifelong behavior of healthy living. It can be as simple as riding a bike, playing tag, jumping rope, or doing chores. The AAP supports the efforts of schools to include more physical activity in the curriculum and encourages parents to support their children's participation.

■ *Physical Activity Guidelines for Americans 2008*

- Children and adolescents should engage in 60 minutes (1 hour) or more of physical activity daily.
- Aerobic: Most of the 60 or more minutes a day should be either moderate- or vigorous-intensity aerobic activity and should include vigorous-intensity physical activity at least three days a week.
- Muscle strengthening: As part of their 60 or more minutes of daily physical activity, children and adolescents should include muscle-strengthening physical activity on at least three days of the week.
- Bone strengthening: As part of their 60 or more minutes of daily physical activity, children and adolescents should include bone-strengthening physical activity on at least three days of the week.

Psychological Support Weight-loss programs that involve parents and other caregivers in treatment report greater success than those without parental involvement. Because obesity in parents and their children tends to be positively correlated, both benefit when parents participate in a weight-loss program. Parental attitudes about food greatly influence children's eating behavior, so it is important that the influence be positive. Otherwise, eating problems may become exacerbated.

Behavioral Changes In contrast to traditional weight-loss programs that focus on *what* to eat, behavioral programs focus on *how* to eat. These techniques involve changing learned habits that lead a child to eat excessively.

Drugs The use of weight-loss drugs to treat obesity in children merits special concern because the long-term effects of these drugs on growth and development have not been studied. The drugs may be used in addition to structured lifestyle changes for carefully selected children or adolescents who are at high risk for severe obesity in adulthood. Only orlistat (see Chapter 7) has been approved for limited use in adolescents aged 12 years and older.[71]

Surgery The use of surgery to treat severe obesity in adults (see Chapter 7) has created interest in its use for adolescents. Limited research shows that, after surgery, extremely obese adolescents lose significant weight and experience improvements in type 2 diabetes and cardiovascular risk factors.[72] The selection criteria for surgery to treat obesity in adolescents (■) are based on recommendations of a panel of pediatricians and surgeons.

■ Surgery may be an option for adolescents who meet the following criteria:

- Have reached physical maturity
- BMI ≥50, or BMI ≥40 with significant weight-related health problems
- Have experienced failure in a formal, six-month weight-loss program
- Are capable of adhering to the long-term lifestyle changes required after surgery

Obesity is prevalent in our society. Because treatment of obesity is frequently unsuccessful, it is most important to prevent its onset. Above all, be sensible in teaching

TABLE 11-6 Examples of Aerobic, Muscle-Strengthening, and Bone-Strengthening Physical Activities for Children and Adolescents

Moderate-to-Vigorous Aerobic Activities	Muscle-Strengthening Activities	Bone-Strengthening Activities
 © Fotokostic/Shutterstock.com	 © Istockphoto.com/Christopher Futcher	 © Irina Fischer/Shutterstock.com
Moderate Active recreation such as hiking, skateboarding, rollerblading Bicycle riding[a] Brisk walking **Vigorous** Active games involving running and chasing, such as tag Bicycle riding[a] Cross-country skiing Jumping rope Martial arts Running Sports such as soccer, ice or field hockey, basketball, swimming, tennis	Games such as tug-of-war Modified push-ups (with knees on the floor) Resistance exercises using body weight, free weights, or resistance bands Rope or tree climbing Sit-ups (curl-ups or crunches) Swinging on playground equipment/bars	Games such as hopscotch Hopping, skipping, jumping Jumping rope Running Sports such as gymnastics, basketball, volleyball, tennis

[a]Some activities, such as bicycling, can be moderate or vigorous, depending on level of effort.

children how to maintain appropriate body weight. Children can easily get the impression that their worth is tied to their body weight. Parents and the media are most influential in shaping self-concept, weight concerns, and dieting practices.[73] Some parents fail to realize that society's ideal of slimness can be perilously close to starvation and that a child encouraged to "diet" cannot obtain the energy and nutrients required for normal growth and development. Even healthy children without diagnosable eating disorders have been observed to limit their growth through "dieting." Weight loss in truly overweight children can be managed without compromising growth, but it should be overseen by a health care professional.

Review Notes

- Childhood obesity has become a major health problem.
- Genetics, energy-dense diets, and physical inactivity play roles in childhood obesity.
- Childhood obesity can impair physical and psychological health.
- The main goal of obesity treatment in children is to improve long-term physical health through permanent healthy lifestyle habits.

MEALTIMES AT HOME

The childhood years are the parents' last chance to influence their children's food choices. Parents who want to promote nutritious choices and healthful habits provide access to nutrient-dense, delicious foods and opportunities for active play at home. Food choices and regular physical activity can not only promote healthy growth but, as mentioned earlier, also help prevent the chronic diseases of later life. Many experts agree that early childhood is the time to put into effect practices that, until recently, were recommended only for adults.

Feeding children requires not only providing a variety of nutritious foods but also nurturing the children's self-esteem and well-being. Parents face a number of challenges in preparing meals that both appeal to their children's tastes and provide needed nutrients. Because the interactions between parents and children can set the stage for lifelong attitudes and habits, a child's preferences should be treated with respect, even when nutrient needs must take precedence.

© Mary Kate Denny/PhotoEdit

Eating is more fun when friends are there.

Honoring Children's Preferences Researchers attempting to explain children's food preferences encounter many contradictions. Children say they like colorful foods yet most often reject green and yellow vegetables while favoring brown peanut butter and white potatoes, apple wedges, and bread. They do like raw vegetables better than cooked ones, though, so it is wise to offer vegetables that are raw or slightly undercooked, crunchy, and bright in color. They should be warm, not hot, because a child's mouth is much more sensitive than an adult's. The flavor should be mild (a child has more taste buds), and smooth foods such as mashed potatoes or pea soup should have no lumps (a child wonders, with some disgust, what the lumps might be).

> **NURSING DIAGNOSIS**
>
> *Readiness for enhanced parenting* applies to clients who have children whose needs (physical and emotional) are met.

Make mealtimes fun for children. Young children like to eat at little tables and to be served little portions of food. They also love to eat with other children and have been observed to stay at the table longer and eat more food when in the company of their peers. Parents who serve food in a relaxed and casual manner, without anxiety, provide an environment in which a child's negative emotions will be minimized.

Avoiding Power Struggles Problems over food often arise during the second or third year, when children begin asserting their independence. Many of these problems stem from the conflict between children's developmental stages and capabilities and parents who, in attempting to do what they think is best for their children, try to control every aspect of eating. Such conflicts can disrupt children's abilities to regulate their own food intakes or to determine their own likes and dislikes. For example, many people share the misconception that children must be persuaded or coerced to try new foods. In fact, the opposite is true. When children are forced to try new foods, especially when offered rewards for eating a particular food, they are less likely to try those foods again than are children who are left to decide for themselves. Similarly, when children are restricted from eating their favorite foods, they are more likely to want those foods. As dietitian and family therapist Ellyn Satter notes, the parent is responsible for *what* the child is offered to eat, but the child is responsible for *how much* and even *whether* to eat.

When introducing new foods at the table, parents are advised to offer them one at a time and only in small amounts at first. The more often a food is presented to a young child, the more likely the child will accept that food.[74] Between 5 and 10 exposures to a new food are necessary before a toddler shows an enhanced preference for the food. Offer the new food at the beginning of the meal, when the child is hungry, and allow the child to make the decision to accept or reject it. Parents have their own inclinations and dislikes; so do children. Never make an issue of food acceptance. Table 11-7 (p. 332) offers tips for feeding picky eaters.

TABLE 11-7 Tips for Feeding Picky Eaters

Get Them Involved

Children are more likely to feel a sense of ownership and may be more interested in trying foods when they participate in:

- meal planning,
- grocery shopping,
- cooking,
- gardening and harvesting the foods they eat.

Be Creative

- Try serving veggies as finger foods with dips or spreads.
- Use cookie cutters to cut fruits and veggies into fun shapes.
- Put healthy snacks in ice cube trays or muffin pans where children can easily reach them and graze on these as they play.
- Serve traditional meals out of order (for example, breakfast for dinner).
- Use healthy foods such as veggies and whole grains in craft projects to help kids become familiar with them and encourage their interest and enthusiasm for these foods.

Enhance Favorite Recipes

- Include sliced or shredded veggies in sauces, casseroles, pancakes, and muffins.
- Serve fruit over cereal, yogurt, or ice cream.
- Bake brownies with black beans or cookies with lentils as an ingredient.

Model and Share

- Be a role model to children by eating healthy foods with them. Offer to share your healthy snack with them, too.
- Sometimes children need to be exposed to a new food multiple times before they develop a taste for the food, so make sure healthy options are always available and don't give up on repeatedly offering foods your child might not seem interested in.
- Encourage your child to taste at least one bite of each food served at a meal.

Respect and Relax

- Remember that it is not uncommon for children to eat sporadically. They have smaller stomachs and therefore are likely to feel full faster and become hungry again not long after a snack or meal.
- Focus on your child's overall weekly intake of food and nutrients rather than daily consumptions.
- If you suspect that your child might not be eating enough to support healthy growth and development, discuss your concerns with your child's doctor. It might be helpful to maintain a food log of everything your child eats over a period of three days to bring along to the doctor appointment.

Source: Based on Mayo Clinic Staff, Children's nutrition: 10 tips for picky eaters, 2011, available at http://www.mayoclinic.com/health/childrens-health/HQ01107.

Choking Prevention Parents must always be alert to the dangers of choking. A choking child is silent, so an adult should be present whenever a child is eating. Make sure the child sits when eating; choking is more likely when a child is running or falling (see p. 303 for a list of foods and nonfood items most likely to cause choking).

Play First Ideally, each meal is preceded, not followed, by the activity the child looks forward to the most. A number of schools have discovered that children eat a much better lunch if it is served after, rather than before, recess. Otherwise, children "hurry up and eat" so that they can go play.

Child Participation Allowing children to help plan and prepare the family's meals provides enjoyable learning experiences and encourages children to eat the foods they have prepared. Vegetables are pretty, especially when fresh, and provide opportunities for children to learn about color, growing things and their seeds, and shapes and textures—all of which are fascinating to young children. Measuring, stirring, decorating, and arranging foods are skills that even a very young child can practice with enjoyment and pride (see Table 11-8).

Snacks Parents may find that their children often snack so much that they aren't hungry at mealtimes. Instead of teaching children *not* to snack, teach them *how* to snack. Provide snacks that are as nutritious as the foods served at mealtime. Snacks

TABLE 11-8 Food Skills and Developmental Milestones of Preschool Children[a]

Food Skills	Developmental Milestones
Age 1 to 2 years • Uses a spoon • Lifts and drinks from a cup • Helps scrub fruits and vegetables, tear lettuce or greens, snap green beans, or dip foods • Can be messy; can be easily distracted	• Large muscles develop • Experiences slowed growth and decreased appetite • Develops likes and dislikes • May suddenly refuse certain foods
Age 3 years • Spears food with a fork • Feeds self independently • Helps wrap, pour, mix, shake, stir, or spread foods • Follows simple instructions	• Medium hand muscles develop • May suddenly refuse certain foods • Begins to request favorite foods • Makes simple either/or food choices
Age 4 years • Uses all utensils and napkin • Helps measure dry ingredients • Learns table manners	• Small finger muscles develop • Influenced by TV, media, and peers • May dislike many mixed dishes
Age 5 years • Measures liquids • Helps grind, grate, and cut (soft foods with dull knife) • Uses hand mixer with supervision	• Fine coordination of fingers and hands develops • Usually accepts food that is available • Eats with minor supervision

[a]These ages are approximate. Healthy, normal children develop at their own pace.

Source: Adapted from MyPlate for Preschoolers, Behavioral Milestones, available at http://www.choosemyplate.gov/preschoolers/healthy-habits/Milestones.pdf.

can even be mealtime foods that are served individually over time, instead of all at once on one plate. When providing snacks to children, think of the food groups and offer such snacks as pieces of cheese, sliced strawberries, cooked baby carrots, and egg salad on whole-wheat crackers (see Table 11-9 on p. 334). Snacks that are easy to prepare should be readily available to children, especially if the children arrive home after school before their parents.

Preventing Dental Caries Children frequently snack on sticky, sugary foods that stay on the teeth and provide an ideal environment for the growth of bacteria that cause dental caries. Teach children to brush and floss after meals, to brush or rinse after eating snacks, to avoid sticky foods, and to select crisp or fibrous foods frequently.

Serving as Role Models In an effort to practice these many tips, parents may overlook perhaps the single most important influence on their children's food habits—themselves. Parents who do not eat oranges should not be surprised when their children refuse to eat oranges. Likewise, parents who dislike the smell of brussels sprouts may not be able to persuade children to try them. Children learn much through imitation. Parents, older siblings, and other caregivers set an irresistible example by sitting with younger children, eating the same foods, and having pleasant conversations during mealtime.

While serving and enjoying food, caregivers can promote both physical and emotional health at every stage of a child's life. They can help their children to develop both a positive self-concept and a positive attitude toward food. If the beginnings are right, children will grow without the conflicts and confusions over foods that lead to nutrition and health problems.

TABLE 11-9 Healthful Snack Ideas—Think Food Groups, Alone and in Combination

Selecting two or more foods from different food groups adds variety and nutrient balance to snacks. The combinations are endless, so be creative.

Grains

Grain products are filling snacks, especially when combined with other foods:

- Cereal with fruit and milk
- Crackers and cheese
- Whole-grain toast with peanut butter
- Popcorn with grated cheese
- Oatmeal raisin cookies with milk

Vegetables

Cut-up fresh, raw vegetables make great snacks alone or in combination with foods from other food groups:

- Celery with peanut butter
- Broccoli, cauliflower, and carrot sticks with a flavored cottage cheese dip

Fruits

Fruits are delicious snacks and can be eaten alone—fresh, dried, or juiced—or combined with other foods:

- Apples and cheese
- Bananas and peanut butter
- Peaches with yogurt
- Raisins mixed with sunflower seeds or nuts

Protein Foods

Meat, poultry, seafood, legumes, soy products, nuts, and seeds add protein to snacks:

- Refried beans with nachos and cheese
- Peanut butter, hummus, or tuna on crackers
- Luncheon meat on whole-grain bread

Milk and Milk Products

Milk can be used as a beverage with any snack, and many other milk products, such as yogurt and cheese, can be eaten alone or with other foods as listed above.

NUTRITION AT SCHOOL

While parents are doing what they can to establish good eating habits in their children at home, others are preparing and serving foods to their children at day care centers and schools. In addition, children begin learning about food and nutrition in the classroom. Meeting the nutrition and education needs of children is critical to supporting their healthy growth and development.[75]

The U.S. government funds programs to provide nutritious, high-quality meals for children at school. Both the School Breakfast Program and the National School Lunch Program provide meals at a reasonable cost to children from families with the financial means to pay. Meals are available free or at reduced cost to children from low-income families.

■ The school breakfast must contain, at a minimum:

- One serving of fluid milk (either unflavored low-fat or flavored or unflavored fat-free)
- One serving of fruit or vegetable (no more than half of the servings may be 100% full-strength juice)
- One to two servings of whole grains; or one serving of whole grains and one serving of meat or meat alternatives

School Breakfast

The School Breakfast Program (■) is available in about 90 percent of the nation's schools that offer school lunch, and 12 million children participate in it.[76] Nevertheless, many children who need the School Breakfast Program either do not have access to the program or do not participate in it.[77] The majority of children who eat school breakfasts are from low-income families. As research results continue to emphasize the positive impact breakfast has on school performance and health, vigorous campaigns to expand school breakfast programs are under way.[78]

School Lunch More than 31 million children receive lunches through the National School Lunch Program—more than half of them free or at a reduced price.[79] School lunches are designed to provide at least a third of the recommendation for energy, protein, vitamin A, vitamin C, iron, and calcium. They must also include specified numbers of servings from each food group. In an effort to help reduce disease risk, all government-funded meals served at schools must follow the *Dietary Guidelines for Americans*. Table 11-10 shows school lunch patterns for children of different ages.

In 2012, the USDA Food and Nutrition Service issued a final rule updating the meal patterns and nutrition standards for school meals. Changes to meals reflected in the

TABLE 11-10 School Lunch Patterns for Different Ages

Food Group	Amount of Food per Week (minimum per day)[a]		
	Elementary School Grades K to 5	Middle School Grades 6 to 8	High School Grades 9 to 12
Fruits (cups)[b]	2½ (½)	2½ (½)	5 (1)
Vegetables (cups)[b]	3¾ (¾)	3¾ (¾)	5 (1)
Dark green[c]	½	½	½
Red/orange[c]	¾	¾	1¼
Beans/peas (legumes)[c]	½	½	½
Starchy[c]	½	½	½
Other[c]	½	½	¾
Additional to reach total[d]	1	1	1½
Grains (oz eq)[e]	8–9 (1)	8–10 (1)	10–12 (2)
Meat or meat alternate (oz eq)	8–10 (1)	9-10 (1)	10–12 (2)
Milk			
Fluid milk (cups)[f]	5 (1)	5 (1)	5 (1)
Other Specifications: Daily amount based on the average for a 5-day week			
kCalorie range[g, h]	550–650	600–700	750–850
Saturated fat (% of total kcalories)[h]	<10	<10	<10
Sodium (mg)[h, i]			
School year 2014–2015	≤1230	≤1360	≤1420
School year 2017–2018	≤935	≤1035	≤1080
School year 2022–2023	≤640	≤710	≤740
Trans fat[h]	Nutrition label or manufacturer specifications must indicate zero grams of *trans* fat per serving.		

[a]Minimum creditable serving is ⅛ cup.
[b]One quarter-cup of dried fruit counts as ½ cup of fruit; 1 cup of leafy greens counts as ½ cup of vegetables. No more than half of the fruit or vegetable offerings may be in the form of juice. All juice must be 100% full-strength.
[c]Larger amounts of these vegetables may be served.
[d]Any vegetable subgroup may be offered to meet the total weekly vegetable requirement.
[e]At least half of the grains offered must be whole grain-rich beginning July 1, 2012 (school year 2012–2013). All grains must be whole grain-rich beginning July 1, 2014 (school year 2014–2015).
[f]Fluid milk must be low-fat (1% milk fat or less, unflavored) or fat-free (unflavored or flavored).
[g]The average daily amount of kcalories for a 5-day school week must be within the range (at least the minimum and no more than the maximum values).
[h]Discretionary sources of kcalories (solid fats and added sugars) may be added to the meal pattern if within the specifications for kcalories, saturated fat, *trans* fat, and sodium. Foods of minimal nutritional value and fluid milk with fat content greater than 1% milk fat are not allowed.
[i]Final sodium specifications are to be reached by school year 2022–2023 or July 1, 2022. Intermediate specifications are established for school year 2014–2015 and 2017–2018.

Sources: U.S. Department of Agriculture 2012; Federal Register 77 (2012): 4110–4111.

rule include greater availability of fruits, vegetables, whole grains, and fat-free and low-fat milk, decreased levels of sodium, saturated fat, and *trans* fat, and guidelines for meeting nutrient needs within specified kcalorie ranges based on age/grade groups for school children. Initial implementation of these new standards began in July of 2012, while additional changes will be phased in, ending with the final goal for reduction in sodium intake required to be in place by school year 2022–2023.[80]

Parents often rely on school lunches to meet a significant part of their children's nutrient needs on school days. Indeed, students who regularly eat school lunches have higher intakes of many nutrients and fiber than students who do not.[81] Children don't always like what they are served, however, and school lunch programs must strike a balance between what children want to eat and what will nourish them and guard their health.

The Academy of Nutrition and Dietetics has set nutrition standards for childcare programs.[82] Among them, meal plans should:

- Be nutritionally adequate and consistent with the *Dietary Guidelines for Americans.*
- Emphasize fresh fruit, fresh and frozen vegetables, whole grains, and low-fat milk and milk products.
- Limit foods and beverages high in energy, added sugars, solid fats, and sodium, and low in vitamins and minerals.
- Provide foods and beverages in quantities and meal patterns appropriate to ensure optimal growth and development.
- Involve parents in planning.
- Provide furniture and eating utensils that are age appropriate and developmentally suitable to encourage children to accept and enjoy mealtime.

In addition, childcare providers can encourage active play for children by creating opportunities for children to engage in both structured and unstructured activity throughout the day.

Competing Influences at School Serving nutritious lunches is only half the battle; students need to eat them, too. Short lunch periods and long lines prevent some students from eating a school lunch and leave others with too little time to complete their meals.[83] Nutrition efforts at schools are also undermined when students can buy what the USDA labels "competitive foods"—meals from fast-food restaurants or à la carte foods such as pizza or snack foods and carbonated beverages from snack bars, school stores, and vending machines.[84] These foods and beverages compete with nutritious school lunches. When students have access to competitive foods, participation in the school lunch program decreases, nutrient intake from lunch declines, and more food is discarded.[85]

Increasingly, school-based nutrition issues are being addressed by legislation. Some states restrict the sale of competitive foods and have higher rates of participation in school meal programs than the national average. Federal legislation mandates that all school districts that participate in the USDA's National School Lunch Program develop and put in place a local wellness policy.[86] By law, wellness policies must:

- Set goals for nutrition education, physical activity, and other school-based activities.
- Establish nutrition guidelines for all foods available on school campuses during the school day.
- Develop a plan to measure policy implementation.

School districts across the nation have made progress toward meeting these goals, but implementation is inconsistent, and, because wellness policies are established locally, a great deal of variety exists among them.[87] Some are well defined and detailed, whereas others are vague.[88] To enhance local wellness policies, standards for competitive foods and beverages served in schools spell out appropriate fat, saturated fat, kcalorie, sugar, and sodium contents. Establishing and implementing these nutrition standards for competitive foods and beverages helps to ensure that all foods served in schools are consistent and comply with the *Dietary Guidelines for Americans.*

Review Notes

- Adults at home and at school need to provide children with nutrient-dense foods and teach them how to make healthful choices. Adults also need to provide ample opportunity for children to be physically active.
- School-based nutrition issues are being addressed by legislation nationwide. Establishing and implementing nutrition standards for all foods available on school campuses helps to ensure that such foods comply with the *Dietary Guidelines for Americans*.

Nutrition during Adolescence

As children pass through **adolescence** on their way to becoming adults, they change in many ways. Their physical changes make their nutrient needs high, and their emotional, intellectual, and social changes make meeting those needs a challenge.

Teenagers make many more choices for themselves than they did as children. They are not fed; they eat. Food choices made during the teen years profoundly affect health, both now and in the future. At the same time, social pressures thrust choices at them: whether to drink alcoholic beverages and whether to develop their bodies to meet extreme ideals of slimness or athletic prowess. Their interest in nutrition—both valid information and misinformation—derives from personal, immediate experiences. They are concerned with how diet can improve their lives now—they try the latest fad diet to fit into a new bathing suit, avoid greasy foods in an effort to clear acne, or eat a plate of pasta to prepare for a big sporting event. In presenting information on the nutrition and health of adolescents, this chapter includes topics of interest to teens.

GROWTH AND DEVELOPMENT DURING ADOLESCENCE

With the onset of adolescence, the steady growth of childhood speeds up abruptly and dramatically, and the growth patterns of females and males become distinct. Hormones direct the intensity and duration of the adolescent growth spurt, profoundly affecting every organ of the body, including the brain. After two to three years of intense growth and a few more at a slower pace, physically mature adults emerge.

In general, the adolescent growth spurt begins at age 10 or 11 for females and at age 12 or 13 for males. It lasts about two and a half years. Before **puberty**, male and female body compositions differ only slightly, but during the adolescent spurt, differences between the genders become apparent in the skeletal system, lean body mass, and fat stores. In females, fat assumes a larger percentage of total body weight, and in males, the lean body mass—principally muscle and bone—increases much more than in females. On average, males grow 8 inches taller, and females, 6 inches taller. Males gain approximately 45 pounds, and females, about 35 pounds.

ENERGY AND NUTRIENT NEEDS

The energy needs of adolescents vary greatly, depending on the current rate of growth, gender, body composition, and physical activity.[89] Boys' energy needs may be especially high; they typically grow faster than girls and, as mentioned, develop a greater proportion of lean body mass. An exceptionally active boy of 15 may need 3500 kcalories or more a day just to maintain his weight. Girls start growing earlier than boys and attain shorter heights and lower weights, so their energy needs peak sooner and decline earlier than those of their male peers. An inactive girl of 15 whose growth is nearly at a standstill may need fewer than 1800 kcalories a day if she is to avoid excessive weight gain. Thus, teenage girls need to pay special attention to being physically active and selecting foods of high nutrient density so that they will meet their nutrient needs without exceeding their energy needs.

adolescence: the period of growth from the beginning of puberty until full maturity. Timing of adolescence varies from person to person.

puberty: the period in life in which a person becomes physically capable of reproduction.

Obesity The insidious problem of obesity becomes ever more apparent in adolescence and often continues into adulthood. Without intervention, overweight teens will face numerous physical and socioeconomic consequences for years to come. The consequences of obesity are so dramatic and our society's attitude toward obese people is so negative that even healthy-weight or underweight teens may perceive a need to lose weight. When taken to extremes, restrictive diets bring dramatic physical consequences of their own, as Nutrition in Practice 7 explains.

Vitamin D Recommendations for most vitamins increase during the teen years (see the tables on the inside front cover). Several of the vitamin recommendations for adolescents are similar to those for adults, including the recently revised recommendations for vitamin D.[90] Vitamin D is essential for bone growth and development. Recent studies of vitamin D status in adolescents show that many are vitamin D deficient; blacks, females, and overweight adolescents are most at risk.[91] Adolescents who do not receive enough vitamin D from fortified foods such as milk and cereals, or from sun exposure each day, may need a supplement.

Iron The need for iron increases during adolescence for both females and males but for different reasons. Iron needs increase for females as they start to menstruate and for males as their lean body mass develops. Hence, the RDA increases at age 14 for both males and females. Because menstruation continues throughout a woman's childbearing years, the RDA for iron remains high for women into late adulthood. For males, the RDA returns to preadolescent values in early adulthood.

In addition, iron needs increase when the adolescent growth spurt begins, whether that occurs before or after age 14. Therefore, boys in a growth spurt need an additional 2.9 milligrams of iron per day above the RDA for their age; girls need an additional 1.1 milligrams per day.

Furthermore, iron recommendations for girls before age 14 do not reflect the iron losses of menstruation, even though the average age of menarche (first menstruation) in the United States is 12.5 years. Therefore, for girls younger than age 14 who have started to menstruate, an additional 2.5 milligrams of iron per day is recommended. Thus, the RDA for iron depends not only on age and gender but also on whether the individual is in a growth spurt or has begun to menstruate, as listed in the margin. (■)

Iron intakes often fail to keep pace with increasing needs, especially for females, who typically consume fewer iron-rich foods such as meat and fewer total kcalories than males. Not surprisingly, iron deficiency is most prevalent among adolescent girls. Iron-deficient children and teens score lower on standardized tests than those who are not iron deficient.

■ Iron RDA for males:
- 9–13 yr (8 mg/day)
- 9–13 yr in growth spurt (10.9 mg/day)
- 14–18 yr (11 mg/day)
- 14–18 yr in growth spurt (13.9 mg/day)

Iron RDA for females:
- 9–13 yr (8 mg/day)
- 9-13 yr in menarche (10.5 mg/day)
- 9–13 yr in menarche and growth spurt (11.6 mg/day)
- 14–18 yr (15 mg/day)
- 14–18 yr in growth spurt (16.1 mg/day)

Calcium Adolescence is a crucial time for bone development, and the requirement for calcium reaches its peak during these years.[92] Unfortunately, many adolescents, especially females, have calcium intakes below recommendations.[93] Low calcium intakes during times of active growth, especially if paired with physical inactivity, can compromise the development of peak bone mass, which is considered the best protection against adolescent fractures and adult osteoporosis. Increasing milk products in the diet to meet calcium recommendations greatly increases bone density.[94] Once again, however, teenage girls are most vulnerable, for their milk—and therefore their calcium—intakes begin to decline at the time when their calcium needs are greatest.[95] Furthermore, women have much greater bone losses than men in later life. In addition to dietary calcium, physical activity makes bones grow stronger. However, because some high schools do not require students to attend physical activity classes, many adolescents must make a point to be physically active during leisure time.

FOOD CHOICES AND HEALTH HABITS

Teenagers like the freedom to come and go as they choose. They eat what they want if it is convenient and if they have the time. With a multitude of afterschool, social, and job activities, they almost inevitably fall into irregular eating habits. At any given time on any given day, a teenager may be skipping a meal, eating a snack, preparing a meal, or consuming food prepared by a parent or restaurant. Adolescents who frequently eat meals with their families, however, eat more fruits, vegetables, grains, and calcium-rich foods and drink fewer soft drinks, than those who seldom eat with their families.[96] They may even smoke, drink alcohol, or abuse drugs less often.[97] Many adolescents begin to skip breakfast on a regular basis, missing out on important nutrients that are not made up at later meals during the day. Compared with those who skip breakfast, teenagers who do eat breakfast have higher intakes of vitamin A, vitamin C, and riboflavin, as well as calcium, iron, and zinc.[98] Teenagers who eat breakfast are therefore more likely to meet their nutrient recommendations.

Breakfast skipping may also lead to weight gain in adolescents. Research shows a dose-response, inverse relationship between breakfast eating and BMI.[99] As adolescents make the transition to adulthood, not only do they skip breakfast more often, but they also eat fast food more often. Both skipping breakfast and eating fast foods lead to weight gain.[100]

Ideally, in light of adolescents' busy schedules and desire for freedom, the adult becomes a **gatekeeper**, controlling the type and availability of food in the teenager's environment. Teenage sons and daughters and their friends should find plenty of nutritious, easy-to-grab food in the refrigerator (meats for sandwiches; low-fat cheeses; fresh, raw vegetables and fruits; fruit juices; and milk) and more in the cabinets (whole-grain breads, peanut butter, nuts, popcorn, and cereal). In many households today, all the adults work outside the home, and teenagers perform some of the gatekeeper's roles, such as shopping for groceries or choosing fast or prepared foods.

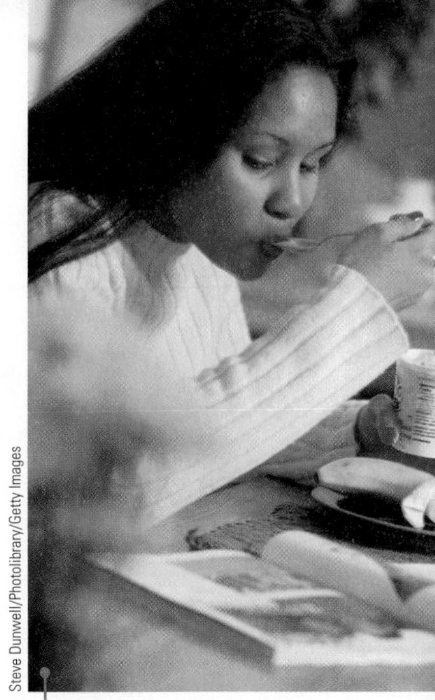

Steve Dunwell/Photolibrary/Getty Images

Nutritious snacks contribute valuable nutrients to an active teen's diet.

Snacks On average, about a fourth of an adolescent's total daily energy intake comes from snacks, which, if chosen carefully, can contribute some of the needed nutrients (see Table 11-9 on p. 334). A survey of more than 5000 adolescents found that those who ate snacks more often were less likely to be overweight or obese and had lower rates of abdominal obesity compared with those who ate snacks less often.[101]

Beverages Most frequently, adolescents drink soft drinks instead of fruit juice or milk with lunch, supper, and snacks. About the only time they select fruit juices is at breakfast. When teens drink milk, they are more likely to consume it with a meal (especially breakfast) than as a snack. Because of their greater food intakes, boys are more likely than girls to drink enough milk to meet their calcium needs.[102]

Soft drinks, when chosen as the primary beverage, may affect bone density, partly because they displace milk from the diet. Over the past three decades, teens (especially girls) have been drinking more soft drinks and less milk. Adolescents who drink soft drinks regularly have a higher energy intake and a lower calcium intake than those who do not.

Soft drinks containing caffeine present a different problem if caffeine intake becomes excessive. Caffeine seems to be relatively harmless, however, when used in moderate doses (the equivalent of fewer than, say, three 12-ounce cola beverages a day). (■) In greater amounts, it can cause the symptoms associated with anxiety—sweating, tenseness, and inability to concentrate.

Eating Away from Home Adolescents eat about one-third of their meals away from home, and their nutritional welfare is enhanced or hindered by the choices they make. A lunch consisting of a hamburger, a chocolate shake, and french fries supplies substantial quantities of many nutrients at a kcalorie cost of about 800, an energy intake some adolescents can afford. When they eat this sort of lunch, teens can adjust their breakfast and dinner choices to include fruits and vegetables for vitamin A, vitamin C, folate, and fiber, and lean meats and legumes for iron and zinc. Fortunately, many fast-food restaurants are offering more nutritious choices than the standard hamburger meal.

Peer Influence Physical maturity and growing independence present adolescents with new choices. The consequences of those choices will influence their health and nutrition status both today and throughout life. Many of the food and health choices

■ Appendix A provides a table of the caffeine contents of beverages, foods, and medications.

gatekeeper: with respect to nutrition, a key person who controls other people's access to foods and thereby exerts a profound impact on their nutrition. Examples are the spouse who buys and cooks the food, the parent who feeds the children, and the caregiver in a day care center.

adolescents make reflect the opinions and actions of their peers. When others perceive milk as "babyish," a teen may choose soft drinks instead; when others skip lunch and hang out in the parking lot, a teen may join in for the camaraderie, regardless of hunger. Some teenagers begin using drugs, alcohol, and tobacco; others wisely refrain. Adults can set up the environment so that nutritious foods are available and can stand by with reliable information and advice about health and nutrition, but the rest is up to the adolescents. Ultimately, they make the choices.

Review Notes

- Nutrient needs rise dramatically as children enter the rapid growth phase of the teen years.
- The busy lifestyles of teenagers add to the challenge of meeting their nutrient needs, especially for iron and calcium.

Assessment of nutrition status in healthy children and adolescents can confirm that development is normal or can catch potential problems early. (■) The Nutrition Assessment Checklist highlights problems to look for when working with children and adolescents.

■ Chapter 13 offers details about nutrition assessment.

Nutrition Assessment Checklist for Children and Adolescents

MEDICAL HISTORY

Check the medical record for:

- ❍ Alcohol, tobacco, or illicit drug abuse
- ❍ Attention deficit/hyperactivity disorder (ADHD)
- ❍ Diabetes or other chronic disorders
- ❍ Eating disorders
- ❍ Food allergies
- ❍ Lactose intolerance
- ❍ Obesity
- ❍ Pregnancy

MEDICATIONS

For children or adolescents being treated with drug therapy for medical conditions, note:

- ❍ Side effects that might reduce food intake or change nutrient needs
- ❍ Proper administration of medication with respect to food intake

DIETARY INTAKE

For all children and adolescents, especially those considered at risk nutritionally, assess the diet for:

- ❍ Total energy
- ❍ Protein
- ❍ Calcium and iron
- ❍ Vitamin A, vitamin C, and folate
- ❍ Fiber

Note the following:

- ❍ Number of days each week a nutritious breakfast is eaten
- ❍ Number of hours the child or teen sleeps each day
- ❍ Number of soft drinks the child or teen drinks each day
- ❍ Number of fast-food meals eaten each day
- ❍ Number and type of snacks eaten each day
- ❍ Type and amount of physical activity
- ❍ Amount of caffeine consumed

ANTHROPOMETRIC DATA

Measure baseline height and weight.

- ❍ Reassess height, weight, and growth patterns at each medical checkup.
- ❍ Note significant obesity or underweight and intervention strategies employed.

LABORATORY TESTS

Monitor the following laboratory tests for children and adolescents:

- ❍ Hemoglobin, hematocrit, or other tests of iron status
- ❍ Blood glucose for children or adolescents with diabetes
- ❍ Blood lead concentrations

PHYSICAL SIGNS

Look for physical signs of:

- ❍ Protein-energy malnutrition
- ❍ Iron deficiency
- ❍ Vitamin A deficiency
- ❍ Vitamin C deficiency
- ❍ Folate deficiency

Self Check

1. Which of the following is a characteristic of iron deficiency in children?
   a. It rarely develops in those with high intakes of milk.
   b. It affects brain function before anemia sets in.
   c. It is a primary factor in hyperactivity.
   d. Milk deficiency enhances mental performance by lowering physical activity level, thereby leading to increased attention span.

2. Three symptoms of lead toxicity are:
   a. diarrhea, irritability, and fatigue.
   b. low blood sugar, hair loss, and skin rash.
   c. increased heart rate, hyperactivity, and dry skin.
   d. bleeding gums, brittle fingernails, and swollen glands.

3. Allergic reactions to foods are more often caused by:
   a. corn, rice, or meats.
   b. eggs, peanuts, or milk.
   c. red meats, milk, or MSG.
   d. seafood, dark greens, or lactose.

4. When introducing new foods to children:
   a. reward children as they try new foods.
   b. offer many choices to encourage variety.
   c. offer one new food at the end of the meal.
   d. offer one new food at the beginning of the meal.

5. Which of the following is *not* true? Children who watch a lot of television are likely to:
   a. become obese.
   b. spend less time being physically active.
   c. learn healthy eating tips from programs.
   d. eat the foods most often advertised on television.

6. Which of the following strategies is *not* effective?
   a. Play first, eat later.
   b. Provide small portions.
   c. Encourage children to help prepare meals.
   d. Use dessert as a reward for eating vegetables.

7. During the growth spurt of adolescence:
   a. females gain more weight than males.
   b. males gain more fat, proportionately, than females.
   c. differences in body composition between males and females become apparent.
   d. similarities in body composition between males and females become apparent.

8. Two nutrients that are usually lacking in adolescents' diets are:
   a. zinc and fat.
   b. iron and calcium.
   c. protein and thiamin.
   d. vitamin A and riboflavin.

9. To help teenagers consume a balanced diet, parents can:
   a. monitor the teens' food intake.
   b. give up—parents can't influence teenagers.
   c. keep the cabinets and refrigerator well stocked.
   d. forbid snacking and insist on regular, well-balanced meals.

10. To balance the day's intake, an adolescent who eats a hamburger, fries, and chocolate shake at lunch might benefit most from a dinner of:
    a. fried chicken, rice, and banana.
    b. ribeye steak, baked potato, and salad.
    c. pork chop, mashed potatoes, and apple juice.
    d. spaghetti with meat sauce, broccoli, and milk.

Answers to these questions are in Appendix H. For more chapter review: Access an interactive eBook, chapter-specific interactive learning tools, including flashcards, quizzes, videos, and more in your Nutrition CourseMate, accessed through CengageBrain.com.

Clinical Applications

1. At two and a half years old, Travis is healthy, though slightly underweight, and headstrong. Travis's mother hovers over him at every meal and insists that he take several bites of every food on his plate, even if he dislikes the food or is not familiar with it. Even though Travis is hungry when he sits down to a meal, his mother's constant urging to get him to eat quickly quells any interest he had in eating. Travis simply folds his arms across his chest, closes his mouth tightly, and refuses to eat any more food. After more begging, pleading, and nagging, Travis's mother becomes angry

and sends him away from the table. Travis is not allowed to snack between meals because his mother is concerned that snacks will ruin his appetite.

- What factors might be contributing to Travis's refusal to eat?
- Travis's mother is concerned about her son's underweight. What strategies would you suggest to help Travis gain weight?
- What advice would you offer Travis's mother to help her improve mealtimes with her son?

2. Loni is a physically inactive, overweight 15-year-old girl who enjoys watching television and playing computer games in her spare time. She typically buys a cola and bag of chips from the vending machine for lunch and then stops by a fast-food restaurant after school for chicken nuggets or a hamburger with french fries. She has noticed that she has gained weight and lacks energy.
   - What nutrients might be excessive or lacking in her current diet plan?
   - What dietary advice would you offer to help Loni look and feel healthier?
   - What would you tell Loni to motivate her to become physically active?

Nutrition on the Net

For further study of the topics in this chapter, access these websites.

- Learn how to care for children and adolescents from the American Academy of Pediatrics and the Canadian Paediatric Society:
  www.aap.org and www.cps.ca
- Download the current growth charts from the Centers for Disease Control and Prevention, and learn about the most recent revision of these charts:
  www.cdc.gov/growthcharts
- Get information on MyPlate for children and preschoolers from the USDA:
  www.choosemyplate.gov
- Get tips for keeping children healthy from the Nemours Foundation:
  www.kidshealth.org
- Visit the National Center for Education in Maternal and Child Health and the National Institute of Child Health and Human Development:
  www.ncemch.org and www.nichd.nih.gov/
- Learn about child nutrition programs:
  www.fns.usda.gov/fns
- Learn how to reduce lead exposure in your home from the U.S. Department of Housing and Urban Development's Office of Healthy Homes and Lead Hazard Control:
  www.hud.gov/lead
- Learn more about food allergies from the American Academy of Allergy, Asthma, and Immunology; the Food Allergy and Anaphylaxis Network; and the International Food Information Council Foundation:
  www.aaaai.org, www.foodallergy.org, and www.foodinsight.org/
- Learn more about hyperactivity from Children and Adults with Attention Deficit/Hyperactivity Disorder:
  www.chadd.org
- Learn how to help children develop positive body images and relate to food in a healthy way:
  www.womenshealth.gov/body-image/kids/index.html
- Visit the Milk Matters section of the National Institute of Child Health and Human Development (NICHD):
  www.nichd.nih.gov/milk/milk.cfm
- Get weight-loss tips for children and adolescents:
  www.shapedown.com
- Under the "Health Issues" menu, select "Conditions" and then "Tobacco" for information on the risks of tobacco use from the American Academy of Pediatrics:
  www.healthychildren.org
- Visit the Smoking & Tobacco Use page of the Centers for Disease Control and Prevention:
  http://cdc.gov/tobacco/

Notes

1. R. P. Heaney, Diet, osteoporosis, and fracture prevention: The totality of the evidence, in A. Bendichand and R. J. Deckelbaum, eds., *Preventive Nutrition: The Comprehensive Guide for Health Professionals*, 4th ed. (New York: Human Press, 2010), pp. 443–469.
2. S. A. Ramsay and coauthors, "Are you done?" Child care providers' verbal communication at mealtimes that reinforce or hinder children's internal cues of hunger and satiation, *Journal of Nutrition Education and Behavior* 42 (2010): 265–270.
3. American Academy of Pediatrics, Policy Statement—Children, adolescents, obesity, and the media, *Pediatrics* 128 (2011): 201–208; V. C. Strasburger, A. B. Jordan, and E. Donnerstein, Health effects of media on children and adolescents, *Pediatrics* 125 (2010): 756–767; Position of the American Dietetic Association: Nutrition guidance for healthy children ages 2 to 11 years, *Journal of the American Dietetic Association* 108 (2008): 1038–1047.
4. J. L. Foltz and coauthors, Population-level intervention strategies and examples for obesity prevention in children, *Annual Review of Nutrition*

32 (2012): 391–415; Position of the American Dietetic Association: Benchmarks for nutrition in child care, *Journal of the American Dietetic Association* 111 (2011): 607–615; N. Larson and coauthors, What role can child-care settings play in obesity prevention? A review of the evidence and call for research efforts, *Journal of the American Dietetic Association* 111 (2011): 1343–1362.

5. Nutritional aspects of vegetarian diets, in R. E. Kleinman, ed., *Pediatric Nutrition Handbook*, 6th ed. (Elk Grove Village, IL: American Academy of Pediatrics, 2009), pp. 201–224.
6. Committee on the Scientific Evaluation of Dietary Reference Intakes, Food and Nutrition Board, Institute of Medicine, *Dietary Reference Intakes for Energy, Carbohydrate, Fiber, Fat, Fatty Acids, Cholesterol, Protein, and Amino Acids* (Washington, DC: National Academies Press, 2005), pp. 265–338.
7. Committee on the Scientific Evaluation of Dietary Reference Intakes, 2005, pp. 339–421.
8. Committee on the Scientific Evaluation of Dietary Reference Intakes, 2005, pp. 769–879.
9. Committee on the Scientific Evaluation of Dietary Reference Intakes, 2005, pp. 422–541.
10. R. D. Baker, F. R. Greer, and the Committee on Nutrition, Clinical Report: Diagnosis and prevention of iron deficiency and iron-deficiency anemia in infants and young children (0–3 years of age), *Pediatrics* 126 (2010): 1040–1050.
11. Expert Panel on Integrated Guidelines for Cardiovascular Health and Risk Reduction in Children and Adolescents, Summary Report, *Pediatrics* 128 (2011): S213–S256.
12. Committee on Dietary Reference Intakes, *Dietary Reference Intakes for Calcium and Vitamin D* (Washington, DC: National Academies Press, 2011), pp. 345–402.
13. C. L.Wagner and F. R. Greer, and the Section on Breastfeeding and Committee on Nutrition, Prevention of rickets and vitamin D deficiency in infants, children, and adolescents, *Pediatrics* 122 (2008): 1142–1152.
14. R. L. Bailey and coauthors, Do dietary supplements improve micronutrient sufficiency in children and adolescents? *Journal of Pediatrics* 161 (2012): 837–842; Feeding the child, in R. E. Kleinman, ed., *Pediatric Nutrition Handbook*, 6th ed. (Elk Grove Village, IL: American Academy of Pediatrics, 2009), pp. 145–174.
15. Bailey and coauthors, 2012; U. Shaikh, R. S. Byrd, and P. Auinger, Vitamin and mineral supplement use by children and adolescents in the 1999–2004 National Health and Nutrition Examination Survey: Relationship with nutrition, food security, physical activity, and health care access, *Archives of Pediatrics and Adolescent Medicine* 163 (2009): 150–157.
16. Position of the American Dietetic Association, 2008.
17. R. R. Briefel and coauthors, The Feeding Infants and Toddlers Study 2008: Study design and methods, *Journal of the American Dietetic Association* 110 (2010): S16–S26; N. F. Butte and coauthors, Nutrient intakes of US infants, toddlers, and preschoolers meet or exceed dietary reference intakes, *Journal of the American Dietetic Association* 110 (2010): S27–S37.
18. M. K. Fox and coauthors, Food consumption patterns of young preschoolers: Are they starting off on the right path? *Journal of the American Dietetic Association* 110 (2010): S52–S59.
19. Butte and coauthors, 2010; M. Vadiveloo, L. Zhu, and P. A. Quatromoni, Diet and physical activity patterns of school-aged children, *Journal of the American Dietetic Association* 109 (2009): 145–151.
20. H. Skouteris and coauthors, Parental influence and obesity prevention in pre-schoolers: A systematic review of interventions, *Obesity Reviews* 12 (2011): 315–328.
21. R. B. Ervin and coauthors, Consumption of added sugar among U.S. children and adolescents, 2005–2008, *NCHS Data Brief* 87 (2012), available at www.cdc.gov/nchs/data/databriefs/db87.htm.
22. Bread for the World, Hunger facts: Domestic, available at www.bread.org/learn/hunger-basics/hunger-facts-domestic.html; A. Coleman-Jensen and coauthors, Household food security in the United States in 2011, *ERS Report Summary*, September 2012, available at www.ers.usda.gov.
23. Position of the American Dietetic Association: Child and adolescent nutrition assistance programs, *Journal of the American Dietetic Association* 110 (2010): 791–799.
24. C. E. Basch, Breakfast and the achievement gap among urban minority youth, *Journal of School Health* 81 (2011): 635–640; S. B. Cooper, S. Bandelow, and M. E. Nevill, Breakfast consumption and cognitive function in adolescent schoolchildren, *Physiology & Behavior* 103 (2011): 431–439; S. P. P. Tin and coauthors, Breakfast skipping and change in body mass index in young children, *International Journal of Obesity* 35 (2011): 899–906; P. R. Deshmukh-Taskar and coauthors, The relationship of breakfast skipping and type of breakfast consumption with nutrient intake and weight status in children and adolescents: The National Health and Nutrition Examination Survey 1999–2006, *Journal of the American Dietetic Association* 110 (2010): 869–878.
25. T. Coppinger and coauthors, Body mass, frequency of eating and breakfast consumption in 9–13-year-olds, *Journal of Human Nutrition and Dietetics* 25 (2012): 43–49; T. V. E. Kral and coauthors, Effects of eating breakfast compared with skipping breakfast on ratings of appetite and intake at subsequent meals in 8- to 10-y-old children, *American Journal of Clinical Nutrition* 93 (2011): 284–291; K. J. Smith and coauthors, Skipping breakfast: Longitudinal associations with cardiometabolic risk factors in the Childhood Determinants of Adult Health Study, *American Journal of Clinical Nutrition* 92 (2010): 1316–1325; L. Dubois and coauthors, Breakfast skipping is associated with differences in meal patterns, macronutrient intakes and overweight among pre-school children, *Public Health Nutrition* 12 (2009): 19–28.
26. K. Kordas, Iron, lead, and children's behavior and cognition, *Annual Review of Nutrition* 30 (2010): 123–148.
27. M. M. Black and coauthors, Iron deficiency and iron-deficiency anemia in the first two years of life: Strategies to prevent loss of developmental potential, *Nutrition Reviews* 69 (2011): S64–S70; M. K. Georgieff, Long-term brain and behavioral consequences of early iron deficiency, *Nutrition Reviews* 69 (2011): S43–S48; C. S. Wang and coauthors, Iron-deficiency anemia in infancy and social emotional development in preschool-aged Chinese children, *Pediatrics* 127 (2011): e927–e933; F. Corapci and coauthors, Longitudinal evaluation of externalizing and internalizing behavior problems following iron deficiency in infancy, *Journal of Pediatric Psychology* 35 (2010): 296–305.
28. Kordas, 2010.
29. K. M. Cecil and coauthors, Decreased brain volume in adults with childhood lead exposure, *PLos Medicine* 27 (2008): e112.
30. Centers for Disease Control and Prevention, Lead, available at www.cdc.gov/nceh/lead/, updated October 23, 2012.
31. Centers for Disease Control and Prevention, Prevention tips, available at www.cdc.gov/nceh/lead/tips.htm, updated June 25, 2012.
32. R. S. Gupta and coauthors, The prevalence, severity, and distribution of childhood food allergy in the United States, *Pediatrics* 128 (2011): e9–e17; Food and Drug Administration, Food allergies: Reducing the risks, January 23, 2009, available at www.fda.gov/consumer/updates/foodallergies012209.html.
33. U.S. Food and Drug Administration, Food allergies: Reducing the risks, January 23, 2009, available at www.fda.gov/consumer/updates/foodallergies012209.html.
34. J. A. Boyce and coauthors, Guidelines for the diagnosis and management of food allergy in the United States: Summary of the NIAID-Sponsored Expert Panel Report, *Journal of the American Dietetic Association* 111 (2011): 17–27.
35. H. N. Cho and coauthors, Nutritional status according to sensitized food allergens in children with atopic dermatitis, *Allergy, Asthma, and Immunology Research* 3 (2010): 53–57.
36. Boyce and coauthors, 2011.
37. Gupta and coauthors, 2011.
38. S. Waserman and W. Watson, Food allergy, *Allergy, Asthma, and Clinical Immunology* 7 (2011): S7.
39. M. Pansare and D. Kamat, Peanut allergy, *Current Opinions in Pediatrics* 22 (2010): 642–646; A. T. Clark and coauthors, Successful oral tolerance induction in severe peanut allergy, *Allergy* 64 (2009): 1218–1220.
40. S. H. Sicherer, T. Mahr, and The Section on Allergy and Immunology, Clinical report—Management of food allergy in the school setting, *Pediatrics* 126 (2010): 1232–1239.
41. U.S. Food and Drug Administration, *Food Facts*, Food allergies: What you need to know, available at www.fda.gov/Food/ResourcesForYou/Consumers/icm079311.htm, updated May 14, 2012; M. M. Pieretti and coauthors, Audit of manufactured products: Use of allergen advisory labels and identification of labeling ambiguities, *Journal of Allergy and Clinical Immunology* 124 (2009): 337–341.
42. L. J. Akinbami and coauthors, Attention deficit hyperactivity disorder among children aged 5–17 years in the United States, 1998–2009, *NCHS Data Brief* 70, August 2011, www.cdc.gov/nchs/data/databriefs/db70.htm.
43. Subcommittee on Attention-Deficit/Hyperactivity Disorder, Steering Committee on Quality Improvement and Management, ADHD: Clinical practice guideline for the diagnosis, evaluation, and treatment of attention-deficit/hyperactivity disorder in children and adolescents, *Pediatrics*

128 (2011): 1007–1022; W. B. Brinkman and coauthors, Parental angst making and revisiting decisions about treatment of attention-deficit/hyperactivity disorder, *Pediatrics* 124 (2009): 580–589.

44. J. G. Millichap and M. M. Yee, The diet factor in attention-deficit/hyper activity disorder, *Pediatrics* 129 (2012): 330–337.
45. R. E. Kleinman and coauthors, A research model for investigating the effects of artificial food colorings on children with ADHD, *Pediatrics* 127 (2011): e1575–1584; T. E. Froehlich and coauthors, Update on environmental risk factors for attention-deficit/hyperactivity disorder, *Current Psychiatry Reports* 13 (2011): 333–344; A. Connolly and coauthors, Pattern of intake of food additives associated with hyperactivity in Irish children and teenagers, *Food Additives and Contaminants: Part A, Chemistry, Analysis, Control, Exposure and Risk Assessment* 27 (2010): 447–456; J. Stevenson and coauthors, The role of histamine degradation gene polymorphisms in moderating the effects of food additives on children's ADHD symptoms, *American Journal of Psychiatry* 167 (2010): 1108–1115.
46. Food and Drug Administration, Food ingredients and colors, available at www.fda.gov, revised April 2010.
47. C. L. Ogden and coauthors, Prevalence of obesity and trends in body mass index among US children and adolescents, 1999–2010, *Journal of the American Medical Association* 307 (2012): 483–490; C. L. Ogden and coauthors, Prevalences of high body mass index in US children and adolescents, 2007–2008, *Journal of the American Medical Association* 303 (2010): 242–249.
48. Expert Panel on Integrated Guidelines for Cardiovascular Health and Risk Reduction in Children and Adolescents, Summary Report, *Pediatrics* 128 (2011): S213–S256; J. J. Reilly, Assessment of obesity in children and adolescents: Synthesis of recent systematic reviews and clinical guidelines, *Journal of Human Nutrition and Dietetics* 23 (2010): 205–211.
49. G. Flores and H. Lin, Factors predicting severe childhood obesity in kindergarteners, *International Journal of Obesity* November 13, 2012. doi: 10.1038/ijo.2012.168.
50. Expert Panel on Integrated Guidelines for Cardiovascular Health and Risk Reduction in Children and Adolescents, 2011; N. S. The and coauthors, Association of adolescent obesity with risk of severe obesity in adulthood, *Journal of the American Medical Association* 304 (2010): 2042–2057; F. M. Biro and M. Wien, Childhood obesity and adult morbidities, *American Journal of Clinical Nutrition* 91 (2010): 1499S–1505S.
51. R. Cooper and coauthors, Associations between parental and offspring adiposity up to midlife: The contribution of adult lifestyle factors in the 1958 British Birth Cohort Study, *American Journal of Clinical Nutrition* 92 (2010): 946–953; K. L. Whitaker and coauthors, Comparing maternal and paternal intergenerational transmission of obesity risk in a large population-based sample, *American Journal of Clinical Nutrition* 91 (2010): 1560–1567; L. Li and coauthors, Intergenerational influences on childhood body mass index: The effect of parental body mass index trajectories, *American Journal of Clinical Nutrition* 89 (2009): 551–557.
52. Biro and Wien, 2010.
53. Centers for Disease Control and Prevention, Overweight and obesity: A growing problem, www.cdc.gov/obesity/childhood/problem.html, updated April 27, 2012; Foltz and coauthors, 2012.
54. S. J. Salvy and coauthors, Influence of parents and friends on children's and adolescents' food intake and food selection, *American Journal of Clinical Nutrition* 93 (2011): 87–92; L. Hall and coauthors, Children's intake of fruit and selected energy-dense nutrient-poor foods is associated with fathers' intake, *Journal of the American Dietetic Association* 111 (2011): 1039–1044; H. A. Raynor and coauthors, The relationship between child and parent food hedonics and parent and child food group intake in children with overweight/obesity, *Journal of the American Dietetic Association* 111 (2011): 425–430; C. Sweetman and coauthors, Characteristics of family mealtimes affecting children's vegetable consumption and liking, *Journal of the American Dietetic Association* 111 (2011): 269–273; S. L. Anzman and coauthors, Parental influence on children's early eating environments and obesity risk: Implications for prevention, *International Journal of Obesity* 34 (2010): 1116–1124; K. J. Gruber and L. A. Haldeman, Using the family to combat childhood and adult obesity, *Preventing Chronic Disease* 6 (2009): A106.
55. K. S. Geller and D. A. Dzewaltowski, Longitudinal and cross-sectional influences on youth fruit and vegetable consumption, *Nutrition Reviews* 67 (2009): 65–76.
56. J. Reedy and S. M. Krebs-Smith, Dietary sources of energy, solid fats, and added sugars among children and adolescents in the United States, *Journal of the American Dietetic Association* 110 (2010): 1477–1484.
57. Ervin and coauthors, 2012; J. A. Welsh and coauthors, Consumption of added sugars is decreasing in the United States, *American Journal of Clinical Nutrition* 94 (2011): 726–734; S. N. Bleich and coauthors, Increasing consumption of sugar-sweetened beverages among US adults: 1988–1994 to 1999–2004, *American Journal of Clinical Nutrition* 89 (2009): 372–381.
58. D. F. Tate and coauthors, Replacing caloric beverages with water or diet beverages for weight loss in adults: Main results of the Choose Healthy Options Consciously Everyday (CHOICE) randomized clinical trial, *American Journal of Clinical Nutrition* 95 (2012): 555–563; V. S. Malik and coauthors, sugar-sweetened beverages, obesity, type 2 diabetes mellitus, and cardiovascular disease risk, *Circulation* 121 (2010): 1356–1364; F. B. Hu and V. S. Malik, Sugar-sweetened beverages and risk of obesity and type 2 diabetes: epidemiologic evidence, *Physiology and Behavior* 100 (2010): 47–54; O. L. Bermudez and X. Gao, Greater consumption of sweetened beverages and added sugars is associated with obesity among US young adults, *Annals of Nutrition and Metabolism* 57 (2010): 211–218; L. Chen and coauthors, Reduction in consumption of sugar-sweetened beverages is associated with weight loss: The PREMIER Trial, *American Journal of Clinical Nutrition* 89 (2009): 1299–1306.
59. U. Ekelund and coauthors, Moderate to vigorous physical activity and sedentary time and caridometabolic risk factors in children and adolescents, *Journal of the American Medical Association* 307 (2012): 704–712.
60. Centers for Disease Control and Prevention, 2012.
61. American Academy of Pediatrics, Council on Communications and Media, Policy statement—Children, adolescents, obesity, and the media, *Pediatrics* 128 (2011): 201–208; D. M. Jackson and coauthors, Increased television viewing is associated with elevated body fatness but not with lower total energy expenditure in children, *American Journal of Clinical Nutrition* 89 (2009): 1031–1036.
62. American Academy of Pediatrics, Council on Communications and Media, 2011; S. B. Sisson and coauthors, TVs in the bedrooms of children: Does it impact health and behavior? *Preventive Medicine* 52 (2011): 104–108.
63. D. J. Anschutz, R. CME Engels, and T. Van Strien, Side effects of television food commercials on concurrent nonadvertised sweet snack food intakes in young children, *American Journal of Clinical Nutrition* 89 (2009): 1328–1333.
64. A. E. Henry and M. Story, Food and beverage brands that market to children and adolescents on the internet: A content analysis of branded web sites, *Journal of Nutrition Education and Behavior* 41 (2009): 353–359.
65. J. P. Chaput and coauthors, Video game playing increases food intake in adolescents: A randomized crossover study, *American Journal of Clinical Nutrition* 93 (2011): 1196–1203; American Academy of Pediatrics, Council on Communications and Media, 2011.
66. K. White, G. Schofield, and A. E. Kilding, Energy expended by boys playing active video games, *Journal of Science and Medicine in Sport* 14 (2011): 130–134.
67. American Academy of Pediatrics, Council on Communications and Media, 2011.
68. W. Tu and coauthors, Intensified effect of adiposity on blood pressure in overweight and obese children, *Hypertension* 58 (2011): 818–824; M. Salvadori and coauthors, Elevated blood pressure in relation to overweight and obesity among children in a rural Canadian community, *Pediatrics* 122 (2008): e821–e827.
69. J. Ohman and coauthors, Early childhood overweight and asthma and allergic sensitization at 8 years of age, *Pediatrics* 129 (2012): 70–76; A. Tirosh and coauthors, Adolescent BMI trajectory and risk of diabetes versus coronary disease, *New England Journal of Medicine* 364 (2011): 1315–1325; J. C. Han, D. A. Lawlor, and S. Y. S. Kimm, Childhood obesity, *Lancet* 375 (2010): 1737–1748; F. M. Biro and M. Wien, Childhood obesity and adult morbidities, *American Journal of Clinical Nutrition* 91 (2010): 1499S–1505S; P. W. Franks and coauthors, Childhood obesity, other cardiovascular risk factors, and premature death, *New England Journal of Medicine* 362 (2010): 458–493; C. J. Lavie, R. V. Milani, and H. O. Ventura, Obesity and cardiovascular disease, *Journal of the American College of Cardiology* 53 (2009): 1925–1933; D. S. Freedman and coauthors, Risk factors and adult body mass index among overweight children: The Bogalusa Heart Study, *Pediatrics* 123 (2009): 750–757.
70. American Heart Association, Overweight in children, 2012, available at http://www.heart.org/HEARTORG/GettingHealthy?Overweight-in-Children_UCM_304054_Article.jsp.
71. A. L. Rogovik and R. D. Goldman, Pharmacologic treatment of pediatric obesity, *Canadian Family Physician* 57 (2011): 195–197.

72. T. H. Inge and coauthors, Reversal of type 2 diabetes mellitus and improvements in cardiovascular risk factors after surgical weight loss in adolescents, *Pediatrics* 123 (2009): 214–222.
73. K. W. Bauer, J. M. Berge, and D. Neumark-Sztainer, The importance of families to adolescents' physical activity and dietary intake, *Adolescent Medicine: State of the Art Reviews* 22 (2011): 601–613; S. L. Anzman and coauthors, Parental influence on children's early eating environments and obesity risk: Implications for prevention, *International Journal of Obesity* 34 (2010): 1116–1124; K. J. Gruber and L. A. Haldeman, Using the family to combat childhood and adult obesity, *Preventing Chronic Disease* 6 (2009): A106.
74. Position of the American Dietetic Association, 2008.
75. Position of the American Dietetic Association, 2011; Position of the American Dietetic Association, School Nutrition Association, and Society for Nutrition Education: Comprehensive school nutrition services, *Journal of the American Dietetic Association* 110 (2010): 1738–1749.
76. The School Breakfast Program, www.fns.usda.gov/cnd/breakfast.htm, updated August 5, 2012; Position of the American Dietetic Association: Local support for nutrition integrity in schools, *Journal of the American Dietetic Association* 110 (2010): 1244–1254.
77. Position of the American Dietetic Association: Local support for nutrition integrity in schools, 2010.
78. M. K. Crepinsek and coauthors, Meals offered and served in US public schools: Do they meet nutrient standards? *Journal of the American Dietetic Association* 109 (2009): S31–S43.
79. National School Lunch Program, www.fns.usda.gov/cnd/lunch/AboutLunch/NSLPFactSheet.pdf, updated August, 2012; Position of the American Dietetic Association: Local support for nutrition integrity in schools, 2010.
80. Food and Nutrition Services (FNS) USDA, Nutrition standards in the National School Lunch and School Breakfast Programs. Final rule, *Federal Register* 77 (2012): 4088–4167.
81. M. Story, The Third School Nutrition Dietary Assessment Study: Findings and policy implications for improving the health of US children, *Journal of the American Dietetic Association* 109 (2009): S7–S13.
82. Position of the American Dietetic Association, 2011.
83. Position of the American Dietetic Association: Local support for nutrition integrity in schools, 2010.
84. Position of the American Dietetic Association: Local support for nutrition integrity in schools, 2010.
85. M. Kakarala, D. R. Keast, and S. Hoerr, Schoolchildren's consumption of competitive foods and beverages, excluding á la carte, *Journal of School Health* 80 (2010): 429–435; Position of the American Dietetic Association: Local support for nutrition integrity in schools, 2010.
86. Position of the American Dietetic Association, School Nutrition Association, and Society for Nutrition Education: Comprehensive school nutrition services, 2010; Position of the American Dietetic Association: Local support for nutrition integrity in schools, 2010.
87. L. R. Turner and F. J. Chaloupka, Student access to competitive foods in elementary schools, trends over time and regional differences, *Archives of Pediatrics & Adolescent Medicine* 166 (2012): 164–169; Centers for Disease Control and Prevention, Availability of less nutritious snack foods and beverages in secondary schools—Selected states, 2002–2008, *Morbidity and Mortality Weekly Report* 58 (2009): 1102–1104.
88. Position of the American Dietetic Association, School Nutrition Association, and Society for Nutrition Education: Comprehensive school nutrition services, 2010.
89. Committee on the Scientific Evaluation of Dietary Reference Intakes, 2005, pp. 177–182.
90. Committee on Dietary Reference Intakes, 2011, pp. 362– 402.
91. S. Saintonge, H. Bang, and L. M. Gerber, Implications of a new definition of vitamin D deficiency in a multiracial US adolescent population: The National Health and Nutrition Examination Survey III, *Pediatrics* 123 (2009): 797–803.
92. Committee on Dietary Reference Intakes, 2011, pp. 35–74.
93. Committee on Dietary Reference Intakes, 2011, pp.458–460; R. L. Bailey and coauthors, Estimation of total usual calcium and vitamin D intakes in the United States, *The Journal of Nutrition* 140 (2010): 817–822.
94. M. Mesias, I. Seiquer, and M. P. Navarro, Calcium nutrition in adolescence, *Critical Reviews in Food Science and Nutrition* 51 (2011): 195–209; L. Esterie and coauthors, Milk, rather than other foods, is associated with vertebral bone mass and circulating IGF-1 in female adolescents, *Osteoporosis International* 20 (2009): 567–575.
95. L. M. Fiorito and coauthors, Girls' early sweetened carbonated beverage intake predicts different patterns of beverage and nutrient intake across childhood and adolescence, *Journal of the American Dietetic Association* 110 (2010): 543–550.
96. A. J. Hammons and B. H. Fiese, Is frequency of shared family meals related to the nutritional health of children and adolescents? *Pediatrics* 127 (2011): e1565–e1574; D. Neumark-Sztainer and coauthors, Family meals and adolescents: What have we learned from Project EAT (Eating Among Teens)? *Public Health Nutrition* 13 (2010): 1113–1121; S. J. Woodruff and R. M. Hanning, Associations between family dinner frequency and specific food behaviors among grade six, seven, and eight students from Ontario and Nova Scotia, *Journal of Adolescent Health* 44 (2009): 431–436.
97. Neumark-Sztainer and coauthors, 2010.
98. P. R. Deshmukh-Taskar and coauthors, The relationship of breakfast skipping and type of breakfast consumption with nutrient intake and weight status in children and adolescents: The National Health and Nutrition Examination Survey 1999–2006, *Journal of the American Dietetic Association* 110 (2011): 869–878; K. J. Smith and coauthors, Skipping breakfast: Longitudinal associations with cardiometabolic risk factors in the Childhood Determinants of Adult Health Study, *American Journal of Clinical Nutrition* 92 (2010): 1316–1325.
99. M. T. Timlin and coauthors, Breakfast eating and weight change in a 5-year prospective analysis of adolescents: Project EAT (Eating Among Teens), *Pediatrics* 121 (2008): e638–e645.
100. C. M. McDonald and coauthors, Overweight is more prevalent than stunting and is associated with socioeconomic status, maternal obesity, and a snacking dietary pattern in school children from Bogota, Colombia, *Journal of Nutrition* 139 (2009): 370–376.
101. D. R. Keast, T. A. Nicklas, and C. E. O'Neil, Snacking is associated with reduced risk of overweight and reduced abdominal obesity in adolescents: National Health and Nutrition Examination survey (NHANES) 1999–2004, *American Journal of Clinical Nutrition* 92 (2010): 428–435.
102. N. D. Brener and coauthors, Beverage consumption among high school students—United States, 2010, *Morbidity and Mortality Weekly Report* 60 (2011): 778–780.

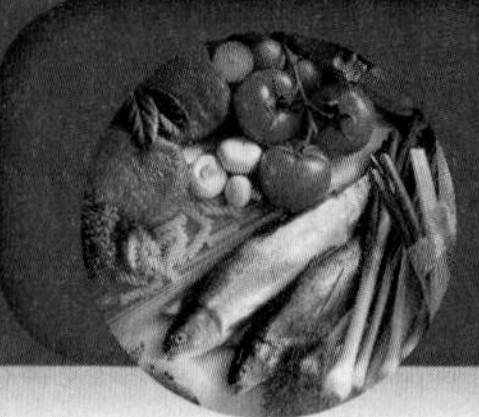

Nutrition in Practice

Childhood Obesity and the Early Development of Chronic Diseases

When people think about the health problems of children and adolescents, they typically think of ear infections, colds, and acne—not heart disease, diabetes, or hypertension. Today, however, unprecedented numbers of U.S. children are being diagnosed with obesity and the serious "adult diseases," such as type 2 diabetes, that accompany overweight.[1] When type 2 diabetes develops before the age of 20, the incidence of diabetic kidney disease and death in middle age increases dramatically, largely because of the long duration of the disease. For children born in the United States in the year 2000, the risk of developing type 2 diabetes sometime in their lives is estimated to be 30 percent for boys and 40 percent for girls. U.S. children are not alone—rapidly rising rates of obesity threaten the health of an alarming number of children around the globe.[2] Without immediate intervention, millions of children are destined to develop type 2 diabetes and hypertension in childhood followed by **cardiovascular disease (CVD)** in early adulthood.[3]

Over the past three decades, researchers have been observing how changes in body weight, blood lipids, blood pressure, and individual behaviors correlate with the development of CVD over time—from infancy to childhood through adolescence and into young adulthood. Some major findings have emerged from this research:

- Changes inside the arteries—changes predictive of CVD—are evident in childhood.
- Obesity in children affects these changes.
- Behaviors that influence the development of obesity and of CVD are learned and begin early in life. These behaviors include overeating, physical inactivity, and cigarette smoking.

This Nutrition in Practice focuses on efforts to prevent childhood obesity, type 2 diabetes, and CVD (see the accompanying glossary for definitions of the relevant terms), but the benefits extend to other obesity-related diseases as well. The years of childhood (ages 2 to 18) are emphasized here, for the earlier in life health-promoting habits become established, the better they will stick.

What about genetics? Do some people inherit the tendency to become obese or develop diabetes or CVD regardless of the lifestyle habits they adopt?

For obesity, as well as for CVD, hypertension, and type 2 diabetes, genetics does not appear to play a *determining* role; that is, a person is not simply destined at birth to develop them. Instead, genetics appears to play a *permissive* role—the potential is inherited and will then develop, if given a push by factors in the environment such as poor diet, sedentary lifestyle, and cigarette smoking.[4] Researchers note that the relationship between genes and the environment is a synergistic one—their combined effects are greater than the sum of their individual effects.

Glossary

atherosclerosis (ATH-er-oh-scler-OH-sis): a type of artery disease characterized by plaques (accumulations of lipid-containing material) on the inner walls of the arteries (see Chapter 21).
- *athero* = porridge or soft
- *scleros* = hard
- *osis* = condition

cardiovascular disease (CVD): a general term for all diseases of the heart and blood vessels. Atherosclerosis is the main cause of CVD. When the arteries that carry blood to the heart muscle become blocked, the heart suffers damage known as coronary heart disease (CHD).
- *cardio* = heart
- *vascular* = blood vessels

fatty streaks: accumulation of cholesterol and other lipids along the walls of the arteries.

plaque (PLACK): an accumulation of fatty deposits, smooth muscle cells, and fibrous connective tissue that develops in the artery walls in atherosclerosis. Plaque associated with atherosclerosis is known as **atheromatous (ATH-er-OH-ma-tus) plaque**.

Many experts agree that preventing or treating obesity in childhood will reduce the rate of chronic diseases in adulthood. Without intervention, most overweight children become overweight adolescents who become overweight adults, and being overweight exacerbates every chronic disease that adults face.[5] Fatty liver, a condition that correlates directly with BMI, was not even recognized in pediatric research until recently. Today, fatty liver disease affects about one in three obese children.[6]

What about events that take place during fetal development—malnutrition, for example? Can they affect a person's tendency to develop diseases later in life?

A theory called *fetal programming,* or *developmental origins of health and disease,* states that maternal malnutrition or other harmful conditions at a critical period of fetal development may have lifelong effects on an individual's pattern of genetic expression and therefore on the tendency to develop obesity and certain diseases.[7] Poor maternal diet or health during pregnancy may alter the infant's bodily functions such as blood pressure, cholesterol metabolism, glucose metabolism, and immune functions that influence disease development.[8] For example, malnutrition during fetal development may encourage metabolic programming to promote nutrient storage to provide a survival advantage in an environment of poor postnatal nutrition. In a postnatal environment of adequate nutrition or overnutrition, however, such adaptations can lead to the development of glucose intolerance, type 2 diabetes, cardiovascular disease, and hypertension.[9]

Why has type 2 diabetes become so prevalent?

In recent years, type 2 diabetes, a chronic disease closely linked with obesity, has been on the rise among children and adolescents as the prevalence of obesity in U.S. youth has increased.[10] Obesity is the most important risk factor for type 2 diabetes—most of the children diagnosed with type 2 diabetes are obese. Most are diagnosed during puberty, but as children become more obese and less active, the disease is appearing in younger and younger children. Type 2 diabetes is most likely to occur in those who are obese and sedentary and have a family history of diabetes.

> **NURSING DIAGNOSIS**
>
> *Sedentary lifestyle* applies to clients who report a lifestyle that is characterized by a low physical activity level.

How does type 2 diabetes develop?

In type 2 diabetes, the body's cells become insulin resistant—that is, the cells become less sensitive to insulin, reducing the amount of glucose entering the cells from the blood. The combination of obesity and insulin resistance produces a cluster of symptoms, including high blood pressure and high blood lipids, which in turn promotes the development of atherosclerosis and the early development of CVD.[11] Other common problems evident by early adulthood include kidney disease, blindness, and miscarriages. The complications of diabetes, especially when encountered at a young age, can shorten life expectancy. Chapter 20 offers a detailed discussion of diabetes.

> **NURSING DIAGNOSIS**
>
> *Risk for unstable blood glucose level* applies to clients who are physically inactive and overweight.

Prevention and treatment of type 2 diabetes depend on weight management, which can be particularly difficult in a young person's world of food advertising, video games, and pocket money for candy bars. The activity and dietary suggestions to help defend against heart disease later in this discussion apply to type 2 diabetes as well.

How does CVD develop, and when does its development begin?

Most CVD involves **atherosclerosis**—the accumulation of cholesterol and other blood lipids along the walls of the arteries. Frequently, atherosclerosis and its complications interfere with the flow of blood to the heart and can lead to coronary heart disease (CHD), which, in turn, raises the likelihood of a heart attack. When atherosclerosis interferes with blood flow to the brain, a stroke can result. Infants are born with healthy, smooth, clear arteries, but within the first decade of life, **fatty streaks** may begin to appear. During adolescence, these fatty streaks may begin to turn into **plaques** (Figure 21-2 in Chapter 21 shows the formation of plaques in atherosclerosis).[12] By early adulthood, the fibrous plaques may begin to calcify and become raised lesions, especially in boys and young men. As the lesions grow more numerous and thicken, the heart disease rate begins to rise, and the rise becomes dramatic at about age 45 in men and 55 in women. From this point on, arterial damage and blockage progress rapidly, and heart attacks and strokes threaten life. In short, the consequences of atherosclerosis, which become apparent only in adulthood, have their beginnings in the first decades of life.[13]

Children with the highest risks of developing heart disease are sedentary and obese; they may also have diabetes, high blood pressure, and high blood cholesterol.[14] In contrast, children with the lowest risks of heart disease are physically active and of normal weight, with low blood pressure and favorable lipid profiles.

Parents do not need to worry about their children's blood cholesterol, do they?

Atherosclerotic lesions reflect blood cholesterol: as blood cholesterol rises, lesion coverage increases. Cholesterol values at birth are similar in all populations; differences

TABLE NP11-1 Cholesterol Values for Children and Adolescents

Disease Risk	Total Cholesterol (mg/dL)	LDL Cholesterol (mg/dL)
Acceptable	<170	<110
Borderline	170–199	110–129
High	≥200	≥130

Note: Adult values appear in Table 21-2 on p. 598.

© Cengage Learning

emerge in early childhood. Standard values for cholesterol screening in children and adolescents are listed in Table NP11-1. Cholesterol concentrations change with age in children and adolescents, however, and are especially variable during puberty.[15] Thus, using a single cut point for all pediatric age groups has limitations.

In general, blood cholesterol tends to rise as dietary saturated fat intakes increase. Blood cholesterol also correlates with childhood obesity, especially abdominal obesity.[16] LDL cholesterol rises with obesity, and HDL declines. These relationships are apparent throughout childhood, and their magnitude increases with age.

Children who are both overweight and have high blood cholesterol are likely to have parents who develop heart disease early. For this reason, selective screening is recommended for children and adolescents of any age who are overweight or obese; those whose parents (or grandparents) have premature (≤55 years of age for men and ≤65 years of age for women) heart disease; those whose parents have elevated blood cholesterol; those who have other risk factors for heart disease such as hypertension, cigarette smoking, or diabetes; and those whose family history is unavailable. Because blood cholesterol in children is a good predictor of adult values, some experts recommend universal screening for all children aged 9 to 11.[17]

Early—but not advanced—atherosclerotic lesions are reversible, making screening and education a high priority. Both those with family histories of heart disease and those with multiple risk factors need intervention.

Is hypertension a concern for children and adolescents?

Pediatricians routinely monitor blood pressure in children and adolescents. High blood pressure may signal an underlying disease or the early onset of hypertension. Childhood hypertension, left untreated, can accelerate the development of atherosclerosis.[18] Diagnosing hypertension in children and adolescents requires consideration of age, gender, and height, and blood pressure cannot be assessed using the simple tables applied to adults.

Like atherosclerosis and high blood cholesterol, hypertension may develop in the first decades of life, especially among obese children, and worsen with time. Children can control their hypertension by participating in regular aerobic activity and by losing weight or maintaining their weight as they grow taller. Restricting dietary sodium also causes an immediate drop in most children's and adolescents' blood pressure.[19]

Regular physical activity lowers risks for heart disease and hypertension in adults; does it do so in children as well?

Yes. Research has also confirmed an association between blood lipids and physical activity in children, similar to that seen in adults. Physically active children have a better lipid profile and lower blood pressure than physically inactive children, and these positive findings often persist into adulthood.[20] The *Physical Activity Guidelines for Americans 2008* recommendations for children and adolescents are listed in Chapter 11 on p. 329.

Just as blood cholesterol and obesity track over the years, so does a child's level of physical activity. Those who are inactive now are likely to still be inactive years later. Similarly, those who are physically active now tend to remain so. Compared with inactive teens, those who are physically active weigh less, smoke less, eat a diet lower in saturated fats, and have better blood lipid profiles. Both obesity and blood cholesterol correlate with the inactive pastime of watching television. The message is clear: physical activity offers numerous health benefits, and children who are active today are most likely to be active for years to come.

Are adult dietary recommendations appropriate for children?

Regardless of family history, all children older than age two should eat a variety of foods and maintain desirable weight (see Table NP11-2). Children (4 to 18 years of age) should receive at least 25 percent and no more than 35 percent of total energy from fat, less than 10 percent from saturated fat, and less than 300 milligrams of cholesterol per day.[21]

Healthy children older than age two can begin the transition to these recommendations by eating fewer foods high in saturated fat and selecting more fruits and vegetables. Healthy meals can occasionally include moderate amounts of a child's favorite food, such as ice cream, even if it is high in saturated fat. A steady diet from the children's menus in some restaurants—which feature chicken nuggets, hot dogs, and french fries—easily exceeds a prudent intake of saturated fat, *trans* fat, and kcalories, however, and invites both nutrient shortages and weight gains.[22] Fortunately, most restaurant chains are changing children's menus to include steamed vegetables, fruit cups, and broiled or grilled chicken—additions welcomed by busy parents who often dine out or purchase take-out foods.

Other fatty foods, such as nuts, vegetable oils, and some varieties of fish such as tuna or salmon, contribute essential fatty acids. Low-fat milk and milk products also deserve special attention in a child's diet for the needed calcium and other nutrients they supply.[23]

Parents and caregivers play a key role in helping children establish healthy eating habits. Balanced meals need to provide lean meat, poultry, fish, and legumes; fruits and vegetables; whole grains; and low-fat milk products. Such meals can provide enough energy and nutrients to support growth and maintain blood cholesterol within a healthy range.

Pediatricians warn parents to avoid extremes. Although intentions may be good, excessive food restriction may create nutrient deficiencies and impair growth. Furthermore, parental control over eating may instigate battles and foster attitudes about foods that can lead to inappropriate eating behaviors.

TABLE NP11-2 American Heart Association Dietary Guidelines and Strategies for Children[a]

- Balance dietary kcalories with physical activity to maintain normal growth.
- Every day, engage in 60 minutes of moderate to vigorous play or physical activity.
- Eat vegetables and fruits daily. Use fresh, frozen, and canned vegetables and fruits and serve at every meal; limit those with added fats, salt, and sugar.
- Limit juice intake to 4 ounces per day.
- Use vegetable oils (canola, soybean, olive, safflower, or other unsaturated oils) and soft margarines low in saturated fat and *trans* fatty acids instead of butter or most other animal fats in the diet.
- Choose whole-grain breads and cereals rather than refined products; read labels and make sure that "whole grain" is the first ingredient.
- Limit or avoid the intake of sugar-sweetened beverages; encourage water.
- Consume low-fat and nonfat milk and milk products daily.
- Include two servings of fish per week, especially fatty fish such as broiled or baked salmon.
- Choose legumes and tofu in place of meat for some meals.
- Choose only lean cuts of meat and reduced-fat meat products; remove the skin from poultry.
- Use less salt, including salt from processed foods. Breads, breakfast cereals, and soups may be high in salt and/or sugar so read food labels and choose high-fiber, low-salt, low-sugar alternatives.
- Limit the intake of high-kcalorie add-ons such as gravy, Alfredo sauce, cream sauce, cheese sauce, and hollandaise sauce.
- Serve age-appropriate portion sizes on/in appropriately sized plates and bowls.

[a]These guidelines are for children two years of age and older.

Source: Adapted from Expert Panel on Integrated Guidelines for Cardiovascular Health and Risk Reduction in Children and Adolescents, Summary report, *Pediatrics* 128 (2011): S213–S256; American Heart Association, S. S. Gidding and coauthors, Dietary recommendations for children and adolescents: A guide for practitioners, *Pediatrics* 117 (2006): 544–559.

Do pediatricians ever prescribe drugs to lower blood cholesterol in children?

Experts agree that children with high blood cholesterol should first be treated with diet. If high blood cholesterol persists despite dietary intervention in children 10 years of age and older, then drugs may be necessary to lower blood cholesterol. Drugs can effectively lower blood cholesterol without interfering with adolescent growth or development.

Can parents or caregivers do anything else to help children reduce their risks of CVD?

Even though the focus of this text is nutrition, another risk factor for heart disease that starts in childhood and carries over into adulthood must also be addressed—cigarette smoking. Each day 3800 young people between the ages of 12 and 17 light up for the first time, and an estimated 1000 become daily cigarette smokers. Among high school students, one in six smokes regularly.[24] Approximately 80 percent of all adult smokers began smoking before the age of 18.

Of those teenagers who continue smoking, half will eventually die of smoking-related causes. Efforts to teach children about the dangers of smoking need to be aggressive. Children are not likely to consider the long-term health consequences of tobacco use. They are more likely to be struck by the immediate health consequences, such as shortness of breath when playing sports, or social consequences, such as having bad breath. Whatever the context, the message to all children and teens should be clear: Don't start smoking. If you've already started, quit.

In conclusion, *adult* heart disease is a major *pediatric* problem. Without intervention, some 60 million children are destined to suffer its consequences within the next 30 years. Optimal prevention efforts focus on children, especially on those who are overweight. Just as young children receive vaccinations against infectious diseases, they need screening for, and education about, chronic diseases. Many health education programs have been implemented in schools around the country. These programs are most effective when they include education in the classroom, heart-healthy meals in the lunchroom, fitness activities on the playground, and parental involvement at home.

Notes

1. C. L. Ogden and coauthors, Prevalence of obesity and trends in body mass index among US children and adolescents, 1999–2010, *Journal of the American Medical Association* 307 (2012): 483–490; A. Tirosh and coauthors, Adolescent BMI trajectory and risk of diabetes versus coronary disease, *New England Journal of Medicine* 364 (2011): 1315–1325; J. C. Han, D. A. Lawlor, and S. Y. S. Kimm, Childhood obesity, *Lancet* 375 (2010): 1737–1748; F. M. Biro and M. Wien, Childhood obesity and adult morbidities, *American Journal of Clinical Nutrition* 91 (2010): 1499S–1505S; P. W. Franks and coauthors, Childhood obesity, other cardiovascular risk factors, and premature death, *New England Journal of Medicine* 362 (2010): 458–493; C. J. Lavie, R. V. Milani, and H. O. Ventura, Obesity and cardiovascular disease, *Journal of the American College of Cardiology* 53 (2009): 1925–1933; D. S. Freedman and coauthors, Risk factors and adult body mass index among overweight children: The Bogalusa Heart Study, *Pediatrics* 123 (2009): 750–757.
2. M. de Onis and coauthors, Global prevalence and trends of overweight and obesity among preschool children, *American Journal of Clinical Nutrition* 92 (2010): 1257–1264; B. M. Popkin, Recent dynamics suggest selected countries catching up to US obesity, *American Journal of Clinical Nutrition* 91 (2010): 284S–288S; C. Bouchard, Childhood obesity: Are genetic differences involved? *American Journal of Clinical Nutrition* 89 (2009): 1494S–1501S.
3. A. L. May, E. V. Kuklina, and P. W. Yoon, Prevalence of cardiovascular disease risk factors among US adolescents, 1999–2008, *Pediatrics* 129 (2012): 1035–1041; C. Friedemann and coauthors, Cardiovascular disease risk in healthy children and its association with body mass index: Systematic review and meta-analysis, *British Medical Journal* 345 (2012): e4759; Expert Panel on Integrated Guidelines for Cardiovascular Health and Risk Reduction in Children and Adolescents, Summary report, *Pediatrics* 128 (2011): S212–S256; M. Juonola and coauthors, Childhood adiposity, adult adiposity, and cardiovascular risk factors, *New England Journal of Medicine* 365 (2011): 1876–1885; Freedman and coauthors, 2009.
4. D. K. Arnett and S. A. Claas, Preventing and controlling hypertension in the era of genomic innovation and environmental transformation, *Journal of the American Medical Association* 308 (2012): 1745–1746; M. Manco and B. Dallapiccola, Genetics of pediatric obesity, *Pediatrics* 130 (2012): 123–133; E. C. Brito and coauthors, Previously associated type 2 diabetes variants may interact with physical activity to modify the risk of impaired glucose regulation and type 2 diabetes, *Diabetes* 58 (2009): 1411–1418; C. Bouchard, Childhood obesity: Are genetics differences involved? *American Journal of Clinical Nutrition* 89 (2009): 1494S–1501S; J. M. Ordovas, Genetic influences on blood lipids and cardiovascular disease risk: Tools for primary prevention, *American Journal of Clinical Nutrition* 89 (2009): 1509S–1517S.
5. M. Juonaloa and coauthors, Childhood adiposity, adult adiposity, and cardiovascular risk factors, *New England Journal of Medicine* 365 (2011): 1876–1885; M. Neovius, J. Sundström, and F. Rasmussen, Combined effects of overweight and smoking late adolescence on subsequent mortality: Nationwide cohort study, *British Medical Journal* 338 (2009): b496.
6. B. G. Koot and coauthors, Lifestyle intervention for non-alcoholic fatty liver disease: Prospective cohort study of its efficacy and factors related to improvement, *Archives of Disease in Childhood* 96 (2011): 669–674.
7. M. L. Gusmão Correia, and coauthors, Developmental origins of health and disease: Experimental and human evidence of fetal programming for metabolic syndrome, *Journal of Human Hypertension* 26 (2012): 405–419; J. L. Tarry-Adkins and S. E. Ozanne, Mechanisms of early life programming: Current knowledge and future directions, *American Journal of Clinical Nutrition* 94 (2011): 1765S–1771S; G. C. Burdge and K. A. Lillycrop, Nutrition, epigenetics, and developmental plasticity: Implications for understanding human disease, *Annual Review of Nutrition* 30 (2010): 315–339; M. E. Symonds, T. Stephenson, and H. Budge, Early determinants of cardiovascular disease: The role of early diet in later blood pressure control, *American Journal of Clinical Nutrition* 89 (2009): 1518S–1522S; S. H. Zeisel, Epigenetic mechanisms for nutrition determinants of later health outcomes, *American Journal of Clinical Nutrition* 89 (2009): 1488S–1493S.
8. J. G. Eriksson, Early growth and coronary heart disease and type 2 diabetes: Findings from the Helsinki Birth Cohort Study (HBCS), *American Journal of Clinical Nutrition* 94 (2011): 1799S–1802S; B. Reusens and coauthors, Maternal malnutrition programs the endocrine pancreas in progeny, *American Journal of Clinical Nutrition* 94 (2011): 1824S–1829S; Tarry-Adkins and Ozanne, 2011; Burdge and Lillycrop, 2010.
9. Tarry-Adkins and Ozanne, 2011.
10. M. Juonala and coauthors, Childhood adiposity, adult adiposity, and cardiovascular risk factors, *New England Journal of Medicine* 365 (2011): 1876–1885; Biro and Wien, 2010; Freedman and coauthors, 2009.
11. Biro and Wien, 2010; E. M. Urbina and coauthors, Youth with obesity and obesity-related type 2 diabetes mellitus demonstrate abnormalities in carotid structure and function, *Circulation* 119 (2009): 2913–2919.
12. W. Insull, The pathology of atherosclerosis: Plaque development and plaque responses to medical treatment, *The American Journal of Medicine* 122 (2009): S3–S14.
13. Friedmann and coauthors, 2012; Expert Panel on Integrated Guidelines for Cardiovascular Health and Risk Reduction in Children and Adolescents, 2011.
14. Centers for Disease Control and Prevention, Prevalence of abnormal lipid levels among youths—United States, 1999–2006, *Morbidity and Mortality Weekly Report* 59 (2010): 29–33.
15. S. R. Daniels, F. R. Greer, and the Committee on Nutrition, Lipid screening and cardiovascular health in childhood, *Pediatrics* 122 (2008): 198–208.
16. Friedemann and coauthors, 2012; May, Kuklina, and Yoon, 2012; Freedman and coauthors, 2009.
17. Expert Panel on Integrated Guidelines for Cardiovascular Health and Risk Reduction in Children and Adolescents, 2011.
18. M. Barton, Childhood obesity: A lifelong health risk, *Acta Pharmacologica Sinica* 2 (2012): 189–193; J. Le and coauthors, "Vascular Age" is advanced in children with atherosclerosis-promoting risk factors, *Circulation Cardiovascular Imaging* 3 (2010): 8–14.
19. S. Stabouli and coauthors, The role of obesity, salt and exercise on blood pressure in children and adolescents, *Expert Review of Cardiovascular Therapy* 9 (2011): 753–761; J. Feber and M. Ahmed, Hypertension in children: New trends and challenges, *Clinical Science* 119 (2010): 151–161.
20. U. Ekelund and coauthors, Moderate to vigorous physical activity and sedentary time and cardiometabolic risk factors in children and adolescents, *Journal of the American Medical Association* 15 (2012): 704–712; N. J. Farpour-Lambert and coauthors, Physical activity reduces systemic blood pressure and improves early markers of atherosclerosis in pre-pubertal obese children, *Journal of the American College of Cardiology* 54 (2009): 2396–2406.
21. Standing Committee on the Scientific Evaluation of Dietary Reference Intakes, Food and Nutrition Board, Institute of Medicine, *Dietary Reference Intakes for Energy, Carbohydrate, Fiber, Fat, Fatty Acids, Cholesterol, Protein, and Amino Acids* (Washington, DC: National Academies Press, 2005), pp. 769–879.
22. L. Johnson and coauthors, Energy-dense, low-fiber, high-fat dietary pattern is associated with increased fatness in childhood, *American Journal of Clinical Nutrition* 87 (2008): 846–854.
23. F. R. Greer, N. F. Krebs, and the Committee on Nutrition, American Academy of Pediatrics, Optimizing bone health and calcium intakes of infants, children, and adolescents, *Pediatrics* 117 (2006): 578–585.
24. Centers for Disease Control and Prevention, Youth and tobacco use, 2009, available at www.cdc.gov/tobacco/data_statistics/fact_sheets/youth_data/tobacco_use/index.htm.

Chapter 12 Nutrition through the Life Span: Later Adulthood

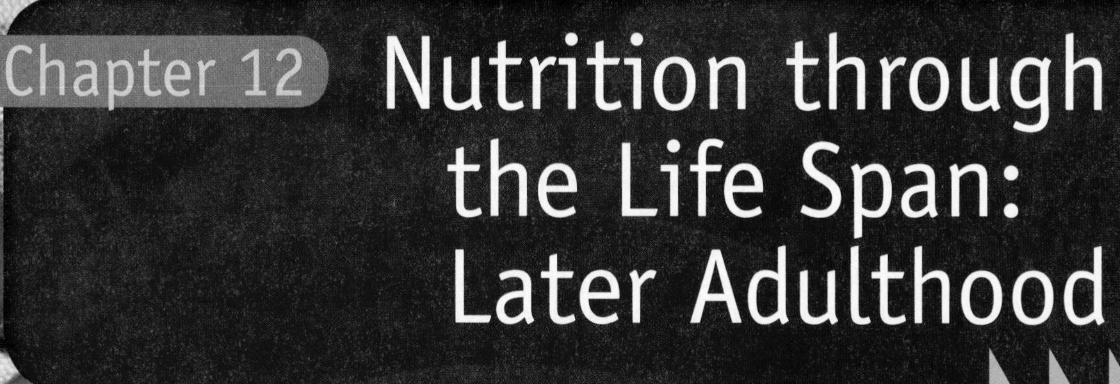

Ulrich Kerth/Getty Images

THE LAST TWO CHAPTERS WERE DEVOTED TO STAGES OF THE LIFE CYCLE that require special nutrition attention: pregnancy, lactation, infancy, childhood, and adolescence. Much of the text before that focused on nutrition to support wellness during adulthood. This chapter describes the special nutrition needs of the later adult years.

The most urgent nutrition need of older people, however, is to have made good food choices in the past! All of life's nutrition choices incur health consequences for the better or for the worse. A single day's intakes of nutrients may exert only a minute effect on body organs and their functions, but, over years and decades, the repeated effects accumulate to have major impacts. This being the case, it is of great importance for everyone, of every age, to pay close attention today to nutrition.

The "graying" of America is a continuing trend.[1] The majority of citizens are now middle aged, and the ratio of old people to young is increasing, as Figure 12-1 shows. Our society uses the arbitrary age of 65 to define the transition point between middle age and old age, but growing "old" happens day by day, with changes occurring gradually over time. (■) Since 1950 the population older than age 65 has almost tripled. Remarkably, the fastest-growing age group has been people older than age 85; since 1950 their numbers have increased sevenfold. The U.S. Bureau of the Census projects that by the year 2040 more than a million Americans will be 100 years old or older.

■ Commonly used age groups:
- Young old (65–74 years)
- Old (75–84 years)
- Oldest old (≥85 years)

life expectancy: the average number of years lived by people in a given society.

life span: the maximum number of years of life attainable by a member of a species.

longevity: long duration of life.

quality of life: a person's perceived physical and mental well-being.

Life expectancy in the United States is 78 years, up from about 47 years in 1900.[2] Advances in medical science—antibiotics and other treatments—are largely responsible for almost doubling the life expectancy in the 20th century. Improved nutrition and an abundant food supply have also contributed to lengthening life expectancy. Ironically, this abundant food supply has also jeopardized the chances of further lengthening life expectancy as obesity rates increase.[3]

The human **life span**, currently estimated at 130 years, is the upper limit of human **longevity**, even given optimal nutrition. With work progressing in medical and genetic technologies, however, the human life span may be extended significantly.

Research in the field of aging is active—and difficult. Researchers are challenged by the diversity of older adults. When older adults experience health problems, it is hard to know whether to attribute these problems to genetics, aging, or other environmental factors such as nutrition. The idea that nutrition can influence the way the human body ages is particularly appealing because diet is a factor that people can control and change.

FIGURE 12-1 The Aging of the U.S. Population

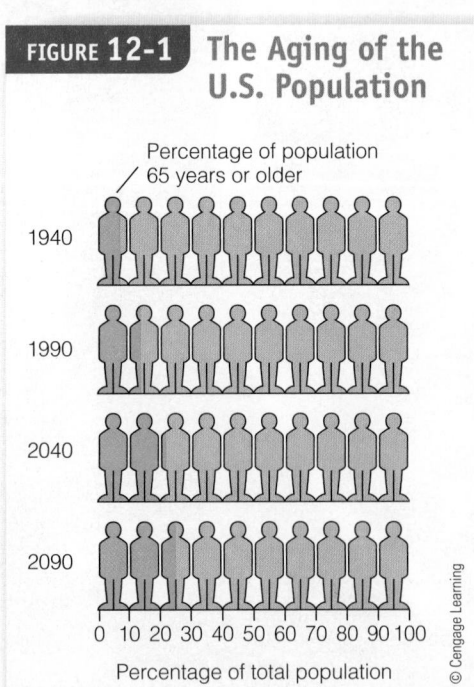

Nutrition and Longevity

What has been learned so far about the effects of nutrition and environment on longevity provides incentive for researchers to keep asking questions about how and why human beings age. Among their questions are:

- To what extent is aging inevitable, and can it be slowed through changes in lifestyle and environment?
- What role does nutrition play in the aging process, and what role can it play in slowing aging?

With respect to the first question, it seems that aging is an inevitable, natural process, programmed into the genes at conception. People can, however, slow the process within the natural limits set by heredity by adopting healthy lifestyle habits such as eating nutritious food and engaging in physical activity. In fact, an estimated 70 to 80 percent of the average person's life expectancy may depend on individual health-related behaviors; genes determine the remaining 20 to 30 percent.

With respect to the second question, good nutrition helps to maintain a healthy body and can therefore ease the aging process in many significant ways.[4] Clearly, nutrition can improve the **quality of life** in the later years.[5]

SLOWING THE AGING PROCESS

One approach researchers use to search out the secret of long life is to study older people. Some people are young for their ages, whereas others are old for their ages. What makes the difference?

Healthy Habits Six lifestyle habits seem to have a profound influence on people's health and therefore on their **physiological age**:[6]

- Eating well-balanced meals (rich in fruits, vegetables, whole grains, poultry, fish, and low-fat milk products)
- Maintaining a healthy body weight
- Engaging in regular physical activity
- Not smoking
- Not using alcohol, or using it in moderation
- Sleeping regularly and adequately

Over the years, the effects of these lifestyle choices accumulate—that is, those who follow all of these practices live longer and have fewer disabilities as they age. They are in better health, even if older in **chronological age**, than people who do not adopt these behaviors. Even though people cannot alter their birth dates, they may be able to add years to, and enhance the quality of, their lives. Physical activity seems to be most influential in preventing or slowing the many changes that many people seem to accept as an inevitable consequence of old age. In other words, physical activity and long life seem to go together.[7]

Physical Activity The many and remarkable benefits of regular physical activity are not limited to the young. Compared with those who are inactive, older adults who are active weigh less; have greater flexibility, more endurance, better balance, and better health; and live longer.[8] They reap additional benefits from various types of activities as well: aerobic activities improve cardiorespiratory endurance, blood pressure, and blood lipid concentrations; moderate endurance activities improve the quality of sleep; and resistance training significantly improves posture and mobility. In fact, regular physical activity is a powerful predictor of a person's mobility in the later years. Mobility, in turn, is closely associated with longevity.[9] Physical activity also increases blood flow to the brain, thereby preserving mental ability, alleviating depression, and supporting independence.[10]

Muscle mass and muscle strength tend to decline with aging, making older people vulnerable to falls and immobility. Falls are a major cause of fear, injury, disability, dependence, and even death among older adults. Regular physical activity tones, firms, and strengthens muscles, helping to improve confidence, reduce the risk of falling, and minimize the risk of injury should a fall occur.

Even without a fall, older adults may become so weak that they can no longer perform life's daily tasks, such as climbing stairs, carrying packages, and opening jars. Resistance training helps older adults to maintain independence by strengthening the muscles needed to perform these tasks.[11] Even in frail, elderly people older than 85 years of age, resistance training has been shown not only to improve balance, muscle strength, and mobility but also to increase energy expenditure and energy intake. This finding highlights another reason to be physically active: a person spending energy on physical activity can afford to eat more food and, with it, more nutrients. People who are committed to an ongoing fitness program can benefit from higher energy and nutrient intakes and still maintain healthy body weights.

Ideally, physical activity should be part of each day's schedule and should be intense enough to prevent muscle atrophy and to speed up the heartbeat and respiration rate. Although aging affects both speed and endurance to some degree, older adults can still train and achieve exceptional performances. Healthy older adults who have not been active can ease into a suitable routine. They can start by walking short distances until they are walking at least 10 minutes continuously and then gradually increase the distance to a 30-minute walk at least five days a week. With persistence, people can achieve great improvements at any age. Table 12-1 (p. 354) provides exercise goals and

NURSING DIAGNOSIS

Impaired physical mobility applies to clients with slowed movement, decreased muscle mass, and decreased muscle strength.

NURSING DIAGNOSIS

Risk for falls applies to clients who are 65 years of age or older, those who have a history of falls, those who live alone, and those who use assistive devices such as a walker or a cane.

physiological age: a person's age as estimated from her or his body's health and probable life expectancy.

chronological age: a person's age in years from his or her date of birth.

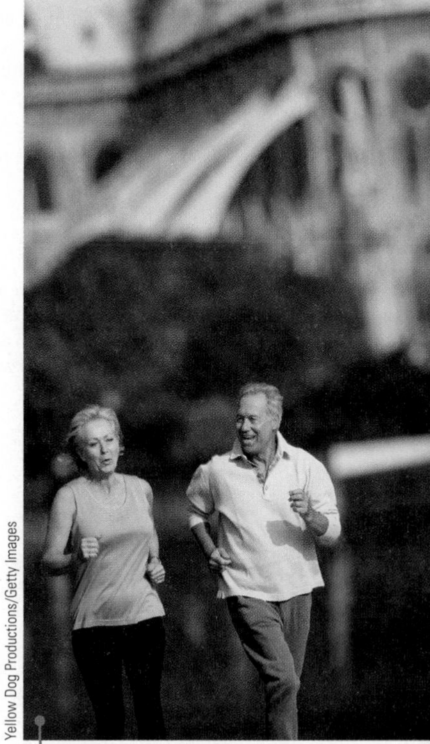

Yellow Dog Productions/Getty Images

Regular physical activity promotes a healthy, independent lifestyle.

TABLE 12-1 Exercise Guidelines for Older Adults

	Aerobic	Strength	Balance	Flexibility
Examples	© Geoff Manasse/Photodisc/PictureQuest	© IT Stock Free/PictureQuest	© IT Stock Free/PictureQuest	© Ron Chapple/Thinkstock/PictureQuest
Start easy and progress gradually	Be active five minutes on most or all days	Using 0- to 2-pound weights, do one set of 8–12 repetitions twice a week	Hold onto table or chair with one hand, then with one finger	Hold stretch for 10 seconds; do each stretch three times
Frequency	At least five days per week of moderate activity or at least four days per week for vigorous activity	At least two (nonconsecutive) days per week	Two to three days each week	At least two days per week, preferably on all days that aerobic or strength activities are performed
Intensity[a]	Moderate, vigorous, or combination	Moderate to high; 10 to 15 repetitions per exercise	—	Moderate
Duration	At least 30 minutes of moderate activity in bouts of at least 10 minutes each or at least 20 minutes of continuous vigorous activity	8 to 10 exercises involving the major muscle groups	—	Stretch major muscle groups for 10–30 seconds, repeating each stretch three to four times
Cautions and comments	Stop if you are breathing so hard you can't talk or if you feel dizziness or chest pain	Breathe out as you contract and in as you relax (do not hold your breath); use smooth, steady movements	Incorporate balance techniques with strength exercises as you progress	Stretch after strength and endurance exercises for 20 minutes, three times a week; use slow, steady movements; bend joints slightly

[a]On a 10-point scale, where sitting = 0 and maximum effort = 10, moderate intensity = 5 to 6 and vigorous intensity = 7 to 8.
Note: Activity recommendations are in addition to routine activities of daily living (such as getting dressed, cooking, and grocery shopping) and moderate activities lasting less than 10 minutes.

Source: C. E. Garber and coauthors, Quantity and quality of exercise for developing and maintaining cardiorespiratory, muscoskeletal, and neuromotor fitness in apparently healthy adults: Guidance for prescribing exercise, *Medicine and Science in Sports and Exercise* 43 (2011): 1334–1359; W. J. Chodzko-Zajko and coauthors, Position Stand of the American College of Sports Medicine: Exercise and physical activity for older adults, *Medicine and Science in Sports and Exercise* 41 (2009):1510–1530.

guidelines for older adults. Relatively few older adults meet these goals. People with medical conditions should check with a physician before beginning an exercise routine, as should sedentary men older than 40 and sedentary women older than 50 who want to participate in a vigorous program.

Restriction of kCalories In their efforts to understand longevity, researchers have not only observed people but have also manipulated influencing factors, such as diet, in animals. This research has produced some interesting and suggestive findings. For example, animals live longer and have fewer age-related diseases when their energy intakes are restricted.[12] These life-prolonging benefits become evident when the diet provides enough food to prevent malnutrition and energy intake of about 70 percent of normal; benefits decline as the age of starting the energy restriction is delayed.

Exactly how energy restriction prolongs life remains unexplained, although gene activity appears to play a key role. Energy restriction in animals prevents alterations in gene expression that are associated with aging. Food restriction may also extend the life span by preventing damaging lipid oxidation, thereby delaying the onset of age-related diseases, such as cancer and atherosclerosis.[13] Other research suggests that energy restriction beneficially alters several aspects of fat cell metabolism, which may play a role in reducing chronic disease risk and extending life. Experiments with food restriction and longevity in animals have *not* suggested any direct applications to human nutrition.

Moderate energy restriction (80 to 90 percent of usual intake) in human beings may be valuable. When people restrict energy intake moderately, body weight, body fat, and blood pressure drop, and blood lipids and insulin response improve—favorable changes for preventing chronic diseases.[14] The reduction in oxidative damage that occurs with energy restriction in animals also occurs in people whose diets include antioxidant nutrients and phytochemicals. Diets such as the Mediterranean diet that include an abundance of fruits, vegetables, olive oil, whole grains, and legumes—with their array of antioxidants and phytochemicals—support good health and long life.[15]

NUTRITION AND DISEASE PREVENTION

Nutrition alone, even if ideal, cannot ensure a long and robust life. Nevertheless, nutrition clearly affects aging and longevity in human beings by way of its role in disease prevention. Among the better-known relationships between nutrition and disease are the following:

- Appropriate energy intake helps prevent *obesity*, *diabetes*, and related *cardiovascular diseases* such as atherosclerosis and hypertension (Chapters 7, 20, and 21) and may influence the development of some forms of *cancer* (Chapter 23).
- Adequate intakes of essential nutrients prevent *deficiency diseases* such as scurvy, goiter, and anemia (Chapters 8 and 9).
- Variety in food intake, as well as ample intakes of certain fruits and vegetables, may be protective against certain types of *cancer* (Chapter 23).
- Moderation in sugar intake helps prevent *dental caries* (Nutrition in Practice 17).
- Appropriate fiber intakes help prevent disorders of the digestive tract such as *constipation*, *diverticulosis*, and possibly *colon cancer* (Chapters 3, 18, and 23).
- Moderate sodium intake and adequate intakes of potassium, calcium, and other minerals help prevent *hypertension* (Chapters 9 and 21).
- An adequate calcium intake throughout life helps protect against *osteoporosis* (Chapter 9).

Other, less well-established links between nutrition and disease are being discovered each day. Research that focuses on how life factors affect aging and disease processes is vital to ensuring that more and more people can look forward to long, healthy lives.

Review Notes

- Life expectancy in the United States increased dramatically in the 20th century.
- Factors that enhance longevity include limited or no alcohol use, regular balanced meals rich in fruits and vegetables, weight control, adequate sleep, abstinence from smoking, and regular physical activity.
- Nutrition alone, even if ideal, cannot guarantee a long and robust life. At the very least, however, nutrition—especially when combined with regular physical activity—can influence aging and longevity in human beings by supporting good health and preventing disease.

Nutrition-Related Concerns during Late Adulthood

Nutrition through the prime years may play a greater role than has been realized in preventing many changes once thought to be inevitable consequences of growing older. The following discussions of cataracts and macular degeneration, arthritis, and the aging brain show that nutrition may provide at least some protection against some of the conditions commonly associated with aging.

CATARACTS AND MACULAR DEGENERATION

Cataracts are age-related thickenings in the lenses of the eye that impair vision. If not surgically removed, they ultimately lead to blindness. Oxidative stress appears to play a significant role in the development of cataracts, and the antioxidant nutrients and phytochemicals in fruits and vegetables may help minimize the damage. A diet high in foods providing ample carotenoids, vitamin C, and vitamin E may be especially important for preventing early onset of cataracts. Also, people who follow the *Dietary Guidelines for Americans* are reported to have fewer cataracts, though cataracts can occur even in well-nourished individuals due to exposure to ultraviolet light, oxidative damage, viral infections, toxic substances, genetic disorders, injury, or other trauma.[16] Most cataracts are vaguely called "senile" cataracts, meaning "caused by aging." In the United States, more than half of all adults age 65 and older have a cataract.

One other diet-related factor may play a role in cataract development: obesity. How obesity may influence the development of cataracts is not known. Risk factors that typically accompany overweight, such as inactivity, diabetes, or hypertension, do not explain the association.[17]

The leading cause of visual loss among older people is **macular degeneration**, a deterioration of the macular region of the eye. As with cataracts, risk factors for age-related macular degeneration include oxidative stress from sunlight. Preventive factors for macular degeneration may include supplements of the omega-3 fatty acid DHA and some B vitamins (folate, vitamin B_6, and vitamin B_{12}), and the carotenoids lutein and zeaxanthin.[18]

ARTHRITIS

Almost 50 million people in the United States have some form of **arthritis**.[19] As the population ages, it is expected that the prevalence will increase to 70 million by 2030. The most common type of arthritis that disables older people is **osteoarthritis**, a painful swelling of the joints. During movement, the ends of bones are normally protected from wear by cartilage and by small sacs of fluid that lubricate the joint. With age, the cartilage sometimes disintegrates, and the joints become malformed and painful to move.

One known connection between osteoarthritis and nutrition is overweight. Weight loss can help overweight people with osteoarthritis, partly because the joints affected are often weight-bearing joints that are stressed and irritated by having to carry excess poundage. Interestingly, though, weight loss often relieves the worst pain of osteoarthritis in the hands as well, even though they are not weight-bearing joints. Jogging and other weight-bearing activities do not worsen osteoarthritis. In fact, both aerobic activity and weight training offer modest improvements in physical performance and pain relief, especially when accompanied by even modest weight loss.[20]

Nutrition quackery to treat arthritis is abundant, but no one universally effective diet for arthritis relief is known. Table 12-2 presents some of the many ineffective dietary treatments for osteoarthritis. Traditional medical intervention for arthritis includes medication and surgery. Two popular supplements for treating osteoarthritis—glucosamine and chondroitin—may alleviate pain and improve mobility, but mixed reports from studies emphasize the need for additional research.[21]

cataracts: clouding of the eye lenses that impairs vision and can lead to blindness.

macular (MACK-you-lar) degeneration: deterioration of the macular area of the eye that can lead to loss of central vision and eventual blindness. The **macula** is a small, oval, yellowish region in the center of the retina that provides the sharp, straight-ahead vision so critical to reading and driving.

arthritis: inflammation of a joint, usually accompanied by pain, swelling, and structural changes.

osteoarthritis: a painful, chronic disease of the joints that occurs when the cushioning cartilage in a joint breaks down; joint structure is usually altered, with loss of function; also called *degenerative arthritis*.

Drugs and supplements used to relieve arthritis can impose nutrition risks; some affect appetite and alter the body's use of nutrients, as Chapter 14 explains.

Another type of arthritis, known as **rheumatoid arthritis**, has a possible link to diet through the immune system. In rheumatoid arthritis, the immune system mistakenly attacks the bone coverings as if they were made of foreign tissue. In some individuals, certain foods, notably a Mediterranean-type diet of fish, vegetables, and olive oil, may moderate the inflammatory responses and provide some relief.[22]

The omega-3 fatty acids commonly found in fish oil reduce joint tenderness and improve mobility in some people with rheumatoid arthritis. The same diet recommended for heart health—one low in saturated fat from meats and milk products and high in omega-3 fats from fish—helps prevent or reduce the inflammation in the joints that makes arthritis so painful.

TABLE 12-2 Ineffective Dietary Strategies for Arthritis

• Alfalfa tea	• Honey
• Aloe vera liquid	• Inositol
• Amino acid supplements	• Kelp
• Blackstrap molasses	• Lecithin
• Burdock root	• Para-amino benzoic acid (PABA)
• Calcium	• Raw liver
• Celery juice	• Superoxide dismutase (SOD)
• Cod liver oil	• Vitamin D
• Copper supplements	• Vitamin megadoses
• Dimethyl sulfoxide (DMSO)	• Watercress
• Fasting	• Yeast
• Fresh fruit	
• Garlic	

THE AGING BRAIN

The brain, like all of the body's organs, responds to both inherited and environmental factors that can enhance or diminish its amazing capacities. (■) One of the challenges researchers face when studying the aging of the human brain is to distinguish among changes caused by normal, age-related, physiological processes; changes caused by diseases; and changes caused by cumulative, extrinsic factors such as diet.

The brain normally changes in some characteristic ways as it ages. For one thing, its blood supply decreases. For another, the number of **neurons**, the brain cells that specialize in transmitting information, diminishes as people age. When the number of nerve cells in one part of the **cerebral cortex** diminishes, hearing and speech are affected. Losses of neurons in other parts of the cortex can impair memory and cognitive function. When the number of neurons in the hindbrain diminishes, balance and posture are affected. Losses of neurons in other parts of the brain affect still other functions.

■ Factors that protect brain function:

- Physical activities
- Intellectual challenges
- Social interactions
- Balanced diet rich in antioxidants

Nutrient Deficiencies and Brain Function

Clinicians now recognize that much of the cognitive loss and forgetfulness generally attributed to aging is due in part to extrinsic, and therefore controllable, factors such as nutrient deficiencies. The ability of neurons to synthesize specific neurotransmitters depends in part on the availability of precursor nutrients that are obtained from the diet. The neurotransmitter serotonin, for example, derives from the amino acid tryptophan. In addition, the enzymes involved in neurotransmitter synthesis require vitamins and minerals to function properly. The fatty acid DHA counteracts the cognitive decline commonly seen in elderly adults.[23] Thus, nutrient deficiencies may contribute to the loss of memory and cognition that some older adults experience.[24] Such losses may be preventable or at least diminished or delayed through diet.

In some instances, the degree of cognitive loss is extensive. Such **senile dementia** may be attributable to a specific disorder such as a brain tumor or Alzheimer's disease.

Alzheimer's Disease

In **Alzheimer's disease**, the most prevalent form of dementia, brain cell death occurs in the areas of the brain that coordinate memory and cognition. Alzheimer's disease afflicts more than five million people in the United States, and that number is expected to triple by the year 2050.[25] Diagnosis of Alzheimer's depends on its characteristic symptoms: the victim gradually loses memory and reasoning ability, the ability to communicate, physical capabilities, and eventually life itself. Table 12-3 (p. 358) compares the signs of Alzheimer's disease with typical age-related changes.

rheumatoid arthritis: a disease of the immune system involving painful inflammation of the joints and related structures.

neurons: nerve cells; the structural and functional units of the nervous system. Neurons initiate and conduct nerve transmissions.

cerebral cortex: the outer surface of the cerebrum, which is the largest part of the brain.

senile dementia: the loss of brain function beyond the normal loss of physical adeptness and memory that occurs with aging.

Alzheimer's disease: a progressive, degenerative disease that attacks the brain and impairs thinking, behavior, and memory.

NURSING DIAGNOSIS

Chronic confusion applies to clients with Alzheimer's disease.

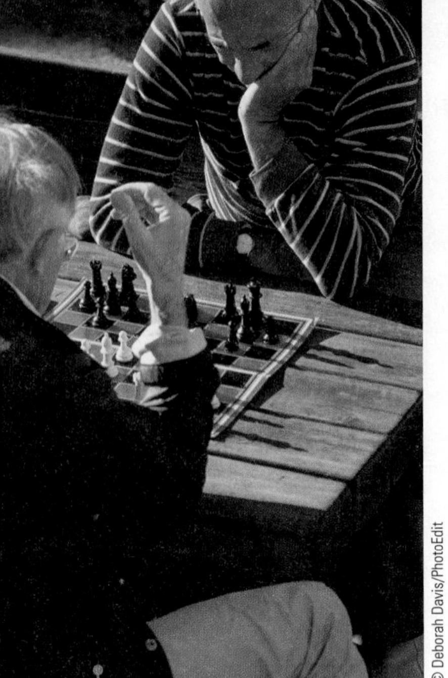

© Deborah Davis/PhotoEdit

Both foods and mental challenges nourish the brain.

TABLE 12-3 Signs of Alzheimer's and Typical Age-Related Changes Compared

Signs of Alzheimer's	Typical Age-Related Changes
Memory loss that disrupts daily life such as asking for the same information repeatedly or asking others to handle tasks of daily living	Forgetting a name or missing an appointment
Challenges in planning or solving problems such as following a recipe or paying monthly bills	Missing a monthly payment or making an error when balancing the checkbook
Difficulty completing familiar tasks at home such as using the microwave, at work such as preparing a report, or at leisure such as playing a game	Needing help recording a television program
Confusion with time or place including current season and location	Not knowing today's date
Trouble understanding visual images and spatial relationships such as judging distances and recognizing self in a mirror	Experiencing visual changes due to cataracts
New problems with words in speaking or writing such as not knowing the name of a common object	Being unable to find the right word to use
Misplacing things and losing the ability to retrace steps such as putting the milk in the closet and having no idea when or where the milk was last seen	Misplacing a pair of glasses or the car keys
Decreased or poor judgment such as giving large sums of money to strangers	Making a bad decision on occasion
Withdrawal from work projects or social activities	Feeling too tired to participate in work, family, or social activities
Changes in mood and personality such as confusion, suspicion, depression, and anxiety, especially when in unfamiliar places or with unfamiliar people	Becoming irritable when routines are disrupted

Source: Adapted from Alzheimer's Association, www.alz.org/alzheimers_disease_10_signs_of_alzheimers.asp.

The primary risk factor for Alzheimer's disease is age, but the exact cause remains unknown.[26] Clearly, genetic factors are involved.[27] Free radicals and oxidative stress also seem to be involved. Nerve cells in the brains of people with Alzheimer's disease show evidence of free-radical attack—damage to DNA, cell membranes, and proteins—and of the minerals that trigger these attacks—iron, copper, zinc, and aluminum.[28] Some research suggests that the antioxidant nutrients can limit free-radical damage and delay or prevent Alzheimer's disease, but more research is needed to confirm this possibility.[29]

Increasing evidence also suggests that overweight and obesity in middle age are associated with dementia, in general, and with Alzheimer's disease, in particular.[30] The possible relationship between obesity and Alzheimer's disease is disturbing given the current obesity epidemic. Efforts to prevent and treat obesity, however, may also help prevent Alzheimer's disease.

senile plaques: clumps of the protein fragment beta-amyloid on the nerve cells, commonly found in the brains of people with Alzheimer's dementia.

neurofibrillary tangles: snarls of the threadlike strands that extend from the nerve cells, commonly found in the brains of people with Alzheimer's dementia.

In Alzheimer's disease, the brain develops **senile plaques** and **neurofibrillary tangles**. Senile plaques are clumps of a protein fragment called beta-amyloid, whereas neurofibrillary tangles are snarls of the fibers that extend from the nerve cells. Both seem to occur in response to oxidative stress.[31] Researchers question whether these characteristics are the cause or the result of Alzheimer's disease.[32] In fact, scientists are unsure whether these plaques and tangles are causing the damage, serving as markers, or even protecting the brain by sequestering the proteins that begin the dementia process. In any case, treatment research focuses on lowering beta-amyloid levels.

Research suggests that cardiovascular disease risk factors such as high blood pressure, diabetes, and elevated levels of homocysteine may be related to the development of Alzheimer's disease.[33] Diets designed to support a healthy heart, which include the omega-3 fatty acids of oily fish, may benefit brain health as well.[34] Similarly, physical activity supports heart health and slows the cognitive decline of Alzheimer's disease.[35]

Treatment for Alzheimer's disease involves providing care to clients and support to their families. Drugs are used to improve or at least to slow the loss of short-term memory and cognition, but they do not cure the disease. Other drugs may be used to control depression, anxiety, and behavior problems.

Maintaining appropriate body weight may be the most important nutrition concern for the person with Alzheimer's disease. Depression and forgetfulness can lead to changes in eating behaviors and poor food intake. Furthermore, changes in the body's weight-regulation system may contribute to weight loss. Perhaps the best that a caregiver can do nutritionally for a person with Alzheimer's disease is to supervise food planning and mealtimes. Providing well-liked and well-balanced meals and snacks in a cheerful atmosphere encourages food consumption. To minimize confusion, offer a few ready-to-eat foods, in bite-size pieces, with seasonings and sauces. To avoid mealtime disruptions, control distractions such as music, television, children, and the telephone.

Review Notes

- Cataracts, age-related macular degeneration, and arthritis afflict millions of older adults, while others face senile dementia and other losses of brain function.
- Oxidative stress plays a significant role in the development of cataracts, and the antioxidant nutrients and phytochemicals in fruits and vegetables may help minimize the damage.
- For some people with rheumatoid arthritis, a Mediterranean-type diet of fish, vegetables, and olive oil may moderate the inflammatory responses and provide some relief.
- Nutrient deficiencies may contribute to the loss of memory and cognition that some older adults experience. Such losses may be preventable or at least diminished or delayed through diet.
- The primary risk factor for Alzheimer's disease is age, but the exact cause remains unknown.

Energy and Nutrient Needs during Late Adulthood

Knowledge about the nutrient needs and nutrition status of older adults has grown considerably in recent years. The Dietary Reference Intakes (DRI) cluster people older than 50 into two age categories—51 to 70 years old and 71 years and older. Research is showing that the nutrition needs of people 51 to 70 years old may be very different from those of people older than 70—a pattern that makes sense considering the wide age span involved.

Setting standards for older people is difficult, though, because individual differences become more pronounced as people grow older. People start out with different genetic predispositions and ways of handling nutrients, and the effects of these differences become magnified with years of unique dietary habits. For example, one person may tend to omit fruits and vegetables from his diet, and by the time he is old, he may have a set of nutrition problems associated with a lack of fiber and antioxidants. Another person who omitted milk and milk products all her life may have nutrition problems related to a lack of calcium. Also, as people age, they may suffer different

TABLE 12-4 Examples of Physical Changes of Aging That Affect Nutrition

Mouth	Tooth loss, gum disease, and reduced salivary output impede chewing and swallowing. Swallowing disorders and choking may become likely. Discomfort and pain associated with eating may reduce food intake.
Digestive tract	Intestines lose muscle strength, resulting in sluggish motility that leads to constipation (see Chapter 18). Stomach inflammation, abnormal bacterial growth, and greatly reduced acid output impair digestion and absorption. Pain may cause food avoidance or reduced intake.
Hormones	For example, the pancreas secretes less insulin and cells become less responsive, causing abnormal glucose metabolism.
Sensory organs	Diminished senses of smell and taste can reduce appetite; diminished sight can make food shopping and preparation difficult.
Body composition	Weight loss and decline in lean body mass lead to lowered energy requirements. May be preventable or reversible through physical activity.
Urinary tract	Increased frequency of urination may limit fluid intake.

chronic diseases and take different medications—both have impacts on nutrient needs. Table 12-4 lists some changes of aging that can affect nutrition. For all these reasons, researchers have difficulty even defining "healthy aging," a prerequisite to developing recommendations that are designed to meet the "needs of practically all healthy persons." Still, some generalizations are valid. The next sections give special attention to a few nutrients of concern.

ENERGY AND ENERGY NUTRIENTS

Energy needs decline with advancing age. As a general rule, adult energy needs decline an estimated 5 percent per decade. One reason is that people usually reduce their physical activity as they age, although they need not do so. Another reason is that basal metabolic rate declines 1 to 2 percent per decade, in part because lean body mass and thyroid hormones diminish. Loss of muscle mass, known as **sarcopenia**, can be significant in the later years (its prevalence is more than 50 percent among those older than 75), and its consequences, dramatic (see Figure 12-2).[36] As skeletal muscle mass diminishes, people lose their ability to move and to maintain balance, making falls likely. The limitations that accompany the loss of muscle mass and strength play a key role in the diminishing health that often accompanies aging.[37] To some extent, however, declines in lean body mass and energy needs may not be entirely inevitable. Optimal nutrition with sufficient protein and regular physical activity, especially resistance training, can help maintain muscle mass and strength and minimize the changes in body composition associated with aging.[38] Physical activity not only increases energy expenditure but, along with sound nutrition, enhances bone density and supports many body functions as well.

sarcopenia (SAR-koh-PEE-nee-ah): age-related loss of skeletal muscle mass, muscle strength, and muscle function.

FIGURE 12-2 Sarcopenia

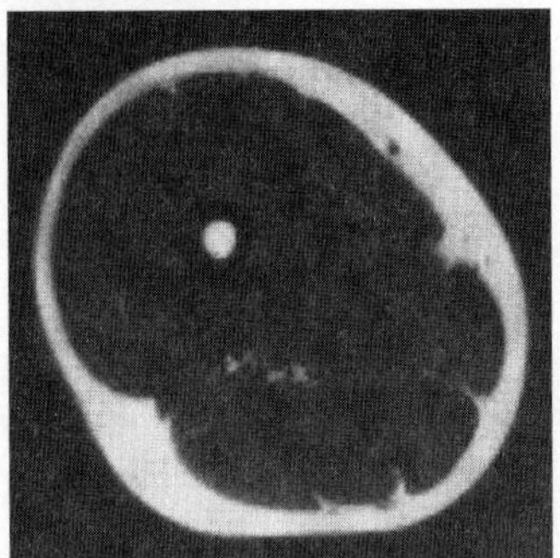

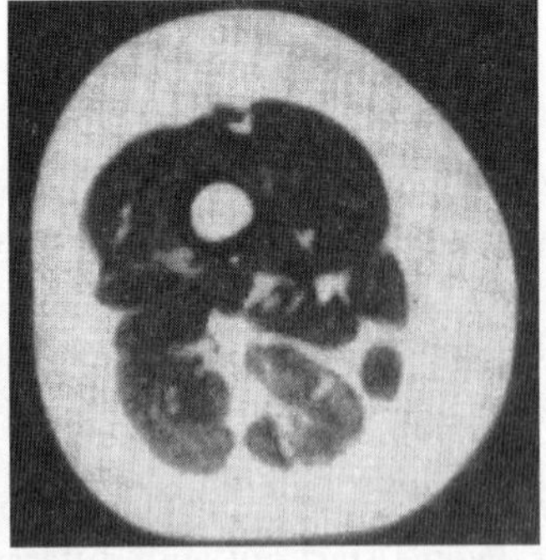

These cross sections of two women's thighs may appear to be about the same size from the outside, but the 20-year-old woman's thigh (left) is dense with muscle tissue. The 64-year-old woman's thigh (right) has lost muscle and gained fat, changes that may be largely preventable with strength-building physical activities.

Courtesy of Dr. William Evans

The lower energy expenditures of many older adults require that they eat less food to maintain their weights. Accordingly, the estimated energy requirements for adults decrease steadily after age 19. Energy intakes typically decline in parallel with needs. Still, many older adults are overweight, indicating that their food intakes do not decline enough to compensate for their reduced energy expenditures. Chapter 6 presents the many health problems that accompany obesity and the BMI guidelines for a healthy body weight (18.5–24.9). These guidelines apply to all adults, regardless of age, but they may be too restrictive for older adults. The importance of body weight in defending against chronic diseases differs for older adults.[39] Being *moderately overweight* may not be harmful. Older adults who are *obese*, however, face serious medical complications and can improve their quality of life with weight loss.[40]

On limited energy allowances, people must select mostly nutrient-dense foods. There is little leeway for added sugars, solid fats, or alcohol. Older adults can follow the USDA Food Patterns (see pp. 18–19), making sure to choose the recommended daily amounts of food from each food group that are appropriate to their energy needs (see Table 1-7 on p. 20).

Protein The protein needs of older adults appear to be about the same as those of younger people. With advancing age, however, people take in fewer total kcalories from food and so may need a greater percentage of kcalories from protein at each meal in order to support a healthy immune system and prevent losing muscle tissue, bone tissue, and other lean body mass.[41]

As energy needs decrease, lower-kcalorie protein sources such as lean tender meats, poultry, fish, boiled eggs, fat-free milk products, and legumes can help hold weight to a healthy level. Underweight or malnourished older adults need the opposite—energy-dense protein sources such as eggs scrambled with margarine, tuna salad with mayonnaise, peanut butter on wheat toast, hearty soups, and milkshakes. Nutrient-fortified supplements in liquid, pudding, cookie, or other forms between meals can also boost energy and nutrient intakes.

Carbohydrate and Fiber As always, abundant carbohydrate is needed to protect protein from being used as an energy source. The recommendation to obtain ample amounts of mostly whole-grain breads, cereals, rice, and pasta holds true for older people. As for younger people, a steady supply of carbohydrate is essential for optimal brain functioning. With age, fiber takes on extra importance for its role in alleviating constipation, a common complaint among older adults and among residents of **health care communities** in particular.

Fruits and vegetables supply soluble fibers and phytochemicals to help ward off chronic diseases, but factors such as transportation problems, limited cooking facilities, and chewing problems limit some elderly people's intakes of fresh fruits and vegetables.[42] Even without such problems, most older adults fail to obtain the recommended daily 25 or more grams of fiber (14 grams per 1000 kcalories).[43] When low fiber intakes are combined with low fluid intakes, inadequate physical activity, and constipating medications, constipation becomes inevitable.

Fat As is true for people of all ages, dietary fat intake needs to be moderate for most older adults—enough to enhance flavors and provide valuable essential fatty acids and other nutrients, but not so much as to raise risks of cancer, atherosclerosis, and other degenerative diseases. The recommendation should not be taken too far; limiting fat too severely may lead to nutrient deficiencies and weight loss—two problems that carry greater health risks in the elderly than overweight.

WATER

Dehydration is a risk for older adults, who may not notice or pay attention to their thirst or who find it difficult and bothersome to get a drink or to get to a bathroom.[44] Older adults who have lost bladder control may be afraid to drink too much water.

NURSING DIAGNOSIS

Risk for deficient fluid volume applies to clients who have deviations affecting fluid intake and access.

health care communities: living environments for people with chronic conditions, functional limitations, or need for supervision or assistance. Health care communities include assisted living facilities, group homes, short-term rehabilitation facilities, skilled nursing facilities, and hospice facilities.

Despite real fluid needs, older people do not seem to feel as thirsty or notice mouth dryness as readily as younger people. Many employees of health care communities such as assisted living facilities say it is hard to persuade their elderly clients to drink enough water and fruit juices.

Total body water decreases as people age, so even mild stresses such as fever or hot weather can precipitate rapid dehydration in older adults. Dehydrated older adults seem to be more susceptible to urinary tract infections, pneumonia, pressure ulcers, confusion, and disorientation.[45] An intake of 9 cups a day of total beverages, including water, is recommended for women; for men, the recommendation is 13 cups a day of total beverages, including water.[46]

VITAMINS AND MINERALS

As research reveals more about how specific vitamins and minerals influence disease prevention and how age-related physiological changes affect nutrient metabolism, optimal intakes of vitamins and minerals for different groups of older adults are being defined. This section highlights the vitamins and minerals of greatest concern to older adults.

Vitamin D Older adults face a greater risk of vitamin D deficiency than younger people do. Vitamin D–fortified milk is the most reliable source of vitamin D, but many older adults drink little or no milk. Further compromising the vitamin D status of many older people, especially those in assisted living facilities or group homes, is their limited exposure to sunlight. Finally, aging reduces the skin's capacity to make vitamin D and the kidneys' ability to convert it to its active form. Not only are older adults not getting enough vitamin D, but they may actually need more to improve both muscle and bone strength. The margin lists vitamin D recommendations to prevent bone loss and to maintain vitamin D status in older people, especially those who engage in minimal outdoor activity.[47] (■)

Vitamin B_{12} The DRI committee recommends that adults aged 51 years and older obtain 2.4 micrograms of vitamin B_{12} daily *and* that vitamin B_{12}–fortified foods (such as fortified cereals) or supplements be used to meet much of the DRI recommended intake.[48] (■) The committee's recommendation reflects the finding that an estimated 10 to 30 percent of adults older than 50 years lose the ability to produce enough stomach acid to make the protein-bound form of vitamin B_{12} available for absorption. Synthetic vitamin B_{12} is reliably absorbed, however. Given the poor cognition, anemia, and other devastating neurological effects associated with a vitamin B_{12} deficiency, an adequate intake is imperative.[49]

One cause of the malabsorption of protein-bound vitamin B_{12} is a condition known as **atrophic gastritis**. An estimated 10 to 30 percent of adults older than 50 have atrophic gastritis.

Folate As is true of vitamin B_{12}, folate intakes of older adults typically fall short of recommendations. (■) The elderly are also more likely to have medical conditions or to take medications that can compromise folate status.

Iron Among the minerals, iron deserves first mention. (■) Iron-deficiency anemia is less common in older adults than in younger people, but it still occurs in some, especially in those with low food energy intakes. Aside from diet, other factors in many older people's lives make iron deficiency likely: chronic blood loss from disease conditions and medicines, and poor iron absorption due to reduced secretion of stomach acid and antacid use. Anyone concerned with older people's nutrition should keep these possibilities in mind.

Zinc Zinc intake is commonly low in older people. Zinc deficiency can depress the appetite and blunt the sense of taste, thereby leading to low food intakes and worsening of zinc status. Zinc deficiency may also increase the likelihood of infectious diseases such as pneumonia.[50] Many medications that older adults commonly use can impair zinc absorption or enhance its excretion and thus lead to deficiency. (■)

■ Vitamin D RDA during late adulthood:
- 15 μg/day (51–70 yr)
- 20 μg/day (>70 yr)

■ Vitamin B_{12} RDA during late adulthood:
- 2.4 μg/day

■ Folate RDA during late adulthood:
- 400 μg/day

■ Iron RDA during late adulthood:
- 8 mg/day

■ Zinc RDA during late adulthood:
- 11 mg/day (men)
- 8 mg/day (women)

atrophic gastritis (a-TRO-fik gas-TRI-tis): a condition characterized by chronic inflammation of the stomach accompanied by a diminished size and functioning of the mucosa and glands.

Calcium The importance of abundant dietary calcium throughout life to protect against osteoporosis has been emphasized throughout this book. The calcium intakes of many people, especially women, in the United States are well below the recommendations. (■) If milk causes stomach discomfort, as many older adults report, then lactose-modified milk or other calcium-rich foods should take its place.

NUTRIENT SUPPLEMENTS FOR OLDER ADULTS

People judge for themselves how to manage their nutrition, and many older adults turn to dietary supplements. Advertisers target older people with appeals to take supplements and eat "health" foods, claiming that these products prevent disease and promote longevity. Quite often those who take supplements are not deficient in the nutrients being supplemented.

Elderly people often benefit from a balanced low-dose vitamin and mineral supplement, however.[51] Such supplements supply many of the needed minerals along with the vitamins often lacking in older people's diets without providing too much of any one nutrient.

Nonetheless, food is still the best source of nutrients for everybody. Supplements are just that—supplements to foods, not substitutes for them. For anyone who is motivated to obtain the best possible health, it is never too late to learn to eat well, become physically active, and adopt other lifestyle changes such as quitting smoking and moderating alcohol use. Table 12-5 summarizes the nutrient concerns of aging. Table 12-6 (p. 364) offers strategies for growing old healthfully.

■ Calcium RDA during late adulthood:

- Men (51–70 yr.): 1000 mg/day
- Women (51–70 yr): 1200 mg/day
- Adults (>70 yr): 1200 mg/day

TABLE 12-5 Summary of Nutrient Concerns of Aging

Nutrient	Effect of Aging	Comments
Water	Lack of thirst and decreased total body water make dehydration likely.	Mild dehydration is a common cause of confusion. Difficulty obtaining water or getting to the bathroom may compound the problem.
Energy	Need decreases as muscle mass decreases (sarcopenia).	Physical activity moderates the decline.
Fiber	Likelihood of constipation increases with low intakes and changes in the GI tract.	Inadequate water intakes and lack of physical activity, along with some medications, compound the problem.
Protein	Needs may stay the same or increase slightly.	Low-fat milk and other high-quality protein foods are appropriate. Low-fat, high-fiber legumes and grains meet both protein and other nutrient needs.
Vitamin B_{12}	Atrophic gastritis is common.	Deficiency causes neurological damage; supplements may be needed.
Vitamin D	Increased likelihood of inadequate intake; skin synthesis declines.	Daily sunlight exposure in moderation or supplements may be beneficial.
Calcium	Intakes may be low; osteoporosis is common.	Stomach discomfort commonly limits milk intake; calcium substitutes or supplements may be needed.
Iron	In women, status improves after menopause; deficiencies are linked to chronic blood losses and low stomach acid output.	Adequate stomach acid is required for absorption; antacid or other medicine use may aggravate iron deficiency; vitamin C and meat increase absorption.
Zinc	Intakes are often inadequate and absorption may be poor, but needs may also increase.	Medications interfere with absorption; deficiency may depress appetite and sense of taste.

TABLE 12-6 Strategies for Growing Old Healthfully

- Choose nutrient-dense foods.
- Be physically active. Walk, run, dance, swim, bike, or row for aerobic activity. Lift weights, do calisthenics, or pursue some other activity to tone, firm, and strengthen muscles. Practice balancing on one foot or doing simple movements with your eyes closed. Modify activities to suit changing abilities and preferences.
- Maintain appropriate body weight.
- Reduce stress—cultivate self-esteem, maintain a positive attitude, manage time wisely, know your limits, practice assertiveness, release tension, and take action.
- For women, discuss with a physician the risks and benefits of estrogen replacement therapy.
- For people who smoke, discuss with a physician strategies and programs to help you quit.
- Expect to enjoy sex, and learn new ways of enhancing it.
- Use alcohol only moderately, if at all; use drugs only as prescribed.
- Take care to prevent accidents.
- Expect good vision and hearing throughout life; obtain glasses and hearing aids if necessary.
- Take care of your teeth; obtain dentures if necessary.
- Be alert to confusion as a disease symptom, and seek diagnosis.
- Take medications as prescribed; see a physician before self-prescribing medicines or herbal remedies and a registered dietitian before self-prescribing supplements.
- Control depression through activities and friendships; seek professional help if necessary.
- Drink six to eight glasses of water every day.
- Practice mental skills. Keep solving math problems and crossword puzzles, playing cards or other games, reading, writing, imagining, and creating.
- Make financial plans early to ensure security.
- Accept change. Work at recovering from losses; make new friends.
- Cultivate spiritual health. Cherish personal values. Make life meaningful.
- Go outside for sunshine and fresh air as often as possible.
- Be socially active—play bridge, join an exercise or dance group, take a class, teach a class, eat with friends, or volunteer time to help others.
- Stay interested in life—pursue a hobby, spend time with grandchildren, take a trip, read, grow a garden, or go to the movies.
- Enjoy life.

THE EFFECTS OF DRUGS ON NUTRIENTS

As people grow older, the use of medicines—from over-the-counter types such as aspirin and laxatives to prescription medications of all kinds—becomes commonplace. Most drugs interact with one or more nutrients in several ways, usually resulting in greater-than-normal needs for these nutrients. Chapter 14 offers a discussion of diet-drug interactions and describes the many reasons why elderly people are vulnerable to such interactions.

The most common drug that can affect nutrition in older people is alcohol. The effects of alcohol on people of all ages are explained in Nutrition in Practice 19.

Review Notes

- Table 12-5 summarizes the nutrient concerns of aging.
- Elderly people often benefit from a balanced low-dose vitamin and mineral supplement. Food is still the best source of nutrients, however.

Food Choices and Eating Habits of Older Adults

To provide any benefit, strategies and interventions to improve a person's nutrition status must be based on knowledge of food preferences and eating patterns. Menus and feeding programs for older adults must take into consideration not only the food likes and dislikes but also the living conditions, economic status, and medical conditions of this diverse group of people. If nutrition intervention is to be successful, it is essential to know what foods people will eat, in what settings they like to eat these foods, and whether they can buy and prepare meals.

Older people are, for the most part, independent, socially sophisticated, mentally lucid, fully participating members of society who report themselves to be happy and healthy. In fact, chronic disabilities among the elderly have declined dramatically in recent years. Older people spend more money per person on foods to eat at home than other age groups and less money on foods eaten away from home. Manufacturers would be wise to cater to the preferences of older adults by providing good-tasting, nutritious foods in easy-to-open, single-serving packages with labels that are easy to read. Such products enable older adults to maintain their independence; most of them want to take care of themselves and need to feel a sense of control and involvement in their own lives. As discussed earlier, another way older adults can take care of themselves is by remaining or becoming physically active. Physical activity helps preserve one's ability to perform daily tasks and so promotes independence.

© Bob Thomas/Stone/Getty Images

Shared meals can brighten the day and enhance the appetite.

INDIVIDUAL PREFERENCES

Familiarity, taste, and health beliefs are most influential on older people's food choices. Eating foods that are familiar, especially ethnic foods that recall family meals and pleasant times, can be comforting. Older adults are choosing poultry and fish, low-fat milk and milk products, and high-fiber breads and grains, indicating that they recognize the importance of diet in supporting good health. Few older adults, however, consume the recommended amounts of milk products.

MEAL SETTING

The food choices and eating habits of older adults are also affected by the changes in lifestyle that often accompany aging in this society. Whether people live alone, with others, or in institutions affects the way they eat. For example, men living alone are most likely to be poorly nourished. Older adults who live alone do not make poorer food choices than those who live with companions; rather, they consume too little food: loneliness is directly related to inadequacies, especially of energy intakes.

DEPRESSION

Another factor affecting food intake and appetite in older people is depression. Loss of appetite and motivation to cook or even to eat frequently accompanies depression. An overwhelming feeling of grief and sadness at the death of a spouse, friend, or family member may leave many people, particularly elderly people, with a feeling of powerlessness to overcome the depression. The support and companionship of family and friends, especially at mealtimes, can help overcome depression and enhance appetite. The Case Study (p. 366) presents a man who has several of these problems. Use the suggestions here, and in the last section of this chapter, to help develop solutions. The Nutrition Assessment Checklist for Older Adults at the end of this chapter helps to pinpoint nutrition-related factors to look for when working with older adults. To determine the risk of malnutrition in older clients, health care providers can keep in mind the characteristics listed in the margin. (■)

■ Risk factors for malnutrition in older adults:

- **D**isease
- **E**ating poorly
- **T**ooth loss or oral pain
- **E**conomic hardship
- **R**educed social contact
- **M**ultiple medications
- **I**nvoluntary weight loss or gain
- **N**eeds assistance with self-care
- **E**lderly person older than 80 years

FOOD ASSISTANCE PROGRAMS

Federally funded programs can provide food and nutrition services for older adults.[52] The Older Americans Act (OAA) provides many different services and support to help older adults remain independent. An integral component of the OAA is the OAA Nutrition Program, which offers services that promote health to older adults. Table 12-7 (p. 366) summarizes food assistance programs available to the elderly.

Case Study: Elderly Man with a Poor Diet

Mr. Brezenoff is 75 years old and lives alone. He has slowly been losing weight since his wife died a year ago. At 5 feet 8 inches tall, he currently weighs 124 pounds. His previous weight was 150 pounds. In talking with Mr. Brezenoff, you realize that he doesn't even like to talk about food, let alone eat it. "My wife always did the cooking before, and I ate well. Now I just don't feel like eating." You manage to find out that he skips breakfast, has soup and bread for lunch, and sometimes eats a cold-cut sandwich or a frozen dinner for supper. He seldom sees friends or relatives. Mr. Brezenoff has also lost several teeth and doesn't eat any raw fruits or vegetables because he finds them hard to chew. He lives on a meager but adequate income.

1. Consult the BMI table (inside back cover), and judge whether Mr. Brezenoff is at a healthy weight. What other assessments might you use to back up your judgment? Is his weight loss significant?
2. What factors are contributing to Mr. Brezenoff's poor food intake? What nutrients are probably deficient in his diet?
3. Look at Mr. Brezenoff as an individual and suggest ways he can improve his diet and his lifestyle.
4. What other aspects of Mr. Brezenoff's physical and mental health should you consider in helping him to improve his food intake?

MEALS FOR SINGLES

Many older adults live alone, and singles of all ages face challenges in purchasing, storing, and preparing food. Large packages of meat and vegetables are often intended for families of four or more, and even a head of lettuce can spoil before one person can use it all. Many singles live in small dwellings and have little storage space for foods. A limited income presents additional obstacles. This section offers suggestions that can help to solve some of the problems singles face, beginning with a note about the dangers of foodborne illness.

Foodborne Illness

The risk of foodborne illness is greater for older adults than for other adults. The consequences of an upset stomach, diarrhea, fever, vomiting, abdominal cramps, and dehydration are oftentimes more severe, sometimes leading to paraly-

TABLE 12-7 Food Assistance Programs for Older Adults

Program
OAA Nutrition Program *Services*: Provides congregate and home-delivered meals to improve older people's nutrition status. Includes transportation to congregate meal sites; shopping assistance; information and referral; and, to some extent, nutrition counseling and education. *Impact*: Improves the nutrient content of high-risk older adults' diets and offers socialization and recreation. Many of the nutrition programs around the country go above and beyond federal requirements of congregate and home meals by offering lunch clubs, ethnic meals, and meals for older homeless people.
Supplemental Nutrition Assistance Program (SNAP) *Services*: Supplements income for low-income households by means of a card similar to a debit card that can be used to purchase food. *Impact*: Serves more as an income supplement for some elderly participants than as a device to improve nutrition status. For other elderly participants, nutrient intakes are higher than those of nonparticipants with similar incomes.
Meals on Wheels *Services*: Delivers meals directly to the homebound elderly; integrated into the meal delivery services provided by the OAA Program. *Impact:* Focuses on filling the need for weekend and holiday meals for homebound elderly people, a service that is limited in the OAA Nutrition Program.
Senior Farmers' Market Nutrition Program *Services*: Provides low-income older adults with coupons that can be exchanged for fresh fruits, vegetables, and herbs at community-supported farmers' markets and roadside stands; administered by the USDA. State agencies may limit sales to specific foods that are locally grown to encourage recipients to support farmers in their own states. *Impact*: Increases fresh fruit and vegetable consumption, provides nutrition information, and even reaches the homebound elderly, a group of people who normally do not have access to farmers' markets.

sis, meningitis, or even death. For these reasons, older adults need to carefully follow the food safety suggestions presented in Nutrition in Practice 2.

Spend Wisely People who have the means to shop and cook for themselves can cut their food bills just by being wise shoppers. Large supermarkets are usually less expensive than convenience stores. A grocery list helps reduce impulse buying, and specials and coupons can save money when the items featured are those that the shopper needs and uses.

Buying the right amount so as not to waste any food is a challenge for people eating alone. They can buy fresh milk in the size best suited for personal needs. Pint-size and even cup-size boxes (■) of milk are available and can be stored unopened on a shelf for up to three months without refrigeration.

Many foods that offer a variety of nutrients for practically pennies have a long shelf life; staples such as rice, pastas, dry powdered milk, and dried legumes can be purchased in bulk and stored for months at room temperature. Other foods that are usually a good buy include whole pieces of cheese rather than sliced or shredded cheese, fresh produce in season, variety meats such as chicken livers, and cereals that require cooking instead of ready-to-serve cereals.

A person who has ample freezing space can buy large packages of meat, such as pork chops, ground beef, or chicken, when they are on sale. Then the meat can be immediately wrapped in individual servings for the freezer. All the individual servings can be put in a bag marked appropriately with the contents and the date.

Frozen vegetables are more economical in large bags than in small boxes. The amount needed can be taken out, and the bag closed tightly with a twist tie or rubber band. If the package is returned quickly to the freezer each time, the vegetables will stay fresh for a long time.

Finally, breads and cereals usually must be purchased in larger quantities. Again the amount needed for a few days can be taken out and the rest stored in the freezer.

People who do not have freezers can ask the grocer to break open a package of wrapped meat and rewrap the portion needed. Similarly, eggs can be purchased by the half-dozen. Eggs do keep for long periods, though, if stored properly in the refrigerator.

Fresh fruits and vegetables can be purchased individually. A person can buy fresh fruit at various stages of ripeness: a ripe one to eat right away, a semiripe one to eat soon after, and a green one to ripen on the windowsill. If vegetables are packaged in large quantities, the grocer can break open the package so that a smaller amount can be purchased. Small cans of fruits and vegetables, even though they are more expensive per unit, are a reasonable alternative, considering that it is expensive to buy a regular-size can and let the unused portion spoil.

Be Creative Creative chefs think of various ways to use foods when only large amounts are available. For example, a head of cauliflower can be divided into thirds. Then one-third is cooked and eaten hot, another third is put into a vinegar and oil marinade for use in a salad, and the last third can be used in a casserole or stew.

A variety of vegetables and meats can be enjoyed stir-fried; inexpensive vegetables such as cabbage, celery, and onion are delicious when crisp cooked in a little oil with herbs or lemon added. Interesting frozen vegetable mixtures are available in larger grocery stores. Cooked, leftover vegetables can be dropped in at the last minute. A bonus of a stir-fried meal is that there is only one pan to wash. Similarly, a microwave oven allows a home chef to use fewer pots and pans. Meals and leftovers can also be frozen or refrigerated in microwavable containers to reheat as needed.

Many frozen dinners or grocery store take-out foods offer nutritious options. Adding a fresh salad, a whole-wheat roll, and a glass of milk can make a nutritionally balanced meal. The "How to" (p. 368) offers time-saving tips to turn convenience foods into nutritious meals.

Also, single people shouldn't hesitate to invite someone to share meals with them whenever there is enough food. It's likely that the person will return the invitation, and both parties will get to enjoy companionship and a meal prepared by others.

■ Boxes of milk that can be stored at room temperature have been exposed to temperatures above those of pasteurization just long enough to sterilize the milk—a process called *ultrahigh temperature (UHT)*.

HOW TO Turn Convenience Foods into Nutritious Meals

These time-saving tips can turn convenience foods into nutritious meals:

- Add extra nutrients and a fresh flavor to canned stews and soups by tossing in some frozen ready-to-use mixed vegetables. Choose vegetables frozen without salty, fatty sauces—prepared foods generally contain enough salt to season the whole dish, including added vegetables.
- Buy frozen vegetables in a bag, toss in a variety of herbs, and use as needed.
- When grilling burgers or chicken, wrap a mixture of frozen broccoli, onion, and carrots in a foil packet with a tablespoon of Italian dressing and grill alongside the meat for seasoned grilled vegetables.
- Use canned fruits in their own juices as desserts. Toss in some frozen berries or peach slices and top with flavored yogurt for an instant fruit salad.
- Prepared rice or noodle dishes are convenient, but those claiming to contain broccoli, spinach, or other vegetables seldom contain enough to qualify as a serving of vegetables. Pump up the nutrient value by adding a half-cup of frozen vegetables per serving of pasta or rice just before cooking.
- Purchase frozen onion, mushroom, and pepper mixtures to embellish jarred spaghetti sauce or small frozen pizzas. Top with parmesan cheese.
- Use frozen shredded potatoes, sold for hash browns, in soups or stews or mix with a handful of shredded reduced-fat cheese or a can of fat-free "cream of anything" soup and bake for a quick and hearty casserole.
- Purchase a bag of triple-washed, ready-to-eat salad and add any or all of the following to make a hearty, healthy salad: a small can of garbanzo beans or a package of frozen edamame (soy beans); a handful of shredded reduced-fat cheese; a hardboiled egg; a handful of toasted almonds or other nuts.
- Open a can of pinto beans or black beans, heat a tablespoon of olive oil, and sauté a little bit of frozen onion and pepper mixture. Add the beans and mash them in the oil and vegetables. Serve the bean mixture in a soft taco shell with shredded reduced-fat cheese, lettuce, and salsa for a tasty, nutritious bean taco.

Review Notes

- Food choices of older adults are affected by health status and changed life circumstances.
- Older people can benefit from both the nutrients provided and the social interaction available at congregate meals. Other government programs deliver meals to those who are homebound.
- With creativity and careful shopping, those living alone can prepare nutritious, inexpensive meals.

Nutrition Assessment Checklist for Older Adults

MEDICAL HISTORY

Check the medical record for:

- ❍ Alcohol abuse
- ❍ Alzheimer's disease or other dementia or confusion
- ❍ Arthritis
- ❍ Cataracts
- ❍ Chronic diseases (cancer, heart disease, hypertension, diabetes)
- ❍ Cigarette, cigar, or pipe smoking; use of other tobacco products
- ❍ Constipation
- ❍ Dehydration
- ❍ Dental disease or tooth loss
- ❍ Depression
- ❍ Inflammation of the stomach (gastritis)
- ❍ Swallowing disorders

MEDICATIONS

For older adults being treated with drug therapy for medical conditions, note:

- ❍ Use of multiple medications—prescription and/or over-the-counter medications such as laxatives and pain relievers
- ❍ Side effects that might reduce food intake or change nutrient needs
- ❍ Proper administration of medication with respect to food intake

- ❍ Malnutrition—is the person's nutrition status questionable even before considering side effects of medications that worsen nutrition status?
- ❍ Diminished mental capacity that might interfere with taking correct medications and doses
- ❍ Dehydration (can alter effects of medications)

DIETARY INTAKE

For all older adults, especially those at risk nutritionally, assess the diet for:

- ❍ Total energy
- ❍ Protein
- ❍ Calcium, iron, and zinc
- ❍ Vitamin B_6, vitamin B_{12}, folate, and vitamin D

Note the following:

- ❍ Number of meals eaten each day
- ❍ Number and ages of people in household
- ❍ Amount of milk consumed each day
- ❍ Type and frequency of outdoor activity
- ❍ Type and frequency of physical activity
- ❍ Financial resources
- ❍ Transportation resources
- ❍ Physical disabilities
- ❍ Mental alertness

ANTHROPOMETRIC DATA

Measure baseline height and weight.

- ❍ Reassess height and weight at each medical checkup.
- ❍ Note significant overweight or underweight, which warrants intervention.
- ❍ Use skinfold measures to reveal altered body composition that may indicate malnutrition and loss of lean tissue.

LABORATORY TESTS

- ❍ Hemoglobin, hematocrit, or other tests of iron status
- ❍ Serum albumin or other measures of protein status
- ❍ Serum folate
- ❍ Serum B_{12}

PHYSICAL SIGNS

Look for physical signs of:

- ❍ Protein-energy malnutrition
- ❍ Iron and zinc deficiency
- ❍ Folate deficiency

Self Check

1. The fastest-growing age group in the United States is:

a. under 21 years of age.
b. 30 to 45 years of age.
c. 50 to 70 years of age.
d. over 85 years of age.

2. Which of the following lifestyle habits can enhance the length and quality of people's lives?

a. Moderate smoking
b. Six hours of sleep daily
c. Regular physical activity
d. Skipping breakfast

3. Which of the following is among the better-known relationships between nutrition and disease prevention?

a. Appropriate fiber intake helps prevent goiter.
b. Moderate sodium intake helps prevent obesity.
c. Moderate sugar intake helps prevent hypertension.
d. Appropriate energy intake helps prevent diabetes and cardiovascular disease.

4. A disease of the immune system that involves painful inflammation of the joints is:

a. sarcopenia.
b. osteoarthritis.
c. senile dementia.
d. rheumatoid arthritis.

5. Examples of low-kcalorie, high-quality protein foods include:

a. cottage cheese, sour cream, and eggs.
b. green and yellow vegetables and citrus fruits.
c. potatoes, rice, pasta, and whole-grain breads.
d. lean meats, poultry, fish, legumes, fat-free milk, and eggs.

6. For malnourished and underweight people, protein- and energy-dense snacks include:

a. fresh fruits and vegetables.
b. yogurt and cottage cheese.
c. whole grains and high-fiber legumes.
d. scrambled eggs and peanut butter on wheat toast.

7. Which of the following does not explain why dehydration is a risk for older adults?

a. They do not seem to feel thirsty.
b. Total body water increases with age.
c. They may find it difficult to get a drink.
d. They may have difficulty swallowing liquids.

8. Inadequate milk intake and limited exposure to sunlight contribute to older adults' risk of:

a. vitamin A deficiency.
b. vitamin D deficiency.
c. riboflavin deficiency.
d. vitamin B_{12} deficiency.

9. Two risk factors for malnutrition in older adults are:
   a. loneliness and multiple medication use.
   b. increased energy needs and lack of fiber.
   c. decreased mineral absorption and antioxidant intake.
   d. high carbohydrate intake and lack of physical activity.

10. Strategies to improve nutrition status when growing old include:
   a. increasing vitamin A intake and exercising 30 minutes daily.
   b. choosing nutrient-dense foods and maintaining appropriate weight.
   c. avoiding high-fiber foods and taking a daily vitamin-mineral supplement.
   d. eating at least one big meal per day and drinking at least 10 glasses of water daily.

Answers to these questions appear in Appendix H. For more chapter review: Access an interactive eBook, chapter-specific interactive learning tools, including flashcards, quizzes, videos, and more in your Nutrition CourseMate, accessed through CengageBrain.com.

Clinical Applications

1. Ms. Hamilton, an 80-year-old woman in excellent health, lives alone, eats a well-balanced diet, enjoys an active social life, and walks every day. Consider the way Ms. Hamilton's health and nutrition status might be affected by the following situations:
   - Many of Ms. Hamilton's friends pass away or move into extended care facilities.
   - Ms. Hamilton falls and breaks her hip.
   - Ms. Hamilton begins to feel isolated and depressed.

   Describe interventions the health care professional can take to help Ms. Hamilton deal with each situation to prevent her from falling into a downward spiral.

Nutrition on the Net

For further study of the topics in this chapter, access these websites.

- Search for "aging," "arthritis," and "Alzheimer's" on the U.S. government health information site:
  www.healthfinder.gov
- Visit the Administration on Aging, part of the Administration for Community Living:
  www.aoa.gov
- Visit the American Geriatrics Society Foundation for Health in Aging:
  www.healthinaging.org
- Visit the National Institute on Aging:
  www.nia.nih.gov
- Visit the American Association of Retired Persons:
  www.aarp.org
- Get nutrition tips for growing older in good health from the Academy of Nutrition and Dietetics:
  www.eatright.org
- Learn more about cataracts and macular degeneration from the National Eye Institute, the Macular Degeneration Partnership, and the American Society of Cataract and Refractive Surgery:
  www.nei.nih.gov, www.amd.org, and www.ascrs.org
- Learn more about arthritis from the Arthritis Society, the Arthritis Foundation, and the National Institute of Arthritis and Musculoskeletal and Skin Diseases:
  www.arthritis.ca, www.arthritis.org, and www.niams.nih.gov
- Learn more about Alzheimer's disease from the NIA Alzheimer's Disease Education and Referral Center and the Alzheimer's Association:
  www.nia.nih.gov/alzheimers and www.alz.org
- Find out about federal government programs designed to help senior citizens maintain good health:
  www.seniors.gov
- Visit the National Council on Aging:
  www.ncoa.org

Notes

1. Position of the Academy of Nutrition and Dietetics, Food and nutrition for older adults: Promoting health and wellness, *Journal of the Academy of Nutrition and Dietetics* 112 (2012): 1255–1277; National Center for Health Statistics, Life expectancy at age 65 years, by sex and race—United States, 2000–2006, *Morbidity and Mortality Weekly Report* 58 (2009): 473.
2. A. M. Minino, Deaths in the United States, 2009, retrieved from www.cdc.gov/nchs/data/databriefs/db64.pdf.
3. E. A. Finkelstein and coauthors, Individual and aggregate years-of-life-lost associated with overweight and obesity, *Obesity* 18 (2010): 333–339; C. J. Lavie and coauthors, Obesity and cardiovascular disease, *Journal of the American College of Cardiology* 53 (2009): 1925–1932.
4. Position of the Academy of Nutrition and Dietetics, 2012.
5. T. Ahmed and N. Haboubi, Assessment and management of nutrition in older people and its importance to health, *Clinical Interventions in Aging* 5 (2010): 207–216.
6. S. Sabia and coauthors, Influence of individual and combined healthy behaviours on successful aging, *Canadian Medical Association Journal* 184 (2012): 1985–1992; E. S. Ford and coauthors, Low-risk lifestyle behaviors and all-cause mortality: Findings from the National Health and Nutrition Examination Survey III Mortality Study, *American Journal of Public Health* 101 (2011): 1922–1929; A. L. Anderson and coauthors, Dietary patterns and survival of older adults, *Journal of the American Dietetic Association* 111 (2011): 84–91.
7. C. P. Wen and coauthors, Minimum amount of physical activity for reduced mortality and extended life expectancy: A prospective cohort study, *Lancet* 378 (2011): 1244–1253; S. W. Farrell and coauthors, Cardiorespiratory fitness, adiposity, and all-cause mortality in women, *Medicine and Science in Sports and Exercise* 42 (2010): 2006–2012; S. Q. Townsend and coauthors, Physical activity at midlife in relation to successful survival in women at age 70 years or older, *Archives of Internal Medicine* 170 (2010): 194–201.
8. W. J. Chodzko and coauthors, American College of Sports Medicine, Position stand: Exercise and physical activity for older adults, *Medicine and Science in Sports and Exercise* 41 (2009): 1510–1530; S. Mandic and coauthors, Characterizing differences in mortality at the low end of the fitness spectrum, *Medicine and Science in Sports and Exercise* 41 (2009): 1573–1579.
9. S. Studenski and coauthors, Gait speed and survival in older adults, *Journal of the American Medical Association* 305 (2011): 50–58.
10. M. O. Melancon , D. Lorrain, and I. J. Dionne, Exercise increases tryptophan availability to the brain in older men age 57–70 years, *Medicine and Science in Sports and Exercise* 44 (2012): 881–887; J. C. Sieverdes and coauthors, Association between leisure time physical activity and depressive symptoms in men, *Medicine and Science in Sports and Exercise* 44 (2012): 260–265; G. F. Bertheussen and coauthors, Associations between physical activity and physical and mental health—A HUNT 3 Study, *Medicine and Science in Sports and Exercise* 43 (2011): 1220–1228; L. D. Baker and coauthors, Effects of aerobic exercise on mild cognitive impairment, *Archives of Neurology* 67 (2010): 71–79; T. Liu-Ambrose and M. G. Donaldson, Exercise and cognition in older adults: Is there a role for resistance training programmes? *British Journal of Sports Medicine* 43 (2009): 25–27.
11. R. Koopman and L. J. van Loon, Aging, exercise, and muscle protein metabolism, *Journal of Applied Physiology,* 106 (2009): 2040–2048; K. A. Martin Ginis and coauthors, Weight training to activities of daily living: Helping older adults make a connection, *Medicine and Science in Sports and Exercise* 38 (2006): 116–121.
12. H. W. Park, Longevity, aging, and caloric restriction: Clive Main McCay and the construction of a multidisciplinary research program, *Historical Studies in the Natural Sciences* 40 (2010): 79–124.
13. R. Pallavi, M. Giogio, and P. G. Pelicci, Insights into the beneficial effect of caloric/dietary restriction for healthy and prolonged life, *Frontiers in Physiology* 3 (2012): 318.
14. L. M. Redman and E. Ravussin, Caloric restriction in humans: Impact on physiological, psychological, and behavioral outcomes, *Antioxidants and Redox Signaling* 14 (2011): 275–287; L. Fontana, The scientific basis of caloric restriction leading to longer life, *Current Opinion in Gastroenterology* 25 (2009): 144–150.
15. F. Sofi and coauthors, Accruing evidence on benefits of adherence to the Mediterranean diet on health: An updated systematic review and meta-analysis, *American Journal of Clinical Nutrition* 92 (2010): 1189–1196.
16. J. A. Mares and coauthors, Healthy diets and the subsequent prevalence of nuclear cataract in women, *Archives of Ophthalmology* 128 (2010): 738–749.
17. L. S. Lim and coauthors, Relation of age-related cataract with obesity and obesity genes in an Asian population, *American Journal of Epidemiology* 169 (2009): 1267–1274.
18. E. J. Johnson, Age-related macular degeneration and antioxidant vitamins: Recent findings, *Current Opinion in Clinical Nutrition and Metabolic Care* 13 (2010): 28–33; W. G. Christen and coauthors, Folic acid, pyridoxine, and cyanocobalamin combination treatment and age-related macular degeneration in women, *Archives of Internal Medicine* 169 (2009): 335–341; J. P. SanGiovanni and coauthors, ω-3 long-chain polyunsaturated fatty acid intake and 12-y incidence of neovascular age-related macular degeneration and central geographic atrophy: AREDS report 30, a prospective cohort study from the Age-Related Eye Disease Study, *American Journal of Clinical Nutrition* 90 (2009): 1601–1607.
19. Y. J. Cheng and coauthors, Prevalence of doctor-diagnosed arthritis and arthritis-attributable activity limitation: United States, 2007–2009, *Morbidity and Mortality Weekly* 59 (2010): 1261–1265.
20. E. M. Ross and C. B. Juhl, Osteoarthritis 2012 year in review: Rehabilitation and outcomes, *Osteoarthritis Cartilage* 20 (2012): 1477–1483; K. R. Vincent and H. K. Vincent, Resistance exercise for knee osteoarthritis, *PM & R* 4 (2012): S45–S52; K. L. Bennell and R. S. Hinman, A review of the clinical evidence for exercise in osteoarthritis of the hip and knee, *Journal of Science and Medicine in Sport* 14 (2011): 4–9.
21. R. L. Ragle and A. D. Sawitzke, Nutraceuticals in the management of osteoarthritis: A critical review, *Drugs and Aging* 29 (2012): 17–31; N. Burdett and J. D. McNeil, Difficulties with assessing the benefit of glucosamine sulphate as treatment for osteoarthritis, *International Journal of Evidence Based Healthcare,* 10 (2012): 222–226; C. T. Vangsness, W. Spiker, and J. Erickson, A review of evidence-based medicine for glucosamine and chondroitin sulfate use in knee osteoarthritis, *Arthroscopy* 25 (2009): 86–94.
22. G. Smedslund and coauthors, Effectiveness and safety of dietary interventions for rheumatoid arthritis: A systematic review of randomized controlled trials, *Journal of the American Dietetic Association* 110 (2010): 727–735; C. Deighton and coauthors, Management of rheumatoid arthritis: Summary of NICE guidance, *British Medical Journal* 338 (2009): b702.
23. N. G. Bazan, M. F. Molina, and W. C. Gordon, Docosahexaenoic acid signalolipidomics in nutrition: Significance in aging, neuroinflammation, macular degeneration, Alzheimer's and other neurodegenerative diseases, *Annual Review of Nutrition* 31 (2011): 321–351.
24. J. G. Walker and coauthors, Oral folic acid and vitamin B-12 supplementation to prevent cognitive decline in community-dwelling older adults with depressive symptoms—The Beyond Ageing Project; A randomized controlled trial, *American Journal of Clinical Nutrition* 95 (2012): 194–203; Bazan, Molina, and Gordon, 2011; A. D. Smith and H. Refsum, Vitamin B-12 and cognition in the elderly, *American Journal of Clinical Nutrition* 89 (2009): 707S–711S; D. M. Lee and coauthors, Association between 25-hydroxyvitamin D levels and cognitive performance in middle-aged and older European men, *Journal of Neurology, Neurosurgery, and Psychiatry* 80 (2009): 722–729.
25. Centers for Disease Control and Prevention, Alzheimer's Disease, retrieved from www.cdc.gov/aging/aginginfo/alzheimers.htm. Site updated February 23, 2011.
26. H. W. Querfurth and F. M. LaFeria, Mechanisms of disease: Alzheimer's disease, *New England Journal of Medicine* 362 (2010): 329–344.
27. J. C. Lambert and coauthors, Genome-wide association study identifies variants at CLU and CR1 associated with Alzheimer's disease, *Nature Genetics* 41 (2009): 1094–1099; A. Burns and S. Lliffe, Clinical review—Alzheimer's disease, *British Medical Journal* 338 (2009): b158.
28. Querfurth and LaFeria, 2010.
29. F. E. Harrison, A critical review of vitamin C for the prevention of age-related decline and Alzheimer's disease, *Journal of Alzheimer's Disease* 29 (2012): 11–26; J. Viña and coauthors, Antioxidant pathways in Alzheimer's disease: Possibilities of intervention, *Current Pharmaceutical Design* 17 (2011): 3861–3864.
30. B. Misiak, J. Leszek, and A. Kiejna, Metabolic syndrome, mild cognitive impairment and Alzheimer's disease—The emerging role of systemic low-grade inflammation and adiposity, *Brain Research Bulletin* 89 (2012): 144–149; S. M. De La Monte, Metabolic derangements mediate cognitive impairment and Alzheimer's disease; Role of peripheral insulin-resistance diseases, *Panminerva Medica* 54 (2012): 171–178; J. A. Luchsinger and D. R. Gustafson, Adiposity and Alzheimer's disease, *Current Opinion in Clinical Nutrition and Metabolic Care* 12 (2009): 15–21.

31. Querfurth and LaFeria, 2010; A. Gella and N. Durany, Oxidative stress in Alzheimer's disease, *Cell Adhesion and Migration* 13 (2009): 88–93.
32. R. J. Castellani and coauthors, Reexamining Alzheimer's disease: Evidence for a protective role for amyloid-beta protein precursor and amyloid-beta, *Journal of Alzheimer's Disease* 18 (2009): 447–452.
33. De La Monte, 2012; Misiak, Leszek, and Kiejan, 2012; J. D. Doecke and coauthors, Blood-based protein biomarkers for diagnosis of Alzheimer Disease, *Archives of Neurology* 2012, doi: 10.1001/archneurol.2012.1282; J. A. Luchsinger and D. R. Gustafson, Adiposity, type 2 diabetes, and Alzheimer's disease, *Journal of Alzheimer's Disease* 16 (2009): 693–704.
34. G. M. Cole and S. A. Frautschy, DHA may prevent age-related dementia, *Journal of Nutrition* 140 (2010): 869–874; Y. Gu and coauthors, Food combination and Alzheimer disease risk: A protective diet, *Archives of Neurology* 67 (2010): 699–706; E. Albanese and coauthors, Dietary fish and meat intake and dementia in Latin America, China, and India: A 10/66 Dementia Research Group population-based study, *American Journal of Clinical Nutrition* 90 (2009): 392–400; C. M. Milte, N. Sinn, and P. R. C. Howe, Polyunsaturated fatty acid status in attention deficit hyperactivity diseorder, depression, and Alzheimer's disease: Towards an omega-3 index for mental health? *Nutrition Reviews* 67 (2009): 573–590; N. Scarmeas and coauthors, Physical activity, diet, and risk of Alzheimer disease, *Journal of the American Medical Association* 302 (2009): 627–637.
35. N. Scarmeas and coauthors, Physical activity, diet, and risk of Alzheimer disease, *Journal of the American Medical Association* 302 (2009): 627–637; T. Liu-Ambrose and M. G. Donaldson, Exercise and cognition in older adults: Is there a role for resistance training programmes? *British Journal of Sports Medicine* 43 (2009): 25–27.
36. Position of the Academy of Nutrition and Dietetics, 2012.
37. Koopman and van Loon, 2009.
38. L. A. Burton and D. Sumukadas, Optimal management of sarcopenia, *Clinical Interventions in Aging* 5 (2010): 217–228; Koopman and van Loon, 2009.
39. Position of the Academy of Nutrition and Dietetics, 2012; J. Cohen-Mansfield and R. Perach, Is there a reversal in the effect of obesity on mortality in old age? *Journal of Aging Research* 2011 (2011): 765071; D. K. Childers and D. B. Allison, The "obesity paradox": A parsimonious explanation for relations among obesity, mortality rate and aging? *International Journal of Obesity* 34 (2010): 1231–1238.
40. Position of the Academy of Nutrition and Dietetics, 2012; L. M. Donini and coauthors, A systematic review of the literature concerning the relationship between obesity and mortality in the elderly, *Journal of Nutrition, Health & Aging* 16 (2012): 89–98.
41. J. E. Morley and coauthors, Nutritional recommendations for the management of sarcopenia, *Journal of the American Medical directors Association* 11 (2010): 391–396; D. K. Layman, Dietary guidelines should reflect new understandings about adult protein needs, *Nutrition and Metabolism* 6 (2009): 12; A. K. Surdykowski and coauthors, Optimizing bone health in older adults: The importance of dietary protein, *Aging Health* 6 (2010): 345–357.
42. Position of the Academy of Nutrition and Dietetics, 2012; R. B. Ervin and B. A. Dye, The effect of functional dentition on Healthy Eating Index scores and nutrient intakes in a nationally representative sample of older adults, *Journal of Public Health Dentistry* 69 (2009): 207–216.
43. D. E. King, A. G. Mainous III, and C. A. Lambourne, Trends in dietary fiber intake in the United states, 1999–2008, *Journal of the Academy of Nutrition and Dietetics* 112 (2012): 642–648; Standing Committee on the Scientific Evaluation of Dietary Reference Intakes, Food and Nutrition Board, Institute of Medicine, *Dietary Reference Intakes for Energy, Carbohydrate, Fiber, Fat, Fatty Acids, Cholesterol, Protein, and Amino Acids* (Washington DC: National Academies Press, 2005), pp. 387–389.
44. Position of the Academy of Nutrition and Dietetics, 2012.
45. Position of the Academy of Nutrition and Dietetics, 2012; Standing Committee on the Scientific Evaluation of Dietary Reference Intakes, Food and Nutrition Board, Institute of Medicine, *Dietary Reference Intakes for Water, Potassium, Sodium, Chloride, and Sulfate* (Washington, DC: National Academies Press, 2005), pp. 118–127.
46. Standing Committee on the Scientific Evaluation of Dietary Reference Intakes, *Dietary Reference Intakes for Water, Potassium, Sodium, Chloride, and Sulfate*, 2005, pp. 149–150.
47. Committee on Dietary Reference Intakes, *Dietary Reference Intakes for Calcium and Vitamin D* (Washington, DC: National Academies Press, 2011), pp. 345–402.
48. Committee on Dietary Reference Intakes, *Dietary Reference Intakes for Thiamin, Riboflavin, Niacin, Vitamin B6, Folate, Vitamin B12, Pantothenic Acid, Biotin, and Choline* (Washington, DC: National Academies Press, 2000), p. 338.
49. A. D. Smith and H. Refsum, Vitamin B-12 and cognition in the elderly, *American Journal of Clinical Nutrition* 89 (2009): 707S–711S.
50. J. B. Barnett, D. H. Hamer, and S. N. Meydani, Low zinc status: A new risk factor for pneumonia in the elderly? *Nutrition Reviews* 68 (2010): 30–37.
51. Position of the Academy of Nutrition and Dietetics, 2012.
52. Position of the Academy of Nutrition and Dietetics, 2012; Position of the American Dietetic Association, American Society for Nutrition, and Society for Nutrition Education: Food and nutrition programs for community-residing older adults, *Journal of the American Dietetic Association* 110 (2010): 463–472.

Nutrition in Practice

Hunger and Community Nutrition

Worldwide, one person in every eight experiences persistent hunger—not the healthy appetite triggered by anticipation of a hearty meal but the painful sensation caused by a lack of food.[1] Hunger deprives a person of the physical and mental energy needed to enjoy a full life and often leads to severe malnutrition and death. In 2011, more than two million children younger than the age of five died as a result of hunger and malnutrition.[2]

In the United States, where most people enjoy a life of relative abundance, 6.8 million households live with **very low food security**—one or more members of these households, many of them children, repeatedly had little or nothing to eat because of a lack of money.[3] Another 11 million households experienced **low food security** or **marginal food security**, somewhat less dire conditions. Given the agricultural bounty and enormous wealth in this country, do these numbers surprise you? The limited or uncertain availability of nutritionally adequate and safe foods is known as **food insecurity** and is a major social problem in our nation today.[4] Table NP12-1 (p. 374) presents questions used in national surveys to identify food insecurity in the United States, and Figure NP12-1 presents the most recent findings. Surveys like these provide crude, but necessary, data to estimate the degree of hunger in this country. The glossary defines related terms.

NURSING DIAGNOSIS

Imbalanced nutrition: Less than body requirements applies to clients who have a lack of food.

FIGURE NP12-1 Food Security of U.S. Households

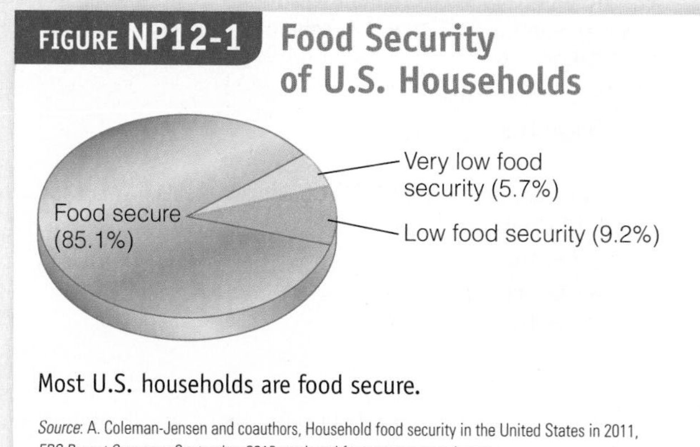

Most U.S. households are food secure.

Source: A. Coleman-Jensen and coauthors, Household food security in the United States in 2011, *ERS Report Summary,* September 2012, retrieved from www.ers.usda.gov.

Glossary

emergency kitchens: programs that provide meals to be eaten on-site; often called *soup kitchens.*

food banks: facilities that collect and distribute food donations to authorized organizations feeding the hungry.

food deserts: urban and rural low-income areas with limited access to affordable and nutritious foods (also defined in Chapter 7).

food insecurity: limited or uncertain access to foods of sufficient quality or quantity to sustain a healthy and active life.

food pantries: community food collection programs that provide groceries to be prepared and eaten at home.

food poverty: hunger occurring when enough food exists in an area but some of the people cannot obtain it because they lack money, are being deprived for political reasons, live in a country at war, or suffer from other problems such as lack of transportation.

food recovery: the collection of wholesome food for distribution to low-income people who are hungry. Four common methods of food recovery are:

- *Field gleaning:* collecting crops from fields that either have already been harvested or are not profitable to harvest.
- *Perishable food rescue or salvage:* collection of perishable produce from wholesalers and markets.
- *Prepared food rescue:* collection of prepared foods from commercial kitchens.
- *Nonperishable food collection:* collection of processed foods from wholesalers and markets.

food security: access to enough food to sustain a healthy and active life.

low food security: a descriptor for households with reduced dietary quality, variety, and desirability but with adequate quantity of food and normal eating patterns. Example: A family whose diet centers on inexpensive, low-nutrient foods such as refined grains, inexpensive meats, sweets, and fats.

marginal food security: a descriptor for households with problems or anxiety at times about accessing adequate food, but without substantial reductions in the quality, variety, or quantity of their food intake. Example: A parent worried that the food purchased would not last until the next paycheck.

very low food security: a descriptor for households that, at times during the year, experienced disrupted eating patterns or reduced food intake of one or more household members because of a lack of money or other resources for food. Example: A family in which one or more members went to bed hungry, lost weight, or didn't eat for a whole day because they did not have enough food.

TABLE NP12-1 U.S. Food Security Survey—Short Form

The food security of an individual or household lies along a continuum. The *existence* of food security can be determined by asking the first six questions below. The *degree* of food insecurity is assessed with an additional four questions that follow. Figure NP12-1 shows the results of the most recent survey.

1. "The food that (I/we) bought just didn't last, and (I/we) didn't have money to get more." Was that often, sometimes, or never true for (you/your household) in the last 12 months?
   [ ] Often true
   [ ] Sometimes true
   [ ] Never true

2. "(I/we) couldn't afford to eat balanced meals." Was that often, sometimes, or never true for (you/your household) in the last 12 months?
   [ ] Often true
   [ ] Sometimes true
   [ ] Never true

3. In the last 12 months, since last (name of current month), did (you/you or other adults in your household) ever cut the size of your meals or skip meals because there wasn't enough money for food?
   [ ] Yes
   [ ] No

If yes, then:

4. How often did this happen—almost every month, some months but not every month, or in only 1 or 2 months?
   [ ] Almost every month
   [ ] Some months but not every month
   [ ] Only 1 or 2 months

5. In the last 12 months, did you ever eat less than you felt you should because there wasn't enough money for food?
   [ ] Yes
   [ ] No

6. In the last 12 months, were you ever hungry but didn't eat because there wasn't enough money for food?
   [ ] Yes
   [ ] No

These additional questions help to determine the severity of food insecurity:

Least severe:
Was this statement often, sometimes, or never true for you in the last 12 months? "We worried whether our food would run out before we got money to buy more."

Somewhat more severe:
Was this statement often, sometimes, or never true for you in the last 12 months? "We couldn't afford to eat balanced meals."

Midrange severity:
In the last 12 months, did you ever cut the size of your meals or skip meals because there wasn't enough money for food?

Most severe:
In the last 12 months, did you ever not eat for a whole day because there wasn't enough money for food?
In the last 12 months, did any of the children ever not eat for a whole day because there wasn't enough money for food?

Source: Economic Research Service, USDA, U.S. Household Food Security Survey Module: Six-Item Short Form, 2008. The Six-Item Short Form and the long survey are available at www.ers.usda.gov/topics/food-nutrition-assistance/food-security-in-the-us/survey-tools.aspx#household.

Why is hunger a problem in developed countries such as the United States where food is abundant?

Hunger has many causes, but in developed countries, the primary cause is **food poverty**. People are hungry not because there is no food nearby to purchase, but because they lack sufficient money with which to buy nutritious food and pay for other necessities, such as housing, clothing, medicines, and utilities. An estimated 15 percent of the population of the United States lives in poverty.[5] Even those above the poverty line may not have **food security**. Physical and mental illnesses and disabilities, sudden job losses, and high living expenses threaten their financial stability. Further contributing to food poverty are other problems such as abuse of alcohol and other drugs; lack of awareness of available food assistance programs; and the reluctance of people, particularly the elderly, to accept what they perceive as "welfare" or "charity." Lack of resources remains the major cause of food poverty in developed countries, and solving this problem would do a lot to relieve hunger.

In the United States, poverty and hunger reach across various segments of society, touching some more than others—notably, single parents living in households with their children, Hispanics, African Americans, and those living in the inner cities. People living in poverty are simply unable to buy sufficient amounts of nourishing foods, even if they are skilled in food shopping. Consequently, their diets tend to be inadequate. For many of the children in these families, school lunch (and breakfast, where available) may be the only nourishment for the day. Otherwise, they go hungry, waiting for an adult to find money for food. Not surprisingly, these children are more likely to have health problems and iron-deficiency anemia than those who eat regularly.[6] They also tend to perform poorly in school and in social situations. For adults, the risk of developing chronic diseases increases.[7] For pregnant women, the risk of developing gestational diabetes more than doubles.[8]

Is it true that hunger and obesity often exist side by side—sometimes within the same household or even the same person?

Ironically, it is true—food insufficiency and obesity often exist side by side.[9] Food insecurity and obesity may logically seem to be mutually exclusive, but research studies consistently show that the highest rates of obesity occur among those living in the greatest poverty and food insecurity.[10]

Low-income urban and rural communities that offer little or no access to affordable nutritious foods—**food deserts** (first mentioned in Chapter 7)—lack access to markets that sell fresh produce.[11] Not surprisingly, people living in food deserts often lack fruits and vegetables in their diets and often fail to meet intake recommendations of the *Dietary Guidelines*.[12] High-fat, high-sugar, refined, energy-dense foods that are readily available in food deserts infamously lack other needed nutrients. Doughnuts, packaged sweet cakes, sugary punches, hamburgers, and french fries fill the stomach, are affordable, are easily obtained at any hour, are easily carried, require no preparation, and taste good. It is not surprising, then, that these high-kcalorie foods edge nutrient-dense staple foods out of the diet.

Economic uncertainty and stress greatly influence the prevalence of obesity.[13] People who are unsure about their next meal may overeat when food or money become available. Interestingly, food insecure people who do *not* participate in food assistance programs have a greater risk of obesity than those who do—it seems that providing a reliable supply of nutritious food may help to prevent obesity among those living with food insecurity.[14]

What U.S. food programs are directed at relieving hunger in the United States?

The Academy of Nutrition and Dietetics calls for aggressive action to bring an end to domestic hunger and to achieve food and nutrition security for all residents of the United States.[15] Many federal and local programs aim to prevent or relieve malnutrition and hunger in the United States.

An extensive network of federal assistance programs provides life-giving food daily to millions of U.S. citizens. An estimated one out of every seven Americans receives food assistance of some kind, at a total cost of almost $90 billion per year.[16] Even so, the programs are not fully successful in preventing hunger, though they do seem to improve the nutrient intakes of those who participate. Programs described in the life cycle chapters include the WIC program for low-income pregnant women, breastfeeding mothers, and their young children (Chapter 10); the school lunch and breakfast programs for children (Chapter 11); and the food assistance programs for older adults such as congregate meals and Meals on Wheels (Chapter 12).

The centerpiece of food programs for low-income people in the United States is the Supplemental Nutrition Assistance Program (SNAP), administered by the U.S. Department of Agriculture (USDA). The USDA issues debit cards through state agencies to households—people who buy and prepare food together. The amount a household receives depends on its size, resources, and income. Recipients may use the cards to purchase food and food-bearing plants and seeds but not to buy tobacco, cleaning items, alcohol, or other nonfood items. The "How to" (p. 376) offers tips for both saving money and preventing food waste.

SNAP is the largest of the federal food assistance programs, both in amount of money spent and in number of people participating. It provides assistance to more than 46 million people at a cost of more than $72 billion per year; about half of the recipients are children.

HOW TO Stretch Food Dollars and Reduce Waste

Eating well on a budget can pose a challenge, but these tips ease the task. For daily menus and recipes for healthy, thrifty meals, visit the USDA Center for Nutrition Policy and Promotion: www.cnpp.usda.gov.

Plan Ahead

- Plan your menus, write grocery lists, and shop only for foods on your list to avoid expensive "impulse" buying.
- Center meals on whole grains, legumes, and vegetables; use smaller quantities of meat, poultry, fish, or eggs.
- Use cooked cereals such as oatmeal instead of ready-to-eat breakfast cereals.
- Cook large quantities when time and money allow; freeze portions for convenient later meals.
- Check for sales and use coupons for products you need; plan meals to take advantage of sale items.

Shop Smart

- Do not shop when hungry.
- Select whole foods instead of convenience foods (raw whole potatoes instead of refrigerated prepared mashed potatoes, for example).
- Try store brands.
- Buy fresh produce in season; buy canned or frozen items at other times.
- Buy large bags of frozen items or dry goods; use as needed and store the remainder.
- Buy fat-free dry milk; mix and refrigerate quantities needed for a day or two. Buy fresh milk by the gallon or half-gallon only if you can use it up before it spoils.
- Buy less expensive cuts of meat, such as beef chuck and pork shoulder roasts; cook with liquid long enough to make the meat tender.
- Buy whole chickens instead of pieces; ask a butcher to show you how to cut them up.
- Frequent discount stores instead of grocery stores for non-food items such as toilet paper and detergent.

Reduce Waste

- Change your thinking from "what do I want to eat" to "what do I have available to eat." You paid for the food you have on hand—so use it up.
- Buy only the amount of fresh foods that you will eat before it spoils.
- Peel away the tough outer layers from stems of asparagus and broccoli; slice and cook the tender stems or add raw to salads.
- Scrub, but don't peel, potatoes before cooking—the skins add color, texture, and nutrients to the dish.
- Before buying food in bulk, plan how to store it properly. If it spoils before use, you'll throw away your savings.
- If your "bargain" bulk food is more than you can use but is still fresh, donate it to your local food bank or homeless shelter. (It won't save you money, but it will provide a wealth of satisfaction.)
- If space permits, compost fruit and vegetable scraps to feed shrubs and other outdoor plants.

Why do health care professionals need to know about food assistance programs?

Health care professionals who work in public health are generally well acquainted with food assistance programs, and often many of their clients receive such assistance. Regardless of the setting in which health care professionals see clients, however, it is important to encourage those who may be having financial problems to talk with a social worker who can assess their eligibility for food assistance programs. The subject of food assistance must be approached in a nonjudgmental and tactful manner—the client may feel uncomfortable about seeking assistance.

Are there other programs aimed at reducing hunger in the United States?

Efforts to resolve the problem of hunger in the United States do not depend solely on federal assistance programs. National **food recovery** programs such as Feeding America coordinate the efforts of **food banks**, **food pantries**, **emergency kitchens**, and homeless shelters that provide food to tens of millions of people a year. Table NP12-2 lists addresses, phone numbers, and websites for Feeding America and other hunger relief organizations.

Feeding the hungry in the United States.

TABLE NP12-2 Hunger-Relief Organizations

Organization	Mission Statement
Bread for the World www.bread.org	Nonpartisan, Christian citizens' movement seeking to influence reform in policies, programs, and conditions that allow hunger and poverty to persist globally.
Catholic Relief Services www.crs.org	Humanitarian service agency assisting the impoverished and disadvantaged through community-based, sustainable development initiatives.
Community Food Security Coalition www.foodsecurity.org	North American coalition working to catalyze food systems that are healthy, sustainable, just, and democratic by building community voice and capacity for change.
Congressional Hunger Center www.hungercenter.org	Bipartisan organization training and inspiring leaders with the intent to end hunger and advocating public policies to create a food-secure world.
Feeding America www.feedingamerica.org	Domestic charity organization providing food assistance through a nationwide network of member food banks and facilitating education to end hunger nationally.
Food and Agriculture Organization (FAO) of the United Nations www.fao.org	International organization leading efforts to achieve food security for all by helping to develop and modernize countries' agriculture, forestry, and fishery practices.
Idealist www.idealist.org	International organization seeking to connect people, organizations, and resources to help build a world where all people can live free and dignified lives.
Oxfam America www.oxfamamerica.org	International relief and development organization aiming to create lasting solutions to poverty, hunger, and injustice.
Pan American Health Organization new.paho.org	International public health agency aiming to strengthen national and local health systems with the purpose of improving the quality of, and lengthening, the lives of the peoples in the Americas.
Society of St. Andrew www.endhunger.org	Ecumenical Christian ministry salvaging and redirecting large amounts of fresh produce to hunger agencies for distribution to the poor.
The Hunger Project www.thp.org	International relief organization attempting to end hunger and poverty by pioneering sustainable, grassroots, women-centered strategies and advocating for their widespread adoption in countries throughout the world.
United Nations Children's Fund (UNICEF) www.unicef.org	International organization advocating for the protection of children's rights, to help meet their basic needs and to expand their opportunities to reach their full potentials.
WhyHunger www.whyhunger.org	Domestic organization supporting and funding community-based organizations intent on empowering individuals and building self-reliance to provide long-term solutions to end hunger and poverty.
World Food Programme www.wfp.org	Food aid branch of the United Nations aiming to prepare for, protect during, and provide assistance after emergencies, as well as reduce hunger and undernutrition.
World Health Organization (WHO) www.who.int	United Nations agency acting as the authority on international public health by influencing policy, setting research agendas, establishing standards, and providing technical support to monitor and assess health trends.

Each year, a tremendous amount of our food supply is wasted in fields, commercial kitchens, grocery stores, and restaurants—enough food to feed millions of people. Food recovery programs collect and distribute good food that would otherwise go to waste. Volunteers might pick corn left in an already harvested field, a grocer might deliver ripe bananas to a local food bank, and a caterer might take leftover chicken salad to a community shelter, for example. All of these efforts help to feed the hungry in the United States.

AP Photo/Jake Schoelkopf

Community-based efforts to feed citizens include food pantries that provide groceries.

What about local efforts and community nutrition programs?

Food recovery programs depend on volunteers. Concerned citizens work through local agencies and churches to feed the hungry. Community-based food pantries provide groceries, and soup kitchens serve prepared meals. Meals often deliver adequate nourishment, but most homeless people receive fewer than one and a half meals a day, so many are still inadequately nourished. Health care professionals can serve as valuable members of community groups seeking to provide food assistance.

Notes

1. *The State of Food Insecurity in the World, 2012, Executive Summary*, retrieved from Food and Agriculture Organization, www.fao.org.
2. World Health Organization, *Children: Reducing mortality,* Fact sheet, September 2012, retrieved from www.who.int/mediacentre/factsheets/fs178/en/index.html.
3. A. Coleman-Jensen and coauthors, *Economic Research Service Report Summary*, September 2012, Household food security in the United States, 2011, retrieved from www.ers.usda.gov/publications/err-economic-research-report/err141.aspx.
4. Position of the American Dietetic Association, Food insecurity in the United States, *Journal of the American Dietetic Association* 110 (2010): 1368–1377.
5. US Census Bureau, *Social, Economic, and Housing Statistics Division: Poverty* (Washington, DC: U.S. Government Printing Office, 2011).
6. H. A. Eicher-Miller and coauthors, Food insecurity is associated with iron deficiency anemia in U.S. adolescents, *American Journal of Clinical Nutrition* 90 (2009): 1358–1371; R. Rose-Jacobs and coauthors, Household food insecurity: Associations with at-risk infant and toddler development, *Pediatrics* 121 (2008): 65–72.
7. H. K. Seligman and D. Schillinger, Hunger and socioeconomic disparities in chronic disease, *New England Journal of Medicine* 363 (2010): 6–9.
8. C. M. Olson, Food insecurity and maternal health during pregnancy, *Journal of the American Dietetic Association* 110 (2010): 690–691.
9. J. C. Eisenmann and coauthors, Is food insecurity related to overweight and obesity in children and adolescents? A summary of sutides, 1995–2009, *Obesity Reviews* 12 (2011): e73–e83.
10. B. M. Popkin, L. S. Adair, and S. W. Ng, Global nutrition transition and the pandemic of obesity in developing countries, *Nutrition Reviews* 70 (2011): 3–21; A. Drewnowski, Obesity, diets, and social inequalities, *Nutrition Reviews* 67 (2009): S36–S39; B. J. Lohman and coauthors, Adolescent overweight and obesity: Links to food insecurity and individual, maternal, and family stressors, *Journal of Adolescent Health* 45 (2009): 230–237; E. Metallinos-Katsaras, B. Sherry, and J. Kallio, Food insecurity is associated with overweight in children younger than 5 years of age, *Journal of the American Dietetic Association* 109 (2009): 1790–1794.
11. A. M. Fretts and coauthors, Associations of processed meat and unprocessed red meat intake with incident diabetes: The Strong Heart Family Study, *American Journal of Clinical Nutrition* 95 (2012): 752–758; H. J. Song and coauthors, Understanding a key feature of urban food stores to develop nutrition intervention, *Journal of Hunger & Environmental Nutrition* 7 (2012): 77–90; Institute of Medicine and National Research Council, *The Public Health Effects of Food Deserts: Workshop Summary* (Washington, DC: The National Academies Press, 2009).
12. B. T. Izumi and coauthors, Associations between neighborhood availability and individual consumption of dark-green and orange vegetables among ethnically diverse adults in Detroit, *Journal of the American Dietetic Association* 111 (2011): 274–279.
13. A. Offer, R. Pechey, and S. Ulijaszek, Obesity under affluence varies by welfare regimes: The effect of fast food, insecurity, and inequality, *Economics and Human Biology* 8 (2010): 297–308.
14. A. Karnik and coauthors, Food insecurity and obesity in New York City primary care clinics, *Medical Care* 49 (2011): 658–661; N. I. Larson and M. T. Story, Food insecurity and weight status among U.S. children and families: A review of the literature, *American Journal of Preventative Medicine* 40 (2011): 166–173.
15. Position of the American Dietetic Association, 2010.
16. USDA Economic Research Service, Economic Bulletin 6-8, The food assistance landscape: 2010 annual report, retrieved from www.ers.usda.gov/media/129642/eib6-8.pdf.

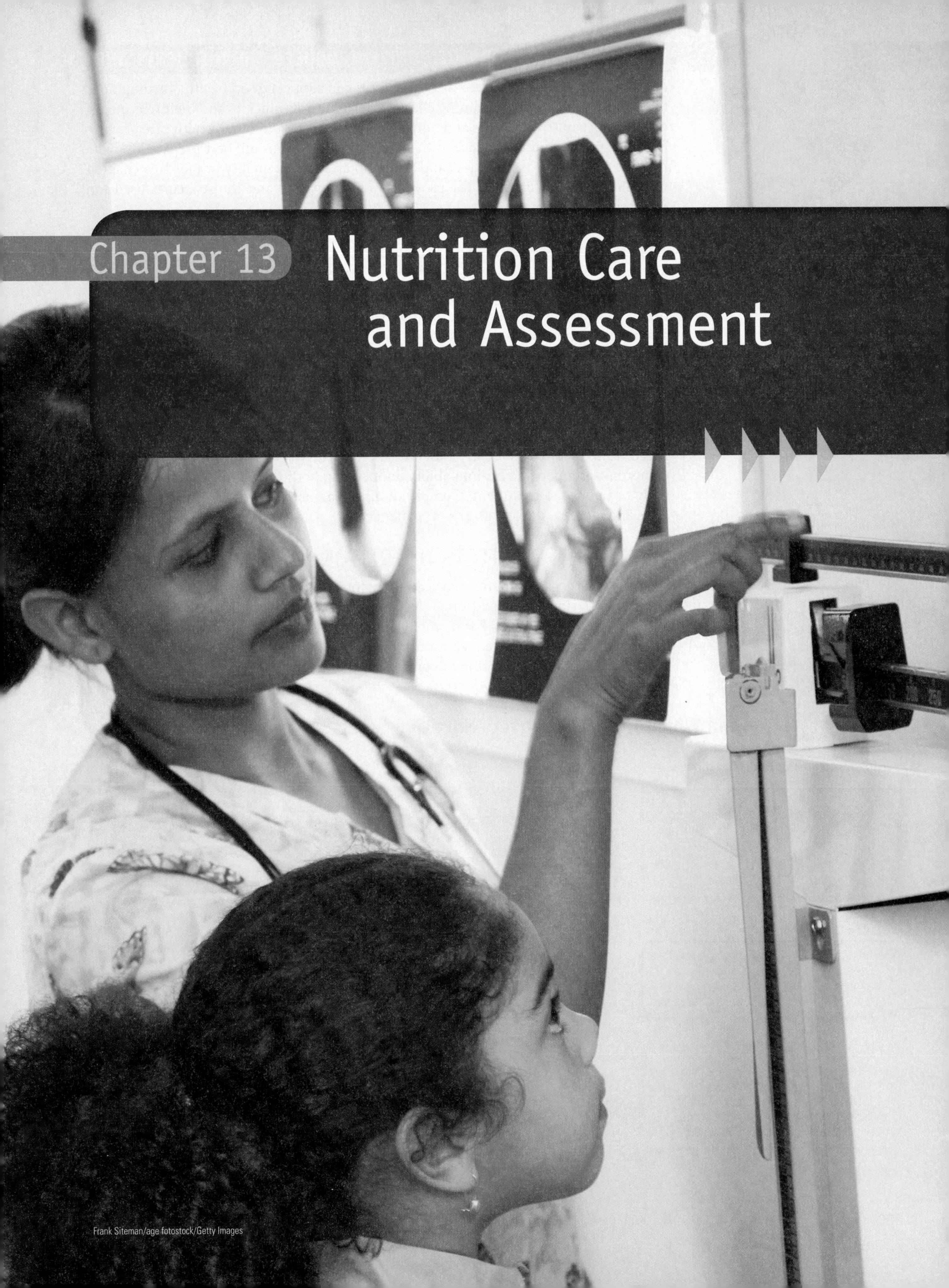

Chapter 13 Nutrition Care and Assessment

Frank Siteman/age fotostock/Getty Images

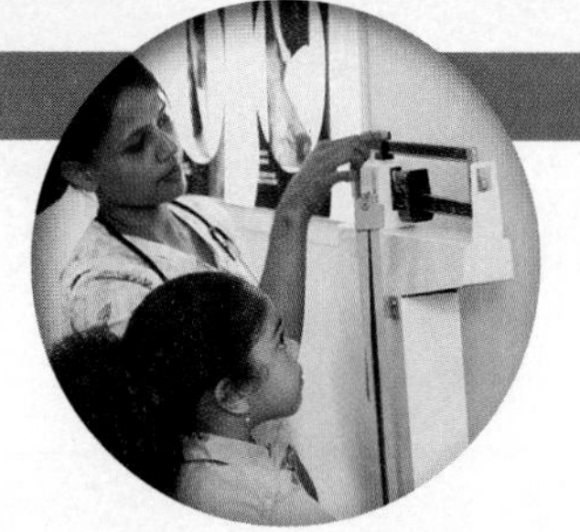

EARLIER CHAPTERS OF THIS BOOK INTRODUCED THE NUTRIENTS AND described how the appropriate dietary choices can support good health. Turning now to clinical nutrition, the remaining chapters describe how illnesses and their treatments can influence nutrition status and nutrient needs. Nurses and other health care providers who consider nutrition care in their treatment decisions are best prepared to help patients (■) recover from illness and maintain an optimal quality of life. This chapter introduces the process used for providing nutrition care and the strategies used for evaluating nutrition status.

Nutrition in Health Care

The busy nurse with many responsibilities may be tempted to put a patient's nutrition needs on the back burner; after all, the benefits of nutrition therapy are not always as obvious or immediate as those of other medical treatments. Correcting nutritional problems, however, may improve both short-term and long-term outcomes of medical treatments and help to prevent complications. Moreover, patients are often concerned about the diet they need to improve their health.

Malnutrition is frequently reported in patients hospitalized with an acute illness. Depending on the patient population, estimates of malnutrition in hospital patients range from 15 to 60 percent.[1] Poor nutrition status weakens immune function and compromises a person's healing ability, influencing both the course of illness and the body's response to treatment. Complications of malnutrition often lengthen hospital stays and increase the overall cost of patient care.

■ Reminder: Nurses may use the term *client* or *patient* when referring to an individual under their care. The clinical chapters emphasize the care of those with serious illnesses; thus, the term *patient* is used throughout these chapters.

anorexia: loss of appetite.

HOW ILLNESS AFFECTS NUTRITION STATUS

Illnesses and their treatments may lead to malnutrition by causing a reduction in food intake, interfering with digestion and absorption, or altering nutrient metabolism and excretion (see Figure 13-1). For example, the nausea caused by some illnesses can diminish appetite and thereby reduce food intake. An inflamed mouth or esophagus may make the physical act of eating uncomfortable. Some medications can cause **anorexia**

FIGURE 13-1 Ways in Which Illness Can Affect Nutrition Status

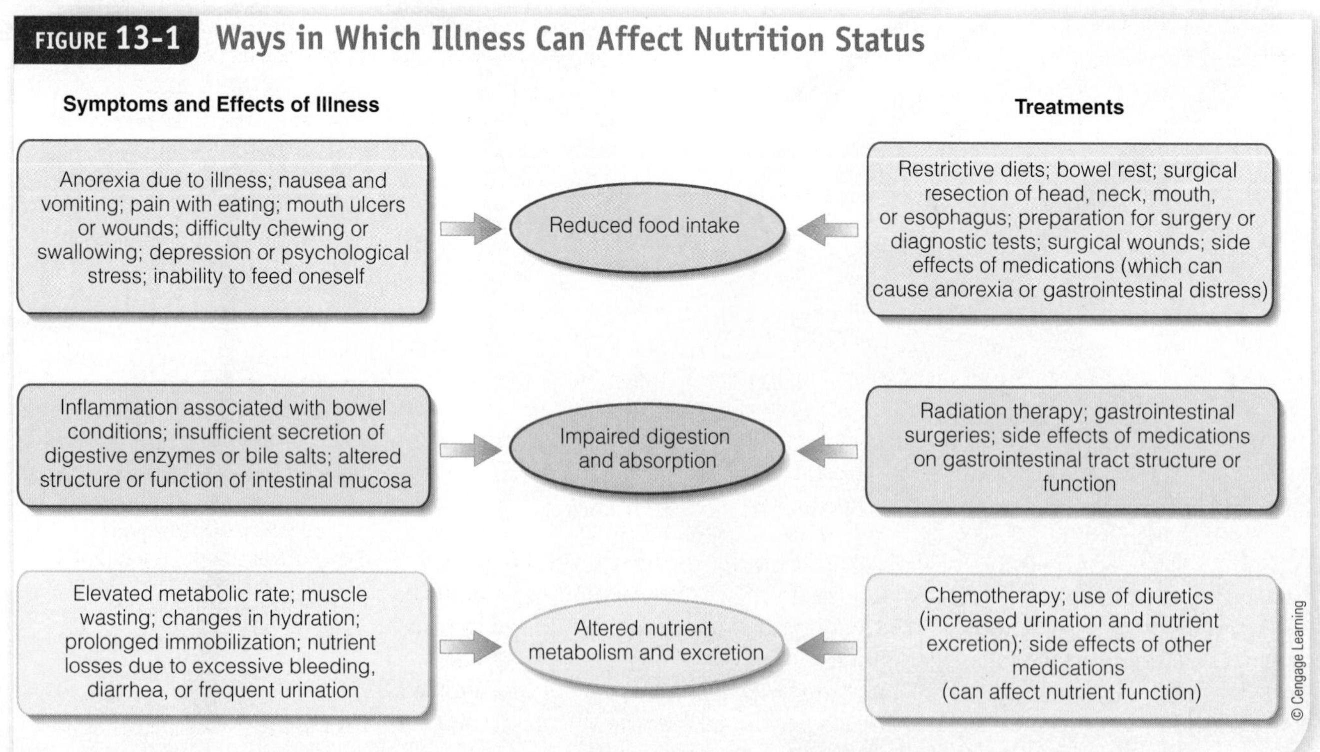

(loss of appetite) or gastrointestinal discomfort or interfere with nutrient function and metabolism. Prolonged bed rest may lead to **pressure sores**, which increase metabolic stress and raise protein and energy needs.

The dietary changes required during an acute illness are usually temporary and can be tailored to accommodate an individual's preferences and lifestyle. Conversely, chronic illnesses (those lasting three months or longer) may require long-term dietary adjustments. For example, diabetes treatment requires lifelong changes in diet and lifestyle that some people may find difficult to adhere to. The challenge for nurses and other health professionals is to help their patients understand the potential benefits of nutrition therapy and accept the dietary changes that can improve their health.

In addition to the direct effects of an illness on nutrition status, the cost of health care can drain financial resources and limit the ability to obtain high-quality foods. Some individuals may lack the space and equipment necessary to store and prepare special meals. Many persons lack the strength and energy required for meal preparation while recovering from illness. Emotional health may suffer as a result of chronic disease or terminal illness, causing a lack of appetite and disinterest in food preparation.

RESPONSIBILITY FOR NUTRITION CARE

Members of the health care team work together to ensure that the nutritional needs of patients are met. The roles of health professionals may vary among different institutions, and their responsibilities may sometimes overlap. Sometimes the patient's nutrition care is incorporated into the medical care plan developed by the entire health care team. Such plans, called **critical pathways**, outline coordinated plans of care for specific medical diagnoses, treatments, or procedures.

Physicians

Physicians are responsible for meeting all of a patient's medical needs, including nutrition. They prescribe **diet orders** and other instructions related to nutrition care, including referrals for nutrition assessment and dietary counseling. Physicians rely on nurses, registered dietitians, and other health professionals to alert them to nutrition problems, suggest strategies for handling nutrition care, and provide nutrition services.

Nurses

Nurses interact closely with patients and thus are in an ideal position to identify people who would benefit from nutrition services. Nurses often screen patients for nutrition problems and may participate in nutrition and dietary assessments. Nurses also provide direct nutrition care, such as encouraging patients to eat, finding practical solutions to food-related problems, recording a patient's food intake, and answering questions about special diets. As members of **nutrition support teams**, nurses are responsible for administering tube and intravenous feedings. In facilities that do not employ registered dietitians, nurses often assume responsibility for much of the nutrition care.

Registered Dietitians

Registered dietitians (■) are food and nutrition experts who are qualified to provide **medical nutrition therapy**. They conduct nutrition and dietary assessments; diagnose nutritional problems; develop, implement, and evaluate **nutrition care plans** (described later); plan and approve menus; and provide dietary counseling and nutrition education services. Registered dietitians may also manage food and cafeteria services in health care institutions.

Registered Dietetic Technicians

Registered dietetic technicians often work in partnership with registered dietitians and assist in the implementation and monitoring of nutrition services. Depending on their background and experience, they may screen patients for nutrition problems, develop menus and recipes, ensure appropriate meal delivery, monitor patients' food choices and intakes, and provide patient education and counseling. Dietetic technicians sometimes supervise foodservice operations and may have roles in purchasing, inventory, quality control, sanitation, or safety.

■ Reminder: A *registered dietitian (RD)* has completed the education and training specified by the Academy of Nutrition and Dietetics (or Dietitians of Canada), including an undergraduate degree in nutrition or dietetics, a supervised internship, and a national registration examination.

pressure sores: regions of skin and tissue that are damaged due to prolonged pressure on the affected area by an external object, such as a bed, wheelchair, or cast; vulnerable areas of the body include buttocks, hips, and heels. Also called *decubitus* (deh-KYU-bih-tus) *ulcers*.

critical pathways: coordinated programs of treatment that merge the care plans of different health practitioners; also called *clinical pathways*.

diet orders: specific instructions concerning dietary management; also called *diet prescriptions* or *nutrition prescriptions*.

nutrition support teams: health care professionals responsible for the provision of nutrients by tube feeding or intravenous infusion.

medical nutrition therapy: nutrition care provided by a registered dietitian; includes assessment of nutrition status, diagnosis of nutrition problems, development of nutrition care plans, and provision of dietary counseling and nutrition education.

nutrition care plans: strategies for meeting an individual's nutritional needs.

Other Health Care Professionals Other heath care professionals who may assist with nutrition care include pharmacists, physical therapists, occupational therapists, speech therapists, nursing assistants, home health care aides, and social workers. These individuals can be instrumental in alerting dietitians or nurses to nutrition problems or may share relevant information about a patient's health status or personal needs.

IDENTIFYING RISK FOR MALNUTRITION

To identify patients who are malnourished or at risk for malnutrition, a **nutrition screening** is conducted within 24 hours of a patient's admission to a hospital or other extended-care facility. A screening may also be included in certain types of outpatient services and community health programs. A nutrition screening involves collecting health-related data that can indicate the presence of protein-energy malnutrition (PEM) (■) or other nutrition problems. The screening should be sensitive enough to identify patients who require nutrition care but simple enough to be completed within 10 to 15 minutes. Usually a nurse, nursing assistant, registered dietitian, or dietetic technician performs and documents the screening.

The information collected in a nutrition screening varies according to the patient population, the type of care offered by the health care facility, and the patient's medical problems. Often included are the admitting diagnosis, physical measurements and laboratory test results obtained during the admission process, and information about diet and health status provided by the patient or caregiver. Table 13-1 lists examples of information that can help to identify individuals at risk of developing malnutrition. A number of screening and assessment tools that use different combinations of these variables have become popular in recent years; one such example is the Subjective Global Assessment, which combines elements of the medical history and physical examination to determine malnutrition risk (see Table 13-2). Briefer screening methods may use just two or three variables; for example, several tools screen for malnutrition risk by evaluating unintentional weight changes and reduced appetite or food intake.[2]

Nursing care plans often include **nursing diagnoses** that suggest the need for nutrition interventions (see Table 13-3). For example, a nursing diagnosis of "impaired swallowing" alerts the nurse to potential problems with food consumption and suggests the need for a modified diet. Such a diagnosis can accompany various medical disorders, including developmental disabilities, diseases involving the esophagus, and neurological conditions. Examples of relevant nursing diagnoses are provided throughout the clinical nutrition chapters.

TABLE 13-1 Criteria for Identifying Malnutrition Risk

Category	Specific Examples
Admission data	Age, medical diagnosis, severity of illness or injury
Anthropometric data	Height and weight, body mass index (BMI), unintentional weight changes, loss of muscle or subcutaneous fat
Functional assessment data	Low hand grip strength, general weakness, impaired mobility
Historical information	History of diabetes, renal disease, or other chronic illness; use of medications that can impair nutrition status; extensive dietary restrictions; food allergies or intolerances; requirement for nutrition support; depression, social isolation, or dementia
Laboratory test results	Blood test results that suggest presence of inflammation (such as low serum protein levels) or anemia
Signs and symptoms	Reduced appetite or food intake, problems that interfere with food intake (such as chewing or swallowing difficulties or nausea and vomiting), localized or general edema, presence of pressure sores

■ Reminder: *Protein-energy malnutrition* is a deficiency of protein and food energy and is characterized by loss of weight and muscle tissue.

nutrition screening: a brief assessment of health-related variables to identify patients who are malnourished or at risk for malnutrition.

nursing diagnoses: clinical judgments about actual or potential health problems that provide the basis for selecting appropriate nursing interventions.

TABLE 13-2 Subjective Global Assessment

The Subjective Global Assessment evaluates a person's risk of malnutrition by ranking key variables of the medical history and physical examination. These variables are each given an A, B, or C rating: A for well nourished, B for potential or mild malnutrition, and C for severe malnutrition. Patients are classified according to the final numbers of A, B, and C ratings.

Medical History

- Body weight changes: percentage change in past six months; weight change in past two weeks
- Dietary changes: suboptimal, low kcalorie, liquid diet, or starvation
- GI symptoms: nausea, diarrhea, vomiting, or anorexia for more than two weeks
- Functional ability: full capacity versus suboptimal, walking versus bedridden
- Degree of disease-related metabolic stress: low, medium, or high

Physical Examination

- Subcutaneous fat loss (triceps or chest)
- Muscle loss (quadriceps or deltoids)
- Ankle edema
- Sacral (lower spine) edema
- Ascites (abdominal edema)

Classification

A: Well nourished: if no significant loss of weight, fat, or muscle tissue and no dietary difficulties, functional impairments, or GI symptoms; also applies to patients with recent weight gain and improved appetite, functioning, or medical prognosis
B: Moderate malnutrition: if 5 to 10 percent weight loss, mild loss of muscle or fat tissue, decreased food intake, and digestive or functional difficulties that impair food intake; the B classification usually applies to patients with an even mix of A, B, and C ratings
C: Severe malnutrition: if more than 10 percent weight loss, severe loss of muscle or fat tissue, edema, multiple GI symptoms, and functional impairments

Sources: R. S. Gibson, *Principles of Nutritional Assessment* (New York: Oxford University Press, 2005), pp. 809–826; A. S. Detsky and coauthors, What is subjective global assessment of nutritional status? *Journal of Parenteral and Enteral Nutrition* 11 (1987): 8–13.

A nutrition screening may lead to a referral for nutrition care. The following section describes the next stage of the process: the method used by dietitians to address nutritional concerns.

THE NUTRITION CARE PROCESS

Registered dietitians use a systematic approach to medical nutrition therapy called the **nutrition care process.** The steps of this process include nutrition assessment, nutrition diagnosis, nutrition intervention, and nutrition monitoring and evaluation, as shown

TABLE 13-3 Nursing Diagnoses with Nutritional Implications

- Chronic confusion
- Chronic pain
- Constipation
- Diarrhea
- Disturbed body image
- Feeding self-care deficit
- Imbalanced nutrition: less than body requirements
- Imbalanced nutrition: more than body requirements
- Impaired dentition
- Impaired oral mucous membrane
- Impaired physical mobility
- Impaired swallowing
- Insufficient breast milk
- Nausea
- Readiness for enhanced nutrition
- Risk for aspiration
- Risk for deficient fluid volume
- Risk for unstable blood glucose level

Source: NANDA International, *Nursing Diagnoses: Definitions and Classification 2012–2014* (Oxford: Wiley-Blackwell, 2012).

nutrition care process: a systematic approach used by dietetics professionals to evaluate and treat nutrition-related problems.

FIGURE 13-2 The Nutrition Care Process

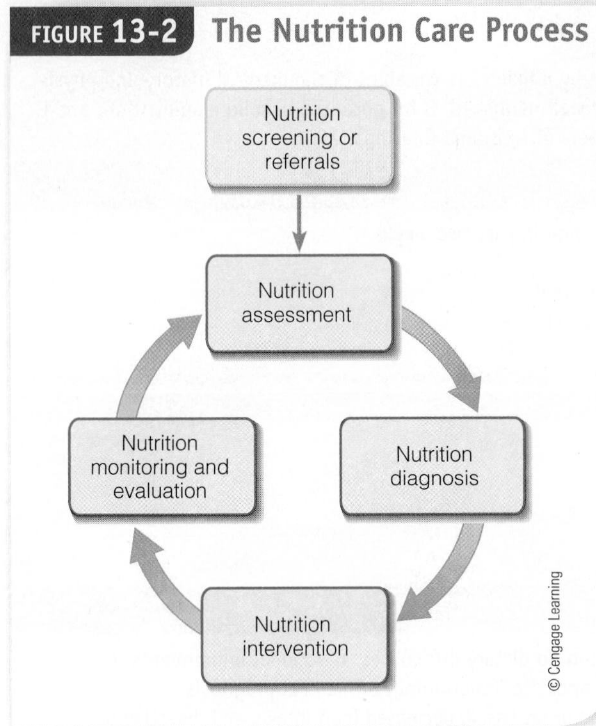

in Figure 13-2.[3] (■) Although the process is easiest to visualize as a series of steps, the steps are often revisited to reassess and revise diagnoses and intervention strategies. Each step of the nutrition care process must be documented in the medical record, providing a record for future reference and facilitating communication among members of the health care team.

Nutrition Assessment Nutrition assessment involves the collection and analysis of health-related data in order to identify specific nutrition problems and their underlying causes. The information may be obtained from the medical record, physical examination, laboratory analyses, medical procedures, an interview with the patient or caregiver, and consultation with other health professionals. The assessment data are used to develop a plan of action to prevent or correct energy or nutrient imbalances, or to determine whether a care plan is working. The second half of this chapter describes the components of nutrition assessment in detail.

Nutrition Diagnosis Each nutrition problem identified by the nutrition assessment receives a separate diagnosis.[4] Nutrition diagnoses, similar to nursing diagnoses, are formatted to include the specific nutrition problem, the etiology or cause, and signs and symptoms that provide evidence of the problem. (■) For example, a nutrition diagnosis might state, "Unintentional weight loss (*the problem*) related to insufficient kcaloric intake (*the etiology or cause*) as evidenced by a 10-pound weight loss (8 percent of body weight) in the past few months (*the sign or symptom*)." Like nursing diagnoses, a nutrition diagnosis can change during the course of an illness.

■ As a comparision, the *nursing process* consists of these steps:

1. Assessment
2. Nursing diagnosis
3. Outcome identification/planning
4. Implementation
5. Evaluation

■ This format is called a *PES statement* because it includes the *P*roblem, the *E*tiology, and the *S*igns and symptoms.

Nutrition Intervention After nutrition problems are identified, the appropriate nutrition care can be planned and implemented. A nutrition intervention may include counseling or education about appropriate dietary and lifestyle practices, a change in medication or other treatment, or adjustments in the meals offered to a hospital patient. To be successful, the intervention should consider an individual's food habits, lifestyle, and other personal factors. Goals are stated in terms of measurable outcomes; for example, goals for an overweight person with diabetes might include improvements in blood glucose levels and body weight. Other goals may be changes in the patient's dietary behaviors and lifestyle; for example, a person with diabetes may learn how to control carbohydrate intake and begin a regular exercise program. Chapter 14 gives additional information about nutrition intervention.

Nutrition Monitoring and Evaluation The effectiveness of the nutrition care plan must be evaluated periodically: the patient's progress should be monitored closely, and updated assessment data or diagnoses may require adjustments in goals or outcome measures. Sometimes a new situation alters nutritional needs; for example, a change in the medical treatment or a new medication may alter a person's tolerance to certain foods. The nutrition care plan must be flexible enough to adapt to the new situation.

If progress is slow or a patient is unwilling or unable to make the suggested changes, the care plan should be redesigned and take into account the reasons why the earlier plan was not successful. The new plan may need to include motivational techniques or additional patient education. If the patient remains unwilling to modify behaviors despite the expected benefits, the health practitioner can try again at a later time when the patient may be more receptive.

NURSING DIAGNOSIS

The nursing diagnosis *readiness for enhanced nutrition* is appropriate for a person who is willing to improve dietary practices.

Review Notes

- Illnesses and their treatments can affect food intake and nutrient needs, leading to malnutrition. In turn, poor nutrition status can influence the course of illness and reduce the effectiveness of medical treatments.
- The combined efforts of each member of the health care team ensure that patients receive optimal nutrition care.
- Nutrition screening identifies individuals who can benefit from nutrition assessment and follow-up care. Nursing diagnoses can also identify potential nutrition problems.
- The nutrition care process includes four interrelated steps: nutrition assessment, nutrition diagnosis, nutrition intervention, and nutrition monitoring and evaluation.

Nutrition Assessment

As described earlier in this chapter, a nutrition assessment provides the information needed for diagnosing nutrition problems and designing a nutrition care plan; follow-up assessments help to determine whether the care plan has been effective. Ideally, the assessment should be sensitive enough to detect subtle nutrition problems and specific enough to identify problem nutrients. The remainder of this chapter describes the types of information and measures that are most often included in a nutrition assessment.

HISTORICAL INFORMATION

Historical information provides valuable clues about nutrition status and nutrient requirements; it also reveals personal preferences that should be considered when developing a nutrition care plan. Table 13-4 summarizes the various types of historical data that contribute to a nutrition assessment. This information can be obtained from the medical record or by interviewing the patient or caregiver.

Medical History

Many medical problems and their treatments can interfere with food intake or require dietary changes; Table 13-5 lists examples. The medical history generally includes the family medical history as well; this information may reveal a person's genetic susceptibilities for diseases that can potentially be prevented with dietary and lifestyle changes.

TABLE 13-4 Historical Information Used in Nutrition Assessment[a]

Medical History	Medication and Supplement History	Personal and Social History	Food and Nutrition History
Age	Prescription drugs	Cognitive abilities	Food intake
Current complaint(s)	Over-the-counter drugs	Cultural/ethnic identity	Food availability
Past medical problems	Dietary and herbal supplements	Educational level	Recent weight changes
Ongoing medical treatments		Employment status	Dietary restrictions
Surgical history		Home/family situation	Food allergies or intolerances
Family medical history		Religious beliefs	Nutrition and health knowledge
Chronic disease risk		Socioeconomic status	Physical activity level and exercise habits
Mental/emotional health status		Use of tobacco, alcohol, or illegal drugs	

[a]Historical information is classified in different ways among medical institutions.

TABLE 13-5 Medical Problems Often Associated with Malnutrition

- Acquired immune deficiency syndrome (AIDS)
- Alcoholism
- Anorexia nervosa or bulimia
- Burns (extensive or severe)
- Cancer and cancer treatments
- Cardiovascular diseases
- Celiac disease
- Chewing or swallowing difficulties
- Chronic kidney disease
- Dementia or other mental illness
- Diabetes mellitus
- Feeding disabilities
- Infections
- Inflammatory bowel diseases
- Liver disease
- Pressure sores
- Surgery (major)
- Vomiting (prolonged or severe)

Medication and Supplement History

A number of medications can have detrimental effects on nutrition status, and some dietary components can alter the absorption or metabolism of drugs. Ingredients in dietary and herbal supplements can also interact with medications. Chapter 14 describes examples of notable diet-drug interactions that may need consideration when planning nutrition care.

Personal and Social History

Personal and social factors can influence food choices as well as a person's ability to manage health and nutrition problems. For example, cultural background or religious beliefs can affect food preferences. Financial concerns may restrict access to health care and nutritious foods. Some individuals may depend on others to prepare or procure food. An individual who lives alone or is depressed may eat poorly or be unable to follow complex dietary instructions. Use of tobacco or illegal drugs may alter food intake or have disruptive effects on health and nutrition status.

Food and Nutrition History

A food and nutrition history (often called a *diet history*) is a detailed account of a person's dietary practices. It includes information about food intake, lifestyle habits, and other factors that may influence food choices, such as food allergies or beliefs about nutrition and health. The procedure often includes an interview about recent food intake (for example, a *24-hour recall*) and a survey about usual food choices (such as a *food frequency questionnaire*). In the hospital setting, direct observation of patients' food intakes is helpful. The following section describes the most common methods of gathering food intake information.

DIETARY ASSESSMENT

Obtaining accurate food intake data is challenging, as the results may vary depending on the individual's memory and honesty and the assessor's skill and training. Each method has its strengths and weaknesses, so best results are obtained by using a combination of approaches. Table 13-6 summarizes the methods most commonly used as well as their most notable advantages and disadvantages.

24-hour dietary recall: a record of foods consumed during the previous day or in the past 24 hours; sometimes modified to include foods consumed in a typical day.

The 24-Hour Dietary Recall

The **24-hour dietary recall** is a guided interview in which an individual recounts all of the foods and beverages consumed during the previous day or in the past 24 hours. The interview includes questions about the times when meals or snacks were eaten, amounts consumed, and ways in which foods were prepared. Food models or measuring cups and spoons are used to help the individual visualize and describe the amounts consumed.

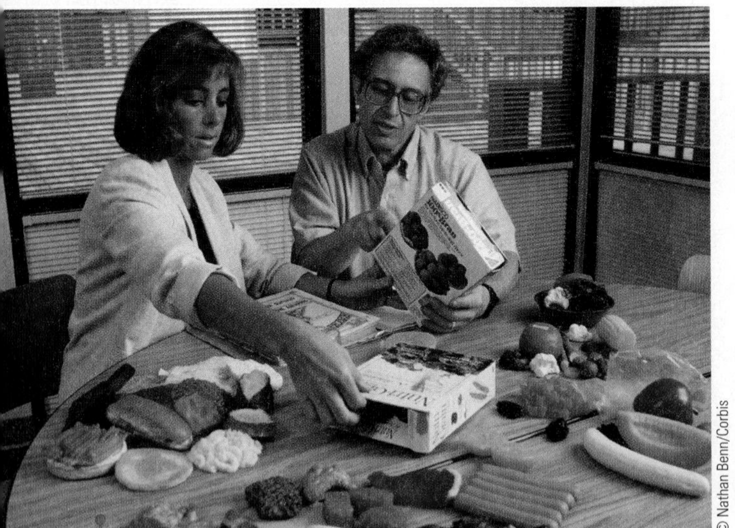

Food models and measuring utensils can help an individual visualize portion sizes.

The *multiple-pass method* is considered the most effective approach for obtaining an accurate list of foods consumed.[5] In this procedure, the interview includes four or five separate passes through the 24-hour period of interest. In the first pass, the respondent provides a "quick list" of foods consumed without prompts by the interviewer. The second pass is conducted to help the respondent remember foods that are often forgotten, such as beverages, bread, additions to foods (such as butter on toast), savory snacks, and sweets. A third and fourth pass elicits additional details about the foods consumed, such as the amounts eaten, preparation methods, and places where foods were obtained or consumed. A final pass is conducted to provide a final opportunity to recall foods and to probe for additional details. The entire multiple-pass interview can be conducted in about 30 to 45 minutes.

TABLE 13-6 Methods for Obtaining Food Intake Data

Method	Description	Advantages	Disadvantages
24-hour dietary recall	Guided interview in which the foods and beverages consumed in a 24-hour period are described in detail.	• Results are not dependent on literacy or educational level of respondent. • Interview occurs after food is consumed, so method does not influence dietary choices. • Results are obtained quickly; method is relatively easy to conduct. • Method does not require reading or writing ability.	• Process relies on memory. • Underestimation and overestimation of food intakes are common. • Food items that cause embarrassment (alcohol, desserts) may be omitted. • Data from a single day cannot accurately represent the respondent's usual intake. • Seasonal variations may not be addressed. • Skill of interviewer affects outcome.
Food frequency questionnaire	Written survey of food consumption during a specific period of time, often a one-year period.	• Process examines long-term food intake, so day-to-day and seasonal variability should not affect results. • Questionnaire is completed after food is consumed, so method does not influence food choices. • Method is inexpensive to administer.	• Process relies on memory. • Food lists often include common foods only. • Serving sizes are often difficult for respondents to evaluate without assistance. • Calculated nutrient intakes may not be accurate. • Food lists for the general population are of limited value in special populations. • Method is not effective for monitoring short-term changes in food intake.
Food record	Written account of food consumed during a specified period, usually several consecutive days. Accuracy is improved by including weights or measures of foods.	• Process does not rely on memory. • Recording foods as they are consumed may improve accuracy of food intake data. • Process is useful for controlling intake because keeping records increases awareness of food choices.	• Recording process itself influences food intake. • Underreporting and portion size errors are common. • Process is time-consuming and burdensome for respondent; requires high degree of motivation. • Method requires literacy and the physical ability to write. • Seasonal changes in diet are not taken into account.
Direct observation	Observation of meal trays or shelf inventories before and after eating; possible only in residential facilities.	• Process does not rely on memory. • Method does not influence food intake. • Method can be used to evaluate the acceptability of a prescribed diet.	• Process is possible only in residential situations. • Method is labor intensive.

After the day's intake is recounted, the interviewer can ask whether the intake that day was fairly typical and, if not, how it varied from the person's usual intake. Recall interviews may be conducted on several nonconsecutive days to get a better representation of a person's usual diet. A disadvantage of the 24-hour recall method is that it does not take into account fluctuations in food intake or seasonal variations. Moreover, food intakes are often underestimated because the process relies on an individual's memory and reporting accuracy.

Food Frequency Questionnaire A **food frequency questionnaire** surveys the foods and beverages regularly consumed during a specific time period. Some questionnaires are qualitative only: food lists contain common foods, organized by food group, with check boxes to indicate frequency of consumption. Other types of questionnaires can collect semiquantitative information by including portion sizes as well. Figure 13-3 shows a sample section of a semiquantitative questionnaire that surveys fruit intake over the previous year. Because the respondent is often asked to estimate food intakes over a one-year period, the results should not be affected by seasonal changes in diet. Conversely, a disadvantage of this method is its inability to determine recent changes in food intake. Another limitation is that the questionnaires typically include only common food items, so the accuracy of food intake data is reduced if an individual consumes atypical foods.

Some shortened food frequency questionnaires focus on food categories relevant to a person's medical condition. For example, a questionnaire designed to evaluate calcium intake may include only milk products, fortified foods, certain fruits and vegetables, and dietary supplements that contain calcium. A computer analysis can then quickly estimate the individual's calcium intake and compare it to recommendations.

food frequency questionnaire: a survey of foods routinely consumed. Some questionnaires ask about the types of food eaten and yield only qualitative information; others include questions about portions consumed and yield semiquantitative data as well.

food record: a detailed log of food eaten during a specified time period, usually several days; also called a *food diary*. A food record may also include information about medications, disease symptoms, and physical activity.

Food Record A **food record** is a written account of foods and beverages consumed during a specified time period, usually several consecutive days. Foods are recorded as they are consumed in order to obtain the most complete and accurate record possible; thus, the process does not rely on memory. A detailed food record includes the types and amounts of foods and beverages consumed, times of consumption, and methods of preparation. It may also include information about medication use, disease symptoms, and physical activity.

The food record can provide valuable information about food intake, as well as a person's response to and compliance with nutrition therapy. Unfortunately, food records require a great deal of time to complete, and people need to be highly motivated to keep accurate records. Another drawback is that the recording process itself may influence food intake. Furthermore, it is difficult to obtain accurate estimates of nutrient intakes in just a few days or even a week due to day-to-day and seasonal variations in food intake.

FIGURE 13-3 Sample Section of a Food Frequency Questionnaire

FRUIT	HOW OFTEN								HOW MUCH			
	Never or less than once per month	1 per mon.	2–3 per mon.	1 per week	2 per week	3–4 per week	5–6 per week	Every day	MEDIUM SERVING	YOUR SERVING SIZE S	M	L
EXAMPLE: Bananas	○	○	○	●	○	○	○	○	1 medium	○ 1/2	● 1	○ 2
Bananas	○	○	○	○	○	○	○	○	1 medium	○ 1/2	○ 1	○ 2
Apples, applesauce	○	○	○	○	○	○	○	○	1 medium or 1/2 cup	○ 1/2	○ 1	○ 2
Oranges (not including juice)	○	○	○	○	○	○	○	○	1 medium	○ 1/2	○ 1	○ 2
Grapefruit (not including juice)	○	○	○	○	○	○	○	○	1/2 medium	○ 1/4	○ 1/2	○ 1
Cantaloupe	○	○	○	○	○	○	○	○	1/4 medium	○ 1/8	○ 1/4	○ 1/2
Peaches, apricots (fresh, in season)	○	○	○	○	○	○	○	○	1 medium	○ 1/2	○ 1	○ 2
Peaches, apricots (canned or dried)	○	○	○	○	○	○	○	○	1 medium or 1/2 cup	○ 1/2	○ 1	○ 2
Prunes, or prune juice	○	○	○	○	○	○	○	○	1/2 cup	○ 1/4	○ 1/2	○ 1
Watermelon (in season)	○	○	○	○	○	○	○	○	1 slice	○ 1/2	○ 1	○ 2
Strawberries, other berries (in season)	○	○	○	○	○	○	○	○	1/2 cup	○ 1/4	○ 1/2	○ 1
Any other fruit, including kiwi, fruit cocktail, grapes, raisins, mangoes	○	○	○	○	○	○	○	○	1/2 cup	○ 1/4	○ 1/2	○ 1

Direct Observation In facilities that serve meals, food intakes can be directly observed and analyzed. This method can also reveal a person's food preferences, changes in appetite, and any problems with a prescribed diet. Although a useful means of discerning patients' intakes, direct observation requires regular and careful documentation and can be labor intensive and costly.

Nurses use direct observation to conduct patients' **kcalorie counts**, which are estimates of the food energy (and often protein) consumed by patients during a single day or several consecutive days. To perform a kcalorie count, the nurse records the dietary items that a patient is given at meals and subtracts the amounts remaining after meals are completed; this procedure allows an analysis of the caloric value of foods and beverages that are actually consumed.

ANTHROPOMETRIC DATA

Measures of body size, known as **anthropometric** measurements, can reveal problems related to both overnutrition and PEM. Height (or length) and weight are the most common anthropometric measures and are used to evaluate growth in children and nutrition status in adults. Other helpful values include skinfold measurements (described in Appendix E) and circumferences of the head, waist, and limbs.

Height (or Length) Poor growth in children can signify malnutrition. In adults, height measurements alone do not reflect current nutrition status but can be used for estimating a person's energy needs or appropriate body weight. Length is measured in infants and children younger than 24 months of age, and height is usually measured in older children and adults. (■) Length can also be measured in adults and children who cannot stand unassisted due to physical or medical reasons. The "How to" describes some standard techniques for measuring length and height.

■ *Length* is measured while a person is recumbent (lying down), whereas *height* is measured while a person is standing upright.

kcalorie counts: estimates of food energy (and often protein) consumed by patients for one or more days.

anthropometric (AN-throw-poe-MEH-trik): related to physical measurements of the human body, such as height, weight, body circumferences, and percentage of body fat.

HOW TO Measure Length and Height

To improve the accuracy of length and height measurements, keep the following in mind:

- Always measure—never ask! Self-reported heights are less accurate than measured heights. If height is not measured, document that the height is self-reported.
- Measure the length of infants and young children by using a measuring board with a fixed headboard and a movable footboard. It generally takes two people to measure length: one person gently holds the infant's head against the headboard; the other straightens the infant's legs and moves the footboard to the bottom of the infant's feet.
- Measure height next to a wall on which a nonstretchable measuring tape or board has been fixed. Ask the person to stand erect without shoes and with heels together. The person's eyes and head should be facing forward, with heels, buttocks, and shoulder blades touching the wall. Place a ruler or other flat, stiff object on the top of the head at a right angle to the wall and carefully note the height measurement. Immediately record length and height measurements to the nearest ⅛ inch or 0.1 centimeter.
- For evaluating growth rate in young children, use the appropriate growth chart (Appendix E) when plotting results. If length is measured, use the growth chart for children between 0 and 36 months; if height is measured, use the chart for individuals between 2 and 20 years.
- Higher values are obtained from supine measurements than from vertical height measurements due to gravity.

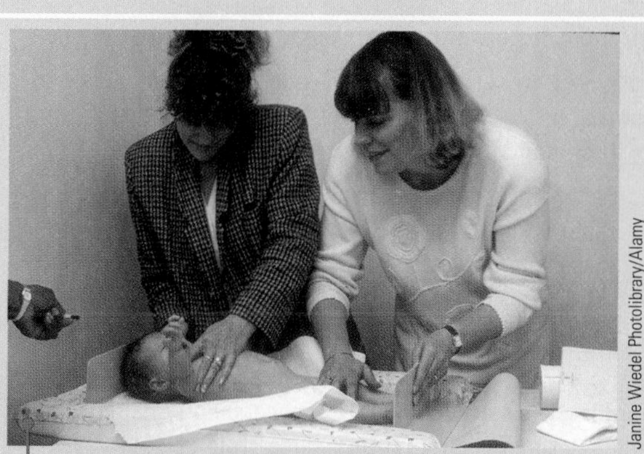

Janine Wiedel Photolibrary/Alamy

It generally takes two people to measure the length of an infant.

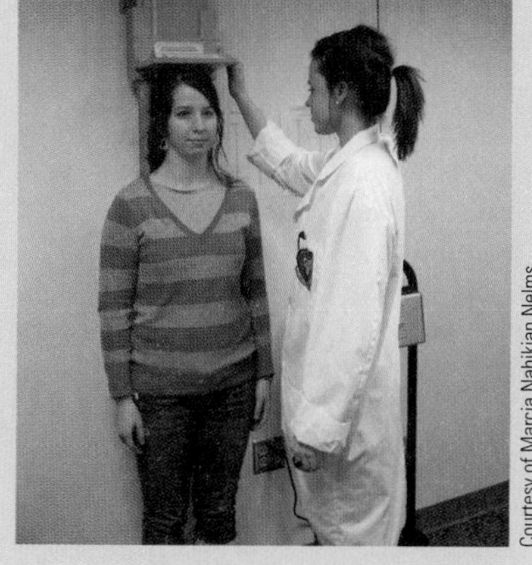

Courtesy of Marcia Nahikian Nelms

Standing erect allows for an accurate height measurement.

In adults who are bedridden or unable to stand, height can be estimated from equations that include either the knee height or the full arm span, both of which correlate well with height.[6] Knee height, which extends from the heel to the top of the knee when the leg is bent at a 90-degree angle, can be measured in either a sitting or supine position with a knee-height caliper; specific formulas are available for different age, gender, and ethnic groups. The full arm span is the distance from the tip of one middle finger to the other while the arms are extended horizontally. In children with disabilities that affect stature, alternative measures of linear growth include the full arm span, lower-leg lengths (knee to heel, similar to the knee height measure), and upper-arm lengths (shoulder to elbow), all of which can be compared with reference percentiles.

Body Weight During clinical care, health care providers monitor body weights closely: weight changes may reflect changes in body water due to illness, and an involuntary weight loss can be a sign of PEM. Body weights can be compared with healthy ranges on height-weight tables and growth charts or used to calculate the body mass index (BMI). (■) A healthy body weight typically falls within a BMI range of 18.5 to 25; thus, an appropriate body weight can usually be estimated by using a BMI table or graph (see the inside back cover of this book).[7] The "How to" includes suggestions for improving the accuracy of weight measurements.

■ Reminder: $BMI = \frac{\text{weight (kg)}}{\text{height (m)}^2}$

Head Circumference A measurement of head circumference helps to assess brain growth and malnutrition in children up to three years of age, although this measure is not necessarily reduced in a malnourished child. Head circumference values can also track brain development in premature and small-for-gestational-age infants. To measure head circumference, the assessor encircles the largest circumference measure of a

HOW TO Measure Weight

Tips for measuring weight include:

- Always measure—never ask! Self-reported weights are often inaccurate. If weight is not measured, document that the weight is self-reported.
- Valid weight measurements require scales that have been carefully maintained, calibrated, and checked for accuracy at regular intervals. Beam balance and electronic scales are the most accurate. Bathroom scales are inaccurate and inappropriate for clinical use.
- Measure an infant's weight with a scale that allows the infant to sit or lie down. The tray should be large enough to support an infant or young child up to 40 pounds, and weight graduations should be in ½-ounce or 10-gram increments. For accurate results, weigh infants without clothes or diapers. Excessive movement by the infant can reduce accuracy.
- Children who can stand are weighed in the same way as adults, using beam balance or electronic scales with platforms large enough for standing comfortably. If repeated weight measurements are needed, each weighing should take place at the same time of day (preferably before breakfast), in the same amount of clothing, after the person has voided, and using the same scale. Record weights to the nearest ¼ pound or 0.1 kilogram.
- Special scales and hospital beds with built-in scales are available for weighing people who are bedridden.

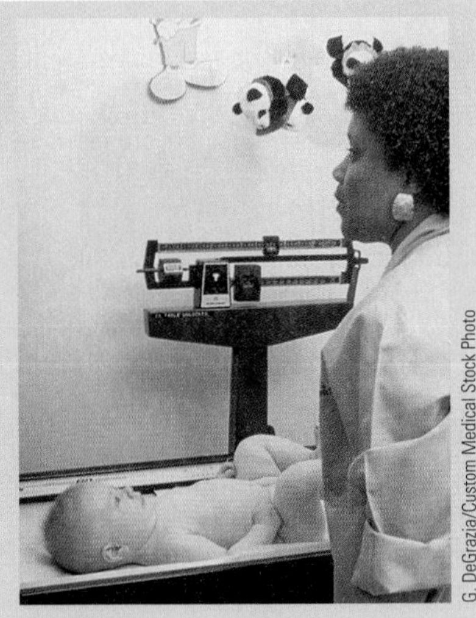

G. DeGrazia/Custom Medical Stock Photo

Infants are weighed on scales that allow them to sit or lie down.

© Image Source/Jupiter Images

Beam balance scales can provide accurate weight measurements in older children and adults.

child's head with a nonstretchable measuring tape: the tape is placed just above the eyebrows and ears, and around the occipital prominence at the back of the head (see the photo). The measurement is read to the nearest ⅛ inch or 0.1 centimeter.

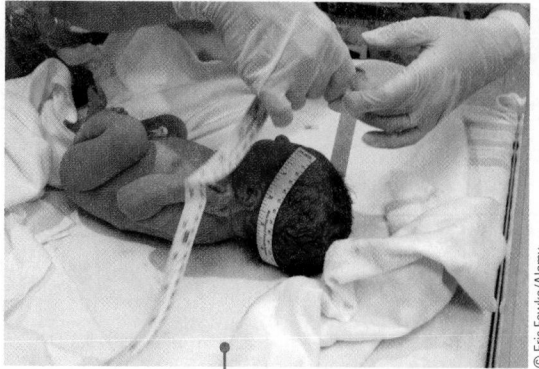

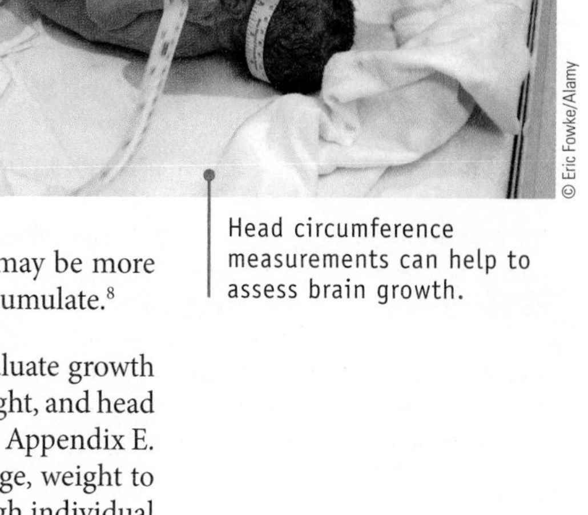
© Eric Fowke/Alamy

Head circumference measurements can help to assess brain growth.

Circumferences of Waist and Limbs Circumferences of the waist and limbs are useful for evaluating body fat and muscle mass, respectively. Waist circumference correlates with intra-abdominal fat and can help in assessing overnutrition. Circumferences of the mid-upper arm, mid-thigh, and mid-calf regions can help in evaluating the effects of illness, aging, and PEM on skeletal muscle tissue. For improved accuracy, circumference measurements are often used together with skinfold measurements to correct for the subcutaneous fat in limbs. In individuals with edema, upper body measurements may be more useful than those in the lower body, where excess body water tends to accumulate.[8]

Anthropometric Assessment in Infants and Children To evaluate growth patterns, the nurse takes periodic measurements of height (or length), weight, and head circumference and plots them on growth charts, such as those provided in Appendix E. The most commonly used growth charts compare height (or length) to age, weight to age, head circumference to age, weight to length, and BMI to age. Although individual growth patterns vary, a child's growth will generally stay at about the same percentile throughout childhood; a sharp drop in a previously steady growth pattern suggests malnutrition. Growth patterns that fall below the 5th percentile may also be cause for concern, although genetic influences must be considered when interpreting low values. Growth charts with BMI-for-age percentiles can be used to assess risk of underweight and overweight in children over two years of age: the 5th and 85th percentiles are used as cutoffs to identify children who may be malnourished or overweight, respectively.[9]

Anthropometric Assessment in Adults To evaluate the nutritional risks associated with illness, clinicians monitor both the total reduction in body weight and the rate of weight loss over time. Weight changes need to be evaluated carefully, however: although unintentional weight loss can indicate malnutrition, weight gain may result from fluid retention (■) rather than recovery of muscle tissue or overnutrition. Moreover, fluid retention can mask the weight loss associated with PEM.

Weight data are often expressed as a percentage of usual body weight (%UBW) or ideal body weight (%IBW). The %UBW is more effective than %IBW for interpreting weight changes that occur in underweight, overweight, or obese individuals. In overweight persons, the %IBW may fail to identify significant weight loss. Conversely, in underweight individuals, the %IBW can overstate the degree of weight loss due to illness. The "How to" (p. 392) describes how to estimate %UBW and %IBW, and Table 13-7 shows how to interpret these values. As noted previously, the rate of weight loss must be considered as well as the amount of loss; for example, an involuntary weight loss of more than 10 percent within a six-month period suggests risk of PEM.[10] Table 13-8 provides additional information about the rates of weight loss that indicate nutritional risk.

Some illnesses discussed in later chapters are associated with losses in muscle tissue that resist nutrition intervention. In older adults, losses in both muscle tissue and

TABLE 13-7 Body Weight and Nutritional Risk

%UBW	%IBW	Nutritional Risk
85–95	80–90	Risk of mild malnutrition
75–84	70–79	Risk of moderate malnutrition
<75	<70	Risk of severe malnutrition

■ Fluid retention often accompanies worsening disease in patients with heart failure, liver cirrhosis, and kidney failure.

HOW TO Estimate and Evaluate %UBW and %IBW

%UBW: To estimate %UBW, compare an individual's current weight with the weight that the person generally maintains:

$$\%UBW = \frac{\text{current weight}}{\text{usual weight}} \times 100$$

For example, if a man loses 32 pounds during illness and his usual weight is 180 pounds, his current weight would be 148 pounds. These values can be incorporated into the previous equation:

$$\%UBW = \frac{148}{180} \times 100 = 82.2\%$$

The man in this example weighs 82.2 percent of his usual weight. A look at Table 13-7 shows that a person who is at 82 percent of UBW may be moderately malnourished.

%IBW: To estimate %IBW, compare an individual's current weight with a reasonable (ideal) weight from a BMI table or other appropriate reference:

$$\%IBW = \frac{\text{current weight}}{\text{ideal weight}} \times 100$$

For example, suppose you wish to calculate the %IBW for a woman who is 5 feet 8 inches tall and weighs 116 pounds. The midpoint of the healthy BMI range is approximately 22, so using a BMI table (as shown on the inside back cover of this book), you estimate that a reasonable weight for this woman would be about 144 pounds:

$$\%IBW = \frac{116}{144} \times 100 = 80.6\%$$

The woman in this example weighs about 80.6 percent of her ideal body weight. A look at Table 13-7 suggests that, at 80.6 percent of IBW, she may be mildly malnourished. Keep in mind that the calculation of "ideal body weight" is somewhat arbitrary because the BMI table and various other references provide a range of weights for individuals of a given height.

height are common even though body weights may remain stable. Thus, clinicians may include skinfold and limb circumference measurements in a nutrition assessment to help them identify changes in body composition that need to be addressed in the treatment plan.

BIOCHEMICAL ANALYSES

Biochemical data provide information about protein-energy nutrition, vitamin and mineral status, fluid and electrolyte balances, and organ function. Most tests are based on analyses of blood and urine samples, which contain proteins, nutrients, and metabolites that reflect nutrition and health status. Repeated measures are more helpful than single values, as serial data can indicate whether a condition is improving or worsening. Table 13-9 lists and describes some common blood tests (■) that may be helpful in a nutrition assessment. Laboratory tests relevant to specific diseases will be discussed in the chapters that follow.

Interpreting laboratory values can be challenging because a number of factors may influence the test results. For example, serum protein values can be affected by fluid imbalances, (■) infections, inflammation, pregnancy, and various other factors. Similarly, serum levels of vitamins and minerals are often poor indicators of nutrient deficiency because the values are affected by multiple variables; therefore, a variety of tests is generally needed to diagnose a nutrition problem. Taken together with other assessment data, however, laboratory test results help to present a clearer picture than is possible to obtain otherwise.

■ Blood test results are reported in terms of either *plasma* or *serum* levels. *Plasma* is the yellow fluid that remains after cells are removed and still contains clotting factors. *Serum* is the fluid remaining after both cells and clotting factors are removed.

■ Fluid retention can cause lab results that are deceptively low. Dehydration may cause lab results to be deceptively high.

■ *Metabolic stress* may be a consequence of infection, injury, illness, inflammation, or surgery.

TABLE 13-8 Rate of Involuntary Weight Loss Associated with Nutritional Risk

% Weight Loss[a]	Time Period
>2%	1 week
>5%	1 month
>7.5%	3 months
>10%	6 months

[a] $\%\text{ weight loss} = \frac{\text{usual weight} - \text{current weight}}{\text{usual weight}} \times 100$

Serum Proteins Serum protein levels can sometimes help in the assessment of protein-energy status, but the levels may fluctuate for other reasons as well.[11] For example, serum proteins are synthesized in the liver, so blood levels of these proteins can reflect liver function. Metabolic stress (■) alters serum proteins because the liver responds by increasing its synthesis of some proteins and reducing the synthesis of others. Values may also be influenced by hydration status, pregnancy, kidney function, zinc status, blood loss, and some medications. Because serum proteins are affected by so many factors, their values must be considered along with other data to evaluate health and nutrition status.

Albumin Albumin is the most abundant serum protein, and its levels are routinely monitored in hospital patients to help gauge the severity of illness.[12] Although many medical conditions influence albumin, it is slow to reflect changes in nutrition status because of its large body pool and slow rate of degradation. (■) In people with chronic PEM, albumin levels remain normal for a long period despite significant protein depletion, and levels fall only after prolonged malnutrition. Likewise, albumin levels increase slowly when malnutrition is treated, so albumin is not a sensitive indicator of effective treatment.

■ In blood tests, the term *half-life* describes the length of time that a substance remains in plasma. The albumin in plasma has a half-life of 14 to 20 days, meaning that half of the amount circulating in plasma is degraded in this time period.

Transferrin Transferrin is an iron-transport protein, and its concentrations respond to iron status, PEM, and various illnesses. Transferrin levels rise as iron status worsens and fall as iron status improves, so using transferrin values to evaluate protein-energy status is difficult if an iron deficiency is also present. Transferrin is degraded more rapidly than albumin, (■) but its levels change relatively slowly in response to nutrition therapy.

■ Transferrin's half-life in plasma is approximately 8 to 10 days.

TABLE 13-9 Routine Laboratory Tests with Nutritional Implications

This table presents a partial listing of some uses of commonly performed lab tests that have implications for nutritional problems.

Laboratory Test	Acceptable Range	Description
Hematology		
Red blood cell (RBC) count	Male: 4.3–5.7 million/μL Female: 3.8–5.1 million/μL	RBC number; helps with anemia diagnosis.
Hemoglobin (Hb)	Male: 13.5–17.5 g/dL Female: 12.0–16.0 g/dL	RBC hemoglobin content; helps with anemia diagnosis.
Hematocrit (Hct)	Male: 39–49% Female: 35–45%	Percent RBC volume in blood; helps with anemia diagnosis.
Mean corpuscular volume (MCV)	80–100 fL	RBC size; helps to distinguish microcytic and macrocytic anemia.
Mean corpuscular hemoglobin concentration (MCHC)	31–37% Hb/cell	RBC Hb concentration; helps with diagnosis of iron-deficiency anemia.
White blood cell (WBC) count	4500–11,000 cells/μL	WBC number; may indicate immune status, infection, or inflammation.
Serum Proteins		
Total protein	6.4–8.3 g/dL	Levels are not highly sensitive or specific to disease; may reflect body protein content, illness, infection, inflammation, changes in hydration or metabolism, pregnancy, or use of certain medications.
Albumin	3.4–4.8 g/dL	Levels may reflect illness or PEM; slow to respond to improvement or worsening of disease.
Transferrin	200–400 mg/dL >60 yr: 180–380 mg/dL	Levels may reflect illness, PEM, or iron deficiency; slightly more sensitive to changes in health status than albumin.
Prealbumin (transthyretin)	10–40 mg/dL	Levels may reflect illness or PEM; more responsive to changes in health status than albumin or transferrin.
C-reactive protein	68–8200 ng/mL	Elevated levels may indicate inflammation or disease.

continued

TABLE 13-9 Routine Laboratory Tests with Nutritional Implications, *continued*

Laboratory Test	Acceptable Range	Description
Serum Enzymes		
Creatine kinase (CK)	Male: 38–174 U/L Female: 26–140 U/L	Different forms are found in the muscle, brain, and heart; elevated levels may indicate a heart attack, brain tissue damage, or skeletal muscle injury.
Lactate dehydrogenase (LDH)	208–378 U/L	Found in many tissues; specific types may be elevated after a heart attack, lung damage, or liver disease.
Alkaline phosphatase	>20 yr: 25–100 U/L	Found in many tissues; often measured to evaluate liver function.
Aspartate aminotransferase (AST, formerly SGOT)	10–30 U/L	Elevated levels may indicate liver disease or liver damage; somewhat increased after muscle injury.
Alanine aminotransferase (ALT, formerly SGPT)	Male: 10–40 U/L Female: 7–35 U/L	Elevated levels may indicate liver disease or liver damage; somewhat increased after muscle injury.
Serum Electrolytes		
Sodium	136–146 mEq/L	Helps with assessment of hydration status or neuromuscular, kidney, and adrenal function.
Potassium	3.5–5.1 mEq/L	Helps with assessment of acid-base balance or kidney function; can also detect potassium imbalances.
Chloride	98–106 mEq/L	Helps with assessment of hydration status or detection of acid-base and electrolyte imbalances.
Other		
Glucose (fasting)[a]	Adult: 74–106 mg/dL >60 yr: 80–115 mg/dL	Helps with diagnosis of glucose intolerance, diabetes mellitus, and hypoglycemia; also used for monitoring diabetes treatment.
Glycated hemoglobin (HbA_{1c})	4.0–6.0% of total Hb	Used for monitoring long-term blood glucose control (approximately 1 to 3 months prior).
Blood urea nitrogen (BUN)	6–20 mg/dL	Primarily used for monitoring kidney function; value altered by liver failure, dehydration, or shock.
Uric acid	Male: 3.5–7.2 mg/dL Female: 2.6–6.0 mg/dL	Used for detection of gout or changes in kidney function; levels affected by age and diet and vary among different ethnic groups.
Creatinine (serum or plasma)	Male: 0.7–1.3 mg/dL Female: 0.6–1.1 mg/dL	Used for monitoring renal function.

[a]Fasting glucose levels that repeatedly exceed 100 mg/dL suggest prediabetes.
Note: μL = microliter; dL = deciliter; fL = femtoliter; ng = nanogram; U/L = units per liter; mEq = milliequivalents.

Source: L. Goldman and A. I. Schafer, coeditors, *Goldman's Cecil Medicine* (Philadelphia: Saunders, 2012).

Prealbumin and Retinol-Binding Protein Levels of prealbumin (also called transthyretin) and retinol-binding protein decrease rapidly during PEM and respond quickly to improved protein intakes. (■) Thus, these proteins are more sensitive than albumin to short-term changes in protein status. Although sometimes used to evaluate malnutrition risk or improvement in nutrition status, they are more expensive to measure than albumin so they are not routinely included during nutrition assessment. Like other serum proteins, their usefulness is somewhat limited because they are affected by a number of different factors, including metabolic stress, zinc deficiency, and various medical conditions.

■ Half-lives of prealbumin and retinol-binding protein are 2 to 3 days and 12 hours, respectively.

PHYSICAL EXAMINATION

As with other assessment methods, interpreting physical signs of malnutrition requires skill and clinical judgment. Most physical signs are nonspecific; they can reflect any of several nutrient deficiencies, as well as conditions unrelated to nutrition. For example, cracked lips may be caused by several B vitamin deficiencies but may also result from sunburn, windburn, or dehydration. Dietary and laboratory data are usually needed as additional evidence to confirm suspected nutrient deficiencies.

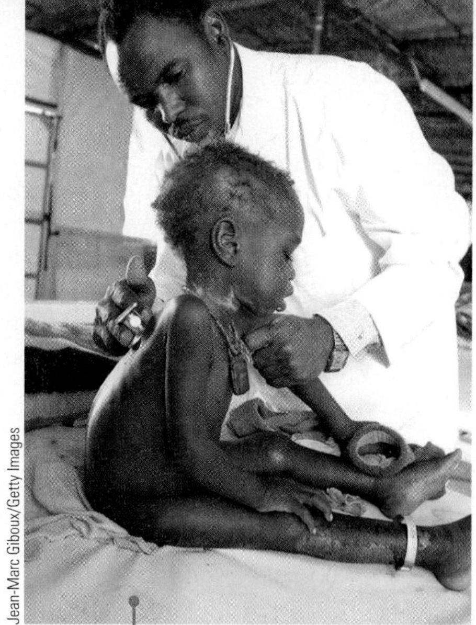
Jean-Marc Giboux/Getty Images

In a child with kwashiorkor, physical signs of malnutrition may include sparse, brittle hair; loss of hair color; a swollen abdomen; and dermatitis.

Clinical Signs of Malnutrition Signs of malnutrition tend to appear most often in parts of the body where cell replacement occurs at a rapid rate, such as the hair, skin, and digestive tract (including the mouth and tongue). Table 13-10 lists some clinical signs of nutrient deficiencies. Many of the symptoms listed occur only in advanced stages of deficiency. Chapters 8 and 9 provide additional examples of clinical signs that develop during nutrient imbalances.

Hydration Status As mentioned previously, fluid imbalances may accompany some illnesses and can also result from the use of certain medications. Thus, recognizing the signs of fluid retention or dehydration is necessary for the correct interpretation of blood test results and the body weight measurement.

TABLE 13-10 Clinical Signs of Nutrient Deficiencies

Body System	Acceptable Appearance	Signs of Malnutrition	Other Causes of Abnormalities
Hair	Shiny, firm in scalp	Dull, brittle, dry, loose; falls out (PEM); corkscrew hair (vitamin C)	Excessive hair bleaching; hair loss from aging, chemotherapy, or radiation therapy
Eyes	Bright; clear; shiny; pink, moist membranes; adjust easily to light	Pale membranes (iron); spots, dryness, night blindness (vitamin A); redness at corners of eyes (B vitamins)	Anemia that is unrelated to nutrition; eye disorders; allergies; aging
Lips	Smooth	Dry, cracked, or with sores in the corners of the lips (B vitamins)	Sunburn, windburn, excessive salivation from ill-fitting dentures or various disorders
Mouth and gums	Oral tissues without lesions, swelling, or bleeding; red tongue; normal sense of taste; teeth without caries; ability to chew and swallow	Bleeding gums (vitamin C); smooth or magenta tongue (B vitamins), poor taste sensation (zinc)	Medications, periodontal disease (poor oral hygiene)
Skin	Smooth, firm, good color	Poor wound healing (PEM, vitamin C, zinc); dry, rough, lack of fat under skin (essential fatty acids, PEM, vitamin A, B vitamins); bruising or bleeding under skin (vitamins C and K); pale (iron)	Poor skin care, diabetes mellitus, aging, medications
Nails	Smooth, firm, uniform, pink	Ridged (PEM); spoon shaped, pale (iron)	—
Other	—	Dementia, peripheral neuropathy (B vitamins); swollen glands at front of neck (PEM, iodine); bowed legs (vitamin D)	Disorders of aging (dementia), diabetes mellitus (peripheral neuropathy)

Fluid retention (also called *edema*) may result from malnutrition, infection, injury, or the use of certain medications. It can be caused by impaired blood circulation or organ dysfunction and frequently accompanies disorders of the heart, blood vessels, liver, kidneys, and lungs. Physical signs of fluid retention include weight gain, facial puffiness, swelling of limbs, abdominal distention, and tight-fitting shoes.

Dehydration can result from vomiting, diarrhea, fever, sweating, excessive urination, blood loss, and skin injuries or burns (due to fluid losses through skin lesions). Risk of dehydration is especially high in older adults, who have a reduced thirst response and various other impairments in fluid regulation.[13] Symptoms of dehydration include thirst, weight loss, dry skin or mouth, reduced skin tension, dark yellow or amber urine, and low urine volume.

Functional Assessment Nutrient deficiencies sometimes impair physiological functions, so health care providers may conduct tests or procedures to help them evaluate some aspects of malnutrition. For example, both PEM and zinc deficiency can depress immunity, which can be evaluated by testing the skin's response to antigens that cause redness and swelling when immune function is adequate. Muscle weakness due to **wasting** (loss of muscle tissue) can be assessed by testing hand-grip strength. Exercise tolerance, which is reduced in heart and lung disorders, may be evaluated using a treadmill or cycle ergometer. The accompanying Case Study can help you review the different components of a nutrition assessment.

wasting: the gradual atrophy (loss) of body tissues; associated with protein-energy malnutrition or chronic illness.

Case Study: Nutrition Screening and Assessment

Lisa Sawrey is an 80-year-old retired businesswoman who has been a widow for 10 years. She uses a walker and has poorly fitting dentures. She was recently admitted to the hospital with pneumonia and also has congestive heart failure and diabetes. She routinely takes several medications to control her blood glucose levels, hypertension, and heart function. In addition to these medications, the physician has recently ordered antibiotics to treat the pneumonia. During an initial nutrition screening, Mrs. Sawrey stated that she had been eating very poorly over the past two weeks. She said that she usually weighs about 125 pounds—a fact that was documented in her medical chart from a previous visit. Although she felt she was losing weight, she didn't know how much weight she may have lost or when she started losing weight. Upon admission to the hospital, Mrs. Sawrey weighed 110 pounds and was 5 feet 3 inches tall. Her serum albumin level was 3.0 grams per deciliter. A physical exam revealed edema, and several other laboratory tests confirmed that she was retaining fluid. As a result of the nutrition screening, Mrs. Sawrey was referred to a nurse for a nutrition assessment.

1. From the brief description provided, which items in Mrs. Sawrey's medical history, personal and social history, and food and nutrition history might alert the nurse that this patient is at risk of malnutrition?
2. Identify a healthy body weight for Mrs. Sawrey, and calculate her %UBW and %IBW. What do the results reveal? What effect does fluid retention have on Mrs. Sawrey's weight?
3. How might fluid retention alter Mrs. Sawrey's serum protein levels? What physical symptoms may have suggested that she was retaining excess fluid?
4. What tools can be used to estimate Mrs. Sawrey's usual food intake? What medical, physical, and personal factors are likely to influence her diet?
5. Describe other types of assessment information that may help the nurse determine whether Mrs. Sawrey should be referred to a registered dietitian.

Review Notes

- Nutrition assessments include historical information, anthropometric data, biochemical analyses, and a physical examination. Historical information includes the medical history, medication and supplement history, personal and social history, and food and nutrition history.
- Health care providers assess food intake using 24-hour dietary recall interviews, food frequency questionnaires, food records, and direct observation.
- Anthropometric measurements help clinicians evaluate growth patterns, overnutrition and undernutrition, and body composition.
- Biochemical analyses help in the assessment of nutrient imbalances but are also influenced by various other medical problems.
- A physical examination allows the assessor to detect signs of nutrient deficiency, fluid imbalances, and functional impairments that are related to nutritional problems.

Self Check

1. Mr. Hom experiences loss of appetite, difficulty swallowing, and mouth pain as a consequence of illness. Mr. Hom is at risk of malnutrition due to:
   a. altered metabolism.
   b. reduced food intake.
   c. altered excretion of nutrients.
   d. altered digestion and absorption.

2. Because of their central role in health care, nurses are well positioned for:
   a. calculating patients' nutrient needs.
   b. providing medical nutrition therapy.
   c. conducting complete nutrition assessments.
   d. identifying patients at risk for malnutrition.

3. Of the following data collected during a nutrition screening, which item does *not* place the person at risk for malnutrition?
   a. Having a health problem that is frequently associated with PEM
   b. Using prescription medications that affect nutrient needs
   c. Residing with a spouse in a middle-income neighborhood
   d. Significantly reducing food intake over the past five or more days

4. The nutrition care process is a systematic approach for:
   a. identifying the nutrient content of foods.
   b. ordering special diets.
   c. conducting nutrition screening.
   d. identifying and meeting the nutrition needs of patients.

5. To conduct complete nutrition assessments, clinicians rely on several sources of information, which include all of the following *except:*
   a. nutrition care plans.
   b. body measurements.
   c. medical, medication, and social histories.
   d. biochemical analyses.

6. Which dietary assessment method does a nurse use to conduct a kcalorie count?
   a. 24-hour recall interview
   b. Food frequency questionnaire
   c. Food record
   d. Direct observation

7. The %UBW of a person who weighs 135 pounds and has a usual body weight of 150 pounds is:
   a. 111 percent.
   b. 90 percent.
   c. 86 percent.
   d. 74 percent.

8. A malnourished, acutely ill patient has just begun to eat after days without significant amounts of food. Which of the following blood test results would change most quickly as the patient's nutrition and health status improves?
   a. Albumin
   b. Transferrin
   c. Serum electrolytes
   d. Retinol-binding protein

9. Which sign of PEM would be unlikely to show up in a physical examination?
   a. Low serum protein levels
   b. Dull, brittle hair
   c. Poor wound healing
   d. Wasting

10. Fluid retention can cause all of the following effects *except:*
    a. weight gain.
    b. facial puffiness.
    c. weight loss.
    d. tight-fitting shoes.

Answers to these questions can be found in Appendix H. For more chapter review: Access an interactive eBook, chapter-specific interactive learning tools, including flashcards, quizzes, videos, and more in your Nutrition CourseMate, accessed through CengageBrain.com.

Clinical Applications

1. Describe the potential nutritional implications of these findings from a patient's medical, personal, and social histories: age 78, lives alone, recently lost spouse, uses a walker, has no natural teeth or dentures, has a history of hypertension and diabetes, uses medications that cause frequent urination.

2. Calculate the %UBW and %IBW for a man who is 5 feet 11 inches tall with a current weight of 150 pounds and a usual body weight of 180 pounds. What additional information do you need to interpret the implications of his weight loss?

3. Nurses and nurses' aides frequently shoulder much of the responsibility for collecting food intake data for kcalorie counts because they typically deliver food trays and snacks and later retrieve them. Why is it important to verify and record both what the patient receives (foods and amounts) and the foods that remain uneaten? When might patients be enlisted in the collection of food intake data, and when might such a course be unwise?

Nutrition on the Net

For further study of topics covered in this chapter, access these websites.

- Learn more about nursing diagnoses at the website of NANDA International:
  www.nanda.org
- Visit the website of the Joint Commission to learn about the accreditation of health care institutions:
  www.jointcommission.org
- Learn about the Mini Nutritional Assessment, a nutrition screening and assessment tool developed by Nestle Nutrition Institute, at this website:
  www.mna-elderly.com
- Analyze your diet by using the "Super Tracker" tool at this website:
  www.choosemyplate.gov

Notes

1. J. V. White and coauthors, Consensus statement of the Academy of Nutrition and Dietetics/American Society for Parenteral and Enteral Nutrition: Characteristics recommended for the identification and documentation of adult malnutrition (undernutrition), *Journal of the American Dietetic Association* 112 (2012): 730–738.
2. P. Charney and M. Marian, Nutrition screening and nutrition assessment, in P. Charney and A. M. Malone, eds., *ADA Pocket Guide to Nutrition Assessment* (Chicago: American Dietetic Association, 2009), pp. 1–19.
3. Writing Group of the Nutrition Care Process/Standardized Language Committee, Nutrition care process and model part I: The 2008 update, *Journal of the American Dietetic Association* 108 (2008): 1113–1117; K. Lacey and E. Pritchett, Nutrition care process and model: ADA adopts road map to quality care and outcomes management, *Journal of the American Dietetic Association* 103 (2003): 1061–1072.
4. Writing Group of the Nutrition Care Process/Standardized Language Committee, 2008.
5. F. E. Thompson and A. F. Subar, Dietary assessment methodology, in A. M. Coulston and C. J. Boushey, eds., *Nutrition in the Prevention and Treatment of Disease* (Burlington, MA: Elsevier Academic Press, 2008), pp. 3–39; A. J. Moshfegh and coauthors, The U.S. Department of Agriculture automated muliple-pass method reduces bias in the collection of energy intakes, *American Journal of Clinical Nutrition* 88 (2008): 324–332.
6. E. Saltzman and M. A. McCrory, Physical assessment of nutritional status, in A. M. Coulston and C. J. Boushey, eds., *Nutrition in the Prevention and Treatment of Disease* (Burlington, MA: Elsevier Academic Press, 2008), pp. 57–73.
7. B. Shah, K. Sucher, and C. B. Hollenbeck, Comparison of ideal body weight equations and published height-weight tables with body mass index tables for healthy adults in the United States, *Nutrition in Clinical Practice* 21 (2006): 312–319.
8. Saltzman and McCrory, 2008.
9. L. Benson, H. J. Baer, and D. C. Kaelber, Trends in the diagnosis of overweight and obesity in children and adolescents: 1999–2007, *Pediatrics* 123 (2009): e153–e158; S. E. Barlow and the Expert Committee, Expert Committee recommendations regarding the prevention, assessment, and treatment of child and adolescent overweight and obesity: Summary report, *Pediatrics* 120 (2007): S164–S192.
10. J. Lefton and A. M. Malone, Anthropometric assessment, in P. Charney and A. M. Malone, eds., *ADA Pocket Guide to Nutrition Assessment* (Chicago: American Dietetic Association, 2009), pp. 154–166.
11. C. W. Thompson, Laboratory assessment, in P. Charney and A. M. Malone, eds., *ADA Pocket Guide to Nutrition Assessment* (Chicago: American Dietetic Association, 2009), pp. 62–153.
12. Thompson, 2009.
13. Standing Committee on the Scientific Evaluation of Dietary Reference Intakes, Food and Nutrition Board, Institute of Medicine, *Dietary Reference Intakes for Water, Potassium, Sodium, Chloride, and Sulfate* (Washington, DC: National Academies Press, 2005), pp. 147–150.

Nutrition in Practice

Nutritional Genomics

Consider this situation: A physician scrapes your cheek to collect a sample of cells and submits the sample to a **genomics** lab. In a short time, you receive a report that reveals your disease susceptibilities and suggests dietary and lifestyle changes that can improve your health. You may even be given a prescription for food choices or a dietary supplement that will best meet your personal nutrient requirements. Unlikely? Perhaps, but these possibilities are being explored by scientists working in the field of **nutritional genomics**, the study of dietary effects on **gene expression**. Recent research suggests that some dietary factors may be more helpful (or more harmful) in people who have particular genetic variations. The promise of nutritional genomics is a custom-designed dietary prescription that fits each person's specific needs. The accompanying glossary defines genomics and related terms.

What is a genome?

Genetic information is encoded in DNA molecules within the nuclei of almost all of the cells in our bodies. Figure NP13-1 shows how the genetic material is organized within the **genome**, the complete set of genetic information within our cells. The DNA molecules are tightly packed along with associated proteins within the 46 **chromosomes**. Segments of a DNA strand that can eventually be translated into proteins are called **genes**. The sequence of **nucleotides** within each gene encodes the amino acid sequence of a particular protein. Scientists estimate that there are between 20,000

FIGURE NP13-1 The Human Genome

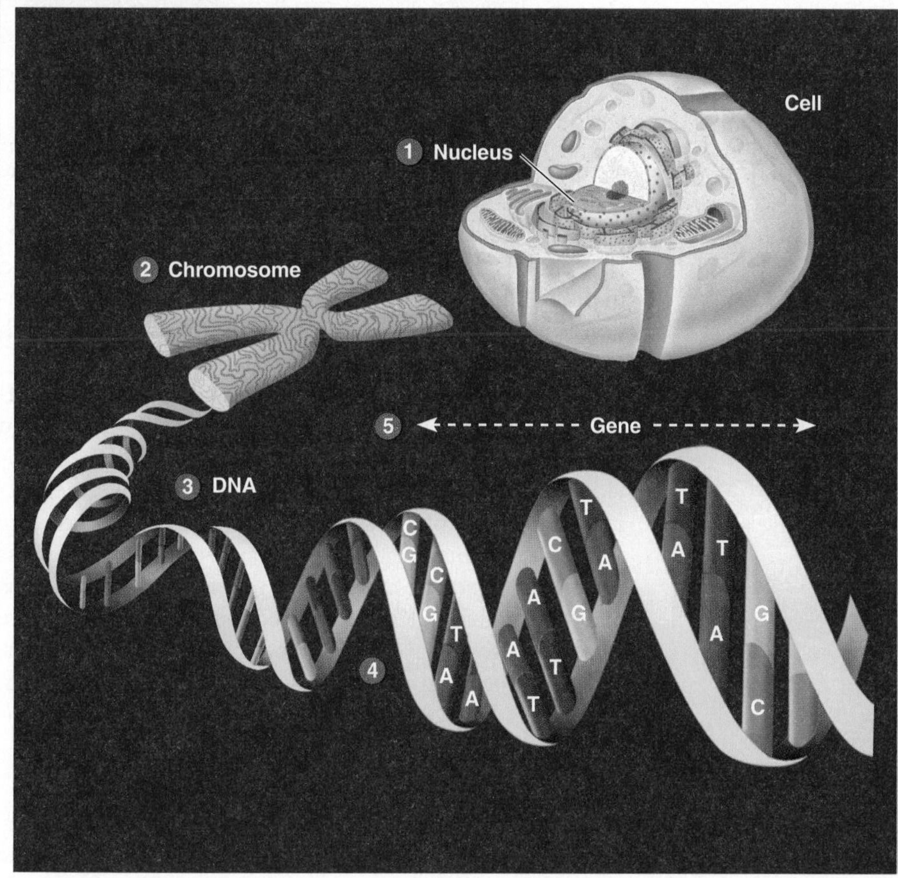

1 The human genome is a complete set of genetic material organized into 46 chromosomes, located within the nucleus of a cell.

2 A chromosome is made of DNA and associated proteins.

3 The double helical structure of a DNA molecule is made up of two long chains of nucleotides. Each nucleotide is composed of a phosphate group, a 5-carbon sugar, and a base.

4 The sequence of nucleotide bases (C, G, A, T) determines the amino acid sequence of proteins. These bases are connected by hydrogen bonding to form base pairs: adenine (A) with thymine (T) and guanine (G) with cytosine (C).

5 A gene is a segment of DNA that includes the information needed to synthesize one or more proteins.

Source: Adapted from "A Primer: From DNA to Life," Human Genome Project, U.S. Department of Energy Office of Science, www.ornl.gov/sci/techresources/Human_Genome/primer_pic.shtml.

Glossary

chromosomes: structures within the nucleus of a cell that contain the cell's DNA and associated proteins.

epigenetics: processes that cause heritable changes in gene expression that are separate from the DNA nucleotide sequence.

gene expression: the process by which a cell converts the genetic code into RNA and protein.

genes: segments of DNA that contain the information needed to make proteins.

genome (JEE-nome): the full complement of genetic material in the chromosomes of a cell.

genomics (jee-NO-miks): the study of genomes.

inherited disorders: medical conditions resulting from genetic defects.

methylation: the addition of methyl ($-CH_3$) groups.

microarray technology: research technology that monitors the expression of thousands of genes simultaneously.

multigene or polygenic: involving a number of genes, rather than a single gene.

noncoding sequences: regions of DNA that do not code for proteins. Some noncoding sequences may have regulatory or structural properties, but most have no known function.

nucleotides: the subunits of DNA and RNA molecules. These compounds—cytosine (C), thymine (T), uracil (U), guanine (G), and adenine (A)—are each composed of a phosphate group, a 5-carbon sugar (ribose), and a nitrogen-containing base. A DNA molecule is made up of two long chains of nucleotides held together by hydrogen bonding between nucleotide bases on opposing strands; each hydrogen-bonded nucleotide couple is called a *base pair.*

nutritional genomics: the study of dietary effects on genetic expression; also known as *nutrigenomics.*

polymorphisms: differences in the DNA sequences among individuals. A **single-nucleotide polymorphism** involves a single nucleotide at a particular area in the DNA strand.

- *poly* = many
- *morph* = form
- *ism* = condition

promoter: a region of DNA involved with gene activation.

transcription factors: proteins that bind DNA at specific sequences to regulate gene expression.

and 25,000 genes in the human genome.[1] However, only a small percentage (1 to 2 percent) of the genome codes for proteins: most DNA consists of **noncoding sequences**, which may help to regulate gene expression or have other functions.

When proteins are made, the information in the DNA sequence is first transcribed (copied) to messenger RNA molecules, which carry the genetic information out of the nucleus. Gene expression can be measured by determining the amounts of messenger RNA in a tissue sample. The expression of thousands of genes can be measured simultaneously using **microarray technology** (see photo).

A DNA microarray allows researchers to monitor the expression of thousands of genes simultaneously.

How did research in nutritional genomics begin?

The recent surge in genomics research grew from the Human Genome Project, a 13-year international effort by industry and government scientists to determine the complete nucleotide sequence of human DNA. Completed in 2003, this project led to enormous advances in the research technologies needed to study genes and genetic variation. Scientists are currently working to identify the individual genes in the genome, the roles of their protein products, the genes and proteins associated with diseases, and the dietary and lifestyle choices that influence the expression of genes involved in disease. In addition, researchers are studying how variations in the noncoding regions of DNA molecules—which regulate gene expression—influence disease risk.[2]

Genetic differences among individuals have been studied for years, as have the specialized dietary therapies used to treat various **inherited disorders**. For example, an individual may inherit a genetic defect that inhibits the normal metabolism of an essential nutrient and may therefore need to consume a diet that contains either more or less of this nutrient. An example of this type of condition is phenylketonuria (PKU), discussed in Nutrition in Practice 16. Genomic research takes this concept a bit further: instead of focusing on alterations in just one or two genes, researchers have been able to study the expression of *multiple* genes.

How do nutrients alter gene expression?

Some nutrients can switch gene expression on or off. The **promoter** region of a gene (a DNA region involved with gene activation) acts as the master switch. A large variety of proteins known as **transcription factors** can bind to areas on the promoter and either enhance or inhibit gene expression. A combination of dietary factors and hormones influences the types of transcription factors that reach the nucleus and their tendency to bind to DNA. Specific examples of how nutrients can influence transcription factors include:

- The transcription factor that enhances the gene expression of enzymes required for cholesterol synthesis enters the nucleus only when the cellular cholesterol content is low.
- The transcription factor that inhibits the expression of ferritin, an iron-storage protein, changes its affinity for DNA based on the iron content of the cell.

Gene expression is also influenced by modifications in the structure of DNA and its packaging in chromosomes. For example, **methylation** of DNA molecules (the addition of methyl groups) alters the expression of numerous genes and depends on both inherited factors and the availability of certain nutrients. In individuals who are susceptible to a disease that is influenced by DNA methylation, an altered diet or nutrient supplementation can potentially reduce the risk of disease.[3] The field of **epigenetics** investigates processes that cause heritable changes in gene expression that are separate from the underlying DNA nucleotide sequence.

How much genetic variation is there among people?

Except for identical twins, no two individuals are genetically identical. However, the variation in the genomes of any two persons is only about 0.5 percent,[4] a difference of approximately one base in every 5000. The most common genetic differences, known as **polymorphisms**, are changes in single nucleotides. A **single-nucleotide polymorphism** may be due to a nucleotide insertion, deletion, or substitution within the DNA molecule. Such variations are significant only if they affect a protein's amino acid sequence in a way that alters the protein's function, or if the change in the DNA molecule changes how a particular gene is regulated.

Genetic variation gives rise to the diversity among human beings—it explains most of the differences in our physical appearances and metabolic characteristics. Along with environmental factors, it also determines our susceptibilities to disease. Diseases affected by a single gene tend to be relatively rare and usually exert their effects early in life. In contrast, common diseases such as heart disease and cancer are influenced by many genes and typically develop over several decades or even longer. In these more complex **multigene**, or **polygenic**, disorders, many genes can contribute to disease risk, but no single gene may be sufficient to cause the disease on its own. In addition, the specific genes that influence a person's susceptibility to disease may vary substantially among different individuals.

What are some examples of single-gene disorders?

Examples of single-gene disorders include phenylketonuria, cystic fibrosis (discussed in Chapter 18), and the iron-overload disease hemochromatosis. Single-gene disorders may seriously disrupt metabolism and often require significant dietary or medical intervention. However, not all single-gene disorders have life-threatening ramifications. For example, lactose intolerance is related to an alteration in the promoter of the lactase gene; the condition may cause gastrointestinal discomfort but is readily managed by simple dietary changes.

How are multigene disorders different from single-gene disorders?

Multigene disorders are usually sensitive to a number of environmental influences, including diet and lifestyle; these environmental factors can directly influence the expression of the genes involved. Multigene disorders tend to develop over many years, so determining genetic susceptibility may allow a person to modify diet and lifestyle appropriately and reduce their disease risk before signs and symptoms appear.

Heart disease is an example of a disease with multiple gene influences.[5] Its many risk factors represent the involvement of an assortment of genes, which affect disparate aspects of physiology and metabolism. Consider that the major risk factors for heart disease include elevated blood cholesterol levels, hypertension, type 2 diabetes, and obesity. The underlying cause of any of these risk factors is rarely known; currently, clinicians screen for the presence of risk factors but not for the reasons why they occur. Should genomic research prove successful, a future assessment might be to identify specific genetic variations that can lead to the development of individual risk factors. For example, tests may determine whether a person's high blood cholesterol levels are due to excessive cholesterol absorption, excessive cholesterol production in the liver, or reduced cholesterol degradation. This information could then guide health care providers to the most appropriate intervention, allowing a better match between treatment recommendations and a person's genetic profile.

Can genomic research be used to explore the differences in nutrient needs among people?

Even though most people apparently can meet their nutrient needs by consuming nutrients at recommended levels, it would be useful to learn more about genetic variations within healthy populations. The techniques that have emerged from genomic research may provide a means for fine-tuning nutrient recommendations for different individuals.[6] Moreover, ideal indicators of nutrient status are still lacking for several of the minerals, such as zinc, magnesium, and chromium. Scientists hope to eventually produce genomic maps that will indicate how various nutrient deficiencies and combinations of deficiencies affect gene expression. These maps may eventually provide data that can help diagnose nutrient deficiencies.

Will knowledge about the human genome substantially change the manner in which health care is provided?

The enthusiasm surrounding genomic research should be put into perspective in terms of both the status of clinical medicine at present and people's willingness to make difficult lifestyle choices. Critics have questioned whether genetic markers for disease are more useful than simple and inexpensive clinical measurements, which reflect both genetic *and* environmental influences. In other words, knowing that a person is genetically predisposed toward high cholesterol levels is not necessarily more useful than knowing the person's actual blood cholesterol level.[7] Furthermore, a person's family history is already a simple genomic tool that indicates a higher genetic risk for certain illnesses. Understanding the family's medical history allows the clinician to use risk-reduction strategies that are appropriate to the culture and literacy of the patient, rather than focusing on genetic risk alone.[8]

Obtaining additional knowledge about disease risk is not necessarily useful unless people are motivated to make serious lifestyle changes. For example, despite the present abundance of disease prevention recommendations, many people seem unwilling or unable to make the changes known to improve health. Researchers have estimated that heart disease and type 2 diabetes are 80 percent and 90 percent preventable, respectively, by changing one's lifestyle to include an appropriate diet, a healthy body weight, and regular exercise, among other factors.[9] Given the difficulty that people have with current recommendations, it is unlikely that they will enthusiastically adopt an even more detailed list of dietary and lifestyle modifications.

What ethical concerns are raised by having extensive knowledge about an individual's genome?

A primary concern with our newfound ability to obtain detailed genetic information is confidentiality: should information about a person's susceptibility to disease be released to others (including other family members at risk) without that person's consent? Concern about privacy issues has led to federal and state legislation that prevents discrimination by group health plans, health insurers, and employers on the basis of an individual's genetic predisposition to disease.[10]

Another consideration is whether genetic testing is always in the best interest of children. Although early knowledge of a child's predisposition to illnesses may be useful for parents who want to provide optimal care, the release of this information could threaten the child's privacy and increase the potential for genetic discrimination in the future. In addition, children at high risk of developing a serious illness may grow up believing that their future choices are limited and may alter their life course to accommodate a disease that may never develop.[11]

Although genomic research has the potential to improve our ability to diagnose and treat disease, it is still unclear how knowledge of the genome will be translated into useful medical treatments. Health care professionals will need to keep informed of the ethical, legal, and social implications of nutritional and medical genomics as this remarkable research continues.

Notes

1. P. Stankiewicz and J. R. Lupski, Gene, genomic, and chromosomal disorders, in L. Goldman and A I. Schafer, *Goldman's Cecil Medicine* (Philadelphia: Saunders, 2012): 187–195.
2. J. Hardy and A. Singleton, Genomewide association studies and human disease, *New England Journal of Medicine* 360 (2009): 1759–1768.
3. S.-W. Choi and S. Friso, Epigenetics: A new bridge between nutrition and health, *Advances in Nutrition* 1 (2010): 8–16; G. C. Burdge and K. A. Lillycrop, Nutrition, epigenetics, and developmental plasticity: Implications for understanding human disease, *Annual Review of Nutrition* 30 (2010): 315–339.
4. G. S. Ginsburg, Applications of molecular technologies to clinical medicine, in L. Goldman and A I. Schafer, *Goldman's Cecil Medicine* (Philadelphia: Saunders, 2012): 199–203.
5. C. J. O'Donnell and E. G. Nabel, Genomics of cardiovascular disease, *New England Journal of Medicine* 365 (2011): 2098–2109.
6. P. J. Stover and M. A. Caudill, Genetic and epigenetic contributions to human nutrition and health: Managing genome–diet interactions, *Journal of the American Dietetic Association* 108 (2008): 1480–1487.
7. W. C. Willett, Balancing life-style and genomics research for disease prevention, *Science* 296 (2002): 695–698.
8. M. J. Khoury and coauthors, Do we need genomic research for the prevention of common diseases with environmental causes? *American Journal of Epidemiology* 161 (2005): 799–805.
9. Khoury and coauthors, 2005.
10. P. R. Reilly and R. M. DeBusk, Ethical and legal issues in nutritional genomics, *Journal of the American Dietetic Association* 108 (2008): 36–40.
11. B. R. Korf, Principles of genetics, in L. Goldman and A I. Schafer, *Goldman's Cecil Medicine* (Philadelphia: Saunders, 2012): 184–187.

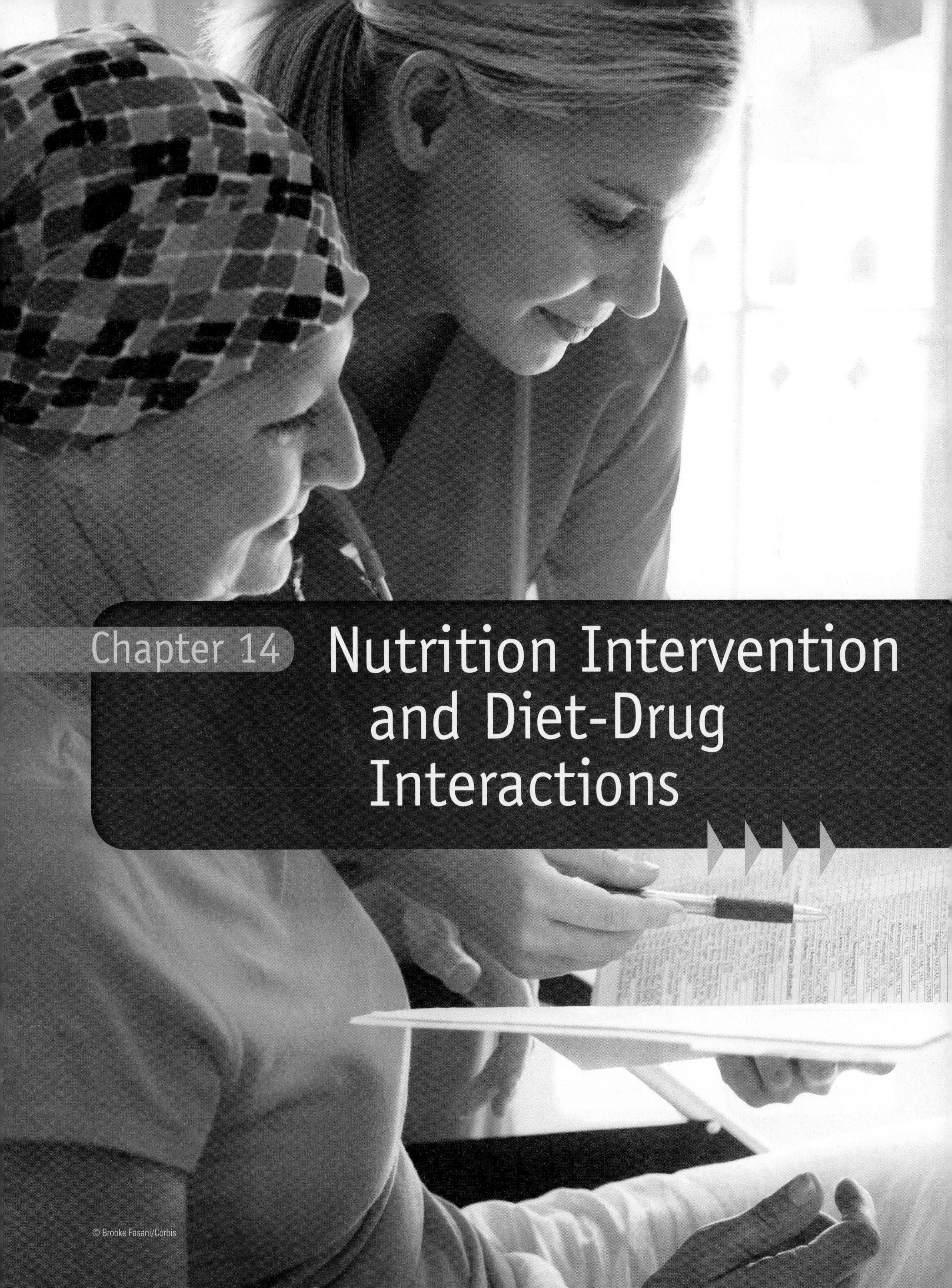

Chapter 14 Nutrition Intervention and Diet-Drug Interactions

THIS CHAPTER DESCRIBES HOW HEALTH CARE PROFESSIONALS ADDRESS nutrition problems and provide nutrition care. A critical part of this process is ensuring that nutritional needs are met, so the chapter also introduces some therapeutic diets and identifies some of the procedures and challenges involved in foodservice delivery. The final section discusses diet-drug interactions that need consideration when health practitioners administer nutrition care.

Implementing Nutrition Care

As explained in Chapter 13, nurses are well positioned to identify patients with nutritional problems. In addition, nursing care plans may include nursing diagnoses that suggest the need for nutritional intervention. This section provides examples of typical nutrition interventions and describes how the nurse can incorporate nutrition care into a care plan.

CARE PLANNING

Once the dietitian or nurse has collected and analyzed assessment information, the next steps of nutrition care can be carried out. Table 14-1 lists examples of nutrition interventions that may be appropriate in either the hospital or outpatient setting; note that some interventions may fall within the scope of nursing practice whereas others require the assistance of other health professionals. Many interventions include a diet order (prescribed by a physician), which provides specific recommendations regarding food, nutrient, or energy intake or feeding method. The care plan may also involve nutrition education or counseling (typically administered by a registered dietitian) to provide the knowledge, skills, and motivation that enable the patient to make dietary and lifestyle changes.

Table 14-2 shows an example of how nutrition care might be incorporated into a nursing care plan. The example describes a patient at risk for protein-energy malnutrition and wasting due to his swallowing difficulty and intolerance to physical activity. After diagnosing his nutrition-related problems, the nurse identifies **expected outcomes** and plans strategies that can correct the problems. This example includes interventions that require referrals to other health practitioners.

expected outcomes: patient-oriented goals that are derived from nursing diagnoses.

TABLE 14-1 Examples of Nutrition Interventions

Intervention	Examples
Food and/or nutrient delivery	Providing appropriate meals, snacks, and dietary supplements Providing specialized nutrition support (tube feedings and parenteral nutrition) Determining the need for feeding assistance or adjustment in feeding environment Managing nutrition-related medication problems
Nutrition education	Providing basic nutrition-related instruction Providing in-depth training to increase dietary knowledge or skills
Nutrition counseling	Helping the individual set priorities and establish goals Motivating the individual to change behaviors Solving problems that interfere with the nutrition care plan
Coordination of nutrition care	Providing referrals or consulting other health professionals or agencies that can assist with treatment Organizing treatments that involve other health professionals or health care facilities Arranging transfer of nutrition care to another professional or location

TABLE 14-2 Incorporating Nutrition Care into the Nursing Care Plan

The following example illustrates how the nurse can address nutrition problems while working through the different phases of the nursing process.

Nursing Process	Nutrition-Related Features
Assessment. The nurse includes nutrition assessment data in the overall assessment database.	The assessment data related to nutrition care are categorized and documented. *Subjective data:* • 74-year-old man with emphysema and dysphagia (difficulty swallowing); lives with wife. • Uses oxygen therapy at home but mostly while sleeping; frequently feels "out of breath" during the day. • Reports coughing while eating; senses that food gets "stuck" in his throat. • Believes he has lost weight recently because his clothes seem looser than usual. Says usual weight is 160 pounds, which is documented in the medical record. • Physically inactive; mostly watches TV during the day. *Objective data:* Height: 5′ 9″; weight: 145 lb; BMI: 21.5; %UBW: 90.6%; albumin: 3.6 g/dL; continually needs to clear throat; exhibits dyspnea (labored breathing) with walking.
Nursing diagnosis. The nurse develops diagnoses that indicate the need for nutrition intervention.	Nursing diagnoses with nutritional implications include: • *Impaired swallowing:* related to neuromuscular impairment. • *Risk for aspiration:* related to ineffective swallowing reflex. • *Imbalanced nutrition: less than body requirements* related to inability to ingest foods. • *Activity intolerance:* related to imbalance between oxygen supply and demand.
Outcome identification. The nurse identifies outcomes associated with improved nutrition status.	Appropriate outcomes include the patient's ability to: • Identify foods and beverages that can be consumed without difficulty. • Consume meals without coughing or aspiration. • Consume adequate amounts of food and beverages to meet nutritional needs. • Avoid further weight loss. • Use oxygen therapy as needed to permit a steady increase in activity.
Planning. The nurse develops strategies to help the patient achieve the expected outcomes.	To achieve the expected outcomes, the nurse plans to: • Educate the patient about body positioning and feeding techniques that can improve his ability to eat without coughing or discomfort. • Provide the patient with information about high-kcalorie and high-protein foods to minimize further weight and lean tissue losses. • Explain the need for daily physical activity, including suggestions that may improve the patient's ability to tolerate additional activity. • Teach the patient to conserve energy while performing activities of daily living. • Provide information about lightweight, portable, supplemental oxygen that can assist the patient during the day. Examples of interventions that would require the involvement of other health professionals include: • The attending physician may need to prescribe dysphagia testing to help determine the most appropriate food plan for the patient. • A dysphagia specialist may need to provide specific feeding recommendations related to the patient's type of dysphagia. • A registered dietitian may need to review and modify the patient's food plan so that he can improve his energy intake and make more appropriate food and beverage selections.
Implementation. The nurse performs the appropriate interventions and refers patient to other health practitioners, as necessary.	
Evaluation. The nurse continually monitors and evaluates the patient's progress in achieving each outcome. The nursing care plan is revised as needed.	

APPROACHES TO NUTRITION CARE

A nutrition care plan often involves significant dietary modifications. To ensure better compliance, the plan must be compatible with the desires and abilities of the person it is designed to help. The challenge is greater if dietary changes are required for extended periods.

Long-Term Nutrition Intervention When long-term changes are necessary, a care plan must take into account the person's current food practices, lifestyle, and degree of motivation. Behavior change is a process that occurs in stages; therefore, more than one consultation is usually necessary. The following approaches may be helpful in implementing long-term dietary changes:[1]

- *Determine the individual's readiness for change.* Some people have little desire to change their food practices, and even those who are willing may not be fully prepared to take the necessary steps. The health practitioner needs to consider a patient's readiness to adopt new dietary behaviors before attempting to implement an ambitious care plan.
- *Emphasize what to eat, rather than what not to eat.* Emphasizing foods to include in the diet, rather than those to restrict, can make dietary changes more appealing. For example, encouraging additional fruits and vegetables is a more attractive message than advising the patient to restrict butter, cream sauces, and ice cream.
- *Suggest only one or two changes at a time.* People are more likely to adopt a nutrition care plan that does not deviate too much from their usual diet. If they succeed in adopting one or two changes, they are more likely to stick to the plan and be open to additional suggestions. Stricter plans may yield quicker results but are useful only for highly motivated people.

Nutrition Education Nutrition education allows patients to learn about the dietary factors that affect their particular medical condition. Ideally, this knowledge can motivate them to change their diet and lifestyle to improve their health status.

A nutrition education program should be tailored to a person's age, level of literacy, and cultural background. Learning style should also be considered: some people learn best by discussion supplemented with written materials, whereas others prefer visual examples, such as food models and measuring devices.[2] Information can be provided in one-on-one sessions or group discussions. The meeting should include an assessment of the person's understanding of the material and commitment to making changes. Follow-up sessions can reveal whether the person has successfully adopted the new plan. For example, a nurse who counsels a woman who is lactose intolerant and hesitant to use milk products might proceed as follows:

- The nurse provides sample menus of a nutritionally adequate diet that limits milk and milk products. Together, the nurse and the woman design menus that consider her food preferences.
- The nurse describes the types and amounts of milk products that would likely be tolerated without causing symptoms and explains how to gradually incorporate these foods into the diet.
- Using diet analysis software, the nurse demonstrates how altering intakes of calcium-containing foods changes a meal's calcium content.
- The nurse explains how to use the Daily Values information on food labels to estimate the calcium content of packaged foods.
- The nurse provides information about the advantages and disadvantages of different calcium supplements.
- The nurse assesses the woman's understanding by having her identify nonmilk products that are high in calcium.

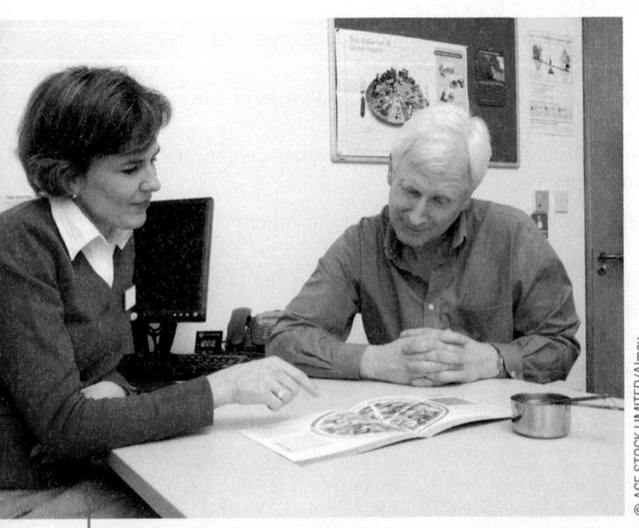

© ACE STOCK LIMITED/Almay

Nutrition counseling requires sensitivity to cultural orientation, educational background, and motivation for change.

Ideally, the nurse would be able to monitor the woman's progress in a subsequent counseling session.

Case Study: Implementing Nutrition Care

Max is an 11-year-old boy who was admitted to the hospital after he passed out while playing with friends. Tests confirm a diagnosis of type 1 diabetes mellitus. Max remains in the hospital for several days until his blood glucose and ketone levels are under control. During this time, he and his family learn about diabetes, the type of diet Max needs to follow, the use of insulin, the monitoring of blood glucose levels, and the required coordination of food intake, insulin, and physical activity. The details of diabetes mellitus are reserved for Chapter 20, but for now you can consider the steps that are necessary for implementing Max's nutrition care.

1. The nurse handling Max's case must identify expected outcomes when writing the care plan. What expected outcomes are appropriate for Max's situation?
2. Given the chronic nature of Max's illness and his age, what approaches should the nurse use when discussing the required meal plan and insulin treatment with Max and his family?
3. What factors need consideration when designing a nutrition education program for Max and his parents?
4. Max will need additional care to learn more about diabetes and to make the adjustments that will allow him to cope with his condition. Why is it important to address follow-up care before Max leaves the hospital?

Follow-up Care For optimal success, the health care professional should monitor the patient's progress and periodically evaluate the effectiveness of the nutrition care plan. Doing so usually involves comparing relevant outcome measures (such as the results of blood tests) with initial values and meeting with the patient to learn whether the plan has been satisfactory from the patient's point of view. Such follow-up efforts can reveal whether the care plan needs to be revised or updated, as is often the case when a person's medical condition changes. The accompanying Case Study provides an opportunity for you to review the implementation of nutrition care.

Review Notes

- Nutrition interventions are designed to correct the nutrition problems associated with illness. The intervention may involve the provision of appropriate foods or meals, nutrition education or counseling, and/or the coordination of nutrition care with other health professionals.
- Using the steps of the nursing process, a nurse can incorporate nutrition care into a nursing care plan.
- A nutrition intervention should take into account a person's food practices, lifestyle, and degree of motivation. Nutrition education should be individualized to accommodate a patient's needs and learning style.
- Nutrition care can be evaluated by reviewing relevant outcome measures of health status and determining the patient's understanding and acceptance of the intervention.

Dietary Modifications

During illness, many patients can meet energy and nutrient needs by following a **regular diet**. Others may require a **modified diet**, which is altered by changing food consistency or texture, nutrient content, or the foods included in the diet. This section introduces several types of dietary modifications and explains their uses during illness.

ENERGY INTAKES IN HOSPITAL PATIENTS

To estimate energy intakes that are appropriate for hospital patients, clinicians typically calculate or measure the resting metabolic rate (RMR) and then adjust the RMR value with "stress factors" that account for medical problems or, in some cases, medical treatments. In ambulatory patients, a factor for activity level may also be applied. Table 14-3 lists examples

regular diet: a diet that includes all foods and meets the nutrient needs of healthy people; may also be called a *standard diet, general diet, normal diet,* or *house diet.*

modified diet: a diet that contains foods altered in texture, consistency, or nutrient content or that includes or omits specific foods; may also be called a *therapeutic diet.*

TABLE 14-3 Selected Equations for Estimating Resting Metabolic Rate (RMR)

Harris–Benedict[a]

Women: RMR = 655.1 + [9.563 × weight (kg)] + [1.85 × height (cm)] − [4.676 × age (years)]

Men: RMR = 66.5 + [13.75 × weight (kg)] + [5.003 × height (cm)] − [6.755 × age (years)]

Mifflin–St. Jeor

Women: RMR = [9.99 × weight (kg)] + [6.25 × height (cm)] − [4.92 × age (years)] − 161

Men: RMR = [9.99 × weight (kg)] + [6.25 × height (cm)] − [4.92 × age (years)] + 5

[a]Although these equations are sometimes used for estimating basal metabolic rate (BMR), they were derived from data measured during resting conditions in most cases.

of RMR equations in common use, and the "How to" presents an example of this method. To obtain more accurate RMR values, clinicians sometimes use **indirect calorimetry**, a procedure that estimates energy expenditure by measuring oxygen consumption and carbon dioxide production during a period of rest.

In critical care patients, energy needs may be considerably higher than normal due to fever, mechanical ventilation, restlessness, or the presence of open wounds. Patients who are critically ill are usually bedridden and inactive, however, so the energy needed for physical activity is minimal. Energy requirements for critical care patients are described in Chapter 16.

MODIFIED DIETS

indirect calorimetry: a procedure that estimates energy expenditure by measuring oxygen consumption and carbon dioxide production during a period of rest.

Diets that contain foods with altered texture and consistency may be advised for individuals with chewing or swallowing difficulties. Diets with modified nutrient or food content may be prescribed to correct malnutrition, relieve disease symptoms, or reduce the risk of developing complications. Patients who have several medical problems may

HOW TO Estimate Appropriate Energy Intakes for Hospital Patients

To estimate an appropriate energy intake for a hospital patient, the health practitioner measures or calculates the patient's resting metabolic rate (RMR) and then applies a "stress factor" to accommodate the additional energy needs imposed by illness. The stress factor 1.25 has been shown to be reasonably accurate for many hospitalized patients; other examples are listed in Table 16-2 on p. 472.

The following example uses the Mifflin–St. Jeor equation (shown in Table 14-3) and the stress factor 1.25 to determine the energy needs of a 57-year-old female patient who is 5 feet 3 inches tall, weighs 115 pounds, and is confined to bed.

Step 1: The patient's weight and height are converted to the units used in the equation:

Weight in kilograms = 115 lb ÷ 2.2 lb/kg = 52.3 kg

Height in centimeters = 63 in × 2.54 cm/in = 160 cm

Step 2: Using the Mifflin–St. Jeor equation for estimating RMR in women:

RMR = [9.99 × weight (kg)] + [6.25 × height (cm)] − [4.92 × age (years)] − 161

= (9.99 × 52.3) + (6.25 × 160) − (4.92 × 57) − 161

= 522 + 1000 − 280 − 161 = 1081 kcal

Step 3: The RMR value is multiplied by the appropriate stress factor:

RMR × stress factor = 1081 × 1.25 = 1351 kcal

Thus, an appropriate energy intake for this patient would be approximately 1351 kcal. Her weight should be monitored to determine if her actual needs are higher or lower.

For a patient who is not confined to bed, an additional activity factor can be applied to accommodate the extra energy needs. For example, if the patient in the example begins limited activity while in the hospital, an activity factor of 1.2 can be multiplied by the results obtained in Step 3:

1351 × activity factor = 1351 × 1.2 = 1621 kcal

The activity factor for a hospitalized patient often falls between 1.1 and 1.4, and it is likely to change as the patient's condition improves.

require a number of dietary modifications. Keep in mind that modified diets should be adjusted to satisfy individual preferences and tolerances and may also need to be altered as a patient's condition changes. Table 14-4 lists examples of modified diets that are often prescribed during illness.[3] Later chapters discuss other types of dietary strategies that may be helpful for treating nutritional problems.

Mechanically Altered Diets Individuals who have difficulty chewing or swallowing may benefit from mechanically altered diets. Chewing difficulties usually result from dental problems. Impaired swallowing, or **dysphagia**, may result from neurological disorders, surgical procedures involving the head and neck, and physiological or anatomical abnormalities that restrict the movement of food within the throat or esophagus. Dysphagia diets are highly individualized because swallowing problems vary in severity and swallowing ability can fluctuate over time. Chapter 17 provides details about the specific diets used for treating dysphagia.

Table 14-5 provides examples of foods included in mechanically altered diets. Although the names for these diets vary, a more restrictive diet may contain mostly

dysphagia: difficulty swallowing.

TABLE 14-4 Examples of Modified Diets

Type of Diet[a]	Description of Diet	Appropriate Uses
Modified Texture and Consistency		
Mechanically altered diets	Contain foods that are modified in texture. Pureed diets include only pureed foods; mechanically altered and soft diets may include solid foods that are mashed, minced, ground, or soft.	Pureed diets are used for people with swallowing difficulty, poor lip and tongue control, or oral hypersensitivity. Mechanically altered and soft diets are appropriate for people with limited chewing ability or certain swallowing impairments.
Blenderized liquid diet	Contains fluids and foods that are blenderized to liquid form.	For people who cannot chew, swallow easily, or tolerate solid foods.
Clear liquid diet	Contains clear fluids or foods that are liquid at room temperature and leave minimal residue in the colon.	For preparation for bowel surgery or colonoscopy, for acute GI disturbances (such as after GI surgeries), or as a transition diet after intravenous feeding. For short-term use only.
Modified Nutrient or Food Content		
Fat-restricted diet	Limits dietary fat to low (<50 g/day) or very low (<25 g/day) intakes.	For people who have certain malabsorptive disorders or symptoms of diarrhea, flatulence, or steatorrhea (fecal fat) resulting from dietary fat intolerance.
Low-fiber diet	Limits dietary fiber; degree of restriction depends on the patient's condition and reason for restriction.	For acute phases of intestinal disorders or to reduce fecal output before surgery. Not recommended for long-term use.
Low-sodium diet	Limits dietary sodium; degree of restriction depends on symptoms and disease severity.	To help lower blood pressure or prevent fluid retention; used in hypertension, congestive heart failure, renal disease, and liver disease.
High-kcalorie, high-protein diet	Contains foods that are kcalorie and protein dense.	Used for patients with high kcalorie and protein requirements (due to cancer, AIDS, burns, trauma, and other conditions); also used to reverse malnutrition, improve nutritional status, or promote weight gain.

[a] Registered dietetians sometimes use the term *nutrition therapy* in place of *diet* when they provide nutrition care; for example, the *fat-restricted diet* may be called *fat-restricted nutrition therapy*.

Source: Academy of Nutrition and Dietetics, *Nutrition Care Manual* (Chicago: Academy of Nutrition and Dietetics, 2012).

TABLE 14-5 Foods Included in Mechanically Altered and Soft Diets

Depending on the feeding problem, a mechanically altered or soft diet may include foods that are pureed, mashed, ground, minced, or soft textured. Foods vary according to tolerance.

Pureed Diets	Mechanically Altered or Soft Diets
Milk products: Milk, smooth yogurt, pudding, custard	**Milk products:** Milk, yogurt with soft fruit, pudding, cottage cheese
Fruits: Pureed fruits and juices without pulp, seeds, skins, or chunks; well-mashed fresh bananas; applesauce	**Fruits:** Canned or cooked fruits without seeds or skin, fruit juices with small amounts of pulp, ripe bananas
Vegetables: Pureed cooked vegetables without seeds, skins, or chunks; mashed potatoes, pureed potatoes with gravy	**Vegetables:** Soft, well-cooked vegetables that are not rubbery or fibrous; well-cooked, moist potatoes
Meats and meat substitutes: Pureed meats; smooth, homogenous soufflés; hummus or other pureed legume spreads	**Meats and meat substitutes:** Ground, minced, or tender meat, poultry, or fish with gravy or sauce; tofu; well-cooked, moist legumes; scrambled or soft-cooked eggs
Breads and cereals: Smooth cooked cereals such as Cream of Wheat, slurried breads and pancakes,[a] pureed rice and pasta	**Breads and cereals:** Cooked cereals or moistened dry cereals with minimal texture, soft pancakes or breads, well-cooked noodles or dumplings in sauce or gravy

[a]Slurried foods are foods mixed with liquid until the consistency is appropriate; they may be gelled and shaped to improve their appearance.

pureed foods (*pureed diet*), whereas a less restrictive diet may include moist, soft-textured foods that easily form a bolus (*mechanically altered diet* or *soft diet*). Diets for people with chewing problems typically include foods that are ground or minced (*ground/minced diet*). Note that the foods used in these diets can overlap, and individual tolerances ultimately determine whether foods are included or excluded.

Blenderized Liquid Diet

Blenderized diets are most often recommended following oral or facial surgeries (for example, jaw wiring). Soft or tender foods that can be blenderized (often with added liquid) are available from all food groups; they include breads and cereals; boiled rice and pasta; cooked vegetables; fresh or cooked fruits without skins or seeds; and cooked, tender meats and fish. Foods that do not blend well should be excluded; these include hard or rubbery foods such as nuts and seeds, coconut, some raw vegetables, dried fruits, hard cheeses, sausages, and frankfurters.

Clear Liquid Diet

Clear liquids, which require minimal digestion and are easily tolerated by the gastrointestinal (GI) tract, are often the foods recommended before some GI procedures (such as GI examinations, X-rays, or surgeries), after GI surgery, or after fasting or intravenous feeding. The **clear liquid diet** consists of clear fluids and foods that are liquid at body temperature and leave little undigested material (called **residue**) in the colon. Permitted foods include clear or pulp-free fruit juices, carbonated beverages, clear meat and vegetable broths (such as consommé and bouillon), fruit-flavored gelatin, fruit ices made from clear juices, frozen juice bars, and plain hard candy. Although the clear liquid diet provides fluid and electrolytes, its nutrient and energy contents are extremely limited. If used for longer than a day or two, this diet should be supplemented with commercially prepared low-residue formulas that provide required nutrients. Figure 14-1 gives an example of a one-day clear liquid menu.

clear liquid diet: a diet that consists of foods that are liquid at body temperature, require minimal digestion, and leave little residue (undigested material) in the colon.

residue: material left in the intestine after digestion; includes mostly dietary fiber and undigested starches and proteins.

Fat-Restricted Diet

Fat restriction may be necessary for reducing the symptoms of fat malabsorption, which often accompanies diseases of the liver, gallbladder, pancreas, and intestines. Fat restriction may also alleviate the symptoms of heartburn. Although the fat intake is occasionally limited to as little as 25 grams daily, it should not be restricted more than necessary because fat is an important source of kcalories. Chapter 18 gives additional information about fat-restricted diets.

Most foods included in a fat-restricted diet provide less than 1 gram of fat per serving. The diet includes fat-free milk products, most breads and cooked grains, fat-free broths and soups, vegetables prepared without fats, most fruits, and fat-free candies and sweets (see Table 18-5 on p. 520). Restricted foods include low-fat and whole-milk products, baked products with added fat (like muffins), and most prepared desserts. Lean meat and meat substitutes are permitted but may be restricted to 4 to 6 ounces per day, depending on the degree of restriction. Some patients with malabsorptive disorders cannot tolerate large amounts of lactose or dietary fiber, so foods that include these substances may also need to be excluded from the diet.

FIGURE 14-1 Menu—Clear Liquid Diet

Sample Menu	
Breakfast	Strained orange juice Flavored gelatin Ginger ale Coffee or tea, sugar
Lunch	Bouillion or consommé Flavored gelatin Frozen juice bars Apple or grape juice Coffee or tea, sugar
Supper	Bouillion or consommé Flavored gelatin Fruit ice Cranberry juice Coffee or tea, sugar
Snacks	Soft drinks Fruit ices Hard candy

Low-Fiber Diet

Fiber restriction is recommended during acute phases of intestinal disorders, when the presence of fiber may exacerbate intestinal discomfort or cause diarrhea or blockages. Low-fiber diets are sometimes used before surgery to minimize fecal volume and after surgery during transition to a regular diet. Long-term fiber restriction is discouraged, however, because it is associated with constipation, diverticulosis, and other problems.

Low-fiber diets often eliminate whole-grain breads and cereals, nuts and nut butters, most fresh fruits (except peeled apples, ripe bananas, and melons), dried fruits, dried beans and peas, and many vegetables (including broccoli, cabbage, corn, onions, peppers, spinach, and winter squash). (■) If required, even greater reductions in colonic residue can be achieved by excluding foods high in resistant starch (see p. 67), milk products that contain significant lactose, and foods that contain fructose or sugar alcohols (such as sorbitol); these foods contribute to colonic residue because some of their nutrients may be poorly digested (such as the lactose in milk) or absorbed (such as sorbitol and fructose).

Low-Sodium Diet

A low-sodium diet can help to prevent or correct fluid retention and may be prescribed for treatment of hypertension, congestive heart failure, kidney disease, and liver disease. The sodium intake that is recommended depends on the illness, the severity of symptoms, and the drug treatment prescribed. In most cases, sodium is restricted to 2000 or 3000 milligrams daily, although more severe restrictions may be used in the hospital setting. Many patients find it difficult to significantly reduce their sodium intake, so while the low-sodium diet is prescribed in an attempt to improve the patient's medical problem, the recommended sodium intake may sometimes exceed the Tolerable Upper Intake Level (UL) for sodium of 2300 milligrams. (■)

Low-sodium diets limit the use of salt when cooking and at the table, eliminate most prepared foods and condiments, and limit consumption of milk and milk products (if excessive). Because so many processed foods are high in sodium, people following a low-sodium diet should check food labels and consume only low-sodium products. Sodium restrictions are difficult to implement on a long-term basis because many people find low-sodium foods unpalatable. Additional information about controlling dietary sodium is provided in Chapters 21 and 22.

High-kCalorie, High-Protein Diet

The high-kcalorie, high-protein diet is used to increase kcalorie and protein intakes in patients who have unusually high requirements or in those who are eating poorly. High-fat foods are added to increase energy intakes; consequently, the diet may exceed 35 percent kcalories from fat. (■) Consuming small, frequent meals and commercial liquid supplements (such as Ensure or Boost) can also help a patient meet increased energy, protein, and other nutrient needs.

■ Specific information about the fiber content of foods can be found in Chapter 3 (p. 81) and Appendix A.

■ The average sodium intake in the United States is approximately 3300 mg per day. The sodium UL was set at 2300 mg to help prevent hypertension.

■ Reminder: Current guidelines recommend a total fat intake between 20 and 35 percent of kcalories, with most fats coming from polyunsaturated and monounsaturated fats.

TABLE 14-6 Foods Included in High-kCalorie, High-Protein Diets

Milk products	Whole milk, half-and-half, cream Milk shakes, eggnog Cheese Ice cream, whipped cream
Fruits	Dried fruit Canned fruit in heavy syrup Avocado
Vegetables	High-kcalorie vegetables such as potatoes, corn, and peas Vegetables prepared with butter, margarine, sour cream, cheese sauces, mayonnaise, or salad dressing Cream of vegetable soups
Meats and high-protein foods	All meats, fish, and poultry, including bacon, frankfurters, and luncheon meats; eggs; beans; tofu Meats that are prepared fried or covered in cream sauces or gravies Protein bars Nuts and seeds, peanut and other nut butters, coconut
Breads and cereals	Granola and dry cereals prepared with whole milk or cream and dried fruit Hot cereals with whole milk or cream, or added fat Pasta, rice, and biscuits with added fat Pancakes, waffles, French toast
Beverages	Fruit juices, fruit smoothies, sweetened beverages Meal replacement drinks Beverages with added protein powders

Examples of foods included in high-kcalorie, high-protein diets are listed in Table 14-6. Some of these foods are high in saturated fat, which is limited in heart-healthy diets. These foods are used liberally in diets for malnourished patients to help correct their immediate nutrition problems—weight loss and muscle wasting. Chapter 23 offers additional suggestions for increasing the kcalorie and protein contents of meals.

VARIATIONS IN THE DIET ORDER

As mentioned previously, the physician has the primary responsibility for prescribing an appropriate diet for a patient in a medical facility. Diet orders must be precise to avoid confusion; for example, a "low-sodium diet" should specify the amount of sodium permitted, as "low sodium" could be interpreted to mean any amount between 500 and 3000 milligrams. The physician often relies on the dietitian or nurse to recommend changes in the diet order when warranted. If a diet order seems inappropriate or outdated, the nurse should alert the physician that a mistake may have been made.

Diet Progression A change in diet as a patient's food tolerance improves is called **diet progression**. For example, the diet order may read, "progress diet from clear liquids to a regular diet as tolerated." In practice, this means that the patient would be given clear beverages initially, and then gradually be provided with other

diet progression: a change in diet as a patient's tolerances permit.

beverages or solid foods that are unlikely to cause discomfort. As another example, the diet may progress from small, frequent feedings to larger meals as tolerance improves. Symptoms such as nausea, vomiting, diarrhea, and gastrointestinal pain suggest intolerance.

Nothing by Mouth (NPO) An order to not give a patient anything at all—food, beverages, or medications—is indicated by NPO, an abbreviation for *non per os,* meaning "nothing by mouth." For example, an order may read "NPO for 24 hours" or "NPO until after X-ray." The NPO order is commonly used during certain acute illnesses or diagnostic tests involving the GI tract.

Alternative Feeding Routes In most cases, patients can meet their nutrient needs by consuming regular foods. Sometimes, however, a person's medical condition makes it difficult to meet nutrient needs orally. In such cases, the physician may order **tube feedings** or **parenteral nutrition**, described in detail in Chapter 15.

- *Tube feedings.* Nutritionally complete formulas can be delivered through a tube placed directly into the stomach or intestine. Tube feedings are preferred to parenteral nutrition if the GI tract is functioning normally.
- *Parenteral nutrition.* A person's medical condition sometimes prohibits the use of the GI tract to deliver nutrients. If the person is malnourished and the GI tract cannot be used for a significant period of time, parenteral nutrition, in which nutrient solutions are supplied intravenously, can meet nutritional needs.

Review Notes

- Dietary modifications prescribed during illness include changes in food texture or consistency, modified energy or nutrient content, or the inclusion or exclusion of certain foods.
- Energy requirements for hospital patients are typically estimated by multiplying a person's resting metabolic rate (RMR) by factors that account for the medical condition, medical treatments, and activity level.
- Mechanically altered diets may be prescribed for people with chewing or swallowing difficulties.
- Clear liquid diets may be recommended before various diagnostic tests or after acute gastrointestinal disturbances, intravenous feedings, or fasts.
- Some medical conditions require the restriction of specific nutrients, such as fat, fiber, or sodium.
- A high-kcalorie, high-protein diet may help to prevent or reverse malnutrition, improve nutrition status, or promote weight gain.
- The diet order must include specific instructions about dietary modifications to avoid confusion. The diet prescribed may change during the course of illness.

Foodservice

The work of a foodservice department may appear deceptively simple: appropriate meals are delivered to patients who need specific types of diets. Behind the scenes, however, a complex system is at work. A foodservice department faces a daily challenge in planning, producing, and delivering hundreds of nutritious meals and accommodating dozens of special diets and food preferences.

When designing menus for modified diets, the dietary and foodservice personnel refer to a **diet manual**, which details the exact foods or preparation methods to

tube feedings: liquid formulas delivered through a tube placed in the stomach or intestine.

parenteral nutrition: the provision of nutrients by vein, bypassing the intestine.

diet manual: a resource that specifies the foods or preparation methods to include or exclude in modified diets and provides sample menus.

Banan Stock/Jupiter Images

Foodservice departments strive to prepare appetizing and nutritious meals and may accommodate dozens of special diets.

FIGURE 14-2 Sample Selective Menu

SODIUM-CONTROLLED	SUNDAY
Lunch	
LF = Low Fat	LSLF = Low Sodium, Low Fat
Meats	
LSLF Baked chicken	LSLF Baked fish (cod)
Starchy Vegetables	
LSLF Rice	LSLF Boiled potatoes
Vegetables	
LSLF Baby carrots	LSLF Green beans
Soup/Salad/Juice	**Dressings**
LSLF Coleslaw	Diet French
LS Chicken broth	Diet Thousand Island
Apple juice	Diet Italian
Tossed salad	
Desserts	
Pears	Fresh fruit
Breads	
Dinner roll	Bran bread
White bread	LS Crackers
Whole wheat bread	
Beverages & Condiments	
Coffee	Sugar
Decaf. coffee	Sugar substitute
Hot tea	Creamer
Decaf. hot tea	Lemon
Iced tea	Herb seasoning
Whole milk	Margarine
2% milk	Diet mustard
Fat-free milk	Diet mayonnaise
No salt	Diet catsup
Name ____________	Room _____

include or exclude in a modified diet. The diet manual may also outline the rationale and indications for use of the diets and include sample menus. The manual may be compiled by the dietetics staff or adopted from another health care facility or a dietetics organization.

FOOD SELECTION

Most hospitals provide **selective menus** from which patients can select their meals (see Figure 14-2). A patient prescribed a modified diet receives menus that include only the foods specified in the hospital's diet manual for that particular diet. The use of selective menus allows patients to choose the foods they prefer and are most likely to eat. An added advantage is that patients who must follow a modified diet can become familiar with the foods permitted on their particular diet.

To improve patient satisfaction with foodservice as well as patients' perceptions of their overall hospital experience, many hospitals are moving toward a room-service, cook-to-order system similar to what an individual might experience in a hotel. In these facilities, the menus list more food choices than usual and include more fresh foods, seasonal ingredients, and local specialties. Entrees are prepared as they are ordered, and food delivery hours have been expanded to better accommodate the patients' varied schedules. A 2008 survey of food and nutrition professionals found that 37 percent of their facilities offered room service.[4]

FOOD SAFETY

Each institution has protocols for handling food products based on the identification of potential hazards and critical control points in food preparation, usually referred to as **Hazard Analysis and Critical Control Points (HACCP).** Generally, a HACCP program addresses food handling, cooking, and storage procedures; cleaning and disinfecting of utensils, surfaces, and equipment; and staff sanitation issues. Personnel involved with preparing or delivering meals need to be aware of the specific HACCP systems at their facility.

IMPROVING FOOD INTAKE

People in hospitals and other medical facilities often lose their appetites as a result of their medical condition, treatment, or emotional distress. Moreover, some medications and other treatments can dramatically alter taste perceptions. Patients may receive meals at specified times, whether they are hungry or not, and often must eat in bed without companionship. Under these types of conditions, eating can become more of a chore than a pleasurable experience. Meals may also be unwelcome if the person is in pain or has been sedated.

To improve food intakes, health professionals should ensure that the patient's room remains calm and quiet during mealtime. Excessive activity, like room maintenance or ward rounds, can distract patients and reduce appetite. If the patient's appetite or sense of taste is affected by illness, the patient can be asked to identify foods that are the most enjoyable. Placing an occasional "surprise" on the tray—a decoration or funny card, for example—may help patients look forward to meals or perk up sagging spirits. The "How to" on p. 415 includes additional suggestions that may help to improve food intake at mealtimes.

selective menus: menus that provide choices in some or all menu categories.

Hazard Analysis and Critical Control Points (HACCP): systems of food or formula preparation that identify food safety hazards and critical control points during foodservice procedures; pronounced *Hah-sip*.

HOW TO Help Hospital Patients Improve Their Food Intakes

1. Empathize with the patient. Show that you understand how difficult eating may be when a person feels too sick to move or too tired to sit up. Help to motivate the patient by explaining how important good nutrition is to recovery.
2. Help patients select the foods they like and mark menus appropriately. When appropriate and permissible, let friends or family members bring favorite foods from outside the hospital.
3. For patients who are weak, suggest foods that require little effort to eat. Eating a roast beef sandwich, for example, requires less effort than cutting and eating a steak. Drinking soup from a cup may be easier than eating it with a spoon.
4. During mealtimes, make sure the patient's room is quiet and has sufficient lighting for viewing the food. See that the room is free of odors that may interfere with the appetite.
5. Help patients prepare for meals. Help them wash their hands and get comfortable, either in bed or in a chair. Adjust the extension table to a comfortable distance and height and make sure it is clean. Take these steps before the tray arrives so that the meal can be served promptly and at the right temperature.
6. When the food cart arrives, check the patient's tray. Confirm that the patient is receiving the right diet, the foods on the tray are those selected from the menu, and the foods look appealing. Order a new tray if the foods are not appropriate.
7. Help with eating, if necessary. Help patients open containers or cut foods, and assist with feeding if patients cannot feed themselves. Encourage patients with little appetite to eat the most nutritious foods first and to drink liquids between meals.
8. Take a positive attitude toward the hospital's food. Never say something like "I couldn't eat this either." Instead, say, "The foodservice department really tries to make foods appetizing. I'm sure we can find a solution."

Review Notes

- Hospital foodservice departments may accommodate the special needs of hundreds of patients daily. Diet manuals specify the foods to include in or exclude from modified diets.
- Many hospitals provide selective menus from which patients can choose meals that are appropriate for their medical condition.
- Foodservice departments must follow the specific HACCP protocols adopted by their institutions to ensure food safety.
- Hospital patients may need assistance at mealtime and encouragement to consume adequate amounts of food.

Diet-Drug Interactions

When working with patients, medical personnel should be alert to possible interactions between drugs and dietary substances. These interactions can raise health care costs and result in serious, and sometimes fatal, complications. Accordingly, health professionals must learn to take steps to prevent or lessen their adverse consequences. Diet-drug interactions (also called *food-drug interactions* or *drug-nutrient interactions*) generally fall into the following categories:

- Drugs may alter food intake by reducing the appetite or by causing complications that make food consumption difficult or unpleasant. Other drugs may increase the appetite and cause weight gain.
- Drugs may alter the absorption, metabolism, or excretion of nutrients. Conversely, nutrients and other food components may alter the absorption, metabolism, and excretion of drugs.
- Some interactions between dietary components and drugs can cause drug toxicity.

Examples of these types of diet-drug interactions are shown in Table 14-7.[5]

TABLE 14-7 Examples of Diet-Drug Interactions

Drugs may alter food intake by:
- Altering the appetite (amphetamines suppress appetite; corticosteroids increase appetite).
- Interfering with taste or smell (amphetamines change taste perceptions).
- Inducing nausea or vomiting (digitalis may do both).
- Interfering with oral function (some antidepressants may cause dry mouth).
- Causing sores or inflammation in the mouth (methotrexate may cause painful mouth ulcers).

Drugs may alter nutrient absorption by:
- Changing the acidity of the digestive tract (antacids may interfere with iron and folate absorption).
- Damaging mucosal cells (cancer chemotherapy may damage mucosal cells).
- Binding to nutrients (bile acid binders bind to fat-soluble vitamins).

Foods and nutrients may alter drug absorption by:
- Stimulating the secretion of gastric acid (the antifungal agent ketoconazole is absorbed better with meals due to increased acid secretion).
- Altering the rate of gastric emptying (drug absorption may be delayed when the drug is taken with food).
- Binding to drugs (calcium binds to tetracycline, reducing the absorption of both substances).
- Competing for absorption sites in the small intestine (dietary amino acids interfere with levodopa absorption).

Drugs and nutrients may interact and alter metabolism by:
- Acting as structural analogs (as do warfarin and vitamin K).
- Using similar enzyme systems (phenobarbital induces liver enzymes that increase the metabolism of folate, vitamin D, and vitamin K).
- Competing for transport on serum proteins (fatty acids and drugs may compete for the same sites on the serum protein albumin).

Drugs may alter nutrient excretion by:
- Altering nutrient reabsorption in the kidneys (some diuretics increase the excretion of sodium and potassium).
- Causing diarrhea or vomiting (diarrhea and vomiting may cause electrolyte losses).

Food substances may alter drug excretion by:
- Inducing the activities of liver enzymes that metabolize drugs, increasing drug excretion (components of charcoal-broiled meats increase the metabolism of warfarin, theophylline, and acetaminophen).

Food substances and drugs may interact and cause drug toxicity by:
- Increasing side effects of the drug (the caffeine in beverages can increase the adverse effects of stimulants).
- Increasing drug action to excessive levels (grapefruit components inhibit the enzymes that degrade certain drugs, increasing drug concentrations in the body).

DRUG EFFECTS ON FOOD INTAKE

Some drugs can make food intake difficult or unpleasant: they may suppress the appetite, cause mouth dryness, alter the sense of taste, lead to inflammation or lesions in the mouth or GI tract, or induce nausea and vomiting. Certain side effects of drugs, including abdominal discomfort, constipation, and diarrhea, may be worsened by food consumption. Medications that cause drowsiness, such as sedatives and some painkillers, can make a person too tired to eat.

Drug complications that reduce food intake are significant only when they continue for a long period. Although many drugs can cause nausea in certain individuals, the nausea often subsides after the first few doses of the medication and therefore has little effect on nutrition status. If side effects persist, other medications may be prescribed to treat them; for example, antinauseants and antiemetics can help to reduce nausea and vomiting and thereby improve food intake.

Some medications stimulate the appetite and encourage weight gain. Unintentional weight gain may result from the use of some antidepressants, antipsychotics, antidiabetic drugs, and corticosteroids (such as prednisone).[6] For some conditions, however, weight gain is desirable. Patients with diseases that cause wasting, such as cancer or AIDS, are sometimes prescribed appetite enhancers such as megestrol acetate (Megace), a progesterone analog, or dronabinol (Marinol), which is derived from the active ingredient in marijuana.

DRUG EFFECTS ON NUTRIENT ABSORPTION

The medications that most often cause widespread nutrient malabsorption are those that upset gastrointestinal function or damage the intestinal mucosa. Antineoplastic and antiretroviral drugs (■) are especially detrimental, although nonsteroidal anti-inflammatory drugs (NSAIDs) and some antibiotics can have similar, though milder, effects. This section describes additional ways in which medications may alter nutrient absorption.

■ *Antineoplastic drugs* combat tumor growth. *Antiretroviral drugs* treat HIV infection.

Drug-Nutrient Binding Some medications bind to nutrients in the GI tract, preventing their absorption. For example, bile acid binders (such as cholestyramine, or Questran), which are used to reduce blood cholesterol levels, may bind to fat-soluble vitamins. Some antibiotics, notably tetracycline and ciprofloxacin (Cipro), bind to the calcium in foods and supplements, reducing the absorption of both the calcium and the antibiotic. Other minerals that may bind to these antibiotics include iron, magnesium, and zinc. Consumers are advised to use dairy products and all mineral supplements at least two hours apart from these medications.

Altered Stomach Acidity Medications that reduce stomach acidity can impair the absorption of vitamin B_{12}, folate, and iron. Examples include antacids, which neutralize stomach acid by acting as weak bases, and antiulcer drugs (such as proton pump inhibitors and H2 blockers), which interfere with acid secretion.

Direct Inhibition Several drugs impede nutrient absorption by interfering with their transport into mucosal cells. For example, the antibiotics trimethoprim (Proloprim) and pyrimethamine (Daraprim) compete with folate for absorption into intestinal cells. The anti-inflammatory medication colchicine, a treatment for gout, inhibits vitamin B_{12} absorption.

Tetra Images/Jupiter Images

Because nurses interact closely with patients, they are often responsible for educating patients about potential diet-drug interactions.

DIETARY EFFECTS ON DRUG ABSORPTION

Major influences on drug absorption include the stomach-emptying rate, the level of acidity in the stomach, and direct interactions with dietary components. The drug's formulation may also influence its absorption. The instructions included with medications typically advise whether food should be included or avoided with use.

Stomach-Emptying Rate Drugs reach the small intestine more quickly when the stomach is empty. Therefore, taking a medication with meals may delay its absorption, although the total amount absorbed may not be lower. As an example, aspirin works faster when taken on an empty stomach, although taking it with food is often encouraged to reduce stomach irritation. A slow drug absorption rate (due to slow stomach emptying) can be a problem if high drug concentrations are needed for effectiveness, as when a hypnotic is taken to induce sleep.

Stomach Acidity Some drugs are better absorbed in an acidic environment, whereas others are absorbed better under more alkaline conditions. For example, reduced stomach acidity (due to secretory disorders or antacid medications) may reduce the absorption of ketoconazole (Nizoral, an antifungal medication) and atazanavir (an antiretroviral medication), but increase the absorption of digoxin (Lanoxin, which treats heart failure) and alendronate (Fosamax, which treats osteoporosis).[7] Some drugs can be damaged by acid and are available in coated forms that resist the stomach's acidity.

Interactions with Dietary Components Some dietary substances can bind to drugs and inhibit their absorption. For example, the phytates (■) in foods can bind to digoxin. High-fiber diets can decrease the absorption of some tricyclic antidepressants due to binding between the fiber and the drugs. As mentioned earlier, minerals can bind to some antibiotics, reducing absorption of both the minerals and the drug.

DRUG EFFECTS ON NUTRIENT METABOLISM

Drugs and nutrients share similar enzyme systems in the small intestine and liver. Consequently, some drugs may enhance or inhibit the activities of enzymes needed for nutrient metabolism. For example, the anticonvulsants phenobarbital and phenytoin increase levels of the liver enzymes that metabolize folate, vitamin D, and vitamin K; therefore, persons using these drugs may require supplements of these vitamins.

The drug methotrexate, which treats cancer (and some inflammatory conditions), acts by interfering with folate metabolism and thus depriving rapidly dividing cancer cells of the folate they need to multiply. Methotrexate resembles folate in structure (see Figure 14-3) and competes with folate for the enzyme that converts folate to its active form. The adverse effects of using methotrexate therefore include symptoms of folate deficiency. These adverse effects can be reduced by using a preactivated form of folate (called leucovorin), which is often prescribed along with methotrexate to ensure that the body's rapidly dividing cells (cells of the digestive tract, skin cells, and red blood cells) receive adequate folate.

Isoniazid (INH) is an antituberculosis drug similar in structure to vitamin B_6. The drug can interfere with both the vitamin's activation and its normal functioning.[8] Because INH must be taken for at least six months to treat infection, vitamin B_6 supplements are often given simultaneously to prevent deficiency.

■ Reminder: *Phytates* are compounds found in many plant foods, including whole grains and legumes. Phytates can bind to minerals and other substances and reduce their absorption.

FIGURE 14-3 Folate and Methotrexate

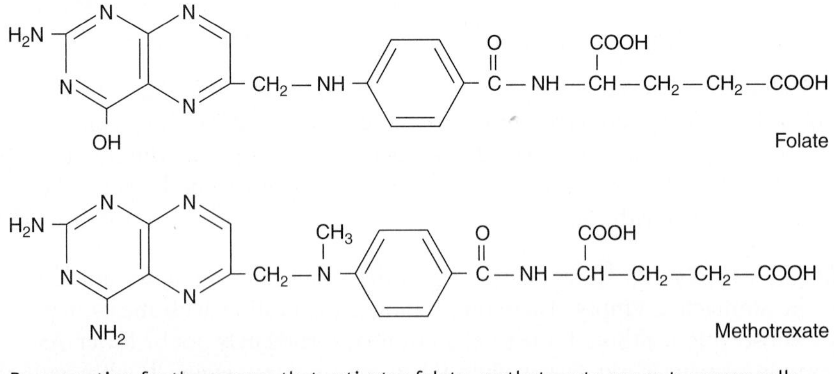

By competing for the enzyme that activates folate, methotrexate prevents cancer cells from obtaining the folate they need to multiply. In the process, normal cells are also deprived of the folate they need.

DIETARY EFFECTS ON DRUG METABOLISM

Some food components alter the activities of enzymes that metabolize drugs or may counteract drug effects in other ways. Compounds in grapefruit juice (or whole grapefruit) have been found to inhibit or inactivate enzymes that metabolize a number of different drugs. As a result of the reduced enzyme action, blood concentrations of the drugs increase, leading to stronger physiological effects. The effect of grapefruit juice can last for a substantial period, possibly as long as 24 to 48 hours after the juice is consumed;[9] thus, the interaction cannot be avoided by separating grapefruit juice consumption from drug administration. Table 14-8 provides examples of drugs that interact with grapefruit juice, as well as some drugs that are unaffected.

A number of dietary substances can alter the activity of the anticoagulant drug warfarin (Coumadin). One important interaction is with vitamin K, which is structurally similar to warfarin. Warfarin acts by blocking the enzyme that activates vitamin K, thereby preventing the synthesis of several blood-clotting factors. (■) The amount of warfarin prescribed is dependent, in part, on how much vitamin K is in the diet. If vitamin K consumption from foods or supplements changes substantially, it can alter the effect of the drug. Individuals using warfarin are advised to consume similar amounts of vitamin K daily to keep warfarin activity stable. The dietary sources highest in vitamin K are green leafy vegetables.

Several popular herbs contain natural compounds that may enhance the activity of warfarin and therefore should be avoided during warfarin treatment. These herbs include St. John's wort, garlic, ginseng, dong quai, danshen, and others.[10]

DRUG EFFECTS ON NUTRIENT EXCRETION

Drugs that increase urine production may reduce nutrient reabsorption in the kidneys, (■) resulting in greater urinary losses of the nutrients. For example, some diuretics can increase losses of calcium, potassium, magnesium, and thiamin; thus, dietary

TABLE 14-8 Examples of Grapefruit Juice–Drug Interactions

Drug Category	Drugs Affected by Grapefruit Juice	Drugs Unaffected by Grapefruit Juice
Cardiovascular drugs	Amiodarone Felodipine Nicardipine	Amlodipine Digoxin Diltiazem
Cholesterol-lowering drugs	Atorvastatin Lovastatin Simvastatin	Fluvastatin Pravastatin Rosuvastatin
Central nervous system drugs	Buspirone Carbamazepine Diazepam	Alprazolam Haloperidol Lorazepam
Anti-infective drugs	Erythromycin Saquinavir	Clarithromycin Quinine
Estrogens	Ethinylestradiol	17-ß-estradiol
Anticoagulants	—	Acenocoumarol Warfarin
Immunosuppressants	Cyclosporine Tacrolimus	Prednisone

■ Reminder: Vitamin K is required for the synthesis of prothrombin and several other blood-clotting proteins.

■ When the kidneys reabsorb a substance, they retain it in the blood. Substances that are not reabsorbed are excreted in urine.

supplements may be necessary to avoid deficiency. Risk of nutrient depletion is higher if multiple drugs with the same effect are used, if kidney function is impaired, or if the medications are used for a long time. Note that some diuretics can cause certain minerals to be retained, rather than excreted.

Corticosteroids, which are used as anti-inflammatory agents and immunosuppressants, promote sodium and water retention and increase urinary potassium excretion.[11] Long-term use of corticosteroids can have multiple adverse effects, which include muscle wasting, bone loss, weight gain, and hyperglycemia, with eventual development of osteoporosis and diabetes.

DIETARY EFFECTS ON DRUG EXCRETION

Inadequate excretion of medications can cause toxicity, whereas excessive losses may reduce the amount available for therapeutic effect. Some food components influence drug excretion by altering the amount reabsorbed in the kidneys. For example, the amount of of lithium (a mood stabilizer) reabsorbed in the kidneys is similar to the amount of sodium that is reabsorbed. Thus, both dehydration and sodium depletion, which promote sodium reabsorption, can result in lithium retention. Similarly, a person with a high sodium intake will excrete more sodium in the urine and, therefore, more lithium. Individuals using lithium are advised to maintain a consistent sodium intake from day to day to maintain stable blood concentrations of lithium.

Urine acidity can affect drug excretion due to the effects of pH on a compound's ionic (chemical) form. The medication quinidine, used to treat arrhythmias, is excreted more readily in acidic urine. Foods or drugs that cause urine to become more alkaline may reduce quinidine excretion and raise blood levels of the medication.

DRUG-NUTRIENT INTERACTIONS AND TOXICITY

Interactions between food components and drugs can cause toxicity or exacerbate a drug's side effects. The combination of tyramine, a food component, and monoamine oxidase (MAO) inhibitors, which treat depression and Parkinson's disease, can be fatal. MAO inhibitors block an enzyme that normally inactivates tyramine, as well as the hormones epinephrine and norepinephrine. When people who take MAO inhibitors consume excessive tyramine, the increased tyramine in the blood can induce a sudden release of stored norepinephrine. This surge in norepinephrine results in severe headaches, rapid heartbeat, and a dangerous rise in blood pressure. For this reason, people taking MAO inhibitors are advised to restrict their intakes of foods rich in tyramine.

Tyramine occurs naturally in foods and is also formed when bacteria degrade the protein in foods. Thus, the tyramine content of a food usually increases when a food ages or spoils. Individuals at risk of tyramine toxicity are advised to buy mainly fresh foods and consume them promptly. Foods that often contain substantial amounts of tyramine are listed in Table 14-9.

TABLE 14-9 Examples of Foods with a High Tyramine Content[a]

- Aged cheeses (cheddar, Gruyère)
- Aged or cured meats (sausage, salami)
- Beer
- Fermented vegetables (sauerkraut, kim chee)
- Fish or shrimp sauce
- Prepared soy foods (miso, tempeh, tofu)
- Soy sauce
- Yeast extracts (Marmite, Vegemite)

[a]The tyramine content of foods depends on storage conditions and processing; thus, the amounts in similar products can vary substantially.

Considering the number of medications available and the many ways in which drugs and dietary substances can interact, it is not surprising that serious side effects are increasingly being recognized. Health professionals should attempt to understand the mechanisms underlying diet-drug interactions, identify them when they occur, and prevent them whenever possible. The accompanying "How to" offers some practical advice about preventing drug-nutrient interactions.

HOW TO Prevent Diet-Drug Interactions

The Joint Commission, an accreditation agency for health care organizations, has recommended that all patients be educated about potential diet-drug interactions. Nurses can help by informing patients of precautions related to medications and watching for signs of problems that may arise.

To prevent diet-drug interactions, first list the types and amounts of over-the-counter drugs, prescription drugs, and dietary supplements that the patient uses on a regular basis. Look up each of these substances in a drug reference and make a note of:

- The appropriate method of administration (twice daily or at bedtime, for example).
- How the drug should be administered with respect to foods, beverages, and specific nutrients (for example, take on an empty stomach, take with food, do not take with milk, or do not drink alcoholic beverages while using the medication).
- How the drug should be used with respect to other medications.
- The side effects that may influence food intake (nausea and vomiting, diarrhea, constipation, or sedation, for example) or nutrient needs (interference with nutrient absorption or metabolism, for example).

A similar process can be used to review the dietary supplements that a person is taking. A reliable reference may list their appropriate uses, possible side effects, and potential interactions with food and medications.

Patients who take multiple medications may need to time their intakes carefully to avoid drug-drug or diet-drug interactions. The nurse can use information from a patient's food and nutrition history (see Chapter 13) to help the patient coordinate meals and drugs so as to avoid interactions.

Some medications have well-known effects on nutrition status. The nurse should remain alert for signs of problems, especially when:

- Nutrition problems are a frequent result of using the medication.
- A patient requires multiple medications.
- The patient is in a high-risk group; for example, a child, a pregnant or lactating woman, an older adult, or a person who is malnourished, abuses alcohol, or has impaired liver or kidney function.
- The patient needs to use the medication for an extended period.

Check with the pharmacist for additional information about drugs and their potential adverse effects.

Review Notes

- Medications can alter food intake and affect the absorption, metabolism, and excretion of nutrients. Components of foods can similarly affect the absorption, metabolism, and excretion of drugs.
- Drugs can alter food intake by increasing or decreasing the appetite, altering the sense of taste, causing GI discomfort, or damaging the lining of the GI tract.
- Drugs can affect nutrient absorption by binding to nutrients, changing stomach acidity, or interfering with nutrient transport into intestinal cells. Dietary substances can influence drug absorption by altering the stomach-emptying rate, changing stomach acidity, or directly binding to the drug.
- Drugs and nutrients may interfere with each other's metabolism because they use similar enzymes in the small intestine and liver.
- Drug-nutrient interactions may cause nutrient losses in urine and alter the urinary excretion of drugs. Some interactions between food components and drugs may result in toxicity.

Self Check

1. A successful nutrition intervention would include a long list of:
   a. dietary changes that the patient should consider making.
   b. foods that the patient should avoid.
   c. appetizing meals and foods that the patient can include in the diet.
   d. reasons why the patient should make dietary changes.

2. Indirect calorimetry is performed by:
   a. estimating the patient's metabolic rate using a predictive equation.
   b. measuring the patient's oxygen consumption and carbon dioxide production.
   c. recording and analyzing the patient's food intake over a 24-hour period.
   d. measuring the patient's body heat losses over a 24-hour period.

3. Mechanically altered diets are often prescribed for individuals with:
   a. disorders of the liver, gallbladder, and pancreas.
   b. unusually high kcalorie and protein requirements.
   c. chewing and swallowing difficulties.
   d. malabsorptive disorders.

4. Which statement regarding the clear liquid diet is true?
   a. Milk and yogurt are acceptable components of the diet.
   b. The diet is low in fiber and residue and therefore is acceptable during acute GI disturbances.
   c. The diet can usually meet all of a patient's nutritional needs.
   d. The diet is usually indicated for patients who are unable to swallow or move their jaws.

5. One disadvantage of the high-kcalorie, high-protein diet is that:
   a. patients often find the diet unpalatable.
   b. the diet contributes a significant amount of colonic residue so it may be poorly tolerated.
   c. the diet can only be continued for a short-term period or nutrient deficiencies would result.
   d. high-fat foods are added to increase energy intakes and therefore the diet may exceed 35 percent kcalories from fat.

6. A health practitioner assisting with a tube feeding is unsure about the length of time an open can of formula can be safely stored. What resource would most likely provide this information?
   a. The manual describing the hospital's HACCP procedures
   b. A diet manual
   c. The diet order
   d. The food label on the formula can

7. Medications that reduce stomach acidity can impair the absorption of:
   a. fat-soluble vitamins.
   b. thiamin and riboflavin.
   c. sodium and potassium.
   d. vitamin B_{12}, folate, and iron.

8. Compounds in grapefruit juice can:
   a. bind to antibiotics, reducing their absorption.
   b. cause excessive drug excretion.
   c. strengthen the effects of certain drugs.
   d. alter acidity in the stomach, impairing drug absorption.

9. Individuals using the medication warfarin must maintain a consistent daily intake of:
   a. vitamin A.
   b. calcium.
   c. sodium.
   d. vitamin K.

10. People who use MAO inhibiters must limit consumption of:
    a. aged cheeses.
    b. whole milk and yogurt.
    c. dark green leafy vegetables.
    d. grapefruit juice.

Answers to these questions appear in Appendix H. For more chapter review: Access an interactive eBook, chapter-specific interactive learning tools, including flashcards, quizzes, videos, and more in your Nutrition CourseMate, accessed through CengageBrain.com.

Clinical Applications

1. David is a 29-year-old male who is 6 feet 2 inches tall and has a usual body weight of 180 pounds. He was admitted to the hospital following an automobile accident and was treated for minor injuries. Using the method described in the "How to" on p. 408, estimate an appropriate energy intake for David using both the Harris-Benedict and Mifflin-St. Jeor equations. Use the stress factor 1.25, with no additional activity factor.

2. A hospital patient with a dysphagia problem has been receiving a soft diet. You notice that she coughs repeatedly while eating meals and eats only small portions of the foods on her plate. You suspect that the diet is inappropriate and that she may need a more restrictive dysphagia diet, but her medical chart doesn't specify the severity of the dysphagia.
   - To address these problems in your care plan, which health professionals would you need to consult, and what information would you require from each of them?
   - The next meal is due within the hour, and you suspect the woman will be intolerant to the foods she is scheduled to receive. You haven't yet heard from the health practitioners you consulted while developing your care plan. What options might you consider in handling this immediate problem?

3. An elderly man in a residential home has been losing weight since his arrival there. He has been taking several medications to treat both a heart problem and a mild case of bronchitis. You notice that he eats only a few bites at mealtimes and seems uninterested in food. Describe several steps you can take to learn whether the medications are interfering with his food intake in some way.

Nutrition on the Net

For further study of topics covered in this chapter, access these websites:

- Visit the website of the Academy of Nutrition and Dietetics to learn more about nutrition services in clinical care: www.eatright.org
- Design patient education materials with information from the MedlinePlus website: www.nlm.nih.gov/medlineplus/
- This Food and Drug Administration website discusses issues of clinical interest and is targeted to health professionals: www.fda.gov/ForHealthProfessionals/
- Visit the home page of the U.S. Food and Drug Administration (FDA), the agency that regulates all drugs in the United States: www.fda.gov
- The FDA provides safety information about drugs and other medical products on the MedWatch website: www.fda.gov/medwatch

Notes

1. K. Glanz, Current theoretical bases for nutrition intervention and their uses, in A. M. Coulston and C. J. Boushey, eds., *Nutrition in the Prevention and Treatment of Disease* (Burlington, MA: Elsevier Academic Press, 2008), pp. 127–138.
2. L. M. Delahanty and J. M. Heins, Tools and techniques to facilitate nutrition intervention, in A. M. Coulston and C. J. Boushey, eds., *Nutrition in the Prevention and Treatment of Disease* (Burlington, MA: Elsevier Academic Press, 2008), pp. 149–167.
3. Academy of Nutrition and Dietetics, *Nutrition Care Manual* (Chicago: Academy of Nutrition and Dietetics, 2012).
4. S. Aase, Hospital foodservice and patient experience: What's new? *Journal of the American Dietetic Association* 111 (2011): 1118–1123.
5. J. I. Boullata and L. M. Hudson, Drug-nutrient interactions: A broad view with implications for practice, *Journal of the Academy of Nutrition and Dietetics* 112 (2012): 506–517; L.-N. Chan, Drug-nutrient interactions, in M. E. Shils and coeditors, eds., *Modern Nutrition in Health and Disease* (Baltimore: Lippincott Williams & Wilkins, 2006), pp. 1539–1553.
6. W. S. Leslie, C. R. Hankey, and M. E. J. Lean, Weight gain as an adverse effect of some commonly prescribed drugs: A systematic review, *QJM: An International Journal of Medicine* 100 (2007): 395–404.
7. E. Lahner and coauthors, Systematic review: Impaired drug absorption related to the co-administration of antisecretory therapy, *Alimentary Pharmacology and Therapeutics* 29 (2009): 1219–1229.
8. P. Preziosi, Isoniazid: Metabolic aspects and toxicological correlates, *Current Drug Metabolism* 8 (2007): 839–851.
9. W. W. McCloskey, K. Zaiken, and R. R. Couris, Clinically significant grapefruit juice-drug interactions, *Nutrition Today* 43 (2008): 19–26.
10. M. L. Chavez, M. A. Jordan, and P. I. Chavez, Evidence-based drug–herbal interactions, *Life Sciences* 78 (2006): 2146–2157.
11. G. W. Cannon, Immunosuppressing drugs including corticosteroids, *Goldman's Cecil Medicine* (Philadelphia: Saunders, 2012), pp. 159–165.

Nutrition in Practice

Complementary and Alternative Therapies

The medical treatments described in the clinical chapters are based on current scientific understanding of human physiology and biochemistry and are generally supported by well-conducted clinical research. This Nutrition in Practice examines therapies that have *not* been scientifically validated and are therefore not currently promoted by conventional medical professionals; these therapies fall into a category called *complementary and alternative medicine (CAM)*. When the therapies are used together with conventional medicine, they are called *complementary*; when used in place of conventional medicine, they are called *alternative*.[1] Note that the term *alternative* may be misleading in that it inappropriately implies that unproven methods of treatment are valid alternatives to conventional treatments.

How popular is CAM in the United States?

Nearly 40 percent of adults in the United States use some form of CAM (excluding the use of prayer).[2] CAM is especially prevalent among persons with chronic pain or debilitating illness; for example, 80 percent of cancer patients reportedly use CAM.[3] Many patients use CAM as an adjunct to conventional medicine—often for symptoms or illnesses that are not sufficiently helped by conventional treatments. CAM therapies remain popular despite the dearth of evidence demonstrating their effectiveness. Reasons for their popularity include consumers' growing interest in self-help measures, the noninvasive nature of many CAM therapies, and the positive interactions consumers have with CAM practitioners.[4]

In response to the enormous popularity of CAM in the United States, in 1998 Congress established the National Center for Complementary and Alternative Medicine (NCCAM), which is now one of the 27 institutes that make up the National Institutes of Health (NIH). NCCAM's missions are to investigate complementary and alternative therapies by funding well-designed scientific studies and to provide authoritative information for consumers and health professionals. If enough evidence is found to support the use of a complementary or alternative therapy, it will likely be incorporated into mainstream medical practice.[5]

Why should mainstream health professionals learn more about CAM?

Due to the enormous consumer interest in trying novel treatments, health professionals need to be familiar with CAM therapies so that they can better communicate with patients regarding their medical care and advise them when an alternative approach conflicts with standard therapy or presents a danger to health. To provide medical students with objective information about CAM, many medical schools in the United States now offer elective courses about alternative forms of treatment. Physicians who practice *integrative medicine* may refer patients for complementary therapies while continuing to provide standard treatments.

What kinds of practices are considered CAM therapies?

CAM encompasses any and all therapies that are not normally part of conventional medicine. Consequently, the list of CAM approaches includes hundreds of advertised therapies purchased and used by consumers. Unfortunately, CAM has become a marketing buzzword and is used by unscrupulous sellers of worthless treatments. NCCAM categorizes CAM therapies as shown in Table NP14-1 and defined in the Glossary of Alternative Therapies on p. 431. Several popular examples are described in this discussion; other examples are discussed on the NCCAM website (http://nccam.nih.gov/health).

In what ways do alternative medical systems differ from conventional medicine?

Alternative medical systems are based on beliefs that lack the scientific basis of the theories underlying conventional medicine. Virtually all of these alternative systems were developed well over 100 years ago, before our bodies' biochemical and physiological processes were well understood. The alternative treatments may appeal to consumers because the interventions are nontechnical and seem nonthreatening. In general, however, the alternative theories and practices remain rooted in the past and have not been updated to include current knowledge. Examples of alternative medical systems include the following:

- **Naturopathic medicine** proposes that a person's natural "life force" can foster self-healing. This life force is allegedly stimulated by certain health-promoting factors and

TABLE NP14-1 Examples of Complementary and Alternative Medicine

Alternative Medical Systems
- Ayurveda
- Homeopathic medicine
- Naturopathic medicine
- Traditional Chinese medicine

Biologically Based Therapies
- Aromatherapy
- Dietary supplements
- Foods and special diets
- Herbal products
- Hormones

Energy Therapies
- Bioelectrical therapies (including electrical and magnetic fields)
- Biofield therapies (including acupuncture, qi gong, and therapeutic touch)

Manipulative and Body-Based Methods
- Chiropractic
- Massage therapy
- Osteopathic manipulation
- Reflexology

Mind-Body Interventions
- Biofeedback
- Faith healing (prayer)
- Meditation
- Mental healing (including hypnotherapy)
- Music, art, and dance therapy

suppressed by excesses and deficiencies. Naturopaths believe that ill health results from an internal disruption rather than from external disease-causing agents. Naturopathic therapies aim to enhance the natural healing powers of the body and may include special diets or fasting, herbal remedies and other dietary supplements, acupuncture, homeopathy, massage, and various other interventions.

- **Homeopathic medicine** is based on the dubious theory that "like cures like." Homeopaths believe that a substance that causes a particular set of symptoms can be used to cure a disease that has similar symptoms. Homeopathic medicines are usually natural substances that are substantially diluted in the belief that dilution increases potency, and most remedies are so extremely diluted that the original substance is no longer present. Homeopaths theorize that even though their remedies no longer contain a diluted substance, they still have powerful healing effects because the water structure is somehow altered during the dilution process used to prepare homeopathic medicines. This theory, however, conflicts with scientific understanding of water structure and properties.
- **Traditional Chinese medicine (TCM)** includes a large number of folk practices that originated in China. TCM is based on the theory that the body has pathways (called *meridians*) that conduct energy (called *qi*; pronounced "chee"). The interrupted flow of qi is believed to cause illness. TCM practices allegedly improve the flow of qi and include acupuncture, qi gong, herbal remedies, dietary practices, and massage. Ironically, the Western approach to managing illness is now the primary system of health care used in China.[6]

What is the theory underlying mind-body interventions?

Mind-body interventions attempt to improve a person's sense of psychological or spiritual well-being despite the presence of illness. The treatments are also used in the hope of reducing stress, dealing with pain, or lowering blood pressure. Some of these therapies have been incorporated into mainstream medicine for stress reduction or relaxation. For example, **biofeedback** training, in which individuals learn to monitor skin temperature, muscle tension, or brain wave activity while practicing relaxation techniques, is frequently taught by behavioral medicine specialists to help patients reduce stress or anxiety. Other techniques to reduce stress and promote relaxation include **meditation**, art and music therapy, and prayer.

The clinical applications of other mind-body therapies are far more questionable. One example is guided **imagery**, in which a person tries to reverse the disease process (for example, shrink a tumor) by using mental pictures. Another example is the use of **faith healing** in place of proven conventional treatments to cure disease.

What are some examples of biologically based therapies?

Biological therapies involve the use of natural products, such as dietary supplements, herbal and plant extracts,

Biofeedback training is a stress reduction and relaxation technique.

and special foods. The most common of these treatments include the use of vitamin and mineral supplements (discussed in Nutrition in Practice 9) and herbal products. Table NP14-2 lists examples of popular herbs, their common uses, and potential adverse effects associated with their use.

How effective are herbal remedies in the treatment of disease?

Despite the popularity of herbal products in the United States, the benefits of their use are uncertain. Although many medicinal herbs contain naturally occurring compounds that exert physiological effects, few herbal products have been rigorously tested, many make unfounded claims, and some may contain contaminants or produce toxic effects.[7] Only a limited number of clinical studies support the traditional uses,* and the results of studies that suggest little or no benefit are rarely publicized by the supplement industry. The NCCAM is currently funding large, controlled trials of several popular herbal treatments in an effort to obtain reliable efficacy and safety data.

TABLE NP14-2 Popular Herbal Products, Their Common Uses, and Adverse Effects

Herb	Scientific Name	Common Uses	Potential Adverse Effects
Black cohosh	*Cimicifuga racemosa*	Relief of menopausal symptoms	Rare; occasional stomach upset, headache, weight gain
Chaparral	*Larrea tridentata*	General tonic, treatment of infection, cancer, and arthritis	Hepatitis, liver failure
Comfrey	*Symphytum officinale*	Wound healing (topical use), treatment of lung and GI disorders	Liver damage
Echinacea	*Echinacea augustifolia, E. pallida, E. purpurea*	Prevention and treatment of upper respiratory infections	Rare; allergic reactions
Feverfew	*Tanacetum parthenium*	Prevention of migraine headache	Mouth and tongue sores, swelling of lips and mouth, stomach upset
Garlic	*Allium sativum*	Reduction of blood clotting, atherosclerosis, blood pressure, and blood cholesterol	Breath and body odor, nausea, hypotension, allergy, excessive bleeding
Ginger	*Zingiber officinale*	Prevention and treatment of nausea and motion sickness	Rare; occasional heartburn
Ginkgo	*Ginkgo biloba*	Treatment of dementia, memory defects, and circulatory impairment	Rare; nausea, stomach upset, diarrhea, allergy, anxiety, insomnia, excessive bleeding
Ginseng	*Panax ginseng, P. quinquefolius*	General tonic, reduction of blood glucose levels	Rare
Kava	*Piper methysticum*	Treatment of anxiety, stress, and insomnia	Rare; stomach upset, restlessness, drowsiness, tremor, headache, allergic skin reactions, occasional hepatitis and liver failure
St. John's wort	*Hypericum perforatum*	Treatment of mild to moderate depression	Skin photosensitivity
Saw palmetto	*Serenoa repens*	Reduction of symptoms associated with enlarged prostate	Rare; abdominal pain, nausea, diarrhea, fatigue, headache, decreased libido
Valerian	*Valeriana officinalis*	Sedation, treatment of insomnia	Rare
Yohimbe	*Pausinystalia yohimbe*	Treatment of erectile dysfunction	Anxiety, dizziness, headache, nausea, rapid heartbeat, hypertension, increased urinary frequency

*For example, some studies suggest that St. John's wort may be effective for treating mild depression, ginger may help to prevent motion sickness, and garlic supplements may improve blood pressure in people with hypertension.

Determining the effectiveness of herbal remedies can be complicated. Herbs contain numerous compounds, and it is often unclear which of these ingredients, if any, might produce the implied beneficial effects. Because the compounds in herbs vary among species and are affected by a plant's growing conditions, different samples of an herb can have different chemical compositions. Even when the active ingredients in an herbal product have been shown to be effective, the dosage suggested on the label may not provide the quantity of active ingredients found to be effective. For example, a consumer group (ConsumerLab.com) tested seven *Ginkgo biloba* products and found that four of the products, when consumed at the recommended dosage, lacked the amounts of the compounds indicated on their labels.[8] Similar problems were found with products containing echinacea, ginseng, St. John's wort, valerian, and various other herbs. In a university study of echinacea preparations, 10 percent of the 59 products tested contained no measurable echinacea, and only 52 percent of the samples contained the variety of echinacea listed on the label.[9]

How safe are the herbal products that are available in the marketplace?

Consumers often assume that because plants are "natural," herbal products must be harmless. Many herbal remedies have toxic effects, however.[10] The most common adverse effects of herbs include diarrhea, nausea, and vomiting. The popular herbs chaparral, comfrey, and kava have caused liver damage. The use of yohimbe (promoted for bodybuilding and erectile dysfunction) has been linked to heart arrhythmias, high blood pressure, anxiety, and seizures. Note that such adverse effects are rarely listed on supplement labels.

Contamination of herbal products is another safety concern. Many products have been found to contain lead and other toxic metals in excessive amounts.[11] Other contaminants frequently found in herbal products include molds, bacteria, and pesticides that have been banned for use on food crops.[12] Adulteration of imported products is a serious concern: several studies found that some herbal products imported from China and Taiwan contained synthetic drugs that were not declared on the label.[13] There have also been reports of illnesses and fatalities occurring from the intentional or accidental substitution of one plant species for another.[14]

Unlike drugs, herbal products do not need FDA approval before they are marketed. According to the Dietary Supplement Health and Education Act (DSHEA) of 1994, the companies that produce or distribute dietary (including herbal) supplements are responsible for determining their safety, yet these companies are not required to provide any evidence or conduct safety studies. If a company receives reports of illness or injury related to the use of its products, it is not required to submit this information to the FDA. In addition, the FDA must show that a supplement is unsafe before it can take action to remove the product from the marketplace.

Is there any risk that herbs may interact with medications like other dietary substances do?

Yes. Like drugs, herbs may either intensify or interfere with the effects of other herbs and drugs, or they may raise the risk of toxicity.[15] For example, garlic, ginkgo, and ginseng may increase the risk of bleeding when used with anticoagulant drugs. St. John's wort has been found to diminish the actions of oral contraceptives, anticoagulants, and other drugs. Unfortunately, information about herb-drug interactions is limited, and much of what is known has been obtained from case studies rather than controlled clinical trials. Table NP14-3 provides some examples of herb-drug interactions.

TABLE NP14-3 Examples of Herb-Drug Interactions

Herb	Drugs	Interaction
Chamomile	Sedatives (barbiturates, benzodiazepines)	May intensify or prolong sedative effects
Echinacea	Immunosuppressant drugs	May suppress drug effects
Feverfew	Anticoagulants, antiplatelet drugs, aspirin	May increase risk of bleeding
Garlic, ginger, ginkgo, ginseng	Anticoagulants, antiplatelet drugs	May increase risk of bleeding
Goldenseal	Anticoagulants	May oppose drug effects, increasing the risk of clot formation
Licorice	Antiarrhythmics, antihypertensives	May oppose drug effects
St. John's wort	Various	May suppress drug effects
Valerian	Sedatives (barbiturates, benzodiazepines)	May intensify or prolong sedative effects

Source: R. S. Porter and J. L. Kaplan, eds., *Merck Manual of Diagnosis and Therapy* (Whitehouse Station, NJ: Merck Sharp and Dohme Corp., 2011): 3421–3432.

Aside from herbal products, have other types of dietary supplements been found to be effective therapies?

As with herbal products, the efficacy and safety of other types of dietary supplements are uncertain. Note that manufacturers are free to market the supplements even if studies fail to show measurable health benefits, and the products are not regulated or tested for safety. Dietary supplements that are currently popular include the following:

- **Hormones.** Some hormones or hormone-like products derived from foods are considered dietary supplements and can be sold over the counter. One example is melatonin, a hormone made by the pineal gland and alleged to correct sleep disorders and prevent jet lag. Another example is the adrenal hormone dehydroepiandrosterone (DHEA), which is promoted to enhance immunity, increase muscle mass, improve memory, and defend against aging.
- **Glucosamine-chondroitin supplements.** Glucosamine and chondroitin are produced in the body and help to maintain joint cartilage. Early studies suggested that supplements containing glucosamine and chondroitin reduced moderate to severe symptoms of osteoarthritis better than a placebo, prompting some physicians to suggest using these supplements for pain relief.[16] Recent studies have cast doubt on the earlier findings, however, and several trials are still in progress.[17]

Is there any evidence supporting the use of various foods or supplements to cleanse or detoxify the body?

No; these practices are useless for the intended purpose. "Cleansing" practices are promoted based on the incorrect idea that the organs, blood, and intestines are "clogged" by toxins and undigested foods and that some types of dietary substances can rid the body of these accumulated toxins and wastes. Consumers who have little or no knowledge about physiology perceive this idea as plausible and are vulnerable to deceptive claims. A common scam is the sale of supplements or recipes for liver and gallbladder "flushes," which are said to cleanse the liver, remove gallstones, or improve medical conditions ranging from allergies to cancer.[18] In many instances, the ingested substances are partially or wholly indigestible and the individual using them passes semisolid materials that resemble gallstones to the untrained eye. Similarly, "colon cleanse" products are combinations of substances such as clay and gelling fibers that cause the users to pass large fecal casts claimed to represent years of accumulated wastes; such wastes are alleged to cause an array of health problems. Note that there is no scientific support for the alleged benefits of products said to "flush" the liver, gallbladder, or colon.

Which alternative practices involve physical manipulation, and how do they work?

Manipulative interventions include physical touch, forceful movement of different parts of the body, and the application of pressure. Some practitioners maintain that special energy fields are manipulated during the physical treatment and that proper energy flow induces healing. The most popular practices include the following:

- **Chiropractic** theory proposes that keeping the nervous system free from obstruction allows the body to heal itself, allegedly because the healing process stems from the brain and is conducted via the spinal cord and nerves to all parts of the body. Chiropractors claim to diagnose illnesses by detecting subluxations in the spine, which are variously described as misaligned vertebrae or pinched nerves that allegedly cause subtle interferences within the nervous system. The main treatment is the adjustment, a manual manipulation that is said to correct a subluxation and restore the body's natural healing ability. Although spinal manipulation has mainly been found to be helpful for improving back pain, many chiropractors still assert that chiropractic can cure disease rather than simply relieve symptoms.[19] For example, many chiropractors promote spinal manipulation to treat infectious diseases and prevent cancer, even though the nervous system and spinal alignment do not play roles in the pathology of these conditions.
- **Massage therapy** is the manipulation of muscle and connective tissue to improve muscle function, reduce pain, or promote relaxation. Massage therapists may also apply heat or cold and give advice about exercises that may improve muscle tone and range of motion. Massage is often integrated into conventional physical therapy, although some massage therapists may incorrectly suggest that massage is a valid treatment for a wide range of medical conditions.

What are the alleged effects of "energy" therapies?

Two categories of therapies involve the alleged curative power of "energy":

- **Biofield therapies** are said to influence the energy that surrounds or pervades the human body, and their proponents claim that an energy therapy can strengthen or restore a person's "energy flow" and induce healing. Acupuncture, qi gong, and therapeutic touch are among the therapies that subscribe to these theories. Note that CAM adherents often use the term *energy* unscientifically and that there is no objective evidence of this sort of energy flow.

- **Bioelectrical** or **bioelectromagnetic therapies** use electric or magnetic fields to allegedly promote healing; for example, magnets have been marketed with claims that they can improve circulation, reduce inflammation, and speed recovery from injuries.

Acupuncture, a component of traditional Chinese medicine, is the most well known of the therapies involving "energy flow." Acupuncture is based on the theory that disease is caused by the disrupted flow of qi through the body, and the treatment allegedly corrects such disruptions and restores health. The practice involves the shallow insertion of stainless steel needles into the skin at designated points on the body, sometimes accompanied by a low-frequency current to produce greater stimulation.

Qi gong is another therapy originating in China that is said to improve the flow of qi within the body. Qi gong masters allegedly cure disease by releasing energy from their body and passing it to the person being treated. Self-help practices include deep breathing, certain types of physical exercise, and concentration and relaxation techniques.

Therapeutic touch is based on the premise that the "healing force" of a practitioner can be used to cure disease. Practitioners claim to identify and correct energy imbalances by passing their hands above a patient's body and transferring "excess energy" to the patient.

Do conventional health practitioners consider CAM treatments to be safe and effective?

As mentioned earlier, CAM treatments are generally excluded from mainstream medical practice because there is no evidence proving that they are effective for treating the diseases and medical conditions for which they are used. Many consumers think otherwise and seem satisfied that these treatments "work." How is this dichotomy to be explained?

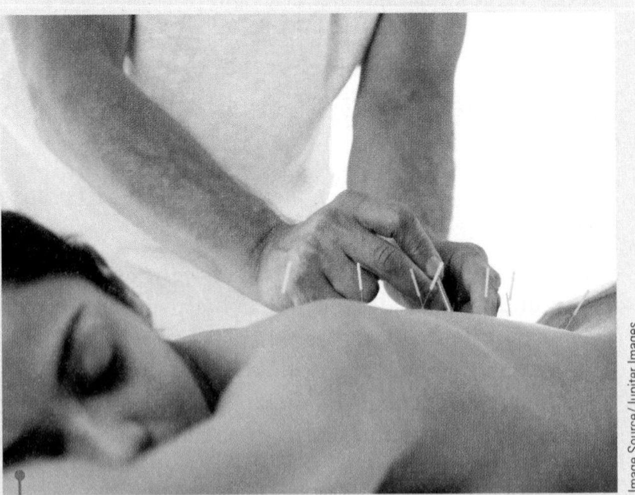

Image Source/Jupiter Images

Acupuncture involves the shallow insertion of stainless steel needles into the skin, sometimes accompanied by a low-frequency current.

Surveys suggest that consumers perceive their visits to CAM therapists as far more pleasant than their visits to conventional health practitioners. As explained earlier, CAM therapists often spend more time with patients, are more attentive, and use less invasive interventions.[20] Self-help measures are encouraged, so the consumer has more control over the treatment. The therapies appear to be more "natural" and to have fewer side effects. Possible explanations for "cures" include the following:

- A person may seem cured because of misdiagnosis; that is, the condition diagnosed by the CAM practitioner may not have actually existed.
- The condition may have been self-limiting, or it may have gone into temporary remission after the treatment.
- Undue credit may be inappropriately assigned to the CAM therapy when the improvement was actually due to a previous or concurrent conventional treatment.
- The placebo effect may have had an influence on the course of disease.

The central question remains: Do the CAM therapies merely make people *feel* better, or do they really *get* better? This question can be answered only by well-controlled research studies.

Are any potential dangers associated with the use of CAM?

One of the attractions of alternative therapies is the assumption that they are safe. Recall, however, the concerns associated with the use of herbal products, which include the potential toxicity of herbal ingredients, product contamination or adulteration, and interactions with conventional medications. Another concern is that use of CAM therapies may delay the use of reliable treatments that have demonstrable benefits.[21] Various reports have described people with treatable medical conditions who suffered permanent disability or death when they were misdiagnosed or improperly treated by CAM practitioners. For example, a rare but well-known risk of spinal cord injury or stroke is associated with a type of cervical manipulation performed by chiropractors.[22] Unfortunately, because most CAM therapies are not regulated or monitored, there are no accurate estimates of their adverse effects.

What should health care professionals do if they think their patients are using CAM?

Health practitioners should be aware when their patients are using CAM therapies that may have consequences influencing the course of a disease and its treatment. Accordingly, it is important to routinely inquire about the use of CAM therapies in a respectful, nonjudgmental manner and to educate patients about the hazards of postponing or stopping conventional treatment. Patients should also be told about potential interactions between conventional treatments and CAM therapies. Some patients may want to learn about differences between evidence-based medical practices and untested CAM theories and may be interested in the integrative medicine options available.

All alternative therapies have one characteristic in common: their effectiveness is, for the most part, unproven.[23] Because patients often choose CAM therapies due to positive interactions with CAM practitioners, health care professionals should realize that empathizing with patients may go a long way toward winning their trust and improving their compliance with therapy. Furthermore, health practitioners should stay informed about unconventional practices by obtaining reliable, objective resources so that they can knowledgeably discuss these options with patients.

Glossary of Alternative Therapies

acupuncture (AK-you-PUNK-chur): a therapy that involves inserting thin needles into the skin at specific anatomical points, allegedly to correct disruptions in the flow of energy within the body.

aromatherapy: inhalation of oil extracts from plants to cure illness or enhance health.

ayurveda: a traditional medical system from India that promotes the use of diet, herbs, meditation, massage, and yoga for preventing and treating illness.

bioelectrical or **bioelectromagnetic therapies:** therapies that involve the unconventional use of electric or magnetic fields to cure illness.

biofeedback: a technique in which individuals are trained to gain voluntary control of certain physiological processes, such as skin temperature or brain wave activity, to help reduce stress and anxiety.

biofield therapies: healing methods based on the belief that illnesses can be cured by manipulating energy fields that purportedly surround and penetrate the body. Examples include *acupuncture, qi gong,* and *therapeutic touch.*

chiropractic (KYE-roh-PRAK-tic): a method of treatment based on the unproven theory that spinal manipulation can restore health.

- A *subluxation* is a misaligned vertebra or other spinal alteration that may cause illness.
- *Adjustment* is the manipulative therapy practiced by chiropractors.

faith healing: the use of prayer or belief in divine intervention to promote healing.

homeopathic (HO-mee-oh-PATH-ic) medicine: a practice based on the theory that "like cures like"; that is, substances believed to cause certain symptoms are prescribed at extremely low concentrations for curing diseases with similar symptoms.

- *homeo* = like
- *pathos* = suffering

hypnotherapy: a technique that uses hypnosis and the power of suggestion to improve health behaviors, relieve pain, and promote healing.

imagery: the use of mental images of things or events to aid relaxation or promote self-healing.

massage therapy: manual manipulation of muscles to reduce tension, increase blood circulation, improve joint mobility, and promote healing of injuries.

meditation: a self-directed technique of calming the mind and relaxing the body.

naturopathic (NAY-chur-oh-PATH-ic) medicine: an approach to health care using practices alleged to enhance the body's natural healing abilities. Treatments may include a variety of alternative therapies including dietary supplements, herbal remedies, exercise, and homeopathy.

osteopathic (OS-tee-oh-PATH-ic) manipulation: a CAM technique performed by a doctor of osteopathy (D.O., or osteopath) that includes deep tissue massage and manipulation of the joints, spine, and soft tissues. A D.O. is a fully trained and licensed medical physician, although osteopathic manipulation has not been proved to be an effective treatment.

qi gong (chee-GUNG): a traditional Chinese system that combines movement, meditation, and breathing techniques and allegedly cures illness by enhancing the flow of qi (energy) within the body.

reflexology: a technique that applies pressure or massage on areas of the hands or feet to allegedly cure disease or relieve pain in other areas of the body; sometimes called *zone therapy.*

therapeutic touch: a technique of passing hands over a patient to purportedly identify energy imbalances and transfer healing power from therapist to patient; also called *laying on of hands.*

traditional Chinese medicine (TCM): an approach to health care based on the concept that illness can be cured by enhancing the flow of qi (energy) within a person's body. Treatments may include herbal therapies, physical exercises, meditation, acupuncture, and remedial massage.

Notes

1. R. S. Porter and J. L. Kaplan, eds., *Merck Manual of Diagnosis and Therapy* (Whitehouse Station, NJ: Merck Sharp and Dohme Corp., 2011), pp. 3411–3420.
2. Porter and Kaplan, 2011; P. M. Barnes, B. Bloom, and R. Nahin, Complementary and alternative medicine use among adults and children: United States, 2007, *CDC National Health Statistics Report* 12 (December 10, 2008): 1–24.
3. M. Charlson, Complementary and alternative medicine, in L. Goldman and A. I. Schafer, eds., *Goldman's Cecil Medicine* (Philadelphia: Saunders, 2012), pp. 177–181.
4. Charlson, 2012; A. M. McCaffrey, G. F. Pugh, and B. B. O'Connor, Understanding patient preference for integrative medical care: Results from patient focus groups, *Journal of General Internal Medicine* 22 (2007): 1500–1505.
5. S. E. Straus, Complementary and alternative medicine, in L. Goldman and D. Ausiello, eds., *Cecil Medicine* (Philadelphia: Saunders, 2008), pp. 206–209.
6. D. R. Haley and coauthors, Five myths of the Chinese health care system, *Health Care Manager* 27 (2008): 147–158; D. Normile, The new face of Chinese medicine, *Science* 299 (2003): 188–190.

7. M. E. Gershwin and coauthors, Public safety and dietary supplementation, *Annals of the New York Academy of Sciences* 1190 (2010): 104–117; U.S. Government Accountability Office (GAO), Herbal dietary supplements: Examples of deceptive or questionable marketing practices and potentially dangerous advice, GAO-10-662T (Washington, DC: May 26, 2010); R. B. van Breemen, H. H. S. Fong, and N. R. Farnsworth, Ensuring the safety of botanical dietary supplements, *American Journal of Clinical Nutrition* 87 (2008): 509S–513S.
8. ConsumerLab.com, Product review: Supplements for memory and cognition enhancement (ginkgo, huperzine A, and acetyl-L-carnitine), available at www.consumerlab.com, posted December 30, 2009; site visited June 25, 2012.
9. C. M. Gilroy and coauthors, Echinacea and truth in labeling, *Archives of Internal Medicine* 163 (2003): 699–704.
10. C. E. Dennehy and C. Tsourounis, Dietary supplements and herbal medications, in B. G. Katzung, ed., *Basic and Clinical Pharmacology* (New York: McGraw-Hill, 2012), pp. 1125–1137.
11. ConsumerLab.com, 2012; Gershwin and coauthors, 2010.
12. V. H. Tournas, E. Katsoudas, and E. J. Miracco, Moulds, yeasts, and aerobic plate counts in ginseng supplements, *International Journal of Food Microbiology* 108 (2006): 178–181; K. S. Leung and coauthors, Systematic evaluation of organochlorine pesticide residues in Chinese materia medica, *Phytotherapy Research* 19 (2005): 514–518.
13. P. A. Cohen and E. Ernst, Safety of herbal supplements: A guide for cardiologists, *Cardiovascular Therapeutics* 28 (2010): 246–253; R. J. Ko, A U.S. perspective on the adverse reactions from traditional Chinese medicines, *Journal of the Chinese Medical Association* 67 (2004): 109–116.
14. M. L. Coghlan and coauthors, Deep sequencing of plant and animal DNA contained within traditional Chinese medicines reveals legality issues and health safety concerns, *PLoS Genetics* 8 (2012): e1002657, available at doi:10.1371/journal.pgen.1002657; van Breemen, Fong, and Farnsworth, 2008.
15. Dennehy and Tsourounis, 2012; Cohen and Ernst, 2010.
16. O. Bruyere and J. Y. Reginster, Glucosamine and chondroitin sulfate as therapeutic agents for knee and hip osteoarthritis, *Drugs and Aging* 24 (2007): 573–580.
17. A. D. Sawitzke and coauthors, The effect of glucosamine and/or chondroitin sulfate on the progression of knee osteoarthritis: A report from the glucosamine/chondroitin arthritis intervention trial, *Arthritis and Rheumatism* 58 (2008): 3183–3191; R. M. Rozendaal and coauthors, Effect of glucosamine sulfate on hip osteoarthritis, *Annals of Internal Medicine* 148 (2008): 268–277; S. Reichenbach and coauthors, Meta-analysis: Chondroitin for osteoarthritis of the knee or hip, *Annals of Internal Medicine* 146 (2007): 580–590.
18. P. Moran, The truth about liver and gallbladder "flushes," available at http://quackwatch.com/01QuackeryRelatedTopics/flushes.html, posted March 9, 2007, site visited June 27, 2012; C. W. Sies and J. Brooker, Could these be gallstones? *Lancet* 365 (2005): 1388–1389.
19. J. C. Keating and coauthors, Subluxation: Dogma or science? *Chiropractic and Osteopathy* 13 (2005): 17–26.
20. McCaffrey, Pugh, and O'Connor, 2007.
21. Porter and Kaplan, 2011.
22. F. C. Albuquerque and coauthors, Craniocervical arterial dissections as sequelae of chiropractic manipulation: Patterns of injury and management, *Journal of Neurosurgery* 115 (2011): 1197–1205; W.-L. Chen and coauthors, Vertebral artery dissection and cerebellar infarction following chiropractic manipulation, *Emergency Medical Journal* 23 (2006): e1, available at doi:10.1136/emj.2004.015636.
23. Barnes, Bloom, and Nahin, 2008.

Chapter 15 Enteral and Parenteral Nutrition Support

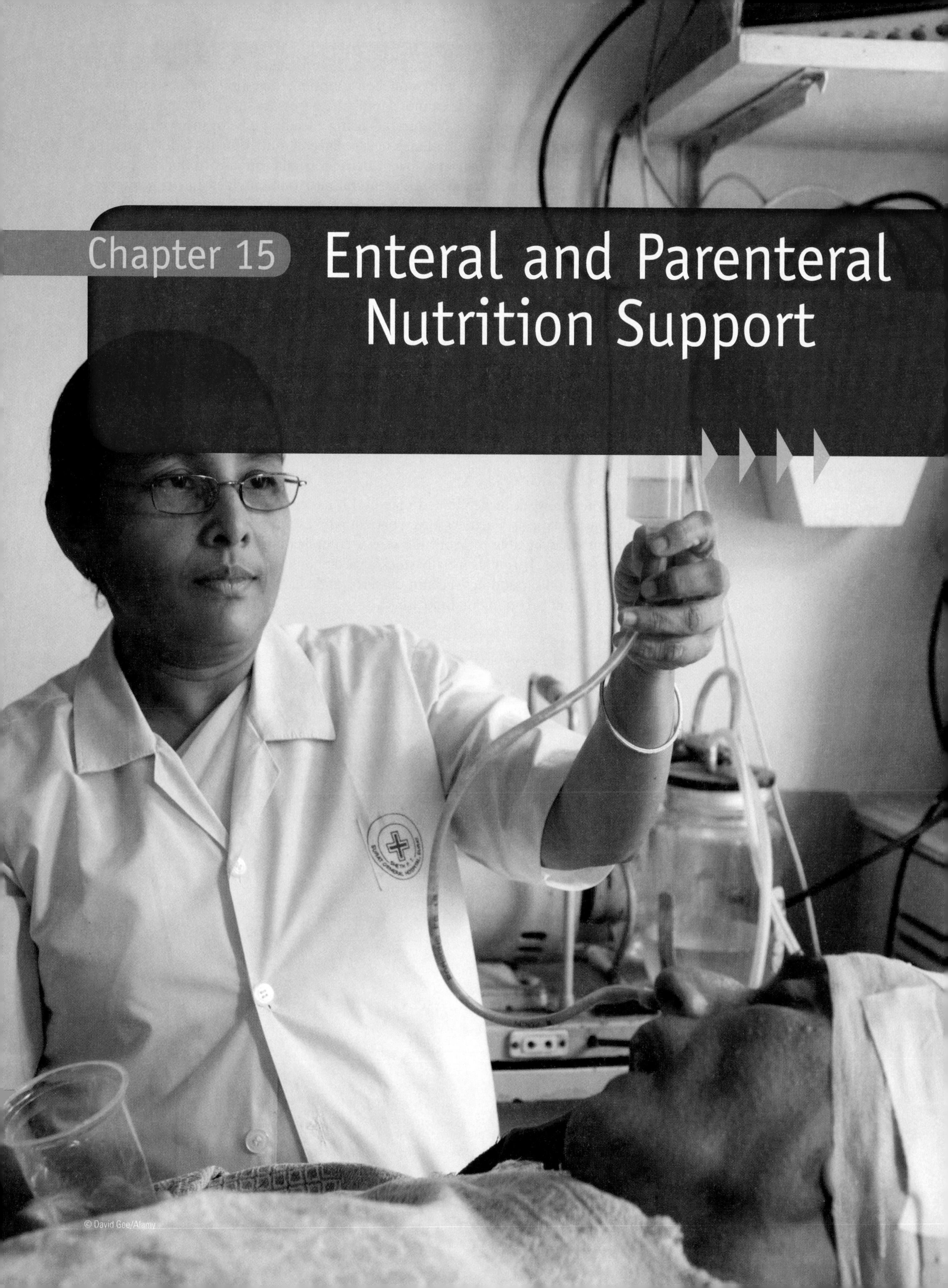

PATIENTS ARE OFTEN TOO SICK TO CONSUME A NORMAL DIET. MOREOVER, some illnesses may interfere with food consumption, digestion, or absorption to such a degree that regular foods cannot supply the necessary nutrients. In such cases, **nutrition support**—the delivery of nutrients using a feeding tube or intravenous infusions—can meet a patient's nutritional needs. **Enteral nutrition** provides nutrients using tube feedings, which deliver nutrient-dense formulas directly to the stomach or small intestine via a thin, flexible tube. **Parenteral nutrition** provides nutrients intravenously to patients who do not have adequate gastrointestinal (GI) function to handle enteral feedings. If the GI tract remains functional, enteral nutrition is preferred over parenteral nutrition because it is associated with fewer infectious complications and is significantly less expensive.[1] Figure 15-1 summarizes the decision-making process for selecting the most appropriate feeding method.

Enteral Nutrition

If gastrointestinal function is normal and a poor appetite is the primary nutrition problem, patients may be able to improve their diets by using oral supplements. If patients are unable to meet their nutrient needs by oral intakes alone, tube feedings can be used to deliver the required nutrients.

ORAL SUPPLEMENTS

Patients who are weak or debilitated may find it easier to consume oral supplements than to consume meals. Furthermore, a patient who can improve nutrition status with supplements may be able to avoid the stress, complications, and expense associated with tube feedings. Hospitals usually stock a variety of nutrient-dense formulas, milkshakes, fruit drinks, puddings, gelatin desserts, and snack bars to provide to patients who are at risk of becoming malnourished.

> ***NURSING DIAGNOSIS***
>
> The nursing diagnosis *imbalanced nutrition: less than body requirements* may be appropriate for a person who needs to use oral supplements to help meet nutrient needs.

FIGURE 15-1 Selecting a Feeding Route

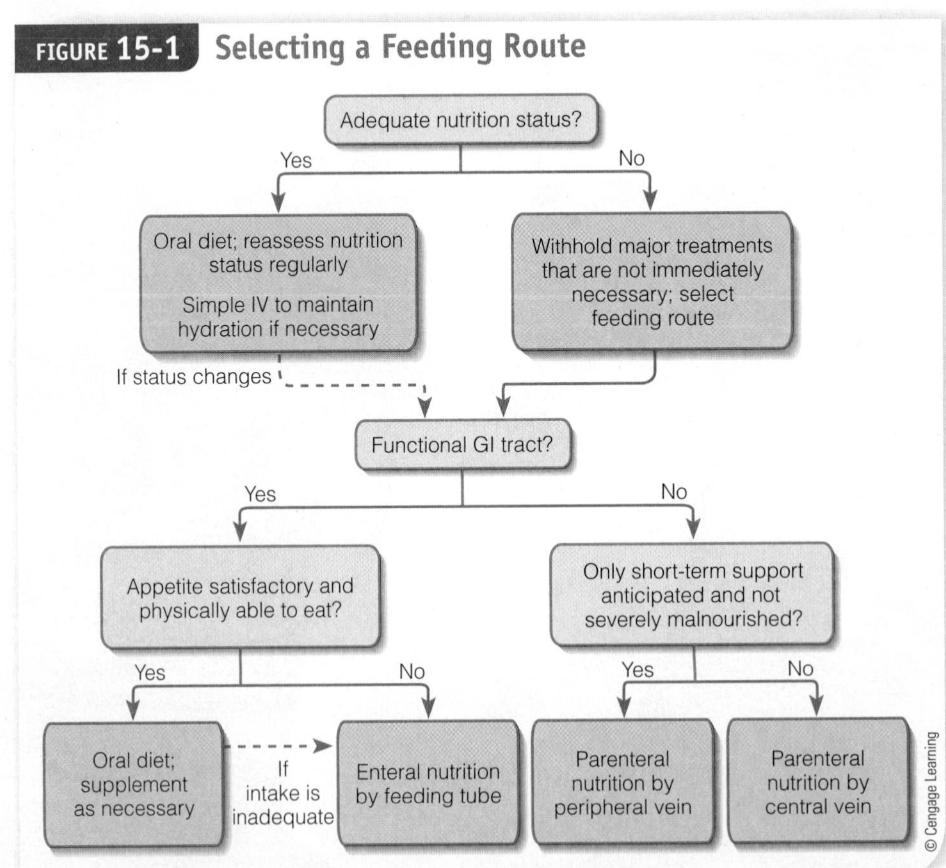

nutrition support: the delivery of nutrients using a feeding tube or intravenous infusions.

enteral (EN-ter-al) nutrition: the provision of nutrients using the GI tract; usually refers to the use of tube feedings.

parenteral (par-EN-ter-al) nutrition: the intravenous provision of nutrients that bypasses the GI tract.

par = beside
entero = intestine

When a patient uses an oral supplement, taste becomes an important consideration. Allowing patients to sample different products and select the ones they prefer helps to promote acceptance. The "How to" offers additional suggestions for helping patients improve their intakes using oral supplements.

Oral supplements are sold in pharmacies and grocery stores for home use; examples of popular liquid supplements include Ensure, Boost, and Carnation Instant Breakfast. These products are sometimes used as convenient meal replacements or supplements by healthy individuals.

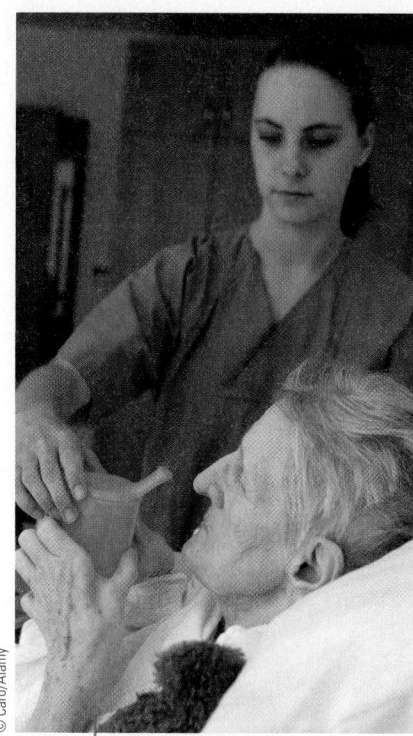

Patients can drink nutrient-dense formulas when they are unable to consume enough food from a regular diet.

CANDIDATES FOR TUBE FEEDINGS

Tube feedings may be recommended for patients at risk of developing protein-energy malnutrition who are unable to consume adequate food and/or oral supplements for several days. The following medical conditions or treatments may indicate the need for tube feedings:

- Severe swallowing disorders
- Impaired motility in the upper GI tract
- GI obstructions and **fistulas** that can be bypassed with a feeding tube
- Certain types of intestinal surgeries
- Little or no appetite for extended periods, especially if the patient is malnourished
- Extremely high nutrient requirements
- Mechanical ventilation
- Mental incapacitation due to confusion, neurological disorders, or coma

Contraindications for tube feedings include severe GI bleeding, high-output fistulas, **intractable** vomiting or diarrhea, complete intestinal obstruction, and severe malabsorption.[2] In addition, several clinical studies have suggested that tube feedings are not always effective in some of the patient populations in which they are routinely used; thus, the decision to use tube feedings should be considered in light of the most recent research evidence.[3]

TUBE FEEDING ROUTES

The feeding route chosen depends on the patient's medical condition, the expected duration of tube feeding, and the potential complications of a particular route. Figure 15-2 illustrates the main feeding routes, and the Glossary of Tube Feeding Routes describes each route.

fistulas (FIST-you-luz): abnormal passages between organs or tissues (or between an internal organ and the body's surface) that permit the passage of fluids or secretions.

intractable: not easily managed or controlled.

HOW TO Help Patients Improve Intakes with Oral Supplements

Patients in hospitals are often quite ill and have poor appetites. Even when a person enjoys an oral supplement, the taste may become monotonous in time. Health practitioners may be able to motivate patients to improve intakes by trying these suggestions:

- Let the patient sample different products that are appropriate for his or her needs, and provide only those that the patient enjoys.
- Serve supplements attractively. For example, a formula offered in a glass on an attractive plate may be more appealing than a formula served from a can with an unfamiliar name.
- Try keeping the formula in an ice bath so that it is cool and refreshing when the patient drinks it. Check with the patient to make sure the colder temperature is suitable.
- If a patient finds the smell of a formula unappealing, it may help to cover the top of the glass with plastic wrap or a lid, leaving just enough room for a straw.
- For patients with little appetite, offer the formula or snack food in small amounts that are easy to tolerate, and serve it more frequently during the day.
- Provide easy access. Keep the supplement close to the patient's bed where it can be reached with little effort and within sight so that the patient is reminded to consume it.
- If the patient stops enjoying a particular product, suggest an alternative. Maintain an updated list of oral supplements that are available at your institution so that you can advise patients about the possible options.

FIGURE 15-2 Tube Feeding Routes

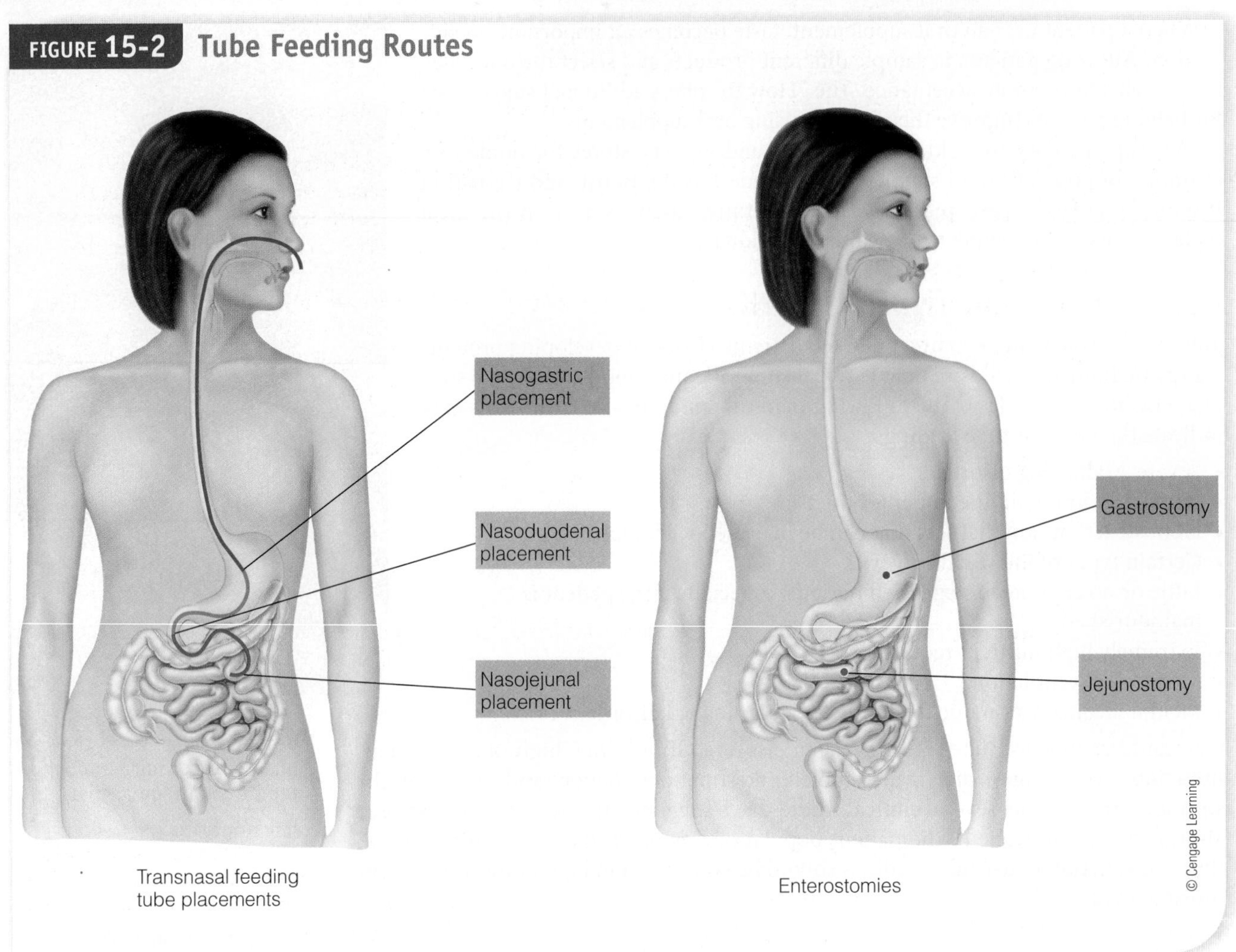

Gastrointestinal Access When a patient is expected to be tube fed for less than four weeks, a **nasogastric** or **nasointestinal** route is generally chosen; for these routes, the feeding tube is passed into the GI tract via the nose. The patient is frequently awake during **transnasal** (through-the-nose) placement of a feeding tube. While the patient is in a slightly upright position with head tilted, the tube is inserted into a nostril and passed into the stomach (nasogastric placement), duodenum (**nasoduodenal** placement), or jejunum (**nasojejunal** placement). If the patient is awake and alert, he or she can swallow water to ease the tube's passage. The final position of the feeding tube tip is verified by abdominal X-ray or other means. In infants, **orogastric** placement, in which the feeding tube is passed into the stomach via the mouth, is sometimes preferred over transnasal routes; this placement allows the infant to breathe more normally during feedings.

When a patient will be tube fed for longer than four weeks or if the nasointestinal route is inaccessible due to an obstruction or other medical reasons, a direct route to the stomach or intestine may be created by passing the tube through an **enterostomy**, an opening in the abdominal wall that leads to the stomach (**gastrostomy**) or jejunum (**jejunostomy**). An enterostomy can be made by either surgical incision or needle puncture.

Selecting a Feeding Route As mentioned, transnasal access is usually preferred when the tube feeding duration is expected to be less than four weeks, and enterostomies are often appropriate when tube feedings are planned for longer periods. Gastric feedings (nasogastric and gastrostomy routes) are preferred whenever possible. These

Glossary of Tube Feeding Routes

For each type of tube placement, the terms are listed in order from the upper to lower organs of the digestive system.

transnasal: a *transnasal feeding tube* is one that is inserted through the nose.

- **nasogastric (NG):** tube is placed into the stomach via the nose.
- **nasointestinal:** tube is placed into the small intestine via the nose; refers to *nasoduodenal* and *nasojejunal* feeding routes.
- **nasoduodenal (ND):** tube is placed into the duodenum via the nose.
- **nasojejunal (NJ):** tube is placed into the jejunum via the nose.

orogastric: tube is inserted into the stomach through the mouth. This method is often used to feed infants because a nasogastric tube may hinder the infant's breathing.

enterostomy (EN-ter-AH-stoe-mee): an opening into the GI tract through the abdominal wall.

- **gastrostomy (gah-STRAH-stoe-mee):** an opening into the stomach through which a feeding tube can be passed. A nonsurgical technique for creating a gastrostomy under local anesthesia is called *percutaneous endoscopic gastrostomy (PEG).*
- **jejunostomy (JEH-ju-NAH-stoe-mee):** an opening into the jejunum through which a feeding tube can be passed. A nonsurgical technique for creating a jejunostomy is called *percutaneous endoscopic jejunostomy (PEJ).* The tube can either be guided into the jejunum via a gastrostomy or passed directly into the jejunum *(direct PEJ).*

feedings are more easily tolerated and less complicated to deliver than intestinal feedings because the stomach controls the rate at which nutrients enter the intestine. Gastric feedings are not possible, however, if patients have gastric obstructions or motility disorders that interfere with the stomach's ability to empty.

Gastric feedings are often avoided in patients at high risk of **aspiration**, a common complication in which fluids enter the lungs, either from the backflow of stomach contents or secretions from the mouth and pharynx. **Aspiration pneumonia**, a lung disease that is sometimes fatal, may result. Aspiration risk is high in patients with esophageal disorders, neuromuscular diseases, and conditions that reduce consciousness or cause dementia. Although health practitioners frequently administer nasointestinal feedings to minimize the likelihood of aspiration, studies have not consistently shown that gastric feedings are associated with a higher pneumonia risk.[4] Table 15-1 summarizes the advantages and disadvantages of the various tube feeding routes.

NURSING DIAGNOSIS

The nursing diagnosis *risk for aspiration* may apply to certain patients using tube feedings.

aspiration: drawing in by suction or breathing; a common complication of enteral feedings in which foreign material enters the lungs, often from GI secretions or the reflux of stomach contents.

aspiration pneumonia: a lung disease resulting from the abnormal entry of foreign material; caused by either bacterial infection or irritation of the lower airways.

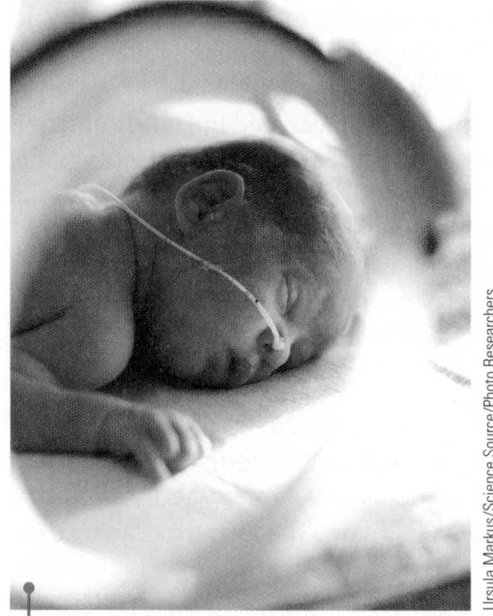

A transnasal feeding tube accesses the GI tract via the nose.

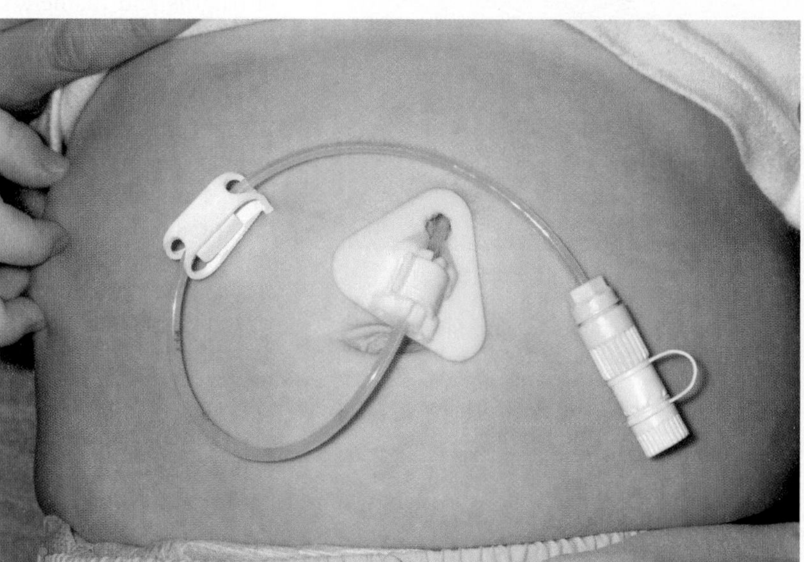

In a gastrostomy, the feeding tube accesses the GI tract through the abdominal wall.

TABLE 15-1 Comparison of Tube Feeding Routes[a]

Insertion Method or Feeding Site	Advantages	Disadvantages
Transnasal	Does not require surgery or incisions for placement; tubes can be placed by a nurse or trained dietitian.	Easy to remove by disoriented patients; long-term use may irritate the nasal passages, throat, and esophagus.
Nasogastric	Easiest to insert and confirm placement; least expensive method; feedings can be given intermittently and without an infusion pump.	Highest risk of aspiration in compromised patients;[b] risk of tube migration to small intestine.
Nasoduodenal and nasojejunal	Lower risk of aspiration in compromised patients;[b] allows for earlier tube feedings than gastric feedings during critical illness; may allow enteral feedings even when obstructions, fistulas, or other medical conditions prevent gastric feedings.	More difficult to insert and confirm placement; risk of tube migration to stomach; feedings require an infusion pump for administration.
Tube enterostomies	Allow the lower esophageal sphincter to remain closed, reducing the risk of aspiration;[b] more comfortable than transnasal insertion for long-term use; site is not visible under clothing.	Tubes must be placed by physician or surgeon; general anesthesia may be required for surgically placed tubes; risk of complications from the insertion procedure; risk of infection at insertion site.
Gastrostomy	Feedings can often be given intermittently and without a pump; easier insertion procedure than a jejunostomy.	Moderate risk of aspiration in high-risk patients.[b]
Jejunostomy	Lowest risk of aspiration;[b] allows for earlier tube feedings than gastrostomy during critical illness; may allow enteral feedings even when obstructions, fistulas, or medical conditions prevent gastric feedings.	Most difficult insertion procedure; most costly method; feedings require an infusion pump for administration.

[a]Relative to other tube-feeding routes. The actual advantages and disadvantages of different insertion procedures depend on the person's medical condition.
[b]The risk of aspiration associated with the different feeding routes is controversial and still under investigation.

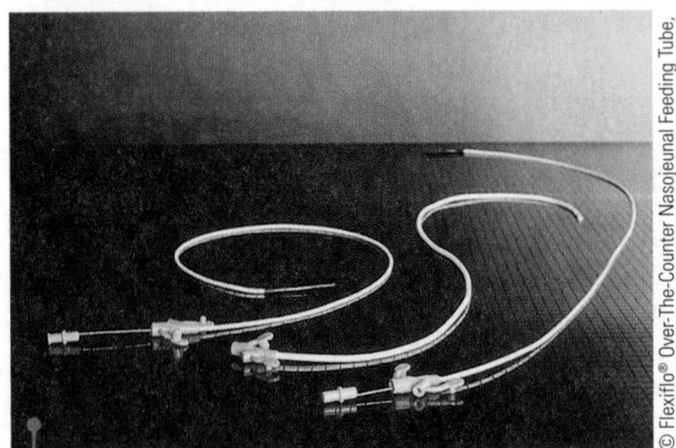

The thin wires protruding from the ends of these feeding tubes are stylets, which stiffen the tubes to ease insertion and are discarded thereafter. The orange Y-connectors provide ports for administering water or medications without disrupting the feeding.

■ 1 French = ⅓ mm
12 French = 12 × ⅓ mm = 4 mm

French units: units of measure that indicate the size of a feeding tube's outer diameter; 1 French unit equals ⅓ millimeter.

Feeding Tubes

Feeding tubes are made from soft, flexible materials (usually silicone, polyurethane, or polyvinyl) and come in a variety of lengths and diameters. The tube selected largely depends on the patient's age and size, the feeding route, and the formula's viscosity. In many cases, the tube selected is the smallest-diameter tube through which the formula will flow without clogging.

The outer diameter of a feeding tube is measured in **French units**, in which each unit equals ⅓ millimeter; thus, a "12 French" feeding tube has a 4-millimeter diameter. (■) The inner diameter depends on the thickness of the tubing material. Double-lumen tubes are also available; these allow a single tube to be used for both intestinal feedings and **gastric decompression**, a procedure in which the stomach contents of patients with motility disorders are removed by suction.

ENTERAL FORMULAS

Most enteral formulas can supply all of an individual's nutrient requirements when consumed in sufficient volume, a necessity for the patient who is using a tube feeding for more than a few days. The formulas can be used alone or provided along with other foods.

Types of Enteral Formulas

Although dozens of enteral formulas are commercially available, most health care institutions stock a limited number.[5] Appendix G lists some examples of enteral formulas as well as the amounts of protein, carbohydrate, fat, and energy they provide. The main types of formulas include the following:

- **Standard formulas**, also called *polymeric formulas*, are provided to individuals who can digest and absorb nutrients without difficulty. They contain intact proteins extracted from milk or soybeans (called **protein isolates**) or a combination of such proteins. The carbohydrate sources include hydrolyzed cornstarch, glucose polymers (such as maltodextrin and corn syrup solids), and sugars. A few commercial formulas, called **blenderized formulas**, are produced from a mixture of whole foods such as chicken, vegetables, fruits, and oil, along with some added vitamins and minerals.
- **Elemental formulas**, also called *hydrolyzed*, *chemically defined*, or *monomeric formulas*, are prescribed for patients who have compromised digestive or absorptive functions. Elemental formulas contain proteins and carbohydrates that have been partially or fully broken down to fragments that require little (if any) digestion. The formulas are often low in fat and may provide fat from **medium-chain triglycerides (MCT)** to ease digestion and absorption.
- **Specialized formulas**, also called *disease-specific* or *specialty formulas*, are designed to meet the specific nutrient needs of patients with particular illnesses. Products have been developed for individuals with liver, kidney, and lung diseases; glucose intolerance; severe wounds; and metabolic stress. Specialized formulas are generally expensive, and their effectiveness is controversial.
- **Modular formulas**, created from individual macronutrient (■) preparations called *modules*, are sometimes prepared for patients who require specific nutrient combinations to treat their illnesses. Vitamin and mineral preparations are also included in these formulas so that they can meet all of a person's nutrient needs. In some cases, one or more modules are added to other enteral formulas to adjust their nutrient composition.

Macronutrient Composition The amounts of protein, carbohydrate, and fat in enteral formulas vary substantially. The protein content of most formulas ranges from 10 to 25 percent of total kcalories;[6] note that protein needs are high in patients with severe metabolic stress, whereas protein restrictions are necessary for patients with chronic kidney disease. Carbohydrate and fat provide most of the energy in enteral formulas; standard formulas generally provide 30 to 60 percent of kcalories from carbohydrate and 10 to 45 percent of kcalories from fat.

Energy Density The energy density of most enteral formulas ranges from 1.0 to 2.0 kcalories per milliliter of fluid. The formulas that have lower energy densities are appropriate for patients with average fluid requirements. Formulas with higher energy densities can meet energy and nutrient needs in a smaller volume of fluid and therefore benefit patients who have high nutrient needs or fluid restrictions. Individuals with high fluid needs can be given a formula with low energy density or be supplied with additional water via the feeding tube or intravenously.

Fiber Content Fiber-containing formulas can be helpful for improving fecal bulk and colonic function, treating diarrhea or constipation, or maintaining blood glucose control. Conversely, fiber-containing formulas are avoided in patients with acute intestinal conditions or pancreatitis and before or after some intestinal examinations and surgeries.

Osmolality **Osmolality** refers to the moles of osmotically active solutes (or *osmoles*) per kilogram of water. (■) An enteral formula with an osmolality similar to that of blood serum (about 300 milliosmoles per kilogram) is an **isotonic formula**, whereas a **hypertonic formula** has an osmolality greater than that of blood serum.

Most enteral formulas have osmolalities between 300 and 700 milliosmoles per kilogram; generally, elemental formulas and nutrient-dense formulas have higher osmolalities than standard formulas. Most people are able to tolerate both isotonic

gastric decompression: the removal of the stomach contents (including swallowed saliva, stomach secretions, and gas) of patients with motility disorders or obstructions that prevent stomach emptying.

standard formulas: enteral formulas that contain mostly intact proteins and polysaccharides; also called *polymeric formulas.*

protein isolates: proteins that have been isolated from foods.

blenderized formulas: enteral formulas that are prepared by using a food blender to mix and puree whole foods.

elemental formulas: enteral formulas that contain proteins and carbohydrates that are partially or fully hydrolyzed; also called *hydrolyzed, chemically defined,* or *monomeric formulas.*

medium-chain triglycerides (MCT): triglycerides that contain fatty acids that are 8 to 10 carbons in length. MCT do not require digestion and can be absorbed in the absence of lipase or bile.

specialized formulas: enteral formulas designed to meet the nutrient needs of patients with specific illnesses; also called *disease-specific* or *specialty formulas.*

modular formulas: enteral formulas prepared in the hospital from *modules* that contain single macronutrients; used for people with unique nutrient needs.

osmolality (OZ-moe-LAL-ih-tee): the concentration of osmotically active solutes in a solution, expressed as milliosmoles (mOsm) per kilogram of solvent.

isotonic formula: a formula with an osmolality similar to that of blood serum (about 300 milliosmoles per kilogram).

iso = equal
tono = pressure

hypertonic formula: a formula with an osmolality greater than that of blood serum.

■ Reminder: The *macronutrients* are proteins, carbohydrates, and fats.

■ Osmotically active solutes affect *osmosis*, the movement of water across semipermeable membranes.

and hypertonic feedings without difficulty. When medications are infused along with enteral feedings, however, the osmotic load increases substantially and may contribute to the diarrhea experienced by many tube-fed patients.

Formula Selection The formula is selected after careful assessment of the patient's medical problems, fluid and nutrition status, and ability to digest and absorb nutrients; some of the factors considered are shown in Figure 15-3. The formula chosen should meet the patient's medical and nutrient needs with the lowest risk of complications and the lowest cost. The vast majority of patients can use standard formulas. A person with a functional, but impaired, GI tract may require an elemental formula. Factors that influence formula selection include:

- *Nutrient and energy needs.* As with patients consuming regular diets, an adjustment in macronutrient and energy intakes may be necessary for tube-fed patients. For example, patients with diabetes may need to control carbohydrate intake, critical care patients may have high protein and energy requirements, and patients with chronic kidney disease may need to limit their intakes of protein and several minerals.

FIGURE 15-3 Selecting a Formula

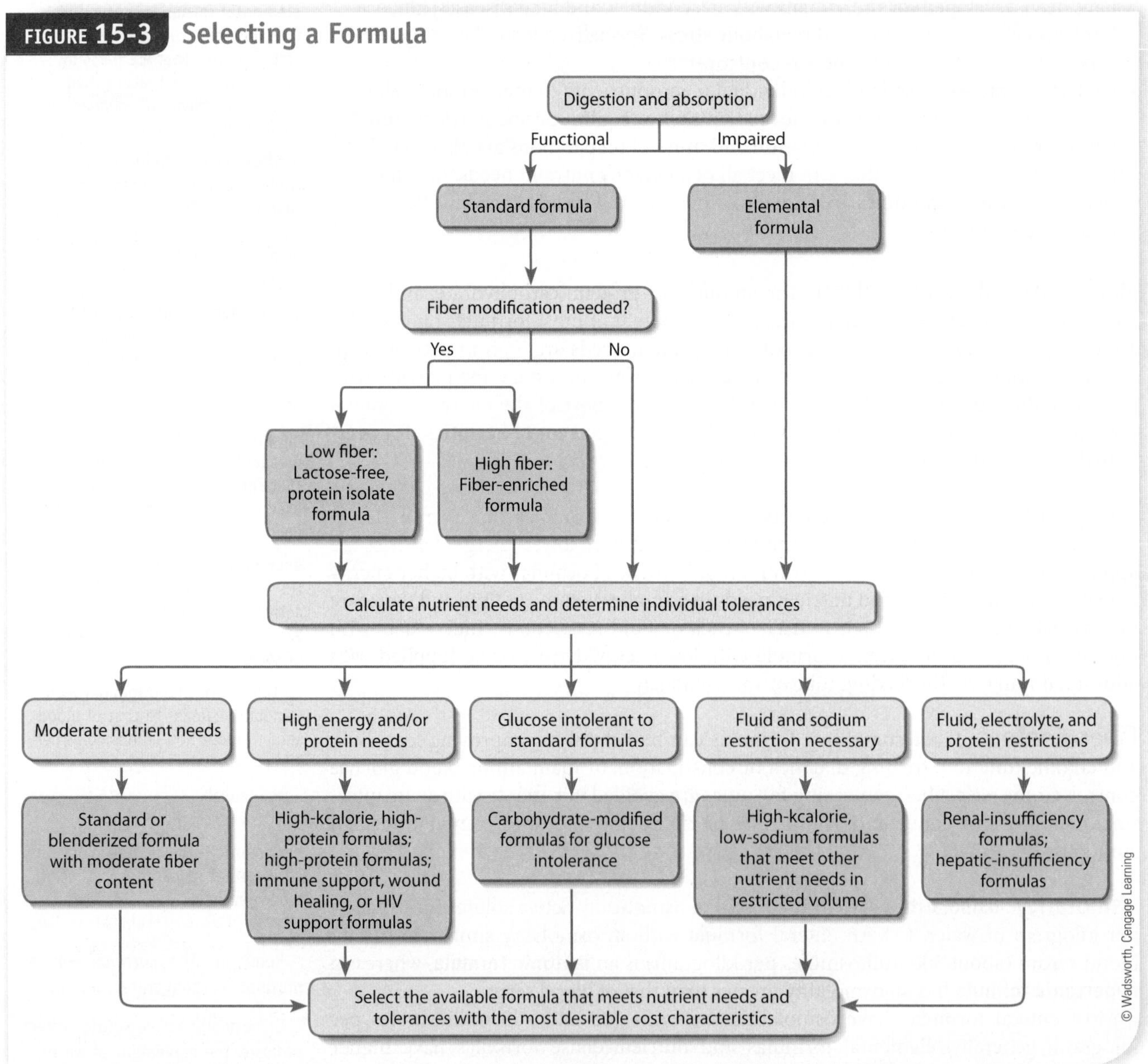

- *Fluid requirements.* High nutrient needs must be met using the volume of formula a patient can tolerate. If fluids are restricted, the formula should have adequate nutrient content and energy density to provide the required nutrients in the volume prescribed.
- *The need for fiber modifications.* The choice of formulas is narrower if fiber intake needs to be high or low. Formulas that provide fiber may be helpful for managing diarrhea, constipation, or hyperglycemia in some patients; other patients may need to avoid fiber due to an increased risk of bowel obstructions.
- *Individual tolerances (food allergies and sensitivities).* Most formulas are lactose free because many patients who need enteral formulas have some degree of lactose intolerance. Many formulas are also gluten free and can accommodate the needs of individuals with celiac disease (gluten sensitivity).

In most cases, formula selection is influenced by availability at the health care facility. The medical staff may initially choose a formula based on the criteria previously mentioned, and then reevaluate the decision according to the patient's response to the formula. Note that few research studies have assessed the effectiveness of the various specialized formulas, so their additional expense may be difficult to justify.

Safe Handling Individuals who are ill or malnourished often have suppressed immune systems, making them vulnerable to infection from foodborne illness. Thus, the personnel involved with preparing or delivering formula—usually individuals in the foodservice department or pharmacy—should be aware of the specific protocols at their facility that prevent formula contamination. (■)

■ Reminder: Health care facilities have protocols for handling food products and formulas based on the potential hazards and critical control points in food preparation, referred to as *Hazard Analysis and Critical Control Points* (*HACCP*).

Formulas are available as *open feeding systems* and *closed feeding systems.* With an **open feeding system**, the formula needs to be transferred from its original packaging to a feeding container. Examples include formulas that are packaged in cans or bottles, concentrates that need to be diluted, and powders that require reconstitution. In a **closed feeding system**, the sterile formula is prepackaged in a container that can be connected directly to a feeding tube. Closed systems are less likely to become contaminated, require less nursing time, and can hang for longer periods of time than

open feeding system: a formula delivery system that requires the formula to be transferred from its original packaging to a feeding container before being administered through the feeding tube.

closed feeding system: a formula delivery system in which the formula comes prepackaged in a container that can be attached directly to the feeding tube for administration.

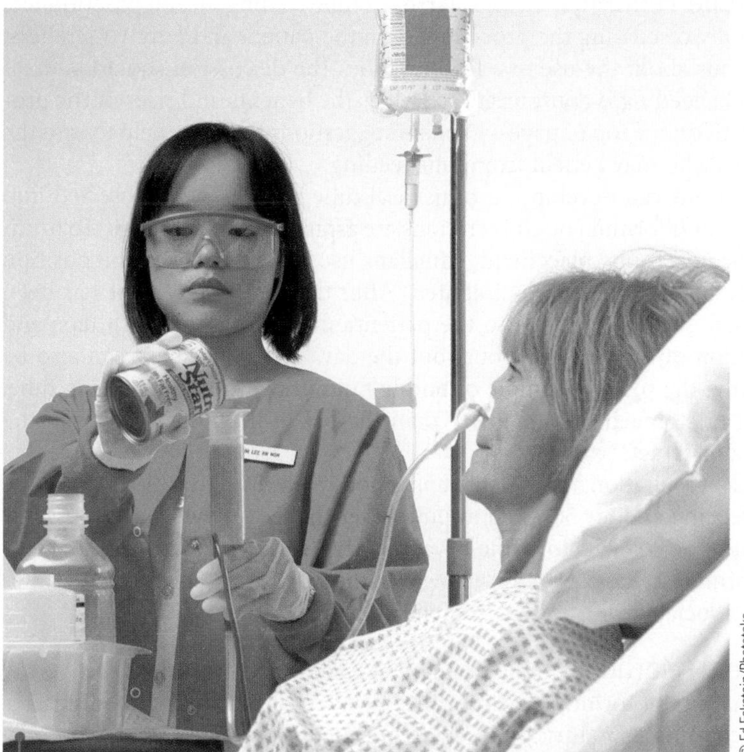

In an open feeding system, the formula is transferred from its original packaging to a feeding container.

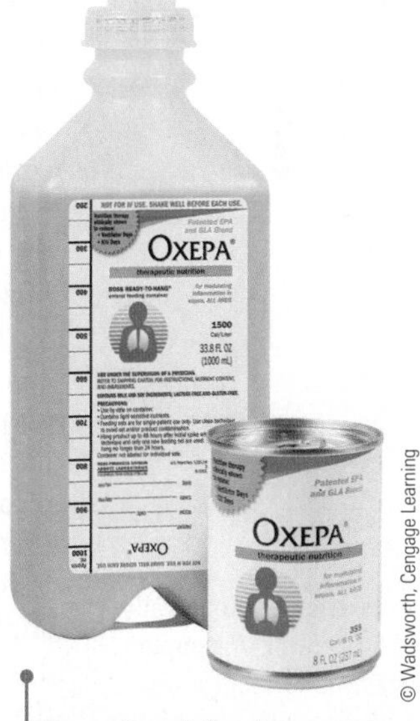

In a closed feeding system, the sterile formula is prepackaged in a container that can be attached directly to a feeding tube, such as the bottle shown on the left. The formula in the can at right can be used in an open feeding system.

open systems. Although closed systems cost more initially, they may be less expensive in the long run because they prevent bacterial contamination and thus avoid the costs of treating infections.

Formula Safety Guidelines After the formula reaches the nursing station, the nursing staff assumes responsibility for its safe handling. Hands should be carefully washed before handling formulas and feeding containers. Some facilities require that nonsterile gloves be worn whenever formulas are handled. The following steps can reduce the risk of formula contamination when using open feeding systems:

- Before opening a can of formula, clean the lid with a disposable alcohol wipe and wash the can opener with detergent and hot water. (Check HACCP protocols for details.) If you do not use the entire can at one feeding, label the can with the date and time it was opened.
- Store opened cans or mixed formulas in clean, closed containers. Refrigerate the unused portion of formula promptly. Discard unlabeled or improperly labeled containers and all opened containers of formula that are not used within 24 hours.
- Hang no more than an 8-hour supply of formula (or a 4-hour supply for newborn infants) when using liquid formula from a can. Formulas prepared from powders or modules should hang no longer than 4 hours. Discard any formula that remains, rinse out the feeding bag and tubing, and add fresh formula to the feeding bag. Use a new feeding container and tubing (except for the feeding tube itself) every 24 hours.[7]

For closed feeding systems, the hang time should be no longer than 24 to 48 hours. Contamination is more likely with the longer time periods.

ADMINISTRATION OF TUBE FEEDINGS

The methods of tube feeding administration vary somewhat from one health care facility to the next. The procedures presented in the following sections are suggested guidelines.

Preparing for Tube Feedings Before starting a tube feeding, health practitioners can ease fears by fully discussing the procedure with the patient and family members, who may feel anxious about the use of a feeding tube. The discussion should address the reasons why tube feeding is appropriate as well as the benefits and risks of the procedure. The "How to" on the top of page 443 offers suggestions that may help to ease the concerns of patients who may benefit from tube feeding.

Serious complications can develop if a transnasal tube is accidentally inserted into the respiratory tract or if formula or GI secretions are aspirated into the lungs. To minimize the risk of incorrect tube placement, clinicians use X-rays to verify the position of the feeding tube before a feeding is initiated. After the tube's placement has been confirmed, the nurse secures the tube to the patient's nose and cheek with tape and monitors the position of the tubing throughout the day. Tube placement can also be monitored by testing the pH of a sample of bodily fluid drawn into the feeding tube; recall that the pH of stomach fluid is lower than the pH of fluid obtained from the intestine or respiratory tract. (■)

■ A fasting gastric sample usually has a pH of 5 or lower. A sample from the intestine or respiratory tract has a pH of about 6 or higher.

To reduce the risk of aspiration, the patient's upper body is elevated to a 30- to 45-degree angle during the feeding and for 30 to 60 minutes after the feeding whenever possible. The addition of blue food coloring to formula was formerly suggested as a means of identifying aspirated formula in lung secretions; however, the practice was discontinued after it was found to be associated with various complications and even deaths.[8]

Formula Delivery Methods A day's nutrient needs can be met by delivering relatively large amounts of formula several times per day (**intermittent feedings**) or smaller amounts continuously during the day (**continuous feedings**). A patient may also start with continuous feedings and gradually transition to intermittent feedings. Each method has specific uses, advantages, and disadvantages.

intermittent feedings: delivery of about 250 to 400 milliliters of formula over 30 to 45 minutes.

continuous feedings: slow delivery of formula at a constant rate over an 8- to 24-hour period.

HOW TO Help Patients Cope with Tube Feedings

Patients may be less apprehensive about tube feedings once they understand the insertion procedure, the expected duration of the tube feeding, and the strategic role that nutrition plays in recovery from disease. The pointers that follow can help health practitioners prepare patients for transnasal tube feedings:

- Allow the patient to see and touch the feeding tube. Understanding that the tube is soft and narrow (only about half the diameter of a pencil) often alleviates anxiety.
- Show the patient how the feeding equipment is attached to the feeding tube, and explain how the feeding will work. For young children, use dolls or stuffed toys to demonstrate tube insertion and feeding procedures.
- Explain that the patient remains fully alert during the procedure and helps to pass the tube by swallowing. A numbing solution sprayed on the back of the throat minimizes discomfort and prevents gagging during the procedure.
- Inform the patient that after the tube has been inserted, most people become accustomed to its presence within a few hours. In most cases, the patient can continue to swallow foods and beverages with the tube in place.

Tube feedings may cause some patients to feel that they have lost control over an important aspect of their lives. They may also feel self-conscious about how the feeding tube looks or feel awkward when moving around with the equipment. A few measures can help:

- Assure the patient that the tube feeding will be temporary, if such assurance is appropriate.
- Involve patients in the decision-making and care process whenever possible. Patients can help to arrange their daily feeding schedules and can perform some of the feeding procedures themselves.
- Show patients how to manipulate the feeding equipment so that they can get out of bed and move around.

Many patients may be relieved to know that they can receive sound nutrition without any effort. As they feel better and begin to eat again, the volume of enteral formula can be reduced and then discontinued when food intake is adequate.

Intermittent feedings are best tolerated when they are delivered into the stomach (not the intestine). Generally, a total of about 250 to 400 milliliters of formula is delivered over 30 to 45 minutes using a gravity drip method or an infusion pump. The exact amount is determined by dividing the required volume of formula into several daily feedings, as shown in the "How to" below. Due to the relatively high volume of formula delivered at one time, intermittent feedings may be difficult for some patients to tolerate, and the risk of aspiration may be higher than with continuous feedings. An advantage of intermittent feedings is that they are similar to the usual pattern of eating and allow the patient freedom of movement between meals.

Rapid delivery of a large volume of formula into the stomach (250 to 500 milliliters over 5 to 15 minutes) is called a **bolus feeding**. This type of feeding may be given every 3 to 4 hours using a syringe. Bolus feedings are convenient for patients and staff because they are rapidly administered, do not require an infusion pump, and allow greater independence for patients. However, bolus feedings can cause abdominal discomfort, nausea, and cramping in some patients, and the risk of aspiration is greater than with other methods of feeding. For these reasons, bolus feedings are used only in patients who are not critically ill.

bolus (BOH-lus) feeding: delivery of about 250 to 500 milliliters of formula over a 5- to 15-minute period.

HOW TO Plan a Tube Feeding Schedule

After selecting a suitable formula, the clinician must determine the volume of formula that meets the patient's nutritional needs. Consider a patient who needs 2000 kcalories daily and is using a standard formula that provides 1.0 kcalorie per milliliter. The total volume of formula required would be 2000 milliliters per day:

$$x \text{ mL} \times 1.0 \text{ kcal/mL} = 2000 \text{ kcal}$$

$$x \text{ mL} = \frac{2000 \text{ kcal}}{1.0 \text{ kcal/mL}} = 2000 \text{ mL}$$

If the patient is to receive intermittent feedings six times a day, he will need about 333 milliliters of formula at each feeding:

$$2000 \text{ mL} \div 6 \text{ feedings} = 333 \text{ mL/feeding}$$

Alternatively, if he is to receive intermittent feedings eight times a day, he will need 250 milliliters (or about one can of ready-to-feed formula) at each feeding:

$$2000 \text{ mL} \div 8 \text{ feedings} = 250 \text{ mL/feeding}$$

If the patient is to receive the formula continuously over 24 hours, he will need about 83 milliliters of formula each hour:

$$2000 \text{ mL} \div 24 \text{ hours} = 83 \text{ mL/hr}$$

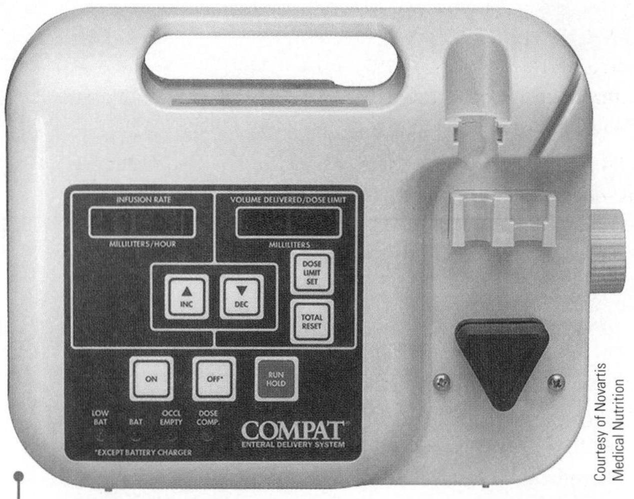

Courtesy of Novartis Medical Nutrition

The delivery of intermittent and continuous feedings can be controlled with an infusion pump.

Continuous feedings are delivered slowly and at a constant rate over a period of 8 to 24 hours. Continuous feedings are used to deliver intestinal feedings and are generally recommended for critically ill patients because the slower delivery rate may be easier to tolerate. Continuous feedings may also be recommended for patients who cannot tolerate intermittent feedings. An infusion pump is required to ensure accurate and steady flow rates; consequently, the feedings can limit the patient's freedom of movement and are also more costly.

Initiating and Advancing Tube Feedings Formula administration techniques vary widely among institutions, so protocols should be reviewed carefully before working with patients. Some general suggestions regarding the delivery of tube feedings in adult patients include the following:

- Formulas are typically provided full-strength, although they may occasionally be diluted if the patient's fluid requirements are high and water needs cannot be met by other means. Formula dilution may sometimes be necessary to improve the flow of a viscous formula.[9]
- Intermittent feedings may start with 60 to 120 milliliters at the initial feeding and be increased by 60 to 120 milliliters at each feeding until the goal volume is reached.[10]
- Continuous feedings can usually start at rates of about 40 to 60 milliliters per hour and be raised by 20 milliliters per hour until the goal rate is reached.[11]
- If the patient cannot tolerate an increased rate of delivery, the feeding rate is slowed until the person adapts. Goal rates can usually be achieved over 24 to 36 hours. In some patients, formula delivery can be started at the goal rate immediately.
- Slower rates of delivery may be better tolerated by critically ill patients, when concentrated formulas are used, or in patients who have undergone an extended period of bowel rest due to surgery, intestinal disease, or the use of parenteral nutrition.[12]

NURSING DIAGNOSIS

The nursing diagnoses *deficient fluid volume* and *excess fluid volume* may apply to patients who are receiving too little or too much water relative to needs.

Meeting Water Needs Although water needs vary, (■) many adults require about 2000 milliliters (about 2 quarts) of water daily. Additional water is required in patients with severe vomiting, diarrhea, fever, excessive sweating, high urine output, fistula drainage, high-output ostomies, blood loss, or open wounds. Fluids may be restricted in persons with kidney, liver, or heart disease.

The water in formulas meets a substantial portion of water needs. Enteral formulas contain about 70 to 85 percent water, or about 700 to 850 milliliters of water per liter of formula. In addition to the water in formulas, water can be provided by flushing water separately through the feeding tube. Water flushes are also conducted to prevent feeding tubes from clogging; the tubes are flushed with about 30 milliliters of warm water about every 4 hours during continuous feedings and before and after each intermittent feeding. The water used for routine flushes should be included when estimating fluid intakes.

■ To estimate fluid requirements in adults and children:

- Adults: allow 30–40 mL/kg; 25–30 mL/kg in adults ≥65 years old
- Children: allow 50–60 mL/kg
- Infants: allow 100–150 mL/kg

Checking the Gastric Residual Volume When a patient receives a gastric feeding, the nurse regularly measures the **gastric residual volume** (the volume of formula and GI secretions remaining in the stomach after feeding) to ensure that the stomach is emptying properly. The gastric residual volume is measured by gently withdrawing the gastric contents through the feeding tube using a syringe, usually before each intermittent feeding and every 4 to 6 hours during continuous feedings. Although opinions vary, some experts recommend that feedings be withheld and an evaluation be conducted if the gastric residual volume exceeds 500 milliliters.[13] If the tendency to accumulate fluids persists, the physician may recommend intestinal feedings or begin drug therapy to improve gastric emptying.

gastric residual volume: the volume of formula and GI secretions remaining in the stomach after a previous feeding.

MEDICATION DELIVERY DURING TUBE FEEDINGS

Patients receiving tube feedings sometimes require one or more medications that need to be delivered through feeding tubes. Because medications can interact with substances in enteral formulas in the same ways that they interact with substances in foods, potential diet-drug interactions must be considered. In addition, some medications may need to be exposed to the acidic stomach environment and thus cannot be administered via an intestinal feeding tube. Medications can also cause feeding tubes to clog. The "How to" provides some guidelines that may help to prevent complications.

Medications and Continuous Feeding

Continuous feedings are ordinarily stopped for 15 minutes before and after medication administration so that the components of enteral formulas do not interfere with the medication's absorption. Some medications may require a longer formula-free interval; for example, feedings need to be stopped for at least one hour before and after administering phenytoin, a medication that controls seizures.[14] In such cases, the formula's delivery rate needs to be increased so that the correct amount of formula can be delivered.

Diarrhea

Medications are a major cause of the diarrhea that frequently accompanies tube feedings. Diarrhea is especially associated with the administration of sorbitol-containing medications, laxatives, and some types of antibiotics.[15] The high osmolality of many liquid medications can also cause diarrhea, so dilution of hypertonic medications may be helpful.

NURSING DIAGNOSIS

The nursing diagnosis *diarrhea* often applies to patients receiving tube feedings.

TUBE FEEDING COMPLICATIONS

Complications are a frequent occurrence during tube feedings. Possible complications include gastrointestinal problems, such as constipation and diarrhea; mechanical problems related to the tube feeding process; and metabolic problems, such as nutrient deficiencies and changes in the body's biochemistry. Examples of the most common complications, along with some preventive and corrective measures, are summarized in Table 15-2.

HOW TO Administer Medications to Patients Receiving Tube Feedings

The pharmacist is your best resource for learning how and when medications can be administered via feeding tubes, especially when you are dealing with an unfamiliar drug. Check with the pharmacist to learn the following:

- Whether a particular medication is known to be incompatible with formulas.
- The proper timing of medication administration to avoid diet-drug interactions.
- For patients using intestinal feedings, whether a medication can be absorbed without exposure to stomach acid.
- Whether a liquid form of a medication is available and, if so, the appropriate dosage of the liquid form.
- If only tablets are available, whether the tablets can be crushed and mixed with water. Enteric-coated and sustained-release medications should not be crushed due to the potential for adverse effects.

In general, it is best to give medications by mouth instead of by tube whenever possible. In some cases, the injectable form of a medication may be the best option. For medications that must be given by feeding tube:

- Do not mix medications with enteral formulas. Do not mix medications together.
- Before administering medications, ensure that the feeding tube is placed correctly, that it is not clogged, and that the gastric residual volume is not excessive.
- Position the patient in a semi-upright position (30 degrees or higher) to prevent aspiration.
- Flush the feeding tube with 30 milliliters of warm water before and after administering a medication. When more than one medication is administered, flush the feeding tube with water between medications.
- Use liquid forms of medications whenever possible. Dilute viscous or hypertonic liquid medications with at least 30 milliliters of water before administering them through the feeding tube.
- If tablets are used, crush tablets to a fine powder and mix with about 30 milliliters of warm water before administering.

TABLE 15-2 Causes and Management of Tube Feeding Complications

Complications	Possible Causes	Preventive/Corrective Measures
Aspiration of formula	Inappropriate tube placement	Ensure correct placement of feeding tube.
	Delayed gastric emptying	Elevate head of bed during and after feeding; decrease formula delivery rate if gastric residual volume is excessive; consider using intestinal feedings in high-risk patients.
	Excessive sedation	Minimize use of medications that cause sedation.
Clogged feeding tube	Excessive formula viscosity	Ensure that tube size is appropriate; flush tubing with water before and after giving formula. Remedies to unclog feeding tubes include flushes with warm water or solutions that contain pancreatic enzymes and sodium bicarbonate; consult pharmacist for more options.
	Improper administration of medications	Use oral, liquid, or injectable medications whenever possible; flush tubing with water before and after a medication is given; avoid mixing medications with formula; dilute thick or sticky liquid medications before administering; crush tablets to a fine powder and mix with water (except enteric-coated or sustained-release medications).
Constipation	Inadequate dietary fiber	Use a formula with appropriate fiber content.
	Dehydration	Provide additional fluids.
	Lack of exercise	Encourage walking and other activities, if appropriate.
	Medication side effect	Consult physician about minimizing or replacing medications that cause constipation.
Diarrhea	Medication intolerance	Dilute hypertonic medications before administering; avoid using poorly tolerated medications.
	Infection in GI tract	Consult physician about specific diagnosis and appropriate treatment.
	Formula contamination	Review safety guidelines for formula preparation and delivery.
	Excessively rapid formula administration	Decrease formula delivery rate or use continuous feedings.
	Lactose or gluten intolerance	Use lactose-free or gluten-free formula in patients with intolerances.
Fluid and electrolyte imbalances	Diarrhea	See items under *Diarrhea*.
	Inappropriate fluid intake or excessive losses	Monitor daily weights, intake and output records, serum electrolyte levels, and clinical signs that indicate dehydration or overhydration; ensure that water intake and formula delivery rates are appropriate.
	Inappropriate insulin, diuretic, or other therapy	Ensure that medication doses are appropriate.
	Inappropriate nutrient intake	Use a formula with appropriate nutrient content; ensure that malnourished patients do not receive excessive nutrients.[a]
Nausea and vomiting, cramps	Delayed stomach emptying	Decrease formula delivery rate or use continuous feedings; halt feeding if gastric residual volume is excessive (>500 mL); evaluate for obstruction; consider use of medications to improve emptying rate.
	Formula intolerance	Ensure that formula is at room temperature, delivery rate is appropriate, and formula odor is not objectionable; consider using formula that is low in fat, low in fiber, or elemental.
	Medication intolerance	Consult physician about replacing medications that are poorly tolerated.
	Response to disease or disease treatment	Consider use of medications that control nausea and vomiting.

[a]An excessive nutrient intake in malnourished patients may cause *refeeding syndrome,* a disorder that can lead to fluid and electrolyte imbalances; see p. 454 for details.

Many complications of tube feeding can be prevented by choosing the most appropriate feeding route, formula, and delivery method. Attention to a patient's primary medical condition and medication use is important as well. Health practitioners responsible for the patient's day-to-day care monitor body weight, hydration status, and results of laboratory tests to detect problems before complications develop.

TRANSITION TO TABLE FOODS

After the patient's condition improves, the volume of formula can be tapered off as the patient gradually shifts to an oral diet. The steps in the transition depend on the patient's medical condition and the type of feeding the patient is receiving. Individuals using continuous feedings are often switched to intermittent feedings initially. In some patients, swallowing function may need to be evaluated before oral feedings begin. Patients receiving elemental formulas may begin the transition by using a standard formula, either orally or via tube feeding. If the patient has not consumed lactose for several weeks, a diet with minimal lactose may be better tolerated. Oral intake should supply about two-thirds of estimated nutrient needs before the tube feeding is discontinued completely.[16] The Case Study allows you to consider the many factors involved in tube feedings.

Review Notes

- Hospitals can provide oral supplements to patients at risk of becoming malnourished; these may include nutrient-dense formulas, milkshakes, fruit drinks, and various snack food items.
- Transnasal feeding routes are preferred for short-term tube feedings, whereas enterostomies are used for longer-term feedings. Gastric feedings are preferred but may be avoided in patients at risk of aspiration.
- Patients may receive standard, elemental, specialized, or modular formulas, which differ in their macronutrient composition, energy density, fiber content, and osmolality. The formula can be delivered intermittently, in bolus feedings, or continuously.
- Formulas can meet a substantial portion of the patient's water requirements, and additional water can be provided by flushing water through the feeding tube.
- Medications should be given separately and accompanied by water flushes to prevent tube clogging.
- Complications of tube feedings can be gastrointestinal, mechanical, or metabolic in nature.

Case Study: Injured Hiker Requiring Enteral Nutrition Support

Sharyn Bartell is a 24-year-old student who suffered multiple fractures when she fell from a cliff while hiking. She has been in the hospital for two weeks and has no appetite. Due to her injuries, she is in traction and is immobile, although the head of her bed can be elevated 45 degrees. Sharyn weighed 140 pounds upon her arrival in the hospital, but she has lost 8 pounds over the course of her hospitalization. The health care team decides that nasoduodenal tube feeding should be instituted before her nutrition status deteriorates further. The standard formula selected for the feeding is lactose free, and Sharyn's nutrient requirements can be met with 2200 milliliters of the formula per day.

1. What steps can be taken to prepare Sharyn for tube feeding? What are some general reasons why nasoduodenal placement of the feeding tube might be preferred over nasogastric placement?
2. The physician's orders specify that the feeding should be given continuously over 18 hours. Using the method shown in the "How to" on p. 443, develop an appropriate tube feeding schedule.
3. Estimate Sharyn's fluid needs using her current weight and the fluid intake range suggested in the margin on p. 444. If Sharyn's formula is 80 percent water, will she receive enough water from the formula? If not, estimate the amount of additional fluid she would need, and explain how this fluid can be provided.
4. Describe precautions that should be taken if Sharyn is to receive medications through the feeding tube.
5. After three days of feeding, Sharyn develops diarrhea. Check Table 15-2 to determine the possible causes. What measures can be taken to correct the diarrhea?

Parenteral Nutrition

The first half of this chapter described how oral supplements and enteral formulas can improve or replace a regular diet. The supplements and formulas cannot be used when intestinal function is inadequate, however, and therefore the ability to meet nutrient needs intravenously is a lifesaving option for critically ill persons. The procedure is costly, however, and is associated with a number of potentially dangerous complications. As previous sections suggested, enteral nutrition support is preferred if the GI tract is functional.

CANDIDATES FOR PARENTERAL NUTRITION

peripheral veins: the small-diameter veins that carry blood from the limbs.

central veins: the large-diameter veins located close to the heart.

peripheral parenteral nutrition (PPN): the infusion of nutrient solutions into peripheral veins, usually a vein in the arm or back of the hand.

phlebitis (fleh-BYE-tiss): inflammation of a vein.

osmolarity: the concentration of osmotically active solutes in a solution, expressed as milliosmoles per liter of solution (mOsm/L). *Osmolality* (mOsm/kg) is an alternative measure used to describe a solution's osmotic properties.

total parenteral nutrition (TPN): the infusion of nutrient solutions into a central vein.

Parenteral nutrition is generally recommended for patients who are unable to digest or absorb nutrients and who are either malnourished or likely to become so (review Figure 15-1). In addition, some medical situations require bowel rest for an extended period due to intestinal inflammation or tissue damage. Thus, patients with the following conditions are often considered candidates for parenteral nutrition:

- Intestinal obstructions or fistulas
- Paralytic ileus (intestinal paralysis)
- Short bowel syndrome (a substantial portion of the small intestine has been removed)
- Intractable vomiting or diarrhea
- Severe gastrointestinal bleeding
- Bone marrow transplants
- Severe malnutrition and intolerance to enteral nutrition

VENOUS ACCESS

Once the decision to use parenteral nutrition has been made, the access site must be selected. The access sites for intravenous feedings fall into two main categories: the **peripheral veins** located in the arms and back of the hands, and the large-diameter **central veins** located near the heart.

Peripheral Parenteral Nutrition

In **peripheral parenteral nutrition (PPN)**, nutrients are delivered using only the peripheral veins. Peripheral veins can be damaged by overly concentrated solutions, however—**phlebitis** may develop, characterized by redness, swelling, and tenderness at the infusion site. To prevent phlebitis, the **osmolarity** of parenteral solutions used for PPN is generally kept below 900 milliosmoles per liter,[17] a concentration that limits the amounts of energy and protein the solution can provide. The "How to" compares the terms *osmolarity* and *osmolality*, both of which can be used to express the osmolar concentration of a solution.

PPN is most often used in patients who require short-term nutrition support (about 7 to 10 days) and do not have high nutrient needs or fluid restrictions. The use of PPN is not possible if the peripheral veins are too weak to tolerate the procedure. In many cases, clinicians must rotate venous access sites to avoid damaging veins.

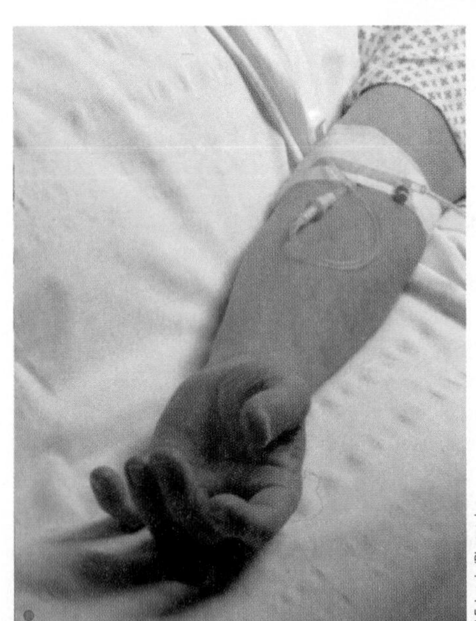

Ed Eckstein/Phototake

The peripheral veins can provide access to the blood for delivery of parenteral solutions.

Total Parenteral Nutrition

Most patients meet their nutrient needs using the larger, central veins, where blood volume is greater and nutrient concentrations do not need to be limited. This method can reliably provide all of a person's nutrient requirements and therefore is called **total parenteral nutrition (TPN)**. Because the central veins carry a large volume of blood, the parenteral solutions are rapidly diluted; thus, patients with high nutrient needs or fluid restrictions can receive the nutrient-dense solutions they require. TPN is also preferred for patients who require long-term parenteral nutrition.

HOW TO Express the Osmolar Concentration of a Solution

The movement of water across biological membranes (called *osmosis*) is influenced by a solution's concentration of *milliosmoles*, the ions and molecules that contribute to water's osmotic pressure. The terms *osmolarity* and *osmolality* are both used to express the concentration of these solutes:

- *Osmolarity* refers to the milliosmoles per liter of solution (mOsm/L).
- *Osmolality* refers to the milliosmoles per kilogram of solvent (mOsm/kg).

The main difference between the terms is that osmolarity refers to a volume of solution that includes the solutes of interest (milliosmoles contained in a liter of the solution), whereas osmolality refers to the solutes separately from the solvent in which they are dissolved (milliosmoles per kilogram of solvent). A second difference is that osmolarity is expressed in terms of the solution's volume, whereas osmolality is expressed in terms of the solvent's weight. Osmolarity and osmolality are roughly equivalent when describing dilute aqueous solutions at room or body temperature; this is because 1 liter of water weighs 1 kilogram, and the solutes contribute little to the volume of the solution.

Osmolarity and osmolality are both used in clinical practice, but they are derived in different ways. Osmolarity is typically calculated using equations that account for the nutrients and electrolytes in biological solutions. Osmolality is usually a measured value obtained using an *osmometer*, a common instrument in hospital laboratories.

There are several ways to access central veins. The tip of a central venous **catheter** can be placed directly into a large-diameter central vein or threaded into a central vein through a peripheral vein (see Figure 15-4). Peripheral insertion of central catheters is less invasive and lower in cost than the insertion of catheters directly into central veins.

PARENTERAL SOLUTIONS

The pharmacies located within health care institutions are often responsible for preparing parenteral solutions. This arrangement is convenient because the pharmacist can customize the solutions to meet patients' nutrient needs and because the solutions have a limited shelf life. The physician typically submits an order form such as the one shown in Figure 15-5 to the pharmacy. Prescriptions for parenteral solutions are highly individualized and may need to be recalculated daily until the patient's condition is stable.

catheter: a thin tube placed within a narrow lumen (such as a blood vessel) or body cavity; can be used to infuse or withdraw fluids or keep a passage open.

FIGURE 15-4 Accessing Central Veins for Total Parenteral Nutrition

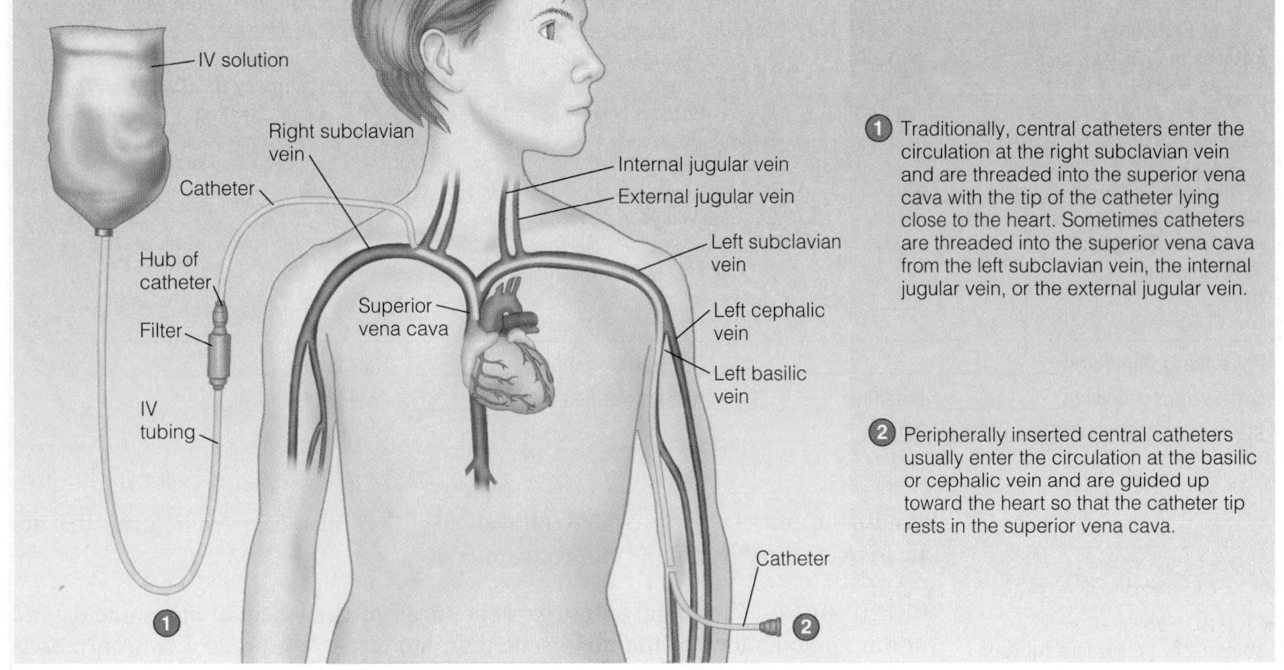
IV solution
Right subclavian vein
Catheter
Internal jugular vein
External jugular vein
Hub of catheter
Left subclavian vein
Superior vena cava
Filter
Left cephalic vein
Left basilic vein
IV tubing
Catheter
1
2
1 Traditionally, central catheters enter the circulation at the right subclavian vein and are threaded into the superior vena cava with the tip of the catheter lying close to the heart. Sometimes catheters are threaded into the superior vena cava from the left subclavian vein, the internal jugular vein, or the external jugular vein.
2 Peripherally inserted central catheters usually enter the circulation at the basilic or cephalic vein and are guided up toward the heart so that the catheter tip rests in the superior vena cava.

FIGURE 15-5 Sample Parenteral Nutrition Order Form

Physician Orders
PARENTERAL NUTRITION (PN) – ADULT

Primary Diagnosis: ______________________ **Ht:** ________ cm **Dosing Wt:** ______ kg

PN Indication: ______________________ **Allergies** ______________________

Instructions: This form must be completed for a new order or continuation of PN and faxed to the Pharmacy by [Insert Time] to receive same day preparation. PN administration begins at [Insert Time]. Contact the Nutrition Support Service at (XXX) XXX-XXXX for additional information.

Administration Route: CVC or PICC *Note: Proper tip placement of the CVC or PICC must be confirmed prior to PN infusion*

Peripheral IV (PIV) (*Final PN Osmolarity ≤ ________ mOsm/L*)

Monitoring: Daily weights, Strict input & output, Bedside glucose monitoring every _____ hours

Na, K, Cl, CO_2, Glucose, BUN, Scr, Mg, PO_4 every ______________

T, Bili, Alk Phos, AST, ALT, Albumin, Triglycerides, Calcium every ______________

Base Solution: Select one — *Parenteral nutrition MUST be administered through a dedicated infusion port and filtered with a 1.2-micron in-line filter at all times. Discard any unused volume after 24 hours.*

PERIPHERAL 2-in-1	CENTRAL 2-in-1	CENTRAL 3-in-1
Dextrose ______ g	Dextrose ______ g	Dextrose ______ g
Amino Acids (*Brand* ________) ______ g	Amino Acids (*Brand* ________) ______ g	Amino Acids (*Brand* ________) ______ g
		Fat Emulsion (*Brand* ________) ______ g
For patients with PIV and established glucose tolerance; Provides _____ kcal; Maximum Rate not to exceed _____ mL/hour	*For patients with CVC or PICC and established glucose tolerance; Provides _____ kcal; Maximum Rate not to exceed _____ mL/hour*	*For patients with CVC or PICC and established glucose/fat emulsion tolerance; Provides _____ kcal; Maximum Rate not to exceed _____ mL/hour*
		Use of additional fat emulsion not required with 3-in-1 base solution

RATE & VOLUME: _____ mL/hour for _____ hours = _____ mL/day
Must specify

or **CYCLIC INFUSION:** _____ mL/hour for _____ hours, then _____ mL/hour for _____ hours = _____ mL/day

Fat Emulsion (*Brand* ____________) ***– via PIV or CVC with 2-in-1 base solutions*** *(Select caloric density & volume)*

10%	250 mL	Infuse at _____ mL/hour over _____ hours	Frequency ______________
20%	500 mL	*(Note: infusions < 4 or > 12 hours not recommended)*	*Discard any unused volume after 12 hours.*

Additives: *(per day)*		***Normal Dosages***
Sodium Chloride	_____ mEq	*1-2 mEq Sodium/kg/day*
as Acetate	_____ mEq	*pH or CO_2 dependent*
as Phosphate	_____ mmol of PO_4	*Consider if hyperkalemic*
Potassium Chloride	_____ mEq	*1-2 mEq Potassium/kg/day*
as Acetate	_____ mEq	*pH or CO_2 dependent*
as Phosphate	_____ mmol of PO_4	*20-40 mmol/day (1 mmol Phos = 1.5 mEq K)*
Calcium **Gluconate**	_____ mEq	*5-15 mEq/day*
Magnesium **Sulfate**	_____ mEq	*8-24 mEq/day*
Adult **Multivitamins**	_____ mL/day	*Contains Vitamin K 150 mcg*
Adult **Trace Elements**	_____ mL/day	*Zn ___ mg, Cu ___ mg, Mn ___ mg, Cr ___ mcg, Se ___ mcg (with normal hepatic function)*
H_2 **Antagonist** ________	_____ mg	*___ mg/day with normal renal function*
Other:		

Additives: *(per day)*
Regular Insulin _____ units
Recommend if hyperglycemic, start with 1 unit for every 10 g of dextrose

***Pharmacy Use Only:* Ca/PO_4 Limit Checked** _____
(Note: Some brands of amino acids contain phosphate)

Physician's Signature: ______________ **Pager Number:** ________ **Date/time:** ________

Orders transcribed by: ________ **Date/time:** ________ **Orders verified by:** ________ **Date/time:** ________

SEND COMPLETED ORDERS TO PHARMACY

Because the nutrients are provided intravenously, they must be given in forms that are safe to inject directly into the bloodstream.

■ A 10 percent amino acid solution supplies 10 g of amino acids per 100 mL of solution.

Amino Acids Parenteral solutions contain all of the essential amino acids and various combinations of the nonessential amino acids. Amino acid concentrations range from 3.5 to 20 percent; (■) the more concentrated solutions are used only for TPN. Just as in regular foods, the amino acids provide 4 kcalories per gram. Disease-specific products are available for patients with liver disease, kidney disease, and metabolic stress, but they are rarely used in practice due to lack of evidence of their benefit.[18]

Carbohydrate Glucose is the main source of energy in parenteral solutions. It is provided in the form dextrose monohydrate, in which each glucose molecule is associated with a single water molecule. Dextrose monohydrate provides 3.4 kcalories per gram, slightly less than pure glucose, which provides 4 kcalories per gram. Dextrose solutions for parenteral nutrition are available in concentrations between 5 and 70 percent. (■) Concentrations greater than 12.5 percent are used only in TPN solutions.[19]

■ A 10 percent dextrose solution provides 10 g of dextrose monohydrate per 100 mL of solution.

In parenteral solutions, the dextrose concentration is indicated by a "D" followed by its concentration in water (W) or normal saline (NS). For example, D5 or D5W indicates that a solution contains 5 percent dextrose in water. Similarly, D5NS means that a solution contains 5 percent dextrose in normal saline.

Lipids Lipid emulsions supply essential fatty acids and are a significant source of energy. Currently, the lipid emulsions available in the United States contain triglycerides from soybean oil,* egg phospholipids to serve as emulsifying agents, and glycerol to make the solutions isotonic. Lipid emulsions are available in 10, 20, and 30 percent solutions, providing 1.1, 2.0, and 3.0 kcalories per milliliter, respectively. Therefore, a 500-milliliter container of 10 percent lipid emulsion would provide 550 kcalories; the same volume of a 20 percent lipid emulsion would provide 1000 kcalories. (■) In the United States, the 30 percent lipid emulsion can be used for preparing mixed parenteral solutions but cannot be directly infused into patients.[20]

■ 500 mL of a 10% lipid emulsion:
500 mL × 1.1 kcal/mL = 550 kcal

500 mL of a 20% lipid emulsion:
500 mL × 2 kcal/mL = 1000 kcal

Lipid emulsions are often provided daily and may supply about 20 to 30 percent of total kcalories. Including lipids as an energy source reduces the need for energy from dextrose and thereby lowers the risk of hyperglycemia in glucose-intolerant patients. Lipid infusions must be restricted in patients with hypertriglyceridemia, however. There is also some concern that lipid emulsions that supply excessive amounts of linoleic acid (possibly including the amount in soybean oil) can suppress some aspects of the immune response.[21]

Fluids and Electrolytes Daily fluid needs usually range from 30 to 40 milliliters per kilogram of body weight in young adults and 25 to 30 milliliters per kilogram of body weight in adults who are 65 years and older, averaging between 1500 and 2500 milliliters for most people. The amounts are adjusted according to daily fluid losses and the results of hydration assessment.

The electrolytes added to parenteral solutions include calcium, magnesium, phosphorus, sodium, potassium, and chloride. The amounts in parenteral solutions differ from DRI values because the nutrients are infused directly into the blood and are not influenced by absorption, as they are when consumed orally. Because electrolyte imbalances can be lethal, electrolyte management by experienced professionals is necessary whenever intravenous therapies are used. Blood tests are administered daily to monitor electrolyte levels until patients have stabilized.

The electrolyte content of parenteral solutions is expressed in *milliequivalents (mEq)*, which are units indicating the number of ionic charges provided by electrolytes. (■) The body's fluids and parenteral solutions are neutral solutions that contain equal numbers of positive and negative charges.

■ Milliequivalents are determined by dividing an ion's molecular weight (MW) by its number of charges. For example:

- For calcium, MW = 40, and the ion has 2 positive charges: 40 ÷ 2 = 20. Thus, 1 mEq of Ca^{++} is equivalent to 20 mg of calcium.
- For sodium, MW = 23, and the ion has 1 positive charge: 23 ÷ 1 = 23. Thus, 1 mEq of Na^{+} is equivalent to 23 mg of sodium.
- 1 mEq of Ca^{++} has the same number of charges as 1 mEq of Na^{+}.

Vitamins and Trace Minerals Commercial multivitamin and trace mineral preparations are added to parenteral solutions to meet micronutrient needs. All of the vitamins are usually included, although a preparation without vitamin K is available for patients using warfarin therapy.[22] (■) The trace minerals usually added to parenteral solutions include chromium, copper, manganese, selenium, and zinc. Iron is excluded

■ Reminder: The anticoagulant warfarin works by interfering with vitamin K's blood-clotting function (see Chapter 14).

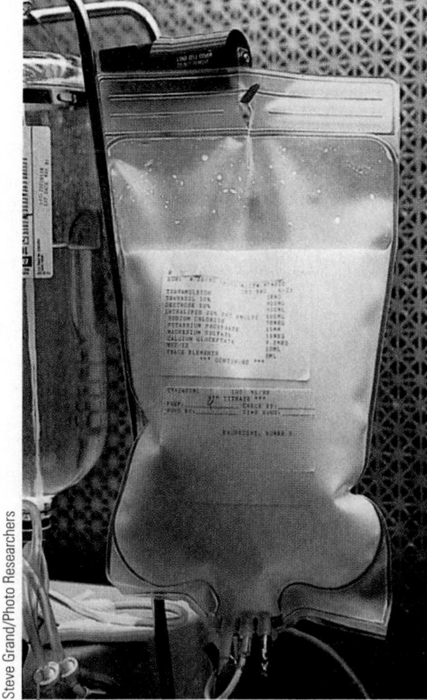

Steve Grand/Photo Researchers

A lipid emulsion gives the parenteral solution a milky white color.

*Outside of the United States, commercially available lipid emulsions may contain soybean oil, medium-chain triglycerides, fish oil, and/or olive oil.

because it can destabilize parenteral solutions that contain lipid emulsions and because the infused iron has had harmful effects in some patients; therefore, special forms of iron need to be injected separately.

Medications To avoid the need for a separate infusion site, medications are occasionally added directly to parenteral solutions or infused through a separate port in the catheter (attached via a Y-connector). The administration of a second solution using a separate port is called a **piggyback**. Insulin, for example, is sometimes added by piggyback to improve glucose tolerance. In practice, few medications are added to parenteral solutions so that potential diet-drug interactions can be avoided.

Parenteral Formulations When a parenteral solution contains dextrose, amino acids, and lipids, it is called a **total nutrient admixture (TNA)** or a **3-in-1 solution**. A **2-in-1 solution** excludes lipids, and the lipid emulsion is administered separately, often by piggyback administration. The administration of TNA solutions is simpler because only one infusion pump is required; however, the addition of lipid emulsion to solutions reduces their stability, a major concern when TNA solutions are compounded. Thus, lipids are usually administered separately when they are not a major energy source and are used only to provide essential fatty acids. The "How to" describes a method for calculating the macronutrient and energy content of a parenteral solution.

Osmolarity Recall that the osmolarity of PPN solutions is limited to 900 milliosmoles per liter because peripheral veins are sensitive to high nutrient concentrations, whereas TPN solutions may be as nutrient dense as necessary. The components of a solution that contribute most to its osmolarity are amino acids, dextrose, and electrolytes; as concentrations of these nutrients increase, the osmolarity of a solution increases. Because lipids contribute little to osmolarity, lipid emulsions are used to increase the energy provided by PPN solutions.

■ Reminder: A *nutrition support team* is a multidisciplinary team of health care professionals who are responsible for the provision of nutrients by tube feeding or intravenous infusion.

piggyback: the administration of a second solution using a separate port in an intravenous catheter.

total nutrient admixture (TNA): a parenteral solution that contains dextrose, amino acids, and lipids; also called a **3-in-1 solution**.

2-in-1 solution: a parenteral solution that contains dextrose and amino acids, but excludes lipids.

ADMINISTERING PARENTERAL NUTRITION

Parenteral nutrition is a complex treatment that requires skills from a variety of disciplines. Many hospitals organize nutrition support teams, (■) consisting of physicians, nurses, dietitians, and pharmacists, that specialize in the provision of both enteral and parenteral nutrition. The nurse, who performs direct patient care, plays a central role in administering and monitoring parenteral infusions.

HOW TO Calculate the Macronutrient and Energy Content of a Parenteral Solution

Suppose a patient is receiving 1.25 liters (1250 milliliters) of a parenteral solution that contains 5 percent amino acids and 30 percent dextrose, supplemented with 250 milliliters of a 20 percent lipid emulsion daily. How many grams of protein and carbohydrate is the person receiving, and what is the total energy intake for the day?

Amino acids:

$$5\% \text{ amino acids} = \frac{5 \text{ g amino acids}}{100 \text{ mL}}$$

$$\frac{5 \text{ g amino acids}}{100 \text{ mL}} \times 1250 \text{ mL} = 62.5 \text{ g of amino acids}$$

$$62.5 \text{ g amino acids} \times 4.0 \text{ kcal/g} = 250 \text{ kcal}$$

Carbohydrate:

$$30\% \text{ dextrose} = \frac{30 \text{ g dextrose}}{100 \text{ mL}}$$

$$\frac{30 \text{ g dextrose}}{100 \text{ mL}} \times 1250 \text{ mL} = 375 \text{ g of dextrose}$$

$$375 \text{ g dextrose} \times 3.4 \text{ kcal/g} = 1275 \text{ kcal}$$

Lipids:

Recall that a 20 percent lipid emulsion provides 2.0 kcalories per milliliter. If the patient is given 250 milliliters of the emulsion:

$$250 \text{ mL} \times 2.0 \text{ kcal/mL} = 500 \text{ kcal}$$

Total energy intake:

$$250 \text{ kcal} + 1275 \text{ kcal} + 500 \text{ kcal} = 2025 \text{ kcal}$$

Insertion and Care of Intravenous Catheters Although skilled nurses can place catheters into peripheral veins, only qualified physicians can insert catheters directly into central veins. Patients may be awake for the procedure and given local anesthesia. Unnecessary apprehension can be avoided by explaining the procedure to the patient beforehand.

Catheter-related problems frequently cause complications (see Table 15-3). Catheters may be improperly positioned or may dislodge after placement. Air can leak into catheters and escape into the bloodstream, obstructing blood flow. Catheters in peripheral veins may cause phlebitis, necessitating reinsertion at an alternate site. A catheter may become clogged from blood clotting or from a buildup of scar tissue around the catheter tip. Catheters are also a leading cause of infection: contamination may be introduced during insertion or may develop at the placement site.

> **NURSING DIAGNOSIS**
>
> The nursing diagnosis *risk for vascular trauma* may apply to patients undergoing intravenous therapy.

To reduce the risk of complications, nurses use aseptic techniques when inserting catheters, changing tubing, or changing a dressing that covers the catheter site. Unusual bleeding or a wet dressing suggests a problem with catheter placement. A change in infusion rate may indicate a clogged catheter. Infection may be indicated by redness or swelling around the catheter site or by an unexplained fever. Routine inspections of equipment and frequent monitoring of patients' symptoms help to minimize the problems associated with catheter use.

Administration of Parenteral Solutions The method used to initiate and advance parenteral nutrition depends on the patient's condition and the potential for complications. One approach is to start the infusion at a slow rate (with a solution that is either full strength or nutrient dilute) and increase the rate gradually over a two- to three-day period. For example, 40 milliliters per hour can be infused during the first 24 hours of administration (supplying 960 milliliters) and the volume increased to the goal rate on the second day. Another method is to give the full volume of a nutrient-dilute solution on the first day and advance nutrient concentrations as tolerated. Solutions can often be started at full volume and full strength unless there is a risk of hyperglycemia or other complications.[23]

Parenteral solutions can be infused continuously over 24 hours (**continuous parenteral nutrition**) or during 8- to 16-hour periods only (**cyclic parenteral nutrition**). Continuous infusions are given to critically ill and malnourished patients who cannot receive adequate nutrients in the shorter time periods. Cyclic infusions are sometimes provided at night so that patients can participate in routine activities during the day. This method is especially suited to patients who require long-term parenteral support or who will be infusing parenteral solutions at home. Patients may begin with continuous parenteral nutrition and transition to cyclic parenteral nutrition as their condition improves.

TABLE 15-3 Potential Complications of Parenteral Nutrition

Catheter-Related	Metabolic
• Air embolism	• Electrolyte imbalances
• Blood clotting at catheter tip	• Gallbladder disease
• Clogging of catheter	• Hyperglycemia, hypoglycemia
• Dislodgment of catheter	• Hypertriglyceridemia
• Improper placement	• Liver disease
• Infection, sepsis	• Metabolic bone disease
• Phlebitis	• Nutrient deficiencies
• Tissue injury	• Refeeding syndrome

continuous parenteral nutrition: continuous administration of parenteral solutions over a 24-hour period.

cyclic parenteral nutrition: administration of parenteral solutions over an 8- to 16-hour period each day.

Regular monitoring can help to prevent complications. The parenteral solution and tubing are checked frequently for signs of contamination. Routine testing of glucose, lipids, and electrolyte levels helps to determine tolerance to solutions. Frequent reassessment of nutrition status may be necessary until a patient has stabilized. Rapid changes in infusion rate are discouraged in some patients due to a risk of developing hyperglycemia or hypoglycemia.

Discontinuing Intravenous Infusions The patient must have adequate GI function before parenteral nutrition can be tapered off and enteral feedings begun. During the transition to oral feedings, a combination of methods is often necessary. Parenteral infusions are usually tapered off at the same time that tube feedings or oral feedings are begun, such that the two methods can together supply the needed nutrients. Clear liquids are generally the first foods offered and include pulp-free fruit juices, soft drinks, and clear broths; small amounts are given initially to determine tolerance. (■) Later feedings include beverages and solid foods that are unlikely to cause discomfort. If gastrointestinal symptoms (such as nausea, vomiting, bloating, or diarrhea) develop, oral feedings are limited in size or frequency until the intestines adapt. Once about 65 to 75 percent of nutrient needs can be provided enterally, the intravenous infusions may be discontinued.

Transitioning to an oral diet is sometimes difficult because a person's appetite remains suppressed for several weeks after parenteral nutrition is terminated. Patients receiving continuous parenteral nutrition may have better appetites during the day if they are switched to nocturnal cyclic infusions before beginning oral intakes.

MANAGING METABOLIC COMPLICATIONS

As discussed previously, the catheters used for intravenous infusions may cause a number of serious complications. This section describes some metabolic complications that may result from parenteral nutrition (review Table 15-3) and some suggestions for managing them.[24]

> **NURSING DIAGNOSIS**
>
> The nursing diagnoses *risk for unstable blood glucose level* and *risk for electrolyte imbalance* often apply to critically ill patients who begin parenteral nutrition.

Hyperglycemia Hyperglycemia (■) most often occurs in patients who are glucose intolerant, receiving excessive energy or dextrose, or undergoing severe metabolic stress. It can be prevented by providing insulin along with parenteral solutions or by restricting the amount of dextrose in parenteral solutions. Dextrose infusions are generally limited to less than 5 milligrams per kilogram of body weight per minute in critically ill adult patients so that the carbohydrate intake does not exceed the maximum glucose oxidation rate.

Hypoglycemia Although uncommon, hypoglycemia sometimes occurs when parenteral nutrition is interrupted or discontinued or if excessive insulin is given. In patients at risk, such as young infants, infusions may be tapered off over several hours before discontinuation. Another option is to infuse a dextrose solution at the same time that parenteral nutrition is interrupted or stopped.

Hypertriglyceridemia Hypertriglyceridemia may develop in critically ill patients who cannot tolerate the amount of lipid emulsion supplied. Patients at risk include those with severe infection, liver disease, kidney failure, or hyperglycemia and those using immunosuppressant or corticosteroid medications. If blood triglyceride levels exceed 500 milligrams per deciliter, lipid infusions should be reduced or stopped.[25]

■ Chapter 14 provides more information about the clear liquid diet.

■ For most patients receiving parenteral nutrition, blood glucose levels should not exceed about 200 mg/dL.

Refeeding Syndrome Severely malnourished patients who are fed aggressively (parenterally or otherwise) may develop **refeeding syndrome**, characterized by electrolyte and fluid imbalances and hyperglycemia. These effects occur because dextrose infusions raise circulating insulin levels, which promote anabolic processes that quickly remove phosphate, potassium, and magnesium from the blood. The altered electrolyte levels can lead to fluid retention and life-threatening changes in organ systems.

refeeding syndrome: a condition that sometimes develops when a severely malnourished person is aggressively fed; characterized by electrolyte and fluid imbalances and hyperglycemia.

Refeeding syndrome generally develops within two weeks of beginning parenteral infusions.[26] The patients at highest risk are those who have experienced chronic malnutrition or substantial weight loss. To prevent refeeding syndrome, health practitioners start parenteral infusions slowly and carefully monitor electrolyte and glucose levels when malnourished patients begin receiving nutrition support.

Liver Disease Fatty liver often results from parenteral nutrition, but it is usually corrected after the parenteral infusions are discontinued. Long-term parenteral nutrition, however, can result in progressive liver disease and may eventually lead to liver failure. To minimize the risk, practitioners are careful to avoid giving the patient excess energy, dextrose, or lipids, which promote fat deposition in the liver. Cyclic infusions may be less problematic than continuous infusions. If appropriate, some oral feedings may be encouraged to reduce the amount of parenteral support necessary.[27]

NURSING DIAGNOSIS

The nursing diagnosis *risk for impaired liver function* may apply to patients who undergo parenteral nutrition.

Gallbladder Disease When parenteral nutrition continues for more than four weeks, sludge (thickened bile) may build up in the gallbladder and eventually lead to gallstone formation. Prevention is sometimes possible by initiating oral feedings or enteral nutrition before problems develop. Patients requiring long-term parenteral nutrition may be given medications to stimulate gallbladder contraction or improve bile flow or may have their gallbladders removed surgically.

Metabolic Bone Disease Long-term parenteral nutrition is associated with reduced bone mineralization and lower bone density, which may be related to altered calcium, phosphorus, magnesium, and sodium metabolism. Inappropriate intakes of vitamin D, vitamin K, and phosphorus have also been implicated. The ideal intervention varies among patients; it may include dietary adjustments, nutrient supplements, medications, and physical activity.[28]

Review Notes

- Peripheral parenteral nutrition is provided to patients who need short-term parenteral support (about 7 to 10 days) and do not have high nutrient needs or fluid restrictions. Total parenteral nutrition supplies nutrient-dense solutions for long-term parenteral support.
- Parenteral solutions include amino acids, dextrose, electrolytes, vitamins, and minerals. Lipid emulsions may be included in the mixture or may be administered separately.
- Parenteral solutions may be initiated gradually or provided at full volume and full strength in selected patients. Critically ill patients may require continuous infusions, whereas healthier patients and long-term users may prefer cyclic infusions.
- Catheters are frequently the cause of complications, which include improper placement or dislodgment, infection, clotting, embolism, and phlebitis.
- Metabolic complications include hyperglycemia, hypoglycemia, hypertriglyceridemia, refeeding syndrome, and diseases affecting the liver, gallbladder, and bone. The Case Study checks your understanding of the concepts introduced in this section.

Case Study: Patient with Intestinal Disease Requiring Parenteral Nutrition

Jason Huang, a 27-year-old man with an inflammatory intestinal disease, underwent a surgical procedure in which a substantial portion of his small intestine was removed. He had received TPN prior to surgery and continued to receive it afterward. After 10 days, tube feeding was begun and initially delivered very small feedings.

1. List some reasons why the nutrition support team initially chose TPN as a means of nutrition support for this patient. How would you explain the need for parenteral nutrition to Jason?
2. Describe the components of a typical TPN solution. Calculate the energy content of 1 liter of a solution that provides 140 grams of dextrose monohydrate, 45 grams of amino acids, and 90 milliliters of 20 percent lipid emulsion. If Jason's energy requirement is 2100 kcalories per day, how many liters of solution will he need each day?
3. Why is it important that Jason begin enteral feedings as soon as possible? Assuming that Jason eventually tolerates a tube feeding, in what ways can the health care team help Jason make the transition from parenteral nutrition to tube feedings? Consider some of the physiological problems that Jason might face when he begins an oral diet.
4. If Jason is unable to meet his nutrient needs orally, he may need to continue tube feeding or TPN at home. As you read through the section on nutrition support at home, consider the factors that would make Jason a good candidate for a home nutrition support program. Consider both the benefits of a proposed program and the problems he could encounter.

Courtesy of Kendall Healthcare

Portable pumps and convenient carrying cases allow freedom of movement for individuals using home nutrition support.

Nutrition Support at Home

Some individuals may require nutrition support—either tube feedings or parenteral nutrition—after a medical condition has stabilized and they no longer require hospital services. For such a person, home nutrition support may be a suitable option. Current medical technology allows for the safe administration of nutrition support in home settings, and insurance coverage often pays a substantial portion of the costs. Medical equipment providers and home infusion companies can provide the supplies, enteral formulas or parenteral solutions, and services necessary for home nutrition care. Most important, patients using these services can continue to receive specialized nutrition care while leading normal lives.

CANDIDATES FOR HOME NUTRITION SUPPORT

Individuals referred for home nutrition support usually need long-term nutrition care for chronic medical conditions. Users of home nutrition services (or their families and other caregivers) must be capable of learning the required procedures and managing any complications that arise. The home should be clean and have adequate storage for formulas or solutions and equipment. The costs should be clearly explained to families who cannot get insurance reimbursement. Candidates for home nutrition support include the following:[29]

- For home enteral nutrition, individuals who have disorders that prevent food from reaching the intestines or interfere with nutrient absorption. Examples include people with head and neck cancers, severe dysphagia, gastric outlet obstructions, and pancreatic or intestinal conditions that cause malabsorption.
- For home parenteral nutrition, individuals who have disorders that severely impede nutrient absorption or interfere with intestinal motility. Examples include people with short bowel syndrome, inflammatory bowel diseases, and intestinal obstructions.

NURSING DIAGNOSIS

The nursing diagnosis *readiness for enhanced self-health management* may be appropriate for individuals who prepare for home nutrition support services.

PLANNING HOME NUTRITION CARE

As with the nutrition support provided in health care facilities, planning for home nutrition care involves decisions about access sites, formulas, and nutrient delivery methods. Users of home services should be involved in the decision making to ensure long-term compliance and satisfaction.

Home Enteral Nutrition Access to the GI tract is possible using either nasal tubes or enterostomies. Although people can learn to place nasogastric tubes themselves, active children and adults often prefer low-profile gastrostomy or jejunostomy tubes, which allow them to lead a more normal lifestyle. Jejunostomy tubes are generally less convenient because the frequent feedings required can interfere with daytime activities.

The advantages and disadvantages associated with the different administration methods should be fully discussed with patients. For gastric feedings, bolus infusions are simplest and can be quickly delivered. If intermittent feedings require slow or reliable delivery rates, infusion pumps may be necessary. Portable pumps can free individuals from the need to infuse formula at home and can also be used when traveling.

The formula chosen for home use is influenced by its cost and availability. Insurance reimbursements do not always include the cost of enteral formulas, which are considered to be "food" products. For this reason, some people choose to prepare simple formulas at home. Blenderizing home-cooked foods is possible, but the foods need to be strained to remove particles and clumps that may obstruct the tube. Closed (ready-to-hang) feeding systems are useful for avoiding contamination risk but are not appropriate for intermittent feedings that require smaller amounts of formula.

Home Parenteral Nutrition

Although both peripheral parenteral nutrition and total parenteral nutrition can be provided at home, long-term therapy requires access to the larger, central veins that are appropriate for TPN. The catheter's exit site is generally placed in a region accessible to the patient. Most people prefer cyclic infusions over continuous infusions and transition to cyclic infusions before discharge from the hospital. Because infusion pumps are required for home TPN, sufficient battery backup is necessary in case electrical service is interrupted. Portable pumps are helpful for individuals who lead active lifestyles.

Parenteral solutions need to be aseptically prepared, and individuals who mix their own solutions must be carefully trained. Ready-made parenteral solutions require refrigeration and are stable for limited periods; for example, 3-in-1 solutions may be stable for only one week when refrigerated.

QUALITY-OF-LIFE ISSUES

Although home nutrition programs can help to improve health and extend life, consumers of these services and their families may struggle with the lifestyle adjustments required. In addition to the economic impact of nutrition support, home feedings are often time consuming and inconvenient. Activities and work schedules must accommodate the feeding schedule. Extra planning is necessary and precautions must be taken when a person wants to travel or participate in sports activities. Explaining one's medical needs to friends and acquaintances may be embarrassing.

Among physical difficulties, people receiving nocturnal feedings often cite disturbed sleep as a major problem. Disruptions may be due to multiple nighttime bathroom visits, noisy infusion pumps, or difficulty finding a comfortable sleeping position when "hooked up." People using parenteral support sometimes prefer infusing solutions during the day to improve their sleeping patterns.

Among social issues, the inability to consume meals with family and friends is often a great concern.[30] Many individuals miss the enjoyment, comfort, and socialization they previously experienced from food and mealtimes. Joining friends at restaurants and attending certain types of social events may become a source of stress for individuals who cannot consume food.

People who depend on nutrition support face many challenges that can affect quality of life. Support groups or counseling resources can help patients cope with the demands of treatment. The Oley Foundation (www.oley.org) is an excellent source of current information and emotional support for individuals who require home nutrition.

Review Notes

- Candidates for home enteral nutrition services have disorders that interfere with swallowing ability, GI motility, or nutrient absorption. Candidates for home parenteral nutrition have disorders that severely impair nutrient absorption or cause intestinal motility problems.
- Patients and caregivers should participate in decisions about access sites, formulas, and nutrient delivery methods. Enteral formulas and parenteral solutions can be purchased or prepared in the home.
- The use of portable pumps may help individuals lead a normal lifestyle. Nevertheless, lifestyle adjustments to nutrition support may be difficult and stressful.

Nutrition Assessment Checklist for People Receiving Enteral Nutrition Support

MEDICAL HISTORY

Check the medical record for medical conditions that:

- ❍ Alter nutrient needs and influence the formula selection
- ❍ Influence the selection of tube placement sites and feeding routes
- ❍ Suggest the length of time that the tube feeding will be needed

Monitor the medical record for complications or risks that may influence the formula selection or delivery technique, including:

- ❍ Aspiration
- ❍ Constipation
- ❍ Fluid and electrolyte imbalances
- ❍ Diarrhea
- ❍ Hyperglycemia
- ❍ Nausea and vomiting
- ❍ Skin irritation

MEDICATIONS

Check medications for those that can cause side effects similar to the complications of tube feeding, such as:

- ❍ Nausea and vomiting
- ❍ Diarrhea
- ❍ Constipation
- ❍ GI discomfort

For medications delivered through the feeding tube, check:

- ❍ Form of medication and possible alternatives
- ❍ Viscosity of liquid medications
- ❍ Potential for diet-drug interactions

DIETARY INTAKE

To assess nutritional adequacy, check to see whether:

- ❍ The formula is appropriate for patient's needs
- ❍ Supplemental water is provided to meet needs
- ❍ The formula is administered as prescribed

ANTHROPOMETRIC DATA

Measure baseline height and weight, and monitor body weight regularly. If weight is not appropriate:

- ❍ Determine whether energy needs have been correctly assessed
- ❍ Check to see if the formula is being delivered as prescribed
- ❍ Check for signs of dehydration or overhydration

LABORATORY TESTS

Check serum and urine tests for signs of:

- ❍ Fluid and electrolyte imbalances
- ❍ Glucose intolerance
- ❍ Improvement or deterioration of the medical condition

PHYSICAL SIGNS

Look for physical signs of:

- ❍ Dehydration or overhydration
- ❍ Delayed gastric emptying
- ❍ Malnutrition

Nutrition Assessment Checklist for People Receiving Parenteral Nutrition Support

MEDICAL HISTORY

Check the medical record for medical conditions that:

- ❍ Prevent the use of enteral nutrition
- ❍ Suggest the time period that parenteral nutrition will be required

Monitor the medical record for complications or risks that may influence the parenteral solution formulation or delivery technique, including:

- ❍ Acid-base imbalances
- ❍ Fluid and electrolyte imbalances
- ❍ Hyperglycemia or hypoglycemia
- ❍ Hypertriglyceridemia
- ❍ Preexisting liver disease
- ❍ Refeeding syndrome

MEDICATIONS

For medications added to the parenteral solution, determine the:

- ❍ Medication's compatibility with the parenteral solution
- ❍ Length of time that the medication can remain stable in solution

For medications infused separately, determine:

- ❍ Length of time that the infusion may need to be stopped
- ❍ Necessary adjustments in parenteral infusions to compensate for medication delivery

DIETARY INTAKE

To assess nutritional adequacy, check to see whether:

- ❍ Patient's nutrient needs were correctly determined
- ❍ Solution is administered as prescribed
- ❍ Infusion pump is operating correctly

ANTHROPOMETRIC DATA

Measure baseline height and weight, and monitor daily weights. If weight is not appropriate:

- ❍ Determine whether energy needs have been correctly assessed
- ❍ Check to see if the parenteral solution is being delivered as prescribed
- ❍ Check for signs of dehydration or overhydration

LABORATORY TESTS

Check serum and urine tests for signs of:

- ❍ Fluid, electrolyte, and acid-base imbalances
- ❍ Hyperglycemia or hypoglycemia
- ❍ Hypertriglyceridemia
- ❍ Abnormal liver function
- ❍ Improvement or deterioration of medical condition

PHYSICAL SIGNS

Routinely monitor the following:

- ❍ Catheter insertion site for signs of infection or inflammation
- ❍ Blood pressure, temperature, pulse, and respiration for signs of fluid, electrolyte, and acid-base imbalances

Look for physical signs of:

- ❍ Dehydration or overhydration
- ❍ Malnutrition

Self Check

1. Of the following, which enteral feeding method would be inappropriate?

a. A bolus feeding via nasogastric tube
b. A bolus feeding via nasojejunal tube
c. An intermittent feeding via gastrostomy
d. A continuous feeding using an infusion pump

2. What would be the most appropriate reason to choose a formula that provides 2 kcalories per milliliter of fluid?

a. The patient has diarrhea.
b. The patient has chronic kidney disease.
c. The patient is unable to tolerate large volumes of fluid.
d. The patient requires a low-residue diet.

3. Compared with intermittent tube feedings, continuous feedings:

a. require an infusion pump.
b. allow greater freedom of movement.
c. are more similar to normal patterns of eating.
d. are associated with more GI side effects.

4. A patient needs 1800 milliliters of formula a day. If the patient is to receive formula intermittently every four hours, how many milliliters of formula will she need at each feeding?

a. 225
b. 300
c. 400
d. 425

5. The nurse using a feeding tube to deliver medications recognizes that:

a. medications given by feeding tube generally do not cause GI complaints.
b. medications can usually be added directly to the feeding container.
c. enteral formulas do not interact with medications in the same way that foods do.
d. thick or sticky liquid medications and crushed tablets can clog feeding tubes.

6. Ideal candidates for parenteral nutrition include all of the following *except:*

a. a patient with complete intestinal obstruction.
b. a person with short bowel syndrome.
c. a patient with intractable diarrhea.
d. a person with a severe swallowing disorder.

7. Iron is typically excluded from parenteral solutions, in part, because:

a. requirements for iron vary substantially from person to person.
b. iron can destabilize solutions that include lipid emulsions.
c. iron restriction is necessary in persons using warfarin therapy.
d. iron promotes fat deposition in the liver.

8. Common complications of parenteral nutrition include all of the following *except:*

a. increased infection risk.
b. fatty liver.
c. diarrhea.
d. gallbladder disease.

9. Refeeding syndrome causes dangerous fluctuations in:
   a. serum electrolytes.
   b. serum liver enzyme levels.
   c. blood triglyceride levels.
   d. ketone bodies.

10. Patients using home parenteral nutrition:
   a. usually prefer continuous rather than cyclic infusions.
   b. are unable to travel or work away from home.
   c. require infusion pumps for use at home.
   d. can only obtain 2-in-1 solutions and so must infuse lipids separately.

Answers to these questions appear in Appendix H. For more chapter review: Access an interactive eBook, chapter-specific interactive learning tools, including flashcards, quizzes, videos, and more in your Nutrition CourseMate, accessed through CengageBrain.com.

Clinical Applications

1. Appendix G provides examples of enteral formulas on the market and lists their energy and macronutrient contents. Select one standard formula and one elemental formula from Tables G-1 and G-2, respectively. For the two formulas you selected, calculate the volume of formula that would meet the energy needs of a patient who requires about 1750 kcalories daily. Use these results in answering the following questions:
   a. What is the amount of protein, carbohydrate, and fat that the patient would obtain in a typical day? Determine the percentages of kcalories that come from carbohydrate and fat. Do these percentages fall within the Acceptable Macronutrient Distribution Ranges described in Chapter 1 (p. 10)?
   b. Tables G-1 and G-2 show the formula volumes that would meet the Reference Daily Intakes (RDI). Would the volumes you obtained meet typical vitamin and mineral needs?

2. A liter of a TPN solution contains 500 milliliters of 50 percent dextrose solution and 500 milliliters of 5 percent amino acid solution. Determine the daily energy and protein intakes of a person who receives 2 liters per day of such a solution. Calculate the average daily energy intake if the person also receives 500 milliliters of a 20 percent fat emulsion three times a week.

3. Consider the clinical, financial, psychological, and social ramifications of using home parenteral nutrition, with no foods allowed by mouth, in answering the following questions:
   a. What would be the advantages of living at home instead of in a hospital or other residential facility? Can you think of some disadvantages?
   b. Think about how you, as the patient, might manage daily infusions: consider the time, cost, and commitment required to maintain the therapy.
   c. If not allowed to consume foods, what possible difficulties might you encounter? How would you handle holidays and special occasions that center around food?

Nutrition on the Net

For further study of topics covered in this chapter, access these websites:

- To learn more about the appropriate uses of enteral and parenteral nutrition, visit the websites of these organizations:
  American Society for Parenteral and Enteral Nutrition: www.nutritioncare.org
  British Association for Parenteral and Enteral Nutrition: www.bapen.org.uk
- Information about enteral formulas is available from these manufacturers' websites:
  Abbott Nutrition: www.abbottnutrition.com
  Nestlé Nutrition: www.nestle-nutrition.com
- To learn about home nutrition support, visit the website of the Oley Foundation, a national nonprofit organization that provides information, outreach services, and emotional support for consumers of home enteral and parenteral services: www.oley.org

Notes

1. K. L. Coughlin, S. I. Austhof, and C. Hamilton, Nutrition support: Indication and efficacy, in A. Skipper, ed., *Dietitian's Handbook of Enteral and Parenteral Nutrition* (Sudbury, MA: Jones and Bartlett Learning, 2012), pp. 22–45.
2. M. Marian and P. Charney, Patient selection and indications for enteral feedings, in P. Charney and A. Malone, eds., *ADA Pocket Guide to Enteral Nutrition* (Chicago: American Dietetic Association, 2006), pp. 1–25.
3. R. L. Koretz, Do data support nutrition support? Part II. Enteral artificial nutrition, *Journal of the American Dietetic Association* 107 (2007): 1374–1380.
4. A. Skipper, Enteral nutrition, in A. Skipper, ed., *Dietitian's Handbook of Enteral and Parenteral Nutrition* (Sudbury, MA: Jones and Bartlett Learning, 2012), pp. 259–280; R. Bankhead and coauthors, A.S.P.E.N. enteral nutrition practice recommendations, *Journal of Parenteral and Enteral Nutrition* 33 (2009): 122–167.
5. Skipper, 2012.
6. Academy of Nutrition and Dietetics, *Nutrition Care Manual* (Chicago: Academy of Nutrition and Dietetics, 2012).
7. Bankhead and coauthors, 2009.
8. K. K. Kattelmann and coauthors, Preliminary evidence for a medical nutrition therapy protocol: Enteral feedings for critically ill patients, *Journal of the American Dietetic Association* 106 (2006): 1226–1241.
9. C. Thompson, Initiation, advancement, and transition of enteral feedings, in P. Charney and A. Malone, eds., *ADA Pocket Guide to Enteral Nutrition* (Chicago: American Dietetic Association, 2006), pp. 123–154.
10. Thompson, 2006; C. R. Parrish, J. Krenitsky, and C. Kusenda, Enteral feeding challenges, in G. Cresci, ed., *Nutrition Support for the Critically Ill Patient: A Guide to Practice* (Boca Raton, FL: Taylor & Francis Group, 2005), pp. 321–340.
11. Academy of Nutrition and Dietetics, 2012; Skipper, Enteral nutrition, 2012.
12. Thompson, 2006.
13. Bankhead and coauthors, 2009; S. A. McClave and coauthors, Guidelines for the provision and assessment of nutrition support therapy in the adult critically ill patient: Society of Critical Care Medicine (SCCM) and American Society for Parenteral and Enteral Nutrition (A.S.P.E.N.), *Journal of Parenteral and Enteral Nutrition* 33 (2009): 277–316.
14. P. D. Wohlt and coauthors, Recommendations for the use of medications with continuous enteral nutrition, *American Journal of Health-Systems Pharmacy* 66 (2009): 1458–1467.
15. M. K. Russell, Monitoring complications of enteral feedings, in P. Charney and A. Malone, eds., *ADA Pocket Guide to Enteral Nutrition* (Chicago: American Dietetic Association, 2006), pp. 155–192.
16. Thompson, 2006.
17. A. Wilmer and G. van den Berghe, Parenteral nutrition, in L. Goldman and A I. Schafer, *Goldman's Cecil Medicine* (Philadelphia: Saunders, 2012): 1394–1397.
18. A. Skipper, Parenteral nutrition, in A. Skipper, ed., *Dietitian's Handbook of Enteral and Parenteral Nutrition* (Sudbury, MA: Jones and Bartlett Learning, 2012), pp. 281–300.
19. M. L. Christensen, Parenteral formulations, in G. Cresci, ed., *Nutrition Support for the Critically Ill Patient: A Guide to Practice* (Boca Raton, FL: Taylor & Francis Group, 2005), pp. 279–302.
20. Academy of Nutrition and Dietetics, 2012.
21. V. W. Vanek and coauthors, A.S.P.E.N. position paper: Clinical role for alternative intravenous fat emulsions, *Nutrition in Clinical Practice* 27 (2012): 150–192.
22. A. M. Malone, Parenteral nutrients and formulations, in P. Charney and A. Malone, eds., *ADA Pocket Guide to Parenteral Nutrition* (Chicago: American Dietetic Association, 2007), pp. 52–63.
23. S. Roberts, Initiation, advancement, and acute complications, in P. Charney and A. Malone, eds., *ADA Pocket Guide to Parenteral Nutrition* (Chicago: American Dietetic Association, 2007), pp. 76–102.
24. M. P. Fuhrman, Complications of long-term parenteral nutrition, in P. Charney and A. Malone, eds., *ADA Pocket Guide to Parenteral Nutrition* (Chicago: American Dietetic Association, 2007), pp. 103–117; Roberts, 2007.
25. Roberts, 2007.
26. Wilmer and van den Berghe, 2012.
27. Roberts, 2007.
28. Fuhrman, 2007.
29. C. Hamilton and T. Austin, Home parenteral nutrition, in P. Charney and A. Malone, eds., *ADA Pocket Guide to Parenteral Nutrition* (Chicago: American Dietetic Association, 2007), pp. 118–146; A. Pattinson and J. Buchholtz, Home enteral nutrition, in P. Charney and A. Malone, eds., *ADA Pocket Guide to Enteral Nutrition* (Chicago: American Dietetic Association, 2006), pp. 193–227.
30. M. F. Winkler, Living with enteral and parenteral nutrition: How food and eating contribute to quality of life, *Journal of the American Dietetic Association* 110 (2010): 169–177.

Nutrition in Practice

Inborn Errors of Metabolism

An **inborn error of metabolism** is an inherited trait, caused by a genetic **mutation**, that results in the absence, deficiency, or dysfunction of a protein that has a critical metabolic role.[1] The severity of the inborn error's effects is ultimately related to the degree of impairment caused by the missing or altered protein. This Nutrition in Practice describes some inborn errors of metabolism and discusses the role of diet in two of these disorders: phenylketonuria and galactosemia. The accompanying glossary defines the relevant terms.

What problems can result from inborn errors of metabolism?

The protein affected by an inborn error may function as an enzyme, receptor, transport protein, or structural protein. When the body fails to make a protein, the functions that depend on that protein are impaired. For example, when an enzyme is missing or malfunctioning in a metabolic pathway that typically converts compound A to compound B, compound A will accumulate and compound B will not be made. The excess of compound A and the lack of compound B may have harmful effects. Furthermore, the imbalances in one pathway may affect other pathways and ultimately cause a number of metabolic and physiologic disturbances. Table NP15-1 lists some examples of inborn errors related to defects in nutrient metabolism.

What role can diet play in treating inborn errors of metabolism?

Medical nutrition therapy is the primary treatment for many inborn errors that involve nutrient metabolism. After the biochemical pathway affected by an inborn error is identified, a health practitioner may be able to manipulate elements of the diet to compensate for deficiencies and excesses. Dietary intervention generally involves restricting substances that cannot be properly metabolized and supplying substances that cannot be produced. Thus, dietary changes may be able to improve outcomes of some inborn errors by:

- Preventing the accumulation of toxic **metabolites**
- Replacing nutrients that are deficient as a result of a defective metabolic pathway
- Providing a diet that supports normal growth and development and maintains health

Glossary

cystic fibrosis: an inherited disorder that affects the transport of chloride across epithelial cell membranes; primarily affects the gastrointestinal and respiratory systems.

galactosemia (ga-LAK-toe-SEE-me-ah): an inherited disorder that affects galactose metabolism. Accumulated galactose causes damage to the liver, kidneys, and brain in untreated patients.

gene therapy: treatment for inherited disorders in which DNA sequences are introduced into the chromosomes of affected cells, prompting the cells to express the protein needed to correct the disease.

genetic counseling: support for families at risk of genetic disorders; involves diagnosis of disease, identification of inheritance patterns within the family, and review of reproductive options.

hemophilia (HE-moh-FEEL-ee-ah): an inherited bleeding disorder characterized by deficiency or malfunction of a plasma protein needed for clotting blood.

inborn error of metabolism: an inherited trait (one that is present at birth) that causes the absence, deficiency, or malfunction of a protein that has a critical metabolic role.

metabolites: products of metabolism; compounds produced by a biochemical pathway.

mutation: a heritable change in the DNA sequence of a gene.

phenylketonuria (FEN-il-KEY-toe-NU-ree-ah) or PKU: an inherited disorder that affects the conversion of the essential amino acid phenylalanine to the amino acid tyrosine. The condition is named after the phenylalanine metabolites—called *phenylketones*—that are excreted in the urine of individuals who have the disorder.

TABLE NP15-1 Nutrition-Related Inborn Errors of Metabolism

Disorder	Affected Nutrient(s) or Substance	Metabolic Defect	Nutritional Treatment
Amino acid metabolism			
Maple syrup urine disease	Branched-chain amino acids (isoleucine, leucine, and valine)	Impaired metabolism of branched-chain amino acids	Restriction of branched-chain amino acids; thiamin supplementation
Phenylketonuria	Phenylalanine	Impaired conversion of phenylalanine to tyrosine	Phenylalanine-restricted diet; tyrosine supplementation
Carbohydrate metabolism			
Galactosemia	Galactose	Impaired conversion of galactose to glucose	Galactose-restricted diet
Glycogen storage disease	Glycogen	Impaired metabolism or transport of glycogen, resulting in glycogen accumulation in tissues	Varies; may require frequent feedings, cornstarch supplementation, high-protein diet
Lipid metabolism			
Carnitine transporter deficiency	Fatty acids	Impaired transport of fatty acids into mitochondria for oxidation	Carnitine supplementation; avoidance of fasting and strenuous exercise
X-adrenoleukodystrophy[a]	Very long-chain fatty acids	Impaired breakdown of very long-chain fatty acids in peroxisomes	Under investigation; limited benefit from restriction of very long-chain fatty acids and supplementation with various fatty acid mixtures[a]
Mineral metabolism			
Hemochromatosis	Iron	Excessive iron absorption (causes iron accumulation)	Avoidance of iron and vitamin C supplements and alcoholic beverages (routine blood draws remove excess iron from the body)
Wilson's disease	Copper	Impaired copper excretion (causes copper accumulation)	Avoidance of copper-rich foods; zinc therapy (reduces copper absorption)

[a]The disease X-adrenoleukodystrophy was featured in the 1992 film *Lorenzo's Oil*.

Successful treatment for an inborn error of metabolism depends on the ability to screen newborns and diagnose metabolic diseases before irreversible damage can occur. After a genetic defect is identified, family members undergo **genetic counseling** to evaluate the likelihood that they may pass on the disorder to future offspring. During counseling, couples may learn about reproductive options such as artificial insemination, *in vitro* fertilization, or prenatal monitoring after conception.

Are there treatments for inborn errors that don't involve dietary changes?

Nondietary therapies can treat some inborn errors of metabolism, although the options are somewhat limited. In some cases, the missing protein is infused; this is the primary means of treating **hemophilia**, caused by deficiency of one of the plasma proteins needed for clotting blood. Drug therapy is the main treatment for some inborn errors, including **cystic fibrosis** (discussed in Chapter 18), which is characterized by a defect that prevents normal chloride transport across cell membranes. Future approaches may include **gene therapy**, a treatment that introduces DNA sequences into the chromosomes of affected cells, prompting the cells to express the protein needed to correct the abnormality.

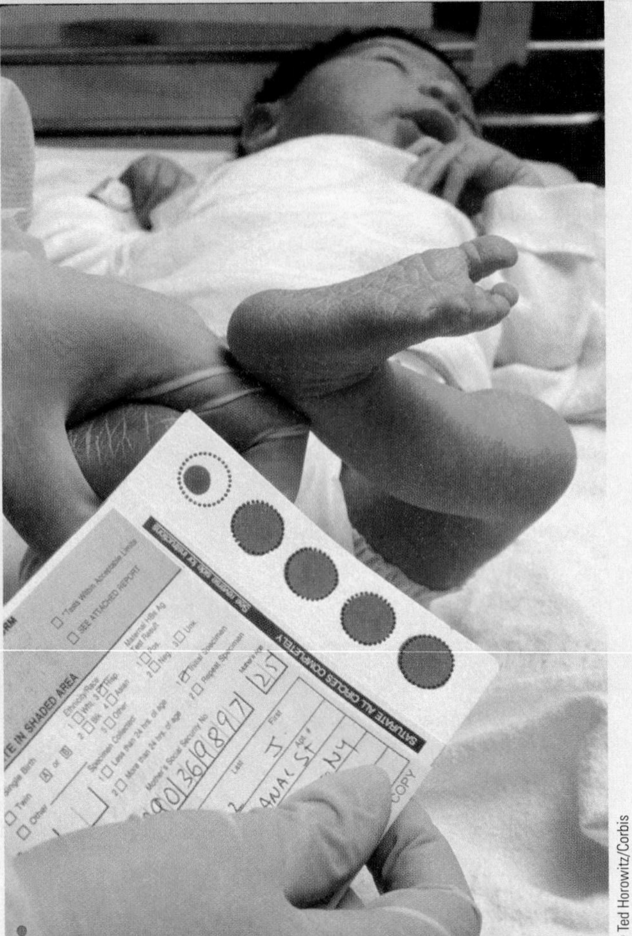

© Ted Horowitz/Corbis

A simple blood test screens newborns for PKU—a common inborn error of metabolism.

What is an example of an inborn error that benefits from dietary treatment?

A classic example is **phenylketonuria (PKU)**, a metabolic disorder that affects amino acid metabolism. PKU occurs in approximately 1 out of every 15,000 births in the United States each year.[2] In PKU, the missing or defective protein is the liver enzyme *phenylalanine hydroxylase*, which converts the essential amino acid phenylalanine to the amino acid tyrosine. This chemical reaction is also the first step in the breakdown of excess phenylalanine. Without the enzyme, phenylalanine accumulates in the blood and tissues and severely damages the developing brain. The impairment in the metabolic pathway also prevents liver synthesis of tyrosine and tyrosine-derived compounds (such as the neurotransmitter epinephrine, the skin pigment melanin, and the hormone thyroxine). Under these conditions, tyrosine becomes essential: the body cannot produce tyrosine, and therefore the diet must supply it.

Although PKU's most debilitating effect is on brain development, other symptoms may manifest if the condition is untreated. Infants with PKU may have poor appetites and grow slowly. They may be irritable or have tremors or seizures. Their bodies and urine may have a musty odor. Their skin may be unusually pale, and they may develop skin rashes. In older children and adults who discontinue treatment, neurological and psychological problems are common. Individuals with elevated phenylalanine levels may exhibit impaired reasoning, a reduced attention span, and poor memory, among other deficits.[3]

How is PKU diagnosed?

PKU must be diagnosed soon after birth so that early treatment can prevent its devastating effects. For this reason, newborns are screened for PKU in all 50 states.[4] A standard blood test for phenylalanine is typically conducted by heel puncture after the infant has consumed several meals containing protein. Abnormal results require further testing.

The screening of newborns for PKU is one of the most common genetic tests in the United States and many other countries. Before widespread newborn screening, infants with PKU demonstrated developmental delays (for example, inability to crawl) by six to nine months of age. By the time parents recognized the problem, the damage was irreversible. Most of the damaging consequences of this disorder are now prevented due to the early detection and treatment of PKU.

What is the treatment for PKU?

The only current treatment for PKU is a diet that restricts phenylalanine and supplies tyrosine so that the blood levels of these amino acids are maintained within safe ranges. Because phenylalanine is an essential amino acid, the diet cannot exclude it completely. Children with PKU need phenylalanine to grow, but they cannot handle excesses without detrimental effects. Therefore, their diets must provide enough phenylalanine to support growth and health, but not so much as to cause harm. The diets must also provide tyrosine, which is an essential nutrient for individuals with PKU. To ensure that blood concentrations of phenylalanine and tyrosine are close to normal, blood tests are performed periodically, and diets are adjusted when necessary. If the dietary treatment is conscientiously followed, it can prevent the symptoms described earlier. Adults with PKU must continue to follow the PKU diet, as well, in order to prevent deterioration in brain function.

What are the main features of the PKU diet?

Central to the PKU diet is the use of an enteral formula that is phenylalanine free yet supplies energy, amino acids, vitamins, and minerals. For infants, the phenylalanine-free formula can be supplemented with measured amounts of breast milk or regular infant formula to provide the phenylalanine needed for growth. Low-phenylalanine formulas are available for infants who must meet all of their nutrient needs with formula. Formula requirements need to be recalculated periodically to accommodate the growing infant's shifting needs for protein, phenylalanine, tyrosine, and energy.

Once food consumption begins, a phenylalanine-free formula supplies the needed amino acids, and foods that contain phenylalanine are carefully monitored. All proteins contain some phenylalanine; therefore, high-protein foods such as meat, fish, poultry, milk, cheese, legumes, and nuts (including peanut butter) are omitted. Fruits, vegetables, and cereals also contain phenylalanine, so only limited amounts are allowed. Low-protein flours and mixes are available for making low-phenylalanine breads, pasta, cakes, and cookies. Foods that do not contain phenylalanine, such as jams, jellies, and most sweeteners, can be used freely. Growth rates and nutrition status are monitored to ensure that the diet is adequate. Note that older children, teens, and adults with PKU should continue to use the phenylalanine-free formulas to help meet their protein and energy needs.

Individuals with PKU should be encouraged to develop creative ways to make their diets enjoyable. The formula can be flavored or combined with fruits or juices to make smoothies or frozen juice bars. Sandwiches can include low-phenylalanine breads and fillings such as mashed bananas or avocados, shredded carrots and olives, or tomato slices with mayonnaise. Children often enjoy creating special recipes with permitted foods to make their choices more varied and to share meals with friends.

How long should a person with PKU continue the dietary treatment?

Lifelong adherence to a phenylalanine-restricted diet is currently recommended for all individuals with PKU, as elevated phenylalanine levels can adversely affect cognitive function at any age. It is especially important that women with PKU maintain safe phenylalanine concentrations during pregnancy. Elevated phenylalanine levels, especially during the first trimester, have been associated with mental retardation and organ malformations in the offspring of PKU mothers who have discontinued dietary treatment.[5]

What is another example of an inborn error that requires dietary changes?

Galactosemia is an example of an inborn error of carbohydrate metabolism. Individuals with galactosemia are deficient in one of the enzymes needed to metabolize galactose, a sugar that is primarily found in milk products (recall that each lactose molecule contains galactose). An accumulation of galactose can cause damage in multiple tissues. Infants with galactosemia who are given milk react with severe vomiting and liver jaundice within days of the initial feeding. Serious liver damage can develop and progress to symptomatic cirrhosis. Other complications may include kidney failure, cataracts, and brain damage. Treatment in the first weeks of life can prevent the most detrimental effects of galactose accumulation, but if treatment is delayed, the damage to the brain is irreversible.[6]

What is the dietary treatment for galactosemia?

The diet for galactosemia is much simpler than the diet for PKU. For one thing, galactose is not an essential nutrient. The galactosemia diet essentially eliminates galactose from the diet and does not need to provide a carefully determined amount of any nutrient, as the PKU diet does. In addition, dietary galactose is primarily obtained from lactose (the milk sugar), so the main focus of dietary treatment is the exclusion of milk and milk products. A number of other foods that contain galactose in substantial amounts, such as organ meats and some legumes, fruits, and vegetables, must also be avoided or restricted. Patients receive food lists that identify the galactose content of common foods.

Infants diagnosed with galactosemia are given lactose-free formulas to meet their nutrient needs. Once a child can consume adequate amounts of regular foods, special formulas are unnecessary. However, care must be taken to ensure that the diet supplies adequate calcium and vitamin D.

How effective is the dietary treatment for galactosemia?

Although the early introduction of a galactose-restricted diet can eliminate the acute toxic effects of galactosemia, complications of the disease may develop despite an individual's compliance with diet therapy. For example, most patients experience delays in speech and language development. Ovarian failure occurs in up to 85 percent of women who have galactosemia.[7] In addition, some evidence suggests that IQ declines as a person with galactosemia ages. The reasons for these long-term complications are not fully understood.

For many inborn errors of metabolism, effective management requires early diagnosis and treatment, as well as control of the environmental factors that may cause toxicity. In some cases, dietary changes are central to treatment and can prevent serious complications. Other inborn errors, however, may not be as easily treated. Future developments in biotechnology may allow gene therapy to assist in the medical treatment of some of these disorders.

Notes

1. L. J. Elsas II, Approach to inborn errors of metabolism, in L. Goldman and A I. Schafer, *Goldman's Cecil Medicine* (Philadelphia: Saunders, 2012), pp. 1340–1346.
2. N. Blau, F. J. van Spronsen, and H. L. Levy, Phenylketonuria, *Lancet* 376 (2010): 1417–1427.
3. Blau, van Spronsen, and Levy, 2010.
4. U.S. Preventive Services Task Force, Screening for phenylketonuria (PKU): U.S. Preventive Services Task Force reaffirmation recommendation, *Annals of Family Medicine* 6 (2008): 166, doi: 10.1370/afm.820.
5. S. D. Cederbaum, Disorders of phenylalanine and tyrosine metabolism, in L. Goldman and D. Ausiello, eds., *Cecil Medicine* (Philadelphia: Saunders, 2008), pp. 1573–1576.
6. L. J. Elsas II, Galactosemia, in L. Goldman and D. Ausiello, eds., *Cecil Medicine* (Philadelphia: Saunders, 2008), pp. 1555–1558.
7. Elsas, 2008.

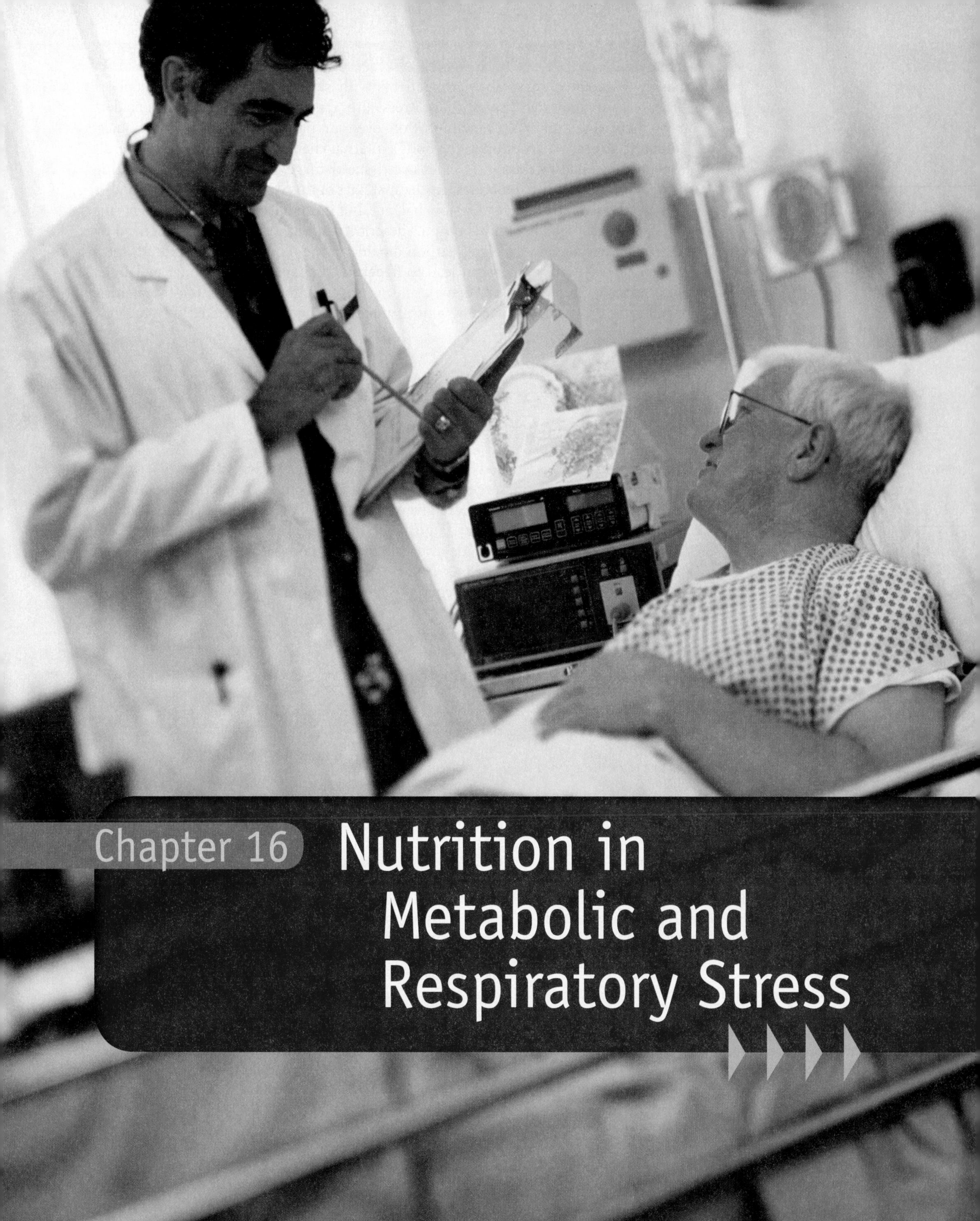

Chapter 16 Nutrition in Metabolic and Respiratory Stress

David Hanover/Getty Images

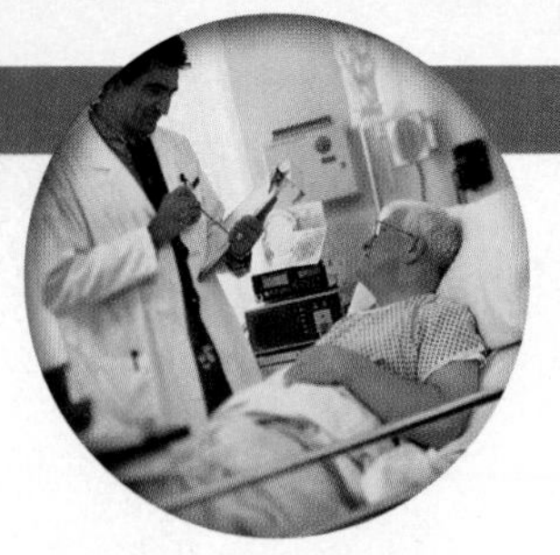

enough to threaten survival. Many patients with severe stress require life support measures and intensive monitoring. Stress also raises nutritional needs considerably—increasing the risk of malnutrition even in previously healthy individuals. **Metabolic stress**, a disruption in the body's internal chemical environment, can result from uncontrolled infections or extensive tissue damage, such as deep, penetrating wounds or multiple broken bones. As the first part of this chapter explains, the body's stress response is an attempt to restore balance, but it can have both helpful and harmful effects. Later sections of this chapter describe **respiratory stress**, characterized by inadequate oxygen and excessive carbon dioxide in the blood and tissues. Both metabolic and respiratory stress can lead to **hypermetabolism** (above-normal metabolic rate), **wasting** (loss of muscle tissue), and, in severe circumstances, life-threatening complications. The accompanying glossary defines some terms related to metabolic stress.

The Body's Responses to Stress and Injury

The **stress response** is the body's *nonspecific* response to a variety of stressors, such as burns, fractures, infection, surgery, and wounds. During stress, the metabolic processes that support immediate survival are given priority, while those of lesser consequence are delayed. Energy is of primary importance, and therefore the energy nutrients are mobilized from storage and made available in the blood. Heart rate and respiration (breathing rate) increase to deliver oxygen and nutrients to cells more quickly, and blood pressure rises. Meanwhile, energy is diverted from processes that are not life

Glossary of Terms Related to Metabolic Stress

acute-phase response: changes in body chemistry resulting from infection, inflammation, or injury; characterized by alterations in serum proteins.

complement: a group of plasma proteins that assist the activities of antibodies.

C-reactive protein: an acute-phase protein often used as an indicator of inflammation; it binds dead or dying cells to activate certain immune responses.

cytokines (SIGH-toe-kines): signaling proteins produced by the body's cells; those produced by white blood cells regulate various aspects of immune function.

eicosanoids (eye-KO-sa-noids): 20-carbon molecules derived from dietary fatty acids that help to regulate blood pressure, blood clotting, and other body functions.

- *eicosa* = twenty

hepcidin: an acute-phase protein involved in the regulation of iron metabolism.

hypermetabolism: a higher-than-normal metabolic rate.

inflammatory response: a group of nonspecific immune responses to infection or injury.

mast cells: cells within connective tissue that produce and release histamine.

metabolic stress: a disruption in the body's chemical environment due to the effects of disease or injury. Metabolic stress is characterized by changes in metabolic rate, heart rate, blood pressure, hormonal status, and nutrient metabolism.

phagocytes (FAG-oh-sites): white blood cells (neutrophils and macrophages) that have the ability to engulf and destroy antigens.

- *phagein* = to eat

respiratory stress: a condition characterized by abnormal oxygen and carbon dioxide levels in body tissues due to abnormal gas exchange between the air and blood.

sepsis: a whole-body inflammatory response caused by infection; characterized by symptoms similar to those of *systemic inflammatory response syndrome* (see definition below).

shock: a severe reduction in blood flow that deprives the body's tissues of oxygen and nutrients; characterized by reduced blood pressure, raised heart and respiratory rates, and muscle weakness.

stress response: the chemical and physical changes that occur within the body during stress.

systemic (sih-STEM-ic): relating to the entire body.

systemic inflammatory response syndrome (SIRS): a whole-body inflammatory response caused by severe illness or trauma; characterized by raised heart and respiratory rates, abnormal white blood cell counts, and fever.

wasting: the breakdown of muscle tissue that results from disease or malnutrition.

sustaining, such as growth, reproduction, and long-term immunity. If stress continues for a long period, interference with these processes begins to cause damage, possibly resulting in growth retardation and illness.

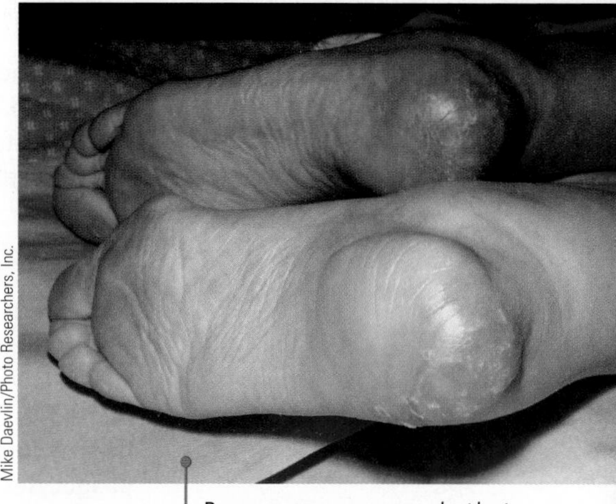
Mike Daevlin/Photo Researchers, Inc.

Pressure sores, wounds that develop when prolonged pressure cuts off blood circulation to the skin and underlying tissues, are a frequent source of metabolic stress in bedridden and wheelchair-bound patients.

HORMONAL RESPONSES TO STRESS

The stress response is mediated by several hormones, which are released into the blood soon after the onset of injury (see Table 16-1).[1] The catecholamines (epinephrine and norepinephrine)—often called the *fight-or-flight hormones*—stimulate heart muscle, raise blood pressure, and increase metabolic rate. Epinephrine also promotes glucagon secretion from the pancreas, prompting the release of nutrients from storage. The steroid hormone cortisol enhances muscle protein degradation, raising amino acid levels in the blood and making amino acids available for conversion to glucose. All of these hormones have similar effects on glucose and fat metabolism, causing the breakdown of glycogen, the production of glucose from amino acids, and the breakdown of triglycerides in adipose tissue. (■) Thus, the combined effects of these hormones contribute to hyperglycemia, which often accompanies critical illness. Two other hormones induced by stress, aldosterone and antidiuretic hormone, help to maintain blood volume by stimulating the kidneys to reabsorb more sodium and water, respectively.

Cortisol's effects can be detrimental when stress is prolonged. In excess, cortisol causes the depletion of protein in muscle, bone, connective tissue, and skin. It impairs wound healing, so high cortisol levels may be especially dangerous for a patient with severe injuries. Because cortisol inhibits protein synthesis, consuming more protein cannot easily reverse tissue losses. Excess cortisol also leads to insulin resistance, contributing to hyperglycemia, and suppresses immune responses, increasing susceptibility to infection. Note that pharmaceutical forms of cortisol (such as *cortisone* and *prednisone*) are common anti-inflammatory medications; their long-term use can cause undesirable side effects such as muscle wasting, thinning of the skin, diabetes, and early osteoporosis.

NURSING DIAGNOSIS

The nursing diagnoses *impaired tissue integrity* and *risk for infection* may apply to a person who undergoes prolonged stress.

TABLE 16-1 Metabolic Effects of Hormones Released during the Stress Response

Hormone	Metabolic Effects
Catecholamines	• Increase in metabolic rate • Glycogen breakdown in liver and muscle • Glucose production from amino acids • Release of fatty acids from adipose tissue • Glucagon secretion from pancreas
Glucagon	• Glycogen breakdown in liver • Glucose production from amino acids • Release of fatty acids from adipose tissue
Cortisol	• Protein degradation • Enhancement of glucagon's action on liver glycogen • Glucose production from amino acids • Release of fatty acids from adipose tissue
Aldosterone	• Sodium reabsorption in kidneys
Antidiuretic hormone	• Water reabsorption in kidneys

■ The catecholamines, glucagon, and cortisol have actions that oppose those of insulin and are therefore referred to as *counterregulatory hormones*.

THE INFLAMMATORY RESPONSE

Cells of the immune system mount a quick, nonspecific response to infection or tissue injury. This so-called **inflammatory response** serves to contain and destroy infectious agents (and their products) and prevent further tissue damage. As in the stress response, there is a delicate balance between a response that protects tissues from further injury and an excessive response that can cause additional damage to tissue.

The Inflammatory Process

The inflammatory response begins with the dilation of arterioles and capillaries at the site of injury, which increases the blood flow to the affected area. The capillaries within the damaged tissue become more permeable, allowing some blood plasma to escape into the tissue and cause local edema (see Figure 16-1). The various changes in blood vessels (■) attract immune cells that can destroy foreign agents and clear cellular debris. Among the first cells to arrive are the **phagocytes**, which slip through gaps between the endothelial cells that form the vessel walls. The phagocytes engulf microorganisms and destroy them with reactive forms of oxygen and hydrolytic enzymes. When inflammation becomes chronic, these normally useful activities of phagocytes can damage healthy tissue.

■ The classic signs of inflammation that accompany altered blood flow are:

- *Swelling*—from the accumulation of fluid at the site of injury.
- *Redness*—from the increase in blood and red blood cells in the injured area.
- *Heat*—from the influx of warm arterial blood.
- *Pain*—from the release of chemical mediators that stimulate pain receptors.

■ *Antihistamines* are medications taken to reduce the effects of histamine.

Mediators of Inflammation

Numerous chemical substances control the inflammatory process. These *mediators* are released from damaged tissue, blood vessel cells, and activated immune cells. Many of them help to regulate more than one step in the process. Histamine, a small molecule similar to an amino acid in structure, is released from granules within **mast cells**, causing vasodilation and capillary permeability. (■) Other compounds that participate in the inflammatory process include **cytokines**, produced by white blood cells (and some other types of cells), and **eicosanoids**, which are derived from dietary fatty acids. Note that most anti-inflammatory medications, including steroidal drugs (such as cortisone and prednisone) and nonsteroidal anti-inflammatory drugs (such as aspirin and ibuprofen), act by blocking eicosanoid synthesis.

Changing dietary fat sources may have subtle effects on the inflammatory process.[2] The major precursor for the eicosanoids is arachidonic acid, which derives from the omega-6 fatty acids in vegetable oils. Some omega-3 fatty acids compete with arachidonic acid and may inhibit the production of the most potent inflammatory mediators. Although health professionals sometimes recommend replacing some of the omega-6

FIGURE 16-1 The Inflammatory Process

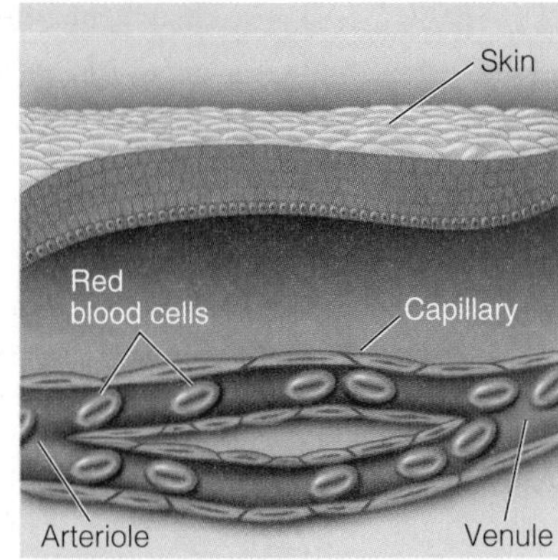

Cells lining the blood vessels lie close together, and normally do not allow the contents to cross into tissue.

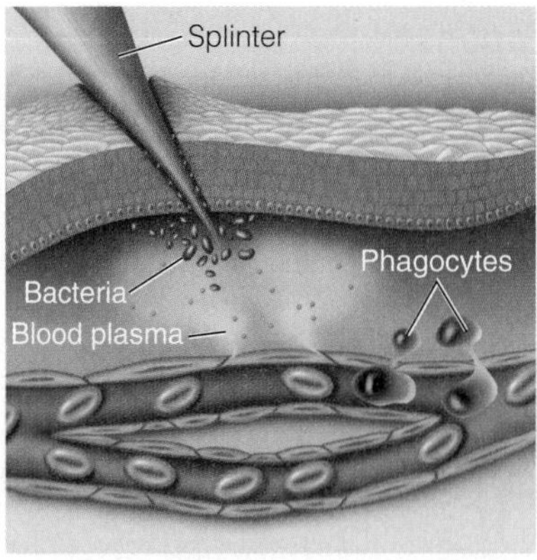

When tissues are damaged, immune cells release histamine, which dilates some blood vessels, increasing blood flow to the damaged area. Fluid leaks out of capillaries (causing swelling), and phagocytes escape between the small gaps in the blood vessel walls.

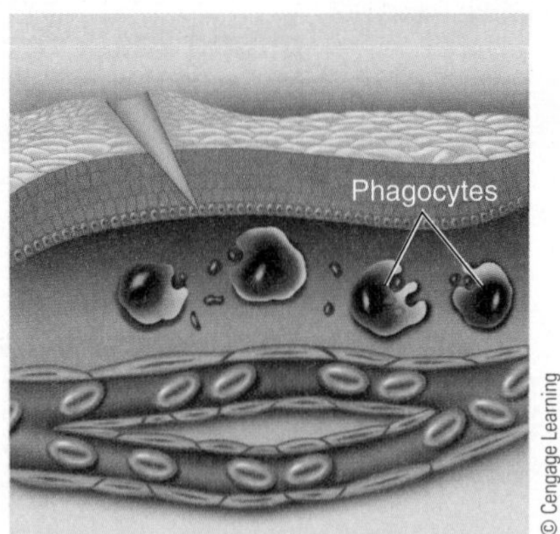

Phagocytes engulf bacteria and disable them with hydrolytic enzymes and reactive forms of oxygen.

fatty acids in the diet with omega-3 fatty acids (■) to reduce inflammation, most clinical studies conducted thus far have not confirmed this benefit.[3]

Systemic Effects of Inflammation In addition to the localized effects described earlier, the cytokines released during acute inflammation produce a number of **systemic** effects, which are collectively known as the **acute-phase response.**[4] Within hours after inflammation, infection, or severe injury, the liver steps up its production of certain plasma proteins (called *acute-phase proteins*), including **C-reactive protein,** (■) **complement, hepcidin,** blood-clotting proteins such as fibrinogen and prothrombin, and others. Conversely, plasma concentrations of albumin, iron, and zinc fall (recall from Chapter 13 that albumin levels are often measured to assess health status). The acute-phase response is accompanied by muscle catabolism to make amino acids available for glucose production, tissue repair, and immune protein synthesis; consequently, negative nitrogen balance (and wasting) frequently results. Other clinical features of the acute-phase response may include fever, an elevated metabolic rate, increased pulse and blood pressure, increased blood neutrophil levels, lethargy, and anorexia.

If inflammation does not resolve, the continued production of pro-inflammatory cytokines may lead to the **systemic inflammatory response syndrome (SIRS)**, which is diagnosed when the patient's symptoms include substantial increases in heart rate and respiratory rate, abnormal white blood cell counts, and/or fever. If these symptoms result from a severe infection, the condition is called **sepsis**. Complications associated with severe cases of SIRS or sepsis include fluid retention and tissue edema, low blood pressure, and impaired blood flow. If the reduction in blood flow is severe enough to deprive the body's tissues of oxygen and nutrients (a condition known as **shock**), multiple organs may fail simultaneously, as discussed in Nutrition in Practice 16.

NURSING DIAGNOSIS

The nursing diagnosis *risk for shock* applies to a person with SIRS or sepsis.

Review Notes

- The stress and inflammatory responses are nonspecific responses to stressors that cause infection and injury.
- The stress response is mediated by the catecholamine hormones, cortisol, and glucagon, which together raise nutrient levels in the blood, stimulate heart rate, raise blood pressure, and increase metabolic rate. Aldosterone and antidiuretic hormone help to maintain adequate blood volume.
- The inflammatory process—mediated by compounds released from damaged tissues, immune cells, and blood vessels—may result in both local and systemic effects.
- Local inflammatory effects include swelling, redness, warmth, and pain in damaged tissue; systemic effects are characterized by changes in acute-phase proteins and increases in body temperature, pulse, blood pressure, metabolic rate, and blood neutrophil levels.
- Persistent, severe inflammation may lead to shock and increases the risk of multiple organ dysfunction.

■ Reminder: Foods high in omega-3 fatty acids include fatty fish, fish oils, flaxseeds, and walnuts.

■ C-reactive protein, the best clinical indicator of the acute-phase response, becomes elevated during many chronic illnesses.

abscesses (AB-sess-es): accumulations of pus.

debridement: the surgical removal of dead, damaged, or contaminated tissue resulting from burns or wounds; helps to prevent infection and hasten healing.

Nutrition Treatment of Acute Stress

As described earlier, an excessive response to metabolic stress can worsen illness and even threaten survival. Therefore, medical personnel must manage both the acute medical condition that initiated stress and the complications that arise as a result of the stress and inflammatory responses. Immediate concerns during severe stress are to restore lost fluids and electrolytes and remove underlying stressors. Thus, initial treatments include administering intravenous solutions to correct fluid and electrolyte imbalances, treating infections, repairing wounds, draining **abscesses** (pus), and removing dead tissue (**debridement**). After stabilization, nutrient needs can be assessed and nutrition therapy provided.

NURSING DIAGNOSIS

Nursing diagnoses that may apply to individuals undergoing metabolic stress include *risk for deficient fluid volume, risk for electrolyte imbalance,* and *risk for unstable blood glucose level.*

DETERMINING NUTRITIONAL REQUIREMENTS

Notable metabolic changes in patients undergoing metabolic stress include hypermetabolism, negative nitrogen balance, insulin resistance, and hyperglycemia. Hypermetabolism and negative nitrogen balance can lead to wasting, which may impair organ function and delay recovery. Hyperglycemia increases the risk of infection, a dangerous problem during critical illness. Therefore, the principal goals of nutrition therapy are to preserve lean (muscle) tissue, maintain immune defenses, and promote healing.

Feeding an acutely stressed patient is often challenging. Underfeeding can worsen negative nitrogen balance and increase lean tissue losses. Overfeeding increases the risks of refeeding syndrome (■) and its associated hyperglycemia. Assessing nutritional needs can be complicated, however, because fluid imbalances prevent accurate weight measurements, and laboratory data may reflect the metabolic alterations of illness rather than the person's nutrition status.

The amounts of protein and energy to provide during acute illness are controversial and still under investigation. Research results have been mixed, in part because various conditions can lead to metabolic stress and each patient's situation is somewhat different. Moreover, protein and energy needs can vary substantially over the course of illness. Clinicians need to closely observe patients' responses to feedings and readjust nutrient intakes as necessary.

Estimating Energy Needs for Acute Stress A common method for determining the energy needs of acutely stressed individuals is to estimate or measure the resting metabolic rate (RMR) and then multiply the result by a stress factor to account for the increased energy requirements of stress and healing. This method was introduced in Chapter 14 (see p. 407 and the "How to" on p. 408); Table 16-2 reviews the use of stress factors and provides examples. Generally, energy needs are increased by fever, mechanical ventilation, restlessness, burns, and the presence of open wounds. Note that hospital patients undergoing acute stress are usually bedridden, so the energy needed for physical activity is minimal.

Some predictive equations used for estimating energy needs include "built-in" stress factors to account for stress, injury, or intensive treatment. Table 16-3 lists examples of

■ Reminder: *Refeeding syndrome* can develop when a severely malnourished person is aggressively fed; it is associated with fluid and electrolyte imbalances and hyperglycemia.

TABLE 16-2 Disease-Specific Stress Factors for Estimating Energy Needs during Metabolic Stress

Method

Step 1. Estimate the energy needed to support the resting metabolic rate (RMR) using indirect calorimetry or a predictive equation (see Table 14-3, p. 408).

Step 2. Multiply the estimated RMR by an appropriate stress factor for acute illness (see the "How to" box on p. 408).

Examples of stress factors[a]

- Intensive care: 1.0 to 1.1
- Cirrhosis (advanced liver disease): 1.2
- Acute kidney injury: 1.3
- Burns (more than 20 percent of body surface): 1.3 to 1.4
- Repletion after acute inflammation: 1.3 to 1.5
- Acute pancreatitis: 1.4 to 1.8

Adjustment for obesity[b]

In some individuals with BMI >30, using an adjusted body weight in RMR equations may help to avoid overfeeding. One method is to use a weight that falls between the person's actual body weight and an ideal body weight (IBW):

$$\text{Adjusted weight} = \text{IBW} + [0.5 \times (\text{Actual body weight} - \text{IBW})]$$

[a]Published values vary; energy intakes should be adjusted if the patient fails to maintain body weight at the energy level provided.
[b]The use of adjusted body weights for determining energy requirements in obese individuals is controversial and under investigation.

Sources: Academy of Nutrition and Dietetics, *Nutrition Care Manual* (Chicago: Academy of Nutrition and Dietetics, 2012); A. Skipper, ed., *Dietitian's Handbook of Enteral and Parenteral Nutrition* (Sudbury, MA: Jones & Bartlett Learning, 2012).

equations used for ventilator-dependent critical care patients and describes the use of the Ireton–Jones equation, which includes multipliers for the presence of trauma and burn injuries. Other equations in current use include factors for other pertinent variables, such as heart rate and respiratory rate.[5]

A quick method for estimating energy needs is to multiply a person's body weight by a factor appropriate for the medical condition. As an example, daily energy needs for patients with acute kidney injury often fall within the range of 20 to 30 kcalories per kilogram of body weight;[6] a patient weighing 160 pounds (72.7 kilograms) may therefore require between 1454 and 2181 kcalories per day. (■) Many critical care patients require between 25 and 30 kcalories per kilogram per day.[7] For critically ill obese patients (BMI > 30), **hypocaloric feedings** may improve patient outcome; the suggested energy intake is 11 to 14 kcalories per kilogram of actual body weight (or 22 to 25 kcalories per kilogram of ideal body weight) daily.[8] (■)

Protein Requirements in Acute Stress

To maintain lean tissue, the protein intakes recommended during acute stress are higher than DRI values. (■) For example, the protein needs of nonobese critically ill patients may range between 1.2 and 2.0 grams per kilogram body weight per day. Obese patients given hypocaloric feedings may require 2.0 to 2.5 grams per kilogram ideal body weight per day to maintain nitrogen balance.[9] Despite high intakes, however, nitrogen balance is difficult to achieve during acute stress because hormonal changes encourage the degradation of body protein. The bed rest required during critical illness also contributes substantially to muscle breakdown.

The amino acids glutamine and arginine are sometimes added to the diets of acutely stressed and immune-compromised patients. Clinical studies have suggested that glutamine supplementation may improve immune function, preserve muscle mass, and reduce mortality rates in critically ill patients.[10] Arginine supplementation may improve the immune responses and nitrogen balances of critically ill and postoperative patients.[11] Although glutamine and arginine are often added to enteral formulas promoted for wound healing and enhanced immunity, their use remains controversial.

■ 72.7 kg × 20 kcal/kg = 1454 kcal
72.7 kg × 30 kcal/kg = 2181 kcal

■ Reminder: Healthy body weights typically fall within the BMI range of 18.5 to 25.

■ Reminder: The protein RDA for adults is 0.8 grams per kilogram body weight.

hypocaloric feedings: reduced-kcalorie feedings that usually include sufficient protein and micronutrients to maintain nitrogen balance and prevent malnutrition; also called *permissive underfeeding*.

TABLE 16-3 Selected Equations for Estimating Energy Needs in Ventilator-Dependent Critical Care Patients

Ireton–Jones[a]

Energy needs (kcal/day) = 1925 + [5 × Wt (kg)] − [10 × Age (yr)] + [281 × Sex] + [292 × Trauma] + [851 × Burn]

where Sex is male (× 1) or female (× 0), Trauma is the presence of physical injury (× 1) or not (× 0), and Burn is the presence of a burn injury (× 1) or not (× 0).

Penn State[a]

Energy needs (kcal/day) = [RMR × 0.85] + [V_E × 33] + [T_{max} × 175] − 6433

where RMR is calculated using the Harris-Benedict equation (see Table 14-3, p. 408), V_E is minute ventilation in liters per minute (reading taken from the ventilator), and T_{max} is the patient's maximum body temperature (in degrees Celsius) in the preceding 24 hours.

Example (Ireton–Jones equation)

Erin is a 27-year-old female patient who weighs 140 pounds (63.6 kilograms). Two days ago, she was severely injured in an automobile accident and is currently being cared for in a critical care unit where she is receiving mechanical ventilation. She did not suffer a burn injury. Using the Ireton–Jones equation shown above, her daily energy needs can be estimated as follows:

Energy needs (kcal/day)
= 1925 + [5 × Wt (kg)] − [10 × Age (yr)] + [281 × Sex] + [292 × Trauma] + [851 × Burn]
= 1925 + (5 × 63.6 kg) − (10 × 27 yr) + (281 × 0) + (292 × 1) + (851 × 0)
= 1925 + 318 − 270 + 0 + 292 + 0 = 2265 kcal

[a]When calculating the energy needs of obese individuals using this equation, the person's actual body weight (not an adjusted body weight) yields satisfactory results.

Carbohydrate and Fat Intakes in Acute Stress The bulk of energy needs are supplied from carbohydrate and fat. Carbohydrate is usually the main source of energy, providing 50 to 60 percent of total energy requirements. (■) In patients with severe hyperglycemia, fat may supply up to 50 percent of kcalories, although high fat intakes may suppress immune function and increase the risk of developing infections and hypertriglyceridemia. Patients with blood triglyceride levels above 300 to 400 milligrams per deciliter may require fat restriction.[12]

Micronutrient Needs in Acute Stress Acutely stressed patients are believed to have increased micronutrient needs, but specific requirements remain unknown.[13] In hypermetabolic patients, the need for B vitamins may be higher to support the increase in energy metabolism. Several micronutrients, including vitamin A, vitamin C, and zinc, have critical roles in immunity and wound healing; thus, their supplementation may speed recovery under certain circumstances. Patients with burns and tissue injuries may have increased requirements for trace minerals due to tissue losses.

The acute-phase response causes a redistribution in the tissue content of some micronutrients that either raises or lowers their blood levels; therefore, micronutrient status is sometimes difficult to interpret. (■) Blood concentrations of trace minerals are monitored in patients receiving parenteral nutrition to ensure that excessive amounts are not given intravenously.

APPROACHES TO NUTRITION CARE IN ACUTE STRESS

As mentioned earlier, the initial care following acute stress focuses on maintaining fluid and electrolyte balances. Simple intravenous solutions often contain dextrose, providing minimal kcalories. Once patients are stable, nutrition support may be necessary if poor appetite, the medical condition, or a medical procedure (such as mechanical ventilation) interferes with food intake. For acutely ill patients with a functional GI tract, early enteral feedings—started in the first 24 to 48 hours after hospitalization—are associated with fewer complications and shorter hospital stays compared with delayed feedings.[14] If enteral nutrition is not possible, malnourished patients may receive parenteral nutrition support soon after admission to the hospital. In previously healthy patients, however, parenteral nutrition support may be withheld during the first seven days of hospitalization to avoid the risk of infectious complications.[15]

Once patients can tolerate oral feedings, a high-kcalorie, high-protein diet is often prescribed, although care must be taken not to overfeed patients who are at risk of developing refeeding syndrome or hyperglycemia. Because meeting protein and energy needs may be difficult, oral supplements are often provided to supplement the diet. Many such formulas have a high nutrient density, and some contain extra amounts of nutrients believed to promote healing or benefit immune function, such as the amino acids arginine and glutamine, omega-3 fatty acids, and the antioxidant nutrients. Nutrient needs should be reassessed frequently as the patient's condition improves. The accompanying Case Study reviews the nutrition care of a patient undergoing acute metabolic stress.

■ Reminder: When parenteral infusions are necessary, dextrose is usually limited to 5 milligrams per kilogram body weight per minute.

■ During the acute-phase response, plasma levels of iron and zinc fall, whereas plasma copper levels rise.

NURSING DIAGNOSIS

The nursing diagnosis *imbalanced nutrition: less than body requirements* may apply to a patient recovering from acute stress.

Case Study: Patient with a Severe Burn

David Bray, a 42-year-old man, has been admitted to intensive care. He suffered a severe burn covering 35 percent of his body when he was trapped inside a burning building. His wife told the nurse that Mr. Bray's height is 6 feet and that he usually weighs about 175 pounds. The physician ordered lab tests, including serum protein concentrations, but the results are not yet available.

1. Identify Mr. Bray's immediate needs after the injury. Describe the initial concerns of the health care team and the measures they might take soon after Mr. Bray's arrival at the hospital.
2. Considering Mr. Bray's condition, what problems might the health care team encounter when they attempt to obtain information that can help them assess his nutrition status? What additional concerns might they have if Mr. Bray was malnourished before he experienced the burn?
3. Estimate Mr. Bray's energy and protein needs (use a protein factor of 2.0 grams per kilogram). What problems may interfere with Mr. Bray's ability to meet his nutrient needs?
4. After Mr. Bray transitions to oral feedings, he is able to obtain only 65 percent of his energy requirements. What other feeding options may be considered?

Review Notes

- In patients hospitalized with acute stress, initial treatments include restoring fluid and electrolyte balances and treating infections and wounds.
- Metabolic stress can result in hypermetabolism, negative nitrogen balance, hyperglycemia, and wasting. The objectives of nutrition care are to preserve muscle tissue, maintain immune defenses, and promote healing.
- To determine energy needs for acute stress, RMR values may be modified using disease-specific stress factors. Another method is the use of simple equations that prescribe specific energy levels per kilogram of body weight.
- Protein recommendations during acute stress are higher than DRI levels to help maintain nitrogen balance, prevent tissue losses, and allow the healing of damaged tissue. Carbohydrates and lipids provide most of the patient's energy needs.
- Micronutrient needs may be increased during acute stress, although specific requirements remain unknown.
- Enteral and parenteral nutrition support or oral supplements may be used to help meet the high nutrient needs of acutely stressed patients.

Nutrition and Respiratory Stress

Some medical problems upset the process of gas exchange between the air and blood and result in respiratory stress, which is characterized by a reduction in the blood's oxygen supply and an increase in carbon dioxide levels. Excessive carbon dioxide in the blood may disturb the breathing pattern enough to interfere with food intake. Moreover, the labored breathing caused by many respiratory disorders entails a higher energy cost than normal breathing does, raising energy needs and increasing carbon dioxide production further. Lung diseases make physical activity difficult and can lead to muscle wasting. Weight loss and malnutrition therefore become dangerous outcomes of some types of respiratory illnesses.

CHRONIC OBSTRUCTIVE PULMONARY DISEASE

Chronic obstructive pulmonary disease (COPD) refers to a group of conditions characterized by the persistent obstruction of airflow through the lungs. Figure 16-2 illustrates the main airways (**bronchi** and **bronchioles**) and air sacs (**alveoli**) of the normal respiratory system, and Figure 16-3 shows how they are altered in COPD. The two main categories of COPD are **chronic bronchitis** and **emphysema**, although most COPD patients display features of both conditions:[16]

- *Chronic bronchitis* is characterized by persistent inflammation and excessive secretions of mucus in the airways of the lungs, which may ultimately thicken and become too narrow for adequate mucus clearance. Chronic bronchitis is diagnosed when a chronic, productive cough persists for at least three months of the year for two consecutive years.
- *Emphysema* is characterized by the breakdown of the lungs' elastic structure and destruction of the walls of the bronchioles and alveoli, changes that significantly reduce the surface area needed for respiration. Emphysema is diagnosed on the basis of clinical signs and the results of lung function tests.

Both chronic bronchitis and emphysema are associated with abnormal levels of oxygen and carbon dioxide in the blood and shortness of breath (**dyspnea**). COPD may eventually lead to respiratory or heart failure, and, together with other chronic respiratory illnesses, ranks as the third leading cause of death in the United States.[17]

COPD is a debilitating condition. Generally, dyspnea worsens as the disease progresses, resulting in dramatic reductions in physical activity and quality of life.

chronic obstructive pulmonary disease (COPD): a group of lung diseases characterized by persistent obstructed airflow through the lungs and airways; includes chronic bronchitis and emphysema.

bronchi (BRON-key), bronchioles (BRON-key-oles): the main airways of the lungs. The singular form of bronchi is *bronchus.*

alveoli (al-VEE-oh-lee): air sacs in the lungs. One air sac is an *alveolus.*

chronic bronchitis (bron-KYE-tis): a lung disorder characterized by persistent inflammation and excessive secretions of mucus in the main airways of the lungs.

emphysema (EM-fih-ZEE-mah): a progressive lung disease characterized by the breakdown of the lungs' elastic structure and destruction of the walls of the bronchioles and alveoli, reducing the surface area involved in respiration.

dyspnea (DISP-nee-ah): shortness of breath.

FIGURE 16-2 The Respiratory System

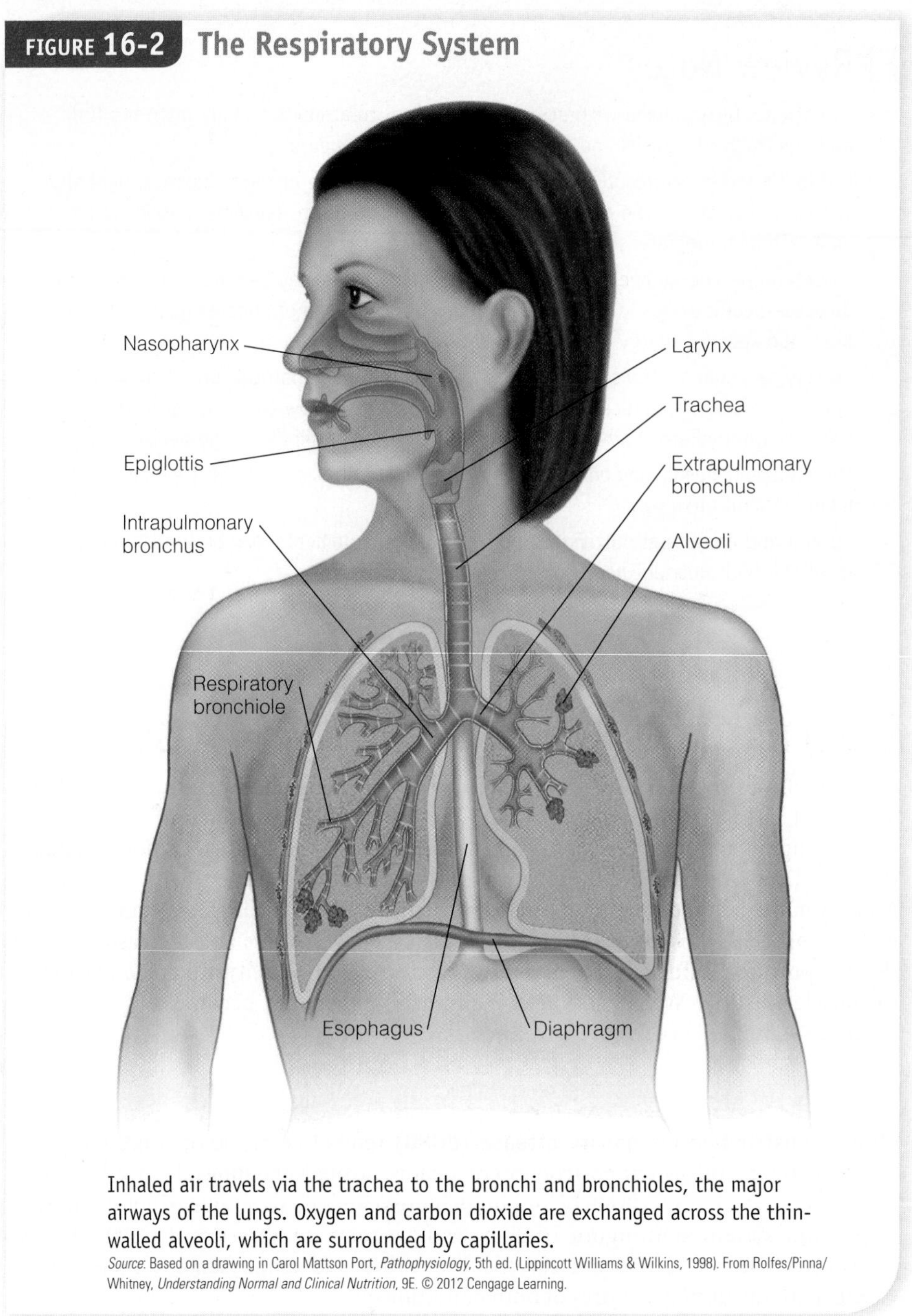

Inhaled air travels via the trachea to the bronchi and bronchioles, the major airways of the lungs. Oxygen and carbon dioxide are exchanged across the thin-walled alveoli, which are surrounded by capillaries.

Source: Based on a drawing in Carol Mattson Port, *Pathophysiology*, 5th ed. (Lippincott Williams & Wilkins, 1998). From Rolfes/Pinna/Whitney, *Understanding Normal and Clinical Nutrition*, 9E. © 2012 Cengage Learning.

> ***NURSING DIAGNOSIS***
>
> Nursing diagnoses that may apply to people with COPD include *impaired gas exchange, ineffective breathing pattern, ineffective airway clearance,* and *activity intolerance.*

Activities of daily living such as bathing or dressing may cause exhaustion or breathlessness. Weight loss and wasting are common in the advanced stages of disease and may result from hypermetabolism, poor food intake, and the actions of various inflammatory proteins.

Causes of COPD Cigarette smoking is the primary risk factor for COPD and is especially damaging when combined with respiratory infections or an occupational exposure to dusts or chemicals. Only a minority of smokers (about 15 percent[18]) develops COPD, however; thus, genetic susceptibility also contributes to its development. Genetic factors are especially likely in patients with early-onset COPD. Alpha-1-antitrypsin deficiency, an inherited disorder, accounts for 1 to 2 percent of COPD cases.[19] Individuals with this defect have inadequate blood levels of a plasma protein (alpha-1-antitrypsin) that normally inhibits the enzymatic breakdown of lung tissue.

FIGURE 16-3 Chronic Obstructive Pulmonary Disease

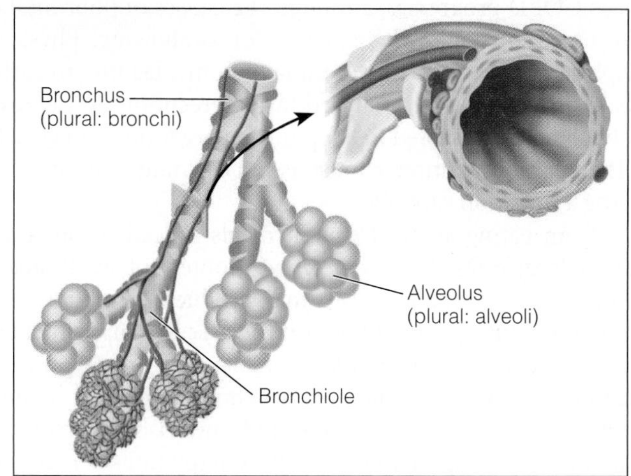

Healthy bronchi provide an open passageway for air. Healthy alveoli permit gas exhange between the air and blood.

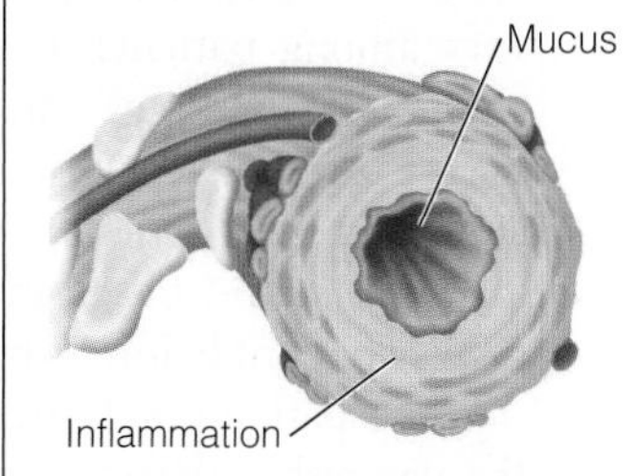

Chronic bronchitis is characterized by inflammation, excessive secretion of mucus, and narrowing of the airways— factors that reduce normal airflow.

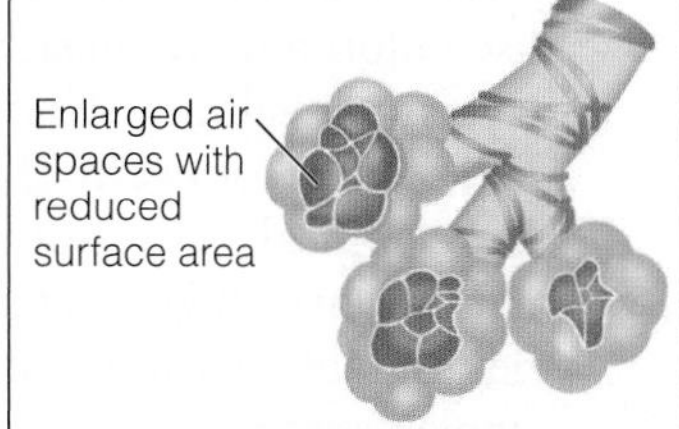

Emphysema is characterized by gradual destruction of the walls separating alveoli and reduced lung elasticity.

Treatment of COPD The primary objectives of COPD treatment are to prevent the disease from progressing and to relieve major symptoms (dyspnea and coughing). Individuals with COPD are encouraged to quit smoking to prevent disease progression and to get vaccinated against influenza and pneumonia to avoid complications. The most frequently prescribed medications are bronchodilators, which improve airflow, and inhaled corticosteroids (anti-inflammatory drugs), which help to relieve symptoms and prevent exacerbations. For people with severe COPD, supplemental oxygen therapy (12 hours daily) can maintain normal oxygen levels in the blood and reduce mortality risk. The Diet-Drug Interactions feature lists nutrition-related effects of the medications used to treat COPD.

Patients who need supplemental oxygen can use lightweight, portable equipment that allows them to move around freely.

Nutrition Therapy for COPD The main goals of nutrition therapy for COPD are to correct malnutrition (which affects up to 60 percent of COPD patients[20]), promote the maintenance of a healthy body weight, and prevent muscle wasting. Energy needs of COPD patients are usually raised due to hypermetabolism (about 20 percent above normal), which results from chronic inflammation and the increased workload of respiratory muscles.[21] Because underweight COPD patients have higher mortality rates, encouraging adequate food intake is generally the main focus of the nutrition care plan. Conversely, excess body weight places an additional strain on the respiratory

DIET-DRUG INTERACTIONS

Check this table for notable nutrition-related effects of the medications discussed in this chapter.

Bronchodilators (albuterol, salmeterol, ipratropium, tiotropium)	**Gastrointestinal effect:** dry mouth (ipratropium, tiotropium)
	Metabolic effect: mild hypokalemia (albuterol, salmeterol)
Corticosteroids (inhaled) (fluticasone, beclomethasone)	**Gastrointestinal effect:** decreased taste sensation
	Metabolic effect: low bone density

system, and so COPD patients who are overweight or obese may benefit from energy restriction and gradual weight reduction.[22]

Food intake often declines as COPD progresses, although the causes of poor intake vary among patients. Dyspnea may interfere with chewing or swallowing. Physical changes in the lungs and diaphragm may reduce abdominal volume, leading to early satiety. Appetite may be affected by medications, altered taste perception, or depression. (■) Some patients may become too disabled to shop or prepare food or may lack adequate support at home. The health practitioner must assess the unique needs of a COPD patient before proposing a nutrition care plan.

Some patients may benefit from eating small, frequent meals spaced throughout the day rather than two or three large ones. The lower energy content of small meals reduces the carbon dioxide load, and the smaller meals may produce less abdominal discomfort and dyspnea. Some individuals may eat better if they receive supplemental oxygen at mealtimes. Consuming adequate fluids should be encouraged to help prevent the secretion of overly thick mucus; however, some patients should consume liquids between meals so as not to interfere with food intake. For undernourished persons, a high-kcalorie, high-protein diet may be helpful, but excessive energy intakes increase the amount of carbon dioxide produced and can increase respiratory stress. Oral supplements may be recommended as between-meal snacks to improve weight gain or endurance, but patients should be cautioned not to consume amounts that reduce energy intake at mealtime.

Pulmonary Formulas Enteral formulas designed for use in COPD provide more kcalories from fat and fewer from carbohydrate than standard formulas. The ratio of carbon dioxide production to oxygen consumption is lower when fat is consumed, so theoretically these formulas should lower respiratory requirements. However, research studies have not confirmed that the reduced-carbohydrate formulas improve clinical outcomes more than moderate energy intakes.[23]

Incorporating an Exercise Program Loss of muscle can be more readily prevented or reversed if the treatment plan includes an effective exercise program. With exercise, patients are likely to see improvements in their strength, endurance, and ability to perform activities of daily living. Both aerobic training and resistance exercise can be beneficial.[24] Some patients may need to increase activity gradually over a period of four to six weeks before reaching exercise goals.[25] The accompanying Case Study can help you review the nutrition care for a patient with COPD.

■ The altered sense of taste in COPD patients may be due to chronic mouth breathing (which causes mouth dryness) or the use of bronchodilators.

RESPIRATORY FAILURE

respiratory failure: a potentially life-threatening condition in which inadequate respiratory function impairs gas exchange between the air and circulating blood, resulting in abnormal levels of tissue gases.

In **respiratory failure**, inadequate respiratory function impairs gas exchange between the air and circulating blood, resulting in abnormal levels of tissue gases that can be life threatening. Any of a large number of conditions that cause lung injury or impair lung function can be the underlying cause of failure. Possible causes of respiratory failure

Case Study: Elderly Man with Emphysema

John Todaro is an 84-year-old man who has emphysema that severely affects both lungs. He is 5 feet 9 inches tall and currently weighs 150 pounds, about 20 pounds less than his weight in earlier years. He lives with a daughter and son-in-law and eats meals with their family. He becomes breathless when eating and when walking around the house, and he feels tired much of the time. A medical clinic recently ordered oxygen therapy for home use, but supplies have not yet arrived. Mr. Todaro's daughter is concerned about her father's recent weight loss and breathlessness.

1. Assess Mr. Todaro's risk of malnutrition, using information from Table 13-7 in Chapter 13 (p. 391). What factors may have contributed to his weight loss?
2. What are possible reasons for Mr. Todaro's difficulty with eating? List some dietary suggestions that may help to improve his appetite and food intake. How might the use of oxygen therapy help?
3. Based on the history given, what factors may account for Mr. Todaro's tiredness? What suggestions would you give Mr. Todaro and his daughter regarding physical activity?

include infection (such as pneumonia or sepsis), physical trauma, aspiration of stomach contents, neuromuscular disorders, smoke inhalation, and airway obstruction.[26]

If an acute lung injury causes enough damage that emergency care is required to restore normal oxygen and carbon dioxide levels, the condition is known as **acute respiratory distress syndrome (ARDS)**. In ARDS, the lungs exhibit extensive inflammation and fluid buildup (called *pulmonary edema*) that interfere with lung ventilation and damage the alveoli. Later stages are associated with a proliferation of lung cells, which results in fibrosis and disrupts lung structure. A dangerous complication of ARDS is the progression to multiple organ dysfunction syndrome, described in Nutrition in Practice 16.

Consequences of Respiratory Failure

Respiratory failure is characterized by severe **hypoxemia** (low oxygen levels in the blood) and **hypercapnia** (excessive carbon dioxide in the blood). The low oxygen content of tissues (**hypoxia**) impedes cellular function and may lead to cell death. Severe hypercapnia can cause **acidosis**, which interferes with normal functioning of the central nervous system. To compensate for respiratory failure, a person breathes more rapidly, and the heart rate increases. The skin may become sweaty and develop a bluish cast (**cyanosis**). Headache, confusion, and drowsiness may occur. Severe cases of respiratory failure can cause heart arrhythmias and, ultimately, coma.

acute respiratory distress syndrome (ARDS): respiratory failure triggered by severe lung injury; a medical emergency that causes dyspnea and pulmonary edema and usually requires mechanical ventilation.

hypoxemia (high-pock-SEE-me-ah): a low level of oxygen in the blood.

hypercapnia (high-per-CAP-nee-ah): excessive carbon dioxide in the blood.

hypoxia (high-POCK-see-ah): a low amount of oxygen in body tissues.

acidosis: acid accumulation in body tissues; depresses the central nervous system and may lead to disorientation and, eventually, coma.

cyanosis (sigh-ah-NOH-sis): a bluish cast in the skin due to the color of deoxygenated hemoglobin. Cyanosis is most evident in individuals with lighter, thinner skin; it is mostly seen on lips, cheeks, and ears and under the nails.

Treatment of Respiratory Failure

The treatment of respiratory failure focuses on supporting lung function and correcting the underlying disorder. Because respiratory failure can be caused by a number of different conditions, treatment plans vary considerably. Individuals with chronic lung disorders may be provided with oxygen therapy via a face mask or nasal tubing to relieve symptoms, whereas patients with ARDS receive mechanical ventilation until they are able to breathe independently. Diuretics may be prescribed to mobilize the fluid that has accumulated in lung tissue. Medications are given to treat infections, keep airways open, or relieve inflammation. Complications are common in ARDS and must be forestalled to prevent multiple organ dysfunction.

Nutrition Therapy for Respiratory Failure

Patients with lung injuries or ARDS are frequently hypermetabolic and/or catabolic and at high risk of muscle wasting. The primary concerns are therefore to supply enough energy and protein to sustain muscle tissue and lung function without overtaxing the compromised respiratory system. Fluid restrictions may be necessary to help correct pulmonary edema. As usual, when nutrition support is necessary, enteral nutrition is preferred over parenteral nutrition. Specific recommendations for respiratory failure include the following:

- *Energy.* Energy needs typically range from 25 to 35 kilocalories per kilgram.[27] The body weight used to determine energy needs may need to be corrected for pulmonary edema. Energy needs can also be estimated using predictive equations such as those described in Tables 16-2 and 16-3. Overfeeding should be avoided because it can cause excessive carbon dioxide production and worsen respiratory function.
- *Protein.* Protein needs are increased in patients with lung inflammation or ARDS. For mild or moderate lung injury, protein recommendations range from 1.0 to 1.5 grams of protein per kilogram of body weight per day. Patients with ARDS may require 1.5 to 2 grams of protein per kilogram of body weight daily.
- *Fluids.* Although most patients have normal fluid requirements, fluid status should be monitored daily to prevent fluid imbalances. Some patients may require fluid restriction to prevent edema in lung tissue, whereas others may become dehydrated due to diuretic therapy, an increase in bronchial secretions, or a low fluid intake. The presence of edema can make it difficult to assess whether a critically ill patient is maintaining weight.

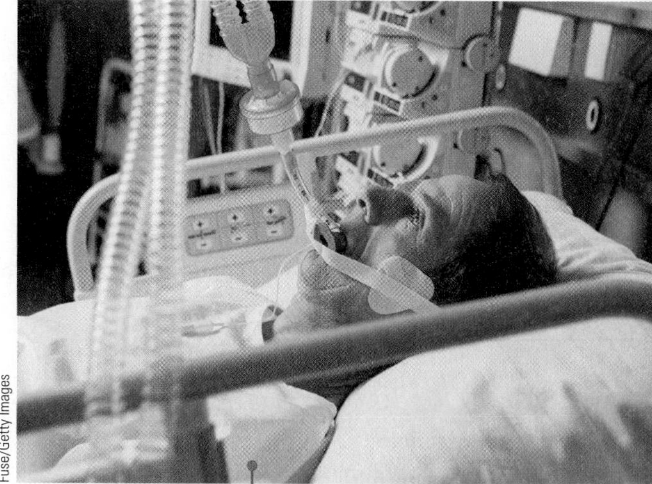

Fuse/Getty Images

Mechanical ventilation controls the rate and amount of oxygen supplied to a person's airways.

Nutrition Support in Respiratory Failure Patients with severe cases of respiratory failure may be unable to eat meals and may require nutrition support. Enteral nutrition is used if the intestine is functional; intestinal feedings may be preferred over gastric feedings because they reduce the risk of aspiration. Patients with acute lung injuries or ARDS may benefit from enteral formulas designed to reduce inflammation and promote healing; these formulas are typically fortified with omega-3 fatty acids and antioxidant nutrients.[28] Nutrient-dense formulas (1.5 to 2.0 kcalories per milliliter) are prescribed for patients with fluid restrictions. If the risk of aspiration is too high to continue enteral feedings, parenteral nutrition support may be considered.

Review Notes

- Respiratory stress can affect body weight, muscle mass, and the normal functioning of all body tissues.
- Chronic obstructive pulmonary diseases (COPDs) are debilitating, progressive illnesses that can lead to malnutrition, muscle wasting, and activity intolerance. The main categories of COPD are chronic bronchitis and emphysema.
- The goals of nutrition therapy for COPD are to improve food intake, maintain proper weight, preserve muscle tissue, and improve exercise endurance.
- Respiratory failure, which is characterized by severe hypoxemia and hypercapnia, can result from conditions that cause lung injury or impair lung function. Acute respiratory distress syndrome (ARDS) is a severe form of respiratory failure that requires emergency care.
- Goals of nutrition therapy for respiratory failure are to supply enough energy and protein to support lung function without burdening the respiratory system. Fluid restrictions may be necessary to reverse pulmonary edema.

Nutrition Assessment Checklist for People Undergoing Metabolic or Respiratory Stress

MEDICAL HISTORY

Check the medical record to determine:

- ❍ Cause of stress
- ❍ Severity of stress
- ❍ Whether any organ system is compromised
- ❍ Whether nutrition support is required

For patients with COPD, check to determine:

- ❍ Degree of breathing difficulty
- ❍ Use of oxygen therapy
- ❍ Activity tolerance

Review the medical record for complications related to underfeeding or overfeeding, such as:

- ❍ Dehydration or fluid overload
- ❍ Electrolyte imbalances
- ❍ Fatty liver
- ❍ Hyperglycemia
- ❍ Hypertriglyceridemia

MEDICATIONS

Record all medications and note:

- ❍ Side effects that may alter food intake or nutrition status

DIETARY INTAKE

If the patient is not meeting nutrition goals:

- ❍ Monitor intakes to ensure that the patient is receiving the diet prescribed.
- ❍ Investigate appetite problems or difficulties with eating.
- ❍ Consider interventions to improve food intake.
- ❍ Consider the need for oral supplements.
- ❍ In patients with COPD, consider problems that may hamper the patient's ability to prepare or consume foods.

ANTHROPOMETRIC DATA

Measure baseline height and weight, and monitor daily weights. Remember that body weight can fluctuate in acutely ill patients who undergo fluid resuscitation. After the patient's weight has stabilized:

- ❍ Reevaluate protein and energy needs.
- ❍ Consider the need to alter the energy prescription to meet weight goals.
- ❍ Nutrient deficiencies
- ❍ Negative nitrogen balance
- ❍ Organ dysfunction or organ function that has normalized

LABORATORY TESTS

Laboratory tests that may be affected by stress and therefore require careful interpretation include:

- ❍ Albumin
- ❍ C-reactive protein
- ❍ Prealbumin
- ❍ Serum iron and zinc
- ❍ Transferrin
- ❍ White blood cell count

Monitor laboratory tests for signs of:

- ❍ Dehydration or fluid overload
- ❍ Electrolyte and acid-base imbalances
- ❍ Hyperglycemia
- ❍ Hypertriglyceridemia

PHYSICAL SIGNS

Regularly assess vital signs, including:

- ❍ Blood pressure
- ❍ Body temperature
- ❍ Pulse
- ❍ Respiratory rate

Look for physical signs of:

- ❍ Protein-energy malnutrition
- ❍ Dehydration or fluid overload
- ❍ Nutrient deficiencies and excesses

Self Check

1. Which of the following metabolic changes accompanies acute stress?
a. Hypoglycemia
b. Reduced heart and respiratory rates
c. Elevated immune responses
d. Muscle protein catabolism

2. After an injury occurs, typical changes in the damaged tissue include all of the following *except:*
a. reduced capillary permeability.
b. increased blood flow.
c. warmth.
d. fluid accumulation.

3. How do nonsteroidal anti-inflammatory drugs suppress inflammatory processes?
a. They interfere with histamine secretion.
b. They inhibit eicosanoid synthesis.
c. They impair the actions of certain cytokines.
d. They inhibit the absorption of omega-6 fatty acids.

4. Examples of acute-phase proteins include all of the following *except:*
a. C-reactive protein.
b. hepcidin.
c. albumin.
d. fibrinogen.

5. Which of the following statements concerning protein and energy recommendations during acute metabolic stress is true?
a. Protein and energy recommendations are similar to those for healthy people.
b. Protein and energy recommendations are reduced because a stressed individual cannot metabolize nutrients normally.
c. Acutely stressed individuals can benefit from as much protein and energy as can be provided.
d. Protein and energy recommendations are increased to minimize muscle tissue losses.

6. Hypocaloric feedings may benefit critical care patients who are:
a. elderly.
b. obese.
c. receiving mechanical ventilation.
d. accumulating fluid.

7. The primary risk factor for COPD is:
a. alpha-1-antitrypsin deficiency.
b. occupational exposure to dusts or chemicals.
c. cigarette smoking.
d. frequent respiratory infections.

8. A primary feature of emphysema is:
a. obstruction within the bronchi.
b. obstruction within the bronchioles.
c. destruction of the walls separating the alveoli.
d. excessive lung elasticity.

9. The weight loss and wasting that often occur in COPD can be caused by:

a. reduced food intake.
b. increased metabolic rate.
c. reduced exercise tolerance.
d. all of the above.

10. Nutrition therapy for a person with respiratory failure includes:

a. careful attention to providing enough, but not too much, energy.
b. a generous fluid intake to facilitate mucus clearance.
c. a high fat intake to help prevent weight loss and wasting.
d. a high carbohydrate intake to increase carbon dioxide production.

Answers to these questions appear in Appendix H. For more chapter review: Access an interactive eBook, chapter-specific interactive learning tools, including flashcards, quizzes, videos, and more in your Nutrition CourseMate, accessed through CengageBrain.com.

Clinical Applications

1. Adam is a 29-year-old male who is 6 feet 2 inches tall and has a usual body weight of 180 pounds. He underwent emergency surgery following a serious injury and is now being cared for in the intensive care unit, where he is receiving mechanical ventilation. Using the Ireton–Jones equation shown in Table 16-3 on p. 473, estimate Adam's energy requirement. Estimate his protein requirement, using the factor 1.5 grams per kilogram of body weight.

2. Ayesha is a 23-year-old law student who was admitted to the hospital following an automobile accident in which she broke several bones and ruptured part of her small intestine. She has been in the hospital for several weeks and has just begun eating table foods. Her brother, who was driving the vehicle, was also seriously injured and nearly lost his life. Aside from the increased nutritional needs imposed by the stress of the accident, discuss how the following factors might interfere with Ayesha's ability to improve her nutrition status:
   - Ayesha's injuries are painful.
   - Ayesha's medications cause drowsiness.
   - Ayesha is depressed.
   - Ayesha is often out of her room for X-rays and other diagnostic tests when the menus and food trays arrive.
   - Ayesha's food intake is sometimes restricted due to the procedures she is undergoing.

   How might these problems be resolved to improve Ayesha's food intake?

Nutrition on the Net

For further study of topics covered in this chapter, access these websites:

- To uncover additional information relevant to critical care, visit these sites:
  American Association of Critical Care Nurses:
  www.aacn.org
  American Society for Parenteral and Enteral Nutrition:
  www.nutritioncare.org
- To learn more about lung diseases, visit these sites:
  American Lung Association:
  www.lungusa.org
  Canadian Lung Association:
  www.lung.ca
  American Thoracic Society:
  www.thoracic.org
  National Heart, Lung and Blood Institute:
  www.nhlbi.nih.gov

Notes

1. P. A. Fitzgerald, Adrenal medulla and paraganglia, in D. G. Gardner and D. Shoback, eds., *Greenspan's Basic & Clinical Endocrinology* (New York: McGraw-Hill, 2011), pp. 345–393; T. B. Carroll and coauthors, Glucocorticoids and adrenal androgens, in D. G. Gardner and D. Shoback, eds., *Greenspan's Basic & Clinical Endocrinology* (New York: McGraw-Hill, 2011), pp. 285–327.
2. P. C. Calder, Mechanisms of action of (n-3) fatty acids, *Journal of Nutrition* 142 (2012): 592S–599S.
3. G. H. Johnson and K. Fritsche, Effect of dietary linoleic acid on markers of inflammation in healthy persons: A systematic review of randomized controlled trials, *Journal of the Academy of Nutrition and Dietetics* 112 (2012): 1029–1041.
4. D. S. Pisetsky, Laboratory testing in the rheumatic diseases, in L. Goldman and A I. Schafer, eds., *Goldman's Cecil Medicine* (Philadelphia: Saunders, 2012): 1651–1656; V. Kumar and coauthors, Acute and chronic inflammation, in V. Kumar and coeditors, *Robbins and Cotran Pathologic Basis of Disease* (Philadelphia: Saunders, 2010), pp. 43–77.
5. D. Frankenfield, Prediction of resting metabolic rate in critically ill adult patients: Results of a systematic review of the evidence, *Journal of the American Dietetic Association* 107 (2007): 1552–1561.
6. M. Kalista-Richards, Acute kidney injury, in A. Skipper, ed., *Dietitian's Handbook of Enteral and Parenteral Nutrition* (Sudbury, MA: Jones & Bartlett Learning, 2012), pp. 157–167.
7. S. A. McClave and coauthors, Guidelines for the provision and assessment of nutrition support therapy in the adult critically ill patient: Society of Critical Care Medicine (SCCM) and American Society for Parenteral and Enteral Nutrition (A.S.P.E.N.), *Journal of Parenteral and Enteral Nutrition* 33 (2009): 277–316.
8. McClave and coauthors, 2009.
9. McClave and coauthors, 2009.
10. McClave and coauthors, 2009.
11. M. Zhou and R. G. Martindale, Arginine in the critical care setting, *Journal of Nutrition* 137 (2007): 1687S–1692S.
12. J. Lefton and P. P. Lopez, Macronutrient requirements: Carbohydrate, protein, and lipid, in G. Cresci, ed., *Nutrition Support for the Critically Ill Patient: A Guide to Practice* (Boca Raton, FL: Taylor & Francis Group, 2005), pp. 99–108.
13. Academy of Nutrition and Dietetics, *Nutrition Care Manual* (Chicago: Academy of Nutrition and Dietetics, 2012).
14. McClave and coauthors, 2009
15. McClave and coauthors, 2009.
16. D. E. Niewoehner, Chronic obstructive pulmonary disease, in L. Goldman and A I. Schafer, eds., *Goldman's Cecil Medicine* (Philadelphia: Saunders, 2012), pp. 537–544; A. N. Husain, The lung, in V. Kumar and coeditors, *Robbins and Cotran Pathologic Basis of Disease* (Philadelphia: Saunders, 2010), pp. 677–737.
17. S. L. Murphy, J. Xu, and K. D. Kochanek, Deaths: Preliminary data for 2010, *National Vital Statistics Reports* 60, no. 4 (Hyattsville, MD: National Center for Health Statistics, 2012).
18. R. S. Porter and J. L. Kaplan, eds., *Merck Manual of Diagnosis and Therapy* (Whitehouse Station, N.J.: Merck Sharp and Dohme Corp., 2011), pp. 1889–1902.
19. Niewoehner, 2012.
20. P. F. Collins, R. J. Stratton, and M. Elia, Nutritional support in chronic obstructive pulmonary disease: A systematic review and meta-analysis, *American Journal of Clinical Nutrition* 95 (2012): 1385–1395.
21. D. A. King, F. Cordova, and S. M. Scharf, Nutritional aspects of chronic obstructive pulmonary disease, *Proceedings of the American Thoracic Society* 5 (2008): 519–523.
22. M. Poulain and coauthors, The effect of obesity on chronic respiratory diseases: Pathophysiology and therapeutic strategies, *Canadian Medical Association Journal* 174 (2006): 1293–1299.
23. McClave and coauthors, 2009; A. Malone, Enteral formula selection, in P. Charney and A. Malone, eds., *ADA Pocket Guide to Enteral Nutrition* (Chicago: American Dietetic Association, 2006), pp. 63–122.
24. W. D. Reid and coauthors, Exercise prescription for hospitalized people with chronic obstructive pulmonary disease and comorbidities: A synthesis of systematic reviews, *International Journal of COPD* 7 (2012): 297–320.
25. Porter and Kaplan, 2011.
26. L. D. Hudson and A. S. Slutsky, Acute respiratory failure, in L. Goldman and A I. Schafer, eds., *Goldman's Cecil Medicine* (Philadelphia: Saunders, 2012), pp. 629–638.
27. Academy of Nutrition and Dietetics, 2012.
28. K. L. Coughlin, S. I. Austhof, and C. Hamilton, Nutrition support: Indication and efficacy, in A. Skipper, ed., *Dietitian's Handbook of Enteral and Parenteral Nutrition* (Sudbury, MA: Jones & Bartlett Learning, 2012), pp. 22–45; McClave and coauthors, 2009.

Nutrition in Practice

Multiple Organ Dysfunction Syndrome

Multiple organ dysfunction syndrome (MODS), also called *multiple organ failure,* is a frequent cause of death in intensive care patients. Described as the progressive dysfunction of two or more of the body's organ systems, MODS most often involves the lungs, kidneys, and liver. MODS is not a disease per se, but rather a late stage of severe illness or injury that results from a severe inflammatory response (discussed in Chapter 16). MODS can be initiated by a number of very different critical illnesses and conditions, including respiratory failure, sepsis, burn injuries, trauma, and pancreatitis. This Nutrition in Practice discusses how MODS develops, the manner in which it is treated, and the importance of its prevention.

How long has MODS been a major clinical problem?

MODS was recognized as a clinical entity only after World War II. Prior to the mid-20th century, patients with severe illnesses or multiple injuries frequently died of shock or circulatory failure. After fluid replacement and blood transfusions became standard treatments, the kidneys became the organs at highest risk, and kidney failure became the most common cause of death. Eventually, physicians learned to better support kidney function by providing appropriate electrolyte solutions and improving urine output. With improved kidney care, the lungs became the most vulnerable organ after severe injury. Improved treatment of respiratory failure eventually led to the current situation: advances in critical care allow patients to survive severe illnesses and injuries, but the body's defenses often overburden organs that were not originally injured.

Why does critical illness sometimes lead to MODS?

As discussed earlier in Chapter 16, injury and infection cause the release of chemical mediators that have systemic (whole-body) effects. A severe, persistent inflammatory response can lead to systemic inflammatory response syndrome (SIRS), which is associated with a constellation of symptoms including fever, increased heart and respiratory rates, and abnormal white blood cell counts. SIRS is a normal adaptive response to a severe insult, but if not reversed quickly enough it can progress to shock, which is characterized by extremely low blood pressure and an inadequate blood supply for the tissues and organs of the body.[1]

As might be expected from a systemic reduction in blood availability, shock can impair numerous organ systems. The abnormal delivery of oxygen and nutrients to tissues and insufficient removal of wastes result in irreversible injury to cells and tissues. Although each organ system is affected differently, ultimately one or more organs may begin to fail. The failure of one organ may place excessive demands on another, causing the second to fail as well. The progression of SIRS to MODS reflects the inability of the body's defenses and medical treatments to counter the detrimental effects of a sustained and potent inflammatory response.

The specific pathophysiology of MODS is poorly understood. Although early reports attempted to link the development of MODS directly to sepsis, sepsis is not present in all cases.[2] Infection often results from impaired immune function and therefore is a frequent consequence of MODS, but it is not necessarily the underlying trigger of organ dysfunction. Recall from this chapter that sepsis gives rise to symptoms identical to those seen in SIRS. Figure NP16-1 illustrates the relationships among SIRS, infection, sepsis, and MODS.

Do organs fail in a specific pattern?

Although the clinical course differs substantially among patients, the sequence of organ dysfunction often follows a similar pattern: first the lungs fail, then the liver, and finally the kidneys, GI tract, or heart. Other organs or systems may also become involved, and each additional failure reduces the likelihood of survival. Table NP16-1 lists the organs and systems most often involved in MODS and the potential consequences of their failure.

Are there any risk factors for MODS?

Epidemiological studies have identified a number of factors that increase risk. For example, people who develop MODS are often older, have multiple or severe injuries, and develop severe infections. Table NP16-2 lists the major risk factors associated with MODS, some of which are discussed below:

- *Age.* Patients who are older than 55 years of age are several times more likely to develop MODS than are younger patients. In elderly patients, the increased risk may be due to the presence of chronic illnesses that directly affect organ function, such as heart disease,

FIGURE NP16-1 Relationships among SIRS, Sepsis, and Multiple Organ Dysfunction Syndrome

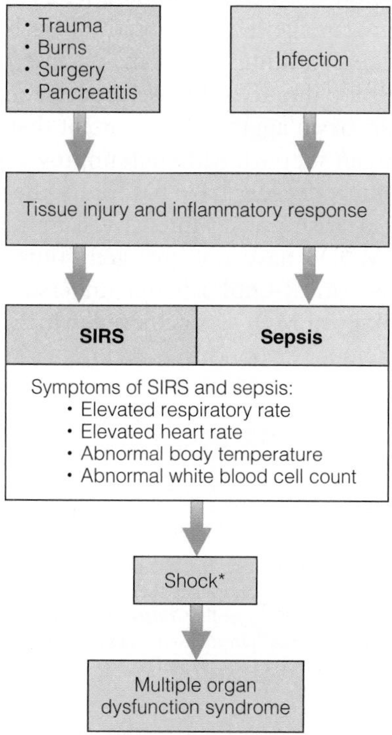

*After critical injury, shock may sometimes precede and be the cause of SIRS.

lung disease, diabetes, or liver damage. Aging also decreases the functional reserve of organs, thereby reducing an older patient's ability to deal with the additional stress that arises during critical illness.

- *Severity of SIRS.* The length of time that SIRS persists is related to the development of MODS. Patients who have SIRS that persists for more than three days are more likely to develop MODS than patients who have SIRS for less than two days.
- *Infection.* Prolonged SIRS can suppress immune function and increase the risk of developing an infection. During hospital stays, critically ill patients often contract pneumonia—the principal infection associated with MODS. The risks of infection and sepsis greatly increase with the use of invasive catheters, which are frequently needed during intensive care to provide oxygen support, intravenous fluid resuscitation, nutrition support, and urine clearance.
- *Blood transfusions.* Blood transfusions are immunosuppressive and may increase a patient's risks of developing infection or sepsis. Blood transfusions frequently have adverse effects that can add further stress; they may cause acute lung injury, allergic reactions, red blood cell hemolysis (breakdown), and other complications.

What is the treatment for MODS?

Once MODS has developed, extensive medical support is needed until the inflammatory response has abated. Unfortunately, aggressive treatments can have damaging effects of their own and may cause further injury to organs that are already weakened by illness. Health practitioners should be aware of the adverse effects of aggressive therapies and remain alert to a patient's responses to treatments. Therapies that are often used to manage MODS include:[3]

- *Lung support.* Mechanical ventilation is used to assist injured lungs and sustain gas exchange.
- *Fluid resuscitation.* Fluids and electrolytes are supplied to restore blood volume and maintain electrolyte balance.
- *Support of heart and blood vessel function.* Medications help to sustain or increase cardiac output and maintain adequate blood pressure.
- *Kidney support.* Hemofiltration or dialysis helps to prevent the buildup of toxic metabolites in blood.
- *Protection against infection.* Antibiotic therapy may reverse or prevent infections.
- *Nutrition support.* Enteral and parenteral nutrition support provide nutrients, help to prevent excessive wasting, and promote recovery.

TABLE NP16-1 Physiological Effects of Organ or System Failure

Organ or System	Effects of Failure
Lungs	Inability to maintain gas exchange
Liver	Altered metabolic processes
Kidneys	Inability to regulate blood volume, maintain electrolytes, remove wastes
Heart	Low cardiac output, low blood pressure, inadequate circulation, shock
GI tract	Impaired digestion and absorption, abnormal bleeding, bacterial translocation
Immune system	Infection, sepsis
Coagulation system	Excessive bleeding or blood clotting
Central nervous system	Decreased perceptions, brain injury, coma

TABLE NP16-2 Factors That Influence Risk of Multiple Organ Dysfunction Syndrome

Age over 55 years
Prior chronic disease
Persistent SIRS
Major infection
Blood transfusions
Severity of tissue injury
Length of time between injury and arrival at hospital
Malnutrition

What can be done to reduce the incidence of MODS?

Because mortality rates for MODS are so high, prevention must be considered at the earliest stages of injury and treatment, before an excessive inflammatory response can cause further damage. Health practitioners have learned to identify the conditions that can increase organ stress whether they are due to a disease process, an inflammatory response, or an aggressive treatment that is intended to provide organ support. Although improvements in care over the past few decades have reduced some of the complications that arise during intensive care, rates of mortality from MODS have not changed. Thus, a focus on prevention is critical until a better understanding of the pathophysiology of MODS is achieved, which may lead to additional therapeutic options.

Notes

1. J. A. Russell, Shock syndromes related to sepsis, in L. Goldman and A I. Schafer, eds., *Goldman's Cecil Medicine* (Philadelphia: Saunders, 2012), pp. 658–666.
2. D. C. Dewar and coauthors, Post-injury multiple organ failure, *Trauma* 13 (2011): 81–91.
3. Russell, 2012; M. H. Oltermann, Systemic inflammatory response and sepsis: A multidisciplinary approach to the nutritional considerations, in G. Cresci, ed., *Nutrition Support for the Critically Ill Patient: A Guide to Practice* (Boca Raton, FL: Taylor & Francis Group, 2005), pp. 565–577.

Chapter 17 Nutrition and Upper Gastrointestinal Disorders

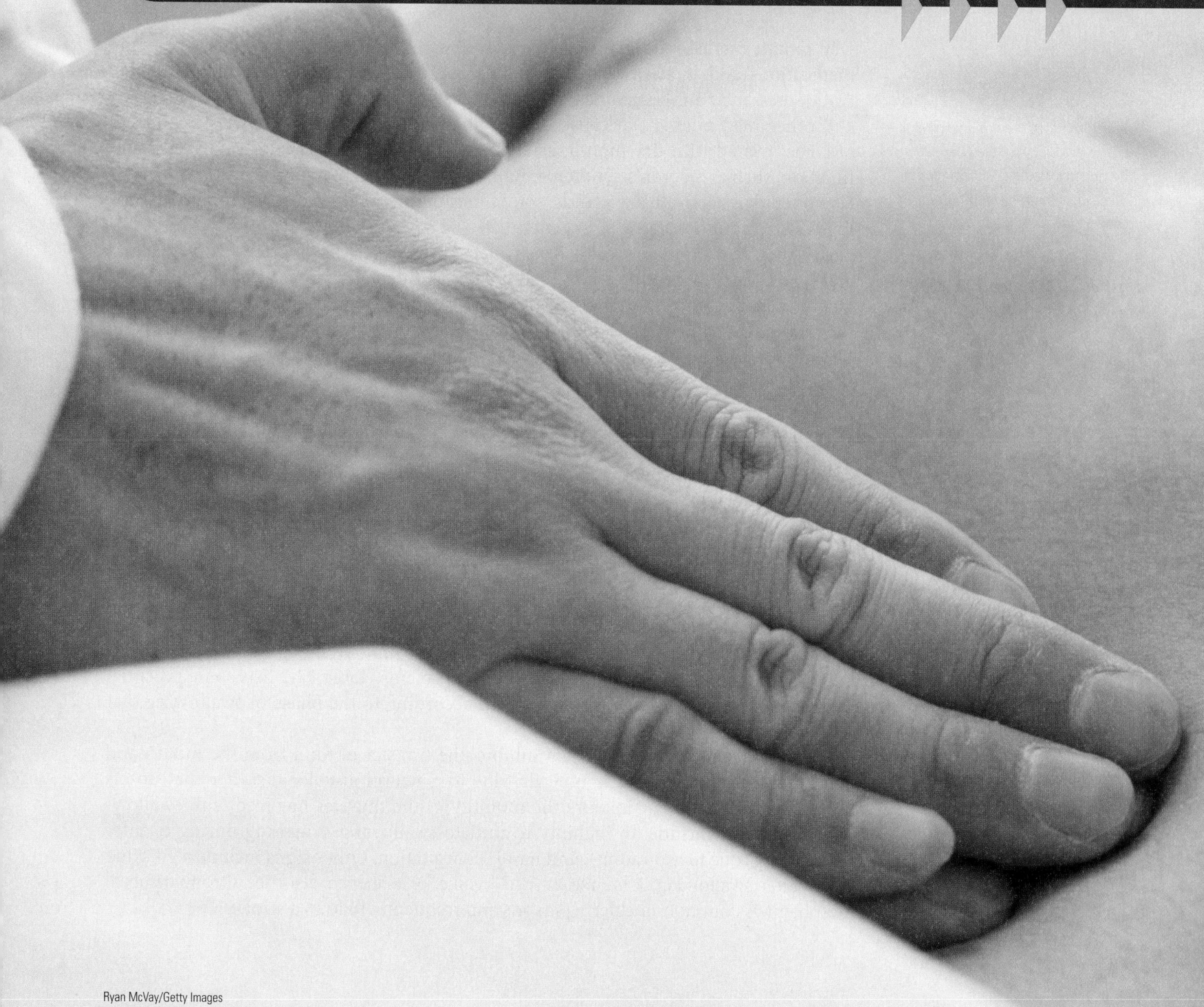

Ryan McVay/Getty Images

GASTROINTESTINAL (GI) ILLNESSES ACCOUNT FOR A SIGNIFICANT fraction of hospital admissions and visits to health practitioners each year. Diagnosis is not always straightforward, however, because many patients with GI complaints exhibit no physical abnormalities. Evaluation therefore requires a detailed review of a patient's symptoms and responses to dietary adjustments. Because GI complications frequently accompany other illnesses, the medical history can sometimes uncover the underlying source of distress. This chapter describes upper GI symptoms and disorders; the next chapter discusses conditions that involve the lower GI tract.

Conditions Affecting the Mouth and Esophagus

Chapter 14 described several types of mechanically altered diets that can be used by individuals who have difficulty chewing and swallowing (see pp. 409–410). This section describes how to relieve the discomfort of dry mouth and examines the causes and treatments of the two most common disorders involving the esophagus: dysphagia (difficulty swallowing) and gastroesophageal reflux disease.

DRY MOUTH

Dry mouth (**xerostomia**), caused by reduced salivary flow, is a side effect of many medications and is associated with a number of diseases and disease treatments. Antidepressants, antihistamines, antihypertensives, antineoplastics, bronchodilators, and other medications can cause dry mouth. Poorly controlled diabetes mellitus is often associated with dry mouth, as are conditions that directly affect salivary gland function, such as **Sjögren's syndrome**. Radiation therapy that treats head and neck cancers often damages salivary glands, sometimes permanently. Excessive mouth breathing is also a common cause of dry mouth.[1]

Dry mouth can impair health in a variety of ways.[2] It can interfere with speaking and swallowing. Mouth infections, bad breath, and dental diseases are more common. Dentures may be uncomfortable to wear, and ulcerations may develop where they contact the mouth. Taste sensation is often diminished, and salty or spicy foods may cause pain. Dry mouth may cause a person to reduce food intake and may thereby increase malnutrition risk. Table 17-1 (p. 489) offers suggestions that can help to manage dry mouth.

NURSING DIAGNOSIS

The nursing diagnosis *impaired oral mucous membrane* is appropriate for a person with dry mouth.

xerostomia (ZEE-roh-STOE-me-ah): dry mouth caused by reduced salivary flow.

xero = dry
stomia = mouth

DYSPHAGIA

The act of swallowing involves multiple processes. In the initial, or **oropharyngeal**, phase of swallowing, muscles in the mouth and tongue propel the bolus of food through the pharynx and into the esophagus. At the same time, tissues of the soft palate prevent food from entering the nasal cavity, and the epiglottis blocks the opening to the trachea to prevent aspiration of food substances or saliva into the lungs. In the second, or **esophageal**, phase of swallowing, peristalsis forces the bolus through the esophagus, and the lower esophageal sphincter relaxes to allow passage of the bolus into the stomach. Due to the many tasks involved in swallowing, dysphagia can result from a number of different physical or neurological problems. Table 17-2 lists some potential causes of dysphagia, which are categorized according to the phase of swallowing that is impaired.[3]

Oropharyngeal dysphagia, which inhibits the transfer of food from the mouth and pharynx to the esophagus, is typically due to a neuromuscular disorder that upsets the swallowing reflex or impairs the mobility of the muscles involved with swallowing. Symptoms include an inability to initiate swallowing, coughing during or after swallowing (due to aspiration), and nasal regurgitation. Other signs include a gurgling noise after swallowing, a hoarse or "wet" voice, or a speech disorder. Oropharyngeal dysphagia is common in elderly persons and frequently follows a stroke.

NURSING DIAGNOSIS

The nursing diagnosis *impaired swallowing* is appropriate for a person with dysphagia.

Sjögren's (SHOW-grenz) syndrome: an autoimmune disease characterized by the destruction of secretory glands, resulting in dry mouth and dry eyes.

oropharyngeal (OR-row-fah-ren-JEE-al): involving the mouth and pharynx.

esophageal (eh-SOF-ah-JEE-al): involving the esophagus.

oropharyngeal dysphagia: an inability to transfer food from the mouth and pharynx to the esophagus; usually caused by a neurological or muscular disorder.

TABLE 17-1 Suggestions for Managing Dry Mouth

Food and Beverage Tips
- Take frequent sips of water or another sugarless beverage.
- Suck on ice cubes or frozen fruit juice bars (unless their coldness causes discomfort).
- Consume foods that have a high fluid content, such as soups, stews, sauces and gravies, yogurt, and pureed fruit.
- Avoid dry foods like toast, chips, and crackers.
- Avoid citrus juices and spicy or salty foods if they cause mouth irritation.

Lifestyle Practices
- Chew sugarless gum to help stimulate salivary flow.
- Avoid caffeine, alcohol, and smoking, which may dry the mouth.
- Use a humidifier during the night.

Saliva Substitutes
- Use over-the-counter saliva substitutes (available as gels, sprays, and tablets), especially just before meals and at bedtime.
- Try rinsing the mouth with a teaspoonful of vegetable oil or softened margarine.

Dental Care
- Pay strict attention to oral hygiene, brushing teeth and flossing at least twice daily. Try to brush immediately after each meal.
- Avoid alcohol- and detergent-containing mouthwashes that may dry and irritate the mouth.
- Ask your dentist about fluoride treatments that help to prevent tooth decay.

Medications
- If dry mouth is caused by a medication, ask your physician about possible alternatives.
- Ask your physician if using a medication to stimulate saliva secretion may be of benefit; examples include cevimeline (Evoxac) and pilocarpine (Salagen).

Esophageal dysphagia interferes with the passage of materials through the esophageal lumen and into the stomach, and is usually caused by an obstruction in the esophagus or a motility disorder. The main symptom is the sensation of food "sticking" in the esophagus after it is swallowed. An obstruction can be caused by a **stricture** (abnormal narrowing), tumor, or compression of the esophagus by surrounding tissues. Whereas an obstruction can prevent the passage of solid foods but may not affect liquids, a motility disorder hinders the passage of both solids and liquids. **Achalasia**, the most common motility disorder, is a degenerative nerve condition affecting the esophagus; it is characterized by impaired peristalsis and incomplete relaxation of the lower esophageal sphincter when swallowing.

NURSING DIAGNOSIS

The nursing diagnoses *imbalanced nutrition: less than body requirements* and *risk for aspiration* often apply to a patient with dysphagia.

Complications of Dysphagia

Health practitioners should be alert to the various complications that may accompany dysphagia. If the condition restricts food consumption, malnutrition and weight loss may occur. Individuals who cannot swallow liquids are at increased risk of dehydration. If aspiration occurs, it may cause choking, airway obstruction, or respiratory infections, including pneumonia. If a person does not have a normal cough reflex, aspiration is more difficult to diagnose and may go unnoticed.

Nutrition Intervention for Dysphagia

To compensate for swallowing difficulties, a person with dysphagia may need to consume foods and beverages that have been physically modified so that they are easier to swallow. Because a wide variety of defects can cause dysphagia, finding the best diet is often a challenge. Furthermore, a person's swallowing ability can fluctuate over time, so the dietary plan needs frequent reassessment.

The National Dysphagia Diet, developed in 2002 by a panel of dietitians, speech and language therapists, and a food scientist, has helped to standardize the nutrition care of dysphagia patients.[4] Table 17-3

TABLE 17-2 Selected Causes of Dysphagia

Oropharyngeal Dysphagia
- Alzheimer's disease (advanced stages)
- Amyotrophic lateral sclerosis (Lou Gehrig's disease)
- Cerebral palsy
- Multiple sclerosis
- Muscular dystrophy
- Myasthenia gravis
- Parkinson's disease
- Poliomyelitis
- Stroke

Esophageal Dysphagia
- Achalasia
- Esophageal cancer
- Esophageal spasm
- External compression (from a tumor, enlarged thyroid gland, or enlarged left atrium)
- Scleroderma
- Strictures (from inflammation, scarring, or a congenital abnormality)

Diamond Crystal Specialty Foods

Can you tell that the foods in this photo are pureed foods shaped with food molds?

presents brief descriptions of the different levels of the diet and some sample meals. After the appropriate level is selected, the diet must be adjusted to suit the person's swallowing abilities and tolerances. In many cases, the most appropriate foods may be determined only by trial and error. A consultation with a swallowing expert, such as a speech and language therapist, is often necessary.

Food Properties and Preparation Foods included in dysphagia diets should have easy-to-manage textures and consistencies. Soft, cohesive foods are easier to swallow than hard or crumbly foods. Moist foods are better tolerated than dry foods. Some foods within a category may be acceptable and others may not; for example, some cookies are soft and tender, whereas others are hard and brittle. Sticky or gummy foods, such as peanut butter and cream cheese, may be difficult to clear from the mouth and throat.

The textures of foods are typically altered to make them easier to swallow. Solid foods are often pureed, mashed, ground, or minced (review Table 14-5 on p. 410). Foods that have more than one texture, such as vegetable soup or cereal with milk, are difficult to manage, so ingredients may be blended to a single consistency with items such as nuts and seeds omitted. Semi-liquid foods such as sauces and gravies may be thickened with food starches (such as cornstarch or potato flakes) during cooking or mixed with commercial food thickeners (■) after cooking until the desired consistency is reached. A variety of pre-thickened food products, including pureed meats, eggs, vegetables, and pasta, are commercially available.

Consuming foods that have a similar consistency can quickly become monotonous. The "How to" offers suggestions for improving the acceptance of pureed and other mechanically altered foods.

■ Most commercial thickening agents are gels or powders made from modified food starches or food gums.

esophageal dysphagia: an inability to move food through the esophagus; usually caused by an obstruction or a motility disorder.

stricture: abnormal narrowing of a passageway; often due to inflammation, scarring, or a congenital abnormality.

achalasia (ack-ah-LAY-zhah): an esophageal disorder characterized by weakened peristalsis and impaired relaxation of the lower esophageal sphincter.

a = without
chalasia = relaxation

Properties of Liquids Thickened liquids are easier to swallow than thin liquids such as water or juice. Table 17-3 describes the four levels of liquid consistencies prescribed for dysphagia patients, referred to as thin, nectarlike, honeylike, and spoon-thick. To increase viscosity, commercial thickeners can be stirred into beverages and other liquid foods, such as soup broths. Some beverages may lose their appeal when thickened; for example, individuals may find thickened coffee and tea unacceptable. Moreover, hydration is more difficult to maintain when a patient has access to only thickened beverages, which are often less acceptable for quenching thirst.

HOW TO Improve Acceptance of Mechanically Altered Foods

Take a moment to think about a meal of pureed or ground foods. A typical dinner of baked chicken, potatoes, carrots, and green beans can look like mounds of differently colored mush. The foods may taste great, but a person may have little appetite before trying a first bite. To improve appetite, be creative when preparing and serving meals:

- Help to stimulate the appetite by preparing favorite foods and foods with pleasant smells. Enliven food flavors with aromatic spices and seasonings.
- Substitute brightly colored vegetables for white vegetables; for example, replace mashed potatoes with mashed sweet potatoes. If serving more than one vegetable, place contrasting colors (such as spinach and carrots) side by side or swirl the two together.
- Shape pureed and ground foods so they resemble traditional dishes; for example, meats can be flattened to form a patty or rounded to resemble meatballs. Use food molds to restore slurried breads and pureed meats to their traditional shapes.
- Try layering ingredients so that the food looks like a fancy casserole or popular hors d'oeuvre. For example, food items can resemble lasagna, moussaka, tamales, or sushi.
- Use attractive plates and silverware to improve the visual appeal of a meal. Colorful garnishes can add interest and eye appeal.

Efforts to improve the appearance of foods can go a long way toward helping people eat nourishing meals and maintain a healthy weight.

TABLE 17-3 National Dysphagia Diet

Level 1: Dysphagia Pureed

Foods should be pureed or well mashed, homogeneous, and cohesive. This diet is for patients with moderate to severe dysphagia and poor oral or chewing ability.

Sample menus:

- *Breakfast:* Cream of Wheat, slurried muffins or pancakes,[a] pureed scrambled eggs, plain or vanilla yogurt, well-mashed bananas, fruit juice without pulp (thickened as needed), coffee or tea (if thin liquids are acceptable).
- *Lunch or dinner:* Pureed tomato soup, slurried crackers, pureed meat or poultry, zucchini soufflé, mashed potatoes with gravy, pureed carrots or green beans, smooth applesauce, pureed peaches, chocolate pudding.

Foods to avoid: Dry breads and cereals, oatmeal, rice, fruit yogurt, cheese (including cottage cheese), peanut butter, nuts and seeds, raw fruits and vegetables, chunky applesauce, fruit preserves with chunks or seeds, tomato sauce with seeds, beverages with pulp, coarsely ground pepper, herbs.

Level 2: Dysphagia Mechanically Altered

Foods should be moist, cohesive, and soft textured and should easily form a bolus. This diet is for patients with mild to moderate dysphagia; some chewing ability is required.

Sample menus:

- *Breakfast:* Moist oatmeal, cornflakes or puffed rice cereal with milk (thickened as needed), moist pancakes or muffins (with butter, margarine, or jam; without nuts or seeds), soft scrambled eggs, cottage cheese, ripe bananas or cooked fruit without skin or seeds, fruit juice (thickened as needed), coffee or tea (if thin liquids are allowed).
- *Lunch or dinner:* Soup with easy-to-chew meat and vegetables; slurried bread or crackers; minced, tender-cooked meat; well-cooked pasta with moist meatballs and meat sauce; baked potato with gravy; soft, tender-cooked vegetables (not fibrous or rubbery); canned peach slices; soft fruit pie (with bottom crust only); soft, smooth chocolate bar.

Foods to avoid: Dry or coarse foods; breads and cereals with nuts, seeds, or dried fruit; frankfurters and sausages; hard-cooked eggs; corn and clam chowders; sandwiches; pizza; sliced cheese; rice; potato skins; French fries; raw vegetables; fibrous, rubbery, or non-tender cooked vegetables such as asparagus, broccoli, brussels sprouts, cabbage, celery, corn, and peas; peanut butter; coconut; nuts and seeds; raw fruit (except banana); cooked fruit with skin or seeds; pineapple; mango; uncooked dried fruit; popcorn; chewy candies (such as caramel or licorice).

Level 3: Dysphagia Advanced

Foods should be moist and in bite-sized pieces when swallowed; foods with mixed textures are included. This diet is for patients with mild dysphagia and adequate chewing ability.

Sample menus:

- *Breakfast:* Cereal with milk, moist pancakes or muffins (with butter, margarine, or jam; without nuts or seeds), poached or scrambled eggs, fruit yogurt, soft fresh fruit (peeled) or berries, coffee or tea (if thin liquids are tolerated).
- *Lunch or dinner:* Chicken noodle soup; moistened crackers or moist bread; thin-sliced tender meat; cheese; moist, soft-cooked potatoes or rice; tender-cooked vegetables; shredded lettuce with dressing; fresh, peeled peach or melon; canned fruit salad; moist chocolate chip cookie (without nuts).

Foods to avoid: Dry or coarse foods; breads and cereals with nuts, seeds, or dried fruit; corn and clam chowders; potato skins; raw vegetables (except shredded lettuce); corn; chunky peanut butter; coconut; nuts and seeds; hard fruit (such as apples or pears); fruit with skin, seeds, or stringy textures (such as mango or pineapple); uncooked dried fruit; fruit leathers; popcorn; chewy candies (such as caramel or licorice).

Liquid Consistencies (only those tolerated are allowed in the diet)

- *Thin:* Watery fluids; may include milk, coffee, tea, juices, carbonated beverages.
- *Nectarlike:* Fluids thicker than water that can be sipped through a straw; may include buttermilk, eggnog, tomato juice, cream soups.
- *Honeylike:* Fluids that can be eaten with a spoon but do not hold their shape; may include honey, some yogurt products, tomato sauce.
- *Spoon-thick:* Thick fluids that must be eaten with a spoon and can hold their shape; may include milk pudding, thickened applesauce.

[a]Slurried foods are starchy foods that have been mixed with liquid to achieve an appropriate consistency; they may be gelled and shaped to improve appearance.

Alternative Feeding Strategies for Dysphagia Some patients may be able to learn alternative feeding techniques to help them compensate for their swallowing problem. For example, changing the position of the head and neck while eating and drinking can minimize some swallowing difficulties. (As an example, cups designed for dysphagia patients allow drinking without tilting the head back.) Individuals with oropharyngeal dysphagia can be taught exercises that strengthen the jaws, tongue, or larynx, or they can learn new methods of swallowing that allow them to consume a normal diet. Speech and language therapists are often responsible for teaching patients these techniques.

GASTROESOPHAGEAL REFLUX DISEASE

Gastroesophageal reflux disease (GERD) is characterized by frequent reflux (backward flow) of the stomach's acidic contents into the esophagus, leading to pain, inflammation, and, possibly, tissue damage. People who suffer from GERD often refer to these symptoms as *heartburn* or *acid indigestion.* The gastroesophageal reflux itself does not necessarily cause symptoms or injury—it occurs occasionally in healthy people and is a problem only if it creates complications and requires lifestyle changes or medical treatment.

gastroesophageal reflux disease (GERD): a condition characterized by the backward flow (reflux) of the stomach's acidic contents into the esophagus.

hiatal hernia: a condition in which the upper portion of the stomach protrudes above the diaphragm; most cases are asymptomatic.

Causes of GERD The lower esophageal sphincter is the main barrier to gastric reflux, so GERD can result if the sphincter muscle is weak or relaxes inappropriately. Other factors that predispose a person to GERD include high stomach pressures and inadequate acid clearance from the esophagus.[5] Conditions associated with high rates of GERD include obesity, pregnancy, and **hiatal hernia**, a condition in which a portion of the stomach protrudes above the diaphragm (see Figure 17-1). During pregnancy,

FIGURE 17-1 The Upper GI Tract, Acid Reflux, and Hiatal Hernia

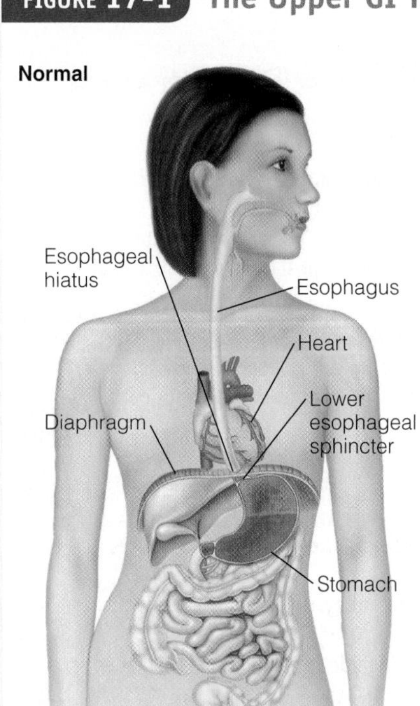

The stomach normally lies below the diaphragm, and the esophagus passes through the esophageal hiatus. The lower esophageal sphincter prevents reflux of stomach contents.

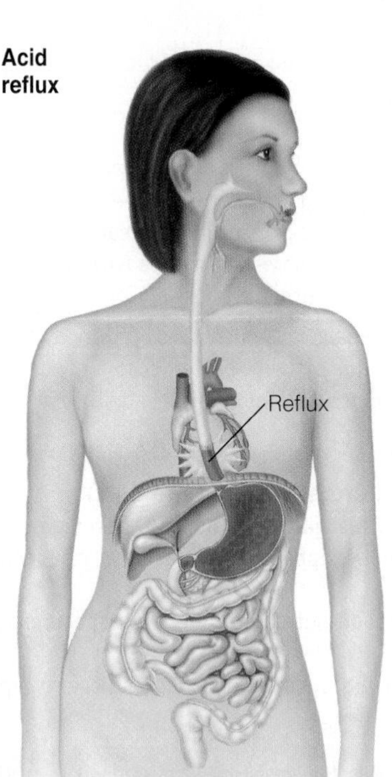

Whenever the pressure in the stomach exceeds the pressure in the esophagus, as can occur with overeating and overdrinking, the chance of reflux increases. The resulting "heartburn" is so-named because it is felt in the area of the heart.

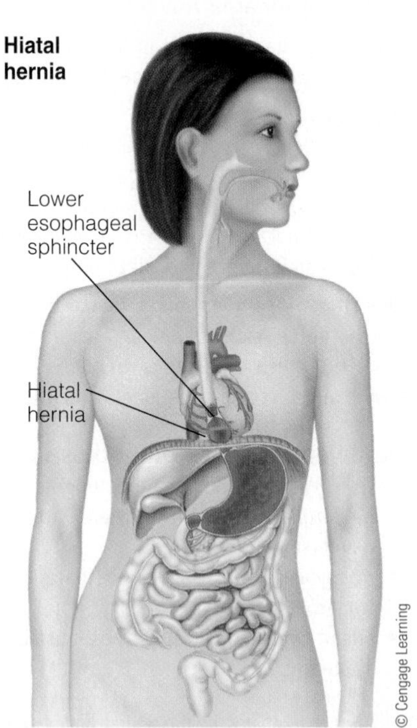

Risk of acid reflux may increase as a consequence of a hiatal hernia. A "sliding" hiatal hernia occurs when part of the stomach, along with the lower esophageal sphincter, rises above the diaphragm.

nearly two-thirds of women report heartburn, which typically worsens during the third trimester.[6] Many medications can increase the risk of reflux, as does the use of nasogastric tubes in tube feedings. Various other conditions or substances can exacerbate GERD by increasing stomach pressures or weakening the lower esophageal sphincter; Table 17-4 lists examples.

Consequences of GERD If gastric acid remains in the esophagus long enough to damage the esophageal lining, the resulting inflammation is called **reflux esophagitis**. Severe and chronic inflammation may lead to esophageal ulcers, with consequent bleeding. Healing and scarring of ulcerated tissue may narrow the inner diameter of the esophagus, causing esophageal stricture. A slowly progressive dysphagia for solid foods sometimes results, and swallowing occasionally becomes painful. Pulmonary disease may develop if gastric contents are aspirated into the lungs. Chronic reflux is also associated with **Barrett's esophagus**, a condition in which damaged esophageal cells are gradually replaced by cells that resemble those in gastric or intestinal tissue; such cellular changes increase the risk of developing esophageal cancer. GERD can also damage tissues in the mouth, pharynx, and larynx, resulting in eroded tooth enamel, sore throat, cough, and laryngitis.[7]

> ***NURSING DIAGNOSIS***
>
> The nursing diagnoses *acute pain* and *risk for aspiration* may apply to a person with GERD.

Treatment of GERD Treatment objectives are to alleviate symptoms and facilitate the healing of damaged tissue. Severe ulcerative disease may require immediate acid-suppressing medication, whereas a mild case may be managed with dietary and lifestyle changes. The "How to" (p. 494) lists lifestyle modifications that may help to prevent the recurrence of gastroesophageal reflux.

Medications that suppress gastric acid secretion help the healing process by reducing the damaging effects of acid on esophageal tissue. **Proton-pump inhibitors** are the most effective of the antisecretory agents and are used both for rapid healing of esophagitis and as a maintenance treatment. Other antisecretory drugs include **histamine-2 receptor blockers** (often referred to as *H2 blockers*) and antacids, which neutralize gastric acid. Antacids are frequently used to relieve occasional heartburn, but they are not necessarily appropriate for GERD because they have only short-term effects and may cause some nutrient deficiencies when used over the long term.

Surgery may be required in severe cases of GERD that are unresponsive to medications and lifestyle changes. In one popular procedure (called *fundoplication*), the upper

TABLE 17-4 Conditions and Substances Associated with Esophageal Reflux

Conditions/Substances That Increase Pressure within the Stomach	Conditions/Substances That Weaken the Lower Esophageal Sphincter
Ascites (abdominal fluid accumulation)	Alcohol
Carbonated beverages	Anticholinergic drugs
Delayed gastric emptying	Antihistamines
Eating large meals	Caffeinated beverages
Lying down after eating	Calcium channel blockers
Obesity	Chocolate
Pregnancy	Cigarette smoking
Wearing tight clothing around the waist or abdomen	Diazepam
	Estrogen, progesterone
	Fatty foods
	Peppermint and spearmint
	Theophylline
	Tricyclic antidepressants

reflux esophagitis: inflammation in the esophagus related to the reflux of acidic stomach contents.

Barrett's esophagus: a condition in which esophageal cells damaged by chronic exposure to stomach acid are replaced by cells that resemble those in the stomach or small intestine, sometimes becoming cancerous.

proton-pump inhibitors: a class of drugs that inhibit the enzyme that pumps hydrogen ions (protons) into the stomach. Examples include omeprazole (Prilosec) and lansoprazole (Prevacid).

histamine-2 receptor blockers: a class of drugs that suppress acid secretion by inhibiting receptors on acid-producing cells; commonly called *H2 blockers*. Examples include cimetidine (Tagamet), ranitidine (Zantac), and famotidine (Pepcid).

HOW TO Manage Gastroesophageal Reflux Disease

Management of GERD may require modifications in diet and lifestyle to minimize discomfort and reduce the recurrence of acid reflux. Recommendations typically include the following:

- Consume only small meals and drink liquids between meals so that the stomach does not become overly distended, which can exert pressure on the lower esophageal sphincter.
- Limit foods that weaken lower esophageal sphincter pressure or increase gastric acid secretion; these include chocolate, fried and fatty foods, spearmint and peppermint, coffee (both caffeinated and decaffeinated), and tea.
- During periods of esophagitis, avoid foods and beverages that may irritate the esophagus, such as citrus fruits and juices, tomato products, garlic, onions, pepper, spicy foods, carbonated beverages, and very hot or very cold foods (depending on individual tolerances).
- Avoid eating bedtime snacks or lying down after meals. Meals should be consumed at least three hours before bedtime.
- Reduce nighttime reflux by elevating the head of the bed on 6-inch blocks, inserting a foam wedge under the mattress, or propping pillows under the head and upper torso.
- Avoid bending over and wearing tight-fitting garments; both can cause pressure in the stomach to increase, heightening the risk of reflux.
- Avoid cigarettes and alcohol; both relax the lower esophageal sphincter.
- Avoid using nonsteroidal anti-inflammatory drugs (NSAIDs) such as aspirin, naproxen, and ibuprofen, which can damage the esophageal mucosa.

Food tolerances among people with GERD can vary markedly. Health practitioners can help patients pinpoint food intolerances by advising them to keep a record of the foods and beverages they consume, as well as any resulting symptoms.

section of the stomach (the fundus) is gathered up around the lower esophagus and sewn in such a way that the lower esophagus and sphincter are surrounded by stomach muscle; this technique increases pressure within the esophagus and fortifies the sphincter muscle. Esophageal strictures are often treated by dilating the esophagus with an inflatable balloon-like device or a fixed-size dilator, or by using surgical approaches. The Case Study will help you to review the treatments available for a patient with GERD.

Review Notes

- Dry mouth can increase the risk of developing oral infections, diminish taste sensation, and lead to reduced food intake. It can be managed with oral hygiene, dietary changes, and saliva substitutes.
- Dysphagia may interfere with food intake and increase the risk of aspiration. Treatment may include dietary adjustments, strengthening exercises, and using different swallowing techniques.
- Gastroesophageal reflux disease (GERD) may lead to inflammation, esophageal ulcers, bleeding, and stricture. Treatment includes the use of acid-suppressing drugs and lifestyle changes.

Case Study: Woman with GERD

Elyssa Rinaldi is a 39-year-old accountant who is 5 feet 4 inches tall and weighs 165 pounds. During a recent physical examination, she mentioned to her physician that she had been feeling fairly well until she began experiencing heartburn, which has progressively become more frequent and painful. The heartburn often occurs after she eats a large meal and is particularly bad after she goes to bed at night. By directly examining the esophageal lumen using an endoscope (a thin, flexible tube equipped with an optical device), the physician found evidence of reflux esophagitis and a slight narrowing throughout the length of the esophagus.

Ms. Rinaldi's medical history does not indicate any significant health problems. During her last physical exam, her physician advised her to stop smoking cigarettes and to lose 20 pounds, but she has not attempted to do either. The nutrition assessment reveals that Ms. Rinaldi is feeling stressed because it is the middle of the tax season. She usually has little time for breakfast, eats a lunch of fast foods while continuing to work at her desk, and eats a large dinner at around 8 p.m. She generally has wine with dinner and another alcoholic beverage later in the evening.

1. Explain to Ms. Rinaldi the meaning of the medical diagnoses *reflux esophagitis* and *esophageal stricture.*
2. From the brief history provided, list the factors and behaviors that increase Ms. Rinaldi's risks of experiencing reflux. What recommendations can you make to help her change these behaviors?
3. What medications might the physician prescribe, and why?

Conditions Affecting the Stomach

Stomach disorders range from occasional bouts of discomfort to severe conditions that require surgery. This section begins with a discussion of *dyspepsia* (often called "indigestion"), the sensation of pain or discomfort in the upper abdomen that occurs after food consumption. More serious stomach conditions that may benefit from dietary adjustments include *gastritis* and *peptic ulcers,* which most often result from bacterial infection or the use of medications that damage the stomach lining.

DYSPEPSIA

Dyspepsia refers to general symptoms of pain or discomfort in the upper abdominal region, which may include stomach pain, gnawing sensations, early satiety, nausea, vomiting, and bloating. These symptoms sometimes indicate the presence of more serious illnesses, such as GERD or peptic ulcer disease. Although about 25 percent of the population experiences dyspepsia, only about half of those affected seek medical attention.[8]

> ***NURSING DIAGNOSIS***
>
> The nursing diagnoses *acute pain* and *nausea* may be appropriate for a person with dyspepsia.

Causes of Dyspepsia

Abdominal symptoms don't always lead to a clear diagnosis. Various medical problems can cause abdominal discomfort, including foodborne illness, GERD, peptic ulcers, gastric motility disorders, gallbladder and pancreatic diseases, and tumors in the upper GI tract. Chronic diseases such as diabetes mellitus, heart disease, and hypothyroidism can sometimes be accompanied by gastric symptoms. Some medications, including aspirin (and other nonsteroidal anti-inflammatory drugs), antibiotics, digitalis, and theophylline, can cause gastrointestinal distress. Some dietary supplements, such as iron and potassium supplements and some herbal products, may cause gastrointestinal problems. Intestinal conditions such as irritable bowel syndrome or lactose intolerance may mimic dyspepsia. Although pinpointing the cause of the symptoms can be difficult, a complete examination is in order if the individual experiences unintentional weight loss, dysphagia, persistent vomiting, GI bleeding, or anemia, which suggest the presence of serious illness.[9]

Potential Food Intolerances

Although many people attribute their symptoms to eating certain foods or spices, controlled studies have been unable to find associations between specific foods and dyspepsia.[10] Substances often reported to cause symptoms include coffee and chili peppers. High-fat meals can slow gastric emptying and thereby exacerbate dyspepsia. Spicy foods may cause some injury to the mucosal lining and exacerbate the pain from a preexisting ulcer. To minimize symptoms, people with dyspepsia are typically advised to consume small, frequent meals; avoid fatty or highly spiced foods; and avoid the specific foods believed to trigger symptoms.[11]

Bloating and Stomach Gas

The feeling of bloating may be caused by excessive gas in the stomach, which accumulates when air is swallowed. Air swallowing often accompanies gum chewing, smoking, rapid eating, drinking carbonated beverages, and using a straw. Omitting these practices generally helps to correct the problem.

NAUSEA AND VOMITING

Nausea and vomiting accompany many illnesses and are common side effects of medications. Although occasional vomiting is not dangerous, prolonged vomiting can cause fluid and electrolyte imbalances and may require medical care. Chronic vomiting can reduce food intake and lead to malnutrition and nutrient deficiencies.

The symptoms that accompany vomiting may give clues about its cause.[12] If abdominal pain is present, a GI disorder or obstruction is usually the cause. If abdominal pain is not present, possible causes of vomiting include medications, foodborne illness, pregnancy, motion sickness, neurological disease, inner ear disorders, hepatitis, and various chronic illnesses.

dyspepsia: symptoms of pain or discomfort in the upper abdominal area, often called *indigestion;* a symptom of illness rather than a disease itself.

dys = bad; impaired
pepsis = digestion

The suffix *-itis* refers to the presence of inflammation in an organ or tissue. Some forms of gastritis are also characterized by tissue destruction (*atrophic gastritis*).

The specific reasons why ulcers develop are not known; only 10 to 15 percent of individuals with chronic *Helicobacter pylori* infection actually develop a peptic ulcer.

intractable vomiting: vomiting that is not easily managed or controlled.

gastritis: inflammation of stomach tissue.

***Helicobacter pylori (H. pylori)*:** a species of bacterium that colonizes gastric mucosa; a primary cause of gastritis and peptic ulcer disease.

hypochlorhydria (HIGH-poe-clor-HIGH-dree-ah): abnormally low gastric acid secretions.

achlorhydria (AY-clor-HIGH-dree-ah): absence of gastric acid secretions.

peptic ulcer: an open sore in the gastrointestinal mucosa; may develop in the esophagus, stomach, or duodenum.

peptic = related to digestion

Treatment of Nausea and Vomiting

Most cases are short lived and require no treatment. When treatment is necessary, the main goal is to find and correct the underlying disorder. Restoring hydration may also be necessary in some individuals. If a medication is the cause, taking it with food may help. If the cause is unknown or the underlying disorder cannot be corrected, medications that suppress nausea and vomiting can be prescribed. People with **intractable vomiting**—severe vomiting that is not easily controlled—may require intravenous nutrition support.

Dietary Interventions

Sometimes nausea can be prevented or improved with dietary measures. To minimize stomach distention, patients should consume small meals and drink beverages between meals rather than during a meal. Dry, starchy foods such as toast, crackers, and pretzels may help to reduce nausea, whereas fatty or spicy foods and foods with strong odors may worsen symptoms. Foods that are cold or at room temperature may be better tolerated than hot foods. Individuals often have strong food aversions when nauseated, and tolerances vary greatly.

GASTRITIS

Gastritis is a general term that refers to inflammation of the stomach mucosa. (■) Acute cases of gastritis typically result from irritating substances or treatments that damage the gastric mucosa, resulting in tissue erosions, ulcers, or hemorrhaging (severe bleeding). Chronic cases may be caused by long-term infections or autoimmune disease and can progress to widespread gastric inflammation and tissue atrophy. Most often, gastritis results from ***Helicobacter pylori*** infection or the use of NSAIDs, which are primary causes of peptic ulcer disease as well.[13] Table 17-5 lists some potential causes of gastritis.

Complications of Gastritis

The extensive tissue damage that sometimes develops in chronic gastritis can disrupt gastric secretory functions. If hydrochloric acid secretions become abnormally low (**hypochlorhydria**) or absent (**achlorhydria**), absorption of nonheme iron and vitamin B_{12} can be impaired, increasing the risk of deficiency. Pernicious anemia, a condition characterized by the autoimmune destruction of stomach cells that produce intrinsic factor, is a late complication of atrophic gastritis that can result in the macrocytic anemia of vitamin B_{12} deficiency (see p. 217).

TABLE 17-5 Potential Causes of Gastritis

Potential Causes of Gastritis
Infection
• Bacterial: *Helicobacter pylori*
• Fungal: *Candida albicans*
• Parasitic: *Anisakis* (nematode infection)
• Viral: *Cytomegalovirus*
Chemical Substances
• Alcohol
• Cancer chemotherapy
• Drugs (especially aspirin and other NSAIDs)
• Ingestion of toxins or corrosive materials
Internal (Bodily) Causes
• Autoimmune disease
• Bile reflux
• Severe stress or sepsis
Miscellaneous
• Foreign bodies
• High salt intake
• Radiation therapy

Dietary Interventions for Gastritis

Dietary recommendations depend on an individual's symptoms. In asymptomatic cases, no dietary adjustments are needed. If pain or discomfort is present, the patient should avoid irritating foods and beverages; these often include alcohol, coffee (including decaffeinated), cola beverages, spicy foods, and fried or fatty foods. If food consumption increases pain or causes nausea and vomiting, food intake should be avoided for 24 to 48 hours to rest the stomach. If hypochlorhydria or achlorhydria is present, supplementation of iron and vitamin B_{12} may be warranted.

PEPTIC ULCER DISEASE

A **peptic ulcer** is an open sore that develops in the GI mucosa when gastric acid and pepsin overwhelm mucosal defenses and destroy mucosal tissue. A primary factor in peptic ulcer development is *H. pylori* infection, which is present in approximately 30 to 60 percent of patients with gastric ulcers and 70 to 90 percent of those with duodenal ulcers.[14] (■) Another major factor is the use of NSAIDs, which have both topical and systemic effects that can damage the GI lining. In rare cases, ulcers may develop from disorders that cause excessive acid secretion. Ulcer risk can be increased by cigarette smoking and psychological stress.[15]

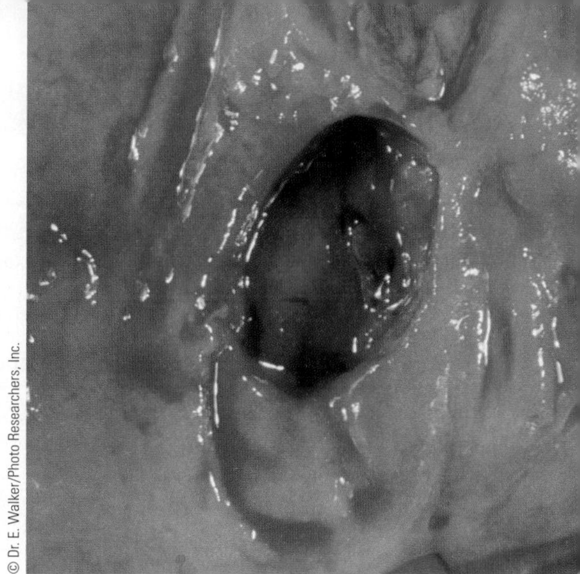

A peptic ulcer, such as the gastric ulcer shown here, damages mucosal tissue and may cause pain and bleeding.

Effects of Psychological Stress Although most ulcers are associated with *H. pylori* infection or NSAID use, about 5 to 20 percent of ulcers develop for other reasons.[16] Psychological stress (■) is not believed to cause ulcers per se, but it has effects on physiological processes and behaviors that may increase a person's vulnerability. The physiological effects of stress vary among individuals but may include hormonal changes that impair immune responses and wound healing, increased secretions of hydrochloric acid and pepsin, and rapid stomach emptying (which increases the acid load in the duodenum). Stress may also lead to behavioral changes, including the increased use of cigarettes, alcohol, and NSAIDs—all potential risk factors for ulcers. Thus, stress may play a contributory role in ulcer development, although its precise effects are not fully understood.

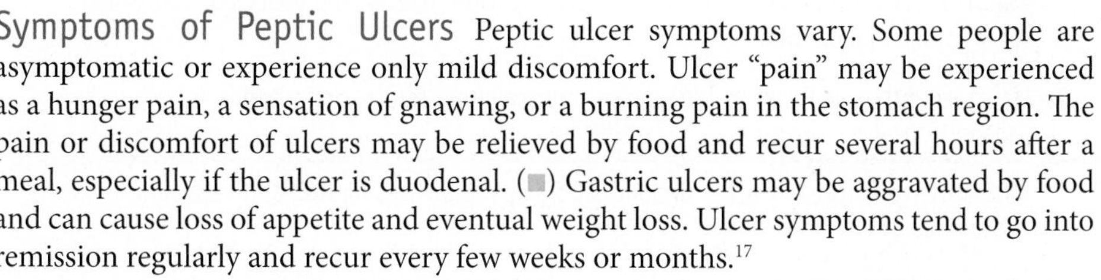

Symptoms of Peptic Ulcers Peptic ulcer symptoms vary. Some people are asymptomatic or experience only mild discomfort. Ulcer "pain" may be experienced as a hunger pain, a sensation of gnawing, or a burning pain in the stomach region. The pain or discomfort of ulcers may be relieved by food and recur several hours after a meal, especially if the ulcer is duodenal. (■) Gastric ulcers may be aggravated by food and can cause loss of appetite and eventual weight loss. Ulcer symptoms tend to go into remission regularly and recur every few weeks or months.[17]

NURSING DIAGNOSES

Nursing diagnoses for people with peptic ulcers may include *acute pain* and *nausea*. For patients with bleeding ulcers, appropriate diagnoses may include *risk for deficient fluid volume, imbalanced nutrition: less than body requirements,* and *fatigue.*

Complications of Peptic Ulcers Peptic ulcers are a major cause of gastrointestinal bleeding, which occurs in up to 15 percent of ulcer cases.[18] Bleeding is a potential cause of death and, if severe, may indicate the need for surgical intervention. Severe bleeding is evidenced by black, tarry stool samples or, occasionally, vomit that resembles coffee grounds. Other serious complications of ulcers include perforations of the stomach or duodenum (sometimes leading directly into the peritoneal cavity), penetration of the ulcer into an adjacent organ, and **gastric outlet obstruction** due to scarring or inflammation.

Drug Therapy for Peptic Ulcers The goals of ulcer treatment are to relieve pain, promote healing, and prevent recurrence. In most cases, treatment requires using a combination of antibiotics to eradicate *H. pylori* infection and/or discontinuing the use of aspirin and other NSAIDs, which can irritate the gastric mucosa and delay healing. The antibiotics used to treat *H. pylori* infection most often include amoxicillin, clarithromycin, metronidazole, and tetracycline. Antisecretory drugs are prescribed to relieve pain and allow healing; these include proton-pump inhibitors, H2 blockers, and antacids (as used in GERD; see the earlier discussion on p. 493). The most frequently prescribed drug regimen is a "triple therapy" that includes two antibiotics and an antisecretory drug. Bismuth preparations (such as Pepto-Bismol) and sucralfate may also help to heal ulcers by coating the gastrointestinal lining and preventing further tissue erosion. See the Diet-Drug Interactions feature on p. 498 for nutrition-related effects of the medications used in ulcer treatment.

Nutrition Care for Peptic Ulcers The goals of nutrition care are to correct nutrient deficiencies, if necessary, and encourage dietary and lifestyle practices that minimize symptoms.[19] Patients should avoid dietary items that increase acid secretion or irritate the gastrointestinal lining; examples include alcohol, coffee and other caffeine-containing beverages, chocolate, and pepper, although individual tolerances vary. Small meals may be better tolerated than large ones. Patients should avoid food consumption for at least two hours before bedtime. Cigarette smoking is discouraged, as it can delay healing and increase the risk of ulcer recurrence. There is no evidence that dietary adjustments can alter the rate of healing or prevent recurrence.[20]

■ *Stress ulcers* occur in about 1 to 2 percent of patients who undergo severe metabolic stress.

■ In the United States, duodenal ulcers are more common than gastric ulcers.

gastric outlet obstruction: an obstruction that prevents the normal emptying of stomach contents into the duodenum.

DIET-DRUG INTERACTIONS

Check this table for notable nutrition-related effects of the medications discussed in this chapter.

Medication	Effects
Antacids (aluminum hydroxide, magnesium hydroxide, calcium carbonate)	**Gastrointestinal effects:** Constipation (aluminum- or calcium-containing antacids), diarrhea (magnesium-containing antacids)
	Dietary interactions: May decrease iron, folate, or vitamin B_{12} absorption
	Metabolic effects: Electrolyte imbalances
Antibiotics (for *H. pylori* infection; include amoxicillin, metronidazole, tetracycline)	**Gastrointestinal effects:** Diarrhea, nausea and vomiting (tetracycline, metronidazole), altered taste sensation (metronidazole)
	Dietary interactions: Avoid alcohol with metronidazole; tetracycline can bind to calcium, iron, magnesium, and zinc, reducing absorption of both the tetracycline and the minerals
Antisecretory drugs (proton-pump inhibitors, H2 blockers)	**Gastrointestinal effects:** Diarrhea, constipation, nausea and vomiting, abdominal pain (proton-pump inhibitors)
	Dietary interactions: May decrease iron, calcium, folate, and vitamin B_{12} absorption
Octreotide	**Gastrointestinal effects:** Abdominal cramps, diarrhea, nausea and vomiting, flatulence
	Dietary interactions: May decrease absorption of fat, fat-soluble vitamins, and vitamin B_{12}
	Metabolic effects: Hyperglycemia, hypothyroidism

Review Notes

- Dyspepsia refers to general symptoms of indigestion such as abdominal pain, nausea, and vomiting. Dietary measures may include avoiding large meals, fatty or spicy foods, and foods that trigger symptoms.
- Gastritis and peptic ulcer disease are most often associated with *Helicobacter pylori* infection, which can be eradicated by antibiotic therapy. NSAID use can promote gastritis and peptic ulcer disease by damaging the mucosal lining.
- Extensive damage to the mucosa may reduce gastric secretions and increase the risks of developing iron and vitamin B_{12} deficiencies.
- Nutrition care for gastritis and peptic ulcer disease includes correcting any nutritional deficiencies that develop and eliminating dietary substances that cause pain or discomfort.

gastrectomy (gah-STREK-ta-mee): the surgical removal of part of the stomach (partial gastrectomy) or the entire stomach (total gastrectomy).

bariatric (BAH-ree-AH-trik) surgery: surgery that treats severe obesity.

baros = weight

Gastric Surgery

Gastric surgery is sometimes necessary for treating stomach cancer, some ulcer complications, and ulcers that are resistant to drug therapy. In recent years, gastric surgeries have also become popular treatments for severe obesity. This section describes **gastrectomy**, the surgery that removes diseased areas of the stomach, and **bariatric surgery**, the type of surgery that treats severe obesity. Because gastric surgeries can interfere with stomach function either temporarily or permanently, patients generally need to make significant dietary adjustments afterward.

GASTRECTOMY

Figure 17-2 illustrates some typical gastrectomy procedures. In a partial gastrectomy, only part of the stomach is removed, and the remaining portion is connected to the duodenum or jejunum. In a total gastrectomy, the surgeon removes the entire stomach and connects the esophagus directly to the small intestine.

Nutrition Care after Gastrectomy

The primary goals of nutrition care after a gastrectomy are to meet the nutritional needs of the postsurgical patient and promote the healing of stomach tissue. Another goal is to prevent discomfort or nutrient deficiencies that may arise due to reduced stomach capacity or altered stomach function. As the next section will describe, some gastric surgeries increase the risk of **dumping syndrome**, a group of symptoms that result when a large amount of food passes rapidly into the small intestine.

Following a gastrectomy, oral intake of fluids and foods is suspended until some healing has occurred, and fluids are supplied intravenously. The first fluids given orally are usually small sips of water, ice chips (melted in the mouth), and broth. Once fluids are tolerated, patients are offered liquid meals (with no sugars) at first, and they usually progress to a soft food diet by the fourth or fifth day after surgery. Tube feedings may be necessary if complications prevent a normal progression to solid foods.[21]

Dietary measures after a gastrectomy are influenced by the size of the remaining stomach, which influences meal size, and the stomach-emptying rate, which affects food tolerances. Initially, the patient is offered small meals and snacks that include only one or two food items; these foods may contain protein (fish, lean meat, eggs, or cheese), fat, and complex carbohydrates (bread, potatoes, and vegetables). Depending on the amount of food tolerated, the patient may require as many as six to eight small meals and snacks per day. The patient should avoid sweets and sugars because they increase osmolarity in the small intestine and potentiate the dumping syndrome (discussed below). Some patients may need to avoid milk products due to lactose intolerance. Soluble fibers may be added to meals to slow stomach emptying and reduce risk of diarrhea. Liquids are restricted during meals (and 30 to 60 minutes before and after meals) due to limited stomach capacity and because liquids can increase the stomach emptying rate. Table 17-6 (p. 500) lists foods that are often permitted or limited in postgastrectomy diets.[22]

FIGURE 17-2 Gastrectomy Procedures

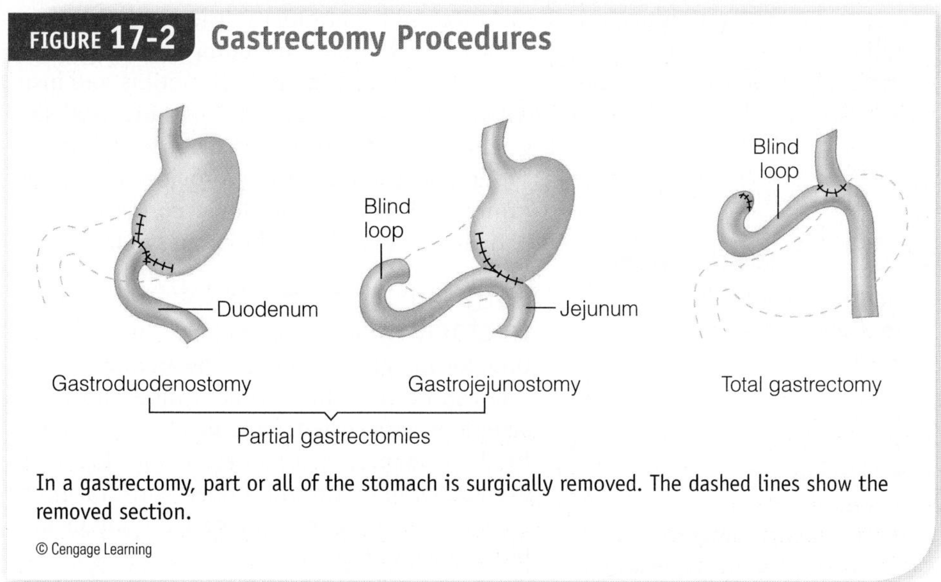

In a gastrectomy, part or all of the stomach is surgically removed. The dashed lines show the removed section.

dumping syndrome: a cluster of symptoms that result from the rapid emptying of an osmotic load from the stomach into the small intestine.

TABLE 17-6 Postgastrectomy Diet

Food Category	Foods Recommended (as tolerated)	Foods to Limit (unless tolerated)
Meat and meat alternates	Lean tender meats, fish, poultry, shellfish, eggs, smooth nut butters	Fried, tough, or chewy meats; frankfurters and sausages; bacon; luncheon meats; dried peas and beans
Milk and milk products	Milk, plain yogurt, mild cheeses	Milk shakes, chocolate milk, sweetened yogurt
Breads and cereals	Breads, crackers, bagels, pasta, and breakfast cereals made from enriched white flour (cereals should contain no added sugars)	Breads and cereals with more than 2 grams of fiber per serving; baked goods with dried fruit, nuts, or seeds; granola; frosted cereals; pastries; doughnuts
Vegetables	Tender-cooked vegetables without peels, skins, or seeds; raw lettuce	Raw vegetables (except lettuce), beets, broccoli, brussels sprouts, cabbage, cauliflower, collard and mustard greens, corn, potato skins
Fruit	Canned fruit without added sugars, bananas, melon	Canned fruit in syrup, raw fruit (except bananas and melons), dried fruit, fruit juices
Beverages	Decaffeinated coffee and tea, beverages sweetened with artificial sweeteners	Caffeinated beverages; alcoholic beverages; fruit drinks or fruit juices; beverages sweetened with sugars, corn syrup, or honey

Dumping Syndrome The dumping syndrome, a common complication of gastrectomy, is characterized by a group of symptoms resulting from rapid gastric emptying. Ordinarily, the pyloric sphincter controls the rate of flow from the stomach into the duodenum. After some types of stomach surgery, the hypertonic gastric contents are no longer regulated and rush into the small intestine more quickly after meals, causing a number of unpleasant effects. Early symptoms can occur within 30 minutes after eating and may include nausea, vomiting, abdominal cramping, diarrhea, lightheadedness, rapid heartbeat, and others (see Table 17-7). These symptoms may be due to a shift of fluid from blood vessels to the intestine that lowers blood volume and increases intestinal distention, and/or the accelerated release of GI hormones that induce strong intestinal contractions. Several hours later, symptoms of hypoglycemia may occur (see Table 17-7) because the unusually large spike in blood glucose following the meal (due to rapid nutrient influx and absorption) can result in an excessive insulin response.

Dietary adjustments can greatly minimize or prevent dumping syndrome. The goals are to limit the amount of food material that reaches the intestine, slow the rate of gastric emptying, and reduce foods that increase hypertonicity. Therefore, meal size is limited, fluids are restricted during meals, and sugars (including milk sugar) are restricted. In some cases, octreotide—a medication that inhibits GI motility and insulin release—may be prescribed to lessen symptoms. The "How to" lists practical suggestions for reducing the occurrence of dumping syndrome. The Case Study provides the opportunity to design a menu for a postgastrectomy patient who is at risk of dumping syndrome.

TABLE 17-7 Symptoms of Dumping Syndrome

Early Dumping Syndrome	Late Dumping Syndrome
Symptoms may begin within 30 minutes after eating. • Abdominal cramps, bloating • Diarrhea • Flushing, sweating • Lightheadedness • Nausea and vomiting • Rapid heartbeat • Weakness, feeling faint	**Symptoms may begin 1 to 3 hours after eating.** • Anxiety • Confusion • Headache, dizziness • Hunger • Palpitations • Sweating • Weakness, feeling faint

Nutrition Problems following a Gastrectomy After a gastrectomy, it may take time for the patient to learn the amount of food that can be consumed without discomfort. The symptoms associated with meals may lead to food avoidance, substantial weight loss, and eventually, malnutrition. Other nutrition problems that may occur after a gastrectomy include the following:[23]

HOW TO Alter the Diet to Reduce Symptoms of Dumping Syndrome

Dietary adjustments can greatly minimize or prevent symptoms of dumping syndrome. The following suggestions may help:

- Eat smaller meals that suit the reduced capacity of the stomach. Increase the number of meals consumed daily so that energy intake is adequate.
- Eat in a relaxed setting. Eat slowly, and chew food thoroughly.
- Include fiber-rich foods in each meal. Adding soluble fibers like pectin or guar gum to meals may help to control symptoms.
- If symptoms of hypoglycemia continue, try including a protein-rich food in each meal.
- Limit the amount of fluid included in meals. Avoid drinking beverages within 30 to 60 minutes before and after meals, but be sure to consume adequate fluid to avoid dehydration.
- Avoid juices, sweetened beverages, and foods that contain high amounts of sugar. Use artificial sweeteners to sweeten beverages and desserts.
- Avoid milk and most milk products, which are high in lactose. Avoid enzyme-treated milk as well, because the breakdown products of lactose (glucose and galactose) can also cause dumping symptoms. Cheese may be better tolerated than milk because its lactose content is low. Make an effort to consume nonmilk calcium sources such as green leafy vegetables, fish with bones, and tofu.
- Avoid carbonated beverages if they cause bloating.
- Avoid foods and beverages that are very hot or very cold, unless tolerated.
- Lie down for 20 to 30 minutes (or longer) after eating to help slow the transit of food to the small intestine. While eating a meal, sit upright.

- *Fat malabsorption.* Fat digestion and absorption may become impaired for a number of reasons after gastrectomy. The accelerated transit of food material may prevent the normal mixing of fat with lipase and bile. If the duodenum has been removed or bypassed, less lipase is available for fat digestion. Bacterial overgrowth, (■) a common consequence of gastric surgeries, can lead to changes in bile acids that upset bile function. The fat malabsorption that results from these changes can eventually cause deficiencies of fat-soluble vitamins and some minerals. Supplemental pancreatic enzymes are sometimes provided to improve fat digestion. Medium-chain triglycerides, which are more easily digested and absorbed, can be used to supply additional fat kcalories.
- *Bone disease.* Osteoporosis and osteomalacia are common outcomes following a gastrectomy. The fat malabsorption described earlier can cause malabsorption of both vitamin D and calcium; (■) furthermore, patients at risk of dumping syndrome may need to avoid milk products, which are among the best sources of these nutrients. Bone density should be monitored during the years following surgery, and supplementation of calcium and vitamin D is often recommended.
- *Anemia.* After a gastrectomy, the reduced secretion of gastric acid and intrinsic factor impairs the absorption of iron and vitamin B_{12}, respectively, often leading to anemia. If the duodenum has been removed or is bypassed, the risk of iron deficiency increases because the duodenum is a major site of iron absorption. Supplementation of both iron and vitamin B_{12} is usually warranted after surgery.

■ Bacterial overgrowth may be a consequence of reduced gastric acid secretions, altered motility of intestinal contents, or changes in intestinal anatomy due to surgical reconstruction. Chapter 18 describes bacterial overgrowth in detail.

■ Fat malabsorption reduces calcium absorption because the negatively charged fatty acids combine with calcium (which is positively charged) and prevent its absorption.

Case Study Nutrition Care after Gastric Surgery

Dirk Hanson, a 58-year-old biology teacher, was admitted to the hospital for gastric surgery after numerous medical treatments failed to manage severe complications related to his peptic ulcer disease. A gastrojejunostomy was performed, and after about 24 hours, Mr. Hanson was able to take small sips of warm water. The health care team anticipates multiple nutrition-related problems and is taking measures to prevent them.

1. Review Figure 17-2 to better understand Mr. Hanson's surgical procedure. Consider the possibilities that he might experience the following symptoms: early satiety, nausea and vomiting, weight loss, dumping syndrome, fat malabsorption, anemia, and bone disease. Explain why each of these conditions may occur.
2. What type of diet will the physician prescribe for Mr. Hanson after he begins eating solid foods? Create a day's worth of menus, using foods from Table 17-6.
3. What advice can you give Mr. Hanson that will help to prevent dumping syndrome? List several foods from each major food group that may cause symptoms of dumping syndrome.

BARIATRIC SURGERY

Bariatric surgery is currently considered the most effective and durable treatment for morbid obesity.[24] (■) Candidates for bariatric surgery are obese individuals who have a body mass index (BMI) greater than 40, or a BMI between 35 and 40 accompanied by severe weight-related problems such as diabetes, hypertension, or debilitating osteoarthritis (a healthy BMI usually falls between 20 and 25). In addition, the patient should have attempted a variety of nonsurgical weight-loss measures—such as dietary adjustments, exercise, medications, and behavior modification—prior to seeking surgery. Bariatric surgery patients should have realistic expectations about the amount of weight they are likely to lose, the diet they will need to follow, and the complications that may ensue. Some types of bariatric surgery can dramatically affect health and nutrition status, and many patients require lifelong management.

> **NURSING DIAGNOSIS**
>
> Nursing diagnoses appropriate for the ideal bariatric surgery candidate may include *readiness for enhanced self-health management* and *readiness for enhanced knowledge.*

Bariatric Surgical Procedures Figure 17-3 illustrates the most popular surgical options for weight reduction. The gastric bypass operation, which accounts for about 88 percent of bariatric surgeries,[25] constructs a small gastric pouch that reduces stomach capacity and thereby restricts meal size. In addition, the gastric pouch is connected directly to the jejunum, resulting in significant nutrient malabsorption because the flow of food bypasses a significant portion of the small intestine. In the gastric banding procedure, a gastric pouch is created using a fluid-filled inflatable band; adjusting the band's fluid level can tighten or loosen the band and alter the size of the opening to the rest of the stomach. A smaller opening slows the pouch-emptying rate and prolongs the sense of fullness after a meal. Whereas the gastric bypass operation is usually permanent, the gastric banding procedure is fully reversible.

Clinical studies suggest that the gastric bypass surgery is more effective than the gastric banding procedure. Although study results vary, gastric bypass patients typically lose about 60 to 70 percent of their excess body weight, whereas those who undergo gastric banding lose about 50 percent of their excess weight.[26]

Nutrition Care after Bariatric Surgery The main objectives of nutrition care after bariatric surgery are to maximize and maintain weight loss, ensure appropriate nutrient intakes, maintain hydration, and avoid complications such as nausea and

■ Bariatric surgery was introduced in Chapter 7 (see p. 173).

FIGURE 17-3 Surgical Procedures for Severe Obesity

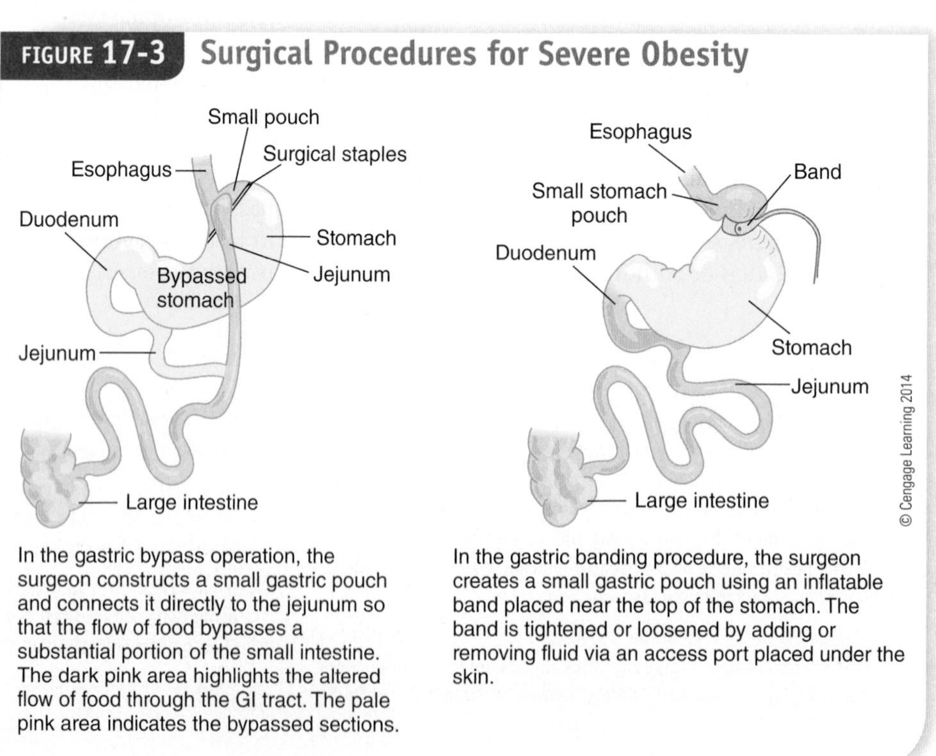

In the gastric bypass operation, the surgeon constructs a small gastric pouch and connects it directly to the jejunum so that the flow of food bypasses a substantial portion of the small intestine. The dark pink area highlights the altered flow of food through the GI tract. The pale pink area indicates the bypassed sections.

In the gastric banding procedure, the surgeon creates a small gastric pouch using an inflatable band placed near the top of the stomach. The band is tightened or loosened by adding or removing fluid via an access port placed under the skin.

vomiting or dumping syndrome.[27] The gastric pouch created by surgery may eventually expand to hold up to one-half cup of food, but its initial capacity is only a few tablespoons. Only sugar-free, noncarbonated clear liquids and low-fat broths are provided during the first two days following bariatric surgery. Afterward, patients consume a liquid diet (high in protein, low in sugars and fat) at first, followed by pureed foods and then solid foods; the diet is advanced as tolerated, usually over a period of four to six weeks. Once the diet progresses to solid foods, patients often consume between three and six small meals per day. Only small portions of food can be consumed at each meal because overeating can stretch the gastric pouch or result in vomiting or regurgitation. Similarly, fluids must be consumed separately from meals to avoid excessive distention. Other dietary recommendations include the following:

- *Protein intake.* Recommendations range from 1.0 to 1.5 grams of protein per kilogram of ideal body weight per day;[28] however, intakes are often lower than recommended. Patients are generally instructed to consume liquid protein supplements regularly and to eat high-protein foods before consuming other foods in a meal.
- *Vitamin and mineral deficiencies.* Bariatric patients have a high risk of developing nutrient deficiencies due to reduced food intake, reduced gastric secretions, and nutrient malabsorption. Supplemental vitamin B_{12}, vitamin D, iron, and calcium are recommended after surgery. A daily multivitamin/mineral supplement ensures that patients meet their needs for other nutrients.
- *Foods to avoid.* Some foods may obstruct the gastric outlet; these include doughy or sticky breads and pasta products; melted cheese; fibrous vegetables such as asparagus and celery; foods with seeds, peels, or skins; nuts; popcorn; and tough, chewy meats.
- *Dumping syndrome.* To avoid symptoms of dumping syndrome, gastric bypass patients must carefully control food portions, avoid foods high in sugars, and consume liquids between meals (review the "How to" on p. 501).

After bariatric surgery, patient education and counseling are critical for weight loss and weight management, and patients also need to learn the elements of a healthy diet. The "How to" includes additional dietary suggestions for patients who have undergone bariatric surgery.

HOW TO Alter Dietary Habits to Achieve and Maintain Weight Loss after Bariatric Surgery

Patients need to learn new dietary habits after bariatric surgery. The following recommendations may help:

- Consume only small portions of food and chew food thoroughly. Use a small spoon, and take small bites. Relax and enjoy the meal, taking at least 15 to 20 minutes to eat.
- Understand that, at first, the appropriate amount of each food served at mealtime may be only a few spoonfuls. Learn to recognize the sensations that occur when the gastric pouch is full. Signs of fullness may include pressure in the stomach region, a slight feeling of nausea, or pain in the upper chest or shoulder.
- To control vomiting, try eating smaller volumes of food, eating more slowly, and avoiding foods known to cause difficulty. Continued vomiting may be a sign that some food choices or amounts are inappropriate.
- Consume food only during designated mealtimes (usually three to six small meals per day), and avoid consuming foods at other times of the day. Snacking throughout the day can become a bad habit that causes weight to be regained.
- Learn to recognize foods that cause problems. Foods that are dry, sticky, or fibrous may be difficult to tolerate during the weeks after surgery.
- Avoid consuming liquids within 30 minutes of mealtime. Avoid high-kcalorie drinks like sweetened soda, milk shakes, and alcoholic beverages. Avoid drinking carbonated beverages or using a straw, as these practices can increase stomach gas and cause bloating.
- Sip water and other beverages throughout the day to obtain sufficient fluids. Aim to consume between 6 and 8 cups of water and other noncaloric beverages daily. Remember that most people meet a significant fraction of their fluid needs by eating food, but a person who has undergone bariatric surgery must limit food intake.
- Engage in regular physical activity. Activity is a valuable aid to weight maintenance and can help to maintain lean tissue while weight is being lost.

Postsurgical Concerns in Gastric Bypass Surgery Common complaints after bariatric surgery include nausea, vomiting, and constipation.[29] Although the cause of these problems varies among patients, dietary noncompliance and inadequate fluid intake are often contributing factors, and improved dietary intakes can help in resolving these conditions. The long-term complications that may develop after gastric bypass surgery are similar to those that arise after gastrectomy and include fat malabsorption, bone disease, and anemia. Rapid weight loss increases a person's risk of developing gallbladder disease; patients at especially high risk sometimes have their gallbladders removed while undergoing bariatric surgery. After weight loss, plastic surgery may be necessary to remove extra skin, especially on the abdomen, buttocks, hips, and thighs.

Review Notes

- Gastric surgeries, which are used to treat cancer, peptic ulcer complications, and obesity, require dietary adjustments after surgery and are associated with complications that may affect nutrition status.
- After a gastrectomy procedure, dietary measures depend on the size of the remaining stomach and the stomach-emptying rate. Common postsurgical complications include fat malabsorption, bone disease, anemia, and dumping syndrome.
- Dumping syndrome, characterized by a group of symptoms caused by rapid gastric emptying, is managed by controlling food portions, avoiding foods high in sugars, and consuming liquids between meals.
- After bariatric surgery, patients must learn to consume appropriate food portions; meet fluid needs; use dietary supplements to prevent nutrient deficiencies; and choose foods that are unlikely to cause abdominal discomfort, vomiting, or dumping syndrome.

Nutrition Assessment Checklist for People with Upper GI Tract Disorders

MEDICAL HISTORY

Check the medical history to uncover conditions or treatments that may:

- ❍ Interfere with chewing or swallowing
- ❍ Lead to dry mouth
- ❍ Lead to dyspepsia, nausea, or vomiting

Check for a medical diagnosis of:

- ❍ Gastritis or peptic ulcer
- ❍ GERD
- ❍ Hiatal hernia
- ❍ Pernicious anemia

For a patient who has undergone gastric surgery, check for the following complications:

- ❍ Anemia
- ❍ Bone disease
- ❍ Dumping syndrome
- ❍ Fat malabsorption

MEDICATIONS

Record all medications and note:

- ❍ Aspirin or NSAID use in patients with gastritis or peptic ulcer disease
- ❍ Medications that may cause dry mouth
- ❍ Medications that may cause nausea and vomiting

To help alleviate nausea, suggest that medications be taken with food, when possible.

DIETARY INTAKE

To devise an acceptable meal plan, obtain:

- ❍ An accurate and thorough record of food intake
- ❍ A thorough record of dietary supplement intake, including both vitamin/mineral supplements and herbal products
- ❍ A record of food allergies and intolerances, as well as food preferences
- ❍ A record of foods that provoke symptoms of GERD, dyspepsia, gastritis, peptic ulcers, or dumping syndrome

For patients on long-term dysphagia diets, monitor appetite, food tolerances, and the variety of foods consumed

ANTHROPOMETRIC DATA

Measure baseline height and weight. Address weight loss early to prevent malnutrition in patients with:

- ❍ Difficulty chewing or swallowing
- ❍ Dumping syndrome

- ❍ Dyspepsia
- ❍ Frequent nausea and vomiting
- ❍ Malabsorption

LABORATORY TESTS

Check laboratory tests for signs of dehydration for patients with:

- ❍ Constipation
- ❍ Dumping syndrome
- ❍ Persistent vomiting

Check laboratory tests for nutrition-related anemia in patients with:

- ❍ Gastritis
- ❍ Long-term use of antisecretory drugs
- ❍ Previous gastric surgeries

PHYSICAL SIGNS

Look for physical signs of:

- ❍ Dehydration—in patients with constipation, dumping syndrome, or persistent vomiting
- ❍ Iron and vitamin B_{12} deficiencies—in patients with hypochlorhydria or achlorhydria

Self Check

1. If a patient with dysphagia has difficulty swallowing solids but can easily swallow liquids:

a. the patient may have achalasia.
b. the problem is most likely a motility disorder.
c. the problem is probably an esophageal obstruction.
d. the patient most likely has oropharyngeal dysphagia.

2. The health practitioner working with a patient with dysphagia should recognize that:

a. highly seasoned foods are often restricted.
b. regular diets do not meet nutrient needs, so supplements are required.
c. pureed foods should be given to minimize the risk of aspiration.
d. the foods allowed are those that can be comfortably and safely chewed and swallowed.

3. Possible consequences of GERD include all of the following *except:*

a. gastric ulcer.
b. reflux esophagitis.
c. dysphagia.
d. Barrett's esophagus.

4. Treating GERD with proton-pump inhibitors is effective because they:

a. provide a protective barrier between gastric acid and the esophageal mucosa.
b. reduce pressure within the stomach.
c. reduce gastric acid secretion.
d. strengthen the lower esophageal sphincter.

5. For the patient with persistent vomiting, the major nutrition-related concern(s) is/are:

a. dehydration and malnutrition.
b. reflux esophagitis.
c. dyspepsia.
d. peptic ulcers.

6. Chronic gastritis may increase risk of:

a. dumping syndrome.
b. bone disease.
c. iron and vitamin B_{12} deficiencies.
d. gallbladder disease.

7. The primary cause of most peptic ulcers is:

a. consumption of spicy foods.
b. psychological stress.
c. smoking cigarettes.
d. *Helicobacter pylori* infection.

8. The main dietary recommendation for patients with gastritis or peptic ulcers is to consume foods that:

a. neutralize stomach acidity.
b. are well tolerated and do not cause discomfort.
c. coat the stomach lining.
d. promote healing of mucosal tissue.

9. Following a gastrectomy, patients can greatly minimize the risk of dumping syndrome by:

a. increasing intake of carbohydrates and avoiding fatty foods.
b. consuming liquids between meals and avoiding foods that supply sugars.
c. consuming a high-fiber diet and minimizing meat and cheese intake.
d. including milk products with each meal to provide protein.

10. After a gastric bypass operation, the patient should be monitored for all of the following conditions *except:*

a. anemia.
b. bone disease.
c. hiatal hernia.
d. fat-soluble vitamin deficiencies.

Answers to these questions appear in Appendix H. For more chapter review: Access an interactive eBook, chapter-specific interactive learning tools, including flashcards, quizzes, videos, and more in your Nutrition CourseMate, accessed through CengageBrain.com.

Clinical Applications

1. Although some individuals require a mechanically altered diet for just a few weeks, others have medical problems that require long-term use of such diets. Consider the difference between working with a person who has had a swallowing problem for years and a person who recently had mouth surgery and is just beginning to eat again.
   - Explain how the needs of these individuals may differ. What nutrition-related problems may develop if a person has been following a restrictive dysphagia diet for several years?
   - Using Table 17-3 and the "How to" on p. 490, create a day's worth of menus for a person who requires long-term use of a pureed dysphagia diet and tolerates only liquids that have a honeylike consistency.

2. Jillian, a 38-year-old woman who is 5 feet 4 inches tall and weighs 227 pounds, has had severe hip and knee osteoarthritis for several years and was recently diagnosed with type 2 diabetes. After trying numerous diet programs without success, she finally visits a bariatric surgeon to learn about the surgical options for treating her obesity.
   - Calculate Jillian's BMI, and explain why or why not Jillian would be a good candidate for bariatric surgery.
   - Should Jillian decide to undergo gastric bypass surgery, she will need to permanently change her dietary habits. Describe the dietary recommendations and nutrition concerns following bariatric surgery. Summarize the measures necessary for preventing vomiting, distention of the gastric pouch, gastric outlet obstruction, and dumping syndrome.
   - Explain why dehydration is a frequent complication following bariatric surgery. What tips can you give Jillian to help her avoid this problem?

Nutrition on the Net

For further study of topics covered in this chapter, access these websites:

- Visit the websites of these organizations to find information that is helpful both for health practitioners and patients with gastrointestinal problems:
  American College of Gastroenterology:
  www.acg.gi.org
  American Gastroenterological Association:
  www.gastro.org
  International Foundation for Functional Gastrointestinal Disorders:
  www.iffgd.org
  National Institute of Diabetes and Digestive and Kidney Diseases, a division of the National Institutes of Health:
  www2.niddk.nih.gov
- Find more information about dysphagia at the Dysphagia Resource Center:
  www.dysphagia.com
- Learn more about *Helicobacter pylori* from the Helicobacter Foundation:
  www.helico.com
- The Consumer Guide to Bariatric Surgery provides general information about bariatric surgeries at this website:
  www.yourbariatricsurgeryguide.com

Notes

1. T. E. Daniels, Diseases of the mouth and salivary glands, in L. Goldman and A I. Schafer, eds., *Goldman's Cecil Medicine* (Philadelphia: Saunders, 2012), pp. 2449–2454; R. S. Porter and J. L. Kaplan, eds., Approach to dental and oral symptoms, *Merck Manual of Diagnosis and Therapy* (Whitehouse Station, NJ: Merck Sharp and Dohme Corp., 2011), pp. 494–516.
2. S. Naguwa and M. E. Gershwin, Sjögren's syndrome, in L. Goldman and A I. Schafer, eds., *Goldman's Cecil Medicine* (Philadelphia: Saunders, 2012), pp. 1713–1716.
3. A. Chaudhury and H. Mashimo, Oropharyngeal and esophageal motility disorders, in N. J. Greenberger, ed., *Current Diagnosis and Treatment: Gastroenterology, Hepatology, and Endoscopy* (New York: McGraw-Hill Companies, 2012), pp. 164–182.
4. The National Dysphagia Diet Task Force, *The National Dysphagia Diet: Standardization for Optimal Care* (Chicago: American Dietetic Association, 2002).
5. J. E. Richter and F. K. Friedenberg, Gastroesophageal reflux disease, in M. Feldman and coeditors, *Sleisenger and Fordtran's Gastrointestinal and Liver Disease* (Philadelphia: Saunders, 2010), pp. 705–726.
6. S. Friedman and J. R. Agrawal, Gastrointestinal and biliary complications of pregnancy, in N. J. Greenberger, ed., *Current Diagnosis and Treatment: Gastroenterology, Hepatology, and Endoscopy* (New York: McGraw-Hill Companies, 2012), pp. 79–104.
7. G. W. Falk and D. A. Katzka, Diseases of the esophagus, in L. Goldman and A I. Schafer, eds., *Goldman's Cecil Medicine* (Philadelphia: Saunders, 2012), pp. 874–886.
8. W. W. Chan and R. Burakoff, Functional (nonulcer) dyspepsia, in N. J. Greenberger, ed., *Current Diagnosis and Treatment: Gastroenterology, Hepatology, and Endoscopy* (New York: McGraw-Hill Companies, 2012), pp. 203–213.
9. Chan and Burakoff, 2012.
10. J. Tack, Dyspepsia, in M. Feldman and coeditors, *Sleisenger and Fordtran's Gastrointestinal and Liver Disease* (Philadelphia: Saunders, 2010), pp. 183–195.
11. Tack, 2010.
12. K. McQuaid, Approach to the patient with gastrointestinal disease, in L. Goldman and A I. Schafer, eds., *Goldman's Cecil Medicine* (Philadelphia: Saunders, 2012), pp. 828–844.
13. E. L. Lee and M. Feldman, Gastritis and gastropathies, in M. Feldman and coeditors, *Sleisenger and Fordtran's Gastrointestinal and Liver Disease* (Philadelphia: Saunders, 2010), pp. 845–860.
14. E. Lew, Peptic ulcer disease, in N. J. Greenberger, ed., *Current Diagnosis and Treatment: Gastroenterology, Hepatology, and Endoscopy* (New York: McGraw-Hill Companies, 2012), pp. 187–197.
15. Lew, 2012.
16. E. J. Kuipers and M. J. Blaser, Acid peptic disease, in L. Goldman and A I. Schafer, eds., *Goldman's Cecil Medicine* (Philadelphia: Saunders, 2012), pp. 886–895.
17. Kuipers and Blaser, 2012.
18. Lew, 2012.
19. Academy of Nutrition and Dietetics, *Nutrition Care Manual* (Chicago: Academy of Nutrition and Dietetics, 2012).
20. R. S. Porter and J. L. Kaplan, eds., Gastritis and peptic ulcer disease, *Merck Manual of Diagnosis and Therapy* (Whitehouse Station, NJ: Merck Sharp and Dohme Corp., 2011), pp. 128–138; F. F. Ferri, *Ferri's Clinical Advisor 2009: Instant Diagnosis and Treatment* (Philadelphia: Mosby, 2009), pp. 693–694.
21. Academy of Nutrition and Dietetics, 2012.
22. Academy of Nutrition and Dietetics, 2012.
23. C. R. Parrish, Post-gastrectomy: Managing the nutrition fall-out, *Nutrition Issues in Gastroenterology* 18 (2004): 63–75.
24. M. K. Robinson and N. J. Greenberger, Treatment of obesity: The impact of bariatric surgery, in N. J. Greenberger, ed., *Current Diagnosis and Treatment: Gastroenterology, Hepatology, and Endoscopy* (New York: McGraw-Hill Companies, 2012), pp. 224–236.
25. G. Woodard and J. Morton, Bariatric surgery, in M. Feldman and coeditors, *Sleisenger and Fordtran's Gastrointestinal and Liver Disease* (Philadelphia: Saunders, 2010), pp. 115–119.
26. M. A. Attiah, Durability of Roux-en-Y gastric bypass surgery: A meta-regression study, *Annals of Surgery* 256 (2012): 251–254; Weight Management Dietetic Practice Group, *ADA Pocket Guide to Bariatric Surgery* (Chicago: American Dietetic Association, 2009), pp. 1–16.
27. Academy of Nutrition and Dietetics, 2012.
28. L. Aills and coauthors, ASMBS allied health nutritional guidelines for the surgical weight loss patient, *Surgery for Obesity and Related Diseases* 4 (2008): S73–S108.
29. Robinson and Greenberger, 2012.

Nutrition in Practice

Nutrition and Oral Health

Various aspects of nutrition and oral health were discussed earlier in this book. Chapters 3 and 9 described the effects of carbohydrate and fluoride, respectively, on the development of dental caries. Chapter 10 explained how the prolonged use of nursing bottles increases the risk of dental caries in babies. This Nutrition in Practice provides additional information about dental caries and introduces other problems related to oral health. The relevant terms are defined in the accompanying glossary.

What is dental caries, and what factors influence its development?

Dental caries is an oral infectious disease that affects the structures and integrity of the teeth. Caries develops when the bacteria that reside in **dental plaque** metabolize dietary carbohydrates and produce acids that attack tooth enamel. If allowed to progress, the decay can penetrate the dentin and destroy other structures that support and maintain the tooth (see Figure NP17-1). The development of dental caries is influenced by the type of carbohydrate consumed, the frequency of carbohydrate intake, the stickiness of the foods that contain carbohydrate, and the availability of saliva to rinse the teeth and neutralize acid. Other factors include oral hygiene, fluoride intake, and the composition of tooth enamel, which together influence a person's susceptibility to caries. Poor nutrition during pregnancy, infancy, or early childhood can impair the development of healthy teeth and increase vulnerability to dental caries.[1] Table NP17-1 lists examples of nutrient deficiencies that influence the development of dental caries.

Which foods promote caries development?

The most **cariogenic** foods are carbohydrate-containing foods that remain in contact with the teeth for prolonged periods, are difficult to clear from the mouth, or are consumed frequently or over an extended time period. Examples include hard candies or lozenges that dissolve slowly in the mouth; sticky or chewy foods such as dried fruit, jelly beans, or chewy bread; starchy snack foods such as pretzels or chips, and sweetened beverages that are repeatedly sipped. These foods can be eaten without provoking tooth decay if they are consumed quickly and removed from tooth surfaces promptly.

Acidic foods and beverages, such as cola drinks, citrus fruits and juices, pickles, and some herbal teas, also contribute to the erosion of tooth enamel.[2] Acidic food items are more cariogenic when consumed after or between meals because they are less likely to be rinsed away by saliva or neutralized by alkaline foods that may be consumed during the meal.

Glossary

cariogenic (KAH-ree-oh-JEN-ik): conducive to development of dental caries.

dental calculus: mineralized dental plaque, often associated with inflammation and bleeding.

dental caries (KAH-reez): infectious disease of the teeth that causes the gradual decay and disintegration of tooth structures.

dental plaque (PLACK): a film of bacteria and bacterial by-products that accumulates on the tooth surface.

gingiva (jin-JYE-va, JIN-jeh-va): the gums.

gingivitis (jin-jeh-VYE-tus): inflammation of the gums, characterized by redness, swelling, and bleeding.

periodontal disease: disease that involves the connective tissue that supports the teeth.

periodontitis: inflammation or degeneration of the tissues that support the teeth.

periodontium: the tissues that support the teeth, including the gums, cementum (bonelike material covering the dentin layer of the tooth), periodontal ligament, and underlying bone.

- *peri* = around, surrounding
- *odont* = tooth

FIGURE NP17-1 Development of Dental Caries

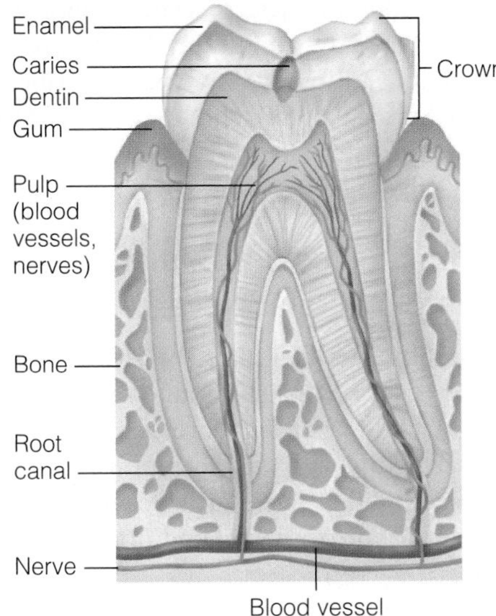

Dental caries begins when acid dissolves the enamel that covers the tooth. If not repaired, the decay may penetrate the dentin and spread into the pulp of the tooth, causing inflammation, abscess, and possible loss of the tooth.

How does saliva protect against caries formation?

In addition to rinsing away the sugars and food particles that remain on the teeth, saliva contains substances (such as proteins, bicarbonate, and phosphates) that dilute and neutralize mouth acidity. Furthermore, saliva contains antimicrobial proteins (immunoglobulins and lysozyme) that defend against bacteria and fungi. Finally, the calcium, phosphate, and fluoride ions in saliva help to prevent dissolution of enamel and promote remineralization. Thus, if salivary secretions are low or absent, the risk of developing dental caries and other dental diseases greatly increases.

Does dental plaque contribute to other dental diseases?

Yes, the deposits of plaque can thicken and lead to other dental problems. As plaque accumulates, it fills with calcium and phosphate, eventually forming **dental calculus**. Calculus develops either at the gum surface or in the crevice between the gum and a tooth; its presence may lead to additional plaque retention. The buildup of plaque and calculus increases the likelihood of infection and subsequent inflammation.

Periodontal disease is the name given to inflammatory conditions that involve the **periodontium**, the structures

Do any foods help to prevent caries?

Yes, foods that stimulate saliva flow, neutralize mouth acidity, or induce the clearance of food particles from the teeth can help to prevent caries formation. Examples include cheese, which increases salivary secretions and contains nutrients that neutralize acid; milk, which reduces mouth acidity due to its nearly neutral pH; and raw vegetables, which require vigorous chewing and therefore stimulate saliva flow. Some foods contain substances that reduce enamel solubility, such as the fluoride in tea and an unidentified compound in cocoa. The sequence in which foods are eaten influences caries development; for example, eating cheese after consuming an acidic fruit dessert can raise plaque pH and reduce caries risk, whereas drinking sugared coffee at the end of the meal can prolong plaque acidity. Chewing sugarless gum after or between meals can help to prevent caries by stimulating salivary flow, pushing saliva into hard-to-reach crevices in and around teeth, and removing food particles from the teeth.[3]

TABLE NP17-1 Nutrient Deficiencies Affecting Development of Dental Caries

Nutrient Deficiency	Effect on Dental Tissues
Protein	Increased solubility of tooth enamel
Vitamin A	Impaired formation of tooth enamel
Vitamin D	Impaired formation of tooth enamel, reduced enamel mineralization
Calcium	Reduced enamel mineralization
Phosphorus	Reduced enamel mineralization
Fluoride	Increased susceptibility to enamel demineralization, reduced protection against decay-causing bacteria

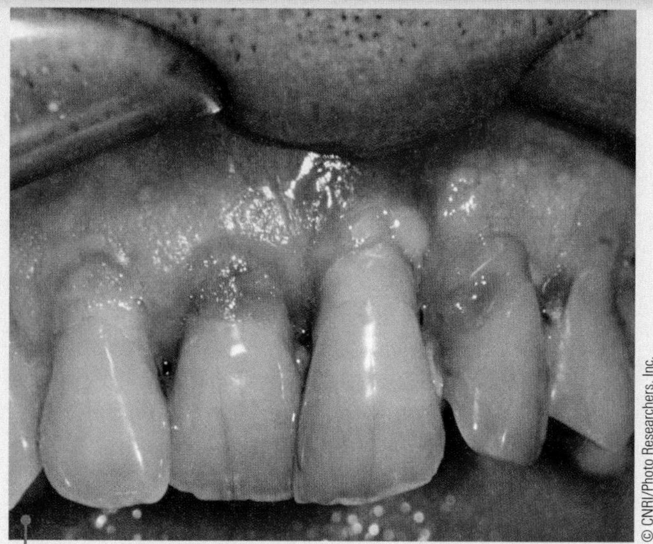
© CNRI/Photo Researchers, Inc.

Periodontal disease destroys the tissues and bones that hold teeth in place.

that support the tooth in its bony socket. The periodontium includes the gums (called **gingiva**), other connective tissues surrounding the tooth, and the bone underneath. Inflammation of the gums, called **gingivitis**, is characterized by redness, bleeding, and swelling of gum tissue. **Periodontitis** is an inflammation of the other tissues surrounding the tooth. As plaque invades the space below the gum line, the combination of toxic bacterial by-products and the body's immune response can damage the tissues holding a tooth in place. Left untreated, the tissues and bone of the peridontium may ultimately be destroyed, leading to permanent tooth loss.

What are the risk factors for periodontal disease, and how is it treated?

Dental plaque is the major risk factor associated with periodontal disease, and the severity of disease is related to the amount of plaque present. Tobacco smoking is another factor, possibly due to its destructive effects on cellular immune responses. The likelihood of developing periodontal disease is increased if a person has a chronic illness that impairs immune status, such as diabetes mellitus or HIV infection. Other risk factors include stress, pregnancy, use of certain medications (including oral contraceptives, antiepileptic drugs, and anticancer drugs), and dental conditions that increase plaque accumulation, such as poorly aligned teeth or ill-fitting bridges.[4] Strategies for reducing risk focus on improving oral hygiene (proper brushing and flossing) and encouraging smoking cessation. Table NP17-2 provides suggestions that may help to decrease the risk of developing oral diseases.

Treatment of periodontal disease depends on the extent of damage. In mild cases, deep cleaning and proper oral hygiene may reverse the condition. Antimicrobial mouth rinses and topical antibiotics may be prescribed to control infection. Surgical approaches may be used to remove plaque or calculus deposits underneath gum tissue or to replace tissues that have been destroyed.

How do chronic illnesses influence a person's risk for dental and oral diseases?

Some chronic illnesses can alter the structure and function of dental tissues, impair immune responses, or cause reduced salivary flow. Furthermore, some medications or treatments can reduce salivary secretions, along with the immune protection that saliva provides. Examples of conditions that contribute to dental and oral diseases include the following:

- *Diabetes mellitus.* Periodontal disease is more prevalent among people with diabetes mellitus, especially those whose diabetes is poorly controlled. People with diabetes often have impaired immune responses and a greater susceptibility to infections. Diabetes also favors the growth of bacteria that tend to infect periodontal tissues. People with diabetes tend to have higher plaque accumulations and dry mouth. In addition, the damaging effects of hyperglycemia weaken the collagen structure of dental tissues, making them more vulnerable to destruction.[5]
- *Human immunodeficiency virus (HIV)/AIDS.* HIV infection is characterized by compromised immunity, and the risk of developing periodontal disease is closely linked to the extent of HIV infection. Those at greatest risk include smokers and patients in the advanced stages of disease. HIV-infected individuals often develop dry mouth as a result of medications or salivary gland dysfunction.[6] In untreated persons, fungal and viral infections are common and may cause burning in the mouth and painful ulcerations.
- *Radiation therapy for oral cancers.* Radiation treatment often causes serious oral and dental complications. Inflammation and tissue damage can be so severe that the radiation treatment may need to be halted or the intensity significantly reduced. Other complications include dry mouth, fungal and viral infections, changes in taste sensation, and tissue and muscle scarring (which often reduces chewing ability). To minimize complications, dental care is often initiated before radiation therapy begins.

Can dental diseases have adverse effects on health beyond their effects on the teeth?

Yes, the bacteria that reside on dental tissues can enter the bloodstream and travel to other tissues; therefore, they may be able to trigger immune responses or cause infections elsewhere in the body. Evidence supports a link between dental bacteria and other medical conditions, including the following:[7]

- *Atherosclerosis and heart disease.* The inflammatory process induced by periodontal pathogens may increase levels of cytokines and other mediators that accelerate the progression of atherosclerosis. In addition, periodontal bacteria may enter into the bloodstream and contribute to the processes of plaque formation or blood clotting.[8]
- *Diabetes mellitus.* The chronic inflammation caused by periodontal disease can exacerbate insulin resistance and provoke events leading to type 2 diabetes. Severe periodontal disease has also been linked to poor glycemic control in persons with diabetes.
- *Respiratory illnesses.* The teeth of hospitalized individuals can become colonized with bacteria that cause respiratory illnesses. In addition, mortality rates associated with pneumonia have been found to be four times higher in individuals with periodontal disease.[9]

Research studies are in progress to confirm cause-and-effect relationships between oral bacteria and the medical conditions described above, as well as the specific mechanisms involved.

Nutrition status and health status have strong influences on oral health. Developing sound eating habits and maintaining good dental hygiene are practices that can promote dental health and possibly reduce the risk of developing other medical problems. Additional studies will help to clarify the complex interactions between dental disease and chronic illnesses.

TABLE NP17-2 Suggestions for Preventing Oral Diseases

Personal hygiene

- Brush your teeth twice daily for at least two minutes with a fluoride toothpaste. Replace your toothbrush every three or four months or whenever the bristles become frayed.
- Floss between your teeth at least once a day.
- Avoid smoking. If you smoke, look into tobacco cessation programs in your area.
- Avoid consuming excessive amounts of alcohol.

Dietary practices

- Reduce your consumption of sugary foods. Avoid consuming sugary and starchy snacks between meals.
- After consuming carbohydrate-containing foods, sugary beverages, or acidic soft drinks or teas, rinse your mouth or clean your teeth.
- If you chew gum, use a sugarless gum.
- Avoid putting an infant or child to bed with a nursing bottle containing anything except plain water.
- In communities that do not provide fluoridated water, provide children at high risk of dental caries with a fluoride supplement.

Professional dental care

- Visit a dentist at least once a year to have your teeth and mouth examined and teeth cleaned. Visit the dentist more often if necessary.
- Ask the dentist if you are a candidate for topical fluoride treatments or tooth sealants, which protect susceptible tooth surfaces.

Notes

1. D. P. DePaola and coauthors, Nutrition and dental medicine, in M. E. Shils and coeditors, *Modern Nutrition in Health and Disease* (Baltimore: Lippincott Williams & Wilkins, 2006), pp. 1152–1178.
2. S. Wongkhantee and coauthors, Effect of acidic food and drinks on surface hardness of enamel, dentine, and tooth-coloured filling materials, *Journal of Dentistry* 34 (2006): 214–220.
3. S. K. Stookey, The effect of saliva on dental caries, *Journal of the American Dental Association* 139 (2008): 11S–17S; DePaola and coauthors, 2006.
4. J. Kim and S. Amar, Periodontal disease and systemic conditions: A bidirectional relationship, *Odontology* 94 (2006): 10–21.
5. American Dietetic Association, Position of the American Dietetic Association: Oral health and nutrition, *Journal of the American Dietetic Association* (2007): 1418–1428; DePaola and coauthors, 2006.
6. J. C. C. Filho and E. M. Giovani, Xerostomy, dental caries and periodontal disease in HIV+ patients, *Brazilian Journal of Infectious Diseases* 13 (2009): 13–17.
7. P. Weidlich and coauthors, Association between periodontal diseases and systemic diseases, *Brazilian Oral Research* 22 (2008): 32–43.
8. P. B. Lockhart and coauthors, Periodontal disease and atherosclerotic vascular disease: Does the evidence support an independent association? *Circulation* 125 (2012): 2520–2544.
9. Weidlich and coauthors, 2008.

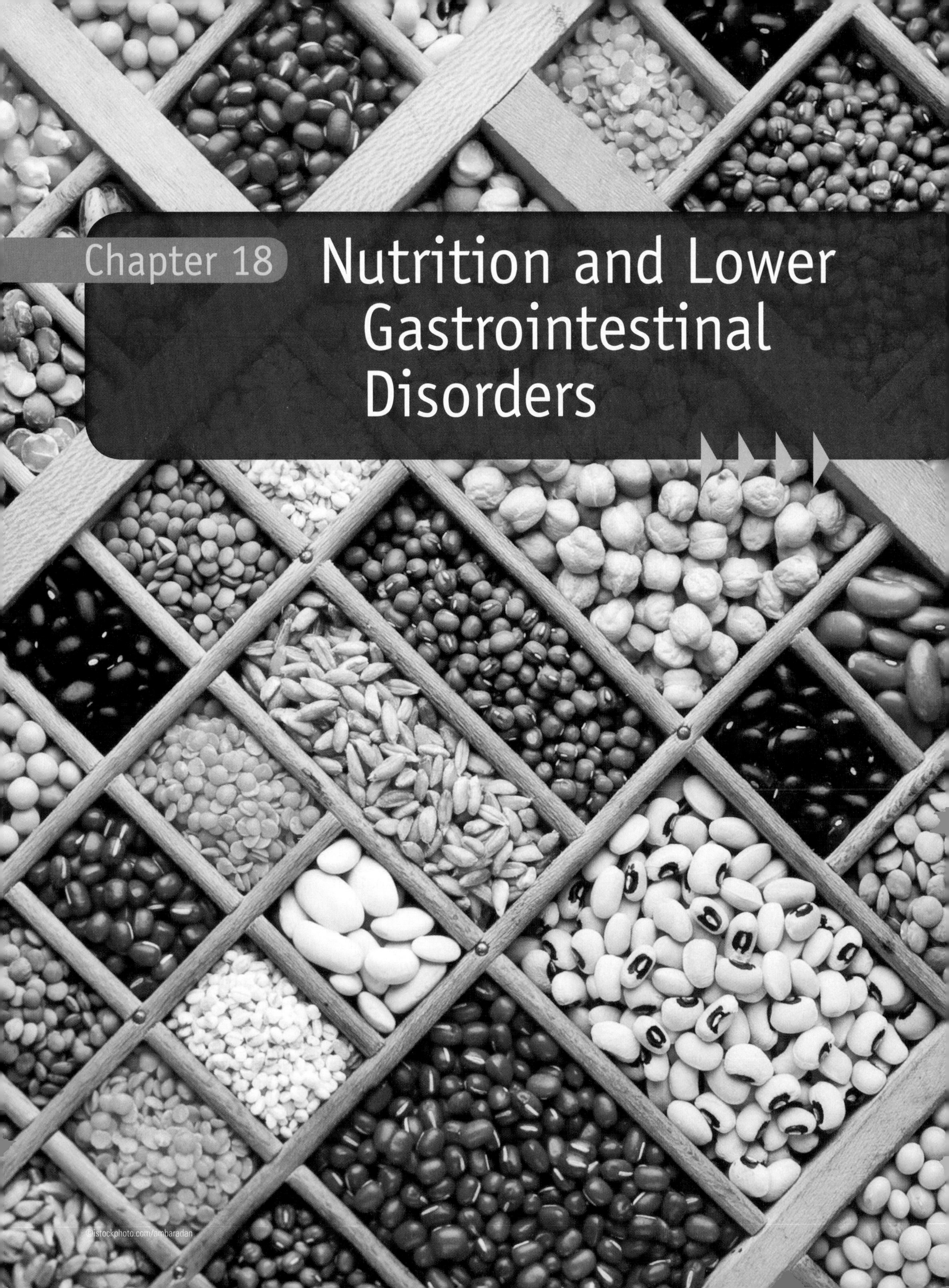

Chapter 18

Nutrition and Lower Gastrointestinal Disorders

BECAUSE MOST DIGESTIVE AND ABSORPTIVE PROCESSES OCCUR IN THE small intestine, disorders of the lower gastrointestinal (GI) tract can interfere substantially with a person's diet and lifestyle. Some of the diets required for these conditions are complicated and difficult to follow, and the foods that are tolerated can vary considerably. In visits with patients, health professionals should ensure that patients understand the nutrition prescription and help to pinpoint difficult foods. They can also suggest ways to make restrictive diets more acceptable.

Common Intestinal Problems

Nearly all people experience occasional intestinal problems, which usually clear up without medical treatment. Intestinal discomfort can sometimes drive a person to seek medical attention, however, and the symptoms may be evidence of a serious intestinal disorder or other illness. The most common intestinal problems and their causes and treatments are discussed below.

CONSTIPATION

A diagnosis of constipation is based, in part, on a defecation frequency of fewer than three bowel movements per week. Other symptoms may include the passage of hard stools and excessive straining during defecation. In some cases, a person's perception of constipation may be due to a mistaken notion of what constitutes "normal" bowel habits, so the person's expectations about bowel function may need to be addressed.

Constipation is much more prevalent among women than men and is a common complaint during pregnancy. The incidence of constipation increases with aging, although older adults (65 years and older) tend to find straining during defecation to be a greater problem than infrequent defecation.[1]

NURSING DIAGNOSIS

The nursing diagnosis *perceived constipation* may apply to people with adequate bowel movements who self-diagnose constipation.

NURSING DIAGNOSIS

The nursing diagnoses *constipation* and *risk for constipation* may apply to persons with medical problems that predispose them to constipation.

Causes of Constipation The risk of constipation is increased in individuals with a low-fiber diet, low food intake, inadequate fluid intake, or low level of physical activity. All of these factors can extend transit time, leading to increased water reabsorption within the colon and dry, hard stools that are difficult to pass. Medical conditions often associated with constipation include diabetes mellitus and hypothyroidism. Neurological conditions such as Parkinson's disease, spinal cord injuries, and multiple sclerosis may cause motor problems that lead to constipation. During pregnancy, women can experience constipation because the enlarged uterus presses against the rectum and colon. Constipation is also a common side effect of several classes of medications and some dietary supplements, including opiate-containing analgesics, tricyclic antidepressants, anticonvulsants, calcium channel blockers, aluminum-containing antacids, and iron and calcium supplements.

■ The fiber DRI for women and men aged 19 to 50 years are 25 and 38 grams, respectively. Recommendations for increasing fiber intake are described on pp. 80–81, and Appendix A includes the fiber values of most common foods.

Treatment of Constipation The primary treatment for constipation is a gradual increase in fiber intake to at least 25 grams per day.[2] (■) High-fiber diets increase stool weight and fecal water content and promote a more rapid transit of materials through the colon. Foods that increase stool weight the most are wheat bran, fruits, and vegetables.[3] Bran intake can be increased by adding bran cereals and whole-wheat bread to the diet or by mixing bran powder with beverages or foods. The transition to a high-fiber diet may be difficult for some people because it can increase intestinal gas, so high-fiber foods should be added gradually, as tolerated. Fiber supplements such as methylcellulose (Citrucel), psyllium (Metamucil, Fiberall), and polycarbophil (a synthetic fiber) are also effective (see Table 18-1); these supplements can be mixed with beverages and taken several times daily. Unlike other fibers, methylcellulose and polycarbophil do not increase intestinal gas.

High-fiber foods promote regular bowel movements.

TABLE 18-1 Laxatives and Bulk-Forming Agents

Laxative Type	Active Ingredients	Product Examples	Method of Action	Cautions
Fiber (bulk-forming agents)	Malt soup extract, methylcellulose, polycarbophil, psyllium	Citrucel, Fiberall, Fiber-Lax, Metamucil	Fiber supplements increase stool weight and aid in the formation of soft, bulky stools. Similar effects are achieved by adding bran to the diet. For mild constipation. Safe for long-term use.	Some fiber supplements may increase flatulence. Psyllium may cause an allergic reaction.
Osmotic laxatives: nonabsorbable salts	Magnesium citrate, magnesium hydroxide	Epsom salts, milk of magnesia, Citromag	Unabsorbed salts attract and retain water in the large intestine and stimulate contractions.	May cause bloating and watery stools or diarrhea. Should be used with caution. Avoid using in renal patients and children.
Osmotic laxatives: nonabsorbable sugars	Lactulose, polyethylene glycol, sorbitol	Cephulac, Chronulac, CoLyte	Unabsorbed sugars attract water to the large intestine and promote softer stools. Must be used for several days to take effect. Safe for long-term use.	May cause flatulence and cramps. Can lose effectiveness over time.
Stimulant or irritant laxatives	Aloe, bisacodyl, cascara, castor oil, senna	Correctol, Dulcolax, Ex-Lax	Act as local irritants to colonic tissue; stimulate peristalsis and mucosal secretions. For moderate-to-severe constipation. Long-term use is discouraged.	Usually given only after milder treatments fail. May alter fluid and electrolyte balances. May lead to laxative dependency.
Stool surfactant agents (stool softeners)	Docusate sodium	Colace, Surfak	Detergent action promotes the mixing of water with stools. Prevents formation of dry, hard stools.	Do not increase stool weight. Limited effectiveness.

Several other measures may also help constipation. Consuming adequate fluid (usually 1.5 to 2 liters daily) may help to increase stool frequency in people who are already consuming a high-fiber diet.[4] An appropriate fluid intake prevents excessive reabsorption of water from the colon, resulting in wetter stools. Adding prunes or prune juice to the diet is often recommended because prunes contain compounds that have a mild laxative effect. Inactive individuals are generally encouraged to increase physical activity, although clinical studies have not confirmed that increasing exercise improves constipation symptoms or stool transit times.[5]

Laxatives Laxatives work by increasing stool weight, increasing the water content of the stool, or stimulating peristaltic contractions. Table 18-1 includes examples of common laxatives and describes their modes of action. Enemas and suppositories (chemicals introduced into the rectum) are also used to promote defecation; they work by distending and stimulating the rectum or by lubricating the stool.

Medical Interventions For patients with severe constipation who do not respond to dietary or laxative treatments, physicians may prescribe medications (called *prokinetic agents*) that stimulate colonic contractions. Physical therapy and biofeedback techniques are sometimes successful in training patients to relax their pelvic muscles more effectively. Surgical interventions are a last resort and include colonic resections and colostomy operations, which are discussed later in the chapter.

TABLE 18-2 Foods That May Increase Intestinal Gas

Foods
Apples
Artichokes
Asparagus
Beer
Broccoli
Brussels sprouts
Cabbage
Carbonated beverages
Carrots
Corn
Dried beans and peas
Fructose-sweetened products
Fruit juices
Green beans
Leeks
Milk products (if lactose intolerant)
Onions
Peanuts
Pears
Turnips
Wheat

INTESTINAL GAS

As mentioned in the previous section, excessive intestinal gas (**flatulence**) may be an unpleasant side effect of consuming a high-fiber diet. Because dietary fibers are not digested, they pass into the colon and are fermented by bacteria, which produce gas as a by-product. (■) Other incompletely digested or poorly absorbed carbohydrates have similar effects; these include fructose, sugar alcohols (sorbitol, xylitol, and mannitol), the indigestible carbohydrates in beans (raffinose and stachyose), and some forms of resistant starch, found in grain products and potatoes. Table 18-2 lists examples of foods commonly associated with excessive gas production, although individual responses vary. Malabsorption disorders (discussed later in this chapter) can cause considerable flatulence because the undigested nutrients can be metabolized by colonic bacteria. Swallowed air that is not expelled by belching may travel to the intestines and be a source of intestinal gas as well.

Many people blame abdominal bloating and pain on excessive gas, but these symptoms do not correlate well with an increase in intestinal gas.[6] In fact, most people who self-diagnose a flatulence problem have no more intestinal gas than others. Some individuals who experience frequent symptoms of abdominal bloating and pain are later diagnosed with irritable bowel syndrome (see pp. 533–534) or dyspepsia (discussed in Chapter 17).

DIARRHEA

Diarrhea is characterized by the passage of frequent, watery stools. In most cases, it lasts for only a day or two and subsides without complication. Severe or persistent diarrhea, however, can cause dehydration and electrolyte imbalances. If chronic, it may lead to weight loss and malnutrition. Diarrhea may be accompanied by other symptoms, such as fever, abdominal cramps, dyspepsia, or bleeding, which help in diagnosing the cause.

Causes of Diarrhea

Diarrhea is a complication of various GI disorders and may also be induced by infections, medications, or dietary substances. It results from inadequate fluid reabsorption in the intestines, sometimes in conjunction with an increase in intestinal secretions.[7] In *osmotic diarrhea*, unabsorbed nutrients or other substances attract water to the colon and increase fecal water content; the usual causes include high intakes of poorly absorbed sugars (such as sorbitol, mannitol, or fructose), lactase deficiency (which causes lactose malabsorption), and ingestion of laxatives that contain magnesium or phosphates. In *secretory diarrhea*, the fluid secreted by the intestines exceeds the amount that can be reabsorbed by intestinal cells. Secretory diarrhea is often due to foodborne illness but can also be caused by intestinal inflammation and irritating chemical substances (such as medications or unabsorbed bile acids). *Motility disorders* may also result in diarrhea because they accelerate the transit of colonic residue, reducing the contact time available for fluid reabsorption.

Acute cases of diarrhea start abruptly and may persist for several weeks; they are frequently caused by viral, bacterial, parasitic, or protozoal infections or occur as a side effect of medications. Chronic diarrhea, which persists for about four weeks or longer,[8] can result from malabsorption disorders, inflammatory diseases, motility disorders, infectious diseases, radiation treatment, and many other conditions. As mentioned in earlier chapters, diarrhea is a frequent complication of tube feedings and it may also occur when oral or enteral feedings are resumed after a period of bowel rest (see Chapter 15).

Medical Treatment of Diarrhea

Correcting the underlying medical problem is the first step in treating diarrhea. For example, antibiotics can be prescribed for treating intestinal infections. If a medication is the cause of diarrhea, a different drug may be prescribed. If certain foods are responsible, they can be omitted from the diet. Bulk-forming agents such as psyllium (Metamucil) or methylcellulose (Citrucel) can help to reduce the liquidity of the stool. If chronic diarrhea does not respond to treatment,

■ Reminder: The *soluble fibers* in foods are more readily fermented in the small intestine than the *insoluble fibers*.

flatulence: the condition of having excessive intestinal gas, which causes abdominal discomfort.

antidiarrheal drugs may be prescribed to slow GI motility or reduce intestinal secretions. Probiotics (■) may be beneficial for certain types of diarrhea (especially infectious diarrhea), but standard treatment protocols have not been developed.[9] People with severe, **intractable** diarrhea sometimes require total parenteral nutrition.

Oral Rehydration Therapy Severe diarrhea requires the replacement of lost fluid and electrolytes. Oral rehydration solutions can be purchased or easily mixed using water, salts, and glucose or sucrose (see the recipe in the margin). (■) The addition of carbohydrate to the solution facilitates sodium and water absorption. Commercial sports drinks are not ideal fluids for rehydration because their sodium content is too low, but they can be used if accompanied by salty snack foods.[10] When diarrhea results in extreme dehydration, intravenous solutions are used to quickly replenish fluid and electrolytes.

Nutrition Therapy for Diarrhea Because diarrhea can develop for numerous reasons, the nutrition prescription depends on the medical diagnosis and severity of the condition. The dietary treatment initially recommended is a low-fiber, low-fat, lactose-free diet.[11] The diet limits foods that contribute to stool volume, such as those with significant amounts of fiber, resistant starch, fructose, sugar alcohols, and lactose (in lactose-intolerant individuals). Fructose and sugar alcohols, which are poorly absorbed, retain fluids in the colon and contribute to osmotic diarrhea. Similarly, milk products may worsen osmotic diarrhea in persons who are lactose intolerant. Avoidance of fatty foods is recommended because they can sometimes aggravate diarrhea. Gas-producing foods (those with poorly digested or malabsorbed carbohydrates) can increase intestinal distention and cause additional discomfort. Patients should avoid caffeinated coffee and tea because caffeine stimulates GI motility and can thereby reduce water reabsorption. Fluid intakes must usually be increased to replace fluid losses. In the treatment of formula-fed infants, apple pectin or banana flakes are sometimes added to formulas to help thicken stool consistency. Table 18-3 lists examples of foods that may worsen diarrhea, although individual tolerances vary.

TABLE 18-3 Foods That May Worsen Diarrhea[a]

Foods to Avoid	Rationale	Selected Examples
High-fiber foods	They increase colonic residue.	Breads and cereals with more than 2 g fiber per serving, fruits and vegetables with peels or skins
Foods with indigestible carbohydrates	They contribute to osmotic diarrhea.	Artichokes, asparagus, brussels sprouts, cabbage, dried beans and peas, fruit, garlic, green beans, leeks, onions, wheat, zucchini
Foods that contain fructose or sugar alcohols	They contribute to osmotic diarrhea.	Dried fruits, fresh fruits (except bananas), fruit juices, fructose-sweetened soft drinks, sugar-free gums and candies
Milk products, if person is lactose intolerant	They contribute to osmotic diarrhea.	Milk and milk products
Gas-producing foods	They increase abdominal discomfort.	Foods with poorly digested or absorbed carbohydrates (including foods listed in the three rows directly above)
Caffeine-containing beverages	They increase intestinal motility.	Coffee, tea, colas, energy drinks

[a]Individual tolerances vary; the foods to avoid are best determined by trial and error.

■ *Probiotics* are live bacteria provided in foods and dietary supplements for the purpose of preventing or treating disease. Nutrition in Practice 18 describes the potential health benefits of probiotics.

■ An oral rehydration solution can be mixed from the following ingredients:

- ½ tsp sodium chloride (table salt)
- ⅓ tsp potassium chloride (salt substitute)
- ¾ tsp sodium bicarbonate (baking soda)
- 1⅓ tbs sugar
- 1 qt water

intractable: not easily managed or controlled.

Review Notes

- The most common intestinal problems include constipation, intestinal gas, and diarrhea.
- Constipation may be caused by a variety of medical problems but generally correlates with low-fiber diets, low food or fluid intakes, and physical inactivity; a high-fiber diet or fiber supplements may be beneficial.
- Intestinal gas is largely produced by colonic bacteria and is usually caused by nutrient malabsorption.
- Diarrhea can result from malabsorption, intestinal infections, motility disorders, or medications; the patient may require oral rehydration therapy to replace fluid and electrolyte losses. A low-fiber, low-fat, lactose-free diet may be initially recommended.

Malabsorption

resection: the surgical removal of part of an organ or body structure.

steatorrhea (stee-AT-or-REE-ah): excessive fat in the stools due to fat malabsorption; characterized by stools that are loose, frothy, and foul smelling due to a high fat content.

steat = fat
rheo = flow

To digest and absorb nutrients, we depend on normal digestive secretions and healthy intestinal mucosa. Malabsorption can therefore be caused by pancreatic disorders that lead to enzyme or bicarbonate deficiencies, disorders that lead to bile deficiency, and inflammatory diseases or medical treatments that damage intestinal tissue. In some cases, the treatment of an intestinal disease requires surgical removal of a section (**resection**) of the small intestine, leaving minimal absorptive capacity in the portion that remains. Table 18-4 lists examples of diseases and treatments that are frequently associated with malabsorption.

Malabsorption rarely involves a single nutrient. When malabsorption is caused by pancreatic enzyme deficiencies, all macronutrients—protein, carbohydrate, and fat—may be affected. When fat is malabsorbed, fat-soluble nutrients and minerals are usually malabsorbed as well. Malabsorption disorders and their treatments can tax nutrition status further by causing complications that alter food intake, raise nutrient needs, and incur additional nutrient losses.

TABLE 18-4 Potential Causes of Malabsorption

Genetic disorders • Enzyme deficiencies
Intestinal disorders • AIDS-related enteropathy • Bacterial overgrowth • Celiac disease • Crohn's disease • Radiation enteritis
Intestinal infections • Giardiasis • Nematode (roundworm) infections
Liver disease (bile insufficiency)
Pancreatic disorders • Chronic pancreatitis • Cystic fibrosis
Surgeries • Gastric or intestinal bypass surgery • Intestinal resection (short bowel syndrome)

FAT MALABSORPTION

Fat is the nutrient most frequently malabsorbed because both digestive enzymes and bile must be present for its digestion. Thus, fat malabsorption often develops when an illness reduces either pancreatic or bile secretions. For example, both pancreatitis and cystic fibrosis can reduce the secretion of pancreatic lipase, whereas severe liver disease can cause bile insufficiency. Motility disorders that accelerate gastric emptying or intestinal transit can cause fat malabsorption because they prevent the normal mixing of dietary fat with lipase and bile. Fat malabsorption can also be caused by conditions or treatments that damage the intestinal mucosa, such as inflammatory bowel diseases, AIDS, and radiation treatments for cancer.

Fat malabsorption is often evidenced by **steatorrhea**, the presence of excessive fat in the stools. Steatorrhea can be evaluated by placing the patient on a high-fat diet (80 to 100 grams per day), performing a 48- to 72-hour stool collection, and measuring the stool's fat content. Healthy individuals generally excrete less than 7 grams of fat per day under these conditions.[12]

Consequences of Fat Malabsorption Fat malabsorption is associated with losses of food energy, essential fatty acids, fat-soluble vitamins, and some minerals (see Figure 18-1). Weight loss may result if the individual does not consume alternative sources of energy. Deficiencies of fat-soluble vitamins and essential fatty acids are common in chronic conditions. Malabsorption of some minerals, including calcium, magnesium, and zinc, often occurs because the minerals form **soaps** with the unabsorbed fatty acids. Calcium deficiency may lead to bone loss, which is further aggravated by the vitamin D deficiency that may be present due to fat malabsorption.

Another consequence of fat malabsorption is an increased risk of kidney stones, which are most often composed of calcium oxalate. (■) The oxalates in foods ordinarily bind to calcium in the small intestine and are excreted in the stool. If calcium instead binds to fatty acids or bile acids, the oxalates are free to be absorbed into the blood and are ultimately excreted in the urine. The risk of developing oxalate stones increases when urinary oxalate levels are high. Kidney stones are discussed further in Chapter 22.

NURSING DIAGNOSIS

The nursing diagnosis *imbalanced nutrition: less than body requirements* is common in people with fat malabsorption.

■ Reminder: *Oxalates* are plant compounds found in green leafy vegetables such as beet greens and spinach. The oxalates can bind to minerals in the GI tract and form complexes that the body cannot absorb.

Nutrition Therapy for Fat Malabsorption If steatorrhea does not improve, a fat-restricted diet may be recommended (see Table 18-5). The diet may help to relieve intestinal symptoms that are aggravated by fat intake (such as diarrhea and flatulence) and reduce vitamin and mineral losses. Because fat is a primary energy source, it should not be restricted more than necessary. Medium-chain triglycerides (MCT), which do not require lipase or bile for digestion and absorption, can be used as an alternative source of dietary fat, although MCT oil does not provide essential fatty acids. Figure 18-2 presents an example of a menu for a fat-restricted diet, and the "How to" on page 521 offers suggestions for following the fat-restricted diet and for using MCT oil.

soaps: chemical compounds formed from fatty acids and positively charged minerals.

FIGURE 18-1 The Consequences of Fat Malabsorption

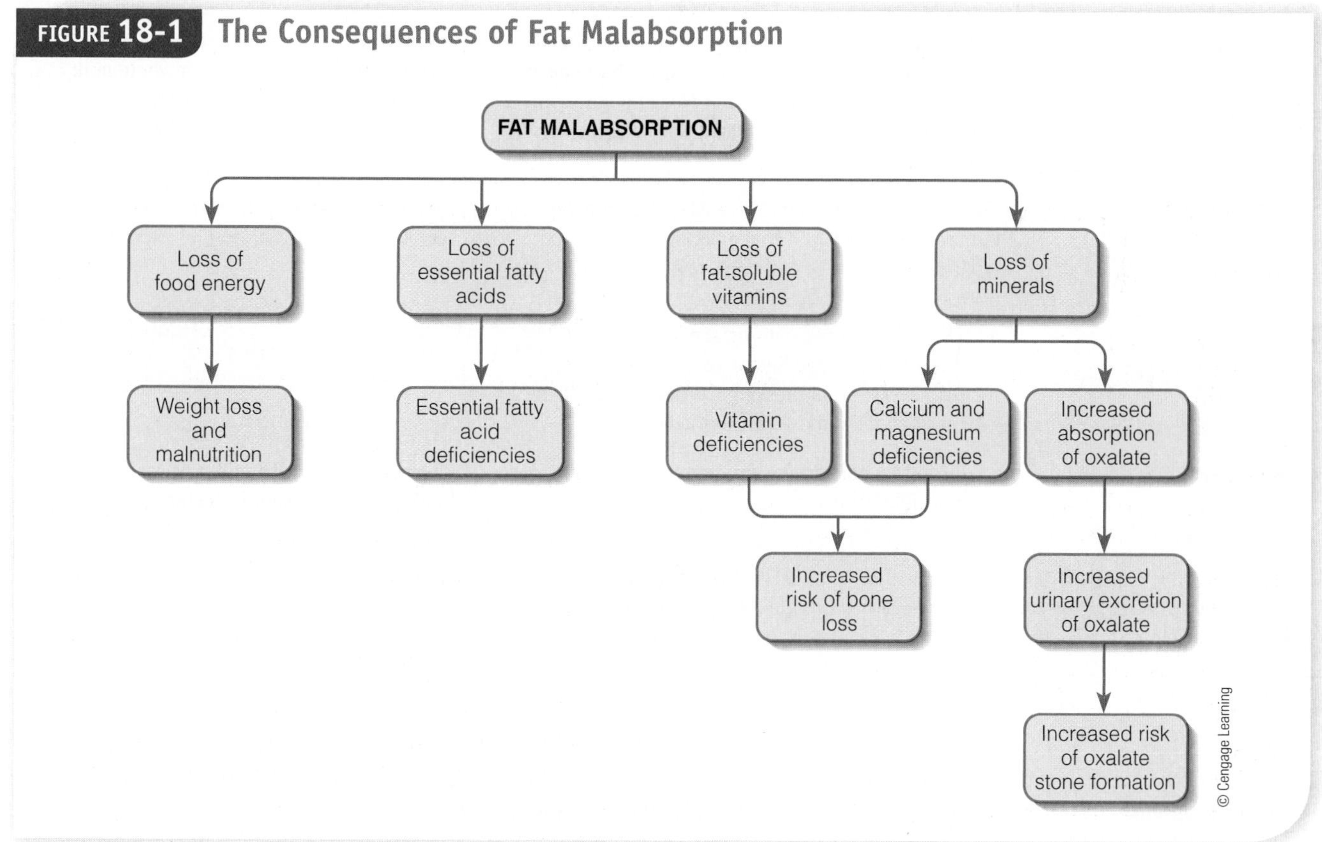

TABLE 18-5 Fat-Restricted Diet

A fat-restricted diet includes mostly low-fat and fat-free foods. For a fat intake of 50 grams per day, limit meats and meat substitutes to 6 ounces per day, and limit fats and oils to 8 teaspoons per day.[a] Foods from other food groups should provide less than 1 gram of fat per serving.

Food Category	Foods Recommended	Foods to Avoid
Meat and meat alternates	Lean meats, fish, and skinless poultry prepared by broiling, roasting, grilling, or boiling; low-fat luncheon meats such as sliced turkey breast; meat alternates such as dried beans or peas; low-fat egg substitutes; egg whites	Meat with visible fat, ground beef (unless extra lean), sausage, bacon, frankfurters, spareribs, duck, tuna packed in oil, whole eggs and egg yolks
Milk and milk products	Fat-free milk, fat-free yogurt, fat-free sour cream substitutes, fat-free half-and-half and cream substitutes, fat-free cheeses. Low-fat milk products can be used in moderation.	Milk products that are not fat free or low fat
Breads, cereals, rice, and pasta	Whole-grain and enriched breads, cooked cereals and most cold breakfast cereals, plain tortillas, bagels, English muffins, fat-free muffins, saltine crackers, graham crackers, pretzels, plain rice, plain noodles and pasta	Biscuits, pancakes, waffles, granola, snack crackers made with fat, cornbread, doughnuts, corn and potato chips, fried rice, buttered or butter-flavored popcorn
Vegetables	All vegetables prepared without added fat	Buttered, creamed, breaded, or fried vegetables; vegetables prepared au gratin style; french-fried potatoes; olives
Fruits	All fruits except avocado	Avocado; fruit dishes prepared with fats, nuts, or coconut
Desserts	Sherbet; fruit ices; flavored gelatin; angel food cake; meringue; fat-free puddings; fat-free bakery products; fat-free ice cream or frozen yogurt; fat-free candies such as marshmallows, jelly beans, and hard candy	Cakes, cookies, pies, and pastries made with fat; puddings made with whole milk or eggs; ice cream; candies made with fat such as caramel and chocolates
Fats and oils	Unsaturated vegetable oils, soft or liquid margarines and spreads, limited amounts of butter or stick margarine (1 teaspoon provides about 3½ to 4½ grams of fat) Each of these foods can replace 1 teaspoon of fat in the amounts specified: 1 tbs salad dressing, 2 tbs low-fat salad dressing, ½ tbs peanut butter, 1 tbs chopped nuts, 2 tbs mashed avocado	Dietary fat that exceeds the amount specified in the nutrition prescription
Beverages	Fruit juices, soft drinks, fat-free milk, coffee, tea, coffee substitutes	Beverages made with milk (unless fat free) or added cream, chocolate milk, eggnog, milk shakes

[a]To achieve a fat intake less than 50 grams, additional reductions may be necessary. For example, for a fat intake of 25 grams per day, limit meats and meat substitutes to 4 ounces per day, and limit fats and oils to 2 teaspoons per day.

BACTERIAL OVERGROWTH

bacterial overgrowth: excessive bacterial colonization of the stomach and small intestine; may be due to low gastric acidity, altered GI motility, mucosal damage, or contamination.

Ordinarily, the GI tract is protected from **bacterial overgrowth** by gastric acid, which destroys bacteria; peristalsis, which flushes bacteria through the small intestine before they multiply; and immunoglobulins secreted into the GI lumen.[13] When bacterial overgrowth does occur, it can lead to fat malabsorption because the bacteria dismantle the bile acids needed for fat emulsification. Deficiencies of the fat-soluble vitamins A, D, and E may eventually develop. The bacteria also produce enzymes and toxins that disturb the intestinal mucosa, destroying some mucosal enzymes (especially lactase)

FIGURE 18-2 Sample Menu—Fat-Restricted Diet

The menu shown here contains approximately 50 grams of fat.

SAMPLE MENU

Breakfast

6 oz orange juice

1 c oatmeal with raisins, nonfat milk, and sugar

1 slice whole-wheat toast with 1 tsp margarine

Coffee with fat-free half and half

Lunch

Turkey breast sandwich (includes 2 slices whole-wheat bread, 2 oz lean turkey breast, 2 tomato slices, lettuce leaf, and 2 tsp mayonnaise)

2 c salad greens with 1 tbs salad dressing

Fruit cup (includes 1 c peaches and ½ c berries) with ½ c orange sherbet

Snack

6 oz nonfat fruit yogurt

6 saltine crackers with 1 tbs peanut butter and ½ tbs honey

Dinner

4 oz cod with sliced lemon and dill

1 slice French bread with 1 tsp butter

1 c steamed rice with herbs and walnut oil (includes ½ tsp oil)

1 c steamed broccoli and carrots with ½ tsp margarine

1 piece angel food cake with fat-free whipped cream

and possibly reducing the absorptive surface area. Some types of bacteria metabolize vitamin B_{12}, reducing its absorption and increasing the risk of deficiency. Although symptoms of bacterial overgrowth are often minor and nonspecific, severe cases may lead to chronic diarrhea, steatorrhea, flatulence, bloating, and weight loss.

Causes of Bacterial Overgrowth

Conditions that impair intestinal motility and allow material to stagnate can greatly increase susceptibility to bacterial overgrowth. For example, in some types of gastric surgery, a portion of the small intestine is bypassed, preventing the flow of material in the bypassed region and allowing bacteria to flourish (see the "blind loop" shown in Figure 17-2 on p. 499). Intestinal motility can also be reduced by strictures, obstructions, and diverticula (protrusions) in the small intestine, as well as by some chronic illnesses, including diabetes mellitus and scleroderma.[14]

HOW TO Follow a Fat-Restricted Diet

For some individuals, fat-restricted diets may be difficult to follow. Fats add flavors, aromas, and textures to foods—characteristics that make foods more enjoyable. Unlike some dietary changes that can be introduced gradually, fat restriction is often implemented immediately, allowing little time for adaptation. These suggestions may help:

- Fat is better tolerated if provided in small portions. Divide the day's allotment into several servings that can be consumed throughout the day.
- Use variety to enhance enjoyment of meals: vary flavors, textures, colors, and seasonings.
- Look for fat-free items when grocery shopping. Incorporate fat-free ingredients when preparing favorite recipes.
- Try fat-free and low-fat condiments to improve the diet's palatability. Experiment with herbs and spices. Instead of butter, use fruit butters on toast. Use butter-flavored granules on vegetables. Replace mayonnaise in sandwiches with spicy mustard. Replace salad dressings with flavored vinegars.
- Avoid products that contain the fat substitute olestra, which may aggravate GI symptoms.

If patients are interested in using medium-chain triglyceride (MCT) oil:

- Explain that MCT products are expensive but that the cost is sometimes covered by medical insurance.
- Advise patients to add MCT oil to the diet gradually. Diarrhea and abdominal cramps may result if too much is used at once. Tolerance to MCT oil may improve in time.
- Advise patients that MCT oil may have an unpleasant taste when used alone. Suggest using MCT oil in recipes as a substitute for regular oil. MCT oil can replace oil in salad dressings, be incorporated into sauces, and be used in cooking or baking. It can also be added to fat-free milk products to make milk shakes.
- Explain that MCT oil should not be used to fry foods because it decomposes at lower temperatures than most cooking oils.

Reduced secretions of gastric acid can also lead to bacterial overgrowth. Possible causes include atrophic gastritis, use of acid-suppressing medications, and some gastrectomy procedures.

Treatment for Bacterial Overgrowth Treatment may include antibiotics to suppress bacterial growth and surgical correction of the anatomical defects that contribute to a motility disorder. Medications may be given to stimulate peristalsis, and acid-suppressing medications should be discontinued. A lactose-restricted diet may reduce flatulence and diarrhea in some individuals.[15] Dietary supplements can help to correct nutrient deficiencies, especially deficiencies of the fat-soluble vitamins, calcium (which combines with malabsorbed fatty acids), and vitamin B_{12}.[16]

LACTOSE INTOLERANCE

Approximately 75 percent of people worldwide have some degree of lactose intolerance, which is caused by the loss or reduction of lactase, the intestinal enzyme that digests the lactose in milk products.[17] Lactose intolerance is especially prevalent among individuals of certain ethnic groups, including Asians, African Americans, Native Americans, Ashkenazi Jews, and Latinos. It can also result from GI disorders that damage the small intestinal mucosa. The primary symptoms of lactose intolerance are diarrhea and increased intestinal gas.

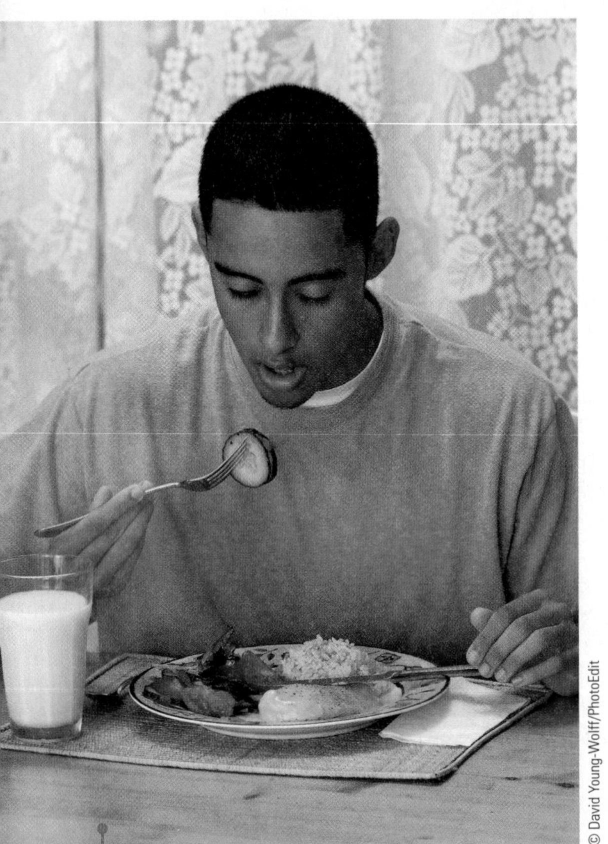

© David Young-Wolff/PhotoEdit

Most people with lactose intolerance can drink milk, especially if they drink it along with other foods and limit the amount they consume at any one time.

Lactose intolerance is rarely serious and is easily managed by simple dietary adjustments. Although people with the condition are sometimes reluctant to consume milk products, clinical studies have found that individuals with lactose intolerance can tolerate up to 2 cups of milk daily without significant symptoms.[18] In addition, the regular consumption of milk products increases the amount of lactose metabolized by intestinal bacteria, which improves lactose tolerance.[19] People who avoid milk for fear of intestinal discomfort can be urged to gradually increase consumption of lactose-containing foods. They may more readily tolerate milk when intake is divided throughout the day and the milk is taken with food. Some people tolerate chocolate milk better than plain milk. Most aged cheeses are well tolerated because they contain little lactose. Yogurts that contain live bacterial cultures are usually acceptable because the bacteria contain lactase, which may aid in lactose digestion. Other options are to add a lactase preparation to milk or to take enzyme tablets before consuming lactose-containing foods. Lactose-free milk is also commercially available.

People who develop lactose intolerance as a result of intestinal illness are often advised to temporarily restrict milk and milk products. Foods that contain lactose can be reintroduced in small amounts once the condition improves. Individuals who restrict milk products should be encouraged to consume alternative food sources of calcium and vitamin D (see Chapters 8 and 9).

Review Notes

- Malabsorption can be caused by reduced digestive secretions, motility disorders, or damaged intestinal mucosa.
- Fat malabsorption, usually indicated by steatorrhea, can cause losses of food energy and deficiencies of essential fatty acids, fat-soluble vitamins, and some minerals. A fat-restricted diet may be prescribed to improve symptoms.
- Bacterial overgrowth can result from conditions that reduce gastric acidity or intestinal motility; it typically causes malabsorption of fat and some essential nutrients.
- Lactose intolerance, caused by lactase deficiency, can be managed by adjusting the amount and timing of lactose consumption.

Conditions Affecting the Pancreas

As mentioned previously, pancreatic disorders can lead to maldigestion and malabsorption due to the impaired secretion of digestive enzymes. (■) This section describes several diseases that disrupt pancreatic function and cause widespread malabsorption.

PANCREATITIS

Pancreatitis is an inflammatory disease of the pancreas. Although mild cases may subside in a few days, other cases can persist for weeks or months. Chronic pancreatitis can lead to irreversible damage to pancreatic tissue and permanent loss of function.

Acute Pancreatitis

In acute pancreatitis, the digestive enyzmes within pancreatic cells become prematurely activated, causing destruction of pancreatic tissue and subsequent inflammation. About 70 to 80 percent of acute cases are caused by gallstones or alcohol abuse; less frequent causes include elevated blood triglyceride levels (greater than 1000 milligrams per deciliter) and various drugs or toxins.[20] Common symptoms include severe abdominal pain, nausea and vomiting, and abdominal distention. In most patients, the condition resolves within a week with no complications. More severe cases may lead to chronic pancreatitis, infection, the systemic inflammatory response syndrome (see p. 471), or multiple organ failure.

NURSING DIAGNOSIS

Nursing diagnoses for people with acute pancreatitis may include *acute pain, nausea, diarrhea,* and *risk for deficient fluid volume.*

Nutrition Therapy for Acute Pancreatitis

The initial treatment for acute pancreatitis is supportive and includes pain control and intravenous hydration. Oral fluids and food are withheld until the patient is pain free and experiences no nausea or vomiting. Afterwards, patients may consume a low-fat liquid diet or small low-fat meals, as tolerated (fat stimulates the pancreas more than other nutrients).[21] In severe pancreatitis, tube feedings may be necessary; either standard formulas or elemental formulas (■) may be used, depending on patient tolerance. Protein and energy needs are high in severe cases due to the catabolic and hypermetabolic effects of inflammation. Patients with acute pancreatitis require nutrient supplements until food intake can meet nutritional needs.

Chronic Pancreatitis

Chronic pancreatitis is characterized by progressive, permanent damage to pancreatic tissue, resulting in the impaired secretion of digestive enzymes and bicarbonate. The condition is usually a result of repeated episodes of acute pancreatitis, which may sometimes be mild enough to go unnoticed. About 70 to 80 percent of chronic pancreatitis cases are caused by excessive alcohol consumption.[22] Cigarette smoking is often a contributing risk factor.

NURSING DIAGNOSIS

The nursing diagnosis *imbalanced nutrition: less than body requirements* may be appropriate for some individuals with chronic pancreatitis.

Most patients with chronic pancreatitis experience persistent abdominal pain, which may worsen with eating and be accompanied by nausea and vomiting. Analgesics or opiate drugs are often needed for pain control. Fat maldigestion and steatorrhea occur in advanced cases (after 5 to 10 years of illness or longer); later, protein and carbohydrate maldigestion may also develop. Food avoidance (due to pain associated with eating) and malabsorption may lead to weight loss and malnutrition. Long-term illness is associated with reduced secretion of insulin and glucagon, and diabetes eventually develops in 40 to 80 percent of patients.[23]

Nutrition Therapy for Chronic Pancreatitis

The objectives of nutrition therapy are to correct malnutrition, reduce malabsorption, and prevent symptom recurrence. Dietary supplements are used to correct nutrient deficiencies, which may be due to malabsorption or to the alcohol abuse that caused the disease. To improve food tolerance, patients should consume small, low-fat meals. They should also avoid alcohol completely and quit smoking cigarettes, as these substances can exacerbate illness and interfere with healing.[24]

■ Reminder: The digestive secretions of the pancreas include bicarbonate and digestive enzymes. The bicarbonate neutralizes the acidic gastric contents that enter the duodenum, and the digestive enzymes break down protein, carbohydrate, and fat.

■ Reminder: An *elemental formula* contains hydrolyzed nutrients that require minimal digestion and are easily absorbed.

enteric coated: refers to medications or enzyme preparations that are coated to withstand stomach acidity and dissolve only at the higher pH of the small intestine.

cystic fibrosis: an inherited disease characterized by the presence of abnormally viscous exocrine secretions; often leads to respiratory illness and pancreatic insufficiency.

NURSING DIAGNOSIS

Nursing diagnoses that may apply to people with cystic fibrosis include *ineffective airway clearance, risk for infection, risk for deficient fluid volume, risk for delayed development, risk for disproportionate growth,* and *imbalanced nutrition: less than body requirements.*

exocrine: pertains to external secretions, such as those of the mucous membranes or the skin. Opposite of *endocrine,* which pertains to hormonal secretions into the blood.

exo = outside
krinein = to secrete

Steatorrhea is usually treated with pancreatic enzyme replacement. Pancreatic enzymes are often **enteric coated** to resist the acidity of the stomach and do not dissolve until the pH is above 5.5. If nonenteric-coated preparations are used, acid-suppressing drugs are also required. Fecal fat concentrations must be monitored to determine if the enzyme treatment has been effective. In some cases, a fat-controlled diet may help to reduce symptoms, and MCT oil can be used as an alternative source of fat kcalories. The Diet-Drug Interactions feature lists nutrition-related side effects of the medications discussed in this chapter.

CYSTIC FIBROSIS

Cystic fibrosis is the most common life-threatening genetic disorder among Caucasians, with an incidence of approximately 1 in 3300 white births.[25] In the United States, about 30,000 people are affected.[26] Cystic fibrosis is characterized by a mutation in the protein that regulates chloride transport across epithelial cell membranes. The abnormality alters the ion concentration and/or viscosity of **exocrine** secretions, causing a broad range of serious complications. Until a few decades ago, few infants born with cystic fibrosis survived to adulthood. Now, with early detection and advances in medical treatment, the median life span has reached 37 years of age, with many patients surviving into their 50s.[27]

Consequences of Cystic Fibrosis

The abnormal glandular secretions that occur in cystic fibrosis ultimately disrupt the functioning of multiple tissues and organs.

DIET-DRUG INTERACTIONS

Check this table for notable nutrition-related effects of the medications discussed in this chapter.

Antidiarrheal drugs	**Gastrointestinal effect:** Constipation
Anti-inflammatory drugs (sulfasalazine, corticosteroids)	**Gastrointestinal effects:** Nausea, vomiting (sulfasalazine)
	Dietary interactions: Sulfasalazine may decrease folate absorption; supplementation is recommended
	Metabolic effects: Anemia (sulfasalazine); fluid retention, hyperglycemia, hypocalcemia, hypokalemia, hypophosphatemia, increased appetite, protein catabolism (corticosteroids)
Antisecretory drugs (proton-pump inhibitors, H2 blockers)	**Gastrointestinal effects:** Diarrhea, constipation, nausea and vomiting, abdominal pain (proton-pump inhibitors)
	Dietary interactions: May decrease iron, calcium, folate, and vitamin B_{12} absorption
Laxatives	**Gastrointestinal effects:** Diarrhea, flatulence, abdominal discomfort
	Metabolic effects: Dehydration, electrolyte imbalances, laxative dependency
Pancreatic enzyme replacements	**Gastrointestinal effects:** Constipation, nausea and vomiting, diarrhea, abdominal cramps, irritation of GI mucosa
	Dietary interactions: May decrease folate and iron absorption
	Metabolic effects: Elevated serum or urinary uric acid levels (with high doses), allergic reactions (rare)

Common complications of cystic fibrosis involve the lungs, pancreas, (■) and sweat glands:

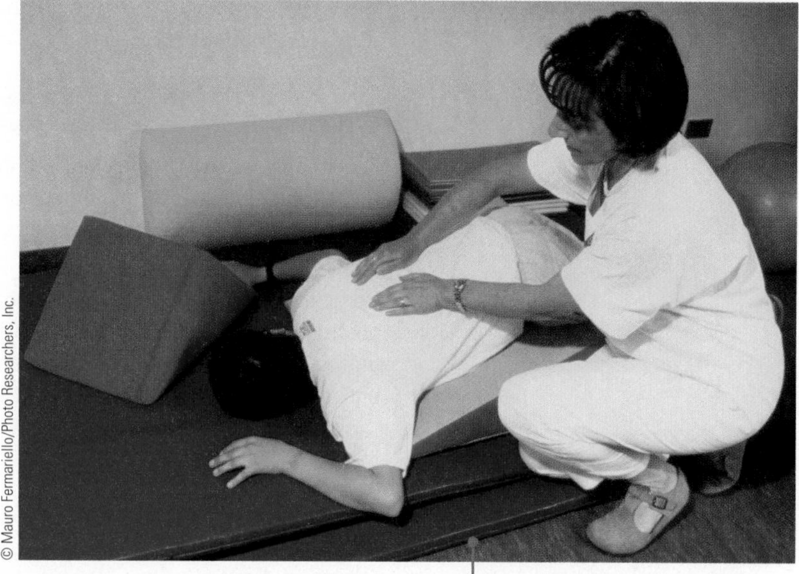
© Mauro Fermariello/Photo Researchers, Inc.

Postural drainage, a type of physical therapy used in treatment of cystic fibrosis, helps to clear the thick, sticky secretions that block airways and increase infection risk.

- *Lung disease.* In cystic fibrosis, the mucus secretions of bronchial tissues are abnormally viscous, leading to airway obstructions and persistent respiratory infections. These problems result in chronic coughing and progressive inflammation in bronchial tissues. The eventual lung damage causes breathing difficulties and lower exercise tolerance. As with other obstructive airway diseases, nutrition status may become impaired due to hypermetabolism, the greater energy cost of labored breathing, and anorexia (loss of appetite). The chronic respiratory infections raise energy needs further. (See Chapter 16 for details about the nutrition problems associated with chronic obstructive lung diseases.)
- *Pancreatic disease.* About 85 to 90 percent of cystic fibrosis patients produce thickened pancreatic secretions that obstruct the pancreatic ducts.[28] The trapped pancreatic enzymes eventually damage pancreatic tissue, leading to progressive atrophy and scarring. Few pancreatic enzymes reach the small intestine, resulting in malabsorption of protein, fat, and fat-soluble vitamins. Other problems that may develop over time include pancreatitis and glucose intolerance or diabetes (due to destruction of the insulin-producing cells).
- *Other complications.* Because cystic fibrosis affects all exocrine secretions, complications typically develop in many other tissues or organs. Salt losses in sweat are usually excessive, increasing the risk of dehydration. Intestinal obstruction is a common problem in newborn infants and may also occur in older patients. Gallbladder and liver diseases may result from bile duct obstructions. Abnormalities in genital tissues cause sterility in men and reduced fertility in women.

Nutrition Therapy for Cystic Fibrosis

Children with cystic fibrosis are chronically undernourished, grow poorly, and have difficulty maintaining normal body weight. Their energy and protein requirements are high due to increased metabolism and nutrient malabsorption. To achieve normal growth and appropriate weight, energy and protein needs may range from 120 to 200 percent of DRI values.[29] (■) Cystic fibrosis patients are typically encouraged to eat high-kcalorie and high-fat foods, eat frequent meals and snacks, and supplement meals with milk shakes or oral dietary supplements. Supplemental tube feedings can help to improve nutrition status if energy intakes are inadequate.

Pancreatic enzyme replacement therapy is a central feature of cystic fibrosis treatment. Supplemental enzymes must be included with every meal or snack. For infants and small children, the contents of capsules are mixed in small amounts of liquid or a soft food (such as applesauce) and fed with a spoon. Enzyme dosages may need to be adjusted if malabsorption continues, as evidenced by poor growth or GI symptoms such as steatorrhea, intestinal gas, or abdominal pain.

The risk of nutrient deficiency depends on the degree of malabsorption. The nutrients of greatest concern include the fat-soluble vitamins, essential fatty acids, and calcium. Multivitamin and fat-soluble vitamin supplements are routinely recommended. The liberal use of table salt and salty foods is encouraged to make up for losses of sodium in sweat.

■ Reminder: The pancreas secretes digestive enzymes and bicarbonate into the digestive tract (*exocrine* secretions) and the hormones insulin and glucagon into the bloodstream (*endocrine* secretions).

■ Reminder: Table 14-6 in Chapter 14 lists foods that are often included in high-kcalorie, high-protein diets.

Review Notes

- Chronic pancreatic disorders can cause widespread maldigestion and malabsorption due to the impaired secretion of digestive enzymes.
- Acute pancreatitis is short lived and does not cause permanent damage, but it requires the withholding of food and liquids until healing has occurred. Chronic pancreatitis leads to digestive enzyme deficiencies and is treated with pancreatic enzyme replacement therapy.
- Cystic fibrosis, a genetic disorder associated with thickened exocrine secretions, causes obstructive lung disease and pancreatic damage. Children with cystic fibrosis have high protein and energy requirements and must use pancreatic enzyme replacement therapy and dietary supplements to reverse malnutrition.

Conditions Affecting the Small Intestine

When the intestinal mucosa is damaged due to inflammation, infection, or other causes, malabsorption is the likely outcome. This section discusses *celiac disease* and the *inflammatory bowel diseases,* which are intestinal illnesses that can damage the intestinal mucosa, and *short bowel syndrome,* the malabsorption disorder that results when a substantial portion of the small intestine is surgically removed.

CELIAC DISEASE

Celiac disease is an immune disorder characterized by an abnormal response to a protein fraction in **wheat gluten** and to related proteins in barley and rye. The reaction to gluten causes severe damage to the intestinal mucosa and subsequent malabsorption. Celiac disease is estimated to affect approximately 1 percent of Caucasian persons in the United States and elsewhere; however, the condition is less prevalent in other ethnic groups.[30]

Consequences of Celiac Disease

The immune reaction to gluten can cause striking changes in intestinal tissue. In affected areas, the villi may be shortened or absent, resulting in a significant reduction in mucosal surface area (and, therefore, in intestinal digestive enzymes). The damage may be restricted to the duodenum or may involve the full length of the small intestine. Individuals with severe disease may malabsorb all nutrients to some degree, especially the macronutrients, calcium, iron, folate, the fat-soluble vitamins, and vitamin B_{12}.[31]

Symptoms of celiac disease include GI disturbances such as diarrhea, steatorrhea, and flatulence. Because lactase deficiency can result from mucosal damage, milk products may exacerbate GI symptoms. Due to nutrient malabsorption, children with celiac disease often exhibit poor growth, low body weights, muscle wasting, and anemia. Adults may develop anemia, bone disorders, neurological symptoms, and fertility problems.[32] Individuals with celiac disease who do not eliminate gluten from the diet are at increased risk of developing intestinal and lymphatic cancers.[33]

Some gluten-sensitive individuals may have few GI symptoms but react to gluten by developing a severe, itchy rash. This condition is called **dermatitis herpetiformis** and requires dietary adjustments similar to those for celiac disease.

Nutrition Therapy for Celiac Disease

The treatment for celiac disease is lifelong adherence to a gluten-free diet. Improvement in symptoms is often evident within several weeks, although mucosal healing can sometimes take years. If lactase deficiency is suspected, patients should avoid lactose-containing foods until the intestine has

NURSING DIAGNOSIS

Nursing diagnoses for people with celiac disease may include *diarrhea, risk for disproportionate growth,* and *imbalanced nutrition: less than body requirements.*

celiac (SEE-lee-ack) disease: an immune disorder characterized by an abnormal response to wheat gluten and related proteins; also called *gluten-sensitive enteropathy* or *celiac sprue.*

wheat gluten (GLU-ten): a family of water-insoluble proteins in wheat; includes the gliadin (GLY-ah-din) fractions that are toxic to persons with celiac disease.

dermatitis herpetiformis (DERM-ah-TYE-tis HER-peh-tih-FOR-mis): a gluten-sensitive disorder characterized by a severe skin rash.

recovered. Dietary supplements can be used to meet micronutrient needs and reverse deficiencies.

The gluten-free diet eliminates foods that contain wheat, barley, and rye (see Table 18-6). Because many foods contain ingredients derived from these grains, foods that are problematic are not always obvious. Even small amounts of gluten may cause symptoms in some people, so patients need to check ingredient lists on food labels carefully. Gluten sources that may be overlooked include beer, caramel coloring, coffee substitutes, communion wafers, imitation meats, malt syrup, medications, salad dressings, soy sauce, and brewer's yeast. Special gluten-free products can be purchased to replace common foods such as bread, pasta, and cereals. Although somewhat expensive, these products increase food choices and allow celiac patients to enjoy foods that would otherwise be forbidden. Patients should also be instructed in food preparation methods that prevent cross-contamination from utensils, cutting boards, and toasters.

TABLE 18-6 Gluten-Free Diet

Food Category	Gluten-Free Choices	Potential Gluten Sources
Meat and meat alternates	Fresh meat, fish, or poultry; shellfish; dried peas and beans; tofu; nuts and seeds; eggs	Luncheon meats, sandwich spreads, meatloaf, meatballs, frankfurters, sausages, poultry injected with broth, imitation meat products, imitation seafood, meat extenders, miso, egg substitutes, dried egg products, dry roasted nuts, peanut butter. *Avoid*: products made with hydrolyzed vegetable protein (HVP), marinades, and soy sauce; breaded foods; foods prepared with cream sauces or gravies.
Milk and milk products	Milk, buttermilk, half-and-half, cream, plain yogurt, cheese, cottage cheese, cream cheese	Chocolate milk, milk shakes, frozen yogurt, flavored yogurt, cheese spreads, cheese sauces. *Avoid*: malted milk, malted milk powders.
Breads, cereals, rice, and pasta	Breads, bakery products, and cereals made with amaranth, arrowroot, buckwheat, corn, flax, hominy grits, millet, potato flour or potato starch, quinoa, rice, sorghum, soybean flour, tapioca, and teff; pasta and noodles made with the grains or starches listed above; corn tacos and corn tortillas	Oatmeal and oat bran (due to contamination), rice crackers, rice cakes, corn cakes. *Avoid:* breads, bakery products, cereals, tortillas, matzo, pastas, and pancake or baking mixes made with wheat, rye, barley, and triticale. *Wheat products* include bulghur, couscous, durum flour, einkorn, emmer, farina, graham flour, kamut, semolina, spelt, wheat bran, and wheat germ. *Barley products* include malt, malt flavoring, and malt extract.
Fruits and vegetables	Any fresh, frozen, or canned fruits or vegetables	French fries from fast-food restaurants, commercial salad dressings, fruit pie fillings, dried fruits (may be dusted with flour). *Avoid:* scalloped potatoes (usually made with wheat flour), creamed vegetables, vegetables dipped in batters.
Desserts	Bakery products made with gluten-free flours, most ice creams, sherbet, sorbet, Italian ices, popsicles, gelatin desserts, egg custards, most chocolate bars, chocolate chips, hard candies, marshmallows, whipped toppings	Some ice creams (especially if made with cookie dough, brownies, nuts, and other added ingredients), icing or frosting, candies and candy bars. *Avoid:* bakery products or doughnuts made with wheat, rye, or barley; puddings made with wheat flour; ice cream or sherbets that contain gluten stabilizers; ice cream cones; licorice.
Beverages	Coffee; tea; cocoa made with pure cocoa powder; soft drinks; wine; distilled alcoholic beverages such as rum, gin, whiskey, and vodka	Instant tea or coffee, coffee substitutes, chocolate drinks, hot cocoa mixes. *Avoid:* beer, ale, lager, malted beverages, cereal beverages, beverages that contain nondairy cream substitutes.

© Polara Studios, Inc.

Gluten-free products help people with celiac disease enjoy a wider variety of foods.

Although most people with celiac disease can safely consume moderate amounts of oats, most oats grown in the United States are contaminated with wheat, barley, or rye. Oats are usually grown in rotation with other grains and may become contaminated during harvesting or processing. However, several oat manufacturers in the United States produce oats in dedicated facilities and test the products to ensure they are gluten free.[34] Individuals who wish to include oats in their diet should be advised to purchase only uncontaminated oats and to limit intakes to the amounts found to be safe (about ½ cup of dry rolled oats or ¼ cup dry steel-cut oats per day).

A gluten-free diet may become monotonous unless care is taken to diversify food choices. The diet can also be a social liability by restricting food choices when individuals eat in restaurants, visit friends, or travel. Nonadherence is common when individuals eat away from home.[35] Nutrition education can help celiac patients learn how to meet their nutrient needs and expand meal options despite dietary constraints. Figure 18-3 shows an example of a menu for a gluten-free diet.

INFLAMMATORY BOWEL DISEASES

Inflammatory bowel diseases are chronic inflammatory illnesses characterized by abnormal immune responses to microbes that inhabit the GI tract.[36] Although both genetic and environmental factors contribute to the development of these diseases, the exact triggers are unknown. Table 18-7 compares the two major forms of inflammatory bowel disease, **Crohn's disease** and **ulcerative colitis**. Crohn's disease usually involves the small intestine and may lead to nutrient malabsorption, whereas ulcerative colitis affects the large intestine, where little nutrient absorption occurs. (■) Both diseases are characterized by periods of active disease interspersed with periods of remission. Nutrient losses can result from tissue damage, bleeding, and diarrhea.

Complications of Crohn's Disease

Crohn's disease may occur in any region of the GI tract, but most cases involve the ileum and/or large intestine. Lesions may develop in different areas in the intestine, with normal tissue separating affected regions (called "skip" lesions). During exacerbations, the inflammation may extend deeply into intestinal tissue and be accompanied by ulcerations, fissures, and **fistulas** (abnormal passages between tissues). Loops of intestine may become matted together. Scar tissue eventually thickens and

■ Although ulcerative colitis affects the large intestine, the condition is included in this section because it is one of the major subtypes of inflammatory bowel disease.

Crohn's disease: an inflammatory bowel disease that usually occurs in the lower portion of the small intestine and the colon. Inflammation may pervade the entire intestinal wall.

ulcerative colitis (ko-LY-tis): an inflammatory bowel disease that involves the rectum and colon; inflammation affects the mucosa and submucosa only.

fistulas (FIST-you-luz): abnormal passages between organs or tissues that permit the passage of fluids or secretions.

FIGURE 18-3 Sample Menu—Gluten-Free Diet

SAMPLE MENU

Breakfast

Orange juice

Gluten-free pancake with maple syrup

Plain yogurt with banana and strawberries

Coffee with half and half

Lunch

Grilled chicken breast with cranberry chutney

Baked potato topped with grated cheddar cheese

Sliced tomato with chopped basil

Raspberry sherbet

Snack

Tortilla chips and guacamole

Hot cocoa (made with cocoa powder)

Dinner

Sauteed catfish with sliced lemon and dill

Wild rice pilaf

Collard greens and garlic sauteed in olive oil

Green salad with oil and vinegar dressing

Vanilla egg custard

© Cengage Learning

TABLE 18-7 Comparison of Crohn's Disease and Ulcerative Colitis

	Crohn's Disease	Ulcerative Colitis
Location of inflammation	About 35% of cases involve the ileum and colon, 28% are in the small intestine only, and 32% are confined to the colon	Confined to the rectum and colon; always involves the rectum but often extends into the colon
Pattern of inflammation	Discrete areas separated by normal tissue ("skip" lesions)	Continuous inflammation throughout the affected region
Depth of damage	Damage throughout all layers of tissue; causes deep fissures that give intestinal tissue a "cobblestone" appearance	Damage primarily in the mucosa and submucosa (layers of intestinal tissue closest to the lumen)
Fistulas	Common	Usually do not occur
Cancer risk	Increased	Greatly increased

stiffens the intestinal wall, narrowing the lumen and possibly causing strictures or obstructions. About 75 percent of patients require surgery within 20 years of diagnosis.[37] Patients with Crohn's disease are also at increased risk of developing intestinal cancers.

Malnutrition may result from poor food intake, malabsorption, nutrient losses (especially of protein) from inflamed tissues, increased needs due to inflammation, and surgical resections that shorten the small intestine. If the ileum is affected, bile acids may become depleted, (■) causing malabsorption of fat, fat-soluble vitamins, calcium, magnesium, and zinc (the minerals bind to the unabsorbed fatty acids). Because the ileum is the site of vitamin B_{12} absorption, deficiency can develop unless the patient is given vitamin B_{12} injections. Anemia may result from bleeding, inadequate absorption of nutrients (iron, folate, and vitamin B_{12}) involved in blood cell formation, or the metabolic effects of chronic illness. Anorexia often develops due to abdominal discomfort and the effects of cytokines produced during the inflammatory process.[38]

Complications of Ulcerative Colitis Ulcerative colitis always involves the rectum and usually extends into the colon. Tissue inflammation is continuous along the length of intestine affected, ending abruptly at the area where healthy tissue begins. The erosion or ulceration affects the mucosa and submucosa only (the layers of intestinal tissue closest to the lumen). In early stages, the mucosa appears reddened and swollen; advanced stages may feature mucosal atrophy, thin colon walls, and, in some cases,

■ Reminder: Most of the bile used during digestion is eventually reabsorbed in the ileum and returned to the liver.

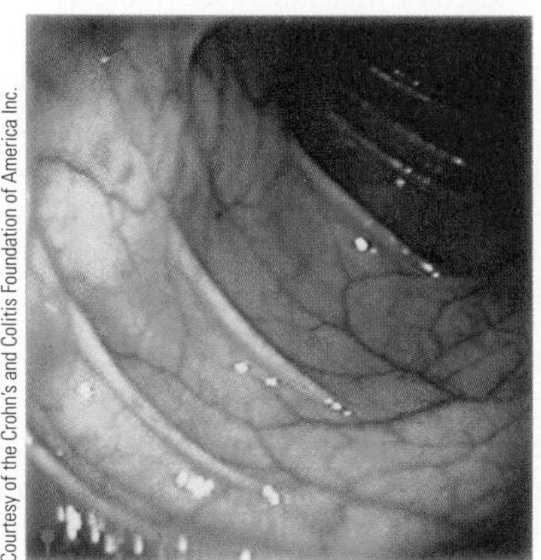
Courtesy of the Crohn's and Colitis Foundation of America Inc.

The healthy colon has a smooth surface with a visible pattern of fine blood vessels.

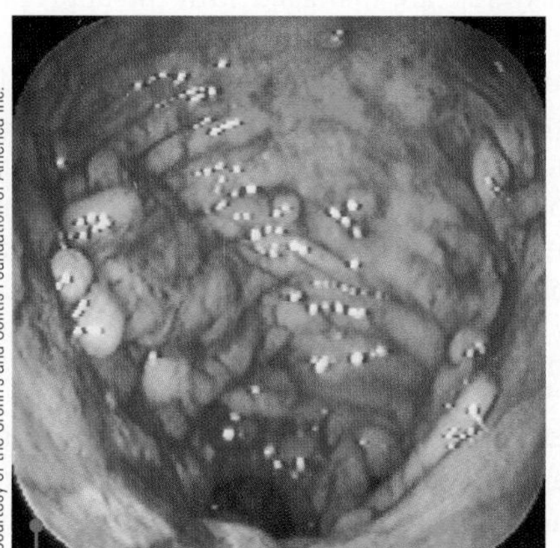
Courtesy of the Crohn's and Colitis Foundation of America Inc.

In Crohn's disease, the mucosa has a "cobblestone" appearance due to deep fissuring in the inflamed mucosal tissue.

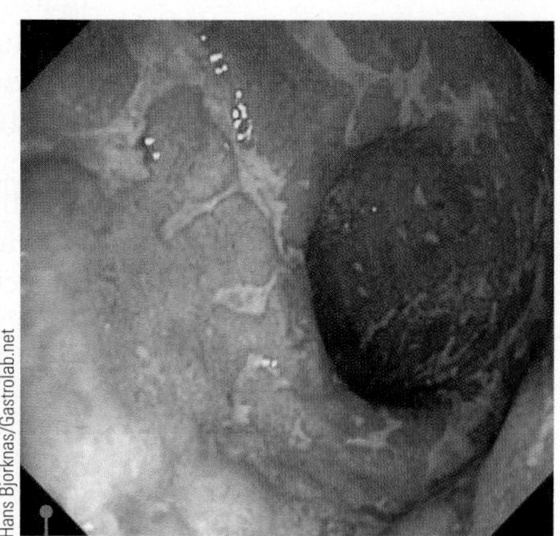
Hans Bjorknas/Gastrolab.net

In ulcerative colitis, the colon appears inflamed and reddened, and ulcers are visible.

NURSING DIAGNOSIS

Nursing diagnoses for people with inflammatory bowel diseases may include *acute pain, nausea, diarrhea, risk for deficient fluid volume,* and *imbalanced nutrition: less than body requirements.*

colon dilation (known as *toxic megacolon*). During active episodes, patients may have frequent, urgent bowel movements that are small in volume and contain blood and mucus. Symptoms vary among patients but may include diarrhea, constipation, rectal bleeding, and abdominal pain.

Although mild disease may cause few complications, weight loss, fever, and weakness are common when most of the colon is involved. Severe disease is often associated with anemia (due to blood loss), dehydration, and electrolyte imbalances. Protein losses from the inflamed tissue can be substantial. A **colectomy** (removal of the colon) is performed in 25 to 40 percent of patients and prevents future recurrence.[39] Colon cancer risk is substantially increased in ulcerative colitis patients.

Drug Treatment of Inflammatory Bowel Diseases

Medications help to control symptoms, reduce inflammation, and minimize complications. The drugs prescribed include antidiarrheal agents, immunosuppressants, anti-inflammatory drugs (usually corticosteroids and salicylates), and antibiotics. Although these medications may allow the patient to achieve and maintain remission, some may cause side effects that are detrimental to nutrition status. The Diet-Drug Interactions feature on p. 524 lists some nutrition-related effects of the medications used in inflammatory bowel diseases.

Nutrition Therapy for Crohn's Disease

Crohn's disease often requires aggressive dietary management because it can lead to protein-energy malnutrition (PEM), nutrient deficiencies, and growth failure in children. Specific dietary measures depend on the functional status of the GI tract and the symptoms and complications that develop; thus, nutrition care varies among patients and throughout the course of illness.

High-kcalorie, high-protein diets may be prescribed to prevent or treat malnutrition or promote healing. Oral supplements may help to increase energy intake and improve weight gain. Vitamin and mineral supplements are usually necessary, especially if nutrient malabsorption is present; nutrients at risk include calcium, iron, magnesium, zinc, folate, vitamin B_{12}, and vitamin D.[40] In some instances, tube feedings are used to supplement the diet or may be the sole means of providing nutrients.

During disease exacerbations, a low-fiber, low-fat diet provided in small, frequent feedings can minimize stool output and reduce symptoms of malabsorption. If diarrhea or flatulence is present, a restricted intake of lactose, fructose, and sorbitol may improve symptoms. Patients with diarrhea should make sure they obtain adequate fluids to prevent dehydration. Individuals with partial obstructions may need to restrict high-fiber foods. Table 18-8 includes examples of the various dietary adjustments that may be beneficial for patients with Crohn's disease.

Nutrition Therapy for Ulcerative Colitis

In most cases, the diet for ulcerative colitis requires few adjustments. As in Crohn's disease, the symptoms and complications that arise are managed with the appropriate dietary measures (see Table 18-8). During disease exacerbations, emphasis is given to restoring fluid and electrolyte balances and correcting deficiencies that result from protein and blood losses; dietary adjustments are based on the extent of bleeding and diarrhea output. Thus, adequate protein, energy, fluid, and electrolytes need to be provided. A low-fiber diet may reduce irritation by minimizing fecal volume. If colon function becomes severely impaired, food and fluids may be withheld and fluids and electrolytes supplied intravenously until colon function is restored.

SHORT BOWEL SYNDROME

The treatment of Crohn's disease, cancers of the small intestine, and other intestinal disorders may include the surgical resection (removal) of a major portion of the small intestine. **Short bowel syndrome** is the malabsorption syndrome that results when the absorptive capacity of the remaining intestine is insufficient for meeting nutritional needs. Without appropriate dietary adjustments, short bowel syndrome can result in fluid and electrolyte imbalances and multiple nutrient deficiencies. Symptoms include diarrhea, steatorrhea, dehydration, weight loss, and growth impairment in children.

colectomy: removal of a portion or all of the colon.

short bowel syndrome: the malabsorption syndrome that follows resection of the small intestine; characterized by inadequate absorptive capacity in the remaining intestine.

TABLE 18-8 Management of Symptoms and Complications in Crohn's Disease

Symptom or Complication	Possible Dietary Measures
Growth failure, weight loss, or muscle wasting	• High-kcalorie, high-protein diet • Oral supplements • Tube feedings
Anorexia or pain with eating	• Small, frequent meals • Oral supplements, as tolerated • Tube feedings if long-term (>5 to 7 days)
Malabsorption	• High-kcalorie diet • Nutrient supplementation
Steatorrhea (fat malabsorption)	• Low-fat diet • Medium-chain triglycerides • Nutrient supplementation
Diarrhea	• Fluid and electrolyte replacement • Nutrient supplementation
Lactose intolerance	• Avoidance of lactose-containing foods
Nutrient deficiencies	• Nutrient-dense diet • Nutrient supplementation
Strictures, partial obstruction, or fistulas	• Low-fiber diet • Liquid supplements
Severe bowel obstruction, high-output fistulas, or severe exacerbations of disease	• Total parenteral nutrition

Figure 18-4 (p. 532) reviews nutrient absorption in the GI tract and describes how absorption is affected by surgical resections. Generally, up to 50 percent of the small intestine can be resected without serious nutritional consequences.[41] More extensive resections lead to generalized malabsorption, and patients may need lifelong parenteral nutrition to supplement oral intakes.

Intestinal Adaptation

After an intestinal resection, the remaining intestine undergoes **intestinal adaptation**, which dramatically improves the intestine's absorptive capacity. Adaptation begins soon after surgery and continues for several years. During this period, the remaining section of intestine develops taller villi and deeper crypts and also grows in length and diameter; these changes dramatically increase the absorptive surface area of the remaining intestine. The ileum has a greater capacity for adaptation than the jejunum; thus, removal of the ileum has more severe consequences than removal of the jejunum. Loss of the ileum permanently disrupts both vitamin B_{12} and bile acid absorption. Depletion of bile acids exacerbates fat malabsorption, and the unabsorbed bile acids irritate the colon walls and can worsen diarrhea. Adaptation is achieved more easily if the colon is present because the colon's resident bacteria can metabolize unabsorbed carbohydrates and produce some usable nutrients. An intact colon also helps to reduce losses of fluids and electrolytes.

Nutrition Therapy for Short Bowel Syndrome

Immediately after a resection, fluids and electrolytes must be supplied intravenously. In the first few weeks after surgery, the fluid losses from diarrhea can be substantial, so appropriate rehydration therapy is critical to recovery. The diarrhea gradually lessens as intestinal adaptation progresses.

intestinal adaptation: physiological changes in the small intestine that increase its absorptive capacity after resection.

FIGURE 18-4 Nutrient Absorption and Consequences of Intestinal Surgeries

WHAT IS ABSORBED

Duodenum/jejunum
- Simple carbohydrates
- Fats
- Amino acids
- Vitamins[a]
- Minerals[a]
- Water

Ileum
- Bile salts
- Vitamin B_{12}
- Water

(Ileum assumes absorptive function of duodenum and jejunum with adaptation)

Colon
- Water
- Electrolytes
- Short-chain fatty acids

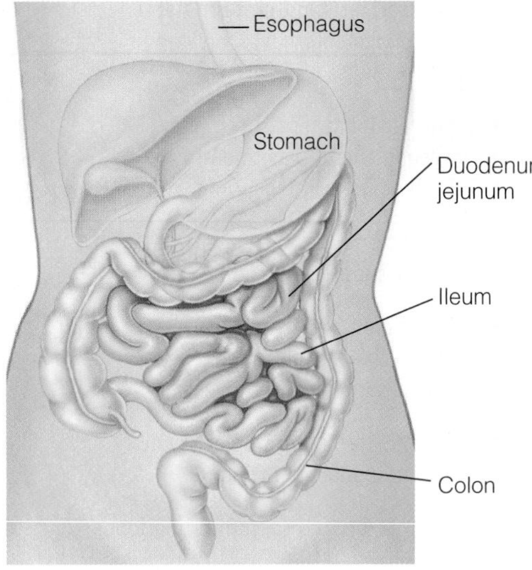

POSSIBLE CONSEQUENCES OF RESECTION

Duodenum/jejunum
- Minimal consequences if the ileum remains intact
- Calcium and iron malabsorption if duodenum resected

Ileum
- Fat malabsorption
- Protein malabsorption
- Malabsorption of fat-soluble vitamins and vitamin B_{12}
- Reduced calcium, magnesium, and zinc absorption
- Fluid losses
- Diarrhea/steatorrhea

Colon
- Fluid and electrolyte losses
- Diarrhea

(Losses are compounded if ileum is also resected)

[a]The absorption of vitamins and minerals begins in the duodenum and continues throughout the length of the small intestine.

Total parenteral nutrition meets nutritional needs after surgery and is gradually replaced by tube feedings and/or oral feedings. To promote intestinal adaptation, (■) the feedings may be started within a week after surgery, after diarrhea subsides somewhat and some bowel function is restored. Initial oral intake may consist of occasional sips of clear, sugar-free liquids, progressing to larger amounts of liquid formulas and then to solid foods, as tolerated. Very small, frequent feedings can utilize the remaining intestine most efficiently. To compensate for malabsorption and reduce the need for parenteral nutrition, a high-kcalorie diet may be encouraged.[42]

The exact diet prescribed for short bowel syndrome depends on the portion of intestine removed, the length of remaining intestine, and whether the colon is still intact; moreover, dietary readjustments may be required as intestinal adaptation progresses.[43] If a high-kcalorie diet is advised and fat is well tolerated, a high-fat, low-carbohydrate diet may help to increase energy intakes. Conversely, a high-complex-carbohydrate, low-fat diet is typically suggested for patients who have an intact colon, because the colon bacteria can metabolize the unabsorbed carbohydrate and produce short-chain fatty acids (which are absorbed in the colon), and the fat restriction may improve steatorrhea. In some cases, medium-chain triglycerides may be added as an energy source.

Dietary choices are tailored according to individual symptoms and tolerances. Some patients may need to drink oral rehydration solutions to stay sufficiently hydrated and obtain adequate electrolytes. Concentrated sweets (which draw fluid into the intestines) should be avoided if they worsen diarrhea. Some patients may be lactose intolerant, but they may minimize symptoms by limiting the amount of milk consumed at one time. Because calcium malabsorption increases the risk of developing kidney stones (see p. 519), a low-oxalate diet may be recommended. Vitamin and mineral supplements can prevent deficiencies from developing due to malabsorption. The accompanying Case Study can help you review the material on Crohn's disease and short bowel syndrome.

■ Adaptation depends on the presence of nutrients in the lumen; therefore, tube and oral feedings are begun as soon as possible after surgery to stimulate the growth of intestinal tissue.

Case Study: Patient with Short Bowel Syndrome

Judi Morel is a 28-year-old economist with an eight-year history of Crohn's disease. Judi is 5 feet 7 inches tall. Three years ago, she underwent a small bowel resection and remained free of active disease for two years. During that time, her symptoms subsided; she was able to tolerate most foods without any problem and gained weight. Ten months ago, Judi experienced a severe flare-up of her Crohn's disease. Since that time, she has lost 15 pounds and currently weighs 118 pounds. She has experienced severe abdominal pain and fatigue that have persisted despite aggressive medical management that included intravenous nutrition. Five days ago, Judi underwent another resection, which left her with 40 percent of healthy small intestine. Her colon is intact. She is experiencing extensive diarrhea.

1. Describe the manifestations of Crohn's disease, and explain why surgery is sometimes performed as part of the treatment. Describe the complications of disease that may affect nutrient needs.
2. Using the BMI table in the back of the book, check the ideal weight range for a person of Judi's height. What nutrition-related concerns are suggested by Judi's recent weight loss? What other nutrition problems did Judi probably experience as a consequence of Crohn's disease?
3. Discuss the complications that may follow an extensive intestinal resection. What factors may affect a person's ability to meet nutrient needs with an oral diet?
4. Describe the dietary progression recommended after an extensive intestinal resection. After Judi is able to eat solid foods, what factors may affect the type of diet that is recommended for her?

Review Notes

- Disorders of the small intestine that cause damage to mucosal tissue, such as celiac disease and Crohn's disease, often result in malabsorption.
- Celiac disease is characterized by an abnormal immune response to wheat gluten and to related proteins in barley and rye. Disease treatment involves lifelong adherence to a gluten-free diet.
- Crohn's disease is an inflammatory illness that can cause extensive damage to all layers of intestinal tissue. Treatment includes medications that suppress inflammation and relieve symptoms, dietary adjustments that reduce symptoms and correct deficiencies, and intestinal resection to remove damaged tissue.
- Ulcerative colitis is an inflammatory disease that damages the mucosa of the rectum and colon. Whereas mild cases may cause few complications, severe cases may require colectomy.
- Short bowel syndrome is a possible consequence of intestinal resection, although intestinal adaptation may improve absorptive capacity over time.

Conditions Affecting the Large Intestine

In the large intestine, the colon moves undigested materials to the rectum and has a central role in maintaining fluid and electrolyte balances. Its bacterial population ferments undigested nutrients and produces short-chain fatty acids and some vitamins that our bodies can absorb and use. This section describes several conditions that may upset the function or structure of the large intestine.

IRRITABLE BOWEL SYNDROME

People with **irritable bowel syndrome** experience chronic and recurring intestinal symptoms that cannot be explained by specific physical abnormalities. The symptoms usually include disturbed defecation (diarrhea and/or constipation), flatulence, and abdominal discomfort or pain; the pain is often aggravated by eating and relieved by defecation. In some patients, symptoms are mild; in others, the disturbances in colonic function can

irritable bowel syndrome: an intestinal disorder of unknown cause that disturbs the functioning of the large intestine; symptoms include abdominal pain, flatulence, diarrhea, and constipation.

NURSING DIAGNOSIS

Nursing diagnoses appropriate for people with irritable bowel syndrome may include *diarrhea, constipation, acute pain, chronic pain,* and *ineffective health maintenance.*

interfere with work and social activities enough to dramatically alter the person's lifestyle and sense of well-being. Irritable bowel syndrome generally occurs in individuals between 20 and 40 years of age and affects twice as many women as men; its prevalence among adults in the United States has been estimated to be between 10 and 20 percent.[44]

Although the causes of irritable bowel syndrome remain elusive, people with the disorder tend to have excessive colonic responses to meals, GI hormones, and psychological stress.[45] Many individuals exhibit hypersensitivity to a normal degree of intestinal distention and feel discomfort when experiencing normal meal transit or typical amounts of intestinal gas. Intestinal motility after meals may be excessive, leading to diarrhea, or be reduced, causing constipation. Some patients show signs of low-grade intestinal inflammation; others may have had a bacterial infection that initiated their GI problems. Diagnosis of irritable bowel syndrome is often difficult because its symptoms are typical of other GI disorders and laboratory tests for the condition are nonexistent.

Treatment of Irritable Bowel Syndrome

Medical treatment of irritable bowel syndrome often includes dietary adjustments, stress management, and behavioral therapies. Medications may be prescribed to manage symptoms but they are not always helpful. The drugs prescribed may include laxatives, antidiarrheal agents, antidepressants, antispasmodics (which reduce pain by relaxing GI muscles), and antibiotics (which alter bacterial populations in the colon).

Nutrition Therapy for Irritable Bowel Syndrome

Although dietary adjustments may reduce symptoms, responses among patients vary considerably. The most common recommendation is to gradually increase fiber intake from food or supplements to relieve constipation and improve stool bulk. However, clinical studies suggest that additional fiber has only marginal effectiveness in improving symptoms and may worsen flatulence.[46] Psyllium supplementation may be helpful for individuals with constipation. Some individuals have fewer symptoms when they consume small, frequent meals instead of larger ones. Foods that aggravate symptoms may include fried or fatty foods, gas-producing foods (review p. 516 and Table 18-2), milk products (not necessarily due to lactose intolerance), wheat products, coffee (with or without caffeine), and alcoholic beverages; however, individual tolerances vary and are best determined by trial and error. Psychological associations have a strong influence on food tolerance, so foods that patients perceive to be problematic should be discussed so that the diet is not restricted unnecessarily. A careful evaluation of the dietary patterns that exacerbate symptoms may uncover the foods and habits most closely associated with intestinal discomfort.

Treatments under investigation for irritable bowel syndrome include peppermint oil,[47] which relaxes smooth muscle, and various types of probiotics.[48] The Case Study can help you apply your knowledge about irritable bowel syndrome to a clinical situation.

Case Study: Patient with Irritable Bowel Syndrome

Hannah Tran is a 22-year-old recent college graduate who began her first professional job in a bank one month ago. As a college student, she occasionally experienced abdominal pain and cramping after eating. She also had frequent bouts of diarrhea and felt somewhat better after bowel movements. Once Hannah began her new job, her symptoms occurred more frequently. At first, she attributed her symptoms to job stress, but when the symptoms continued for several months, she decided to see her physician. After taking a careful history and conducting tests to rule out other bowel disorders, the physician diagnosed irritable bowel syndrome. The physician prescribed bulk-forming fibers and advised Hannah to keep a record of her food intake and symptoms for one week. Hannah was then referred to a dietitian for a review of her dietary record. The dietitian noticed that Hannah routinely drank several cups of coffee in the morning and had large meals for lunch and dinner. Hannah often ate out in Mexican restaurants and favored spicy foods, refried beans, and fatty desserts. Between meals, she snacked on low-carbohydrate foods sweetened with sugar alcohols and drank several cans of soda daily. Her dietary fiber intake, however, totaled only about 13 grams daily.

1. Describe the characteristics of irritable bowel syndrome to Hannah, and indicate the role that stress might play in her illness.
2. Explain how the record of food intake and symptoms might be helpful in devising an appropriate dietary plan for Hannah. Could any of the foods that are currently in Hannah's diet be aggravating her symptoms?
3. What dietary measures may benefit individuals with irritable bowel syndrome? What problems might some of the dietary changes cause?

DIVERTICULAR DISEASE OF THE COLON

Diverticulosis refers to the presence of pebble-sized herniations (outpockets) in the intestinal mucosa, known as diverticula (see Figure 18-5). The prevalence of diverticulosis increases with age, occurring in 50 to 65 percent of 80-year-old individuals.[49] Most people with diverticulosis are symptom free and remain unaware of the condition until a complication develops.

Although the cause of diverticulosis is unclear, changes in connective tissue proteins that occur with aging may contribute to its development.[50] Epidemiological studies and controlled studies in rats have suggested that low-fiber diets may increase risk due to fiber's influence on transit time, stool volume, and intraluminal pressures;[51] however, some recent clinical studies were unable to find an association between low-fiber intakes and the incidence of diverticulosis.[52] Similarly, a relationship between diverticulosis and other proposed risk factors, such as high meat intake, constipation, and physical inactivity, has not been confirmed by reliable studies.

FIGURE 18-5 Diverticula in the Colon

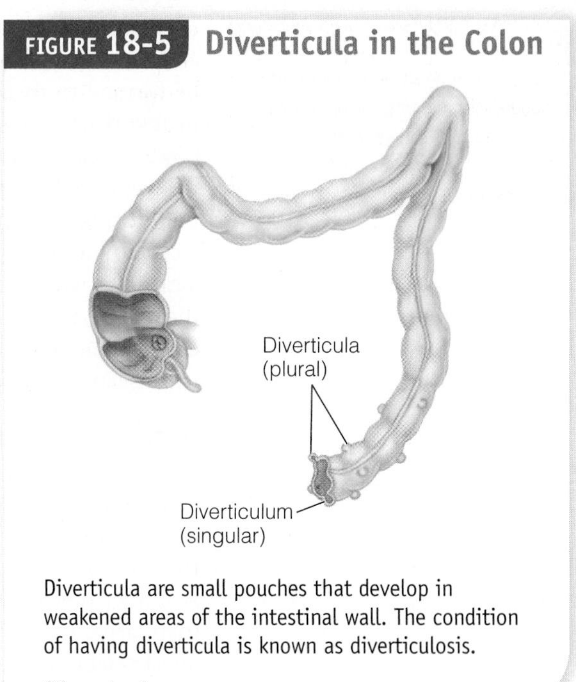

Diverticula are small pouches that develop in weakened areas of the intestinal wall. The condition of having diverticula is known as diverticulosis.

Diverticulitis

Inflammation or infection sometimes develops in the area around a diverticulum. This condition, called **diverticulitis**, is the most common complication of diverticulosis, affecting 10 to 25 percent of individuals with the condition.[53] It is thought to result from erosion of the diverticular wall due to high intraluminal pressure or particulate matter, leading to inflammation and eventually a microperforation that causes subsequent infection. If the infection spreads to adjacent organs, fistulas may develop. Less frequently, the infection spreads to the peritoneal cavity, causing life-threatening illness. Symptoms of diverticulitis may include persistent abdominal pain, tenderness in the affected area, fever, constipation, and diarrhea. Anorexia, nausea, and vomiting may also occur.

> **NURSING DIAGNOSIS**
>
> Nursing diagnoses that may apply to people with diverticulitis include *acute pain, constipation, diarrhea,* and *risk for deficient fluid volume.*

diverticulosis (DYE-ver-tic-you-LOH-sis): an intestinal condition characterized by the presence of small herniations (called diverticula) in the intestinal wall.

diverticulitis (DYE-ver-tic-you-LYE-tis): inflammation or infection involving diverticula.

Treatment for Diverticular Disease

Medical treatment for diverticulosis is necessary only if symptoms develop. Patients are often advised to increase fiber intake to relieve constipation and other symptoms, although high-fiber diets have not been shown to reverse diverticulosis or prevent disease progression.[54] Fiber should be increased gradually to ensure tolerance; the emphasis should be on insoluble fiber sources such as wheat bran, whole-grain products, fruits, and vegetables. Bulk-forming agents, such as psyllium, can help to increase fiber intake if food sources are insufficient. Avoiding nuts, seeds, and popcorn is sometimes suggested to prevent complications, but no evidence is available to justify the recommendation.[55]

Patients with diverticulitis may need antibiotics to treat infections and, possibly, pain-control medications. In mild cases, a clear liquid diet may be advised initially, with progression to solid foods as tolerated. In severe cases, bowel rest is necessary (oral fluids and food are withheld), and fluids are provided intravenously. Afterwards, an oral diet is gradually reintroduced as the condition improves, beginning with clear liquids and progressing to a low-fiber diet until inflammation and bleeding subside.[56] After recovery, a high-fiber diet is recommended to help prevent symptom recurrence. Surgical interventions are sometimes necessary to treat complications of diverticulitis and may include removal of the affected portion of the colon.

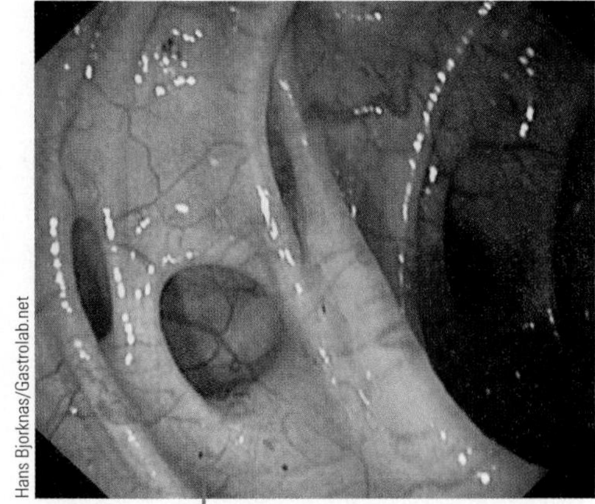

Diverticula are frequently seen during a colonoscopy, a procedure that uses a flexible, lighted tube to examine the inside of the colon.

COLOSTOMIES AND ILEOSTOMIES

An *ostomy* is a surgically created opening (called a **stoma**) in the abdominal wall through which dietary wastes can be eliminated. Whereas a permanent ostomy is necessary after a partial or total colectomy, a temporary ostomy is sometimes constructed to bypass the colon after injury or extensive surgery. To create the stoma, the cut end

NURSING DIAGNOSIS

Nursing diagnoses appropriate for people with ostomies may include *disturbed body image*, *risk for diarrhea*, *risk for deficient fluid volume*, and *risk for impaired skin integrity*.

of the remaining segment of functional intestine is routed through an opening in the abdominal wall and stitched in place so that it empties to the exterior. The stoma can be formed from a section of the colon (**colostomy**) or ileum (**ileostomy**), as shown in Figure 18-6. Conditions that may require these procedures include inflammatory bowel diseases, diverticulitis, and colorectal cancers.

To collect wastes, a disposable bag is affixed to the skin around the stoma and emptied during the day as needed. Alternatively, an interior pouch can be surgically constructed behind the stoma using intestinal tissue, and the pouch can be emptied with a catheter when convenient. Stool consistency varies according to the length of colon that is functional. If a small portion of the colon is absent or bypassed, the stools may continue to be semisolid. If the entire colon has been removed or is bypassed, the ability to reabsorb fluid and electrolytes is substantially reduced, and the output is liquid.

Nutrition Care for Patients with Ostomies

The nutrition care after an ostomy depends on the length of colon removed and the portion of ileum that remains, so dietary adjustments are individualized according to the surgical procedure and symptoms that develop afterward. Following surgery, the diet progresses from clear liquids that are low in sugars to a regular meal plan, as tolerated. To reduce stool output, a low-fiber diet may be recommended. Small, frequent meals may be more acceptable than larger ones. To determine food tolerances, patients should try small amounts of questionable foods and assess their effects; a food that causes problems can be tried again later. Appropriate fluid and electrolyte intakes should be encouraged when a large portion of the colon has been removed.

People with ileostomies need to chew thoroughly to ensure that foods are adequately digested and to prevent obstructions, a common complication due to the small diameter of the ileal lumen. Foods high in insoluble fibers are sometimes avoided because they reduce transit time, may cause obstructions, and increase stool output. To replace electrolyte losses, patients are encouraged to use salt liberally and to ingest beverages with added electrolytes (such as sports drinks and oral rehydration beverages), if necessary. If a large portion of the ileum has been removed, fat malabsorption may occur due to bile acid depletion, and vitamin B_{12} injections may be required.

FIGURE 18-6 Colostomy and Ileostomy

Colostomy

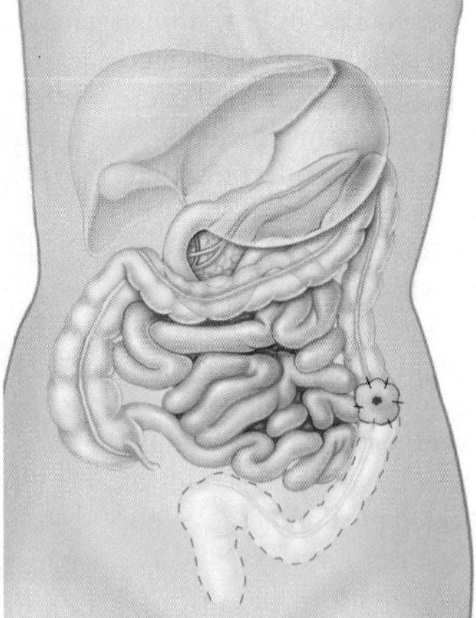

In a colostomy, a portion of the colon is removed or bypassed, and the stoma is formed from the remaining section of functional colon.

Ileostomy

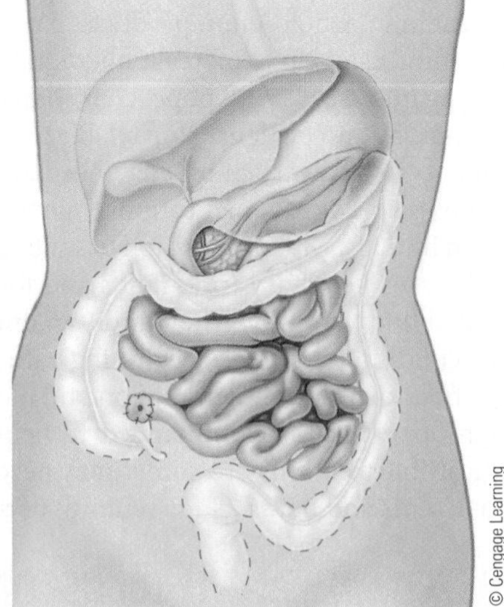

In an ileostomy, the entire colon is removed or bypassed, and the stoma is formed from the ileum.

stoma (STOE-ma): a surgically created opening in a body tissue or organ.

colostomy (co-LAH-stoe-me): a surgical passage through the abdominal wall into the colon.

ileostomy (ill-ee-AH-stoe-me): a surgical passage through the abdominal wall into the ileum.

Dietary concerns after colostomies depend on the length of colon remaining. Most patients have no dietary restrictions and can return to a regular diet. Patient concerns may include stool odors, excessive gas production, and diarrhea. If a large portion of colon was removed, recommendations may be similar to those given to ileostomy patients.

Obstructions As mentioned, foods that are incompletely digested can cause obstructions, a primary concern of ileostomy patients. Although these patients can consume almost any food that is cut into small pieces and carefully chewed, the following foods may cause difficulty: celery, coconut, corn, mushrooms, peas, raw cabbage (for example, in coleslaw), salad greens, dried fruit, unpeeled fresh fuits, pineapple, nuts, seeds, popcorn, frankfurters, sausages, and tough, chewy meats.

Reducing Gas and Odors Persons with ostomies are often concerned about foods that may increase gas production or cause strong odors. Foods that may cause excessive gas include those listed in Table 18-2 on p. 516; practices that increase gas formation include smoking, gum or tobacco chewing, using drinking straws, drinking carbonated beverages, and eating quickly. Foods that sometimes produce unpleasant odors include asparagus, beer, broccoli, brussels sprouts, cabbage, cauliflower, dried beans and peas, eggs, fish, garlic, and onions. Foods that may help to reduce odors include buttermilk, cranberry juice, parsley, and yogurt.

Diarrhea Examples of foods that may aggravate diarrhea were listed in Table 18-3 on p. 517. Foods and dietary substances that may thicken stool include applesauce, banana (or banana flakes), cheese, pasta, pectin, potatoes, smooth peanut butter, tapioca, and white rice. What works may differ for each individual, however, and is best determined by trial and error.

Review Notes

- Irritable bowel syndrome is characterized by chronic, recurring intestinal symptoms such as diarrhea and/or constipation, abdominal pain, and flatulence. Although the causes are unknown, the disorder is influenced by food intake, stress, and psychological factors.
- Diverticulosis is often asymptomatic until complications develop; its prevalence increases with advancing age. Diverticulitis, which involves inflammation or infection of the diverticula, may require medical treatment and temporary bowel rest.
- Colostomies and ileostomies are surgically created openings in the abdominal wall using the cut end of the colon or ileum. Fluid and electrolyte requirements are greater after an ostomy if colon function is reduced or absent.

Nutrition Assessment Checklist for People with Lower GI Tract Disorders

MEDICAL HISTORY

Check the medical record for diseases that:

- ❍ Cause chronic GI symptoms, such as irritable bowel syndrome or ulcerative colitis
- ❍ Interfere with pancreatic enzyme secretion, such as chronic pancreatitis or cystic fibrosis
- ❍ Interfere with nutrient absorption, such as Crohn's disease or celiac disease

Check for surgical procedures involving the lower GI tract, such as:

- ❍ Intestinal resections or bypass surgeries
- ❍ Ileostomy
- ❍ Colostomy

Check for the following symptoms or complications:

- ❍ Anemia
- ❍ Bacterial overgrowth

- Bone disease
- Constipation
- Diarrhea, dehydration
- Fistulas
- Lactose intolerance
- Nutrient deficiencies
- Obstructions
- Oxalate kidney stones
- Poor growth, in children
- Steatorrhea

MEDICATIONS

Check for medications or dietary supplements that may:

- Cause constipation or diarrhea
- Interfere with food intake by causing nausea, vomiting, abdominal discomfort, dry mouth, or drowsiness
- Alter appetite or nutrient needs

DIETARY INTAKE

Note the following problems, and contact the dietitian if you suspect difficulties such as:

- Poor appetite or food intake
- Food intolerances
- Inadequate fiber intake, in patients with constipation
- Inadequate fluid intake
- Malabsorbed carbohydrates, in patients with diarrhea

ANTHROPOMETRIC DATA

Measure baseline height and weight. Address weight loss early to prevent malnutrition in patients with:

- Severe or persistent diarrhea
- Nutrient malabsorption

LABORATORY TESTS

Check laboratory tests for signs of dehydration, electrolyte imbalances, nutrient deficiencies, and anemia in patients with:

- Severe or persistent diarrhea
- Nutrient malabsorption
- Intestinal resections

PHYSICAL SIGNS

Look for physical signs of:

- Dehydration
- Essential fatty acid and fat-soluble vitamin deficiencies
- Folate and vitamin B_{12} deficiencies
- Mineral deficiencies
- Protein-energy malnutrition

Self Check

1. The health practitioner advising an elderly patient with constipation encourages the patient to:

a. consume a low-fat diet low in sodium.
b. consume a high-protein diet rich in calcium.
c. eliminate gas-forming foods from the diet.
d. gradually add high-fiber foods to the diet.

2. Osmotic diarrhea often results from:

a. excessive colonic contractions.
b. excessive fluid secretion by the intestines.
c. nutrient malabsorption.
d. viral, bacterial, or protozoal infections.

3. Nutrition problems that may result from fat malabsorption include all of the following *except:*

a. weight loss.
b. essential amino acid deficiencies.
c. bone loss.
d. oxalate kidney stones.

4. Common nutrition problems associated with bacterial overgrowth in the stomach and small intestine include:

a. sensitivity to gluten.
b. fat malabsorption and vitamin B_{12} deficiency.
c. constipation.
d. permanent loss of digestive enzymes.

5. Chronic pancreatitis and cystic fibrosis are both treated with:

a. intestinal resection.
b. postural drainage.
c. enzyme replacement therapy.
d. stool softeners.

6. A person with celiac disease must avoid products containing:

a. wheat, barley, and rye.
b. lactose.
c. excessive fat.
d. corn, rice, and millet.

7. Despite the intestine's capacity for adaptation, removal of which section of the intestine will likely result in fat malabsorption and multiple nutrient deficiencies?
   a. Duodenum
   b. Jejunum
   c. Ileum
   d. Colon

8. Symptoms of irritable bowel syndrome most often include:
   a. nausea and vomiting.
   b. weight loss and malnutrition.
   c. strong odors and obstructions.
   d. constipation, diarrhea, and flatulence.

9. Diverticulosis is most often associated with:
   a. aging.
   b. inadequate exercise.
   c. intestinal surgery.
   d. a high-fiber diet.

10. After an ileostomy, the most serious concern is that:
   a. the diet is too restrictive to meet nutrient needs.
   b. waste disposal causes frequent daily interruptions.
   c. fluid restrictions prevent patients from drinking beverages freely.
   d. incompletely digested foods may cause obstructions.

Answers to these questions appear in Appendix H. For more chapter review: Access an interactive eBook, chapter-specific interactive learning tools, including flashcards, quizzes, videos, and more in your Nutrition CourseMate, accessed through CengageBrain.com.

Clinical Applications

1. A health practitioner working with a patient with a constipation problem provides him with detailed information about a high-fiber diet. At a follow-up appointment, the patient reports no change in symptoms. His food diary for that day shows that he consumed an omelet and toast for breakfast and a sandwich with juice for lunch.
   - Considering these two meals only, what additional information would help the health practitioner evaluate the man's compliance with the diet he was given?
   - Review the discussion about fiber in Chapter 3, and create a one-day menu that provides the DRI for fiber for an adult male, using the fiber values listed in Appendix A.

2. Using Table 18-5 on p. 520 as a guide, plan a day's menus for a diet containing approximately 50 grams of fat. Take care to make the meals both palatable and nutritious. How can these menus be improved using the suggestions in the "How to" on p. 521?

3. As stated in this chapter, treatment of celiac disease is deceptively simple—eliminate wheat, barley, and rye, and possibly oats. Remaining on a gluten-free diet is more challenging than it appears, however.
   - Randomly select 10 of your favorite snack and convenience foods. Take a trip to the grocery store, and check the labels of the products you selected to see if they would be allowed on a gluten-free diet. Keep in mind that the labels may not list all offending ingredients.
   - Find acceptable substitutes for the products that are not allowed, either by substituting other foods or by checking for gluten-free products in the grocery store. If you have access to the Internet, you may want to investigate websites that advertise gluten-free products to get an idea of what's available.

Nutrition on the Net

For further study of topics covered in this chapter, access these websites:

- Visit the websites of these organizations to find information that is helpful both for health practitioners and patients with gastrointestinal problems:
  American College of Gastroenterology:
  gi.org
  American Gastroenterological Association:
  www.gastro.org
  National Institute of Diabetes and Digestive and Kidney Diseases, a division of the National Institutes of Health:
  www2.niddk.nih.gov
- Find additional information about cystic fibrosis at the website of the Cystic Fibrosis Foundation:
  www.cff.org/home

- Find more information about celiac disease by visiting these websites:
  Celiac Disease Foundation:
  www.celiac.org
  Celiac Sprue Association:
  www.csaceliacs.org
  Gluten Intolerance Group:
  www.gluten.net
- Learn more about inflammatory bowel diseases at the website of the Crohn's and Colitis Foundation of America:
  www.ccfa.org

Notes

1. A. J. Lembo and S. P. Ullman, Constipation, in M. Feldman, L. S. Friedman, and L. J. Brandt, eds., *Sleisenger and Fordtran's Gastrointestinal and Liver Disease* (Philadelphia: Saunders, 2010), pp. 259–284.
2. Academy of Nutrition and Dietetics, *Nutrition Care Manual* (Chicago: American Academy of Nutrition and Dietetics, 2012); C. A. Ternent and coauthors, Practice parameters for the evaluation and management of constipation, *Diseases of the Colon and Rectum* 50 (2007): 2013–2022.
3. Standing Committee on the Scientific Evaluation of Dietary Reference Intakes, Food and Nutrition Board, Institute of Medicine, *Dietary Reference Intakes for Energy, Carbohydrate, Fiber, Fat, Fatty Acids, Cholesterol, Protein, and Amino Acids* (Washington, DC: National Academies Press, 2002).
4. Ternent and coauthors, 2007.
5. Lembo and Ullman, 2010.
6. F. Azpiroz and M. D. Levitt, Intestinal gas, in M. Feldman, L. S. Friedman, and L. J. Brandt, eds., *Sleisenger and Fordtran's Gastrointestinal and Liver Disease* (Philadelphia: Saunders, 2010), pp. 233–240.
7. L. R. Schiller and J. H. Sellin, Diarrhea, in M. Feldman, L. S. Friedman, and L. J. Brandt, eds., *Sleisenger and Fordtran's Gastrointestinal and Liver Disease* (Philadelphia: Saunders, 2010), pp. 211–232.
8. Schiller and Sellin, 2010.
9. D. Wolvers and coauthors, Guidance for substantiating the evidence for beneficial effects of probiotics: Prevention and management of infections by probiotics, *Journal of Nutrition* 140 (2010): 698S–712S.
10. Schiller and Sellin, 2010.
11. Academy of Nutrition and Dietetics, 2012.
12. J. S. Trier, Intestinal malabsorption, in N. J. Greenberger, ed., *Current Diagnosis and Treatment: Gastroenterology, Hepatology, and Endoscopy* (New York: McGraw-Hill Companies, 2012), pp. 237–257.
13. Trier, 2012.
14. S. O'Mahony and F. Shanahan, Enteric microbiota and small intestinal bacterial overgrowth, in M. Feldman, L. S. Friedman, and L. J. Brandt, eds., *Sleisenger and Fordtran's Gastrointestinal and Liver Disease* (Philadelphia: Saunders, 2010), pp. 1769–1778.
15. J. K. DiBaise, Nutritional consequences of small intestinal bacterial overgrowth, *Practical Gastroenterology* 32 (December 2008): 15–28.
16. C. E. Semrad, Approach to the patient with diarrhea and malabsorption, in L. Goldman and A I. Schafer, eds., *Goldman's Cecil Medicine* (Philadelphia: Saunders, 2012), pp. 895–913.
17. S. Hertzler and coauthors, Nutrient considerations in lactose intolerance, in A. M. Coulston and C. J. Boushey, eds., *Nutrition in the Prevention and Treatment of Disease* (Burlington, MA: Elsevier, 2008), pp. 755–770.
18. Hertzler and coauthors, 2008.
19. A. Szilagyi and coauthors, Differential impact of lactose/lactase phenotype on colonic microflora, *Canadian Journal of Gastroenterology* 24 (2010): 373–379.
20. B. U. Wu, D. L. Conwell, and P. A. Banks, Acute pancreatitis, in N. J. Greenberger, ed., *Current Diagnosis and Treatment: Gastroenterology, Hepatology, and Endoscopy* (New York: McGraw-Hill Companies, 2012), pp. 311–318.
21. Wu, Conwell, and Banks, 2012; Academy of Nutrition and Dietetics, 2012.
22. C. E. Forsmark, Pancreatitis, in L. Goldman and A I. Schafer, eds., *Goldman's Cecil Medicine* (Philadelphia: Saunders, 2012), pp. 937–944.
23. C. E. Forsmark, Chronic pancreatitis, in M. Feldman, L. S. Friedman, and L. J. Brandt, eds., *Sleisenger and Fordtran's Gastrointestinal and Liver Disease* (Philadelphia: Saunders, 2010), pp. 985–1015.
24. Wu, Conwell, and Banks, 2012; Academy of Nutrition and Dietetics, 2012.
25. B. J. Rosenstein, Cystic fibrosis, in R. S. Porter and J. L. Kaplan, eds., *Merck Manual of Diagnosis and Therapy* (Whitehouse Station, NJ: Merck Sharp and Dohme Corp., 2011), pp. 2881–2886.
26. F. J. Accurso, Cystic fibrosis, in L. Goldman and A I. Schafer, eds., *Goldman's Cecil Medicine* (Philadelphia: Saunders, 2012), pp. 544–548.
27. Accurso, 2012; M. P. Boyle, Adult cystic fibrosis, *Journal of the American Medical Association* 298 (2007): 1787–1793.
28. Rosenstein, 2011.
29. Academy of Nutrition and Dietetics, 2012.
30. A. Rubio-Tapia and coauthors, Prevalence of celiac disease in the United States, *American Journal of Gastroenterology* 107 (2012): 1538–1544; Semrad, 2012.
31. M. M. Niewinski, Advances in celiac disease and gluten-free diet, *Journal of the American Dietetic Association* (2008): 661–672.
32. R. J. Farrell and C. P. Kelly, Celiac disease and refractory celiac disease, in M. Feldman, L. S. Friedman, and L. J. Brandt, eds., *Sleisenger and Fordtran's Gastrointestinal and Liver Disease* (Philadelphia: Saunders, 2010), pp. 1797–1820.
33. Semrad, 2012.
34. N. Raymond, J. Heap, and S. Case, The gluten-free diet: An update for health professionals, *Practical Gastroenterology* 30 (September 2006): 67–92.
35. Raymond, Heap, and Case, 2006.
36. R. S. Blumberg and S. B. Snapper, Inflammatory bowel disease: Immunologic considerations and therapeutic implications, in N. J. Greenberger, ed., *Current Diagnosis and Treatment: Gastroenterology, Hepatology, and Endoscopy* (New York: McGraw-Hill Companies, 2012), pp. 12–22.
37. B. E. Sands and C. A. Siegel, Crohn's disease, in M. Feldman, L. S. Friedman, and L. J. Brandt, eds., *Sleisenger and Fordtran's Gastrointestinal and Liver Disease* (Philadelphia: Saunders, 2010), pp. 1941–1973.
38. Sands and Siegel, 2010.
39. J. L. Irani, A. Ramsanahie, and R. Bleday, Inflammatory bowel disease: Surgical considerations, in N. J. Greenberger, ed., *Current Diagnosis and Treatment: Gastroenterology, Hepatology, and Endoscopy* (New York: McGraw-Hill Companies, 2012), pp. 36–46.
40. Academy of Nutrition and Dietetics, 2012.
41. Trier, 2012.
42. A. L. Buchman, Short bowel syndrome, in M. Feldman, L. S. Friedman, and L. J. Brandt, eds., *Sleisenger and Fordtran's Gastrointestinal and Liver Disease* (Philadelphia: Saunders, 2010), pp. 1779–1795.
43. Buchman, 2010; C. R. Parrish, The clinician's guide to short bowel syndrome, *Practical Gastroenterology* 31 (September 2005): 67–106.
44. S. Friedman, Irritable bowel syndrome, in N. J. Greenberger, ed., *Current Diagnosis and Treatment: Gastroenterology, Hepatology, and Endoscopy* (New York: McGraw-Hill Companies, 2012), pp. 297–309.
45. Friedman, 2012; N. J. Talley, Irritable bowel syndrome, in M. Feldman, L. S. Friedman, and L. J. Brandt, eds., *Sleisenger and Fordtran's Gastrointestinal and Liver Disease* (Philadelphia: Saunders, 2010), pp. 2091–2104.
46. Talley, 2010; W. D. Heizer, S. Southern, and S. McGovern, The role of diet in symptoms of irritable bowel syndrome in adults: A narrative review, *Journal of the American Dietetic Association* 109 (2009): 1204–1214; A. Sanjeevi and D. F. Kirby, The role of food and dietary intervention in the irritable bowel syndrome, *Practical Gastroenterology* 32 (July 2008): 33–42.
47. Heizer, Southern, and McGovern, 2009.

48. P. Moayyedi and coauthors, The efficacy of probiotics in the treatment of irritable bowel syndrome: A systematic review, *Gut* 59 (2010): 325–332.
49. A. C. Travis and R. S. Blumberg, Diverticular disease of the colon, in N. J. Greenberger, ed., *Current Diagnosis and Treatment: Gastroenterology, Hepatology, and Endoscopy* (New York: McGraw-Hill Companies, 2012), pp. 259–272.
50. Travis and Blumberg, 2012; J. M. Fox and N. H. Stollman, Diverticular disease of the colon, in M. Feldman, L. S. Friedman, and L. J. Brandt, eds., *Sleisenger and Fordtran's Gastrointestinal and Liver Disease* (Philadelphia: Saunders, 2010), pp. 2073–2089.
51. Travis and Blumberg, 2012; Fox and Stollman, 2010.
52. A. F. Peery and coauthors, A high-fiber diet does not protect against asymptomatic diverticulosis, *Gastroenterology* 142 (2012): 266–272; L. L. Strate, Lifestyle factors and the course of diverticular disease, *Digestive Diseases* 30 (2012): 35–45.
53. Travis and Blumberg, 2012.
54. C. Ünlü and coauthors, A systematic review of high-fibre dietary therapy in diverticular disease, *International Journal of Colorectal Disease* 27 (2012): 419–427.
55. S. Tarleton and J. K. DiBaise, Low-residue diet in diverticular disease: Putting an end to a myth, *Nutrition in Clinical Practice* 26 (2011): 137–142.
56. Academy of Nutrition and Dietetics, 2012.

Nutrition in Practice

Probiotics and Intestinal Health

Soon after birth, the warm, nutrient-rich environment within the gastrointestinal tract is colonized by a wide variety of bacterial species. In fact, the approximately 10 trillion bacterial cells inhabiting our bodies (**flora**) represent more than 90 percent of all our cells. Most bacterial cells reside in our colon, which harbors 400 to 500 different species.[1] Although the exact composition of intestinal bacteria varies among individuals, the pattern within an individual tends to remain constant over time, fluctuating somewhat due to age, illness, antibiotic treatment, and, to some extent, dietary factors. Table NP18-1 lists the predominant types of bacteria that colonize the human intestines, and Table NP18-2 shows how the bacterial populations vary within different regions of the GI tract.

Over the past several decades, nutritional scientists and microbiologists have tried to determine whether **probiotics**—live, **nonpathogenic** microorganisms supplied in sufficient numbers to possibly benefit our health—can be useful for preventing or treating various medical conditions. Although the diseases of interest include gastrointestinal disorders, researchers have also been studying the effects of probiotics on cancer, immune system disorders, and other illnesses. This Nutrition in Practice discusses some of the research and explains some of the issues involved in selecting and consuming probiotics. The accompanying glossary defines some relevant terms.

TABLE NP18-1 Intestinal Flora

Predominant Types	Subdominant Types
Bacteroides	Enterobacteria
Bifidobacteria	Enterococci
Clostridia	Escherichia
Eubacteria	Klebsiella
Peptococci	Lactobacilli
Peptostreptococci	Micrococci
Ruminococci	Staphylococci

TABLE NP18-2 Bacterial Populations in the Gastrointestinal Tract

Organ	Total Bacteria (per mL of contents)
Stomach, duodenum	10 to 1000
Jejunum, ileum	10^4 to 10^8
Colon	10^{10} to 10^{12}

How do our intestinal bacteria influence health?

Intestinal bacteria can benefit our health in a number of different ways. First, the bacteria degrade much of our undigested or unabsorbed dietary carbohydrate, including dietary fibers, starch that is resistant to digestion, and poorly absorbed sugars and sugar alcohols. In turn, the bacteria produce some vitamins, as well as short-chain fatty acids that our colonic epithelial cells and other body cells can use as an energy source. Intestinal bacteria also assist in the development and maintenance of mucosal tissue, protect intestinal tissue from **pathogenic** bacteria, and stimulate immune defenses in mucosal cells and other body tissues.[2]

Certain nondigestible substances in food, called **prebiotics**, can stimulate the growth or activity of resident bacteria within the large intestine. Prebiotics include some of the carbohydrates found in artichokes, asparagus, bananas, chicory root, garlic, Jerusalem artichokes, leeks, onions, and other foods.[3] Because the intestinal bacteria that degrade these substances produce gas as a by-product, people who consume high amounts of these foods may experience more flatulence than usual.

Why are certain types of bacteria considered "probiotic"?

For microbes to be "probiotic"—that is, beneficial to health—they must be nonpathogenic when consumed. They must survive their transit through the digestive tract; therefore, they must be resistant to destruction by stomach acid, bile, and other digestive substances. They should be able to alter the intestinal environment in some way that is beneficial to the human host, either by producing antimicrobial substances, altering immune defenses, metabolizing undigested foodstuffs, or protecting the intestinal walls.

Glossary

bacterial translocation: movement of bacteria across the intestinal mucosa, allowing access to body tissues.
flora: the bacteria that normally reside in a person's body.
nonpathogenic: not capable of causing disease.
pathogenic: capable of causing disease.
prebiotics: indigestible substances in foods that stimulate the growth of nonpathogenic bacteria within the large intestine.
probiotics: live microorganisms provided in foods and dietary supplements for the purpose of preventing or treating disease.

Probiotic bacteria must be consumed in large amounts—between 100 million and 100 billion live bacteria per day—to survive in sufficient numbers to influence the bacterial populations in the large intestine; a serving of yogurt usually provides these amounts. Carefully controlled studies have not found that probiotic bacteria actually *colonize* the intestine, however, as they are no longer detected in fecal or intestinal samples once ingestion of the probiotic product stops.[4] Note that only a few different types of bacteria are used in foods, and the relatively small amounts consumed cannot compete with the huge populations that normally populate our digestive tract.

What types of medical problems are helped by probiotics?

Although the results of research studies vary, probiotics may help to prevent and treat some gastric and intestinal disorders (including inflammatory bowel diseases and irritable bowel syndrome), alter susceptibility to food allergens and alleviate some allergy symptoms, and improve the availability and digestibility of various nutrients. Other potential benefits include improved immune responses, reduced symptoms of lactose intolerance, and reduced cancer risk.[5]

Much of the research investigating probiotics and intestinal illness has focused on the prevention and treatment of infectious diarrhea. For example, controlled trials have suggested that certain strains of probiotic bacteria may shorten the duration of diarrhea caused by rotavirus infection in infants and children, decrease the incidence of traveler's diarrhea in tourists visiting high-risk areas, and prevent the recurrence of infectious diarrhea in hospitalized patients.[6] In studies of children and adults using antibiotics, some strains of probiotic bacteria have been shown to reduce the incidence and duration of antibiotic-associated diarrhea. As another example, some studies have suggested that probiotic treatment may help to reduce the recurrence of *pouchitis,* an inflammation of the surgical pouch created in patients who have had an ileostomy or colostomy.[7]

Despite promising research results thus far, there are no clear conclusions about the appropriate probiotic doses or durations of treatment for many of these conditions. Moreover, the beneficial effects of one bacterial strain cannot be extrapolated to other strains of the same species, as different strains can have contrasting effects.[8] Thus, individuals who decide to consume probiotic-containing foods and supplements to benefit their health cannot be certain that the substances they use will help their condition. At best, probiotics should be considered an adjunct therapy rather than a primary treatment for an illness.

What are the main dietary sources of probiotics?

Probiotics are provided mainly by fermented foods. In the United States, yogurt and acidophilus milk are produced using various species of lactobacilli and bifidobacteria, although the species are chosen for their ability to produce desirable food products rather than their potential health benefits. In Europe and Asia, food products containing probiotic bacteria include yogurt, milk, ice cream, oatmeal gruel, and soft drinks. Although lactobacilli are used to produce various other fermented food products, such as sauerkraut, pickles, brined olives, and sausages, these foods retain few, if any, live bacteria after they undergo typical food processing methods.[9]

A number of companies market probiotic supplements, which are available in capsules, tablets, and powders.[10] Because probiotic products contain living organisms, storage conditions may affect viability—heat, moisture, and oxygen can reduce survival times—and therefore consumers should check the expiration date before purchasing a product. When a consumer group (ConsumerLab.com)

Various species of *Lactobacillus* are used in the production of fermented food products, such as the foods shown in this photo.

tested 12 probiotic supplements, they found that 2 of the products contained substantially fewer live bacteria than was claimed on the label.[11] Thus, there is no guarantee that a dietary supplement will contain the numbers of microbes expected.

Are there any potential problems associated with the use of probiotics?

Yes. One major concern is the possibility that probiotic bacteria may cause infection in immune-compromised individuals. Various species of probiotic bacteria, including *Lactobacillus* species, have been isolated from the infection sites of severely ill individuals who were consuming the probiotic.[12] Risk is increased by the use of antibiotic therapy (which reduces intestinal flora populations), illnesses or medications that suppress immunity, and illnesses that increase risk of **bacterial translocation** (including inflammatory bowel illnesses and intestinal infections). Care should be taken to inquire about probiotic use in these patients.

Other safety concerns are related to the lack of industry standards for probiotics in foods and supplements: the concentrations and strains of probiotic organisms in foods may vary substantially.[13] Thus, a consumer who wishes to try probiotics would find it difficult to determine how much of a product to consume in order to achieve the desired effect.

The microbes that inhabit the GI tract have critical roles in maintaining the integrity of intestinal tissues and influence health in various other ways. Preliminary research suggests that altering our bacterial populations by consuming probiotics or prebiotics may help to improve our defenses against certain illnesses. Additional studies are needed to verify the beneficial effects of probiotics and prebiotics and to develop standard protocols that can be used for treating illness.

Notes

1. S. O'Mahony and F. Shanahan, Enteric microbiota and small intestinal bacterial overgrowth, in M. Feldman, L. S. Friedman, and L. J. Brandt, eds., *Sleisenger and Fordtran's Gastrointestinal and Liver Disease* (Philadelphia: Saunders, 2010), pp. 1769–1778.
2. O'Mahony and Shanahan, 2010.
3. E. M. M. Quigley, Prebiotics and probiotics: Their role in the management of gastrointestinal disorders in adults, *Nutrition in Clinical Practice* 27 (2012): 195–200.
4. B. Corthésy, H. R. Gaskins, and A. Mercenier, Cross-talk between probiotic bacteria and the host immune system, *Journal of Nutrition* 137 (2007): 781S–790S.
5. G. T. Rijkers and coauthors, Guidance for substantiating the evidence for beneficial effects of probiotics: Current status and recommendations for future research, *Journal of Nutrition* 140 (2010): 671S–676S; S. Rabot and coauthors, Guidance for substantiating the evidence for beneficial effects of probiotics: Impact of probiotics on digestive system metabolism, *Journal of Nutrition* 140 (2010): 677S–689S.
6. D. Wolvers and coauthors, Guidance for substantiating the evidence for beneficial effects of probiotics: Prevention and management of infections by probiotics, *Journal of Nutrition* 140 (2010): 698S–712S; L. V. McFarland, Probiotics and diarrhea, *Annals of Nutrition and Metabolism* 57 (suppl. 1) (2010): 10–11.
7. D. Haller and coauthors, Guidance for substantiating the evidence for beneficial effects of probiotics: Probiotics in chronic inflammatory bowel disease and the functional disorder irritable bowel syndrome, *Journal of Nutrition* 140 (2010): 690S–697S.
8. Quigley, 2012.
9. L. C. Douglas and M. E. Sanders, Probiotics and prebiotics in dietetics practice, *Journal of the American Dietetic Association* 108 (2008): 510–521.
10. L. E. Morrow, V. Gogineni, and M. A. Malesker, Probiotics in the intensive care unit, *Nutrition in Clinical Practice* 27 (2012): 235–241.
11. ConsumerLab.com, Product review: Probiotics for adults, children, and pets, available at www.consumerlab.com; site visited September 1, 2012.
12. K. Whelan and C. E. Myers, Safety of probiotics in patients receiving nutritional support: A systematic review of case reports, randomized controlled trials, and nonrandomized trials, *American Journal of Clinical Nutrition* 91 (2010): 687–703.
13. E. R. Farnworth, The evidence to support health claims for probiotics, *Journal of Nutrition* 138 (2008): 1250S–1254S.

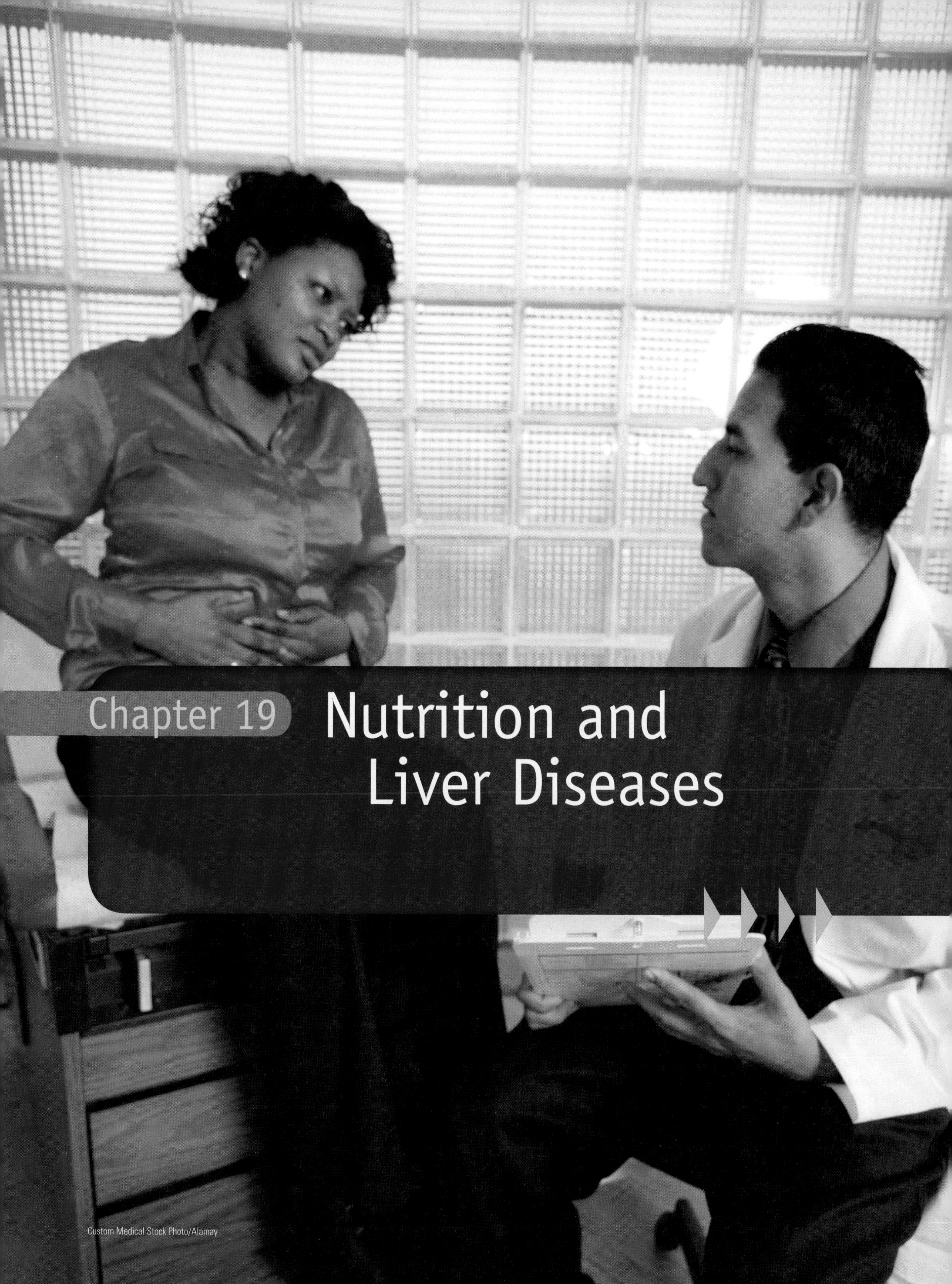

Chapter 19 Nutrition and Liver Diseases

Custom Medical Stock Photo/Alamay

THE LIVER IS THE MOST METABOLICALLY ACTIVE ORGAN IN THE BODY.

It plays a central role in processing, storing, and redistributing the nutrients provided by the foods we eat. The liver produces most of the proteins that circulate in plasma, including albumin, blood clotting proteins, and transport proteins; it also produces the bile that emulsifies fat during digestion. In addition, the liver detoxifies drugs and alcohol and processes excess nitrogen so that it can be safely excreted as urea. If damage or disease hinders the liver's ability to perform its various functions, the effects on health and nutrition status can be profound.

Liver disease progresses slowly. Its primary symptom, fatigue, often goes unnoticed. Other symptoms may be so mild that complications develop before liver disease is diagnosed. Once liver disease is recognized, preserving the remaining liver function becomes a concern because the liver can regenerate some healthy tissue, improving the prognosis. Preventing additional damage is the principal means of avoiding liver failure or transplantation.

Fatty Liver and Hepatitis

Fatty liver and hepatitis are the two most common disorders affecting the liver. Although both conditions may be mild and are usually reversible, each may progress to more serious illness and eventually cause liver damage.

FATTY LIVER

Fatty liver is an accumulation of fat in liver tissue. It develops when the amount of fat produced in the liver or picked up from the blood exceeds the amount the liver can use or export to the blood via lipoproteins. Fatty liver may be caused by defects in metabolism, exposure to various drugs and toxins, or excessive alcohol ingestion (see Nutrition in Practice 19). In cases unrelated to alcohol, insulin resistance (■) is the primary risk factor; thus, fatty liver frequently accompanies diabetes mellitus, metabolic syndrome, and obesity.[1] Other causes of fatty liver include genetic disorders, long-term total parenteral nutrition, and protein-energy malnutrition. Fatty liver is estimated to affect 20 percent or more of the adult population in the United States.[2]

Consequences of Fatty Liver In many individuals, fatty liver is asymptomatic and causes no harm. In other cases, it may be associated with inflammation (**steatohepatitis**), liver enlargement (**hepatomegaly**), and fatigue. If liver damage and scarring develop, fatty liver may progress to cirrhosis (discussed in a later section), liver failure, or liver cancer.[3]

Fatty liver is a frequent cause of abnormal liver enzyme levels in the blood. Laboratory findings may include elevated blood concentrations of the liver enzymes ALT and AST, (■) as well as increased levels of triglycerides, cholesterol, and glucose. Table 19-2 on p. 549 provides normal ranges for these liver enzymes.

Treatment of Fatty Liver The usual treatment for fatty liver is to eliminate the factors that cause it. For example, if fatty liver is due to alcohol abuse or drug treatment, it may improve after the patient discontinues use of the substance. In patients with elevated blood lipids, fatty liver may improve after blood lipid levels are lowered. An appropriate treatment for obese or diabetic patients might be weight reduction, increased physical activity, or medications that improve insulin sensitivity. Rapid weight loss should be discouraged, however, because it may accelerate the progression of liver disease.[4] Note that lifestyle modifications are not always successful in reversing fatty liver, especially in patients who lack the usual risk factors.

HEPATITIS

Hepatitis, a condition of liver inflammation, results from damage to liver tissue. Most often, the damage is caused by infection with specific viruses, designated by the letters A, B, C, D, and E. Other causes include excessive alcohol intake, exposure to some

■ Reminder: Insulin resistance is the reduced sensitivity to insulin in liver, muscle, and adipose cells.

■ The liver enzymes ALT (alanine aminotransferase) and AST (aspartate aminotransferase) are involved in amino acid catabolism.

fatty liver: an accumulation of fat in liver tissue; also called *hepatic steatosis*.

steatohepatitis (STEE-ah-to-HEP-ah-TYE-tis): liver inflammation that is associated with fatty liver.

hepatomegaly (HEP-ah-toe-MEG-ah-lee): enlargement of the liver.

hepatitis (hep-ah-TYE-tis): inflammation of the liver.

drugs and toxic chemicals, and fatty liver disease. A number of herbal products are reported to cause hepatitis; they include chaparral, germander, ma huang, jin bu huan, kava kava, kombucha, senna, and skullcap.[5] Less common causes of hepatitis include infection with other viruses and autoimmune diseases.

Viral Hepatitis In the United States, acute hepatitis (■) is most often caused by infection with hepatitis virus A, B, or C (see Table 19-1). Specific features of these viruses include the following:

- *Hepatitis A virus* (HAV) is primarily spread via fecal-oral transmission, which usually involves the ingestion of foods or beverages that have been contaminated with fecal material. Outbreaks of HAV infection are often associated with floods and other natural disasters, when inadequately treated sewage contaminates water supplies. Less frequently, HAV infection is contracted by blood transfusion. Vaccinations against HAV are recommended for food handlers, child care workers, travelers to regions where the virus is endemic, recipients of blood products, illicit drug users, and persons with unsafe sexual practices.[6] HAV infection usually resolves within a few months and does not cause chronic illness or permanent liver damage.
- *Hepatitis B virus* (HBV) is transmitted by infected blood or needles, by sexual contact with an infected person, or from mother to infant during childbirth. A major global health concern, HBV has infected one-third of the world population, although chronic illness develops in less than 10 percent of cases.[7] Vaccinations are currently recommended for newborn infants and children, health care workers, recipients of blood products, dialysis patients, sexually active adults, and users of injected drugs.
- *Hepatitis C virus* (HCV) is spread via infected blood or needles but is not readily spread by sexual contact or childbirth. Most HCV cases progress to chronic illness, and currently HCV infection is the most common cause of chronic liver disease in the United States.[8] No vaccine is available to protect against HCV infection. Preventive measures include blood donor screening, viral inactivation of blood products, infection control practices in health care settings, and risk reduction counseling to high-risk individuals.[9]

■ Hepatitis is considered *acute* if it lasts less than six months, whereas *chronic* cases are those that last six months or longer.

■ Jaundice results when liver dysfunction impairs the metabolism of bilirubin, a breakdown product of hemoglobin that is normally eliminated in bile. Accumulation of bilirubin in the bloodstream leads to yellow discoloration of tissues.

jaundice (JAWN-dis): yellow discoloration of the skin and eyes due to an accumulation of bilirubin, a breakdown product of hemoglobin that normally exits the body via bile secretions.

Symptoms of Hepatitis The effects of hepatitis depend on the cause and severity of the condition. Individuals with mild or chronic cases are often asymptomatic. The onset of acute hepatitis may be accompanied by fatigue, malaise, nausea, anorexia, and pain in the liver area. The liver is often slightly enlarged and tender. **Jaundice** (yellow discoloration of tissues) may develop, causing yellowing of the skin, urine, and sclera (whites of the eyes). (■) Other symptoms of hepatitis may include fever, muscle weakness, joint pain, and skin rashes. Serum levels of the liver enzymes ALT and AST are typically elevated. Chronic hepatitis can cause complications that are typical of liver cirrhosis and may lead to liver cancer.

NURSING DIAGNOSIS

Nursing diagnoses appropriate for people with hepatitis may include *fatigue, risk for deficient fluid volume (due to vomiting and diarrhea), imbalanced nutrition: less than body requirements, acute pain,* and *risk for impaired liver function.*

TABLE 19-1 Features of Hepatitis Viruses

Hepatitis Virus	Major Mode of Transmission	New Cases (United States, 2010)	Chronic Disease Rate (% of cases)	Chronic Cases (United States, 2010)	Vaccine
A	Fecal-oral	17,000	None	0	Available
B	Bloodborne, sexual transmission	35,000	Newborn infants: 90% Children (1 to 5 years): 30–50% Adults: 5%	0.8–1.4 million	Available
C	Bloodborne	17,000	75–85%	2.7–3.9 million	None

Note: There are fewer new cases of hepatitis C virus (HCV) infection than of hepatitis B virus (HBV) infection each year, but more HCV cases become chronic. Therefore, there are more HCV carriers than HBV carriers.

Source: Centers for Disease Control and Prevention, *Viral hepatitis surveillance: United States, 2010* (Atlanta: U.S. Department of Health and Human Services, 2012).

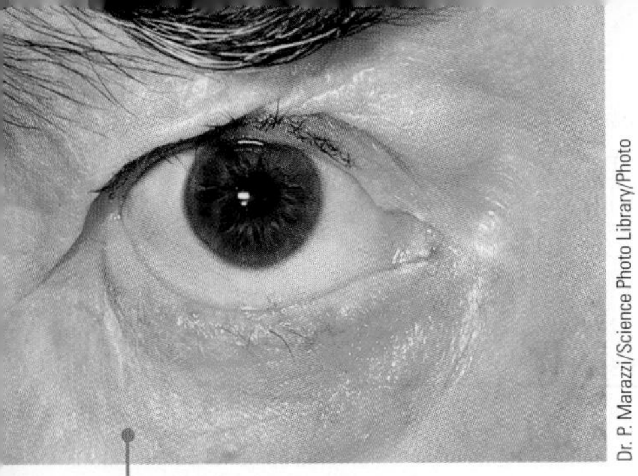
Dr. P. Marazzi/Science Photo Library/Photo Researchers, Inc.

Jaundice is a yellow discoloration of the tissues that is most easily seen in the sclera.

Treatment of Hepatitis Hepatitis is treated with supportive care, such as bed rest (if necessary) and an appropriate diet. Hepatitis patients should avoid substances that irritate the liver, such as alcohol, drugs, and dietary supplements that cause liver damage. Hepatitis A infection usually resolves without the use of medications. Antiviral agents may be used to treat HBV and HCV infections; examples include lamivudine and ribavirin, which block viral replication, and interferon alfa, which both inhibits viral replication and enhances immune responses.[10] Nonviral forms of hepatitis may be treated with anti-inflammatory and immunosuppressant drugs. Hospitalization is not required for hepatitis unless other medical conditions or complications hamper recovery.

Nutrition Therapy for Hepatitis Nutrition care varies according to a patient's symptoms and nutrition status. Most individuals require no dietary changes. Those with anorexia or abdominal discomfort may find small, frequent meals easier to tolerate. Malnourished individuals need to consume adequate protein and energy to replenish nutrient stores; the diet should include about 1.0 to 1.2 grams of protein per kilogram of body weight each day.[11] A low-fat diet, with fat limited to less than 30 percent of total kcalories, may be necessary for those with steatorrhea. Patients with persistent vomiting may require fluid and electrolyte replacement. Oral nutritional supplements can be helpful for improving nutrient intakes.

Review Notes

- Fatty liver can result from metabolic defects, exposure to some drugs and toxins, or excessive alcohol intake. Insulin resistance is a primary risk factor for fatty liver; thus, the condition often accompanies diabetes mellitus, metabolic syndrome, and obesity.
- Hepatitis is frequently caused by viral infection but may also result from alcohol abuse, drug toxicity, fatty liver, and other causes. Hepatitis B and C viruses may lead to chronic hepatitis.
- Fatty liver may be treated by eliminating the factors that cause it. Treatment of hepatitis involves supportive care, such as bed rest, elimination of liver toxins, and dietary measures that maintain or improve nutrition status.

Cirrhosis

Cirrhosis is a late stage of chronic liver disease. Long-term liver disease gradually destroys liver tissue, leading to scarring (fibrosis) in some regions and small areas of regenerated, healthy tissue in others. As the disease progresses, the scarring becomes more extensive, leaving fewer areas of healthy tissue. A cirrhotic liver is often shrunken and has an irregular, nodular appearance. Cirrhosis is characterized by impaired liver function and may eventually result in liver failure. Together, chronic liver disease and cirrhosis rank as the 12th leading cause of death in the United States.[12]

The chief causes of cirrhosis in the United States are chronic hepatitis C infection and alcoholic liver disease, followed by nonalcoholic fatty liver disease and chronic hepatitis B infection.[13] Additional causes include other types of chronic hepatitis, drug-induced liver injury, some inherited metabolic disorders, and bile duct blockages, which cause bile acids to accumulate to toxic levels in the liver.

NURSING DIAGNOSIS

Nursing diagnoses for people with cirrhosis may include *fatigue, nausea, ineffective protection, risk for impaired skin integrity, chronic pain,* and *imbalanced nutrition: less than body requirements.*

cirrhosis (sih-ROE-sis): an advanced stage of liver disease in which extensive scarring replaces healthy liver tissue, causing impaired liver function and liver failure.

CONSEQUENCES OF CIRRHOSIS

Many patients with liver disease remain asymptomatic for years. Because liver damage progresses slowly, the effects of chronic liver disease may be subtle at first. Initial symptoms are usually nonspecific and may include fatigue, malaise, anorexia, and

© Martin Rotker/Phototake

Normal liver tissue is smooth and has a regular texture.

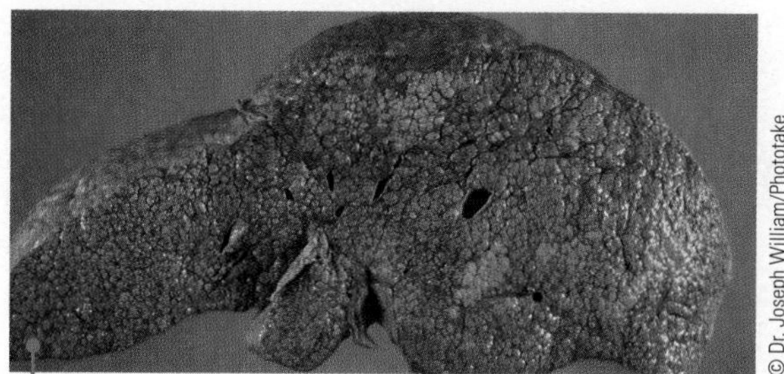

© Dr. Joseph William/Phototake

A cirrhotic liver has an irregular, nodular appearance. The nodules represent clusters of regenerating cells within the damaged liver tissue.

weight loss. Later, the decline in liver function may lead to metabolic disturbances: patients may develop anemia, bruise easily, and be more susceptible to infections. If bile obstruction occurs, jaundice, fat malabsorption, and **pruritis** (itchy skin) are likely. The physical changes in liver tissue may interfere with blood flow, causing fluid to accumulate in blood vessels and body tissues. Advanced cirrhosis can disrupt kidney, lung, and brain function. Figure 19-1 illustrates some common clinical effects of liver cirrhosis, and later sections describe some of these complications in more detail.

Table 19-2 lists laboratory tests that are used to monitor the extent of liver damage. Serum liver enzyme levels are elevated in liver disease because the injured liver tissue releases the enzymes into the bloodstream. Serum levels of bilirubin may be elevated if the liver is too damaged to process it or if bile ducts are blocked and prevent its excretion. Reduced synthesis of plasma proteins in the liver lowers albumin levels and extends blood-clotting time. Liver damage also impairs the conversion of ammonia to urea, causing ammonia levels in the blood to rise.

■ Reminder: The *portal vein* is the large blood vessel that carries nutrient-rich blood from the GI tract to the liver.

pruritis: itchy skin.

Portal Hypertension A large volume of blood normally flows through the liver. The portal vein (■) and hepatic artery together supply approximately 1.5 liters (about 1.5 quarts) of blood each

FIGURE 19-1 Clinical Effects of Liver Cirrhosis

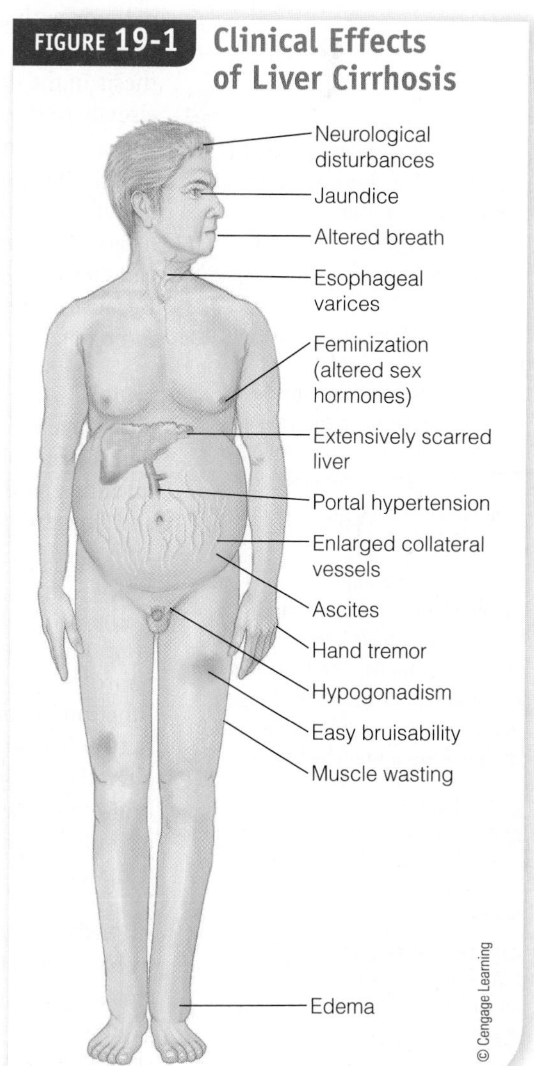

© Cengage Learning

TABLE 19-2 Laboratory Tests for Evaluation of Liver Disease

Laboratory Test	Normal Ranges (serum)	Values in Liver Disease
Alanine aminotransferase (ALT)	Male: 10–40 U/L Female: 7–35 U/L	Elevated
Albumin	3.4–4.8 g/dL	Decreased
Alkaline phosphatase	25–100 U/L	Normal or elevated
Ammonia	15–45 µg N/dL	Elevated
Aspartate aminotransferase (AST)	10–30 U/L	Elevated
Bilirubin (total)	0.3–1.2 mg/dL	Elevated
Blood urea nitrogen (BUN)	6–20 mg/dL	Normal or decreased
Prothrombin time[a]	11–15 seconds	Prolonged

[a]The test for prothrombin time evaluates the clotting ability of blood.
Note: U/L = units per liter; dL = deciliter; µg = micrograms; N = nitrogen

Source: L. Goldman and A. I. Schafer, eds., *Goldman's Cecil Medicine* (Philadelphia: Saunders, 2012).

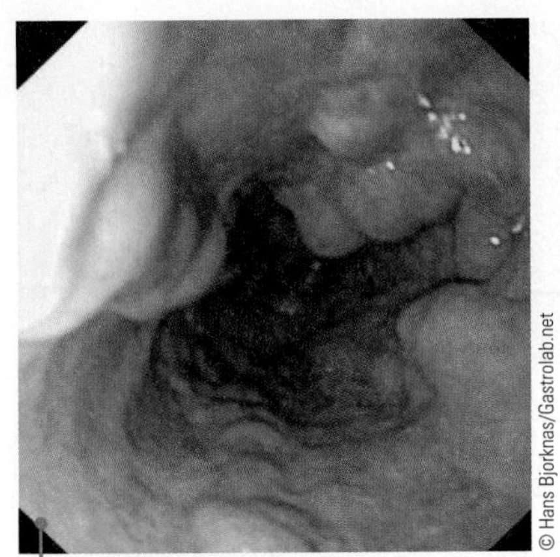

© Hans Bjorknas/Gastrolab.net

Esophageal varices, such as the one shown here, may protrude into the lumen and be vulnerable to rupture and bleeding.

minute to the extensive network of vessels in the liver. The scarred tissue of a cirrhotic liver impedes the flow of blood, three-fourths of which is supplied by the portal vein.[14] In addition, the restricted blood flow within the liver stimulates the release of vasodilators (such as nitric oxide) in nearby arteries, leading to a greater volume of portal blood. The increased portal blood coupled with resistance to blood flow within the liver causes a rise in blood pressure within the portal vein, called **portal hypertension**.

Collateral Vessels and Gastroesophageal Varices When blood flow through the portal vein is impeded, the blood is forced backward into the veins that normally supply the portal vein with blood. The blood is then diverted to the systemic circulation via **collateral vessels**, which develop and expand throughout the gastrointestinal (GI) tract and in regions near the abdominal wall. As portal pressure builds, some of these collaterals can become enlarged and engorged with blood, forming abnormally dilated vessels called **varices** (see the photo). The varices that develop in the esophagus (*esophageal varices*) and stomach (*gastric varices*) are vulnerable to rupture because they have thin walls and often bulge into the lumen. If ruptured, they can cause massive bleeding that is sometimes fatal. The blood loss is exacerbated by the liver's reduced production of blood-clotting factors.

Ascites Within 10 years of disease onset, about 50 percent of cirrhosis patients develop **ascites**, a large accumulation of fluid in the abdominal cavity. The development of ascites indicates that liver damage has reached a critical stage, as half of patients with ascites die within 5 years.[15] Ascites is primarily a consequence of portal hypertension, sodium and water retention in the kidneys, and reduced albumin synthesis in the diseased liver. As a result of portal hypertension, the distorted blood flow elsewhere in the body alters kidney function, leading to sodium and water retention and an accumulation of body fluid. Elevated pressure within the liver's small blood vessels (**sinusoids**) causes fluid to leak into lymphatic vessels and, ultimately, the abdominal cavity. The movement of water into the abdomen is exacerbated by low levels of serum albumin, a protein that helps to retain fluid in blood vessels. Ascites can cause abdominal discomfort and early satiety, which contribute to malnutrition. Because ascites can raise body weight considerably, weight changes may be difficult to interpret.

Hepatic Encephalopathy Advanced liver disease often leads to **hepatic encephalopathy**, a disorder characterized by abnormal neurological functioning. Signs of hepatic encephalopathy include adverse changes in personality, behavior, mood, mental ability, and motor functions (see Table 19-3). At worst, amnesia, unresponsiveness, and **hepatic coma** may develop. Although hepatic encephalopathy is fully reversible with medical treatment, the prognosis is poor when it progresses to the advanced stages.

The exact causes of hepatic encephalopathy remain elusive, although elevated blood ammonia levels are thought to play a key role in its development due to ammonia's neurotoxicity. Other substances that may accumulate in brain tissue and disturb brain function include sulfur compounds, naturally occurring benzodiazepines, short-chain fatty acids, and manganese.[16] Another theory is that the brain's neurotransmitters are altered by an increased ratio of aromatic amino acids to branched-chain amino acids in brain tissue, a result of disordered amino acid metabolism in the liver. (■) Most likely, a combination of metabolic abnormalities contributes to the disruption in neurological functioning.

Elevated Blood Ammonia Levels Much of the body's free ammonia is produced by bacterial action on unabsorbed dietary protein in the colon. Normally, the liver extracts this ammonia from portal blood and converts it to urea, which is then excreted by the kidneys. In advanced liver disease, the liver is unable to process the ammonia sufficiently. In addition, much of the ammonia-laden blood bypasses the liver by way of collateral vessels and reaches the general blood circulation, causing a substantial increase in the

NURSING DIAGNOSIS

Nursing diagnoses appropriate for people with esophageal or gastric varices include *ineffective protection* and *risk for bleeding.*

NURSING DIAGNOSIS

Nursing diagnoses appropriate for people with ascites may include *impaired comfort* and *imbalanced nutrition: less than body requirements.*

NURSING DIAGNOSIS

Nursing diagnoses appropriate for people with hepatic encephalopathy may include *chronic confusion, impaired memory, disturbed sleep pattern,* and *adult failure to thrive.*

■ The aromatic amino acids—phenylalanine, tyrosine, and tryptophan—have carbon rings in their side groups. The branched-chain amino acids are leucine, isoleucine, and valine; their side groups have a branched structure.

portal hypertension: elevated blood pressure in the portal vein due to obstructed blood flow through the liver and greater inflow of portal blood.

collateral vessels: blood vessels that enlarge or newly form to allow an alternative pathway for diverted blood.

varices (VAH-rih-seez): abnormally dilated blood vessels (singular: *varix*).

ammonia that reaches brain tissue. Although ammonia levels do not correlate well with the degree of neurological impairment in hepatic encephalopathy, ammonia-reducing medications can successfully reverse the neurological symptoms.[17]

Malnutrition and Wasting Most patients with cirrhosis develop protein-energy malnutrition (PEM) and experience some degree of wasting. Malnutrition is usually caused by a combination of factors (see Table 19-4). Patients may consume less food due to reduced appetite, GI symptoms, early satiety associated with ascites, or fatigue. If the diet is restricted in sodium (to treat ascites), foods may seem unpalatable. Fat malabsorption is common due to reduced bile availability, which may lead to steatorrhea and deficiencies of the fat-soluble vitamins and some minerals. Additional nutrient losses may result from diarrhea, vomiting, and GI bleeding. If cirrhosis is a consequence of alcohol abuse, multiple nutrient deficiencies may be present.

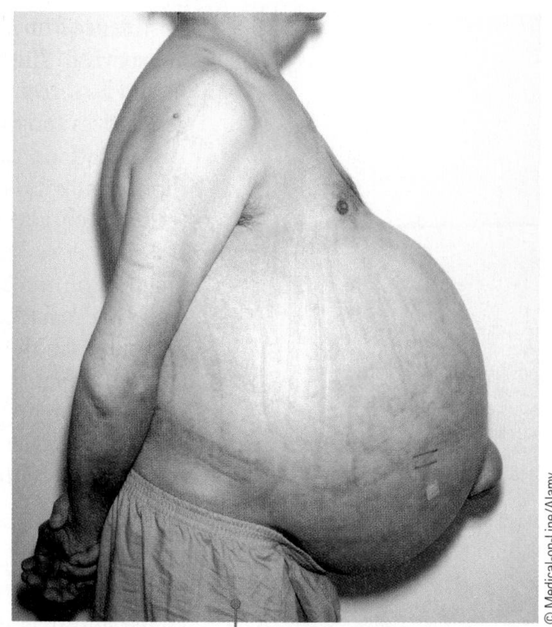

© Medical-on-Line/Alamy

Ascites can be caused by various illnesses, but cirrhosis is the underlying cause in most patients with the condition.

TREATMENT OF CIRRHOSIS

Medical treatment for cirrhosis aims to correct the underlying cause of disease and prevent or treat complications. Supportive care, including an appropriate diet and avoidance of liver toxins, promotes recovery and helps to prevent further damage.

TABLE 19-3 Stages of Hepatic Encephalopathy

Early Stages	Middle Stages	Later Stages
• Personality changes • Short attention span • Depression, irritability • Lack of coordination • Tremor • Sleep disturbances	• Mood and behavior changes • Disorientation • Lethargy • Slurred speech • Pronounced tremor • Changes in sleep–wake cycle	• Confusion, amnesia • Somnolence to semi-stupor • Involuntary eye movements • Muscular rigidity • Abnormal reflexes • Coma

TABLE 19-4 Possible Causes of Malnutrition in Liver Disease

Mechanism	Examples
Reduced nutrient intake	Abdominal discomfort, altered mental status, altered taste sensation, anorexia, early satiety (due to ascites), effects of medications (including GI disturbances and taste changes), fasting for medical procedures, fatigue, nausea and vomiting, dietary restrictions
Malabsorption or nutrient losses	Diarrhea, effects of medications (including malabsorption and nutrient losses from diuretic use), fat malabsorption (due to reduced bile flow), GI bleeding, vomiting
Altered metabolism or increased nutrient needs	Hypermetabolism, impaired protein synthesis, infections or inflammation, muscle catabolism, reduced nutrient storage and metabolism in the liver

ascites (ah-SIGH-teez): an abnormal accumulation of fluid in the abdominal cavity.

sinusoids: the small, capillary-like passages that carry blood through liver tissue.

hepatic encephalopathy (en-sef-ah-LOP-ah-thie): a neurological complication of advanced liver disease that is characterized by changes in personality, mood, behavior, mental ability, and motor functions.

encephalo = brain
pathy = disease

hepatic coma: loss of consciousness resulting from severe liver disease.

Abstinence from alcohol is critical for preserving the remaining liver function and extending survival. Antiviral medications may be prescribed to treat viral infections. Patients should be screened and treated for life-threatening complications, such as gastroesophageal varices and liver cancer. Liver transplantation may be necessary in advanced cirrhosis.

Medications can effectively treat many of the complications that accompany cirrhosis. Individuals with portal hypertension and varices may be given propranolol (Inderal) or octreotide (Sandostatin), which reduce portal blood pressure and bleeding risk. Diuretics can help to control portal hypertension and ascites; common examples include spironolactone (Aldactone) and furosemide (Lasix). Lactulose, a nonabsorbable disaccharide, treats hepatic encephalopathy by reducing ammonia production and absorption in the colon. The antibiotic rifaximin is an alternative treatment for elevated ammonia that works by altering bacterial populations. To stimulate the appetite and promote weight gain, megestrol acetate (Megace) or dronabinol (Marinol) may be prescribed. The Diet-Drug Interactions feature lists potential nutritional problems associated with these medications.

NUTRITION THERAPY FOR CIRRHOSIS

Nutrition care for cirrhosis is customized to each patient's needs, which vary considerably and depend on the accompanying complications. Common problems include

DIET-DRUG INTERACTIONS

Check this table for notable nutrition-related effects of the medications discussed in this chapter.

Appetite stimulants (megestrol acetate, dronabinol)	**Gastrointestinal effects:** Nausea, vomiting, diarrhea
	Dietary interaction: Dronabinol potentiates the effects of alcohol
	Metabolic effect: Hyperglycemia (megestrol acetate)
Diuretics (furosemide, spironolactone)	**Gastrointestinal effects:** Dry mouth, anorexia, decreased taste perception
	Dietary interactions: Furosemide's bioavailability is reduced when taken with food; licorice root may interfere with the effects of diuretics
	Metabolic effects: Fluid and electrolyte imbalances,[a] hyperglycemia, hyperlipidemias, thiamin deficiency (furosemide), elevated uric acid levels (furosemide)
Immunosuppressants (cyclosporine, tacrolimus)	**Gastrointestinal effects:** Nausea, vomiting, abdominal discomfort, diarrhea, constipation, anorexia
	Dietary interactions: Limit alcohol intake due to the potential for toxic effects; the bioavailability of tacrolimus is reduced when the drug is taken with food; grapefruit juice can raise serum concentrations of these drugs to toxic levels
	Metabolic effects: Electrolyte imbalances, hyperglycemia, hyperlipidemias, anemia
Lactulose	**Gastrointestinal effects:** Nausea, vomiting, diarrhea, flatulence
	Dietary interactions: Calcium and magnesium supplements may reduce the effectiveness of lactulose
	Metabolic effects: Fluid and electrolyte imbalances

[a]*Furosemide* is a potassium-wasting diuretic; patients should increase intakes of potassium-rich foods. *Spironolactone* is a potassium-sparing diuretic; patients should avoid supplemental potassium and potassium-containing salt substitutes.

PEM and muscle wasting; therefore, protein and energy intakes must be sufficient for maintaining nitrogen balance. Dietary substances that may cause additional liver injury should be avoided; examples include alcohol, some herbal supplements, and vitamin or mineral megadoses. If esophageal varices are present, a soft diet (■) may be prescribed to reduce risk of bleeding. Table 19-5 lists the general dietary guidelines for cirrhosis, which are discussed in the sections that follow.

Energy Daily energy needs for patients with cirrhosis often fall within the range of 35 to 40 kilocalories per kilogram of dry body weight (weight without ascites);[18] energy intakes can be started within this range and adjusted as necessary to achieve a healthy body weight. Requirements may be higher for patients with hypermetabolism, infection, nutrient malabsorption, or recent unintentional weight loss. A value for dry weight can be obtained after diuretic therapy or after a medical procedure that directly removes excess abdominal fluid.

Many patients with cirrhosis have difficulty consuming enough food to achieve good nutrition status. Some individuals may find four to six small meals easier to tolerate than three large meals each day. Oral supplements, including liquid formulas and energy bars, can help to improve energy intakes. The "How to" on p. 554 offers additional suggestions that can help a patient meet energy needs.

■ Reminder: A *soft diet* contains only moist, soft-textured foods; see Table 14-5 in Ch. 14 (p. 410) for details.

TABLE 19-5 Nutrition Therapy for Liver Cirrhosis

Energy
- Energy needs may range from 35 to 40 kcal/kg body weight per day; the energy intake can be started within this range and adjusted as necessary.
- Energy requirements may be higher in patients with hypermetabolism, infection, malabsorption, or malnutrition. Energy requirements may be lower in patients who would benefit from weight loss.
- In patients with ascites, use the estimated dry body weight when calculating nutrition needs.

Meal frequency
- To improve food intake, patients should consume small meals four to six times daily.

Protein
- Provide 0.8 to 1.2 g protein/kg dry body weight per day to maintain or improve nitrogen balance.
- Some patients may require 1.2 to 1.5 g protein/kg dry body weight per day.
- In patients with hepatic encephalopathy, the protein intake should be spread throughout the day; protein restriction is not recommended as it may worsen malnutrition.

Carbohydrate
- No carbohydrate restrictions unless the patient has insulin resistance or diabetes.
- Persons with insulin resistance or diabetes should monitor carbohydrate intakes and consume a diet that maintains blood glucose control.

Fat
- No fat restrictions unless fat malabsorption is present.
- If fat is malabsorbed, restrict fat to 30% of total kcalories or as necessary to control steatorrhea; use medium-chain triglycerides (MCT) to increase kcalories.

Sodium and fluid
- Restrict sodium as necessary to control ascites; 2000 mg sodium per day is adequate restriction in most cases.
- If ascites is accompanied by low serum sodium levels (less than 128 mEq/L), restrict fluids to 1200 to 1500 mL per day. In severe cases (sodium level less than 125 mEq/L), restrict fluids to 1000 to 1200 mL per day.

Vitamins and minerals
- Ensure adequate intake from diet or supplements based on individual needs.

HOW TO Help the Cirrhosis Patient Eat Enough Food

Individuals with cirrhosis often have difficulty consuming enough food to prevent malnutrition and its consequences. Ascites and GI symptoms such as nausea and vomiting may interfere with food intake. Fatigue may cause a lack of interest in food preparation. Sodium restrictions may make foods unpalatable. To improve food intake:

- If nutrient restrictions are necessary, make sure the patient fully understands how to modify the diet so that food intake is not restricted unnecessarily. Provide lists of acceptable foods and menus. Explain how recipes can be altered so that favorite foods can still be incorporated into the diet.
- Suggest between-meal snacks during the day and a snack at bedtime. A liquid supplement like Boost or Ensure can substitute for a snack and requires no preparation. Snacks should not be consumed within two hours of meals, or they may reduce appetite at mealtime.
- If the patient has little appetite or is quickly satiated, suggest foods that are higher in food energy, such as whole milk instead of reduced-fat milk or canned fruit that is packed in heavy syrup instead of fruit juice. Suggest that beverages be consumed separately from meals.
- Recommend energy boosters. Cream sauces and gravies can add kcalories to entrées. Fruit juices and fruit nectars can substitute for drinking water. The following additions can boost the energy content of meals:
  - Sour cream and butter—on vegetables and potatoes
  - Mayonnaise—in sandwiches and salads
  - Half-and-half and light cream—in soups and on cereals
  - Hard-cooked eggs—in casseroles and meat loaf
  - Cheese—in salads and casseroles and melted on steamed vegetables
  - Peanut butter, nut butters, and cream cheese—on crackers or celery and in milk shakes
  - Chopped nuts—in salads, cooked cereals, and bakery products

Sodium-restricted diets are recommended for treating ascites and other medical conditions, including kidney and heart disorders. The "How to" on p. 612 (Chapter 21) offers suggestions to help patients implement sodium restrictions. To improve the palatability of low-sodium meals:

- Suggest that patients replace salt with strong-flavored herbs and spices such as chili powder, coriander, cumin, curry powder, garlic, ginger, lemon, mint, and parsley.
- Advise patients to check food labels to learn the sodium content of packaged foods. Similar products may be available that are lower in sodium. (Persons using potassium-sparing diuretics should be cautioned to avoid salt substitutes that replace sodium with potassium.)

Offer support and encouragement to the patient with cirrhosis. Significant weight loss is less likely to occur if dietary advice is provided before problems progress.

Protein To maintain or improve nitrogen balance, the protein recommendation is 0.8 to 1.2 grams of protein per kilogram of body weight per day based on dry weight or an appropriate weight for height.[19] (■) Some patients may need to consume higher amounts of protein (1.2 to 1.5 grams of protein per kilogram body weight daily) to prevent wasting or improve protein status.[20] In patients with hepatic encephalopathy, the protein intake should be spread throughout the day so that only modest amounts are consumed at each meal. Protein restriction is not helpful because an inadequate protein intake can worsen malnutrition and wasting.

In an attempt to normalize plasma amino acid ratios and possibly improve the mental status of patients with hepatic encephalopathy, some health care providers may prescribe enteral formulas enriched with branched-chain amino acids. Clinical studies testing the use of these formulas have yielded mixed results, however, and their use is recommended only in patients who do not respond to conventional treatment.[21]

Carbohydrate and Fat Carbohydrate provides a substantial proportion of energy needs. Many patients with cirrhosis are insulin resistant, however, and require medications or insulin to manage their hyperglycemia. These individuals should follow the dietary guidelines for diabetes: monitor carbohydrate intakes and consume a diet that maintains blood glucose levels within a normal range (see Chapter 20). Carbohydrate intakes should be fairly consistent from day to day for improved blood glucose control.

Fat provides both energy and essential fatty acids. In patients with fat malabsorption, fat intake may be restricted to less than 30 percent of total kcalories or as necessary to control steatorrhea. Medium-chain triglycerides (MCT) may be used to provide additional energy, although essential fatty acids cannot be obtained from MCT oils and may need to be supplemented. Severe steatorrhea warrants supplementation of the fat-soluble vitamins, calcium, magnesium, and zinc (see Chapter 18).

■ Reminder: The protein RDA for healthy adults is 0.8 g/kg.

Sodium and Fluid Patients with ascites are typically advised to restrict sodium. Because ascites is partly caused by sodium and water retention in the kidneys, treatment usually includes both sodium restriction (to no more than 2000 milligrams of sodium per day) and diuretic therapy to promote fluid loss. (■) Potassium intake should be monitored if a potassium-wasting diuretic (such as furosemide) is used.

Many patients find low-sodium diets unpalatable, so some health practitioners may allow a more liberal sodium intake and depend on diuretics to remove excess fluid. If patients do not respond to sodium restriction and diuretic therapy, fluid may be removed from the abdomen by surgical puncture (**paracentesis**) or may be diverted to the bloodstream using a catheter (**peritoneovenous shunt**).

Fluid restriction (■) may be necessary when ascites is accompanied by a low concentration of serum sodium. If the sodium level falls below 128 milliequivalents per liter, the fluid intake should be limited to 1200 to 1500 milliliters daily; with a sodium level below 125 milliequivalents per liter, fluids should be restricted to 1000 to 1200 milliliters per day.[22]

Vitamins and Minerals Vitamin and mineral deficiencies are common in patients with cirrhosis due to the effects of illness, disease complications, or the alcohol abuse that may have induced liver disease. Thus, nutrient supplementation is often necessary. If steatorrhea is present, fat-soluble nutrients can be provided in water-soluble forms. Patients with esophageal varices may find it easier to ingest supplements in liquid form.

Food Safety Because people with cirrhosis are susceptible to infections, they should avoid foods that may increase risk of foodborne illness, such as unpasteurized milk products; undercooked meat, fish, poultry, and eggs; unwashed fruits and vegetables; raw vegetable sprouts; and unpasteurized juices.[23] Nutrition in Practice 2 provides additional information about safe food practices (pp. 53–62).

Enteral and Parenteral Nutrition Support In patients who are unable to consume enough food, tube feedings may be infused overnight as a supplement to oral intakes or may replace oral feedings entirely. Although standard formulas are often appropriate, an energy-dense, moderate-protein, low-electrolyte formula may be necessary for patients with ascites or fluid restrictions. In patients with esophageal varices, the feeding tube should be as narrow and flexible as possible to prevent rupture and bleeding. Parenteral nutrition support should be considered for patients who are unable to tolerate enteral feedings due to intestinal obstruction, gastrointestinal bleeding, or uncontrollable vomiting. To avoid excessive fluid delivery, patients with ascites typically require concentrated parenteral solutions, which are infused into central veins. The Case Study can help you apply your knowledge of cirrhosis to a clinical situation.

■ Table 22-1 (p. 627) and the "How to" on p. 612 (Chapter 21) provide information about restricting dietary sodium.

■ The "How to" on p. 637 (Chapter 22) describes a method for implementing fluid restriction.

paracentesis (pah-rah-sen-TEE-sis): a surgical puncture of a body cavity with an aspirator to draw out excess fluid.

peritoneovenous (PEH-rih-toe-NEE-oh-VEE-nus) shunt: a surgical passage created between the peritoneum and the jugular vein to divert fluid and relieve ascites. The peritoneum is the membrane that surrounds the abdominal cavity.

Case Study: Man with Cirrhosis

Lenny Levitt, a 49-year-old carpenter, has just been diagnosed with cirrhosis, which is a consequence of his alcohol abuse over the past 25 years. Although he understands that he has an alcohol problem and recently entered an alcohol rehabilitation program, he is still drinking. At 5 feet 8 inches tall, Mr. Levitt, who formerly weighed 160 pounds, now weighs 130 pounds. According to family members, he is showing signs of mental deterioration, such as forgetfulness and an inability to concentrate. He is jaundiced and appears thin, although his abdomen is distended with ascites. Laboratory findings indicate elevated serum concentrations of AST, ALT, and ammonia; reduced albumin levels; and hyperglycemia.

1. Do Mr. Levitt's laboratory values suggest liver disease? Compare the results of his laboratory tests with the values shown in Table 19-2.
2. From the limited information available, evaluate Mr. Levitt's nutrition status. What medical problem makes it difficult to interpret his present weight? Describe the development of that type of problem in liver disease, and explain how the diet is usually adjusted for such a patient.
3. Estimate Mr. Levitt's energy and protein needs. Describe the general diet you might recommend for him. What suggestions do you have for increasing his energy intake?
4. Explain the significance of Mr. Levitt's elevated blood ammonia levels. What signs may indicate that he is undergoing mental decline?
5. Describe each of the following complications of liver disease: portal hypertension, jaundice, and gastroesophageal varices. What complication may result if the esophageal varices are not treated?

Review Notes

- Liver cirrhosis is characterized by extensive fibrosis and progressive liver dysfunction. The primary causes of cirrhosis in the United States are hepatitis C infection and alcoholic liver disease.
- Symptoms of cirrhosis include fatigue, GI disturbances, anorexia, and weight loss; eventually, patients may bruise easily and be more susceptible to infections.
- Complications of cirrhosis include portal hypertension, gastroesophageal varices, ascites, hepatic encephalopathy, and malnutrition.
- Treatment of cirrhosis is highly individualized and depends on the accompanying symptoms and complications. Both drug therapies and dietary adjustments are usually necessary. If warranted, the diet may need to be restricted in fat, sodium, or fluids.

Liver Transplantation

Acute or chronic liver disease can lead to liver failure, in which case liver transplantation is the only remaining treatment option. The most common illnesses that precede liver transplantation are chronic hepatitis C infection and alcoholic liver disease, which account for about 50 percent of liver transplant cases.[24] The five-year survival rate among transplant recipients ranges from 54 to 81 percent, depending on the cause of illness.[25] Complications such as ascites and hepatic encephalopathy worsen the prognosis.

Nutrition Status of Transplant Patients As mentioned earlier, advanced liver disease is usually associated with malnutrition, which can increase the risk of complications following a liver transplant. Evaluating nutrition status in transplant candidates can be difficult, however, because liver dysfunction and malnutrition often have similar metabolic effects. If fluid retention is present, it can mask weight loss and alter anthropometric and laboratory values. Correcting malnutrition prior to transplant surgery can help speed recovery after the surgery.

Posttransplantation Concerns The immediate concerns following a transplant are organ rejection and infection. Immunosuppressive drugs, including prednisone, cyclosporine, and tacrolimus, help to reduce the immune responses that cause rejection, but they also raise the risk of infection. Infections are a potential cause of death following a liver transplant; therefore, antibiotics and antiviral medications are prescribed to reduce infection risk.

NURSING DIAGNOSIS

Nursing diagnoses for people with liver transplants may include *ineffective protection, risk for infection,* and *imbalanced nutrition: less than body requirements.*

Immunosuppressive drugs can affect nutrition status in numerous ways. Gastrointestinal side effects include nausea, vomiting, diarrhea, abdominal pain, and mouth sores. Some medications may alter appetite and taste perception. Some of the drugs may cause hyperglycemia or outright diabetes, which may need to be controlled with insulin. Electrolyte and fluid imbalances are common. Other possible effects include hypertension, hyperlipidemias, kidney toxicity, protein catabolism, and increased osteoporosis risk.[26]

Protein and energy requirements are increased after transplantation due to the stress of surgery. High-kcalorie, high-protein snacks and oral supplements can help the transplant patient meet postsurgical needs. Vitamin and mineral supplementation is also an integral part of nutrition care. To help transplant patients avoid developing foodborne illnesses, health practitioners should provide information about food safety measures, such as cooking foods adequately, washing fresh produce, and avoiding foods that may be contaminated. See Nutrition in Practice 2 (pp. 53–62) for additional information on food safety.

Review Notes

- Liver transplantation has improved the long-term outlook for patients with advanced liver disease. Transplant patients are typically malnourished and may have medical problems that affect transplant success.
- Due to the potential for organ rejection, immunosuppressive drugs are prescribed following surgery. Use of these drugs increases the risk of infection, and the drugs have side effects that can impair nutrition status and general health.

Nutrition Assessment Checklist for People with Disorders of the Liver

MEDICAL HISTORY

Check the medical record to determine:

- ❍ Type of liver disorder
- ❍ Cause of the liver disorder
- ❍ If the patient has received a liver transplant

Review the medical record for complications that may alter nutritional needs, including:

- ❍ Abdominal pain
- ❍ Anemia
- ❍ Ascites
- ❍ Esophageal varices
- ❍ Hepatic encephalopathy
- ❍ Impaired kidney or lung function
- ❍ Infections
- ❍ Insulin resistance or diabetes mellitus
- ❍ Malabsorption
- ❍ Malnutrition
- ❍ Pancreatitis

MEDICATIONS

In patients with liver dysfunction, the risk of diet-drug interactions is high because most drugs are metabolized in the liver. Risk of interactions is intensified for patients with:

- ❍ Ascites (medications may take a long time to reach the liver)
- ❍ Renal failure (medications often undergo further metabolism in the kidneys and are excreted in the urine)
- ❍ Malnutrition
- ❍ Multiple prescriptions
- ❍ Long-term medication use

DIETARY INTAKE

For patients with fatty liver, pay special attention to:

- ❍ Energy intake, if the patient is overweight or malnourished, has diabetes, or is receiving total parenteral nutrition
- ❍ Carbohydrate intake, if the patient has diabetes or is receiving total parenteral nutrition
- ❍ Alcohol abuse

For patients with hepatitis, cirrhosis, or ascites:

- ❍ Check appetite.
- ❍ Ensure that energy and nutrient intakes are adequate.
- ❍ Determine whether alcohol is being consumed.
- ❍ Determine whether sodium or fluid restriction is warranted.
- ❍ Base energy intakes on desirable weight or an estimated dry weight to avoid overfeeding.

ANTHROPOMETRIC DATA

Take baseline height and weight measurements, and monitor weight regularly. For patients with ascites and edema:

- ❍ Monitor weight changes to evaluate the degree of fluid retention.
- ❍ Remember that the patient may be malnourished and weight may be deceptively high.

LABORATORY TESTS

Note that albumin and serum proteins are often reduced in people with liver disease and are not appropriate indicators of nutrition status. Review the following laboratory test results to assess liver function:

- ❍ Albumin
- ❍ Alkaline phosphatase
- ❍ ALT and AST
- ❍ Ammonia
- ❍ Bilirubin
- ❍ Prothrombin time

Check laboratory test results for complications associated with liver failure, including:

- ❍ Anemia
- ❍ Decreased renal function
- ❍ Fluid retention
- ❍ Hyperglycemia

PHYSICAL SIGNS

Look for physical signs of:

- ❍ Fluid retention (ascites and edema)
- ❍ PEM (muscle wasting and unintentional weight loss)
- ❍ Nutrient deficiencies

Self Check

1. In cases of fatty liver that are unrelated to excessive alcohol intakes, the primary risk factor is:

a. following a high-protein diet.
b. use of illicit drugs.
c. following a high-fat diet.
d. insulin resistance.

2. Which of the following statements about hepatitis is true?

a. Chronic hepatitis can progress to cirrhosis.
b. Whatever the cause of hepatitis, symptoms are typically severe.
c. Vaccines are available to protect against hepatitis A, B, and C viruses.
d. HCV infection is usually spread through contaminated foods and water.

3. Which blood test values often indicate the presence of liver disease?

a. Elevated albumin levels
b. Elevated liver enzyme levels
c. Reduced bilirubin levels
d. Reduced alkaline phosphatase levels

4. Esophageal varices are a dangerous complication of liver disease primarily because they:

a. interfere with food intake.
b. divert blood flow from the GI tract.
c. can lead to massive bleeding.
d. contribute to hepatic encephalopathy.

5. A complication of cirrhosis that contributes to the development of ascites is:

a. portal hypertension.
b. increased sodium excretion by the kidneys.
c. elevated serum albumin levels.
d. bile obstruction.

6. Lactulose, a nonabsorbable disaccharide, is generally prescribed to patients who develop which complication of cirrhosis?

a. Portal hypertension
b. Hepatic encephalopathy
c. Bile insufficiency
d. Ascites

7. With respect to protein intake, patients with hepatic encephalopathy should:

a. consume a high-protein diet.
b. restrict protein intake to avoid elevations in serum ammonia.
c. spread protein intake evenly throughout the day.
d. use formulas enriched with aromatic amino acids to meet their protein needs.

8. People with ascites must often restrict dietary intake of:

a. fat.
b. protein.
c. sugars.
d. sodium.

9. After a liver transplant, a major health concern is:

a. jaundice.
b. bile insufficiency.
c. infection.
d. hepatic coma.

10. Dietary concerns after a liver transplant include all of the following *except:*

a. severe protein restrictions that are difficult to adhere to.
b. increased risk of foodborne illness.
c. gastrointestinal side effects of medications.
d. altered appetite and taste perception from medications.

Answers to these questions appear in Appendix H. For more chapter review: Access an interactive eBook, chapter-specific interactive learning tools, including flashcards, quizzes, videos, and more in your Nutrition CourseMate, accessed through CengageBrain.com.

Clinical Applications

1. Vijaya Reddy is a college student who visited relatives near her parents' birthplace in Anantapur, India, during summer vacation. Although her relatives provided boiled or purified water at their home, they occasionally took Vijaya to local restaurants, where she drank tap water. Several weeks after Vijaya returned home, she developed flu-like symptoms and started feeling extremely tired. She also experienced upper abdominal pain and felt nauseous after meals. After her roommate told her that her eyes and skin appeared yellow, she knew something was definitely wrong. A physician at the student health center diagnosed hepatitis.
   - Which type of hepatitis did Vijaya most likely have?
   - What additional symptoms may develop? Is Vijaya's condition likely to become chronic?
   - What medical treatment is suggested for Vijaya's condition? Describe the dietary modifications that may be necessary in some cases.

2. As discussed in the section on cirrhosis, many patients develop protein-energy malnutrition and wasting during the course of illness. Review Table 19-4 to find examples of problems that may lead to malnutrition. Select three nutrition or medical problems (from the "Examples" column), and discuss the complications of liver disease that may cause the problems you selected. What dietary or medical treatments can help in managing these problems?

Nutrition on the Net

For further study of topics covered in this chapter, access these websites:

- To obtain additional information about liver diseases, visit the American Liver Foundation and the Canadian Liver Foundation:
  www.liverfoundation.org and www.liver.ca
- The websites of the following organizations include information that is helpful for both health practitioners and patients with liver diseases:
  American College of Gastroenterology:
  gi.org
  American Gastroenterological Association:
  www.gastro.org
- Learn more about hepatitis by visiting the Hepatitis Foundation International:
  www.hepfi.org
- To uncover more information about liver transplants, search the CenterSpan Transplant News Network:
  www.centerspan.org

Notes

1. D. E. Cohen and F. A. Anania, Nonalcoholic fatty liver disease, in N. J. Greenberger, ed., *Current Diagnosis and Treatment: Gastroenterology, Hepatology, and Endoscopy* (New York: McGraw-Hill Companies, 2012), pp. 502–508.
2. Cohen and Anania, 2012.
3. A. E. Reid, Nonalcoholic fatty liver disease, in M. Feldman, L. S. Friedman, and L. J. Brandt, eds., *Sleisenger and Fordtran's Gastrointestinal and Liver Disease* (Philadelphia: Saunders, 2010), pp. 1401–1411.
4. Reid, 2010.
5. J. H. Lewis, Liver disease caused by anesthetics, toxins, and herbal preparations, in M. Feldman, L. S. Friedman, and L. J. Brandt, eds., *Sleisenger and Fordtran's Gastrointestinal and Liver Disease* (Philadelphia: Saunders, 2010), pp. 1447–1459.
6. A. Rutherford and J. L. Dienstag, Viral hepatitis, in N. J. Greenberger, ed., *Current Diagnosis and Treatment: Gastroenterology, Hepatology, and Endoscopy* (New York: McGraw-Hill Companies, 2012), pp. 449–475.
7. J. M. Crawford and C. Liu, Liver and biliary tract, in V. Kumar and coeditors, *Robbins and Cotran Pathologic Basis of Disease* (Philadelphia: Saunders, 2010), pp. 833–890.
8. Rutherford and Dienstag, 2012.
9. Centers for Disease Control and Prevention, *Viral hepatitis surveillance: United States, 2010* (Atlanta: U.S. Department of Health and Human Services, 2012).
10. S. Safrin, Antiviral agents, in B. G. Katzung, ed., *Basic and Clinical Pharmacology* (New York: McGraw-Hill/Lange, 2012), pp. 861–890.
11. Academy of Nutrition and Dietetics, *Nutrition Care Manual* (Chicago: Academy of Nutrition and Dietetics, 2012).
12. S. L. Murphy, J. Xu, and K. D. Kochanek, Deaths: Preliminary data for 2010, *National Vital Statistics Reports* 60 (January 2012).
13. G. Garcia-Tsao, Cirrhosis and its sequelae, in L. Goldman and A I. Schafer, eds., *Goldman's Cecil Medicine* (Philadelphia: Saunders, 2012), pp. 999–1007.
14. V. H. Shah and P. S. Kamath, Portal hypertension and gastrointestinal bleeding, in M. Feldman, L. S. Friedman, and L. J. Brandt, eds., *Sleisenger and Fordtran's Gastrointestinal and Liver Disease* (Philadelphia: Saunders, 2010), pp. 1489–1516.
15. N. J. Greenberger, Ascites and spontaneous bacterial peritonitis, in N. J. Greenberger, ed., *Current Diagnosis and Treatment: Gastroenterology, Hepatology, and Endoscopy* (New York: McGraw-Hill Companies, 2012), pp. 514–521.

16. N. J. Greenberger, Portal systemic encephalopathy and hepatic encephalopathy, in N. J. Greenberger, ed., *Current Diagnosis and Treatment: Gastroenterology, Hepatology, and Endoscopy* (New York: McGraw-Hill Companies, 2012), pp. 509–521.
17. Greenberger, Portal systemic encephalopathy and hepatic encephalopathy, 2012.
18. A. O'Brien and R. Williams, Nutrition in end-stage liver disease: Principles and practice, *Gastroenterology* 134 (2008): 1729–1740; M. Plauth and coauthors, ESPEN guidelines on enteral nutrition: Liver disease, *Clinical Nutrition* 25 (2006): 285–294.
19. Academy of Nutrition and Dietetics, 2012.
20. O'Brien and Williams, 2008.
21. P. Caruana and N. Shah, Hepatic encephalopathy: Are NH_4 levels and protein restriction obsolete? *Practical Gastroenterology* (May 2011): 6–18; S. A. McClave and coauthors, Guidelines for the provision and assessment of nutrition support therapy in the adult critically ill patient: Society of Critical Care Medicine (SCCM) and American Society for Parenteral and Enteral Nutrition (A.S.P.E.N.), *Journal of Parenteral and Enteral Nutrition* 33 (2009): 277–316.
22. Academy of Nutrition and Dietetics, 2012.
23. Academy of Nutrition and Dietetics, 2012.
24. E. B. Keeffe, Hepatic failure and liver transplantation, in L. Goldman and A I. Schafer, eds., *Goldman's Cecil Medicine* (Philadelphia: Saunders, 2012), pp. 1007–1011.
25. Keeffe, 2012.
26. A. A. Qamar, Liver transplantation, in N. J. Greenberger, ed., *Current Diagnosis and Treatment: Gastroenterology, Hepatology, and Endoscopy* (New York: McGraw-Hill Companies, 2012), pp. 555–566.

Nutrition in Practice

Alcohol in Health and Disease

As Chapter 19 described, excessive alcohol ingestion is a primary cause of liver disease. Alcohol can be toxic to other organs as well, including the brain, gastrointestinal (GI) tract, and pancreas. In addition, **alcohol abuse** can lead to a number of nutrient deficiencies. Moderate use of alcohol, however, has been associated with various health benefits, such as lower heart disease risk in middle-aged and older adults. This Nutrition in Practice discusses current recommendations concerning alcohol and the health problems and benefits associated with its use. The Glossary defines the relevant terms.

What are the current dietary guidelines for alcohol consumption?

For those who choose to drink alcoholic beverages, the *Dietary Guidelines for Americans* recommend that women and men limit their average daily intakes of alcohol to one drink and two drinks per day, respectively. In addition, women should avoid consuming more than three drinks in a single day, whereas men should consume no more than four drinks in one day.[1] One **drink** is defined as 12 ounces of beer, 5 ounces of wine, 10 ounces of wine cooler, or 1½ ounces of 80 proof distilled spirits such as gin, rum, vodka, and whiskey. Some individuals should not consume alcohol at all, such as pregnant and lactating women, women of childbearing age who may become pregnant, children and adolescents, people using medications that can interact with alcohol, and people who are unable to voluntarily restrict their alcohol intake. Alcohol should also be avoided by anyone who is involved in an activity that requires attention or coordination, such as driving or operating machinery.

About 64 percent of adults in the United States drink alcoholic beverages.[2] The prevalence of alcohol abuse is estimated to be between 7.4 and 9.7 percent in the general population.

What happens to alcohol in the body?

Recall that alcohol is a source of food energy, providing 7 kcalories per gram. Although a small amount of alcohol is metabolized in the stomach, most of the alcohol consumed is quickly absorbed in the stomach or small intestine and passes readily into the body's cells, where alcohol concentrations reach levels similar to those in the blood. The liver is the site of most alcohol metabolism, although its ability to metabolize alcohol is somewhat limited: the average adult metabolizes only 7 to 10 grams of alcohol per hour, about the amount in one drink.[3] The main product of alcohol metabolism is **acetate**, which can be used as a source of energy by most tissues.

Because there is no storage pool for alcohol and it can be toxic to cells, the metabolism of alcohol in the liver takes priority over that of other substances. Thus, alcohol suppresses the breakdown of fat for energy, leading to fat accumulation in the liver and the increased release of triglyceride-carrying lipoproteins (VLDL). An excessive alcohol intake inhibits both the storage of glycogen and the liver's production of glucose between meals. Heavy drinkers are at risk of developing hypoglycemia, an effect that is accentuated in people with diabetes who use insulin or medications to reduce glucose levels. Alcohol ingestion can also impair the synthesis and secretion of a number of liver proteins, including albumin and fibrinogen (a blood-clotting protein).

Glossary

acetaldehyde (ah-set-AL-deh-hide): an intermediate in alcohol metabolism that, in excess, can damage the body's tissues and interfere with cellular functions.

acetate (AH-seh-tate): the product of alcohol metabolism that is used as a source of energy by many tissues in the body.

alcohol abuse: the continued use of alcohol despite the development of social, legal, or health problems. Individuals who abuse alcohol are called *alcoholics.*

alcoholic liver disease: liver disease that is related to excessive alcohol consumption. Disorders that may develop include fatty liver, hepatitis, and cirrhosis, which may lead to liver failure.

drink: an alcoholic beverage that contains about half an ounce of pure alcohol. One drink is equivalent to 12 ounces of beer, 5 ounces of wine, 10 ounces of wine cooler, or 1½ ounces of 80 proof distilled spirits.

Each of the amounts shown is equivalent to one drink.

How is alcohol toxic to cells?

Alcohol can damage cells either directly, or indirectly via the metabolite **acetaldehyde** that is created during alcohol metabolism. High alcohol concentrations change the structure of cell membranes, increasing membrane permeability. Cell membrane proteins may become damaged, interfering with the transport of substances across the membrane. The presence of alcohol also activates inflammatory responses in cells, which can eventually lead to cell death and tissue damage. Acetaldehyde can cause numerous adverse effects: it binds to proteins and interferes with their functions, alters immune responses, causes oxidative damage, inhibits DNA repair, and prevents the formation of microtubules within cells.

How does alcohol affect brain function?

Alcohol acts as a central nervous system depressant. It can cause sedation, slow reaction times, and relieve anxiety. In excess, it impairs judgment, reduces inhibitions, and impairs speech and motor functions. Extremely high blood alcohol levels can lead to coma, respiratory depression, and death.

Chronic heavy drinking can cause certain types of neurological damage. The most common abnormality is injury to the peripheral nerves, which is evidenced by tingling in the hands and feet, lack of muscular coordination, and changes in a person's manner of walking. Visual impairments, such as blurred vision and optic nerve degeneration, can also occur.

Chronic drinkers who are forced to discontinue their use of alcohol may experience withdrawal symptoms, which include anxiety, irritability, and palpitations. In severe cases, symptoms may include nausea and vomiting, delirium, and seizures.

What are some other long-term consequences of drinking too much alcohol?

Alcoholic liver disease is the most common complication of alcohol abuse, occurring in about 15 to 30 percent of chronic heavy drinkers.[4] The disease develops in a manner similar to the disease progression described in Chapter 19: fatty liver, hepatitis, and, eventually, cirrhosis and liver failure. In addition to liver damage, alcohol can cause damage to the GI tract, pancreas, and heart (see Figure NP19-1). In the GI tract, alcohol's eventual effects may include gastritis, ulcers, nutrient malabsorption, diarrhea, and GI cancers. Chronic alcohol ingestion frequently injures pancreatic tissue and may lead to chronic pancreatitis. Heavy drinking is also associated with heart arrhythmias, impaired heart muscle contractility, and hypertension.

What are the effects of excessive alcohol consumption on nutrition status?

An excessive alcohol intake can cause multiple nutrient deficiencies. Because alcohol supplies 7 kcalories per gram, it displaces other energy sources along with the essential nutrients such foods would provide. As mentioned earlier, alcohol can cause nutrient malabsorption as a result of direct damage to the GI mucosa. Alcohol also interferes with the way the body processes nutrients. Examples of common deficiencies in persons who abuse alcohol include:[5]

- *Vitamin A.* Alcohol and vitamin A are metabolized by similar enzymes, so heavy drinking—which induces the enzymes that break down alcohol—increases the degradation of vitamin A.
- *Thiamin.* Alcohol abuse is the most common cause of thiamin deficiency in the United States. Heavy drinking is associated with low thiamin intakes, and alcohol ingestion dramatically reduces thiamin absorption.
- *Folate.* Alcohol reduces the absorption of folate in the small intestine and increases folate degradation. Because folate has a central role in cell division (and cells of the GI tract are rapidly dividing cells), folate deficiency contributes to the malabsorption of other nutrients as well.

Can alcohol have disruptive effects on the metabolism of medications?

Yes. Alcohol consumption activates enzymes that metabolize certain drugs, so alcohol ingestion can alter drug metabolism. For example, the consumption of several drinks daily activates enzymes that convert acetaminophen (Tylenol) to chemicals that are toxic to the liver; thus, alcohol drinkers should avoid combining alcohol

FIGURE NP19-1 Effects of Excessive Alcohol Intake on Organ Systems

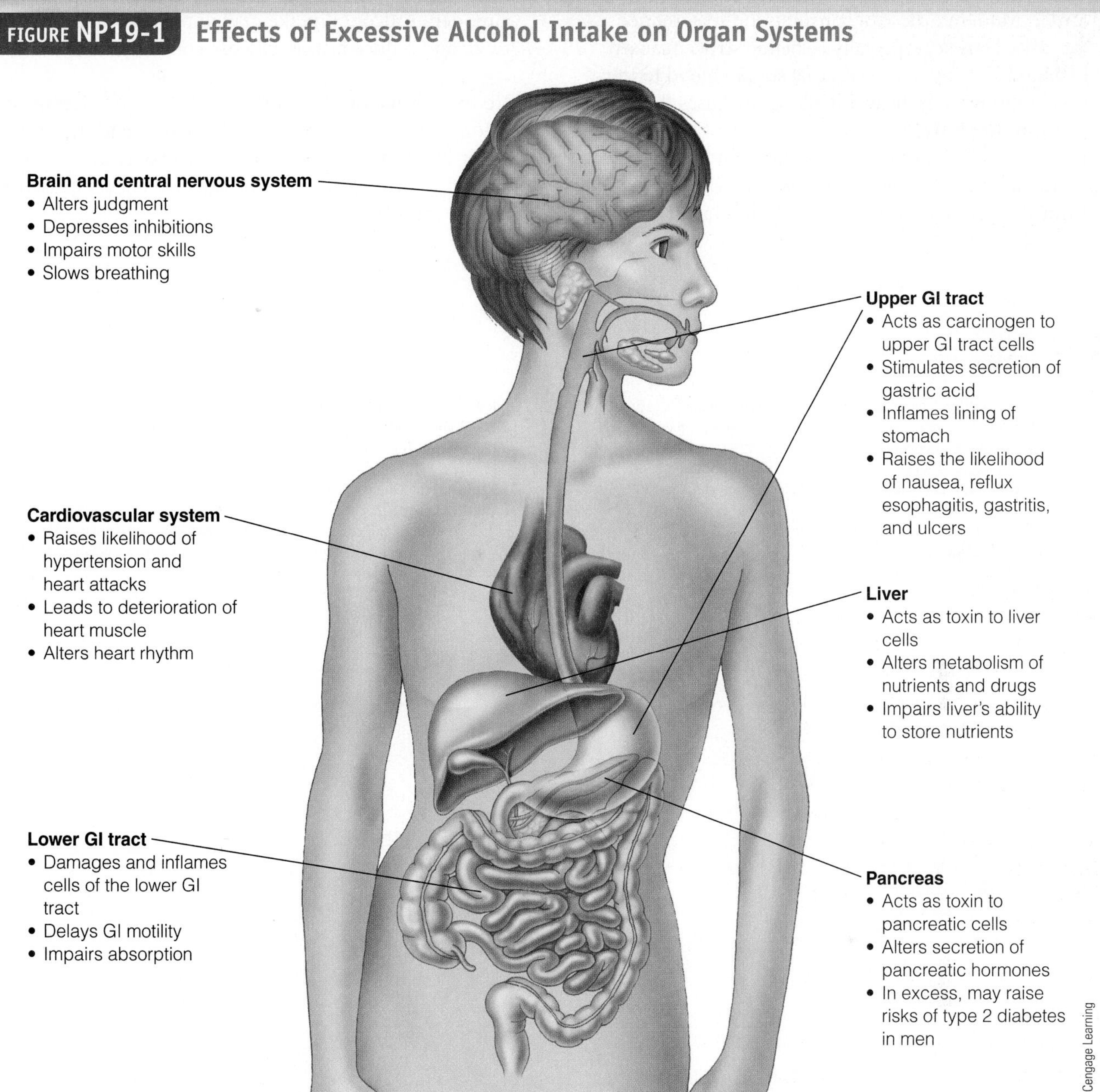

and acetaminophen. In contrast, impaired drug metabolism is likely when the drugs and alcohol require similar enzymes for their metabolism. As an example, when alcohol is ingested at the same time that some sedative drugs are taken, drug metabolism is delayed until the alcohol is degraded, prolonging the drugs' effects.[6]

Does alcohol have any beneficial effects on health?

Yes. Epidemiological studies in Western countries have consistently shown that a moderate alcohol intake reduces total mortality among middle-aged and older men and women. In addition, researchers have learned that moderate alcohol drinking can help reduce risks of developing the following chronic illnesses:[7]

- *Heart disease.* Individuals with a light to moderate intake of alcohol have a lower risk of heart disease than nondrinkers. Alcohol helps to protect against the development of atherosclerosis, increases levels of high-density lipoproteins (HDLs), and reduces the tendency for blood clotting. Its protective effects are seen mainly in older persons who have one or more classic risk factors for heart disease.[8]

- *Stroke.* Moderate alcohol consumption may reduce the risk of stroke, especially ischemic stroke (caused by blood clotting in arteries that supply blood to the brain). Conversely, heavy drinking increases the risk of all forms of stroke.
- *Diabetes.* The risk of type 2 diabetes may be lower in moderate drinkers than in individuals who abstain from alcohol. In addition, some studies have found that a moderate alcohol intake may improve insulin sensitivity and reduce fasting glucose concentrations.

Note that unlike other recommendations for disease prevention, medical personnel rarely recommend that a nondrinker begin drinking to reduce disease risk, due to the potential for addiction and the adverse effects caused by excessive intakes.

Notes

1. *Report of the Dietary Guidelines Advisory Committee on the Dietary Guidelines for Americans, 2010*, available at http://www.cnpp.usda.gov/DGAs2010-DGACReport.htm; site visited September 14, 2012.
2. P. G. O'Connor, Alcohol abuse and dependence, in L. Goldman and A I. Schafer, eds., *Goldman's Cecil Medicine* (Philadelphia: Saunders, 2012), pp. 146–153.
3. S. B. Masters, The alcohols, in B. G. Katzung, ed., *Basic and Clinical Pharmacology* (New York: McGraw-Hill Companies, 2012), pp. 389–401.
4. Masters, 2012.
5. P. M. Suter, Alcohol: Its role in nutrition and health, in J. W. Erdman, I. A. Macdonald, and S. H. Zeisel, eds., *Present Knowledge in Nutrition* (Ames, IA: Wiley-Blackwell, 2012): pp. 912–938.
6. Masters, 2012.
7. Report of the Dietary Guidelines Advisory Committee on the Dietary Guidelines for Americans, 2010.
8. A. Lichtenstein and coauthors, Diet and lifestyle recommendations revision 2006: A scientific statement from the American Heart Association Nutrition Committee, *Circulation* 114 (2006): 82–96.

Chapter 20 Nutrition and Diabetes Mellitus

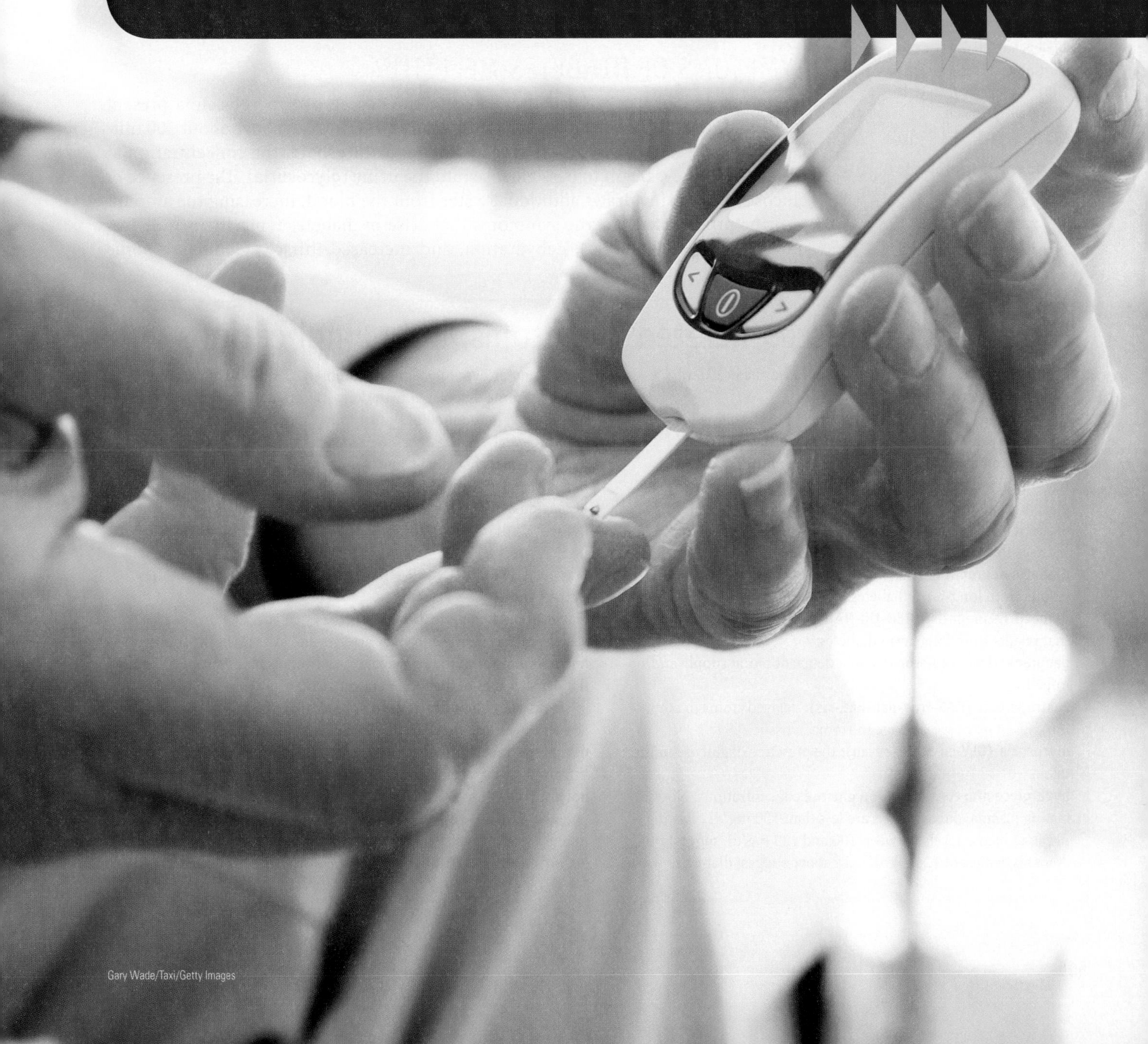

Gary Wade/Taxi/Getty Images

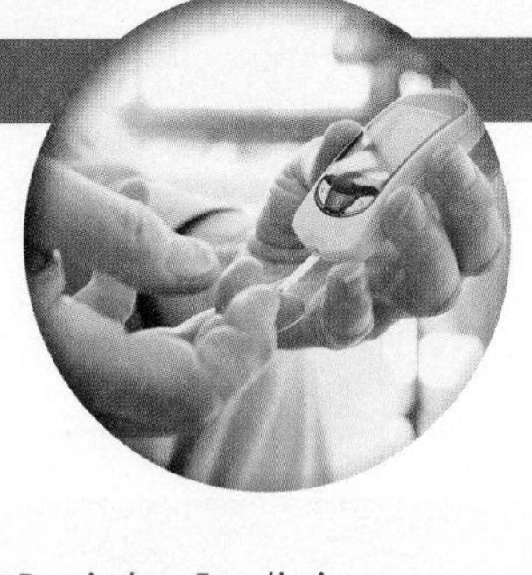

THE INCIDENCE OF DIABETES MELLITUS IS STEADILY INCREASING IN THE United States and many other countries. It now affects an estimated 11.3 percent of adults aged 20 and older in the United States, or about 26 million people.[1] About 27 percent of persons with diabetes are unaware that they have it,[2] a danger because its damaging effects often occur before symptoms develop. Diabetes ranks seventh among the leading causes of death in the United States. It also contributes to the development of other life-threatening diseases, including heart disease and kidney failure, which are discussed in the two chapters that follow. The glossary below defines diabetes-related symptoms and complications.

■ Reminder: *Insulin* is a pancreatic hormone that regulates blood glucose concentrations. Its actions are countered mainly by the hormone *glucagon*.

diabetes (DYE-ah-BEE-teez) mellitus: a group of metabolic disorders characterized by hyperglycemia and disordered insulin metabolism.

diabetes = siphon (in Greek), referring to the excessive passage of urine that is characteristic of untreated diabetes

mellitus = sweet, honeylike

renal threshold: the blood concentration of a substance that exceeds the kidneys' capacity for reabsorption, causing the substance to be passed into the urine.

Overview of Diabetes Mellitus

The term **diabetes mellitus** refers to metabolic disorders characterized by elevated blood glucose concentrations and disordered insulin metabolism. People with diabetes may be unable to produce sufficient insulin or use insulin effectively, or they may have both types of abnormalities. (■) These impairments result in defective glucose uptake and utilization in muscle and adipose cells and unrestrained glucose production in the liver. The result is **hyperglycemia**, a marked elevation in blood glucose levels that can ultimately cause damage to blood vessels, nerves, and tissues.

SYMPTOMS OF DIABETES MELLITUS

Symptoms of diabetes are usually related to the degree of hyperglycemia present (see Table 20-1). When the plasma glucose concentration rises above about 200 milligrams per deciliter (mg/dL), it exceeds the **renal threshold**, the concentration at which the kidneys begin to pass glucose into the urine (**glycosuria**). The presence of glucose in the urine draws additional water from the blood, increasing the amount of urine produced. Thus, the symptoms that arise in diabetes typically include frequent urination (**polyuria**), dehydration, and increased thirst (**polydipsia**). Some

Glossary of Diabetes-Related Symptoms and Complications

acetone breath: a distinctive fruity odor on the breath of a person with ketosis.
claudication (CLAW-dih-KAY-shun): pain in the legs while walking; usually due to an inadequate supply of blood to muscles.
diabetic coma: a coma that occurs in uncontrolled diabetes; may be due to diabetic ketoacidosis, the hyperosmolar hyperglycemic syndrome, or severe hypoglycemia.
diabetic nephropathy (neh-FRAH-pah-thee): damage to the kidneys that results from long-term diabetes.
diabetic neuropathy (nur-RAH-pah-thee): nerve damage that results from long-term diabetes.
diabetic retinopathy (REH-tih-NAH-pah-thee): retinal damage that results from long-term diabetes.
gangrene: death of tissue due to a deficient blood supply and/or infection.
gastroparesis (GAS-troe-pah-REE-sis): delayed stomach emptying caused by nerve damage in stomach tissue.
glycosuria (GLY-co-SOOR-ee-ah): the presence of glucose in the urine.
hyperglycemia: elevated blood glucose concentrations. Normal fasting plasma glucose levels are less than 100 mg/dL. Fasting plasma glucose levels between 100 and 125 mg/dL suggest prediabetes; values of 126 mg/dL and above suggest diabetes.
hyperosmolar hyperglycemic syndrome: a condition of extreme hyperglycemia associated with dehydration, hyperosmolar blood, and altered mental status; sometimes called the *hyperosmolar hyperglycemic nonketotic state.*
hypoglycemia: abnormally low blood glucose levels. In diabetes, hypoglycemia is treated when plasma glucose falls below 70 mg/dL.
ketoacidosis (KEY-toe-ass-ih-DOE-sis): an acidosis (lowering of blood pH) that results from the excessive production of ketone bodies.
ketonuria (KEY-toe-NOOR-ee-ah): the presence of ketone bodies in the urine.
macrovascular complications: disorders that affect large blood vessels, including the coronary arteries and arteries of the limbs.
microalbuminuria: the presence of albumin (a blood protein) in the urine, a sign of diabetic nephropathy.
microvascular complications: disorders that affect small blood vessels, including those in the retina and kidneys.
peripheral vascular disease: disorders characterized by impaired blood circulation in the limbs.
polydipsia (POL-ee-DIP-see-ah): excessive thirst.
polyphagia (POL-ee-FAY-jee-ah): excessive appetite or food intake.
polyuria (POL-ee-YOOR-ree-ah): excessive urine production.

people lose weight and have an increased appetite (**polyphagia**) as a result of the nutrient depletion that occurs when insulin is deficient. Another potential consequence of hyperglycemia is blurred vision, caused by the exposure of eye tissues to hyperosmolar fluids. (■) Increased infections are common in individuals with diabetes and may be due to weakened immune responses and impaired circulation. In some cases, constant fatigue is the only symptom and may be related to altered energy metabolism, dehydration, or other effects of the disease.

TABLE 20-1 Symptoms of Diabetes Mellitus

- Frequent urination (polyuria)
- Dehydration, dry mouth
- Increased thirst (polydipsia)
- Weight loss
- Increased hunger (polyphagia)
- Blurred vision
- Increased infections
- Fatigue

DIAGNOSIS OF DIABETES MELLITUS

The diagnosis of diabetes is based primarily on plasma glucose levels, which can be measured under fasting conditions (■) or at random times during the day. In some cases, an **oral glucose tolerance test** is given: the individual ingests a 75-gram glucose load, and plasma glucose is measured at one or more time intervals following glucose ingestion. **Glycated hemoglobin (HbA_{1c})** levels, which reflect hemoglobin's exposure to glucose over a two- to three-month period, are an indirect assessment of blood glucose levels. The following criteria are currently used to diagnose diabetes:

- The plasma glucose concentration of a blood sample obtained at a random time during the day (without regard to food intake) is 200 mg/dL or higher, and classic symptoms of hyperglycemia (such as polyuria, polydipsia, and unexplained weight loss) are present.
- The plasma glucose concentration is 126 mg/dL or higher after a fast of at least eight hours.
- The plasma glucose concentration measured two hours after a 75-gram glucose load is 200 mg/dL or higher.
- The HbA_{1c} level is 6.5 percent or higher.

Overt symptoms of hyperglycemia help to confirm the diagnosis. Otherwise, a diagnosis of diabetes is confirmed only if a subsequent test yields similar results.

The term **prediabetes** pertains to individuals who have blood glucose levels between normal and diabetic, that is, between 100 and 125 mg/dL when fasting or between 140 and 199 mg/dL when measured two hours after ingesting a 75-gram glucose load. HbA_{1c} levels between 5.7 and 6.4 percent also suggest prediabetes. Although people with prediabetes are usually asymptomatic, they are at increased risk of developing diabetes and cardiovascular diseases.[3] Prediabetes is currently estimated to affect approximately 35 percent of adults in the United States[4] and 23 percent of adolescents aged 12 to 19 years,[5] and it is especially prevalent among those who are overweight or obese.

NURSING DIAGNOSIS

Nursing diagnoses for people with diabetes may include *risk for unstable blood glucose level, risk for deficient fluid volume, risk for infection,* and *adult failure to thrive.*

■ Reminder: *Osmolarity* refers to the concentration of osmotically active particles in solution. Hyperglycemia causes the body's fluids to become *hyperosmolar,* meaning that they have an abnormally high osmolarity.

■ Normal fasting plasma glucose levels are approximately 75 to 100 mg/dL (published values vary).

oral glucose tolerance test: a test that evaluates a person's ability to tolerate an oral glucose load.

glycated hemoglobin (HbA_{1c}): hemoglobin that has nonenzymatically attached to glucose; the level of HbA_{1c} in blood helps to diagnose diabetes and evaluate long-term glycemic control. Also called *glycosylated hemoglobin*.

prediabetes: the condition in which blood glucose levels are higher than normal (fasting plasma glucose between 100 and 125 mg/dL) but not high enough to be diagnosed as diabetes.

TYPES OF DIABETES MELLITUS

Table 20-2 (p. 568) lists features of the two main types of diabetes, type 1 and type 2 diabetes. Pregnancy can lead to abnormal glucose tolerance and the condition known as *gestational diabetes* (discussed later in this chapter), which often resolves after pregnancy but is a risk factor for type 2 diabetes. Diabetes can also be caused by medical conditions that damage the pancreas or interfere with insulin function.

Type 1 Diabetes

Type 1 diabetes accounts for about 5 to 10 percent of diabetes cases. It is usually caused by **autoimmune** destruction of the pancreatic beta cells, which produce and secrete insulin. By the time symptoms develop, the damage to the beta cells has progressed so far that insulin must be supplied exogenously, most often by injection. Although the reason for the autoimmune attack is usually unknown, environmental toxins or infections are likely triggers. People with type 1 diabetes often have a genetic susceptibility for the disorder and are at increased risk of developing other autoimmune diseases.

type 1 diabetes: the type of diabetes characterized by absolute insulin deficiency that usually results from autoimmune destruction of pancreatic beta cells.

autoimmune: an immune response directed against the body's own tissues.

auto = self

TABLE 20-2 Features of Type 1 and Type 2 Diabetes Mellitus

Feature	Type 1 Diabetes	Type 2 Diabetes
Prevalence in diabetic population	5–10 percent of cases	90–95 percent of cases
Age of onset	<30 years	>40 years[a]
Associated conditions	Autoimmune diseases, viral infection, inherited factors	Obesity, aging, inactivity, inherited factors
Major defect	Destruction of pancreatic beta cells; insulin deficiency	Insulin resistance; insulin deficiency relative to needs
Insulin secretion	Little or none	Varies; may be normal, increased, or decreased
Requirement for insulin therapy	Always	Sometimes
Former names	Juvenile-onset diabetes Insulin-dependent diabetes	Adult-onset diabetes Noninsulin-dependent diabetes

[a]Incidence of type 2 diabetes is increasing in children and adolescents; in more than 90 percent of these cases, it is associated with overweight or obesity and a family history of type 2 diabetes.

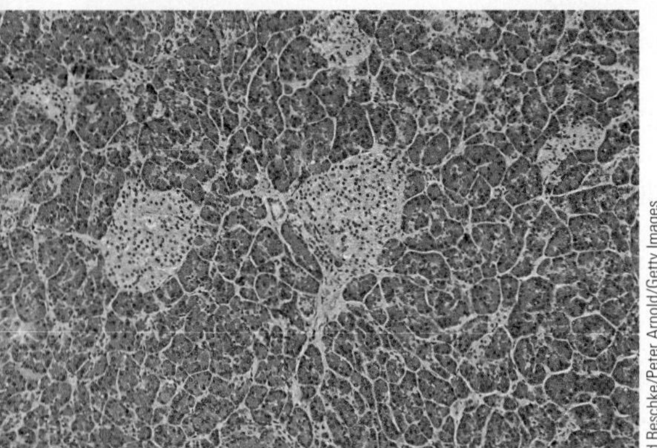

Cross-sections of the pancreas reveal small clusters of cells known as the islets of Langerhans; these regions contain the beta cells that produce insulin.

Type 1 diabetes usually develops during childhood or adolescence, and symptoms may appear abruptly in previously healthy children.[6] Classic symptoms are polyuria, polydipsia, weight loss, and weakness or fatigue. **Ketoacidosis**—acidosis due to the excessive production of ketone bodies—is sometimes the first sign of disease. (■) Disease onset tends to be more gradual in individuals who develop type 1 diabetes in later years. Blood tests that detect antibodies to insulin, pancreatic islet cells, and pancreatic enzymes can confirm the diagnosis and help to predict development of the disease in close relatives.

■ Reminder: *Ketone bodies* are products of fat metabolism that are produced in the liver; they accumulate in tissues when fatty acids are released in abnormally high amounts from adipose tissue.

Type 2 Diabetes

Type 2 diabetes is the most prevalent form of diabetes, accounting for 90 to 95 percent of cases. It is often asymptomatic for many years before diagnosis. The defect in type 2 diabetes is **insulin resistance**, the reduced sensitivity to insulin in muscle, adipose, and liver cells, coupled with relative insulin deficiency, the lack of sufficient insulin to manage glucose effectively. Normally, the pancreatic beta cells secrete more insulin to compensate for insulin resistance. In type 2 diabetes, insulin levels are often abnormally high (**hyperinsulinemia**), but the additional insulin is insufficient to compensate for its diminished effect in cells. Thus, the hyperglycemia that develops in type 2 diabetes represents a mismatch between the amount of insulin required and the amount produced by beta cells. Beta cell function tends to worsen over time in people with type 2 diabetes, and insulin production gradually declines as the condition progresses.[7]

Although the precise causes of type 2 diabetes are unknown, risk is substantially increased by obesity (especially abdominal obesity), aging, and physical inactivity. An estimated 80 percent of individuals with type 2 diabetes are obese, and obesity itself can directly cause some degree of insulin resistance.[8] (■) Prevalence increases with age and approaches 27 percent in persons 65 years of age or older; however, many of these cases remain undiagnosed.[9] Inherited factors strongly influence risk; type 2 diabetes is more common in certain ethnic populations, including African Americans, Asian Americans, Hispanic Americans, Mexican Americans, Native Americans, and Pacific Islanders.

■ Nutrition in Practice 20 provides information about the relationship between obesity and insulin resistance.

type 2 diabetes: the type of diabetes characterized by insulin resistance coupled with insufficient insulin secretion.

insulin resistance: reduced sensitivity to insulin in muscle, adipose, and liver cells.

hyperinsulinemia: abnormally high levels of insulin in the blood.

Type 2 Diabetes in Children and Adolescents

Although most cases of type 2 diabetes are diagnosed in individuals older than 40 years of age, children and teenagers who are overweight or obese or have a family history of diabetes are at increased risk.

Because type 2 diabetes is frequently asymptomatic, it is generally identified in youths only when high-risk groups are screened for the disease.

Increased rates of both type 1 and type 2 diabetes have been documented in children in past decades and correlate with the rise in childhood obesity. Type 1 and type 2 diabetes are sometimes difficult to distinguish in children, however, and a few studies suggest that some children diagnosed with type 1 diabetes may actually have had type 2 diabetes.[10] Type 2 diabetes is still extremely rare in children; for example, its estimated incidence in 10- to 19-year-old African American and Native American youths—the groups at highest risk—is 19 and 32 cases per 100,000 individuals per year, respectively.[11] Its increasing prevalence, however, indicates that routine screening and diabetes prevention programs may be important safeguards for children at risk.

PREVENTION OF TYPE 2 DIABETES MELLITUS

Clinical studies suggest that lifestyle changes can prevent or delay the incidence of type 2 diabetes in individuals at risk. In the Diabetes Prevention Program, a multicenter trial of 3234 adults with impaired glucose tolerance, dietary changes and increased physical activity led to a 58 percent reduction in diabetes incidence over a 3-year period.[12] Other studies have shown that lifestyle modifications can reduce diabetes risk for as long as 10 to 20 years.[13] Based on the results of these studies, guidelines for diabetes prevention include the following strategies:*

- *Weight management.* A sustained weight loss of 5 to 10 percent of body weight is recommended for overweight and obese individuals. If weight loss cannot be achieved, healthy eating behaviors should be encouraged to prevent additional weight gain.
- *Active lifestyle.* At least 150 minutes of moderate physical activity, such as brisk walking, is recommended weekly.
- *Dietary modifications.* An increased intake of whole grains and dietary fiber has been associated with a reduced risk for type 2 diabetes. Individuals who are overweight or obese should decrease their intake of dietary fat to avoid consuming excessive energy.
- *Regular monitoring.* Individuals at risk should be monitored each year to check for the possible development of type 2 diabetes. If necessary, they can be provided with additional counseling, education, or resources.

ACUTE COMPLICATIONS OF DIABETES MELLITUS

Untreated diabetes may result in life-threatening complications. Insulin deficiency can cause significant disturbances in energy metabolism, and severe hyperglycemia can lead to dehydration and electrolyte imbalances. In treated diabetes, hypoglycemia (low blood glucose) is a possible complication of inappropriate disease management. Figure 20-1 (p. 570) presents an overview of some of the acute effects of insulin insufficiency on energy metabolism.

Diabetic Ketoacidosis in Type 1 Diabetes

A severe lack of insulin causes diabetic ketoacidosis. Without insulin, glucagon's effects become more pronounced, leading to the unrestrained breakdown of the triglycerides in adipose tissue and the protein in muscle. As a result, an increased supply of fatty acids and amino acids arrives in the liver, fueling the production of ketone bodies and glucose. Ketone bodies, which are acidic, can reach dangerously high levels in the bloodstream (ketoacidosis) and spill into the urine (**ketonuria**). Blood pH typically falls below 7.30 (blood pH normally ranges between 7.35 and 7.45). Blood glucose concentrations usually exceed 250 mg/dL and rise above 1000 mg/dL in severe cases. The main features of diabetic ketoacidosis therefore include severe ketosis (abnormally high levels of ketone bodies), acidosis, and hyperglycemia.

*The antidiabetic medication metformin may be beneficial for preventing diabetes in very high-risk individuals, such as those who are very obese, have severe or worsening hyperglycemia, or have a history of gestational diabetes.

FIGURE 20-1 **Acute Effects of Insulin Insufficiency**

The effects of insulin insufficiency can be grouped according to its effects on carbohydrate, protein, and fat metabolism.

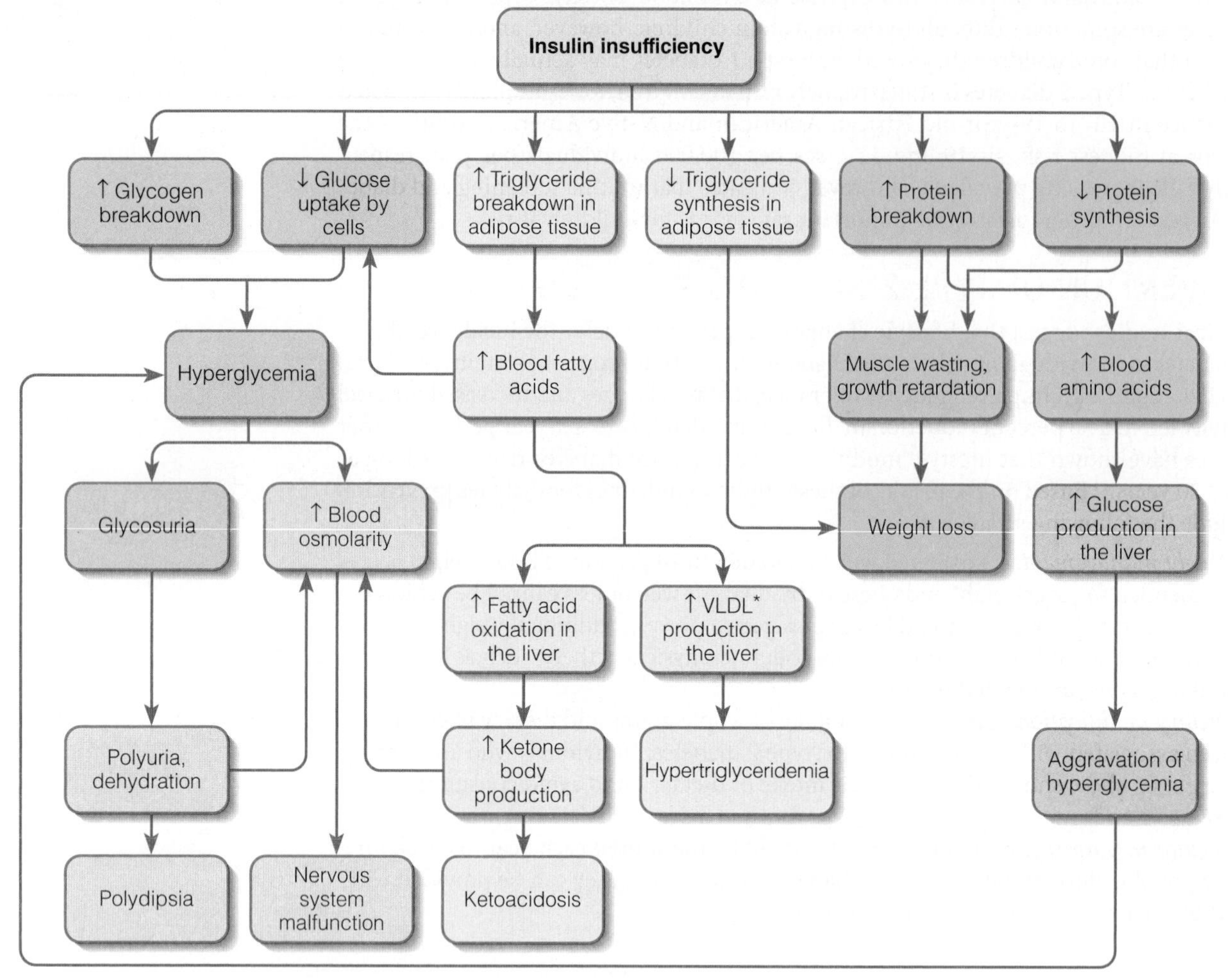

*Very low density lipoproteins: these lipoproteins transport triglycerides from the liver to other tissues.

Patients with ketoacidosis may exhibit symptoms of both acidosis and dehydration. Acidosis is partially corrected by exhalation of carbon dioxide, so rapid or deep breathing is characteristic. (■) Ketone accumulation is sometimes evident by a fruity odor on a person's breath (**acetone breath**). Significant fluid loss (polyuria) accompanies the hyperglycemia, lowering blood volume and blood pressure and depleting electrolytes. In response, patients may demonstrate marked fatigue, lethargy, nausea, and vomiting. Mental state may vary from alert to comatose (**diabetic coma**). (■) Treatment of diabetic ketoacidosis includes insulin therapy to correct the hyperglycemia, intravenous fluid and electrolyte replacement, and, in some cases, bicarbonate therapy to treat acidosis.

Diabetic ketoacidosis is sometimes the earliest sign that leads to diagnosis of type 1 diabetes, but more often it results from inappropriate diabetes treatment (such as missed insulin injections), illness or infection, alcohol abuse, or other physiological stressors. The condition usually develops quickly, within one to two days. Mortality rates are generally less than 5 percent but may exceed 20 percent among elderly individuals with other complications or illnesses.[14] Although diabetic ketoacidosis can occur in type 2 diabetes—usually due to severe stressors such as infection, trauma, or surgery—it rarely develops because even relatively low insulin concentrations are able to suppress ketone body production.

■ Bicarbonate is a buffer in the blood that corrects acidosis. The acid (H^+) combines with bicarbonate (HCO_3^-) to form carbonic acid (H_2CO_3), which breaks down to water (H_2O) and carbon dioxide (CO_2). The carbon dioxide is then exhaled.

■ Diabetic coma was a frequent cause of death before insulin was routinely used to manage diabetes.

Hyperosmolar Hyperglycemic Syndrome in Type 2 Diabetes The **hyperosmolar hyperglycemic syndrome** is a condition of severe hyperglycemia and dehydration that develops in the absence of significant ketosis. As mentioned earlier, the hyperglycemia that develops in poorly controlled diabetes leads to polyuria, which results in substantial fluid and electrolyte losses. In the hyperosmolar hyperglycemic syndrome, patients are unable to recognize thirst or adequately replace fluids due to age, illness, sedation, or incapacity. The profound dehydration that eventually develops exacerbates the rise in blood glucose levels, which often exceed 600 mg/dL and may climb above 1000 mg/dL. Blood plasma may become so hyperosmolar as to cause neurological abnormalities, such as confusion, speech or vision impairments, muscle weakness, abnormal reflexes, and seizures; about 10 percent of patients lapse into coma.[15] Treatment includes intravenous fluid and electrolyte replacement and insulin therapy.

The hyperosmolar hyperglycemic syndrome is sometimes the first sign of type 2 diabetes in older persons. It is usually precipitated by an infection, serious illness, or drug treatment that alters insulin action or secretion. Unlike diabetic ketoacidosis, the condition often evolves slowly, over one week or longer; the absence of clinical symptoms can delay its diagnosis. The mortality rate may be as high as 20 percent, in part because the condition occurs more often in older patients with cardiovascular disease or other major illnesses.[16]

NURSING DIAGNOSIS

Nursing diagnoses for people with diabetic ketoacidosis or the hyperosmolar hyperglycemic syndrome may include *ineffective self-health management, deficient fluid volume, risk for electrolyte imbalance,* and *acute confusion.*

Hypoglycemia **Hypoglycemia**, or low blood glucose, is the most frequent complication of type 1 diabetes and may occur in type 2 diabetes as well. It is due to the inappropriate management of diabetes rather than the disease itself, and is usually caused by excessive dosages of insulin or antidiabetic drugs, prolonged exercise, skipped or delayed meals, inadequate food intake, or the consumption of alcohol without food. Hypoglycemia is the most frequent cause of coma in insulin-treated patients and is believed to account for 3 to 4 percent of deaths in this population.[17]

Symptoms of hypoglycemia include sweating, heart palpitations, shakiness, hunger, weakness, dizziness, and irritability. Mental confusion may prevent a person from recognizing the problem and taking such corrective action as ingesting glucose tablets, juice, or candy. If hypoglycemia occurs during the night, patients may be completely unaware of its presence.

CHRONIC COMPLICATIONS OF DIABETES MELLITUS

Prolonged exposure to high glucose concentrations can damage cells and tissues. Glucose nonenzymatically combines with proteins, producing molecules that eventually break down to form reactive compounds known as **advanced glycation end products (AGEs)**; in diabetes, these AGEs accumulate to such high levels that they alter the structures of proteins and stimulate metabolic pathways that are damaging to tissues. In addition, excessive glucose promotes the production and accumulation of sorbitol, which increases oxidative stress within cells and causes cellular injury.

Chronic complications of diabetes typically involve the large blood vessels (**macrovascular complications**), smaller vessels such as arterioles and capillaries (**microvascular complications**), and the nervous system (**diabetic neuropathy**). Other tissues adversely affected include the lens of the eye and the skin; cataracts, glaucoma, and various types of skin lesions sometimes develop. Infections are common in diabetes, a possible consequence of hyperglycemia, impaired circulation, and/or depressed immune responses. Many of these complications appear 15 to 20 years after diabetes onset.[18] In individuals with type 2 diabetes, complications often develop before diabetes is diagnosed.

NURSING DIAGNOSIS

Nursing diagnoses for persons with chronic complications of diabetes may include *ineffective peripheral tissue perfusion, risk for infection, risk for impaired skin integrity,* and *risk for injury.*

Macrovascular Complications The damage caused by diabetes accelerates the development of atherosclerosis in the arteries of the heart, brain, and limbs. Cardiovascular diseases are the leading cause of death in people with diabetes, accounting for up to 70 percent of deaths.[19] Moreover, type 2 diabetes is frequently accompanied by multiple risk factors for cardiovascular disease, including hypertension and blood lipid abnormalities. (■) People with diabetes also have increased tendencies for thrombosis (blood

■ The *metabolic syndrome,* a cluster of symptoms often associated with insulin resistance (including hyperglycemia, hypertension, and abnormal blood lipids), substantially increases heart disease risk (see Nutrition in Practice 20).

advanced glycation end products (AGEs): reactive compounds formed after glucose combines with protein; AGEs can damage tissues and lead to diabetic complications.

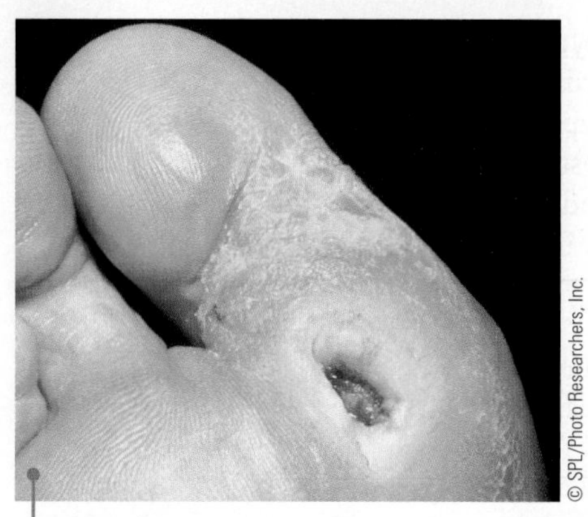

© SPL/Photo Researchers, Inc.

Foot ulcers are a common complication of diabetes because blood circulation is impaired, which slows healing, and nerve damage dampens foot pain, delaying recognition and treatment of cuts and bruises.

clot formation) and abnormal ventricle function, both of which can worsen the clinical course of heart disease.

Peripheral vascular disease (impaired blood circulation in the limbs) increases the risk of **claudication** (pain while walking) and contributes to the development of foot ulcers (see the photo). Left untreated, foot ulcers can lead to **gangrene** (tissue death), and some patients require foot amputation, a major cause of disability in individuals with diabetes.

Microvascular Complications Long-term diabetes is associated with detrimental changes in capillary structure and function, such as the thickening of basement membranes, growth of fibrous tissue (scarring), increased capillary permeability, and proliferation of vessels that function abnormally. The primary microvascular complications involve the retina of the eye and the kidneys.

In **diabetic retinopathy**, the weakened capillaries of the retina leak fluid, lipids, or blood, causing local edema or hemorrhaging. The defective blood flow also leads to damage and scarring within retinal tissue. New blood vessels eventually form, but they are fragile and bleed easily, releasing blood and proteins that obscure vision. The retinal changes usually occur after an individual has had diabetes for many years. Up to 80 percent of diabetes patients develop retinopathy 15 to 20 years after diabetes onset.[20]

In **diabetic nephropathy**, damage to the kidneys' specialized capillaries prevents adequate blood filtration, resulting in abnormal urinary protein losses (**microalbuminuria**). As the kidney damage worsens, urine production decreases and nitrogenous wastes accumulate in the blood; eventually, the individual requires dialysis (artificial filtration of blood) to survive. Because the kidneys normally regulate blood volume and blood pressure, inadequate kidney function leads to high blood pressure in patients with nephropathy. Kidney failure eventually develops in about 30 to 40 percent of patients with type 1 diabetes and up to 20 percent of those with type 2 diabetes.[21] (Chapter 22 provides details about the progression and treatment of chronic kidney disease.)

Diabetic Neuropathy The extent of neuropathy (nerve damage) that develops in diabetes depends on the severity and duration of hyperglycemia. Symptoms of neuropathy vary and may be experienced as deep pain or burning in the legs and feet, weakness of the arms and legs, or numbness and tingling in the hands and feet. Pain and cramping, especially in the legs, are often severe during the night and may interrupt sleep. Neuropathy also contributes to the development of foot ulcers because cuts and bruises may go unnoticed until wounds are severe. Other manifestations of neuropathy include sweating abnormalities, disturbances in bladder and bowel function, sexual dysfunction, constipation, and delayed stomach emptying (**gastroparesis**). Neuropathy occurs in about 50 percent of diabetes cases.[22]

Review Notes

- Diabetes mellitus is a chronic condition characterized by inadequate insulin secretion and/or impaired insulin action. Diagnosis is based on indicators of hyperglycemia.
- In type 1 diabetes, the pancreas secretes little or no insulin, and insulin therapy is necessary for survival. Type 2 diabetes is characterized by insulin resistance coupled with relative insulin deficiency; disease risk is increased by obesity, aging, and physical inactivity.
- Acute complications of diabetes include diabetic ketoacidosis, in which hyperglycemia is accompanied by ketosis and acidosis, and the hyperosmolar hyperglycemic syndrome, characterized by severe hyperglycemia, dehydration, and possible mental impairments. Another acute complication, hypoglycemia, is usually a consequence of inappropriate disease management.
- Chronic complications of diabetes include macrovascular disorders such as cardiovascular and peripheral vascular diseases, microvascular conditions such as retinopathy and nephropathy, and neuropathy.

Treatment of Diabetes Mellitus

Diabetes is a chronic and progressive illness that requires lifelong treatment. Managing blood glucose levels is a delicate balancing act that involves meal planning, proper timing of medications, and physical exercise. Frequent adjustments in treatment are often necessary to establish good **glycemic** control. Individuals with type 1 diabetes require insulin therapy for survival. Type 2 diabetes is initially treated with nutrition therapy and exercise, but most patients eventually need antidiabetic medications or insulin. Diabetes management becomes even more difficult once complications develop. Although the health care team must determine the appropriate therapy, the individual with diabetes ultimately assumes much of the responsibility for treatment and therefore requires education in self-management of the disease.

TREATMENT GOALS

The main goal of diabetes treatment is to maintain blood glucose levels within a desirable range to prevent or reduce the risk of complications. Several multicenter clinical trials have shown that intensive diabetes treatment, which keeps blood glucose levels tightly controlled, can greatly reduce the incidence and severity of diabetic retinopathy, nephropathy, and neuropathy.[23] (■) Therefore, maintenance of near-normal glucose levels has become the fundamental objective of diabetes care plans. Other goals of treatment include maintaining healthy blood lipid concentrations, controlling blood pressure, and managing weight—measures that can help to prevent or delay diabetes complications as well. Table 20-3 provides examples of the general differences between conventional and intensive therapies for type 1 diabetes. For type 2 diabetes, intensive therapy involves the addition of medications or insulin to standard dietary and lifestyle modifications. Note that intensive therapy is recommended only if the benefits of therapy outweigh the potential risks.*

Diabetes education provides an individual with the knowledge and skills necessary to implement treatment. The primary instructor is often a **Certified Diabetes Educator (CDE)**, a health care professional (often a nurse or dietitian) who has specialized knowledge about diabetes treatment and the health education process. To manage

TABLE 20-3 Comparison of Conventional and Intensive Therapies for Type 1 Diabetes

	Conventional Therapy	Intensive Therapy
Blood glucose monitoring	Monitored daily	Monitored at least three times daily
Insulin therapy	One or two daily injections; no daily adjustments	Three or more daily injections or use of an external insulin pump; dosage adjusted according to the results of blood glucose monitoring and expected carbohydrate intake
Advantages	Fewer incidences of severe hypoglycemia; less weight gain	Delayed progression of retinopathy, nephropathy, and neuropathy
Disadvantages	More rapid progression of retinopathy, nephropathy, and neuropathy	Twofold to threefold increase in severe hypoglycemia; weight gain; increased risk of becoming overweight

*Intensive treatment may be inappropriate for individuals with limited life expectancies or a history of hypoglycemia and middle-aged or older adults with previous heart disease or multiple heart disease risk factors.

■ Studies that evaluated the benefits of intensive treatment include the Diabetes Control and Complications Trial and the United Kingdom Prospective Diabetes Study.

glycemic (gly-SEE-mic): pertaining to blood glucose.

Certified Diabetes Educator (CDE): a health care professional who specializes in diabetes management education; certification is obtained from the National Certification Board for Diabetes Educators.

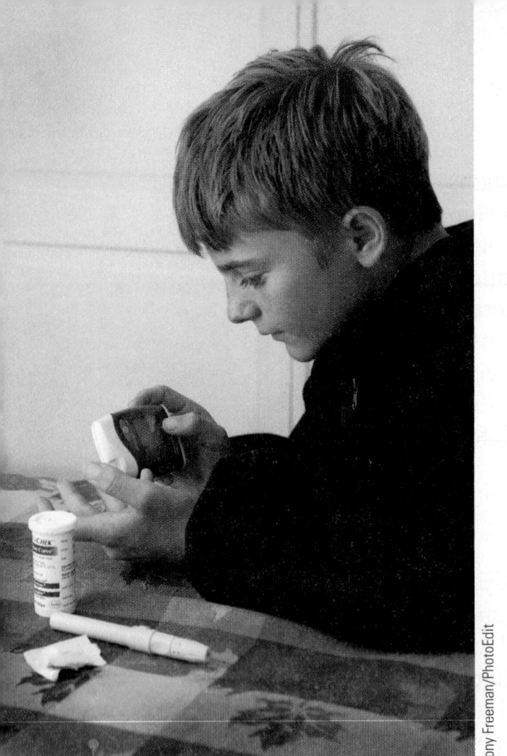

Self-monitoring of blood glucose involves applying a drop of blood from a finger prick to a chemically treated paper strip, which is then analyzed for glucose.

diabetes, patients need to learn about appropriate meal planning, medication administration, blood glucose monitoring, weight management, appropriate physical activity, and prevention and treatment of diabetic complications.

EVALUATING DIABETES TREATMENT

Diabetes treatment is largely evaluated by monitoring glycemic status. Good glycemic control requires frequent home monitoring of blood glucose levels using a glucose meter, referred to as **self-monitoring of blood glucose**. (■) Glucose testing provides valuable feedback when the patient adjusts food intake, medications, and physical activity and is helpful for preventing hypoglycemia. Ideally, patients with type 1 diabetes should test blood glucose levels three or more times daily—and more frequently when therapy is adjusted. Some patients may achieve better glycemic control by also using a **continuous glucose monitoring** system, which measures tissue glucose levels every few minutes using a tiny sensor placed under the skin. Although self-monitoring of blood glucose is also useful in type 2 diabetes, the recommended frequency varies according to the specific needs of individual patients.

Long-Term Glycemic Control Health care providers periodically evaluate long-term glycemic control by measuring HbA_{1c} levels. The glucose in blood freely enters red blood cells and attaches to hemoglobin molecules in direct proportion to the amount of glucose present. Because the life span of red blood cells averages 120 days, the percentage of HbA_{1c} reflects glycemic control over the preceding two to three months (the average age of circulating red blood cells). The goal of diabetes treatment is usually an HbA_{1c} value less than 7 percent,[24] (■) but the percentage is often markedly higher in people with diabetes, even those who are maintaining near-normal blood glucose levels. Less stringent HbA_{1c} goals may be suitable for certain patients, including those with limited life expectancy, advanced diabetic complications, or a history of severe hypoglycemia.

The **fructosamine test** is sometimes conducted to determine glycemic control over the preceding two- to three-week period. This test determines the nonenzymatic glycation of serum proteins (primarily albumin), which have a shorter half-life than hemoglobin. Most often, the fructosamine test is used to evaluate recent adjustments in diabetes treatment or glycemic control during pregnancy. The test cannot be used if the patient has a liver or kidney disease that lowers serum protein levels.

Monitoring for Long-Term Complications Individuals with diabetes are routinely monitored for signs of long-term complications. Blood pressure is measured at each checkup. Annual lipid screening is suggested for most adult patients. Routine checks for urinary protein (microalbuminuria) can determine if nephropathy has developed. Physical examinations generally screen for signs of retinopathy, neuropathy, and foot problems.

Ketone Testing Ketone testing, which checks for the development of ketoacidosis, should be performed if symptoms are present or if risk has increased due to acute illness, stress, or pregnancy. Both blood and urine tests are available for home use, although the blood tests are generally more reliable. Ketone testing is most useful for patients who have type 1 diabetes or gestational diabetes. Individuals with type 2 diabetes may produce excessive ketone bodies when severely stressed by infection or trauma.

NUTRITION THERAPY: DIETARY RECOMMENDATIONS

Nutrition therapy can both improve glycemic control and slow the progression of diabetic complications. As always, the nutrition care plan must consider personal preferences and lifestyle habits. In addition, dietary intakes must be modified to accommodate growth, lifestyle changes, aging, and any complications that develop. Although all members of the diabetes care team should understand the principles of dietary treatment, a registered dietitian is best suited to design and implement the nutrition therapy

■ Goals for glycemic control:

- Fasting glucose: 70–130 mg/dL
- One to two hours after mealtime: <180 mg/dL
- Bedtime glucose: 90–150 mg/dL

■ In people without diabetes, HbA_{1c} is typically less than 6 percent of total hemoglobin.

self-monitoring of blood glucose: home monitoring of blood glucose levels using a glucose meter.

continuous glucose monitoring: continuous monitoring of tissue glucose levels using a small sensor placed under the skin.

fructosamine test: a measurement of glycated serum proteins that analyzes glycemic control over the preceding two to three weeks; also known as the *glycated albumin test* or the *glycated serum protein test*.

provided to diabetes patients. This section presents the dietary recommendations for diabetes; a later section describes meal-planning strategies.

Macronutrient Intakes The recommended macronutrient distribution (percent of kcalories from carbohydrate, fat, and protein) depends on food preferences and metabolic factors (for example, insulin sensitivity, blood lipid levels, and kidney function).[25] Intakes suggested for the general population (■) are often used as a guideline. Day-to-day consistency in carbohydrate intake is associated with better glycemic control, unless the patient is undergoing intensive insulin therapy that matches insulin doses to mealtime carbohydrate intakes.[26]

Total Carbohydrate Intake The amount of carbohydrate consumed has the greatest influence on blood glucose levels after meals—the more grams of carbohydrate ingested, the greater the glycemic response. The carbohydrate recommendation is based in part on the person's metabolic needs (which are related to the type of diabetes, degree of glucose tolerance, and blood lipid levels), the type of insulin or other medications used to manage the diabetes, and individual preferences. Low-carbohydrate diets, which restrict carbohydrate intake to less than 130 grams per day, are not recommended.

Carbohydrate Sources Different carbohydrate-containing foods have different effects on blood glucose levels; for example, consuming a portion of white rice causes blood glucose to increase more than would a similar portion of barley. A food's glycemic effect is influenced by the type of carbohydrate in a food, the food's fiber content, the preparation method, the other foods included in a meal, and individual tolerances. (■) A food's glycemic effect is not usually a primary consideration when treating diabetes, however, because clinical studies investigating the potential benefits of low-glycemic index diets on glycemic control have had mixed results.[27] Nonetheless, high-fiber, minimally processed foods—which typically have more moderate effects on blood glucose than do highly processed, starchy foods—are among the foods frequently recommended for persons with diabetes.

Sugars A common misperception is that people with diabetes need to avoid sugar and sugar-containing foods. In reality, table sugar (sucrose), made up of glucose and fructose, has a lower glycemic effect than starch. Because moderate consumption of sugar has not been shown to adversely affect glycemic control,[28] sugar recommendations for people with diabetes are similar to those for the general population, which suggest minimizing foods and beverages that contain added sugars. However, sugars and sugary foods must be counted as part of the daily carbohydrate allowance.

Although fructose has a minimal glycemic effect, its use as an added sweetener is not advised because excessive dietary fructose may adversely affect blood lipid levels. (Note that it is not necessary to avoid the naturally occurring fructose in fruits and vegetables.) Sugar alcohols (such as sorbitol and maltitol) have lower glycemic effects than glucose or sucrose, but their use has not been found to significantly improve long-term glycemic control. Artificial sweeteners (such as aspartame, saccharin, and sucralose) contain no digestible carbohydrate and can be safely used in place of sugar.

Fiber Fiber recommendations for individuals with diabetes are similar to those for the general population; (■) thus, people with diabetes are encouraged to include fiber-rich foods such as legumes, whole-grain cereals, fruits, and vegetables in their diet. Although some studies have suggested that very high fiber intakes (44 to 50 grams per day) may improve glycemic control, many individuals have difficulty enjoying or tolerating such large amounts of fiber.[29]

Dietary Fat Because people with diabetes are at high risk of developing cardiovascular diseases, guidelines for dietary fat are similar to those suggested for other persons at risk: saturated fat should be less than 7 percent of total kcalories, *trans* fat should be minimized, and cholesterol intake should be limited to less than 200 milligrams daily.[30] Dietary strategies for cardiovascular disease are discussed further in Chapter 21.

■
- Carbohydrate: 45 to 65% of kcalories
- Fat: 20 to 35% of kcalories
- Protein: 10 to 35% of kcalories

■ Reminder: The *glycemic index* ranks foods according to their glycemic effect (see Nutrition in Practice 3). The website www.glycemicindex.com provides glycemic index values for a wide variety of common foods.

■ Fiber DRIs for adult women and men range from 21 to 38 g; check the DRI table on the inside front cover of this text for specific values.

Protein In the United States, the average protein intake is about 15 percent of the energy intake. Although small, short-term studies have suggested that protein intakes above 20 percent of kcalories may improve glycemic control, increase satiety, and help with weight loss, the long-term effects of such diets on diabetes management are unknown.[31] In addition, high protein intakes are discouraged because they may be detrimental to kidney function in patients with nephropathy.

Alcohol Use in Diabetes Guidelines for alcohol intake are similar to those for the general population, which recommend that women and men limit their average daily intakes of alcohol to one drink and two drinks per day, respectively. (■) In addition, individuals using insulin or medications that promote insulin secretion should consume food when they ingest alcoholic beverages to avoid hypoglycemia (alcohol can cause hypoglycemia by interfering with glucose production in the liver). Conversely, an excessive alcohol intake (three or more drinks per day) can worsen hyperglycemia and raise triglyceride levels in susceptible persons. People who should avoid alcohol include pregnant women and individuals with advanced neuropathy, abnormally high triglyceride levels, or a history of alcohol abuse.[32]

Micronutrients Micronutrient recommendations for people with diabetes are the same as for the general population. Vitamin and mineral supplementation is not recommended unless nutrient deficiencies develop; those at risk include the elderly, pregnant or lactating women, strict vegetarians, and individuals on kcalorie-restricted diets.

Although various micronutrients have been tested for their potential benefits in managing diabetes, results have not been promising. Studies testing the use of chromium supplements for improving glycemic control have had inconsistent results, and supplementation is not currently recommended.[33] Similarly, clinical trials testing the effectiveness of vitamin C, vitamin E, and beta-carotene supplements for reducing the oxidative stress associated with diabetes or preventing complications have not found any benefit, and some data suggested possible harm.[34]

Body Weight in Type 2 Diabetes Because excessive body fat can worsen insulin resistance, weight loss is recommended for overweight or obese individuals who have diabetes.[35] Even moderate weight loss (5 percent of body weight) can help to improve glycemic control, blood lipid levels, and blood pressure. Weight loss is most beneficial early in the course of diabetes, before insulin secretion has diminished.

NUTRITION THERAPY: MEAL-PLANNING STRATEGIES

Dietitians provide a number of meal-planning strategies to help people with diabetes maintain glycemic control. These strategies emphasize control of carbohydrate intake and portion sizes. Initial dietary instructions may include guidelines for maintaining a healthy diet, improving blood lipids, and reducing cardiovascular risk factors. Sample menus that include commonly eaten foods can help to illustrate general principles.

Carbohydrate Counting Carbohydrate-counting techniques are simpler and more flexible than other menu-planning approaches and are widely used for planning diabetes diets. Carbohydrate counting works as follows: After an interview in which the dietitian learns about the patient's usual food intake and calculates nutrient and energy needs, the patient is given a daily carbohydrate allowance, divided into a pattern of meals and snacks according to individual preferences. The carbohydrate allowance can be expressed in grams or as the number of carbohydrate portions allowed per meal (see Table 20-4, p. 577). The user of the plan need only be concerned about meeting carbohydrate goals and can select from any of the carbohydrate-containing food groups when planning meals (see Table 20-5 and Figure 20-2). Although encouraged to make healthy food choices, the individual has the freedom to choose the foods desired at each meal without risking loss of glycemic control. Some people may also need guidance about noncarbohydrate foods to help them choose a healthy diet that improves blood lipids or energy intakes. The "How to" on pp. 577–578 shows how to implement carbohydrate counting in clinical practice.

■ Reminder: One drink is equivalent to 12 oz of beer, 5 oz of wine, or 1½ oz of 80 proof distilled spirits such as gin, rum, vodka, and whiskey.

HOW TO Use Carbohydrate Counting in Clinical Practice

1. The first step in basic carbohydrate counting is to determine an appropriate carbohydrate allowance and suitable distribution pattern; an example is shown in Table 20-4. To ensure that the carbohydrate level is acceptable to the person using the plan, a nutrition assessment can help the dietitian estimate the person's usual energy and carbohydrate intakes and food habits. Frequent monitoring of blood glucose levels can help to determine whether additional carbohydrate restriction would be helpful.

The example given in Table 20-4 illustrates a meal pattern for a person consuming 2000 kcalories daily with a carbohydrate allowance of 50 percent of kcalories. This is calculated as follows:

$$2000 \text{ kcal} \times 50\% = 1000 \text{ kcal of carbohydrate}$$

$$\frac{1000 \text{ kcal carbohydrate}}{4 \text{ kcal/g carbohydrate}} = 250 \text{ g carbohydrate/day}$$

$$\frac{250 \text{ g carbohydrate}}{15 \text{ g/1 carbohydrate portion}} = 16.7 \text{ carbohydrate portions/day}$$

2. The distribution of carbohydrates among meals and snacks is based on both individual preferences and metabolic needs. In type 1 diabetes, the insulin regimen must coordinate with the individual's dietary and lifestyle choices. People using conventional insulin therapy must maintain a consistent carbohydrate intake from day to day to match their particular insulin prescription, whereas those using intensive therapy can alter insulin dosages when carbohydrate intakes change. People with type 2 diabetes are encouraged to develop dietary patterns that suit their lifestyle and medication schedules. For all types of diabetes, the carbohydrate recommendation may need to be altered periodically to improve blood glucose control.

3. Carbohydrate counting can be done in one of two ways:
   - Count the grams of carbohydrate provided by foods.
   - Count carbohydrate portions, expressed in terms of servings that contain approximately 15 grams each.

Success with carbohydrate counting requires knowledge about the food sources of carbohydrates and an understanding of portion control. As shown in Table 20-5, food selections that contain about 15 grams of carbohydrate are interchangeable.

continued

TABLE 20-4 Sample Carbohydrate Distribution for a 2000-kCalorie Diet

	Carbohydrate Allowance	
Meals	Grams	Portions[a]
Breakfast	60	4
Lunch	60	4
Afternoon snack	30	2
Dinner	75	5
Evening snack	30	2
Totals	**255 g**	**17**

Note: The carbohydrate allowance in this example is approximately 50 percent of total kcalories.
[a]1 portion = 15 g carbohydrate = 1 portion of starchy food, milk, or fruit.

TABLE 20-5 Carbohydrate-Containing Food Groups and Sample Portion Sizes

Bread, cereal, rice, and pasta: 1 portion = 15 g carbohydrate

1 slice of bread or 1 tortilla
½ English muffin
¾ c unsweetened, ready-to-eat cereal
½ c cooked oatmeal
⅓ c cooked rice or pasta

Starchy vegetables: 1 portion = 15 g carbohydrate

1 small (3 oz) potato
½ c canned or frozen corn
½ c cooked beans
1 c winter squash, cubed

Fruit: 1 portion = 15 g carbohydrate

1 medium apple, orange, or peach
1 small banana
¾ c blueberries or chopped pineapple
½ c apple or orange juice

Milk products: 1 portion = 12 g carbohydrate; may be rounded up to 15 g for ease in counting carbohydrate portions

1 c milk (whole, low-fat, or fat-free)
1 c buttermilk
6 oz plain yogurt

Sweets and desserts: Carbohydrate content varies; portions listed contain approximately 15 g

½ c ice cream
2 sandwich cookies (with cream filling)
½ frosted cupcake
1 granola bar (1 oz)
1 tbs honey

Nonstarchy vegetables: 1 portion = 3–6 g carbohydrate; 3 servings are equivalent to 1 carbohydrate portion; can be disregarded if fewer than 3 servings are consumed

½ c cooked cauliflower
½ c cooked cabbage, collards, or kale
½ c cooked okra
½ c diced or raw tomatoes

Note: Unprocessed meats, fish, and poultry contain negligible amounts of carbohydrate.

HOW TO Use Carbohydrate Counting in Clinical Practice, *continued*

The portions of foods that contain 15 grams may vary substantially, however, even among foods in a single food group. Accurate carbohydrate counting often requires instruction and practice in portion control using measuring cups, spoons, and a food scale. Food lists that indicate the carbohydrate contents of common foods are available from the American Diabetes Association and the Academy of Nutrition and Dietetics; these are helpful resources for learning carbohydrate-counting methods.

When using packaged foods, individuals should check the Nutrition Facts panel of food labels to find the carbohydrate content of a serving. If the fiber content is greater than 5 grams per serving, it should be subtracted from the *Total Carbohydrate* value, as fiber does not contribute to blood glucose (some health practitioners may suggest subtracting only half of the grams of fiber). If the sugar alcohol content is greater than 5 grams per serving, half of the grams of sugar alcohol can be subtracted from the *Total Carbohydrate* value.

4. Once they have learned the basic carbohydrate counting method, individuals can select whatever foods they wish as long as they do not exceed their carbohydrate goals. Figure 20-2 shows a day's menu that provides the carbohydrate allowance shown in Table 20-4. Although carbohydrate counting focuses on a single macronutrient, people using this technique should be encouraged to follow a healthy eating plan that meets other dietary objectives as well.

FIGURE 20-2 Translating Carbohydrate Portions into a Day's Meals

SAMPLE MENU

	Carbohydrate Portions
Breakfast:	
Carbohydrate goal = 4 portions or 60 g.	
¾ c unsweetened, ready-to-eat cereal	1
½ c low-fat milk	½
1 scrambled egg	—
1 slice whole-wheat toast (with margarine or butter)	1
6 oz orange juice	1½
Coffee (without milk or sugar)	—
Lunch:	
Carbohydrate goal = 4 portions or 60 g.	
1 tuna salad sandwich (includes 2 slices whole-grain bread, mayonnaise)	2
6 oz yogurt (plain) with ¾ c blueberries and artificial sweetener	2
Diet cola	—

	Carbohydrate Portions
Afternoon snack:	
Carbohydrate goal = 2 portions or 30 g.	
2 sandwich cookies	1
1 c low-fat millk	1
Dinner:	
Carbohydrate goal = 5 portions or 75 g.	
4 oz grilled steak	—
1 small baked potato (with margarine or butter)	1
Corn on cob, 1 large ear	2
½ c steamed collard greens[a]]	
1 c sliced, raw tomatoes[a]]	1
½ c ice cream	1
Evening snack:	
Carbohydrate goal = 2 portions or 30 g.	
1 medium apple	1
1 oz granola bar	1

[a]Three servings of nonstarchy vegetables are equivalent to 1 carbohydrate portion.

Carbohydrate counting is taught at different levels of complexity depending on a person's needs and abilities. The basic carbohydrate-counting method just described can be helpful for most people, although it requires a consistent carbohydrate intake from day to day to match the medication or insulin regimen. Advanced carbohydrate counting allows more flexibility but is best suited for patients using intensive insulin therapy. With this method, a person can determine the specific dose of insulin needed to cover the amount of carbohydrate consumed at a meal. The person is then free to choose the types and portions of food desired without sacrificing glycemic control. Advanced carbohydrate counting requires some training and should be attempted only after an individual has mastered more basic methods.

Exchange Lists for Meal Planning The exchange list system is an alternative meal-planning method, although it may be more difficult for patients to learn than carbohydrate counting. The exchange system sorts foods according to their proportions of carbohydrate, fat, and protein so that each item in a food group (or "exchange list") has a similar macronutrient and energy content (see pp. C-1 to C-2). Thus, any food on a list can be exchanged, or traded, for any other food on the same list without affecting the macronutrient balance in a day's meals. Although the exchange list system can be helpful for individuals who want to maintain a diet with specific percentages of protein, carbohydrate, and fat, it offers no advantages for maintaining glycemic control and is less flexible than carbohydrate counting. This system of meal planning is described further in Appendix C (Appendix B for Canadians).

The exchange lists can be helpful resources for individuals using carbohydrate-counting methods because the portions in the exchange lists are interchangeable with the portions used in carbohydrate counting. Foods listed in the starch, fruit, and milk lists, for example, are equivalent to carbohydrate portions, as each item contains approximately 15 grams of carbohydrate (see pp. C-3 to C-6; note that the carbohydrate in the milk exchanges can be rounded up to 15 grams). In the list labeled "Sweets, Desserts, and Other Carbohydrates" (pp. C-7 to C-8), the carbohydrate portions are indicated in the far-right column.

INSULIN THERAPY

Insulin therapy is necessary for individuals who cannot produce enough insulin to meet their metabolic needs. It is therefore required by people with type 1 diabetes and those with type 2 diabetes who cannot maintain glycemic control with medications, diet, and exercise. The pancreas normally secretes insulin in relatively low amounts between meals and during the night (called *basal insulin*) and in much higher amounts when meals are ingested. Ideally, the insulin treatment should reproduce the natural pattern of insulin secretion as closely as possible.

Insulin Preparations The forms of insulin that are commercially available differ by their onset of activity, timing of peak activity, and duration of effects. Table 20-6 and Figure 20-3 (p. 580) show how insulin preparations are classified: they may be rapid acting (lispro, aspart, and glulisine), short acting (regular), intermediate acting (NPH), or long acting (glargine and detemir), thereby allowing substantial flexibility in establishing a suitable insulin regimen.[36] The rapid- and short-acting insulins are typically used at mealtimes, whereas the intermediate- and long-acting insulins provide basal insulin for the periods between meals and during the night. Thus, mixtures of several types of insulin can produce greater glycemic control than any one type alone. Several premixed formulations are also available; examples are listed in Table 20-6.

■ Because insulin is a protein, it would be destroyed by digestive processes if taken orally.

subcutaneous (sub-cue-TAY-nee-us): beneath the skin.

syringes: devices used for injecting medications. A syringe consists of a hypodermic needle attached to a hollow tube with a plunger inside.

Insulin Delivery Insulin is most often administered by **subcutaneous** injection, either self-administered or provided by caregivers. (■) Disposable **syringes**, which are filled from vials that contain multiple doses of insulin, are the most common devices used for injecting insulin. Another option is to use insulin pens, injection devices that resemble permanent marking pens. Disposable insulin pens are prefilled with insulin and used one time only, whereas reusable pens can be fitted with prefilled insulin cartridges and replaceable needles. Some individuals use insulin pumps, computerized devices that infuse insulin through thin, flexible tubing that remains in the skin; the pump can be attached to a belt or kept in a pocket. Some of the newer insulin pumps include built-in continuous glucose monitoring systems.

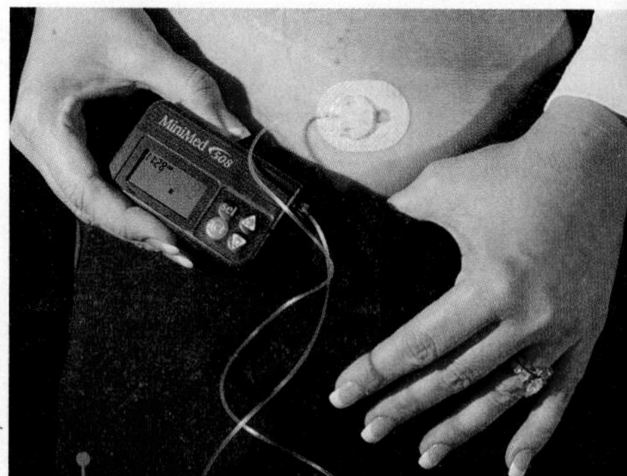

Courtesy of Medtronic

An external insulin pump delivers a low dosage of insulin continuously and bolus doses at mealtimes.

Insulin Regimen for Type 1 Diabetes Type 1 diabetes is best managed with intensive insulin therapy, which involves multiple daily injections of several types of insulin or the use of an insulin pump. Insulin pumps are usually programmed to deliver low amounts of rapid-acting insulin continuously (to meet basal insulin needs) and bolus doses of rapid-acting insulin at mealtimes. In persons who inject insulin,

TABLE 20-6 Insulin Preparations

Form of Insulin	Common Preparations	Onset of Action	Peak Action	Duration of Action
Rapid acting	Lispro (Humalog) Aspart (Novolog) Glulisine (Apidra)	5–15 minutes	60–90 minutes	3–5 hours
Short acting	Regular	30 minutes	2–3 hours	5–8 hours
Intermediate acting	NPH	2–4 hours	6–10 hours	10–16 hours
Long acting	Glargine (Lantus) Detemir (Levemir)	1–2 hours	Steady effects	24 hours
Insulin mixtures (with sample ratios)	NPH/regular (70:30) NPL (modified lispro)/ lispro (50:50)	Variable; depends on formulation	Variable; depends on formulation	Variable; depends on formulation

intermediate- or long-acting insulin meets basal insulin needs, and rapid- or short-acting insulin is injected before meals. (■) At least three or more daily injections are required for good glycemic control. Simpler regimens involve twice-daily injections of a mixture of intermediate- and short-acting insulin. Regimens that include three or more injections allow for greater flexibility in carbohydrate intake and meal timing. With fewer injections, the timing of both meals and injections must be similar from day to day to avoid periods of insulin deficiency or excess.

■ Rapid-acting insulin begins working within 15 minutes, so it can be injected right before a meal. Short-acting insulin requires a half-hour wait before the meal can begin.

A person using intensive therapy must learn to accurately determine the amount of insulin to inject before each meal. The amount required depends on the premeal blood glucose level, the carbohydrate content of the meal, and the person's body weight and sensitivity to insulin. To determine insulin sensitivity, the individual keeps careful records of food intake, insulin dosages, and blood glucose levels. Eventually, these records are analyzed by medical personnel to determine the appropriate **carbohydrate-to-insulin ratio** for that individual, which assists in calculating insulin doses at mealtime. Intensive therapy allows for substantial variation in food intake and lifestyle, but it requires frequent testing of blood glucose levels and a good understanding of carbohydrate counting.

carbohydrate-to-insulin ratio: the amount of carbohydrate that can be handled per unit of insulin; on average, every 15 grams of carbohydrate requires about 1 unit of rapid- or short-acting insulin.

FIGURE 20-3 Effects of Insulin Preparations

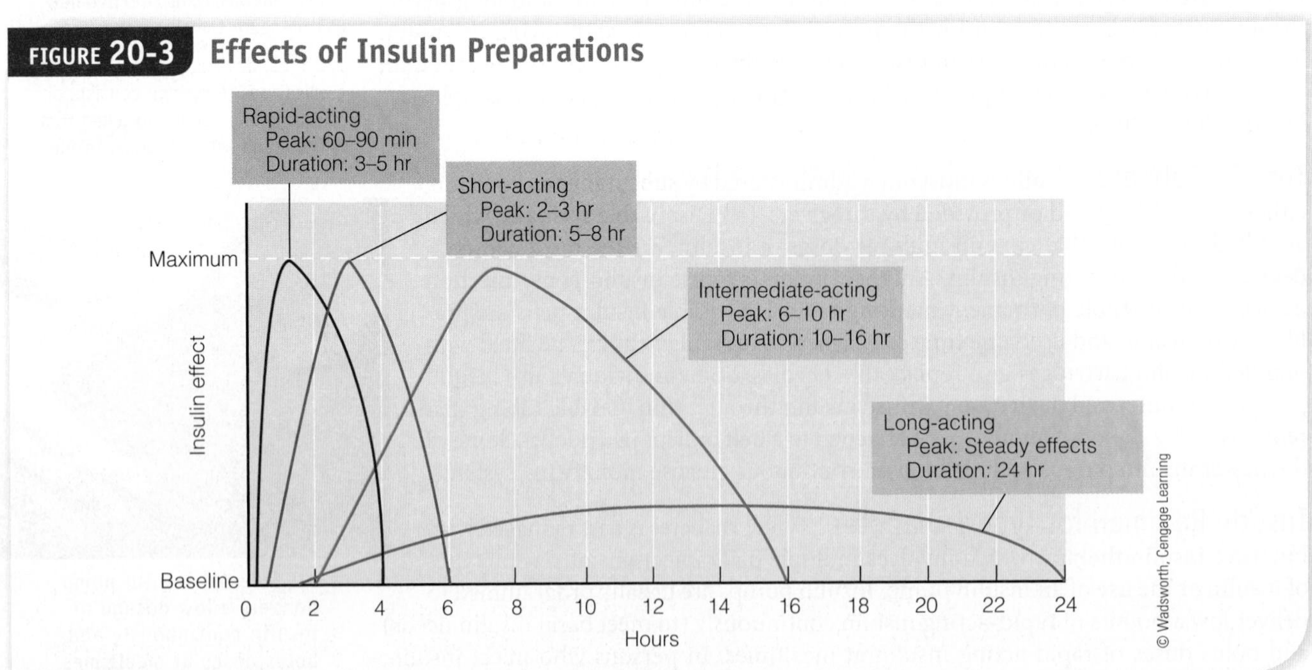

After insulin therapy is initiated, persons with type 1 diabetes may experience a temporary remission of disease symptoms and a reduced need for insulin, known as the *honeymoon period*. The remission is due to a temporary improvement in pancreatic beta-cell function and may last for several weeks or months. It is important to anticipate this period of remission to avoid insulin excess. In all cases, the honeymoon period eventually ends, and the patient must reinstate full insulin treatment.

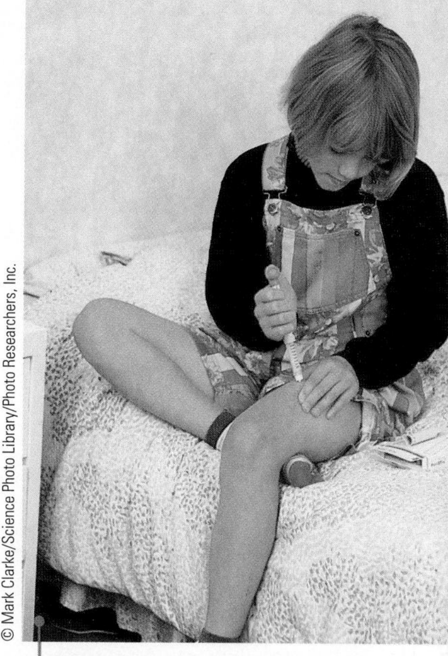

Children often become adept at administering the insulin they require.

Insulin Regimen for Type 2 Diabetes About 30 percent of people diagnosed with type 2 diabetes can benefit from insulin therapy.[37] Although initial treatment of type 2 diabetes may involve diet therapy, physical activity, and oral antidiabetic medications, long-term results with these treatments are often disappointing. As the disease progresses, pancreatic function worsens, and many individuals require insulin therapy to maintain glycemic control.

Many possible regimens can be used to control type 2 diabetes. Some persons may be treated with insulin alone, whereas others may use insulin in combination with other antidiabetic drugs. Many patients need only one or two daily injections. Some regimens involve a mixture of intermediate-acting and rapid- or short-acting insulin in the morning and an injection of intermediate- or long-acting insulin at dinner or before bedtime. In other cases, only a single injection of intermediate- or long-acting insulin may be needed at bedtime.[38] Doses and timing are adjusted according to the results of blood glucose self-monitoring.

Insulin Therapy and Hypoglycemia Hypoglycemia is the most common complication of insulin treatment, although it may also result from the use of some oral antidiabetic drugs. It most often results from intensive insulin therapy because the attempt to attain near-normal blood glucose levels increases the risk of overtreatment.

Hypoglycemia can be corrected with the immediate intake of glucose or a carbohydrate-containing food. Usually, 15 to 20 grams of carbohydrate (■) can relieve hypoglycemia in about 15 minutes, although patients should retest their blood glucose levels after 15 minutes in case additional treatment is necessary. Foods that provide pure glucose yield a better response than foods that contain other sugars, such as sucrose or fructose. Individuals using insulin are usually advised to carry glucose tablets or a source of carbohydrate that can be readily ingested. After blood glucose normalizes, patients should consume a meal or snack to prevent recurrence. Individuals at risk of severe hypoglycemia are often given prescriptions for the hormone glucagon, which can be injected by caregivers in case of unconsciousness.

■ Each of the following sources provides approximately 15 g of carbohydrate:

- Glucose tablets: 4 tablets
- Table sugar: 4 tsp
- Maple syrup: 4 tsp
- Orange juice: ½ c
- Jelly beans: 15 small

Insulin Therapy and Weight Gain Weight gain is sometimes an unintentional side effect of insulin therapy, especially in individuals undergoing intensive insulin treatment. Although the exact causes of the weight gain are unclear, it may partly be due to insulin's stimulatory effect on fat synthesis. Patients may be able to avoid weight gain by reducing the ratio of basal to mealtime insulin and improving carbohydrate-counting skills to obtain better estimates of mealtime insulin requirements.[39] Concerns about weight should not discourage the use of intensive therapy, which is associated with longer life expectancy and fewer complications than occur with conventional therapy.

Fasting Hyperglycemia Insulin therapy must sometimes be adjusted to prevent **fasting hyperglycemia**, which typically develops in the early morning after an overnight fast of at least eight hours. The usual cause is a waning of insulin action during the night due to insufficient insulin. A second possibility, known as the **dawn phenomenon**, is an increase of blood glucose in the morning due to the early morning secretion of growth hormone, which reduces insulin sensitivity. Less frequently, fasting hyperglycemia develops as a result of nighttime hypoglycemia, which causes the secretion of hormones that stimulate glucose production; the resulting condition is known as **rebound hyperglycemia** (also called the **Somogyi effect**). Whatever the cause, fasting hyperglycemia can be treated by adjusting the dosage or formulation of insulin administered in the evening.[40]

fasting hyperglycemia: hyperglycemia that typically develops in the early morning after an overnight fast of at least eight hours.

dawn phenomenon: morning hyperglycemia that is caused by the early-morning release of growth hormone, which reduces insulin sensitivity.

rebound hyperglycemia: hyperglycemia that results from the release of counterregulatory hormones following nighttime hypoglycemia; also called the **Somogyi effect**.

ANTIDIABETIC DRUGS

Treatment of type 2 diabetes often requires the use of oral medications and injectable drugs other than insulin. These drugs can improve hyperglycemia by several modes of action: they can stimulate insulin secretion, suppress glucagon secretion, increase insulin sensitivity, improve glucose utilization in tissues, reduce glucose production in the liver, delay stomach emptying, or delay carbohydrate digestion and absorption. Treatment may involve the use of a single medication (monotherapy) or a combination of several medications (combination therapy). By utilizing several mechanisms at once, combination therapy achieves more rapid and sustained glycemic control than is possible with monotherapy. Table 20-7 lists examples of antidiabetic drugs, and the Diet-Drug Interactions feature lists some of their nutrition-related effects. Because medications cannot replace the benefits offered by dietary adjustments and physical activity, persons with diabetes should be advised to continue both.

PHYSICAL ACTIVITY AND DIABETES MANAGEMENT

Regular physical activity can improve glycemic control considerably and is therefore a central feature of diabetes management. Physical activity also benefits other aspects of health, including cardiovascular risk factors and body weight. People with diabetes are advised to perform at least 150 minutes of moderate-intensity activity per week, spread over at least three days. In addition, individuals with type 2 diabetes should engage in resistance exercise at least twice weekly unless contraindicated by a medical condition that increases risk of injury.[41] Both aerobic and resistance exercise can improve insulin sensitivity.

TABLE 20-7 Antidiabetic Drugs

Drug Category	Common Examples	Mode of Action
Alpha-glucosidase inhibitors	Acarbose (Precose) Miglitol (Glyset)	Delay carbohydrate digestion and absorption
Amylin analogs (injected)	Pramlintide (Symlin)	Suppress glucagon secretion, delay stomach emptying, increase satiety
Biguanides	Metformin (Glucophage)	Inhibit liver glucose production, improve glucose utilization
Bile acid sequestrants	Colesevelam (Welchol)	Unknown; may inhibit liver glucose production
Dipeptidyl peptidase 4 (DPP-4) inhibitors	Saxagliptin (Onglyza) Sitagliptin (Januvia)	Improve insulin secretion, suppress glucagon secretion, delay stomach emptying
Dopamine D2 receptor agonists	Bromocriptine (Cycloset)	Increase insulin sensitivity
GLP-1 receptor agonists (injected)	Exenatide (Byetta) Liraglutide (Victoza)	Improve insulin secretion, suppress glucagon secretion, delay stomach emptying, increase satiety
Meglitinides	Nateglinide (Starlix) Repaglinide (Prandin)	Stimulate insulin secretion by the pancreas
Sulfonylureas	Glipizide (Glucotrol) Glyburide (Diabeta)	Stimulate insulin secretion by the pancreas
Thiazolidinediones	Pioglitazone (Actos) Rosiglitazone (Avandia)[a]	Increase insulin sensitivity

[a]Use is restricted in the United States and various other countries due to the increased risk of heart attack or heart failure in persons using the drug.

DIET-DRUG INTERACTIONS

Check this table for notable nutrition-related effects of the medications discussed in this chapter.

Drug	Effects
Alpha-glucosidase inhibitors	**Gastrointestinal effects:** Flatulence, abdominal cramps, diarrhea
	Metabolic effects: May decrease blood concentrations of calcium and vitamin B_6
Biguanides (metformin)	**Gastrointestinal effects:** Metallic taste, nausea, vomiting, anorexia, flatulence, abdominal cramps, diarrhea
	Dietary interaction: Excessive alcohol intake may cause lactic acidosis, which requires emergency treatment
	Metabolic effects: Decreased folate and vitamin B_{12} absorption, which may lead to deficiency
Meglitinides	**Metabolic effects:** Hypoglycemia, weight gain
Sulfonylureas	**Gastrointestinal effects:** Nausea, abdominal cramps, diarrhea, constipation
	Dietary interaction: Alcohol may delay drug absorption and prolong hypoglycemia (if hypoglycemia occurs)
	Metabolic effects: Hypoglycemia, weight gain, allergic skin reactions
Thiazolidinediones	**Metabolic effects:** Weight gain, fluid retention, edema, anemia, decreased bone density and increased risk of fractures (women)

Medical Evaluation before Exercise Before a person with diabetes begins a new exercise program, a medical evaluation should screen for problems that may be aggravated by certain activities. Complications involving the heart and blood vessels, eyes, kidneys, feet, and nervous system may limit the types of activity recommended. For individuals with a low level of fitness who have been relatively inactive, only mild or moderate exercise may be prescribed at first; a short walk at a comfortable pace may be the first activity suggested. People with severe retinopathy should avoid vigorous aerobic or resistance exercise, which may lead to retinal detachment and damage to eye tissue. Individuals with peripheral neuropathy should ensure that they wear proper footwear (■) during exercise; those with a foot injury or open sore should avoid weight-bearing activity. To prevent dehydration, proper hydration should be encouraged before and during exercise.

Maintaining Glycemic Control People who do not have diabetes maintain blood glucose levels during physical activity because their insulin levels drop and secretions of glucagon and epinephrine increase, promoting glucose production in the liver. In people who use insulin or medications that induce insulin secretion, however, blood glucose levels fall during activity because the insulin promotes rapid consumption of glucose by exercising muscles and blocks glucose synthesis by the liver. For this reason, insulin should not be injected immediately before exercise because it can lead to hypoglycemia, and medications that promote insulin secretion may require dosage adjustments due to exercise.

Individuals who use insulin or medications that induce insulin secretion should check blood glucose levels both before and after an activity. If blood glucose is below 100 mg/dL, carbohydrate should be consumed before the exercise begins.[42] Additional carbohydrate may be needed during or after prolonged activity or even several hours after the activity is completed. Individuals with type 1 diabetes who have ketosis should avoid vigorous activity, which increases ketone body production and can worsen the ketosis.

■ The use of protective foot gear, such as gel soles and socks that prevent blisters, can help to prevent foot trauma during exercise.

SICK-DAY MANAGEMENT

Illness, infection, or injury can cause hormonal changes that raise blood glucose levels and greatly increase the risk of developing diabetic ketoacidosis or the hyperosmolar hyperglycemic syndrome. During illness, patients with diabetes should measure blood glucose and ketone levels several times daily. They should continue to use antidiabetic drugs, including insulin, as prescribed; adjustments in dosages may be necessary if hyperglycemia persists. If appetite is poor, patients should select easy-to-manage foods and beverages that provide the prescribed amount of carbohydrate at each meal. To prevent dehydration, especially if vomiting or diarrhea is present, patients should make sure they consume adequate amounts of liquids throughout the day.

> ***NURSING DIAGNOSIS***
>
> The nursing diagnosis *risk for unstable blood glucose level* may apply to an individual with diabetes who has an illness or injury.

Review Notes

- Diabetes treatment includes nutrition therapy, the use of insulin or other antidiabetic medications, and appropriate physical activity. Glycemic control is evaluated by monitoring blood glucose levels and glycated hemoglobin.
- The quantity of carbohydrate consumed has the greatest influence on blood glucose levels after meals. The total amount of carbohydrate ingested is more important than the type of carbohydrate consumed.
- Carbohydrate counting is widely used in menu planning and can be taught at different levels of complexity, depending on individual needs and abilities.
- Insulin therapy is required for patients who cannot produce sufficient insulin and may be used in both type 1 and type 2 diabetes. Antidiabetic drugs prescribed for type 2 diabetes can improve insulin secretion and effectiveness, suppress glucagon secretion, reduce glucose production by the liver, and delay carbohydrate absorption.
- Physical activity improves glycemic status and enhances various aspects of general health. Illness may worsen glycemic control and often necessitates adjustments in medications and careful attention to dietary and fluid requirements. The Case Study provides an opportunity to review the factors that influence treatment of type 1 diabetes.

Case Study: Child with Type 1 Diabetes

Nora is a 12-year-old girl who was diagnosed with type 1 diabetes two years ago. She practices intensive therapy and has had the support of her parents and an excellent diabetes management team. With their help, Nora has been able to assume the bulk of the responsibility for her diabetes care and has managed to control her blood glucose remarkably well. In the past few months, however, Nora has been complaining bitterly about the impositions diabetes has placed on her life and her interactions with friends. Sometimes she refuses to monitor her blood glucose levels, and she has skipped insulin injections a few times. Recently, Nora was admitted to the emergency room complaining of fever, nausea, vomiting, and intense thirst. The physician noted that Nora was confused and lethargic. A urine test was positive for ketones, and her blood glucose levels were 400 mg/dL. The diagnosis was diabetic ketoacidosis.

1. Describe the metabolic events that lead to ketoacidosis. Were Nora's symptoms and laboratory tests consistent with the diagnosis?
2. Review Table 20-3, and consider the advantages and disadvantages that intensive therapy might have for Nora. Describe the complications associated with long-term diabetes.
3. Discuss how Nora's age might influence her ability to cope with and manage her diabetes. Why might she feel that diabetes is disrupting her life? List suggestions that may help. How would you explain the importance of glycemic control to a 12-year-old girl?

Diabetes Management in Pregnancy

Women with diabetes face new challenges during pregnancy. Due to hormonal changes, pregnancy increases insulin resistance and the body's need for insulin, so maintaining glycemic control may be more difficult. In addition, 4 to 14 percent of nondiabetic women in the United States develop gestational diabetes (the prevalence depends on the patient population).[43] (■) Women with gestational diabetes are at greater risk of developing type 2 diabetes later in life, and their children are at increased risk of developing obesity and type 2 diabetes as they enter adulthood.

A pregnancy complicated by diabetes increases health risks for both mother and fetus. Uncontrolled diabetes is linked with increased incidences of miscarriage, birth defects, and fetal deaths. Newborns are more likely to suffer from respiratory distress and to develop metabolic problems such as hypoglycemia, jaundice, and hypocalcemia. Women with diabetes often deliver babies with **macrosomia** (abnormally large bodies), which makes delivery more difficult and can result in birth trauma or the need for a cesarean section. Macrosomia results because maternal hyperglycemia induces excessive insulin production by the fetal pancreas, which stimulates growth and fat deposition.[44]

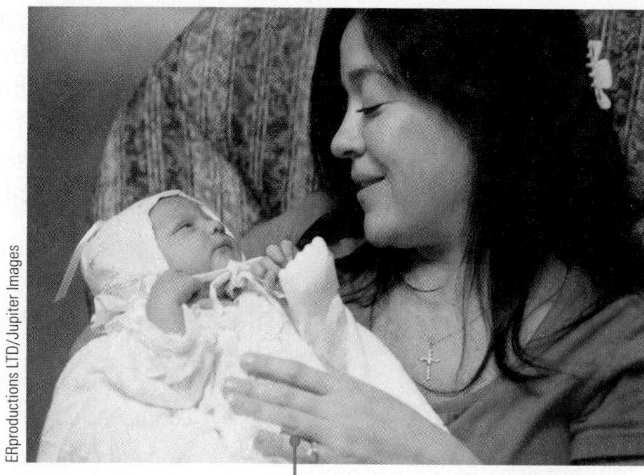
ERproductions LTD/Jupiter Images

Glycemic control during pregnancy offers the best chance of a safe delivery and a healthy infant.

PREGNANCY IN TYPE 1 OR TYPE 2 DIABETES

Women with diabetes who achieve glycemic control at conception and during the first trimester of their pregnancy substantially reduce the risks of birth defects and spontaneous abortion. For this reason, it is recommended that women contemplating pregnancy receive preconception care to avoid the complications associated with poorly controlled diabetes. Maintaining glycemic control during the second and third trimesters can minimize the risks of macrosomia and morbidity in newborn infants.

Women with type 1 diabetes require intensive insulin therapy during pregnancy. Insulin adjustments may be necessary every few weeks due to changes in insulin sensitivity. Patients with type 2 diabetes are usually switched from their usual medications to insulin therapy to prevent possible toxicity to the fetus. Although metformin and the sulfonylurea glyburide may be safe to use at conception and during early pregnancy in pregnant women with type 2 diabetes, research data are limited in this population so physicians may be reluctant to prescribe the drugs.[45]

Nutrient requirements during pregnancy are similar for women with and without diabetes. In women with diabetes, however, carbohydrate intakes must be balanced with insulin treatment and physical activity to avoid hypoglycemia and hyperglycemia. To help with this goal, women should consume meals and snacks at similar times each day, and select carbohydrate sources that facilitate glucose control after meals, such as whole grains, fruits, and vegetables.[46] An evening snack is usually required to prevent overnight hypoglycemia and ketosis. When insulin dosages are adjusted, the diabetic woman will need to modify her carbohydrate intake as well.

NURSING DIAGNOSIS

Nursing diagnoses for diabetic pregnancies may include *risk for disturbed maternal/fetal dyad; risk for delayed development: fetal;* and *risk for disproportionate growth: fetal.*

GESTATIONAL DIABETES

Risk of gestational diabetes is highest in women who have a family history of diabetes, are obese, are in a high-risk ethnic group (African American, Asian American, Hispanic American, Native American, or Pacific Islander), or have previously given birth to an infant weighing more than 9 pounds. To ensure that appropriate treatment is offered, physicians routinely test all women for gestational diabetes between 24 and 28 weeks of gestation. In high-risk women, screening should begin prior to pregnancy or soon after conception. Even mild hyperglycemia can have adverse effects on a developing fetus and may lead to complications during pregnancy.[47]

Women with gestational diabetes who are overweight or obese may need to adjust their energy intakes during pregnancy. Although adequate energy is needed for fetal development, a modest kcaloric reduction (about 30 percent less than total energy needs) may improve glycemic control without increasing the risk of ketosis.[48] Restricting carbohydrate to 40 to 45 percent of total energy intake may improve blood glucose levels after meals. Carbohydrate is usually poorly tolerated in the morning; therefore, restricting carbohydrate (to about 30 grams) at breakfast is often necessary.[49] The remaining carbohydrate intake should be spaced throughout the day in several

■ Gestational diabetes was introduced in Chapter 10.

macrosomia (MAK-roh-SOH-mee-ah): the condition of having an abnormally large body; in infants, refers to birth weights of 4000 grams (8 pounds 13 ounces) and above.

meals and snacks, including an evening snack to prevent ketosis during the night. Regular aerobic activity is often recommended because it can help to improve glycemic control. Women who fail to achieve glycemic goals by diet and exercise alone may need to use insulin or an antidiabetic drug that is safe during pregnancy; glyburide or metformin may be prescribed as an alternative to insulin for some patients. The Case Study reviews the connections between gestational diabetes and type 2 diabetes.

Review Notes

- Careful management of blood glucose levels before and during pregnancy may prevent complications in mother and infant.
- Women with diabetes who become pregnant may need to adjust their insulin therapy or medications. Meals and snacks should be consumed at similar times each day; an evening snack may help to prevent overnight ketosis.
- Women with gestational diabetes may need to restrict energy and carbohydrate intakes to maintain appropriate glucose levels. Insulin or an antidiabetic drug may be prescribed to help them maintain glycemic control.

Case Study: Woman with Type 2 Diabetes

Teresa Cordova is a 41-year-old Mexican American woman recently diagnosed with type 2 diabetes. Mrs. Cordova developed gestational diabetes while she was pregnant with her second child. Her blood glucose levels returned to normal following pregnancy, and she was advised to get regular checkups, maintain a desirable weight, and engage in regular physical activity. Although she reports that she does not overeat and that she exercises regularly, she has been unable to maintain a healthy weight. At 5 feet 3 inches tall, Mrs. Cordova currently weighs 155 pounds. She has decided to lose weight and join a gym because she is concerned about the long-term effects of diabetes and the possibility that she may need insulin injections. She is also concerned about her husband and children because they are overweight and not very active. The physician refers Mrs. Cordova to a dietitian to help her plan a diet.

1. What factors in Mrs. Cordova's medical history increase her risk for diabetes? Are her husband and children also at risk?
2. Describe the general characteristics of a diet and exercise program that would be appropriate for Mrs. Cordova. How might weight loss and physical activity benefit her diabetes?
3. If Mrs. Cordova is unable to control her blood glucose with diet and physical activity, what treatment might be suggested? Explain to Mrs. Cordova why she would probably not require insulin at this time.
4. What dietary and lifestyle changes may help to prevent diabetes in Mrs. Cordova's husband and children?

Nutrition Assessment Checklist for People with Diabetes

MEDICAL HISTORY

Check the medical record to determine:

- ❍ Type of diabetes
- ❍ Duration of diabetes
- ❍ Acute and chronic complications
- ❍ Conditions, including pregnancy, that may alter treatment

MEDICATIONS

For people with preexisting diabetes who use antidiabetic drugs (including insulin), note:

- ❍ Type of medication
- ❍ Administration schedule

Check for use of other medications, including:

- ❍ Medications that affect blood glucose levels
- ❍ Cholesterol- and triglyceride-lowering medications
- ❍ Antihypertensive medications

DIETARY INTAKE

To devise an acceptable meal plan and coordinate medications, obtain:

- ❍ An accurate and thorough record of food intake and meal patterns
- ❍ An account of usual physical activities

At medical checkups, reassess the person's ability to:

- ❍ Maintain an appropriate carbohydrate intake
- ❍ Maintain an appropriate energy intake
- ❍ Monitor blood glucose levels at home
- ❍ Adjust insulin and diet to accommodate sick days
- ❍ Use appropriate foods to treat hypoglycemia

ANTHROPOMETRIC DATA

Take accurate baseline height and weight measurements as a basis for:

- ❍ Appropriate energy intake
- ❍ Initial insulin therapy

Periodically reassess height and weight for children and weight for adults and pregnant women to ensure that the meal plan provides an appropriate energy intake.

LABORATORY TESTS

Monitor the success of diabetes treatment using these tests:

- ❍ Blood lipid concentrations
- ❍ Blood or urinary ketones
- ❍ Glycated hemoglobin
- ❍ Urinary protein (microalbuminuria)

PHYSICAL SIGNS

Look for physical signs of:

- ❍ Dehydration, especially in older adults
- ❍ Foot ulcers
- ❍ Nerve damage
- ❍ Vision problems

Self Check

1. Which of the following is characteristic of type 1 diabetes?

a. Abdominal obesity increases risk.
b. The pancreas makes little or no insulin.
c. It is the predominant form of diabetes.
d. It often arises during pregnancy.

2. Which of the following is true about type 2 diabetes?

a. It is usually an autoimmune disease.
b. The pancreas makes little or no insulin.
c. Diabetic ketoacidosis is a common complication.
d. Chronic complications may develop before it is diagnosed.

3. Most chronic complications associated with diabetes result from:

a. altered kidney function.
b. infections that deplete nutrient reserves.
c. weight gain and hypertension.
d. damage to blood vessels and nerves.

4. Long-term glycemic control is usually evaluated by:

a. self-monitoring of blood glucose.
b. testing urinary ketone levels.
c. measuring glycated hemoglobin.
d. testing urinary protein levels (microalbuminuria).

5. Regarding dietary carbohydrate, a patient with diabetes should be most concerned about:

a. consuming the correct quantity of carbohydrate at each meal or snack.
b. consuming the correct proportion of sugars, starches, and fiber in meals.
c. avoiding added sugars and kcaloric sweeteners.
d. choosing meals with ideal proportions of protein, carbohydrate, and fat.

6. Which of the following is true regarding the general use of alcohol in diabetes?

a. A serving of alcohol is considered part of the carbohydrate allowance.
b. Alcohol contributes to hyperglycemia and should be avoided completely.
c. Alcohol can cause hypoglycemia and should therefore be consumed with food if patients use insulin or medications that stimulate insulin secretion.
d. Patients can use alcohol in unlimited quantities unless they are pregnant.

7. The meal-planning strategy best suited to all people with diabetes is:

a. carbohydrate counting.
b. the exchange list system.
c. following menus and recipes provided by a registered dietitian.
d. the approach that best helps the patient control blood glucose levels.

8. A patient using intensive insulin therapy is likely to follow a regimen that involves:

a. twice-daily injections that combine short-, intermediate-, and long-acting insulin in each injection.
b. a mixture of intermediate- and long-acting insulin injected between meals.
c. multiple daily injections that supply basal insulin and precise insulin doses at each meal.
d. the use of both insulin and oral antidiabetic agents.

9. In a person who has previously maintained good glycemic control, hyperglycemia can be precipitated by:
   a. infections or illnesses.
   b. chronic alcohol ingestion.
   c. undertreatment of hypoglycemia.
   d. prolonged exercise.

10. Which dietary modification may be helpful for women with gestational diabetes?
   a. Consuming most of the day's carbohydrate allotment in the morning
   b. Restricting carbohydrate to about 30 grams at breakfast
   c. Avoiding food intake after dinner
   d. Reducing energy intake to about 50 percent of the calculated requirement

Answers to these questions appear in Appendix H. For more chapter review: Access an interactive eBook, chapter-specific interactive learning tools, including flashcards, quizzes, videos, and more in your Nutrition CourseMate, accessed through CengageBrain.com.

Clinical Applications

1. Using the carbohydrate-counting method described in the "How to" on pp. 577–578, determine an appropriate carbohydrate intake (in both grams and portions) for a man with type 2 diabetes who requires approximately 2600 kcalories daily. Assume he would benefit from a carbohydrate allowance that is 50 percent of his energy intake. Using information from Tables 20-4 and 20-5, develop a one-day sample menu that is likely to meet his carbohydrate goals. Use the exchange lists in Appendix C to find additional examples of foods to include in your menu.

2. Take a trip to a pharmacy or use information from an online drugstore to price these items: blood glucose meter, test strips for the glucose meter selected, lancets, insulin, and syringes. Determine the approximate cost of insulin and syringes for a person who uses 12 units of short-acting insulin (regular) and 18 units of intermediate-acting insulin (NPH) taken in three injections daily (thus, the insulin requirement is 30 units per day). Also estimate the cost of testing blood glucose three times daily. Approximately how much would these supplies cost per month?

Nutrition on the Net

For further study of topics covered in this chapter, access these websites:

- Visit the American Diabetes Association and the Joslin Diabetes Center to find information on a wide range of topics related to diabetes: www.diabetes.org and www.joslin.org
- Comprehensive and reliable information about diabetes for both health practitioners and consumers is available from the National Institute of Diabetes and Digestive and Kidney Diseases and the Centers for Disease Control and Prevention: www2.niddk.nih.gov and www.cdc.gov/diabetes
- Find out how to become a diabetes educator by visiting the website of the American Association of Diabetes Educators: www.diabeteseducator.org

Notes

1. Centers for Disease Control and Prevention, National diabetes fact sheet: National estimates and general information on diabetes and prediabetes in the United States, 2011 (Atlanta, GA: U.S. Department of Health and Human Services, Centers for Disease Control and Prevention, 2011).
2. Centers for Disease Control and Prevention, 2011.
3. American Diabetes Association, Diagnosis and classification of diabetes mellitus, *Diabetes Care* 35 (2012): S64–S71.
4. Centers for Disease Control and Prevention, 2011.
5. A. L. May, E. V. Kuklina, and P. W. Yoon, Prevalence of cardiovascular disease risk factors among U.S. adolescents, 1999–2008, *Pediatrics* 129 (2012): 1035–1041.
6. S. E. Inzucchi and R. S. Sherwin, Type 1 diabetes mellitus, in L. Goldman and A I. Schafer, eds., *Goldman's Cecil Medicine* (Philadelphia: Saunders, 2012), pp. 1475–1489.
7. U. Masharani and M. S. German, Pancreatic hormones and diabetes mellitus, in D. G. Gardner and D. Shoback, eds., *Greenspan's Basic and Clinical Endocrinology* (New York: McGraw-Hill/Lange, 2011), pp. 573–655.
8. A. Maitra, The endocrine system, in V. Kumar and coeditors, *Robbins and Cotran Pathologic Basis of Disease* (Philadelphia: Saunders, 2010), pp. 1097–1164.
9. Centers for Disease Control and Prevention, 2011.
10. R. B. Lipton, Incidence of diabetes in children and youth—tracking a moving target, *Journal of the American Medical Association* 297 (2007): 2760–2762.
11. Centers for Disease Control and Prevention, 2011.
12. W. C. Knowler and coauthors, Reduction in the incidence of type 2 diabetes with lifestyle intervention or metformin, *New England Journal of Medicine* 346 (2002): 393–403.

13. American Diabetes Association, Standards of medical care in diabetes—2012, *Diabetes Care* 35 (2012): S11–S63.
14. Masharani and German, 2011.
15. S. E. Inzucchi and R. S. Sherwin, Type 2 diabetes mellitus, in L. Goldman and A I. Schafer, eds., *Goldman's Cecil Medicine* (Philadelphia: Saunders, 2012), pp. 1489–1499.
16. D. G. Gardner, Endocrine emergencies, in D. G. Gardner and D. Shoback, eds., *Greenspan's Basic and Clinical Endocrinology* (New York: McGraw-Hill/Lange, 2011), pp. 763–786.
17. Inzucchi and Sherwin, Type 1 diabetes mellitus, 2012.
18. Maitra, The endocrine system, 2010.
19. Inzucchi and Sherwin, Type 2 diabetes mellitus, 2012.
20. Maitra, The endocrine system, 2010.
21. Masharani and German, 2011.
22. Masharani and German, 2011.
23. American Diabetes Association, Implications of the United Kingdom Prospective Diabetes Study, *Diabetes Care* 21 (1998): 2180–2184; Diabetes Control and Complications Trial Research Group, The effect of intensive treatment of diabetes on the development and progression of long-term complications in insulin-dependent diabetes mellitus, *New England Journal of Medicine* 329 (1993): 977–986.
24. American Diabetes Association, Standards of medical care in diabetes—2012, 2012.
25. American Diabetes Association, Standards of medical care in diabetes—2012, 2012; American Diabetes Association, Nutrition recommendations and interventions for diabetes, *Diabetes Care* 31 (2008): S61–S78.
26. M. J. Franz and coauthors, The evidence for medical nutrition therapy for type 1 and type 2 diabetes in adults, *Journal of the American Dietetic Association* 110 (2010): 1852–1889.
27. Franz and coauthors, 2010.
28. Franz and coauthors, 2010; American Diabetes Association, 2008.
29. Franz and coauthors, 2010.
30. American Diabetes Association, 2008.
31. Franz and coauthors, 2010; American Diabetes Association, 2008.
32. American Diabetes Association, 2008.
33. American Diabetes Association, Standards of medical care in diabetes—2012, 2012.
34. American Diabetes Association, Standards of medical care in diabetes—2012, 2012.
35. American Diabetes Association, Standards of medical care in diabetes—2012, 2012; Franz and coauthors, 2010.
36. M. S. N. Kennedy, Pancreatic hormones and antidiabetic drugs, in B. G. Katzung, ed., *Basic and Clinical Pharmacology* (New York: McGraw-Hill/Lange, 2012), pp. 743–768; Masharani and German, 2011.
37. Kennedy, 2012.
38. Inzucchi and Sherwin, Type 2 diabetes mellitus, 2012; Kennedy, 2012.
39. R. J. Brown and coauthors, Uncoupling intensive insulin therapy from weight gain and hypoglycemia in type 1 diabetes, *Diabetes Technology and Therapeutics* 13 (2011): 457–460.
40. Masharani and German, 2011.
41. American Diabetes Association, Standards of medical care in diabetes—2012, 2012.
42. American Diabetes Association, Standards of medical care in diabetes—2012, 2012.
43. K. Rosene-Montella, Common medical problems in pregnancy, in L. Goldman and A I. Schafer, eds., *Goldman's Cecil Medicine* (Philadelphia: Saunders, 2012), pp. 1555–1565.
44. A. Maitra, Diseases of infancy and childhood, in V. Kumar and coeditors, *Robbins and Cotran Pathologic Basis of Disease* (Philadelphia: Saunders, 2010), pp. 447–483.
45. Masharani and German, 2011.
46. J. L. Kitzmiller and coauthors, Managing preexisting diabetes for pregnancy, *Diabetes Care* 31 (2008): 1060–1079.
47. Z. Hussain and L. Jovanovic, Nutritional strategies in pregestational, gestational, and postpartum diabetic patients, in J. I. Mechanick and E. M. Brett, eds., *Nutritional Strategies for the Diabetic and Prediabetic Patient* (Boca Raton, FL: CRC Press, 2006), pp. 133–148.
48. American Diabetes Association, 2008.
49. Academy of Nutrition and Dietetics, *Nutrition Care Manual* (Chicago: Academy of Nutrition and Dietetics, 2012).

Nutrition in Practice

The Metabolic Syndrome

Chapter 20 described how insulin resistance—a reduced sensitivity to insulin in muscle, adipose, and liver cells—can contribute to hyperglycemia and hyperinsulinemia and, eventually, to type 2 diabetes. Insulin resistance is also a central feature of several other conditions, including the **metabolic syndrome**, a cluster of metabolic abnormalities that are associated with increased risk of developing cardiovascular diseases (CVD) and type 2 diabetes. This Nutrition in Practice describes how the metabolic syndrome is diagnosed, how and why it might develop, its potential consequences, and current treatment approaches. The Glossary defines the relevant terms.

How is the metabolic syndrome diagnosed, and how common is it in the United States?

Table NP20-1 lists the laboratory values used to identify the metabolic syndrome, which is diagnosed when at least three of the following disorders are present: hyperglycemia, abdominal obesity, **hypertriglyceridemia** (elevated blood triglyceride levels), reduced high-density lipoprotein (HDL) cholesterol levels, and hypertension (high blood pressure). An estimated 34 percent of adults in the United States may meet the criteria for the metabolic syndrome.[1] As Figure NP20-1 shows, prevalence increases with age. Risk also varies among ethnic groups: Mexican Americans have the highest incidence of the metabolic syndrome in the United States, with an overall prevalence of about 37 percent.[2]

Because the disorders that identify the metabolic syndrome are considered independent risk factors for heart disease or diabetes, some medical experts have questioned whether the diagnosis of metabolic syndrome is a useful one.[3] The main benefit of grouping the disorders may be to guide clinical management of these interrelated metabolic problems.[4] However, some studies indicate that heart disease risk actually varies substantially among individuals with the metabolic syndrome, suggesting that further screening is needed to identify those who may benefit from aggressive treatment.[5]

What causes the metabolic syndrome?

Both genetic and environmental factors probably contribute to the development of the metabolic syndrome.[6] However, the close relationship between abdominal obesity and insulin resistance suggests that the current obesity crisis may be largely responsible for its high prevalence. Visceral fat is thought to induce a number of metabolic

TABLE NP20-1 Features of the Metabolic Syndrome

Metabolic syndrome is diagnosed when three or more of the following abnormalities are present.

Measure	Diagnostic Cut Point
Hyperglycemia	Fasting plasma glucose ≥ 100 mg/dL
Abdominal obesity	Waist circumference >40 in. in men, >35 in. in women
Hypertriglyceridemia	VLDLs ≥150 mg/dL
Reduced HDL cholesterol	HDLs <40 mg/dL in men, <50 mg/dL in women
Hypertension	Blood pressure ≥130/85 mm Hg

Glossary

adiponectin (AH-dih-poe-NECK-tin): a hormone produced by adipose cells that improves insulin sensitivity.

fibrinogen (fye-BRIN-oh-jen): a liver protein that promotes blood clot formation.

hypertriglyceridemia (HYE-per-try-gliss-er-rye-DEE-me-ah): high blood triglyceride levels. Blood triglycerides are carried in *very-low-density lipoproteins (VLDL).*

metabolic syndrome: a cluster of interrelated disorders, including abdominal obesity, insulin resistance, high blood pressure, and abnormal blood lipids, which together increase risk of diabetes and cardiovascular disease; also known as *insulin resistance syndrome* or *syndrome X.*

plasminogen activator inhibitor-1: a protein that promotes blood clotting by inhibiting blood clot degradation within blood vessels.

resistin (re-ZIST-in): a hormone produced by adipose cells that induces insulin resistance.

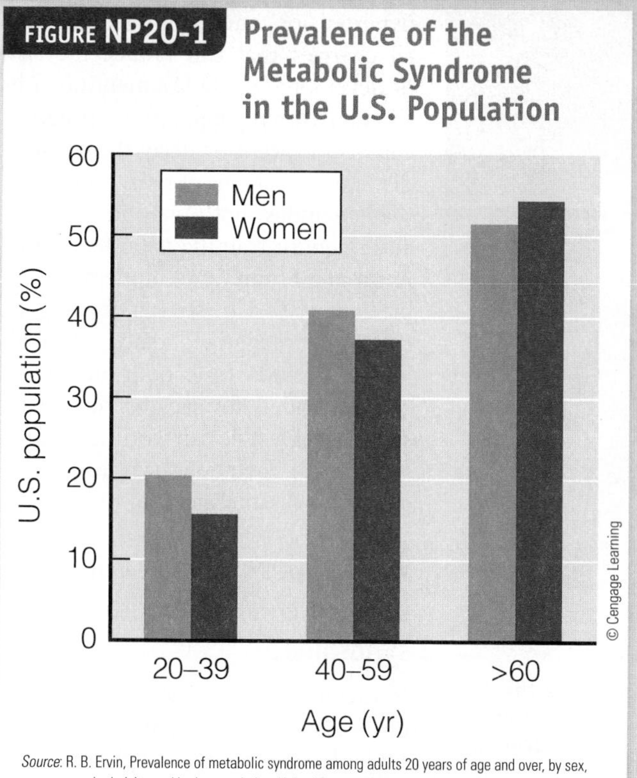

FIGURE NP20-1 Prevalence of the Metabolic Syndrome in the U.S. Population

Source: R. B. Ervin, Prevalence of metabolic syndrome among adults 20 years of age and over, by sex, age, race and ethnicity, and body mass index: United States, 2003-2006, *National Health Statistics Reports*, no. 13 (Hyattsville, MD: National Center for Health Statistics, 2009).

changes that promote insulin resistance, which then leads to hyperglycemia and other abnormalities. Note that overweight individuals (BMI from 25 to 29.9) have about a sixfold increased risk of developing the metabolic syndrome compared with underweight and normal-weight persons (BMI less than 25).[7]

How does obesity lead to insulin resistance?

Various theories have been proposed to explain the relationship between obesity and insulin resistance. Some research indicates that the enlarged adipose cells of obese individuals have a reduced capacity to store triglyceride and may also increase their release of fatty acids into the bloodstream, resulting in the abnormal accumulation of triglycerides in the muscle and liver; the high fat content of these tissues may alter cellular responses to insulin and result in insulin resistance.[8] In addition, obesity causes adipose cells to alter the hormones and proteins they release into the blood, promoting a state of insulin resistance.[9] For example, obesity is associated with the reduced secretion of **adiponectin**, a hormone that promotes insulin sensitivity and glucose tolerance. Conversely, the adipose cells release larger amounts of the hormone **resistin**, which promotes insulin resistance. The enlarged adipose cells also activate local macrophages (immune cells), which secrete a number of cytokines (signaling proteins) that induce inflammation and blood coagulation. The inflammatory process activates multiple metabolic changes that antagonize insulin responsiveness.

Can obesity lead to other problems related to the metabolic syndrome?

Abdominal obesity is frequently associated with blood lipid abnormalities. Because the insulin-resistant adipose cells release more fatty acids into the blood, the liver must accelerate its production of very-low-density lipoproteins (VLDL), and hypertriglyeridemia develops. Excessive body fatness is also associated with elevated low-density lipoprotein (LDL) cholesterol levels and reduced HDL levels.

Obesity also heightens the risk of developing hypertension, although the underlying mechanisms are poorly understood. The elevated blood pressure may be due, in part, to the expanded blood volume that develops in obesity.[10] In addition, the insulin resistance and hyperinsulinemia that accompany obesity may stimulate the sympathetic nervous system, altering hormonal secretions and blood vessel activity in ways that favor a rise in blood pressure.[11] Hyperinsulinemia also promotes sodium reabsorption in the kidneys, resulting in fluid retention and increased blood volume.

How does the metabolic syndrome contribute to cardiovascular disease risk?

The disorders that characterize the metabolic syndrome—obesity, lipid abnormalities, and hypertension—are all independent risk factors for CVD. In addition, the condition is often associated with blood vessel dysfunction and the tendency to form blood clots, characteristics that favor the development of atherosclerosis and raise the risks of heart attack or stroke.[12] For example, individuals with the metabolic syndrome exhibit reduced production of the vasodilator *nitric oxide* and increased secretion of the vasoconstrictor *endothelin-1*—changes that enhance vasoconstriction and stimulate the release of proinflammatory cytokines. These cytokines release factors that increase blood vessel permeability, recruit immune cells, and increase oxidative stress, thereby promoting atherosclerosis. Blood vessel inflammation, obesity, and insulin resistance may all promote increased production of procoagulant proteins such as **fibrinogen** and **plasminogen activator inhibitor-1**.[13] Individuals with the metabolic syndrome are also at increased risk of developing diabetes, which is another major risk factor for CVD.

What is the usual treatment for the metabolic syndrome?

The main treatment goals are to correct the abnormalities that increase CVD and diabetes risk. In most individuals, a combination of weight loss and physical activity can improve insulin resistance, blood pressure, and blood lipid levels.[14] Even a moderate weight loss (about 7 percent of body weight) can improve symptoms, although many people find this difficult to achieve. Additional dietary strategies depend on a patient's specific symptoms. If dietary and lifestyle modifications are not successful, medications may be prescribed. Because effective treatment requires lifelong commitment, health care providers should work with patients to develop a treatment plan that they are willing to adopt.

What dietary strategies, other than weight loss, are suggested for people with the metabolic syndrome?

In individuals with hypertriglyceridemia, the general recommendation is to reduce intake of added sugars and refined grain products (soda, juices, white bread, sweetened cereal, and desserts) and increase servings of whole grains and foods high in fiber (whole-wheat bread, oatmeal, legumes, fruits, and vegetables). In some people, carbohydrate restriction may help to reduce blood triglyceride levels and improve hyperglycemia. Including fish in the diet each week may also improve triglyceride levels. Individuals with hypertension are encouraged to reduce sodium intake and increase consumption of fruits and vegetables and low-fat milk products. A diet low in saturated fat, *trans* fats, and cholesterol can help to reduce LDL cholesterol levels. Chapter 21 describes additional dietary modifications that may reduce CVD risk.

Regular exercise can reduce the risks of developing the metabolic syndrome, cardiovascular diseases, and type 2 diabetes.

Why is physical activity recommended for people with the metabolic syndrome?

Regular physical activity helps with weight management and may also improve blood lipid concentrations, hypertension, and insulin resistance—all changes that can reduce the risk of developing CVD. As mentioned in Chapter 20, a regular exercise program can also prevent or delay the onset of diabetes in persons at risk. A program that includes both aerobic exercise and strength training is best. A minimum of 30 minutes of moderate aerobic activity (brisk walking, jogging, or cycling) is suggested daily, although longer periods (one hour daily) are recommended for weight control. A sedentary lifestyle can worsen the progression of metabolic syndrome and should be discouraged.

What types of medications are used to treat the metabolic syndrome?

If dietary and lifestyle changes are unsuccessful, medications may be prescribed to correct hypertriglyceridemia and hypertension (Chapter 21 provides details). At present, antidiabetic drugs are not routinely used to treat insulin resistance in patients with the metabolic syndrome due to insufficient evidence that the drugs can improve long-term outcomes better than lifestyle changes.

As explained in this Nutrition in Practice, the metabolic syndrome consists of a cluster of interrelated disorders that may increase the risk of developing CVD and type 2 diabetes. Whereas most of the features of the metabolic syndrome are individual risk factors for CVD, in combination they may raise risk twofold to threefold. Treatment of the metabolic syndrome emphasizes dietary and lifestyle changes.

Notes

1. V. L. Roger and coauthors, Heart disease and stroke statistics—2012 update: A report from the American Heart Association, *Circulation* 125 (2012): e2–e220; R. B. Ervin, Prevalence of metabolic syndrome among adults 20 years of age and over, by sex, age, race and ethnicity, and body mass index: United States, 2003–2006, *National Health Statistics Reports*, no. 13 (Hyattsville, MD: National Center for Health Statistics, 2009).
2. Ervin, 2009.
3. A. M. Kanaya and C. Vaisse, Obesity, in D. G. Gardner and D. Shoback, eds., *Greenspan's Basic and Clinical Endocrinology* (New York: McGraw-Hill/Lange, 2011), pp. 699–709.
4. U. Masharani and M. S. German, Pancreatic hormones and diabetes mellitus, in D. G. Gardner and D. Shoback, eds., *Greenspan's Basic and Clinical Endocrinology* (New York: McGraw-Hill/Lange, 2011), pp. 573–655.
5. S. Malik and coauthors, Impact of subclinical atherosclerosis on cardiovascular disease events in individuals with metabolic syndrome and diabetes, *Diabetes Care* 34 (2011): 2285–2290.
6. Z. T. Bloomgarden, The 6th Annual World Congress on the Insulin Resistance Syndrome, *Diabetes Care* 32 (2009): e104–e111.
7. Ervin, 2009.
8. G. H. Goossens, The role of adipose tissue dysfunction in the pathogenesis of obesity-related insulin resistance, *Physiology and Behavior* 94 (2008): 208–218.
9. Kanaya and Vaisse, 2011; A. Maitra, The endocrine system, in V. Kumar and coeditors, *Robbins and Cotran Pathologic Basis of Disease* (Philadelphia: Saunders, 2010), pp. 1097–1164.
10. R. G. Victor, Arterial hypertension, in L. Goldman and A I. Schafer, eds., *Goldman's Cecil Medicine* (Philadelphia: Saunders, 2012): 373–389.
11. L. Duvnjak, T. Bulum, and Z. Metelko, Hypertension and the metabolic syndrome, *Diabetologia Croatica* 37 (2008): 83–89.
12. B. B. Dokken, The pathophysiology of cardiovascular disease and diabetes: Beyond blood pressure and lipids, *Diabetes Spectrum* 21 (2008): 160–165.
13. E. Kassi and coauthors, Metabolic syndrome: Definitions and controversies, *BMC Medicine* 9 (May 5, 2011), available at DOI:10.1186/1741-7015-9-48; Dokken, 2008.
14. F. Magkos and coauthors, Management of the metabolic syndrome and type 2 diabetes through lifestyle modification, *Annual Review of Nutrition* 29 (2009): 223–256.

Chapter 21 Nutrition and Disorders of the Heart and Blood Vessels

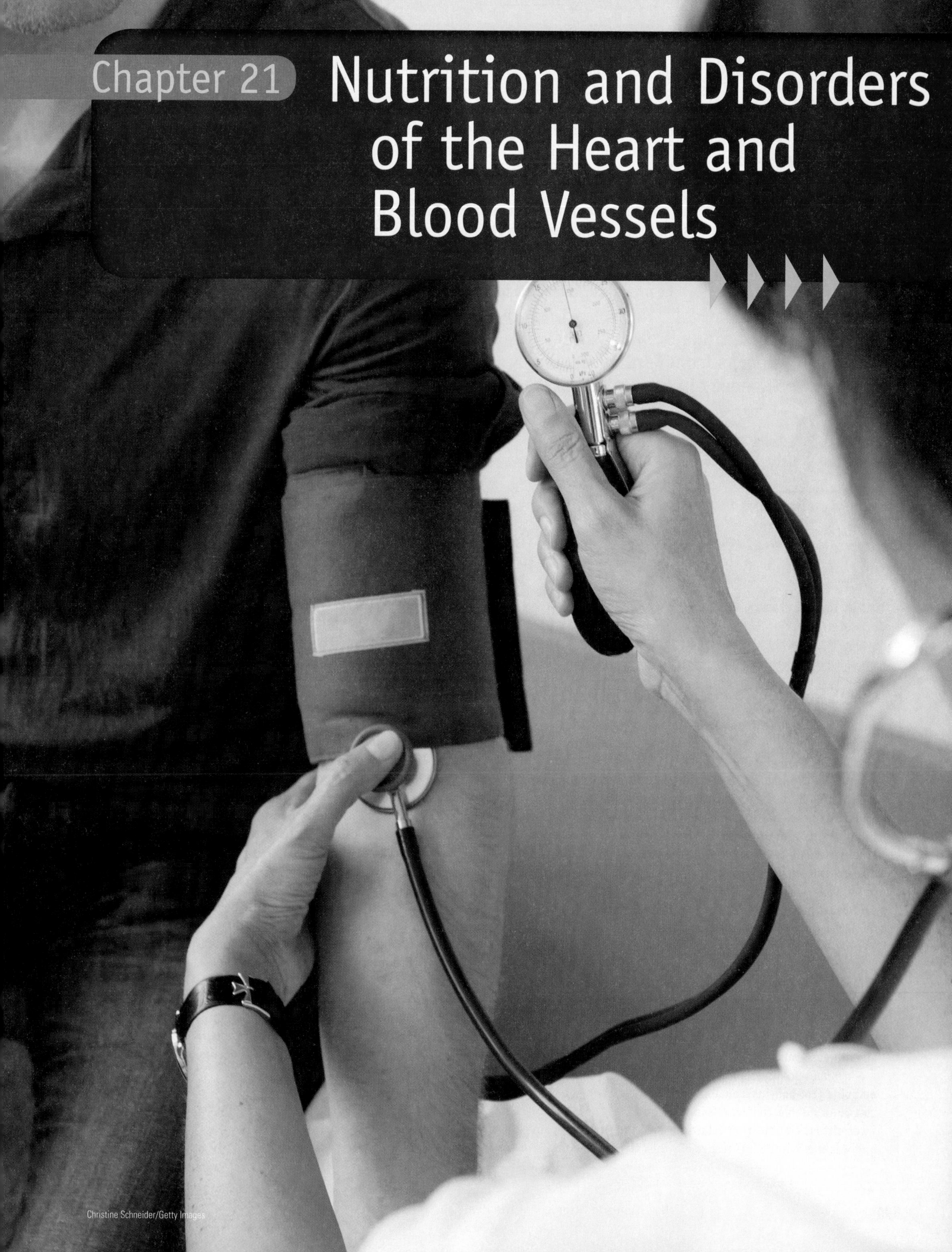

Christine Schneider/Getty Images

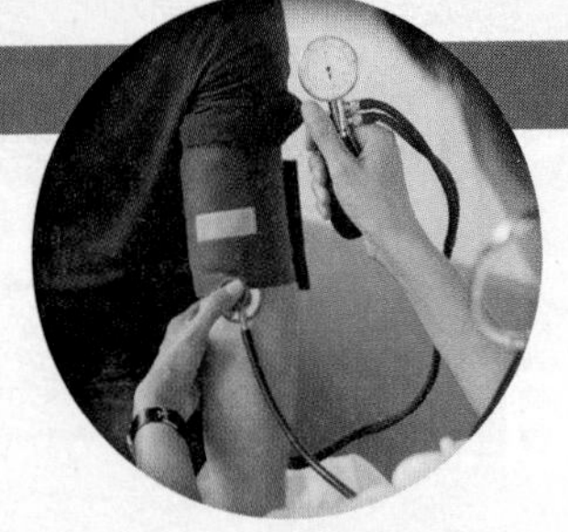

CARDIOVASCULAR DISEASE (CVD), A GROUP OF DISORDERS INVOLVING the heart and blood vessels, is responsible for approximately 33 percent of deaths in the United States.[1] The various types of CVD claim about as many lives as the next three leading causes of death combined. Although many people assume that heart conditions are men's diseases, more women than men die from CVD each year. Furthermore, CVD is a global health issue; it is the leading cause of death worldwide.[2] Figure 21-1 shows the percentages of deaths in the United States resulting from all types of CVD.

■ Atherosclerosis is the most common form of *arteriosclerosis,* a more general term for arterial diseases that are characterized by abnormally thickened walls and lost elasticity.

The most common form of CVD is **coronary heart disease (CHD)**, which is usually caused by **atherosclerosis** in the coronary arteries that supply blood to the heart muscle. If atherosclerosis restricts blood flow in these arteries, the resulting deprivation of oxygen and nutrients can destroy heart tissue and cause a **myocardial infarction**—a **heart attack**. When the blood supply to brain tissue is blocked, a **stroke** occurs. Both heart attack and stroke may result in disablement or death. This chapter describes these and other cardiovascular disorders. The accompanying glossary defines common terms related to CVD.

FIGURE 21-1 Percentage Breakdown of Deaths from Cardiovascular Diseases in the United States

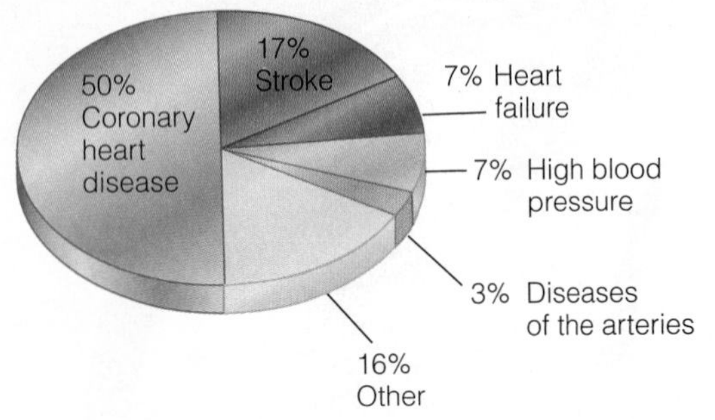

Source: V. L. Roger and coauthors, Heart disease and stroke statistics—2012 update: A report from the American Heart Association, *Circulation* 125 (2012): e2–e220.

Atherosclerosis

In atherosclerosis, (■) the artery walls become progressively thickened due to an accumulation of fatty deposits, fibrous connective tissue, and smooth muscle cells, collectively known as **plaque**. Atherosclerosis initially arises in response to minimal but chronic injuries that damage the inner arterial wall. The resulting inflammatory response (■) attracts immune cells and increases the permeability of the vessel wall. Low-density lipoproteins (LDLs) slip under the thin layer of endothelial

Glossary of Terms Related to Cardiovascular Diseases

aneurysm (AN-you-rih-zum): an abnormal enlargement or bulging of a blood vessel (usually an artery) caused by weakness in the blood vessel wall.

angina (an-JYE-nah or AN-ji-nah) pectoris: a condition caused by ischemia in the heart muscle that results in discomfort or dull pain in the chest region. The pain often radiates to the left shoulder and arm or to the back, neck, and lower jaw.

atherosclerosis (ATH-er-oh-scler-OH-sis): a type of artery disease characterized by accumulations of fatty material on the inner walls of arteries.

cardiovascular disease (CVD): a group of disorders involving the heart and blood vessels.

- *cardio* = heart
- *vascular* = blood vessels

coronary heart disease (CHD): a chronic, progressive disease characterized by obstructed blood flow in the coronary arteries; also called *coronary artery disease.*

embolism (EM-boh-lizm): the obstruction of a blood vessel by an embolus, causing sudden tissue death.

- *embol* = to insert, plug

embolus (EM-boh-lus): an abnormal particle, such as a blood clot or air bubble, that travels in the blood.

intermittent claudication (claw-dih-KAY-shun): severe pain and weakness in the legs (especially the calves) caused by inadequate blood supply to the muscles; it usually occurs with walking and subsides during rest.

ischemia (iss-KEE-mee-a): inadequate blood supply within a tissue due to obstructed blood flow.

myocardial (MY-oh-CAR-dee-al) infarction (in-FARK-shun): death of heart muscle caused by a sudden reduction in coronary blood flow; also called a **heart attack** or *cardiac arrest.*

- *myo* = muscle
- *cardial* = heart
- *infarct* = tissue death

plaque (PLACK): an accumulation of fatty deposits, smooth muscle cells, and fibrous connective tissue in blood vessels.

stroke: a sudden injury to brain tissue resulting from impaired blood flow through an artery that supplies blood to the brain; also called a *cerebrovascular accident.*

- *cerebro* = brain

thrombosis (throm-BOH-sis): the formation or presence of a blood clot in blood vessels. A *coronary thrombosis* occurs in a coronary artery, and a *cerebral thrombosis* occurs in an artery that supplies blood to the brain.

- *thrombo* = clot

thrombus: a blood clot formed within a blood vessel that remains attached to its place of origin.

(blood vessel) cells; eventually, additional lipids, connective tissue, and calcium accumulate and the plaque thickens. Atherosclerosis often develops in regions where the arteries branch or bend because the blood flow is disturbed in those areas (see Figure 21-2); it begins to develop as early as childhood or adolescence and usually progresses over several decades before symptoms develop.

FIGURE 21-2 Plaque Formation in Atherosclerosis

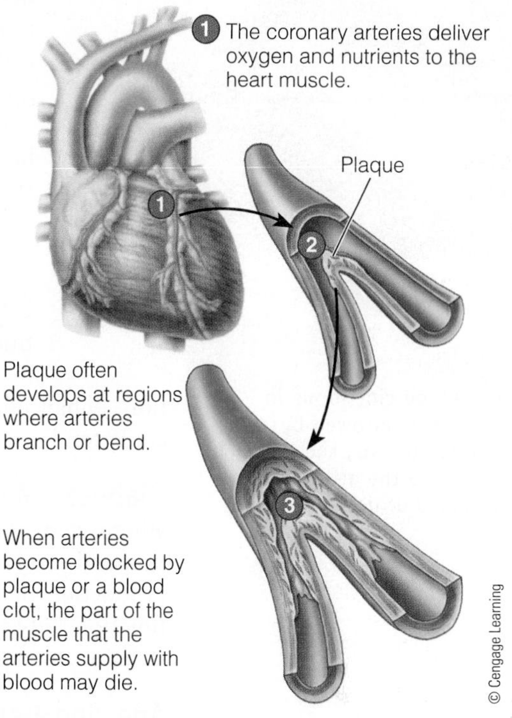

1 The coronary arteries deliver oxygen and nutrients to the heart muscle.
Plaque
2 Plaque often develops at regions where arteries branch or bend.
3 When arteries become blocked by plaque or a blood clot, the part of the muscle that the arteries supply with blood may die.
© Cengage Learning

CONSEQUENCES OF ATHEROSCLEROSIS

As atherosclerosis worsens, it may eventually narrow the lumen of an artery and interfere with blood flow. Some types of plaque are highly susceptible to rupture, which promotes blood clotting within the artery (**thrombosis**). A blood clot (**thrombus**) may enlarge in time and ultimately obstruct blood flow. A portion of a clot can also break free (**embolus**) and travel through the circulatory system until it lodges in a narrowed artery and shuts off blood flow to the surrounding tissue (**embolism**). Most complications of atherosclerosis result from the deficiency of blood and oxygen within the tissue served by an obstructed artery (**ischemia**).

Atherosclerosis can affect almost any organ or tissue in the body and, accordingly, is a major cause of disability or death. Obstructed blood flow in the coronary arteries can cause pain or discomfort in the chest and surrounding regions (**angina pectoris**) or lead to a heart attack. As mentioned earlier, obstructed blood flow to the brain can cause injury to or destruction of brain tissue, or a stroke. Impaired blood circulation in the legs can cause fatigue and pain while walking, known as **intermittent claudication**. Blockage of the arteries that supply the kidneys can result in kidney disease or even acute kidney injury.

Atherosclerosis is the most common cause of an **aneurysm**—the abnormal dilation of a blood vessel. Plaque can weaken the blood vessel wall, and eventually the pressure of blood flow can cause the damaged region to stretch and balloon outward. Aneurysms can rupture and lead to massive bleeding and death, particularly when a large vessel such as the aorta is affected. In the arteries of the brain, an aneurysm may lead to bleeding within the brain, a coma, or a stroke.

CAUSES OF ATHEROSCLEROSIS

The factors that initiate atherosclerosis either cause direct damage to the artery wall or allow lipid materials to penetrate its surface. Factors that generally worsen atherosclerosis are those that lead to plaque rupture or blood coagulation. The development of advanced atherosclerosis is a long-term process that involves recurrent plaque rupture, thrombosis, and healing at sites in the artery wall.[3]

■ C-reactive protein, an acute-phase protein secreted during the inflammatory response, is associated with increased heart disease risk (see Chapter 16).

Shear Stress/Hypertension

The stress of blood flow along artery walls—called *shear stress*—can cause physical damage to arteries.[4] (■) Hypertension (high blood pressure) intensifies the stress of blood flow on arterial tissue, provoking a low-grade inflammatory state that may stimulate plaque formation or progression.[5]

■ *Shear stress* is a stress that occurs sideways against a surface rather than perpendicular to a surface.

Abnormal Blood Lipids

When levels are high, LDLs (■) are actively taken up and retained in susceptible regions in the artery wall. Elevated VLDL (very-low-density lipoprotein) levels promote **atherogenic** responses in immune cells and endothelial cells.[6] Because high-density lipoproteins (HDLs) remove cholesterol from circulation and contain proteins that inhibit inflammation, LDL oxidation, and plaque accumulation, low HDL levels can contribute to the development of atherosclerosis as well.[7]

■ Reminder: LDLs transport cholesterol in the blood, whereas VLDLs transport triglycerides. In clinical practice, VLDLs are commonly referred to as *blood triglycerides*.

atherogenic: able to initiate or promote atherosclerosis.

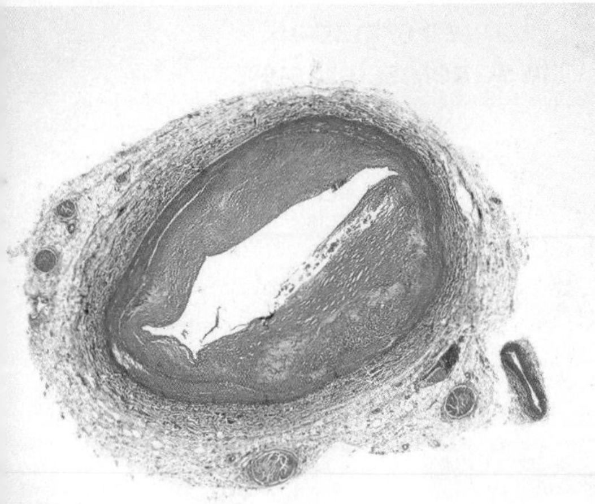

Dr. Gladden/Visuals Unlimited/Corbis

If a blood clot forms in an artery narrowed by atherosclerosis, the tissue served by the artery may be damaged or destroyed.

LDLs vary in size and density, and these LDL subtypes have differing effects on heart disease risk. The smallest, most dense LDLs can slip into artery walls easily and are therefore more atherogenic than the larger, less dense LDLs.[8] In addition, people who have small, dense LDLs frequently have elevated VLDL and low HDL levels. This lipoprotein profile is especially prevalent in individuals with the metabolic syndrome and type 2 diabetes. (■)

Elevated concentrations of a variant form of LDL called *lipoprotein(a)* have been found to speed the progression of atherosclerosis and to raise the risk of various types of CVD. Lipoprotein(a) levels are primarily genetically determined and are influenced to only a minor degree by age and environmental factors.

Cigarette Smoking Chemicals in cigarette smoke (including nicotine) are toxic to endothelial cells, and the resulting damage contributes to arterial injury. Other effects of smoking include chronic inflammation, enhanced blood coagulation, increased LDL cholesterol, and decreased HDL cholesterol—all effects that can promote the progression of atherosclerosis.[9]

Diabetes Mellitus Diabetes can both initiate and accelerate the development of atherosclerosis. Chronic hyperglycemia leads to the accumulation of advanced glycation end products (■), which promote inflammation and oxidative stress, induce the production of compounds that favor plaque progression, and disturb blood vessel function.[10] By various other mechanisms, diabetes increases tendencies for vasoconstriction, blood clotting, and plaque rupture.[11]

Age and Gender Advancing age promotes atherosclerosis due to the cumulative exposure to risk factors and the degeneration of arterial cells with age. Risk of atherosclerosis increases substantially in men aged 45 or older and women aged 55 or older. After menopause, women's risk increases, in part, because the reduction in estrogen levels has unfavorable effects on lipoprotein levels and arterial function.[12] (■) Levels of the amino acid homocysteine, which may upset endothelial cell function, rise with age and are generally higher in men; however, researchers have not determined whether elevated homocysteine levels directly contribute to the disease process or are merely an indicator of abnormal metabolism.[13] (■)

■ Disorders characterized by abnormal levels of blood lipids are called *hyperlipidemias* or *dyslipidemias*.

■ Reminder: *Advanced glycation end products* are reactive compounds formed after glucose nonenzymatically attaches to proteins (see Chapter 20).

■ Estrogen replacement therapy after menopause has mixed effects on heart disease risk; it can improve endothelial function and reduce LDL levels, but it also promotes blood clotting.

■ Blood homocysteine levels are influenced by intakes of folate, vitamin B_{12}, and vitamin B_6.

Review Notes

- Atherosclerosis, characterized by the buildup of arterial plaque, can lead to complications such as angina pectoris, heart attack, stroke, intermittent claudication, kidney disease, and aneurysms.
- Plaque develops in response to long-term, chronic inflammation at susceptible sites in artery walls. Leading causes of plaque formation and progression include shear stress, hypertension, elevated LDL and VLDL levels, cigarette smoking, diabetes, and aging.

Coronary Heart Disease

Coronary heart disease (CHD), also called *coronary artery disease*, is the most common type of cardiovascular disease.[14] As discussed earlier, CHD is characterized by impaired blood flow through the coronary arteries, which may lead to angina pectoris, heart attack, or even sudden death. CHD is most often caused by atherosclerosis but occasionally results from a **spasm** or inflammatory condition that causes narrowing of the coronary arteries.

spasm: a sudden, forceful, and involuntary muscle contraction.

SYMPTOMS OF CORONARY HEART DISEASE

Symptoms of CHD usually arise only after many years of disease progression. In angina pectoris and heart attacks, pain or discomfort most often occurs in the chest region and may be perceived as a feeling of heaviness, constriction, or squeezing; the pain may radiate to the left neck and shoulder, arms, back, or jaw.[15] In angina pectoris, the symptoms are often triggered by exertion and subside with rest; in a heart attack, the pain may be severe, last longer, and occur without exertion. Other symptoms of CHD include shortness of breath, unusual weakness or fatigue, lightheadedness or dizziness, nausea, vomiting, and lower abdominal discomfort. Women are more likely than men to have a heart condition (or even a heart attack) that is unaccompanied by chest pain or acute symptoms.

> ***NURSING DIAGNOSIS***
>
> Nursing diagnoses for individuals with CHD may include *risk for decreased cardiac tissue perfusion, acute pain,* and *activity intolerance.*

EVALUATING RISK FOR CORONARY HEART DISEASE

Because CHD develops over many years, prevention should begin well before symptoms appear. Population studies have suggested that about 90 percent of people with CHD have at least one of the four classic risk factors: smoking, high LDL cholesterol, hypertension, and diabetes.[16] These and other major risk factors for CHD are listed in Table 21-1; most of the risk factors listed can be modified by changes in diet and lifestyle.

CHD Risk Assessment Risk assessment requires several key laboratory measures (see Table 21-2) and a thorough medical history. A lipoprotein profile (also called a *blood lipid profile*), which includes measures of total cholesterol, LDL and HDL cholesterol, and blood triglycerides, should be obtained every five years starting at 20 years of age. Some clinicians may use the ratio of total cholesterol to HDL cholesterol or LDL cholesterol to HDL cholesterol to help assess CHD risk. In persons with high blood triglycerides, the non-HDL cholesterol level (total cholesterol minus HDLs) may be more accurate than the LDL level for predicting CHD risk.[17] Overweight and obesity predispose to CHD, particularly when abdominal obesity is present. Hypertension is a major risk factor, and the risk for CHD increases as blood pressure increases. Finally, cigarette smoking and the presence of diabetes strongly contribute to CHD risk. The "How to" on p. 599 presents a screening method for assessing a person's 10-year risk of developing CHD that includes some of these risk factors.

In some individuals, a CHD risk assessment may include tests that provide additional detail about blood lipids or that indicate the presence of atherosclerosis.[18] Blood lipid status is sometimes evaluated by measuring LDL and HDL subclasses, the LDL particle number, lipoprotein(a) levels, or levels of proteins or enzymes associated with lipoproteins (especially *apolipoprotein B*, a component of LDLs, lipoprotein(a), and VLDLs). Atherosclerosis may be evaluated using the *coronary calcium score*, a computed tomography (CT) scan that analyzes the calcium content of plaque in the coronary arteries.

Blood Cholesterol Levels and CHD Risk Once a person's level of risk has been identified, much of the treatment focuses on lowering LDL cholesterol. Elevated LDL levels are directly related to the development of atherosclerosis, and clinical studies have confirmed that LDL-lowering treatments can successfully reduce CHD mortality rates.[19] CHD is seldom seen in populations that maintain desirable LDL levels.

TABLE 21-1 Risk Factors for CHD

Major Nonmodifiable Risk Factors
- Advancing age (men ≥ 45 years; women ≥ 55 years)
- Male gender
- Family history of heart disease

Major Modifiable Risk Factors
- High LDL cholesterol
- High blood triglyceride (VLDL) levels
- Low HDL cholesterol
- Hypertension (high blood pressure)
- Diabetes mellitus
- Obesity (especially abdominal obesity)
- Physical inactivity
- Cigarette smoking
- Alcohol overconsumption (≥3 drinks per day)
- An atherogenic diet (includes high saturated fat, cholesterol, and *trans* fat intakes; low fruit and vegetable intakes)

Note: Risk factors highlighted in yellow have relationships with diet.

Sources: M. J. Klag, Epidemiology of cardiovascular disease, in L. Goldman and A. I. Schafer, eds., *Goldman's Cecil Medicine* (Philadelphia: Saunders, 2012), pp. 256–260; V. L. Roger and coauthors, Heart disease and stroke statistics—2012 update: A report from the American Heart Association, *Circulation* 125 (2012): e2–e220.

TABLE 21-2 Laboratory Measures for CHD Risk Assessment

Clinical Measures	Desirable	Borderline Risk	High Risk
Total blood cholesterol (mg/dL)	<200	200–239	≥240
LDL cholesterol (mg/dL)	<100[a]	130–159	160–189[b]
HDL cholesterol (mg/dL)[c]	≥60	Men: 40–59 Women: 50–59	Men: <40 Women: <50
Triglycerides, fasting (mg/dL)	<150	150–199	200–499[d]
Body mass index (BMI)[e]	18.5–24.9	25–29.9	≥30
Blood pressure (systolic and diastolic pressure)	<120/<80	120–139/80–89[f]	≥140/≥90[g]

[a]LDL levels of 100–129 mg/dL indicate a near-optimal level; <70 mg/dL is a desirable goal for very high-risk persons.
[b]LDL levels ≥190 mg/dL indicate a very high risk.
[c]To estimate non-HDL cholesterol, subtract the HDL level from the total cholesterol level; non-HDL cholesterol risk levels are 30 mg/dL higher than the LDL risk levels.
[d]Triglyceride levels ≥500 mg/dL indicate a very high risk.
[e]Body mass index (BMI) was defined in Chapter 6; BMI standards are found on the inside back cover.
[f]These values indicate prehypertension for most individuals.
[g]These values indicate stage one hypertension; ≥160/≥100 indicates stage two hypertension. Physicians use these classifications to determine medical treatment.

As mentioned earlier, HDLs help to protect against atherosclerosis, and low HDL levels often coexist with other lipid abnormalities; thus, a low HDL value is highly predictive of CHD risk. In addition, low HDL levels are usually associated with other CHD risk factors, such as obesity, smoking, inactivity, and insulin resistance. Although having adequate HDLs is beneficial, however, high HDL levels do not necessarily confer additional benefit.[20] Moreover, chronic inflammation—which is often associated with atherosclerosis, diabetes, and various other conditions—may lead to HDL dysfunction and increased CHD risk despite the presence of high HDL concentrations.[21]

THERAPEUTIC LIFESTYLE CHANGES FOR LOWERING CHD RISK

People who have CHD or multiple risk factors for CHD are often advised to make dietary and lifestyle changes before considering drug treatment. This approach to risk reduction, known as Therapeutic Lifestyle Changes (TLC), is summarized in Table 21-3. (■) The main features of the TLC plan include a blood cholesterol-lowering diet, regular physical activity, and weight reduction. If the recommendations are followed carefully, substantial progress may be seen after six weeks. People with a high risk of CHD should try to lower LDL cholesterol with at least a three-month trial of TLC before starting drug therapy. This section describes the elements of the TLC plan in detail.

NURSING DIAGNOSIS

The nursing diagnosis *readiness for enhanced self-health management* applies to people who wish to alter their health behaviors to reduce risk of CHD.

Saturated Fat

Of the dietary lipids, saturated fat has the strongest effect on blood cholesterol levels, and replacing saturated fat with monounsaturated and polyunsaturated fats can generally lower LDL levels. The TLC recommendation is to consume less than 7 percent of total kcalories as saturated fat. The average saturated fat intake in the United States is about 11 percent of total kcalories consumed.[22]

For most people, cutting down on saturated fat involves more than just switching from butter to vegetable oil, as the main sources of saturated fat in the United States are full-fat cheeses, pizza, grain-based and dairy-based desserts, chicken dishes, sausages, frankfurters, bacon, and ribs.[23] Thus, choosing fat-free or low-fat milk products,

■ For the most current guidelines related to CHD risk reduction, access this website: www.nhlbi.nih.gov/guidelines.

HOW TO Assess a Person's Risk of Heart Disease

This assessment estimates a person's 10-year risk for experiencing a major coronary event associated with CHD, such as a heart attack.[a] A high score does not mean that the person *will* have a heart attack, but it warns of the possibility and suggests the need to consult a physician. To use this tool, you need to know a person's age, total and HDL cholesterol levels, and blood pressure.

Age (years):

	Men	Women
20–34	−9	−7
35–39	−4	−3
40–44	0	0
45–49	3	3
50–54	6	6
55–59	8	8
60–64	10	10
65–69	11	12
70–74	12	14
75–79	13	16

HDL (mg/dL):

	Men	Women
≥60	−1	−1
50–59	0	0
40–49	1	1
<40	2	2

Systolic Blood Pressure (mm Hg):

	Untreated		Treated	
	Men	Women	Men	Women
<120	0	0	0	0
120–129	0	1	1	3
130–139	1	2	2	4
140–159	1	3	2	5
≥160	2	4	3	6

Total Cholesterol (mg/dL):

	Age 20–39		Age 40–49		Age 50–59		Age 60–69		Age 70–79	
	Men	Women	Men	Women	Men	Women	Men	Women	Men	Women
<160	0	0	0	0	0	0	0	0	0	0
160–199	4	4	3	3	2	2	1	1	0	1
200–239	7	8	5	6	3	4	1	2	0	1
240–279	9	11	6	8	4	5	2	3	1	2
≥280	11	13	8	10	5	7	3	4	1	2
Smoking (any cigarette smoking in the past month):										
Smoker	8	9	5	7	3	4	1	2	1	1
Nonsmoker	0	0	0	0	0	0	0	0	0	0

Scoring Heart Disease Risk

Add up the total points: ___. Using the table at the right, find the total in the first column for the appropriate gender, and then check the second column to learn the percentage risk of developing severe CHD within the next 10 years. A person's risk for an acute coronary event (such as a heart attack) may be identified as low (<10%), moderate (10–20%), or high (>20 %). Treatment strategies vary according to a person's risk category.

Men		Women	
Total	Risk (%)	Total	Risk (%)
<0	<1	<9	<1
0–4	1	9–12	1
5–6	2	13–14	2
7	3	15	3
8	4	16	4
9	5	17	5
10	6	18	6
11	8	19	8
12	10	20	11
13	12	21	14
14	16	22	17
15	20	23	22
16	25	24	27
≥17	≥30	≥25	≥30

[a]A link to the electronic version of this assessment is available on the ATP III page of the National Heart, Lung, and Blood Institute's website (www.nhlbi.nih.gov/guidelines/cholesterol).

Source: Adapted from Expert Panel on Detection, Evaluation, and Treatment of High Blood Cholesterol in Adults (Adult Treatment Panel III), *Third Report of the National Cholesterol Education Program (NCEP)*, NIH publication no. 02-5216 (Bethesda, MD: National Heart, Lung, and Blood Institute, 2002), section III.

TABLE 21-3 Reducing Risk of CHD with Therapeutic Lifestyle Changes

Dietary Strategies

- Limit saturated fat to less than 7 percent of total kcalories and cholesterol to less than 200 milligrams per day. Maintaining a fat intake that is 25 to 35 percent of total kcalories may help with this goal.
- Replace saturated fats with unsaturated fats from fish, vegetable oils, and nuts or with carbohydrates from whole grains, legumes, fruits, and vegetables.
- Avoid food products that contain *trans* fats. The *trans* fat content in packaged foods is shown on the Nutrition Facts panel.
- Choose foods high in soluble fibers, including oats, barley, legumes, and fruit. Food supplements that contain psyllium seed husks can be used to help lower LDL cholesterol levels.
- Regularly consume food products that contain added plant sterols or stanols.
- To reduce blood pressure, limit sodium intake to <1500 milligrams per day and choose a diet that is high in fruits and vegetables, low-fat milk products, nuts, and whole grains.
- Fish can be consumed regularly as part of a CHD risk-reduction diet.
- If alcohol is consumed, it should be limited to one drink daily for women and two drinks daily for men.

Lifestyle Choices

- Physical activity: Engage in a minimum of 150 minutes per week of moderate-intensity activity or 75 minutes per week of vigorous activity, or an equivalent combination.
- Smoking cessation: Exposure to any form of tobacco smoke should be minimized.

Weight Reduction

- In overweight or obese individuals, weight reduction may improve other CHD risk factors. The general goals of a weight-management program should be to prevent weight gain, reduce body weight, and maintain a lower body weight over the long term.
- The initial goal of a weight-loss program should be to lose no more than 5 to 10 percent of the original body weight.

selecting lean meats or fish, and avoiding certain types of desserts are usually more effective ways of reducing saturated fat.

Replacing saturated fat with carbohydrate can also lower LDL cholesterol, but such a change may raise blood triglyceride (VLDL) levels. The effect on blood triglycerides can be minimized by limiting added sugars and including fiber-rich foods; ideally, the diet should include generous amounts of whole grains, legumes, fruits, and vegetables. The TLC diet recommends a carbohydrate intake in the range of 50 to 60 percent of total kcalories.

Polyunsaturated and Monounsaturated Fat As described in the previous section, replacing saturated fat with either monounsaturated or polyunsaturated fat helps to lower LDL levels. A switch to polyunsaturated fat tends to have the greater effect, but it also promotes a slight reduction in HDL cholesterol.[24] Other concerns are that high intakes of polyunsaturated fat may increase inflammation within the body (■) or contribute to oxidative stress. (■) Therefore, TLC guidelines limit polyunsaturated fat to 10 percent of total kcalories, whereas up to 20 percent of kcalories from monounsaturated fat are allowed. Note that most polyunsaturated fat in the diet consists of omega-6 fatty acids such as linoleic acid; omega-3 fatty acids may have beneficial effects on heart disease risk, as described in a later section.

Total Fat For people with substantial saturated fat intakes, limiting total fat may indirectly reduce saturated fat. Therefore, the TLC recommendation for total fat is an intake of 25 to 35 percent of kcalories. Individuals with elevated blood triglycerides may benefit from achieving a fat intake at the upper end of this range (30 to 35 percent) so that their carbohydrate intakes are not excessive. Fat intakes higher than 35 percent of kcalories are discouraged because they may promote weight gain in some people.

■ Reminder: Polyunsaturated fatty acids are precursors for the eicosanoids, which mediate inflammation (see pp. 470–471).

■ Polyunsaturated fats are more susceptible to oxidation than saturated fats, although the clinical significance of consuming high amounts is still unknown.

***Trans* Fat** *Trans* fats can raise LDL cholesterol levels, and when they replace saturated fats in the diet (as when stick margarine replaces butter), they may also reduce HDL cholesterol levels. Furthermore, *trans* fats may raise CHD risk by promoting inflammation and endothelial dysfunction.[25] The TLC recommendation is to keep *trans* fat intake as low as possible.

Most sources of *trans* fats are products made with partially hydrogenated vegetable oils; examples include baked goods such as crackers, cookies, and doughnuts; snack foods such as potato chips and corn chips; and fried foods such as French fries and fried chicken. In the past few years, food manufacturers have reformulated many food products so that they contain little or no *trans* fat. In some cases, unfortunately, the *trans* fats have been replaced with saturated fat sources, so consumers should read labels carefully to avoid both types of cholesterol-raising fats.

Dietary Cholesterol Dietary cholesterol has a minimal effect on LDL cholesterol levels, and the effects of altering intake vary among individuals. Moreover, cholesterol's influence on CHD risk is somewhat unclear: although some research studies have found a relationship between dietary cholesterol and CHD risk, others have not.[26] The TLC recommendation is a cholesterol intake of less than 200 milligrams per day, a level that may reduce LDL levels in some individuals by about 3 to 5 percent.[27] Cholesterol intakes of women and men in the United States average about 261 and 333 milligrams per day, respectively.[28] Eggs contribute about one-quarter of the cholesterol in the U.S. diet, followed by chicken, beef, and cheese.[29]

The effect of eggs on CHD risk is controversial. While egg intakes have not been linked to CHD risk in healthy populations,[30] a number of observational studies have found an association in persons with diabetes.[31] Furthermore, a study that evaluated the effects of egg yolk intake on carotid artery plaque found that the individuals (both with and without diabetes) who had the highest egg intakes had the highest amounts of plaque.[32] Thus, the number of eggs to include in a heart-healthy diet is undecided, and different guidelines may be necessary for healthy and high-risk populations.

Soluble Fibers Soluble, viscous fibers can reduce LDL cholesterol levels by inhibiting cholesterol and bile absorption in the small intestine and reducing cholesterol synthesis in the liver. Good sources of soluble fibers include oats, barley, legumes, and fruits. The soluble fiber in psyllium seed husks, frequently used to treat constipation, is effective for lowering cholesterol levels when used as a dietary supplement.

Plant Sterols and Stanols Foods or supplements that contain significant amounts of plant sterols or plant stanols can help to lower LDL cholesterol levels. (■) Plant sterols or stanols are added to various food products, such as margarine and orange juice, or supplied in dietary supplements. These plant compounds work by interfering with cholesterol and bile absorption. About 2 grams of plant sterols daily (provided by 2 to 2½ tablespoons of sterol-enriched margarines) can lower LDL cholesterol by up to 10 percent.[33]

Sodium and Potassium Intakes Excessive dietary sodium may raise blood pressure, whereas potassium can help to lower blood pressure. A low-sodium diet that contains generous amounts of fruits and vegetables, low-fat milk products, nuts, and whole grains has been found to substantially reduce blood pressure, largely due to the diet's content of potassium and several other minerals that have blood pressure–lowering effects. This diet (the *DASH Eating Plan*) and other factors that influence blood pressure are discussed in a later section (see pp. 611–612).

Fish and Omega-3 Fatty Acids The omega-3 fatty acids in fish, known as EPA and DHA (see p. 96), may benefit individuals at risk of CHD by suppressing inflammation, lowering blood triglyceride levels, reducing blood clotting, and stabilizing heart rhythm. In addition, including fish in the diet can reduce CHD risk because fish is low in saturated fat and often replaces meat dishes that contain saturated fat. The American Heart Association recommends consuming two or more servings of fish per week, with an emphasis on fatty fish.[34] Of note, the use of fish oil supplements has not been shown to reduce heart attacks or heart disease-related deaths in most clinical trials.[35]

Alcohol Light to moderate consumption of alcohol—from beer, wine, or liquor—has favorable effects on atherosclerosis, HDL cholesterol levels, blood-clotting activity, insulin resistance, and overall CHD risk.[36] Consumption should be limited to one

■ Plant sterols are extracted from soybeans and pine tree oils and are hydrogenated to produce the plant stanols that are added to commercial products.

© Ronnie Kaufman/Corbis

Regular aerobic exercise can strengthen the cardiovascular system, promote weight loss, reduce blood pressure, and improve blood glucose and lipid levels.

drink daily for women and two for men, however, because higher intakes may promote plaque formation and increase blood triglyceride levels and blood pressure. Because alcohol consumption increases risk of various cancers and may have other detrimental effects on health (see Nutrition in Practice 19), nondrinkers are not encouraged to start drinking in an effort to decrease their risk for CHD.

Regular Physical Activity Regular aerobic activity reverses a number of risk factors for CHD: it can lower blood triglycerides, raise HDL levels, lower blood pressure, promote weight loss, improve insulin sensitivity, strengthen heart muscle, and increase coronary artery size and tone. Current guidelines suggest a minimum of 150 minutes per week of moderate-intensity activity or 75 minutes per week of vigorous activity, or an equivalent combination.[37] Activities that use large muscle groups have the greatest benefits; such activities include brisk walking, running, swimming, cycling, stair-stepping, and cross-country skiing. If preferred, physical activity can be divided into several sessions during the day.

Smoking Cessation Cigarette smoking is a major risk factor for CHD as well as for other types of cardiovascular disease. (■) In addition to promoting atherosclerosis, cigarette smoking decreases the oxygen-carrying capacity of the blood, raises the heart rate, inhibits vasodilation, reduces exercise tolerance, and promotes blood clotting, among other effects.[38] Smoking just one or two cigarettes daily—even low-tar, low-nicotine cigarettes—increases CHD risk. Quitting smoking improves CHD risk quickly; the incidence of CHD drops to levels near those of nonsmokers within three years.[39] Currently, about 21 percent of men and 18 percent of women in the United States are cigarette smokers.[40]

Weight Reduction Obesity—especially abdominal obesity—is often associated with a number of metabolic abnormalities that increase CHD risk, such as insulin resistance, hypertension, elevated blood triglycerides, low HDL levels, and reduced LDL size. (■) In addition, the adipose tissue of obese individuals produces various types of inflammatory mediators and blood clotting factors, raising the risks of both atherosclerosis and heart attack. Obesity also strains the heart and blood vessels because **cardiac output** is greater, resulting in a greater workload for the left ventricle, which pumps blood to the major arteries.[41]

In persons who are obese, weight reduction can improve such CHD risk factors as hypertension, blood lipid abnormalities, and insulin resistance. However, individuals should focus on weight reduction only after they have adopted other dietary measures to lower LDL cholesterol. This approach ensures that LDL reduction is given priority and that the individual does not receive a multitude of dietary suggestions at one time. The initial goal of a weight-loss program should be a loss of no more than 5 to 10 percent of a person's original body weight.[42] For some, avoiding additional weight gain may be a desirable starting point.

■ Cigar and pipe smoking can also increase the risk of CHD, but the risk may not be as high because the smoke is less likely to be inhaled.

■ The pattern of metabolic complications associated with obesity characterizes the metabolic syndrome, which is described in Nutrition in Practice 20.

Successful Adherence to Lifestyle Changes Adopting multiple lifestyle changes at once is challenging. Health practitioners can help to motivate patients by explaining the reasons for each change, setting obtainable goals, and providing practical suggestions. In some individuals, high LDL cholesterol levels may persist despite adherence to a TLC program, and drug therapy may be the only effective treatment. Review Table 21-3 (p. 600) for a summary of the recommendations discussed in this section. The "How to" offers suggestions for implementing a heart-healthy diet.

LIFESTYLE CHANGES FOR HYPERTRIGLYCERIDEMIA

Hypertriglyceridemia (elevated blood triglycerides) (■) affects nearly one-third of adults in the United States.[43] It is common in people with diabetes mellitus, obesity, and the metabolic syndrome and may also result from other disorders. Elevated blood

■ Blood triglycerides:
- Borderline high: 150–199 mg/dL
- High: 200–499 mg/dL
- Very high: ≥500 mg/dL

cardiac output: the volume of blood pumped by the heart within a specified period of time.

HOW TO Implement a Heart-Healthy Diet

For some people, following a heart-healthy diet may require significant changes in food choices. It is often easier to adopt a new diet if only a few changes are made at a time. Discussing positive choices (what to eat) first, rather than negative ones (what not to eat), may improve compliance. These suggestions can help patients implement their diet:

Breads, Cereals, and Pasta

- Choose whole-grain breads and cereals. Make sure the first ingredient on bread and cereal labels is "whole wheat flour" rather than "enriched wheat flour." Consume oats and barley regularly, as they are good sources of soluble fibers.
- Bakery products and snack foods often contain *trans* fats. Choose products whose labels list 0 grams of *trans* fat on the Nutrition Facts panel; the ingredient lists should not include any "partially hydrogenated vegetable oil," the main source of *trans* fatty acids.
- Avoid products that contain tropical oils (coconut, palm, or palm kernel oil), which are high in saturated fat.

Fruits and Vegetables

- Incorporate at least one or two servings of fruits and vegetables into each meal. Keep the refrigerator stocked with a variety of colorful fruits and vegetables (baby carrots, blueberries, grapes) that can be eaten when the urge to nibble arises.
- Check food labels on canned products carefully. Canned vegetables (especially tomato-based products) are often high in sodium. Fruits that are canned in juice are higher in nutrient density than those canned in syrup.
- Avoid French fries from fast-food restaurants, which are often prepared with *trans* fats. Restrict high-sodium foods such as pickles, olives, sauerkraut, and kimchee.

Lunch and Dinner Entrées

- Limit meat, fish, and poultry intake to 5 ounces per day. Plan to eat fish twice a week, preferably fatty fish such as salmon, tuna, and mackerel.
- Select lean cuts of beef, such as sirloin tip and round steak; lean cuts of pork, such as loin chops and tenderloin; and skinless poultry pieces. Trim visible fat before cooking.
- Select extra-lean ground meat and drain well after cooking. Use lean ground turkey, without skin added, in place of ground beef.
- Prepare pasta and vegetable stir-fry dishes several times weekly to help reduce meat intake and increase vegetable intake. Use soybean products and other legumes as sources of protein.
- Limit cholesterol-rich organ meats (liver, brain, sweetbreads) and shrimp. Limit intake of whole eggs to two per week, as the yolks are high in cholesterol (about 185 milligrams per yolk). Replace whole eggs in recipes with egg whites or commercial egg substitutes or similar reduced-cholesterol products.
- Restrict these high-sodium foods: cured or smoked meats such as beef jerky, bologna, corned beef, frankfurters, ham, luncheon meats, salt pork, and sausage; salty or smoked fish such as anchovies, caviar, salted or dried cod, herring, and smoked salmon; and canned, frozen, or packaged soups, sauces, and entrées.

Milk Products

- Select fat-free or low-fat milk products only. Use yogurt or fat-free sour cream to make dips or salad dressings. Substitute evaporated fat-free milk for heavy cream.
- Restrict foods high in saturated fat or sodium, such as cheese, processed cheeses, and ice cream or other milk-based desserts.

Fats and Oils

- Prepare salad dressings and other foods with vegetable oils rich in omega-3 fatty acids, such as canola, soybean, flaxseed, and walnut oils. Select other unsaturated vegetable oils, such as corn, olive, peanut, safflower, sesame, and sunflower oils, instead of saturated fat sources such as butter and lard.
- Select margarines that indicate 0 grams of *trans* fat on the Nutrition Facts panel; these products should contain little or no "partially hydrogenated vegetable oil." Tub margarines are less likely to contain *trans* fat than stick margarines. To help lower LDL cholesterol levels, use margarines with added plant sterols or stanols.
- Add unsalted nuts or avocados to meals to make them more appetizing; these foods are good sources of unsaturated fats.
- Avoid tropical oils (coconut, palm, or palm kernel oil), which are high in saturated fat.

Spices and Seasonings

- Use salt only at the end of cooking, and you will need to add much less. Use salt substitutes at the table.
- Spices and herbs can improve food flavor without adding sodium. Try using more garlic, ginger, basil, curry or chili powder, cumin, pepper, lemon, mint, oregano, rosemary, and thyme.
- Check the sodium content on food labels. Flavorings and sauces that are usually high in sodium include bouillon cubes, soy sauce, steak and barbecue sauces, relishes, mustard, and catsup.

Snacks and Desserts

- Select low-sodium and low–saturated fat snacks such as unsalted pretzels and nuts, plain popcorn, and unsalted chips and crackers. Check labels on snack foods and desserts to ensure that they do not include *trans* fats.
- Enjoy angel food cake, which is made without egg yolks and added fat. Select low-fat frozen desserts such as sherbet, sorbet, fruit bars, and some low-fat ice creams.
- Choose canned or dried fruits and crunchy raw vegetables to boost fruit and vegetable intake.

triglycerides may coexist with elevated LDL cholesterol or occur separately. Whereas mild or moderate hypertriglyceridemia is often associated with increased risk of CHD, severe hypertriglyceridemia can cause additional complications, including fatty deposits in the skin and soft tissues and acute pancreatitis.[44]

Nutrition Therapy for Hypertriglyceridemia Dietary and lifestyle changes can improve most cases of mild hypertriglyceridemia.[45] Excessive weight gain and an inactive lifestyle may both raise triglyceride levels. Dietary factors that increase triglyceride levels include high intakes of alcohol and refined carbohydrates; sucrose and fructose are the carbohydrates with the strongest effect. Thus, controlling body weight, becoming physically active, restricting alcohol, and limiting intakes of refined carbohydrates (especially sweetened beverages and food items made with white flour and added sugars) are basic treatments for hypertriglyceridemia. As mentioned earlier, high triglyceride levels are often associated with low HDL levels, and the lifestyle changes listed here are likely to improve HDL levels as well.

Severe Hypertriglyceridemia Extreme elevations in blood triglycerides are usually caused by genetic mutations that upset lipoprotein metabolism. In addition to dietary and lifestyle changes, medications are usually necessary for lowering blood triglyceride levels above 500 milligrams per deciliter. If blood triglycerides exceed 1000 milligrams per deciliter, a very low-fat diet, providing less than 15 percent of kcalories from fat, may be required.[46] Patients must also eliminate consumption of alcoholic beverages.

Fish Oil Supplements and Hypertriglyceridemia Fish oil supplements are sometimes recommended for treating hypertriglyceridemia. Clinical trials suggest that a daily intake of 2 to 4 grams of EPA and DHA (combined) may lower triglyceride levels by 10 to 50 percent.[47] Although fish oil supplements may be effective for reducing blood triglyceride levels, research studies have not shown that their use in hypertriglyceridemia patients can improve cardiovascular disease outcomes.[48] In addition, fish oil therapy should be monitored by a physician due to the potential for adverse effects.

VITAMIN SUPPLEMENTATION AND CHD RISK

Patients are often interested in the potential benefits of using certain types of dietary supplements for reducing CHD risk, particularly B vitamin and antioxidant supplements. Most clinical trials have not been able to confirm any benefits from using these supplements, as described in this section.

B Vitamin Supplements and Homocysteine As mentioned earlier, elevated blood homocysteine is a risk factor for CHD, but whether homocysteine itself is directly damaging or is simply an indicator of other abnormalities remains unknown. Although increased intakes of folate, vitamin B_6, and vitamin B_{12} can lower homocysteine levels, clinical trials have not demonstrated that supplementation with these vitamins can reduce the incidence of heart attacks in those at risk.[49] Hence, B vitamin supplements are not currently recommended for patients at risk for CHD.

Antioxidant Supplements Because oxidative stress promotes atherosclerosis, researchers have hypothesized that antioxidant supplementation may inhibit atherosclerosis progression and reduce CHD risk. Several epidemiological studies suggested that antioxidant-rich diets can protect against CHD, but because persons who consume such diets usually maintain a healthy lifestyle and body weight as well, it has been difficult to determine whether the antioxidants were responsible for the effect. Most studies that have tested supplementation with single antioxidants (such as vitamins C or E) or combinations of antioxidants have produced weak or inconsistent results, and several studies suggested possible harm.[50] Until more data are available, the use of antioxidant supplements is not recommended for heart disease prevention.

DRUG THERAPIES FOR CHD PREVENTION

Individuals who cannot reach LDL goals with dietary and lifestyle changes alone may be prescribed one or more medications.[51] The most common drugs prescribed are *statins* (such as Lipitor or Zocor), which reduce cholesterol synthesis in the liver. *Bile acid sequestrants* (such as Colestid or Questran) reduce LDL cholesterol levels by interfering with bile acid reabsorption in the small intestine. For lowering triglyceride levels and increasing HDL, both *fibrates* (such as Lopid) and *nicotinic acid* (a form of niacin) are effective; nicotinic acid can also reduce LDL and lipoprotein(a) levels. Individuals using these medications should continue their dietary and lifestyle modifications so that they can use the minimum effective doses of the drugs they require.

In addition to lipid-lowering medications, some people may require drugs that suppress blood clotting (such as anticoagulants and aspirin) or reduce blood pressure. Nitroglycerin (a vasodilator) may be given to alleviate angina as needed. Some medications may affect nutrition status or food intake (see the Diet-Drug Interactions feature); the interactions can be even more complicated when multiple medications are used.

TREATMENT OF HEART ATTACK

As explained earlier, a heart attack occurs when the blood supply to heart muscle is blocked, causing damage or death to heart tissue. If the blood flow within the artery is restored quickly, the heart muscle may be saved; if not, the muscle tissue dies. Drug therapies given immediately after a heart attack may include thrombolytic drugs (clot-busting drugs), anticoagulants, aspirin, painkillers, and medications that regulate heart rhythm and reduce blood pressure.[52] Patients are not given food or beverages, except for sips of water or clear liquids, until their condition stabilizes. Once able to eat, they are initially offered small portions of foods that are low in sodium, saturated fat, and cholesterol. The sodium restriction helps to limit fluid retention but may be lifted after several days if the patient shows no signs of heart failure.

A heart attack patient needs to regain strength and learn strategies that can reduce the risk of a future heart attack; such strategies are similar to the Therapeutic Lifestyle Changes (TLC) described earlier. Thus, the cardiac rehabilitation programs in hospitals and outpatient clinics include exercise therapy, instruction about heart-healthy food choices, help with smoking cessation, and medication counseling. The programs often last several months. Home-based rehabilitation programs are also beneficial, but they are more limited in scope and lack the benefit of group interaction.

NURSING DIAGNOSIS

Nursing diagnoses for persons who experience a heart attack may include *decreased cardiac output, risk for decreased cardiac tissue perfusion, acute pain,* and *activity intolerance.*

Review Notes

- Long-term CHD management emphasizes risk reduction. Modifiable risk factors include elevated LDL and triglyceride levels, low HDL levels, hypertension, diabetes, obesity, an inactive lifestyle, cigarette smoking, and various dietary factors.
- Dietary and lifestyle changes can help patients reduce LDL levels and eliminate other risk factors. Dietary recommendations are to reduce saturated fat, *trans* fats, and cholesterol; increase soluble fiber; and incorporate plant sterols or stanols and fish into the diet. Other recommendations include regular physical activity, smoking cessation, and weight reduction.
- Treatment for mild hypertriglyceridemia emphasizes weight control, regular physical activity, avoiding a high carbohydrate intake (especially foods with added sugars), and alcohol restriction. Severe hypertriglyceridemia requires drug therapies and dietary fat restriction.
- Medications given after a heart attack suppress blood clotting, regulate heart rhythm, and reduce blood pressure. To reduce the risk of a future heart attack, patients must learn strategies similar to the TLC approach.

DIET-DRUG INTERACTIONS

Check this table for notable nutrition-related effects of the medications discussed in this chapter.

Anticoagulants (warfarin)	**Dietary interactions:** Warfarin requires a consistent vitamin K intake to maintain effectiveness. Drug effects may be enhanced with supplementation of vitamin E, fish oils, garlic, ginkgo, and glucosamine. Drug effects may be reduced with coenzyme Q, St. John's wort, and green tea. Avoid alcohol.
Antihypertensives	
Calcium channel blockers	**Gastrointestinal effects:** Nausea, GI discomfort, flatulence, constipation, diarrhea
	Dietary interactions: Avoid herbal supplements that contain natural licorice. Avoid grapefruit juice, which may enhance drug effects (depends on specific drug used). Avoid alcohol.
	Metabolic effects: Edema, flushing
ACE inhibitors[a]	**Gastrointestinal effects:** Reduced taste sensation
	Dietary interactions: Food intake and certain mineral supplements may interfere with absorption (depends on specific drug used). Avoid herbal supplements that contain natural licorice.
	Metabolic effects: Elevated serum potassium levels
Antilipimics	
Statins	**Gastrointestinal effects:** Constipation, flatulence, GI discomfort
	Dietary interactions: Avoid grapefruit juice and red yeast rice, which may enhance drug effects, and St. John's wort, which may reduce drug effects (interactions depend on specific drug used).
	Metabolic effects: Elevated serum liver enzymes
Bile acid sequestrants	**Gastrointestinal effects:** Constipation, flatulence, GI discomfort
	Dietary interactions: May reduce absorption of fat, fat-soluble vitamins, and some minerals.
	Metabolic effects: Electrolyte imbalances, nutrient deficiencies
Nicotinic acid	**Gastrointestinal effects:** GI discomfort (unless taken with milk or food), nausea, diarrhea, flatulence
	Dietary interactions: Alcoholic beverages may increase side effects.
	Metabolic effects: Elevated serum liver enzymes, elevated uric acid levels, hyperglycemia, flushing
Digoxin	**Gastrointestinal effects:** Anorexia, nausea, vomiting, diarrhea
	Dietary interactions: Antacids or magnesium supplements can reduce drug absorption. St. John's wort may reduce drug efficacy.
	Metabolic effects: Electrolyte imbalances
Diuretics (furosemide, spironolactone)	**Gastrointestinal effects:** Dry mouth, anorexia, decreased taste perception
	Dietary interactions: Furosemide's bioavailability is reduced when taken with food. Licorice root may interfere with the effects of diuretics.
	Metabolic effects: Fluid and electrolyte imbalances,[b] hyperglycemia, hyperlipidemias, thiamin deficiency (furosemide), elevated uric acid levels (furosemide)

[a]ACE is an abbreviation for *angiotensin-converting enzyme*. An ACE inhibitor interferes with the conversion of angiotensin I to angiotensin II, a peptide that helps to regulate blood pressure.

[b]*Furosemide* is a potassium-wasting diuretic; patients should increase intakes of potassium-rich foods. *Spironolactone* is a potassium-sparing diuretic; patients should avoid supplemental potassium and potassium-containing salt substitutes.

Stroke

Stroke is the fourth most common cause of death in the United States and a leading cause of long-term disability in adults.[53] About 87 percent of strokes are **ischemic strokes**, caused by the obstruction of blood flow to brain tissue. **Hemorrhagic strokes** occur in 13 percent of cases and result from bleeding within the brain, which damages brain tissue.[54] Most ischemic strokes are a result of ruptured atherosclerotic plaque and subsequent blood clot formation, but an embolism (■) may also cause a stroke. Hemorrhagic strokes often result from the rupture of a blood vessel that has been weakened by atherosclerosis and chronic hypertension. Hemorrhagic strokes are generally more deadly: about 38 percent result in death within 30 days.[55]

Strokes that occur suddenly and are short-lived (lasting several minutes to several hours) are called **transient ischemic attacks (TIAs)**. These brief strokes are a warning sign that a more severe stroke may follow, and they need to be evaluated and treated quickly.[56] TIAs typically cause short-term neurological symptoms, such as confusion, slurred speech, numbness, paralysis, or difficulty speaking. Treatment includes the use of aspirin and other drugs that inhibit blood clotting.

STROKE PREVENTION

Stroke is largely preventable by recognizing its risk factors and making lifestyle choices that reduce risk. Many of the risk factors are similar to those for heart disease; they include hypertension, elevated LDL cholesterol, diabetes mellitus, cigarette smoking, and a history of cardiovascular disease. Medications that suppress blood clotting reduce the risk of ischemic stroke, especially in people who have suffered a first stroke or a transient ischemic attack. The drugs typically prescribed include antiplatelet drugs (including aspirin) or anticoagulants such as warfarin (Coumadin). (■) Anticoagulant therapy requires regular follow-up and occasional adjustments in dosage to prevent excessive bleeding.

STROKE MANAGEMENT

The effects of a stroke vary according to the area of the brain that has been injured. Body movements, senses, and speech are often impaired, and one side of the body may be weakened or paralyzed. Early diagnosis and treatment are necessary to preserve brain tissue and minimize long-term disability. Ideally, thrombolytic (clot-busting) drugs should be used within 4½ hours following an ischemic stroke to restore blood flow and prevent further brain damage.[57] After patients have stabilized, they are usually started on medications that help to prevent stroke recurrence or complications, including anticoagulants or antiplatelet drugs, antihypertensives, and blood lipid-lowering drugs.

Rehabilitation programs typically start as soon as possible after stabilization. Patients must be evaluated for neurological deficits, sensory loss, mobility impairments, bowel and bladder function, communication ability, and psychological problems. Rehabilitation services often include physical therapy, occupational therapy, speech and language pathology, and kinesiotherapy (training to improve strength and mobility).

The focus of nutrition care is to help patients maintain nutrition status and overall health despite the disabilities caused by the stroke. Some patients may need to learn about dietary treatments that improve blood lipid levels and blood pressure. The initial assessment should determine the nature of the patient's self-feeding difficulty (if any) and the adjustments required for appropriate food intake. Dysphagia (difficulty swallowing) is a frequent complication of stroke and is associated with a poorer prognosis. (■) Difficulty with speech prevents patients from communicating food preferences or describing the problems they may be having with eating. Coordination problems can make it hard for patients to grasp utensils or bring food from table to mouth. In some cases, tube feedings may be necessary until the patient has regained these skills. Nutrition in Practice 21 describes additional options for feeding people with disabilities such as those that follow stroke.

NURSING DIAGNOSIS

Nursing diagnoses for stroke patients may include *impaired physical mobility, impaired swallowing, risk for aspiration, impaired verbal communication,* and *feeding self-care deficit.*

■ Reminder: An *embolism* is the obstruction of a blood vessel by a traveling blood clot or air bubble (an *embolus*).

■ Warfarin acts by interfering with vitamin K's blood-clotting function (see Chapter 14).

■ Reminder: Chapter 17 describes the nutrition care of patients with dysphagia.

ischemic strokes: strokes caused by the obstruction of blood flow to brain tissue.

hemorrhagic strokes: strokes caused by bleeding within the brain, which destroys or compresses brain tissue.

transient ischemic attacks (TIAs): brief ischemic strokes that cause short-term neurological symptoms.

Review Notes

- The two major types of strokes, ischemic and hemorrhagic stroke, may be a consequence of atherosclerosis, hypertension, or both. Transient ischemic attacks, which are short-lived ischemic strokes, are a warning sign that a more severe stroke may follow.
- Strokes are largely preventable by reversing modifiable risk factors, such as hypertension, cigarette smoking, diabetes mellitus, and elevated LDL cholesterol.
- Treatment of a stroke includes the use of anticlotting drugs such as antiplatelet drugs and anticoagulants. Rehabilitation services evaluate the extent of neurological and functional impairment caused by a stroke and provide the therapy patients need to regain lost function. A patient who has had a major stroke may have problems eating normally due to lack of coordination or difficulty swallowing.

Hypertension

Hypertension (high blood pressure) affects about one-third of adults in the United States.[58] Prevalence is especially high among African Americans, who develop hypertension earlier in life and sustain higher average blood pressures throughout their lives than other ethnic groups. An estimated 22 percent of people with hypertension are unaware that they have it.[59] (■)

Although people cannot feel the physical effects of hypertension, it is a primary risk factor for atherosclerosis and cardiovascular diseases. Elevated blood pressure forces the heart to work harder to eject blood into the arteries; this effort weakens heart muscle and increases the risk of developing heart arrhythmias, heart failure, and even sudden death. Hypertension is also a primary cause of stroke and kidney failure, and reducing blood pressure can dramatically reduce the incidence of these diseases.

■ Blood pressure is measured both when heart muscle contracts (*systolic* blood pressure) and when it relaxes (*diastolic* blood pressure). Measurements are expressed as millimeters of mercury (mm Hg).

	Systolic	Diastolic
• Desirable	<120	<80
• Prehypertension	120–139	80–89
• Hypertension	≥140	≥90

FACTORS THAT INFLUENCE BLOOD PRESSURE

Although the underlying causes of most cases of hypertension are not fully understood, much is known about the physiological factors that affect blood pressure, the force exerted by the blood on artery walls. As shown in Figure 21-3, blood pressure depends on the volume of blood pumped by the heart (cardiac output) and the resistance the blood encounters in the arterioles (peripheral resistance). (■) When either cardiac output or peripheral resistance increases, blood pressure rises. Cardiac output is raised when heart rate or blood volume increases; peripheral resistance is affected mostly by the diameters of the arterioles and blood viscosity. Blood pressure is therefore influenced by the nervous system, which regulates heart muscle contractions and arteriole diameters, and hormonal signals, which may cause fluid retention or blood vessel constriction. The kidneys also play a role in regulating blood pressure by controlling the secretion of the hormones involved in vasoconstriction and retention of sodium and water.

■ The equation describing this relationship is blood pressure (BP) = cardiac output (CO) × peripheral resistance (PR).

FACTORS THAT CONTRIBUTE TO HYPERTENSION

In 90 to 95 percent of hypertension cases, the cause is unknown (called **primary** or **essential hypertension**).[60] In other cases, hypertension is caused by a known physical or metabolic disorder (**secondary hypertension**), such as an abnormality in an organ or hormone involved in blood pressure regulation. For example, conditions characterized by the narrowing of renal arteries often result in the increased production of proteins and hormones that stimulate water retention and vasoconstriction, thereby raising blood pressure. A number of hormonal disorders and medications may also cause secondary hypertension.

primary hypertension: hypertension with an unknown cause; also known as **essential hypertension.**

secondary hypertension: hypertension that results from a known physiological abnormality.

FIGURE 21-3 **Determinants of Blood Pressure**

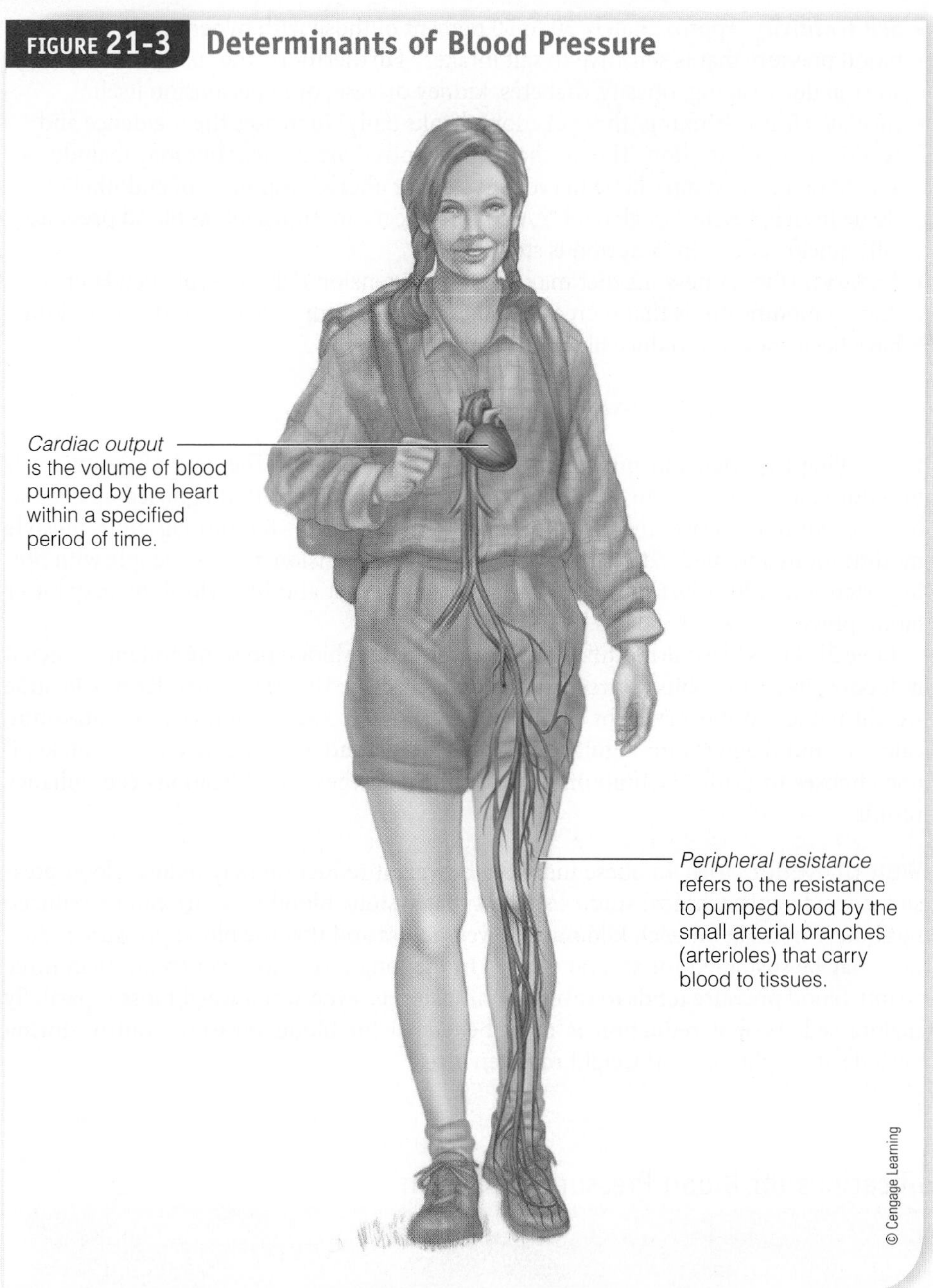

Cardiac output is the volume of blood pumped by the heart within a specified period of time.
Peripheral resistance refers to the resistance to pumped blood by the small arterial branches (arterioles) that carry blood to tissues.
© Cengage Learning

A number of risk factors for hypertension have been identified. These include the following:

- *Aging.* Hypertension risk increases with age. About two-thirds of persons older than 65 years have hypertension.[61] Moreover, individuals who have normal blood pressure at age 55 still have a 90 percent risk of developing high blood pressure during their lifetimes.[62]
- *Genetic factors.* Risk of hypertension is similar among family members. It is also more prevalent and severe in certain ethnic groups; for example, the prevalence in African American adults is about 41 percent, compared with a prevalence of about 28 percent in whites and Mexican Americans.[63]
- *Obesity.* Hypertension risk increases as body fatness increases. Numerous clinical studies have confirmed a strong relationship between excess body fat and increased blood pressure.[64] Obesity raises blood pressure, in part, by increasing blood volume, activating the sympathetic nervous system, and stimulating processes that cause blood vessel constriction.

NURSING DIAGNOSIS

Nursing diagnoses for some individuals with hypertension may include *imbalanced nutrition: more than body requirements* and *ineffective self health management.*

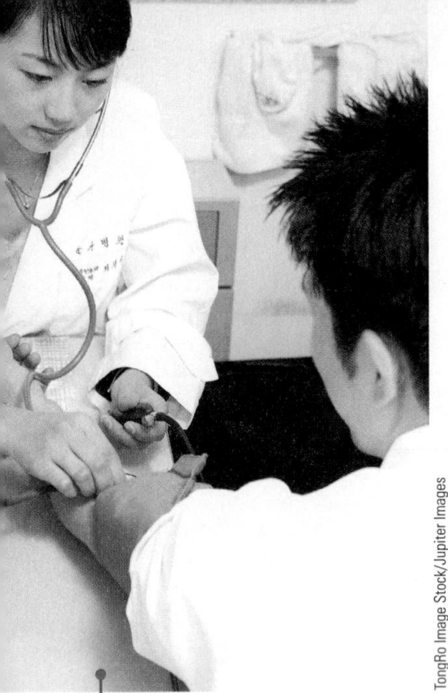
TongRo Image Stock/Jupiter Images

Screening people for hypertension is a first step toward early detection and prevention of complications.

- *Salt sensitivity.* Approximately 30 to 50 percent of those with hypertension have blood pressure that is sensitive to salt intake.[65] Furthermore, the salt sensitivity may worsen due to aging, obesity, diabetes, kidney disease, or hypertension itself.[66]
- *Alcohol.* Heavy drinking (three or more drinks daily) increases the incidence and severity of hypertension. The mechanisms involved are unclear but may include activation of the sympathetic nervous system or altered responses of endothelial tissue in the presence of alcohol.[67] Alcohol's effects are transient, as blood pressure falls quickly after consumption is stopped.
- *Dietary factors.* A person's diet may affect hypertension risk. As explained later, dietary modifications that increase intakes of potassium, calcium, and magnesium have been shown to reduce blood pressure.

TREATMENT OF HYPERTENSION

Controlling hypertension improves CVD risk considerably. The goal of treatment is to reduce blood pressure to <140/<90 mm Hg; for people with diagnosed CHD, diabetes, or kidney disease, the blood pressure goal is <130/<80 mm Hg. Both lifestyle modifications and medications are used to treat hypertension. (■) For people with prehypertension (120–139/80–89 mm Hg), changes in diet and lifestyle alone may lower blood pressure to a normal level.

Table 21-4 lists lifestyle modifications that can reduce blood pressure and the expected reduction in systolic blood pressure for each change. The recommendations include weight reduction if overweight or obese; a diet low in sodium and rich in potassium, calcium, and magnesium; regular physical activity; and a moderate alcohol intake, if one chooses to drink.[68] Combining two or more of these modifications can enhance results.

■ For updated guidelines related to hypertension treatment, access this website: www.nhlbi.nih.gov/guidelines.

Weight Reduction

In obese individuals, weight reduction may reduce blood pressure considerably. Clinical studies suggest that systolic blood pressure can be reduced by about 1 mm Hg for each kilogram of weight loss and that the blood pressure reduction may be sustained for several years.[69] In the long term, however (more than three years), blood pressure tends to revert to initial levels, even when weight loss is partially maintained. Weight reduction is most beneficial for blood pressure control during periods of weight loss and weight maintenance.[70]

TABLE 21-4 Lifestyle Modifications for Blood Pressure Reduction

Modification	Recommendation	Expected Reduction in Systolic Blood Pressure
Weight reduction	Maintain healthy body weight (BMI below 25).	5–20 mm Hg/10 kg weight loss
DASH Eating Plan[a]	Adopt a diet rich in fruits, vegetables, and low-fat milk products with reduced saturated fat intake.	8–14 mm Hg
Sodium restriction	Reduce dietary sodium intake to less than 1500 milligrams sodium (less than 4 grams salt) per day.[b]	5–10 mm Hg
Physical activity	Perform aerobic physical activity for at least 30 minutes per day, most days of the week.	4–9 mm Hg
Moderate alcohol consumption	Men: Limit to two drinks per day. Women and lighter-weight men: Limit to one drink per day.	2–4 mm Hg

[a]The DASH Eating Plan was tested in a study called *D*ietary *A*pproaches to *S*top *H*ypertension.
[b]The American Heart Association recommends an intake of <1500 mg sodium per day for the entire U.S. population.

Sources: P. K. Whelton and coauthors, Sodium, blood pressure, and cardiovascular disease: Further evidence supporting the American Heart Association sodium reduction recommendations, *Circulation* 126 (2012): 2880–2889; *Seventh Report of the Joint National Committee on Prevention, Detection, Evaluation, and Treatment of High Blood Pressure (JNC 7)*, NIH publication no. 03-5231 (Bethesda, MD: National Institutes of Health, National Heart, Lung, and Blood Institute, and National High Blood Pressure Education Program, May 2003).

Dietary Approaches for Reducing Blood Pressure Several research studies have shown that a significant reduction in blood pressure can be achieved by following a diet that emphasizes fruits, vegetables, and low-fat dairy products and includes whole grains, poultry, fish, and nuts.[71] The diet tested in these studies, now known as the *DASH Eating Plan,* (■) provides more fiber, potassium, magnesium, and calcium than the typical American diet. The diet also limits red meat, sweets, sugar-containing beverages, saturated fat (to 7 percent of kcalories), and cholesterol (to 150 milligrams per day), so it is beneficial for reducing CHD risk as well. During the eight-week study period when hypertensive subjects consumed the DASH diet, their systolic blood pressures fell by 11.4 mm Hg more than the blood pressures of subjects who remained on the standard American control diet.[72] The DASH Eating Plan, shown in Table 21-5, is a dietary pattern that meets the goals specified in the *Dietary Guidelines for Americans 2010* report.[73]

The DASH Eating Plan is even more effective when accompanied by a low sodium intake. In a research study that tested the blood pressure–lowering effects of the DASH dietary pattern in combination with sodium restriction, the best results were achieved when sodium was reduced to 1500 milligrams daily—a level much lower than the

■ The DASH Eating Plan is based on the test diet used in a study called *D*ietary *A*pproaches to *S*top *H*ypertension.

TABLE 21-5 The DASH Eating Plan

Food Group	Recommended Servings for Different Energy Intakes (*servings per day except as noted*)			
	1600 kcal	2000 kcal	2600 kcal	3100 kcal
Grains and grain products[a] (1 serving = 1 slice bread, 1 oz dry cereal,[b] or ½ c cooked rice, pasta, or cereal)	6	6–8	10–11	12–13
Vegetables (1 serving = ½ c cooked vegetables, 1 c raw leafy vegetables, or ½ c vegetable juice)	3–4	4–5	5–6	6
Fruits (1 serving = 1 medium fruit; ½ c fresh, frozen, or canned fruit; ¼ c dried fruit; or ½ c fruit juice)	4	4–5	5–6	6
Milk products (low fat or fat free) (1 serving = 1 c milk or yogurt, or 1½ oz cheese)	2–3	2–3	3	3–4
Meat, poultry, and fish (1 serving = 1 oz cooked lean meat, poultry, or fish; or 1 egg)	3–4 oz or less	6 oz or less	6 oz or less	6–9 oz or less
Nuts, seeds, and legumes (1 serving = ⅓ c nuts, 2 tbs peanut butter, 2 tbs seeds, or ½ c cooked dry beans or peas)	3–4 per week	4–5 per week	1	1
Fats and oils (1 serving = 1 tsp vegetable oil or soft margarine, ½ tbs mayonnaise, or 1 tbs salad dressing)	2	2–3	3	4
Sweets and added sugars (1 serving = 1 tbs sugar, jelly, or jam; ½ c sorbet; or 1 c lemonade)	3 or less per week	5 or less per week	≤ 2	≤ 2

[a]Whole grains are recommended for most servings consumed.
[b]One ounce of dry cereal may be equivalent to ½ to 1¼ cups, depending on the cereal. Check the food label for the portion size.

Source: U.S. Department of Agriculture and U.S. Department of Health and Human Services, *Dietary Guidelines for Americans, 2010* (Washington, DC: U.S. Government Printing Office, 2010).

amounts typically consumed in the United States (average daily sodium intakes for men and women are 3760 milligrams and 2828 milligrams, respectively).[74]

Sodium restriction by itself can have a modest blood pressure–lowering effect (review Table 21-4), but some people are more responsive than others. Although a low-sodium diet may improve blood pressure to some extent, it should be combined with other lifestyle modifications for greater effect. The "How to" lists practical suggestions for restricting sodium intake; additional detail is provided in Table 22-1 on p. 627.

- *ACE inhibitors* interfere with the production of angiotensin II, a peptide that regulates blood pressure.
- *Beta-blockers* reduce heart rate and cardiac output.
- *Calcium channel blockers* inhibit vasoconstriction.

Drug Therapies for Reducing Blood Pressure

People with hypertension usually require two or more medications to meet their blood pressure goals. Using a combination of drugs with different modes of action can reduce the doses of each drug needed and minimize side effects. Most treatments include diuretics, which lower blood pressure by reducing blood volume. Other medications prescribed include angiotensin-converting enzyme (ACE) inhibitors, beta-blockers, and calcium channel blockers; (■) these drugs are also used to treat various heart conditions. Drug dosages may need regular adjustment until the blood pressure goal is reached.

Review Notes

- About one in three persons in the United States has hypertension, which increases the risk of developing CHD, stroke, heart failure, and kidney failure.
- Blood pressure is elevated by factors that increase blood volume, heart rate, or resistance to blood flow. Although the underlying cause of most hypertension cases is unknown, risk factors include aging, family history, ethnicity, obesity, and various dietary factors.
- Treatment of hypertension usually includes a combination of lifestyle modifications (such as adopting the DASH Eating Plan, restricting sodium, and increasing physical activity) and drug therapies. The Case Study provides an opportunity to review the risk factors and treatments for CHD and hypertension.

HOW TO Reduce Sodium Intake

- Select fresh, unprocessed foods. Packaged foods, canned goods, and frozen meals are often high in sodium.
- Do not use salt at the table or while cooking. Salt substitutes may be useful for some people. Salt substitutes often contain potassium, however, and are not appropriate for people using diuretics that promote potassium retention in the blood.
- Avoid eating in fast-food restaurants; most menu choices are very high in sodium.
- Check food labels. The labeling term *low sodium* is a better guide than the terms *reduced sodium* (contains 25 percent less sodium than the regular product) or *light in sodium* (contains 50 percent less sodium). To be labeled *low sodium,* a food product must contain less than 140 milligrams of sodium per serving. Keep your sodium goal in mind when you read labels.
- Recognize the high-sodium foods in each food category, and purchase only unsalted or low-sodium varieties of these products if they are available. High-sodium foods include the following:
  - Snack foods made with added salt, such as tortilla chips, popcorn, and nuts.
  - Processed meats, such as ham, corned beef, bologna, salami, sausage, bacon, frankfurters, and pastrami.
  - Processed fish, such as salted fish and canned fish.
  - Tomato-based products, such as tomato sauce, tomato juice, pizza, canned tomatoes, and catsup.
  - Canned soups and broths; note that even reduced-sodium varieties may contain excessive sodium.
  - Cheese, such as cottage cheese, American cheese, and Parmesan and most other hard cheeses.
  - Bakery products made with baking powder or baking soda (sodium bicarbonate), such as cakes, cookies, doughnuts, and muffins.
  - Condiments and relishes, such as bouillon cubes, olives, and pickled vegetables.
  - Flavoring sauces, such as soy sauce, barbecue sauce, and steak sauce.
- Check for the word *sodium* on medication labels. Sodium is often an ingredient in some types of antacids and laxatives.

Case Study: Patient with Cardiovascular Disease

Robert Reid, a 48-year-old African American computer programmer, is 5 feet 9 inches tall and weighs 240 pounds. He sits for long hours at work and is too tired to exercise when he gets home at night. His meals usually include fatty meats, eggs, and cheese, and he likes dairy desserts such as pudding and ice cream. He has a family history of CHD and hypertension. His recent laboratory tests show that his blood pressure is 160/100 mm Hg, and his LDL and HDL levels are 160 mg/dL and 35 mg/dL, respectively. He smokes a pack of cigarettes each day and usually has two glasses of wine at both lunch and dinner.

1. Identify Mr. Reid's major risk factors for CHD and hypertension. Which can be modified? What complications might occur if he delays treatment for his blood lipids and blood pressure?
2. What dietary changes would you recommend that could help to improve Mr. Reid's blood pressure and his LDL cholesterol? Explain the rationale for each dietary change. Prepare a day's menu for Mr. Reid using the DASH Eating Plan as an outline for your choices.
3. What other laboratory tests or measurements would you need to better assess Mr. Reid's condition? Why?
4. Describe several benefits that Mr. Reid might obtain from a program that includes weight reduction and regular physical activity. Explain why the use of alcohol can be both a protective and a damaging lifestyle habit.
5. Assuming that Mr. Reid does not make any changes in his diet and lifestyle and suffers a heart attack, identify the elements of a cardiac rehabilitation program that would be critical for his long-term survival.

Heart Failure

Heart failure, also called *congestive heart failure,* is characterized by the heart's inability to pump adequate blood, resulting in inadequate blood delivery and a buildup of fluids in the veins and tissues. Heart failure has various causes, but it is often a consequence of chronic disorders that create extra work for the heart muscle, such as hypertension or CHD. To accommodate the extra workload, the heart enlarges or pumps faster or harder, but it eventually may weaken enough to fail completely. Heart failure develops most often in older adults and the elderly: the majority of cases occur in individuals 65 years or older.[75]

CONSEQUENCES OF HEART FAILURE

The symptoms and consequences of heart failure depend on the side of the heart that fails. The right side of the heart normally receives blood from the peripheral tissues and pumps the blood to the lungs. With impaired pumping, blood backs up in the peripheral tissues and abdominal organs. Fluid may accumulate in the lower extremities and in the liver and abdomen, causing chest pain, difficulty with digestion and absorption, and swelling in the legs, ankles, and feet. In contrast, the left side of the heart receives blood from the lungs and pumps it to the peripheral tissues. A weakened left heart can cause a buildup of fluid in the lungs (called *pulmonary edema*), resulting in extreme shortness of breath and limited oxygen for activity; in severe cases, it can lead to respiratory failure. With inadequate blood flow, the functions of various organs, such as the liver and kidneys, may become impaired. The effects of heart failure also depend on the severity of illness: mild cases may be asymptomatic, but severe cases may cause considerable damage to health.

Heart failure often affects a person's food intake and level of physical activity. In persons with abdominal bloating and liver enlargement, pain and discomfort may worsen with meals. Limb weakness and fatigue can limit physical activity. End-stage heart failure is often accompanied by **cardiac cachexia**, a condition of severe malnutrition characterized by significant weight loss and tissue wasting. Cardiac cachexia may develop due to increased levels of cytokines that promote catabolism, elevated metabolic rate, loss of appetite, reduced food intake, and malabsorption.[76] The resultant weakness further lowers the person's strength, functional capacity, and activity levels.

NURSING DIAGNOSIS

Nursing diagnoses appropriate for patients with heart failure may include *decreased cardiac output, excess fluid volume, impaired gas exchange, activity intolerance,* and *imbalanced nutrition: less than body requirements.*

heart failure: a condition with various causes that is characterized by the heart's inability to pump adequate blood to the body's cells, resulting in fluid accumulation in the tissues; also called *congestive heart failure.*

cardiac cachexia: a condition of severe malnutrition that develops in heart failure patients; characterized by weight loss and tissue wasting.

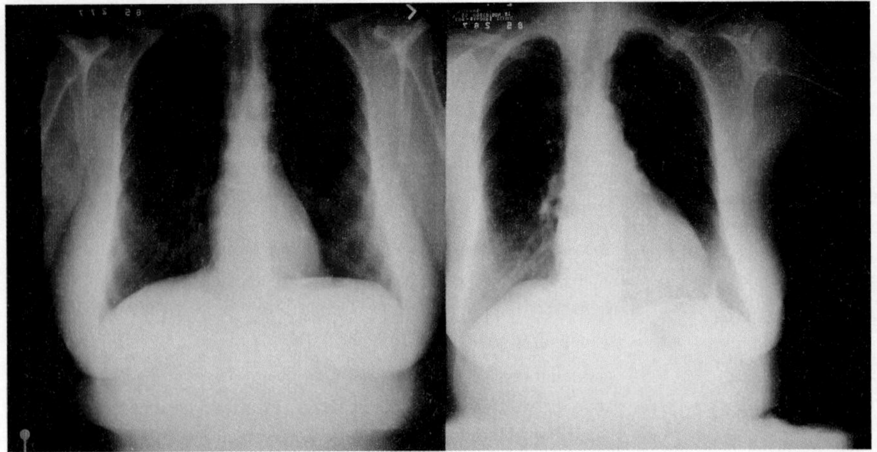

© Susan Leavines/Photo Researchers, Inc.

An overburdened heart enlarges in an effort to supply blood to the body's tissues.

MEDICAL MANAGEMENT OF HEART FAILURE

Heart failure is a chronic, progressive illness that may require frequent hospitalizations. Many patients face a combination of debilitating symptoms, complex treatments, and an uncertain outcome. Important goals of medical therapy are to slow disease progression and enhance the patient's quality of life.

The specific treatment for heart failure depends on the nature and severity of the illness. Medications help to manage fluid retention and improve heart function. Dietary sodium and fluid restrictions can help to prevent fluid accumulation. Vaccinations for influenza and pneumonia reduce the risk of developing respiratory infections. Treatment of CHD risk factors, such as hypertension and lipid disorders, can help to slow disease progression. Heart failure patients are also encouraged to participate in exercise programs to avoid becoming physically disabled and to improve endurance.

Drug Therapies for Heart Failure The medications prescribed for heart failure include diuretics, ACE inhibitors, angiotensin receptor blockers, beta-blockers, vasodilators, and digitalis.[77] The diuretics are given to reverse or prevent fluid retention. The patient must monitor fluid fluctuations with daily weight measurements and can make small adjustments in the diuretic dose as needed. The other drugs listed help to improve heart and blood vessel functioning and blood flow.

Nutrition Therapy for Heart Failure The main dietary recommendation for heart failure is a sodium restriction of 2000 milligrams or less daily to reduce the likelihood of fluid retention. (■) In patients with persistent or recurrent fluid retention, fluid intakes may be restricted to 2 liters per day or less.[78] Individuals who have difficulty eating due to abdominal or chest pain may tolerate small, frequent meals better than large meals.

Other Dietary Recommendations Patients with heart failure may be prone to constipation due to diuretic use and reduced physical activity. Maintaining an adequate fiber intake can help to minimize constipation problems. Because alcohol consumption may worsen heart function, some patients may need to restrict or avoid alcoholic beverages. Patients on diuretic therapy or restricted diets may benefit from daily multivitamin-mineral supplementation.[79]

■ The recommendation to limit sodium intake to 2000 mg is not very restrictive. The sodium recommendation for individuals at risk of hypertension is 1500 mg daily.

Cardiac Cachexia No known therapies can reverse cardiac cachexia, and prognosis is poor. For some patients, liquid supplements, tube feedings, or parenteral nutrition support can be supportive additions to treatment.

Review Notes

- Heart failure is usually a chronic, progressive condition that results from other cardiovascular illnesses.
- In heart failure, the heart is unable to pump adequate blood to tissues. Consequences may include fluid accumulation in the veins, lungs, and other organs and impaired organ function.
- Treatment of heart failure includes drug therapies that reduce fluid accumulation and improve heart function. Nutrition therapy may include sodium, fluid, and alcohol restrictions and nutrient supplementation.

Nutrition Assessment Checklist for People with Cardiovascular Diseases

MEDICAL HISTORY

Check the medical record for a diagnosis of:

- ❍ Coronary heart disease
- ❍ Stroke
- ❍ Hypertension
- ❍ Heart failure

Review the medical record for complications related to cardiovascular diseases:

- ❍ Heart attack
- ❍ Transient ischemic attack
- ❍ Cardiac cachexia

Note risk factors for CHD or stroke that are related to diet, including:

- ❍ Elevated LDL or triglyceride levels
- ❍ Obesity or overweight
- ❍ Diabetes
- ❍ Hypertension

MEDICATIONS

For patients using drug treatments for cardiovascular diseases, note:

- ❍ Side effects that may alter food intake
- ❍ Medications that may interact with grapefruit juice
- ❍ Use of warfarin, which requires a consistent vitamin K intake
- ❍ Use of diuretics or other drugs associated with potassium imbalances
- ❍ Potential diet-drug or herb-drug interactions

DIETARY INTAKE

For patients with CHD, a previous stroke, or hypertension, assess the diet for:

- ❍ Energy intake
- ❍ Saturated fat, *trans* fat, cholesterol, and sodium content
- ❍ Soluble fiber and plant sterol or plant stanol content
- ❍ Intake of whole grains, fruits, vegetables, legumes, and nuts
- ❍ Alcohol content

For patients with complications resulting from cardiovascular diseases:

- ❍ Check physical disabilities that may interfere with food preparation or consumption following a stroke.
- ❍ Check adequacy of food and nutrient intake in patients with heart failure.

ANTHROPOMETRIC DATA

Measure baseline height and weight, and reassess weight at each medical checkup. Note whether patients are meeting weight goals, including:

- ❍ Weight loss or maintenance in patients who are overweight
- ❍ Weight maintenance in patients with advanced heart failure

Remember that weight may be deceptively high in people who are retaining fluids, especially individuals with heart failure.

LABORATORY TESTS

Monitor the following laboratory tests in people with cardiovascular diseases:

- ❍ LDL cholesterol, blood triglycerides, and HDL cholesterol
- ❍ Blood glucose in patients with diabetes
- ❍ Serum potassium in patients using diuretics, antihypertensive medications, or digoxin
- ❍ Blood-clotting times in patients using anticoagulants
- ❍ Indicators of fluid retention in patients with heart failure

PHYSICAL SIGNS

Blood pressure measurement is routine in physical exams but is especially important for people who:

- ❍ Have cardiovascular diseases
- ❍ Have experienced a heart attack or stroke
- ❍ Have risk factors for CHD or hypertension

Look for signs of:

- ❍ Potassium imbalances (muscle weakness, numbness and tingling, irregular heartbeat) in those using diuretics, antihypertensive medications, or digoxin
- ❍ Fluid overload in patients with heart failure

Self Check

1. Ischemia in the coronary arteries is a frequent cause of:
   a. angina pectoris.
   b. hemorrhagic stroke.
   c. aneurysm.
   d. hypertension.

2. Risk factors for atherosclerosis include all of the following *except:*
   a. smoking.
   b. hypertension.
   c. diabetes mellitus.
   d. elevated HDL cholesterol.

3. To screen for risk of coronary heart disease, which laboratory assessment should be conducted every five years in individuals who are 20 years and older?
   a. Coronary calcium score
   b. Lipoprotein profile
   c. Blood pressure check
   d. LDL particle number

4. The dietary lipids with the strongest LDL cholesterol–raising effects are:
   a. monounsaturated fats.
   b. polyunsaturated fats.
   c. saturated fats.
   d. plant sterols.

5. The fatty acids EPA and DHA may help to improve heart disease risk due to their beneficial effects on:
   a. LDL and HDL levels.
   b. blood triglyceride levels and blood-clotting activity.
   c. cardiac output and heart rate.
   d. LDL size and particle number.

6. Patients with mild hypertriglyceridemia may improve their triglyceride levels by:
   a. reducing sodium intake.
   b. consuming moderate amounts of alcohol.
   c. limiting intakes of refined carbohydrates.
   d. reducing cholesterol intake.

7. Which medications reduce cholesterol synthesis in the liver?
   a. Bile acid sequestrants
   b. Fibrates
   c. ACE inhibitors
   d. Statins

8. Hemorrhagic stroke:
   a. is the most common type of stroke.
   b. results from obstructed blood flow within brain tissue.
   c. comes on suddenly and usually lasts for up to 30 minutes.
   d. results from bleeding within the brain, which damages brain tissue.

9. Hypertensive patients can benefit from all of the following dietary and lifestyle modifications *except:*
   a. including fat-free or low-fat milk products in the diet.
   b. reducing total fat intake.
   c. consuming generous amounts of fruits, vegetables, legumes, and nuts.
   d. reducing sodium intake.

10. Nutrition therapy for a patient with heart failure usually includes:
    a. weight loss.
    b. reducing total fat intake.
    c. sodium restriction.
    d. cholesterol restriction.

Answers to these questions appear in Appendix H. For more chapter review: Access an interactive eBook, chapter-specific interactive learning tools, including flashcards, quizzes, videos, and more in your Nutrition CourseMate, accessed through CengageBrain.com.

Clinical Applications

1. List risk factors for coronary heart disease, and identify possible interrelationships among the factors. For example, a woman over 55 years of age is also at risk for diabetes; a person with diabetes is more likely to have hypertension.

2. Review the DASH Eating Plan shown in Table 21-5. As the chapter describes, the DASH dietary pattern is helpful for lowering blood pressure and for reducing CHD risk as well.
   - List elements of the DASH Eating Plan that are consistent with the TLC recommendations.
   - Suggest ways in which a person following the DASH Eating Plan might accomplish the following additional dietary modifications: consume a higher percentage of fat from monounsaturated sources, reduce intake of *trans* fats, and include EPA/DHA and plant sterols in the diet.

Nutrition on the Net

For further study of topics covered in this chapter, access these websites:

- To search for additional information about cardiovascular diseases, obtain the American Heart Association dietary recommendations, or find links to other relevant materials, visit the website of the American Heart Association: www.heart.org
- To learn about improving health care and life expectancy of ethnic minority populations at high risk for cardiovascular diseases, visit the website of the International Society on Hypertension in Blacks: www.ishib.org
- Information about cardiovascular diseases, the DASH Eating Plan, and implementation of heart-healthy diets is available at the websites of the National Heart, Lung, and Blood Institute and the Heart and Stroke Foundation of Canada: www.nhlbi.nih.gov and www.heartandstroke.ca
- Learn more about the prevention and treatment of stroke at the website of the National Stroke Association: www.stroke.org

Notes

1. V. L. Roger and coauthors, Heart disease and stroke statistics—2012 update: A report from the American Heart Association, *Circulation* 125 (2012): e2–e220.
2. World Health Organization, *World Health Statistics 2012* (Geneva: World Health Organization, 2012), available at www.who.int/gho/publications/world_health_statistics/EN_WHS2012_Full.pdf, site visited November 12, 2012.
3. R. N. Mitchell and F. J. Schoen, Blood vessels, in V. Kumar and coeditors, eds., *Robbins and Cotran Pathologic Basis of Disease* (Philadelphia: Saunders, 2010), pp. 487–528; W. Insull, The pathology of atherosclerosis: Plaque development and plaque responses to medical treatment, *American Journal of Medicine* 122 (2009): S3–S14.
4. C. Cheng and coauthors, Atherosclerotic lesion size and vulnerability are determined by patterns of fluid shear stress, *Circulation* 113 (2006): 2744–2753.
5. V. Cachofeiro and coauthors, Inflammation: A link between hypertension and atherosclerosis, *Current Hypertension Reviews* 5 (2009): 40–48.
6. M. Miller and coauthors, Triglycerides and cardiovascular disease: A scientific statement from the American Heart Association, *Circulation* 123 (2011): 2292–2333.
7. K. Mahdy Ali and coauthors, Cardiovascular disease risk reduction by raising HDL cholesterol—Current therapies and future opportunities, *British Journal of Pharmacology* 167 (2012): 1177–1194; D. Kothapalli and coauthors, Cardiovascular protection by apoE and apoE-HDL linked to suppression of ECM gene expression and arterial stiffening, *Cell Reports* 2 (2012): 1–13.
8. G. K. Hansson and A. Hamsten, Atherosclerosis, thrombosis, and vascular biology, in L. Goldman and A. I. Schafer, eds., *Goldman's Cecil Medicine* (Philadelphia: Saunders, 2012), pp. 409–412; K. Musunuru, Atherogenic dyslipidemia: Cardiovascular risk and dietary intervention, *Lipids* 45 (2010): 907–914.
9. N. L. Benowitz, Tobacco, in L. Goldman and D. Ausiello, eds., *Cecil Medicine* (Philadelphia: Saunders, 2008), pp. 162–166.
10. E. Galkina and K. Ley, Immune and inflammatory mechanisms of atherosclerosis, *Annual Review of Immunology* 27 (2009): 165–197.
11. C. E. Tabit and coauthors, Endothelial dysfunction in diabetes mellitus: Molecular mechanisms and clinical implications, *Reviews in Endocrine & Metabolic Disorders* 11 (2010): 61–74; G. Orasanu and J. Plutzky, The pathologic continuum of diabetic vascular disease, *Journal of the American College of Cardiology* 53 (2009): S35–S42.
12. M. A. Maturana, M. C. Irigoyen, and P. M. Spritzer, Menopause, estrogens, and endothelial dysfunction: Current concepts, *Clinics* 62 (2007): 77–86.
13. K. L. Schalinske and A. L. Smazal, Homocysteine imbalance: A pathological metabolic marker, *Advances in Nutrition* 3 (2012): 755–762; K. S. McCully, Homocysteine, vitamins, and vascular disease prevention, *American Journal of Clinical Nutrition* 86 (2007): 1563S–1568S.
14. Roger and coauthors, 2012.
15. W. E. Boden, Angina pectoris and stable ischemic heart disease, in L. Goldman and A. I. Schafer, eds., *Goldman's Cecil Medicine* (Philadelphia: Saunders, 2012), pp. 412–425.
16. Roger and coauthors, 2012.
17. P. S. Jellinger and coauthors, American Association of Clinical Endocrinologists' guidelines for management of dyslipidemia and prevention of atherosclerosis, *Endocrine Practice* 18 (2012): 269–293; J. S. Rana and coauthors, The role of non-HDL cholesterol in risk stratification for coronary artery disease. *Current Atherosclerosis Reports* 14 (2012): 130–134.
18. Jellinger and coauthors, 2012; M. H. Davidson and coauthors, Clinical utility of inflammatory markers and advanced lipoprotein testing: Advice from an expert panel of lipid specialists, *Journal of Clinical Lipidology* 5 (2011): 338–367.
19. C. F. Semenkovich, Disorders of lipid metabolism, in L. Goldman and A. I. Schafer, eds., *Goldman's Cecil Medicine* (Philadelphia: Saunders, 2012), pp. 1346–1354.
20. B. F. Voight and coauthors, Plasma HDL cholesterol and risk of myocardial infarction: A Mendelian randomization study, *Lancet* 380 (2012): 572–580.
21. Mahdy Ali and coauthors, 2012; J. P. Corsetti and coauthors, Inflammation reduces HDL protection against primary cardiac risk, *European Journal of Clinical Investigation* 40 (2010): 483–489.
22. U.S. Department of Agriculture, Agricultural Research Service, Nutrient intakes from food: Mean amounts consumed per individual, by gender and age, *What We Eat in America, NHANES 2009–2010* (2012), available at www.ars.usda.gov/ba/bhnrc/fsrg, site visited November 20, 2012.
23. U.S. Department of Agriculture and U.S. Department of Health and Human Services, *Dietary Guidelines for Americans, 2010* (Washington, DC: U.S. Government Printing Office, 2010).
24. Jellinger and coauthors, 2012.
25. D. Mozaffarian, A. Aro, and W. C. Williett, Health effects of trans-fatty acids: Experimental and observational evidence, *European Journal of Clinical Nutrition* 63 (2009): S5–S21.
26. M. L. Fernandez, Rethinking dietary cholesterol, *Current Opinion in Clinical Nutrition and Metabolic Care* 15 (2012): 117–121; J. D. Spence, D. J. A. Jenkins, and J. Davignon, Dietary cholesterol and egg yolks: Not for patients at risk of vascular disease, *Canadian Journal of Cardiology* 26 (2010): e336–e339.
27. M. M. Kanter and coauthors, Exploring the factors that affect blood cholesterol and heart disease risk: Is dietary cholesterol as bad for you as history leads us to believe? *Advances in Nutrition* 3 (2012): 711–717.
28. U.S. Department of Agriculture, Agricultural Research Service, 2012.
29. National Cancer Institute, Applied Research Program, *Sources of Cholesterol among the U.S. Population, 2005–06*, available at http://riskfactor.cancer.gov/diet/foodsources/cholesterol, site visited November 20, 2012.
30. Fernandez, 2012.
31. D. K. Houston and coauthors, Dietary fat and cholesterol and risk of cardiovascular disease in older adults: The Health ABC study, *Nutrition, Metabolism & Cardiovascular Diseases* 21 (2011): 430–437; Spence, Jenkins, and Davignon, 2010.
32. J. D. Spence, D. J. A. Jenkins, and J. Davignon, Egg yolk consumption and carotid plaque, *Atherosclerosis* 224 (2012): 469–473.
33. J. Plat and coauthors, Progress and prospective of plant sterol and plant stanol research: Report of the Maastricht meeting, *Atherosclerosis* 225 (2012): 521–533.

34. D. M. Lloyd-Jones and coauthors, Defining and setting national goals for cardiovascular health promotion and disease reduction: The American Heart Association's Strategic Impact Goal through 2020 and beyond, *Circulation* 121 (2010): 586–613.
35. E. C. Rizos and coauthors, Association between omega-3 fatty acid supplementation and risk of major cardiovascular disease events, *Journal of the American Medical Association* 308 (2012): 1024–1033; S. M. Kwak and coauthors, Efficacy of omega-3 fatty acid supplements (eicosapentaenoic acid and docosahexaenoic acid) in the secondary prevention of cardiovascular disease, *Archives of Internal Medicine* 172 (2012): 686–694.
36. S. E. Brien and coauthors, Effect of alcohol consumption on biological markers associated with risk of coronary heart disease: Systematic review and meta-analysis of interventional studies, *British Medical Journal* 342 (2011): doi:10.1136/bmj.d636; P. E. Ronksley and coauthors, Association of alcohol consumption with selected cardiovascular disease outcomes: A systematic review and meta-analysis, *British Medical Journal* 342 (2011): doi:10.1136/bmj.d671.
37. Lloyd-Jones and coauthors, 2010.
38. Benowitz, 2008.
39. M. J. Klag, Epidemiology of cardiovascular disease, in L. Goldman and A. I. Schafer, eds., *Goldman's Cecil Medicine* (Philadelphia: Saunders, 2012), pp. 256–260.
40. Roger and coauthors, 2012.
41. A. M. Kanaya and C. Vaisse, Obesity, in D. G. Gardner and D. Shoback, eds., *Greenspan's Basic and Clinical Endocrinology* (New York: McGraw-Hill/Lange, 2011), pp. 699–709.
42. H. M. Seagle, H. R. Wyatt, and J. O. Hill, Obesity: Overview of treatments and interventions, in A. M. Coulston, C. J. Boushey, and M. G. Ferruzzi, eds., *Nutrition in the Prevention and Treatment of Disease* (London: Academic Press/Elsevier, 2013), pp. 445–464.
43. Miller and coauthors, 2011.
44. M. J. Malloy and J. P. Kane, Disorders of lipoprotein metabolism, in D. G. Gardner and D. Shoback, eds., *Greenspan's Basic and Clinical Endocrinology* (New York: McGraw-Hill/Lange, 2011), pp. 675–698.
45. L. Berglund and coauthors, Evaluation and treatment of hypertriglyceridemia, *Journal of Clinical Endocrinology and Metabolism* 97 (2012): 2969–2989; Miller and coauthors, 2011.
46. R. C. Oh and J. B. Lanier, Management of hypertriglyceridemia, *American Family Physician* 75 (2007): 1365–1371.
47. Berglund and coauthors, 2012.
48. Berglund and coauthors, 2012; Miller and coauthors, 2011.
49. M. J. Jardine and coauthors, The effect of folic acid based homocysteine lowering on cardiovascular events in people with kidney disease: Systematic review and meta-analysis, *British Medical Journal* 344 (2012): doi: http://dx.doi.org/10.1136/bmj.e3533; R. Clarke and coauthors, Homocysteine and vascular disease: Review of published results of the homocysteine-lowering trials, *Journal of Inherited Metabolic Disease* 34 (2011): 83–91.
50. S. D. Holligan and coauthors, Atherosclerotic cardiovascular disease, in J. W. Erdman, I. A. Macdonald, and S. H. Zeisel, eds., *Present Knowledge in Nutrition* (Ames, IA: Wiley-Blackwell, 2012), pp. 745–805.
51. M. J. Malloy and J. P. Kane, Agents used in dyslipidemia, in B. G. Katzung, ed., *Basic and Clinical Pharmacology* (New York: McGraw-Hill/Lange, 2012), pp. 619–633.
52. J. L. Anderson, ST segment elevation acute myocardial infarction and complications of myocardial infarction, in L. Goldman and A. I. Schafer, eds., *Goldman's Cecil Medicine* (Philadelphia: Saunders, 2012), pp. 434–448.
53. Roger and coauthors, 2012.
54. Roger and coauthors, 2012.
55. Roger and coauthors, 2012.
56. J. A. Zivin, Ischemic cerebrovascular disease, in L. Goldman and A. I. Schafer, eds., *Goldman's Cecil Medicine* (Philadelphia: Saunders, 2012), pp. 2310–2320.
57. Zivin, 2012.
58. Roger and coauthors, 2012.
59. Roger and coauthors, 2012.
60. R. G. Victor, Arterial hypertension, in L. Goldman and A. I. Schafer, eds., *Goldman's Cecil Medicine* (Philadelphia: Saunders, 2012), pp. 373–389.
61. G. L. Bakris, Arterial hypertension, in R. S. Porter and J. L. Kaplan, eds., *Merck Manual of Diagnosis and Therapy* (Whitehouse Station, NJ: Merck Sharp and Dohme Corp., 2011), pp. 2065–2081.
62. Bakris, 2011.
63. Roger and coauthors, 2012; C. D. Fryar and coauthors, Trends in nutrient intakes and chronic health conditions among Mexican-American adults, a 25-year profile: United States, 1982–2006. *National Health Statistics Reports* 50 (Mar. 28, 2012).
64. J. A. N. Dorresteijn, F. L. J. Visseren, and W. Spiering, Mechanisms linking obesity to hypertension, *Obesity Reviews* 13 (2012): 17–26.
65. V. Savica, G. Bellinghieri, and J. D. Kopple, The effect of nutrition on blood pressure, *Annual Review of Nutrition* 30 (2010): 365–401.
66. P. K. Whelton and coauthors, Sodium, blood pressure, and cardiovascular disease: Further evidence supporting the American Heart Association sodium reduction recommendations, *Circulation* 126 (2012): 2880–2889; Savica, Bellinghieri, and Kopple, 2010.
67. Victor, 2012; G. Soardo and coauthors, Effects of alcohol withdrawal on blood pressure in hypertensive heavy drinkers, *Journal of Hypertension* 24 (2006): 1493–1498.
68. Whelton and coauthors, 2012; National High Blood Pressure Education Program/National Institutes of Health, *The Seventh Report of the Joint National Committee on Prevention, Detection, Evaluation, and Treatment of High Blood Pressure (JNC 7)*, NIH publication no. 03-5233 (Bethesda, MD: National Heart, Lung, and Blood Institute, 2003).
69. L. H. Kuller, Weight loss and reduction of blood pressure and hypertension, *Hypertension* 54 (2009): 700–701; L. Aucott and coauthors, Long-term weight loss from lifestyle intervention benefits blood pressure?: A systematic review, *Hypertension* 54 (2009): 756–762.
70. Savica, Bellinghieri, and Kopple, 2010; Kuller, 2009; Aucott and coauthors, 2009.
71. F. M. Sacks and coauthors, Effects on blood pressure of reduced dietary sodium and the Dietary Approaches to Stop Hypertension (DASH) diet, *New England Journal of Medicine* 344 (2001): 3–10; L. J. Appel and coauthors, A clinical trial on the effects of dietary patterns on blood pressure, *New England Journal of Medicine* 336 (1997): 1117–1124.
72. Appel and coauthors, 1997.
73. U.S. Department of Agriculture and U.S. Department of Health and Human Services, 2010.
74. Centers for Disease Control and Prevention, Vital Signs: Food categories contributing the most to sodium consumption—United States, 2007–2008, *Morbidity and Mortality Weekly Report* 61 (2012): 1–7.
75. B. M. Massie, Heart failure: Pathophysiology and diagnosis, in L. Goldman and A. I. Schafer, eds., *Goldman's Cecil Medicine* (Philadelphia: Saunders, 2012), pp. 295–303.
76. Massie, 2012.
77. B. G. Katzung, Drugs used in heart failure, in B. G. Katzung, ed., *Basic and Clinical Pharmacology* (New York: McGraw-Hill/Lange, 2012), pp. 211–225.
78. Academy of Nutrition and Dietetics, *Nutrition Care Manual* (Chicago: Academy of Nutrition and Dietetics, 2012); J. J. V. McMurray and M. A. Pfeffer, Heart failure: Management and prognosis, in L. Goldman and A. I. Schafer, eds., *Goldman's Cecil Medicine* (Philadelphia: Saunders, 2012), pp. 303–318.
79. J. Lindenfeld and coauthors, HFSA 1010 comprehensive heart failure practice guideline, *Journal of Cardiac Failure* 16 (2010): 475–539.

Nutrition in Practice

Helping People with Feeding Disabilities

Chapter 21 referred to difficulties following a stroke that can interfere with the ability to eat independently. This Nutrition in Practice discusses the problems faced by individuals who must cope with various disabilities that interfere with the process of eating, including those that interfere with chewing, swallowing, or bringing food to the mouth. These obstacles can arise at any time during a person's life and from any number of causes. An infant may be born with a physical impairment such as cleft palate; an adolescent may lose motor control following injuries sustained in an automobile accident; an older adult may struggle with the pain of arthritis or the mental deterioration of dementia. Table NP21-1 lists some of the conditions that may lead to feeding problems.

In what ways can disabilities impair a person's ability to eat?

Eating and drinking require a considerable number of individual coordinated motions. Consider an infant learning the skills required for feeding: each step—sitting, grasping cups and utensils, bringing food to the mouth, biting, chewing, and swallowing—requires coordinated movements. An injury or disability that interferes with any of these movements can lead to feeding problems and inadequate food intake. Total food intake is often significantly reduced when individuals with inefficient motor function take a long time to eat.[1] Difficulties that affect procurement of food, such as the inability to drive, walk, or carry groceries, can also lower food intake and lead to malnutrition and weight loss.

TABLE NP21-1 Conditions That May Lead to Feeding Problems

The following conditions may lead to feeding problems by interfering with a person's ability to suck, bite, chew, swallow, or coordinate hand-to-mouth movements.

- Accidents
- Amputations
- Arthritis
- Birth defects
- Brain tumors
- Cerebral palsy
- Cleft palate
- Down syndrome
- Head injuries
- Huntington's chorea
- Language, visual, or hearing impairments
- Multiple sclerosis
- Muscle weakness
- Muscular dystrophy
- Parkinson's disease
- Polio
- Spinal cord injuries
- Stroke

Can disabilities alter a person's energy needs?

Yes, certain disabilities can either increase or decrease energy requirements.[2] Disabilities that affect muscle tension or mobility can reduce physical activity and, consequently, energy requirements. Other disabilities, such as certain forms of cerebral palsy, cause involuntary muscle activity that raises energy requirements. Loss of a limb due to amputation reduces energy needs in proportion to the weight and metabolism represented by the missing limb, but energy needs may be greater if an individual increases activity to compensate for the loss, such as by propelling a wheelchair. Because the effects of disabilities are often unpredictable, the health care practitioner may find it difficult to assess energy requirements until weight gain or loss has occurred.

Overweight and obesity often accompany conditions that limit mobility or result in short stature; examples include Down syndrome and spina bifida. Obesity may also develop because the family or caregiver provides an inappropriate amount of food, sometimes out of sympathy for the disabled individual. In these cases, the health practitioner may need to counsel the family or caregiver about appropriate food choices and portion sizes.

Which health professionals typically work with people who have feeding problems?

Evaluating and treating feeding problems often involve the joint efforts of health care professionals from a variety of disciplines, including nurses, dietitians, occupational and physical therapists, speech-language pathologists, and dentists. Together, these professionals evaluate each patient's dietary needs and assess abilities to chew, sip, swallow, grasp utensils, use utensils to pick up foods, and bring foods from the plate to the mouth. A speech-language pathologist most often evaluates chewing and swallowing abilities and trains patients to use lips, tongue, and throat for eating and speaking. An occupational therapist

Adaptive feeding equipment can help patients with feeding disabilities gain independence.

can demonstrate alternative feeding strategies, including changes in body position that improve feeding, techniques for handling utensils and food, and use of special feeding devices.

Direct observation of a patient during mealtimes allows health care professionals to assess current eating behaviors, demonstrate feeding techniques, monitor the patient's and caregiver's understanding of the techniques, and evaluate how well the care plan is working. To illustrate, consider a child with a feeding problem caused by hypersensitivity to oral stimulation. The health care professional may start by teaching the caregiver to gently stroke the child's face with a hand, washcloth, or soft toy. Once the child tolerates touch on less sensitive areas of the face, the health care professional may encourage the caregiver to slowly begin to rub the child's lips, gums, palate, and tongue. With time, the child may be better able to tolerate the presence of food in the mouth. Examples of other strategies that can help feeding problems are listed in Table NP21-2.

TABLE NP21-2 Interventions for Feeding-Related Problems

Inability to Suck

- Use squeeze bottles, which do not require sucking, to express liquids into the mouth.
- Place a spoon on the center of the tongue, and apply downward pressure to stimulate sucking.
- Apply rhythmic, slow strokes on the tongue to alter tongue position and improve the sucking response.

Inability to Chew

- Place foods between teeth to promote chewing.
- Improve chewing skills with foods of different textures; for example, fruit leathers stimulate jaw movements but dissolve quickly enough to minimize choking.
- Provide soft foods that require minimal chewing or are easily chewed.

Inability to Swallow

- Provide thickened liquids, pureed foods, and moist foods that form boluses easily.
- Provide cold formulas, frozen fruit juice bars, and ice; cold substances promote swallowing movements by the tongue and soft palate.
- Make sure the patient's jaw and lips are closed to facilitate swallowing action.
- Correct posture and head position if they interfere with swallowing ability.

Inability to Grasp or Coordinate Movements

- Provide utensils that have modified handles, or are smaller or larger as necessary.
- Encourage the use of hands for feeding if utensils are difficult to maneuver.
- Provide plates with food guards to prevent spilling.
- Supply clothing protection.

Impaired Vision

- Place foods (meats, vegetables) in similar locations on the plate at each meal.
- Provide plates with food guards to prevent spilling.

© Cengage Learning

Can special equipment be used to help people with certain feeding difficulties?

Yes. Adaptive feeding devices can make a remarkable difference in a person's ability to eat independently. Figure NP21-1 shows a few of the many special feeding devices that are available and describes their uses. Other examples of adaptive equipment include specialized chairs to improve posture, bolsters inserted under arms to improve elbow stability, and raised trays or eating surfaces to simplify hand-to-mouth movements.

Sometimes, despite the best efforts of all involved, a patient is unable to consume enough food by mouth. In these cases, tube feedings can help to improve nutrition status. Tube feedings are typically recommended for patients with severe dysphagia (difficulty swallowing), aspiration pneumonia, recurrent malnutrition, or failure to thrive.[3]

In what ways can feeding difficulties affect family life?

Mealtimes are a critical time for social interaction, and therefore individuals with feeding problems may encounter emotional and social problems if they are unable to participate. Children may fail to develop social skills, whereas adults may miss the social stimulation that mealtimes provide. Individuals should be encouraged to sit with family and friends during meals so that they are not deprived of the social and cultural aspects of eating.

The responsibility of caring for a person with a feeding problem can frequently overwhelm a caregiver.[4] Caring for

FIGURE NP21-1 Examples of Adaptive Feeding Devices

Utensils

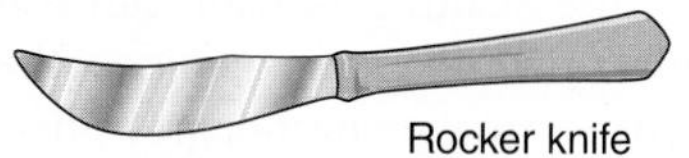

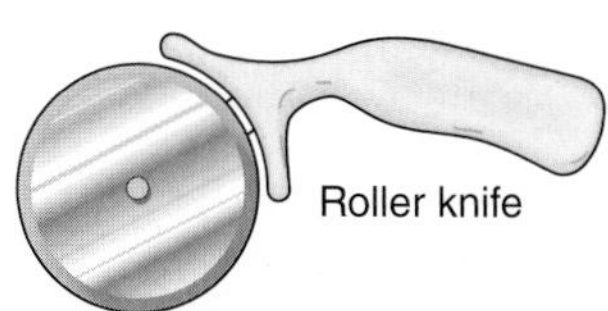

People with only one arm or hand may have difficulty cutting foods and may appreciate using a *rocker knife* or a *roller knife.*

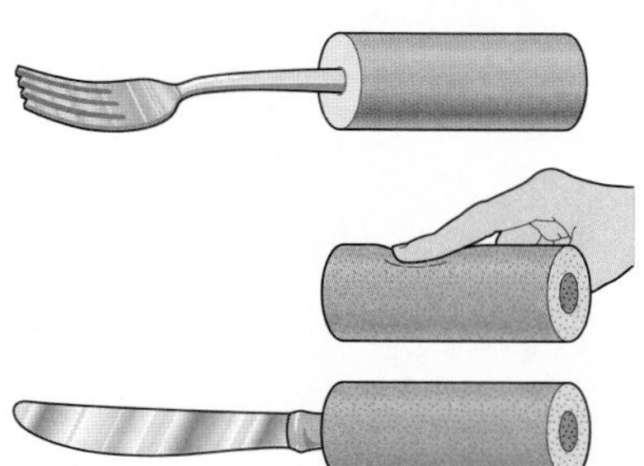

People with a limited range of motion can feed themselves better when they use *flatware with built-up handles.*

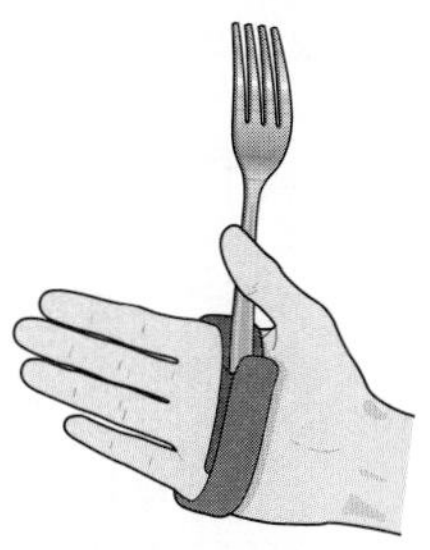

People with extreme muscle weakness may be able to eat with a *utensil holder.*

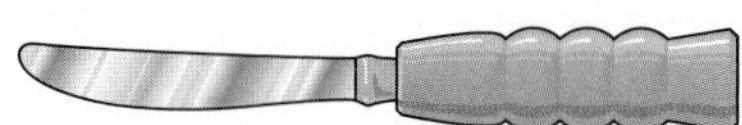

For people with tremors, spasticity, and uneven jerky movements, *weighted utensils* can aid the feeding process.

Battery-powered feeding machines enable people with severe limitations to eat with less assistance from others.

Plates

People who have limited dexterity and difficulty maneuvering food find *scoop dishes* or *food guards* useful.

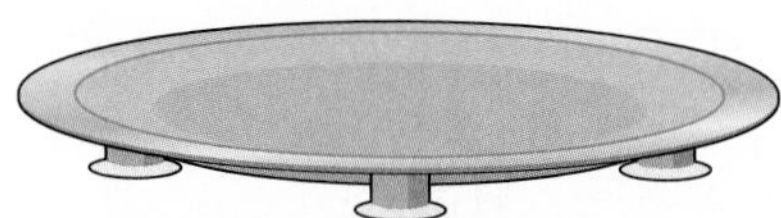

People with uncontrolled or excessive movements might move dishes around while eating and may benefit from using *unbreakable dishes with suction cups.*

Cups

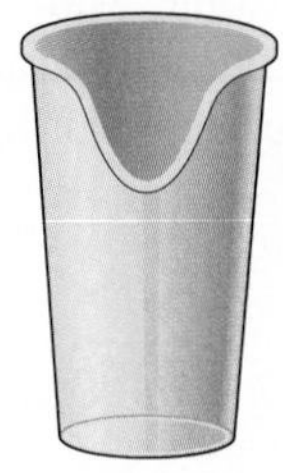

People with limited neck motion can use a *cutout plastic cup*.

Two-handed cups enable people with moderate muscle weakness to lift a cup with two hands.

People with uncontrolled or excessive movements might prefer to drink liquids from a *covered cup* or glass with a *slotted opening* or *spout*.

A soft, flexible long plastic straw may also ease the task of drinking.

a person with disabilities requires time and patience—and many new therapies to be learned and administered. The caregiver may spend many hours preparing special foods, monitoring the use of adaptive feeding equipment, and helping with feedings. Moreover, a person with disabilities may need help with other tasks as well, and all may require a considerable amount of time. In many cases, a caregiver receives little or no assistance. These conditions may lead to strained interactions between caregiver and patient and cause stress and frustration. Psychologists can offer counseling to patients or caregivers to help them adjust, and all members of the health care team can offer emotional support and practical suggestions to ease caregivers' responsibilities and frustrations.

Successful therapy for people with feeding disabilities requires the involvement of many health care professionals and depends on accurate identification of impaired feeding skills and determination of appropriate interventions. Ideally, with training, people with disabilities attain total independence—they are able to prepare, serve, and eat nutritionally adequate food daily without help. In some cases, these goals can be met with the help of caregivers. The combined efforts of the health care team can support both patients and caregivers in enhancing quality of life and in achieving independence to the greatest degree possible.

Notes

1. Position of the Canadian Paediatric Society: Nutrition in neurologically impaired children, *Paediatrics and Child Health* 14 (2009): 395–401.
2. Position of the American Dietetic Association: Providing nutrition services for people with developmental disabilities and special health care needs, *Journal of the American Dietetic Association* 110 (2010): 296–307.
3. Position of the American Dietetic Association, 2010.
4. P. Raina and coauthors, The health and well-being of caregivers of children with cerebral palsy, *Pediatrics* 115 (2005): e626–e636.

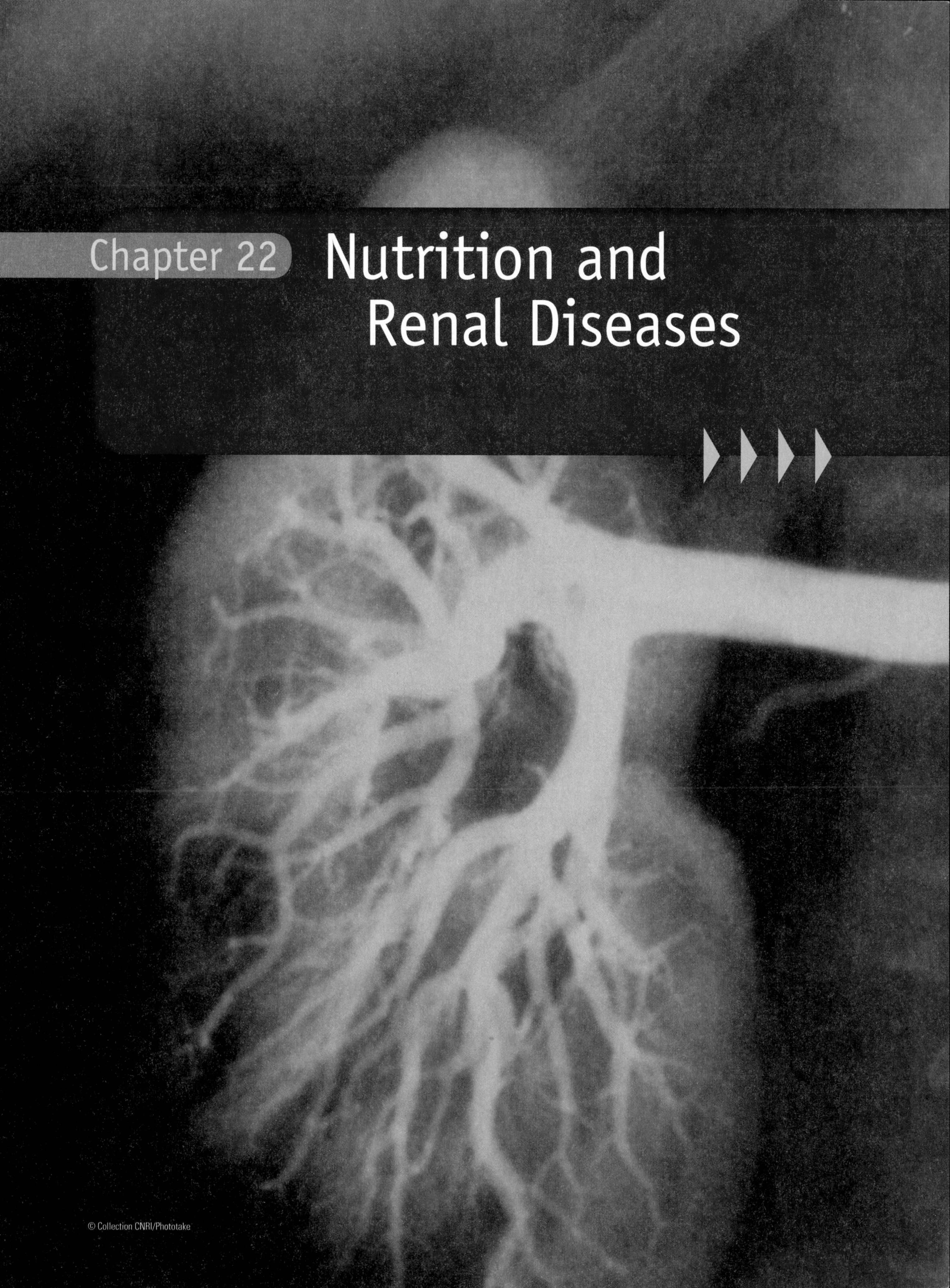

Chapter 22 Nutrition and Renal Diseases

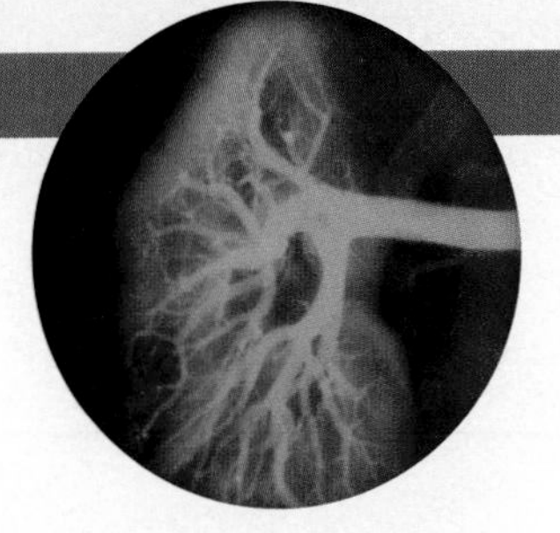

THE TWO KIDNEYS SIT JUST ABOVE THE WAIST ON EACH SIDE OF THE spinal column. As part of the urinary system, they are responsible for filtering the blood and removing excess fluid and wastes for elimination in urine. Because the kidneys are so proficient at this task, disturbances in body fluids that result from water intake, physical activity, and metabolism are normally corrected within hours.

Figure 22-1 shows the kidneys' placement and structure and one of their functional units, the **nephron**. Within each nephron, the **glomerulus**, a ball-shaped tuft of capillaries, serves as a gateway through which the blood components must pass to form **filtrate**. The glomerulus and surrounding **Bowman's capsule** function like a sieve, retaining blood cells and most plasma proteins in the blood while allowing fluid and small solutes to enter the nephron's system of **tubules**. As the filtrate passes through the tubules, its composition continuously changes as some of its components are reabsorbed and returned to the blood via capillaries surrounding each tubule. Eventually, the remaining filtrate enters a duct shared by several nephrons, and additional water is reabsorbed to form the final urine product. By filtering the blood and forming urine, the kidneys regulate the extracellular fluid volume and osmolarity, electrolyte concentrations, and acid-base balance. They also excrete metabolic waste products such as urea and creatinine, as well as various drugs and toxins. Other roles of the kidneys include the following:

- Secretion of the enzyme *renin,* which helps to regulate blood pressure
- Production of the hormone **erythropoietin**, which stimulates the production of red blood cells in the bone marrow
- Conversion of vitamin D to its active form, thereby helping to regulate calcium balance and bone formation

Subsequent sections of this chapter explain how **renal** diseases can interfere with the kidneys' various functions and severely disrupt health.

nephron (NEF-ron): the functional unit of the kidneys, consisting of a glomerulus and tubules.

nephros = kidney

glomerulus (gloh-MEHR-yoo-lus): a tuft of capillaries within the nephron that filters water and solutes from the blood as urine production begins (plural: *glomeruli*).

filtrate: the substances that pass through the glomerulus and travel through the nephron's tubules, eventually forming urine.

Bowman's (BOE-minz) capsule: a cuplike component of the nephron that surrounds the glomerulus and collects the filtrate that is passed to the tubules.

FIGURE 22-1 The Kidneys and Nephron Function

Kidney
Ureter
Pelvis
Bladder
Kidney, sectioned to show location of nephrons
Renal artery
Renal vein
A nephron (a working unit of the kidney). Each kidney contains about one million nephrons.
Blood vessel
Glomerulus
Capillaries of glomerulus
Tubule
To the body
To the bladder
1 Blood flows into the glomerulus, and some of its fluid, with dissolved substances, is absorbed into the tubule.
2 Then the fluid and substances needed by the body are returned to the blood in vessels alongside the tubule.
3 The tubule passes waste materials on to the bladder.

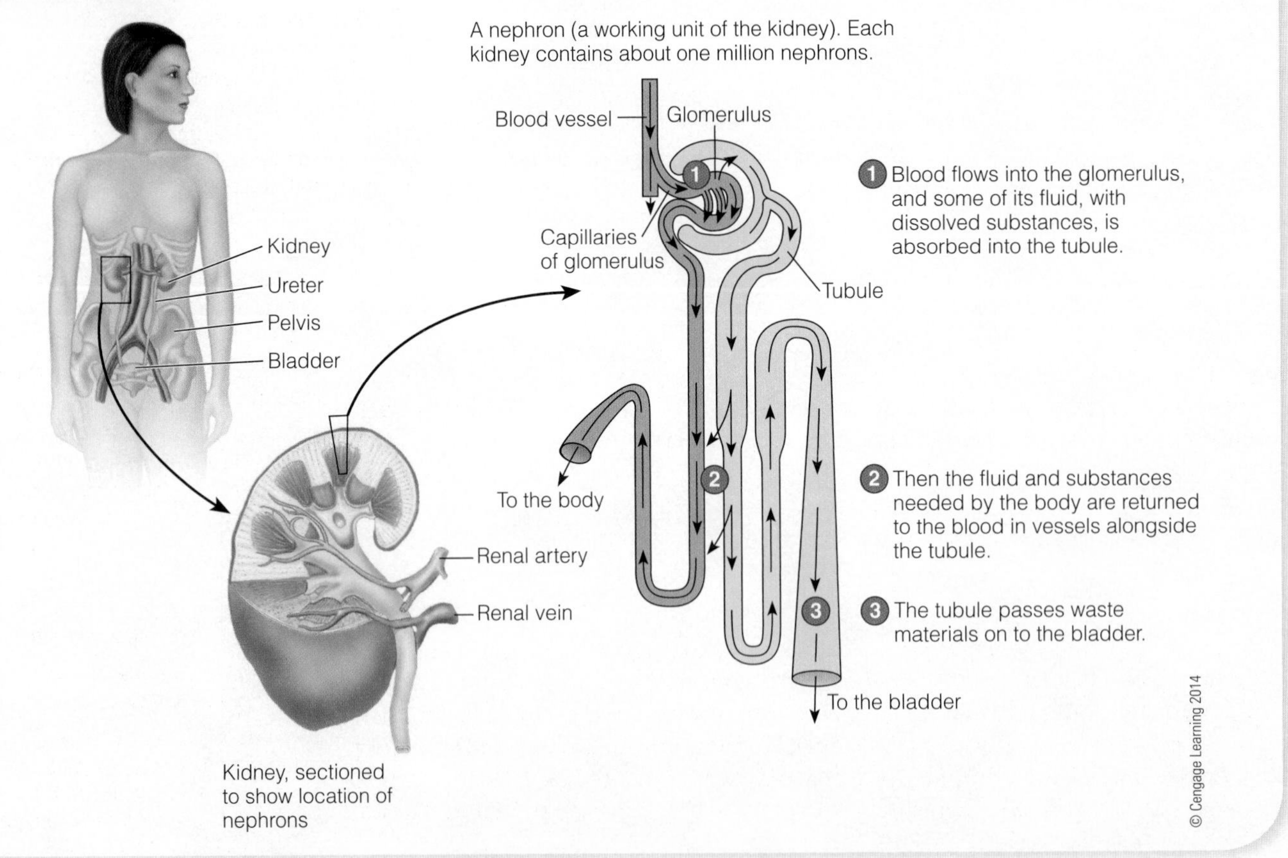

The Nephrotic Syndrome

The **nephrotic syndrome** is not a specific disease; rather, the term refers to a syndrome caused by significant urinary protein losses (**proteinuria**) that result from severe glomerular damage. The condition arises because damage to the glomeruli increases their permeability to plasma proteins, allowing the proteins to escape into the urine. The loss of plasma proteins (typically more than 3½ grams per day) causes serious consequences, including edema, blood lipid abnormalities, blood coagulation disorders, and infections. In some cases, the nephrotic syndrome can progress to renal failure.

Causes of the nephrotic syndrome include glomerular disorders, diabetic nephropathy, immunological and hereditary diseases, infections (involving the kidneys or elsewhere in the body), chemical damage (from medications or illicit drugs), and some cancers.[1] Depending on the underlying condition, some patients may experience one or more relapses and require additional treatment to prevent proteinuria from recurring.

CONSEQUENCES OF THE NEPHROTIC SYNDROME

In the nephrotic syndrome, urinary protein losses generally average about 8 grams daily.[2] The liver attempts to compensate for these losses by increasing its synthesis of various plasma proteins, but some of the proteins are produced in excessive amounts. The imbalance in plasma protein concentrations contributes to a number of complications.

NURSING DIAGNOSIS

The nursing diagnoses excess *fluid volume*, *risk for infection*, and *imbalanced nutrition: less than body requirements* may apply to people with the nephrotic syndrome.

Edema

Albumin is the most abundant plasma protein, and it is the protein with the most significant urinary losses as well. The **hypoalbuminemia** characteristic of the nephrotic syndrome contributes to a fluid shift from blood plasma to the interstitial spaces and, thus, edema. (■) Impaired sodium excretion also contributes to edema: the nephrotic kidney tends to reabsorb sodium in greater amounts than usual, causing sodium and water retention within the body.[3]

■ Reminder: Plasma proteins, such as albumin, help to maintain fluid balance within the blood.

Blood Lipid and Blood Clotting Abnormalities

Individuals with the nephrotic syndrome frequently have elevated levels of low-density lipoproteins (LDL), very-low-density lipoproteins (VLDL), and the more atherogenic LDL variant known as lipoprotein(a). Furthermore, the risk of blood clotting is increased due to urinary losses of proteins that inhibit blood clotting and elevated levels of plasma proteins that favor clotting. The blood clotting abnormalities increase the risk of **deep vein thrombosis** and similar disorders.

Other Effects of the Nephrotic Syndrome

The proteins lost in urine include immunoglobulins (antibodies) and vitamin D–binding protein. Depletion of immunoglobulins increases susceptibility to infection. Loss of vitamin D–binding protein results in lower vitamin D and calcium levels and increases the risk of rickets in children. Patients with the nephrotic syndrome frequently develop protein-energy malnutrition (PEM) and muscle wasting from the continued proteinuria. Figure 22-2 (p. 626) summarizes the effects of urinary protein losses in the nephrotic syndrome.

TREATMENT OF THE NEPHROTIC SYNDROME

Medical treatment of the nephrotic syndrome requires diagnosis and management of the underlying disorder responsible for the proteinuria. Complications are managed with medications and nutrition therapy. The drugs prescribed may include diuretics, ACE inhibitors (which reduce protein losses), lipid-lowering drugs, anti-inflammatory drugs (usually corticosteroids, such as prednisone), and immunosuppressants (such as cyclosporine).[4] Nutrition therapy can help to prevent PEM, correct lipid abnormalities, and alleviate edema.

tubules: tubelike structures of the nephron that process filtrate during urine production. The tubules are surrounded by capillaries that reabsorb substances retained by tubule cells.

erythropoietin (eh-RITH-ro-POY-eh-tin): a hormone made by the kidneys that stimulates red blood cell production.

renal (REE-nal): pertaining to the kidneys.

nephrotic (neh-FROT-ik) syndrome: a syndrome associated with disorders that cause severe glomerular damage, resulting in significant urinary protein losses.

proteinuria (PRO-teen-NOO-ree-ah): loss of protein, mostly albumin, in the urine; also known as *albuminuria*.

hypoalbuminemia: low plasma albumin concentrations.

deep vein thrombosis: formation of a stationary blood clot (thrombus) in a deep vein, usually in the leg, which causes inflammation, pain, and swelling and is potentially fatal.

FIGURE 22-2 Consequences of Urinary Protein Losses in the Nephrotic Syndrome

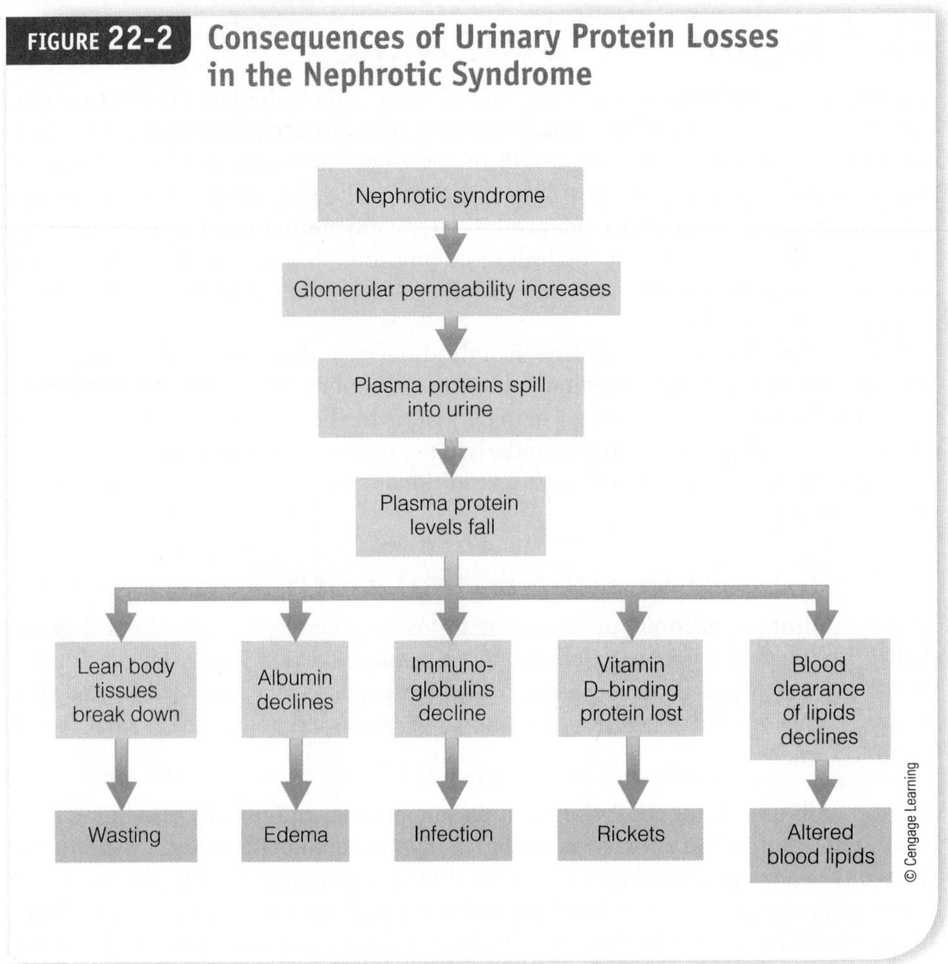

Protein and Energy Meeting protein and energy needs helps to minimize losses of muscle tissue. High-protein diets are not advised, however, because they can exacerbate urinary protein losses and result in further damage to the kidneys. Instead, the protein intake should fall between 0.8 and 1.0 gram per kilogram of body weight per day; (■) at least half of the protein consumed should come from high-quality sources, such as milk products, meat, fish, poultry, eggs, and soy products.[5] An adequate energy intake (about 35 kcalories per kilogram of body weight daily) sustains weight and spares protein. Weight loss or infections suggest the need for additional kcalories.

Lipids As Chapter 21 explains, a diet low in saturated fat, *trans* fats, cholesterol, and refined sugars helps to control elevated LDL and VLDL levels. Dietary measures are usually inadequate for controlling blood lipids, however, so physicians usually prescribe lipid-lowering medications as well. In some cases, treating the underlying cause of nephrotic syndrome is sufficient for correcting the lipid disorders.[6]

Sodium and Potassium A low-sodium diet can help to control edema; therefore, sodium intake is often limited to 1 to 2 grams daily.[7] Table 22-1 provides guidelines for following a diet restricted to 2 grams of sodium. If diuretics prescribed for the edema cause potassium losses, patients are encouraged to select foods rich in potassium (see Chapter 9).

Vitamins and Minerals Multivitamin/mineral supplementation can help patients avoid nutrient deficiencies; nutrients at risk include vitamin B_6, vitamin B_{12}, folate, iron, copper, and zinc.[8] (■) To reduce risk of bone loss and rickets, calcium (about 1000 to 1500 milligrams per day) and vitamin D supplements are also advised.

■ The protein RDA for adults is 0.8 g/kg body weight.

■ Nutrient deficiencies may develop if the carrier proteins for nutrients are lost in the urine.

TABLE 22-1 Low-Sodium Diet

General Guidelines

About 75 percent of the sodium in a typical diet comes from processed foods, about 10 percent from unprocessed natural foods, and about 15 percent from table salt. With this in mind:

- Whenever possible, select fresh foods, which are usually low in sodium.
- Select frozen and canned food products that have been prepared without added salt.
- Avoid adding salt to foods while cooking or at the table.
- When dining in restaurants, ask that meals be prepared without salt.

Sodium in Foods

All foods contain sodium, but some contain more than others. Use the information in the table below to plan meals that are low in sodium.

Food Group	Serving Size	Sodium per Serving (mg)
Milk products	1 cup milk or yogurt; 1 oz hard cheese (cheddar, Swiss, jack)	150–200
	Avoid: buttermilk, cottage cheese, cheese spreads, prepared cheeses (such as American cheese)	
Meat, fish, poultry, and eggs	3 oz fresh meat, fish, or poultry; 1 large egg	60
	Avoid: luncheon meats, corned beef, salt pork, sausage, frankfurters, bacon, canned meats or fish, fresh meats prepared with injected broth	
Fruits and vegetables	½ cup fresh vegetables, ½ cup fresh or frozen fruit, 6 oz fruit juice; 6 oz tomato or vegetable juice without added salt	10–20
	Avoid: pickled vegetables, olives, tomato or vegetable juices with added salt; dried fruits with added sodium sulfite	
Breads and cereals	½–⅔ cup dry or cooked cereal without added salt, ½ cup cooked rice or pasta	0–10
	1 slice bread, 1 roll or tortilla, ½–⅔ cup dry or cooked cereal prepared with salt	150
	Avoid: pancakes, waffles, muffins, biscuits, and quick breads made with baking powder or baking soda; instant and ready-to-eat cereals with >175 mg sodium; salted snack foods	
Condiments	Unsalted butter; low-sodium salad dressings, mayonnaise, sauces, and gravies; low-sodium catsup, mustard, and hot sauces; garlic or onion powders without added salt; lemon or lime juice; vinegar	Varies; check labels
	Avoid: commercial salad dressings, gravy or soup mixes, barbeque sauces, soy sauce, steak sauces, spices or herb products made with salt, bouillon, meat tenderizers, monosodium glutamate	

A Sample Diet Restricted to 2 Grams (2000 mg) of Sodium

Using the guidelines presented here, an individual can develop a variety of sample menus. A possible plan for a day might look like this:

Food Group	Sodium (mg)
Meat, 6 oz (2 servings × 60 mg)	120
Milk, 3 c (3 servings × 150 mg)	450
Fruit, 2 servings	negligible
Vegetables, 3 servings	45
Whole-grain bread, 4 slices (4 × 150 mg)	600
Salt, ¼ tsp (used lightly at meals)	600
Total	**1815**

Individuals can use the remainder of the sodium allowance for whatever foods they desire. The sodium content of most foods can be determined by reading food labels or using food composition tables (such as Appendix A). See additional information about reducing sodium intake in Chapter 21 (p. 612).

Review Notes

- The nephrotic syndrome is characterized by urinary protein losses due to glomerular damage. Complications include edema, lipid and blood clotting abnormalities, infections, and PEM.
- Medications treat the underlying cause of proteinuria and manage complications. The diet should provide sufficient protein and energy to maintain health, but patients should avoid consuming excess protein. Other dietary adjustments can help to correct edema, lipid disorders, and nutrient deficiencies.

Acute Kidney Injury

In **acute kidney injury**, kidney function deteriorates rapidly, over hours or days. The loss of kidney function reduces urine output and allows nitrogenous wastes to build up in blood. The degree of renal dysfunction varies from mild to severe. With prompt treatment, acute kidney injury is often reversible, although mortality rates are high, ranging from 30 to 80 percent when kidney damage is involved.[9] Most cases of acute kidney injury develop in the hospital, occurring in 5 to 7 percent of hospitalized patients.[10]

acute kidney injury: the rapid decline of kidney function over a period of hours or days; potentially a cause of acute renal failure.

oliguria (OL-lih-GOO-ree-ah): an abnormally low amount of urine, often less than 400 mL/day.

anuria (ah-NOO-ree-ah): the absence of urine; clinically identified as urine output less than 50 mL/day.

CAUSES OF ACUTE KIDNEY INJURY

Many different disorders can lead to acute kidney injury, and it often develops as a consequence of severe illness, injury, or surgery. To aid in diagnosis and treatment, its causes are commonly classified as prerenal, intrarenal, or postrenal. *Prerenal* factors are those that cause a sudden reduction in blood flow to the kidneys; they often involve a severe stressor such as heart failure, shock, or blood loss. Factors that damage kidney tissue, such as infections, toxicants, drugs, or direct trauma, are classified as *intrarenal* causes of acute kidney injury. *Postrenal* factors are those that prevent excretion of urine due to urinary tract obstructions. Table 22-2 provides examples of specific disorders that may cause acute kidney injury.

NURSING DIAGNOSIS

Nursing diagnoses for people with acute kidney injury may include *impaired urinary elimination, excess fluid volume, risk for infection, decreased cardiac output*, and *fatigue*.

CONSEQUENCES OF ACUTE KIDNEY INJURY

A decline in renal function alters the composition of blood and urine. The kidneys become unable to regulate the levels of electrolytes, acid, and nitrogenous wastes in the blood. Urine may be diminished in quantity (**oliguria**) or absent (**anuria**), leading to fluid retention. Diagnosis is often a complex task because the clinical effects can be subtle and vary according to the underlying cause of disease.

TABLE 22-2 Causes of Acute Kidney Injury

Prerenal Factors (60 to 70% of cases)	Intrarenal Factors (25 to 40% of cases)	Postrenal Factors (5 to 10% of cases)
• **Low blood volume or pressure:** hemorrhage, burns, sepsis or shock, anaphylactic reactions, nephrotic syndrome, gastrointestinal losses, diuretics, antihypertensive medications • **Renal artery disorders:** blood clots or emboli, stenosis, aneurysm, trauma • **Heart disorders:** heart failure, heart attack, arrhythmias	• **Vascular disorders:** sickle-cell disease, diabetes mellitus, transfusion reactions • **Obstructions (within kidney):** inflammation, tumors, stones, scar tissue • **Renal injury:** infections, environmental contaminants, drugs, medications, *E. coli* food poisoning	• **Obstructions (ureter or bladder):** strictures, tumors, stones, trauma • **Prostate disorders:** cancer or hyperplasia • **Renal vein thrombosis** • **Bladder disorders:** neurological conditions, bladder rupture • **Pregnancy**

Fluid and Electrolyte Imbalances About half of patients with acute kidney injury experience oliguria, producing less than 400 milliliters of urine per day (normal urine volume is 1000 to 1500 milliliters daily).[11] The reduced excretion of fluids and electrolytes results in sodium retention and elevated levels of potassium, phosphate, and magnesium in the blood. Elevated potassium levels (**hyperkalemia**) are of particular concern because potassium imbalances can alter heart rhythm and lead to heart failure. Elevated serum phosphate levels (**hyperphosphatemia**) (■) promote excessive secretion of parathyroid hormone, which leads to losses of bone calcium. Due to the sodium retention and reduced urine production, edema is a common symptom of acute kidney injury and may be apparent as puffiness in the face and hands and swelling of the feet and ankles.

Uremia As a result of impaired kidney function, nitrogen-containing compounds and various other waste products may accumulate in the blood—a condition often referred to as **uremia**. The clinical outcome, called the *uremic syndrome,* includes a cluster of disorders caused by impairments in multiple body systems. Although the clinical effects vary among patients, complications may include hormonal imbalances, electrolyte and acid-base imbalances, disturbed heart and gastrointestinal functioning, neuromuscular disturbances, and depressed immunity, among other abnormalities. The uremic syndrome is described in more detail later in this chapter.

TREATMENT OF ACUTE KIDNEY INJURY

Treatment of acute kidney injury involves a combination of drug therapy, **dialysis**, (■) and nutrition therapy to restore fluid and electrolyte balances and minimize blood concentrations of toxic waste products. Both medical care and dietary measures are highly individualized to suit each patient's needs. Correcting the underlying illness is necessary to prevent further damage to the kidneys.

In oliguric patients (those with reduced urine production), recovery from kidney injury sometimes begins with a period of **diuresis**, in which large amounts of fluid (up to 3 liters daily) are excreted.[12] Because tubular function is minimal at this stage, electrolytes may not be sufficiently reabsorbed; consequently, both fluid depletion and electrolyte losses become a concern. Patients with this pattern of recovery (generally those with tubular injury) require close monitoring in case they need fluid and electrolyte replacement.

Drug Treatment in Acute Kidney Injury Because kidney function is required for drug excretion, patients may need to use lower doses of their usual medications to compensate for limited urine output. Conversely, dialysis treatment may increase losses of some drugs, and doses may need to be increased. Drugs that are **nephrotoxic** (including some antibiotics and nonsteroidal anti-inflammatory drugs) must be avoided until kidney function improves.

Medications prescribed for acute kidney injury depend on the cause of illness and complications that develop. Inflammatory conditions may require treatment with immunosuppressants. Edema is treated with diuretics; furosemide (Lasix) is the usual choice. Patients with hyperkalemia may be given potassium-exchange resins that bind potassium in the gastrointestinal tract and reduce its absorption. Rapid correction of hyperkalemia may require the use of insulin, which drives extracellular potassium into cells; glucose must be coadministered to prevent hypoglycemia. To reduce serum phosphorus levels, phosphate binders may be provided with meals to prevent phosphorus absorption. If acidosis is present, bicarbonate may be administered orally or intravenously. (■)

Energy and Protein Acute kidney injury is often associated with other critical illnesses, resulting in hypermetabolism and a prolonged state of catabolism and muscle wasting. Therefore, sufficient energy and protein must be ingested to preserve muscle

■ To measure serum phosphate levels, the phosphorus content of the blood is analyzed; thus, the terms *serum phosphate* and *serum phosphorus* are often used interchangeably.

■ Nutrition in Practice 22 describes common dialysis procedures, including *continuous renal replacement therapy,* the approach usually used for treating acute kidney injury.

■ Reminder: Bicarbonate corrects acidosis because the acid (H^+) combines with bicarbonate (HCO_3^-) to form carbonic acid (H_2CO_3), which breaks down to water (H_2O) and carbon dioxide (CO_2); the carbon dioxide is then exhaled.

hyperkalemia (HIGH-per-ka-LEE-me-ah): elevated serum potassium levels.

hyperphosphatemia (HIGH-per-fos-fa-TEE-me-ah): elevated serum phosphate levels.

uremia (you-REE-me-ah): the accumulation of nitrogenous and various other waste products in the blood (literally, "urine in the blood"); often associated with disorders that reflect impairments in multiple body systems. The term may also be used to indicate the toxic state that results when wastes are retained in the blood.

dialysis (dye-AH-lih-sis): a treatment that removes wastes and excess fluid from the blood after the kidneys have stopped functioning.

diuresis (DYE-uh-REE-sis): increased urine production.

nephrotoxic: toxic to the kidneys.

mass. Initially, the patient may be provided with 25 to 35 kcalories per kilogram of body weight per day, while body weight is monitored to ensure that energy intake is adequate.[13]

Protein contributes nitrogen, increasing the kidneys' workload, but intake should be sufficient to maintain nitrogen balance (to the extent possible) and prevent additional wasting. Protein recommendations are influenced by kidney function, the degree of catabolism, and the use of dialysis (dialysis removes nitrogenous wastes). For noncatabolic patients who are not treated with dialysis, protein intakes should be limited to 0.8 to 1.2 grams per kilogram body weight per day.[14] Higher intakes (1.2 to 1.5 grams per kilogram daily) may be recommended if kidney function improves, the patient is catabolic, or the treatment includes dialysis. Patients who must receive additional amounts of protein (such as those with burns or large wounds) require more frequent dialysis to accommodate the nitrogen load.[15]

Fluids Health practitioners can assess fluid status by monitoring weight fluctuations, blood pressure, pulse rates, and the appearance of the skin and mucous membranes. Another method is to measure serum sodium concentrations: a low level of sodium often indicates excessive fluid intake, and a high level of sodium suggests inadequate fluid intake.

Fluid balance must be restored in patients who are either overhydrated or dehydrated. Thereafter, fluid needs can be estimated by measuring urine output and adding about 500 milliliters to account for the water lost from skin, lungs, and perspiration. An individual with fever, vomiting, or diarrhea requires additional fluid. Patients undergoing dialysis can ingest fluids more freely.

Electrolytes Serum electrolyte levels are monitored closely to determine appropriate electrolyte intakes. Depending on the results of laboratory tests and the clinical assessment, restrictions may be necessary for potassium (2000 to 3000 milligrams per day), phosphorus (8 to 15 milligrams per kilogram body weight per day), and sodium (2000 to 3000 milligrams per day).[16] Patients undergoing dialysis may be allowed more liberal intakes. As mentioned previously, oliguric patients who experience diuresis at the beginning of the recovery period may need electrolyte replacement to compensate for urinary losses.

Enteral and Parenteral Nutrition Some patients need nutrition support to obtain adequate energy and nutrients. Enteral support (tube feeding) is preferred over parenteral nutrition because it is less likely to cause infection and sepsis. Enteral formulas for patients with acute kidney injury are more kcalorically dense and may have lower protein and electrolyte concentrations than standard formulas. Total parenteral nutrition is necessary only if patients are severely malnourished or cannot consume food or tolerate tube feedings for more than 14 days.[17]

Review Notes

- Acute kidney injury is characterized by a rapid decline in kidney function, causing a buildup of fluid, electrolytes, and nitrogenous wastes in the blood. Causes may involve prerenal, intrarenal, or postrenal factors.
- Acute kidney injury may cause fluid and electrolyte imbalances and uremia. If hyperkalemia develops, it can alter heart rhythm and lead to heart failure.
- Acute kidney injury is treated with medications, dialysis, and dietary modifications. The accompanying Case Study checks your understanding of acute kidney injury.

■ The kidneys' ability to function despite loss of nephrons is referred to as *renal reserve.*

chronic kidney disease: kidney disease characterized by gradual, irreversible deterioration of the kidneys; also known as *chronic renal failure.*

Chronic Kidney Disease

Unlike acute kidney injury, in which kidney function declines suddenly and rapidly, **chronic kidney disease** is characterized by gradual, irreversible deterioration. Because the kidneys have a large functional reserve, (■) the disease typically progresses over

Case Study: Woman with Acute Kidney Injury

Catherine Garber is a 42-year-old office manager admitted to the hospital's intensive care unit. She was first seen in the emergency room with severe edema, headache, nausea and vomiting, and a rapid heart rate. She reported an inability to pass more than minimal amounts of urine in the past two days. Her son, who drove her to the emergency room, reported that she had missed work for several days and seemed confused and unusually tired. Laboratory tests revealed elevated serum creatinine, BUN, and potassium levels. After learning from her medical history that Mrs. Garber had begun taking penicillin earlier in the week, the physician diagnosed acute kidney injury, probably caused by a reaction to the medication. Mrs. Garber is 5 feet 3 inches tall and weighs 125 pounds.

1. Describe the probable reason for Mrs. Garber's inability to produce urine. Is her reaction to penicillin considered a prerenal, intrarenal, or postrenal cause of kidney injury? Give examples of other medical problems that can cause acute kidney injury.
2. What medications can the physician prescribe to treat Mrs. Garber's edema and hyperkalemia? What recommendation is likely regarding her continued use of penicillin?
3. What concerns should be kept in mind when determining Mrs. Garber's energy, protein, fluid, and electrolyte needs during acute kidney injury? How would dialysis treatment alter recommendations?
4. After treatment begins, Mrs. Garber suddenly begins producing copious amounts of urine. How may this development alter dietary treatment?

many years without causing symptoms. Patients are typically diagnosed late in the course of illness, after most kidney function has been lost.[18]

The most common causes of chronic kidney disease are diabetes mellitus and hypertension, which are estimated to cause 45 and 27 percent of cases, respectively.[19] Other conditions that lead to chronic kidney disease include inflammatory, immunological, and hereditary diseases that directly involve the kidneys. Chronic kidney disease affects approximately 13 percent of the U.S. population.[20]

Reminder: *Creatinine* is a waste product of creatine, a nitrogen-containing compound in muscle cells.

Reminder: *Aldosterone* promotes sodium (and therefore water) retention and potassium excretion.

end-stage renal disease: an advanced stage of chronic kidney disease in which dialysis or a kidney transplant is necessary to sustain life.

CONSEQUENCES OF CHRONIC KIDNEY DISEASE

In the early stages of chronic kidney disease, the nephrons compensate by enlarging so that they can handle the extra workload. As the nephrons deteriorate, however, there is additional work for the remaining nephrons. The overburdened nephrons continue to degenerate until finally the kidneys are unable to function adequately, resulting in kidney failure. Once the extent of kidney damage necessitates active treatment—either dialysis or a kidney transplant—the condition is classified as **end-stage renal disease**. Without intervention at this stage, an individual cannot survive. Table 22-3 lists common clinical effects of the early and advanced stages of chronic kidney disease. Many symptoms of chronic kidney disease are nonspecific, which may delay diagnosis of the condition.

Renal disease is evaluated using the **glomerular filtration rate (GFR)**, the rate at which the kidneys form filtrate. The GFR can be estimated using predictive equations that are based on serum creatinine levels, (■) age, gender, race, and body size. Table 22-4 (p. 632) shows how chronic kidney disease is classified according to estimated GFR. Other laboratory measures used to assess kidney function include urinary protein levels, BUN, and the ratio of albumin to creatinine in a urine sample.

TABLE 22-3 Clinical Effects of Chronic Kidney Disease

Early Stages
- Anorexia
- Exercise intolerance
- Fatigue
- Headache
- Hypercoagulation
- Hypertension
- Proteinuria, hematuria (blood in urine)

Advanced Stages
- Anemia, bleeding tendency
- Cardiovascular disease
- Confusion, mental impairments
- Electrolyte imbalances
- Fluid retention, edema
- Hormonal abnormalities
- Itching
- Metabolic acidosis
- Nausea and vomiting
- Peripheral neuropathy
- Protein-energy malnutrition
- Reduced immunity
- Renal osteodystrophy

Altered Electrolytes and Hormones

As the GFR falls, the increased activity of the remaining nephrons is often sufficient to maintain electrolyte excretion. Therefore, fluid and electrolyte disturbances may not develop until the third or fourth stage of chronic kidney disease. A number of hormonal adaptations also help to regulate electrolyte levels, but these changes may cause complications of their own. The increased secretion of aldosterone (■) helps to prevent increases in serum potassium

TABLE 22-4 Evaluation of Chronic Kidney Disease

Stage of Disease	Description	GFR[a] (mL/min per 1.73 m²)
1	Kidney damage with normal or increased GFR	≥90
2	Kidney damage with mildly decreased GFR	60–89
3	Moderately decreased GFR	30–59
4	Severely decreased GFR	15–29
5	Kidney failure	<15 (or undergoing dialysis)

[a]Glomerular filtration rate, or GFR, is estimated from the Modification of Diet in Renal Disease study equation and is based on age, gender, race, and calibration for serum creatinine. Normal GFR is approximately 125 mL/min.

Source: A. S. Levey and coauthors, National Kidney Foundation practice guidelines for chronic kidney disease: Evaluation, classification, and stratification, *Annals of Internal Medicine* 139 (2003): 137–147.

■ Reminder: *Parathyroid hormone* helps to regulate serum concentrations of calcium and phosphorus. Elevated parathyroid hormone stimulates bone turnover and the release of calcium from bone into blood.

■ The *Subjective Global Assessment*, described in Chapter 13 (Table 13-2 on p. 383), is sometimes used to assess PEM risk in patients with chronic kidney disease.

NURSING DIAGNOSIS

Nursing diagnoses for people with chronic kidney disease may include *impaired urinary elimination, excess fluid volume, risk for infection, decreased cardiac output, fatigue,* and *imbalanced nutrition: less than body requirements.*

glomerular filtration rate (GFR): the rate at which filtrate is formed within the kidneys, normally about 125 mL/min.

renal osteodystrophy: a bone disorder that develops in patients with chronic kidney disease as a result of increased secretion of parathyroid hormone, reduced serum calcium, acidosis, and impaired vitamin D activation by the kidneys.

uremic syndrome: the cluster of disorders caused by inadequate kidney function; complications include fluid, electrolyte, and hormonal imbalances; altered heart function; neuromuscular disturbances; and other metabolic derangements.

but contributes to fluid overload and the development of hypertension (in patients who were not previously hypertensive). Increased secretion of parathyroid hormone (■) helps to prevent elevations in serum phosphorus but contributes to bone loss and the development of **renal osteodystrophy**, a bone disorder common in renal patients. Electrolyte imbalances are likely when GFR becomes extremely low (less than 5 milliliters per minute), when hormonal adaptations are inadequate, or when intakes of water and electrolytes are either very restricted or excessive.

Because the kidneys are responsible for maintaining acid-base balance, acidosis often develops in chronic kidney disease. Although usually mild, the acidosis exacerbates renal bone disease because compounds in bone (for example, protein and phosphates) are released to buffer the acid in blood.

Uremic Syndrome

Uremia usually develops during the final stages of chronic renal failure, when the GFR falls below about 15 milliliters per minute.[21] As mentioned previously, the numerous complications that result from uremia are collectively known as the **uremic syndrome**. Clinical effects may include the following:[22]

- *Hormonal imbalances.* Diseased kidneys are unable to produce erythropoietin, causing anemia. Reduced production of active vitamin D contributes to bone disease. Altered levels of various other hormones may upset growth, reproductive function (menstruation, sperm production), and blood glucose regulation.
- *Altered heart function/increased heart disease risk.* Fluid and electrolyte imbalances result in hypertension, arrhythmias, and eventual heart muscle enlargement. Excessive parathyroid hormone secretion leads to calcification of arteries and heart tissue. Patients with uremia are at increased risk of stroke, heart attack, and heart failure.
- *Neuromuscular disturbances.* Initial symptoms may be mild, and include malaise, irritability, and altered thought processes. Later effects include muscle cramping, restless leg syndrome, sensory deficits, tremor, and seizures.
- *Other effects.* Defects in platelet function and clotting factors prolong bleeding time and contribute to bruising, gastrointestinal bleeding, and anemia. Skin changes include increased pigmentation and severe pruritus (itchiness). Patients with uremia typically have suppressed immune responses and are at high risk of developing infections.

Protein-Energy Malnutrition

Patients with chronic kidney disease often develop PEM and wasting. (■) Anorexia is thought to contribute to the poor food intake of kidney patients and may result from hormonal disturbances, nausea and vomiting, restrictive diets, uremia, and medications. Nutrient losses also contribute to malnutrition and may be a consequence of vomiting, diarrhea, gastrointestinal bleeding, and dialysis. In addition, many of the illnesses that lead to chronic kidney disease induce a catabolic state that contributes to protein losses.

TREATMENT OF CHRONIC KIDNEY DISEASE

The goals of treatment for patients with chronic kidney disease are to slow disease progression and prevent or alleviate symptoms. Dietary measures help to prevent PEM and weight loss. Once kidney disease reaches the final stages, dialysis or a kidney transplant is necessary to sustain life.

Drug Therapy for Chronic Kidney Disease Medications help to control some of the complications associated with chronic kidney disease. Treatment of hypertension is critical for slowing disease progression and reducing cardiovascular disease risk; thus, antihypertensive drugs are usually prescribed (see Chapter 21). Some antihypertensive drugs (such as ACE inhibitors) can reduce proteinuria, helping to prevent additional kidney damage. Anemia is usually treated by injection or intravenous administration of erythropoietin (epoetin). Other common drug treatments include phosphate binders (taken with food) to reduce serum phosphorus levels, sodium bicarbonate to reverse acidosis, and cholesterol-lowering medications. Supplementation with active vitamin D (called *calcitriol)* helps to raise serum calcium and reduce parathyroid hormone levels.

Dialysis Dialysis replaces kidney function by removing excess fluid and wastes from the blood. In **hemodialysis**, the blood is circulated through a **dialyzer** (artificial kidney), where it is bathed by a **dialysate**, a solution that selectively removes fluid and wastes. In **peritoneal dialysis**, the dialysate is infused into a person's peritoneal cavity, and blood is filtered by the peritoneum (the membrane that surrounds the abdominal cavity). After several hours, the dialysate is drained, removing unneeded fluid and wastes. Nutrition in Practice 22 provides additional information about dialysis.

Nutrition Therapy for Chronic Kidney Disease The patient's diet strongly influences disease progression, the development of complications, and serum levels of nitrogenous wastes and electrolytes. Because the dietary measures for chronic kidney disease are complex and nutrient needs change frequently during the course of illness, a dietitian who specializes in renal disease is best suited to provide nutrition therapy. Table 22-5 (p. 634) summarizes the general dietary guidelines for patients in different stages of illness. Because patients' needs vary considerably, actual recommendations should be based on the results of a careful and complete nutrition assessment.

Energy The energy intake should be high enough to allow patients to maintain a healthy weight and to prevent wasting. Foods and beverages with high energy density (■) are typically recommended. Malnourished patients may require oral supplements or tube feedings to maintain weight. (See the "How to" on p. 657 in Chapter 23 for suggestions for increasing the energy content of meals.)

The dialysate used in peritoneal dialysis contains glucose in order to draw fluid from the blood to the peritoneal cavity by osmosis; about 40 to 50 percent of this glucose is absorbed.[23] The kcalories from glucose (as many as 800 kcalories daily) must be included in estimates of energy intake. Weight gain is sometimes a problem when peritoneal dialysis continues for a long period.[24]

Protein A low-protein diet may be prescribed to reduce the amount of nitrogenous waste produced. Furthermore, low-protein diets supply less phosphorus than high-protein diets, reducing the risks associated with hyperphosphatemia. Because renal patients often develop PEM, however, their diet must provide enough protein to meet needs and prevent wasting. During the later stages of kidney disease, recommended protein intakes typically fall between 0.6 and 0.8 grams per kilogram of body weight per day; the amount suggested is generally reduced as the disease progresses.[25] At least 50 percent of the protein consumed should come from high-quality protein sources (such as eggs, milk products, meat, poultry, fish, and soybeans) to ensure that the patient consumes adequate amounts of the essential amino acids. Low-protein breads, pastas, and other grain-based products are commercially available to help renal patients improve energy intakes without increasing protein consumption.

■ Reminder: Foods with high energy density contain a high number of kcalories per unit weight; these foods are generally high in fat and low in water content.

hemodialysis (HE-moe-dye-AL-ih-sis): a treatment that removes fluids and wastes from the blood by passing the blood through a dialyzer.

dialyzer (DYE-ah-LYE-zer): a machine used in hemodialysis to filter the blood; also called an *artificial kidney*.

dialysate (dye-AL-ih-sate): the solution used in dialysis to draw wastes and fluids from the blood.

peritoneal (PEH-rih-toe-NEE-al) dialysis: a treatment that removes fluids and wastes from the blood by using the body's peritoneal membrane as a filter.

TABLE 22-5 Dietary Guidelines for Chronic Kidney Disease[a]

Nutrient	Predialysis[b]	Hemodialysis	Peritoneal Dialysis
Energy (kcal/kg body weight)	35 for <60 years old 30–35 for ≥60 years old	35 for <60 years old 30–35 for ≥60 years old	35 for <60 years old 30–35 for ≥60 years old (total energy intake includes kcalories absorbed from the dialysate)
Protein (g/kg body weight)	GFR >50 mL/min: >0.8 GFR <50 mL/min: 0.6–0.8 (≥50% high-quality proteins)	≥1.2 (≥50% high-quality proteins)	≥1.2–1.3 (≥50% high-quality proteins)
Fat	As necessary to maintain a healthy lipid profile	As necessary to maintain a healthy lipid profile	As necessary to maintain a healthy lipid profile
Fluid (mL/day)	Unrestricted if urine output is normal	500–1000 plus daily urine output	As necessary to maintain fluid balance
Sodium (mg/day)	Varies; usually <2400	2000–3000	3000–4000
Potassium (mg/day)	GFR >60 mL/min: unrestricted GFR <60 mL/min: <2400	2000–3000; adjust according to serum potassium levels	3000–4000; adjust according to serum potassium levels
Phosphorus (mg/day)	GFR >60 mL/min: usually unrestricted GFR <60 mL/min: 800–1000 if serum phosphorus or parathyroid hormone is elevated	800–1000 if serum phosphorus or parathyroid hormone is elevated	800–1000 if serum phosphorus or parathyroid hormone is elevated
Calcium (mg/day)	GFR >60 mL/min: DRI levels GFR <60 mL/min: ≤2000 from diet and medications	≤2000 from diet and medications	≤2000 from diet and medications

[a]Values listed in this table apply to adults; recommendations for children vary with age.
[b]Predialysis guidelines apply to patients in stages 1 through 4; by stage 5, either hemodialysis or peritoneal dialysis is necessary.

Sources: Academy of Nutrition and Dietetics, *Nutrition Care Manual* (Chicago: Academy of Nutrition and Dietetics, 2012); D. J. Goldstein-Fuchs and C. M. Goeddeke-Merickel, Nutrition and kidney disease, in A. Greenberg, ed., *Primer on Kidney Diseases* (Philadelphia: Saunders, 2009), pp. 479–486.

Because of the high risk of wasting and compliance difficulties associated with low-protein diets, some dietitians suggest that patients consume higher amounts of protein to preserve health.[26] Once dialysis has begun, protein restrictions can be relaxed because dialysis removes nitrogenous wastes and results in some amino acid losses as well.

Lipids To control elevated blood lipids and reduce heart disease risk, patients with chronic kidney disease are advised to restrict their intakes of saturated fat, *trans* fat, and cholesterol. Although patients are often encouraged to consume high-fat foods to improve their energy intakes, the foods they select should provide mostly unsaturated fats. Good choices include nuts and seeds, oil-based salad dressings, mayonnaise, (■) avocados, and soybean products (see Nutrition in Practice 4 for additional suggestions).

■ Reminder: Most salad dressings and mayonnaise products are made with polyunsaturated or monounsaturated vegetable oils.

Sodium and Fluids As kidney disease progresses, patients excrete less urine and cannot handle normal intakes of sodium and fluids. Recommendations depend on the total urine output, changes in body weight and blood pressure, and serum sodium levels. A rise in body weight and blood pressure suggests that the person is retaining sodium and fluid; conversely, declines in these measurements indicate fluid loss. Most people with kidney disease tend to retain sodium and may benefit from mild restriction; less frequently, a patient may have a salt-wasting condition that requires additional dietary sodium.

Fluids are not restricted until urine output decreases. For a person who is neither dehydrated nor overhydrated, the daily fluid intake should match the daily urine output. (Obligatory water losses—from skin and lungs—are replaced by the water contained in the solid foods that are consumed.) Once a person is on dialysis, sodium and fluid intakes should be controlled so that only about 2 pounds of water weight are gained daily—this excess fluid is then removed during the next dialysis treatment. Patients on fluid-restricted diets should be advised that foods such as flavored gelatin, soups, fruit ices, and frozen fruit juice bars contribute to the fluid allowance.

© Craig M. Moore

In a renal diet, at least half of the protein consumed should be from high-quality protein sources such as eggs, milk, meat, poultry, and fish.

Potassium Most patients can handle typical intakes of potassium in the early stages of disease. Potassium restrictions are advised for patients who have hyperkalemia, have diabetic nephropathy (which increases risk of hyperkalemia), or are in later stages of illness. Conversely, potassium supplementation may be necessary for persons using potassium-wasting diuretics.

Dialysis patients must control potassium intakes to prevent hyperkalemia or, more rarely, **hypokalemia**. Restriction is necessary for people treated with hemodialysis, whereas those undergoing peritoneal dialysis can consume potassium more freely. Recommended intakes are based on serum potassium levels, renal function, medications, and the dialysis procedure used.

All fresh foods provide potassium, but some fruits and vegetables contain such high amounts that that some patients must restrict intakes. Table 22-6 shows the potassium content of some common fruits and vegetables. Foods in other food groups may be

TABLE 22-6 Potassium Guide—Fruits and Vegetables

This table lists common fruits and vegetables according to their potassium content. One serving is ½ cup raw fruit or cooked vegetable unless otherwise noted. Keep in mind that the portion size may determine how a food is categorized. Check Appendix A for additional information about the potassium content of foods.

High Potassium (>250 mg per serving)	Medium Potassium (150–250 mg per serving)	Low Potassium (<150 mg per serving)
Avocado	Apple (1 medium)	Blueberries
Banana	Apricots (2 whole)	Cabbage
Beets	Asparagus	Carrots (1 medium)
Chard	Broccoli	Cauliflower
Dates (3 whole)	Cantaloupe	Cucumbers
Nectarine (1 small)	Celery	Eggplant
Orange (1 medium)	Corn	Grapes
Parsnips	Grapefruit (½ fruit)	Green beans
Potatoes	Honeydew melon	Green pepper
Pumpkin	Kale	Lettuce (4 leaves, raw)
Raisins	Peach (1 small)	Onions (1 small)
Spinach	Pear (1 medium)	Plum (1 small)
Sweet potatoes	Peas	Strawberries
Tomato	Zucchini	Watermelon

hypokalemia (HIGH-po-ka-LEE-me-ah): low serum potassium levels.

© Craig M. Moore

People on a renal diet can consume most fruits and vegetables in limited amounts.

high in potassium as well; examples include dried beans, fish, milk and milk products, molasses, nuts and nut butters, and wheat bran. Note that salt substitutes and other low-sodium products often contain potassium chloride, which people on a potassium-restricted diet should avoid. Appendix A provides additional information about the potassium content of common foods.

Phosphorus, Calcium, and Vitamin D To minimize the risk of bone disease, serum phosphorus and calcium levels are monitored in kidney disease patients, and laboratory values help to guide recommendations. Elevated serum phosphorus levels indicate the need for dietary phosphorus restriction and, if necessary, the use of phosphate binders (taken with meals). Because many phosphate binders are calcium salts, patients are at risk of developing **hypercalcemia** in response to simultaneous calcium and vitamin D supplementation. As mentioned previously, vitamin D supplementation is standard treatment for many renal patients, (■) although the amount prescribed depends on the serum levels of calcium, phosphorus, and parathyroid hormone.

High-protein foods are also high in phosphorus, so the protein-restricted diets consumed by predialysis patients curb phosphorus intakes as well. After dialysis treatments begin and protein intakes are liberalized, phosphate binders become essential for phosphorus control. Because foods that are rich in calcium (such as milk and milk products) are usually high in phosphorus and are therefore restricted, patients may rely on calcium supplements (or calcium-based phosphate binders) to meet their calcium needs. Table 22-7 lists examples of foods that are high in phosphorus.

■ Reminder: Diseased kidneys are unable to produce activated vitamin D, which normally regulates calcium absorption and helps to maintain serum calcium levels.

hypercalcemia (HIGH-per-kal-SEE-me-ah): elevated serum calcium levels.

intradialytic parenteral nutrition: the infusion of nutrients during hemodialysis, often providing amino acids, dextrose, lipids, and some trace minerals.

Vitamins and Minerals The restrictive renal diet interferes with vitamin and mineral intakes, increasing the risk of deficiencies. In addition, patients treated with dialysis lose water-soluble vitamins and some trace minerals into the dialysate. Multivitamin supplements are typically recommended for all patients with chronic kidney disease. Supplements prescribed for dialysis patients typically supply generous amounts of folate and vitamin B_6—0.8 to 1 milligram and 10 milligrams per day, respectively—along with recommended amounts of the other water-soluble vitamins.[27] Supplemental vitamin C should be limited to 100 milligrams per day because excessive intakes can contribute to kidney stone formation in individuals at risk (see p. 641). Vitamin A supplements are not recommended because vitamin A levels tend to rise as kidney function worsens.

Iron deficiency is common in hemodialysis patients and may be due to inadequate erythropoietin, gastrointestinal bleeding, reduced iron absorption, or blood losses associated with the dialysis treatment.[28] Intravenous administration of iron, in conjunction with erythropoietin therapy, is more effective than oral iron supplementation for improving iron status.

TABLE 22-7 Foods High in Phosphorus[a]

- Barley
- Bran (oat, wheat)
- Buckwheat groats
- Bulgur
- Canned iced teas
- Canned lemonade
- Coconut
- Cola beverages
- Cornmeal
- Couscous
- Dried peas and beans
- Fish
- Milk products
- Nuts and seeds
- Organ meats
- Peanut butter
- Processed meats
- Soybeans, tofu

[a]For a complete list, visit the USDA's Nutrient Database at http://ndb.nal.usda.gov/. Click on "Nutrient lists," and then find the list of foods sorted in descending order by phosphorus content (click on the letter "W" to the right of the word "Phosphorus").

Enteral and Parenteral Nutrition Nutrition support is sometimes necessary for renal patients who cannot consume adequate amounts of food. The enteral formulas suitable for patients with chronic kidney disease are more kcalorically dense and have lower protein and electrolyte concentrations than standard formulas. **Intradialytic parenteral nutrition** is an option for supplying supplemental nutrients to dialysis patients; this technique combines parenteral infusions with hemodialysis treatments. An advantage of this approach is that the volume of parenteral solution infused can be simultaneously removed (recall that fluid intake is controlled in dialysis patients). However, clinical studies have not shown intradialytic parenteral nutrition to be more successful than oral supplementation in improving the nutrition status of malnourished dialysis patients.[29] Currently, the technique is used mainly in patients with PEM who have not responded well to oral supplements.[30]

Dietary Compliance Adhering to a renal diet is probably the most difficult aspect of treatment for patients with chronic kidney disease. These patients often require extensive counseling once multiple dietary restrictions become necessary. Depending on the stage of illness and the patient's laboratory values, the renal diet may limit protein, fluids, sodium, potassium, and phosphorus, thereby affecting food selections from all major food groups. In addition, adjustments in nutrient intake are required as the disease progresses. If the kidney disease was caused by diabetes, patients must also continue the dietary changes necessary for controlling blood glucose levels. Because renal diets have so many restrictions, patient compliance is often a problem. The "How to" provides suggestions to help patients comply with renal diets, and Table 22-8 on p. 638 shows an example of a one-day menu that includes some of these restrictions. The accompanying Case Study (p. 638) allows you to apply your knowledge about chronic kidney disease and hemodialysis.

KIDNEY TRANSPLANTS

A preferred alternative to dialysis in patients with end-stage renal disease is kidney transplantation.[31] A successful kidney transplant restores kidney function, allows a more liberal diet, and frees the patient from routine dialysis. Given the choice, many patients would prefer transplants, but the demand for suitable kidneys far exceeds the supply. Other barriers to transplantation include advanced age, poor health, and financial difficulties. Approximately 30 percent of patients with end-stage renal disease receive a kidney transplant.[32]

Immunosuppressive Drug Therapy To prevent tissue rejection following transplant surgery, patients require high doses of immunosuppressive drugs such as corticosteroids, cyclosporine, tacrolimus, and azathioprine. These drugs have multiple effects that can alter nutrition status, including nausea, vomiting, diarrhea, glucose intolerance, altered blood lipids, fluid retention, hypertension, and increased risk of infection. Because immunosuppressive drug therapy increases the risk of foodborne infection,

NURSING DIAGNOSIS

The nursing diagnoses *risk for infection* and *ineffective protection* may apply to persons with kidney transplants.

HOW TO Help Patients Comply with a Renal Diet

Patients with renal disease and their caregivers face considerable challenges as they learn to manage a renal diet. The following suggestions may help:

1. *To keep track of fluid intake:*
   - Fill a container with an amount of water equal to your total fluid allowance. Each time you consume a liquid food or beverage, discard an equivalent amount of water from the container. The amount remaining in the container will show you how much fluid you have left for the day.
   - Be sure to save enough fluid to take medications.
2. *To help control thirst:*
   - Chew gum or suck hard candy.
   - Suck on frozen grapes.
   - Freeze allowed beverages to a semisolid state so that they take longer to consume. Or, fill an ice-cube tray with your favorite fruit-flavored beverage, and suck on flavored ice cubes during the day.
   - Add lemon juice or crumpled mint leaves to water to make it more refreshing.
   - Gargle with refrigerated mouthwash.
3. *To increase the energy content of meals:*
   - Add extra margarine or a flavored oil to rice, noodles, breads, crackers, and cooked vegetables. Add extra salad dressing or mayonnaise to salads.
   - Add nondairy whipped toppings to desserts.
   - Include fried foods in your diet.
4. *To include more of your favorite vegetables in meals:*
   - Consult your nurse or dietitian to learn whether you can safely use the process of leaching to remove some of the potassium from vegetables.
   - To leach potassium from vegetables: Cut the vegetables into 1/8-inch slices and rinse. Soak the vegetables in a large amount of warm water for two hours—about 10 parts of water to 1 part of vegetables. Rinse vegetables well. Boil vegetables using 5 parts of water to 1 part of vegetables.
5. *To prevent the diet from becoming monotonous:*
   - Experiment with new combinations of allowed foods.
   - Substitute nondairy products for milk products. Nondairy products, which are lower in protein, phosphorus, and potassium, can substitute for milk and add energy to the diet.
   - Add flavor to foods by seasoning with garlic, onion, chili powder, curry powder, oregano, mint, basil, parsley, pepper, or lemon juice.
   - Consult a nurse or dietitian when you want to eat restricted foods. Many restricted foods can be used occasionally and in small amounts if the menu is carefully adjusted.

TABLE 22-8 Chronic Kidney Disease—One-Day Menu

The menu below provides 2028 kcalories, 46 g protein, 784 mg phosphorus, 2190 mg potassium, and 1510 mg sodium.[a] The energy and protein content would be appropriate for a 135-pound predialysis patient.

Breakfast

- Corn flakes with milk (1 cup cereal, ½ cup whole milk)
- Apricot nectar (1 cup)
- Caffé latte (brewed coffee, 2 tsp sugar, ½ cup cream substitute)

Lunch

- Turkey sandwich (2 slices white bread, 1½ oz dark meat, 5 slices cucumber, 1 tbs mayonnaise)
- Grape juice (1 cup)
- Orange sherbet (½ cup)

Dinner

- Spaghetti with tomato sauce (1 cup cooked spaghetti, ½ cup bottled tomato sauce, ½ tbs grated cheese)
- Green beans with olive oil (1 cup cooked green beans, 1 tbs olive oil)
- Biscuit with margarine (2½-inch biscuit, ½ tbs margarine)
- Baked apple with nondairy sour cream (1 large apple, ¼ cup nondairy sour cream)

[a]Energy and nutrient values were obtained from the USDA National Nutrient Database for Standard Reference: http://ndb.nal.usda.gov/

food safety guidelines should be provided to patients and caregivers. The Diet-Drug Interactions feature summarizes the nutrition-related effects of immunosuppressants and other drugs mentioned in this chapter.

Nutrition Therapy after Kidney Transplant After patients recover from transplant surgery, most nutrients can be consumed at levels recommended for the general population. The protein recommendation is similar to that of the general population, about 0.8 to 1.0 gram per kilogram of body weight per day.[33] Patients should attempt to maintain a healthy body weight and consume a diet that reduces their risk for cardiovascular diseases.

For most transplant patients, the side effects of drugs are the primary reason that dietary adjustments may be required. Although sodium, potassium, phosphorus, and fluid intakes are usually liberalized following a transplant, serum electrolyte levels must be monitored because some drug therapies can cause electrolyte imbalances or fluid retention. If corticosteroids are used as immunosuppressants, calcium supplementation

Case Study: Man with Chronic Kidney Disease

Thomas Stone is a 55-year-old banker who developed chronic kidney disease as a result of hypertension. His condition was discovered several years ago, when routine laboratory tests revealed elevated serum creatinine and BUN levels. Since then, he has been taking antihypertensive medications and restricting dietary sodium, but he reported difficulty following the low-protein diet that was also prescribed. Mr. Stone recently visited his doctor with complaints of low urine output and reduced sensation in his hands and feet. He also reported feeling drowsy at work and mentioned that he was bruising more than usual. The examination revealed a 9-pound weight gain since his last visit and swelling in his ankles and feet. Tests revealed that his GFR had fallen to 10 milliliters per minute. Mr. Stone is 5 feet 8 inches tall and normally weighs 160 pounds.

1. Explain how chronic kidney disease progresses. What happens to GFR, serum creatinine levels, and BUN as renal function declines?
2. Describe the clinical effects you would expect during the final stage of disease, when kidney failure develops. Explain the significance of each of Mr. Stone's physical complaints.
3. Explain why a low-sodium, low-protein diet was prescribed for Mr. Stone at a former visit. What energy and protein intakes were probably recommended at that time?
4. The physician determines that Mr. Stone's kidney disease has reached the final stage and prescribes hemodialysis. How will dialysis alter Mr. Stone's diet? Calculate his new protein recommendation, and compare it to the amount of protein recommended before dialysis. What other changes in nutrient intake may be necessary?

DIET-DRUG INTERACTIONS

Check this table for notable nutrition-related effects of the medications discussed in this chapter.

Medication	Effects
Immunosuppressants (cyclosporine, tacrolimus)	**Gastrointestinal effects:** Nausea, vomiting, abdominal discomfort, diarrhea, constipation, anorexia
	Dietary interactions: Limit alcohol intake due to the potential for toxic effects; the bioavailability of tacrolimus is reduced when the drug is taken with food; grapefruit juice can raise serum concentrations of these drugs to toxic levels
	Metabolic effects: Electrolyte imbalances, hyperglycemia, hyperlipidemias, anemia
Immunosuppressants (corticosteroids)	**Metabolic effects:** Fluid retention, hyperglycemia, hypocalcemia, hypokalemia, hypophosphatemia, increased appetite, protein catabolism
Phosphate binders (calcium-based)	**Gastrointestinal effect:** Constipation
	Metabolic effects: Electrolyte imbalances
Potassium-exchange resins (sodium polystyrene sulfonate)	**Gastrointestinal effects:** Anorexia, constipation
	Dietary interactions: Calcium and magnesium supplements must be taken separately
	Metabolic effects: Fluid and sodium retention, hypokalemia, hypocalcemia, hypomagnesemia
Potassium citrate	**Gastrointestinal effects:** Nausea, vomiting, abdominal pain, diarrhea
	Metabolic effect: Hyperkalemia

is recommended because the medication increases urinary calcium losses. If drug treatment leads to hyperglycemia, patients should limit intakes of refined carbohydrates and concentrated sweets; for some individuals, oral medications or insulin therapy may be necessary. As noted earlier, patients must carefully follow food safety guidelines to avoid foodborne illness.

Review Notes

- Chronic kidney disease causes gradual loss of kidney function and often results from long-standing diabetes mellitus or hypertension.
- Depending on the stage of chronic kidney disease, complications may include fluid and electrolyte imbalances, hypertension, renal osteodystrophy, mental impairments, bleeding abnormalities, anemia, increased risk for cardiovascular disease, and reduced immunity.
- Treatment of chronic kidney disease can slow disease progression and correct complications, and includes drug therapies, dialysis, and nutrition therapy. Dietary measures may feature a low-protein diet, controlled fluid and sodium intakes, phosphorus restrictions, and calcium and vitamin D supplementation; potassium restrictions are usually necessary after dialysis treatment begins.
- Kidney transplantation in patients with chronic kidney disease restores renal function and liberalizes dietary restrictions.

Kidney Stones

Approximately 12 percent of men and 6 percent of women in the United States develop one or more **kidney stones** during their lifetimes.[34] A kidney stone is a crystalline mass that forms within the urinary tract. Although stones are often asymptomatic, their passage can cause severe pain or block the urinary tract. Stones tend to recur but can be prevented with dietary measures and medical treatment.

kidney stones: crystalline masses that form in the urinary tract; also called *renal calculi* or *nephrolithiasis*.

© Nathan Griffith/Corbis

The most common type of kidney stone is composed of calcium oxalate crystals, as shown here. Kidney stones may be as small as a bread crumb or as large as a golf ball.

FORMATION OF KIDNEY STONES

Kidney stones develop when stone constituents become concentrated in urine, allowing crystals to form and grow. About 70 percent of kidney stones are made up primarily of calcium oxalate. Less commonly, stones are composed of calcium phosphate, uric acid, the amino acid cystine, or magnesium ammonium phosphate (the latter are known as *struvite* stones). Factors that predispose an individual to stone formation include the following:

- *Dehydration* or *low urine volume,* which promotes the crystallization of minerals and other compounds in urine.
- *Obstruction,* which prevents the flow of urine and encourages salt precipitation.
- *Urine acidity,* which affects the dissolution of urinary constituents. Some stones form more readily in acidic urine, whereas others form in alkaline urine.
- *Metabolic factors,* which affect the presence of compounds that either promote or inhibit crystal growth.
- *Renal disease,* which is associated with calcification of tissues and phosphate accumulation.

The most common types of kidney stones are described in this section.

Calcium Oxalate Stones

The most common abnormality in people with calcium oxalate stones is **hypercalciuria** (elevated urinary calcium levels). Hypercalciuria can result from excessive calcium absorption, impaired calcium reabsorption in kidney tubules, or elevated serum levels of parathyroid hormone or vitamin D. However, some people with calcium oxalate stones excrete normal amounts of calcium in the urine, and the reason they form stones is unknown.

Elevated urinary oxalate levels, or **hyperoxaluria**, also promote the formation of calcium oxalate crystals. Oxalate is a normal product of metabolism that readily binds to calcium. Hyperoxaluria may reflect an increase in the body's synthesis of oxalate or increased absorption from dietary sources. (■)

■ Reminder: Fat malabsorption promotes oxalate absorption, increasing the risk of forming calcium oxalate stones (see Chapter 18).

Uric Acid Stones

Uric acid stones develop when the urine is abnormally acidic, contains excessive uric acid, or both. These stones are frequently associated with **gout**, a metabolic disorder characterized by elevated uric acid levels in the blood and urine. A diet rich in **purines** also contributes to high uric acid levels; purines are abundant in animal proteins (meat, poultry, seafood) and degrade to uric acid in the body. In addition, a high intake of animal protein increases urine acidity, which promotes the crystallization of uric acid.

Cystine and Struvite Stones

Cystine stones can form in people with the inherited disorder **cystinuria**, in which the renal tubules are unable to reabsorb the amino acid cystine. The abnormality results in abnormally high concentrations of cystine in the urine, leading to subsequent crystallization and stone formation. **Struvite** stones, composed primarily of magnesium ammonium phosphate, form in alkaline urine; the urinary pH is sometimes elevated due to the bacterial degradation of urea to ammonia. Struvite stones can accompany chronic urinary infections or disorders that interfere with urinary flow.

CONSEQUENCES OF KIDNEY STONES

In most cases, kidney stones do not pose serious medical problems. Small stones can readily pass through the ureters and out of the body with minimal treatment.

Renal Colic

A stone passing through the ureter can produce severe, stabbing pain, called **renal colic**. The pain can be severe enough to cause nausea and vomiting and sometimes requires medication. Blood may appear in the urine (**hematuria**) as a result of damage to the kidney or ureter lining.

Urinary Tract Complications

Depending on the location of the stone, symptoms may include urination urgency, frequent urination, or inability to urinate. Stones that are unable to pass through the ureter can cause a urinary tract obstruction and possibly lead to infection.

hypercalciuria (HIGH-per-kal-see-YOO-ree-ah): elevated urinary calcium levels.

hyperoxaluria (HIGH-per-ox-ah-LOO-ree-ah): elevated urinary oxalate levels.

gout (GOWT): a metabolic disorder characterized by elevated uric acid levels in the blood and urine and the deposition of uric acid in and around the joints, causing acute joint inflammation.

purines (PYOO-reens): products of nucleotide metabolism that degrade to uric acid.

cystinuria (SIS-tin-NOO-ree-ah): an inherited disorder characterized by the elevated urinary excretion of several amino acids, including cystine.

struvite (STROO-vite): crystals of magnesium ammonium phosphate.

renal colic: the intense pain that occurs when a kidney stone passes through the ureter; the pain typically begins in the back and intensifies as the stone travels toward the bladder.

hematuria (HE-mah-TOO-ree-ah): blood in the urine.

PREVENTION AND TREATMENT OF KIDNEY STONES

Solutes are less likely to crystallize and form stones in dilute urine. Therefore, people who form kidney stones are advised to drink 12 to 16 cups of fluids daily to maintain urine volumes of at least 2½ liters per day.[35] Additional fluid may be needed in hot weather or if an individual is extremely active. For some patients, dietary modifications, medications, or surgical stone removal may be necessary.

NURSING DIAGNOSIS

Nursing diagnoses for patients with kidney stones may include *acute pain, risk for infection,* and *impaired urinary elimination.*

Calcium Oxalate Stones Most dietary measures and drug treatments for calcium oxalate stones aim to reduce urinary calcium and oxalate levels. Dietary measures may include adjustments in calcium, oxalate, protein, and sodium intakes.[36] Patients should consume adequate calcium from food sources (about 800 and 1200 milligrams per day for men and women, respectively) because dietary calcium combines with oxalate in the intestines, reducing oxalate absorption and helping to control hyperoxaluria. (■) Conversely, low-calcium diets promote oxalate absorption and higher urinary oxalate levels. Some individuals with hyperoxaluria may benefit from dietary oxalate restriction (see Table 22-9). High protein and sodium intakes increase urinary calcium excretion, so moderate protein consumption (0.8 to 1.0 gram per kilogram of body weight per day) and a controlled sodium intake (no more than 3450 milligrams daily) are also advised. Vitamin C intakes should not exceed the RDA (90 and 75 milligrams for men and women, respectively) because vitamin C degrades to oxalate in the body.[37] Medications used to prevent calcium oxalate stones include thiazide diuretics, which reduce urinary calcium levels; potassium citrate (a base), which inhibits crystal formation; and allopurinol (Zyloprim), which reduces uric acid production in the body and may have other effects.[38]

James Darell/Getty Images

Drinking plenty of water throughout the day is the most important measure for preventing kidney stones.

Uric Acid Stones Diets restricted in purines may help to control urinary uric acid levels. Because all animal proteins contain purines, strict dietary control over a long period may be difficult to achieve. In addition, the benefits of purine restriction are unclear. The drug treatments used for uric acid stones include allopurinol to reduce uric acid levels and potassium citrate to reduce urine acidity.

■ Because calcium supplements can elevate urinary calcium levels, they are not as helpful as food sources of calcium.

Cystine and Struvite Stones High fluid intakes may prevent the formation of cystine stones in some patients, whereas other individuals require drug therapy to

TABLE 22-9 Foods High in Oxalate

Vegetables	Fruits	Other
Beets*	Apricots, dried	Barley
Chard	Blackberries	Buckwheat
Collard greens	Blueberries	Chocolate*
Dried beans	Currants, red	Cocoa
Eggplant	Figs	Cornmeal, grits
Escarole	Grapes, Concord	Miso
Green beans	Kiwi	Nuts, nut butters*
Kale	Lemon peel	Peanut butter*
Leeks	Oranges, orange peel	Sesame seeds, tahini
Mustard greens	Raspberries	Soybean products
Parsley	Rhubarb*	Tea*
Spinach*	Strawberries*	Wheat bran*
Sweet potatoes		Whole-wheat flour

Note: The oxalate content of many foods has not been analyzed, and few studies have been conducted to determine which foods raise urinary oxalate levels.

*The foods marked with an asterisk have been documented to raise urinary oxalate levels and should be avoided by people who form calcium oxalate stones.

reduce cystine production in the body. Medications frequently prescribed include penicillamine (Cuprimine) and tiopronin (Thiola), which increase the solubility of cystine, and potassium citrate, which reduces urine acidity. For preventing struvite stones, preventing or promptly treating urinary tract infections is a central strategy.

Medical Treatment for Kidney Stones Medical treatment may be necessary for a kidney stone that is too large to pass, blocks urine flow, or causes severe pain or bleeding. Medications that relax the ureter and increase urine flow may be given to facilitate stone passage. Some kidney stones can be fragmented into pieces that are small enough to pass in the urine; the most common method is *extracorporeal shock wave lithotripsy*, a procedure that uses high-amplitude sound waves to degrade the stone. Surgical methods that involve physical removal of kidney stones have a higher success rate but are also more invasive.

Review Notes

- Kidney stones form when stone constituents crystallize in urine. Complications include renal colic, difficulty with urination, and obstruction.
- Kidney stones may be prevented by maintaining urine volumes of at least 2½ liters daily. Dietary measures include the consumption of appropriate amounts of calcium, oxalates, protein, sodium, and purines.
- Symptomatic kidney stones are sometimes treated with medications that facilitate stone passage or surgeries that fragment or remove stones.

Nutrition Assessment Checklist for People with Kidney Diseases

MEDICAL HISTORY

Check the medical record to determine:

- ❍ Degree of kidney function
- ❍ Cause of the nephrotic syndrome or kidney disease
- ❍ Type of dialysis, if appropriate
- ❍ Whether the patient has received a kidney transplant
- ❍ Type of kidney stone

Review the medical record for complications that may alter nutritional needs:

- ❍ Anemia
- ❍ Diabetes mellitus
- ❍ Edema or oliguria
- ❍ Hyperlipidemia
- ❍ Hypertension
- ❍ Metabolic stress or infection
- ❍ Protein-energy malnutrition

MEDICATIONS

Assess risks for medication-related malnutrition related to:

- ❍ Long-term use of medications
- ❍ Multiple medication use, especially if medications affect nutrition status

For all patients with kidney diseases, note:

- ❍ Whether medications or supplements contain electrolytes that must be controlled
- ❍ Use of drugs or herbs that may be toxic to the kidneys

DIETARY INTAKE

For patients with the nephrotic syndrome, kidney disease, or kidney transplant, assess intakes of:

- ❍ Protein and energy
- ❍ Fluid
- ❍ Vitamins, especially vitamin D
- ❍ Minerals, especially calcium, phosphorus, iron, and electrolytes

For patients with kidney stones or a history of kidney stones:

- ❍ Stress the need to drink plenty of fluids throughout the day.
- ❍ Assess intake of calcium, oxalate, sodium, protein, purines, or vitamin C, as appropriate for the type of stone.

ANTHROPOMETRIC DATA

Take accurate baseline height and weight measurements. Keep in mind that:

- ❍ Fluid retention due to the nephrotic syndrome or kidney failure can mask malnutrition.
- ❍ For dialysis patients, the weight measured immediately after the dialysis treatment (called the *dry weight*) most accurately reflects the person's true weight. Rapid weight gain between dialysis treatments reflects fluid retention. If fluid retention is excessive, review fluid intake to determine if the patient understands and is complying with diet recommendations.

LABORATORY TESTS

Note that serum protein levels are often low in patients with nephrotic syndrome or advanced kidney disease.

Review the following laboratory test results to assess the degree of kidney function and response to treatments:

- Blood urea nitrogen (BUN)
- Creatinine
- Glomerular filtration rate (GFR)
- Serum electrolytes
- Urinary protein

Check laboratory test results for complications associated with kidney disease, including:

- Anemia
- Hyperglycemia
- Hyperlipidemia
- Hyperparathyroidism (related to bone disease)

PHYSICAL SIGNS

For patients with nephrotic syndrome or kidney disease, look for physical signs of:

- Bone disease
- Dehydration or fluid retention
- Hyperkalemia
- Iron deficiency
- Uremia

Self Check

1. Which of the following is *not* a function of the kidneys?
   a. Activation of vitamin K
   b. Maintenance of acid-base balance
   c. Elimination of metabolic waste products
   d. Maintenance of fluid and electrolyte balances

2. The nephrotic syndrome frequently results in:
   a. the uremic syndrome.
   b. oliguria.
   c. edema.
   d. renal colic.

3. Dietary recommendations for patients with the nephrotic syndrome include:
   a. a high protein intake.
   b. sodium restriction.
   c. potassium and phosphorus restrictions.
   d. fluid restriction.

4. What is a common treatment for hyperkalemia?
   a. Eliminating potassium from the diet
   b. Using potassium-wasting diuretics to increase potassium losses
   c. Increasing fluid intake to maintain a high urine volume
   d. Using potassium-exchange resins, which bind potassium in the GI tract

5. Fluid requirements for oliguric patients are estimated by adding about ____ milliliters to the volume of urine output.
   a. 100
   b. 300
   c. 500
   d. 750

6. What is the most common cause of chronic kidney disease?
   a. Diabetes mellitus
   b. Hypertension
   c. Autoimmune disease
   d. Exposure to toxins

7. A person with chronic kidney disease who has been following a renal diet for several years begins hemodialysis treatment. An appropriate dietary adjustment would be to:
   a. reduce protein intake.
   b. consume protein more liberally.
   c. increase intakes of sodium and water.
   d. consume potassium and phosphorus more liberally.

8. Which of the following nutrients may be unintentionally restricted when a patient restricts phosphorus intake?
   a. Fluid
   b. Calcium
   c. Potassium
   d. Sodium

9. Most kidney stones are made primarily from:
   a. struvite.
   b. uric acid.
   c. calcium oxalate.
   d. cystine.

10. Treatment for all kidney stones includes:
   a. dietary oxalate restriction.
   b. dietary protein restriction.
   c. vitamin C supplementation.
   d. a fluid intake that maintains a urine volume of at least 2½ liters per day.

Answers to these questions appear in Appendix H. For more chapter review: Access an interactive eBook, chapter-specific interactive learning tools, including flashcards, quizzes, videos, and more in your Nutrition CourseMate, accessed through CengageBrain.com.

Clinical Applications

1. A person with chronic kidney disease may need multiple medications to control disease progression and treat symptoms and complications. For people with diabetes and hyperlipidemias who develop chronic kidney disease, medications might include insulin, oral hypoglycemic drugs, antihypertensives, diuretics, lipid-lowering medications, and phosphate binders. Review the nutrition-related side effects of these medications. Describe the ways in which these medications may make it harder for people to maintain nutrition status.
2. Because the diet for chronic kidney disease is so restrictive, patients often find the diet difficult to manage and maintain over the long term. Review the suggestions in the "How to" on p. 637. Can you think of additional suggestions that may help? List some ideas that may help patients adjust to each of the different aspects of their renal diets.

Nutrition on the Net

For further study of topics covered in this chapter, access these websites:

- To search for specific topics related to kidney diseases, dialysis, and kidney transplants, visit these sites:
  Kidney Foundation of Canada:
  www.kidney.ca
  National Institute of Diabetes and Digestive and Kidney Diseases:
  www2.niddk.nih.gov
  National Kidney Foundation:
  www.kidney.org
- To find materials for patients with kidney diseases, visit the American Association of Kidney Patients:
  www.aakp.org
- To find more information about kidney stones, visit the Oxalosis and Hyperoxaluria Foundation:
  www.ohf.org
- To see photographs of kidney stones, visit the website of the Louis C. Herring and Company Laboratory:
  www.herringlab.com

Notes

1. G. B. Appel and J. Radhakrishnan, Glomerular disorders and nephrotic syndromes, in L. Goldman and A. I. Schafer, eds., *Goldman's Cecil Medicine* (Philadelphia: Saunders, 2012), pp. 761–771.
2. B. R. Don and G. A. Kaysen, Proteinuria and nephrotic syndrome, in R. W. Schrier, ed., *Renal and Electrolyte Disorders* (Philadelphia: Lippincott Williams & Wilkins, 2010), pp. 519–558.
3. Don and Kaysen, 2010.
4. Appel and Radhakrishnan, 2012; Don and Kaysen, 2010.
5. Academy of Nutrition and Dietetics, *Nutrition Care Manual* (Chicago: Academy of Nutrition and Dietetics, 2012); Don and Kaysen, 2010.
6. Don and Kaysen, 2010.
7. Academy of Nutrition and Dietetics, 2012.
8. Academy of Nutrition and Dietetics, 2012; G. M. Podda and coauthors, Abnormalities of homocysteine and B vitamins in the nephrotic syndrome, *Thrombosis Research* 120 (2007): 647–652.
9. B. A. Molitoris, Acute kidney injury, in L. Goldman and A. I. Schafer, eds., *Goldman's Cecil Medicine* (Philadelphia: Saunders, 2012), pp. 756–761.
10. Molitoris, 2012.
11. R. W. Schrier and C. L. Edelstein, Acute kidney injury: Pathogenesis, diagnosis, and management, in R. W. Schrier, ed., *Renal and Electrolyte Disorders* (Philadelphia: Lippincott Williams & Wilkins, 2010), pp. 325–388.
12. C. E. Alpers, The kidney, in V. Kumar and coeditors, *Robbins and Cotran Pathologic Basis of Disease* (Philadelphia: Saunders, 2010), pp 905–969.
13. Academy of Nutrition and Dietetics, 2012; J. M. Gervasio, W. P. Garmon, and M. R. Holowatyj, Nutrition support in acute kidney injury, *Nutrition in Clinical Practice* 26 (2011): 374–381.
14. Academy of Nutrition and Dietetics, 2012.
15. J. Krenitsky and M. H. Rosner, Nutritional support for patients with acute kidney injury: How much protein is enough or too much? *Practical Gastroenterology* (June 2011): 28–42.
16. Academy of Nutrition and Dietetics, 2012.
17. Schrier and Edelstein, 2010.
18. M. Chonchol and L. Chan, Chronic kidney disease: Manifestations and pathogenesis, in R. W. Schrier, ed., *Renal and Electrolyte Disorders* (Philadelphia: Lippincott Williams & Wilkins, 2010), pp. 389–425.
19. W. E. Mitch, Chronic kidney disease, in L. Goldman and A. I. Schafer, eds., *Goldman's Cecil Medicine* (Philadelphia: Saunders, 2012), pp. 810–818.
20. Mitch, 2012.
21. G. T. Obrador, Chronic renal failure and the uremic syndrome, in E. V. Lerma and coeditors, *Current Diagnosis and Treatment: Nephrology and Hypertension* (New York: McGraw-Hill/Lange, 2009), pp. 149–154.
22. Chonchol and Chan, 2010; A. Schieppati, R. Pisoni, and G. Remuzzi, Pathophysiology of chronic kidney disease, in A. Greenberg, ed., *Primer on Kidney Diseases* (Philadelphia: Saunders, 2009), pp. 422–445.
23. J. Podel and coauthors, Glucose absorption in acute peritoneal dialysis, *Journal of Renal Nutrition* 10 (2000): 93–97.
24. A. J. Hutchison and A. Vardhan, Peritoneal dialysis, in A. Greenberg, ed., *Primer on Kidney Diseases* (Philadelphia: Saunders, 2009), pp. 459–471.
25. Academy of Nutrition and Dietetics, 2012; D. J. Goldstein-Fuchs and C. M. Goeddeke-Merickel, Nutrition and kidney disease, in A. Greenberg, ed., *Primer on Kidney Diseases* (Philadelphia: Saunders, 2009), pp. 478–486.
26. J. A. Beto and V. K. Bansal, Medical nutrition therapy in chronic kidney failure: Integrating clinical practice guidelines, *Journal of the American Dietetic Association* 104 (2004): 404–409.
27. Goldstein-Fuchs and Goeddeke-Merickel, 2009.
28. N. Tolkoff-Rubin, Treatment of irreversible renal failure, in L. Goldman and A. I. Schafer, eds., *Goldman's Cecil Medicine* (Philadelphia: Saunders, 2012), pp. 818–826.
29. N. J. Cano and coauthors, Intradialytic parenteral nutrition does not improve survival in malnourished hemodialysis patients: A 2-year multicenter, prospective, randomized study, *Journal of the American Society of Nephrology* 18 (2007): 2583–2591.
30. T. G. Axelsson, M. Chmielewski, and B. Lindholm, Kidney disease, in J. W. Erdman, I. A. Macdonald, and S. H. Zeisel, eds., *Present Knowledge in Nutrition* (Ames: IA: Wiley-Blackwell, 2012), pp. 874–888.
31. Tolkoff-Rubin, 2012.
32. S. Beddhu, Outcome of end-stage renal disease therapies, in A. Greenberg, ed., *Primer on Kidney Diseases* (Philadelphia: Saunders, 2009), pp. 472–477.
33. Academy of Nutrition and Dietetics, 2012.
34. G. C. Curhan, Nephrolithiasis, in L. Goldman and A. I. Schafer, eds., *Goldman's Cecil Medicine* (Philadelphia: Saunders, 2012), pp. 789–794.
35. Academy of Nutrition and Dietetics, 2012.
36. Academy of Nutrition and Dietetics, 2012.
37. J. Lamarche and coauthors, Vitamin C-induced oxalate nephropathy, *International Journal of Nephrology* 2011 (2011), available at DOI:10.4061/2011/146927.
38. Curhan, 2012.

Nutrition in Practice

Dialysis

Although there is no perfect substitute for one's own kidneys, dialysis offers a life-sustaining treatment option for people with chronic kidney disease who develop renal failure. Dialysis can serve as a long-term treatment or as a temporary measure to sustain life until a suitable kidney donor can be found. Dialysis can also restore fluid and electrolyte balances in patients with acute kidney injury. Clinicians who routinely work with renal patients should understand how dialysis procedures work. This Nutrition in Practice describes the process of dialysis and outlines the various types of procedures that are available. The accompanying glossary (p. 646) defines the relevant terms.

How does dialysis work?

Dialysis removes excess fluids and wastes from the blood by employing the processes of **diffusion**, **osmosis**, and **ultrafiltration** (see Figure NP22-1). The dialysate, a solution similar in composition to normal blood plasma, is carried through a compartment beside a **semipermeable membrane**; the person's blood flows in the opposite direction along the other side of the membrane. The semipermeable membrane acts like a filter: small molecules such as urea and glucose can pass through microscopic pores in the membrane, whereas large molecules are unable to cross.

In *hemodialysis,* the tiny tubes that carry blood through the dialyzer are made of materials that serve as semipermeable membranes. In *peritoneal dialysis,* the body's peritoneal membrane, rich with blood vessels, is used to filter the blood.

How are solutes separated from the blood in dialysis?

The chemical composition of the dialysate affects the movement of solutes across the semipermeable membrane. When the concentration of a substance is lower in the dialysate than in the blood, the substance—provided it can cross the membrane—will diffuse out of the blood. For example, the goal is to remove as much as possible of the waste product urea from the blood, so the dialysate contains no urea. For many other solutes, the dialysate is adjusted so that only excesses will be removed. Potassium can be removed from the blood, for example, by providing a dialysate that has a lower concentration of potassium than is found in the person's blood. The dialysate must contain some potassium, however; otherwise, the blood potassium would fall too low.

The dialysate can also be used to add needed components back into the blood. For a person with acidosis, for example, bases such as bicarbonate are added to the dialysate; the bases then move by diffusion into the blood to alleviate the acidosis.

FIGURE NP22-1 Diffusion, Osmosis, and Ultrafiltration

Diffusion

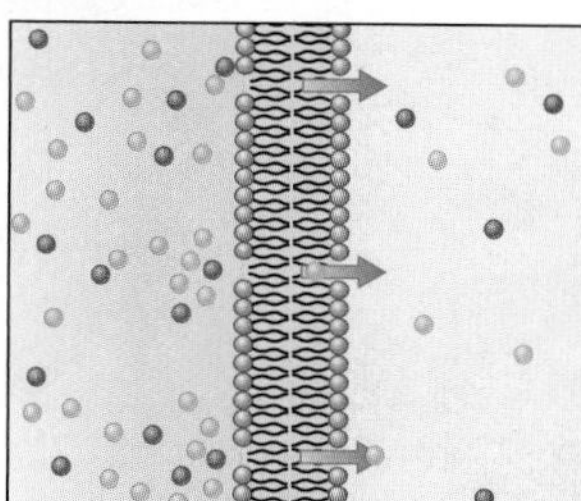

Small molecules (electrolytes and waste products) move from an area of high concentration to an area of low concentration by diffusion.

Osmosis

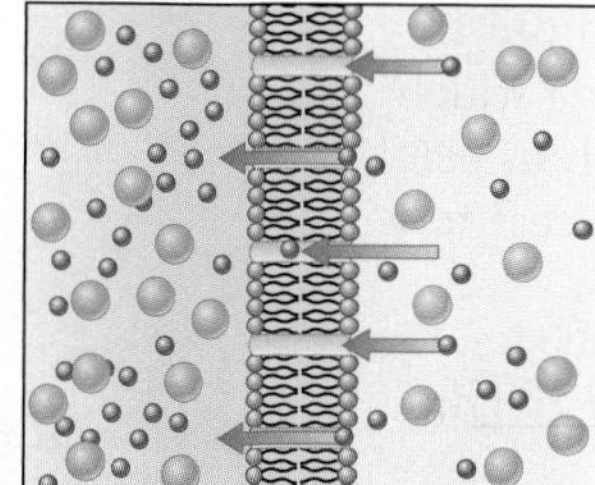

Water moves from an area of high water concentration to an area of low water concentration. In other words, water moves toward the side where solutes are more concentrated.

Ultrafiltration

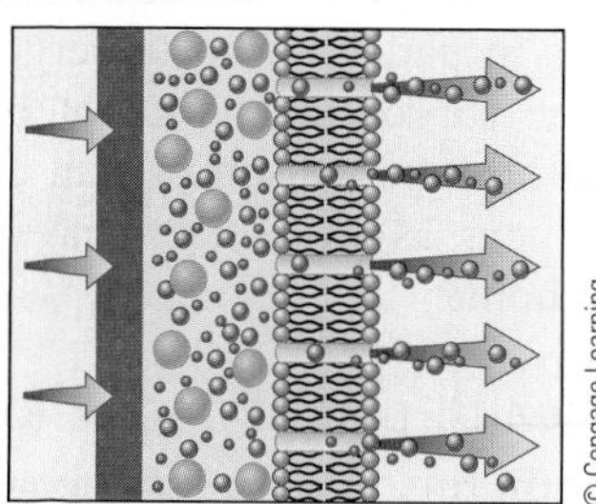

Pressure squeezes water and small molecules through the pores of a semipermeable membrane during ultrafiltration.

Glossary

continuous ambulatory peritoneal dialysis (CAPD): the most common method of peritoneal dialysis; involves frequent exchanges of dialysate, which remains in the peritoneal cavity throughout the day.
continuous renal replacement therapy (CRRT): a slow, continuous method of removing solutes and/or fluids from blood by gently pumping blood across a filtration membrane over a prolonged time period.
diffusion: movement of solutes from an area of high concentration to one of low concentration.
hemofiltration: removal of fluid and solutes by pumping blood across a membrane; no osmotic gradients are created during the process.
oncotic pressure: the pressure exerted by fluid on one side of a membrane as a result of osmosis.
osmosis: movement of water across a membrane toward the side where solutes are more concentrated.
peritonitis: inflammation of the peritoneal membrane.
pressure gradient: the change in pressure over a given distance. In dialysis, a pressure gradient is created between the blood and the dialysate.
semipermeable membrane: a membrane that allows some, but not all, particles to pass through.
ultrafiltration: removal of fluids and solutes from blood by using pressure to transfer the blood across a semipermeable membrane.
urea kinetic modeling: a method of determining the adequacy of dialysis treatment by calculating the urea clearance from blood.

How is fluid removed from the blood?

Because albumin and other plasma proteins are so adept at retaining fluids in blood, osmosis alone is not an efficient process for removing fluid. In hemodialysis, a **pressure gradient** is created between the blood and the dialysate. Most modern dialyzers produce *positive* pressure in the blood compartment and *negative* pressure in the dialysate compartment, establishing a pressure gradient that "pushes" water (and accompanying solutes) through the pores of the membrane. This process, called ultrafiltration, relies on pumps to establish an appropriate flow rate between the blood and the dialysate.

How often do patients require hemodialysis, and how long do treatments last?

Most patients undergo hemodialysis three times weekly and the treatments last three to five hours. Other options include short daily dialysis, performed for about two hours each day, and daily nocturnal hemodialysis, in which dialysis is done at home five or six nights a week while the patient is sleeping.[1] Although some studies have reported improved outcomes in patients who undergo more frequent dialysis, these approaches have not been widely adopted.[2] Note that most patients must visit dialysis centers to obtain treatment, as few patients have access to a dialysis machine at home.

How does the health practitioner know if the dialysis treatment has been effective?

A number of methods have been devised for gauging the adequacy of dialysis treatment. The most common method is **urea kinetic modeling**, a technique that evaluates the amount of urea cleared from the blood. The formula used most often is Kt/V, where K is the amount of urea cleared, t is the time spent on dialysis, and V is the blood volume. The value obtained indicates whether the patient has undergone sufficient dialysis; the goal is a Kt/V result of approximately 1.2. Because technical data (such as dialyzer clearance data, blood flow rate, and dialysate flow rate) need to be incorporated into the calculation, the computation is usually done by computer analysis. Current treatment guidelines recommend that hemodialysis adequacy be evaluated at least monthly, or more often if problems develop or patients are noncompliant.[3]

Are any complications associated with hemodialysis?

Yes. Although lifesaving, hemodialysis is associated with a substantial number of complications.[4] Problems at the vascular access site include infections and blood clotting. Hypotension can develop while blood is circulated through the dialyzer. Muscle cramping often occurs during the procedure, especially in the hands, legs, and feet.

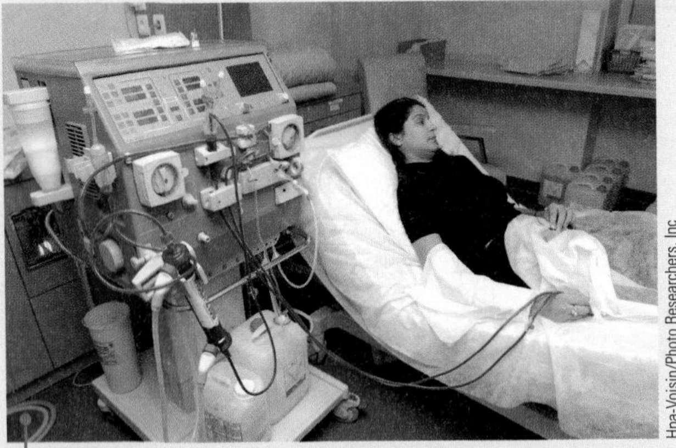

During hemodialysis, blood passes through a dialyzer where wastes are extracted, and the cleansed blood is returned to the body.

Blood losses can worsen anemia, which is already severe in two-thirds of patients beginning hemodialysis treatment.[5] Patients may also experience headaches, weakness, nausea, vomiting, restlessness, and agitation. Many patients experience extreme fatigue after a hemodialysis treatment, and some may require rest or sleep.

How does peritoneal dialysis work?

In peritoneal dialysis, the peritoneal membrane surrounding the abdominal organs serves as a semipermeable membrane. The dialysate is infused into a catheter that empties into the peritoneal space—the space within the abdomen near the intestines (see Figure NP22-2). In the most common procedure, **continuous ambulatory peritoneal dialysis (CAPD)**, the dialysate remains in the peritoneal cavity for four to six hours, after which it is drained and replaced with fresh dialysate (about 2 to 3 liters in adults). Generally, the dialysate solution is exchanged four times daily and requires only about 30 minutes to drain and replace.

Because a pressure gradient cannot be created in the peritoneal cavity as it can in a dialyzer, the glucose concentration in the dialysate must be high enough to create sufficient **oncotic pressure** to draw fluid from the blood. As indicated in Chapter 22, a substantial amount of glucose can be absorbed into the patient's blood and may contribute to weight gain over time. The high glucose load may also cause hyperglycemia and hypertriglyceridemia in some patients.

What are the advantages and disadvantages of peritoneal dialysis?

Peritoneal dialysis offers a number of advantages over hemodialysis: vascular access is not required, dietary restrictions are fewer, and the procedure can be scheduled when convenient. The most common complication is infection, which can occur at the catheter site or within the peritoneal cavity (**peritonitis**). Other problems that may arise include blood clotting in the catheter, catheter migration, and abdominal hernia due to the dialysate volume.

What are the features of continuous renal replacement therapy?

In people with acute kidney injury, **continuous renal replacement therapy (CRRT)** removes fluids and wastes. CRRT uses the process of **hemofiltration**, in which blood is gently pumped across a filtration membrane over a prolonged time period. (This process differs from dialysis treatments that rely on the diffusion of wastes across a membrane into the dialysate.) Either a pump or the patient's own blood pressure moves the blood across the

FIGURE NP22-2 Peritoneal Dialysis

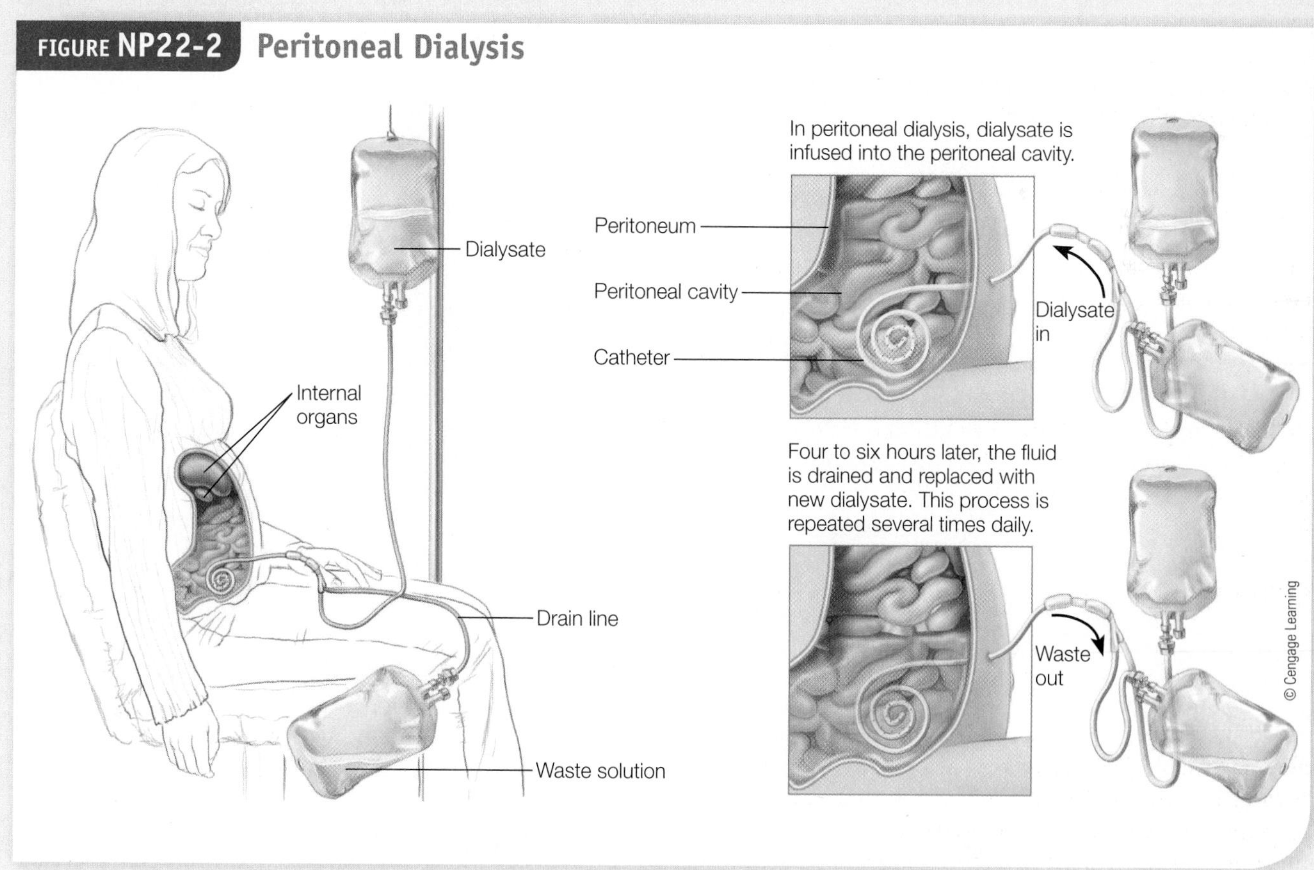

Dialysate
Internal organs
Drain line
Waste solution
Peritoneum
Peritoneal cavity
Catheter
In peritoneal dialysis, dialysate is infused into the peritoneal cavity.
Dialysate in
Four to six hours later, the fluid is drained and replaced with new dialysate. This process is repeated several times daily.
Waste out
© Cengage Learning

membrane. The procedure can be used to remove fluids, solutes, or both. Some patients require fluid replacement during the procedure to maintain adequate blood volume, so hydration status must be closely monitored.

The use of CRRT is advantageous in acute care situations because it corrects imbalances without causing sudden shifts in blood volume, which are poorly tolerated in acute care patients. In addition, replacement fluids can include parenteral feedings without upsetting fluid balance. Complications include clotting problems, damage to arteries, and inadequate blood flow rates in hypotensive patients.

Dialysis and CRRT help to remove the wastes and fluids that are normally removed by healthy kidneys. Although these procedures cannot restore the kidneys' hormonal functions, they provide a lifesaving means of alleviating symptoms of uremia, hypertension, and edema.

Notes

1. J. I. McMillan, Renal replacement therapy, in R. S. Porter and J. L. Kaplan, eds., *Merck Manual of Diagnosis and Therapy* (Whitehouse Station, NJ: Merck Sharp and Dohme Corp., 2011), pp. 2447–2453.
2. E. D. Weinhandl and coauthors, Survival in daily home hemodialysis and matched thrice-weekly in-center hemodialysis patients, *Journal of the American Society of Nephrology* 23 (2012): 895–904; P. Susantitaphong and coauthors, Effect of frequent or extended hemodialysis on cardiovascular parameters: A meta-analysis, *American Journal of Kidney Diseases* 59 (2012): 689–699.
3. National Kidney Foundation, K/DOQI clinical practice guidelines for hemodialysis adequacy: Update 2006, www.kidney.org/PROFESSIONALS/kdoqi/guideline_upHD_PD_VA/hd_guide2.htm; accessed October 20, 2012.
4. N. Tolkoff-Rubin, Treatment of irreversible renal failure, in L. Goldman and A. I. Schafer, eds., *Goldman's Cecil Medicine* (Philadelphia: Saunders, 2012), pp. 818–826.
5. Tolkoff-Rubin, 2012.

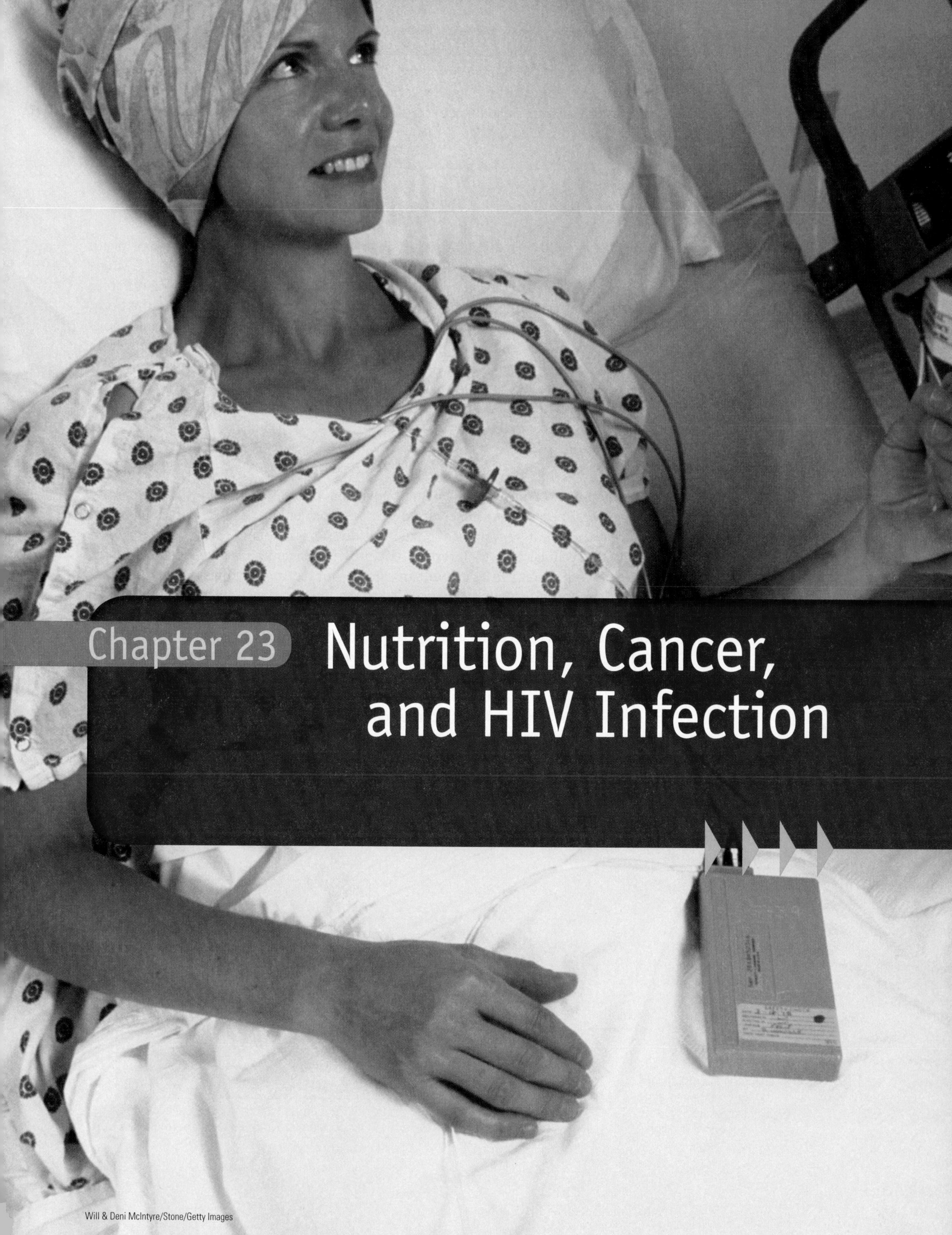

Chapter 23 Nutrition, Cancer, and HIV Infection

Will & Deni McIntyre/Stone/Getty Images

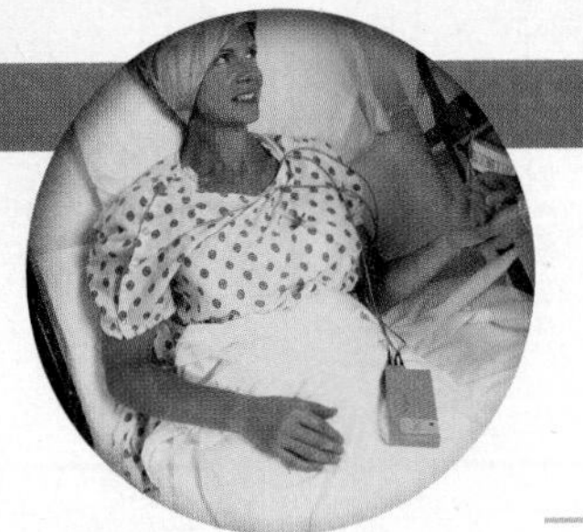

ALTHOUGH THEY ARE DISTINCT DISORDERS, FROM A NUTRITIONAL standpoint **cancers** and **human immunodeficiency virus (HIV)** share some similarities. Both disorders have debilitating effects that influence nutritional needs, and both can lead to severe wasting in advanced cases. These illnesses require nutrition therapy that is highly individualized based on the symptoms manifested and the tissues or organs involved.

Cancer

Cancer, the growth of **malignant** tissue, is the second most common cause of death in the United States, ranking just below cardiovascular disease. Cancer is not a single disorder, however; there are many different kinds of malignant growths. The different types of cancer have different characteristics, occur in different locations in the body, take different courses, and require different treatments. (■) Whereas an isolated, non-spreading type of skin cancer may be removed in a physician's office with no effect on nutrition status, advanced cancers—especially those of the gastrointestinal (GI) tract and pancreas—can seriously impair nutrition status. In the United States, the most common cancers are breast cancer (in women), prostate cancer (in men), lung cancer, and colorectal cancers.[1]

■ Cancers are classified by the tissues or cells from which they develop:

- *Adenocarcinomas* (ADD-eh-no-CAR-sih-NO-muz) arise from glandular tissues.
- *Carcinomas* (CAR-sih-NO-muz) arise from epithelial tissues.
- *Leukemias* (loo-KEY-mee-uz) arise from white blood cell precursors.
- *Lymphomas* (lim-FOE-muz) arise from lymphoid tissue.
- *Melanomas* (MEL-ah-NO-muz) arise from pigmented skin cells.
- *Myelomas* (MY-ah-LOE-muz) arise from plasma cells in the bone marrow.
- *Sarcomas* (sar-KO-muz) arise from connective tissues, such as muscle or bone.

HOW CANCER DEVELOPS

The development of cancer, called **carcinogenesis**, often proceeds slowly and continues for several decades. A cancer usually arises from either genetic mutations or the altered expression of genes that regulate cell division in a single cell.[2] These changes may promote cellular growth, interfere with growth restraint, or prevent cellular death. The affected cell thereby loses its built-in capacity for halting cell division and produces daughter cells with the same genetic defects. As the abnormal mass of cells, called a **tumor** (or *neoplasm*), grows, (■) a network of blood vessels forms to supply the tumor with the nutrients it needs to support its growth. The tumor can disrupt the functioning of the normal tissue around it, and some tumor cells may **metastasize**, spreading to other regions in the body. In leukemia (cancer affecting the white blood cells), the abnormal cells do not form a tumor; they accumulate in the blood and other tissues. Figure 23-1 illustrates the steps in cancer development.

■ An abnormal cellular mass that is noncancerous is called a *benign* tumor.

cancers: diseases characterized by the uncontrolled growth of abnormal cells, which can destroy adjacent tissues and spread to other areas of the body via lymph or blood.

The reasons why cancers develop are numerous and varied. Vulnerability to cancer is sometimes inherited, as when a person is born with a genetic defect that alters DNA structure, function, or repair. Certain metabolic processes may initiate carcinogenesis, as when chronic inflammation increases the rate of cell division and the risk of a damaging mutation. More often, cancers are caused by interactions between a person's

FIGURE 23-1 Cancer Development

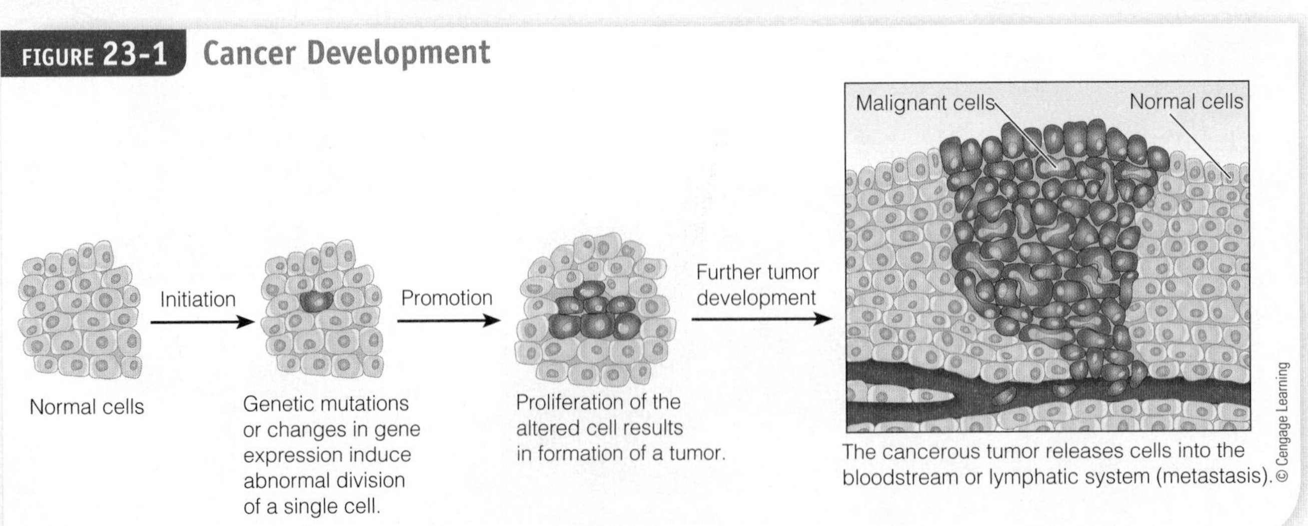
Initiation
Promotion
Further tumor development
Malignant cells
Normal cells
Normal cells
Genetic mutations or changes in gene expression induce abnormal division of a single cell.
Proliferation of the altered cell results in formation of a tumor.
The cancerous tumor releases cells into the bloodstream or lymphatic system (metastasis).
© Cengage Learning

genes and the environment. Exposure to cancer-causing substances, or **carcinogens**, may induce genetic mutations that lead to cancer; other substances may stimulate division or proliferation of the altered cells. Table 23-1 provides examples of environmental factors that increase cancer risk.

NUTRITION AND CANCER RISK

Like other environmental factors, diet and lifestyle strongly influence cancer risk. Various food components can alter processes of DNA repair, gene expression, or cell differentiation in ways that affect cancer development.[3] Moreover, certain food compounds can directly damage DNA, alter the metabolism of carcinogens by liver enzymes, or inhibit carcinogen formation in the body. Energy balance and growth rates may influence cancer risk by modifying rates of cell division (and therefore, the risks of damaging mutations) as well as levels of hormones that regulate cell growth. Table 23-2 lists examples of nutrition-related factors that may increase or decrease the risk of developing cancer.

human immunodeficiency virus (HIV): the virus that causes acquired immune deficiency syndrome (AIDS). HIV destroys immune cells and progressively impedes the body's ability to fight infections and certain cancers.

malignant (ma-LIG-nent): describes a cancerous cell or tumor, which can injure healthy tissue and spread cancer to other regions of the body.

carcinogenesis (CAR-sin-oh-JEN-eh-sis): the process of cancer development.

tumor: an abnormal tissue mass that has no physiological function; also called a *neoplasm* (NEE-oh-plazm).

metastasize (meh-TAS-tah-size): to spread from one part of the body to another; refers to cancer cells.

carcinogens (CAR-sin-oh-jenz or car-SIN-oh-jenz): substances that can cause cancer (the adjective is *carcinogenic*).

Nutrition and Increased Cancer Risk

As shown in Table 23-2, obesity is a risk factor for a number of different cancers, including some relatively common cancers such as colon cancer and postmenopausal breast cancer. Obesity increases cancer risk, in part, by altering levels of hormones that influence cell growth, such as the sex hormones, insulin, and several types of growth factors.[4] For example, in the case of breast cancer in postmenopausal women, the hormone estrogen is likely involved: obese women have higher estrogen levels than lean women do because adipose tissue is the primary source of estrogen after menopause. The increase in circulating estrogen may create an environment that encourages carcinogenesis in breast tissue.

Alcohol consumption correlates strongly with cancers of the head and neck, colon, rectum, and breast. For head and neck cancers, the risk is multiplied when alcohol drinkers also smoke tobacco.[5] In addition, alcohol abuse may damage the liver and

TABLE 23-1 Environmental Factors That Increase Cancer Risk

Environmental Factors	Cancer Sites
Aflatoxins (toxins in moldy peanuts or grains)	Liver
Arsenic	Skin, lung, bladder, kidney
Asbestos[a]	Lung, pleura, peritoneum
Chromium (hexavalent) compounds	Nasal cavity, lung
Estrogen replacement therapy	Endometrium, breast
Immunosuppressant drugs	Lymphoid tissues, liver
Infection with *Helicobacter pylori*	Stomach
Infection with hepatitis B and hepatitis C viruses	Liver
Infection with human papillomavirus (HPV)	Cervix, vulva, vagina, penis, anus, oropharynx
Ionizing radiation (X-rays, radon, radioactive isotopes, and other sources)	White blood cells (leukemia), thyroid, lung, bladder, breast, bone, liver
Tobacco[b]	Lip, oral cavity, nasal cavity, pharynx, larynx, lung, esophagus, stomach, colon, rectum, liver, pancreas, kidney, bladder
Ultraviolet radiation (sun exposure)	Skin

[a]Risk is greatly increased in cigarette smokers.
[b]A combined exposure to tobacco and alcohol multiplies the risks of developing cancers of the oral cavity, pharynx, larynx, and esophagus.

Sources: M. J. Thun and A. Jemel, Epidemiology of cancer, in L. Goldman and A. I. Schafer, eds., *Goldman's Cecil Medicine* (Philadelphia: Saunders, 2012), pp. 1177–1182; World Cancer Research Fund/American Institute for Cancer Research, *Food, Nutrition, Physical Activity, and the Prevention of Cancer: A Global Perspective* (Washington, DC: American Institute for Cancer Research, 2007), pp. 157–171.

TABLE 23-2 Nutrition-Related Factors That Influence Cancer Risk

Nutrition-Related Factors[a]	Cancer Sites
Factors that may increase cancer risk	
Obesity	Esophagus, colon, rectum, pancreas, gallbladder, kidney, breast (postmenopausal), endometrium
Red meat, processed meats	Colon, rectum
Salted and salt-preserved foods	Stomach
Beta-carotene supplements	Lung[b]
High-calcium diets (more than 1500 mg daily)	Prostate
Alcohol[c]	Mouth, pharynx, larynx, esophagus, colon, rectum, liver, breast
Low level of physical activity[d]	Colon, breast (postmenopausal), endometrium
Factors that may decrease cancer risk	
Fruits and nonstarchy vegetables	Lung, mouth, pharynx, larynx, esophagus, stomach
Carotenoid-containing foods	Lung, mouth, pharynx, larynx, esophagus
Tomato products	Prostate
Allium vegetables (onion, garlic)	Stomach, colon, rectum
Vitamin C–containing foods	Esophagus
Folate-containing foods	Pancreas
Fiber-containing foods	Colon, rectum
Milk and calcium supplements	Colon, rectum
High level of physical activity[d]	Colon, breast (postmenopausal), endometrium

[a]Altered cancer risk is associated with high intakes of the dietary substances listed.
[b]Cancer risk is increased in tobacco smokers and may not apply to other groups.
[c]A combined exposure to alcohol and tobacco multiplies the risks of developing cancers of the oral cavity, pharynx, larynx, and esophagus.
[d]Physical activity may influence cancer risk by altering body fatness, intestinal transit time, insulin sensitivity, hormone levels, enzyme activities, or immune responses.

Source: World Cancer Research Fund/American Institute for Cancer Research, *Food, Nutrition, Physical Activity, and the Prevention of Cancer: A Global Perspective* (Washington, DC: American Institute for Cancer Research, 2007).

instigate the development of liver cancer. These findings illustrate why the potential benefits of moderate alcohol consumption on cardiovascular disease risk must be weighed against the potential dangers.

Food preparation methods are responsible for producing certain types of carcinogens. Cooking meat, poultry, and fish at high temperatures (by frying or broiling, for example) causes carcinogens to form in these foods.[6] Carcinogens also accompany the smoke that adheres to food during grilling, and they are present in the charred surfaces of grilled meat and fish. (■) However, the cancer risk from eating such foods is unclear because the biological actions of these carcinogens are modulated by other dietary components, including compounds in vegetables and other plant foods. In several population studies, consumption of well-cooked meats was linked to cancers of the colon, rectum, breast, and prostate.[7]

■ To minimize carcinogen formation during cooking:

- Marinate meats before cooking.
- Use lower-heat options such as roasting, stewing, or microwaving.
- Choose lean meats for grilling, and take care not to blacken surfaces.
- To reduce smoke formation, prevent fat from dripping onto the heat source.

Nutrition and Decreased Cancer Risk The consumption of fruits and vegetables may provide some benefits in protecting against the development of cancer (see Table 23-2). Fruits and vegetables contain both nutrients and phytochemicals with antioxidant activity, and these substances may prevent or reduce the oxidative reactions in cells that cause DNA damage. Phytochemicals may also help to inhibit carcinogen production in the body, enhance immune responses that protect against cancer development, or promote enzyme reactions that inactivate carcinogens.[8] The B vitamin folate, which is provided by certain fruits and vegetables, plays roles in DNA synthesis and repair; thus, inadequate folate intakes may allow DNA damage to accumulate. Fruits and vegetables also contribute dietary fiber, which may help to protect against colon and rectal cancers by diluting potential carcinogens in fecal matter and accelerating their removal from the GI tract. Table 23-3 summarizes the dietary and lifestyle practices that may help to reduce the risk of developing cancer.

Polara Studios, Inc.

Cruciferous vegetables, such as cauliflower, broccoli, and brussels sprouts, contain nutrients and phytochemicals that inhibit cancer development.

CONSEQUENCES OF CANCER

Once cancer develops, its consequences depend on the location of the cancer, its severity, and the treatment. The complications that develop are often due to a tumor's impingement on surrounding tissues. Nonspecific effects of cancer include anorexia, malaise, weight loss, night sweats, and fever.[9] During the early stages, many cancers produce no symptoms, and the person may be unaware of the threat to health.

Wasting Associated with Cancer

Anorexia, muscle wasting, weight loss, anemia, and fatigue typify **cancer cachexia**, a condition of severe malnutrition that develops in up to 50 percent of cancer patients.[10] Without adequate energy and nutrients, the body is poorly equipped to maintain organ function, support immune defenses, and mend damaged tissues. An involuntary weight loss of more than 10 percent, which indicates significant malnutrition, is cause for concern;[11] care must be taken not to overlook unintentional weight loss in patients who are overweight or obese. Cachexia is directly responsible for as many as 20 percent of cancer deaths.[12]

cancer cachexia (ka-KEK-see-ah): a wasting syndrome associated with cancer that is characterized by anorexia, muscle wasting, weight loss, and fatigue.

TABLE 23-3 Guidelines for Reducing Cancer Risk

Achieve and maintain a healthy body weight throughout life. • Be as lean as possible within the normal range of body weight for your height. • Avoid weight gain and increases in waist circumference throughout adulthood.
Be physically active as part of everyday life. • For adults: engage in moderate activity (equivalent to brisk walking) for at least 30 minutes each day; increase duration or intensity of physical activity as fitness improves. • For children and adolescents: engage in moderate to vigorous activity for at least 60 minutes each day. • Limit sedentary habits such as watching television.
Choose a healthy diet that emphasizes plant sources. • Limit consumption of energy-dense foods (>225 kcal/100 g food) and sugary drinks that contribute to weight gain. • Consume relatively unprocessed grains and/or legumes with every meal. Choose whole-grain products instead of processed (refined) grains. • Consume at least 2½ cups of nonstarchy vegetables and fruits every day.
Limit consumption of foods that may increase cancer risk. • Limit consumption of red meats (beef, pork, or lamb) to 18 ounces per week. • Limit consumption of processed meats (those preserved by smoking, curing, or salting). • Avoid salt-preserved, salted, and salty foods. • Avoid moldy grains and legumes.
Limit consumption of alcoholic beverages. • For women: drink no more than one drink daily. • For men: drink no more than two drinks daily.
Aim to meet nutritional needs through the diet. • Obtain necessary nutrients from the diet. Dietary supplements are not recommended for cancer prevention, and they may have unexpected adverse effects.
Avoid using tobacco in any form.

Sources: L. H. Kushi and coauthors, American Cancer Society guidelines on nutrition and physical activity for cancer prevention: Reducing the risk of cancer with healthy food choices and physical activity, *CA: A Cancer Journal for Clinicians* 62 (2012): 30–67; World Cancer Research Fund/American Institute for Cancer Research, *Food, Nutrition, Physical Activity, and the Prevention of Cancer: A Global Perspective* (Washington, DC: American Institute for Cancer Research, 2007).

NURSING DIAGNOSIS

The nursing diagnoses *imbalanced nutrition: less than body requirements, adult failure to thrive,* and *ineffective protection* may apply to those with cancer cachexia.

Many factors play a role in the wasting associated with cancer. Cytokines, (■) released by both tumor cells and immune cells, induce an inflammatory and catabolic state. The combined effects of a poor appetite, accelerated and abnormal metabolism, and the diversion of nutrients to support tumor growth result in a lower supply of energy and nutrients at a time when demands are high. Appetite and food intake are further disturbed by the effects of treatments and medications prescribed for cancer patients. Unlike in starvation, nutrition intervention alone is unable to reverse cachexia.[13]

Metabolic Changes The metabolic changes that arise in cancer exacerbate the wasting described in the previous section.[14] Cancer patients exhibit an increased rate of protein turnover, (■) but reduced muscle protein synthesis. Muscle contributes amino acids for glucose production, further depleting the body's supply of protein. Triglyceride breakdown increases, elevating serum lipids. Many patients develop insulin resistance. These metabolic abnormalities help to explain why people with cancer fail to regain lean tissue or maintain healthy body weights even when they are consuming adequate energy and nutrients.

NURSING DIAGNOSIS

The nursing diagnoses *nausea, chronic pain, fatigue, death anxiety, impaired oral mucous membrane, impaired swallowing, diarrhea,* and *constipation* may be associated with poor food intake in people with cancer.

Anorexia and Reduced Food Intake Anorexia is a major contributor to the wasting associated with cancer. Some factors that contribute to anorexia or otherwise reduce food intake include:

- *Chronic nausea and early satiety.* People with cancer frequently experience nausea and a premature feeling of fullness after eating small amounts of food.
- *Fatigue.* People with cancer may tire easily and lack the energy to prepare and eat meals. Once cachexia develops, these tasks become even more difficult.
- *Pain.* People in pain may have little interest in eating, particularly if eating makes the pain worse.
- *Mental stress.* A cancer diagnosis can cause distress, anxiety, and depression, all of which may reduce appetite. Facing and undergoing cancer treatments induce additional psychological stress.
- *Gastrointestinal obstructions.* A tumor may partially or completely obstruct a portion of the GI tract, causing complications such as nausea and vomiting, early satiety, delayed gastric emptying, and bacterial overgrowth. Some patients with obstructions are unable to tolerate oral diets.
- *Effects of cancer therapies.* Chemotherapy and radiation treatments for cancer frequently have side effects that make food consumption difficult, such as nausea, vomiting, dry mouth, altered taste perceptions, food aversions, mouth sores, inflammation of the mouth and esophagus, difficulty swallowing, abdominal pain or discomfort, diarrhea, and constipation.

TREATMENTS FOR CANCER

The primary medical treatments for cancer—surgery, chemotherapy, radiation therapy, or any combination of the three—aim to remove cancer cells, prevent further tumor growth, and alleviate symptoms.[15] The likelihood of effective treatment is highest with early detection and intervention. Because treatment decisions are difficult and cancer therapies have considerable side effects, patients rely on health care providers to help them make informed decisions.

■ The cytokines that induce cachexia include tumor necrosis factor-α, interleukin-1, interleukin-6, and γ-interferon.

■ Reminder: *Protein turnover* refers to the continuous degradation and synthesis of the body's proteins.

Surgery Surgery is performed to remove tumors, determine the extent of cancer, and protect nearby tissues. Often, surgery must be followed by other cancer treatments to prevent the growth of new tumors. The acute metabolic stress caused by surgery raises protein and energy needs and can exacerbate wasting. Surgery also contributes to pain, fatigue, and anorexia, all of which can reduce food intake at a time when nutritional needs are substantial. Blood loss contributes to nutrient losses and further exacerbates malnutrition. Some surgeries can have long-term effects on nutrition status (see Table 23-4).

chemotherapy: the use of drugs to arrest or destroy cancer cells; these drugs are called *antineoplastic agents.*

Chemotherapy **Chemotherapy** relies on the use of drugs to treat cancer and is used to inhibit tumor growth, shrink tumors before surgery, and prevent or eradicate metastasis. Some cancer drugs interfere with the process of cell division; (■) others sterilize cells that are in a resting phase and are not actively dividing. Unfortunately, most of these drugs have toxic effects on normal cells as well and are especially damaging to rapidly dividing cells, such as those of the GI tract, skin, and bone marrow. The bone marrow damage can suppress the production of red blood cells (causing anemia) and white blood cells (causing **neutropenia**). Some of the newer drugs target properties specific to cancer cells and are better tolerated by the body's tissues. Table 23-5 (p. 656) describes some nutrition-related side effects that may result from chemotherapy.

TABLE 23-4 Nutrition-Related Side Effects of Cancer Surgeries

Head and Neck Surgeries
Aspiration
Dry or sore mouth
Reduced chewing or swallowing ability
Reduced sense of taste or smell
Esophageal Resection
Acid reflux
Altered gastric motility
Reduced swallowing ability
Gastric Resection
Dumping syndrome
Early satiety
Inadequate gastric acid secretion
Malabsorption of iron, folate, and vitamin B_{12}
Intestinal Resection
Bile insufficiency
Diarrhea
Fluid and electrolyte imbalances
General malabsorption
Pancreatic Resection
Diabetes mellitus
General malabsorption

Radiation Therapy

Radiation therapy treats cancer by bombarding cancer cells with X-rays, gamma rays, or various atomic particles. These treatments generate reactive forms of oxygen, such as superoxide and hydroxyl radicals, which can damage cellular DNA and cause cell death. Newer techniques are able to focus radiation directly at tumors and minimize damage to nearby tissues. An advantage of radiation therapy over surgery is that it can shrink tumors while preserving organ structure and function. Compared with chemotherapy, radiation therapy is better able to target specific regions of the body, rather than involving all body cells. Nonetheless, radiation therapy can damage healthy tissues and sometimes has long-term detrimental effects on nutrition status (see Table 23-5). Radiation to the head and neck area can damage the salivary glands and taste buds, causing inflammation, dry mouth, and a reduced sense of taste; in severe cases, damage may be permanent. Radiation treatment in the lower abdominal area can cause **radiation enteritis**, an inflammatory condition of the small intestine that causes nausea, vomiting, malabsorption, and diarrhea; the condition may persist for months or years and lead to chronic malabsorption in some individuals.

Hematopoietic Stem Cell Transplantation

Hematopoietic stem cell transplantation replaces the blood-forming stem cells that have been destroyed by high-dose chemotherapy or radiation therapy. These procedures may be used to treat leukemias, lymphomas, and multiple myeloma.[16] If possible, the stem cells are collected from the patient's bone marrow or circulating blood before chemotherapy or radiation treatment begins so that it is not necessary to find a separate donor. If another person's cells are used, the patient must take immunosuppressant drugs to prevent **tissue rejection**.

The treatments required for stem cell transplantation can have a substantial impact on food intake and nutrition status. The high-dose chemotherapy or radiation therapy preceding the transplant and the immunosuppressant drugs often required afterward can impair immune function substantially and increase the risk of infection and foodborne illness. Other common complications include anorexia, nausea, vomiting, dry mouth, altered taste sensations, inflamed mucous membranes, malabsorption, and diarrhea. Patients are often unable to consume adequate food during or after the procedures and usually require nutrition support.

Biological Therapies

Newer therapies for cancer include the use of biological molecules that stimulate immune responses against cancer cells (also called *immunotherapy*). These substances include antibodies, cytokines, and other proteins that strengthen the body's immune defenses, enable the destruction of cancer cells, or interfere with cancer development in some way. Although side effects vary, many of these treatments can cause anorexia, GI symptoms, and general discomfort, reducing a person's ability or desire to consume adequate amounts of food.

■ One drug that inhibits cell division is *methotrexate,* which closely resembles the B vitamin folate (see Figure 14-3 on p. 418). Folate is required for cell division because it is needed for DNA synthesis. Methotrexate works by blocking activity of the enzyme that converts folate to its active form.

NURSING DIAGNOSIS

Nursing diagnoses that may apply to people with stem cell transplantation include *nausea, diarrhea, imbalanced nutrition: less than body requirements, ineffective protection, risk for infection,* and *impaired oral mucous membrane.*

neutropenia: a low white blood cell (neutrophil) count, which increases susceptibility to infection.

radiation therapy: the use of X-rays, gamma rays, or atomic particles to destroy cancer cells.

TABLE 23-5 Nutrition-Related Side Effects of Chemotherapy and Radiation Therapy

	Reduced Nutrient Intake	Accelerated Nutrient Losses	Altered Metabolism
Chemotherapy	Abdominal pain Anorexia Mouth ulcers Nausea and vomiting Reduced taste sensation	Diarrhea Gastrointestinal inflammation Malabsorption Vomiting	Anemia, neutropenia Fluid and electrolyte imbalances as a consequence of vomiting, diarrhea, or malabsorption Hyperglycemia Interference with vitamins or body compounds Negative nitrogen and micronutrient balances Secondary effects of malnutrition, infection, or inflammation
Radiation therapy	Anorexia Damage to teeth, jaws, or salivary glands Dysphagia Esophagitis Mouth sores Nausea and vomiting Reduced salivary secretions Reduced taste sensation	Blood loss from intestine and bladder Diarrhea Fistulas Intestinal obstructions Malabsorption Radiation enteritis Vomiting	Fluid and electrolyte imbalances as a consequence of vomiting, diarrhea, or malabsorption Secondary effects of malnutrition, infection, or inflammation

Medications to Combat Anorexia and Wasting Medications prescribed to stimulate the appetite and promote weight gain include megestrol acetate (Megace), a synthetic compound similar in structure to the hormone progesterone, and dronabinol (Marinol), which resembles the psychoactive ingredient in marijuana and stimulates the appetite at doses that have minimal mental effects. Antiemetic drugs, which control nausea and vomiting, are typically coadministered with chemotherapeutic drugs to improve appetite and food intake. Under investigation are medications that promote muscle protein synthesis, induce the secretion of growth hormone or growth factors, or reduce catabolism.[17]

■ Reminder: *Complementary and alternative medicine (CAM)* refers to health care practices that have not been proved to be effective and consequently are not included as part of conventional treatment (see Nutrition in Practice 14).

Alternative Therapies Many patients turn to *complementary and alternative medicine (CAM)* (■) to assist them in their fight against cancer. Patients may use CAM because they wish to have more control over their treatments or because they are concerned about the effectiveness of conventional approaches. Although few abandon conventional medicine, up to 80 percent of cancer patients combine one or more CAM approaches with standard treatment.[18] Many patients do not discuss their use of CAM with physicians.

Multivitamin and herbal supplements are among the most frequently used CAM therapies. Although many supplements can be used without risk, some may have adverse effects or interfere with conventional treatments. Use of the herb St. John's wort, for example, can reduce the effectiveness of some anticancer drugs.[19] As another example, some studies suggest that antioxidant supplements interfere with chemotherapy and radiation treatments.[20] Most research suggests that the use of dietary supplements (including multivitamin supplements) is unable to improve outcomes or survival after a cancer diagnosis and may actually increase mortality rates.[21]

radiation enteritis: inflammation of intestinal tissue caused by radiation therapy.

hematopoietic stem cell transplantation: transplantation of the stem cells that produce red blood cells and white blood cells; the stem cells are obtained from the bone marrow *(bone marrow transplantation)* or circulating blood.

haima = blood
poiesis = to make

tissue rejection: destruction of donor tissue by the recipient's immune system, which recognizes the donor cells as foreign.

NUTRITION THERAPY FOR CANCER

The goals of nutrition therapy for cancer patients are to maintain a healthy weight, preserve muscle tissue, prevent or correct nutrient deficiencies, and provide a diet that patients can tolerate and enjoy despite the complications of illness.[22] Appropriate nutrition care helps patients preserve their strength and improves recovery after stressful cancer treatments. Moreover, malnourished cancer patients develop more complications and have shorter survival times than patients who maintain good nutrition status.

Because there are many forms of cancer and a variety of potential treatments, nutritional needs among cancer patients vary considerably. Furthermore, a person's needs may change at different stages of illness. Patients should be screened for malnutrition when cancer is diagnosed and be reassessed during the treatment and recovery periods.

Protein and Energy For patients at risk of weight loss and wasting, the focus of nutrition care is to ensure appropriate intakes of protein and energy. Protein requirements are often between 1.0 and 1.6 grams per kilogram of body weight daily.[23] Energy needs may range from 25 to 35 kcalories per kilogram of body weight per day for most patients, with higher amounts recommended for patients with severe stress, hypermetabolism, and various other complications.[24] Patients who cannot eat adequate food may be able to meet their needs by supplementing the diet with nutrient-dense oral supplements. The "How to" provides suggestions that can help to increase the energy and protein content of meals. (■)

Although weight loss is a problem for many cancer patients, breast cancer patients often gain weight.[25] The weight gain occurs during the first two years after breast cancer diagnosis and is associated with an increase in total body fat. By discussing weight maintenance soon after diagnosis and encouraging physical activity, health practitioners can help patients avoid unnecessary weight gain.

■ Reminder: The high-kcalorie, high-protein diet, which is appropriate for some individuals with cancer, is described in Chapter 14 (see pp. 411–412).

HOW TO Increase kCalories and Protein in Meals

To increase the energy content of a meal, try these suggestions:

- *Meats.* Choose high-fat meats instead of lean meats. Sauté or pan-fry meat instead of baking or roasting it, and use sauces or gravies liberally. Sprinkle bacon bits or sausage pieces on vegetable dishes.
- *Cheese.* Include cheese slices or cream cheese in sandwiches made with luncheon meats. Spread cream cheese on raw vegetables, toast, and crackers or mix into dishes that contain chopped fruit.
- *Half-and-half and cream.* Replace milk or water with half-and-half or cream in soups, sauces, hot chocolate, desserts, mashed potatoes, and cold cereals. Use sour cream or cream sauces on potato dishes, vegetable dishes, and soups. Add whipped cream to fruit salads and desserts.
- *Breads and cereals.* Choose high-fat grain products such as granola, pancakes, waffles, French toast, and biscuits. Prepare hot cereals with whole milk or cream, or added fat.
- *Fruits.* Mash avocados to make guacamole, or use mashed avocado as a sandwich spread. Add chopped dried fruits to salads and baked goods. Snack on dried fruits between meals.
- *Nuts.* Add chopped nuts to pasta dishes, stir-fried vegetables, fruit salads, and green salads. Use nut meats in baked products. Spread nut butters on bread and crackers.
- *Butter or margarine.* Melt on pasta, potatoes, rice, and cooked vegetables. Add to hot cereals, casseroles, and soups. Spread liberally on bread, crackers, and rolls.
- *Mayonnaise or salad dressings.* Add to pasta, tuna, and potato salads. Use as dressings for raw or cooked vegetables.
- *Beverages.* Replace water and non-kcaloric beverages with sweetened drinks, fruit juices, and milk shakes. Drink whole milk instead of low-fat or nonfat milk. Add strawberry or chocolate syrup to plain milk to boost kcalories.

These suggestions can help to add protein to a meal:

- *Meats.* Add small chunks of meat to soups, egg dishes, casseroles, bean dishes, and pasta sauces. Add minced meats to vegetable dishes. Add chunks of cooked chicken or turkey to salads.
- *Eggs.* Add raw eggs when preparing casseroles, meatballs, and hamburgers. Add chopped hard-cooked eggs to salads, vegetable dishes, sandwich fillings, and pasta and potato salads.
- *Cheese.* Melt on burgers, meat loaf, cooked vegetables, scrambled eggs, casseroles, and potatoes. Add cottage cheese to casseroles, egg dishes, pasta recipes, and salad dressings. Grate hard cheeses and sprinkle on soups, salads, and cooked vegetable dishes. Avoid using reduced-fat cheeses.
- *Milk.* Use in place of water when preparing cereals and soups. Use cream sauces (which are made with milk) to flavor vegetable and pasta dishes.
- *Powdered milk (use full-fat milk powder if available).* Add to recipes that include milk. Dissolve extra milk powder into milk-containing beverages. Stir into hot cereals, potato dishes, casseroles, and sauces. Add to scrambled eggs, hamburgers, and meat loaf.
- *Protein supplements.* Snack on protein bars between meals. Add protein powders to beverages and shakes. Drink meal replacement formulas, such as Ensure or Boost, instead of juices or soda.

Managing Symptoms and Complications A thorough nutrition assessment often uncovers specific problems or symptoms that interfere with food consumption. Table 23-6 lists dietary considerations related to cancers affecting different sites in the body. The "How to" describes dietary strategies that may alleviate symptoms and improve food intake. Patients' responses to these strategies can vary considerably, and in some cases a number of adjustments may be necessary.

TABLE 23-6 Dietary Considerations for Specific Cancers

Cancer Sites	Common Complications[a]	Possible Dietary Measures
Brain and nervous system	Chewing or swallowing difficulty, headache, altered taste or smell sensation, difficulty feeding oneself	Mechanically altered diet, use of adaptive feeding devices (see Nutrition in Practice 21)
Head and neck[b]	Chewing or swallowing difficulty, aspiration, inflamed mucosa, dry mouth, altered taste or smell sensation	Tube feeding, mechanically altered diet
Esophagus	Swallowing difficulty, aspiration, obstruction, acid reflux, inflamed mucosa	Tube feeding, mechanically altered diet
Stomach	Anorexia, early satiety, reduced secretion of gastric acid and intrinsic factor, delayed stomach emptying, dumping syndrome, malabsorption, nutrient deficiencies	Tube feeding (for obstruction or unmanageable dumping syndrome); postgastrectomy diet; small, frequent meals; limited sugars and insoluble fibers (see Chapter 17); nutrient supplementation
Intestine	Inflamed mucosa, bacterial overgrowth, obstruction, lactose intolerance, general malabsorption, bile insufficiency, nutrient deficiencies, short bowel syndrome (if resected), altered bowel function, fluid and electrolyte imbalances	Tube feeding or total parenteral nutrition for obstruction, enteritis, or short bowel syndrome; fat- and lactose-restricted diet (see Chapter 18); nutrient supplementation
Pancreas	Reduced secretion of digestive enzymes, bile insufficiency, general malabsorption, nutrient deficiencies, hyperglycemia	Fat-restricted diet; enzyme replacement (see Chapter 18); small, frequent meals; carbohydrate-controlled diet (Chapter 20); nutrient supplementation

[a]Actual complications depend on the exact location of the cancer and the specific methods used for treating the cancer.
[b]Includes cancers of the salivary glands, oral and nasal cavities, pharynx, and larynx.

HOW TO Help Patients Handle Food-Related Problems

In people with cancer or HIV infection, various complications can interfere with food intake. Health care providers can try to identify a patient's specific problems and offer appropriate solutions. Not every suggestion will work for each person; encourage patients to experiment and find strategies that work best.

I just don't have an appetite.

- Eat small meals and snacks at regular times each day.
- Eat the largest meal at the time of day when you feel the best.
- Include nutrient-dense foods in meals, and consume them before other foods.
- Indulge in favorite foods throughout the day. Serve foods attractively.
- Avoid drinking large amounts of liquids before or with meals.
- Eat in a pleasant and relaxed environment. Eat with family and friends when possible.
- Listen to your favorite music or enjoy a TV or radio program while you eat.
- Ask your doctor about appetite-enhancing medications.

I am too tired to fix meals and eat.

- Let family members and friends prepare food for you.
- Obtain foods that are easy to prepare and easy to eat, such as sandwiches, frozen dinners, take-out meals from restaurants, instant breakfast drinks, liquid formulas, and energy bars.

HOW TO Help Patients Handle Food-Related Problems, *continued*

- Find time to rest before you attempt to prepare a large meal.
- Prepare soups, stews, and casserole dishes in sufficient quantity to provide enough for several meals, so that you will have enough to eat at times when you are too tired to cook.

Foods just don't taste right.

- Brush your teeth or use mouthwash before you eat.
- Consume foods chilled or at room temperature. Use plastic, rather than metal, eating utensils.
- Choose eggs, fish, poultry, and milk products instead of meats.
- Experiment with sauces, seasonings, herbs, spices, and sweeteners to improve food taste and flavor.
- Save your favorite foods for times when you are not feeling nauseated.

I am nauseated a lot of the time, and sometimes I need to vomit.

- Consume liquids throughout the day to replace fluids.
- If you become nauseated from chemotherapy treatments, avoid eating for at least two hours before treatments.
- Consume your largest meal at a time when you are least likely to feel nauseous.
- Try consuming smaller meals, and eat slowly. Experiment with foods to see if some foods cause nausea more than others.
- Avoid foods and meals that have strong odors or are fatty, greasy, or gas forming.

I have problems chewing and swallowing food.

- Experiment with food consistencies to find the ones you can manage best. Thin liquids, dry foods, and sticky foods (such as peanut butter) are often difficult to swallow.
- Add sauces and gravies to dry foods.
- Drink fluids during meals to ease chewing and swallowing.
- Try using a straw to drink liquids. Experiment with beverage thickeners if you cannot tolerate thin beverages.
- Tilt your head forward and backward to see if you can swallow more easily when your head is positioned differently.

I have sores in my mouth, and they hurt when I eat.

- Try eating chilled or frozen foods; they are often soothing.
- Try soft foods such as ice cream, milk shakes, bananas, applesauce, mashed potatoes, cottage cheese, and macaroni and cheese. Mix dry foods with sauces or gravies.
- Cut foods into smaller pieces, so they are less likely to irritate the mouth.
- Avoid foods that irritate mouth sores, such as citrus fruits and juices, tomatoes and tomato-based products, spicy foods, foods that are very salty, foods with seeds (such as poppy seeds and sesame seeds) that can scrape the sores, and coarse foods such as raw vegetables, crackers, corn chips, and toast.
- Ask your doctor about using a local anesthetic solution such as lidocaine before eating to reduce pain.
- Use a straw for drinking liquids, in order to bypass the sores.

My mouth is really dry.

- Rinse your mouth with warm salt water or mouthwash frequently. Avoid using mouthwash that contains alcohol.
- Drink small amounts of liquid frequently between meals.
- Ask your doctor or pharmacist about medications or saliva substitutes that can help a dry mouth condition.
- Use sour candy or chewing gum to stimulate the flow of saliva.
- Sip fluids frequently while eating. Add broth, sauces, gravies, mayonnaise, butter, or margarine to dry foods.
- Make sure you brush your teeth and floss regularly to prevent tooth decay and oral infections.

I am having trouble with constipation.

- Drink plenty of fluids. Try warm fluids, especially in the morning.
- Eat whole-grain breads and cereals, nuts, fresh fruits and vegetables, prunes, and prune juice. Avoid refined carbohydrate foods such as white bread, white rice, and pasta.
- Engage in physical activity regularly.
- Try an over-the-counter bulk-forming agent, such as methylcellulose (Citrucel), psyllium (Metamucil or Fiberall), or polycarbophil (Fiber-Lax).

I am having trouble with diarrhea.

- To avoid dehydration, drink plenty of fluids throughout the day. Salty broths and soups, diluted fruit juices, and sports drinks are good choices. Avoid caffeine- and alcohol-containing beverages. For severe diarrhea, try oral rehydration formulas that are commercially prepared.
- Avoid foods and beverages that increase gas, such as legumes, onions, vegetables of the cabbage family, foods that contain sorbitol or mannitol, and carbonated beverages.
- Try using lactase enzyme replacements when you use milk products in case you are experiencing lactose intolerance. Yogurt and aged cheeses may be easier to tolerate than milk and fresh cheeses.
- Avoid fatty foods if you are fat intolerant. Try reducing your intake of whole-grain breads and cereals if they worsen the diarrhea.
- Eat small, frequent meals instead of large ones. Try consuming cool or lukewarm foods instead of very cold or hot foods.
- Ask your doctor about using a bulk-forming agent or antidiarrheal medication.

Low-Microbial Diet

Patients with suppressed immunity or neutropenia may be prescribed a **low-microbial diet** (also called a *neutropenic diet*), which includes only foods that are unlikely to be contaminated with bacteria or other microbes.[26] Generally, patients should consume only well-cooked meats and eggs, pasteurized milk products, well-washed fruits and vegetables, and shelf-stable packaged foods. Foods that must be avoided include unwashed raw fruits and vegetables; unpasteurized juices and milk products; undercooked meat, poultry, and eggs; leftover luncheon meats and meat spreads; leftover foods that have not been adequately reheated; and foods from salad bars or street vendors. In addition, patients should be instructed to follow safe food-handling practices to minimize the risk of foodborne illness (see Nutrition in Practice 2).

Enteral and Parenteral Nutrition Support

Tube feedings or parenteral nutrition may be needed by patients who have long-term or permanent gastrointestinal impairment or are experiencing complications that interfere with food intake.[27] For example, many patients undergoing radiation therapy for head and neck cancers have dysphagia and require long-term tube feeding. Parenteral nutrition is reserved for patients who have inadequate GI function, such as those with severe radiation enteritis. Whenever possible, enteral nutrition is strongly preferred over parenteral nutrition, to preserve GI function and avoid infection.

low-microbial diet: a diet that contains foods that are unlikely to be contaminated with bacteria and other microbes.

Review Notes

- Cancer arises from mutations in the genes that control cell division. Some dietary substances promote carcinogenesis, whereas others may help to prevent cancer.
- Cancer's effects on nutrition status depend on the type of cancer a person has, its severity, and the methods used to treat the cancer. Cancer cachexia is a frequent complication of cancer and may be a consequence of anorexia, altered metabolism, and responses to cancer treatment.
- Medical treatments for most cancers include surgery, chemotherapy, and/or radiation therapy, which remove cancer cells, prevent tumor growth, and alleviate symptoms. Nutrition therapy aims to minimize weight loss and wasting, correct deficiencies, and manage complications that impair food intake. The Case Study allows you to apply information about nutrition and cancer to a clinical situation.

Case Study: Woman with Cancer

Jeanne Monroy is a 58-year-old public relations consultant who was recently diagnosed with colon cancer after a routine colonoscopy, a procedure in which the colon is examined using a flexible tube attached to an optical device. Mrs. Monroy is scheduled to have surgery to remove the segment of colon that contains the tumor and to determine if the cancer has spread to the surrounding lymph nodes and, possibly, other organs. The nurse completing the nutrition assessment finds that Mrs. Monroy is 5 feet 5 inches tall and weighs 178 pounds. Mrs. Monroy usually spends most of the day sitting and has little time to engage in recreational exercise. Her diet typically includes red meat at both lunch and dinner, and she consumes one or two glasses of wine with both meals. She eats two or three servings of fruits and vegetables each day, although she does not like green leafy vegetables very much. She rarely drinks milk or consumes milk products.

1. Review Table 23-2 on p. 652, and describe the factors in Mrs. Monroy's diet and lifestyle that may have contributed to the development of colon cancer.
2. What symptoms and complications may arise after colon surgery and impair nutrition status? If the cancer team decides that Mrs. Monroy needs follow-up chemotherapy, how might the chemotherapy affect her nutrition status?
3. If Mrs. Monroy is unresponsive to treatment and her cancer progresses, she may develop cancer cachexia. Describe this syndrome, its causes, and its consequences.
4. Provide suggestions that may help Mrs. Monroy handle the following problems should they develop: poor appetite, fatigue, taste alterations, nausea and vomiting, chewing and swallowing difficulties, mouth sores, dry mouth, diarrhea, constipation, and weight loss.

HIV Infection

Perhaps the most infamous infectious disease today is **acquired immune deficiency syndrome (AIDS)**. AIDS develops from infection with human immunodeficiency virus (HIV), which attacks the immune system and disables a person's defenses against other diseases, including infections and certain cancers. Then these diseases—which would produce mild, if any, illness in people with healthy immune systems—destroy health and life. In the 30-plus years since AIDS has been identified, it has caused over 25 million deaths worldwide.[28]

Although the global incidence of HIV infection has been declining in recent years, its prevalence continues to be high in sub-Saharan Africa, where it affects 5 percent of the adult population (see Table 23-7).[29] Fortunately, remarkable progress has been made in understanding and treating HIV infection. Access to antiretroviral drugs continues to increase throughout the world, reducing AIDS-related deaths and the risk of HIV transmission.

PREVENTION OF HIV INFECTION

As there is still no cure for AIDS, the best course is prevention. HIV is most often sexually transmitted and can be spread by direct contact with contaminated body fluids, such as blood, semen, vaginal secretions, and breast milk. Because many people remain symptom-free during the early stages of infection, they may not realize that they can pass the infection to others. To reduce the spread of HIV infection, individuals at risk (see Table 23-8, p. 662) are encouraged to undergo testing. A blood test can usually detect HIV antibodies within several months after exposure and, often, after two or three weeks. An estimated 21 percent of persons in the United States who have HIV infection are unaware that they are infected.[30]

CONSEQUENCES OF HIV INFECTION

HIV infection destroys immune cells that have a protein called CD4 on their surfaces.[31] The cells most affected are the **helper T cells**, (■) also called *CD4+ T cells* because the presence of CD4 is a primary characteristic. HIV is able to enter the helper T cells and induce them to produce additional copies of the virus, thus perpetuating and exacerbating the infection. Other cells that have the CD4 protein (and are infected by HIV) include tissue macrophages and certain cells of the central nervous system. Early symptoms of HIV infection are nonspecific and may include fever, sore throat, malaise, swollen lymph nodes, skin rashes, muscle and joint pain, and diarrhea. After these symptoms subside, many people remain symptom-free for 5 to 10 years or even longer. If the HIV infection is not treated, however, the depletion of T cells eventually increases the person's susceptibility to **opportunistic infections**—that is, infections caused by microorganisms that normally do not cause disease in healthy individuals.

■ *T cells* are lymphocytes that develop in the thymus gland. The other lymphocytes are the *B cells* (which develop in bone marrow) and *natural killer cells.*

TABLE 23-7 The HIV and AIDS Epidemic at a Glance, 2010

Stage of Epidemic	World	Sub-Saharan Africa	Asia	North America
Individuals living with HIV infection	34,000,000	22,900,000	4,800,000	1,300,000
Individuals newly infected with HIV	2,700,000	1,900,000	360,000	58,000
AIDS-related deaths	1,800,000	1,200,000	310,000	20,000

Source: World Health Organization, *Global HIV/AIDS Response: Epidemic Update and Health Sector Progress towards Universal Access: Progress Report 2011*, available at http://whqlibdoc.who.int/publications/2011/9789241502986_eng.pdf; site visited November 2, 2012.

acquired immune deficiency syndrome (AIDS): the late stage of illness caused by infection with the human immunodeficiency virus (HIV); characterized by severe damage to immune function.

helper T cells: lymphocytes that have a specific protein called CD4 on their surfaces and therefore are also known as *CD4+ T cells;* these are the cells most affected in HIV infection.

opportunistic infections: infections from microorganisms that normally do not cause disease in healthy people but are damaging to persons with compromised immune function.

TABLE 23-8 Risk Factors for HIV Infection

- History of receiving blood transfusions or blood components before 1985
- Infant born to mother with HIV infection
- Intravenous drug use in which syringes are shared among users
- Sexual contact with intravenous drug users, prostitutes, or individuals with a history of HIV or other sexually transmitted diseases
- Sexual contact with multiple partners
- Unsafe sexual practices

NURSING DIAGNOSIS

Nursing diagnoses appropriate for those with HIV infection may include *ineffective protection, risk for infection, impaired oral mucous membrane, risk for impaired skin integrity, chronic pain, diarrhea, risk for deficient fluid volume, fatigue, imbalanced nutrition: less than body requirements, hopelessness,* and *death anxiety.*

The term *AIDS* applies to the advanced stages of HIV infection, in which the inability to fight illness allows a number of serious diseases and complications to develop; such **AIDS-defining illnesses** include severe infections, certain cancers, and wasting of muscle tissue. Without treatment, AIDS develops in 26 to 36 percent of HIV-infected persons within 7 years.[32] Health practitioners evaluate disease progression by measuring the concentrations of helper T cells and circulating virus (called the *viral load*) and by monitoring clinical symptoms. Although current drug therapies dramatically slow the progression of HIV infection, the drugs' side effects may make it difficult for patients to adhere to treatments, as discussed in several of the following sections.

AIDS-defining illnesses: diseases and complications associated with the later stages of an HIV infection, including wasting, recurrent bacterial pneumonia, opportunistic infections, and certain cancers.

lipodystrophy (LIP-oh-DIS-tro-fee): abnormalities in body fat and fat metabolism that may result from drug treatments for HIV infection. The accumulation of abdominal fat is sometimes called *protease paunch.*

buffalo hump: the accumulation of fatty tissue at the base of the neck.

lipomas (lih-POE-muz): benign tumors composed of fatty tissue.

candidiasis: a fungal infection on the mucous membranes of the oral cavity and elsewhere; usually caused by *Candida albicans.*

herpes simplex virus: a common virus that can cause blisterlike lesions on the lips and in the mouth.

Lipodystrophy

Many patients who use drug therapies to suppress HIV infection develop abnormalities in body fat and fat metabolism known as **lipodystrophy.**[33] Patients may lose fat from the face and extremities, accumulate abdominal fat, or both. Also observed are breast enlargement (in both men and women), fat accumulation at the base of the neck (called a **buffalo hump**), and benign growths composed of fat tissue (called **lipomas**). The changes in body composition are often disfiguring and may cause physical discomfort; moreover, patients often develop hypertriglyceridemia, elevated low-density lipoprotein (LDL) cholesterol levels, reduced high-density lipoprotein (HDL) cholesterol levels, insulin resistance, and hyperinsulinemia. The specific reasons for the development of lipodystrophy are unclear.

Weight Loss and Wasting

Even with effective treatment of HIV infection, weight loss and wasting are ongoing problems for many HIV-infected patients. *HIV-associated wasting* is diagnosed in patients who unintentionally lose 5 percent of body weight within 3 months, 7.5 percent within 6 months, or 10 percent within 12 months.[34] The wasting has been linked with accelerated disease progression, reduced strength, and fatigue. In the later stages of AIDS, the wasting is severe and increases the risk of death. Much as in cancer, wasting associated with HIV infection has many causes: anorexia and inadequate food intake, altered metabolism, malabsorption, chronic diarrhea, and diet-drug interactions.

Anorexia and Reduced Food Intake

Inadequate food intake is a key factor in the development of wasting. Poor food intake may result from various factors, including the following:

- *Emotional distress, pain, and fatigue.* The physical and social problems that accompany chronic illness may cause fear, anxiety, and depression, which contribute to anorexia. Pain and fatigue, which may be associated with some disease complications, can cause anorexia and difficulty with eating.
- *Oral infections.* The oral infections associated with HIV infection can cause discomfort and interfere with food consumption. Common infections include **candidiasis** and **herpes simplex virus** infection. Candidiasis (commonly called *thrush*) can cause mouth pain, dysphagia (difficulty swallowing), and altered taste sensation; infection with herpes simplex virus may cause painful lesions around the lips and in the mouth.
- *Respiratory disorders.* Respiratory infections, including pneumonia and tuberculosis, are common in people with HIV infection. Symptoms may include chest pain, shortness of breath, and cough, which interfere with eating and contribute to anorexia.

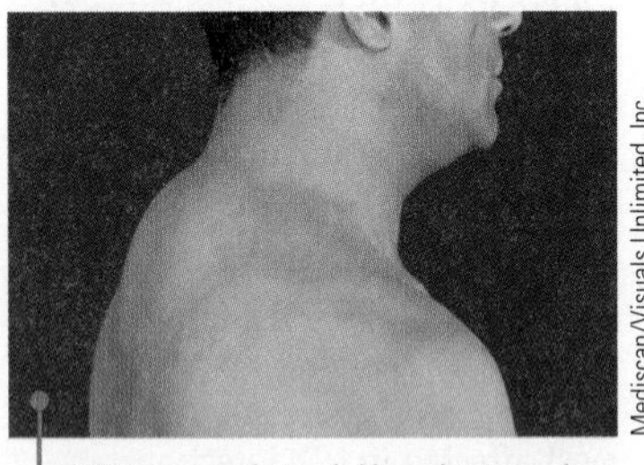

HIV-associated lipodystrophy is sometimes evident by the accumulation of fatty tissue at the base of the neck, referred to as *buffalo hump.*

- *Cancer.* As described earlier in this chapter, cancer leads to anorexia for numerous reasons. In addition, **Kaposi's sarcoma**, a type of cancer frequently associated with HIV infection, can cause lesions in the mouth and throat that make eating painful.
- *Medications.* The medications given to treat HIV infection, other infections, and cancer often cause anorexia, nausea and vomiting, altered taste sensation, food aversions, and diarrhea.

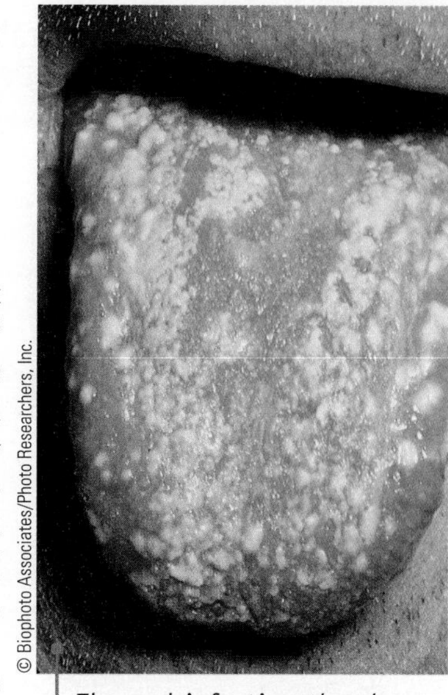

The oral infection *thrush* is easily identified by the characteristic milky white patches that appear on the tongue.

GI Tract Complications Complications of HIV infection involving the GI tract may result from opportunistic infections, medications, or the HIV infection itself.[35] In addition to the oral infections described previously, infections may develop in the esophagus, stomach, and intestines. Medications that treat viral, parasitic, and fungal infections in the GI tract can contribute to bacterial overgrowth. Furthermore, many patients develop nausea, vomiting, and diarrhea from the medications used to suppress HIV. As a result of these multiple problems, HIV-infected patients using standard treatments face an extremely high risk of malnutrition due to the combination of GI discomfort, bacterial overgrowth, malabsorption, and nutrient losses from vomiting, steatorrhea, and diarrhea.

Patients in the advanced stages of HIV infection often develop pathological changes in the small intestine referred to as *AIDS enteropathy*.[36] The condition is characterized by villus atrophy and blunting, intestinal cell losses, and inflammation. The result is a substantial reduction in the intestinal absorptive area, causing malabsorption, diarrhea, and weight loss.

Neurological Complications Neurological complications may be a consequence of HIV infection, immune suppression (causing cancers and infections that target brain tissue), or the medications used to treat HIV infection.[37] Clinical features include mild to severe dementia, muscle weakness and gait disturbances, and pain, numbness, and tingling in the legs and feet. Neurological impairments are usually more pronounced in the advanced stages of AIDS.

Other Complications Patients with HIV infection can develop anemia due to nutrient malabsorption, blood loss, disturbed bone marrow function, medication side effects, or the chronic illness itself. (■) HIV infection may also lead to skin disorders (rashes, infections, and cancers), eye disorders (retinal infection or detachment), kidney diseases (nephrotic syndrome or chronic kidney disease), and coronary heart disease.

TREATMENTS FOR HIV INFECTION

Although there is no cure for HIV infection, treatments can help to slow its progression, reduce complications, and alleviate pain. The standard drug treatment for suppressing HIV infection is a combination of at least three antiretroviral drugs.[38] Table 23-9 lists the major drug categories included in antiretroviral therapy and describes the drugs' modes of action. These antiretroviral agents have multiple adverse effects that make their long-term use difficult to tolerate. In addition to the GI effects discussed previously, side effects include skin rashes, headache, anemia, tingling and numbness, hepatitis, pancreatitis, and kidney stones. Thus, although antiretroviral therapy has improved the life span and quality of life for many patients, the drug regimens are difficult to adhere to and cause complications that require continual management.

In addition to antiretroviral drugs, adjunct drug therapies may be necessary to prevent or treat infections, treat HIV-associated cancers, or manage other complications that arise over the course of illness.[39] Medications are often prescribed to treat vomiting, anorexia, diarrhea, pain, blood lipid abnormalities, or glucose intolerance. The Diet-Drug Interactions feature (p. 664) summarizes the nutrition-related effects of some of the antiretroviral agents and other drugs mentioned in this chapter.

■ Anemia can develop during chronic illness due to the altered distribution of iron in tissues and reduced blood cell synthesis, among other abnormalities.

Kaposi's (kah-POH-seez) sarcoma: a common cancer in HIV-infected persons that is characterized by lesions in the skin, lungs, and GI tract.

TABLE 23-9 Antiretroviral Drugs for Treatment of HIV Infection

Category	Examples	Mode of Action
CCR5 antagonists	Maraviroc	CCR5 antagonists prevent HIV from entering cells by blocking a membrane receptor on the host cell.
Fusion inhibitors	Enfuvirtide	Fusion inhibitors prevent HIV from entering cells by binding a viral protein needed for its entry.
Integrase inhibitors	Raltegravir	Integrase inhibitors impair the function of HIV's integrase enzyme, which incorporates viral DNA into the host cell's genome.
Non-nucleoside reverse transcriptase inhibitors (NNRTI)	Delavirdine, efavirenz, nevirapine	NNRTI bind active sites on HIV's reverse transcriptase enzyme, blocking the ability of HIV to produce DNA copies of its genetic material.
Nucleoside reverse transcriptase inhibitors (NRTI)	Didanosine, lamivudine, zidovudine	As analogs of the nucleosides needed for DNA synthesis, NRTI impair the ability of HIV's reverse transcriptase enzyme to produce usable copies of DNA.
Protease inhibitors (PI)	Saquinavir, ritonavir, indinavir	PI inhibit HIV's protease enzyme, which cleaves HIV's gene products into usable structural proteins.

Sources: U.S. Department of Health and Human Services, Panel on Antiretroviral Guidelines for Adult and Adolescents, *Guidelines for the Use of Antiretroviral Agents in HIV-1-Infected Adults and Adolescents*, pp. F1–F27 (updated March 27, 2012), available at http://www.aidsinfo.nih.gov/contentfiles/lvguidelines/adultandadolescentgl.pdf; site visited November 3, 2012; S. Safrin, Antiviral agents, in B. G. Katzung, ed., *Basic and Clinical Pharmacology* (New York: Lange/McGraw-Hill, 2012), pp. 861–890.

Control of Anorexia and Wasting Anabolic hormones, appetite stimulants, and regular physical activity have been successful in reversing unintentional weight loss and increasing muscle mass in HIV-infected patients. Testosterone and human growth hormone have demonstrated positive effects on muscle tissue, especially in combination with resistance training. A regular program of resistance exercise improves muscle mass and strength and corrects some of the metabolic abnormalities (altered blood lipids and insulin resistance) that are common in HIV-infected patients. The medications megestrol acetate and dronabinol (described on p. 656) are sometimes prescribed to stimulate appetite and improve weight gain, although much of the weight increase is attributable to a gain of fat rather than lean tissue.[40]

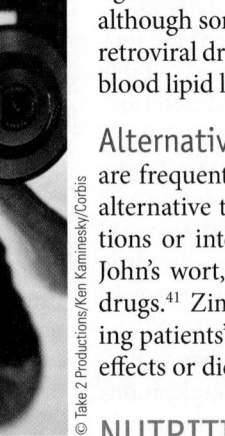

Resistance training can help a person with HIV infection maintain muscle mass and strength.

Control of Lipodystrophy Treatment strategies for lipodystrophy are under investigation. Both aerobic activity and resistance training may help to reduce abdominal fat, although some patients opt for cosmetic surgery. Patients may be given alternative antiretroviral drugs to alleviate symptoms. Medications may be prescribed to treat abnormal blood lipid levels and insulin resistance.

Alternative Therapies Like cancer patients, people with HIV infection and AIDS are frequently tempted to try unconventional methods of treatment. Although many alternative therapies are harmless, some may have side effects that worsen complications or interfere with treatment. For example, herbal preparations that contain St. John's wort, echinacea, or garlic may reduce the effectiveness of some antiretroviral drugs.[41] Zinc megadoses may increase the progression of HIV infection.[42] Monitoring patients' use of dietary supplements is essential to reduce the likelihood of adverse effects or diet-drug interactions.

NUTRITION THERAPY FOR HIV INFECTION

HIV-infected individuals must learn how to maintain body weight and muscle mass, prevent malnutrition, and cope with nutrition-related side effects of medications. Therefore, nutrition assessment and counseling should begin soon after a patient is

DIET-DRUG INTERACTIONS

Check this table for notable nutrition-related effects of the medications discussed in this chapter.

Drug	Effects
Appetite stimulants (megestrol acetate, dronabinol)	**Gastrointestinal effects:** Nausea, vomiting, diarrhea
	Dietary interaction: Dronabinol potentiates the effects of alcohol
	Metabolic effect: Hyperglycemia (megestrol acetate)
Didanosine	**Gastrointestinal effects:** Nausea, vomiting, anorexia, dry mouth, altered taste sensation, abdominal pain, diarrhea
	Dietary interactions: Take medication either one-half hour before or two hours after a meal; avoid magnesium supplements and alcohol while taking drug
	Metabolic effects: Pancreatitis; anemia; increased serum uric acid and liver enzyme levels
Methotrexate	**Gastrointestinal effects:** Nausea, vomiting, gingivitis, diarrhea
	Dietary interaction: Reduces folate absorption
	Metabolic effects: Liver toxicity, increased serum uric acid levels, anemia
Ritonavir	**Gastrointestinal effects:** Nausea, vomiting, altered taste sensation, anorexia, diarrhea
	Dietary interaction: Must be taken with food; avoid using alcohol while taking drug
	Metabolic effects: Hyperglycemia, liver toxicity, jaundice, hyperlipidemias (especially hypertriglyceridemia)
Zidovudine	**Gastrointestinal effects:** Nausea, vomiting, anorexia
	Dietary interactions: Do not take with a high-fat meal, which may decrease drug absorption
	Metabolic effects: Insulin resistance, diabetes, anemia, hyperlipidemias

Note: Most antiretroviral drugs that treat HIV infection have gastrointestinal and metabolic side effects; only a few are listed here as examples.

diagnosed with HIV infection. The initial assessment should include an evaluation of body weight and body composition. Follow-up measurements may indicate the need to adjust dietary recommendations and drug therapies.

Weight Management Since the development of successful drug therapies for HIV infection, obesity and overweight have become more prevalent than wasting among HIV-infected individuals in the United States.[43] Because excessive body weight can increase risks for cardiovascular disease and diabetes, moderate weight loss is recommended for patients with HIV infection who are overweight or obese.

Individuals who experience weight loss and wasting may benefit from a high-kcalorie, high-protein diet. Daily energy needs may range from 30 to 40 kcalories per kilogram body weight, and protein requirements may be as high as 1.2 to 2.0 grams per kilogram body weight per day.[44] If food consumption is difficult, small, frequent feedings may be better tolerated than several large meals. The addition of nutrient-dense snacks, protein or energy bars, and oral supplements can improve intakes. Liquid formulas may be useful for the person who is too tired to eat or prepare meals. Review the "How to" on p. 657 for additional suggestions for adding energy and protein to the diet.

Metabolic Complications As mentioned, individuals who are using antiretroviral drugs frequently develop insulin resistance and elevated triglyceride and LDL cholesterol levels. Treating these problems often requires both medications and dietary adjustments. Patients should be advised to achieve or maintain a desirable weight, replace saturated fats with monounsaturated and polyunsaturated fats, increase fiber intake, and limit intakes of *trans*-fatty acids, cholesterol, added sugars, and alcohol. Regular physical activity can improve both insulin resistance and blood lipid levels. (■) If problems persist, alternative antiretroviral medications may be prescribed in an attempt to improve the metabolic abnormalities.

Vitamins and Minerals Vitamin and mineral needs of people with HIV infections are highly variable, and little information is available concerning specific needs. Because nutrient deficiencies are likely to result from reduced food intake, malabsorption, diet-drug interactions, and nutrient losses, multivitamin-mineral supplements are usually recommended. Patients should be cautioned to maintain intakes that are close to DRI recommendations, however, due to the risk of adverse interactions between excessive amounts of certain nutrients and antiretroviral drugs.[45]

Symptom Management The discomfort associated with antiretroviral therapy, opportunistic GI infections, and symptoms of malabsorption can make food consumption difficult, and problems such as vomiting and diarrhea contribute to fluid and electrolyte losses. The "How to" pp. 658–659 describes measures for improving food and fluid intakes in individuals with these problems.

Food Safety The depressed immunity of people with HIV infections places them at extremely high risk of developing foodborne infections. Health practitioners should caution patients about their high susceptibility to foodborne illness and provide detailed instructions about the safe handling and preparation of foods; for some individuals, a low-microbial diet may be suggested (see p. 660). Water can also be a source of foodborne illness and is a common cause of **cryptosporidiosis** in HIV-infected individuals. In places where water quality is questionable, patients should consult their local health departments to determine whether the tap water is safe to drink. If not, or to take additional safety measures, water used for drinking and making ice cubes should be boiled for one minute.

Enteral and Parenteral Nutrition Support In later stages of illness, people with HIV infections may be unable to consume enough food and may need aggressive nutrition support. Tube feedings are preferred whenever the GI tract is functional; they can be provided at night to supplement oral diets consumed during the day. Parenteral nutrition is reserved for patients who are unable to tolerate enteral nutrition, such as those with GI obstructions that prevent food intake. For individuals with severe malabsorption, orally administered hydrolyzed formulas containing medium-chain triglycerides may be as effective as parenteral nutrition for reversing weight loss and wasting.[46] For either type of nutrition support, careful measures are necessary to avoid bacterial contamination of nutrient formulas and feeding equipment.

■ Additional suggestions for managing insulin resistance and hyperlipidemias are available in Chapters 20 and 21, respectively.

cryptosporidiosis (KRIP-toe-spor-ih-dee-OH-sis): a foodborne illness caused by the parasite *Cryptosporidium parvum*.

Review Notes

- By attacking immune cells, HIV causes progressive damage to immune function and may eventually lead to AIDS.
- Improved drug therapies have slowed the progression of HIV infection; however, the drugs may promote the development of lipodystrophy, characterized by body fat redistribution, abnormal lipid levels, and insulin resistance.
- HIV infection may lead to weight loss and wasting, anorexia, and various complications that affect food intake. Dietary adjustments, resistance training, and medications can help patients maintain their weight and prevent wasting.
- People with HIV infection must pay strict attention to food safety guidelines to prevent foodborne illnesses. The Case Study provides an opportunity to review the nutritional concerns of a person with HIV infection.

Man with HIV Infection

Three years ago, George Judd, a 37-year-old financial planner, sought medical help when he began feeling run-down and developed a painful white fungal infection over his mouth and tongue. The presence of thrush, recent weight loss, and anemia alerted Mr. Judd's physician to the possibility of an HIV infection. When Mr. Judd tested positive for HIV, he and his family and friends were devastated by the news, but those close to him have remained supportive. During the three years since Mr. Judd began antiretroviral drug therapy, he has maintained his weight but has also developed lipodystrophy and hypertriglyceridemia. Mr. Judd is 6 feet tall and currently weighs 185 pounds. He occasionally develops diarrhea and sometimes anorexia.

1. Describe lipodystrophy, and discuss its typical pattern in people who have an HIV infection. What adjustments in treatment and lifestyle may be helpful for Mr. Judd?
2. Describe an appropriate diet for Mr. Judd. What strategies may improve his problems with diarrhea and anorexia? Suggest reasons why diarrhea and anorexia may develop in people with HIV infections.
3. Explain why an HIV infection can lead to wasting as the disease progresses to the later stages. What recommendations may be helpful for maintaining weight and health if wasting becomes a problem?

Nutrition Assessment Checklist for People with Cancer or HIV Infections

MEDICAL HISTORY

Check the medical record to determine:

- ❍ Type and stage of cancer
- ❍ Stage of HIV infection

Review the medical record for complications that may alter nutrition therapy, including:

- ❍ Altered organ function
- ❍ Altered taste perception
- ❍ Anorexia
- ❍ Dry mouth and oral infections
- ❍ GI symptoms and infections
- ❍ Hyperlipidemias
- ❍ Insulin resistance
- ❍ Malnutrition and wasting

MEDICATIONS

For patients with cancer or HIV infections:

- ❍ Check medications to identify potential diet-drug interactions.
- ❍ Recommend the use of anti-nauseants, if needed.
- ❍ Ask about the use of dietary supplements, including herbal products.

For cancer patients who require chemotherapy:

- ❍ Recommend strategies to prevent food aversions.
- ❍ Offer suggestions for managing drug-related complications.

For HIV-infected patients using antiretroviral drug therapy:

- ❍ Remind patients that some drugs are better absorbed with foods and that others must be taken on an empty stomach.
- ❍ Help patients work out a medication schedule that suits their lifestyle and is timed appropriately in regard to food intake.
- ❍ Offer suggestions for managing drug-related complications.

DIETARY INTAKE

For patients with poor food intakes and weight loss:

- ❍ Determine the reasons for reduced food intake.
- ❍ Offer appropriate suggestions to improve food intake.
- ❍ Provide interventions before weight loss progresses too far.

For patients with HIV infections who experience weight gain, elevated triglyceride or LDL cholesterol levels, or hyperglycemia:

- ❍ Assess the diet for energy, total fat, types of fat, carbohydrates, fiber, and sugars.
- ❍ For patients with hyperlipidemias, recommend a diet low in saturated fat, *trans*-fatty acids, cholesterol, and sugars.
- ❍ For patients with hyperglycemia, recommend a consistent carbohydrate intake at meals and snacks that emphasizes complex carbohydrates and limits concentrated sweets.
- ❍ Recommend regular physical activity for weight control and for improving blood lipid levels and insulin resistance.

ANTHROPOMETRIC DATA

Take baseline height and weight measurements, monitor weight regularly, and suggest dietary adjustments for weight maintenance, if necessary. Remember that body composition may change without affecting body weight. Perform baseline and periodic body composition measurements in HIV-infected patients who are using antiretroviral drug therapy.

LABORATORY TESTS

Note that albumin and other serum proteins may be reduced in patients with cancer or HIV infections, especially in those experiencing wasting. Check laboratory tests for indications of:

- Anemia
- Dehydration
- Elevated LDL cholesterol levels
- Elevated triglyceride levels
- Hyperglycemia

For patients with HIV infections, evaluate disease progression by checking:

- Helper T cell counts
- Viral load

PHYSICAL SIGNS

Look for physical signs of:

- Dehydration (especially for patients with fever, vomiting, or diarrhea)
- Kaposi's sarcoma
- Oral infections
- Protein-energy malnutrition and wasting

Self Check

1. Which of these dietary substances may help to protect against cancer?
a. Alcoholic beverages
b. Well-cooked meats, poultry, and fish
c. Animal fats
d. Compounds in fruits and vegetables

2. The metabolic changes that often accompany cancer include all of the following *except:*
a. increased triglyceride breakdown.
b. increased protein turnover.
c. increased muscle protein synthesis.
d. insulin resistance.

3. An advantage of radiation therapy over chemotherapy is that:
a. radiation is not damaging to rapidly dividing cells.
b. side effects of radiation therapy do not include malnutrition.
c. radiation can be directed toward the regions affected by cancer.
d. the radiation used is too weak to damage GI tissues.

4. Although many cancer patients lose weight, which type of cancer is often associated with weight gain?
a. Kidney cancer
b. Breast cancer
c. Colon cancer
d. Lung cancer

5. Which food below should be avoided by a patient consuming a low-microbial diet?
a. Baked potato
b. Pasteurized yogurt
c. Banana
d. Leftover luncheon meat

6. HIV can enter and destroy:
a. epithelial cells.
b. helper T cells.
c. liver cells.
d. intestinal cells.

7. In patients who develop lipodystrophy, which abnormality is unlikely?
a. Increased abdominal fat
b. Increased fat in the arms and legs
c. Fat accumulation at the base of the neck
d. Hypertriglyceridemia

8. In people with HIV infection, mouth sores may be caused by all of the following, *except:*
a. cryptosporidiosis.
b. Kaposi's sarcoma.
c. herpes simplex virus.
d. candidiasis.

9. Megestrol acetate and dronabinol are:
a. medications used to promote weight gain.
b. protease inhibitors that fight HIV infection.
c. medications that treat common opportunistic infections.
d. anabolic hormones that promote gain of muscle tissue.

10. To prevent cryptosporidiosis, a person with HIV infection may need to:
a. wash hands carefully before meals.
b. avoid consuming undercooked meat, poultry, and eggs.
c. consume a high-kcalorie, high-protein diet.
d. boil drinking water for one minute.

Answers to these questions appear in Appendix H. For more chapter review: Access an interactive eBook, chapter-specific interactive learning tools, including flashcards, quizzes, videos, and more in your Nutrition CourseMate, accessed through CengageBrain.com.

Clinical Applications

1. Consider the nutrition problems that may develop in a 36-year-old woman with a malignant brain tumor that affects her ability to move the right side of her body (including the tongue) and to speak coherently. She is taking a pain medication that makes her nauseated and sleepy. Her expected survival time is only about six months.
   - If she is right-handed, how might her impairment interfere with eating? What suggestions do you have for overcoming this problem?
   - How might her nutrition status be affected by her inability to communicate effectively? What suggestions may help?
   - In what ways might the pain medication she is taking affect her nutrition status?

2. Various types of chronic conditions can lead to weight loss and wasting. For some of these conditions, such as Crohn's disease and celiac disease (Chapter 18), diet is a cornerstone of treatment. For others, such as cancer and HIV infection, nutrition plays a supportive role. What determines whether nutrition plays a primary role or a supportive role in the treatment of disease?

Nutrition on the Net

For further study of topics covered in this chapter, access these websites:

- To learn more about cancer, including risk factors, screening, detection, treatments (including nutrition), and support networks, visit these sites:
  American Association for Cancer Research:
  www.aacr.org
  American Cancer Society:
  www.cancer.org
  American Institute for Cancer Research:
  www.aicr.org
  National Cancer Institute:
  www.cancer.gov
  OncoLink:
  www.oncolink.org
- For more information about nutrition and cancer prevention, check the resources provided at these websites:
  American Cancer Society:
  www.cancer.org/acs/groups/cid/documents/webcontent/002577-pdf.pdf
  American Institute for Cancer Research:
  www.aicr.org/reduce-your-cancer-risk/recommendations-for-cancer-prevention/
- To find additional information about HIV infection and AIDS, visit these sites:
  AIDSinfo, an information service provided by the U.S. Department of Health and Human Services:
  aidsinfo.nih.gov
  AIDS.gov, a resource provided by various departments and agencies of the federal government:
  aids.gov
  The Body:
  www.thebody.com
  UCSF Center for HIV Information:
  hivinsite.ucsf.edu
- To review information about safe food handling provided by U.S. government agencies, visit this site:
  www.foodsafety.gov

Notes

1. National Center for Health Statistics, *Health, United States, 2011: With Special Feature on Socioeconomic Status and Health* (Hyattsville, MD: U.S. Department of Health and Human Services, 2012), pp. 181–183.
2. H. T. Lynch and C. Richard Boland, Cancer genetics, in L. Goldman and A. I. Schafer, eds., *Goldman's Cecil Medicine* (Philadelphia: Saunders, 2012), pp. 1182–1184, J. A. Moscow and K. H. Cowan, Biology of cancer, in L. Goldman and A. I. Schafer, eds., *Goldman's Cecil Medicine* (Philadelphia: Saunders, 2012), pp. 1184–1187.
3. J. L. Freudenheim, Nutrition and genetic factors in carcinogenesis, in A. M. Coulston, C. J. Boushey, and M. G. Ferruzzi, eds., *Nutrition in the Prevention and Treatment of Disease* (London: Academic Press/Elsevier, 2013), pp. 645–656.
4. K. Robien, C. L. Rock, and W. Demark-Wahnefried, Nutrition and cancers of the breast, endometrium, and ovary, in A. M. Coulston, C. J. Boushey, and M. G. Ferruzzi, eds., *Nutrition in the Prevention and Treatment of Disease* (London: Academic Press/Elsevier, 2013), pp. 657–672.
5. World Cancer Research Fund/American Institute for Cancer Research, *Food, Nutrition, Physical Activity, and the Prevention of Cancer: A Global Perspective* (Washington, DC: American Institute for Cancer Research, 2007).
6. R. J. Turesky and L. Le Marchand, Metabolism and biomarkers of heterocyclic aromatic amines in molecular epidemiology studies: Lessons learned from aromatic amines, *Chemical Research in Toxicology* 24 (2011): 1169–1214.
7. W. Zheng and S.-A. Lee, Well-done meat intake, heterocyclic amine exposure, and cancer risk, *Nutrition and Cancer* 61 (2009): 437–446.
8. World Cancer Research Fund/American Institute for Cancer Research, 2007.

9. H. S. Rugo, Paraneoplastic syndromes and other non-neoplastic effects of cancer, in L. Goldman and A. I. Schafer, eds., *Goldman's Cecil Medicine* (Philadelphia: Saunders, 2012), pp. 1192–1200.
10. V. C. Vaughan, P. Martin, and P. A. Lewandowski, Cancer cachexia: Impact, mechanisms and emerging treatments, *Journal of Cachexia, Sarcopenia, and Muscle* 3 (2012). doi: 10.1007/s13539-012-0087-1; M. J. Tisdale, Mechanisms of cancer cachexia, *Physiological Reviews* 89 (2009): 381–410.
11. K. C. H. Fearon, Cancer cachexia: Developing multimodal therapy for a multidimensional problem, *European Journal of Cancer* 44 (2008): 1124–1132.
12. Vaughan, Martin, and Lewandowski, 2012; Fearon, 2008.
13. S. Dodson and coauthors, Muscle wasting in cancer cachexia: Clinical implications, diagnosis, and emerging treatment strategies, *Annual Review of Medicine* 62 (2011): 8.1–8.15.
14. Vaughan, Martin, and Lewandowski, 2012; K. C. H. Fearon, The 2011 ESPEN Arvid Wretlind lecture: Cancer cachexia: The potential impact of translational research on patient-focused outcomes, *Clinical Nutrition* 31 (2012): 577–582; L. G. Melstrom and coauthors, Mechanisms of skeletal muscle degradation and its therapy in cancer cachexia, *Histology and Histopathology* 22 (2007): 805–814.
15. M. C. Perry, Approach to the patient with cancer, in L. Goldman and A. I. Schafer, eds., *Goldman's Cecil Medicine* (Philadelphia: Saunders, 2012), pp. 1164–1177.
16. J. M. Vose, Hematopoietic stem cell transplantation, in L. Goldman and A. I. Schafer, eds., *Goldman's Cecil Medicine* (Philadelphia: Saunders, 2012), pp. 1158–1162.
17. S. Dodson and coauthors, Muscle wasting in cancer cachexia: Clinical implications, diagnosis, and emerging treatment strategies, *Annual Review of Medicine* 62 (2011): 8.1–8.15.
18. M. Charlson, Complementary and alternative medicine, in L. Goldman and A. I. Schafer, eds., *Goldman's Cecil Medicine* (Philadelphia: Saunders, 2012), pp. 177–181.
19. A. Agins, *ADA Quick Guide to Drug-Supplement Interactions* (Chicago: American Dietetic Association, 2011), pp. 19–20.
20. Agins, 2011; M. L. Heaney and coauthors, Vitamin C antagonizes the cytotoxic effects of antineoplastic drugs, *Cancer Research* 68 (2008): 8031–8038.
21. C. L. Rock and coauthors, Nutrition and physical activity guidelines for cancer survivors, *CA: A Cancer Journal for Clinicians* 62 (2012): 242–274.
22. Rock and coauthors, 2012.
23. J. Doley, HIV and cancer, in A. Skipper, ed., *Dietitian's Handbook of Enteral and Parenteral Nutrition* (Sudbury, MA: Jones and Bartlett Learning, 2012), pp. 199–216.
24. Academy of Nutrition and Dietetics, *Nutrition Care Manual* (Chicago: Academy of Nutrition and Dietetics, 2012).
25. Rock and coauthors, 2012; N. Saquib and coauthors, Weight gain and recovery of pre-cancer weight after breast cancer treatments: Evidence from the women's healthy eating and living (WHEL) study, *Breast Cancer Research and Treatment* 105 (2007): 177–186; G. Makari-Judson, C. H. Judson, and W. C. Mertens, Longitudinal patterns of weight gain after breast cancer diagnosis: Observations beyond the first year, *Breast Journal* 13 (2007): 258–265.
26. Academy of Nutrition and Dietetics, 2012.
27. Doley, 2012.
28. C. W. Dieffenbach and A. S. Fauci, Thirty years of HIV and AIDS: Future challenges and opportunities, *Annals of Internal Medicine* 154 (2011): 766–771.
29. World Health Organization, *Global HIV/AIDS Response: Epidemic Update and Health Sector Progress towards Universal Access: Progress Report 2011*, available at http://whqlibdoc.who.int/publications/2011/9789241502986_eng.pdf; site visited November 2, 2012.
30. J. Prejean and coauthors, Estimated HIV incidence in the United States, 2006–2009, *PLoS One* 6 (2011). doi: 10.1371/journal.pone.0017502
31. G. M. Shaw, Biology of human immunodeficiency viruses, in L. Goldman and A. I. Schafer, eds., *Goldman's Cecil Medicine* (Philadelphia: Saunders, 2012), pp. 2177–2181.
32. Shaw, 2012.
33. H. Masur, L. Healey, and C. Hadigan, Treatment of human immunodeficiency virus infection and acquired immunodeficiency syndrome, in L. Goldman and A. I. Schafer, eds., *Goldman's Cecil Medicine* (Philadelphia: Saunders, 2012), pp. 2185–2196.
34. J. L. Gerrior, Unintentional weight loss and wasting in HIV infection, in K. M. Hendricks, K. R. Dong, and J. L. Gerrior, eds., *Nutrition Management of HIV and AIDS* (Chicago: American Dietetic Association, 2009), pp. 41–55.
35. T. A. Knox and C. Wanke, Gastrointestinal manifestations of HIV and AIDS, in L. Goldman and A. I. Schafer, eds., *Goldman's Cecil Medicine* (Philadelphia: Saunders, 2012), pp. 2196–2199.
36. F. Maingat and coauthors, Inflammation and epithelial cell injury in AIDS enteropathy: Involvement of endoplasmic reticulum stress, *FASEB Journal* 25 (2011): 2211–2220, doi: 10.1096/fj.10-175992
37. J. R. Berger and A. Nath, Neurologic complications of human immunodeficiency virus infection, in L. Goldman and A. I. Schafer, eds., *Goldman's Cecil Medicine* (Philadelphia: Saunders, 2012), pp. 2218–2222.
38. U.S. Department of Health and Human Services, Panel on Antiretroviral Guidelines for Adult and Adolescents, *Guidelines for the Use of Antiretroviral Agents in HIV-1-Infected Adults and Adolescents*, pp. F1–F27 (updated March 27, 2012), available at http://www.aidsinfo.nih.gov/contentfiles/lvguidelines/adultandadolescentgl.pdf; site visited November 3, 2012; S. Safrin, Antiviral agents, in B. G. Katzung, ed., *Basic and Clinical Pharmacology* (New York: Lange/McGraw-Hill, 2012), pp. 861–890.
39. Masur, Healey, and Hadigan, 2012.
40. Gerrior, 2009; K. Mulligan and coauthors, Testosterone supplementation of megestrol therapy does not enhance lean tissue accrual in men with human immunodeficiency virus-associated weight loss: A randomized, double-blind, placebo-controlled, multicenter trial, *Journal of Clinical Endocrinology and Metabolism* 92 (2007): 563–570.
41. C. E. Dennehy and C. Tsourounis, Dietary supplements and herbal medications, in B. G. Katzung, ed., *Basic and Clinical Pharmacology* (New York: The McGraw-Hill Companies, Inc., 2012), pp. 1125–1137.
42. Doley, 2012.
43. L. Vining, General nutrition issues for healthy living with HIV infection, in K. M. Hendricks, K. R. Dong, and J. L. Gerrior, eds., *Nutrition Management of HIV and AIDS* (Chicago: American Dietetic Association, 2009), pp. 23–40.
44. Gerrior, 2009.
45. Doley, 2012; A. Howard, K. M. Hendricks, and J. Dwyer, Dietary supplement use in HIV infection, in K. M. Hendricks, K. R. Dong, and J. L. Gerrior, eds., *Nutrition Management of HIV and AIDS* (Chicago: American Dietetic Association, 2009), pp. 109–127.
46. Doley, 2012.

Nutrition in Practice

Ethical Issues in Nutrition Care

As with other medical technologies, the availability of specialized nutrition support forces health care professionals and members of our society to face difficult **ethical** issues. When medical treatments prolong life by merely delaying death, the lifetime that remains may be of extremely low quality. This Nutrition in Practice examines the ethical dilemmas that clinicians must face when dealing with patients in critical care. The accompanying glossary defines the relevant terms.

If providing care can do little to promote recovery, is it appropriate to withhold or withdraw treatment?

In attempting to answer questions such as these, health professionals must consider the following ethical principles:[1]

- A patient has the right to make decisions concerning his or her own well-being (**patient autonomy**), even if refusing treatment could result in death. It is generally accepted that a patient's preferences should take precedence over the desires of others.
- A patient should be fully informed of a treatment's benefits and risks in a fair and honest manner (**disclosure**). A patient's acceptance of a treatment that has been adequately disclosed is considered **informed consent**.
- A patient must have the mental capacity to make appropriate health care decisions (**decision-making capacity**). If a patient is incapable of doing so, a person designated by the patient should serve as a **surrogate** decision maker.
- The potential benefits (**beneficence**) of any treatment should outweigh its potential harm (**maleficence**).
- Health care providers must determine whether the provision of health care to one patient would unfairly limit the care of other patients (**distributive justice**).

Although these principles may seem simple and logical, it is often difficult to determine the appropriate action to

Glossary

advance health care directive: written or oral instruction regarding one's preferences for medical treatment to be used in the event of becoming incapacitated; also called an *advance medical directive* or a *living will.*

beneficence (be-NEF-eh-sense): the act of performing beneficial services rather than harmful ones.

cardiopulmonary resuscitation (CPR): life-sustaining treatment that supplies oxygen and restores a person's ability to breathe and pump blood.

decision-making capacity: the ability to understand pertinent information and make appropriate decisions; known as *decision-making competency* within the legal system.

defibrillation: life-sustaining treatment in which an electronic device is used to shock the heart and reestablish a pattern of normal contractions. Defibrillation is used when the heart has arrhythmias or has experienced arrest.

dialysis: life-sustaining treatment in which a patient's blood is filtered using selective diffusion through a semipermeable membrane; substitutes for kidney function.

disclosure: the act of revealing pertinent information. For example, clinicians should accurately describe proposed tests and procedures, their benefits and risks, and alternative approaches.

distributive justice: the equitable distribution of resources.

do-not-resuscitate (DNR) order: a request by a patient or surrogate to withhold cardiopulmonary resuscitation.

durable power of attorney: a legal document (sometimes called a *health care proxy)* that gives legal authority to another *(a health care agent)* to make medical decisions in the event of incapacitation.

ethical: in accordance with accepted principles of right and wrong.

futile: describes medical care that will not improve the medical circumstances of a patient.

health care agent: a person given legal authority to make medical decisions for another in the event of incapacitation.

informed consent: a patient's or caregiver's agreement to undergo a treatment that has been adequately disclosed. Persons must be mentally competent in order to make the decision.

maleficence (mah-LEF-eh-sense): the act of doing evil or harm.

mechanical ventilation: life-sustaining treatment in which a mechanical ventilator is used to substitute for a patient's failing lungs.

patient autonomy: a principle of self-determination, such that patients (or surrogate decision makers) are free to choose the medical interventions that are acceptable to them, even if they choose to refuse interventions that may extend their lives.

persistent vegetative state: a vegetative mental state resulting from brain injury that persists for at least one month. Individuals lose awareness and the ability to think but retain noncognitive brain functions, such as motor reflexes and normal sleep patterns.

surrogate: a substitute; a person who takes the place of another.

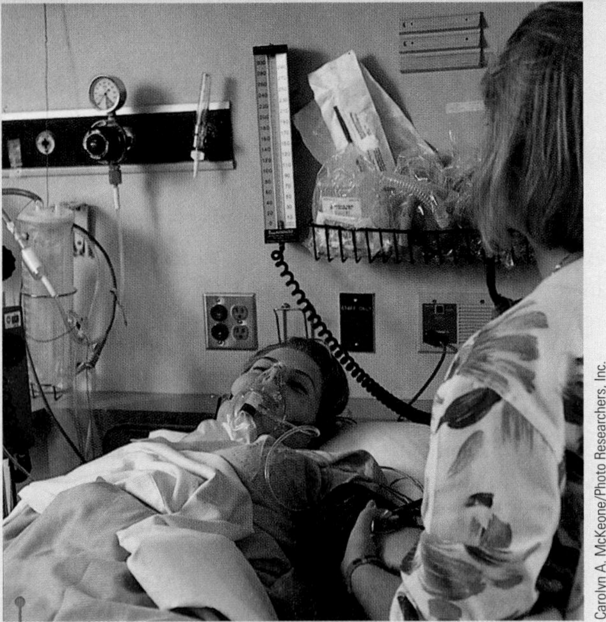
Carolyn A. McKeone/Photo Researchers, Inc.

Is it ever morally or legally appropriate to withhold or withdraw nutrition support?

take during intensive care. When clinicians and families disagree, the courts may be asked to decide.

What kinds of treatments can help to sustain a patient's life?

Nutrition support and hydration are both considered life-sustaining treatments because withholding or withdrawing either can result in death. Other life-sustaining treatments include **cardiopulmonary resuscitation (CPR)**, which supplies oxygen and restores a person's ability to breathe and pump blood; **defibrillation**, in which an electronic device shocks the heart and reestablishes normal contractions; **mechanical ventilation**, which substitutes for lung function; and **dialysis**, which substitutes for kidney function.[2]

Do patients have a right to life-sustaining treatments?

Although life-sustaining treatments are readily provided to patients who have a reasonable chance of recovering from illness, it may be difficult to determine the best course of action for patients who are dying or who are unlikely to regain consciousness. Under such circumstances, such treatments may be considered **futile** because they are unable to improve the outcome of disease or increase the patient's comfort and well-being.[3] If patients or caregivers demand a treatment that health practitioners have determined to be useless, a legal resolution may be required. Conversely, medical personnel may find it objectionable to withdraw life support when they know that the inevitable consequence will be the patient's death.

How have the courts resolved conflicts involving nutrition support?

One of the landmark cases involving nutrition support concerned Nancy Cruzan, who suffered permanent and irreversible brain damage after a car crash in 1983 when she was 26 years of age.[4] After she had been in a **persistent vegetative state** for five years, her parents requested permission to discontinue tube feeding, but hospital staff refused to honor the request, and the matter was taken to court. The Missouri Supreme Court determined that Nancy had never definitively stated her "right to die" wishes and that her parents were unable to make such a request for her. The court also stated that preserving life, no matter what its quality, should take precedence over all other considerations. Nancy's parents appealed the ruling, but in 1990, the U.S. Supreme Court upheld the Missouri Supreme Court in a five-to-four decision. Three witnesses were eventually found who could testify that Nancy would not desire life-sustaining treatment under the circumstances, and the Court finally granted permission to remove the feeding tube. This case illustrates the importance of having an **advance health care directive** (discussed in a later section) that specifies one's preferences for medical treatment in the event of incapacitation.

In a more recent case, the spouse and parents of a patient in a persistent vegetative state fought a 10-year legal battle over her medical care. In 1990, at the age of 25, Terri Schiavo suffered a full cardiac arrest.[5] She initially fell into a coma, but her condition evolved into a persistent vegetative state that was considered irreversible. Despite the neurologists' diagnosis and a series of computed tomography (CT) and magnetic resonance imaging (MRI) scans showing extensive brain atrophy, her parents maintained that she was minimally conscious and could improve somewhat with rigorous treatment. Her husband, who was legally responsible for her care, insisted that she would never have wanted to be kept alive in a vegetative state. Like Nancy Cruzan, Terri had never expressed her wishes in an advance directive.

In 1998, Terri's husband filed a petition to have her feeding tube removed, and a Florida court approved the motion in February 2000. Although Terri's parents appealed, an appeals court affirmed the decision, and the Florida Supreme Court declined to review the case. In April 2001, Terri's physicians removed her feeding tube, but within days, a federal circuit court judge ordered it to be reinserted and reopened the case. Eventually, the various motions filed by the parents were dismissed and Terri's feeding tube was removed for the second time in October 2003. Within days, the Florida legislature passed a bill known as *Terri's Law* that gave the governor the authority to intervene, and Governor Jeb Bush ordered the feeding tube reinserted. A year later, Florida's Supreme Court declared Terri's Law to be unconstitutional. Although the governor appealed the decision, his appeal was rejected in January 2005. Terri's feeding tube was removed for the third time in March 2005.

Despite emergency petitions by her parents and an attempt by the U.S. Congress to have her case reconsidered, the courts refused to grant a restraining order, and Terri died 13 days after her feeding tube was removed.

How can people ensure that their wishes will be considered in the event that they become incapacitated?

A person can declare preferences about medical treatments in an advance health care directive, sometimes called a *living will.* These directives include instructions about life-sustaining procedures that a person does or does not want. The documents are incorporated into the medical record and updated when appropriate; they take effect only if a physician determines that a patient lacks the ability to understand and make decisions about available treatments. If a person's preferences are unknown, decisions are based on a patient's best interests as determined by a caregiver or family member.[6]

Another important directive is a **durable power of attorney** (sometimes called a *health care proxy*), in which another person (a **health care agent**) is appointed to act as decision maker in the event of incapacitation. The agent should understand one's medical preferences and be absolutely trustworthy. If an agent is given comprehensive power to supervise care, he or she may make decisions about medical staff, health care facilities, and medical procedures.

Laws regarding advance directives vary from state to state. In some states, nutrition and hydration are not considered life-sustaining treatments, and a person's instructions about them may need to be indicated separately. Some states restrict the use of advance directives to terminal illness or disallow them during pregnancy. Generally, advance directives created in one state are honored in another.

How does a "do-not-resuscitate" order differ from other advance directives?

A **do-not-resuscitate (DNR) order** is frequently used to withhold CPR in the event of cardiopulmonary arrest, which occurs too suddenly for deliberate decision making. A DNR order is written in the medical record as other directives are, but it does not exclude the use of other life-prolonging measures. A DNR order is most often used in patients with serious illnesses or advanced age. Some institutions allow a physician to write a DNR order for a patient who has a poor prognosis, but the physician must inform the patient or surrogate if this is done.

Have advance directives changed the way that medical care is provided?

Not really. Despite the availability of advance directives, only about 47 percent of people in the United States have completed one.[7] Furthermore, patients' preferences often change as their medical conditions evolve or they learn more about their prognosis.[8] In addition, advance directives are sometimes too general or vague to guide specific treatment decisions.

Physicians must often provide patient care before they have a chance to discuss treatments with patients or caregivers. In many cases, life-sustaining treatments are begun without the prior knowledge of patients or their decision makers, or the treatments continue even if patients want them stopped. Patients who are fully aware of treatment options and clearly state their preferences are more likely to be successful at obtaining the care they desire.

What resources are available to individuals who have difficulty making decisions about life-sustaining medical treatments?

Many medical institutions have ethics committees that meet regularly to update patient care policies pertaining to end-of-life treatments; these committees provide guidelines to help families and hospital staff who face difficult treatment decisions. Medical staff may also provide referrals for hospice care to dying patients who prefer comfort and palliation over invasive procedures in their final days.

Notes

1. E. J. Emanuel, Bioethics in the practice of medicine, in L. Goldman and A. I. Schafer, eds., *Goldman's Cecil Medicine* (Philadelphia: Saunders, 2012), pp. 4–9; J. M. Luce and D. B. White, A history of ethics and law in the intensive care unit, *Critical Care Clinics* 25 (2009): 221–237.
2. Luce and White, 2009.
3. Emanuel, 2012; J. M. Luce, End-of-life decision making in the intensive care unit, *American Journal of Respiratory and Critical Care Medicine* 182 (2010): 6–11.
4. J. O. Maillet, Position of the American Dietetic Association: Ethical and legal issues in nutrition, hydration, and feeding, *Journal of the American Dietetic Association* 108 (2008): 873–882.
5. R. Cranford, Facts, lies, and videotapes: The permanent vegetative state and the sad case of Terri Schiavo, *Journal of Law, Medicine, and Ethics* 33 (2005): 363–372.
6. L. Snyder and C. Leffler, Position paper: Ethics manual, *Annals of Internal Medicine* (2005): 560–582.
7. Emanuel, 2012.
8. Luce and White, 2009.

appendixes

Contents

appendix A Table of Food Composition

CONTENTS

This edition of the table of food composition includes a wide variety of foods. It is updated with each edition to reflect current nutrient data for foods, to remove outdated foods, and to add foods that are new to the marketplace.* The nutrient database for this appendix is compiled from a variety of sources, including the USDA Nutrient Database and manufacturers' data. The USDA database provides data for a wider variety of foods and nutrients than other sources. Because laboratory analysis for each nutrient can be quite costly, manufacturers tend to provide data only for those nutrients mandated on food labels. Consequently, data for their foods are often incomplete; any missing information on this table is designated as a dash. Keep in mind that a dash means only that the information is unknown and should not be interpreted as a zero. A zero means that the nutrient is not present in the food.

Whenever using nutrient data, remember that many factors influence the nutrient contents of foods. These factors include the mineral content of the soil, the diet fed to the animal or the fertilizer used on the plant, the season of harvest, the method of processing, the length and method of storage, the method of cooking, the method of analysis, and the moisture content of the sample analyzed. With so many influencing factors, users should view nutrient data as a close approximation of the actual amount.

- ***Fats*** Total fats, as well as the breakdown of total fats to saturated, monounsaturated, and polyunsaturated fats, are listed in the table. The fatty acids seldom add up to the total in part due to rounding but also because values may include some non-fatty acids, such as glycerol, phosphate, or sterols. *Trans*-fatty acids are not listed separately in this edition because newer hydrogenated fats generally add less than 0.5 g *trans* fat to a serving of food, an amount often reported as 0.
- ***Vitamin A, Vitamin E, and Folate*** In keeping with the RDA for vitamin A, vitamin E, and folate, this appendix presents data for vitamin A in micrograms (µg) RAE; vitamin E data in milligrams (mg) alpha-tocopherol, listed in the table as Vit E (mg α); and folate as micrograms (µg) DFE, listed in the table as Fola (µg).
- ***Bioavailability*** Keep in mind that the availability of nutrients from foods depends not only on the quantity provided by a food, but also on the amount absorbed and used by the body—the bioavailability. The bioavailability of folate from fortified foods, for example, is greater than from naturally occurring sources. Similarly, the body can make niacin from the amino acid tryptophan, but niacin values in this table (and most databases) report preformed niacin only. Chapter 8 provides conversion factors and additional details.
- ***Using the Table*** The foods and beverages in this table are organized into several categories, which are listed at the head of each right-hand page. Page numbers are provided, and each group is color-coded to make it easier to find individual foods.
- ***Caffeine Sources*** Caffeine occurs in several plants, including the familiar coffee bean, the tea leaf, and the cocoa bean from which chocolate is made. Most human societies use caffeine regularly, most often in beverages, for its stimulant effect and flavor. Caffeine contents of beverages vary depending on the plants they are made from, the climates and soils where the plants are grown, the grind or cut size, the method and duration of brewing, and the amounts served. The accompanying table shows that, in general, a cup of coffee contains the most caffeine; a cup of tea, less

*This food composition table has been prepared by Cengage Learning. The nutritional data are supplied by Axxya Systems.

than half as much; and cocoa or chocolate, less still. As for cola beverages, they are made from kola nuts, which contain caffeine, but most of their caffeine is added, using the purified compound obtained from decaffeinated coffee beans. The FDA lists caffeine as a multipurpose GRAS substance (■) that may be added to foods and beverages. Drug manufacturers use caffeine in many products.

■ A GRAS substance is one that is "generally recognized as safe."

TABLE Caffeine Content of Selected Beverages, Foods, and Medications

Beverages and Foods	Serving Size	Average (mg)
Coffee		
Brewed	8 oz	95
Decaffeinated	8 oz	2
Instant	8 oz	64
Tea		
Brewed, green	8 oz	30
Brewed, herbal	8 oz	0
Brewed, leaf or bag	8 oz	47
Instant	8 oz	26
Lipton Brisk iced tea	12 oz	7
Nestea Cool iced tea	12 oz	12
Snapple iced tea (all flavors)	16 oz	42
Soft Drinks		
A&W Creme Soda	12 oz	29
Barq's Root Beer	12 oz	18
Coca-Cola	12 oz	30
Dr. Pepper, Mr. Pibb, Sunkist Orange	12 oz	36
A&W Root Beer, club soda, Fresca, ginger ale, 7-Up, Sierra Mist, Sprite, Squirt, tonic water, caffeine-free soft drinks	12 oz	0
Mello Yello	12 oz	51
Mountain Dew	12 oz	45
Pepsi	12 oz	32
Energy Drinks		
5-hour Energy	2 oz	138
AMP, Red Bull	8-8.3 oz	80
Monster	16 oz	160
No Fear, Rock Star	16 oz	174
Wired X344	16 oz	344
Other Beverages		
Chocolate milk or hot cocoa	8 oz	5
Starbucks Frappuccino Mocha	9.5 oz	72
Starbucks Frappuccino Vanilla	9.5 oz	64
Yoohoo chocolate drink	9 oz	3

Beverages and Foods	Serving Size	Average (mg)
Candies		
Baker's chocolate	1 oz	26
Dark chocolate covered coffee beans	1 oz	235
Dark chocolate, semisweet	1 oz	18
Milk chocolate	1 oz	6
Milk chocolate covered coffee beans	1 oz	224
White chocolate	1 oz	0
Foods		
Frozen yogurt, Ben & Jerry's coffee fudge	1 cup	85
Frozen yogurt, Häagen-Dazs coffee	1 cup	40
Ice cream, Starbucks coffee	1 cup	50
Ice cream, Starbucks Frappuccino bar	1 bar	15
Yogurt, Dannon coffee flavored	1 cup	45

Drugs	Serving Size	Average (mg)
Cold Remedies		
Coryban-D, Dristan	1 tablet	30
Diuretics		
Aqua-Ban	1 tablet	100
Pre-Mens Forte	1 tablet	100
Pain Relievers		
Anacin, BC Fast Pain Reliever	1 tablet	32
Excedrin, Midol, Midol Max Strength	1 tablet	65
Stimulants		
Awake, NoDoz	1 tablet	100
Awake Maximum Strength, Caffedrine, NoDoz Maximum Strength, Stay Awake, Vivarin	1 tablet	200
Weight-Control Aids		
Dexatrim	1 tablet	200

NOTE: The FDA suggests a maximum of 65 milligrams per 12-ounce cola beverage but does not regulate the caffeine contents of other beverages. Because products change, contact the manufacturer for an update on products you use regularly.

APPENDIX A

(Computer code is for Cengage Diet Analysis program)
(For purposes of calculations, use "0" for t, <1, <.1, <.01, etc.)

TABLE A-1 Table of Food Composition

DA+ Code	Food Description	Quantity	Measure	Wt (g)	H_2O (g)	Ener (kcal)	Prot (g)	Carb (g)	Fiber (g)	Fat (g)	Fat Breakdown (g)		
											Sat	Mono	Poly
Breads, Baked Goods, Cakes, Cookies, Crackers, Chips, Pies													
	Bagels												
8534	Cinnamon and raisin	1	item(s)	71	22.7	194	7.0	39.2	1.6	1.2	0.2	0.1	0.5
14395	Multi-grain	1	item(s)	61	—	170	6.0	35.0	1.0	1.5	0.5	0.1	0.4
8538	Oat bran	1	item(s)	71	23.4	181	7.6	37.8	2.6	0.9	0.1	0.2	0.3
4910	Plain, enriched	1	item(s)	71	25.8	182	7.1	35.9	1.6	1.2	0.3	0.4	0.5
4911	Plain, enriched, toasted	1	item(s)	66	18.7	190	7.4	37.7	1.7	1.1	0.2	0.3	0.6
	Biscuits												
25008	Biscuits	1	item(s)	41	15.8	121	2.6	16.4	0.5	4.9	1.4	1.4	1.8
16729	Scone	1	item(s)	42	11.5	148	3.8	19.1	0.6	6.2	2.0	2.5	1.3
25166	Wheat biscuits	1	item(s)	55	21.0	162	3.6	21.9	1.4	6.7	1.9	1.9	2.5
	Bread												
325	Boston brown, canned	1	slice(s)	45	21.2	88	2.3	19.5	2.1	0.7	0.1	0.1	0.3
8716	Bread sticks, plain	4	item(s)	24	1.5	99	2.9	16.4	0.7	2.3	0.3	0.9	0.9
25176	Cornbread	1	piece(s)	55	25.9	141	4.7	18.3	0.9	5.4	2.1	1.4	1.5
327	Cracked wheat	1	slice(s)	25	9.0	65	2.2	12.4	1.4	1.0	0.2	0.5	0.2
9079	Croutons, plain	¼	cup(s)	8	0.4	31	0.9	5.5	0.4	0.5	0.1	0.2	0.1
8582	Egg	1	slice(s)	40	13.9	113	3.8	19.1	0.9	2.4	0.6	0.9	0.4
8585	Egg, toasted	1	slice(s)	37	10.5	117	3.9	19.5	0.9	2.4	0.6	1.1	0.4
329	French	1	slice(s)	32	8.9	92	3.8	18.1	0.8	0.6	0.2	0.1	0.3
8591	French, toasted	1	slice(s)	23	4.7	73	3.0	14.2	0.7	0.5	0.1	0.1	0.2
42096	Indian fry, made with lard (Navajo)	3	ounce(s)	85	26.9	281	5.7	41.0	—	10.4	3.9	3.8	0.9
332	Italian	1	slice(s)	30	10.7	81	2.6	15.0	0.8	1.1	0.3	0.2	0.4
1393	Mixed grain	1	slice(s)	26	9.6	69	3.5	11.3	1.9	1.1	0.2	0.2	0.5
8604	Mixed grain, toasted	1	slice(s)	24	7.6	69	3.5	11.3	1.9	1.1	0.2	0.2	0.5
8605	Oat bran	1	slice(s)	30	13.2	71	3.1	11.9	1.4	1.3	0.2	0.5	0.5
8608	Oat bran, toasted	1	slice(s)	27	10.4	70	3.1	11.8	1.3	1.3	0.2	0.5	0.5
8609	Oatmeal	1	slice(s)	27	9.9	73	2.3	13.1	1.1	1.2	0.2	0.4	0.5
8613	Oatmeal, toasted	1	slice(s)	25	7.8	73	2.3	13.2	1.1	1.2	0.2	0.4	0.5
1409	Pita	1	item(s)	60	19.3	165	5.5	33.4	1.3	0.7	0.1	0.1	0.3
7905	Pita, whole wheat	1	item(s)	64	19.6	170	6.3	35.2	4.7	1.7	0.3	0.2	0.7
338	Pumpernickel	1	slice(s)	32	12.1	80	2.8	15.2	2.1	1.0	0.1	0.3	0.4
334	Raisin, enriched	1	slice(s)	26	8.7	71	2.1	13.6	1.1	1.1	0.3	0.6	0.2
8625	Raisin, toasted	1	slice(s)	24	6.7	71	2.1	13.7	1.1	1.2	0.3	0.6	0.2
10168	Rice, white, gluten free, wheat free	1	slice(s)	38	—	130	1.0	18.0	0.5	6.0	0	—	—
8653	Rye	1	slice(s)	32	11.9	83	2.7	15.5	1.9	1.1	0.2	0.4	0.3
8654	Rye, toasted	1	slice(s)	29	9.0	82	2.7	15.4	1.9	1.0	0.2	0.4	0.3
336	Rye, light	1	slice(s)	25	9.3	65	2.0	12.0	1.6	1.0	0.2	0.3	0.3
8588	Sourdough	1	slice(s)	25	7.0	72	2.9	14.1	0.6	0.5	0.1	0.1	0.2
8592	Sourdough, toasted	1	slice(s)	23	4.7	73	3.0	14.2	0.7	0.5	0.1	0.1	0.2
491	Submarine or hoagie roll	1	item(s)	135	40.6	400	11.0	72.0	3.8	8.0	1.8	3.0	2.2
8596	Vienna, toasted	1	slice(s)	23	4.7	73	3.0	14.2	0.7	0.5	0.1	0.1	0.2
8670	Wheat	1	slice(s)	25	8.9	67	2.7	11.9	0.9	0.9	0.2	0.2	0.4
8671	Wheat, toasted	1	slice(s)	23	5.6	72	3.0	12.8	1.1	1.0	0.2	0.2	0.4
340	White	1	slice(s)	25	9.1	67	1.9	12.7	0.6	0.8	0.2	0.2	0.3
1395	Whole wheat	1	slice(s)	46	15.0	128	3.9	23.6	2.8	2.5	0.4	0.5	1.4
	Cakes												
386	Angel food, prepared from mix	1	piece(s)	50	16.5	129	3.1	29.4	0.1	0.2	0	0	0.1
8772	Butter pound, ready to eat, commercially prepared	1	slice(s)	75	18.5	291	4.1	36.6	0.4	14.9	8.7	4.4	0.8
28517	Carrot	1	slice(s)	131	56.6	339	4.8	56.5	1.9	11.1	1.0	5.7	3.8
4931	Chocolate with chocolate icing, commercially prepared	1	slice(s)	64	14.7	235	2.6	34.9	1.8	10.5	3.1	5.6	1.2
8756	Chocolate, prepared from mix	1	slice(s)	95	23.2	352	5.0	50.7	1.5	14.3	5.2	5.7	2.6
393	Devil's food cupcake with chocolate frosting	1	item(s)	35	8.4	120	2.0	20.0	0.7	4.0	1.8	1.6	0.6
8757	Fruitcake, ready to eat, commercially prepared	1	piece(s)	43	10.9	139	1.2	26.5	1.6	3.9	0.5	1.8	1.4
1397	Pineapple upside down, prepared from mix	1	slice(s)	115	37.1	367	4.0	58.1	0.9	13.9	3.4	6.0	3.8
411	Sponge, prepared from mix	1	slice(s)	63	18.5	187	4.6	36.4	0.3	2.7	0.8	1.0	0.4
8817	White with coconut frosting, prepared from mix	1	slice(s)	112	23.2	399	4.9	70.8	1.1	11.5	4.4	4.1	2.4
8819	Yellow with chocolate frosting, ready to eat, commercially prepared	1	slice(s)	64	14.0	243	2.4	35.5	1.2	11.1	3.0	6.1	1.4

PAGE KEY: A-4 = Breads/Baked Goods A-10 = Cereal/Rice/Pasta A-14 = Fruit A-18 = Vegetables/Legumes A-28 = Nuts/Seeds A-30 = Vegetarian A-32 = Dairy A-40 = Eggs A-40 = Seafood A-42 = Meats A-46 = Poultry A-46 = Processed Meats A-48 = Beverages A-52 = Fats/Oils A-54 = Sweets A-56 = Spices/Condiments/Sauces A-60 = Mixed Foods/Soups/Sandwiches A-64 = Fast Food A-84 = Convenience A-86 = Baby Foods

Chol (mg)	Calc (mg)	Iron (mg)	Magn (mg)	Pota (mg)	Sodi (mg)	Zinc (mg)	Vit A (µg)	Thia (mg)	Vit E (mg α)	Ribo (mg)	Niac (mg)	Vit B_6 (mg)	Fola (µg)	Vit C (mg)	Vit B_{12} (µg)	Sele (µg)
0	13	2.69	19.9	105.1	228.6	0.80	14.9	0.27	0.22	0.19	2.18	0.04	123.5	0.5	0	22.0
0	60	1.08	—	—	310.0	—	0	—	—	—	—	—	—	0	—	—
0	9	2.18	22.0	81.7	360.0	0.63	0.7	0.23	0.23	0.24	2.10	0.03	95.1	0.1	0	24.3
0	63	4.29	15.6	53.3	318.1	1.34	0	0.42	0.07	0.18	2.82	0.04	160.5	0.7	0	16.2
0	65	2.97	15.8	56.1	316.8	0.85	0	0.39	0.07	0.17	2.88	0.04	134.0	0	0	16.6
0	38	0.94	6.0	47.4	206.0	0.20	—	0.16	0.01	0.12	1.20	0.01	46.8	0.1	0.1	7.1
49	79	1.35	7.1	48.7	277.2	0.29	64.7	0.14	0.42	0.15	1.19	0.02	49.1	0	0.1	10.9
0	57	1.21	16.1	81.0	321.1	0.42	—	0.19	0.01	0.14	1.65	0.03	63.2	0.1	0.1	0
0	32	0.94	28.4	143.1	284.0	0.22	11.3	0.01	0.14	0.05	0.50	0.03	6.3	0	0	9.9
0	5	1.02	7.7	29.8	157.7	0.21	0	0.14	0.24	0.13	1.26	0.01	61.0	0	0	9.0
21	94	0.91	10.5	71.5	209.8	0.48	—	0.14	0.32	0.15	1.03	0.04	78.7	1.7	0.2	6.2
0	11	0.70	13.0	44.3	134.5	0.31	0	0.09	—	0.06	0.92	0.08	19.0	0	0	6.3
0	6	0.30	2.3	9.3	52.4	0.06	0	0.04	—	0.02	0.40	0.00	15.7	0	0	2.8
20	37	1.21	7.6	46.0	196.8	0.31	25.2	0.17	0.10	0.17	1.93	0.02	52.0	0	0	12.0
21	38	1.23	7.8	46.6	199.8	0.31	25.5	0.14	0.10	0.16	1.77	0.02	47.7	0	0	12.2
0	14	1.16	9.0	41.0	208.0	0.29	0	0.13	0.05	0.09	1.52	0.03	73.6	0.1	0	8.7
0	11	0.89	7.1	32.2	165.6	0.24	0	0.10	0.04	0.09	1.24	0.02	49.9	0	0	6.8
6	48	3.43	15.3	65.5	279.8	0.29	0	0.36	0.00	0.18	3.91	0.03	166.7	—	0	15.8
0	23	0.88	8.1	33.0	175.2	0.25	0	0.14	0.08	0.08	1.31	0.01	91.2	0	0	8.2
0	27	0.65	20.3	59.8	109.2	0.44	0	0.07	0.09	0.03	1.05	0.06	19.5	0	0	8.6
0	27	0.65	20.4	60.0	109.7	0.44	0	0.06	0.10	0.03	1.05	0.07	16.8	0	0	8.6
0	20	0.93	10.5	44.1	122.1	0.26	0.6	0.15	0.13	0.10	1.44	0.02	36.0	0	0	9.0
0	19	0.92	9.2	33.2	121.0	0.28	0.5	0.12	0.13	0.09	1.29	0.01	28.1	0	0	8.9
0	18	0.72	10.0	38.3	161.7	0.27	1.4	0.10	0.13	0.06	0.84	0.01	23.5	0	0	6.6
0	18	0.74	10.3	38.5	162.8	0.28	1.3	0.09	0.13	0.06	0.77	0.02	18.7	0.1	0	6.7
0	52	1.57	15.6	72.0	321.6	0.50	0	0.35	0.18	0.19	2.77	0.02	99.0	0	0	16.3
0	10	1.95	44.2	108.8	340.5	0.97	0	0.21	0.39	0.05	1.81	0.17	22.4	0	0	28.2
0	22	0.91	17.3	66.6	214.7	0.47	0	0.10	0.13	0.09	0.98	0.04	40.0	0	0	7.8
0	17	0.75	6.8	59.0	101.4	0.18	0	0.08	0.07	0.10	0.90	0.01	40.6	0	0	5.2
0	17	0.76	6.7	59.0	101.8	0.19	0	0.07	0.07	0.09	0.81	0.02	35.5	0.1	0	5.2
0	100	1.08	—	—	140	—	—	0.15	—	0.10	1.20	—	47.5	0	—	—
0	23	0.90	12.8	53.1	211.2	0.36	0	0.13	0.10	0.10	1.21	0.02	48.3	0.1	0	9.9
0	23	0.89	12.5	53.1	210.3	0.36	0	0.11	0.10	0.09	1.09	0.02	42.9	0.1	0	9.9
0	20	0.70	3.9	51.0	175.0	0.18	0	0.10	—	0.08	0.80	0.01	5.3	0	0	8.0
0	11	0.91	7.0	32.0	162.5	0.23	0	0.11	0.05	0.07	1.19	0.03	57.5	0.1	0	6.8
0	11	0.89	7.1	32.2	165.6	0.24	0	0.10	0.04	0.09	1.24	0.02	49.9	0	0	6.8
0	100	3.80	—	128.0	683.0	—	0	0.54	—	0.33	4.50	0.04	—	0	—	42.0
0	11	0.89	7.1	32.2	165.6	0.24	0	0.10	0.04	0.09	1.24	0.02	49.9	0	0	6.8
0	36	0.87	12.0	46.0	130.3	0.30	0	0.09	0.05	0.08	1.30	0.03	24.8	0.1	0	7.2
0	38	0.94	13.6	51.3	140.5	0.34	0	0.10	0.06	0.09	1.44	0.04	23.0	0	0	7.7
0	38	0.94	5.8	25.0	170.3	0.19	0	0.11	0.06	0.08	1.10	0.02	42.8	0	0	4.3
0	15	1.42	37.3	144.4	159.2	0.69	0	0.13	0.35	0.10	1.83	0.09	35.9	0	0	17.8
0	42	0.11	4.0	67.5	254.5	0.06	0	0.04	0.01	0.10	0.08	0.00	14.5	0	0	7.7
166	26	1.03	8.3	89.3	298.5	0.34	111.8	0.10	—	0.17	0.98	0.03	46.5	0	0.2	6.6
0	65	2.18	23.0	279.6	367.7	0.44	—	0.25	0.01	0.19	1.73	0.10	98.2	4.6	0	14.7
27	28	1.40	21.8	128.0	213.8	0.44	16.6	0.01	0.62	0.08	0.36	0.02	14.7	0.1	0.1	2.1
55	57	1.53	30.4	133.0	299.3	0.65	38.0	0.13	—	0.20	1.08	0.03	37.1	0.2	0.2	11.3
19	21	0.70	—	46.0	92.0	—	—	0.04	—	0.05	0.30	—	8.1	0	—	2.0
2	14	0.89	6.9	65.8	116.1	0.11	3.0	0.02	0.38	0.04	0.34	0.02	13.8	0.2	0	0.9
25	138	1.70	15.0	128.8	366.9	0.35	71.3	0.17	—	0.17	1.36	0.03	44.9	1.4	0.1	10.8
107	26	0.99	5.7	88.8	143.6	0.37	48.5	0.10	—	0.19	0.75	0.03	33.4	0	0.2	11.7
1	101	1.29	13.4	110.9	318.1	0.37	13.4	0.14	0.13	0.21	1.19	0.03	57.12	0.1	0.1	12.0
35	24	1.33	19.2	113.9	215.7	0.39	21.1	0.07	—	0.10	0.79	0.02	20.5	0	0.1	2.2

TABLE A-1 Table of Food Composition *(continued)*

(Computer code is for Cengage Diet Analysis program)
(For purposes of calculations, use "0" for t, <1, <.1, <.01, etc.)

											Fat Breakdown (g)		
DA+ Code	Food Description	Quantity	Measure	Wt (g)	H_2O (g)	Ener (kcal)	Prot (g)	Carb (g)	Fiber (g)	Fat (g)	Sat	Mono	Poly
Breads, Baked Goods, Cakes, Cookies, Crackers, Chips, Pies—*continued*													
8822	Yellow with vanilla frosting, ready to eat, commercially prepared	1	slice(s)	64	14.1	239	2.2	37.6	0.2	9.3	1.5	3.9	3.3
	Snack cakes												
8791	Chocolate snack cake, creme filled, with frosting	1	item(s)	50	9.3	200	1.8	30.2	1.6	8.0	2.4	4.3	0.9
25010	Cinnamon coffee cake	1	piece(s)	72	22.6	231	3.6	35.8	0.7	8.3	2.2	2.6	3.0
16777	Funnel cake	1	item(s)	90	37.6	276	7.3	29.1	0.9	14.4	2.7	4.7	6.1
8794	Sponge snack cake, creme filled	1	item(s)	43	8.6	155	1.3	27.2	0.2	4.8	1.1	1.7	1.4
	Snacks, chips, pretzels												
29428	Bagel chips, plain	3	item(s)	29	—	130	3.0	19.0	1.0	4.5	0.5	—	—
29429	Bagel chips, toasted onion	3	item(s)	29	—	130	4.0	20.0	1.0	4.5	0.5	—	—
38192	Chex traditional snack mix	1	cup(s)	45	—	197	3.0	33.3	1.5	6.1	0.8	—	—
654	Potato chips, salted	1	ounce(s)	28	0.6	155	1.9	14.1	1.2	10.6	3.1	2.8	3.5
8816	Potato chips, unsalted	1	ounce(s)	28	0.5	152	2.0	15.0	1.4	9.8	3.1	2.8	3.5
5096	Pretzels, plain, hard, twists	5	item(s)	30	1.0	114	2.7	23.8	1.0	1.1	0.2	0.4	0.4
4632	Pretzels, whole wheat	1	ounce(s)	28	1.1	103	3.1	23.0	2.2	0.7	0.2	0.3	0.2
4641	Tortilla chips, plain	6	item(s)	11	0.2	53	0.8	7.1	0.6	2.5	0.3	0.8	0.5
	Cookies												
8859	Animal crackers	12	item(s)	30	1.2	134	2.1	22.2	0.3	4.1	1.0	2.3	0.6
8876	Brownie, prepared from mix	1	item(s)	24	3.0	112	1.5	12.0	0.5	7.0	1.8	2.6	2.3
25207	Chocolate chip cookies	1	item(s)	30	3.7	140	2.0	16.2	0.6	7.9	2.1	3.3	2.1
8915	Chocolate sandwich cookie with extra creme filling	1	item(s)	13	0.2	65	0.6	8.9	0.4	3.2	0.7	2.1	0.3
14145	Fig Newtons cookies	1	item(s)	16	—	55	0.5	11.0	0.5	1.3	0	—	—
8920	Fortune cookie	1	item(s)	8	0.6	30	0.3	6.7	0.1	0.2	0.1	0.1	0
25208	Oatmeal cookies	1	item(s)	69	12.3	234	5.7	45.1	3.1	4.2	0.7	1.3	1.8
25213	Peanut butter cookies	1	item(s)	35	4.1	163	4.2	16.9	0.9	9.2	1.7	4.7	2.3
33095	Sugar cookies	1	item(s)	16	4.1	61	1.1	7.4	0.1	3.0	0.6	1.3	0.9
9002	Vanilla sandwich cookie with creme filling	1	item(s)	10	0.2	48	0.5	7.2	0.2	2.0	0.3	0.8	0.8
	Crackers												
9012	Cheese cracker sandwich with peanut butter	4	item(s)	28	0.9	139	3.5	15.9	1.0	7.0	1.2	3.6	1.4
9008	Cheese crackers (mini)	30	item(s)	30	0.9	151	3.0	17.5	0.7	7.6	2.8	3.6	0.7
33362	Cheese crackers, low sodium	1	serving(s)	30	0.9	151	3.0	17.5	0.7	7.6	2.9	3.6	0.7
8928	Honey graham crackers	4	item(s)	28	1.2	118	1.9	21.5	0.8	2.8	0.4	1.1	1.1
9016	Matzo crackers, plain	1	item(s)	28	1.2	112	2.8	23.8	0.9	0.4	0.1	0	0.2
9024	Melba toast	3	item(s)	15	0.8	59	1.8	11.5	0.9	0.5	0.1	0.1	0.2
9028	Melba toast, rye	3	item(s)	15	0.7	58	1.7	11.6	1.2	0.5	0.1	0.1	0.2
14189	Ritz crackers	5	item(s)	16	0.5	80	1.0	10.0	0	4.0	1.0	—	—
9014	Rye crispbread crackers	1	item(s)	10	0.6	37	0.8	8.2	1.7	0.1	0	0	0.1
9040	Rye wafer	1	item(s)	11	0.6	37	1.1	8.8	2.5	0.1	0	0	0
432	Saltine crackers	5	item(s)	15	0.8	64	1.4	10.6	0.5	1.7	0.2	1.1	0.2
9046	Saltine crackers, low salt	5	item(s)	15	0.6	65	1.4	10.7	0.5	1.8	0.4	1.0	0.3
9052	Snack cracker sandwich with cheese filling	4	item(s)	28	1.1	134	2.6	17.3	0.5	5.9	1.7	3.2	0.7
9054	Snack cracker sandwich with peanut butter filling	4	item(s)	28	0.8	138	3.2	16.3	0.6	6.9	1.4	3.9	1.3
9048	Snack crackers, round	10	item(s)	30	1.1	151	2.2	18.3	0.5	7.6	1.1	3.2	2.9
9050	Snack crackers, round, low salt	10	item(s)	30	1.1	151	2.2	18.3	0.5	7.6	1.1	3.2	2.9
9044	Soda crackers	5	tem(s)	15	0.8	64	1.4	10.6	0.5	1.7	0.2	1.1	0.2
9059	Wheat cracker sandwich with cheese filling	4	item(s)	28	0.9	139	2.7	16.3	0.9	7.0	1.2	2.9	2.6
9061	Wheat cracker sandwich with peanut butter filling	4	item(s)	28	1.0	139	3.8	15.1	1.2	7.5	1.3	3.3	2.5
9055	Wheat crackers	10	item(s)	30	0.9	142	2.6	19.5	1.4	6.2	1.6	3.4	0.8
9057	Wheat crackers, low salt	10	item(s)	30	0.9	142	2.6	19.5	1.4	6.2	1.6	3.4	0.8
9022	Whole wheat crackers	7	item(s)	28	0.8	124	2.5	19.2	2.9	4.8	1.0	1.6	1.8
	Pastry												
16754	Apple fritter	1	item(s)	17	6.4	61	1.0	5.5	0.2	3.9	0.9	1.7	1.1
41565	Cinnamon rolls with icing, refrigerated dough	1	serving(s)	44	12.3	145	2.0	23.0	0.5	5.0	1.5	—	—
4945	Croissant, butter	1	item(s)	57	13.2	231	4.7	26.1	1.5	12.0	6.6	3.1	0.6
9096	Danish, nut	1	item(s)	65	13.3	280	4.6	29.7	1.3	16.4	3.8	8.9	2.8
9115	Doughnut with creme filling	1	item(s)	85	32.5	307	5.4	25.5	0.7	20.8	4.6	10.3	2.6

PAGE KEY: A-4 = Breads/Baked Goods A-10 = Cereal/Rice/Pasta A-14 = Fruit A-18 = Vegetables/Legumes A-28 = Nuts/Seeds A-30 = Vegetarian A-32 = Dairy A-40 = Eggs A-40 = Seafood A-42 = Meats A-46 = Poultry A-46 = Processed Meats A-48 = Beverages A-52 = Fats/Oils A-54 = Sweets A-56 = Spices/Condiments/Sauces A-60 = Mixed Foods/Soups/Sandwiches A-64 = Fast Food A-84 = Convenience A-86 = Baby Foods

Chol (mg)	Calc (mg)	Iron (mg)	Magn (mg)	Pota (mg)	Sodi (mg)	Zinc (mg)	Vit A (µg)	Thia (mg)	Vit E (mg α)	Ribo (mg)	Niac (mg)	Vit B_6 (mg)	Fola (µg)	Vit C (mg)	Vit B_{12} (µg)	Sele (µg)
35	40	0.68	3.8	33.9	220.2	0.16	12.2	0.06	—	0.04	0.32	0.01	25.6	0	0.1	3.5
0	58	1.80	18.0	88.0	194.5	0.52	0.5	0.01	0.54	0.03	0.46	0.07	17.5	1.0	0	1.7
26	55	1.36	9.9	91.9	277.6	0.30	—	0.17	0.23	0.16	1.29	0.02	66.1	0.3	0.1	9.6
62	126	1.90	16.2	152.1	269.1	0.65	49.5	0.23	1.54	0.32	1.86	0.04	75.6	0	0.3	17.7
7	19	0.54	3.4	37.0	155.1	0.12	2.1	0.06	0.50	0.05	0.52	0.01	23.0	0	0	1.3
0	0	0.72	—	45.0	70.0	—	0	—	—	—	—	—	—	0	0	—
0	0	0.72	—	50.0	300.0	—	0	—	—	—	—	—	—	0	0	—
0	0	0.55	—	75.8	621.2	—	0	0.09	—	0.05	1.21	—	38.5	0	—	—
0	7	0.45	19.8	465.5	148.8	0.67	0	0.01	1.91	0.06	1.18	0.20	21.3	5.3	0	2.3
0	7	0.46	19.0	361.5	2.3	0.30	0	0.04	2.58	0.05	1.08	0.18	12.8	8.8	0	2.3
0	11	1.29	10.5	43.8	514.5	0.25	0	0.13	0.10	0.18	1.57	0.03	86.0	0	0	1.7
0	8	0.76	8.5	121.9	57.6	0.17	0	0.12	—	0.08	1.85	0.07	15.3	0.3	0	—
0	19	0.25	15.8	23.2	45.5	0.26	0	0.00	0.46	0.01	0.13	0.02	2.2	0	0	0.7
0	13	0.82	5.4	30.0	117.9	0.19	0	0.10	0.03	0.09	1.04	0.01	49.5	0	0	2.1
18	14	0.44	12.7	42.2	82.3	0.23	42.2	0.03	—	0.05	0.24	0.02	9.4	0.1	0	2.8
13	11	0.69	12.4	62.1	108.8	0.24	—	0.08	0.54	0.06	0.87	0.01	14.1	0	0	4.1
0	2	1.01	4.7	17.8	45.6	0.10	0	0.02	0.25	0.02	0.25	0.00	9.0	0	0	1.1
0	10	0.36	—	—	57.5	—	0	—	—	—	—	—	—	0	—	—
0	1	0.12	0.6	3.3	21.9	0.01	0.1	0.01	0.00	0.01	0.15	0.00	8.4	0	0	0.2
0	26	1.93	48.8	176.7	311.1	1.42	—	0.26	0.23	0.13	1.35	0.09	65.6	0.3	0	17.4
13	27	0.65	21.1	112.8	154.1	0.46	—	0.08	0.73	0.09	1.85	0.05	35.0	0.1	0.1	4.8
18	5	0.30	1.7	12.2	49.4	0.08	—	0.04	0.28	0.05	0.31	0.01	13.1	0	0	3.1
0	3	0.22	1.4	9.1	34.9	0.04	0	0.02	0.16	0.02	0.27	0.00	8.2	0	0	0.3
0	14	0.76	15.7	61.0	198.8	0.29	0.3	0.15	0.66	0.08	1.63	0.04	39.8	0	0.1	2.3
4	45	1.43	10.8	43.5	298.5	0.33	8.7	0.17	0.01	0.12	1.40	0.16	72.3	0	0.1	2.6
4	45	1.43	10.8	31.8	137.4	0.33	5.1	0.17	0.09	0.12	1.40	0.16	40.2	0	0.1	2.6
0	7	1.04	8.4	37.8	169.4	0.22	0	0.06	0.09	0.08	1.15	0.01	18.5	0	0	2.9
0	4	0.89	7.1	31.8	0.6	0.19	0	0.11	0.01	0.08	1.10	0.03	4.8	0	0	10.5
0	14	0.55	8.9	30.3	124.4	0.30	0	0.06	0.06	0.04	0.61	0.01	29.0	0	0	5.2
0	12	0.55	5.9	29.0	134.9	0.20	0	0.07	—	0.04	0.70	0.01	19.4	0	0	5.8
0	20	0.72	—	10.0	135.0	—	—	—	—	—	—	—	—	0	—	—
0	3	0.24	7.8	31.9	26.4	0.23	0	0.02	0.08	0.01	0.10	0.02	6.5	0	0	3.7
0	4	0.65	13.3	54.5	87.3	0.30	0	0.04	0.08	0.03	0.17	0.03	5.0	0	0	2.6
0	10	0.84	3.3	23.1	160.8	0.12	0	0.01	0.14	0.06	0.78	0.01	33.2	0	0	1.5
0	18	0.81	4.1	108.6	95.4	0.11	0	0.08	0.01	0.06	0.78	0.01	33.2	0	0	2.9
1	72	0.66	10.1	120.1	392.3	0.17	4.8	0.12	0.06	0.19	1.05	0.01	44.8	0	0	6.0
0	23	0.77	15.4	60.2	201.0	0.31	0.3	0.13	0.57	0.07	1.71	0.04	34.2	0	0	3.0
0	36	1.08	8.1	39.9	254.1	0.20	0	0.12	0.60	0.10	1.21	0.01	55.8	0	0	2.0
0	36	1.08	8.1	106.5	111.9	0.20	0	0.12	0.60	0.10	1.21	0.01	55.8	0	0	2.0
0	10	0.84	3.3	23.1	160.8	0.12	0	0.01	0.14	0.06	0.78	0.01	33.2	0	0	1.5
2	57	0.73	15.1	85.7	255.6	0.24	4.8	0.10	—	0.12	0.89	0.07	26.9	0.4	0	6.8
0	48	0.74	10.6	83.2	226.0	0.23	0	0.10	—	0.08	1.64	0.03	26.0	0	0	6.1
0	15	1.32	18.6	54.9	238.5	0.48	0	0.15	0.15	0.09	1.48	0.04	56.1	0	0	1.9
0	15	1.32	18.6	60.9	84.9	0.48	0	0.15	0.15	0.09	1.48	0.04	21.6	0	0	10.1
0	14	0.86	27.7	83.2	184.5	0.60	0	0.05	0.24	0.02	1.26	0.05	7.8	0	0	4.1
14	9	0.26	2.2	22.4	6.8	0.09	7.1	0.03	0.07	0.04	0.23	0.01	9.2	0.2	0.1	2.6
0	—	0.72	—	—	340.1	—	0	—	—	—	—	—	—	—	—	—
38	21	1.15	9.1	67.3	424.1	0.42	117.4	0.22	0.47	0.13	1.24	0.03	74.1	0.1	0.1	12.9
30	61	1.17	20.8	61.8	236.0	0.56	5.9	0.14	0.53	0.15	1.49	0.06	79.3	1.1	0.1	9.2
20	21	1.55	17.0	68.0	262.7	0.68	9.4	0.28	0.24	0.12	1.90	0.05	92.7	0	0.1	9.2

TABLE A-1 Table of Food Composition *(continued)*

(Computer code is for Cengage Diet Analysis program)
(For purposes of calculations, use "0" for t, <1, <.1, <.01, etc.)

DA+ Code	Food Description	Quantity	Measure	Wt (g)	H_2O (g)	Ener (kcal)	Prot (g)	Carb (g)	Fiber (g)	Fat (g)	Fat Breakdown (g) Sat	Mono	Poly
Breads, Baked Goods, Cakes, Cookies, Crackers, Chips, Pies—*continued*													
9117	Doughnut with jelly filling	1	item(s)	85	30.3	289	5.0	33.2	0.8	15.9	4.1	8.7	2.0
4947	Doughnut, cake	1	item(s)	47	9.8	198	2.4	23.4	0.7	10.8	1.7	4.4	3.7
9105	Doughnut, cake, chocolate glazed	1	item(s)	42	6.8	175	1.9	24.1	0.9	8.4	2.2	4.7	1.0
437	Doughnut, glazed	1	item(s)	60	15.2	242	3.8	26.6	0.7	13.7	3.5	7.7	1.7
10617	Toaster pastry, brown sugar cinnamon	1	item(s)	50	5.3	210	3.0	35.0	1.0	6.0	1.0	4.0	1.0
30928	Toaster pastry, cream cheese	1	item(s)	54	—	200	3.0	23.0	0	11.0	4.5	—	—
	Muffins												
25015	Blueberry	1	item(s)	63	29.7	160	3.4	23.0	0.8	6.0	0.9	1.5	3.3
9189	Corn, ready to eat	1	item(s)	57	18.6	174	3.4	29.0	1.9	4.8	0.8	1.2	1.8
9121	English muffin, plain, enriched	1	item(s)	57	24.0	134	4.4	26.2	1.5	1.0	0.1	0.2	0.5
29582	English muffin, toasted	1	item(s)	50	18.6	128	4.2	25.0	1.5	1.0	0.1	0.2	0.5
9145	English muffin, wheat	1	item(s)	57	24.1	127	5.0	25.5	2.6	1.1	0.2	0.2	0.5
8894	Oat bran	1	item(s)	57	20.0	154	4.0	27.5	2.6	4.2	0.6	1.0	2.4
	Granola bars												
38161	Kudos milk chocolate granola bars w/fruit and nuts	1	item(s)	28	—	90	2.0	15.0	1.0	3.0	1.0	—	—
38196	Nature Valley banana nut crunchy granola bars	2	item(s)	42	—	190	4.0	28.0	2.0	7.0	1.0	—	—
38187	Nature Valley fruit 'n' nut trail mix bar	1	item(s)	35	—	140	3.0	25.0	2.0	4.0	0.5	—	—
1383	Plain, hard	1	item(s)	25	1.0	115	2.5	15.8	1.3	4.9	0.6	1.1	3.0
4606	Plain, soft	1	item(s)	28	1.8	126	2.1	19.1	1.3	4.9	2.1	1.1	1.5
	Pies												
454	Apple pie, prepared from home recipe	1	slice(s)	155	73.3	411	3.7	57.5	2.3	19.4	4.7	8.4	5.2
470	Pecan pie, prepared from home recipe	1	slice(s)	122	23.8	503	6.0	63.7	—	27.1	4.9	13.6	7.0
33356	Pie crust mix, prepared, baked	1	slice(s)	20	2.1	100	1.3	10.1	0.4	6.1	1.5	3.5	0.8
9007	Pie crust, ready to bake, frozen, enriched, baked	1	slice(s)	16	1.8	82	0.7	7.9	0.2	5.2	1.7	2.5	0.6
472	Pumpkin pie, prepared from home recipe	1	slice(s)	155	90.7	316	7.0	40.9	—	14.4	4.9	5.7	2.8
	Rolls												
8555	Crescent dinner roll	1	item(s)	28	9.7	78	2.7	13.8	0.6	1.2	0.3	0.3	0.6
489	Hamburger roll or bun, plain	1	item(s)	43	14.9	120	4.1	21.3	0.9	1.9	0.5	0.5	0.8
490	Hard roll	1	item(s)	57	17.7	167	5.6	30.0	1.3	2.5	0.3	0.6	1.0
5127	Kaiser roll	1	item(s)	57	17.7	167	5.6	30.0	1.3	2.5	0.3	0.6	1.0
5130	Whole wheat roll or bun	1	item(s)	28	9.4	75	2.5	14.5	2.1	1.3	0.2	0.3	0.6
	Sport bars												
37026	Balance original chocolate bar	1	item(s)	50	—	200	14.0	22.0	0.5	6.0	3.5	—	—
37024	Balance original peanut butter bar	1	item(s)	50	—	200	14.0	22.0	1.0	6.0	2.5	—	—
36580	Clif Bar chocolate brownie energy bar	1	item(s)	68	—	240	10.0	45.0	5.0	4.5	1.5	—	—
36583	Clif Bar crunchy peanut butter energy bar	1	item(s)	68	—	250	12.0	40.0	5.0	6.0	1.5	—	—
36589	Clif Luna Nutz over Chocolate energy bar	1	item(s)	48	—	180	10.0	25.0	3.0	4.5	2.5	—	—
12005	PowerBar apple cinnamon	1	item(s)	65	—	230	9.0	45.0	3.0	2.5	0.5	1.5	0.5
16078	PowerBar banana	1	item(s)	65	—	230	9.0	45.0	3.0	2.5	0.5	1.0	0.5
16080	PowerBar chocolate	1	item(s)	65	6.4	230	10.0	45.0	3.0	2.0	0.5	0.5	1.0
29092	PowerBar peanut butter	1	item(s)	65	—	240	10.0	45.0	3.0	3.5	0.5	—	—
	Tortillas												
1391	Corn tortillas, soft	1	item(s)	26	11.9	57	1.5	11.6	1.6	0.7	0.1	0.2	0.4
1669	Flour tortilla	1	item(s)	32	9.7	100	2.7	16.4	1.0	2.5	0.6	1.2	0.5
	Pancakes, waffles												
8926	Pancakes, blueberry, prepared from recipe	3	item(s)	114	60.6	253	7.0	33.1	0.8	10.5	2.3	2.6	4.7
5037	Pancakes, prepared from mix with egg and milk	3	item(s)	114	60.3	249	8.9	32.9	2.1	8.8	2.3	2.4	3.3
1390	Taco shells, hard	1	item(s)	13	1.0	62	0.9	8.3	0.6	2.8	0.6	1.6	0.5
30311	Waffle, 100% whole grain	1	item(s)	75	32.3	200	6.9	25.0	1.9	8.4	2.3	3.3	2.1
9219	Waffle, plain, frozen, toasted	2	item(s)	66	20.2	206	4.7	32.5	1.6	6.3	1.1	3.2	1.5
500	Waffle, plain, prepared from recipe	1	item(s)	75	31.5	218	5.9	24.7	1.7	10.6	2.1	2.6	5.1

PAGE KEY: A-4 = Breads/Baked Goods A-10 = Cereal/Rice/Pasta A-14 = Fruit A-18 = Vegetables/Legumes A-28 = Nuts/Seeds A-30 = Vegetarian A-32 = Dairy A-40 = Eggs A-40 = Seafood A-42 = Meats A-46 = Poultry A-46 = Processed Meats A-48 = Beverages A-52 = Fats/Oils A-54 = Sweets A-56 = Spices/Condiments/Sauces A-60 = Mixed Foods/Soups/Sandwiches A-64 = Fast Food A-84 = Convenience A-86 = Baby Foods

Chol (mg)	Calc (mg)	Iron (mg)	Magn (mg)	Pota (mg)	Sodi (mg)	Zinc (mg)	Vit A (μg)	Thia (mg)	Vit E (mg α)	Ribo (mg)	Niac (mg)	Vit B_6 (mg)	Fola (μg)	Vit C (mg)	Vit B_{12} (μg)	Sele (μg)
22	21	1.49	17.0	67.2	249.1	0.63	14.5	0.26	0.36	0.12	1.81	0.08	88.4	0	0.2	10.6
17	21	0.91	9.4	59.7	256.6	0.25	17.9	0.10	0.90	0.11	0.87	0.02	32.9	0.1	0.1	4.4
24	89	0.95	14.3	44.5	142.8	0.23	5.0	0.01	0.08	0.02	0.19	0.01	27.3	0	0	1.7
4	26	0.36	13.2	64.8	205.2	0.46	2.4	0.53	—	0.04	0.39	0.03	13.2	0.1	0.1	5.0
0	0	1.80	—	70.0	190.0	—	—	0.15	—	0.17	2.00	0.20	21.0	0	0	—
10	100	1.80	—	—	220.0	—	—	0.15	—	0.17	2.00	—	21.0	0	0.6	—
20	56	1.02	7.8	70.2	289.4	0.28	—	0.17	0.75	0.15	1.25	0.02	62.5	0.4	0.1	8.8
15	42	1.60	18.2	39.3	297.0	0.30	29.6	0.15	0.45	0.18	1.16	0.04	63.8	0	0.1	8.7
0	30	1.42	12.0	74.7	264.5	0.39	0	0.25	—	0.16	2.21	0.02	57.0	0	0	—
0	95	1.36	11.0	71.5	252.0	0.38	0	0.19	0.16	0.14	1.90	0.02	62.5	0.1	0	13.5
0	101	1.63	21.1	106.0	217.7	0.61	0	0.24	0.25	0.16	1.91	0.05	46.7	0	0	16.6
0	36	2.39	89.5	289.0	224.0	1.04	0	0.14	0.37	0.05	0.23	0.09	79.2	0	0	6.3
0	200	0.36	—	—	60.0	—	0	—	—	—	—	—	—	0	0	—
0	20	1.08	—	120.0	160.0	—	0	—	—	—	—	—	—	0	—	—
0	0	0.00	—	—	95.0	—	0	—	—	—	—	—	—	0	—	—
0	15	0.72	23.8	82.3	72.0	0.50	0	0.06	—	0.03	0.39	0.02	5.8	0.2	0	4.0
0	30	0.72	21.0	92.3	79.0	0.42	0	0.08	—	0.04	0.14	0.02	6.8	0	0.1	4.6
0	11	1.73	10.9	122.5	327.1	0.29	17.1	0.22	—	0.16	1.90	0.05	58.9	2.6	0	12.1
106	39	1.80	31.7	162.3	319.6	1.24	100.0	0.22	—	0.22	1.03	0.07	41.5	0.2	0.2	14.6
0	12	0.43	3.0	12.4	145.8	0.07	0	0.06	—	0.03	0.47	0.01	22.2	0	0	4.4
0	3	0.36	2.9	17.6	103.5	0.05	0	0.04	0.42	0.06	0.39	0.01	16.3	0	0	0.5
65	146	1.96	29.5	288.3	348.8	0.71	660.3	0.14	—	0.31	1.21	0.07	43.4	2.6	0.1	11.0
0	39	0.93	5.9	26.3	134.1	0.18	0	0.11	0.02	0.09	1.16	0.02	47.6	0	0.1	5.5
0	59	1.42	9.0	40.4	206.0	0.28	0	0.17	0.03	0.13	1.78	0.03	73.5	0	0.1	8.4
0	54	1.87	15.4	61.6	310.1	0.53	0	0.27	0.23	0.19	2.41	0.02	86.1	0	0	22.3
0	54	1.86	15.4	61.6	310.1	0.53	0	0.27	0.23	0.19	2.41	0.01	86.1	0	0	22.3
0	30	0.69	24.1	77.1	135.5	0.57	0	0.07	0.26	0.04	1.04	0.06	8.5	0	0	14.0
3	100	4.50	40.0	160.0	180.0	3.75	—	0.37	—	0.42	5.00	0.50	102.0	60.0	1.5	17.5
3	100	4.50	40.0	130.0	230.0	3.75	—	0.37	—	0.42	5.00	0.50	102.0	60.0	1.5	17.5
0	250	4.50	100.0	370.0	150.0	3.00	—	0.37	—	0.25	3.00	0.40	80.0	60.0	0.9	14.0
0	250	4.50	100.0	230.0	250.0	3.00	—	0.37	—	0.25	3.00	0.40	80.0	60.0	0.9	14.0
0	350	5.40	80.0	190.0	190.0	5.25	—	1.20	—	1.36	16.00	2.00	400.0	60.0	6.0	24.5
0	300	6.30	140.0	125.0	100.0	5.25	—	1.50	—	1.70	20.00	2.00	400.0	60.0	6.0	—
0	300	6.30	140.0	190.0	100.0	5.25	0	1.50	—	1.70	20.00	2.00	400.0	60.0	6.0	—
0	300	6.30	140.0	200.0	95.0	5.25	0	1.50	—	1.70	20.00	2.00	400.0	60.0	6.0	5.1
0	300	6.30	140.0	130.0	120.0	5.25	0	1.50	—	1.70	20.00	2.00	400.0	60.0	6.0	—
0	21	0.32	18.7	48.4	11.7	0.34	0	0.02	0.07	0.02	0.39	0.06	1.3	0	0	1.6
0	41	1.06	7.0	49.6	203.5	0.17	0	0.17	0.06	0.08	1.14	0.01	64.3	0	0	7.1
64	235	1.96	18.2	157.3	469.7	0.61	57.0	0.22	—	0.31	1.73	0.05	60.4	2.5	0.2	16.0
81	245	1.48	25.1	226.9	575.7	0.85	82.1	0.22	—	0.35	1.40	0.12	61.6	0.7	0.4	—
0	13	0.25	11.3	29.7	51.7	0.21	0.1	0.03	0.09	0.01	0.25	0.03	11.1	0	0	0.6
71	194	1.60	28.5	171.0	371.3	0.87	48.8	0.15	0.32	0.25	1.47	0.08	38.3	0	0.4	20.0
10	203	4.56	15.8	95.0	481.8	0.35	262.7	0.34	0.64	0.46	5.86	0.68	78.5	0	1.9	8.3
52	191	1.73	14.3	119.3	383.3	0.51	48.8	0.19	—	0.26	1.55	0.04	51.0	0.3	0.2	34.7

TABLE A-1 Table of Food Composition *(continued)*

(Computer code is for Cengage Diet Analysis program)
(For purposes of calculations, use "0" for t, <1, <.1, <.01, etc.)

DA+ Code	Food Description	Quantity	Measure	Wt (g)	H_2O (g)	Ener (kcal)	Prot (g)	Carb (g)	Fiber (g)	Fat (g)	Fat Breakdown (g) Sat	Mono	Poly
Cereal, Flour, Grain, Pasta, Noodles, Popcorn													
	Grain												
2861	Amaranth, dry	½	cup(s)	98	9.6	365	14.1	64.5	9.1	6.3	1.6	1.4	2.8
1953	Barley, pearled, cooked	½	cup(s)	79	54.0	97	1.8	22.2	3.0	0.3	0.1	0	0.2
1956	Buckwheat groats, cooked, roasted	½	cup(s)	84	63.5	77	2.8	16.8	2.3	0.5	0.1	0.2	0.2
1957	Bulgur, cooked	½	cup(s)	91	70.8	76	2.8	16.9	4.1	0.2	0	0	0.1
1963	Couscous, cooked	½	cup(s)	79	57.0	88	3.0	18.2	1.1	0.1	0	0	0.1
1967	Millet, cooked	½	cup(s)	120	85.7	143	4.2	28.4	1.6	1.2	0.2	0.2	0.6
1969	Oat bran, dry	½	cup(s)	47	3.1	116	8.1	31.1	7.2	3.3	0.6	1.1	1.3
1972	Quinoa, dry	½	cup(s)	85	11.3	313	12.0	54.5	5.9	5.2	0.6	1.4	2.8
	Rice												
129	Brown, long grain, cooked	½	cup(s)	98	71.3	108	2.5	22.4	1.8	0.9	0.2	0.3	0.3
2863	Brown, medium grain, cooked	½	cup(s)	98	71.1	109	2.3	22.9	1.8	0.8	0.2	0.3	0.3
37488	Jasmine, saffroned, cooked	½	cup(s)	280	—	340	8.0	78.0	0	0	0	0	0
30280	Pilaf, cooked	½	cup(s)	103	74.0	129	2.1	22.2	0.6	3.3	0.6	1.5	1.0
28066	Spanish, cooked	½	cup(s)	244	184.2	241	5.7	50.2	3.3	1.9	0.4	0.6	0.7
2867	White glutinous, cooked	½	cup(s)	87	66.7	84	1.8	18.3	0.9	0.2	0	0.1	0.1
484	White, long grain, boiled	½	cup(s)	79	54.1	103	2.1	22.3	0.3	0.2	0.1	0.1	0.1
482	White, long grain, enriched, instant, boiled	½	cup(s)	83	59.4	97	1.8	20.7	0.5	0.4	0	0.1	0
486	White, long grain, enriched, parboiled, cooked	½	cup(s)	79	55.6	97	2.3	20.6	0.7	0.3	0.1	0.1	0.1
1194	Wild brown, cooked	½	cup(s)	82	60.6	83	3.3	17.5	1.5	0.3	0	0	0.2
	Flour and grain fractions												
505	All purpose flour, self-rising, enriched	½	cup(s)	63	6.6	221	6.2	46.4	1.7	0.6	0.1	0	0.2
503	All purpose flour, white, bleached, enriched	½	cup(s)	63	7.4	228	6.4	47.7	1.7	0.6	0.1	0	0.2
1643	Barley flour	½	cup(s)	56	5.5	198	4.2	44.7	2.1	0.8	0.2	0.1	0.4
383	Buckwheat flour, whole groat	½	cup(s)	60	6.7	201	7.6	42.3	6.0	1.9	0.4	0.6	0.6
504	Cake wheat flour, enriched	½	cup(s)	69	8.6	248	5.6	53.5	1.2	0.6	0.1	0.1	0.3
426	Cornmeal, degermed, enriched	½	cup(s)	69	7.8	255	5.0	54.6	2.8	1.2	0.1	0.2	0.5
424	Cornmeal, yellow whole grain	½	cup(s)	61	6.2	221	4.9	46.9	4.4	2.2	0.3	0.6	1.0
1978	Dark rye flour	½	cup(s)	64	7.1	207	9.0	44.0	14.5	1.7	0.2	0.2	0.8
1644	Masa corn flour, enriched	½	cup(s)	57	5.1	208	5.3	43.5	5.5	2.1	0.3	0.6	1.0
1976	Rice flour, brown	½	cup(s)	79	9.4	287	5.7	60.4	3.6	2.2	0.4	0.8	0.8
1645	Rice flour, white	½	cup(s)	79	9.4	289	4.7	63.3	1.9	1.1	0.3	0.3	0.3
1980	Semolina, enriched	½	cup(s)	84	10.6	301	10.6	60.8	3.2	0.9	0.1	0.1	0.4
2827	Soy flour, raw	½	cup(s)	42	2.2	185	14.7	14.9	4.1	8.8	1.3	1.9	4.9
1990	Wheat germ, crude	2	tablespoon(s)	14	1.6	52	3.3	7.4	1.9	1.4	0.2	0.2	0.9
506	Whole wheat flour	½	cup(s)	60	6.2	203	8.2	43.5	7.3	1.1	0.2	0.1	0.5
	Breakfast bars												
39230	Atkins Morning Start apple crisp breakfast bar	1	item(s)	37	—	170	11.0	12.0	6.0	9.0	4.0	—	—
10571	Nutri-Grain apple cinnamon cereal bar	1	item(s)	37	—	140	2.0	27.0	1.0	3.0	0.5	2.0	0.5
10647	Nutri-Grain blueberry cereal bar	1	item(s)	37	5.4	140	2.0	27.0	1.0	3.0	0.5	2.0	0.5
10648	Nutri-Grain raspberry cereal bar	1	item(s)	37	5.4	140	2.0	27.0	1.0	3.0	0.5	2.0	0.5
10649	Nutri-Grain strawberry cereal bar	1	item(s)	37	5.4	140	2.0	27.0	1.0	3.0	0.5	2.0	0.5
	Breakfast cereals, hot												
1260	Cream of Wheat, instant, prepared	½	cup(s)	121	—	388	12.9	73.3	4.3	0	0	0	0
365	Farina, enriched, cooked w/water and salt	½	cup(s)	117	102.4	56	1.7	12.2	0.3	0.1	0	0	0
363	Grits, white corn, regular and quick, enriched, cooked w/water and salt	½	cup(s)	121	103.3	71	1.7	15.6	0.4	0.2	0	0.1	0.1
8636	Grits, yellow corn, regular and quick, enriched, cooked w/salt	½	cup(s)	121	103.3	71	1.7	15.6	0.4	0.2	0	0.1	0.1
8657	Oatmeal, cooked w/water	½	cup(s)	117	97.8	83	3.0	14.0	2.0	1.8	0.4	0.5	0.7
5500	Oatmeal, maple and brown sugar, instant, prepared	1	item(s)	198	150.2	200	4.8	40.4	2.4	2.2	0.4	0.7	0.8
5510	Oatmeal, ready to serve, packet, prepared	1	item(s)	186	158.7	112	4.1	19.8	2.7	2.0	0.4	0.7	0.8
	Breakfast cereals, ready to eat												
1197	All-Bran	1	cup(s)	62	1.3	160	8.1	46.0	18.2	2.0	0.4	0.4	1.3
1200	All-Bran Buds	1	cup(s)	91	2.7	212	6.4	72.7	39.1	1.9	0.4	0.5	1.2

PAGE KEY: A-4 = Breads/Baked Goods A-10 = Cereal/Rice/Pasta A-14 = Fruit A-18 = Vegetables/Legumes A-28 = Nuts/Seeds A-30 = Vegetarian A-32 = Dairy A-40 = Eggs A-40 = Seafood A-42 = Meats A-46 = Poultry A-46 = Processed Meats A-48 = Beverages A-52 = Fats/Oils A-54 = Sweets A-56 = Spices/Condiments/Sauces A-60 = Mixed Foods/Soups/Sandwiches A-64 = Fast Food A-84 = Convenience A-86 = Baby Foods

Chol (mg)	Calc (mg)	Iron (mg)	Magn (mg)	Pota (mg)	Sodi (mg)	Zinc (mg)	Vit A (µg)	Thia (mg)	Vit E (mg α)	Ribo (mg)	Niac (mg)	Vit B_6 (mg)	Fola (µg)	Vit C (mg)	Vit B_{12} (µg)	Sele (µg)
0	149	7.40	259.3	356.8	20.5	3.10	0	0.06	—	0.20	1.24	0.20	47.8	4.1	0	—
0	9	1.04	17.3	73.0	2.4	0.64	0	0.06	0.01	0.04	1.61	0.09	12.6	0	0	6.8
0	6	0.67	42.8	73.9	3.4	0.51	0	0.03	0.07	0.03	0.79	0.06	11.8	0	0	1.8
0	9	0.87	29.1	61.9	4.6	0.51	0	0.05	0.01	0.02	0.91	0.07	16.4	0	0	0.5
0	6	0.30	6.3	45.5	3.9	0.20	0	0.05	0.10	0.02	0.77	0.04	11.8	0	0	21.6
0	4	0.75	52.8	74.4	2.4	1.09	0	0.12	0.02	0.09	1.59	0.13	22.8	0	0	1.1
0	27	2.54	110.5	266.0	1.9	1.46	0	0.55	0.47	0.10	0.43	0.07	24.4	0	0	21.2
0	40	3.88	167.4	478.5	4.2	2.62	0.8	0.30	2.06	0.26	1.28	0.40	156.4	0	0	7.2
0	10	0.41	41.9	41.9	4.9	0.61	0	0.09	0.02	0.02	1.49	0.14	3.9	0	0	9.6
0	10	0.51	42.9	77.0	1.0	0.60	0	0.09	—	0.01	1.29	0.14	3.9	0	0	38.0
0	—	2.16	—	—	780.0	—	—	—	—	—	—	—	—	—	—	—
0	11	1.16	9.3	54.6	390.4	0.37	33.0	0.13	0.28	0.02	1.23	0.06	73.1	0.4	0	4.3
0	37	1.52	95.4	330.5	97.1	1.40	—	0.27	0.12	0.05	3.24	0.38	89.9	22.6	0	14.3
0	2	0.12	4.4	8.7	4.4	0.35	0	0.01	0.03	0.01	0.25	0.02	0.9	0	0	4.9
0	8	0.94	9.5	27.7	0.8	0.38	0	0.12	0.03	0.01	1.16	0.07	76.6	0	0	5.9
0	7	1.46	4.1	7.4	3.3	0.40	0	0.06	0.01	0.01	1.43	0.04	97.4	0	0	4.0
0	15	1.43	7.1	44.2	1.6	0.29	0	0.16	0.01	0.01	1.82	0.12	107.4	0	0	7.3
0	2	0.49	26.2	82.8	2.5	1.09	0	0.04	0.19	0.07	1.05	0.11	21.3	0	0	0.7
0	211	2.90	11.9	77.5	793.7	0.38	0	0.42	0.02	0.24	3.64	0.02	193.4	0	0	21.5
0	9	2.90	13.7	66.9	1.2	0.42	0	0.48	0.02	0.30	3.68	0.02	183.3	0	0	21.2
0	16	0.70	45.4	185.9	4.5	1.04	0	0.06	—	0.02	2.56	0.14	4.5	0	0	2.0
0	25	2.42	150.6	346.2	6.6	1.86	0	0.24	0.18	0.10	3.68	0.34	32.4	0	0	3.4
0	10	5.01	11.0	71.9	1.4	0.42	0	0.61	0.01	0.29	4.65	0.02	194.6	0	0	3.4
0	2	2.98	24.1	104.9	4.8	0.48	7.6	0.42	0.10	0.28	3.66	0.12	231.1	0	0	8.0
0	4	2.10	77.5	175.1	21.3	1.10	6.7	0.22	0.24	0.12	2.20	0.18	15.2	0	0	9.4
0	36	4.12	158.7	467.2	0.6	3.58	0.6	0.20	0.90	0.16	2.72	0.28	21.1	0	0	22.8
0	80	4.10	62.7	169.9	2.8	1.00	0	0.80	0.08	0.42	5.60	0.20	190.9	0	0	8.5
0	9	1.56	88.5	228.3	6.3	1.92	0	0.34	0.94	0.06	5.00	0.58	12.6	0	0	—
0	8	0.26	27.6	60.0	0	0.62	0	0.10	0.08	0.02	2.04	0.34	3.2	0	0	11.9
0	14	3.64	39.2	155.3	0.8	0.86	0	0.66	0.20	0.46	5.00	0.08	219.2	0	0	74.6
0	87	2.70	182.0	1067.0	5.5	1.65	2.5	0.24	0.82	0.48	1.83	0.18	146.4	0	0	3.2
0	6	0.90	34.4	128.2	1.7	1.76	0	0.27	—	0.07	0.97	0.18	40.4	0	0	11.4
0	20	2.32	82.8	243.0	3.0	1.74	0	0.26	0.48	0.12	3.82	0.20	26.4	0	0	42.4
0	200	—	—	90.0	70.0	—	—	0.22	—	0.25	3.00	—	—	9.0	—	—
0	200	1.80	8.0	75.0	110.0	1.50	—	0.37	—	0.42	5.00	0.50	40.0	0	—	—
0	200	1.80	8.0	75.0	110.0	1.50	—	0.37	—	0.42	5.00	0.50	40.0	0	0	—
0	200	1.80	8.0	70.0	110.0	1.50	—	0.37	—	0.42	5.00	0.50	40.0	0	0	—
0	200	1.80	8.0	55.0	110.0	1.50	—	0.37	—	0.42	5.00	0.50	40.0	0	0	—
0	862	34.91	21.4	150.8	732.6	0.86	—	1.59	—	1.47	21.55	2.15	122.2	0	0	—
0	5	0.58	2.3	15.1	383.3	0.09	0	0.07	0.01	0.05	0.57	0.01	139.2	0	0	10.6
0	4	0.73	6.1	25.4	269.8	0.08	0	0.10	0.02	0.07	0.87	0.03	46.0	0	0	3.8
0	4	0.73	6.1	25.4	269.8	0.08	2.4	0.10	0.02	0.07	0.87	0.03	44.8	0	0	3.3
0	11	1.05	31.6	81.9	4.7	1.17	0	0.09	0.09	0.02	0.26	0.01	7.0	0	0	6.3
0	26	6.83	49.9	126.4	403.5	1.03	0	1.02	—	0.05	1.56	0.30	42.2	0	0	11.1
0	21	3.96	44.7	112.4	240.9	0.92	0	0.60	—	0.04	0.77	0.18	18.7	0	0	3.8
0	241	10.90	224.4	632.4	150.0	3.00	300.1	1.40	—	1.68	9.16	7.44	1362.8	12.4	12.0	5.8
0	57	13.64	186.4	909.1	614.5	4.55	464.5	1.09	1.42	1.27	15.45	6.09	2054.8	18.2	18.2	26.3

TABLE A-1 Table of Food Composition *(continued)*

(Computer code is for Cengage Diet Analysis program)
(For purposes of calculations, use "0" for t, <1, <.1, <.01, etc.)

DA+ Code	Food Description	Quantity	Measure	Wt (g)	H_2O (g)	Ener (kcal)	Prot (g)	Carb (g)	Fiber (g)	Fat (g)	Fat Breakdown (g) Sat	Mono	Poly
Cereal, Flour, Grain, Pasta, Noodles, Popcorn—*continued*													
1199	Apple Jacks	1	cup(s)	33	0.9	130	1.0	30.0	0.5	0.5	0	—	—
1204	Cap'n Crunch	1	cup(s)	36	0.9	147	1.3	30.7	1.3	2.0	0.5	0.4	0.3
1205	Cap'n Crunch Crunchberries	1	cup(s)	35	0.9	133	1.3	29.3	1.3	2.0	0.5	0.4	0.3
1206	Cheerios	1	cup(s)	30	1.0	110	3.0	22.0	3.0	2.0	0	0.5	0.5
3415	Cocoa Puffs	1	cup(s)	30	0.6	120	1.0	26.0	0.2	1.0	—	—	—
1207	Cocoa Rice Krispies	1	cup(s)	41	1.0	160	1.3	36.0	1.3	1.3	0.7	0	0
5522	Complete wheat bran flakes	1	cup(s)	39	1.4	120	4.0	30.7	6.7	0.7	—	—	—
1211	Corn Flakes	1	cup(s)	28	0.9	100	2.0	24.0	1.0	0	0	0	0
1247	Corn Pops	1	cup(s)	31	0.9	120	1.0	28.0	0.3	0	0	0	0
1937	Cracklin' Oat Bran	1	cup(s)	65	2.3	267	5.3	46.7	8.0	9.3	4.0	4.7	1.3
1220	Froot Loops	1	cup(s)	32	0.8	120	1.0	28.0	1.0	1.0	0.5	0	0
38214	Frosted Cheerios	1	cup(s)	37	—	149	2.5	31.1	1.2	1.2	—	—	—
372	Frosted Flakes	1	cup(s)	41	1.1	160	1.3	37.3	1.3	0	0	0	0
38215	Frosted Mini Chex	1	cup(s)	40	—	147	1.3	36.0	0	0	0	0	0
10268	Frosted Mini-Wheats	1	cup(s)	59	3.1	208	5.8	47.4	5.8	1.2	0	0	0.6
38216	Frosted Wheaties	1	cup(s)	40	—	147	1.3	36.0	0.3	0	0	0	0
1223	Granola, prepared	½	cup(s)	61	3.3	298	9.1	32.5	5.5	14.7	2.5	5.8	5.6
2415	Honey Bunches of Oats honey roasted	1	cup(s)	40	0.9	160	2.7	33.3	1.3	2.0	0.7	1.2	0.1
1227	Honey Nut Cheerios	1	cup(s)	37	0.9	149	3.7	29.9	2.5	1.9	0	0.6	0.6
2424	Honeycomb	1	cup(s)	22	0.3	83	1.5	19.5	0.8	0.4	0	—	—
10286	Kashi whole grain puffs	1	cup(s)	19	—	70	2.0	15.0	1.0	0.5	0	—	—
41142	Kellogg's Mueslix	1	cup(s)	83	7.2	298	7.6	60.8	6.1	4.6	0.7	2.4	1.5
1231	Kix	1	cup(s)	24	0.5	96	1.6	20.8	0.8	0.4	—	—	—
30569	Life	1	cup(s)	43	1.7	160	4.0	33.3	2.7	2.0	0.3	0.6	0.6
1233	Lucky Charms	1	cup(s)	24	0.6	96	1.6	20.0	0.8	0.8	—	—	—
38220	Multi Grain Cheerios	1	cup(s)	30	—	110	3.0	24.0	3.0	1.0	—	—	—
1201	Multi-Bran Chex	1	cup(s)	63	1.3	216	4.3	52.9	8.6	1.6	0	0	0.5
13633	Post Bran Flakes	1	cup(s)	40	1.5	133	4.0	32.0	6.7	0.7	0	—	—
1241	Product 19	1	cup(s)	30	1.0	100	2.0	25.0	1.0	0	0	0	0
32432	Puffed rice, fortified	1	cup(s)	14	0.4	56	0.9	12.6	0.2	0.1	0	—	—
32433	Puffed wheat, fortified	1	cup(s)	12	0.4	44	1.8	9.6	0.5	0.1	0	—	—
13334	Quaker 100% natural granola oats and honey	½	cup(s)	48	—	220	5.0	31.0	3.0	9.0	3.8	4.1	1.2
13335	Quaker 100% natural granola oats, honey, and raisins	½	cup(s)	51	—	230	5.0	34.0	3.0	9.0	3.6	3.8	1.1
2420	Raisin Bran	1	cup(s)	59	5.0	190	4.0	46.0	8.0	1.0	0	0.1	0.4
1244	Rice Chex	1	cup(s)	31	0.8	120	2.0	27.0	0.3	0	0	0	0
1245	Rice Krispies	1	cup(s)	26	0.8	96	1.6	23.2	0	0	0	0	0
5593	Shredded Wheat	1	cup(s)	49	0.4	177	5.8	40.9	6.9	1.1	0.1	0	0.2
1248	Smacks	1	cup(s)	36	1.1	133	2.7	32.0	1.3	0.7	—	—	—
1246	Special K	1	cup(s)	31	0.9	110	7.0	22.0	0.5	0	0	0	0
3428	Total corn flakes	1	cup(s)	23	0.6	83	1.5	18.0	0.6	0	0	0	0
1253	Total whole grain	1	cup(s)	40	1.1	147	2.7	30.7	4.0	1.3	—	—	—
1254	Trix	1	cup(s)	30	0.6	120	1.0	27.0	1.0	1.0	—	—	—
382	Wheat germ, toasted	2	tablespoon(s)	14	0.8	54	4.1	7.0	2.1	1.5	0.3	0.2	0.9
1257	Wheaties	1	cup(s)	36	1.2	132	3.6	28.8	3.6	1.2	—	—	—
	Pasta, noodles												
449	Chinese chow mein noodles, cooked	½	cup(s)	23	0.2	119	1.9	12.9	0.9	6.9	1.0	1.7	3.9
1995	Corn pasta, cooked	½	cup(s)	70	47.8	88	1.8	19.5	3.4	0.5	0.1	0.1	0.2
448	Egg noodles, enriched, cooked	½	cup(s)	80	54.2	110	3.6	20.1	1.0	1.7	0.3	0.5	0.4
1563	Egg noodles, spinach, enriched, cooked	½	cup(s)	80	54.8	106	4.0	19.4	1.8	1.3	0.3	0.4	0.3
440	Macaroni, enriched, cooked	½	cup(s)	70	43.5	111	4.1	21.6	1.3	0.7	0.1	0.1	0.2
2000	Macaroni, tricolor vegetable, enriched, cooked	½	cup(s)	67	45.8	86	3.0	17.8	2.9	0.1	0	0	0
1996	Plain pasta, fresh-refrigerated, cooked	½	cup(s)	64	43.9	84	3.3	16.0	—	0.7	0.1	0.1	0.3
1725	Ramen noodles, cooked	½	cup(s)	114	94.5	104	3.0	15.4	1.0	4.3	0.2	0.2	0.2
2878	Soba noodles, cooked	½	cup(s)	95	69.4	94	4.8	20.4	—	0.1	0	0	0
2879	Somen noodles, cooked	½	cup(s)	88	59.8	115	3.5	24.2	—	0.2	0	0	0.1
493	Spaghetti, al dente, cooked	½	cup(s)	65	41.6	95	3.5	19.5	1.0	0.5	0.1	0.1	0.2
2884	Spaghetti, whole wheat, cooked	½	cup(s)	70	47.0	87	3.7	18.6	3.2	0.4	0.1	0.1	0.1
	Popcorn												
476	Air popped	1	cup(s)	8	0.3	31	1.0	6.2	1.2	0.4	0	0.1	0.2
4619	Caramel	1	cup(s)	35	1.0	152	1.3	27.8	1.8	4.5	1.3	1.0	1.6

PAGE KEY: A-4 = Breads/Baked Goods A-10 = Cereal/Rice/Pasta A-14 = Fruit A-18 = Vegetables/Legumes A-28 = Nuts/Seeds A-30 = Vegetarian A-32 = Dairy A-40 = Eggs A-40 = Seafood A-42 = Meats A-46 = Poultry A-46 = Processed Meats A-48 = Beverages A-52 = Fats/Oils A-54 = Sweets A-56 = Spices/Condiments/Sauces A-60 = Mixed Foods/Soups/Sandwiches A-64 = Fast Food A-84 = Convenience A-86 = Baby Foods

Chol (mg)	Calc (mg)	Iron (mg)	Magn (mg)	Pota (mg)	Sodi (mg)	Zinc (mg)	Vit A (μg)	Thia (mg)	Vit E (mg α)	Ribo (mg)	Niac (mg)	Vit B_6 (mg)	Fola (μg)	Vit C (mg)	Vit B_{12} (μg)	Sele (μg)
0	0	4.50	8.0	30.0	130.0	1.50	150.2	0.37	—	0.42	5.00	0.50	196.0	15.0	1.5	2.4
0	5	6.80	20.0	73.3	266.7	5.00	2.5	0.51	—	0.57	6.68	0.67	946.8	0	0	6.7
0	7	6.53	18.7	73.3	240.0	5.13	2.4	0.51	—	0.57	6.68	0.67	910.7	0	0	6.7
0	100	8.10	40.0	95.0	280.0	3.75	150.3	0.37	—	0.42	5.00	0.50	493.2	6.0	1.5	11.3
0	100	4.50	8.0	50.0	170.0	3.75	0	0.37	—	0.42	5.00	0.50	165.9	6.0	1.5	2.0
0	53	6.00	10.7	66.7	253.3	2.00	200.1	0.49	—	0.56	6.67	0.67	442.8	20.0	2.0	5.8
0	0	24.00	53.3	226.7	280.0	20.00	300.1	2.00	—	2.27	26.67	2.67	909.1	80.0	8.0	4.1
0	0	8.10	3.4	25.0	200.0	0.16	149.8	0.37	—	0.42	5.00	0.50	221.8	6.0	1.5	1.4
0	0	1.80	2.5	25.0	120.0	1.50	150.0	0.37	—	0.42	5.00	0.50	174.7	6.0	1.5	2.0
0	27	2.40	80.0	293.3	200.0	2.00	299.9	0.49	—	0.56	6.67	0.67	217.1	20.0	2.0	14.4
0	0	4.50	8.0	35.0	150.0	1.50	150.1	0.37	—	0.42	5.00	0.50	166.1	15.0	1.5	2.3
0	124	5.60	19.9	68.4	261.3	4.67	—	0.46	—	0.52	6.22	0.62	444.4	7.5	1.9	—
0	0	6.00	3.7	26.7	200.0	0.20	200.1	0.49	—	0.56	6.67	0.67	260.4	8.0	2.0	1.8
0	133	12.00	—	33.3	266.7	4.00	—	0.49	—	0.56	6.67	0.67	266.7	8.0	2.0	—
0	0	16.66	69.4	196.7	5.8	1.74	0	0.43	—	0.49	5.78	0.58	193.5	0	1.7	2.4
0	133	10.80	0	46.7	266.7	10.00	—	1.00	—	1.13	13.33	1.33	901.2	8.0	4.0	—
0	48	2.58	106.8	329.4	15.3	2.45	0.6	0.44	6.77	0.17	1.30	0.17	50.0	0.7	0	17.0
0	0	10.80	21.3	0	253.3	0.40	—	0.49	—	0.56	6.67	0.67	549.6	0	2.0	—
0	124	5.60	39.8	112.0	336.0	4.67	—	0.46	—	0.52	6.22	0.62	444.4	7.5	1.9	8.8
0	0	2.03	6.0	26.3	165.4	1.13	—	0.28	—	0.32	3.74	0.37	126.1	0	1.1	—
0	0	0.36	—	60.0	0	—	0	0.03	—	0.03	0.80	0.00	—	0	—	—
0	48	6.83	74.2	363.3	257.5	5.67	136.7	0.67	6.00	0.67	8.33	3.08	1030.0	0.3	9.2	14.4
0	120	6.48	6.4	28.0	216.0	3.00	120.2	0.30	—	0.34	4.00	0.40	317.8	4.8	1.2	4.8
0	149	11.87	41.3	120.0	213.3	5.33	0.9	0.53	—	0.60	7.12	0.71	607.6	0	0	10.7
0	80	3.60	12.8	48.0	168.0	3.00	—	0.30	—	0.34	4.00	0.40	300.7	4.8	1.2	4.8
0	100	18.00	24.0	85.0	200.0	15.00	—	1.50	—	1.70	20.00	2.00	699.3	15.0	6.0	—
0	108	17.50	64.8	237.7	410.6	4.05	171.1	0.40	—	0.45	5.40	0.54	893.3	6.5	1.6	4.9
0	0	10.80	80.0	266.7	280.0	2.00	—	0.49	—	0.56	6.67	0.67	453.3	0	2.0	—
0	0	18.00	16.0	50.0	210.0	15.00	225.3	1.50	—	1.70	20.00	2.00	675.9	60.0	6.0	3.6
0	1	4.43	3.5	15.8	0.4	0.14	0	0.36	—	0.25	4.94	0.01	2.7	0	0	1.5
0	3	3.80	17.4	41.8	0.5	0.28	0	0.31	—	0.21	4.23	0.02	3.8	0	0	14.8
0	61	1.20	51.0	220.0	20.0	1.05	0.5	0.13	—	0.12	0.82	0.07	16.8	0.2	0.1	8.3
0	59	1.20	49.0	250.0	20.0	0.99	0.5	0.13	—	0.12	0.80	0.08	15.8	0.4	0.1	8.8
0	20	10.80	80.0	360.0	360.0	2.25	—	0.37	—	0.42	5.00	0.50	248.4	0	2.1	—
0	100	9.00	9.3	35.0	290.0	3.75	—	0.37	—	0.42	5.00	0.50	389.4	6.0	1.5	1.2
0	0	1.44	12.8	32.0	256.0	0.48	120.1	0.30	—	0.34	4.80	0.40	237.4	4.8	1.2	4.1
0	18	2.90	60.3	179.3	1.1	1.37	0	0.14	—	0.12	3.47	0.18	21.1	0	0	2.0
0	0	0.48	10.7	53.3	66.7	0.40	200.2	0.49	—	0.56	6.67	0.67	224.6	8.0	2.0	17.5
0	0	8.10	16.0	60.0	220.0	0.90	225.1	0.52	—	0.59	7.00	2.00	675.8	21.0	6.0	7.0
0	752	13.53	0	22.6	157.9	11.28	112.8	1.13	22.56	1.28	15.04	1.50	518.2	45.1	4.5	1.2
0	1333	24.00	32.0	120.0	253.3	20.00	200.4	2.00	31.32	2.27	26.67	2.67	901.2	80.0	8.0	1.9
0	100	4.50	0	15.0	190.0	3.75	150.3	0.37	—	0.42	5.00	0.50	155.4	6.0	1.5	6.0
0	6	1.28	45.2	133.8	0.6	2.35	0.7	0.23	2.25	0.11	0.79	0.13	49.7	0.8	0	9.2
0	24	9.72	38.4	126.0	264.0	9.00	180.4	0.90	—	1.02	12.00	1.20	403.6	7.2	3.6	1.7
0	5	1.06	11.7	27.0	98.8	0.31	0	0.13	0.78	0.09	1.33	0.02	31.1	0	0	9.7
0	1	0.18	25.2	21.7	0	0.44	2.1	0.04	—	0.02	0.39	0.04	4.2	0	0	2.0
23	10	1.17	16.8	30.4	4.0	0.52	4.8	0.23	0.13	0.11	1.66	0.03	110.4	0	0.1	19.1
26	15	0.87	19.2	29.6	9.6	0.50	8.0	0.19	0.46	0.09	1.17	0.09	75.2	0	0.1	17.4
0	5	0.90	12.6	30.8	0.7	0.36	0	0.19	0.04	0.10	1.18	0.03	83.3	0	0	18.5
0	7	0.33	12.7	20.8	4.0	0.30	3.4	0.08	0.14	0.04	0.72	0.02	71.0	0	0	13.3
21	4	0.73	11.5	15.4	3.8	0.36	3.8	0.13	—	0.10	0.64	0.02	66.6	0	0.1	—
18	9	0.89	8.5	34.5	414.5	0.30	—	0.08	—	0.04	0.71	0.03	4.0	0.1	0	—
0	4	0.45	8.5	33.2	57.0	0.11	0	0.09	—	0.02	0.48	0.03	6.6	0	0	—
0	7	0.45	1.8	25.5	141.7	0.19	0	0.01	—	0.03	0.08	0.01	1.8	0	0	—
0	7	1.00	12.4	51.5	0.5	0.35	0	0.12	0.04	0.07	0.90	0.04	77.4	0	0	40.0
0	11	0.74	21.0	30.8	2.1	0.57	0	0.08	0.21	0.03	0.50	0.06	3.5	0	0	18.1
0	1	0.25	11.5	26.3	0.6	0.25	0.8	0.01	0.02	0.01	0.18	0.01	2.5	0	0	0
2	15	0.61	12.3	38.4	72.5	0.20	0.7	0.02	0.42	0.02	0.77	0.01	1.8	0	0	1.3

TABLE A-1 Table of Food Composition *(continued)*

(Computer code is for Cengage Diet Analysis program)
(For purposes of calculations, use "0" for t, <1, <.1, <.01, etc.)

DA+ Code	Food Description	Quantity	Measure	Wt (g)	H_2O (g)	Ener (kcal)	Prot (g)	Carb (g)	Fiber (g)	Fat (g)	Fat Breakdown (g)		
											Sat	Mono	Poly
Cereal, Flour, Grain, Pasta, Noodles, Popcorn—*continued*													
4620	Cheese flavored	1	cup(s)	36	0.9	188	3.3	18.4	3.5	11.8	2.3	3.5	5.5
477	Popped in oil	1	cup(s)	11	0.1	64	0.8	5.0	0.9	4.8	0.8	1.1	2.6
Fruit and Fruit Juices													
	Apples												
952	Juice, prepared from frozen concentrate	½	cup(s)	120	105.0	56	0.2	13.8	0.1	0.1	0	0	0
225	Juice, unsweetened, canned	½	cup(s)	124	109.0	58	0.1	14.5	0.1	0.1	0	0	0
224	Slices	½	cup(s)	55	47.1	29	0.1	7.6	1.3	0.1	0	0	0
946	Slices without skin, boiled	½	cup(s)	86	73.1	45	0.2	11.7	2.1	0.3	0	0	0.1
223	Raw medium, with peel	1	item(s)	138	118.1	72	0.4	19.1	3.3	0.2	0	0	0.1
948	Dried, sulfured	¼	cup(s)	22	6.8	52	0.2	14.2	1.9	0.1	0	0	0
226	Applesauce, sweetened, canned	½	cup(s)	128	101.5	97	0.2	25.4	1.5	0.2	0	0	0.1
227	Applesauce, unsweetened, canned	½	cup(s)	122	107.8	52	0.2	13.8	1.5	0.1	0	0	0
38492	Crabapples	1	item(s)	35	27.6	27	0.1	7.0	0.9	0.1	0	0	0
	Apricot												
228	Fresh without pits	4	item(s)	140	120.9	67	2.0	15.6	2.8	0.5	0	0.2	0.1
229	Halves with skin, canned in heavy syrup	½	cup(s)	129	100.1	107	0.7	27.7	2.1	0.1	0	0	0
230	Halves, dried, sulfured	¼	cup(s)	33	10.1	79	1.1	20.6	2.4	0.2	0	0	0
	Avocado												
233	California, whole, without skin or pit	½	cup(s)	115	83.2	192	2.2	9.9	7.8	17.7	2.4	11.3	2.1
234	Florida, whole, without skin or pit	½	cup(s)	115	90.6	138	2.5	9.0	6.4	11.5	2.2	6.3	1.9
2998	Pureed	⅛	cup(s)	28	20.2	44	0.5	2.4	1.8	4.0	0.6	2.7	0.5
	Banana												
4580	Dried chips	¼	cup(s)	55	2.4	285	1.3	32.1	4.2	18.5	15.9	1.1	0.3
235	Fresh whole, without peel	1	item(s)	118	88.4	105	1.3	27.0	3.1	0.4	0.1	0	0.1
	Blackberries												
237	Raw	½	cup(s)	72	63.5	31	1.0	6.9	3.8	0.4	0	0	0.2
958	Unsweetened, frozen	½	cup(s)	76	62.1	48	0.9	11.8	3.8	0.3	0	0	0.2
	Blueberries												
959	Canned in heavy syrup	½	cup(s)	128	98.3	113	0.8	28.2	2.0	0.4	0	0.1	0.2
238	Raw	½	cup(s)	73	61.1	41	0.5	10.5	1.7	0.2	0	0	0.1
960	Unsweetened, frozen	½	cup(s)	78	67.1	40	0.3	9.4	2.1	0.5	0	0.1	0.2
	Boysenberries												
961	Canned in heavy syrup	½	cup(s)	128	97.6	113	1.3	28.6	3.3	0.2	0	0	0.1
962	Unsweetened, frozen	½	cup(s)	66	56.7	33	0.7	8.0	3.5	0.2	0	0	0.1
35576	**Breadfruit**	1	item(s)	384	271.3	396	4.1	104.1	18.8	0.9	0.2	0.1	0.3
	Cherries												
967	Sour red, canned in water	½	cup(s)	122	109.7	44	0.9	10.9	1.3	0.1	0	0	0
3000	Sour red, raw	½	cup(s)	78	66.8	39	0.8	9.4	1.2	0.2	0.1	0.1	0.1
3004	Sweet, canned in heavy syrup	½	cup(s)	127	98.2	105	0.8	26.9	1.9	0.2	0	0.1	0.1
969	Sweet, canned in water	½	cup(s)	124	107.9	57	1.0	14.6	1.9	0.2	0	0	0
240	Sweet, raw	½	cup(s)	73	59.6	46	0.8	11.6	1.5	0.1	0	0	0
	Cranberries												
3007	Chopped, raw	½	cup(s)	55	47.9	25	0.2	6.7	2.5	0.1	0	0	0
1717	Cranberry apple juice drink	½	cup(s)	123	102.6	77	0	19.4	0	0.1	0	0	0.1
1638	Cranberry juice cocktail	½	cup(s)	127	109.0	68	0	17.1	0	0.1	0	0	0.1
241	Cranberry juice cocktail, low calorie, with saccharin	½	cup(s)	119	112.8	23	0	5.5	0	0	0	0	0
242	Cranberry sauce, sweetened, canned	¼	cup(s)	69	42.0	105	0.1	26.9	0.7	0.1	0	0	0
	Dates												
244	Domestic, chopped	¼	cup(s)	45	9.1	125	1.1	33.4	3.6	0.2	0	0	0
243	Domestic, whole	¼	cup(s)	45	9.1	125	1.1	33.4	3.6	0.2	0	0	0
	Figs												
975	Canned in heavy syrup	½	cup(s)	130	98.8	114	0.5	29.7	2.8	0.1	0	0	0.1
974	Canned in water	½	cup(s)	124	105.7	66	0.5	17.3	2.7	0.1	0	0	0.1
973	Raw, medium	2	item(s)	100	79.1	74	0.7	19.2	2.9	0.3	0.1	0.1	0.1
	Fruit cocktail and salad												
245	Fruit cocktail, canned in heavy syrup	½	cup(s)	124	99.7	91	0.5	23.4	1.2	0.1	0	0	0
978	Fruit cocktail, canned in juice	½	cup(s)	119	103.6	55	0.5	14.1	1.2	0	0	0	0
977	Fruit cocktail, canned in water	½	cup(s)	119	107.6	38	0.5	10.1	1.2	0.1	0	0	0
979	Fruit salad, canned in water	½	cup(s)	123	112.1	37	0.4	9.6	1.2	0.1	0	0	0

PAGE KEY: A-4 = Breads/Baked Goods A-10 = Cereal/Rice/Pasta A-14 = Fruit A-18 = Vegetables/Legumes A-28 = Nuts/Seeds A-30 = Vegetarian A-32 = Dairy A-40 = Eggs A-40 = Seafood A-42 = Meats A-46 = Poultry A-46 = Processed Meats A-48 = Beverages A-52 = Fats/Oils A-54 = Sweets A-56 = Spices/Condiments/Sauces A-60 = Mixed Foods/Soups/Sandwiches A-64 = Fast Food A-84 = Convenience A-86 = Baby Foods

Chol (mg)	Calc (mg)	Iron (mg)	Magn (mg)	Pota (mg)	Sodi (mg)	Zinc (mg)	Vit A (µg)	Thia (mg)	Vit E (mg α)	Ribo (mg)	Niac (mg)	Vit B_6 (mg)	Fola (µg)	Vit C (mg)	Vit B_{12} (µg)	Sele (µg)
4	40	0.79	32.5	93.2	317.4	0.71	13.6	0.04	—	0.08	0.52	0.08	3.9	0.2	0.2	4.3
0	0	0.22	8.7	20.0	116.4	0.34	0.9	0.01	0.27	0.00	0.13	0.01	2.8	0	0	0.2
0	7	0.31	6.0	150.6	8.4	0.05	0	0.00	0.01	0.02	0.05	0.04	0	0.7	0	0.1
0	9	0.46	3.7	147.6	3.7	0.04	0	0.03	0.01	0.02	0.12	0.04	0	1.1	0	0.1
0	3	0.06	2.7	58.8	0.5	0.02	1.6	0.01	0.10	0.01	0.05	0.02	1.6	2.5	0	0
0	4	0.16	2.6	75.2	0.9	0.03	1.7	0.01	0.04	0.01	0.08	0.04	0.9	0.2	0	0.3
0	8	0.16	6.9	147.7	1.4	0.05	4.1	0.02	0.24	0.03	0.12	0.05	4.1	6.3	0	0
0	3	0.30	3.4	96.8	18.7	0.04	0	0.00	0.11	0.03	0.20	0.03	0	0.8	0	0.3
0	5	0.44	3.8	77.8	3.8	0.05	1.3	0.01	0.26	0.03	0.24	0.03	1.3	2.2	0	0.4
0	4	0.14	3.7	91.5	2.4	0.03	1.2	0.01	0.25	0.03	0.22	0.03	1.2	1.5	0	0.4
0	6	0.12	2.5	67.9	0.4	—	0.7	0.01	0.20	0.01	0.03	—	2.0	2.8	0	—
0	18	0.54	14.0	362.6	1.4	0.28	134.4	0.04	1.24	0.05	0.84	0.07	12.6	14.0	0	0.1
0	12	0.38	9.0	180.6	5.2	0.14	80.0	0.02	0.77	0.02	0.48	0.07	2.6	4.0	0	0.1
0	18	0.87	10.5	381.5	3.3	0.12	59.1	0.00	1.42	0.02	0.85	0.05	3.3	0.3	0	0.7
0	15	0.66	33.3	583.0	9.2	0.78	8.0	0.08	2.23	0.16	2.19	0.31	102.3	10.1	0	0.4
0	12	0.19	27.6	403.6	2.3	0.45	8.0	0.02	3.03	0.04	0.76	0.08	40.3	20.0	0	—
0	3	0.14	8.0	134.1	1.9	0.17	1.9	0.01	0.57	0.03	0.48	0.07	22.4	2.8	0	0.1
0	10	0.69	41.8	294.8	3.3	0.40	2.2	0.04	0.13	0.01	0.39	0.14	7.7	3.5	0	0.8
0	6	0.30	31.9	422.4	1.2	0.17	3.5	0.03	0.11	0.08	0.78	0.43	23.6	10.3	0	1.2
0	21	0.45	14.4	116.6	0.7	0.38	7.9	0.01	0.84	0.02	0.47	0.02	18.0	15.1	0	0.3
0	22	0.60	16.6	105.7	0.8	0.19	4.5	0.02	0.88	0.03	0.91	0.05	25.7	2.3	0	0.3
0	6	0.42	5.1	51.2	3.8	0.09	2.6	0.04	0.49	0.07	0.14	0.05	2.6	1.4	0	0.1
0	4	0.20	4.4	55.8	0.7	0.12	2.2	0.03	0.41	0.03	0.30	0.04	4.4	7.0	0	0.1
0	6	0.14	3.9	41.9	0.8	0.05	1.6	0.03	0.37	0.03	0.40	0.05	5.4	1.9	0	0.1
0	23	0.55	14.1	115.2	3.8	0.24	2.6	0.03	—	0.04	0.29	0.05	43.5	7.9	0	0.5
0	18	0.56	10.6	91.7	0.7	0.15	2.0	0.04	0.57	0.02	0.51	0.04	41.6	2.0	0	0.1
0	65	2.07	96.0	1881.6	7.7	0.46	0	0.42	0.38	0.11	3.45	0.38	53.8	111.4	0	2.3
0	13	1.67	7.3	119.6	8.5	0.09	46.4	0.02	0.28	0.05	0.22	0.05	9.8	2.6	0	0
0	12	0.25	7.0	134.1	2.3	0.08	49.6	0.02	0.05	0.03	0.31	0.03	6.2	7.8	0	0
0	11	0.44	11.4	183.4	3.8	0.12	10.1	0.02	0.29	0.05	0.50	0.03	5.1	4.6	0	0
0	14	0.45	11.2	162.4	1.2	0.10	9.9	0.03	0.29	0.05	0.51	0.04	5.0	2.7	0	0
0	9	0.26	8.0	161.0	0	0.05	2.2	0.02	0.05	0.02	0.11	0.04	2.9	5.1	0	0
0	4	0.13	3.3	46.8	1.1	0.05	1.7	0.01	0.66	0.01	0.05	0.03	0.6	7.3	0	0.1
0	4	0.09	1.2	20.8	2.5	0.02	0	0.00	0.15	0.00	0.00	0.00	0	48.4	0	0
0	4	0.13	1.3	17.7	2.5	0.04	0	0.00	0.28	0.00	0.05	0.00	0	53.5	0	0.3
0	11	0.05	2.4	29.6	3.6	0.02	0	0.00	0.06	0.00	0.00	0.00	0	38.2	0	0
0	3	0.15	2.1	18.0	20.1	0.03	1.4	0.01	0.57	0.01	0.06	0.01	0.7	1.4	0	0.2
0	17	0.45	19.1	291.9	0.9	0.12	0	0.02	0.02	0.02	0.56	0.07	8.5	0.2	0	1.3
0	17	0.45	19.1	291.9	0.9	0.12	0	0.02	0.02	0.02	0.56	0.07	8.5	0.2	0	1.3
0	35	0.36	13.0	128.2	1.3	0.14	2.6	0.03	0.16	0.05	0.55	0.09	2.6	1.3	0	0.3
0	35	0.36	12.4	127.7	1.2	0.15	2.5	0.03	0.10	0.05	0.55	0.09	2.5	1.2	0	0.1
0	35	0.36	17.0	232.0	1.0	0.14	7.0	0.06	0.10	0.04	0.40	0.10	6.0	2.0	0	0.2
0	7	0.36	6.2	109.1	7.4	0.09	12.4	0.02	0.49	0.02	0.46	0.06	3.7	2.4	0	0.6
0	9	0.25	8.3	112.6	4.7	0.11	17.8	0.01	0.47	0.02	0.48	0.06	3.6	3.2	0	0.6
0	6	0.30	8.3	111.4	4.7	0.11	15.4	0.02	0.47	0.01	0.43	0.06	3.6	2.5	0	0.6
0	9	0.37	6.1	95.6	3.7	0.10	27.0	0.02	—	0.03	0.46	0.04	3.7	2.3	0	1.0

TABLE A-1 Table of Food Composition *(continued)*

(Computer code is for Cengage Diet Analysis program)
(For purposes of calculations, use "0" for t, <1, <.1, <.01, etc.)

DA+ Code	Food Description	Quantity	Measure	Wt (g)	H_2O (g)	Ener (kcal)	Prot (g)	Carb (g)	Fiber (g)	Fat (g)	Fat Breakdown (g) Sat	Mono	Poly
Fruit and Fruit Juices—*continued*													
	Gooseberries												
982	Canned in light syrup	½	cup(s)	126	100.9	92	0.8	23.6	3.0	0.3	0	0	0.1
981	Raw	½	cup(s)	75	65.9	33	0.7	7.6	3.2	0.4	0	0	0.2
	Grapefruit												
251	Juice, pink, sweetened, canned	½	cup(s)	125	109.1	57	0.7	13.9	0.1	0.1	0	0	0
249	Juice, white	½	cup(s)	124	111.2	48	0.6	11.4	0.1	0.1	0	0	0
3022	Pink or red, raw	½	cup(s)	114	100.8	48	0.9	12.2	1.8	0.2	0	0	0
248	Sections, canned in light syrup	½	cup(s)	127	106.2	76	0.7	19.6	0.5	0.1	0	0	0
983	Sections, canned in water	½	cup(s)	122	109.6	44	0.7	11.2	0.5	0.1	0	0	0
247	White, raw	½	cup(s)	115	104.0	38	0.8	9.7	1.3	0.1	0	0	0
	Grapes												
255	American, slip skin	½	cup(s)	46	37.4	31	0.3	7.9	0.4	0.2	0.1	0	0
256	European, red or green, adherent skin	½	cup(s)	76	60.8	52	0.5	13.7	0.7	0.1	0	0	0
3159	Grape juice drink, canned	½	cup(s)	125	106.6	71	0	18.2	0.1	0	0	0	0
259	Grape juice, sweetened, with added vitamin C, prepared from frozen concentrate	½	cup(s)	125	108.6	64	0.2	15.9	0.1	0.1	0	0	0
3060	Raisins, seeded, packed	¼	cup(s)	41	6.8	122	1.0	32.4	2.8	0.2	0.1	0	0.1
987	**Guava, raw**	1	item(s)	55	44.4	37	1.4	7.9	3.0	0.5	0.2	0	0.2
35593	**Guavas, strawberry**	1	item(s)	6	4.8	4	0	1.0	0.3	0	0	0	0
3027	**Jackfruit**	½	cup(s)	83	60.4	78	1.2	19.8	1.3	0.2	0	0	0.1
990	**Kiwi fruit or Chinese gooseberries**	1	item(s)	76	63.1	46	0.9	11.1	2.3	0.4	0	0	0.2
	Lemon												
262	Juice	1	tablespoon(s)	15	13.8	4	0.1	1.3	0.1	0	0	0	0
993	Peel	1	teaspoon(s)	2	1.6	1	0	0.3	0.2	0	0	0	0
992	Raw	1	item(s)	108	94.4	22	1.3	11.6	5.1	0.3	0	0	0.1
	Lime												
269	Juice	1	tablespoon(s)	15	14.0	4	0.1	1.3	0.1	0	0	0	0
994	Raw	1	item(s)	67	59.1	20	0.5	7.1	1.9	0.1	0	0	0
995	**Loganberries, frozen**	½	cup(s)	74	62.2	40	1.1	9.6	3.9	0.2	0	0	0.1
	Mandarin orange												
1038	Canned in juice	½	cup(s)	125	111.4	46	0.8	11.9	0.9	0	0	0	0
1039	Canned in light syrup	½	cup(s)	126	104.7	77	0.6	20.4	0.9	0.1	0	0	0
999	**Mango**	½	cup(s)	83	67.4	54	0.4	14.0	1.5	0.2	0.1	0.1	0
1005	**Nectarine, raw, sliced**	½	cup(s)	69	60.4	30	0.7	7.3	1.2	0.2	0	0.1	0.1
	Melons												
271	Cantaloupe	½	cup(s)	80	72.1	27	0.7	6.5	0.7	0.1	0	0	0.1
1000	Casaba melon	½	cup(s)	85	78.1	24	0.9	5.6	0.8	0.1	0	0	0
272	Honeydew	½	cup(s)	89	79.5	32	0.5	8.0	0.7	0.1	0	0	0
318	Watermelon	½	cup(s)	76	69.5	23	0.5	5.7	0.3	0.1	0	0	0
	Orange												
14412	Juice with calcium and vitamin D	½	cup(s)	120	—	55	1.0	13.0	0	0	0	0	0
29630	Juice, fresh squeezed	½	cup(s)	124	109.5	56	0.9	12.9	0.2	0.2	0	0	0
14411	Juice, not from concentrate	½	cup(s)	120	—	55	1.0	13.0	0	0	0	0	0
278	Juice, unsweetened, prepared from frozen concentrate	½	cup(s)	125	109.7	56	0.8	13.4	0.2	0.1	0	0	0
3040	Peel	1	teaspoon(s)	2	1.5	2	0	0.5	0.2	0	0	0	0
273	Raw	1	item(s)	131	113.6	62	1.2	15.4	3.1	0.2	0	0	0
274	Sections	½	cup(s)	90	78.1	42	0.8	10.6	2.2	0.1	0	0	0
	Papaya, raw												
16830	Dried, strips	2	item(s)	46	12.0	119	1.9	29.9	5.5	0.4	0.1	0.1	0.1
282	Papaya	½	cup(s)	70	62.2	27	0.4	6.9	1.3	0.1	0	0	0
35640	**Passion fruit, purple**	1	item(s)	18	13.1	17	0.4	4.2	1.9	0.1	0	0	0.1
	Peach												
285	Halves, canned in heavy syrup	½	cup(s)	131	103.9	97	0.6	26.1	1.7	0.1	0	0	0.1
286	Halves, canned in water	½	cup(s)	122	113.6	29	0.5	7.5	1.6	0.1	0	0	0
290	Slices, sweetened, frozen	½	cup(s)	125	93.4	118	0.8	30.0	2.3	0.2	0	0.1	0.1
283	Raw, medium	1	item(s)	150	133.3	59	1.4	14.3	2.3	0.4	0	0.1	0.1
	Pear												
8672	Asian	1	item(s)	122	107.7	51	0.6	13.0	4.4	0.3	0	0.1	0.1
293	D'Anjou	1	item(s)	200	168.0	120	1.0	30.0	5.2	1.0	0	0.2	0.2
294	Halves, canned in heavy syrup	½	cup(s)	133	106.9	98	0.3	25.5	2.1	0.2	0	0	0
1012	Halves, canned in juice	½	cup(s)	124	107.2	62	0.4	16.0	2.0	0.1	0	0	0
291	Raw	1	item(s)	166	139.0	96	0.6	25.7	5.1	0.2	0	0	0

PAGE KEY: A-4 = Breads/Baked Goods A-10 = Cereal/Rice/Pasta A-14 = Fruit A-18 = Vegetables/Legumes A-28 = Nuts/Seeds A-30 = Vegetarian A-32 = Dairy A-40 = Eggs A-40 = Seafood A-42 = Meats A-46 = Poultry A-46 = Processed Meats A-48 = Beverages A-52 = Fats/Oils A-54 = Sweets A-56 = Spices/Condiments/Sauces A-60 = Mixed Foods/Soups/Sandwiches A-64 = Fast Food A-84 = Convenience A-86 = Baby Foods

Chol (mg)	Calc (mg)	Iron (mg)	Magn (mg)	Pota (mg)	Sodi (mg)	Zinc (mg)	Vit A (μg)	Thia (mg)	Vit E (mg α)	Ribo (mg)	Niac (mg)	Vit B_6 (mg)	Fola (μg)	Vit C (mg)	Vit B_{12} (μg)	Sele (μg)
0	20	0.42	7.6	97.0	2.5	0.14	8.8	0.03	—	0.07	0.19	0.02	3.8	12.6	0	0.5
0	19	0.23	7.5	148.5	0.8	0.09	11.3	0.03	0.28	0.02	0.23	0.06	4.5	20.8	0	0.5
0	10	0.45	12.5	202.2	2.5	0.08	0	0.05	0.05	0.03	0.40	0.03	12.5	33.6	0	0.1
0	11	0.25	14.8	200.1	1.2	0.06	1.2	0.05	0.27	0.02	0.25	0.05	12.4	46.9	0	0.1
0	25	0.09	10.3	154.5	0	0.07	66.4	0.04	0.14	0.03	0.23	0.06	14.9	35.7	0	0.1
0	18	0.50	12.7	163.8	2.5	0.10	0	0.04	0.11	0.02	0.30	0.02	11.4	27.1	0	1.1
0	18	0.50	12.2	161.0	2.4	0.11	0	0.05	0.11	0.03	0.30	0.02	11.0	26.6	0	1.1
0	14	0.07	10.4	170.2	0	0.08	2.3	0.04	0.15	0.02	0.30	0.05	11.5	38.3	0	1.6
0	6	0.13	2.3	87.9	0.9	0.02	2.3	0.04	0.09	0.02	0.14	0.05	1.8	1.8	0	0
0	8	0.27	5.3	144.2	1.5	0.05	2.3	0.05	0.14	0.05	0.14	0.07	1.5	8.2	0	0.1
0	9	0.16	7.5	41.3	11.3	0.04	0	0.28	0.00	0.44	0.18	0.04	1.3	33.1	0	0.1
0	5	0.13	5.0	26.3	2.5	0.05	0	0.02	0.00	0.03	0.16	0.05	1.3	29.9	0	0.1
0	12	1.06	12.4	340.3	11.6	0.07	0	0.04	—	0.07	0.46	0.07	1.2	2.2	0	0.2
0	10	0.14	12.1	229.4	1.1	0.12	17.1	0.03	0.40	0.02	0.59	0.06	27.0	125.6	0	0.3
0	1	0.01	1.0	17.5	2.2	—	0.3	0.00	—	0.00	0.03	0.00	—	2.2	0	
0	28	0.49	30.5	250.0	2.5	0.35	12.4	0.02	—	0.09	0.33	0.09	11.5	5.5	0	0.5
0	26	0.23	12.9	237.1	2.3	0.10	3.0	0.02	1.11	0.01	0.25	0.04	19.0	70.5	0	0.2
0	1	0.00	0.9	18.9	0.2	0.01	0.2	0.00	0.02	0.00	0.02	0.01	2.0	7.0	0	0
0	3	0.01	0.3	3.2	0.1	0.01	0.1	0.00	0.01	0.00	0.01	0.00	0.3	2.6	0	0
0	66	0.75	13.0	156.6	3.2	0.10	2.2	0.05	—	0.04	0.21	0.11	—	83.2	0	1.0
0	2	0.02	1.2	18.0	0.3	0.01	0.3	0.00	0.03	0.00	0.02	0.01	1.5	4.6	0	0
0	22	0.40	4.0	68.3	1.3	0.07	1.3	0.02	0.14	0.01	0.13	0.02	5.4	19.5	0	0.3
0	19	0.47	15.4	106.6	0.7	0.25	1.5	0.04	0.64	0.03	0.62	0.05	19.1	11.2	0	0.1
0	14	0.34	13.7	165.6	6.2	0.64	53.5	0.10	0.12	0.04	0.55	0.05	6.2	42.6	0	0.5
0	9	0.47	10.1	98.3	7.6	0.30	52.9	0.07	0.13	0.06	0.56	0.05	6.3	24.9	0	0.5
0	8	0.10	7.4	128.7	1.7	0.03	31.4	0.05	0.92	0.04	0.48	0.11	11.6	22.8	0	0.5
0	4	0.19	6.2	138.7	0	0.12	11.7	0.02	0.53	0.02	0.78	0.02	3.5	3.7	0	0
0	7	0.17	9.6	213.6	12.8	0.14	135.2	0.03	0.04	0.01	0.59	0.05	16.8	29.4	0	0.3
0	9	0.29	9.4	154.7	7.7	0.06	0	0.01	0.04	0.03	0.20	0.14	6.8	18.5	0	0.3
0	5	0.15	8.8	201.8	15.9	0.07	2.7	0.03	0.01	0.01	0.37	0.07	16.8	15.9	0	0.6
0	5	0.18	7.6	85.1	0.8	0.07	21.3	0.02	0.04	0.01	0.13	0.03	2.3	6.2	0	0.3
0	175	0.00	12.0	225.0	0	—	0	0.08	—	0.03	0.40	0.06	30.0	36.0	0	—
0	14	0.25	13.6	248.0	1.2	0.06	12.4	0.11	0.05	0.04	0.50	0.05	37.2	62.0	0	0.1
0	10	0.00	12.5	225.0	0	0.06	0	0.08	—	0.03	0.40	0.06	30.0	36.0	0	0.1
0	11	0.12	12.5	236.6	1.2	0.06	6.2	0.10	0.25	0.02	0.25	0.06	54.8	48.4	0	0.1
0	3	0.01	0.4	4.2	0.1	0.01	0.4	0.00	0.01	0.00	0.01	0.00	0.6	2.7	0	0
0	52	0.13	13.1	237.1	0	0.09	14.4	0.11	0.23	0.05	0.36	0.07	39.3	69.7	0	0.7
0	36	0.09	9.0	162.9	0	0.06	9.9	0.07	0.16	0.03	0.25	0.05	27.0	47.9	0	0.4
0	73	0.30	30.4	782.9	9.2	0.21	83.7	0.06	2.22	0.08	0.93	0.05	58.0	37.7	0	1.8
0	17	0.07	7.0	179.9	2.1	0.05	38.5	0.02	0.51	0.02	0.24	0.01	26.6	43.3	0	0.4
0	2	0.28	5.2	62.6	5.0	0.01	11.5	0.00	0.00	0.02	0.27	0.01	2.5	5.4	0	0.1
0	4	0.35	6.6	120.5	7.9	0.11	22.3	0.01	0.64	0.03	0.80	0.02	3.9	3.7	0	0.4
0	2	0.39	6.1	120.8	3.7	0.11	32.9	0.01	0.59	0.02	0.63	0.02	3.7	3.5	0	0.4
0	4	0.46	6.3	162.5	7.5	0.06	17.5	0.01	0.77	0.04	0.81	0.02	3.8	117.8	0	0.5
0	9	0.37	13.5	285.0	0	0.25	24.0	0.03	1.09	0.04	1.20	0.03	6.0	9.9	0	0.2
0	5	0.00	9.8	147.6	0	0.02	0	0.01	0.14	0.01	0.26	0.02	9.8	4.6	0	0.1
0	22	0.50	12.0	250.0	0	0.24	—	0.04	1.00	0.08	0.20	0.03	14.6	8.0	0	1.0
0	7	0.29	5.3	86.5	6.7	0.10	0	0.01	0.10	0.02	0.32	0.01	1.3	1.5	0	0
0	11	0.36	8.7	119.0	5.0	0.11	0	0.01	0.10	0.01	0.25	0.02	1.2	2.0	0	0
0	15	0.28	11.6	197.5	1.7	0.16	1.7	0.02	0.19	0.04	0.26	0.04	11.6	7.0	0	0.2

TABLE A-1 Table of Food Composition *(continued)*

(Computer code is for Cengage Diet Analysis program)
(For purposes of calculations, use "0" for t, <1, <.1, <.01, etc.)

DA+ Code	Food Description	Quantity	Measure	Wt (g)	H_2O (g)	Ener (kcal)	Prot (g)	Carb (g)	Fiber (g)	Fat (g)	Fat Breakdown (g)		
											Sat	Mono	Poly
Fruit and Fruit Juices—*continued*													
1017	**Persimmon**	1	item(s)	25	16.1	32	0.2	8.4	—	0.1	0	0	0
	Pineapple												
3053	Canned in extra heavy syrup	½	cup(s)	130	101.0	108	0.4	28.0	1.0	0.1	0	0	0
1019	Canned in juice	½	cup(s)	125	104.0	75	0.5	19.5	1.0	0.1	0	0	0
296	Canned in light syrup	½	cup(s)	126	108.0	66	0.5	16.9	1.0	0.2	0	0	0.1
1018	Canned in water	½	cup(s)	123	111.7	39	0.5	10.2	1.0	0.1	0	0	0
299	Juice, unsweetened, canned	½	cup(s)	125	108.0	66	0.5	16.1	0.3	0.2	0	0	0.1
295	Raw, diced	½	cup(s)	78	66.7	39	0.4	10.2	1.1	0.1	0	0	0
1024	**Plantain, cooked**	½	cup(s)	77	51.8	89	0.6	24.0	1.8	0.1	0.1	0	0
300	**Plum, raw, large**	1	item(s)	66	57.6	30	0.5	7.5	0.9	0.2	0	0.1	0
1027	**Pomegranate**	1	item(s)	154	124.7	105	1.5	26.4	0.9	0.5	0.1	0.1	0.1
	Prunes												
5644	Dried	2	item(s)	17	5.2	40	0.4	10.7	1.2	0.1	0	0	0
305	Dried, stewed	½	cup(s)	124	86.5	133	1.2	34.8	3.8	0.2	0	0.1	0
306	Juice, canned	1	cup(s)	256	208.0	182	1.6	44.7	2.6	0.1	0	0.1	0
	Raspberries												
309	Raw	½	cup(s)	62	52.7	32	0.7	7.3	4.0	0.4	0	0	0.2
310	Red, sweetened, frozen	½	cup(s)	125	90.9	129	0.9	32.7	5.5	0.2	0	0	0.1
311	**Rhubarb, cooked with sugar**	½	cup(s)	120	81.5	140	0.5	37.5	2.7	0.1	0	0	0.1
	Strawberries												
313	Raw	½	cup(s)	72	65.5	23	0.5	5.5	1.4	0.2	0	0	0.1
315	Sweetened, frozen, thawed	½	cup(s)	128	99.5	99	0.7	26.8	2.4	0.2	0	0	0.1
16828	**Tangelo**	1	item(s)	95	82.4	45	0.9	11.2	2.3	0.1	0	0	0
	Tangerine												
1040	Juice	½	cup(s)	124	109.8	53	0.6	12.5	0.2	0.2	0	0	0
316	Raw	1	item(s)	88	74.9	47	0.7	11.7	1.6	0.3	0	0.1	0.1
Vegetables, Legumes													
	Amaranth												
1043	Leaves, boiled, drained	½	cup(s)	66	60.4	14	1.4	2.7	—	0.1	0	0	0.1
1042	Leaves, raw	1	cup(s)	28	25.7	6	0.7	1.1	—	0.1	0	0	0
8683	**Arugula leaves, raw**	1	cup(s)	20	18.3	5	0.5	0.7	0.3	0.1	0	0	0.1
	Artichoke												
1044	Boiled, drained	1	item(s)	120	100.9	64	3.5	14.3	10.3	0.4	0.1	0	0.2
2885	Hearts, boiled, drained	½	cup(s)	84	70.6	45	2.4	10.0	7.2	0.3	0.1	0	0.1
	Asparagus												
566	Boiled, drained	½	cup(s)	90	83.4	20	2.2	3.7	1.8	0.2	0	0	0.1
568	Canned, drained	½	cup(s)	121	113.7	23	2.6	3.0	1.9	0.8	0.2	0	0.3
565	Tips, frozen, boiled, drained	½	cup(s)	90	84.7	16	2.7	1.7	1.4	0.4	0.1	0	0.2
	Bamboo shoots												
1048	Boiled, drained	½	cup(s)	60	57.6	7	0.9	1.2	0.6	0.1	0	0	0.1
1049	Canned, drained	½	cup(s)	66	61.8	12	1.1	2.1	0.9	0.3	0.1	0	0.1
	Beans												
1801	Adzuki beans, boiled	½	cup(s)	115	76.2	147	8.6	28.5	8.4	0.1	0	—	—
511	Baked beans with franks, canned	½	cup(s)	130	89.8	184	8.7	19.9	8.9	8.5	3.0	3.7	1.1
513	Baked beans with pork in sweet sauce, canned	½	cup(s)	127	89.3	142	6.7	26.7	5.3	1.8	0.6	0.6	0.5
512	Baked beans with pork in tomato sauce, canned	½	cup(s)	127	93.0	119	6.5	23.6	5.1	1.2	0.5	0.7	0.3
1805	Black beans, boiled	½	cup(s)	86	56.5	114	7.6	20.4	7.5	0.5	0.1	0	0.2
14597	Chickpeas, garbanzo beans or bengal gram, boiled	½	cup(s)	82	49.4	134	7.3	22.5	6.2	2.1	0.2	0.5	0.9
569	Fordhook lima beans, frozen, boiled, drained	½	cup(s)	85	62.0	88	5.2	16.4	4.9	0.3	0.1	0	0.1
1806	French beans, boiled	½	cup(s)	89	58.9	114	6.2	21.3	8.3	0.7	0.1	0	0.4
2773	Great northern beans, boiled	½	cup(s)	89	61.1	104	7.4	18.7	6.2	0.4	0.1	0	0.2
2736	Hyacinth beans, boiled, drained	½	cup(s)	44	37.8	22	1.3	4.0	—	0.1	0.1	0.1	0
570	Lima beans, baby, frozen, boiled, drained	½	cup(s)	90	65.1	95	6.0	17.5	5.4	0.3	0.1	0	0.1
515	Lima beans, boiled, drained	½	cup(s)	85	57.1	105	5.8	20.1	4.5	0.3	0.1	0	0.1
579	Mung beans, sprouted, boiled, drained	½	cup(s)	62	57.9	13	1.3	2.6	0.5	0.1	0	0	0
510	Navy beans, boiled	½	cup(s)	91	58.1	127	7.5	23.7	9.6	0.6	0.1	0.1	0.4
32816	Pinto beans, boiled, drained, no salt added	½	cup(s)	63	58.8	14	1.2	2.6	—	0.2	0	0	0.1
1052	Pinto beans, frozen, boiled, drained	½	cup(s)	47	27.3	76	4.4	14.5	4.0	0.2	0	0	0.1

PAGE KEY: A-4 = Breads/Baked Goods A-10 = Cereal/Rice/Pasta A-14 = Fruit A-18 = Vegetables/Legumes A-28 = Nuts/Seeds A-30 = Vegetarian A-32 = Dairy A-40 = Eggs A-40 = Seafood A-42 = Meats A-46 = Poultry A-46 = Processed Meats A-48 = Beverages A-52 = Fats/Oils A-54 = Sweets A-56 = Spices/Condiments/Sauces A-60 = Mixed Foods/Soups/Sandwiches A-64 = Fast Food A-84 = Convenience A-86 = Baby Foods

Chol (mg)	Calc (mg)	Iron (mg)	Magn (mg)	Pota (mg)	Sodi (mg)	Zinc (mg)	Vit A (µg)	Thia (mg)	Vit E (mg α)	Ribo (mg)	Niac (mg)	Vit B_6 (mg)	Fola (µg)	Vit C (mg)	Vit B_{12} (µg)	Sele (µg)
0	7	0.62	—	77.5	0.3	—	0	—	—	—	—	—	—	16.5	0	—
0	18	0.49	19.5	132.6	1.3	0.14	1.3	0.11	—	0.03	0.36	0.09	6.5	9.5	0	—
0	17	0.35	17.4	151.9	1.2	0.12	2.5	0.12	0.01	0.02	0.35	0.09	6.2	11.8	0	0.5
0	18	0.49	20.2	132.3	1.3	0.15	2.5	0.11	0.01	0.03	0.36	0.09	6.3	9.5	0	0.5
0	18	0.49	22.1	156.2	1.2	0.15	2.5	0.11	0.01	0.03	0.37	0.09	6.2	9.5	0	0.5
0	16	0.39	15.0	162.5	2.5	0.14	0	0.07	0.03	0.03	0.25	0.13	22.5	12.5	0	0.1
0	10	0.22	9.3	84.5	0.8	0.09	2.3	0.06	0.02	0.03	0.39	0.09	14.0	37.0	0	0.1
0	2	0.45	24.6	358.1	3.9	0.10	34.7	0.04	0.10	0.04	0.58	0.19	20.0	8.4	0	1.1
0	4	0.11	4.6	103.6	0	0.06	11.2	0.02	0.17	0.02	0.27	0.02	3.3	6.3	0	0
0	5	0.46	4.6	398.9	4.6	0.18	7.7	0.04	0.92	0.04	0.46	0.16	9.2	9.4	0	0.9
0	7	0.16	6.9	123.0	0.3	0.07	6.6	0.01	0.07	0.03	0.32	0.03	0.7	0.1	0	0
0	24	0.51	22.3	398.0	1.2	0.24	21.1	0.03	0.24	0.12	0.90	0.27	0	3.6	0	0.1
0	31	3.02	35.8	706.6	10.2	0.53	0	0.04	0.30	0.17	2.01	0.55	0	10.5	0	1.5
0	15	0.42	13.5	92.9	0.6	0.26	1.2	0.02	0.54	0.02	0.37	0.03	12.9	16.1	0	0.1
0	19	0.81	16.3	142.5	1.3	0.22	3.8	0.02	0.90	0.05	0.28	0.04	32.5	20.6	0	0.4
0	174	0.25	16.2	115.0	1.0	—	—	0.02	—	0.03	0.25	—	—	4.0	0	—
0	12	0.30	9.4	110.2	0.7	0.10	0.7	0.02	0.21	0.02	0.28	0.03	17.3	42.3	0	0.3
0	14	0.59	7.7	125.0	1.3	0.06	1.3	0.01	0.30	0.09	0.37	0.03	5.1	50.4	0	0.9
0	38	0.09	9.5	172.0	0	0.06	10.5	0.08	0.17	0.03	0.26	0.05	28.5	50.5	0	0.5
0	22	0.25	9.9	219.8	1.2	0.04	16.1	0.07	0.16	0.02	0.12	0.05	6.2	38.3	0	0.1
0	33	0.13	10.6	146.1	1.8	0.06	29.9	0.05	0.18	0.03	0.33	0.07	14.1	23.5	0	0.1
0	138	1.49	36.3	423.1	13.9	0.58	91.7	0.01	—	0.09	0.37	0.12	37.6	27.1	0	0.6
0	60	0.65	15.4	171.1	5.6	0.25	40.9	0.01	—	0.04	0.18	0.05	23.8	12.1	0	0.3
0	32	0.29	9.4	73.8	5.4	0.09	23.8	0.01	0.09	0.02	0.06	0.01	19.4	3.0	0	0.1
0	25	0.73	50.4	343.2	72.0	0.48	1.2	0.06	0.22	0.10	1.33	0.09	106.8	8.9	0	0.2
0	18	0.51	35.3	240.2	50.4	0.33	0.8	0.04	0.16	0.07	0.93	0.06	74.8	6.2	0	0.2
0	21	0.81	12.6	201.6	12.6	0.54	45.0	0.14	1.35	0.12	0.97	0.07	134.1	6.9	0	5.5
0	19	2.21	12.1	208.1	347.3	0.48	49.6	0.07	1.47	0.12	1.15	0.13	116.2	22.3	0	2.1
0	16	0.50	9.0	154.8	2.7	0.36	36.0	0.05	1.08	0.09	0.93	0.01	121.5	22.0	0	3.5
0	7	0.14	1.8	319.8	2.4	0.28	0	0.01	—	0.03	0.18	0.06	1.2	0	0	0.2
0	5	0.21	2.6	52.4	4.6	0.43	0.7	0.02	0.41	0.02	0.09	0.09	2.0	0.7	0	0.3
0	32	2.30	59.8	611.8	9.2	2.03	0	0.13	—	0.07	0.82	0.11	139.2	0	0	1.4
8	62	2.24	36.3	304.3	556.9	2.42	5.2	0.08	0.21	0.07	1.17	0.06	38.9	3.0	0.4	8.4
9	75	2.08	41.7	326.4	422.5	1.73	0	0.05	0.03	0.07	0.44	0.07	10.1	3.5	0	6.3
9	71	4.09	43.0	373.2	552.8	6.93	5.1	0.06	0.12	0.05	0.62	0.08	19.0	3.8	0	5.9
0	23	1.80	60.2	305.3	0.9	0.96	0	0.21	—	0.05	0.43	0.05	128.1	0	0	1.0
0	40	2.36	39.4	238.6	5.7	1.25	0.8	0.09	0.28	0.05	0.43	0.11	141.0	1.1	0	3.0
0	26	1.54	35.7	258.4	58.7	0.62	8.5	0.06	0.24	0.05	0.90	0.10	17.9	10.9	0	0.5
0	56	0.95	49.6	327.5	5.3	0.56	0	0.11	—	0.05	0.48	0.09	66.4	1.1	0	1.1
0	60	1.88	44.3	346.0	1.8	0.77	0	0.14	—	0.05	0.60	0.10	90.3	1.2	0	3.6
0	18	0.33	18.3	114.0	0.9	0.16	3.0	0.02	—	0.03	0.20	0.01	20.4	2.2	0	0.7
0	25	1.76	50.4	369.9	26.1	0.49	7.2	0.06	0.57	0.04	0.69	0.10	14.4	5.2	0	1.5
0	27	2.08	62.9	484.5	14.5	0.67	12.8	0.11	0.11	0.08	0.88	0.16	22.1	8.6	0	1.7
0	7	0.40	8.7	62.6	6.2	0.29	0.6	0.03	0.04	0.06	0.51	0.03	18.0	7.1	0	0.4
—	63	2.14	48.2	354.0	0	0.93	0	0.21	0.01	0.06	0.59	0.12	127.4	0.8	0	2.6
0	9	0.41	11.3	61.7	32.1	0.10	0	0.04	—	0.03	0.45	0.03	18.3	3.8	0	0.4
0	24	1.27	25.4	303.6	39.0	0.32	0	0.12	—	0.05	0.29	0.09	16.0	0.3	0	0.7

TABLE A-1 Table of Food Composition *(continued)*

(Computer code is for Cengage Diet Analysis program)
(For purposes of calculations, use "0" for t, <1, <.1, <.01, etc.)

DA+ Code	Food Description	Quantity	Measure	Wt (g)	H_2O (g)	Ener (kcal)	Prot (g)	Carb (g)	Fiber (g)	Fat (g)	Fat Breakdown (g) Sat	Mono	Poly
	Vegetables, Legumes—*continued*												
514	Red kidney beans, canned	½	cup(s)	128	99.0	108	6.7	19.9	6.9	0.5	0.1	0.2	0.2
1810	Refried beans, canned	½	cup(s)	127	96.1	119	6.9	19.6	6.7	1.6	0.6	0.7	0.2
1053	Shell beans, canned	½	cup(s)	123	111.1	37	2.2	7.6	4.2	0.2	0	0	0.1
1670	Soybeans, boiled	½	cup(s)	86	53.8	149	14.3	8.5	5.2	7.7	1.1	1.7	4.4
1108	Soybeans, green, boiled, drained	½	cup(s)	90	61.7	127	11.1	9.9	3.8	5.8	0.7	1.1	2.7
1807	White beans, small, boiled	½	cup(s)	90	56.6	127	8.0	23.1	9.3	0.6	0.1	0.1	0.2
575	Yellow snap, string or wax beans, boiled, drained	½	cup(s)	63	55.8	22	1.2	4.9	2.1	0.2	0	0	0.1
576	Yellow snap, string or wax beans, frozen, boiled, drained	½	cup(s)	68	61.7	19	1.0	4.4	2.0	0.1	0	0	0.1
	Beets												
584	Beet greens, boiled, drained	½	cup(s)	72	64.2	19	1.9	3.9	2.1	0.1	0	0	0.1
2730	Pickled, canned with liquid	½	cup(s)	114	92.9	74	0.9	18.5	3.0	0.1	0	0	0
581	Sliced, boiled, drained	½	cup(s)	85	74.0	37	1.4	8.5	1.7	0.2	0	0	0.1
583	Sliced, canned, drained	½	cup(s)	85	77.3	26	0.8	6.1	1.5	0.1	0	0	0
580	Whole, boiled, drained	2	item(s)	100	87.1	44	1.7	10.0	2.0	0.2	0	0	0.1
585	**Cowpeas or black-eyed peas, boiled, drained**	½	cup(s)	83	62.3	80	2.6	16.8	4.1	0.3	0.1	0	0.1
	Broccoli												
588	Chopped, boiled, drained	½	cup(s)	78	69.6	27	1.9	5.6	2.6	0.3	0.1	0	0.1
590	Frozen, chopped, boiled, drained	½	cup(s)	92	83.5	26	2.9	4.9	2.8	0.1	0	0	0.1
587	Raw, chopped	½	cup(s)	46	40.6	15	1.3	3.0	1.2	0.2	0	0	0
16848	**Broccoflower, raw, chopped**	½	cup(s)	32	28.7	10	0.9	1.9	1.0	0.1	0	0	0
	Brussels sprouts												
591	Boiled, drained	½	cup(s)	78	69.3	28	2.0	5.5	2.0	0.4	0.1	0	0.2
592	Frozen, boiled, drained	½	cup(s)	78	67.2	33	2.8	6.4	3.2	0.3	0.1	0	0.2
	Cabbage												
595	Boiled, drained, no salt added	1	cup(s)	150	138.8	35	1.9	8.3	2.8	0.1	0	0	0
35611	Chinese (pak choi or bok choy), boiled with salt, drained	1	cup(s)	170	162.4	20	2.6	3.0	1.7	0.3	0	0	0.1
16869	Kim chee	1	cup(s)	150	137.5	32	2.5	6.1	1.8	0.3	0	0	0.2
594	Raw, shredded	1	cup(s)	70	64.5	17	0.9	4.1	1.7	0.1	0	0	0
596	Red, shredded, raw	1	cup(s)	70	63.3	22	1.0	5.2	1.5	0.1	0	0	0.1
597	Savoy, shredded, raw	1	cup(s)	70	63.7	19	1.4	4.3	2.2	0.1	0	0	0
35417	**Capers**	1	teaspoon(s)	4	—	2	0	0	0	0	0	0	0
	Carrots												
8691	Baby, raw	8	item(s)	80	72.3	28	0.5	6.6	2.3	0.1	0	0	0.1
601	Grated	½	cup(s)	55	48.6	23	0.5	5.3	1.5	0.1	0	0	0.1
1055	Juice, canned	½	cup(s)	118	104.9	47	1.1	11.0	0.9	0.2	0	0	0.1
600	Raw	½	cup(s)	61	53.9	25	0.6	5.8	1.7	0.1	0	0	0.1
602	Sliced, boiled, drained	½	cup(s)	78	70.3	27	0.6	6.4	2.3	0.1	0	0	0.1
32725	**Cassava or manioc**	½	cup(s)	103	61.5	165	1.4	39.2	1.9	0.3	0.1	0.1	0
	Cauliflower												
606	Boiled, drained	½	cup(s)	62	57.7	14	1.1	2.5	1.4	0.3	0	0	0.1
607	Frozen, boiled, drained	½	cup(s)	90	84.6	17	1.4	3.4	2.4	0.2	0	0	0.1
605	Raw, chopped	½	cup(s)	50	46.0	13	1.0	2.6	1.2	0	0	0	0
	Celery												
609	Diced	½	cup(s)	51	48.2	8	0.3	1.5	0.8	0.1	0	0	0
608	Stalk	2	item(s)	80	76.3	13	0.6	2.4	1.3	0.1	0	0	0.1
	Chard												
1057	Swiss chard, boiled, drained	½	cup(s)	88	81.1	18	1.6	3.6	1.8	0.1	0	0	0
1056	Swiss chard, raw	1	cup(s)	36	33.4	7	0.6	1.3	0.6	0.1	0	0	0
	Collard greens												
610	Boiled, drained	½	cup(s)	95	87.3	25	2.0	4.7	2.7	0.3	0	0	0.2
611	Frozen, chopped, boiled, drained	½	cup(s)	85	75.2	31	2.5	6.0	2.4	0.3	0.1	0	0.2
	Corn												
29614	Yellow corn, fresh, cooked	1	item(s)	100	69.2	107	3.3	25.0	2.8	1.3	0.2	0.4	0.6
615	Yellow creamed sweet corn, canned	½	cup(s)	128	100.8	92	2.2	23.2	1.5	0.5	0.1	0.2	0.3
612	Yellow sweet corn, boiled, drained	½	cup(s)	82	57.0	89	2.7	20.6	2.3	1.1	0.2	0.3	0.5

PAGE KEY: A-4 = Breads/Baked Goods A-10 = Cereal/Rice/Pasta A-14 = Fruit A-18 = Vegetables/Legumes A-28 = Nuts/Seeds A-30 = Vegetarian A-32 = Dairy A-40 = Eggs A-40 = Seafood A-42 = Meats A-46 = Poultry A-46 = Processed Meats A-48 = Beverages A-52 = Fats/Oils A-54 = Sweets A-56 = Spices/Condiments/Sauces A-60 = Mixed Foods/Soups/Sandwiches A-64 = Fast Food A-84 = Convenience A-86 = Baby Foods

Chol (mg)	Calc (mg)	Iron (mg)	Magn (mg)	Pota (mg)	Sodi (mg)	Zinc (mg)	Vit A (µg)	Thia (mg)	Vit E (mg α)	Ribo (mg)	Niac (mg)	Vit B_6 (mg)	Fola (µg)	Vit C (mg)	Vit B_{12} (µg)	Sele (µg)
0	32	1.62	35.8	327.7	330.2	2.09	0	0.13	0.02	0.11	0.57	0.10	25.6	1.4	0	0.6
10	44	2.10	41.7	337.8	378.2	1.48	0	0.03	0.00	0.02	0.39	0.18	13.9	7.6	0	1.6
0	36	1.21	18.4	133.5	409.2	0.33	13.5	0.04	0.04	0.07	0.25	0.06	22.1	3.8	0	2.6
0	88	4.42	74.0	442.9	0.9	0.98	0	0.13	0.30	0.24	0.34	0.20	46.4	1.5	0	6.3
0	131	2.25	54.0	485.1	12.6	0.82	7.2	0.23	—	0.14	1.13	0.05	99.9	15.3	0	1.3
0	65	2.54	60.9	414.4	1.8	0.97	0	0.21	—	0.05	0.24	0.11	122.6	0	0	1.2
0	29	0.80	15.6	186.9	1.9	0.23	2.5	0.05	0.28	0.06	0.38	0.04	20.6	6.1	0	0.3
0	33	0.59	16.2	85.1	6.1	0.32	4.1	0.02	0.03	0.06	0.26	0.04	15.5	2.8	0	0.3
0	82	1.36	49.0	654.5	173.5	0.36	275.8	0.08	1.30	0.20	0.35	0.09	10.1	17.9	0	0.6
0	12	0.46	17.0	168.0	299.6	0.29	1.1	0.01	—	0.05	0.28	0.05	30.6	2.6	0	1.1
0	14	0.67	19.6	259.3	65.5	0.30	1.7	0.02	0.03	0.03	0.28	0.05	68.0	3.1	0	0.6
0	13	1.54	14.5	125.8	164.9	0.17	0.9	0.01	0.02	0.03	0.13	0.04	25.5	3.5	0	0.4
0	16	0.79	23.0	305.0	77.0	0.35	2.0	0.02	0.04	0.04	0.33	0.06	80.0	3.6	0	0.7
0	106	0.92	42.9	344.9	3.3	0.85	33.0	0.08	0.18	0.12	1.15	0.05	104.8	1.8	0	2.1
0	31	0.52	16.4	228.5	32.0	0.35	60.1	0.04	1.13	0.09	0.43	0.15	84.2	50.6	0	1.2
0	30	0.56	12.0	130.6	10.1	0.25	46.9	0.05	1.21	0.07	0.42	0.12	51.5	36.9	0	0.6
0	21	0.33	9.6	143.8	15.0	0.19	14.1	0.03	0.36	0.05	0.29	0.08	28.7	40.6	0	1.1
0	11	0.23	6.4	96.0	7.4	0.20	2.6	0.02	0.01	0.03	0.23	0.07	18.2	28.2	0	0.2
0	28	0.93	15.6	247.3	16.4	0.25	30.4	0.08	0.33	0.06	0.47	0.13	46.8	48.4	0	1.2
0	20	0.37	14.0	224.8	11.6	0.18	35.7	0.08	0.39	0.08	0.41	0.22	78.3	35.4	0	0.5
0	72	0.24	22.5	294.0	12.0	0.30	6.0	0.08	0.20	0.04	0.36	0.16	45.0	56.2	0	0.9
0	158	1.76	18.7	630.7	459.0	0.28	360.4	0.04	0.14	0.10	0.72	0.28	69.7	44.2	0	0.7
0	144	1.26	27.0	379.5	996.0	0.36	288.0	0.06	0.36	0.10	0.80	0.32	88.5	79.6	0	1.5
0	28	0.33	8.4	119.0	12.6	0.12	3.5	0.04	0.10	0.02	0.16	0.08	30.1	25.6	0	0.2
0	31	0.56	11.2	170.1	18.9	0.15	39.2	0.04	0.07	0.05	0.29	0.14	12.6	39.9	0	0.4
0	24	0.28	19.6	161.0	19.6	0.18	35.0	0.05	0.11	0.02	0.21	0.13	56.0	21.7	0	0.6
0	0	0.00	—	—	140	—	0	—	—	—	—	—	—	0	—	—
0	26	0.71	8.0	189.6	62.4	0.13	552.0	0.02	—	0.02	0.44	0.08	21.6	2.1	0	0.7
0	18	0.16	6.6	176.0	37.9	0.13	459.2	0.03	0.36	0.03	0.54	0.07	10.4	3.2	0	0.1
0	28	0.54	16.5	344.6	34.2	0.21	1128.1	0.11	1.37	0.07	0.46	0.26	4.7	10.0	0	0.7
0	20	0.18	7.3	195.2	42.1	0.15	509.4	0.04	0.40	0.04	0.60	0.08	11.6	3.6	0	0.1
0	23	0.26	7.8	183.3	45.2	0.15	664.6	0.05	0.80	0.03	0.50	0.11	10.9	2.8	0	0.5
0	16	0.27	21.6	279.1	14.4	0.35	1.0	0.08	0.19	0.04	0.87	0.09	27.8	21.2	0	0.7
0	10	0.19	5.6	88.0	9.3	0.10	0.6	0.02	0.04	0.03	0.25	0.10	27.3	27.5	0	0.4
0	15	0.36	8.1	125.1	16.2	0.11	0	0.03	0.05	0.04	0.27	0.07	36.9	28.2	0	0.5
0	11	0.22	7.5	151.5	15.0	0.14	0.5	0.03	0.04	0.03	0.26	0.11	28.5	23.2	0	0.3
0	20	0.10	5.6	131.3	40.4	0.07	11.1	0.01	0.14	0.03	0.16	0.04	18.2	1.6	0	0.2
0	32	0.16	8.8	208.0	64.0	0.10	17.6	0.01	0.21	0.04	0.25	0.05	28.8	2.5	0	0.3
0	51	1.98	75.3	480.4	156.6	0.29	267.8	0.03	1.65	0.08	0.32	0.07	7.9	15.8	0	0.8
0	18	0.64	29.2	136.4	76.7	0.13	110.2	0.01	0.68	0.03	0.14	0.03	5.0	10.8	0	0.3
0	133	1.10	19.0	110.2	15.2	0.21	385.7	0.03	0.83	0.10	0.54	0.12	88.4	17.3	0	0.5
0	179	0.95	25.5	213.4	42.5	0.22	488.8	0.04	1.06	0.09	0.54	0.09	64.6	22.4	0	1.3
0	2	0.61	32.0	248.0	242.0	0.48	13.0	0.20	0.09	0.07	1.60	0.06	46.0	6.2	0	0.2
0	4	0.48	21.8	171.5	364.8	0.67	5.1	0.03	0.09	0.06	1.22	0.08	57.6	5.9	0	0.5
0	2	0.36	21.3	173.8	0	0.50	10.7	0.17	0.07	0.05	1.32	0.04	37.7	5.1	0	0.2

TABLE A-1 Table of Food Composition *(continued)*

(Computer code is for Cengage Diet Analysis program)
(For purposes of calculations, use "0" for t, <1, <.1, <.01, etc.)

DA+ Code	Food Description	Quantity	Measure	Wt (g)	H_2O (g)	Ener (kcal)	Prot (g)	Carb (g)	Fiber (g)	Fat (g)	Fat Breakdown (g) Sat	Mono	Poly
	Vegetables, Legumes—*continued*												
614	Yellow sweet corn, frozen, boiled, drained	½	cup(s)	82	63.2	66	2.1	15.8	2.0	0.5	0.1	0.2	0.3
618	**Cucumber**	¼	item(s)	75	71.7	11	0.5	2.7	0.4	0.1	0	0	0
16870	**Cucumber, kim chee**	½	cup(s)	75	68.1	16	0.8	3.6	1.1	0.1	0	0	0
	Dandelion greens												
620	Chopped, boiled, drained	½	cup(s)	53	47.1	17	1.1	3.4	1.5	0.3	0.1	0	0.1
2734	Raw	1	cup(s)	55	47.1	25	1.5	5.1	1.9	0.4	0.1	0	0.2
1066	**Eggplant, boiled, drained**	½	cup(s)	50	44.4	17	0.4	4.3	1.2	0.1	0	0	0
621	**Endive or escarole, chopped, raw**	1	cup(s)	50	46.9	8	0.6	1.7	1.5	0.1	0	0	0
8784	**Jicama or yambean**	½	cup(s)	65	116.5	49	0.9	11.4	6.3	0.1	0	0	0.1
	Kale												
623	Frozen, chopped, boiled, drained	½	cup(s)	65	58.8	20	1.8	3.4	1.3	0.3	0	0	0.2
29313	Raw	1	cup(s)	67	56.6	33	2.2	6.7	1.3	0.5	0.1	0	0.2
	Kohlrabi												
1072	Boiled, drained	½	cup(s)	83	74.5	24	1.5	5.5	0.9	0.1	0	0	0
1071	Raw	1	cup(s)	135	122.9	36	2.3	8.4	4.9	0.1	0	0	0.1
	Leeks												
1074	Boiled, drained	½	cup(s)	52	47.2	16	0.4	4.0	0.5	0.1	0	0	0
1073	Raw	1	cup(s)	89	73.9	54	1.3	12.6	1.6	0.3	0	0	0.1
	Lentils												
522	Boiled	¼	cup(s)	50	34.5	57	4.5	10.0	3.9	0.2	0	0	0.1
1075	Sprouted	1	cup(s)	77	51.9	82	6.9	17.0	—	0.4	0	0.1	0.2
	Lettuce												
625	Butterhead leaves	11	piece(s)	83	78.9	11	1.1	1.8	0.9	0.2	0	0	0.1
624	Butterhead, Boston or Bibb	1	cup(s)	55	52.6	7	0.7	1.2	0.6	0.1	0	0	0.1
626	Iceberg	1	cup(s)	55	52.6	8	0.5	1.6	0.7	0.1	0	0	0
628	Iceberg, chopped	1	cup(s)	55	52.6	8	0.5	1.6	0.7	0.1	0	0	0
629	Looseleaf	1	cup(s)	36	34.2	5	0.5	1.0	0.5	0.1	0	0	0
1665	Romaine, shredded	1	cup(s)	56	53.0	10	0.7	1.8	1.2	0.2	0	0	0.1
	Mushrooms												
15585	Crimini (about 6)	3	ounce(s)	85	—	28	3.7	2.8	1.9	0	0	0	0
8700	Enoki	30	item(s)	90	79.7	40	2.3	6.9	2.4	0.3	0	0	0.1
1079	Mushrooms, boiled, drained	½	cup(s)	78	71.0	22	1.7	4.1	1.7	0.4	0	0	0.1
1080	Mushrooms, canned, drained	½	cup(s)	78	71.0	20	1.5	4.0	1.9	0.2	0	0	0.1
630	Mushrooms, raw	½	cup(s)	48	44.4	11	1.5	1.6	0.5	0.2	0	0	0.1
15587	Portabella, raw	1	item(s)	84	—	30	3.0	3.9	3.0	0	0	0	0
2743	Shiitake, cooked	½	cup(s)	73	60.5	41	1.1	10.4	1.5	0.2	0	0.1	0
	Mustard greens												
2744	Frozen, boiled, drained	½	cup(s)	75	70.4	14	1.7	2.3	2.1	0.2	0	0.1	0
29319	Raw	1	cup(s)	56	50.8	15	1.5	2.7	1.8	0.1	0	0	0
	Okra												
16866	Batter coated, fried	11	piece(s)	83	55.6	156	2.1	12.7	2.0	11.2	1.5	3.7	5.5
32742	Frozen, boiled, drained, no salt added	½	cup(s)	92	83.8	26	1.9	5.3	2.6	0.3	0.1	0	0.1
632	Sliced, boiled, drained	½	cup(s)	80	74.1	18	1.5	3.6	2.0	0.2	0	0	0
	Onions												
635	Chopped, boiled, drained	½	cup(s)	105	92.2	46	1.4	10.7	1.5	0.2	0	0	0.1
2748	Frozen, boiled, drained	½	cup(s)	106	97.8	30	0.8	7.0	1.9	0.1	0	0	0
1081	Onion rings, breaded and pan fried, frozen, heated	10	piece(s)	71	20.2	289	3.8	27.1	0.9	19.0	6.1	7.7	3.6
633	Raw, chopped	½	cup(s)	80	71.3	32	0.9	7.5	1.4	0.1	0	0	0
16850	Red onions, sliced, raw	½	cup(s)	57	50.7	24	0.5	5.8	0.8	0	0	0	0
636	Scallions, green or spring onions	2	item(s)	30	26.9	10	0.5	2.2	0.8	0.1	0	0	0
16860	**Palm hearts, cooked**	½	cup(s)	73	50.7	84	2.0	18.7	1.1	0.1	0	0	0.1
637	**Parsley, chopped**	1	tablespoon(s)	4	3.3	1	0.1	0.2	0.1	0	0	0	0
638	**Parsnips, sliced, boiled, drained**	½	cup(s)	78	62.6	55	1.0	13.3	2.8	0.2	0	0.1	0
	Peas												
639	Green peas, canned, drained	½	cup(s)	85	69.4	59	3.8	10.7	3.5	0.3	0.1	0	0.1
641	Green peas, frozen, boiled, drained	½	cup(s)	80	63.6	62	4.1	11.4	4.4	0.2	0	0	0.1
35694	Pea pods, boiled with salt, drained	½	cup(s)	80	71.1	32	2.6	5.2	2.2	0.2	0	0	0.1

PAGE KEY: A-4 = Breads/Baked Goods A-10 = Cereal/Rice/Pasta A-14 = Fruit A-18 = Vegetables/Legumes A-28 = Nuts/Seeds A-30 = Vegetarian A-32 = Dairy A-40 = Eggs A-40 = Seafood A-42 = Meats A-46 = Poultry A-46 = Processed Meats A-48 = Beverages A-52 = Fats/Oils A-54 = Sweets A-56 = Spices/Condiments/Sauces A-60 = Mixed Foods/Soups/Sandwiches A-64 = Fast Food A-84 = Convenience A-86 = Baby Foods

Chol (mg)	Calc (mg)	Iron (mg)	Magn (mg)	Pota (mg)	Sodi (mg)	Zinc (mg)	Vit A (µg)	Thia (mg)	Vit E (mg α)	Ribo (mg)	Niac (mg)	Vit B_6 (mg)	Fola (µg)	Vit C (mg)	Vit B_{12} (µg)	Sele (µg)
0	2	0.38	23.0	191.1	0.8	0.51	8.2	0.02	0.05	0.05	1.07	0.08	28.7	2.9	0	0.6
0	12	0.20	9.8	110.6	1.5	0.14	3.8	0.01	0.01	0.01	0.07	0.03	5.3	2.1	0	0.2
0	7	3.61	6.0	87.8	765.8	0.38	—	0.02	—	0.02	0.34	0.08	17.3	2.6	0	—
0	74	0.95	12.6	121.8	23.1	0.15	179.6	0.07	1.28	0.09	0.27	0.08	6.8	9.5	0	0.2
0	103	1.70	19.8	218.3	41.8	0.22	279.4	0.10	1.89	0.14	0.44	0.13	14.8	19.2	0	0.3
0	3	0.12	5.4	60.9	0.5	0.06	1.0	0.04	0.20	0.01	0.30	0.04	6.9	0.6	0	0
0	26	0.41	7.5	157.0	11.0	0.39	54.0	0.04	0.22	0.03	0.20	0.01	71.0	3.2	0	0.1
0	16	0.78	15.5	194.0	5.2	0.20	1.3	0.02	0.59	0.04	0.25	0.05	15.5	26.1	0	0.9
0	90	0.61	11.7	208.7	9.8	0.11	477.8	0.02	0.59	0.07	0.43	0.05	9.1	16.4	0	0.6
0	90	1.14	22.8	299.5	28.8	0.29	515.2	0.07	—	0.08	0.66	0.18	19.4	80.4	0	0.6
0	21	0.33	15.7	280.5	17.3	0.26	1.7	0.03	0.43	0.02	0.32	0.13	9.9	44.6	0	0.7
0	32	0.54	25.7	472.5	27.0	0.04	2.7	0.06	0.64	0.02	0.54	0.20	21.6	83.7	0	0.9
0	16	0.56	7.3	45.2	5.2	0.02	1.0	0.01	—	0.01	0.10	0.04	12.5	2.2	0	0.3
0	53	1.86	24.9	160.2	17.8	0.10	73.9	0.05	0.81	0.02	0.35	0.20	57.0	10.7	0	0.9
0	9	1.65	17.8	182.7	1.0	0.63	0	0.08	0.05	0.04	0.52	0.09	89.6	0.7	0	1.4
0	19	2.47	28.5	247.9	8.5	1.16	1.5	0.17	—	0.09	0.86	0.14	77.0	12.7	0	0.5
0	29	1.02	10.7	196.4	4.1	0.16	137.0	0.04	0.14	0.05	0.29	0.06	60.2	3.1	0	0.5
0	19	0.68	7.1	130.9	2.7	0.11	91.3	0.03	0.09	0.03	0.19	0.04	40.1	2.0	0	0.3
0	10	0.22	3.8	77.5	5.5	0.08	13.7	0.02	0.09	0.01	0.07	0.02	15.9	1.5	0	0.1
0	10	0.22	3.8	77.5	5.5	0.08	13.7	0.02	0.09	0.01	0.07	0.02	15.9	1.5	0	0.1
0	13	0.31	4.7	69.8	10.1	0.06	133.2	0.02	0.10	0.02	0.13	0.03	13.7	6.5	0	0.2
0	18	0.54	7.8	138.3	4.5	0.13	162.4	0.04	0.07	0.03	0.17	0.04	76.2	13.4	0	0.2
0	0	0.67	—	—	32.6	—	0	—	—	—	—	—	—	0	0	—
0	1	0.98	14.4	331.2	2.7	0.54	0	0.16	0.01	0.14	5.31	0.07	46.8	0	0	2.0
0	5	1.35	9.4	277.7	1.6	0.67	0	0.05	0.01	0.23	3.47	0.07	14.0	3.1	0	9.3
0	9	0.61	11.7	100.6	331.5	0.56	0	0.06	0.01	0.01	1.24	0.04	9.4	0	0	3.2
0	1	0.24	4.3	152.6	2.4	0.25	0	0.04	0.01	0.19	1.73	0.05	7.7	1.0	0	4.5
0	39	0.35	—	—	9.9	—	0	—	—	—	—	—	—	0	0	—
0	2	0.31	10.2	84.8	2.9	0.96	0	0.02	0.02	0.12	1.08	0.11	15.2	0.2	0	18.0
0	76	0.84	9.8	104.3	18.8	0.15	265.5	0.03	1.01	0.04	0.19	0.08	52.5	10.4	0	0.5
0	58	0.81	17.9	198.2	14.0	0.11	294.0	0.04	1.12	0.06	0.45	0.10	104.7	39.2	0	0.5
2	54	1.13	32.2	170.8	109.7	0.44	14.0	0.16	1.50	0.12	1.29	0.11	43.7	9.2	0	3.6
0	88	0.61	46.9	215.3	2.8	0.57	15.6	0.09	0.29	0.11	0.72	0.04	134.3	11.2	0	0.6
0	62	0.22	28.8	108.0	4.8	0.34	11.2	0.10	0.21	0.04	0.69	0.15	36.8	13.0	0	0.3
0	23	0.24	11.5	174.3	3.1	0.21	0	0.03	0.02	0.02	0.17	0.12	15.7	5.5	0	0.6
0	17	0.32	6.4	114.5	12.7	0.06	0	0.02	0.01	0.02	0.14	0.06	13.8	2.8	0	0.4
0	22	1.20	13.5	91.6	266.3	0.29	7.8	0.19	—	0.09	2.56	0.05	73.1	1.0	0	2.5
0	18	0.16	8.0	116.8	3.2	0.13	0	0.03	0.01	0.02	0.09	0.09	15.2	5.9	0	0.4
0	13	0.10	5.7	82.4	1.7	0.09	0	0.02	0.01	0.01	0.04	0.08	10.9	3.7	0	0.3
0	22	0.44	6.0	82.8	4.8	0.11	15.0	0.01	0.16	0.02	0.15	0.01	19.2	5.6	0	0.2
0	13	1.23	7.3	1318.4	10.2	2.72	2.2	0.03	0.36	0.12	0.62	0.53	14.6	5.0	0	0.5
0	5	0.23	1.9	21.1	2.1	0.04	16.0	0.00	0.02	0.00	0.05	0.00	5.8	5.1	0	0
0	29	0.45	22.6	286.3	7.8	0.20	0	0.06	0.78	0.04	0.56	0.07	45.2	10.1	0	1.3
0	17	0.80	14.5	147.1	214.2	0.60	23.0	0.10	0.02	0.06	0.62	0.05	37.4	8.2	0	1.4
0	19	1.21	17.6	88.0	57.6	0.53	84.0	0.22	0.02	0.08	1.18	0.09	47.2	7.9	0	0.8
0	34	1.57	20.8	192.0	192.0	0.29	41.6	0.10	0.31	0.06	0.43	0.11	23.2	38.3	0	0.6

TABLE A-1 Table of Food Composition *(continued)*

(Computer code is for Cengage Diet Analysis program)
(For purposes of calculations, use "0" for t, <1, <.1, <.01, etc.)

DA+ Code	Food Description	Quantity	Measure	Wt (g)	H_2O (g)	Ener (kcal)	Prot (g)	Carb (g)	Fiber (g)	Fat (g)	Fat Breakdown (g) Sat	Mono	Poly
Vegetables, Legumes—*continued*													
1082	Peas and carrots, canned with liquid	½	cup(s)	128	112.4	48	2.8	10.8	2.6	0.3	0.1	0	0.2
1083	Peas and carrots, frozen, boiled, drained	½	cup(s)	80	68.6	38	2.5	8.1	2.5	0.3	0.1	0	0.2
2750	Snow or sugar peas, frozen, boiled, drained	½	cup(s)	80	69.3	42	2.8	7.2	2.5	0.3	0.1	0	0.1
640	Snow or sugar peas, raw	½	cup(s)	32	28.0	13	0.9	2.4	0.8	0.1	0	0	0
29324	Split peas, sprouted	½	cup(s)	60	37.4	77	5.3	16.9	—	0.4	0.1	0	0.2
	Peppers												
644	Green bell or sweet, boiled, drained	½	cup(s)	68	62.5	19	0.6	4.6	0.8	0.1	0	0	0.1
643	Green bell or sweet, raw	½	cup(s)	75	69.9	15	0.6	3.5	1.3	0.1	0	0	0
1664	Green hot chili	1	item(s)	45	39.5	18	0.9	4.3	0.7	0.1	0	0	0
1663	Green hot chili, canned with liquid	½	cup(s)	68	62.9	14	0.6	3.5	0.9	0.1	0	0	0
1086	Jalapeno, canned with liquid	½	cup(s)	68	60.4	18	0.6	3.2	1.8	0.6	0.1	0	0.3
8703	Yellow bell or sweet	1	item(s)	186	171.2	50	1.9	11.8	1.7	0.4	0.1	0	0.2
1087	**Poi**	½	cup(s)	120	86.0	134	0.5	32.7	0.5	0.2	0	0	0.1
	Potatoes												
1090	Au gratin mix, prepared with water, whole milk and butter	½	cup(s)	124	97.7	115	2.8	15.9	1.1	5.1	3.2	1.5	0.2
1089	Au gratin, prepared with butter	½	cup(s)	123	90.7	162	6.2	13.8	2.2	9.3	5.8	2.6	0.3
5791	Baked, flesh and skin	1	item(s)	202	151.3	188	5.1	42.7	4.4	0.3	0.1	0	0.1
645	Baked, flesh only	½	cup(s)	61	46.0	57	1.2	13.1	0.9	0.1	0	0	0
1088	Baked, skin only	1	item(s)	58	27.4	115	2.5	26.7	4.6	0.1	0	0	0
5795	Boiled in skin, flesh only, drained	1	item(s)	136	104.7	118	2.5	27.4	2.1	0.1	0	0	0.1
5794	Boiled, drained, skin and flesh	1	item(s)	150	115.9	129	2.9	29.8	2.5	0.2	0	0	0.1
647	Boiled, flesh only	½	cup(s)	78	60.4	67	1.3	15.6	1.4	0.1	0	0	0
648	French fried, deep fried, prepared from raw	14	item(s)	70	32.8	187	2.7	23.5	2.9	9.5	1.9	4.2	3.0
649	French fried, frozen, heated	14	item(s)	70	43.7	94	1.9	19.4	2.0	3.7	0.7	2.3	0.2
1091	Hashed brown	½	cup(s)	78	36.9	207	2.3	27.4	2.5	9.8	1.5	4.1	3.7
652	Mashed with margarine and whole milk	½	cup(s)	105	79.0	119	2.1	17.7	1.6	4.4	1.0	2.0	1.2
653	Mashed, prepared from dehydrated granules with milk, water, and margarine	½	cup(s)	105	79.8	122	2.3	16.9	1.4	5.0	1.3	2.1	1.4
2759	Microwaved	1	item(s)	202	145.5	212	4.9	49.0	4.6	0.2	0.1	0	0.1
2760	Microwaved in skin, flesh only	½	cup(s)	78	57.1	78	1.6	18.1	1.2	0.1	0	0	0
5804	Microwaved, skin only	1	item(s)	58	36.8	77	2.5	17.2	4.2	0.1	0	0	0
1097	Potato puffs, frozen, heated	½	cup(s)	64	38.2	122	1.3	17.8	1.6	5.5	1.2	3.9	0.3
1094	Scalloped mix, prepared with water, whole milk and butter	½	cup(s)	124	98.4	116	2.6	15.9	1.4	5.3	3.3	1.5	0.2
1093	Scalloped, prepared with butter	½	cup(s)	123	99.2	108	3.5	13.2	2.3	4.5	2.8	1.3	0.2
	Pumpkin												
1773	Boiled, drained	½	cup(s)	123	114.8	25	0.9	6.0	1.3	0.1	0	0	0
656	Canned	½	cup(s)	123	110.2	42	1.3	9.9	3.6	0.3	0.2	0	0
	Radicchio												
8731	Leaves, raw	1	cup(s)	40	37.3	9	0.6	1.8	0.4	0.1	0	0	0
2498	Raw	1	cup(s)	40	37.3	9	0.6	1.8	0.4	0.1	0	0	0
657	**Radishes**	6	item(s)	27	25.7	4	0.2	0.9	0.4	0	0	0	0
1099	**Rutabaga, boiled, drained**	½	cup(s)	85	75.5	33	1.1	7.4	1.5	0.2	0	0	0.1
658	**Sauerkraut, canned**	½	cup(s)	118	109.2	22	1.1	5.1	3.4	0.2	0	0	0.1
	Seaweed												
1102	Kelp	½	cup(s)	40	32.6	17	0.6	3.8	0.5	0.2	0.1	0	0
1104	Spirulina, dried	½	cup(s)	8	0.4	22	4.3	1.8	0.3	0.6	0.2	0.1	0.2
1106	**Shallots**	3	tablespoon(s)	30	23.9	22	0.8	5.0	—	0	0	0	0
	Soybeans												
1670	Boiled	½	cup(s)	86	53.8	149	14.3	8.5	5.2	7.7	1.1	1.7	4.4
2825	Dry roasted	½	cup(s)	86	0.7	388	34.0	28.1	7.0	18.6	2.7	4.1	10.5
2824	Roasted, salted	½	cup(s)	86	1.7	405	30.3	28.9	15.2	21.8	3.2	4.8	12.3
8739	Sprouted, stir fried	½	cup(s)	63	42.3	79	8.2	5.9	0.5	4.5	0.6	1.0	2.5
	Soy products												
1813	Soy milk	1	cup(s)	240	211.3	130	7.8	15.1	1.4	4.2	0.5	1.0	2.3
2838	Tofu, dried, frozen (koyadofu)	3	ounce(s)	85	4.9	408	40.8	12.4	6.1	25.8	3.7	5.7	14.6

PAGE KEY: A-4 = Breads/Baked Goods A-10 = Cereal/Rice/Pasta A-14 = Fruit A-18 = Vegetables/Legumes A-28 = Nuts/Seeds A-30 = Vegetarian A-32 = Dairy A-40 = Eggs A-40 = Seafood A-42 = Meats A-46 = Poultry A-46 = Processed Meats A-48 = Beverages A-52 = Fats/Oils A-54 = Sweets A-56 = Spices/Condiments/Sauces A-60 = Mixed Foods/Soups/Sandwiches A-64 = Fast Food A-84 = Convenience A-86 = Baby Foods

Chol (mg)	Calc (mg)	Iron (mg)	Magn (mg)	Pota (mg)	Sodi (mg)	Zinc (mg)	Vit A (μg)	Thia (mg)	Vit E (mg α)	Ribo (mg)	Niac (mg)	Vit B_6 (mg)	Fola (μg)	Vit C (mg)	Vit B_{12} (μg)	Sele (μg)
0	29	0.96	17.9	127.5	331.5	0.74	368.5	0.09	—	0.07	0.74	0.11	23.0	8.4	0	1.1
0	18	0.75	12.8	126.4	54.4	0.36	380.8	0.18	0.41	0.05	0.92	0.07	20.8	6.5	0	0.9
0	47	1.92	22.4	173.6	4.0	0.39	52.8	0.05	0.37	0.09	0.45	0.13	28.0	17.6	0	0.6
0	14	0.65	7.6	63.0	1.3	0.08	17.0	0.04	0.12	0.02	0.19	0.05	13.2	18.9	0	0.2
0	22	1.34	33.6	228.6	12.0	0.62	4.8	0.12	—	0.08	1.84	0.14	86.4	6.2	0	0.4
0	6	0.31	6.8	112.9	1.4	0.08	15.6	0.04	0.34	0.02	0.32	0.15	10.9	50.6	0	0.2
0	7	0.25	7.5	130.4	2.2	0.09	13.4	0.04	0.27	0.02	0.35	0.16	7.5	59.9	0	0
0	8	0.54	11.3	153.0	3.2	0.13	26.6	0.04	0.31	0.04	0.42	0.12	10.4	109.1	0	0.2
0	5	0.34	9.5	127.2	797.6	0.10	24.5	0.01	0.46	0.02	0.54	0.10	6.8	46.2	0	0.2
0	16	1.28	10.2	131.2	1136.3	0.23	57.8	0.03	0.47	0.03	0.27	0.13	9.5	6.8	0	0.3
0	20	0.85	22.3	394.3	3.7	0.31	18.6	0.05	—	0.04	1.65	0.31	48.4	341.3	0	0.6
0	19	1.06	28.8	219.6	14.4	0.26	3.6	0.16	2.76	0.05	1.32	0.33	25.2	4.8	0	0.8
19	103	0.39	18.6	271.0	543.3	0.29	64.4	0.02	—	0.10	1.16	0.05	8.7	3.8	0	3.3
28	146	0.78	24.5	485.1	530.4	0.85	78.4	0.08	—	0.14	1.22	0.21	16.0	12.1	0	3.3
0	30	2.18	56.6	1080.7	20.2	0.72	2.0	0.12	0.08	0.09	2.84	0.62	56.6	19.4	0	0.8
0	3	0.21	15.3	238.5	3.1	0.18	0	0.06	0.02	0.01	0.85	0.18	5.5	7.8	0	0.2
0	20	4.08	24.9	332.3	12.2	0.28	0.6	0.07	0.02	0.06	1.77	0.35	12.8	7.8	0	0.4
0	7	0.42	29.9	515.4	5.4	0.40	0	0.14	0.01	0.02	1.95	0.40	13.6	17.7	0	0.4
0	13	1.27	34.1	572.0	7.4	0.46	0	0.14	0.01	0.03	2.13	0.44	15.0	18.4	0	—
0	6	0.24	15.6	255.8	3.9	0.21	0	0.07	0.01	0.01	1.02	0.21	7.0	5.8	0	0.2
0	16	1.05	30.8	567.0	8.4	0.39	0	0.08	0.09	0.03	1.34	0.37	16.1	21.2	0	0.4
0	8	0.51	18.2	315.7	271.6	0.26	0	0.09	0.07	0.02	1.55	0.12	19.6	9.3	0	0.1
0	11	0.43	27.3	449.3	266.8	0.37	0	0.13	0.01	0.03	1.80	0.37	12.5	10.1	0	0.4
1	23	0.27	19.9	344.4	349.6	0.31	43.0	0.09	0.44	0.04	1.23	0.25	9.4	11.0	0.1	0.8
2	36	0.21	21.0	164.8	179.5	0.26	49.3	0.09	0.53	0.09	0.90	0.16	8.4	6.8	0.1	5.9
0	22	2.50	54.5	902.9	16.2	0.72	0	0.24	—	0.06	3.46	0.69	24.2	30.5	0	0.8
0	4	0.31	19.4	319.0	5.4	0.25	0	0.10	—	0.01	1.26	0.25	9.3	11.7	0	0.3
0	27	3.44	21.5	377.0	9.3	0.29	0	0.04	0.01	0.04	1.28	0.28	9.9	8.9	0	0.3
0	9	0.41	10.9	199.7	307.2	0.21	0	0.08	0.15	0.02	0.97	0.08	9.0	4.0	0	0.4
14	45	0.47	17.4	252.2	423.7	0.31	43.5	0.02	—	0.06	1.28	0.05	12.4	4.1	0	2.0
15	70	0.70	23.3	463.1	410.4	0.49	0	0.08	—	0.11	1.29	0.22	16.0	13.0	0	2.0
0	18	0.69	11.0	281.8	1.2	0.28	306.3	0.03	0.98	0.09	0.50	0.05	11.0	5.8	0	0.2
0	32	1.70	28.2	252.4	6.1	0.20	953.1	0.02	1.29	0.06	0.45	0.06	14.7	5.1	0	0.5
0	8	0.23	5.2	120.8	8.8	0.25	0.4	0.01	0.90	0.01	0.10	0.02	24.0	3.2	0	0.4
0	8	0.23	5.2	120.8	8.8	0.25	0.4	0.01	0.90	0.01	0.10	0.02	24.0	3.2	0	0.4
0	7	0.09	2.7	62.9	10.5	0.07	0	0.00	0.00	0.01	0.06	0.01	6.8	4.0	0	0.2
0	41	0.45	19.6	277.1	17.0	0.30	0	0.07	0.27	0.04	0.61	0.09	12.8	16.0	0	0.6
0	35	1.73	15.3	200.6	780.0	0.22	1.2	0.03	0.17	0.03	0.17	0.15	28.3	17.3	0	0.7
0	67	1.12	48.4	35.6	93.2	0.48	2.4	0.02	0.32	0.04	0.16	0.00	72.0	1.2	0	0.3
0	9	2.14	14.6	102.2	78.6	0.15	2.2	0.18	0.38	0.28	0.96	0.03	7.1	0.8	0	0.5
0	11	0.36	6.3	100.2	3.6	0.12	18.0	0.02	—	0.01	0.06	0.09	10.2	2.4	—	0.4
0	88	4.42	74.0	442.9	0.9	0.98	0	0.13	0.30	0.24	0.34	0.20	46.4	1.5	0	6.3
0	120	3.39	196.1	1173.0	1.7	4.10	0	0.36	—	0.64	0.90	0.19	176.3	4.0	0	16.6
0	119	3.35	124.7	1264.2	140.2	2.70	8.6	0.08	0.78	0.12	1.21	0.17	181.5	1.9	0	16.4
0	52	0.25	60.4	356.6	8.8	1.32	0.6	0.26	—	0.12	0.69	0.10	79.9	7.5	0	0.4
0	60	1.53	60.0	283.2	122.4	0.28	0	0.14	0.26	0.16	1.23	0.18	43.2	0	0	11.5
0	310	8.27	50.2	17.0	5.1	4.16	22.1	0.42	—	0.27	1.01	0.24	78.2	0.6	0	46.2

TABLE A-1 Table of Food Composition *(continued)*

(Computer code is for Cengage Diet Analysis program)
(For purposes of calculations, use "0" for t, <1, <.1, <.01, etc.)

DA+ Code	Food Description	Quantity	Measure	Wt (g)	H_2O (g)	Ener (kcal)	Prot (g)	Carb (g)	Fiber (g)	Fat (g)	Fat Breakdown (g) Sat	Mono	Poly
	Vegetables, Legumes—*continued*												
13844	Tofu, extra firm	3	ounce(s)	85	—	86	8.6	2.2	1.1	4.3	0.5	0.9	2.8
13843	Tofu, firm	3	ounce(s)	85	—	75	7.5	2.2	0.5	3.2	0	0.9	2.3
1816	Tofu, firm, with calcium sulfate and magnesium chloride (nigari)	3	ounce(s)	85	72.2	60	7.0	1.4	0.8	3.5	0.7	1.0	1.5
1817	Tofu, fried	3	ounce(s)	85	43.0	230	14.6	8.9	3.3	17.2	2.5	3.8	9.7
13841	Tofu, silken	3	ounce(s)	85	—	42	3.7	1.9	0	2.3	0.5	—	—
13842	Tofu, soft	3	ounce(s)	85	—	65	6.5	1.1	0.5	3.2	0.5	1.1	2.2
1671	Tofu, soft, with calcium sulfate and magnesium chloride (nigari)	3	ounce(s)	85	74.2	52	5.6	1.5	0.2	3.1	0.5	0.7	1.8
	Spinach												
663	Canned, drained	½	cup(s)	107	98.2	25	3.0	3.6	2.6	0.5	0.1	0	0.2
660	Chopped, boiled, drained	½	cup(s)	90	82.1	21	2.7	3.4	2.2	0.2	0	0	0.1
661	Chopped, frozen, boiled, drained	½	cup(s)	95	84.5	32	3.8	4.6	3.5	0.8	0.1	0	0.4
662	Leaf, frozen, boiled, drained	½	cup(s)	95	84.5	32	3.8	4.6	3.5	0.8	0.1	0	0.4
659	Raw, chopped	1	cup(s)	30	27.4	7	0.9	1.1	0.7	0.1	0	0	0
8470	Trimmed leaves	1	cup(s)	32	27.5	3	0.9	0	2.8	0.1	—	—	—
	Squash												
1662	Acorn winter, baked	½	cup(s)	103	85.0	57	1.1	14.9	4.5	0.1	0	0	0.1
29702	Acorn winter, boiled, mashed	½	cup(s)	123	109.9	42	0.8	10.8	3.2	0.1	0	0	0
29451	Butternut, frozen, boiled	½	cup(s)	122	106.9	47	1.5	12.2	1.8	0.1	0	0	0
1661	Butternut winter, baked	½	cup(s)	102	89.5	41	0.9	10.7	3.4	0.1	0	0	0
32773	Butternut winter, frozen, boiled, mashed, no salt added	½	cup(s)	121	106.4	47	1.5	12.2	—	0.1	0	0	0
29700	Crookneck and straightneck summer, boiled, drained	½	cup(s)	65	60.9	12	0.6	2.6	1.2	0.1	0	0	0.1
29703	Hubbard winter, baked	½	cup(s)	102	86.8	51	2.5	11.0	—	0.6	0.1	0	0.3
1660	Hubbard winter, boiled, mashed	½	cup(s)	118	107.5	35	1.7	7.6	3.4	0.4	0.1	0	0.2
29704	Spaghetti winter, boiled, drained, or baked	½	cup(s)	78	71.5	21	0.5	5.0	1.1	0.2	0	0	0.1
664	Summer, all varieties, sliced, boiled, drained	½	cup(s)	90	84.3	18	0.8	3.9	1.3	0.3	0.1	0	0.1
665	Winter, all varieties, baked, mashed	½	cup(s)	103	91.4	38	0.9	9.1	2.9	0.4	0.1	0	0.2
1112	Zucchini summer, boiled, drained	½	cup(s)	90	85.3	14	0.6	3.5	1.3	0	0	0	0
1113	Zucchini summer, frozen, boiled, drained	½	cup(s)	112	105.6	19	1.3	4.0	1.4	0.1	0	0	0.1
	Sweet potatoes												
666	Baked, peeled	½	cup(s)	100	75.8	90	2.0	20.7	3.3	0.2	0	0	0.1
667	Boiled, mashed	½	cup(s)	164	131.4	125	2.2	29.1	4.1	0.2	0.1	0	0.1
668	Candied, home recipe	½	cup(s)	91	61.1	132	0.8	25.4	2.2	3.0	1.2	0.6	0.1
670	Canned, vacuum pack	½	cup(s)	100	76.0	91	1.7	21.1	1.8	0.2	0	0	0.1
2765	Frozen, baked	½	cup(s)	88	64.5	88	1.5	20.5	1.6	0.1	0	0	0
1136	Yams, baked or boiled, drained	½	cup(s)	68	47.7	79	1.0	18.7	2.7	0.1	0	0	0
32785	**Taro shoots, cooked, no salt added**	½	cup(s)	70	66.7	10	0.5	2.2	—	0.1	0	0	0
	Tomatillo												
8774	Raw	2	item(s)	68	62.3	22	0.7	4.0	1.3	0.7	0.1	0.1	0.3
8777	Raw, chopped	½	cup(s)	66	60.5	21	0.6	3.9	1.3	0.7	0.1	0.1	0.3
	Tomato												
16846	Cherry, fresh	5	item(s)	85	80.3	15	0.7	3.3	1.0	0.2	0	0	0.1
671	Fresh, ripe, red	1	item(s)	123	116.2	22	1.1	4.8	1.5	0.2	0	0	0.1
675	Juice, canned	½	cup(s)	122	114.1	21	0.9	5.2	0.5	0.1	0	0	0
75	Juice, no salt added	½	cup(s)	122	114.1	21	0.9	5.2	0.5	0.1	0	0	0
1699	Paste, canned	2	tablespoon(s)	33	24.1	27	1.4	6.2	1.3	0.2	0	0	0.1
1700	Puree, canned	¼	cup(s)	63	54.9	24	1.0	5.6	1.2	0.1	0	0	0.1
1118	Red, boiled	½	cup(s)	120	113.2	22	1.1	4.8	0.8	0.1	0	0	0.1
3952	Red, diced	½	cup(s)	90	85.1	16	0.8	3.5	1.1	0.2	0	0	0.1
1120	Red, stewed, canned	½	cup(s)	128	116.7	33	1.2	7.9	1.3	0.2	0	0	0.1
1125	Sauce, canned	¼	cup(s)	61	55.6	15	0.8	3.3	0.9	0.1	0	0	0
8778	Sun dried	½	cup(s)	27	3.9	70	3.8	15.1	3.3	0.8	0.1	0.1	0.3
8783	Sun dried in oil, drained	¼	cup(s)	28	14.8	59	1.4	6.4	1.6	3.9	0.5	2.4	0.6
	Turnips												
678	Turnip greens, chopped, boiled, drained	½	cup(s)	72	67.1	14	0.8	3.1	2.5	0.2	0	0	0.1

PAGE KEY: A-4 = Breads/Baked Goods A-10 = Cereal/Rice/Pasta A-14 = Fruit A-18 = Vegetables/Legumes A-28 = Nuts/Seeds A-30 = Vegetarian A-32 = Dairy A-40 = Eggs A-40 = Seafood A-42 = Meats A-46 = Poultry A-46 = Processed Meats A-48 = Beverages A-52 = Fats/Oils A-54 = Sweets A-56 = Spices/Condiments/Sauces A-60 = Mixed Foods/Soups/Sandwiches A-64 = Fast Food A-84 = Convenience A-86 = Baby Foods

Chol (mg)	Calc (mg)	Iron (mg)	Magn (mg)	Pota (mg)	Sodi (mg)	Zinc (mg)	Vit A (μg)	Thia (mg)	Vit E (mg α)	Ribo (mg)	Niac (mg)	Vit B_6 (mg)	Fola (μg)	Vit C (mg)	Vit B_{12} (μg)	Sele (μg)
0	65	1.16	84.1	—	0	—	0	—	—	—	—	—	—	0	0	—
0	108	1.16	56.1	—	0	—	0	—	—	—	—	—	—	0	0	—
0	171	1.36	31.5	125.9	10.2	0.70	0	0.05	0.01	0.05	0.08	0.06	16.2	0.2	0	8.4
0	316	4.14	51.0	124.2	13.6	1.69	0.9	0.14	0.03	0.04	0.08	0.08	23.0	0	0	24.2
0	56	0.34	33.1	—	4.7	—	0	—	—	—	—	—	—	0	1.7	—
0	108	1.16	35.5	—	0	—	0	—	—	—	—	—	—	0	1.9	—
0	94	0.94	23.0	102.1	6.8	0.54	0	0.04	0.01	0.03	0.45	0.04	37.4	0.2	0	7.6
0	136	2.45	81.3	370.2	28.9	0.48	524.3	0.02	2.08	0.14	0.41	0.11	104.8	15.3	0	1.5
0	122	3.21	78.3	419.4	63.0	0.68	471.6	0.08	1.87	0.21	0.44	0.21	131.4	8.8	0	1.4
0	145	1.86	77.9	286.9	92.2	0.46	572.9	0.07	3.36	0.16	0.41	0.12	115.0	2.1	0	5.2
0	145	1.86	77.9	286.9	92.2	0.46	572.9	0.07	3.36	0.16	0.41	0.12	115.0	2.1	0	5.2
0	30	0.81	23.7	167.4	23.7	0.16	140.7	0.02	0.61	0.06	0.22	0.06	58.2	8.4	0	0.3
0	25	2.13	25.5	134.1	38.0	0.18	—	0.03	—	0.05	0.18	0.07	0	7.5	0	—
0	45	0.95	44.1	447.9	4.1	0.17	21.5	0.17	—	0.01	0.90	0.19	19.5	11.1	0	0.7
0	32	0.68	31.9	322.2	3.7	0.13	50.2	0.12	—	0.01	0.65	0.14	13.5	8.0	0	0.5
0	23	0.70	10.9	161.9	2.4	0.14	203.3	0.06	0.14	0.05	0.56	0.08	19.5	4.3	0	0.6
0	42	0.61	29.6	289.6	4.1	0.13	569.1	0.07	1.31	0.01	0.99	0.12	19.4	15.4	0	0.5
0	23	0.70	10.9	161.2	2.4	0.14	202.4	0.06	—	0.05	0.56	0.08	19.4	4.2	0	0.6
0	14	0.31	13.6	137.1	1.3	0.19	5.2	0.03	—	0.02	0.29	0.07	14.9	5.4	0	0.1
0	17	0.48	22.4	365.1	8.2	0.15	308.0	0.07	—	0.04	0.57	0.17	16.3	9.7	0	0.6
0	12	0.33	15.3	252.5	5.9	0.11	236.0	0.05	0.14	0.03	0.39	0.12	11.8	7.7	0	0.4
0	16	0.26	8.5	90.7	14.0	0.15	4.7	0.02	0.09	0.01	0.62	0.07	6.2	2.7	0	0.2
0	24	0.32	21.6	172.8	0.9	0.35	9.9	0.04	0.12	0.03	0.46	0.05	18.0	5.0	0	0.2
0	23	0.45	13.3	247.0	1.0	0.23	267.5	0.02	0.12	0.07	0.51	0.17	20.5	9.8	0	0.4
0	12	0.32	19.8	227.7	2.7	0.16	50.4	0.04	0.11	0.04	0.39	0.07	15.3	4.1	0	0.2
0	19	0.54	14.5	216.3	2.2	0.22	10	0.05	0.13	0.04	0.43	0.05	8.9	4.1	0	0.2
0	38	0.69	27.0	475.0	36.0	0.32	961.0	0.10	0.71	0.10	1.48	0.28	6.0	19.6	0	0.2
0	44	1.18	29.5	377.2	44.3	0.33	1290.7	0.09	1.54	0.08	0.88	0.27	9.8	21.0	0	0.3
7	24	1.03	10.0	172.6	63.9	0.13	0	0.01	—	0.03	0.36	0.03	10.0	6.1	0	0.7
0	22	0.89	22.0	312.0	53.0	0.18	399.0	0.04	1.00	0.06	0.74	0.19	17.0	26.4	0	0.7
0	31	0.47	18.4	330.1	7.0	0.26	913.3	0.05	0.67	0.04	0.49	0.16	19.3	8.0	0	0.5
0	10	0.35	12.2	455.6	5.4	0.13	4.1	0.06	0.23	0.01	0.37	0.15	10.9	8.2	0	0.5
0	10	0.28	5.6	240.8	1.4	0.37	2.1	0.02	—	0.03	0.56	0.07	2.1	13.2	0	0.7
0	5	0.42	13.6	182.2	0.7	0.15	4.1	0.03	0.25	0.02	1.25	0.03	4.8	8.0	0	0.3
0	5	0.41	13.2	176.9	0.7	0.15	4.0	0.03	0.25	0.02	1.22	0.04	4.6	7.7	0	0.3
0	9	0.22	9.4	201.5	4.3	0.14	35.7	0.03	0.45	0.01	0.50	0.06	12.8	10.8	0	0
0	12	0.33	13.5	291.5	6.2	0.20	51.7	0.04	0.66	0.02	0.73	0.09	18.5	15.6	0	0
0	12	0.52	13.4	278.2	326.8	0.18	27.9	0.06	0.39	0.04	0.82	0.14	24.3	22.2	0	0.4
0	12	0.52	13.4	278.2	12.2	0.18	27.9	0.06	0.39	0.04	0.82	0.14	24.3	22.2	0	0.4
0	12	0.97	13.8	332.6	259.1	0.20	24.9	0.02	1.41	0.05	1.00	0.07	3.9	7.2	0	1.7
0	11	1.11	14.4	274.4	249.4	0.22	16.3	0.01	1.23	0.05	0.91	0.07	6.9	6.6	0	0.4
0	13	0.82	10.8	261.6	13.2	0.17	28.8	0.04	0.67	0.03	0.64	0.10	15.6	27.4	0	0.6
0	9	0.24	9.9	213.3	4.5	0.15	37.8	0.03	0.48	0.01	0.53	0.07	13.5	11.4	0	0
0	43	1.70	15.3	263.9	281.8	0.22	11.5	0.06	1.06	0.04	0.91	0.02	6.4	10.1	0	0.8
0	8	0.62	9.8	201.9	319.6	0.12	10.4	0.01	0.87	0.04	0.59	0.06	6.7	4.3	0	0.1
0	30	2.45	52.4	925.3	565.7	0.53	11.9	0.14	0.00	0.13	2.44	0.09	18.4	10.6	0	1.5
0	13	0.73	22.3	430.4	73.2	0.21	17.6	0.05	—	0.10	0.99	0.08	6.3	28.0	0	0.8
0	99	0.58	15.8	146.2	20.9	0.10	274.3	0.03	1.35	0.05	0.30	0.13	85.0	19.7	0	0.6

TABLE A-1 Table of Food Composition *(continued)*

(Computer code is for Cengage Diet Analysis program)
(For purposes of calculations, use "0" for t, <1, <.1, <.01, etc.)

DA+ Code	Food Description	Quantity	Measure	Wt (g)	H_2O (g)	Ener (kcal)	Prot (g)	Carb (g)	Fiber (g)	Fat (g)	Fat Breakdown (g) Sat	Mono	Poly
Vegetables, Legumes—*continued*													
679	Turnip greens, frozen, chopped, boiled, drained	½	cup(s)	82	74.1	24	2.7	4.1	2.8	0.3	0.1	0	0.1
677	Turnips, cubed, boiled, drained	½	cup(s)	78	73.0	17	0.6	3.9	1.6	0.1	0	0	0
	Vegetables, mixed												
1132	Canned, drained	½	cup(s)	82	70.9	40	2.1	7.5	2.4	0.2	0	0	0.1
680	Frozen, boiled, drained	½	cup(s)	91	75.7	59	2.6	11.9	4.0	0.1	0	0	0.1
7489	V8 100% vegetable juice	½	cup(s)	120	—	25	1.0	5.0	1.0	0	0	0	0
7490	V8 low sodium vegetable juice	½	cup(s)	120	—	25	0	6.5	1.0	0	0	0	0
7491	V8 spicy hot vegetable juice	½	cup(s)	120	—	25	1.0	5.0	0.5	0	0	0	0
	Water chestnuts												
31073	Sliced, drained	½	cup(s)	75	70.0	20	0	5.0	1.0	0	0	0	0
31087	Whole	½	cup(s)	75	70.0	20	0	5.0	1.0	0	0	0	0
1135	**Watercress**	1	cup(s)	34	32.3	4	0.8	0.4	0.2	0	0	0	0
Nuts, Seeds, and Products													
	Almonds												
32940	Almond butter with salt added	1	tablespoon(s)	16	0.2	101	2.4	3.4	0.6	9.5	0.9	6.1	2.0
1137	Almond butter, no salt added	1	tablespoon(s)	16	0.2	101	2.4	3.4	0.6	9.5	0.9	6.1	2.0
32886	Blanched	¼	cup(s)	36	1.6	211	8.0	7.2	3.8	18.3	1.4	11.7	4.4
32887	Dry roasted, no salt added	¼	cup(s)	35	0.9	206	7.6	6.7	4.1	18.2	1.4	11.6	4.4
29724	Dry roasted, salted	¼	cup(s)	35	0.9	206	7.6	6.7	4.1	18.2	1.4	11.6	4.4
29725	Oil roasted, salted	¼	cup(s)	39	1.1	238	8.3	6.9	4.1	21.7	1.7	13.7	5.3
508	Slivered	¼	cup(s)	27	1.3	155	5.7	5.9	3.3	13.3	1.0	8.3	3.3
1138	**Beechnuts, dried**	¼	cup(s)	57	3.8	328	3.5	19.1	5.3	28.5	3.3	12.5	11.4
517	**Brazil nuts, dried, unblanched**	¼	cup(s)	35	1.2	230	5.0	4.3	2.6	23.3	5.3	8.6	7.2
1166	**Breadfruit seeds, roasted**	¼	cup(s)	57	28.3	118	3.5	22.8	3.4	1.5	0.4	0.2	0.8
1139	**Butternuts, dried**	¼	cup(s)	30	1.0	184	7.5	3.6	1.4	17.1	0.4	3.1	12.8
	Cashews												
32931	Cashew butter with salt added	1	tablespoon(s)	16	0.5	94	2.8	4.4	0.3	7.9	1.6	4.7	1.3
32889	Cashew butter, no salt added	1	tablespoon(s)	16	0.5	94	2.8	4.4	0.3	7.9	1.6	4.7	1.3
1140	Dry roasted	¼	cup(s)	34	0.6	197	5.2	11.2	1.0	15.9	3.1	9.4	2.7
518	Oil roasted	¼	cup(s)	32	1.1	187	5.4	9.6	1.1	15.4	2.7	8.4	2.8
	Coconut, shredded												
32896	Dried, not sweetened	¼	cup(s)	23	0.7	152	1.6	5.4	3.8	14.9	13.2	0.6	0.2
1153	Dried, shredded, sweetened	¼	cup(s)	23	2.9	116	0.7	11.1	1.0	8.3	7.3	0.4	0.1
520	Shredded	¼	cup(s)	20	9.4	71	0.7	3.0	1.8	6.7	5.9	0.3	0.1
	Chestnuts												
1152	Chinese, roasted	¼	cup(s)	36	14.6	87	1.6	19.0	—	0.4	0.1	0.2	0.1
32895	European, boiled and steamed	¼	cup(s)	46	31.3	60	0.9	12.8	—	0.6	0.1	0.2	0.2
32911	European, roasted	¼	cup(s)	36	14.5	88	1.1	18.9	1.8	0.8	0.1	0.3	0.3
32922	Japanese, boiled and steamed	¼	cup(s)	36	31.0	20	0.3	4.5	—	0.1	0	0	0
32923	Japanese, roasted	¼	cup(s)	36	18.1	73	1.1	16.4	—	0.3	0	0.1	0.1
4958	**Flax seeds or linseeds**	¼	cup(s)	43	3.3	225	8.4	12.3	11.9	17.7	1.7	3.2	12.6
32904	**Ginkgo nuts, dried**	¼	cup(s)	39	4.8	136	4.0	28.3	—	0.8	0.1	0.3	0.3
	Hazelnuts or filberts												
32901	Blanched	¼	cup(s)	30	1.7	189	4.1	5.1	3.3	18.3	1.4	14.5	1.7
32902	Dry roasted, no salt added	¼	cup(s)	30	0.8	194	4.5	5.3	2.8	18.7	1.3	14.0	2.5
1156	**Hickory nuts, dried**	¼	cup(s)	30	0.8	197	3.8	5.5	1.9	19.3	2.1	9.8	6.6
	Macadamias												
32905	Dry roasted, no salt added	¼	cup(s)	34	0.5	241	2.6	4.5	2.7	25.5	4.0	19.9	0.5
32932	Dry roasted, with salt added	¼	cup(s)	34	0.5	240	2.6	4.3	2.7	25.5	4.0	19.9	0.5
1157	Raw	¼	cup(s)	34	0.5	241	2.6	4.6	2.9	25.4	4.0	19.7	0.5
	Mixed nuts												
1159	With peanuts, dry roasted	¼	cup(s)	34	0.6	203	5.9	8.7	3.1	17.6	2.4	10.8	3.7
32933	With peanuts, dry roasted, with salt added	¼	cup(s)	34	0.6	203	5.9	8.7	3.1	17.6	2.4	10.8	3.7
32906	Without peanuts, oil roasted, no salt added	¼	cup(s)	36	1.1	221	5.6	8.0	2.0	20.2	3.3	11.9	4.1
	Peanuts												
2807	Dry roasted	¼	cup(s)	37	0.6	214	8.6	7.9	2.9	18.1	2.5	9.0	5.7
2806	Dry roasted, salted	¼	cup(s)	37	0.6	214	8.6	7.9	2.9	18.1	2.5	9.0	5.7
1763	Oil roasted, salted	¼	cup(s)	36	0.5	216	10.1	5.5	3.4	18.9	3.1	9.4	5.5
1884	Peanut butter, chunky	1	tablespoon(s)	16	0.2	94	3.8	3.5	1.3	8.0	1.3	3.9	2.4
30303	Peanut butter, low sodium	1	tablespoon(s)	16	0.2	95	4.0	3.1	0.9	8.2	1.8	3.9	2.2
30305	Peanut butter, reduced fat	1	tablespoon(s)	18	0.2	94	4.7	6.4	0.9	6.1	1.3	2.9	1.8

PAGE KEY: A-4 = Breads/Baked Goods A-10 = Cereal/Rice/Pasta A-14 = Fruit A-18 = Vegetables/Legumes A-28 = Nuts/Seeds A-30 = Vegetarian A-32 = Dairy A-40 = Eggs A-40 = Seafood A-42 = Meats A-46 = Poultry A-46 = Processed Meats A-48 = Beverages A-52 = Fats/Oils A-54 = Sweets A-56 = Spices/Condiments/Sauces A-60 = Mixed Foods/Soups/Sandwiches A-64 = Fast Food A-84 = Convenience A-86 = Baby Foods

Chol (mg)	Calc (mg)	Iron (mg)	Magn (mg)	Pota (mg)	Sodi (mg)	Zinc (mg)	Vit A (µg)	Thia (mg)	Vit E (mg α)	Ribo (mg)	Niac (mg)	Vit B_6 (mg)	Fola (µg)	Vit C (mg)	Vit B_{12} (µg)	Sele (µg)
0	125	1.59	21.3	183.7	12.3	0.34	441.2	0.04	2.18	0.06	0.38	0.06	32.0	17.9	0	1.0
0	26	0.14	7.0	138.1	12.5	0.09	0	0.02	0.02	0.02	0.23	0.05	7.0	9.0	0	0.2
0	22	0.86	13.0	237.2	121.4	0.33	475.1	0.04	0.24	0.04	0.47	0.06	19.6	4.1	0	0.2
0	23	0.74	20.0	153.8	31.9	0.44	194.7	0.06	0.34	0.10	0.77	0.06	17.3	2.9	0	0.3
0	20	0.36	12.9	260.0	310.0	0.24	100.0	0.05	—	0.03	0.87	0.17	—	30.0	0	—
0	20	0.36	—	450.0	70.0	—	100.0	0.02	—	0.02	0.75	—	—	30.0	0	—
0	20	0.36	12.9	240.0	360.0	0.24	50.0	0.05	—	0.03	0.88	0.17	—	15.0	0	—
0	7	0.00	—	—	5.0	—	0	—	—	—	—	—	—	2.0	—	—
0	7	0.00	—	—	5.0	—	0	—	—	—	—	—	—	2.0	—	—
0	41	0.06	7.1	112.2	13.9	0.03	54.4	0.03	0.34	0.04	0.06	0.04	3.1	14.6	0	0.3
0	43	0.59	48.5	121.3	72.0	0.49	0	0.02	4.16	0.10	0.46	0.01	10.4	0.1	0	0.8
0	43	0.59	48.5	121.3	1.8	0.48	0	0.02	—	0.09	0.46	0.01	10.4	0.1	0	—
0	78	1.34	99.7	249.0	10.2	1.13	0	0.07	8.95	0.20	1.32	0.04	10.9	0	0	1.0
0	92	1.55	98.7	257.4	0.3	1.22	0	0.02	8.97	0.29	1.32	0.04	11.4	0	0	1.0
0	92	1.55	98.7	257.4	117.0	1.22	0	0.02	8.97	0.29	1.32	0.04	11.4	0	0	1.0
0	114	1.44	107.5	274.4	133.1	1.20	0	0.03	10.19	0.30	1.43	0.04	10.6	0	0	1.1
0	71	1.00	72.4	190.4	0.3	0.83	0	0.05	7.07	0.27	0.91	0.03	13.5	0	0	0.7
0	1	1.39	0	579.7	21.7	0.20	0	0.16	—	0.20	0.48	0.38	64.4	8.8	0	4.0
0	56	0.85	131.6	230.7	1.1	1.42	0	0.21	2.00	0.01	0.10	0.03	7.7	0.2	0	671.0
0	49	0.50	35.3	616.7	15.9	0.58	8.5	0.22	—	0.12	4.20	0.22	33.6	4.3	0	8.0
0	16	1.21	71.1	126.3	0.3	0.94	1.8	0.12	—	0.04	0.31	0.17	19.8	1.0	0	5.2
0	7	0.81	41.3	87.4	98.2	0.83	0	0.05	0.15	0.03	0.26	0.04	10.9	0	0	1.8
0	7	0.81	41.3	87.4	2.4	0.83	0	0.05	—	0.03	0.26	0.04	10.9	0	0	1.8
0	15	2.06	89.1	193.5	5.5	1.92	0	0.07	0.32	0.07	0.48	0.09	23.6	0	0	4.0
0	14	1.95	88.0	203.8	4.2	1.73	0	0.12	0.30	0.07	0.56	0.10	8.1	0.1	0	6.5
0	6	0.76	20.7	125.2	8.5	0.46	0	0.01	0.10	0.02	0.13	0.07	2.1	0.3	0	4.3
0	3	0.45	11.6	78.4	60.9	0.42	0	0.01	0.09	0.00	0.11	0.06	1.9	0.2	0	3.9
0	3	0.48	6.4	71.2	4.0	0.21	0	0.01	0.04	0.00	0.11	0.01	5.2	0.7	0	2.0
0	7	0.54	32.6	173.0	1.4	0.33	0	0.05	—	0.03	0.54	0.15	26.1	13.9	0	2.6
0	21	0.80	24.8	328.9	12.4	0.11	0.5	0.06	—	0.03	0.32	0.10	17.5	12.3	0	—
0	10	0.32	11.8	211.6	0.7	0.20	0.4	0.08	0.18	0.05	0.48	0.18	25.0	9.3	0	0.4
0	4	0.19	6.5	42.8	1.8	0.14	0.4	0.04	—	0.01	0.19	0.03	6.1	3.4	0	—
0	13	0.75	23.2	154.8	6.9	0.51	1.4	0.15	—	—	0.24	0.14	21.4	10.1	0	—
0	142	2.13	156.1	354.0	11.9	1.83	0	0.06	0.14	0.06	0.59	0.39	118.4	0.5	0	2.3
0	8	0.62	20.7	390.2	5.1	0.26	21.5	0.17	—	0.07	4.58	0.25	41.4	11.4	0	—
0	45	0.98	48.0	197.4	0	0.66	0.6	0.14	5.25	0.03	0.46	0.17	23.4	0.6	0	1.2
0	37	1.31	51.9	226.5	0	0.74	0.9	0.10	4.58	0.03	0.61	0.18	26.4	1.1	0	1.2
0	18	0.64	51.9	130.8	0.3	1.29	2.1	0.26	—	0.04	0.27	0.06	12.0	0.6	0	2.4
0	23	0.88	39.5	121.6	1.3	0.43	0	0.23	0.19	0.02	0.76	0.12	3.4	0.2	0	3.9
0	23	0.88	39.5	121.6	88.8	0.43	0	0.23	0.19	0.02	0.76	0.12	3.4	0.2	0	3.9
0	28	1.24	43.6	123.3	1.7	0.44	0	0.40	0.18	0.05	0.83	0.09	3.7	0.4	0	1.2
0	24	1.27	77.1	204.5	4.1	1.30	0.3	0.07	—	0.07	1.61	0.10	17.1	0.1	0	1.0
0	24	1.26	77.1	204.5	229.1	1.30	0	0.06	3.74	0.06	1.61	0.10	17.1	0.1	0	2.6
0	38	0.92	90.4	195.8	4.0	1.67	0.4	0.18	—	0.17	0.70	0.06	20.2	0.2	0	—
0	20	0.82	64.2	240.2	2.2	1.20	0	0.16	2.52	0.03	4.93	0.09	52.9	0	0	2.7
0	20	0.82	64.2	240.2	296.7	1.20	0	0.16	2.84	0.03	4.93	0.09	52.9	0	0	2.7
0	22	0.54	63.4	261.4	115.2	1.18	0	0.03	2.49	0.03	4.97	0.16	43.2	0.3	0	1.2
0	7	0.30	25.6	119.2	77.8	0.45	0	0.02	1.01	0.02	2.19	0.07	14.7	0	0	1.3
0	6	0.29	25.4	107.0	2.7	0.47	0	0.01	1.23	0.02	2.14	0.07	11.8	0	0	1.2
0	6	0.34	30.6	120.4	97.2	0.50	0	0.05	1.20	0.01	2.63	0.06	10.8	0	0	1.4

TABLE A-1 Table of Food Composition *(continued)*

(Computer code is for Cengage Diet Analysis program)
(For purposes of calculations, use "0" for t, <1, <.1, <.01, etc.)

DA+ Code	Food Description	Quantity	Measure	Wt (g)	H_2O (g)	Ener (kcal)	Prot (g)	Carb (g)	Fiber (g)	Fat (g)	Fat Breakdown (g)		
											Sat	Mono	Poly
Nuts, Seeds, and Products—*continued*													
524	Peanut butter, smooth	1	tablespoon(s)	16	0.3	94	4.0	3.1	1.0	8.1	1.7	3.9	2.3
2804	Raw	¼	cup(s)	37	2.4	207	9.4	5.9	3.1	18.0	2.5	8.9	5.7
	Pecans												
32907	Dry roasted, no salt added	¼	cup(s)	28	0.3	198	2.6	3.8	2.6	20.7	1.8	12.3	5.7
32936	Dry roasted, with salt added	¼	cup(s)	27	0.3	192	2.6	3.7	2.5	20.0	1.7	11.9	5.6
1162	Oil roasted	¼	cup(s)	28	0.3	197	2.5	3.6	2.6	20.7	2.0	11.3	6.5
526	Raw	¼	cup(s)	27	1.0	188	2.5	3.8	2.6	19.6	1.7	11.1	5.9
12973	**Pine nuts or pignolia, dried**	1	tablespoon(s)	9	0.2	58	1.2	1.1	0.3	5.9	0.4	1.6	2.9
	Pistachios												
1164	Dry roasted	¼	cup(s)	31	0.6	176	6.6	8.5	3.2	14.1	1.7	7.4	4.3
32938	Dry roasted, with salt added	¼	cup(s)	32	0.6	182	6.8	8.6	3.3	14.7	1.8	7.7	4.4
1167	**Pumpkin or squash seeds, roasted**	¼	cup(s)	57	4.0	296	18.7	7.6	2.2	23.9	4.5	7.4	10.9
	Sesame												
32912	Sesame butter paste	1	tablespoon(s)	16	0.3	94	2.9	3.8	0.9	8.1	1.1	3.1	3.6
32941	Tahini or sesame butter	1	tablespoon(s)	15	0.5	89	2.6	3.2	0.7	8.0	1.1	3.0	3.5
1169	Whole, roasted, toasted	3	tablespoon(s)	10	0.3	54	1.6	2.4	1.3	4.6	0.6	1.7	2.0
	Soy nuts												
34173	Deep sea salted	¼	cup(s)	28	—	119	11.9	8.9	4.9	4.0	1.0	—	—
34174	Unsalted	¼	cup(s)	28	—	119	11.9	8.9	4.9	4.0	0	—	—
	Sunflower seeds												
528	Kernels, dried	1	tablespoon(s)	9	0.4	53	1.9	1.8	0.8	4.6	0.4	1.7	2.1
29721	Kernels, dry roasted, salted	1	tablespoon(s)	8	0.1	47	1.5	1.9	0.7	4.0	0.4	0.8	2.6
29723	Kernels, toasted, salted	1	tablespoon(s)	8	0.1	52	1.4	1.7	1.0	4.8	0.5	0.9	3.1
32928	Sunflower seed butter with salt added	1	tablespoon(s)	16	0.2	93	3.1	4.4	—	7.6	0.8	1.5	5.0
	Trail mix												
4646	Trail mix	¼	cup(s)	38	3.5	173	5.2	16.8	2.0	11.0	2.1	4.7	3.6
4647	Trail mix with chocolate chips	¼	cup(s)	38	2.5	182	5.3	16.8	—	12.0	2.3	5.1	4.2
4648	Tropical trail mix	¼	cup(s)	35	3.2	142	2.2	23.0	—	6.0	3.0	0.9	1.8
	Walnuts												
529	Dried black, chopped	¼	cup(s)	31	1.4	193	7.5	3.1	2.1	18.4	1.1	4.7	11.0
531	English or Persian	¼	cup(s)	29	1.2	191	4.5	4.0	2.0	19.1	1.8	2.6	13.8
Vegetarian Foods													
	Prepared												
34222	Brown rice and tofu stir-fry (vegan)	8	ounce(s)	227	244.4	302	16.5	18.0	3.2	21.0	1.7	4.7	13.4
34368	Cheese enchilada casserole (lacto)	8	ounce(s)	227	80.3	385	16.6	38.4	4.1	17.8	9.5	6.1	1.1
34247	Five bean casserole (vegan)	8	ounce(s)	227	175.8	178	5.9	26.6	6.0	5.8	1.1	2.5	1.9
34261	Lentil stew (vegan)	8	ounce(s)	227	227.9	188	11.5	35.9	11.0	0.7	0.1	0.1	0.3
34397	Macaroni and cheese (lacto)	8	ounce(s)	227	352.1	391	18.1	37.1	1.0	18.7	9.8	6.0	1.8
34238	Steamed rice and vegetables (vegan)	8	ounce(s)	227	222.9	587	11.2	87.9	5.8	23.1	4.1	8.7	9.1
34308	Tofu rice burgers (ovo-lacto)	1	piece(s)	218	77.6	435	22.4	68.6	5.6	8.4	1.7	2.4	3.5
34276	Vegan spinach enchiladas (vegan)	1	piece(s)	82	59.2	93	4.9	14.5	1.8	2.4	0.3	0.6	1.3
34243	Vegetable chow mein (vegan)	8	ounce(s)	227	163.3	166	6.5	22.1	2.0	6.4	0.7	2.7	2.5
34454	Vegetable lasagna (lacto)	8	ounce(s)	227	178.9	208	13.7	29.9	2.6	4.1	2.3	1.1	0.3
34339	Vegetable marinara (vegan)	8	ounce(s)	252	200.7	104	3.0	16.7	1.4	3.1	0.4	1.4	1.0
34356	Vegetable rice casserole (lacto)	8	ounce(s)	227	178.9	238	9.7	24.4	4.0	12.5	4.9	3.5	3.1
34311	Vegetable strudel (ovo-lacto)	8	ounce(s)	227	63.1	478	12.0	32.4	2.5	33.8	11.5	16.7	3.9
34371	Vegetable taco (lacto)	1	item(s)	85	46.5	117	4.2	13.6	2.9	5.6	2.1	1.9	1.3
34282	Vegetarian chili (vegan)	8	ounce(s)	227	191.4	115	5.6	21.4	7.1	1.5	0.2	0.3	0.7
34367	Vegetarian vegetable soup (vegan)	8	ounce(s)	227	257.9	111	3.2	16.0	3.2	5.0	1.0	2.1	1.6
	Boca burger												
32067	All American flamed grilled patty	1	item(s)	71	—	90	14.0	4.0	3.0	3.0	1.0	—	—
32074	Boca chik'n nuggets	4	item(s)	87	—	180	14.0	17.0	3.0	7.0	1.0	—	—
32075	Boca meatless ground burger	½	cup(s)	57	—	60	13.0	6.0	3.0	0.5	0	—	—
32072	Breakfast links	2	item(s)	45	—	70	8.0	5.0	2.0	3.0	0.5	—	—
32071	Breakfast patties	1	item(s)	38	—	60	7.0	5.0	2.0	2.5	0	—	—
35780	Cheeseburger meatless burger patty	1	item(s)	71	—	100	12.0	5.0	3.0	5.0	1.5	—	—
33958	Original meatless chik'n patties	1	item(s)	71	—	160	11.0	15.0	2.0	6.0	1.0	—	—

PAGE KEY: A-4 = Breads/Baked Goods A-10 = Cereal/Rice/Pasta A-14 = Fruit A-18 = Vegetables/Legumes A-28 = Nuts/Seeds A-30 = Vegetarian A-32 = Dairy A-40 = Eggs A-40 = Seafood A-42 = Meats A-46 = Poultry A-46 = Processed Meats A-48 = Beverages A-52 = Fats/Oils A-54 = Sweets A-56 = Spices/Condiments/Sauces A-60 = Mixed Foods/Soups/Sandwiches A-64 = Fast Food A-84 = Convenience A-86 = Baby Foods

Chol (mg)	Calc (mg)	Iron (mg)	Magn (mg)	Pota (mg)	Sodi (mg)	Zinc (mg)	Vit A (µg)	Thia (mg)	Vit E (mg α)	Ribo (mg)	Niac (mg)	Vit B_6 (mg)	Fola (µg)	Vit C (mg)	Vit B_{12} (µg)	Sele (µg)
0	7	0.30	24.6	103.8	73.4	0.47	0	0.01	1.44	0.02	2.14	0.09	11.8	0	0	0.9
0	34	1.67	61.3	257.3	6.6	1.19	0	0.23	3.04	0.04	4.40	0.12	87.6	0	0	2.6
0	20	0.78	36.8	118.3	0.3	1.41	1.9	0.12	0.35	0.03	0.32	0.05	4.5	0.2	0	1.1
0	19	0.75	35.6	114.5	103.4	1.36	1.9	0.11	0.34	0.03	0.31	0.05	4.3	0.2	0	1.1
0	18	0.68	33.3	107.8	0.3	1.23	1.4	0.13	0.70	0.03	0.33	0.05	4.1	0.2	0	1.7
0	19	0.69	33.0	111.7	0	1.23	0.8	0.18	0.38	0.04	0.32	0.06	6.0	0.3	0	1.0
0	1	0.47	21.6	51.3	0.2	0.55	0.1	0.03	0.80	0.02	0.37	0.01	2.9	0.1	0	0.1
0	34	1.29	36.9	320.4	3.1	0.71	4.0	0.26	0.59	0.05	0.44	0.39	15.4	0.7	0	2.9
0	35	1.34	38.4	333.4	129.6	0.73	4.2	0.26	0.61	0.05	0.45	0.40	16.0	0.7	0	3.0
0	24	8.48	303.0	457.4	10.2	4.22	10.8	0.12	0.00	0.18	0.99	0.05	32.3	1.0	0	3.2
0	154	3.07	57.9	93.1	1.9	1.17	0.5	0.04	—	0.03	1.07	0.13	16.0	0	0	0.9
0	21	0.66	14.3	68.9	5.3	0.69	0.5	0.24	—	0.02	0.85	0.02	14.7	0.6	0	0.3
0	94	1.40	33.8	45.1	1.0	0.68	0	0.07	—	0.02	0.43	0.07	9.3	0	0	0.5
0	59	1.07	—	—	148.1	—	0	—	—	—	—	—	—	0	—	—
0	59	1.07	—	—	9.9	—	0	—	—	—	—	—	—	0	—	—
0	7	0.47	29.3	58.1	0.8	0.45	0.3	0.13	2.99	0.03	0.75	0.12	20.4	0.1	0	4.8
0	6	0.30	10.3	68.0	32.8	0.42	0	0.01	2.09	0.02	0.56	0.06	19.0	0.1	0	6.3
0	5	0.57	10.8	41.1	51.3	0.44	0	0.03	—	0.02	0.35	0.07	19.9	0.1	0	5.2
0	20	0.76	59.0	11.5	83.2	0.85	0.5	0.05	—	0.05	0.85	0.13	37.9	0.4	0	—
0	29	1.14	59.3	256.9	85.9	1.20	0.4	0.17	—	0.07	1.76	0.11	26.6	0.5	0	—
2	41	1.27	60.4	243.0	45.4	1.17	0.8	0.15	—	0.08	1.65	0.09	24.4	0.5	0	—
0	20	0.92	33.6	248.2	3.5	0.41	0.7	0.15	—	0.04	0.51	0.11	14.7	2.7	0	—
0	19	0.97	62.8	163.4	0.6	1.05	0.6	0.01	0.56	0.04	0.14	0.18	9.7	0.5	0	5.3
0	29	0.85	46.2	129.0	0.6	0.90	0.3	0.10	0.20	0.04	0.32	0.15	28.7	0.4	0	1.4
0	353	6.34	118.3	501.4	142.2	2.03	—	0.23	0.07	0.14	1.49	0.36	51.8	24.8	0	14.8
39	441	2.44	34.6	191.2	1139.7	1.84	—	0.31	0.05	0.35	2.23	0.11	118.3	20.4	0.4	20.0
0	48	1.78	40.8	364.1	613.6	0.61	—	0.10	0.52	0.07	0.93	0.11	64.5	8.3	0	3.3
0	34	3.23	50.0	548.8	436.5	1.42	—	0.24	0.14	0.16	2.31	0.29	202.7	26.4	0	12.1
43	415	1.71	45.4	267.8	1641.0	2.32	—	0.32	0.27	0.48	2.18	0.13	162.7	0.9	0.8	33.3
0	91	3.31	153.1	810.1	3117.8	2.04	—	0.37	3.03	0.21	6.16	0.64	70.1	35.2	0	18.8
52	467	9.01	89.7	455.6	2449.5	2.06	—	0.27	0.12	0.26	3.43	0.29	167.7	2.0	0.1	43.0
0	117	1.13	40.4	170.5	134.2	0.68	—	0.07	—	0.07	0.53	0.10	20.3	1.8	0	5.1
0	189	3.70	28.0	310.3	372.7	0.76	—	0.13	0.05	0.11	1.43	0.14	76.8	8.0	0	6.5
10	176	1.86	41.9	470.0	759.4	1.14	—	0.26	0.05	0.25	2.49	0.22	124.5	19.0	0.4	21.8
0	17	0.94	19.1	189.9	439.6	0.42	—	0.15	0.55	0.08	1.36	0.12	88.4	23.5	0	10.8
17	190	1.28	29.3	414.2	626.0	1.24	—	0.16	0.35	0.29	2.00	0.19	154.8	56.0	0.2	5.8
29	200	2.15	24.5	181.0	512.1	1.24	—	0.28	0.20	0.31	2.88	0.11	111.4	17.4	0.2	19.7
7	77	0.88	26.3	174.1	280.7	0.59	—	0.08	0.04	0.06	0.49	0.08	38.7	4.6	0	3.0
0	65	1.98	41.0	543.1	390.7	0.74	—	0.14	0.15	0.10	1.31	0.18	47.7	20.3	0	4.4
0	46	1.87	34.9	550.3	729.5	0.56	—	0.13	0.55	0.09	1.99	0.27	49.9	29.9	0	1.4
5	150	1.80	—	—	280.0	—	0	—	—	—	—	—	—	0	—	—
0	40	1.44	—	—	500.0	—	—	—	—	—	—	—	—	0	—	—
0	60	1.80	—	—	270.0	—	0	—	—	—	—	—	—	0	—	—
0	20	1.44	—	—	330.0	—	0	—	—	—	—	—	—	0	—	—
0	20	1.08	—	—	280.0	—	0	—	—	—	—	—	—	0	—	—
5	80	1.80	—	—	360.0	—	—	—	—	—	—	—	—	0	—	—
0	40	1.80	—	—	430.0	—	—	—	—	—	—	—	—	0	—	—

TABLE A-1 Table of Food Composition *(continued)*

(Computer code is for Cengage Diet Analysis program)
(For purposes of calculations, use "0" for t, <1, <.1, <.01, etc.)

DA+ Code	Food Description	Quantity	Measure	Wt (g)	H_2O (g)	Ener (kcal)	Prot (g)	Carb (g)	Fiber (g)	Fat (g)	Fat Breakdown (g) Sat	Mono	Poly
Vegetarian Foods—*continued*													
32066	Original patty	1	item(s)	71	—	70	13.0	6.0	4.0	0.5	0	—	—
32068	Roasted garlic patty	1	item(s)	71	—	70	12.0	6.0	4.0	1.5	0	—	—
37814	Roasted onion meatless burger patty	1	item(s)	71	—	70	11.0	7.0	4.0	1.0	0	—	—
	Gardenburger												
37810	BBQ chik'n with sauce	1	item(s)	142	—	250	14.0	30.0	5.0	8.0	1.0	—	—
39661	Black bean burger	1	item(s)	71	—	80	8.0	11.0	4.0	2.0	0	—	—
39666	Buffalo chik'n wing	3	item(s)	95	—	180	9.0	8.0	5.0	12.0	1.5	—	—
39665	Country fried chicken with creamy pepper gravy	1	item(s)	142	—	190	9.0	16.0	2.0	9.0	1.0	—	—
37808	Flamed grilled chik'n	1	item(s)	71	—	100	13.0	5.0	3.0	2.5	0	—	—
37803	Garden vegan	1	item(s)	71	—	100	10.0	12.0	2.0	1.0	—	—	—
39663	Homestyle classic burger	1	item(s)	71	—	110	12.0	6.0	4.0	5.0	0.5	—	—
37807	Meatless breakfast sausage	1	item(s)	43	—	50	5.0	2.0	2.0	3.5	0	—	—
37809	Meatless meatballs	6	item(s)	85	—	110	12.0	8.0	4.0	4.5	1.0	—	—
37806	Meatless riblets with sauce	1	item(s)	142	—	160	17.0	11.0	4.0	5.0	0	—	—
29913	Original	1	item(s)	71	—	90	10.0	8.0	3.0	2.0	0.5	—	—
39662	Sun-dried tomato basil burger	1	item(s)	71	—	80	10.0	11.0	3.0	1.5	0.5	—	—
29915	Veggie medley	1	item(s)	71	—	90	9.0	11.0	4.0	2.0	0	—	—
	Loma Linda												
9311	Big franks, canned	1	item(s)	51	—	110	11.0	3.0	2.0	6.0	1.0	1.5	3.5
9323	Fried chik'n with gravy	2	piece(s)	80	45.9	150	12.0	5.0	2.0	10	1.5	2.5	5.0
9326	Linketts, canned	1	item(s)	35	21.0	70	7.0	1.0	1.0	4.0	0.5	1.0	2.5
9336	Redi-Burger patties, canned	1	slice(s)	85	50.5	120	18.0	7.0	4.0	2.5	0.5	0.5	1.5
9350	Swiss Stake pattie with gravy, frozen	1	piece(s)	92	65.7	130	9.0	9.0	3.0	6.0	1.0	1.5	3.5
9354	Tender Rounds meatball substitute, canned in gravy	6	piece(s)	80	53.9	120	13.0	6.0	1.0	4.5	0.5	1.5	2.5
	Morningstar Farms												
33707	America's Original Veggie Dog links	1	item(s)	57	—	80	11.0	6.0	1.0	0.5	0	—	—
9362	Better'n Eggs egg substitute	¼	cup(s)	57	50.3	20	5.0	0	0	0	0	0	0
9371	Breakfast bacon strips	2	item(s)	16	6.8	60	2.0	2.0	0.5	4.5	0.5	1.0	3.0
9368	Breakfast sausage links	2	item(s)	45	26.8	80	9.0	3.0	2.0	3.0	0.5	1.5	1.0
33705	Chik'n nuggets	4	piece(s)	86	—	190	12.0	18.0	2.0	7.0	1.0	2.0	4.0
11587	Chik patties	1	item(s)	71	36.3	150	9.0	16.0	2.0	6.0	1.0	1.5	2.5
2531	Garden veggie patties	1	item(s)	67	40.1	100	10.0	9.0	4.0	2.5	0.5	0.5	1.5
33702	Spicy black bean veggie burger	1	item(s)	78	—	140	12.0	15.0	3.0	4.0	0.5	1.0	2.5
9412	Vegetarian chili, canned	1	cup(s)	230	172.6	180	16.0	25.0	10.0	1.5	0.5	0.5	0.5
	Worthington												
9424	Chili, canned	1	cup(s)	230	167.0	280	24.0	25.0	8.0	10.0	1.5	1.5	7.0
9436	Diced chik, canned	¼	cup(s)	55	42.7	50	9.0	2.0	1.0	0	0	0	0
9440	Dinner roast, frozen	1	slice(s)	85	53.2	180	14.0	6.0	3.0	11.0	1.5	4.5	5.0
9420	Meatless chicken slices, frozen	3	slice(s)	57	38.9	90	9.0	2.0	0.5	4.5	1.0	1.0	2.5
36702	Meatless chicken style roll, frozen	1	slice(s)	55	—	90	9.0	2.0	1.0	4.5	1.0	1.0	2.5
9428	Meatless corned beef, sliced, frozen	3	slice(s)	57	31.2	140	10.0	5.0	0	9.0	1.0	2.0	5.0
9470	Meatless salami, sliced, frozen	3	slice(s)	57	32.4	120	12.0	3.0	2.0	7.0	1.0	1.0	5.0
9480	Meatless smoked turkey, sliced	3	slice(s)	57	—	140	10.0	4.0	0	9.0	1.5	2.0	5.0
9462	Prosage links	2	item(s)	45	26.8	80	9.0	3.0	2.0	3.0	0.5	0.5	2.0
9484	Stakelets patty beef steak substitute, frozen	1	piece(s)	71	41.5	150	14.0	7.0	2.0	7.0	1.0	2.5	3.5
9486	Stripples bacon substitute	2	item(s)	16	6.8	60	2.0	2.0	0.5	4.5	0.5	1.0	3.0
9496	Vegetable Skallops meat substitute, canned	½	cup(s)	85	—	90	17.0	4.0	3.0	1.0	0	0	0.5
Dairy													
	Cheese												
1433	Blue, crumbled	1	ounce(s)	28	12.0	100	6.1	0.7	0	8.1	5.3	2.2	0.2
884	Brick	1	ounce(s)	28	11.7	105	6.6	0.8	0	8.4	5.3	2.4	0.2
885	Brie	1	ounce(s)	28	13.7	95	5.9	0.1	0	7.8	4.9	2.3	0.2
34821	Camembert	1	ounce(s)	28	14.7	85	5.6	0.1	0	6.9	4.3	2.0	0.2
5	Cheddar, shredded	¼	cup(s)	28	10.4	114	7.0	0.4	0	9.4	6.0	2.7	0.3
888	Cheddar or colby	1	ounce(s)	28	10.8	112	6.7	0.7	0	9.1	5.7	2.6	0.3
32096	Cheddar or colby, low fat	1	ounce(s)	28	17.9	49	6.9	0.5	0	2.0	1.2	0.6	0.1
889	Edam	1	ounce(s)	28	11.8	101	7.1	0.4	0	7.9	5.0	2.3	0.2

PAGE KEY: A-4 = Breads/Baked Goods A-10 = Cereal/Rice/Pasta A-14 = Fruit A-18 = Vegetables/Legumes A-28 = Nuts/Seeds A-30 = Vegetarian A-32 = Dairy A-40 = Eggs A-40 = Seafood A-42 = Meats A-46 = Poultry A-46 = Processed Meats A-48 = Beverages A-52 = Fats/Oils A-54 = Sweets A-56 = Spices/Condiments/Sauces A-60 = Mixed Foods/Soups/Sandwiches A-64 = Fast Food A-84 = Convenience A-86 = Baby Foods

Chol (mg)	Calc (mg)	Iron (mg)	Magn (mg)	Pota (mg)	Sodi (mg)	Zinc (mg)	Vit A (μg)	Thia (mg)	Vit E (mg α)	Ribo (mg)	Niac (mg)	Vit B_6 (mg)	Fola (μg)	Vit C (mg)	Vit B_{12} (μg)	Sele (μg)
0	60	1.80	—	—	280.0	—	0	—	—	—	—	—	—	0	—	—
0	60	1.80	—	—	370.0	—	0	—	—	—	—	—	—	0	—	—
0	100	2.70	—	—	300.0	—	—	—	—	—	—	—	—	0	—	—
0	150	1.08	—	—	890.0	—	—	—	—	—	—	—	—	0	—	—
0	40	1.44	—	—	330.0	—	—	—	—	—	—	—	—	0	—	—
0	40	0.72	—	—	1000.0	—	—	—	—	—	—	—	—	0	—	—
5	40	1.44	—	—	550.0	—	—	—	—	—	—	—	—	0	—	—
0	60	3.60	—	—	360.0	—	—	—	—	—	—	—	—	0	—	—
0	40	4.50	—	—	230.0	—	—	—	—	—	—	—	—	0	—	—
0	80	1.44	—	—	380.0	—	—	—	—	—	—	—	—	0	—	—
0	20	0.72	—	—	120.0	—	—	—	—	—	—	—	—	0	—	—
0	60	1.80	—	—	400.0	—	—	—	—	—	—	—	—	0	—	—
0	60	1.80	—	—	720.0	—	—	—	—	—	—	—	—	3.6	—	—
0	80	1.08	30.4	193.4	490.0	0.89	—	0.10	—	0.15	1.08	0.08	10.1	1.2	0.1	7.0
5	60	1.44	—	—	260.0	—	—	—	—	—	—	—	—	3.6	—	—
0	40	1.44	27.0	182.0	290.0	0.46	—	0.07	—	0.08	0.90	0.09	10.6	9.0	0	4.0
0	0	0.77	—	50.0	220.0	—	0	0.22	—	0.10	2.00	0.70	—	0	2.4	—
0	20	1.80	—	70.0	430.0	0.33	0	1.05	—	0.34	4.00	0.30	—	0	2.4	—
0	0	0.36	—	20.0	160.0	0.46	0	0.12	—	0.20	0.80	0.16	—	0	0.9	—
0	0	1.06	—	140.0	450.0	—	0	0.15	—	0.25	4.00	0.40	—	0	1.2	—
0	0	0.72	—	200.0	430.0	—	0	0.45	—	0.25	10.00	1.00	—	0	5.4	—
0	20	1.08	—	80.0	340.0	0.66	0	0.75	—	0.17	2.00	0.16	—	0	1.2	—
0	0	0.72	—	60.0	580.0	—	0	—	—	—	—	—	—	0	—	—
0	20	0.72	—	75.0	90.0	0.60	37.5	0.03	—	0.34	0.00	0.08	24.0	—	0.6	—
0	0	0.36	—	15.0	220.0	0.05	0	0.75	—	0.04	0.40	0.07	—	0	0.2	—
0	0	1.80	—	50.0	300.0	—	0	0.37	—	0.17	7.00	0.50	—	0	3.0	—
0	20	2.70	—	320.0	490.0	—	0	0.52	—	0.25	5.00	0.30	—	0	1.5	—
0	0	1.80	—	210.0	540.0	—	0	1.80	—	0.17	2.00	0.20	—	0	1.2	—
0	40	0.72	—	180.0	350.0	—	—	—	—	—	—	—	—	0	—	—
0	40	1.80	—	320.0	470.0	—	0	—	—	—	0.00	—	—	0	—	—
0	40	3.60	—	660.0	900.0	—	—	—	—	—	—	—	—	0	—	—
0	40	3.60	—	330.0	1130.0	—	0	0.30	—	0.13	2.00	0.70	—	0	1.5	—
0	0	1.08	—	100.0	220.0	0.24	0	0.06	—	0.10	4.00	0.08	—	0	0.2	—
0	20	1.80	—	120.0	580.0	0.64	0	1.80	—	0.25	6.00	0.60	—	0	1.5	—
0	250	1.80	—	250.0	250.0	0.26	0	0.37	—	0.13	4.00	0.30	—	0	1.8	—
0	100	1.08	—	240.0	240.0	—	0	0.37	—	0.13	4.00	0.30	—	0	1.8	—
0	0	1.80	—	130.0	460.0	0.26	0	0.45	—	0.17	5.00	0.30	—	0	1.8	—
0	0	1.08	—	95.0	800.0	0.30	0	0.75	—	0.17	4.00	0.20	—	0	0.6	—
0	60	2.70	—	60.0	450.0	0.23	0	1.80	—	0.17	6.00	0.40	—	0	3.0	—
0	0	1.44	—	50.0	320.0	0.36	0	1.80	—	0.17	2.00	0.30	—	0	3.0	—
0	40	1.08	—	130.0	480.0	0.50	0	1.20	—	0.13	3.00	0.30	—	0	1.5	—
0	0	0.36	—	15.0	220.0	0.05	0	0.75	—	0.03	0.40	0.08	—	0	0.2	—
0	0	0.36	—	10.0	390.0	0.67	0	0.03	—	0.03	0.00	0.01	—	0	0	—
21	150	0.08	6.5	72.6	395.5	0.75	56.1	0.01	0.07	0.10	0.28	0.04	10.2	0	0.3	4.1
27	191	0.12	6.8	38.6	158.8	0.73	82.8	0.00	0.07	0.10	0.03	0.01	5.7	0	0.4	4.1
28	52	0.14	5.7	43.1	178.3	0.67	49.3	0.02	0.06	0.14	0.10	0.06	18.4	0	0.5	4.1
20	110	0.09	5.7	53.0	238.7	0.67	68.3	0.01	0.06	0.14	0.18	0.06	17.6	0	0.4	4.1
30	204	0.19	7.9	27.7	175.4	0.87	74.9	0.01	0.08	0.10	0.02	0.02	5.1	0	0.2	3.9
27	194	0.21	7.4	36.0	171.2	0.87	74.8	0.00	0.07	0.10	0.02	0.02	5.1	0	0.2	4.1
6	118	0.11	4.5	18.7	173.5	0.51	17.0	0.00	0.01	0.06	0.01	0.01	3.1	0	0.1	4.1
25	207	0.12	8.5	53.3	273.6	1.06	68.9	0.01	0.06	0.11	0.02	0.02	4.5	0	0.4	4.1

TABLE A-1 Table of Food Composition *(continued)*

(Computer code is for Cengage Diet Analysis program)
(For purposes of calculations, use "0" for t, <1, <.1, <.01, etc.)

DA+ Code	Food Description	Quantity	Measure	Wt (g)	H_2O (g)	Ener (kcal)	Prot (g)	Carb (g)	Fiber (g)	Fat (g)	Fat Breakdown (g) Sat	Mono	Poly
Dairy—*continued*													
890	Feta	1	ounce(s)	28	15.7	75	4.0	1.2	0	6.0	4.2	1.3	0.2
891	Fontina	1	ounce(s)	28	10.8	110	7.3	0.4	0	8.8	5.4	2.5	0.5
8527	Goat cheese, soft	1	ounce(s)	28	17.2	76	5.3	0.3	0	6.0	4.1	1.4	0.1
893	Gouda	1	ounce(s)	28	11.8	101	7.1	0.6	0	7.8	5.0	2.2	0.2
894	Gruyere	1	ounce(s)	28	9.4	117	8.5	0.1	0	9.2	5.4	2.8	0.5
895	Limburger	1	ounce(s)	28	13.7	93	5.7	0.1	0	7.7	4.7	2.4	0.1
896	Monterey jack	1	ounce(s)	28	11.6	106	6.9	0.2	0	8.6	5.4	2.5	0.3
13	Mozzarella, part skim milk	1	ounce(s)	28	15.2	72	6.9	0.8	0	4.5	2.9	1.3	0.1
12	Mozzarella, whole milk	1	ounce(s)	28	14.2	85	6.3	0.6	0	6.3	3.7	1.9	0.2
897	Muenster	1	ounce(s)	28	11.8	104	6.6	0.3	0	8.5	5.4	2.5	0.2
898	Neufchatel	1	ounce(s)	28	17.6	74	2.8	0.8	0	6.6	4.2	1.9	0.2
14	Parmesan, grated	1	tablespoon(s)	5	1.0	22	1.9	0.2	0	1.4	0.9	0.4	0.1
17	Provolone	1	ounce(s)	28	11.6	100	7.3	0.6	0	7.5	4.8	2.1	0.2
19	Ricotta, part skim milk	¼	cup(s)	62	45.8	85	7.0	3.2	0	4.9	3.0	1.4	0.2
18	Ricotta, whole milk	¼	cup(s)	62	44.1	107	6.9	1.9	0	8.0	5.1	2.2	0.2
20	Romano	1	tablespoon(s)	5	1.5	19	1.6	0.2	0	1.3	0.9	0.4	0
900	Roquefort	1	ounce(s)	28	11.2	105	6.1	0.6	0	8.7	5.5	2.4	0.4
21	Swiss	1	ounce(s)	28	10.5	108	7.6	1.5	0	7.9	5.0	2.1	0.3
	Imitation cheese												
42245	Imitation American cheddar cheese	1	ounce(s)	28	15.1	68	4.7	3.3	0	4.0	2.5	1.2	0.1
53914	Imitation cheddar	1	ounce(s)	28	15.1	68	4.7	3.3	0	4.0	2.5	1.2	0.1
	Cottage cheese												
9	Low fat, 1% fat	½	cup(s)	113	93.2	81	14.0	3.1	0	1.2	0.7	0.3	0
8	Low fat, 2% fat	½	cup(s)	113	89.6	102	15.5	4.1	0	2.2	1.4	0.6	0.1
	Cream cheese												
11	Cream cheese	2	tablespoon(s)	29	15.6	101	2.2	0.8	0	10.1	6.4	2.9	0.4
17366	Fat-free cream cheese	2	tablespoon(s)	30	22.7	29	4.3	1.7	0	0.4	0.3	0.1	0
10438	Tofutti Better than Cream Cheese	2	tablespoon(s)	30	—	80	1.0	1.0	0	8.0	2.0	—	6.0
	Processed cheese												
24	American cheese food, processed	1	ounce(s)	28	12.3	94	5.2	2.2	0	7.1	4.2	2.0	0.3
25	American cheese spread, processed	1	ounce(s)	28	13.5	82	4.7	2.5	0	6.0	3.8	1.8	0.2
22	American cheese, processed	1	ounce(s)	28	11.1	106	6.3	0.5	0	8.9	5.6	2.5	0.3
9110	Kraft deluxe singles pasteurized process American cheese	1	ounce(s)	28	—	108	5.4	0	0	9.5	5.4	—	—
23	Swiss cheese, processed	1	ounce(s)	28	12.0	95	7.0	0.6	0	7.1	4.5	2.0	0.2
	Soy cheese												
10437	Galaxy Foods vegan grated parmesan cheese alternative	1	tablespoon(s)	8	—	23	3.0	1.5	0	0	0	0	0
10430	Nu Tofu cheddar flavored cheese alternative	1	ounce(s)	28	—	70	6.0	1.0	0	4.0	0.5	2.5	1.0
	Cream												
26	Half and half cream	1	tablespoon(s)	15	12.1	20	0.4	0.6	0	1.7	1.1	0.5	0.1
32	Heavy whipping cream, liquid	1	tablespoon(s)	15	8.7	52	0.3	0.4	0	5.6	3.5	1.6	0.2
28	Light coffee or table cream, liquid	1	tablespoon(s)	15	11.1	29	0.4	0.5	0	2.9	1.8	0.8	0.1
30	Light whipping cream, liquid	1	tablespoon(s)	15	9.5	44	0.3	0.4	0	4.6	2.9	1.4	0.1
34	Whipped cream topping, pressurized	1	tablespoon(s)	3	1.8	8	0.1	0.4	0	0.7	0.4	0.2	0
	Sour cream												
30556	Fat-free sour cream	2	tablespoon(s)	32	25.8	24	1.0	5.0	0	0	0	0	0
36	Sour cream	2	tablespoon(s)	24	17.0	51	0.8	1.0	0	5.0	3.1	1.5	0.2
	Imitation cream												
3659	Coffeemate nondairy creamer, liquid	1	tablespoon(s)	15	—	20	0	2.0	0	1.0	0	0.5	0
40	Cream substitute, powder	1	teaspoon(s)	2	0	11	0.1	1.1	0	0.7	0.7	0	0
904	Imitation sour cream	2	tablespoon(s)	29	20.5	60	0.7	1.9	0	5.6	5.1	0.2	0
35972	Nondairy coffee whitener, liquid, frozen	1	tablespoon(s)	15	11.7	21	0.2	1.7	0	1.5	0.3	1.1	0
35976	Nondairy dessert topping, frozen	1	tablespoon(s)	5	2.4	15	0.1	1.1	0	1.2	1.0	0.1	0
35975	Nondairy dessert topping, pressurized	1	tablespoon(s)	4	2.7	12	0	0.7	0	1.0	0.8	0.1	0

PAGE KEY: A-4 = Breads/Baked Goods A-10 = Cereal/Rice/Pasta A-14 = Fruit A-18 = Vegetables/Legumes A-28 = Nuts/Seeds A-30 = Vegetarian A-32 = Dairy A-40 = Eggs A-40 = Seafood A-42 = Meats A-46 = Poultry A-46 = Processed Meats A-48 = Beverages A-52 = Fats/Oils A-54 = Sweets A-56 = Spices/Condiments/Sauces A-60 = Mixed Foods/Soups/Sandwiches A-64 = Fast Food A-84 = Convenience A-86 = Baby Foods

Chol (mg)	Calc (mg)	Iron (mg)	Magn (mg)	Pota (mg)	Sodi (mg)	Zinc (mg)	Vit A (µg)	Thia (mg)	Vit E (mg α)	Ribo (mg)	Niac (mg)	Vit B_6 (mg)	Fola (µg)	Vit C (mg)	Vit B_{12} (µg)	Sele (µg)
25	140	0.18	5.4	17.6	316.4	0.81	35.4	0.04	0.05	0.23	0.28	0.12	9.1	0	0.5	4.3
33	156	0.06	4.0	18.1	226.8	0.99	74.0	0.01	0.07	0.05	0.04	0.02	1.7	0	0.5	4.1
13	40	0.53	4.5	7.4	104.3	0.26	81.6	0.02	0.05	0.10	0.12	0.07	3.4	0	0.1	0.8
32	198	0.06	8.2	34.3	232.2	1.10	46.8	0.01	0.06	0.09	0.01	0.02	6.0	0	0.4	4.1
31	287	0.04	10.2	23.0	95.3	1.10	76.8	0.01	0.07	0.07	0.03	0.02	2.8	0	0.5	4.1
26	141	0.03	6.0	36.3	226.8	0.59	96.4	0.02	0.06	0.14	0.04	0.02	16.4	0	0.3	4.1
25	211	0.20	7.7	23.0	152.0	0.85	56.1	0.00	0.07	0.11	0.02	0.02	5.1	0	0.2	4.1
18	222	0.06	6.5	23.8	175.5	0.78	36.0	0.01	0.04	0.08	0.03	0.02	2.6	0	0.2	4.1
22	143	0.12	5.7	21.5	177.8	0.82	50.7	0.01	0.05	0.08	0.02	0.01	2.0	0	0.6	4.8
27	203	0.11	7.7	38.0	178.0	0.79	84.5	0.00	0.07	0.09	0.02	0.01	3.4	0	0.4	4.1
22	21	0.07	2.3	32.3	113.1	0.14	84.5	0.00	—	0.05	0.03	0.01	3.1	0	0.1	0.9
4	55	0.04	1.9	6.3	76.5	0.19	6.0	0.00	0.01	0.02	0.01	0.00	0.5	0	0.1	0.9
20	214	0.14	7.9	39.1	248.3	0.91	66.9	0.01	0.06	0.09	0.04	0.02	2.8	0	0.4	4.1
19	167	0.27	9.2	76.9	76.9	0.82	65.8	0.01	0.04	0.11	0.04	0.01	8.0	0	0.2	10.3
31	127	0.23	6.8	64.6	51.7	0.71	73.8	0.01	0.06	0.12	0.06	0.02	7.4	0	0.2	8.9
5	53	0.03	2.1	4.3	60	0.12	4.8	0.00	0.01	0.01	0.00	0.00	0.4	0	0.1	0.7
26	188	0.15	8.5	25.8	512.9	0.59	83.3	0.01	—	0.16	0.20	0.03	13.9	0	0.2	4.1
26	224	0.05	10.8	21.8	54.4	1.23	62.4	0.01	0.10	0.08	0.02	0.02	1.7	0	0.9	5.2
10	159	0.08	8.2	68.6	381.3	0.73	32.3	0.01	0.07	0.12	0.03	0.03	2.0	0	0.1	4.3
10	159	0.09	8.2	68.6	381.3	0.73	32.3	0.01	0.07	0.12	0.04	0.03	2.0	0	0.1	4.3
5	69	0.15	5.7	97.2	458.8	0.42	12.4	0.02	0.01	0.18	0.14	0.07	13.6	0	0.7	10.2
9	78	0.18	6.8	108.5	458.8	0.47	23.7	0.02	0.02	0.20	0.16	0.08	14.7	0	0.8	11.5
32	23	0.34	1.7	34.5	85.8	0.15	106.1	0.01	0.08	0.05	0.02	0.01	3.8	0	0.1	0.7
2	56	0.05	4.2	48.9	163.5	0.26	83.7	0.01	0.00	0.05	0.04	0.01	11.1	0	0.2	1.5
0	0	0.00	—	—	135.0	—	0	—	—	—	—	—	—	0	—	—
23	162	0.16	8.8	82.5	358.6	0.90	57.0	0.01	0.06	0.14	0.04	0.02	2.0	0	0.4	4.6
16	159	0.09	8.2	68.6	381.3	0.73	49.0	0.01	0.05	0.12	0.03	0.03	2.0	0	0.1	3.2
27	156	0.05	7.7	47.9	422.1	0.80	72.0	0.01	0.07	0.10	0.02	0.02	2.3	0	0.2	4.1
27	338	0.00	0	33.8	459.0	1.22	114.0	—	—	0.14	—	—	—	0	0.2	—
24	219	0.17	8.2	61.2	388.4	1.02	56.1	0.00	0.09	0.07	0.01	0.01	1.7	0	0.3	4.5
0	60	0.00	—	75.0	97.5	—	—	—	—	—	—	—	—	—	—	—
0	200	0.36	—	—	190.0	—	—	—	—	—	—	—	—	0	—	—
6	16	0.01	1.5	19.5	6.2	0.08	14.6	0.01	0.05	0.02	0.01	0.01	0.5	0.1	0	0.3
21	10	0.00	1.1	11.3	5.7	0.03	61.7	0.00	0.15	0.01	0.01	0.00	0.6	0.1	0	0.1
10	14	0.01	1.4	18.3	6.0	0.04	27.2	0.01	0.08	0.02	0.01	0.01	0.3	0.1	0	0.1
17	10	0.00	1.1	14.6	5.1	0.03	41.9	0.00	0.13	0.01	0.01	0.00	0.6	0.1	0	0.1
2	3	0.00	0.3	4.4	3.9	0.01	5.6	0.00	0.01	0.00	0.00	0.00	0.1	0	0	0
3	40	0.00	3.2	41.3	45.1	0.16	23.4	0.01	0.00	0.04	0.02	0.01	3.5	0	0.1	1.7
11	28	0.01	2.6	34.6	12.7	0.06	42.5	0.01	0.14	0.03	0.01	0.00	2.6	0.2	0.1	0.5
0	0	0.00	—	30.0	0	—	0	0.01	—	0.01	0.20	—	—	0	—	—
0	0	0.02	0.1	16.2	3.6	0.01	0	0.00	0.01	0.00	0.00	0.00	0	0	0	0
0	1	0.11	1.7	46.3	29.3	0.34	0	0.00	0.21	0.00	0.00	0.00	0	0	0	0.7
0	1	0.00	0	28.9	12.0	0.00	0.2	0.00	0.12	0.00	0.00	0.00	0	0	0	0.2
0	0	0.00	0.1	0.9	1.2	0.00	0.3	0.00	0.05	0.00	0.00	0.00	0	0	0	0.1
0	0	0.00	0	0.8	2.8	0.00	0.2	0.00	0.04	0.00	0.00	0.00	0	0	0	0.1

TABLE A-1 Table of Food Composition *(continued)*

(Computer code is for Cengage Diet Analysis program)
(For purposes of calculations, use "0" for t, <1, <.1, <.01, etc.)

DA+ Code	Food Description	Quantity	Measure	Wt (g)	H_2O (g)	Ener (kcal)	Prot (g)	Carb (g)	Fiber (g)	Fat (g)	Fat Breakdown (g) Sat	Mono	Poly
Dairy—*continued*													
	Fluid milk												
60	Buttermilk, low fat	1	cup(s)	245	220.8	98	8.1	11.7	0	2.2	1.3	0.6	0.1
54	Low fat, 1%	1	cup(s)	244	219.4	102	8.2	12.2	0	2.4	1.5	0.7	0.1
55	Low fat, 1%, with nonfat milk solids	1	cup(s)	245	220.0	105	8.5	12.2	0	2.4	1.5	0.7	0.1
57	Nonfat, skim or fat free	1	cup(s)	245	222.6	83	8.3	12.2	0	0.2	0.1	0.1	0
58	Nonfat, skim or fat free with nonfat milk solids	1	cup(s)	245	221.4	91	8.7	12.3	0	0.6	0.4	0.2	0
51	Reduced fat, 2%	1	cup(s)	244	218.0	122	8.1	11.4	0	4.8	3.1	1.4	0.2
52	Reduced fat, 2%, with nonfat milk solids	1	cup(s)	245	217.7	125	8.5	12.2	0	4.7	2.9	1.4	0.2
50	Whole, 3.3%	1	cup(s)	244	215.5	146	7.9	11.0	0	7.9	4.6	2.0	0.5
	Canned milk												
62	Nonfat or skim evaporated	2	tablespoon(s)	32	25.3	25	2.4	3.6	0	0.1	0	0	0
63	Sweetened condensed	2	tablespoon(s)	38	10.4	123	3.0	20.8	0	3.3	2.1	0.9	0.1
61	Whole evaporated	2	tablespoon(s)	32	23.3	42	2.1	3.2	0	2.4	1.4	0.7	0.1
	Dried milk												
64	Buttermilk	¼	cup(s)	30	0.9	117	10.4	14.9	0	1.8	1.1	0.5	0.1
65	Instant nonfat with added vitamin A	¼	cup(s)	17	0.7	61	6.0	8.9	0	0.1	0.1	0	0
5234	Skim milk powder	¼	cup(s)	17	0.7	62	6.1	9.1	0	0.1	0.1	0	0
907	Whole dry milk	¼	cup(s)	32	0.8	159	8.4	12.3	0	8.5	5.4	2.5	0.2
909	**Goat milk**	1	cup(s)	244	212.4	168	8.7	10.9	0	10.1	6.5	2.7	0.4
	Chocolate milk												
33155	Chocolate syrup, prepared with milk	1	cup(s)	282	227.0	254	8.7	36.0	0.8	8.3	4.7	2.1	0.5
33184	Cocoa mix with aspartame, added sodium and vitamin A, no added calcium or phosphorus, prepared with water	1	cup(s)	192	177.4	56	2.3	10.8	1.2	0.4	0.3	0.1	0
908	Hot cocoa, prepared with milk	1	cup(s)	250	206.4	193	8.8	26.6	2.5	5.8	3.6	1.7	0.1
69	Low fat	1	cup(s)	250	211.3	158	8.1	26.1	1.3	2.5	1.5	0.8	0.1
68	Reduced fat	1	cup(s)	250	205.4	190	7.5	30.3	1.8	4.8	2.9	1.1	0.2
67	Whole	1	cup(s)	250	205.8	208	7.9	25.9	2.0	8.5	5.3	2.5	0.3
70	**Eggnog**	1	cup(s)	254	188.9	343	9.7	34.4	0	19.0	11.3	5.7	0.9
	Breakfast drinks												
10093	Carnation Instant Breakfast classic chocolate malt, prepared with skim milk, no sugar added	1	cup(s)	243	—	142	11.1	21.3	0.7	1.3	0.7	—	—
10092	Carnation Instant Breakfast classic French vanilla, prepared with skim milk, no sugar added	1	cup(s)	273	—	150	12.9	24.0	0	0.4	0.4	—	—
10094	Carnation Instant Breakfast stawberry sensation, prepared with skim milk, no sugar added	1	cup(s)	243	—	142	11.1	21.3	0	0.4	0.4	—	—
10091	Carnation Instant Breakfast strawberry sensation, prepared with skim milk	1	cup(s)	273	—	220	12.5	38.8	0	0.4	0.4	—	—
1417	Ovaltine rich chocolate flavor, prepared with skim milk	1	cup(s)	258	—	170	8.5	31.0	0	0	0	0	0
8539	**Malted milk, chocolate mix, fortified, prepared with milk**	1	cup(s)	265	215.8	223	8.9	28.9	1.1	8.6	5.0	2.2	0.5
	Milkshakes												
73	Chocolate	1	cup(s)	227	164.0	270	6.9	48.1	0.7	6.1	3.8	1.8	0.2
3163	Strawberry	1	cup(s)	226	167.8	256	7.7	42.8	0.9	6.3	3.9	—	—
74	Vanilla	1	cup(s)	227	169.2	254	8.8	40.3	0	6.9	4.3	2.0	0.3
	Ice cream												
4776	Chocolate	½	cup(s)	66	36.8	143	2.5	18.6	0.8	7.3	4.5	2.1	0.3
12137	Chocolate fudge, no sugar added	½	cup(s)	71	—	100	3.0	16.0	2.0	3.0	1.5	—	—
16514	Chocolate, soft serve	½	cup(s)	87	49.9	177	3.2	24.1	0.7	8.4	5.2	2.4	0.3
16523	Sherbet, all flavors	½	cup(s)	97	63.8	139	1.1	29.3	3.2	1.9	1.1	0.5	0.1
4778	Strawberry	½	cup(s)	66	39.6	127	2.1	18.2	0.6	5.5	3.4	—	—
76	Vanilla	½	cup(s)	72	43.9	145	2.5	17.0	0.5	7.9	4.9	2.1	0.3
12146	Vanilla chocolate swirl, fat-free, no sugar added	½	cup(s)	71	—	100	3.0	14.0	2.0	3.0	2.0	—	—
82	Vanilla, light	½	cup(s)	76	48.3	125	3.6	19.6	0.2	3.7	2.2	1.0	0.2
78	Vanilla, light, soft serve	½	cup(s)	88	61.2	111	4.3	19.2	0	2.3	1.4	0.7	0.1

PAGE KEY: A-4 = Breads/Baked Goods A-10 = Cereal/Rice/Pasta A-14 = Fruit A-18 = Vegetables/Legumes A-28 = Nuts/Seeds A-30 = Vegetarian A-32 = Dairy A-40 = Eggs A-40 = Seafood A-42 = Meats A-46 = Poultry A-46 = Processed Meats A-48 = Beverages A-52 = Fats/Oils A-54 = Sweets A-56 = Spices/Condiments/Sauces A-60 = Mixed Foods/Soups/Sandwiches A-64 = Fast Food A-84 = Convenience A-86 = Baby Foods

Chol (mg)	Calc (mg)	Iron (mg)	Magn (mg)	Pota (mg)	Sodi (mg)	Zinc (mg)	Vit A (µg)	Thia (mg)	Vit E (mg α)	Ribo (mg)	Niac (mg)	Vit B_6 (mg)	Fola (µg)	Vit C (mg)	Vit B_{12} (µg)	Sele (µg)
10	284	0.12	27.0	370.0	257.3	1.02	17.2	0.08	0.12	0.37	0.14	0.08	12.3	2.5	0.5	4.9
12	290	0.07	26.8	366.0	107.4	1.02	141.5	0.04	0.02	0.45	0.22	0.09	12.2	0	1.1	8.1
10	314	0.12	34.3	396.9	127.4	0.98	144.6	0.09	—	0.42	0.22	0.11	12.3	2.5	0.9	5.6
5	306	0.07	27.0	382.2	102.9	1.02	149.5	0.11	0.02	0.44	0.23	0.09	12.3	0	1.3	7.6
5	316	0.12	36.8	419.0	129.9	1.00	149.5	0.10	0.00	0.42	0.22	0.11	12.3	2.5	1.0	5.4
20	285	0.07	26.8	366.0	100.0	1.04	134.2	0.09	0.07	0.45	0.22	0.09	12.2	0.5	1.1	6.1
20	314	0.12	34.3	396.9	127.4	0.98	137.2	0.09	—	0.42	0.22	0.11	12.3	2.5	0.9	5.6
24	276	0.07	24.4	348.9	97.6	0.97	68.3	0.10	0.14	0.44	0.26	0.08	12.2	0	1.1	9.0
1	93	0.09	8.6	105.9	36.7	0.28	37.6	0.01	0.00	0.09	0.05	0.01	2.9	0.4	0.1	0.8
13	109	0.07	9.9	141.9	48.6	0.36	28.3	0.03	0.06	0.16	0.08	0.02	4.2	1.0	0.2	5.7
9	82	0.06	7.6	95.4	33.4	0.24	20.5	0.01	0.04	0.10	0.06	0.01	2.5	0.6	0.1	0.7
21	359	0.09	33.3	482.5	156.7	1.21	14.9	0.11	0.03	0.48	0.27	0.10	14.2	1.7	1.2	6.2
3	209	0.05	19.9	289.9	93.3	0.75	120.5	0.07	0.00	0.30	0.15	0.06	8.5	1.0	0.7	4.6
3	214	0.05	20.3	296.0	95.3	0.76	123.1	0.07	0.00	0.30	0.15	0.06	8.7	1.0	0.7	4.7
31	292	0.15	27.2	425.6	118.7	1.06	82.2	0.09	0.15	0.38	0.20	0.09	11.8	2.8	1.0	5.2
27	327	0.12	34.2	497.8	122.0	0.73	139.1	0.11	0.17	0.33	0.67	0.11	2.4	3.2	0.2	3.4
25	251	0.90	50.8	408.9	132.5	1.21	70.5	0.11	0.14	0.46	0.38	0.09	14.1	0	1.1	9.6
0	92	0.74	32.6	405.1	138.2	0.51	0	0.04	0.00	0.20	0.16	0.04	1.9	0	0.2	2.5
20	263	1.20	57.5	492.5	110.0	1.57	127.5	0.09	0.07	0.45	0.33	0.10	12.5	0.5	1.1	6.8
8	288	0.60	32.5	425.0	152.5	1.02	145.0	0.09	0.05	0.41	0.31	0.10	12.5	2.3	0.9	4.8
20	273	0.60	35.0	422.5	165.0	0.97	160.0	0.11	0.10	0.45	0.41	0.06	5.0	0	0.8	8.5
30	280	0.60	32.5	417.5	150.0	1.02	65.0	0.09	0.15	0.40	0.31	0.10	12.5	2.3	0.8	4.8
150	330	0.50	48.3	419.1	137.2	1.16	116.8	0.08	0.50	0.48	0.26	0.12	2.5	3.8	1.1	10.7
9	444	4.00	88.9	631.1	195.6	3.38	—	0.33	—	0.45	4.44	0.44	4.0	26.7	1.3	8.0
9	500	4.50	100.0	665.0	192.0	3.75	—	0.37	—	0.51	5.00	0.49	100.0	30.0	1.5	9.0
9	444	4.00	88.9	568.9	186.7	3.38	—	0.33	—	0.45	4.44	0.44	88.9	26.7	1.3	8.0
9	500	4.47	100.0	665.0	288.0	3.75	—	0.37	—	0.51	5.07	0.50	100.0	30.0	1.5	8.8
5	350	3.60	100.0	—	270.0	3.75	—	0.37	—	—	4.00	0.40	—	12.0	1.2	—
27	339	3.76	45.1	577.7	230.6	1.16	903.7	0.75	0.15	1.31	11.08	1.01	13.3	31.8	1.1	12.5
25	300	0.70	36.4	508.9	252.2	1.09	40.9	0.10	0.11	0.50	0.28	0.05	11.4	0	0.7	4.3
25	256	0.24	29.4	412.0	187.9	0.81	58.9	0.10	—	0.44	0.39	0.10	6.8	1.8	0.7	4.8
27	332	0.22	27.3	415.8	215.8	0.88	56.8	0.06	0.11	0.44	0.33	0.09	15.9	0	1.2	5.2
22	72	0.61	19.1	164.3	50.2	0.38	77.9	0.02	0.19	0.12	0.14	0.03	10.6	0.5	0.2	1.7
10	100	0.36	—	—	65.0	—	—	—	—	—	—	—	—	0	—	—
22	103	0.32	19.0	192.0	43.3	0.45	66.6	0.03	0.22	0.13	0.11	0.03	4.3	0.5	0.3	2.5
0	52	0.13	7.7	92.6	44.4	0.46	9.7	0.02	0.02	0.08	0.07	0.02	6.8	5.6	0.1	1.3
19	79	0.13	9.2	124.1	39.6	0.22	63.4	0.03	—	0.16	0.11	0.03	7.9	5.1	0.2	1.3
32	92	0.06	10.1	143.3	57.6	0.49	85.0	0.03	0.21	0.17	0.08	0.03	3.6	0.4	0.3	1.3
10	100	0.00	—	—	65.0	—	—	—	—	—	—	—	—	0	—	—
21	122	0.14	10.6	158.1	56.2	0.55	97.3	0.04	0.09	0.19	0.10	0.03	4.6	0.9	0.4	1.5
11	138	0.05	12.3	194.5	61.6	0.46	25.5	0.04	0.05	0.17	0.10	0.04	4.4	0.8	0.4	3.2

TABLE A-1 Table of Food Composition *(continued)*

(Computer code is for Cengage Diet Analysis program)
(For purposes of calculations, use "0" for t, <1, <.1, <.01, etc.)

DA+ Code	Food Description	Quantity	Measure	Wt (g)	H_2O (g)	Ener (kcal)	Prot (g)	Carb (g)	Fiber (g)	Fat (g)	Fat Breakdown (g) Sat	Mono	Poly
Dairy—*continued*													
	Soy desserts												
10694	Tofutti low fat vanilla fudge nondairy frozen dessert	½	cup(s)	70	—	140	2.0	24.0	0	4.0	1.0	—	—
15721	Tofutti premium chocolate supreme nondairy frozen dessert	½	cup(s)	70	—	180	3.0	18.0	0	11.0	2.0	—	—
15720	Tofutti premium vanilla nondairy frozen dessert	½	cup(s)	70	—	190	2.0	20.0	0	11.0	2.0	—	—
	Ice milk												
16517	Chocolate	½	cup(s)	66	42.9	94	2.8	16.9	0.3	2.1	1.3	0.6	0.1
16516	Flavored, not chocolate	½	cup(s)	66	41.4	108	3.5	17.5	0.2	2.6	1.7	0.6	0.1
	Pudding												
25032	Chocolate	½	cup(s)	144	109.7	155	5.1	22.7	0.7	5.4	3.1	1.7	0.2
1923	Chocolate, sugar free, prepared with 2% milk	½	cup(s)	133	—	100	5.0	14.0	0.3	3.0	1.5	—	—
1722	Rice	½	cup(s)	113	75.6	151	4.1	29.9	0.5	1.9	1.1	0.5	0.1
4747	Tapioca, ready to eat	1	item(s)	142	102.0	185	2.8	30.8	0	5.5	1.4	3.6	0.1
25031	Vanilla	½	cup(s)	136	109.7	116	4.7	17.6	0	2.8	1.6	0.9	0.2
1924	Vanilla, sugar free, prepared with 2% milk	½	cup(s)	133	—	90	4.0	12.0	0.2	2.0	1.5	—	—
	Frozen yogurt												
4785	Chocolate, soft serve	½	cup(s)	72	45.9	115	2.9	17.9	1.6	4.3	2.6	1.3	0.2
1747	Fruit varieties	½	cup(s)	113	80.5	144	3.4	24.4	0	4.1	2.6	1.1	0.1
4786	Vanilla, soft serve	½	cup(s)	72	47.0	117	2.9	17.4	0	4.0	2.5	1.1	0.2
	Milk substitutes												
	Lactose free												
16081	Fat-free, calcium fortified [milk]	1	cup(s)	240	—	80	8.0	13.0	0	0	0	0	0
36486	Low fat milk	1	cup(s)	240	—	110	8.0	13.0	0	2.5	1.5	—	—
36487	Reduced fat milk	1	cup(s)	240	—	130	8.0	12.0	0	5.0	3.0	—	—
36488	Whole milk	1	cup(s)	240	—	150	8.0	12.0	0	8.0	5.0	—	—
	Rice												
10083	Rice Dream carob rice beverage	1	cup(s)	240	—	150	1.0	32.0	0	2.5	0	—	—
17089	Rice Dream original rice beverage, enriched	1	cup(s)	240	—	120	1.0	25.0	0	2.0	0	—	—
10087	Rice Dream vanilla enriched rice beverage	1	cup(s)	240	—	130	1.0	28.0	0	2.0	0	—	—
	Soy												
34750	Soy Dream chocolate enriched soy beverage	1	cup(s)	240	—	210	7.0	37.0	1.0	3.5	0.5	—	—
34749	Soy Dream vanilla enriched soy beverage	1	cup(s)	240	—	150	7.0	22.0	0	4.0	0.5	—	—
13840	Vitasoy light chocolate soymilk	1	cup(s)	240	—	100	4.0	17.0	0	2.0	0.5	0.5	1.0
13839	Vitasoy light vanilla soymilk	1	cup(s)	240	—	70	4.0	10.0	0	2.0	0.5	0.5	1.0
13836	Vitasoy rich chocolate soymilk	1	cup(s)	240	—	160	7.0	24.0	1.0	4.0	0.5	1.0	2.5
13835	Vitasoy vanilla delite soymilk	1	cup(s)	240	—	120	7.0	13.0	1.0	4.0	0.5	1.0	2.5
	Yogurt												
3615	Custard style, fruit flavors	6	ounce(s)	170	127.1	190	7.0	32.0	0	3.5	2.0	—	—
3617	Custard style, vanilla	6	ounce(s)	170	134.1	190	7.0	32.0	0	3.5	2.0	0.9	0.1
32101	Fruit, low fat	1	cup(s)	245	184.5	243	9.8	45.7	0	2.8	1.8	0.8	0.1
29638	Fruit, nonfat, sweetened with low-calorie sweetener	1	cup(s)	241	208.3	123	10.6	19.4	1.2	0.4	0.2	0.1	0
93	Plain, low fat	1	cup(s)	245	208.4	154	12.9	17.2	0	3.8	2.5	1.0	0.1
94	Plain, nonfat	1	cup(s)	245	208.8	137	14.0	18.8	0	0.4	0.3	0.1	0
32100	Vanilla, low fat	1	cup(s)	245	193.6	208	12.1	33.8	0	3.1	2.0	0.8	0.1
5242	Yogurt beverage	1	cup(s)	245	199.8	172	6.2	32.8	0	2.2	1.4	0.6	0.1
38202	Yogurt smoothie, nonfat, all flavors	1	item(s)	325	—	290	10.0	60.0	6.0	0	0	0	0
	Soy yogurt												
34617	Stonyfield Farm O'Soy strawberry-peach pack organic cultured soy yogurt	1	item(s)	113	—	100	5.0	16.0	3.0	2.0	0	—	—
34616	Stonyfield Farm O'Soy vanilla organic cultured soy yogurt	1	item(s)	170	—	150	7.0	26.0	4.0	2.0	0	—	—
10453	White Wave plain silk cultured soy yogurt	8	ounce(s)	227	—	140	5.0	22.0	1.0	3.0	0.5	—	—

PAGE KEY: A-4 = Breads/Baked Goods A-10 = Cereal/Rice/Pasta A-14 = Fruit A-18 = Vegetables/Legumes A-28 = Nuts/Seeds A-30 = Vegetarian A-32 = Dairy A-40 = Eggs A-40 = Seafood A-42 = Meats A-46 = Poultry A-46 = Processed Meats A-48 = Beverages A-52 = Fats/Oils A-54 = Sweets A-56 = Spices/Condiments/Sauces A-60 = Mixed Foods/Soups/Sandwiches A-64 = Fast Food A-84 = Convenience A-86 = Baby Foods

Chol (mg)	Calc (mg)	Iron (mg)	Magn (mg)	Pota (mg)	Sodi (mg)	Zinc (mg)	Vit A (µg)	Thia (mg)	Vit E (mg α)	Ribo (mg)	Niac (mg)	Vit B_6 (mg)	Fola (µg)	Vit C (mg)	Vit B_{12} (µg)	Sele (µg)
0	0	0.00	—	8.0	90.0	—	0	—	—	—	—	—	—	0	—	—
0	0	0.00	—	7.0	180.0	—	0	—	—	—	—	—	—	0	—	—
0	0	0.00	—	2.0	210.0	—	0	—	—	—	—	—	—	0	—	—
6	94	0.15	13.1	155.2	40.6	0.36	15.7	0.03	0.05	0.11	0.08	0.02	3.9	0.5	0.3	2.2
16	76	0.05	9.2	136.2	48.5	0.47	90.4	0.02	0.05	0.11	0.06	0.01	3.3	0.1	0.2	1.3
35	149	0.46	31.3	226.7	137.0	0.71	—	0.05	0.00	0.22	0.15	0.06	8.3	1.2	0.5	4.9
10	150	0.72	—	330.0	310.0	—	—	0.06	—	0.26	—	—	—	0	—	—
7	113	0.28	15.8	201.4	66.4	0.52	41.6	0.03	0.05	0.17	0.34	0.06	4.5	0.2	0.2	4.8
1	101	0.15	8.5	130.6	205.9	0.31	0	0.03	0.21	0.13	0.09	0.03	4.3	0.4	0.3	0
35	146	0.17	17.2	188.9	136.4	0.52	—	0.04	0.00	0.22	0.10	0.05	8.0	1.2	0.5	4.6
10	150	0.00	—	190.0	380.0	—	—	0.03	—	0.17	—	—	—	0	—	—
4	106	0.90	19.4	187.9	70.6	0.35	31.7	0.02	—	0.15	0.22	0.05	7.9	0.2	0.2	1.7
15	113	0.52	11.3	176.3	71.2	0.31	55.4	0.04	0.10	0.20	0.07	0.04	4.5	0.8	0.1	2.1
1	103	0.21	10.1	151.9	62.6	0.30	42.5	0.02	0.07	0.16	0.20	0.05	4.3	0.6	0.2	2.4
3	500	0.00	—	—	125.0	—	100.0	—	—	—	—	—	—	0	0	—
10	300	0.00	—	—	125.0	—	100.0	—	—	—	—	—	—	0	—	—
20	300	0.00	—	—	125.0	—	98.2	—	—	—	—	—	—	0	—	—
35	300	0.00	—	—	125.0	—	58.1	—	—	—	—	—	—	0	—	—
0	20	0.72	—	82.5	100.0	—	—	—	—	—	—	—	—	1.2	—	—
0	300	0.00	13.3	60.0	90.0	0.24	—	0.06	—	0.00	0.84	0.07	—	0	1.5	—
0	300	0.00	—	53.0	90.0	—	—	—	—	—	—	—	—	0	1.5	—
0	300	1.80	60.0	350.0	160.0	0.60	33.3	0.15	—	0.06	0.80	0.12	60.0	0	3.0	—
0	300	1.80	40.0	260.0	140.0	0.60	33.3	0.15	—	0.06	0.80	0.12	60.0	0	3.0	—
0	300	0.72	24.0	200.0	140.0	0.90	—	0.09	—	0.34	—	—	24.0	0	0.9	—
0	300	0.72	24.0	200.0	120.0	0.90	—	0.09	—	0.34	—	—	24.0	0	0.9	—
0	300	1.08	40.0	320.0	150.0	0.90	—	0.15	—	0.34	—	—	60.0	0	0.9	—
0	40	0.72	—	320.0	115.0	—	0	—	—	—	—	—	—	0	—	—
15	300	0.00	16.0	310.0	100.0	—	—	—	—	0.25	—	—	—	0	—	—
15	300	0.00	16.0	310.0	100.0	—	—	—	—	0.25	—	—	—	0	—	—
12	338	0.14	31.9	433.7	129.9	1.64	27.0	0.08	0.04	0.39	0.21	0.09	22.1	1.5	1.1	6.9
5	369	0.62	41.0	549.5	139.8	1.83	4.8	0.10	0.16	0.44	0.49	0.10	31.3	26.5	1.1	7.0
15	448	0.19	41.7	573.3	171.5	2.18	34.3	0.10	0.07	0.52	0.27	0.12	27.0	2.0	1.4	8.1
5	488	0.22	46.6	624.8	188.7	2.37	4.9	0.11	0.00	0.57	0.30	0.13	29.4	2.2	1.5	8.8
12	419	0.17	39.2	536.6	161.7	2.03	29.4	0.10	0.04	0.49	0.26	0.11	27.0	2.0	1.3	12.0
13	260	0.22	39.2	399.4	98.0	1.10	14.7	0.11	0.00	0.51	0.30	0.14	29.4	2.1	1.5	—
5	300	2.70	100.0	580.0	290.0	2.25	—	0.37	—	0.42	5.00	0.50	100.0	15.0	1.5	—
0	100	1.08	24.0	5.0	20.0	—	0	0.22	—	0.10	—	0.04	—	0	0	—
0	150	1.44	40.0	15.0	40.0	—	—	0.30	—	0.13	—	0.08	—	0	0	—
0	400	1.44	—	0	30.0	—	0	—	—	—	—	—	—	0	—	—

TABLE A-1 Table of Food Composition *(continued)*

(Computer code is for Cengage Diet Analysis program)
(For purposes of calculations, use "0" for t, <1, <.1, <.01, etc.)

DA+ Code	Food Description	Quantity	Measure	Wt (g)	H_2O (g)	Ener (kcal)	Prot (g)	Carb (g)	Fiber (g)	Fat (g)	Fat Breakdown (g)		
											Sat	Mono	Poly
Eggs													
	Eggs												
99	Fried	1	item(s)	46	31.8	90	6.3	0.4	0	7.0	2.0	2.9	1.2
100	Hard boiled	1	item(s)	50	37.3	78	6.3	0.6	0	5.3	1.6	2.0	0.7
101	Poached	1	item(s)	50	37.8	71	6.3	0.4	0	5.0	1.5	1.9	0.7
97	Raw, white	1	item(s)	33	28.9	16	3.6	0.2	0	0.1	0	0	0
96	Raw, whole	1	item(s)	50	37.9	72	6.3	0.4	0	5.0	1.5	1.9	0.7
98	Raw, yolk	1	item(s)	17	8.9	54	2.7	0.6	0	4.5	1.6	2.0	0.7
102	Scrambled, prepared with milk and butter	2	item(s)	122	89.2	204	13.5	2.7	0	14.9	4.5	5.8	2.6
	Egg substitute												
4028	Egg Beaters	¼	cup(s)	61	—	30	6.0	1.0	0	0	0	0	0
920	Frozen	¼	cup(s)	60	43.9	96	6.8	1.9	0	6.7	1.2	1.5	3.7
918	Liquid	¼	cup(s)	63	51.9	53	7.5	0.4	0	2.1	0.4	0.6	1.0
Seafood													
	Cod												
6040	Atlantic cod or scrod, baked or broiled	3	ounce(s)	85	64.6	89	19.4	0	0	0.7	0.1	0.1	0.2
1573	Atlantic cod, cooked, dry heat	3	ounce(s)	85	64.6	89	19.4	0	0	0.7	0.1	0.1	0.2
2905	**Eel, raw**	3	ounce(s)	85	58.0	156	15.7	0	0	9.9	2.0	6.1	0.8
	Fish fillets												
25079	Baked	3	ounce(s)	84	79.9	99	21.7	0	0	0.7	0.1	0.1	0.3
8615	Batter coated or breaded, fried	3	ounce(s)	85	45.6	197	12.5	14.4	0.4	10.5	2.4	2.2	5.3
25082	Broiled fish steaks	3	ounce(s)	85	68.1	128	24.2	0	0	2.6	0.4	0.9	0.8
25083	Poached fish steaks	3	ounce(s)	85	67.1	111	21.1	0	0	2.3	0.3	0.8	0.7
25084	Steamed	3	ounce(s)	85	72.2	79	17.2	0	0	0.6	0.1	0.1	0.2
25089	**Flounder, baked**	3	ounce(s)	85	64.4	113	14.8	0.4	0.1	5.5	1.1	2.2	1.4
1825	**Grouper, cooked, dry heat**	3	ounce(s)	85	62.4	100	21.1	0	0	1.1	0.3	0.2	0.3
	Haddock												
6049	Baked or broiled	3	ounce(s)	85	63.2	95	20.6	0	0	0.8	0.1	0.1	0.3
1578	Cooked, dry heat	3	ounce(s)	85	63.1	95	20.6	0	0	0.8	0.1	0.1	0.3
1886	**Halibut, Atlantic and Pacific, cooked, dry heat**	3	ounce(s)	85	61.0	119	22.7	0	0	2.5	0.4	0.8	0.8
1582	**Herring, Atlantic, pickled**	4	piece(s)	60	33.1	157	8.5	5.8	0	10.8	1.4	7.2	1.0
1587	**Jack mackerel, solids, canned, drained**	2	ounce(s)	57	39.2	88	13.1	0	0	3.6	1.1	1.3	0.9
8580	**Octopus, common, cooked, moist heat**	3	ounce(s)	85	51.5	139	25.4	3.7	0	1.8	0.4	0.3	0.4
1831	**Perch, mixed species, cooked, dry heat**	3	ounce(s)	85	62.3	100	21.1	0	0	1.0	0.2	0.2	0.4
1592	**Pacific rockfish, cooked, dry heat**	3	ounce(s)	85	62.4	103	20.4	0	0	1.7	0.4	0.4	0.5
	Salmon												
2938	Coho, farmed, raw	3	ounce(s)	85	59.9	136	18.1	0	0	6.5	1.5	2.8	1.6
1594	Broiled or baked with butter	3	ounce(s)	85	53.9	155	23.0	0	0	6.3	1.2	2.3	2.3
29727	Smoked chinook (lox)	2	ounce(s)	57	40.8	66	10.4	0	0	2.4	0.5	1.1	0.6
154	**Sardine, Atlantic with bones, canned in oil**	3	ounce(s)	85	50.7	177	20.9	0	0	9.7	1.3	3.3	4.4
	Scallops												
155	Mixed species, breaded, fried	3	item(s)	47	27.2	100	8.4	4.7	—	5.1	1.2	2.1	1.3
1599	Steamed	3	ounce(s)	85	64.8	90	13.8	2.0	0	2.6	0.4	1.0	0.8
1839	**Snapper, mixed species, cooked, dry heat**	3	ounce(s)	85	59.8	109	22.4	0	0	1.5	0.3	0.3	0.5
	Squid												
1868	Mixed species, fried	3	ounce(s)	85	54.9	149	15.3	6.6	0	6.4	1.6	2.3	1.8
16617	Steamed or boiled	3	ounce(s)	85	63.3	89	15.2	3.0	0	1.3	0.4	0.1	0.5
1570	**Striped bass, cooked, dry heat**	3	ounce(s)	85	62.4	105	19.3	0	0	2.5	0.6	0.7	0.9
1601	**Sturgeon, steamed**	3	ounce(s)	85	59.4	111	17.0	0	0	4.3	1.0	2.0	0.7
1840	**Surimi, formed**	3	ounce(s)	85	64.9	84	12.9	5.8	0	0.8	0.2	0.1	0.4
1842	**Swordfish, cooked, dry heat**	3	ounce(s)	85	58.5	132	21.6	0	0	4.4	1.2	1.7	1.0
1846	**Tuna, yellowfin or ahi, raw**	3	ounce(s)	85	60.4	92	19.9	0	0	0.8	0.2	0.1	0.2
	Tuna, canned												
159	Light, canned in oil, drained	2	ounce(s)	57	33.9	112	16.5	0	0	4.6	0.9	1.7	1.6
355	Light, canned in water, drained	2	ounce(s)	57	42.2	66	14.5	0	0	0.5	0.1	0.1	0.2
33211	Light, no salt, canned in oil, drained	2	ounce(s)	57	33.9	112	16.5	0	0	4.7	0.9	1.7	1.6

PAGE KEY: A-4 = Breads/Baked Goods A-10 = Cereal/Rice/Pasta A-14 = Fruit A-18 = Vegetables/Legumes A-28 = Nuts/Seeds A-30 = Vegetarian A-32 = Dairy A-40 = Eggs A-40 = Seafood A-42 = Meats A-46 = Poultry A-46 = Processed Meats A-48 = Beverages A-52 = Fats/Oils A-54 = Sweets A-56 = Spices/Condiments/Sauces A-60 = Mixed Foods/Soups/Sandwiches A-64 = Fast Food A-84 = Convenience A-86 = Baby Foods

Chol (mg)	Calc (mg)	Iron (mg)	Magn (mg)	Pota (mg)	Sodi (mg)	Zinc (mg)	Vit A (μg)	Thia (mg)	Vit E (mg α)	Ribo (mg)	Niac (mg)	Vit B_6 (mg)	Fola (μg)	Vit C (mg)	Vit B_{12} (μg)	Sele (μg)
210	27	0.91	6.0	67.6	93.8	0.55	91.1	0.03	0.56	0.23	0.03	0.07	23.5	0	0.6	15.7
212	25	0.59	5.0	63.0	62.0	0.52	84.5	0.03	0.51	0.25	0.03	0.06	22.0	0	0.6	15.4
211	27	0.91	6.0	66.5	147.0	0.55	69.5	0.02	0.48	0.20	0.03	0.06	17.5	0	0.6	15.8
0	2	0.02	3.6	53.8	54.8	0.01	0	0.00	0.00	0.14	0.03	0.00	1.3	0	0	6.6
212	27	0.91	6.0	67.0	70.0	0.55	70.0	0.03	0.48	0.23	0.03	0.07	23.5	0	0.6	15.9
210	22	0.46	0.9	18.5	8.2	0.39	64.8	0.03	0.43	0.09	0.00	0.06	24.8	0	0.3	9.5
429	87	1.46	14.6	168.4	341.6	1.22	174.5	0.06	1.33	0.53	0.09	0.14	36.6	0.2	0.9	27.5
0	20	1.08	4.0	85.0	115.0	0.60	112.5	0.15	—	0.85	0.20	0.08	60.0	0	1.2	—
1	44	1.18	9.0	127.8	119.4	0.58	6.6	0.07	0.95	0.23	0.08	0.08	9.6	0.3	0.2	24.8
1	33	1.32	5.6	207.1	111.1	0.82	11.3	0.07	0.17	0.19	0.07	0.00	9.4	0	0.2	15.6
47	12	0.41	35.7	207.5	66.3	0.49	11.9	0.07	0.68	0.06	2.13	0.24	6.8	0.8	0.9	32.0
47	12	0.41	35.7	207.5	66.3	0.49	11.9	0.07	0.68	0.06	2.13	0.24	6.8	0.9	0.9	32.0
107	17	0.42	17.0	231.3	43.4	1.37	887.0	0.13	3.40	0.03	2.97	0.05	12.8	1.5	2.6	5.5
44	8	0.31	29.1	489.0	86.1	0.48	—	0.02	—	0.05	2.47	0.46	8.1	3.0	1.0	44.3
29	15	1.79	20.4	272.2	452.5	0.37	9.4	0.09	—	0.09	1.78	0.08	17.0	0	0.9	7.7
37	55	0.97	96.7	524.3	62.9	0.49	—	0.05	—	0.08	6.47	0.36	12.6	0	1.2	42.5
32	48	0.85	84.0	455.6	54.7	0.42	—	0.05	—	0.07	5.92	0.33	11.5	0	1.1	37.0
41	12	0.29	24.7	319.3	41.7	0.34	—	0.06	—	0.06	1.89	0.21	6.1	0.8	0.8	32.0
44	19	0.34	47.3	224.7	280.2	0.20	—	0.06	0.40	0.07	2.02	0.18	7.4	2.8	1.6	33.5
40	18	0.96	31.5	404.0	45.1	0.43	42.5	0.06	—	0.01	0.32	0.29	8.5	0	0.6	39.8
63	36	1.15	42.5	339.4	74.0	0.40	16.2	0.03	0.42	0.03	3.94	0.29	6.8	0	1.2	34.4
63	36	1.14	42.5	339.3	74.0	0.40	16.2	0.03	—	0.03	3.93	0.29	11.1	0	1.2	34.4
35	51	0.91	91.0	489.9	58.7	0.45	45.9	0.05	—	0.07	6.05	0.33	11.9	0	1.2	39.8
8	46	0.73	4.8	41.4	522.0	0.31	154.8	0.02	1.02	0.08	1.98	0.10	1.2	0	2.6	35.1
45	137	1.15	21.0	110.0	214.9	0.57	73.7	0.02	0.58	0.12	3.50	0.11	2.8	0.5	3.9	21.4
82	90	8.11	51.0	535.8	391.2	2.85	76.5	0.04	1.02	0.06	3.21	0.55	20.4	6.8	30.6	76.2
98	87	0.98	32.3	292.6	67.2	1.21	8.5	0.06	—	0.10	1.61	0.11	5.1	1.4	1.9	13.7
37	10	0.45	28.9	442.3	65.5	0.45	60.4	0.03	1.32	0.07	3.33	0.22	8.5	0	1.0	39.8
43	10	0.29	26.4	382.7	40.0	0.36	47.6	0.08	—	0.09	5.79	0.56	11.1	0.9	2.3	10.7
40	15	1.02	26.9	376.6	98.6	0.56	—	0.13	1.14	0.05	8.33	0.18	4.2	1.8	2.3	41.0
13	6	0.48	10.2	99.2	1134.0	0.17	14.7	0.01	—	0.05	2.67	0.15	1.1	0	1.8	21.6
121	325	2.48	33.2	337.6	429.5	1.10	27.2	0.04	1.70	0.18	4.43	0.14	10.2	0	7.6	44.8
28	20	0.38	27.4	154.8	215.8	0.49	10.7	0.02	—	0.05	0.70	0.06	23.3	1.1	0.6	12.5
27	20	0.22	45.9	238.0	358.7	0.78	32.3	0.01	0.16	0.05	0.84	0.11	10.2	2.0	1.1	18.2
40	34	0.20	31.5	444.0	48.5	0.37	29.8	0.04	—	0.00	0.29	0.39	5.1	1.4	3.0	41.7
221	33	0.85	32.3	237.3	260.3	1.48	9.4	0.04	—	0.39	2.21	0.04	11.9	3.6	1.0	44.1
227	31	0.62	28.9	192.1	356.2	1.49	8.5	0.01	1.17	0.32	1.69	0.04	3.4	3.2	1.0	43.7
88	16	0.91	43.4	279.0	74.8	0.43	26.4	0.09	—	0.03	2.17	0.29	8.5	0	3.8	39.8
63	11	0.59	29.8	239.7	388.5	0.35	198.9	0.06	0.52	0.07	8.30	0.19	14.5	0	2.2	13.3
26	8	0.22	36.6	95.3	121.6	0.28	17.0	0.01	0.53	0.01	0.18	0.02	1.7	0	1.4	23.9
43	5	0.88	28.9	313.8	97.8	1.25	34.9	0.03	—	0.09	10.02	0.32	1.7	0.9	1.7	52.5
38	14	0.62	42.5	377.6	31.5	0.44	15.3	0.37	0.42	0.04	8.33	0.77	1.7	0.8	0.4	31.0
10	7	0.79	17.6	117.3	200.6	0.51	13.0	0.02	0.49	0.07	7.03	0.06	2.8	0	1.2	43.1
17	6	0.87	15.3	134.3	191.5	0.43	9.6	0.01	0.19	0.04	7.52	0.19	2.3	0	1.7	45.6
10	7	0.78	17.6	117.4	28.3	0.51	0	0.02	—	0.06	7.03	0.06	2.8	0	1.2	43.1

TABLE A-1 Table of Food Composition *(continued)*

(Computer code is for Cengage Diet Analysis program)
(For purposes of calculations, use "0" for t, <1, <.1, <.01, etc.)

DA+ Code	Food Description	Quantity	Measure	Wt (g)	H_2O (g)	Ener (kcal)	Prot (g)	Carb (g)	Fiber (g)	Fat (g)	Fat Breakdown (g)		
											Sat	Mono	Poly
Seafood—*continued*													
33212	Light, no salt, canned in water, drained	2	ounce(s)	57	42.6	66	14.5	0	0	0.5	0.1	0.1	0.2
2961	White, canned in oil, drained	2	ounce(s)	57	36.3	105	15.0	0	0	4.6	0.7	1.8	1.7
351	White, canned in water, drained	2	ounce(s)	57	41.5	73	13.4	0	0	1.7	0.4	0.4	0.6
33213	White, no salt, canned in oil, drained	2	ounce(s)	57	36.3	105	15.0	0	0	4.6	0.9	1.4	1.9
33214	White, no salt, canned in water, drained	2	ounce(s)	57	42.0	73	13.4	0	0	1.7	0.4	0.4	0.6
	Yellowtail												
8548	Mixed species, cooked, dry heat	3	ounce(s)	85	57.3	159	25.2	0	0	5.7	1.4	2.2	1.5
2970	Mixed species, raw	2	ounce(s)	57	42.2	83	13.1	0	0	3.0	0.7	1.1	0.8
	Shellfish, meat only												
1857	Abalone, mixed species, fried	3	ounce(s)	85	51.1	161	16.7	9.4	0	5.8	1.4	2.3	1.4
16618	Abalone, steamed or poached	3	ounce(s)	85	40.7	177	28.8	10.1	0	1.3	0.3	0.2	0.2
	Crab												
1851	Blue crab, canned	2	ounce(s)	57	43.2	56	11.6	0	0	0.7	0.1	0.1	0.2
1852	Blue crab, cooked, moist heat	3	ounce(s)	85	65.9	87	17.2	0	0	1.5	0.2	0.2	0.6
8562	Dungeness crab, cooked, moist heat	3	ounce(s)	85	62.3	94	19.0	0.8	0	1.1	0.1	0.2	0.3
1860	**Clams, cooked, moist heat**	3	ounce(s)	85	54.1	126	21.7	4.4	0	1.7	0.2	0.1	0.5
1853	**Crayfish, farmed, cooked, moist heat**	3	ounce(s)	85	68.7	74	14.9	0	0	1.1	0.2	0.2	0.4
	Oysters												
8720	Baked or broiled	3	ounce(s)	85	68.6	89	5.6	3.2	0	5.8	1.3	2.1	1.9
152	Eastern, farmed, raw	3	ounce(s)	85	73.3	50	4.4	4.7	0	1.3	0.4	0.1	0.5
8715	Eastern, wild, cooked, moist heat	3	ounce(s)	85	59.8	117	12.0	6.7	0	4.2	1.3	0.5	1.6
8584	Pacific, cooked, moist heat	3	ounce(s)	85	54.5	139	16.1	8.4	0	3.9	0.9	0.7	1.5
1865	Pacific, raw	3	ounce(s)	85	69.8	69	8.0	4.2	0	2.0	0.4	0.3	0.8
1854	**Lobster, northern, cooked, moist heat**	3	ounce(s)	85	64.7	83	17.4	1.1	0	0.5	0.1	0.1	0.1
1862	**Mussel, blue, cooked, moist heat**	3	ounce(s)	85	52.0	146	20.2	6.3	0	3.8	0.7	0.9	1.0
	Shrimp												
158	Mixed species, breaded, fried	3	ounce(s)	85	44.9	206	18.2	9.8	0.3	10.4	1.8	3.2	4.3
1855	Mixed species, cooked, moist heat	3	ounce(s)	85	65.7	84	17.8	0	0	0.9	0.2	0.2	0.4
Beef, Lamb, Pork													
	Beef												
4450	Breakfast strips, cooked	2	slice(s)	23	5.9	101	7.1	0.3	0	7.8	3.2	3.8	0.4
174	Corned beef, canned	3	ounce(s)	85	49.1	213	23.0	0	0	12.7	5.3	5.1	0.5
33147	Cured, thin siced	2	ounce(s)	57	32.9	100	15.9	3.2	0	2.2	0.9	1.0	0.1
4581	Jerky	1	ounce(s)	28	6.6	116	9.4	3.1	0.5	7.3	3.1	3.2	0.3
	Ground beef												
5898	Lean, broiled, medium	3	ounce(s)	85	50.4	202	21.6	0	0	12.2	4.8	5.3	0.4
5899	Lean, broiled, well done	3	ounce(s)	85	48.4	214	23.8	0	0	12.5	5.0	5.7	0.3
5914	Regular, broiled, medium	3	ounce(s)	85	46.1	246	20.5	0	0	17.6	6.9	7.7	0.6
5915	Regular, broiled, well done	3	ounce(s)	85	43.8	259	21.6	0	0	18.4	7.5	8.5	0.5
	Beef rib												
4241	Rib, small end, separable lean, 0″ fat, broiled	3	ounce(s)	85	53.2	164	25.0	0	0	6.4	2.4	2.6	0.2
4183	Rib, whole, lean and fat, ¼″ fat, roasted	3	ounce(s)	85	39.0	320	18.9	0	0	26.6	10.7	11.4	0.9
	Beef roast												
16981	Bottom round, choice, separable lean and fat, ⅛″ fat, braised	3	ounce(s)	85	46.2	216	27.9	0	0	10.7	4.1	4.6	0.4
16979	Bottom round, separable lean and fat, ⅛″ fat, roasted	3	ounce(s)	85	52.4	185	22.5	0	0	9.9	3.8	4.2	0.4
16924	Chuck, arm pot roast, separable lean and fat, ⅛″ fat, braised	3	ounce(s)	85	42.9	257	25.6	0	0	16.3	6.5	7.0	0.6
16930	Chuck, blade roast, separable lean and fat, ⅛″ fat, braised	3	ounce(s)	85	40.5	290	22.8	0	0	21.4	8.5	9.2	0.8
5853	Chuck, blade roast, separable lean, 0″ trim, pot roasted	3	ounce(s)	85	47.4	202	26.4	0	0	9.9	3.9	4.3	0.3
4296	Eye of round, choice, separable lean, 0″ fat, roasted	3	ounce(s)	85	56.5	138	24.4	0	0	3.7	1.3	1.5	0.1
16989	Eye of round, separable lean and fat, ⅛″ fat, roasted	3	ounce(s)	85	52.2	180	24.2	0	0	8.5	3.2	3.6	0.3

PAGE KEY: A-4 = Breads/Baked Goods A-10 = Cereal/Rice/Pasta A-14 = Fruit A-18 = Vegetables/Legumes A-28 = Nuts/Seeds A-30 = Vegetarian A-32 = Dairy A-40 = Eggs A-40 = Seafood A-42 = Meats A-46 = Poultry A-46 = Processed Meats A-48 = Beverages A-52 = Fats/Oils A-54 = Sweets A-56 = Spices/Condiments/Sauces A-60 = Mixed Foods/Soups/Sandwiches A-64 = Fast Food A-84 = Convenience A-86 = Baby Foods

Chol (mg)	Calc (mg)	Iron (mg)	Magn (mg)	Pota (mg)	Sodi (mg)	Zinc (mg)	Vit A (µg)	Thia (mg)	Vit E (mg α)	Ribo (mg)	Niac (mg)	Vit B_6 (mg)	Fola (µg)	Vit C (mg)	Vit B_{12} (µg)	Sele (µg)
17	6	0.86	15.3	134.4	28.3	0.43	0	0.01	—	0.04	7.52	0.19	2.3	0	1.7	45.6
18	2	0.36	19.3	188.8	224.5	0.26	2.8	0.01	1.30	0.04	6.63	0.24	2.8	0	1.2	34.1
24	8	0.55	18.7	134.3	213.6	0.27	3.4	0.00	0.48	0.02	3.28	0.12	1.1	0	0.7	37.2
18	2	0.36	19.3	188.8	28.3	0.26	0	0.01	—	0.04	6.63	0.24	2.8	0	1.2	34.1
24	8	0.54	18.7	134.4	28.3	0.27	3.4	0.00	—	0.02	3.28	0.12	1.1	0	0.7	37.3
60	25	0.53	32.3	457.6	42.5	0.56	26.4	0.14	—	0.04	7.41	0.15	3.4	2.5	1.1	39.8
31	13	0.28	17.0	238.1	22.1	0.29	16.4	0.08	—	0.02	3.86	0.09	2.3	1.6	0.7	20.7
80	31	3.23	47.6	241.5	502.6	0.80	1.7	0.18	—	0.11	1.61	0.12	11.9	1.5	0.6	44.1
144	50	4.84	68.9	295.0	980.1	1.38	3.4	0.28	6.74	0.12	1.89	0.21	6.0	2.6	0.7	75.6
50	57	0.47	22.1	212.1	188.8	2.27	1.1	0.04	1.04	0.04	0.77	0.08	24.4	1.5	0.3	18.0
85	88	0.77	28.1	275.6	237.3	3.58	1.7	0.08	1.56	0.04	2.80	0.15	43.4	2.8	6.2	34.2
65	50	0.36	49.3	347.0	321.5	4.65	26.4	0.04	—	0.17	3.08	0.14	35.7	3.1	8.8	40.5
57	78	23.78	15.3	534.1	95.3	2.32	145.4	0.12	—	0.36	2.85	0.09	24.7	18.8	84.1	54.4
117	43	0.94	28.1	202.4	82.5	1.25	12.8	0.03	—	0.06	1.41	0.11	9.4	0.4	2.6	29.1
43	36	5.30	37.4	125.0	403.8	72.22	60.4	0.07	0.98	0.06	1.04	0.04	7.7	2.8	14.7	50.7
21	37	4.91	28.1	105.4	151.3	32.23	6.8	0.08	—	0.05	1.07	0.05	15.3	4.0	13.8	54.1
89	77	10.19	80.8	239.0	358.9	154.45	45.9	0.16	—	0.15	2.11	0.10	11.9	5.1	29.8	60.9
85	14	7.82	37.4	256.8	180.3	28.27	124.2	0.10	0.72	0.37	3.07	0.07	12.8	10.9	24.5	131.0
43	7	4.34	18.7	142.9	90.1	14.13	68.9	0.05	—	0.20	1.70	0.04	8.5	6.8	13.6	65.5
61	52	0.33	29.8	299.4	323.2	2.48	22.1	0.01	0.85	0.05	0.91	0.06	9.4	0	2.6	36.3
48	28	5.71	31.5	227.9	313.8	2.27	77.4	0.25	—	0.35	2.55	0.08	64.6	11.6	20.4	76.2
150	57	1.07	34.0	191.3	292.4	1.17	0	0.11	—	0.11	2.60	0.08	33.2	1.3	1.6	35.4
166	33	2.62	28.9	154.8	190.5	1.32	57.8	0.02	1.17	0.02	2.20	0.10	3.4	1.9	1.3	33.7
27	2	0.71	6.1	93.1	509.2	1.44	0	0.02	0.06	0.05	1.46	0.07	1.8	0	0.8	6.1
73	10	1.76	11.9	115.7	855.6	3.03	0	0.01	0.12	0.12	2.06	0.11	7.7	0	1.4	36.5
23	6	1.53	10.8	243.2	815.9	2.25	0	0.04	0.00	0.10	2.98	0.19	6.2	0	1.5	16.0
14	6	1.53	14.5	169.2	627.4	2.29	0	0.04	0.13	0.04	0.49	0.05	38.0	0	0.3	3.0
58	6	2.00	17.9	266.2	59.5	4.63	0	0.05	—	0.23	4.21	0.23	7.6	0	1.8	16.0
69	12	2.21	18.4	250.0	62.4	5.86	0	0.08	—	0.23	5.10	0.16	9.4	0	1.7	19.0
62	9	2.07	17.0	248.3	70.6	4.40	0	0.02	—	0.16	4.90	0.23	7.6	0	2.5	16.2
71	12	2.30	18.5	242.4	72.4	5.18	0	0.08	—	0.23	4.93	0.17	8.5	0	1.6	18.0
65	16	1.59	21.3	319.8	51.9	4.64	0	0.06	0.34	0.12	7.15	0.53	8.5	0	1.4	29.2
72	9	1.96	16.2	251.7	53.6	4.45	0	0.06	—	0.14	2.85	0.19	6.0	0	2.1	18.7
68	6	2.29	17.9	223.7	35.7	4.59	0	0.05	0.41	0.15	5.05	0.36	8.5	0	1.7	29.3
64	5	1.83	14.5	182.0	29.8	3.76	0	0.05	0.34	0.12	3.92	0.29	6.8	0	1.3	23.0
67	14	2.15	17.0	205.8	42.5	5.93	0	0.05	0.45	0.15	3.63	0.25	7.7	0	1.9	24.1
88	11	2.66	16.2	198.2	55.3	7.15	0	0.06	0.17	0.20	2.06	0.22	4.3	0	1.9	20.9
73	11	3.12	19.6	223.7	60.4	8.73	0	0.06	—	0.23	2.27	0.24	5.1	0	2.1	22.7
49	5	2.16	16.2	200.7	32.3	4.28	0	0.05	0.30	0.15	4.69	0.34	8.5	0	1.4	28.0
54	5	1.98	15.3	193.1	31.5	3.95	0	0.05	0.34	0.13	4.37	0.31	7.7	0	1.5	25.2

TABLE A-1 Table of Food Composition *(continued)*

(Computer code is for Cengage Diet Analysis program)
(For purposes of calculations, use "0" for t, <1, <.1, <.01, etc.)

DA+ Code	Food Description	Quantity	Measure	Wt (g)	H_2O (g)	Ener (kcal)	Prot (g)	Carb (g)	Fiber (g)	Fat (g)	Fat Breakdown (g)		
											Sat	Mono	Poly
Beef, Lamb, Pork—*continued*													
	Beef steak												
4348	Short loin, t-bone steak, lean and fat, ¼″ fat, broiled	3	ounce(s)	85	43.2	274	19.4	0	0	21.2	8.3	9.6	0.8
4349	Short loin, t-bone steak, lean, ¼″ fat, broiled	3	ounce(s)	85	52.3	174	22.8	0	0	8.5	3.1	4.2	0.3
4360	Top loin, prime, lean and fat, ¼″ fat, broiled	3	ounce(s)	85	42.7	275	21.6	0	0	20.3	8.2	8.6	0.7
	Beef variety												
188	Liver, pan fried	3	ounce(s)	85	52.7	149	22.6	4.4	0	4.0	1.3	0.5	0.5
4447	Tongue, simmered	3	ounce(s)	85	49.2	242	16.4	0	0	19.0	6.9	8.6	0.6
	Lamb chop												
3275	Loin, domestic, lean and fat, ¼″ fat, broiled	3	ounce(s)	85	43.9	269	21.4	0	0	19.6	8.4	8.3	1.4
	Lamb leg												
3264	Domestic, lean and fat, ¼″ fat, cooked	3	ounce(s)	85	45.7	250	20.9	0	0	17.8	7.5	7.5	1.3
	Lamb rib												
182	Domestic, lean and fat, ¼″ fat, broiled	3	ounce(s)	85	40.0	307	18.8	0	0	25.2	10.8	10.3	2.0
183	Domestic, lean, ¼″ fat, broiled	3	ounce(s)	85	50.0	200	23.6	0	0	11.0	4.0	4.4	1.0
	Lamb shoulder												
186	Shoulder, arm and blade, domestic, choice, lean and fat, ¼″ fat, roasted	3	ounce(s)	85	47.8	235	19.1	0	0	17.0	7.2	6.9	1.4
187	Shoulder, arm and blade, domestic, choice, lean, ¼″ fat, roasted	3	ounce(s)	85	53.8	173	21.2	0	0	9.2	3.5	3.7	0.8
3287	Shoulder, arm, domestic, lean and fat, ¼″ fat, braised	3	ounce(s)	85	37.6	294	25.8	0	0	20.4	8.4	8.7	1.5
3290	Shoulder, arm, domestic, lean, ¼″ fat, braised	3	ounce(s)	85	41.9	237	30.2	0	0	12.0	4.3	5.2	0.8
	Lamb variety												
3375	Brain, pan fried	3	ounce(s)	85	51.6	232	14.4	0	0	18.9	4.8	3.4	1.9
3406	Tongue, braised	3	ounce(s)	85	49.2	234	18.3	0	0	17.2	6.7	8.5	1.1
	Pork, cured												
29229	Bacon, Canadian style, cured	2	ounce(s)	57	37.9	89	11.7	1.0	0	4.0	1.3	1.8	0.4
161	Bacon, cured, broiled, pan fried or roasted	2	slice(s)	16	2.0	87	5.9	0.2	0	6.7	2.2	3.0	0.7
35422	Breakfast strips, cured, cooked	3	slice(s)	34	9.2	156	9.8	0.4	0	12.5	4.3	5.6	1.9
189	Ham, cured, boneless, 11% fat, roasted	3	ounce(s)	85	54.9	151	19.2	0	0	7.7	2.7	3.8	1.2
29215	Ham, cured, extra lean, 4% fat, canned	2	2 ounce(s)	57	41.7	68	10.5	0	0	2.6	0.9	1.3	0.2
1316	Ham, cured, extra lean, 5% fat, roasted	3	ounce(s)	85	57.6	123	17.8	1.3	0	4.7	1.5	2.2	0.5
16561	Ham, smoked or cured, lean, cooked	1	slice(s)	42	27.6	66	10.5	0	0	2.3	0.8	1.1	0.3
	Pork chop												
32671	Loin, blade, chops, lean and fat, pan fried	3	ounce(s)	85	42.5	291	18.3	0	0	23.6	8.6	10	2.6
32672	Loin, center cut, chops, lean and fat, pan fried	3	ounce(s)	85	45.1	236	25.4	0	0	14.1	5.1	6.0	1.6
32682	Loin, center rib, chops, boneless, lean and fat, braised	3	ounce(s)	85	49.5	217	22.4	0	0	13.4	5.2	6.1	1.1
32603	Loin, center rib, chops, lean, broiled	3	ounce(s)	85	55.4	158	21.9	0	0	7.1	2.4	3.0	0.8
32478	Loin, whole, lean and fat, braised	3	ounce(s)	85	49.6	203	23.2	0	0	11.6	4.3	5.2	1.0
32481	Loin, whole, lean, braised	3	ounce(s)	85	52.2	174	24.3	0	0	7.8	2.9	3.5	0.6
	Pork leg or ham												
32471	Pork leg or ham, rump portion, lean and fat, roasted	3	ounce(s)	85	48.3	214	24.6	0	0	12.1	4.5	5.4	1.2
32468	Pork leg or ham, whole, lean and fat, roasted	3	ounce(s)	85	46.8	232	22.8	0	0	15.0	5.5	6.7	1.4
	Pork ribs												
32693	Loin, country style, lean and fat, roasted	3	ounce(s)	85	43.3	279	19.9	0	0	21.6	7.8	9.4	1.7
32696	Loin, country style, lean, roasted	3	ounce(s)	85	49.5	210	22.6	0	0	12.6	4.5	5.5	0.9

PAGE KEY: A-4 = Breads/Baked Goods A-10 = Cereal/Rice/Pasta A-14 = Fruit A-18 = Vegetables/Legumes A-28 = Nuts/Seeds A-30 = Vegetarian A-32 = Dairy A-40 = Eggs A-40 = Seafood A-42 = Meats A-46 = Poultry A-46 = Processed Meats A-48 = Beverages A-52 = Fats/Oils A-54 = Sweets A-56 = Spices/Condiments/Sauces A-60 = Mixed Foods/Soups/Sandwiches A-64 = Fast Food A-84 = Convenience A-86 = Baby Foods

Chol (mg)	Calc (mg)	Iron (mg)	Magn (mg)	Pota (mg)	Sodi (mg)	Zinc (mg)	Vit A (μg)	Thia (mg)	Vit E (mg α)	Ribo (mg)	Niac (mg)	Vit B_6 (mg)	Fola (μg)	Vit C (mg)	Vit B_{12} (μg)	Sele (μg)
58	7	2.56	17.9	233.9	57.8	3.56	0	0.07	0.18	0.17	3.29	0.27	6.0	0	1.8	10.0
50	5	3.11	22.1	278.1	65.5	4.34	0	0.09	0.11	0.21	3.93	0.33	6.8	0	1.9	8.5
67	8	1.88	19.6	294.3	53.6	3.85	0	0.06	—	0.15	3.96	0.31	6.0	0	1.6	19.5
324	5	5.24	18.7	298.5	65.5	4.44	6586.3	0.15	0.39	2.91	14.86	0.87	221.1	0.6	70.7	27.9
112	4	2.22	12.8	156.5	55.3	3.47	0	0.01	0.25	0.25	2.96	0.13	6.0	1.1	2.7	11.2
85	17	1.53	20.4	278.1	65.5	2.96	0	0.08	0.11	0.21	6.03	0.11	15.3	0	2.1	23.3
82	14	1.59	19.6	263.7	61.2	3.79	0	0.08	0.11	0.21	5.66	0.11	15.3	0	2.2	22.5
84	16	1.59	19.6	229.5	64.6	3.40	0	0.07	0.10	0.18	5.95	0.09	11.9	0	2.2	20.3
77	14	1.87	24.7	266.1	72.3	4.47	0	0.08	0.15	0.21	5.56	0.12	17.9	0	2.2	26.4
78	17	1.67	19.6	213.4	56.1	4.44	0	0.07	0.11	0.20	5.22	0.11	17.9	0	2.2	22.3
74	16	1.81	21.3	225.3	57.8	5.13	0	0.07	0.15	0.22	4.89	0.12	21.3	0	2.3	24.2
102	21	2.03	22.1	260.3	61.2	5.17	0	0.06	0.12	0.21	5.66	0.09	15.3	0	2.2	31.6
103	22	2.29	24.7	287.5	64.6	6.20	0	0.06	0.15	0.23	5.38	0.11	18.7	0	2.3	32.1
2130	18	1.73	18.7	304.5	133.5	1.70	0	0.14	—	0.31	3.87	0.19	6.0	19.6	20.5	10.2
161	9	2.23	13.6	134.4	57.0	2.54	0	0.06	—	0.35	3.13	0.14	2.6	6.0	5.4	23.8
28	5	0.38	9.6	195.0	798.9	0.78	0	0.42	0.11	0.09	3.53	0.22	2.3	0	0.4	14.2
18	2	0.22	5.3	90.4	369.6	0.56	1.8	0.06	0.04	0.04	1.76	0.04	0.3	0	0.2	9.9
36	5	0.67	8.8	158.4	713.7	1.25	0	0.25	0.08	0.12	2.58	0.11	1.4	0	0.6	8.4
50	7	1.13	18.7	347.7	1275.0	2.09	0	0.62	0.26	0.28	5.22	0.26	2.6	0	0.6	16.8
22	3	0.53	9.6	206.4	711.6	1.09	0	0.47	0.09	0.13	3.00	0.25	3.4	0	0.5	8.2
45	7	1.25	11.9	244.1	1023.1	2.44	0	0.64	0.21	0.17	3.42	0.34	2.6	0	0.6	16.6
23	3	0.39	9.2	132.7	557.3	1.07	0	0.28	0.10	0.10	2.10	0.19	1.7	0	0.3	10.7
72	26	0.74	17.9	282.4	57.0	2.71	1.7	0.52	0.17	0.25	3.35	0.28	3.4	0.5	0.7	29.7
78	23	0.77	24.7	361.5	68.0	1.96	1.7	0.96	0.21	0.25	4.76	0.39	5.1	0.9	0.6	33.2
62	4	0.78	14.5	329.1	34.0	1.76	1.7	0.44	—	0.20	3.66	0.26	3.4	0.3	0.4	28.4
56	22	0.57	21.3	291.7	48.5	1.91	0	0.48	0.08	0.18	6.68	0.57	0	0	0.4	38.6
68	18	0.91	16.2	318.1	40.8	2.02	1.7	0.53	0.20	0.21	3.75	0.31	2.6	0.5	0.5	38.5
67	15	0.96	17.0	329.1	42.5	2.10	1.7	0.56	0.17	0.22	3.90	0.32	3.4	0.5	0.5	41.0
82	10	0.89	23.0	318.1	52.7	2.39	2.6	0.63	0.18	0.28	3.95	0.26	2.6	0.2	0.6	39.8
80	12	0.85	18.7	299.4	51.0	2.51	2.6	0.54	0.18	0.26	3.89	0.34	8.5	0.3	0.6	38.5
78	21	0.90	19.6	292.6	44.2	2.00	2.6	0.75	—	0.29	3.67	0.37	4.3	0.3	0.7	31.6
79	25	1.09	20.4	296.8	24.7	3.24	1.7	0.48	—	0.29	3.96	0.37	4.3	0.3	0.7	36.0

TABLE A-1 Table of Food Composition *(continued)*

(Computer code is for Cengage Diet Analysis program)
(For purposes of calculations, use "0" for t, <1, <.1, <.01, etc.)

DA+ Code	Food Description	Quantity	Measure	Wt (g)	H_2O (g)	Ener (kcal)	Prot (g)	Carb (g)	Fiber (g)	Fat (g)	Fat Breakdown (g) Sat	Mono	Poly
Beef, Lamb, Pork—*continued*													
	Pork shoulder												
32626	Shoulder, arm picnic, lean and fat, roasted	3	ounce(s)	85	44.3	270	20.0	0	0	20.4	7.5	9.1	2.0
32629	Shoulder, arm picnic, lean, roasted	3	ounce(s)	85	51.3	194	22.7	0	0	10.7	3.7	5.1	1.0
	Rabbit												
3366	Domesticated, roasted	3	ounce(s)	85	51.5	168	24.7	0	0	6.8	2.0	1.8	1.3
3367	Domesticated, stewed	3	ounce(s)	85	50.0	175	25.8	0	0	7.2	2.1	1.9	1.4
	Veal												
3391	Liver, braised	3	ounce(s)	85	50.9	163	24.2	3.2	0	5.3	1.7	1.0	0.9
3319	Rib, lean only, roasted	3	ounce(s)	85	55.0	151	21.9	0	0	6.3	1.8	2.3	0.6
1732	Deer or venison, roasted	3	ounce(s)	85	55.5	134	25.7	0	0	2.7	1.1	0.7	0.5
Poultry													
	Chicken												
29562	Flaked, canned	2	ounce(s)	57	39.3	97	10.3	0.1	0	5.8	1.6	2.3	1.3
	Chicken, fried												
29632	Breast, meat only, breaded, baked or fried	3	ounce(s)	85	44.3	193	25.3	6.9	0.2	6.6	1.6	2.7	1.7
35327	Broiler breast, meat only, fried	3	ounce(s)	85	51.2	159	28.4	0.4	0	4.0	1.1	1.5	0.9
36413	Broiler breast, meat and skin, flour coated, fried	3	ounce(s)	85	48.1	189	27.1	1.4	0.1	7.5	2.1	3.0	1.7
36414	Broiler drumstick, meat and skin, flour coated, fried	3	ounce(s)	85	48.2	208	22.9	1.4	0.1	11.7	3.1	4.6	2.7
35389	Broiler drumstick, meat only, fried	3	ounce(s)	85	52.9	166	24.3	0	0	6.9	1.8	2.5	1.7
35406	Broiler leg, meat only, fried	3	ounce(s)	85	51.5	177	24.1	0.6	0	7.9	2.1	2.9	1.9
35484	Broiler wing, meat only, fried	3	ounce(s)	85	50.9	179	25.6	0	0	7.8	2.1	2.6	1.8
29580	Patty, fillet or tenders, breaded, cooked	3	ounce(s)	85	40.2	256	14.5	12.2	0	16.5	3.7	8.4	3.7
	Chicken, roasted, meat only												
35409	Broiler leg, meat only, roasted	3	ounce(s)	85	55.0	162	23.0	0	0	7.2	1.9	2.6	1.7
35486	Broiler wing, meat only, roasted	3	ounce(s)	85	53.4	173	25.9	0	0	6.9	1.9	2.2	1.5
35138	Roasting chicken, dark meat, meat only, roasted	3	ounce(s)	85	57.0	151	19.8	0	0	7.4	2.1	2.8	1.7
35136	Roasting chicken, light meat, meat only, roasted	3	ounce(s)	85	57.7	130	23.1	0	0	3.5	0.9	1.3	0.8
35132	Roasting chicken, meat only, roasted	3	ounce(s)	85	57.3	142	21.3	0	0	5.6	1.5	2.1	1.3
	Chicken, stewed												
1268	Gizzard, simmered	3	ounce(s)	85	57.8	124	25.8	0	0	2.3	0.6	0.4	0.3
1270	Liver, simmered	3	ounce(s)	85	56.8	142	20.8	0.7	0	5.5	1.8	1.2	1.7
3174	Meat only, stewed	3	ounce(s)	85	56.8	151	23.2	0	0	5.7	1.6	2.0	1.3
	Duck												
1286	Domesticated, meat and skin, roasted	3	ounce(s)	85	44.1	287	16.2	0	0	24.1	8.2	11.0	3.1
1287	Domesticated, meat only, roasted	3	ounce(s)	85	54.6	171	20.0	0	0	9.5	3.5	3.1	1.2
	Goose												
35507	Domesticated, meat and skin, roasted	3	ounce(s)	85	44.2	259	21.4	0	0	18.6	5.8	8.7	2.1
35524	Domesticated, meat only, roasted	3	ounce(s)	85	48.7	202	24.6	0	0	10.8	3.9	3.7	1.3
1297	Liver pate, smoked, canned	4	tablespoon(s)	52	19.3	240	5.9	2.4	0	22.8	7.5	13.3	0.4
	Turkey												
3256	Ground turkey, cooked	3	ounce(s)	85	50.5	200	23.3	0	0	11.2	2.9	4.2	2.7
3263	Patty, batter coated, breaded, fried	1	item(s)	94	46.7	266	13.2	14.8	0.5	16.9	4.4	7.0	4.4
219	Roasted, dark meat, meat only	3	ounce(s)	85	53.7	159	24.3	0	0	6.1	2.1	1.4	1.8
222	Roasted, fryer roaster breast, meat only	3	ounce(s)	85	58.2	115	25.6	0	0	0.6	0.2	0.1	0.2
220	Roasted, light meat, meat only	3	ounce(s)	85	56.4	134	25.4	0	0	2.7	0.9	0.5	0.7
1303	Turkey roll, light and dark meat	2	slice(s)	57	39.8	84	10.3	1.2	0	4.0	1.2	1.3	1.0
1302	Turkey roll, light meat	2	slice(s)	57	42.5	56	8.4	2.9	0	0.9	0.2	0.2	0.1
Processed Meats													
	Beef												
1331	Corned beef loaf, jellied, sliced	2	slice(s)	57	39.2	87	13.0	0	0	3.5	1.5	1.5	0.2

PAGE KEY: A-4 = Breads/Baked Goods A-10 = Cereal/Rice/Pasta A-14 = Fruit A-18 = Vegetables/Legumes A-28 = Nuts/Seeds A-30 = Vegetarian A-32 = Dairy A-40 = Eggs A-40 = Seafood A-42 = Meats A-46 = Poultry A-46 = Processed Meats A-48 = Beverages A-52 = Fats/Oils A-54 = Sweets A-56 = Spices/Condiments/Sauces A-60 = Mixed Foods/Soups/Sandwiches A-64 = Fast Food A-84 = Convenience A-86 = Baby Foods

Chol (mg)	Calc (mg)	Iron (mg)	Magn (mg)	Pota (mg)	Sodi (mg)	Zinc (mg)	Vit A (μg)	Thia (mg)	Vit E (mg α)	Ribo (mg)	Niac (mg)	Vit B_6 (mg)	Fola (μg)	Vit C (mg)	Vit B_{12} (μg)	Sele (μg)
80	16	1.00	14.5	276.4	59.5	2.93	1.7	0.44	—	0.25	3.33	0.29	3.4	0.2	0.6	28.6
81	8	1.20	17.0	298.5	68.0	3.46	1.7	0.49	—	0.30	3.66	0.34	4.3	0.3	0.7	32.7
70	16	1.93	17.9	325.7	40.0	1.93	0	0.07	—	0.17	7.17	0.40	9.4	0	7.1	32.7
73	17	2.01	17.0	255.1	31.5	2.01	0	0.05	0.37	0.14	6.09	0.28	7.7	0	5.5	32.7
435	5	4.34	17.0	279.8	66.3	9.55	8026	0.15	0.57	2.43	11.18	0.78	281.5	0.9	72.0	16.4
98	10	0.81	20.4	264.5	82.5	3.81	0	0.05	0.30	0.24	6.37	0.23	11.9	0	1.3	9.4
95	6	3.80	20.4	284.9	45.9	2.33	0	0.15	—	0.51	5.70	—	—	0	—	11.0
35	8	0.89	6.8	147.4	408.2	0.79	19.3	0.01	—	0.07	3.58	0.19	2.3	0	0.2	—
67	19	1.05	24.7	222.6	450.2	0.84	—	0.08	—	0.09	10.97	0.46	4.3	0	0.3	—
77	14	0.96	26.4	234.7	67.2	0.91	6.0	0.06	0.35	0.10	12.57	0.54	3.4	0	0.3	22.3
76	14	1.01	25.5	220.3	64.6	0.93	12.8	0.06	0.39	0.11	11.68	0.49	6.0	0	0.3	20.3
77	10	1.13	19.6	194.8	75.7	2.45	21.3	0.06	0.65	0.19	5.13	0.29	9.4	0	0.3	15.6
80	10	1.12	20.4	211.8	81.6	2.73	15.3	0.06	—	0.20	5.22	0.33	7.7	0	0.3	16.7
84	11	1.19	21.3	216.0	81.6	2.53	17.0	0.07	0.38	0.21	5.68	0.33	7.7	0	0.3	16.0
71	13	0.96	17.9	176.9	77.4	1.80	15.3	0.03	0.40	0.10	6.15	0.50	3.4	0	0.3	21.6
49	11	0.75	19.6	244.8	411.4	0.79	4.3	0.09	1.04	0.12	5.99	0.24	35.7	0	0.2	13.9
80	10	1.11	20.4	205.8	77.4	2.43	16.2	0.06	0.22	0.19	5.37	0.31	6.8	0	0.3	18.8
72	14	0.98	17.9	178.6	78.2	1.82	15.3	0.03	0.22	0.10	6.21	0.50	3.4	0	0.3	21.0
64	9	1.13	17.0	190.5	80.8	1.81	13.6	0.05	—	0.16	4.87	0.26	6.0	0	0.2	16.7
64	11	0.91	19.6	200.7	43.4	0.66	6.8	0.05	0.22	0.07	8.90	0.45	2.6	0	0.3	21.9
64	10	1.02	17.9	194.8	63.8	1.29	10.2	0.05	—	0.12	6.70	0.34	4.3	0	0.2	20.9
315	14	2.71	2.6	152.2	47.6	3.75	0	0.02	0.17	0.17	2.65	0.06	4.3	0	0.9	35.0
479	9	9.89	21.3	223.7	64.6	3.38	3385.8	0.24	0.69	1.69	9.39	0.64	491.6	23.7	14.3	70.1
71	12	0.99	17.9	153.1	59.5	1.69	12.8	0.04	0.22	0.13	5.20	0.22	5.1	0	0.2	17.8
71	9	2.29	13.6	173.5	50.2	1.58	53.6	0.14	0.59	0.22	4.10	0.15	5.1	0	0.3	17.0
76	10	2.29	17.0	214.3	55.3	2.21	19.6	0.22	0.59	0.39	4.33	0.21	8.5	0	0.3	19.1
77	11	2.40	18.7	279.8	59.5	2.22	17.9	0.06	1.47	0.27	3.54	0.31	1.7	0	0.3	18.5
82	12	2.44	21.3	330.0	64.6	2.69	10.2	0.07	—	0.33	3.47	0.39	10.2	0	0.4	21.7
78	36	2.86	6.8	71.8	362.4	0.47	520.5	0.04	—	0.15	1.30	0.03	31.2	0	4.9	22.9
87	21	1.64	20.4	229.6	91.0	2.43	0	0.04	0.28	0.14	4.09	0.33	6.0	0	0.3	31.6
71	13	2.06	14.1	258.5	752.0	1.35	9.4	0.09	0.87	0.17	2.16	0.18	57.3	0	0.2	20.8
72	27	1.98	20.4	246.6	67.2	3.79	0	0.05	0.54	0.21	3.10	0.30	7.7	0	0.3	34.8
71	10	1.30	24.7	248.3	44.2	1.48	0	0.03	0.07	0.11	6.37	0.47	5.1	0	0.3	27.3
59	16	1.14	23.8	259.4	54.4	1.73	0	0.05	0.07	0.11	5.81	0.45	5.1	0	0.3	27.3
31	18	0.76	10.2	153.1	332.3	1.13	0	0.05	0.19	0.16	2.72	0.15	2.8	0	0.1	16.6
19	4	0.21	10.8	242.1	590.8	0.50	0	0.01	0.07	0.08	4.05	0.23	2.3	0	0.2	7.4
27	6	1.15	6.2	57.3	540.4	2.31	0	0.00	—	0.06	0.99	0.06	4.5	0	0.7	9.8

TABLE A-1 Table of Food Composition *(continued)*

(Computer code is for Cengage Diet Analysis program)
(For purposes of calculations, use "0" for t, <1, <.1, <.01, etc.)

DA+ Code	Food Description	Quantity	Measure	Wt (g)	H_2O (g)	Ener (kcal)	Prot (g)	Carb (g)	Fiber (g)	Fat (g)	Fat Breakdown (g) Sat	Mono	Poly
Processed Meats—*continued*													
	Bologna												
13459	Beef	1	slice(s)	28	15.1	90	3.0	1.0	0	8.0	3.5	4.3	0.3
13461	Light, made with pork and chicken	1	slice(s)	28	18.2	60	3.0	2.0	0	4.0	1.0	2.0	0.4
13458	Made with chicken and pork	1	slice(s)	28	15.0	90	3.0	1.0	0	8.0	3.0	4.1	1.1
13565	Turkey bologna	1	slice(s)	28	19.0	50	3.0	1.0	0	4.0	1.0	1.1	1.0
	Chicken												
7125	Breast, smoked	1	slice(s)	10	—	10	1.8	0.3	0	0.2	0	—	—
	Ham												
7127	Deli-sliced, honey	1	slice(s)	10	—	10	1.7	0.3	0	0.3	0.1	—	—
7126	Deli-sliced, smoked	1	slice(s)	10	—	10	1.7	0.2	0	0.3	0.1	—	—
8614	**Beef and pork mortadella, sliced**	2	slice(s)	46	24.1	143	7.5	1.4	0	11.7	4.4	5.2	1.4
1323	**Pork olive loaf**	2	slice(s)	57	33.1	133	6.7	5.2	0	9.4	3.3	4.5	1.1
1324	**Pork pickle and pimento loaf**	2	slice(s)	57	34.2	128	6.4	4.8	0.9	9.1	3.0	4.0	1.6
	Sausages and frankfurters												
37296	Beerwurst beef, beer salami (bierwurst)	1	slice(s)	29	16.6	74	4.1	1.2	0	5.7	2.5	2.7	0.2
37257	Beerwurst pork, beer salami	1	slice(s)	21	12.9	50	3.0	0.4	0	4.0	1.3	1.9	0.5
35338	Berliner, pork and beef	1	ounce(s)	28	17.3	65	4.3	0.7	0	4.9	1.7	2.3	0.4
37298	Bratwurst pork, cooked	1	piece(s)	74	42.3	181	10.4	1.9	0	14.3	5.1	6.7	1.5
37299	Braunschweiger pork liver sausage	1	slice(s)	15	8.2	51	2.0	0.3	0	4.5	1.5	2.1	0.5
1329	Cheesefurter or cheese smokie, beef and pork	1	item(s)	43	22.6	141	6.1	0.6	0	12.5	4.5	5.9	1.3
1330	Chorizo, beef and pork	2	ounce(s)	57	18.1	258	13.7	1.1	0	21.7	8.2	10.4	2.0
8600	Frankfurter, beef	1	item(s)	45	23.4	149	5.1	1.8	0	13.3	5.3	6.4	0.5
202	Frankfurter, beef and pork	1	item(s)	45	25.2	137	5.2	0.8	0	12.4	4.8	6.2	1.2
1293	Frankfurter, chicken	1	item(s)	45	28.1	100	7.0	1.2	0.2	7.3	1.7	2.7	1.7
3261	Frankfurter, turkey	1	item(s)	45	28.3	100	5.5	1.7	0	7.8	1.8	2.6	1.8
37275	Italian sausage, pork, cooked	1	item(s)	68	32.0	234	13.0	2.9	0.1	18.6	6.5	8.1	2.2
37307	Kielbasa or kolbassa, pork and beef	1	slice(s)	30	18.5	67	5.0	1.0	0	4.7	1.7	2.2	0.5
1333	Knockwurst or knackwurst, beef and pork	2	ounce(s)	57	31.4	174	6.3	1.8	0	15.7	5.8	7.3	1.7
37285	Pepperoni, beef and pork	1	slice(s)	11	3.4	51	2.2	0.4	0.2	4.4	1.8	2.1	0.3
37313	Polish sausage, pork	1	slice(s)	21	11.4	60	2.8	0.7	0	5.0	1.8	2.3	0.5
206	Salami, beef, cooked, sliced	2	slice(s)	52	31.2	136	6.5	1.0	0	11.5	5.1	5.5	0.5
37272	Salami, pork, dry or hard	1	slice(s)	13	4.6	52	2.9	0.2	0	4.3	1.5	2.0	0.5
40987	Sausage, turkey, cooked	2	ounce(s)	57	36.9	111	13.5	0	0	5.9	1.3	1.7	1.5
8620	Smoked sausage, beef and pork	2	ounce(s)	57	30.6	181	6.8	1.4	0	16.3	5.5	6.9	2.2
8619	Smoked sausage, pork	2	ounce(s)	57	32.0	178	6.8	1.2	0	16.0	5.3	6.4	2.1
37273	Smoked sausage, pork link	1	piece(s)	76	29.8	295	16.8	1.6	0	24.0	8.6	11.1	2.8
1336	Summer sausage, thuringer, or cervelat, beef and pork	2	ounce(s)	57	25.6	205	9.9	1.9	0	17.3	6.5	7.4	0.7
37294	Vienna sausage, cocktail, beef and pork, canned	1	piece(s)	16	10.4	37	1.7	0.4	0	3.1	1.1	1.5	0.2
	Spreads												
1318	Ham salad spread	¼	cup(s)	60	37.6	130	5.2	6.4	0	9.3	3.0	4.3	1.6
32419	Pork and beef sandwich spread	4	tablespoon(s)	60	36.2	141	4.6	7.2	0.1	10.4	3.6	4.6	1.5
	Turkey												
13604	Breast, fat free, oven roasted	1	slice(s)	28	—	25	4.0	1.0	0	0	0	0	0
13606	Breast, hickory smoked fat free	1	slice(s)	28	—	25	4.0	1.0	0	0	0	0	0
16049	Breast, hickory smoked slices	1	slice(s)	56	—	50	11.0	1.0	0	0	0	0	0
16047	Breast, honey roasted slices	1	slice(s)	56	—	60	11.0	3.0	0	0	0	0	0
16048	Breast, oven roasted slices	1	slice(s)	56	—	50	11.0	1.0	0	0	0	0	0
7124	Breast, oven roasted	1	slice(s)	10	—	10	1.8	0.3	0	0.1	0	0	0
13567	Turkey ham, 10% water added	2	slice(s)	56	40.9	70	10.0	2.0	0	3.0	0	0.4	0.6
37270	Turkey pastrami	1	slice(s)	28	20.3	35	4.6	1.0	0	1.2	0.3	0.4	0.3
3262	Turkey salami	2	slice(s)	57	39.1	98	10.9	0.9	0.1	5.2	1.6	1.8	1.4
37318	Turkey salami, cooked	1	slice(s)	28	20.4	43	4.3	0.1	0	2.7	0.8	0.9	0.7
Beverages													
	Beer												
866	Ale, mild	12	fluid ounce(s)	360	332.3	148	1.1	13.3	0.4	0	0	0	0
686	Beer	12	fluid ounce(s)	356	327.7	153	1.6	12.7	0	0	0	0	0
16886	Beer, non alcoholic	12	fluid ounce(s)	360	328.1	133	0.8	29.0	0	0.4	0.1	0	0.2

PAGE KEY: A-4 = Breads/Baked Goods A-10 = Cereal/Rice/Pasta A-14 = Fruit A-18 = Vegetables/Legumes A-28 = Nuts/Seeds A-30 = Vegetarian A-32 = Dairy A-40 = Eggs A-40 = Seafood A-42 = Meats A-46 = Poultry A-46 = Processed Meats A-48 = Beverages A-52 = Fats/Oils A-54 = Sweets A-56 = Spices/Condiments/Sauces A-60 = Mixed Foods/Soups/Sandwiches A-64 = Fast Food A-84 = Convenience A-86 = Baby Foods

Chol (mg)	Calc (mg)	Iron (mg)	Magn (mg)	Pota (mg)	Sodi (mg)	Zinc (mg)	Vit A (µg)	Thia (mg)	Vit E (mg α)	Ribo (mg)	Niac (mg)	Vit B_6 (mg)	Fola (µg)	Vit C (mg)	Vit B_{12} (µg)	Sele (µg)
20	0	0.36	3.9	47.0	310.0	0.56	0	0.01	—	0.03	0.67	0.04	3.6	0	0.4	—
20	40	0.36	5.6	45.6	300.0	0.45	0	—	—	—	—	—	—	0	—	—
30	20	0.36	5.9	43.1	300.0	0.39	0	—	—	—	—	—	—	0	—	—
20	40	0.36	6.2	42.6	270.0	0.51	0	—	—	—	—	—	—	0	—	—
4	0	0.00	—	—	100.0	—	0	—	—	—	—	—	—	0	—	—
4	0	0.12	—	—	100.0	—	0	—	—	—	—	—	—	0.6	—	—
4	0	0.12	—	—	103.3	—	0	—	—	—	—	—	—	0.6	—	—
26	8	0.64	5.1	75.0	573.2	0.96	0	0.05	0.10	0.07	1.23	0.06	1.4	0	0.7	10.4
22	62	0.30	10.8	168.7	842.9	0.78	34.1	0.16	0.14	0.14	1.04	0.13	1.1	0	0.7	9.3
33	62	0.75	19.3	210.7	740.7	0.95	44.3	0.22	0.22	0.06	1.41	0.23	21.0	4.4	0.3	4.5
18	3	0.44	3.5	66.5	264.9	0.71	0	0.02	0.05	0.03	0.98	0.04	0.9	0	0.6	4.7
12	2	0.15	2.7	53.3	261.0	0.36	0	0.11	0.03	0.04	0.68	0.07	0.6	0	0.2	4.4
13	3	0.32	4.3	80.2	367.7	0.70	0	0.10	—	0.06	0.88	0.05	1.4	0	0.8	4.0
44	33	0.95	11.1	156.9	412.2	1.70	0	0.37	0.01	0.13	2.36	0.15	1.5	0.7	0.7	15.7
24	1	1.42	1.7	27.5	131.5	0.42	641.0	0.03	0.05	0.23	1.27	0.05	6.7	0	3.1	8.8
29	25	0.46	5.6	88.6	465.3	0.96	20.2	0.10	0.10	0.06	1.24	0.05	1.3	0	0.7	6.8
50	5	0.90	10.2	225.7	700.2	1.93	0	0.35	0.12	0.17	2.90	0.30	1.1	0	1.1	12.0
24	6	0.67	6.3	70.2	513.0	1.10	0	0.01	0.09	0.06	1.06	0.04	2.3	0	0.8	3.7
23	5	0.51	4.5	75.2	504.0	0.82	8.1	0.09	0.11	0.05	1.18	0.05	1.8	0	0.6	6.2
43	33	0.52	9.0	90.9	379.8	0.50	0	0.02	0.09	0.11	2.10	0.14	3.2	0	0.2	10.4
35	67	0.66	6.3	176.4	485.1	0.82	0	0.01	0.27	0.08	1.65	0.06	4.1	0	0.4	6.8
39	14	0.97	12.2	206.7	820.8	1.62	6.8	0.42	0.17	0.15	2.83	0.22	3.4	0.1	0.9	15.0
20	13	0.44	4.9	84.4	283.0	0.61	0	0.06	0.06	0.06	0.87	0.05	1.5	0	0.5	5.4
34	6	0.37	6.2	112.8	527.3	0.94	0	0.19	0.32	0.07	1.55	0.09	1.1	0	0.7	7.7
13	2	0.15	2.0	34.7	196.7	0.30	0	0.05	0.00	0.02	0.59	0.04	0.7	0.1	0.2	2.4
15	2	0.29	2.9	37.3	199.3	0.40	0	0.10	0.04	0.03	0.71	0.03	0.4	0.2	0.2	3.7
37	3	1.14	6.8	97.8	592.8	0.92	0	0.04	0.08	0.08	1.68	0.08	1.0	0	1.6	7.6
10	2	0.16	2.8	48.4	289.3	0.53	0	0.11	0.02	0.04	0.71	0.07	0.3	0	0.4	3.3
52	12	0.84	11.9	169.0	377.1	2.19	7.4	0.04	0.10	0.14	3.24	0.18	3.4	0.4	0.7	0
33	7	0.42	7.4	101.5	516.5	0.71	7.4	0.10	0.07	0.06	1.66	0.09	1.1	0	0.3	0
35	6	0.33	6.2	273.9	468.9	0.74	0	0.12	0.14	0.10	1.59	0.10	0.6	0	0.4	10.4
52	23	0.87	14.4	254.6	1136.6	2.13	0	0.53	0.18	0.19	3.43	0.26	3.8	1.5	1.2	16.4
42	5	1.15	7.9	147.4	737.1	1.45	0	0.08	0.12	0.18	2.44	0.14	1.1	9.4	3.1	11.5
14	2	0.14	1.1	16.2	155.0	0.25	0	0.01	0.03	0.01	0.25	0.01	0.6	0	0.2	2.7
22	5	0.35	6.0	90.0	547.2	0.66	0	0.26	1.04	0.07	1.25	0.09	0.6	0	0.5	10.7
23	7	0.47	4.8	66.0	607.8	0.61	15.6	0.10	1.04	0.08	1.03	0.07	1.2	0	0.7	5.8
10	0	0.00	—	—	340.0	—	0	—	—	—	—	—	—	0	—	—
10	0	0.00	—	—	300.0	—	0	—	—	—	—	—	—	0	—	—
25	0	0.72	—	—	720.0	—	0	—	—	—	—	—	—	0	—	—
20	0	0.72	—	—	660.0	—	0	—	—	—	—	—	—	0	—	—
20	0	0.72	—	—	660.0	—	0	—	—	—	—	—	—	0	—	—
4	0	0.06	—	—	103.3	—	0	—	—	—	—	—	—	0	—	—
40	0	0.72	12.3	162.4	700.0	1.44	0	—	—	—	—	—	—	0	—	—
19	3	1.19	4.0	97.8	278.1	0.61	1.1	0.01	0.06	0.07	1.00	0.07	1.4	4.6	0.1	4.6
43	23	0.70	12.5	122.5	569.3	1.31	1.1	0.24	0.13	0.17	2.25	0.24	5.7	0	0.6	15.0
22	11	0.35	6.2	61.2	284.6	0.65	0.6	0.12	0.06	0.08	1.12	0.12	2.8	0	0.3	7.5
0	18	0.07	21.6	90.0	14.4	0.03	0	0.03	0.00	0.10	1.62	0.18	21.6	0	0.1	2.5
0	14	0.07	21.4	96.2	14.3	0.03	0	0.01	0.00	0.08	1.82	0.16	21.4	0	0.1	2.1
0	25	0.21	25.2	28.8	46.8	0.07	—	0.07	0.00	0.18	3.99	0.10	50.4	1.8	0.1	4.3

TABLE A-1 Table of Food Composition *(continued)*

(Computer code is for Cengage Diet Analysis program)
(For purposes of calculations, use "0" for t, <1, <.1, <.01, etc.)

DA+ Code	Food Description	Quantity	Measure	Wt (g)	H_2O (g)	Ener (kcal)	Prot (g)	Carb (g)	Fiber (g)	Fat (g)	Fat Breakdown (g) Sat	Mono	Poly
Beverages—*continued*													
31609	Bud Light beer	12	fluid ounce(s)	355	335.5	110	0.9	6.6	0	0	0	0	0
31608	Budweiser beer	12	fluid ounce(s)	355	327.7	145	1.3	10.6	0	0	0	0	0
869	Light beer	12	fluid ounce(s)	354	335.9	103	0.9	5.8	0	0	0	0	0
31613	Michelob beer	12	fluid ounce(s)	355	323.4	155	1.3	13.3	0	0	0	0	0
31614	Michelob Light beer	12	fluid ounce(s)	355	329.8	134	1.1	11.7	0	0	0	0	0
	Gin, rum, vodka, whiskey												
857	Distilled alcohol, 100 proof	1	fluid ounce(s)	28	16.0	82	0	0	0	0	0	0	0
687	Distilled alcohol, 80 proof	1	fluid ounce(s)	28	18.5	64	0	0	0	0	0	0	0
688	Distilled alcohol, 86 proof	1	fluid ounce(s)	28	17.8	70	0	0	0	0	0	0	0
689	Distilled alcohol, 90 proof	1	fluid ounce(s)	28	17.3	73	0	0	0	0	0	0	0
856	Distilled alcohol, 94 proof	1	fluid ounce(s)	28	16.8	76	0	0	0	0	0	0	0
	Liqueurs												
33187	Coffee liqueur, 53 proof	1	fluid ounce(s)	35	10.8	113	0	16.3	0	0.1	0	0	0
3142	Coffee liqueur, 63 proof	1	fluid ounce(s)	35	14.4	107	0	11.2	0	0.1	0	0	0
736	Cordials, 54 proof	1	fluid ounce(s)	30	8.9	106	0	13.3	0	0.1	0	0	0
	Wine												
861	California red wine	5	fluid ounce(s)	150	133.4	125	0.3	3.7	0	0	0	0	0
858	Domestic champagne	5	fluid ounce(s)	150	—	105	0.3	3.8	0	0	0	0	0
690	Sweet dessert wine	5	fluid ounce(s)	147	103.7	235	0.3	20.1	0	0	0	0	0
1481	White wine	5	fluid ounce(s)	148	128.1	121	0.1	3.8	0	0	0	0	0
1811	Wine cooler	10	fluid ounce(s)	300	267.4	159	0.3	20.2	0	0.1	0	0	0
	Carbonated												
31898	7 Up	12	fluid ounce(s)	360	321.0	140	0	39.0	0	0	0	0	0
692	Club soda	12	fluid ounce(s)	355	354.8	0	0	0	0	0	0	0	0
12010	Coca-Cola Classic cola soda	12	fluid ounce(s)	360	319.4	146	0	40.5	0	0	0	0	0
693	Cola	12	fluid ounce(s)	368	332.7	136	0.3	35.2	0	0.1	0	0	0
2391	Cola or pepper-type soda, low calorie with saccharin	12	fluid ounce(s)	355	354.5	0	0	0.3	0	0	0	0	0
9522	Cola soda, decaffeinated	12	fluid ounce(s)	372	333.4	153	0	39.3	0	0	0	0	0
9524	Cola, decaffeinated, low calorie with aspartame	12	fluid ounce(s)	355	354.3	4	0.4	0.5	0	0	0	0	0
1415	Cola, low calorie with aspartame	12	fluid ounce(s)	355	353.6	7	0.4	1.0	0	0.1	0	0	0
1412	Cream soda	12	fluid ounce(s)	371	321.5	189	0	49.3	0	0	0	0	0
31899	Diet 7 Up	12	fluid ounce(s)	360	—	0	0	0	0	0	0	0	0
12031	Diet Coke cola soda	12	fluid ounce(s)	360	—	2	0	0.2	0	0	0	0	0
29392	Diet Mountain Dew soda	12	fluid ounce(s)	360	—	0	0	0	0	0	0	0	0
29389	Diet Pepsi cola soda	12	fluid ounce(s)	360	—	0	0	0	0	0	0	0	0
12034	Diet Sprite soda	12	fluid ounce(s)	360	—	4	0	0	0	0	0	0	0
695	Ginger ale	12	fluid ounce(s)	366	333.9	124	0	32.1	0	0	0	0	0
694	Grape soda	12	fluid ounce(s)	372	330.3	160	0	41.7	0	0	0	0	0
1876	Lemon lime soda	12	fluid ounce(s)	368	330.8	147	0.2	37.4	0	0.1	0	0	0
29391	Mountain Dew soda	12	fluid ounce(s)	360	314.0	170	0	46.0	0	0	0	0	0
3145	Orange soda	12	fluid ounce(s)	372	325.9	179	0	45.8	0	0	0	0	0
1414	Pepper-type soda	12	fluid ounce(s)	368	329.3	151	0	38.3	0	0.4	0.3	0	0
29388	Pepsi regular cola soda	12	fluid ounce(s)	360	318.9	150	0	41.0	0	0	0	0	0
696	Root beer	12	fluid ounce(s)	370	330.0	152	0	39.2	0	0	0	0	0
12044	Sprite soda	12	fluid ounce(s)	360	321.0	144	0	39.0	0	0	0	0	0
	Coffee												
731	Brewed	8	fluid ounce(s)	237	235.6	2	0.3	0	0	0	0	0	0
9520	Brewed, decaffeinated	8	fluid ounce(s)	237	234.3	5	0.3	1.0	0	0	0	0	0
16882	Cappuccino	8	fluid ounce(s)	240	224.8	79	4.1	5.8	0.2	4.9	2.3	1.0	0.2
16883	Cappuccino, decaffeinated	8	fluid ounce(s)	240	224.8	79	4.1	5.8	0.2	4.9	2.3	1.0	0.2
16880	Espresso	8	fluid ounce(s)	237	231.8	21	0	3.6	0	0.4	0.2	0	0.2
16881	Espresso, decaffeinated	8	fluid ounce(s)	237	231.8	21	0	3.6	0	0.4	0.2	0	0.2
732	Instant, prepared	8	fluid ounce(s)	239	236.5	5	0.2	0.8	0	0	0	0	0
	Fruit drinks												
29357	Crystal Light sugar-free lemonade drink	8	fluid ounce(s)	240	—	5	0	0	0	0	0	0	0
6012	Fruit punch drink with added vitamin C, canned	8	fluid ounce(s)	248	218.2	117	0	29.7	0.5	0	0	0	0
31143	Gatorade Thirst Quencher, all flavors	8	fluid ounce(s)	240	—	50	0	14.0	0	0	0	0	0
260	Grape drink, canned	8	fluid ounce(s)	250	210.5	153	0	39.4	0	0	0	0	0
17372	Kool-Aid (lemonade/punch/fruit drink)	8	fluid ounce(s)	248	220.0	108	0.1	27.8	0.2	0	0	0	0

PAGE KEY: A-4 = Breads/Baked Goods A-10 = Cereal/Rice/Pasta A-14 = Fruit A-18 = Vegetables/Legumes A-28 = Nuts/Seeds A-30 = Vegetarian A-32 = Dairy A-40 = Eggs A-40 = Seafood A-42 = Meats A-46 = Poultry A-46 = Processed Meats A-48 = Beverages A-52 = Fats/Oils A-54 = Sweets A-56 = Spices/Condiments/Sauces A-60 = Mixed Foods/Soups/Sandwiches A-64 = Fast Food A-84 = Convenience A-86 = Baby Foods

Chol (mg)	Calc (mg)	Iron (mg)	Magn (mg)	Pota (mg)	Sodi (mg)	Zinc (mg)	Vit A (μg)	Thia (mg)	Vit E (mg α)	Ribo (mg)	Niac (mg)	Vit B_6 (mg)	Fola (μg)	Vit C (mg)	Vit B_{12} (μg)	Sele (μg)
0	18	0.14	17.8	63.9	9.0	0.10	0	0.03	—	0.10	1.39	0.12	14.6	0	0	4.0
0	18	0.10	21.3	88.8	9.0	0.07	0	0.02	—	0.09	1.60	0.17	21.3	0	0.1	4.0
0	14	0.10	17.7	74.3	14.2	0.03	0	0.01	0.00	0.05	1.38	0.12	21.2	0	0.1	1.4
0	18	0.10	21.3	88.8	9.0	0.07	0	0.02	—	0.09	1.60	0.17	21.3	0	0.1	4.0
0	18	0.14	17.8	63.9	9.0	0.10	0	0.03	—	0.10	1.39	0.12	14.6	0	0	4.0
0	0	0.01	0	0.6	0.3	0.01	0	0.00	—	0.00	0.00	0.00	0	0	0	0
0	0	0.01	0	0.6	0.3	0.01	0	0.00	0.00	0.00	0.00	0.00	0	0	0	0
0	0	0.01	0	0.6	0.3	0.01	0	0.00	0.00	0.00	0.00	0.00	0	0	0	0
0	0	0.01	0	0.6	0.3	0.01	0	0.00	0.00	0.00	0.00	0.00	0	0	0	0
0	0	0.01	0	0.6	0.3	0.01	0	0.00	—	0.00	0.00	0.00	0	0	0	0
0	0	0.02	1.0	10.4	2.8	0.01	0	0.00	0.00	0.00	0.05	0.00	0	0	0	0.1
0	0	0.02	1.0	10.4	2.8	0.01	0	0.00	—	0.00	0.05	0.00	0	0	0	0.1
0	0	0.02	0.6	4.5	2.1	0.01	0	0.00	0.00	0.00	0.02	0.00	0	0	0	0.1
0	12	1.43	16.2	170.6	15.0	0.14	0	0.01	0.00	0.04	0.11	0.05	1.5	0	0	—
0	—	—	—	—	—	—	—	—	—	—	—	—	—	—	0	—
0	12	0.34	13.2	135.4	13.2	0.10	0	0.01	0.00	0.01	0.30	0.00	0	0	0	0.7
0	13	0.39	14.8	104.7	7.4	0.18	0	0.01	0.00	0.01	0.15	0.06	1.5	0	0	0.1
0	18	0.75	15.0	129.0	24.0	0.18	—	0.01	0.03	0.03	0.13	0.03	3.0	5.4	0	0.6
0	—	—	—	0.6	75.0	—	—	—	—	—	—	—	—	—	—	—
0	18	0.03	3.5	7.1	74.6	0.35	0	0.00	0.00	0.00	0.00	0.00	0	0	0	0
0	—	—	—	0	49.5	—	0	—	—	—	—	—	—	0	—	—
0	7	0.41	0	7.4	14.7	0.06	0	0.00	0.00	0.00	0.00	0.00	0	0	0	0.4
0	14	0.06	3.5	14.2	56.8	0.11	0	0.00	0.00	0.00	0.00	0.00	0	0	0	0.3
0	7	0.08	0	11.2	14.9	0.03	0	0.00	0.00	0.00	0.00	0.00	0	0	0	0.4
0	11	0.06	0	24.9	14.2	0.03	0	0.02	0.00	0.08	0.00	0.00	0	0	0	0.3
0	11	0.39	3.5	28.4	28.4	0.03	0	0.02	0.00	0.08	0.00	0.00	0	0	0	0
0	19	0.18	3.7	3.7	44.5	0.26	0	0.00	0.00	0.00	0.00	0.00	0	0	0	0
0	—	—	—	77.0	45.0	—	—	—	—	—	—	—	—	—	—	—
0	—	—	—	18.0	42.0	—	0	—	—	—	—	—	—	0	—	—
0	—	—	—	70.0	35.0	—	—	—	—	—	—	—	—	—	—	—
0	—	—	—	30.0	35.0	—	—	—	—	—	—	—	—	—	—	—
0	—	—	—	109.5	36.0	—	0	—	—	—	—	—	—	0	—	—
0	11	0.65	3.7	3.7	25.6	0.18	0	0.00	0.00	0.00	0.00	0.00	0	0	0	0.4
0	11	0.29	3.7	3.7	55.8	0.26	0	0.00	—	0.00	0.00	0.00	0	0	0	0
0	7	0.41	3.7	3.7	33.2	0.14	0	0.00	0.00	0.00	0.05	0.00	0	0	0	0
0	—	—	—	0	70.0	—	—	—	—	—	—	—	—	—	—	—
0	19	0.21	3.7	7.4	44.6	0.36	0	0.00	—	0.00	0.00	0.00	0	0	0	0
0	11	0.14	0	3.7	36.8	0.14	0	0.00	—	0.00	0.00	0.00	0	0	0	0.4
0	—	—	—	0	35.0	—	—	—	—	—	—	—	—	—	—	—
0	18	0.18	3.7	3.7	48.0	0.26	0	0.00	0.00	0.00	0.00	0.00	0	0	0	0.4
0	—	—	—	0	70.5	—	0	—	—	—	—	—	—	0	—	—
0	5	0.02	7.1	116.1	4.7	0.04	0	0.03	0.02	0.18	0.45	0.00	4.7	0	0	0
0	7	0.14	11.8	108.9	4.7	0.00	0	0.00	0.00	0.03	0.66	0.00	0	0	0	0.5
12	144	0.19	14.4	232.8	50.4	0.50	33.6	0.04	0.09	0.27	0.13	0.04	7.2	0	0.4	4.6
12	144	0.19	14.4	232.8	50.4	0.50	33.6	0.04	0.09	0.27	0.13	0.04	7.2	0	0.4	4.6
0	5	0.30	189.6	272.6	33.2	0.11	0	0.00	0.04	0.42	12.34	0.00	2.4	0.5	0	0
0	5	0.30	189.6	272.6	33.2	0.11	0	0.00	0.04	0.42	12.34	0.00	2.4	0.5	0	0
0	10	0.09	9.5	71.6	9.5	0.01	0	0.00	0.00	0.00	0.56	0.00	0	0	0	0.2
0	0	0.00	—	160.0	40.0	—	0	—	—	—	—	—	—	0	—	—
0	20	0.22	7.4	62.0	94.2	0.02	5.0	0.05	0.04	0.05	0.05	0.02	9.9	89.3	0	0.5
0	0	0.00	—	30.0	110.0	—	0	—	—	—	—	—	—	0	—	—
0	130	0.17	2.5	30.0	40.0	0.30	0	0.00	0.00	0.01	0.02	0.01	0	78.5	0	0.3
0	14	0.45	5.0	49.6	31.0	0.19	—	0.03	—	0.05	0.04	0.01	4.3	41.6	0	1.0

TABLE A-1 Table of Food Composition *(continued)*

(Computer code is for Cengage Diet Analysis program)
(For purposes of calculations, use "0" for t, <1, <.1, <.01, etc.)

DA+ Code	Food Description	Quantity	Measure	Wt (g)	H_2O (g)	Ener (kcal)	Prot (g)	Carb (g)	Fiber (g)	Fat (g)	Fat Breakdown (g) Sat	Mono	Poly
Beverages—*continued*													
17225	Kool-Aid sugar free, low calorie tropical punch drink mix, prepared	8	fluid ounce(s)	240	—	5	0	0	0	0	0	0	0
266	Lemonade, prepared from frozen concentrate	8	fluid ounce(s)	248	221.6	99	0.2	25.8	0	0.1	0	0	0
268	Limeade, prepared from frozen concentrate	8	fluid ounce(s)	247	212.6	128	0	34.1	0	0	0	0	0
14266	Odwalla strawberry C monster smoothie blend	8	fluid ounce(s)	240	—	160	2.0	38.0	0	0	0	0	0
10080	Odwalla strawberry lemonade quencher	8	fluid ounce(s)	240	—	110	0	28.0	0	0	0	0	0
10099	Snapple fruit punch fruit drink	8	fluid ounce(s)	240	—	110	0	29.0	0	0	0	0	0
10096	Snapple kiwi strawberry fruit drink	8	fluid ounce(s)	240	211.2	110	0	28.0	0	0	0	0	0
	Slim Fast ready-to-drink shake												
16054	French vanilla ready to drink shake	11	fluid ounce(s)	325	—	220	10.0	40.0	5.0	2.5	0.5	1.5	0.5
40447	Optima rich chocolate royal ready-to-drink shake	11	fluid ounce(s)	330	—	180	10.0	24.0	5.0	5.0	1.0	3.5	0.5
16055	Strawberries n cream ready to drink shake	11	fluid ounce(s)	325	—	220	10.0	40.0	5.0	2.5	0.5	1.5	0.5
	Tea												
33179	Decaffeinated, prepared	8	fluid ounce(s)	237	236.3	2	0	0.7	0	0	0	0	0
1877	Herbal, prepared	8	fluid ounce(s)	237	236.1	2	0	0.5	0	0	0	0	0
735	Instant tea mix, lemon flavored with sugar, prepared	8	fluid ounce(s)	259	236.2	91	0	22.3	0.3	0.2	0	0	0
734	Instant tea mix, unsweetened, prepared	8	fluid ounce(s)	237	236.1	2	0.1	0.4	0	0	0	0	0
733	Tea, prepared	8	fluid ounce(s)	237	236.3	2	0	0.7	0	0	0	0	0
	Water												
1413	Mineral water, carbonated	8	fluid ounce(s)	237	236.8	0	0	0	0	0	0	0	0
33183	Poland spring water, bottled	8	fluid ounce(s)	237	237.0	0	0	0	0	0	0	0	0
1821	Tap water	8	fluid ounce(s)	237	236.8	0	0	0	0	0	0	0	0
1879	Tonic water	8	fluid ounce(s)	244	222.3	83	0	21.5	0	0	0	0	0
Fats and Oils													
	Butter												
104	Butter	1	tablespoon(s)	14	2.3	102	0.1	0	0	11.5	7.3	3.0	0.4
2522	Butter Buds, dry butter substitute	1	teaspoon(s)	2	—	5	0	2.0	0	0	0	0	0
921	Unsalted	1	tablespoon(s)	14	2.5	102	0.1	0	0	11.5	7.3	3.0	0.4
107	Whipped	1	tablespoon(s)	9	1.5	67	0.1	0	0	7.6	4.7	2.2	0.3
944	Whipped, unsalted	1	tablespoon(s)	11	2.0	82	0.1	0	0	9.2	5.9	2.4	0.3
	Fats, cooking												
2671	Beef tallow, semisolid	1	tablespoon(s)	13	0	115	0	0	0	12.8	6.4	5.4	0.5
922	Chicken fat	1	tablespoon(s)	13	0	115	0	0	0	12.8	3.8	5.7	2.7
5454	Household shortening with vegetable oil	1	tablespoon(s)	13	0	115	0	0	0	13.0	3.4	5.5	2.7
111	Lard	1	tablespoon(s)	13	0	115	0	0	0	12.8	5.0	5.8	1.4
	Margarine												
114	Margarine	1	tablespoon(s)	14	2.3	101	0	0.1	0	11.4	2.1	5.5	3.4
5439	Soft	1	tablespoon(s)	14	2.3	103	0.1	0.1	0	11.6	1.7	4.4	2.1
32329	Soft, unsalted, with hydrogenated soybean and cottonseed oils	1	tablespoon(s)	14	2.5	101	0.1	0.1	0	11.3	2.0	5.4	3.5
928	Unsalted	1	tablespoon(s)	14	2.6	101	0.1	0.1	0	11.3	2.1	5.2	3.5
119	Whipped	1	tablespoon(s)	9	1.5	64	0.1	0.1	0	7.2	1.2	3.2	2.5
	Spreads												
54657	I Can't Believe It's Not Butter!, tub, soya oil (non-hydrogenated)	1	tablespoon(s)	14	2.3	103	0.1	0.1	0	11.6	2.8	2.0	5.1
2708	Mayonnaise with soybean and safflower oils	1	tablespoon(s)	14	2.1	99	0.2	0.4	0	11.0	1.2	1.8	7.6
16157	Promise vegetable oil spread, stick	1	tablespoon(s)	14	4.2	90	0	0	0	10.0	2.5	2.0	4.0
	Oils												
2681	Canola	1	tablespoon(s)	14	0	120	0	0	0	13.6	1.0	8.6	3.8
120	Corn	1	tablespoon(s)	14	0	120	0	0	0	13.6	1.8	3.8	7.4
122	Olive	1	tablespoon(s)	14	0	119	0	0	0	13.5	1.9	9.9	1.4

PAGE KEY: A-4 = Breads/Baked Goods A-10 = Cereal/Rice/Pasta A-14 = Fruit A-18 = Vegetables/Legumes A-28 = Nuts/Seeds A-30 = Vegetarian A-32 = Dairy A-40 = Eggs A-40 = Seafood A-42 = Meats A-46 = Poultry A-46 = Processed Meats A-48 = Beverages A-52 = Fats/Oils A-54 = Sweets A-56 = Spices/Condiments/Sauces A-60 = Mixed Foods/Soups/Sandwiches A-64 = Fast Food A-84 = Convenience A-86 = Baby Foods

Chol (mg)	Calc (mg)	Iron (mg)	Magn (mg)	Pota (mg)	Sodi (mg)	Zinc (mg)	Vit A (µg)	Thia (mg)	Vit E (mg α)	Ribo (mg)	Niac (mg)	Vit B_6 (mg)	Fola (µg)	Vit C (mg)	Vit B_{12} (µg)	Sele (µg)
0	0	0.00	—	10.1	10.1	—	0	—	—	—	—	—	—	6.0	—	—
0	10	0.39	5.0	37.2	9.9	0.05	0	0.01	0.02	0.05	0.04	0.01	2.5	9.7	0	0.2
0	5	0.00	4.9	24.7	7.4	0.02	0	0.01	0.00	0.01	0.02	0.01	2.5	7.7	0	0.2
0	20	0.72	—	0	20.0	—	0	—	—	—	—	—	—	600.0	0	—
0	0	0.00	—	70.0	10.0	—	0	—	—	—	—	—	—	54.0	0	—
0	0	0.00	—	20.0	10.0	—	0	—	—	—	—	—	—	0	0	—
0	0	0.00	—	40.0	10.0	—	0	—	—	—	—	—	—	0	0	—
5	400	2.70	140.0	600.0	220.0	2.25	—	0.52	—	0.59	7.00	0.70	120.0	60.0	2.1	17.5
5	1000	2.70	140.0	600.0	220.0	2.25	—	0.52	—	0.59	7.00	0.70	120.0	30.0	2.1	17.5
5	400	2.70	140.0	600.0	220.0	2.25	—	0.52	—	0.59	7.00	0.70	120.0	60.0	2.1	17.5
0	0	0.04	7.1	87.7	7.1	0.04	0	0.00	0.00	0.03	0.00	0.00	11.9	0	0	0
0	5	0.18	2.4	21.3	2.4	0.09	0	0.02	0.00	0.01	0.00	0.00	2.4	0	0	0
0	5	0.05	2.6	38.9	5.2	0.02	0	0.00	0.00	0.00	0.02	0.00	0	0	0	0.3
0	7	0.02	4.7	42.7	9.5	0.02	0	0.00	0.00	0.01	0.07	0.00	0	0	0	0
0	0	0.04	7.1	87.7	7.1	0.04	0	0.00	0.00	0.03	0.00	0.00	11.9	0	0	0
0	33	0.00	0	0	2.4	0.00	0	0.00	—	0.00	0.00	0.00	0	0	0	0
0	2	0.02	2.4	0	2.4	0.00	0	0.00	—	0.00	0.00	0.00	0	0	0	0
0	7	0.00	2.4	2.4	7.1	0.00	0	0.00	0.00	0.00	0.00	0.00	0	0	0	0
0	2	0.02	0	0	29.3	0.24	0	0.00	0.00	0.00	0.00	0.00	0	0	0	0
31	3	0.00	0.3	3.4	81.8	0.01	97.1	0.00	0.32	0.01	0.01	0.00	0.4	0	0	0.1
0	0	0.00	0	1.6	120.0	0.00	0	0.00	0.00	0.00	0.00	0.00	0	0	0	—
31	3	0.00	0.3	3.4	1.6	0.01	97.1	0.00	0.32	0.01	0.01	0.00	0.4	0	0	0.1
21	2	0.01	0.2	2.4	77.7	0.01	64.3	0.00	0.21	0.00	0.00	0.00	0.3	0	0	0.1
25	3	0.00	0.2	2.7	1.3	0.01	78.0	0.00	0.26	0.00	0.00	0.00	0.3	0	0	0.1
14	0	0.00	0	0	0	0.00	0	0.00	0.34	0.00	0.00	0.00	0	0	0	0
11	0	0.00	0	0	0	0.00	0	0.00	0.34	0.00	0.00	0.00	0	0	0	0
0	0	0.00	0	0	0	0.00	0	0.00	—	0.00	0.00	0.00	0	0	0	—
12	0	0.00	0	0	0	0.01	0	0.00	0.07	0.00	0.00	0.00	0	0	0	0
0	4	0.01	0.4	5.9	133.0	0.00	115.5	0.00	1.26	0.01	0.00	0.00	0.1	0	0	0
0	4	0.00	0.3	5.5	155.4	0.00	142.7	0.00	1.00	0.00	0.00	0.00	0.1	0	0	0
0	4	0.00	0.3	5.4	3.9	0.00	103.1	0.00	0.98	0.00	0.00	0.00	0.1	0	0	0
0	2	0.00	0.3	3.5	0.3	0.00	115.5	0.00	1.80	0.00	0.00	0.00	0.1	0	0	0
0	2	0.00	0.2	3.4	97.1	0.00	73.7	0.00	0.45	0.00	0.00	0.00	0.1	0	0	0
0	4	0.00	0.3	5.5	155.3	0.00	142.6	0.00	0.72	0.00	0.00	0.00	0.1	0	0	0
8	2	0.06	0.1	4.7	78.4	0.01	11.6	0.00	3.03	0.00	0.00	0.08	1.1	0	0	0.2
0	10	0.18	—	8.7	90.0	—	—	0.00	—	0.00	0.00	—	—	0.6	—	
0	0	0.00	0	0	0	0.00	0	0.00	2.37	0.00	0.00	0.00	0	0	0	0
0	0	0.00	0	0	0	0.00	0	0.00	1.94	0.00	0.00	0.00	0	0	0	0
0	0	0.07	0	0.1	0.3	0.00	0	0.00	1.93	0.00	0.00	0.00	0	0	0	0

TABLE A-1 Table of Food Composition *(continued)*

(Computer code is for Cengage Diet Analysis program)
(For purposes of calculations, use "0" for t, <1, <.1, <.01, etc.)

DA+ Code	Food Description	Quantity	Measure	Wt (g)	H_2O (g)	Ener (kcal)	Prot (g)	Carb (g)	Fiber (g)	Fat (g)	Fat Breakdown (g) Sat	Mono	Poly
Fats and Oils—*continued*													
124	Peanut	1	tablespoon(s)	14	0	119	0	0	0	13.5	2.3	6.2	4.3
2693	Safflower	1	tablespoon(s)	14	0	120	0	0	0	13.6	0.8	10.2	2.0
923	Sesame	1	tablespoon(s)	14	0	120	0	0	0	13.6	1.9	5.4	5.7
128	Soybean, hydrogenated	1	tablespoon(s)	14	0	120	0	0	0	13.6	2.0	5.8	5.1
130	Soybean, with soybean and cottonseed oil	1	tablespoon(s)	14	0	120	0	0	0	13.6	2.4	4.0	6.5
2700	Sunflower	1	tablespoon(s)	14	0	120	0	0	0	13.6	1.8	6.3	5.0
357	**Pam original no stick cooking spray**	1	serving(s)	0	0.2	0	0	0	0	0	0	0	0
	Salad dressing												
132	Blue cheese	2	tablespoon(s)	30	9.7	151	1.4	2.2	0	15.7	3.0	3.7	8.3
133	Blue cheese, low calorie	2	tablespoon(s)	32	25.4	32	1.6	0.9	0	2.3	0.8	0.6	0.8
1764	Caesar	2	tablespoon(s)	30	10.3	158	0.4	0.9	0	17.3	2.6	4.1	9.9
29654	Creamy, reduced calorie, fat-free, cholesterol-free, sour cream and/or buttermilk and oil	2	tablespoon(s)	32	23.9	34	0.4	6.4	0	0.9	0.2	0.2	0.5
29617	Creamy, reduced calorie, sour cream and/or buttermilk and oil	2	tablespoon(s)	30	22.2	48	0.5	2.1	0	4.2	0.6	1.0	2.4
134	French	2	tablespoon(s)	32	11.7	146	0.2	5.0	0	14.3	1.8	2.7	6.7
135	French, low fat	2	tablespoon(s)	32	17.4	74	0.2	9.4	0.4	4.3	0.4	1.9	1.6
136	Italian	2	tablespoon(s)	29	16.6	86	0.1	3.1	0	8.3	1.3	1.9	3.8
137	Italian, diet	2	tablespoon(s)	30	25.4	23	0.1	1.4	0	1.9	0.1	0.7	0.5
139	Mayonnaise-type	2	tablespoon(s)	29	11.7	115	0.3	7.0	0	9.8	1.4	2.6	5.3
942	Oil and vinegar	2	tablespoon(s)	32	15.2	144	0	0.8	0	16.0	2.9	4.7	7.7
1765	Ranch	2	tablespoon(s)	30	11.6	146	0.1	1.6	0	15.8	2.3	5.2	7.6
3666	Ranch, reduced calorie	2	tablespoon(s)	30	20.5	62	0.1	2.2	0	6.1	1.1	1.8	2.9
940	Russian	2	tablespoon(s)	30	11.6	107	0.5	9.3	0.7	7.8	1.2	1.8	4.4
939	Russian, low calorie	2	tablespoon(s)	32	20.8	45	0.2	8.8	0.1	1.3	0.2	0.3	0.7
941	Sesame seed	2	tablespoon(s)	30	11.8	133	0.9	2.6	0.3	13.6	1.9	3.6	7.5
142	Thousand Island	2	tablespoon(s)	32	14.9	118	0.3	4.7	0.3	11.2	1.6	2.5	5.8
143	Thousand Island, low calorie	2	tablespoon(s)	30	18.2	61	0.3	6.7	0.4	3.9	0.2	1.9	0.8
	Sandwich spreads												
138	Mayonnaise with soybean oil	1	tablespoon(s)	14	2.1	99	0.1	0.4	0	11.0	1.6	2.7	5.8
140	Mayonnaise, low calorie	1	tablespoon(s)	16	10.0	37	0	2.6	0	3.1	0.5	0.7	1.7
141	Tartar sauce	2	tablespoon(s)	28	8.7	144	0.3	4.1	0.1	14.4	2.2	3.8	7.7
Sweets													
4799	**Butterscotch or caramel topping**	2	tablespoon(s)	41	13.1	103	0.6	27.0	0.4	0	0	0	0
	Candy												
1786	Almond Joy candy bar	1	item(s)	45	4.3	220	2.0	27.0	2.0	12.0	8.0	3.3	0.7
1785	Bit-O-Honey candy	6	item(s)	40	—	190	1.0	39.0	0	3.5	2.5	—	—
33375	Butterscotch candy	2	piece(s)	12	0.6	47	0	10.8	0	0.4	0.2	0.1	0
1701	Chewing gum, stick	1	item(s)	3	0.1	7	0	2.0	0.1	0	0	0	0
33378	Chocolate fudge with nuts, prepared	2	piece(s)	38	2.9	175	1.7	25.8	1.0	7.2	2.5	1.5	2.9
1787	Jelly beans	15	item(s)	43	2.7	159	0	39.8	0.1	0	0	0	0
1784	Kit Kat wafer bar	1	item(s)	42	0.8	210	3.0	27.0	0.5	11.0	7.0	3.5	0.3
4674	Krackel candy bar	1	item(s)	41	0.6	210	2.0	28.0	0.5	10.0	6.0	3.9	0.4
4934	Licorice	4	piece(s)	44	7.3	154	1.1	35.1	0	1.0	0	0.1	0
1780	Life Savers candy	1	item(s)	2	—	8	0	2.0	0	0	0	0	0
1790	Lollipop	1	item(s)	28	—	108	0	28.0	0	0	0	0	0
4679	M & Ms peanut chocolate candy, small bag	1	item(s)	49	0.9	250	5.0	30.0	2.0	13.0	5.0	5.4	2.1
1781	M & Ms plain chocolate candy, small bag	1	item(s)	48	0.8	240	2.0	34.0	1.0	10.0	6.0	3.3	0.3
4673	Milk chocolate bar, Symphony	1	item(s)	91	0.9	483	7.7	52.8	1.5	27.8	16.7	7.2	0.6
1783	Milky Way bar	1	item(s)	58	3.7	270	2.0	41.0	1.0	10.0	5.0	3.5	0.3
1788	Peanut brittle	1½	ounce(s)	43	0.3	207	3.2	30.3	1.1	8.1	1.8	3.4	1.9
1789	Reese's peanut butter cups	2	piece(s)	51	0.8	280	6.0	19.0	2.0	15.5	6.0	7.2	2.7
4689	Reese's pieces candy, small bag	1	item(s)	43	1.1	220	5.0	26.0	1.0	11.0	7.0	0.9	0.4
33399	Semisweet chocolate candy, made with butter	½	ounce(s)	14	0.1	68	0.6	9.0	0.8	4.2	2.5	1.4	0.1
1782	Snickers bar	1	item(s)	59	3.2	280	4.0	35.0	1.0	14.0	5.0	6.1	2.9
4694	Special Dark chocolate bar	1	item(s)	41	0.4	220	2.0	25.0	3.0	12.0	8.0	4.6	0.4
4695	Starburst fruit chews, original fruits	1	package(s)	59	3.9	240	0	48.0	0	5.0	1.0	2.1	1.8

PAGE KEY: A-4 = Breads/Baked Goods A-10 = Cereal/Rice/Pasta A-14 = Fruit A-18 = Vegetables/Legumes A-28 = Nuts/Seeds A-30 = Vegetarian A-32 = Dairy A-40 = Eggs A-40 = Seafood A-42 = Meats A-46 = Poultry A-46 = Processed Meats A-48 = Beverages A-52 = Fats/Oils A-54 = Sweets A-56 = Spices/Condiments/Sauces A-60 = Mixed Foods/Soups/Sandwiches A-64 = Fast Food A-84 = Convenience A-86 = Baby Foods

Chol (mg)	Calc (mg)	Iron (mg)	Magn (mg)	Pota (mg)	Sodi (mg)	Zinc (mg)	Vit A (µg)	Thia (mg)	Vit E (mg α)	Ribo (mg)	Niac (mg)	Vit B_6 (mg)	Fola (µg)	Vit C (mg)	Vit B_{12} (µg)	Sele (µg)
0	0	0.00	0	0	0	0.00	0	0.00	2.11	0.00	0.00	0.00	0	0	0	0
0	0	0.00	0	0	0	0.00	0	0.00	4.63	0.00	0.00	0.00	0	0	0	0
0	0	0.00	0	0	0	0.00	0	0.00	0.19	0.00	0.00	0.00	0	0	0	0
0	0	0.00	0	0	0	0.00	0	0.00	1.10	0.00	0.00	0.00	0	0	0	0
0	0	0.00	0	0	0	0.00	0	0.00	1.64	0.00	0.00	0.00	0	0	0	0
0	0	0.00	0	0	0	0.00	0	0.00	5.58	0.00	0.00	0.00	0	0	0	0
0	0	0.00	0	0.3	1.5	0.01	0.1	0.00	0.00	0.00	0.00	0.00	0	0	0	0
5	24	0.06	0	11.1	328.2	0.08	20.1	0.00	1.80	0.03	0.03	0.01	7.8	0.6	0.1	0.3
0	28	0.16	2.2	1.6	384.0	0.08	—	0.01	0.08	0.03	0.01	0.01	1.0	0.1	0.1	0.5
1	7	0.05	0.6	8.7	323.4	0.03	0.6	0.00	1.56	0.00	0.01	0.00	0.9	0	0	0.5
0	12	0.08	1.6	42.6	320.0	0.05	0.3	0.00	0.21	0.01	0.01	0.01	1.9	0	0	0.5
0	2	0.03	0.6	10.8	306.9	0.01	—	0.00	0.71	0.00	0.01	0.01	0	0.1	0	0.5
0	8	0.25	1.6	21.4	267.5	0.09	7.4	0.01	1.60	0.01	0.06	0.00	0	0	0	0
0	4	0.27	2.6	34.2	257.3	0.06	8.6	0.01	0.09	0.01	0.14	0.01	0.6	0	0	0.5
0	2	0.18	0.9	14.1	486.3	0.03	0.6	0.00	1.47	0.01	0.00	0.01	0	0	0	0.6
2	3	0.19	1.2	25.5	409.8	0.05	0.3	0.00	0.06	0.00	0.00	0.02	0	0	0	2.4
8	4	0.05	0.6	2.6	209.0	0.05	6.2	0.00	0.60	0.01	0.00	0.01	1.8	0	0.1	0.5
0	0	0.00	0	2.6	0.3	0.00	0	0.00	1.46	0.00	0.00	0.00	0	0	0	0.5
1	4	0.03	1.2	8.4	354.0	0.01	5.4	0.00	1.84	0.01	0.00	0.00	0.3	0.1	0	0.1
0	5	0.01	1.5	8.4	413.7	0.01	0.9	0.00	0.72	0.01	0.00	0.00	0.3	0.1	0	0.1
0	6	0.20	3.0	51.9	282.3	0.06	13.2	0.01	0.98	0.01	0.16	0.02	1.5	1.4	0	0.5
2	6	0.18	0	50.2	277.8	0.02	0.6	0.00	0.12	0.00	0.00	0.00	1.0	1.9	0	0.5
0	6	0.18	0	47.1	300.0	0.02	0.6	0.00	1.50	0.00	0.00	0.00	0	0	0	0.5
8	5	0.37	2.6	34.2	276.2	0.08	4.5	0.46	1.28	0.01	0.13	0.00	0	0	0	0.5
0	5	0.27	2.1	60.6	249.3	0.05	4.8	0.01	0.30	0.01	0.13	0.00	0	0	0	0
5	1	0.03	0.1	1.7	78.4	0.02	11.2	0.01	0.72	0.01	0.00	0.08	0.7	0	0	0.2
4	0	0.00	0	1.6	79.5	0.01	0	0.00	0.32	0.00	0.00	0.00	0	0	0	0.3
8	6	0.20	0.8	10.1	191.5	0.05	20.2	0.00	0.97	0.00	0.01	0.07	2.0	0.1	0.1	0.5
0	22	0.08	2.9	34.4	143.1	0.07	11.1	0.01	—	0.03	0.01	0.01	0.8	0.1	0	0
0	18	0.33	30.3	126.5	65.0	0.36	0	0.01	—	0.06	0.21	—	—	0	—	—
0	20	0.00	—	—	150.0	—	0	—	—	—	—	—	—	0	—	—
1	0	0.00	0	0.4	46.9	0.01	3.4	0.00	0.01	0.00	0.00	0.00	0	0	0	0.1
0	0	0.00	0	0.1	0	0.00	0	0.00	0.00	0.00	0.00	0.00	0	0	0	0
5	22	0.74	20.9	69.5	14.8	0.54	14.4	0.02	0.09	0.03	0.12	0.03	6.1	0.1	0	1.1
0	1	0.05	0.9	15.7	21.3	0.02	0	0.00	0.00	0.01	0.00	0.00	0	0	0	0.5
3	60	0.36	16.4	126.0	30.0	0.51	0	0.07	—	0.22	1.07	0.05	59.6	0	0.1	2.0
3	40	0.36	—	168.8	50.0	—	0	—	—	—	—	—	—	0	—	—
0	0	0.22	2.6	28.2	126.3	0.07	0	0.01	0.07	0.01	0.04	0.00	0	0	0	—
0	0	0.00	—	0	0	—	0	0.00	—	0.00	0.00	—	—	0	—	0
0	0	0.00	—	—	10.8	—	0	0.00	—	0.00	0.00	—	—	0	—	1.0
5	40	0.36	36.5	170.6	25.0	1.13	14.8	0.03	—	0.06	1.60	0.04	17.3	0.6	0.1	1.9
5	40	0.36	19.6	127.4	30.0	0.46	14.8	0.02	—	0.06	0.10	0.01	2.9	0.6	0.1	1.4
22	228	0.82	61.0	398.6	91.9	1.00	0	0.06	—	0.25	0.14	0.10	10.9	2.0	0.4	—
5	60	0.18	19.8	140.1	95.0	0.41	15.1	0.02	—	0.06	0.20	0.02	5.8	0.6	0.2	3.3
5	11	0.51	17.9	71.4	189.2	0.37	16.6	0.05	1.08	0.01	1.12	0.03	19.6	0	0	1.1
3	40	0.72	45.4	217.4	180.0	0.93	0	0.12	—	0.08	2.35	0.07	28.1	0	0.1	2.3
0	20	0.00	18.9	169.9	80.0	0.32	0	0.04	—	0.06	1.22	0.03	12.0	0	0.1	0.8
3	5	0.44	16.3	51.7	1.6	0.23	0.4	0.01	—	0.01	0.06	0.01	0.4	0	0	0.5
5	40	0.36	42.3	—	140.0	1.37	15.3	0.03	—	0.06	1.60	0.05	23.5	0.6	0.1	2.7
0	0	1.80	45.5	136.0	50.0	0.59	0	0.01	—	0.02	0.16	0.01	0.8	0	0	1.2
0	10	0.18	0.6	1.2	0	0.00	—	0.00	—	0.00	0.00	0.00	0	30	0	0.5

TABLE A-1 Table of Food Composition *(continued)*

(Computer code is for Cengage Diet Analysis program)
(For purposes of calculations, use "0" for t, <1, <.1, <.01, etc.)

DA+ Code	Food Description	Quantity	Measure	Wt (g)	H_2O (g)	Ener (kcal)	Prot (g)	Carb (g)	Fiber (g)	Fat (g)	Fat Breakdown (g) Sat	Mono	Poly
Sweets—*continued*													
4698	Taffy	3	piece(s)	45	2.2	179	0	41.2	0	1.5	0.9	0.4	0.1
4699	Three Musketeers bar	1	item(s)	60	3.5	260	2.0	46.0	1.0	8.0	4.5	2.6	0.3
4702	Twix caramel cookie bars	2	item(s)	58	2.4	280	3.0	37.0	1.0	14.0	5.0	7.7	0.5
4705	York peppermint pattie	1	item(s)	39	3.9	160	0.5	32.0	0.5	3.0	1.5	1.2	0.1
	Frosting, icing												
4760	Chocolate frosting, ready to eat	2	tablespoon(s)	31	5.2	122	0.3	19.4	0.3	5.4	1.7	2.8	0.6
4771	Creamy vanilla frosting, ready to eat	2	tablespoon(s)	28	4.2	117	0	19.0	0	4.5	0.8	1.4	2.2
17291	Dec-A-Cake variety pack candy decoration	1	teaspoon(s)	4	—	15	0	3.0	0	0.5	0	—	—
536	White icing	2	tablespoon(s)	40	3.6	162	0.1	31.8	0	4.2	0.8	2.0	1.2
	Gelatin												
13697	Gelatin snack, all flavors	1	item(s)	99	96.8	70	1.0	17.0	0	0	0	0	0
2616	Sugar free, low calorie mixed fruit gelatin mix, prepared	½	cup(s)	121	—	10	1.0	0	0	0	0	0	0
548	**Honey**	1	tablespoon(s)	21	3.6	64	0.1	17.3	0	0	0	0	0
	Jams, jellies												
550	Jam or preserves	1	tablespoon(s)	20	6.1	56	0.1	13.8	0.2	0	0	0	0
42199	Jams, preserves, dietetic, all flavors, w/sodium saccharin	1	tablespoon(s)	14	6.4	18	0	7.5	0.4	0	0	0	0
552	Jelly	1	tablespoon(s)	21	6.3	56	0	14.7	0.2	0	0	0	0
545	**Marshmallows**	4	item(s)	29	4.7	92	0.5	23.4	0	0.1	0	0	0
4800	**Marshmallow cream topping**	2	tablespoon(s)	40	7.9	129	0.3	31.6	0	0.1	0	0	0
555	**Molasses**	1	tablespoon(s)	20	4.4	58	0	14.9	0	0	0	0	0
4780	**Popsicle or ice pop**	1	item(s)	59	47.5	47	0	11.3	0	0.1	0	0	0
	Sugar												
559	Brown sugar, packed	1	teaspoon(s)	5	0.1	17	0	4.5	0	0	0	0	0
563	Powdered sugar, sifted	⅓	cup(s)	33	0.1	130	0	33.2	0	0	0	0	0
561	White granulated sugar	1	teaspoon(s)	4	0	16	0	4.2	0	0	0	0	0
	Sugar substitute												
1760	Equal sweetener, packet size	1	item(s)	1	—	0	0	0.9	0	0	0	0	0
13029	Splenda granular no calorie sweetener	1	teaspoon(s)	1	—	0	0	0.5	0	0	0	0	0
1759	Sweet N Low sugar substitute, packet	1	item(s)	1	0.1	4	0	0.5	0	0	0	0	0
	Syrup												
3148	Chocolate syrup	2	tablespoon(s)	38	11.6	105	0.8	24.4	1.0	0.4	0.2	0.1	0
29676	Maple syrup	¼	cup(s)	80	25.7	209	0	53.7	0	0.2	0	0.1	0.1
4795	Pancake syrup	¼	cup(s)	80	30.4	187	0	49.2	0	0	0	0	0
Spices, Condiments, Sauces													
	Spices												
807	Allspice, ground	1	teaspoon(s)	2	0.2	5	0.1	1.4	0.4	0.2	0	0	0
1171	Anise seeds	1	teaspoon(s)	2	0.2	7	0.4	1.1	0.3	0.3	0	0.2	0.1
729	Bakers' yeast, active	1	teaspoon(s)	4	0.3	12	1.5	1.5	0.8	0.2	0	0.1	0
683	Baking powder, double acting with phosphate	1	teaspoon(s)	5	0.2	2	0	1.1	0	0	0	0	0
1611	Baking soda	1	teaspoon(s)	5	0	0	0	0	0	0	0	0	0
8552	Basil	1	teaspoon(s)	1	0.8	0	0	0	0	0	0	0	0
34959	Basil, fresh	1	piece(s)	1	0.5	0	0	0	0	0	0	0	0
808	Basil, ground	1	teaspoon(s)	1	0.1	4	0.2	0.9	0.6	0.1	0	0	0
809	Bay leaf	1	teaspoon(s)	1	0	2	0	0.5	0.2	0.1	0	0	0
11720	Betel leaves	1	ounce(s)	28	—	17	1.8	2.4	0	0	—	—	—
730	Brewers' yeast	1	teaspoon(s)	3	0.1	8	1.0	1.0	0.8	0	0	0	0
11710	Capers	1	teaspoon(s)	5	—	0	0	0	0	0	0	0	0
1172	Caraway seeds	1	teaspoon(s)	2	0.2	7	0.4	1.0	0.8	0.3	0	0.2	0.1
1173	Celery seeds	1	teaspoon(s)	2	0.1	8	0.4	0.8	0.2	0.5	0	0.3	0.1
1174	Chervil, dried	1	teaspoon(s)	1	0	1	0.1	0.3	0.1	0	0	0	0
810	Chili powder	1	teaspoon(s)	3	0.2	8	0.3	1.4	0.9	0.4	0.1	0.1	0.2
8553	Chives, chopped	1	teaspoon(s)	1	0.9	0	0	0	0	0	0	0	0
51420	Cilantro (coriander)	1	teaspoon(s)	0	0.3	0	0	0	0	0	0	0	0
811	Cinnamon, ground	1	teaspoon(s)	2	0.2	6	0.1	1.9	1.2	0	0	0	0
812	Cloves, ground	1	teaspoon(s)	2	0.1	7	0.1	1.3	0.7	0.4	0.1	0	0.1
1175	Coriander leaf, dried	1	teaspoon(s)	1	0	2	0.1	0.3	0.1	0	0	0	0
1176	Coriander seeds	1	teaspoon(s)	2	0.2	5	0.2	1.0	0.8	0.3	0	0.2	0
1706	Cornstarch	1	tablespoon(s)	8	0.7	30	0	7.3	0.1	0	0	0	0

Chol (mg)	Calc (mg)	Iron (mg)	Magn (mg)	Pota (mg)	Sodi (mg)	Zinc (mg)	Vit A (μg)	Thia (mg)	Vit E (mg α)	Ribo (mg)	Niac (mg)	Vit B_6 (mg)	Fola (μg)	Vit C (mg)	Vit B_{12} (μg)	Sele (μg)
4	4	0.00	0	1.4	23.4	0.09	12.2	0.01	0.04	0.01	0.00	0.00	0	0	0	0.3
5	20	0.36	17.5	80.3	110.0	0.33	14.5	0.01	—	0.03	0.20	0.01	0	0.6	0.1	1.5
5	40	0.36	18.5	116.8	115.0	0.45	15.0	0.09	—	0.13	0.69	0.01	13.9	0.6	0.1	1.2
0	0	0.33	23.4	66.1	10.0	0.28	0	0.01	—	0.03	0.31	0.01	1.5	0	0	—
0	2	0.44	6.4	60.0	56.1	0.09	0	0.00	0.48	0.00	0.03	0.00	0.3	0	0	0.2
0	1	0.04	0.3	9.5	51.5	0.01	0	0.00	0.43	0.08	0.06	0.00	2.2	0	0	0
0	0	0.00	—	—	15.0	—	0	—	—	—	—	—	—	0	—	—
0	4	0.01	0.4	5.6	76.4	0.01	44.4	0.00	0.32	0.01	0.00	0.00	0	0	0	0.3
0	0	0.00	—	0	40.0	—	0	—	—	—	—	—	—	0	—	—
0	0	0.00	0	0	50.0	0.00	0	0.00	0.00	0.00	0.00	0.00	0	0	0	—
0	1	0.08	0.4	10.9	0.8	0.04	0	0.00	0.00	0.01	0.02	0.01	0.4	0.1	0	0.2
0	4	0.10	0.8	15.4	6.4	0.01	0	0.00	0.02	0.02	0.01	0.00	2.2	1.8	0	0.4
0	1	0.56	0.7	9.7	0	0.01	0	0.00	0.01	0.00	0.00	0.00	1.3	0	0	0.2
0	1	0.04	1.3	11.3	6.3	0.01	0	0.00	0.00	0.01	0.01	0.00	0.4	0.2	0	0.1
0	1	0.06	0.6	1.4	23.0	0.01	0	0.00	0.00	0.00	0.02	0.00	0.3	0	0	0.5
0	1	0.08	0.8	2.0	32.0	0.01	0	0.00	0.00	0.00	0.03	0.00	0.4	0	0	0.7
0	41	0.94	48.4	292.8	7.4	0.05	0	0.01	0.00	0.00	0.18	0.13	0	0	0	3.6
0	0	0.31	0.6	8.9	4.1	0.08	0	0.00	0.00	0.00	0.00	0.00	0	0.4	0	0.1
0	4	0.03	0.4	6.1	1.3	0.00	0	0.00	0.00	0.00	0.01	0.00	0	0	0	0.1
0	0	0.01	0	0.7	0.3	0.00	0	0.00	0.00	0.00	0.00	0.00	0	0	0	0.2
0	0	0.00	0	0.1	0	0.00	0	0.00	0.00	0.00	0.00	0.00	0	0	0	0
0	0	0.00	0	0	0	0.00	0	0.00	0.00	0.00	0.00	0.00	0	0	0	0
0	0	0.00	—	—	0	—	—	0.00	—	0.00	0.00	—	—	0	0	—
0	0	0.00	—	—	0	—	0	—	0.00	—	—	—	—	0	—	—
0	5	0.79	24.4	84.0	27.0	0.27	0	0.00	0.01	0.01	0.12	0.00	0.8	0.1	0	0.5
0	54	0.96	11.2	163.2	7.2	3.32	0	0.01	0.00	0.01	0.02	0.00	0	0	0	0.5
0	2	0.02	1.6	12.0	65.6	0.06	0	0.01	0.00	0.01	0.00	0.00	0	0	0	0
0	13	0.13	2.6	19.8	1.5	0.01	0.5	0.00	—	0.00	0.05	0.00	0.7	0.7	0	0.1
0	14	0.77	3.6	30.3	0.3	0.11	0.3	0.01	—	0.01	0.06	0.01	0.2	0.4	0	0.1
0	3	0.66	3.9	80.0	2.0	0.25	0	0.09	0.00	0.21	1.59	0.06	93.6	0	0	1.0
0	339	0.51	1.8	0.2	363.1	0.00	0	0.00	0.00	0.00	0.00	0.00	0	0	0	0
0	0	0.00	0	0	1258.6	0.00	0	0.00	0.00	0.00	0.00	0.00	0	0	0	0
0	2	0.02	0.6	2.6	0	0.01	2.3	0.00	0.01	0.00	0.01	0.00	0.6	0.2	0	0
0	1	0.01	0.4	2.3	0	0.00	1.3	0.00	—	0.00	0.00	0.00	0.3	0.1	0	0
0	30	0.58	5.9	48.1	0.5	0.08	6.6	0.00	0.10	0.00	0.09	0.03	3.8	0.9	0	0
0	5	0.25	0.7	3.2	0.1	0.02	1.9	0.00	—	0.00	0.01	0.01	1.1	0.3	0	0
0	110	2.29	—	155.9	2.0	—	—	0.04	—	0.07	0.19	—	—	0.9	0	—
0	6	0.46	6.1	50.7	3.3	0.21	0	0.41	—	0.11	1.00	0.06	104.3	0	0	0
0	—	—	—	—	105.0	—	—	—	—	—	—	—	—	—	0	—
0	14	0.34	5.4	28.4	0.4	0.11	0.4	0.01	0.05	0.01	0.07	0.01	0.2	0.4	0	0.3
0	35	0.89	8.8	28.0	3.2	0.13	0.1	0.01	0.02	0.01	0.06	0.01	0.2	0.3	0	0.2
0	8	0.19	0.8	28.4	0.5	0.05	1.8	0.00	—	0.00	0.03	0.01	1.6	0.3	0	0.2
0	7	0.37	4.4	49.8	26.3	0.07	38.6	0.01	0.75	0.02	0.20	0.09	2.6	1.7	0	0.2
0	1	0.01	0.4	3.0	0	0.01	2.2	0.00	0.00	0.00	0.01	0.00	1.1	0.6	0	0
0	0	0.01	0.1	1.7	0.2	0.00	1.1	0.00	0.01	0.00	0.00	0.00	0.2	0.1	0	0
0	23	0.19	1.4	9.9	0.2	0.04	0.3	0.00	0.05	0.00	0.03	0.00	0.1	0.1	0	0.1
0	14	0.18	5.5	23.1	5.1	0.02	0.6	0.00	0.17	0.01	0.03	0.01	2.0	1.7	0	0.1
0	7	0.25	4.2	26.8	1.3	0.02	1.8	0.01	0.01	0.01	0.06	0.00	1.6	3.4	0	0.2
0	13	0.29	5.9	22.8	0.6	0.08	0	0.00	—	0.01	0.03	—	0	0.4	0	0.5
0	0	0.03	0.2	0.2	0.7	0.01	0	0.00	0.00	0.00	0.00	0.00	0	0	0	0.2

TABLE A-1 Table of Food Composition *(continued)*

(Computer code is for Cengage Diet Analysis program)
(For purposes of calculations, use "0" for t, <1, <.1, <.01, etc.)

DA+ Code	Food Description	Quantity	Measure	Wt (g)	H_2O (g)	Ener (kcal)	Prot (g)	Carb (g)	Fiber (g)	Fat (g)	Fat Breakdown (g) Sat	Mono	Poly
Spices, Condiments, Sauces—*continued*													
1177	Cumin seeds	1	teaspoon(s)	2	0.2	8	0.4	0.9	0.2	0.5	0	0.3	0.1
11729	Cumin, ground	1	teaspoon(s)	5	—	11	0.4	0.8	0.8	0.4	—	—	—
1178	Curry powder	1	teaspoon(s)	2	0.2	7	0.3	1.2	0.7	0.3	0	0.1	0.1
1179	Dill seeds	1	teaspoon(s)	2	0.2	6	0.3	1.2	0.4	0.3	0	0.2	0
1180	Dill weed, dried	1	teaspoon(s)	1	0.1	3	0.2	0.6	0.1	0	0	0	0
34949	Dill weed, fresh	5	piece(s)	1	0.9	0	0	0.1	0	0	0	0	0
4949	Fennel leaves, fresh	1	teaspoon(s)	1	0.9	0	0	0.1	0	0	—	—	—
1181	Fennel seeds	1	teaspoon(s)	2	0.2	7	0.3	1.0	0.8	0.3	0	0.2	0
1182	Fenugreek seeds	1	teaspoon(s)	4	0.3	12	0.9	2.2	0.9	0.2	0.1	—	—
11733	Garam masala, powder	1	ounce(s)	28	—	107	4.4	12.8	0	4.3	—	—	—
1067	Garlic clove	1	item(s)	3	1.8	4	0.2	1.0	0.1	0	0	0	0
813	Garlic powder	1	teaspoon(s)	3	0.2	9	0.5	2.0	0.3	0	0	0	0
1068	Ginger root	2	teaspoon(s)	4	3.1	3	0.1	0.7	0.1	0	0	0	0
1183	Ginger, ground	1	teaspoon(s)	2	0.2	6	0.2	1.3	0.2	0.1	0	0	0
35497	Leeks, bulb and lower-leaf, freeze-dried	¼	cup(s)	1	0	3	0.1	0.6	0.1	0	0	0	0
1184	Mace, ground	1	teaspoon(s)	2	0.1	8	0.1	0.9	0.3	0.6	0.2	0.2	0.1
1185	Marjoram, dried	1	teaspoon(s)	1	0	2	0.1	0.4	0.2	0	0	0	0
1186	Mustard seeds, yellow	1	teaspoon(s)	3	0.2	15	0.8	1.2	0.5	0.9	0	0.7	0.2
814	Nutmeg, ground	1	teaspoon(s)	2	0.1	12	0.1	1.1	0.5	0.8	0.6	0.1	0
2747	Onion flakes, dehydrated	1	teaspoon(s)	2	0.1	6	0.1	1.4	0.2	0	0	0	0
1187	Onion powder	1	teaspoon(s)	2	0.1	7	0.2	1.7	0.1	0	0	0	0
815	Oregano, ground	1	teaspoon(s)	2	0.1	5	0.2	1.0	0.6	0.2	0	0	0.1
816	Paprika	1	teaspoon(s)	2	0.2	6	0.3	1.2	0.8	0.3	0	0	0.2
817	Parsley, dried	1	teaspoon(s)	0	0	1	0.1	0.2	0.1	0	0	0	0
818	Pepper, black	1	teaspoon(s)	2	0.2	5	0.2	1.4	0.6	0.1	0	0	0
819	Pepper, cayenne	1	teaspoon(s)	2	0.1	6	0.2	1.0	0.5	0.3	0.1	0	0.2
1188	Pepper, white	1	teaspoon(s)	2	0.3	7	0.3	1.6	0.6	0.1	0	0	0
1189	Poppy seeds	1	teaspoon(s)	3	0.2	15	0.5	0.7	0.3	1.3	0.1	0.2	0.9
1190	Poultry seasoning	1	teaspoon(s)	2	0.1	5	0.1	1.0	0.2	0.1	0	0	0
1191	Pumpkin pie spice, powder	1	teaspoon(s)	2	0.1	6	0.1	1.2	0.3	0.2	0.1	0	0
1192	Rosemary, dried	1	teaspoon(s)	1	0.1	4	0.1	0.8	0.5	0.2	0.1	0	0
11723	Rosemary, fresh	1	teaspoon(s)	1	0.5	1	0	0.1	0.1	0	0	0	0
2722	Saffron powder	1	teaspoon(s)	1	0.1	2	0.1	0.5	0	0	0	0	0
11724	Sage	1	teaspoon(s)	1	—	1	0	0.1	0	0	—	—	—
1193	Sage, ground	1	teaspoon(s)	1	0.1	2	0.1	0.4	0.3	0.1	0	0	0
30189	Salt substitute	¼	teaspoon(s)	1	—	0	0	0	0	0	0	0	0
30190	Salt substitute, seasoned	¼	teaspoon(s)	1	—	1	0	0.1	0	0	0	—	—
822	Salt, table	¼	teaspoon(s)	2	0	0	0	0	0	0	0	0	0
1194	Savory, ground	1	teaspoon(s)	1	0.1	4	0.1	1.0	0.6	0.1	0	—	—
820	Sesame seed kernels, toasted	1	teaspoon(s)	3	0.1	15	0.5	0.7	0.5	1.3	0.2	0.5	0.6
11725	Sorrel	1	teaspoon(s)	3	—	1	0.1	0.1	0	0	0	—	—
11721	Spearmint	1	teaspoon(s)	2	1.6	1	0.1	0.2	0.1	0	0	0	0
35498	Sweet green peppers, freeze-dried	¼	cup(s)	2	0	5	0.3	1.1	0.3	0	0	0	0
11726	Tamarind leaves	1	ounce(s)	28	—	33	1.6	5.2	0	0.6	—	—	—
11727	Tarragon	1	ounce(s)	28	—	14	1.0	1.8	0	0.3	—	—	—
1195	Tarragon, ground	1	teaspoon(s)	2	0.1	5	0.4	0.8	0.1	0.1	0	0	0.1
11728	Thyme, fresh	1	teaspoon(s)	1	0.5	1	0	0.2	0.1	0	0	0	0
821	Thyme, ground	1	teaspoon(s)	1	0.1	4	0.1	0.9	0.5	0.1	0	0	0
1196	Turmeric, ground	1	teaspoon(s)	2	0.3	8	0.2	1.4	0.5	0.2	0.1	0	0
11995	Wasabi	1	tablespoon(s)	14	10.7	10	0.7	2.3	0.2	0	—	—	—
	Condiments												
674	Catsup or ketchup	1	tablespoon(s)	15	10.4	15	0.3	3.8	0	0	0	0	0
703	Dill pickle	1	ounce(s)	28	26.7	3	0.2	0.7	0.3	0	0	0	0
138	Mayonnaise with soybean oil	1	tablespoon(s)	14	2.1	99	0.1	0.4	0	11.0	1.6	2.7	5.8
140	Mayonnaise, low calorie	1	tablespoon(s)	16	10.0	37	0	2.6	0	3.1	0.5	0.7	1.7
1682	Mustard, brown	1	teaspoon(s)	5	4.1	5	0.3	0.3	0	0.3	—	—	—
700	Mustard, yellow	1	teaspoon(s)	5	4.1	3	0.2	0.3	0.2	0.2	0	0.1	0
706	Sweet pickle relish	1	tablespoon(s)	15	9.3	20	0.1	5.3	0.2	0.1	0	0	0
141	Tartar sauce	2	tablespoon(s)	28	8.7	144	0.3	4.1	0.1	14.4	2.2	3.8	7.7
	Sauces												
685	Barbecue sauce	2	tablespoon(s)	31	18.9	47	0	11.3	0.2	0.1	0	0	0.1
834	Cheese sauce	¼	cup(s)	63	44.4	110	4.2	4.3	0.3	8.4	3.8	2.4	1.6
32123	Chili enchilada sauce, green	2	tablespoon(s)	57	53.0	15	0.6	3.1	0.7	0.3	0	0	0.1
32122	Chili enchilada sauce, red	2	tablespoon(s)	32	24.5	27	1.1	5.0	2.1	0.8	0.1	0	0.4

PAGE KEY: A-4 = Breads/Baked Goods A-10 = Cereal/Rice/Pasta A-14 = Fruit A-18 = Vegetables/Legumes A-28 = Nuts/Seeds A-30 = Vegetarian A-32 = Dairy A-40 = Eggs A-40 = Seafood A-42 = Meats A-46 = Poultry A-46 = Processed Meats A-48 = Beverages A-52 = Fats/Oils A-54 = Sweets A-56 = Spices/Condiments/Sauces A-60 = Mixed Foods/Soups/Sandwiches A-64 = Fast Food A-84 = Convenience A-86 = Baby Foods

Chol (mg)	Calc (mg)	Iron (mg)	Magn (mg)	Pota (mg)	Sodi (mg)	Zinc (mg)	Vit A (µg)	Thia (mg)	Vit E (mg α)	Ribo (mg)	Niac (mg)	Vit B_6 (mg)	Fola (µg)	Vit C (mg)	Vit B_{12} (µg)	Sele (µg)
0	20	1.39	7.7	37.5	3.5	0.10	1.3	0.01	0.07	0.01	0.09	0.01	0.2	0.2	0	0.1
0	20	—	—	43.6	4.8	—	—	—	—	—	—	—	—	—	—	—
0	10	0.59	5.1	30.9	1.0	0.08	1.0	0.01	0.44	0.01	0.06	0.02	3.1	0.2	0	0.3
0	32	0.34	5.4	24.9	0.4	0.10	0.1	0.01	—	0.01	0.05	0.01	0.2	0.4	0	0.3
0	18	0.48	4.5	33.1	2.1	0.03	2.9	0.00	—	0.00	0.02	0.01	1.5	0.5	0	—
0	2	0.06	0.6	7.4	0.6	0.01	3.9	0.00	0.01	0.00	0.01	0.00	1.5	0.9	0	—
0	1	0.02	—	4.0	0.1	—	—	0.00	—	0.00	0.01	0.00	—	0.3	0	—
0	24	0.37	7.7	33.9	1.8	0.07	0.1	0.01	—	0.01	0.12	0.01	—	0.4	0	—
0	7	1.24	7.1	28.5	2.5	0.09	0.1	0.01	—	0.01	0.06	0.02	2.1	0.1	0	0.2
0	215	9.24	93.6	411.1	27.5	1.07	—	0.09	—	0.09	0.70	—	0	0	0	—
0	5	0.05	0.8	12.0	0.5	0.03	0	0.01	0.00	0.00	0.02	0.03	0.1	0.9	0	0.4
0	2	0.07	1.6	30.8	0.7	0.07	0	0.01	0.01	0.00	0.01	0.08	0.1	0.5	0	1.1
0	1	0.02	1.7	16.6	0.5	0.01	0	0.00	0.01	0.00	0.02	0.01	0.4	0.2	0	0
0	2	0.20	3.3	24.2	0.6	0.08	0.1	0.00	0.32	0.00	0.09	0.01	0.7	0.1	0	0.7
0	3	0.06	1.3	19.2	0.3	0.01	0.1	0.01	—	0.00	0.02	0.01	2.9	0.9	0	0
0	4	0.23	2.8	7.9	1.4	0.03	0.7	0.01	—	0.01	0.02	0.00	1.3	0.4	0	0
0	12	0.49	2.1	9.1	0.5	0.02	2.4	0.00	0.01	0.00	0.02	0.01	1.6	0.3	0	0
0	17	0.32	9.8	22.5	0.2	0.18	0.1	0.01	0.09	0.01	0.26	0.01	2.5	0.1	0	4.4
0	4	0.06	4.0	7.7	0.4	0.04	0.1	0.01	0.00	0.00	0.02	0.00	1.7	0.1	0	0
0	4	0.02	1.5	27.1	0.4	0.03	0	0.01	0.00	0.00	0.01	0.02	2.8	1.3	0	0.1
0	8	0.05	2.6	19.8	1.1	0.04	0	0.01	0.01	0.00	0.01	0.02	3.5	0.3	0	0
0	24	0.66	4.1	25.0	0.2	0.06	5.2	0.01	0.28	0.01	0.09	0.01	4.1	0.8	0	0.1
0	4	0.49	3.9	49.2	0.7	0.08	55.4	0.01	0.62	0.03	0.32	0.08	2.2	1.5	0	0.1
0	4	0.29	0.7	11.4	1.4	0.01	1.5	0.00	0.02	0.00	0.02	0.00	0.5	0.4	0	0.1
0	9	0.60	4.1	26.4	0.9	0.03	0.3	0.00	0.01	0.01	0.02	0.01	0.2	0.4	0	0.1
0	3	0.14	2.7	36.3	0.5	0.04	37.5	0.01	0.53	0.01	0.15	0.04	1.9	1.4	0	0.2
0	6	0.34	2.2	1.8	0.1	0.02	0	0.00	—	0.00	0.01	0.00	0.2	0.5	0	0.1
0	41	0.26	9.3	19.6	0.6	0.28	0	0.02	0.03	0.01	0.02	0.01	1.6	0.1	0	0
0	15	0.53	3.4	10.3	0.4	0.04	2.0	0.00	0.02	0.00	0.04	0.02	2.1	0.2	0	0.1
0	12	0.33	2.3	11.3	0.9	0.04	0.2	0.00	0.01	0.00	0.03	0.01	0.9	0.4	0	0.2
0	15	0.35	2.6	11.5	0.6	0.03	1.9	0.01	—	0.01	0.01	0.02	3.7	0.7	0	0.1
0	2	0.04	0.6	4.7	0.2	0.01	1.0	0.00	—	0.00	0.01	0.00	0.8	0.2	0	—
0	1	0.07	1.8	12.1	1.0	0.01	0.2	0.00	—	0.00	0.01	0.01	0.7	0.6	0	0
0	4	—	1.1	2.7	0	0.01	—	0.00	—	—	—	—	—	—	0	—
0	12	0.19	3.0	7.5	0.1	0.03	2.1	0.01	0.05	0.00	0.04	0.01	1.9	0.2	0	0
0	7	0.00	0	603.6	0.1	—	0	—	—	—	—	—	—	0	—	—
0	0	0.00	—	476.3	0.1	—	0	—	—	—	—	—	—	0	—	—
0	0	0.01	0	0.1	581.4	0.00	0	0.00	0.00	0.00	0.00	0.00	0	0	0	0
0	30	0.53	5.3	14.7	0.3	0.06	3.6	0.01	—	—	0.05	0.02	—	0.7	0	0.1
0	3	0.21	9.2	10.8	1.0	0.27	0.1	0.03	0.01	0.01	0.15	0.00	2.6	0	0	0
0	—	—	—	—	0.1	—	—	—	—	—	—	—	—	—	—	—
0	4	0.22	1.2	8.7	0.6	0.02	3.9	0.00	—	0.00	0.01	0.00	2.0	0.3	0	—
0	2	0.16	3.0	50.7	3.1	0.03	4.5	0.01	0.06	0.01	0.11	0.03	3.7	30.4	0	0.1
0	85	1.48	20.2	—	—	—	—	0.06	—	0.02	1.16	—	—	0.9	0	—
0	48	—	14.5	128.1	2.6	0.17	—	0.04	—	—	—	—	—	0.6	0	—
0	18	0.51	5.6	48.3	1.0	0.06	3.4	0.00	—	0.02	0.14	0.03	4.4	0.8	0	0.1
0	3	0.14	1.3	4.9	0.1	0.01	1.9	0.00	—	0.00	0.01	0.00	0.4	1.3	0	—
0	26	1.73	3.1	11.4	0.8	0.08	2.7	0.01	0.10	0.01	0.06	0.01	3.8	0.7	0	0.1
0	4	0.91	4.2	55.6	0.8	0.09	0	0.00	0.06	0.01	0.11	0.04	0.9	0.6	0	0.1
0	13	0.11	—	—	—	—	—	0.02	—	0.01	0.07	—	—	11.2	0	—
0	3	0.07	2.9	57.3	167.1	0.03	7.1	0.00	0.21	0.02	0.21	0.02	1.5	2.3	0	0
0	12	0.10	2.0	26.1	248.1	0.03	2.6	0.01	0.02	0.01	0.03	0.01	0.3	0.2	0	0
5	1	0.03	0.1	1.7	78.4	0.02	11.2	0.00	0.72	0.01	0.00	0.08	0.7	0	0	0.2
4	0	0.00	0	1.6	79.5	0.01	0	0.00	0.32	0.00	0.00	0.00	0	0	0	0.3
0	6	0.09	1.0	6.8	68.1	0.01	0	0.00	0.09	0.00	0.01	0.00	0.2	0.1	0	—
0	3	0.07	2.5	6.9	56.8	0.03	0.2	0.01	0.01	0.00	0.02	0.00	0.4	0.1	0	1.6
0	0	0.13	0.8	3.8	121.7	0.02	9.2	0.00	0.08	0.01	0.03	0.00	0.2	0.2	0	0
8	6	0.20	0.8	10.1	191.5	0.05	20.2	0.00	0.97	0.00	0.01	0.07	2.0	0.1	0.1	0.5
0	4	0.06	3.8	65.0	349.7	0.04	3.8	0.00	0.20	0.01	0.15	0.01	0.6	0.2	0	0.4
18	116	0.13	5.7	18.9	521.6	0.61	50.4	0.00	—	0.07	0.01	0.01	2.5	0.3	0.1	2.0
0	5	0.36	9.5	125.7	61.9	0.11	—	0.02	0.00	0.02	0.63	0.06	5.7	43.9	0	0
0	7	1.05	11.1	231.3	113.8	0.14	—	0.01	0.00	0.21	0.61	0.34	6.6	0.3	0	0.3

TABLE A-1 Table of Food Composition *(continued)*

(Computer code is for Cengage Diet Analysis program)
(For purposes of calculations, use "0" for t, <1, <.1, <.01, etc.)

DA+ Code	Food Description	Quantity	Measure	Wt (g)	H_2O (g)	Ener (kcal)	Prot (g)	Carb (g)	Fiber (g)	Fat (g)	Fat Breakdown (g) Sat	Mono	Poly
Spices, Condiments, Sauces—*continued*													
29688	Hoisin sauce	1	tablespoon(s)	16	7.1	35	0.5	7.1	0.4	0.5	0.1	0.2	0.3
1641	Horseradish sauce, prepared	1	teaspoon(s)	5	3.3	10	0.1	0.2	0	1.0	0.6	0.3	0
16670	Mole poblano sauce	½	cup(s)	133	102.7	156	5.3	11.4	2.7	11.3	2.6	5.1	3.0
29689	Oyster sauce	1	tablespoon(s)	16	12.8	8	0.2	1.7	0	0	0	0	0
1655	Pepper sauce or Tabasco	1	teaspoon(s)	5	4.8	1	0.1	0	0	0	0	0	0
347	Salsa	2	tablespoon(s)	32	28.8	9	0.5	2.0	0.5	0.1	0	0	0
52206	Soy sauce, tamari	1	tablespoon(s)	18	12.0	11	1.9	1.0	0.1	0	0	0	0
839	Sweet and sour sauce	2	tablespoon(s)	39	29.8	37	0.1	9.1	0.1	0	0	0	0
1613	Teriyaki sauce	1	tablespoon(s)	18	12.2	16	1.1	2.8	0	0	0	0	0
25294	Tomato sauce	½	cup(s)	150	132.8	63	2.2	11.9	2.6	1.8	0.2	0.4	0.9
728	White sauce, medium	¼	cup(s)	63	46.8	92	2.4	5.7	0.1	6.7	1.8	2.8	1.8
1654	Worcestershire sauce	1	teaspoon(s)	6	4.5	4	0	1.1	0	0	0	0	0
	Vinegar												
30853	Balsamic	1	tablespoon(s)	15	—	10	0	2.0	0	0	0	0	0
727	Cider	1	tablespoon(s)	15	14.0	3	0	0.1	0	0	0	0	0
1673	Distilled	1	tablespoon(s)	15	14.3	2	0	0.8	0	0	0	0	0
12948	Tarragon	1	tablespoon(s)	15	13.8	2	0	0.1	0	0	0	0	0
Mixed Foods, Soups, Sandwiches													
	Mixed dishes												
16652	Almond chicken	1	cup(s)	242	186.8	281	21.8	15.8	3.4	14.7	1.8	6.3	5.6
25224	Barbecued chicken	1	serving(s)	177	99.3	327	27.1	15.7	0.5	17.1	4.8	6.8	3.8
25227	Bean burrito	1	item(s)	149	81.8	326	16.1	33.0	5.6	14.8	8.3	4.7	0.9
9516	Beef and vegetable fajita	1	item(s)	223	143.9	397	22.4	35.3	3.1	18.0	5.9	8.0	2.5
16796	Beef or pork egg roll	2	item(s)	128	85.2	225	9.9	18.4	1.4	12.4	2.9	6.0	2.6
177	Beef stew with vegetables, prepared	1	cup(s)	245	201.0	220	16.0	15.0	3.2	11.0	4.4	4.5	0.5
30233	Beef stroganoff with noodles	1	cup(s)	256	190.1	343	19.7	22.8	1.5	19.1	7.4	5.7	4.4
16651	Cashew chicken	1	cup(s)	242	186.8	281	21.8	15.8	3.4	14.7	1.8	6.3	5.6
30274	Cheese pizza with vegetables, thin crust	2	slice(s)	140	76.6	298	12.7	35.4	2.5	12.0	4.9	4.7	1.6
30330	Cheese quesadilla	1	item(s)	54	18.3	190	7.7	15.3	1.0	10.8	5.2	3.6	1.3
215	Chicken and noodles, prepared	1	cup(s)	240	170.0	365	22.0	26.0	1.3	18.0	5.1	7.1	3.9
30239	Chicken and vegetables with broccoli, onion, bamboo shoots in soy based sauce	1	cup(s)	162	125.5	180	15.8	9.3	1.8	8.6	1.7	3.0	3.1
25093	Chicken cacciatore	1	cup(s)	244	175.7	284	29.9	5.7	1.3	15.3	4.3	6.2	3.3
28020	Chicken fried turkey steak	3	ounce(s)	492	276.2	706	77.1	68.7	3.6	12.0	3.4	2.9	3.9
218	Chicken pot pie	1	cup(s)	252	154.6	542	22.6	41.4	3.5	31.3	9.8	12.5	7.1
30240	Chicken teriyaki	1	cup(s)	244	158.3	364	51.0	15.2	0.7	7.0	1.8	2.0	1.7
25119	Chicken waldorf salad	½	cup(s)	100	67.2	179	14.0	6.8	1.0	10.8	1.8	3.1	5.2
25099	Chili con carne	¾	cup(s)	215	174.4	198	13.7	21.4	7.5	6.9	2.5	2.8	0.5
1062	Coleslaw	¾	cup(s)	90	73.4	70	1.2	11.2	1.4	2.3	0.3	0.6	1.2
1574	Crab cakes, from blue crab	1	item(s)	60	42.6	93	12.1	0.3	0	4.5	0.9	1.7	1.4
32144	Enchiladas with green chili sauce (enchiladas verdes)	1	item(s)	144	103.8	207	9.3	17.6	2.6	11.7	6.4	3.6	1.0
2793	Falafel patty	3	item(s)	51	17.7	170	6.8	16.2	—	9.1	1.2	5.2	2.1
28546	Fettuccine alfredo	1	cup(s)	244	88.7	279	13.1	46.1	1.4	4.2	2.2	1.0	0.4
32146	Flautas	3	item(s)	162	78.0	438	24.9	36.3	4.1	21.6	8.2	8.8	2.3
29629	Fried rice with meat or poultry	1	cup(s)	198	128.5	333	12.3	41.8	1.4	12.3	2.2	3.5	5.7
16649	General Tso chicken	1	cup(s)	146	91.0	296	18.7	16.4	0.9	17.0	4.0	6.3	5.3
1826	Green salad	¾	cup(s)	104	98.9	17	1.3	3.3	2.2	0.1	0	0	0
1814	Hummus	½	cup(s)	123	79.8	218	6.0	24.7	4.9	10.6	1.4	6.0	2.6
16650	Kung pao chicken	1	cup(s)	162	87.2	434	28.8	11.7	2.3	30.6	5.2	13.9	9.7
16622	Lamb curry	1	cup(s)	236	187.9	257	28.2	3.7	0.9	13.8	3.9	4.9	3.3
25253	Lasagna with ground beef	1	cup(s)	237	158.4	284	16.9	22.3	2.4	14.5	7.5	4.9	0.8
442	Macaroni and cheese, prepared	1	cup(s)	200	122.3	390	14.9	40.6	1.6	18.6	7.9	6.4	2.9
29637	Meat filled ravioli with tomato or meat sauce, canned	1	cup(s)	251	198.7	208	7.8	36.5	1.3	3.7	1.5	1.4	0.3
25105	Meat loaf	1	slice(s)	115	84.5	245	17.0	6.6	0.4	16.0	6.1	6.9	0.9
16646	Moo shi pork	1	cup(s)	151	76.8	512	18.9	5.3	0.6	46.4	6.9	15.8	21.2
16788	Nachos with beef, beans, cheese, tomatoes and onions	1	serving(s)	551	253.5	1576	59.1	137.5	20.4	90.8	32.6	41.9	9.4
6116	Pepperoni pizza	2	slice(s)	142	66.1	362	20.2	39.7	2.9	13.9	4.5	6.3	2.3
29601	Pizza with meat and vegetables, thin crust	2	slice(s)	158	81.4	386	16.5	36.8	2.7	19.1	7.7	8.1	2.2
655	Potato salad	½	cup(s)	125	95.0	179	3.4	14.0	1.6	10.3	1.8	3.1	4.7

PAGE KEY: A-4 = Breads/Baked Goods A-10 = Cereal/Rice/Pasta A-14 = Fruit A-18 = Vegetables/Legumes A-28 = Nuts/Seeds A-30 = Vegetarian A-32 = Dairy A-40 = Eggs A-40 = Seafood A-42 = Meats A-46 = Poultry A-46 = Processed Meats A-48 = Beverages A-52 = Fats/Oils A-54 = Sweets A-56 = Spices/Condiments/Sauces A-60 = Mixed Foods/Soups/Sandwiches A-64 = Fast Food A-84 = Convenience A-86 = Baby Foods

Chol (mg)	Calc (mg)	Iron (mg)	Magn (mg)	Pota (mg)	Sodi (mg)	Zinc (mg)	Vit A (μg)	Thia (mg)	Vit E (mg α)	Ribo (mg)	Niac (mg)	Vit B_6 (mg)	Fola (μg)	Vit C (mg)	Vit B_{12} (μg)	Sele (μg)
0	5	0.16	3.8	19.0	258.4	0.05	0	0.00	0.04	0.03	0.18	0.01	3.7	0.1	0	0.3
2	5	0.00	0.5	6.7	14.6	0.01	8.0	0.00	0.02	0.01	0.00	0.00	0.5	0.1	0	0.1
1	38	1.81	58.3	280.9	304.8	1.15	13.3	0.06	1.72	0.08	1.84	0.09	15.9	3.4	0.1	1.1
0	5	0.02	0.6	8.6	437.3	0.01	0	0.00	0.00	0.02	0.23	0.00	2.4	0	0.1	0.7
0	1	0.05	0.6	6.4	31.7	0.01	4.1	0.00	0.00	0.00	0.01	0.01	0.1	0.2	0	0
0	9	0.14	4.8	95.0	192.0	0.11	4.8	0.01	0.37	0.01	0.02	0.05	1.3	0.6	0	0.3
0	4	0.43	7.3	38.7	1018.9	0.08	0	0.01	0.00	0.03	0.72	0.04	3.3	0	0	0.1
0	5	0.20	1.2	8.2	97.5	0.01	0	0.00	—	0.01	0.11	0.03	0.2	0	0	—
0	5	0.30	11.0	40.5	689.9	0.01	0	0.01	0.00	0.01	0.22	0.01	1.4	0	0	0.2
0	23	1.24	28.9	536.8	268.6	0.36	—	0.08	0.52	0.08	1.64	0.20	23.2	32.0	0	1.0
4	74	0.20	8.8	97.5	221.3	0.25	—	0.04	—	0.11	0.25	0.02	3.1	0.5	0.2	—
0	6	0.30	0.7	45.4	55.6	0.01	0.3	0.00	0.00	0.01	0.03	0.00	0.5	0.7	0	0
0	0	0.00	—	—	0	—	0	—	—	—	—	—	—	0	—	—
0	1	0.03	0.7	10.9	0.7	0.01	0	0.00	0.00	0.00	0.00	0.00	0	0	0	0
0	1	0.09	0	2.3	0.1	0.00	0	0.00	0.00	0.00	0.00	0.00	0	0	0	5.0
0	0	0.07	—	2.3	0.7	—	—	0.07	—	0.07	0.07	—	—	0.3	0	—
41	68	1.86	58.1	539.7	510.6	1.50	31.5	0.07	4.11	0.22	9.57	0.43	26.6	5.1	0.3	13.6
120	26	1.70	32.4	419.7	500.9	2.67	—	0.09	0.01	0.24	6.87	0.40	15.0	7.9	0.3	19.5
38	333	3.01	52.5	447.6	510.6	1.98	—	0.28	0.01	0.30	1.92	0.19	134.4	8.2	0.3	15.9
45	85	3.65	37.9	475.0	756.0	3.52	17.8	0.38	0.80	0.29	5.33	0.39	102.6	23.4	2.1	28.3
74	31	1.68	20.5	248.3	547.8	0.89	25.6	0.32	1.28	0.24	2.55	0.18	38.4	4.0	0.3	17.5
71	29	2.90	—	613.0	292.0	—	—	0.15	0.51	0.17	4.70	—	—	17.0	0	15.0
74	69	3.25	35.8	391.7	816.6	3.63	69.1	0.21	1.25	0.30	3.80	0.21	69.1	1.3	1.8	27.9
41	68	1.86	58.1	539.7	510.6	1.50	31.5	0.07	4.11	0.22	9.57	0.43	26.6	5.1	0.3	13.6
17	249	2.78	28.0	294.0	739.2	1.42	47.6	0.29	1.05	0.33	2.84	0.14	89.6	15.3	0.4	18.6
23	190	1.04	13.5	75.6	469.3	0.86	58.3	0.11	0.43	0.15	0.89	0.02	32.4	2.4	0.1	9.2
103	26	2.20	—	149.0	600.0	—	—	0.05	—	0.17	4.30	—	75.5	0	—	29.0
42	28	1.19	22.7	299.7	620.5	1.32	81.0	0.07	1.11	0.14	5.28	0.36	16.2	22.5	0.2	12.0
109	47	1.97	40.0	489.3	492.1	2.13	—	0.11	0.00	0.20	9.81	0.57	25.3	14.0	0.3	22.6
156	423	8.79	110.3	1182.9	880.4	6.18	—	0.72	0.00	1.05	20.16	1.20	180.1	2.7	1.3	97.6
68	66	3.32	37.8	390.6	652.7	1.94	259.6	0.39	1.05	0.39	7.25	0.23	113.4	10.3	0.2	27.0
156	51	3.26	68.3	588.0	3208.6	3.75	31.7	0.15	0.58	0.36	16.68	0.88	24.4	2.0	0.5	36.1
42	20	0.82	23.9	202.5	246.5	1.13	—	0.05	0.62	0.09	4.06	0.25	15.8	2.5	0.2	10.7
27	42	2.83	50.6	636.8	864.8	2.36	—	0.15	0.01	0.22	3.18	0.19	58.1	10.3	0.6	7.3
7	41	0.53	9.0	162.9	20.7	0.18	47.7	0.06	—	0.05	0.24	0.11	24.3	29.4	0	0.6
90	63	0.64	19.8	194.4	198.0	2.45	34.2	0.05	—	0.04	1.74	0.10	36.6	1.7	3.6	24.4
27	266	1.07	38.5	251.4	276.3	1.26	—	0.07	0.02	0.16	1.27	0.17	44.6	59.3	0.2	6.0
0	28	1.74	41.8	298.4	149.9	0.76	0.5	0.07	—	0.08	0.53	0.06	47.4	0.8	0	0.5
9	218	1.83	38.4	163.9	472.9	1.24	—	0.41	0.00	0.35	2.85	0.09	225.0	1.6	0.4	38.4
73	146	2.66	61.3	222.9	885.7	3.43	0	0.10	0.10	0.16	3.00	0.26	95.7	0	1.2	36.7
103	38	2.77	33.7	196.0	833.6	1.34	41.6	0.33	1.60	0.18	4.17	0.27	146.5	3.4	0.3	22.0
66	26	1.46	23.4	248.2	849.7	1.40	29.2	0.10	1.62	0.18	6.28	0.28	23.4	12.0	0.2	19.9
0	13	0.65	11.4	178.0	26.9	0.21	59.0	0.03	—	0.05	0.56	0.08	38.3	24.0	0	0.4
0	60	1.91	35.7	212.8	297.7	1.34	0	0.10	0.92	0.06	0.49	0.49	72.6	9.7	0	3.0
65	50	1.96	63.2	427.7	907.2	1.50	38.9	0.15	4.32	0.14	13.22	0.58	42.1	7.5	0.3	23.0
90	38	2.95	40.1	493.2	495.6	6.60	—	0.08	1.29	0.28	8.03	0.21	28.3	1.4	2.9	30.4
68	233	2.22	40.1	420.1	433.6	2.70	—	0.21	0.21	0.29	3.06	0.22	91.1	15.0	0.8	21.4
34	310	2.06	40.0	258.0	784.0	2.06	180.0	0.27	0.72	0.43	2.18	0.08	100.0	0	0.5	30.6
15	35	2.10	20.1	283.6	1352.9	1.28	27.6	0.19	0.70	0.16	2.77	0.14	60.2	21.6	0.4	13.3
85	59	1.87	21.8	300.8	411.7	3.40	—	0.08	0.00	0.27	3.72	0.13	18.7	0.9	1.6	17.9
172	32	1.57	25.7	333.7	1052.5	1.82	49.8	0.49	5.39	0.36	2.88	0.31	21.1	8.0	0.8	30.0
154	948	7.32	242.4	1201.2	1862.4	10.68	259.0	0.29	7.71	0.81	6.39	1.09	148.8	16.0	2.6	44.1
28	129	1.87	17.0	305.3	533.9	1.03	105.1	0.26	—	0.46	6.09	0.11	76.7	3.3	0.4	26.1
36	258	3.14	31.6	352.3	971.7	2.02	49.0	0.37	1.13	0.37	3.77	0.19	91.6	15.6	0.6	22.8
85	24	0.81	18.8	317.5	661.3	0.38	40.0	0.09	—	0.07	1.11	0.17	8.8	12.5	0	5.1

TABLE A-1 Table of Food Composition *(continued)*

(Computer code is for Cengage Diet Analysis program)
(For purposes of calculations, use "0" for t, <1, <.1, <.01, etc.)

DA+ Code	Food Description	Quantity	Measure	Wt (g)	H_2O (g)	Ener (kcal)	Prot (g)	Carb (g)	Fiber (g)	Fat (g)	Fat Breakdown (g) Sat	Mono	Poly
Mixed Foods, Soups, Sandwiches—*continued*													
25109	Salisbury steaks with mushroom sauce	1	serving(s)	135	101.8	251	17.1	9.3	0.5	15.5	6.0	6.7	0.8
16637	Shrimp creole with rice	1	cup(s)	243	176.6	309	27.0	27.7	1.2	9.2	1.7	3.6	2.9
497	Spaghetti and meatballs with tomato sauce, prepared	1	cup(s)	248	174.0	330	19.0	39.0	2.7	12.0	3.9	4.4	2.2
28585	Spicy thai noodles (pad thai)	8	ounce(s)	227	73.3	221	8.9	35.7	3.0	6.4	0.8	3.3	1.8
33073	Stir fried pork and vegetables with rice	1	cup(s)	235	173.6	348	15.4	33.5	1.9	16.3	5.6	6.9	2.6
28588	Stuffed shells	2½	item(s)	249	157.5	243	15.0	28.0	2.5	8.1	3.1	3.0	1.3
16821	Sushi with egg in seaweed	6	piece(s)	156	116.5	190	8.9	20.5	0.3	7.9	2.2	3.2	1.5
16819	Sushi with vegetables and fish	6	piece(s)	156	101.6	218	8.4	43.7	1.7	0.6	0.2	0.1	0.2
16820	Sushi with vegetables in seaweed	6	piece(s)	156	110.3	183	3.4	40.6	0.8	0.4	0.1	0.1	0.1
25266	Sweet and sour pork	¾	cup(s)	249	205.9	265	29.2	17.1	1.0	8.1	2.6	3.5	1.5
16824	Tabouli, tabbouleh or tabuli	1	cup(s)	160	123.7	198	2.6	15.9	3.7	14.9	2.0	10.9	1.6
25276	Three bean salad	½	cup(s)	99	82.2	95	1.9	9.7	2.6	5.9	0.8	1.4	3.5
160	Tuna salad	½	cup(s)	103	64.7	192	16.4	9.6	0	9.5	1.6	3.0	4.2
25241	Turkey and noodles	1	cup(s)	319	228.5	270	24.0	21.2	1.0	9.2	2.4	3.5	2.3
16794	Vegetable egg roll	2	item(s)	128	89.8	201	5.1	19.5	1.7	11.6	2.5	5.7	2.6
16818	Vegetable sushi, no fish	6	piece(s)	156	99.0	226	4.8	49.9	2.0	0.4	0.1	0.1	0.1
	Sandwiches												
1744	Bacon, lettuce and tomato with mayonnaise	1	item(s)	164	97.2	341	11.6	34.2	2.3	17.6	3.8	5.5	6.7
30287	Bologna and cheese with margarine	1	item(s)	111	45.6	345	13.4	29.3	1.2	19.3	8.1	7.0	2.4
30286	Bologna with margarine	1	item(s)	83	33.6	251	8.1	27.3	1.2	12.1	3.7	5.0	2.1
16546	Cheese	1	item(s)	83	31.0	261	9.1	27.6	1.2	12.7	5.4	4.2	2.1
8789	Cheeseburger, large, plain	1	item(s)	185	78.9	564	32.0	38.5	2.6	31.5	12.5	10.2	1.0
8624	Cheeseburger, large, with bacon, vegetables, and condiments	1	item(s)	195	91.4	550	30.8	36.8	2.5	30.9	11.9	10.6	1.3
1745	Club with bacon, chicken, tomato, lettuce, and mayonnaise	1	item(s)	246	137.5	546	31.0	48.9	3.0	24.5	5.3	7.5	9.4
1908	Cold cut submarine with cheese and vegetables	1	item(s)	228	131.8	456	21.8	51.0	2.0	18.6	6.8	8.2	2.3
30247	Corned beef	1	item(s)	130	74.9	265	18.2	25.3	1.6	9.6	3.6	3.5	1.0
25283	Egg salad	1	item(s)	126	72.1	278	10.7	28.0	1.4	13.5	2.9	4.2	5.0
16686	Fried egg	1	item(s)	96	49.7	226	10.0	26.2	1.2	8.6	2.3	3.2	1.9
16547	Grilled cheese	1	item(s)	83	27.5	291	9.2	27.9	1.2	15.8	6.0	5.7	3.0
16659	Gyro with onion and tomato	1	item(s)	105	68.3	163	12.0	20.0	1.1	3.5	1.3	1.3	0.5
1906	Ham and cheese	1	item(s)	146	74.2	352	20.7	33.3	2.0	15.5	6.4	6.7	1.4
31890	Ham with mayonnaise	1	item(s)	112	56.3	271	13.0	27.9	1.9	11.6	2.8	4.0	4.0
756	Hamburger, double patty, large, with condiments and vegetables	1	item(s)	226	121.5	540	34.3	40.3	—	26.6	10.5	10.3	2.8
8793	Hamburger, large, plain	1	item(s)	137	57.7	426	22.6	31.7	1.5	22.9	8.4	9.9	2.1
8795	Hamburger, large, with vegetables and condiments	1	item(s)	218	121.4	512	25.8	40.0	3.1	27.4	10.4	11.4	2.2
25134	Hot chicken salad	1	item(s)	98	48.4	242	15.2	23.8	1.3	9.2	2.9	2.5	3.0
25133	Hot turkey salad	1	item(s)	98	50.1	224	15.6	23.8	1.3	6.9	2.3	1.6	2.5
1411	Hotdog with bun, plain	1	item(s)	98	52.9	242	10.4	18.0	1.6	14.5	5.1	6.9	1.7
30249	Pastrami	1	item(s)	134	71.2	328	13.4	27.8	1.6	17.7	6.1	8.3	1.2
16701	Peanut butter	1	item(s)	93	23.6	345	12.2	37.6	3.3	17.4	3.4	7.7	5.2
30306	Peanut butter and jelly	1	item(s)	93	24.2	330	10.3	41.9	2.9	14.7	2.9	6.5	4.4
1909	Roast beef submarine with mayonnaise and vegetables	1	item(s)	216	127.4	410	28.6	44.3	—	13.0	7.1	1.8	2.6
1910	Roast beef, plain	1	item(s)	139	67.6	346	21.5	33.4	1.2	13.8	3.6	6.8	1.7
1907	Steak with mayonnaise and vegetables	1	item(s)	204	104.2	459	30.3	52.0	2.3	14.1	3.8	5.3	3.3
25288	Tuna salad	1	item(s)	179	102.2	415	24.5	28.4	1.6	22.4	3.5	6.2	11.4
30283	Turkey submarine with cheese, lettuce, tomato, and mayonnaise	1	item(s)	277	168.0	529	30.4	49.4	3.0	22.8	6.8	6.0	8.6
31891	Turkey with mayonnaise	1	item(s)	143	74.5	329	28.7	26.4	1.3	11.2	2.6	2.6	4.8
	Soups												
25296	Bean	1	cup(s)	301	253.1	191	13.8	29.0	6.5	2.3	0.7	0.8	0.5
711	Bean with pork, condensed, prepared with water	1	cup(s)	253	215.9	159	7.3	21.0	7.3	5.5	1.4	2.0	1.7

PAGE KEY: A-4 = Breads/Baked Goods A-10 = Cereal/Rice/Pasta A-14 = Fruit A-18 = Vegetables/Legumes A-28 = Nuts/Seeds A-30 = Vegetarian A-32 = Dairy A-40 = Eggs A-40 = Seafood A-42 = Meats A-46 = Poultry A-46 = Processed Meats A-48 = Beverages A-52 = Fats/Oils A-54 = Sweets A-56 = Spices/Condiments/Sauces A-60 = Mixed Foods/Soups/Sandwiches A-64 = Fast Food A-84 = Convenience A-86 = Baby Foods

Chol (mg)	Calc (mg)	Iron (mg)	Magn (mg)	Pota (mg)	Sodi (mg)	Zinc (mg)	Vit A (μg)	Thia (mg)	Vit E (mg α)	Ribo (mg)	Niac (mg)	Vit B_6 (mg)	Fola (μg)	Vit C (mg)	Vit B_{12} (μg)	Sele (μg)
60	74	1.94	23.8	314.5	360.5	3.45	—	0.10	0.00	0.27	3.95	0.13	20.8	0.7	1.6	17.4
180	102	4.68	63.2	413.1	330.5	1.72	94.8	0.29	2.06	0.11	4.75	0.21	121.5	12.9	1.2	49.3
89	124	3.70	—	665.0	1009.0	—	81.5	0.25	—	0.30	4.00	—	—	22.0	—	22.0
37	31	1.56	49.4	181.3	591.6	1.05	—	0.18	0.35	0.13	1.82	0.17	55.8	22.6	0.1	3.2
46	38	2.71	33.0	396.9	569.5	2.08	—	0.51	0.38	0.20	5.07	0.30	162.4	18.8	0.4	22.8
30	188	2.26	49.4	403.0	471.5	1.41	—	0.26	0.00	0.26	3.83	0.24	161.2	17.9	0.2	28.9
214	45	1.84	18.7	135.7	463.3	0.98	106.1	0.13	0.67	0.28	1.35	0.13	82.7	1.9	0.7	20.3
11	23	2.15	25.0	202.8	340.1	0.78	45.2	0.26	0.24	0.07	2.76	0.14	121.7	3.6	0.3	13.9
0	20	1.54	18.7	96.7	152.9	0.68	25.0	0.19	0.12	0.03	1.86	0.13	118.6	2.3	0	9.8
74	40	1.76	35.6	619.9	621.8	2.53	—	0.81	0.20	0.37	6.69	0.66	14.7	11.9	0.7	49.6
0	30	1.21	35.2	249.6	796.8	0.48	54.4	0.07	2.43	0.04	1.11	0.11	30.4	26.1	0	0.5
0	26	0.96	15.5	144.8	224.2	0.30	—	0.02	0.88	0.04	0.26	0.04	32.2	10.0	0	2.7
13	17	1.02	19.5	182.5	412.1	0.57	24.6	0.03	—	0.07	6.86	0.08	8.2	2.3	1.2	42.2
77	69	2.56	33.2	400.8	577.1	2.51	—	0.23	0.28	0.30	6.41	0.29	109.5	1.4	1.1	33.4
60	29	1.65	17.9	193.3	549.1	0.48	25.6	0.15	1.28	0.20	1.59	0.09	46.1	5.5	0.2	11.3
0	23	2.38	21.8	157.6	369.7	0.82	48.4	0.28	0.15	0.05	2.44	0.12	135.7	3.7	0	8.1
21	79	2.36	27.9	351.0	944.6	1.08	44.3	0.32	1.16	0.24	4.36	0.21	105.0	9.7	0.2	27.1
40	258	2.38	24.4	215.3	941.3	1.88	102.1	0.31	0.55	0.35	2.97	0.14	91.0	0.2	0.8	20.4
17	100	2.24	16.6	138.6	579.3	1.02	44.8	0.29	0.49	0.21	2.92	0.12	88.8	0.2	0.5	16.0
22	233	2.04	19.9	127.0	733.7	1.24	97.1	0.25	0.47	0.30	2.25	0.05	88.8	0	0.3	13.1
104	309	4.47	44.4	401.5	986.1	5.75	0	0.35	—	0.77	8.26	0.49	129.48	0	2.8	38.9
98	267	4.03	44.9	464.1	1314.3	5.20	0	0.33	—	0.67	8.25	0.47	122.9	1.4	2.4	6.6
71	157	4.57	46.7	464.9	1087.3	1.82	41.8	0.54	1.52	0.40	12.82	0.61	172.2	6.4	0.4	42.3
36	189	2.50	68.4	394.4	1650.7	2.57	70.7	1.00	—	0.79	5.49	0.13	109.4	12.3	1.1	30.8
46	81	3.04	19.5	127.4	1206.4	2.26	2.6	0.23	0.20	0.24	3.42	0.11	88.4	0.3	0.9	31.2
219	85	2.25	18.8	159.0	423.1	0.87	—	0.27	0.12	0.43	2.06	0.15	112.1	0.9	0.6	29.9
206	104	2.79	17.3	117.1	438.7	0.92	89.3	0.26	0.66	0.40	2.26	0.10	110.4	0	0.6	24.2
22	235	2.05	19.9	128.7	763.6	1.26	129.5	0.19	0.72	0.28	2.05	0.05	58.1	0	0.2	13.2
28	47	1.77	22.1	218.4	235.2	2.33	9.5	0.23	0.26	0.20	3.12	0.13	63.0	3.2	0.9	18.3
58	130	3.24	16.1	290.5	770.9	1.37	96.4	0.30	0.29	0.48	2.68	0.20	78.8	2.8	0.5	23.1
34	91	2.47	23.5	210.6	1097.6	1.13	5.6	0.57	0.50	0.26	3.80	0.25	90.7	2.2	0.2	20.2
122	102	5.85	49.7	569.5	791.0	5.67	0	0.36	—	0.38	7.57	0.54	110.7	1.1	4.1	25.5
71	74	3.57	27.4	267.2	474.0	4.11	0	0.28	—	0.28	6.24	0.23	80.8	0	2.1	27.1
87	96	4.92	43.6	479.6	824.0	4.88	0	0.41	—	0.37	7.28	0.32	115.5	2.6	2.4	33.6
39	115	1.88	20.0	172.4	505.1	1.19	—	0.23	0.28	0.22	4.84	0.19	79.5	0.5	0.2	19.9
37	114	1.99	21.3	189.0	494.5	1.07	—	0.22	0.28	0.20	4.26	0.22	79.6	0.5	0.2	23.4
44	24	2.31	12.7	143.1	670.3	1.98	0	0.23	—	0.27	3.64	0.04	60.7	0.1	0.5	26.0
51	80	3.02	22.8	182.2	1364.1	2.70	2.7	0.28	0.26	0.26	4.97	0.14	89.8	0.3	1.0	14.3
0	110	2.92	66.0	226.0	580.3	1.32	0	0.31	2.39	0.23	6.72	0.18	130.2	0	0	13.2
0	94	2.50	56.7	198.1	492.9	1.12	—	0.26	2.01	0.20	5.66	0.15	110.7	0.1	0	11.2
73	41	2.80	67.0	330.5	844.6	4.38	30.2	0.41	—	0.41	5.96	0.32	88.6	5.6	1.8	25.7
51	54	4.22	30.6	315.5	792.3	3.39	11.1	0.37	—	0.30	5.86	0.26	68.1	2.1	1.2	29.2
73	92	5.16	49.0	524.3	797.6	4.52	0	0.40	—	0.36	7.30	0.36	128.5	5.5	1.6	42.0
59	78	2.97	35.9	316.3	724.7	1.02	—	0.27	0.34	0.26	12.07	0.46	99.6	1.8	2.4	76.9
64	307	4.59	49.9	534.6	1759.0	2.74	74.8	0.49	1.19	0.65	3.91	0.31	171.7	10.5	0.6	42.1
67	100	3.46	34.3	304.6	564.9	3.00	5.7	0.28	0.74	0.32	6.84	0.46	94.4	0	0.3	40
5	79	3.05	61.8	588.8	689.0	1.41	—	0.27	0.02	0.15	3.63	0.23	140.1	3.6	0.2	7.9
3	78	1.89	43.0	371.9	883.0	0.96	43.0	0.08	1.08	0.03	0.52	0.03	30.4	1.5	0	7.8

TABLE A-1 Table of Food Composition *(continued)*

(Computer code is for Cengage Diet Analysis program)
(For purposes of calculations, use "0" for t, <1, <.1, <.01, etc.)

DA+ Code	Food Description	Quantity	Measure	Wt (g)	H_2O (g)	Ener (kcal)	Prot (g)	Carb (g)	Fiber (g)	Fat (g)	Fat Breakdown (g) Sat	Mono	Poly
Mixed Foods, Soups, Sandwiches—*continued*													
713	Beef noodle, condensed, prepared with water	1	cup(s)	244	224.9	83	4.7	8.7	0.7	3.0	1.1	1.2	0.5
825	Cheese, condensed, prepared with milk	1	cup(s)	251	206.9	231	9.5	16.2	1.0	14.6	9.1	4.1	0.5
826	Chicken broth, condensed, prepared with water	1	cup(s)	244	234.1	39	4.9	0.9	0	1.4	0.4	0.6	0.3
25297	Chicken noodle soup	1	cup(s)	286	258.4	117	10.8	10.9	0.9	2.9	0.8	1.1	0.7
827	Chicken noodle, condensed, prepared with water	1	cup(s)	241	226.1	60	3.1	7.1	0.5	2.3	0.6	1.0	0.6
724	Chicken noodle, dehydrated, prepared with water	1	cup(s)	252	237.3	58	2.1	9.2	0.3	1.4	0.3	0.5	0.4
823	Cream of asparagus, condensed, prepared with milk	1	cup(s)	248	213.3	161	6.3	16.4	0.7	8.2	3.3	2.1	2.2
824	Cream of celery, condensed, prepared with milk	1	cup(s)	248	214.4	164	5.7	14.5	0.7	9.7	3.9	2.5	2.7
708	Cream of chicken, condensed, prepared with milk	1	cup(s)	248	210.4	191	7.5	15.0	0.2	11.5	4.6	4.5	1.6
715	Cream of chicken, condensed, prepared with water	1	cup(s)	244	221.1	117	3.4	9.3	0.2	7.4	2.1	3.3	1.5
709	Cream of mushroom, condensed, prepared with milk	1	cup(s)	248	215.0	166	6.2	14.0	0	9.6	3.3	2.0	1.8
716	Cream of mushroom, condensed, prepared with water	1	cup(s)	244	224.6	102	1.9	8.0	0	7.0	1.6	1.3	1.7
25298	Cream of vegetable	1	cup(s)	285	250.7	165	7.2	15.2	1.9	8.6	1.6	4.6	1.9
16689	Egg drop	1	cup(s)	244	228.9	73	7.5	1.1	0	3.8	1.1	1.5	0.6
25138	Golden squash	1	cup(s)	258	223.9	145	7.6	20.4	0.4	4.1	0.8	2.2	0.9
16663	Hot and sour	1	cup(s)	244	209.7	161	15.0	5.4	0.5	7.9	2.7	3.4	1.1
28054	Lentil chowder	1	cup(s)	244	202.8	153	11.4	27.7	12.6	0.5	0.1	0.1	0.2
28560	Macaroni and bean	1	cup(s)	246	138.8	146	5.8	22.9	5.1	3.7	0.5	2.2	0.6
714	Manhattan clam chowder, condensed, prepared with water	1	cup(s)	244	225.1	73	2.1	11.6	1.5	2.1	0.4	0.4	1.2
28561	Minestrone	1	cup(s)	241	185.4	103	4.5	16.8	4.8	2.3	0.3	1.4	0.4
717	Minestrone, condensed, prepared with water	1	cup(s)	241	220.1	82	4.3	11.2	1.0	2.5	0.6	0.7	1.1
28038	Mushroom and wild rice	1	cup(s)	244	199.7	86	4.7	13.2	1.7	0.3	0	0	0.2
828	New England clam chowder, condensed, prepared with milk	1	cup(s)	248	212.2	151	8.0	18.4	0.7	5.0	2.1	0.7	0.6
28036	New England style clam chowder	1	cup(s)	244	227.5	61	3.8	8.8	1.8	0.2	0.1	0	0
28566	Old country pasta	1	cup(s)	252	183.3	146	6.5	18.3	3.6	4.5	2.0	2.4	0.9
725	Onion, dehydrated, prepared with water	1	cup(s)	246	235.7	30	0.8	6.8	0.7	0	0	0	0
16667	Shrimp gumbo	1	cup(s)	244	207.2	166	9.5	18.2	2.4	6.7	1.3	2.9	2.0
28037	Southwestern corn chowder	1	cup(s)	244	217.8	98	4.9	17.0	2.4	0.5	0.1	0.1	0.2
30282	Soybean (miso)	1	cup(s)	240	218.6	84	6.0	8.0	1.9	3.4	0.6	1.1	1.4
25140	Split pea	1	cup(s)	165	119.7	72	4.5	16.0	1.6	0.3	0.1	0	0.2
718	Split pea with ham, condensed, prepared with water	1	cup(s)	253	206.9	190	10.3	28.0	2.3	4.4	1.8	1.8	0.6
726	Tomato vegetable, dehydrated, prepared with water	1	cup(s)	253	238.4	56	2.0	10.2	0.8	0.9	0.4	0.3	0.1
710	Tomato, condensed, prepared with milk	1	cup(s)	248	213.4	136	6.2	22.0	1.5	3.2	1.8	0.9	0.3
719	Tomato, condensed, prepared with water	1	cup(s)	244	223.0	73	1.9	16.0	1.5	0.7	0.2	0.2	0.2
28595	Turkey noodle	1	cup(s)	244	216.9	114	8.1	15.1	1.9	2.4	0.3	1.1	0.7
28051	Turkey vegetable	1	cup(s)	244	220.8	96	12.2	6.6	2.0	1.1	0.3	0.2	0.3
25141	Vegetable	1	cup(s)	252	228.1	82	5.2	16.5	4.5	0.3	0	0	0.1
720	Vegetable beef, condensed, prepared with water	1	cup(s)	244	224.0	76	5.4	9.9	2.0	1.9	0.8	0.8	0.1
28598	Vegetable gumbo	1	cup(s)	252	184.5	170	4.4	28.9	3.6	4.7	0.7	3.2	0.5
721	Vegetarian vegetable, condensed, prepared with water	1	cup(s)	241	222.7	67	2.1	11.8	0.7	1.9	0.3	0.8	0.7
Fast Food													
	Arby's												
36094	Au jus sauce	1	serving(s)	85	—	43	1.0	7.0	0	1.3	0.4	—	—
751	Beef 'n cheddar sandwich	1	item(s)	195	—	445	22.0	44.0	2.0	21.0	6.0	—	—
9279	Cheddar curly fries	1	serving(s)	198	—	631	8.0	73.0	7.0	37.4	6.8	—	—

PAGE KEY: A-4 = Breads/Baked Goods A-10 = Cereal/Rice/Pasta A-14 = Fruit A-18 = Vegetables/Legumes A-28 = Nuts/Seeds A-30 = Vegetarian A-32 = Dairy A-40 = Eggs A-40 = Seafood A-42 = Meats A-46 = Poultry A-46 = Processed Meats A-48 = Beverages A-52 = Fats/Oils A-54 = Sweets A-56 = Spices/Condiments/Sauces A-60 = Mixed Foods/Soups/Sandwiches A-64 = Fast Food A-84 = Convenience A-86 = Baby Foods

Chol (mg)	Calc (mg)	Iron (mg)	Magn (mg)	Pota (mg)	Sodi (mg)	Zinc (mg)	Vit A (μg)	Thia (mg)	Vit E (mg α)	Ribo (mg)	Niac (mg)	Vit B_6 (mg)	Fola (μg)	Vit C (mg)	Vit B_{12} (μg)	Sele (μg)
5	20	1.07	7.3	97.6	929.6	1.51	12.2	0.06	1.22	0.05	1.03	0.03	29.3	0.5	0.2	7.3
48	289	0.80	20.1	341.4	1019.1	0.67	358.9	0.06	—	0.33	0.50	0.07	10.0	1.3	0.4	7.0
0	10	0.51	2.4	209.8	775.9	0.24	0	0.01	0.04	0.07	3.34	0.02	4.9	0	0.2	0
24	25	1.38	16.4	340.1	774.5	0.77	—	0.15	0.02	0.16	5.57	0.13	37.2	1.8	0.3	10.2
12	14	1.59	9.6	53.0	638.7	0.38	26.5	0.13	0.07	0.10	1.30	0.04	28.9	0	0	11.6
10	5	0.50	7.6	32.8	577.1	0.20	2.5	0.20	0.12	0.07	1.08	0.02	27.7	0	0.1	9.6
22	174	0.86	19.8	359.6	1041.6	0.91	62.0	0.10	—	0.27	0.88	0.06	29.8	4.0	0.5	8.0
32	186	0.69	22.3	310.0	1009.4	0.19	114.1	0.07	—	0.24	0.43	0.06	7.4	1.5	0.5	4.7
27	181	0.67	17.4	272.8	1046.6	0.67	178.6	0.07	—	0.25	0.92	0.06	7.4	1.2	0.5	8.0
10	34	0.61	2.4	87.8	985.8	0.63	163.5	0.02	—	0.06	0.82	0.01	2.4	0.2	0.1	7.0
10	164	1.36	19.8	267.8	823.4	0.79	81.8	0.10	1.01	0.29	0.62	0.05	7.4	0.2	0.6	6.0
0	17	1.31	4.9	73.2	775.9	0.24	9.8	0.05	0.97	0.05	0.50	0.00	2.4	0	0	2.9
1	80	1.20	17.5	340.7	787.9	0.56	—	0.12	1.05	0.18	3.32	0.12	39.5	10.7	0.3	4.3
102	22	0.75	4.9	219.6	729.6	0.48	41.5	0.02	0.29	0.19	3.02	0.05	14.6	0	0.5	7.6
4	262	0.78	42.4	542.2	515.6	0.88	—	0.16	0.52	0.30	1.14	0.16	32.7	12.5	0.7	6.0
34	29	1.24	19.5	373.3	1561.6	1.43	—	0.26	0.12	0.24	4.96	0.20	14.6	0.5	0.4	19.3
0	48	4.38	59.3	626.2	26.7	1.57	—	0.24	0.06	0.12	1.87	0.32	192.7	16.1	0	3.5
0	59	1.90	32.4	275.9	531.0	0.51	—	0.16	0.37	0.12	1.44	0.10	92.3	9.1	0	8.8
2	27	1.56	9.8	180.6	551.4	0.87	48.8	0.02	1.22	0.03	0.77	0.09	9.8	3.9	3.9	9.0
0	62	1.70	29.9	287.5	442.7	0.42	—	0.09	0.23	0.09	0.70	0.07	62.0	13.3	0	3.5
2	34	0.91	7.2	313.3	911.0	0.74	118.1	0.05	—	0.04	0.94	0.09	50.6	1.2	0	8.0
0	30	1.38	27.3	376.9	283.8	1.00	—	0.06	0.07	0.23	3.21	0.14	27.3	3.7	0.1	4.7
17	169	3.00	29.8	456.3	887.8	0.99	91.8	0.20	0.54	0.43	1.96	0.17	22.3	5.2	11.9	10.9
3	89	1.30	29.2	503.7	256.8	0.52	—	0.06	0.02	0.10	1.30	0.15	24.2	11.8	3.0	3.8
5	57	2.43	51.9	500.4	355.7	0.78	—	0.21	0.01	0.14	2.64	0.20	105.2	22.0	0.1	10.3
0	22	0.12	9.8	76.3	851.2	0.12	0	0.03	0.02	0.03	0.15	0.06	0	0.2	0	0.5
51	105	2.85	48.8	461.2	441.6	0.90	80.5	0.18	1.90	0.12	2.52	0.19	102.5	17.6	0.3	14.9
1	83	1.03	26.3	434.4	217.4	0.57	—	0.08	0.09	0.13	1.81	0.21	32.1	39.6	0.2	1.7
0	65	1.87	36.0	362.4	988.8	0.86	232.8	0.06	0.96	0.16	2.61	0.15	57.6	4.6	0.2	1.0
0	28	1.26	29.8	328.1	602.4	0.52	—	0.10	0.00	0.07	1.50	0.16	49.5	8.1	0	0.7
8	23	2.27	48.1	399.7	1006.9	1.31	22.8	0.14	—	0.07	1.47	0.06	2.5	1.5	0.3	8.0
0	20	0.60	10.1	169.5	334.0	0.20	10.1	0.06	0.43	0.09	1.26	0.06	12.7	3.0	0.1	2.0
10	166	1.36	29.8	466.2	711.8	0.84	94.2	0.09	0.44	0.31	1.35	0.15	5.0	15.6	0.6	9.2
0	20	1.31	17.1	273.3	663.7	0.29	24.4	0.04	0.41	0.07	1.23	0.10	0	15.4	0	6.1
26	28	1.40	24.1	223.8	395.9	0.75	—	0.21	0.01	0.12	2.85	0.15	65.1	7.1	0.1	13.7
21	38	1.48	25.0	423.4	348.7	0.99	—	0.09	0.01	0.09	3.64	0.27	23.9	10.6	0.2	10.3
0	40	2.08	39.1	681.5	670.3	0.67	—	0.16	0.00	0.09	2.70	0.26	36.3	22.4	0	2.2
5	20	1.09	7.3	168.4	773.5	1.51	190.3	0.03	0.58	0.04	1.00	0.07	9.8	2.4	0.3	2.7
0	56	1.85	38.9	360.2	518.4	0.64	—	0.18	0.64	0.08	1.77	0.17	107.5	21.9	0	4.1
0	24	1.06	7.2	207.3	814.6	0.45	171.1	0.05	1.39	0.04	0.90	0.05	9.6	1.4	0	4.3
0	0	—	—	—	1510.0	—	—	—	—	—	—	—	—	—	—	—
51	80	3.96	—	—	1274.0	—	—	—	—	—	—	—	—	1.8	—	—
0	80	3.24	—	—	1476.0	—	—	—	—	—	—	—	—	9.6	—	—

TABLE A-1 Table of Food Composition *(continued)*

(Computer code is for Cengage Diet Analysis program)
(For purposes of calculations, use "0" for t, <1, <.1, <.01, etc.)

DA+ Code	Food Description	Quantity	Measure	Wt (g)	H_2O (g)	Ener (kcal)	Prot (g)	Carb (g)	Fiber (g)	Fat (g)	Fat Breakdown (g) Sat	Mono	Poly
Fast Food—*continued*													
34770	Chicken breast fillet sandwich, grilled	1	item(s)	233	—	414	32.0	36.0	3.0	17.0	3.0	—	—
36131	Chocolate shake, regular	1	serving(s)	397	—	507	13.0	83.0	0	13.0	8.0	—	—
36045	Curly fries, large size	1	serving(s)	198	—	631	8.0	73.0	7.0	37.0	7.0	—	—
36044	Curly fries, medium size	1	serving(s)	128	—	406	5.0	47.0	5.0	24.0	4.0	—	—
752	Ham 'n cheese sandwich	1	item(s)	167	—	304	23.0	35.0	1.0	7.0	2.0	—	—
36048	Homestyle fries, large size	1	serving(s)	213	—	566	6.0	82.0	6.0	37.0	7.0	—	—
36047	Homestyle fries, medium size	1	serving(s)	142	—	377	4.0	55.0	4.0	25.0	4.0	—	—
33465	Homestyle fries, small size	1	serving(s)	113	—	302	3.0	44.0	3.0	20.0	4.0	—	—
9249	Junior roast beef sandwich	1	item(s)	125	—	272	16.0	34.0	2.0	10.0	4.0	—	—
9251	Large roast beef sandwich	1	item(s)	281	—	547	42.0	41.0	3.0	28.0	12.0	—	—
39640	Market Fresh chicken salad with pecans sandwich	1	item(s)	322	—	769	30.0	79.0	9.0	39.0	10.0	—	—
39641	Market Fresh Martha's Vineyard salad, without dressing	1	serving(s)	330	—	277	26.0	24.0	5.0	8.0	4.0	—	—
34769	Market Fresh roast turkey and Swiss sandwich	1	serving(s)	359	—	725	45.0	75.0	5.0	30.0	8.0	—	—
9267	Market Fresh roast turkey ranch and bacon sandwich	1	serving(s)	382	—	834	49.0	75.0	5.0	38.0	11.0	—	—
39642	Market Fresh Santa Fe salad, without dressing	1	serving(s)	372	—	499	30.0	42.0	7.0	23.0	8.0	—	—
39650	Market Fresh Southwest chicken wrap	1	serving(s)	251	—	567	36.0	42.0	4.0	29.0	9.0	—	—
37021	Market Fresh Ultimate BLT sandwich	1	item(s)	294	—	779	23.0	75.0	6.0	45.0	11.0	—	—
750	Roast beef sandwich, regular	1	item(s)	154	—	320	21.0	34.0	2.0	14.0	5.0	—	—
36132	Strawberry shake, regular	1	serving(s)	397	—	498	13.0	81.0	0	13.0	8.0	—	—
2009	Super roast beef sandwich	1	item(s)	198	—	398	21.0	40.0	2.0	19.0	6.0	—	—
36130	Vanilla shake, regular	1	serving(s)	369	—	437	13.0	66.0	0	13.0	8.0	—	—
	Auntie Anne's												
35371	Cheese dipping sauce	1	serving(s)	35	—	100	3.0	4.0	0	8.0	4.0	—	—
35353	Cinnamon sugar soft pretzel	1	item(s)	120	—	350	9.0	74.0	2.0	2.0	0	—	—
35354	Cinnamon sugar soft pretzel with butter	1	item(s)	120	—	450	8.0	83.0	3.0	9.0	5.0	—	—
35372	Marinara dipping sauce	1	serving(s)	35	—	10	0	4.0	0	0	0	0	0
35357	Original soft pretzel	1	serving(s)	120	—	340	10.0	72.0	3.0	1.0	0	—	—
35358	Original soft pretzel with butter	1	item(s)	120	—	370	10.0	72.0	3.0	4.0	2.0	—	—
35359	Parmesan herb soft pretzel	1	item(s)	120	—	390	11.0	74.0	4.0	5.0	2.5	—	—
35360	Parmesan herb soft pretzel with butter	1	item(s)	120	—	440	10.0	72.0	9.0	13.0	7.0	—	—
35361	Sesame soft pretzel	1	item(s)	120	—	350	11.0	63.0	3.0	6.0	1.0	—	—
35362	Sesame soft pretzel with butter	1	item(s)	120	—	410	12.0	64.0	7.0	12.0	4.0	—	—
35364	Sour cream and onion soft pretzel	1	item(s)	120	—	310	9.0	66.0	2.0	1.0	0	—	—
35366	Sour cream and onion soft pretzel with butter	1	item(s)	120	—	340	9.0	66.0	2.0	5.0	3.0	—	—
35373	Sweet mustard dipping sauce	1	serving(s)	35	—	60	0.5	8.0	0	1.5	1.0	—	—
35367	Whole wheat soft pretzel	1	item(s)	120	—	350	11.0	72.0	7.0	1.5	0	—	—
35368	Whole wheat soft pretzel with butter	1	item(s)	120	—	370	11.0	72.0	7.0	4.5	1.5	—	—
	Boston Market												
34978	Butternut squash	¾	cup(s)	143	—	140	2.0	25.0	2.0	4.5	3.0	—	—
35006	Caesar side salad	1	serving(s)	71	—	40	3.0	3.0	1.0	20.0	2.0	—	—
35013	Chicken Carver sandwich with cheese and sauce	1	item(s)	321	—	700	44.0	68.0	3.0	29.0	7.0	—	—
34979	Chicken gravy	4	ounce(s)	113	—	15	1.0	4.0	0	0.5	0	—	—
35053	Chicken noodle soup	¾	cup(s)	283	—	180	13.0	16.0	1.0	7.0	2.0	—	—
34973	Chicken pot pie	1	item(s)	425	—	800	29.0	59.0	4.0	49.0	18.0	—	—
35054	Chicken tortilla soup with toppings	¾	cup(s)	227	—	340	12.0	24.0	1.0	22.0	7.0	—	—
35007	Cole slaw	¾	cup(s)	125	—	170	2.0	21.0	2.0	9.0	2.0	—	—
35057	Cornbread	1	item(s)	45	—	130	1.0	21.0	0	3.5	1.0	—	—
34980	Creamed spinach	¾	cup(s)	191	—	280	9.0	12.0	4.0	23.0	15.0	—	—
34998	Fresh vegetable stuffing	1	cup(s)	136	—	190	3.0	25.0	2.0	8.0	1.0	—	—
34991	Garlic dill new potatoes	¾	cup(s)	156	—	140	3.0	24.0	3.0	3.0	1.0	—	—
34983	Green bean casserole	¾	cup(s)	170	—	60	2.0	9.0	2.0	2.0	1.0	—	—
34982	Green beans	¾	cup(s)	91	—	60	2.0	7.0	3.0	3.5	1.5	—	—

PAGE KEY: A-4 = Breads/Baked Goods A-10 = Cereal/Rice/Pasta A-14 = Fruit A-18 = Vegetables/Legumes A-28 = Nuts/Seeds A-30 = Vegetarian A-32 = Dairy A-40 = Eggs A-40 = Seafood A-42 = Meats A-46 = Poultry A-46 = Processed Meats A-48 = Beverages A-52 = Fats/Oils A-54 = Sweets A-56 = Spices/Condiments/Sauces A-60 = Mixed Foods/Soups/Sandwiches A-64 = Fast Food A-84 = Convenience A-86 = Baby Foods

Chol (mg)	Calc (mg)	Iron (mg)	Magn (mg)	Pota (mg)	Sodi (mg)	Zinc (mg)	Vit A (μg)	Thia (mg)	Vit E (mg α)	Ribo (mg)	Niac (mg)	Vit B_6 (mg)	Fola (μg)	Vit C (mg)	Vit B_{12} (μg)	Sele (μg)
9	90	3.06	—	—	913.0	—	—	—	—	—	—	—	—	10.8	—	—
34	510	0.54	—	—	357.0	—	—	—	—	—	—	—	—	5.4	—	—
0	80	3.24	—	—	1476.0	—	—	—	—	—	—	—	—	9.6	—	—
0	50	1.98	—	—	949.0	—	—	—	—	—	—	—	—	6.0	—	—
35	160	2.70	—	—	1420.0	—	—	—	—	—	—	—	—	1.2	—	—
0	50	1.62	—	—	1029.0	—	—	—	—	—	—	—	—	12.6	—	—
0	30	1.08	—	—	686.0	—	—	—	—	—	—	—	—	8.4	—	—
0	30	0.90	—	—	549.0	—	—	—	—	—	—	—	—	6.6	—	—
29	60	3.06	—	—	740.0	—	0	—	—	—	—	—	—	0	—	—
102	70	6.30	—	—	1869.0	—	0	—	—	—	—	—	—	0.6	—	—
74	180	4.32	—	—	1240.0	—	—	—	—	—	—	—	—	30.0	—	—
72	200	1.62	—	—	454.0	—	—	—	—	—	—	—	—	33.6	—	—
91	360	5.22	—	—	1788.0	—	—	—	—	—	—	—	—	10.2	—	—
109	330	5.40	—	—	2258.0	—	—	—	—	—	—	—	—	11.4	—	—
59	420	3.60	—	—	1231.0	—	—	—	—	—	—	—	—	36.6	—	—
88	240	4.50	—	—	1451.0	—	—	—	—	—	—	—	—	7.8	—	—
51	170	4.68	—	—	1571.0	—	—	—	—	—	—	—	—	16.8	—	—
44	60	3.60	—	—	953.0	—	0	—	—	—	—	—	—	0	—	—
34	510	0.72	—	—	363.0	—	—	—	—	—	—	—	—	6.6	—	—
44	70	3.78	—	—	1060.0	—	—	—	—	—	—	—	—	6.0	—	—
34	510	0.36	—	—	350.0	—	—	—	—	—	—	—	—	5.4	—	—
10	100	0.00	—	—	510.0	—	—	—	—	—	—	—	—	0	—	—
0	20	1.98	—	—	410.0	—	0	—	—	—	—	—	—	0	—	—
25	30	2.34	—	—	430.0	—	—	—	—	—	—	—	—	0	—	—
0	0	0.00	—	—	180.0	—	0	—	—	—	—	—	—	0	—	—
0	30	2.34	—	—	900.0	—	0	—	—	—	—	—	—	0	—	—
10	30	2.16	—	—	930.0	—	—	—	—	—	—	—	—	0	—	—
10	80	1.80	—	—	780.0	—	—	—	—	—	—	—	—	1.2	—	—
30	60	1.80	—	—	660.0	—	—	—	—	—	—	—	—	1.2	—	—
0	20	2.88	—	—	840.0	—	0	—	—	—	—	—	—	0	—	—
15	20	2.70	—	—	860.0	—	—	—	—	—	—	—	—	0	—	—
0	30	1.98	—	—	920.0	—	—	—	—	—	—	—	—	0	—	—
10	40	2.16	—	—	930.0	—	—	—	—	—	—	—	—	0	—	—
40	0	0.00	—	—	120.0	—	0	—	—	—	—	—	—	0	—	—
0	30	1.98	—	—	1100.0	—	0	—	—	—	—	—	—	0	—	—
10	30	2.34	—	—	1120.0	—	—	—	—	—	—	—	—	0	—	—
10	59	0.80	—	—	35.0	—	—	—	—	—	—	—	—	22.2	—	—
0	60	0.43	—	—	75.0	—	—	—	—	—	—	—	—	5.4	—	—
90	211	2.85	—	—	1560.0	—	—	—	—	—	—	—	—	15.8	—	—
0	0	0.00	—	—	570.0	—	0	—	—	—	—	—	—	0	—	—
55	0	1.07	—	—	220.0	—	—	—	—	—	—	—	—	1.8	—	—
115	40	4.50	—	—	800.0	—	—	—	—	—	—	—	—	1.2	—	—
45	123	1.32	—	—	1310.0	—	—	—	—	—	—	—	—	18.4	—	—
10	41	0.48	—	—	270.0	—	—	—	—	—	—	—	—	24.5	—	—
5	0	0.71	—	—	220.0	—	0	—	—	—	—	—	—	0	—	—
70	264	2.84	—	—	580.0	—	—	—	—	—	—	—	—	9.5	—	—
0	41	1.48	—	—	580.0	—	—	—	—	—	—	—	—	2.5	—	—
0	0	0.85	—	—	120.0	—	0	—	—	—	—	—	—	14.3	—	—
5	20	0.72	—	—	620.0	—	—	—	—	—	—	—	—	2.4	—	—
0	43	0.38	—	—	180.0	—	—	—	—	—	—	—	—	5.1	—	—

TABLE A-1 Table of Food Composition *(continued)*

(Computer code is for Cengage Diet Analysis program)
(For purposes of calculations, use "0" for t, <1, <.1, <.01, etc.)

DA+ Code	Food Description	Quantity	Measure	Wt (g)	H_2O (g)	Ener (kcal)	Prot (g)	Carb (g)	Fiber (g)	Fat (g)	Fat Breakdown (g) Sat	Mono	Poly
Fast Food—*continued*													
34984	Homestyle mashed potatoes	¾	cup(s)	221	—	210	4.0	29.0	3.0	9.0	6.0	—	—
34985	Homestyle mashed potatoes and gravy	1	cup(s)	334	—	225	5.0	33.0	3.0	9.5	6.0	—	—
34988	Hot cinnamon apples	¾	cup(s)	145	—	210	0	47.0	3.0	3.0	0	—	—
34989	Macaroni and cheese	¾	cup(s)	221	—	330	14.0	39.0	1.0	12.0	7.0	—	—
51193	Market chopped salad with dressing	1	item(s)	563	—	580	11.0	31.0	9.0	48.0	9.0	—	—
34970	Meatloaf	1	serving(s)	218	—	480	29.0	23.0	2.0	33.0	13.0	—	—
39383	Nestle Toll House chocolate chip cookie	1	item(s)	78	—	370	4.0	49.0	2.0	19.0	9.0	—	—
34965	Quarter chicken, dark meat, no skin	1	item(s)	134	—	260	30.0	2.0	0	13.0	4.0	—	—
34966	Quarter chicken, dark meat, with skin	1	item(s)	149	—	280	31.0	3.0	0	15.0	4.5	—	—
34963	Quarter chicken, white meat, no skin or wing	1	item(s)	173	—	250	41.0	4.0	0	8.0	2.5	—	—
34964	Quarter chicken, white meat, with skin and wing	1	item(s)	110	—	330	50.0	3.0	0	12.0	4.0	—	—
34968	Roasted turkey breast	5	ounce(s)	142	—	180	38.0	0	0	3.0	1.0	—	—
35011	Seasonal fresh fruit salad	1	serving(s)	142	—	60	1.0	15.0	1.0	0	0	0	0
51192	Spinach with garlic butter sauce	1	serving(s)	170	—	130	5.0	9.0	5.0	9.0	6.0	—	—
34969	Spiral sliced holiday ham	8	ounce(s)	227	—	450	40.0	13.0	0	26.0	10.0	—	—
35003	Steamed vegetables	1	cup(s)	136	—	50	2.0	8.0	3.0	2.0	0	—	—
35005	Sweet corn	¾	cup(s)	176	—	170	6.0	37.0	2.0	4.0	1.0	—	—
35004	Sweet potato casserole	¾	cup(s)	198	—	460	4.0	77.0	3.0	17.0	6.0	—	—
	Burger King												
29731	Biscuit with sausage, egg, and cheese	1	item(s)	191	—	610	20.0	33.0	1.0	45.0	15.0	—	—
14249	Cheeseburger	1	item(s)	133	—	330	17.0	31.0	1.0	16.0	7.0	—	—
14251	Chicken sandwich	1	item(s)	219	—	660	24.0	52.0	4.0	40.0	8.0	—	—
3808	Chicken Tenders, 8 pieces	1	serving(s)	123	—	340	19.0	21.0	0.5	20.0	5.0	—	—
14259	Chocolate shake, small	1	item(s)	315	—	470	8.0	75.0	1.0	14.0	9.0	—	—
29732	Croissanwich with sausage and cheese	1	item(s)	106	37.2	370	14.0	23.0	0.5	25.0	9.0	12.7	3.3
14261	Croissanwich with sausage, egg, and cheese	1	item(s)	159	71.4	470	19.0	26.0	0.5	32.0	11.0	15.8	6.1
3809	Double cheeseburger	1	item(s)	189	—	500	30.0	31.0	1.0	29.0	14.0	—	—
14244	Double Whopper sandwich	1	item(s)	373	—	900	47.0	51.0	3.0	57.0	19.0	—	—
14245	Double Whopper with cheese sandwich	1	item(s)	398	—	990	52.0	52.0	3.0	64.0	24.0	—	—
14250	Fish Filet sandwich	1	item(s)	250	—	630	24.0	67.0	4.0	30.0	6.0	—	—
14255	French fries, medium, salted	1	serving(s)	116	—	360	4.0	41.0	4.0	20.0	4.5	—	—
14262	French toast sticks, 5 pieces	1	serving(s)	112	37.6	390	6.0	46.0	2.0	20.0	4.5	10.6	2.9
14248	Hamburger	1	item(s)	121	—	290	15.0	30.0	1.0	12.0	4.5	—	—
14263	Hash brown rounds, small	1	serving(s)	75	27.1	230	2.0	23.0	2.0	15.0	4.0	—	—
14256	Onion rings, medium	1	serving(s)	91	—	320	4.0	40.0	3.0	16.0	4.0	—	—
39000	Tendercrisp chicken sandwich	1	item(s)	286	—	780	25.0	73.0	4.0	43.0	8.0	—	—
37514	TenderGrill chicken sandwich	1	item(s)	258	—	450	37.0	53.0	4.0	10.0	2.0	—	—
14258	Vanilla shake, small	1	item(s)	296	—	400	8.0	57.0	0	15.0	9.0	—	—
1736	Whopper sandwich	1	item(s)	290	—	670	28.0	51.0	3.0	39.0	11.0	—	—
14243	Whopper with cheese sandwich	1	item(s)	315	—	760	33.0	52.0	3.0	47.0	16.0	—	—
	Carl's Jr												
33962	Carl's bacon Swiss crispy chicken sandwich	1	item(s)	268	—	750	31.0	91.0	—	28.0	28.0	—	—
10801	Carl's Catch fish sandwich	1	item(s)	215	—	560	19.0	58.0	2.0	27.0	7.0	—	1.9
10862	Carl's Famous Star hamburger	1	item(s)	254	—	590	24.0	50.0	3.0	32.0	9.0	—	—
10785	Charbroiled chicken club sandwich	1	item(s)	270	—	550	42.0	43.0	4.0	23.0	7.0	—	2.9
10866	Charbroiled chicken salad	1	item(s)	437	—	330	34.0	17.0	5.0	7.0	4.0	—	1.0
10855	Charbroiled Santa Fe chicken sandwich	1	item(s)	266	—	610	38.0	43.0	4.0	32.0	8.0	—	—
10790	Chicken stars, 6 pieces	1	serving(s)	85	—	260	13.0	14.0	1.0	16.0	4.0	—	1.6
34864	Chocolate shake, small	1	serving(s)	595	—	540	15.0	98.0	0	11.0	7.0	—	—
10797	Crisscut fries	1	serving(s)	139	—	410	5.0	43.0	4.0	24.0	5.0	—	—
10799	Double Western Bacon cheeseburger	1	item(s)	308	—	920	51.0	65.0	2.0	50	21.0	—	6.6

PAGE KEY: A-4 = Breads/Baked Goods A-10 = Cereal/Rice/Pasta A-14 = Fruit A-18 = Vegetables/Legumes A-28 = Nuts/Seeds A-30 = Vegetarian A-32 = Dairy A-40 = Eggs A-40 = Seafood A-42 = Meats A-46 = Poultry A-46 = Processed Meats A-48 = Beverages A-52 = Fats/Oils A-54 = Sweets A-56 = Spices/Condiments/Sauces A-60 = Mixed Foods/Soups/Sandwiches A-64 = Fast Food A-84 = Convenience A-86 = Baby Foods

Chol (mg)	Calc (mg)	Iron (mg)	Magn (mg)	Pota (mg)	Sodi (mg)	Zinc (mg)	Vit A (µg)	Thia (mg)	Vit E (mg α)	Ribo (mg)	Niac (mg)	Vit B_6 (mg)	Fola (µg)	Vit C (mg)	Vit B_{12} (µg)	Sele (µg)
25	51	0.46	—	—	660.0	—	—	—	—	—	—	—	—	19.2	—	—
25	100	0.59	—	—	1230.0	—	—	—	—	—	—	—	—	24.9	—	—
0	16	0.28	—	—	15.0	—	—	—	—	—	—	—	—	0	—	—
30	345	1.65	—	—	1290.0	—	—	—	—	—	—	—	—	0	—	—
10	—	—	—	—	2010.0	—	—	—	—	—	—	—	—	—	—	—
125	140	3.77	—	—	970.0	—	—	—	—	—	—	—	—	1.8	—	—
20	0	1.32	—	—	340.0	—	—	—	—	—	—	—	—	0	—	—
155	0	1.52	—	—	260.0	—	0	—	—	—	—	—	—	0	—	—
155	0	2.14	—	—	660.0	—	0	—	—	—	—	—	—	0	—	—
125	0	0.89	—	—	480.0	—	0	—	—	—	—	—	—	0	—	—
165	0	0.78	—	—	960.0	—	0	—	—	—	—	—	—	0	—	—
70	20	1.80	—	—	620.0	—	0	—	—	—	—	—	—	0	—	—
0	16	0.29	—	—	20.0	—	—	—	—	—	—	—	—	29.5	—	—
20	—	—	—	—	200.0	—	—	—	—	—	—	—	—	—	—	—
140	0	1.73	—	—	2230.0	—	0	—	—	—	—	—	—	0	—	—
0	53	0.46	—	—	45.0	—	—	—	—	—	—	—	—	24.0	—	—
0	0	0.43	—	—	95.0	—	—	—	—	—	—	—	—	5.8	—	—
20	44	1.18	—	—	210.0	—	—	—	—	—	—	—	—	9.8	—	—
210	250	2.70	—	—	1620.0	—	89.9	—	—	—	—	—	—	0	—	—
55	150	2.70	—	—	780.0	—	—	0.24	—	0.31	4.17	—	—	1.2	—	—
70	64	2.89	—	—	1440.0	—	—	0.50	—	0.32	10.29	—	—	0	—	—
55	20	0.72	—	—	960.0	—	—	0.14	—	0.11	10.93	—	—	0	—	—
55	333	0.79	—	—	350.0	—	—	0.11	—	0.61	0.26	—	—	2.7	0	—
50	99	1.78	20.1	217.3	810.0	1.51	—	0.34	1.03	0.33	4.33	—	—	0	0.6	22.2
180	146	2.63	28.6	313.2	1060.0	2.08	—	0.38	1.66	0.51	4.72	0.28	—	0	1.1	38.0
105	250	4.50	—	—	1030.0	—	—	0.26	—	0.44	6.37	—	—	1.2	—	—
175	150	8.07	—	—	1090.0	—	—	0.39	—	0.59	11.05	—	—	9.0	—	—
195	299	8.08	—	—	1520.0	—	—	0.39	—	0.66	11.03	—	—	9.0	—	—
60	101	3.62	—	—	1380.0	—	—	—	—	—	—	—	—	3.6	—	—
0	20	0.71	—	—	590.0	—	0	0.15	—	0.48	2.30	—	—	8.9	—	—
0	60	1.80	21.3	124.3	440.0	0.57	—	0.31	0.98	0.19	2.88	0.05	—	0	0	13.7
40	80	2.70	—	—	560.0	—	—	0.25	—	0.28	4.25	—	—	1.2	—	—
0	0	0.36	—	—	450.0	—	0	0.11	0.83	0.06	1.35	0.17	—	1.2	—	—
0	100	0.00	—	—	460.0	—	0	0.14	—	0.09	2.32	—	—	0	—	—
75	79	4.43	—	—	1730.0	—	—	—	—	—	—	—	—	8.9	—	—
75	57	6.82	—	—	1210.0	—	—	—	—	—	—	—	—	5.7	—	—
60	348	0.00	—	—	240.0	—	—	0.11	—	0.63	0.21	—	—	2.4	0	—
51	100	5.38	—	—	1020.0	—	—	0.38	—	0.43	7.30	—	—	9.0	—	—
115	249	5.38	—	—	1450.0	—	—	0.38	—	0.51	7.28	—	—	9.0	—	—
80	200	5.40	—	—	1900.0	—	—	—	—	—	—	—	—	2.4	—	—
80	150	2.70	—	—	990.0	—	60.0	—	—	—	—	—	—	2.4	—	—
70	100	4.50	—	—	910.0	—	—	—	—	—	—	—	—	6.0	—	—
95	200	3.60	—	—	1330.0	—	—	—	—	—	—	—	—	9.0	—	—
75	200	1.80	—	—	880.0	—	—	—	—	—	—	—	—	30.0	—	—
100	200	3.60	—	—	1440.0	—	—	—	—	—	—	—	—	9.0	—	—
35	19	1.02	—	—	470.0	—	0	—	—	—	—	—	—	0	—	—
45	600	1.08	—	—	360.0	—	0	—	—	—	—	—	—	0	—	—
0	20	1.80	—	—	950.0	—	0	—	—	—	—	—	—	12.0	—	—
155	300	7.20	—	—	1730.0	—	—	—	—	—	—	—	—	1.2	—	—

TABLE A-1 Table of Food Composition *(continued)*

(Computer code is for Cengage Diet Analysis program)
(For purposes of calculations, use "0" for t, <1, <.1, <.01, etc.)

DA+ Code	Food Description	Quantity	Measure	Wt (g)	H_2O (g)	Ener (kcal)	Prot (g)	Carb (g)	Fiber (g)	Fat (g)	Fat Breakdown (g) Sat	Mono	Poly
Fast Food—*continued*													
14238	French fries, small	1	serving(s)	92	—	290	5.0	37.0	3.0	14.0	3.0	—	—
10798	French toast dips without syrup, 5 pieces	1	serving(s)	155	—	370	8.0	49.0	0	17.0	5.0	—	1.4
10802	Onion rings	1	serving(s)	128	—	440	7.0	53.0	3.0	22.0	5.0	—	0.8
34858	Spicy chicken sandwich	1	item(s)	198	—	480	14.0	48.0	2.0	26.0	5.0	—	—
34867	Strawberry shake, small	1	serving(s)	595	—	520	14.0	93.0	0	11.0	7.0	—	—
10865	Super Star hamburger	1	item(s)	348	—	790	41.0	52.0	3.0	47.0	14.0	—	—
38925	The Six Dollar burger	1	item(s)	429	—	1010	40.0	60.0	3.0	66.0	26.0	—	—
10818	Vanilla shake, small	1	item(s)	398	—	314	10.0	51.5	0	7.4	4.7	—	—
10770	Western Bacon cheeseburger	1	item(s)	225	—	660	32.0	64.0	2.0	30.0	12.0	—	4.8
	Chick Fil-A												
38746	Biscuit with bacon, egg, and cheese	1	item(s)	163	—	470	18.0	39.0	1.0	26.0	9.0	—	—
38747	Biscuit with egg	1	item(s)	135	—	350	11.0	38.0	1.0	16.0	4.5	—	—
38748	Biscuit with egg and cheese	1	item(s)	149	—	400	14.0	38.0	1.0	21.0	7.0	—	—
38753	Biscuit with gravy	1	item(s)	192	—	330	5.0	43.0	1.0	15.0	4.0	—	—
38752	Biscuit with sausage, egg, and cheese	1	item(s)	212	—	620	22.0	39.0	2.0	42.0	14.0	—	—
38771	Carrot and raisin salad	1	item(s)	113	—	170	1.0	28.0	2.0	6.0	1.0	—	—
38761	Chargrilled chicken Cool Wrap	1	item(s)	245	—	390	29.0	54.0	3.0	7.0	3.0	—	—
38766	Chargrilled chicken garden salad	1	item(s)	275	—	180	22.0	9.0	3.0	6.0	3.0	—	—
38758	Chargrilled chicken sandwich	1	item(s)	193	—	270	28.0	33.0	3.0	3.5	1.0	—	—
38742	Chicken biscuit	1	item(s)	145	—	420	18.0	44.0	2.0	19.0	4.5	—	—
38743	Chicken biscuit with cheese	1	item(s)	159	—	470	21.0	45.0	2.0	23.0	8.0	—	—
38762	Chicken Caesar Cool Wrap	1	item(s)	227	—	460	36.0	52.0	3.0	10.0	6.0	—	—
38757	Chicken deluxe sandwich	1	item(s)	208	—	420	28.0	39.0	2.0	16.0	3.5	—	—
38764	Chicken salad sandwich on wheat bun	1	item(s)	153	—	350	20.0	32.0	5.0	15.0	3.0	—	—
38756	Chicken sandwich	1	item(s)	170	—	410	28.0	38.0	1.0	16.0	3.5	—	—
38768	Chick-n-Strip salad	1	item(s)	327	—	400	34.0	21.0	4.0	20.0	6.0	—	—
38763	Chick-n-Strips	4	item(s)	127	—	300	28.0	14.0	1.0	15.0	2.5	—	—
38770	Cole slaw	1	item(s)	128	—	260	2.0	17.0	2.0	21.0	3.5	—	—
38776	Diet lemonade, small	1	cup(s)	255	—	25	0	5.0	0	0	0	0	0
38755	Hashbrowns	1	serving(s)	84	—	260	2.0	25.0	3.0	17.0	3.5	—	—
38765	Hearty breast of chicken soup	1	cup(s)	241	—	140	8.0	18.0	1.0	3.5	1.0	—	—
38741	Hot buttered biscuit	1	item(s)	79	—	270	4.0	38.0	1.0	12.0	3.0	—	—
38778	IceDream, small cone	1	item(s)	135	—	160	4.0	28.0	0	4.0	2.0	—	—
38774	IceDream, small cup	1	serving(s)	227	—	240	6.0	41.0	0	6.0	3.5	—	—
38775	Lemonade, small	1	cup(s)	255	—	170	0	41.0	0	0.5	0	—	—
38777	Nuggets	8	item(s)	113	—	260	26.0	12.0	0.5	12.0	2.5	—	—
38769	Side salad	1	item(s)	108	—	60	3.0	4.0	2.0	3.0	1.5	—	—
38767	Southwest chargrilled salad	1	item(s)	303	—	240	25.0	17.0	5.0	8.0	3.5	—	—
40481	Spicy chicken cool wrap	1	serving(s)	230	—	380	30.0	52.0	3.0	6.0	3.0	—	—
38772	Waffle potato fries, small, salted	1	serving(s)	85	—	270	3.0	34.0	4.0	13.0	3.0	—	—
	Cinnabon												
39572	Caramellata Chill w/whipped cream	16	fluid ounce(s)	480	—	406	10.0	61.0	0	14.0	8.0	—	—
39571	Cinnabon Bites	1	serving(s)	149	—	510	8.0	77.0	2.0	19.0	5.0	—	—
39570	Cinnabon Stix	5	item(s)	85	—	379	6.0	41.0	1.0	21.0	6.0	—	—
39567	Classic roll	1	item(s)	221	—	813	15.0	117.0	4.0	32.0	8.0	—	—
39568	Minibon	1	item(s)	92	—	339	6.0	49.0	2.0	13.0	3.0	—	—
39573	Mochalatta Chill w/whipped cream	16	fluid ounce(s)	480	—	362	9.0	55.0	0	13.0	8.0	—	—
39569	Pecanbon	1	item(s)	272	—	1100	16.0	141.0	8.0	56.0	10.0	—	—
	Dairy Queen												
1466	Banana split	1	item(s)	369	—	510	8.0	96.0	3.0	12.0	8.0	—	—
38552	Brownie Earthquake®	1	serving(s)	304	—	740	10.0	112.0	0	27.0	16.0	—	—
38561	Chocolate chip cookie dough blizzard,® small	1	item(s)	319	—	720	12.0	105.0	0	28.0	14.0	—	—
1464	Chocolate malt, small	1	item(s)	418	—	640	15.0	111.0	1.0	16.0	11.0	—	—
38541	Chocolate shake, small	1	item(s)	397	—	560	13.0	93.0	1.0	15.0	10.0	—	—
17257	Chocolate soft serve	½	cup(s)	94	—	150	4.0	22.0	0	5.0	3.5	—	—
1463	Chocolate sundae, small	1	item(s)	163	—	280	5.0	49.0	0	7.0	4.5	—	—

PAGE KEY: A-4 = Breads/Baked Goods A-10 = Cereal/Rice/Pasta A-14 = Fruit A-18 = Vegetables/Legumes A-28 = Nuts/Seeds A-30 = Vegetarian A-32 = Dairy A-40 = Eggs A-40 = Seafood A-42 = Meats A-46 = Poultry A-46 = Processed Meats A-48 = Beverages A-52 = Fats/Oils A-54 = Sweets A-56 = Spices/Condiments/Sauces A-60 = Mixed Foods/Soups/Sandwiches A-64 = Fast Food A-84 = Convenience A-86 = Baby Foods

Chol (mg)	Calc (mg)	Iron (mg)	Magn (mg)	Pota (mg)	Sodi (mg)	Zinc (mg)	Vit A (µg)	Thia (mg)	Vit E (mg α)	Ribo (mg)	Niac (mg)	Vit B_6 (mg)	Fola (µg)	Vit C (mg)	Vit B_{12} (µg)	Sele (µg)
0	0	1.08	—	—	170.0	—	0	—	—	—	—	—	—	21.0	—	—
3	0	0.00	—	—	470.0	—	0	0.25	—	0.23	2.00	—	—	0	—	—
0	20	0.72	—	—	700.0	—	0	—	—	—	—	—	—	3.6	—	—
40	100	3.60	—	—	1220.0	—	—	—	—	—	—	—	—	6.0	—	—
45	600	0.00	—	—	340.0	—	0	—	—	—	—	—	—	0	—	—
130	100	7.20	—	—	980.0	—	—	—	—	—	—	—	—	9.0	—	—
145	279	4.29	—	—	1960.0	—	—	—	—	—	—	—	—	16.7	—	—
30	401	0.00	—	—	234.0	—	0	—	—	—	—	—	—	0	—	—
85	200	5.40	—	—	1410.0	—	60.0	—	—	—	—	—	—	1.2	—	—
270	150	2.70	—	—	1190.0	—	—	—	—	—	—	—	—	0	—	—
240	80	2.70	—	—	740.0	—	—	—	—	—	—	—	—	0	—	—
255	150	2.70	—	—	970.0	—	—	—	—	—	—	—	—	0	—	—
5	60	1.80	—	—	930.0	—	0	—	—	—	—	—	—	0	—	—
300	200	3.60	—	—	1360.0	—	—	—	—	—	—	—	—	0	—	—
10	40	0.36	—	—	110.0	—	—	—	—	—	—	—	—	4.8	—	—
65	200	3.60	—	—	1020.0	—	—	—	—	—	—	—	—	6.0	—	—
65	150	0.72	—	—	620.0	—	—	—	—	—	—	—	—	30	—	—
65	80	2.70	—	—	940.0	—	—	—	—	—	—	—	—	6.0	—	—
35	60	2.70	—	—	1270.0	—	0	—	—	—	—	—	—	0	—	—
50	150	2.70	—	—	1500.0	—	—	—	—	—	—	—	—	0	—	—
80	500	3.60	—	—	1350.0	—	—	—	—	—	—	—	—	1.2	—	—
60	100	2.70	—	—	1300.0	—	—	—	—	—	—	—	—	2.4	—	—
65	150	1.80	—	—	880.0	—	—	—	—	—	—	—	—	0	—	—
60	100	2.70	—	—	1300.0	—	—	—	—	—	—	—	—	0	—	—
80	150	1.44	—	—	1070.0	—	—	—	—	—	—	—	—	6.0	—	—
65	40	1.44	—	—	940.0	—	—	—	—	—	—	—	—	0	—	—
25	60	0.36	—	—	220.0	—	—	—	—	—	—	—	—	36.0	—	—
0	0	0.36	—	—	5.0	—	0	—	—	—	—	—	—	15.0	—	—
5	20	0.72	—	—	380.0	—	—	—	—	—	—	—	—	0	—	—
25	40	1.08	—	—	900.0	—	—	—	—	—	—	—	—	0	—	—
0	60	1.80	—	—	660.0	—	0	—	—	—	—	—	—	0	—	—
15	100	0.36	—	—	80.0	—	—	—	—	—	—	—	—	0	—	—
25	200	0.36	—	—	105.0	—	—	—	—	—	—	—	—	0	—	—
0	0	0.36	—	—	10.0	—	0	—	—	—	—	—	—	15.0	—	—
70	40	1.08	—	—	1090.0	—	0	—	—	—	—	—	—	0	—	—
10	100	0.00	—	—	75.0	—	—	—	—	—	—	—	—	15.0	—	—
60	200	1.08	—	—	770.0	—	—	—	—	—	—	—	—	24.0	—	—
60	200	3.60	—	—	1090.0	—	—	—	—	—	—	—	—	3.6	—	—
0	20	1.08	—	—	115.0	—	0	—	—	—	—	—	—	1.2	—	—
46	—	—	—	—	187.0	—	—	—	—	—	—	—	—	—	—	—
35	—	—	—	—	530.0	—	—	—	—	—	—	—	—	—	—	—
16	—	—	—	—	413.0	—	—	—	—	—	—	—	—	—	—	—
67	—	—	—	—	801.0	—	—	—	—	—	—	—	—	—	—	—
27	—	—	—	—	337.0	—	—	—	—	—	—	—	—	—	—	—
46	—	—	—	—	252.0	—	—	—	—	—	—	—	—	—	—	—
63	—	—	—	—	600.0	—	—	—	—	—	—	—	—	—	—	—
30	250	1.80	—	—	180.0	—	—	—	—	—	—	—	—	15.0	—	—
50	250	1.80	—	—	350.0	—	—	—	—	—	—	—	—	0	—	—
50	350	2.70	—	—	370.0	—	—	—	—	—	—	—	—	1.2	—	—
55	450	1.80	—	—	340.0	—	—	—	—	—	—	—	—	2.4	—	—
50	450	1.44	—	—	280.0	—	—	—	—	—	—	—	—	2.4	—	—
15	100	0.72	—	—	75.0	—	—	—	—	—	—	—	—	0	—	—
20	200	1.08	—	—	140.0	—	—	—	—	—	—	—	—	0	—	—

TABLE A-1 Table of Food Composition *(continued)*

(Computer code is for Cengage Diet Analysis program)
(For purposes of calculations, use "0" for t, <1, <.1, <.01, etc.)

DA+ Code	Food Description	Quantity	Measure	Wt (g)	H_2O (g)	Ener (kcal)	Prot (g)	Carb (g)	Fiber (g)	Fat (g)	Fat Breakdown (g) Sat	Mono	Poly
Fast Food—*continued*													
1462	Dipped cone, small	1	item(s)	156	—	340	6.0	42.0	1.0	17.0	9.0	4.0	3.0
38555	Oreo cookies blizzard, small	1	item(s)	283	—	570	11.0	83.0	0.5	21.0	10.0	—	—
38547	Royal Treats Peanut Buster® Parfait	1	item(s)	305	—	730	16.0	99.0	2.0	31.0	17.0	—	—
17256	Vanilla soft serve	½	cup(s)	94	—	140	3.0	22.0	0	4.5	3.0	—	—
	Domino's												
31606	Barbeque buffalo wings	1	item(s)	25	—	50	6.0	2.0	0	2.5	0.5	—	—
31604	Breadsticks	1	item(s)	30	—	115	2.0	12.0	0	6.3	1.1	—	—
37551	Buffalo Chicken Kickers	1	item(s)	24	—	47	4.0	3.0	0	2.0	0.5	—	—
37548	CinnaStix	1	item(s)	30	—	123	2.0	15.0	1.0	6.1	1.1	—	—
37549	Dot, cinnamon	1	item(s)	28	7.6	99	1.9	14.9	0.7	3.7	0.7	—	—
31605	Double cheesy bread	1	item(s)	35	—	123	4.0	13.0	0	6.5	1.9	—	—
31607	Hot buffalo wings	1	item(s)	25	—	45	5.0	1.0	0	2.5	0.5	—	—
	Domino's Classic hand tossed pizza												
31573	America's favorite feast, 12″	1	slice(s)	102	—	257	10.0	29.0	2.0	11.5	4.5	—	—
31574	America's favorite feast, 14″	1	slice(s)	141	—	353	14.0	39.0	2.0	16.0	6.0	—	—
37543	Bacon cheeseburger feast, 12″	1	slice(s)	99	—	273	12.0	28.0	2.0	13.0	5.5	—	—
37545	Bacon cheeseburger feast, 14″	1	slice(s)	137	—	379	17.0	38.0	2.0	18.0	8.0	—	—
37546	Barbeque feast, 12″	1	slice(s)	96	—	252	11.0	31.0	1.0	10.0	4.5	—	—
37547	Barbeque feast, 14″	1	slice(s)	131	—	344	14.0	43.0	2.0	13.5	6.0	—	—
31569	Cheese, 12″	1	slice(s)	55	—	160	6.0	28.0	1.0	3.0	1.0	—	—
31570	Cheese, 14″	1	slice(s)	75	—	220	8.0	38.0	2.0	4.0	1.0	—	—
37538	Deluxe feast, 12″	1	slice(s)	201	101.8	465	19.5	57.4	3.5	18.2	7.7	—	—
37540	Deluxe feast, 14″	1	slice(s)	273	138.4	627	26.4	78.3	4.7	24.1	10.2	—	—
31685	Deluxe, 12″	1	slice(s)	100	—	234	9.0	29.0	2.0	9.5	3.5	—	—
31694	Deluxe, 14″	1	slice(s)	136	—	316	13.0	39.0	2.0	12.5	5.0	—	—
31686	Extravaganzza, 12″	1	slice(s)	122	—	289	13.0	30.0	2.0	14.0	5.5	—	—
31695	Extravaganzza, 14″	1	slice(s)	165	—	388	17.0	40.0	3.0	18.5	7.5	—	—
31575	Hawaiian feast, 12″	1	slice(s)	102	—	223	10.0	30.0	2.0	8.0	3.5	—	—
31576	Hawaiian feast, 14″	1	slice(s)	141	—	309	14.0	41.0	2.0	11.0	4.5	—	—
31687	Meatzza, 12″	1	slice(s)	108	—	281	13.0	29.0	2.0	13.5	5.5	—	—
31696	Meatzza, 14″	1	slice(s)	146	—	378	17.0	39.0	2.0	18.0	7.5	—	—
31571	Pepperoni feast, extra pepperoni and cheese, 12″	1	slice(s)	98	—	265	11.0	28.0	2.0	12.5	5.0	—	—
31572	Pepperoni feast, extra pepperoni and cheese, 14″	1	slice(s)	135	—	363	16.0	39.0	2.0	17.0	7.0	—	—
31577	Vegi feast, 12″	1	slice(s)	102	—	218	9.0	29.0	2.0	8.0	3.5	—	—
31578	Vegi feast, 14″	1	slice(s)	139	—	300	13.0	40.0	3.0	11.0	4.5	—	—
	Domino's thin crust pizza												
31583	America's favorite, 12″	1	slice(s)	72	—	208	8.0	15.0	1.0	13.5	5.0	—	—
31584	America's favorite, 14″	1	slice(s)	100	—	285	11.0	20.0	2.0	18.5	7.0	—	—
31579	Cheese, 12″	1	slice(s)	49	—	137	5.0	14.0	1.0	7.0	2.5	—	—
31580	Cheese, 14″	1	slice(s)	68	27.0	214	8.8	19.0	1.4	11.4	4.6	2.9	2.5
31688	Deluxe, 12″	1	slice(s)	70	—	185	7.0	15.0	1.0	11.5	4.0	—	—
31697	Deluxe, 14″	1	slice(s)	94	—	248	10.0	20.0	2.0	15.0	5.5	—	—
31689	Extravaganzza, 12″	1	slice(s)	92	—	240	11.0	16.0	1.0	15.5	6.0	—	—
31698	Extravaganzza, 14″	1	slice(s)	123	—	320	14.0	21.0	2.0	20.5	8.0	—	—
31585	Hawaiian, 12″	1	slice(s)	71	—	174	8.0	16.0	1.0	9.5	3.5	—	—
31586	Hawaiian, 14″	1	slice(s)	100	—	240	11.0	21.0	2.0	13.0	5.0	—	—
31690	Meatzza, 12″	1	slice(s)	78	—	232	11.0	15.0	1.0	15.0	6.0	—	—
31699	Meatzza, 14″	1	slice(s)	104	—	310	14.0	20.0	2.0	20	8.0	—	—
31581	Pepperoni, extra pepperoni and cheese, 12″″	1	slice(s)	68	—	216	9.0	14.0	1.0	14.0	5.5	—	—
31582	Pepperoni, extra pepperoni and cheese, 14″	1	slice(s)	93	—	295	13.0	20.0	1.0	19.0	7.5	—	—
31587	Vegi, 12″	1	slice(s)	71	—	168	7.0	15.0	1.0	9.5	3.5	—	—
31588	Vegi, 14″	1	slice(s)	97	—	231	10.0	21.0	2.0	13.5	5.0	—	—
	Domino's Ultimate deep dish pizza												
31596	America's favorite, 12″	1	slice(s)	115	—	309	12.0	29.0	2.0	17.0	6.0	—	—
31702	America's favorite, 14″	1	slice(s)	162	—	433	17.0	42.0	3.0	23.5	8.0	—	—
31590	Cheese, 12″	1	slice(s)	90	—	238	9.0	28.0	2.0	11.0	3.5	—	—
31591	Cheese, 14″	1	slice(s)	128	53.9	351	14.5	41.0	2.9	13.2	5.2	3.8	2.5
31589	Cheese, 6″	1	item(s)	215	—	598	22.9	68.4	3.9	27.6	9.9	—	—
31691	Deluxe, 12″	1	slice(s)	122	—	287	11.0	29.0	2.0	15.0	5.0	—	—

PAGE KEY: A-4 = Breads/Baked Goods A-10 = Cereal/Rice/Pasta A-14 = Fruit A-18 = Vegetables/Legumes A-28 = Nuts/Seeds A-30 = Vegetarian A-32 = Dairy A-40 = Eggs A-40 = Seafood A-42 = Meats A-46 = Poultry A-46 = Processed Meats A-48 = Beverages A-52 = Fats/Oils A-54 = Sweets A-56 = Spices/Condiments/Sauces A-60 = Mixed Foods/Soups/Sandwiches A-64 = Fast Food A-84 = Convenience A-86 = Baby Foods

Chol (mg)	Calc (mg)	Iron (mg)	Magn (mg)	Pota (mg)	Sodi (mg)	Zinc (mg)	Vit A (μg)	Thia (mg)	Vit E (mg α)	Ribo (mg)	Niac (mg)	Vit B_6 (mg)	Fola (μg)	Vit C (mg)	Vit B_{12} (μg)	Sele (μg)
20	200	1.08	—	—	130.0	—	—	—	—	—	—	—	—	1.2	—	—
40	350	2.70	—	—	430.0	—	—	—	—	—	—	—	—	1.2	—	—
35	300	1.80	—	—	400.0	—	—	—	—	—	—	—	—	1.2	—	—
15	150	0.72	—	—	70.0	—	—	—	—	—	—	—	—	0	—	—
26	10	0.36	—	—	175.5	—	—	—	—	—	—	—	—	0	—	—
0	0	0.72	—	—	122.1	—	—	—	—	—	—	—	—	0	—	—
9	0	0.00	—	—	162.5	—	—	—	—	—	—	—	—	0	—	—
0	0	0.72	—	—	111.4	—	—	—	—	—	—	—	—	0	—	—
0	6	0.59	—	—	85.7	—	—	—	—	—	—	—	—	0	—	—
6	40	0.72	—	—	162.3	—	—	—	—	—	—	—	—	0	—	—
26	10	0.36	—	—	254.5	—	—	—	—	—	—	—	—	1.2	—	—
																—
22	100	1.80	—	—	625.5	—	—	—	—	—	—	—	—	0.6	—	—
31	140	2.52	—	—	865.5	—	—	—	—	—	—	—	—	0.6	—	—
27	140	1.80	—	—	634.0	—	—	—	—	—	—	—	—	0	—	—
38	190	2.52	—	—	900.0	—	—	—	—	—	—	—	—	0	—	—
20	140	1.62	—	—	600.0	—	—	—	—	—	—	—	—	0.6	—	—
27	190	2.16	—	—	831.5	—	—	—	—	—	—	—	—	0.6	—	—
0	0	1.80	—	—	110.0	—	0	—	—	—	—	—	—	0	—	—
0	0	2.70	—	—	150.0	—	0	—	—	—	—	—	—	0	—	—
40	199	3.56	—	—	1063.1	—	—	—	—	—	—	—	—	1.4	—	—
53	276	4.84	—	—	1432.2	—	—	—	—	—	—	—	—	1.8	—	—
17	100	1.80	—	—	541.5	—	—	—	—	—	—	—	—	0.6	—	—
23	130	2.34	—	—	728.5	—	—	—	—	—	—	—	—	1.2	—	—
28	140	1.98	—	—	764.0	—	—	—	—	—	—	—	—	0.6	—	—
37	190	2.70	—	—	1014.0	—	—	—	—	—	—	—	—	1.2	—	—
16	130	1.62	—	—	546.5	—	—	—	—	—	—	—	—	1.2	—	—
23	180	2.34	—	—	765.0	—	—	—	—	—	—	—	—	1.2	—	—
28	130	1.80	—	—	739.5	—	—	—	—	—	—	—	—	0	—	—
37	190	2.52	—	—	983.5	—	—	—	—	—	—	—	—	0	—	—
24	130	1.62	—	—	670.0	—	70.9	—	—	—	—	—	—	0	—	—
33	180	2.34	—	—	920.0	—	104.7	—	—	—	—	—	—	0	—	—
13	130	1.62	—	—	489.0	—	—	—	—	—	—	—	—	0.6	—	—
18	180	2.34	—	—	678.0	—	—	—	—	—	—	—	—	0.6	—	—
23	100	0.90	—	—	533.0	—	—	—	—	—	—	—	—	2.4	—	—
32	140	1.26	—	—	736.5	—	—	—	—	—	—	—	—	3.0	—	—
10	90	0.54	—	—	292.5	—	60.0	—	—	—	—	—	—	1.8	—	—
14	151	0.48	17.7	125.1	338.0	0.02	64.6	0.05	1.01	0.07	0.69	—	—	2.4	0.5	24.1
19	100	0.90	—	—	449.0	—	—	—	—	—	—	—	—	2.4	—	—
24	130	1.08	—	—	601.0	—	—	—	—	—	—	—	—	3.6	—	—
29	140	1.08	—	—	671.5	—	—	—	—	—	—	—	—	2.4	—	—
38	190	1.44	—	—	886.5	—	—	—	—	—	—	—	—	3.6	—	—
17	130	0.72	—	—	454.0	—	—	—	—	—	—	—	—	3.0	—	—
24	180	0.90	—	—	637.5	—	—	—	—	—	—	—	—	3.6	—	—
29	140	0.90	—	—	647.0	—	—	—	—	—	—	—	—	1.8	—	—
38	190	1.26	—	—	865.5	—	—	—	—	—	—	—	—	2.4	—	—
26	130	0.72	—	—	577.0	—	80.0	—	—	—	—	—	—	1.8	—	—
35	80	1.08	—	—	792.5	—	105.8	—	—	—	—	—	—	2.4	—	—
14	130	0.72	—	—	396.5	—	—	—	—	—	—	—	—	2.4	—	—
19	180	1.08	—	—	550.5	—	—	—	—	—	—	—	—	3.0	—	—
25	120	2.34	—	—	796.5	—	—	—	—	—	—	—	—	0.6	—	—
34	170	3.24	—	—	1110.0	—	—	—	—	—	—	—	—	0.6	—	—
11	110	1.98	—	—	555.5	—	70.0	—	—	—	—	—	—	0	—	—
18	189	3.78	32.0	209.9	718.1	1.75	99.8	0.29	1.13	0.31	5.44	—	—	0	0.6	45.6
36	295	4.67	—	—	1341.4	—	174.0	—	—	—	—	—	—	0.5	—	—
20	120	2.16	—	—	712.0	—	—	—	—	—	—	—	—	1.2	—	—

TABLE A-1 Table of Food Composition *(continued)*

(Computer code is for Cengage Diet Analysis program)
(For purposes of calculations, use "0" for t, <1, <.1, <.01, etc.)

DA+ Code	Food Description	Quantity	Measure	Wt (g)	H_2O (g)	Ener (kcal)	Prot (g)	Carb (g)	Fiber (g)	Fat (g)	Fat Breakdown (g) Sat	Mono	Poly
Fast Food—*continued*													
31700	Deluxe, 14″	1	slice(s)	156	—	396	15.0	42.0	3.0	20.0	7.0	—	—
31692	Extravaganzza, 12″	1	slice(s)	136	—	341	14.0	30.0	2.0	19.0	7.0	—	—
31701	Extravaganzza, 14″	1	slice(s)	186	—	468	20.0	43.0	3.0	25.5	9.5	—	—
31599	Hawaiian, 12″	1	slice(s)	114	—	275	12.0	30.0	2.0	13.0	5.0	—	—
31600	Hawaiian, 14″	1	slice(s)	162	—	389	17.0	43.0	3.0	18.0	6.5	—	—
31693	Meatzza, 12″	1	slice(s)	121	—	333	14.0	29.0	2.0	19.0	7.0	—	—
31703	Meatzza, 14″	1	slice(s)	167	—	458	19.0	42.0	3.0	25.0	9.5	—	—
31593	Pepperoni, extra pepperoni and cheese, 12″	1	slice(s)	110	—	317	13.0	29.0	2.0	17.5	6.5	—	—
31594	Pepperoni, extra pepperoni and cheese, 14″	1	slice(s)	155	—	443	18.0	42.0	3.0	24.0	9.0	—	—
31602	Vegi, 12″	1	slice(s)	114	—	270	11.0	30.0	2.0	13.5	5.0	—	—
31603	Vegi, 14″	1	slice(s)	159	—	380	15.0	43.0	3.0	18.0	6.5	—	—
31598	With ham and pineapple tidbits, 6″	1	item(s)	430	—	619	25.2	69.9	4.0	28.3	10.2	—	—
31595	With Italian sausage, 6″	1	item(s)	430	—	642	24.8	69.6	4.2	31.1	11.3	—	—
31592	With pepperoni, 6″	1	item(s)	430	—	647	25.1	68.5	3.9	32.0	11.7	—	—
31601	With vegetables, 6″	1	item(s)	430	—	619	23.4	70.8	4.6	28.7	10.1	—	—
	In-n-Out Burger												
34391	Cheeseburger with mustard and ketchup	1	serving(s)	268	—	400	22.0	41.0	3.0	18.0	9.0	—	—
34374	Cheeseburger	1	serving(s)	268	—	480	22.0	39.0	3.0	27.0	10.0	—	—
34390	Cheeseburger, lettuce leaves instead of buns	1	serving(s)	300	—	330	18.0	11.0	3.0	25.0	9.0	—	—
34377	Chocolate shake	1	serving(s)	425	—	690	9.0	83.0	0	36.0	24.0	—	—
34375	Double-Double cheeseburger	1	serving(s)	330	—	670	37.0	39.0	3.0	41.0	18.0	—	—
34393	Double-Double cheeseburger with mustard and ketchup	1	serving(s)	330	—	590	37.0	41.0	3.0	32.0	17.0	—	—
34392	Double-Double cheeseburger, lettuce leaves instead of buns	1	serving(s)	362	—	520	33.0	11.0	3.0	39.0	17.0	—	—
34376	French fries	1	serving(s)	125	—	400	7.0	54.0	2.0	18.0	5.0	—	—
34373	Hamburger	1	item(s)	243	—	390	16.0	39.0	3.0	19.0	5.0	—	—
34389	Hamburger with mustard and ketchup	1	serving(s)	243	—	310	16.0	41.0	3.0	10.0	4.0	—	—
34388	Hamburger, lettuce leaves instead of buns	1	serving(s)	275	—	240	13.0	11.0	3.0	17.0	4.0	—	—
34379	Strawberry shake	1	serving(s)	425	—	690	9.0	91.0	0	33.0	22.0	—	—
34378	Vanilla shake	1	serving(s)	425	—	680	9.0	78.0	0	37.0	25.0	—	—
	Jack in the Box												
30392	Bacon ultimate cheeseburger	1	item(s)	338	—	1090	46.0	53.0	2.0	77.0	30.0	—	—
1740	Breakfast Jack	1	item(s)	125	—	290	17.0	29.0	1.0	12.0	4.5	—	—
14074	Cheeseburger	1	item(s)	131	—	350	18.0	31.0	1.0	17.0	8.0	—	—
14106	Chicken breast strips, 4 pieces	1	serving(s)	201	—	500	35.0	36.0	3.0	25.0	6.0	—	—
37241	Chicken club salad, plain, without salad dressing	1	serving(s)	431	—	300	27.0	13.0	4.0	15.0	6.0	—	—
14064	Chicken sandwich	1	item(s)	145	—	400	15.0	38.0	2.0	21.0	4.5	—	—
14111	Chocolate ice cream shake, small	1	serving(s)	414	—	880	14.0	107.0	1.0	45.0	31.0	—	—
14073	Hamburger	1	item(s)	118	—	310	16.0	30.0	1.0	14.0	6.0	—	—
14090	Hash browns	1	serving(s)	57	—	150	1.0	13.0	2.0	10.0	2.5	—	—
14072	Jack's Spicy Chicken sandwich	1	item(s)	270	—	620	25.0	61.0	4.0	31.0	6.0	—	—
1468	Jumbo Jack hamburger	1	item(s)	261	—	600	21.0	51.0	3.0	35.0	12.0	—	—
1469	Jumbo Jack hamburger with cheese	1	item(s)	286	—	690	25.0	54.0	3.0	42.0	16.0	—	—
14099	Natural cut french fries, large	1	serving(s)	196	—	530	8.0	69.0	5.0	25.0	6.0	—	—
14098	Natural cut french fries, medium	1	serving(s)	133	—	360	5.0	47.0	4.0	17.0	4.0	—	—
1470	Onion rings	1	serving(s)	119	—	500	6.0	51.0	3.0	30.0	6.0	—	—
33141	Sausage, egg, and cheese biscuit	1	item(s)	234	—	740	27.0	35.0	2.0	55.0	17.0	—	—
14095	Seasoned curly fries, medium	1	serving(s)	125	—	400	6.0	45.0	5.0	23.0	5.0	—	—
14077	Sourdough Jack	1	item(s)	245	—	710	27.0	36.0	3.0	51.0	18.0	—	—
37249	Southwest chicken salad, plain, without salad dressing	1	serving(s)	488	—	300	24.0	29.0	7.0	11.0	5.0	—	—
14112	Strawberry ice cream shake, small	1	serving(s)	417	—	880	13.0	105.0	0	44.0	31.0	—	—

PAGE KEY: A-4 = Breads/Baked Goods A-10 = Cereal/Rice/Pasta A-14 = Fruit A-18 = Vegetables/Legumes A-28 = Nuts/Seeds A-30 = Vegetarian A-32 = Dairy A-40 = Eggs A-40 = Seafood A-42 = Meats A-46 = Poultry A-46 = Processed Meats A-48 = Beverages A-52 = Fats/Oils A-54 = Sweets A-56 = Spices/Condiments/Sauces A-60 = Mixed Foods/Soups/Sandwiches A-64 = Fast Food A-84 = Convenience A-86 = Baby Foods

Chol (mg)	Calc (mg)	Iron (mg)	Magn (mg)	Pota (mg)	Sodi (mg)	Zinc (mg)	Vit A (μg)	Thia (mg)	Vit E (mg α)	Ribo (mg)	Niac (mg)	Vit B_6 (mg)	Fola (μg)	Vit C (mg)	Vit B_{12} (μg)	Sele (μg)
26	170	3.06	—	—	974.5	—	—	—	—	—	—	—	—	1.2	—	—
31	160	2.52	—	—	934.5	—	—	—	—	—	—	—	—	1.2	—	—
40	220	3.42	—	—	1260.0	—	—	—	—	—	—	—	—	1.2	—	—
19	150	1.98	—	—	717.0	—	—	—	—	—	—	—	—	1.2	—	—
26	210	2.88	—	—	1011.0	—	—	—	—	—	—	—	—	1.8	—	—
31	160	2.34	—	—	910.5	—	—	—	—	—	—	—	—	0	—	—
40	220	3.24	—	—	1230.0	—	—	—	—	—	—	—	—	0.6	—	—
27	150	2.16	—	—	840.5	—	86.5	—	—	—	—	—	—	0	—	—
37	220	3.06	—	—	1166.0	—	115.4	—	—	—	—	—	—	0.6	—	—
15	150	2.16	—	—	659.5	—	—	—	—	—	—	—	—	0.6	—	—
21	220	3.06	—	—	924.0	—	—	—	—	—	—	—	—	1.2	—	—
43	298	4.84	—	—	1497.8	—	—	—	—	—	—	—	—	1.5	—	—
45	302	4.89	—	—	1478.1	—	—	—	—	—	—	—	—	0.6	—	—
47	299	4.81	—	—	1523.7	—	167.9	—	—	—	—	—	—	0.6	—	—
36	307	5.10	—	—	1472.5	—	—	—	—	—	—	—	—	4.7	—	—
60	200	3.60	—	—	1080.0	—	—	—	—	—	—	—	—	12.0	—	—
60	200	3.60	—	—	1000.0	—	—	—	—	—	—	—	—	9.0	—	—
60	200	2.70	—	—	720.0	—	—	—	—	—	—	—	—	12.0	—	—
95	300	0.72	—	—	350.0	—	—	—	—	—	—	—	—	0	—	—
120	350	5.40	—	—	1440.0	—	—	—	—	—	—	—	—	9.0	—	—
115	350	5.40	—	—	1520.0	—	—	—	—	—	—	—	—	12.0	—	—
120	350	4.50	—	—	1160.0	—	—	—	—	—	—	—	—	12.0	—	—
0	20	1.80	—	—	245.0	—	0	—	—	—	—	—	—	0	—	—
40	40	3.60	—	—	650.0	—	—	—	—	—	—	—	—	9.0	—	—
35	40	3.60	—	—	730.0	—	—	—	—	—	—	—	—	12.0	—	—
40	40	2.70	—	—	370.0	—	—	—	—	—	—	—	—	12.0	—	—
85	300	0.00	—	—	280.0	—	—	—	—	—	—	—	—	0	—	—
90	300	0.00	—	—	390.0	—	—	—	—	—	—	—	—	0	—	—
140	308	7.38	—	540.0	2040.0	—	—	—	—	—	—	—	—	0.6	—	—
220	145	3.48	—	210.0	760.0	—	—	—	—	—	—	—	—	3.5	—	—
50	151	3.61	—	270.0	790.0	—	40.2	—	—	—	—	—	—	0	—	—
80	18	1.60	—	530.0	1260.0	—	—	—	—	—	—	—	—	1.1	—	—
65	280	3.35	—	560.0	880.0	—	—	—	—	—	—	—	—	50.4	—	—
35	100	2.70	—	240.0	730.0	—	—	—	—	—	—	—	—	4.8	—	—
135	460	0.47	—	840.0	330.0	—	—	—	—	—	—	—	—	0	—	—
40	100	3.60	—	250.0	600.0	—	0	—	—	—	—	—	—	0	—	—
0	10	0.18	—	190.0	230.0	—	0	—	—	—	—	—	—	0	—	—
50	150	1.80	—	450.0	1100.0	—	—	—	—	—	—	—	—	9.0	—	—
45	164	4.92	—	380.0	940.0	—	—	—	—	—	—	—	—	9.8	—	—
70	234	4.20	—	410.0	1310.0	—	—	—	—	—	—	—	—	8.4	—	—
0	20	1.42	—	1240.0	870.0	—	0	—	—	—	—	—	—	8.9	—	—
0	19	1.01	—	840.0	590.0	—	0	—	—	—	—	—	—	5.6	—	—
0	40	2.70	—	140.0	420.0	—	40.0	—	—	—	—	—	—	18.0	—	—
280	88	2.36	—	310.0	1430.0	—	—	—	—	—	—	—	—	0	—	—
0	40	1.80	—	580.0	890.0	—	—	—	—	—	—	—	—	0	—	—
75	200	4.50	—	430.0	1230.0	—	—	—	—	—	—	—	—	9.0	—	—
55	274	4.10	—	670.0	860.0	—	—	—	—	—	—	—	—	43.8	—	—
135	466	0.00	—	750.0	290.0	—	—	—	—	—	—	—	—	0	—	—

TABLE A-1 Table of Food Composition *(continued)*

(Computer code is for Cengage Diet Analysis program)
(For purposes of calculations, use "0" for t, <1, <.1, <.01, etc.)

DA+ Code	Food Description	Quantity	Measure	Wt (g)	H_2O (g)	Ener (kcal)	Prot (g)	Carb (g)	Fiber (g)	Fat (g)	Fat Breakdown (g) Sat	Mono	Poly
Fast Food—*continued*													
14078	Ultimate cheeseburger	1	item(s)	323	—	1010	40.0	53.0	2.0	71.0	28.0	—	—
14110	Vanilla ice cream shake, small	1	serving(s)	379	—	790	13.0	83.0	0	44.0	31.0	—	—
	Jamba Juice												
31645	Aloha Pineapple smoothie	24	fluid ounce(s)	730	—	500	8.0	117.0	4.0	1.5	1.0	—	—
31646	Banana Berry smoothie	24	fluid ounce(s)	719	—	480	5.0	112.0	4.0	1.0	0	—	—
31656	Berry Lime Sublime smoothie	24	fluid ounce(s)	728	—	460	3.0	106.0	5.0	2.0	1.0	—	—
31647	Carribean Passion smoothie	24	fluid ounce(s)	730	—	440	4.0	102.0	4.0	2.0	1.0	—	—
38422	Carrot juice	16	fluid ounce(s)	472	—	100	3.0	23.0	0	0.5	0	—	—
31648	Chocolate Moo'd smoothie	24	fluid ounce(s)	634	—	720	17.0	148.0	3.0	8.0	5.0	—	—
31649	Citrus Squeeze smoothie	24	fluid ounce(s)	727	—	470	5.0	110.0	4.0	2.0	1.0	—	—
31651	Coldbuster smoothie	24	fluid ounce(s)	724	—	430	5.0	100.0	5.0	2.5	1.0	—	—
31652	Cranberry Craze smoothie	24	fluid ounce(s)	793	—	460	6.0	104.0	4.0	0.5	0	—	—
31654	Jamba Powerboost smoothie	24	fluid ounce(s)	738	—	440	6.0	105.0	6.0	1.0	0	—	—
38423	Lemonade	16	fluid ounce(s)	483	—	300	1.0	75.0	0	0	0	0	0
31657	Mango-a-go-go smoothie	24	fluid ounce(s)	690	—	440	3.0	104.0	4.0	1.5	0.5	—	—
38424	Orange juice, freshly squeezed	16	fluid ounce(s)	496	—	220	3.0	52.0	0.5	1.0	0	—	—
38426	Orange/carrot juice	16	fluid ounce(s)	484	—	160	3.0	37.0	0	1.0	0	—	—
31660	Orange-a-peel smoothie	24	fluid ounce(s)	726	—	440	8.0	102.0	5.0	1.5	0	—	—
31662	Peach Pleasure smoothie	24	fluid ounce(s)	720	—	460	4.0	108.0	4.0	2.0	1.0	—	—
31665	Protein Berry Pizzaz smoothie	24	fluid ounce(s)	710	—	440	20.0	92.0	5.0	1.5	0	—	—
31668	Razzmatazz smoothie	24	fluid ounce(s)	730	—	480	3.0	112.0	4.0	2.0	1.0	—	—
31669	Strawberries Wild smoothie	24	fluid ounce(s)	725	—	450	6.0	105.0	4.0	0.5	0	—	—
38421	Strawberry Tsunami smoothie	24	fluid ounce(s)	740	—	530	4.0	128.0	4.0	2.0	1.0	—	—
38427	Vibrant C juice	16	fluid ounce(s)	448	—	210	2.0	50.0	1.0	0	0	0	0
38428	Wheatgrass juice, freshly squeezed	1	ounce(s)	28	—	5	0.5	1.0	0	0	0	0	0
	Kentucky Fried Chicken (KFC)												
31850	BBQ baked beans	1	serving(s)	136	—	220	8.0	45.0	7.0	1.0	0	—	—
31853	Biscuit	1	item(s)	57	—	220	4.0	24.0	1.0	11.0	2.5	—	—
51223	Boneless Fiery Buffalo Wings	6	item(s)	211	—	530	30.0	44.0	3.0	26.0	5.0	—	—
39386	Boneless Honey BBQ Wings	6	item(s)	213	—	570	30.0	54.0	5.0	26.0	5.0	—	—
51224	Boneless Sweet & Spicy Wings	6	item(s)	203	—	550	30.0	50.0	3.0	26.0	5.0	—	—
31851	Cole slaw	1	serving(s)	130	—	180	1.0	22.0	3.0	10.0	1.5	—	—
31842	Colonel's Crispy Strips	3	item(s)	151	—	370	28.0	17.0	1.0	20.0	4.0	—	—
31849	Corn on the cob	1	item(s)	162	—	150	5.0	26.0	7.0	3.0	1.0	—	—
51221	Double Crunch sandwich	1	item(s)	213	—	520	27.0	39.0	3.0	29.0	5.0	—	—
3761	Extra Crispy chicken, breast	1	item(s)	162	—	370	33.0	10.0	2.0	22.0	5.0	—	—
3762	Extra Crispy chicken, drumstick	1	item(s)	60	—	150	12.0	4.0	0	10.0	2.5	—	—
3763	Extra Crispy chicken, thigh	1	item(s)	114	—	290	17.0	16.0	1.0	18.0	4.0	—	—
3764	Extra Crispy chicken, whole wing	1	item(s)	52	—	150	11.0	11.0	1.0	7.0	1.5	—	—
51218	Famous Bowls mashed potatoes with gravy	1	serving(s)	531	—	720	26.0	79.0	6.0	34.0	9.0	—	—
51219	Famous Bowls rice with gravy	1	serving(s)	384	—	610	25.0	67.0	5.0	27.0	8.0	—	—
31841	Honey BBQ chicken sandwich	1	item(s)	147	—	290	23.0	40.0	2.0	4.0	1.0	—	—
31833	Honey BBQ wing pieces	6	item(s)	157	—	460	27.0	26.0	3.0	27.0	6.0	—	—
10859	Hot wings pieces	6	piece(s)	134	—	450	26.0	19.0	2.0	30.0	7.0	—	—
42382	KFC Snacker sandwich	1	serving(s)	119	—	320	14.0	29.0	2.0	17.0	3.0	—	—
31848	Macaroni and cheese	1	serving(s)	136	—	180	8.0	18.0	0	8.0	3.5	—	—
31847	Mashed potatoes with gravy	1	serving(s)	151	—	140	2.0	20.0	1.0	5.0	1.0	—	—
10825	Original Recipe chicken, breast	1	item(s)	161	—	340	38.0	9.0	2.0	17.0	4.0	—	—
10826	Original Recipe chicken, drumstick	1	item(s)	59	—	140	13.0	3.0	0	8.0	2.0	—	—
10827	Original Recipe chicken, thigh	1	item(s)	126	—	350	19.0	7.0	1.0	27.0	7.0	—	—
10828	Original Recipe chicken, whole wing	1	item(s)	47	—	140	10.0	4.0	0	9.0	2.0	—	—
51222	Oven roasted Twister chicken wrap	1	item(s)	269	—	520	30.0	46.0	4.0	23.0	3.5	—	—
31844	Popcorn chicken, small or individual	1	item(s)	114	—	370	19.0	21.0	2.0	24.0	4.5	—	—
31852	Potato salad	1	serving(s)	128	—	180	2.0	22.0	2.0	9.0	1.5	—	—
10845	Potato wedges, small	1	serving(s)	102	—	250	4.0	32.0	3.0	12.0	2.0	—	—
31839	Tender Roast chicken sandwich with sauce	1	item(s)	236	—	430	37.0	29.0	2.0	18.0	3.5	—	—

PAGE KEY: A-4 = Breads/Baked Goods A-10 = Cereal/Rice/Pasta A-14 = Fruit A-18 = Vegetables/Legumes A-28 = Nuts/Seeds A-30 = Vegetarian A-32 = Dairy A-40 = Eggs A-40 = Seafood A-42 = Meats A-46 = Poultry A-46 = Processed Meats A-48 = Beverages A-52 = Fats/Oils A-54 = Sweets A-56 = Spices/Condiments/Sauces A-60 = Mixed Foods/Soups/Sandwiches A-64 = Fast Food A-84 = Convenience A-86 = Baby Foods

Chol (mg)	Calc (mg)	Iron (mg)	Magn (mg)	Pota (mg)	Sodi (mg)	Zinc (mg)	Vit A (μg)	Thia (mg)	Vit E (mg α)	Ribo (mg)	Niac (mg)	Vit B_6 (mg)	Fola (μg)	Vit C (mg)	Vit B_{12} (μg)	Sele (μg)
125	308	7.39	—	480.0	1580.0	—	—	—	—	—	—	—	—	0.6	—	—
135	532	0.00	—	750.0	280.0	—	—	—	—	—	—	—	—	0	—	—
5	200	1.80	60.0	1000.0	30.0	0.30	—	0.37	—	0.34	2.00	0.60	60.0	102.0	0	1.4
0	200	1.44	40.0	1010.0	115.0	0.60	—	0.09	—	0.25	0.80	0.70	24.0	15.0	0.2	1.4
5	200	1.80	16.0	510.0	35.0	0.30	—	0.06	—	0.25	6.00	0.70	140.0	54.0	0	1.4
5	100	1.80	24.0	810.0	60.0	0.30	—	0.09	—	0.25	5.00	0.50	100.0	78.0	0	1.4
0	150	2.70	80.0	1030.0	250.0	0.90	—	0.52	—	0.25	5.00	0.70	80.0	18.0	0	5.6
30	500	1.08	60.0	810.0	380.0	1.50	—	0.22	—	0.76	0.40	0.16	16.0	6.0	1.5	4.2
5	100	1.80	80.0	1170.0	35.0	0.30	—	0.37	—	0.34	1.90	0.60	100.0	180.0	0	1.4
5	100	1.08	60.0	1260.0	35.0	16.50	—	0.37	—	0.34	3.00	0.40	121.5	1302.0	0	1.4
0	250	1.44	16.0	500.0	50.0	0.30	—	0.03	—	0.25	5.00	0.60	120.0	54.0	0	1.4
0	1200	1.80	480.0	1070.0	45.0	16.50	—	5.55	—	6.12	68.00	7.40	640.0	288.0	10.8	77.0
0	20	0.00	8.0	200.0	10.0	0.00	—	0.03	—	0.17	14.00	1.80	320.0	36.0	0	0
5	100	1.08	24.0	780.0	50.0	0.30	—	0.15	—	0.25	5.00	0.70	120.0	72.0	0	1.4
0	60	1.08	60.0	990.0	0	0.30	—	0.45	—	0.13	2.00	0.20	160.0	246.0	0	0
0	100	1.80	60.0	1010.0	125.0	0.60	—	0.45	—	0.25	3.00	0.50	120.0	132.0	0	2.8
0	250	1.80	80.0	1380.0	160.0	0.90	—	0.45	—	0.42	2.00	0.50	140.0	240.0	0.6	1.4
5	100	0.72	32.0	740.0	60.0	0.30	—	0.06	—	0.25	4.00	0.60	80.0	18.0	0	1.4
0	1100	2.62	60.0	650.0	240.0	0.58	—	0.08	—	0.17	1.20	0.70	58.3	60.0	0	5.6
5	150	1.80	32.0	810.0	70.0	0.30	—	0.09	—	0.34	6.00	1.00	160.0	60.0	0	1.4
5	250	1.80	40.0	1050.0	180.0	0.90	—	0.12	—	0.34	0.80	0.40	40.0	60.0	0.6	1.4
5	100	1.08	24.0	480.0	10.0	0.30	—	0.06	—	0.34	14.00	1.80	320.0	90.0	0	1.4
0	20	1.08	40.0	720.0	0	0.30	—	0.30	—	0.10	1.60	0.40	80.0	678.0	0	0
0	0	1.80	8.0	80.0	0	0.00	0	0.03	—	0.03	0.40	0.04	16.0	3.6	0	2.8
0	100	2.70	—	—	730.0	—	—	—	—	—	—	—	—	1.2	—	—
0	40	1.80	—	—	640.0	—	—	—	—	—	—	—	—	0	—	—
65	40	1.80	—	—	2670.0	—	—	—	—	—	—	—	—	1.2	—	—
65	40	1.80	—	—	2210.0	—	—	—	—	—	—	—	—	1.2	—	—
65	60	1.80	—	—	2000.0	—	—	—	—	—	—	—	—	1.2	—	—
5	40	0.72	—	—	270.0	—	—	—	—	—	—	—	—	12.0	—	—
65	40	1.44	—	—	1220.0	—	0	—	—	—	—	—	—	1.2	—	—
0	60	1.08	—	—	10.0	—	—	—	—	—	—	—	—	6.0	—	—
55	100	2.70	—	—	1220.0	—	—	—	—	—	—	—	—	6.0	—	—
85	20	2.70	—	—	1020.0	—	—	—	—	—	—	—	—	1.2	—	—
55	0	1.44	—	—	300.0	—	0	—	—	—	—	—	—	0	—	—
95	20	2.70	—	—	700.0	—	—	—	—	—	—	—	—	—	—	—
45	20	1.08	—	—	340.0	—	—	—	—	—	—	—	—	0	—	—
35	200	5.40	—	—	2330.0	—	—	—	—	—	—	—	—	6.0	—	—
35	200	4.50	—	—	2130.0	—	—	—	—	—	—	—	—	6.0	—	—
60	80	2.70	—	—	710.0	—	—	—	—	—	—	—	—	2.4	—	—
140	40	1.80	—	—	970.0	—	—	—	—	—	—	—	—	21.0	—	—
115	40	1.44	—	—	990.0	—	—	—	—	—	—	—	—	1.2	—	—
25	60	2.70	—	—	690.0	—	—	—	—	—	—	—	—	2.4	—	—
15	150	0.72	—	—	800.0	—	—	—	—	—	—	—	—	1.2	—	—
0	40	1.44	—	—	560.0	—	—	—	—	—	—	—	—	1.2	—	—
135	20	2.70	—	—	960.0	—	—	—	—	—	—	—	—	6.0	—	—
70	20	1.08	—	—	340.0	—	—	—	—	—	—	—	—	0	—	—
110	20	2.70	—	—	870.0	—	—	—	—	—	—	—	—	1.2	—	—
50	20	1.44	—	—	350.0	—	0	—	—	—	—	—	—	1.2	—	—
60	40	6.30	—	—	1380.0	—	—	—	—	—	—	—	—	15.0	—	—
25	40	1.80	—	—	1110.0	—	0	—	—	—	—	—	—	0	—	—
5	0	0.36	—	—	470.0	—	—	—	—	—	—	—	—	6.0	—	—
0	20	1.08	—	—	700.0	—	0	—	—	—	—	—	—	0	—	—
80	80	2.70	—	—	1180.0	—	—	—	—	—	—	—	—	9.0	—	—

TABLE A-1 Table of Food Composition *(continued)*

(Computer code is for Cengage Diet Analysis program)
(For purposes of calculations, use "0" for t, <1, <.1, <.01, etc.)

DA+ Code	Food Description	Quantity	Measure	Wt (g)	H_2O (g)	Ener (kcal)	Prot (g)	Carb (g)	Fiber (g)	Fat (g)	Fat Breakdown (g) Sat	Mono	Poly
Fast Food—*continued*													
	Long John Silver												
39392	Baked cod	1	serving(s)	101	—	120	22.0	1.0	0	4.5	1.0	—	—
3777	Batter dipped fish sandwich	1	item(s)	177	—	470	18.0	48.0	3.0	23.0	5.0	—	—
37568	Battered fish	1	item(s)	92	—	260	12.0	17.0	0.5	16.0	4.0	—	—
37569	Breaded clams	1	serving(s)	85	—	240	8.0	22.0	1.0	13.0	2.0	—	—
37566	Chicken plank	1	item(s)	52	—	140	8.0	9.0	0.5	8.0	2.0	—	—
39404	Clam chowder	1	item(s)	227	—	220	9.0	23.0	0	10.0	4.0	—	—
39398	Cocktail sauce	1	ounce(s)	28	—	25	0	6.0	0	0	0	0	0
3770	Coleslaw	1	serving(s)	113	—	200	1.0	15.0	3.0	15.0	2.5	1.8	4.1
39400	French fries, large	1	item(s)	142	—	390	4.0	56.0	5.0	17.0	4.0	—	—
3774	Fries, regular	1	serving(s)	85	—	230	3.0	34.0	3.0	10.0	2.5	—	—
3779	Hushpuppy	1	piece(s)	23	—	60	1.0	9.0	1.0	2.5	0.5	—	—
3781	Shrimp, batter-dipped, 1 piece	1	piece(s)	14	—	45	2.0	3.0	0	3.0	1.0	—	—
39399	Tartar sauce	1	ounce(s)	28	—	100	0	4.0	0	9.0	1.5	—	—
39395	Ultimate Fish sandwich	1	item(s)	199	—	530	21.0	49.0	3.0	28.0	8.0	—	—
	McDonald's												
50828	Asian salad with grilled chicken	1	item(s)	362	—	290	31.0	23.0	6.0	10.0	1.0	—	—
2247	Barbecue sauce	1	item(s)	28	—	45	0	11.0	0	0	0	0	0
737	Big Mac hamburger	1	item(s)	219	—	560	25.0	47.0	3.0	30.0	10.0	—	—
29777	Caesar salad dressing	1	package(s)	44	—	150	1.0	5.0	0	13.0	2.5	—	—
38391	Caesar salad with grilled chicken, no dressing	1	serving(s)	278	230.6	181	26.4	10.5	3.1	6.0	2.9	1.7	0.8
38393	Caesar salad without chicken, no dressing	1	serving(s)	190	170.4	84	6.0	8.1	3.0	3.9	2.2	0.9	0.3
738	Cheeseburger	1	item(s)	119	—	310	15.0	35.0	1.0	12.0	6.0	—	—
29775	Chicken McGrill sandwich	1	item(s)	213	—	400	27.0	38.0	3.0	16.0	3.0	—	—
1873	Chicken McNuggets, 6 piece	1	serving(s)	96	—	250	15.0	15.0	0	15.0	3.0	—	—
3792	Chicken McNuggets, 4 piece	1	serving(s)	64	—	170	10.0	10.0	0	10.0	2.0	—	—
29774	Crispy chicken sandwich	1	item(s)	232	121.8	500	27.0	63.0	3.0	16.0	3.0	5.7	7.4
743	Egg McMuffin	1	item(s)	139	76.8	300	17.0	30.0	2.0	12.0	4.5	3.8	2.5
742	Filet-O-Fish sandwich	1	item(s)	141	—	400	14.0	42.0	1.0	18.0	4.0	—	—
2257	French fries, large	1	serving(s)	170	—	570	6.0	70.0	7.0	30.0	6.0	—	—
1872	French fries, small	1	serving(s)	74	—	250	2.0	30.0	3.0	13.0	2.5	—	—
33822	Fruit 'n Yogurt Parfait	1	item(s)	149	111.2	160	4.0	31.0	1.0	2.0	1.0	0.2	0.1
739	Hamburger	1	item(s)	105	—	260	13.0	33.0	1.0	9.0	3.5	—	—
2003	Hash browns	1	item(s)	53	—	140	1.0	15.0	2.0	8.0	1.5	—	—
2249	Honey sauce	1	item(s)	14	—	50	0	12.0	0	0	0	0	0
38397	Newman's Own creamy caesar salad dressing	1	item(s)	59	32.5	190	2.0	4.0	0	18.0	3.5	4.6	9.6
38398	Newman's Own low fat balsamic vinaigrette salad dressing	1	item(s)	44	29.1	40	0	4.0	0	3.0	0	1.0	1.2
38399	Newman's Own ranch salad dressing	1	item(s)	59	30.1	170	1.0	9.0	0	15.0	2.5	9.0	3.7
1874	Plain Hotcakes with syrup and margarine	3	item(s)	221	—	600	9.0	102.0	2.0	17.0	4.0	—	—
740	Quarter Pounder hamburger	1	item(s)	171	—	420	24.0	40.0	3.0	18.0	7.0	—	—
741	Quarter Pounder hamburger with cheese	1	item(s)	199	—	510	29.0	43.0	3.0	25.0	12.0	—	—
2005	Sausage McMuffin with egg	1	item(s)	165	82.4	450	20.0	31.0	2.0	27.0	10.0	10.9	4.6
50831	Side salad	1	item(s)	87	—	20	1.0	4.0	1.0	0	0	0	0
	Pizza Hut												
39009	Hot chicken wings	2	item(s)	57	—	110	11.0	1.0	0	6.0	2.0	—	—
14025	Meat Lovers hand tossed pizza	1	slice(s)	118	—	300	15.0	29.0	2.0	13.0	6.0	—	—
14026	Meat Lovers pan pizza	1	slice(s)	123	—	340	15.0	29.0	2.0	19.0	7.0	—	—
31009	Meat Lovers stuffed crust pizza	1	slice(s)	169	—	450	21.0	43.0	3.0	21.0	10.0	—	—
14024	Meat Lovers thin 'n crispy pizza	1	slice(s)	98	—	270	13.0	21.0	2.0	14.0	6.0	—	—
14031	Pepperoni Lovers hand tossed pizza	1	slice(s)	113	—	300	15.0	30.0	2.0	13.0	7.0	—	—
14032	Pepperoni Lovers pan pizza	1	slice(s)	118	—	340	15.0	29.0	2.0	19.0	7.0	—	—
31011	Pepperoni Lovers stuffed crust pizza	1	slice(s)	163	—	420	21.0	43.0	3.0	19.0	10.0	—	—
14030	Pepperoni Lovers thin 'n crispy pizza	1	slice(s)	92	—	260	13.0	21.0	2.0	14.0	7.0	—	—
10834	Personal Pan pepperoni pizza	1	slice(s)	61	—	170	7.0	18.0	0.5	8.0	3.0	—	—
10842	Personal Pan supreme pizza	1	slice(s)	77	—	190	8.0	19.0	1.0	9.0	3.5	—	—

PAGE KEY: A-4 = Breads/Baked Goods A-10 = Cereal/Rice/Pasta A-14 = Fruit A-18 = Vegetables/Legumes A-28 = Nuts/Seeds A-30 = Vegetarian A-32 = Dairy A-40 = Eggs A-40 = Seafood A-42 = Meats A-46 = Poultry A-46 = Processed Meats A-48 = Beverages A-52 = Fats/Oils A-54 = Sweets A-56 = Spices/Condiments/Sauces A-60 = Mixed Foods/Soups/Sandwiches A-64 = Fast Food A-84 = Convenience A-86 = Baby Foods

Chol (mg)	Calc (mg)	Iron (mg)	Magn (mg)	Pota (mg)	Sodi (mg)	Zinc (mg)	Vit A (µg)	Thia (mg)	Vit E (mg α)	Ribo (mg)	Niac (mg)	Vit B_6 (mg)	Fola (µg)	Vit C (mg)	Vit B_{12} (µg)	Sele (µg)
90	20	0.72	—	—	240.0	—	—	—	—	—	—	—	—	0	—	—
45	60	2.70	—	—	1210.0	—	—	—	—	—	—	—	—	2.4	—	—
35	20	0.72	—	—	790.0	—	—	—	—	—	—	—	—	4.8	—	—
10	20	1.08	—	—	1110.0	—	0	—	—	—	—	—	—	0	—	—
20	0	0.72	—	—	480.0	—	0	—	—	—	—	—	—	2.4	—	—
25	150	0.72	—	—	810.0	—	—	—	—	—	—	—	—	0	—	—
0	0	0.00	—	—	250.0	—	—	—	—	—	—	—	—	0	—	—
20	40	0.36	—	222.7	340.0	0.70	—	0.07	—	0.08	2.34	—	—	18.0	—	—
0	0	0.00	—	—	580.0	—	0	—	—	—	—	—	—	24.0	—	—
0	0	0.00	—	370.0	350.0	0.30	0	0.09	—	0.01	1.60	—	—	15.0	—	—
0	20	0.36	—	—	200.0	—	0	—	—	—	—	—	—	0	—	—
15	0	0.00	—	—	160.0	—	0	—	—	—	—	—	—	1.2	—	—
15	0	0.00	—	—	250.0	—	0	—	—	—	—	—	—	0	—	—
60	150	2.70	—	—	1400.0	—	—	—	—	—	—	—	—	4.8	—	—
65	150	3.60	—	—	890.0	—	—	—	—	—	—	—	—	54.0	—	—
0	0	0.00	—	55.0	260.0	—	—	—	—	—	—	—	—	0	—	—
80	250	4.50	—	400.0	1010.0	—	—	—	—	—	—	—	—	1.2	—	—
10	40	0.18	—	30.0	400.0	—	—	—	—	—	—	—	—	0.6	—	—
67	178	1.77	—	708.9	767.3	—	—	0.15	—	0.19	10.62	—	127.9	29.2	0.2	—
10	163	1.15	17.1	410.4	157.7	—	—	0.08	—	0.07	0.40	—	102.6	26.8	0	0.4
40	200	2.70	—	240.0	740.0	—	60.0	—	—	—	—	—	—	1.2	—	—
70	150	2.70	—	510.0	1010.0	—	—	—	—	—	—	—	—	6.0	—	—
35	20	0.72	—	240.0	670.0	—	—	—	—	—	—	—	—	1.2	—	—
25	0	0.36	—	160.0	450.0	—	—	—	—	—	—	—	—	1.2	—	—
60	80	3.60	62.6	526.6	1380.0	1.53	41.8	0.46	2.27	0.39	12.85	—	94.2	6.0	0.4	—
230	300	2.70	26.4	218.2	860.0	1.59	—	0.36	0.82	0.51	4.31	0.20	109.8	1.2	0.9	—
40	150	1.80	—	250.0	640.0	—	36.2	—	—	—	—	—	—	0	—	—
0	20	1.80	—	—	330.0	—	0	—	—	—	—	—	—	9.0	—	—
0	20	0.72	—	—	140.0	—	0	—	—	—	—	—	—	3.6	—	—
5	150	0.67	20.9	248.8	85.0	0.53	0	0.06	—	0.17	0.35	—	19.4	9.0	0.3	—
30	150	2.70	—	210.0	530.0	—	5.0	—	—	—	—	—	—	1.2	—	—
0	0	0.36	—	210.0	290.0	—	0	—	—	—	—	—	—	1.2	—	—
0	0	0.00	—	0	0	—	0	—	—	—	—	—	—	0	—	—
20	61	0.00	3.0	16.0	500.0	0.20	—	0.01	15.43	0.02	0.01	0.64	2.4	0	0.1	0.1
0	4	0.00	1.3	8.8	730.0	0.01	—	0.00	0.00	0.00	0.00	0.00	0	2.4	0	0
0	40	0.00	1.8	70.4	530.0	0.03	0	0.01	—	0.08	0.01	0.02	0.6	0	0	0.2
20	150	2.70	—	280.0	620.0	—	—	—	—	—	—	—	—	0	—	—
70	150	4.50	—	390.0	730.0	—	10.0	—	—	—	—	—	—	1.2	—	—
95	300	4.50	—	440.0	1150.0	—	100.0	—	—	—	—	—	—	1.2	—	—
255	300	3.60	29.7	282.2	950.0	2.01	—	0.43	0.82	0.56	4.83	0.24	—	0	1.2	—
0	20	0.72	—	—	10.0	—	—	—	—	—	—	—	—	15.0	—	—
70	0	0.36	—	—	450.0	—	—	—	—	—	—	—	—	0	—	—
35	150	1.80	—	—	760.0	—	—	—	—	—	—	—	—	6.0	—	—
35	150	2.70	—	—	750.0	—	—	—	—	—	—	—	—	6.0	—	—
55	250	2.70	—	—	1250.0	—	—	—	—	—	—	—	—	9.0	—	—
35	150	1.44	—	—	740.0	—	—	—	—	—	—	—	—	6.0	—	—
40	200	1.80	—	—	710.0	—	57.7	—	—	—	—	—	—	2.4	—	—
40	200	2.70	—	—	700.0	—	57.7	—	—	—	—	—	—	2.4	—	—
55	300	2.70	—	—	1120.0	—	—	—	—	—	—	—	—	3.6	—	—
40	200	1.44	—	—	690.0	—	58.0	—	—	—	—	—	—	2.4	—	—
15	80	1.44	—	—	340.0	—	38.5	—	—	—	—	—	—	1.4	—	—
20	80	1.86	—	—	420.0	—	—	—	—	—	—	—	—	3.6	—	—

TABLE A-1 Table of Food Composition *(continued)*

(Computer code is for Cengage Diet Analysis program)
(For purposes of calculations, use "0" for t, <1, <.1, <.01, etc.)

DA+ Code	Food Description	Quantity	Measure	Wt (g)	H_2O (g)	Ener (kcal)	Prot (g)	Carb (g)	Fiber (g)	Fat (g)	Fat Breakdown (g) Sat	Mono	Poly
Fast Food—*continued*													
39013	Personal Pan Veggie Lovers pizza	1	slice(s)	69	—	150	6.0	19.0	1.0	6.0	2.0	—	—
14028	Veggie Lovers hand tossed pizza	1	slice(s)	118	—	220	10.0	31.0	2.0	6.0	3.0	—	—
14029	Veggie Lovers pan pizza	1	slice(s)	119	—	260	10.0	30.0	2.0	12.0	4.0	—	—
31010	Veggie Lovers stuffed crust pizza	1	slice(s)	172	—	360	16.0	45.0	3.0	14.0	7.0	—	—
14027	Veggie Lovers thin 'n crispy pizza	1	slice(s)	101	—	180	8.0	23.0	2.0	7.0	3.0	—	—
39012	Wing blue cheese dipping sauce	1	item(s)	43	—	230	2.0	2.0	0	24.0	5.0	—	—
39011	Wing ranch dipping sauce	1	item(s)	43	—	210	0.5	4.0	0	22.0	3.5	—	—
	Starbucks												
38052	Cappuccino, tall	12	fluid ounce(s)	360	—	120	7.0	10.0	0	6.0	4.0	—	—
38053	Cappuccino, tall nonfat	12	fluid ounce(s)	360	—	80	7.0	11.0	0	0	0	0	0
38054	Cappuccino, tall soymilk	12	fluid ounce(s)	360	—	100	5.0	13.0	0.5	2.5	0	—	—
38059	Cinnamon spice mocha, tall nonfat w/o whipped cream	12	fluid ounce(s)	360	—	170	11.0	32.0	0	0.5	—	—	—
38057	Cinnamon spice mocha, tall w/whipped cream	12	fluid ounce(s)	360	—	320	10.0	31.0	0	17.0	11.0	—	—
38051	Espresso, single shot	1	fluid ounce(s)	30	—	5	0	1.0	0	0	0	0	0
38088	Flavored syrup, 1 pump	1	serving(s)	10	—	20	0	5.0	0	0	0	0	0
32562	Frappuccino bottled coffee drink, mocha	9½	fluid ounce(s)	298	—	190	6.0	39.0	3.0	3.0	2.0	—	—
32561	Frappuccino coffee drink, all bottled flavors	9½	fluid ounce(s)	281	—	190	7.0	35.0	0	3.5	2.5	—	—
38073	Frappuccino, mocha	12	fluid ounce(s)	360	—	220	5.0	44.0	0	3.0	1.5	—	—
38067	Frappuccino, tall caramel w/o whipped cream	12	fluid ounce(s)	360	—	210	4.0	43.0	0	2.5	1.5	—	—
38070	Frappuccino, tall coffee	12	fluid ounce(s)	360	—	190	4.0	38.0	0	2.5	1.5	—	—
39894	Frappuccino, tall coffee, light blend	12	fluid ounce(s)	360	—	110	5.0	22.0	2.0	1.0	0	—	—
38071	Frappuccino, tall espresso	12	fluid ounce(s)	360	—	160	4.0	33.0	0	2.0	1.5	—	—
39897	Frappuccino, tall mocha, light blend	12	fluid ounce(s)	360	—	140	5.0	28.0	3.0	1.5	0	—	—
39887	Frappuccino, tall Strawberries and Creme, w/o whipped cream	12	fluid ounce(s)	360	—	330	10.0	65.0	0	3.5	1.0	—	—
38063	Frappuccino, tall Tazo chai creme w/o whipped cream	12	fluid ounce(s)	360	—	280	10.0	52.0	0	3.5	1.0	—	—
38066	Frappuccino, tall Tazoberry	12	fluid ounce(s)	360	—	140	0.5	36.0	0.5	0	0	0	0
38065	Frappuccino, tall Tazoberry Crème	12	fluid ounce(s)	360	—	240	4.0	54.0	0.5	1.0	0	—	—
38080	Frappuccino, tall vanilla w/o whipped cream	12	fluid ounce(s)	360	—	270	10.0	51.0	0	3.5	1.0	—	—
39898	Frappuccino, tall white chocolate mocha, light blend	12	fluid ounce(s)	360	—	160	6.0	32.0	2.0	2.0	1.0	—	—
38074	Frappuccino, tall white chocolate w/o whipped cream	12	fluid ounce(s)	360	—	240	5.0	48.0	0	3.5	2.5	—	—
39883	Java Chip Frappuccino, tall w/o whipped cream	12	fluid ounce(s)	360	—	270	5.0	51.0	1.0	7.0	4.5	—	—
33111	Latte, tall w/nonfat milk	12	fluid ounce(s)	360	335.3	120	12.0	18.0	0	0	0	0	0
33112	Latte, tall w/whole milk	12	fluid ounce(s)	360	—	200	11.0	16.0	0	11.0	7.0	—	—
33109	Macchiato, tall caramel w/nonfat milk	12	fluid ounce(s)	360	—	170	11.0	30.0	0	1.0	0	—	—
33110	Macchiato, tall caramel w/whole milk	12	fluid ounce(s)	360	—	240	10.0	28.0	0	10.0	6.0	—	—
33107	Mocha coffee drink, tall nonfat, w/o whipped cream	12	fluid ounce(s)	360	—	170	11.0	33.0	1.0	1.5	0	—	—
38089	Mocha syrup	1	serving(s)	17	—	25	1.0	6.0	0	0.5	0	—	—
33108	Mocha, tall mocha w/whole milk	12	fluid ounce(s)	360	—	310	10.0	32.0	1.0	17.0	10.0	—	—
38042	Steamed apple cider, tall	12	fluid ounce(s)	360	—	180	0	45.0	0	0	0	0	0
38087	Tazo chai black tea, soymilk, tall	12	fluid ounce(s)	360	—	190	4.0	39.0	0.5	2.0	0	—	—
38084	Tazo chai black tea, tall	12	fluid ounce(s)	360	—	210	6.0	36.0	0	5.0	3.5	—	—
38083	Tazo chai black tea, tall nonfat	12	fluid ounce(s)	360	—	170	6.0	37.0	0	0	0	0	0
38076	Tazo iced tea, tall	12	fluid ounce(s)	360	—	60	0	16.0	0	0	0	0	0
38077	Tazo tea, grande lemonade	16	fluid ounce(s)	480	—	120	0	31.0	0	0	0	0	0
38045	Vanilla crème steamed nonfat milk, tall w/whipped cream	12	fluid ounce(s)	360	—	260	11.0	33.0	0	8.0	5.0	—	—
38046	Vanilla crème steamed soymilk, tall w/whipped cream	12	fluid ounce(s)	360	—	300	8.0	37.0	1.0	12.0	6.0	—	—
38044	Vanilla crème steamed whole milk, tall w/whipped cream	12	fluid ounce(s)	360	—	330	10.0	31.0	0	18.0	11.0	—	—

PAGE KEY: A-4 = Breads/Baked Goods A-10 = Cereal/Rice/Pasta A-14 = Fruit A-18 = Vegetables/Legumes A-28 = Nuts/Seeds A-30 = Vegetarian A-32 = Dairy A-40 = Eggs A-40 = Seafood A-42 = Meats A-46 = Poultry A-46 = Processed Meats A-48 = Beverages A-52 = Fats/Oils A-54 = Sweets A-56 = Spices/Condiments/Sauces A-60 = Mixed Foods/Soups/Sandwiches A-64 = Fast Food A-84 = Convenience A-86 = Baby Foods

Chol (mg)	Calc (mg)	Iron (mg)	Magn (mg)	Pota (mg)	Sodi (mg)	Zinc (mg)	Vit A (µg)	Thia (mg)	Vit E (mg α)	Ribo (mg)	Niac (mg)	Vit B_6 (mg)	Fola (µg)	Vit C (mg)	Vit B_{12} (µg)	Sele (µg)
10	80	1.80	—	—	280.0	—	—	—	—	—	—	—	—	3.6	—	—
15	150	1.80	—	—	490.0	—	—	—	—	—	—	—	—	9.0	—	—
15	150	2.70	—	—	470.0	—	—	—	—	—	—	—	—	9.0	—	—
35	250	2.70	—	—	980.0	—	—	—	—	—	—	—	—	9.0	—	—
15	150	1.44	—	—	480.0	—	—	—	—	—	—	—	—	9.0	—	—
25	20	0.00	—	—	550.0	—	0	—	—	—	—	—	—	0	—	—
10	0	0.00	—	—	340.0	—	0	—	—	—	—	—	—	0	—	—
25	250	0.00	—	—	95.0	—	—	—	—	—	—	—	—	1.2	0	—
3	200	0.00	—	—	100.0	—	—	—	—	—	—	—	—	0	0	—
0	250	0.72	—	—	75.0	—	—	—	—	—	—	—	—	0	0	—
5	300	0.72	—	—	150.0	—	—	—	—	—	—	—	—	0	0	—
70	350	1.08	—	—	140.0	—	—	—	—	—	—	—	—	2.4	0	—
0	0	0.00	—	—	0	—	0	—	—	—	—	—	—	0	0	—
0	0	0.00	—	—	0	—	0	—	—	—	—	—	—	0	0	—
12	219	1.08	—	530.0	110.0	—	—	—	—	—	—	—	—	0	—	—
15	250	0.36	—	510.0	105.0	—	—	—	—	—	—	—	—	0	—	—
10	150	0.72	—	—	180.0	—	—	—	—	—	—	—	—	0	0	—
10	150	0.00	—	—	180.0	—	—	—	—	—	—	—	—	0	0	—
10	150	0.00	—	—	180.0	—	—	—	—	—	—	—	—	0	0	—
0	150	0.00	—	—	220.0	—	—	—	—	—	—	—	—	0	—	—
10	100	0.00	—	—	160.0	—	—	—	—	—	—	—	—	0	0	—
0	150	0.72	—	—	220.0	—	—	—	—	—	—	—	—	0	—	—
3	350	0.00	—	—	270.0	—	—	—	—	—	—	—	—	21.0	—	—
3	350	0.00	—	—	270.0	—	—	—	—	—	—	—	—	3.6	0	—
0	0	0.00	—	—	30.0	—	0	—	—	—	—	—	—	0	0	—
0	150	0.00	—	—	125.0	—	0	—	—	—	—	—	—	1.2	0	—
3	350	0.00	—	—	370.0	—	—	—	—	—	—	—	—	3.6	0	—
3	150	0.00	—	—	250.0	—	—	—	—	—	—	—	—	0	—	—
10	150	0.00	—	—	210.0	—	—	—	—	—	—	—	—	0	0	—
10	150	1.44	—	—	220.0	—	—	—	—	—	—	—	—	0	—	—
5	350	0.00	39.8	—	170.0	1.35	—	0.12	—	0.47	0.36	0.13	17.5	0	1.3	—
45	400	0.00	46.6	—	160.0	1.28	—	0.12	—	0.54	0.34	0.14	16.8	2.4	1.2	—
5	300	0.00	—	—	160.0	—	—	—	—	—	—	—	—	1.2	—	—
30	300	0.00	—	—	135.0	—	—	—	—	—	—	—	—	2.4	—	—
5	300	2.70	—	—	135.0	—	—	—	—	—	—	—	—	0	—	—
0	0	0.72	—	—	0	—	0	—	—	—	—	—	—	0	0	—
55	300	2.70	—	—	115.0	—	—	—	—	—	—	—	—	0	—	—
0	0	1.08	—	—	15.0	—	0	—	—	—	—	—	—	0	0	—
0	200	0.72	—	—	70.0	—	—	—	—	—	—	—	—	0	0	—
20	200	0.36	—	—	85.0	—	—	—	—	—	—	—	—	1.2	0	—
5	200	0.36	—	—	95.0	—	—	—	—	—	—	—	—	0	0	—
0	0	0.00	—	—	0	—	0	—	—	—	—	—	—	0	0	—
0	0	0.00	—	—	15.0	—	0	—	—	—	—	—	—	4.8	0	—
35	350	0.00	—	—	170.0	—	—	—	—	—	—	—	—	0	0	—
30	400	1.44	—	—	130.0	—	—	—	—	—	—	—	—	0	0	—
65	350	0.00	—	—	140.0	—	—	—	—	—	—	—	—	0	0	—

TABLE A-1 Table of Food Composition *(continued)*

(Computer code is for Cengage Diet Analysis program)
(For purposes of calculations, use "0" for t, <1, <.1, <.01, etc.)

DA+ Code	Food Description	Quantity	Measure	Wt (g)	H_2O (g)	Ener (kcal)	Prot (g)	Carb (g)	Fiber (g)	Fat (g)	Fat Breakdown (g) Sat	Mono	Poly
Fast Food—*continued*													
38090	Whipped cream	1	serving(s)	27	—	100	0	2.0	0	9.0	6.0	—	—
38062	White chocolate mocha, tall nonfat w/o whipped cream	12	fluid ounce(s)	360	—	260	12.0	45.0	0	4.0	3.0	—	—
38061	White chocolate mocha, tall w/ whipped cream	12	fluid ounce(s)	360	—	410	11.0	44.0	0	20.0	13.0	—	—
38048	White hot chocolate, tall nonfat w/o whipped cream	12	fluid ounce(s)	360	—	300	15.0	51.0	0	4.5	3.5	—	—
38050	White hot chocolate, tall soymilk w/whipped cream	12	fluid ounce(s)	360	—	420	11.0	56.0	1.0	16.0	9.0	—	—
38047	White hot chocolate, tall w/ whipped cream	12	fluid ounce(s)	360	—	460	13.0	50.0	0	22.0	15.0	—	—
	Subway												
15842	Cheese steak sandwich, 6", wheat bread	1	item(s)	250	—	360	24.0	47.0	5.0	10.0	4.5	—	—
40478	Chicken and bacon ranch sandwich, 6", white or wheat bread	1	serving(s)	297	—	540	36.0	47.0	5.0	25.0	10.0	—	—
38622	Chicken and bacon ranch wrap with cheese	1	item(s)	257	—	440	41.0	18.0	9.0	27.0	10.0	—	—
32045	Chocolate chip cookie	1	item(s)	45	—	210	2.0	30.0	1.0	10.0	6.0	—	—
32048	Chocolate chip M&M cookie	1	item(s)	45	—	210	2.0	32.0	0.5	10.0	5.0	—	—
32049	Chocolate chunk cookie	1	item(s)	45	—	220	2.0	30.0	0.5	10.0	5.0	—	—
4024	Classic Italian B.M.T. sandwich, 6", white bread	1	item(s)	236	—	440	22.0	45.0	2.0	21.0	8.5	—	—
15838	Classic tuna sandwich, 6", wheat bread	1	item(s)	250	—	530	22.0	45.0	4.0	31.0	7.0	—	—
15837	Classic tuna sandwich, 6", white bread	1	item(s)	243	—	520	21.0	43.0	2.0	31.0	7.5	—	—
16397	Club salad, no dressing and croutons	1	item(s)	412	—	160	18.0	15.0	4.0	4.0	1.5	—	—
3422	Club sandwich, 6", white bread	1	item(s)	250	—	310	23.0	45.0	2.0	6.0	2.5	—	—
4030	Cold cut combo sandwich, 6", white bread	1	item(s)	242	—	400	20.0	45.0	2.0	17.0	7.5	—	—
34030	Ham and egg breakfast sandwich	1	item(s)	142	—	310	16.0	35.0	3.0	13.0	3.5	—	—
3885	Ham sandwich, 6", white bread	1	item(s)	238	—	310	17.0	52.0	2.0	5.0	2.0	—	—
3888	Meatball marinara sandwich, 6", wheat bread	1	item(s)	377	—	560	24.0	63.0	7.0	24.0	11.0	—	—
4651	Meatball sandwich, 6", white bread	1	item(s)	370	—	550	23.0	61.0	5.0	24.0	11.5	—	—
15839	Melt sandwich, 6", white bread	1	item(s)	260	—	410	25.0	47.0	4.0	15.0	5.0	—	—
32046	Oatmeal raisin cookie	1	item(s)	45	—	200	3.0	30.0	1.0	8.0	4.0	—	—
16379	Oven-roasted chicken breast sandwich, 6", wheat bread	1	item(s)	238	—	330	24.0	48.0	5.0	5.0	1.5	—	—
32047	Peanut butter cookie	1	item(s)	45	—	220	4.0	26.0	1.0	12.0	5.0	—	—
4655	Roast beef sandwich, 6", wheat bread	1	item(s)	224	—	290	19.0	45.0	4.0	5.0	2.0	—	—
3957	Roast beef sandwich, 6", white bread	1	item(s)	217	—	280	18.0	43.0	2.0	5.0	2.5	—	—
16378	Roasted chicken breast, 6", white bread	1	item(s)	231	—	320	23.0	46.0	3.0	5.0	2.0	—	—
34028	Southwest steak and cheese sandwich, 6", Italian bread	1	item(s)	271	—	450	24.0	48.0	6.0	20.0	6.0	—	—
4032	Spicy Italian sandwich, 6", white bread	1	item(s)	220	—	470	20.0	43.0	2.0	25.0	9.5	—	—
4031	Steak and cheese sandwich, 6", white bread	1	item(s)	243	—	350	23.0	45.0	3.0	10.0	5.0	—	—
32050	Sugar cookie	1	item(s)	45	—	220	2.0	28.0	0.5	12.0	6.0	—	—
40477	Sweet onion chicken teriyaki sandwich, 6", white or wheat bread	1	serving(s)	281	—	370	26.0	59.0	4.0	5.0	1.5	—	—
38623	Turkey breast and bacon melt wrap with chipotle sauce	1	item(s)	228	—	380	31.0	20.0	9.0	24.0	7.0	—	—
15834	Turkey breast and ham sandwich, 6", white bread	1	item(s)	227	—	280	19.0	45.0	2.0	5.0	2.0	—	—
16376	Turkey breast sandwich, 6", white bread	1	item(s)	217	—	270	17.0	44.0	2.0	4.5	2.0	—	—

PAGE KEY: A-4 = Breads/Baked Goods A-10 = Cereal/Rice/Pasta A-14 = Fruit A-18 = Vegetables/Legumes A-28 = Nuts/Seeds A-30 = Vegetarian A-32 = Dairy A-40 = Eggs A-40 = Seafood A-42 = Meats A-46 = Poultry A-46 = Processed Meats A-48 = Beverages A-52 = Fats/Oils A-54 = Sweets A-56 = Spices/Condiments/Sauces A-60 = Mixed Foods/Soups/Sandwiches A-64 = Fast Food A-84 = Convenience A-86 = Baby Foods

Chol (mg)	Calc (mg)	Iron (mg)	Magn (mg)	Pota (mg)	Sodi (mg)	Zinc (mg)	Vit A (μg)	Thia (mg)	Vit E (mg α)	Ribo (mg)	Niac (mg)	Vit B_6 (mg)	Fola (μg)	Vit C (mg)	Vit B_{12} (μg)	Sele (μg)
40	0	0.00	—	—	10.0	—	—	—	—	—	—	—	—	0	0	—
5	400	0.00	—	—	210.0	—	—	—	—	—	—	—	—	0	0	—
70	400	0.00	—	—	210.0	—	—	—	—	—	—	—	—	2.4	0	—
10	450	0.00	—	—	250.0	—	—	—	—	—	—	—	—	0	0	—
35	500	1.44	—	—	210.0	—	—	—	—	—	—	—	—	0	0	—
75	500	0.00	—	—	250.0	—	—	—	—	—	—	—	—	3.6	0	—
35	150	8.10	—	—	1090.0	—	—	—	—	—	—	—	—	18.0	—	—
90	250	4.50	—	—	1400.0	—	—	—	—	—	—	—	—	21.0	—	—
90	300	2.70	—	—	1680.0	—	—	—	—	—	—	—	—	9.0	—	—
15	0	1.08	—	—	150.0	—	—	—	—	—	—	—	—	0	—	—
10	20	1.00	—	—	100.0	—	—	—	—	—	—	—	—	0	—	—
10	0	1.00	—	—	100.0	—	—	—	—	—	—	—	—	0	—	—
55	150	2.70	—	—	1770.0	—	—	—	—	—	—	—	—	16.8	—	—
45	100	5.40	—	—	1030.0	—	—	—	—	—	—	—	—	21.0	—	—
45	100	3.60	—	—	1010.0	—	—	—	—	—	—	—	—	16.8	—	—
35	60	3.60	—	—	880.0	—	—	—	—	—	—	—	—	30.0	—	—
35	60	3.60	—	—	1290.0	—	—	—	—	—	—	—	—	13.8	—	—
60	150	3.60	—	—	1530.0	—	—	—	—	—	—	—	—	16.8	—	—
190	80	4.50	—	—	720.0	—	66.7	—	—	—	—	—	—	3.6	—	—
25	60	2.70	—	—	1375.0	—	—	—	—	—	—	—	—	13.8	—	—
45	200	7.20	—	—	1610.0	—	—	—	—	—	—	—	—	36.0	—	—
45	200	5.40	—	—	1590.0	—	—	—	—	—	—	—	—	31.8	—	—
45	150	5.40	—	—	1720.0	—	—	—	—	—	—	—	—	24.0	—	—
15	20	1.08	—	—	170.0	—	—	—	—	—	—	—	—	0	—	—
45	60	4.50	—	—	1020.0	—	—	—	—	—	—	—	—	18.0	—	—
15	20	0.72	—	—	200.0	—	—	—	—	—	—	—	—	0	—	—
20	60	6.30	—	—	920.0	—	—	—	—	—	—	—	—	18.0	—	—
20	60	4.50	—	—	900.0	—	—	—	—	—	—	—	—	13.8	—	—
45	60	2.70	—	—	1000.0	—	—	—	—	—	—	—	—	13.8	—	—
45	150	8.10	—	—	1310.0	—	—	—	—	—	—	—	—	21.0	—	—
55	60	2.70	—	—	1650.0	—	—	—	—	—	—	—	—	16.8	—	—
35	150	6.30	—	—	1070.0	—	—	—	—	—	—	—	—	13.8	—	—
15	0	0.72	—	—	140.0	—	—	—	—	—	—	—	—	0	—	—
50	80	4.50	—	—	1220.0	—	—	—	—	—	—	—	—	24.0	—	—
50	200	2.70	—	—	1780.0	—	—	—	—	—	—	—	—	6.0	—	—
25	60	2.70	—	—	1210.0	—	—	—	—	—	—	—	—	13.8	—	—
20	60	2.70	—	—	1000.0	—	—	—	—	—	—	—	—	13.8	—	—

TABLE A-1 Table of Food Composition *(continued)*

(Computer code is for Cengage Diet Analysis program)
(For purposes of calculations, use "0" for t, <1, <.1, <.01, etc.)

DA+ Code	Food Description	Quantity	Measure	Wt (g)	H_2O (g)	Ener (kcal)	Prot (g)	Carb (g)	Fiber (g)	Fat (g)	Fat Breakdown (g) Sat	Mono	Poly
Fast Food—*continued*													
15841	Veggie Delite sandwich, 6", wheat bread	1	item(s)	167	—	230	9.0	44.0	4.0	3.0	1.0	—	—
16375	Veggie Delite, 6", white bread	1	item(s)	160	—	220	8.0	42.0	2.0	3.0	1.5	—	—
32051	White chip macadamia nut cookie	1	item(s)	45	—	220	2.0	29.0	0.5	11.0	5.0	—	—
	Taco Bell												
29906	7-Layer burrito	1	item(s)	283	—	490	17.0	65.0	9.0	18.0	7.0	—	—
744	Bean burrito	1	item(s)	198	—	340	13.0	54.0	8.0	9.0	3.5	—	—
749	Beef burrito supreme	1	item(s)	248	—	410	17.0	51.0	7.0	17.0	8.0	—	—
33417	Beef Chalupa Supreme	1	item(s)	153	—	380	14.0	30.0	3.0	23.0	7.0	—	—
34474	Beef Gordita Baja	1	item(s)	153	—	340	13.0	29.0	4.0	19.0	5.0	—	—
29910	Beef Gordita Supreme	1	item(s)	153	—	310	14.0	29.0	3.0	16.0	6.0	—	—
2014	Beef soft taco	1	item(s)	99	—	200	10.0	21.0	3.0	9.0	4.0	—	—
10860	Beef soft taco supreme	1	item(s)	135	—	250	11.0	23.0	3.0	13.0	6.0	—	—
34472	Chicken burrito supreme	1	item(s)	248	—	390	20.0	49.0	6.0	13.0	6.0	—	—
33418	Chicken Chalupa Supreme	1	item(s)	153	—	360	17.0	29.0	2.0	20.0	5.0	—	—
34475	Chicken Gordita Baja	1	item(s)	153	—	320	17.0	28.0	3.0	16.0	3.5	—	—
29909	Chicken quesadilla	1	item(s)	184	—	520	28.0	40.0	3.0	28.0	12.0	—	—
29907	Chili cheese burrito	1	item(s)	156	—	390	16.0	40.0	3.0	18.0	9.0	—	—
10794	Cinnamon twists	1	serving(s)	35	—	170	1.0	26.0	1.0	7.0	0	—	—
29911	Grilled chicken Gordita Supreme	1	item(s)	153	—	290	17.0	28.0	2.0	12.0	5.0	—	—
14463	Grilled chicken soft taco	1	item(s)	99	—	190	14.0	19.0	1.0	6.0	2.5	—	—
29912	Grilled Steak Gordita Supreme	1	item(s)	153	—	290	15.0	28.0	2.0	13.0	5.0	—	—
29904	Grilled steak soft taco	1	item(s)	128	—	270	12.0	20.0	2.0	16.0	4.5	—	—
29905	Grilled steak soft taco supreme	1	item(s)	135	—	235	13.0	21.0	1.0	11.0	6.0	—	—
2021	Mexican pizza	1	serving(s)	216	—	530	20.0	42.0	7.0	30.0	8.0	—	—
29894	Mexican rice	1	serving(s)	131	—	170	6.0	23.0	1.0	11.0	3.0	—	—
10772	Meximelt	1	serving(s)	128	—	280	15.0	22.0	3.0	14.0	7.0	—	—
2011	Nachos	1	serving(s)	99	—	330	4.0	32.0	2.0	21.0	3.5	—	—
2012	Nachos Bellgrande	1	serving(s)	308	—	770	19.0	77.0	12.0	44.0	9.0	—	—
2023	Pintos 'n cheese	1	serving(s)	128	—	150	9.0	19.0	7.0	6.0	3.0	—	—
34473	Steak burrito supreme	1	item(s)	248	—	380	18.0	49.0	6.0	14.0	7.0	—	—
33419	Steak Chalupa Supreme	1	item(s)	153	—	360	15.0	28.0	2.0	21.0	6.0	—	—
747	Taco	1	item(s)	78	—	170	8.0	13.0	3.0	10.0	3.5	—	—
2015	Taco salad with salsa, with shell	1	serving(s)	548	—	840	30.0	80.0	15.0	45.0	11.0	—	—
14459	Taco supreme	1	item(s)	113	—	210	9.0	15.0	3.0	13.0	6.0	—	—
748	Tostada	1	item(s)	170	—	230	11.0	27.0	7.0	10.0	3.5	—	—
Convenience Meals													
	Banquet												
29961	Barbeque chicken meal	1	item(s)	281	—	330	16.0	37.0	2.0	13.0	3.0	—	—
14788	Boneless white fried chicken meal	1	item(s)	286	—	310	10.0	21.0	4.0	20.0	5.0	—	—
29960	Fish sticks meal	1	item(s)	207	—	470	13.0	58.0	1.0	20.0	3.5	—	—
29957	Lasagna with meat sauce meal	1	item(s)	312	—	320	15.0	46.0	7.0	9.0	4.0	—	—
14777	Macaroni and cheese meal	1	item(s)	340	—	420	15.0	57.0	5.0	14.0	8.0	—	—
1741	Meatloaf meal	1	item(s)	269	—	240	14.0	20.0	4.0	11.0	4.0	—	—
39418	Pepperoni pizza meal	1	item(s)	191	—	480	11.0	56.0	5.0	23.0	8.0	—	—
33759	Roasted white turkey meal	1	item(s)	255	—	230	14.0	30.0	5.0	6.0	2.0	—	—
1743	Salisbury steak meal	1	item(s)	269	196.9	380	12.0	28.0	3.0	24.0	12.0	—	—
	Budget Gourmet												
1914	Cheese manicotti with meat sauce entrée	1	item(s)	284	194.0	420	18.0	38.0	4.0	22.0	11.0	6.0	1.3
1915	Chicken with fettucini entrée	1	item(s)	284	—	380	20.0	33.0	3.0	19.0	10.0	—	—
3986	Light beef stroganoff entrée	1	item(s)	248	177.0	290	20.0	32.0	3.0	7.0	4.0	—	—
3996	Light sirloin of beef in herb sauce entrée	1	item(s)	269	214.0	260	19.0	30.0	5.0	7.0	4.0	2.3	0.3
3987	Light vegetable lasagna entrée	1	item(s)	298	227.0	290	15.0	36.0	4.8	9.0	1.8	0.9	0.6
	Healthy Choice												
9425	Cheese French bread pizza	1	item(s)	170	—	340	22.0	51.0	5.0	5.0	1.5	—	—
9306	Chicken enchilada suprema meal	1	item(s)	320	251.5	360	13.0	59.0	8.0	7.0	3.0	2.0	2.0
3821	Familiar Favorites lasagna bake with meat sauce entrée	1	item(s)	255	—	270	13.0	38.0	4.0	7.0	2.5	—	—

PAGE KEY: A-4 = Breads/Baked Goods A-10 = Cereal/Rice/Pasta A-14 = Fruit A-18 = Vegetables/Legumes A-28 = Nuts/Seeds A-30 = Vegetarian A-32 = Dairy A-40 = Eggs A-40 = Seafood A-42 = Meats A-46 = Poultry A-46 = Processed Meats A-48 = Beverages A-52 = Fats/Oils A-54 = Sweets A-56 = Spices/Condiments/Sauces A-60 = Mixed Foods/Soups/Sandwiches A-64 = Fast Food A-84 = Convenience A-86 = Baby Foods

Chol (mg)	Calc (mg)	Iron (mg)	Magn (mg)	Pota (mg)	Sodi (mg)	Zinc (mg)	Vit A (µg)	Thia (mg)	Vit E (mg α)	Ribo (mg)	Niac (mg)	Vit B_6 (mg)	Fola (µg)	Vit C (mg)	Vit B_{12} (µg)	Sele (µg)
0	60	4.50	—	—	520.0	—	—	—	—	—	—	—	—	18.0	—	—
0	60	2.70	—	—	500.0	—	—	—	—	—	—	—	—	13.8	—	—
15	20	0.72	—	—	160.0	—	—	—	—	—	—	—	—	0	—	—
25	250	5.40	—	—	1350.0	—	—	—	—	—	—	—	—	15.0	—	—
5	200	4.50	—	—	1190.0	—	5.9	—	—	—	—	—	—	4.8	—	—
40	200	4.50	—	—	1340.0	—	9.9	—	—	—	—	—	—	6.0	—	—
40	150	2.70	—	—	620.0	—	—	—	—	—	—	—	—	3.6	—	—
35	100	2.70	—	—	780.0	—	—	—	—	—	—	—	—	2.4	—	—
40	150	2.70	—	—	620.0	—	—	—	—	—	—	—	—	3.6	—	—
25	100	1.80	—	—	630.0	—	—	—	—	—	—	—	—	1.2	—	—
40	150	2.70	—	—	650.0	—	—	—	—	—	—	—	—	3.6	—	—
45	200	4.50	—	—	1360.0	—	—	—	—	—	—	—	—	9.0	—	—
45	100	2.70	—	—	650.0	—	—	—	—	—	—	—	—	4.8	—	—
40	100	1.80	—	—	800.0	—	—	—	—	—	—	—	—	3.6	—	—
75	450	3.60	—	—	1420.0	—	—	—	—	—	—	—	—	1.2	—	—
40	300	1.80	—	—	1080.0	—	—	—	—	—	—	—	—	0	—	—
0	0	0.37	—	—	200.0	—	0	—	—	—	—	—	—	0	—	—
45	150	1.80	—	—	650.0	—	—	—	—	—	—	—	—	4.8	—	—
30	100	1.08	—	—	550.0	—	14.6	—	—	—	—	—	—	1.2	—	—
40	100	2.70	—	—	530.0	—	—	—	—	—	—	—	—	3.6	—	—
35	100	2.70	—	—	660.0	—	—	—	—	—	—	—	—	3.6	—	—
35	120	1.44	—	—	565.0	—	29.2	—	—	—	—	—	—	3.6	—	—
40	350	3.60	—	—	1000.0	—	—	—	—	—	—	—	—	4.8	—	—
15	100	1.44	—	—	790.0	—	—	—	—	—	—	—	—	3.6	—	—
40	250	2.70	—	—	880.0	—	—	—	—	—	—	—	—	2.4	—	—
3	80	0.71	—	—	530.0	—	0	—	—	—	—	—	—	0	—	—
35	200	3.60	—	—	1280.0	—	—	—	—	—	—	—	—	4.8	—	—
15	150	1.44	—	—	670.0	—	—	—	—	—	—	—	—	3.6	—	—
35	200	4.50	—	—	1250.0	—	9.9	—	—	—	—	—	—	9.0	—	—
40	100	2.70	—	—	530.0	—	—	—	—	—	—	—	—	3.6	—	—
25	80	1.08	—	—	350.0	—	—	—	—	—	—	—	—	1.2	—	—
65	450	7.20	—	—	1780.0	—	—	—	—	—	—	—	—	12.0	—	—
40	100	1.08	—	—	370.0	—	—	—	—	—	—	—	—	3.6	—	—
15	200	1.80	—	—	730.0	—	—	—	—	—	—	—	—	4.8	—	—
																—
50	40	1.08	—	—	1210.0	—	0	—	—	—	—	—	—	4.8	—	—
45	80	1.44	—	—	1200.0	—	—	—	—	—	—	—	—	18.0	—	—
55	20	1.44	—	—	710.0	—	—	—	—	—	—	—	—	0	—	—
20	100	2.70	—	—	1170.0	—	—	—	—	—	—	—	—	0	—	—
20	150	1.44	—	—	1330.0	—	0	—	—	—	—	—	—	0	—	—
30	0	1.80	—	—	1040.0	—	0	—	—	—	—	—	—	0	—	—
35	150	1.80	—	—	870.0	—	0	—	—	—	—	—	—	0	—	—
25	60	1.80	—	—	1070.0	—	—	—	—	—	—	—	—	3.6	—	—
60	40	1.44	—	—	1140.0	—	0	—	—	—	—	—	—	0	—	—
85	300	2.70	45.4	484.0	810.0	2.29	—	0.45	—	0.51	4.00	0.22	30.7	0	0.7	—
85	100	2.70	—	—	810.0	—	—	0.15	—	0.42	6.00	—	—	0	—	—
35	40	1.80	38.9	280.0	580.0	4.71	—	0.17	—	0.36	4.28	0.27	18.9	2.4	2.5	—
30	40	1.80	57.7	540.0	850.0	4.81	—	0.15	—	0.29	5.53	0.37	38.4	6.0	1.6	—
15	283	3.03	78.5	420.0	780.0	1.39	—	0.22	—	0.45	3.13	0.32	74.8	59.1	0.2	—
10	350	3.60	—	—	600.0	—	—	—	—	—	—	—	—	0	—	—
30	40	1.44	—	—	580.0	—	—	—	—	—	—	—	—	3.6	—	—
20	100	1.80	—	—	600.0	—	—	—	—	—	—	—	—	0	—	—

TABLE A-1 Table of Food Composition *(continued)*

(Computer code is for Cengage Diet Analysis program)
(For purposes of calculations, use "0" for t, <1, <.1, <.01, etc.)

DA+ Code	Food Description	Quantity	Measure	Wt (g)	H_2O (g)	Ener (kcal)	Prot (g)	Carb (g)	Fiber (g)	Fat (g)	Fat Breakdown (g) Sat	Mono	Poly
Convenience Meals—*continued*													
13744	Familiar Favorites sesame chicken with vegetables and rice entrée	1	item(s)	255	—	260	17.0	34.0	4.0	6.0	2.0	2.0	2.0
9316	Lemon pepper fish meal	1	item(s)	303	—	280	11.0	49.0	5.0	5.0	2.0	1.0	2.0
9322	Traditional salisbury steak meal	1	item(s)	354	250.3	360	23.0	45.0	5.0	9.0	3.5	4.0	1.0
9359	Traditional turkey breasts meal	1	item(s)	298	—	330	21.0	50.0	4.0	5.0	2.0	1.5	1.5
	Stouffers												
2313	Cheese French bread pizza	1	serving(s)	294	—	380	15.0	43.0	3.0	16.0	6.0	—	—
11138	Cheese manicotti with tomato sauce entrée	1	item(s)	255	—	360	18.0	41.0	2.0	14.0	6.0	—	—
2366	Chicken pot pie entrée	1	item(s)	284	—	740	23.0	56.0	4.0	47.0	18.0	12.4	10.5
11116	Homestyle baked chicken breast with mashed potatoes and gravy entrée	1	item(s)	252	—	270	21.0	21.0	2.0	11.0	3.5	—	—
11146	Homestyle beef pot roast and potatoes entrée	1	item(s)	252	—	260	16.0	24.0	3.0	11.0	4.0	—	—
11152	Homestyle roast turkey breast with stuffing and mashed potatoes entrée	1	item(s)	273	—	290	16.0	30.0	2.0	12.0	3.5	—	—
11043	Lean Cuisine Comfort Classics baked chicken and whipped potatoes and stuffing entrée	1	item(s)	245	—	240	15.0	34.0	3.0	4.5	1.0	2.0	1.0
11046	Lean Cuisine Comfort Classics honey mustard chicken with rice pilaf entrée	1	item(s)	227	—	250	17.0	37.0	1.0	4.0	1.0	1.0	1.0
9479	Lean Cuisine Deluxe French bread pizza	1	item(s)	174	—	310	16.0	44.0	3.0	9.0	3.5	0.5	0.5
360	Lean Cuisine One Dish Favorites chicken chow mein with rice	1	item(s)	255	—	190	13.0	29.0	2.0	2.5	0.5	1.0	0.5
11054	Lean Cuisine One Dish Favorites chicken enchilada Suiza with Mexican-style rice	1	serving(s)	255	—	270	10.0	47.0	3.0	4.5	2.0	1.5	1.0
9467	Lean Cuisine One Dish Favorites fettucini alfredo entrée	1	item(s)	262	—	270	13.0	39.0	2.0	7.0	3.5	2.0	1.0
11055	Lean Cuisine One Dish Favorites lasagna with meat sauce entrée	1	item(s)	298	—	320	19.0	44.0	4.0	7.0	3.0	2.0	0.5
	Weight Watchers												
11164	Smart Ones chicken enchiladas suiza entrée	1	item(s)	255	—	340	12.0	38.0	3.0	10.0	4.5	—	—
39763	Smart Ones chicken oriental entrée	1	item(s)	255	—	230	15.0	34.0	3.0	4.5	1.0	—	—
11187	Smart Ones pepperoni pizza	1	item(s)	198	—	400	22.0	58.0	4.0	9.0	3.0	—	—
39765	Smart Ones spaghetti bolognese entrée	1	item(s)	326	—	280	17.0	43.0	5.0	5.0	2.0	—	—
31512	Smart Ones spicy szechuan style vegetables and chicken	1	item(s)	255	—	220	11.0	34.0	4.0	5.0	1.0	—	—
Baby Foods													
787	Apple juice	4	fluid ounce(s)	127	111.6	60	0	14.8	0.1	0.1	0	0	0
778	Applesauce, strained	4	tablespoon(s)	64	56.7	26	0.1	6.9	1.1	0.1	0	0	0
779	Bananas with tapioca, strained	4	tablespoon(s)	60	50.4	34	0.2	9.2	1.0	0	0	0	0
604	Carrots, strained	4	tablespoon(s)	56	51.7	15	0.4	3.4	1.0	0.1	0	0	0
770	Chicken noodle dinner, strained	4	tablespoon(s)	64	54.8	42	1.7	5.8	1.3	1.3	0.4	0.5	0.3
801	Green beans, strained	4	tablespoon(s)	60	55.1	16	0.7	3.8	1.3	0.1	0	0	0
910	Human milk, mature	2	fluid ounce(s)	62	53.9	43	0.6	4.2	0	2.7	1.2	1.0	0.3
760	Mixed cereal, prepared with whole milk	4	ounce(s)	113	84.6	128	5.4	18.0	1.5	4.0	2.2	1.2	0.4
772	Mixed vegetable dinner, strained	2	ounce(s)	57	50.3	23	0.7	5.4	0.8	0	—	—	0
762	Rice cereal, prepared with whole milk	4	ounce(s)	113	84.6	130	4.4	18.9	0.1	4.1	2.6	1.0	0.2
758	Teething biscuits	1	item(s)	11	0.7	44	1.0	8.6	0.2	0.6	0.2	0.2	0.1

PAGE KEY: A-4 = Breads/Baked Goods A-10 = Cereal/Rice/Pasta A-14 = Fruit A-18 = Vegetables/Legumes A-28 = Nuts/Seeds A-30 = Vegetarian A-32 = Dairy A-40 = Eggs A-40 = Seafood A-42 = Meats A-46 = Poultry A-46 = Processed Meats A-48 = Beverages A-52 = Fats/Oils A-54 = Sweets A-56 = Spices/Condiments/Sauces A-60 = Mixed Foods/Soups/Sandwiches A-64 = Fast Food A-84 = Convenience A-86 = Baby Foods

Chol (mg)	Calc (mg)	Iron (mg)	Magn (mg)	Pota (mg)	Sodi (mg)	Zinc (mg)	Vit A (µg)	Thia (mg)	Vit E (mg α)	Ribo (mg)	Niac (mg)	Vit B_6 (mg)	Fola (µg)	Vit C (mg)	Vit B_{12} (µg)	Sele (µg)
35	18	0.72	—	—	580.0	—	—	—	—	—	—	—	—	12.0	—	—
35	20	0.36	—	—	580.0	—	—	—	—	—	—	—	—	30.0	—	—
45	80	2.70	—	—	580.0	—	—	—	—	—	—	—	—	21.0	—	—
35	40	1.80	—	—	600.0	—	—	—	—	—	—	—	—	0	—	—
30	200	1.80	—	230.0	660.0	—	—	—	—	—	—	—	—	2.4	—	—
70	250	1.44	—	550.0	920.0	—	—	—	—	—	—	—	—	6.0	—	—
65	150	2.70	—	—	1170.0	—	—	—	—	—	—	—	—	2.4	—	—
55	20	0.72	—	490.0	770.0	—	0	—	—	—	—	—	—	0	—	—
35	20	1.80	—	800.0	960.0	—	—	—	—	—	—	—	—	6.0	—	—
45	40	1.08	—	490.0	970.0	—	—	—	—	—	—	—	—	3.6	—	—
25	40	1.16	—	500.0	650.0	—	—	—	—	—	—	—	—	3.6	—	—
30	64	0.38	—	370.0	650.0	—	—	—	—	—	—	—	—	0	—	—
20	150	2.70	—	300.0	700.0	—	—	—	—	—	—	—	—	15.0	—	—
25	40	0.72	—	380.0	650.0	—	—	—	—	—	—	—	—	2.4	—	—
20	150	0.72	—	350.0	510.0	—	—	—	—	—	—	—	—	2.4	—	—
15	200	0.72	—	290.0	690.0	—	0	—	—	—	—	—	—	0	—	—
30	250	1.47	—	610.0	690.0	—	—	—	—	—	—	—	—	2.4	—	—
40	200	0.72	—	—	800.0	—	—	—	—	—	—	—	—	2.4	—	—
35	40	0.72	—	—	790.0	—	—	—	—	—	—	—	—	6.0	—	—
15	200	1.08	—	401.0	700.0	—	69.1	—	—	—	—	—	—	4.8	—	—
15	150	3.60	—	—	670.0	—	—	—	—	—	—	—	—	9.0	—	—
10	40	1.44	—	—	890.0	—	—	—	—	—	—	—	—	0	—	—
0	5	0.72	3.8	115.4	3.8	0.03	1.3	0.01	0.76	0.02	0.10	0.03	0	73.4	0	0.1
0	3	0.12	1.9	45.4	1.3	0.01	0.6	0.01	0.36	0.02	0.04	0.02	1.3	24.5	0	0.2
0	3	0.12	6.0	52.8	5.4	0.04	1.2	0.01	0.36	0.02	0.08	0.04	3.6	10.0	0	0.4
0	12	0.20	5.0	109.8	20.7	0.08	320.9	0.01	0.29	0.02	0.25	0.04	8.4	3.2	0	0.1
10	17	0.40	9.0	89.0	14.7	0.32	70.4	0.03	0.12	0.04	0.44	0.04	8.3	0	0	2.4
0	23	0.40	12.0	87.6	3.0	0.12	10.8	0.02	0.04	0.04	0.20	0.02	14.4	0.2	0	0
9	20	0.02	1.8	31.4	10.5	0.10	37.6	0.01	0.04	0.02	0.10	0.01	3.1	3.1	0	1.1
12	249	11.82	30.6	225.7	53.3	0.80	28.4	0.49	—	0.65	6.54	0.07	10.2	1.4	0.3	
—	12	0.18	6.2	68.6	4.5	0.08	77.1	0.01	—	0.02	0.28	0.04	4.5	1.6	0	0.4
12	271	13.82	51.0	215.5	52.2	0.72	24.9	0.52	—	0.56	5.90	0.12	6.8	1.4	0.3	4.0
0	11	0.39	3.9	35.5	28.4	0.10	3.1	0.02	0.02	0.05	0.47	0.01	7.6	1.0	0	2.6

appendix B WHO: Nutrition Recommendations Canada: Guidelines and Meal Planning

CONTENTS

This appendix presents nutrition recommendations from the World Health Organization (WHO) and details for Canadians on the *Eating Well with Canada's Food Guide* and the *Beyond the Basics* meal-planning system.

Nutrition Recommendations from WHO

The World Health Organization (WHO) has assessed the relationships between diet and the development of chronic diseases. Its recommendations include:

- Energy: sufficient to support growth, physical activity, and a healthy body weight (BMI between 18.5 and 24.9) and to avoid weight gain greater than 11 pounds (5 kilograms) during adult life
- Total fat: 15 to 30 percent of total energy
- Saturated fatty acids: <10 percent of total energy
- Polyunsaturated fatty acids: 6 to 10 percent of total energy
- Omega-6 polyunsaturated fatty acids: 5 to 8 percent of total energy
- Omega-3 polyunsaturated fatty acids: 1 to 2 percent of total energy
- *Trans*-fatty acids: <1 percent of total energy
- Total carbohydrate: 55 to 75 percent of total energy
- Sugars: <10 percent of total energy
- Protein: 10 to 15 percent of total energy
- Cholesterol: <300 mg per day
- Salt (sodium): <5 g salt per day (<2 g sodium per day), appropriately iodized
- Fruits and vegetables: ≥400 g per day (about 1 pound)
- Total dietary fiber: >25 g per day from foods
- Physical activity: one hour of moderate-intensity activity, such as walking, on most days of the week

Eating Well with Canada's Food Guide

Figure B-1 presents the 2007 *Eating Well with Canada's Food Guide*. Additional publications, which are available from Health Canada (■) through its website, provide many more details.

■ Search for "Canada's food guide" at Health Canada: www.hc-sc.gc.ca

FIGURE B-1 Eating Well with Canada's Food Guide

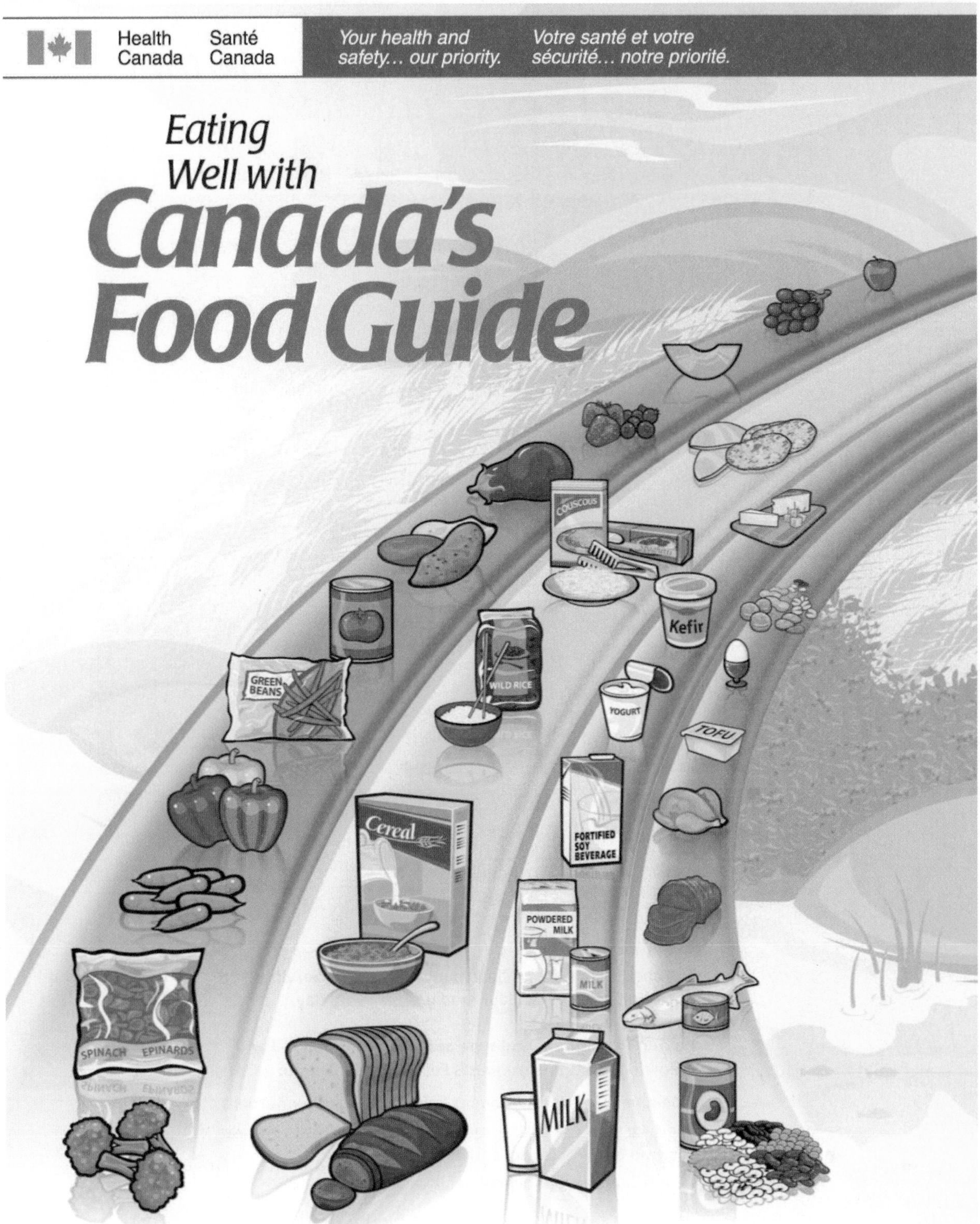

Health Canada Santé Canada

Your health and safety... our priority. Votre santé et votre sécurité... notre priorité.

Eating Well with Canada's Food Guide

COUSCOUS

Kefir

GREEN BEANS

WILD RICE

YOGURT

TOFU

Cereal

FORTIFIED SOY BEVERAGE

POWDERED MILK

MILK

SPINACH EPINARDS

MILK

Canada

FIGURE B-1 Eating Well with Canada's Food Guide (*Continued*)

Recommended Number of Food Guide Servings per Day

	Children			Teens		Adults			
Age in Years	2-3	4-8	9-13	14-18		19-50		51+	
Sex	Girls and Boys			Females	Males	Females	Males	Females	Males
Vegetables and Fruit	4	5	6	7	8	7-8	8-10	7	7
Grain Products	3	4	6	6	7	6-7	8	6	7
Milk and Alternatives	2	2	3-4	3-4	3-4	2	2	3	3
Meat and Alternatives	1	1	1-2	2	3	2	3	2	3

The chart above shows how many Food Guide Servings you need from each of the four food groups every day.

Having the amount and type of food recommended and following the tips in *Canada's Food Guide* will help:

- Meet your needs for vitamins, minerals and other nutrients.
- Reduce your risk of obesity, type 2 diabetes, heart disease, certain types of cancer and osteoporosis.
- Contribute to your overall health and vitality.

FIGURE B-1 Eating Well with Canada's Food Guide (*Continued*)

What is One Food Guide Serving?
Look at the examples below.

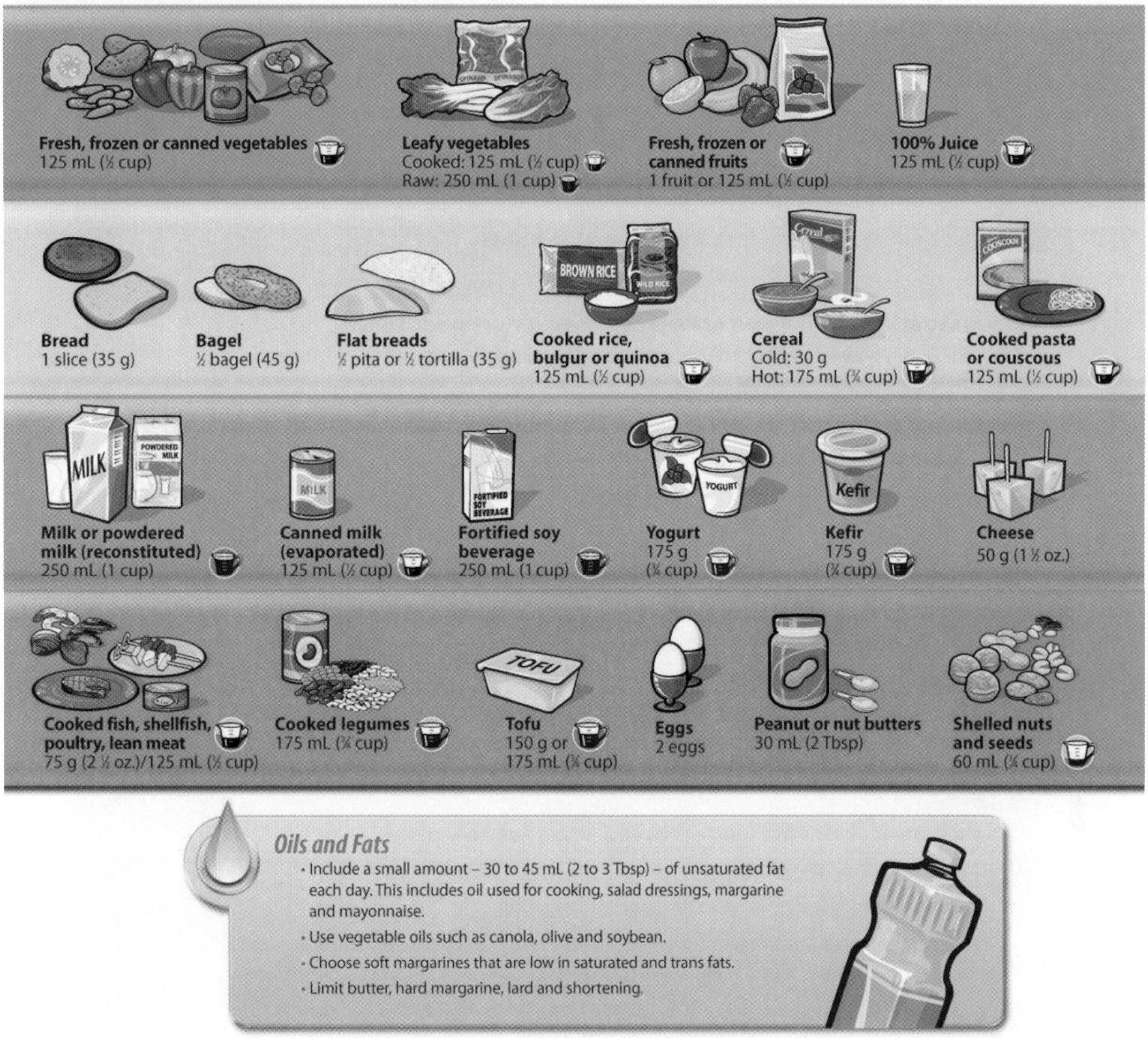

Fresh, frozen or canned vegetables
125 mL (½ cup)

Leafy vegetables
Cooked: 125 mL (½ cup)
Raw: 250 mL (1 cup)

Fresh, frozen or canned fruits
1 fruit or 125 mL (½ cup)

100% Juice
125 mL (½ cup)

Bread
1 slice (35 g)

Bagel
½ bagel (45 g)

Flat breads
½ pita or ½ tortilla (35 g)

Cooked rice, bulgur or quinoa
125 mL (½ cup)

Cereal
Cold: 30 g
Hot: 175 mL (¾ cup)

Cooked pasta or couscous
125 mL (½ cup)

Milk or powdered milk (reconstituted)
250 mL (1 cup)

Canned milk (evaporated)
125 mL (½ cup)

Fortified soy beverage
250 mL (1 cup)

Yogurt
175 g
(¾ cup)

Kefir
175 g
(¾ cup)

Cheese
50 g (1 ½ oz.)

Cooked fish, shellfish, poultry, lean meat
75 g (2 ½ oz.)/125 mL (½ cup)

Cooked legumes
175 mL (¾ cup)

Tofu
150 g or
175 mL (¾ cup)

Eggs
2 eggs

Peanut or nut butters
30 mL (2 Tbsp)

Shelled nuts and seeds
60 mL (¼ cup)

Oils and Fats

- Include a small amount – 30 to 45 mL (2 to 3 Tbsp) – of unsaturated fat each day. This includes oil used for cooking, salad dressings, margarine and mayonnaise.
- Use vegetable oils such as canola, olive and soybean.
- Choose soft margarines that are low in saturated and trans fats.
- Limit butter, hard margarine, lard and shortening.

FIGURE B-1 Eating Well with Canada's Food Guide (*Continued*)

Make each Food Guide Serving count...
wherever you are – at home, at school, at work or when eating out!

- **Eat at least one dark green and one orange vegetable each day.**
  - Go for dark green vegetables such as broccoli, romaine lettuce and spinach.
  - Go for orange vegetables such as carrots, sweet potatoes and winter squash.
- **Choose vegetables and fruit prepared with little or no added fat, sugar or salt.**
  - Enjoy vegetables steamed, baked or stir-fried instead of deep-fried.
- **Have vegetables and fruit more often than juice.**

- **Make at least half of your grain products whole grain each day.**
  - Eat a variety of whole grains such as barley, brown rice, oats, quinoa and wild rice.
  - Enjoy whole grain breads, oatmeal or whole wheat pasta.
- **Choose grain products that are lower in fat, sugar or salt.**
  - Compare the Nutrition Facts table on labels to make wise choices.
  - Enjoy the true taste of grain products. When adding sauces or spreads, use small amounts.

- **Drink skim, 1%, or 2% milk each day.**
  - Have 500 mL (2 cups) of milk every day for adequate vitamin D.
  - Drink fortified soy beverages if you do not drink milk.
- **Select lower fat milk alternatives.**
  - Compare the Nutrition Facts table on yogurts or cheeses to make wise choices.

- **Have meat alternatives such as beans, lentils and tofu often.**
- **Eat at least two Food Guide Servings of fish each week.***
  - Choose fish such as char, herring, mackerel, salmon, sardines and trout.
- **Select lean meat and alternatives prepared with little or no added fat or salt.**
  - Trim the visible fat from meats. Remove the skin on poultry.
  - Use cooking methods such as roasting, baking or poaching that require little or no added fat.
  - If you eat luncheon meats, sausages or prepackaged meats, choose those lower in salt (sodium) and fat.

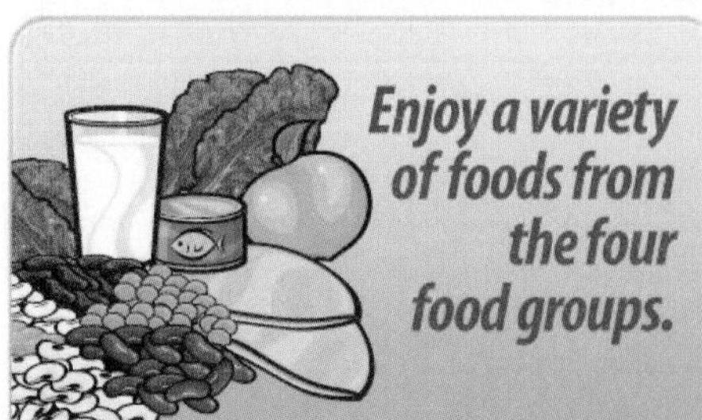

Enjoy a variety of foods from the four food groups.

Satisfy your thirst with water!

Drink water regularly. It's a calorie-free way to quench your thirst. Drink more water in hot weather or when you are very active.

* Health Canada provides advice for limiting exposure to mercury from certain types of fish. Refer to www.hc-sc.gc.ca for the latest information.

FIGURE B-1 Eating Well with Canada's Food Guide (*Continued*)

Advice for different ages and stages...

Children

Following *Canada's Food Guide* helps children grow and thrive.

Young children have small appetites and need calories for growth and development.

- Serve small nutritious meals and snacks each day.
- Do not restrict nutritious foods because of their fat content. Offer a variety of foods from the four food groups.
- Most of all... be a good role model.

Women of childbearing age

All women who could become pregnant and those who are pregnant or breastfeeding need a multivitamin containing **folic acid** every day. Pregnant women need to ensure that their multivitamin also contains **iron**. A health care professional can help you find the multivitamin that's right for you.

Pregnant and breastfeeding women need more calories. Include an extra 2 to 3 Food Guide Servings each day.

Here are two examples:

- Have fruit and yogurt for a snack, or
- Have an extra slice of toast at breakfast and an extra glass of milk at supper.

Men and women over 50

The need for **vitamin D** increases after the age of 50.

In addition to following *Canada's Food Guide*, everyone over the age of 50 should take a daily vitamin D supplement of 10 μg (400 IU).

How do I count Food Guide Servings in a meal?

Here is an example:

Vegetable and beef stir-fry with rice, a glass of milk and an apple for dessert		
250 mL (1 cup) mixed broccoli, carrot and sweet red pepper	=	2 **Vegetables and Fruit** Food Guide Servings
75 g (2 ½ oz.) lean beef	=	1 **Meat and Alternatives** Food Guide Serving
250 mL (1 cup) brown rice	=	2 **Grain Products** Food Guide Servings
5 mL (1 tsp) canola oil	=	part of your **Oils and Fats** intake for the day
250 mL (1 cup) 1% milk	=	1 **Milk and Alternatives** Food Guide Serving
1 apple	=	1 **Vegetables and Fruit** Food Guide Serving

FIGURE B-1 Eating Well with Canada's Food Guide (*Continued*)

Eat well and be active today and every day!

The benefits of eating well and being active include:

- Better overall health.
- Lower risk of disease.
- A healthy body weight.
- Feeling and looking better.
- More energy.
- Stronger muscles and bones.

Be active

To be active every day is a step towards better health and a healthy body weight.

Canada's Physical Activity Guide recommends building 30 to 60 minutes of moderate physical activity into daily life for adults and at least 90 minutes a day for children and youth. You don't have to do it all at once. Add it up in periods of at least 10 minutes at a time for adults and five minutes at a time for children and youth.

Start slowly and build up.

Eat well

Another important step towards better health and a healthy body weight is to follow *Canada's Food Guide* by:

- Eating the recommended amount and type of food each day.
- Limiting foods and beverages high in calories, fat, sugar or salt (sodium) such as cakes and pastries, chocolate and candies, cookies and granola bars, doughnuts and muffins, ice cream and frozen desserts, french fries, potato chips, nachos and other salty snacks, alcohol, fruit flavoured drinks, soft drinks, sports and energy drinks, and sweetened hot or cold drinks.

Read the label

- Compare the Nutrition Facts table on food labels to choose products that contain less fat, saturated fat, trans fat, sugar and sodium.
- Keep in mind that the calories and nutrients listed are for the amount of food found at the top of the Nutrition Facts table.

Nutrition Facts
Per 0 mL (0 g)

Amount	% Daily Value
Calories 0	
Fat 0 g	0 %
Saturates 0 g + Trans 0 g	0 %
Cholesterol 0 mg	
Sodium 0 mg	0 %
Carbohydrate 0 g	0 %
Fibre 0 g	0 %
Sugars 0 g	
Protein 0 g	
Vitamin A 0 %	Vitamin C 0 %
Calcium 0 %	Iron 0 %

Limit trans fat

When a Nutrition Facts table is not available, ask for nutrition information to choose foods lower in trans and saturated fats.

Take a step today...

✓ Have breakfast every day. It may help control your hunger later in the day.

✓ Walk wherever you can – get off the bus early, use the stairs.

✓ Benefit from eating vegetables and fruit at all meals and as snacks.

✓ Spend less time being inactive such as watching TV or playing computer games.

✓ Request nutrition information about menu items when eating out to help you make healthier choices.

✓ Enjoy eating with family and friends!

✓ Take time to eat and savour every bite!

For more information, interactive tools, or additional copies visit Canada's Food Guide on-line at:
www.hc-sc.gc.ca

or contact:

Publications
Health Canada
Ottawa, Ontario K1A 0K9
E-Mail: publications@hc-sc.gc.ca
Tel.: 1-866-225-0709
Fax: (613) 941-5366
TTY: 1-800-267-1245

Également disponible en français sous le titre : Bien manger avec le Guide alimentaire canadien

This publication can be made available on request on diskette, large print, audio-cassette and braille.

 HC Pub.: 4651 Cat.: H164-38/1-2007E ISBN: 0-662-44467-1

Beyond the Basics: Meal Planning for Healthy Eating, Diabetes Prevention and Management

Beyond the Basics: Meal Planning for Healthy Eating, Diabetes Prevention and Management is Canada's system of meal planning.[1] Similar to the U.S. exchange system, *Beyond the Basics* sorts foods into groups and defines portion sizes to help people manage their blood glucose and maintain a healthy weight. Because foods that contain carbohydrate raise blood glucose, the food groups are organized into two sections—those that contain carbohydrate (presented in Table B-1) and those that contain little or no carbohydrate (shown in Table B-2). One portion from any of the food groups listed in Table B-1 provides about 15 grams of available carbohydrate (total carbohydrate minus fiber) and counts as one carbohydrate choice. Within each group, foods are identified as those to "choose more often" (generally higher in vitamins, minerals, and fiber) and those to "choose less often" (generally higher in sugar, saturated fat, or *trans* fat).

[1]The tables for the Canadian meal planning system are adapted from *Beyond the Basics: Meal Planning for Healthy Eating, Diabetes Prevention and Management,* copyright 2005, with permission of the Canadian Diabetes Association. Additional information is available from www.diabetes.ca.

TABLE B-1 Food Groups that Contain Carbohydrate

KEY

● Choose more often
▲ Choose less often

1 serving = 15 g carbohydrate or 1 carbohydrate choice	
Food	**Measure**
Grains and starches: 15 g carbohydrate, 3 g protein, 0 g fat, 286 kJ (68 kcal)	
▲ Bagel, large (11 cm, 4.5″)	1/4
▲ Bagel, small (8 cm, 3″)	1/2
▲ Bannock, fried	1.5″ × 2.5″
● Bannock, whole grain baked	1.5″ × 2.5″
● Barley, pearled, cooked	125 mL (1/2 c)
▲ Bread, white	1 slice (30 g)
● Bread, whole grain	1 slice (30 g)
● Bulghur, cooked	125 mL (1/2 c)
▲ Bun, hamburger or hotdog	1/2
▲ Cereal, flaked unsweetened	125 mL (1/2 c)
● Cream of Wheat, cooked	175 mL (3/4 c)
● Red River, cooked	125 mL (1/2 c)
● Chapati (15 cm, 6″)	1 (44 g)
● Corn	125 mL (1/2 c)
● Couscous, cooked	125 mL (1/2 c)
▲ Crackers, soda type	7
▲ Croutons	175 mL (3/4 c)
● English muffin, whole grain	1/2 (28 g)
▲ French fries	10 (50 g)
● Millet, cooked	1/3 c (75 mL)
▲ Naan bread (15 cm, 6″)	1/4
▲ Pancake (10 cm, 4″)	1
● Pasta, cooked	125 mL (1/2 c)
▲ Pita bread, white (15 cm, 6″)	1/2
● Pita bread, whole wheat (15 cm, 6″)	1/2
▲ Pizza crust (30 cm, 12″)	1/12 (90 g)

(continued)

Source: *Beyond the Basics: Meal Planning for Healthy Eating, Diabetes Prevention and Management*, copyright 2005, Canadian Diabetes Association. Additional information is available from www.diabetes.ca

KEY

● Choose more often
▲ Choose less often

TABLE B-1 Food Groups that Contain Carbohydrate (*continued*)

Food	Measure
Grains and starches: 15 g carbohydrate, 3 g protein, 0 g fat, 286 kJ (68 kcal)—continued	
● Plantain, cooked, mashed	⅓ c (75 mL)
● Potatoes, boiled or baked	½ medium (84 g)
● Rice, white or brown, cooked	⅓ c (75 mL)
● Roti (15 cm, 6″)	1 (44 g)
● Soup, thick, chunky type	250 mL (1 c)
● Sweet potato, mashed	⅓ c (75 mL)
▲ Taco shells (13 cm, 5″)	2 (17 g)
● Tortilla, wheat flour (25 cm, 10″)	½
▲ Waffle (medium)	1 (39 g)
Fruits: 15 g carbohydrate, 1–2 g protein, 0 g fat, 269 kJ (64 kcal)	
● Apple	1 medium (138 g)
● Applesauce, unsweetened	125 mL (½ c)
● Banana	1 small or ½ large
● Blackberries	500 mL (2 c)
● Cherries	15 (102 g)
● Fruit, canned in juice	125 mL (½ c)
▲ Fruit, dried	60 mL (¼ c)
● Grapefruit	1 small
● Grapes	15 or ½ c (80 g)
● Kiwi	2 medium (150 g)
▲ Juice	125 mL (½ c)
● Mango	½ medium
● Melon	250 mL (1 c)
● Orange	1 medium
● Other berries	250 mL (1 c)
● Pear	1 medium
● Pineapple	175 mL (¾ c)
● Plum	2 medium
● Raspberries	500 mL (2 c)
● Strawberries	500 mL (2 c)
Milk and alternatives: 15 g carbohydrate, 7 g protein, variable fat, 386–651 kJ (92–155 kcal)	
● Evaporated milk, canned	125 mL (½ c)
● Milk, fluid	250 mL (1 c)
● Milk powder, skim	60 mL (4 tbs)
● Soy beverage, flavoured	125 mL (½ c)
● Soy beverage, plain	250 mL (1 c)
● Soy yogourt, flavoured	75 mL (⅓ c)
● Yogourt, nonfat, plain	175 mL (¾ c)
● Yogourt, skim, artificially sweetened	250 mL (1 c)
Other choices (sweet foods and snacks): 15 g carbohydrate, variable protein and fat	
▲ Brownies, unfrosted	5 cm × 5 cm (2″ × 2″)
▲ Cake, unfrosted	5 cm × 5 cm (2″ × 2″)
▲ Cookies, arrowroot or gingersnap	3–4
▲ Jam, jelly, marmalade	15 mL (1 tbs)

(*continued*)

TABLE B-1 Food Groups that Contain Carbohydrate (*continued*)

Food	Measure
Other choices (sweet foods and snacks): 15 g carbohydrate, variable protein and fat—continued	
● Milk pudding, skim, no sugar added	125 mL (1/2 c)
▲ Muffin, plain	1 small (45 g)
▲ Oatmeal granola bar	1 (28 g)
● Popcorn, low fat, air popped	750 mL (3 c)
▲ Pretzels, low fat, large	7
▲ Pretzels, low fat, sticks	30
▲ Sugar, white	15 mL (1 tbs, 3 tsp, or 3 packets)
▲ Syrup, honey, molasses	1 tbs (15 mL)

KEY

● Choose more often
▲ Choose less often

TABLE B-2 Food Groups that Contain Little or No Carbohydrate

Food	Measure
Vegetables: To encourage consumption, most vegetables are considered "free"	
▲ Artichokes, Jerusalem[a]	
● Asparagus	
● Beans, yellow or green	
● Bean sprouts	
● Beets	
● Broccoli	
● Cabbage	
● Carrots	
● Cauliflower	
● Celery	
● Cucumber	
● Eggplant	
● Kale	
● Leeks	
● Mushrooms	
● Okra	
● Onions	
▲ Parsnips[a]	
▲ Peas[a]	
● Peppers	
● Rutabagas	
● Salad vegetables	
▲ Squash, Hubbard, pumpkin, spaghetti	
▲ Squash, acorn[a], butternut[a]	
● Tomatoes, fresh	
● Tomatoes, canned, regular	
▲ Tomatoes, canned, stewed[a]	
● Turnips	
	(*continued*)

[a]These vegetables contain enough carbohydrate to be counted as one carbohydrate choice (15 g of available carbohydrate) when the portion size eaten is 1 cup (250 mL) or more.

Source: *Beyond the Basics: Meal Planning for Healthy Eating, Diabetes Prevention and Management*, copyright 2005, Canadian Diabetes Association. Additional information is available from www.diabetes.ca

KEY

● Choose more often
▲ Choose less often

TABLE B-2 Food Groups that Contain Little or No Carbohydrate (*continued*)

Food	Measure
Meat and alternatives: 0 g carbohydrate, 7 g protein, 3–5 g fat, 307 kJ (73 kcal)	
● Cheese, skim (<7% milk fat)	2.5 cm × 2.5 cm × 2.5 cm (1″ × 1″ × 1″)
● Cheese, light (<20% milk fat)	2.5 cm × 2.5 cm × 2.5 cm (1″ × 1″ × 1″)
▲ Cheese, regular (≥21% milk fat)	2.5 cm × 2.5 cm × 2.5 cm (1″ × 1″ × 1″)
● Cottage cheese (1–2% milk fat)	60 mL (¼ c)
● Egg	1 medium-large
● Fish, canned in oil or water	60 mL (¼ c)
● Fish, fresh or frozen, cooked	30 g (1 oz)
● Hummus	90 g (⅓ c)
● Legumes, cooked	125 mL (½ c)
● Meat, game, cooked	30 g (1 slice)
● Meat, ground, lean or extra lean, cooked	30 g (2 tbs)
● Meat, lean, cooked	30 g (1 slice)
● Meat, organ or tripe, cooked	30 g (1 slice)
● Meat, prepared, low fat	30 g (1–3 slices)
▲ Meat, prepared, regular fat	30 g (1–3 slices)
▲ Meat, regular, cooked	30 g (1–3 slices)
● Peameal/back bacon, cooked	30 g (1–2 slices)
● Poultry, ground, lean, cooked	30 g (2 tbs)
● Poultry, skinless, cooked	30 g (1 slice)
▲ Poultry/wings, skin on, cooked	45 g (2)
● Shellfish, cooked	30 g (3 medium)
● Tofu (soybean)	½ block (100 g)
● Vegetarian meat alternatives	30 g (1 oz)
Fats: 0 g carbohydrate, 0 g protein, 5 g fat, 189 kJ (45 kcal)	
▲ Avocado	34 g (⅙)
▲ Bacon	30 g (1 slice)
▲ Butter	5 mL (1 tsp)
▲ Cheese, spreadable	15 mL (1 tbs)
▲ Margarine, non-hydrogenated, regular	5 mL (1 tsp)
▲ Mayonnaise, light	15 mL (1 tbs)
▲ Nuts	15 mL (1 tbs)
▲ Oil, canola or olive	5 mL (1 tsp)
▲ Salad dressing, regular	5 mL (1 tsp)
▲ Seeds	15 g (1 tbs)
▲ Tahini	8 mL (½ tbs)
Extras: <5 g carbohydrate, 84 kJ (20 kcal)	
Broth	
Coffee	
Herbs and spices	
Ketchup	
Mustard	
Relish	
Sugar-free gelatin	
Sugar-free soft drinks	
Tea	

appendix C Exchange Lists for Diabetes

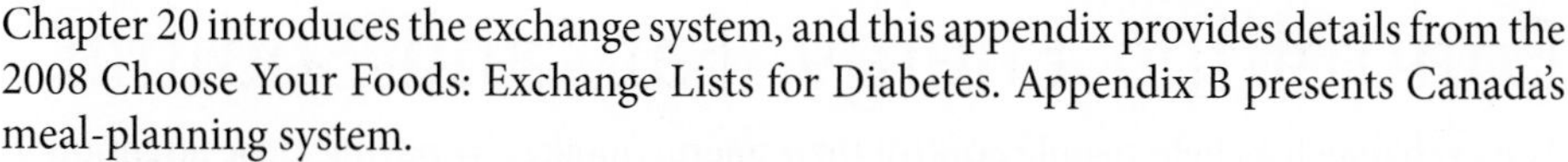

Chapter 20 introduces the exchange system, and this appendix provides details from the 2008 Choose Your Foods: Exchange Lists for Diabetes. Appendix B presents Canada's meal-planning system.

Exchange lists can help people with diabetes to manage their blood glucose levels by controlling the amount and kinds of carbohydrates they consume. These lists can also help in planning diets for weight management by controlling kcalorie and fat intake.

CONTENTS

The Exchange System

The exchange system sorts foods into groups by their proportions of carbohydrate, fat, and protein (Table C-1 on p. C-2). These groups may be organized into several exchange lists of foods (Tables C-2 through C-12 on pp. C-3–C-16). For example, the carbohydrate group includes these exchange lists:

- Starch
- Fruits
- Milk (fat-free, reduced-fat, and whole)
- Sweets, Desserts, and Other Carbohydrates
- Nonstarchy Vegetables

Then any food on a list can be "exchanged" for any other on that same list. Another group for alcohol has been included as a reminder that these beverages often deliver substantial carbohydrate and kcalories, and therefore warrant their own list.

Serving Sizes

The serving sizes have been carefully adjusted and defined so that a serving of any food on a given list provides roughly the same amount of carbohydrate, fat, and protein, and, therefore, total energy. Any food on a list can thus be exchanged, or traded, for any other food on the same list without significantly affecting the diet's energy-nutrient balance or total kcalories. For example, a person may select 17 small grapes or ½ large grapefruit as one fruit exchange, and either choice would provide roughly 15 grams of carbohydrate and 60 kcalories. A whole grapefruit, however, would count as 2 fruit exchanges.

To apply the system successfully, users must become familiar with the specified serving sizes. A convenient way to remember the serving sizes and energy values is to keep in mind a typical item from each list (review Table C-1).

The Foods on the Lists

Foods do not always appear on the exchange list where you might first expect to find them. They are grouped according to their energy-nutrient contents rather than by their source (such as milks), their outward appearance, or their vitamin and mineral contents. For example, cheeses are grouped with meats (not milk) because, like meats, cheeses contribute energy from protein and fat but provide negligible carbohydrate.

For similar reasons, starchy vegetables such as corn, green peas, and potatoes are found on the Starch list with breads and cereals, not with the vegetables. Likewise, bacon is grouped with the fats and oils, not with the meats.

Diet planners learn to view mixtures of foods, such as casseroles and soups, as combinations of foods from different exchange lists. They also learn to interpret food labels with the exchange system in mind.

Controlling Energy, Fat, and Sodium

The exchange lists help people control their energy intakes by paying close attention to serving sizes. People wanting to lose weight can limit foods from the Sweets, Desserts, and Other Carbohydrates and Fats lists, and they might choose to avoid the Alcohol list altogether. The Free Foods list provide low-kcalorie choices.

By assigning items like bacon to the Fats list, the exchange lists alert consumers to foods that are unexpectedly high in fat. Even the Starch list specifies which grain products contain added fat (such as biscuits, cornbread, and waffles) by marking them with a symbol to indicate added fat (the symbols are explained in the table keys). In addition, the exchange lists encourage users to think of fat-free milk as milk and of whole milk as milk with added fat, and to think of lean meats as meats and of medium-fat and high-fat meats as meats with added fat. To that end, foods on the milk and meat lists are separated into categories based on their fat contents (review Table C-1). The Milk list is subdivided for fat-free, reduced fat, and whole; the meat list is subdivided for lean, medium fat, and high fat. The meat list also includes plant-based proteins, which tend to be rich in fiber. Notice that many of these foods (p. C-11) bear the symbol for "high fiber."

People wanting to control the sodium in their diets can begin by eliminating any foods bearing the "high sodium" symbol. In most cases, the symbol identifies foods that, in one serving, provide 480 milligrams or more of sodium. Foods on the Combination Foods or Fast Foods lists that bear the symbol provide more than 600 milligrams of

TABLE C-1 The Food Lists

Lists	Typical Item/Portion Size	Carbohydrate (g)	Protein (g)	Fat (g)	Energy[a] (kcal)
Carbohydrates					
Starch[b]	1 slice bread	15	0–3	0–1	80
Fruits	1 small apple	15	—	—	60
Milk					
Fat-free, low-fat, 1%	1 c fat-free milk	12	8	0–3	100
Reduced-fat, 2%	1 c reduced-fat milk	12	8	5	120
Whole	1 c whole milk	12	8	8	160
Sweets, desserts, and other carbohydrates[c]	2 small cookies	15	varies	varies	varies
Nonstarchy vegetables	½ c cooked carrots	5	2	—	25
Meat and Meat Substitutes					
Lean	1 oz chicken (no skin)	—	7	0–3	45
Medium-fat	1 oz ground beef	—	7	4–7	75
High-fat	1 oz pork sausage	—	7	8+	100
Plant-based proteins	½ c tofu	varies	7	varies	varies
Fats	1 tsp butter	—	—	5	45
Alcohol	12 oz beer	varies	—	—	100

[a]The energy value for each exchange list represents an approximate average for the group and does not reflect the precise number of grams of carbohydrate, protein, and fat. For example, a slice of bread contains 15 grams of carbohydrate (60 kcalories), 3 grams protein (12 kcalories), and a little fat—rounded to 80 kcalories for ease in calculating. A ½ cup of vegetables (not including starchy vegetables) contains 5 grams carbohydrate (20 kcalories) and 2 grams protein (8 more), which has been rounded down to 25 kcalories.

[b]The Starch list includes cereals, grains, breads, crackers, snacks, starchy vegetables (such as corn, peas, and potatoes), and legumes (dried beans, peas, and lentils).

[c]The Sweets, Desserts, and Other Carbohydrates list includes foods that contain added sugars and fats such as sodas, candy, cakes, cookies, doughnuts, ice cream, pudding, syrup, and frozen yogurt.

Source: *Exchange Lists* taken from *Choose Your Foods: Exchange Lists for Diabetes.* Copyright © 2008 Academy of Nutrition and Dietetics. Adapted and reprinted with permission.

sodium. Other foods may also contribute substantially to sodium (consult Chapter 9 for details).

Planning a Healthy Diet

To obtain a daily variety of foods that provide healthful amounts of carbohydrate, protein, and fat, as well as vitamins, minerals, and fiber, the meal plan for adults and teenagers should include at least:

- Two to three servings of nonstarchy vegetables
- Two servings of fruits
- Six servings of grains (at least three of whole grains), beans, and starchy vegetables
- Two servings of low-fat or fat-free milk
- About 6 ounces of meat or meat substitutes
- *Small* amounts of fat and sugar

The actual amounts are determined by age, gender, activity levels, and other factors that influence energy needs.

TABLE C-2 Starch

The Starch list includes bread, cereals and grains, starchy vegetables, crackers and snacks, and legumes (dried beans, peas, and lentils). 1 starch choice = 15 grams carbohydrate, 0–3 grams protein, 0–1 grams fat, and 80 kcalories.

NOTE: In general, one starch exchange is ½ cup cooked cereal, grain, or starchy vegetable; ⅓ cup cooked rice or pasta; 1 ounce of bread product; ¾ ounce to 1 ounce of most snack foods.

Bread

Food	Serving Size
Bagel, large (about 4 oz)	¼ (1 oz)
▽ Biscuit, 2½ inches across	1
Bread	
☺ reduced-kcalorie	2 slices (1½ oz)
white, whole-grain, pumpernickel, rye, unfrosted raisin	1 slice (1 oz)
Chapatti, small, 6 inches across	1
▽ Cornbread, 1¾ inch cube	1 (1½ oz)
English muffin	½
Hot dog bun or hamburger bun	½ (1 oz)
Naan, 8 inches by 2 inches	¼
Pancake, 4 inches across, ¼ inch thick	1
Pita, 6 inches across	½
Roll, plain, small	1 (1 oz)
▽ Stuffing, bread	⅓ cup
▽ Taco shell, 5 inches across	2
Tortilla, corn, 6 inches across	1
Tortilla, flour, 6 inches across	1
Tortilla, flour, 10 inches across	⅓
▽ Waffle, 4-inch square or 4 inches across	1

Cereals and Grains

Food	Serving Size
Barley, cooked	⅓ cup
Bran, dry	
☺ oat	¼ cup
☺ wheat	½ cup
☺ Bulgur (cooked)	½ cup
Cereals	
☺ bran	½ cup
cooked (oats, oatmeal)	½ cup
puffed	1½ cups
shredded wheat, plain	½ cup
sugar-coated	½ cup
unsweetened, ready-to-eat	¾ cup
Couscous	⅓ cup
Granola	
low-fat	¼ cup
▽ regular	¼ cup
Grits, cooked	½ cup
Kasha	½ cup
Millet, cooked	⅓ cup

(*continued*)

KEY

☺ = More than 3 grams of dietary fiber per serving.

▽ = Extra fat, or prepared with added fat. (Count as 1 starch + 1 fat.)

🧂 = 480 milligrams or more of sodium per serving.

Source: *Exchange Lists* taken from *Choose Your Foods: Exchange Lists for Diabetes.* Copyright © 2008 Academy of Nutrition and Dietetics. Adapted and reprinted with permission.

TABLE C-2 Starch (*continued*)

Cereals and Grains—continued

Food	Serving Size
Muesli	¼ cup
Pasta, cooked	⅓ cup
Polenta, cooked	⅓ cup
Quinoa, cooked	⅓ cup
Rice, white or brown, cooked	⅓ cup
Tabbouleh (tabouli), prepared	½ cup
Wheat germ, dry	3 Tbsp
Wild rice, cooked	½ cup

Starchy Vegetables

Food	Serving Size
Cassava	⅓ cup
Corn	½ cup
on cob, large	½ cob (5 oz)
☺ Hominy, canned	¾ cup
☺ Mixed vegetables with corn, peas, or pasta	1 cup
☺ Parsnips	½ cup
☺ Peas, green	½ cup
Plantain, ripe	⅓ cup
Potato	
baked with skin	¼ large (3 oz)
boiled, all kinds	½ cup or ½ medium (3 oz)
▽ mashed, with milk and fat	½ cup
french fried (oven-baked)[a]	1 cup (2 oz)
☺ Pumpkin, canned, no sugar added	1 cup
Spaghetti/pasta sauce	½ cup
☺ Squash, winter (acorn, butternut)	1 cup
☺ Succotash	½ cup
Yam, sweet potato, plain	½ cup

Crackers and Snacks[b]

Food	Serving Size
Animal crackers	8
Crackers	
▽ round-butter type	6
saltine-type	6
▽ sandwich-style, cheese or peanut butter filling	3
▽ whole-wheat regular	2–5 (¾ oz)
☺ whole-wheat lower fat or crispbreads	2–5 (¾ oz)
Graham cracker, 2½-inch square	3
Matzoh	¾ oz
Melba toast, about 2-inch by 4-inch piece	4
Oyster crackers	20
Popcorn	3 cups
▽ ☺ with butter	3 cups
☺ no fat added	3 cups
☺ lower fat	3 cups
Pretzels	¾ oz
Rice cakes, 4 inches across	2
Snack chips	
fat-free or baked (tortilla, potato), baked pita chips	15–20 (¾ oz)
▽ regular (tortilla, potato)	9–13 (¾ oz)

Beans, Peas, and Lentils[c]

The choices on this list count as 1 starch + 1 lean meat.

Food	Serving Size
☺ Baked beans	⅓ cup
☺ Beans, cooked (black, garbanzo, kidney, lima, navy, pinto, white)	½ cup
☺ Lentils, cooked (brown, green, yellow)	½ cup
☺ Peas, cooked (black-eyed, split)	½ cup
🧂 ☺ Refried beans, canned	½ cup

KEY

☺ = More than 3 grams of dietary fiber per serving.

▽ = Extra fat, or prepared with added fat. (Count as 1 starch + 1 fat.)

🧂 = 480 milligrams or more of sodium per serving.

[a]Restaurant-style french fries are on the Fast Foods list.
[b]For other snacks, see the Sweets, Desserts, and Other Carbohydrates list. For a quick estimate of serving size, an open handful is equal to about 1 cup or 1 to 2 ounces of snack food.
[c]Beans, peas, and lentils are also found on the Meat and Meat Substitutes list.

TABLE C-3 Fruits

Fruit[a]

The Fruits list includes fresh, frozen, canned, and dried fruits and fruit juices. 1 fruit choice = 15 grams carbohydrate, 0 grams protein, 0 grams fat, and 60 kcalories.

NOTE: In general, one fruit exchange is ½ cup canned or fresh fruit or unsweetened fruit juice; 1 small fresh fruit (4 ounces); 2 tablespoons dried fruit.

Food	Serving Size
Apple, unpeeled, small	1 (4 oz)
Apples, dried	4 rings
Applesauce, unsweetened	½ cup
Apricots	
canned	½ cup
dried	8 halves
☺ fresh	4 whole (5½ oz)
Banana, extra small	1 (4 oz)
☺ Blackberries	¾ cup
Blueberries	¾ cup
Cantaloupe, small	⅓ melon or 1 cup cubed (11 oz)
Cherries	
sweet, canned	½ cup
sweet fresh	12 (3 oz)
Dates	3
Dried fruits (blueberries, cherries, cranberries, mixed fruit, raisins)	2 Tbsp
Figs	
dried	1½
☺ fresh	1½ large or 2 medium (3½ oz)
Fruit cocktail	½ cup
Grapefruit	
large	½ (11 oz)
sections, canned	¾ cup
Grapes, small	17 (3 oz)
Honeydew melon	1 slice or 1 cup cubed (10 oz)
☺ Kiwi	1 (3½ oz)
Mandarin oranges, canned	¾ cup
Mango, small	½ (5½ oz) or ½ cup
Nectarine, small	1 (5 oz)
☺ Orange, small	1 (6½ oz)
Papaya	½ or 1 cup cubed (8 oz)
Peaches	
canned	½ cup
fresh, medium	1 (6 oz)
Pears	
canned	½ cup
fresh, large	½ (4 oz)
Pineapple	
canned	½ cup
fresh	¾ cup
Plums	
canned	½ cup
dried (prunes)	3
small	2 (5 oz)
☺ Raspberries	1 cup
☺ Strawberries	1¼ cup whole berries
☺ Tangerines, small	2 (8 oz)
Watermelon	1 slice or 1¼ cups cubes (13½ oz)

Fruit Juice

Food	Serving Size
Apple juice/cider	½ cup
Fruit juice blends, 100% juice	⅓ cup
Grape juice	⅓ cup
Grapefruit juice	½ cup
Orange juice	½ cup
Pineapple juice	½ cup
Prune juice	⅓ cup

KEY

☺ = More than 3 grams of dietary fiber per serving.

⚠ = Extra fat, or prepared with added fat. (Count as 1 starch + 1 fat.)

🧂 = 480 milligrams or more of sodium per serving.

[a]The weight listed includes skin, core, seeds, and rind.

Source: *Exchange Lists* taken from *Choose Your Foods: Exchange Lists for Diabetes.* Copyright © 2008 Academy of Nutrition and Dietetics. Adapted and reprinted with permission.

TABLE C-4 Milk

The Milk list groups milks and yogurts based on the amount of fat they have (fat-free/low fat, reduced fat, and whole). Cheeses are found on the Meat and Meat Substitutes list and cream and other dairy fats are found on the Fats list.

NOTE: In general, one milk choice is 1 cup (8 fluid ounces or ½ pint) milk or yogurt.

Milk and Yogurts

Food	Serving Size
Fat-free or low-fat (1%)	
1 fat-free/low-fat milk choice = 12 g carbohydrate, 8 g protein, 0–3 g fat, and 100 kcal.	
Milk, buttermilk, acidophilus milk, Lactaid	1 cup
Evaporated milk	½ cup
Yogurt, plain or flavored with an artificial sweetener	⅔ cup (6 oz)
Reduced-fat (2%)	
1 reduced-fat milk choice = 12 g carbohydrate, 8 g protein, 5 g fat, and 120 kcal.	
Milk, acidophilus milk, kefir, Lactaid	1 cup
Yogurt, plain	⅔ cup (6 oz)
Whole	
1 whole milk choice = 12 g carbohydrate, 8 g protein, 8 g fat, and 160 kcal.	
Milk, buttermilk, goat's milk	1 cup
Evaporated milk	½ cup
Yogurt, plain	8 oz

Dairy-Like Foods

Food	Serving Size	Count as
Chocolate milk		
fat-free	1 cup	1 fat-free milk + 1 carbohydrate
whole	1 cup	1 whole milk + 1 carbohydrate
Eggnog, whole milk	½ cup	1 carbohydrate + 2 fats
Rice drink		
flavored, low fat	1 cup	2 carbohydrates
plain, fat-free	1 cup	1 carbohydrate
Smoothies, flavored, regular	10 oz	1 fat-free milk + 2½ carbohydrates
Soy milk		
light	1 cup	1 carbohydrate + ½ fat
regular, plain	1 cup	1 carbohydrate + 1 fat
Yogurt		
and juice blends	1 cup	1 fat-free milk + 1 carbohydrate
low carbohydrate (less than 6 grams carbohydrate per choice)	⅔ cup (6 oz)	½ fat-free milk
with fruit, low-fat	⅔ cup (6 oz)	1 fat-free milk + 1 carbohydrate

Source: *Exchange Lists* taken from *Choose Your Foods: Exchange Lists for Diabetes.*

TABLE C-5 Sweets, Desserts, and Other Carbohydrates

1 other carbohydrate choice = 15 grams carbohydrate, variable grams protein, variable grams fat, and variable kcalories.

NOTE: In general, one choice from this list can substitute for foods on the Starch, Fruits, or Milk lists.

Beverages, Soda, and Energy/Sports Drinks

Food	Serving Size	Count as
Cranberry juice cocktail	½ cup	1 carbohydrate
Energy drink	1 can (8.3 oz)	2 carbohydrates
Fruit drink or lemonade	1 cup (8 oz)	2 carbohydrates

(*continued*)

TABLE C-5 Sweets, Desserts, and Other Carbohydrates (*continued*)

Food	Serving Size	Count as
Beverages, Soda, and Energy/Sports Drinks—continued		
Hot chocolate		
regular	1 envelope added to 8 oz water	1 carbohydrate + 1 fat
sugar-free or light	1 envelope added to 8 oz water	1 carbohydrate
Soft drink (soda), regular	1 can (12 oz)	2½ carbohydrates
Sports drink	1 cup (8 oz)	1 carbohydrate
Brownies, Cake, Cookies, Gelatin, Pie, and Pudding		
Food	**Serving Size**	**Count as**
Brownie, small, unfrosted	1¼-inch square, ⅞ inch high (about 1 oz)	1 carbohydrate + 1 fat
Cake		
angel food, unfrosted	1/12 of cake (about 2 oz)	2 carbohydrates
frosted	2-inch square (about 2 oz)	2 carbohydrates + 1 fat
unfrosted	2-inch square (about 2 oz)	1 carbohydrate + 1 fat
Cookies		
chocolate chip	2 cookies (2¼ inches across)	1 carbohydrate + 2 fats
gingersnap	3 cookies	1 carbohydrate
sandwich, with crème filling	2 small (about ⅔ oz)	1 carbohydrate + 1 fat
sugar-free	3 small or 1 large (¾–1 oz)	1 carbohydrate + 1–2 fats
vanilla wafer	5 cookies	1 carbohydrate + 1 fat
Cupcake, frosted	1 small (about 1¾ oz)	2 carbohydrates + 1–1½ fats
Fruit cobbler	½ cup (3½ oz)	3 carbohydrates + 1 fat
Gelatin, regular	½ cup	1 carbohydrate
Pie		
commercially prepared fruit, 2 crusts	⅙ of 8-inch pie	3 carbohydrates + 2 fats
pumpkin or custard	⅛ of 8-inch pie	1½ carbohydrates + 1½ fats
Pudding		
regular (made with reduced-fat milk)	½ cup	2 carbohydrates
sugar-free or sugar- and fat-free (made with fat-free milk)	½ cup	1 carbohydrate
Candy, Spreads, Sweets, Sweeteners, Syrups, and Toppings		
Food	**Serving Size**	**Count as**
Candy bar, chocolate/peanut	2 “fun size” bars (1 oz)	1½ carbohydrates + 1½ fats
Candy, hard	3 pieces	1 carbohydrate
Chocolate “kisses”	5 pieces	1 carbohydrate + 1 fat
Coffee creamer		
dry, flavored	4 tsp	½ carbohydrate + ½ fat
liquid, flavored	2 Tbsp	1 carbohydrate
Fruit snacks, chewy (pureed fruit concentrate)	1 roll (¾ oz)	1 carbohydrate
Fruit spreads, 100% fruit	1½ Tbsp	1 carbohydrate
Honey	1 Tbsp	1 carbohydrate
Jam or jelly, regular	1 Tbsp	1 carbohydrate
Sugar	1 Tbsp	1 carbohydrate
Syrup		
chocolate	2 Tbsp	2 carbohydrates
light (pancake type)	2 Tbsp	1 carbohydrate
regular (pancake type)	1 Tbsp	1 carbohydrate

(*continued*)

Source: *Exchange Lists* taken from *Choose Your Foods: Exchange Lists for Diabetes*.

TABLE C-5 Sweets, Desserts, and Other Carbohydrates (*continued*)

Condiments and Sauces[a]

Food	Serving Size	Count as
Barbeque sauce	3 Tbsp	1 carbohydrate
Cranberry sauce, jellied	1/4 cup	1 1/2 carbohydrates
[salt shaker] Gravy, canned or bottled	1/2 cup	1/2 carbohydrate + 1/2 fat
Salad dressing, fat-free, low-fat, cream-based	3 Tbsp	1 carbohydrate
Sweet and sour sauce	3 Tbsp	1 carbohydrate

Doughnuts, Muffins, Pastries, and Sweet Breads

Food	Serving Size	Count as
Banana nut bread	1-inch slice (1 oz)	2 carbohydrates + 1 fat
Doughnut		
cake, plain	1 medium (1 1/2 oz)	1 1/2 carbohydrates + 2 fats
yeast type, glazed	3 3/4 inches across (2 oz)	2 carbohydrates + 2 fats
Muffin (4 oz)	1/4 muffin (1 oz)	1 carbohydrate + 1/2 fat
Sweet roll or Danish	1 (2 1/2 oz)	2 1/2 carbohydrates + 2 fats

Frozen Bars, Frozen Desserts, Frozen Yogurt, and Ice Cream

Food	Serving Size	Count as
Frozen pops	1	1/2 carbohydrate
Fruit juice bars, frozen, 100% juice	1 bar (3 oz)	1 carbohydrate
Ice cream		
fat-free	1/2 cup	1 1/2 carbohydrates
light	1/2 cup	1 carbohydrate + 1 fat
no sugar added	1/2 cup	1 carbohydrate + 1 fat
regular	1/2 cup	1 carbohydrate + 2 fats
Sherbet, sorbet	1/2 cup	2 carbohydrates
Yogurt, frozen		
fat-free	1/3 cup	1 carbohydrate
regular	1/2 cup	1 carbohydrate + 0–1 fat

Granola Bars, Meal Replacement Bars/Shakes, and Trail Mix

Food	Serving Size	Count as
Granola or snack bar, regular or low-fat	1 bar (1 oz)	1 1/2 carbohydrates
Meal replacement bar	1 bar (1 1/3 oz)	1 1/2 carbohydrates + 0–1 fat
Meal replacement bar	1 bar (2 oz)	2 carbohydrates + 1 fat
Meal replacement shake, reduced kcalorie	1 can (10–11 oz)	1 1/2 carbohydrates + 0–1 fat
Trail mix		
candy/nut-based	1 oz	1 carbohydrate + 2 fats
dried fruit-based	1 oz	1 carbohydrate + 1 fat

KEY

[salt shaker] = 480 milligrams or more of sodium per serving.

[a]You can also check the Fats list and Free Foods list for other condiments.

TABLE C-6 Nonstarchy Vegetables

The Nonstarchy Vegetables list includes vegetables that have few grams of carbohydrates or kcalories; starchy vegetables are found on the Starch list. 1 nonstarchy vegetable choice = 5 grams carbohydrate, 2 grams protein, 0 grams fat, and 25 kcalories.

NOTE: In general, one nonstarchy vegetable choice is ½ cup cooked vegetables or vegetable juice or 1 cup raw vegetables. Count 3 cups of raw vegetables or 1½ cups of cooked vegetables as one carbohydrate choice.

Nonstarchy Vegetables[a]	
Amaranth or Chinese spinach	Kohlrabi
Artichoke	Leeks
Artichoke hearts	Mixed vegetables (without corn, peas, or pasta)
Asparagus	Mung bean sprouts
Baby corn	Mushrooms, all kinds, fresh
Bamboo shoots	Okra
Beans (green, wax, Italian)	Onions
Bean sprouts	Oriental radish or daikon
Beets	Pea pods
🧂 Borscht	☺ Peppers (all varieties)
Broccoli	Radishes
☺ Brussels sprouts	Rutabaga
Cabbage (green, bok choy, Chinese)	🧂 Sauerkraut
☺ Carrots	Soybean sprouts
Cauliflower	Spinach
Celery	Squash (summer, crookneck, zucchini)
☺ Chayote	Sugar pea snaps
Coleslaw, packaged, no dressing	☺ Swiss chard
Cucumber	Tomato
Eggplant	Tomatoes, canned
Gourds (bitter, bottle, luffa, bitter melon)	🧂 Tomato sauce
Green onions or scallions	🧂 Tomato/vegetable juice
Greens (collard, kale, mustard, turnip)	Turnips
Hearts of palm	Water chestnuts
Jicama	Yard-long beans

KEY

☺ = More than 3 grams of dietary fiber per serving.

🧂 = 480 milligrams or more of sodium per serving.

[a]Salad greens (like chicory, endive, escarole, lettuce, romaine, spinach, arugula, radicchio, watercress) are on the Free Foods list.

TABLE C-7 Meat and Meat Substitutes

The Meat and Meat Substitutes list groups foods based on the amount of fat they have (lean meat, medium-fat meat, high-fat meat, and plant-based proteins).

Lean Meats and Meat Substitutes

1 lean meat choice = 0 grams carbohydrate, 7 grams protein, 0–3 grams fat, and 100 kcalories.

Food	Amount
Beef: Select or Choice grades trimmed of fat: ground round, roast (chuck, rib, rump), round, sirloin, steak (cubed, flank, porterhouse, T-bone), tenderloin	1 oz
Beef jerky	1 oz
Cheeses with 3 grams of fat or less per oz	1 oz
Cottage cheese	¼ cup
Egg substitutes, plain	¼ cup
Egg whites	2
Fish, fresh or frozen, plain: catfish, cod, flounder, haddock, halibut, orange roughy, salmon, tilapia, trout, tuna	1 oz
Fish, smoked: herring or salmon (lox)	1 oz
Game: buffalo, ostrich, rabbit, venison	1 oz
Hot dog with 3 grams of fat or less per oz (8 dogs per 14 oz package) *Note: May be high in carbohydrate.*	1
Lamb: chop, leg, or roast	1 oz
Organ meats: heart, kidney, liver *Note: May be high in cholesterol.*	1 oz
Oysters, fresh or frozen	6 medium
Pork, lean	
Canadian bacon	1 oz
rib or loin chop/roast, ham, tenderloin	1 oz
Poultry, without skin: Cornish hen, chicken, domestic duck or goose (well-drained of fat), turkey	1 oz
Processed sandwich meats with 3 grams of fat or less per oz: chipped beef, deli thin-sliced meats, turkey ham, turkey kielbasa, turkey pastrami	1 oz
Salmon, canned	1 oz
Sardines, canned	2 medium
Sausage with 3 grams of fat or less per oz	1 oz
Shellfish: clams, crab, imitation shellfish, lobster, scallops, shrimp	1 oz
Tuna, canned in water or oil, drained	1 oz
Veal, lean chop, roast	1 oz

Medium-Fat Meat and Meat Substitutes

1 medium-fat meat choice = 0 grams carbohydrate, 7 grams protein, 4–7 grams fat, and 130 kcalories.

Food	Amount
Beef: corned beef, ground beef, meatloaf, Prime grades trimmed of fat (prime rib), short ribs, tongue	1 oz
Cheeses with 4–7 grams of fat per oz: feta, mozzarella, pasteurized processed cheese spread, reduced-fat cheeses, string	1 oz
Egg *Note: High in cholesterol, so limit to 3 per week.*	1
Fish, any fried product	1 oz
Lamb: ground, rib roast	1 oz
Pork: cutlet, shoulder roast	1 oz
Poultry: chicken with skin; dove, pheasant, wild duck, or goose; fried chicken; ground turkey	1 oz
Ricotta cheese	2 oz or ¼ cup
Sausage with 4–7 grams of fat per oz	1 oz
Veal, cutlet (no breading)	1 oz

High-Fat Meat and Meat Substitutes

1 high-fat meat choice = 0 grams carbohydrate, 7 grams protein, 8+ grams fat, and 150 kcalories. These foods are high in saturated fat, cholesterol, and kcalories and may raise blood cholesterol levels if eaten on a regular basis. Try to eat 3 or fewer servings from this group per week.

Food	Amount
Bacon	
pork	2 slices (16 slices per lb or 1 oz each, before cooking)
turkey	3 slices (½ oz each before cooking)
Cheese, regular: American, bleu, brie, cheddar, hard goat, Monterey jack, queso, and Swiss	1 oz
Hot dog: beef, pork, or combination (10 per lb-sized package)	1
Hot dog: turkey or chicken (10 per lb-sized package)	1
Pork: ground, sausage, spareribs	1 oz
Processed sandwich meats with 8 grams of fat or more per oz: bologna, pastrami, hard salami	1 oz
Sausage with 8 grams fat or more per oz: bratwurst, chorizo, Italian, knockwurst, Polish, smoked, summer	1 oz

(continued)

Source: *Exchange Lists* taken from *Choose Your Foods: Exchange Lists for Diabetes*. Copyright © 2008 Academy of Nutrition and Dietetics. Adapted and reprinted with permission.

TABLE C-7 Meat and Meat Substitutes (*continued*)

Plant-Based Proteins

1 plant-based protein choice = variable grams carbohydrate, 7 grams protein, variable grams fat, and variable kcalories.
Because carbohydrate content varies among plant-based proteins, you should read the food label.

Food	Serving Size	Count as
"Bacon" strips, soy-based	3 strips	1 medium-fat meat
☺ Baked beans	1/3 cup	1 starch + 1 lean meat
☺ Beans, cooked: black, garbanzo, kidney, lima, navy, pinto, white[a]	1/2 cup	1 starch + 1 lean meat
☺ "Beef" or "sausage" crumbles, soy-based	2 oz	1/2 carbohydrate + 1 lean meat
"Chicken" nuggets, soy-based	2 nuggets (1 1/2 oz)	1/2 carbohydrate + 1 medium-fat meat
☺ Edamame	1/2 cup	1/2 carbohydrate + 1 lean meat
Falafel (spiced chickpea and wheat patties)	3 patties (about 2 inches across)	1 carbohydrate + 1 high-fat meat
Hot dog, soy-based	1 (1 1/2 oz)	1/2 carbohydrate + 1 lean meat
☺ Hummus	1/3 cup	1 carbohydrate + 1 high-fat meat
☺ Lentils, brown, green, or yellow	1/2 cup	1 carbohydrate + 1 lean meat
☺ Meatless burger, soy-based	3 oz	1/2 carbohydrate + 2 lean meats
☺ Meatless burger, vegetable- and starch-based	1 patty (about 2 1/2 oz)	1 carbohydrate + 2 lean meats
Nut spreads: almond butter, cashew butter, peanut butter, soy nut butter	1 Tbsp	1 high-fat meat
☺ Peas, cooked: black-eyed and split peas	1/2 cup	1 starch + 1 lean meat
🧂 ☺ Refried beans, canned	1/2 cup	1 starch + 1 lean meat
"Sausage" patties, soy-based	1 (1v oz)	1 medium-fat meat
Soy nuts, unsalted	3/4 oz	1/2 carbohydrate + 1 medium-fat meat
Tempeh	1/4 cup	1 medium-fat meat
Tofu	4 oz (1/2 cup)	1 medium-fat meat
Tofu, light	4 oz (1/2 cup)	1 lean meat

KEY

☺ = More than 3 grams of dietary fiber per serving.

▽! = Extra fat, or prepared with added fat. (Add an additional fat choice to this food.)

🧂 = 480 milligrams or more of sodium per serving (based on the sodium content of a typical 3-oz serving of meat, unless 1 or 2 oz is the normal serving size).

[a]Beans, peas, and lentils are also found on the Starch list; nut butters in smaller amounts are found in the Fats list.

TABLE C-8 Fats

Fats and oils have mixtures of unsaturated (polyunsaturated and monounsaturated) and saturated fats. Foods on the Fats list are grouped together based on the major type of fat they contain. 1 fat choice = 0 grams carbohydrate, 0 grams protein, 5 grams fat, and 45 kcalories.

NOTE: In general, one fat exchange is 1 teaspoon of regular margarine, vegetable oil, or butter; 1 tablespoon of regular salad dressing.

When used in large amounts, bacon and peanut butter are counted as high-fat meat choices (see Meat and Meat Substitutes list). Fat-free salad dressings are found on the Sweets, Desserts, and Other Carbohydrates list. Fat-free products such as margarines, salad dressings, mayonnaise, sour cream, and cream cheese are found on the Free Foods list.

Monounsaturated Fats

Food	Serving Size
Avocado, medium	2 Tbsp (1 oz)
Nut butters (*trans* fat-free): almond butter, cashew butter, peanut butter (smooth or crunchy)	1½ tsp
Nuts	
almonds	6 nuts
Brazil	2 nuts
cashews	6 nuts
filberts (hazelnuts)	5 nuts
macadamia	3 nuts
mixed (50% peanuts)	6 nuts
peanuts	10 nuts
pecans	4 halves
pistachios	16 nuts
Oil: canola, olive, peanut	1 tsp
Olives	
black (ripe)	8 large
green, stuffed	10 large

Polyunsaturated Fats

Food	Serving Size
Margarine: lower-fat spread (30%–50% vegetable oil, *trans* fat-free)	1 Tbsp
Margarine: stick, tub (*trans* fat-free) or squeeze (*trans* fat-free)	1 tsp
Mayonnaise	
reduced-fat	1 Tbsp
regular	1 tsp
Mayonnaise-style salad dressing	
reduced-fat	1 Tbsp
regular	2 tsp
Nuts	
Pignolia (pine nuts)	1 Tbsp
walnuts, English	4 halves
Oil: corn, cottonseed, flaxseed, grape seed, safflower, soybean, sunflower	1 tsp
Oil: made from soybean and canola oil—Enova	1 tsp
Plant stanol esters	
light	1 Tbsp
regular	2 tsp

Polyunsaturated Fats—continued

Food	Serving Size
Salad dressing	
🧂 reduced-fat *Note: May be high in carbohydrate.*	2 Tbsp
🧂 regular	1 Tbsp
Seeds	
flaxseed, whole	1 Tbsp
pumpkin, sunflower	1 Tbsp
sesame seeds	1 Tbsp
Tahini or sesame paste	2 tsp

Saturated Fats

Food	Serving Size
Bacon, cooked, regular or turkey	1 slice
Butter	
reduced-fat	1 Tbsp
stick	1 tsp
whipped	2 tsp
Butter blends made with oil	
reduced-fat or light	1 Tbsp
regular	1½ tsp
Chitterlings, boiled	2 Tbsp (½ oz)
Coconut, sweetened, shredded	2 Tbsp
Coconut milk	
light	⅓ cup
regular	1½ Tbsp
Cream	
half and half	2 Tbsp
heavy	1 Tbsp
light	1½ Tbsp
whipped	2 Tbsp
whipped, pressurized	¼ cup
Cream cheese	
reduced-fat	1½ Tbsp (¾ oz)
regular	1 Tbsp (½ oz)
Lard	1 tsp
Oil: coconut, palm, palm kernel	1 tsp
Salt pork	¼ oz
Shortening, solid	1 tsp
Sour cream	
reduced-fat or light	3 Tbsp
regular	2 Tbsp

KEY

🧂 = 480 milligrams or more of sodium per serving.

TABLE C-9 Free Foods

A "free" food is any food or drink choice that has less than 20 kcalories and 5 grams or less of carbohydrate per serving.

- Most foods on this list should be limited to 3 servings (as listed here) per day. Spread out the servings throughout the day. If you eat all 3 servings at once, it could raise your blood glucose level.
- Food and drink choices listed here without a serving size can be eaten whenever you like.

Low Carbohydrate Foods

Food	Serving Size
Cabbage, raw	½ cup
Candy, hard (regular or sugar-free)	1 piece
Carrots, cauliflower, or green beans, cooked	¼ cup
Cranberries, sweetened with sugar substitute	½ cup
Cucumber, sliced	½ cup
Gelatin	
dessert, sugar-free	
unflavored	
Gum	
Jam or jelly, light or no sugar added	2 tsp
Rhubarb, sweetened with sugar substitute	½ cup
Salad greens	
Sugar substitutes (artificial sweeteners)	
Syrup, sugar-free	2 Tbsp

Modified Fat Foods with Carbohydrate

Food	Serving Size
Cream cheese, fat-free	1 Tbsp (½ oz)
Creamers	
nondairy, liquid	1 Tbsp
nondairy, powdered	2 tsp
Margarine spread	
fat-free	1 Tbsp
reduced-fat	1 tsp
Mayonnaise	
fat-free	1 Tbsp
reduced-fat	1 tsp
Mayonnaise-style salad dressing	
fat-free	1 Tbsp
reduced-fat	1 tsp
Salad dressing	
fat-free or low-fat	1 Tbsp
fat-free, Italian	2 Tbsp
Sour cream, fat-free or reduced-fat	1 Tbsp
Whipped topping	
light or fat-free	2 Tbsp
regular	1 Tbsp

Condiments

Food	Serving Size
Barbecue sauce	2 tsp
Catsup (ketchup)	1 Tbsp
Honey mustard	1 Tbsp
Horseradish	
Lemon juice	
Miso	1½ tsp
Mustard	
Parmesan cheese, freshly grated	1 Tbsp
Pickle relish	1 Tbsp
Pickles	
[sodium] dill	1½ medium
sweet, bread and butter	2 slices
sweet, gherkin	¾ oz
Salsa	¼ cup
[sodium] Soy sauce, light or regular	1 Tbsp
Sweet and sour sauce	2 tsp
Sweet chili sauce	2 tsp
Taco sauce	1 Tbsp
Vinegar	
Yogurt, any type	2 Tbsp

Drinks/Mixes

Any food on the list—without a serving size listed—can be consumed in any moderate amount.

- [sodium] Bouillon, broth, consommé
- Bouillon or broth, low-sodium
- Carbonated or mineral water
- Club soda
- Cocoa powder, unsweetened (1 Tbsp)
- Coffee, unsweetened or with sugar substitute
- Diet soft drinks, sugar-free
- Drink mixes, sugar-free
- Tea, unsweetened or with sugar substitute
- Tonic water, diet
- Water
- Water, flavored, carbohydrate free

Seasonings

Any food on this list can be consumed in any moderate amount.

- Flavoring extracts (for example, vanilla, almond, peppermint)
- Garlic
- Herbs, fresh or dried
- Nonstick cooking spray
- Pimento
- Spices
- Hot pepper sauce
- Wine, used in cooking
- Worcestershire sauce

KEY

[sodium] = 480 milligrams or more of sodium per serving.

Source: *Exchange Lists* taken from *Choose Your Foods: Exchange Lists for Diabetes.* Copyright © 2008 Academy of Nutrition and Dietetics. Adapted and reprinted with permission.

TABLE C-10 Combination Foods

Many foods are eaten in various combinations, such as casseroles. Because "combination" foods do not fit into any one choice list, this list of choices provides some typical combination foods.

Entrees

Food	Serving Size	Count as
🧂 Casserole type (tuna noodle, lasagna, spaghetti with meatballs, chili with beans, macaroni and cheese)	1 cup (8 oz)	2 carbohydrates + 2 medium-fat meats
🧂 Stews (beef/other meats and vegetables)	1 cup (8 oz)	1 carbohydrate + 1 medium-fat meat + 0–3 fats
Tuna salad or chicken salad	½ cup (3½ oz)	½ carbohydrate + 2 lean meats + 1 fat

Frozen Meals/Entrees

Food	Serving Size	Count as
🧂 ☺ Burrito (beef and bean)	1 (5 oz)	3 carbohydrates + 1 lean meat + 2 fats
🧂 Dinner-type meal	generally 14–17 oz	3 carbohydrates + 3 medium-fat meats + 3 fats
🧂 Entrée or meal with less than 340 kcalories	about 8–11 oz	2–3 carbohydrates + 1–2 lean meats
Pizza		
🧂 cheese/vegetarian, thin crust	¼ of a 12 inch (4½–5 oz)	2 carbohydrates + 2 medium-fat meats
🧂 meat topping, thin crust	¼ of a 12 inch (5 oz)	2 carbohydrates + 2 medium-fat meats + 1½ fats
🧂 Pocket sandwich	1 (4½ oz)	3 carbohydrates + 1 lean meat + 1–2 fats
🧂 Pot pie	1 (7 oz)	2½ carbohydrates + 1 medium-fat meat + 3 fats

Salads (Deli-Style)

Food	Serving Size	Count as
Coleslaw	½ cup	1 carbohydrate + 1½ fats
Macaroni/pasta salad	½ cup	2 carbohydrates + 3 fats
🧂 Potato salad	½ cup	1½–2 carbohydrates + 1–2 fats

Soups

Food	Serving Size	Count as
🧂 Bean, lentil, or split pea	1 cup	1 carbohydrate + 1 lean meat
🧂 Chowder (made with milk)	1 cup (8 oz)	1 carbohydrate + 1 lean meat + 1½ fats
🧂 Cream (made with water)	1 cup (8 oz)	1 carbohydrate + 1 fat
🧂 Instant	6 oz prepared	1 carbohydrate
🧂 with beans or lentils	8 oz prepared	2½ carbohydrates + 1 lean meat
🧂 Miso soup	1 cup	½ carbohydrate + 1 fat
🧂 Oriental noodle	1 cup	2 carbohydrates + 2 fats
Rice (congee)	1 cup	1 carbohydrate
🧂 Tomato (made with water)	1 cup (8 oz)	1 carbohydrate
🧂 Vegetable beef, chicken noodle, or other broth-type	1 cup (8 oz)	1 carbohydrate

KEY

☺ = More than 3 grams of dietary fiber per serving.

⚠ = Extra fat, or prepared with added fat.

🧂 = 600 milligrams or more of sodium per serving (for combination food main dishes/meals).

Source: *Exchange Lists* taken from *Choose Your Foods: Exchange Lists for Diabetes.*

TABLE C-11 Fast Foods

The choices in the Fast Foods list are not specific fast-food meals or items, but are estimates based on popular foods. Ask the restaurant or check its website for nutrition information about your favorite fast foods.

Breakfast Sandwiches

Food	Serving Size	Count as
🧂 Egg, cheese, meat, English muffin	1 sandwich	2 carbohydrates + 2 medium-fat meats
🧂 Sausage biscuit sandwich	1 sandwich	2 carbohydrates + 2 high-fat meats + 3½ fats

Main Dishes/Entrees

Food	Serving Size	Count as
🧂 ☺ Burrito (beef and beans)	1 (about 8 oz)	3 carbohydrates + 3 medium-fat meats + 3 fats
🧂 Chicken breast, breaded and fried	1 (about 5 oz)	1 carbohydrate + 4 medium-fat meats
Chicken drumstick, breaded and fried	1 (about 2 oz)	2 medium-fat meats
🧂 Chicken nuggets	6 (about 3½ oz)	1 carbohydrate + 2 medium-fat meats + 1 fat
🧂 Chicken thigh, breaded and fried	1 (about 4 oz)	½ carbohydrate + 3 medium-fat meats + 1½ fats
🧂 Chicken wings, hot	6 (5 oz)	5 medium-fat meats + 1½ fats

Oriental

Food	Serving Size	Count as
🧂 Beef/chicken/shrimp with vegetables in sauce	1 cup (about 5 oz)	1 carbohydrate + 1 lean meat + 1 fat
🧂 Egg roll, meat	1 (about 3 oz)	1 carbohydrate + 1 lean meat + 1 fat
Fried rice, meatless	½ cup	1½ carbohydrates + 1½ fats
🧂 Meat and sweet sauce (orange chicken)	1 cup	3 carbohydrates + 3 medium-fat meats + 2 fats
🧂 ☺ Noodles and vegetables in sauce (chow mein, lo mein)	1 cup	2 carbohydrates + 1 fat

Pizza

Food	Serving Size	Count as
Pizza		
🧂 cheese, pepperoni, regular crust	⅛ of a 14 inch (about 4 oz)	2½ carbohydrates + 1 medium-fat meat + 1½ fats
🧂 cheese/vegetarian, thin crust	¼ of a 12 inch (about 6 oz)	2½ carbohydrates + 2 medium-fat meats + 1½ fats

Sandwiches

Food	Serving Size	Count as
🧂 Chicken sandwich, grilled	1	3 carbohydrates + 4 lean meats
🧂 Chicken sandwich, crispy	1	3½ carbohydrates + 3 medium-fat meats + 1 fat
Fish sandwich with tartar sauce	1	2½ carbohydrates + 2 medium-fat meats + 2 fats
Hamburger		
🧂 large with cheese	1	2½ carbohydrates + 4 medium-fat meats + 1 fat
regular	1	2 carbohydrates + 1 medium-fat meat + 1 fat
🧂 Hot dog with bun	1	1 carbohydrate + 1 high-fat meat + 1 fat
Submarine sandwich		
🧂 less than 6 grams fat	6-inch sub	3 carbohydrates + 2 lean meats
🧂 regular	6-inch sub	3½ carbohydrates + 2 medium-fat meats + 1 fat
Taco, hard or soft shell (meat and cheese)	1 small	1 carbohydrate + 1 medium-fat meat + 1½ fats

(*continued*)

KEY

☺ = More than 3 grams of dietary fiber per serving.

⚠ = Extra fat, or prepared with added fat.

🧂 = 600 milligrams or more of sodium per serving (for fast-food main dishes/meals).

Source: *Exchange Lists* taken from *Choose Your Foods: Exchange Lists for Diabetes.* Copyright © 2008 Academy of Nutrition and Dietetics. Adapted and reprinted with permission.

TABLE C-11 Fast Foods (*continued*)

Salads

Food	Serving Size	Count as
🧂 ☺ Salad, main dish (grilled chicken type, no dressing or croutons)		1 carbohydrate + 4 lean meats
Salad, side, no dressing or cheese	Small (about 5 oz)	1 vegetable

Sides/Appetizers

Food	Serving Size	Count as
▽ French fries, restaurant style	small	3 carbohydrates + 3 fats
	medium	4 carbohydrates + 4 fats
	large	5 carbohydrates + 6 fats
🧂 Nachos with cheese	small (about 4½ oz)	2½ carbohydrates + 4 fats
🧂 Onion rings	1 serving (about 3 oz)	2½ carbohydrates + 3 fats

Desserts

Food	Serving Size	Count as
Milkshake, any flavor	12 oz	6 carbohydrates + 2 fats
Soft-serve ice cream cone	1 small	2½ carbohydrates + 1 fat

KEY

☺ = More than 3 grams of dietary fiber per serving.

▽ = Extra fat, or prepared with added fat.

🧂 = 600 milligrams or more of sodium per serving (for fast-food main dishes/meals).

TABLE C-12 Alcohol

1 alcohol equivalent = variable grams carbohydrate, 0 grams protein, 0 grams fat, and 100 kcalories.

NOTE: In general, one alcohol choice (½ ounce absolute alcohol) has about 100 kcalories. For those who choose to drink alcohol, guidelines suggest limiting alcohol intake to 1 drink or less per day for women, and 2 drinks or less per day for men. To reduce your risk of low blood glucose (hypoglycemia), especially if you take insulin or a diabetes pill that increases insulin, always drink alcohol with food. While alcohol, by itself, does not directly affect blood glucose, be aware of the carbohydrate (for example, in mixed drinks, beer, and wine) that may raise your blood glucose.

Alcoholic Beverage	Serving Size	Count as
Beer		
light (4.2%)	12 fl oz	1 alcohol equivalent + ½ carbohydrate
regular (4.9%)	12 fl oz	1 alcohol equivalent + 1 carbohydrate
Distilled spirits: vodka, rum, gin, whiskey, 80 or 86 proof	1½ fl oz	1 alcohol equivalent
Liqueur, coffee (53 proof)	1 fl oz	1 alcohol equivalent + 1 carbohydrate
Sake	1 fl oz	½ alcohol equivalent
Wine		
dessert (sherry)	3½ fl oz	1 alcohol equivalent + 1 carbohydrate
dry, red or white (10%)	5 fl oz	1 alcohol equivalent

Source: *Exchange Lists* taken from *Choose Your Foods: Exchange Lists for Diabetes*.

appendix D Physical Activity and Energy Requirements

Chapter 6 described how to calculate your estimated energy requirements by using an equation that accounts for your gender, age, weight, height, and physical activity level. This appendix presents tables that provide a shortcut to estimating total energy requirements, as developed by the *Dietary Guidelines for Americans 2010*, and based on the equations of the Committee on Dietary Reference Intakes.

Table D-1 describes activity levels for three groups of people: sedentary, moderately active, or active. Once you have found an activity level that approximates your own, find your daily kcalorie need in Table D-2.

CONTENTS

TABLE D-1 Sedentary, Moderately Active, and Active People

Sedentary	A lifestyle that includes only the light physical activity associated with typical day-to-day life.
Moderately active	A lifestyle that includes physical activity equivalent to walking about 1.5 to 3 miles per day at 3 to 4 miles per hour in addition to the light physical activity associated with typical day-to-day life.
Active	A lifestyle that includes physical activity equivalent to walking more than 3 miles per day at 3 to 4 miles per hour in addition to the light physical activity associated with typical day-to-day life.

TABLE D-2 Estimated kCalorie Needs per Day by Age, Gender, and Physical Activity Level (Detailed)

Estimated amounts of kcalories needed to maintain kcalorie balance for various gender and age groups at three different levels of physical activity. The estimates are rounded to the nearest 200 kcalories. An individual's kcalorie needs may be higher or lower than these average estimates.[a]

	Gender/Activity Level					
Age (years)	Male/ Sedentary	Male/ Moderately Active	Male/Active	Female[b]/ Sedentary	Female[b]/ Moderately Active	Female[b]/ Active
2	1000	1000	1000	1000	1000	1000
3	1200	1400	1400	1000	1200	1400
4	1200	1400	1600	1200	1400	1400
5	1200	1400	1600	1200	1400	1600
6	1400	1600	1800	1200	1400	1600
7	1400	1600	1800	1200	1600	1800
8	1400	1600	2000	1400	1600	1800
9	1600	1800	2000	1400	1600	1800
10	1600	1800	2200	1400	1800	2000

continued

TABLE D-2 Estimated kCalorie Needs per Day by Age, Gender, and Physical Activity Level (Detailed) (*continued*)

Age (years)	Gender/Activity Level					
	Male/ Sedentary	Male/ Moderately Active	Male/Active	Female[b]/ Sedentary	Female[b]/ Moderately Active	Female[b]/ Active
11	1800	2000	2200	1600	1800	2000
12	1800	2200	2400	1600	2000	2200
13	2000	2200	2600	1600	2000	2200
14	2000	2400	2800	1800	2000	2400
15	2200	2600	3000	1800	2000	2400
16-18	2400	2800	3200	1800	2000	2400
19-20	2600	2800	3000	2000	2200	2400
21-25	2400	2800	3000	2000	2200	2400
26-30	2400	2600	3000	1800	2000	2400
31-35	2400	2600	3000	1800	2000	2200
36-40	2400	2600	2800	1800	2000	2200
41-45	2200	2600	2800	1800	2000	2200
46-50	2200	2400	2800	1800	2000	2200
51-55	2200	2400	2800	1600	1800	2200
56-60	2200	2400	2600	1600	1800	2200
61-65	2000	2400	2600	1600	1800	2000
66-75	2000	2200	2600	1600	1800	2000
76+	2000	2200	2400	1600	1800	2000

[a]Based on Estimated Energy Requirements (EER) equations, using reference heights (average) and reference weights (healthy) for each age-gender group. For children and adolescents, reference height and weight vary. For adults the reference man is 5 feet 10 inches tall and weighs 154 pounds. The reference woman is 5 feet 4 inches tall and weighs 126 pounds. EER equations are from the Institute of Medicine. *Dietary Reference Intakes for Energy, Carbohydrate, Fiber, Fat, Fatty Acids, Cholesterol, Protein, and Amino Acids*. Washington, DC: The National Academies Press; 2002.

[b]Estimates for females do not include women who are pregnant or breastfeeding.

Source: U.S. Department of Agriculture and U.S. Department of Health and Human Services, *Dietary Guidelines for Americans 2010*, www.dietaryguidelines.gov.

appendix E Nutrition Assessment: Supplemental Information

Chapter 13 describes data from nutrition assessments that help health professionals evaluate patients' nutrition status and nutrient needs. This appendix provides additional information that may be useful for complete assessments.

Weight Gain during Pregnancy

Chapter 10 describes desirable weight-gain patterns during pregnancy. Figure E-1 shows prenatal weight-gain grids, which are used to plot the rate of weight gain during pregnancy.

FIGURE E-1 **Recommended Prenatal Weight Gain Based on Prepregnancy Weight**

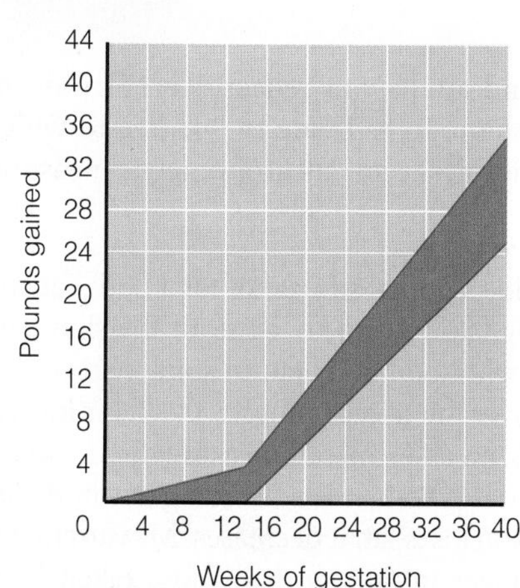

Normal-weight women should gain about $3\frac{1}{2}$ pounds in the first trimester and just under 1 pound/week thereafter, achieving a total gain of 25 to 35 pounds by term.

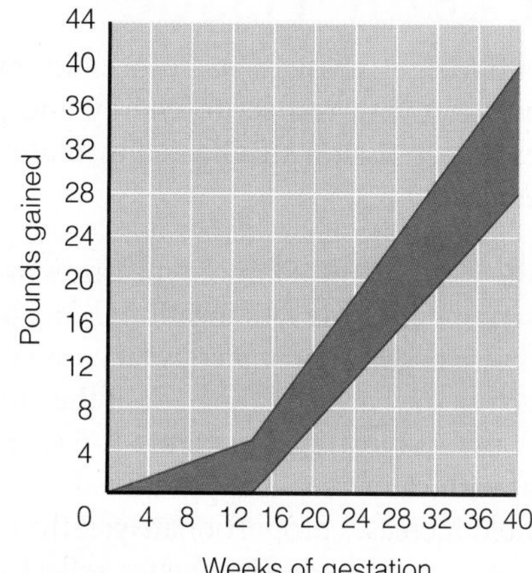

Underweight women should gain about 5 pounds in the first trimester and just over 1 pound/week thereafter, achieving a total gain of 28 to 40 pounds by term.

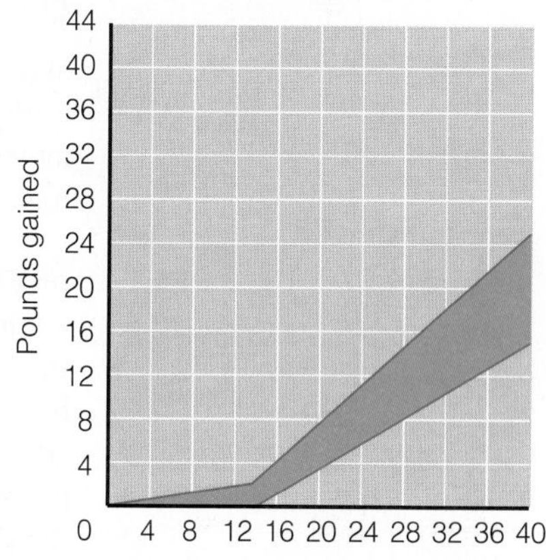

Overweight women should gain about 2 pounds in the first trimester and $\frac{2}{3}$ pound/week thereafter, achieving a total gain of 15 to 25 pounds.

Growth Charts

Health professionals evaluate physical development by monitoring growth rates of children and comparing these rates with those on standard growth charts. Standard charts compare length or height to age, weight to age, weight to length, head circumference to age, and body mass index (BMI) (■) to age. Although individual growth patterns vary, a child's growth curve will generally stay at about the same percentile throughout childhood. In children whose growth has been retarded, nutrition rehabilitation will ideally cause height and weight to increase to higher percentiles. In overweight children, the goal is for weight to remain stable as height increases, until weight becomes appropriate for height.

■ Reminder: The *body mass index (BMI)* is an index of a person's weight in relation to height, determined by dividing the weight in kilograms by the square of the height in meters:

$$\text{BMI} = \frac{\text{Weight (kg)}}{\text{Height (m)}^2}$$

To evaluate growth in infants, an assessor uses a chart such as those in Figures E-2 through E-5. (■) For example, the assessor follows these steps to determine the weight percentile:

- Select the appropriate chart based on age and gender.
- Locate the child's age along the horizontal axis on the bottom of the chart.
- Locate the child's weight in pounds or kilograms along the vertical axis.
- Mark the chart where the age and weight lines intersect, and read off the percentile.

For other measures, the assessor follows a similar procedure, using the appropriate chart. (When length is measured, use the chart for birth to 36 months; when height is measured, use the chart for 2 to 20 years.) Once all of the measures are plotted on growth charts, a skilled clinician can begin to interpret the data. Ideally, the height, weight, and head circumference should be in roughly the same percentile.

Head circumference is generally measured in children under two years of age. Since the brain grows rapidly before birth and during early infancy, extreme and chronic malnutrition during these times can impair brain development, curtailing the number of brain cells and reducing head circumference. Nonnutritional factors, such as certain disorders and genetic variation, can also influence head circumference.

Measures of Body Fat and Lean Tissue

Significant weight changes in both children and adults can reflect overnutrition or undernutrition with respect to energy and protein. To estimate the degree to which fat stores or lean tissues are affected by malnutrition, several anthropometric measurements are useful.

Skinfold Measures Skinfold measures provide a good estimate of total body fat and a fair assessment of the fat's location. Most body fat lies directly beneath the skin, and the thickness of this subcutaneous fat correlates with total body fat. In some parts of the body, such as the back and the back of the arm over the triceps muscle, this fat is loosely attached. (■) As illustrated in Figure E-6 (p. E-7), an assessor can measure the thickness of the fat with calipers that apply a fixed amount of pressure. If a person gains body fat, the skinfold increases proportionately; if the person loses fat, it decreases. Measurements taken from central-body sites better reflect changes in fatness than those taken from upper sites (arm and back). Because subcutaneous fat may be thicker in one area than in another, skinfold measurements are often taken at three or four different places on the body (including upper-, central-, and lower-body sites); the sum of these measures is then compared to standard values. In some situations, the triceps skinfold measurement alone may be used because it is easily accessible. Triceps skinfold measures greater than 15 millimeters in men or 25 millimeters in women suggest excessive body fat.

Waist Circumference Chapter 6 explains how fat distribution correlates with health risks and mentions that the waist circumference is a valuable indicator of abdominal fat. To measure waist circumference, the assessor places a nonstretchable tape around the person's body, crossing just above the upper hip bones and making sure that the tape remains on a level horizontal plane on all sides (see Figure E-7, p. E-7). The tape is tightened slightly, but without compressing the skin.

Waist-to-Hip Ratio The waist-to-hip ratio assesses abdominal obesity, but it offers no advantage over the waist circumference alone. To calculate the waist-to-hip ratio, divide the waistline measurement by the hip measurement. (■) In general, women with a waist-to-hip ratio of 0.8 or greater and men with a waist-to-hip ratio of 0.9 or greater have an increased risk of developing diabetes and cardiovascular diseases.

■ Additional growth charts are available at www.cdc.gov/growthcharts.

■ Common sites for skinfold measures:

- Triceps
- Biceps
- Subscapular (below shoulder blade)
- Suprailiac (above hip bone)
- Abdomen
- Upper thigh

■ The calculation of waist-to-hip ratio in a woman with a 28-inch waist and 38-inch hips is 28 ÷ 38 = 0.74.

FIGURE E-2 Length-for-Age and Weight-for-Age Percentiles

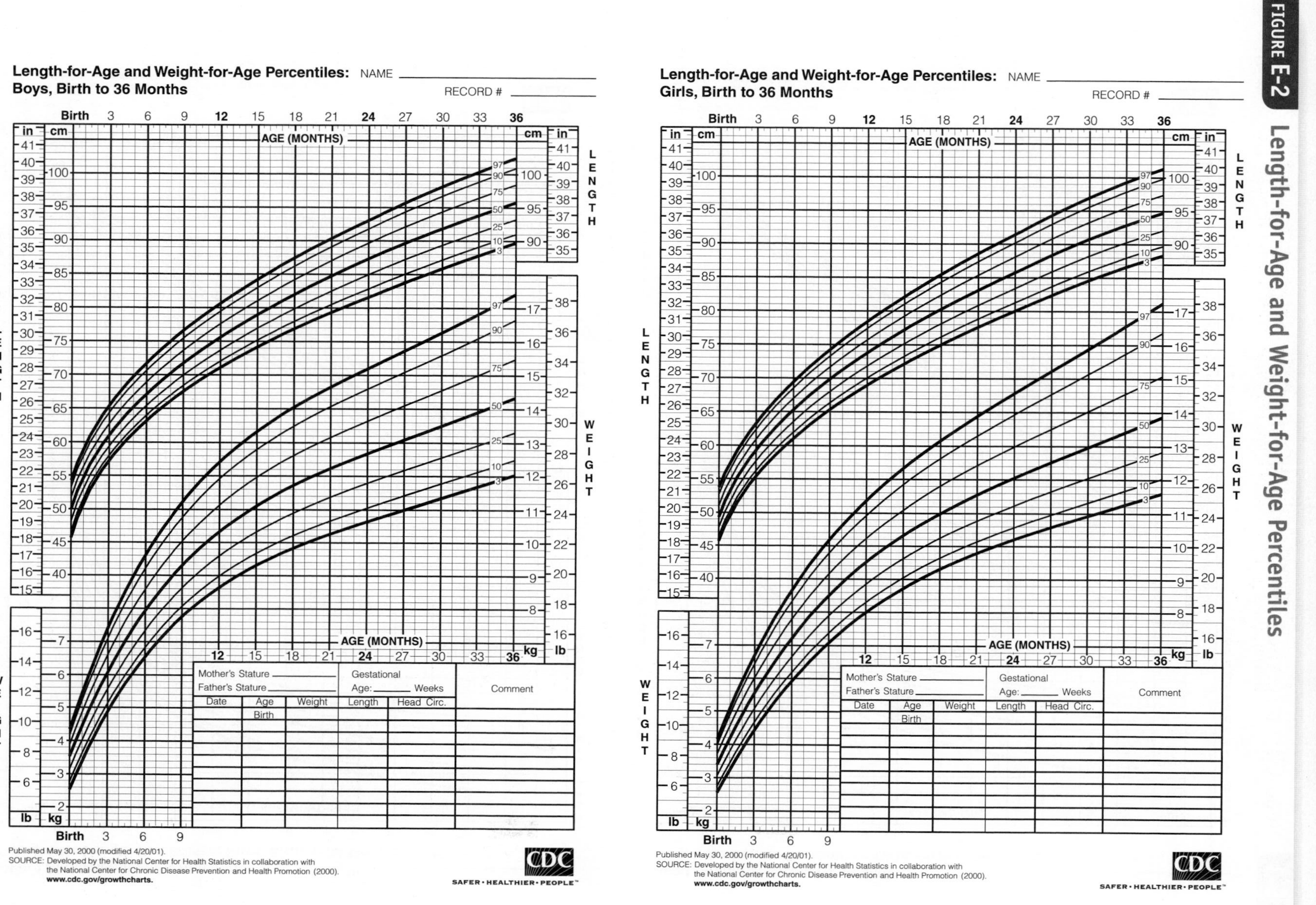

FIGURE E-3 Head Circumference-for-Age and Weight-for-Length Percentiles

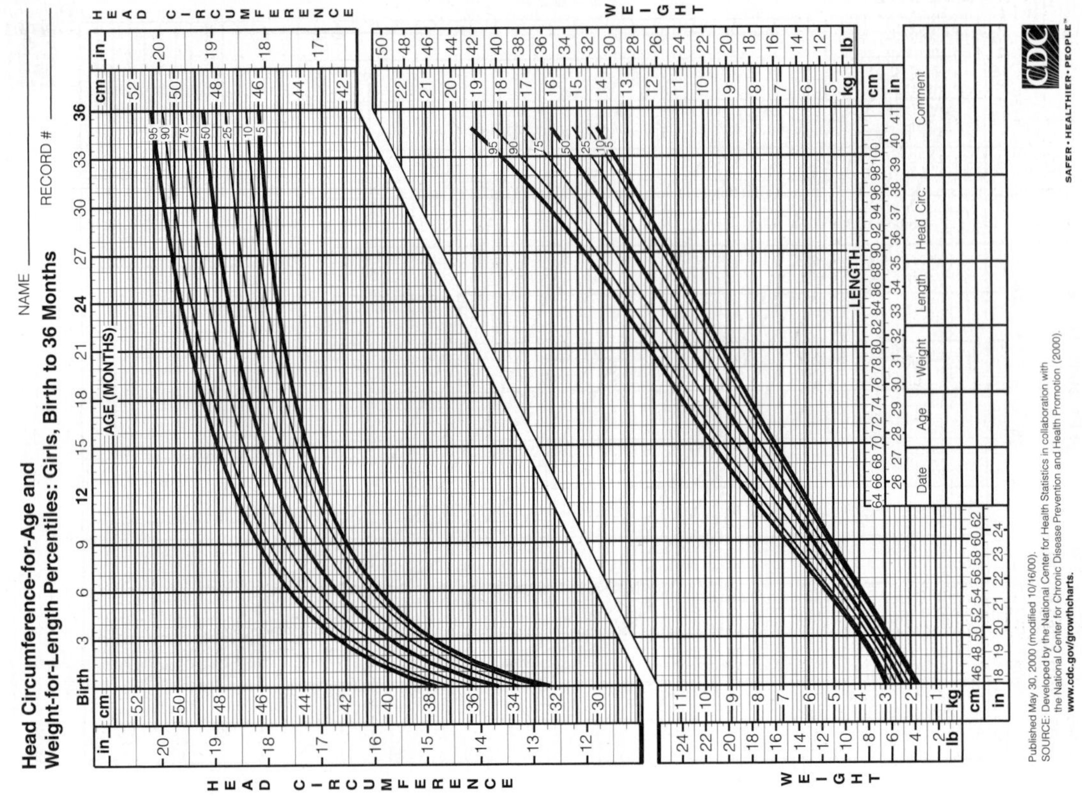

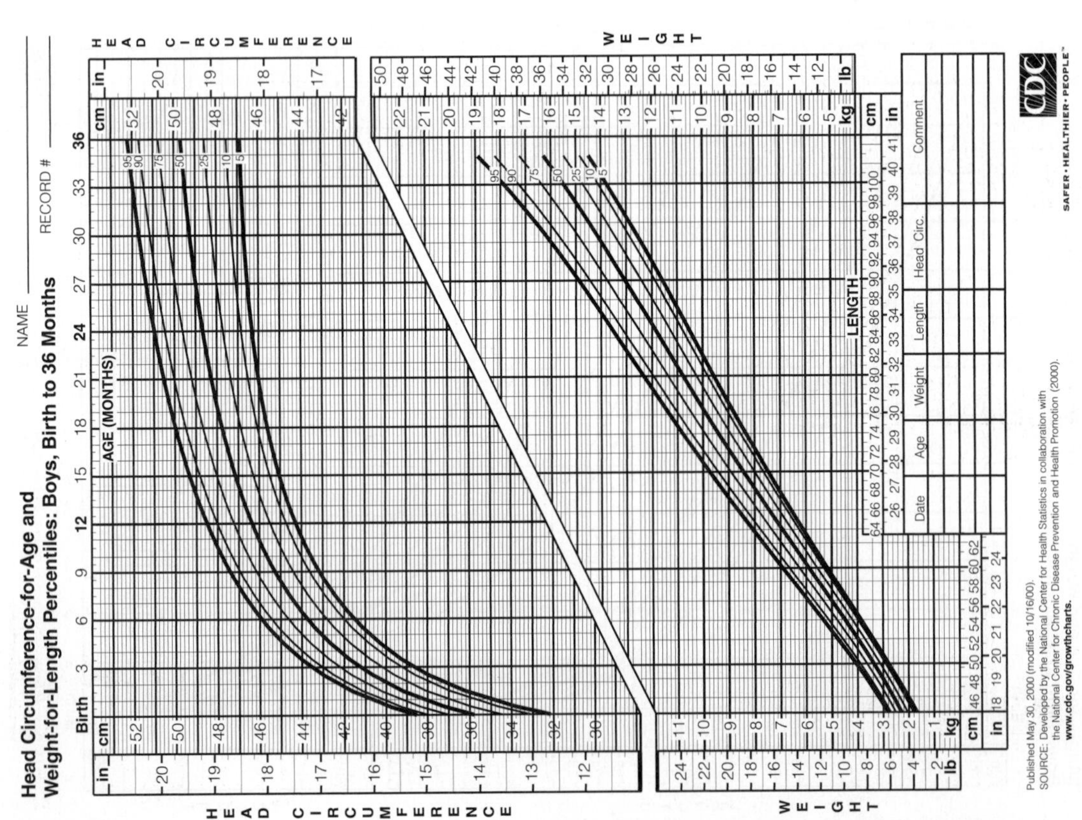

FIGURE E-4 Stature-for-Age and Weight-for-Age Percentiles

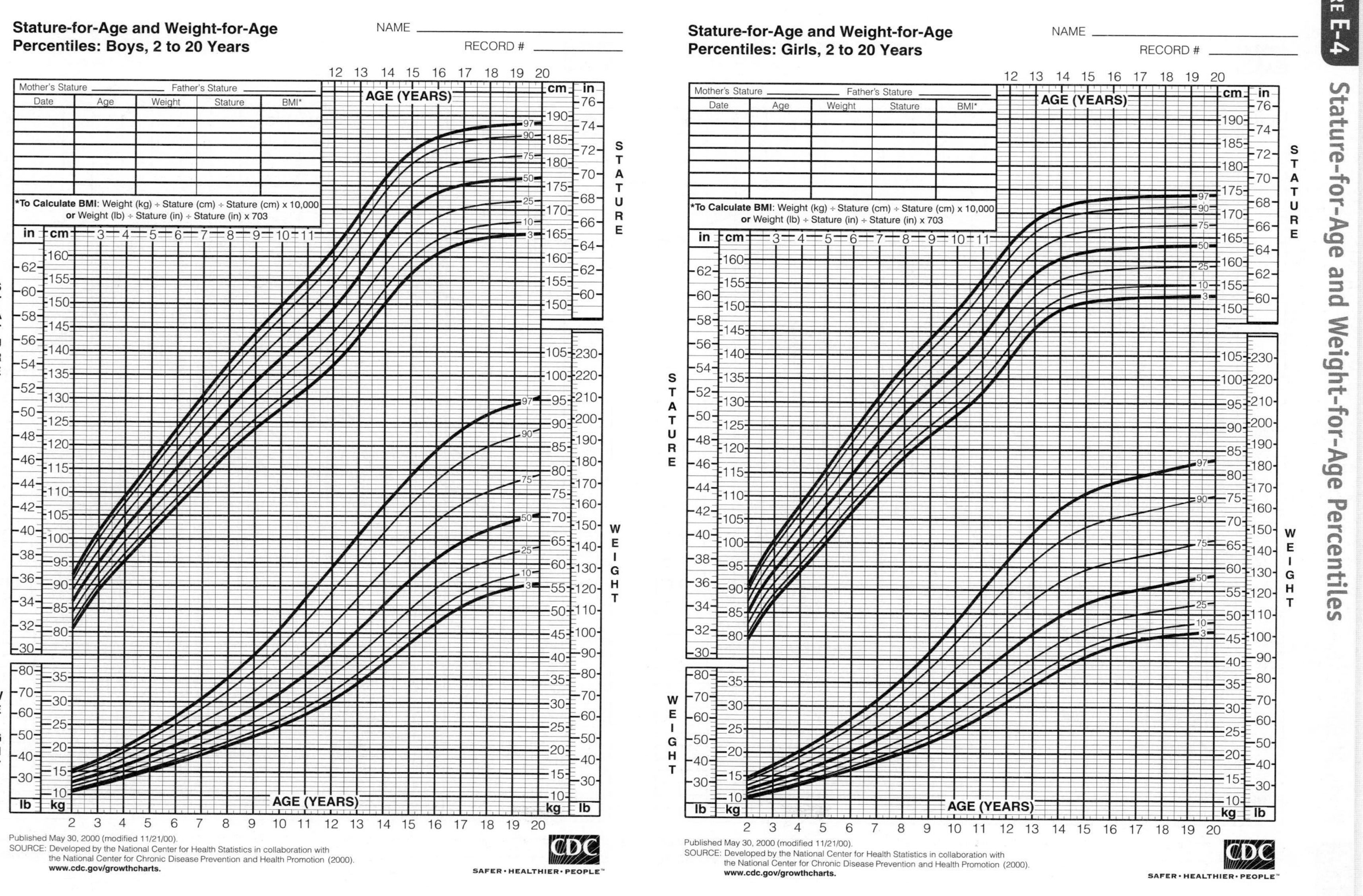

FIGURE E-5 Body Mass Index-for-Age Percentiles

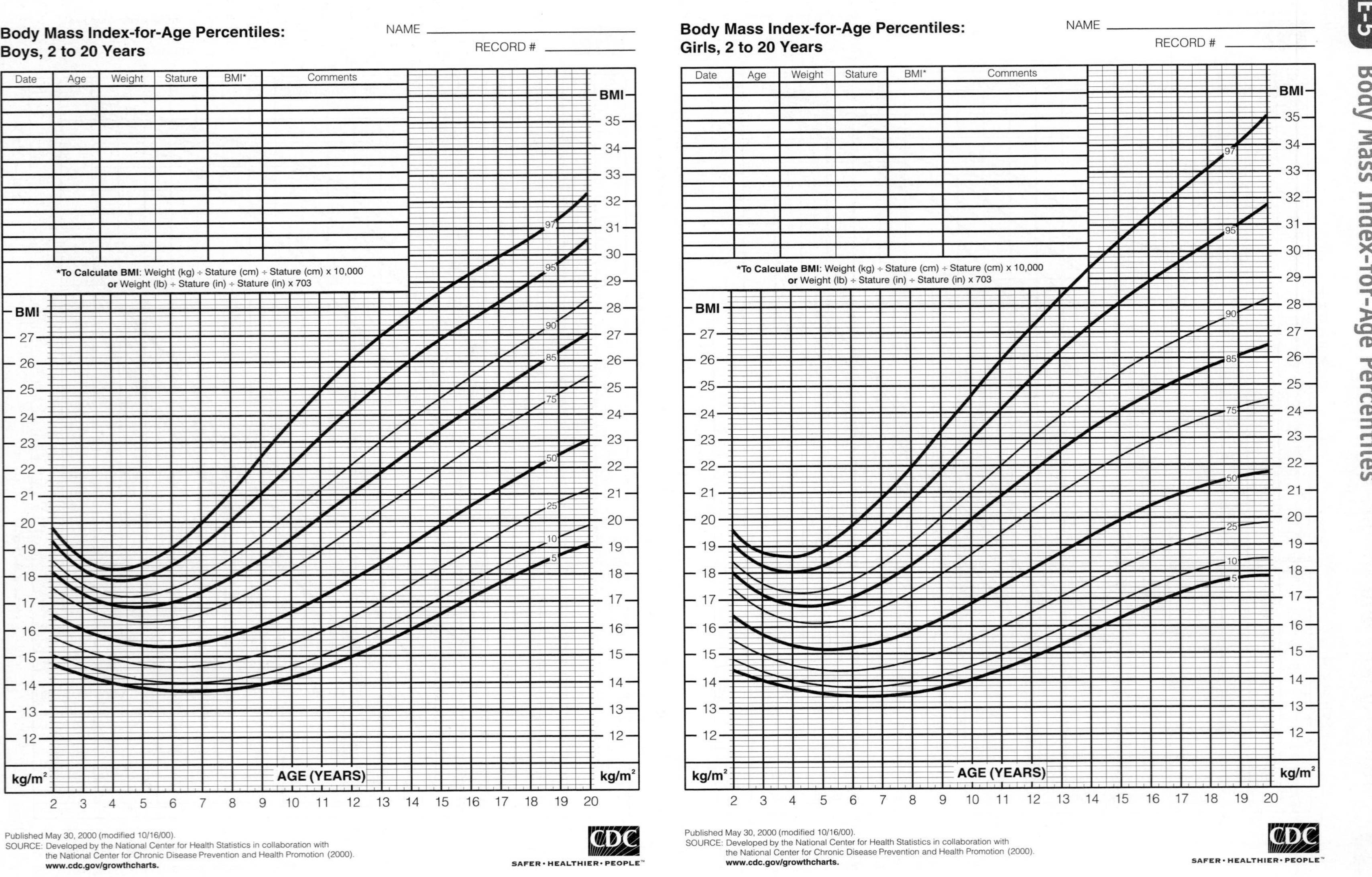

FIGURE E-6 How to Measure the Triceps Skinfold

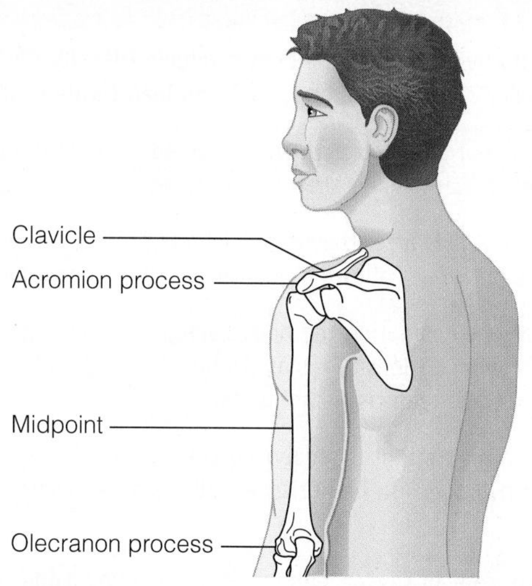

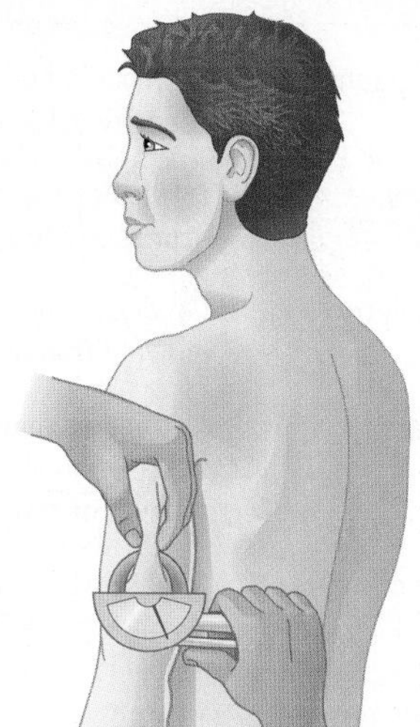

A. Find the midpoint of the arm:
   1. Ask the subject to bend his or her arm at the elbow and lay the hand across the stomach. (If he or she is right-handed, measure the left arm, and vice versa.)
   2. Feel the shoulder to locate the acromion process. It helps to slide your fingers along the clavicle to find the acromion process. The olecranon process is the tip of the elbow.
   3. Place a measuring tape from the acromion process to the tip of the elbow. Divide this measurement by 2 and mark the midpoint of the arm with a pen.

B. Measure the skinfold:
   1. Ask the subject to let his or her arm hang loosely to the side.
   2. Grasp a fold of skin and subcutaneous fat between the thumb and forefinger slightly above the midpoint mark. Gently pull the skin away from the underlying muscle. (This step takes a lot of practice. If you want to be sure you don't have muscle as well as fat, ask the subject to contract and relax the muscle. You should be able to feel if you are pinching muscle.)
   3. Place the calipers over the skinfold at the midpoint mark, and read the measurement to the nearest 1.0 millimeter in two to three seconds. (If using plastic calipers, align pressure lines, and read the measurement to the nearest 1.0 millimeter in two to three seconds.)
   4. Repeat steps 2 and 3 twice more. Add the three readings, and then divide by 3 to find the average.

Hydrodensitometry To estimate body density using hydrodensitometry, the person is weighed twice—first on land and then again when submerged in water. Underwater weighing usually generates a good estimate of body fat and is useful in research, although the technique has drawbacks: it requires bulky, expensive, and nonportable equipment. Furthermore, submerging some people in water (especially those who are very young, very old, ill, or fearful) is difficult and not well tolerated.

Bioelectrical Impedance To measure body fat using the bioelectrical impedance method, a very-low-intensity electrical current is briefly sent through the body by way of electrodes placed on the wrist and ankle. Fat impedes the flow of electricity; thus, the magnitude of the current is influenced by the body fat content. Recent food intake and hydration status can influence results. Like other anthropometric methods, bioelectrical impedance requires standardized procedures and calibrated instruments.

A number of other methods are sometimes used to estimate the body's content of body fat. Table E-1 (p. E-8) describes common techniques often used in the clinical or research setting.

FIGURE E-7 How to Measure Waist Circumference

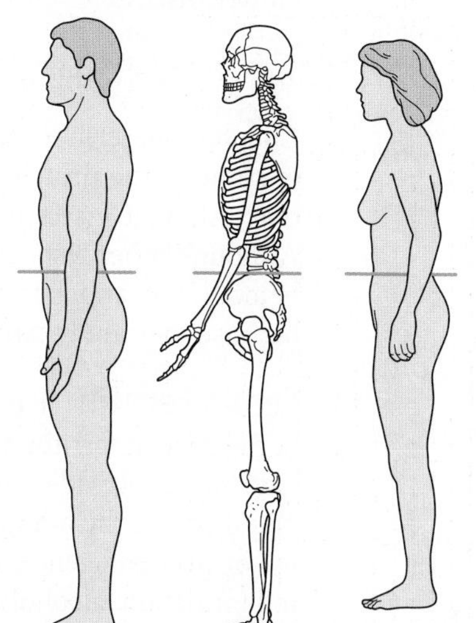

Place the measuring tape around the waist just above the bony crest of the hip. The tape runs parallel to the floor and is snug (but does not compress the skin). The measurement is taken at the end of normal expiration.

Source: National Institutes of Health Obesity Education Initiative, *Clinical Guidelines on the Identification, Evaluation, and Treatment of Overweight and Obesity in Adults* (Washington, D.C.: U.S. Department of Health and Human Services, 1998), p. 59.

TABLE E-1 Selected Methods for Estimating Body Fat Content

Method	Description
Air-displacement plethysmography (Bod Pod®)	Estimates body density by measuring the body's volume (density = mass/volume); the density value allows derivation of the body's fat and lean tissue contents
Bioelectrical impedance assay	Measures the magnitude of an electrical current passed through the body; electrical conductivity is higher in lean tissues than in fat tissue
Dual energy X-ray absorptiometry	Analyzes the change in X-rays after they contact body tissues; fat and lean tissues have different effects on X-rays, allowing quantification of the various tissues
Hydrodensitometry (underwater weighing)	Estimates body density by comparing the body's weight on land and in water or by measuring the body's volume (density = mass/volume); the density value allows derivation of the body's fat and lean tissue contents
Isotope dilution—deuterated water	Measures total body water content by analyzing the dilution of heavy water (water with a heavy form of hydrogen) in body tissues; allows an estimate of lean tissue and body fat content
Skinfolds	Estimates subcutaneous fat in several regions of the body by using calipers to measure skinfold thicknesses
Ultrasound	Estimates subcutaneous fat in several regions of the body by using ultrasound to measure skinfold thicknesses

Nutritional Anemias

Anemia, a symptom of a wide variety of nutrition- and nonnutrition-related disorders, is characterized by the reduced oxygen-carrying capacity of blood. Iron, folate, and vitamin B_{12} deficiencies—caused by inadequate intake, poor absorption, or abnormal metabolism of these nutrients—are the most common causes of nutritional anemias. Table E-2 lists laboratory tests that distinguish among the various nutrition-related anemias. Some nonnutrition-related causes of anemia include massive blood loss, infections, hereditary blood disorders such as sickle-cell anemia, and chronic liver or kidney disease.

ASSESSMENT OF IRON STATUS

Chapter 9 describes the progression of iron deficiency in detail, (■) as well as the roles of some of the proteins involved in iron metabolism. This section describes the various tests that assess iron status, and Table E-3 (p. E-10) provides acceptable values. Although other tests are more specific for detecting the early stages of iron deficiency, hemoglobin and hematocrit are most often used to detect iron-deficiency anemia because they are inexpensive and easily measured.

Serum Ferritin

In the initial stage of iron deficiency, iron stores diminish. Iron is stored in the protein ferritin, which is located in the liver, spleen, and bone marrow. Serum ferritin values provide a noninvasive estimate of iron stores, because the ferritin levels in blood reflect the amounts stored in the tissues. Serum ferritin is not a reliable indicator of iron deficiency, however, because its concentrations are increased by infection, inflammation, alcohol consumption, and liver disease.

■ Reminder: Iron deficiency progresses as follows:

1. Iron stores diminish
2. Transport iron decreases
3. Hemoglobin production falls

Serum Iron and Total Iron-Binding Capacity (TIBC)

Early stages of iron deficiency are characterized by reduced levels of serum iron, which represent the amount of iron bound to transferrin, the iron transport protein. Total iron-binding capacity (TIBC) is a measure of the total amount of iron that the transferrin in blood can carry;

TABLE E-2 Laboratory Tests Useful for Evaluating Nutrition-Related Anemias

Test or Test Result	What It Reflects
For Anemia (general)	
Hemoglobin (Hg)	Total amount of hemoglobin in the red blood cells (RBCs)
Hematocrit (Hct)	Percentage of RBCs in the total blood volume
Red blood cell (RBC) count	Number of RBCs
Mean corpuscular volume (MCV)	RBC size; helps to determine if anemia is microcytic (iron deficiency) or macrocytic (folate or vitamin B_{12} deficiency)
Mean corpuscular hemoglobin concentration (MCHC)	Hemoglobin concentration within the average RBC; helps to determine if anemia is hypochromic (iron deficiency) or normochromic (folate or vitamin B_{12} deficiency)
Bone marrow aspiration	The manufacture of blood cells in different developmental states
For Iron-Deficiency Anemia	
↓ Serum ferritin	Early deficiency state with depleted iron stores
↓ Transferrin saturation	Progressing deficiency state with diminished transport iron
↑ Erythrocyte protoporphyrin	Later deficiency state with limited hemoglobin production
For Folate-Deficiency Anemia	
↓ Serum folate	Progressing deficiency state
↓ RBC folate	Later deficiency state
For Vitamin B_{12}-Deficiency Anemia	
↓ Serum vitamin B_{12}	Progressing deficiency state
↑ Serum methylmalonic acid	Vitamin B_{12} deficiency
Schilling test	Adequacy of vitamin B_{12} absorption

thus, it is an indirect measure of the transferrin content of blood. During iron deficiency, the liver produces more transferrin in an effort to increase iron transport capacity, and therefore iron depletion is characterized by an increase in TIBC. TIBC reflects liver function as well as changes in iron metabolism.

Transferrin Saturation The percentage of transferrin that is saturated with iron is an indirect measure derived from the serum iron and total iron-binding capacity measures, as follows:

$$\%\ \text{Transferrin saturation} = \frac{\text{serum iron}}{\text{total iron-binding capacity}} \times 100$$

During iron deficiency, transferrin saturation decreases. The transferrin saturation value is a useful indicator of iron status because it includes information about both the iron and transferrin contents of blood.

Erythrocyte Protoporphyrin The iron-containing molecule in hemoglobin is heme, which is formed from iron and protoporphyrin. Protoporphyrin accumulates in the blood when iron supplies are inadequate for the formation of heme. However, levels of protoporphyrin may increase when hemoglobin synthesis is impaired for other reasons, such as lead poisoning or inflammation.

TABLE E-3 Criteria for Assessing Iron Status

Laboratory Test	Acceptable Values	Effect of Iron Deficiency
Serum ferritin	Male: 20–250 ng/mL Female: 10–120 ng/mL	Lower than normal
Serum iron	Male: 60–175 μg/dL Female: 50–170 μg/dL	Lower than normal
Total iron-binding capacity	250–450 μg/dL	Higher than normal
Transferrin saturation	Male: 20–50% Female: 15–50%	Lower than normal
Erythrocyte protoporphyrin	<70 μg/dL red blood cells	Higher than normal
Hemoglobin (Hb), whole blood	Male: 13.5–17.5 g/dL Female: 12.0–16.0 g/dL	Lower than normal
Hematocrit (Hct)	Male: 39–49% Female: 35–45%	Lower than normal
Mean corpuscular volume (MCV)	80–100 fL	Lower than normal

Note: ng = nanogram, μg = microgram, dL = deciliter, nmol = nanomole, fL= femtoliter

Source: L. Goldman and A. I. Schafer, eds., *Goldman's Cecil Medicine* (Philadelphia: Saunders, 2012), pp. 2558–2569.

Hemoglobin When iron stores become depleted, hemoglobin production is impaired, and symptoms of anemia may eventually develop. Hemoglobin's usefulness in evaluating iron status is limited, however, because hemoglobin concentrations drop fairly late in the development of iron deficiency, and other nutrient deficiencies and medical conditions can also alter hemoglobin concentrations.

Hematocrit The hematocrit is the percentage of the total blood volume occupied by red blood cells. To measure the hematocrit, a clinician spins the blood samples in a centrifuge to separate the red blood cells from the plasma. Low values indicate a reduced number or size of red blood cells. Although this test is not specific for iron status, it can help to detect the presence of iron-deficiency anemia.

Mean Corpuscular Volume (MCV) The hematocrit value divided by the red blood cell count provides a measure of the average size of a red blood cell, referred to as the mean corpuscular volume (MCV). This measure helps to classify the type of anemia that is present. In iron deficiency, the red blood cells are smaller than average (microcytic cells).

ASSESSMENT OF FOLATE AND VITAMIN B_{12} STATUS

Folate deficiency and vitamin B_{12} deficiency present a similar clinical picture—an anemia characterized by abnormally large, misshapen, and immature red blood cells (megaloblastic cells). Distinguishing between folate and vitamin B_{12} deficiency is essential, however, because their treatments differ. Giving folate to a person with vitamin B_{12} deficiency improves many of the test results indicative of vitamin B_{12} deficiency, but this would be a dangerous treatment because vitamin B_{12} deficiency causes nerve damage that folate cannot correct. Thus, inappropriate folate administration masks vitamin B_{12}-deficiency anemia, and nerve damage worsens. For this reason, it is critical to determine whether an anemia characterized by macrocytic cells results from a folate deficiency or from a vitamin B_{12} deficiency. Several of the following assessment measures help to make this distinction.

Mean Corpuscular Volume (MCV) As previously mentioned, MCV is a measure of red blood cell size. In folate and vitamin B_{12} deficiencies, the red blood cells are larger than average, or macrocytic. Macrocytic cells are not necessarily indicative of nutrient deficiency, however, as they may also result from a high alcohol intake, liver disease, and various medications.

Serum Folate and Vitamin B_{12} Levels Analyses of serum folate and vitamin B_{12} levels are usually among the first tests conducted to determine the cause of macrocytic red blood cells. The presence of low serum levels of either nutrient is consistent with a deficiency of that nutrient, whereas adequate levels can help to rule out deficiency. Folate levels are not a specific measure of folate status, however; they may increase after folate consumption, and decrease due to alcohol consumption, pregnancy, or use of anticonvulsants. The folate level in red blood cells—called erythrocyte folate—correlates well with folate stores and can help to diagnose folate deficiency, but this test is not available at all institutions. Table E-4 shows acceptable ranges for tests used for assessing folate and vitamin B_{12} status.

Methylmalonic Acid and Homocysteine Levels To determine whether a nutrient deficiency is present, clinicians can measure the levels of substances that accumulate when the functions of that nutrient are impaired. For example, blood levels of the amino acid homocysteine are usually increased by both folate and vitamin B_{12} deficiency because both nutrients are needed for its metabolism. Methylmalonic acid, a breakdown product of several amino acids, requires vitamin B_{12} for its metabolism; hence, serum levels increase as a result of vitamin B_{12} deficiency. Because methylmalonic acid levels are not influenced by folate status, this measure is useful in distinguishing between folate and vitamin B_{12} deficiency.

Schilling Test As Chapter 8 explains, vitamin B_{12} deficiency most often results from malabsorption, not poor intake. The Schilling test can help to diagnose malabsorption of vitamin B_{12}: after the patient takes an oral dose of radioactive vitamin B_{12}, a urine test determines whether the vitamin B_{12} was absorbed. The Schilling test is rarely performed at present due to the difficulty in obtaining the chemical reagents needed for the test.

TABLE E-4 Criteria for Assessing Folate and Vitamin B_{12} Status

Laboratory Test	Acceptable Range	Effect of Folate or Vitamin B_{12} Deficiency
Serum folate	3–16 ng/mL	Reduced in folate deficiency
Erythrocyte folate	140–628 ng/mL packed cells	Reduced in folate deficiency
Serum vitamin B_{12}	200–835 pg/mL	Reduced in vitamin B_{12} deficiency
Serum methylmalonic acid	150–370 nmol/L	Increased in vitamin B_{12} deficiency
Plasma homocysteine	Male: 4–16 µmol/L Female: 3–14 µmol/L	Increased in folate or vitamin B_{12} deficiency

Note: ng = nanogram, pg = picogram, nmol = nanomole, µmol = micromole

Source: L. Goldman and A. I. Schafer, eds., *Goldman's Cecil Medicine* (Philadelphia: Saunders, 2012), pp. 2558–2569; R. S. Porter and J. L. Kaplan, eds., *Merck Manual of Diagnosis and Therapy* (Whitehouse Station, NJ: Merck Sharp and Dohme Corp., 2011), pp. 3491–3502.

Antibodies to Intrinsic Factor The presence of serum antibodies for intrinsic factor can help to confirm a diagnosis of pernicious anemia, an autoimmune disease characterized by destruction of the cells that produce intrinsic factor (a protein required for vitamin B_{12} absorption; see Chapter 8). Serum antibodies to the parietal cells that produce and release intrinsic factor may also indicate pernicious anemia, but these antibodies may be present in various other conditions as well.

Cautions about Nutrition Assessment

The tests outlined in this appendix yield information that becomes meaningful only when they are conducted and interpreted by a skilled clinician. Potential sources of error may be introduced at any step, from the collection of samples to the analysis and reporting of data. Equipment must be regularly calibrated to ensure accuracy of measurements. In addition, the assessor must keep in mind that few tests may be specific to the nutrient of interest alone, and lab results may reflect physiological processes other than the ones being tested. Furthermore, because many tests are not sensitive enough to detect the early stages of deficiency, follow-up testing is often necessary to identify a nutrition problem.

appendix F Aids to Calculation

Many mathematical problems have been worked out in the "How to" sections of the text. These pages provide additional help and examples.

Conversion Factors

A *conversion factor* is a numerical factor—expressed as a ratio—that can convert a quantity expressed in one unit to another unit; for example, the factor may be used to convert pounds to kilograms or feet to inches. To create a conversion factor, an equality (such as *1 kilogram = 2.2 pounds*) is expressed as a fraction:

$$\frac{1 \text{ kg}}{2.2 \text{ lb}} \text{ and } \frac{2.2 \text{ lb}}{1 \text{ kg}}.$$

Because a conversion factor has a value of *1* (the value in the numerator is equal to the value in the denominator), it can be used as a multiplier to change the *unit* of measure without changing the *value* of the measurement. To convert the units of a measurement, the fraction must have the desired unit in the numerator, as the unit in the denominator will cancel out the original unit.

Example 1 Convert the weight of 130 pounds to kilograms.

- Multiply 130 pounds by a conversion factor (fraction) that includes both pounds and kilograms and is arranged so that the desired unit (kilograms) is in the numerator:

$$130 \text{ lb} \times \frac{1 \text{ kg}}{2.2 \text{ lb}} = \frac{130 \text{ kg}}{2.2} = 59 \text{ kg}.$$

As this example shows, the unit for *pounds* cancels out and the unit for *kilograms* remains in the solution.

Example 2 The food label on a bottle of apple juice shows the contents in both fluid ounces and liters. How many liters are contained in a bottle that holds 64 fluid ounces?

- Multiply 64 ounces by a conversion factor that includes both fluid ounces and liters, with the desired unit (liters) in the numerator:

$$64 \text{ fl oz} \times \frac{1 \text{ L}}{33.8 \text{ fl oz}} = \frac{64 \text{ L}}{33.8} = 1.89 \text{ L}.$$

Percentages

A percentage expresses a fraction that has 100 in the denominator; for example, the term *50 percent* is equivalent to the fraction 50/100. Similar to other fractions, percentages are used to express a *proportion* of the *whole*; therefore, the units in the numerator and denominator must be similar. Any fraction can be expressed in hundredths and converted to a percentage by dividing the numerator by the denominator and multiplying by 100:

$$1/4 = 0.25.$$
$$0.25 \times 100 = 25\%.$$

Example 3 Suppose your energy intake for the day is 2000 kcalories (kcal) and your recommended energy intake is 2400 kcalories. What percent of the recommended energy intake did you consume?

- Divide your intake by the recommended intake:

  2000 kcal (your intake) ÷ 2400 kcal (recommended intake) = 0.83.

- Multiply by 100 to express the decimal as a percentage:

  0.83 × 100 = 83%

Example 4 A percentage can also be greater than 100. Suppose your intake of vitamin C is 120 milligrams and your RDA (male) is 90 milligrams. What percent of the RDA for vitamin C did you consume?

120 mg (your intake) ÷ 90 mg (RDA) = 1.33.
1.33 × 100 = 133%.

Weights and Measures

LENGTH

1 meter (m) = 39 in
1 centimeter (cm) = 0.4 in
1 inch (in) = 2.5 cm
1 foot (ft) = 30 cm

TEMPERATURE

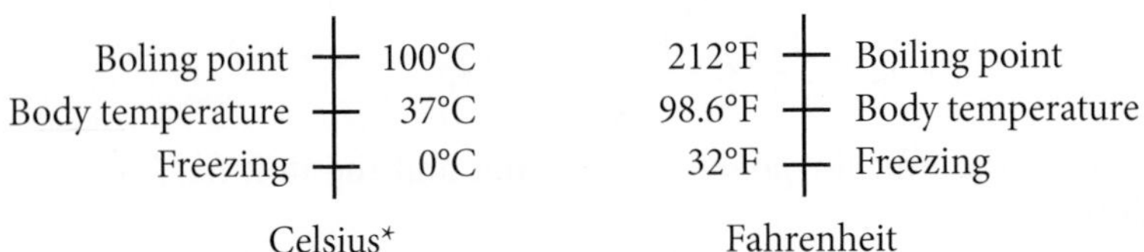

- To find degrees Fahrenheit (°F) when you know degrees Celsius (°C), multiply by 1.8 and then add 32.
- To find degrees Celsius (°C) when you know degrees Fahrenheit (°F), subtract 32 and then multiply by 0.56.

VOLUME

1 liter (L) = 1000 mL, 0.26 gal, 1.06 qt, 2.1 pt, or 33.8 fl oz
1 milliliter (mL) = 1/1000 L or 0.03 fl oz
1 gallon (gal) = 128 oz, 8 c, or 3.8 L
1 quart (qt) = 32 fl oz, 4 c, or 0.95 L
1 pint (pt) = 16 fl oz, 2 c, or 0.47 L
1 cup (c) = 8 fl oz, 16 tbs, about 250 mL, or 0.25 L
1 ounce (oz) = 30 mL
1 tablespoon (tbs) = 3 tsp or 15 mL
1 teaspoon (tsp) = 5 mL

*Also known as *centigrade*.

WEIGHT

1 kilogram (kg) = 1000 g or 2.2 lb
1 gram (g) = 1/1000 kg, 1000 mg, or 0.035 oz
1 milligram (mg) = 1/1000 g or 1000 µg
1 microgram (µg) = 1/1000 mg
1 pound (lb) = 16 oz, 454 g, or 0.45 kg
1 ounce (oz) = about 28 g

ENERGY

1 kilojoule (kJ) = 0.24 kcal
1 megajoule (MJ) = 240 kcal
1 kcalorie (kcal) = 4.2 kJ
1 g carbohydrate = 4 kcal = 17 kJ
1 g fat = 9 kcal = 37 kJ
1 g protein = 4 kcal = 17 kJ
1 g alcohol = 7 kcal = 29 kJ

appendix G Enteral Formulas

The large number of enteral formulas available allows patients to meet a wide variety of medical needs. The first step in narrowing the choice of formulas is to determine the patient's ability to digest and absorb nutrients. Table G-1 on pp. G through G-2 lists examples of standard formulas for patients who can adequately digest and absorb nutrients, and Table G-2 on p. G-3 provides examples of elemental formulas for patients with a limited ability to digest or absorb nutrients. Each formula is listed only once, although a formula may have more than one use. A high-protein formula, for example, may also be a fiber-containing formula. Tables G-3 and G-4 on p. G-4 list modules that can be used to prepare modular formulas or enhance enteral formulas.

The information shown in this appendix reflects the literature provided by manufacturers and does not suggest endorsement by the authors. Manufacturers frequently add new formulas, discontinue old ones, and change formula compositions. Consult the manufacturers' literature and websites for updates and additional examples of enteral formulas.[a] The following products are listed in this appendix:

- *Abbott Nutrition:* Glucerna 1.0 Cal, Jevity 1 Cal, Jevity 1.5 Cal, Nepro with Carb Steady, Osmolite 1 Cal, Oxepa, Pivot 1.5 Cal, Polycose, Promote, Promote with Fiber, Pulmocare, Suplena with Carb Steady, Vital 1.0 Cal
- *Nestlé Nutrition*: Beneprotein, Compleat, Compleat Pediatric, Diabetisource AC, Fibersource HN, Impact, Impact Peptide 1.5, Impact Glutamine, Isosource HN, MCT Oil, Microlipid, Novasource Renal, Nutren 1.0, Nutren 1.0 Fiber, Nutren 1.5, Nutren 2.0, Nutren Glytrol, Nutren Junior, Nutren Pulmonary, Nutren Replete, Nutren Replete Fiber, NutriHep, Peptamen, Peptamen Junior, Vivonex Pediatric, Vivonex T.E.N.

[a]Sources for the information in this appendix: Abbott Nutrition, www.abbottnutrition.com; Nestlé Nutrition, www.nestle-nutrition.com.

TABLE G-1 Standard Formulas

Product	Volume to Meet 100% RDI[a] (mL)	Energy (kcal/mL)	Protein or Amino Acids (g/L)	Carbohydrate (g/L)	Fat (g/L)	Notes
Lactose-Free Standard Formulas						
Compleat	1400	1.06	48	132	40	Blenderized formula, 6 g fiber/L
Nutren 1.0	1500	1.00	40	128	38	25% fat from MCT
Osmolite 1 Cal	1321	1.06	44	144	35	20% fat from MCT
Lactose-Free, Fiber-Enhanced Formulas						
Jevity 1 Cal	1321	1.06	44	155	35	14 g fiber/L
Nutren 1.0 Fiber	1500	1.00	40	128	38	14 g fiber/L
Promote with Fiber	1000	1.00	63	138	28	14 g fiber/L

[a]RDI = Reference Daily Intakes, which are labeling standards for vitamins, minerals, and protein. Consuming 100 percent of the RDI will meet the nutrient needs of most people using the product.

continued

TABLE G-1 Standard Formulas, *continued*

Product	Volume to Meet 100% RDI[a] (mL)	Energy (kcal/mL)	Protein or Amino Acids (g/L)	Carbohydrate (g/L)	Fat (g/L)	Notes
Lactose-Free, High-kCalorie Formulas						
Jevity 1.5 Cal	1000	1.50	64	216	50	22 g fiber/L
Nutren 1.5	1000	1.50	60	168	68	50% fat from MCT
Nutren 2.0	750	2.00	80	196	104	75% fat from MCT
Lactose-Free, High-Protein Formulas						
Fibersource HN	1250	1.20	54	160	39	20% fat from MCT, 10 g fiber/L
Isosource HN	1165	1.20	54	160	39	20% fat from MCT, low residue
Promote	1000	1.00	63	130	26	19% fat from MCT, low residue
Specialized Formulas: Pediatric (1 to 10 years)						
Compleat Pediatric	Varies[b]	1.06	48	132	40	Blenderized formula, 6 g fiber/L
Nutren Junior	Varies[b]	1.00	30	110	50	20% fat from MCT
Specialized Formulas: Glucose Intolerance						
Diabetisource AC	1250	1.20	60	100	59	36% kcal from carbohydrate, 15 g fiber/L
Glucerna 1.0 Cal	1420	1.00	42	96	54	34% kcal from carbohydrate, 14 g fiber/L
Nutren Glytrol	1500	1.00	45	100	48	40% kcal from carbohydrate, 15 g fiber/L
Specialized Formulas: Immune System Support						
Impact	1500	1.00	56	132	28	Enriched with arginine, nucleotides, and omega-3 fatty acids
Impact Peptide 1.5	1000	1.50	94	140	64	Same as above
Impact Glutamine	1000	1.30	78	148	43	Same as above, and enriched with glutamine
Specialized Formulas: Chronic Kidney Disease (CKD)						
Nepro with Carb Steady	944	1.80	81	161	96	Low in potassium and phosphorus; to be used after dialysis has been instituted

[a]RDI = Reference Daily Intakes, which are labeling standards for vitamins, minerals, and protein. Consuming 100 percent of the RDI will meet the nutrient needs of most people using the product.

[b]Depends on age of child

continued

TABLE G-1 Standard Formulas, *continued*

Product	Volume to Meet 100% RDI[a] (mL)	Energy (kcal/mL)	Protein or Amino Acids (g/L)	Carbohydrate (g/L)	Fat (g/L)	Notes
Novasource Renal	1000	2.00	91	183	100	Low in electrolytes; to be used after dialysis has been instituted
Suplena with Carb Steady	1000	1.80	45	196	96	Low in protein and electrolytes; for patients with CKD (stage 3 or 4)
Specialized Formulas: Respiratory Insufficiency						
Nutren Pulmonary	1000	1.50	68	100	95	56% kcal from fat, 40% fat from MCT
Oxepa	946	1.50	63	105	94	Enriched with omega-3 fatty acids and antioxidants; for mechanically ventilated patients
Pulmocare	947	1.50	63	106	93	55% kcal from fat, 20% fat from MCT, enriched with antioxidant nutrients
Specialized Formulas: Wound Healing						
Nutren Replete	1000	1.00	62	112	34	Enhanced with vitamins and minerals; for patients recovering from surgery, burns, or pressure ulcers
Nutren Replete Fiber	1000	1.00	62	112	34	Same as above; 14 g fiber/L

Note: MCT = Medium-chain triglycerides

[a]RDI = Reference Daily Intakes, which are labeling standards for vitamins, minerals, and protein. Consuming 100 percent of the RDI will meet the nutrient needs of most people using the product.

[b]Depends on age of child

TABLE G-2 Elemental Formulas

Product	Volume to Meet 100% RDI[a] (mL)	Energy (kcal/mL)	Protein or Amino Acids (g/L)	Carbohydrate (g/L)	Fat (g/L)	Notes
Specialized Elemental Formula: Hepatic Insufficiency						
NutriHep	1000	1.50	40	290	21	Free amino acids, high in branched-chain amino acids and low in aromatic amino acids
Specialized Elemental Formula: Immune System Support						
Pivot 1.5 Cal	1500	1.50	94	172	51	Enriched with arginine, glutamine, omega-3 fatty acids, and antioxidant nutrients
Specialized Elemental Formulas: Malabsorption						
Peptamen	1500	1.00	40	128	39	70% fat from MCT
Vital 1.0 Cal	1422	1.00	40	130	38	Enhanced with prebiotics and antioxidants
Vivonex T.E.N.	2000	1.00	38	204	3	Powder form, 100% free amino acids, very low fat
Specialized Elemental Formulas: Pediatric (1 to 10 years)						
Peptamen Junior	Varies[b]	1.00	30	136	38	60% fat from MCT
Vivonex Pediatric	Varies[b]	0.80	24	126	23	Powder form, 100% free amino acids

Note: MCT = Medium-chain triglycerides

[a]RDI = Reference Daily Intakes, which are labeling standards for vitamins, minerals, and protein. Consuming 100 percent of the RDI will meet the nutrient needs of most people using the product.

[b]Depends on age of child

TABLE G-3 Protein and Carbohydrate Modules

Product	Major Ingredient	Energy (kcal/g)	Nutrient Content (g/100 g)
Beneprotein	Whey protein powder	3.6	86 g protein
Polycose	Hydrolyzed cornstarch (powder)	3.8	94 g carbohydrate

TABLE G-4 Fat Modules

Product	Major Ingredient	Energy (kcal/mL)	Fat Content (g/100 mL)
MCT Oil	Coconut and/or palm kernel oil	7.7	93
Microlipid	Safflower oil	4.5	50

appendix H Answers to Self Check Questions

CHAPTER 1:
1. d, 2. d, 3. c, 4. d, 5. c, 6. b, 7. c, 8. b, 9. c, 10. c

CHAPTER 2:
1. b, 2. a, 3. d, 4. a, 5. c, 6. a, 7. d, 8. c, 9. d, 10. d

CHAPTER 3:
1. b, 2. b, 3. c, 4. a, 5. a, 6. a, 7. b, 8. a, 9. d, 10. c

CHAPTER 4:
1. b, 2. c, 3. a, 4. d, 5. a, 6. a, 7. d, 8. d, 9. b, 10. a

CHAPTER 5:
1. c, 2. b, 3. d, 4. d, 5. a, 6. d, 7. c, 8. b, 9. c, 10. d

CHAPTER 6:
1. b, 2. c, 3. d, 4. d, 5. b, 6. a, 7. a, 8. b, 9. c, 10. c

CHAPTER 7:
1. b, 2. a, 3. a, 4. c, 5. b, 6. c, 7. d, 8. b, 9. c, 10. b

CHAPTER 8:
1. d, 2. d, 3. c, 4. a, 5. d, 6. d, 7. b, 8. c, 9. c, 10. a

CHAPTER 9:
1. d, 2. b, 3. a, 4. d, 5. c, 6. d, 7. a, 8. d, 9. d, 10. b

CHAPTER 10:
1. a, 2. d, 3. c, 4. c, 5. b, 6. d, 7. c, 8. d, 9. d, 10. c

CHAPTER 11:
1. b, 2. a, 3. b, 4. d, 5. c, 6. d, 7. c, 8. b, 9. c, 10. d

CHAPTER 12:
1. d, 2. c, 3. d, 4. d, 5. d, 6. d, 7. b, 8. b, 9. a, 10. b

CHAPTER 13:
1. b, 2. d, 3. c, 4. d, 5. a, 6. d, 7. b, 8. d, 9. a, 10. c

CHAPTER 14:
1. c, 2. b, 3. c, 4. b, 5. d, 6. a, 7. d, 8. c, 9. d, 10. a

CHAPTER 15:
1. b, 2. c, 3. a, 4. b, 5. d, 6. d, 7. b, 8. c, 9. a, 10. c

CHAPTER 16:
1. d, 2. a, 3. b, 4. c, 5. d, 6. b, 7. c, 8. c, 9. d, 10. a

CHAPTER 17:
1. c, 2. d, 3. a, 4. c, 5. a, 6. c, 7. d, 8. b, 9. b, 10. c

CHAPTER 18:
1. d, 2. c, 3. b, 4. b, 5. c, 6. a, 7. c, 8. d, 9. a, 10. d

CHAPTER 19:
1. d, 2. a, 3. b, 4. c, 5. a, 6. b, 7. c, 8. d, 9. c, 10. a

CHAPTER 20:
1. b, 2. d, 3. d, 4. c, 5. a, 6. c, 7. d, 8. c, 9. a, 10. b

CHAPTER 21:
1. a, 2. d, 3. b, 4. c, 5. b, 6. c, 7. d, 8. d, 9. b, 10. c

CHAPTER 22:
1. a, 2. c, 3. b, 4. d, 5. c, 6. a, 7. b, 8. b, 9. c, 10. d

CHAPTER 23:
1. d, 2. c, 3. c, 4. b, 5. d, 6. b, 7. b, 8. a, 9. a, 10. d

Glossary

2-in-1 solution: a parenteral solution that contains dextrose and amino acids, but excludes lipids.

24-hour dietary recall: a record of foods consumed during the previous day or in the past 24 hours; sometimes modified to include foods consumed in a typical day.

abscesses (AB-sess-es): accumulations of pus.

Acceptable Daily Intake (ADI): the amount of a nonnutritive sweetener that individuals can safely consume each day over the course of a lifetime without adverse effect. It includes a 100-fold safety factor.

Acceptable Macronutrient Distribution Ranges (AMDR): ranges of intakes for the energy-yielding nutrients that provide adequate energy and nutrients and reduce the risk of chronic disease.

achalasia (ack-ah-LAY-zhah): an esophageal disorder characterized by weakened peristalsis and impaired relaxation of the lower esophageal sphincter.

achlorhydria (AY-clor-HIGH-dree-ah): absence of gastric acid secretions.

acid-base balance: the balance maintained between acid and base concentrations in the blood and body fluids.

acidosis: acid accumulation in body tissues; depresses the central nervous system and may lead to disorientation and, eventually, coma.

acids: compounds that release hydrogen ions in a solution.

acquired immune deficiency syndrome (AIDS): the late stage of illness caused by infection with the human immunodeficiency virus (HIV); characterized by severe damage to immune function.

acute kidney injury: the rapid decline of kidney function over a period of hours or days; potentially a cause of acute renal failure.

acute respiratory distress syndrome (ARDS): respiratory failure triggered by severe lung injury; a medical emergency that causes dyspnea and pulmonary edema and usually requires mechanical ventilation.

added sugars: sugars, syrups, and other kcaloric sweeteners that are added to foods during processing or preparation or at the table. Added sugars do not include the naturally occurring sugars found in fruits and milk products.

adequacy: the characteristic of a diet that provides all the essential nutrients, fiber, and energy necessary to maintain health and body weight.

Adequate Intakes (AI): a set of values that are used as guides for nutrient intakes when scientific evidence is insufficient to determine an RDA.

adipokines (AD-ih-poh-kynz): protein hormones made and released by adipose tissue (fat) cells.

adipose tissue: the body's fat, which consists of masses of fat-storing cells called adipose cells.

adolescence: the period of growth from the beginning of puberty until full maturity. Timing of adolescence varies from person to person.

advanced glycation end products (AGEs): reactive compounds formed after glucose combines with protein; AGEs can damage tissues and lead to diabetic complications.

adverse reactions: unusual responses to food (including intolerances and allergies).

aerobic physical activity: activity in which the body's large muscles move in a rhythmic manner for a sustained period of time. Aerobic activity, also called *endurance activity*, improves cardiorespiratory fitness. Brisk walking, running, swimming, and bicycling are examples.

AIDS-defining illnesses: diseases and complications associated with the later stages of an HIV infection, including wasting, recurrent bacterial pneumonia, opportunistic infections, and certain cancers.

alcohol-related birth defects (ARBD): a condition caused by prenatal alcohol exposure that is diagnosed when there is a history of substantial, regular maternal alcohol intake or heavy episodic drinking and birth defects known to be associated with alcohol exposure.

alcohol-related neurodevelopmental disorder (ARND): a condition caused by prenatal alcohol exposure that is diagnosed when there is a confirmed history of substantial, regular maternal alcohol intake or heavy episodic drinking and behavioral, cognitive, or central nervous system abnormalities known to be associated with alcohol exposure.

aldosterone (al-DOS-ter-own): a hormone secreted by the adrenal glands that stimulates the reabsorption of sodium by the kidneys; also regulates chloride and potassium concentrations.

alkalosis: excessive base in the blood and body fluids.

alpha-lactalbumin (lackt-AL-byoo-min): the chief protein in human breast milk, as **casein** (CAY-seen) is the chief protein in cow's milk.

alveoli (al-VEE-oh-lee): air sacs in the lungs. One air sac is an *alveolus*.

Alzheimer's disease: a progressive, degenerative disease that attacks the brain and impairs thinking, behavior, and memory.

amino (a-MEEN-oh) acids: building blocks of protein. Each contains an amino group, an acid group, a hydrogen atom, and a distinctive side group, all attached to a central carbon atom.

amniotic (am-nee-OTT-ic) sac: the "bag of waters" in the uterus in which the fetus floats.

anaphylactic (AN-ah-feh-LAC-tic) shock: a life-threatening whole-body allergic reaction to an offending substance.

anencephaly (AN-en-SEF-a-lee): an uncommon and always fatal type of neural tube defect; characterized by the absence of a brain.

anorexia: loss of appetite.

anthropometric (AN-throw-poe-MEH-trik): related to physical measurements of the human body, such as height, weight, body circumferences, and percentage of body fat.

antibodies: large proteins of the blood and body fluids, produced in response to invasion of the body by unfamiliar molecules (mostly proteins) called *antigens*. Antibodies inactivate the invaders and so protect the body.

antidiuretic hormone (ADH): a hormone released by the pituitary gland in response to high salt concentrations in the blood. The kidneys respond by reabsorbing water.

antigens: substances that elicit the formation of antibodies or an inflammation reaction from the immune system. A bacterium, a virus, a toxin, and a protein in food that causes allergy are all examples of antigens.

antioxidant (anti-OX-ih-dant): a compound that protects other compounds from oxygen by itself reacting with oxygen. Antioxidant food additives are preservatives that delay or prevent rancidity of foods and other damage to food caused by oxygen. (See also: *dietary antioxidants.*)

anuria (ah-NOO-ree-ah): the absence of urine; clinically identified as urine output less than 50 mL/day.

appetite: the psychological desire to eat; a learned motivation that is experienced as a pleasant sensation that accompanies the sight, smell, or thought of appealing foods.

artery: a vessel that carries blood away from the heart.

arthritis: inflammation of a joint, usually accompanied by pain, swelling, and structural changes.

artificial fats: zero-energy fat replacers that are chemically synthesized to mimic the sensory and cooking qualities of naturally occurring fats but are totally or partially resistant to digestion.

ascites (ah-SIGH-teez): an abnormal accumulation of fluid in the abdominal cavity.

ascorbic acid: one of the two active forms of vitamin C. Many people refer to vitamin C by this name.

aspiration: drawing in by suction or breathing; a common complication of enteral feedings in which foreign material enters the lungs, often from GI secretions or the reflux of stomach contents.

aspiration pneumonia: a lung disease resulting from the abnormal entry of foreign material; caused by either bacterial infection or irritation of the lower airways.

atherogenic: able to initiate or promote atherosclerosis.

atherosclerosis (ATH-er-oh-scler-OH-sis): a type of artery disease characterized by accumulations of fatty material on the inner walls of arteries.

atrophic gastritis (a-TRO-fik gas-TRI-tis): a condition characterized by chronic inflammation of the stomach accompanied by a diminished size and functioning of the mucosa and glands.

autoimmune: an immune response directed against the body's own tissues.

bacterial overgrowth: excessive bacterial colonization of the stomach and small intestine; may be due to low gastric acidity, altered GI motility, mucosal damage, or contamination.

balance: the dietary characteristic of providing foods in proportion to one another and in proportion to the body's needs.

bariatric (BAH-ree-AH-trik) surgery: surgery that treats severe obesity.

Barrett's esophagus: a condition in which esophageal cells damaged by chronic exposure to stomach acid are replaced by cells that resemble those in the stomach or small intestine, sometimes becoming cancerous.

basal metabolic rate (BMR): the rate of energy use for metabolism under specified conditions: after a 12-hour fast and restful sleep, without any physical activity or emotional excitement, and in a comfortable setting. It is usually expressed as kcalories per kilogram of body weight per hour.

basal metabolism: the energy needed to maintain life when a person is at complete digestive, physical, and emotional rest. Basal metabolism is normally the largest part of a person's daily energy expenditure.

bases: compounds that accept hydrogen ions in a solution.

behavior modification: the changing of behavior by the manipulation of *antecedents* (cues or environmental factors that trigger behavior), the behavior itself, and *consequences* (the penalties or rewards attached to behavior).

beriberi: the thiamin-deficiency disease; characterized by loss of sensation in the hands and feet, muscular weakness, advancing paralysis, and abnormal heart action.

beta-carotene: a vitamin A precursor made by plants and stored in human fat tissue; an orange pigment.

BHA, BHT: preservatives commonly used to slow the development of "off" flavors, odors, and color changes caused by oxidation.

bifidus (BIFF-id-us, by-FEED-us) factors: factors in colostrum and breast milk that favor the growth of the "friendly" bacterium *Lactobacillus* (lack-toe-ba-SILL-us) *bifidus* in the infant's intestinal tract. These bacteria prevent other, less desirable intestinal inhabitants from flourishing.

binders: chemical compounds in foods that combine with nutrients (especially minerals) to form complexes the body cannot absorb. Examples include *phytates* and *oxalates*.

bioactive food components: compounds in foods (either nutrients or phytochemicals) that alter physiological processes in the body.

bioavailability: the rate and extent to which a nutrient is absorbed and used.

blastocyst (BLASS-toe-sist): the developmental stage of the zygote when it is about five days old and ready for implantation.

blenderized formulas: enteral formulas that are prepared by using a food blender to mix and puree whole foods.

body mass index (BMI): an index of a person's weight in relation to height; determined by dividing the weight (in kilograms) by the square of the height (in meters).

bolus (BOH-lus): the portion of food swallowed at one time.

bolus (BOH-lus) feeding: delivery of about 250 to 500 milliliters of formula over a 5- to 15-minute period.

Bowman's (BOE-minz) capsule: a cuplike component of the nephron that surrounds the glomerulus and collects the filtrate that is passed to the tubules.

bronchi (BRON-key), bronchioles (BRON-key-oles): the main airways of the lungs. The singular form of bronchi is *bronchus*.

buffalo hump: the accumulation of fatty tissue at the base of the neck.

buffers: compounds that can reversibly combine with hydrogen ions to help keep a solution's acidity or alkalinity constant.

built environment: the buildings, roads, utilities, homes, fixtures, parks, and all other manufactured entities that form the physical characteristics of a community.

calories: units in which energy is measured. (See also: *kilocalories*.)

cancer cachexia (ka-KEK-see-ah): a wasting syndrome associated with cancer that is characterized by anorexia, muscle wasting, weight loss, and fatigue.

cancers: diseases characterized by the uncontrolled growth of abnormal cells, which can destroy adjacent tissues and spread to other areas of the body via lymph or blood.

candidiasis: a fungal infection on the mucous membranes of the oral cavity and elsewhere; usually caused by *Candida albicans*.

capillaries: small vessels that branch from an artery. Capillaries connect arteries to veins. Oxygen, nutrients, and waste materials are exchanged across capillary walls.

carbohydrate-to-insulin ratio: the amount of carbohydrate that can be handled per unit of insulin; on average, every 15 grams of carbohydrate requires about 1 unit of rapid- or short-acting insulin.

carbohydrates: energy nutrients composed of monosaccharides.

carcinogenesis (CAR-sin-oh-JEN-eh-sis): the process of cancer development.

carcinogens (CAR-sin-oh-jenz or **car-SIN-oh-jenz):** substances that can cause cancer (the adjective is *carcinogenic*).

cardiac cachexia: a condition of severe malnutrition that develops in heart failure patients; characterized by weight loss and tissue wasting.

cardiac output: the volume of blood pumped by the heart within a specified period of time.

cardiovascular disease (CVD): a group of disorders involving the heart and blood vessels.

cataracts: clouding of the eye lenses that impairs vision and can lead to blindness.

cathartics (ca-THART-ics): strong laxatives.

catheter: a thin tube placed within a narrow lumen (such as a blood vessel) or body cavity; can be used to infuse or withdraw fluids or keep a passage open.

celiac (SEE-lee-ack) disease: an immune disorder characterized by an abnormal response to wheat gluten and related proteins; also called *gluten-sensitive enteropathy* or *celiac sprue.*

central obesity: excess fat around the trunk of the body; also called *abdominal fat* or *upper-body fat.*

central veins: the large-diameter veins located close to the heart.

cerebral cortex: the outer surface of the cerebrum, which is the largest part of the brain.

Certified Diabetes Educator (CDE): a health care professional who specializes in diabetes management education; certification is obtained from the National Certification Board for Diabetes Educators.

cesarean (see-ZAIR-ee-un) section: surgical childbirth, in which the infant is taken through an incision in the woman's abdomen.

chemotherapy: the use of drugs to arrest or destroy cancer cells; these drugs are called *antineoplastic agents.*

choline: a nonessential nutrient that can be made in the body from an amino acid.

chronic bronchitis (bron-KYE-tis): a lung disorder characterized by persistent inflammation and excessive secretions of mucus in the main airways of the lungs.

chronic diseases: diseases characterized by slow progression, long duration, and degeneration of body organs due in part to such personal lifestyle elements as poor food choices, smoking, alcohol use, and lack of physical activity.

chronic hypertension: in pregnant women, hypertension that is present and documented before pregnancy; in women whose prepregnancy blood pressure is unknown, the presence of sustained hypertension before 20 weeks of gestation.

chronic kidney disease: kidney disease characterized by gradual, irreversible deterioration of the kidneys; also known as *chronic renal failure.*

chronic malnutrition: malnutrition caused by long-term food deprivation; characterized in children by short height for age (stunting).

chronic obstructive pulmonary disease (COPD): a group of lung diseases characterized by persistent obstructed airflow through the lungs and airways; includes chronic bronchitis and emphysema.

chronological age: a person's age in years from his or her date of birth.

chylomicrons (kye-lo-MY-crons): the lipoproteins that transport lipids from the intestinal cells into the body. The cells of the body remove the lipids they need from the chylomicrons, leaving chylomicron remnants to be picked up by the liver cells.

chyme (KIME): the semiliquid mass of partly digested food expelled by the stomach into the duodenum (the top portion of the small intestine).

cirrhosis (sih-ROE-sis): an advanced stage of liver disease in which extensive scarring replaces healthy liver tissue, causing impaired liver function and liver failure.

clear liquid diet: a diet that consists of foods that are liquid at body temperature, require minimal digestion, and leave little residue (undigested material) in the colon.

clinically severe obesity: a BMI of 40 or greater, or a BMI of 35 or greater with one or more serious conditions such as hypertension. Another term used to describe the same condition is *morbid obesity.*

closed feeding system: a formula delivery system in which the formula comes prepackaged in a container that can be attached directly to the feeding tube for administration.

coenzyme (co-EN-zime): a small molecule that works with an enzyme to promote the enzyme's activity. Many coenzymes contain B vitamins as part of their structure.

cognitive skills: as taught in behavior therapy, changes to conscious thoughts with the goal of improving adherence to lifestyle modifications; examples are problem-solving skills or the correction of false negative thoughts, termed *cognitive restructuring.*

colectomy: removal of a portion or all of the colon.

collagen: the characteristic protein of connective tissue.

collateral vessels: blood vessels that enlarge or newly form to allow an alternative pathway for diverted blood.

colostomy (co-LAH-stoe-me): a surgical passage through the abdominal wall into the colon.

colostrum (co-LAHS-trum): a milk-like secretion from the breasts that is rich in protective factors. Colostrum is present during the first day or so after delivery, before milk appears.

complementary proteins: two or more proteins whose amino acid assortments complement each other in such a way that the essential amino acids missing from one are supplied by the other.

conditionally essential amino acid: an amino acid that is normally nonessential but must be supplied by the diet in special circumstances when the need for it becomes greater than the body's ability to produce it.

conjugated linoleic acid: a collective term for several fatty acids that have the same chemical formulas as linoleic acid but with different configurations.

continuous feedings: slow delivery of formula at a constant rate over an 8- to 24-hour period.

continuous glucose monitoring: continuous monitoring of tissue glucose levels using a small sensor placed under the skin.

continuous parenteral nutrition: continuous administration of parenteral solutions over a 24-hour period.

cornea (KOR-nee-uh): the hard, transparent membrane covering the outside of the eye.

cretinism (CREE-tin-ism): an iodine-deficiency disease characterized by mental and physical retardation.

critical pathways: coordinated programs of treatment that merge the care plans of different health practitioners; also called *clinical pathways.*

critical period: a finite period during development in which certain events occur that will have irreversible effects on later developmental stages; usually a period of rapid cell division.

Crohn's disease: an inflammatory bowel disease that usually occurs in the lower portion of the small intestine and the colon. Inflammation may pervade the entire intestinal wall.

cryptosporidiosis (KRIP-toe-spor-ih-dee-OH-sis): a foodborne illness caused by the parasite *Cryptosporidium parvum.*

cultural competence: an awareness and acceptance of one's own and others' cultures combined with the skills needed to interact effectively with people of diverse cultures.

cyanosis (sigh-ah-NOH-sis): a bluish cast in the skin due to the color of deoxygenated hemoglobin. Cyanosis is most evident in individuals with lighter, thinner skin; it is mostly seen on lips, cheeks, and ears and under the nails.

cyclic parenteral nutrition: administration of parenteral solutions over an 8- to 16-hour period each day.

cystic fibrosis: an inherited disease characterized by the presence of abnormally viscous exocrine secretions; often leads to respiratory illness and pancreatic insufficiency.

cystinuria (SIS-tin-NOO-ree-ah): an inherited disorder characterized by the elevated urinary excretion of several amino acids, including cystine.

Daily Values: reference values developed by the FDA specifically for use on food labels.

dawn phenomenon: morning hyperglycemia that is caused by the early-morning release of growth hormone, which reduces insulin sensitivity.

debridement: the surgical removal of dead, damaged, or contaminated tissue resulting from burns or wounds; helps to prevent infection and hasten healing.

deep vein thrombosis: formation of a stationary blood clot (thrombus) in a deep vein, usually in the leg, which causes inflammation, pain, and swelling, and is potentially fatal.

deficient: in regard to nutrient intake, describes the amount below which almost all healthy people can be expected, over time, to experience deficiency symptoms.

dehydration: the loss of water from the body that occurs when water output exceeds water input. The symptoms progress rapidly from thirst, to weakness, to exhaustion and delirium, and end in death if not corrected.

denaturation (dee-nay-cher-AY-shun): the change in a protein's shape brought about by heat, acid, or other agents. Past a certain point, denaturation is irreversible.

dental caries: the gradual decay and disintegration of a tooth.

dermatitis herpetiformis (DERM-ah-TYE-tis HER-peh-tih-FOR-mis): a gluten-sensitive disorder characterized by a severe skin rash.

diabetes (DYE-ah-BEE-teez) mellitus: a group of metabolic disorders characterized by hyperglycemia and disordered insulin metabolism.

dialysate (dye-AL-ih-sate): the solution used in dialysis to draw wastes and fluids from the blood.

dialysis (dye-AH-lih-sis): a treatment that removes wastes and excess fluid from the blood after the kidneys have stopped functioning.

dialyzer (DYE-ah-LYE-zer): a machine used in hemodialysis to filter the blood; also called an *artificial kidney.*

diet manual: a resource that specifies the foods or preparation methods to include or exclude in modified diets and provides sample menus.

diet orders: specific instructions concerning dietary management; also called *diet prescriptions* or *nutrition prescriptions.*

diet progression: a change in diet as a patient's tolerances permit.

dietary antioxidants: compounds typically found in plant foods that significantly decrease the adverse effects of oxidation on living tissues. The major antioxidant vitamins are vitamin E, vitamin C, and beta-carotene.

dietary fibers: a general term denoting in plant foods the polysaccharides cellulose, hemicellulose, pectins, gums, and mucilages, as well as the nonpolysaccharide lignins, which are not digested by human digestive enzymes, although some are digested by GI tract bacteria.

dietary folate equivalents (DFE): the amount of folate available to the body from naturally occurring sources, fortified foods, and supplements, accounting for differences in bioavailability from each source.

Dietary Reference Intakes (DRI): a set of values for the dietary nutrient intakes of healthy people in the United States and Canada. These values are used for planning and assessing diets.

dietary supplements: products that are added to the diet and contain any of the following ingredients: a vitamin; a mineral; an herb or other botanical; an amino acid; a metabolite; a constituent; or an extract.

differentiation: the development of specific functions different from those of the original.

digestion: the process by which complex food particles are broken down to smaller absorbable particles.

digestive system: all the organs and glands associated with the ingestion and digestion of food.

dipeptide: two amino acids bonded together.

disaccharides (dye-SACK-uh-rides): pairs of sugar units bonded together.

discretionary kcalories: the kcalories remaining in a person's energy allowance after consuming enough nutrient-dense foods to meet all nutrient needs for a day.

diuresis (DYE-uh-REE-sis): increased urine production.

diuretics (dye-yoo-RET-ics): medications that promote the excretion of water through the kidneys. Not all diuretics increase the urinary loss of potassium. Some, called potassium-sparing diuretics, are less likely to result in a potassium deficiency (see Chapter 21).

diverticulitis (DYE-ver-tic-you-LYE-tis): inflammation or infection involving diverticula.

diverticulosis (DYE-ver-tic-you-LOH-sis): an intestinal condition characterized by the presence of small herniations (called diverticula) in the intestinal wall.

dumping syndrome: a cluster of symptoms that result from the rapid emptying of an osmotic load from the stomach into the small intestine.

dyspepsia: symptoms of pain or discomfort in the upper abdominal area, often called *indigestion*; a symptom of illness rather than a disease itself.

dysphagia: difficulty swallowing.

dyspnea (DISP-nee-ah): shortness of breath.

eating pattern: customary intake of foods and beverages over time.

eclampsia (eh-KLAMP-see-ah): a severe complication during pregnancy in which seizures occur.

edema (eh-DEEM-uh): the swelling of body tissue caused by leakage of fluid from the blood vessels and accumulation of the fluid in the interstitial spaces.

electrolyte: a salt that dissolves in water and dissociates into charged particles called ions.

electrolyte solutions: solutions that can conduct electricity.

elemental formulas: enteral formulas that contain proteins and carbohydrates that are partially or fully hydrolyzed; also called *hydrolyzed, chemically defined*, or *monomeric formulas.*

embryo (EM-bree-oh): the developing infant from two to eight weeks after conception.

emphysema (EM-fih-ZEE-mah): a progressive lung disease characterized by the breakdown of the lungs' elastic structure and destruction of the walls of the bronchioles and alveoli, reducing the surface area involved in respiration.

empty-kcalorie foods: a popular term used to denote foods that contribute energy but lack protein, vitamins, and minerals.

emulsifier: a substance that mixes with both fat and water and that disperses the fat in the water, forming an emulsion.

end-stage renal disease: an advanced stage of chronic kidney disease in which dialysis or a kidney transplant is necessary to sustain life.

energy density: a measure of the energy a food provides relative to the amount of food (kcalories per gram).

energy-yielding nutrients: the nutrients that break down to yield energy the body can use. The three energy-yielding nutrients are carbohydrate, protein, and fat.

enrichment: the addition to a food of nutrients to meet a specified standard. In the case of refined bread or cereal, five nutrients have been added: thiamin, riboflavin, niacin, and folate in amounts approximately equivalent to, or higher than, those originally present and iron in amounts to alleviate the prevalence of iron-deficiency anemia.

enteral (EN-ter-al) nutrition: the provision of nutrients using the GI tract; usually refers to the use of tube feedings.

enteric coated: refers to medications or enzyme preparations that are coated to withstand stomach acidity and dissolve only at the higher pH of the small intestine.

environmental tobacco smoke (ETS): the combination of exhaled smoke (mainstream smoke) and smoke from lighted cigarettes, pipes, or cigars (sidestream smoke) that enters the air and may be inhaled by other people.

enzymes: protein catalysts. A catalyst is a compound that facilitates chemical reactions without itself being changed in the process.

EPA, DHA: omega-3 fatty acids made from linolenic acid. The full name for EPA is *eicosapentaenoic* (EYE-cosa-PENTA-ee-NO-ick) *acid*. The full name for DHA is *docosahexaenoic* (DOE-cosa-HEXA-ee-NO-ick) *acid*.

epinephrine: one of the stress hormones secreted whenever emergency action is needed; prescribed therapeutically to relax the bronchioles during allergy or asthma attacks.

epithelial (ep-i-THEE-lee-ul) cells: cells on the surface of the skin and mucous membranes.

epithelial tissue: tissue composing the layers of the body that serve as selective barriers between the body's interior and the environment (examples are the cornea, the skin, the respiratory lining, and the lining of the digestive tract).

erythrocyte (er-REETH-ro-cite) hemolysis (he-MOLL-uh-sis): rupture of the red blood cells, caused by vitamin E deficiency.

erythrocyte protoporphyrin (PRO-toh-PORE-fe-rin): a precursor to hemoglobin.

erythropoietin (eh-RITH-ro-POY-eh-tin): a hormone made by the kidneys that stimulates red blood cell production.

esophageal (eh-SOF-ah-JEE-al): involving the esophagus.

esophageal dysphagia: an inability to move food through the esophagus; usually caused by an obstruction or a motility disorder.

essential amino acids: amino acids that the body cannot synthesize in amounts sufficient to meet physiological need. Nine amino acids are known to be essential for human adults: histidine, isoleucine, leucine, lysine, methionine, phenylalanine, threonine, tryptophan, and valine.

essential fatty acids: fatty acids that the body requires but cannot make and so must be obtained through the diet.

essential nutrients: nutrients a person must obtain from food because the body cannot make them for itself in sufficient quantities to meet physiological needs.

Estimated Average Requirements (EAR): the average daily nutrient intake levels estimated to meet the requirements of half of the healthy individuals in a given age and gender group; used in nutrition research and policymaking and as the basis on which RDA values are set.

Estimated Energy Requirement (EER): the dietary energy intake level that is predicted to maintain energy balance in a healthy adult of a defined age, gender, weight, and physical activity level consistent with good health.

ethnic diets: foodways and cuisines typical of national origins, races, cultural heritages, or geographic locations.

exclusive breastfeeding: an infant's consumption of human milk with no supplementation of any type (no water, no juice, no nonhuman milk, and no foods) except for vitamins, minerals, and medications.

exocrine: pertains to external secretions, such as those of the mucous membranes or the skin. Opposite of *endocrine*, which pertains to hormonal secretions into the blood.

expected outcomes: patient-oriented goals that are derived from nursing diagnoses.

extracellular fluid: fluid residing outside the cells; includes the fluid between the cells (*interstitial fluid*), plasma, and the water of structures such as the skin and bones. Extracellular fluid accounts for about one-third of the body's water.

fasting hyperglycemia: hyperglycemia that typically develops in the early morning after an overnight fast of at least eight hours.

fat replacers: ingredients that replace some or all of the functions of fat in foods and may or may not provide energy.

fats: lipids that are solid at room temperature (70°F or 21°C).

fatty acids: organic compounds composed of a chain of carbon atoms with hydrogen atoms attached and an acid group at one end.

fatty liver: an accumulation of fat in liver tissue; also called *hepatic steatosis.*

fermentation: the anaerobic (without oxygen) breakdown of carbohydrates by microorganisms that releases small organic compounds along with carbon dioxide and energy.

fertility: the capacity of a woman to produce a normal ovum periodically and of a man to produce normal sperm; the ability to reproduce.

fetal alcohol spectrum disorders (FASD): a spectrum of physical, behavioral, and cognitive disabilities caused by prenatal alcohol exposure.

fetal alcohol syndrome (FAS): the cluster of symptoms seen in an infant or child whose mother consumed excessive alcohol during her pregnancy. FAS includes, but is not limited to, brain damage, growth retardation, mental retardation, and facial abnormalities.

fetus (FEET-us): the developing infant from eight weeks after conception until its birth.

filtrate: the substances that pass through the glomerulus and travel through the nephron's tubules, eventually forming urine.

fistulas (FIST-you-luz): abnormal passages between organs or tissues (or between an internal organ and the body's surface) that permit the passage of fluids or secretions.

fitness: the characteristics that enable the body to perform physical activity; more broadly, the ability to meet routine physical demands with enough reserve energy to rise to a physical challenge; or the body's ability to withstand stress of all kinds.

flatulence: the condition of having excessive intestinal gas, which causes abdominal discomfort.

fluid and electrolyte balance: maintenance of the necessary amounts and types of fluid and minerals in each compartment of the body fluids.

fluorapatite (floor-APP-uh-tite): the stabilized form of bone and tooth crystal, in which fluoride has replaced the hydroxy portion of hydroxyapatite.

fluorosis (floor-OH-sis): mottling of the tooth enamel from ingestion of too much fluoride during tooth development.

follicle (FOLL-i-cul): a group of cells in the skin from which a hair grows.

food allergy: an adverse reaction to food that involves an immune response; also called *food-hypersensitivity reactions.*

food aversions: strong desires to avoid particular foods.

food cravings: deep longings for particular foods.

food deserts: urban and rural low-income areas with limited access to affordable and nutritious foods.

food frequency questionnaire: a survey of foods routinely consumed. Some questionnaires ask about the types of food eaten and yield only qualitative information; others include questions about portions consumed and yield semiquantitative data as well.

food group plan: a diet-planning tool that sorts foods into groups based on nutrient content and then specifies that people should eat certain amounts of food from each group.

food intolerances: adverse reactions to foods or food additives that do not involve the immune system.

food record: a detailed log of food eaten during a specified time period, usually several days; also called a *food diary.* A food record may also include information about medications, disease symptoms, and physical activity.

foodways: the eating habits and culinary practices of a people, region, or historical period.

fortification: the addition to a food of nutrients that were either not originally present or present in insignificant amounts. Fortification can be used to correct or prevent a widespread nutrient deficiency, to balance the total nutrient profile of a food, or to restore nutrients lost in processing.

free radicals: highly reactive chemical forms that can cause destructive changes in nearby compounds, sometimes setting up a chain reaction.

French units: units of measure that indicate the size of a feeding tube's outer diameter; 1 French unit equals ⅓ millimeter.

fructosamine test: a measurement of glycated serum proteins that analyzes glycemic control over the preceding two to three weeks; also known as the *glycated albumin test* or the *glycated serum protein test.*

fructose: a monosaccharide; sometimes known as *fruit sugar.*

functional foods: whole or modified foods that contain bioactive food components believed to provide health benefits, such as reduced disease risks, beyond the benefits that their nutrients contribute. All whole foods are functional in some ways because they provide at least some needed substances, but certain foods stand out as rich sources of bioactive food components.

galactose: a monosaccharide; part of the disaccharide lactose.

gastrectomy (gah-STREK-ta-mee): the surgical removal of part of the stomach (partial gastrectomy) or the entire stomach (total gastrectomy).

gastric banding: a surgical means of producing weight loss by restricting stomach size with a constricting band.

gastric bypass: surgery that restricts stomach size and reroutes food from the stomach to the lower part of the small intestine; creates a chronic, lifelong state of malabsorption by preventing normal digestion and absorption of nutrients.

gastric decompression: the removal of the stomach contents (including swallowed saliva, stomach secretions, and gas) of patients with motility disorders or obstructions that prevent stomach emptying.

gastric outlet obstruction: an obstruction that prevents the normal emptying of stomach contents into the duodenum.

gastric residual volume: the volume of formula and GI secretions remaining in the stomach after a previous feeding.

gastritis: inflammation of stomach tissue.

gastroesophageal reflux disease (GERD): a condition characterized by the backward flow (reflux) of the stomach's acidic contents into the esophagus.

gastrointestinal (GI) tract: the digestive tract. The principal organs are the stomach and intestines.

gastrointestinal motility: spontaneous motion in the digestive tract accomplished by involuntary muscular contractions.

gatekeeper: with respect to nutrition, a key person who controls other people's access to foods and thereby exerts a profound impact on their nutrition. Examples are the spouse who buys and cooks the food, the parent who feeds the children, and the caregiver in a day-care center.

gestation: the period of about 40 weeks (three trimesters) from conception to birth; the term of a pregnancy.

gestational diabetes: glucose intolerance with first onset or first recognition during pregnancy.

gestational hypertension: high blood pressure that develops in the second half of pregnancy and usually resolves after childbirth.

ghrelin (GRELL-in): a hormone produced primarily by the stomach cells. It signals the hypothalamus of the brain to stimulate appetite and food intake.

glomerular filtration rate (GFR): the rate at which filtrate is formed within the kidneys, normally about 125 mL/min.

glomerulus (gloh-MEHR-yoo-lus): a tuft of capillaries within the nephron that filters water and solutes from the blood as urine production begins (plural: *glomeruli*).

glucagon (GLOO-ka-gon): a hormone that is secreted by special cells in the pancreas in response to low blood glucose concentration and elicits release of glucose from storage.

glucose: a monosaccharide; the sugar common to all disaccharides and polysaccharides; also called *blood sugar* or *dextrose*.

glycated hemoglobin (HbA_{1c}): hemoglobin that has nonenzymatically attached to glucose; the level of HbA_{1c} in blood helps to diagnose diabetes and evaluate long-term glycemic control. Also called *glycosylated hemoglobin*.

glycemic (gly-SEE-mic): pertaining to blood glucose.

glycemic index: a method of classifying foods according to their potential for raising blood glucose.

glycemic response: the extent to which a food raises the blood glucose concentration and elicits an insulin response.

glycerol (GLISS-er-ol): an organic compound, three carbons long, that can form the backbone of triglycerides and phospholipids.

glycogen (GLY-co-gen): a polysaccharide composed of glucose, made and stored by liver and muscle tissues of human beings and animals as a storage form of glucose. Glycogen is not a significant food source of carbohydrate and is not counted as one of the polysaccharides in foods.

goiter (GOY-ter): an enlargement of the thyroid gland due to an iodine deficiency, malfunction of the gland, or overconsumption of a thyroid antagonist. Goiter caused by iodine deficiency is sometimes called *simple goiter*.

goitrogen (GOY-troh-jen): a substance that enlarges the thyroid gland and causes *toxic goiter*. Goitrogens occur naturally in such foods as cabbage, kale, brussels sprouts, cauliflower, broccoli, and kohlrabi.

gout (GOWT): a metabolic disorder characterized by elevated uric acid levels in the blood and urine and the deposition of uric acid in and around the joints, causing acute joint inflammation.

Hazard Analysis and Critical Control Points (HACCP): systems of food or formula preparation that identify food safety hazards and critical control points during foodservice procedures; pronounced *hassip*.

health: a range of states with physical, mental, emotional, spiritual, and social components. At a minimum, health means freedom from physical disease, mental disturbances, emotional distress, spiritual discontent, social maladjustment, and other negative states. At a maximum, health means *wellness*.

health care communities: living environments for people with chronic conditions, functional limitations, or need for supervision or assistance. Health care communities include assisted living facilities, group homes, short-term rehabilitation facilities, skilled nursing facilities, and hospice facilities.

health claims: statements that characterize the relationship between a nutrient or other substance in food and a disease or health-related condition.

Healthy People: a national public health initiative under the jurisdiction of the U.S. Department of Health and Human Services (DHHS) that identifies the most significant preventable threats to health and focuses efforts toward eliminating them.

heart failure: a condition with various causes that is characterized by the heart's inability to pump adequate blood to the body's cells, resulting in fluid accumulation in the tissues; also called *congestive heart failure*.

***Helicobacter pylori (H. pylori)*:** a species of bacterium that colonizes gastric mucosa; a primary cause of gastritis and peptic ulcer disease.

helper T cells: lymphocytes that have a specific protein called CD4 on their surfaces and therefore are also known as *CD4+ T cells*; these are the cells most affected in HIV infection.

hematocrit (hee-MAT-oh-crit): measurement of the volume of the red blood cells packed by centrifuge in a given volume of blood.

hematopoietic stem cell transplantation: transplantation of the stem cells that produce red blood cells and white blood cells; the stem cells are obtained from the bone marrow (*bone marrow transplantation*) or circulating blood.

hematuria (HE-mah-TOO-ree-ah): blood in the urine.

hemochromatosis (heem-oh-crome-a-TOH-sis): iron overload characterized by deposits of iron-containing pigment in many tissues, with tissue damage. Hemochromatosis is usually caused by a hereditary defect in iron absorption.

hemodialysis (HE-moe-dye-AL-ih-sis): a treatment that removes fluids and wastes from the blood by passing the blood through a dialyzer.

hemoglobin: the oxygen-carrying protein of the red blood cells.

hemolytic (HE-moh-LIT-ick) anemia: the condition of having too few red blood cells as a result of erythrocyte hemolysis.

hemorrhagic (hem-oh-RAJ-ik) disease: the vitamin K–deficiency disease in which blood fails to clot.

hemorrhagic strokes: strokes caused by bleeding within the brain, which destroys or compresses brain tissue.

hepatic coma: loss of consciousness resulting from severe liver disease.

hepatic encephalopathy (en-sef-ah-LOP-ah-thie): a neurological complication of advanced liver disease that is characterized by changes in personality, mood, behavior, mental ability, and motor functions.

hepatic portal vein: the vein that collects blood from the GI tract and conducts it to capillaries in the liver.

hepatic vein: the vein that collects blood from the liver capillaries and returns it to the heart.

hepatitis (hep-ah-TYE-tis): inflammation of the liver.

hepatomegaly (HEP-ah-toe-MEG-ah-lee): enlargement of the liver.

hepcidin (HEP-sid-in): an acute-phase protein involved in the regulation of iron metabolism; specifically, a hormone secreted by the liver in response to elevated blood iron. Hepcidin reduces iron's absorption from the intestine and its release from storage.

herpes simplex virus: a common virus that can cause blisterlike lesions on the lips and in the mouth.

hiatal hernia: a condition in which the upper portion of the stomach protrudes above the diaphragm; most cases are asymptomatic.

high-density lipoproteins (HDL): the type of lipoproteins that transport cholesterol back to the liver from peripheral cells; composed primarily of protein.

high-fructose corn syrup: a widely used commercial kcaloric sweetener made by adding enzymes to cornstarch to convert a portion of its glucose molecules into sweet-tasting fructose.

high-quality proteins: dietary proteins containing all the essential amino acids in relatively the same amounts that human beings require. They may also contain nonessential amino acids.

high-risk pregnancy: a pregnancy characterized by risk factors that make it likely the birth will be surrounded by problems such as premature delivery, difficult birth, retarded growth, birth defects, and early infant death.

histamine-2 receptor blockers: a class of drugs that suppress acid secretion by inhibiting receptors on acid-producing cells; commonly called *H2 blockers*. Examples include cimetidine (Tagamet), ranitidine (Zantac), and famotidine (Pepcid).

homeostasis (HOME-ee-oh-STAY-sis): the maintenance of constant internal conditions (such as chemistry, temperature, and blood pressure) by the body's control system.

hormones: chemical messengers. Hormones are secreted by a variety of glands in the body in response to altered conditions. Each travels to one or more target tissues or organs and elicits specific responses to restore normal conditions.

human immunodeficiency virus (HIV): the virus that causes acquired immune deficiency syndrome (AIDS). HIV destroys immune cells and progressively impedes the body's ability to fight infections and certain cancers.

hunger: the physiological need to eat, experienced as a drive to obtain food; an unpleasant sensation that demands relief.

hydrogenation (high-dro-gen-AY-shun): a chemical process by which hydrogen atoms are added to monounsaturated or polyunsaturated fats to reduce the number of double bonds, making the fats more saturated (solid) and more resistant to oxidation (protecting against rancidity). Hydrogenation produces *trans*-fatty acids.

hyperactivity: inattentive and impulsive behavior that is more frequent and severe than is typical of others of a similar age; professionally called **attention deficit/hyperactivity disorder (ADHD)**.

hypercalcemia (HIGH-per-kal-SEE-me-ah): elevated serum calcium levels.

hypercalciuria (HIGH-per-kal-see-YOO-ree-ah): elevated urinary calcium levels.

hypercapnia (high-per-CAP-nee-ah): excessive carbon dioxide in the blood.

hyperinsulinemia: abnormally high levels of insulin in the blood.

hyperkalemia (HIGH-per-ka-LEE-me-ah): elevated serum potassium levels.

hyperoxaluria (HIGH-per-ox-ah-LOO-ree-ah): elevated urinary oxalate levels.

hyperphosphatemia (HIGH-per-fos-fa-TEE-me-ah): elevated serum phosphate levels.

hypertension: high blood pressure.

hypertonic formula: a formula with an osmolality greater than that of blood serum.

hypoalbuminemia: low plasma albumin concentrations.

hypocaloric feedings: reduced-kcalorie feedings that usually include sufficient protein and micronutrients to maintain nitrogen balance and prevent malnutrition; also called *permissive underfeeding*.

hypochlorhydria (HIGH-poe-clor-HIGH-dree-ah): abnormally low gastric acid secretions.

hypokalemia (HIGH-po-ka-LEE-me-ah): low serum potassium levels.

hyponatremia (HIGH-po-na-TREE-me-ah): a decreased concentration of sodium in the blood.

hypothalamus (high-po-THAL-ah-mus): a brain center that controls activities such as maintenance of water balance, regulation of body temperature, and control of appetite.

hypoxemia (high-pock-SEE-me-ah): a low level of oxygen in the blood.

hypoxia (high-POCK-see-ah): a low amount of oxygen in body tissues.

ileostomy (ill-ee-AH-stoe-me): a surgical passage through the abdominal wall into the ileum.

immunity: the body's ability to defend itself against diseases.

implantation: the stage of development in which the blastocyst embeds itself in the wall of the uterus and begins to develop; occurs during the first two weeks after conception.

indirect calorimetry: a procedure that estimates energy expenditure by measuring oxygen consumption and carbon dioxide production during a period of rest.

inflammation: an immunological response to cellular injury characterized by an increase in white blood cells.

inorganic: not containing carbon or pertaining to living things.

insoluble fibers: the tough, fibrous structures of fruits, vegetables, and grains; indigestible food components that do not dissolve in water.

insulin: a hormone secreted by the pancreas in response to (among other things) high blood glucose. It promotes cellular glucose uptake for use or storage.

insulin resistance: the condition in which a normal amount of insulin produces a subnormal effect in muscle, adipose, and liver cells, resulting in an elevated fasting glucose; a metabolic consequence of obesity that precedes type 2 diabetes.

intermittent feedings: delivery of about 250 to 400 milliliters of formula over 30 to 45 minutes.

intestinal adaptation: physiological changes in the small intestine that increase its absorptive capacity after resection.

intestinal flora: the bacterial inhabitants of the GI tract.

intractable: not easily managed or controlled.

intractable vomiting: vomiting that is not easily managed or controlled.

intradialytic parenteral nutrition: the infusion of nutrients during hemodialysis, often providing amino acids, dextrose, lipids, and some trace minerals.

intrinsic factor: a substance secreted by the stomach cells that binds with vitamin B_{12} in the small intestine to aid in the absorption of vitamin B_{12}.

iron deficiency: the condition of having depleted iron stores.

iron overload: toxicity from excess iron.

iron-deficiency anemia: a blood iron deficiency characterized by small, pale red blood cells; also called *microcytic hypochromic anemia.*

irritable bowel syndrome: an intestinal disorder of unknown cause that disturbs the functioning of the large intestine; symptoms include abdominal pain, flatulence, diarrhea, and constipation.

ischemic strokes: strokes caused by the obstruction of blood flow to brain tissue.

isotonic formula: a formula with an osmolality similar to that of blood serum (about 300 milliosmoles per kilogram).

jaundice (JAWN-dis): yellow discoloration of the skin and eyes due to an accumulation of bilirubin, a breakdown product of hemoglobin that normally exits the body via bile secretions.

Kaposi's (kah-POH-seez) sarcoma: a common cancer in HIV-infected persons that is characterized by lesions in the skin, lungs, and GI tract.

kcalorie (energy) control: management of food energy intake.

kcalorie counts: estimates of food energy (and often, protein) consumed by patients for one or more days.

keratin (KERR-uh-tin): a water-insoluble protein; the normal protein of hair and nails.

ketones (KEY-tones): acidic, water-soluble compounds produced by the liver during the breakdown of fat when carbohydrate is not available; technically known as *ketone bodies.*

kidney stones: crystalline masses that form in the urinary tract; also called *renal calculi* or *nephrolithiasis.*

kilocalorie: the unit in which food energy is measured (1000 calories equal 1 kilocalorie), abbreviated *kcalories* or *kcal.* One kcalorie is the amount of heat necessary to raise the temperature of 1 kilogram (kg) of water 1°C. The scientific use of the term *kcalorie* is the same as the popular use of the term *calorie.*

kwashiorkor (kwash-ee-OR-core or **kwash-ee-or-CORE):** severe malnutrition characterized by failure to grow and develop, edema, changes in the pigmentation of the hair and skin, fatty liver, anemia, and apathy.

lactadherin (lack-tad-HAIR-in): a protein in breast milk that attacks diarrhea-causing viruses.

lactation: production and secretion of breast milk for the purpose of nourishing an infant.

lactoferrin (lack-toe-FERR-in): protein in breast milk that binds iron and keeps it from supporting the growth of the infant's intestinal bacteria.

lactose: a disaccharide composed of glucose and galactose; commonly known as *milk sugar.*

laparoscopic weight-loss surgery: a weight-loss surgery procedure in which surgeons gain access to the abdomen via several small incisions. A tiny video camera is inserted through one of the incisions and surgical instruments through the others. The surgeons watch their work on a large-screen monitor.

lapses: periods of returning to old habits.

lecithins: one type of phospholipid.

legumes (lay-GYOOMS, LEG-yooms): plants of the bean and pea family with seeds that are rich in protein compared with other plant-derived foods.

leptin: a hormone produced by fat cells under the direction of the (*ob*) gene. It decreases appetite and increases energy expenditure.

life expectancy: the average number of years lived by people in a given society.

life span: the maximum number of years of life attainable by a member of a species.

limiting amino acid: an essential amino acid that is present in dietary protein in the shortest supply relative to the amount needed for protein synthesis in the body.

linoleic acid, linolenic acid: polyunsaturated fatty acids that are essential for human beings.

lipids: a family of compounds that includes triglycerides (fats and oils), phospholipids, and sterols. Lipids are characterized by their insolubility in water.

lipodystrophy (LIP-oh-DIS-tro-fee): abnormalities in body fat and fat metabolism that may result from drug treatments for HIV infection. The accumulation of abdominal fat is sometimes called *protease paunch.*

lipomas (lih-POE-muz): benign tumors composed of fatty tissue.

lipoprotein lipase (LPL): an enzyme mounted on the surface of fat cells (and other cells) that hydrolyzes triglycerides in the blood into fatty acids and glycerol for absorption into the cells. There they are metabolized or reassembled for storage.

lipoproteins: clusters of lipids associated with proteins that serve as transport vehicles for lipids in the lymph and blood.

listeriosis: a serious foodborne infection that can cause severe brain infection or death in a fetus or newborn; caused by the bacterium *Listeria monocytogenes*, which is found in soil and water.

longevity: long duration of life.

low birthweight (LBW): a birthweight less than 5½ lb (2500 g); indicates probable poor health in the newborn and poor nutrition status of the mother during pregnancy. Optimal birthweight for a full-term infant is 6.8 to 7.9 lb (about 3100 to 3600 g).

low-density lipoproteins (LDL): the type of lipoproteins derived from VLDL as cells remove triglycerides from them. LDL carry cholesterol and triglycerides from the liver to the cells of the body and are composed primarily of cholesterol.

low-microbial diet: a diet that contains foods that are unlikely to be contaminated with bacteria and other microbes.

low-risk pregnancy: a pregnancy characterized by factors that make it likely the birth will be normal and the infant healthy.

lymph (LIMF): the body fluid found in lymphatic vessels. Lymph consists of all the constituents of blood except red blood cells.

lymphatic system: a loosely organized system of vessels and ducts that conveys the products of digestion toward the heart.

macrosomia (MAK-roh-SOH-mee-ah): the condition of having an abnormally large body; in infants, refers to birth weights of 4000 grams (8 pounds 13 ounces) and above.

macular (MACK-you-lar) degeneration: deterioration of the macular area of the eye that can lead to loss of central vision and eventual blindness. The *macula* is a small, oval, yellowish region in the center of the retina that provides the sharp, straight-ahead vision so critical to reading and driving.

major minerals: essential mineral nutrients required in the adult diet in amounts greater than 100 milligrams per day.

malignant (ma-LIG-nent): describes a cancerous cell or tumor, which can injure healthy tissue and spread cancer to other regions of the body.

malnutrition: any condition caused by deficient or excess energy or nutrient intake or by an imbalance of nutrients.

maltose: a disaccharide composed of two glucose units; sometimes known as *malt sugar*.

marasmus (ma-RAZZ-mus): severe malnutrition characterized by poor growth, dramatic weight loss, loss of body fat and muscle, and apathy.

medical nutrition therapy: nutrition care provided by a registered dietitian; includes assessment of nutrition status, diagnosis of nutrition problems, development of nutrition care plans, and provision of dietary counseling and nutrition education.

medium-chain triglycerides (MCT): triglycerides that contain fatty acids that are 8 to 10 carbons in length. MCT do not require digestion and can be absorbed in the absence of lipase or bile.

metastasize (meh-TAS-tah-size): to spread from one part of the body to another; refers to cancer cells.

microalbuminuria: the presence of albumin (a blood protein) in the urine, a sign of diabetic nephropathy.

microvilli (MY-cro-VILL-ee or **MY-cro-VILL-eye):** tiny, hairlike projections on each cell of every villus that can trap nutrient particles and transport them into the cells. The singular form is **microvillus**.

moderate-intensity physical activity: physical activity that requires some increase in breathing and/or heart rate and expends 3.5 to 7 kcalories per minute. Walking at a speed of 3 to 4.5 miles per hour (about 15 to 20 minutes to walk one mile) is an example.

moderation: the provision of enough, but not too much, of a substance.

modified diet: a diet that contains foods altered in texture, consistency, or nutrient content or that includes or omits specific foods; may also be called a *therapeutic diet.*

modular formulas: enteral formulas prepared in the hospital from *modules* that contain single macronutrients; used for people with unique nutrient needs.

molybdenum (mo-LIB-duh-num): a trace element.

monoglycerides: molecules of glycerol with one fatty acid attached. A molecule of glycerol with two fatty acids attached is a *diglyceride.*

monosaccharides (mon-oh-SACK-uh-rides): single sugar units.

monounsaturated fatty acid (MUFA): a fatty acid that has one point of unsaturation; for example, the oleic acid found in olive oil.

mucous membrane: membrane composed of mucus-secreting cells that lines the surfaces of body tissues.

myoglobin: the oxygen-carrying protein of the muscle cells.

naturally occurring sugars: sugars that are not added to a food but are present as its original constituents, such as the sugars of fruit or milk.

nephron (NEF-ron): the functional unit of the kidneys, consisting of a glomerulus and tubules.

nephrotic (neh-FROT-ik) syndrome: a syndrome associated with disorders that cause severe glomerular damage, resulting in significant urinary protein losses.

nephrotoxic: toxic to the kidneys.

neural tube: the embryonic tissue that later forms the brain and spinal cord.

neural tube defects (NTD): malformations of the brain, spinal cord, or both that occur during embryonic development and result in birth defects that may cause lifelong disability or death.

neurofibrillary tangles: snarls of the threadlike strands that extend from the nerve cells, commonly found in the brains of people with Alzheimer's dementia.

neurons: nerve cells; the structural and functional units of the nervous system. Neurons initiate and conduct nerve transmissions.

neutropenia: a low white blood cell (neutrophil) count, which increases susceptibility to infection.

niacin equivalents (NE): the amount of niacin present in food, including the niacin that can theoretically be made from tryptophan, its precursor, present in the food.

night blindness: the slow recovery of vision after exposure to flashes of bright light at night; an early symptom of vitamin A deficiency.

nitrogen balance: the amount of nitrogen consumed (N in) as compared with the amount of nitrogen excreted (N out) in a given period of time. The laboratory scientist can estimate the protein in a sample of food, body tissue, or excreta by measuring the nitrogen in it.

nonessential amino acids: amino acids that the body can synthesize.

nonnutritive sweeteners: synthetic or natural food additives that offer sweet flavor but with negligible or no calories per serving; also called *artificial sweeteners, intense sweeteners, noncaloric sweeteners,* and *very low-calorie sweeteners.*

nursing bottle tooth decay: extensive tooth decay due to prolonged tooth contact with formula, milk, fruit juice, or other carbohydrate-rich liquid offered to an infant in a bottle.

nursing diagnoses: clinical judgments about actual or potential health problems that provide the basis for selecting appropriate nursing interventions.

nutrient claims: statements that characterize the quantity of a nutrient in a food.

nutrient density: a measure of the nutrients a food provides relative to the energy it provides. The more nutrients and the fewer kcalories, the higher the nutrient density.

nutrient profiling: ranking foods based on their nutrient composition.

nutrients: substances obtained from food and used in the body to provide energy and structural materials and to serve as regulating agents to promote growth, maintenance, and repair. Nutrients may also reduce the risks of some diseases.

nutrition: the science of foods and the nutrients and other substances they contain, and of their ingestion, digestion, absorption, transport, metabolism, interaction, storage, and excretion. A broader definition includes the study of the environment and of human behavior as it relates to these processes.

nutrition care plans: strategies for meeting an individual's nutritional needs.

nutrition care process: a systematic approach used by dietetics professionals to evaluate and treat nutrition-related problems.

nutrition screening: a brief assessment of health-related variables to identify patients who are malnourished or at risk for malnutrition.

nutrition support: the delivery of nutrients using a feeding tube or intravenous infusions.

nutrition support teams: health care professionals responsible for the provision of nutrients by tube feeding or intravenous infusion.

nutritive sweeteners: sweeteners that yield energy, including both the sugars and the sugar alcohols.

obese: having too much body fat with adverse health effects; BMI 30 or more.

oils: lipids that are liquid at room temperature (70°F or 21°C).

olestra: a synthetic fat made from sucrose and fatty acids that provides zero kcalories per gram; also known as *sucrose polyester*.

oliguria (OL-lih-GOO-ree-ah): an abnormally low amount of urine, often less than 400 mL/day.

omega-3 fatty acids: polyunsaturated fatty acids in which the endmost double bond is three carbons back from the end of the carbon chain; relatively newly recognized as important in nutrition. Linolenic acid is an example.

omega-6 fatty acid: a polyunsaturated fatty acid with its endmost double bond six carbons back from the end of its carbon chain; long recognized as important in nutrition. Linoleic acid is an example.

open feeding system: a formula delivery system that requires the formula to be transferred from its original packaging to a feeding container before being administered through the feeding tube.

opportunistic infections: infections from microorganisms that normally do not cause disease in healthy people but are damaging to persons with compromised immune function.

opsin (OP-sin): the protein portion of the visual pigment molecule.

oral glucose tolerance test: a test that evaluates a person's ability to tolerate an oral glucose load.

organic: carbon containing. The four organic nutrients are carbohydrate, fat, protein, and vitamins.

oropharyngeal (OR-row-fah-ren-JEE-al): involving the mouth and pharynx.

oropharyngeal dysphagia: an inability to transfer food from the mouth and pharynx to the esophagus; usually caused by a neurological or muscular disorder.

osmolality (OZ-moe-LAL-ih-tee): the concentration of osmotically active solutes in a solution, expressed as milliosmoles (mOsm) per kilogram of solvent.

osmolarity: the concentration of osmotically active solutes in a solution, expressed as milliosmoles per liter of solution (mOsm/L). *Osmolality* (mOsm/kg) is an alternative measure used to describe a solution's osmotic properties.

osteoarthritis: a painful, chronic disease of the joints that occurs when the cushioning cartilage in a joint breaks down; joint structure is usually altered, with loss of function; also called *degenerative arthritis*.

osteomalacia (os-tee-oh-mal-AY-shuh): a bone disease characterized by softening of the bones. Symptoms include bending of the spine and bowing of the legs. The disease occurs most often in adults with renal failure or malabsorption disorders.

osteoporosis (os-tee-oh-pore-OH-sis): literally, porous bones; reduced density of the bones, also known as *adult bone loss*.

overnutrition: overconsumption of food energy or nutrients sufficient to cause disease or increased susceptibility to disease; a form of malnutrition.

overweight: body weight greater than the weight range that is considered healthy; BMI 25.0 to 29.9.

ovum (OH-vum): the female reproductive cell, capable of developing into a new organism upon fertilization; commonly referred to as an egg.

oxidation (OKS-ee-day-shun): the process of a substance combining with oxygen.

paracentesis (pah-rah-sen-TEE-sis): a surgical puncture of a body cavity with an aspirator to draw out excess fluid.

parenteral (par-EN-ter-al) nutrition: the intravenous provision of nutrients that bypasses the GI tract.

pellagra (pell-AY-gra): the niacin-deficiency disease. Symptoms include the "4 Ds": diarrhea, dermatitis, dementia, and, ultimately, death.

peptic ulcer: an open sore in the gastrointestinal mucosa; may develop in the esophagus, stomach, or duodenum.

peripheral parenteral nutrition (PPN): the infusion of nutrient solutions into peripheral veins, usually a vein in the arm or back of the hand.

peripheral veins: the small-diameter veins that carry blood from the limbs.

peristalsis (peri-STALL-sis): successive waves of involuntary muscular contractions passing along the walls of the GI tract that push the contents along.

peritoneal (PEH-rih-toe-NEE-al) dialysis: a treatment that removes fluids and wastes from the blood by using the body's peritoneal membrane as a filter.

peritoneovenous (PEH-rih-toe-NEE-oh-VEE-nus) shunt: a surgical passage created between the peritoneum and the jugular vein to divert fluid and relieve ascites. The peritoneum is the membrane that surrounds the abdominal cavity.

pernicious (per-NISH-us) anemia: a blood disorder that reflects a vitamin B_{12} deficiency caused by lack of intrinsic factor and characterized by large, immature red blood cells and damage to the nervous system (*pernicious* means "highly injurious or destructive").

pH: the concentration of hydrogen ions. The lower the pH, the stronger the acid. Thus, pH 2 is a strong acid; pH 6 is a weak acid; pH 7 is neutral; and a pH above 7 is alkaline.

phlebitis (fleh-BYE-tiss): inflammation of a vein.

phospholipids: one of the three main classes of lipids; compounds that are similar to triglycerides but have *choline* (or another compound) and a phosphorus-containing acid in place of one of the fatty acids.

physiological age: a person's age as estimated from her or his body's health and probable life expectancy.

phytates: nonnutrient components of grains, legumes, and seeds. Phytates can bind minerals such as iron, zinc, calcium, and magnesium in insoluble complexes in the intestine, and the body excretes them unused.

phytochemicals (FIGH-toe-CHEM-ih-cals): compounds in plants that confer color, taste, and other characteristics. Some phytochemicals are bioactive food components in functional foods. Nutrition in Practice 8 provides details.

pica (PIE-ka): a craving for nonfood substances; also known as *geophagia* (jee-oh-FAY-jee-uh) when referring to clay-eating behavior.

piggyback: the administration of a second solution using a separate port in an intravenous catheter.

pigment: a molecule capable of absorbing certain wavelengths of light so that it reflects only those that we perceive as a certain color.

pituitary (pit-TOO-ih-tary) gland: in the brain, the "king gland" that regulates the operation of many other glands.

placenta (pla-SEN-tuh): an organ that develops inside the uterus early in pregnancy, in which maternal and fetal blood circulate in close proximity and exchange materials. The fetus receives nutrients and oxygen across the placenta; the mother's blood picks up carbon dioxide and other waste materials to be excreted via her lungs and kidneys.

polypeptide: 10 or more amino acids bonded together. An intermediate strand of between 4 and 10 amino acids is an *oligopeptide*.

polysaccharides: long chains of monosaccharide units arranged as starch, glycogen, or fiber.

polyunsaturated fatty acids (PUFA): fatty acids with two or more points of unsaturation. For example, linoleic acid has two such points, and linolenic acid has three. Thus, polyunsaturated *fat* is composed of triglycerides containing a high percentage of PUFA.

portal hypertension: elevated blood pressure in the portal vein due to obstructed blood flow through the liver and greater inflow of portal blood.

portion size: the quantity of food served or eaten at one meal or snack; *not* a standard amount.

precursors: compounds that can be converted into other compounds; with regard to vitamins, compounds that can be converted into active vitamins; also known as *provitamins*.

prediabetes: the condition in which blood glucose levels are higher than normal but not high enough to be diagnosed as diabetes (fasting plasma glucose between 100 and 125 mg/dL).

preeclampsia (PRE-ee-KLAMP-see-ah): a condition characterized by hypertension and protein in the urine.

preformed vitamin A: vitamin A in its active form.

prehypertension: blood pressure values that predict hypertension.

prenatal supplements: nutrient supplements specifically designed to provide the nutrients needed during pregnancy, particularly folate, iron, and calcium, without excesses or unneeded constituents.

pressure sores: regions of skin and tissue that are damaged due to prolonged pressure on the affected area by an external object, such as a bed, wheelchair, or cast; vulnerable areas of the body include buttocks, hips, and heels. Also called *decubitus* (deh-KYU-bih-tus) ulcers.

primary hypertension: hypertension with an unknown cause; also known as **essential hypertension**.

protein digestibility: a measure of the amount of amino acids absorbed from a given protein intake.

protein isolates: proteins that have been isolated from foods.

protein turnover: the continuous breakdown and synthesis of body proteins involving the recycling of amino acids.

proteins: compounds made from strands of amino acids composed of carbon, hydrogen, oxygen, and nitrogen atoms. Some amino acids also contain sulfur atoms.

proteinuria (PRO-teen-NOO-ree-ah): loss of protein, mostly albumin, in the urine; also known as *albuminuria*.

proton-pump inhibitors: a class of drugs that inhibit the enzyme that pumps hydrogen ions (protons) into the stomach. Examples include omeprazole (Prilosec) and lansoprazole (Prevacid).

pruritis: itchy skin.

puberty: the period in life in which a person becomes physically capable of reproduction.

purines (PYOO-reens): products of nucleotide metabolism that degrade to uric acid.

quality of life: a person's perceived physical and mental well-being.

radiation enteritis: inflammation of intestinal tissue caused by radiation therapy.

radiation therapy: the use of X-rays, gamma rays, or atomic particles to destroy cancer cells.

rancid: the term used to describe fats when they have deteriorated, usually by oxidation. Rancid fats often have an "off" odor.

rebound hyperglycemia: hyperglycemia that results from the release of counterregulatory hormones following nighttime hypoglycemia; also called the *Somogyi effect*.

Recommended Dietary Allowances (RDA): a set of values reflecting the average daily amounts of nutrients considered adequate to meet the known nutrient needs of practically all healthy people in a particular life stage and gender group; a goal for dietary intake by individuals.

refeeding syndrome: a condition that sometimes develops when a severely malnourished person is aggressively fed; characterized by electrolyte and fluid imbalances and hyperglycemia.

refined grain: a product from which the bran, germ, and husk have been removed, leaving only the endosperm.

reflux esophagitis: inflammation in the esophagus related to the reflux of acidic stomach contents.

regular diet: a diet that includes all foods and meets the nutrient needs of healthy people; may also be called a *standard diet*, *general diet*, *normal diet*, or *house diet*.

renal (REE-nal): pertaining to the kidneys.

renal colic: the intense pain that occurs when a kidney stone passes through the ureter; the pain typically begins in the back and intensifies as the stone travels toward the bladder.

renal osteodystrophy: a bone disorder that develops in patients with chronic kidney disease as a result of increased secretion of parathyroid hormone, reduced serum calcium, acidosis, and impaired vitamin D activation by the kidneys.

renal threshold: the blood concentration of a substance that exceeds the kidneys' capacity for reabsorption, causing the substance to be passed into the urine.

requirement: the lowest continuing intake of a nutrient that will maintain a specified criterion of adequacy.

resection: the surgical removal of part of an organ or body structure.

residue: material left in the intestine after digestion; includes mostly dietary fiber and undigested starches and proteins.

resistance training: the use of free weights or weight machines to provide resistance for developing muscle strength, power, and endurance; also called *weight training*. A person's own body weight may also be used to provide resistance as when a person does push-ups, pull-ups, or abdominal crunches.

resistant starches: starches that escape digestion and absorption in the small intestine of healthy people.

respiratory failure: a potentially life-threatening condition in which inadequate respiratory function impairs gas exchange between the air and circulating blood, resulting in abnormal levels of tissue gases.

resting metabolic rate (RMR): a measure of the energy use of a person at rest in a comfortable setting—similar to the BMR but with less stringent criteria for recent food intake and physical activity. Consequently, the RMR is slightly higher than the BMR.

retina (RET-in-uh): the layer of light-sensitive nerve cells lining the back of the inside of the eye; consists of rods and cones.

retinol activity equivalents (RAE): a measure of vitamin A activity; the amount of retinol that the body will derive from a food containing preformed retinol or its precursor beta-carotene.

retinol-binding protein (RBP): the specific protein responsible for transporting retinol.

rheumatoid arthritis: a disease of the immune system involving painful inflammation of the joints and related structures.

rhodopsin (ro-DOP-sin): a light-sensitive pigment of the retina; contains the retinal form of vitamin A and the protein opsin.

rickets: the vitamin D–deficiency disease in children.

salts: compounds composed of charged particles (ions). An example of a salt is potassium chloride (K^+Cl^-).

sarcopenia (SAR-koh-PEE-nee-ah): age-related loss of skeletal muscle mass, muscle strength, and muscle function.

satiation (say-she-AY-shun): the feeling of satisfaction and fullness that occurs during a meal and halts eating. Satiation determines how much food is consumed during a meal.

satiety: the feeling of fullness and satisfaction that occurs after a meal and inhibits eating until the next meal. Satiety determines how much time passes between meals.

saturated fatty acid: a fatty acid carrying the maximum possible number of hydrogen atoms (having no points of unsaturation).

screen time: sedentary time spent using an electronic device, such as a television, computer, or video game player.

scurvy: the vitamin C–deficiency disease.

secondary hypertension: hypertension that results from a known physiological abnormality.

segmentation: a periodic squeezing or partitioning of the intestine by its circular muscles that both mixes and slowly pushes the contents along.

selective menus: menus that provide choices in some or all menu categories.

self-efficacy: a person's belief in his or her ability to succeed in an undertaking.

self-monitoring of blood glucose: home monitoring of blood glucose levels using a glucose meter.

senile dementia: the loss of brain function beyond the normal loss of physical adeptness and memory that occurs with aging.

senile plaques: clumps of the protein fragment beta-amyloid on the nerve cells, commonly found in the brains of people with Alzheimer's dementia.

serving size: a standardized quantity of a food; such information allows comparisons when reading food labels and consistency when following the *Dietary Guidelines*.

set-point theory: the theory that the body tends to maintain a certain weight by means of its own internal controls.

severe acute malnutrition (SAM): malnutrition caused by recent severe food restriction; characterized in children by underweight for height (wasting).

short bowel syndrome: the malabsorption syndrome that follows resection of the small intestine; characterized by inadequate absorptive capacity in the remaining intestine.

sinusoids: the small, capillary-like passages that carry blood through liver tissue.

Sjögren's (SHOW-grenz) syndrome: an autoimmune disease characterized by the destruction of secretory glands, resulting in dry mouth and dry eyes.

skinfold measure: a clinical estimate of total body fatness in which the thickness of a fold of skin on the back of the arm (over the triceps muscle), below the shoulder blade (subscapular), or in other places is measured with a caliper.

soaps: chemical compounds formed from fatty acids and positively charged minerals.

solid fats: fats that are not usually liquid at room temperature; commonly found in most foods derived from animals and vegetable oils that have been hydrogenated. Solid fats typically contain more saturated and *trans* fats than most oils.

soluble fibers: indigestible food components that readily dissolve in water and often impart gummy or gel-like characteristics to foods. An example is pectin from fruit, which is used to thicken jellies.

spasm: a sudden, forceful, and involuntary muscle contraction.

Special Supplemental Nutrition Program for Women, Infants, and Children (WIC): a high-quality, cost-effective health care and nutrition services program administered by the U.S. Department of Agriculture for low-income women, infants, and children who are nutritionally at risk. WIC provides supplemental foods, nutrition education, and referrals to health care and other social services.

specialized formulas: enteral formulas designed to meet the nutrient needs of patients with specific illnesses; also called *disease-specific* or *specialty formulas*.

spina (SPY-nah) bifida (BIFF-ih-dah): one of the most common types of neural tube defects; characterized by the incomplete closure of the spinal cord and its bony encasement.

standard formulas: enteral formulas that contain mostly intact proteins and polysaccharides; also called *polymeric formulas*.

starch: a plant polysaccharide composed of glucose and digestible by human beings.

steatohepatitis (STEE-ah-to-HEP-ah-TYE-tis): liver inflammation that is associated with fatty liver.

steatorrhea (stee-AT-or-REE-ah): excessive fat in the stools due to fat malabsorption; characterized by stools that are loose, frothy, and foul smelling due to a high fat content.

sterile: free of microorganisms such as bacteria.

steroids (STARE-oids): medications used to reduce tissue inflammation, to suppress the immune response, or to replace certain steroid hormones in people who cannot synthesize them.

sterols: one of the main classes of lipids; includes cholesterol, vitamin D, and the sex hormones (such as testosterone).

stoma (STOE-ma): a surgically created opening in a body tissue or organ.

stricture: abnormal narrowing of a passageway; often due to inflammation, scarring, or a congenital abnormality.

structure-function claims: statements that describe how a product may affect a structure or function of the body; for example, "calcium builds strong bones." Structure-function claims do not require FDA authorization.

struvite (STROO-vite): crystals of magnesium ammonium phosphate.

subcutaneous (sub-cue-TAY-nee-us): beneath the skin.

subcutaneous fat: fat stored directly under the skin.

sucrose: a disaccharide composed of glucose and fructose; commonly known as *table sugar*, *beet sugar*, or *cane sugar*.

sugar alcohols: sugarlike compounds in the chemical family *alcohol* derived from fruits or manufactured from carbohydrates; sugar alcohols are absorbed more slowly than other sugars, are metabolized differently, and do not elevate the risk of dental caries. Examples are maltitol, mannitol, sorbitol, isomalt, lactitol, and xylitol.

syringes: devices used for injecting medications. A syringe consists of a hypodermic needle attached to a hollow tube with a plunger inside.

tannins: compounds in tea (especially black tea) and coffee that bind iron.

teratogenic (ter-AT-oh-jen-ik): causing abnormal fetal development and birth defects.

thermic effect of food: an estimation of the energy required to process food (digest, absorb, transport, metabolize, and store ingested nutrients).

tissue rejection: destruction of donor tissue by the recipient's immune system, which recognizes the donor cells as foreign.

tocopherol (tuh-KOFF-er-ol): a general term for several chemically related compounds, one of which has vitamin E activity.

Tolerable Upper Intake Levels (UL): a set of values reflecting the highest average daily nutrient intake levels that are likely to pose no risk of toxicity to almost all healthy individuals in a particular life stage and gender group. As intake increases above the UL, the potential risk of adverse health effects increases.

tolerance level: the maximum amount of residue permitted in a food when a pesticide is used according to the label directions.

total nutrient admixture (TNA): a parenteral solution that contains dextrose, amino acids, and lipids; also called a **3-in-1 solution**.

total parenteral nutrition (TPN): the infusion of nutrient solutions into a central vein.

trace minerals: essential mineral nutrients required in the adult diet in amounts less than 100 milligrams per day.

***trans*-fatty acids:** fatty acids in which the hydrogen atoms next to the double bond are on opposite sides of the carbon chain.

transferrin (trans-FERR-in): the body's iron-carrying protein.

transient ischemic attacks (TIAs): brief ischemic strokes that cause short-term neurological symptoms.

triglycerides (try-GLISS-er-rides): one of the main classes of lipids; the chief form of fat in foods and the major storage form of fat in the body; composed of glycerol with three fatty acids attached.

trimester: a period representing one-third of the term of gestation. A trimester is about 13 to 14 weeks.

tripeptide: three amino acids bonded together.

tube feedings: liquid formulas delivered through a tube placed in the stomach or intestine.

tubules: tubelike structures of the nephron that process filtrate during urine production. The tubules are surrounded by capillaries that reabsorb substances retained by tubule cells.

tumor: an abnormal tissue mass that has no physiological function; also called a *neoplasm* (NEE-oh-plazm).

type 1 diabetes: the type of diabetes characterized by absolute insulin deficiency that usually results from autoimmune destruction of pancreatic beta cells.

type 2 diabetes: the type of diabetes that accounts for 90 to 95 percent of diabetes cases and usually results from insulin resistance coupled with insufficient insulin secretion.

ulcerative colitis (ko-LY-tis): an inflammatory bowel disease that involves the rectum and colon; inflammation affects the mucosa and submucosa only.

umbilical (um-BIL-ih-cul) cord: the ropelike structure through which the fetus's veins and arteries reach the placenta; the route of nourishment and oxygen into the fetus and the route of waste disposal from the fetus.

undernutrition: underconsumption of food energy or nutrients severe enough to cause disease or increased susceptibility to disease; a form of malnutrition.

underweight: body weight lower than the weight range that is considered healthy; BMI below 18.5.

unsaturated fatty acid: a fatty acid with one or more points of unsaturation where hydrogen atoms are missing (includes monounsaturated and polyunsaturated fatty acids).

urea (yoo-REE-uh): the principal nitrogen-excretion product of protein metabolism; generated mostly by removal of amine groups from unneeded amino acids or from the sacrifice of amino acids to meet a need for energy.

uremia (you-REE-me-ah): the accumulation of nitrogenous and various other waste products in the blood (literally, "urine in the blood"); often associated with disorders that reflect impairments in multiple body systems. The term may also be used to indicate the toxic state that results when wastes are retained in the blood.

uremic syndrome: the cluster of disorders caused by inadequate kidney function; complications include fluid, electrolyte, and hormonal imbalances; altered heart function; neuromuscular disturbances; and other metabolic derangements.

USDA Food Patterns: the USDA's food group plan for ensuring dietary adequacy that assigns foods to five major food groups.

uterus (YOO-ter-us): the womb, the muscular organ within which the infant develops before birth.

varices (VAH-rih-seez): abnormally dilated blood vessels (singular: *varix*).

variety (dietary): consumption of a wide selection of foods within and among the major food groups (the opposite of monotony).

vein: a vessel that carries blood back to the heart.

very-low-density lipoproteins (VLDL): the type of lipoproteins made primarily by liver cells to transport lipids to various tissues in the body; composed primarily of triglycerides.

vigorous-intensity physical activity: physical activity that requires a large increase in breathing and/or heart rate and expends more than 7 kcalories per minute. Walking at a very brisk pace (>4.5 miles per hour) or running at a pace of at least 5 miles per hour are examples.

villi (VILL-ee or VILL-eye): fingerlike projections from the folds of the small intestine. The singular form is **villus**.

visceral fat: fat stored within the abdominal cavity in association with the internal abdominal organs, as opposed to fat stored directly under the skin (subcutaneous fat); also called *intra-abdominal fat.*

viscous: having a gel-like consistency.

vitamin A: a fat-soluble vitamin. Its three chemical forms are *retinol* (the alcohol form), *retinal* (the aldehyde form), and *retinoic acid* (the acid form).

vitamins: essential, nonkcaloric, organic nutrients needed in tiny amounts in the diet.

voluntary activities: the component of a person's daily energy expenditure that involves conscious and deliberate muscular work: walking, lifting, climbing, and other physical activities. Voluntary activities normally require less energy in a day than basal metabolism does.

waist circumference: a measurement used to assess a person's abdominal fat.

wasting: the gradual atrophy (loss) of body tissues; associated with protein-energy malnutrition or chronic illness.

water balance: the balance between water intake and water excretion that keeps the body's water content constant.

water intoxication: the rare condition in which body water contents are too high. The symptoms may include confusion, convulsion, coma, and even death in extreme cases.

wean: to gradually replace breast milk with infant formula or other foods appropriate to an infant's diet.

wellness: maximum well-being; the top range of health states; the goal of the person who strives toward realizing his or her full potential physically, mentally, emotionally, spiritually, and socially.

wheat gluten (GLU-ten): a family of water-insoluble proteins in wheat; includes the gliadin (GLY-ah-din) fractions that are toxic to persons with celiac disease.

xerostomia (ZEE-roh-STOE-me-ah): dry mouth caused by reduced salivary flow.

zygote (ZY-goat): the product of the union of ovum and sperm; a fertilized ovum.

Index

This index lists primarily topics that received significant mention in the text. Inclusive pages (for example, 53–56) indicate major discussions; pages in **bold** refer to defined terms; pages in *italics* refer to figures, diagrams, or chemical structures; pages followed by "t" refer to tables.